International Conference on Modern Trends in Manufacturing Technologies and Equipment (ICMTMTE 2017)

MATEC Web of Conferences Volume 129 (2017)

Sevastopol, Russia
11 - 15 September 2017

Part 1 of 2

Editors:

Sergey Bratan
Sergey Borbatyuk
Sergey Leonov
Stanislav Roshchupkin

ISBN: 978-1-5108-5189-4

Printed from e-media with permission by:

Curran Associates, Inc.
57 Morehouse Lane
Red Hook, NY 12571

Some format issues inherent in the e-media version may also appear in this print version.

Printed by Curran Associates, Inc. (2017)

For additional information, please contact EDP Sciences – Web of Conferences at the address below.

EDP Sciences – Web of Conferences
17, Avenue du Hoggar
Parc d'Activité de Courtabœuf
BP 112
F-91944 Les Ulis Cedex A
France

Phone: +33 (0) 1 69 18 75 75
Fax: +33 (0) 1 69 28 84 91

contact-edps@webofconferences.org

Additional copies of this publication are available from:

Curran Associates, Inc.
57 Morehouse Lane
Red Hook, NY 12571 USA
Phone: 845-758-0400
Fax: 845-758-2633
Email: curran@proceedings.com
Web: www.proceedings.com

TABLE OF CONTENTS

PART 1

PART 2

ICMTMTE 2017

PREFACE

The International Conference on Modern Trends in Manufacturing Technologies and Equipment 2017 (ICMTME 2017) was held in Sevastopol, Russia from 11 to 15 September 2017 and was organized by the Sevastopol State University, National University of Science and Technology «MISIS», Polzunov Altai State Technical University, Inlink Ltd. and International Union of Machine Builders.

ICMTMTE 2017 was organized to provide scientists and industry experts with the opportunity to summarize the latest developments in manufacturing technologies, materials science, manufacturing equipment and tools, to discuss future developments in the new technique and technologies, to build up the network and partnerships for the future collaboration.

The conference was supported financially by the Russian Foundation for Basic Research (grant № 17-08-20448).

The conferees represented different countries – Russia, Ukraine, Kazakhstan, Armenia, Turkey, Germany, China, Taiwan, Vietnam. The proceedings of ICMTMTE 2017 consists of 221 selected papers, which were presented orally at the conference or via poster sessions.

We express our deepest gratitude to all members of the scientific committee, invited speakers, reviewers, sponsors, participants and members of the ICMTMTE 2017 for their valuable contribution to the successful organization of ICMTMTE 2017. We also would like to express our special thanks to Isaline Augusto and all staff from EDP Sciences for their great support to ICMTMTE 2017.

With best wishes,

Prof. Sergey Bratan, General Chair of ICMTMTE 2017

Dr. Stanislav Roshchupkin, Secretary of ICMTMTE 2017

CONFERENCE COMMITTEES

Conference Chair
Prof. Sergey Bratan (Sevastopol State University, Sevastopol, RUS)

Conference Co-Chair
Prof. Sergey Gorbatyuk (National University of Science and Technology «MISIS», Moscow, RUS)
Prof. Sergey Leonov (Polzunov Altai State Technical University, Barnaul, RUS)

International Program Committee
Prof. U. Abdulgazis (Crimean. Eng. and Ped. Un., RUS)
Prof. T. Duyun (Belgorod St. Technol. Un., RUS)
Prof. A. Mikhaylov (Donetsk National Tech. Un., UKR)
Prof. G. Bahadirov (IMSS named after Urazbaeva, UZB)
Prof. A. Buchacz (Silesian Technical University, POL)
Prof. N. Danijarov (Karaganda St. Tech. Un., KAZ)
Prof. Y. Novoselov (Sevastopol State Un., RUS)
Prof. A. Kharchenko (Sevastopol State Un., RUS)
Prof. A. Bohonskiy (Sevastopol State Un., RUS)
Prof. N. Pokintelitsa (Sevastopol State Un., RUS)
Prof. R. Giniavichus (Vilnius Gediminas Tech. Un., LTU)
Assoc. Prof. A. Giutuni (Tunis Tech. University, TUN)
Prof. M. Heyfetc (Nat. Academy of Sc. of Belarus, BLR)
Prof. A. Handozhko (Bryansk St. Tech. University, RUS)
Assoc. Prof. G-J. Jasper (Plymouth University, GBR)
Prof. R. Kasper (OVG University Magdeburg, DEU)
Prof. S. Khristoforian (St. Eng.Un. of Armenia, ARM)
Prof. G. Barsukov (Orel State University, RUS)
Prof. O. Pruteanu (Tech. Un. "Gh. Asachi", ROU)
Prof. M. Radovanovic (Tech. Un. of Nis, SRB)
Prof. A. Toca (Technical University Moldova, MDA)
Prof. R. Turmanidze (Tech. Un. of Georgia, GEO)
Assoc. Prof. K. Monkova (Tech. University Kosice, SVK)
Prof. A. Popov (Omsk State Technical University, RUS)
Prof. Z. Sharifov (Azerbaijan Technical. University, AZE)
Prof. A. Suslov (Moscow State Industrial University, RUS)
Prof. V. Gusev (Donetsk National Tech. Un., UKR)
Prof. F. Yakubov (Crimean. Eng. and Ped. Un., RUS)
Prof. A. Malikov (Tula State University, RUS)
Prof. V. Nosenko (Volgograd State Tech. Un. RUS)
Prof. A. Korotkov (Kuzbas State Tech. Un., RUS)
Prof. A. Kozlov (Lipetsk State Tech. Un. RUS)

Organizing Committee
Assoc. Prof. P. Novikov (Sevastopol State Un., RUS)
Assoc. Prof. S. Sazonov (Sevastopol State Un., RUS)
Assoc. Prof. D. Sidorov (Sevastopol State Un., RUS)
Assoc. Prof. V. Bogutskiy (Sevastopol State Un., RUS)
Assoc. Prof. A. Tarakhovskiy (Sevastopol State Un., RUS)
Assoc. Prof. L. Shron (Sevastopol State Un., RUS)
M.Sc. A. Kolesov (Sevastopol State Un., RUS)

Conference Secretary
Assoc. Prof. S. Roshchupkin (Sevastopol State Un., RUS)

СЕВАСТОПОЛЬСКИЙ
ГОСУДАРСТВЕННЫЙ
УНИВЕРСИТЕТ

edp sciences

MATEC Web of Conferences

Statement of Peer review

In submitting conference proceedings to MATEC Web of Conferences, I certify to the Publisher that I adhere to the **Policy on Publishing Integrity** of the journal in order to safeguard good scientific practice in publishing.

1. All articles have been subjected to peer review administered by the proceedings editors.
2. Reviews have been conducted by expert referees, who have been requested to provide unbiased and constructive comments aimed, whenever possible, at improving the work.
3. Proceedings editors have taken all reasonable steps to ensure the quality of the materials they publish and their decision to accept or reject a paper for publication has been based only on the merits of the work and the relevance to the journal.

Title, date and place of the conference

International Conference on Modern Trends in Manufacturing Technologies and Equipment 2017 (ICMTME 2017), September 11-15, 2017, Sevastopol, Russia

Proceedings editor(s):

Prof. Sergey Bratan, Prof. Sergey Gorbatyuk, Prof. Sergey Leonov, Dr. Stanislav Roshchupkin

Date and editor's signature

09.10. 2017

MATEC Web of Conferences **129**, 01001 (2017) DOI: 10.1051/matecconf/201712901001
ICMTMTE 2017

The combined finishing of geometrically-complex parts

Vladislav Smolentsev[1], *Oleg* Kirillov[1,*], and *Vasily* Kotukov[1]

[1]Voronezh State Technical University, 394026, Moskovsky Prospekt, 14, Voronezh, Russian Federation

Abstract. The paper describes the combined hydroabrasive dimensional processing, which involves the application of an electric field of metal parts with complex volumetric profiles, requiring jet control with reference to 5-6 coordinates. One should also consider the reciprocal impact of technological parameters upon formation of parts machined. The mechanism of dimensional material removal has been developed, which is the basis for working out the informational technology of a combined process. The objects of study are adjoining parts of spatially complex surfaces, like gas-turbine engine blades, or axial and radial wheels of turbopump units, as well as the processes of unsoldering in cooled parts of combustion chambers, jet nozzles and complex-shaped part sections with limited tool access into the processing zone, including those made from materials of low machinability. The combined process of hydroabrasive finishing has been developed, which involves the induction of electric current for geometrically-complex surfaces of any geometry, including in areas, inaccessible or hardly accessible for a metal-cutting tool when manufacturing parts from alloys of low machinability.

1 Introduction

The dimensional machining of geometrically-complex parts encounters great difficulties in formation of precision transitional zones. These include the rounded edges of gas-turbine engine blades with high twist angles, where the applicable string method of machining by abrasive belts makes it difficult to copy the spatially variable profile with a tool. The more precise production of such sections by narrow strips results in drastic decrease in performance, accuracy of parts and tool durability.

Apart from that, the applied methods of manufacturing edges and coupling zones of turbopump unit blades are rather labour-intensive and unstable, since these zones are hardly accessible for an instrument, requiring the unproductive manual work to ensure the desired profile precision. This hinders the fulfillment of the necessary tactical, technical and operational requirements for new-generation transport equipment.

Jet methods, including the combined hydroabrasive machining, make it possible to switch from the string method, involving the consecutive formation of separate part zones (mainly, accessible for a tool), to pointwise remote shaping of surfaces of any profile. The presence of an electric field provides the opportunity to fully exclude impregnation and perform the finishing operations on parts of any geometry, including those with areas hidden within a product.

The modern designs of high-tech transport equipment contain a great number of high-precision elements[1, 2, 3], having no direct access by a metal-cutting tool, and the development of new technological methods and devices for their finishing considerably expands the possibilities of manufacturers of perspective machine -building products. It has required the scientific rationale and the assessment of capability for dimensional hydroabrasive machining of geometrically-complex metal parts, using the accumulated knowledge of combined processes with ensuring the technological parameters of finishing for geometrically-complex surfaces, including those having a spatial geometry.

2 The state of studies in the area under consideration

The analysis of available publications [3, 4] on the subject of the article has shown that the issues related to combined hydroabrasive finishing are insufficiently addressed in the domestic and translated technical literature, though the principal possibility of effective implementation of the method is shown by scientists and specialisits. Therefore, it is quite reasonable to set a goal of implementing new technological processes and devices, the emergence of which provides the possibility of finishing geometrically-complex surfaces of metal parts. At the same time, the current scientific groundwork [5-11] is sufficient for developing new methods and devices, ensuring the attainment of the stated goal. It demands the theoretical studies on developing the mathematical apparatus for establishing the relationship of the profiles of any processing complexity with calculated and designated parameters of the combined process.

* Corresponding author: kirillov.olli@yandex.ru

3 The mechanism of dimensional material removal using the combined method

With reference to finishing, the application of an electric field helps to ensure the required accuracy of the part profile and the quality of the surface layer. It controls the boundary of the jet cross-section and the electric modes of combined machining, eliminates the impregnation (particularly, viscous and also heat-resistant materials), and reduces the surface roughness.

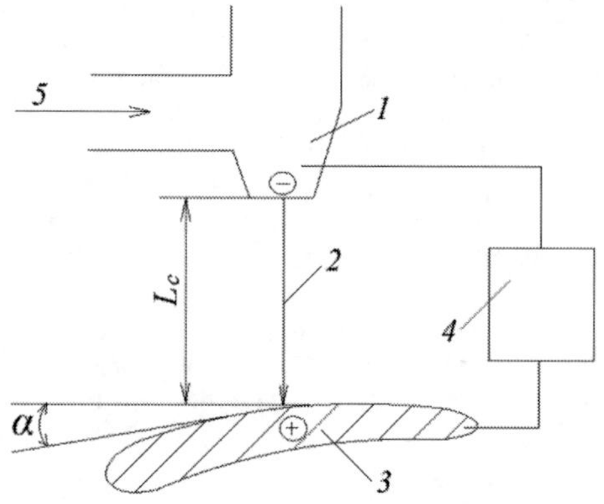

Fig. 1. The scheme of combined hydroabrasive machining: 1 – the nozzle for operating medium supply; 2 – the jet of operating medium; 3 – the machined part (anode); 4 – the source of direct current; 5 – the channel for current-conducting pellets and fluid (if necessary); L_c – the jet length; *a* – the angle of jet attack.

The figure 1 presents the scheme of combined hydroabrasive machining, in which the research is focused at parameters of an electric field and a jet.

The length of the undisturbed part of a jet with clear boundaries, ensuring the required accuracy of coupling between the geometrically-complex surfaces, is calculated using the formula:

$$L_н = K_н d_г \left(1 + \frac{1}{\sqrt{\frac{D_c \gamma_c \sigma_c}{\mu_c}}} \right) \sqrt{\frac{v_c^2 D_c \gamma_c}{\sigma_c}}, \quad (1)$$

where K_n – the numerical coefficient, considering the properties of the operating medium; d_g – the equivalent diameter of abrasive pellets, mm; D_c - the diameter of the injector nozzle, mm; γ_c – the average density of the operating medium in a jet, kg/m^3; σ_c – the surface tension coefficient of the operating medium, N/m; μ_c – the dynamic viscosity coefficient of the operating medium, Pa/s; v_c – the average speed of the operating medium in a jet, m/s.

The condition for dimensional machining is

$$Lc < Ln \quad (2)$$

The distance from the injector nozzle section to the workpiece machining zone is determined by the drawing of the part.

To regulate the dimensional material removal by a jet with hard abrasive, it is necessary to set a half-angle of jet spray, which depends on the jet pressure P_c

$$P_c = \sqrt{(Q_c \gamma_c \vec{v}_1)^2 + (Q_c \gamma_c \vec{v}_2)^2} \cdot \sin \beta, \quad (3)$$

where Q_c – the volume flow of the operating medium, m^3/s; γ_c – the density of the medium, g/cm^3; v_1 – the jet speed before contact with the surface machined; v_2 – the flow speed after leaving the processing zone; β – the angle of an attack, grad.

To ensure the conditions for getting finishing parameters, it is necessary to set the voltage U, applied to the jet, and the current I, required for anode process flow.

$$U = \left\{ \frac{j_{кр} L_c}{1 - \frac{3\varphi_c}{2 + \varphi_c - 0{,}394\varphi_c^{10/3}}} \right\} \chi_0^{-1} \quad (4)$$

where $j_{кр}$ – the critical value of current density, beyond which the anodic removal of metal is possible; L_c – the distance from the nozzle section to the processing point; φ_c – the level of filling the electrolyte with current-conducting particles; χ_0 – the specific electric conductivity of the operating medium before the start of machining (as measured).

The current I, supplied to the processing zone at an equivalent pellet radius is calculated using the formula:

$$I = \left(\frac{U}{2\pi k \rho_1 r} + \frac{U}{\pi \rho_2 r(\pi + 2k) + 2R_n} \right) n \frac{L_c}{L_0}, \quad (5)$$

where U – electrode voltage, V; k – the coefficient, taking account of the ratio between the specific electrolyte resistance and the same parameter for pellet material; p_2 – the specific resistance of the pellet material; $n = f(r,l)$ – the number of pellets along the length L_c; r – the radius of abrasive pellets; R_n – the transfer resistance of current flow from the electrolyte to the pellet; L_c – the distance between the nozzle section and the processing point; L_0 – the distance from the nozzle section to the workpiece along the jet axis during the adjustment.

We assume, that the speed of allowance removal (v_n) under the impact of abrasive pellets depends on parameters of an electric field and the distance:

$$v_n = K_э v_н L_c^{0,3} \quad (6)$$

where $K_э$ is the coefficient, taking account of the time of allowance removal speed growth during stripping the modified layer through anodic workpiece dissolution in the processing zone. For steel parts, it is equal to 1,3-1,5; $v_н$ – the speed of allowance removal in the processed workpiece zone, closest to the nozzle; it is calculated in the same way as the law of anodic dissolution for stationary electrodes.

Taking account of the regularities of allowance removal speed, it is possible to set the time, required to obtain the finished part profile with set parameters of accuracy and surface layer quality.

4 The system of jet control during multi-axis movement of the machining zone

The mechanism of allowance removal, discussed above, makes it possible to design the automated system of jet movement, ensuring the required accuracy during finishing of adjoining surfaces of arbitrary shape. The system relies on the mathematical description of the movement of a jet, in which its boundaries are located on the processed section, and the variable material removal is ensured by the change of jet position (as a rule, by the angle of attack), provided the required quality of workpiece surface layer is ensured. In case of small width of the coupling section (edge), the jet axis must be located on the symmetry line of an adjoining profile with coordinates which can also be set by points X_{ij}, Y_{ij}, Z_{ij}.

To ensure the coupling of spatial surfaces, the law of allowance removal speed change, depending on the angle of attack, is used as a control parameter. The machining of an arbitrary surface by a jet is possible with real-time control of all movements.

When a jet moves along L-L (fig.2), it is required to control its vector with reference to minimal number of control parameters.

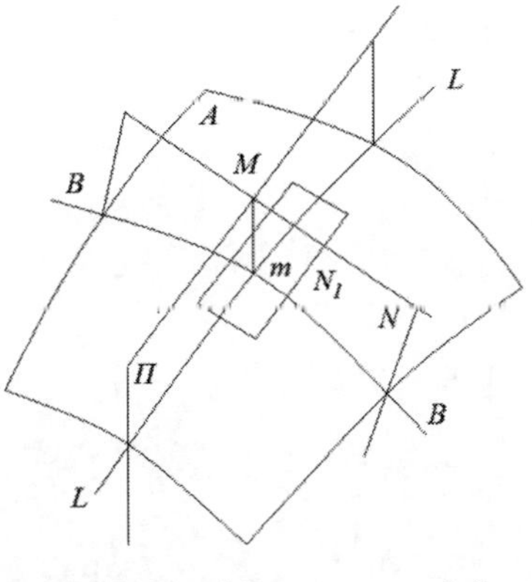

Fig. 2. The scheme of jet movement during geometrically-complex surface machining:
m– the point of jet contact with a processed zone; *M*– the nozzle section for the supply of operating medium; *L-L*, *B-B* – the generatrices of the processed surface; *N*, *P*, *N*ı – the planes in the processing zone

The methodology is developed for calculating the pitch of linear shift and finding the coordinates of a processing point with account of the jet radius R_c in the contact point.

$$\Delta X = \left[1 - \frac{R_c^2 xx \cos\alpha}{\left(1+z_x'^2\right)^{3/2}} + \frac{R_c^2 \alpha_x' z_x' \sin\alpha}{\left(1+z_x'^2\right)^{1/2}} \right] \Delta x; \tag{7}$$

$$\Delta Y = R_c(\theta)\alpha_x' \cos\alpha\Delta x; \tag{8}$$

$$\Delta Z = \left[z_x' + \frac{R_c z' x z_{xx}'' \cos\alpha}{\left(1+z_x'^2\right)^{3/2}} + \frac{R_c \alpha_x' \sin\alpha}{\left(1+z_x'^2\right)^{1/2}} \right] \Delta x \tag{9}$$

where $\acute{\alpha}$ – the slope angle of the processed surface with respect to the jet axis; x', y', z'– the change of jet axis coordinates due to variable speed of contact point movement along the line with a curvilinear generatric (the first derivative); x'', y'', z'' – the change of jet axis coordinates due to variable acceleration during machining of surface sections of various curvature (the second derivative with respect to the axis specified); z''_{xx}– indicates the movement along the axis x, with variable speed and acceleration.

The machining of the surface, specified by coordinates, is more complicated. These include the engine blades with highly twisted edges.

Provided the angle of tangent slope towards the part profile is constant, then, by analogy with mechanical processing of such surfaces, the position of a point on the processed part zone can be represented as:

$$X_{i,j} = \frac{x_{i,j} R_c \cos\alpha_{i,j}}{\alpha_{i+1,j} - \alpha_{i,j}} \left(\sqrt{1+\alpha^2{}_{i+1,j}} - \sqrt{1+\alpha^2{}_{i,j}} \right) \tag{10}$$

where α_{i+1} – the angle on the subsequent line.

$$Y_{i,j} = y_i; \tag{11}$$

$$Z_{i,j} = \alpha_i x_{i,j} + y_j tg\alpha_{i+} + D_{i,j} \tag{12}$$

where $D_{i,j}$ is jet divergence at the nozzle section along the coordinates; $d_{i,j}$ is the divergence in a processing point, shifting along the line with unit vectors.

With account of regularities discovered, the analysis of possible trajectory movements of a jet during combined hydroabrasive finishing has been conducted, using the examples of geometrically-complex surfaces, in which a standard unit is a profile of the airfoil edge and the transfer blade zone of jet and rocket engines, as well as other blade-type machines. The analysis has shown that the reviewed process is most efficient for products with limited tool access, where the jet method has significant advantage.

5 Conclusion

The study confirms the possibility of using available and cheap, or current-conducting abrasives in combined finishing, not previously implemented in hydroabrasive machining [12, 13]. The stable accuracy of 6-8 quality degrees has been achieved, given the roughness of surface layer for steels is less than 1 μm, which ensures high-

DOI: 10.1051/matecconf/201712901001

quality finishing of parts and transfer zones with profiles of any spatial geometry.

References

1. The RF Patent № 2465994 MPK B23H5/00. *The method of eliminating the impregnation effect during hydroabrasive separation of viscous metals* / Patent pending 2011105507: claimed 14.02.2010: published 20.08.2012 The Bulletin № **31**.
2. The RF Patent № 2455132 MPK B23H. *The method of hydroabrasive separation of viscous metals* / Patent pending 2010149206: claimed 30.11.2010: published 10.07.2012. The Bulletin № **19**.
3. The RF Patent №2536128 МПК B23H. *The method of combine separation of metals* / Patent pending 2012116013: claimed 19.04.2012: published. 20.12.2014 . The Bulletin № **35**
4. G.V.Barsukov, A.V.Mikheev, *The determination of hydroabrasive cutting productivity with account of abrasive grain characteristics*, Journal of Engineering. The reference book, № **1**, PP.9-14 (2008)
5. B.A. Burnashov, *The determination of cutting strength during cutting of material by a high-pressure ice-water jet*, The Bulletin of Bryansk State Technical University, № **4**, PP. 17 – 20 (2008)
6. E.V.Goncharov, V.P.Smolentsev, O.N.Kirillov, *The combined separation of metals in the electric field*, The fundamental and applied problems of engineering and technology. № **2-5** (292), PP. 10 – 15 (2012)
7. D.G. Moriss, M. Mazmy, C. Noseda, *Creep resistance in a new alloy based on Fe3Al*, Scr. Met et Mater, № **2** – 31, PP. 173 – 178 (1994)
8. V.K. Sikka, S. Viswanathan, C.G. McKaamey, *Development and commercialization status of Fe3Al – based intermetallic alloys,* Struct. Intermetallics: Champion, PP. 26 – 30 (1993)
9. V.P.Smolentsev, E.V.Goncharov, *The calculation of technological modes of hydroabrasive separation of materials with application of an electrical field,* The Bulletin of Voronezh State Technical University, Volume 8, № **4**, PP. 130-133 (2012)
10. Y.S.Stepanov, G.V.Barsukov, E.G.Alyushin, *The modern technologies of hydro- and hydroabrasive workpiece machining*, Science-based technologies in machine-construction, № **6**, PP. 11-17 (2012)
11. Y.S.Stepanov, A.P.Cherepenko, M.A.Burnashov, *The technology of cutting the technical textile materials and their deriviative compositions, applied in transport machine-construction*, The Reference. The Journal of Engineering, № **1**, PP. 3 – 6 (1999)
12. Y.S.Stepanov, A.N.Kachanov, MA.Burnashov, *The methodology of calculating the technological parameters of cutting non-metal materials by ice-water tools*, Polzunovsky Vestnik, № **1-2**, PP. 276-277 (2009)
13. A.G. Suslov, B.M.Bazrov, V.F.Bezyazychnyi et al, *Science-based technologies in machine-construction,* (Moscow: Mashinostroenie, 2012)

Adhesive wear mechanism under combined electric diamond grinding

Vyacheslav Popov[1,*], *Pavel* Arkhipov[1], and *Daniel* Rychkov[1]

[1]Bratsk State University, 40 Makarenko, Bratsk, 665709, Russian Federation

Abstract. The article provides a scientific substantiation of loading of metal-bond diamond grinding wheels and describes the mechanism of contact interaction (interlocking) of wheels with tool steel as well as its general properties having an influence on combined electric diamond grinding efficiency. The study concluded that a loaded layer can be formed in a few stages different by nature. It is known, that one of the causes of grinding degradation is a continuous loading of active grits (abrasive grinding tool) by workpiece chips. It all affects the diamond grinding wheels efficiency and grinding ability with a result in increase of tool pressure, contact temperature and wheels specific removal rate. Science has partially identified some various methods to minimize grinding wheel loading, however, as to loading of metal-bond diamond grinding wheels the search is still in progress. Therefore, research people have to state, that in spite of the fact that the wheels made of cubic boron nitride are of little use as applied to ceramic, ultrahard, hard-alloyed hard-to-machine and nano-materials of the time, but manufactures have to apply cubic boron nitride wheels wherein diamond ones preferable.

1 Introduction

Loading is an adhesion of grinding dust (chips out of workpieces, tools, lubricoolant, etc.) to the active grits during the abrasive machining [1–7]. A dressing process is needed to remove worn grains, to disclose new and sharp grains and to restore the geometry of cutting face of the wheel [8–18]. Currently, they have ascertained the nature of chip adhesion as adhesive and diffusive one. In order to provide its scientific justification it is needed to widen and deepen the scope of investigation: from macro and through micro to an atomic level [19]. We developed a theory of wheel loading [20, 21]. It has been placed beyond a doubt that a loading layer has a complex, mechanic and physicochemical nature and consists of workpiece components, wheel bond, secondary compounds formed as a result of adhesive and diffusion phenomenon, chemical reaction including recrystallization. Mathematical modeling of contact interaction was carried out on a micro level. The necessary prerequisites were created for modelling (simulation) objects on an atomic level [22].

2 Experimental investigation procedure

Investigation procedure of diamond grinding wheels (DGW) surface after combined electric diamond grinding (CEDG) was aimed to identify a radial component of cutting pressure, P_y (N/cm^2); cutting power efficiency, N_{ef} (W). Artificial thermocouple was used to take contact temperature in the grinding area. A scanning electron microscopy, SEM (Carl Zeiss EVO®50 XVP) was used to get images of metalic bond DGW surfaces. Microrelief parameters were identifired by an optical interferometry (Zygo NewView™7300). Chemical composition of DGW surfaces was identifired by an Energy-Dispersive X-ray Analysis (Oxford Instruments) and a phase structure of DGW surfaces was recieved with the help of X-ray structural analysis (ARL X'TRA (Thermo Fisher Scientific)) [20].

3 Investigation of relationship (dependence) of radial cutting pressure and cutting power efficiency

Investigation of dependence of a radial cutting pressure (Fig. 1) and a cutting power efficiency (Fig. 2) from CEDG operating modes reviels that machining force (cutting pressure) at diamond grinding as applied to high-speed steel (HSS) is 3–5 times higher than when grinding hard alloys due to an intence diamond-steel interaction (interworking). Therfore, by the nature of tool-load change, one can inderectly judge some phisical phenomena occured in the cutting area. Under certain electrical (i_{ed}, i_{et}) and mechanical (S, V, t) machining conditions, one can achieve the following goals: set up a self-sharpening mode on DGW; block its loading; stabilize the cutting power at its minimum power value.

4 Investigation of temperature in contact zone

* Corresponding author: berkutoff@rambler.ru

DOI: 10.1051/matecconf/201712901002

Investigation of the contact temperature in the cutting zone (Fig. 3) shows that a thermal conduction of HSS is poor and get hot very fast during grinding. High temperature appears to be an accelerator in the process of plastic deformation, formation of a defect layer and some other contact actions [23–25], as well as it intencifies crack formation along the tool nose (cutting edge), with its further failure.

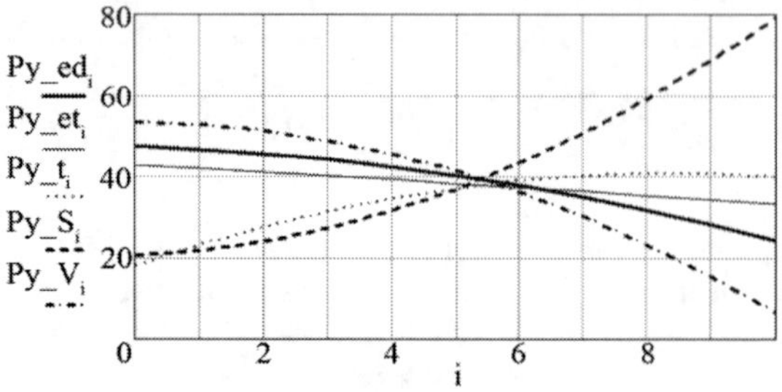

Fig. 1. Graphical relationship (dependence) of radial cutting pressure P_y on CEDG operating modes

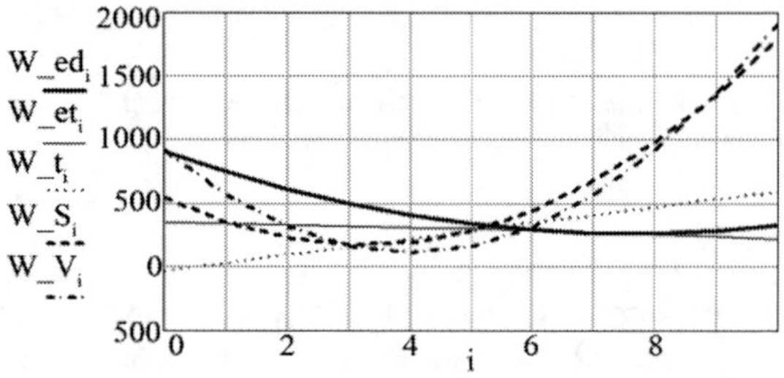

Fig. 2. Graphical relationship (dependence) of cutting power efficiency W on CEDG operating modes

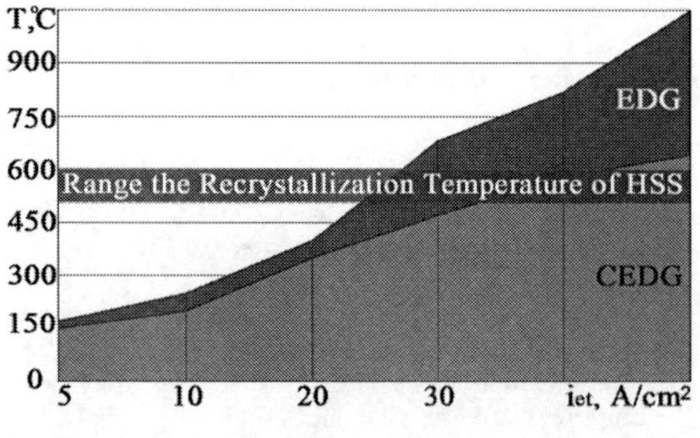

Fig. 3. Temperature in the Contact Zone

Therefore, it is practically impossible to apply a common way of machinning to hardened HSS, taking into account that the temperature in the contact zone can be about 1200°C. At CEDG and under the recommended conditions the temperature appears 150–180°C only that does not bring to recrystallization of HSS components (recrystallization temperature for components: Fe 450°C, Cr 850–1075°C, W 1200°C, V 700–800°C, Mo 850–1220°C). This is a good showing [26–29], so far as high hardness of the tool made of HSS remains its quality upon heating up to 500–600°C. On the whole, a tool being sharpened under CEDG will remain cold when the heat load falls only on the thin upper layer which is being removed while machining.

5 Energy-Dispersive X-ray analysis (Oxford Instruments)

Energy-Dispersive X-ray Analysis of chemical composition of the loaded layer (Fig. 4) revealed that a diffraction pattern has planty of weak reflections that is typical of inorganic matters containing essential amount of different phases. In addition, poor quality of diffraction patterns at the statistical accuracy of 5 sec. indicates a high content of crystal defects in the present patterns [30]. Spectral analysis of surface X-Ray patterns was carried out and decoded. (Fig. 5, b). Components of Fe, Al, Cu, Zn, Mg, O, C in the patterns were found according to the results of the chemical analysis.

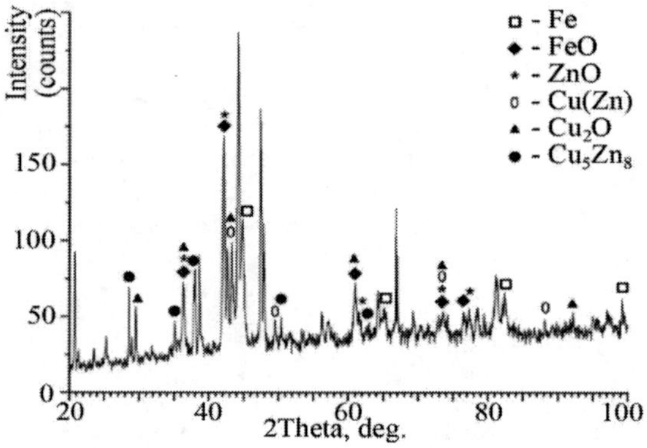

Fig. 4. Analysis of the DGW surface diffraction pattern after CEDG HSS

It should be observed that during the formation of a loaded layer, as well as a defect one when grinding without lubricoolant, the heat factor is of primary importance. Exactly heat appears to be the key initiator of physical and chemical phenomena and reactions. In each case when machining and grinding different materials with different abrasives, there can arise different adhesion mechanisms which are more relevant to a specific thermodynamic, phase or kinetic process (Fig. 5).

Mechanical adhesion (Fig. 5, a).

At the very beginning of grinding, when grits just begin scratching agaist the material, then a mechanical adhesion interlocking of chips between grits and metal-bond DGW takes place. Thence arise surface residual compressive stresses. It makes possible to minimize loading and save high cutting abilities of DGW by lubricoolant feed speed up or by reducing the depth of cut, besides the correct selection of grinding tool specification.

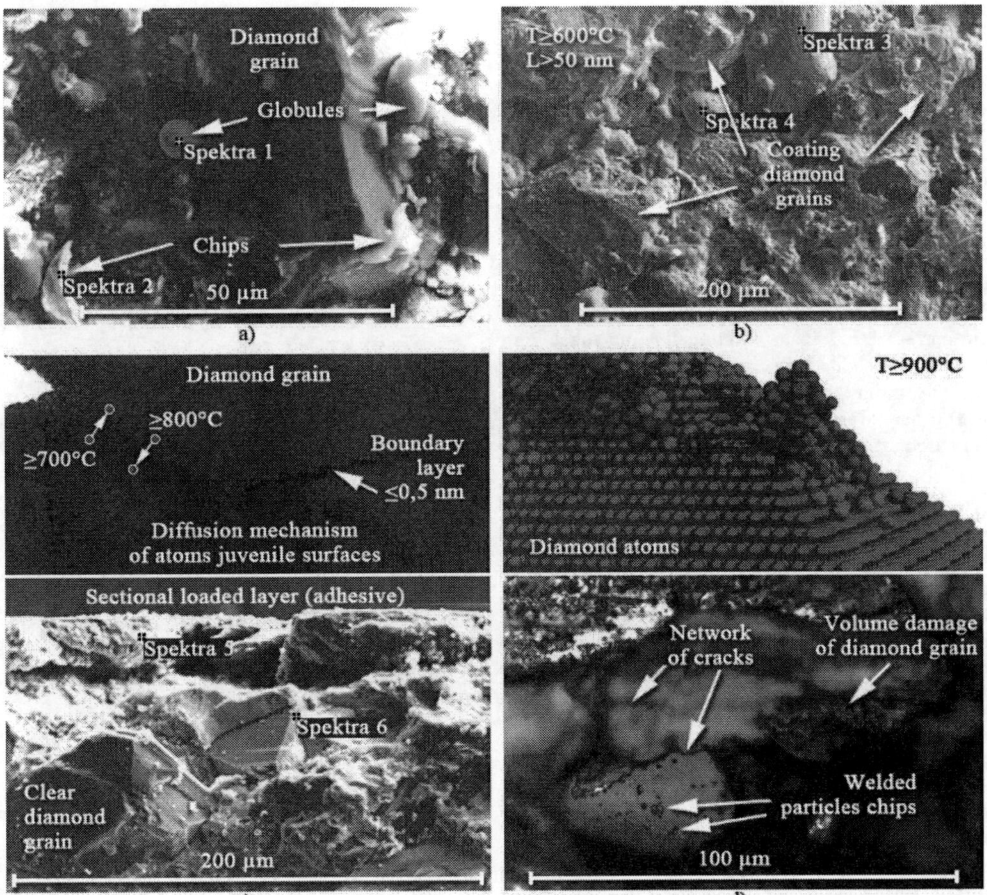

Fig. 5. Mechanism of contact interaction under CEDG: a) mechanical adhesion; b) molecular adhesion; c) chemical adhesion; d) electrical adhesion

Grinding wear of grits, typical to mechanical adhesion, lays in localized microdistruction of cutting grits and thier chipping due to tearing of a sticking stuff off periodically. As a rule, the rate of loading is higher when cutting plastic materials and it tends to tearing right up to the bond level when cutting some breakable materials. This affects the grinding material in a form of negative changes in the surface layer due to ecrease of cutting force. The risk of thermal damages to the surface increases significantly.

Adsoptive or molecular adhesion (Fig. 5, b).

Chip gradual accumulation results in its packing between grits and in DGW bonds that leads to a pressure boost. This makes an adsoption to be diffused all over the active surface, which in turn goes with a heat release. Surface residual compressive stresses are generated creating favourable conditions for a phisical adsoption inside cracks of plastically deformed material (chips). Worn sports on the surface of grits are progressing; a constant friction involves grits in an additional air oxidation that results in chips of greater size.

Molecular adhesion process can be compensated by use of emulsions and oils, creation of inert atmosphere or adding a special active admixture to a lubricoolant. The literature , in this stage of loading, recommends to replace a diamond abrasive (grit) with an el'bor borazon, silicon carbide or more porous abrasive tool.

Simultaneous increase in cutting tool force (Fig. 1), cutting power (Fig. 2) and temperature (Fig. 3) in the contact zone results in formation of burnings on the machining surface. Carbide grits start to grow making the machining surface more scored and rough.

Chemical adhesion (Fig. 5, c).

Further increase in pressure and temperature due to a significant contact area extension is the reason of the formation of an interfacial (interphase) contact (Fig. 4). The phase transformation brings to the compressive or streching surface stresses, subject to a gain or reduction of the transfered material phase volume. By this time both loaded and machining surfaces are juvenile. Thermodynamic preconditions arise for a Van-der-Vaals interaction, i.e. chemical bonding, that is more appropriate for the materials of the same chemical nature. The chemical adhesion compound, since its formation, can save its integrity and high mechanical properties in spite of external action or corrosive medium. From now on, the ready-built bond is the most difficult to destroy in usual way. The only way to remove a loaded layer is a mechanical tear of chemical bonds by various technique of DGW trimming or a combined machining approach. As a result of the chemical adhesion a DGW wear encreases significantly and whole grits have been torn out yet. Some irreversible structural changes occur in the machining material which are considered a waste.

Electrical adhesion (Fig. 5, d).

If the above mentioned conditions should be observed during the machining process thence arises simultanious plastical deformation both in the machining surface and the loading layer. This deformation includes all three types: mechanical,themal and phase one. But if grinding is carried out by a combined method, with electric current bringing in additionally into machining zone, then wheel loading might be as well due to sticking of grinding dust and grits. Temperature increase might be observed in the DGW bonds that results in grits cracking along ruptures under the heat impact.

6 Results and discussion

In terms of the mentioned above as well as of results of complex research of CEDG tool steel, it seams possible to create such conditions which be able to prevent DGW loading and widen process possibility of their application in elctrical and chemical treatment. It will allow to reach a new and competitive production level [31].

7 Conclusions

As a result of analitical and experimental research, an urgent scientific issue has been addressed. This up to date goal is of great economic worth that makes it possible to discover qualitative patterns of relationship of matal-bond DGW loading mechanism subject to the specified thermodynamic phase and kinetic properties of grinding.

Formation of a loaded layer on the surfaces of DGW under impact of high temperature, pressure and interatomic interaction has been well-grounded and corroborated by methods of spectral, X-Ray energo-dispersive analysis and raster electronic microscopy. The chemical composition of a loaded layer has been identified with the help of analyses mentioned above. The nature of DGW loading has been identified as adhesive-diffusive which takes place at grinding tool iron-carbon steel at the expense of carbon diffusion. As a result there is a breakable overstrengthened structure with a heightened microhardnees. All this affect the upper layer and decrease its mechanical properties.

The reported study was funded by RFBR according to the research project No. 16-38-00123 мол_а.

References

1. W.B. Rowe, 2nd ed. Oxford: Elsevier, (2014)
2. J. Badger, S. Murphy, G.E. O'Donnell, AMR, **126–128,** 597–602, (2010)
3. S. Malkin, Guo Changsheng, New York, Industrial Press, (2007)
4. M.J. Jackson, J.P. Davim, Springer US, (2011)
5. I.D. Marinescu, M.P. Hitchiner, E. Uhlmann, W.B. Rowe, I. Inasaki, CRC Press, (2016)
6. R.I. King, R.S. Hahn, Springer Science & Business Media, (2012)
7. H.G. Sachsel, C.A.E. Precision, Xlibris Corporation, (2010)
8. M. Schöpf, I. Beltrami, M. Boccadoro, D. Kramer, CIRP AMT, **50 (1),** 125–128, (2001)
9. E. Brinksmeier, Y. Mutlugunes, F. Klocke, J.C. Aurich, P. Shore, H. Ohmori, CIRP AMT, **59 (2),** 652–671, (2010)
10. A.G. Mamalis, M. Horvath, A.I. Grabchenko, J. MPT, **97 (1-3)**, 120–125, (2000)
11. J.B.J.W. Hegeman, J.Th.M. De Hosson, G. De With, Wear, **248 (1-2),** 187–196, (2001)
12. C.C. Chang, A.Z. Szeri, Wear, **216 (1),** 77–86, (1998)
13. B. Zhang, X.L. Zheng, H. Tokura, M. Yoshikawa, J. MPT, **132 (1-3)**, 353–364, (2003)
14. H. Hamdi, H. Zahouani, J.M. Bergheau, J. MPT, **147 (3),** 277–285, (2004)
15. P. Durgumahanti, V. Singh, R.P. Venkateswara, Inter. J. MTM, **50 (3),** 231–240, (2010)
16. Z. Zhang, F. Huo, Y. Wu, H. Huang, Inter. J. MTM, **51 (1),** 18–24, (2011)
17. X. Chen, W.B. Rowe, R. Cai, Inter. J. MTM, **42 (5)** 585–593, (2002)
18. J. Webster, M. Tricard, CIRP AMT, **53 (2),** 597–617, (2004)
19. V.Y. Popov, A.S. Yanyushkin, Eas. Eur. Scien. J. **2,** 301–310, (2014)
20. V.Yu. Popov, A.S. Yanyushkin, Y.I. Zamashchikov, AMM, **799–800,** 291–298, (2015)
21. D.V. Lobanov, P.V. Arkhipov, A.S. Yanyushkin, V.Yu. Skeeba, IOP Conf. Ser.: MSE, **142,** 012081, (2016)
22. V.Y. Popov, A.N. Khlystov, A.V. Bondin, CRM, **8 (1),** 161–172, (2016)
23. V. Ivancivsky, K. Parts, V. Popov, AMM, **788,** 129–135, (2015)
24. V.Yu. Skeeba, V.V. Ivancivsky, N.V. Martyushev, D.V. Lobanov, N.V. Vakhrushev, A.K. Zhigulev, KEM, **712,** 105–111, (2016)
25. V.Yu. Skeeba, V.V. Ivancivsky, A.V. Kutyshkin, K.A. Parts, IOP Conf. Ser.: MSE, **126,** 012016, (2016)
26. Hossam Halfa, J. MMCE, **1,** 257–270, (2013)
27. V.Yu. Skeeba, V.V. Ivancivsky, D.V. Lobanov, A.K. Zhigulev and P.Yu. Skeeba, IOP Conf. Ser.: MSE, **125,** 012031, (2016)
28. L.A. Bokhoeva, V.E. Rogov, A.S. Chermoshentseva, D.V. Lobanov, IOP Conf. Ser.: MSE, **142,** 012077, (2016)

29. V.V. Ivancivsky, V.Yu. Skeeba, I.A. Bataev, D.V. Lobanov, N.V. Martyushev, O.V. Sakha, I.V. Khlebova, IOP Conf. Ser.: MSE, **156,** 012025, (2016)
30. J.H. Chen, D. Bernaerts, J.W. Seo, G. van Tendeloo, H. Kagi, Mag. Let., **77 (3),** 135–140, (1998)
31. A.S. Yanyushkin, D.V. Lobanov, D.A. Rychkov, AMM. **770**, 739–743, (2015)

 DOI: 10.1051/matecconf/201712901003

Defects in diamonds as the basic adhesion grinding

Vyacheslav Popov[1,*], *Daniel* Rychkov[1], and *Pavel* Arkhipov[1]

[1]Bratsk State University, 665709, 40 Makarenko, Bratsk, Russian Federation

Abstract. Defects arising in diamond crystals during their synthesis, give rise to the formation of more complex defects that arise in the operational process of diamonds themselves. In case of external action these defects have immediate strong effect on operating characteristics of devices, products or tools containing diamonds. When external action these defects and have immediate strong effect on production characteristics of devices, products or instruments containing diamonds. This is particularly manifested in the grinding of tough, ductile materials based on aluminum, magnesium or titanium. Developed on the basis of modern digital technology and molecular dynamics visualization method allows to show it at the atomic level. The problems of practical application of the combined electric diamond grinding are examined. Experiments were conducted on the grinding tool materials with metal-bonded diamond grinding wheels.

1 Introduction

The basic parameter which identifies quantitative, qualitative and price characteristics of engineering projects of the time is a possibility to carry out a virtual experiment. This is less expensive but more effective and safe against a physical one. One of the key criteria of rapid prototypical industry (RP) is a possibility to visualize a model in material. Modeling by the molecular dynamics method (MD) provides an opportunity to visualize an atomic structure of the material in the model. RP models make possible to carry out some special experiments, which are not available in the real-world, while MD models provide for experiments at the atomic structure of the material being created.

Recent years have revealed a stable interest tendency of manufactures to diamond grinding wheels [1]. High-tech diamond tools, in spite of their crashproof properties, given to lose their free-cutting features during grinding because of diamond crystals wear out or their adhesion loading.

Study of the process in the contact zone, description of reasons of diamond grinding wheels (DGW) efficiency loss using a computer technique for modeling of the structure and material properties at cutting-make the study to be fundamental.

The study concluded that a loaded layer could be formed in a few stages different by nature and which were corroborated with an up-to-date scientific investigation of adhesion [2–7]. In each case in machining and grinding different materials with different abrasives, different adhesion mechanisms, which are more relevant to a specific thermodynamic, phase or kinetic process can dominate over there [8–18]. In order to expand the mechanism of adhesion, it is necessary to carry out comprehensive analyses in terms of material science, mechanical engineering technology, chemistry and solid-state physics by atomic visualization of juvenile surfaces [19, 20]. To this effect it is necessary to take into account the specific character of artificial diamond synthesis, which was practically neglected by previous researchers.

However, it is very important to know what internal and surface defects have diamond crystals, which are used for fabrication of modern grinding tools.

At present metal-synthetic diamond crystals are used for abrasive tools. Such crystals can be formed under high pressure and temperature in the metal melt with dissolved carbon therein. This is attributed to the fact that a growth medium on the basis of metal catalyst, e.g. like nickel, chromium or cobalt is well able to dissolve graphite. All side phases in the form of inclusions can penetrate into crystals in the reaction chamber [17–21].

Thus, nearly each diamond has nanodimensional structural imperfections and defects after synthesis. All researchers, whose investigations have been dedicated to the issues of synthetic diamond, seem to be unanimous in the idea, that defects occurrence in diamonds has a significant influence on their physical properties (mechanical, magnetic, electrical, etc.), particularly on their hardnees, and specifically on their heat resistance after synthesis [22].

So far as diamond properties are defined by the defects, the operating characteristics of diamond tool, containing diamonds, depend directly on them. The tool experiences some kind of external influence (pressure, temperature, oxidation, electrical current, etc.) during its grinding and hence, the behaviour of diamond in the tool depends on this influence as well. There is evidence that modelling of cutting operation, considering main defects which are inherent in synthetic diamonds will be more accurate both from a fundamental point of view and a computer modelling of solid-state physics objects.

* Corresponding author: berkutoff@rambler.ru

2 Influence of point defects (microflaw)

Nitrogen is the main inclusion of the diamond-type crystal lattice which initiates formation of lot of point defects [19, 22]. Thus, single vacancy defects (Fig. 1, a), single substituent and interstitial atoms (Fig. 1, b, c), could be passed into multi-vacancies clusters.

So far as substituents belong to a different chemical element and of a different shape, crystal lattice can be widened or compressed under the influence of such transitions. Stored in the lattice energy, being released during grinding, might bring to formation of cracks inside diamond and dicrease its strength properties.

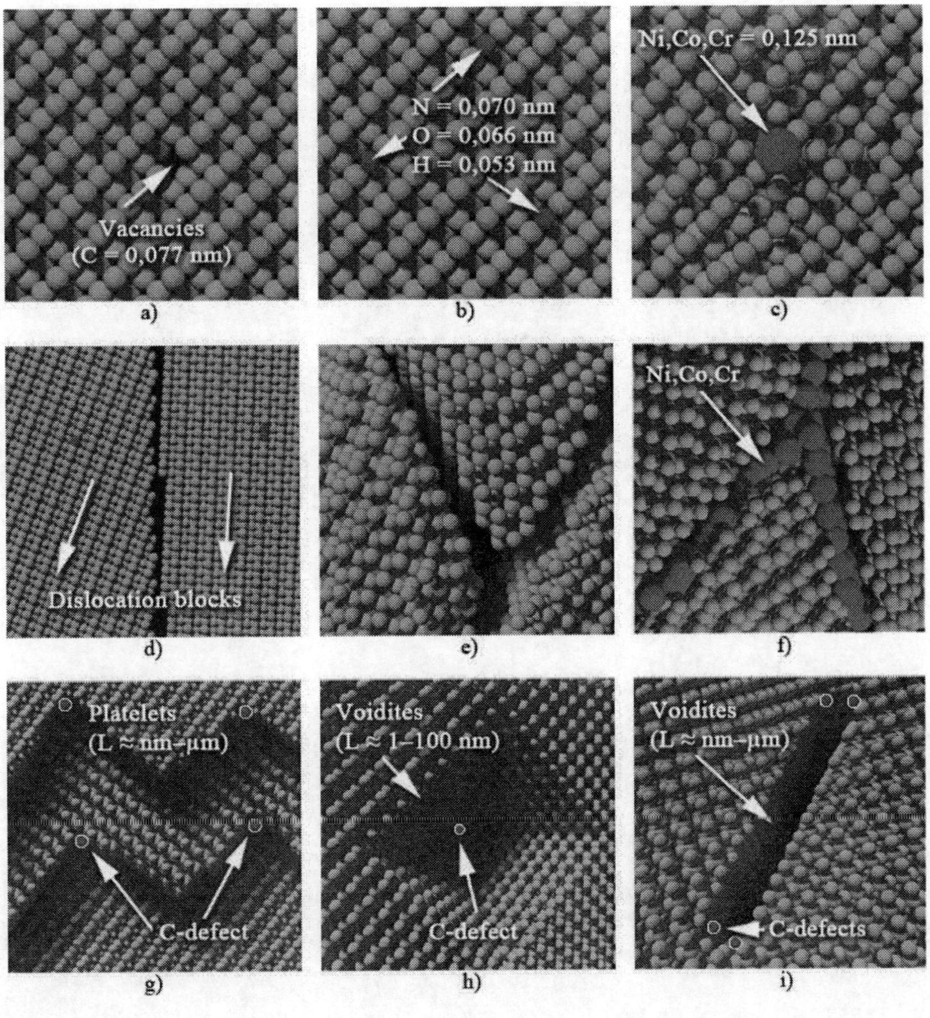

Fig. 1. Modelling of synthetic diamond defects, with respect to the crystal morphology. Point defects (microflows): a) vacancies; b) substituents (C-defect); c) interstitial atoms; d) dislocations. Three-dimensional defects (macrodefects): e) interblock cracks; f) metal contamination; j) platelets defects; h) octahendra voidites; i) planar voidites.

3 Influence of three-dimensional defects (macrodefects)

There is evidence that adhesion of grinded off material to the grinding wheel bond, or so to say loading itself, mostly occurs due to some defects of crystal surfaces. As a rule, they are represented in the form of crystal lattice blocks (Fig. 1, d). Cracks (Fig. 1, e) and cracks with metal contamination (Fig. 1, f) can be located both on the crystal surface and inside diamond, having the extent from some nanometers to a few micrometers. The defects under the total influence of high temperature and increased cutting force interact with grinding surface on an atomic level. Example, diffusion of carbon out of the machining surface to the diamond crystal can be observed at temperature above 700°C, but diffusion of carbon out of the diamond crystal to the machining surface can be observed at temperature above 800°C.

The greater amont of metal contamination the higher oxidation rate of diamond crystals. Under electrical diamond grinding they can interact with electrolyte, and thus dicrease cutting properties of wheel grit.

DOI: 10.1051/matecconf/201712901003

So far as point defects can originate three-dimensional ones, there are both platelets (Fig. 1, g) and planar (Fig. 1, h, i) defects in the crystal. Three-dimensional defects as well as point ones can be located on the crystal surface and inside diamond grains, therefore diamond crystal fracture directly depends upon an external cutting force and heat process.

Platelets defects (Fig. 1, g) are formed out of carbon interstitial atoms with nitrogen impurity [19, 22].

Nano-sized voidites can be of length from one nanometer (Fig. 1, h) to a few micrometers (Fig. 1, i) and located between grains of a similar orientation at their boundaries or along dislocation lines [19, 22].

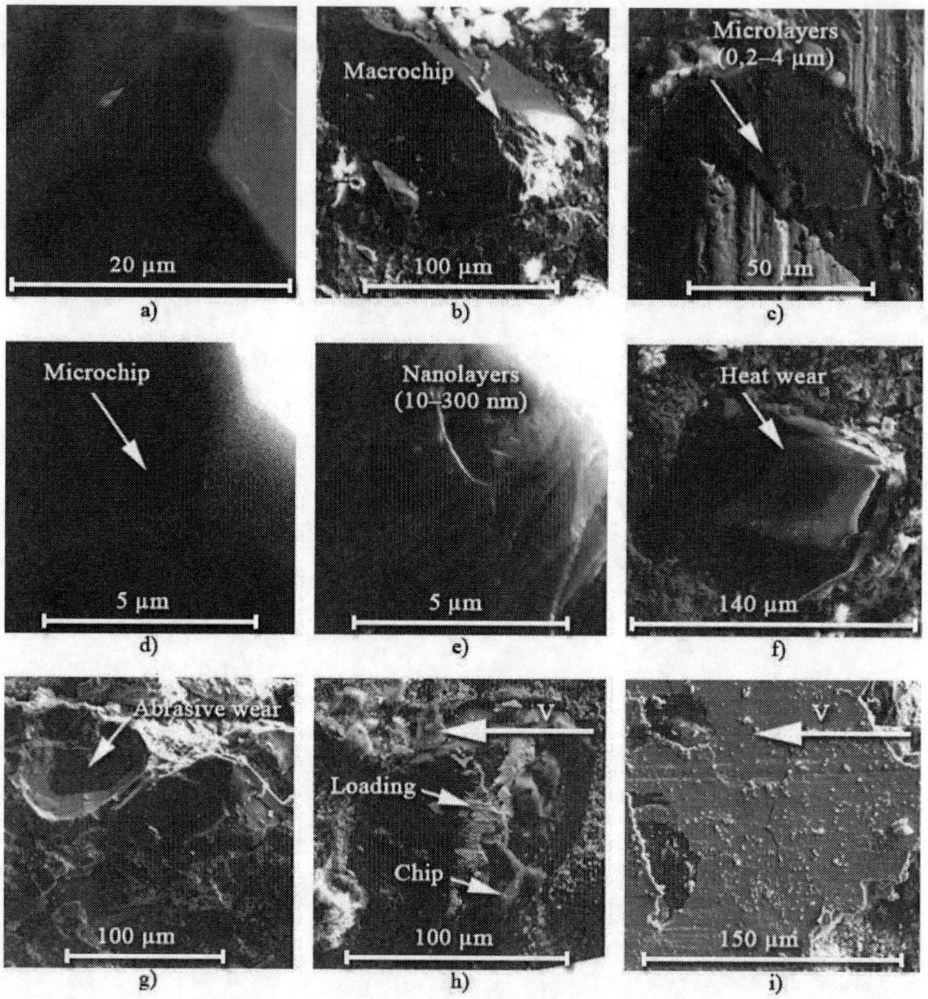

Fig. 2. Typical diamond wear (crystal face failure) under grinding: a) cutting diamond face before grinding; b) three-demencional failure (volume fracture) of the diamond top in form of macrochip; c) cutting face wear along microlayer growth; d) three-demencional failure (volume fracture) of the diamond top in form of microchip; e) cutting face wear along nanolayer growth; f) fretting of cutting face resulting in formation of a smooth wear surface; g) diamong breakoff out of wheel bond; h) loading of the cutting face; i) totally loaded diamonds.

4 Effect of point and three-dimensional defects on diamond wear

In grinding the most crystals (grit) of the grinding wheel will save their shape and integrity. (Fig. 2, a). However, it would suggest some point defects and three-demensional ones, presenting in crystals, caused their definite wear out. The three-demensional rapture of the diamond's top occurs in form of macrochip (Fig. 2, b) or microchip (Fig. 2, d) along tiny (micro) cracks on the surface of crystal. It should be noted that a diamond crystal consists of nanolayers, so to say growth stages, which have the size of from 10 to 300 nm (Fig. 2, e). These nanolayers are subsequently formed into microlayers having the dimension form 0,2 to 4 µm (Fig. 2, c). It is remarkable that such a devision of wear to micro and macro levels might be observed at rupture of diamond cutting surfaces, which are in direct contact with a machining surface.

It is reckoned [1, 3–12, 17, 18, 23], that depending on the machining mode and physico-chemical properties of the work surface, diamond wear process pattern can be devided into: abrasive (Fig. 2, g), adhesion (Fig. 2, h), diffusion and heat wears (Fig. 2, f). But the comprehensive study of the wheel surface has revealed [19, 21, 24] that, with no dependence on the machining condition, crystals of all types of wear can be observed on the level above the wheel bond (Fig. 2, i). Obviously, it is because defects give rise to nonuniform stress in a crystal. Deformation occurs during grinding and initiates converging of point defects (C-defects), but grinded-off chips, being ingressed into cracks, expand them and provoke one or another type of wear.

Hence, it is clear that the most diamond defects come from availability of only one foreign atom (N, H, O, Ni, Cr, Co, etc.), or lack of one carbon atom in a crystal lattice. Surely the defects can develope further under high temperature and pressure in the cutting zone [24–28]. It seems plausible that synthesis of defect-free diamonds might be one of the promising ways to prevent loading of metal-bonded grinding wheels [21, 29–31].

However, while further work is under way, by now there is a possibility to reduce a loading impact significantly by the following steps:

- create a preventive film on the wheel surface;
- replace standard electrolyte with alternative lubricoolant;
- develope and implement new advanced designs of grinding wheels;
- active wheel dressing during grinding in accordance with the technology of electro-diamond machining;
- combination of above-mentioned methods.

5 Conclusions

The presence of different defects in diamonds has a strong influence on their physical properties and abrasive tool performance, that in turn, has a bad effect on the workpiece quality.

After synthesis cracks occuring on a crystal face or inside crystal as well as cracks filled up with metal inclusion give rise to adhesion-diffusion loading of the abrasion layer with workpiece chips.

Up-to-date additive technology provide designers with new technological solutions and oppotunities at modelling of materials by molecular dynamics method.

The reported study was funded by RFBR according to the research project No. 16-38-00123 мол_а.

References

1. W.B. Rowe, 2nd ed. Oxford: Elsevier, (2014)
2. J. Badger, S. Murphy, G.E. O'Donnell, AMR, **126–128** (2010) 597–602
3. S. Malkin, Guo Changsheng, New York, Industrial Press, (2007)
4. M.J. Jackson, J.P. Davim, Springer US, (2011)
5. I.D. Marinescu, M.P. Hitchiner, E. Uhlmann, W.B. Rowe, I. Inasaki, CRC Press, (2016)
6. R.I. King, R.S. Hahn, Springer Science & Business Media, (2012)
7. H.G. Sachsel, C.A.E. Precision, Xlibris Corporation, (2010)
8. M. Schöpf, I. Beltrami, M. Boccadoro, D. Kramer, CIRP AMT, **50 (1),** 125–128 (2001)
9. E. Brinksmeier, Y. Mutlugunes, F. Klocke, J.C. Aurich, P. Shore, H. Ohmori, CIRP AMT, **59 (2),** 652–671 (2010)
10. A.G. Mamalis, M. Horvath, A.I. Grabchenko, J. MPT, **97 (1-3),** 120–125 (2000)
11. J.B.J.W. Hegeman, J.Th.M. De Hosson, G. De With, Wear, **248 (1-2),** 187–196 (2001)
12. C.C. Chang, A.Z. Szeri, Wear, **216 (1)**, 77–86 (1998)
13. B. Zhang, X.L. Zheng, H. Tokura, M. Yoshikawa, J. MPT, **132 (1-3)**, 353–364 (2003)
14. H. Hamdi, H. Zahouani, J.M. Bergheau, J. MPT, **147 (3),** 277–285(2004)
15. P. Durgumahanti, V. Singh, R.P. Venkateswara, Inter. J. MTM, **50 (3),** 231–240 (2010)
16. Z. Zhang, F. Huo, Y. Wu, H. Huang, Inter. J. MTM, **51 (1),** 18–24 (2011)
17. X. Chen, W.B. Rowe, R. Cai, Inter. J. MTM, **42 (5),** 585–593 (2002)
18. J. Webster, M. Tricard, CIRP AMT, **53 (2),** 597–617 (2004)
19. V.Y. Popov, A.S. Yanyushkin, A.N. Khlystov, Metal Working and Material Science, **4,** 16–23 (2016)
20. V.Y. Popov, A.N. Khlystov, A.V. Bondin, CRM, **8 (1)** 161–172, (2016)
21. V.Yu. Popov, A.S. Yanyushkin, Y.I. Zamashchikov, AMM, **799–800**, 291–298 (2015)
22. J.H. Chen, D. Bernaerts, J.W. Seo, G. van Tendeloo, H. Kagi, Mag. Let., **77 (3),** 135–140 (1998)
23. L.A. Bokhoeva, V.E. Rogov, A.S. Chermoshentseva, D.V. Lobanov, IOP Conf. Ser.: MSE, **142**, 012077 (2016)
24. V. Ivancivsky, K. Parts, V. Popov, AMM, **788,** 129–135 (2015)
25. V.Yu. Skeeba, V.V. Ivancivsky, N.V. Martyushev, D.V. Lobanov, N.V. Vakhrushev, A.K. Zhigulev, KEM, **712,** 105–111 (2016)
26. V.Yu. Skeeba, V.V. Ivancivsky, A.V. Kutyshkin, K.A. Parts, IOP Conf. Ser.: MSE, **126,** 012016 (2016)
27. Hossam Halfa, J. MMCE, **1,** 257–270 (2013)
28. D.V. Lobanov, P.V. Arkhipov, A.S. Yanyushkin, V.Yu. Skeeba, IOP Conf. Ser.: MSE, **142,** 012081(2016)
29. V.Yu. Skeeba, V.V. Ivancivsky, D.V. Lobanov, A.K. Zhigulev and P.Yu. Skeeba, IOP Conf. Ser.: MSE, **125**, 012031(2016)

DOI: 10.1051/matecconf/201712901003

30. V.V. Ivancivsky, V.Yu. Skeeba, I.A. Bataev, D.V. Lobanov, N.V. Martyushev, O.V. Sakha, I.V. Khlebova, IOP Conf. Ser.: MSE, **156,** 012025 (2016)
31. A.S. Yanyushkin, D.A. Rychkov, D.V. Lobanov, Procedia Engineering **150,** 942–947 (2016)

DOI: 10.1051/matecconf/201712901004

Providing dismountable rivet bonded joints through the use of hot-melt adhesives

Natalia Baurova[1,*], *Alexander* Anoprienko[1], and *Yulia* Romanova[1]

[1]Moscow State Automobile and Road Technical University (MADI), 125319, Moscow, Leningradskii pr.64, Russia;

Abstract. The article presents the comparative analysis of adhesives used for obtaining rivet bonded joints. The dismantling characteristics of adhesive rivet bonded joints of different types are considered. It is shown that the application of hot-melt adhesives proved to be a far more advanced technique used for the materials being joined.

Today aluminum alloys and composites (carbon-filled plastics) are widely used in mechanical engineering. Joining parts from aluminum alloys or carbon-filled plastics to steel parts of the bearing frame is extremely difficult or impossible to perform with the use of traditional welding. For this reason the riveting technique has been increasingly used in mechanical engineering in recent years.

In the process of riveting, the protruding end of the shank is upset by hammer blows to form the closing head of a required shape and size [1]. This technology is more than 2000 years old. During excavations archeologists often discover various objects in which this joining technique was applied. At the beginning of the 19th century riveting was used as one of the main methods for producing permanent joints. The Eiffel Tower in Paris, for instance, is assembled directly with the application of rivets. However, the development of welding technology considerably reduced the sphere of applications of riveted joints. However, when manufacturing areas dealing with permanent joints started to use dissimilar materials, riveting moved to the foreground again.

In recent years riveted joints have also found use in the assembly of dissimilar materials or materials with poor weldability, heat-treated or finished precision components for which heating is prohibited, and as a result it is impossible to apply welding [2-3]. In certain cases riveting is the only method for producing joints.

The main disadvantages of riveted joints are high labour-intensive costs, high consumption of metal, weakness of the working cross-section and the effect of stress concentration caused by the holes required for rivets. The strength characteristics of riveted joints can be improved by adding glue (rivet bonding). The idea of rivet bonding is that the technique combines the two independent operations – «bonding» and «riveting». In mechanical engineering the process of making uncured rivet bonded joints has become the most widespread method. The core of this technological process is that rivets are fitted after the adhesive is applied, but before it is cured. [4].

Epoxy adhesives have been traditionally used to make joints [5]. The main application advantages of epoxy adhesives are good joint sealing, simplicity and low cost of this method, good corrosion resistance, no need for heating and lack of stress concentration in the joint area.

The drawbacks of epoxy materials include high requirements imposed on the quality of surface preparation, a long time of curing, toxicity and difficult dismounting. The high strength of glue and lack of thermoplasticity require additional power load and extra tools which definitely complicates dismantling. Furthermore, the faying surfaces of the bonded parts may be damaged while dismounting.

An alternative type of adhesives applied to create joints is hot-melt adhesives that are a form of thermoplastics [6-7]. Unlike epoxies that are thermosetting resins, hot-melt adhesives are reversible. Initially solid adhesives get melted and become viscous-flow at elevated temperatures. Once adhesives cool, they revert to the original solid state. This method provides a strong bond line. Depending on a brand, bonding temperatures range between 100 - 200°C [7].

The basic requirements for hot-melt adhesives are a great bond strength and environmental resistance [8]. All that prevents the joint from altering its characteristics while in operation.

The advantages of hot-melt adhesives applied together with rivet bonded joints are that they are easily removed and provide increased strength and stiffness to structures under static, dynamic and vibration loads. This is because the structure distortion (e.g. in the three-point bending) is not accompanied by load-bearing points moving with respect to each other [9]. All these features increase the service life of rivet bonded joints.

Dismounting of rivet bonded joints with hot melt adhesives is ensured by the fact that thermoplastic hot-melt adhesives go to a liquid state and lose their

*Corresponding author: nbaurova@mail.ru

tackiness when heated [10]. Table 1 shows the labour coefficient for assembly and dismounting of rivet joints, rivet bonded joints with epoxy adhesives and rivet bonded joints with hot-melt adhesives.

Table 1. The labour coefficient for assembly / dismounting of joints

Type of joint	The labour coefficient, %	
	Assembly	**Dismounting**
Rivet joint	100	100
rivet bonded joint with epoxy adhesive	160	180
rivet bonded joint with hot-melt adhesive	160	130

The sequence of operations on dismounting rivet bonded joints with hot-melt adhesive looks as follows: first, the bond line is cleaned, any dirt or paint, if there is any, is removed from the bond line. Then, rivets are drilled out (like during the process of rivet joint dismounting). After that the bond line is being heated with the hot air gun until the hot-melt adhesive gets melted (from 100 to 200°C). When heated the solid adhesive becomes viscous-flow and joint surfaces get separated without any application of load. The rest of molten adhesive is removed from the plate surface with cloth. The result is a clean and tidy surface suitable for repeated bonding or riveting.

When producing rivet bonded joints with epoxy adhesive it is also necessary to observe the microclimatic parameters of a working area. High humidity causes curing to decelerate or even stop [5] whereas hot-melt adhesives are moisture resistant [10].

Thus, we can draw a conclusion that using thermoplastic hot-melt adhesives is one of the most efficient ways to manufacture rivet bonded joints.

Hybrid joints combining adhesive bonding with rivet joints can generally be created in one of three ways:

- the sequential method (installing rivets after the adhesive layer has hardened);
- the «fixing» method (installing rivets before adhesives have cured);
- the injection method (joining parts with a rivet joint before injecting the adhesive into the gap between the components).

Each of these technologies has advantages and disadvantages (table 2). The choice of technology is based on the conditions of a particular production. If the parts are assembled for single-unit production, then the first and third technologies are the most effective. The application of the second technology is preferable for mass production. This is due to the fact that this technology does not require any changes in riveting modes and the equipment earlier used for riveting can be applied.

The «fixing» method (when joining the components together is followed by the hardening of the adhesive layer) is widely used for automotive body repairs.

This alternative technology provides a higher level of joint efficiency compared to adhesive bonding, riveting or the technology when rivets are installed after the adhesive has cured. The «fixing» method also enables to use technologically recent adhesives providing a strong adhesive layer.

The principle drawbacks of the method are the rivet tool contamination caused by excessive uncured adhesive and the increased drill wear [3].

Table 2. Comparison of different methods of creating rivet bonded joints

Methods of creating rivet bonded joints	Advantages	Drawbacks
The «fixing» method: riveting before adhesive curing	- high quality of the adhesive layer; - a wide range of adhesives	- contamination of drills by excessive uncured adhesive
The sequential method: riveting after adhesive curing	- uniform thickness of the adhesive layer; - no clamping is required; - a wide range of adhesives	- potential damage of the adhesive layer while rivet; - the increased wear of drills
The injection method: riveting before adhesive injecting	-high quality of the adhesive layer	- a special tool is required to apply the adhesive; - the increased assembling time required for adhesive hardening; - the limited range of adhesives

The quality of rivet bonded joints depends to a large extent on a rivet and adhesive type chosen. Being placed like weld points with a certain span the rivet holes represent further stress raisers. Therefore, each type of joint requires a particular type of rivet.

The hybrid joints used in the automobile industry and machine building are mainly based on two rivet types: blind and metal piercing rivets. Typically, metal piercing rivets are called «self-piercing» due to the fact that their installation does not require any pre-drilled holes. Self-piercing rivet bonded joint is shown in fig.1. However, they can be installed on a flat surface only (with no high spots), e.g. in door openings or window screens. While setting such rivets the protrusion is formed on one side

DOI: 10.1051/matecconf/201712901004

only. At the same time the rivet 'cap' is positioned close to the part surface that allows it to be almost imperceptible after painting. The main disadvantage of self-piercing rivets is that their installation requires special equipment whereas blind rivets are installed with traditional riveting hammers.

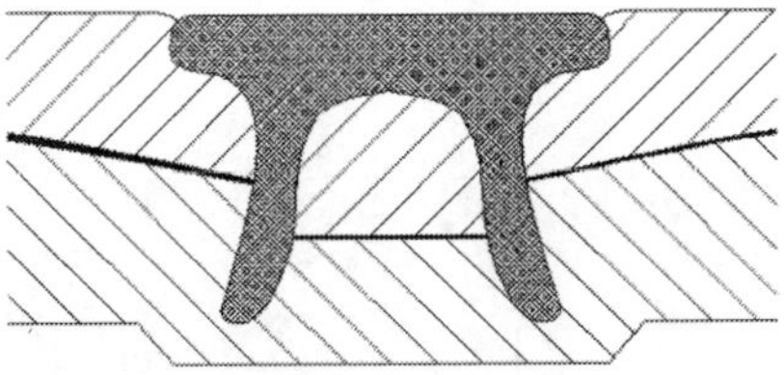

Fig.1. Self-piercing rivet bonded joint.

Blind rivets require pre-drilled holes. Therefore, the use of such rivets is undesirable for the assembly of large machine parts with numerous holes because there is a high risk of the holes' mismatch caused by the displacement of plates relative to each other. Blind rivet bonded joint is shown in fig.2.

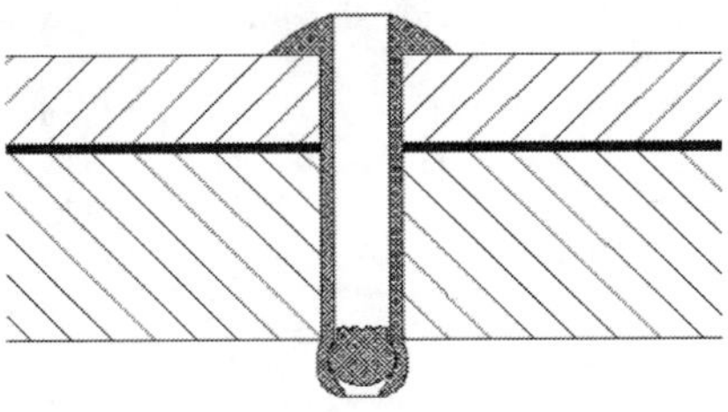

Fig.2. Blind rivet bonded joint.

Rivet joints comply with the condition according to which any mechanical loads are taken by rivets. However, rivet bonded joints do not satisfy it, for the adhesive applied shares load stresses with the rivet [5]. The long bond line results in the redistribution of stresses. Therefore, the span between rivets can be increased.

The type of adhesive depends on a particular method applied to create a rivet bonded joint. Riveting before adhesive curing allows film, phenol-formaldehyde and hot or cold cure epoxy adhesives to be used because in comparison to other methods of rivet bonding this one provides no restrictions on stress-strain behavior of adhesive (i.e. on its brittleness). When a rivet bonded joint is subjected to loadings the adhesive layer takes a large part of stresses thereby relieving rivets and reducing the strain of plates. For this reason, cold cure epoxy adhesives with encapsulants are a more perspective material used for automotive body repairs [8-9]. Fig.3 shows the behavior of epoxy adhesive with encapsulants under shear stress 50% of failure load.

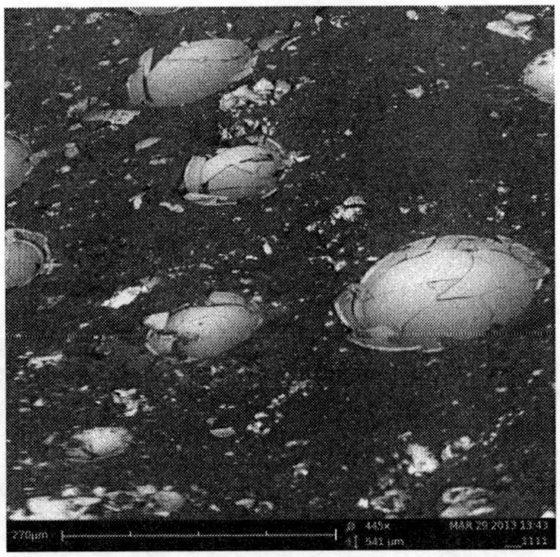

Fig.3. Epoxy adhesive with encapsulants under shear stress 50% of failure load.

The analysis of strength characteristics of different rivet bonded joints is given in accordance with industry standards GOST 1.00872-77 [10]. Some tests were performed to determine the maximum shear load of rivet bonded joints. The test samples of rivet and rivet bonded joints comply with the requirements of GOST 6996-66. The test results are presented in Table 3. Once the tests have been conducted, mean failure load is calculated.

Table 3. Test results of the calculations of maximum shear load

Test subject	Max. shear load, kgf (N)
Blind riveted joint	245 (2402)
blind rivet bonded joint	246 (2413)
Self-piercing riveted joint	251 (2462)
self-piercing rivet bonded joint	301 (2952)

The tests have shown that self-piercing rivet bonded joints provide the greatest shear strength whereas adding adhesives to blind riveted joints has not affected their strength. If rivet bonding of plates with different thickness is required, the strength of such joints is to be somewhat poorer compared to that provided by plates of uniform thickness. To reduce the negative effect of uneven gauge it is necessary to place the closing head on the side of a thicker plate. Thus, the use of modern rivets and adhesives allows the adhesive layer to take a greater part of stresses that, in turn, leads to the extended service life of joints and the reduced material deformation.

The fault cause analysis and possible preventive measures (remedies) are shown in Table 4.

Table 4. Typical faults in rivet bonded joints, causes and remedies

№ п/п	Fault	Cause	Remedy
1.	Deflection of joining plates	Excessive pressure on plates. Diameter of rivet shaft is larger than diameter of hole (the rivet is set forcefully)	Check the amount of force applied while setting the rivet. Check hole dimensions correspond to the diameter of the rivet shank
2.	Metal swell under the rivet heads	Riveting was performed without upsetting the rivet spots by straining tie	Upset the rivet spots by straining tie
3.	Hollows on joining plates and rivet heads	The wrong choice of tool, careless work	Use dedicated upsetting
4.	Incomplete closing head	The wrong choice of rivet type (the rivet shank is shorter than it is supposed to be)	Calculate the optimum length of the rivet shank and check presorting of rivets for those of required length has been performed
5.	Deformation (bend) of rivet shaft	The rivet shank of excessive length or misalignment of rivet axes and the bucking tool	Check the geometry of rivets corresponds to the calculations made
6.	Shank flattening	Clearance gap between joined parts	Check the geometry and proper bearing of parts
7.	Gapping between the closing head and the die head	The rivet head not tightly set	Check the rivet head is set tight. Remove any contamination as well as facings out of holes
8.	Mismatch between the closing head and the die head	Displacement of parts before or during the process of riveting. Angular position of drill while in operation	Provide visual inspection during the process of drilling. Make sure parts to be riveted are fixed and unmoved
9.	Misalignment of rivets	Non-uniform thickness of bond line, displacement of parts while adhesive bonding	Provide pre-spacing of rivet holes, make sure parts to be riveted are fixed and unmoved
10.	Starved spots	Initial poor fit of parts (increased clearance between joining parts). Air inclusions and voids in bond line.	Check the geometry and firm adherence of parts. Check the thickness and uniformity of adhesive layer. Comply with adhesive bonding technology
11.	Low adhesive strength	Contaminated surfaces of plates to be bonded. failure of cooling modes for adhesive	Check pretreatment procedures are correct. Observe the dedicated temperature modes and provide accurate time control while hot-melt heating or cooling

The quality control test takes an important part in the creation of rivet bonded joints. At an early stage it is monitored by visual inspection that there are no such faults as:

- undue local deformation of joining parts including their deflection while setting the rivet and metal swell under the rivet heads (this fault mostly occurs while riveting metal sheet 5 mm and thicker); hollows on joining plates and rivet heads.
- incomplete closing head;
- deformation (bend) and flattening of the rivet shank;
- gapping between the closing head and the die head;

DOI: 10.1051/matecconf/201712901004

-mismatch between the closing head and the die head along with misalignment of rivets;

- starved spots or peeling (low adhesive strength).

Thus, the tests conducted allow us to draw a conclusion that the adhesive used in a rivet bonded joint of high quality not only protects the inner surfaces of substrates from corrosion but also seals the joint off. It was found that rivet bonded joints provide a greater shear strength (by 15-35 %), even tear (by 20 %) and uneven tear (by 25-40 %) than corresponding rivet joints [1, 5, 9]. Furthermore, rivet bonded joints show better strength characteristics under dead-load stresses and vibration strain than similar rivet joints.

References

1. N. Lapina, N. Baurova, Polym. Sci., Ser. D, **9**(3), 326-330 (2016).
2. V. Zorin, Mechaniz. Stroitelst, **77**(10), 43-46 (2016).
3. N. Baurova, V. Zorin., V. Prikhodko, Polym. Sci., Ser. D, **9**(4), 402-406 (2016).
4. A. Konoplin, Remont. Vosstan. Moderniz, **8**, 37-39 (2016).
5. A. Konoplin, Vsematerial. Encicloped. Sprav., **7**, 10-14 (2015).
6. G. Malysheva, M. Shablygin, T. Guzeva, Fibre Chemistry, **47**(2), 85-88 (2015).
7. A.Borodulin, A. Marycheva, G. Malysheva, Glass Physics and Chemistry. **41**(6), 660-664 (2015)
8. A. Sergeyev, R. Turusov, N. Baurova, A. Kuperman, Mechan. of Comp. Mat. **51**(3), 321-332 (2015)
9. V. Nelyub, Fibre Chemistry. **47**(1), 40-42 (2015)
10. V. Zorin, Mechaniz. Stroitelst., **77**(1), 12-15 (2016)

DOI: 10.1051/matecconf/201712901005

Intensification of grinding technology and methods for studying the structure of fibrous materials in the process of its production

Aidar Saetshin[1,*], *Zimfira* Valishina[1], *Evgenie* Matuhin[2], and *Anatoliy* Kostochko[1]

[1]Kazan National Research Technological University, Kazan
[2]FKP "Kazan State Gunpowder Plant", Kazan

Abstract. One of the most important indicators of the suitability of new types of fibrous materials for processing in products with specified properties is the degree of grinding. A detailed analysis of the influence of various factors on the quality of grinding of fibrous materials was carried out. The information is necessary for predicting the properties of new types of materials and controlling the technological process for obtaining fibrous materials with improved and controlled characteristics. The advantages of a modern method for analyzing the properties of fibers in the process of obtaining new types of materials are shown, which makes it possible to evaluate the active fiber surface and its fractional composition.

The grinding phase plays a significant role in producing the nitrate ester of cellulose [1-3].

Control process principles of grinding the different types of nitrocellulose (NC) are based on abstract content about processes of fibrous materials grinding. It is set on the basis of systematic researches and the revealed dependence connecting the fractional composition of NC with technological parameters of grinding process and structural-morphological properties of NC, rate of plastification and the quality of NC compositions [4-6].

The main purpose in this phase is to obtain the NC fibers with the necessary geometrical sizes, that provide fast invasion of wash liquids in fibers and removal from channels the encapsulated acids remained after previous stages of stabilization process. When grinding fibers in the water, the change of their structural properties is happened. Therefore the term "grinding" doesn't reflect the essence of technical and technological processes in installations like cylinder engine, conical and other mills.

The existing technological processes and mechanism don't allow to obtain fibrous materials, in particular the nitrate esters of cellulose homogeneous in fractional composition [3].

The problem of increasing the homogeneity and stability of properties according to solubility, molecular weight of NC is relevant in connection with the expansion of the scope and tightening of requirements on new types of nitrate esters of cellulose (table 2). The increase of NC quality characteristics is possible due to the optimization of the technological process or modernization of production.

The main experimental data on investigation of grinding mechanism of NC fibrous in working parts of knife machine are described in this paper. Also presented the principle of fiber processing in various grinding mechanisms and issues related to the analysis of NC morphological structure with the help of modern analyses method of fiber properties.

For NC grinding are currently used the conical mills (RK-01) and disk mills DMK (disk mill Kuzmina), differing by the principle of grinding. When grinding in conical mill, the fiber is "become ground" and in the result the NC has "ground" surface. When grinding in the DMK the fiber is generally cut. The Impact on the fiber during grinding effects characteristics of products [9-10].

The impact of grinding method on proper-ties of obtained NC has been confirmed in case of the study of thermo mechanical properties and kinetics of nitrate ester cellulose swelling on the basis of cotton and wood cellulose (table 1).

The presented results of the study in the table 2 show that the grinding method effects on technical and technological NC properties, affecting the quality on received NC compositions.

Table 1. The impact of grinding method on physical, chemical and technological characteristics of cellulose nitrates, received from various types of raw materials [4].

cellulose	Type of mill	Nitrogen content, ml NO/g	Vis cosi ty,° E	Sieve analysis (sieve 0,16 %)	Degree of swellin g;ΔV
Cotton cellulose	RK-01	191.4	2.0	1.0	0.50
	DMK	191.8	2.5	2.3	0.35
Bleached sulfite	RK-01	189.4	2.2	3.1	0.38
	DMK	190.2	2.0	2.7	0.40

Insufficient knowledge regarding the mechanism of grinding NC, especially in the way of impact of driven elements on fiber, and also the current mechanization of grinding phase, doesn't allow to prepare the fiber for plastification [7-8].

*Corresponding author: aidar.saetschin@yandex.ru

DOI: 10.1051/matecconf/201712901005

By grinding NC with high degree substitution in DMK, the fractional composition depends on the way of input and removing of NC water suspension from machine. By grinding NC on the scheme "under pressure", the aver-age length of fiber is 0.168 mm, bulk mass is 108 kg/m³. By grinding on the scheme "free flow", the average length of fiber is 0.131 mm, bulk mass is 128 kg/m³. By grinding in conical mill RK-01 due to the multiple passes through crushing knives the average length of fiber is 0.207 mm, bulk mass is approximately 108 kg/m³.

By grinding in DMK on the scheme "under pressure", by increasing the suspension concentrates from 4 to 8%, the fraction content with the size of fiber 0.16 mm is increasing from 0 to 27%. This is because the increasing of suspension concentrate of NC products leads to the increasing of NC stay period in mechanism through speed reduction of mass flow. For this purpose the probability of fiber breakthrough across knives-edges reduces. At the same time the NC fractional composition is directly linked to suspension concentrate. The higher is the concentration, the greater the degree of NC grinding.

Table 2. Physico-chemical characteristics of analyzed nitrates of cellulose.

Indicator name	NC	
	№1	№2
Raw material input	Cotton cellulose	
Nitrogen content, %	13,1	11,8-12,5
Degree of grinding: sieve residue 016,%	Max 7	Max 9

The fractional composition of NC №1 и №2 (table 2) is established in DMK (figure 1) [11] depending on handling time and load intensity. The figure 1 shows that by increasing the grinding time and DMK engine load, there a significant reduction of fraction with the length of fiber more than 0.5 mm. There is also a notable increase in fractional composition due to the grinding in DMK under engine load more than 180 A.

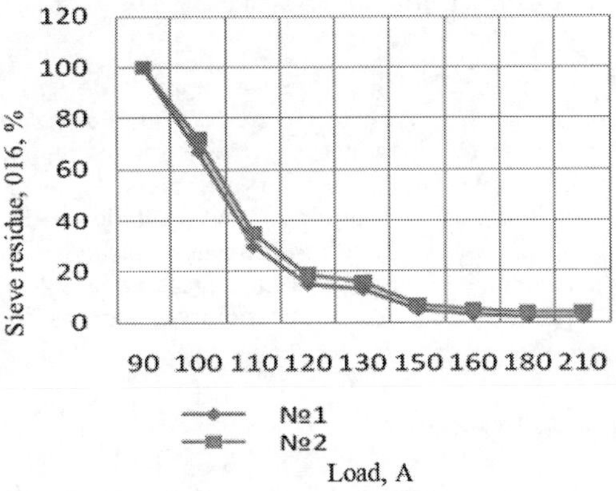

Fig.1. The quality of cellulose nitrate grinding by the method of sieve residue 0.16 when grinding in DMK.

On the basis of the above, it may be concluded that it is necessary to use a special mechanism that would not only cut fibers, but brush them for receiving required parameters in the clarity of grinding during the NC grinding [12].

The degree of grinding alongside with viscosity, solubility, degree of etherification and molecular weight distribution is the one of the most important characteristics of quality that determines technological properties of NC and nitrate cellulose compositions (table 3).

Table 3. Comparative characteristic of physico-chemical and structural properties of NC with analog.

Indicator name	sampleNC №2C[x]	New type of NC	analog
		NC sample № 1-01	
Nitrogen content, %	13.21	13.2	13.17-13.3
Viscosity, mPas	15.7	57	45-60
Bet, m²/g	8.71	7.3	
solubilityalcohol/ether, %	5.1	28.3	28-35
Solubility in alcohol, %	1.0	1.3	Max 3
Grinding degree: Fibre volume, ml/10g	108	98	Max 95

[x)] For NC grinding (Sample № 2C) the conical mill RK-01 is used

To date, it is necessary to give the correct assessment to the degree of NC grinding. Until recently there has been a lack of reliable method of quality assessment for NC grinding. The current methods (volumetric method, "artillery sample", sieve method) don't provide with complete information on grinding quality and don't allow to define the main factors, affecting on the process of NC qualities formation with improved characteristics. Atthepresenttimethenewmethodisused [13-15].

The use of new method of NC morphological structure analyses is necessary for establishing the connection among particle sizes, structure and qualities of nitrate cellulose products.

This method allows to analyze samples with the provision of data on the fiber sizes (length, width), grain characteristics, coarseness, fiber contortion and other parameters of morphological structure of fiber and grains.

The principle of measurement is following: high resolution camera captures halftone images from which the analyzer calculates the different properties of fiber with the help of images analyses. The image of sample fibers is created in a measuring cell with depth of sharpness of 0.5 mm according to the standard ISO 16505-2. The length of fiber is measured along axis of fiber symmetry. Due to this the accuracy of measurements of the contorted and retorted fibers is ensured. Besides, for the calculations of deformations the projection length of fiber is measured (the linear distance among fiber points which are in the longest distance one from another).

The results of analyses are presented in the tables (4,5,6,7,8) and figures (2,3,4,5).

DOI: 10.1051/matecconf/201712901005

Table 4. Distribution of fractions to length, width and mass for NC № 1-01 on the basis of modern analyses method.

Fractional composition, mm	Length, %	Mass, %	Width, %
1: 0-0.2	27.8	25.1	16.1
2: 0.2-0.6	54.7	56.3	17.4
3: 0.6-1.2	16.3	17.3	17.8
4: 1.2-2.0	1.2	1.3	17.6
5: 2.0-3.2	0.1	0.1	19.0

Table 5. The results of sample processing NC №1-01.

Coarseness, mg/m	0.098
Twist, %	22.94
Flaked grains A, %	35.22
Laminar grains B, %	5.79
Grains in percentage from weighted average, %	67.05
Width of fiber, mkm	17.5

Table 6. The average length of fiber for sample NC №1-01.

Average length of fiber	Value
Lc(n) Arithmeticaverageoffiberlength, mm	0.179
Lc(l) Weightedaveragetofiberlength, mm	0.381
Lc(w) Weightedaveragetofibermass, mm	0.574

Table 7. Distributions of fractions to length, width and mass for NC № 2Con the basis of modern analyses method.

Fractional composition, mm	Length, %	Mass, %	Width, %
1: 0-0.2	45.8	41.4	16.9
2: 0.2-0.6	48.3	52.7	18.7
3: 0.6-1.2	5.6	5.7	18.3
4: 1.2-2.0	0.3	0.2	16.3

Table 8. The results of sample processingNC №2C.

Coarseness, mg/m	0.110
Twist, %	21.39
Flaked grains A, %	52.17
Laminar grains B, %	4.23
Grains in percentage from weighted average, %	78.61
Fiber width, mkm	16.8

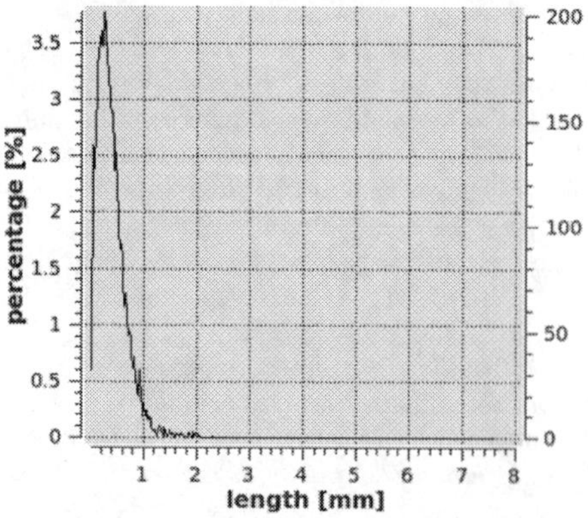

Fig. 2. Weighed average distribution to grain length of sample № 1-01.

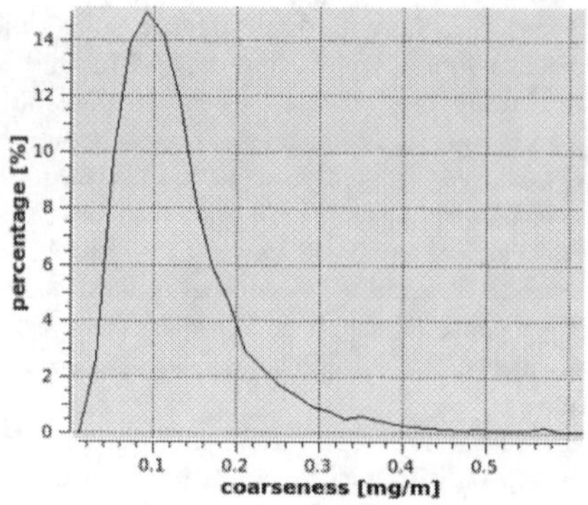

Fig. 3. Distribution to coarseness for sample NC № 1-01.

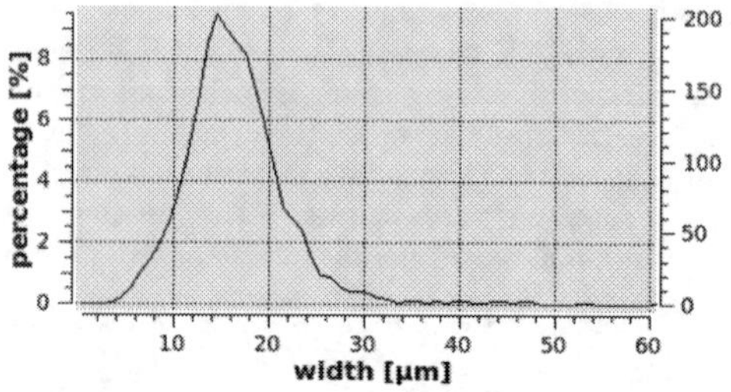

Fig.4. Distribution to width of fibre for sample NC №2C.

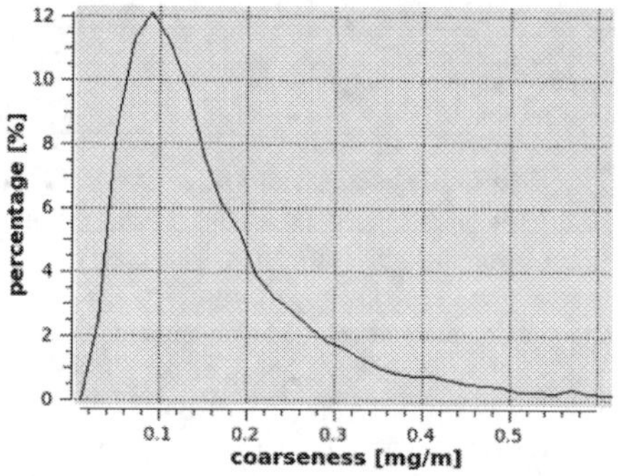

Fig. 5. Distribution to coarseness for sample NC № 2C.

Consequently, the use of appropriate grinding mechanism will allow to receive NC with homogeneous properties by gravimetric density, structure, active area,fractional composition, polymerization degree, speed of plasticizer absorption and nitrate cellulose compositions with improved mechanical characteristics.

The modern method is characterized by accuracy, informativity and there is a possibility to use it as express method.

References

1. V. I. Gindich, L.V. Zabelin, *Production of cellulosenitrates. Technology and equipment* (1984)
2. A.V.Kostochko, B.M. Kazban, *Powders, solid rocket fuel and its propert ies* (2015)
3. Z.T. Valishina, A.A. Saetshin, V.G. Borbuzanov, N.Kh. Giniyatov, E.L. Matukhin, A.V. Kostochko,

Structure and properties of a new type of cellulose nitrate with improved characteristics, Ammunition, Scientific and technical collection of the State Scientific Center of the Federal State Unitary Enterprise "TsNIIKhMim. DI. Mendeleev", **2**, 31 (2016)

4. G.M. Marchenko, M. Zabelin, *Production of nitrates of cellulose. Physical and chemical foundations of production and processing of cellulose nitrates* (1988)
5. V.I. Kovalenko, O.V. Mikhailov, G.M. Khrapkovsky, *Cellulose nitrate: molecular-structural heterogeneity* (2003)
6. V.M. Lebedeva, A.S. Semochkin, D.M. Kaganov, *Intensification of the process of dehydration of cellulose nitrates on centrifuges of periodic action,* Proceedings All-Russian. Conf. RCTUthem. DIMendeleev, 452 (2010)
7. V.I.Gindich, *Production of cellulose nitrates and of acids regeneration* (1995)
8. Z.T. Valishina, A.A. Saetshin, V.G. Borbuzanov, E.L. Matukhin, A.V. Kostochko, *Structure and properties of high-quality dispersed cellulose nitrates*, Materials of the VIII All-Russian Conf., , 157 (2016)
9. A.A. Saetshin, Z.T. Valishina, E.L. Matukhin, I.I.Malov, G.N. Galiullina, A.V. Kostochko, *Method for Investigating the Morphological Structure of Powder Celluloses,* Bulletin of the Kazan Technological University, **18**, 103 (2016)
10. A.A. Saetshin, A.Yu. Nizami, F.T. Valishina, F.T.Yusupov, V.G. Borbuzanov, E.L. Matukhin, Z.T. A.V. Kostochko, *Equipment and Materials, Technologies and equipment for pulverizing of cellulose-containing material,* Materials of the International Scientific and Technical Conference "Innovative Engineering Technologies, **1**, 154 (2016)
11. F.T. Yusupov, A.A. Saetshin, V.G. Borbuzanov, E.L. Matukhin, Z.T. Valishina, A.V. Kostochko, *Improvement of technology and equipment for the modification of cellulosic materials (separation, aggregation, conditioning,* Materials of the International Scientific and Technical Conference "Innovative machine-building technologies, equipment and materials, **1**, 234 (2016)
12. A.A. Saetshin, Z.T. Valishina, E.L. Matukhin, A.V. Kostochko, *A modern method for analyzing the properties of cellulose-containing materials,* Ecological problems of economic subjects: a collection of the VI International Scientific and Practical Conference, 77 (2016)
13. A.A. Saetshin, Z.T. Valishina, E.L. Matukhin, A.V. Kostochko, *A modern method for investigating the properties of high-quality dispersed celluloses and nitrate esters of cellulose,* Collection of abstracts of reports "Modern problems of special technical chemistry, 30 (2016)
14. A.A. Saetshin, Z.T. Valishina, E.L. Matukhin, G.N. Galiullina, I.I. Malov, A.V. Kostochko, *Reagents and materials for construction, operation and repair of oil, gas and gas condensate wells: production, properties and experience of application,* Materials of the XX International Scientific and Practical Conf., 106 (2016)
15. Z.T. Valishina, A.V. Kostochko, O.T. Shipin, R.F. Gatina, Yu.M. Mikhailov, *Structure and properties of new types of nitrate esters,* Vestnik Kazanskogo tekhnol.un-ta, **9**, 281 (2010)

DOI: 10.1051/matecconf/201712901006

Selection of direct drives for tool and workpiece rotation for gear-hobbing machines of new generation

Vladimir Novikov[1,*], *Alexey* Isaev[1], and *Viktor* Bushuev[1]

[1]Sub-department "Machine tools" of MSTU "STANKIN", 127055 Moscow, Russia

Abstract. Selection of direct drives for tool and workpiece rotation based on their computed torques for gear-hobbing machines of new generation is considered. Significant influence of torsional stiffness of electromechanical drive system, their tuning and peculiarities of operation in the continuous tracking mode on the drives' maximum torque is shown.

1 Features of gear-hobbing machines of new generation

Modern machine tools of new generation are built, as a rule, on the basis of mechatronic modules for realization of shape-generating motions, and the interrelated rotations of the tool and workpiece are implemented by means of direct drives mounted on the corresponding spindles. In this case, both design and operational advantages are achieved.

This paper deals with selection of the torque values of the drives for the rotation of cutting tool and workpiece in the new generation gear-milling machine ⌀200 mod. 5320F4 with accuracy P class is considered. The machine tool is designed at MSTU "STANKIN" and manufactured by the machine-tool plant "SASTA" JSC (Sasovo, Ryazan region).

When creating new generation machine tools with main performance indicators depending on the electromechanical properties of drives, a number of completely new and serious problems arise, including the choice of the force-torque parameters of the drives.

Let us consider the main distinctive features of direct drives of gear-hobbing machines affecting the their torques:

1. The lack of intermediate gears in the short generating train of the machine allows to achieve the maximum possible mechanical torsional stiffness.

2. Dynamic characteristics of the drive depended on its tuning affect the stiffness (the ratio of torque variation to the angle of rotation variation) and the maximum component of the cutting torque per revolution of the hob.

3. The drives are continuously working in the tracking mode, providing a constant gear ratio between the hob and the workpiece, so apart from the cutting torque, they consume the torque for acceleration (acceleration and braking) of the motors.

The experiments were carried out on a prototype model of the machine Ø200mm mod. 5320F4 at the MSTU "STANKIN". Steel workpieces of gear wheels with different number of teeth and modules (m = 3;4;6) were machined with worm hobs. The gear hobbing was carried out mainly in one pass (at the full height of the tooth) with feed rate ranging from 1 to 4 mm/rev. During the work, measurement of torques at the table and the tool, the rotation speeds and spindle rotation angles (machine sensors), cyclic errors of the generating train were conducted.

2 Study of the cutting torques during roughing

Experimental investigations of cutting forces and torques during rough machining on traditional machines have shown that per single table revolution the torque changes approximately according to the sinusoidal law with frequencies equal to the number of cut teeth Z and the $Z{\cdot}Z_0$ product, where Z_0 is the number of flutes in the worm hob [1,2,3,4]. The maximum torque on the worm hob is about 2 times higher than the average.

Figure 1 shows typical graphs of the torque on the drives when roughing the gear with the following parameters: m=4; Z=36; β=0^0, with the following cutting parameters: cutting depth t=9mm; cutting speed V_{HOB}=24 m/min; feed rate S=3,6 mm/rev. M_{MAX}, M_{AVR}, M_{CONST} - maximum, average and constant torque on the drives; $M_{MAX(AVR)}$ is the average of the maximum (peak) values of the torques per revolution of the hob (according to the number of flutes).

* Corresponding author: 9bobah@gmail.com

DOI: 10.1051/matecconf/201712901006

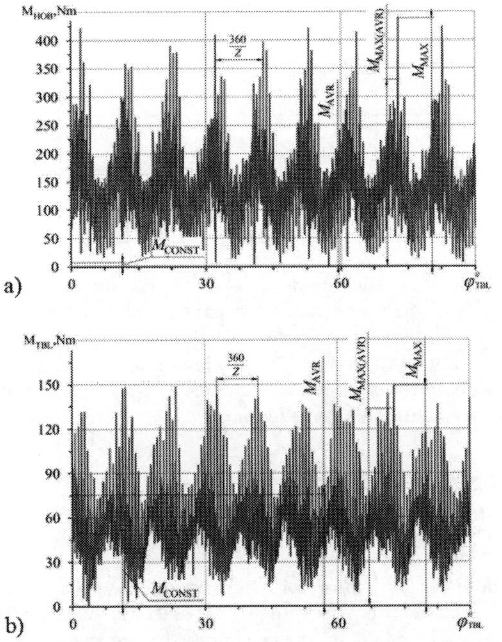

Fig.1 The graphs of the moments of the worm hob (a) and the table (b) when roughing the tooth wheel with the adjustment of the drives to the maximum passband for 0.5 revolutions of the table.

The torque has 18 peaks per 0.5 revolutions of the table, according to the number of gear teeth (Z=36). At the maximum cross section of the removed chips, the maximum torque M_{MAX} appears. The torque at the hob is changing as at machines with mechanical generating train, except for the increased value of the maximum to average torque ratio. Important parameters when choosing drives of hobbing machines are: mean peak torques $M_{MAX(AVR)}$ per revolution of the hob (according to the number of flutes) and the maximum torque M_{MAX}

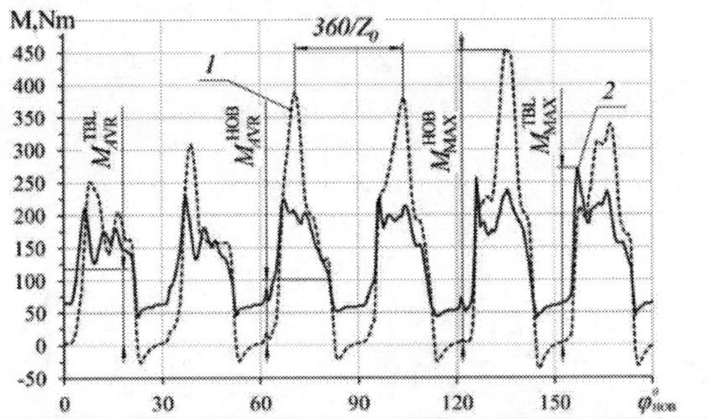

Fig.2 Torques of the hob (1) and the table (2) for rough cutting when rotating the hob at 180^0

Figure 2 shows the torques of the M_{HOB} worm hob and the M_{TBL} table when machining a steel helical wheel under cutting parameters close to the maximum achievable ones at this machine, namely: m=6; Z=29; β=20^0; t=2,25·m=13,5mm; V_{HOB}=24m/min; S=1.5mm/rev; Z_0=12. The graphs show 6 maximum peaks of torque at 180^0 (according to the number of flutes).

The maximum torque at the hob $M_{MAX.HOB}$=455Nm has exceeded the average $M_{AVR.HOB}$ =150Nm ~ 3 times and limited the further increase of cutting parameters for the permissible torque on the engine ($M_{MAX(NGN.HOB)}$=500Nm). The maximum torque at the table $M_{MAX.TBL}$=270Nm has exceeded the average $M_{AVR.TBL}$=160Nm in 1.7 times and does not limit the possibility of further increase in cutting parameters ($M_{MAX(NGN.TBL)}$=716Nm).

Table 1 shows the average experimental data on the analysis of torques on the drives, obtained from the roughing of steel workpieces with modules 3,4 and 6 mm, with different cutting parameters.

Table 1. Average torques at the drives during roughing operations.

	M_{MAX}, Nm	$M_{MAX(AVR)}$, Nm	M_{AVR}, Nm	K_M, Nm	K_{AVR}, Nm
hob	300	200	100	3	2
table	200	175	120	1,7	1,5

Where $K_M = M_{MAX} / M_{AVR}$;
$K_{AVR} = M_{MAX(AVR)} / M_{AVR}$;

The maximum torque at the hob (M_{MAX}) exceeds the average torque by approximately 3 times (2.8 to 3.2 times depending on cutting parametes). The average values of peak torques per revolution of the hob ($M_{MAX(AVR)}$) exceed the average torque at the hob by 2 times, and by 1.5 times at the workpiece.

The results obtained for increasing the maximum torque by 3 times relative to the nominal differ from the results obtained on machines with traditional relatively non-rigid kinematics, where the maximum torque exceeds the average by approximately 2 times [1,2,3,4].

Therefore, to test the suggested hypothesis of increasing the maximum torque (M_{MAX}) due to an increase in the torsional rigidity of the generating train, experiments were conducted with different tuning of the drives which affects the change in the "electrical" component of the torsional stiffness. The following values of the bandwidth of the drives for speed were achieved, depending on the settings: for the hob f=55-410Hz; for the table f=16-175Hz. The work of the drives was studied under conditionally called "good" tuning (f_{HOB}=410Гц; f_{TBL}=175Гц) and "bad" tuning (f_{HOB}=55Hz; f_{TBL} = 16Hz).

3 Influence of the drive control settings on the cutting torque

Figure 3 shows the graphs of the torques on the drives with the "bad" setting (at the cutting parameters shown in Fig. 1), within 0.25 turn of the hob, where 1, 2 and 3 are the torques caused by the motor acceleration, the total torque (dotted line) and the cutting torque,

Table 2. Effect of the quality of the drive settings on their moments.

Tuning quality		Bandwidth f, Hz	M_{MAX}, Nm	Mmax,Nm	M_{ABR}, Nm	M_{CONST}, Nm	$\Delta M_{\varepsilon(MAX)}$, Nm	M_{MAX}/M_{AVR}	$M_{MAX}/M_{MAX(AVR)}$
"good" & "good"	hob	410	440	250	175	50	37	2,51	1,76
	table	175	170	120	75	25	16	2,26	1,41
"bad" & "bad"	hob	55	335	245	150	30	18	2,2	1,37
	table	16	135	105	70	18	10	1,93	1,29

respectively. Since the HSS worm hob used in the experiment has 14 flutes (Z_0 = 14), the graphs show 4 maximum peak torques at the angle of 90 degree. In addition to the cutting torque, the torques $M\varepsilon$ caused by the acceleration of the engines are shown, which have about 10 peaks between the two cuts of the worm hob.

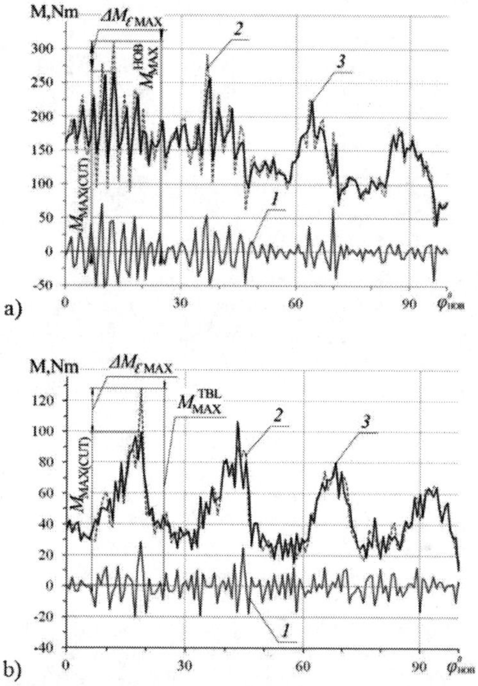

Fig.3 Torques at the hob (a) and the worktable (b) for roughing the wheel, taking into account the cutting, acceleration and friction moments.

The total torques of the drive is total (for the example of the maximum torque):

$M_{MAX} = M_{MAX(CUT)} + \Delta M_{\varepsilon(MAX)}$, where

$M_{MAX(CUT)}$ is the maximum cutting torque, which includes the constant component of the M_{CONST} torque, which depends on the friction in the spindle bearings;

$\Delta M_{\varepsilon(MAX)}$ is the acceleration moment;

Acceleration moment (Fig. 1) can both increase and decrease the maximum drive torque during the cutting process.

The influence of the quality of the adjustment of the drives on their moments and torques during the machining of the wheels in Fig. 3 is illustrated by the data in Table 2.

It follows from Table 2, that all the moments and torques described in this table have increased with the improvement of the quality of the settings and the bandwidth of the drives. The maximum torque M_{MAX} at the hob and the table has increased ~ by 25-30% with improvement of the drive tuning.

4 Study of drives rigidity

Let us define the "electric" torsional stiffness of the drives $C_E = \Delta M/\Delta\varphi$ equal to the ratio of the torque variation ΔM to the drive torsional deformation variation $\Delta\varphi$ in the area of one cut of the hob (see Fig. 4), i.e. at its turn on an angle $\varphi = 360/Z_0$ (similar to the rigidity in the mechanical system).

Fig. 4 shows, as an example, the graphs of the torque variation ΔM of the worm hob drive and the corresponding variation of the worm hob rotation angle $\Delta\varphi$ and the table for their "bad" adjustment and for the cutting parameters shown in Fig. 1.

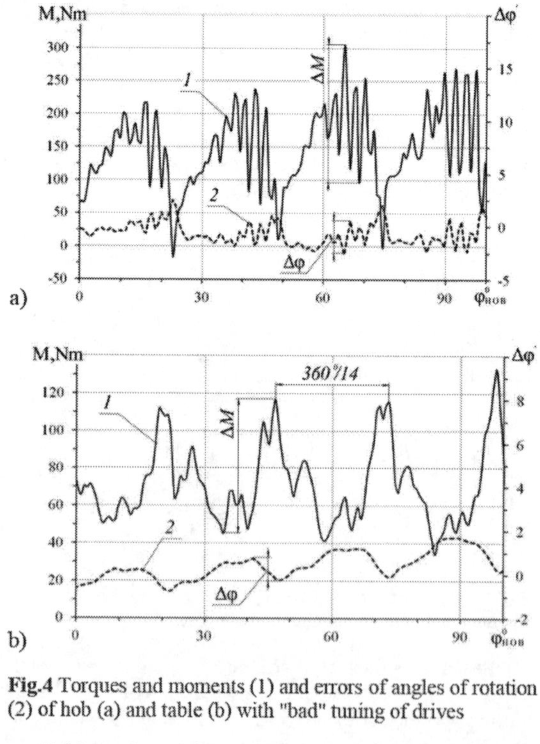

Fig.4 Torques and moments (1) and errors of angles of rotation (2) of hob (a) and table (b) with "bad" tuning of drives

Table 3 shows the average values of the stiffness of the drives C_{TS} and their "electrical" C_E and "mechanical"

DOI: 10.1051/matecconf/201712901006

C_M components for the rough machining of the same workpiece with different drive tuning.

Table 3. Rigidity of drives with different settings.

drive	tuning quality	C_E, Nm/rad	C_M, Nm/rad	C_{TS}, Nm/rad
hob	bad	144460	75670	49660
table	bad	151340	670700	123480
hob	good	189170	75670	54050
table	good	213250	670700	161800

"Electrical" stiffness was determined by the readings of the angular spindle position sensor. "Mechanical" stiffness was computed at the sections of the kinematic chains from the cutting zone to the angular position sensor. Tangential contact deformation of the joints of the pedestal with the table and the workpiece with the worktable were also taken into account in addition to the rigidity of the spindles, cotters and keys.

Table 3 has shown that the rigidity of the drives C_{TS} depends on their adjustment and it is increasing from "bad" tuning to "good" one by approximately 9 and 30% for the hob and table, respectively, as the maximum M_{MAX} moments on drives 30 and 26% increase (See table 2). The "electric" stiffness has exceeded the "mechanical" rigidity by about 2.5 times at the hob but it is 3 times lower than "mechanical" stiffness at the table. During the calculation, contact deformations of the gears and keyed joints, as well as torsional deformations of the shafts were taken into account.

Thus, the main reason for maximum torque increase is a significant increase of electromechanical stiffness of drives of the generating train in machines with direct drives, in comparison with traditional machines. The aggregated calculation of torsional stiffness of the kinematic chain of the generating train of a traditional machine of the similar size has shown that the unit stiffness of hob is approximately by an order of magnitude less than the drive stiffness of the new generation machine.

An important feature the new generation machines' drives is their continuous operation in the tracking mode, which makes it necessary to take into account the acceleration and braking moments of the engine.

Those moments were determined using the formula $M_\varepsilon = I \cdot \varepsilon$, where I is the moment of inertia of the rotating masses, and ε is acceleration computed by differentiating the speed on the worm hob and the table.

In Fig. 5, the graphs of moments for roughing the gear are given as an example; m=6; Z=50; β=16^0; t=12mm; S=1mm/rev; Z_0=12, where: 1,2,3 - acceleration moment of the engine, total torque and moment from cutting (dotted line), respectively.

Thus, we have obtained the following specifications:

-for the hob:

I_{HOB} = 0,2 kg·m^2; ε_{MAX}=650rad/s^2; ε_{AVR}=175rad/s^2;

-for the table:

I_{TBL} = 2,4kg·m^2; ε_{MAX} =40rad/s^2; ε_{AVR} =24rad/s^2;

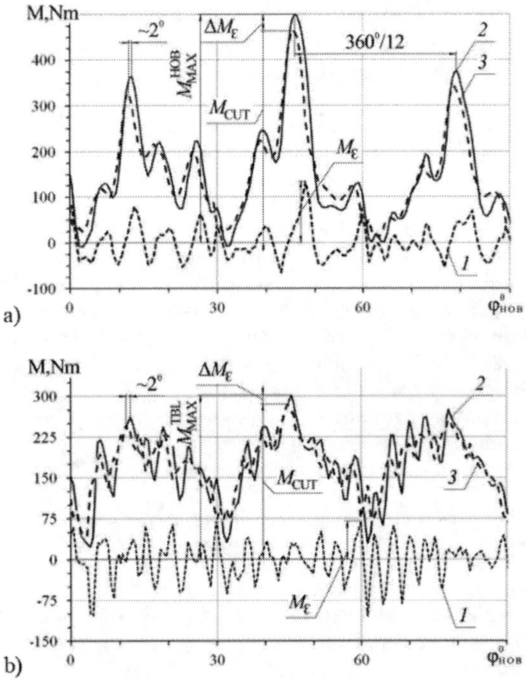

Fig.5 Torques at the hob (a) and worktable (b) when roughing the wheel

The maximum accelerating moment M_ε coincides in phase (with a small lag of ~ 2^0) with the maximum cutting torque M_{CUT} at the worm hob. This is the case for all peak torques for the turn of the hob, since it is the cutting torque that causes the speed to change. In the cases considered (when machining wheels with m = 3;4;6), the motor torque exceeds the cutting torque peaks: for the hob - ΔM_ε =7÷10%; For the table - ΔM_ε=8÷26%.

The most strongly the accelerating moments affect the general torques of drives during finishing, since in some cases they become comparable with the cutting torques.

Analysis of the influence of the cutting process, under different cutting parameters and workpiece parameters, on acceleration torque M_ε, made it possible to establish accelerations for roughing and finishing operations:

The average value of the acceleration peaks (in terms of the number of flutes Z_0) was: on the hob - $\varepsilon_{AVR(HOB)}$ = 160 Rad/s^2 (varied from 70 to 400 Rad/s^2); on the table - $\varepsilon_{AVR(TBL)}$=17Rad/s^2 (varied from 5 to 40 Rad/s^2);

The maximum values are $\varepsilon_{MAX(HOB)}$=345 Rad/s^2 (varied from 180 to 650 Rad/s^2); $\varepsilon_{MAX(TBL)}$=27Rad/s^2 (varied from 8 to 50 Rad/s^2);

At finishing operations, the acceleration is less than about 3 times that for roughing operations.

The conducted experiments showed that accelerations at the drives of the hob and at the table practically do not influence the average torque of the drives, but increase its maximum component M_{MAX} with coefficients $K_{\varepsilon(HOB)}$ ~ 1,1 for the hob and $K_{\varepsilon(TBL)}$ ~ 1,15 for the table.

5 Algorithm for selecting torques of the main motions motors of new generation gear hobbing machines

Based on the results of the research, the following algorithm is proposed for selecting the torques of drives for rotating hob and workpiece:

1. Taking into account the characteristic parameters of the tools and machined wheels, using the recommended (close to maximal) cutting parameters (module, feed, depth, etc.), according to existing formulas from the literature determine the required average torque at the hob. For increased productivity, reference values of feed rate can be increased by at least 1,5 times (if these are not limited by the cutting process itself), since in the new generation machines the cutting parameters are limited as a rule by the torques of the drives.

The average torque of the worktable is determined by:

$$M_{AVR} \geq (F_Z \cdot \sin\beta + F_X \cdot \cos\beta) \cdot d_{WKPC}/2 + M_{FR}, \quad (1)$$

Where F_Z is the average value of the tangential component of the cutting force;

F_X is a axial (along the axis of the hob) component of the cutting force (in the first approximation can be ignored);

β is the maximum angle of inclination of the tooth;

d_{WKPC} is the maximum diameter of the workpiece;

M_{FR} is the moment of friction in the table supports;

2. According to the calculated average torque, the M_{AVR} is determined:

The maximum torque at the engine:

$$M_{MAX} = M_{AVR} \cdot K_M \cdot K_\varepsilon \quad (2)$$

The moment limiting the mode of operation $M_{S6\,(40\%)}$:

$$M_{MAX(AVR)} = M_{AVR} \cdot K_{AVR} \cdot K_\varepsilon \quad (3)$$

For hob:

$K_{M(HOB)} = 3$; $K_{AVR(HOB)} = 2$; $K_{\varepsilon\,(HOB)} = 1{,}1$;

For the table:

$K_{M(TBL)} = 1{,}7$; $K_{AVR(TBL)} = 1{,}5$; $K_{\varepsilon(TBL)} = 1{,}15$;

3. The computed values of the torques are then compared with drive specifications from catalogues, taking into account that:

$$\begin{aligned} &M_{RT} \geq M_{AVR}; \\ &M_{NGN.S6(40\%)} \geq M_{MAX(AVR)} \\ &M_{MAX(NGN)} > M_{MAX}; \end{aligned} \quad (4)$$

6 Conclusions

-The maximum moments and torques arising in the direct drives of the hob and workpiece in the gear-hobbing machines of new generation depend on the parameters of the cutting process, as well as on the torsional stiffness of the drives and the quality of their tuning, and exceed the maximum moments and torques of machines with mechanical generating train by approximately 1.5 times.

-it is recommended to select the drive motors on the average calculated cutting torque, determined using the known methods. The maximum permitted motor torque shall exceed the computed average torque by at least 3 times at the hob and 1.7 times at the workpiece.

-Moments in the drives arising from accelerations during their operation in a continuous tracking mode mainly affect the work at finishing operations.

This work was financially supported by the Ministry of Education and Science of Russian Federation in the framework of the state task in the field of scientific activity of MSTU «STANKIN» № 9.1372.2017/4.6.

References

1. Ya.I. Adam, G.G. Ovumyan, *Spravochnik zuboreza[Gear cutter guide]* (Moscow, Mashinostroenie, 1983)
2. V.V. Bushuev, S.P. Nalyotov, *Tyazhelye zuboobrabatyvayushchie stanki.[Heavy gearhobbing machines]* 2nd revised and enlarged edition (Moscow: Mashinostroenie, 1986)
3. V.N. Bashkirov, A.A. Serov, *Opredelenie secheniya sreza pri zubofrezerovanii cilindricheskih kolyos chervyachnoj frezoj.[Determination of the cross section of cut in the gear hobbing of cylindrical wheels with a hob cutter]* (Stanki i instrument, 1984)
4. Ali M. Abood. *Dynamic Analysis Of The Cutting Forces In Gear Hobbing* (University of Newcastle, UK., 2002)
5. Vl.Vas. Bushuev., Vik.Val. Bushuev, V.A. Novikov. *Kinematicheskaya tochnost' zubofrezernyh stankov novogo pokoleniya[Kinematic accuracy of the new generation gear hobbing machines]*, Vestnik MGTU "STANKIN" **№3(38)** (2016).

Modeling of induction heating equipment for the vulcanization process

Sergey Karpushkin[1,*], *Aleksey* Glebov[1], and *Sergey* Karpov[1]

[1]Tambov State Technical University, 392000 Tambov, Sovetskaya 106, Russia

Abstract. The paper explores the problem of induction heating of press equipment for the vulcanization of rubber products. We describe a mathematical model of induction heating of press plates in the three-dimensional formulation given the automated stabilization of temperature, using positional and PID-control. The solution results are compared with the industrial experimental data.

1 Introduction

Hydraulic presses with heating plates, maintaining a temperature of 250 °C on the working surface for the vulcanization of rubber products and a temperature to 550 °C for treating metal products with high level of output, are often used to implement the processes of heating and pressure treatment of products. The resistive heaters or inductors can be used as heating elements.

The advantages of induction heating include high power density, reliability and durability of heaters. However, with an increase in the working temperature of the plate surface, the inductors capacity decreases a lot. In such cases, it is possible to provide the desired rate of heating by increasing the frequency of the supply voltage.

The capacity of resistive heating elements is essentially independent of the plate temperature. The disadvantage this type of heaters is a limited service life (as the wire is used it will gradually oxidize, forming areas of local overheating which will eventually lead to its destruction). In this regard, in the vulcanization process, preference is given to induction heating plates with current industrial frequency (50 Hz).

2 Statement of the problem

To reduce material costs at the stages of design, construction and operation of the equipment, improve its process characteristics it is necessary to optimize the design of the heating plate and to obtain the desired configuration of the temperature field on the work surface.

Mathematical modeling of induction heating processes is associated with the formalization of the related non-stationary electromagnetic and thermal processes in the volume of plates.

Low-frequency fields can only be effectively used for volumetric heating of ferromagnetic materials. The nonlinear dependence of magnetic field induction on its strength in ferromagnetic bodies considerably complicates the calculation of heat release. The existing methods for calculating the induction heating equipment have the following disadvantages:

1) the use of engineering calculation methods [1] based on the solution of Maxwell's linear equations in one-dimensional formulation, using empirical data does not allow for production and analysis of the specific heat distribution in ferromagnetic bodies of a complex shape;

2) the use of simplified mathematical models of induction heating process do not take into account the non-linearity characteristics of ferromagnetic materials [2,3] and can lead to significant errors in calculating the heat generation by eddy currents;

3) the approximation of the inductor geometry by the bodies of canonical shape [4] leads to large errors in most practical problems.

In this regard, it is necessary to develop effective approaches to mathematical description of electromagnetic processes and solving the systems of equations of mathematical models, which, on the one hand, take into account all the features of the processes in question, and on the other hand, do not lead to an excessive amount of computation.

3 Mathematical model of induction heating of a single plate of the press

A mathematical model for the induction heating process of a hydraulic single-plate press includes the formula to calculate the heat generation by eddy currents and the equation for heat distribution in the volume of plate given the external heat loss.

Mathematical modeling of eddy current fields is based on the allocation of two regions in the volume of the plate and the surrounding airspace:

Ω_1 is the region of eddy currents with non-zero electrical conductivity;

Ω_2 is the region without eddy currents, which might contain external source currents.

* Corresponding author: karp@mail.gaps.tstu.ru

When considering the problem of induction heating of press plates the region Ω1 is represented by the plate material, and the region Ω2 comprises inductors and surrounding airspace.

We make the following assumptions:

1) Maxwell's displacement current is absent (we consider low-frequency fields in a conducting medium);

2) the phenomenon of magnetic hysteresis is absent (high power electromagnetic fields are used) [1,5];

3) properties of materials are isotropic.

Then for the region Ω1 differential equations of the electromagnetic field (Maxwell's equations) are of the form:

$$\mathrm{rot}\mathbf{H} - \gamma\mathbf{E} = 0\,; \tag{1}$$

$$\mathrm{rot}\mathbf{E} + \frac{\partial \mathbf{B}}{\partial t} = 0\,; \tag{2}$$

$$\mathrm{div}\mathbf{B} = 0\,, \tag{3}$$

where **H** is magnetic field strength, A/m; **E** is electric field strength, W/m; γ is specific electric conductivity, $\mathrm{Ohm}^{-1}\mathrm{m}^{-1}$; **B** is magnetic induction, T; t is time, s.

For the region Ω_2:

$$\mathrm{rot}\mathbf{H} = \mathbf{J}^{ext}\,; \tag{4}$$

$$\mathrm{div}\mathbf{B} = 0\,, \tag{5}$$

where $\mathbf{J}^{ext}$ is current density supplied from an external source, $\mathrm{A/m^2}$.

Boundary conditions for the contact area of regions Ω_1 and Ω_2 are used as single-valued conditions for the variables H, E, and B, as well as on the remote edge [6].

To solve Maxwell's equations we introduced the following auxiliary functions: magnetic vector potential **A** (Wb/m) and electric scalar potential V (W), which are connected with the vectors **B** and **E** by the following relationships:

$$\mathbf{B} = \mathrm{rot}\mathbf{A}\,, \tag{6}$$

$$\mathbf{E} = -\frac{\partial \mathbf{A}}{\partial t} - \mathrm{grad}V\,. \tag{7}$$

Then the system of equations (1) - (3) can be written as equation

$$\mathrm{rot}\left(\mu^{-1}\mathrm{rot}\mathbf{A}\right) + \gamma\frac{\partial \mathbf{A}}{\partial t} + \gamma\,\mathrm{grad}V = 0\,, \tag{8}$$

where μ is absolute magnetic permeability, H/m.

Similarly, we can write the system of equations (4) - (5):

$$\mathrm{rot}\left(\mu^{-1}\mathrm{rot}\mathbf{A}\right) = \mathbf{J}^{ext}\,. \tag{9}$$

From (7) it follows that the electric scalar potential V is determined up to a constant. To resolve the uncertainty, is enough to set the value of the potential V at a point of the computational domain. According to (6) magnetic vector potential **A** is determined up to an arbitrary gradient of a scalar function. The uniqueness of solution is achieved by calibration. At present, for the calculation of the eddy current fields the Coulomb gauge is most commonly used [7]

$$\mathrm{div}\mathbf{A} = 0\,. \tag{10}$$

From the computational perspective, it is extremely difficult to ensure that field **A** is divergence-free under non-permanent magnetic permeability [8]. For the first time, the problem of satisfying the Coulomb gauge condition in a three-dimensional formulation was successfully solved by the authors in [6], who modified the differential equations (8) and (9), adding to their left parts the penalty function $-\mathrm{grad}\left(\mu^{-1}\mathrm{div}\mathbf{A}\right)$, and formulated the boundary conditions to ensure a unique solution.

Then for the region Ω1 modified equations will have the form:

$$\mathrm{rot}\left(\mu^{-1}\mathrm{rot}\mathbf{A}\right) - \mathrm{grad}\left(\mu^{-1}\mathrm{div}\mathbf{A}\right) + \gamma\frac{\partial \mathbf{A}}{\partial t} + \gamma\,\mathrm{grad}V = 0\,; \tag{11}$$

$$\mathrm{div}\left(-\gamma\frac{\partial \mathbf{A}}{\partial t} - \gamma\,\mathrm{grad}V\right) = 0\,. \tag{12}$$

For the region Ω_2:

$$\mathrm{rot}\left(\mu^{-1}\mathrm{rot}\mathbf{A}\right) - \mathrm{grad}\left(\mu^{-1}\mathrm{div}\mathbf{A}\right) = \mathbf{J}^{ext}\,. \tag{13}$$

Boundary conditions on the remote edge (the condition of zero for the normal component of the magnetic induction):

$$\mathbf{n} \times \mathbf{A} = 0\,; \tag{14}$$

$$\mu^{-1}\mathrm{div}\mathbf{A} = 0\,; \tag{15}$$

where **n** is a unit normal vector to the surface of the boundary.

Boundary conditions at the interface of the regions Ω_1 and Ω_2:

$$\mathbf{A}_1 = \mathbf{A}_2\,; \tag{16}$$

$$\mu_1^{-1}\mathrm{rot}\mathbf{A}_1 \times \mathbf{n}_1 + \mu_2^{-1}\mathrm{rot}\mathbf{A}_2 \times \mathbf{n}_2 = 0\,; \tag{17}$$

$$\mu_1^{-1}\mathrm{div}\mathbf{A}_1 - \mu_2^{-1}\mathrm{div}\mathbf{A}_2 = 0\,; \tag{18}$$

$$\mathbf{n}\cdot\left(-\gamma\frac{\partial \mathbf{A}}{\partial t} - \gamma\,\mathrm{grad}V\right) = 0\,. \tag{19}$$

where indices 1 and 2 show that characteristics belong to the corresponding region.

In [6] it is proved that the system of equations and boundary conditions (11)-(19) is equivalent to the system of Maxwell's differential equations and provides automatic execution of the Coulomb gauge (10).

To calculate the heat release in the volume of heating plate the total current density is required:

$$\mathbf{J} = \mathbf{J}^{ind} + \mathbf{J}^{ext}\,, \tag{20}$$

DOI: 10.1051/matecconf/201712901007

where $\mathbf{J}^{ind}$ is density of eddy (induced) currents (A/m²) found by the expression

$$\mathbf{J}^{ind} = -\gamma\frac{\partial \mathbf{A}}{\partial t} - \gamma \text{grad} V . \quad (21)$$

The vector of external source current density is determined in the general case by the value of the current by inductors and their structural characteristics:

$$\mathbf{J}^{ext} = f\left(I_i, \mathbf{G}^{(i)}\right), i = 1,\ldots,n_{ind} , \quad (22)$$

where I_i is current passing through the i-th inductor, A; $\mathbf{G}^{(i)}$ is vector of design characteristics of the i-th inductor; n_{ind} is the number of inductors in the heating plate.

The module of average current density of the inductor can be calculated by the formula:

$$J_i^{ext} = \frac{\omega_i I_i}{b_i h_i}, i = 1,\ldots,n_{ind} , \quad (23)$$

where ω_i is the number of winding turns of the i-th inductor; b_i, h_i are the width and depth of the slot cross-section for the i-th inductor, respectively, m.

The current direction is expressed by the basis vectors of the coordinate system, and depends on the geometry of the inductor. Structural characteristics of the most common inductors of rectangular shape can be expressed in the form of the following vector:

$$\mathbf{G}^{(i)} = \left\{d_i, \omega_i, x_i^c, y_i^c, l_i, s_i, b_i, h_i\right\}, \quad (24)$$

where d_i is wire diameter, m; x^c_i, y^c_i are coordinates of the center of the i-th inductor, m; l_i, s_i are its length and width, m.

We note that the external source currents and the eddy currents do not spatially overlap, and the equation (20) the union of regions Ω1 and Ω 2, is necessary for the correct record of relations characterizing the distribution of heat in the volume of plates.

Thus, to calculate the field of eddy currents in each calculation point of the region Ω1 it is necessary to find the 4 unknowns (3 components of the magnetic vector potential **A** and the value of the scalar electric potential V), in the region Ω2 it is necessary to find the 3 unknowns (vector **A**). As noted above, the problem of determining the unknowns is significantly complicated by the non-linearity of magnetic properties of the material to be heated. When an external current source is sinusoidal and we assume the constancy of magnetic permeability of ferromagnetic material of the plate, the expressions (11)-(13) are simplified to linear equations in quasi-stationary complex representation:

$$-\mu^{-1}\Delta\dot{\mathbf{A}} + j2\pi f\gamma\dot{\mathbf{A}} + \gamma \text{grad}\dot{V} = 0 ; \quad (11')$$

$$\text{div}\left(-j2\pi f\gamma\dot{\mathbf{A}} - \gamma \text{grad}\dot{V}\right) = 0 ; \quad (12')$$

$$-\mu^{-1}\Delta\dot{\mathbf{A}} = \dot{\mathbf{J}}_{ext} , \quad (13')$$

where j is an imaginary unit ($j^2 = -1$).

Similarly, boundary conditions (14) - (19) are simplified.

Computing time spent to solve the linear equations (11′)-(13′) are reduced by approximately one order, the main difficulty is related to the replacement of the magnetization curve with the linear dependence. To describe the process of heat distribution in a plate the non-stationary heat equation, taking into account the internal heat sources is used:

$$\frac{\partial T}{\partial t} = a\nabla^2 T + \frac{q}{c\rho} , \quad (25)$$

where $T = T(x,y,z,t)$ is temperature in the point of the volume of the plate with coordinates (x,y,z) at the time point t, °C; $q = q(x,y,z)$ is specific heat generation in this point, W/m³; $a = \lambda/(c\rho)$ is coefficient of thermal diffusivity of the plate material, m²/s; c, ρ, λ are its specific heat capacity (J/(kg·K)), desnity (kg/m³) and heat conductivity coefficient (W/(m·K)), respectively.

The average specific heat generation in the plate for the steady state mode is determined by the Joule- Lenz's law:

$$q = f \int_{t_{st}}^{t_{st}+\frac{1}{f}} \frac{J^2}{\gamma} dt , \quad (26)$$

where f is the current frequency, Hz; J is the total current density module, see (20); t_{st} is conditional stabilization time of electromagnetic processes, after which the changes in the amplitude values of the current density can be neglected, s.

When using linear equations of the electromagnetic field (11′)-(13′) heat generation is calculated by the formula:

$$q = \frac{|J|^2}{2\gamma} , \quad (26')$$

where $|J| = \sqrt{\text{Re}\,\dot{J}^2 + \text{Im}\,\dot{J}^2}$ is the complex amplitude current density module.

The initial condition for the solution of equation (25) is

$$T(x, y, z, 0) = T_0 , \quad (27)$$

where T_0 is the ambient air temperature, °C.

Heat transfer from the working surface, the cover and the ends of the heating plate is described by boundary conditions of the third kind:

$$-\lambda\left.\frac{\partial T}{\partial n}\right|_{\Omega_{pl,r}} = \alpha_r\left(T_r - T_0\right), r = 1,\ldots,6, \quad (28)$$

where $\Omega_{pl,r}$ is the r-th surface of the heating plate, m²; α_r is the heat transfer coefficient from the r-th surface of the plate, W/(m²·K), found by the criterion equation according to [9]; T_r is the average temperature of the r-th surface of the plate (°C).

We used a system of finite-element analysis of ANSYS to solve the equation (25) under conditions (11)-(13), (26)-(28) and under conditions (11')-(13') (26'), (27), (28). Fig. 1 shows the results of the solution of equation (25) under conditions (11')-(13') (26'), (27), (28) for an industrial induction heating plate sized 500×410 mm with four identical heaters with a total capacity of 5.5 kW heated from the initial temperature T_0 = 12° C to the operating temperature of 170 ° C. Fig. 1 also shows the results of the industrial experiment conducted at JSC "ARTI-Zavod", Tambov to study the process of heating the plate using several thermocouples: in the corners and in the center of the work surface (see [10]). The experimental data for the control thermocouple are shown as rectangular markers, and those for the central thermocouple are shown as triangular markers.

The maximum deviation of the calculation results from the experimental data was 6.1° C and 5.3° C (at time 180 s) for the control thermocouple and central thermocouple, respectively. The average error for the heating period of 0-1968 plate was 4.6% and 4.9% for the control thermocouple and the central thermocouple, respectively. We note that with the heating of the plate, at time of 1470 s, the deviation did not exceed 2% and 3% for the control thermocouple and the central thermocouple, respectively.

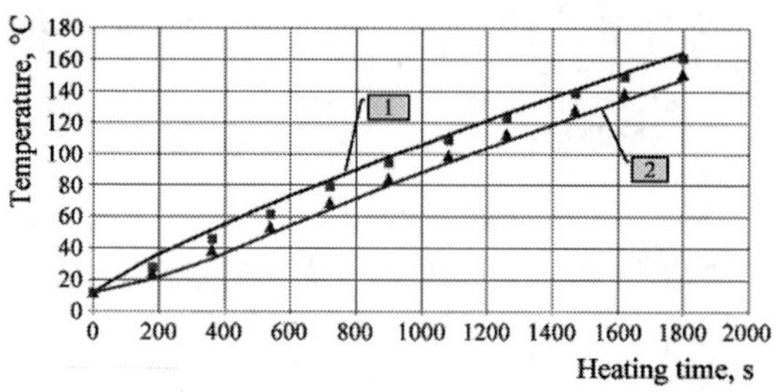

1 – temperature in the point of control thermocouple localization, 2 – temperature in the point of central thermocouple localization

Fig.1. Comparing the results of the plate temperature throughout its heating and the experimental data

The mode of stabilization of a given operating temperature on the surface of the induction heating plate can be achieved by using a two-position controller (heaters on or off). To calculate the temperature field of the plate, we propose the following expression:

$$I_i(t) = \frac{I_i}{2}\left(1 + (-1)^{I_{SW}(t)+1}\right),\ i = 1,\ldots,n_{in,} \qquad (29)$$

where $I_{SW}(t) = \begin{cases} 1, \text{если } T_s(t) < T_d \\ 0, \text{если } T_s(t) > T_u \end{cases}$ is the heaters condition ("1" – ON, "0" – OFF); $T_s(t)$ is the plate temperature in the point of control thermocouple localization, °C; T_d, T_u are lower and upper response thresholds of the controller, respectively, °C.

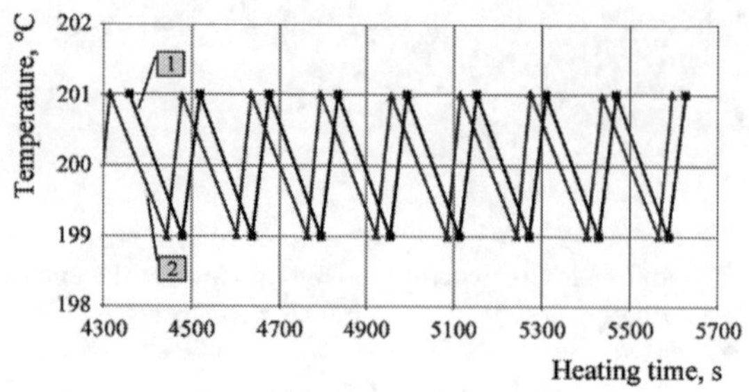

Fig. 2. Comparing the calculation results of the plate temperature in a period of stabilization with the experimental data

Fig. 2 shows a comparison of design temperatures in the place of the control thermocouple localization on the industrial induction heating plate (broken line 2) and the experimental data (broken line 1) in the automatic stabilization mode of the working surface temperature. As can be seen, a graph showing the results of calculation differs from the graph showing the experimental data only in phase: the backlog of the calculated curve from the experimental one is on average 34 s. For these time intervals, this error is negligible.

4 Results and discussion

We described a mathematical model of the induction heating process and automatic stabilization of temperature on the working surface of a hydraulic single-plate press. The model takes into account the non-linearity magnetic characteristics of the plate material and can be applied to a three-dimensional analysis of the eddy current fields.

Calculation of temperature fields of induction heating plates with a certain distribution of heat generation by the eddy currents does not present any difficulties. The main difficulties arise at the stage of solving the three-dimensional nonlinear equations of the electromagnetic field.

It is shown that in the case of constant magnetic permeability the electromagnetic field equations are greatly simplified. Nevertheless, the direct use of the simplified model when calculating the eddy currents in the ferromagnetic plate is not possible. In this regard, it remains topical to develop the relevant methods for calculating magnetic permeability equivalent to the magnetization curve by the criterion of active power.

The comparison of the calculation results with the industrial experimental data brought us to the conclusion about the adequacy of the proposed mathematical models to the real electromagnetic and thermal processes in the volume of the induction heating plate of a hydraulic press.

This work was financially supported by Ministry of Education and Science of the Russian Federation within the basic part (project 8.7082.2017/8.9).

DOI: 10.1051/matecconf/201712901007

References

1. V. Rudnev, D. Loveless, R. Cook, M. Black, *Handbook of Induction Heating* (CRC Press Reference, 2002)
2. K.H. Cho, International Journal of Thermal Sciences, **60**, 195-204 (2012)
3. A.O. Glebov, S.V. Karpov, S.V. Karpushkin. Autom. Remote Control, **75**(6), 1120-1129 (2014)
4. A. Canova, F. Dughiero, F. Fasolo, M. Forzan, F. Freschi, L. Giaccone, M. Repetto, IEEE Transactions on Magnetics, **45**(3), 1855-1858 (2009)
5. F. Bay, V. Labbe, Y. Favennec, J.L. Chenot, International Journal for Numerical Methods in Engineering, **58**(6), 839-867 (2003)
6. O. Bíró, K. Preis, IEEE Transactions on Magnetics. **25**(4), 3145-3159 (1989)
7. A.A. Rodríguez, A. Valli, *Eddy Current Approximation of Maxwell Equations. Theory, algorithms and applications* (Springer Science & Business Media, 2010)
8. L. Heyun, T. Liming, L. Udpa, Y. Sun IEEE Transactions on Magnetics, **35**(3), 1354-1357 (1999)
9. P.G. Romankov, V.F. Frolov, O.M. Flisjuk, M.I. Kurochkina, *Metody rascheta processov i apparatov himicheskoj tehnologii* [Calculation methods for the processes and devices of chemical technology] SPb: Himija, 1998. (in Russian)
10. S.V. Karpushkin, S.V. Karpov, A.O. Glebov. Transactions Tambov STU, **17**(1), 110-120 (2011)

Design optimization of heating plates for hydraulic presses

Evgeniy Malygin[1], *Mikhail* Krasnyanskiy[1], and *Vladimir* Mokrozub[1,*]

[1]Tambov State Technical University, 392000, Sovetskaya, 106, Tambov, Russia

Abstract. The paper describes the problem of design optimization of heating plates for hydraulic presses used for heat treatment of rubber, plastics, and metals. The problem involves finding the number and capacity of heaters or the heating steam pressure, the appropriate configuration of slots for heaters or steam supply channels to ensure the required heating rate of the plate to the operating temperature, and the permissible level of non-uniformity of the temperature field of its working surface. The adequacy of the mathematical description of the thermal process in heating plates for presses is confirmed by comparing the results of the solution of mathematical models with the data of the industrial experiments. The results of solving the problem for induction heating plates and resistive heating plates are presented.

1 Introduction

Heating plates are the main elements of heating systems for hydraulic presses used for heat treatment of rubber, plastics, metals and alloys. The selection of the plate design depends on the type of heating, the number and capacity of heaters (inductors or RTDs), configuration of slots, and in the case of steam heating - the required steam pressure and configuration of steam channels. The most widespread method of heating plates of industrial presses at operating temperatures up to 250 ° C is induction, because induction heaters are easy to manufacture, economical and durable. At high operating temperatures, the use of induction heating is necessitated by the use of electric current generators with a frequency of 50 Hz; thus, the use of resistive heaters is considered to be more efficient (RTDs). Regarding the use of saturated steam for heating of the plates, this method is not as energy efficient as the induction heating, as described in [1], and it has fewer possibilities for forming working temperature fields of desired configuration on the working surfaces of the plate. In addition, it is more complicated to implement.

The present paper addresses the issues of selecting optimum parameters for heating plates designs, and modeling of their temperature fields.

2 Statement of the Problem

As a rule, design and engineering departments of industrial enterprises have to solve the problem of maintaining the desired degree of non-uniformity of temperature fields of working surfaces of single uninsulated heating plates of designed presses рабочих (from ±5 to ±1°C). It is assumed that the use of such plates would ensure the uniform heating of press molds and workpieces and, as a result, it would increase the possibility of manufacturing a wide range of products. Therefore, design optimization of heating plates for presses involves determining the number and power of heaters or pressure of the heating steam, the appropriate configuration of slots to accommodate the heaters or channels to supply the heating steam. This will ensure the required rate of heating the plate up to operating temperature, and permissible level of non-uniformity of temperature field of its working surface.

It is proposed to use the difference between the maximum and minimum operating temperatures of the surface of the plate as the optimality criterion:

$$\Delta = \max_{x,y} T(x, y, w, \tau_h) - \min_{x,y} T(x, y, w, \tau_h), \quad (1)$$

where $\delta \le x \le (l-\delta)$; $\delta \le y \le (s-\delta)$; l, s is the length and width of the plate, respectively, m; δ is the width of the plate edge, which is not included in the working surface, m; $w = h$ is the plate height, m, if the working surface is its upper surface, and if the working surface is its lower surface, then $w = 0$; τ_z is the predetermined duration of heating the working surface of the plate from the ambient temperature T_0 to the predetermined temperature T_z, sec.

It is necessary to find the number of heaters or steam channels n, the necessary capacity of each of them, the configuration of the steam channels or slots to accommodate heaters $\mathbf{G}^{(i)}$, at which the function (1) reaches a minimum value and the following restrictions are fulfilled.

1. Temperature difference on the working surface of the plate:

$$\left|T(x, y, w, \tau_z) - T_z\right| \le \Delta T_z \;\; \forall (x, y) \in S_p \,, \quad (2)$$

where ΔT_z is the predetermined values of the permissible temperature difference on the working surface of the

* Corresponding author: mokr@mail.gaps.tstu.ru

plate, °C; S_p is a set of points positions on the working surface of the plate (as a point (0,0) is taken the lower left corner of the plate).

2. geometric dimensions of slots for heaters or steam channels and their position in the plate:

$$\mathbf{G}_{\min}^{(i)} \le \mathbf{G}^{(i)} < \mathbf{G}_{\max}^{(i)}, i = 1,\ldots,n , \qquad (3)$$

where $\mathbf{G}^{(i)}$ is a vector of slot parameters of the i-th heater or steam channel.

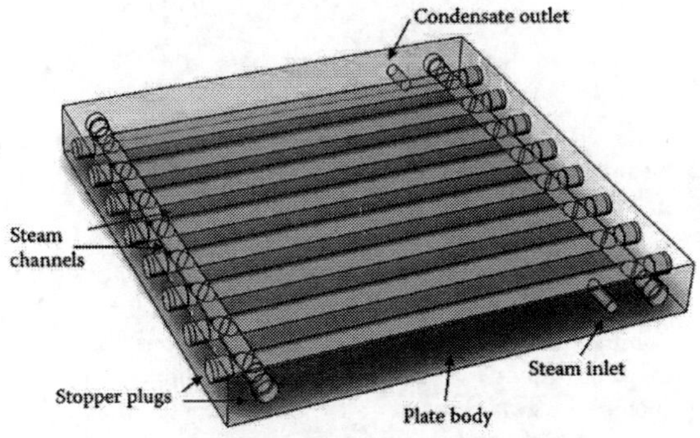

a – steam heating plate

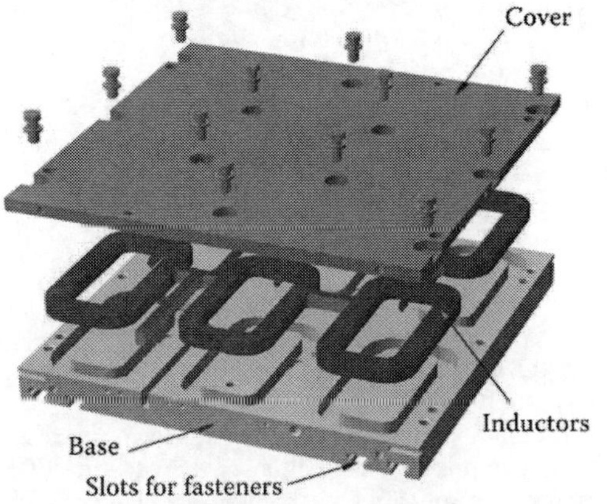

b – induction heating plate

Fig.1. Designs of heating plates for presses

Configuration of steam channels of industrial heating plates generally corresponds to the one shown in Fig.1a, therefore, the components of the vectors $\mathbf{G}^{(i)}$, $i = 1,\ldots,n$ are the inner diameter of the channel (dk_i) and the distance between its axis and the axes of adjacent channels (on the left – bl_i and on the right – br_i), or the adjacent channel axis and the edge of the working surface of the plate.

Induction heaters of industrial heating plates are typically wire coils in the shape of rectangles, as shown in Fig. 1b. In such a case, the components of the vector $\mathbf{G}^{(i)}$ are the depth (g_i), and the width (b_i) of the slot of the i-th inductor, the length (l_i) and the width (s_i) of the inductor, x-coordinate (xc_i) and y-coordinate (yc_i) of its geometric center, the values of which have to ensure the fulfillment of restrictions on the distance between the slots of adjacent inductors, between the inductor slot and edges of the working surface of the plate.

Resistive heaters (RTDs) are a wire made of a material with high electrical resistance (nichrom, fechral), coiled into a spiral. The shape of slots can be arbitrary; therefore, the components of vector $\mathbf{G}^{(i)}$ for the resistive heater are the depth (g_i) and the width (b_i) of the slot cross section, as well as the bending radius of its axis (R_i), the value of which is limited by the restriction on the bending radius of the spiral: the adjacent turns shall be external, see. Fig. 2.

The minimum permissible bending radius of the slot axis R_{min} is determined by the condition

$$h_0 > 2d, \qquad (4)$$

where $h_0 = h - \dfrac{hD}{2R}$ is distance between turns of spiral on the inner radius, m, as shown in Fig. 2; d is the heater wire diameter, m; h is the winding pitch of the straight spiral, m; D is the winding diameter of the spiral on the midline, m; R is the bending radius of the spiral along the axial line, m.

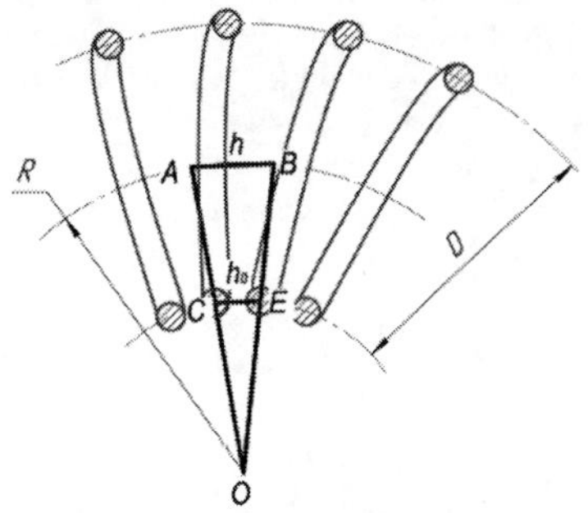

Fig. 2. Sectional vies of the bending spiral fragment

In addition, the position of slots for RTDs in the heating plate is limited by the restriction

$$V_i \cap V_j = \varnothing, \; i, j = 1,...,n, \; i \ne j , \qquad (5)$$

where V_i is the volume of the i-th slot of the heater with the inclusion of the surrounding area defined by restrictions on the distance between the slots of adjacent heaters, the heater slot and the boundaries of the working surface of the plate (m^3).

3. The length of the wire of the induction or resistive heater. For the induction heater the restriction on the number of coils (nv_i) of the wire of a given diameter given the insulation (di_i) is as follows:

$$nv_i \le \frac{4 \cdot b_i \cdot g_i \cdot \chi_i}{\pi \cdot di_i^2}, i = 1,...,n , \qquad (6)$$

where χ_i is the slot fill factor.

For the resistive heater:

$$\gamma_i \frac{U_i^2 \pi d_i^2}{4P_i} \le L_i \le \frac{Lp_i}{t_i}\sqrt{(\pi D_i)^2 + t_i^2}, \; i = 1,...,n , \qquad (7)$$

where L_i is the wire length of the i-th heater, m; D_i, t_i is the average diameter and the wire winding step of the i-th heater, m; Lp_i is the slot length to accommodate the i-ro heater, m;

$d_i = \sqrt[3]{\dfrac{4Q_i^2}{\gamma_i U_i^2 W_i \pi^2}}$ is the required diameter of its of wire, m, see in [2]; γ_i is specific electric conductivity of the heater material, ohm^{-1}m^{-1}; U_i is voltage applied to it, V; Q_i is rated power of the i-th heater, W; W_i is its surface power, W/m^2, see in [3].

4. The total capacity, which provides a desired heating rate of the plates and the heat losses are compensated by the ambient air:

$$\sum_{i=1}^{n} Q_i \geq \frac{l \cdot s \cdot h \cdot \rho}{\tau_z} \int_{T_0}^{T_z} c(T)dT + \sum_{r=1}^{6} \alpha_r S_{pl,r}(T_r - T_0), \quad (8)$$

where c, ρ are specific heat capacity, J·kg^{-1}·K^{-1}, and density, kg/m^3 of the plate material, respectively; $S_{pl,r}$ is the area of the r-th surface of the heating plate: working surface, the cover, the side ends, m^2; α_r is the heat transfer coefficient from the r-th surface of the plate, W·m^{-2}K^{-1}, see in [4]; T_r is the average temperature of the r-th surface of the plate, °C, for a period of time $[0,\tau_z]$.

We note that power generated in the time period $[0,\tau_z]$ in each of the plate channels of steam heating can be found according to [4]:

$$Q_i = \frac{G_i r(t_n)}{\tau_z}, \; i = 1,...,n, \quad (9)$$

where G_i is steam flow through the i-th channel, kg/s; $r(t_n)$ is specific heat of steam condensation at a temperature t_n, J/kg.

To determine the temperature field of the working surface of the plate it is necessary to solve the heat equation with internal heat sources [5]:

$$\frac{\partial T}{\partial \tau} = a\nabla^2 T + \frac{q}{c\rho}, \quad (10)$$

where $T = T(x,y,z,\tau)$ is the temperature in the poing ot the volume plate with coordinates (x,y,z) at the moment of time τ, °C; $q = q(x,y,z,T_{av})$ is the specific heat generation is this point, W/m^3;

$T_{av} = T_{av}(\tau) = \dfrac{1}{h \cdot s \cdot l} \cdot \int_0^h \int_0^s \int_0^l T(x,y,z,\tau)dxdydz$ is the average temperature, °C, of the plate in the moment of time τ; $a = \lambda/(c\rho)$ is the coefficient of temperature conductivity of the plate material, m^2/s; λ is its heat conductivity coefficient, W·m^{-1}K^{-1}.

The initial condition for the solution of equation (9)

$$T(x,y,z,0) = T_0. \quad (11)$$

Heat transfer from the external surfaces of the non-insulated heating plate is described by the boundary conditions of the third kind:

$$-\lambda \left. \frac{\partial T}{\partial n} \right|_{S_{pl,r}} = \alpha_r (T_r - T_0), \; r = 1,...,6. \quad (12)$$

The method for determining the intensity of internal heat release in the plate $q(x,y,z,T_{av})$ depends on the method of heating. For the steam heating plate, see Fig. 1a, if we assume identity and uniformity of heat release throughout the volume of each steam channel, we obtain:

$$q(x,y,z,T_{av}) = \begin{cases} \dfrac{4 \cdot \alpha_n \cdot (t_n - T_{av})}{dk_i}, \text{ if } (x,y,z) \in vk_i, i = 1,...,n \\ 0, \text{otherise} \end{cases} \quad (13)$$

where α_n is coefficient of heat transfer of condensing steam in horizontal cylindrical channels, W·m^{-2}K^{-1}, see in [4]; t_n is the steam condensation temperature, °C; vk_i is the volume of the i-th channel of plate, m^3.

For induction heating plate, see Fig. 1b, if we assume that heat release of each inductor is uniform in the slot volume designed for it, and the characteristics of the material volume to be filled with the inductor coil and special paste correspond to the characteristics of the material plate, we obtain:

$$q(x,y,z) = \begin{cases} q_{av0i}, \text{ if } (x,y,z) \in v_i, i = 1,...,n \\ 0, \text{otherwise} \end{cases}, \quad (14)$$

where q_{av0i} is the average specific power of the i-th inductor for the period $[0,\tau_z]$, W/m^3; v_i is the volume for the released power of the i-th inductor. m^3.

Consequently

$$q_{av0i} = \frac{Q_{avi}}{2b_i \cdot g_i (s_i + l_i - 2b_i)}, \quad (15)$$

where Q_{avi} is the average power of the i-th inductor for the period $[0,\tau_z]$, W, see in [6].

Rounded corners of inductors shown in Fig. 1b, greatly complicate the solution of the equation (9), and therefore we make an assumption about their absence, with the total length of the rectangular inductor equal to the length of the real one.

For the resistive heating plate, if we assume uniform heat release in the volume of the slot where the heater is accommodated, then

$$q(x,y,z) = \begin{cases} \dfrac{Q_i}{v_i}, \text{if } (x,y,z) \in v_i, i = 1,...,n, \\ 0, \text{otherwise}, \end{cases} \quad (16)$$

To solve equation (10) under the conditions (11), (12), (13) or (14), (15) or (16) we used the system of finite element analysis ANSYS [7]. The adequacy of the proposed mathematical description of the industrial processes of induction heating plates and resistive heating plated was confirmed by comparing the calculation results with the results of experiments conducted at JSC "ARTI-Zavod", Tambov, see Fig. 3a, and JSC "Zavod Tambovpolimermash", see Fig. 3b.

DOI: 10.1051/matecconf/201712901008

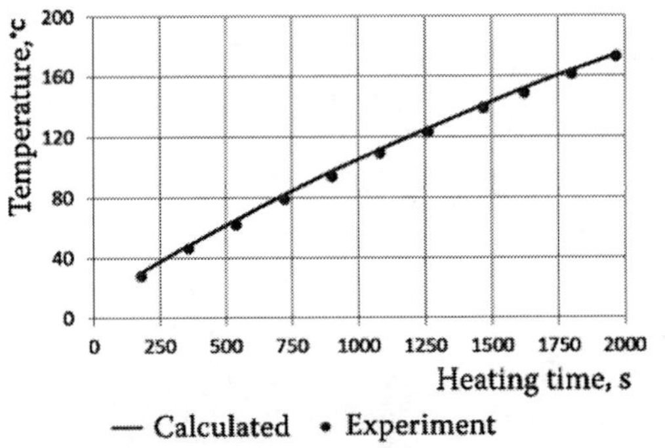

a – induction heating plate

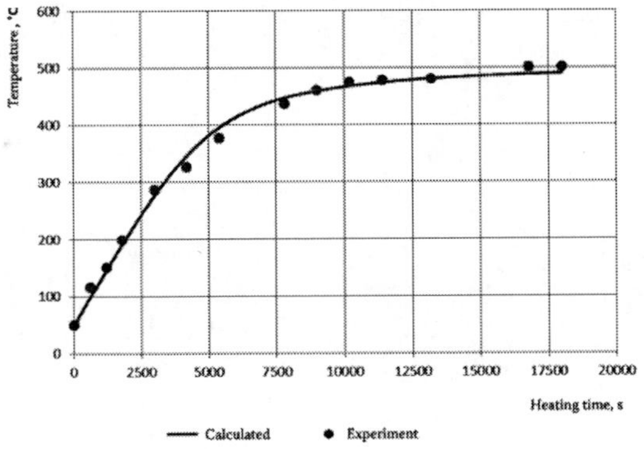

b – resistive heating plate

Fig. 3. Comparing the results of the solution of equation (10) with the experimental data

Fig. 3a shows the calculations and measurements of the temperature field at the location of the control thermocouple of the plate sized 500×410 mm with four rectangular inductors during its heating. The average absolute deviation of calculation results from the experimental data was 2.6 ° C; the relative deviation was 0.7%. For more details about the conditions and the results of this experiment, see [8]. Fig. 3b shows a comparison of the calculation results and measurements of the temperature in the center of the working surface of the resistive heating plate sized 800×800 mm and maintaining an operating temperature of 550° C. The average deviation of the calculated values and the experimental data was 8.9 ° C or 1.5%.

To solve the problem of minimizing the criterion (1) under the conditions (2), (3), (6), (8) (induction heating), conditions (2) - (5) (7) (8) (resistive heating) and conditions (2), (3) (8) (9) (steam heating) an algorithm based on the use of methods of computing experiment planning was developed [9]. The application of this algorithm to two industrial induction heating plates (rectangular heaters shown in Fig. 4a and concentric heaters) made it possible to optimize dimensions, arrangement and values of the heaters power, see in [9] and to find the power of resistive spiral heaters and configuration of the slots to accommodate them in the heating plate, maintaining an operating temperature of 550 ° C (Fig. 4b).

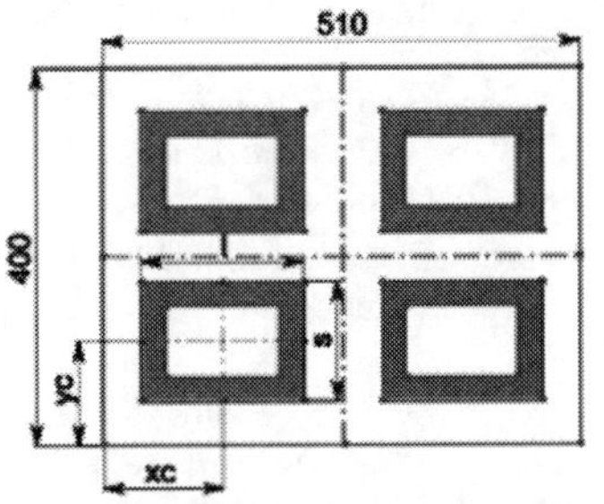

a – induction heating plate

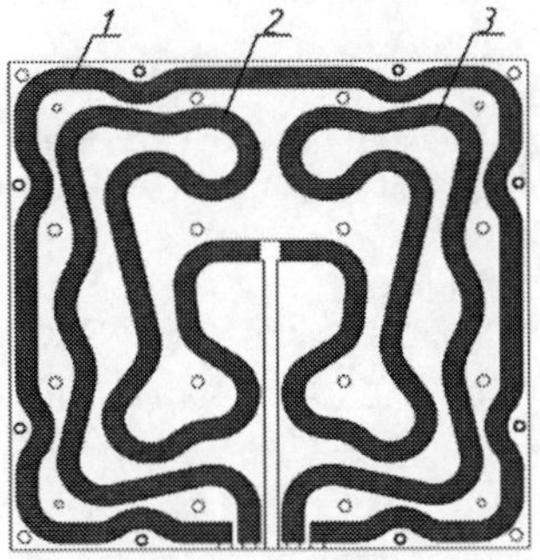

b – resistive heating plate

Fig. 4. Heating plates, for which the optimization problem has been solved

3 Results and Discussion

Thus, the statement of the optimization problem of design parameters for the heating plates of hydraulic presses involves the search for the minimum of the criterion (1) under conditions:
- for induction heating plates – (2), (3), (6), (8);
- for resistive heating plates – (2)-(5) (7), (8)
- for steam heating plates – (2), (3), (8), (9).

The solution of equation (10) of heat distribution in the volume of the plate with internal heat sources under conditions (10), (11) and conditions (13), (14) and (15) or (16) was performed using a system of finite-element analysis ANSYS. The adequacy of the mathematical description of heat distribution in the plate volume was confirmed by the results of industrial experiments.

The problem of design optimization of heating plates as solved using the method of computational experiment planning. This method was successfully used to optimize industrial designs of induction heating and resistive heating plates.

This work was financially supported by Ministry of Education and Science of the Russian Federation within the basic part (project 8.7082.2017/8.9).

References

1. S.V. Karpushkin, S.V. Karpov, A.O. Glebov. Chem. Petr. Eng., **51(5)**, 388-395 (2015)

 DOI: 10.1051/matecconf/201712901008

2. I.A. Fel'dman, M.B. Gutman, G.K. Rubin, N.I. Shadrich, *Raschet i konstruirovanie nagrevatelej jelektropechej soprotivlenija [Calculation and design of electric resistance heaters]* (M.-L.: Jenergija, 1966) (in Russian)
3. E.N. Malyigin, S.V. Karpushkin, A.O. Glebov, S.V. Karpov, Rus. Eng. Res., **36(9)**, 708-712 (2016)
4. P.G. Romankov, V.F. Frolov, O.M. Flisjuk, M.I. Kurochkina, *Metody rascheta processov i apparatov himicheskoj tehnologii [Methods for calculating the processes and devices of chemical technology]* (SPb.: Himija, 1998) (in Russian)
5. N.S. Koshljakov, Je.B. Gliner, M.M. Smirnov, Uravnenija v chastnyh proizvodnyh matematicheskoj fiziki [Partial differential equations of mathematical physics] (M.: Vysshaja shkola, 1970) (in Russian)
6. A.O. Glebov, S.V. Karpov, S.V. Karpushkin, Autom. Remote Control, **75(6)**, 1120-1129 (2014)
7. K.A. Basov, *ANSYS: Spravochnik pol'zovatelja [ANSYS: User Manual]*. (M.: DMK Press, 2005) (in Russian)
8. S.V. Karpushkin, S.V. Karpov, A.O. Glebov, Transactions Tambov STU, **17(1)**, 110-120 (2011)
9. A.V. Ostroukh, A.O. Glebov, S.V. Karpov, S.V. Karpushkin, M.N. Krasnyanskiy, Am. J. Appl. Sci., **11(6)**, 939-946 (2014)

DOI: 10.1051/matecconf/201712901009

An information model of a hydraulic press and its representation in the automated system

Evgeny Malygin[1], *Vladimir* Mokrozub[1,*], and *Vladimir* Nemtinov[1]

[1]Tambov State Technical University, 392000, Sovetskaya, 106, Tambov, Russia

Abstract. The paper describes the information model of a hydraulic press and its representation in the automated system. The proposed model makes it possible to design typical technical objects with minimal participation of a design engineer. The model allows determining the structure of a technical object, finding the dimensions of the elements and developing the working drawings.

1 Introduction

Technical object (TO) is a system that consists of the elements in a determined relationship with each other. TO and its elements have characteristics that ensure performance of functions for which is TO is designed.

Design of technical objects has two main directions: exploratory and standardized. The exploratory design results in the original design, which claims to obtain a patent. The standardized design results in the creation of TO assembled from standard elements. The paper deals with the development of an automated information system (AIS) intended for the design of hydraulic presses.

The AIS is based on the information model of the press (IMP) designed to produce documentation.

Currently, ISO 10303 standards regulate the way of presenting the information about the product, [1] proposes to use polychromatic sets to describe the properties of products, [2] provides methods of structural synthesis based on various graph models, [3] considers the presentation and design of product family, [4] describes methods of drawing up group specifications.

The above publications give little idea of how to make a description of the object in compliance with the technical specification (TS) so that it would be sufficient to produce the working draft of the TO with minimum human involvement and a full set of design drawings for parts and assembly units. Below we consider the information model of the press (IMP) and its representation in the information system that makes it possible to produce a working draft of the press according to the technical specification (TS).

2 The structure of the information model of the press

The initial data for the design of the press are the compressive force; the size of plates; the number of platens; the distance between the plates.

The AIS includes the following steps of obtaining embodiment (s) of the TO design: determining the TO structure (composition and element types), determining the dimensions of the elements, preparation of the working drawings.

According to the given steps the information model of a technical object (IM TO) can be described as follows:

$$M =< E, M^S, M^P, M^R >, \text{Fig. 1.}$$

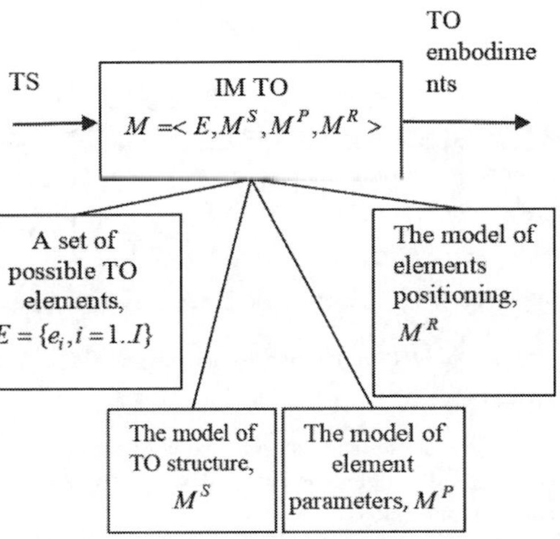

Fig. 1. The IM TO structure.

A set of possible TO elements. Each element of the TO (e_i) has a set of properties (parameters) $P^i = \{p_j^{\,i}, j = 1..J^i\}$, and each property has a set of possible values $Z^{ij} = \{z_k^{\,ij}, k = 1..K^{ij}\}$. The examples of element properties include the element type, geometry and specifications, material, etc.

* Corresponding author: mokr@mail.gaps.tstu.ru

The model of TO structure (M^S) must provide for:

- the selection of the elements which are required for the objet to perform its functions (e.g. if the object has the function of heating, it must have an element or elements providing heating);
- the selection of the type of element (e.g. if there is a heating plate and the operating temperature is more than 250 C°, then the selected heating elements are RTDs).

The basis of M^S is AND-OR tree described in some detail, for example, in [1]. A fragment of AND-OR tree for the hydraulic press is shown in Fig. 2. AND-OR tree specifies possible structures for the TO. To find the structure and elements satisfying the TS it is necessary to have the rules to select these elements. These rules can be divided into two classes:

- the rules to determine the presence of the element in the TO. For example, if heating is required, the heating device must be present. These rules allow searching by the nodes labelled "AND", i.e. to form the structure of the TO.
- the rules to determine the type of the element. The rules allow searching by the nodes labelled "OR".

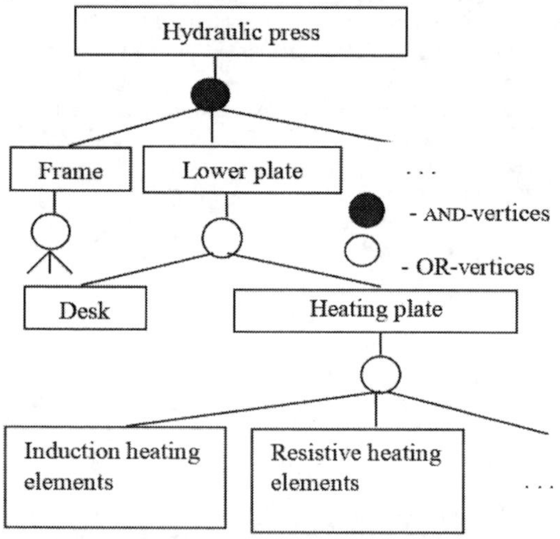

Fig. 2. A fragment of AND-OR tree for a hydralic press

The model of element parameters M^P can be presented in the form $M^P =< M_1^P, M_2^P >$, where M_1^P is a set of standard dimensions of elements, M_2^P is a set of conditions (equations, rules), connecting element dimensions.

Standard elements of the TO in practice are represented in GOST, OST, STP in tables of dimensions (tables of bolts, flanges) which have one or more dimensions determining all the other (for example, for the hydraulic cylinder it is the diameter of a piston)

Dimensions of elements are the required parameters you need to know for the production of working drawings of parts and assembly units. Their number is great, but in practice (with the given structure of the TO), and knowing the value of a small number of determining dimensions (DD) of the TO elements, one can calculate all the others. For example, for the hydraulic press, the DD is the length and the width of the heating plate and the number of platens of the press.

The dimensions of the TO elements can be divided into two types:

a) dimensions that depends on the dimensions of other elements of the TO, single parameters [1];

b) dimensions that depend on the other dimensions of the same element, unitary parameters [1].

The relationship of type b) dimensions of elements is represented as a table of standard dimensions. To represent the dependency of type a) one can use a variety of software tools, such as Mathcad.

Design engineers give preference to Mathcad as they can make and modify the expressions connecting the dimensions of type a).

The model of elements positioning. After identifying all parameters of the elements we can construct a 3D model of the TO (to assemble the elements), i.e. clearly define the position of elements relative to each other. Modern graphic editors allow building parametric 3D models of assemblies. However, their practical use in the design of a new TO is difficult, because you first need to build a lot of options of parametric 3D models, which are determined by AND-OR tree. It is preferable to make a mathematical description of the designed object, which will allow for a particular kind of structure to build its 3D model automatically. This description is called a model of elements positioning.

The formal presentation of the positioning model is as follows:

$$M^R =< E^{3D}, S, Y^R >,$$

where E^{3D} is a set of 3D models elements; S is a register of types of interfaces between the basic axes, edges and faces of the elements; Y^R is the rules governing the interface between basic geometrical parameters of elements.

Any solid-state geometric object is characterized by a plurality of axes, edges, surfaces, which can be called the basic geometric elements (BGEs). The BGE interface types may be as follows:

s_1 is parallelism (or concurrency) of planes or axes;

s_2 is concentricity;

s_3 is intersection angle relative to each other (including perpendicularity);

s_4 is tapping;

s_5 is symmetry, etc.

When positioning the TO elements in space, firstly, we must identify the BGEs (axis edges, edges) of each element of the TO. Secondly, we must define the interfaces between the BGEs of the connected elements of the TO. By doing this, we obtain:

$$Y^R = \bigcap_t (S_k{}^t, p_{in}, p_{jm}, x_t),$$

where t is the interface number, S_k^t is the interface type ($k=1$ is parallelism, $k=2$ is concentricity, etc.), p_{in} is the n-th BGE of the i-th element of the TO, p_{jm} is the m-th BRE of the j-th element of the TO, x_t is the parameter characterizing the interface t (the angle at the intersection of elements, the distance between the planes in the case of parallelism, etc.)

3 Representation of the information model in the relational database

In a relational database, a set of possible elements of the TO $E=\{e_i, i=1..I\}$ can be realized by the technology OBJECT-PROPERTY-VALUE (Fig. 3)

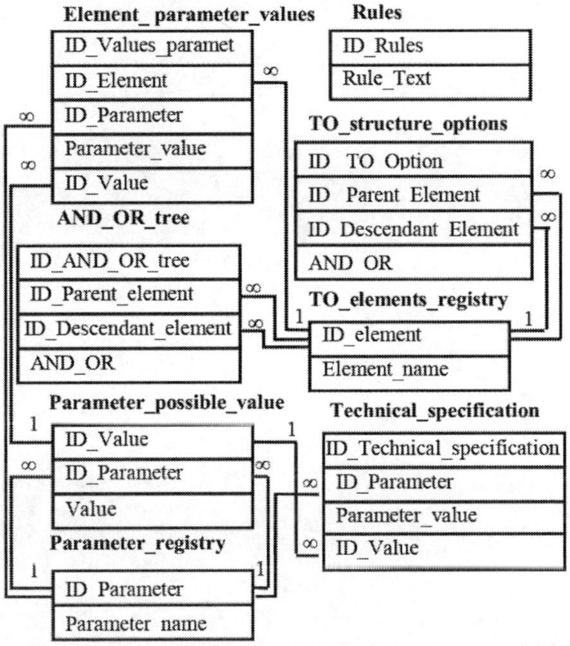

Fig. 3. The ER-diagram of a set of the TO possible elements and the model determining the TO structure

A set of the TO possible elements and their parameters is given in the Tables: "TO_ elements_registry", "Parameters_registry". "Parameter_oossible_values" and "Element_ parameters_values". For example, an element called "Hydraulic cylinder" is stored in the Table "TO_elements_registy", a parameter called "Piston diameter" is stored in the Table "Parameters_registry" and the "Parameter value" of 400 is stored in the Table "Element_ parameters_values". The values of some parameters are a bounded set (e.g., the "Body_Material" can take on values from a finite list of steel). The lists of possible values of the parameters are in the Table 'Parameters_possible_values".

For representation of an AND-OR tree in a relational database, we use a hierarchical structure provided in [5]. In such a case, the Table "TO_elements_registry" contains both physical elements and element classes (e.g., bottom, unit flange, etc.). Obviously, classes can have parameters typical of the class elements. The Table "Technical_specification" (Fig. 3) contains the requirements to the TO specifies in the Table "Parameters_registry". The Table "TO_structure_options" contains the result of combining the rules of selecting the elements from the Table "Rules" and the Tables "AND-OR_tree", "Element_ parameters_valuess" and "Technical_specification".

We consider one of the possible options for the rules in the Table "Rules". Assuming that for the parameter "Working temperature" ID_Parameter = 50, for the element "Induction heating elements" ID_Element = 15. Then the rule "If the operating temperature is less than 250 ° C, the heating element must be induction type" in the form of a SQL statement is as follows:

SELECT ID_Element, Name_element FROM Registry_Elements_TO PE WHERE PE.ID_Element=15 AND EXISTS (SELECT * FROM Technical_specification TS WHERE TS.ID_Paremeter=50 and TS.Value_parameter <250)

One can start checking rules in MS-SQL (T-SQL) using EXECUTE statement or SP_EXECUTESQL system procedure; the latter is preferable, as it allows for a dynamic substitution of parameters.

For representation of standard dimensions of elements in the relational database, one can use the following methods:

1) add dimensions of elements to the table "Parameters_registry" and fill in the Table "Element_ parameters_values" for each element (the Table "TO_elements_ registry");
2) for each element of the TO create a table containing element dimensions and add a field with the name of the this table to the Table "TO_elements_ registry";
3) create a wide table (for example, a table with a number of fields up to 100-200), which contains dimensions of all standard elements (one line corresponds to one physical standard element, one column corresponds to one dimension);
4) store information about dimensions in the form of the text string of the type "name_dimension =value_dimension", e.g., "D1 = 1000, D2 = 500, S = 8".
5) use combinations of the above listed methods.

All these methods have advantages and disadvantages. For example, when using method 2, we have lots of small tables; when using method 3, hard disk space is used inefficiently; when using method 4, it is necessary to make additional efforts to cut the actual value of the dimension out of the line.

The ER-Diagram of the positioning model is shown in Fig. 4.

To build the working drawings of the TO with minimal participation of the design engineer, one must have a library of parametric 3D models and 2D drawings of the elements. Legends of parametric dimensions must match the legends in the model of element parameters determination M^P. Here, we define the BGEs, which are used in the positioning model M^R.

Dimensions in the model of element parameters determination can be designated as follows: unique_name_element.parametric_dimension.

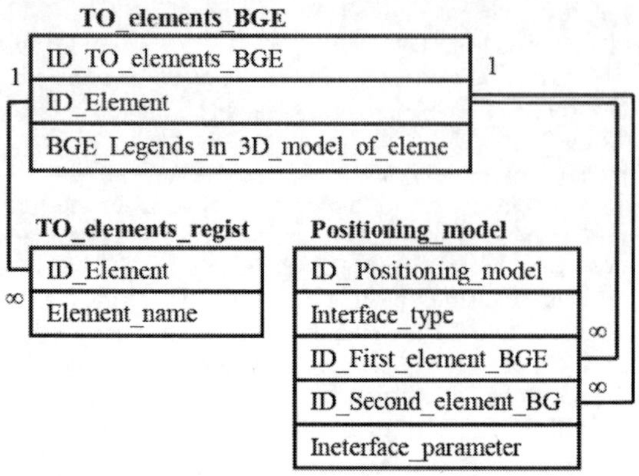

Fig. 4. The ER-diagram of the positioning model

Modern graphic editors (Compass, T-Flex, Inventor) have a programmer interface to create software for the following functions:
- reconstructing 3D models and 2D drawings of the TO elements from the library of standard elements using dimensions from M^P;
- constructing assembly units from the resulting elements of 3D models on the basis of the positioning model M^R.

4 Results and discussion

We considered the TO IM for the hydraulic press and its representation in the automated information system. The created IM consist of the registry of elements, the structure model, the parameters model, the TO positioning model. For representation of the IM in the automated system, we used the relational database. We described the structure of the relational database for the storage and processing of all elements of the IM.

The proposed structure of the IM and methods of its representation make it possible to create intelligent automated system for technical documentation generation with minimal human participation at different stages of the TO construction [6].

This work was financially supported by Ministry of Education and Science of the Russian Federation within the basic part (project 8.7082.2017/8.9).

References

1. V.V. Pavlov. *Strukturnoe modelirovanie v CALS-tehnologijah [Structural modeling in CALS-technologies]* (M.: Nauka, 2006) (in Russian)
2. G.S. Ivanova. *Sposoby predstavlenija strukturnyh modelej [Methods of representation of structural models],* Nauka i obrazovanie Jel. № FS 77-30569, № 1 (2007) (in Russian)
3. V.M. Tret'jakov, *Osnovy metodiki proektirovanija semejstva izdelij [Fundamentals of product family design techniques],* Avtomatizacija i sovremennye tehnologii. **№ 2,** pp.25-33 (2004) (in Russian)
4. V.G. Mokrozub, *Predstavlenie struktury izdelij v reljacionnoj baze dannyh [Representation of product structure in the relational database],* Informacionnye tehnologii, **№11**, pp. 11-13 (2008) (in Russian)
5. V.G. Mokrozub, K.D. Manuilov, V.V. Gorshkov, T.S. Gorshkova, Chem. Petr. Eng., **51**(9-10), pp. 613-618 (2015)
6. V.G. Mokrozub, V.A. Nemtinov, Chem. Petr. Eng., **51**(7), pp. 487-492 (2015)

The perfection of the construction of a combined cutting tool on the basis of the results of mathematical modelling of working cutting processes in RecurDyn

Vladimir Poddubny[1], *Yuri* Kryazhev[1], and *Eugene* Gorin[1,*]

[1]Altai State Technical University (ASTU), 46 Lenin St., Barnaul, 656038, Russia

Abstract. As the title implies the article describes how to optimize the construction of a combined cutting tool on the example of developed design of the face milling cutter with regulable rigidity of damping elements in order to improve the vibration resistance of the cutting process. RecurDyn is proposed, which is widely used for creating models of different mechanical systems, their analysis and optimization of construction, uses the ideology of visual object-oriented programming and computer research of volume solid-state models. Much attention is given to the description of the mechanical and mathematical model of the face milling cutter in RecurDyn and the results of mathematical modeling of the face milling cutter with damping elements, consisting of individual elements, with the possibility of program controlling its operation in the process of cutting. The applying of RecurDyn made it possible to carry out a complex assessment of influence of separate elements of a design of the combined cutting tool on quantitative and qualitative parameters of milling process and to define optimal values of the input and output parameters of technological process of machining for various damping elements.

1 The modeling of mechanical systems in RecyrDun

One of the most important stages in the design of cutting tools is the development of their mathematical models and subsequent modeling for a preliminary assessment of the quality of functioning, in order to determine the degree of influence of the produced structural or technological changes on the qualitative and quantitative parameters of the cutting process.

In RecyrDun the ideology of visual object-oriented programming is used. This software is designed to create models of various mechanical systems, their analysis and optimization.

The core module of the RecyrDunProfessional consists of Body, Joint, Force and Contact.Body makes it possible to include solid bodies of correct geometric shape in the structure (ellipsoid, cylinder, parallelepiped, torus, cone). Mass geometric characteristics are determined by body sizes and material densities, which are set by the user. Joint contains a set of connections providing various movements of the connected system bodies relatively each other. Force allows to describe the various forces acting on the bodies of a mechanical system (the elastic force of Spring with a given rigidity and damping, the force of Axial, acting in the given direction and described by the analytical expression, and many others). The great advantage of RecyrDun in comparison with the most known programs for modeling multibody mechanical systems is the possibility of describing contacts between the bodies of the system without an analytical description of this interaction. Contact gives us this possibility.The interface of contact interaction allows setting the direction of force interaction, stiffness, and damping. Another advantage of RecyrDun is the possibility to include the deforming finite element bodies in the structure of solid bodies system, to make the contact between solid and deformed bodies. It is provided by Flexible.

It is possible to import 3D models from other software (CATIA, Parasolid, SolidWorks).

2 The improving the vibration resistance of the cutting process by using combined cutting tools with regular rigidity

The vibrations of the technological system, which occur during the cutting process have a great influence on the accuracy of machining and the firmness of the cutting tool. When the fluctuations of the cutting tool appear relatively to processed surface its firmness decreases and there are some cross-section errors on the workpiece, a sinuosity develops and the roughness of the processed surface increases. At the same time the productivity decreases, because of the increase in idle time of the equipment, connected with the replacement and adjustment of the tool. The vibrations generated by machining after reaching certain amplitude values are further strengthened due to the instability of the cutting forces at incision and pushing of the cutting tool, as well as by changing the true cutting angles with relative oscillations of the system elements, which leads to changes in the cutting force and the movements caused by it. At small rigidity of technological system, these sources

* Corresponding author: evgenijgorin@gmail.com

DOI: 10.1051/matecconf/201712901010

can not only enhance the oscillations that have arisen, but also become the reason of their emergence. Due to the harmful effects of the vibrations of the technological system on the accuracy, productivity and firmness of the cutting tool at the design of technological processes, the construction of machining attachments and the cutting tool, the task of complete elimination of danger of vibrations or reducing them to the minimum acceptable level is set. The most effective means of struggling vibrations and enhancing the vibration resistance of a technological system is the regulation of rigidity of all elements of a dynamic system, where the most important link in this system is "detail-tool". The degree of stability of the technological system determines its ability to dissipate the energy introduced by the external action, and can be estimated from the rapidity of attenuation of deviations in transition processes [1]. The greater the degree of stability in this particular component of the dynamic system, the faster the decay of transition process of incision of the cutting teeth, the less deviation in the set cutting process.

In this article we consider the reducing of vibrational loads on technological system by absorbing the energy of the oscillating motion of the face milling cutter and controlling the vibration resistance of the cutting tool by increasing its damping properties. At the department of "Technology of Mechanical Engineering" Altai State Technical University the construction of a face mill is developed (a figure 1), in which rubber inserts (damping elements) are used to regulate its rigidity, the number of which can vary depending on the required cutting conditions and the natural frequency of the cutting tool (patent No. 2559078). The main advantage of the proposed method is its wide universality and, as a result, the possibility of its application not only for mills, but also for other types of cutting tools. Damping elements that regulate the rigidity of the face milling cutter can be selected by material and design, taking into account the dynamic characteristics of the technological system, which increases the efficiency of its use.

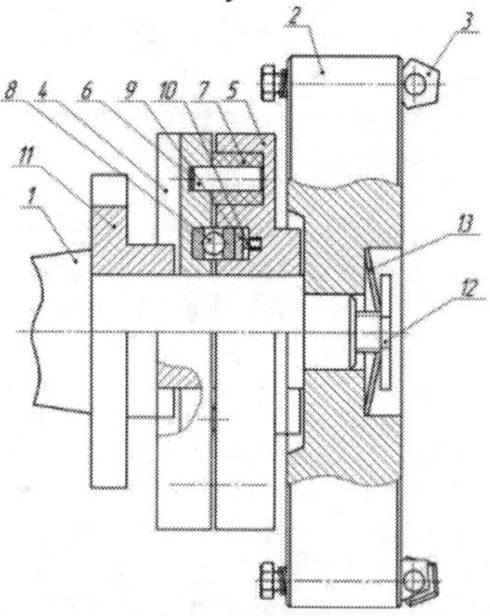

Fig. 1. Combined face mill with damping elements. 1 – mandrel; 2- milling cutter body; 3 - cutting plates; 4 - top ring of knot of damping; 5 – low ring of knot of damping; 6 – pin; 7 - elastic damping elements; 8 - thrust bearing; 9 - distance washer; 10 - elastic figured washer; 11 – lead; 12 - adjusting screw; 13 - spring washer

3 Mathematical model of calculation of vibration characteristics for the processes of face milling

The tool (milling cutter) is represented as a single-mass model with 5 degrees of freedom, which corresponds to the given mass-rigid characteristics.

Figure 2 shows the scheme of the tool and degree of freedom of its dynamic model: x, y - linear movements relative to the principal axes ox, oy; Ψ_x, Ψ_y - angular movements in the planesxoz и yoz; θ – amount of twist; ω – angular speed of rotation of the tool; D – diameter of a mill; S_Z – feed per tooth; φ – mill angle of rotation; φ_P – central working corner of a mill; φ_{en} - mill angle of entry.

The system of the differential equations of process:

$$\begin{cases} m_X \dfrac{\partial^2 x}{\partial t^2} + \eta_X \dfrac{\partial x}{\partial t} + C_X x = P_X(x, y, \psi_X, \psi_Y, \theta, t) \\ m_Y \dfrac{\partial^2 y}{\partial t^2} + \eta_Y \dfrac{\partial y}{\partial t} + C_Y x = P_Y(x, y, \psi_X, \psi_Y, \theta, t) \\ I_X \dfrac{\partial^2 \psi_X}{\partial t^2} + \eta_{\psi_X} \dfrac{\partial \psi_X}{\partial t} + C_{\psi_X} \psi_X = M_X(x, y, \psi_X, \psi_Y, \theta, t) \\ I_Y \dfrac{\partial^2 \psi_Y}{\partial t^2} + \eta_{\psi_Y} \dfrac{\partial \psi_Y}{\partial t} + C_{\psi_Y} \psi_Y = M_Y(x, y, \psi_X, \psi_Y, \theta, t) \\ I_\theta \dfrac{\partial^2 \theta}{\partial t^2} + \eta_\theta \dfrac{\partial \theta}{\partial t} + C_\theta \theta = M_\theta(x, y, \psi_X, \psi_Y, \theta, t) \end{cases} \quad (1)$$

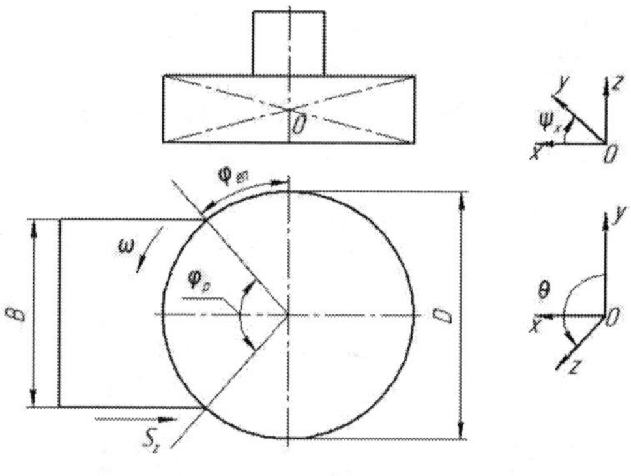

Fig. 2. Scheme of operation of the tool and degree of freedom of dynamic model.

where m_X, m_Y– given masses for the directionsox, oy; I_x, I_y, I_θ– given moments of inertia of the tool relative to the main axes ox, oy, oz ; η_x, η_y, $\eta_{\psi x}$, $\eta_{\psi y}$, η_θ– damping ratio; C_x, C_y, $C_{\psi x}$, $C_{\psi y}$, C_θ– given rigidity; P_x, P_y, M_x, M_y, M_θ– according to effort in the directionox, oyand external moments relative to the main axes ox, oy, oz; t – time.

The interaction between the tool and the workpiece during milling is replaced by the effect on the cutting force tool system: axial– P_O, radial – P_R, tangential – P_τ. Depending on the type of material and processing conditions, the following relationships are obtained from the experimental results for cutting forces:

$$P_R/P_\tau = 0{,}3 \div 0{,}4; P_O/P_\tau = 0{,}4 \div 0{,}45, \quad (2)$$

DOI: 10.1051/matecconf/201712901010

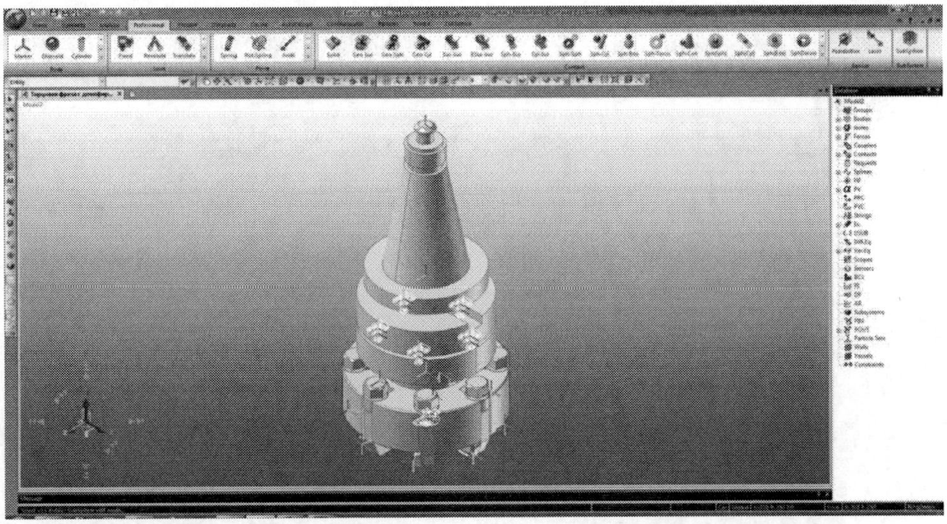

Fig. 3. RecyrDun program window with imported solid model of face milling cutter with damping elements.

4 Results of modeling of working processes of cutting by a face mill with dampening elements

To work in RecyrDun a solid model of the face mill should be used, created in Solidworks Version 2016.

The face milling process was simulated with the following initial data:D=120 mm – diameter of a mill; z=8 – number of teeth; t=1 mm – cutting depth; B=85 mm – cutting width; S_z=0,1 mm/tooth – feed; n=500 rpm – frequency of rotation of a mill; v=188,4 m/m - milling speed. Within the conducted researches process of cutting was modelled with use of the damping elements from rubber (HL 3.1) and steel (St45). Characteristics of knot of damping - radial rigidity and damping were set in baseParametric Value List (Figure 4).

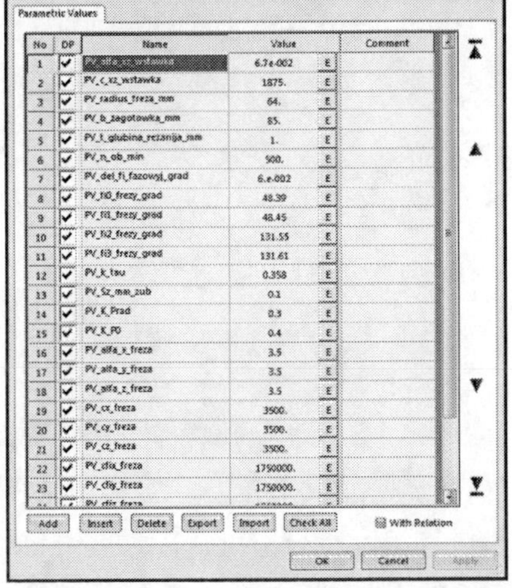

Fig. 4. Initial parameters of modeling of process of milling in RecyrDun

The research in Analisis made it possible to simulate the dynamic state of the combined face mill in the process of cutting steel workpiece of specified sizes and physical and mechanical properties and to carry out a dynamic analysis for various parameters of the milling regime.In this case, RecurDyn, taking into account the connections between the elements, develops equations that connect the reaction with the property of the material, limit and load. After this RecurDyn ordering the equations in a big system of the joint algebraic equations, finds unknown.

During the research Dynamic/Kinematic Analysis graphs of vibration movements along the Z-axis of the combined face milling cutter were obtained while milling the steel workpiece with various damping elements. Figure 5 shows an example of a comparison of the Z-axis oscillation curves for face milling for inserts of rubber and steel.

As can be seen from the graphs shown in Fig. 6, the magnitude of the vibration movements along the Z-axisfor milling with damping elements made of steel is greater than the value of the vibration movements for milling with damping elements made of rubber by 20-30% during the processing time. In this case, the milling process with damping elements made of rubber is stable, since the swing amplitude along the Z axis is stable.

The results of the simulation in RecyrDun correlate with the results of the experimental work carried out on the vertical 5-axis Mazak Variaxis 630 machining center. The workpiece from 30HGSN2A-VD was processed by a projected face milling cutter on the cutting modes: t=1 mm; - cutting depth; S_z=0,1 mm/tooth – feed; n=500 rpm – frequency of rotation of a mill; v=188,4 m/m – milling speed.

On the studied parameters of cutting modesvalue of a roughness of the processed surface in the Ra parameter has decreased by 19,2% for a combined face mill with the damping elements from rubber in comparison with a roughness of the processed surface worked by a mill with the damping elements made of steel. In this case, the vibration movement during the modeling in RecyrDun

DOI: 10.1051/matecconf/201712901010

according to standard deviations and swings with a high accuracy is related to the roughness of the processed surfaces obtained during the experimental studies. Thus, in the process of face milling with a mill with rubber inserts, the oscillations of the cutting tool are reduced, which decrease the roughness of the processed surface in the same ratio as in the modeling in RecyrDun.

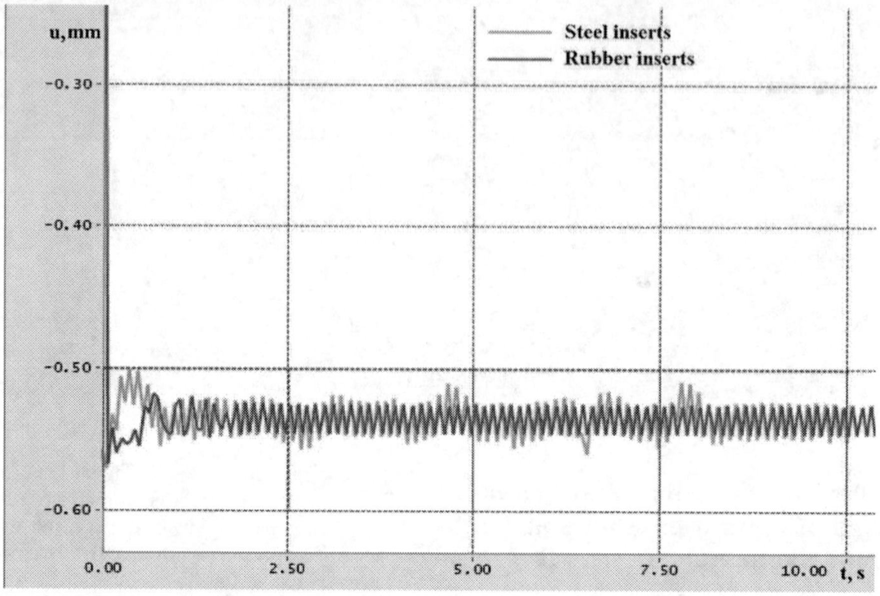

Fig. 5. Comparison of the vibration movements graphs along the Z-axis of the combined face milling cutter with steel and rubber damping elements for modeling face milling in RecyrDun

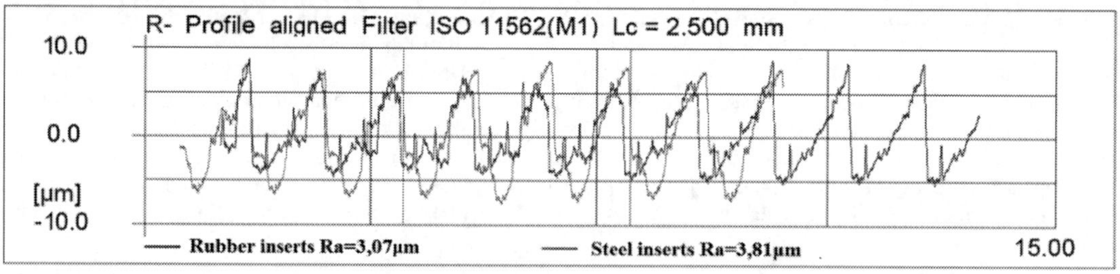

Fig. 6. Oscillograms of measurement on profilograph-profilometer "HOMMEL TESTER W55" roughness Ra of the processed surface when milling with a combined face milling cutter on cutting modes: t=1mm, SM=0,1 mm/tooth, V= 188,4 m/m

References

1. E. U. Tatarkin, U. A. Kryazhev, E. S. Ognevenko, *New aspects of the application of the method of acoustic emission in engineering and metalworking* (ASTU, 2015)
2. V. I. Poddubny, U. A. Kryazhev, E. U. Gorin, *Development of a 3-D model of a dynamic cutting process with a face mill in RecurDyn*. Actual problems in mechanical engineering: **III** International Scientific and Practical Conference (2016)
3. *RecurDyn. Professional Multi Flexible Body Dynamics Simulation Software* [Electronic resource]. URL: http://functionbay.co.kr/pdf/brochure_e_web.pdf

The ultrasonic grinding process temperature field study

Alexander Unyanin[1] and *Albert* Khusainov[2,*]

[1]Ulyanovsk State Technical University, Department of Machine Building, 432027, Sev. Venec, 32, Ulyanovsk, Russia
[2] Ulyanovsk State University, Department of Physics and Engineering, 432017, L. Tolstoy, 42, Ulyanovsk, Russia

Abstract. The use of ultrasonic energy is a promising way to improve the efficiency of the grinding process. However, the analytical study of local temperatures during ultrasonic grinding was not performed. In the process of research, physical and mathematical models have been developed for calculating the temperature field during grinding, taking into account the change of the kinematics of microcutting by abrasive grains and the change of the mechanical characteristics of the workpiece material when ultrasonic vibrations are applied. The models take into account that the parameters, characterizing the workpiece material resistance to dispersing, and the thermophysical properties of abrasive grain, workpiece, chips and external environment, depend on the temperature. The modeling was performed on the basis of a simultaneous solution of the thermal conductivity differential equations, written for each interacting object. For equation calculation the finite-element method was used. The methodology and software for the temperature field calculation have been developed. The temperature modeling results are shown. The effect of the vibration amplitude and phase on the local temperatures in the area of abrasive grain, chips and workpiece contact, including the grain and the workpiece contact time, has been determined. The factors affecting the local temperatures have been determined. During ultrasonic activation the workpiece temperature is lowered by 10 %, and the local temperatures in the abrasive grain, workpiece and chips contact area are lowered by 30 %.

1 Introduction

Grinding wheel performance improvement and treated parts quality improvement while ultrasonic grinding are largely due to this process thermal intensity change [1-3]. However, the analytical study of the temperature field while ultrasonic grinding was not performed.

The thermal processes in grinding are studied sufficiently well [4-6], but in most analytical studies the grinding wheel and the workpiece contact area represents a solid thermal source and the average contact temperatures are calculated. Local temperatures must be known in order to assess the grinding wheel working conditions and the quality parameters of the grinded workpiece thin surface layer [7-9].

The aim of the study is to determine the ultrasonic vibration parameters affecting the temperature field of the grinding process.

For achieving the aim the following problems were solved:

– physical and mathematical models development for the temperature field calculation;

– numerical modeling methodology and software development for the temperature field calculation;

– temperatures numerical modeling and the analysis of the obtained results.

2 Study Methodology

Consideration has been given to heat release in the area of the workpiece material deformation with the abrasive grain and in the areas of grain contact with chips *NE* and the workpiece *EV* (the densities of the heat release sources are q_{∂}, q_{1T}, q_{2T}, respectively) (Figure 1). It was assumed that the source of q_{∂} density was evenly distributed, and the sources of q_{1T} and q_{2T} density were distributed by antisymmetric normal and exponential law, respectively [9-11]. The components of the microcutting force, the power and the thermal sources density were calculated by the analytic dependences [9, 10]. For example, the heat release density in the grain and the workpiece contact area at the distance of x from the point E (see Figure 1) may be calculated as follows

$$q_{2T}(x) = q_{2T} \cdot \exp[-k_0 \cdot x]; \qquad (1)$$

$$q_{2T} = \frac{W_{2T} \cdot k_0}{l_2 \cdot \exp[-k_0 \cdot l_2]}; \; W_{2T} = \frac{\mu_o \cdot \tau_s \cdot f_t \cdot V_K}{3\mu_s}, \qquad (2)$$

where k_0 is coefficient, m^{-1}; f_t is a surface area of the grain friction against the workpiece, m^2; μ_s is an internal friction coefficient [12]; l_2 is a size of the site dullness on the grain, m; τ_s are shear stresses, Pa [12-14]; μ_0 is a coefficient of the grain friction against the workpiece; V_K is a working speed of the grinding wheel, m/s; W_{2T} is a heat release power, W.

* Corresponding author: a.s.khusainov@gmail.com

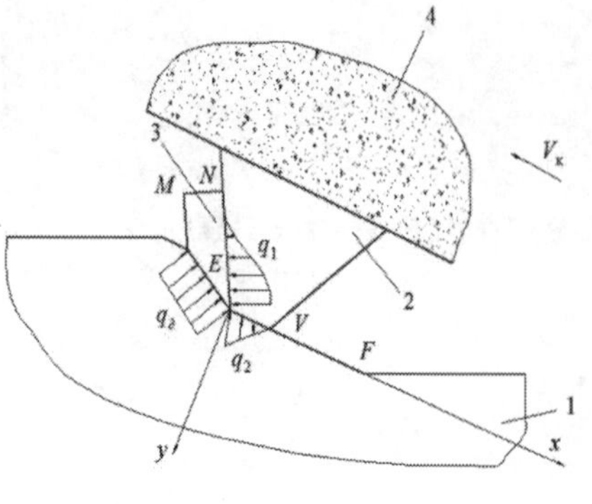

Fig. 1. The scheme for the heat exchange calculation in the grain and the workpiece contact area: 1 – workpiece; 2 – grain; 3 – chips; 4 – grinding wheel

The abrasive grain is presented as a truncated pyramid, the chips and the workpiece are presented as the cuboids [5, 9]. To improve the accuracy of the results, the dependences of the parameters characterizing the workpiece material resistance to dispersing [14, 15] are taken into account, and the dependences of the thermophysical properties of all the interacting objects (including the external environment objects) on the temperature are also considered. The modeling was performed on the basis of a simultaneous solution of the thermal conductivity differential equations, written for each interacting object, and considering the speed of their relative displacement (of the grain in relation to the workpiece and of the chips in relation to the grain).

To consider the temperature field formation as a result of pile-up of thermal pulses from separate grains, displacement (see Figure 1) of a series of the grains successively contacting with the workpiece was modeled.

When flat grinding without ultrasonic vibrations the depth of the cutting grain penetration into the workpiece material depending on the distance l, covered by the grain on the trajectory of contact with the workpiece (Curve 1 in Figure 2), is described by the dependency

$$a(l) = a_{\text{k}} + A \cdot l, \tag{3}$$

where a_{k} is a critical microcutting depth, m; A is a coefficient [9, 10].

When there is the workpiece vibration in the direction which is perpendicular to its processed surface (see Figure 2), the depth of the cutting grain penetration into the workpiece may be presented as follows

$$a_1(l) = a(l) + A_{\text{u}} \cdot sin(\omega \frac{l}{V_{\text{K}}} + \varphi), \tag{4}$$

where A_{u} is the amplitude of the workpiece vibration in the direction which is perpendicular to the processed surface, m; ω is a vibrations cyclic (circular) frequency, rad/c; φ is an ultrasonic vibration phase, rad.

When ultrasonic vibrations take place on a part of the trajectory of contact with the workpiece, the grind will have a greater penetration depth than that without vibrations; on the other part of a contact the penetration depth is less (see Figure 2).

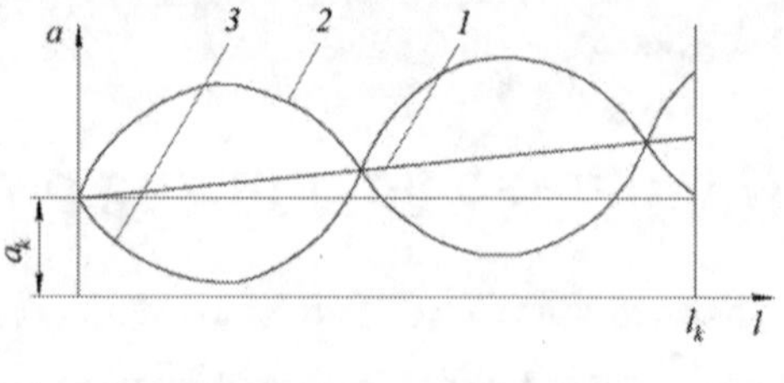

Fig. 2. Dependence of the depth of the grain penetration into the workpiece a on the distance l covered by the grain on the contact trajectory: 1 – without vibrations; 2, 3 – with the workpiece vibration with the phase φ = 0° and 180°, respectively

The workpiece vibration with ultrasonic frequency results in a qualitative change of the process of its material dispersing by the abrasive grains of the grinding wheel. Ultrasonic activation is accompanied by the workpiece material ultimate strength and fluidity reduction and the reduction of its ability for hardening in the deformation process [1].

The average sheer stresses by sheer area, defining the grinding wheel grains microcutting force,

$$\tau_s = \frac{\sigma_{\text{B}} \cdot 1,558 \cdot e^{2,34 \cdot 10^{-3} \cdot T}}{\sqrt{3}}, \tag{5}$$

where σ_{B} is a workpiece material temporary resistance, Pa [14, 16]; T is a workpiece material deformed layer temperature, K. The ultrasonic activation of the workpiece results in the reduction of σ_{B} by 10 … 15 % [1].

The thermal conduction differential equations of the objects (grain, workpiece, grinding wheel and chips) were solved in conjunction with the general boundary conditions in the contact area, which allows to consider the thermal flows distribution between the objects. For example, in the limits of a contact of the workpiece 1 with the grain 2 (see Figure 1) the surface interaction conditions are described by the following boundary conditions:

$$\frac{\partial T_1}{\partial y} = -\frac{q'_{2T}}{\lambda_1};\ \frac{\partial T_2}{\partial y} = -\frac{q''_{2T}}{\lambda_2};\ T_1 = T_2; \tag{6}$$

$$q'_{2T} + q''_{2T} = q_{2T}, \tag{7}$$

where T_1 and T_2 are the workpiece and the grain temperatures in the contact area; λ_1, λ_2 are the workpiece and the grain materials conductivity coefficients, W/(mK); q'_{2T}, q''_{2T} are the density of the thermal flows into the workpiece and the grain, respectively.

Heat exchange at the boundaries of the objects, which contact with the process liquid and air, is set in the form of Newton-Richmann law describing the convective heat exchange process as

$$-\lambda_z \frac{\partial T_z}{\partial n} = \alpha_j \cdot (T_z - T_0), \quad (8)$$

where z is an object number (z = 1, 2, 3, 4 for the workpiece, the grain, the chips and the grinding wheel, respectively); j is a surface number; α_j is a heat transfer coefficient from j surface, W/(m^2·K).

The finite-element method was used for the conductivity differential equations calculation [17]. Numerical procedure on the basis of the conductivity equations discrete analogs is implemented in the original programs.

The adequacy to the real conditions of the physical and mathematical models adopted while temperatures calculation was tested by comparing the average contact temperature experimental values while ultrasonic grinding of the workpieces made of stainless steel 12X18H9T with the calculation values. The difference between the calculation and experimental values does not exceed 26 %, that indicates the possibility of using the method for the thermophysical analysis of the grinding process.

For conducting of the experimental studies an experimental plant, based on the face-grinding machine and equipped with the ultrasonic device, has been developed [18, 19]. The workpiece is one of the vibration system links and is attached between the vibrations source and the reflector. The total length of the vibrations source, the workpiece and the support is equal to the ultrasonic vibration wave length, that is why the workpiece has a maximum vibrations amplitude. The ultrasonic vibration minimum energy losses and the workpiece vibrations high amplitude are also achieved by the adequate contact of the vibrations source with the workpiece that is provided while attaching it.

Experimentally, it has been established that, with the vibrations amplitude in the direction perpendicular to the workpiece processed surface A_u exceeding 3 µm, the grinding wheel wear intensity and the height parameters of the processed surface microgeometry increase. Therefore, while modeling and experimental temperature study the amplitude A_u was limited to 3 µm.

The numerical temperature modeling was performed by the following source data: multipass flat grinding with the grinding wheel face; the workpiece material is steel 40X and 12X18H9T; the grinding wheel grains material is Brown Aluminium Oxide 24A and borazon; Brown Aluminium Oxide graininess is *F*60 (25-th); the grinding wheel operating speed V_K = 35 ... 70 m/s; traverse speed V_S = 10 m/min; grinding depth t = 0,01 mm. The ultrasonic activation of the workpiece with various amplitudes and phases was modeled with the ultrasonic vibration frequency of 18600 Hz. The local temperatures on the grain and the workpiece contact site *EV* and the grain and the chips contact site *NE*, and also the workpiece surface layers temperature were recorded.

3 Study Results and Discussion

The results of modeling when processing the workpieces made of steel 40X with Brown Aluminium Oxide wheel, are presented below. At the vibrations amplitude A_u = 1 µm the temperature T_2 in the area of the grain and the workpiece contact was lower by 8 ... 9 %, the temperature T_2' in the area of the grain and the chips contact was lower by 8 ... 14 %. With ultrasonic vibrations with the amplitude A_u = 2 µm the local temperatures decreased to a greater extent: the temperature T_2 decreased by 9 ... 14 %, T_2' decreased by 17 ... 31 % (Figures 3, 4).

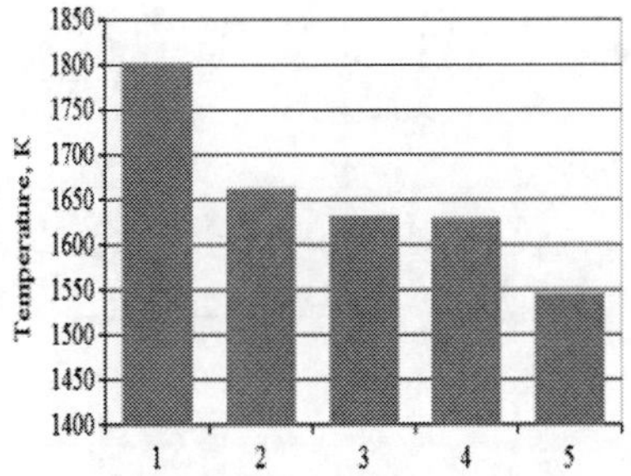

Fig. 3. The temperature T_2 in the area of the grain and the workpiece contact: 1 – A_u = 0 (without ultrasonic vibration); 2 – A_u = 1 µm, φ = 0°; 3 – A_u = 1 µm, φ = 180°; 4 – A_u = 2 µm, φ = 0°; 5 – A_u = 2 µm, φ = 180°.

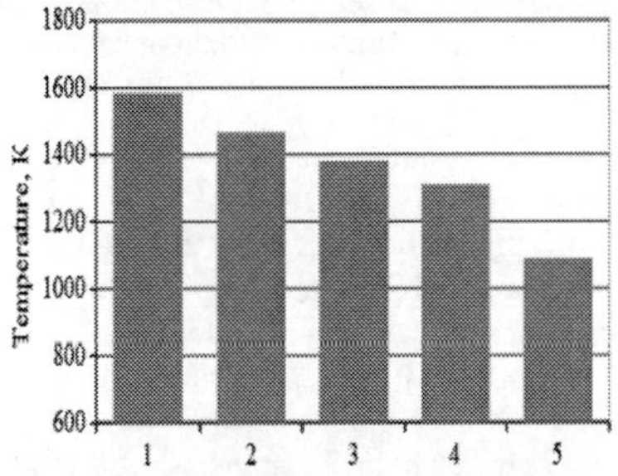

Fig. 4. The temperature T_2' in the area of the grain and the chips contact: 1 – A_u = 0 (without ultrasonic vibration); 2 – A_u = 1 µm, φ = 0°; 3 – A_u = 1 µm, φ = 180°; 4 – A_u = 2 µm, φ = 0°; 5 – A_u = 2 µm, φ = 180°.

Local temperatures are significantly dependent on the ultrasonic vibration phase, with lower temperatures being recorded during the phase φ = 180°. The obtained result may be explained by the fact that during this phase the minimum length of the trajectory, where the grain performs the workpiece microcutting, takes place. In this case, the grain does not perform microcutting at the greater trajectory part, that causes the process thermal power intensity decrease. The maximum cutting trajectory length is obtained with φ = 0°; therefore, at this trajectory the high temperatures are recorded. The microcutting trajectories lengths and the local temperatures with the phases 90° and 270° have intermediate values.

The temperature difference during various phases is as more as the ultrasonic vibration amplitude is higher. In the case of A_u = 1 µm, the difference in temperatures T_2 with the phases φ = 0° and φ = 180° is 2 %, in temperatures T_2' is 6 %; at A_u = 2 µm this difference is 5 and 15 %, respectively (see Figures 3, 4).

In the case of grinding without ultrasonic vibrations, the local temperatures increase with the increase of the workpiece and the grain contact duration (Figures 5, 6, Curve 1). With ultrasonic activation with the phase φ = 0° in the initial contact period the grain penetration depth into the workpiece is higher, and at the trajectory end is lower than without ultrasonic activation (see Figure 2). Accordingly, the local temperatures in the initial period of contact with the workpiece are slightly lower than at A_u = 0 (Curves 1 and 2 in Figures 5, 6). At the final stage of the grain contact with the workpiece the temperature difference is significantly higher, as the penetration depth is lower.

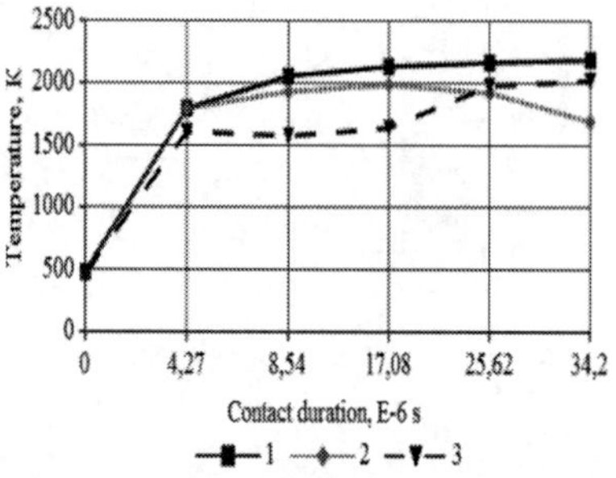

Fig. 5. Dependence of the temperature T_2 in the area of the grain and the workpiece contact on the contact duration: 1 – A_u = 0; 2 – A_u = 2 µm, φ = 0°; 3 – A_u = 2 µm, φ = 180°.

With the vibrations phase φ = 180° in the initial period of contact both the depth of the grain penetration into the workpiece and the local temperatures are lower than they are while grinding without ultrasonic vibrations (Curves 1 and 3 in Figures 5, 6). In particular, at the time moment of the grain contact with the workpiece $\tau = 8{,}54 \cdot 10^{-6}$ s the temperature T_2 is lower by 23 %, T_2' is lower by 3.25 times. At the final stage the depth of the grain penetration into the workpiece with ultrasonic vibrations with the phase φ = 180° increases, so, the temperatures also increase; however, at this stage the temperatures are lower than they are when grinding without ultrasonic vibrations.

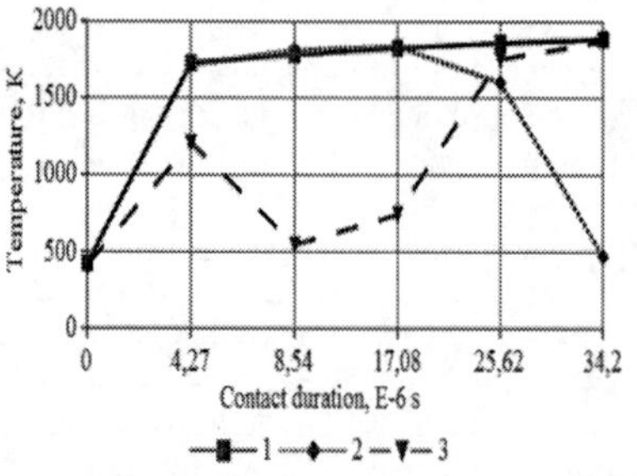

Fig. 6. The dependence of the temperature T_2' in the grain and the chips contact area on the contact duration: 1 – A_u = 0; 2 – A_u = 2 µm, φ = 0°; 3 – A_u = 2 µm, φ = 180°.

On the site of the grain contact with the workpiece the temperature is unevenly distributed (Figure 7). The maximum temperature is recorded at a distance of 5 ... 10 µm from the top of the grain (point *E* in Figure 1).

The average workpiece temperature at a distance of 2 and 100 µm from its surface and the workpiece material deformed layer temperature with ultrasonic vibrations were lower by 9 ... 11 %, whereas the vibrations amplitude only slightly affects these temperatures (Figure 8). Ultrasonic vibrations are more likely to affect the temperature in the workpiece surface layers than the temperature at a distance from this surface (Figure 9).

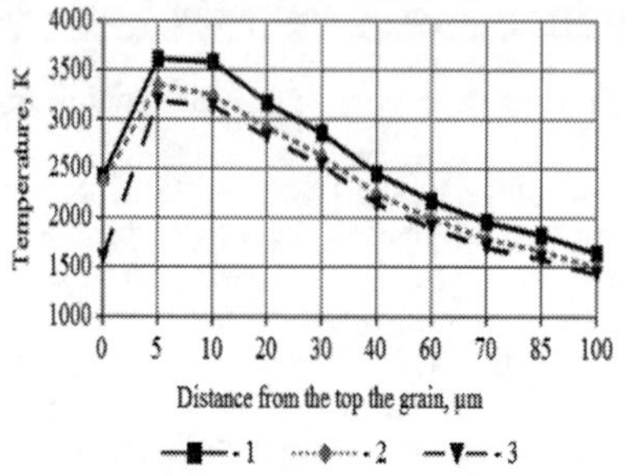

Fig. 7. The temperature distribution on the site of the grain contact with the workpiece: 1 – A_u = 0; 2 – A_u = 2 µm, φ = 0°; 3 – A_u = 2 µm, φ = 180°.

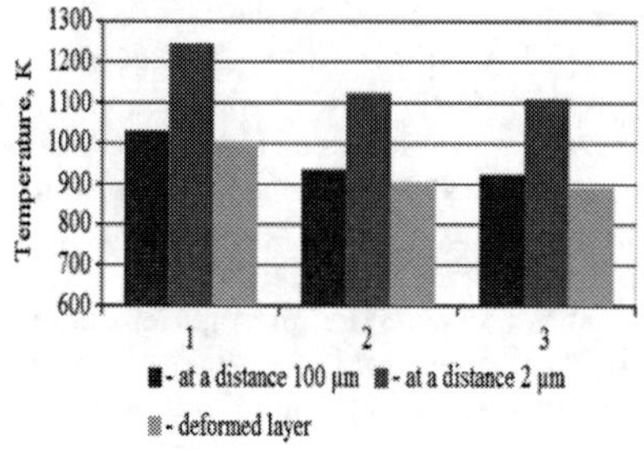

Fig. 8. The workpiece temperature: 1 – A_u = 0; 2 – A_u = 1 µm; 3 – A_u = 2 µm.

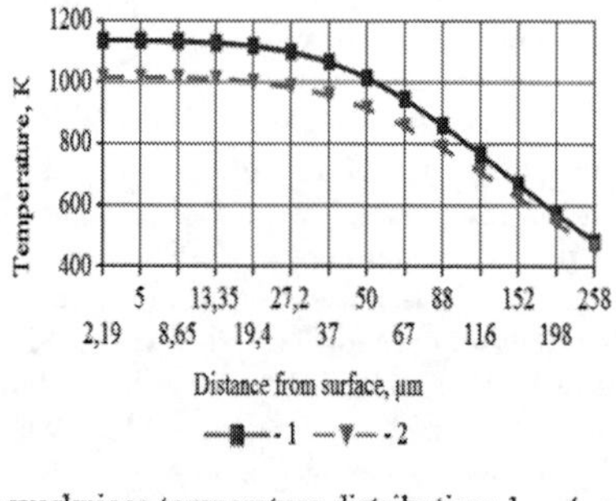

Fig. 9. The workpiece temperature distribution: 1 – A_u = 0; 2 – A_u = 2 µm.

Using the grinding wheels made of borazon provides lower local temperatures (by 1.5 times) and workpiece temperatures (more than twice), compared to using the grinding wheels made of the Brown Aluminium Oxide at the same mode. The reason is the higher borazon thermal conductivity coefficient, the smaller size of the site dullness on the grains and the greater distance between the cutting grains, thus the workpiece surface layer gets cold to lower temperatures.

When the operating speed of the borazon grinding wheel increases due to the heat release sources power increase, the local temperatures also increase. When grinding without ultrasonic vibrations with the speed increase from 35 to 70 m/s, the temperature in the area of the grain contact with the workpiece and with the chips increases by 13 and 26 %, respectively. With ultrasonic vibrations with the amplitude of 2 μm not exceeding the depth of the grain penetration into the workpiece (the grain does not lose the contact with the workpiece during ultrasonic activation), the temperatures in the area of the grain contact with the workpiece and the chips were lower than those without ultrasonic vibrations by 12 and 15 % at the grinding wheel speed of 35 m/s, and by 13 and 17 % at the speed of 70 m/s. Consequently, with the grinding wheel speed increase, the ultrasonic vibrations provide the slightly greater temperature drop. With ultrasonic vibrations with the amplitude of 3 μm exceeding the depth of the grain penetration into the workpiece (the grain loses the contact), the temperatures were reduced by almost 2 times.

4 Main results and conclusions

1. The ultrasonic vibrations during grinding allow to reduce the temperature in the workpiece surface layers to 10 %, the local temperatures in the area of the grain contact with the workpiece and the chips – to 13 and 30 %, respectively.
2. The temperature decrease is mainly affected by change of the stressed and strained state in the processing area due the workpiece material strength specifications reduction and the change of the microcutting kinematics during ultrasonic vibrations;
3. The effect of the ultrasonic vibration amplitude and phase on the temperature field and the effect of ultrasonic vibration parameters on the local temperatures are established, depending on the duration of the grain and the workpiece contact.

References

1. M. F. Vologin, V. V. Kalashnikov, M. S. Nerubai, B. L. Shtrikov, *Application of ultrasound and explosion during processing and assembly* (2002)
2. C. Guo, S. Malkin, Trans. ASME. J. Eng. Ind., **117 (4)**, 571 – 577 (1995)
3. J. Ja, J. Kim, Ju. Sun, J. Jang, Trans. ASME. J. Manuf. Sci and Eng., **121 (3)**, 378 – 384 (1999)
4. D. G. Evseev, A. N. Salnikov, *Physical basis of the grinding process* (1978)
5. A. N. Reznikov, L. A. Reznikov, *Thermal processes in technological systems* (1990)
6. A. V. Yakimov, P. T. Slobodnyak, A. V. Usov, *Thermophysics of mechanical processing* (1991)
7. A. N. Unyanin, STIN, **1**, 28 – 32 (2006)
8. L. V. Khudobin, A. N. Unyanin, Russian Engineering Research, **28 (9)**, 913 (2008)
9. L. V. Khudobin, A. N. Unyanin *Minimizing the loading of grinding wheels* (2007)
10. A. N. Unyanin, Izvestiya vuzov. Mashinostroenie, **6**, 41 – 50 (2006)
11. A. N. Unyanin, Russian Engineering Research, **29 (6)**, 630 - 632 (2009)
12. V. V. Efimov, *Model of grinding process using coolant* (1992)
13. L. N. Filimonov, *Resistance of grinding wheels* (1973)
14. S. N. Korchak, *Performance of grinding process of steel parts* (1974)
15. A. A. Diakonov, Izvestiya vuzov. Mashinostroenie, **7**, 60 – 62 (2007)
16. V. A. Nosenko, *Grinding of adhesion-active metals* (2000)
17. D. Shi, *Numerical methods in heat transfer problems* (1988)
18. A. N. Unyanin, I. Yu. Terekhin, Patent RU 2418671 RF, IPC B 24 B 1 / 04, B 24 B 5 / 04, **14** (2011)
19. A. N. Unyanin, A. S. Khusainov, Procedia Engineering, **150**, 1000 – 1006 (2016).

 DOI: 10.1051/matecconf/201712901012

Automation of the early stages of plating lines design

Vladimir Nemtinov[1,*], *Nikita* Bolshakov[1], and *Yulia* Nemtinova[1]

[1] Tambov State Technical University, 392000 Sovetskaya st. 106,Tambov, Russia

Abstract. In this paper we propose the procedure of automated selection of technological schemes for plating lines with specific operating conditions and criteria set by the customers using the theory of expert systems construction.

1 Introduction

Nowadays, many branches of national economy use plating to impart specific properties to components or final products. The main plating processes are protective and decorative coating. Companies working in the field of plating techniques are increasingly receiving orders for coating of small series or single details, but design of new plating lines requires substantial amount of time.

In order to significantly reduce the time for design of plating lines with customer-specified characteristics of end products, it is necessary to create an information system for automated solution of this task. Development of such systems allows: significant reduction in the use of supplies, replacing routine manual calculations with the software complex, visualization of all possible coating variants with final optimality ratings.

Analysis of literature sources in the field of plating lines design [1 - 11] has shown that no attention is paid to the automated selection of plating processes.

In this regard, the purpose of this work is to develop a procedure for automated synthesis of the structure of plating technological processes, which is implemented at the initial stage of plating lines design taking into account specific operating conditions of the final product and criteria set by the consumer.

In this case, the main input characteristics includes: type of detail, type of metal, purpose of coating, operating conditions.

The most advanced method for solving this problem is application of expert systems [1]. In order to choose an optimal variant of plating processes' structure, it is necessary to clearly define evaluation criteria, which are broken down into groups: costs, manufacturability, reliability and safety. In this case, it is recommended to use a comprehensive assessment, which is based on distribution of weights between these groups taking into account specific conditions.

2 Problem statement for technological scheme selection

In this paper, we propose the following statement of the problem for technological scheme selection for coating of metal products: it is required to find an optimal type of coating depending on consumer's requirements, which ensures the following equation:

$$k_{opt} = \arg\min_{k \in K} Q_1(k) \tag{1}$$

Since a multicriteria approach is used for selecting the optimal variant of technological scheme's structure, it is necessary to choose normalizing methods for the set of criteria and their ranking, as well as the method of multicriteria choice. In this paper, the optimality criterion Q_1 is the sum of weighted relative losses of the following criteria: criterion of discounted costs for plating stages implementation, criterion of reliability of system's functioning, criterion of manufacturability and criterion of safety [2].

The integral criterion Q_1 is equal to

$$Q_1(k) = \sum_{i=1}^{4} (\rho_i \omega_{1i}(k)) \cdot \tag{2}$$

Where ρ_1, ρ_2, ρ_3, ρ_4- weight coefficients.

$$\rho = \{\rho_i\} = \left\{ \rho_i : \rho_i > 0, i = 1,\ldots,4, \sum_{i=1}^{4} \rho_i = 1 \right\}, \tag{3}$$

where $\rho_i \omega_{1i}(k)$ - weighted losses of the i-th criterion; $\omega_{1i}(k) = \omega_{1i}(F_{1i}(k))$, $(i = 1,\ldots,4, k \in K)$- monotonous functions that transform each objective function $F_{1i}(k)$ to a dimensionless form; $F_{11}(k)$ - economic criterion, which includes agregate discounted costs on implementation of the system; $F_{12}(k)$ - criterion of reliability of system's functioning; $F_{13}(k)$, $F_{14}(k)$ – criterion of manufacturability and criterion of safety, respectively. Notably, we should obtain minimum for the function $F_{11}(k)$ and maximum for the functions $F_{12}(k)$, $F_{13}(k)$, $F_{14}(k)$.

Functions $\omega_{1i}(k)$ $(k \in K)$ are set as follows:

* Corresponding author: nemtinov@mail.gaps.tstu.ru

 DOI: 10.1051/matecconf/201712901012

$$\omega_{11}(k)=\frac{F_{11}(k)-F_{11}^{0}}{F_{11}^{\max}-F_{11}^{0}},\quad \omega_{12}(k)=\frac{F_{12}^{0}-F_{12}(k)}{F_{12}^{0}-F_{12}^{\min}}, \tag{4}$$

$$\omega_{13}(k)=\frac{F_{13}^{0}-F_{13}(k)}{F_{13}^{0}-F_{13}^{\min}},\quad \omega_{14}(k)=\frac{F_{14}^{0}-F_{14}(k)}{F_{14}^{0}-F_{14}^{\min}}.$$

Where $F_{11}^{\max}$ - maximal value of the minimizing function $F_{11}(k)(k\in K)$ on the set of acceptable values K; $F_{12}^{\min}$, $F_{13}^{\min}$, $F_{14}^{\min}$ - minimal values of the maximizing functions $F_{12}(k)$, $F_{13}(k)$, $F_{14}(k)$ $(k\in K)$ on the set of acceptable values K; F_{11}^{0}, F_{12}^{0}, F_{13}^{0}, F_{14}^{0} - optimal values of objective functions $F_{11}(k)$, $F_{12}(k)$, $F_{13}(k)$, $F_{14}(k)$ $(k\in K)$ on the set of acceptable values K. The values $\omega_{1i}(k)$ $(i=1,\dots,4,\ k\in K)$ are within the limits from 0 to 1.

Since in this paper the full enumeration method is used for searching of the optimal solution, it is not necessary to achieve the equality of weighted losses of $\rho_i\omega_{1i}(k)$.

In order to choose a single solution to the problem, it is necessary to specify weight coefficients $\rho_i(i=\overline{1,4})$ that satisfy the equation (3) and reflect relative importance of the objective functions $F_{11}(k)$, $F_{12}(k)$, $F_{13}(k)$, $F_{14}(k)$ $(k\in K)$. In this case the most effective methods are ranking and assigning of points [3].

When creating a database of plating methods, implementation of each stage is estimated (agregately) in terms of discounted costs. This economic criterion does not give the exact costs estimates, since at this stage there is only information about the stages of plating, on the basis of which it is possible estimate approximate costs of a particular techological scheme with the help of expert estimates [4]. The components of the criterion $F_{11}(k)$ are:

$$F_{11}=\sum_{i=1}^{N_k}S_{ik},\ k\in K. \tag{5}$$

Reliability is defined as the property of equipment to perform its specific functions, keeping in time the initial values of performance indicators within the specified limits, corresponding to the modes and operational conditions, maintenance and repair of the equipment:

$$F_{12}=\max_{K}\prod_{i=1}^{N_k}Pt_{ik},\ k\in K. \tag{6}$$

Manufacturability of the process is its convenience and ease, the ability to perform the process (i.e. to obtain the required result) with the least labor expenditure^

$$F_{13}=\max_{K}\prod_{i=1}^{N_k}Te_{ik},\ k\in K. \tag{7}$$

Safety criterion is determined on the basis of indicators of the object's elements, which allow calculating the probability of fire and explosion hazards caused by the equipment:

$$F_{14}=\max_{K}\prod_{i=1}^{N_k}(1-Pb_{ik}),\ k\in K. \tag{8}$$

Where S_{ik}, Pt_{ik}, Te_{ik}, Pb_{ik} are the values of costs, manufacturability, reliability and safety for the i-th stage of the k–th variant of the technological scheme.

Criteria values are set by the experts for specific design conditions.

With the help of decision-making mechanism, it is possible to find a standard combination of technological stages that provide the required coating thickness at minimal cost.

3 Information-logical model of decision-making support

For a formalized descriptions of information arrays for solving of the problem (1) - (8), it is requred to create a structured database. Data structure for the studied subject area can be displayed by a set of information-logical models (ILM) and production models (PM) for decision-makig support. When constructing such models, data normalization requirements should be provided, and the models themselves should be presented in a canonical form.

In general, the ILM for decision-making support of plating stages selection with subsequent selection of appropriate hardware represents the union of data sets and relationships between them, written in the form of rules:

$$\mathrm{M}=\langle D,P\rangle,\ D=(d_1,\dots,d_i,\dots,d_N),\ P=(p_1,\dots,p_j,\dots,p_S)$$

where M - ILM operator, $d_1,\dots,d_N$ - ILM data set, $p_1,\dots,p_S$ - set of rules.

In turn, the rules for the model are constructed as follows: if ... (conditions are met), then ... (implementation of the result).

$$S_{\text{detail}}=\{s_{\text{detail},1},\dots,s_{\text{detail},i},\dots,s_{\text{detail},I}\};$$

$$PR_{\text{detail}}=\{pr_{\text{detail},1},\dots,pr_{\text{detail},i},\dots,pr_{\text{detail},I}\},\ i=\overline{1,I}.$$

In order to obtain the variants of plating, it is necessary to compile several databases of reference materials. A fragment of one of them is given in Table 1.

Further, queries to the database are formed as special conditional rules. Examples of the rules for selecting the type and thickness of coating for a deatil with a non-special form made of medium alloy carbon steel to be used under light operating conditions for corrosion protection will look as follows :

Rule 1: If for $s_{\text{detail},1}$ = "carbon steel" it is required to satisfy the properties $s_{\text{detail},6}$ = "protective coating" and $s_{\text{detail},10}$ = "thickness of coating 0-0,5 μm" and operating conditions $s_{\text{detail},18}$ = "to be used in normal conditions", than the plating method is $pr_{\text{detail},4}$ = "zinc coating".

Rule 2: If for $s_{\text{detail},1}$ = "carbon steel" it is required to satisfy the properties $s_{\text{detail},7}$ = "protective and decorative coating" and $s_{\text{detail},10}$ = "thickness of coating 0-0,5 μm" and operating conditions $s_{\text{detail},18}$ = "to be used in normal conditions", than the plating method is $pr_{detail,4}$ = "cadmium coating".

Rule 3: If for $s_{detal,2}$ = "corrosion-resisting steel" it is required to satisfy the properties $s_{detail,6}$ = "protective coating" and $s_{detail,10}$ = " thickness of coating 0-0,5 μm" and operating conditions $s_{detail,18}$ = "to be used in normal conditions", than plating method is $pr_{detail,4}$ = "nickel coating".

Rule 35: If for $s_{detail,1}$= " medium alloy carbon steel" and $s_{detail,5}$= "protective coating" and $s_{detail,11}$= "to be used in light condtions", than the type and thickness of coating is $pr_{detail,1}$= "zinc coating of 6 μm " and $pr_{detail,2}$= "nickel coating of 15 μm".

Table 1. Database fragment of coating types

Coating metal	Type	Operating conditions			
		light	mid	hard	extra
Zn	Corrosion Protection	6	15	15	15
Nk	Corrosion protection	15	27	-	-
Cu+Sl	Reduce the transient resistance	9;3	24;9	30;9	-
Cr	Friction	18	36	48	60
Nk+Sn	Soldering	9;3	9;18	-	-
...	...	...	...	...	...

For each type of coating a separate database is created containing a set of standard technological operations of applying metal coating on details.

For example, for the input parameters specified above, the consumer will receive the types of metals that can cover a detail based on his requirements, their minimum thickness and the corresponding number of technological processes. For a specific example of input characteristics described above, the customer will see the chain of technological processes. A fragment of technological stages of zinc and nickel coating is given in the Table 2.

Each of the possible schemes is evaluated by an expert according to F_{11} - F_{14} criteria. Thus, the consumer will receive visualized variants of processing a detail with calculated priority value for each of them. This will allow the user to select the best type of processing and to start designing technological processes with the exact constructive and regime parameters of the equipment shortly afterwards.

The fastest, most visualized and convenient way to create an information system is the PHP programming language with a DBMS in the Internet - MySQL. This method will not only provide a visual representation of information, but will also allow the users to work with the information system remotely and to integrate various modules into it.

The procedure of automated synthesis of technological schemes of plating stages with the use of expert systems theory suggested in this paper allows users to choose the most suitable technological schemes of plating lines for specific operating conditions and consumer's criteria in the cases of high dimensionality of possible variants. Such a procedure can be used for the subsequent design of technological processes of plating of metal details with determination of precise regime and design parameters of the equipment.

Table 2. Database fragment of coating stages

Zinc coating	Nickel coating
1. Chemical degreasing	1. Chemical degreasing
2. Electrochemical degreasing: cathode and anode	2. Electrochemical degreasing: cathode and anode
3. Rinsing with warm flowing water.	3. Rinsing with warm flowing water.
4. Rinsing with cold flowing water.	4. Rinsing with cold flowing water.
5. Etching.	5. Etching.
6. Activation.	6. Activation.
7. Rinsing with cold flowing water.	7. Rinsing with cold flowing water.
8. Zinc coating.	8. Nickel coating.
9. Rinsing with warm flowing water.	9. Rinsing with warm flowing water.
10. Two-stage rinsing with cold flowing water.	10. Two-stage rinsing with cold flowing water
11. Clearing.	11. Drying.
12. Etching in chrome pickle.	
13. Rinsing with cold non-flowing water. (trapping).	
14. Rinsing with cold flowing water.	
15. Drying.	

Approbation of the proposed approach to selection of the technological scheme's structure for plating lines was carried out on the example of details with the following characteristics: material – carbon steel; purpose of details – to be used in industry, in contact with air, one-time contact with water (moisture, rain); purpose of coating – to implement protective properties against corrosion; required coating thickness is 10 μm; operating conditions of the coated part – normal climatic conditions. As a result of solving the problem of technological scheme selection zinc coating scheme was chosen (see Table 2).

4 Results and discussion

The procedure of automated synthesis of technological schemes for plating stages with the use of expert systems theory suggested in this paper allows users to choose the most suitable scheme for specific operating conditions and consumer's criteria.

References

1. T. Tanaka, J. Ohwi, L.V. Litvintseva, S.V. Ulyanov, Soft computing. **L(2),** 88-95 (1997).
2. E.N. Malygin, V.A. Nemtinov, Yu.V. Nemtinova, Theor. Found. Chem. Eng. 2003. **37(№ 6),** 613-618 (2003).
3. Seong-Joo Kim, Woo -Kyoung Choi, and Hong-Tae M.Gh. Negoita et al., (Eds.): KES 2004, LNAI 3215, P. 589 (2004)
4. V.G. Mokrozub, V.A. Nemtinov, Chemical and petroleum engineering **51(7-8)**, 31-35 (2015).
5. W.M. Cheung, P.G. Maropoulos, International Journal of Computer Applications in Technology, **V. 37(3-4)**, 182-193 (2010).
6. J.M. Wang, L. Zhang, Y.B. Liu, X.N. Mo, G.Q. Ren, Computer Integrated Manufacturing Systems, CIMS. **16(10)**, 2017-2023 (2010).
7. V.G. Mokrozub, K.D. Manuilov, V.V. Gorshkov, T.S. Gorshkova, Chemical and Petroleum Engineering **51(9-10)**, 613- 617 (2016).
8. V.A. Nemtinov, Yu.V. Nemtinova, Journal of Computer and Systems Sciences International **44(3)**, 389- 397 (2005).
9. Yu.V. Litovka, D.S. Solovjev, Transactions of the Tambov State Technical University **18(1)**, 128-135 (2015) (In Russ., abstract in Eng.)
10. Yu.V. Litovka, V.Z. Cao, Transactions of the Tambov State Technical University **22(1)**, 68-74 (2016) (In Russ., abstract in Eng.)
11. V.V. Konkina, D.S. Solovev, Yu.V. Litovka, Transactions of the Tambov State Technical University, **21(2),** 248-256 (2015) (In Russ., abstract in Eng.)

 DOI: 10.1051/matecconf/201712901013

The capability of pulsed laser radiation for cutting band saws hardening

Evgeny Marinin[1,*], *Denis* Sergeev[1], and *Nadezhda* Marinina[2]

[1]FSBEI HE «VyatSU», FTED, 36, Moskovskaya street, Kirov city, Kirov region, 610000, Russian Federation
[2]FSBEI HE «VyatSU», FL, 36, Moskovskaya street, Kirov city, Kirov region, 610000, Russian Federation

Abstract. The article deals with the possibilities of pulsed laser radiation for hardening the band saws. The regimes of pulsed laser hardening the band saws of 1 mm thick made of tool steel 9CrV are grounded theoretically and experimentally tested. Selected and justified modes of treatment harden in the autohardening mode without additional heat removal. The results of the experimental research of microhardness are presented and formed as a result of processing of the microstructure. Selected modes increase the microhardness of the surface to 8500 MPa and form ultra highly dispersed structure in the surface layer characterized by high resistance to abrasion.

1 Work actuality

Band saw machines are widely used at sawing wood species. Also band saw machines for metal and contour band saw technology for metal cutting are more and more used due to its advantages: high productivity; high quality and accuracy of cutting; tool durability; low energy consumption; convenience and simplicity of service.

The band saw machines operate with the fine band saw max. 1-1,5 mm. thick To achieve optimum conditions when working with certain groups of materials and blanks form it is necessary to define exactly the parameters of the saw blade: the tooth spacing, the shape and the saw wrest, the tensile force and the operation modes of band saw machine.

Saws welded in a ring (loop) of a certain length are used in band saw machines. Welding of saws in the loops is performed by melting with the specialized welding units of butt-steam contact welding. It is a thermomechanical process of formation of metal permanent connections, when local heating of the welded parts entails plastic deformation, developing under the influece of the compressive force.

The band saws from carbon steels and bimetal – with spring base and teeth of high speed steel or hard alloy are used for modern band saw technologies of metals processing. The variety of constructive options of band saws and the use of optimal modes of cutting provide high productivity and reduce the cost of technological processing at the band saw machines compared to other methods of spacer cutting blanks.

Band saws made of qualitative high carbon tool steel is designed for economical cutting of various alloyed and non-alloyed steels; but according to its technical and operational indicators they are inferior to the bimetal blades.

Bimetal blades with the tooth of high speed steel are the most widely used in modern sawing technology. This steel has high hardness and red hardness. Specially selected spring steel of the saw foundations gives it strength and durability. The reliable electron-beam welded joint with the cutting teeth allows to use these saws in the most difficult production conditions.

The short-term operation of each tooth during the cutting, the normal cooling and lubrication determine the relatively long-term service of the saw.

Each instrument has a certain performance and durability relation. The durability of the saw tooth and the durability of the blade to the cyclic alternating loads appear from the movement of the saw in the tape mechanism of the band saw.

To extent the service life of the saw it is necessary to run a saw with a gradual increase of the tooth load. It should also be noted that the running of a band saw is always performed with the light processing materials, regardless of what material is used after the grinding of the the band saw.

To achieve the optimal conditions when working with certain material groups and the shape of the blanks it is necessary to select the parameters of the saw blade accurately: the pitch, the shape, the saw tooth setting and the operating modes of the band saw machine – the speed and the feed of the band saw.

There is a problem of choice of band saw blade parameters in the process of production equipment with the band saws to ensure the most appropriate results [1]. Analyzing the available information we conclude that the resistance of band saws depends mostly on the combination of the following factors [2]:

* Corresponding author: e.marrini@gmail.com

DOI: 10.1051/matecconf/201712901013

1. The accuracy of selection of settings for cutting modes: cutting speed (m/min) and productivity (cm^2/min).
2. The correct selection of the blade parameters (type and size, material, brand, shape of the tooth form, tooth spacing) depends on the shape and material of the cut blanks.
3. The material grade and the shape of processed blanks and the quality of the blanks material.
4. The design features of the band saw equipment.
5. Technical state of the band saw equipment and its independent units (unit of tension, pulleys and bearings, carbide and roller guides, the uniformity and the smooth lowering of the sawing frame).
6. The quality and concentration of the used cutting coolant.
7. The quality of the run-in new band blade.

The main reasons for premature failure of the band saws [2]:

1. Punching out (knocking out) of the teeth. This is due to the following reasons: the saw's pitch is too small or the saw's pitch is too large; the blanks are not securely attached; the saw's speed is too slow resulting in excessive plunging; too high supply pressure resulting in excessive sawing into the material; weak tension of the saw leads to slippage; slipping (stopping) of the saw under the load, resulting in excessive cutting of the saw into the material, the cleaning brush of the saws does not work or is worn out.
2. The cracks of the teeth grooves. It is caused by: the difficult movement of the saw in the guides and pulleys due to the contamination of the pulleys or the reduction of the gap in the guides; the gap between the guides is too large; the guides are too far from the blank, the side guides clamp the saw in the tooth grooves area; the slightly clamped side guides lead to the tilt of the saw; the wrong tension of the saw.
3. The cracks of the back of the saw. The reasons of it may: wear of the upper support bearing in the guides, high supply pressure, wear of the side guides, the blade is pressed against the pulley shoulder.
4. Beating (vibration) of the saw. It can be caused by: the curved weld seam, the wrong pitch of the saw, the absence of teeth (broken), too low or high delivery pressure.
5. The premature blunting of the teeth. It can be caused by: too high saw speed for this material, too small or too large pitch of the saw, saw blade are not parallel to the feed direction, defects at the side guides, poorly fixed or worn guides.

The performance and durability [3] are the criteria of tools quality. They are determined by the material properties of blade and methods of processing and hardening.

There are many methods of hardening to increase durability of the cutting tools: rigging tool teeth with the carbide plates; electrocontact hardening and tempering of the teeth in the high-frequency current field; the use of thermochemical or thermomechanical treatment technologies, the method of electric spark machining and others [4].

There is a method of hardening the narrow band saws [5] by the elastic-plastic deformation of a saw blade creating a pre-curved condition and compressive stress on the saw surface. It leads to decrease of amplitude values of stresses in the saw during the operation [6].

Thus, the problem of hardening the cutting band saws is undoubtedly the task of today.

The aim of this work is to determine the possibility of hardening the saw blades with the laser hardening by pulsed solid state laser, analytical selection of hardening modes and their experimental approbation.

2 Theoretical part

Rather promising method of hardening the material of band saws is laser hardening. This kind of hardening treatment is studied properly but is not widespread nowadays. Laser hardening hardens both structural and tool steels [7-9]. The ultra-fine structure characterized by high strength and plasticity is formed as the result of properly chosen modes of such hardening treatment on the surface of impact [9,10]. This combination of mechanical characteristics is required for band saws. Also the laser treatment allows to create compressive stress in the surface layer [9], that affects positively the wear resistance of band saws [5].

The choice of laser hardening modes of the band saw at the last stage of preparation for work is an interesting challenge at the turn of science and industry. The following requirements is fulfilled when performing the laser hardening:

1. preserving the original surface roughness;
2. ensuring the maximum depth of the hardened zone;
3. reducing the heat affected area having reduced mechanical strength;
4. the possibility of technological reproducibility of the process.

These requirements mean that the laser hardening is made without melting the surface with a minimal energy expenditure into hardened saw and the cooling is made by the heat removal into the material. The option of heat removal from the product during laser treatment limits the technological reproducibility of the process and increases the cost of treatment and reduces productivity.

The installations generating pulsed laser radiation provide a minimal energy expenditure in the surface layer and technological reproducibility. Such equipment reduces significantly the heat input to the workpiece by the pulse energy selection, its time and pulses frequency.

3 The methodology of the experiment

The samples of band saws of 1 mm. thick were subjected to the laser hardening. The samples material is tool steel 9CrV subjected to bulk oil hardening from 850 C and tempering at 250 C. This type of pre-heat treatment get higher values of mechanical characteristics by laser hardening [11]. The installation HTF-100 is used as the source of laser radiation.

The time of the laser pulse is found at the heat affected area 200 μm max. and technological equipment capability:

$$h_t = 2 \cdot \sqrt{a \cdot \tau_i} \quad (1)$$

where h_t – depth of heat penetration during the laser pulse, m; τ_i – time of laser pulse, sec; a – heat diffusivity of material, m²/s determined by the formula:

$$a = \frac{\lambda}{\rho c} \quad (2)$$

λ – heat diffusivity of steel 9CrV, λ = 52,3 W/m K [12], ρ – steel 9CrV density, ρ = 7790 kg/m³ [12], c – heat capacity of steel 9CrV, c = 486 J/kg·K [12].

For experimental work the value τ_i = 0.5 ms was taken.

In case of lack of time dependence of the laser radiation intensity and time temperature distribution into the material is described by the following function [13]:

$$T(z) := 2 \cdot \frac{A}{\lambda} \cdot q \cdot \sqrt{a \cdot \tau_i} \cdot \left(ierfc \frac{z}{2\sqrt{a \cdot \tau_i}} \right) \quad (3)$$

where z – the coordinate counted inward from the surface, μm, A – absorption coefficient of laser radiation (A = 0.35), q – the power density of laser radiation, W/m²,$ierfc$ – the function of probability multiple integral [14].

The tabular value of the function probability multiple integral was used in calculations. The required energy is calculated according to the formation of a hardened area, 50 μm min. The result was the choice of the following power values of the laser pulse of 1 J and 1.2 J. Graphical results of the temperature modeling of the material at a point of time τ_i = 0.5 ms that meets the requirements is shown in fig. 1. The diameter of the focused laser radiation is d = 1 mm.

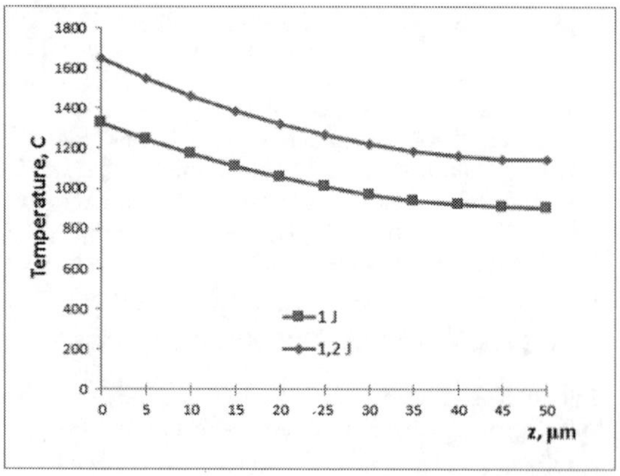

Fig. 1. Temperature distribution deep into at the point of time.

Laser heating is characterized by a high temperature gradient and considerable nonequilibrium transformations. It shifts the phase transformation to higher temperatures [11]. Therefore the estimated surface temperature is chosen within the range 1100-15000 C.

Metallographic studies are carried out with the microscope Neophot-21, the microhardness determination are made with the microthermometry PMT-3.

4 Experimental results and discussion

The microstructure formed in the surface layer shown in fig. 2.

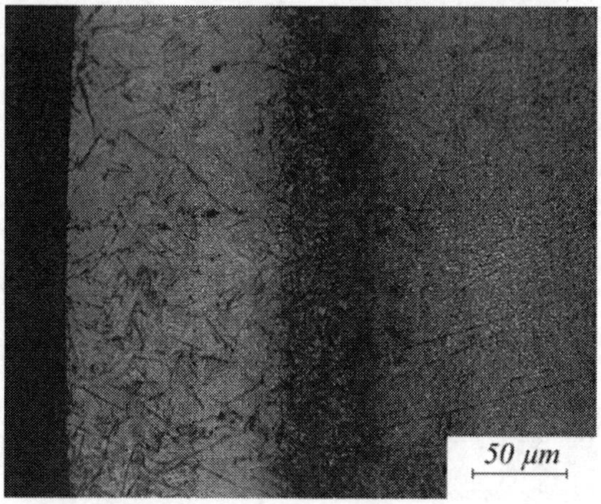

Fig. 2. Microstructure of the treated area.

Microhardness measurements were made deep into samples. Obtained values are shown in fig. 3.

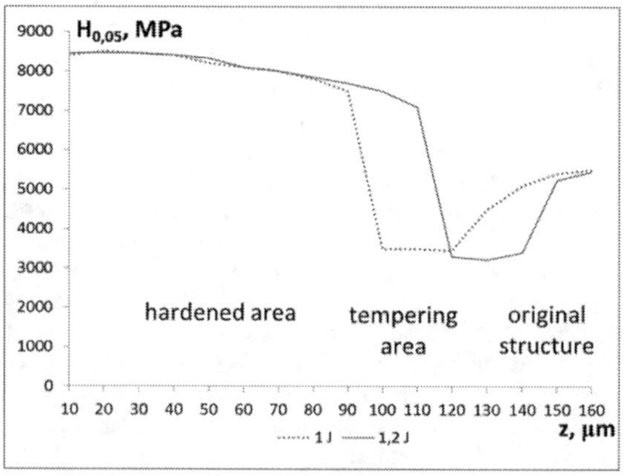

Fig. 3. Microhardness deep into the sample.

As a result of laser hardening two characteristic areas are formed: the area of hardening consisting of martensite, residual austenite and cementite and tempering area, which is a mixture of perlite of various dispersity.

The maximum value of the microhardness is to 8500 MPa. The hardened area when energy pulse used is 1 J, averaged 90 μm, the pulse energy 1.2 J – 110 μm.

The laser processing of band saws with pulse energy of 2; 2.5; and 3 J was made to control the design theoretical model. The melt bath is formed after the use of such modes on the surface of the tool, as the result dendritic cast structure is formed (figs. 4, 5). It should be noted that the wider heat affected area was formed. The maximum value of the microhardness is 6900 MPa. In comparison with pulse energy of 1.2 J there was no the increase of the hardened area. This is due to the low thickness of the band saw preventing the high speed of cooling by increasing the impact energy. To ensure hardening in the autohardening mode heat is removed; it limits significantly the technological reproducibility of the process.

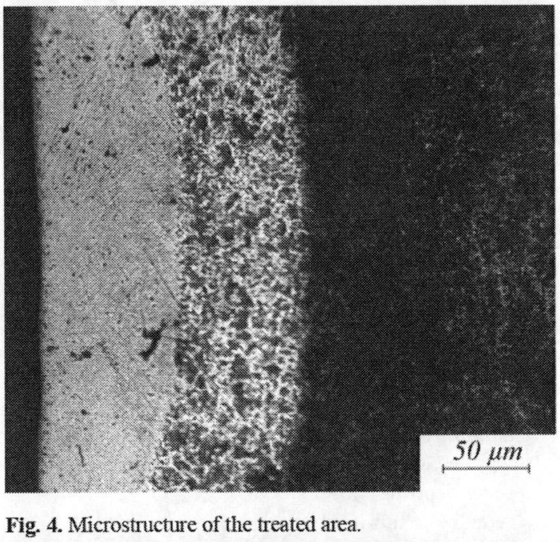

Fig. 4. Microstructure of the treated area.

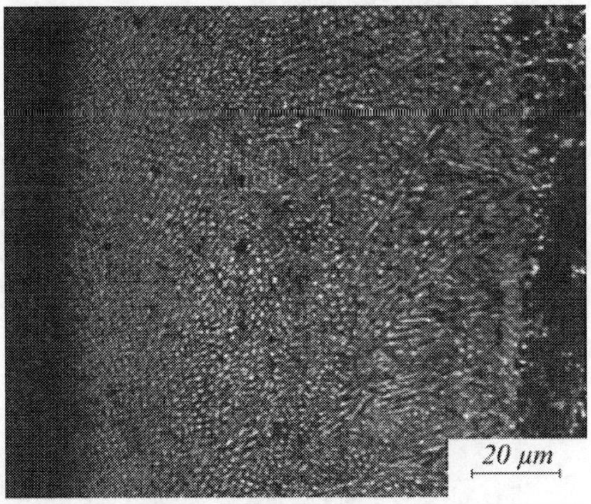

Fig. 5. Dendritic cast structure.

5 Conclusions

1. It is shown the fundamental capability of improving the strength performance of band saws made of tool steel 9CrV by means of pulsed laser radiation.

2. The modes of the pulsed laser radiation are selected analytically and confirmed experimentally. They increase the microhardness of the samples of 1 mm thick to 8500 MPa without additional heat removal during processing.

3. Hardening by pulsed laser radiation is used to determine the band saws areas. There laser processing increases to the maximum the durability of the saw blade.

References

1. D. G. Sergeev, *All-Russian annual scientific and technical conference "Society, science, innovation"*, p. 1163-1167 (2016)
2. D. G. Sergeev, M. N. Yurkin *All-Russian annual scientific and technical conference "Society, science, innovation"*, p. 1280-1283 (2012)
3. V. P. Ivanovskiy, D. Y. Kovaleva Complex criterion of quality tools. *Actual tendences of scientific researches of the XXI century: theory and practice*, **3**, p. 385-388 (2015)
4. Y. I. Mulin, O. V. Kazannikov, V. D. Vlasenko Hardening of cutting tools of bandsaw machines with formation on their working surfaces of functional covers with electro spark processing. *The Bulletin of the Pacific State University*, p. 91-98 (2011)
5. A. F. Dulevich *The works of BGTU. Series. II, Forest and woodworking industry*, **XVII**, p. 331-333 (2009)
6. S. V. Kiselev, A. V. Blokhin, A. F. Dulevich Band saws with high durability. *Actual problems of forest complex*, **46**, p. 153-155 (2016)
7. N. N. Rykalin, A. A. Uglov, N. V. Zuev and others *Laser and electron ray processing of materials* (1985)
8. A. A. Uglov, V. M. Makhtunov, T. P. Shmyreva, I. Y.Smurov, M. B. Ignatiev, Phisics and chemistry of materials treatment, **5** (1986)
9. A. N. Safonov, A. G. Grigoriants, *Laser methods of heat treatment in machine building* (1986)
10. E. A. Marinin, M. A. Surovcev *All-Russian annual scientific and technical conference "Society, science, innovation"*, p. 1266-1267 (2014)
11. G. N. Gavrilov, *Engineering and mastering of technologies of surface heat hardening and cladding of metallic materials by laser emission* (2000)
12. A. S. Okhotin, R. P. Borovikova, T. V. Nechaeva, A. S. Pushkarevskiy, *The heat conductivity of solid bodies* (1984)
13. N. N. Rykalin, *Laser and electron ray processing of materials* (1985)
14. J. F. Asmus, S. F. Baker, *Record of 10th Symposium on Electron, Ion and Laser Beam Technology (L. Marton, ed.)*, p. 241-246 (1969)

DOI: 10.1051/matecconf/201712901014

Modelling of destructive ability of water-ice-jet while machine processing of machine elements

Mikhail Burnashov[1,*], *Alexander* Prezhbilov[1], and *Elena* Stepanova[1]

[1] Orel State University of I.S. Turgenev, Orel, Russian Federation

Abstract. This paper represents the classification of the most common contaminants, appearing on the surfaces of machine elements after a long-term service.The existing well-known surface cleaning methods are described and analyzed in the framework of this paper. The article is intended to provide the reader with an understanding of the process of cleaning and removing contamination from machine elements surface by means of water-ice-jet with preprepared beforehand particles, as well as the process of water-ice-jet formation. The paper deals with the description of such advantages of this method as low costs, wastelessness, high quality of the surface, undergoing processing, minimization of harmful impact upon environment and eco-friendliness, which makes it differ radically from formerly known methods. The scheme of interection between the surface and ice particle is represented. A thermo-physical model of destruction of contaminants by means of a water-ice-jet cleaning technology was developed on its basis. The thermo-physical model allows us to make setting of processing mode and the parameters of water-ice-jet scientifically substantiated and well-grounded.

1 Introduction

The hydro jet technologies are widely used in industry. They are necessary for cuttingmaterials, cleaning contaminants off the surfaces of machine elements [1-20].

2 Problem description

During an operation of machine elements contaminants (which differ in kind, composition, properties and the strength of the adhesion to the surfaces of machine elements) are being formed on their outer and inner surfaces. These contaminants reduce the durability of machine elements, as well as their service life and resistance to corrosion.

When repairs are carried out, contaminants undermine labour productivity, make accuracy of monitoring and defectation of machine elements worse, reduce repair quality and resource of repaired machines and their elements.

The most widespread causes of contaminative coating formation include an emulsive and oil slick, an introduction of the contaminants from the environment, thermal decomposition of oils, an oxydation of metal surfaces, burnt-on (sticking) sand, oil remains, scale etc.

Contaminants on the objects under repairs are divided into the following types due to their chemical composition:

• organic (oil and fat deposit, varnish coating films, lubricants);
• non-organic (scale, road mud, corrosion products);
• mixed (carbon deposit, varnish, grease, industrial contaminants).

Contamination of plant items, assembly units and machine elements include:

• outer deposit,
• remains of lubricative materials,
• carbon deposit (building-up),
• corrosion products,
• scale,
• remains of old paint coatings.

The most common and widespread contaminants are remains of fuel-lubricative materials and products of their transformation. Drastic changes of lubricative materials, occuring during use of machines, are caused by the processes of 'aging'-oxydation and polymerization. The following types may be singled out:

1. an incomplete combustion of fuel,
2. oxydation products,
3. products of destruction of hydrocarbons,
4. polymerization products,
5. products of condensation and coagulation of hydrocarbon and heteroorganic compound,
6. corrosion products and biodamage of metals in the medium of remains of fuel lubricative materials.

According to up-to-date state of the field under discussion the industrial processes must be environment-friendly, energy-efficient and meet industry needs and evolving governmental regulations. These demands are also obligatory for cleaning for the removal of contaminants from surfaces.

Water-jet \ cryo-jet methods are considered to be highly promising and universal methods of cleaning among existing ones.

*Corresponding author: stker@rambler.ru

At the same time, a hydroabrasive cleaning with sand and similar materials, performing the function of an abrasive, has a wide range of drawbacks. Let's mention some of them:

- the abrasive complicates cleaning of inner and inaccessible surfaces;
- roughening of clean surfaces;
- the necessity of utilization of the abrasive, contaminated during the blasting process;
- a high cost of an abrasive material and its delivery.

Therefore, renewal and replacement of the abrasive material is absolutely necessary for a wide introduction of hydroabrasive technology of cleaning of machine elements.

Furthermore, water-ice jet cleaning technology is not only an ecologically clean but also a cost-saving method.

Ice particles possess properties of hard particles, which allow them to remove a great deal of contaminants during accelerating to higher speeds and avoid causing damage to a base of surfaces, undergoing cleaning. Use of water and ice makes regeneration of clean media easier.

In comparison with other methods of cleaning, water-ice jet technology obtains some advantages:
1 minimization of harmful impact upon environment;
2 a low-cost cleaning materials;
3 an absence of the necessity to transport and keep large amounts of the abrasive material;
4 a possibility of a closed and waste-free cycle;
5 a reduction in an abrasive influence on material;
6 the abrasive does not block up peep-holes of machine elements joints;
7 an absence of dust in the cleaning process;
8 a high durability of an instrument.

A water-ice jet formation for cleaning machine elements surfaces is run in the following way: a highly pressured stream, having a diametre ranging from 0,15 to 0,3 mm, is being mixed with entrained stream of ice particles according to an ejection principle. As a result, a two-phased structure, called a water-ice jet, is formed. Use of ice particles, prepared beforehand, is preferable and optimal for this process. The size of particles varies from 1,5 to 2 mm (Fig. 1).

Fig. 1. The picture of an ice particle

Mechanism of coating destruction, caused by water-ice jet, is determined by a multiphase character of highspeed flow (water-ice particles) (Fig. 2).Having spent part of its energy on an acceleration of ice particles, water jet creates voltage, equal to its strength, in the material, which is undergoing destruction. At the same time, ice particles are making a full impact on the material, causing microcracks. They concentrate stresses, which enhances cleaning efficiency in general.

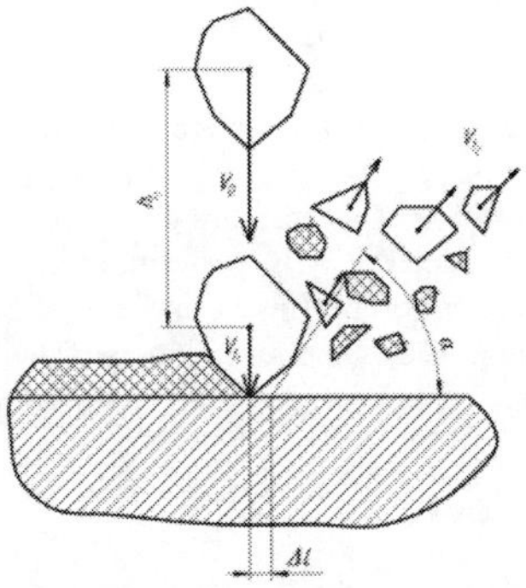

Fig.2. The scheme of an impact on removable coating of an ice particle and its destruction.

Economic efficiency is achieved in the process of deriving granulated ice by means of crushing ice lumps in industrial ice generators. During further mechanic crushing necessary fraction is separated from ice crumb. After production of working granulated ice with a help of this method, the model of an ice particle was created in the programming environment Compass-3D (Fig. 3).

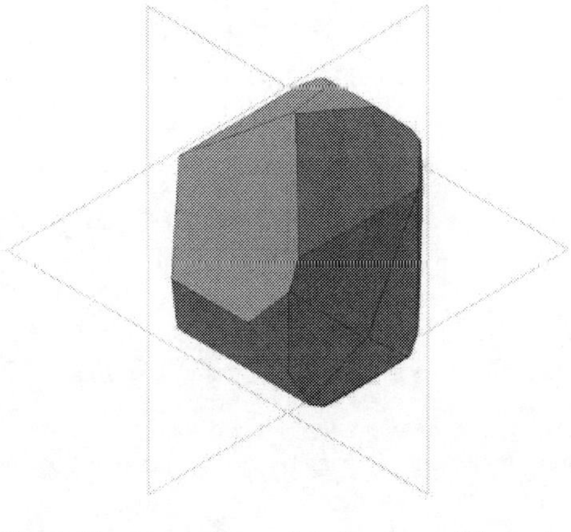

Fig. 3. The scene of the model of an ice particle (an isometric projection)

3 Thermo-physical model of destruction

While being formed, water-ice jet may undergo the following phases:
1. A highly pressured water jet bursts from a stream-forming checkerwork, creating an area of low pressure in a plenum (mixing) chamber, where ice particles are being entrained from the supplying canal.
2. Water-jet-ice particles are entrained by a water flow and mix with it, causing an intensive heat exchange.

3. Mixture of water and ice particles is being accelerated in a collimator.

Energy is generated in the closed system 'an ice particle-a destructed material-machine elements' by means of impulsive force and interaction and is being transformed into a heat stream and a work of cutting. The definition of a fraction of energy, spent on heating an ice particle, accompanied by creation of local zones of reversed phase transfer (transmission) (RPT) of ice and heating of material, causes a particular interest.

Contaminating coatings have a low capacity for heat transfer, therefore, temperature of a contact surface 'an ice particle-material'will determine mechanic and thermal physical properties of material of coating.

The scheme of an interaction between the surface and an ice particle is represented in fig.4.

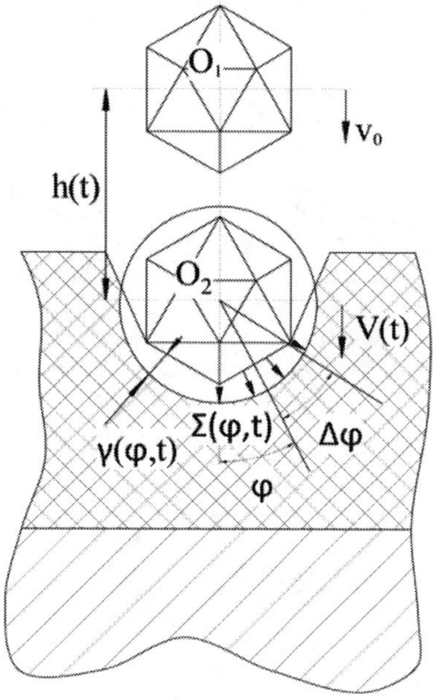

Fig. 4.A schematic design and calculation of the process of interaction of a water jet and ice particles with coating

An equation of an ice particle motion in the material prepared to be removed may be written in the following form:

$$m(t)\frac{dV}{dt} =$$

$$= \frac{1}{2}C_x\rho_c(T)S(h)V^2(t) - \sigma_{conp}(t)S(h) - F_{mp}(t,\phi); \quad (1)$$

where $m(t)$is a current value of the mass of an ice particle, taking into considerationRPT on the contact surface; $V(t)$ is a current value of the speed of penetration; $C_x$is a drag coefficient of an ice particle; $S(h,\varphi)$is a current value of a projection of section area of an ice particle, taking into account depth of penetration and presence of RPT on the elementary area within the solid angle $\Delta\varphi$:

$$S(h,\phi) = \pi R^2(t,\phi)\cdot\left[\frac{h(t)}{R(t,\phi)}H(R-h) + H(h-R)\right] \quad (2)$$

where$h(t)$, $R(t)$are the functions of depth of penetration and the current radius of an ice particle, taking into consideration RPT respectively; $H(...)$ isthe Heaviside step function.

The first component of the right part of an expression (1) characterizes the loss of kinetic energy of an ice particle, when interacting with the material. The second component reflects the resistance of material to destructive impact, and the third component is the friction force, influencing the particle, while penetrating into the removable material.

Friction force will depend on availability of liquid phase, which is relevant to the scheme of motion of an ice particle in the material with phase transitions on the contact surface:

$$F_{mp}(t,\phi) =$$

$$= f_V(t)\sigma_r(t,\phi)S(h,\phi)\cdot H\left[dA(t,\phi) - (q + C_V\Delta T_1)dm(t,\phi)\right] +$$

$$+2\mu_1\frac{V(t,\phi)}{\gamma(t,\phi)}S(h,\phi)\cdot H\left[(q + C_V\Delta T_1)dm(t,\phi) + dA(t,\phi)\right]; \quad (3)$$

$$\sigma_x(t,\phi) = \frac{1}{2}\rho_c(t)V^2(t,\phi);$$

$$\sigma_r(t,\phi) = \frac{\upsilon_л}{1-\upsilon_л}\sigma_x(t,\phi); \quad (4)$$

where $(q + C_V\Delta T_1)dm(t,\varphi)$ isa value of the energy necessary for the transformation of the solid phase of the spherical layer by the mass dm in liquid state; $\mu_1$isa dynamic viscosity coefficient of liquid phase; γ is a size of a gap between ice particle surface and a profile of a cavity in an obstacle. The value of γ in bounds of a discrete angular coordinate $\Delta\varphi$is defined by the mass of the solid phase, limited by the surface in the form of a spherical layer, 'lost' in each particular discrete moment of time. $f_V(T)$ is a function of a speed coefficient of friction; $\sigma_x(t,\varphi)$ is a value of an axial tension on the contact surface of an ice particle with an obstacle, generally equal to a pressure of a velocity (dynamic, kinetic-energy) head; $\sigma_r(t,\varphi)$ is a value of radial voltage on the contact surface of an ice particle with an obstacle.

The work of the resistance forces, determining the intensity of the heat of the contact area, is represented with a help of the following sum:

$$A(t,\varphi) = A_1(t,\varphi) + A_2(t,\varphi) + A_3(t,\varphi), \quad (5)$$

where $A_1(t,\varphi)$, $A_2(t,\varphi)$, $A_3(t,\varphi)$is a work of inertial constituent of force of resistance, work of a friction force and work of forces of plastic resistance to the deformation of coating respectively.

Change of the size of an ice particle, when cleaning within an angular coordinate φ, is given according to the following dependency:

$$d(t,\phi) = 2 \cdot \sqrt[3]{\frac{3}{4\pi\rho_{Л}}\left(m(t) - \int_{0}^{\frac{\pi}{2}} m_{Л}(\phi)d\phi\right)}; \quad (6)$$

where $\rho_{л}$ is a density of an ice particle; $m_{л}(\varphi)$ is a mass of an ice particle, «lost»,

$$A(t,\varphi)=Q(t,\varphi) \quad (7)$$

where $Q(t,\varphi)$ is a the thermal energy of the system of the contact of bodies.

In accordance with the condition of heat transfer through boundary surface of two bodies in contact in case of an ideal thermal contact [4] and the law of conservation of energy, an amount of ice, having been transmitted into an ice particle, and a respective change of its mass may be determined in a following way:

$$\begin{aligned} dQ_1(t,\phi) &= \frac{\upsilon_M}{\upsilon_Л} dA(t,\phi); \\ m(t) &= m(t-\tau) - \int_{0}^{\phi_K} \frac{Q_1(t,\phi)}{q + C_V^M \Delta T_1} d\phi, \end{aligned} \quad (8)$$

where ΔT_1 is a range of temperature change of the contact layer of an ice particle, determining the RPT moment, ($\Delta T_1 = T_{OФП} - T_{01}$; $T_{OФП}$ is a temperature of RPT; T_{01} is an initial temperature of an ice particle; m_0 is an initial mass of an ice particle.

Taking into consideration the fact that a (described in this paper) single act of an interaction of an ice particle with the material of coating, suffering damage, is similar to others, given a particular similarity coefficient, it becomes possible to calculate the percentage of energy, spent on heating of an ice particle, which is accompanied by appearance of local areas of a reverse phase tranfer (transmission) (RPT) of ice and on heating of the material.

4 Conclusion

A received thermo-physical model allows to develop and, having been proved scientifically, to appoint the modes of water-ice cleaning:
- water jet pressure;
- a diameter of a jet forming nozzle;
- the dimensions of ice particles;
- the temperature of storage of ice pellets;
- the distance from the nozzle to the surface;
- feeding speed of a nozzle apparatus during processing.

References

1. M.C. Kurco, R.F. Chadwick, Technika (Suisse), **vol. 21, no 3,** 2079 - 2083(1972)
2. B-L Liu, L-H Liu. Papers presented at the 14th Internationals on Jetting Technology. Held in Brugge, Belgium, 21-23 September, 203-210
3. M. Hashish, C.M. Dunsky, Papers presented at the 14th Internationals on Jetting Technology. Held in Brugge, Belgium, 21-23 September, 329-343
4. H. Kiyohashi, K. Handa. Proceedings of the International Symposium on New Applications of Waterjet Technology, Ishinomaki, Japan, 51-60 (1999)
5. H. Liu, T. Butler. Processing of the new applications of water jet technology. Isinomaki, Japan, 51 – 60 (1999)
6. *Water jet cutting: a production tool.* Machining and Production Engeneering.,**vol. 141, no 3631.** 18-19 (1983)
7. D.E. Hegland, Automation, **vol 22, no 8,** 40 – 44 (1975)
8. K.F. Hensen, E.C. La-Brush, Trans. ASME, **B. 97, no 3,** 1067 – 1073 (1975)
9. Vatteroch polymer-nyttskarwertyg. *Nord Emball,* **vol 41, no 2,** 25-26 (1975)
10. A.Y. Pushkarev, K.A. Golovin, A.B. Zabin, Y.I. Yeruhimivich, Proceeding of the International Symposium on New Applications of Water Jet Technology. Oktober 19-21, Ishinomaki, Japan. Ishinomaki Senshu Universiti, 421-425 (1999)
11. T.J. Louis, 5th American Waterjet Conference, Toronto, Canada, August, 145 – 168 (1989)
12. M. Hessling, 9th International Symposium on Jet CuttingTechnology, Sendai, Japan, October, **paper G3,** 357 - 376 (1988)
13. A.A. El-Domiaty, M.A. Shabara, A.A. Abdel-Rahman, A.K. Al-Sabeeh, Int. J. Adv. Manuf. Technol, **vol12, no 4,** 255-265. (1996)
14. P. Etchells. Aerosp. Compos, and Mater, **vol 3, no 1,** 40 – 43 (1991)
15. P.D. Kovscek, C.D. Taylor, E.D. Thimons, Rept. Invest / BurMines US Dep. Inter., **no 9201,** 1-10 (1988)
16. C. Merchant Howard, D. Chalupuik James, Noise Countr. Eng. J., **vol 29, no 3,** 85-89 (1987)
17. Anon, High Pressure Water Jet Systems, **Part 2, no. 4,** 20 –23 (1993)
18. Yazici, Sina Abrasive Jet Cutting and Drilling Rock, Ph.D. Dissertation in Mining Engineering, University of Missouri - Rolla, Rolla, Missouri, 203 p. (1989)
19. K. Faber, H. Oweinah, 10th International Symposium on Jet Cutting Technology, Amsterdam, **25,** 365 - 384(1990)
20. E.J. Chalmers, 6th American Water Jet Conference, Houston, Texas, 345 - 354(1991)

Research the quality of the caprolon turning with ceramic cutting tools

Oleg Erenkov[1,*], *Alexander* Ivakhnenko[2], and *Victor* Protasev[3]

[1]Pacific National University, 680035 Khabarovsk, Russia
[2]South-West State University, 305040 Kursk, Russia
[3]Tula State University, 300041Tula, Russia

Abstract. The aim of this study is the comparative analysis of caprolon turning with replaceable polyhedral plates made from different materials namely one plate made from wolfram and cobalt alloy VC8 another from ceramic material tungsten free carbide KNT16. The basis of the experimental setup is laboratory research stand STD.201-2 which is used to study the technological system dynamics and the cutting forces registration at turning. The stand functionally consists of the head-tool holder, the interface unit, the sensors connecting module and PC. It is experimentally proved that caprolon turning both with the VC8 alloy cutting plate and the plate made from ceramic material KNT16 is a stable process. This is evidenced by the slight changes values of cutting force main component wherein the range of changes in the case of turning with the ceramic plate is smaller in comparison with the turning option with wolfram and cobalt alloy plate for the selected processing mode.

1 Introduction

The surface layer condition for a component made of polymer material after machining is evaluated by the physic mechanical properties of these materials and also geometric characteristics of the machined surface. The main geometric characteristics of the surface of a turned component are roughness of shape accuracy.

An increase in roughness and the indices of surface shape deviation for a component from roundness, cylindricity, and rectilinearity arises during machining normally due to a high level of vibrations of the engineering lathe–attachment–tool–component system, unevenness and deformation of the machined material, nonuniformity of machining tolerances.

Turning of polymer materials on the basis of traditional engineering solutions does not in all cases provide the required level of geometric characteristics for a machined surface, since the properties of these materials differ markedly from those of traditional structural materials (steel, cast iron, other alloys).

The correct choice of cutting tool material and a given power parameters in the treatment process is one of the technical solutions, the implementation of which will improve the quality of production of products from polymeric composite materials as well as lowering their cost. The cutting tools on the base of so-called "hard" alloys VC6 and VC8 traditionally are used for the treatment of polymeric composite materials, including the caprolon. At the present time the ceramic cutting tool is actively used for the processing of metal materials and alloys [1,2]. Cutting ceramics are characterized by high hardness, including heating, wear resistance, chemical inertness to most metals during the cutting process. The complex of these properties ceramic is significantly superior to conventional cutting tool materials such as high speed steel and hard alloys [3,4,5].

The comparative analysis of the turning of caprolon with the use of the replaceable multifaceted plates made of different material is the aim of the study in this work.

The scientific novelty of the work consists in the obtained results proving the possibility and expediency of using a ceramic tool based on titanium carbonitride with additives of nickel and molybdenum for turning the caprolon, as evidenced by the values of the oscillation of the cutting force and the character of chip formation.

2 The methodology of the experimental studies

Caprolon was chosen as a study material. Caprolon is the structural polymer widely used in engineering because of its antifriction properties, wear resistance, high mechanical strength and good ability in cutting processes. The experimental setup scheme is presented on Figure 1.The basis of the experimental setup is a laboratory research stand STD.201-2, which is used to study the dynamics of the technological system machine-fixture – tool - work piece and registering cutting forces at turning. Functionally the stand consists of a head-tool holder, the interface module connection of sensors, a personal computer. The laboratory stand measurements range of dynamic loads is from 0 to 15000 *N*. The most important component of the cutting force P_z was chosen

* Corresponding author: Erenkov@list.ru

for the comparative analysis of the turning process which determines the quality and performance of the of turning process in generally [6].

The composite cutters were used for turning. The cutters are equipped by many-sided replaceable plates made from different tool materials. One plate made from hard alloy VC8 and another plate from ceramic material KNT16 which is a tungsten free alloy. Alloy KNT16 produced by OOO Virial (Russia, St. Petersburg) and has the following chemical composition: titanium carbonitride - 74%, nickel - 19.5%, molybdenum - 6.5%.

The selected plates are used for turning of hard and churlish materials which include caprolon. Turning of experimental specimens was carried in a universal turning-screw cutting lathe 16K20F3S47. The geometric parameters and materials of the cutting tool, and also the cutting and feed rates were adopted on the basis of previous studies [6] and kept constant during this series of experiments.

Evaluation of shape deviation for machined component surfaces was performed using a digital measuring unit Rondcom 41C for such indices as non-cylindricity (including non-rectilinearity) of the generating line, and non-roundness.

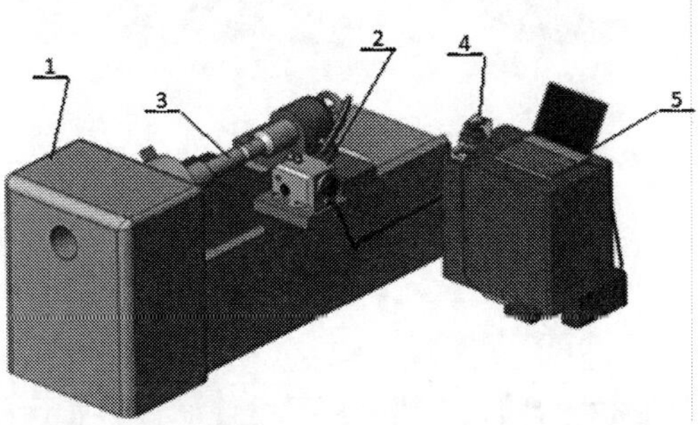

Fig. 1. Experimental set-up:1 – lathe; 2 – head of the tool holder; 3 – work-peace; 4 – the interface unit; 5 – PC.

Component surface roughness was monitored by means of a TR 200 profilometer. As input parameters for surface roughness according to GOST 2789-73 the average deviations were selected for profile R_a, height of profile unevenness for ten points R_z, greatest height of profile unevenness R_{max}, average step of profile unevenness S_m. In addition, in accordance with the international standard ISO 4288 the following surface roughness parameters were determined: distance from the tip of the greatest projection of the profile to the central line R_p; distance from the bottom of the greatest depression of the profile to the central line R_m; profile asymmetry S_k. According to conditions of ISO 4288, the profile with favorable values of parameter S_k has a clear high peaks, that differ from the average value. A surface with negative values of parameter S_k has clear deep depressions in a smooth profile plateau. In less clear cases, parameter S_k approaches zero. If $S_k > 1.5$, then this means that the component surface does not have a simple shape, and it is probably impossible to characterize surface layer quality adequately according to parameters R_a and R_z.

The research of caprolon chip structure obtained at different variants of turning was carried out with an electron microscope Primo Star.

Thus, experimental studies contained the following types of research such as turning blanks by researched plates at the selected cutting conditions, the registration of the cutting forces, chip selection and study its structure, analysis of the results.

3 Experimental study and discussion of results

The results of experimental studies on turning caprolon work pieces by cutting tools made from VC8and KNT16 alloys are graphically showed on Figure 2. Analysis of the presented data allows us to conclude that for all variants of turning the change of cutting force P_z occurs in small intervals. The maximum and minimum values of cutting force P_z investigated at different treatment options are summarized in Table 1. As follows from the data in Table 1 the overall range of the cutting force P_z values variation is from 2.6 to 5.7%. Such assessment is compelling evidence that the process of caprolon pieces turning by selected plates under the above cutting parameters is stable. The cutting process stability in turn is a necessary condition for obtaining the high quality of the machined work piece surface [7].

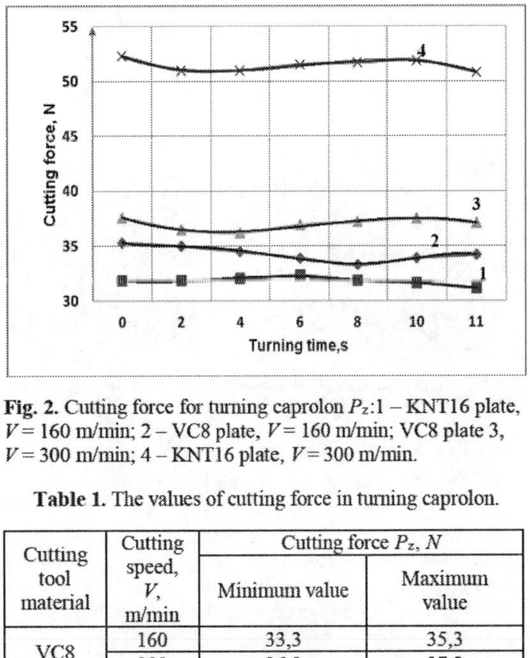

Fig. 2. Cutting force for turning caprolon P_z:1 – KNT16 plate, V = 160 m/min; 2 – VC8 plate, V = 160 m/min; VC8 plate 3, V = 300 m/min; 4 – KNT16 plate, V = 300 m/min.

Table 1. The values of cutting force in turning caprolon.

Cutting tool material	Cutting speed, V, m/min	Cutting force P_z, N	
		Minimum value	Maximum value
VC8	160	33,3	35,3
	300	36,2	37,2
KNT16	160	50,6	51,9
	300	31,3	32,3

The highest values of cutting force P_z are observed when caprolon is turning by ceramic plate at cutting speed 300 m/min. This effect can be explained by the fact that with an increase of turning speed from 160 to 300 m/min cutting temperature increase occurs in the

 DOI: 10.1051/matecconf/201712901015

cutting zone [8]. Tungsten free KNT16 solid alloy has a lower thermal conductivity as compared with the VC carbide and therefore the thermal load caused by the temperature increase in the cutting zone, accumulates mostly in the surface layer of work piece material.

The material becomes more viscous and its destruction occurs by plastic deformation. At the same time respectively the specific work of cutting is increased. The cutting specific work is spent on cutting layer separation and the formation of the chip by means of shear stresses which are the reason for the increase of the cutting force P_z. This assumption is confirmed by electron micrographs of caprolon chips presented on Fig.3 and Fig.4.

The analysis of chip micrographs showed that the nature of the chip formation at caprolon turning with the specified speeds for the two plates is practically the same. This is evidenced by the fact of the identical type continuous shearing chip formation also the nature of the shear elements regularity distribution o and their geometric parameters.

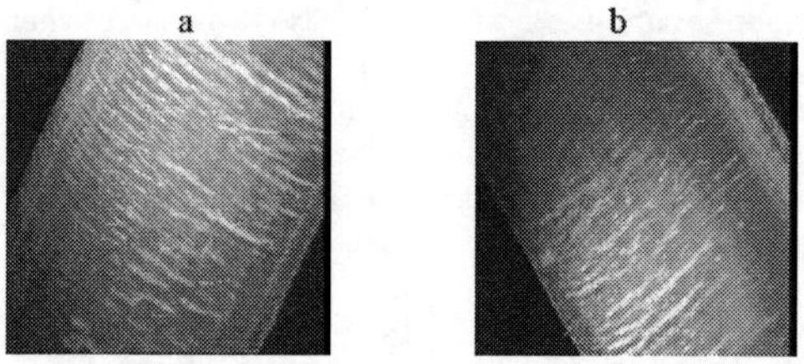

Fig. 3. Caprolon chips obtained by turning at cutting speed of 150 m / min: a - VK8 alloy plate; b - KNT16 alloy plate.

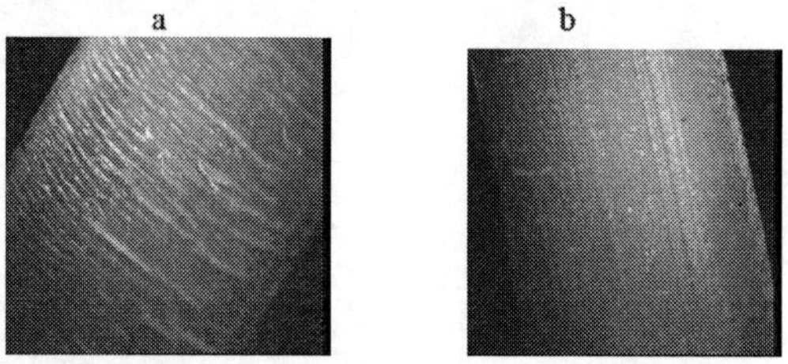

Fig. 4. Chips caprolon obtained by turning at cutting speed of 300 m/min: a - VK8 alloy plate; b - KNT16 alloy plate.

An exception is the option of caprolon turning with KNT16 alloy plate at a speed of 300 m/min. The shear elements of the chips have so small geometric parameters that chips can surely be attributed to the continuous drain. The continuous drain chip forming during caprolon turning continuous also shows the stability of the cutting process and ensure high quality of the machined work piece surface [6].

Numerical values of the test roughness parameters, provided in Table 2, confirm that use of KNT16 alloy plate makes it possible to obtain subsequently a higher quality component surface layer compared with the version of traditional turning with VC8 alloy blade. Here the value of parameter S_k confirms the validity of evaluating roughness according to known parameters R_a and R_z.

Table 2. The roughness values of caprolon treated surface.

Machining type (at a speed 300 m/min)	Roughness parameters, μm					
	R_a	R_z	R_p	R_m	S_m	S_k
Turning with KNT16 alloy	2.5	7.37	4.17	5.63	0.15	–0.8
Turning with VC8 alloy	6.6	20.2	14.8	23.1	0.5	–0.6

The deviations of component surface shape such as roundness, cylindricity, and rectilinearity were evaluated with the aim of more detailed comparative analysis. The obtained parameters are presented correspondingly in Figs. 5, 6, and 7.

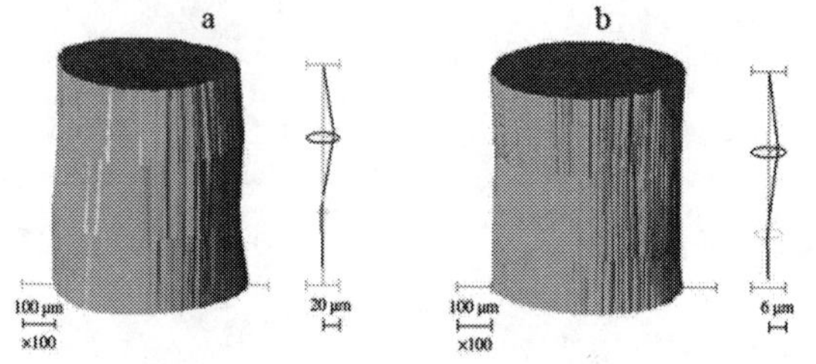

Fig. 5. Deviations from cylindricity after caprolon machining: a) VC8 alloy plate; b) KNT16 alloy plate.

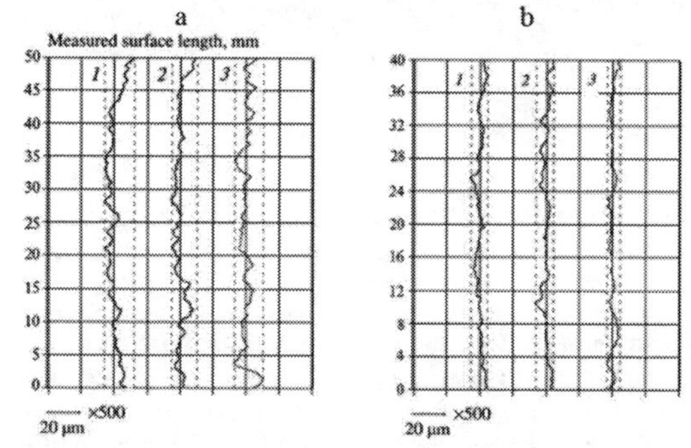

Fig. 6. Deviations from generatrix rectilinearity after caprolon machining: a) VC8 alloy plate; b) KNT16 alloy plate.

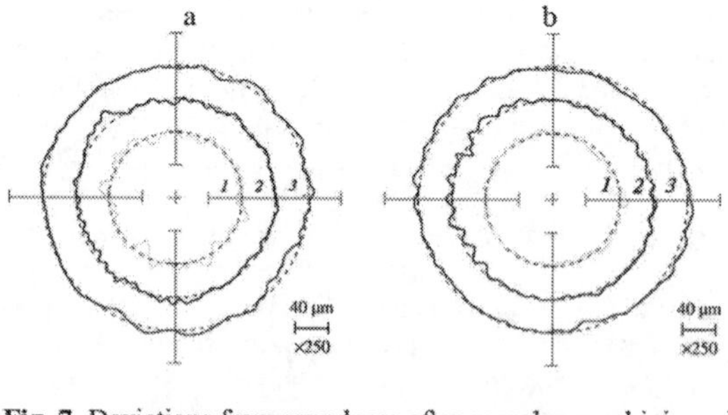

Fig. 7. Deviations from roundness after caprolon machining: a) VC8 alloy plate; b) KNT16 alloy plate.

DOI: 10.1051/matecconf/201712901015

According to data for non-cylindricity the caprolon turning with KNT16 alloy plate leads to a reduction in non-cylindricity by approximately more than three times. For example, whereas in the case of caprolon turning with VC8 alloy blade the maximum deviation from cylindricity is 20 μm, after using the KNT16 alloy plate this deviation does not exceed 6 μm (Fig. 5).

Values of component generatrix non-rectilinearity obtained with the test versions of machining have very little difference and for example 28 μm at caprolon turning with VC8 alloy blade and 15 μm after turning with the KNT16 alloy plate (see Fig. 6).

On the basis of the data obtained (see Fig. 7) for the deviation of component roundness it has been established that use KNT16 alloy plate for caprolon turning promotes some reduction in deviation from roundness. After caprolon turning by the VC8 alloy blade the deviation from roundness is 21 μm and after treatment with KNT16 alloy plate it is 19 μm.

4 Results of production tests

We consider the results of production tests for sliding bearing supports made from caprolon and installed in a screw feeder device drive shaft bearing unit for rubber technology production components.

The aim of the test is to study the effect of machined surface quality for bearing supports made of caprolon on the operating efficiency of a bearing unit for this device.

Two sliding supports were tested in succession. The seating surface of support of No.1 was machined by the cutting the KNT16 alloy plate and turning of support No. 2 the billet was made by VC8 alloy blade. After installing bearings within a block for each support made of caprolon they were operated for 120 hours after which vibration diagnostics were performed.

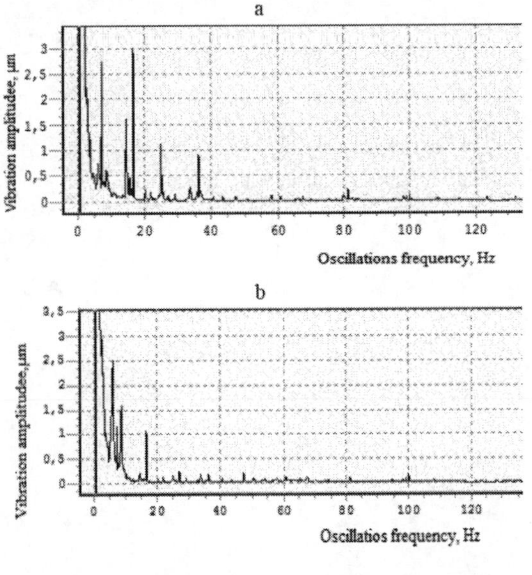

Fig. 8. Vibration amplitude: a - support No.1; b - support No.2.

Vibration diagnostics of bearings was carried out according to a standard [9] using the SD-21 instrument for vibration values control. The output estimates of support operating efficiency were vibration amplitude *A*, μm; vibration speed of oscillations, *V*, mm/sec; vibration acceleration of oscillations, *a*, mm/sec^2.

Analysis of data obtained shows that the use of support No.1 manufactured by caprolon turning with KNT16 alloy plate provides more efficient operation of a bearing assembly and the device as a whole. This is indicated by the lower values of recorded vibration parameters compared with similar values for support No.2 (see Fig. 8).

5 Conclusions

So the following conclusions can be formulated on the basis of the research results:

- experimentally proved that turning process is stable both at caprolon cutting with VC8 alloy blade and caprolon cutting with plate from ceramic material KNT16. The turning process stability is confirmed by the slight values changes of cutting force main component P_z the while the change range in the case of caprolon turning with ceramic plate is smaller in comparison with the change range at caprolon turning with VC8 alloy blade for the selected processing modes;
- established experimentally that caprolon machining is advisable to use many-sided replaceable plates made from KNT16 while at a speed of cut of 300m/min the use of the ceramic plate is more preferable compared with a plate from VC8 alloy as evidenced by the type of received chips;
- the use of many-sided replaceable plates for polymeric materials turning allows expanding the range of instruments used and increasing the efficiency of the overall turning process;
- the caprolon turning with cutter equipped with KNT16 alloy plate leads to a marked reduction in such indices for surface shape deviation, as non-roundness and non-cylindricity, and there is almost no effect on non-rectilinearity;
- results of production tests entirely confirm the expediency of using caprolon turning by ceramic cutting tools made since realization of this approach finally made it possible to provide a reduction in the vibration level of industrial devices.

References

1. B.N. Arzamasov, *Materials Science* (M.: Publishing house of the Moscow Higher Technical School Bauman, 2001)
2. F. Klocke, *Manufacturing Processes 1: Cutting* (Berlin: Springer-Verlag., 2011)
3. A.S. Vereshchaka, *Cutting materials* (M.: Higher School Publishing., 2009)
4. G. Smith, *Cutting Tool Technology* (Industrial Handbook, London: Springer-Verlag London Limited, 2008)

5. P. Davim, *Machining of Hard Materials* (London: Springer-Verlag London Limited, 2011)
6. O.Yu. Erenkov, S.A. Kovalchuk, A.V. Gavrilova, *Combined method of plastic work piece machining based on a pretreatment mechanical down*, Rare metals, Vol.**26**, Spec. Issue, pp.20-24 (2007)
7. O.Yu. Erenkov, A.G. Ivakhnenko, M.V. Radchenko, *Oscillatory process of production systems during turning of caprolon blanks*, Chemical and Petroleum Engineering, Vol. **49**, Issue 5–6, pp. 411-417 (2013)
8. O.Yu. Erenkov, E.G. Kalita, E.O. Ivakhnenko, *Study of the Effect of Cutting Regimes on Chip Formation*, Chemical and Petroleum Engineering, Vol.**50**, Issue 3-4, pp. 273-276 (2014)
9. GOST R ISO 7919-3–99. Monitoring the State of Machines from the Results of Measuring Vibrations on Rotating Shafts, Izd. Standartov, Moscow (1999)

DOI: 10.1051/matecconf/201712901016

Electroerosion formation and technology of cast iron coatings on aluminum alloys

Vladislav P. Smolentsev [1], *Oleg N.* Fedonin[2], and *Eugene V.* Smolentsev [1,*]

[1]Voronezh State Technical University, 14, Moskovsky Ave., Voronezh, 394026, Russia
[2]Bryansk State Technical University, 7, 50-letiya Octyabrya Blvd, Bryansk, 241035, Russia

Abstract. At present in the course of designing basic production parts and industrial equipment designers pay more and more attention to aluminum alloys having a number of properties compared favorably with other materials. In particular, technological aluminum tool electrodes without coating in the presence of products of processing with alkali in the composition of operation environment are being destroyed at the expense of intensified material dissolution. It is shown in the paper that the method offered by the authors and covered by the patents on cast iron coating of products made of aluminum alloys, allows obtaining on a product surface the layers with high adhesion durability ensuring a high protection against destruction in the friction units including operation in hostile environment. Thereupon, aluminum, as compared with iron-based alloys used at manufacturing technological equipment for electrical methods of processing, has a high electrical and thermal conduction, its application will allow achieving considerable energy-saving in the course of parts production. A procedure for the design of a technological process of qualitative cast iron coatings upon aluminum tool electrodes and parts of basic production used in different branches of mechanical engineering is developed.

1 Introduction

In industry there has been developing a new direction of operation characteristics increase in products by means of a surface layer modification including the metal coatings corresponding to the requirements of new equipment designers. In such a way, in aerospace industry one tries to combine in one object of production different operational requirements which were considered earlier to be unrealizable [10-13]. One of these objects represents the parts made of light aluminum alloys having high wear-resistance, resistance to hostile environment, for example, due to cast iron coatings. In recent years encouraging results on electroerosion coatings made of materials with a high melting point (for example, cast iron) are obtained. For manufacturing a qualitative layer it was necessary that an inter-electrode gap control should be used where an electrode vibrates along the axis with the frequency of 50-120 Hz. At the same time, low-frequency oscillations of the tool must be coordinated with the process of coating as without reciprocal tool motions an aluminum alloy melts under coating applied and the part geometry is destroyed.

2 Mechanism of coating

Of great importance at the definition of quality parameters in the coating applied (for example, hardness, coating wear-resistance, adhesion properties and others) is a depth of the surface layer changed in which thermal and diffusion processes take place [1, 2], defining the heat and mass transfer (when applying cast iron coatings upon an aluminum basis the thermal changes and also the doping of aluminum with carbon, iron and other elements are possible). The diagram of the process of the surface layer formation and metal coating is presented in Fig. 1.

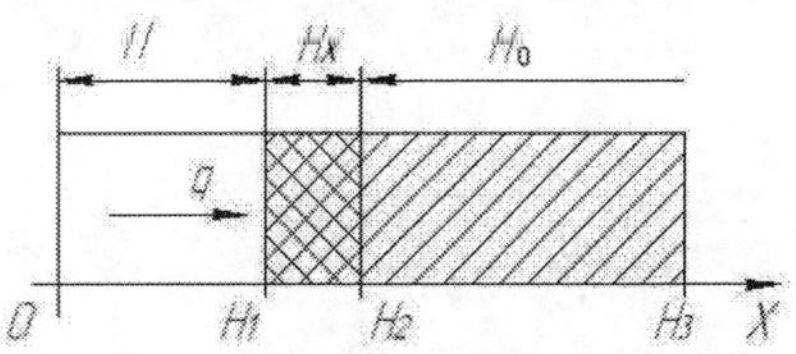

Fig. 1. The structure of the coating layer: H – layer of metal applied; H_x – molten layer; H_0 – blank metal; q – heat flow coming into surface layers of the area of blank coating.

The Figure shows a dynamics of coating formation. A high-temperature metal drop (for example, cast iron) torn off from the anode and moving through a discharge channel falls onto heat-resisting oxide film on the surface of metal processed with a low melting point the initial temperature of which is T_0. Without flux presence after the contact of a drop with a coating area, a film heating and a blank sub-surface layer reflow occur. As a result, a melt layer of variable quantity $H_x(\tau)$, which disturbs blank accuracy and surface layer quality, is formed. It is supposed that the process should proceed at the vibration

* Corresponding author: smolentsev.rabota@gmail.com

DOI: 10.1051/matecconf/201712901016

of a tool electrode with the frequency recommended, for example, in the work [3].

A new method [4] is offered for the cast iron coatings with the surface preparation in the course of the combined processing by means of the chemical removal of a high-temperature oxide film preventing the production of qualitative cast iron layers [5]. At the heart of this method lie the heat- and mass transfer and heat insulator removal from the surface of a blank during the coating application.

In the study of the process of the heat- and mass transfer during the alloying and coating

the dynamics of melting boundary motions into material of a blank and a possibility of the element diffusion of metal coated in the melt bath and a coating and also changes of diffusion coefficients and thermal physical characteristics in the discharge period were taken into account.

At that, the following assumptions were accepted:

– the pressure in the discharge channel is accepted to be equal to the atmospheric one that corresponds to reality only at the initial stage of the discharge. Inasmuch as the duration of each discharge does not exceed some micro-seconds, then such an assumption is quite substantiated;

– as a result of the breakdown of the inter-electrode gap, plasma is not formed, though it concerns only separate stages of the pulse. Here, at the expense of the mass transfer, plasma moves away from the coating applied and cannot affect substantially the process;

– the blank surface melting is carried out without taking into account the evaporation and metal atmospheric emission from the surface of heating;

– the problem solution was carried out in the Cartesian coordinate system the beginning of which coincides with the external boundary of the layer of the coating applied.

$$\frac{\partial C_1}{\partial \tau} = \frac{\partial}{\partial x}\left(D_1 \frac{\partial C_1}{\partial x}\right), \quad 0 \le x \le H_1, \tag{1}$$

$$c_1\rho_1 \frac{\partial T_1}{\partial \tau} = \frac{\partial}{\partial x}\left(\lambda_1 \frac{\partial T_1}{\partial x}\right), \quad 0 \le x \le H_1, \tag{2}$$

$$c_2\rho_2 \frac{\partial T_2}{\partial \tau} = \frac{\partial}{\partial x}\left(\lambda_2 \frac{\partial T_2}{\partial x}\right) + \frac{Q_{mel} \cdot W}{H_2}, \quad H_1 \le x \le H_3, \tag{3}$$

$$\frac{\partial C_2}{\partial \tau} = \frac{\partial}{\partial x}\left(D_2 \frac{\partial C_2}{\partial x}\right), \quad H_1 \le x \le H_2, \tag{4}$$

where c_1, c_2 – heat capacity; ρ_1, ρ_2 – density; T_1, T_2 – temperature; λ_1, λ_2 – thermal-conductivity coefficients; τ – time; x – coordinate; H_1, H_2, H_3 – thickness of the layer coated, that of the molten one and a layer of the main metal accordingly; C_1, C_2 – mass concentration of anode elements in the coating and in molten metal; D_1, D_2 – diffusion coefficients; Q_{mel} – melting heat of an aluminum alloy; $W=\rho_2 V$ – mass speed of melting; V – linear melting rate. Indices 1 and 2 correspond to metal of the anode and that of the cathode (of a blank).

Diffusion coefficients D_1 and D_2 are defined through the exponential dependence on temperature [6]

$$D_1(T) = A_1 \exp\left(-\frac{E_1}{RT_1}\right), \tag{5}$$

$$D_2(T) = A_2 \exp\left(-\frac{E_2}{RT_2}\right), \tag{6}$$

where A_1 и A_2 – pre-exponential multipliers (factors of the metal diffusion in a coating and a base); E_1 and E_2 – activation energy; R – gas constant.

Initial conditions (at τ=0):

$$T = T_0, \tag{7}$$

$$C = C_{01} \text{ at } 0 \le x \le H_1, \tag{8}$$

$$C = C_{02} \text{ at } H_1 \le x \le H_3, \tag{9}$$

$$H_x = 0, \tag{10}$$

where T_0 – initial temperature; C_{01} and C_{02} initial concentration of anode elements in a coating and in melt (micro-bath); H_x – molten layer thickness.

Boundary conditions:

$$\frac{\partial T_1}{\partial x} = 0 \text{ at } x=0, \tag{11}$$

$$\lambda_2 \frac{\partial T_2}{\partial x} = q + q_* \text{ at } x = H_1,\ T_2 = T_{mel}, \tag{12}$$

$$\frac{\partial T_2}{\partial x} = 0 \text{ at } x = H_3, \tag{13}$$

$$C = C_1 \text{ at } x = 0, \tag{14}$$

$$D_1 \frac{\partial C_1}{\partial x} = D_2 \frac{\partial C_2}{\partial x} \text{ at } x = H_1 \tag{15}$$

$$\frac{\partial C_2}{\partial x} = 0 \text{ at } x = H_2, \tag{16}$$

The expression (12) is a boundary condition of the second kind which describes body heating by two sources of heat.

The first source – part of heat from electro-spark discharges coming into surface layers of the blank worked – q.

The second source q_* – heat transferred by metal applied of the electrode.

At that q and q_* – are defined through the formulae

$$q = \frac{P_{av}}{S} \cdot K, \tag{17}$$

$$q_* = L_{fe} \cdot m_K. \tag{18}$$

Here S – electrode cross-sectional area (under condition that the diameter of the discharge channel is close to the diameter of an electrode operation part which corresponds to a tapered end of the tool taking into account its wear); L_{fe} – latent heat of the electrode material melting; $m_к$ – mass of the drop of material deposited on a blank during one pulse.

The average power of the generator (P_{av}) estimated through average current (I_{av}) and voltage (U_{av})

$$P_{av} = U_{av} I_{av} \tag{19}$$

Factor K in (17) characterizes the part of pulse energy consumed for heating a blank surface:

$$K = 1 - \frac{2}{p+1}, \tag{20}$$

where p – coefficient of thermal activity depending upon the thermal conductivity, heat capacity and density of electrode materials:

$$p = \sqrt{\frac{\lambda_1 c_1 \rho_1}{\lambda_2 c_2 \rho_2}} \tag{21}$$

The system of obtained non-linear differential equations of thermal conductivity and diffusion for coating metal and metal for a blank with non-linear boundary conditions on the surface of the environment division was being solved through the method of finite differences.

The numerical investigations were carried out in the range of the parameters of outer effects and characteristics of the transfer [6, 7] typical for the modes of electroerosion metal coatings.

To solve the system of equations it is necessary to know the thickness of the molten layer, H_x the numerical value of which is calculated at any point of time through the formula:

$$H_x(\tau) = h\left(i(\tau) + \frac{T_{i-1}(\tau) - T_{mel}}{T_{i-1}(\tau) - T_i(\tau)}\right), \tag{22}$$

where h – pitch of a differential grid by x; i – number of a mesh point in which a temperature value achieved a metal melting temperature at a certain point of time; T_{i-1}, T_i – temperature value at the same point of time in (i - 1) and i-th points of a grid; T_{mel} – melting temperature of a metal basis.

By (23) it is possible to calculate a linear melting rate:

$$V = \frac{H_x(\tau + \Delta\tau) - H_x(\tau)}{\Delta\tau}, \tag{23}$$

where $H_x(\tau + \Delta\tau)$ and $H_x(\tau)$ – thickness of a molten layer at the point of time under consideration and at a previous point of time; $\Delta\tau$ – pitch of a differential grid in time.

In the course of the fulfillment of numerical computations the following parameter values were used [6]: c_1 = 500 J/(kg ^{0}K); c_2 = 250 J/(kg ^{0}K); ρ_1 = 7100 kg/m^3; ρ_2 = 2800 kg/m^3; λ_1 = (46÷50) W/m·^{0}K; λ_2 = 140 W/m·^{0}K; T_{mel} = 930 ^{0}K; T_0 = 300 ^{0}K; S = (2÷4) mm^2; L_{fe} = 217,72 kJ/kg; Q_{mel} = 390 J/g; E_1 = 4806 kJ/mol; E_2 = 125 kJ/mol; C_{01} = (92,8÷94,6) % (Fe); C_{02} = (90,5÷92,9) %.

P_{av} and m_K – defined through a computational method.

3 Technological process designing

The analyzed mechanism of a surface layer formation allows carrying out the computations of basic parameters of machining modes, taking into account the peculiarities of the processes of coating under the flux layer and developing the recommendations for the material selection for manufacturing a part and a tool electrode shown in [5]. On the basis of this a procedure for the design of a technological process of cast iron coating upon aluminum alloys under a flux layer by means of an electroerosion method is developed.

The paper reports the consideration of two basic stages of the process designed [7]: the preparatory stage and the production one, including the part coating process consisting mainly of operations of the coating layer formation by the electroerosion method.

In the preparatory phase the following is carried out:

1) the selection substantiation of aluminum alloy parts, requiring cast iron coating or that of other materials with a high melting point, ensuring the required operation properties of the product;

2) the assessment of possibility to achieve the required technological indices calculated, according to the dependences shown at the description of the mechanism of coating formation; here some options are possible:

a) basic operation requirements are obtainable and a technological process should be designed for its realization;

b) at the current moment, some requirements cannot be obtained, but for the rest of indices a new process is efficient technically and economically. Here it should be expedient to introduce some precise definitions into a technical specification or technical requirements in a detail drawing and to design a technological process;

c) the specified requirements to a product are still unattainable and one has to pass to other kinds of materials and technological processes;

3) at the same stage, the improvement of the manufacturability of the part design is carried out in the course of which it is necessary to take into account: the character of production, dimensions of parts, physical-engineering properties of aluminum alloys, their peculiarities in behavior in the course of coating application, an error value of a real size, the specified technical requirements to coatings;

4) (if necessary) one improves the geometry of a part, taking into account further machining, calculates machining allowance and obtainable thickness of a qualitative layer obtained;

5) the choice and preparation of means for the technological equipment are carried out: as a rule, all devices used in the course of parts production may be chosen from the number of those used under conditions of a repetition work (at small dimensions of parts there is no need in the selection of handling machinery, which simplifies considerably equipment design and reduces its manufacturing cost);

6) the geometry and material are chosen for a tool electrode. For coating it is recommended to use a rod tool electrode made of cast iron of SCh20 kind. The application of cast iron surface layers for the purpose of

DOI: 10.1051/matecconf/201712901016

corrosion protection is most efficient for electrodes for electrical-chemical treatment in alkaline medium and in environment containing chloride compounds in which aluminum alloys are destroyed quickly. The operation properties of such coatings, in addition to physical, mechanical and chemical characteristics of electrode material used, depend upon the reasonableness of a mode choice and technological process perfection [5]. At the same time, besides structure, properties and chemical composition of material of the coating obtained a significant criterion is smoothness of the coating, its thickness, a diffusion layer depth ensuring a durable adhesion of the coating with the basis and so on [5];

7) the working environment should be chosen and its composition is to be defined. The compatibility of the electrode material and working environment is a significant factor defining not only the efficiency of the process, but also a possibility of its fulfillment. Authors of many works emphasized the dependence of the character of the anode material application on cathode of the chemical interaction of electrode material with elements of inter-electrode environment [8], [9]. It is offered to use as a working medium the flux of FS-71 type having activating properties at coating upon aluminum alloys by the electroerosion method. In the course of processing, the flux enters the reaction with a surface oxide film Al_2O_3 destroys it, dissolves partially and removes it as a suspension cleaning, in such a way, the surface of a blank against a film, which not only prevents a stable process, but is often a main reason in the destruction of a cathode surface;

8) the modes of electroerosion coating are established [5;7], the quantity of layers essential for the formation of a coating with the quality required is defined. It is recommended to apply not less than three coating layers. At that it should be taken into account that the application of the second layer causes roughness increase, but increases heat-resistance and smoothness of the layer formed.

9) the formation of the technological process and in accordance with this, the manufacturing (the second stage of designing) of aluminum alloy parts with cast iron coating is carried out.

On the basis of the route mentioned above a procedure for the technological process designing taking into account the specificity of the coating under consideration which includes parts control on the whole and surfaces subject to processing in particular is formed. A possibility of processing is formed, a real value of the parameter after working is defined, a conformance to an error value is checked up.

At the second (manufacturing) stage the preparation of a blank surface for coating is carried out. When preparing a surface for coating there should be removed stains of dirt, paint, oils and fats. The operation is to be carried out by mechanical purification, rinse, degreasing and etching. An area joining the surface worked with the width of (25 ÷ 30) mm is subject to cleaning. At that the application of abrasive tools for this purpose is unacceptable as crumbled abrasive particles introducing inculcating into a surface can form various inclusions affecting adversely the quality of a coating performed.

Degreasing is carried out with the aid of organic solvents: gasoline, acetone. Parts degreased are washed out in lukewarm water (temperature not low than 298 K) and rubbed clean with a stiff bristle brush. Etching should be carried out in 15-% aqueous solution of nitric acid at the temperature of (333 ÷ 338) K during (2 ÷ 5) min or in 30 % aqueous solution of nitric acid at the temperature of (293 ÷ 298) K during (1 ÷ 2) min to neutralize remains of alkali, after that the parts should be washed out in lukewarm water.

Further, the improvement of modes is fulfilled on the equipment and manufacturing a sample with the aid of which the control of coating quality and the assessment of process efficiency is carried out.

On the basis of the results obtained in the course of the fulfillment of control operations the correction of modes and working a basic work-pieces batch are carried out.

The computations and experiments performed have shown that at electroerosion cast iron coating (material SCh-20) on aluminum alloy (Material D16) a rational combination of modes and conditions of treatment carrying out are the following:

- A_{pulse}= (0,1 ÷ 0,5) J;
- U_{av} = (50 ÷ 60) V;
- $I_{short\ circuit}$ = (3 ÷ 5) A;
- τ_{pulse} = (10^2 ÷ 10^3) μs
- feed speed of "electrode-tool" V_e = (50 ÷ 200) mm/min;
- number of layers n = 3;

Coating is carried out during the period of time essential for the formation of the coating with the thickness required, T_0 = (3 ÷ 5) min/cm^2.

4 Application of coatings for production of technological equipment and parts of basic production

A general scheme of the operation sequence in the technological process is realized at manufacturing a tool electrode made of aluminum alloy; an assembly drawing and a drawing of the aluminum flange with coating are shown in Fig. 2: *a*) general view: 1 – aluminum alloy flange with cast iron coating for the tool electrode mounting, the fulfillment of current supply and electrolyte supply (place "a"); 2 – transition mandrel; 3 – electrically insulating dielectric coating; 4 – working element of the aluminum alloy tool electrode with cast iron coating; *b*) external view and basic dimensions of a flange.

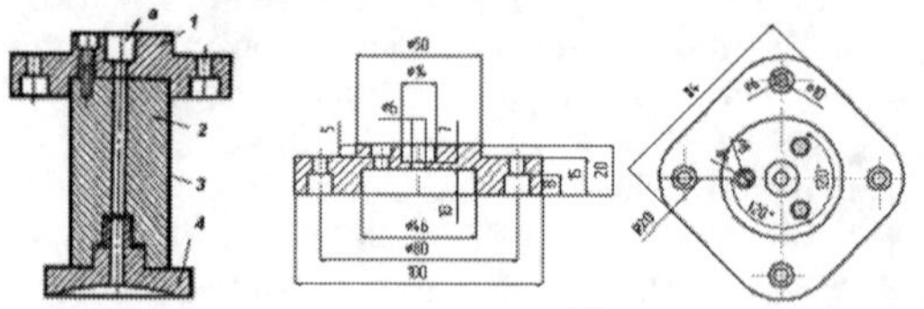

a) b)

Fig. 2. The composite tool electrode for the combined erosion-chemical treatment

The increased working capacity of the coatings mentioned is ensured by the absence of a wedging effect under conditions of liquid friction and also by a specific profile characterized by a round micro-profile affecting in a positive way a load distribution.

The properties mentioned above and also a high durability of such coatings under the influence of electric current of high density and an electrolyte flow make their application possible as protective ones at manufacturing baths, cisterns, current leads and other elements of technological equipment operating in the modes of cavitation liquid flow at a long term of operation, for example, at combined erosion-chemical treatment.

The application of the developed method of coating for the corrosion resistance increase of surfaces in parts subjected to the impact of hostile working medium is also promising. The external view of some parts is presented in Fig. 3.

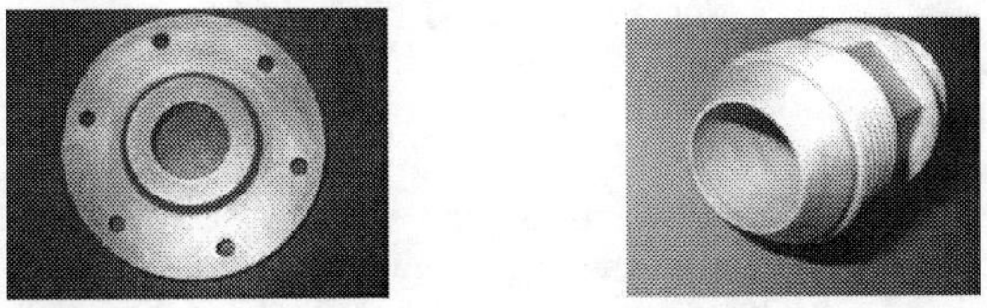

a) b)

Fig. 3. The external view of aluminum alloy parts which are to be coated with cast iron under a flux layer: *a* - flange; *b* - connection

Before coating, the parts (Fig.3) made of material VD17 are subjected to machining, the accuracy control of geometrical parameters and the mutual location of surfaces, after that the layers of cast iron coating SCh20 with the general thickness, 0,4+0,1mm are applied on the surface of the side of the mating areas. After that the surface coated is ground with allowance removal of 0.2-0.22mm and the part becomes resistant to wear and durable to hostile medium impact.

5 Conclusion

The physical model and the mechanism of electroerosion coating are formed taking into account the necessity of low-temperature oscillations of a tool electrode, the specificity of the formation of high temperature surface layers on aluminum alloys under a layer of chemically active flux.

The system of indices essential for designing technological processes [7] for high temperature coatings on the surface of parts made of aluminum alloys by the electroerosion method is developed.

It is shown that the application of parts with a coating allows attaining the efficient protection against wear and hostile medium impact at the reduction of product mass, material electrical conductivity increase at the expense of its substitution and decreasing more than twice power consumption at its supply for processing.

References

1. A.V. Perova, *Proceedings of OrelSTU*, **4(272)**, 8 (2008)
2. V.P. Smolentsev, O.N. Kirillov, S.V. Dultsev, M.V. Shchipanov, *Patent 2318637 (The Russian Federation), Bulletin 7 (2008).*
3. G.P. Teslin, *Soldering with Vibration Application* (L., LDNTP, 1967)
4. V.P. Smolentsev, A.V. Grebenshchikov, A.V. Perova, B.I. Omigov, *Patent 2405662 (The Russian Federation),* Bulletin **34** (2010)
5. A.V. Perova, V.P. Smolentsev, A.S. Gribanov, *Aerospace Techniques*, 2008.
6. A.P. Eichis, B.Ya. Temkina, *Technology of Surface Treatment of Aluminum and Its Alloys* (M.: Machgiz, 1963)
7. E.V. Smolentsev, *Design of Electrical and Combined Methods of Treatment* (M., Mechanical Engineering, 2005)
8. O.N. Krakhmalyov, D.I.Petreshin, O.N. Fedonin, *Modern problems of theory of machines*, **4**, 78 (2016)
9. O.N. Krakhmalyov, D.I.Petreshin, O.N. Fedonin, IOP: *Conference Series. Materials Science and Engineering*, 124 (2015)
10. A.V. Khandozhko, D.I. Petreshin, O.N. Fedonin, V.B. Protasiev, N.A. Bizyakina, D.Yu. Shukin, IOP: *Conference Series. Materials Science and Engineering,* **124** (2015)
11. D.I. Petreshin, A.V. Khandozhko, O.N. Fedonin, IOP: *Conference Series. Materials Science and Engineering,* **124** (2015)
12. S.V. Safonov, E.V. Smolentsev, O.N. Fedonin, IOP: *Conference Series. Materials Science and Engineering,* **124** (2015)
13. V.P. Smolentsev, S.V. Safonov, E.V. Smolentsev, O.N. Fedonin, IOP: *Conference Series. Materials Science and Engineering,* **124** (2015)

On the estimation of feasibility of engineering services use by machine-building enterprises

Ramil Mavlioutov[1,*], *Ekaterina* Butenko[1], and *Regina* Menyailova[1]

[1]Volgograd State Technical University, 400005 Volgograd, the Russian Federation

Abstract. The article highlights the main advantages of engineering services in the implementation of the project. An approach to the estimation of the engineering console as an additional income received by the customer company as a result of engaging an engineering company is proposed. This approach allows to evaluate the validity of the use of engineering services both by the machine-building enterprises themselves as potential customers and engineering companies.

1 Introduction

The imperative of the development of modern economy is the permanent accelerating growth of the level of science-intensive production, goods, works and services. As a result, the importance of engineering activity is growing in the context of the economic relations of industrial enterprises. In the first place, this trend is affected by the machine-building enterprises. The theory and practice of engineering business does not provide for a single category of “engineering service”. Within the framework of this article, an engineering service is understood as a complex of professional services of a commercial nature provided by engineers at all stages of the technological chain of creating a new facility or developing equipment (converting new scientific knowledge into a technical product or a new industrial facility). This complex can include both the whole range of services and individual components [1]. The engineering company is understood as a professional intermediary uniting fundamental and applied science, design and research institutes and manufacturing enterprises, integrating and coordinating the work within the framework of the implementation of new technological projects [2].

The current situation is characterized by the emergence of the practice of applying advanced approaches to managing the operation of engineering companies in Russia. There have been significant changes in the sphere of increasing trade in complex types of equipment against the background of the increasing globalization of economic relations. This necessitates special knowledge for solving both technological and organizational issues. This is the guarantee of the competitiveness of machine-building enterprises in the field of growing struggle for an ever more demanding consumer [3]. At the same time, among the peculiarities of the current situation, first of all, it is necessary to allocate a small motivation in machine-building enterprises to use the services of engineering companies due to low efficiency and low possibility of raising it. There is a contradiction of high input costs for attracting an engineering company with the expected impact.

Due to the lack of a formalized approach that allows to assess the estimation of feasibility of engineering services use within the framework of this article we are going to present the results of our developments on this subject. For this purpose, the study is aimed at solving problems: clarifying the conceptual instrument in terms of determining the economic effect of engaging the engineering company; highlighting the main advantages of engaging an engineering company; determining the impact of engineering services of each of the identified advantages on the amount of revenue of the customer company, as well as their total impact.

A number of researchers note the high potential of engaging engineering companies [4-9]. Here are its main components.

1. Reduction of project implementation deadlines.
2. End-to-end technological responsibility. Within the traditional approach, there is an accumulation of deviations from technological requirements, which leads to a drop in the company's income.
3. High quality, genetics of technical solutions.
4. Reducing the cost of risks. The engineering company plans in advance measures to prevent and mitigate the consequences of risks.
5. Application of project financing. Financial institutions interact on the basis of the most favourable conditions (lowering interest rates, shifting from collateral financing to project one). Long-term contracts, encourage counterparties to cooperate with the use of all kinds of discounts.

In order to identify the income received by the contracting company as a result of engaging the engineering company, we propose the use of the concept

* Corresponding author: ramil-2002@mail.ru

DOI: 10.1051/matecconf/201712901017

of an engineering console. (abridged from English consolidated annuity).

The appearance of the engineering console is due to the following main factors: application of the competencies inherent in the engineering company, throughout the life cycle of the implemented project; the unique nature of the competencies used by the engineering company; priority of suppliers and contractors with a high degree of contractual terms. The cumulative effect of the mutual influence of these factors creates a positive impact on the profitability of the implemented project and, therefore, the investment attractiveness.

Due to the fact that projects involving the engineering company tend to have a significant implementation period, measured in several years, the time factor is mandatory [10]. Net present value (NPV) is the most widely used indicator to estimate the projects taking into account the time factor. Taking into account our definition of the engineering console category (hereinafter - EC), as well as the specifics of its formation, we propose in this article to calculate the EC as the difference between the NPV of the project involving the engineering company (hereinafter - the NPV_E) and the NPV of the project excluding its involvement (hereinafter - NPV_0).

$$EC = NPV_E - NPV_0 \tag{1}$$

The value of the net present value of the project is determined by the well-known formula:

$$NPV = \sum_{i=1}^{n} \frac{NCF_i}{(1+I)^i} - \sum_{i=1}^{n} \frac{Inv_i}{(1+I)^i}, \tag{2}$$

where NCF_i is the net cash flow of the i-period of the investment project,

Inv_i are investments, attached to the i -period of the investment project,

I is the cost of capital attracted for the implementation of the investment project,

n is the period of implementation of the investment project (the calculation level).

To concretize and implement the subsequent analysis it is expedient to present the formula (2) in the form:

$$NPV = \sum_{i=1}^{n} \frac{(1-IT_i)(P_iQ_i - VC_i - FC_i)}{(1+I)^i} - \sum_{i=1}^{n} \frac{(1-IT_i)(MP_i + OE_i)}{(1+I)^i} - \sum_{i=1}^{n} \frac{Inv_i}{(1+I)^i}, \tag{3}$$

where IT_i is the corporate income tax rate,

P_i is the unit price (monetary units a piece)

Q_i is the sales volume in physical terms,

VC_i are variable costs in the cost of production,

FC_i are constant costs in the cost of production,

MP_i are interest payments within the framework of loan servicing,

OE_i are other expenses attributable to the financial result of the organization.

World experience of functioning of engineering companies allows to formulate the advantages provided by their attraction:

1. Reduction of the project implementation period;
2. Risk reduction;
3. Generation of engineering solutions;
4. The high level of business reputation of the engineering company is correlated to the company-customer.

It is necessary to consider in detail the impact of the engagement of the engineering company on the components of the NPV.

2 Reduction of the project implementation period

The use of services of engineering companies ensures the combination of the implementation of a number of processes, which provides an opportunity to reduce the duration of the pre-investment stage and the investment stage. Together, this enables the customer company to begin generating profits at an early stage through an early entry into the operational phase [11]. From this we can conclude that nE < n0. If we assume that the other conditions are equal, the investment project with a shorter implementation period, has a higher NPV. Here it is possible to identify the emergence of the engineering console, due to the reduction in the duration of the investment project.

3 Risk reduction

In the process of estimation NPV, one of the determining factors is the value of the capital cost involved in the implementation of the investment project. The implementation of any investment project implies the existence of a temporary lag between the application of resources and the resulting output, which may result in a change in the deadlines of the project. This causes the emergence of a risk category. Accounting for the risks that accompany the investment project, transpose the investors' preferences. Due to the fact that investors objectively seek to minimize risks, a project with lower risks will be more preferable. The consequence of the latter one is the rise in price of capital as the risks of the implemented project increase [12]. At the same time, risks have a direct impact on the elements of the calculation of the NPV. Thus, we think that is not correct to offset risks solely on the cost of capital. In this regard, it seems necessary to classify the risks accompanying the implementation of the investment project and which are taken into account in the estimation of the NPV. A common feature is to divide the risks of the investment project by sources of their origin into internal and external ones [13].

With reference to an industrial enterprise, the following risks are included in the number of accompanying internal risks:

the risk of inconsistency in the qualifications of personnel in those jobs that they are engaged in. It causes the potential growth of fixed and variable

expenses as a consequence of the adoption of incorrect, untimely and unjustified management decisions, poor quality of the products. We denote it as R_{st}.
production risk due to technological processes. This risk is caused by the possible failure of production equipment due to improper maintenance, operation or negligence. The consequence of this is the growth of fixed expenses, as well as the other ones attributable to the financial result. We denote this risk as R_t.
risk of providing the enterprise with financial resources. It is caused by fluctuations in the value of money, as a consequence of inflationary processes, changes in the Central Bank's discount rate and the exchange rate of the national currency relatively to foreign currencies. As a result, it is possible to depreciate the available funds at the disposal of the enterprise, as well as increase the cost of attracting external financing. The consequences of this are the increase of variable expenses, fixed expenses, as well as interest payments within the framework of loan servicing and other expenses attributable to the financial result. We denote this risk as R_m.

We include the following risks in the risks of external order accompanying the investment project:
risk of disruption in the supply of materials. This risk is caused by possible fraud, the lack of technical ability to produce the necessary materials, the supplier's decision to change the terms of delivery (up to the termination of the contract), accidents in transport. As a result, fixed expenses, as well as other expenses attributable to the financial result, are growing. We denote this risk as R_s.
risk of competition. It is caused by the possible entry of products to the market by competing enterprises, which largely satisfies the needs of consumers, as well as the use of uncivilized forms of competition (misinformation of consumers, compromising goods, pressure on suppliers and banks, etc.). This risk leads to a decrease in prices and sales. Denote this risk as R_r.
risk of force majeure. Because of their nature, unforeseen events of this kind (natural disasters, wars, epidemics) generate other expenses attributable to the financial result. Denote this risk as R_f.
risk of deterioration of market conditions. This risk is generated by problems in the sales sphere: incorrect selection of the sales market, incomplete information about competitors, slow introduction of innovations into production. This risk carries the risk of slowing the pace of sales, accompanied by a drop in selling prices. Denote this risk as R_c.

These risks affect not only the implementation of the forecast of net cash flow for a certain period of the investment project, but also in determining the cost of capital involved in the implementation of the investment project, I. Therefore, we should cite the risk of raising the cost of capital, which is denoted as R_I.

So, relying on the formula (3), the total influence exerted on the NPV by the described risks can be represented in next formula (4). Further, let us single out the risk factors characteristic for the components of this formula (4):

RC_1 is the risk factor for increasing the cost of capital I;

RC_2 is the risk factor for reducing the unit price P_i;

RC_3 is the risk factor for decreasing sales Q_i;

RC_4 is the risk factor for the growth of variable expenses in the cost of production VC_i;

RC_5 is the risk factor for increasing fixed expenses in the cost of production FC_i;

RC_6 is the risk factor for the growth of interest payments within the framework of the MP_i loan service;

RC_7 is the risk factor for the increase in other expenses attributable to the financial result of the organization, OE_i. So, the total influence exerted on the NPV can be represented as:

$$
\begin{aligned}
NPV = &\sum_{i=1}^{n} \frac{(1-IT_i)(R_r R_c P_i Q_i)}{(1+R_I I)^i} - \\
&-\sum_{i=1}^{n} \frac{(1-IT_i)(R_{st} R_m VC_i + R_{st} R_t R_m R_s FC_i)}{(1+R_I I)^i} - \\
&-\sum_{i=1}^{n} \frac{(1-IT_i)(R_m MP_i + R_t R_m R_s R_f OE_i)}{(1+R_I I)^i} - \\
&-\sum_{i=1}^{n} \frac{Inv_i}{(1+R_I I)^i}
\end{aligned}
\tag{4}
$$

By means of the allocated risk factors, formula (4) can be transformed into:

$$
\begin{aligned}
NPV = &\\
= &\sum_{i=1}^{n} \frac{(1-IT_i)((1-RC_1)(1-RC_2)P_i Q_i)}{(1+(1+RC_1)I)^i} - \\
&-\sum_{i=1}^{n} \frac{(1-IT_i)((1-RC_4)VC_i)}{(1+(1+RC_1)I)^i} - \\
&-\sum_{i=1}^{n} \frac{(1-IT_i)((1+RC_5)FC_i)}{(1+(1+RC_1)I)^i} - \\
&-\sum_{i=1}^{n} \frac{(1-IT_i)((1+RC_6)MP_i)}{(1+(1+RC_1)I)^i} - \\
&-\sum_{i=1}^{n} \frac{(1-IT_i)((1+RC_7)OE_i)}{(1+(1+RC_1)I)^i} - \\
&-\sum_{i=1}^{n} \frac{Inv_i}{(1+(1+RC_1)I)^i}
\end{aligned}
\tag{5}
$$

Taking in the investment project of the engineering company allows to reduce the risks accompanying its implementation. This is due to the presence of a significant level of competence in engineering companies, based on practical experience. In addition, a positive contribution is made by the possession of engineering companies by a systematic array of data on potential risks, the degree of their destructive influence and the possibility of prevention [14, 15]. Here, it is possible to identify the emergence of an engineering console, caused by a reduction in the risks of an investment project.

4 Generation of engineering solutions

At the investment stage and the operational stage of the project implementation, the possession of an engineering company with competencies and accumulated knowledge allows it to choose the optimal solutions in terms of the choice of technologies, equipment, achieving the specified quality at minimum expenses from the point of technical and economic characteristics [16]. This is manifested in the following: an increase in the unit price and sales volume, accompanied by a decrease in the cost price in terms of variable expenses (at the investment stage); decrease in the amount of investments applied (at the investment stage).

Accounting for the effect of generating engineering solutions by introducing in the formula (3) the effect of the generation of engineering solutions C_g allows us to provide a general effect on the amount of net present value as:

$$NPV = \sum_{i=1}^{n}\frac{(1-IT_i)((1+C_g)P_iQ_i)}{(1+I)^i} - \sum_{i=1}^{n}\frac{(1-IT_i)((1-C_g)VC_i+FC_i)}{(1+I)^i} - \sum_{i=1}^{n}\frac{(1-IT_i)(MP_i+OE_i)}{(1+I)^i} - \sum_{i=1}^{n}\frac{(1-C_g)Inv_i}{(1+I)^i} \quad (6)$$

Here it is possible to identify the emergence of an engineering console, caused by the generation of engineering solutions to the investment project.

5 The high level of business reputation of the engineering company is correlated to the company-customer

The participation of an engineering company with a business reputation and established relationships with credit institutions provides the opportunity to reduce the cost of financing the project. The reduction of the cost of capital involved in the implementation of the investment project will lead to decrease of interest payments within the framework of debt servicing [17]. The introduction in the formula (3) of the image coefficient of the engineering company C_{img} allows to determine the impact on the NPV of this advantage of the engineering company:

$$NPV = \sum_{i=1}^{n}\frac{(1-IT_i)(P_iQ_i-VC_i-FC_i)}{(1+(1-C_{img})I)^i} - \sum_{i=1}^{n}\frac{(1-IT_i)((1-C_{img})MP_i+OE_i)}{(1+(1-C_{img})I)^i} - \sum_{i=1}^{n}\frac{Inv_i}{(1+I)^i} \quad (7)$$

Here you can identify the origin of the engineering console, due to the high level of business reputation of the engineering company.

The cumulative effect exerted on the net present value of the investment project by four advantages of involving in its implementation an engineering company can be represented as:

$$NPV = \sum_{i=1}^{n}\frac{(1-IT_i)(1+C_g)(1-RC_1)(1-RC_2)P_iQ_i}{(1+(1-C_{img})(1+RC_1)I)^i} - \sum_{i=1}^{n}\frac{(1-IT_i)(1-C_g)(1-RC_4)VC_i)}{(1+(1-C_{img})(1+RC_1)I)^i} - \sum_{i=1}^{n}\frac{(1-IT_i)(1+RC_5)FC_i}{(1+(1-C_{img})(1+RC_1)I)^i} - \sum_{i=1}^{n}\frac{(1-IT_i)(1-C_{img})(1+RC_6)MP_i}{(1+(1-C_{img})(1+RC_1)I)^i} - \sum_{i=1}^{n}\frac{(1-IT_i)(1+RC_7)OE_i}{(1+(1-C_{img})(1+RC_1)I)^i} - \sum_{i=1}^{n}\frac{(1-C_g)Inv_i}{(1+(1+RC_1)I)^i} \quad (8)$$

We decided that the engineering console is defined as the difference between the net present value of the investment project with the involvement of the engineering company (NPV_E) and the net present value of the investment project without involving the engineering company (NPV_0). Since the value of NPV_0 does not change, it is necessary to maximize the NPV_E, to maximize the size of the engineering console:

$$NPV_g = \sum_{i=1}^{n}\frac{(1-IT_i)(1+C_g)(1-RC_1)(1-RC_2)P_iQ_i}{(1+(1-C_{img})(1+RC_1)I)^i} - \sum_{i=1}^{n}\frac{(1-IT_i)(1-C_g)(1-RC_4)VC_i)}{(1+(1-C_{img})(1+RC_1)I)^i} - \sum_{i=1}^{n}\frac{(1-IT_i)(1+RC_5)FC_i}{(1+(1-C_{img})(1+RC_1)I)^i} - \sum_{i=1}^{n}\frac{(1-IT_i)(1-C_{img})(1+RC_6)MP_i}{(1+(1-C_{img})(1+RC_1)I)^i} - \sum_{i=1}^{n}\frac{(1-IT_i)(1+RC_7)OE_i}{(1+(1-C_{img})(1+RC_1)I)^i} - \sum_{i=1}^{n}\frac{(1-C_g)Inv_i}{(1+(1+RC_1)I)^i} \to \max \quad (9)$$

when the constraint is satisfied: $0 \le i \le n$.

So, regardless of expenses caused by the involvement of an engineering company in the implementation of an investment project, there is a positive economic effect, which should be denoted as an engineering console. We have proposed an approach to the definition of an engineering console, which takes into account four advantages for an engineering enterprise from engaging an engineering company: shortening the duration of the project; risk reduction; generation of engineering solutions; high level of business reputation of the engineering company is correlated to the company-customer. The proposed approach can be used in the practice of the functioning of engineering enterprises in the process of making managerial decisions. In addition, the approach is applicable to the activities of the engineering companies themselves as a tool to promote their services. Along with this, the results of the study can be applied in developing approaches to assessing the feasibility of using engineering services by industrial enterprises of various industry affiliations.

References

1. A.Yu. Glebanova, A.S. Semenov and A.V. Kashirin, *Engineering as a special competence in creating a new product: current state and trends,* Journal of Economy and entrepreneurship, **56**, 843-847 (2015) (in Russian)
2. G.L. Ragatz, R.B. Handfield and T.V. Scannell, *Success factors for integrating suppliers into new product development*, Journal of Product Innovation Management, **14**, 190-202 (1997)
3. K. Clarke, *Can engineers lead again?* Engineering, **2**, 19-20 (2016)
4. P.V. Grigor'ev and Yu.S. Kan, *Optimal control of the investment portfolio with respect to the quantile criterion,* Automation and Remote Control, **65**, 319-336 (2004)
5. L. Burneika, *Managing changes in product configurations*, Solid State Phenomena, **144**, 208-213 (2009)
6. Z.A. Mingaleva, *Structural modernization of economy and innovation development*, World Applied Sciences Journal, **20**, 1313-1316 (2012)
7. W. Aarseth, A. Rolstadås and B. Andersen, *Managing organizational challenges in global projects*, International Journal of managing Projects in Business, **7**, 103-132 (2013)
8. K.V. Aikhel, *Business engineering – managing for sustainable development of enterprises,* Science Time, **2**, 24-28 (2016) (in Russian)
9. P.V. Krush and I.V. Makaliuk, *Study of internal factors influence on transaction expenses of engineering companies*, Naukovyi Visnyk Natsionalnoho Hirnychoho Universytetu, **1**, 134-143 (2014)
10. E.-H. Ng and N.G. Beruvides, *Multiple internal rate of return revisited: frequency of occurrences*, Engeeniring Economist, **60**, 75-87 (2015)
11. O.O. Orlov, *Problems of innovation projects' evaluation in mechanical engineering,* Actual Problems of Economics, **163**, 43-51 (2015)
12. N.V. Usmanova and N.A. Orlova, *The role of transaction costs in risk management of investment projects*, Journal of Advanced Research in Law and Economics, **7**, 1226-1233 (2016)
13. B. Yermekbayeva and R. Doszhan, *Specifics of risk management for small and medium business enterprises,* World Applied Sciences Journal, **27**, 774-778 (2013)
14. P. Li P, Ch. Li and J. Wan, *The comparison and selection of investment projects: a new standatdized risk measure,* Engeeniring Economist, **48**, 127 (2003)
15. Y. Wang and M. Li, *The role of internal audit in engineering project risk management*, Procedia Engineering, **24**, 689-694 (2011)
16. A.E. Sklyarov and A.Yu. Glebanova, *Innovative activity of high-technology companies as assessment and forecasting object*, Proceedings of the Voronezh State University of Engineering Technologies, **3**, 296-299 (2016) (in Russian)
17. Y.P. Anisimov, Y.V. Zhuravlyov and A. A. Shvets, *Analysis of efficiency of innovative projects,* Proceedings of the Voronezh State University of Engineering Technologies, **3**, 180-185 (2012) (in Russian)

MATEC Web of Conferences **129**, 01018 (2017) DOI: 10.1051/matecconf/201712901018
ICMTMTE 2017

Prospects for the development of the theory of designing optimal cycles of machining in a multidimensional space of control parameters

Aleksandra Akintseva[1,*] and *Pavel* Pereverzev[2]

[1] South Ural State University, Department of Engineering and Technology, 454080 Chelyabinsk, Lenin Avenue 76, Russian Federation
[2]South Ural State University, Department of Automated Engineering Technology,76, Lenin Avenue, Chelyabinsk, 454080, Russian Federation

Abstract. In the engineering industry there are still no domestic systems of designing efficient automatic cycles for CNC machines because of the lack of a theory of their design. The article describes the approach to the development of the designing theory of optimal automatic cycles, covering all types of machining operations. The basis of this theory is a mathematical model for controlling the process of metal removal, adequately describing the process of metal removal using the whole complex of control parameters as with all possible combinations of the main technological factors as and in the wide ranges of their allowable variation. The dynamic programming method (DPM) is used as a mathematical method of optimization; it can be used for complex optimization of the processing cycle in the multidimensional space of control parameters with any number of constraints verified at each step of a directed enumeration of calculation options. As a result, it becomes possible to optimize an unlimited number of control parameters for all types of machining operations performed on CNC machines

1 Introduction

One of the main problems of modern automated mechanical engineering is the low productivity of operations of billets machining on CNC machines, especially in serial and small-scale production when there are no test billets. The reason for this problem is the lack of design systems for optimal stepwise processing cycles used in control programs for CNC machines. The solution of the problem above is connected with great difficulties, because the ideal variant involves simultaneous optimization of two dozen of control parameters. For example, in case of internal grinding operation with three-stage machining cycle it`s necessary to set six feeds for each stage of the cycle (three radial and three axial) and three values of the part of the removed allowance at each stage. As for mode parameters, the rotation frequency of the circle and the billet must be also set. Parameters of the circle characterization and technological adjustment also refer to the control parameters. Therefore, when designing the operation, the technologist sets: grain material, bundle material, graininess, hardness and structure of the circle, geometric parameters of the circle, overtravel and other parameters. And all these parameters above should be optimal and they should be selected simultaneously. It also should be noted that the choice of the optimum parameters of the stepwise cycle should be made by the technologist considering various technological limitations inherent in the projected machining operation, namely: limitations on the requirements of the operational sketch (tolerances of diameter, shape and location, required roughness and hardness of the machined surface), machine limitations (ranges of feed, rotational frequencies and drive powers, achievable processing accuracy, rigidity of the technological system), limitations on the active control device (number of stages, measurement error), limitations on the cutting tool (design, geometry, strength, wear resistance), limitations on the batch of the processed billets (allowance fluctuation, blunting of the cutting tool), etc.

In fact, in the absence of automated systems for designing machining cycles for CNC machines the technologist must simultaneously set dozens of optimal control parameters considering various technological limitations. It`s a complex scientific and technical task which is naturally beyond the power of any technologist. Therefore, in real production technologists set processing cycles based on existing production statistics and outdated general machine-building standards for cutting modes of the 90-ies of the edition, intended more for rationing operations [1,2, etc.]. It should be noted that the standards are based on the domestic production statistics and reflect the average level of cutting regimes in the country in the 80s at the then existing machine tool park.

Therefore, regulatory recommendations are neither optimal nor effective for modern CNC machines and they do not provide high-performance processing. According to the regulations, at first the technologist

* Corresponding author: popovaav@susu.ru

selects the parameters of the cutting tool on the basis of some recommendations. Then, depending on the range of diameters and processing lengths, the technologist selects the normative base cycle and the rotational speed of the part for fixed basic regulatory conditions. If the actual conditions in production differ from normative ones then the technologist must somehow make a correction of the regime parameters of the cycle or cutting tool. The results of such changes in the cycle and parameters of the cutting tools are poorly predictable due to the lack of verification models. Since such a normative cycle carries a high risk of getting a reject, for the guarantee, the technologist additionally underestimates the performance parameters while reducing the cycle productivity.

As a result, optimization of machining cycles for CNC machines is simply impossible in such production conditions. This situation is of a mass nature in engineering. Therefore, CNC machines are used inefficiently and machining operations are performed according to our estimation with a performance which is understated up to two times.

It should be noted that there is insufficient attention given to the problems of the theory of designing optimal machining cycles for CNC machines. An overwhelming number of researchers are studying the particular aspects of the problem of designing optimal control trajectories [3-8, etc.]. The greatest attention is paid to modeling the cutting force and modeling the technological limitations (on accuracy, roughness, defect layer, wear and tool persistence), affecting the performance of operations for particular conditions. The results of these studies are particular recommendations for partial optimization of cutting modes for a limited number of constraints in a narrow range of their variation. There are very few works aimed at developing the theory of constructing optimal control programs for CNC machine tools that simultaneously consider a set of technological limitations (on parameters of the billet, machined surface, machine, cutting tools, machine control system, unstable conditions of production organization and course of the processing process).

Thus, the domestic mechanical engineering has to solve actual problem of increasing the productivity of machining operations performed on CNC machines.

2 Development of the automated design system of optimal cycles of machining in a multidimensional space of control parameters

The theory of optimal step cycles designing has been developed to solve the problem above and to increase the effective use of CNC machines in machining operations based on the following three components:

1. Mathematical model that adequately describes the process of metal removal using the entire complex of control parameters, as with all possible combinations of basic technological factors as and in wide ranges of their permissible variation. For example, on fig. 1 three-stage cycle with circular plunge grinding is shown. The program feed S_{Π} is set from the machine control panel. The change of S_{Π} in the three-stage cycle during the grinding process is performed according to the commands of the active control device (ACD) depending on the remaining part of the allowance. The actual feed S_{Φ}, which occurs when grinding the machined surface (fig. 1) and a result of which the cutting force arises and varies depending on the rigidity of the technological system and other numerous parameters. The S_{Φ} value is controlled by a stepwise switching of the program feed S_{Π}.

2. Mathematical models of the set of technological constraints described above. Interrelation between the stepwise cycle of the radial feed and technological constraints complex is shown on fig. 1. Most of the limitations cut off the range of allowable values of the actual feed S_{Φ}. The limitation on the range of feeds on the machine cuts off the range of allowable values of the program feed S_{Π}.

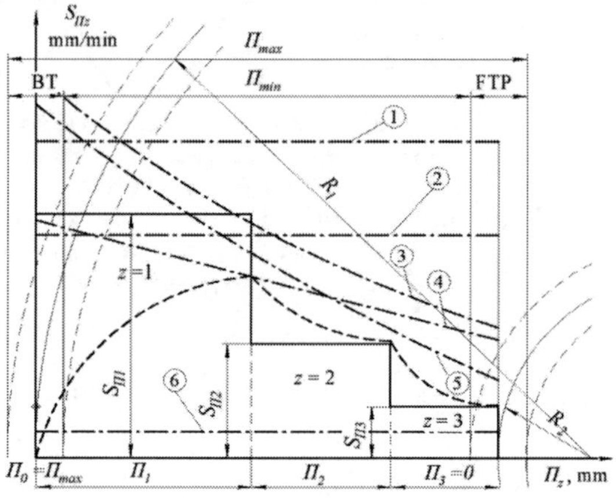

Fig. 1. Range of allowable values of the actual feeds of the three-stage circle plunge grinding: SΦz – actual feed, mm/min; 1 – limitation on the maximum feed speed; 2 – limitation on the circle falling; 3 – limitation on the dimensional accuracy; 4 – limitation on the burn; 5 – roughness limitation; 6 – limitation on the maximum feed speed; BT – billet tolerance; FPT – finished part tolerance;

3. Mathematical method of optimization based on the dynamic programming method (DPM). The process of optimizing the processing cycle is to maximally approximate the trajectory of the S_{Φ} to the active limit of constraints by stepwise switching of the program feed of the S_{Π} (fig. 1). The more the number of stages of S_{Π} switching, the closer the trajectory can be approximated to the S_{Φ} limit constraints and the higher performance of processing cycle can be achieved. Thus, the «stepless» cycle, i.e. cycle with an infinite number of steps will be the most productive. However, in practice, cycles which have no more than five steps are used because the addition of each new step increases cycle productivity by a fraction of a percent. In addition, as the number of steps increases the complexity of setting up the process of processing and its realization increase. Therefore the ACD has a limited number of feed switching stages. Since it is necessary to simultaneously optimize several

DOI: 10.1051/matecconf/201712901018

dozens of control parameters (listed above), a new method of applying the DPM has been developed for their optimization; it allows designing optimal step cycles in a multidimensional space of control parameters. The advantage of this new technique is its insensitivity to the number of optimized control parameters and the lack of requirements for their properties by differentiability, nonlinearity and complexity of their mathematical models. For example, the optimized cycle of radial fed for round external grinding is a stepwise linear function (fig. 1), which can not be described in the form of a single formula and this function does not have the property of differentiability. In addition, the new optimization technique is also insensitive to the number and properties of the technology constraint models and does not require previous creation of the control parameters limits. For example, the roughness constraint of the processed surface is specified and valid only for the finite point of the cycle (i.e. the roughness should be obtained at the end of the machining cycle). Therefore, this limitation can not be calculated and carried out in advance throughout the processing cycle, although it is conventionally drawn in the graph fig. 1. Similarly, the constraint on the processing accuracy also acts only at the end of the cycle and it can not be calculated and drawn on the graph throughout the cycle.

Creation of the computer-aided design system is suggested to lead by the following typical algorithm:

1) creation of a mathematical model for controlling the process of metal removal and control of a complex of control parameters, productivity of the machining operation;

2) creation of mathematical models of control parameters constraints;

3) adaptation of the dynamic programming method for optimization of control cycles in a multidimensional space of control parameters;

4) development of software for the design system of optimal processing cycles.

Approaches to stages 1 and 2 are described in detail in a number of works on the example of round external and internal grinding [9-10, etc.]. Let us consider the third stage in more detail as it is the most difficult and little studied. Optimization of the processing cycle is carried out with the help of DPM in analogy with the optimization of the transport task when it is necessary to find the optimal route on a branched network of roads from the starting point to the final [11].

As the DPM refers to discrete methods then the partial optimization of the cycle for two control parameters (program feed and distribution of allowance for the cycle stages) is carried out on the coordinate grid «Program feed $S_П$ – Allowance $П$» (fig. 2). To obtain the coordinate grid the axes of the graph are divided into a plurality of discrete values (discrete). The procedure for optimizing the DPM is carried out by the coordinate grid notes. Therefore, the smaller the grid cell (i.e. the more discrete) the higher the accuracy of the optimal solution but the higher the complexity of the optimization due to the large number of nodes. If the optimization of the cycle is carried out without considering the limitations on the number of permissible feed switching steps then, as a result of optimization, we get the optimal «stepless» (code name) cycle because of a huge number of discrete a stepless cycle is obtained; this cycle due to the discreteness of DPM will have many small steps the number of which depends only on the discreteness of the coordinate grid. On fig. 2 the stepless cycle on the coordinate grid with the number of discrete 10x20 is shown. If the optimization of the cycle is carried out considering the limitation on the number of permissible feed switching stages, then for example, for a three-step cycle we obtain the optimal three-step cycle as a result of optimization (fig. 2), in which the program feeds and allowances for each step are optimal.

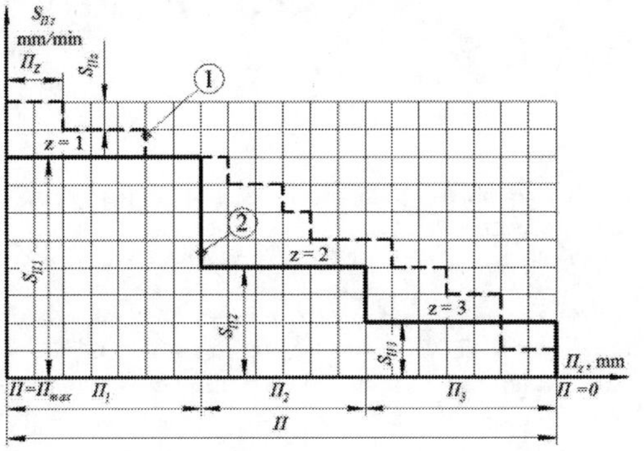

Fig. 2. Coordinate grid «Program feed SП – Allowance П» with a number of discrete 10x20: 1 – «stepless» cycle; 2 – three-step cycle;

First let us consider the simplest case when the limitation on the number of switching steps is not considered. In accordance with the DPM optimization procedure in all grid notes (fig. 3) the information cell with the coordinates $[n, m]$ is placed, where n – discrete number of allowance, m – discrete number of program feed. In the process of optimizing DPM between two grid notes it is necessary to specify the travel time τ_x, necessary to move from one node to another. To know the travel time between two nodes the process of metal removal is simulated; it is equal to the allowance discrete at the feed levels depending on the coordinates of the nodes between which the stroke is made. As several moves can be made to one node the best one is selected from all possible moves which ensures the minimum accumulated time. The optimal (minimal) time of reaching the state $T_{n,m}$ is written to the information cell and in the variable m^* the coordinate of the node from which the optimal move is made is memorized (fig. 3). After applying the standard DPM procedure – reverse move, we get the optimal trajectory of «stepless» cycle (fig. 2), in which the values of the program feed $S_П$ and corresponding to them discrete allowances $П_z$ are optimal.

 DOI: 10.1051/matecconf/201712901018

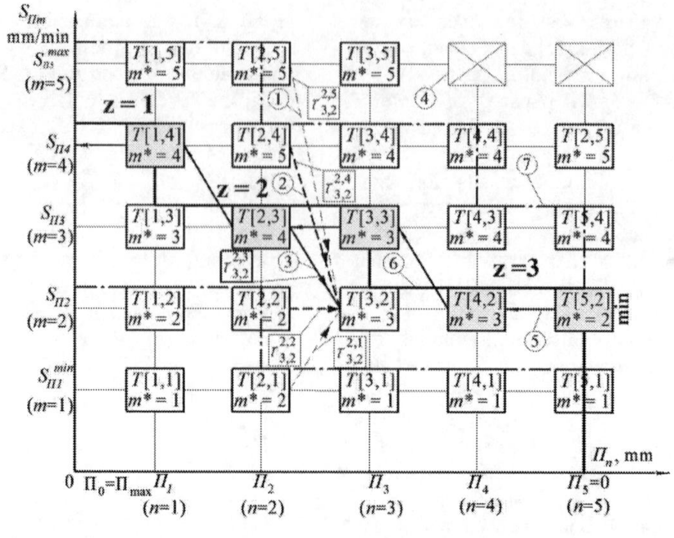

Fig. 3. Coordinate grid for the optimization of the round plunge grinding cycle by dynamic programming method: T [m, n] – optimal time of reaching the state $T_{n,m}$; 1 – competing moves; 2 – admissible moves; 3 – optimal move; 4 – information cells eliminated from the calculation; 5 – reverse move procedure; 6 – trajectory of optimal control; 7 – possible cycles;

As the «stepless» cycles are not used in production practice let us consider the optimization of the cycle considering the limitation on the number of cycle stages. In this case, the structure of the information cell is complicated because the number of stage is added. An example of the information cell for the three-step cycle is given on (fig. 4).

z	$T_{n,m,z}$	$m^*(n,m,z)$
z_{max}	$T_{n,m,z_{max}}$	$m^*(n,m,z_{max})$
...	...	...
$z+1$	$T_{n,m,z+1}$	$m^*(n,m,z+1)$
z	$T_{n,m,z}$	$m^*(n,m,z)$
$z-1$	$T_{n,m,z-1}$	$m^*(n,m,z-1)$
...	...	...
1	$T_{n,m,1}$	$m^*(n,m,1)$

Fig. 4. Structure and content of the information cell at optimization considering the limitation on the number of cycle stages: z – stage number of the cycle; $T_{n,m,z}$ – optimal (minimal) time of reaching the state; m^* – memorized value of the coordinate grid note from which the optimal move is made.

In the process of optimization considering the limitation on the permissible number of program feed switching the following principle should be observed: if a move with program feed level is made, then the stage number z increases by one, otherwise the stage number z does not change (fig. 5). This principle remains in the case of reverse move [12-13].

After the DPM reverse move the partially optimal three-step cycle will be obtained (fig. 2), in which only two control parameters are optimized: program feed and distribution of the removed allowance part along the cycle stages.

Let us consider an even more complicated case when another control parameter is optimized. For example, let us optimize the processing cycles for two feeds (radial and axial) depending on the remaining part of the allowance. Optimization is carried out on the coordinate grid « Program feed $S_П$ – Allowance $П$ ». Thus, two control parameters are considered in the DPM system. The third control parameter – program axial feed V_{Soc} will be considered in the information cell structure (fig. 6).

V_{Socb}	z	$T_{n,m,b,z}$	$m^*_{(n,m,b,z)}$	$b^*_{(n,m,b,z)}$
V_{Soc2} (b=2)	z=3	$T_{n,m,2,3}$	$m^*_{(n,m,2,3)}$	$b^*_{(n,m,2,3)}$
	z=2	$T_{n,m,2,2}$	$m^*_{(n,m,2,2)}$	$b^*_{(n,m,2,2)}$
	z=1	$T_{n,m,2,1}$	$m^*_{(n,m,2,1)}$	$b^*_{(n,m,2,1)}$
V_{Soc1} (b=1)	z=3	$T_{n,m,1,3}$	$m^*_{(n,m,1,3)}$	$b^*_{(n,m,1,3)}$
	z=2	$T_{n,m,1,2}$	$m^*_{(n,m,1,2)}$	$b^*_{(n,m,1,2)}$
	z=1	$T_{n,m,1,1}$	$m^*_{(n,m,1,1)}$	$b^*_{(n,m,1,1)}$

Fig. 6. Structure of the information cell at the cycles optimization for two feeds depending on the part of the removed allowance.

Optimization of control parameters and finding the trajectory is carried out according to the same principles as shown on fig. 5. The result of the optimization will be two optimal trajectories for the radial and axial feeds with the same number of stages and distribution of the allowance in the stages. Similarly, any number of optimized control parameters can be added including parameters of the circle characteristics and technological adjustments. As a result it becomes possible to optimize an unlimited number of cycle control parameters if there is an adequate model of the process of controlling metal removal productivity.

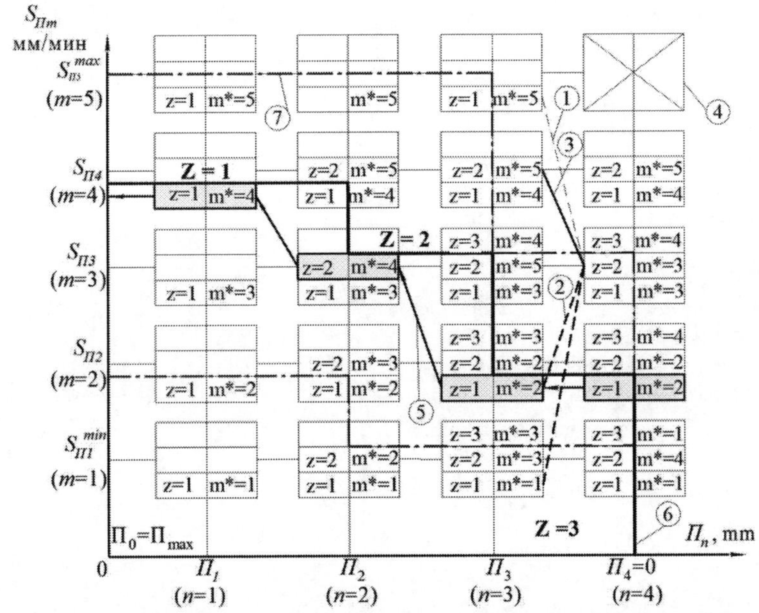

Fig. 5. Reverse move procedure considering the limitation for the admissible number of program feed switching: 1 – competing moves; 2 – admissible moves; 3 – optimal move; 4 – information cells eliminated from the calculation; 5 – reverse move procedure; 6 – trajectory of optimal control; 7 – possible cycles.

The solution to the problem of productivity increasing for all types of machining operations performed on CNC machines is beyond the power of one department because for all types of operations it is necessary to develop processes models and constraint models with subsequent adaptation of the DPM for optimization of the control cycles in the multidimensional space of control parameters. It is necessary to make a lot of analysis which will require modern laboratories with test benches for all types of operations. The solution of such a problem is possible only within the framework of the implementation of a single large project on the basis of a constantly operating technology center in which technologists, programmers, specialists in CNC systems, scientists with knowledge of various types of processing and related areas of scientific knowledge should participate. With the processes and constraints models it is possible to start creating software for optimization the processing cycles and control programs. Software development consists of known typical stages:

– analysis of requirements for the created software to determine the key goals and objectives of the final product and formation of the technical task;

– software design, choice of tools and operating system;

– coding software on the basis of the chosen programming language;

– testing and debugging of software allowing to eliminate programming errors and achieve the ultimate goal — full-featured work of the developed program;

– implementation of the software in operation and its further support is the final stage of the development.

3 Conclusion

1) developed theory of optimal automatic cycles designing:

– covers all types of machining operations (including turning, milling, grinding, drilling) performed on CNC machines;

– based on the development of the DPM application for the complex optimization of the processing cycle in the multidimensional space of control parameters with any number of constraints verified at each step of a directed search of variants calculation.

2) software structure for optimizing DPM cycles is united for different types of machining;

3) in general for the entire automated engineering industry it is necessary to develop over a hundred techniques and software products for the design of optimal cycles what will raise the machining productivity in the machine-building industry by two or more times on CNC machines;

4) task of developing hundreds of techniques and software products for designing optimal cycles is beyond the power of a one team. Its solution requires mobilization of all scientific and technological potential of the country under the management of a national technology center with a developed network of research laboratories.

References

1. *Engineering industry standards of time and the cutting modes for rationing of the works carried out on the*

universal and multi-purpose machines with numerical program control. Part II. Standards of the cutting modes (1990)

2. D.V. Ardashev, D.E. Anelchik, G. I. Butorin and etc., *Modes of cutting for work performed on grinding and finishing machines with manual control and semi-automatic machines: reference book* (2007)

3. G.B. Lurye, *Optimization of a cycle of grinding on the basis of adaptive management* (1979)

4. V.N. Mikhel'kevich, *Automatic Control of Grinding* (1975)

5. Y.K. Novoselov, *Dynamics of surface formation during abrasive processing* (2012)

6. S.S. Saykin, P.M. Salov, T.G. Vinogradova, Innovation in the educational process **8**, 61 (2010)

7. A.A. Nikolaenko, Technology of machine building **5**, 57 (2011)

8. D.Yu. Pimenov, Journal of Friction and Wear **1**, 643 (2015)

9. P.P. Pereverzev, A.V. Akintseva, Russian Engineering Research **4**, 23 (2016)

10. P.P. Pereverzev, A.V. Akintseva, Bulletin of the South Ural State University. Ser. Mashinostroenie **2**, 61 (2016)

11. R. Bellman, *Dynamic programming* (1960)

12. P.P. Pereverzev, A.V. Akintseva, Russian Engineering Research **1**, 11 (2017)

13. P.P. Pereverzev, A.V. Akintseva, Russian Engineering Research **2**, **13** (2017)

Complex optimization of parameters for controlling the cycle of internal grinding by the method of dynamic programming

Aleksandra Akintseva[1,*] and *Pavel* Pereverzev[2]

[1] South Ural State University, Department of Engineering and Technology, 454080 Chelyabinsk, Lenin Avenue 76, Russian Federationa
[2]South Ural State University, Department of Automated Engineering Technology,76, Lenin Avenue, Chelyabinsk, 454080, Russian Federationa

Abstract. The technique of designing the optimal cycles of intragrinding processing which allows performing complex optimization of control parameters is presented. As a mathematical optimization technique the method of dynamic programming is used; it allows to take into account any number of optimized control parameters and constraints of the objective function. In this case, this optimization method is not sensitive to the type of mathematical models of the process and the constraints of the objective function. Complex optimization can simultaneously optimize the parameters of the grinding cycle (the number of cycle stages, feed and allowance for the stages), the cutting speed (billet and circle speed of rotation), parameters of the circle characteristics (hardness, granularity, structure, material of grains and bundle) and parameters of technological adjustments (diameter and height of the circle, circle overtravel).

1 Introduction

Internal grinding is practically the only method of hole final processing which ensures fulfillment of high requirements for accuracy and quality of treated surface. The fulfillment of the drawing requirements for accuracy and quality in the process of internal grinding depends on a variety of technological factors most of which can be set at the beginning of the cycle or directly controlled during the allowance removal process (control parameters of the cycle). Recommendations for the selection of a particular parameter of cycle controlling presented in various normative references [1,2,etc.], have a number of significant drawbacks: 1) they based on statistical data from the 1900s and have a wide range; 2) they are designed to calculate the main operating time on machine tools with manual control for the convenience of normalizing operations; 3) they do not take into account the whole complex of technological factors (for example, when choosing the width of the circle only the length of the hole is taken into account); 4) there are no recommendations regarding the choice of the cycle structure, escape of the mandrel, etc. Therefore, enterprises are forced to select parameters of controlling of the cycle manually, leaning against the saved up experience. The cycles selected in this way require additional inputs of production resources; they are not only suboptimal but also not rational because in order to fulfill the requirements of the drawing on accuracy and quality underestimated radial feed rates are assigned. As a result the production capacity of modern CNC machines is used only by 40-60%. Often the productivity of these machines is lower than that of universal machines, especially for small batches of parts.

In the currently known methods of cycles designing [3-6 and etc.] the control of one cycle parameter (most often a radial feed), i.e. complex optimization of cycle control parameters is not carried out. Separately, it should be noted that in the vast majority of techniques optimization of the cycle structure itself is not considered. The number of stages of the cycle is assumed to be a constant value independent of the initial processing conditions. Most of the cycle design techniques do not use the mathematical optimization method. As a result, the cycles designed according to these methods are rational (they do not guarantee a minimum processing time) and do not ensure compliance with the drawing requirements for accuracy due to the absence of one of the most important and mandatory models of productivity limitations in accuracy of processing operations, which should include not only the limitations associated with the tolerance of diametrical size but also the tolerances of the shape and disposition of the surfaces. It should be noted that the methods of cycles designing do not take into account changes in the processing conditions (blunting of the circle grains, fluctuation of allowance and initial accuracy of the treated surface in a batch of the billets, etc.), what makes them unsuitable for serial production conditions.

2 Methodology of complex design of internal grinding optimal cycles

* Corresponding author: popovaav@susu.ru

DOI: 10.1051/matecconf/201712901019

The solution to the problem described above is the complex method of designing of internal grinding optimal cycles, developed by us, which allows performing simultaneous optimization of the control parameters complex of cycle control.

The control parameters of the grinding cycle can be divided into four main groups. The first group includes structural components of the cycle (as the cycle control parameters), namely the number of stages of the cycle, the distribution of the allowance for the steps of the cycle. The second group includes cutting modes – radial and axial feeds, number of billet rotations, and the speed of the circle. The third group includes parameters of the tool – characteristics of the grinding circle (number of graininess, hardening degree, etc.). The fourth group is quite extensive and includes adjustment parameters (type of reverse stroke – working or idle, diameter and height of the circle, value of the circle overtravel, escape of the mandrel, etc.)

Let us calculate the number of control parameters on examples of three- and five-step cycles because the number of control parameters depends on the cycle structure (tab.1).

It is seen from Table. 1 that the number of control parameters is reckoned in tens. Due to the absence of the complex optimization method of all control parameters for grinding operations fulfilled on CNC machines it becomes clear why in the production conditions cycle is designed manually by the following chain: at first, the parameters of the circle are selected from the specifications of the drawing. Then the machine, cutting modes and parameters of technological adjustment are assigned. The whole process of the operation designing is conducted by the technologist manually from the ranges often intuitively and on the basis of existing experience with an underestimation of the cutting conditions to a level at which the quality of the treated surface is guaranteed. In this situation the design of high-performance operations on CNC machines is impossible on the scale of the entire automated production of the country. Therefore the development of the technique for complex optimization of control parameters for internal grinding operations and the creation of CAD for CNC machines on its basis is an actual task.

Due to the presence of elastic displacements in the technological system and the inertia of the moving masses the actually removed allowance is not equal to the program feed. Therefore the optimization method uses a model of process control (the model of allowance removal), which reflects the relation between the program feed, actually removed allowance, elastic deformations of the system with all possible combinations of the main technological factors and in all ranges of their allowable variation [9].

The task of cycle optimization for processing of a specific batch of billets requires searching of deterministic regime parameters of the cycle with unstable values of a number of technological limitations and processing conditions.

Table 1. Calculation of the control parameters number for step cycles of internal grinding

Group of control parameters	Name of the control parameter	The number of control parameters for a stepped cycle		
		2 stages	3 stages	5 stages
Structural components of the cycle	Number of cycle stages	1	1	1
	Distribution of allowance for the cycle stages	2	3	5
Cutting modes	Radial feed on the steps of the cycle	2	3	5
	Axial feed on the steps of the cycle	2	3	5
	Billet rotational speed	1	1	1
	Circle rotational speed	1	1	1
Characteristics of the grinding circle	Grain material	1	1	1
	Bundle material	1	1	1
	Graininess	1	1	1
	Hardness	1	1	1
	Structure	1	1	1
Process adjustment parameters	Circle diameter	1	1	1
	Circle height	1	1	1
	Circle overtravel	1	1	1
	Type of reverse motion	1	1	1
	TOTAL	18	21	27

This formulation of the optimization problem is the most complex and relates to the field of non-linear programming. Analysis of discrete optimization methods showed that the dynamic programming method (DPM) is the most suitable for cycle's optimization [7]. In Table. 2 a comparative analysis of the DPM speed and exhaustive search for a different number of optimized parameters. The use of DPM is caused by the fact that this method does not require preliminary construction of the area boundary of feasible constraints and is not sensitive to the properties of control models and constraints. The cycle optimization is carried out by the DPM in analogy with the optimization of the transport problem which requires an optimal route on a branched network of roads

DOI: 10.1051/matecconf/201712901019

from the starting point to the final [7]. With regard to optimizing of processing cycle, the road map is similar to the «process state variants» for different feeds on different parts of the allowance. In our case the starting point is the initial state of the process (billet parameters, tool characteristics, etc.); the final point is the final state of the process (parameters of the finished part for accuracy and quality, main time, etc.). As a result the problem of searching a cycle having a minimal basic time among various possible variants of cycles arises.

Table 2. Comparative labor input of optimization methods for a different number of optimized parameters

	Exhaustive search	DPM
Two optimized parameters		
Formula for calculating the number of variants	$K_B = n^m$	$K_B = n\, m^2$
$n = 2$, $m = 2$	4	8
$n = 5$, $m = 5$	3125	125
$n = 10$, $m = 100$	10^{10}	1000
Five optimized parameters		
Formula for calculating the number of variants	$K_B = (((n^m)^k)^v)^d$	$K_B = n\, m^2\, k\, v\, d$
$n = 2$, $m = 2$, $k = 2$, $v = 2$, $d = 2$	65 536	64
$n = 5$, $m = 5$, $k = 3$, $v = 5$, $d = 3$	$2\ 10^{+157}$	$5{,}625\ 10^3$
$n = 10$, $m = 100$, $k = 5$, $v - 10$, $d - 5$	10^{5000}	$2{,}5\ 10^7$

DPM does not limit the number of optimized control parameters. As a result it becomes possible not only to design the structure of the optimal cycle but also to carry out a complex optimization of the cycle control parameters of all four groups. An increase in the number of optimized parameters leads to a complication of the optimization process (an increase in possible variants of cycle structures, etc.). For example, if the cycle is optimized for only one feed, then the optimization procedure is performed on a flat coordinate grid (Fig. 1). If the cycles for two or more control parameters are optimized, then the optimization procedure is performed in a multidimensional matrix. When optimizing two control parameters – radial (n) and axial feeds (b) three-dimensional optimization matrix is used (Fig. 2 and 3). It is necessary to perform the convolution of the three-dimensional matrix on the plane to carry out the DPM optimization procedure. As a result, the information cell located at the nodes of the coordinate grid will become more complicated due to the addition of a new optimized parameter – axial feed. In the article [8] an example of optimization of two control parameters for the cycle – axial and radial feeds with the help of DPM. The information cell shown in Fig. 4 and 5 allows optimizing the following control parameters: radial feed (n), axial feed speed (b), allowance (m), diameter (d) and height (T) of the circle, number of cycle stages (z).

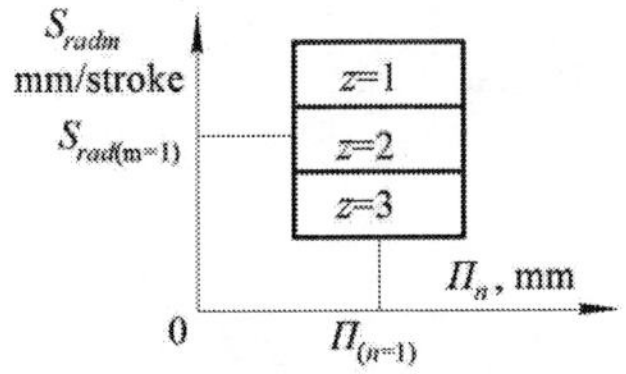

Fig. 1. «Information cell» in the coordinate system «radial feed, allowance», used to optimize two parameters

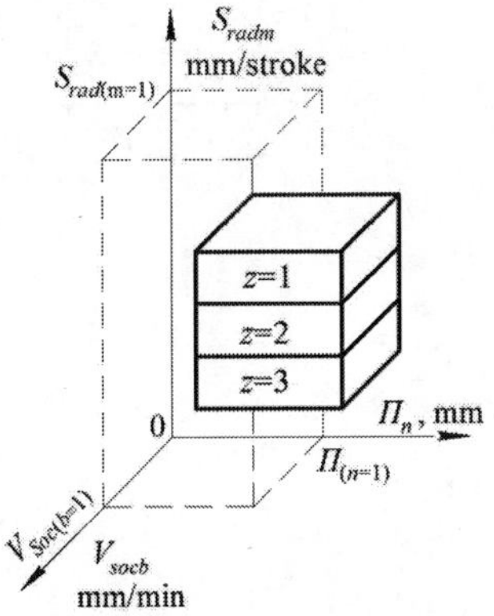

Fig. 2. «Information cell» in the coordinate system «radial feed, allowance», used to optimize three parameters (three-dimensional representation)

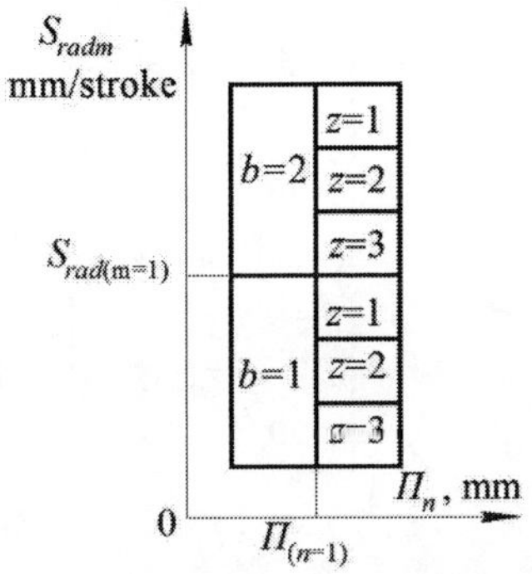

Fig. 3. «Information cell» in the coordinate system «radial feed, allowance», used to optimize three parameters (two-dimensional representation)

Thus, the DPM does not limit either the number of optimized control parameters or the number of constraints of the objective function. Constraints imposed throughout the cycle can be divided into two groups. The first group includes constraints imposed during the whole grinding cycle (on the falling of the circle, drive power, admissible range of feeds, etc.). The second group includes constraints imposed at the end of the cycle (accuracy [10], grinding without burn, roughness of the treated surface, etc.).

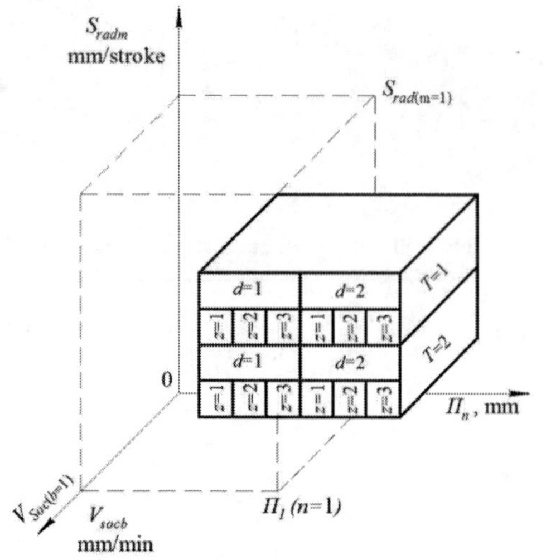

Fig. 4. «Information cell» in the coordinate system «radial feed, allowance», used to optimize five parameters (three-dimensional representation)

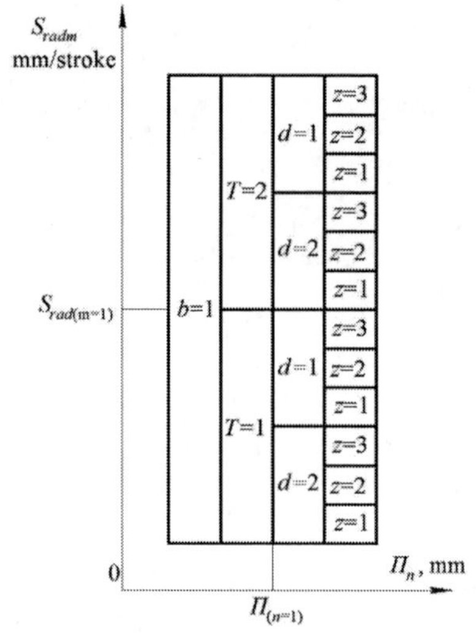

Fig. 5. «Information cell» in the coordinate system «radial feed, allowance», used to optimize five parameters (two-dimensional representation)

It should be noted separately that in order to increase the stability of the accuracy and quality of the treated surface to the effect of variable technological factors, verification of the objective function constraints of the DPM should be carried out in two groups of limiting combinations of the values of technological processing conditions. The imposition of constraints of the objective function for each competing stroke is carried out in two variants with the use of both favorable values (minimum values of the bluntness degree of the circle, the initial radial runout of the billet, allowance, maximum diameter of the circle) and unfavorable factors (maximum bluntness of the circle, minimum values of the initial radial runout of the billet, allowance, diameter of the circle).

3 Conclusion

1. Despite the emergence of new high-performance CNC machine tools in domestic engineering industry the problem connected with the existing low productivity of operations performed on these machines has not yet been solved. The reason for this is the lack of reference literature, CAD and engineering techniques allowing designing cycles for CNC machines which meet the requirements of modern automated production.
2. Developed technique of complex optimization of control parameters for internal grinding operations allows carrying out the simultaneous optimization of an unlimited number of cycle control parameters (number of cycle stages, feed and allowance for the stages), cutting speed (speed of the billet and the circle), parameters of the circle (hardness, graininess, structure, grain and bundle material) and parameters of technological adjustments (diameter and height of the circle, circle overtravel).

References

1. D.V. Ardashev, D.E. Anelchik, G. I. Butorin and etc., *Modes of cutting for work performed on grinding and finishing machines with manual control and semi-automatic machines: reference book* (2007)
2. A.N. Malova and etc., *Handbook of the technologist of the machine builder: reference book* (1972)
3. S.S. Saykin, P.M. Salov, T.G. Vinogradova, Innovation in the educational process **8**, 85 (2010)
4. Y.K. Novoselov, *Dynamics of surface formation during abrasive processing* (2012)
5. V.N Mikhelkevich, *Automatic grinding control* (1975)
6. A.A. Nikolaenko, Technology of machine building. **5**, 57 (2011)
7. R. Bellman, *Dynamic programming* (1960)
8. P.P. Pereverzev, A.V. Akintseva, Procedia Engineering **129**, 121 (2015)
9. P.P. Pereverzev, A.V. Akintseva, Russian Engineering Research **10**, 888 (2016)
10. P.P. Perverzev, A.V. Akintseva, Bulletin of the SUSU. Ser. Mashinostroenie **3**, 44 (2016)

Influence of technological factors of blade processing on the forming of the defect layer

Julius L. Tchigirinsky[1,*] and *Trong Ngo* Quang[1]

[1] "Technology of Machinery" Department, Volgograd State Technical University, 28, Lenin's av., Volgograd, Russia, 400005

Abstract. Identified the interrelationships between the depth of the surface layer with modified physical-mechanical properties (defective or work-hardened or fill hard layer), the parameters of cutting process, and the properties of materials of tool and workpiece. Justified the necessity of research the processes of formation the properties of the machined surface from the standpoint of the thermo-physics of the cutting process.

The formation of subsurface deformation is an attribute of the wedge cutting process. The subsurface deformation significantly affects on the resistance to corrosion and other performance properties of the product. A number of researchers [2, 3, 5] notes that in the surface layer, in addition to changes in physical and mechanical characteristics of the material, such as microhardness, strength and ductility may also change in the crystal structure. As a rule, the condition of the subsurface deformation is evaluated by two basic indicators: the depth and degree of hardening. The degree of hardening is defined [2] as the change in the microhardness of the surface layer in relation to the original material.

In the studies of A. D. Makarov, L. G. Odintsova and other authors [2, 3, 12] identified the main reasons for the formation of the layer with modified physical and mechanical properties – the combined effects of cutting forces and increased temperature induced during a machining. At research by N. V. Talantov [4] and his students proved that the formation of the deformed layer is greatly influenced by the friction at the rake surface and rear surface of the tool, forces and temperature processes in the primary shear zone, temperature and diffusion processes of the "blank-tool" and "chip-tool". All of these factors are the result of the influence of independent technological conditions of a metal cutting process: physical-mechanical properties of materials of the tool and the workpiece, the method of forming, cutting conditions (cutting speed, depth of cut, feed) and the geometric parameters of the cutting wedge.

When considering the microstructure of the depth of the defective layer (Fig. 1), it is possible to distinguish three main zones. In subsurface layer (A), the workpiece material is maximally loaded with both force and temperature influences. As a consequence, the change of the microstructure in this layer is clearly pronounced, in comparison with the deeper layers of the deformed layer (B) and in comparison with the core of the workpiece material. This is evident from a comparison of the size and shape of the crystals in Fig. 1 – considerable differences in size and shape of the crystals, and clearly shows the direction of the impact cutting forces.

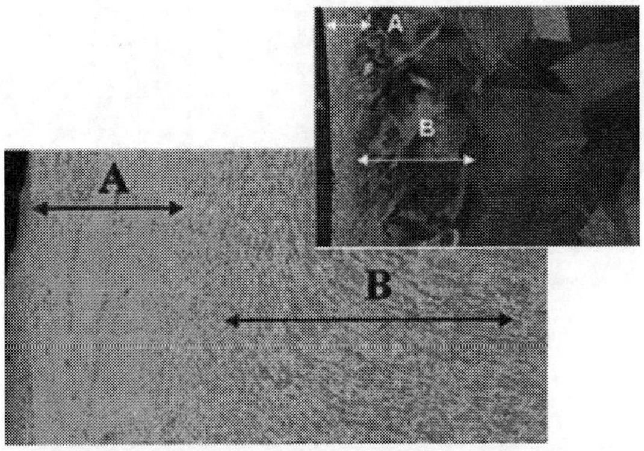

Fig. 1. Microstructure of subsurface deformation [9,11]

The change of the microstructure in the layer (A) is significant, because force is enhanced by high temperatures on the surface finish. In layer (B), the level of temperature-force influence is not so significant. As a consequence, the change of the microstructure of the material and the size of the crystals is less pronounced than in the layer (A), which indicates a smaller degree of plastic deformation of the material. The intensity of deformation and notable changes in the microstructure gradually decrease to the equilibrium state of the core material of the workpiece.

In some cases, when a certain combination of physical and mechanical properties of the contact pairs "tool-workpiece" and cutting parameters, a layer with reduced mechanical properties can be formed on the surface finish [11]. This phenomenon is observed as a result of high cutting speeds with a low thermal conductivity of cutting tools, Such conditions promote intensive heating of the surface finish, which ultimately can lead to the appearance of microcracks on the surface of the product and reduce its performance properties [11]. From the analysis, it follows that the temperature-force conditions

* Corresponding author: Julio-Tchigirinsky@yandex.ru

 DOI: 10.1051/matecconf/201712901020

of processing are critical in the study of properties and regularities of formation of the defect layer.
Classification of the factors affecting the process of formation of the defect layer is shown in Fig. 2.

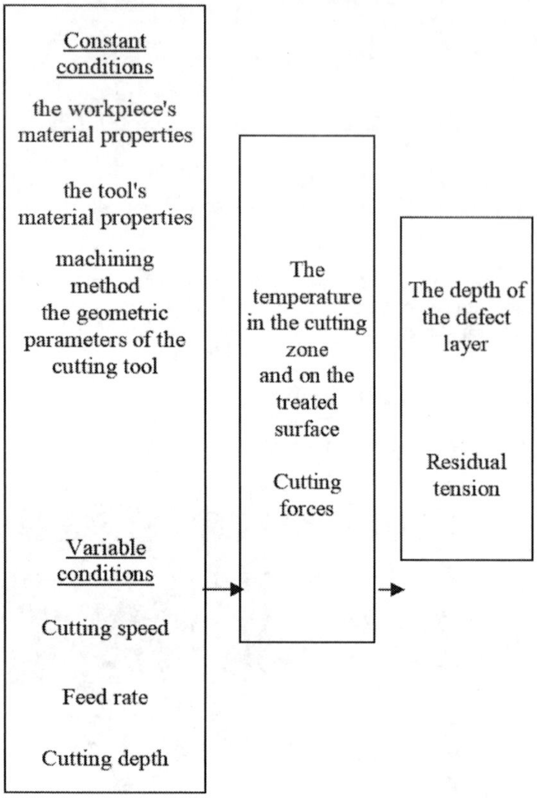

Fig. 2. Parameters that affect the depth of the defective layer

Investigations [2, 12] have established the influence of the cutting speed and tool wear on the depth of the hardened layer, the degree of influence wear is much greater. This is confirmed by the studies of N. V. Talantov [4] and the conclusion about the influence of cutting speed on the regularity of the flow of force and temperature processes in the processing zone and, ultimately, the intensity of the tool wear. Experimentally [2] determined the relationship between cutting speed and depth (hc) of the defect layer, and the degree (N %) of deformation (Fig. 3). In the low cutting speeds (Section I) the depth and degree of hardening decrease monotonically with increasing cutting speed. This is due to the increasing temperature [2], a gradual change in the form of a contact interaction, and the transition from cutting with the build-up [4]. To the formation of stagnant zones (Section II) on the rake surface. Continue to increase the cutting speed, due to the combined effects of the processes of strain hardening and temperature softening, the character of the depth and degree deformation of the curves changes, a plastic contac is formed on the rake surface (Section III). The increase in the intensity of tool wear and contributes to the growth of the chamfer of the flank wear of the tool, i.e., to increase the frictional forces and temperatures.

The results of experimental studies and conclusions of A. D. Makarov [2] that defines the section II (cutting speed 20...50 m/min) as the range of optimal temperatures are in good agreement with the concept of "transition cutting speeds" [4] in which there is a change of types of contact interaction.
The higher degree of influence of tool wear, compared with the cutting speed, is explained by the fact that tool wear, as "explanatory" factor defines the total influence of cutting speed and temperature-deformation processes in the defomed chip.

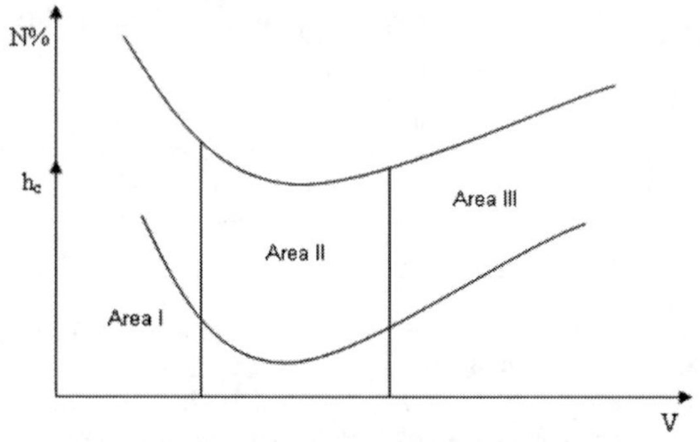

Fig. 3. Effect of cutting speed on depth and degree of defective layer

The effect of the feed on surface layer shown in Fig. 4. Curve (I) shows the patterns of change in the depth and degree of deformation in the range of cutting speeds corresponding to the optimal temperature [2] (Section II, Fig. 3).

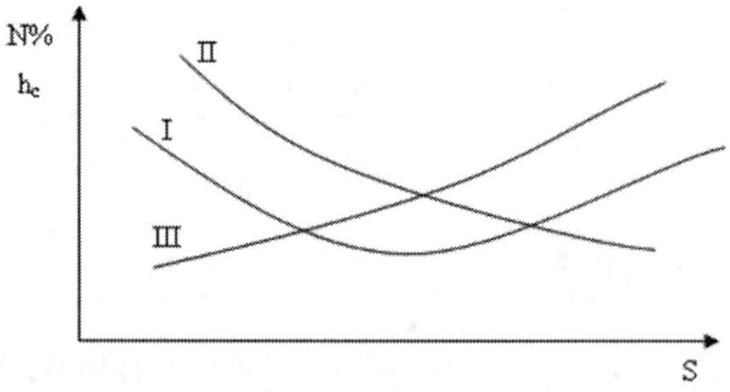

Fig. 4. Effect of depth feed and degree of defective layer

Curve (II) corresponds to the range of low cutting speeds – the total tension state of the surface layer decreases with increasing feed rate. Curve (III) determines the influence of feed rate on the depth and degree of deformation of the surface layer in the range of high cutting speeds (Section III, Fig. 3), corresponding to the plastic contact of the tool and the chip.
The decrease the rake angle γ, the dearance angle α and increasing the radius of the rounding edge of the cutting edge ρ increases the effect of friction forces at the rake and flank surfaces and the tip of the tool, increasing the shear angle β, which ultimately increases the depth of the defect layer.

The two mechanisms for the formation of the hardened layer on the surface finish is indicated by studies [6] aimed to study the effect of wear-resistant coatings: the depth and degree of deformation in the surface layer are determined by the friction on contact surfaces and thermo-physical characteristics of the tool. The change of friction coefficient on the rake surface of the tool leads to the change of the chip reduction coefficient, which affects the intensity of the temperature in the cutting zone. Also, the intensity of heat release depends on the thermal conductivity of the tool material. The coefficient of friction at the flank surface of the tool determines the degree of stresses in the contact zone and, consequently, the degree of deformation of the hardened layer. Studies [2] show that there are clear differences in the depth and degree of deformation when machining with tools made of hard alloys with different thermal conductivities – WC 92-Co 8, TiC 5-WC 85-Co 10, TiC 14-WC 78-Co 8.

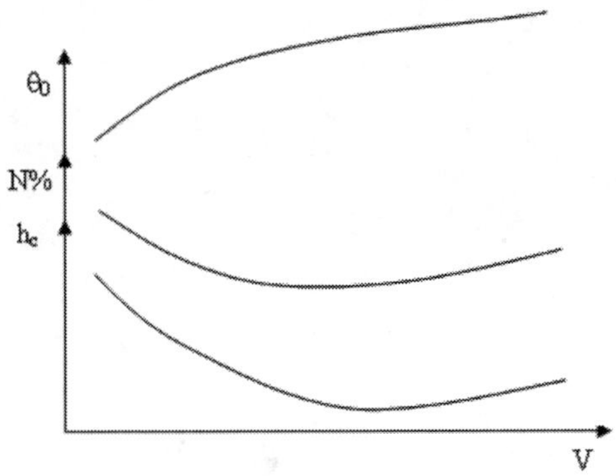

Fig. 5. Effect of cutting speed on depth and degree of defective layer and average contact temperature [2]

The results of investigations of the state of the surface layer in the processing of heat-resistant deformable alloys shown (Fig. 5) that the depth and the degree of deformation reaches the minimum values in the range of medium (Fig. 3, section II) cutting speeds, while the temperature dependence in the contact zone is monotonous.

Given that the thermophysical characteristics of steels change [1, 2, 10] depending on the degree of deformation, it is possible to say that the regularities of formation of the surface layer are not unique. Determination of the relationship between the characteristics of the hardened surface layer only from the elements of cutting conditions can lead to incorrect conclusions.

The analysis allows to make preliminary conclusions about the importance of the effect of thermal processes in the contact zone "tool-workpiece" on the properties of the surface layer of machined parts, i.e., depending on the depth and degree of deformation of the hardened layer from the thermophysical characteristics of structural and tool materials.

To check the formulated hypotheses, a multivariate regression analysis [8] of the results of experimental [9, 11, 12, 13] studies of the milling end mills was carried out. As the parameter of the analyzed state of the surface layer of articles made of titanium (Ti-6Al-4V) alloy, the arithmetic mean deviation of the microroughness Ra is considered.

For contact pair titanium alloy (Ti-6Al-4V) – is a single-carbide with wear-resistant coating (TiN mill end ϕ 12 mm); cutting speed V = [120/ 160/ 200] m/min, depth of cut t = [0,20/ 0,35/ 0,5] mm, feed S= [0,10/ 0,12/ 0,14] mm/tooth – results of statistical analysis are presented in Table. 1.

Table 1. The degree of influence of technological factors on the height of microroughness (Ti-6Al-4V-TiN)

Parameter Specification model	Model specification		
	Linear	Power	Exponential
Non-correlated (independent) factors			
Random factors (uncertainty)	-0,196	0,004	0,004
Cutting depths (*t*)	0,063	0,028	0,027
Feed rate (*Sz*)	0,634	0,609	0,602
Cutting speed (*V*)	-0,503	-0,552	-0,551
Pseudo-factors, taking into account the interactions between independent factors			
Cutting depths – feed rate (cross-sectional area of the cut)	-0,114	0,096	-0,117
Cutting depths – cutting speed	-0,169	0,002	0,006
Feed rate – cutting speed	0,077	-0,081	0,093
Error estimation and model adequacy			
Standard deviation	3,532	3,532	3,532
Fisher's criterion	5,548	13,049	9,124
Model adequacy, %	99,90	99,90	99,90
Relative error, %	3,62	1,31	2,50

The coefficients of the regression models (Table. 1) built in the normalized coordinate space with allowance for the implementation of a standard Gauss-Markov, allow to judge about the degree of influence of explanatory variables on the behavior of the response function. For analysis of the selected three models, the most often considered in technology, specifications. Based on an assessment of modeling error adopted power model.

Positioning the three most important explanatory variables (technological factors, the arguments of the model) in descending order of the absolute values of the regression coefficients of the normalized model, we get the following range: feed, cutting speed, chip thickness. From the above discussion, it follows that the formation of the defect layer is the result of a direct effect of temperature-force processes in the contact zone "tool-workpiece", the depth of the defect layer are determined by the combined effects of input processing parameters [7]: cutting conditions, machining method, parameters of the cutting tool, physical-mechanical properties of contacting materials of the tool and the workpiece.

DOI: 10.1051/matecconf/201712901020

Conclusion

Analysis of the results of the regression analysis allows: First, confirm the traditional view of the geometric nature of the microrelief of the surface finish – the most significant factor is the feed rate;

Second, to make a conclusion about the significant influence of thermal processes in the contact zone "tool-workpiece" on the state of the surface layer, because even for geometric characteristics (Ra) of the surface layer is important effect of the size of the heat source;

Thirdly, to formulate directions for further research: a more subtle directional study of the temperature-force processes in the shear zone in the contact zone "tool-workpiece" to build a physical model of the formation of modified physical-mechanical properties of the surface layer and the determination of quantitative characteristics of the complex interrelationships between the state of the surface layer with processing conditions to control the performance properties of the surfaces finish.

References

1. A. M. Rozenberg, A. N. Eremin, *Elements of the theory of the process of metal cutting* (Moscow. – Sverdlovsk : Machgiz, 1956)
2. A. D. Makarov, *Optimization of cutting processes* (Moscow : Machinery Publishing, 1976)
3. L. G. Odintsov, *Hardening and finishing the parts of machines by surface plastic deformation : Handbook* (Moscow : Machinery Publishing, 1987)
4. N. V. Talantov, *Fizicheskie osnovy protsessa rezaniya, iznashivaniya i razrusheniya instrumenta (Physical Principles of Cutting and Tool Wear and Failure)* (Moscow : Machinery Publisher, 1992)
5. M. E. Blanter, *The metallography and heat treatment* (Moscow: Machgiz, 1963)
6. V, P. Tabakov, A. V. Chihranov, *Wear-resistant coating of cutting tools operating under conditions of continuous cutting* (Ulyanovsk : UlSTU, 2007)
7. A. I. Khaimovitch, A. V. Kuznetcov, *Analytical modeling of power mode high-speed milling of material with viscoplastic hardening*, Vestnik of Samara University. Aerospace engineering, technology and engineering, № **5-2 (36)**, P. 167-172 (2012)
8. Ju. L. Tchigirinsky, *Mathematical methods for process control of machining : monography* / Volgograd State Technical University. – Volgograd. – 139 p. (2010)
9. G. A. Oosthuizen, K. Nunco, P. J. T. Conradie, D. M. Dimitrov, *The effect of cutting parameters on surface integrity in milling Ti6Al4V*, South African Journal of Industrial Engineering. Vol **27** (**4**), p. 115-123. (2016)
10. I. N. Kozachukhnenko, A. A. Bondarev, D. V. Krainev, Yu. N. Polyanchikov, H. V. Khodjamberdyev, *The justification to improve the surface quality when machining with pre-hardening heat-resistant steel 30CrMo4 from the position of the dislocation view*, Bull. VSTU, no. **1 (156)**, p. 6-19 (2015)
11. Pete Crawforth. *Towards a Micromechanistic Understanding of Imparted Subsurface Deformation During Machining of Titanium Alloys* (University of Sheffield, 2014)
12. Zhou Jinming, Volodymr Bushlya, Ru Lin Peng, Jan-Eric Stahl, *Identification of Subsurface Deformation in Machining of Inconel 718*, Trans Tech Publications, Switzerland, Vols. **117-119**, pp 1681-1688 (2011)
13. Liu Chunjing, Tang Dunbing, He Hua, Chen Xingqiang, *Prediction of surface roughness for end milling titanium alloy using modified particle swarm optimization LS-VM*, Transactions of Nanjing University of Aeronautics & Astronautics, Vol. **30** No.**1**, p. 53-61 (2013)

Influence of value of the criterion of blunting on the selection of the brand of hard alloy and optimum cutting speed when turning austenitic steel

Andrew A. Lipatov [1,*] and *Julius* L. Tchigirinsky[1]

[1] Volgograd State Technical University, 28, Lenin's av., Volgograd, Russia, 400005

Abstract. A mechanisms of wear carbide tool when turning of austenitic steel 18-8 by cutters from firm hard-alloys of various groups (WC-Co, TiC-WC-Co, TiC-TaC-WC-Co) were studied. During wear resistant tests it is established, that the prevalence of one of the two mechanisms of wear (of adhesion-fatigue or diffusional) depends on the brand of hard alloy. It is shown, that *in* machining by titanium-containing carbide tool the intensity of the growth of the wear platform on the back surface of the tool as wear changes. This is due to the smooth transition from the predominance of adhesion-fatigue wear to prevalence of diffusional wear. Therefore, the intensity of wear should be considered as current, depending on the value wear platform, and value of the criterion of blunting is influence on the selection of the brand of hard-alloy and optimum cutting speed.

In theory and practice of cutting of metals much attention is pay to the selection of the most effective brand of hard alloy and to definition of optimal cutting speed. The urgency of this problem increases in the case of cutting of austenitic steels and other materials that are hard to machining.

The appointment of a hard alloy is produced mainly on the basis of the machined material and the type of machining (the machining type is qualitatively to fix – roughing, semi-finishing and finishing). Optimum cutting speed, i.e. the speed corresponding to the maximum path length of cut up to a specified criterion blunting (permissible value wear) tool, is considered as a fixed, unchanging from the beginning of cutting to blunting the tool under given conditions (machined and tool materials, modes of cutting, tool geometry, and so on).

It has been established that a wide range of cutting speeds in machining austenitic steel at the wear platform at the rear surface of tool prevails plastic contact – complete adhesion of the machining material to the tool, with plastic flow of the machined material [1]. However, adhesive-fatigue wear is retained to high cutting speeds and hence high temperatures at the wear platform, – especially for tools made of titanium-containing hard-alloys of TiC-Wc-Co and TiC-TaC-WC-Co. This is associated with the unstable chip formation seen in cutting austenitic steel, which leads to cyclic fluctuation of the forces and thermal loads on the wear area of the tool's rear surface [2, 3]. It influence on the peculiarities of tool wear. It is important to establish this peculiarities.

The endurance-tests carried out by the longitudinal turning austenitic steel 18-10 (Fe and $C \leq 0{,}12$ %, Cr 18 %, Ni 10 %, $Ti \leq 0{,}8$ %), showed that when choosing a brand of hard alloy and value of optimal cutting speed should take into account the criterion of blunting the tool (in most cases – allowable width h_{re} of the wear platform on the rear surface).

In our experiments we used cutters with mechanically fixed plates of various hard alloys – WC 94-Co 6, TiC 15-WC 79-Co 6 and TiC 8-TaC 12-WC 71-Co 9. The speed v of cutting was changed from 15 to 180 m/min at supply 0.3 mm/turn and cutting depth 1.5 mm. The basic geometric parameters of cutters: front angle $\gamma = 0^o$; rear angle $\alpha = 10^o$; primary plane angle $\varphi = 45^o$; tip radius r = 0.3 mm. As an indicator for durability used resultant path length L to blunting (h_{re} = 0.3 mm).

The obtained dependences of the total path length L traveled to achieve of wear h_{re} = 0.3 mm, from speed v is presented in Fig. 1.

Monotonously decreasing character of the L(v) dependence for WC 94-Co 6 alloy (curve 1 in Fig. 1) is connected with the known fact – low resistant of tungsten carbide by diffusional dissolution in machined steels. Therefore, in the entire investigated speed range dominated diffusional wear, the intensity of which is sharply in-creased with increasing of the speed and temperature of cutting. Therefore, WC 94-Co 6 alloy cutters the best results only at low v (30 and 45 m/min; at speed v = 15 m/min wear was so low that the test is not fully drawn) are obtained.

At higher speeds (60 and 90 m/min) the highest wear resistance showed cutters of TiC 8-TaC 12-WC 71-Co 9. For them dependence L(v) has a maximum near v = 60 m/min (curve 2 in Fig. 1), which also explains the well-known [4, 5] fact: the increase of the cutting speed for titanium hard alloys has replaced the predominant mechanism of wear – adhesive-fatigue wear to diffusional wear.

Testing of TiC 15-WC 79-Co 6 cutters, in any speed not managed to bring to the criterion of blunting h_{re} = 0.3 mm. The cutters was premature to removing due to the

* Corresponding author: leeandrej@yandex.ru

dominant wear of the tip of the cutter, wear of the secondary cutting edge and a corresponding catastrophic deterioration in the quality of the processed surface. So the L(v) dependence for

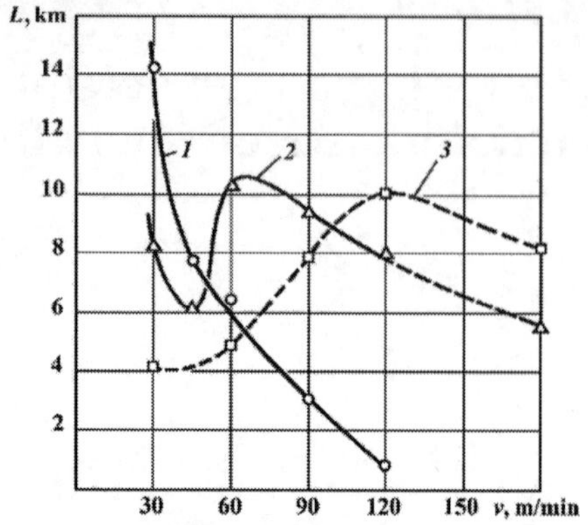

Fig. 1. Dependence of the total path length L to blunting (h_{re} = 0.3 mm) from the speed v of cutting for compared hard-alloys
WC 94-Co 6 (1),
TiC 8-TaC 12-WC 71-Co 9 (2)
and TiC 15-WC 79-Co 6 (3)

TiC 15-WC 79-Co 6 (curve 3), which also has extreme character, was obtained by extrapolation of source endurance curves to h_{re} = 0.3 mm (also the values of total path length L for TiC 8-TaC 12-WC 71-Co 9 at speeds from 120 to 180 m/min) and is shown in Fig. 1 as the dashed line.

Thus, in the practically applicable range of cutting speeds when turning of austenitic steel, best results showed the tool of TiC 8-TaC 12-WC 71-Co 9. However, analysis of initial endurance curves h_{re}(L) (here h_{re} – the current value of wear, and L is the path length from the beginning of test) showed that this is true not always. On Fig. 2 summarizes the dependencies h_{re}(L) for cutters of comparable hard alloys at v = 60 m/min (as per Fig. 1 the best is hard alloy TiC 8-TaC 12-WC 71-Co 9, and this speed is optimal for him).

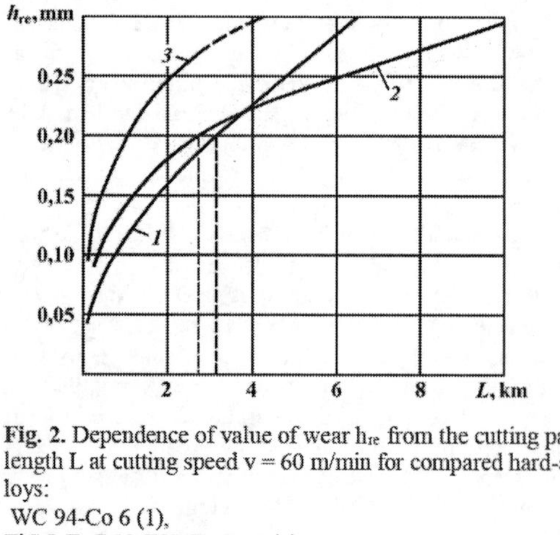

Fig. 2. Dependence of value of wear h_{re} from the cutting path length L at cutting speed v = 60 m/min for compared hard-alloys:
WC 94-Co 6 (1),
TiC 8-TaC 12-WC 71-Co 9 (2)
and TiC 15-WC 79-Co 6 (3)

From Fig. 2 it is seen that in all length of the path L the greatest wear h_{re} showed a fragile, sensitive adhesive-fatigue wear TiC 15-WC 79-Co 6 hard-alloy (curve 3). Moreover, the predominance of this mechanism of wear is saved to high speeds and temperatures of cutting [3] due to the instability of chip formation, inherent to cutting austenitic steels [6, 7] (it is impossible not to notice, that the unfortunate title of the articles [6, 7] is the result of the editor's error; initial author's version was «Instability of Chip Formation and the Dependence of the Thermal Conductivity of Machined Material on Temperature»).

Curves h_{re}(L) for hard-alloys WC 94-Co 6 and TiC 8-TaC 12-WC 71-Co 9 (respectively, 1 and 2 in Fig. 2) are intersect. Therefore, if the criterion of blunting was made lower value h_{re}, such as 0.2 mm, the path of cut up blunting the show WC 94-Co 6 alloy cutter – 3.15 km vs. 2.7 km by TiC 8-TaC 12-WC 71-Co 9 (see builds on Fig. 2, made by the vertical dashed lines).

Intersection curves is explained by the change of the prevailing wear TiC 8-TaC 12-WC 71-Co 9 alloy not only with the growth of v, but at cutting fixed-value – due to blunting the tool and a corresponding increase in temperature on a wear platform [8]. For sharp cutter temperature at the wear platform is small, prevails adhesive-fatigue wear mechanism, generating total high wear intensity. With a blunting of cutter temperature at the wear platform growing and intensity of the adhesive-fatigue wear decreases, i.e. the instrument starts working in the prevalence of dif-fusion wear, resistance which TiC 8-TaC 12-WC 71-Co 9 alloy much higher than that of WC 94-Co 6. There-fore, the slope of the curve h_{re}(L) for TiC 8-TaC 12-WC 71-Co 9 (2) becomes smaller than for WC 94-Co 6 (1). As a result, to achieve of the criterion of blunting h_{re} = 0.3 mm cutter TiC 8-TaC 12-WC 71-Co 9 alloy is considerably larger total path length of cutting (10.3 km, compared 6.4 km from WC 94-Co 6).

This result is rather unexpected – in the appointment of a lesser magnitude criterion of blunting, which is characteristic of the transition from semi-finishing to the finishing machining, the best result was a show WC 94-Co 6 hard-alloy, the scope is closer to the semi-finishing processed than TiC 8-TaC 12-WC 71-Co 9 (flexural strength 1.5 GPa by WC 94-Co 6 versus 1.3 GPa at TiC 8-TaC 12-WC 71-Co 9 [9]). In addition, hard-alloy WC 94-Co 6 belongs to the group of application K (processing of cast-iron, nonferrous metals, etc.) and not to group M (processing of high-alloyed steels), as TiC 8-TaC 12-WC 71-Co 9 [5]. Thus, the criterion of blunting can essentially (and non-obviously) influence the choice of the brand of hard-alloy. It becomes clear why in single and small batch production in the processing of small non-rigid parts made of high-alloyed steels (when you want to minimize cutting forces and often reface the tool) are often used WC-Co cutters.

Not less interesting results were obtained when comparing the dependencies h_{re}(L) obtained by TiC 8-TaC 12-WC 71-Co 9 cutters at machining with different cutting speeds v (Fig. 3).

As can be seen from Fig. 3a, for the alloy WC 94-Co 6 curves are positioned so that for a more lesser cutting speed (providing a increase total path length to blunting) curve h_{re}(L) is located entirely below, than a similar curve

 DOI: 10.1051/matecconf/201712901021

for less bigger speed. This can be considered a convincing evidence of the prevalence of diffusion wear not only in the entire investigated range of speeds (curve 1 in Fig. 1), but at all value of h_{re} (in Fig. 3a we see, that at all value of the criterion blunting with increase v path length steadily decreases).

Different picture is observed for the alloy TiC 8-TaC 12-WC 71-Co 9 (see Fig. 3b): curves intersect. Therefore, the value of the optimal speed becomes a dependent by value of the criterion of blunting. For example, when assigning a value of criterion blunting the h_{re} = 0.3 mm up to speed 60 m/min (curve 2) bigger total path length than v = 120 m/min (curve 3). If you take a smaller value of the criterion of blunting, for example h_{re} = 0.2 mm (when performing for finishing operations or

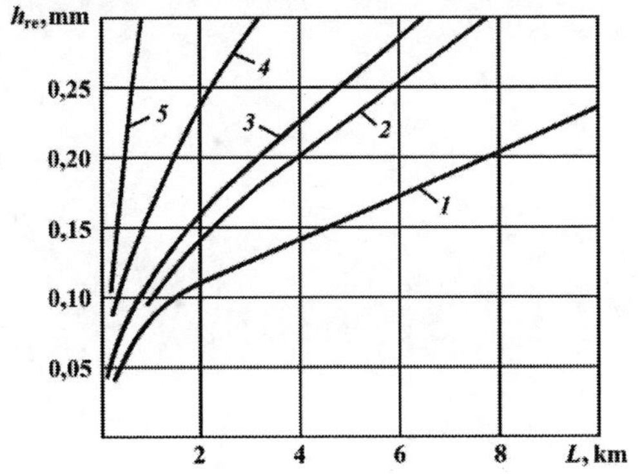

a)

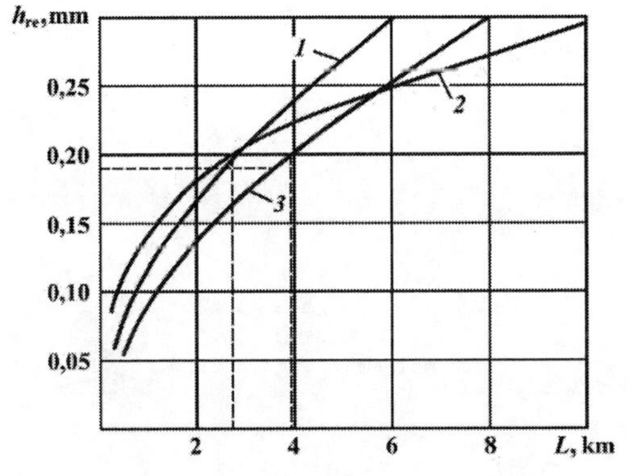

b)

Fig. 3. Dependence of the value of wear h_{re} from the path length L at different cutting speeds v in turning 18-10 steel by hard-alloy cutters WC 94-Co 6 (a: 1 – v = 30 m/min, 2 – v = 45 m/min, 3 – v = 60 m/min, 4 – v = 90 m/min, 5 – v = 120 m/min) and TiC 8-TaC 12-WC 71-Co 9 (b: 1 – v = 45 m/min, 2 – v = 60 m/min, 3 – v = 120 m/min)

for small non-rigid details), the more profitable it would be a speed of 120 m/min (L = 3.9 km vs. 2.7 km to 60 m/min – see made a vertical dashed straight lines building in Fig. 3b). Interbreeding of the curves in Fig. 3b is explained by the change of the prevailing mechanism of wear with increasing h_{re}. At most, this manifested for v = 60 m/min (curve 2; this curve is already appeared in Fig. 2, and comments are given above). When cutting with v = 120 m/min at small wear platform intensity of adhesion-fatigue wear was less than at v = 60 m/min, is due to the higher of the contact temperature. Or another, already prevailed diffusion wear, the intensity of which was less than the intensity of adhesion-fatigue

wear at v = 60 m/min. Therefore, wear platform at v = 120 m/min grew slower, and to achieve h_{re} = 0.2 mm cutter makes bigger path length. However, with further increase h_{re} intensity of the prevailing diffusion wear for v = 120 m/min changes slightly. At the same time, at speed v = 60 m/min transition to the dominance of the diffusion wear, the intensity of which is much less than at cutting with v = 120 m/min, due to the lower temperature. As a result, to achieve the value criterion of blunting h_{re} = 0.3 mm cutter makes a longer total path length already at v = 60 m/min.

Thus, the value of optimum cutting speeds also can depend from criterion of blunting, and the term «optimum cutting speed» needs some correction. In the conventional sense of the optimal speed is mean some originally established cutting speed, that unchanged from the beginning of cutting to blunting the tool and provides the maximal path length of cutting (the maximum number of de-tails when machined to blunting of tool for production). At that, the wear rate defined as the ratio of value of the criterion of blunting to the resulting path length of cutting ($J = h_{re}/L$). Results of conducted research suggest the advisability of introducing a concept of current intensity of wear: $J_c = dh_{re}/dL$.

As we see in Fig. 3b, for small values of the h_{re} value J_c is minimal for v = 120 m/min, but for large values of h_{re} – for v = 60 m/min. Therefore optimum cutting speed is best defined not as a fixed, but as current. The current optimal speed corresponds to the minimum value of current intensity of wear J_c, which in turn depends from the distance traveled cutting and degree of blunting the tool. At continuously changing the cutting speed in the course of machining on the criterion of minimizing the current intensity of wear can be achieved significantly bigger the resulting path length to blunting than using a fixed speed. For example, the machining of 18-10 steel with hard-alloy cutter TiC 8-TaC 12-WC 71-Co 9 can start at speed v = 120 m/min and machine to value of blunting $h_{re} \approx 0.19$ mm. At this value h_{re} are equalized current intensity of wear for speeds of 120 and 60 m/min (tangent lines to curves 2 and 3 in Fig. 3, b in the points of intersection of the horizontal dashed line corresponding to h_{re} = 0.19 mm, have the same degree of slope). If after reaching this value of wear reduce speed up to 60 m/min, the path length of cutting to blunting h_{re} = 0.3 mm may increase to 11,5 km (this is easy to verify by doing simple graphic manipulation with curves 2 and 3 in Fig. 3b).

The possibility to increase tool life in such a way is confirmed by experiment. Even a single step reduction of the cutting speed from 120 to 60 m/min, made at reaching the value h_{re} = 0.2 mm, the path length to blunting increased by 15 % (at the value of criterion blunting h_{re} = 0.3 mm) compared with the path length at a constant speed of 60 m/min.

When using adaptive systems to continuously adjust the speed of cutting at blunting of cutting tool effect may be even higher.

The results obtained make it possible to improve the efficiency of machining austenitic steels by a more informed choice of the brand of hard-alloy and by correct cutting speed as wear of the tool.

In addition, when turning austenitic steels is of great importance the correct assignment of the front angle of the instrument, providing a different effect on the prevailing mechanism and wear rate of tools made of hard-alloys of different groups [10]. For resistant to adhesion-fatigue wear cutters from WC 94-Co 6 increase to angle γ leads to a decrease of the contact temperature and decrease of wear rate due to the suppression of the diffusion component. Hard-alloy TiC 15-WC 79-Co 6 good resistance to diffusion wear, but fragile in terms of the adhesion-fatigue effect. Therefore, the influence of front angle on wear resistance is the opposite of TiC 15-WC 79-Co 6, and to increase the resistance necessary to reduce the angle γ. For TiC 8-TaC 12-WC 71-Co 9 dependence of the resistance of the angle γ is more complex. At small cutting speeds to increase the resistance of the cutting angle should be reduced, and when the cutting speed – increase [10].

References

1. A. A. Lipatov, *Influence of the Temperature Dependence of the Thermal Conductivity of the Machined Material on the Contact Interaction and the Type of Chip Formation*, Stanki Instrum., No. **8**, p. 37–40 (2006)
2. A. A. Lipatov, *Influence of Chip Formation and the Wear of a Hard-Alloy Tool in Cutting Austenitic Steel*, Russian Engineering Research, Vol. 28, No. **6**, p. 17–19 (2008)
3. A. A. Lipatov, *Influence of the Instability of Chip Formation and Peculiarities of Contact Interaction on the Wear Platform Rear Surface of Hard-Alloy Tool on its Wear in Turning Austenitic Steel*, Bull, VSTU, issue **8**, no. **13 (100)**, p. 30–34 (2012)
4. T. N. Loladze. *Prochnost' i iznosostoikost' rezhushchego instrumenta (Strength and Wear Resistance of Cutting Tools)* (Moscow : Mashinostroenie, 1982)
5. N. V. Talantov, *Fizicheskie osnovy protsessa rezaniya, iznashivaniya i razrusheniya instrumenta (Physical Principles of Cutting and Tool Wear and Failure)* (Moscow : Machinery Publisher, 1992)
6. A. A. Lipatov, *Influence of Chip Formation on the Temperature Dependence of the Thermal Conductivity of Machined Material*, Stanki Instrum. No. **10**. p. 23–25 (2012)
7. A. A. Lipatov, *Influence of Chip Formation on the Temperature Dependence of the Blank's Thermal Conductivity*, Russian Engineering Research, Vol. 33, No. **5**, p. 299–301 (2013)
8. A. A. Lipatov, Yu. L. Chigirinnskii, S. I. Kormilitsyn, *Temperature-power Char-acteristics of Contact Interaction in Cutting Austenitic Steel*, Stanki Instrum., no. **7**, p. 21–25 (2013)
9. M. F. Bazhenov, S. G. Baychman, D. G. Karpachev, *Tverdye cplavy: Spravochnik (Hard-Alloys: a Handbook)* (Moscow : Metallurgiya, 1978)
10. A. A. Lipatov, Yu. L. Chigirinnskii, *Influence of the Front Angle on the Wear Carbide Tool in Turning Austenitic Steel in Terms of the Instability of the Chip Formation*, Bull. VSTU, issue **6**, no. **12 (72)**, p. 26–28 (2010)

DOI: 10.1051/matecconf/201712901022

Problems of modernization and technical re-equipment of Russian machine-building enterprises

Olga Antonyan[1], *Elena* Karpushko[1], and *Anna* Solovyeva[1,*]

[1]Volgograd State Technical University, Institute of Architecture and Construction, 400074 Volgograd. ul. Akademicheskaya, 1, Russia

Abstract. Article substantiates the need for modernization and technical re-equipment in Russian Federation. The goal of the research and tasks for achieving this goal has been determined. The existing conditions for development of machine building complex of Russian Federation in recent history and dynamics of economic indicators that characterize it in recent years are considered. Trends in domestic machine building industry are revealed, which directly characterize the slowdown in the rate of country's economic functioning. Urgency of modernization and technical re-equipment of machine-building, its upgrading on the basis of wide introduction of the latest scientific and technical achievements and advanced technologies, is underlined, as without advancing development of machine building, it is impossible to solve problems of technological safety and problems relating to social sphere and standard of living of country's population. Particular attention is paid to the main problems and tasks of modernization, technical re-equipment of engineering enterprises. It is noted that at present stage the demand for modernization of Russian engineering is obvious. The results of the research show how urgent it is to change the forms and methods of production organization, re-equipment of technological capacities, and introduction of modern management systems. However, in order to correctly formulate the main priorities and directions of modernization, an in-depth analysis is needed.

1 Introduction

Nowadays, industry modernization, its technical re-equipment is based on the widespread introduction of the latest scientific and technological achievements and advanced technologies which are a top priority for Russian economy. One of the most important tasks is accelerated development of machine building, an industry that provides not only modernization of entire national economy, but also contributes to the increase in qualification level of population [4].

Today, fundamental modernization of production and use of modern technologies, materials and equipment is a necessary condition for increasing the competitiveness of products and efficiency of work of most domestic enterprises. Therefore, the subject of the study, devoted to the problems of modernization and technical re-equipment of Russian machine-building enterprises is relevant today.

The purpose of the study is to examine the problems of modernization and technical re-equipment of Russian engineering enterprises and to search for strategic directions for their development. To achieve this goal, it is necessary to:
- give an analysis of the results of machine-building complex development of Russian Federation;
- define the main tasks of modernization and technical re-equipment of machine-building enterprises;
- develop directions for improving the processes of modernization and technical re-equipment at the enterprises of the machine-building industry in Russia.

During the problem's studying process, works of Russian and foreign scientists, and own developments in this direction were used.

2 Results of development of machine-building complex in Russia

2.1 Conditions for development of machine building complex in recent history of Russia

Development of machine building in Russia during 1990-2016 is indicative and reflects all country events related to privatization, economy decline of 1990s, series of financial crises and increase in production in early 2000s. All this led to the loss of competitive positions in the world market and growth of imports under category of «machinery, equipment and vehicles» [2]. Nowadays, enterprises of this industry are under considerable pressure from outside, connected with economic sanctions from European Union and United States. Also, depreciation of the national currency imposes additional restrictions on the development of the considered sector. Existing conditions from one country can be regarded as threats to the existence of machine-building in our country; on the other hand, there are additional

* Corresponding author: solovan@yandex.ru

 DOI: 10.1051/matecconf/201712901022

opportunities and competitive advantages in comparison with Western producers.

2.2 Indicators of machine-building complex development for 2016

The year of 2016 was difficult for the entire Russian machine-building industry. Shipbuilding industry has shown the need to activate fundamental and exploratory research to create promising marine equipment for various purposes. For development of shipbuilding industry, the federal targeted and state programs are being implemented in the country, such as «Development of shipbuilding and engineering for development of offshore fields for 2013-2030» and «Development of defense industry complex of Russian Federation for 2016-2025».

The year 2016 was not easy for motor-car industry. The production of passenger cars in 2016 decreased by 27.7%, which was mainly due to a decrease in personal income and investment activity, as well as a reduction in consumer lending and leasing operations due to higher interest rates on loans. Volume of domestic car market for the specified period also decreased by 35.7%. The production of trucks in 2016 decreased by 14%, which is also related to the relative saturation of this market and a reduction in the freight turnover of road transport [12].

At the end of 2016, there was also a decline in the agricultural machinery industry, which amounted to 14.3%.

In general, in 2016, the dynamics of production in the machine building complex of Russia deteriorated. The total production in the machine-building industries decreased by 10.3% as compared to the same period. The current decline in production in engineering in 2016 was predicted, as this industry was the most sensitive to instability in the investment climate.

3 Modernization and technical re-equipment of engineering enterprises as a decisive factor for increasing competitiveness

3.1 Prerequisites for modernization and technical re-equipment of machine-building industry

The implementation of state programs to accelerate the pace of economic development of the country is possible only if the system of modernization of domestic machine-building complex is accelerated, which determines the main factors for development of productive forces in the medium and long term. As numerous researches of scientists show, without advancing development of machine building it is impossible to solve problems of technological safety and problems related to the social sphere and standard of living of the population. In recent years, negative trends in domestic engineering have become noticeable, which directly characterize the slowdown in the pace of country's economy functioning. It is necessary to note first of all:

1. Decrease in output volume of science-intensive products, creation of which is associated with the use of scientific and technological progress.
2. Slow formation and development of innovative potential due to low level of innovative activity.
3. Reducing competitiveness of machine-building enterprises.
4. Significant long-term aging of all elements of fixed capital (and, above all, active part of the basic production assets of machine-building enterprises).
5. Incompleteness of the implementation of conversion programs for many large enterprises in the country.
6. Slowing down the pace of lower costs for the production of industrial products in engineering.
7. Weakening of functional-technological and economic ties in the system of «metallurgy and machine building».
8. Investment failure to carry out works related to the fundamental modernization of the material and technical base of engineering enterprises.
9. Insufficiently effective application of «double» technologies in machine building, which allows to increase (according to the data of domestic and foreign machine-building enterprises) the effectiveness of production operations without significant investment.
10. Aging of labor in many machine-building centers of the country.
11. Complexity of solving the organizational and economic problems arising in connection with the need to carry out activities related to import substitution in the leading sub-branches of machine building in Russia.

In modern conditions, a special role is assigned to solving the problems of increasing efficiency of machine building - locomotive for development of the entire industrial complex of the country. Over the past 15 years, the pace of development of this complex has significantly decreased due to a number of reasons, and primarily because of the underestimation of the significance of the implementation of the new wave of industrialization in the leading sectors of the country's economy. Studies show that the development of industry in any country is determined by the production potential of machine building and metalworking.

3.2 The main problems and tasks of modernization and technical re-equipment of engineering enterprises

Nowadays, modernization and technical re-equipment of machine building, its improvement on the basis of widespread introduction of the latest scientific and technical achievements and advanced technologies is an urgent need. One of the most important tasks is the acceleration of the development of machine building, an industry that provides not only modernization of the entire national economy, but also contributes to the increase in the qualification level of the population [4].

It should be noted that Russia's machine building industry is distinguished by a very high degree of production concentration. Prospects for modernization of enterprises directly related to the development of its product markets, stability and efficiency of economic ties with major consumers [17].

However, it must be understood that access to financial resources is by no means the only condition for the implementation of modernization projects. Thus, in some large machine-building enterprises it is possible to observe how financial and other results of large-scale investment projects do not correspond to the declared ones, time for transition to new technologies is delayed by several years [1].

In addition, at each stage of the investment project of modernization and technical re-equipment, various organizational and managerial problems arise related to the lack of automation and standardization of processes, lack of skills for working with standard software for some responsible executors, need for clarifications and adjustments, lengthy process of providing services of necessary information and coordination of documentation in connection with a large number of matching persons, etc.

As a result, almost half of the time from the whole process of delivery and putting into operation occupies work with documentation, besides the most part from this half takes time, required for document's accommodation.

One of the reasons for this state of affairs is linear-functional structure of enterprise management, which stimulates functional isolation and reduces effectiveness of interactions between individual participants in end-to-end processes. All this leads to non-compliance with the timing of the implementation of modernization measures and availability of undeveloped and rolling investments from year to year, which in turn requires additional time and labor to update target specifications and other necessary documentation.

Modernization in the framework of engineering business groups is carried out through the following activities:

- purchase of Russian-made equipment;
- purchase of second-hand imported equipment;
- purchase of new imported equipment, technical level of which corresponds to the world one.

Modernization through the acquisition of Russian equipment has its positive and negative sides. On the one hand, domestic equipment is easy to operate and cheap. Its use stimulates the development of machine and machine building industries, thereby gradually reducing the dependence of domestic industry from import. On the other hand, the acquisition of Russian equipment, technical level of which is lower than the world level, creates the prerequisites for a future technological backwardness of enterprises.

The acquisition of used imported assets is widespread among Russian machine-building enterprises due to their lower cost compared to new equipment, on the one hand, and relatively high reliability, on the other hand. Usually, their wear and tear is low, as in developed countries the intervals between replacement of assets for new, constantly improving models are constantly reduced.

Negative side of modernization through the use of foreign equipment is increase of dependence of engineering enterprises on imports.

Usually Russian companies implement several types of measures to update fixed capital in parallel. Thus, certain advanced imported equipment units are purchased for operations that critically affect quality and cost of production.

Less advanced assets are acquired to perform most of the operations of main and auxiliary production. Usually such assets are built into technological chains, mainly consisting of domestic elements.

As for the quantitative aspect of investments, mainly projects with relatively low reported costs and short payback periods are being implemented. Such projects, as a rule, are aimed at maintaining a stable volume of production or at a slight increase in production, reduction in the cost of production, as well as implementation of activities prescribed by safety regulations, social and environmental regulations.

Projects with high adjusted costs and long payback periods for creation of new and radical changes in existing production complexes (full-scale modernization of production) are rare.

While development of modernization measures, in particular when developing the concept of a new production (or reconstruction of an existing one), a new technology or its partial improvement, along with the solution of technical and technological tasks, optimization of movement and stocks in the supply of production, minimization of the duration of the production cycle, minimization of losses and downtime, modernization of the entire technological chain of the production cycle.

The general tasks of modernizing the production of innovative industrial enterprises can be structured into several main groups:

- technical modernization (renewal of fixed capital due to its physical or moral depreciation);
- technological modernization (introduction of innovative technologies on the basis of own developments or using various forms of technology transfer);
- product modernization (expansion of the range of previously manufactured products through their modification or cardinal technological upgrade, development of innovative products);
- modernization of the management system (improvement of the enterprise management system. For example, projects to introduce new information technologies, to remove inefficient capacities and non-core productions for outsourcing, translation of the reporting system to international standards, etc.);
- modernization of the personnel management system of the enterprise (changes in the structure of the personnel and its qualitative level due to the transformation of the hiring-selection system, carrying out programs of in-service training and retraining of personnel).

3.3 Potential directions for modernization and technical re-equipment improvement of engineering enterprises

At this stage, the need for Russian engineering in the modernization is obvious. The results of the research show how urgent it is to change the forms and methods of organizing production, re-equipping technological capacities, introducing more modern management systems. However, in order to properly formulate the main priorities and directions of modernization, an in-depth analysis is needed.

According to the observations of Russian researchers, large enterprises often implement large-scale investment projects, practical significance and financial results of which differ significantly from the stated goals, the result of technical re-equipment is ineffective equipment, capabilities of which are far from being used completely, and the transition to new technologies is delayed by several years.

Analysis of the expediency of modernization of the machine-building enterprise is represented in the form of a comprehensive, step-by-step and element-wise research, purpose of which is:

- determining the level of sufficiency of capacity development for carrying out modernization activities;
- identification of possible barriers / obstacles in the course of modernization and development of measures to eliminate them;
- definition of priority areas for modernization.

Modernization projects, as a rule, are implemented with the aim of increasing production volumes and reducing the cost of production (through the use of advanced technologies and materials, optimizing the number of production personnel, saving energy resources, freeing production spaces, etc.), developing new products and improving the quality of already existing [18]. The values of potential effects for machine-building enterprises with an obsolete equipment park are presented in Table. 1.

Table 1. Potential effects of modernization

<table>
<tr><th>Main effects</th><th>During updating of equipment park</th><th>During using of old equipment park with improvement of technology and use of modern tools and equipment</th></tr>
<tr><td>Reduction of electricity costs</td><td rowspan="3">in 3-4 times</td><td rowspan="3">in 1,2 - 1,5 times</td></tr>
<tr><td>Reduction of production space</td></tr>
<tr><td>Reduction of production personnel</td></tr>
<tr><td>Cost Reduction</td><td>up to 50% (taking into account the cost of new equipment)</td><td>up to 20%</td></tr>
</table>

Consideration of modernization problems and technical re-equipment at the machine-building enterprises allows suggesting the main directions for its improvement:

1) provision of the company's units with personnel in accordance with the calculation of the number and required qualifications;

2) implementation of a training program using the capabilities of a corporate training center, cooperation with educational institutions of cities;

3) preservation and development of intellectual, scientific and technical potential of the enterprise;

4) creation of motivation system, development and professional growth of employees, aimed at maximum professional and personal self-realization of employees;

5) improving corporate culture and increasing social attractiveness of the enterprise.

4 Conclusion

Thus, the main task of industrial policy at the present stage is technological modernization of production and increasing competitiveness of products of machine-building enterprises. One of the most important factors ensuring the solution of this problem is qualitative and quantitative composition of used means of production.

Future will be determined by whether the industries will develop on the basis of the achievements of Russian science and technology, or it will be limited only to the transfer of foreign productions to Russia, as a territory with a relatively cheap labor force and with «not the first freshness» technologies.

Modernization of Russian machine building industry no longer needs to be delayed. Today, the radical modernization of production facilities and the use of modern technologies, materials and equipment are a prerequisite for increasing the competitiveness of products and the efficiency of the work of most domestic enterprises.

References

1. O.N. Antonyan, E.N. Karpushko, A.S. Solovyeva, 16th International Scientific Conference Globalization and Its Socio-Economic Constructions **1**, 50-58 (2016)
2. N. I. Aksenova; E. A. Prikhodko, BK 2: Political sciences, law, finance, economics and tourism conference proceedings **III**, 483-490, (2016)
3. C. Azani, R. Khorramshahgol, 9th World Multi-Conference on Systemics, Cybernetics and Informatics **1**, 298-303 (2005)
4. S.P Baranenko, K.D. Busygin, Actual problems of the humanities and natural sciences **9**, 94-102 (2014).
5. J. Bakowski, Sustainable building and refurbishment for next generations, 35-38 (2013)

6. L. Boronina, D. Ronzhina, 7th international conference of education, research and innovation, 5921-5928 (2014)
7. A. Costea, N Rachieru, R. Nita, 13th international conference modern technologies, quality and innovation, 203-206 (2009)
8. Z.V. Danilovich, Z.V. Viktorovich, B.V. Vasilievich, Asian Social Science **10(24)**, 28-34 (2014)
9. L.D. Gitelman, M.V Kozhevnikov, Economy of Region **13(1),** 204-215 (2017)
10. N.V. Kalenskaya, O.V. Pratchenko, Asian Social Science **11(11)**, 357-363 (2015)
11. N Kalenskaya, First global conference on contemporary issues in education, 203-208 (2015)
12. V.V. Kobzev, M.K. Izmailov, Organizer of production **25(1),** 69-83 (2017).
13. E.A. Kolesnik, III international scientific symposium on lifelong wellbeing in the world **19**, 336-343 (2017)
14. Z. Mingaleva, H. Danilina, Life Science Journal **11(8)**, 211-215 (2014)
15. M.L. Melnikov, L.A. Gracheva, Gornyi Zhurnal **6**, 65-67 (2014)
16. B. Melnikas, 7th international scientific conference business and management, 416-425 (2012)
17. I.I. Novakivskyi, N.S. Lushchak, V.Z. Borbulevych, Marketing and management of innovations **3**, 30-245 (2016)
18. Yu.K. Salnikova, Vestnik of the South-Russian State Technical University **5**, 54-59 (2016).
19. A. Schoenmann, J. Greitemann, G. Reinhart, Sixth international conference on changeable, agile, reconfigurable and virtual production **52**, 198-203 (2016)
20. Y. Vertakova, Y. Polozhentseva, M. Klevtsova, Innovation management, development sustainability, and competitive economic growth **I – VII,** 2168-2178 (2016)

DOI: 10.1051/matecconf/201712901023

Research of vibration resistance of non-rigid shafts turning with various technological set-ups

Sergey L. Vasilevykh [1,*], and *Evgeniy S.* Shelihov [2]

[1]Vyatka State University, Kirov, Russian Federation
[2] Orenburg State University, Orenburg, Russian Federation

Abstract. The article considers the definition of the stability range of a dynamic system for turning non-rigid shafts with different technological set-ups: standard and developed ones; they are improved as a result of this research. The topicality of the study is due to the fact that processing such parts is associated with significant difficulties caused by deformation of the workpiece under the cutting force as well as occurrence of vibration of the part during processing, they are so intense and in practice they force to significantly reduce the cutting regime, recur to multiple-pass operation, lead to premature deterioration of the cutter, as a result, reduce the productivity of machining shafts on metal-cutting machines. In this connection, the purpose of the present research is to determine the boundaries of the stability regions with intensive turning of non-rigid shafts. In the article the basic theoretical principles of construction of a mathematical system focused on the process of non-free cutting of a dynamic machine are justified. By means of the developed mathematical model interrelations are established and legitimacies of influence of various technological set-ups on stability of the dynamic system of the machine-tool-device-tool-blank are revealed. The conducted researches allow to more objectively represent difficult processes that occur in a closed dynamic system of a machine.

1 Introduction

The components with rotation bodies account for 30 % of all the machine components. It is the most labour consuming to make the details characterized by fragility: all kinds of shafts, torsion bars, shafts, stems, leading cylinders, nonrigid shafts, etc.

Non-rigid shafts are the ones with length-to-diameter (L to D) rate is over 12 (L/D >12). Due to fragility of the processed non-rigid shaft the technological system machine-device-tool-blank is pliable to the influence of external transverse force and dynamic factors of the cutting process [1].

Processing such components is quite difficult due to the fact that a processed component gets deformed during cutting and to vibration in the component during processing. These factors can be so strong that the cutting regime is to be reduced or multiple-pass operation is to be used, they cause decrease in firmness and operational life of the cutting instrument. [2-7]. Vibrations are not wanted at finish process with cutting at shall depth, when non-vibration movement of the component and the cutter in the cutting zone can lead to defective goods.

The problem of vibration is topical at metal-processing with the help of CNC machines, as it leads not just to decrease in processing accuracy: vibrations in the cutting zone can also lead to accelerated amortization of the machine.

At the same time uncontrolled mechanical oscillations with a comparatively big amplitude are a limiting factor for increasing productivity of cutting process [8,9]. And oscillations appear due to presence and mutual influence of technological cutting conditions, external perturbing forces and parameters of the elastic system of a manufacturing lathe. That's why a decrease in efficiency of non-rigid shafts processing depends mainly on providing the conditions of processing stability.

Nowadays stability is understood according to the experience of the technologist. The existing methods do not provide accuracy, as the mathematical models used for its calculations are oversimplified. Coping with this research issue successfully depends not only on the traditional ways, but also on availability of necessary mathematical models that are able to describe the interdependence of vibrations of the elastic system of the machine and the dynamic process of cutting.

Thus the aim of the research is to work out a mathematical model by meas of which it is possible to calculate the limits of stability zones at linear turning of non-rigid shafts for different kinds of technological equipment of a turning lathe.

The tasks are the following:

1. To work out the methods of researching the mathematical model as for vibration-resistance at linear turning of non-rigid shafts;

* Corresponding author: shelevgen@mail.ru

2. To get a characteristic equation in order to realize the mathematical model of vibration-resistance;

3. To determine the stability zones of the dynamic system under research at lathe work of non-rigid shafts with different technological equipment.

2 Materials and Methods

In the research materials and methods of cutting theory, structural resistance, mathematical modeling were used. The research of influence of various technological equipment of turning lathes on vibration-resistance was fulfilled by the methods of serial presence detection of the parameters of the worked out mathematical model.

2.1 Mathematical modeling of vibration-resistance of linear turning of non-rigid shafts

To calculate the resistance of the dynamic system in question let us consider the analytical model (Pic. 1).

Let us assume that the shaft vibrates in plane that goes through the symmetry axis of the processed shaft.

In the cutting area of the analytical model there is a support attachment group of the processed shaft.

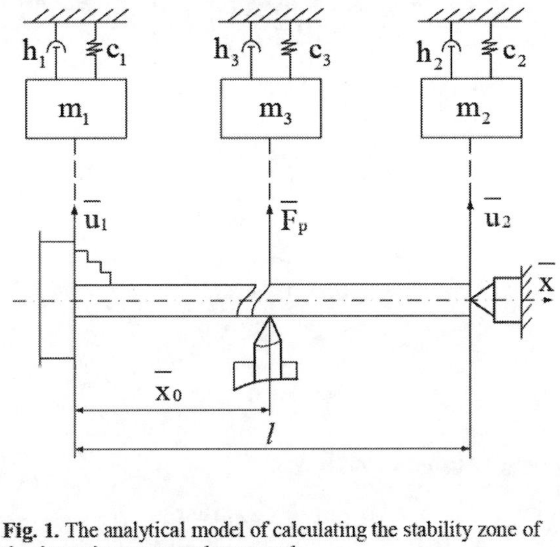

Fig. 1. The analytical model of calculating the stability zone of the dynamic system under research

The analytical model shows: equivalent mass m_i, damping h_i, rigidity of bearings in the sections Ci, where i = 1...3, shaft length l, coordinate of the cutter application $\overline{x}_0$, cutting power $\overline{F}_p$, angle of flexure of the processed shaft $\overline{U}(\overline{x},\overline{t})$.

The analytical model embraces all the mentioned kinds of technological defects (Table 1). Researching the first and the fourth kinds of technological defects the bearer 3 is not taken into account.

Below one can see the method of researching stability of the mathematical model.

The equation of free oscillations of a non-rigid shaft is as follows:

$$EJ\frac{\partial^4 \overline{U_i}}{\partial \overline{x}^4} + m\frac{\partial^2 \overline{U_i}}{\partial \overline{t}^2} = 0, \qquad (1)$$

where EJ – rigidity of the processed shaft at the flexure; $\overline{U}$ – the flexure of the processed shaft, m – mass of a unit of shaft length.

In the cutting process the cutter excites a perturbing force which is applied in some intermediate section of the shaft dividing the length in two parts.

Thus it is necessary to determine eight constants from the four conditions at the ends of the shaft and four conditions of connection.

The conditions of connection are the following:

$$\begin{aligned} &\overline{U}_1(\overline{x}_0,\overline{t}) = \overline{U}_2(0,\overline{t}), \\ &\overline{U}_1'(\overline{x}_0,\overline{t}) = \overline{U}_2'(0,\overline{t}), \\ &\overline{U}_1''(\overline{x}_0,\overline{t}) = \overline{U}_2''(0,\overline{t}), \end{aligned} \qquad (2)$$

$$EJ\left(\overline{U}_1'''(\overline{x}_0,\overline{t}) - \overline{U}_2'''(0,\overline{t})\right) =$$

$$= \overline{F}_p - \left(m_3\overline{U}_2''(0,\overline{t}) + h_3\overline{U}_2'(0,\overline{t}) + C_3\overline{U}_2(0,\overline{t})\right)$$

where U1 means a flexure on the part from the left end of the shaft ($x = 0$) to the cutter ($x = \overline{x}_0$), i. e. at the interval $\left[0, \overline{x}_0\right]$, and U_2 means a flexure on the part from the cutter to the right end of the shaft, i. e. at the interval $\left[\overline{x}_0, l\right]$, m_3, h_3, C_3 parameters which characterize the outfit in the cutting zone.

The cutting power is expressed by the equation:

$$\overline{F}_p = K^{(1)}\overline{U}_1(\overline{x}_0,\overline{t}) - K^{(2)}\overline{U}_2(\overline{x}_0,\overline{t}-\overline{\tau}), \qquad (3)$$

where τ is a rotation period of the shaft (lag time); $K^{(1)}$, $K^{(2)}$ is dynamic characteristics of cutting of the first and second type.

The first three equations in conditions of connection mean continuity of a bending curve of the shaft (flexure), the angle of flexure of the shaft's profile, and the moment of flection in the force point of cutting power, accordingly. The fourth equation means the jump of function of transverse force in the given profile by the amount characterized by the following equation:

$$\overline{F}_p - \left(m_3\overline{U}_2(0,t) + h_3\overline{U}_2(0,t) + C_3\overline{U}_2(0,\overline{t})\right), \qquad (4)$$

The boundary data are the following:

$$\overline{U}_1''(0,t) = 0,$$

$$-EJ\overline{U}_1'''(0,t) =$$

$$=m_1\overline{U}_1''(0,\bar{t})+h_1\overline{U}_1'(0,\bar{t})+C_1\overline{U}_1(0,\bar{t}),$$

$$\overline{U}_2''(l-\bar{x}_0,\bar{t})=0, \tag{5}$$

$$EJ\overline{U}_2'''(l-\bar{x}_0,\bar{t})=m_2\overline{U}_2''(l-\bar{x}_0,\bar{t})+$$

$$+h_2\overline{U}_2'(l-\bar{x}_0,\bar{t})+C_2\overline{U}_2(l-\bar{x}_0,\bar{t}).$$

To bring the system (1–4) to dimensionless form let us consider the following change of variables:

$$x=\frac{\bar{x}}{l-\bar{x}_0},\ U=\frac{\overline{U}}{T},\ t=\frac{\bar{t}}{\tau}.$$

Let us introduce:

$$A=\frac{EJ}{(l-\bar{x}_0)^4},\quad B=\frac{m}{\tau^2},\quad \alpha=\sqrt{\frac{A}{B}},$$

$$m_{1,2,3}=\frac{m_{1,2,3}}{A(l-\bar{x}_0)\tau^2},\quad C_{1,2,3}=\frac{C_{1,2,3}}{A(l-\bar{x}_0)},$$

$$h_{1,2,3}=\frac{h_{1,2,3}}{A(l-\bar{x}_0)\tau},$$

$$P\div\frac{d}{dt},\quad q\div\frac{d}{dx},$$

$$V_{1,2,3}^{(p)}=(m_{1,2,3}p^2+h_{1,2,3}p+C_{1,2,3}).$$

Using Laplase transformation to the system of equation at zero initial condition let us pay attention to the figuring class according to variables x and t, step by step. With the equation (1) we obtain:

$$U(q,p)=\frac{q^4}{q^4+4\alpha^4}\cdot U(o,p)+$$

$$+\frac{q^3}{q^4+4\alpha^4}\cdot U'(o,p)+\frac{q^2}{q^4+4\alpha^4}\cdot U''(o,p)+ \tag{6}$$

$$+\frac{q}{q^4+4\alpha^4}\cdot U'''(o,p),$$

$$4\alpha=\frac{m(l-\bar{x}_0)^4}{EJ\cdot\tau^2}\cdot p^2.$$

Let us express the original of the function U(q,p) according to the variable q as follows:

$$U(x,p)=Y_1(\alpha,x)\cdot U(o,p)+$$

$$+\frac{Y_2(\alpha,x)}{\alpha}\cdot U(o,p)+ \tag{7}$$

$$+\frac{Y_3(\alpha,x)}{\alpha^2}\cdot U''(o,p)+\frac{Y_4(\alpha,x)}{\alpha^3}\cdot U'''(o,p).$$

where $Y_1(t)$, $Y_2(t)$, $Y_3(t)$ $Y_4(t)$ are Krylov functions.

Putting the equation (6) in limited conditions and in conjugating conditions, and using the properties of Krylov functions $Y_1(0) = 1$, $Y_2(0) = Y_3(0) = Y_4(0)$, we shall get a linear system of algebraic equations in relation with $U_1(o,p)$, $U_1'(o,p)$, $U_1''(o,p)$, $U_1'''(o,p)$, $U_2(o,p)$, $U_2'(o,p)$, $U_2''(o,p)$, $U_2'''(o,p)$ as follows:

$$U_1''=0,$$

$$U_1'''+U_1V_1=0,$$

$$-4\alpha^2Y_3(\alpha)U_2-4\alpha^2Y_4(\alpha)U_2'+$$

$$+\frac{Y_2(\alpha)}{\alpha}U_2'''=0,$$

$$\left[4\alpha^3Y_3(\alpha)+Y_1(\alpha)V_2\right]\cdot U_2+$$

$$+\left[4\alpha^2Y_3(\alpha)+\frac{Y_2(\alpha)V_2}{\alpha}\right]\cdot U_2'+$$

$$+\left[4\alpha^2Y_4(\alpha)+\frac{Y_3(\alpha)V_2}{\alpha^2}\right]\cdot U_2''+$$

$$+\left[-Y_1(\alpha)+\frac{Y_4(\alpha)V_2}{\alpha^3}\right]\cdot U_2'''=0,$$

$$Y_1(\alpha,x_0)\cdot U_1+\frac{Y_2(\alpha,x)}{\alpha}\cdot U_1'+$$

$$+\frac{Y_4(\alpha,x_0)}{\alpha^3}\cdot U_1'''-U_2=0,$$

$$-4\alpha Y_4(\alpha,x_0)\cdot U_1+Y_1(\alpha,x_0)\cdot U_1'+$$

$$+\frac{Y_3(\alpha,x_0)}{\alpha^2}\cdot U_1''-U_2'=0,$$

$$-4\alpha^2Y_3(\alpha,x_0)\cdot U_1-4\alpha Y_4(\alpha,x_0)\cdot U_1'+ \tag{8}$$

$$+\frac{Y_2(\alpha,x_0)}{\alpha}\cdot U_1'''-U_2''=0,$$

$$-4\alpha^3Y_2(\alpha,x_0)\cdot U_1-4\alpha^2Y_3(\alpha,x_0)\cdot U_1'+$$

$$+Y_1(\alpha,x_0)\cdot U_1''+V_3U_2-U_2'''=F_p$$

Fp is determined this way:

$$F_p=\left[\begin{array}{c}K_2^{(1)}\cdot\sin^2\beta+0{,}5K_3^{(1)}\cdot\cos2\beta-\\ -\left(K_2^{(2)}\cdot\sin^2\beta+0{,}5K_3^{(2)}\cdot\cos2\beta\right)\cdot e^{-Pt}\end{array}\right]\times$$

$$\times HU_1(x_0,p),$$

here $K_{2,3}^{(1),(2)}$ means the given dynamic characteristics of cutting (DCC); H – means tap depth at which the DCC were taken; β – the angle of slope of vibration direction of the processed shaft at taking DCC characteristics.

Substituting the results obtained using the system of equations according to Cramer rule for $U_1(o,p)$, $U_1'(o,p)$, $U_1''(o,p)$, $U_1'''(o,p)$ into the equation (7) we can find the characteristic equation for the dynamic system under research:

$$1-\frac{\Phi(p,x_0)}{\Delta(p,x_0)}\cdot F_p=0, \qquad (9)$$

where Δ(p, x₀) means a characteristic equation for an open-loop system.

$$\Phi(p,x_0)=Y_1(\alpha,x_0)\cdot A_1+ +\frac{Y_2(\alpha,x_0)}{\alpha}\cdot A_2+\frac{Y_4(\alpha,x_0)}{\alpha^3}\cdot A_3, \qquad (10)$$

with A1, A2, A3 as algebraic additions of the corresponding elements of the matrix got from the system of equations (7). The research is described in the works [10-15].

2.2 The research of stability zones for different kinds of equipment of a turning lathe

At lathe work with non-rigid shafts, in order to research the influence of different kinds of technological equipment of a turning lathe on vibration-resistance, it is the most convenient and economical to use the method of successive parameters identification of the worked out mathematical model which is limited by the process of restricted oblique cutting.

Choosing W = -1/H as the complex parameter on the basis of which the curve of D-division is drawn, it is possible to find the depth cutting limit for different outfit. The calculations presented are made with the help of a complex of programs where the method of stability calculation in "the minor" is used according to the following algorithm.

1. For calculating the transfer function for each value of angular rate on a complex plane (ReW, JmW) a hodograph of the transfer function of the researched system is made (Pic. 2).

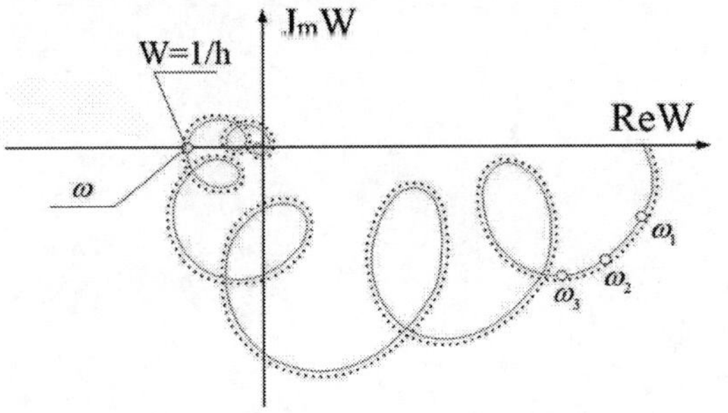

Fig.2 The curve of D-division for determinig the stability area "in the minor"

2. The points of crossings of the hodograph with the real axis ReW are stated, to do that the imaginary parts *W(p)* are compared in sign at different values of angular rate ω and $\omega + K$ following each other, where K is a spacing. They are compared until the moment when the imaginary parts have different signs. For the point of hodograph and ReW crossing the arithmetical average of the real parts of the function in the corresponding points is taken.

3. Having found all the crossing points of the hodograph and the real axis ReW one should choose the point that is the most remote from the central point of coordinates. It is the boundary of the stability area in "the minor", and the corresponding value of cyclic frequency is the frequency of excitation of author-oscillations.

The value of the limit cutting depth for stability in "the minor" which is calculated for all the kinds of the outfit under researched (Table 1).

In the table: 1 means processing with the help of a lathe self-centering chuck with contraction by the rear turning centre; 2 means processing in the centres with driving equipment; 3 means processing in a lathe self-centering chuck with cam readjustment with contractor by the rear turning centre with readjustment; 4 means processing in the leading riffled centre with contraction by the rear rotating centre with readjustment; 5 means the same as 4 but using a steady vibration suppressor which is moving along the axis of the processed component.

Table 1. The data of the limit cutting depth which correspond stability "in the minor"

The variant of outfit	The limit cutting depth (mm)	
	modeling	experiment
1	0,138	0,15
2	0,124	0,1
3	0,164	0,15
4	0,09	0,1
5	4,61	5

3 Conclusion

The following results were obtained in the research:

1. Methods of stating the stability zone of a dynamic system at sharpening non-rigid shafts with different outfit were worked out.

2. A characteristic equation for calculating the limit cutting depth by means of ECM was got.

3. By means of the worked-out mathematical model the boundaries of stability zones for different types of outfits of a lathe machine were stated.

The methods of researching vibration-resistance of linear turning of non-rigid shafts let reflect complex dynamic processes in a closed-loop system of a lathe machine more objectively. With the help of a mathematical model it is possible to determine the boundaries of the stability zones at intensive lathe processing of non-rigid shafts, to find out the connection between stability zones, the technological system machine-device-tool-blank, and the limit cutting depth with different outfits of a lathe machine. On the basis of the analysis it is possible to give important practical recommendations.

References

1. V.A. Kudinov, *Machine Dynamics* (Moscow, 1967)
2. S.L. Vasilevykh, L.A. Vasilevykh, *System technology* (Dnepropetrovsk, 2009)
3. S.L. Vasilevykh, V.E. Saitov, *Features of processing of nonrigid shafts,* Modern science-intensive technologies № **11**, pp. 67-68 (2012)
4. S.L Vasilevykh, V.E. Saitov, *Technological means for ensuring the vibration resistance of the process of turning non-rigid shafts,* Modern high technology, №**3**, pp. 7-10 (2014)
5. S.L. Vasilevykh, *All-Russian annual scientific and practical conference Society, Science, Innovations: Sat. Materials, Vyatka State University* (Kirov: VyatGU, 2014), pp. 1556-1560
6. X. Han, H. Ouyang, M. Wang, Y. Mao, N. Hassan, *Selfexcited vibration of workpieces in a turning process,* Proceedings of the Institution of Mechanical Engineers, Part C: Journal of Mechanical Engineering Science, № **226(8s),** pp. 1958-1970 (2012)
DOI: 10.1177/0954406211435880
7. V.N. Poduraev, *Machining with vibrations* (Moscow, 1970)
8. C.F. Bisu, P. Darnis, A. Gérard, J-Y. K'nevez, *Displacements analysis of self-excited vibrations in turning,* The International Journal of Advanced Manufacturing Technology **44(1s)**, pp. 1-16 (2008) DOI: 10.1007/s00170-008-1815-8
9. C.F. Bisu, A. Gérard, J-Y K'nevez, R. Laheurte, O. Cahuc, *Self-excited vibrations in turning: Forces torsor analysis.* The International Journal of Advanced Manufacturing Technology **44**, pp. 447-462 (2009) DOI: 10.1007/s00170-008-1850-5
10. S.L. Vasilevykh, A.V. Ivanaysky, *The choice of the design model of the elastic system of the machine with various technological equipment.* Technology of mechanical engineering, № **7**, pp. 23-26 (2013)
11. S.L Vasilevykh, V.E. Saitov, *Research of vibrostability in plane turning of non-rigid shafts* (Saarbruken, Germany, LAMBERT Academic Publiahing, 2012)
12. Yu.I. Gorodetsky, V.Ya. Prodiu, *Experimental determination of the dynamic characteristics of the cutting process with the aid of a computer*, Machines and tools **6**, pp. 25-27. (1980),
13. Yu.I. Gorodetsky, A.S. Budankov, V.N. Komarov, *On a system of the experimental study of the dynamics in the process of cutting metals.* Problems of mechanical engineering and reliability of machines, *RAS,* **1**, pp. 80-86 (2008)
14. S.L. Vasilevykh, A.P. Galkin, Yu.I. Gorodetsky, *Mathematical modeling and the study of the stability of non-rigid plane turning of shafts with different technological support,* Univer collection of scientific papers Mathematical Modeling and Optimization, pp. 42-50 (1991)
15. O.A.Yamnikova, *Construction of mathematical models of non-rigid shaft vibrations in cutting,* STIN **1**, pp. 18-21 (2003)

DOI: 10.1051/matecconf/201712901024

Method of increasing durability of thin-walled structural elements

Vladimir Mironov[1,2], *Olga* Lukashuk[2,*], and *Dmitriy* Ogorelkov[2]

[1]Institute of Engineering Science, RAS (Ural Branch), Yekaterinburg, Russia
[2]Ural Federal University, Yekaterinburg, Russia

Abstract. The article analyzes how efficient damping introduced into designs of structural elements by means of such light fillers as foam plastics could be. It cites the results of computing experiments designed to assess how the number of significant free vibration cycles depends on the width of a beam wall and the height of a filler layer. The prediction is made of a substantial increase in durability of the machines which operate in a dynamic impulse-loading mode.

1 Introduction

The problem (common for many branches of technology) of extending the life of a product is dealt with using different approaches. Most often, they are about modifying that surface layer of the product where fatigue cracks tend to appear [1]. For thin-walled structural elements, such approaches have limited application.

On the other hand, the increase in durability could be achieved by decreasing dynamics of machines and constructions – for example, by means of different vibration dampers [2]. In building and aircraft construction, those goals have long and quite successfully been realized through the usage of three-layer plates and shells [3].

Box beams and thin-walled closed-profile rods are widely utilized in bearing metallic structures of mining and transporting machines. During their normal operation, such machines are subject to a wide range of regular or random dynamic loads.

The decrease in dynamics helps to extend the life of such machines and improve the working conditions for their operators. The usage of such light-density fillers as foam plastics – though leading to some insignificant increase in product weight – provide a substantial gain in the protection of personnel (from noise and harmful vibration) and structures (from corrosion). But the main purpose of such a filler capable of absorbing the energy of vibrations is to reduce their number and amplitude when external dynamic loads are applied.

The goal of the work is to make preliminary assessment on practicability of using artificial low-modulus elastic materials to damp thin-walled elements of metallic structures. The task is realized via conducting computing experiments in the environment of the SolidWorks software.

Estimating bending stresses turns out to be a crucial factor in assessing strength and durability of most elements pertaining to bearing structures of machines. Therefore, the research is limited to free vibration of beams subject to bending. To achieve that specific goal, the following tasks were realized:
- calculating the number of significant damped free vibration cycles;
- parametrically analyzing a cantilever beam;
- parametrically analyzing an overhead crane bridge girder.

2 Calculation of the number of significant free vibration cycles for a beam

Evaluative calculations justify analyzing vibrations of a beam as a material point of a reduced mass m_p on a spring of an equivalent rigidity c_e. Free vibrations of such a point are described by homogeneous differential equations of the second order [4]. Neglecting resistance forces, we have

$$\ddot{x} + p^2 x = 0, \tag{1}$$

where $p^2 = c_e / m_p$ – square of angular frequency, c_e – rigidity of a steel beam, m_p – reduced rigidity of a beam without a filler. The solution of equation (1) describes free harmonic vibrations of a frequency $\omega = p / 2\pi$ and a period $T = 1/\omega$.

When a system is affected by low resistance forces proportional to a first degree speed of motion $R = \alpha x$ – which is the case, for example, for building structures – we have

* Corresponding author: oldim96@mail.ru

$$\ddot{x} + 2n\dot{x} + p_1^2 x = 0\,, \tag{2}$$

where $n = \alpha / 2m_{pr}$, α – factor of resistance, m_{pr} – reduced mass of a beam with a filler, $p_1^2 = c_{er} / m_{pr}$ – square of angular frequency for a beam with a filler, c_{er} – equivalent rigidity of a beam with a filler. The solution of equation (2) describes damped vibrations of a frequency $\omega_1 = p_1 / 2\pi$ and a conditional period $T_1 = 2\pi / p_1$. The angular frequency of free damped vibrations is determined by the expression

$$p_1 = \sqrt{p^2 - n^2}\,. \tag{3}$$

Starting conditions to determine the integration constants in equations (1) and (2) are set to be the same as the ones realized in the follow-up experiment: for $t = 0$, we have $x = x_0$; $\dot{x}_0 = 0$.

The solution of equation (2) describes, for $n < p_1$, damped beam vibrations with a gradually diminishing amplitude. The envelopes of the process curve are determined by the functions

$$x = \pm x_0 e^{-nt}\,. \tag{4}$$

Significant vibration amplitudes, regardless of an initial displacement x_0, are estimated by the relation $x / x_0 \geq 0{,}1$. Then, the corresponding time t_* could be determined through formulas (3) and (4)

$$t_* = -\frac{\ln 0{,}1}{n} = -\frac{\ln 0{,}1}{\sqrt{p^2 - p_1^2}}\,. \tag{5}$$

The predicted number of significant cycles N is estimated by

$$N = \frac{t_*}{T_1} = -\frac{p_1 \ln 0{,}1}{2\pi\sqrt{p^2 - p_1^2}} = -\frac{\omega_1 \ln 0{,}1}{2\pi\sqrt{\omega^2 - \omega_1^2}} \tag{6}$$

and could be determined for known frequencies p and p_1. Modern computer systems allow to find natural frequencies ω, ω_1 for thin-walled elements of any structure.

Based on the results of energy approach to assessment of damping properties of a structure, we select the width S of the wall and the height h of the foam plastic layer in a beam as variable parameters.

Frequency analysis for a cantilever beam in its initial state and with a foam plastic filler is made using the SolidWorks software environment [5].

Estimating the damping effect which foam plastics produce using formula (6), suppose that ω is the natural frequency of a beam without a filler, and ω_1 is the one of a beam with a filler.

3 Parametric analysis of test cantilever beams

For the purposes of conducting the experiment, it was decided to use steel test beams of l = 600 *mm* in length, b = 25 *mm* in box-section width, H = 70 *mm* in height and S = 1,5 *mm* in wall thickness. To carry out the frequency analysis, a solid model for a box beam and an assembly model filled with a foam plastic material were built. Lowest vibration frequencies of beams in their initial state or with the foam plastic filler were determined for the interval $S \in [0{,}2;1{,}5]$ *mm*.

The results of the numerical experiment are presented in Fig. 1 as relations between the number of significant vibration cycles N and the thickness of the filled-beam wall S.

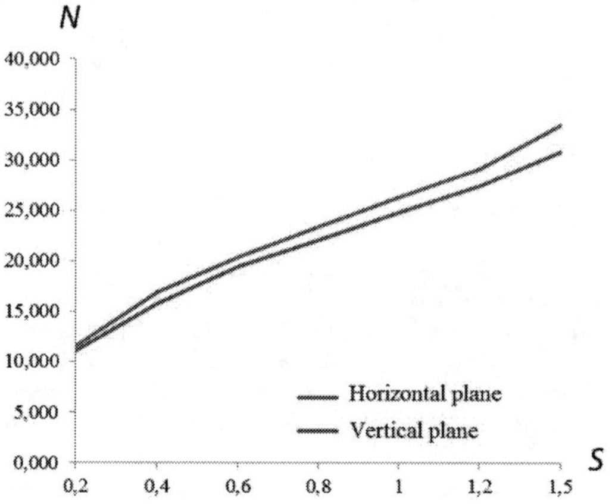

Fig. 1. The number of significant vibrations in relation to the thickness of the beam section wall.

Reference sources cite data on logarithmical decrements of metallic bridge structures in the range of $\delta = 0{,}05...0{,}12$. For the given significance level of 0,1, the number of free vibrations is within 49-19 cycles. In accordance to the graphs in Fig. 1, the number of vibrations is 33-11 cycles, which is approximately 30 % lower than in the case of the values for the box beam. This result should be considered as a significant one. The obtained data on the predicted number of cycles for a filled beam pass validation tests.

4 Parametric analysis of an overhead crane bridge girder

Calculations for the design of an overhead bridge crane of 20 tons in lifting capacity and 19,5 meters in span are made using the procedure described in the tutorial [6]. On the condition of structural rigidity, the following cross-sectional dimensions were chosen for the bridge girder: H = 1,1 *m*, b = 0,6 *m*, S = 9 *mm*.

Fig. 2 shows the form of bending vibrations for the girder assumed to be fastened on specific selected conditions.

 DOI: 10.1051/matecconf/201712901024

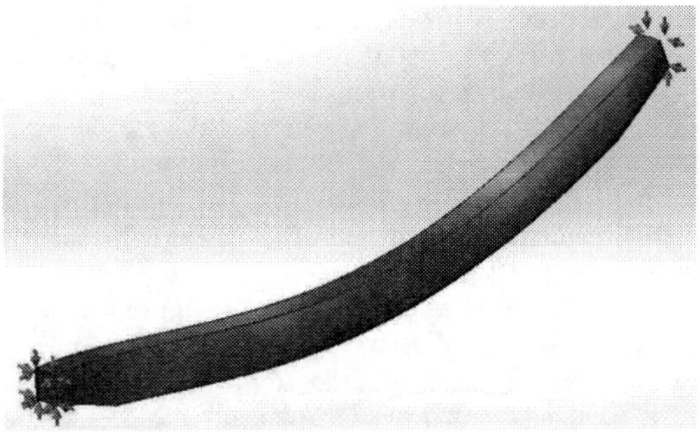

Fig. 2. The form of bending vibrations for the bridge girder of the crane.

The logarithmical decrement of vibrations δ is related to the factor of damping n – which is determined from expression (3) – as $\delta = T_1 n$. Its values are often given in reference sources. Tables 1 and 2 provide the decrement values predicted for bridge girders with various heights of the filler layer *H*.

Table 1. Logarithmical decrement of free damped vibrations δ in the horizontal plane.

Wall thickness, *mm*	Filler layer height, *mm*				
	220	440	660	880	1100
4	0,158	0,176	0,190	0,207	0,231
5	0,134	0,152	0,168	0,182	0,207
6	0,116	0,134	0,151	0,168	0,189
7	0,091	0,109	0,128	0,145	0,174
8	0,070	0,094	0,115	0,129	0,160
9	0,053	0,081	0,100	0,113	0,146
10	0,049	0,074	0,094	0,109	0,137

Table 2. Logarithmical decrement of free damped vibrations δ in the vertical plane.

Wall thickness, *mm*	Filler layer height, *mm*				
	220	440	660	880	1100
4	0,124	0,173	0,203	0,233	0,251
5	0,099	0,145	0,173	0,197	0,222
6	0,089	0,125	0,152	0,177	0,194
7	0,081	0,127	0,149	0,169	0,184
8	0,076	0,111	0,137	0,153	0,171
9	0,067	0,100	0,122	0,143	0,158
10	0,065	0,098	0,118	0,133	0,152

Fig. 3 shows graphs for the decrement change in relation to the parameters selected. The damping capacity of a structure decrease with an increase in the wall thickness and a decrease in the filler layer height.

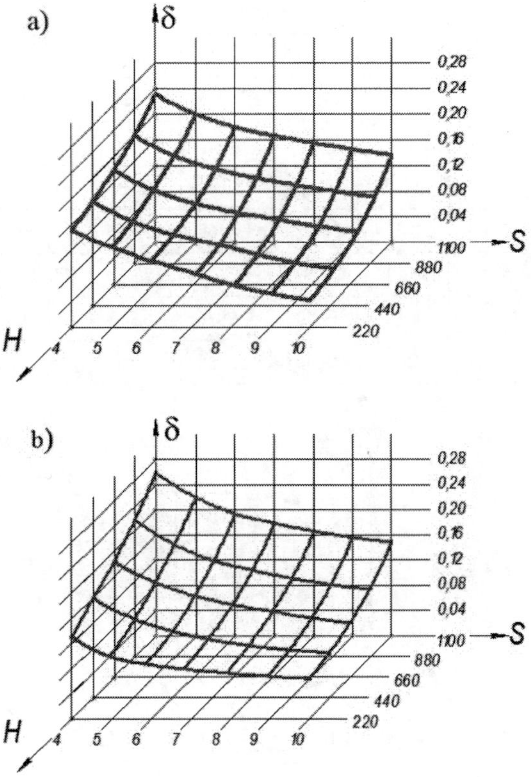

Fig. 3. Relation of the logarithmical decrement of damping to the wall thickness and the foam plastic layer height in the horizontal a) and vertical b) planes.

Similarly, the number of significant cycles also decreases (Tables 3 and 4). Graphs for the relation of the number of cycles to the girder wall thickness and the damping layer height for the given intervals are provided by Fig. 4.

Table 3. The number of significant cycles *N* in the horizontal plane.

Wall thickness, *mm*	Filler layer height, *mm*				
	220	440	660	880	1100
4	14,604	13,093	12,096	11,122	9,964
5	17,121	15,175	13,717	12,634	11,125
6	19,876	17,152	15,200	13,731	12,162
7	25,287	21,059	17,951	15,852	13,246
8	32,899	24,509	19,946	17,791	14,435
9	43,417	28,508	23,063	20,380	15,729
10	47,454	31,003	24,495	21,045	16,764

DOI: 10.1051/matecconf/201712901024

Table 4. The number of significant cycles *N* in the vertical plane.

Wall thickness, *mm*	Filler layer height, *mm*				
	220	440	660	880	1100
4	18,515	13,093	11,365	9,883	9,157
5	23,239	15,175	13,302	11,705	10,382
6	25,777	17,152	15,184	12,995	11,863
7	28,429	21,059	15,468	13,643	12,499
8	30,130	24,509	16,773	15,064	13,487
9	34,401	28,508	18,876	16,158	14,530
10	35,625	31,003	19,588	17,303	15,110

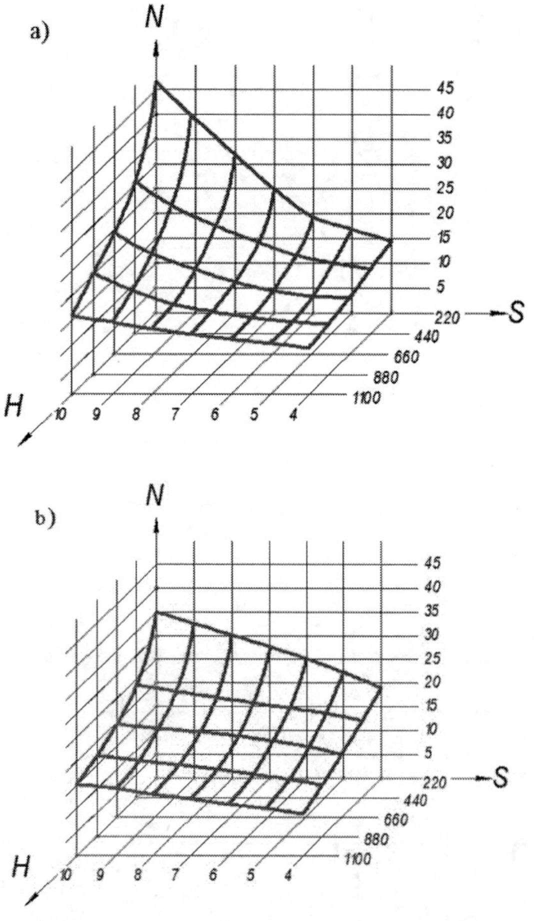

Fig. 4. The number of significant vibrations for the bridge girder in the horizontal a) and vertical b) planes.

Both for the laboratory-tested beam in Fig. 1 and the bridge girder in Fig. 4, the prediction is made of getting major damping effect from using light low-modulus foam plastics as fillers. Thus, for the calculated wall thickness of 9 *mm*, an increase in the filler layer height led to a decrease in the number of the bridge girder vibrations from 43 to 16 in the horizontal plane and from 34 to 14 in the vertical one (see Tables 3 and 4). The filler does not impact the selection of the girder section, which is determined by the conditions of structural strength and rigidity. But it does improve the recommendation given on the maximum possible decrease in the wall thickness.

The decrease in the number of significant cycles predicted for metallic structures damped by means of low-modulus foam plastics would provide the basis for a very efficient method of extending the life of machines. And, particularly, it is true for mining, lifting, carrying, road-building machines, which are subject, during their operation, to a wide range of dynamic loads. A more founded conclusion could be provided by conducting tests on steel beams under repeatedly varied impulse loading.

Conclusion

Computing experiments revealed only a possibility in principle of extending the life of machine parts damped by a filler. It is necessary to check directly in experiments the fatigue properties of the filler, its adhesion to metal. It would be useful to further study the layout of positioning the filler within a structure in order to find a rational solution. Assessment of the project cost is also needed to define optimization problems. At the same time, the provided results showed that such a method of damping has good prospects in terms of increasing the durability of structural elements.

References

1. A.M. Shhipachev, V.S. Muhin, Izv. vuzov. Aviacionnaja tehnika, **1**, 76 (2002)
2. Ya.L. Liberman, K.Yu. Letnev, *Automated Vibration Elimination for the Load Lifted or Lowered on a Flexible Suspension* (UrFU, Ekaterinburg, 2012)
3. V.V. Bolotin, Yu.N. Novichkov, *Mechanics of Multilayer Structures* (Mashinostroenie, Moscow, 1980)
4. V.I. Smirnov, *Course on Higher Mathematics* (Nauka, Moscow, 1974, **2**)
5. A.A. Aljamovskij, A.A. Sobachkin, E.V. Odincova, etc., *Computer-aided Design in SolidWorks for Engineering Purposes* (BHV-Peterburg, Saint Petersburg, 2005)
6. V.E. Dus'e, Yu.V. Navarskij, V.P. Zhegul'skij, *Calculation and Design of Bridge Crane Metallic Structures* (UGTU-UPI, Ekaterinburg, 2007)

 DOI: 10.1051/matecconf/201712901025

Neural network management of technological systems at the finish operations

Irina Ovsyanikova[1,*] and *Aleksandr* Tarapanov [1]

[1]Orel State University of I.S. Turgenev, Orel, Rassian Federation

Abstract. In the article, problems of forming systems management on final operations by means of the neural networks providing high quality of functioning in conditions of uncertainty (insufficiency) of information about the control object and the external environment are considered. It is emphasized that neural networks define the most informative set of the diagnosing attributes and establish a strict correlation between a process condition and results of indirect measurements. It is also shown that the networks containing nonlinear components can be used for approximation of nonlinear mapping with any desirable degree of accuracy; parameters of neural networks can be adjusted in real time considering the input/output data. Advantages of back-propagation neural networks at management of process systems of final operations in order to ensure high quality of providing products are given.

1 Introduction

In modern mechanical engineering, the considerable part of output product is the multiproduct and small-lot manufacture. High requirements are made to the final transactions influencing the quality of the products. It predetermines ensuring of high reliability and fast adaptation of machines to constantly changing modes of functioning and the nomenclature of the details.

The solution to the common problem, and its practical embodiment in the integrated automated (integrated programmed) production is possible only when all separate tasks are implemented.

Regardless of the presence or the absence of information preprocessing (selection of the main attributes, conversion to different form in the new parametric space, etc.), the signal of transformation of process system will represent the vector in any parametric space. Then this vector is compared to the vectors used at the training stage. The majority of neural network models uses described transaction.

2 Problem description

Input and current information about the state of process system and perturbations of the environment becomes the main factor for neural network (NN) management in order to develop the algorithm of training/management.

By manipulating the range of the input signal, eventually, it is possible to receive NN with optimum properties and parameters.

Despite the numerous works devoted to the theoretical modeling and pilot studies of machines, state of the technology equipment with alternative system positions cannot be completely predicted. Therefore, the quality of processed details can only be stated. Virtually all the serial equipment is provided only with local communications that causes low "controllability" and "observability" by output criteria. The main feedback on output parameters is implemented in rare cases [1].

Indirect assessment of technological process is extremely important as it is almost only one information base of the automated equipment adjusting in the course its operation when the impacts and parameters are constantly changing.

The reference to the system managed by an artificial intelligence is connected with "illegibility" of the considered objects. At the same time, the illegibility of images, representations and conceptions is introduced into formal models by different methods.

Selection from the signal of images familiar to the system and its subsequent representation by neurons is performed on the input-output level. Input vectors are fuzzy through both training and recognition, i.e. there is the small dispersion of the vectors belonging to one class. In this context, the neuronet performing this operation shall have the certain capability to statistical averaging [2, 3, 4].

3 Formal model of electroerosive manufacture

The generalizing capability of back-propagation network allows to obtain the correct output at management of process systems of metalworking even with incomplete or partly incorrect input vector. Also the network has the remarkable property to successfully process the noisy, distorted or partially damaged information [5 - 7].

Final operations of technological process almost completely determine the accuracy and quality of the surface of the produced products. In the single-piece and small-lot production, electron discharge machining

*Corresponding author: ovsynikova2012@yandex.ru

(EDM) is widely used as a final transaction at the production of such high-precision products as compression moulds and stamps, which are in turn define the quality of the details.

The key objective of increasing the efficiency and quality of electron discharge machining (EDM) is to find the optimum modes of the accompanying processes, determine the linkages between the input and output processing parameters in order to reduce the wear of the tool - electrode (ET).

Because of the small duration and high power of single discharge, the metal layer on the processed section melts, the major part of power is the share of destruction of the surface layer. Change of energy of single discharge is defined by the form of the impulse and current-voltage characteristic of the pulse generator.

Technological indicators of processing, namely, the roughness Rz, microns, productivity M, mm3/min., depend on the working pulse-repetition rate f, kHz, porosity q, current intensity I, A, voltage U, V. Height of microroughnesses of Rz, microns, shall not exceed 20 microns with the maximum performance considering thermal-physical properties of materials ET and the processed detail.

The temperature of the working environment in the processing zone influences the physical state of the interelectrode distance (ID). These thermal processes are connected with the performance of EDM, the quality of the processed surface and wear of ET. When the temperature of working environment increases, its heat capacity and viscosity, as well as its cooling capability and the possibility of capture and carrying out of the products of the erosion from ID decreases.

All above-mentioned factors indicate the possibility of management or improvement of performance and quality of the surfaces, processed by the electron discharge machining method due to increasing the zone of power costs on electrodischarge destruction by the method of the parameters of impulses optimization.

The management systems based on different methods of management which can be divided into three main groups are applied to obtain necessary parameters of process of EDM:

- feed control EI is carried out by means of the closed simple autonomous circuit of management;
- program control - provides installation of parameters of the processing or the choice of empirical or theoretical parameters of processing;
- adaptive control - carries out a continuous complex process evaluation of processing and installation of parameters during the process of EDM.

The different combinations of the described systems are used. The role of adaptive management system consists in the identification and elimination of all deviations of process from its normal course [8, 9].

One of the important indicators of EDM-process of EEO is process stability. Stability is considered as a capability to keep during the certain period of time the continuous process of metal removal and evacuation of products of the erosion, despite the emergence of the accidental or natural disturbance influences in the system.

For determination the desired combinations of the parameters providing the most productive mode, it is necessary to know the optimum value of the duration or porosity of the impulse with different pulse-repetition rate for certain data of working areas and currents. Thus, for each mode it is necessary to determine the optimum value of porosity [8].

The sequence of choosing the mode of electron discharge machining for copy-piercing machine is as follows:

- determination of model of the machine and pulse generator, determination of material and parameters of ET-quality, type of dielectric liquid and its filtration degree, polarity of electrodes connection;
- the form of digit pulse is adjusting by taking into account the material of the detail and the wearing depth of ET, the form and the required roughness of the processed surface;
- parameters of impulses are set;
- admission controller for receiving working impulses with the certain pulse-repetition rate frequency is configured;
- rate and injection pressure of working liquid are set; frequency of relaxation and vibration of electrodes is determined depending on the value of working current and the wearing depth of ET;
- electrode gaps are checked and changes are made to the tool sizes, processing time is defined.

System synergistic approach [7] is the most promising direction in the complex studying of electric discharge processes and the mechanism of formation of the surface layer of blank. According to the approach, process of electron discharge machining should be considered as a system of dissipative processes. The majority of technological processes of EDM cannot be presented in the form of simple systems, therefore, the approach based on the synthesis the component is applied for formalization of processes of the processing zone.

The main principle of synthetic information modeling is the "black box" principle, when external functioning of the system is modelled merely informational. Set of the dissipative processes occurring in the processing zone should be consider as the "black box"; estimated values of the dependent parameters of EDM, such as tool wear rate, slag, temperature in the processing zone, etc., are the output data.

Information modeling is applied to create the management model allowing to create the process environment and to virtually manage it, avoiding the influence of the arbitrary factors present at the practical experiment. Special "miscalibration" parameters are introduced into the model in order to calculate the probabilistic assessments. The main objective of information modeling is to ensure required output values of the EDM parameters; and the indirect problem is to minimize the influence of arbitrary factors.

In the literature [2-9] the proofs that there is the neural network (NN) for any algorithm are given. NN allows to model nonlinear processes, to work with noisy data and to adapt to any external influences.

The model of the management system is based on the fully connected neural networks, since the NN have the

DOI: 10.1051/matecconf/201712901025

biggest number of connectivity under the same number of neurons; that in turn increases the data capacity of the network. Fully connected network configuration is the most universal and does not require experiments with connection diagrams for each task. In case of network emulation on the average computer, fully connected network has significant features in the functioning speed, simplicity of implementation without affecting the quality of trainability.

A possibility of application of neural networks for solving the management problem is based on the fact that NN consisted of two layers and having a random large number of nodes in hidden layer can approximate any function of real numbers with the desired accuracy. Therefore NN with one hidden layer can be used to solve management tasks.

Adaptive control NN are designed so that the output signal of the managed model corresponds to the output signal of previously defined model, which has required characteristics. Structurally NN has at least two inputs, and one or several outputs. The output signal represents the weighted sum of input signals, and this amount depends on the values and relations between input signals. The following parameters are supplied into the system: porosity q, current intensity I, voltage U (which define the impulse form); consumption and power liquid circulation pressure; frequency of the relaxation and vibration of electrodes; interelectrode gaps and sizes of the tool; processing time. Output parameters are unloading V of metal and the received height Rz of microroughnesses.

The system should be asymptotically stable, the managed system tracks the reference model with the zero error; transition processes at the stage of adaptive control or the training management have the guaranteed limits.

Management based on the training principle means the gradual accumulation of experience of system behavior at a high degree of uncertainty of its initial states. Improvement of system functioning occurs by the results of system behavior. Nature of experience accumulation during the learning is manifold, for example, accumulated experience can be positive or negative, systematized or accidental, own or introduced, simulation (artificial) or natural, etc. However, all accumulation methods have a common feature, which is gradual separation of knowledge domain from all set of ignorance. Therefore, in the theory of the taught systems this feature is reflected in rather quickly developing direction connected with the development of automatic systems of classification or image identification.

Objective complexity of training, both with and without encouragement, is that the classes from the nearest points of the managed space are separable. In other words, intersections of classes when the same points belong to different classes are possible.

The main objective is to determine the most informative set of the diagnostic attributes and to establish the correlation between the state of forming process and results of indirect measurements.

The majority of the known models is based on the regression or correlation dependences obtained by statistical processing of experimental data.

At any recognition algorithm, there has always been the procedure of comparison of the input signal with the standards, which are available in memory.

It is necessary to clarify the parameters of network after defining its general structure. The parameters of the network are the amount of input and output neurons, the number of neurons of Kokhonen and Grossberg layers, and values of synaptic connections.

Back-propagation network (BP) works with the continuous or binary vectors consisting of zero and units. As a result of learning, input vectors are associated with the output data. When the network is trained, supplying the input images results in the output data. The correct output can be obtained when the input is incomplete or a partially incorrect. The hypersurface received as a result of the functioning of BP network, provides the possibility of forecasting because of the continuity principle. It is clear that the forecasting will be more precise inside the hyperparallelepiped of training images; an error will be greater at the solution of the extrapolation problem.

The BP network consists of two layers of neurons (fig. 1): Kokhonen's layer and Grossberg's layer. Neurons of Kokhonen's layer work by the principle "the winner takes it all" in the mode of functioning (recognition), defining the cluster to which the input image belongs. Then the output star of the Grossberg's layer reproduces the corresponding image on the outputs of the network by a signal of the winner neuron in Kokhonen's layer.

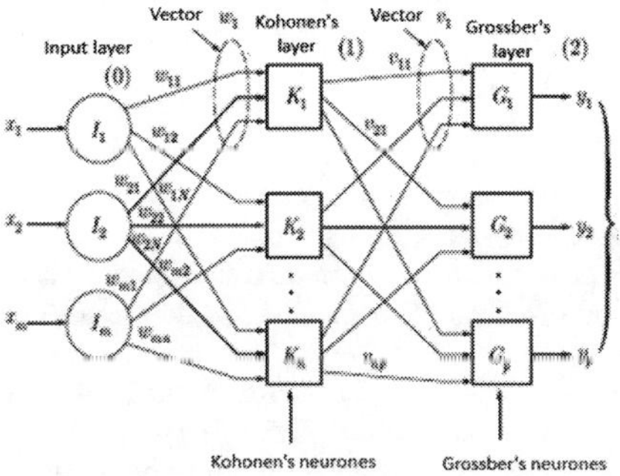

Fig. 1. Back-propagation network

Like other networks, the BP-network works in two modes: training and working. In the first case, weighted coefficients are adjusted with the vectors X and Y as input; in the second mode Y or X are the input of already trained network, and both X and Y are the output. In comparison with the different networks, back-propagation network is trained much quicker, but the accuracy of result is less.

Researches of the last years [5 - 11] showed that the parallel configuration of neuronets provides increased speed of calculations. The networks containing nonlinear components can be used for approximation of nonlinear images with any desirable accuracy degree.

Training by the algorithm of error back-propagation assumes direct and the return passes over all layers of network. At the forward pass the input vector is supplied on the input layer of neural network, then it extends from one layer to another. This resulted in generating of the set of output signals, which is the practical reaction of network to this input image.

During the forward pass all synaptic weights of network are fixed. During the backward pass all synaptic weights are adjusted according to the rule of error correction, namely: the actual output of network is subtracted from the desirable, therefore the signal of error is generated. Subsequently, this signal distributes in the direction, which is opposite to the direction of synaptic connections. Synaptic weights are adjusted to ensure the maximum approach of the output signal of network to the desirable [4, 11].

Presence of the layer of the matrix of coefficients allows the neuronet to weigh input signals, using the set of signal transmission routes by summing up them on with the its own weights. As a result, each combination of all input signals changing in some set ranges will correspond to the value of the output signal.

Nowadays, the number of different architecture of neural network management systems construction are well developed and widely used. Generation of adequate control signal for management of dynamics of the control object states is performed by neural net controller.

State transition is performed by the optimum trajectory; algorithm of error back-propagation, which is one of methods of training of the multilayer neural networks of direct distribution, also called multilayer perceptron, is used as the training algorithm. This iterative gradient algorithm is used in order to minimize the error of the multilayer perceptron work and to receive the acceptable output [8].

To date, large market of neural network products is formed. The modern level of development of microelectronics allows to produce the neurochips consisting of very large number of the simple elements capable to perform only arithmetic transactions. The vast majority of products is presented in the form of the modeling software. The leading companies develop specialized neurochips or neuroboards in the form of add-on devices to usual computers [4, 5, 9, 10].

The hardware neurosystems are rather expensive and primarily intended for professional use.

It is possible to use NeuroShell 2.0 for the research purposes. The advantage of this program is compatibility with the popular packet of data management Microsoft Excel that makes the product convenient for mass use. The nAPL programming language, problem-oriented for the neuronet, programming environment in C++ and the UNIX-compatible operating system allow to develop neural network systems of EDM management.

The main increment of productivity is provided due to application of parallel analog multiplication of input signals by weight coefficients.

The combination of concepts, methods and mathematic of the modern nonlinear management theory with the theory of the learning artificial neural networks gives great opportunities for structural synthesis of complex dynamic management systems of EDM, allowing to receive the effective combination of performance and parameters of quality of the processed surface.

4 Conclusions

1. Including the neural networks in the management system will allow to distinguish essentially new properties: the system will obtain the capability to understand and study the processes and operating conditions, accumulating the gained knowledge and to intentionally use it for improvement of quality characteristics.
2. The integration of modern information technologies into technological process allow to increase processing accuracy. New management systems including all complex of modern equipment guarantee repetition of the programmed geometry with the high precision rate.
3. Neural networks are designed so that the output signal of the managed technological system corresponds to the output signal of previously defined model, which has required quality characteristics.
4. In comparison with the different networks, back-propagation network is trained much quicker, but the accuracy of result is less. The capability of network allows to obtain the correct output at management of process systems of metalworking even with incomplete or partly incorrect input vector. In addition, the network has remarkable property to successfully process the noisy, distorted or partially damaged information.

References

1. B.M. Brzhozovskij, M.B. Brovkova, V.V.Martynov *Opredelenie parametrov kachestva izdelij pri mehanoobrabotke v real'nom vremeni na osnove nejrosetevyh algoritmov* // Nelinejnaja di¬namika i prikladnaja sinergetika: Materialy Mezhdunar. konf. - Komsomol'sk-na-Amure, pp.62-65. (2003)
2. L.N. Jasnickij *Vvedenie v iskusstvennyj intellekt : ucheb. posobie dlja stud. vyssh. ucheb. zavedenij L. N. Jasnickij.* -- 2-e izd., ispr. -- M. : Izdatel'skij centr «Akademija», p.176 (2008)
3. T.Kohonen *Self-organization and associative memory. Series in Information Sciences*, **vol. 8**. Berlin: Springer Verlag p.82 (1984)
4. Hajkin Sajmon. *Nejronnye seti. Polnyj kurs* 2-e izd., ispr.: Per. s angl. – M.: OOO «I. D. Vil'jams», p.1104 (2006)
5. R. Dorf, R. Bishop *Sovremennye sistemy upravlenija* /Per. s angl. B. I. Kopylova. – M.: Laboratorija bazovyh znanij, p.832 (2009)
6. A.Galushkin *Nejronnye seti. Osnovy teorii* – M.: Gorjachaja Linija - Telekom, p.496 (2012)
7. I.V. Ovsyanikova, *Upravlenie parametrami processa formirovanija slozhnyh poverhnostej.* /I.V. Ovsyanikova, A.S. Tarapanov, V.S. Shorkin «Fundamental'nye i prikladnye problemy tehniki i tehnologii» **№2** , Gosuniversitet – UNPK, Orel, p.56-60 (2012)
8. I.V.Ovsyanikova *Informacionnoe modelirovanie adaptivnogo upravlenija jelektrojerozionnoj obrabotkoj* // I.V. Ovsyanikova, E.A. Sychev, A.S.

Tarapanov «Fundamental'nye i prikladnye problemy tehniki i tehnologii» **№5**, Gosuniversitet – UNPK, Orel, p.103-107 (2013)

9. Stuart J. Russell & Peter *Norvig Artificial Intelligence: A Modern Approach* (2nd ed.), Upper Saddle River, New Jersey: Prentice Hall p.460 (2003)
10. Ryshard Tadeusevich, Barbara Borovik, Tomash Gonchazh, Bartosh Lepper Jelementarnoe vvedenie v tehnologiju nejronnyh setej s primerami programm / Perevod I. D. Rudinskogo. — M.: Gorjachaja linija - Telekom, 408 p. (2011)
11. Rutkovskij Leshek *Metody i tehnologii iskusstvennogo intellekta* Per. s pol'sk. I. D. Rudinskogo Izdatel'stvo: Gorjachaja linija – Telekom, 520p. (2010)

Scanning the welded seams of titanium alloys by using subminiature eddy current transducers

Sergey Dmitriev[1], *Vladimir* Malikov[1,*], and *Anatoly* Sagalakov[1]

[1]Altai State University, Faculty physics and technology, 65604 Barnaul, Russia

Abstract. The eddy current transducer (ECT) of the transformer type is used to construct a sensor for investigating titanium sheets joined by a welded joint. The characteristics of the ECT are presented. The measurement method for controlling the occurrence of defects in the welded joints in titanium alloys is described. The experimental results obtained by the ECT on two welded titanium sheets are presented. The depth of penetration of the field of eddy currents into the investigated object is determined and the dependences describing the response of the ECT at different depths of the defect are outlined. The relationships can be used for evaluating the quality of welded joints and determining the reliability of welding.

1 Introduction

It is well-known that many inspection methods describe the violation of the suitability of an object for further application. Therefore, they can be used only for checking individual items in a production batch [1]. Alternative methods of non-destructive eddy current inspection can be used for testing all components so that the quality can be determined accurately.

The problem of evaluating the quality of welding and welded joints in different metals has been discussed many times in different scientific journals. For example, in [2], experiments were carried out to evaluate the relative strength of a welded joint and a parent metal. In [3], the authors propose a new method of evaluating the parameters of the quality of welded joints based on the effect of a high-speed waterjet on the surface of the inspected object.

An urgent problem of non-destructive inspection in engineering is the determination of the quality of welded joints in titanium alloys. Titanium and its joints are used in large volumes in aircraft construction, shipbuilding, chemical engineering and important machine components. Commercial titanium is used for the manufacture of components designed for operation in corrosive media. However, low-quality welded joints in titanium may result in the fracture of components produced from this material. The application of eddy current flaw detection for evaluating the quality of welded joints in titanium alloys is an important task requiring urgent solution. Since the eddy current inspection method is not sensitive to non-conducting paint layers, it can be used for the diagnostics of components with lacquer and varnish coatings. The currently available eddy current transducers (ECTs) can be used to determine the quality of welded joints with a high degree of accuracy [4]. The experiments carried out in this work show that the method can be used for evaluating the homogeneity of material of a welded joint.

One of the well-known shortcomings of eddy current flaw inspection is the inspection of the quality of thin surface layers of electrically conducting materials. The design of the ECT, the small dimensions and the special form of the core allow defects to be found at depths of up to 5 mm.

2 Description of the system

We have developed a superminiature eddy-current transformer for local monitoring of the physical parameters in titanium-alloy plates and weld seams [5]. In contrast to existing sensors, it permits local measurements with sections measuring a few microns, to depths of the order of 5 mm. The electrical conductivity of the alloy is directly measurable, and its distribution over the sample surface and thickness may be readily established. The eddy-current method is based on the dependence of the current magnitude and distribution on the geometric and electromagnetic parameters of the sample and on the mutual position of the sensor and the sample. The basic informative parameter β_0 is a generalized characteristic of the object, the eddy-current sensor, and the frequency of the electromagnetic field [5]. We need to develop an adequate model of the response of a eddy-current transformers of plate type, which are sensitive to many variables and permit reproduction of the voltage hodograph at small values of β_0. We have plotted hodographs illustrating the influence of various parameters on the induced voltage on the basis of the proposed model [5]. The eddy-current transformer is connected to the sound board of a

* Corresponding author: osys11@gmail.com

DOI: 10.1051/matecconf/201712901026

computer with special software that controls the voltage at the transformers exciting winding and also reads the voltage at the measuring winding (in arbitrary units). On the basis of preliminary calibration, these readings are converted to values of the electrical conductivity. The exciting winding (diameter D_1 = 0.12–0.13 mm) of the superminiature eddy-current transformer consists of ten turns of copper wire (cross-sectional area 5 μm^2). The measuring winding (diameter 0.05–0.08 mm) consists of 130 turns of copper wire (cross-sectional area 20 μm^2). To minimize the influence of the exciting winding on the final signal, the circuit includes a compensation winding that consists of 20 turns of copper wire (crosssectional area 5 μm^2), connected to the measuring winding in such a way that the voltage of the exciting winding is subtracted from the result. The windings are wrapped around a core of 2000 NMZ ferrite (relative magnetic permeability μ_{max} = 500 or else of 81NMA alloy annealed by a special method (if greater localization of the magnetic field is required). The core consists of a tetrahedral pyramid (height 1 mm),with a square base (sides of 0.2 mm). The measuring winding rests on the points of the pyramid, which improves the localization of the magnetic field.

The eddy-current transducer (Figure 1) is a transformer with measuring (1), exciting (2), and compensation (3) windings and a magnetic circuit 4, which is located inside the cylindrical platform 5 with tracks that are cut on the external side for windings. The platform is impregnated with a compound 6 at a temperature of 200°C to prevent the disintegration of the windings when the ferrite screen 7, which is intended for the localization of the electromagnetic field on the tested object, is put in place. From the outside the transducer is contained in a corundum washer 8, which protects the core 4 from contacting the tested object.

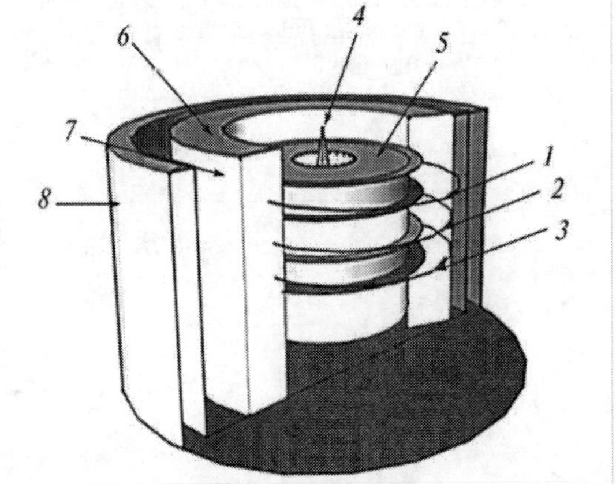

Fig. 1. Scheme of eddy current transducer

Such transformers permit effective localization of the magnetic field, so that defects as small as 250 μm may be detected. In addition, the magnetic field penetrates into the sample to a considerable depth when working at relatively low frequencies. The corresponding software is written in the language C++ for the Windows operating system. By means of the Windows mixer subsystem, the software controls the voltage applied to the exciting winding, thereby specifying the amplitude and frequency of the sinusoidal digital signal from the virtual generator.

3 Operation of the measuring system

The digital signal from the virtual generator is sent to the input of the digital–analog converter of the sound card. The analog signal then passes through a power amplifier to the exciting winding of the eddy current transformer. The sinusoidal signal creates an electromagnetic field, which, on interacting with the sample, induces an emf in the measuring winding of the eddy-current transformer. This voltage is sent through a preamp to the microphone input of the sound card and then to the input of the analog–digital converter in the sound card. The resulting digital signal is sent to the analysis and control module of the software. This module determines the magnitude of the digital signal (in conventional units) corresponding to the voltage U_m at the measuring winding.

The computer's sound card permits variation in frequency of the electromagnetic field created by the exciting winding within the range 100–2000 Hz in the course of scanning.

4 Results and discussion

A set of measurements was performed on VT1-0 technical titanium plates that were connected by welded seams to demonstrate the appropriateness of the proposed unit for the determination of the quality of welded seams. The plate thickness was 5 mm and the width of the welded seam was 5 mm. Before the measurements were started, the transducer was calibrated. The calibration consisted of the determination of the induced voltage from the flawfree sector. The measured characteristic was the voltage that is induced by the field of eddy currents that arise in the tested object. In this experiment a sector for calibration was selected on an intentionally flawfree plate that was manufactured from an identical titanium plate. The calibration was performed at different frequencies. The frequency was varied from 500 to 2000 Hz with a 100Hz step. The subsequent scanning was carried out by displacing the transducer along or across the welded seam or across the flaw area. During the experiments it was determined that the optimal frequency range of the electromagnetic field of the exciting winding for studying titanium is approximately 1500 Hz.

The scanning lengthwise of the weld seam surface was conducted in order to determine the weld seam homogeneity. Herewith, it should be noted that any significant change in the signal amplitude wasn't pointed out. The results of the experiment are presented in Fig. 2.

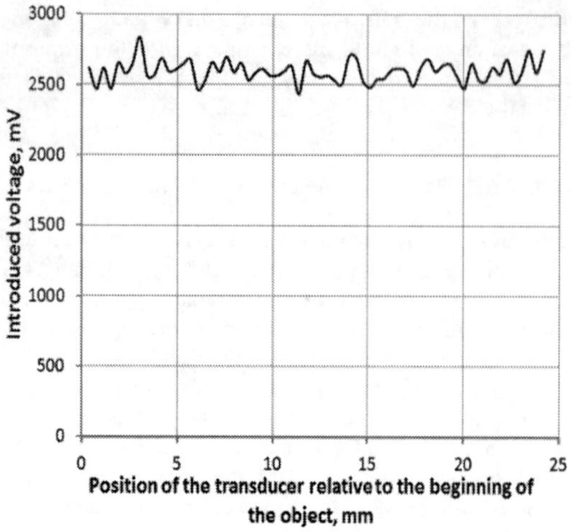

Fig.2. The response value during the scanning lengthwise of the welded seam.

The results of the experiment say for the relatively homogeneous structure of the weld seam but don't provide us with any information regarding the quality of the seam itself. Judging by the results of the given experiment, it could be concluded about the equal distribution of the weld seam defects or its zero-defect quality only.

The scanning was conducted crosswise of the weld seam in the next experiment. The length of the weld seam was 150 mm. The seam was divided into 30 sections, 5 mm each, so that the signal from the seam itself as well as the one coming directly from the slabs was withdrawn. The dependences received were averaged. The results of the experiment are presented in Fig. 3,4.

In the specimen given, the influence of a deficient weld seam on the inserted voltage is clearly traced by a significant drop in the signal amplitude in the weld seam area in comparison with the slabs area.

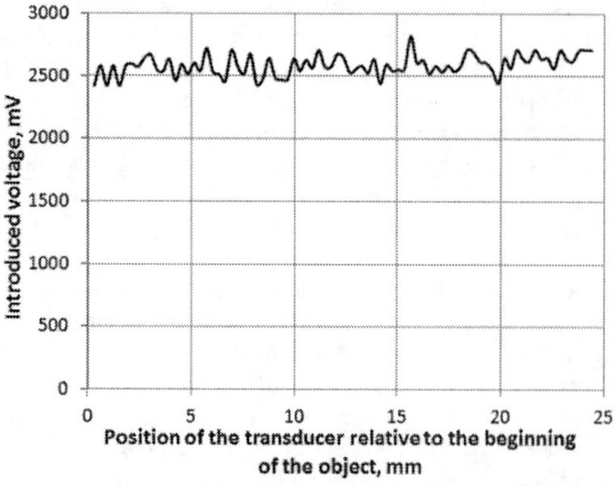

Fig. 3. The response value during the scanning lengthwise of the welded seam Specimen №1

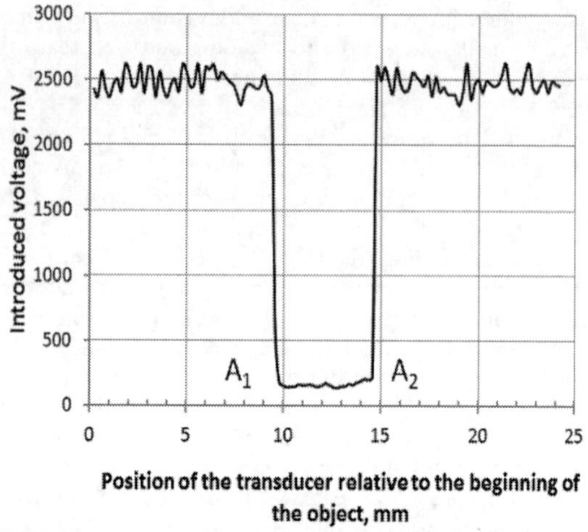

Fig. 4. The response value during the scanning of the welded seam: (a) sample no. 1, A1–A2 are the boundaries of the welded seam, and (b) sample no. 2.

The conclusion about the weld seam of Specimen 1 low quality was made based on the experiment conducted. The low quality of the welding was confirmed directly by the weld seam cut as well. Specimen 2 scanning showed the absence of the signal amplitude excursion within the frame of the weld seam. The weld seam of Specimen 2 cut showed the high quality of welding.

Additional experiments were carried out in simulation of a similar amplitude drop. In these experiments, two titanium sheets of the same thickness were placed in tight contact and subsequently the interface was scanned with a scanning frequency of 1600 Hz. The results of the experiment are presented in Fig. 5. These experiments yielded dependence identical to that shown in Fig. 4 in the region of the welded joint in the specimen. The signal amplitude in the region of the interface decreased by more than an order of magnitude in comparison with the signal amplitude from the sheets.

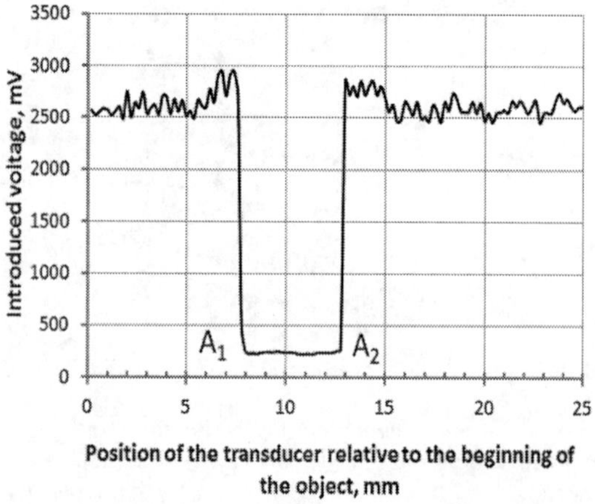

Fig. 5. The response value during the scanning in the area of two slabs junction. A1 – A2 correspond to the edges of the area where the junction presence influences the inserted voltage.

DOI: 10.1051/matecconf/201712901026

4 Conclusions

The results of the experiments conducted showed a high efficiency of the developed method in the search for defects of the titanium slabs weld seams and in the evaluation of the welding quality. Thanks to the change of the eddy-current transducer signal amplitude in the area of the weld seam, we succeeded in defining the edges of the weld seam exactly, its low quality being obvious due to a steep drop in the signal amplitude. Likewise, the dependence, received as a result of Specimen 2, characterized by a high welding quality, scanning, showed the absence of any significant signal amplitude changes.

So, the analysis of the eddy-current transducer response can be used for the evaluation of the performed welding quality.

References

1. S. Dmitriev, A Sagalakov, Polz. Vest. **2**, 45 (2010)
2. V. Bulychev, R. Latypov, Svar. Proiz. **6**, 120 (2012)
3. M. Abashin, Svar. Proiz. **9**, 57 (2014)
4. V. Polyakov, V. Malikov, A. Ishlov, S. Dmitriev, Adv Mater Res. **880**, 105 (2014)
5. S. Dmitriev, L. Dmitrieva, V. Malikov, A. Sagalakov, IOP Conf. S.: Mat. Sci. Eng. **116**, 1 (2015)

DOI: 10.1051/matecconf/201712901027

Quantitative evaluation of the effectiveness of best available technologies of form-shaping

Leonid Shvartsburg[1,*], *Elena* Butrimova[1], and *Olga* Yagolnitser[1]

[1] MSTU "STANKIN", department "Engineering Ecology and Life safety", 127055 Moscow Vadkovsky lane, 3a, Russia

Abstract. The article examines the issues of creation of the best available technologies of form-shaping and the methodology of quantitative evaluation of their effectiveness. It is shown that creation of a unified methodology for generation of best available technologies of form-shaping and quantitative evaluation of the effectiveness of these technologies is possible on the basis of energy-related representation of technological processes. In this case, the technological process is represented as two processes – the process of converting electrical energy into mechanical energy and the process of transfer of mechanical energy to the processing zone. This approach allows to create unified methodologies regardless of the wide variety of technological processes of form-shaping, conditions and parameters of their implementation. It is shown that for each particular implementation of technological processes of form-shaping, energy losses during conversion and transfer of energy and increased consumption of current due to "underload" of process equipment determine the environment friendliness of each specific implementation of the technological process. The main directions for improvement of efficiency of the best available technologies of form-shaping are indicated.

1 Problem definition

Technological processes of form-shaping are among the most important types of human economic activities; their implementation in a large measure determines the economic potential of any country, defines the competitiveness of products. For this reason, major attention is paid to improving the quality of these technological processes. This also applies to implementation of advanced technologies of form-shaping and information technologies as well as to improvement of automation facilities, improvement of tools quality and other innovations [1-9].

The most important quality indices of technological processes are the indices that characterize their impact on the environment and on humans manifested through consumption and wastes (fig.1).

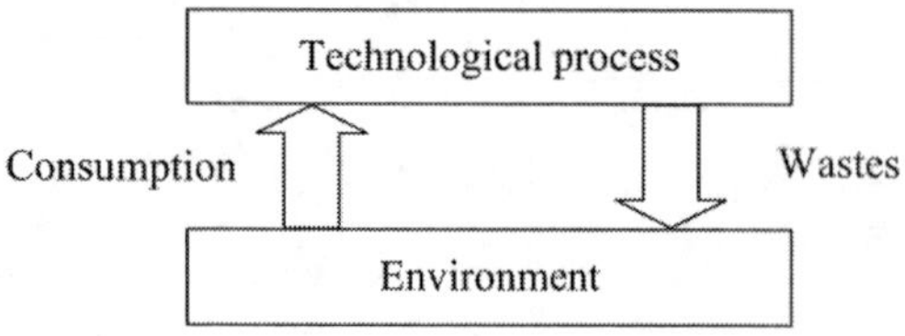

Fig. 1. Impact of technological processes on the environment and humans.

It should be noted that the products also are a waste, merely delayed in time. Therefore, environmentally invariant technological processes of form-shaping do not exist.

It should also be noted that technological processes of form-shaping are characterized by large diversity of types and kinds (milling, grinding, etc.), by technical characteristics of equipment on which they are implemented, by conditions of their implementation. Besides, their implementation requires extra-wide range of regulation, reaching values of 10,000:1 and higher, and ultra-small error.

For this reason, the problem of reducing the impact of form-shaping technological processes on the environment and humans in course of specific implementation of these processes is a major problem facing Russia's machine-building industry; solving this problem is aimed at improving the quality of these processes as well as at increasing their competitiveness.

2 Analysis of recent achievements in the study area

An important factor in reducing the impact of technological processes of form-shaping on the environment and humans is the creation of technologies of form-shaping as best available technologies (BAT).

BAT is a technology for manufacture of products (goods), performance of works, rendering of services determined on the basis of achievements of science and engineering and the best combination of criteria for

* Corresponding author: lesh@stankin.ru

achieving the aims of environment protection, on condition of availability of technical feasibility of its application.

These criteria include:

- the lowest level of negative impact on the environment per unit of time or per volume of manufactured products;
- economic efficiency of its implementation and operation;
- application of resource-saving and energy-saving methods;
- duration of implementation;
- industrial implementation at 2 or more facilities.

Currently the regulatory framework in Russia in the field of BAT implementation includes Federal laws, resolutions and orders of the Government of the Russian Federation, orders of the Ministry of industry and trade, Ministry of natural resources and environment and Rosstandart issued during the period 2014-2016. Rosstandart is the Federal executive body authorized by the Government of the Russian Federation dated 23.12.2014, No. 1458 for attesting a technology as best available technology.

As noted at the International forum of technological development TECHNOPROM 2014, implementation of BAT in various branches of economic activity is comprehensive and includes three main components - industrial policy, ecological policy and improvement of the system of state regulation based on BAT.

Form-shaping technologies do not relate to technologies that require urgent transfer to BAT, as opposed, for example, to power-generation technologies or minerals extraction technologies. However, widespread character of these technologies and concentration in industrialized regions predetermine the necessity for proactive steps towards their transition to BAT.

3 Purpose and objectives of the work

The purpose of this work is the creation of a unified methodology for generation of form-shaping BAT and quantitative evaluation of the effectiveness of these technologies for each specific implementation of technological processes, taking into account the state of technological equipment, tools and conditions for the implementation of these processes.

The objective of this work is the substantiation of the interrelation of the energy-related representation of technological processes of form-shaping with their negative impact on the environment and humans.

4 Main part

Creation of a unified methodology for generation of form-shaping BAT and quantitative evaluation of the effectiveness of these technologies across wide variety of technological processes and conditions of their implementation can be provided through the energy-related representation of these processes [10].

In this case, the technological process of form-shaping may be represented as two processes - the process of converting electrical energy into mechanical energy and the process of transfer of mechanical energy to the processing zone.

The first process is performed by electric motors (AC or DC motors), and the second process – by the kinematics of the technological equipment.

As any physical processes, the processes of energy conversion and transmission involve losses that are characterized by the losses (ΔP_{conv} and ΔP_{trans}) of active component of consumed power. These losses form the negative impact of technological processes on the environment and on humans [11].

Indeed, for the formation of negative impact on the environment and on humans, energy is needed – electromagnetic energy (electromagnetic pollution), heat energy (heat wastes), mechanical energy (vibration wastes, solid wastes), etc. Therefore, the energy of the negative impact on the environment and on humans is determined by the power losses during the implementation of these technological processes, because the energy source is only the energy supplied to the technological equipment.

For this reason, the active component of power consumed from the source exceeds the cutting power calculated by the technologist and applied to the motor shaft, and this excess, which is caused by losses, determines the negative impact of technological processes of form-shaping on the environment and on humans.

It should also be noted that when induction motors are used in technological equipment as converters of electrical energy into mechanical energy, it is necessary to take measures for providing the maximum value of power factor (cosφ). Low value of this coefficient is explained by the fact that, in the vast majority of cases, the equipment is operating at underload, i.e. at conditions when the cutting power reduced to the motor shaft is substantially smaller than the rated capacity of the motor (P_n). This leads to emergence of substantial reactive component of the consumed power, and hence to increase of the value of total consumed power, what generates additional negative load on the environment [12].

Thus, these two factors - energy losses and low power factor - determine the negative impact of technological processes of form-shaping on the environment and humans during their implementation, and also the excessive consumption of energy (physically this means that the technological process is implemented at excessive values of consumption current).

In addition it should be noted that reduction of consumption of energy determined by S value has a considerable environmental impact in itself, because electricity production is one of the types of economic activities that have the greatest negative impact on the environment.

To achieve the goals of environmental protection, what is required for creation of the best available technologies, it is necessary to fulfill two conditions.

Firstly, it is necessary to reduce losses of active power during implementation of the technological process of form-shaping, and secondly, to improve the power factor.

So, the generation of form-shaping BAT is resolved into minimizing the full power (S) consumed during implementation of these technologies by reducing the cutting power (P_{cut}) applied to the motor shaft through efficiency of the machine (η) and energy losses (ΔP_{conv}, ΔP_{trans}), and by improving the power factor ($\cos\varphi$) (1).

$$S = \frac{(P_{cut}/\eta) + \Delta P_{conv} + \Delta P_{trans}}{\cos\varphi} \qquad (1)$$

To a considerable extent this corresponds to BAT for ensuring energy efficiency presented in the European Climate Change Programme.

Based on the stated above, the diagram of generation of form-shaping BAT may be represented as shown in fig. 2.

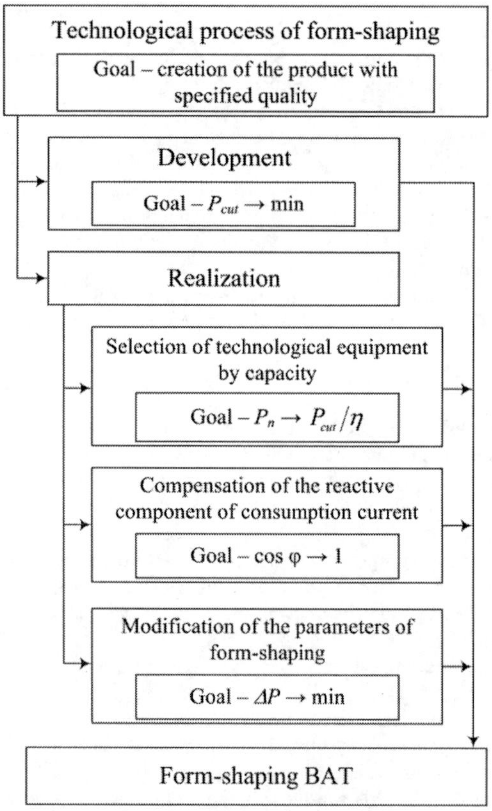

Fig. 3. Diagram of generation of form-shaping BAT.

All stated above applies equally to technological processes implemented on general-purpose equipment and CNC equipment, additive and digital technologies, nanotechnologies, to formation of lean production.

It should be added that after solving the technical issues of generation of form-shaping BAT, for inclusion of these technologies in BAT reference book it is necessary to carry out several organizational and organizationally-technical measures established by regulatory documents in this area of activities.

Hence, for generation of form-shaping BAT, the following set of measures should be **recommended**.

1. During development of the technological process, basing on state-of-the-art technologies, innovations, resource-saving and energy-saving, tool management and metrological support, in order to achieve BAT goal, the process engineer creates the technological process providing minimal cutting power of machining steps.

2. In course of implementation of technological processes of form-shaping, it is necessary to select the equipment by capacity of electric motors, achieving the value of cutting power reduced to the motor shaft as close as possible to motor rated power (to avoid « underload » of the machine as much as possible). Besides, operation of motor at idle should be minimized.

3. In case of « underload » of the machine during implementation of technological processes, it is necessary to install capacitors to compensate the reactive component of the consumption current, or install special energy-efficient motors in the technological equipment.

4. Reduction of energy losses during energy conversion and transmission to the processing zone is achieved by changing the parameters of the technological process of form-shaping. However, it may cause change of static and dynamic characteristics, stability of the system. All this may result in change of quality indices of the technological processes; if the quality indices decrease, the optimization problem must be solved.

Using this approach to creation of best available technologies of form-shaping, the quantitative evaluation of their effectiveness may be performed basing on the ratio of power required for cutting process and calculated by the technologist (P_{cut}), and consumed power measured (S) for the specific implementation of the technological process of form-shaping. This ratio should be called the integrated environmental index of quality of technological processes I, because it takes into account the integral environmental impact of these processes due to energy losses (ΔP_{conv} and ΔP_{trans}) and machine « underload » (cos φ is small) (2).

$$I = \frac{P_{cut}/\eta}{S} = \frac{(P_{cut}/\eta)\cdot\cos\varphi}{(P_{cut}/\eta) + \Delta P_{conv} + \Delta P_{trans}} \qquad (2)$$

5 Conclusions

1. During generation of a form-shaping BAT, it is necessary to use energy analysis of technological processes, allowing the creation a unified methodology for the quantitative evaluation of the effectiveness of these technologies through comparison of cutting power with full consumed power.

2. Energy losses during implementation of technological processes and excessive consumption of current determine the aggregate effect of the specific implementation of the technological process on the environment and humans.

3. This approach enables the quantitative assessment of effectiveness of general-purpose technologies as well as of digital and additive technologies, and facilitates the transition to lean machine-building production.

All of the above applies equally to the technological processes implemented on general-purpose equipment and CNC equipment, to additive and digital technologies, nanotechnologies, to formation of lean production.

Therefore, the introduction of best available technologies of form-shaping is one of the most important basic directions for increase of competitiveness of machine-building, because it is aimed at increasing of resource intensity, reducing of energy consumption of technological processes, at preventing adverse impacts on the environment and on humans, taking into account the particular features of the specific branch of industry.

References

1. S.B. Egorov,, A.V. Kapitanov,, V.G. Mitrofanov, L.E. Shvartsburg, N.A. Ivanova, S.A. Ryabov, *Modern digital manufacturing technical support centers,* Mathematics Education Open Access, **11, 7,** 2213-2225 (2016)
2. A.V. Kapitanov, *Special Characteristics of the Multi-product Manufacturing,* Procedia Engineering **150,** 832-836 (2016)
3. L. Shvartsburg, N. Ivanova, S. Ryabov, T. Zaborowski, *Chemical contaminations in a process of polishing with an implementation of liquid LCTS,* Life science journal **11(10s),** 228-230 (2014)
4. L. E. Shvartsburg, E.V. Butrimova, N.V. Drozdova, *Possibilities and adaptation of software product MS Visio for visualization of the ecological information,* Practical and Educational-Methodical JOURNAL LIFE SAFETY **10,** 35-37 (2011)
5. E.V. Butrimova, N.V. Drozdova, *CALS-technologies in solving the problems of monitoring of energy pollutions,*VESTNIK MSTU «STANKIN» **3,** 43-45 (2008)
6. M. Yu. Khudoshina, O.V. Butrimova, *Development methodology of information retrieval system for environmentally sound selection of lubricant cooling technological fluids in mechanical engineering,* Technology of oil and gas **1,** 54-59 (2014)
7. M. Yu. Khudoshina, O.V. Butrimova, *Stage of conceptual design of the database of coolant-cutting technological means, systems for their application and disposal,* VESTNIK MSTU «STANKIN» **1,** 150-154 (2010)
8. S.N. Grigoriev, M.P. Kozochkin, F.S. Sabirov, A.A. Kutin, *Diagnostic systems as basis for technological improvement,* Procedia CIRP. **1,** 599-604 (2012)
9. V.A. Grechishnikov, P.V. Domnin, V.A. Kosarev, Yu.E. Petukhov, V.B. Romanov, B.E. Sedov, *Shaping by means of complex cutting tools,* Russian Engineering Research **34 , 7,** 461-465 (2014)
10. S.B. Egorov, A.V. Kapitanov,, V.G. Mitrofanov, L.E. Shvartsburg, N.A. Ivanova, S.A. Ryabov, *Formation of the integral ecological quality index of the technological processes in machine building based on their energy efficiency,* International Journal of Environmental and Science Education **11, 11,** 4065-4078 (2016)
11. L. E. Shvartsburg, E.V. Butrimova, N.V. Drozdova, *Experimental research of distribution vibroacoustics factors in the environment for forecasting of their levels in the certain point of space,* Practical and Educational-Methodical JOURNAL LIFE SAFETY **2,** 27-30 (2012)
12. S. I. Gvozdkova, L. E. Shvartsburg, *Minimization of energy losses by increasing the power factor,* VESTNIK MSTU «STANKIN» **2,** 32-36 (2012)

Engineering support of production of high-precision component assemblies

Alexandr Nazaryev [1,*] and *Petr* Bochkarev [1]

[1]Yuri Gagarin State Technical University of Saratov, 410054 Saratov 77 Politechnicheskaya street, Russia

Abstract. In the present paper the discipline of the holistic approach (complex of design procedures) which provide the effectiveness of the realization of the assembly operations is considered. The effective realization of the assembly operations is based on the links between the process design of the manufacturing of the high-precision products and the process design of the assembly production of the high-precision products.

Today there is an objective of the engineering output of the high-precision products within a short time and by the short runs before the manufacturing. The required precision is achieved with the traditional methods of manufacturing. They are: the complication of the processing equipment and the applying of the information technologies and measurement technologies [1]. By now, however, the functional precision of the high-precision products (the devices and the machines) has surpassed the precision of the machining equipment and the assembly equipment almost by a decade. This leads to the problem of the achieving of the specified quality and accuracy of the high-precision products during their manufacture [2]. In general, it is solved by the expensive methods, such as the numerous overhauling, reprocessing and amendment of the construction. Moreover, especially at the assembly production stage, the influence of the unknown and out-of-control reasons creates the considerable difficulties and causes the unpredictable variation of the specified operational factors [2-4].

The existing methods of the achievement of the assembling precision (fig. 1) are not multipurpose and they depend on the nature of the manufacture and the design features of the products. For instance, the selective assembly method provides the required quality of the assembly of the products from the elements with the precision of the manufacturing which is considerably lower than is needed. It provides a low costs of the manufacture, but it is effective only within the high-volume manufacturing or the mass serial production. Beyond that generally a capability to use the methods such as the selective assembly, the specific adaptation and the adjustment is not forecasted in frames of the assembly of the high-precision products. Appreciably more often the manufacturing manners to achieve the required precision (also called honing processes) are used, what influences negatively on the continuance and the value of the assembly cycle.

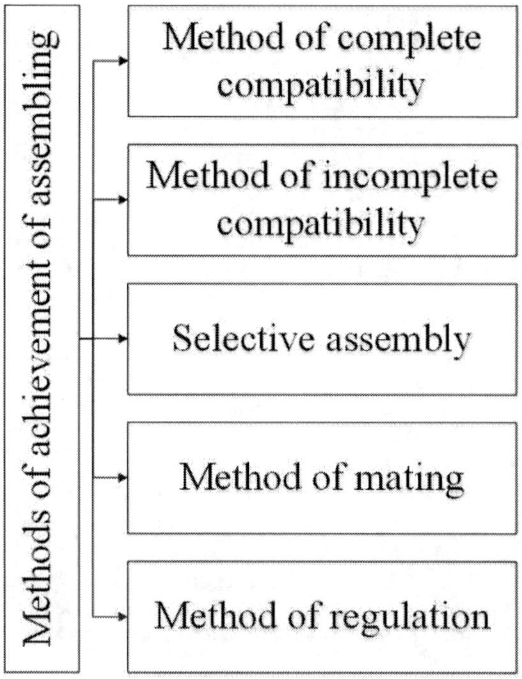

Fig. 1. Methods of the achievement of the assembling

The short terms and the high quality of the process design are the basic requirements of the effective work of the machine-building enterprises. For this very reason it is necessary to conduct researches for the development of the holistic approach to the solving of the machine-building issues. Specially the issues, which are applicable to the assembly of the high-precision products [3-4]. Now there are just the segmental solutions for some specific tasks of these issues, such as the upgrading of the quality and the precision of the production, the reduction of the material costs, the adoption of the

* Corresponding author: alex121989@mail.ru

resource-saving technologies and so on. But the unifying vision of the system meeting the real requirements does not exist [2].

There is provided an approach (complex of design procedures) established by the linkage between the process design of the manufacturing and the process design of the assembly production meeting the built-in requirements. This linkage is shown at the fig. 2 below. Such approach allows to choose the optimal manufacturing methods of the processing of the units meeting the requirements of the following assembling depending on the real manufacturing situation. In turn, this makes possible to reduce labour inputs, production time and working cost, to improve quality and precision of the high precision products. And to reduce labour inputs and production time during the process design.

The scheme of the complex includes the following procedures:

- Assemblage of the high precision products requirements analysis;
- Analysis of the probable manufacturing methods of the processing of items of the high precision products;
- Choice of the efficient manufacturing methods on the ground of analysis.

The scheme of the complex in general terms is given at the fig. 3.

The front-end analysis and the design of the efficient manufacturing methods within the system of automated sequencing of manufacturing methods (SASMM) [5-7] include a number of stages.

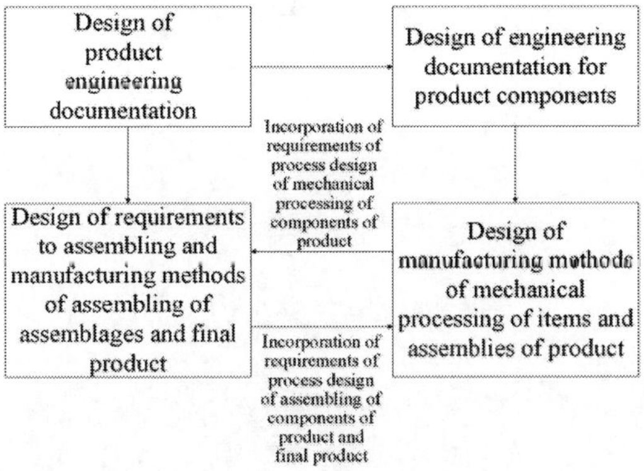

Fig. 2. Linkage between process design of manufacturing and process design of assembly production meeting built-in requirements

The dimensional design analysis of the assembling product and the items, used during the assemblage, and dimensional manufacturing analysis of the variety of the presumable manufacturing methods designed in SASMM are held at the initial stages. The source data for the dimensional design analysis incudes:

- The assembling product engineering documentation;
- The engineering documentation for the items, used during the assemblage.

On the ground of the source data the dimensional design analysis of the assembling product and the items, used during the assemblage is conducted. In the context of this analysis the critical requests of assemblage and items could be established.

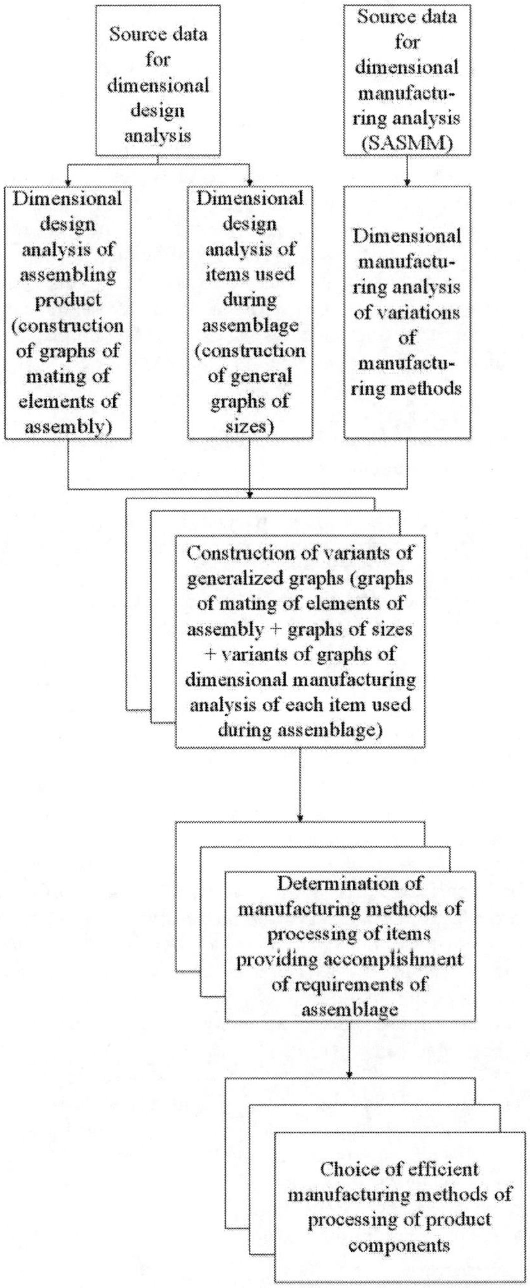

Fig. 3. Scheme of complex of design procedures

A graph with nodes showing elementary surfaces and edges showing their dimensional assignments represents the dimensional assignments of the mechanical-engineering items (assembling products). The graphs of the mating of elements of assembly and the graphs of the sizes are built up in consequence of the dimensional design analysis. The mathematical model of representation and analysis of items and assemblage was

DOI: 10.1051/matecconf/201712901028

used for automatisation of the stage. The graphs of the mating of elements of assembly are the source data for this model [8].

Today the design of manufacturing methods is impossible without technologists. This leads to human approach to making decisions at the process design stage following by reduction of effectiveness of the manufacturing systems during the realization of manufacturing methods. Beyond that the design of manufacturing methods with manual methods in the context of multiproduct manufacture renders an analysis of the limited number of variants at the separate designing stages possible. The loss of quality of process engineering solutions and extension of terms are consequence of it. The system of automated sequencing of manufacturing methods is one of the approaches which make it possible to make process design automatic. This system provides the concurrent design of manufacturing methods for the target set of items during the period under consideration with account for an objective manufacturing situation [5-7].

The source data for dimensional manufacturing analysis is set of options of manufacturing methods of production of items used during the assemblage which was received in SASMM.

The dimensional manufacturing analysis consists of construction of graphs of possible manufacturing methods of production of items. The graph of possible manufacturing methods of production of items represents the combining of two trebles, which are called "derived" and "original". The derived treble represents the machining dimensions. The original treble represents the engineering dimensions and dimensions of allowances. This way, graph of the manufacturing method allows to represent the geometric structure of manufacturing method of the processing in coded format. It also could be considered as a mathematical model of the manufacturing method of the processing. Using of such mathematical model allows to choose the optimal manufacturing methods of the processing of the units meeting the requirements of precision of manufacturing and the requirements of assembly of the product in the highest degree.

The item, called "Bush" (fig. 4), and the designing of precision involving (two modifications of the manufacturing methods) the use of the theory of graphs are shown at the fig. 5 ("a" and "b"). Where "A" means the engineering dimensions, "T" means the machining dimensions, "S" means the ingoing size and "Z" means the allowances for machining.

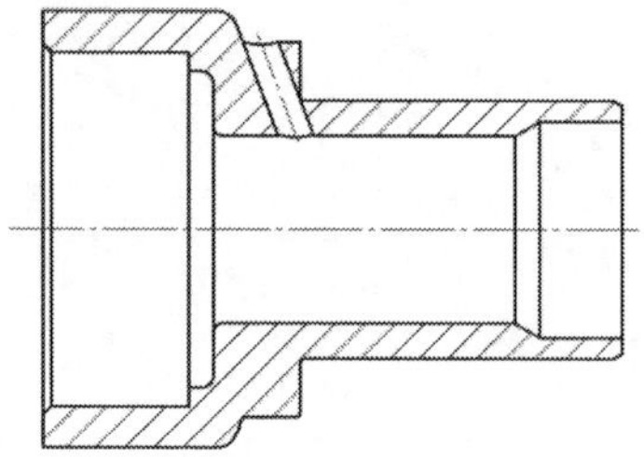

Fig. 4. Item "Bush"

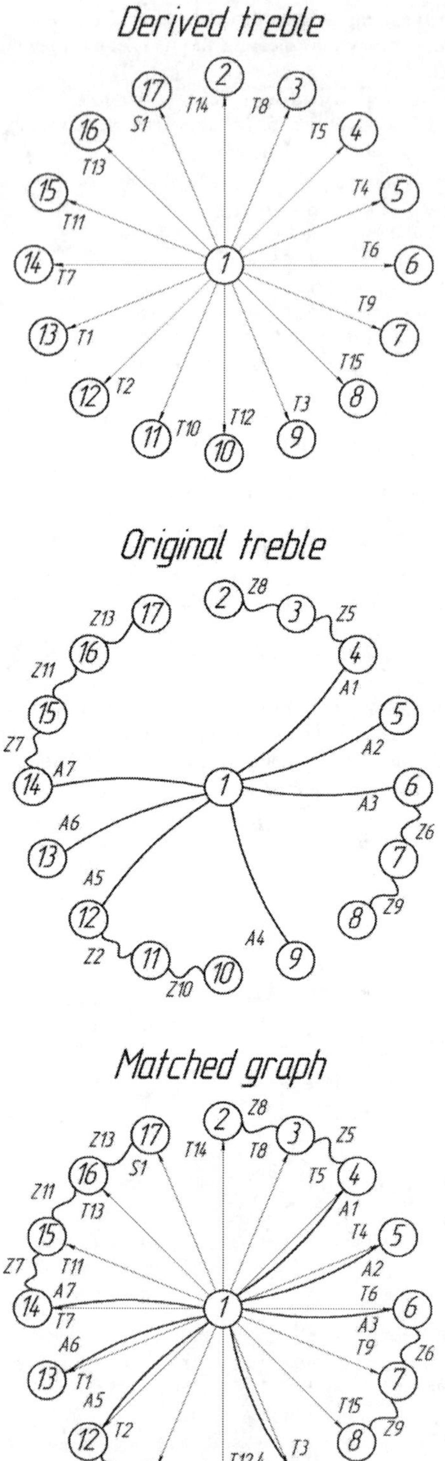

Fig 5.a. Designing of precision for manufacturing method №1 for item "Bush"

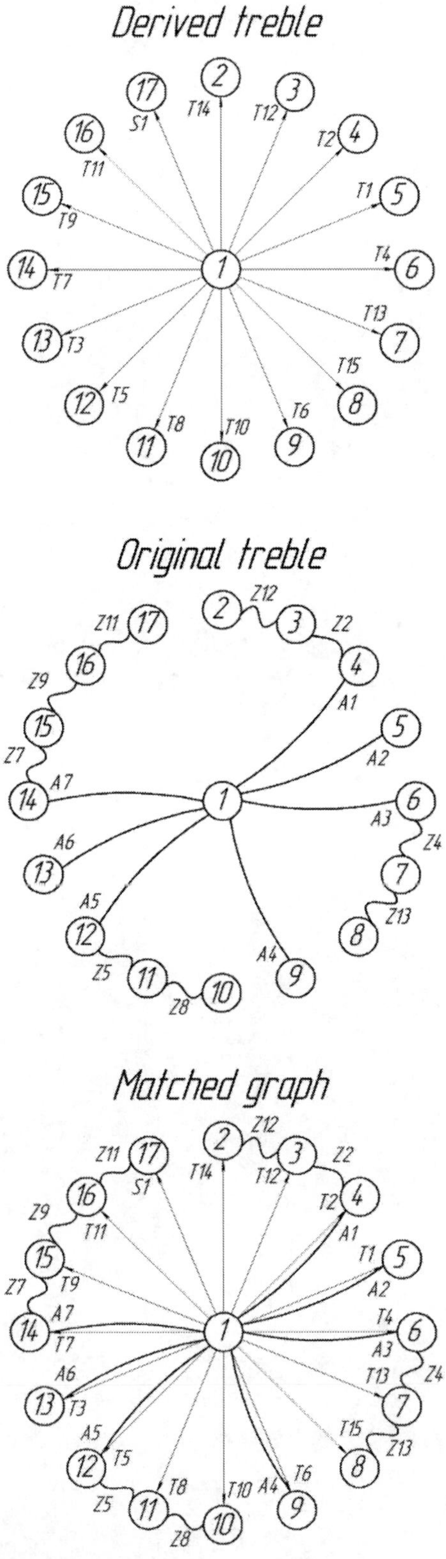

Fig 5.b. Designing of precision for manufacturing method №2 for item "Bush".

The graph of the manufacturing method and the dimension chains are the source data for the approach used for the automatisation of the designing of the manufacturing dimension chains. This approach involves two matrices, one of them is the matrix of nominal dimensions and another is the matrix of allowances. The method in question has some advantages, such as lack of necessity to explore the increasing dimension chain loops and the reducing dimension chain loops and to set up the equations of designing. The other advantage is that it is necessary to construct just the dimension chain and two adjacency matrices (the matrix of allowances and the matrix of nominal dimensions) in order to calculate all sizes and all their allowances. The device also doesn't require any special enumeration for surfaces and allows to automatize the designing of the form deflection circuits and the position deflection circuits. And the other plus point is that the device applies to dimensional analysis of the manufacturing methods of items like "a body of rotation" and more complicated products [9-10].

At the final stage we should choose the set of manufacturing methods of production of items meeting the requirements of assembly. And then it is necessary to define the modification of the efficient manufacturing method with account for an objective manufacturing situation.

The conclusion

In the present paper the scheme of complex of design procedures providing the effective accomplishment of the assembly operations which is based on the links between the process design of the manufacturing of the high-precision products meeting the built-in requirements and the process design of the assembly production of the high-precision products meeting the built-in requirements is represented. In addition, there are the possible solutions for the complex of design procedures analyzed. The realization of this method in SASMM is relevant because the complex of design procedures allows to consider the objective manufacturing situation and to choose the efficient manufacturing methods of the processing of items with account for requirements of assembly. As a result of that it should be possible to reduce labour inputs, production time and working cost, to improve quality and precision of the high precision products. And also it should be possible to reduce labour inputs and production time during the process design.

References

1. B.S. Padun, Tool engineering, **8**, 12-15 (2014)
2. A.N. Semenov, Tools and technologies, **21 – 22**, 122-124 (2004)
3. A.G. Suslov, A.M. Dal'skij, *Scientific fundamentals of technology of mechanical*

DOI: 10.1051/matecconf/201712901028

engineering (Mashinostroenie Publ., Moscow, 2002)

4. B.M. Bazrov, *Fundamentals of technology of mechanical engineering* (Mashinostroenie Publ., Moscow, 2005)
5. P.Yu. Bochkaryov, Technology of Mechanical Engineering, **1**, 10-14 (2002)
6. S.G. Mitin, P.Yu. Bochkaryov, Science intensive technologies in mechanical engineering, **11 (41)**, 44-47 (2014)
7. A.V. Nazaryev, P.Yu. Bochkaryov, Science intensive technologies in mechanical engineering, **12**, 28-34 (2016)
8. F.V. Grechnikov, S.F. Tlustenko, Vestnik SSAU, **3-4**, 38-43 (2011)
9. G. Thimm, G.A. Britton, S.C. Fok, International Journal of Advanced Manufacturing Technology, 261-271 (2004)
10. A.V. Muholzoev, Vestnik YUUrGU. Ser. Mechanical Engineering Industry, **3**, 48–55 (2015)

Problems of ineffective modernization at industrial enterprises and ways of their solution

Dmitry Kiselev[1,*], *Yuri* Popov[1], and *Olga* Kanygina[1]

[1]Volgograd State Technical University (VSTU), Russia

Abstract. The article considers the reasons of inefficient modernization of industrial enterprises. The above statistics confirm the need for a qualitative transformation of technological processes and approaches in the organization and management of production enterprises. Primarily this is due to technical and moral condition of fixed assets and the level of innovation in enterprises. The authors identified several major causes of the problem (inefficient) industrial modernization, which also explains the weak and the dynamics of modernization processes in Russia. The problem is due to how ineffective the actions of the heads of enterprises and factors of external environment. The authors see the solution of the mentioned problems in several directions related to the activities of state bodies and improvement actions of the management of enterprises.

1 Introduction

For many years in Russia at different levels raise issues related to a significant lag in the technological level of the production base of domestic industrial enterprises, not only from world-class companies, but also from the requirements of the current demand in the markets of the country. The need for radical change repeatedly discussed at various levels of government and the scientific community, and, as a rule, in the process of re-industrialization of the industry or the modernization of some of its segments.

In most cases, the products produced by domestic enterprises, is uncompetitive in many respects, including fixed – price and quality, and their relationship. Also note that the domestic enterprises give little importance to systems adapted promotion and service products, accounting for a considerable share in a competitive advantage in industrial markets. This makes many domestic manufacturers to occupy a worthy place in the markets of not only international level, but also across the country.

The ongoing modernization of industrial enterprises is often ineffective, because not achieved the intended system purpose and/or upgrade costs are more predictable because of flaws in the planned activities, which lowers the expected financial result.

2 The analysis of statistical data

In general, the activity of industrial enterprises is characterized by low levels of revenue and profit, which are much lower than expected and unit cost often exceeds all reasonable limits. Official statistics [1] based on cost of production and selling products per 1 rub. of output in the Russian Federation as a whole senses the presence of profits of industrial enterprises, but there is a negative trend in 2015, while for some types of activities the same, or close to 1 rub., or exceeds it, i.e. enterprises operating at a loss.

Table 1. Dynamics of production costs and selling products.

The cost of production and selling products per 1 ruble of the manufactured products, (rub)	2015	2016
machine tools	0,935	0,899
lifting and handling equipment	0,943	0,911
woodworking machinery	0,938	1,023
machines and equipment for mining and construction	0,996	1,065
forging equipment	1,096	0,732
machines for animal husbandry	0,997	0,981
other machines, not included in other groupings	1,031	1,118

The volume of shipped goods of own production in the manufacturing sector has a constant growth trend over the past 7 years the increase amounted to 236%

* Corresponding author: kiselevde7@yandex.ru

(2016 - 33 898,1 billion rubles – in 2009. – 14 351,9 billion rubles), including: manufacture of machinery and equipment 182%; the manufacture of electrical and optical equipment – 251%, production of vehicles and equipment 280%. However, if we turn to the production volumes of industrial products (machines, equipment, vehicles, and equipment) in real terms, in the period 2012-2014. (official data on 2015 and 2016 missing) recorded a decrease of production volumes.

Modernization of production in the first place should be directed to the fixed assets of enterprises. In the domestic industry condition of capital assets requires the acceleration of the pace of modernization, since statistics [1] shows that the average age of machinery and equipment is 11.2 g., and in manufacturing of 12.1 g. the Degree of deterioration reaches of 47.1 %, and the proportion of completely worn-out Fund – of 15.8%, including machinery and equipment of 24.5% in the manufacturing sector, 15% and 20.5%, respectively.

In addition to technical state of assets of enterprises, the importance of the modernization is orientation to the innovative development. However, only a minority of businesses seeking to introduce technological innovations in production. In 2015, they accounted for 8.3% of all organizations that have an innovative activity. And this despite the fact that the cost of innovation has increased from 2010 in 3 times (from 400 803,8 million to 1 200 363,8 million) [1].

If we consider the costs of domestic enterprises for R&d in % of GDP, the rate is 1.24 percent, while in the leading countries it is equal to 2,5 %-4,5 %, and in the "second tier" - 1,5 % - 2,5 % , while the absolute size of r&d spending in Russia is significantly lower.

Table 2. Structure of R&d funding in the world [2].

Country	R&d expenses, $ million. USA	The share of sectors in funding R&d costs, %			
		Business	State	Foreign investors	Other sources
USA	456 977	60,9	27,7	4,5	6,9
China	333 521,6	74,6	21,1	0,9	3,4
Japan	162 347,2	75,5	17,3	0,5	6,7
South Korea	68 051,5	75,7	23,8	0,3	1,1
France	57 986,8	55,4	35,0	7,6	2,0
UK	41 743,4	46,6	27,0	20,7	5,7
Russia	36 614	28,2	67,6	3,0	1,2
Poland	8 114	30,0	61,0	5,6	3,4

We add that the world economic forum brought the indicator "the ability of companies to borrow and adapt technologies", according to which Russia is on the 41 place among 133 countries. This suggests that Russian enterprises for various reasons do not tend to invest not only in the development and introduction of own technologies and innovations, but is even difficult to borrow advanced technology from their Western colleagues. In the end we have a system, "focused on imitation, and not to create radical innovations and new technologies" [3]. All this points to the weakness of governance in domestic enterprises, which are in most cases unable to become leaders in their industry, and just trying to follow the technological leaders of the West. Details of the reasons for the decline in investment flows to the science highlighted in [4], including: lack of demand from the business sector on the results of scientific activities, the lack of near term returns from basic research in the field of science, devoted to the economic problems, the lack of clear financial guidelines, etc.

It is obvious that in order to claim the role of leader, there must be sufficient volume of all types of resources including financial and intellectual. Here we must again regret to say that, on the one hand, or financial-credit institution, nor the state contributes to obtaining cheap long-term funds on a substantial development and/or modernization of production facilities and other enterprises are experiencing an acute shortage of personnel that could ensure the leading position of domestic producers and developers.

Often businesses are investing heavily in the modernization of production and management systems, however, do not get the expected results. The question arises, how justified is in these cases, the ongoing modernization involving replacement or improvement of fixed assets and production technologies, retrain staff and change suppliers with the introduction of modern principles and approaches in the organization and management of production. All of this requires substantial financial investments, some of which are at the expense of public funds, often not giving the expected return. In the opinion of the authors, there are several most important reasons why most of the modernization programmers do not give the expected effect.

3 Problems of ineffective modernization

The invalidity of the modernization programs. "The immediate motive of modernization is a mismatch between resources and the level of skills of the enterprise" [5]. Modernization, as any action within the enterprise cannot be undertaken for its own sake, and must rely on a well-designed purpose that provides an understanding of what she is. This understanding is based on data about what characteristics will have the market in which the company operates. The availability of these data is ensured by several factors.

First, it is a fully developed and multidimensional statistical data, which should be easy access. Unlike Western developed countries, the available statistical database in Russia more than scarce. The average entrepreneur who does not have access to the resources of government, practically deprived of the opportunity to assess the state of the market in which it operates, especially at the regional level. Accordingly, there is no opportunity to assess the degree of saturation of a particular market niches and customer satisfaction.

Second, is the formation of forecasts of development of the market situation. The lack of access to statistical data deprives managers of the ability to generate objective forecasts using statistical models. In addition to quantitative projections, there is a scarcity of qualitative projections generated by the experts. If you still have the opportunity to find a quality expert forecasts at the Federal level, the forecasts of the regional level is practically absent. It simultaneously points to the lack of appropriate specialists (experts), and lack of demand for their services.

Thirdly, it is marketing research, which unlike even the best statistical data show characteristics of the buyers, the nuances of their behavior and their problems and intentions. Analysis of the market of such services in the city of Volgograd showed that the demand for marketing research is tends to zero. On the one hand, this is due to the fact that business leaders do not realize the entire benefit of marketing research, and on the other, what do marketing companies do not have enough specialists who could show the benefits of such research through qualitative work.

The weakness of the above factors determines the fact that the leaders of organizations and entrepreneurs are unable to develop a full and reasonable strategy that would ensure their success in the target market. Lack of strategy leads to lack of comprehension when making decisions about upgrading, because executives do not understand why it is performed, and what results they can expect. As a consequence, current actions, and related financial investments ineffective. This problem is especially evident in small and medium-sized enterprises, which usually do not have the capacity to have on staff specialists in strategic analysis and strategy development for financial reasons.

The lack of planning or weak. Observations show that the organization of small and medium format in the management process almost does not involve planning. This is due to such factors as burdensome expenses on the collection and analysis of information; the difficulty of developing internal standards in planning; the downplaying of the role of planning, the desire to work intuitively, not knowing the types and kinds of planning.

Large companies, given the scale of their activities, are forced to carry out planning, including strategic, tactical or limited. We should note that the quality of the development of plans varies significantly: some companies develop complete plans, others take their planning to the formal procedures and often limited to a plan to sell [6].

One of the issues is inefficient modernization of large enterprises, with a network of offices and branches is inefficient budgeting system, aimed at the development of the budget in a timely manner, often leading to irrational allocation of resources across time and across types of work in economic activity and in the decrease of efficiency of expenses on modernization of production facilities.

Insufficient level of qualification of the staff. Speaking of qualifications, here it is necessary to involve all categories of employees.

First, it is the managers who in most cases do not possess sufficient knowledge and skills in the development of the strategy. In the vast majority of cases the documents in the title which is the word "strategy", they are not, it is at best a long-term plan. A big role plays the effectiveness of the management responsible for the strategic and tactical planning and management. In relation to the ongoing enterprise modernization requires particular attention to innovative management, involving design, forecasting, effective decision-making.

Secondly, it engineering staff, which, based on the latest scientific discoveries and development needs by relating their own ideas and projects with forecasted characteristics of the demand, to form a product, the characteristics of which are demanded by target markets in the foreseeable future, with reference to and which needs to be modernized. Again, by analyzing the activity of the domestic enterprises, we can see that in many cases new products is not a good enough copy of what has already been developed by foreign companies. In those cases when it is possible to create world-class products, it still is not always in demand. Moreover, such products are often not reached the stage of mass production. Real gains have mainly to say in cases where the products are manufactured under the order. In other words, domestic engineers, as a rule, it does not show the ability to anticipate the future; to go for bold decisions; to create products that can become a reference point for foreign businesses to find creative solutions to resolve certain problems, etc. To some extent, the assistance in that they could assist marketing experts, studying the problem from consumers, but in the vast majority of cases, this category of workers, the management of the enterprise driving all marketing activities in the area of product promotion, which reduces the professional level of marketers.

Thirdly, it is the direct perpetrators – production workers, which to some extent depends on the quality of products. Despite the fact that the primary responsibility for this lies with the leadership, a lack of skilled workers, low production and other factors do not allow to produce competitive products.

The lack of long-term development goals. The goal of modern managers is primarily in the area of minimizing costs to maximize profit in a short period, all efforts are focused on using the market to achieve short-term goals. Often the management is following the path of least resistance, reducing first costs for personnel and materials. Enterprise development, including modernization, is not correlated with the specified target setting. The modernization includes the transformation of fixed assets, of the logistics system of the enterprise processes enterprise management requires a strategic and long-term decisions and longer payback period.

The lack of a productive mechanism for assessing the appropriateness and effectiveness of public investment. We have already mentioned that a large part of innovation is the initiative, and therefore with the support of the state. It is obvious that the government allocates adequate financial resources. However, as practice shows, there are no clearly proven and effective mechanisms to assess the appropriateness of investing public funds so that they bring good returns at least in the long term. Thus, we can figuratively say that the state plays in a losing of the lottery, which is unacceptable, given the sources of state revenues and existing obligations.

Obviously, there must be some structure that will carry out a comprehensive market assessment of the proposed projects. It should be recognized that such organizations exist, for example "Agency for strategic initiatives". But, just look at the website of the Agency to understand that it from the entire set required for such structure functions performs only one – the transfer of public money into projects. Indeed, this organization does not help either to develop efficient projects, nor to evaluate their market or social desirability, for which he was subjected to sharp criticism by the Russian President at the end of 2016.

The sanctions restrictions. The complicated geopolitical situation around Russia, has led to significant restrictions in the supply of new technologies and equipment to domestic enterprises and finished industrial products from abroad and made it difficult for exports. And those businesses that are leaders in technological development, and those who try to use the strategy of following the leaders (imitation), currently face important constraints in the implementation of the modernization process. Difficult situation and fragmentation of relations in science and industry do not allow to compensate for these limitations through the manufacture and supply of domestic technologies and equipment. It should be noted that the issue of revitalization of the industrial enterprises is largely a matter of the time required for design and development (modernization) of production capacities, setting up a system of promotion and service of products, the required return on investment. The current economic and geopolitical situation increases uncertainty and leads to significant risks for investors.

4 Direction in solving the problems identified

1. Strengthening the activities of the state in creating a favorable information environment for conducting business. This includes not only better quality of the statistical base available to entrepreneurs and business leaders, but also the encouragement, including financial aspects, creation of research institutes, centers and units as on a commercial basis, and research-based organizations. Currently there is an acute shortage of professionals who can effectively collect information, process it and generate a reasonably accurate projections and estimates. Therefore, at the initial stage, emphasis should be placed on the training of such experts
2. To create the conditions necessary qualification and quality management leaders of the organization. At present one should speak not just about the absence of market relations in Russia, but also about the growing monopolization of markets, both at Federal and regional levels. You should make every effort to promote competition, which will create a greater number of businesses. Those that will be more competitive, will become the centers of attraction of investments; qualified personnel; the attention of buyers, suppliers and marketing intermediaries, etc. The leaders of the companies that have to deal with the outflow of all these resources will be forced to look for ways to improve the skills and quality of management. In this situation, significant importance will be the adaptation of training programs in various training centers of the needs of the market.
3. Reorientation ineffective reform of higher education by absorption (dissolution) of educational institutions to increase the effectiveness of their training [7]. In this case, you must change the parameters for assessing the effectiveness of training. The main criterion for evaluating the educational institution should be the quality of training of graduates, its demand in the labour-market, and in an additional – additional criteria, such as number of scientific papers, often not reflecting the level of the teacher and his work; the amount earned by teachers, financial resources, etc. Accordingly, should change and qualifications of the teaching staff, which appropriately should be subject to direct (appraisal fee, etc.) and indirect (students ' knowledge) assessment. Negative testing experience of students in the framework of the procedures of attestation of Universities showed that one should look for other ways of inspection. It should be particularly noted that the growth of these requirements is meaningless without a real increase in the level of motivation and stimulation of workers of educational institutions.
4. Integration of industry and science. The experience of world leaders, proves the impossibility of sustainable and dynamic development of the economy and the competitiveness of enterprises without deep integration of industry and science. The success of individual domestic enterprises do not give opportunities to talk about systemic change in the whole industry. The responsibility for such integration falls not so much on the state, which must actively influence both sides of the process as on the businesses themselves that need to pay more attention to the qualification of its employees and cooperation with scientific and research organizations.
5. The main impetus in the implementation of the previous two paragraphs can specify the state. It is appropriate to talk about the partnership of business, educational and research organizations and the state, which can be seen in the various system interactions [8, 9, 10 11]: clusters, public-private partnerships or concessions, priority development areas, etc. we emphasize that the role of the state should not be passive, i.e. reduced to some guarantees, promises, preferences, etc., but at the same time and should not violate the laws of the market economy, to encourage competition. You need to create a stable and favorable institutional environment for business, ready to observe the laws and to bear their share of social responsibility, which is necessary to ensure strict observance of laws after thorough revision to ensure realism and to corruption [12].
6. The creation of new effective structures and improvement of existing organizations that can provide real assistance to entrepreneurs and businesses in the development of various projects with elements of innovation and modernization and to conduct a comprehensive evaluation of the proposed projects in terms of the viability and profitability based on global-

DOI: 10.1051/matecconf/201712901029

level indicators. Observations show that, in the majority of cases the leaders are in dire need of not only effective consulting support, but also teaching business and management in market conditions. Almost always mismanagement and the collapse of the organizations (except illegal action) are the result of ignorance of basic management theories and techniques. Here the great role of the organization in the likeness of the chambers of Commerce, business incubators, etc.

7. Systematizing and improving the effectiveness of corporate planning, including strategic, tactical and operational cycles, but with a primary focus on the development of strategic plans. While the main focus should be on strategic and marketing studies and market forecasting. This will generate a more informed and therefore effective strategies and, thereby, to increase the level of decisions, including in the field of modernization of production, which will be more focused – with a focus on specific investment opportunities in the market.

5 Conclusion

The state of the economy and industry in Russia continues to be at a high enough level. Despite the breakthroughs in a very narrow spheres of activity, the competitiveness of most enterprises and their products is low. Assessing the reasons for this situation, many note the high level of physical and moral depreciation of fixed assets of domestic enterprises and the substantial technological gap due to insufficient investment in science, both at the country level and at the level of individual enterprises.

At the same time, there are many examples where significant investments in innovative activity, inextricably linked with the modernization of production, did not bring the results expected. Products updated its productive and technological base, were often as not in demand as the products made on the basis of the old production technologies.

The reasons for this situation due to deficiencies in the activities of the various stakeholders in the economic and managerial relations. Businesses that are upgrading, are not sufficient to base their decisions on statistical data, expert forecasts, and marketing research. Insufficient level of qualification of managers and employees does not allow to formulate long-term and strategic goals, and develop plans and innovative products designed to meet the needs of the market. The state by providing funding for the most significant innovative and modernization projects, is weak control over the reasonability of their usage, which leads to low efficiency of such investment. The weak interaction of scientific and educational institutions and the industrial complex to reduce the possibilities for a qualitative transformation of production processes and marketing.

The most important areas of focus to remedy this situation, according to the authors, should be considered a significant improvement in the information base for making informed decisions when upgrading. Increasing the level of training of managerial and engineering personnel of enterprises should be a priority in solving the considered problems. It is necessary to improve the activities of educational institutions for the training of personnel with adequate qualifications and research organizations to enhance research and the integration of their work with industrial enterprises. Appropriate and more active state involvement in significant projects to build conditions for business development through comprehensive support and forming partnerships.

References

1. Federal State Statistics Service: http://www.gks.ru
2. National Science Board. 2016. Arlington, VA: National Science Foundation Science and Engineering Indicators 2016 (NSB-2016-1), 4/45-46 (2016).
3. Ministry of Economic Development of the Russian Federation, The strategy of innovative development of the Russian Federation for the period till 2020, 16 (2011)
4. A.A. Tarasyev, I.A. Krivenko, M.S. Pecherkina, T.O. Kashina, Economy of Region, **12(1)**, 303-314 (2016)
5. G.B. Kleiner, Economy of Region, **13(1)**, 13-24 (2017)
6. D. Kiselev, Y. Popov, O Kanygina, *Globalization and Its Socio-Economic Consequences*, 16th International Scientific Conference, **II**, 884-891 (2016)
7. Ye.V. Balatsky, Journal of the new economic association, **4(24)**, 111-140 (2014)
8. 8 S. Zemtsov, V. Barinova, A. Pankratov, E. Kutsenko, Foresight and STI Governance, **10(3)**, 34–52 (2016)
9. V. Abashkin, A. Boyarov, E. Kutsenko, Foresight-Russia, **6(3)**, 16-27 (2012)
10. O.A. Romanova, Economy of Region, **1**, 43 -56 (2014)
11. N.V. Kashina, Economy of Region, **12(2)**, 569-585 (2016)
12. G.A.Yavlinsky, Journal of the new economic association, **2(26)**, 173-178 (2015)

 DOI: 10.1051/matecconf/201712901030

Development of the mechanical engineering complex on the basis of the improvement of large and small businesses relations

Svetlana Sokolova[1,*], *Mikhail* Belyaev[1], and *Victoria* Batmanova[2]

[1]The department of economics and management of projects in the construction industry, Volgograd State Technical University, Volgograd, Russia
[2]Department of world and regional Economics, Volgograd State University, Volgograd, Russia

Abstract. Condition, pace and character of the development of the mechanical engineering complex is in many aspects a crucial factor for the social and economic situation of any country. The development of market relations, changes of the conditions of doing business encourage the enterprises to search new managerial methods and to improve the interaction forms. In this respect the display of the peculiarities of the interaction of large machine engineering enterprises and small business in this sphere and also the assessment of the relationship of their development is an important and crucial issue under modern conditions. The most widely spread forms of the cooperation of large scale mechanical engineering enterprises and small businesses of the industry are: outsourcing, franchising, leasing, subcontracting, venture financing, creation of regional forms of the cooperation of large and small firms. However cooperation processes of large scale and small entrepreneurship in Russia are not properly developed. The authors determine the factors hindering the growth of the machine building industry, suggest the recommendations for the development of the large scale enterprises and small business in the industry, substantiate the role of the government in this process. Besides the mechanism of the state support of the development of small business is described.

1 Introduction

In the modern context the level of the social and economic development of a country in many respects depends on the condition of such an important industrial complex as mechanical engineering and machine manufacturing industry. At different times the problems of the development of machine manufacturing were studied by Borodachev N.A., Dvigubskiy I.A., Dyachenko P.E., Kashirina A.I., Linnik Yu.V., Matalin A.A., Mitrofanov S.P., Sokolovskiy A.P., Teame I.A., Yukhin A.I. and others. At present moment in Russia different levels of government pay attention to the development of the machine engineering complex. In the frame of the current governmental program "Development of the industry and the increase its competitiveness" the largest part of subordinate programs (heavy machine construction, energy machine building, agricultural machine construction, transport machine construction, machine construction on specialized industries and so on) in any way are directed on the development of the machine construction industry. In compliance with this program the main directions of the support and the development of the mechanical engineering complex are the modernization of the capital funds used in the technological process; encouragement of scientific research and developments directed on the creation of new technologies and materials; competitivity growth and so on. However despite the measures taken by the Russian government more recently the decrease of the index of machine and equipment production [1, p. 43-45], the reduction of the dynamics of the investments into the fixed capital are observed [1, p. 133-134]. The stagnation in this sector of economy encourages the search of new methods and forms of the support and the development of machine manufacturing [2-21].

In the authors' point of view the reserves of the efficiency growth of the machine engineering complex can be connected with the cooperation of the machine manufacturing firms with relatively smaller enterprises. Such cooperation can be both advantageous for both large scale and small enterprises. In the modern context the level of the development of small entrepreneurship is one of the key indicators characterizing the stability and maturity of a society. In economically developed countries of the world in the structure of the GDP the share of small enterprises exceeds the percent of the large scale enterprises and makes up 70% or more. In our country this index equals only 20%. In this respect the encouragement of the small entrepreneurship, the detection of the interaction of large scale machine construction enterprises and small firms, the

* Corresponding author: svetalekssok@yandex.ru

improvement of their cooperation is especially important and vibrant problem today.

The goal of the research is the study of the peculiarities of the cooperation and the possibilities of the improvement of the interaction of large scale enterprises of machine construction industry and small enterprises.

2 Peculiarities of the cooperation of large scale mechanical construction enterprises and small entrepreneurship

The practical experience of the industrial development in the world confirms the expediency of the cooperation of large scale and small enterprises. Today this process has the fastest tempo in the countries of the East. Large mechanical construction enterprises often place orders of details and components at small enterprises located as a rule not far from the large one. This fact contributes to an increase of sustainability and stability of small business and also allows the large scale machine construction enterprises solving more complicated, serious and responsible problems. As a result the synergetic effect arises. The character and the intensity of the cooperation depends first of all on the industry where the enterprise functions, level of the development and the structure of the regional economy and the country as a whole.

The need of cooperation of small enterprises with large scale businesses depends considerably on the sphere of their activity. The largest share of small firms provides different services. Today in Russia the most widely spread spheres of work of small enterprises are car and consumer equipment repair (39,7 %), gross sales (24,7 %), real estate transactions, rent and services provision (20,4 %), construction (11,0 %), retail sales (10,6 %), manufacturing (9,5 %).

The analysis showed that at the present time the quantity of small enterprises in our country is several times lower than the number of large scale enterprises and this fact contradicts the international practice. This tendency is typical of all the territory of the Russian Federation. The calculated indices of the ratio of small firms to large scale enterprises and also indices of industrial production in the Southern federal district are given in the tables 1, 2.

In Russia the cooperation is most widely spread and is typical of science and scientific services, construction and industry where the share of cooperating enterprises makes up 71 %, 50 % и 49 % correspondingly [1]. In general in our country the processes of the interaction of large scale and small enterprises are not spread in a sufficient way. However under the conditions of the uncertainty and of the crisis the interaction of large scale enterprises and small business is expected to increase considerably the efficiency of use of time, labour, finance, material and information resources. At the same moment it is possible to transfer temporarily resources from one actor to another and in both directions.

Table 1. Index of ratio of small firms to large scale enterprises in the Southern federal district of Russia.

Index	2012	2013	2014	2015	2016
The republic of Adygeya	0,51	0,51	0,53	0,53	0,54
The republic of Kalmykiya	0,12	0,18	0,20	0,25	0,33
Krasnodar kray	0,40	0,44	0,44	0,46	0,48
Astrakhan oblast (region)	0,51	0,50	0,53	0,55	0,56
Volgograd oblast (region)	0,51	0,51	0,52	0,52	0,52
Rostov oblast (region)	0,64	0,62	0,61	0,59	0,57

Table 2. Index of machine manufacturing in the Southern federal district of Russia.

Index	2012	2013	2014	2015	2016
The republic of Adygeya	108,4	100,2	121,5	103,8	111,7
The republic of Kalmykiya	99,8	116,9	85,0	109,1	102,5
Krasnodar kray	107,6	102,6	105,3	102,5	104,3
Astrakhan oblast (region)	109,4	103,0	92,6	100,3	98,6
Volgograd oblast (region)	106,8	101,7	104,5	102,2	103,3
Rostov oblast (region)	110,5	106,8	106,1	161,4	113,6

Under modern economic conditions in Russia often arise situations when small enterprises can provide the load of industrial capacities of large machine construction enterprises. Small firms usually have intellectual property, innovative technologies but do not have necessary equipment and other fixed assets helping make money on the use of the innovations. Such a redistribution of resources allows improving the financial situation of large machine construction enterprises which get the order. But at the same time such a "reverse" interaction can lead to an increase of costs and the appreciation of the production due to singular contracts and refusal from serial production. If in large scale machine engineering enterprises there isn't well timed technical modernization in long term period, they will lose competitive advantages and even can become bankrupts.

Today the main forms of the cooperation of large scale and small enterprises in Russia are outsourcing, franchising, leasing, sub contracts, venture financing, creation of regional forms of cooperation of large scale and small enterprises. An efficient implementation of the forms of the interaction of enterprises mentioned above is possible only at an active participation from the part of the state as a regulator providing the creation of necessary institutional conditions.

3 Assessment of cooperation of large scale enterprises and small business (using the statistics of the Southern federal district of Russia)

The condition of large scale and small enterprises at the level of a region can be assessed on the basis of qualitative and quantitative criteria. The authors supposed that there exists the interaction of the condition of large scale enterprises and small businesses. The correlation and regression analysis was used for the calculation of this relation. At the same time the condition of the industry of the Southern federal district for ten years (2006–2015) was studied. The analysis of the following indices characterizing the small entrepreneurship were taken:

- quantity of small firms per 1000 of working age population;
- turnover of small firms per 1000 of regional inhabitants;
- share of the working population at small firms in the total number of the employed population of a region and others.

In order to assess the condition of large scale industrial business the following parameters were used :

- volume of industrial production per 1000 of regional population;
- number of on going large scale industrial enterprises per 1000 of working age population ;
- volume of investments into the fixed capital per 1000 inhabitants of a region;
- share of population working at large industrial enterprises in the total number of employed population of a region and so on.

In the result of the analysis the correlaton between various indices was established. It was determined that the largest correlation is observed between the volume of industrial production per 1000 inhabitants of a region and the quantity of objects of small enterpreneurship per 1000 of woking age population. The results of the correlation and regression analysis of these indices are presented in the table 3.

Thus, between the indices characterizing the large scale and small business there exists a correlation that can be ranged from an obvious one to a strong one. The authors suppose that a very strong relation between the indices in the Adygeya republic is explained first of all by a large number of small enterprises in the sphere of services and client treatment that assure the needs of large scale business. The reason for a weak correlation in some regions (Krasnodar kray, Astrakhan region) is the structure of regional economy, big share of small enterprises working independently and which are self sufficient and which do not need active cooperation with large scale business at the modern stage of their development.

Table 3. Assessment of proximity of the relation of indices of large scale enterprises and small business in the Southern federal district of Russia.

Index	Coefficient of correlation	Regression equation	Standard error of regression
Southern federal district of Russia	0,97	$y = 0{,}02x + 98095$	2637,52
The republic of Adygeya	0,95	$y = 0{,}01x + 2846$	56,98
The republic of Kalmykiya	0,63	$y = -0{,}03x + 2097$	336,38
Krasnodar kray	0,43	$y = 0{,}003x + 2097$	2220,14
Astrakhan oblast (region)	-0,23	$y = -0{,}002x + 9494$	460,53
Volgograd oblast (region)	0,52	$y = 0{,}01x + 21872$	1373,35
Rostov oblast (region)	0,82	$y = 0{,}07x - 1364$	6278,31

For a qualitative assessment of the correlation proximity the Cheddock scale was used. The systematic estimation of the coefficient of paired correlation by means of the comparison of its absolute value with a critical index at the importance level α, correpsonds to 0,05, degree of freedom f, equal to 8. If $r_{расч} \geq r_{крит}$ so with the given degree of possibility (95%) it can be said that between the analyzed numerous aggregates there exists a linear relation. The importance of the coefficient of paired correlation was also studied on the basis the Student's t-test for a large volume of selection

4 Foundations of the development of large scale enterprises and small business

The analysis of the scientific literature and materials presented in the mass media allowed showing factors limiting the growth of machine construction industry in Russia and the development of the cooperation between the large scale and small enterprises showed the following:

- insufficient demand for the production of the enterprises in the domestic market;
- insufficient demand for the production of the enterprises in the foreign market;
- competitive imports;
- high taxation level;
- wear and absence of equipment;
- uncertainty of the economic situation;
- high interest rates for commercial credit;
- insufficiency of finance;
- insufficiency of qualified workers;

DOI: 10.1051/matecconf/201712901030

- absence or imperfectness of normative and legal base.

The analysis made showed that the biggest problems in the modern context are connected with a high level of taxation (54 %), insufficient demand for the production of an enterprise in the domestic market (53 %) and the uncertainty of the economic situation (39 %).

Under such conditions it is imporant to implement the governmental support of industrial enterprises and especially of small firms. When the authorities are inert the stagnation can make serious problems for the economy. The governmental support of small enterprises has a serious importance for the development of the interaction of large scale enterprises and small business. The program of the development of the small entrepreneurship is expected to set goals, tasks and ways of their achievement, basic indices, to foresee a complex state support of the given sector of the economy. The mechanism of the state support of small entrepreneurship is shown in figure 1.

It is worth mentioning that in the present context large scale machine construction industry is becoming more and more automated and less labour intensive what leads to the release of people. They need to have the opportunity of doing the business that gives profit. Unfortunately the culture of individual entrepreneurship is not developed considerably and its increase allows forming a civilized cooperation of large scale enterprises and small business and allows augmenting considerably the resource potential of a country.

5 Conclusion

The contemporary copperation of large scale machine construction enterprises and small business has its peculiarities and problems. Most often the cooperation is realized in the form of an outsourcing, franchising, leasing, sub contracting, venture financing and others. There exists a stable relation between the condition of large scale enterprises and small business.

For the intensification of the production of Russia the state is expected to implement a resonable and efficient support of large scale enterprises and small firms. A special attention should be paid to the mechanism of the state support of the small business including the organizational, finanical, informational and consulting components.

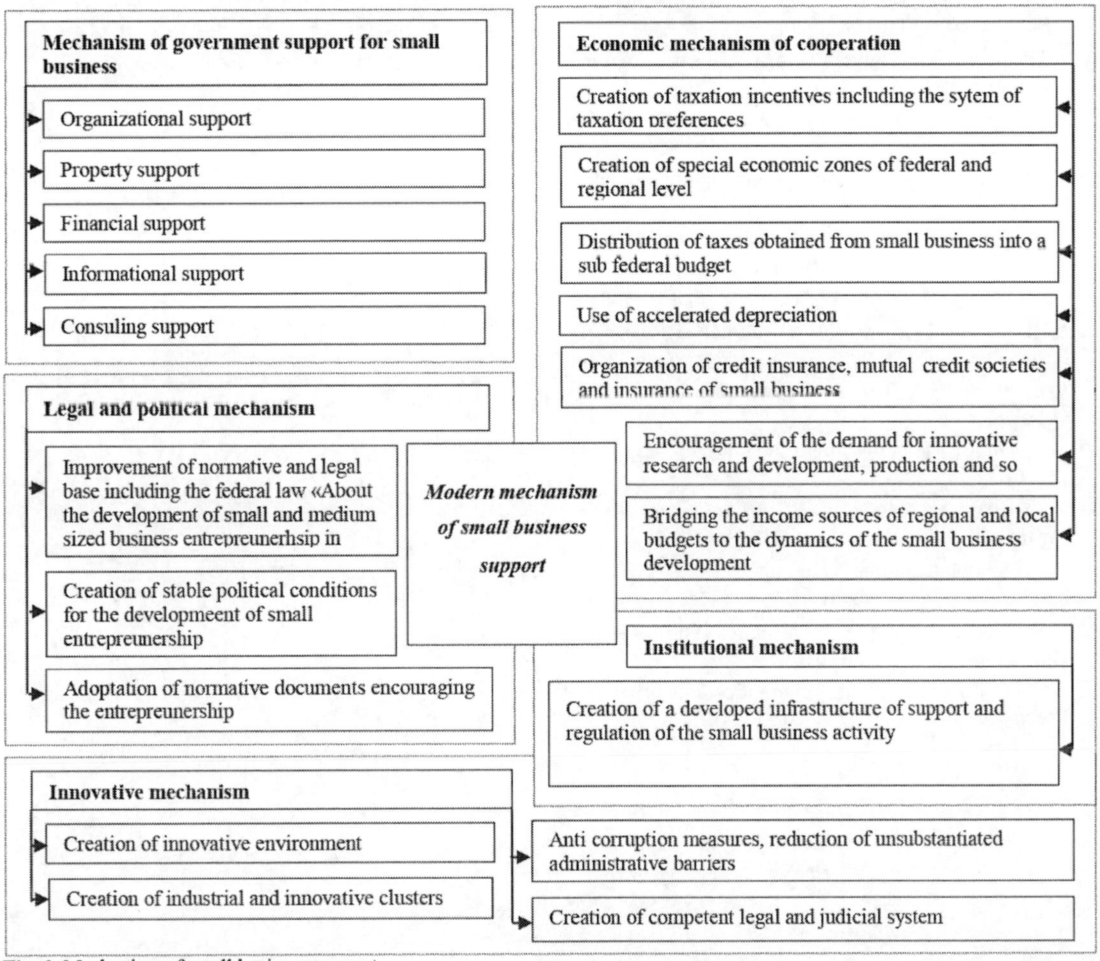

Fig. 1. Mechanism of small business support.

In the authors' opinion the development of the cooperation between large scale enterprises and small firms is especially important under modern economic conditions as it provides absolutely new level for the use of resouces based on the efficiency and thrift.

References

1. *Industrial production in Russia*. 347 (2016)
2. A. Shleifer, R.W. Vishny, Journal of Political Economy, **95**, 461 (1986)
3. A.D. Evans, Studies in Comparative International Development, **51**, 456 (2016)
4. B.S. Min, Journal of Business Ethics, 1 (2016)
5. C. Wickert, Business and Society, **55**, 792 (2016)
6. D. Palia, Journal of Empirical Legal Studies, **13**, 756 (2016)
7. D.H. Kim, M.O. Sambou, M.S. Jung, Empirical evidence from Korea. Sustainability, **8**, 1119 (2016)
8. G. Kaplanoglou, V.T. Rapanos, N. Daskalakis, European Journal of Law and Economics, 1 (2016)
9. G.M. Olevsky, World Economy and International Relations, **60**, 17 (2016)
10. I. Distinguin, C. Rugemintwari, R. Tacneng, World Development, **84**, 18 (2016)
11. K. Ding, T. Gokan, X. Zhu, Annals of Regional Science, 1 (2016)
12. K.G. Palepu, Journal of Accounting and Economics, **8**, 3 (1986)
13. L. Hou, L. Xue, S.N. Bui, W. Kettinger, Information Technology and Management, **17**, 379 (2016)
14. M.C. Jensen, The Journal of Finance, **48**, 831 (1993)
15. M.C. Jensen, W.H. Meckling, Journal of Financial Economics, **3**, 305 (1976)
16. O.E. Williamson, The Yale Law Journal, **93**, 1197 (1984)
17. Q. Guo, C. He, D. Li, Urban Studies, **53**, 2584 (2016)
18. R. Morck, A. Shleifer, R.W. Vishny, Journal of Financial Economics, **20**, 293 (1988)
19. R. Tresca, International Journal of Social Ecology and Sustainable Development, **7**, 35 (2016)
20. R.F. Malaquias, Y. Hwang, Information Development, **32**, 1613 (2016)
21. S. Mayadunne, S. Park, International Journal of Production Economics, **182**, 519 (2016)

DOI: 10.1051/matecconf/201712901031

Nonlinear phenomena in a high–power heating unit with pulse modulated control

Yuriy Alexsandrovich Gol'tsov[1,*], *Alexander Stepanovich* Kizhuk[1], and *Vasiliy Grigorievich* Rubanov[1]

[1]Department of Engineering Cybernetics, Belgorod State Technological University n.a. V.G. Shukhov, Kostyukov Str., 46, 308012, Belgorod, Russia

Abstract. In this paper we describe the transitions from regular periodic mode to quasiperiodicity that can be observed in a pulse–width modulated control system for a high–power heating unit. The behavior of such a system can be described by a set of two coupled nonautonomous differential equations with discontinuous right–hand sides. We reduce the investigation of this system to the studying of a two–dimensional piecewise–smooth map. We demonstrate how a closed invariant curve associated with quasiperiodic dynamics can arise from a stable periodic motion through a Neimark–Sacker bifurcation. The paper also considers a variety of interesting nonlinear phenomena, including phase–locking modes, multistablilty and hysteretic transitions.

1 Introduction

The technology of monocrystals growth is a controlled crystallization process, during which the quality of a growing crystal is determined by the accuracy of controlling phase transition conditions [1, 2]. When growing a sapphire crystal it is necessary to provide the temperature change in the crucible from 25°C to 2050°C with a certain degree of its increase and decrease.

In this way, it is required an automatic control system with the function of software setting of the temperature variation in the crucible with the required accuracy.

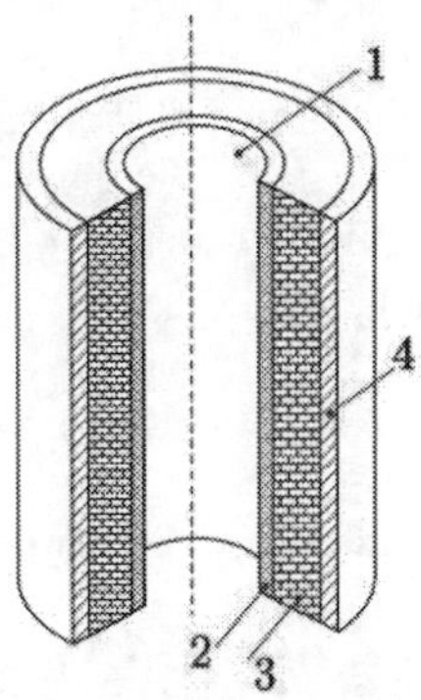

Fig. 1. Schematic diagram of the heating unit.

Figure 1 shows a schematic diagram of the considered heating unit [3]. The heating unit includes the following areas: internal furnace space denoted by number 1, filled with air or gas; nichrome electric heater (denoted by 2), uniformly located in the inner layer of the lining (denoted by number 3) which is produced of magnesite bricks. The outer lining layer 4 is constructed of mineral wool that is in a cylindrical beaker made of galvanized steel.

The geometric shape of the furnace is a limited cylinder, with the lining on its top and bottom. The heat transfer in the first two areas is performed only due to thermal conductivity, as the materials of which the furnace lining and the heater are made, are solid and opaque.

In order to perform the synthesis of the control law using a classical approximation method based on the experimental transient behavior of the heating unit we have defined the following transfer function [3, 4]

$$W(s)\frac{K}{(T_1\cdot s+1)(T_2\cdot s+1)}, \qquad (1)$$

where K is the transfer factor of the plant, and T_1, T_2 are its time constants.

In the modern industry the temperature regulators with thyristor converters are widely used.

However, the use of converters of this type greatly distorts the input current. This causes the appearance of non–sinusoidal modes in the supply network.

In order to eliminate these regimes and to obtain a fast response, an accurate control, and a higher efficiency we have developed a heating unit control system [3, 4] based on a high–frequency switching power electronics converter with pulse–width modulation [5].

At the same time, with real conditions of operation smaller or larger changes of parameters always take place. Such variations often lead to the loss of stability of the normal operation regime and to the appearance of complex dynamical behaviors, including subharmonic, quasiperiodic or chaotic oscillations [6–8].

* Corresponding author: umin@mail.ru

The appearance of such oscillatory modes leads not only to the reduction of the control accuracy by several orders, but also to the sudden breakdown of the technological equipment.

The purpose of the present paper is the numerical study of bifurcation phenomena in the pulse modulated control system of a high–power heating unit.

The paper is organized as follows. In sections 2 and 3 we describe a model of the heating unit with pulse–width modulated control. The behavior of such a system may be represented by a two–dimensional piecewise–smooth set of nonautonomous differential equations. We reduce the investigation of this system to studying the dynamics of a two–dimensional piecewise–smooth mapping.

In section 4, we discuss the bifurcation phenomena occurring in the considered system. The obtained results are summarized in Conclusions.

2 Continuous – time model

The dynamics of the heating unit with pulse width modulated control, a continuous linear part which is described by the transfer function (1), can be represented by the following differential equation:

$$T_1T_2\frac{d^2x}{dt^2}+(T_1+T_2)\frac{dx}{dt}+x=KU\varphi(\varepsilon), \qquad (2)$$

where x is the temperature in the heating unit, ε and $\varphi(\varepsilon)$ are the input (error) and output signals of the pulse–width modulator.

Let us introduce the state variables

$$x=x_1,\ \frac{dx}{dt}=x_2.$$

Then, the closed–loop system expressed by (2) can now be rewritten as a set of two coupled ordinary differential equations:

$$\frac{dx_1}{dt}=x_2,$$

$$\frac{dx_2}{dt}=-\frac{1}{T_1T_2}x_1-\frac{T_1+T_2}{T_1T_2}x_2+\frac{KU}{T_1T_2}\varphi(\varepsilon). \qquad (3)$$

The switching function $\varphi(\varepsilon)$ (output signal of the modulator) is defined as:

$$\varphi(\varepsilon)=\begin{cases}1, & k\cdot T_0<t<k\cdot T_0+\tau_k;\\ 0, & k\cdot T_0+\tau_k<t<(k+1)\cdot T_0;\end{cases}$$

$$k=0,1,2,...$$

Here T_0 is the period of the ramp signal and τ_k is the pulse duration within the time interval $k\cdot T<t<(k+1)\cdot T_0)$.

In the present analysis we shall consider a system with the proportional feedback corrector and pulse–width modulation of the first type (see, for example, [6]).

In such a system, depending on the error signal ε the value of τ_k is determined as:

$$\tau_k=\begin{cases}0, & \varepsilon<0;\\ T_0, & \varepsilon>V_s;\\ \dfrac{T_0\varepsilon}{V_s}, & 0\le\varepsilon\le V_s,\end{cases}$$

where

$$\varepsilon=\alpha(V_{ref}-\beta x_1(kT_0)).$$

Here α and $Vref$ are the amplification constant and reference signal. Vs represents the amplitude of the rump signal and β is the sensitivity of the temperature sensor. The mathematical model (3) of such a system may be cast into the form:

$$\frac{dX}{dt}=\begin{cases}AX+B, & if\quad \xi(t,X)>0,\\ AX, & if\quad \xi(t,X)<0,\end{cases} \qquad (5)$$

with

$$X=\begin{bmatrix}x_1\\x_2\end{bmatrix},\ A=\begin{bmatrix}0 & 1\\ -\dfrac{1}{T_1T_2} & -\dfrac{T_2+T_2}{T_1T_2}\end{bmatrix},\ B=\begin{bmatrix}0\\ \dfrac{KU}{T_1T_2}\end{bmatrix}$$

and

$$\xi(t,X)=V_{ref}-\beta\cdot x_1(k\cdot T_0)-\frac{V_s}{\alpha}(t/T_0-\lfloor t/T_0\rfloor).$$

Here the number $\lfloor t/T_0\rfloor$ denotes the largest integer not greater than t/T_0 (i.e., the integer part, or floor, of t/T_0). The scalar function $\xi(t,X)$ determines the switching manifold

$$\Sigma=\{(t,X):\xi(t,X)=0\}$$

that separates the state space into two different regions

$$\Sigma_+=\{(t,X):\xi(t,X)>0\},$$

$$\Sigma_-=\{(t,X):\xi(t,X)<0\}$$

in which the dynamic behavior of the system (5) is governed by different vector fields. It easy to show, that any solution of (5) will intersect the switching manifold (or discontinuity boundary) Σ transversely (see [8]).

3 Discrete time map

The investigation of the dynamical system (5) has been reduced to the study of the two–dimensional stroboscopic map.

Within the domain

$$kT_0 < t < t_k,\ t_k = kT_0 + \tau_k,$$

the system (5) has the form

$$\frac{dX}{dt} = AX + B, \quad X(k \cdot T_0) = X_k$$

with solution

$$X(t) = \exp(A(t - k \cdot T_0)[X_k + D] - D.$$

At the switching time $t = t_k$:

$$X(t_k) = \exp(A(t_k - k \cdot T_0)[X_k + D] - D,$$

$$D = A^{-1}B.$$

In the subsequent time interval

$$t_k < t < (k+1)T_0,$$

system (5) takes the form

$$\frac{dX}{dt} = AX,$$

with the solution

$$X(t) = \exp(A(t - t_k))X(t_k).$$

In this way, for $t = (k+1) \cdot T_0$

$$X_{k+1} = \exp(A((k+1) \cdot T_0 - t_k))X(t_k), \qquad (6)$$

$$X_{k+1} = X((k+1) \cdot T_0).$$

Substituting the expression for $X(t_k)$ into (6), we obtain the two–dimensional piecewise–smooth map [6]

$$X_{k+1} = \exp(A \cdot T_0)[.X_k + D] - \exp(T_0 - \tau_k)D, \qquad (7)$$

$$k = 0, 1, 2, 3, \ldots$$

Here τ_k

$$\tau_k = \begin{cases} 0, & \varepsilon_k < 0; \\ T_0, & \varepsilon_k > V_s; \\ \dfrac{T_0 \varepsilon_k}{V_s}, & 0 \le \varepsilon_k \le V_s; \end{cases}$$

with

$$\varepsilon_k = \alpha(V_{ref} - CX_k),$$

where $C = (\beta, 0)$ is a row such that $CB = 0$.

In the following simulations, we shall use: $T_1T_2 = 10240\ s^2$; $T1+T_2 = 352$ s; $K = 327.8$ C°/V; $T_0 = 10$ s; $U = 24$ V; $\beta = 0.01$ V/C°; $Vs = 5$ V; $\alpha > 0$; $Vref = 5$ V.

The gain factor α and the input voltage U_0 will be used as the main control parameters. The period T of periodic motion is an integer multiple to the rump period T_0, i.e. $T = mT_0$, $m = 1, 2, \ldots$. We will refer to this type motion as a period–*m cycle* or *m–cycle*.

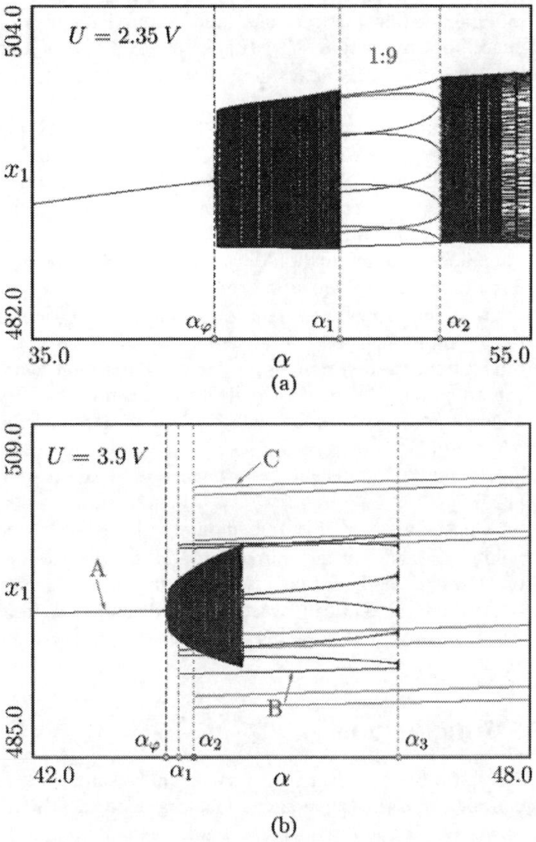

Fig. 2. (a) Birth of the closed invariant curve associated with quasiperiodic dynamics from the stable 1–cycle through a Neimark–Sacker bifurcation at the point α_φ. (b) One–dimensional bifurcation diagram for $U = 3.9$ V. The diagram illustrates a form of multistability in which stable 7 – and 8 – cycles coexist with a stable closed invariant curve.

4 Bifurcation analysis

Depending on the parameter values, a variety of different bifurcation scenarios is observed.

Figure 2(a) illustrates the appearance of the quasiperiodic orbit form a stable 1 – cycle. For the values of the corrector gain factor $\alpha < \alpha_\varphi$ the map (6) shows a single stable 1 – cycle. As α increases, the 1 – cycle undergoes a Neimark–Sacker bifurcation at the point $\alpha = \alpha_\varphi$ in which the stable focus becomes an unstable one. The loss of stability of the period – 1 cycle gives rise to a stable closed invariant curve, associated with quasiperiodic dynamics.

The asymptotic behavior of the system (7) on the invariant curve depends on the rotation number. If the rotation number is irrational, the invariant curve is densely filled with points of the orbit of map (7), and the dynamics is quasiperiodic. When the rotation number is rational, the closed invariant curve contains a pair of cycles, one of which is stable, while the other is a saddle. The attracting closed invariant curve includes two cycles, a saddle and a stable node, and the saddle–node connection composed of the unstable manifolds of the saddle cycle. Parameters regions corresponding to rational rotation numbers are called Arnold tongues or phase–locking windows [9, 10]. Figure 2(a) shows also the 1:9 phase–locking window. Here α_1 and α_2 are the saddle–node bifurcation points.

The bifurcation behavior of the system (7) is not completely covered by the cases described above. As illustrated in Fig. 2(b) for higher values of the input voltage U more complicated nonlinear phenomena are possible, including multistablilty and hysteretic transitions. The bifurcation diagram in Fig. 2(b) contains three branches denoted as *A*, *B* and *C*.

When the parameter α increases, the stable 1 – cycle belonging to the branch *A* undergoes a Neimark–Sacker bifurcation. In this bifurcation, a closed invariant curve corresponding to the quasiperiodic dynamics softly appears and the 1 – cycle becomes an unstable focus. With further increase in the value α, the system enters the region of multistability through a fold bifurcation [11, 12].

On the part of this region that falls between the points α_2 and α_3 the stable 7 – and 8 – cycles (see branches *B* and *C)* coexist with the quasiperiodic attractor. Here one can observe a hard transition from period – 7 cycle to a stable 8 – cycle or quasiperiodic attractor.

5 Conclusions

The purpose of this paper was to study some of the complex dynamic phenomena that can arise in a high–power heating unit with pulse width modulated control. The mathematical model of this system was represented as a two–dimensional set of nonautonomous differential equations with discontinuous right–hand sides.

The first step in our investigation was to reduce this system to a two–dimensional piecewise–smooth map (7). We showed, how the high–power heating unit with pulse–width modulated control can exhibit both a transition to quasiperiodic dynamics through a Neimark–Sacker bifurcation and a variety of interesting nonlinear phenomena, including phase–locking modes, multistablilty and hysteretic transitions.

In this paper, we discussed only the first steps towards understanding of the observed nonlinear phenomena.

Further development of these investigations is important both from the theoretical and applied points of view, as it will provide a theoretical foundation for design and control of a wide class of pulse–modulated control systems.

The authors would like to thank Prof. Zh.T. Zhusubaliyev for his careful reading of and constructive suggestions to the first draft of this manuscript.

Research is carried out with the financial support of The Ministry of Education and Science of the Russian Federation within the Public contract project 2.1396.2017/PCh.

References

1. H. J. Scheel, T. Fukuda, *Crystal growth technology* (John Wiley & Sons, 2003)
2. H. J. Scheel, P. Capper, *Crystal growth technology: From fundamentals and simulation to large–scale production* (John Wiley & Sons, 2008)
3. V.G.Rubanov, A.S. Kizhuk, Y.A. Gol`tsov, Int. J. Soft Comp. **10,** (2015)
4. Y.A. Gol`tsov, A.S. Kizhuk, V.G.Rubanov, Int. J. Pharm. Techn. **12,** (2016)
5. M.H. Rashid, F.L. Luo, *Power Electronics Handbook* (Elsevier Science, 2011)
6. Zh.T. Zhusubaliyev, E. Mosekilde, *Bifurcations and chaos in piecewise–smooth dynamical system* (World Scientific, Singapore, 2003)
7. S.Banerjee, G. C. Verghese, *Nonlinear phenomena in power electronics* (IEEE Press, New York, 2001)
8. Di. M Bernardo, C. J. Budd, A. R. Champneys, P. Kowalczyk, *Piecewise–smooth dynamical systems: Theory and applications* (Springer–Verlag, New York, 2008)
9. J.M.T Thompson, H.B. Stewart, *Nonlinear dynamics and chaos* (Wiley, Chichester, 1986)
10. Yu.A. Kuznetsov, *Elements of applied bifurcation theory* (Springer–Verlag, New York, 2004)
11. Zh.T. Zhusubaliyev, E. Mosekilde, O.O Yanochkina, IEEE Trans. Power Electron. **26**, 4 (2011)
12. Zh.T. Zhusubaliyev, E. Mosekilde, Math. Comput. Simul. **109** (2015)

MATEC Web of Conferences 129, 01032 (2017) DOI: 10.1051/matecconf/201712901032
ICMTMTE 2017

Monitoring of the stress-strain state of the surface layer of a part in the cutting process using vibroacoustic diagnostics

Dmitry Allenov[1,*], *Mikhail* Kozochkin[2], and *Irina* Andryushchenko[1]

[1]Russian University of friendship of people, 117198, Miklukho-Maklaya 6, Moscow, Russia
[2]Moscow State University of Technology "STANKIN", 127055, Vadkovsky per. 1, Moscow, Russia

Abstract. The state of the surface layer after mechanical process influences on the performance characteristics of the parts: corrosion and fatigue resistance, dimensional stability during the operating time. The wear of tools can have a decisive influence on the surface deformation. The present paper is devoted to monitoring the influence of wear on the surface layer using the analysis of vibration signals.

1 Introduction

Currently in engineering, many parts have to work in conditions of high temperatures, in contact with corrosive media, in the presence of high static and dynamic stresses. All these factors lead to the appearance of various defects on the part surface. Most often, the appearance of such defects occurs in a thin, near-surface layer of parts, which is largely formed when the appropriate processing technology is realized. Practice established a close relationship between the properties of the surface layer of parts and their performance properties, including strength and durability. There are many reasons for this connection [10]. In stressed parts, the surface layer is exposed to more severe conditions; risks and irregularities on the surface are places of formation and development of fatigue cracks; the presence of dislocations and microcracks of an atomic scale at high degrees of deformation promotes the formation of untreated microcracks prone to spreading; when parts are used in conditions of high temperatures, strength-free parts without deformation hardening prevail; large degrees of deformation reduce the parts resistance to corrosion-mechanical failure and reduce the resistance of protective coatings, and also affect the relaxation rate of the surface layer properties and its weakening.

Since the set of properties of the processed surface characterize the quality of the surface layer of the processed part, or the quality of the treatment, many researchers all around the world have studied the effects of cutting regimes and the geometry of the cutting tool on these properties. It was found that the properties of the surface layer begin to form in the zone of advanced hardening before the cutting edge of the tool. The development of a dislocation structure in the form of slip bands, an increase in the dislocation density lead to hardening of the surface layer [3,10,13,14]. The intensity of hardening of the surface layer, determined by the density of dislocations, depends on the level of the applied load and the action time. In turn, the dislocation density is closely related to the increase in the hardness of the hardened material and the degree of its hardening. In the literature, it is noted that with increasing hardening of the treated surface, the heterogeneity of the properties of the surface layer also increases, even if all processing conditions are constant. The heterogeneity of the surface layer is manifested in the nonuniform distribution of point defects, surface areas are characterized by different chemical properties. A micro relief is formed on the surface with irregularly shaped dimples and projections, with a random distribution of height and pitch of microroughness. This leads to the appearance in the surface layer of places with increased tendency to cracking, corrosion, adhesion and abrasion [13]. In the process of chip separation, the thickness of the layer jammed by the cutting edge changes randomly, creating the prerequisites for the nonuniform hardening of the surface layer and the fluctuation of the microroughness parameters.

A latent energy of deformation accumulated by the surface layer as a result of hardening can serve as an integral characteristic of the stress-strain state [2,13]. This energy is accumulated as a result of elastic distortions of the crystal lattice caused by an increase in the dislocation density. The lower the latent energy of deformation in the surface layer creates the higher fatigue resistance of a part operating at ordinary temperature. When working in conditions of elevated temperatures, this effect is particularly pronounced.

An important consequence of the instability of the physical-mechanical state of the surface layer after cutting is the violation of dimensional stability of parts after some time after treatment. This problem is relevant not only for thin-walled products, but also for precise details.

An experimental assessment of the influence of technological parameters on the accumulation of hidden

* Corresponding author: allenov_dg@rudn.university

surface energy has shown that there is an ambiguous effect of the same parameters on the level of latent energy when processing various materials. It has been experimentally established that the wear of the cutting edge of tools affects the accumulation of the latent energy of deformation at times to a greater extent than the cutting regimes [13]. The effect of tool wear on the level of hidden energy is further complicated by the fact that there is no identical evaluation of the very nature of wear. The adopted wear control on the back face does not guarantee its fair evaluation with respect to the influence on the accumulation of the hidden energy of deformation of the surface layer of the part. It is shown in [9] that wear forms occur when the tool with a greater faceted wear on the back side creates less surface deformations than a tool with less wear and tear on the back side. It suggests that widespread periodic monitoring of radial wear can not guarantee the stability of the surface layer of the part, since radial wear can be recalculated into wear at the back edge, but the uncertainty of the force impact on the surface of the part will remain large due to the accidental nature of the fractures. The rounding radius of the cutting edge has only a statistical connection with the wear facet at the back edge. In addition, modeling the destruction of the cutting edge in the form of a portion of a cylindrical surface is often a very remote approximation to the real form [4]. The purpose of this research is to create an integrated system for ensuring the quality of the surface layer during blade processing by controlling indirect parameters.

2 Monitoring of the tool wear state

It is known that the destruction of the cutting edge, which causes intensive deformations of the treated surface, leads to an increase in the energy consumed, corresponding to an increase in cutting forces and the amount of heat released. The control of cutting forces can be carried out by changing the active power consumed by the drives on the machine. However, in the final treatment regimes, the method does not always give satisfactory accuracy. This is especially noticeable for finishing with a large ratio of the engine power used and the power used for cutting [7]. This situation is typical for processing centers, where both roughing and finishing of parts from different materials and different configurations are carried out. Temperature control in the cutting zone is a difficult task even for laboratory conditions. The connection between the vibroacoustic (VA) activity of the cutting process and wear of the cutting tool is widely noted in the technical literature. VA signals are relatively easy to control with the help of accelerometers, but the disadvantages of VA diagnostic methods include the difficulty of isolating useful information from the set of parameters of VA signals accompanying cutting. The most simple parameters of VA signals, which are easily obtained in digital or analog form, are the effective amplitudes in the allocated frequency bands. That is, the distortion of the shape of the VA signal spectrum can be used as a diagnostic indication of the change in the state of the cutting tool. Distortion of the shape of the spectrum assumes a change in the ratio of effective amplitudes for different frequency ranges. The increase in heat release in the area of contact between the surfaces of the tool and the material being processed leads to a change in the mechanical properties in the friction pair, for example, to a decrease in the hardness of the material being processed. It was shown in [6, 8] that the amplitude of the high-frequency VA signal in the frictional contact in the first approximation with increasing hardness of the softer element of the friction pair increases linearly. Thus, with increasing temperature (and a drop in the mechanical characteristics of the material being processed) in frictional contact under conditions of adhesive friction, the high-frequency components in the spectrum of the VA signal should decrease. On the other hand, the growth of the load in the contact tool with the machined surface causes additional deformation of the entire technological system and increases its potential energy [5], making it less stable when chip elements are allocated. The situation can be aggravated by a decrease in the bearing capacity of the cutting surface with an excessive increase in the contact temperature, which forms a ripple of the surface and instability of the properties of the surface layer. When separating the chips, the balance of forces in the process system is violated, leading to the emergence of potential energy, the relaxation of elastic deformations with the appearance of oscillations at the natural frequencies of the technological system. According to the above, it follows that with an increase in the potential energy determined by the cutting forces, the oscillations in the technological system in the region of comparatively low frequencies will also grow. This allowed us to use the ratio of the effective amplitudes (Kf) of the vibration signal in the region of high and low frequencies as a diagnostic feature. Fig. 1 shows examples of the spectra of VA signals from the accelerometer that accompanied the turning of a roller of steel 45 with a depth of 0.5 mm, a feed of 0.15 mm / rev, with a cutting speed of 47 m / min. The spectra are presented to illustrate extreme situations in the experiment: for cutting an acute plate and a plate that caused the greatest deformation of the surface layer. On the inset in the corner of Fig. 1 shows the change in the diagnostic parameter K_f for five plates with different wear of the cutting edge, the number of which increased with increasing influence on the strain rate of the surface layer.

Fig. 1 shows clearly that when the worn-out No. 5 plate was used, the amplitudes in the low-frequency region (0.7-1.5 kHz) increased in the BA signal spectrum, but in the high-frequency region (4-15 kHz) there was a significant decrease in amplitudes

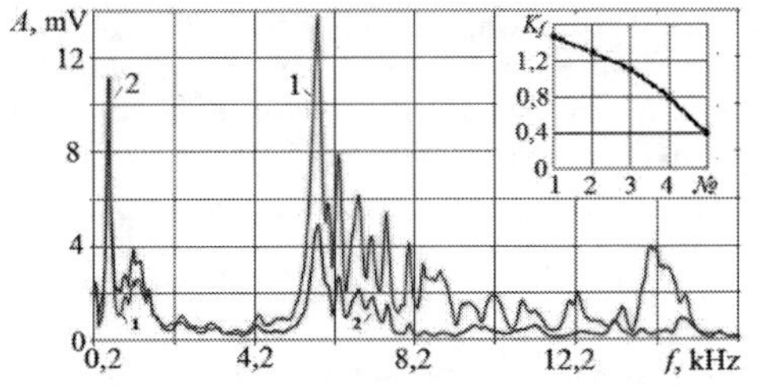

Fig. 1. Spectra of VA signals for sharp (spectrum 1) and blunt (spectrum 2) plates, on the inset change in the values of the parameter Kf with increasing wear influence on the deformation of the surface layer (plate numbers are given).

If we take the ratio Kf of the effective amplitudes for the indicated high-frequency and low-frequency ranges, then when working with an sharp plate Kf = 1.5, and when working with plate No. 5, the ratio Kf = 0.4. The inset in Fig. 1 shows the graph of the change in Kf with the increase in the number of all the cutting inserts used in the experiments. The values of Kf decrease monotonically in accordance with the increase in the intensity of surface deformations caused by the degradation of the cutting edge. At the same time, wear was formally measured according to the width "h" of the chamfer of wear along the rear face: 1 - h1 = 0 mm; 2 - h2 = 0.6 mm; 3 - h3 = 0.68 mm; 4 - h4 = 1.1 mm; 5 - h5 = 0.8 mm. It can be seen that the largest chamfer of wear along the back face was at sample No. 4, and the smallest value of Kf was at plate number 5. This is explained by the fact that on the wear face there were random stains, which determined deformations of the treated surface.

3 Monitoring of the stress-strain state

Measurements of the deformations of surfaces treated with the five plates presented made it possible to construct the dependence of the parameter Kf on the intensity of deformation of the surface of the parts εint. To measure the intensity of deformations, we used the experimental method of fission grids applied to the side surfaces of a billet with a mesh size of 50 μm [16]. In Fig. 2 is a graph of this relationship. On the chart, it is possible to single out sections of run-in, normal wear and catastrophic wear.

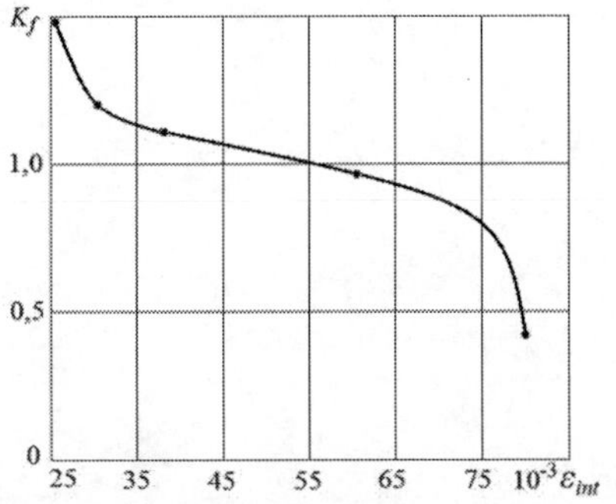

Fig. 2. The parameter change of Kf with increasing strain intensity of εint of the Surface layer.

Kf parameter can be used to monitor of tool wear effect on the intensity of deformation of the surface layer, but with constant processing modes.

When changing the processing modes, Kf values may vary, causing a change in the criteria for the failure of the tool.

Diagnostic parameter search, which was sensitive to a change in the instrument state, but varied little with the processing modes changes, is also a significant goal of ongoing research.

Practice has shown that the VA signals and Kf parameter depend on the cutting speed to the greatest extent.

Fig. 3 shows the dependence of Kf on the cutting speed for an acute and worn tool when sharpening an alloy NiCr20TiAl.

Being a tool with wear values, Kf change five times when changing the cutting speed from 15 to 50 m / min.

Fig. 3 shows the change graph of qf, defined as the value relation of Kf with an acute tool to the value of Kf with a blunted instrument.

It is obvious that the cutting speed of qf varies within 20% in the entire range, with real speed variations relative to the optimum value for the alloy NiCr20TiAl, qf changes are no more than 5%.

Thus, knowing Kf values for different cutting regimes for an acute tool, it is possible to monitor qf value in a wide range of variation of the cutting conditions, with a single failure criterion to have [qf].

For the case shown in Fig. 3 the failure criterion [qf] = 2, which means Kf decrease is allowed from the initial level of 2 times.

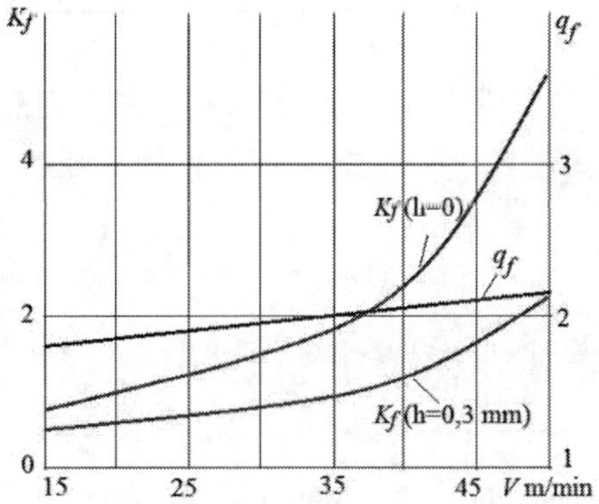

Fig. 3. Variation in Kf and qf values for the cutting speed variation during the sharpening the alloy NiCr20TiAl.

It is important to note that intense self-oscillations may occur in the process of cutting with the processing conditions change, under which Kf behavior may be violated.

At high requirements of surface quality, work with self-oscillations should be eliminated.

The detection of modes with self-oscillations can be done using the VA signals analysis [8].

4 Conclusion

There is a close connection between the latent surface energy, the dislocation density and the number of microcracks with the deformation intensity of the surface layer during the blade machining of the parts.

For parts with high requirements for dimensional, corrosion and fatigue stability, it is necessary to monitor the instrument state changes affecting the deformation of the surface layer of machined parts in automated production environments.

It was shown that this function could be executed by the VA signals analysis accompanying cutting.

As a diagnostic parameter, good prospects for use were established at the ratio of the effective amplitudes of the VA signals taken in the high-frequency and low-frequency ranges, which reflect the shape distortion of this signal spectrum

Further VA monitoring can be obtained for monitoring operations related to nano- cutting [15], where the state of the surface layer plays a decisive role.

References

1. A. Samraj, S. Sayeed, J. Emerson Raja, J. Hossen, A. Rahman, *Dynamic clustering estimation of tool flank wear in turning process using SVD models of the emitted sound signals*, World Academy of Science, Engineering and Technology **80,** pp. 1322–1326 (2011)
2. V.F. Bezjazychnyj, B.M. Drapkin, M.A. Prokofiev, M.V. Timofeev, *Research of stored energy of deformation by metal during indentation by sphere indenter* [Issledovanie zapasennoi metallom energii deformatsii pri vdavlianii charovogo indentora], Zavodskaya labortoriya. Diagnostika materialov [Factory laboratory. Material diagnostics], no.**4**, t.**71**, pp.32-35 (2005)
3. V.F. Bezjazychnyj, M.A. Prokofiev, A.N. Sutyagin, *Technological Assurance of Wear Resistance of Surface Layer of Machine Parts After Turning*, Russian Internet Journal of Industrial Engineering, Vol. **2**, no. **4**, pp. 42-46. (2014) (http://indust-engineering.ru/)
4. I.R. Dobrovinskiy, Yu.T. Medvedik., M.Yu. Medvedik, V.S. Chapaev, *Povolzhskiy region. Tekhnicheskie nauki,* Izvestiya vysshikh uchebnykh zavedeniy. [University proceedings. Volga region. Engineering sciences], no. **2 (26)**, pp. 59-70 (2013)
5. M.P. Kozochkin, *Influence of Machine_Tool Dynamics on the Vibration in Cutting. Russian Engineering Research*, Vol. **34**, No. **9**, pp. 573–577 (2014)
6. M.P. Kozochkin, *Nonlinear cutting dynamics. Russian Engineering Research,* **32 (4)**, (2012 (13)
7. M.P. Kozochkin, A.R. Maslov, and A.N. Porvatov, *Information-measurement and control systems for force and vibroacoustic parameters,* Measurement Techniques, Vol. **58**, No. **8**, pp. 839-844 (2015)
8. M.P. Kozochkin, A.N. Porvatov, *Effect of Adhesion Bonds in Friction Contact on Vibroacoustic Signal and Autooscillations,* Journal of Friction and Wear, Vol. **35**, No. **5**, pp. 389–395 (2014)
9. M.P. Kozochkin, M.A. Volosova, D.G. Allenov *Effect of wear of tool cutting edge on detail surface layer deformation and parameters of vibro-acoustic signals,* Materials Science Forum, Vol. **876**, pp 50-58 (2016)
10. V.S. Muhin, *Poverhnost: tehnologicheskie aspekty prochnosty detaley GTD.* (Moscow: Nauka, 2005)
11. S. Orhan, A. Osman, Er, N. Camus-cu, E. Aslan, *Tool wear evaluation by vibration analysis during end milling of AISI D3 cold work tool steel with 35 HRC hardness,* NDT&E International **40**, pp. 121–126 (2007)
12. A.S. Sidorov, *Monitoring i prognozirovanie iznosa rezhushchego instrumenta v mekhatronnykh stanochnykh sistemakh [Monitoring and forecasting tool wear in mechatronic machine systems], Abstract of dissertation for the degree of Candidate of Technical Sciences* (Ufa, 2007)
13. V.K. Starkov, *Dislokatsionnye predstavleniya o rezanii metallov [Dislocation representations about metal cutting]* (Mashinostroenie, Moscow, 1979)
14. A.G. Suslov, *Technologicheskoe obespechenie parametrov sostoyaniya poverchnostnogo sloya [Technological assurance of surface layer quality parameters]* (Mashinostroenie, Moscow, 1987)
15. Q. Wang, Q. Bai, J. Chen, Y. Sun, Y. Guo, Y. Liang, *Subsurface defects structural evolution in nano-cutting of single crystal copper,* Appl, Surf Sci.; **344,** pp. 38–46 (2015)
16. Werkstüchklassifizierung und auswahl für ein flexibles//Fertigungssystem Madrich. Oberkochen H. «Werkstatts technik». 71. №8. P. 485-489 (1981)

DOI: 10.1051/matecconf/201712901033

Technological system self-tuning when milling

Aleksandr Kozlov[1], *Evgenij* Kiryuschenko[2,*], and *Andrej* Kirichek[3]

[1]Lipetsk State Technical University,398600, 30, Moskovskaya Str.,Lipetsk, Russia
[2]PJ-SCo. 'Novolipetsky Metallurgical Complex', 398040, 2, Metallurgists' Square, Lipetsk, Russia
[3] Bryansk State Technical University, 241000,7, 50 let Octyabrya Blvd, Bryansk, Russia

Abstract. This paper reports the dynamics of the technological system during the milling of complex parts and the investigations of possibility in efficiency increase of a cutting process at the expense of conditions creation for a system self-tuning at which a general level of oscillations will not decrease in it. This can be achieved with a special damping device located on the machine screw gears supports. In this case, if the period of the wave on the cutting surface does not coincide with the period of natural oscillations, then there will be no resonance, and the level of oscillations will be much smaller.

The modern technology of multiaxes milling allows manufacturing parts with various forms, but the productivity of such a process has a number of limitations. [1] Such limitations may be requirements to accuracy assurance in a form, parameters of worked surface roughness, and in some cases, - rigidity of a tool and a technological system on the whole [2].

The impossibility to increase rigidity considerably causes the oscillations appearance when cutting which affects in a negative way upon quality and productivity of processing.

As a rule there is an approach to the problem of milling efficiency increase from the standpoint of vibrations decrease. A basic problem at the development of methods for milling efficiency increase is an adequate presentation of basic regularities in the excitation of forced oscillations and self-oscillations as factors defining machining accuracy [3-5].

A technological system is considered as a mechanical oscillation system with the contour of some masses connected between each other with elastic and dissipative elements (Fig. 1).

The vector equation of the motion of the i-th mass:

$$m_i \cdot \overline{a}_i = \overline{F}_i + \overline{P}_i - \overline{F}_{i+1}, \qquad (1)$$

Pi – external force influence upon the i-th mass, H; Fi – force causing elastic damping on the i-th unit, H; m – mass of the unit, kg; a – unit acceleration, mm/s2.

$$F_i = C_i \cdot (X'_{i-1} - X'_i) + h_i \cdot (v_{i+1} - v_i), \qquad (2)$$

$Fu_i = -m_i \cdot a_i$ – inertial force on the i-th unit, H; X'i – vibrodisplacement of the i-th mass, mm; vi, – vibrovelocity of the i-th mass, mm/s. [6]
Vibration velocity is a velocity of an oscillating mass displacement. From the standpoint of vibro-diagnostics it is a universal parameter which allows comparing different oscillating processes. With its aid it is possible to determine a correlation of an oscillating process with cutting forces changes.

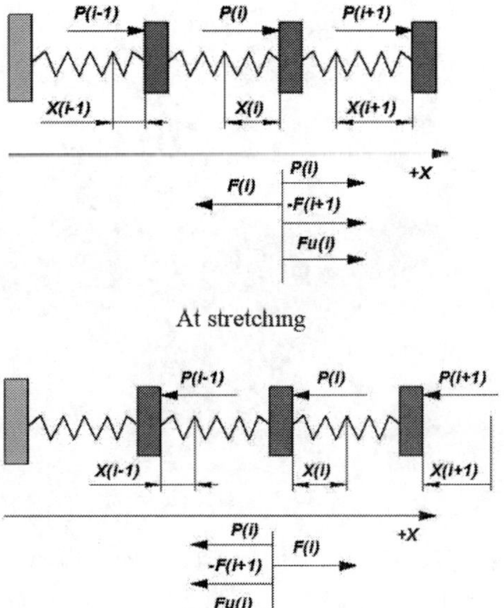

Fig. 1. Design scheme of an elastic system contour

It is well-known [7- 9] that when milling the primary oscillations are formed with secondary resonant forced oscillations. In such a way, the vibrations of the technological system are a sum of some oscillating processes. Hence, on the basis of the analysis of experimental data there is obtained a dependence

* Corresponding author: kiryuschenko_ev@nlmk.com

describing changes in vibro-velocity of a milling machine spindle:

$$v = A_1 \cdot e^{-0,1S}\cos(f_1 \cdot S + \varphi_{11}) + \\ + A_2 \cdot e^{-0,1n}\sin(f_2 \cdot n + \varphi_{12}) + v_0, \quad (3)$$

v – vibrovelocityinthesystem, mm/s; A_1andA_2 – amplitudesof corresponding oscillating processes, mm; S – feed, mm/min; n – rotation frequency, rpm; f – frequency of corresponding oscillations, Hz; φ – difference of oscillation phases in the corresponding direction, rad; v_0 – vibrovelocity out of a cutting area, mm/s.

The peculiarity of a milling process is an interrupted character of cutting which may be characterized by lag time. Lag time corresponds to the value of phase difference this is the same time interval, but measured in a phase (id est, in the angles of the phase):

$$\varphi_{21x} = \omega_x \cdot \tau_{21} = \omega_x \cdot (k+i) \cdot T_x = (k+i) \cdot 2\pi \ (4)$$

φ_{21x} –difference of phase oscillations in the system on X axisduring two adjacent teeth motion through the given point in the area, rad; ω_x –angular frequency of oscillations on X axis, rad/s.

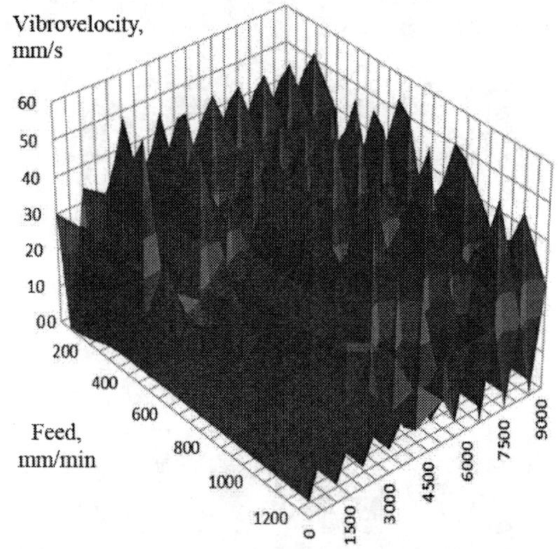

Fig. 2. Machining parameter impact upon vibrovelocity

If we consider a trace on the surface of cutting as a driving force with the frequency p,then the phase of forced oscillations of the system must lag behind the phase of a disturbing force for the value:

$$2\pi \cdot i_p = -\arctan\frac{2 \cdot \varepsilon \cdot \gamma}{1-\gamma^2}, \quad (5)$$

ε – dimensionless factor of attenuation; γ–ratio of the frequency of the driving force $p$to the own frequencyf_x.

During the analysis of vibrovelocity changes (Fig.2) on the basis of equations (3) and (4) there is observed a soft excitation ofregenerative oscillations. It appears from this that the system steady at cutting on a trace but with a small coefficient of rigidity may become unsteady at cutting along a trace when a rigidity coefficient will exceed a certain value. At the equality of own and driving frequencies the phase displacement does not depend upon ε may be equal only to $-\pi/2$. Strictly speaking, this value of the phase displacement corresponds to conditions of resonance occurrence [10].

In such a way, the increase of oscillations level at disturbances on a trail is determined, in the first place, by resonance and near-resonance phenomena caused by a coincidence of the own frequency of the system with the frequency of the wave on the surface of cutting. The results obtained have also shown that the parameter affecting the level of oscillations is a phase displacement at the initial moment of trail cutting. At that a minimum level of oscillations occurs at the initial displacement equal to $\pi/2$, and a maximum one – at $-\pi/2$. At that, if the initial displacement ofphases differs from $-\pi/2$, then a part of oscillations energy is consumed for the system tuning, and oscillations are attenuated rather quickly. If the initial displacement of phases is equal to $-\pi/2$ or close to this value, then the system is set up immediately to the essential displacement of phases and in the system arise the established oscillations. In consequence of the existence of friction in the system such oscillations are damped gradually and damping time depends upon the initial displacement of phases.

It is possible to eliminate resonant oscillations with large amplitudes, if a damping device is introduced into the structure of metal-cutting equipment. As an object with a damper is a system with two degrees of freedom along with damping at a certain frequency there are appeared two new resonances corresponding to the own frequencies of the two-degree system (Fig. 3). In this case a damping device is an additional body with the mass of m_2 connected with the object through a spring with the rigidity of c_2 and a damper with the coefficient of viscous resistance, h.

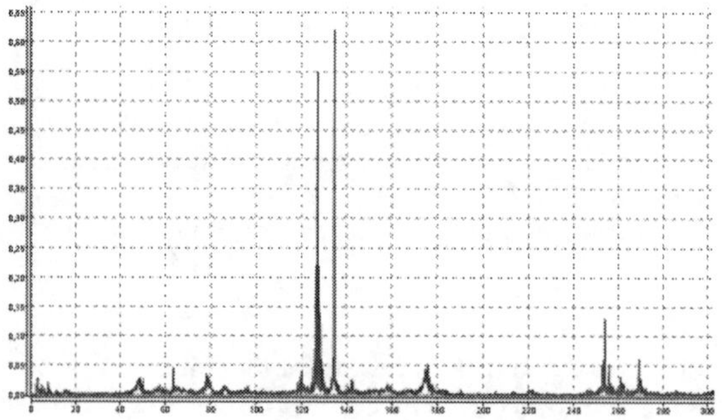

Fig. 3. Oscillation spectrum (value of spindle vibrovelocity in mm/s at corresponding frequencies, Hz) of portable milling equipment

In the presence of a damper a complete oscillations damping becomes impossible as it affects in addition a technological system. [11] But a damper allows obtaining the limited amplitude of system oscillations in all the frequency range at an efficient selection of damper parameters (Fig. 4). This phenomenon may be called a self-tuning of the oscillation system [12].

Differential equations of the system with a damper become:

$$m_1\ddot{x}_1 + h\dot{x}_1 + (c_1 + c_2)x_1 - h\dot{x}_2 - c_2x_2 = F_i,(6)$$

$$m_2\ddot{x}_2 - h\dot{x}_1 - c_2x_1 + h\dot{x}_2 + c_2x_2 = 0$$

m_1 – system mass, kg; m_2 – damper mass, kg; c_1 – system rigidity, c_2 – damper rigidity, h – coefficient of viscous resistance.

Let us add to these equations some dimensionless parameters: β, η, δ.

$$\beta = \frac{m_2}{m_1}, \quad \delta = \frac{\sqrt{c_2/m_2}}{\omega_0}, \quad \eta = \frac{h}{2m_2\omega_0},$$

ω_0 – eigen frequency of machine oscillations, 1/s; n_2 – partial frequency of the damper, 1/s.

Then, after transforms we obtain an expression for the dimensionless amplitude of damper oscillations:

$$A_1 = \sqrt{\frac{(\delta^2 - \gamma^2)^2 + 4\eta^2\gamma^2}{\theta^2 + 4\eta^2\gamma^2(1 - \gamma^2 - \beta\gamma^2)^2}}, \qquad (7)$$

where$\theta = (1 - \gamma^2)(\delta^2 - \gamma^2) - \beta\delta^2\gamma^2$.

The expression (7) obtained allows calculating the amplitude of oscillations for a damping device for the achievement by a technological system a self-tuning state.

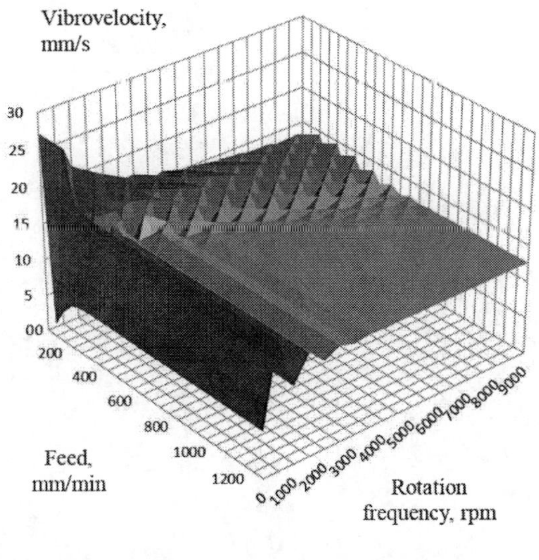

Fig. 4. Theoretical limitation of values of the spindle vibrovelocity in milling equipment at damping

In such a way, the more power for tuning is consumed, the less it becomes for oscillations excitations. To reduce a regenerative effect it is possible by means of the system tuning from resonance with a trail. If a wave period on the surface of cutting does not coincide with the period of eigen oscillations, then resonance will not appear and the level of oscillations will be considerably lesser.

In such a way, at the phase displacement for -π/2 occurs a system self-tuning as a result of which a part of oscillation processes is consumed and a general level of oscillations will drop. The more energy is consumed for tuning, the lesser it is for oscillations excitation. It results from this that one of the efficient ways for regenerative oscillations damping may be the formation of such conditions for its operation at which it should be necessary for it to carry out constantly such a retuning. It may be achieved with the aid of a special damping device located on bearings of machine spiral gears. At that if the period of the wave on the surface of cutting does not coincide with the period of eigen oscillations, then resonance will not appear and the level of oscillations will be much lesser.

References

1. R. Radulescu, *A General Cutting Process Model for High Speed Machining Dynamic and Thermal Considerations: PhD Thesis* (University of Illinois at Urbana-Champaign, 1993)
2. V.M. Svinin, *Study of in multi-blade processing regenerative self-oscillations research,* Metal Machining **3(28)**, p. 28–30 (2005)
3. V.I. Gu-zeev, V.V. Batuev, *Taking into account the effect of traces of the previous treatment on the accuracy of milling of space-complex surfaces*, Progressive technologies in mechanical engineering: Sat. Sci. Tr., Cheliabinsk: Izd-voIuUrGU, p. 115-117 (2003)
4. V.M. Svinin, *Self-organization of secondary auto-oscillations during blade machining*, STIN, **1**, p.7-13 (2006)
5. Sastry Sridhar, Kapoor Shiv G., DeVor Richard E., Dullerud Geir E. ASME J. Eng. Indus. **123**, p. 753–756 (2001)
6. A.M. Kozlov, E.V. Kiryushchenko, S.F. Kuznetsov, *The vibration dampening methodfor portable milling equipment*, Fundamental and Applied Problems in Engineering and Technology **5**, p. 91-97 (2014)
7. Yu.I. Gorodetsky, S.N. Strebulyaev, Yu.E. Maiorova, *Dynamic system self-oscillations investigation for a milling machine with a nonlinear element*, Bulletin of Scientific Technical Development. National Technological Group **9(25)**, p. 91–95 (2005)
8. S.A.Vasin, *Prediction of Tool Vibration Resistance at Turning and Milling* (Moscow: Mechanical Engineering, 2006)
9. A.M. Kozlov, G.E. Malyutin, *The finishing milling productivity Increasing of complex shape concave surfaces on CNC machines*, Bulletin of Mechanical Engineering **12**, p. 71-75 (2014)
10. E.V. Kiriushchenko, *Modeling the dynamics of the technological system in milling*, Bulletin of Voronezh State Technical University, **8(10)**, p. 87- 93 (2012)
11. G.E. Maliutin, *Determination of the cutting forces for a complete finishing of the concave surfaces of a*

complex shape with spherical milling cutters on CNC machines, Fundamental and Applied Problems in Engineering and Technology **4(306)**, p. 74-81 (2014)

12. A.M. Kozlov, E.V. Kiryushchenko, *Optimal processing modes selection technique for vibration velocity minimization of during milling,* Current Fundamental and Applied Investigations, **1(4)** p. 32-38 (2012)

 DOI: 10.1051/matecconf/201712901034

Effect of changing polarity of graphite tool/ Hadfield steel workpiece couple on machining performances in die sinking EDM

Hacı Bekir Özerkan[1,*]

[1] Gazi University Vocational School of Technical Sciences, Department of Machine and Metal Technology, 06374 Ankara, Turkey

Abstract. In this study, machining performance ouput parameters such as machined surface roughness (SR), material removal rate (MRR), tool wear rate (TWR), were experimentally examined and analyzed with the diversifying and changing machining parameters in (EDM). The processing parameters (input par.) of this research are stated as tool material, peak current (I), pulse duration (t_{on}) and pulse interval (t_{off}). The experimental machinings were put into practice by using Hadfield steel workpiece (prismatic) and cylindrical graphite electrodes with kerosene dielectric at different machining current, polarity and pulse time settings. The experiments have shown that the type of tool material, polarity (direct polarity forms higher MRR, SR and TWR), current (high current lowers TWR and enhances MRR, TWR) and pulse on time (t_{on}=48□s is critical threshold value for MRR and TWR) were influential on machining performance in electrical discharge machining.

1 Introduction

The electrical discharge machining (EDM) is widely applied, practiced and important manufacturing technique, in die and mold making industry to fabricate multifaceted and three dimensional complicated geometries made of electrical conductive materials carry out roughy and/or mirror-like finishing operations optionally [1]. This technique process was predicate on removing material from an untreated raw metal part by way of sequential random electrical discharges between tool and the work piece in dielectric environment [2]. The electrode is approached to the work piece until the initial former sparks occur since therefore the operating voltage creates the required and sufficient energy to ionize the dielectric fluid [3]. In this technique, main EDM performance parameters are known as electrode material, work piece materials, dielectric and the electrical parameters consists of polarity, pulse on/off time, current and voltage. Ho and Newman remarked that EDM research topics could be arranged into three basic titles, these are measurement and determining of machining performance under different machining conditions, the influences of machining inputs and designation and fabricating of tools. They educed that the main and effective machining performance outputs are attributed to the MRR, tool wear and machined surface quality [1]. One of the most important electrical parameters of EDM is polarity [4].For briefly enlightening of electrical erosion mentality, and the morphological, metallurgical and textural ultimate surface characteristics, it's always necessary to study the effect of polarity which is defined as reverse polarity (electrode positive) and straight polarity (electrode negative) [5,6].

From the previous and published literature, main machining performance characteristics (MRR, TWR, SR) of difficult to cut Hadfield steel are still infrequent in die sinker EDM with graphite tools. Hence, it's believed that more studies on difficult to hard metals could be optimize the process and allow more feasible ways to increase the overall productivity of machinability by using graphite electrodes. The main purpose of this experimental research is to obtain the effect of machining outputs such as MRR, RWR and SR (R_a). And also this paper describes the wear behaviour of graphite tools with both polarity cases when machining Hadfield steel. The results will be utilized to choose optimal tool and work material arrangement.

2 Experimental design and procedure

The experimental machining tests were performed on Furkan die- sinking EDM (Fig. 1)which is energized by a 100 A iso-pulse type generator. Commercial pure kerosene was selected and used as dielectric during the experimental machinings and lateral flushing system was used to remove metal debris away.

Fig. 1. ZNC Die-sinker EDM machine used in experiments.

* Corresponding author: ozerkan@gazi.edu.tr

Table 1 depicts the chemical composition of the Hadfield steel and the dimensions of the prismatic workpiece samples are 30mmX40mmX10mm. Both workpiece surfaces were ground.

Table 1. Hadfield steel's chemical composition

Hadfield Steel			
C	**Si**	**Mn**	**Cr**
1,08	0,62	13,6	0,72
Mo	**Ni**	**Cu**	**Fe**
0,26	0,3	0,18	83,2

It's known that the first rule of choosing the tool material is the internal structure, toughness and hardness of workpiece materials. In EDM of steels; copper, copper-tungsten and graphite electrodes are most widely used. Due to the difficult cutting of Hadfield steel with conventional methods, hoping that we can get good results by the best conductive graphite. The electrodes of 20 mm diameter were used for machining Hadfield steel machinability experiments. The physical properties of graphite and in order to compare to copper are given in **Table 2**.

Table 2. Physical properties and comparision of graphite and copper

	Graphite	Copper
Density (g/cm^3)	1,8111	8,904
Melting point (°C)	3350	1083
Electrical resistivity (Ω.m)	2.5×10^{-6} ~5×10^{-6}	1.68×10^{-8}
Hardness	HB 10	HB 100

Pulse on time of 3, 6, 12, 24, 48, 99, 201 μs, pulse off time of 3, 6, 12, 24 μs, discharge currents of 6, 12.5, 25, 50 A, and lateral flushing pressures of 0.5 bar were used. The other machining parameters; flushing pressure, an open circuit voltage of 80 V, machining depth of 1 mm were kept stationary during the machinings. In this study, 28 different machining with reverse polarity (electrode positive) and 28 different machining with direct (straight) polarity (electrode negative) were experimentalized. Each experiment under same machining condition was carried out twice and average value of the two results was used. For obtaining the lowest TWR, highest MRR and desired SR, the effects of I (A), t_{on} (μs) and polarity exchange were investigated and evaluated. Surface roughness measurements of machined area were performed by portable TR200 stylus device. Cut-off and evaluation length were selected from standard tables as 0.8mm and 4mm, respectively. Ra, Rz(DIN) and Sm roughness parameters were measured and recorded simultaneously. Five measurements were taken from five different zones of machined area and Ra is used for composing SR assesment graphics. The five SR measurements averaged and used in the study.

3 Results and discussion

The main machining performance outputs namely as; material removal rate (MRR-mm^3/min), relative wear ratio (%) and surface roughness (R_a-μm) specified and evaluated according to the selected and studied machining performance inputs such as type of tool polarity, machining current and pulse-on time. MRR depicts the volume of removed workpiece per machining time. Eroding of both tool and workpiece is known as the fundamental law of EDM. Electrode wear rate (TWR) is defined as the ratio of weared weight of the tool to the machining time. And also relative electrode wear (%) is expressed as the percentage ratio of TWR to the eroded material volume from the work piece (MRR).

Thermo-physical features of tool material specify the talent to machining energy of the EDM and remove metals. For generating a spark, peak current is flowed through the gap after dielectric is broken down. At this point, electrons runaway from the tool for coming into collison with molecules of the dielectric fluid. Hereby dielectric is vaporized and energy plasma channel is generated permitting the spark to materialise and embody. However, graphite tools are able to emit deionizer electrons at much lower temperatures and the time required to form the energy channel is considerably less. Consequently graphite starts forming spark more fastly, intensely and strongly, resulting in explicitly higher metal removal rates. Figure 2 shows the effect of polarity, t_{on} and I variaties on MRR. It is seen obviously that MRR increases with the increase in current with both reverse polarity and direct polarity. Also the inclination of the curves clarifies that MRR increases meaningfully with the increase in current with direct polarity as compared to reverse polarity.

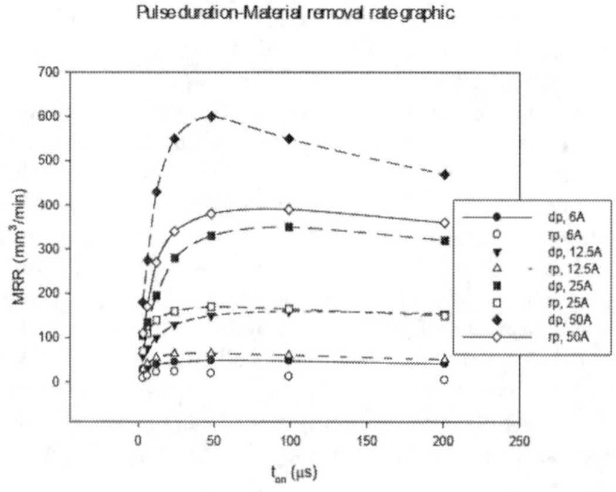

Fig. 2. MRR (mm^3/min) – t_{on}(μs) variation.

With an increase in current the discharge spark energy increases providing higher MRR. In direct polarity case more positively charged ions are emitted from the tools which strike on the work piece. Due to higher mass the momentum of these positively charged ions is greater than that of negatively charged electrons resulting in more eroded material from the workpiece in comparison with the electrodes. Maximum MRR is obtained at 50 amps current with negative polarity. And also the effect of pulse-on time on MRR is shown in Figure 2. MRR is increasing stably with increase in the level of the pulse-on time and apparent that longer than 48μs pulse durations MRR decreases due to the bad, unstable, unbalanced spark discharges because of inadequate and short pulse-off times. Sparking energy is related with the MRR and directly proportional to the pulse on time (t_{on}). However, alonger pulse duration forms and brings on uneffective long arcing that increases dielectric temperature and reduces MRR.

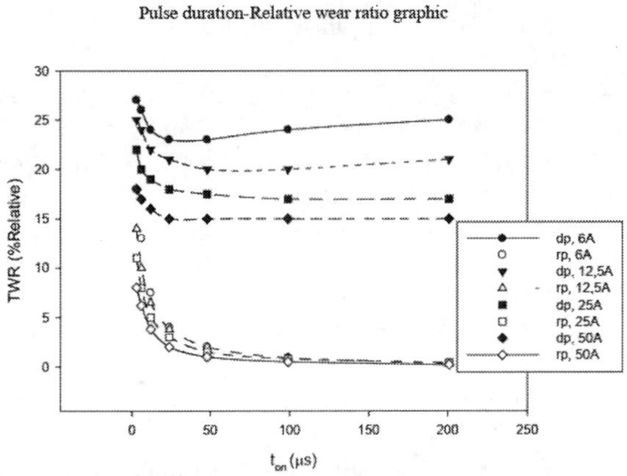

Fig. 3 TWR (% Relative) – t_{on}(μs) variation.

Fig. 3 illustrates the variation of relative wear according to pulse on time and current. There is a clear decrease in relative wear (%) with increase in pulse on time from (3-48 μs) as tool wear rate decrease and material removal rate increases in both polarity conditions. TWR decreases with t_{on} time continuously in a non- linear manner. But increase in t_{on} time from 48-201 μs is provided moderate decrease in tool wear. Reduction in tool wear may be attributed to spark energy is directly proportional with t_{on}. So that discharge energy increases with increase in t_{on} but after definite t_{on} leads to spreading of the discharges. By the reason of this spreading reduction of heat transfer to the tool and increasing of heat transfer the work piece and dielectric occurs which leads to less tool wear and high material removal rate. And also additionally, when pulse duration and cuurent increases relative electrode wear decreases because of graphite tool composes of extra carbon element right along with carbon layer composes from deionization of kerosene adhered the machining area of electrode tool and becomes thick enough. These two affects prevent and reduce the further tool wear and also relative tool wear, as a consequence of that at lower pulse duration the relative electrode wear is more and at higher pulse duration it would be lowered.

And also from Figure 3 it is also seen that as the current increases from low to medium-high levels respectively, the relative electrode wear diminishes. Graphite, can be used on high currents without much loss and deformation in reverse polarization.

At higher t_{on} carbon layer becomes more thicker and it inhibits tool wear and thus at 200μs pulse duration the value of relative electrode wear decreases majorly as compared to the value at medium current level. So that increasing of current at low pulse duration levels relative electrode wear (%) decreases. And also it is seen that relative electrode wear curves for direct and reverse polarity track the same decreasing trend. Lower relative electrode wear is achieved with the reverse polarity at all current and pulse duration values. It is known that at reverse polarity 40% thermal energy is generated at anode and 25% thermal energy is generated at cathode (Xia,*et al.*) [7] and its opposite is true for direct polarity (25 % heat is generated at work piece and 40% heat is generated at tool). Lowest relative electrode wear is achieved at high current at high pulse duration with direct polarity and its value is 0.2%. During the experiments ratio of relative electrode wear (TWR at reverse polarity/ TWR at direct polarity) ranges from 1.5-75 on an average reverse polarity gives 3.7-5.1 times more relative electrode wear as compared to direct polarity. Minimum TWR is obtained at 50 amps current with reverse polarity. Tool wear is dependent on intensity of spark enegy and tool materials type. High electrical conductivitity and melting point (difficult to cut materials) of any metal behaves more resistable to wear because of requiring large amount of energy to melt. Typically used tool materials perform different wearing behaviours with different workpiece materials and polarity combinations. Consequently reverse polarity is more suitable for getting lower relative graphite wears. The surface roughness results at direct and reverse polarity are shown in Figure 4 at different current and pulse duration levels. It is seen obviously that Ra increases with the increase of pulse duration and current despite this at direct polarity case R_a values are higher than the reverse polarity condition. This tendency is opposite to the general trend of increasing surface roughness with the reverse polarity. This observation is supported by Han, *et al.* [8] and Lin *et al.* [9,10]. This effect may be because of higher electrical conductivity of graphite electrodes which produces high intensity impact spark forces that strikes on the work piece with the increase of I and t_{on} at direct polarity. Thus this higher impact forces cavitated and formed more deep and wide discharge craters. Consequently the surface integrity improves as the discharge energy and spark impact force decreases but with graphite tool and direct polarity is occured opposite situation that produces rougher surfaces. From the curves of surface roughness at reverse polarity it is seen that as the pulse duration and current increases the surface roughness decreases. This surface structure is harmonic and agreeable at both polarity with the tool wear graphs. Generation of large craters due to higher sparking energy and also electrical conductivity and melting point of tool material ends up with higher tool wear and big roughness values with higher t_{on} and I. Thus, for achieving better surface finishings it necessitates low I and t_{on} with graphite tools at reverse polarity. From the experimental results it is seen that direct polarity gives almost 0.8-1.8 times higher surface roughness as compared to reverse polarity.

The results show that direct polarity (– tool) is more acceptable, attractive and effective about that MRR is higher, in response to this TWR is lower when using positive tool. It can not clearly say any better performance obtained with tool/workpiece couple as cathode or anode. Direct polarity is suitable for both roughing operations. Good MRR could be achieved in rough machining with negative polarity tools. MRR is 25-81% higher than for machining with positive tool. So that with direct polarity metal removal rate and tool wear is higher as compared to reverse polarity with graphite tools. Polarity is very important for MRR, electrode wear and surface roughness.

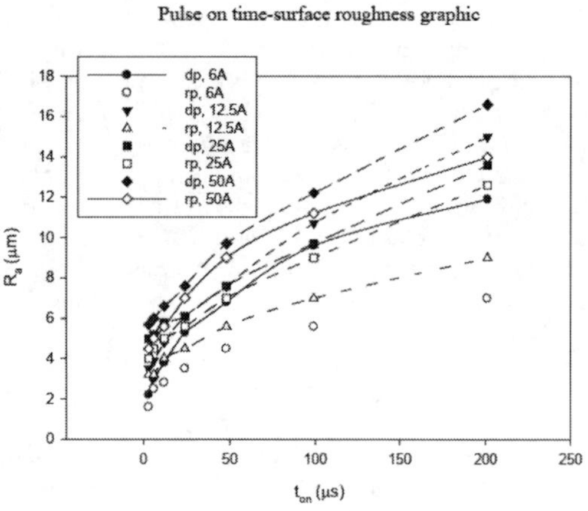

Fig. 4. R_a (µm)–t_{on}(µs) variation.

4 Conclusion

An inclusive experimental study has been carried out for researching the influences of main processing parameters on optimal shaping performance of Hadfield steel in EDM. The machining inputs are tool material, polarity and pulse on time, machining outputs are MRR, TWR (%) and surface roughness. The succinct, brief observations may be summarized from this study.

• MRR increases together with both direct and reverse polarity associated with increasings in current. The slope of the curve shows that MRR increases drastically and progressively in full value current increasings with direct polarity as compared to reverse polarity. It's explicit that longer than 48□s (can be accepted as threshold value) pulse durations MRR decreases due to the bad, unstable, unbalanced sparks.

• There is an evidence decrease in relative wear (%) with increase in pulse on time from (3-48 µs) as TWR decrease and lower relative TWR (%) is achieved with the reverse polarity at all current and pulse on time values. Furthermore machining current increasings of at low pulse on time levels relative electrode wear (%) decreases exaggeratively. Minimum TWR is obtained at maximum current with reverse polarity.

• The effect of polarity was interesting by means of reverse polarity produced shallow and small craters however direct polarity led to larger craters. Roughness of machined surfaces increases with the increase in current and ton with both direct and reverse polarity.

It is important the type of material to be processed, higher power levels and reverse polarity must be used for difficult to cut hard metals with conventional machining processes. The working polarity plays an important role effecting the material removal rate and the properties of the machined surface. Choosing the right parameters is critical for different electrode and workpiece materials for successful operations.

References

1. K.H. Ho, S.T. Newman, Int. J. of Machine Tools & Man. **43** (2003)
2. C.J. Luis, I. Puertas, G. Villa, J. of Materials Processing Tech. **164–165**: 889–896 (2005)
3. B. Bojorquez, R.T. Marloth. O.S. Es-Said, Eng. Failure Analysis, **9** (2002)
4. H. El-Hofy, Advanced Machining Processes: Nontraditional and Hybrid Machining Processes, (McGraw-Hill Companies, 2005)
5. C.H. Che Haron, et al., Journal of Materials Processing Technology **87** (2001)
6. C. Cogun, S. Akaslan, KSME International Journal. **16** (2002)
7. H. Xia, M. Kunieda, N. Nishiwaki, Int. J. of Electrical Mach., **1** (1996)
8. F. Han, Y. Yamada, T. Kawakami, M. Kunieda, Precision Engineering **30** (2006)
9. Y.C. Lin, B.H. Yan, F.Y. Huang, Int. J. of Adv.Man. Tech. **18** (2001a)
10. Y.C. Lin, B.H. Yan, F.Y. Huang, J.l of Mat. Processing Tech. **115** (2001b)

DOI: 10.1051/matecconf/201712901035

Increasing the capabilities of machining attachments by means of ferromagnetic liquids

Alexander Alexandrovich Boldyrev[1,*], *Alexander Ivanovich* Boldyrev[1], and *Oleg Nikolaevich* Fedonin[2]

[1] Voronezh state technical university, 14 Moscow Avenue, 394026, Voronezh, Russia
[2] Bryansk State Technical University, 7, 50-letiya Octyabrya Blvd, Bryansk, 241035, Russia

Abstract. The article is devoted to application of ferromagnetic liquids in machining attachments of technical process for producing machine components. Application of such liquids allows to simplify equipment and decrease its cost, to increase accuracy of workpieces treatment by elimination of additional processing of workpieces base surfaces. It is proposed to set the workpiece not on supports of machine-tools but on the layer of rheological liquid hardened in outer magnetic field. At the same time contact pressure of rheological medium on side surface provides slip resistance caused by the forces appearing during treatment. The article contains calculated dependences for definition of forces holding the workpiece on machine table, being fully or partially submerged into ferromagnetic medium during mechanical processing. Criterion for applicability of ferromagnetic liquid at workpiece setting during treatment is proposed. On the ground of achieved per samples experimental verification of formulated statements and calculated dependences the authors worked out recommendations for selection of rheological media in liquid state for workpiece setting and holding taking into account frequent changes of working objects. Development and running of such machining attachments is efficient in modern industry that is characterized by low serial production and wide assortment of produced goods during finishing mechanical, electrical and combined work, where scientifically grounded application of ferromagnetic liquids increases production profitability by 1,5-2 times.

1 Introduction

Modern industry is characterized by low serial production, wide assortment of produced goods, and consequently by a great deal of quite difficult machining attachments including the appliances for installation and work holding. It is known that the cost of machining attachments may amount to 90 % of production prime cost. That is why design and manufacture of such appliances exercise significant influence over production cost of goods manufacture. For reduction of expenses on preproduction it is effectually to use multi-operated equipment instead of special tools.

2 Relevance. Scientific significance of the question

In machine-building we use different types of machining attachments: magnetic, cryogenic, etc. However there are some severe restrictions in this sphere. Accurate location reference surfaces, workpieces and fixturing components made of magnetic alloys, powerful magnetic elements are required for magnetic catch mechanisms. For frozen liquids (based on water) you need accurate bases, powerful energy-consuming refrigerating systems, complex modernization of equipment for accumulation, feed, and regeneration of the liquid, etc.

The analysis of performed tasks allowed to make preliminary conclusions that it is possible to use setting in ferromagnetic liquid for different types of treatment. Analytical models and correspondences for cutting forces definition available in literature, can be used for design of machining attachments based on rheological media. However it is necessary to estimate the forces adequative for reliable and accurate workpiece positioning and prove the possibility of magnetic liquid use in typical production methods of high-variety manufacturing [1-5].

Manufacture transition to the system of individual orders required scientific substantiation of rheological media use for simplification and cheapening of machining attachments, increasing workpiece accuracy, deletion of additional treatment of workpiece reference surfaces. Their application accelerated conversion and initiation of competitive products (including cost attractiveness) what is relevant to machine-building.

3 Setting of the problem

The research task is development of flexible method of workpieces setting and justification of field for its

* Corresponding author: alexboldyrev@yandex.ru

DOI: 10.1051/matecconf/201712901035

rational application with the use of ferromagnetic liquids properties and managing their state by outer magnetic field for creating bases without production of specialized supports.

Our research of workpiece setting in rheological liquid was based on operational hypothesis according to which change of magnetic field operating upon rheological medium allows to reach viscosity value providing permanent position of work area relative to the tool, appliance, or basic components of equipment [7]. At the same time the provided holding force is suitable for achievement of accurate and reliable workpiece position at the finishing stage of treatment. Development of this hypothesis allowed to create new treatment methods and means for their materialization [7-12].

4 The theoretical part

Conceptually new approach involves setting workpiece on the layer of ferromagnetic liquid hardened in outer field instead of machine-tool supports. Contact pressure of rheological medium on side surface provides slip resistance during processing [7, 13].

The scheme showing method of workpiece positioning in rheological medium is given in Fig. 1.

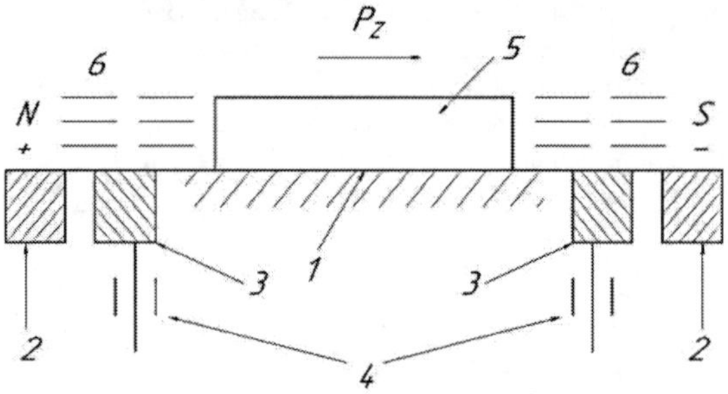

Fig. 1. Scheme of workpiece positioning on magnetic table: 1 – magnetic table; 2 и 3 – magnets; 4 – controllers of magnetic field force; 5 – workpiece; 6 – rheological ferromagnetic liquid

For the workpiece holding on the table it is necessary to create the force (F_H) of standard pressure on table surface determining the stability of position for the workpiece basic parts and pillars with magnetic liquid at complete submergence of the workpiece or its parts into the rheological medium.

$$F_H = (\mu_0 m_z \Delta H) S_e, \tag{1}$$

where μ_0 – coefficient of magnetic inductivity in vacuum (μ_0=4π·10-7 H/m); m_z – magnetic moment of particle; ΔH – gradient of magnetic intensity, kA/m; S_e – workpiece surface area (plane view).

Resistance of the force of workpiece treatment during cutting (F_C) will be composed of the constituent (F_T) for friction of contact areas, pressed by the workpiece mass (m) and friction at standard influence of F_H, but also of the resistance (F_e) of thick ferromagnetic liquid in the field of magnetic influence on it, taking into account the surface area (S_b) of workpiece part, contacting with rheological liquid by the side opposite to the influence of cutting force (Fig. 2).

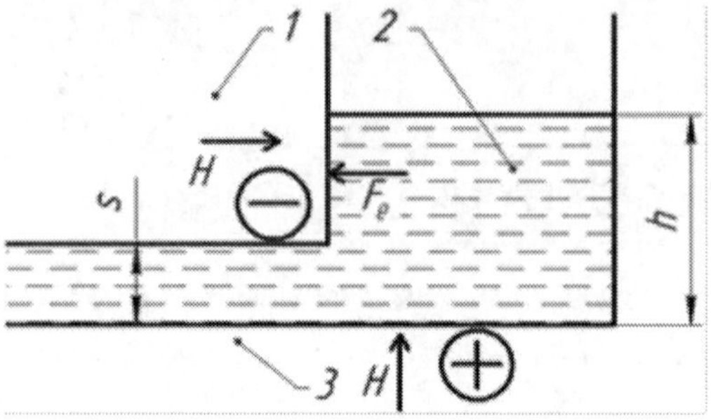

Fig. 2. The forces acting on the workpiece on the machine table: 1 – workpiece; 2 – rheological ferromagnetic liquid; 3 – machine table

$$F_C = F_T + F_e > K_P F_z, \tag{2}$$

where F_z – treatment force; K_P – coefficient considering dynamic and other impacts on workpiece during processing.

Frictional force of contact areas:

$$F_T = K_T(F_H + F_d), \tag{3}$$

where F_d – pressing force of worpiece mass (m); K_T – surfaces friction coefficient.

$$F_T = K_T(\mu_0 m_z \Delta H S_e + gm). \tag{4}$$

Friction force (F_e) of thick medium (ferromagnetic liquid) depends on its density. Density may be taken equal to the force at the beginning of elastic material (rheological medium) flowing. This force can be found by means of maximum tangential stress τ_{max}:

$$F_e = \tau_{max} / S_b, \tag{5}$$

where $\tau_{max} = \{1 - \mu\}\{1 - \tau arctg[1/\tau] - 3/[4(1+\tau^2)]\}/2$; μ – medium viscosity (for ferromagnetic liquid); $\tau = z/(2l)$; z – clearance between surfaces of contact bodies (maximum viscosity layer thickness); l – workpiece width.

Subsequently

$$\tau_{max} = 0{,}31 F_0 (\text{Hertz's task solution}), \tag{6}$$

$$F_e = F_H(S_b) = \mu_0 m_z \Delta H S_b, \tag{7}$$

$$F_C = K_T \mu_0 m_z \Delta H S_e + K_T gm + \mu_0 m_z \Delta H S_b = \\ = \mu_0 m_z \Delta H (K_T S_e + S_b) + K_T gm. \tag{8}$$

Criterion for applicability of ferromagnetic liquid for work setting during its treatment:

$$\Delta\rho' VgH \le kT, \tag{9}$$

where $\Delta\rho'$ – difference in material density of particles and base (liquid); V – particle volume; g – gravitation constant near the ground; H – magnetic liquid layer height; kT – thermal-motion energy of particles (k – Boltzmann's constant; T – temperature of magnetic liquid).

At the known contact surface area of workpiece with the table S_e minimum requirement for layer height of magnetic liquid (H_{min}) at workpiece width l and length l_g.

$$F_C = \mu_0 m_z \Delta H(K_T l l_g + lH) + K_T g \rho_{wp} l l_g H, \quad (10)$$

where ρ_{wp} – density of workpiece material.

Whence H_{min} is determined by numerical methods.

Workpiece stability criterion:

$$H_{min} \leq H_{wp}, \quad (11)$$

where H_{wp} – height of workpiece.

5 Practical significance

Experimental verification for formulated statements and calculated dependences was achieved per samples [6]. Research results also led to definition of optimum concentration of magnetic phase in rheological liquid with setting workpieces made of different materials. For magnetic materials we recommend the rheological media based on machine grease of high viscosity with 20 % part concentration by volume, this increases holding force as compared to traditional fixation on magnetic table: 41-56 % in shear and 20-36 % under separation. For non-magnetic parts specific holding forces determined with the help of experimental plant are approximately (3-3,5)·10^5 N/m^2 in shear and (4-6)·10^4 N/m^2 under separation. The rated workpiece holding forces were tested on plain grinder at treatment under standard conditions. With due consideration of reserve coefficient the necessary forces were provided, the setting base was maintained. We proved the availability of non-magnetic workpiece holding forces to be approximately (12-15)·10^5 N/m^2 in shear and (2-3)·10^5 N/m^2 under separation.

Application of the research results allowed to speed up of the industrial technological cycle in operating high-variety manufacturing due to labour intensity reduction at pre-operation of workpieces setting bases and in some cases due to removal of grippers and reduction in expenditures on machining attachments production. Moreover a number of fixtures and tools having adaptive interaction with rheological medium were created, they allowed to simplify and cheapen their construction, reduce the costs of storing and check out of machining attachments.

6 Conclusions

The research of methods for workpiece setting and holding in rheological liquid allowed to work out scientifically grounded recommendations for the reasonable field of rheological liquid application in high-variety manufacture, their use is desirable when receiving high accuracy surfaces on rough workpiece bases during finishing work, particularly in operations with physicotechnical processes.

References

1. V.F. Aleshin, A.Yu. Kolobov etc., *The use of magnetic fluids in the renovation of engineering objects* (Moscow, MSTU, 2010)
2. A.A. Boldyrev, V.P. Smolentsev, *Requirements to the magnetic fluid , used in tools, jigs and fixtures*, Rybinsk State Aviation Technological University Journal, № **2** (**23**), pp. 127-131 (2012)
3. D.V. Orlov, Y.O. Michalev etc., *Magnetic fluids in mechanical* (Moscow, Engineering, 1993)
4. D.V. Orlov, Y.I. Stradomsky etc., *Influence of the geometric dimensions of the work area on the magnetic-seal held them pressure drop* (Ivanovo, Ivanovo Energy Institute, 1987)
5. A.G. Suslov, B.M. Bazrov, V.F. Bezyazychny etc., *High technologies in mechanical engineering*, (Moscow, Engineering, 2012)
6. V.P. Smolentsev, A.A. Boldyrev, *The mechanism of the process control installation of individual workpieces in rheological fluid*, Fundamental and applied problems of engineering and technology, № **2/3(292)**, pp. 56-60 (2012)
7. Russian patent 2312000, A.S. Revin, A.V. Lisicyn, V.P. Smolentsev, *Method of attachment parts of the non-magnetic material, and device for its realization*, (2007)
8. Russian patent 2191664, V.P. Smolentsev, G.P. Smolentsev, V.Y. Sklokin, *Device for electrochemical and mechanical treatment*, (2002)
9. Russian patent 2216437, V.P. Smolentsev, K.M. Gazizulin, *Electrochemical machining method*, (2003)
10. Russian patent 2229966, V.P. Smolentsev, G.P. Smolentsev, A.V. Levin, *The electrode tool for electrochemical marking and its manufacturing method*, (2002)
11. Russian patent 2457934, V.P. Smolentsev, A.A. Boldyrev, G.N. Klimova, I.U. Kuznecov, *Forming method for abrasive tool working zone*, (2012)
12. Yu.M. Baron, *Technology of abrasive processing in a magnetic field* (Leningrad, Engineering, 1975)
13. E.Ya. Blum, M.M. Mayorov, etc., *Magnetic liquids*, (Riga, Knowledge, 1989)

Modelling of cutting block cut surface at faceted surfaces machining using planetary gear set

Vadim Kuts[1,*], *Mikhail* Razumov[1], and *Nikolay* Kalutskikh[1]

[1]Federal State Budgetary Educational Institution of Higher Education "Southwest State University", SWSU International Office, 94, 50 let Oktyabrya, Kursk, Russia

Abstract. The article resolved the relevant problem to assign process parameters for polygonal turning machining ensuring the component has its pre-set characteristics. To accomplish this, the authors modelled the machining process and described the cut surface formed by the cutter block edge during machining.

Today cylindrical components with multi-faceted surfaces have found application in various areas of the economy, in particular: machine construction, instrument engineering, robotics, agriculture, mining and even the medical industry [1].

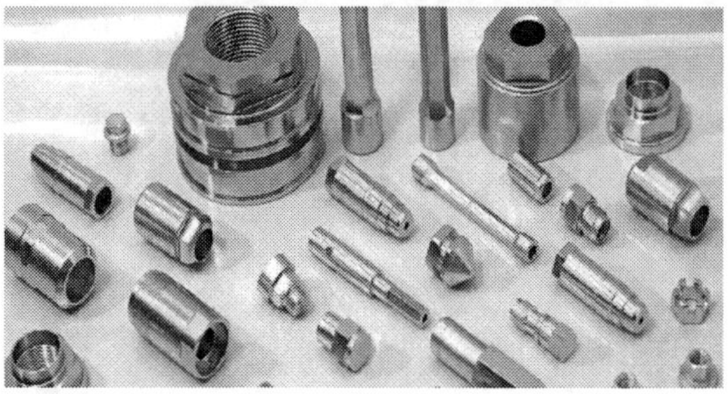

Fig 1. Examples of details with polyhedral surfaces

The conducted analysis of the production methods for faceted surfaces helped to choose the planetary motion method of the turning machine cutting edges. This method works as follows: the planetary gear set assigns a complex curve trajectory (trochoid) to the cutting tool point Fig.1 [1].

To determine the planetary gear set structural parameters, the authors resolved the problem of the rectangle side approximation error calculation with the prolonged trochoid section [1]. However, to assign the machining process parameters the component has its pre-set characteristics, it is necessary to model the machining process, namely, to make a cut surface model.

Fig. 2. Planetary Gear Set

Generally, the cut surface is described by the formula

$$Q_i(\varphi(t), \theta(t) z(t), S) == {}^6A(\varphi(t))\, {}^2A(R-r) \cdot {}^6A(-\theta(t)) \cdot r_{ui}(S) \quad (1)$$

Where

$\varphi(t) = \frac{\pi n}{30} \cdot t$; $\theta(t) = \frac{R}{r} \cdot \frac{\pi n}{30} \cdot t$; $z(t) = \frac{S_{мин}}{60} \cdot t$, $S_{мин}$ – feed per minute, N – number of the cutting block rotations in relation to the workpiece.

Let us build a model for cutting block edges. Fig. 3 presents a square indexable insert.

---

[*] Corresponding author: kuc-vadim@yandex.ru

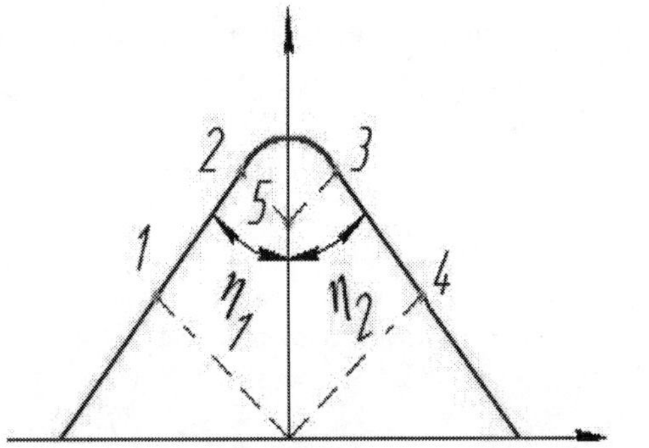

Fig. 3. Design model for the indexable insert cutting edge points: 1,2,3,3 5 – reference points; η1 η2 – angles between the X axis and the insert surface

Let us determine the generating points coordinates
Point 1 coordinates:

$$x_1 = 0{,}5d \cdot \sin\eta_1 \,;\; y_1 = 0{,}5d \cdot \sin\eta_1$$

Point 2 coordinates:

$$x_2 = \frac{\sqrt{2}}{2} d \cdot \cos\eta_1 \,;\; y_2 = -r \cdot \sin\eta_1$$

Point 3 coordinates:

$$x_3 = x_2 \,;\; y_2 = -y_2$$

Point 4 coordinates:

$$x_4 = x_1 \,;\; y_4 = -y_1$$

Point 5 coordinates:

$$x_5 = \frac{\sqrt{2}}{2} d - r \cdot \sqrt{2} \,;\; y_5 = 0$$

$$r_{\Pi} = [x(S), y(S), z(S), 1]^T$$

where

$$y(S) = \begin{cases} y_1; S \le 0 \\ y_1 + S; 0 < S \le y_4 - y_1 \\ y_u; S > y_u - y_1 \end{cases} \quad (2)$$

$$x(S) = \begin{cases} x_1; S \le 0 \\ (x_2 - x_1)\cdot(y(S) - y_1)/(y_2 - y_1) + x_1; 0 < S \le y_2 - y_1 \\ x_5 + \sqrt{r^2 - (y(S) - y_5)^2}; y_2 - y < S \le y_3 - y_1 \\ (x_4 - x_3)\cdot(y(S) - y_3)/(y_4 - y_3) + x_3; y_3 - y_1 < S \le y_4 \\ x_4; S < y_4 \end{cases}$$

$$N_{\Pi}(S) = \left[-\bar{j}\cdot\bar{r}_s(S)\cdot\sin\gamma_{\Pi}; \bar{i}\cdot\bar{r}_s(S)\cdot\sin\gamma_{\Pi}; -\cos\gamma_{\Pi}; 0\right]^T$$

$$N_3(S) = \left[-\bar{j}\cdot\bar{r}_s(S)\cdot\cos\alpha_{\Pi}; \bar{i}\cdot\bar{r}_s(S)\cdot\cos\alpha_{\Pi}; \sin\alpha_{\Pi}; 0\right]^T$$

Further, let us set the indexable insert parameters; the transition to the coordinates system 1 is the first action

$$^{01}A = {}^2A(-d)\cdot{}^6A\left(\frac{\pi}{2}\right)\cdot{}^5A\left(-\frac{\pi}{2}\right) \quad (3)$$

Rotation for the angle λ is the second action

$$^{12}A = {}^6A(-\lambda) \quad (4)$$

After that, let us set the insert for the angle

$$\gamma_g = \gamma_K - \gamma_{\Pi K}$$

$$^{23}A = {}^4A(-\gamma_g) \quad (5)$$

Move to the rotation center

$$^{34}A = {}^2A(-S_{\Pi})\cdot{}^3A(-r_{\Pi}) \quad (6)$$

If $\varphi \ne \varphi_2$, set the angle φ

$$^{45}A = {}^5A(\varphi) \quad (7)$$

$$^{56}A = {}^6A(-(0{,}5d + m - r)) \quad (8)$$

$$^{67}A = {}^6A\left(\frac{\pi}{2}\right)\cdot{}^2A\left(-\frac{\pi}{2}\right) \quad (9)$$

The setting matrix calculation in the coordinates system
Reference data:
$\bar{j}_{21}$ - vector $\bar{j}_2 = [0,1,0,0]^T$ presented in the coordinates system x_1, y_1, z_1
$\bar{k}_{21}$ - vector $\bar{k}_2 = [0,0,1,0]^T$ presented in the coordinates system x_1, y_1, z_1
$\bar{r}_{21}$ - vector defining the position of the coordinates x_2, y_2, z_2 origin within the system x_1, y_1, z_1

The A_{21} matrix elements are determined by the system

$$\begin{cases} \bar{j}_{21} = A_{21}\cdot\bar{j}_2 \\ \bar{k}_{21} = A_{21}\cdot\bar{k}_2 \\ [\bar{j}_{21}\times\bar{k}_{21}] = A_{21}\cdot\bar{i}_2 \\ \bar{r}_{21} = A_{21}\cdot e^4 \end{cases} \quad (10)$$

where $\bar{i}_2 = (1,0,0,0)^T$

$$A_{21} = \begin{bmatrix} \bar{i}_1\cdot[\bar{j}_1\times\bar{k}_{21}] & \bar{i}_1\cdot\bar{j}_{21} & \bar{i}_1\cdot\bar{k}_{21} & \bar{i}_1\cdot\bar{r}_{21} \\ \bar{j}_1\cdot[\bar{j}_1\times\bar{k}_{21}] & \bar{j}_1\cdot\bar{j}_{21} & \bar{j}_1\cdot\bar{k}_{21} & \bar{j}_1\cdot\bar{r}_{21} \\ \bar{k}_1\cdot[\bar{j}_1\times\bar{k}_{21}] & \bar{k}_1\cdot\bar{j}_{21} & \bar{k}_1\cdot\bar{k}_{21} & \bar{k}_1\cdot\bar{r}_{21} \\ 0 & 0 & 0 & 1 \end{bmatrix} \quad (11)$$

The method for calculation of the residual ridges and roughness parameters

1. Define the point on the cutting edge which contacts with the component surface (shape-generating point) by solving the following equation

$$\bar{j}\cdot\bar{Q}(S,t) = 0 \quad (12)$$

In relation to the S edge parameter at t=0

2. Calculate the A_W matrix elements for setting the sectional plane coordinates system under the set parameters

$$[0,0,-1,0]^T,[-1,0,0,0]^T,Q(S,t) \tag{13}$$

Determine the coordinates for the A point

$$r_A = A_W \cdot Q(S,t) \tag{14}$$

Determine the coordinates for the B point

$$r_B = A_W \cdot Q(S,t+T_{об}) \tag{15}$$

Determine the coordinates for the C point by solving the equation system

$$\begin{cases} \overline{i}_w \cdot \overline{A}_w \cdot Q(S_1,t_1) = \overline{i}_w \cdot \overline{A}_w \cdot Q(S_2,t_2) \\ \overline{j}_w \cdot \overline{A}_w \cdot Q(S_1,t_1) = \overline{j}_w \cdot \overline{A}_w \cdot Q(S_2,t_2) \\ \overline{k}_w \cdot A_w Q(S_1,t_1) = 0 \\ \overline{k}_w \cdot A_w Q(S_2,t_2) = 0 \\ 0 \le S_1 \le S_{m4} \\ 0 \le S_2 \le S_{m4} \\ t - \frac{T_{об}}{4} \le t_1 \le t + \frac{T_{об}}{4} \\ t + T_{об} - \frac{T_{об}}{4} \le t_2 \le t + T_{об} + \frac{T_{об}}{4} \end{cases} \tag{16}$$

In relation to the S_1, S_2, t_1, t_2. parameter

Hence $r_c = A_W \cdot Q(S_1,t_1)$

Assessment for the layers being cut

1) Define the normal vector towards the cut surface at the cut-in moment t

$$\overline{N}_{pi}(S,t) = [\overline{Q}_{is} \cdot (S,t) \times \overline{Q}_{it} \cdot (S,t)] \cdot$$
$$sign\begin{Bmatrix} -[\overline{Q}_{is} \cdot (S,t) \times \overline{Q}_{it} \cdot (S,t)] \cdot^6 A(\varphi(t)) \cdot^2 \\ A(R-r) \cdot^6 A(Q(t)) \cdot \overline{N}_{zi}(S) \end{Bmatrix}$$

Let the A_j transition matrix be defined in the system of x_j, y_j, z_j coordinates with the origin in the $\overline{Q}_i(S,t)$ point and the y_j axis along the $\overline{Q}_{is}(S,t)$ vector and the z_j axis along the $\overline{N}_{pi}(S,t)$ vector

Define the cross point of the z_j axis with the cut surface of the i cutter shaped during the previous cutting block rotation

$$\begin{cases} \overline{i}_j \cdot \overline{A_j \cdot Q_i(S^*, t+\Delta t)} = 0 \\ \overline{j}_j \cdot \overline{A_j \cdot Q_i(S^*, t+\Delta t)} = 0 \\ 0 \le S \le S_n \\ t - T_{об} - \frac{T_{об}}{N_r} < \Delta t < t - T_{об} + \frac{T_{об}}{N_r} \end{cases} \tag{17}$$

The system is solved in relation to the S^* and Δt parameters

Define the thickness $Q_1(S,t)$

$$Q_1(S,t) = F_1(S,t) \cdot \overline{k}_i \cdot \overline{A_j \cdot Q_i(S^*, t+\Delta t)} \tag{18}$$

Let the thickness value of the layer being cut limited by the workpiece material be defined as follows:

Solve the equation system in relation to the U and V parameters

$$\begin{cases} \overline{i}_j \cdot \overline{A_j \cdot PR_z(U,V)} = 0 \\ \overline{j}_j \cdot \overline{A_j \cdot PR_z(U,V)} = 0 \\ 0 \le U \le 2\pi \\ 0 \le V \le H_{заг} \end{cases} \tag{19}$$

where $H_{заг}$ - workpiece thickness.

2) Calculate the thickness $a_2(S,t)$

$$a_2(S,t) = F_1(S,t) \cdot \overline{k}_j \cdot \overline{A_j \cdot QPR \cdot (U,V)} \tag{20}$$

The thickness for the layer being cut is calculated as

$$a(S,t) = \min(a_1(S,t), a_2(S,t)) \tag{21}$$

Basing upon the model built for the cut surface, it is possible to calculate assessment parameters for the multi-faceted surfaces machining with further determining of the rational values for the polygonal turning parameters.

References

1. M.S. Razumov, *Enhancing Performance of Multi-faceted Outer Surfaces Shape Generation by means of Planetary Gear Set: the author's abstract of the thesis*, Candidate in Engineering Science (Kursk, 2011)
2. M.S. Razumov, A.N Grechukhin., A.O. Gladyshkin, *Calculation of the blank twist angle at processing with planetary turning*, Proceedings of the Institution of Mechanical Engineers, Part C: Journal of Mechanical Engineering Science, **230 (18)**, pp. 3298-3303 (2016) DOI: 10.1177/0954406215608657
3. M.S. Razumov , A.O. Gladyshkin, A.I. Pykhtin, *Evaluation of rational quantity of profile moment-transferring joints sides* / [Procjena racionalne količine stranica profilnih spojeva za prijenos okretnog momenta], Source of the Document Tehnicki Vjesnik **23 (2)**, pp. 575-577 (2016) DOI: 10.17559/TV-20140614032348
4. M. Razumov, A. Grechukhin, A. Pykhtin, *Determination of Shape Errors during Polygonal Turning of Polyhedrons with an Odd Number of Faces* / Procedia Engineering **150**, PP 844-848 (2016) DOI: 10.1016/j.proeng.2016.07.130

5. A.I. Barbot'ko, M.S. Razumov, A.I.Pykhtin, P.A. Ponkratov, *Selecting the cutting edge of a shaping tool*, Russian Engineering Research 35 (5), pp. 344-346 (2015) DOI: 10,3103 / S1068798X15050068
6. M. Razumov, *Form error calculation during polygonal sharpening of polyhedrons with even number of sides*, Metallurgical and Mining Industry **7 (1)**, pp. 66-69
7. M.S. Razumov, *Technological parameters determination of faced rolls combined treatment* / Metallurgical and Mining Industry, **7 (2)**, pp. 178-182 (2015)
8. M.S. Razumov, A.O. Gladyshkin, E.I. Yatsun, A.A. Zinovkin, O.S. Zubkova, *Selection of rational geometrical parameters of profile of moment transmitted connections*, Metallurgical and Mining Industry **7 (8)**, pp. 512-516 (2015)
9. M. Razumov, A. Gladyshkin, V. Kassikhin, A. Pykhtin, E. Skripkina, *Stress calculation of moment transmitting roll with profile on the base of Reuleaux triangle*, Metallurgical and Mining Industry **7 (8)**, pp. 512-516 (2014)
10. M.S. Razumov, A.I. Pykhtin, A.V. Maslennikov, *Error of polyhedral profiles in single-cutter shaping by means of a planetary mechanism,* Russian Engineering Research **32 (9-10)**, pp. 681-684 (2012) DOI 10,3103 / S1068798X12070180
11. A.I. Barbot'ko, A.V. Maslennikov, M.S. Razumov, *Attachment for clamping bushes when shaping their polyhedral external surfaces*, Russian Engineering Research, **31(12)**, pp. 1253-12542011 DOI: 10.3103/S1068798X11120069
12. Y.A. Maksimenko, V.V. Kuts, *Machining by a disk mill of variable radius*, Russian Engineering Research, Volume **34**, Issue **12**, , Pages 785-788 (2014)
13. V.V. Kuts, I.V. Kucheryaev, *Machine-tool adjustment in machining the housings of composite mills*, Russian Engineering Research, Volume **28**, Issue **10**, Pages 1007-1009 (2008)
14. V.V. Kuts, *The profile distortion value calculation for shaped to be machined surface at the CAD/CAM - development for assembled shaped mills*, Avtomatizatsiya i Sovremennye Tekhnologii, Issue **11**, Pages 5-7 (2004)
15. S.G. Emel'yanov, V.V. Kuts, M.S. Merzhoeva, *Graphoanalytical design method of combined core drills equipped with rB eplaceable and indexing cutting inserts*, Avtomatizatsiya i Sovremennye Tekhnologii, Issue **11**, Pages 19-23 (2003)
16. Y.A. Maksimenko, V.V. Kuts, *Analysis of changes in error processing rk-profile shaft with teeth resharpening cutters with radial constructive feed*, Spravochnik. Inzhenernyi zhurnal, pp.008-012 (2014)
17. O. Kuts, A. Gorokhov, A. Prokofiev, *Profile simulation of rated workable surface in thin-walled part with small angle of taper and generating surfaces of compound milling cutters*, International Conference on Mechanical Engineering, Automation and Control Systems (MEACS), Tomsk, pp. 1-4 (2015) doi: 10.1109/MEACS.2015.7414900
18. L.G. Vainer, A.G. Ivakhnenko, I.V. Karabanov, *Systematic Dynamic Factors in the Shaping of Opposite End Surfaces*, Russian Engineering Research **5**, pp. 340–345 (2014)
19. A.V. Kirichek, A.G. Ivakhnenko, E.O. Ivakhnenko, A.Y. Altukhov, *Geometric Accuracy of the Machines with Strut-Type Structures*, International Journal of Applied Engineering Research **21**, pp. 995–995 (2014)
20. O. Kuts, A. Gorokhov, A. Prokofiev, *Assessment of error in position and orientation of replaceable polyhedral plates in disk milling cutters for profile shafts machining*, 2015 International Conference on Mechanical Engineering, Automation and Control Systems (MEACS), Tomsk, pp. 1-5 (2015) doi: 10.1109/MEACS.2015.7414899
21. A.I. Timchenko, *Tekhnologiya izgotovleniya detaley profil'nykh besshponochnykh soedineniy* [Manufacturing technology of keyless connection of profile pieces] (Moscow, VNIITEMR, 1988)
22. S.G. Emel'yanov, A.O. Gladyshkin, M.S. Razumov, S.F. Yatsun, *Automatization of work preparation of profile rolls*, Izvestiya YuZGU **1(1)**, 113-116. (2012)
23. A.I. Barbot'ko, M.S. Razumov, Lathing of polyhedrons with even number of sides, Vestnik mashinostroeniya, No**1**, p. 46-48 (2010)
24. A.I. Barbot'ko, P.A. Ponkratov i M.S. Razumov, *Metod generatornoy obrabotki profil'nykh valov dolbyakami* [Method of generating processing of grooved rolls by form cutters], V Mezhdunarodnaya nauchno-tekhnicheskaya konferentsiya «Mashinostroenie – osnova tekhnologicheskogo razvitiya Rossii (TM-2013)». Kursk, YuZGU, p.p. 440-442 (2013)
25. V.N. Kassikhin, M.S. Razumov, A.O. Gladyshkin, N.E. Bykovskaya, *Automation of durability calculation for shaft torsion,* News of Southwest state university, Equipment and technologies, SWSU, No. **2**, p. **1**, pp. 179-181 (2012)
26. S.G. Emelyanov, M.S. Razumov, S.A. Chevychelov, A.N. Grechukhin, A.O. Gladyshkin, *Method of forming profile sections wagons springrailway*, Proceedings of the Southwestern State University, Series: Technology **2**, 23-28 (2013)
27. A.I. Barbotko, M.S. Razumov, A.O. Gladyshkin, P.A. Ponkratov, O.V. Eskova, O.V Ovchinkin, *Automated system for the implementation of graphic solutions power function*, Proceedings of the Southwestern State University **1,** 90-94 (2013)
28. Barbot'ko A and Gladyshkin A. *The basis of theory of mathematical modelling: text-book* (St. Oskol: ''TNT''Ltd, 2009)

29. V. Vorona, I. Shchurov, *Working kinematic front angle in turning a sinusoidal linear surface*, Russian Engineering Research **5**, 469-472 (2012)
30. V. Vorona, S. Chinenov, *Rational rear cutter angles for multipassturning of a complex sinusoidal contour surface*, Russian Engineering Research **11**, 1156-1157 (2011)
31. A. Herberg, *Modeling of self-induced vibrations that occur during the machining process of casting patterns with the use of the fuzzy neural networks method*, Archives of Metallurgy and Materials **3**, 871-875 (2013)
32. S. Wiewiorowska, *Analysis of the influence of drawing processparameters on the mechanical properties of trip-structure steelwires*. Archives of Metallurgy and Materials **2**, 573-578 (2013)
33. Z. Muskalski, *Selected problems from the high-carbon steel wire drawing theory and technology*, Archives of Metallurgy and Materials **2**, 527-535 (2014)

 DOI: 10.1051/matecconf/201712901037

Optical diagnostics of selective laser melting and monitoring of single-track formation

Andrey Gusarov[1], *Dmitriy* Kotoban[1,*], and *Ivan* Zhirnov[1]

[1] Moscow State University of Technology "STANKIN", 127994, Vadkovskii per. 1, Moscow, Russia

Abstract. The article presents the optical diagnostics results of the selective laser melting process of single-track production. The track defects detection (such as balling effect, powder free zone formation, sparking) was shown, as well as the visualization of the independent particles consolidation in a solid track. The metal evaporation and the formation of the melt pool specific gas dynamic conditions were considered as important physical phenomena. The velocities of the particle emission from the melt pool, the rate of their involvement, and the velocity of the gas flow were estimated. The results make it possible to evaluate the kinetics of mass transfer under selective laser melting process. The surface thermal field of the laser-irradiated zone strongly influences the material qualitative characteristics after selective laser melting. The results becomes the basis for the development of optical monitoring and diagnostic systems for laser additive manufacturing processes based on the melt pool temperature online controlling.

1 Introduction

The first strategy of scientific and technological development of Russia consisting in the transition to the new industries is directly related to additive manufacturing i.e. to the technologies of the metal, ceramics or polymers parts 3D printing. Modern additive technologies are the new paradigm for the new quality industrial product development and manufacturing [1, 2]. The rearward face of the extensive development of additive manufacturing is its limitations, such as: build material limitations, low reproducibility and low productivity [3]. The most common and prospective additive manufacturing technologies (selective laser or electron beam melting, 3D laser cladding) do not provide either sufficient process online control or quality control of the material [4].

The basis for active control systems developing is the results of the optical diagnostics of selective laser melting (SLM) process [5, 6]. The previous works showed actual results: the real (not brightness) temperature estimating method was found [7, 8], the process parameters influence on the track formation was evaluated [9], and the external conditions influence on the process was determined [10]. The laser-irradiated area visualization and numerical simulation provides considering all physical factors and phenomena during the laser processing [11, 12].

Optical diagnostics involves the use of a high-speed CCD camera allows visual monitoring of fast processes with high resolution, an infrared camera allows analyzing the thermal cycle of the treatment zone, and a pyrometer measures the average temperature of a small area of the treatment zone [13, 14]. A serious problem in the way of wide additive production technologies industrialization is the need for an integrated approach to the additive technologies qualification for industrial manufacturing. The present work is devoted to the combination of accumulated knowledge and new scientific research in the field of control of track formation, process kinetics, and evaluation of the temperature of the melt pool.

2 Materials and methods

Optical diagnostics was carried out on an experimental laboratory-scale selective laser melting machine developed in the laboratory of innovative additive technologies of MSTU "STANKIN" according to the scheme on the Fig. 1. The energy source was the fiber laser LK-200V (IPG-Photonics, Russia) with a maximum power is 200 W, the radiation wavelength is 1070 nm. The focal beam diameter is 100 μm, and the spot transverse energy distribution is similar to the Gaussian.

As a measuring device, a high-speed (HS) camera of the visible range Photron SA5 (USA) was used with a frequency of 10 kHz (frames / s) (exposure ≈0.1 ms) and a spatial resolution of 0.1px / μm, as well as infrared (IR) FLIR Phoenix RDAC camera (Flir, USA) (working wavelength range is 1100-1700 nm) with a frequency of 340 Hz (exposure time: 3 ms) and a resolution of 0.87 pixels / micron. The survey frequency and resolution was limited by the capacity of the data bus. To exclude the laser radiation exposure of the HS-camera CCD-matrix, a notch-filter cuts off the wavelength of 1064 ± 20 nm was used. Optical instruments were installed in

* Corresponding author: kotobandv@mail.ru

 DOI: 10.1051/matecconf/201712901037

the working chamber at an angle of at least 45 ° to the surface of the substrate.

The austenitic stainless steel type AISI321 (12H18N9T, JSC Polema, Russia) with powder fraction of 0-40 μm and a molybdenum powder Mo 99,5% (Amdry 313X, Oerlicon Metco) with powder fraction of 30-75 μm were used as the initial powders. As a substrate, the 321 steel plates of 4 mm were used in the work. Steel powder elemental composition is % mass (Fe – balance, C - 0.1, Si - 0.68, Mn -0.9, Ti - 0.92, Cr – 17.1, Ni – 9.7, S – 0.01, P – 0.03, H2O – 0.022).. The powders were produced by gas atomization and presented almost the spherical shape and homogeneous fraction distribution. The powder were sieved with 0.050 mesh (mesh cell is 50 μm) and dried in the vacuum oven at 120 °C for 2 hours. The substrate was sandblasted and the surface was cleaned with alcohol.

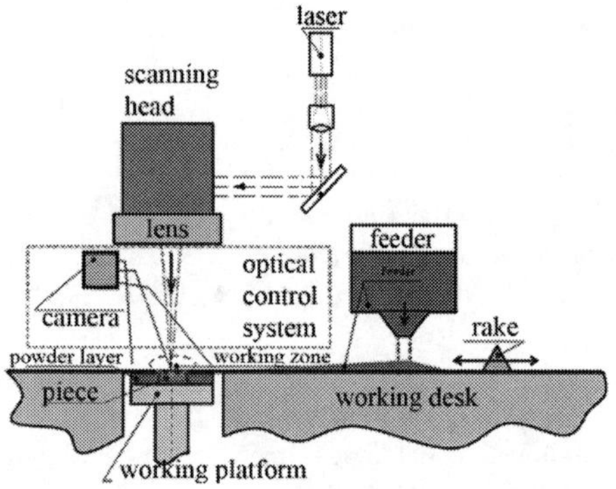

Fig. 1. Optical diagnostics equipment scheme coupled with SLM unit.

First SLM optical diagnostics experiments were carried out with 321 steel powder, which was uniformly leveled on the substrate with thickness of 100 μm (the controlling was provided by automatic leveling system). The laser beam pass the lines of 15 mm in length on the powder surface using different process parameters: laser radiation power of 30 W, beam scanning speed of 50-300 mm/s. The visible and thermography captures were saved and analyzed.

The second stage experiments were carried out with molybdenum powder with the following process parameters: powder leveling thickness was 100 μm, the laser power of 100 and 170 W, the beam scanning speed of 50, 75 and 100 mm/s.

The third step consists in the similar investigation with steel powder with the laser power of 100 W, the beam scanning speed of 50, 75 and 100 mm/s. The step objective was to found the difference of steel and molybdenum process.

The final stage was carried out without powder leveling by 321 steel substrate scanning with scanning speed of 20 and 200 mm/s and laser power of 170 W.

For each stage of experiments, the IR and visible range data were collected, saved and analyzed. The picture analysis consists the particle tracking velocity method.

3 Results and discussion

The results of the first study showed the degradation of the track (by single spheres, the balling effect (see Fig. 2), and the presence of a changing zone without a powder surrounding the built track (see Fig. 3). Obviously, the formation of single spheres in the track results from the reduction of the specific energy on the track unit length. The IR-camera (Fig. 2) data shows at lower power, the maximum temperature remains virtually unchanged, but the temperature distribution changes significantly. Therefore pyrometry (where the signal value is averaged by the matrix) shows qualitative changes in the workspace, but a study is almost impossible with this type of optical mean. Meanwhile, the uniformity of the molten bath can be estimated by applying the methods of digital filtering and coordinate measurement, as well as by calculating of the longitudinal track interruptions number.

Formation of a powder-free zone almost always follows the SLM process. The video images showed powder particles movement directed to the center of the laser-irradiated zone during powder heating and consolidation into a monolithic track. The movement id directly influenced by the applied specific energy

For the steel powder, the consolidation proved to be so fast that it did not possible to deep investigation of the consolidation mechanism. Obviously, the consolidation time could characterize this process [15]. For standard metals and alloys, the melting and fusion time was less than 1 ms apparently due to their fluidity in liquid state.

For the second stage of the experiments, the single tracks were obtained after laser scanning of the molybdenum powder. The choice of molybdenum was proved by its extremely high melting point (Tm = 2620 °C). Optical diagnostics of the process showed a clear picture of evaporation from the laser exposure zone (Fig. 4). Due to the high melting point of the used powder, the single powder particles located outside the melt pool are not fused among themselves and are in motion induced by the evaporation-driven gas flux. The single frame in Fig. 4a shows a powder-free zone appears during the motion of particles adjacent to the laser exposure zone (the melt pool). These particles do not have metallurgical contact with the substrate and are freely displaced by the gas dynamical forces. Serial images overlaying (Figure 4b) allows determining the particles motion path. As it is shown at the image, all moving particles on the substrate are directed to the center of the melt pool. Further, some non-molten particles or liquid particles released from the melt pool move up and out by the evaporation-driven gas flux. Evidently, the process intrinsic evaporation stream is directed perpendicular to the melt pool surface forms a zone of reduced pressure in the region adjacent to the melt pool, and the resulting gas stream involves the powder particles. Similarly, the third stage of work with steel powder was carried out. Using serial images when observing the motion of the same single particles, the velocities of their displacement and the value of the driving force could be calculated (Table 1, 2).

 DOI: 10.1051/matecconf/201712901037

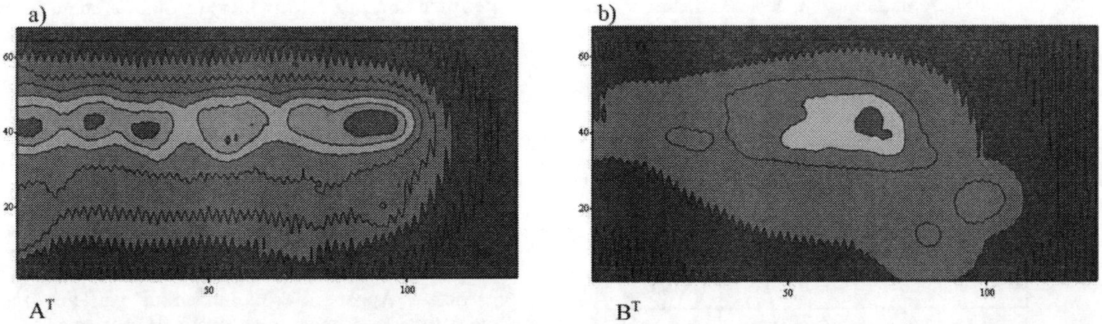

Fig. 2. Distribution of IR-camera signal for unstable track (a) and stable track (b). The process parameters: laser radiation power of 30 W, scanning speed of (a) 300 mm/s, (b) 100 mm/s.

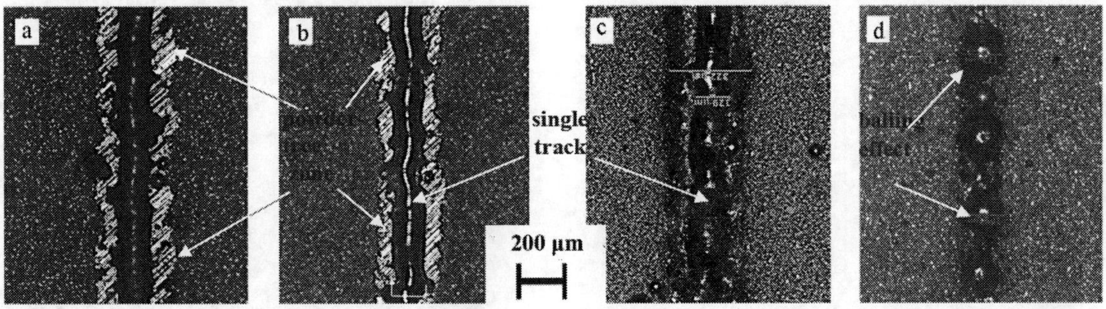

Fig. 3. Melted and solidified under different process parameters tracks from the steel powder (top view, the powder is not cleaned, optical microscopy is used): a) laser power is 30 W, scanning speed is 100 mm/s, b) 30 W, 150 mm/s, c) 30 W, 200 mm/s, d) 30 W, 300 mm/s.

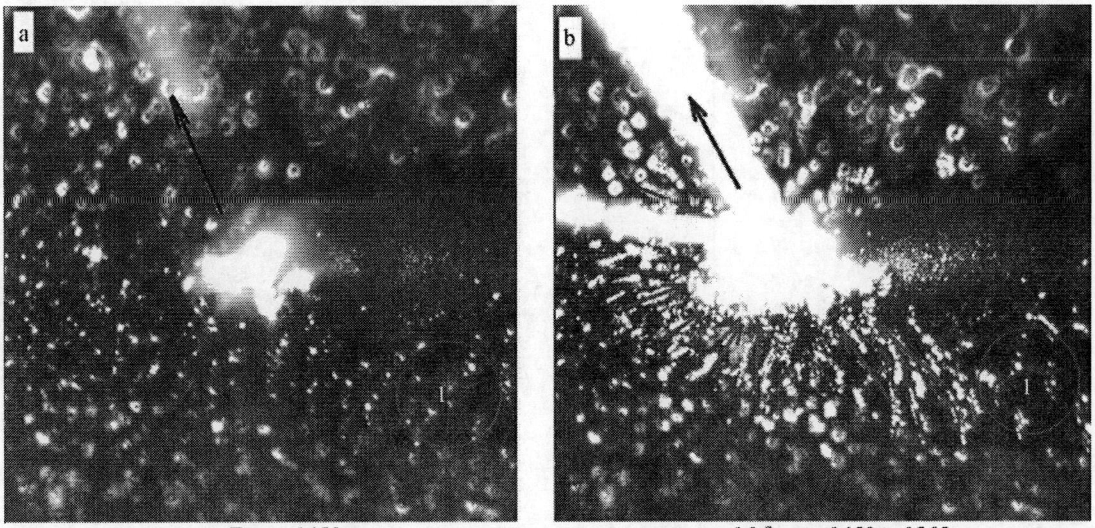

Frame 1450 s 16 frames 1450 to 1565 s

Fig. 4. The visualization obtained from the HS-camera in SLM process of molybdenum powder: a) single frame, b) overlapped series of 16 frames, 1- forming powder-free zone.

Table 1 presents the calculated parameter values for both powders that move out from the laser exposure zone. The data was provided by image treatment using the specific automatized calculation software with PTV.

Table. 1. Parameters characterized the powder particles movement in gas dynamic flux in SLM process out from the laser exposure zone

	321 steel	Mo
Sphere radii, m	$45 \cdot 10^{-6}$	$53 \cdot 10^{-6}$
Volume, m^3	$3,82 \cdot 10^{-13}$	$62,3 \cdot 10^{-14}$
Density, kg/m^3	7760	10220
Mass, kg	$2961 \cdot 10^{-6}$	$6370 \cdot 10^{-6}$
Exposure time, s	$3 \cdot 10^{-4}$	$4 \cdot 10^{-4}$
Path distance, m	$159 \cdot 10^{-6}$	$636 \cdot 10^{-6}$
Particle velocity, m/s	0,53	1,59
Acceleration, m/s^2	1767	3975
Lifting force, N	$5,23 \cdot 10^{-6}$	$25,3 \cdot 10^{-6}$

Table 2. Parameters characterized the powder particles movement in gas dynamic flux in SLM process in the center of laser exposure zone

	321 steel	Mo
Sphere radii, m	$27,8 \cdot 10^{-6}$	$23 \cdot 10^{-6}$
Volume, m^3	$8,95 \cdot 10^{-14}$	$5,09 \cdot 10^{-14}$
Density, kg/m^3	7760	10220
Mass, kg	$694 \cdot 10^{-6}$	$521 \cdot 10^{-6}$
Exposure time, s	$2 \cdot 10^{-4}$	$12 \cdot 10^{-4}$
Path distance, m	$59 \cdot 10^{-6}$	$146 \cdot 10^{-6}$
Particle velocity, m/s	0,30	0,12
Acceleration, m/s^2	1475	101
Lifting force, N	$1,02 \cdot 10^{-6}$	$5,28 \cdot 10^{-8}$

(Particle tracking velocimetry) method. It could be noted that for more dense particles of molybdenum powder, the values of the lifting force are much higher than for steel powder particles.

Table. 2 contains data on the calculation of the velocity of the particles moving toward the center of the melt pool. For the less dense particles of the 321 steel powder, the entrainment velocity into the melt pool is higher than for the molybdenum powder particles. This does not agree with the data in Table. 1, where the velocity of molybdenum particles is higher than for steel particles. Apparently, molybdenum particles require more time to accelerate (because their inertia is higher due to mass), and thus could develop a greater speed. More detailed information could be obtained from numerical simulation. At present, it is obvious that the driving force for the movement of particles is the evaporation flux from the laser exposure zone perpendicular to the melt pool surface.

The experiments on the visualization and calculation of the velocity of the gas evaporation flux passed without using a powder on a 321 steel surface. The reason was to eliminate the particles ejected from the melt pool due to the flood exposure. The other one was to verify the evaporation flux velocity is higher than the particles one.

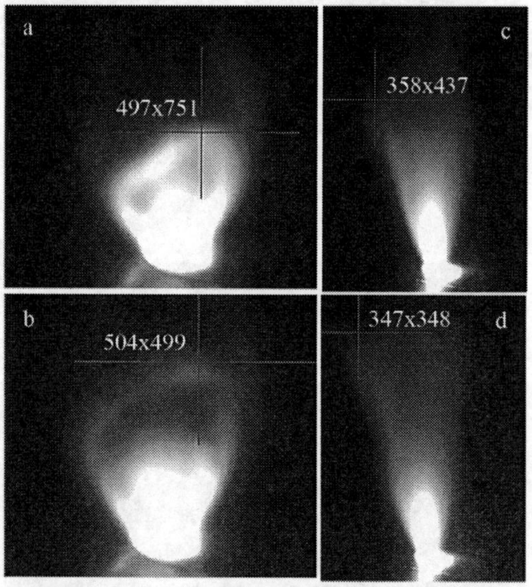

Fig. 5. Image series of the 321 steel plate melting Process parameters: laser power is 170 W, a, b) scanning speed is 20 mm/s, c, d) scanning speed is 300 mm/s

Table. 3. Vapor velocity calculation for scanning speed of 20 mm/s

	x	y	Moving distance, px.	Moving distance, m	Exposure time, s	Vapor velocity, m/s	Average	Std. dev.
1	501	595	26	0,00026	0,0001	3	3	1,14
	509	570						
2	405	649	20	0,00020	0,0001	2		
	400	630						
3	497	571	28	0,00028	0,0001	3		
	505	544						
4	497	571	72	0,00072	0,0002	4		
	504	499						
5	158	483	49	0,00049	0,0001	5		
	153	434						

Table. 4. Vapor velocity calculation for scanning speed of 300 mm/s

	x	y	Moving distance, px.	Moving distance, m	Exposure time, s	Vapor velocity, m/s	Average	Std. dev.
1	654	428	79	0,00079	0,0001	8	9	0,93
	648	349						
2	679	402	81	0,00081	0,0001	8		
	692	322						
3	602	433	99	0,00100	0,0001	10		
	612	334						
4	358	437	90	0,00090	0,0001	9		
	347	348						

It proves that the motion of the powder particles in SLM process is driven by the evaporation gas flux as well (the other reason apparently is melt pool dynamics). At this stage of the experiments, the melting of the substrate was carried out without a powder (Fig. 5). When calculating the speed of vapor, the image scale of 10 μm/px was used. Table. 3 and 4 presents the calculated velocity of the gas flux during scanning of the 321 steel substrate at a speed of 20 mm/s and 300 mm/s. The average value of the vapor velocity for several points in the image is 3 m/s. The value is much higher than the beam scanning speed and single particle velocities. The standard deviations contain high values relative to the mean value, which apparently could depend on the particles sampling being studied in each case. Due to the fact that the flux velocity at different points will be various, mean values with deviations are representative qualitatively.

It should be noted that as the scanning speed increases, the velocity of the evaporation gas flux from the melt pool increases significantly. Apparently, this is due to the decrease in the cross section of the flow passage, while the flow rate hold constant.

The mechanism of the powder-free zone formation, therefore, consists in the formation of a gas-dynamic flux in the laser exposure zone induced by the draft (stack effect) by evaporation flux from the melt pool surface. At the same time, the powder particles are involved along with the gas flux, incorporated into the melt pool or being thrown out in the form of sparks. Evaporation flux, apparently, directly depends on the temperature field on the surface of the melt pool, i.e. it is related to the process parameters. Simple conclusions show that the evaporation, coupled with the hydrodynamic behavior of the melt pool under the influence of laser radiation, form limitations in the technology of laser layer-by-layer 3D synthesis of the parts, as well as in other laser technologies. A similar result was obtained in [16], where the authors found similar effects (similar to Bernoulli effect) and described the medium gas dynamics behavior in the SLM process. The parameters having a significant influence on physical processes in the melting zone must be controlled. As shown, apparently the temperature, its distribution or the initial signal (or brightness temperature) of the IR-camera or pyrometer can be used for contactless process control, based on the developed criteria. The problem of estimating the actual values of the melt pool surface temperatures become particularly difficult due to the specifics of the techniques and the direct dependence of the signal on the radiation and signal propagation conditions from the treatment zone. In addition, the question of the calibration of optical elements in the complete set of the optical path of the SLM machine is currently being solved.

4 Conclusion

The use of diagnostic tools as the HS-camera and the IR-camera allows studying the fast processes occurring in the laser exposure zone, diagnosing the reasons of the particles motion leading to the formation of non-uniform distribution of the melted material volume. Some of these processes are unfavorable and can affect the quality of the part, made from the superposition of single tracks stacked next to each other.

The results of the study showed the kinetics of particles consolidation into a monolithic track and a gas-dynamic mechanism of the powder-free zone formation. In the course of the study, the velocities of the particles involved in the melt pool and particles ejected from the treatment zone were calculated. The value of the velocity of the evaporation gas flux, which is the cause of the motion of particles in the SLM process, is established. The methods and limitations of gas flow control are shown.

This work has been financed by the sources of Russian Science Foundation (grant agreement №15-19-00254 from 18.05.2015).

References

1. S. N. Grigoriev, T.V. Tarasova, Met. Sci. Heat Treat. **57**(9-10), 579-584 (2016)
2. F.Domniţa, H. Rotaru, MATEC Web of Conferences **94** (2017)
3. M. Baumers, P. Dickens, C. Tuck, R. Hague, Technol. Forecast. Soc. **102**, (2016)
4. J.-P. Kruth, L. Froyen, J. Van Vaerenbergh, P. Mercelis, M. Rombouts, B.Lauwers, J Mat Proc Tech **149**, 1, (2004)
5. I. Smurov, IV International WLT-Conference on Lasers in Manufacturing, 537-546 (2007)
6. M. Doubenskaia, M. Pavlov, Y. Chivel, Key Engineering Materials **437**, 458-461 (2010)
7. M. A. Doubenskaia, I. V. Zhirnov, V. I. Teleshevskiy, P. Bertrand, I. Y. Smurov, Mater. Sci. Forum **834**, 93-102 (2015)
8. P. Y. Peretyagin, I. V. Zhirnov, Y. G. Vladimirov, T. V. Tarasova, A. A. Okunkova, Russian Engineering Research **35**, 6, 473-476 (2015)
9. I. Zhirnov, C. Protasov, D. Kotoban, A.V. Gusarov, T. Tarasova, J. Therm. Spray Technol. **26**, 648, 1-13. (2017)
10. M. Doubenskaia, D. Kotoban, I. Zhirnov, Mech. Ind. **17**, 7, 707 (2016)
11. I. Zhirnov, I. Yadroitsava, I. Yadroitsev Mater. Sci. Forum. **828**, 474-481 (2015)
12. I. O. Kovaleva, S. N. Grigoriev, A. V. Gusarov, Phys. Proc. **56**, 421-428 (2014)
13. P. Lott, H. Schleifenbaum, W. Meiners, K. Wissenbach, C. Hinke, J.Bültmann, Phys. Proc. **12**, 683-690 (2011)
14. J.-P. Kruth, P. Mercelis, J. Van Vaerenbergh, T. Craeghs, Proc. 3rd int. conf. adv. res. virt. and rapid prototyping, 521-527 (2007)
15. C. E. Protasov, R. S. Khmyrov, S. N. Grigoriev, A. V. Gusarov Int. J. Heat Mass Transfer **104**, 665-674 (2017)
16. M. J. Matthews, G. Guss, S. A. Khairallah, A. M. Rubenchik, P. J. Depond, W. E. King, Acta Mater. **114**, 33-42 (2016)

 DOI: 10.1051/matecconf/201712901038

Increasing of the carbide cutting tool life by developing the multilayer coatings

Vladimir Tabakov[1], *Aleksey* Chikhranov[1,*], and *Sergei* Sizov[1]

[1]Federal State Budgetary Educational Institution of Higher Education Ulyanovsk State University, Ulyanovsk, Russian Federation
[2]Joint-Stock Company Ulyanovsk Design Bureau of Instrumentation, Ulyanovsk, Russian Federation

Abstract. The paper presents the results of studying the structural parameters and mechanical properties of multi-element coatings based on-TiZrN, TiZrAlN, TiZrCrN, and TiNbAlN. The phase composition, the structural parameters, and the mechanical properties of such coatings are shown to depend on the design of cathode evaporators and the machine layout at the moment of application. Compared to TiN coating, multi-element coatings have higher residual compressive stress, smaller CSR, higher microdeformations of the crystal lattice, better mechanical properties, yet weaker adhesion to the tool base. Based on studying their structural parameters and mechanical properties, we have designed double-layer coatings. Use of such multi-layer coatings prolongs the tool life of carbide plates compared to TiN coating.

1 Introduction

Multi-layer coatings are increasingly used to improve the efficiency of cutters [1-4]. The efficiency of such coatings depends on the mechanical properties of individual layers with specific functions [5, 6]. The mechanical properties of coatings depend on their deposition. Changing the layout of the machine and the design of cathodes has a significant impact on the mechanical properties of coatings [7-13]. Therefore, by changing the deposition conditions, we can adjust the properties of multi-layer coatings.

This research is to improve the efficiency of carbide tools by adjusting the layer-specific composition of multi-layer coatings.

2 Methodology of Experiment

Durable coatings were applied to MK8 carbide plates using a Bulat-6 machine. The chemical composition of such coatings was identified by quantitative X-ray analysis on a MAP-4 machine with ZAF adjustments being factored in. The parameters of the coating structure (lattice spacing a, half-width of the X-ray diffractive line β_{111}) and the residual compressive stresses σ_o were studied by means of a ДРОН-3M diffractometer; relative microdeformations of the crystal lattice $\Delta a/a$ and the coherent scattering regions (CSR) D were calculated as in [14]. Microhardness $H_м$, first-kind elasticity modulus E, fracture toughness $K_{IC\Pi}$ were determined as in [15]. The adhesion strength was evaluated on the bases of the stratification coefficient calculated using a TK-2M hardness tester as in [1]. The efficiency of MK8 carbide plates was evaluated based on the tool life when turning workpieces made of 38ХГН steel.

3 Methodology of Experiment

The multi-layer coating architecture has been selected based on their application in continuous turning as recommended in [2, 6, 16]. According to those papers, the upper layer should create favorable interaction of machined and tool materials at the front surface of the cutter to lower equivalent stresses and stabilize the shape of the cutting wedge. At the same time, in order to contain the fracturing processes, it is desirable that this layer has high mechanical properties as well as high level of its own residual compressive stress that contribute to the generation of higher normal compressive stresses during the cutting. As shown in [2, 4], multi-element coatings, i.e. three-element coatings based on modified titanium nitride, are the best when it comes to meeting such requirements. Beside the above requirements, coatings should have strong adhesion to the tool base; single-element coatings are adhesion-stronger than their multi-element counterparts [1, 2].

Based on the above, we studied TiN, TiZrN, TiZrAlN, TiZrCrN and TiNbAlN which were then used as layers in multi-layer coatings. TiZrN coatings were applied by two cathodes of titanium and zirconium located opposite to each other; TiZrAlN was applied by the same cathodes plus a cathode made of a titan-aluminum alloy. TiNbAlN was applied by two titanium cathodes plus a titan-niobium-aluminum cathode; TiZrCrN was applied by composite cathodes: two titanium cathodes with zirconium inserts and one titanium cathode with a chromium insert. Table 1 presents the chemical composition of the coatings.

* Corresponding author: chihranov@mail.ru

Table 1. Chemical composition of coatings.

Coating	Elements in the coating at. %				
	Ti	Zr	Cr	Nb	Al
TiZrN	57.2	42.8	-	-	-
TiZrAlN	57.5	39.75	-	-	2.8
TiNbAlN	83.4	-	-	14.5	2.1
TiZrCrN	75.4	13.4	11.2	-	-

TiZrN and TiZrAlN are biphasic coatings. TiZrN diffraction pattern in shown in Figure 1a. As shown there, there are two diffraction peaks that are identified as TiN (111) and ZrN (111), both having a cubic crystal lattice. At an angle of 2θ=35.2 grad. between them, there is a diffraction peak identifiable as solid TiZrN solution (111). According [6, 15], such coatings have a multi-layer structure where the multi-layers of separate phases and the solid solution alternate to improve the mechanical properties. TiNbAlN and TiZrCrN are uniphasic coatings. The diffraction patterns of these coatings (see Figure 1b) only show one high-intensity diffraction peak identifiable as TiN (111). Being uniphasic proves the homogeneity of such coatings and lack of multi-layer structure.

Table 2 presents the results of studying the structural parameters of multi-element coatings compated to TiN.

Table 2. Structural parameters and mechanical properties.

Coating	TiN	TiZrN	TiZrAlN	TiNbAlN	TiZrCrN
a, nm	0.4335	0.4356	0.4344	0.4309	0.4279
β_{111}, deg	0.40	0.82	1.06	0.70	0.59
σ_o, MPa	-775	-1352	-1487	-1875	-1422
D, nm	26.0	11.63	8.90	13.70	16.56
$\Delta a/a$ 10^{-3}	4.8	10.6	13.8	9.1	7.5
H_μ, GPa	24.2	36.0	39.8	37.6	36.8
E, GPa	315	385	424	402	405
K_{IC}, MPa·m$^{1/2}$	8.24	8.73	12.32	11.73	10.35
K_O	0	0.08	0.61	0.13	1.57
H_μ^3/E^2, GPa	0.143	0.315	0.351	0.329	0.304

It is found out that compared to TiN, multi-element coatings have a greated X-ray line half-width, greater residual compressive stresses, smaller CSR, and greater relative microdeformations of the crystal lattice. Comparing bi- and tri-element coatings shows that the latter have higher residual compressive stresses, whereas TiNbAlN has the greatest value of σ_o.

The half-width of the X-ray line depends on the design of cathodes and the layout of the machine used to apply the coating. β_{111} is at highest for TiZrAlN applied with solid titanium and zirconium cathodes; such coatings have a multi-layer stucture. The absence of thereof in TiZrCrN is what results in lower β_{111}. These data accord to what is described in [6, 16]. Multi-layer TiZrAlN coatings also have the smallest SCR *D* and the greatest value of $\Delta a/a$.

Multi-element coatings have higher mechanical properties compated to TiN, see Table 2. Their microhardness is 1.48-1.64 times higher, and the modulus of elasticity is 1.2-1.35 times higher depending on the composition. Tri-element coatings have higher microhardness, modulus of elasticity, and fracture toughness, but lower adhesion strength compared to bi-element coatings. TiZrCrN have the least adhesion strength, whereas its stratification coefficient is considerably higher than that of other three-element coatings. Multi-element coatings do not differ significantly in terms of resistance to plastic deformation (the ratio H_μ^3/E^2). These data are in line with what is described in [6, 16], which is due to the effect of various mechanisms strengthening the material upon their condensation.

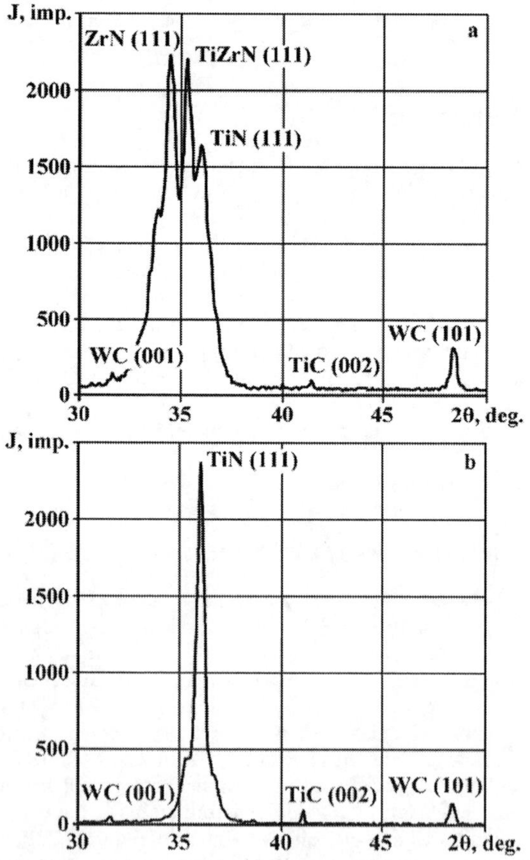

Fig. 1. Fragments of TiZrN (a) and TiNbAlN (b) diffraction patterns

Analyzing the results of studies, we note that TiZrN coatings have the greatest adhesion strength, whereas TiNbAlN and TiZrAlN boast the best combination of mechanical properties and residual stresses to generate higher normal compressive stress during cutting.

Based on these data and the principle of how multi-layer coatings are formed for continuous cutting, we have developed double-layer coatings where the lower level is TiZrN and the upper level is TiNbAlN, TiZrAlN, TiZrCrN. The efficiency thereof was tested by turning 38ХГН steel workpieces. Figure 2 presents the results.

It has been found out that use of our bi-layer coatings prolongs the tool life of carbide plates 1.8-2.8 times compared to TiN (the exact value depends on the cutting parameters and the material machined). In such terms, the most efficient coating is the double-layer with a TiNbAlN upper layer, the least efficient is that with TiZrCrN upper layer. TiZrN-TiZrCrN being less efficient is due to the lower adhesion strength of the lower layer as well as due to lower mechanical properties and residual compressive stresses characteristic of coatings made by composite cathodes [1, 2].

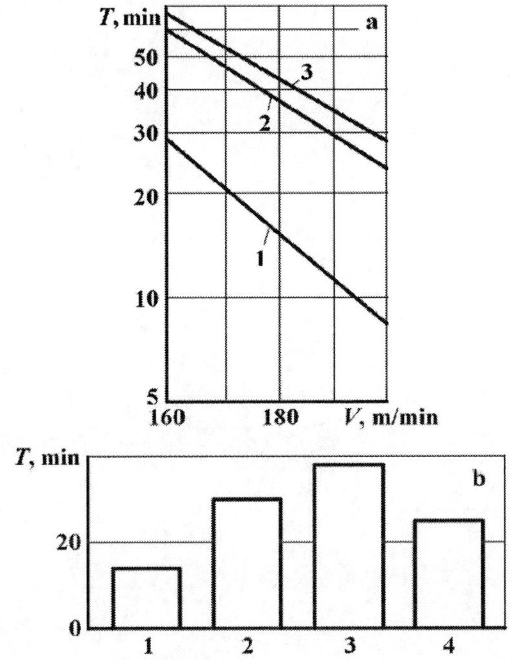

Fig. 2. The dependence of the tool life from the cutting speed (a) and the MK8 plate tool life diagram (b) when turning 38ХГН steel workpieces: a - S=0.15 mm/rot., t=0.5 mm; b - V=140 m/min, S=0.3 mm/rot., t=0.5 mm; 1-TiN, 2 - TiZrN-TiZrAlN, 3 - TiZrN-TiNbAlN, 4 - TiZrN-TiZrCrN

4 Conclusion

Multi-layer coatings improve the tool life of carbide tools compared to single-layer coatings. The most efficient are the coatings where the upper layer gives a favorable combination of mechanical properties and residual compressive stresses while the lower layer immediately adjacent to the tool has sufficient adhesion strength.

References

1. A.S. Vereshchaka, A.A. Vereshchaka, A.I. Bulycheva, *Multi-layer Nanodispersive Coatings for Cutters, (Mnogosloynye nanodispersnye pokrytiya dlya rezhushchego instrumenta)*, Strengthening Technologies and Coatings, (Uprochnyayushchiye tekhnologii i pokrytiya), **5**, 33-39, (2014)
2. V.P. Tabakov, *Generating Durable Ion-Plasma Coatings for Cutters, (Formirovaniye iznosostoykikh ionno-plazmennykh pokryty rezhushchego instrumenta)* (Mashinostroyeniye, 2008)
3. A.S. Vereshchaka, *Efficiency of Cutters with Durable Coatings, (Rabotosposobnost rezhushchego instrumenta s iznosostoykimi pokrytiyami)* (Mashinostroyeniye, 1993)
4. V.P. Tabakov, A.V. Chihranov, *Improving the Efficiency of Carbide Tools by Directed Selection of Durable Coating Composition Parameters, (Povysheniye rabotosposobnosti tverdosplavnogo instrumenta putem napravlennogo vybora parametrov sostava iznosostoykogo pokrytiya)*, STIN, **3**, 14-18, (2016)
5. D.I. Sagitov, *Use of Durable Ion Plasma Coatings in Threading, (Primeneniye ionno-plazmennykh iznosostoykikh pokryty na operatsiyakh narezaniya rezby)*, High-Tech in Mechanical Engineering: Proceedings of the All-Russian Research and Technology Web-Conference, 87-89, (2015)
6. A.A. Vereshchaka, A.S. Vereshchaka, S.N. Grigoryev, *Multi-layer Composite Nano-structured Coatings for Tools Used in Demanding Operations, (Mnogosloyno-kompozitsionnye nanostrukturirovannye pokrytiya dlya rezhushchikh instrumentov, rabotayushchikh v tyazhelykh usloviyakh)*, Strengthening Technologies and Coatings, (Uprochnyayushchiye tekhnologii i pokrytiya), **12**, 3-11, (2012)
7. V.S. Goncharov, S.S. Goncharov, N.G. Babenko, *Developing the Physical Model of a Durable Tool Complex, (Razrabotka fizicheskoy modeli iznosostoykogo kompleksa rezhushchego instrumenta)*, Book of Proceedings of the First International Research and Practice Conference On the Research, Development, and Application of High-Tech in Industry, (Issledovaniye, razrabotka i primeneniye vysokikh tekhnology v promyshlennosti), 249-254, (2005)
8. M.Sh. Migranov, *Efficiency of Metal Cutting by Tools with Durable Multi-layer Coating (Effektivnost rezaniya metallov pri ispolzovanii instrumenta s mnogosloynym iznosostoykim pokrytiyem)*, Mechanics and Physics of Proccess on Surfaces and in the Contact of Solids and Parts of Technological and Power Equipment, (Mekhanika i fizika protsessov na poverkhnosti i v kontakte tverdykh tel, detaley tekhnologicheskogo i energeticheskogo oborudovaniya), **6**, 139-144, (2013)

9. V.N. Gadalov, V.G. Salnikov, D.N. Romanenko, I.V. Shirin, *Durable Coatings for Cutters, (Iznosostoykiye pokrytiya dlya rezhushchego instrumenta)*, Mechanical Engineering Technology, (Tekhnologiya mashinostroyeniya), **7**, 26-28, (2011)
10. M. E. Maxims, Durable Coatings and Their *Application to Improve the Performance of Cutters, (Iznosostoykiye pokrytiya i ikh primeneniye v povyshenii ekspluatatsionnykh kharakteristik rezhushchego instrumenta)*, Bulletin of the Moscow State Technical University MAMI, **4 (22)**, Vol. **2**, 27-34, (2014)
11. G.S. Fox-Rabinovich, S.C. Veldhuis, A.L. Kovalev et al, *Design and performance of AlTiN and TiAlCrN PVD coatings for machining of hard to cut materials*, Surface and coatings technology, **204 (4)**, 489-496 (2009)
12. M.A. Volosova, V.D. Gurin, *Influence of vacuum-plasma nitride coatings on contact processes and a mechanism of wear of working surfaces of high-speed steel cutting tool at interrupted cutting*, Journal of Friction and Wear, **34 (3)**, 183-189, (2013)
13. I.V. Blinkov, O.A. Volkhonskii, V.N. Anikin et al, *Nanostructured wear-resistant coatings based on multicomponent nitrides and produced by vacuum-arc ion-plasma deposition*, Protection of Metals and Physical Chemistry of Surfaces, **48 (6)**, 649-655, (2012)
14. S.S. Gorelik, L.N. Rastorguyev, Yu.A. Skakov, *X-ray Structure and Electron-Optical Analysis of Metals, (Rentgenostrukturny i elektronnooptichesky analiz metallov)* (Metallurgizdat, 1970)
15. V.P. Tabakov, A.V. Chihranov, *Determining the Mechanical Properties of Durable Ion Plasma Coatings Based on TiNi, (Opredeleniye mekhanicheskikh kharakteristik iznosostoykikh ionno-plazmennykh pokryty na osnove nitrida titana)*, Bulletin of Samara RAS Research Center, **12 (4)**, 292-297, (2010)
16. V.P. Tabakov, S.N. Grigoryev, A.S. Vereshchaka, *Principles of Generating and Applying Durable Cutter Coatings, (Printsipy formirovaniya i tekhnologii naneseniya iznosostoykikh pokryty rezhushchego instrumenta)* (Ulyanovsk State Technical University, 2012)
17. B.Ya. Mokritskiy, *Controlling the Efficiency of Tools by Means of Coating Application, (Upravleniye rabotosposobnostyu instrumenta pri nanesenii pokryty)*, STIN, **11**, 11-15, (2010)

DOI: 10.1051/matecconf/201712901039

Design of the tool for periodic not evolvent profiles

Roman Anisimov[1,*], and *Alexander* Tarapanov[1]

[1]Orel State University named after I.S. Turgenev, Komsomolskaya street, 95, Orel, 302026, Russia

Abstract. The new approach to profiling of the tool for processing of parts with periodic not evolvent profiles are considered in the article The discriminatory analysis of periodic profiles including repetition of profile both in the plane of perpendicular axis of part, and in the plane of passing along part of axis is offered. In the basis of the offered profiling method the idea of space shaping by rated surface of product of tool surface lies. The big advantage of the offered approach in profiling is its combination with the analysis of parameters of process of engineering work. It allows to predict the accuracy and surface quality of product with not evolvent periodic profile. While using the offered approach the pinion cutter for processing of wheels with internal triangular teeths and mill for processing of the screw of the counter of consumption of liquid, complex profile of which consists of several formings, have been received

1 Relevance

In modern mechanical engineering, the products containing complex periodic not evolvent profiles which are widely used to torque transfer in set of mechanisms of machines and devices. Variants of form of profiles vary widely. In addition to traditional evolvent profile (cogwheels with evolvent teeths, connections with the help of involute splines) the triangular, cycloidal, trapezoidal and other profiles are also widely used. (fig. 1) It is caused by constructive purpose of certain mechanism and advantages of each of profiles. For example for cycloidal profile it is the possibility of essential change of contact ratio, reduction of quantity of teeths on gear wheel, without possible cutting, reduction of sliding speed of profiles; for triangular one it is the transfer of big torsion torques at thin-walled plugs and hollow shafts; for spindle gearing it is bigger gear ratio in planetary reducer at rather high efficiency and compactness of transfer. [1,2]

The great variety of helicoids of different profile is applied in products of the machine-building direction. The main advantages of screw mechanisms are the constant gear ratio, the possibility of obtaining large axial forces, the ease of achievement of self-braking, high accuracy of movement, reliability and durability. Helicoids (fig. 1) can also be attributed to periodic profiles on a number of grounds, such as the presence of a complex profile, which is repeated with a certain step, but unlike the gear and spline teeth and joints, it is repeated in the plane passing along the axis of the product. In these products forming may consist of a set of straight lines and curves of the second or more orders of magnitude. The products with a helical surface are also received by the method of running-in which can be both with axis of gravity and without it. When processing helicoids the probability of undercut of profile of product is high that it is necessary to predict and to take into account when designing the tool. The kinematics of processing of helicoids is much more difficult than toothed and slit ones which complicates the process of profiling of the tool and increases the terms of its design. [3] One of the most complex and labor-consuming tasks of tool production is profiling of the tool for processing of periodic not evolvent profiles. The peculiarity of the design of the tool is caused by the lack of unambiguous compliance between the profile of the tool and the profile of the forming surface in any of sections. Even more complex problem is the solution of a number of the related to profiling tasks of optimization of receiving the profile of the tool and machining process. For example, the definition of the admissible errors of the installation of the tool on the machine, the definition of the admissible tool wear. The particularly important this question becomes in the process of designing of the instrument for products with a helical surface. [4,5,6]

At the same time the conditions of a rigid competition require reducing the time for project works. Therefore, the fundamental criterion for new profiling methods should be the ability to create computer-aided engineering systems. The existing profiling methods, which differ in degree of formalization and mathematical description, are not always fully amenable to full automation due to inherent contradictions in them. The poor use of computer-aided design tool in domestic engineering affects the quality and cost of the final product seriously. [4,5,6]

The optimal solution for automation of project works are interactive software systems in which the possibility of active interaction of the designed system with the person is laid.

* Corresponding author: roman.anisim@gmail.com

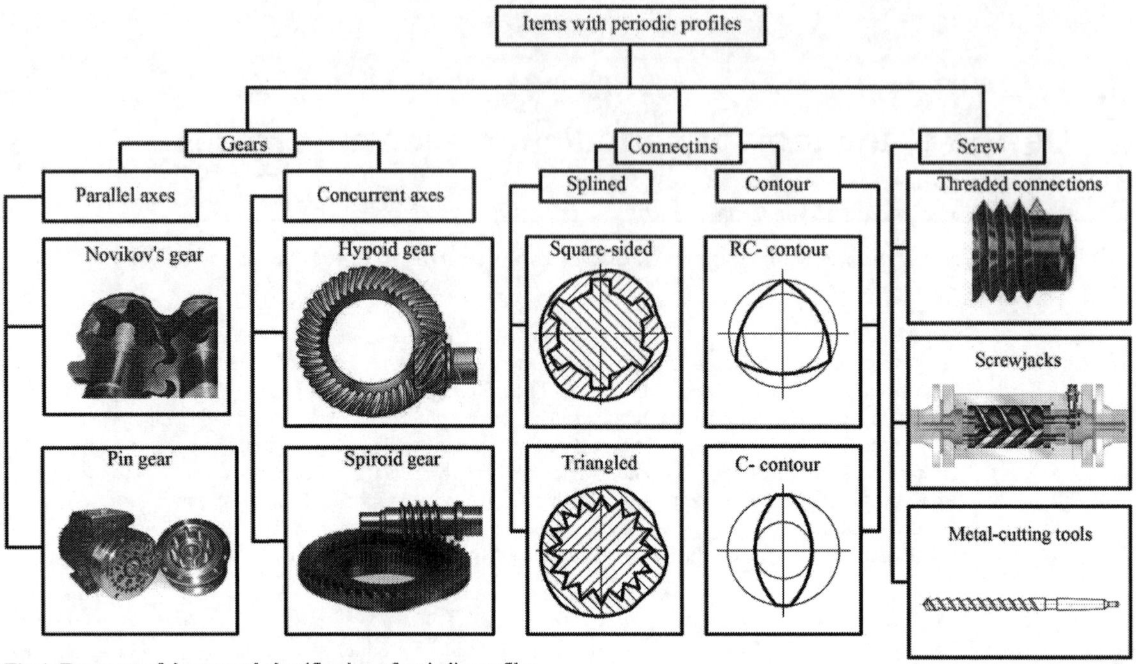

Fig 1. Fragment of the general classification of periodic profiles.

2 A method of profiling of the tool for periodic not evolvent profiles

The machining process of complex periodic structures is characterized by a large number of different processing methods and their variants, the complexity of profiling of the cutting edges of the tool, high requirements to surface quality and to kinematic accuracy and errors of a profile even at the preliminary stage of formation.

In the integrated analysis method, the mathematical model of processing by cutting includes the description of the movement of the elementary site of the edge of the cutting tool in the space of formation, the calculation of kinematic changes of face and back angles, the thickness of the cut-off layer, the determination of the components of cutting forces, the wear of the elementary site of the cutting edge, the determination of the errors of machining and the figures of the quality of surface.[7,8]

The diversity to a complex method of the analysis of the shaping of periodic profiles gives that circumstance that each microelement of the studied space has at the same time geometrical and physicomechanical properties of a part and of a tool, possesses functional and space assignment and a certain time span, contains the framework of the model used is the ability to control the shaping in accordance with the specified objectives and constraints. The qualitative indicators of a rated surface, the cutting modes, parameters of the unit of the tool, the time of processing and the cost value can be refered to the latter. [7,8,9]

The method is based on the description of the process of shaping, or processing of a tool by not evolvent periodic surface, where the directing lines of a rated surface act as the cutting edges. For each site of a periodic not evolvent profile there is the thickness of the cut-off layer close to zero at the point N (fig. 2). For each posion of a hollow of a wheel on a certain elementary site of an initial profile of a hollow the point with thickness close to zero will be one. This point can be considered belonging to the treated surface and the tool surface, and its coordinates are adapted coordinates of the tool profile. [8]

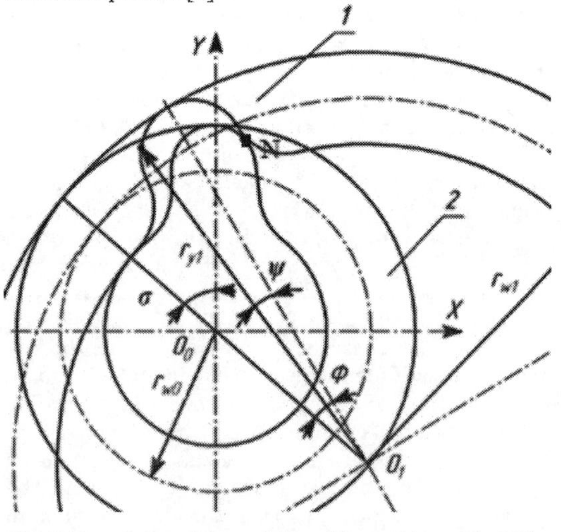

Fig. 2. The scheme of determining of the position of the hollow of the wheel 1 relative to the tooth shaping cutter 2 when profiling.

The schematic of the formation of the generating surface is reduced to the output expression for each coordinate that defines the position of each point of not evolvent periodic surface at any moment in time of formation. In this case, all movements are performed by the nominal surface periodic not evolvent profile. This scheme of shaping is constructed by sequential transformation of the original expression that defines the initial profile of the nominal surface. In the process of creating the schema of shaping there is a need to change the direction of the axes or move the center of the coordinates. This is accomplished through the rotation of one of the planes and the movement the whole system along the axes. The transformations used to build a scheme of forming are divided into 2 types. (table 1). [10,11]

Table 1. Transformations for creation of the scheme of shaping

Turn, rotation	Translational movement, linear shift
$x_n = f(x_{n-1}; y_{n-1}; z_{n-1}; \psi_n)$	$x_n = f(x_{n-1}; \psi_n)$
$y_n = f(x_{n-1}; y_{n-1}; z_{n-1}; \psi_n)$	$y_n = f(y_{n-1}; \psi_n)$
$z_n = f(x_{n-1}; y_{n-1}; z_{n-1}; \psi_n)$	$z_n = f(z_{n-1}; \psi_n)$

where x_{n-1}, y_{n-1}, z_{n-1}- the function equation received from the previous transformation;

ψ_n- value of the parameter corresponding to the current transformation.

The index n denotes the sequence number of the conversion. The direction of rotation in each case is determined by a parameter that can increase or decrease.

The input data of system are: the upper bound and the lower bound of sets ψ_1, $\psi_2,...,\psi_n$ the direction of change (increase or reduction).

Transformations at such operations differ only in the fact that the parameter for creation has the fixed value.

This approach just allows mathematically to provide practically any kinematically difficult scheme, and also to use it for creation of software systems.

Not evolvent periodic profile is set in the edge section. The forming of all profiles applied in mechanical engineering is described by widespread algebraic curves (a straight line, a circle arch, an epicycloid, spirals, an evolvent, etc.) (table 2). The use of such task of forming allows to find any quantity of points of the line of contact of a tool and rated surfaces. [10,11]

The profile received by the results of the offered technique is used further for the solution of a direct task - forecasting of parameters of processing by the tool of a rated surface. In the course of the solution of a direct task they receive the values of components of force of cutting which are then used in the basis of obtaining values of accuracy and roughness of the product. In the mathematical apparatus of the solution of a direct task the rigidity of technology system and the deformations arising in it, their influence on accuracy and roughness of the product are considered

Consecutive application of the offered method of profiling of the tool for not evolvent periodic profiles in combination with the mathematical apparatus solving a direct problem allows you to make adjustments to the tool profile taking into account elastic deformations of the technological system. It will lead to the improvement of the quality of the made products - to decrease roughness and to increase accuracy

Table 2. Examples of the algebraic curves in a parametrical form used for creation of the forming line in the edge section of different rated surfaces

Scheme of curve plotting	Equation of a curve in a parametrical look
1	2
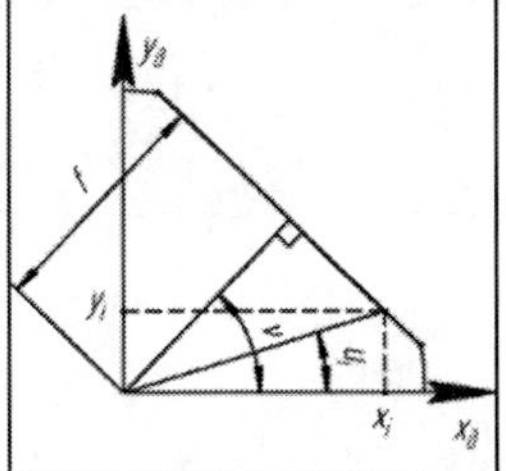	Straight line: $x_i = \dfrac{f}{\cos(u_i - \nu)} \cdot \cos(u_i)$ $y_i = \dfrac{f}{\cos(u_i - \nu)} \cdot \sin(u_i)$
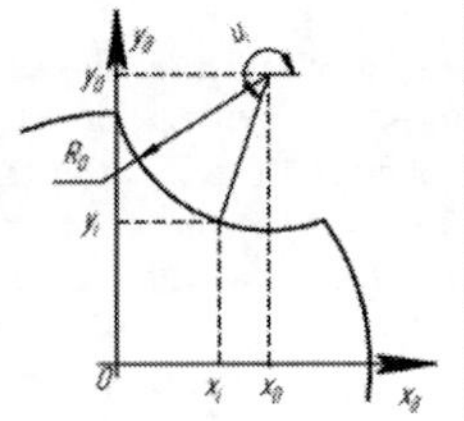	Circle arch: $x_i = x_0 + R_0 \times \cos(u_i)$ $y_i = y_0 + R_0 \times \sin(u_i)$
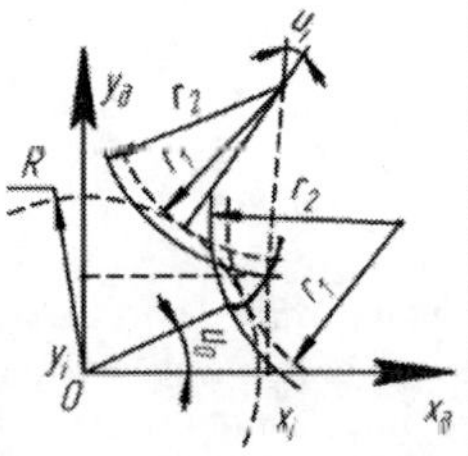	Epicycloid: $x_i = (R + r_1)\cos(u_i + u_0) - -r_2 \cos\left[\left(\dfrac{R+r_1}{r_1}\right)(u_i + u_0)\right]$ $y_i = (R + r_1)\sin(u_i + u_0) - -r_2 \sin\left[\left(\dfrac{R+r_1}{r_1}\right)(u_i + u_0)\right]$
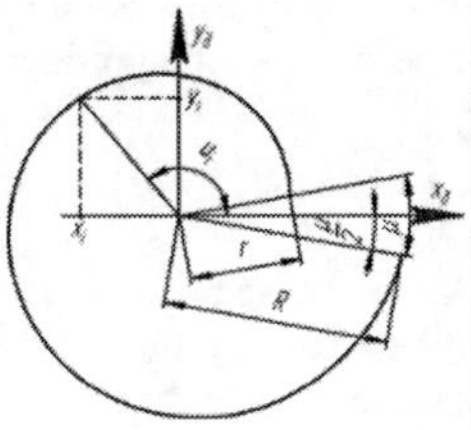	Archimedean spiral: $x_i = \dfrac{\begin{pmatrix} r(2\pi - \mu) + (R - r) \cdot \\ \cdot \left(u_i - \dfrac{\mu}{2}\right) \end{pmatrix} \cdot \cos(u_i)}{2\pi - \mu}$ $y_i = \dfrac{\begin{pmatrix} r(2\pi - \mu) - (R - r) \cdot \\ \cdot \left(u_i - \dfrac{\mu}{2}\right) \end{pmatrix} \cdot \sin(u_i)}{2\pi - \mu}$
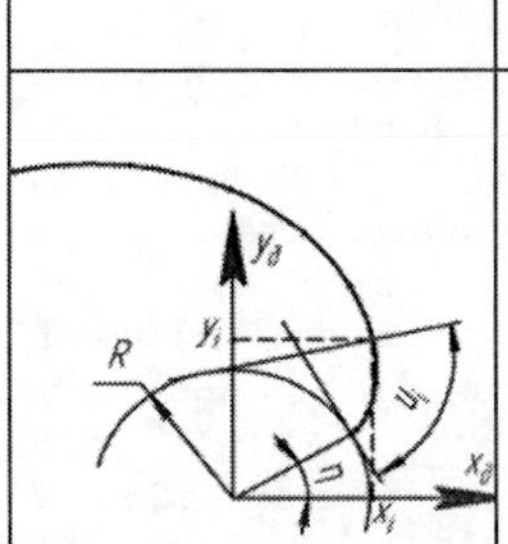	Evolvent: $x_i = R\begin{pmatrix} \cos(u_i - \eta) + u_i \cdot \\ \cdot \sin(u_i - \eta) \end{pmatrix}$ $y_i = R\begin{pmatrix} \sin(u_i - \eta) - u_i \cdot \\ \cdot \cos(u_i - \eta) \end{pmatrix}$

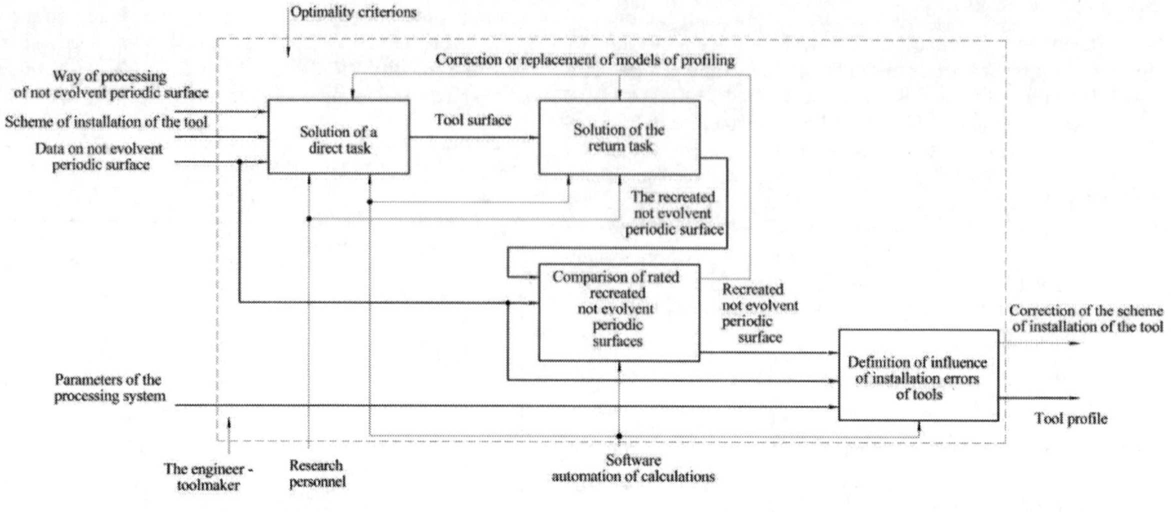

Fig 3. IDEF-0 chart of process of profiling of the tool.

For a more visual demonstration of the stages of profiling with the subsequent forecasting of parameters of the processing of periodic not evolvent profiles the chart of process of profiling of the tool is presented on fig. 3. The process of profiling of the tool is combined with the analysis of parameters of process of cutting that allows to improve the accuracy and surface quality of the products with periodic not evolvent profiles. [12, 13]

The creation of the profile of the rated tool surface and its correction, definition of the modes of processing, the number of passes, the material of the tool and the model of the machine at which it can be reached, the precision and qualitative characteristics of the processed products set by designers of machine-building productions with periodic not evolvent profiles is a consequence of implementation of this chart. This makes it possible to reduce the normal time for the existing technological process, as well as to determine the optimal values of output parameters for the new technological process of processing of the product.

3 Conclusions

A new approach to profiling which is based on the spatial process of forming the nominal surface by the tool one with the further analysis of parameters of processing by cutting is offered The profiling method uses the mathematical apparatus adaptable for creating CAD programs. As a result the combination of making and rated surfaces is visually controlled with an accuracy of 10^{-6} mm. It is collectively with the following analysis of processing allows to find the optimal combination of the tool profile , the parameters of the nominal surface of the settings of the tool and parameters of the treatment process.

The proposed approach provides the reduction in financial and time costs of design of the tool and the increase in the accuracy and performance, significantly reducing the implementation of the time of technological processes. In the environment of mathematical and software methods the pinion cutters for a wheel with internal triangular teeths and mills for the liquid consumption counter screw were profiled (fig. 4, fig. 5) [14].

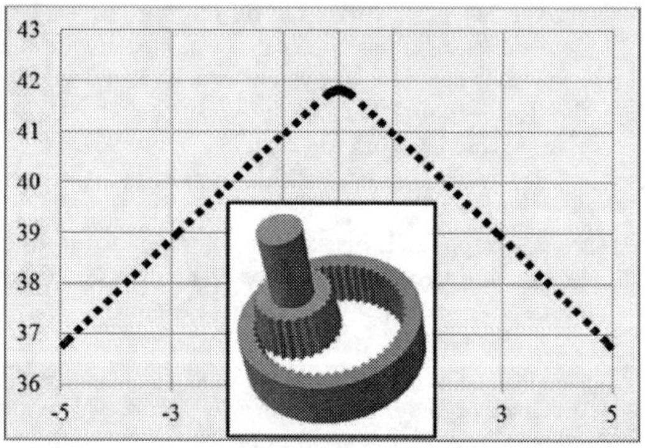

Fig. 4. Results of profiling of the pinion cutter for a triangular profile.

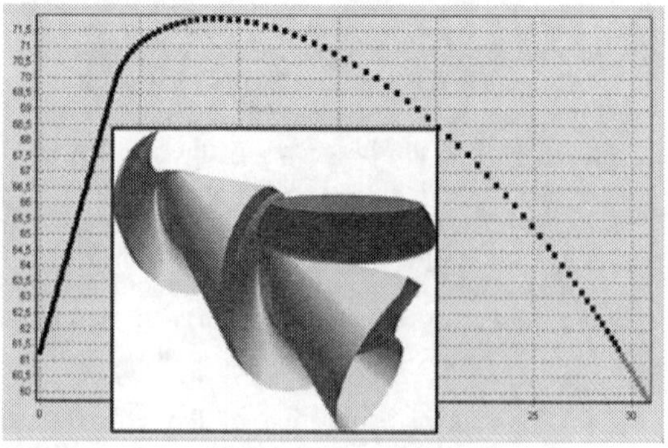

Fig. 5. Results of profiling of the mill for the liquid consumption counter screw.

References

1. R.V. Anisimov, A.S. Tarapanov, *Modern technologies and the equipment for a gear shaping of wheels with internal teeths*, Regional scientific and practical conference of young scientists and graduate students «Scientific capacity of Orlovschina in modernization of an industrial complex of the small cities of Russia», p.4-8 (2010)
2. K D Bouzakis, *Manufacturing of cylindrical gears by generating cutting processes, A critical synthesis of analysis methods,* CIRP Annals-Manufacturing Technology, 57 p. 676-696 (2008)
3. S.I. Brusov, M.V. Zhuplov, A.S. Tarapanov, *The description of an initial helicoid when profiling screw-cutting tools,* News Oryol GTU. Fundamental and applied problems of the equipment and technology **№3** p., 18–23 (2007)
4. R.V. Anisimov, *Improvement of quality of details of vehicles with internal teeths of not evolvent profile,* Topical issues of innovative development of a transport complex. Materials of the International scientific and practical conference, 91-95 (2011)
5. G Figliolini, P Rea, *Effects of the design parameters of involute gears generated by rack-cutters* International Gear Conference 294 (2014)
6. M.V. Zhuplov, *Interactive profiling of disk mills for processing of compound half-closed helicoids on the basis of three-dimensional display of the scheme of profiling,* News Oryol GTU. Fundamental and applied problems of the equipment and technology **№ 2-3/274** (560) p.24-29 (2009)
7. M.V. Zhuplov, S.I. Brusov, A.S. Tarapanov, *An integrated approach in profiling and a research of processing of helicoids,* Problems of quality of machines and their competitiveness: materials of the 6th international scientific and technical conference p. 269–270 (2008)
8. R.V. Anisimov, S. I. Mironov, *Mathematical display of schemes of cutting and determination of thickness of the cut-off layer when processing not evolvent profiles of cogwheels with internal teeths* Fundamental and applied problems of the equipment and technology **№2/3(280),** p.16-21 (2010)
9. K Nagamura, K Ikejo, F G Tutulan, *Design and performance of gear pumps with a non involute tooth profile* Proceedings of the Institution of Mechanical Engineers, Part B: J of Engineering Manufacture. **218,** 699-711 (2004)
10. R.V. Anisimov, *Profiling of gear cutting for wheels with internal not evolvent teeths,* Fundamental and application-oriented problems of upgrade of the modern mechanical engineering and metallurgy: The collection of scientific works of the international scientific and technical conference devoted to the 50 anniversary of department of technology of mechanical engineering of LGTU, 11-15 (2012)
11. M.V. Zhuplov, S.I. Brusov, A.S. Tarapanov, *The description of an initial helicoid when profiling screw-cutting tools,* News Oryol GTU. Fundamental and applied problems of the equipment and technology **№3** p. 18–23 (2007)
12. R.V. Anisimov , A.S. Tarapanov, *Forecasting of a roughness of processing of cogwheels with internal teeths of not evolvent profile,* High technologies in mechanical engineering: Materials of the International scientific and technical conference p. 46-50 (2012)
13. F L Litvin, I Gonzalez-Perez, A Fuentes, K Hayasaka, *Design and investigation of gear drives with non-circular gears applied for speed variation and generation of functions,* Computer methods in applied mechanics and engineering **45,** 3783-3802 (2008)
14. S.I. Brusov, M.V. Zhuplov, A.S. Tarapanov, *Option of definition of a profile of a disk mill for processing of a helicoid on the basis of space mathematical display of the scheme of cutting,* High technologies in mechanical engineering: theses of reports of the All-Russian scientific and technical Internet conference with the international participation p.135-140 (2008)

Modern problems of energy efficiency programs implementation at the enterprises of mechanical engineering

Natalia Borisova[1,*], *Alexander* Borisov[2], and *Oleg* Dushko[3]

[1]The department of management and urban development and construction, Volgograd State Technical University, Volgograd, Russia
[2]The department of economics and project management in construction, Volgograd State Technical University, Volgograd, Russia
[3] The faculty of distance education, Volgograd State Technical University, Volgograd, Russia

Abstract. Russia is the one of the most wasteful countries in the world in using of fuel and energy resources. Energy-saving potential is huge enough. Thereby energy efficiency programs implementation at the enterprises of mechanical engineering is the one of priority tasks of modern economics of Russia and requires radical measures of increasing the efficiency of its implementation. The purpose of Author's search is scientific and practical research of modern problems of energy efficiency programs implementation at the enterprises of mechanical engineering. In achieving of this purpose the following tasks were solved by authors: justified relevance and necessity of implementation of energy-saving policies at the enterprises of mechanical engineering in the regions of Russia; indentified the most significant problems preventing effective implementation of energy-saving program at the enterprises in the regions of Russia; proposed possible directions solving reveal problems in sphere of energy-saving programs implementation at the enterprises in the regions of Russia. In presented research Author's collective focuses on problems of energy audit using in practice as the main instrument for working-out of energy-saving programs at the enterprises of mechanical engineering and the its subsequent implementation. The subject of Author's research is the factors defining energy-saving implementation at the enterprises of mechanical engineering, the object is the enterprises of this one. Methodological scientific and practical research is based on a complex approach with using of methods of comparative, analytical and logical analysis.

1 Introduction

The main reason motivating of energy-saving is the factor exhaustibility of fuel and energy resources. The idea of energy-saving arose in the end of the 20th century during the period of energy crisis most of the world's advanced countries have sought to improve the efficiency and use new alternative and renewable energy sources [1; 8; 10; 12; 13; 17; 19].

In new economical facilities and for many years energy-saving is the fundamental factor of effective functioning and perspective development of many enterprises of industry in particular the enterprises of mechanical engineering of many leading countries including Russia and its regions [7; 9; 14; 15; 16; 20].

Practical implementation of energy-saving potential is the higher priority of Russian's federal and regional economic policy because in new economical conditions every percent of fuel and energy resources saving used in production activity of the industrial enterprises in particular the enterprises of mechanical engineering may give increase of national income above 0,35% [2]. The main purpose of evergy-saving policy in mechanical engineering is cost price decline of manufactured products, increase of its competitiveness and also increase of efficiency of fuel and energy's using.

2 Relevance and problem situation

Today Russia is a energy-surplus country, efficiency of energy resources using at the rate is about 30% that is above 2/3 of energy is wasted ineffectively. This situation worsens every year, if the measures of available potential of energy-saving's using weren't applied in our country, the onset of energy crisis would begin. In 2015 energy intensity indicators decreased by 5,61% to the level of 2008. Despite the positive dynamics, it's very difficult to achieve the result without sound energy-saving policy specified "in energy country's policy until 2020" (decrease in energy intensity of GDP per 40%). [2].

Multiplicative effect is necessary for energy efficiency at the expense of organizational, legal, technical, technological, economical and other measures aimed at volume reduction of used fuel and energy resources while maintaining the corresponding beneficial effect by its using. The most priority directions of energy-saving policy are industrial sector and also sub-sectors. Industrial enterprises are the biggest users of energy resources: they account up to 50% of energy consumption in the country. Wherein the biggest fuel and energy resources losses are observed also at

* Corresponding author: borisovani06@mail.ru

industrial sectors (47%). It attests about huge energy-saving's potential. [4].

Mechanical engineering is a leading sector of Russia's economic and its main system-forming element defining industrial potential and defensive capacity's condition, sustainable functioning of all industrial sectors and filling the consumer market. Mechanical engineering is the one of the most capacious sectors of Russian industry including production every possible machinery, equipment and fittings. The etreprises's production of mechanical engineering plays a big role at scientific and technical progress's achievements implementation in all national economy's spheres of our country. The share of mechanical engineering accounts about 20% amount of issue of all Russian manufacturing industries. [4].

Analysis of energy-using's levels at mechanical engineering sector shows that the sector has significant reserves for increasing energy efficiency. At the present stage mechanical engineering is characterized by wide using of all kinds of energy at the expense of different energy resources: electric power, steam, hot water, liquid, gaseous and solid fuel. The share of electric power accounts about 24%, fuel – 38%, heat power – 38%. 10-15% of consumed energy is expended for technological purposes in mechanical engineering. The electrification's level in mechanical engineering is higher than in industry generally at that more than half the electricity is used in mechanical processes. Specific gravity of electric and technological processes exceeds 50% on some mechanical engineering enterprises. [5].

Of all consumed energy resources at the mechanical engineering enterprises are expended about 30% for technological processes and about 70% for CHPP, boiler rooms, ventilation, lighting, production of compressed air, intrafactory transport and other auxiliary needs. [5]. In its turn, moral and physical deterioration of enterprises's fixed assets (equipment and buildings) is a result of constant consumtion's increas of different kinds of energy in many mechanical engineering's enterprises of Russian regions. We give practical example to justify our conclusions:

- OJSC «Russian machines» is a large Russian diversified holding company iincluded in group «Basic element» and unifying 27 enterprises of mechanical engineering (automotive, automotive components, raliway engineering, aircraft construction, agricultural and construction machinery) in 12 Russian regions. It presented general data about value its energy resources during 2009. In 2009 total cost of energy resources of OJSC «Russian machines» was 5,2 billion rubles, to the end of 2013 their value increased to 9 billion rubles. With a rise in value of energy resources was a rise in share of energy consumption in total cost production. So in 2009 energy consumption accounted for 3,4% of total cost production and to the end of 2013 the share of energy consumption increased to 6%. Eventually using of morally and physically obsolete technologies of power supply at the factories of corporation «Russian machines» (boiler equipment) was the result of constant increasing of energy consumption. According to set of the rules «Boiler settings» (set of the rules 89.13330.2012 Boiler settings) approximate lifetime of this equipment is less than 10 years. In 2013 boiler equipment's age structure at the factories of corporation «Russian machines» was introduced in the following way [6]:

- 12% of equipment is less than 10 years old;
- 14% of equipment is more than 10 years old;
- 16% of equipment is more than 20 years old;
- 47% of equipment is more than 30 years old;
- 10% of equipment is more than 40 years old.

By this way 88% of boiler equipment of OJSC «Russian machines» is exceeded standard operating life that's why energy consumption and total cost production increase.

Leading energy-saving experts affirm that in boiler rooms used at the enterprises of mechanical engineering are ineffective water treatment systems so it leads to scale formation in boilers and heat exchanger and also to decline of economy. With sedimentation of water heating boiler the boilers's efficiency falls to 60% in 2-3 years of exploitation in bad water treatment conditions. [2].

Besides increased equipment deterioration, many mechanical engineering enterprises have high moral and physical deterioration of their industrial buildings (CHPP, boiler rooms) that also promotes rising fuel and energy resources costs, rising prime cost of engineering products and decline of competitiveness. Many regional enterprises at mechanical engineering were created during planned ecomony concerning their building's age is more than 30-40 years old. As a rule most of industrial funds of Russian regions weren't capital and current refurbished that's why it got old and worn out. Authors consider that the one of main reasons of untimely undustrial building's capital and current repairs is a change of political and economic regim in the country leading to reduction in management efficiency of their production and also to limitation of financial resources. Practice shows that industrial buildings have big physical deterioration consume more than 70% of electric power, 49,8% of natural gas, 40% of fossil energy resources [2].

Irrational using of fuel and energy recourses inflicts significant damage for mechanical engineering sector. In its turn rising energy prices, limitation of energy resources made very actual problem of energy-saving for the Russian enterprises of mechanical engineering and its regions.

3 Research and formulation of the problems of energy efficiency programs implementation

The legislative framework of energy-saving policy is formed beginning from 1996. So a number of legislative acts about energy-saving was accepted by the State Duma of the Russian Federation : the Federal law from 1996 №28-FL «About energy-saving», the target program «Russia's energy-saving from 1998 to 2005» from 1998 №80, «Russia's energy strategy to 2020» approved by the government order of the Russian

Federation №1234-GO from 2003. Subprogram «Energy-saving in energy intensive sectors» of the Federal target program «Russia's energy-saving» was elaborated and recommended for practical implementation in 2000.

Energy audit or energy survey is the main source of information about the maximum losses of fuel and energy resources and the main instrument for development of energy-saving programs and their subsequent implementation on practice. Energy audit is a interconnected complex of technical, managerial, organizational, economical and other measures including information's collection and processing about using of fuel and energy resources for getting reliable data about their volume of using, indicators of energy efficiency; potential identification of energy-saving and possibilities of indicators of energy efficiency's increasing with submission of Acquired data in energy passport [2].

Energy passport is formed according to the results of energy survey which contains information about the volume and structure of production and consumption of fuel and energy resources at an enterprise, the volume and structure of their losses and also may be given recommendations about using of concrete energy-saving technologies taking into account the specifics of production activity of an investigated enterprise. [2].

From the moment of output and adoption of many regional and federal programs in energy-saving field has passed more than 15-20 years however massive development of these programs didn't happen. Energy audit experts say that top-brass of mechanical engineering's enterprises ignore the necessity of energy audit, energy-saving programs elaboration and implementation because of lack of sufficient financial resources. Also not all mechanical engineering's enterprises leaders don't recognize the necessity of energy audit for their enterprises. Such attitude of mechanical engineering's enterprises leaders to energy-saving is mostly defined the share of energy consumption in total cost mechanical engineering production. And if in chemical industry the share of energy consumption may reach 40%, then for mechanical engineering this number is from 6% to 15%. As a result energy-saving's issues at many mechanical engineering enterprises aren't included into the list of strategic directions of their development, they are located in the introduction of only the main power engineer of an enterprise [3].

However it should be noted that some enterprises of mechanical engineering are paying active attention to energy-saving at their industry and getting sufficient economic effect from it. The bright example of energy-saving policy at mechanical engineering sector is OJSC «United aircraft corporation» (OJSC «UAC»). Average cost share of aircraft enterprises OJSC «UAK» is 7% for energy resources. But relatively small indicator of main products energy intensity didn't intervene OJSC «UAK» specialists to elaborate corporate policy in energy-saving field which presents constant monitoring of energy efficiency program status of their aircraft factories, promotion of proven solutions on other enterprises of corporation, analysis of energy tariffs, staff motivation. As a result energy-saving corporate program decreased energy consumption per unit of production significantly (all told per 40% in 2012 and also per 30% in 2015) and allowed saving money to 3 billion rubles every year. From 2011 to 2015 energy costs for OJSC «UAK» could increase from 3,16 to 8,2 billion rubles because of volume production increasing and increasing of tariffs for energy resources. But due to energy efficiency program implementation those costs didn't exceed 4,8 billion rubles according to estimates of OJSC «UAK» specialists [3].

Energy passport may be considered as an instrument of support for management decision-making in energy-saving and increasing of mechanical engineering's energy efficiency field. The drafting of full energy passport is possible only according to results of a comprehensive energy analysis of all its sites (in particular by kinds of activity). But currently some problems are connected with energy audit organization and implementation. According to the Federal law №261-FL of the Russian Federation from 23.11.2009 «About energy-saving and increasing of energy efficiency and also about amending certain legislative acts of the Russian Federation» energy analysis isn't compulsory for enterprises with the exception of the following organizations:

- government departments and local governments endowed with the rights of legal entities;
- organizations with state participation or local government;
- organizations which the total costs of all kinds of fuel, fuel oil, thermal energy, coal, electric energy are more than 10 million rubles per annum [2].

In conclusion, many enterprises of different sectors of economy are situated in the list of organizations which must conduct energy audit so it's a positive factor. However there are some problems which exist in practice of conducting energy audit caused by accepted the Federal law №261-FL of Russian Federation from 23.11.2009. So many leading energy audit specialists consider that prescribed conditions by this document don't apply in reality. Let's learn these problems in detail.

The one of the main problems is incomprehensible form and use of energy passport. Also there are no clear and unified requirements for its formalization. No one knows how energy passport must look like, what it must contain and what instance must affirm it. Energy specialists say that the law doesn't prescribe clear rules for conducting of energy audit, it just presents some requirements and terms within which the leaders of corresponding enterprises must make energy passport. As a result specialists research energy efficiency potential of industrial enterprises according to old forms which elaborated in 2000's because of lack of clear recommendations for energy audit conducting. Experienced energy audit specialists think that another important problem is almost impossible to find a competent auditor. In current market conditions there are more than energy audit organizations function emerging during popularization of this topic in our country. But only hundreds of existing energy audit organizations match self-regulating organizations in energy audit field. The current problem is that the standard course duration about energy auditing fundamentals lasts only 72 hours. It's impossible to teach a future energy audit specialist during this time. As a result less than 100 energy audit companies in Russia can offer qualified services. For this reason in rare happenings energy efficiency analysis is conducted economically competently [1].

Also very important problem is that there is no unified methodology about energy audit conducting at the enterprises of mechanical engineering in Russia wherein its algorithm would be prescribed clearly. No methodology serving as a standard for comparison analysis results of investigated enterprise. However we need to remember that energy audit conducting will give nothing if don't invest in energy-saving technologies for the enterprises of mechanical engineering [1]. It's necessary to get experience of foreign companies offering their own modern energy-saving technologies and clear methods about energy audit conducting for the enterprises of mechanical engineering [7; 9; 15; 16; 18]

It should be noted that experience of foreign countries prove that energy audit in our country needs governmental support within the framework of energy-saving of mechanical engineering enterprises as complex process. For example, in Finland general directions of energy audit work come from the Ministry of Labor and Finland's economy. State Center for Economics, Ecology and Transport provides subsidies for an enterprise in the end of work when an auditor made a report and checked the quality of work with a customer. There are different models of energy audit for every sphere of a client's enterprise. If in Russia energy audit is given to the business and self-regulating organizations, in Finland the government is interested at the energy audit results because it allocates funds for expertise of projects. Besides Finnish auditors take a report about energy audit, Russian auditors take an energy passport. [1; 11].

According to the experience of Finland energy audit is a complex approach during the analysis of an enterprise, it's not just solution of technical tasks but also economic tasks and also tasks about production development based on the introduction of specialized technologies, preservation of the environment –reduction of emissions and negative impact on it. For Russia Finland's experience may be estimated first and foremost that they have data of comparison of energy-saving potential which theoretically grounded by energy auditors at technical report and actual potential of energy-saving realized those years in their country. The accuracy of theoretical forecasts of Finnish energy auditors is about 70%. [1; 11].

At the present time energy audit secured by law and its official status is confirmed formally. But on practice concept of energy audit exists only in law, its real function is reduced to formality and ambiguity.

In this way on the surface are the key problems in energy-saving programs implementation for the enterprises of mechanical engineering through energy audit caused by the Federal law №261-FL from 23.11.2009 on practice. We formulate them briefly and clearly:

- lack of clear rules and any budget for energy-saving programs implementation prescribed by law;
- lack of unified form of energy passport and clear requirements about its formalization for the mechanical engineering enterprises;
- deficiency of competent energy audit specialists able to correctly identify reserves on possible saving of energy resources for the enterprises of mechanical engineering;
- lack of unified and clear methods for energy audit conducting for the enterprises of mechanical engineering at the federal and regional levels;
- lack of investment projects about elaboration of energy-saving technology for mechanical engineering enterprises;
- lack of regional service, centers for energy audit organization and implementation at the enterprises of mechanical engineering.

At the moment only some Russian mechanical engineering enterprises conduct energy audit, elaborate and implement energy-saving measures. And also some enterprises of mechanical engineering try to decide their problems about energy efficiency by way of elaboration of energy-saving programs according to the energy audit results which collide with similar problems described above by the authors. We give practical example to justify our conclusions.

So described above by the authors OJSC «Russian machines» started to solve its problems in a complex manner about energy efficiency in several stages because of constant increasing of energy consumption and its share in prime cost of engineering products and also elevated morally and physically deterioration of its fixed assets in particular equipment in 2014. The basic stage of problem solving about energy efficiency was energy audit within the framework of the following ways: conducted energy audit for all 27 enterprises; elaborated energy efficiency programs; the corporate enterprises are provided with registered energy passports; implemented measures within the framework of approved programs and energy efficiency increasing through the conclusion of energy service contracts. But the leaders of OJSC «Russian machines» collided with the following problems during energy audit [6]:

1. The expertise process of energy passports isn't worked out and formalized finally at self-regulating organizations with subsequent registration in the Ministry of energy. Formed energy passports aren't registered that doesn't allow to have feedback for improvement of the methodology of energy audit during work at other enterprises;

2. The search of competent auditors is complicated because of lack of such specialists with profile basic education and experience. With a high demand for staff of this direction supply of personnel is very low both in quality and quantity;

3. No methodology for energy audit so it complicates starting phase of activity.

The following stage of problem solving about energy efficiency of OJSC «Russian machines» will be a transition to professional management of energy infrastructure for reduction of costs on it and purchased energy resources of corporate enterprises and also elaboration of energy management optimization program with accumulation of experience and increasing of business capitalization to gain access to cheap financing. The final stage of problem solving about energy efficiency in a complex manner of OJSC «Russian machines» will become the creation of its own generation based on energy settings working in the mode of combined generation of electric and thermal energy by building a mini-cogeneration plant at the corporate enterprises to provide baseline load coverage. As a result synergy of all described directions presupposes achieving the maximum energy-saving effect, reduction in cost and increasing of competitiveness of production of OJSC «Russian machines». [6].

4 The basic directions of the decision of the put problems

In our country attitude to using of energy resources, fuel and energy resources should be called wasteful and this is more a tradition than a temporary. Unfortunately economy of all types of energy is represented as the requirement of supervisory authorities for us. At the same time the bases of careful attitude to all types of energy are laid down in childhood abroad [8; 12].

We agree with the opinion of energy audit experts that it's necessary to live tens of years in Russia to change consciousness of fellow citizens and attitude to the concept of «economical consumption of energy resources». Now our mindset makes us act quite differently even if law is called up to limit the rights traditionally. Also many energy audit experts think that creation of the federal and regional coordination centers is necessary to improve the system of work organization on energy audit. Representatives of executive and legislative branches, energy public organizations, self-regulating organizations and various political parties should be included in these centers.

In this way according to the leading energy audit expert's opinion and their recommendations we formulated the basic directions about energy efficiency program implementation of the regional enterprises of mechanical engineering on practice:

- creation of the federal and regional coordination centers which organize work on integrated energy audit for the enterprises of mechanical engineering according to which elaborate priority measures for energy-saving and its practical implementation;
- elaboration of unified methodology on the federal level and also clear mothodologies of energy audit, energy-saving technologies, various models of energy audit conducting;
- elaboration unified form of energy passport on the federal level for the enterprises of mechanical engineering and also elaboration of clear requrements to its formalization;
- financial governmental support for energy audit conducting of regional enterprises of mechanical engineering;
- re-attestation of specialists of energy audit companies with the aim of the compliance of their competence with the provided energy audit services for the enterprises of mechanical engineering and other industrial sectors;
- change of criteria (more stringent) of the licensing of energy audit companies provided energy audit services for the enterprises of mechanical engineering and other industrial sectors;
- public and private investment projects on development of energy-saving technologies for the enterprises of mechanical engineering;
- change of the awareness of necessity energy-saving meausres implementaion by way of intensification of outreach work on energy-saving by radio, television, mass media for the leaders of the mechanical engineering enterprises and population in general.

5 Conclusion

Energy-saving has to become the priority direction of activity for the enterprises of mechanical engineering at the moment. Active energy-saving policy is a necessary condition of increasing of production competitiveness and acceleration of scientific and technical progress in mechanical engineering industrial sectors. Also energy-saving will allow contributing to improvement to the environment.

Interest and involvement of all participants in production including executive authorities of each region of Russia are necessary condition for successful energy-saving policy implementation in mechanical engineering sector. It's possible to increase the availability of means of accounting and automatic regulation of production processes and using of fuel and energy resources and also more efficient to keep records and control of energy consumption only when maximum mobilization of used energy resources and optimal return of subjects of energy-saving policy will be implemented.

References

1. N. Borisova, A. Borisov, and E.Viprickaya, Economy and business, **12-3**, 412 (2015)
2. V. Bushuev, Energy policy, **1**, 3 (2014)
3. M. Dubinsky, Energy-saving, **3**, 32 (2010)
4. E. Korosteleva, Young scientist, 1(8), 140 (2011)
5. N. Lazutkina, Modern problems of science and education, 2, (2013)
6. E. Razumny, Statements, **4179** (2016)
7. R. Simanjuntak, P. Dupuis, L. Canale, N. Sinisuka, & G. Zissis, *IEEE Industry Application Society Annual Meeting*. (2014)
8. X. Zhao, H. Li, L. Wu, & Y. Qi, Energy Policy, **66**, 170, (2014)

9. V. Blass, C. Corbett, M. Delmas, & S. Muthulingam, Energy, **65**, 560 (2014)
10. A. Sclafani, *ASME 2010 4th International Conference on Energy Sustainability,* 1. (2010)
11. L. Okkonen, & N. Suhonen, Energy Policy, **38**, 3443 (2010)
12. T. Fleiter, J. Schleich, & P. Ravivanpong, Energy Policy, **51**, 863 (2012)
13. S. G.Al-Ghamdi, & M. M. Bilec, Environmental Science & Technology, **49,** 4048 (2015)
14. P. Sun, & P. Nie, Renewable Energy, **74,** 255 (2015).
15. M. Marques, & R. Neves-Silva, Journal of Cleaner Production, **88,** 105 (2015)
16. A. Al-Mofleh, S. Taib, & H. Al-Gulman, *2008 5th International Symposium on Mechatronics and Its Applications*. (2008)
17. W. Zhang, S. Liu, N. Li, H. Xie, & X. Li, Renewable and Sustainable Energy Reviews, **49**, 395 (2015)
18. A. G. Engelen, & J. S. Collins, *2010 43rd Hawaii International Conference on System Sciences*. (2010).
19. M. Abdel-Wahab, & D. Ali, International Journal of Green Energy, **10(2),** 137 (2013)
20. A. Mansikkasalo, & P. Soderholm, *2012 9th International Conference on the European Energy Market* (2012)

DOI: 10.1051/matecconf/201712901041

Calculation method and simulation of work of the ring elastic compensator for sheet-forming

Volodymyr Kukhar[1], *Elena* Balalayeva[1,*], and *Oleg* Nesterov[1]

[1]SHEI "Pryazovskyi State Technical University", Universytets'ka, 7, Mariupol, 87500, Ukraine

Abstract. The design of the elastic ring compensator of «press-and-die" system errors, the method and software for calculating its geometric parameters are given in the article. The developed method is based on the account of unevenness of the radial and altitudinal deformation (barrel-type shape) of the elastic working element made of polyurethane. A specific example of computer implementation of the developed mathematical model for calculating the sizes of such a compensator for steel-forming operations on an open-crank press is made. Approbation of the compensator was carried out in industrial conditions.

1 Introduction

The accuracy of "press-and-die" systems are affected from misalignment of the slide axis to press table surface in the loaded state, from distortion in the gaps between press and slide guides, as well as cumulative errors of the punching unit. [1-4]. Mechanical and elastic compensators are widely used to reduce errors in the "press-and-die" system of open-crank presses and ring compensators of elastomeric materials has not been studied enough. This makes it difficult to calculate their design parameters.

2 Analysis of literature and the problem formulation

There are two basic approaches to eliminate the influence of deformation of the C-shaped type frame on the alignment of "press-and-die" system. The first approach is the closure of open C-shaped type frames with using the ties [5], which greatly reduces its deformation. A second approach for solving the skew problem is to eliminate the misalignment between the upper and lower basing plates with various additional devices or elements (fig. 1, a). These devices or parts are called "compensators". A further alternative approach to reducing of errors of the "press-and-die" system can be die-free [6] and impression-free [7-8] methods of metal-forming.

The main purpose of installing a skew compensator in forging equipment is to ensure the transfer of the operating force from basic die elements, which have lost alignment as a result of deformation of the frame, to coaxial (parallel) basic die surfaces (fig. 1, a).

The style for the names is First Name, typed in italic 10-point Times, then Last Name, typed in 10-point Times, with a comma after all except the last author, which is separated by comma + "and". Do not use academic titles.

Affiliations of authors should be typed in 9-point Times. They should be preceded by a numerical superscript corresponding to the same superscript after the name of the author concerned. Please ensure that affiliations are as full and complete as possible and include the country.

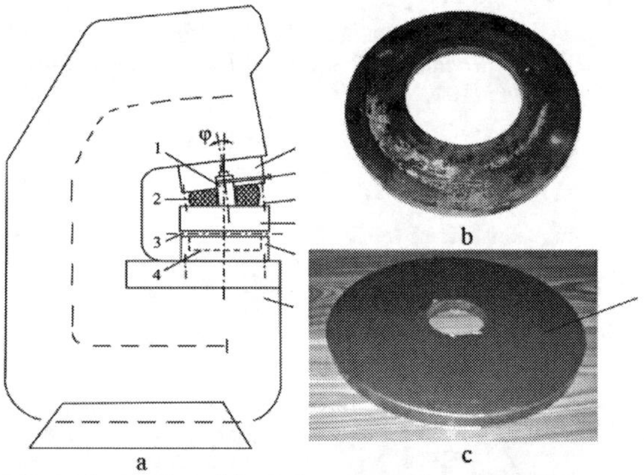

Fig. 1. The opening of the C-shaped press frame (a) and the design of the elastic ring compensator (b), and its upgrading design (c): 1 – shank; 2 – compensator; 3 – die line connector; 4 – circuit of a formed product; 5 – slider; 6 – fixing pin; 7 – fasteners; 8 – upper plate; 9 –lower plate; 10 – frame; φ – skew angle of the slider

A promising direction to reduce of distortions in the "press-and-die" system is the using of different constructions of compensators, which are made, for example, of different brands of polyurethane. The choice of polyurethane is complicated by wide range of functional properties of the polyurethanes from various manufacturers. Recommendations for optimization of this choice are given in [9]. The method of calculating the optimum design parameters is discussed in [10]. Also, an automated method of calculation for the reconfigurable polyurethane compensators, taking into account the errors of "press-and-die" system, is proposed [11].

* Corresponding author: balalaevaeu@gmail.com

Polyurethane elements are widely used for compensation of power and geometric deviations in metal forming technologies [10-13].

The method of fastening the upper plate to the press slider through an elastic compensator (fig. 1, a) is considered in [10, 11]. The example of the design of the ring elastic compensator is shown on fig. 1, b. However, the work of the ring compensator can be accompanied by cracking of the elastic plate because during the influence of deformation force the material can flow into non-technological openings, forming regions of increased concentration of stresses and unregulated deformations.

To solve this problem, the variant of the upgrading design of the compensator (fig. 1, c) was developed, according to which the ring elastic compensator is placed between two ground metal plates with a central hole and connected to them using an adhesive bond.

The disadvantage of the above-described method is the lack of optimal bonding conditions (the glue is applied to the entire surface area of the contact surfaces of the protective plates and the elastic element), which should be determined on the basis of the structural and technological features of the ring compensator and the physical and mechanical characteristics of the glue.

3 Purpose of the work

The purpose of this article is to develop a mathematical model and study the deformations of the ring elastic compensator for modeling its shaping under the conditions of the "press-and-die" system work with slider skewing, as well as the calculation of tangential stresses in the adhesive bond with protective plates to determine the optimum gluing parameters, which will help to reduce adhesive consumption, tangential stresses, increase the reliability of fastening the compensator and improve the working conditions.

4 Research material

The compression pre-tests were carried with different shape samples from polyurethane brand SKU-PFL-100 (СКУ-ПФЛ-100, 100 Shore hardness). The approximating dependence between the pressures during compression of polyurethane q and the degree of upsetting $\varepsilon = (H_0 - h) / H_0$ (where H_0 is the initial height of the compensator, h is the height of the deformed compensator) for $\varepsilon \leq 0.3$ is received as the result of studying the influence of the shape of the compensator on its elastic properties [10, 11]:

$$q = 52\varepsilon + 1.92 . \quad (1)$$

A ring polyurethane element (fig. 2) with an outer radius R, an inner radius R_n and a height H_0 was considered. The support area of the compensator was determinated as $F = \pi(R^2 - R_n^2)/4$. Misalignment of the slide at an angle φ causes the uneven deformation along the height of the compensator (fig. 2). Moreover, the maximum of angle deviation should not exceed $\varphi \leq \mathrm{arctg}(H_0/2R)$.

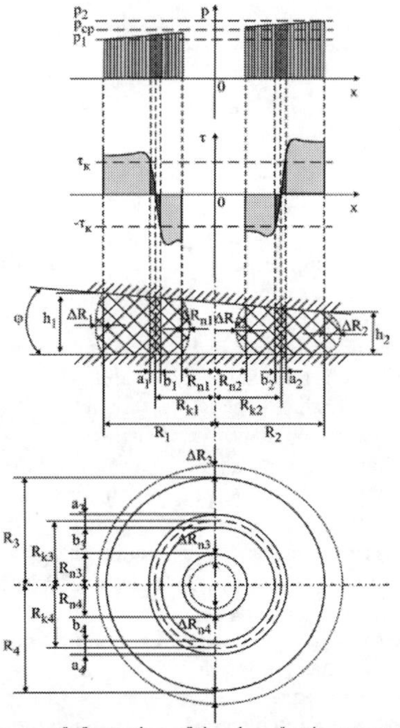

Fig. 2. Uneven deformation of the ring elastic compensator

The upsetting of the ring elastic compensator is accompanied by a simultaneous increase in the outer radius R and a decrease in the inner radius R_n. The changes in the radii ΔR and ΔR_n by analogy with the flow of a deformed ring metal preform at four characteristic points were considered. Without considering the shear strain, we had: $\Delta R_3 = \Delta R_4$ и $\Delta R_{n3} = \Delta R_{n4}$. Based on the accepted assumptions, only three characteristic points were considered.

The deformation of the ring elastic compensator is characterized by the presence of a neutral flow line of the material. The boundary of the material flow is the surface defined by the radius R_k. The contact tangential stress τ at points with radius R_k is equal to zero. At the same time, its average value was determined as:

$$\tau = \psi\tau_s ; \quad \tau_s = q/\sqrt{3} ; \quad \psi = \mu + \frac{1}{8} \cdot \frac{R - R_n}{H_0} \cdot (1 - \mu) \cdot \sqrt{\mu} , \quad (2)$$

where μ is the coefficient of friction.

The radius of the critical surface R_k, as well as the change in the outer radius ΔR and the inner radius ΔR_n at the i-point not taking into account the barrel shape and taking into account the unevenness of the radial deformation along the height were determined by the methods [I.Ya. Tarnovsky, 1963; L.A. Shofman, 1961]. However, preliminary calculations showed a significant difference in the values of R_k which showed the need to refine the dependence to determine the dimensions of the ring element with respect to barrel shape:

$$R_{k_i} = 63.25 \sqrt{R^2 \cdot R_n^2 \frac{R^2 - R_n^2}{8000R^2 \cdot R_n^2 \cdot \ln\frac{R}{R_n} + 1599h_i^2\left(R^2 - R_n^2\right)}} . \quad (3)$$

 DOI: 10.1051/matecconf/201712901041

The change in the outer radius ΔR and the outer radius ΔR_n at the i point was calculated as:

$$\Delta R_i = 0.375 \cdot R \cdot (H_0 - h_i) / H_0 \cdot \left(1 - R_{k_i}^2 / R^2\right);$$

$$\Delta R_{n_i} = -0.375 \cdot R_n \cdot (H_0 - h_i) / H_0 \cdot \left(1 - R_{k_i}^2 / R_n^2\right). \quad (4)$$

The compression pressure p_i at the upsetting of the ring polyurethane compensator is distributed unevenly along its diameter, so the calculation was applied for each deformable part on the four sides of the ring compensator taking into account the unevenness of deformation of the elastic element.

In order to achieve optimum bonding characteristics, the application of glue (diagum FL, diagum P, diaflex) [14] should be carried out only on areas limited by the shear stress zone τ with values not exceeding the critical shear stresses τ_k (fig. 2). According to the proposed method, a number of diagrams of tangential stresses are constructed depending on the geometric parameters of the compensator and zones are defined for which $\tau < \tau_k$ (fig. 2). The tangential stress τ in the adhesive layer of the polyurethane ring compound with metal protective plates is calculated by the method of Lame, described, for example, in [15].

The dependence of the distances from the flow separation line R_k to the outer edge (a) and the inner edge (b) of the compensator from three factors (S/D, S/H_0, ε) was established using the method of experiment planning (2^3):

$$a = 10.283 + 0.034\frac{S}{D} + 0.46\frac{S}{H_0} - 1.109\varepsilon + 0.135\frac{S}{D}\cdot\frac{S}{H_0} -$$

$$-0.073\frac{S}{H_0}\cdot\varepsilon - 0.034\frac{S}{D}\cdot\varepsilon - 0.015\frac{S}{D}\cdot\frac{S}{H_0}\cdot\varepsilon; \quad b = 0.929a \,.(6)$$

where S is the wall thickness of the compensator; D is outer diameter of the compensator.

The special software is written on the base of the developed calculation method. This software allows to calculate the dimensions of the ring elastic compensator after the upsetting with and without taking into account the barrel shape and also determine the bonding areas of the ring compensator with protective plates. The research of the work of the elastic ring compensator with parameters R = 130 mm, R_n = 32.5 mm, H_0 = 13 mm (fig. 3, a) was carried out. The angle φ = 0.1 rad. and the force P = 0.2 MN (fig. 3, a). According to the results of simulation (fig. 3, b), it is advisable to apply the adhesive to the areas of the compensator surface limited by distances a_1 = 14.01 mm and a_2 = 13.94 mm to the outer edge, b_1 = 13.01 mm and b_2 = 12.946 mm to the inner edge of the neutral line, R_{k1} = 74.95 mm and R_{k2} = 74.99 mm; the width of the adhesive layer from the maximum deformation side is 27.02 mm and from the minimal deformation side is 26.9 mm. As a result, the consumption of the adhesive compound was reduced by 72 %.

The upgrated design of the ring elastic compensator was tested during solving of the problem of increasing the accuracy of the dimensions of the stamped part "Bottom" of the washing machine "Donbass" (fig. 4) in the conditions of the enterprise JSC "Elektrobytpribor" (Mariupol).

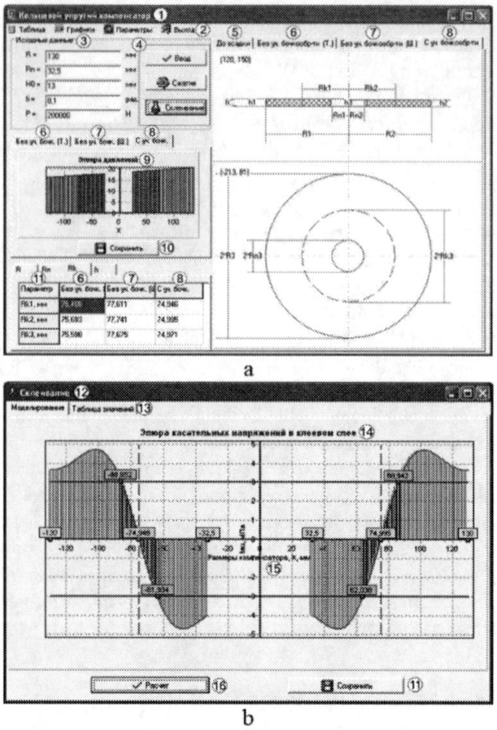

Fig. 3. Software interface for modeling of the work of the ring elastic compensator taking into accountthe barrel shape (a) and plotting the shear stresses in the adhesive layer (b):
1 – The ring elastic compensator ; 2 – Table, Graphics, Parametrs, Exit ; 3 – Initial data; 4 – Input, Upsetting, Gluing; 5 – Before upsetting; 6 – Without taking into account the barrel shape (by I.Ya. Tarnovsky); 7 – Without taking into account the barrel shape (by L.A. Shofman); 8 – With taking into account the barrel shape; 9 – Diagram of the compression pressure; 10 – Save; 11 – Parametrs; 12 – Gluing; 13 – Simulation, Table of values; 14 – Diagram of the tangential stresses in the adhesive layer; 15 – Compensator sizes; 16 – Calculation

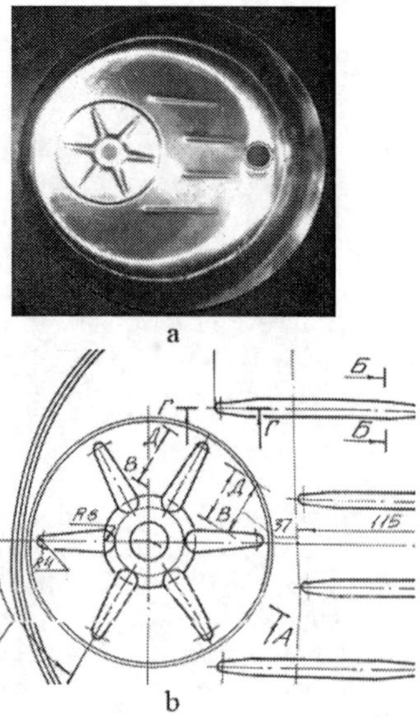

Fig. 4. The bottom of the washing machine "Donbass" (a) and a fragment of its drawing (b)

 DOI: 10.1051/matecconf/201712901041

The compensator in the form of a flat ring made of polyurethane of SKU-PFL-100 grade had an outer diameter of 260 mm, an inner diameter of 65 mm and a height of 13 mm (fig. 1, b); The calculated thickness of two ground metal plates (fig. 1, c) was 0.8 mm (steel X18N9T). The operation was carried out on a single-crank press of a single action with a C-shaped frame with a nominal force of 1 MN (model KE 2130A).

The depth of molded ribs of activator jack (steel X18H9T) was measured from the front and rear of the press with depth gauge at six points on the extreme edges of the axis. Measurements were carried out using the compensator and without it. The upgraded design of the compensator made it possible to reduce the difference in depth of the ribs of the opening under the activator by 64-73 %.

5 Conclusions

Based on the performed analysis of the ring elastic compensator work under conditions of the "press-and-die" system work with slider skewing, its design, methodology and software for calculating its sizes after the upsetting were developed. This made it possible to evaluate the slider skew on the shaping of the elastic element, to determine the dimensions of the areas for bonding the elastic compensator to the protective plates, and to minimize shearing stresses in the adhesive layer by 72% and to improve the quality of the finished metal products from the sheet by 64-73%.

References

1. R.L. Norton, *Machine Design. An Integrated Approach* (Upper Saddle River, NJ: Pearson Prentice Hal, 2006)
2. D.A. Smith, *Why Press Slide out of Parallel Problems Affect Part Quality and Available Tonnage* (Dearborn, MI: Society of Manufacturing Engineers, 1988)
3. *Slide Out-of-Parallel Problems: Why they affect part quality and available tonnage* (The Fabricator, The Fabricator's and Manufacturer's Association, International, Rockford, Illinois, 1990)
4. A. Dasgupta, M. Pecht, *Material Failure Mechanisms and Damage Models*, IEEE Transactions on Reliability, **40 (5)** (1991)
5. M. Cirek, V. Kubec, *Analysis of energy consumption of spindle presses,* Tehnički vjesnik, **13 (1,2)**, (2006)
6. V.V. Kukhar, Metallurgical and Mining Industry, **6**, 122–132 (2015)
7. V. Kukhar, V. Burko, A. Prysiazhnyi, E. Balalayeva, M. Nyhnibeda, Eastern-European Journal Of Enterprise Technologies, **3 (7(81))**, 53-61 (2016)
8. V. Kukhar, V. Artiukh, O. Serduik, E. Balalayeva, Procedia Engineering, **165**, 1693–1704 (2016)
9. V. Artiukh, S. Karlushin, E. Sorochan, Procedia Engineering, **117**, 938–944 (2015)
10. V. Kukhar, E. Balalayeva, O. Tuzenko, V. Burko, Multidisciplinary Journal of Research in Engineering and Technology, **2(3)**, 593–604
11. E.Yu. Balalayeva, V.V. Kukhar, O.V. Hrushko, HCTL Open STL, **6** (2014)
12. Firas M.F. Al-Quran, M.E. Matarneh, V.G. Artukh, Research Journal of Applied Sciences, Engineering and Technology, **4 (11)**, 1585–1589 (2012)
13. Nabeel S. Gharaibeh, Mohammed I. Matarneh, V.G. Artyukh, Research Journal of Applied Sciences, Engineering and Technology, **8 (12)**, 1461–1464 (2014)
14. A.A. Ishenko, G.V. Artiukh, Protection of Metallurgical Machines from Breakdown, **2**, 159–166 (1997)
15. S.G. Kushnarenko, *Metal forming in engineering* (Kharkov, Osnova, 1990)

The relationship of cutting force with hole quality in drilling process of AISI H13 steel

İsmail Tekaüt[1,*], *Halil* Demir[2] *and Ulvi* Şeker[3]

[1] Gazi University, Vocational School of Technical Sciences, Ankara, Turkey
[2] Karabük University, Faculty of Technology, Department of Manufacturing Engineering, Karabük, Turkey
[3] Gazi University, Faculty of Technology, Department of Manufacturing Engineering, Ankara, Turkey

Abstract. The harmony of the drilling machine-cutting tool-work piece is very important for producing the machine part with the ideal dimensions. For this purpose in this study, the effect of cutting forces on hole quality (surface roughness, diameter deviation and circular deviation) was investigated by 14 mm diameter uncoated and (AlCrN monolayer) coated carbide drills for drilling AISI H13 hot work tool steel on vertical machining center. Four different cutting speeds (60, 75, 90 and 108 m / min) and three different feed rates (0.15, 0.20 and 0.25 mm / rev) were used in the experiments. Cutting forces have been found to be effective in improving hole quality. Better hole quality has obtained with coated drills than uncoated drills in experiments. It has been observed that coated drills have the effect of improving the hole quality due to the operation with less cutting force and better chip evacuation.

1 Introduction

Drilling is very important in machining. Since a larger drilled hole than the desired dimension is cause to junk of the part or the small hole from desired dimension need to a second process as reaming. These processes lead to time lost and increased cost. Therefore, It is very important to produce holes in desired quality in one go. In studies in literature; It is emphasized that cutting force increases with increase of feed rate and decreases with increase of cutting speed in drilling process. They observed that magnitude of the cutting forces had a negative effect on hole quality. Literature review indicates that cutting parameters and accordingly cutting forces are important factors to hole quality in drilling process [1-11]. Therefore, İn this experimental study, AISI H13 steel is drilled at four cutting speeds (60, 75, 90 and 108 m/min) and three feed rates (0.15, 0.20 and 0.25 mm/rev) using uncoated and AlCrN coated solid carbide tools with 14 mm diameter. The relationship of cutting force with hole quality in drilling process of AISI H13 steel are investigated.

2 Material and method

The AISI H13 hot work tool with hardness of 90 HRB (184 HB) was drilled 14 mm in diameter with uncoated and multilayer coated (AlCrN monolayer) solid helical carbide drill. The drilling depth is 3D and the cutting parameters were determined by choosing a starting value in the range of the cutting parameters recommended for the drills selected in the manufacturer's catalog (KARCAN). The initial value of 60 m/min was reproduced in accordance with the ISO standard for Standard Numbers (R10, R20 and R40), starting from the condition that at least four different cutting speeds should be selected for each cutting condition specified in the life test standards. Experiments were carried out on these conditions using four different cutting speeds (60, 75, 90, 108 m/min) and three different feeds (0.15, 0.20 and 0.25 mm/rev). The specimens prepared to be 25 mm in outer diameter and 35 mm in hole length were drilled at Johnford VMC- 550 brand CNC vertical machining center in Gazi University Technical Education Faculty, Department of Mechanical Education. In the experiments, quartz crystal based Kistler 9272 type dynamometer and Kistler Type 5070 amplifier, which have the same measuring capacity of three shear force components (Fx, Fy, Fz) and momentum (Mz) were used. Average surface roughness, gross deviation (within tolerances) and circularity deviation of the machined surfaces were measured in order to determine the quality of holes obtained from the experimental studies depending on the input parameters. A needle-tipped Mahr Perthometer M1 device with a portable surface roughness device was used to measure the average surface roughness (Ra) values on the machined surfaces. DEA GLOBAL CLASSIC 07.10.07 three-dimensional coordinate measuring instrument was used in determining of diameter and circularity deviation values. For each sample, the deviation values were determined by taking the arithmetic mean of the values obtained in first measurement from 30 points below 2 mm, in the second measurement from 30 points below 4 mm, in the third measurement from 30 points below 6 mm below from the bottom of the part. Arithmetic averages of the

* Corresponding author: ismailt@gazi.edu.tr

 DOI: 10.1051/matecconf/201712901042

values obtained by measuring the circularity 2, 4 and 6 mm below the surface of the part were calculated to obtain the circularity deviation values.

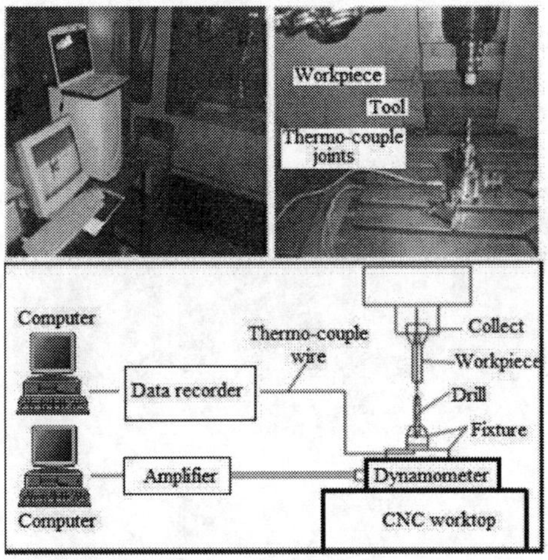

Fig. 1. Experimental setup

3 Results and discussion

The feed force, cutting force and moment values obtained by drilling the AISI H13 steel with uncoated and coated carbide drills at different feed rates and cutting speed combinations are plotted and shown in Fig. 2.

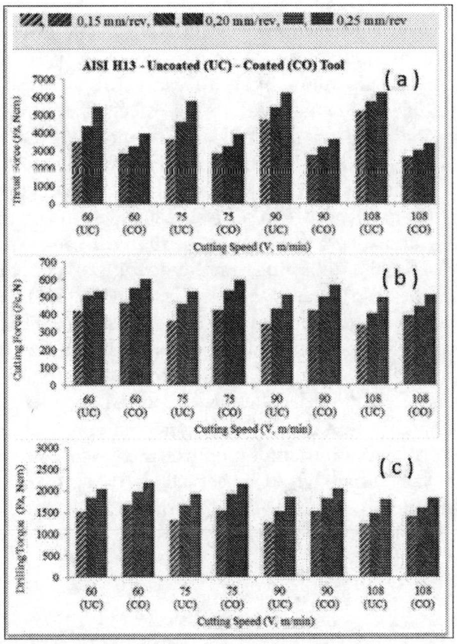

Fig. 2. Cutting forces depending on feed rate and cutting speed.

When examining of the graphs in Figure 2, both feed forces and moment values increased with increasing feed rate in all drilling experiments with uncoated and coated drills. This is attributed to the increase in the amount of advancement and the increase in the amount of drill cutting edges to increase the unit force in order to move faster along the perforation axis per unit of time and to raise the unit volume. Increasing chip volume with increasing feed rate caused the drill to use more cutting force to pull the increased chip volume from the material, resulting in larger moments generated by larger forces (Fig 2.a-Fig. 2.c).

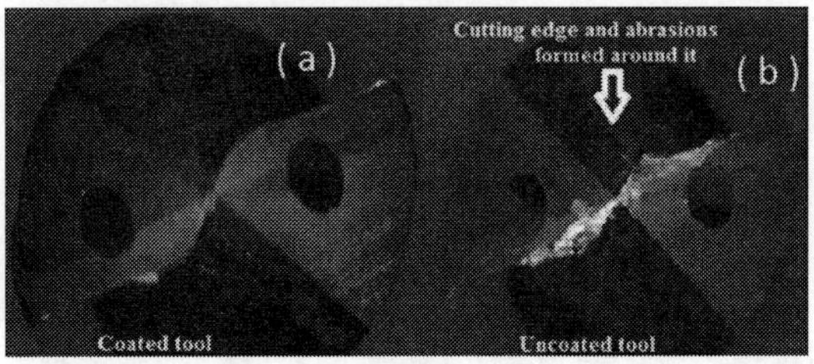

Fig. 3. Coated and uncoated drill

When the graphs are evaluated in terms of cutting speed, it was seen that the increase of the cutting speed in the drilling of the AISI H13 steel with uncoated drills caused to increase in the feed forces and decrease in the moment values. In drilling operations with veneer drills, the increase in cutting speed results in a decrease in both feed forces and moment values. In drilling operations with coated drills, the increase in cutting speed results in a decrease in both feed forces and moment values. This can be attributed to increased wear on the uncoated drills (see Figure 3.a) with increased cutting speed, and higher performance due to the coating material properties on coated drills. In processes with coated drills, increased temperature with increasing cutting speed [12, 13] caused to the deformation of the chip with less force and the facilitated chip evacuation led to a reduction in feed forces. The reduction in the feed forces and the cutting forces resulted in a reduction in moment values (see Figure 3.a - Figure 3.c). As the cutting speed increases, the cutting force and moment values decrease while the feed force increases in the case of uncoated drills. This situation can be explained by the increased wear in the end zone of uncoated drills (see Figure 3.b). Because of the increase in cutting speed, the wear on the drill has increased in the processes with uncoated drills. Due to the increased wear, the drill used more feed force to feed along the hole axis and caused to increase in this force. Increased wear with increased cutting speed made it difficult for the uncoated drill to feed along the hole axis and the drill's distance to the drill hole along the drill axis was further shortened. The shortening of the length that the drill was going to take at unit time has caused the reduction in the amount of chips to be removed. The reduction in chip volume caused a reduction in the cutting forces and so moments, as opposed to the feed forces. This is evidenced by the fact that in drilling with coated drills the decreasing feed force with decreasing cutting speed is less and less prone to decreasing than uncoated drills, while the cutting force obtained by operations with coated drills are higher than those obtained with uncoated drills.

The roughness values occurred on the hole surface after drilling the AISI H13 steel with uncoated and coated drills are shown in Fig. 4 by converting the deviations from diameter and circularity into graphs. When examining the graphs, the increase in feed rate in both drill types led to an increase in surface roughness. This situation can be attributed to the increase in the feed forces with the increase in the feed rate (see Figure 2). Increasing cutting forces had a negative effect on surface quality. When evaluating in terms of cutting speed, increase of the cutting speed improved the hole surface quality obtained in the processes with coated drills, while it causes the deterioration in hole surface quality of same conditions with uncoated drills. This can be explained by the fact that the increase in cutting speed accelerates the wear in uncoated drills, the high performance of the coated drill in the case of coated drills, and the low friction coefficient of the coating material. Increased cutting speed in the case of coated drills improves surface roughness, while in the case of uncoated drills, acceleration of the wear at the drill bit with increasing cutting speed was resulted in a rougher surface (Fig. 4.a). When Figure 4.b is examined, the deviation values of diameter increase with the increase of the feed rate in all experimental results. This can be explained by an increase in the chip volume with increasing the feed rate and so increase in the cutting forces. The increase in the cutting forces acting on drill with increasing feed rate caused in drill axis more deviation from the hole axis in the drilling process. In the case of both uncoated and coated drills, the increase of the cutting speed led to the increase of the deviation values. This can be explained by more deviation of the drill axis from the hole axis with the result of the increase of the machine-tool vibration with increasing cutting speed and. The deviations obtained from the uncoated drills are higher than those obtained with the drilled drills. This can be explained by increased wear in the uncoated drill bit area parallel to the cutting speed increase with reducing the centering feature of the drill (see Figure 2.b). The reduction of the centering property of the drill led to more deviation of drill axis from the hole axis with both the increasing feed force and the increased machine-tool vibration together with the cutting speed. Not losing the centering feature of coated drills (see Figure 3.a.), the quick release of chip and the production with lower cutting forces were resulted in a lower vibration amplitude. Therefore, lesser deviation values were recorded for coating drillings than uncoated drills.

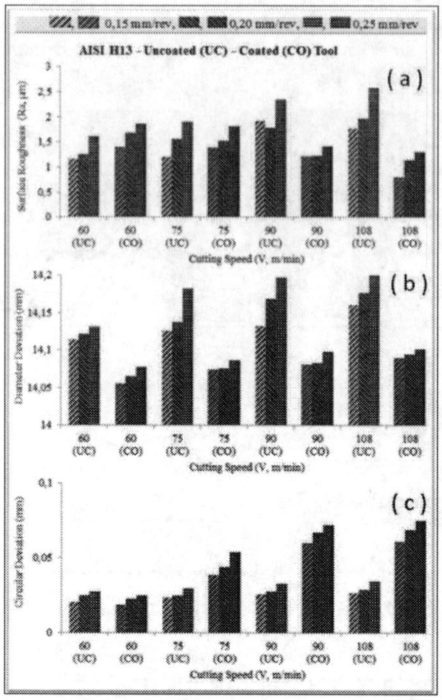

Fig.4. Cutting speed hole quality relations

In all experiments with uncoated and coated drills, the increase in the feed rate caused an increase in the deviations from the circularity. This can be attributed to increasing cutting forces as a result of the increasing feed rate. This is because the vibrations between the tool and the workpiece during drilling affect the drill bit and the higher cutting forces acting on the drill during the machining are more likely to cause the drill vibration amplitude to increase due to the effect of the tool-lathe vibration, causing the drill axis to move further away from the hole axis led to the increase in deviation from the circularity with the deviation of the diameter. The increase in shear rate in all test results led to an increase in deviation from circularity. As the drill rotation speed increases, the vibration amplitude of the system also increases and causes an increase in deviation from circularity. The fact that the deviations from circularity of holes obtained in uncoated drills are lower than that of coated drills can be explained by the fact that the cutting forces of coated drills (tangential force) are greater than that of uncoated drills (see Figure 3.a and Figure 3.c). The drill vibration amplitude has increased because the high tangential forces observed in the coated drills contribute to the drill-lathe vibration as a compelling force during chip removing.

4 Conclusions

The following results were obtained in the drilling process of AISI H13 steel with uncoated and (AlCrN monolayer) coated carbide drills.

• The increase in the feed rate in all experiments led to an increase in cutting forces. This is attributed to the increased chip volume with increasing feed rate.
• Increase in cutting speed leads to an increase in the feed force in uncoated drills, which causes a decrease in cutting force and moment. This caused the drill bit to expand too much force to move along the hole axis in the result of wear on the drill bit together with the cutting speed. However, due to wear, the reduction of the chip volume removed by unit time caused to more decrease of cutting force and moment.
• The cutting speed increase has resulted in a reduction in cutting forces in all experiments with coated drills. This is attributed to the acceleration of the chip evacuation along with the increase of the cutting speed and to the deformation of the chip with less force.
• Increase in the feed rate led to increasing of surface roughness, deviations from the diameter and circularity. This condition is attributed to increasing cutting forces.
• Increase in cutting speed causes increase of roughness values in uncoated drills, but decrease in coated drills. This can be attributed to increased wear in uncoated drills, and to rapid chip removal in coated drills.
• The increase in cutting speed causes to increase in the deviation from both the diameter and circularity. This can be attributed to an increase in the lathe-tool vibration with increasing cutting speed.
• In drilling operations with uncoated tools, the deviation values from diameter are greater than those with drilled drills. This can be explained by the reduction of the centering feature of the drill due to the wear on the tip of the uncoated drill.
• The values of deviation from the circularity in the operations with the coated drills are higher than those of the uncoated drills. This can be explained by the increase in vibration amplitude of the drill.

References

1. D. Umbrello, S. Rizzuti, J.C. Outeiro, R. Shivpuri, R. M'Saoubi, J. of Mat. Processing Tech., 199 (2008)
2. M. Usta, Master Thesis, G.U. Institute of Science, Ankara, (2010)
3. O. İşbilir, E. Ghassemieha, Procedia Eng., 10 (2011)
4. E. İnçal, MSc Thesis, Y.T.Ü., İstanbul, (2007)
5. T. Kıvak, U. Şeker, International Iron & Steel Symp., Karabük, (2012)
6. A. Çakır, M. Nalbant, A. Duran, H.B.Ulaş, 3. National Chip Manuf. Int. Symp, Ankara, (2012)
7. N. Tosun, C. Özek, G.Tosun, 11. International Machine Design and Manuf. Cong., Antalya (2004)
8. İ. Özkul, B. Buldum, A. Akkurt, Pamukkale University, Journal of Engineering Sciences, 19 (2013)
9. T. Kıvak, A. Çiçek, Y. Turgut, İ. Uygur, E. Ekici, 6th Int. Advanced Techn. Symp. (IATS'11), Elazığ (2011)
10. G. Meral, M. Sarıkaya, H. Dilipak, Erciyes Üniversitesi Fen Bilimleri Enstitüsü Dergisi, 27 (2011)
11. A.D. Jayal, A.K. Balaji, R. Sesek, Journal of Mat. Pro. Techn. 9 (2007)
12. M.C. Çakır, Vipaş A.Ş., Bursa, (1999)
13. W.B. Palmer, P.L.B. Oxley, Proc. Inst. Mech. Engrs., 173 (1959)

Determination of the profile of the worm-type tool

Aleksandr Golovko[1,*]

[1]Kazan Federal University, pr.Syuyumbike, d.10A, Naberezhnye Chelny, 423812 ,Russian Federation

Abstract. The cutting wedge of disk shaver at gear shaving has not fully satisfactory geometry: clearance The cutting wedge of the disk shaver when gear shaving is not fully geometrically satisfactory. The clearance on cutting edges are practically equal to zero. We would suggest this is the reason that the disk shavers machining of gear-wheels with large allowance is not possible. The construction of a worm-type tool is known also for clean machining of points of involute of gear-wheels made with continuous spiral cutting edges. The difference of this construction from the construction of N. Chemborisov is that a cutting wedge has the structurally made clearance α (Authors: N. Smorkalov, V.Skripin, V.Ptitcin, IU.Belugin). Polishing of side of coil is the most responsible and labor intensive operation when making a worm-type tool with the continuous spiral cutting edges of construction of N.Smorkalov and other.Dependences allowing to define the coordinates of key points of axial section of side of coil of the worm-type tool were obtained. These coordinates will allow to define the parameters of setting of abrasive disc in relation to the worm-type tool.

One of the variants of the machining of gear teeth, used in engineering, involves the use of gear shaving as a semi-finishing operation before thermal machining and gear honing [1-5].

Gear shaving pertains to the methods of machining on the principle of so-called "free" rolling, in which there is no rigid kinematic connection between the tool and the gear wheel. This feature of the pro gear shaving process positively influences its dynamics, since there are no additional sources of vibrations inherent in the chain of rolling. At the same time, the absence of a rigid kinematic connection between the tool and the wheel leads to a significant decrease in the radial runout of the gear ring, the difference in adjacent circumferential steps, the accumulated error of the circumferential pitch at the angle of 180 °, no oscillation of the measuring interaxial distance after the gear shaving occurs [6-13].

The cutting wedge of the disk shaver during gear shaving has a not quite satisfactory geometry: the clearance angels on the cutting edges are almost equal to zero. For this reason, the machining of gear wheels by rotary gear shaving tool with large allowances for machining (for example, after pre-rolling the teeth) is impossible. It should also be taken into account the high laboriousness of making gear shaving tools, the piece-time for gear grinding reaches 5 ... 6 hours.

For clean treatment of points of gear-wheels can be used multishank tools allowing to use technology of Power Skiving (Fig. 1).

Fig. 1. Scheme of technology of Power Skiving.

However for using this technology the special expensive equipment is needed.

There are known the design of tools for finishing the teeth of involute gears with spiral-screw cutting edges made on one side of the turns [14, 15]. The gears are machined with the coordinated rotation of the tool and the workpiece around their axes and moving them relative to each other in the axial direction of the workpiece, the processing of the right and left side of the profile of the teeth possible only after the wheel is reinstalled.

The drawback of these tools is the complexity of the design and manufacturing technology. A worm-type tool for finishing involute gear teeth is known with a continuous helical cutting edge, while its helical surface has a rectangular profile [16] [Fig. 2].

* Corresponding author: a.n.gol@yandex.ru

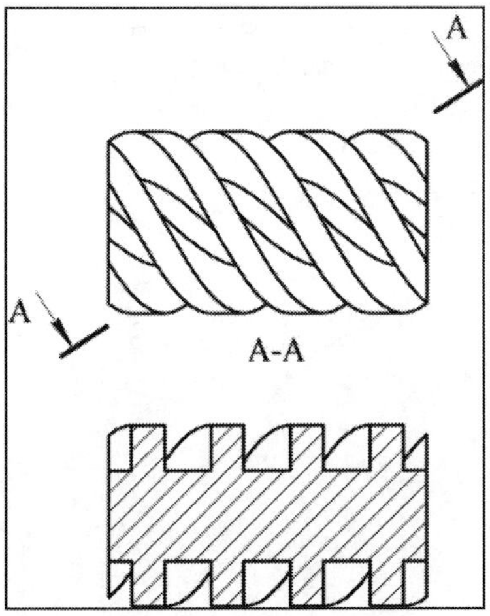

Fig. 2. Construction of the worm-type tool with spiral cutting edges (Authors: N. Chemborisov, R. Khisamutdinov, V. Avdeev).

The disadvantage of this tool is that the cutting wedge during its operation has a not quite satisfactory geometry: the clearance at the cutting edges is practically zero.

Also known is a worm-type tool for finishing the teeth of involute gears made with continuous screw cutting edges, different from the design of N. Chemborisov and others [16] in that the cutting wedge has a structurally executed rear angle α [17] (Fig. 3).

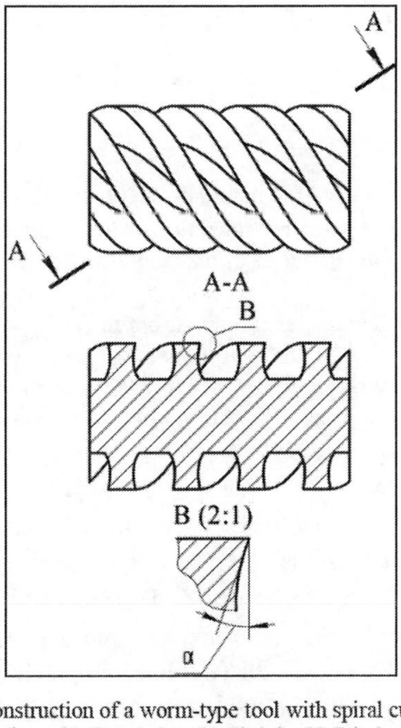

Fig. 3. Construction of a worm-type tool with spiral cutting edges (Authors: N. Smorkalov, V. Skripin, V. Ptitcin, IU. Belugin).

The machining is carried out with the coordinated rotation of the tool 1 and the wheel 2 (Fig.4,5,6,7).

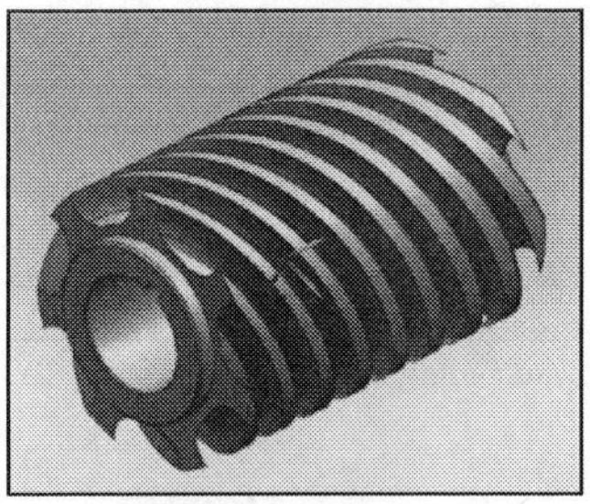

Fig.4. 3D model of a worm-type tool.

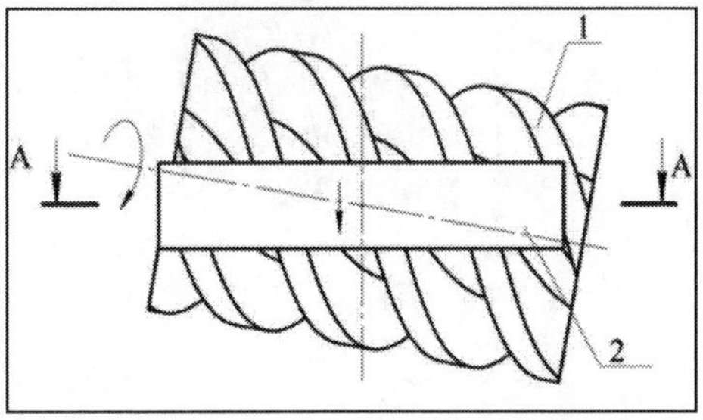

Fig. 5. Scheme of work of a worm-type tool with spiral cutting edges.

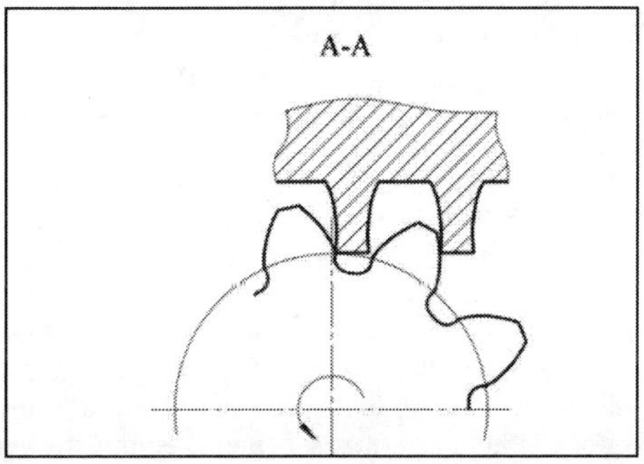

Fig. 6. Scheme of work of a worm-type tool with spiral cutting edges.

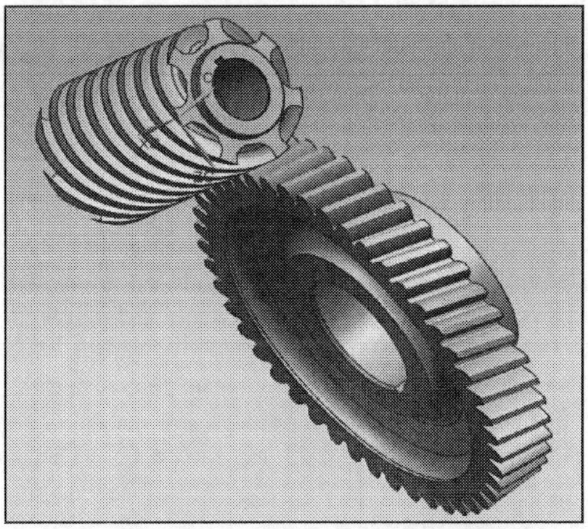

Fig.7. Visualization of adjustment of the worm-type tool on the machine in the module "Assembly" of the NX system.

The cutting wedge element has a front surface coinciding with the outer surface of the tool and moves in relative motion from the head of the wheel to the leg.

The most important and time-consuming operation for the production of a worm-type tool with solid screw cutting edges is the grinding of the lateral surface of the winding. The lateral surface profile is specified initially in a section perpendicular to the screw cutting edge (Fig. 8).

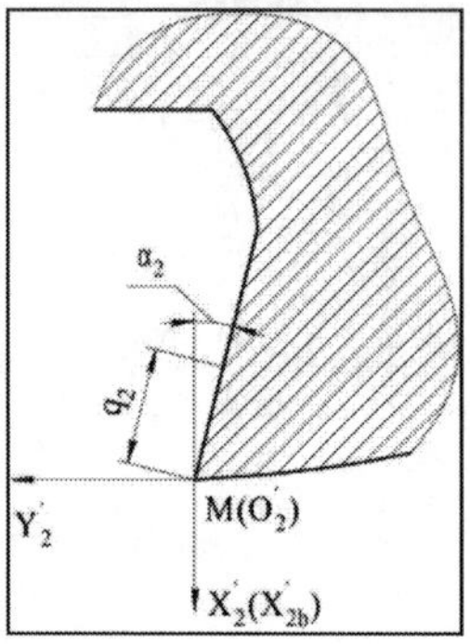

Fig. 8. The normal section of the profile coil of the worm-type tool.

The working section of the profile, which acts as the back surface of the cutting wedge, is formed in the form of a straight line segment. It is convenient to construct the solution of the inverse problem by comparing a theoretically accurate profile in the axial section and a profile obtained by numerical simulation.

This method is the most acceptable, since the use of a combination of the axial section of the worm-type tool and the cross-sections of the rotation surfaces (in this case cylindrical surfaces) considered in [18] makes it possible to obtain the L_s (line of shaping) in the form of a straight line parallel to the axis of the worm tool type. In order to implement this scheme, it is necessary at the first stage to develop a methodology for calculating the coordinates of the nodal points of the profile of the lateral surface of the screw of the worm-type tool in the axial section.

We introduce a coordinate system $X_2Y_2Z_2$ rigidly connected with a worm-type tool (Fig. 9), whose axis Z_2 coincides with the axis X_2 of the worm-type tool, and the initial point of the screw cutting edge is located on the axis. Index 2 of the specified coordinate system is related to the fact that the worm-type tool in this case serves as a detail. With the current point M of the screw cutting edge, we connect the origin of the coordinate system, $X_2'Y_2'Z_2'$ in which we describe a straight section of the profile of the lateral surface of a worm-type tool. The axis Z_2' of the coordinate system coincides with the tangent to the helical cutting edge and makes an angle λ_2 with the end plane of the worm-type tool. The angle of rise of the screw cutting edge can be calculated using the well-known expression:

$$tg\lambda_2 = \frac{p_{z2}}{2\pi r_{a2}}, \qquad (1)$$

Where p_{z2} - the axial movement of the helical cutting edge of the worm-type tool,

r_{a2} - the radius of the circumference of the worm-type tool protrusions.

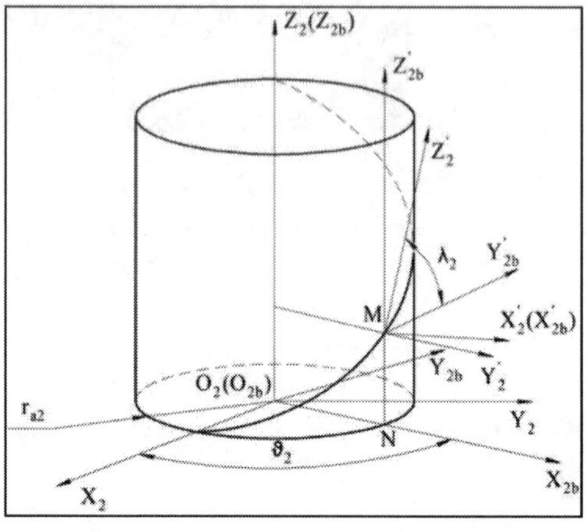

Fig. 9. Transition scheme from the system $X_2'Y_2'Z_2'$ to the system $X_2Y_2Z_2$.

The coordinate axis X_2' lies in the end plane of the worm-type tool and crosses its axis. Thus, the coordinate plane $X_2'O_2'Y_2'$ coincides with the normal cross section. Equations of the rectilinear portion of the lateral surface of the coil of the worm-type tool can be written in the following form (Fig. 8):

$$\begin{aligned} x_2' &= -q_2 \cos\alpha_2 \\ y_2' &= -q_2 \sin\alpha_2 , \\ z_2' &= 0 \end{aligned} \qquad (2)$$

Where q_2 - the parameter determining the position of the point under consideration on the generator,

α_2 - the rear angle of the cutting wedge of the worm-type tool.

To calculate the coordinates of the node points of the axial section of the worm-type tool, it is necessary to make a transition from the coordinate system $X_2'Y_2'Z_2'$ to the system $X_2Y_2Z_2$. We make this transition in three steps (Fig. 9):

a) rotate the coordinate system $X_2'Y_2'Z_2'$ around the axis X_2' by an angle equal to $\pi/2-\lambda_2$, then the coordinate system $X_2'Y_2'Z_2'$ will take a position $X_{2в}'Y_{2в}'Z_{2в}'$, while the axis $Z_{2в}'$ is parallel Z_2;

b) we transfer the origin of coordinates O_2' in O_2 due to its displacement in the direction of the axis $X_{2в}'$ by an amount r_{a2} and in the direction of the axis $Z_{2в}'$ by the value of the segment $MN = p_2\vartheta_2$, where p_2 - the parameter of the screw cutting edge. In this case, the system $X_{2в}'Y_{2в}'Z_{2в}'$ will move to the coordinate system $X_{2в}Y_{2в}Z_{2в}$;

DOI: 10.1051/matecconf/201712901043

c) rotate the coordinate system $X_{2в}Y_{2в}Z_{2в}$ around the axis Z_2 by an angle ϑ_2 , then the coordinate system $X_{2в}Y_{2в}Z_{2в}$ will take the position $X_2Y_2Z_2$.

Conversion of coordinates is realized by means of the following matrix equality:

$$r_2 = M_{22в}M_{2в2в'}M_{2в'2'}r_{2'}\,, \quad (3)$$

Where r_2 and $r_{2'}$ - the column matrices of the corresponding radii of the vectors of the same point in the systems $X_2Y_2Z_2$ and $X_2'Y_2'Z_2'$;

$M_{2в'2'}$ - matrix of transition from system $X_2'Y_2'Z_2'$ to system $X_{2в}'Y_{2в}'Z_{2в}'$;

$M_{2в2в'}$ - matrix of transition from system $X_{2в}'Y_{2в}'Z_{2в}'$ to system $X_{2в}Y_{2в}Z_{2в}$;

$M_{22в}$ - matrix of transition from system $X_{2в}Y_{2в}Z_{2в}$ to system $X_2Y_2Z_2$.

In the coordinate form, taking into account expression (3), the transition from the coordinate system $X_2'Y_2'Z_2'$ to the coordinate system $X_2Y_2Z_2$ will be determined by equations that can be written:

$$x_2 = (x_2' + r_{a2})\cos\vartheta_2 - (y_2'\sin\lambda_2 + z_2'\cos\lambda_2)\sin\vartheta_2$$

$$y_2 = (x_2' + r_{a2})\sin\vartheta_2 + (y_2'\sin\lambda_2 + z_2'\cos\lambda_2)\cos\vartheta_2 \quad (4)$$

$$z_2 = -y_2'\cos\lambda_2 + z_2'\sin\lambda_2 + p_2\vartheta_2$$

Equating the expression x_2 for zero, we obtain the equation:

$$(x_2' + r_{a2})\cos\vartheta_2 - (y_2'\sin\lambda_2 + z_2'\cos\lambda_2)\sin\vartheta_2 = 0 \quad (5)$$

Substituting for this equation the expressions for x_2' , y_2' and z_2' from (2) after a simple transformation we obtain the equation from which we can obtain the formula for calculating the angle ϑ_2 :

$$\vartheta_2 = arctg\left(\frac{q_2\cos\alpha_2 - r_{a2}}{q_2\sin\alpha_2\sin\lambda_2}\right). \quad (6)$$

If, for a fixed value q_2 , we calculate ϑ_2 by formula (6) and substitute the obtained value ϑ_2 into expressions (4) for y_2 and z_2 , then it is possible to determine the coordinates of the node point of the axial section of the lateral surface of the screw of the worm-type tool.

The obtained coordinates of the node points of the axial section of the lateral surface of the turn of the worm-type tool allow to determine the parameters for setting the grinding wheel with respect to the worm-type tool. These adjusting parameters will make it possible to carry out the most labor-intensive operation of manufacturing this tool, namely, grinding the lateral surface of the winding.

References

1. A.A. Matalin, *Manufacturing engineering* (Lan, St. Petersburg, 2010)
2. M.F. Pashkevich, *Manufacturing engineering. Course and degree design* (Grevtsov's publishing house, Minsk, 2010)
3. V.L. Kulygin, *Manufacturing engineering* (BASTET, Moscow, 2011)
4. O.V. Taratynov, *Manufacturing engineering. Design bases on the computer* (FORUM, Moscow, 2011)
5. I.S. Ivanov, *Manufacturing engineering* (INFRA-M, Moscow, 2012)
6. 6. V.A. Arshinov, G.A. Alekseev, *Cutting of metals and the cutting tool* (Mechanical engineering, Moscow, 1968)
7. P.R. Rodin, *Metal-cutting tools* (Vishcha shkola, Kiev, 1974)
8. I.A. Ordinartsev, *Reference book of the toolmaker* (Mechanical engineering, Leningrad, 1987)
9. G.N. Sakharov, O.B. Arbuzov, Yu.L. Borovoy, *Metal-cutting tools* (Mechanical engineering, Moscow, 1989)
10. O.I. Averyanov, V.V. Klepikov, *Cutting tool* (MGIU, Moscow, 2007)
11. A.A. Ryzhkin, K.G. Shuchev, A.G. Skhirtladze, *Cutting tool* (Phoenix, Rostov-on-Don, 2009)
12. E.E. Feldstein, M.A. Korniyevich, *Cutting tool. Operation* (INFRA-M, Moscow, 2012)
13. D.V. Kozhevnikov, V.A. Grechishnikov, S.V. Kirsanov, S.N. Grigoriev, *Cutting tool* (Mechanical engineering, Moscow, 2014)
14. Inventor's certificate USSR, no. 1106609, 1984
15. Inventor's certificate USSR, no. 1004030, 1983
16. Patent RF, no. 103771, 2011
17. Patent RF, no. 2005013, 1993
18. N.V. Smorkalov, *Numerical modeling of surfaces when processing by cutting* (Publishing house of the Kama state polytechnical institute, Naberezhnye Chelny, 2003)

Method for measuring thermo-emf of a "tool-workpiece" natural thermocouple in chip forming machining

Alexander S. Sergeev[1,*], *Zhanna S.* Tikhonova[1], and *Tatiana V.* Uvarova[1]

[1]Volgograde State Technical University, Linine Ave. 28, Volgograde 400005, Russia

Abstract. The article presents a new method and a device for measuring the signal of thermoelectromotive force of a "tool – workpiece" natural thermocouple in the chip forming machining during the turning process. It is proposed to use a device of stationary current-collecting connection for fixing and measuring thermo-electromotive force signal, which allows to eliminate the influence of the transient resistance for existing constructions of movable current collectors. This circumstance makes it possible to reduce the measurement error of the signal and at the same time to increase its reliability. The review and analysis of the most similar constructive solutions have been carried out and there are also experimental data values which have been obtained when comparing the proposed device and the prototype.

The processes of edge cutting machining of the workpiece with tool material are accompanied by the thermoelectromotive force (thermo-emf) generated by "tool – workpiece" natural thermocouple. It is known that the variation of the thermo-emf signal is observed due to the change in the work of the electrons liberation from the hard alloy and the processed workpiece. The value of the contact component for the thermo-emf signal of the tool-workpiece pair will be determined from the condition of equalizing the electrons electrochemical potentials. This fact gives the opportunity to use thermo-emf signal of a "tool - workpiece" as a diagnostic parameter for evaluating physico-mechanical properties of various combinations of "tool-workpiece" contact pairs for subsequent assignment and optimization of the cutting modes while processing the same workpiece with the given tool. And for this purpose, thermo-emf signal should be obtained on strictly defined test run modes, which are the same for all groups of processed and tool materials.

Using the informative ability of the cutting thermo-emf signal to ensure the reliability for the automated determination of the edge cutting modes is presented in many scientific works, in particular in [1-5].

In addition to the occurrence of a parasitic component in the signal and the need for electrical insulation of the tool and the steel workpiece, the main disadvantage of thermo-emf signal application is the dependence of the transient resistance and hence, the value of the thermo-emf on the structure and type of the current collectors installed on the machine tool spindle units. Thus, the main way to increase the reliability level of the thermo-emf signal, and consequently, the reliability of the automated determination of cutting modes, is the development of a method for picking up the thermo-emf signal without the use of movable current collectors of the contact type. Let us consider the best known methods for measuring the cutting thermo-emf signal.

There is a method for measuring the cutting EMF [6], in which a tool or a workpiece is isolated from the machine tool mass, then the tool and the workpiece are brought into complete electrical circuit and the cutting EMF generated on the contact surfaces of the tool is registered. The whole working section of the tool, excluding the zones in which the cutting EMF is measured, should be coated with a dielectric wear-resistant layer.

This well-known method has some disadvantages. When the tool is isolated from the machine tool mass it is necessary to spread dielectric wear-resistant coating over all the cutting tool elements involved in the machining but it greatly increases the production cost and reduces its productivity.

In the case of the workpiece insulation from the machine tool mass, this method involves the use of a current collector on the side of the machine spindle unit, which reduces the reliability of the measured cutting EMF due to electrical losses at the coupling sites between the current collector contacts and the bearing part of the machine spindle during its rotation at different frequencies.

Another well-known method for measuring the cutting EMF is described in the work [7]. This method provides the opportunity of measuring the variable component in the cutting EMF signal, which is picked up by a current collector device, while isolating the cutting tool and the workpiece from the machine tool mass.

The drawback of this method [7] is that the use of the current collector device reduces the reliability of the cutting EMF signal and also reduces the equipment

* Corresponding author: sergeevapp@mail.ru

 DOI: 10.1051/matecconf/201712901044

versatility when this method of EMF measurement is industrially introduced.

The most similar procedure for the same purpose is a method for measuring the thermo-emf of a "tool-workpiece" natural thermocouple [8]. It involves preliminary thermo-emf measurement of a "tool-workpiece" natural thermocouple in chip forming machining in the speed range above the build-up forming zone at idle speed of the machine, while measuring the resistance of the oil films of the machine bearings. At the resistance value equal to or more than 100 ohms, it is necessary to record a useful thermo-emf signal of operating cutting modes using a peak detector as a filter for separating parasitic thermo-emf, also fixing the cutting torque which is characterized by the onset of stable chip formation.

The disadvantage of the given method [8] is that it requires measuring the resistance of the oil films in the machine bearings, which is quite unstable and fluctuates over a wide range, but the method does not specify how to register the useful thermo-emf signal if the resistance of the oil films of the machine bearings is less than 100 ohms. In addition, this method involves the use of a graphite current collector on the side of the machine spindle unit, which reduces the reliability of the measured thermo-emf signal due to electrical losses at the coupling sites between the current collector contacts and the measuring flange of the machine spindle.

In this regard, the authors propose a new technical solution to reduce an error in measuring the useful thermo-emf signal by including an isolated tool and a workpiece in a complete electrical circuit without using intermediate members for fixing the electrical signal (current collectors of various designs).

The technical result that can be obtained by the implementation of the proposed technical solution consists in the accuracy measurement increase of the useful thermo-emf signal of a natural thermocouple, which enhances technological application of the informative ability of the thermo-emf signal in a "tool-workpiece" contact pair.

The given technical result is achieved by the fact that in order to increase the measurement accuracy of the useful thermo-emf signal that is produced during the cutting operation, the measurement of this signal should be made with respect to the insulated non-conductive shims of the cutting tool and the grounded body of the machine tool. The workpiece relative to the machine body is not additionally isolated, and the current collector in the measurement scheme is not applied.

In the proposed method of measuring the thermo-emf of a "tool-workpiece" natural thermocouple we should use complete electrical insulation of the tool by non-conductive shims rather than the parts of oil films in the machine bearings. It allows to significantly reduce the distortions of the useful thermo-emf signal due to constant dielectric properties of non-current conducting shims, in contrast to the changing resistance level of machine oil films [8]. The stiffness of the MFTW system embedded in the machine tool design and the level of vibration resistance must be maintained.

In this case, a mobile current collector (see prototype [8]) is not used, and a peak detector (amplitude rectifier) is directly connected to the grounded body of the machine tool, thus reducing the measurement error of the thermo-emf signal by eliminating electrical losses at the coupling sites between the current collector contacts and the measuring flange of the machine tool spindle.

The main point of the technical solution is explained by the scheme for the thermo-emf measurement of a natural thermocouple on a lathe (Figure 1). The workpiece 1 is fixed in the jaw-chuck of the machine tool. The cutting is carried out in cantilever or with the pressing by the rear center support. 3. The cutter 2 is fixed in the tool holder and is additionally electrically isolated by non-conductive shims 4. Metal cutting is performed in the speed range above the build-up forming zone and the thermo-emf of the "tool-workpiece" natural thermocouple is measured during the chip forming machining by a millivoltmeter 6 with the help of the peak detector circuit (amplitude rectifier) 5 connected on one side to the cutting tool body 2, and on the other side to the grounded body of the machine tool.

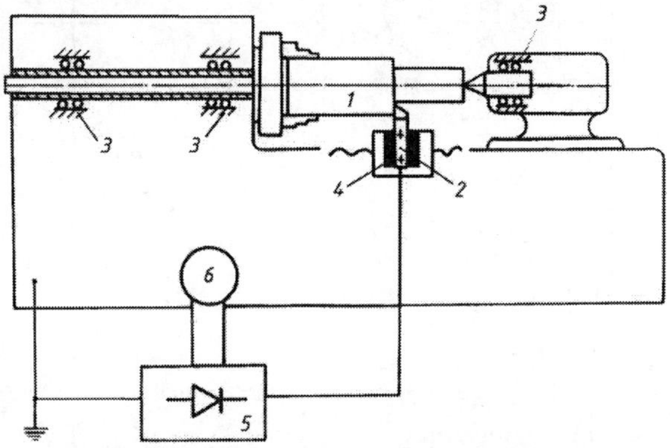

Fig. 1. Thermo-emf signal measuring scheme.

The process is carried out in the following way. Preliminary, at a selected speed corresponding to the range above the build-up forming zone, a metal workpiece fixed directly to the machine is processed with a tool fastened with non-conductive shims. A peak detector (amplitude rectifier) connected in the circuit between the recording device and the thermo-emf source of a natural thermocouple is used as a filter for clearing the parasitic thermo-emf interference. The registration of the useful thermo-emf signal of a natural thermocouple is carried out from the initial moment of cutting until the emergence of a stable constant value on the recording instrument for 4-5 seconds. The storage period of the registered level of the useful thermo-emf signal is determined by condenser capacity which is mounted at the output of the peak detector.

Table 1 shows the results of the experimental verification of the proposed method for measuring the thermo-emf of a "tool-workpiece" natural thermocouple when metal machining on a 1K62 lathe with a carbide tool, at the cutting speed range above the build-up forming zone, using the electrical insulation of the tool.

Table 1. The results of the experimental verification and comparison with the prototype.

№	ISO system	Tool brand and type		Reference thermo-emf signal measurement *E*, mV	Thermo-emf signal measurement of a natural thermocouple									
		Alloy grade, replaceable faceted inserts, RFI (Machinability group)	Coating type		Method (prototype) [8]						Proposed method			
					Tool is insulated, current collector is used						Tool is insulated, current collector is not used			
					Workpiece supported as a cantilever without insulation		Workpiece is fixed with the support of the rear center				Workpiece supported as a cantilever		Workpiece is fixed with the support of the rear center	
							Workpiece is not insulated		Workpiece is insulated					
					E_{ab}, mV	δ, %	E_{ab}, mV	δ, %	E_{bd}, mV	δ, %	E_{bc}, mV	δ, %	E_{bc}, mV	δ, %
1	SNMA 120408	BK6 (K20)	-	8,5	7,9	7,1	8,2	3,5	7,7	9,4	8,3	2,4	8,4	1,2
2	SNMA 120408	BK6 (K20)	TiN	12,2	12,8	-4,9	12	1,6	12,5	-2,5	12,2	0	12,1	0,8
3	TCMW 110204	BK8 (M30)	-	7,4	7,2	2,7	7,7	-4,1	7,1	4,05	7,2	2,7	7,5	-1,4
4	WCMT 080408	T15K6 (P10)	-	11,9	12,2	-2,5	12,1	-1,7	12,2	-2,5	11,8	0,8	11,8	0,8
5	WCMT 080408	T15K6 (P10)	TiN	12,7	11,6	8,7	11,8	7,1	11,6	8,7	12,4	2,4	12,5	1,6
6	WCMT 080408	T15K6 (P10)	TiCN	12,2	11,9	2,5	11,8	3,3	11,8	3,3	12,2	0,0	12,1	0,8
7	SCMT 120408	TH20 (P01)	-	7,4	7,6	-2,7	7,6	-2,7	7,2	2,7	7,3	1,4	7,5	-1,4
8	SCMT 120408	TH20 (P01)	TiAlN	7,5	7,2	4,0	7,3	2,7	7,3	2,7	7,4	1,3	7,4	1,3
9	SPUN 120308	T5K10 (P30)	-	14,6	14,7	-0,7	14,7	-0,7	14,8	-1,4	14,4	1,4	14,4	1,4
10	SPUN 120308	T5K10 (P30)	-	15,4	15,4	0,0	15,9	-3,2	15,3	0,6	15,2	1,3	15,3	0,6
11	CNMG 120408-PM 4225	GC 4225	TiN	10,7	11,5	-7,5	10,8	-0,9	11	-2,8	10,5	1,9	10,8	-0,9
12	CNMG 120408-PM 4225	GC 4225	TiN	11,6	11,4	1,7	11,8	-1,7	11,2	3,4	11,9	-2,6	11,7	-0,9
13	TPUN 110304	TT7K12 (P40)	-	9,2	9,7	-5,4	9,5	-3,3	8,7	5,4	9	2,2	9,1	1,1
14	TCMT 16T308E-UR	T30K4 (P01)	TiN	9,5	9,4	1,1	9,2	3,2	9,2	3,2	9,6	-1,1	9,6	-1,1
Average value of the relative error δ_{av}, %					3,3		2,8		3,5		1,5		0,9	

It also presents the comparative analysis of the given method with the prototype [8]. The following are the cemented carbide tools used for the experiment: BK6, BK8, T5K10, T15K6, T30K4, TH20, TT7K12. The cemented carbide cutting tools come under the Russian Standard (GOST) 2209-69, which are bolted or fixed mechanically to the tool holder. Under the International Standard Organization's classifications, the tools listed above have equivalents in the ranges that consist of K20, M30, P10, P30, P01 and P40 (ISO 2004). The measurement and fixation of the thermo-emf signal used for the relative comparison of the readings was carried out according to the scheme in figure 2. The workpiece (disk) 1 is fixed with a nut 7 on a special mandrel which is isolated from the chuck cams 6 with non-conductive shims 2. The diameter of the workpiece 1 is about 300 mm, which allows to use the minimum spindle speed 9 and directly register the signal from the shielded cable 4. The cutting tool 3 is also isolated from the body of the lathe tool holder 15 by means of non-current conducting shims. The signal is transferred from the workpiece 1 through the nut 7 and the shielded cable 4 to one of the digital millivoltmeter inputs, the second input of the digital millivoltmeter accepts the signal from the cutting tool 3 via the shielded cable 16. The complete electrical circuit of the natural thermocouple is completely isolated from the machine body by means of non-current conductive shims on the side of the cutting tool 3 and the workpiece 1. In this case the thermo-emf signal (see figure 2) is transmitted from the workpiece 1 through the bearing assemblies 5, 11, 12 of the machine spindle unit, the gearing of the transmission 8, 14, the change gear bracket, the feed shaft (or the feedscrew)

13 and the machine rear center (in the case of rear center support machining) to the machine body 10 due to the formation of electrically conductive oil films in the conjunctions spots of the mating units of the machine during the mechanical energy transfer [9].

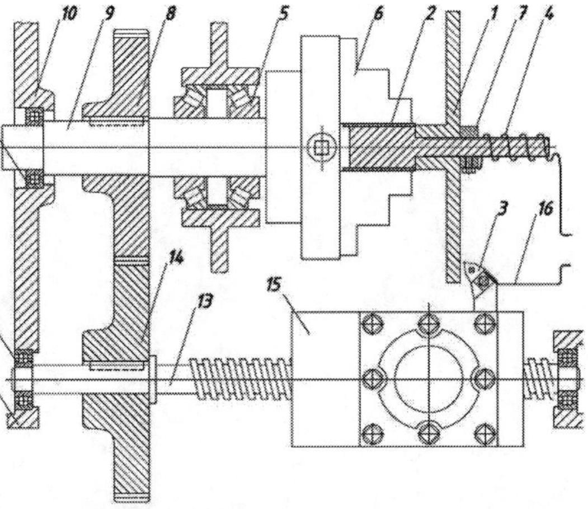

Fig. 2. Measurement and fixation scheme of the thermo-emf signal used for relative comparison of the indications.

The process of generating useful thermo-emf signal was recorded by means of a digital millivoltmeter. The processed metal is 40X steel, the tool is a replaceable faceted non-re-sharpened hard alloy inserts of various grades, type of coating and geometry (see Table 1).

The error in measuring the thermo-emf signal was estimated from the average value of the relative error δav. Fig. 3 shows the comparison of the indicators average values of the relative measuring error in the thermo-emf signal between the prototype and the method proposed by the authors.

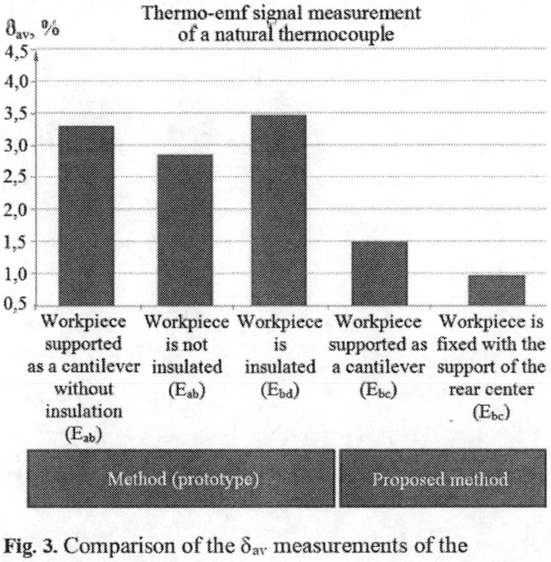

Fig. 3. Comparison of the δ_{av} measurements of the thermo-emf signal.

Figure 4 shows the location of the measurement points (cold junction points) of the thermo-emf value for a "processed workpiece-tool" natural thermocouple in chip forming machining. In Figure 4 and Table 1, the measured values of the thermo-emf are represented in the form, for example, E_{ab} = 12.1 V, which means that the signal of 12.1 V amplitude was measured between point a (current collector output) and point b (cutting edge of the cutting tool).

Thus, it is possible to register the reference value of the thermo-emf signal which is used in the further comparison of the proposed method and the prototype (see Table 1).

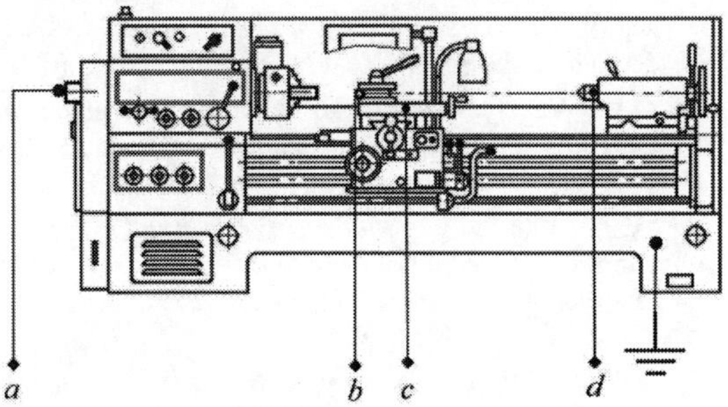

Fig. 4. Arrangement scheme of measurement points for the thermo-emf value.

The proposed method provides accurate measurement of the useful thermo-emf signal of a natural thermocouple and can be applied in the automatic thermo-emf measurement mode, for example, programmatically on CNC machines where the time of measurement, storage and reset of the signal can be set by a timer. The implementation of the proposed method allows to intensify the measurement accuracy of the useful thermo-emf signal of a natural thermocouple (see Table 1), which increases the reliability and enhances the technological possibility of using the informative ability of the thermo-emf signal of the "tool-workpiece" contact pair [1-5].

References

1. A.L. Plotnikov, A. S. Sergeev, N. G. Zaytseva, *Use of the informative ability of a natural thermocouple signal to ensure the reliability of automated determination of edge cutting machining*. Science based technologies in machine building, № **5**, pp. 35-40 (2013)
2. A.L. Plotnikov, V.A. Kamayev, A.S. Sergeyev, T.V. Uvarova, *Modernisation of universal lathes and their operating systems in order to increase their work efficiency,* News of VSTU. Ser. Actual problems of the control, computer technology and computer science in technical systems, Volgograd, № **11 (190)**, pp. 61-66 (2016)
3. A.L. Plotnikov, A.S. Sergeyev, T.V. Uvarova, N.G. Zaytseva, *Cutting force calculation and back calculating of the tool life for three stages of*

turning corrosion-resistant steels, Certificate of the state registration of the software program № 2015619947, 17.09.15 RF, IPC (no). VSTU (2015)

4. E.G. Krylov, A.U. Gorelova, A.S. Sergeev, *The way to control the condition of the cutting edges of prefabricated multiblade tools*. Patent RF, no. 2496629 IPC, B23Q17/09, VSTU (2013)
5. A.L. Plotnikov, A.S. Sergeev, Zh.S. Tikhonova, *Pecularities of the use of cutting EMF signal in automated machine tool manufacturing*, Science based technologies in machine building, № **6**, pp. 21-27 (2016)
6. V.I. Aleksandrov, A.G. Borodayev, E.I. Glinkin, *Cutting EMF signal measurement method*, Patent RF, no. 2149745, IPC 7 V23 Q17 / 09, TSTU (2000)
7. S.V. Vasil'yev, *EMF and cutting temperature. Machine tools and instruments*, №**10**, pp. 20-22 (1980)
8. A.L. Plotnikov, *Measuring method for the thermoelectric power of a tool – workpiece natural thermocouple*, Patent RF, no. 2117557, IPC 7 V23 V25 / 06, VSTU (1999)
9. Y. A.Mit'kin, *Electrophysical characteristics of liquid dielectrics containing impurities: tutorial* (Ivanovo State Power University, Ivanovo, 2002)

Influence of the surface layer characteristics on the regularities of the cutting process

Dmitriy V. Krainev[1,*], *Maria Yu.* Polyanchikova[1], and *Alexander A.* Bondarev[1]

[1]Volgograde State Technical University, Linine Ave. 28, Volgograde 400005, Russia

Abstract. The article considers the influence of the surface layer characteristics on the regularities of the cutting process and the formation of the quality of the surface machined. This effect has been confirmed by the study results of the combined cutting method with advanced plastic deformation (APD). The work estimates the impact of the change in the surface layer properties on the forces and temperature of cutting, stability of the chip formation and quality parameters of the surface machined.

The rise in the tech level specifies the challenges of increasing the effectiveness of the parts production and formation of the required performance characteristics of their surfaces. It is noteworthy that at present, inadequate attention is paid to all the parameters of the machine part surface both in construction, process design, and production. Machining considers for the most part the geometrical parameters and surface roughness. To improve the surface performance parameters such as the surface structure, hardness, wear resistance, etc., some special operations are assigned, including the strengthening treatment, honing finishing operations and heat treatment of various types. However, in certain cases, the desired surface parameters can be achieved by less number of operations and at less cost.

Besides, the concept of machining design, in our view, requires some updating and improvements. Despite the continuing technological advance and potential of certain branches of machine-building industry, which are the basis for any industrial production (first of all, machine tool building, metallurgy, cutting tool production, etc.), for upward movement, they cannot fully ensure the necessary stability of all technological system parameters in course of machining operations. So, this fact causes the parameter instability of the surface machined, and hence the performance instability of the manufactured parts, assemblies, mechanisms, and machines.

The machining procedure of a part can be considered as a sequence of operations on a workpiece. Each operation to some extent changes a set of the surface parameters of the workpiece machined, with the processing result to depend not only on the type and mode of exposure and the tool applied, but also on the previous state and the surface layer of the workpiece. The processing result depended not only on the type and modes of action and the tool applied, but also on the previous state of the workpiece and the surface layer. The effectiveness of the cutting process was considerably influenced by the properties of the material being processed, the workpiece and the cutting tool. The indicated properties given in the reference literature and certificates for tool and construction materials were averaged and allowed for a wide range of characteristics [1, 2]. This significantly reduced the reliability and effectiveness of the technological preparation of production and also affected the stability of the processing results.

For a better understanding the essence of this issue, let's look at the physics of the cutting process. Cutting is a complex and specific form of plastic deformation and fracture, when high rates of deformation combine with extremely high pressures on the surface of the material processed, high temperatures and a gradient of stresses and deformations.

In edge cutting machining, at the shear moment, an elastic and then plastic deformation of the metal occurred at a certain point, while the tool undergoes elastic deformation. Hardening of the cutter-contacting layers of the material processed is accompanied by the release of heat, which is distributed into the workpiece itself, chips, cutting tools and the environment. The heat release intensity which depends on a set of parameters, including the characteristics of the material processed, tool, etc., and has a considerable effect on the nature of the cutting process. When the softening temperatures [3] are reached, the metal brakes down along the cutting line in the direction of the cutting wedge motion.

At the same time, due to the forward going deformation wave, the metal is reinforced again (the area of advanced hardening), and hardening of the metal under the surface machined also occurs due to the effect of the cutting edge on this area. So, there are permanent phenomena of hardening and softening of elementary metal layers.

Thus, the cutting process involves the interaction of two mutually opposite processes, i.e. the strain hardening and thermal softening.

* Corresponding author: krainevdv@mail.ru

DOI: 10.1051/matecconf/201712901045

Since the strain hardening does not affect the entire volume of the workpiece, but primarily the surface layer, we can say that the maximum processing efficiency is impossible to achieve without taking into account the influence of the surface layer characteristics on the cutting process and the quality of the surface machined.

The significance of this effect has been confirmed by the results of studies of the combined cutting method with advanced plastic deformation (APD) [4, 5, 6].

When using the APD, the metal, which had a lower plasticity than that in the case of traditional cutting, came into the area of the chip formation and hardening of the "chip—cutter" contact interaction zone. In addition, there was a weakening of the stress level in the primary deformation zone due to a decrease in the work to bring the volume of the treated body to fracture along the cutting line that helped to reduce the tangential component of the cutting force. The presented changes weakened the spread of the hardening zone of the height and length contact plastic deformations and also caused a decrease in the level of hardening of the secondary deformations zone and, consequently, resistance to the deformation. This was proved experimentally by the established fact of microhardness reduction on the chip roots (Fig. 1) in the APD turning.

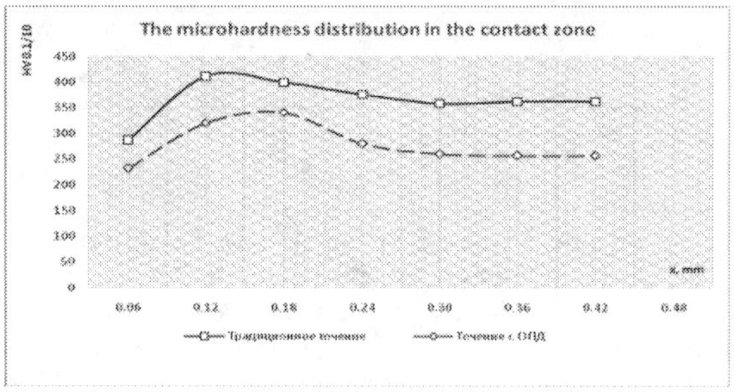

Fig. 1. The microhardness distribution in the contact zone on the front surface of the cutting edge at a distance of 10 μm from the front surface in traditional turning and the APD turning (30XMA steel, hard alloy BK6; t=1mm; v=90m/min; so = 0.256 mm/rev. In the APD turning, the CAPD is 2).

To determine the mode of the APD application, the characteristic of the combined processing, i.e. "the ADP coefficient," was used [4, 5]:

$$C_{APD} = \frac{h_{hard}}{t}, \tag{1}$$

where h_{hard} is the depth of the hardened layer resulted from the APD phase, mm;
t is the depth of cutting, mm.

The reduction of resistance to the plastic deformation when moving volumes of the material processed through the zones of the chip formation and contact interaction of chips with a cutting wedge determined the decrease in the heat release intensity.

The cutting temperature was measured by the artificial thermocouple method for the processing modes and conditions (Fig. 1). A reduction in the power of the heat source in the cutting zone was registered by 17% in the APD processing compared with the conventional turning. The reduction in the power of the heat source in the APD cutting was also established based on the measurements analysis of the natural thermocouple thermo-EMF (Fig. 2).

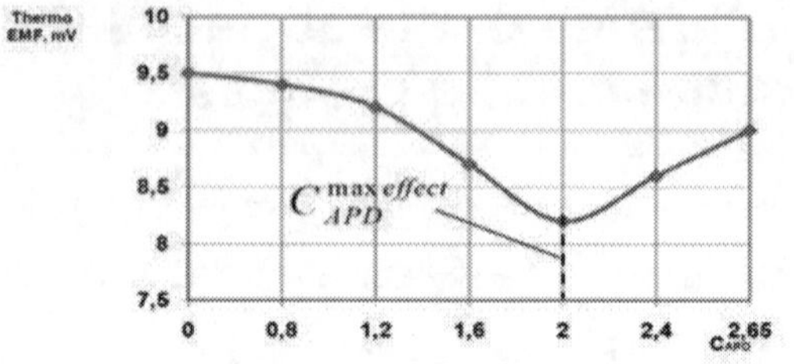

Fig. 2. The quantities of the thermoelectromotive force signal of the cutting process in traditional turning and the APD turning for different CAPD values (20X13 steel is T15K6 hard alloy, t = 1 mm, v = 90 m/min, so = 0.256 mm/rev. CAPD=0 in traditional turning).

Analysis of the experimental data on the measurement of the thermo-EMF of the natural thermocouple confirmed the decrease in the cutting temperature in the APD turning (Table 1).

Table 1. The thermo-EMF signal values of the cutting process in traditional and APD turning, mV

Processing modes and conditions	Traditional turning E, mV	ADP turning E, mV
20X13 steel is T15K6 hard alloy; t=1mm;v=90 m/min; s_o=0.256 mm/rev. In APD turning, the C_{APD}=2	9.5	8.2
14X17H2 steel is BK6hard alloy; t=0.5 mm; v=135 m/min; s_o=0.166 mm/rev. In APD turning, the C_{APD}=1.6	12.4	10.9

The nature of the chip formation process was determined not only by the heat release intensity, but also by the laws of the heat distribution in the cutting zone. The intensity and direction of the heat fluxes in the cutting zone depended on the thermal conductivity λ of the contacting materials. The application of the APD in this case had an effect on the thermal conductivity of the metal of the layer cut (Table 2).

Table 2. The values of the integral thermal conductivity of the initial and deformed material

	Material	Traditional turning	ADP turning ($C_{ADP\ max.\ efficiency.}$)
Heat-conduction λ_{total} *(W/(m*K))*	20X13 steel	21.7	11.17
	30XMA steel	35.08	31.07

The decrease in the amount of the work, which was required to remove the material layer and decrease in the integral sum of the stresses, acting in the zone of advanced hardening, contributed to a decrease in the value of the response applied to the front surface of the tool, i.e., the tangential component of the P_Z cutting force when using the APD, on average by 15-20 % (Fig. 3).

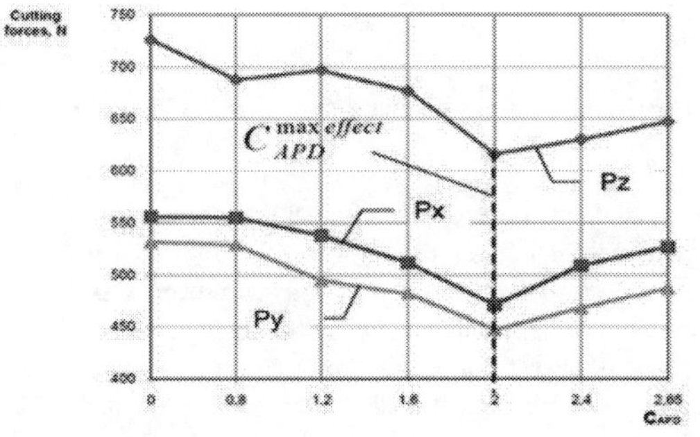

Fig. 3. The values of the tangential PZ, radial PY and axial PX components of the cutting force for traditional turning and APD turning for different values of the CAPD (20X13 steel is T15K6 hard alloy; t = 1 mm; v = 90 m/min; so = 0.256 mm/rev. CAPD=0 in traditional turning.

The APD in turning ensures the reduction of the roughness parameters of the surface machined in comparison with the traditional turning (Fig. 4).

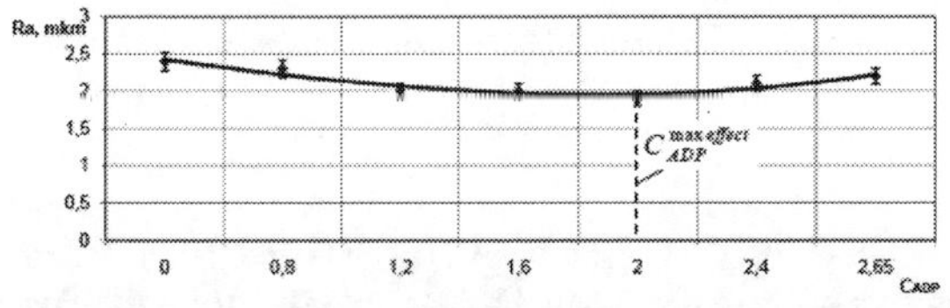

Fig. 4. Dependence of the mean arithmetic deviation of the *Ra* profile of the surface machined on the C_{APD} value (20X13 steel is T15K6 hard alloy; v_p = 90 m/min; s_o = 0.256 mm/rev.; t_p = 1 mm; the C_{APD} = 0 in traditional turning).

In the APD turning, the size of the metal separation center along the cutting line decreased due to the stresses reduction in the cutting zone, and the destruction proceeded more stably, including due to the decrease in the plastic deformation instability under conditions of cyclic chip formation. In addition, the APD helped to reduce the values of the *Ra, Rmax, Rz, Sm* parameters of the microrelief by reducing the wear of the blades.

Under the influence of the APD, the characteristics of the surface layer (the deformation degree and thermal conductivity) changed that affected the cutting process.

This action reduced the cutting forces and temperature, and the chip formation stability enhanced. This led to a change in the quality parameters of the surface machined, i.e., improvement of the surface micro-profile and, as a consequence, to an increase in the performance characteristics of the surface of the workpiece. Thus, it can be concluded that the characteristics of the surface layer have a significant effect on the regularities of the cutting process and the quality of the surface machined. However, up to the present time, theoretical relationships between the parameters of the surface layer, cutting process characteristics and performance properties of the parts have not been established. These dependences would allow solving the problem of technological support of the performance properties specified. The management of the technological processes to ensure the required state of the surface layer is based mainly on the accumulated production experience and results of the laboratory studies presented in the form of reference material and particular empirical formulas. This requires a theoretical generalization of the accumulated knowledge, establishment of theoretical and experimental dependencies of the characteristics of the surface layer, processing methods and regimes and the development of the control approach to the technological heredity in order to create optimal technological parts processing that provides the required quality of the surface machined, production level and minimum cost.

References

1. A.L. Plotnikov, *Management of the cutting modes on lathes with CNC* (RPK Politekhnik, Volgograd, 2003)
2. Ju.L. Tchigirinsky, *Mathematical methods of the control of the machining processes* (VSTU, Volgograd, 2010)
3. N.V. Talantov, *Physics of the processes of cutting, wear and tear of tools* (Mechanical Engineering, Moscow, 1992)
4. V.M. Yaroslavtsev, Bulletin of Bauman MSTU **1**, 119 (2015)
5. D.V. Kraynev, Yu.N. Polyanchikov, A.A. Bondarev, *Improving the turning efficiency of wrought steels and alloys with advanced plastic deformationm*, (VSTU, Volgograd, 2015)
6. P.A. Norchenko, D.V. Kraynev, I.N. Kozachuhnenko, Procedia Engineering. **150**, 821 (2016)

Methodology of multiaxial machines formats volumetric accuracy comparative evaluation

Svetlana Lukina[1,*], and *Igor* Manaenkov[1]

[1]Moscow polytechnic university, 38 B.Semyonovskaya st., Moscow 107023, Russian Federation

Abstract. The main characteristic of machine tool system, providing the most efficient multi-axis processing, is machine volumetric accuracy, which may be demonstrated in the machine working space. The development of methods and means of controlling volumetric accuracy is a very important engineering and process task. Solving this issue may ensure competitive ability of the domestic fixtures and tools for multi-axis form making during the treatment with the help of power tools. Early project stage is the most important one, as the success of the machine project itself highly depends on the right choice of arrangement. In this research, we presented design scheme of machine-tool unit, stated its initial and boundary conditions, and also there was developed a model of comparative assessment of elastic displacement in the working space. Volumetric accuracy is formed by means of inaccuracy space field of the processing system, which occur in the formation zone under the influence of agitation, which may be determined by simulation and analytical modeling of machine tool system. Within the framework of this research, there was developed the method of comparative assessment of volumetric accuracy of multi-axis machine compositions with not less than 5 interpolate axis.

1 Introduction

The modern stage of technological system development for high-technology treatment is characterized by increase of competitiveness in the sector of produced manufacturing equipment. It is shown in raised demands for quality, reliability and multitask functionality of technological machines, which are caused by appearance and study of new high-precision and high-production technologies [1].

Development of methods and means for formation precision control is a very important project and technological task for multi-axis treatment of details with curved surfaces [7,8]. Solving this issue may ensure competitive ability of the domestic machine-tools and increase effectiveness of multi-axis form making.

Low level volumetric accuracy of machine tool systems and the absence of multifaceted approach for ensuring the multi-axis form making during the stage of preparing and introduction of new constructive decisions for multi-axis machine tools are the reasons for the lack of effectiveness in multi-axis formation, which is certainly an engineering and manufacturing problem. [2].

The main characteristic of the machine tool system, which ensures most effective multi-axis treatment is the machine volumetric accuracy which is displayed in its working space. [6,10,14]

The aim of this research is the creation of the method of comparative assessment of multi-axis machine compositions volumetric accuracy (with not less than 5 interpolate axis) in order to find the best constructive decisions "for technological tasks" considering certain manufacturing conditions.

The development of methods and means of controlling volumetric accuracy is a very important engineering and process task.

2 Principal provisions of comparative assessment methods for volumetric accuracy of multi-axis machine tools

Volumetric accuracy of metal-working machine means integral characteristics showing its ability to provide certain form-making accuracy level as well as the technological system error law within the limits of working space [3].

The basis of the multi-axis machine volumetric accuracy modelling method is formed by the following approaches and assumptions:

1. Multi-axis machine working space is formed with the help of linear axis motion lengths with overlapping of sectors from rotary kinematic pairs rotation movements. Rotary kinematic pairs which do not change coordinate position of the instrument and work piece coordinate system are not considered in configurations.

2. Multi-axis machine tool system may be decomposed into simple elements – units with kinematic pairs ("joints") with its own constructive volume. Each kinematic pair consists of two joint elements – movable and immovable, construction of which depends on certain project synthesis principles, which are numerically formalized by the system of restrictions.

* Corresponding author: lukina_sv@mail.ru

DOI: 10.1051/matecconf/201712901046

3. Machine-tool unit models are the basis for the formation of volumetric accuracy in the working space. It is supposed that each machine tool unit has position deviations due to perturbance, and they are transferred to working space for further integration considering machine n-degrees of movement, distributed on kinematic branches of the instrument (IB) and work piece (WB). As per the typical scheme for unit fixation by means of guide surfaces, machine-tool movable units are considered as fixed-end beams on condition that the area from load point to fixing point (joint) is non-deformable [4].

4. Within the limits of the working space all the functional properties of the machine tools are represented, and are expressed by means of its working characteristic: geometric and kinematic accuracy, hardness, heat and vibration resistance, fast action etc, which are subject to calculation and analytical or experimental identification during the assessment of the technical level of the machine being produced.

5. Volumetric accuracy of the multi-axis machine is formed by the integration of the set of characteristics of the resulting error, on the whole set of points in the working space. Level of working space scanning detalization by means of multitude of points may be determined antecedently in correspondence with the aims and tasks of the certain research. Average machine performance parameters are formed on the basis of the value data in the central point of the working space.

3 Implementation assessment method for multi-axis machines and its graphic interpretation

Geometric image of the synthesized machine constructive arrangement consists of the combination of movable unit separate constructions, which are located in the global coordinates respective to working space, according to the earlier chosen scheme of movable unit degree distribution on machine branches.

In accordance with [13] it is necessary to carry out a synthesis of machine constructive arrangements. Each arrangement corresponds to the certain code which characterizes the distribution of the kinematic mobilities on machine branches (IB and WB) relative to stationary (O).

Each movable unit (rotary or translation) is characterized by the set of constructive parameters expressing its ability to ensure necessary coordinate movability and resilience to loads and perturbances. (fig. 1).

The last one is shown through the size of guidelines and contact characteristics of joint elements. Besides, the units has all the necessary movement parameters, this setting the required form and machine working space size. Joint failures or distortion which may occur in the joints of the movable units are the result of the whole set of perturbances (forced, heat, technological manufacturing and assembly defects or others).

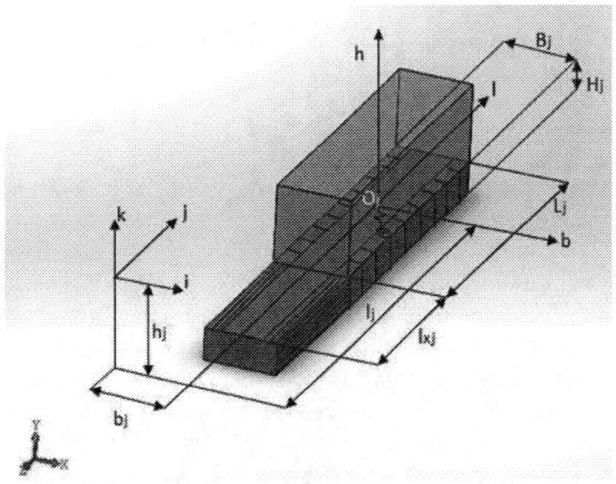

Fig. 1. Machine unit model with translation movement, implementing the joint "kinematic couple P"

Failures in the joints influence the location of machine end units (instrument and workpiece) from the nominal state during formation. In order to assess the degree of influence of each unit failures on to the accuracy of treatment, they should be integrated or differentiated along the working space, estimating the instability. This may be carried out by the procedure of separate joints discrepancy transference into the treatment points in the working space.

While setting initial and boundary conditions of the machine unit calculation scheme the following conditions and allowances are taken into account:

- unit form persistence condition as for absolutely solid body, which is considered during the calculation of joints contact flexibility considering unit form persistence, located from the joint to formation point;
- elasticity condition, as per which machine movable unit fixed on the guidelines has certain elasticity as per Hooke's law [9] and joint's non-disclosure during tipping moment from the cutting forces and the weight of movable unit;
- stability condition, according to which joint area inertia moments are opposed to tipping and twisting moments from the loads which move, push and perturb due to the cutting force and weight;
- cutting force is applied in the working space calculation point (action point), and weight is applied in the centre of unit mass;
- during the joint displacement, there within the working space may occur deviations from the location of the joints carrying instrument and work piece, which influence geometric, static and dynamic machine accuracy.

For the volumetric accuracy analytical assessment of the multi-axis machine Euler laws were used, which allow to form multiple resultant displacements within the working space, caused by cutting force and the weight of movable units. So, for example, the developed model of elastic displacement within the working space along the coordinate axis i, j, k (x-y-z) for the unit moving along the axis i of the machine in the coordinate system is as follows:

 DOI: 10.1051/matecconf/201712901046

$$\delta_{ii}=\frac{P_i\cdot r_{ij}^2-P_j\cdot l_i\cdot r_{ij}}{L_i^3\cdot R_{ik}}+\frac{P_i\cdot r_{ik}^2-P_k\cdot l_i\cdot r_{ik}}{L_i^3\cdot R_{ij}}\rightarrow min \quad (1)$$

where δ_{ii}, δ_{ij}, δ_{ik} - displacement of the calculated point p along the axis i, caused by rotation of the units moving along the axis i, j, k; P_i, P_j, P_k - cutting force components which influence the instrument and are directed along the axis i, j, k; L_i, B_k, R_{ij}, R_{ik}, R_{jk} и R_{kj} - sizes of guiding units moving along the axis i, j, k, and parallel to axis j и k; l_i и r_j, - escape (along the length, width and height of the unit) from the treatment point p to the unit elastic centre measured along the axis i , j , and k;

Displacement along the coordinate axis i (δij) and j (δik) for the units moving relative to coordinate axis j and k, are determined in the same way.

During the modelling there was used superposition principle, according to which resultant displacement in the working point of the working space is determined by means of transferring all the discrepancies from the whole set into this point. [9,12]. Displacement, characterising machine arrangement accuracy as a whole, was determined by the following expression:

$$\delta_\Sigma=\sqrt{\delta_{\Sigma x}{}^2+\delta_{\Sigma y}{}^2+\delta_{\Sigma z}{}^2} \quad (2)$$

where $\delta_{\Sigma x}=\delta_x(IB)-\delta_x(WB)$, $\delta_{\Sigma y}=\delta_y(IB)-\delta_y(WB)$, $\delta_{\Sigma z}=\delta_z(IB)-\delta_z(WB)$.

Here $\delta_x(IB)$, $\delta_y(IB)$, $\delta_z(IB)$ - displacements of the calculated point of the instrument branch along the axis x,y,z , $\delta_x(WB)$, $\delta_y(WB)$, $\delta_z(WB)$. - displacements of the calculated point of the workpiece branch along the axis x,y,z .

Calculated scanning of the working space discrete points allows to determine average displacement along the working zone, as well as their spreading.

For the correct comparison of the arrangement variants there must be assured certain conditions of comparability due to the statement of equal guideline sizes in the correspondent units, as well as equal passage length along the axis on all the arrangements. Also there was considered constructive condition of joint associativity which ensures stability of machine operation as a whole [11,12,15].

Developed method of the volumetric accuracy comparative assessment of the multi-axis machines includes the following stages:

1) Creation of the initial geometric 3D-image of the machine;

2) Geometric synthesis of alternative variants of arrangement;

3) Volumetric accuracy calculation for the variants of arrangement.

4) Comparative analysis of the results, and choosing the most appropriate machine arrangement as per the volumetric accuracy.

On the pictures (2) and (3) there is an example of visualization of 3D-geometric synthesis of the arrangement (fig. 2) and volumetric accuracy assessment (fig. 3).

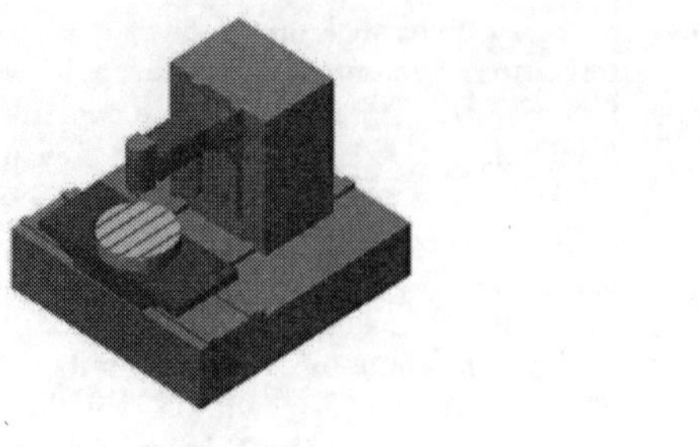

Fig.2. Arrangement 1 (RyTzTy0TxRz)

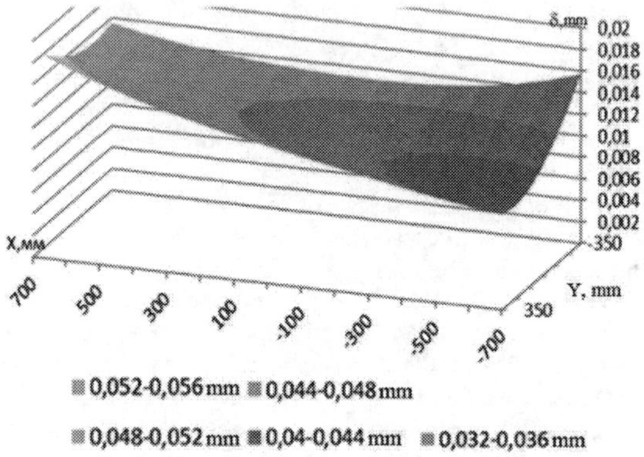

Fig. 3. Machine volumetric accuracy formation (arrangement 1)

Developed method was implemented in Visual Basic.NET and MS Excel. The priority of solving this applied research task was registered by Certificate of computer program registration № 2013661646 [5].

The results of comparative assessment allow to forecast achievable accuracy of multi-axis treatment and choose the best project decision for the machine as per the criterion of volumetric accuracy.

4 Conclusion

Formalized results allow to set the connections between construction layout factors and volumetric accuracy output parameters of the multi-axis machine. This may allow to control machine volumetric accuracy on the early project stages while choosing the composition and parametrization.

Visualization of the calculation method is represented by the synthesis of parametric 3D-geometric image of machine arrangement, and formation of the volumetric accuracy graphics in the working space discretized by the multitude of design points.

The analysis of received results allows to justify the best constructive decisions in correspondence with stated technological requirements and restrictions, and also to make recommendations on different use of machine arrangement for various manufacturing conditions.

A further development of the study is the adaptation of the proposed new methodology for technologically conditioned synthesis of multi-axis machines in high-tech engineering industries for the manufacture of constructively complex parts of the aviation and defense industries.

References

1. S.V. Lukina, I.V. Manaenkov, *Tehnologicheskij sintez mehatronnyh stanochnyh sistem dlja mnogokoordinatnoj obrabotki//* In the collection: XXVI Mezhdunarodnaja Innovacionno-orientirovannaja konferencija molodyh uchjonyh i studentov MIKMUS-2014. Works of a conference. Russian Academy of Sciences; RFFI; Office of power, mechanical engineering, mechanics and management processes; Institute of engineering science of A. A. Blagonravov of RAS. p. 428-431. (2015)
2. S.V. Lukina, S.N. Ivannikov, I.V. Manaenkov, *Povyshenie jeffektivnosti mnogokoordinatnoj obrabotki putem optimizacii komponovki mnogoosevoj stanochnoj sistemy*. News of the MAMI Moscow state technical university, **№ 2 (20)**, p. 34-37. (2014)
3. Ju.D. Vragov, *Analiz komponovok metallorezhushhih stankov/ Osnovy komponetiki* (M.: Mashinostroenie, 1978)
4. D. N. Reshetov, V. N. Portman, *Tochnost' metallorezhushhih stankov* (M.: Mashinostroenie, 1986)
5. S.V. Lukina, I.V. Manaenkov, Certificate on the state registration of the computer program of Rospatent № 2013661646 «*Sistema avtomatizirovannogo sinteza sredstv osnashhenija mnogoosevoj obrabotki*» (2013)
6. S.V. Lukina, *Prognosticheskoe modelirovanie proektnyh innovacionnyh reshenij po konfiguracii sredstv osnashhenija vysokotehnologichnyh proizvodstv*, Innovacii. **№ 8 (202)**, p. 68-71 (2015)
7. V.M. Makarov, S.V. Lukina, *Naukoemkij inzhiniring v zadachah tehperevooruzhenija*. RITM: Remont. Innovacii. Tehnologii. Modernizacija **№ 8 (8),** p. 16-20 (2013).
8. S.V. Lukina, *Metodika optimizacii proizvodstvennoj dejatel'nosti promyshlennogo predprijatija na osnove kompleksa prognosticheskih modelej formirovanija i vybora proektnyh innovacionnyh reshenij v oblasti vysokotehnologichnyh proizvodstv*. Vestnik MSTU Stankin, **№ 1 (32),** p. 125-129 (2015)
9. S.V. Lukina, S.N. Ivannikov, M.V.Krutjakova, I.V. Manaenkov, *Tehnologicheskij sintez mehatronnyh stanochnyh sistem dlja mnogoosevoj obrabotki*, Izvestija Moskovskogo gosudarstvennogo tehnicheskogo universiteta MAMI, **T. 2. № 1 (15)**, p. 46-51 (2013)
10. S.V. Lukina, *Razrabotka kompleksa prognosticheskih modelej formirovanija i vybora proektnyh innovacionnyh reshenij v oblasti vysokotehnologichnyh proizvodstv,* Aktual'nye problemy v mashinostroenii **№ 2,** p. 451-456 (2015)
11. S.V. Lukina, S.N. Ivannikov, I.V. Manaenkov, *Metodika formirovanija i vybora optimal'noj konfiguracii formoobrazujushhej sistemy mnogokoordinatnoj obrabotki*, News of the MAMI Moscow state technical university **T. 2, № 2,** p. 237-242 (2013)
12. V.M. Makarov, S.V. Lukina, *Upravlenie konfiguraciej izdelij dvojnogo naznachenija*. Glavnyj mehanik **№ 3,** p. 28-34 (2015)
13. V.M. Makarov, S.V Lukina, *Programmy razvitija - jekspertnyj podhod*. RITM: Remont. Innovacii. Tehnologii. Modernizacija **№ 9 (97),** p. 19-23 (2014)
14. V.M. Makarov, S.V. Lukina, P.A. Lebed' *Imitacionnoe modelirovanie v zadachah tehnologicheskogo inzhiniringa*. RITM: Remont. Innovacii. Tehnologii. Modernizacija **№ 2 (70),** p. 16-22 (2012)
15. S.V. Lukina, *Avtomatizacija procedur formirovanija i vybora strukturnyh komponovok sbornyh rezhushhih instrumentov na jetape tehnicheskoj podgotovki proizvodstva,* Vestnik Saratovskogo gosudarstvennogo tehnicheskogo universiteta, **T. 3, № 1 (57),** p. 241-247 (2011)

DOI: 10.1051/matecconf/201712901047

Phase-chronometric measuring systems for the provision of technological processes and diagnostics of the production of electromechanical engineering systems

Ekaterina Tumakova[1,*], and *Aleksandr* Komshin[1]

[1]Bauman MSTU, Machine building technology Department, 105005 Moscow, Russia

Abstract. In paper the main problems and objectives assessment of the current technical condition of the machine-building equipment are considered. Modern measuring systems used in engineering analysis. The paper considers a phase-chronometric information technology-metrological support for the evaluation of the technical condition of the synchronous electromechanical systems on the example of turbine CHP. Analysis of the main problems in the diagnosis of electromechanical systems is given. Phase-chronometric method as a basis for building a new system of diagnosis of electromechanical systems reviewed. The paper describes the main elements of technology, assessment of the economic effects of its introduction in the industry.

The quality of repair of the main and auxiliary equipment at the majority of thermal power plants (TPPs) and combined heat and power plants (CHP plants) has significantly deteriorated over the past few years. Qualification of specialists in the field of diagnostics of these objects continues to fall every year. At the same time, it is difficult to estimate the damages caused by the failure of the turbine due to an accident. In addition to the economic losses that arise in the enterprise itself, which generates electricity, significant losses are borne by enterprises and organizations that have not received this energy. In addition, it is necessary to take into account the repair costs for the restoration of equipment that has suffered an accident, depending on its scale.

An accident or failure of the main or auxiliary equipment of a TPP or CHP plant can lead to significant losses, and in some cases, catastrophic consequences in a number of ways:

• Underdevelopment of planned energy by a particular power plant can lead to losses associated with higher energy costs due to the connection of energy sources with lower efficiency.

• The exclusion of a part of the power plant's equipment leads to a rise in the cost of electricity production due to a decrease in the equipment utilization rate.

• Repair costs for restoring equipment that has been damaged (depends on the scale of the accident).

• Damage to enterprises that have not received energy, as a result of an accident or a lack of the necessary capacity reserve for power generating stations.

• Damage to the population caused by a sudden stoppage of public electric transport, interruptions in mass events, disconnection of household appliances, etc.

Serious consequences for long breaks in the heat supply of houses and businesses. [1]

At present, a number of diagnostic methods are known for technical diagnostics of turbine units, the purposes of which are: monitoring of technical condition, finding a location and determining the causes of failures (failures) and forecasting technical condition. The most widely used diagnostic methods based on vibration (vibration spectrum analysis, high-frequency vibration - "shock pulse method", vibration estimation, vibration velocity, vibration acceleration, etc.). [2]

To effectively solve the problems of creating and operating turbine units of a CHP, it is necessary to use methods and diagnostic tools that have a single principle of presenting metrological and information support for the investigated object at all stages of the life cycle. This possibility is provided by the phase-chronometric method, developed at the Bauman MSTU. [3, 4]

As an important feature of the phase-chronometric method, it should be noted that it belongs to the phase measurement methods. [5] When studying the devices of cyclic action, the operation of the phase-chronometric system is based on recording the moment of passage of a certain value of the phase of the cycle. Information on the technical state of the object is contained in variations in the duration of time intervals corresponding to the passage of the intervals (quanta) of the phase.

The use of the phase-chronometric method opens up new possibilities. Phase-chronometric diagnostic systems have a higher metrological level: the phase study provides fundamentally new information, and chronometry makes it possible to improve accuracy.

Accuracy, speed and completeness of the obtained characteristics of the turbine unit in combination with mathematical models provide both control of the

* Corresponding author: tumakova_ekaterina91@mail.ru

development of emergency situations and control of the residual resource.

In order to create an effective diagnostic system, it is necessary to perform computational mathematical modeling along with measuring the controlled parameters in order to identify the object under study with its operating modes. Including simulating critical defects and emergency operating modes, this cannot be realized experimentally. [6]

In this work, a mathematical model was developed that should ensure the connection between theory and practice. The input and output parameters of the mathematical model have the same format as the measurement data. The parameters studied during the experiment are physical-mechanical and structural-technological parameters of the object.

The model performs a diagnostic role, namely, defects or deviations from the norm are laid in it, and the response to them of the system is analyzed. As a result, defect detection algorithms are formed in real experimental data obtained by the time interval measurement system. [7]

To construct a mathematical model of a functioning electromechanical system, the maximum possible approximation of the equations to be modeled in a model to a specific object is necessary. The model includes equations (blocks of programs) describing the individual elements of the investigated object (excitation, control system, etc.). For the elements of the object that require special attention (the place of the most frequent breakdowns), the model moves from their general description to the particular one. [8]

1 Measurement and control of torsional oscillations of shafting

Opportunities are opened to address current day-to-day tasks of operating heat and power engineering, such as:

- prevention of damage to rotors of steam turbines;
- diagnostics of crack formation in rotors of turbines and generators under operating conditions;
- measurement and control of torsional oscillations of shaft lines for the purpose of recording accumulated damage;
- creation and introduction of stationary and portable monitoring systems for torsional oscillations of shafts of turbine units, etc.

An example of the chronogram of the turbine unit turbine rotation and the natural frequency spectrum of the torsional oscillations of the turbo unit are shown in Figures 1 – 4. [9]

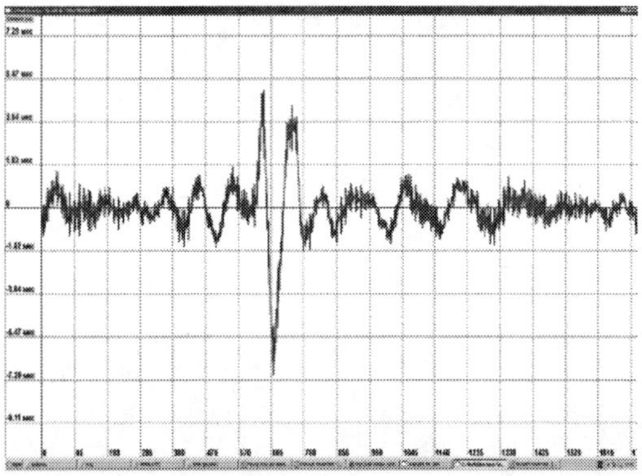

Fig. 1. Chronogram of rotation of turbo unit TVB-200-2-K-200-130.

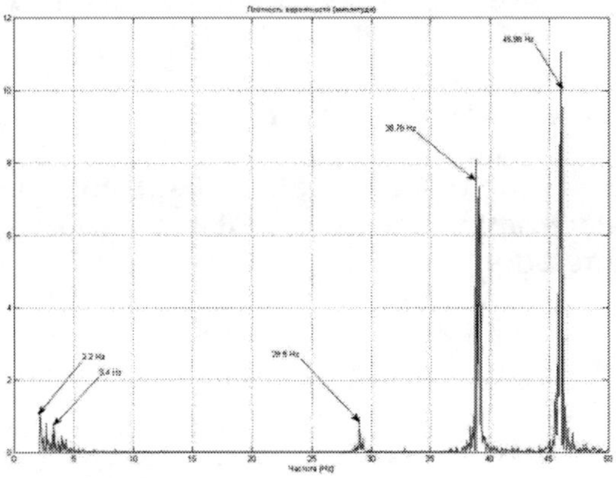

Fig. 2. Spectrum of natural frequencies of torsional vibrations of turbine TIA-220-2-K-200-130.

Here the following main frequencies are distinguished: 2,2 Hz, 3,4 Hz, 28,9 Hz 38,78 Hz and 45,98 Hz.

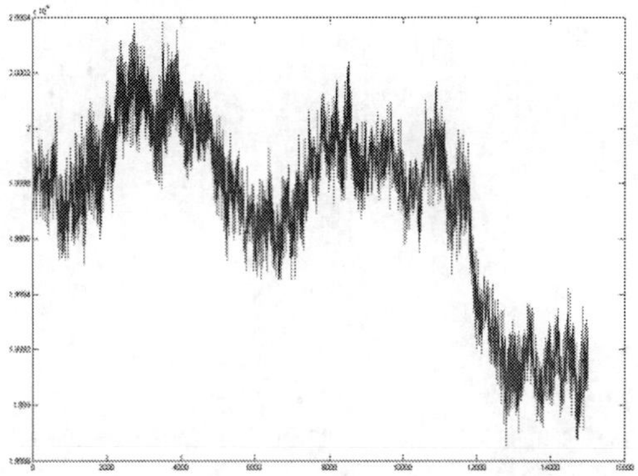

Fig. 3. Chronogram of rotation of turbo unit TVB-320-2UZ-T-250 / 300-240-2.

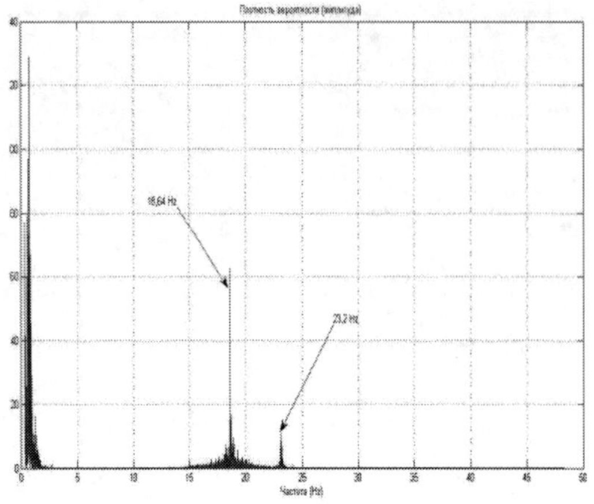

Fig. 4. Spectrum of natural frequencies of torsional vibrations of turbine TVB-320-2UZ-T-250 / 300-240-2.

Here the following main frequencies are distinguished: 18,64 Hz и 23,2 Hz.

2 Diagnostics of elements of electromechanical systems during operation

In the process of operation one of the key challenges is the diagnosis of technical condition of functioning of the object.

Consider the example of the detection and localization of a malfunction of the turbogenerator. In Fig. 5 shows the chronogram of rotation of the defective turbo generator TVV-320-2У3-T-250/300-240-2 where in the process (without stop machine) discovered the sparking brushes of the brush unit.

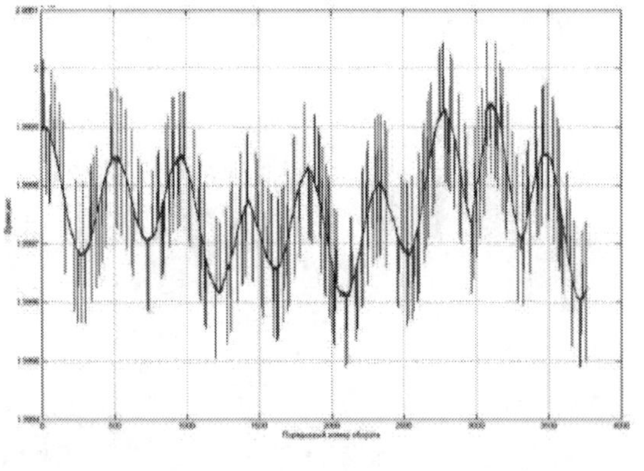

Fig. 5. The chronogram of rotation of the turbo generator TVV-320-2У3-T-250/300-240-2 reflecting the diagnosed defect – sparking brushes of the brush unit.

This defect is manifested as follows: observed periodic bursts of high amplitude on the chronograms of rotation of the turbine. The reason for the fault can be [10]:

- unbalance of brushes or their incorrect installation;
- weak contact of brushes to the collector;
- poor contact in the pick-up rings;
- irregular arrangement of the main and additional poles.

Thus, in the process of operation is determined by the defect is localized in the design of the entire turbine unit, is determined by the cause of the malfunction, i.e. solved a direct problem diagnostic: monitoring of the technical state of the object, evaluation of technical condition and forecasting of a residual resource of elements or turbine unit as a whole. [9]

3 Economic efficiency of the introduction of phase-chronometric systems in heat power engineering

To estimate the magnitude of the economic effect and the appropriateness of introducing the phase-chronometric method in power engineering, calculations were made based on the following initial data for thermal power plants:

- The cost of installing a new TA (T-250-300) at the station is 1500 million rubles. This number is obtained on the basis of averaging of open tender information, since there is no standard cost, and the specific cost is specified depending on the features of the station and the operating conditions of the repair organization.
- The service life of the turbine unit can be extended to 50 years, there are data on the appointment of an individual resource to turbines for such a period by the decision of technical experts, although there have been cases of turbines and longer periods.
- The project lifetime is 50 years.
- For the normal level of project costs (100%), the value obtained during operation of the turbine set by the standard is accepted, where the service life is 25 years, and the repair interval is 4 years. During the project, 2 turbo units are required - at the beginning of operation and after 25 years.
- When extending the service life of the turbine unit, a period of 50 years is assigned - 1 turbine unit is required throughout the project.
- Carrying out of major repairs (MR) it is estimated in 32 million roubles.
- Average repair (AR) is estimated at 7 million roubles.
- Carrying out of the current repair (CR) is estimated at 4 million roubles.

The economic effect of the introduction of information-metrological phase-chronometric technology is calculated. In the end, if the effect of extending the resource will be reached by half, that is, in the amount of 15 million roubles. For one unit, the result of the project in the projected prices (discounted at a rate of 11%) will be as follows:

- net reduced income - 970 million roubles.
- Payback period - 4 years.

Economic calculations were carried out jointly with the department of IBM "Industrial logistics" MSTU.

The result shows, that the project has a good reserve on size of received economy at prolongation of a resource of a turbo generator. [9]

DOI: 10.1051/matecconf/201712901047

Diagnosis of the technical state of electromechanical systems is relevant for various areas of modern industry, such as nuclear power, oil and gas transmission, heat supply, water supply, etc. [8]

The development of electromechanical systems and their diagnostics is an integral part of the modern world. Recently, the use of low-power electric cars has increased significantly - in automation and computer equipment. The development and creation of new types of generators is becoming increasingly important every year. So, since 2012 in the United States, work is under way to create a new type of deep-water electric generator that can transform into electricity the energy of stable deep currents - a source of clean energy. [11]

This work was supported as part of the research 9.4968.2017/BP for the implementation of the initiative scientific projects basic part of state task.

References

1. A.D. Trukhniy, B.V. Lomakin, *Thermal steam turbines and turbine units: a textbook for high schools* (MEU, 2002)
2. O.S. Vakhromeev, R.T. Karimov, A.I. Nadeev, *Modern methods of diagnostics of electromechanical systems* (2005)
3. Zroychikov [and others], Heat Power Engineer. **6** (2005)
4. M.I. Kiselev [and others], Heat Power Engineer. **11** (2006)
5. A.B. Syritskii, Meas. Techn. **6** (2016)
6. K.S. Ermakov, E.V. Tumakova, Sciencee.& Educ. **10** (2015)
7. D.D. Boldasov, K.G. Potapov, A.B. Syritskii, Russ. Eng. Research **8** (2016)
8. E.V. Tumakova, Meas. Techn. **6** (2016)
9. E.V. Tumakova, A.S. Komshin, IIKYSS (2014)
10. A.S. Komshin, O.V. Medvedeva, Meas. Techn. **5** (2014)
11. E.V. Tumakova, A.S. Komshin, The Fut. Eng. Of Russ. (2015)

MATEC Web of Conferences 129, 01048 (2017) DOI: 10.1051/matecconf/201712901048
ICMTMTE 2017

The use of measuring phase-chronometric systems in the production of cyclic aggregates of aircraft

Ekaterina Metelkina[1,*] and *Aleksandr* Komshin[1]

[1]Bauman MSTU, Machine building technology Department, 105005 Moscow, Russia

Abstract. The paper gives an assessment of the current technical condition of the machine-building industry as a whole in the aviation industry. Its weak sides are revealed. Based on the obtained data and the evaluation of existing methods for diagnosing cyclic mechanisms, an approach of the phase-chronometric method based on precise measurements of time intervals is proposed. A functional diagram of the information-measuring system is presented. The results of modeling the real and ideal operating cycle of the gear-boxes are analyzed.

1 Introduction

Wear of productive assets in Russia reaches critical values, but their renovation is rather slow. At the same time, the requirements to ensure their reliability only increase. In the park of operating machines and mechanisms, objects close to the exhaustion of their resources, hence potentially hazardous and requiring replacement, are preserved. Indicators reflecting the prospects for the use of a particular machine building object are determined by the level of working off of a design, manufacturing quality and information-metrological support. [1]

Helicopter construction in the 21st century is gaining momentum. The need for this type of air transport is obvious, due to its characteristics: maneuvering in urban conditions, hovering over the terrain, transportation of people and goods to hard-to-reach places, etc. Helicopters are necessary in medicine, the army, transport companies, the Ministry for Emergency Situations, the Ministry of Internal Affairs. In the general structure of the fleet of aircraft used to carry out air operations, helicopters occupy 54%. Since this type of technology is science-intensive, to maintain in a state of readiness, operating organizations require significant funds and a competently constructed operating system.

Since the first launch of the MI-1 helicopter in 1952, there have been cardinal changes in the field of Russian helicopter construction. Despite the fact that work is now actively being carried out to create perspective rotorcraft, research and development work are carried out, a large proportion of all helicopter engineering activities are still aimed at modernizing existing models and types of equipment.

Moral and physical wear and tear is inherent in the Russian fleet of vessels, besides flight characteristics and consumer properties do not correspond to modern market requirements. All this significantly affects to the safety of flights. The issue of establishing the resources and terms of service of civil aviation equipment and the transition from operation over the established resource to operation according to the current technical condition is also acute.[2] In the world practice these problems are solved by equipping the helicopters with modern hardware and new technical means. In addition to it fault diagnosis systems are being developed.

The presence of a defect and its development to a certain extent do not immediately lead to catastrophic consequences. This indicates the need to detect faults at a time as close to their generation as possible in order to take measures to repair or replace the faulty unit to prevent a possible accident. To solve this problem, special diagnostic systems are being developed that allow early detection of faults in units and assemblies. Their presence on the aircraft is a necessary prerequisite for the maintenance of "by state".

The sources of the initial data for the diagnostic systems are the signals carrying information about the technical state of the units and aggregates. These signals can have a different physical nature, it is important only that they directly or indirectly reflect the state of the object.

The fault diagnosis systems that are currently in use differ in the principles of operation. The most universal is vibration diagnostics. The essence of this diagnostic method consists in the periodic removal of vibration information, the determination of the values of diagnostic parameters, the comparison of current values with thresholds, and decision-making when exceeding them.

Progressive wear of the teeth of the gear wheels of the gearbox leads to a significant increase in the value of the overall vibration level. The appearance of fatigue damage on individual teeth (chipping, cleavage, crack at the base, failure) causes an increase in the values of the

* Corresponding author: katya2110m@yandex.ru

DOI: 10.1051/matecconf/201712901048

total level of the harmonic components of the speed of the driving and driven gears.

Using this method, a large number of parameters can be diagnosed without resorting to dismantling the controlled object itself.

However, the vibration parameters are sensitive to the operating conditions of the equipment, which leads to the impossibility of comparing the results of measurements obtained at large time intervals.

The processes of friction and wear occurring in the units of gearboxes are accompanied by the loss of the metal of the conjugate parts in the form of wear products, which are carried away from the friction zone by the lubricating oil washing these parts. Hence a fundamentally different from the previous approach to diagnostics is appeared - diagnostics of the content of wear products in lubricating oil.

Varieties of this method for controlling the content of wear products in lubricating oil are: the method of spectral analysis of oil; Method of mass spectrometry; Ferrographic method; Radiometric method; X-ray spectral analysis of oil, etc.

In addition to these, in practice, there is used an approach to assess of the technical condition of the gearboxes, based on measuring the temperature of the oil, the increase of which indicates an increased load or an insufficient level of oil in the gearbox.

From the standpoint of the operation of engineering facilities, there is an importance of the frequency of diagnostic features in conditions of continuous operation. In such conditions, it is necessary to emphasize the significant technical advantage of the approach, which is based on precision phase-chronometric measurements forming a time series reflecting the influence of the work process on the measured physical quantity-the time interval corresponding to different measurement intervals of the working cycle.[3] Variations of time intervals characterize the features of changing the parameters of the working cycle, the mathematical processing of which allows you to evaluate the effect of design changes on output parameters.

2 The application of the phase-chronometric method to the diagnostics of gearboxes

This method can be used to ensure the operation of the cyclic mechanisms of aviation equipment, namely, helicopter gearboxes. The diagnostic system based on the phase-chronometric method is based on measuring the kinematic parameters of the operating cycle of the mechanism with phase quantization and time sampling and is able to provide a relative measurement error of not more than $5 \cdot 10^{-4}$% under operating conditions.[4]

The functional diagram of the phase-chronometric system is shown in Figure 1.

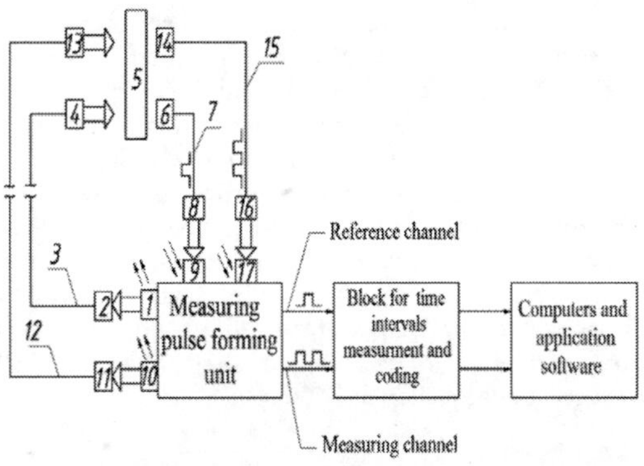

Fig. 1. Functional diagram of the phase-chronometric system.

The following designations are adopted for the reference and measuring channels, respectively: 1, 10 - Radiating diode; 2, 11 - Input collimator; 3, 12 - Optical cable; 4, 13 - Output collimator; 5 - Measuring disc; 6, 14 - Input collimator; 7, 15 - Optical cable; 8, 16 - Output collimator; 9, 17 – Photodiode.

The phase-chronometric system is an information-measuring complex. It includes an angular position sensor, a system for collecting information and determining time intervals, a personal computer, applied and special software. The system for collecting information and determining the time intervals is a board that integrally combines a highly stable pulse generator, a scale converter, one or more counters, a measuring pulse driver, and a chronocomparator.[5]

All these elements can be implemented in a microcontroller, the board of which contains an ADC.

The principle of operation of the photovoltaic system consists in the sequential recording of the duration of whole and multiple periods of output shaft rotation. The luminous flux is created by the primary photoelectric converter and the disk with grooves. The resulting light flux is transferred to the photoconverters of the electronic unit, where a measuring pulse is formed. Then the signal on the communication lines goes to the specialized computer board, where the pulses of the master quartz generator are counted. Due to it, precise measurements of the time intervals between pairs of consecutive pulses are possible. By comparing the obtained time characteristics and the results of mathematical modeling, one can judge the technical state of the diagnostic object.

3 Analysis of data obtained by using the phase-chronometric method

When applying the phase-chronometric method, the results of measurements are arrays of experimental data from successive time series that reflect the effect of the operation of the mechanism on time intervals. Mathematical processing of experimental data, for example, spectral analysis, allows for a general and element-by-element analysis of the operation of the mechanism.

DOI: 10.1051/matecconf/201712901048

To make phase-chronometric diagnostics, it is necessary, first of all, to analyze and implement the work cycle of the product. After this, in detail consider the dynamics and kinematics of the interaction of details and parts of the product to form a mathematical model.

A consistent technique for processing the already acquired measurement information consists of several stages, the first of which is the acquisition and analysis of primary information in the form of rotation chronograms (mathematical modeling). Figure 2 shows the chronogram of the output shaft rotation of the gearbox at a frequency of 7600 rpm.

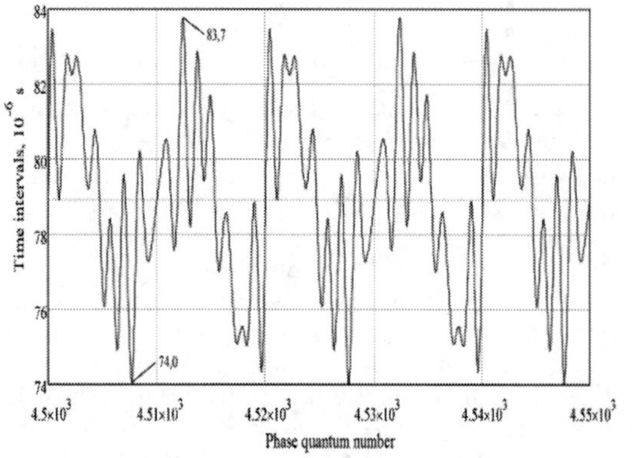

Fig. 2. Chronogram of rotation of the output shaft at 7600 rpm.

The spectral analysis (Fig. 3) showed that the natural frequencies of the mechanism correspond to 50.635 Hz; 163.384 Hz; 382.159 Hz; 658.614 Hz; 988.827 Hz; 1004.601 Hz.

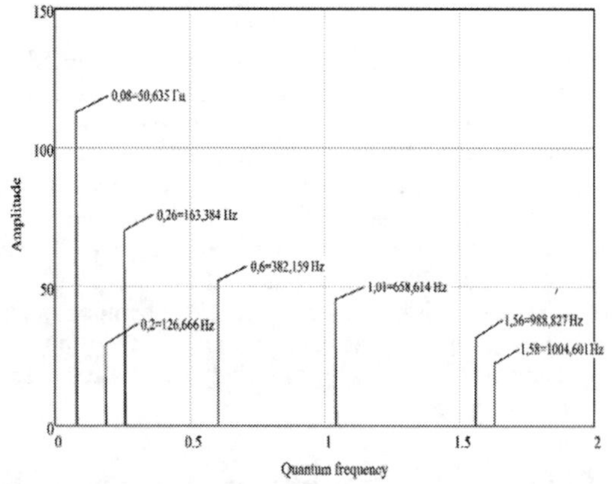

Fig. 3. Spectral analysis of the signal at 7600 rpm.

In the real process of operation and operation, there are effects associated with vibrations in the mechanism, with regime parameters, with uneven engine operation, the effect of assembly errors, temperature deformations, and the degradation of the properties of structural materials and the imbalance of the system resulting from the operation of the mechanism. All these components affect the real chronogram of the output shaft rotation. The real chronogram is shown in Figure 4.

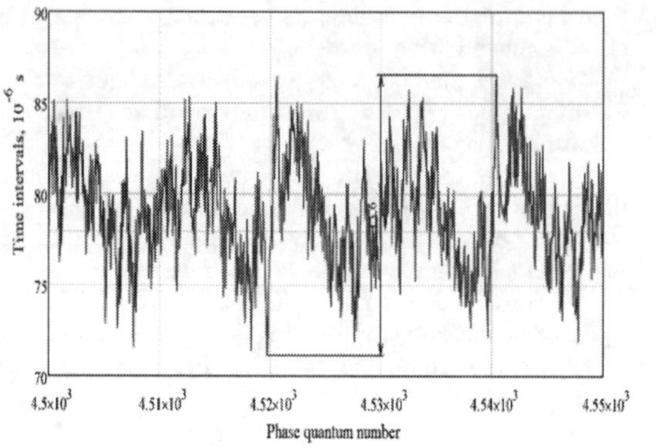

Fig. 4. Real chronogram of rotation of the output shaft at a frequency of 7600 rpm.

The spectrum of the real signal is shown in Figure 5.

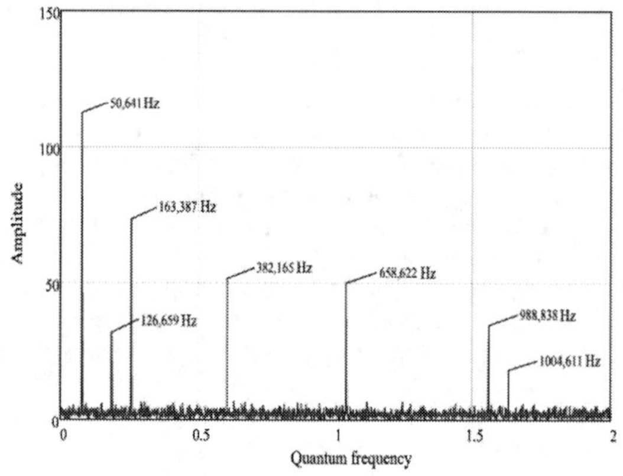

Fig. 5. Spectrum of real signal.

Comparison of the values of the natural frequencies of the mechanism, obtained as a result of mathematical and experimental modeling, shows that during operation, there is the change of own frequencies of the mechanism in comparison with the initial state. By these changes, one can judge the state of the mechanism itself. A comparison of these values is presented in Table 1.

Table 1. Comparison of the values of the natural frequencies of the mechanism.

The value of the natural frequencies of the mechanism, Hz		Relative error of natural frequencies, $\Delta\varpi$
The result of mathematical modeling	The result of processing the experimental data	
36,682	36,696	$3{,}8.10^{-2}$%
91,603	91,613	$1{,}1.10^{-2}$%
218,430	218,445	$6{,}8.10^{-3}$%

Table 1. Continued.

The value of the natural frequencies of the mechanism, Hz		Relative error of natural frequencies, $\Delta\varpi$
The result of mathematical modeling	The result of processing the experimental data	
476,671	476,684	2,7.10^{-3}%
608,511	608,519	1,3.10^{-3}%
715,530	-	-

Experimental studies on operating technical objects have shown that precise chronometric control of the working cycle phase ensures the detection of qualitative repetitive characteristics and individual quantitative parameters of the object under study.[6] All this testifies to the advisability of using measuring phase-chrochronometric systems in the conditions of production of cyclic aggregates of aviation equipment, which will make it possible to obtain measurement information on a fundamentally new level and in the future will allow to go from operation to a specified resource for operation in accordance with technical condition.

References

1. M.I. Kiselev, V.I. Pronyakin, Y.V. Chivilev, Mir Izm. **4** (2007)
2. E.D. Metelkina, IIKYSS (2016)
3. A.S. Komshin, Stand.& Qual. **12** (2015)
4. M.I. Kiselev, Stand.& Qual. **10** (2013)
5. K.G. Potapov, *Research and development of a method and means for assessing the current technical condition of the main drives of turning equipment based on the phase-chronometric approach* (BMSTU, 2015)
6. M.I. Kiselev [and others], High Techn. **9** (2016)

Phasechronometric measuring technologies application for turning tool wear monitoring

Anthony Syritsky[1,*]

[1]Bauman MSTU, Machine building technology Department, 105005 Moscow, Russia

Abstract. The article briefly describes the results of creating a phasechronometric system for monitoring the state of a turning tool. The realization of the method in practice is shown. The composition of the original measuring unit is described. The results of theoretical and experimental studies are presented. The working capacity of the new method in industrial conditions is proved. The proposed method makes it possible to monitor the wear of the lathe with high accuracy during machining.

In the key of manufacturing informatization it becomes an important issue to obtain information about all the key options of work metalworking machines. Lack of information and let one but very important parameter during processing will mean reducing the effect of gaining momentum convergence of information technology and metalworking [1,2]. One of such important parameters are the information about the technical condition of the machine components and the actual state of the least reliable processing element of the system - the cutting tool. It is an assessment tool wear directly in the treatment process with a view to the timely replacement and diagnosis of rolling lathes to evade the system of preventive maintenance.

In this study as a tool to obtain more information about the parameters of the work lathes considered phasechranometric method [3], which is based on the continuous measurement of the time intervals between the phases of the machine cycle or machine (Fig .1).

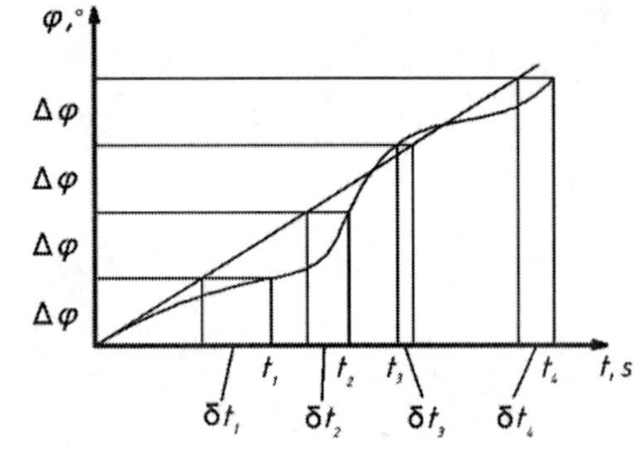

Fig. 1. Phasechronometric measurements principle.

In this method, time intervals $\delta t_1 ... \delta t_i$, the corresponding rotation (displacement) of the vehicle chassis undergoing a cycle are measured, in this case the machine spindle at a certain angle Δφ (cycle phase). Experience about phasechronometric on other machines shows that the measurement of time intervals of the machine spindle revolution in the processing phase, we can obtain information about the operation of the machine. Information on the operation of all machine components, including the state of the tool is included in the chronogram of rotation, which is a graphic reflection of the values of the time series $\delta t_1 ... \delta t_i$ (chronogram formation circuit is shown in Figure 2). Real chronograms are shown in Figure 3-4. Figure 4 shows the chronogram for turning a new cutter. Figure 5 shows the chronogram for turning with a tool with wear VB = 400μm

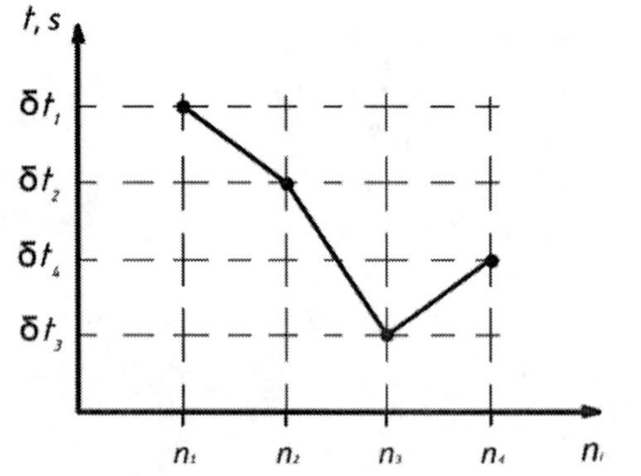

Fig. 2. Chronogram formation.

* Corresponding author: syritsky@yandex.ru

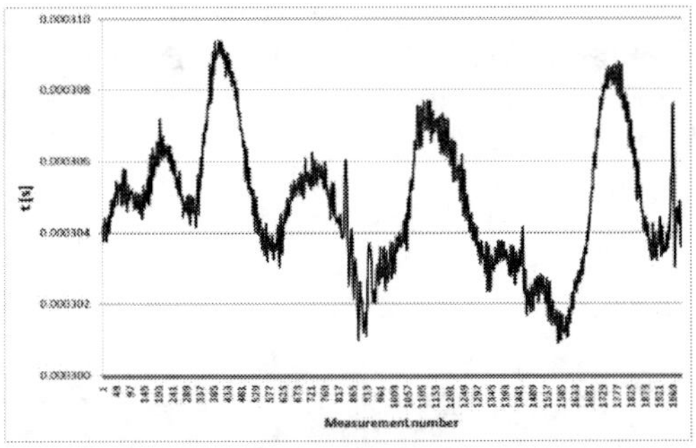

Fig. 3. Chronogram for turning a new cutter.

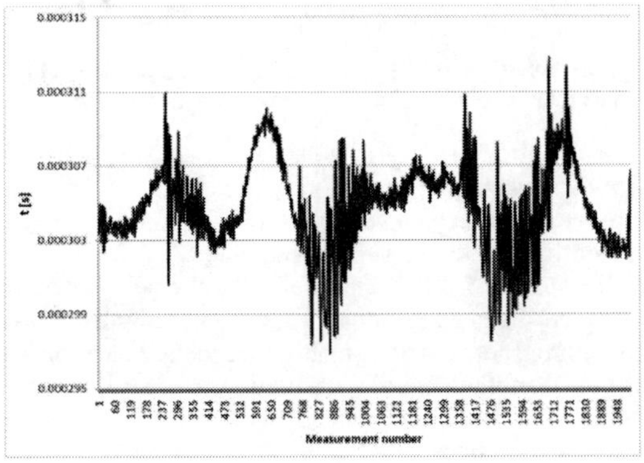

Fig. 4. Chronogram for turning with a tool with wear VB=400 μm

This approach is radically different from those that are used at present in mechanical amplitude techniques (vibration diagnostics and so on.) firstly by accuracy of the obtaining measurement information. Due to the fact that at the moment the frequency and time is possible to measure with the highest accuracy, phasechronometric systems are characterized by a relative error of measurement at the level of 10-2 - 10-4%. Measuring channels used to date diagnostic systems do not reach this level of precision.

With regard to the lathes phasechronometric system consists of three main elements (Figure 1): a primary converter, programmable logic integrated circuit module (FPGAs) and embedded computer.

On the spindle is set as transducer or incremental encoder, or information element (disc with grooves) in conjunction with inductive or optical sensor [10,11]. An example of installing a sensor on a lathe is shown in Figures 5, 6.

Fig. 5. General view of the angle sensor mounting

Fig. 6. Installation of the angle sensor on the lathe machine

The converter generates a sinusoidal analog signal U (t), which is then transmitted to the FPGA for processing where the signal is converted to digital form for transmission to the computer. FPGA module in turn consist of several blocks (Fig .7).

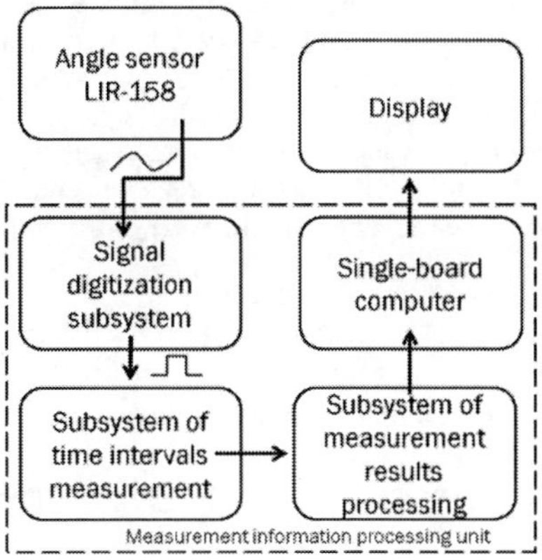

Fig. 7. Measurement unit scheme.

The output of the comparator goes to eliminating bounce circuit which registers clearly establish the level of high-frequency vibrations (bounce) during the transition from the state of the log. 0 in the log. 1 and vice versa. The basic unit of measurement system is formed by the timer and reset circuitry, and represents a real-time clock for measuring time intervals. Because of the different work time FPGA and embedded computer (FPGA works in real-time, embedded computer - no) required a buffer, which is used to synchronize threads. Through a protocol driver from the measured time intervals are generated measuring data packets which are transmitted to the embedded computer for further processing, visualization, and software compatibility with the corresponding industrial equipment. Embedded Systems is a single-board computer with Linux Lubuntu 13.04 operating system. Software platform consists of two parts: a driver for receiving and transmitting data from the bus and the user application to signal processing and imaging. Next, the data as a function of the node status information fed to the machine operator system.

Special interest in matters of the development phasechronometric system-turning tool of state of the cutting edge causes the problem of determining the blunt tool criteria. To build an effective monitoring system requires the determination of a parameter that varies over

DOI: 10.1051/matecconf/201712901049

time with wear. That option change of the relative proportion of high-frequency oscillations in the spectrum A (f) rotation chronogram (Figure 8) was selected.

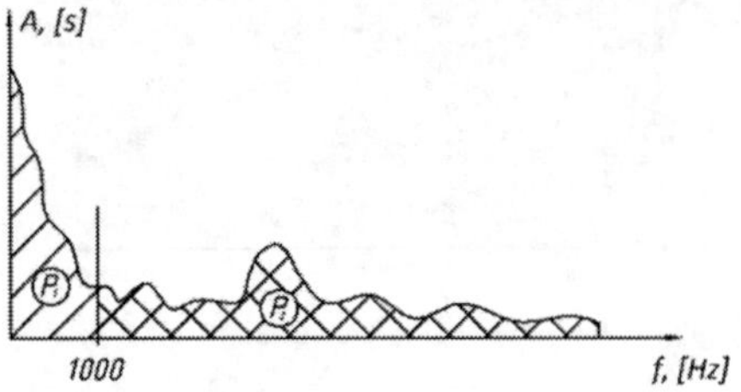

Fig. 8. The relative proportion of high-frequency oscillations in the rotation chronogram spectrum.

It is known that fluctuations of up to 1000 Hz, for this type of machines includes information about uneven work machine motor, toothing defects, lubrication. It is known that fluctuations connected with the wear of the turning tool is high-frequency. Therefore, dividing the power spectrum of the field of high-frequency oscillations (region P_2) to a total capacity of the oscillation spectrum (area P_1 + region P_2), you can get a percentage of Ψ that increase is correlated with the wear of the cutting edge:

$$\Psi = \frac{P_2}{P_1 + P_2} \cdot 100\%$$

where P_1 - total capacity of the oscillation spectrum, W;

P_2 - spectrum power of the field of high-frequency vibrations, W.

Experimental research has focused on the confirmation of the above hypothesis. Their implementation consisted in processing a cylindrical workpiece surface diameter of 70 mm made of steel 45 in the machine 16K20 and recording chronometric data during cutting. Depth of cut t = 1 mm, feed S = 0,2 mm, spindle speed of the machine n = 1000 rev / min. Tool final state was evaluated by measuring the VB parameter on a precision video-measuring machine Mitutoyo Quick Vision Apex Pro. Flank wear was 500 micrometers. The results are presented in figure 9.

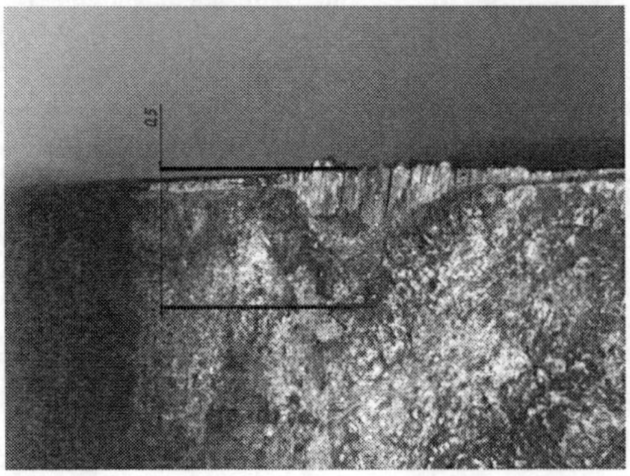

Fig. 9. Flank wear, mm.

Consistently through equal treatment intervals of time were constructed assess changes in the relative proportion of high-frequency oscillations in the spectrum chronogram Ψ rotation. The results are shown in Figure 10:

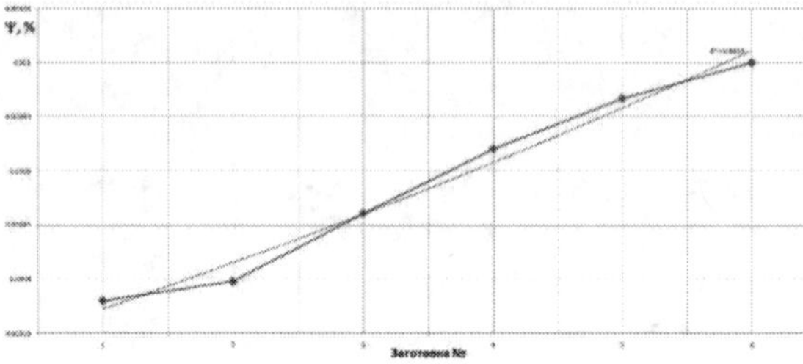

Fig. 10. Assesment of changes in the relative proportion of high-frequency vibrations in the range of rotation chronogram.

As seen in Figure 10, a straight trend line increases with increasing tool wear, so at the moment this method of processing the received measurement information can be identified as the most perspective.

Because of the analysis of the experimental dependence shown in Figure 10, it was concluded that as the cutting tool wear power high-frequency components of the machine spindle oscillation increases relative power of low-frequency components. This can be taken as a basis for choosing a blunt cutting tool test.

The results will form the basis of the general concept phasechronometric system that can be built into the design of machine tools. The functions of the system are the diagnostics of technical condition of machine components and cutting tool wear monitoring

References

1. J. Lee, B. Bagheri, H. Kao, E. Lapira, Manuf. Leadership Journ., **02,** (2015).
2. D.T. Matt, E. Rauch, Industriemanagement **01,** (2015)
3. M. Kiselev and V. Pronyakin, Meas. Techn. **9,** (2001)
4. A.B. Syritskii, Meas. Techn. **6,** (2016)
5. A.S. Komshin, Meas. Techn. **8,** (2013)
6. M.I. Kiselev [and others], Meas. Techn. **9,** (2011)
7. K. Potapov, A. Syritskii, Pribory. **5,** (2014)
8. D. Boldasov, N. Lazarev, A. Syritskii, Pribory. **10,** (2015)

DOI: 10.1051/matecconf/201712901050

The additive technology for obtaining thermal storages based on nanomodified materials

Alexander Shchegolkov[1,*], *Pavel* Borovskikh[2], and *Larisa* Shirshova[3]

[1]Tambov State Technical University, 392000, Tambov, Russia
[2] Martin-Luther-Universität Halle-Wittenberg, Germany
[3]Peoples Friendship University of Russia, 117198, Moscow, Russia

Abstract. The approaches for the implementation of an additive technology for obtaining heat accumulators are considered. The implementation of the provided technology can be realized on the standard 3D printers, which are aimed to obtain plastic materials. However, the software of the printers has to include adjusting analytical expressions, which take into account thermophysical properties of the heat-retaining materials. The analytical expressions have been derived by solving a mathematical model. The mathematical model contains the main data on thermophysical and mechanical and physical properties of the nanomodified material. These properties of the nanomodified material are defined during the experimental studies.

1 Introduction

An increase in efficiency of renewable energies is linked to the application of new innovative technologies and materials. Heat-retaining materials can be also used for an improvement of the element base of photovoltaics. It corresponds to the heat-retaining materials, which are capable for storing thermal energy of a phase transition [1-3]. In this case - for photovoltaics devices - the method of application is associated with an introduction of new functional capabilities (accumulation of thermal energy) and the stabilization of temperature condition. This type of the application allows obtaining both electric and thermal energy by use of the same device. In this case both economic and energy efficiencies increase. However, materials - capable to accumulate heat of a phase transition - lose their shape that is they melt and transit from a solid state into a liquid one. Thus, it hinders their application and leads to a need to use the auxiliary bodies and heat-exchanging elements. There is number of studies, in which potential improvements of such materials with a phase transition are researched, as for example an improvement of paraffin by introducing carbon nanotubes and grapheme into it [4-11]. The primary objective, which is set by authors of this paper, is related to the improvement of thermophysical properties, especially heat conduction. Other properties, which help to stabilize the shape, are not studied. Therein the relevant research goal is to study the issues of the nanomodification of paraffin and composite materials based on it in order to be able to apply them into solar power systems. Moreover, the aim is to obtain finished products, which have stable mechanical and physical properties in a wide temperature range and are suitable for using them in the additive (3D) technologies [2,3]. It is necessary to obtain a material with stable properties and to develop a technique for mathematical modeling in order to implement the additive technologies with a use of the nanomodified materials. The analytical functions will be derived from the mathematical model in order to use them in software for 3D printers.

2 Additive technology

Standard 3D printers, which are used for producing plastic products, can be utilized to implement the additive technology for obtaining heat storages based on nanomodified materials. There are number of techniques of the application of 3D technology for a creation of sophisticated products, which have channels for liquid [4,5]. The Fused Deposition Modeling (FDM) is employed to produce the thermal storage. A printing file of the STL format is created after the calculations - related to the modes of the thermal storage – have been performed. This allows segmenting the virtual model into layers. An expendable material in this printing technology is a composite polymer thread, which heats up to a melting temperature when it enters the extruder and is laid in doses through a nozzle on the working surface of 3D printer. Thus, a thermal storage - the form-lines and geometry of which precisely correspond to the virtual prototype - is being created layer by layer.

However, it is necessary to consider higher thermal capacity of heat-retaining materials and their mechanical and physical properties while creating heat-retaining modules. These properties are considered in specialized software. Analytical expressions - which take into account properties of heat-retaining material - are required for this purpose. These analytical expressions can be derived using approaches related to a solution of partial differential equations [14]. In the framework of the mathematical modeling it is necessary to know

* Corresponding author: Energynano@yandex.ru

mechanical and physical properties of the nanomodified materials - planned to be applied – as well as distinctive features of these materials under conditions of various temperature regimes.

3.1 Synthesis of samples of the nanomodified polymeric composites

The samples - containing P-2 paraffin (Russia) and polyethylene (low pressure (PND) PE80B-275) (Russia) – were prepared for the research. The multi-layered cylindrical carbon nanotubes of the" Taunit M" type – produced by JSC Nanotechcentre, Tambov, Russia –were employed as a modifying additive agent (Fig. 1) [12,13].

Fig. 1. SEM-image of the carbon nanotubes of the "Taunit" type

3.2 The technique of a modification of a composite based on paraffin and polyethylene by carbon nanotubes

The modification was carried out in 2 stages:
1) introducing carbon nanotubes – "M" into paraffin;
2) introducing a mixture of paraffin and carbon nanotubes into polyethylene by means of extrusion.

At the first stage the carbon nanotubes were placed into the silicone tank of the volume of 0,002 m3 and afterwards paraffin "P2" heated to 80°C was added. This mixture was processed by ultrasound with a frequency of 22,5 kHz within 1 hour while stirring it by means of a mechanical stirrer (200 rpm) and keeping temperature of 70°C. As a result, black nontransparent material has been produced. Afterwards, the material was granulated by the press-granulator. The size of granules was in the range of 1 to 8 mm.

The obtained granules were mixed (with mass ratio of 1:1) with polyethylene granules. Furthermore, these granules were placed into the hopper of the extruder. The extruder operates with a driving motor of 1,5 kW and a heater of 2 kW. The produced material has been granulated after the extrusion.

3.3 Methods for studying samples of the nanomodified polymeric composites

The studies of thermophysical properties were conducted by means of "IT - λ – 400" device (measurement of heat conductivity, fig. 2 and 3) and "IT - S– 400" device (measurement of thermal capacity) within a monotonous warming mode with an average speed of 0,2°C/s under adiabatic conditions.

The study of the density of granules was conducted by immersing them in distilled water and afterwards calculating the density with a consideration of displacement volume. The determination of bulk density was conducted by placing granules into the transparent vessel, shaking it within 2-3 minutes and then measuring the occupied volume.

The study of a temperature dependence of the volume was conducted in a measuring cylinder with volume of 0,001 m3. This measuring cylinder was put into a transparent cylinder of bigger volume (0,01 m3).

The free space between them was filled with the distilled water. The temperature of water was varied in increments of 2 °C.

The study of durability was conducted by using a hydraulic press. The study of durability was conducted by using a hydraulic press. The granules were placed in a space between metal sheets.

4 Results of the analysis of the nanomodified composite

Figure 2 shows the results of a study on temperature dependence of thermal capacity of three materials: a composite (the nanomodified paraffin and polyethylene); paraffin modified by carbon nanotubes-"M"; and pure paraffin.

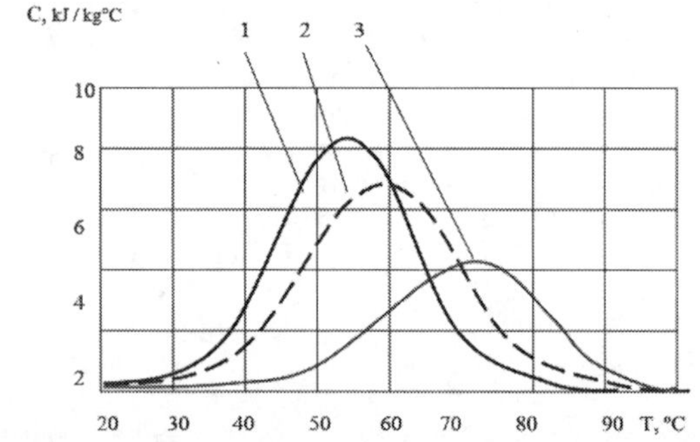

Fig. 2. Temperature dependence of thermal capacity of the nanomodified composite for 3D technology: 1 – paraffin; 2 nanomodified paraffin; 2 – paraffin + polyethylene (carbon nanotubes-"M").

Figure 3 shows the results of a study on temperature dependence of heat conductivity of composites modified by carbon nanotubes as well as pure paraffin.

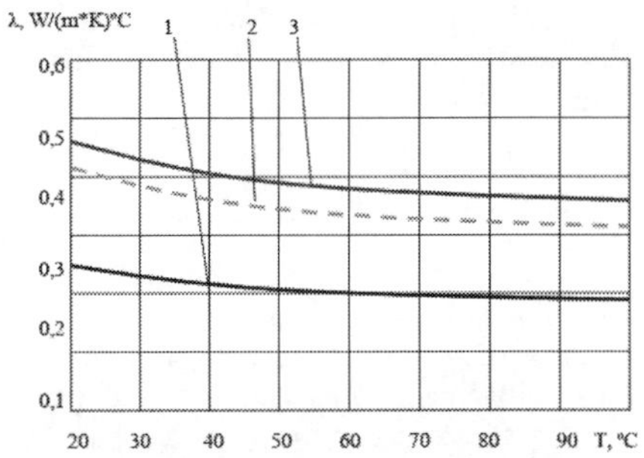

Fig. 3. Temperature dependence of heat conductivity of the nanomodified composite for 3D technology: 1 – paraffin; 2 - paraffin + carbon nanotubes-"M", 2 – paraffin + polyethylene (carbon nanotubes-"M").

In the figure 4 the temperature dependence of volume changes of composite materials is presented.

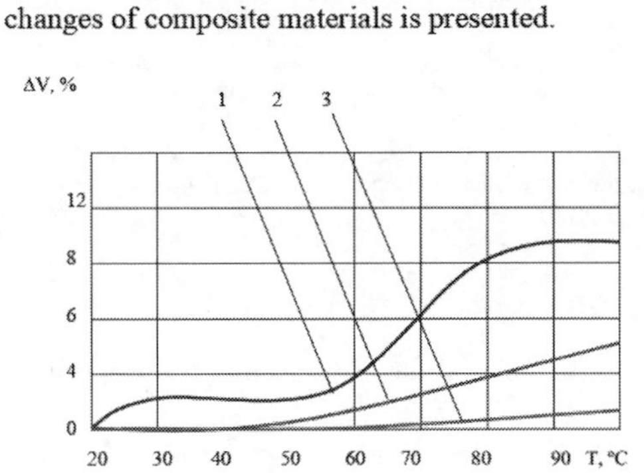

Fig. 4. Temperature dependence of volume changes of the nanomodified composite for 3D technology: 1 – paraffin; 2 - nanomodified paraffin (carbon nanotubes-"M"); 3 – paraffin + polyethylene (carbon nanotubes-"M").

Mechanical and physical properties of the composite paraffin + polyethylene (carbon nanotubes-"M") for 3D technology are presented in the table 1

Table 1. Mechanical and physical properties.

№	Mechanical and physical properties	paraffin + polyethylene (carbon nanotubes-"M").
1	bulk density, kg/m^3	7
2	real density, kg/m^3	1,2
3	Durability at compression of granules, MPa	0,15

5 The technique for deriving analytical functions for 3D technology

Within a framework of the additive technology it is necessary to consider the following situation: there is an overlaying of the material on a photo cell. From the perspective of physical model adaptation the photo cell is considered as an unbounded two-layer sheet with a functional source.

Problem of heat conductivity for an unbounded two-layer sheet with a functional source:

$$\frac{\partial t_i(x_i,\tau)}{\partial \tau}=a_i\frac{\partial^2 t_i(x_i,\tau)}{\partial x_i^2}+\frac{q_i(x_i,\tau)}{c_i\rho_i};\quad i=1,2;\quad 0\le x_i\le R_i \tag{1}$$

$$t_i(x_i,0)=f_i(x_i); \tag{2}$$

$$\lambda_1\frac{\partial t_1(0,\tau)}{\partial x_1}-\alpha_1\left(t_1(0,\tau)-t_{c1}\right)=0; \tag{3}$$

$$\lambda_2\frac{\partial t_2(R_2,\tau)}{\partial x_2}+\alpha_2\left(t_2(R_2,\tau)-t_{c2}\right)=0; \tag{4}$$

$$\lambda_1\frac{\partial t_1(R_1,\tau)}{\partial x_1}=\lambda_2\frac{\partial t_2(0,\tau)}{\partial x_2}; \tag{5}$$

$$t_1(R_1,\tau)=t_2(0,\tau). \tag{6}$$

Here, x – coordinate of thickness of a sheet, m;

τ – time, s;

$t_i(x_i,\tau)$ – temperature field of the *i*-th sheet layer, °C;

R_i – thickness of the *i*-th sheet layer, m;

a_i – thermal diffusivity of a material of the *i*-th sheet layer, m^2/s;

$f_i(x_i)$ – initial temperature profile of the *i*-th sheet layer, °C;

λ_i, c_i – respectively heat conductivity, W/(m°C), and thermal capacity, J/(kg°C), of a material of the *i*-th sheet layer;

α_i – coefficient of convection heat transfer on a surface of the *i*-th sheet layer, W/(m^2°C);

t_i – environment temperature from the perspective of the *i*-th sheet layer, °C.

The solution of the equations (1) – (6) is derived by the method of final integral transformations of the "x" coordinate [14].

The solution is the following:

$$t_i(x_i,\tau)=P_i(x_i)+\sum_{n=1}^{\infty}\frac{T_n(\tau)}{N_n}S_n(x_i); \tag{7}$$

$$\bar{t}_i(\tau)=Q_i(x_i)+\sum_{n=1}^{\infty}\frac{T_n(\tau)}{N_n}Z_{i,n}; \tag{8}$$

$$P_1(x_1)=\left(\frac{x_1}{\lambda_1}+\frac{1}{\alpha_1}\right)W+t_{c1}; \tag{9}$$

$$P_2(x_2)=\left(\frac{x_2}{\lambda_2}+\frac{R_1}{\lambda_1}+\frac{1}{\alpha_1}\right)W+t_{c2}; \tag{10}$$

$$Q_1(x_1)=\left(\frac{R_1}{2\lambda_1}+\frac{1}{\alpha_1}\right)W+t_{c1}; \tag{11}$$

$$Q_2(x_2)=\left(\frac{R_2}{2\lambda_2}+\frac{R_1}{\lambda_1}+\frac{1}{\alpha_1}\right)W+t_{c2}; \tag{12}$$

$$W=\frac{t_{c2}-t_{c1}}{\frac{1}{\alpha_2}+\frac{R_2}{\lambda_2}+\frac{R_1}{\lambda_1}+\frac{1}{\alpha_1}}; \tag{13}$$

$$Z_{i,n}=\frac{\sqrt{a_i}}{\mu_n R_i}\left(\cos(\varphi_{i,n})-\cos\left(\frac{\mu_n R}{\sqrt{a_i}}+\varphi_{i,n}\right)\right); \quad (14)$$

$$T_n(\tau)=\exp(-\mu_n^2\tau)\left(D_n+\int_0^{\tau}G(\mu_n,\tau)\exp(\mu_n^2\tau)d\tau\right); \quad (15)$$

$$D_n=\sum_{i=1}^{2}\frac{\lambda_i}{a_i}C_i\int_0^{R_i}\left(f_i(x_i)-P_i(x_i)\right)\sin\left(\frac{\mu_n}{\sqrt{a_i}}x_i+\varphi_{i,n}\right)dx_i; \quad (16)$$

$$G(\mu_n,\tau)=\sum_{i=1}^{2}\frac{\lambda_i}{a_i}\int_0^{R_i}\frac{q_i(x_i,\tau)}{c_i\rho_i}\sin\left(\frac{\mu_n}{\sqrt{a_i}}x_i+\varphi_{i,n}\right)dx_i. \quad (17)$$

μ_n - the *n-th* consecutive positive root of the equation

$$\frac{\lambda_1\sqrt{a_2}}{\lambda_2\sqrt{a_1}}\cos\left(\frac{\mu}{\sqrt{a_1}}R_1+\varphi_1\right)\sin(\varphi_2)+\sin\left(\frac{\mu}{\sqrt{a_1}}R_1+\varphi_1\right)\cos(\varphi_2)=0; \quad (18)$$

$$\varphi_1=arctg\left(\frac{\lambda_1\mu}{\alpha_1\sqrt{a_1}}\right); \quad (19)$$

$$\varphi_2=-R_2\frac{\mu}{\sqrt{a_2}}-arctg\left(\frac{\lambda_2\mu}{\alpha_2\sqrt{a_2}}\right); \quad (20)$$

$$S_n(x_i)=C_{i,n}\sin\left(\frac{\mu_n}{\sqrt{a_i}}x_i+\varphi_{i,n}\right); \quad (21)$$

$$C_{1,n}=1;\ C_{2,n}=\frac{\sin\left(\frac{\mu_n}{\sqrt{a_{12}}}R_1+\varphi_{1,n}\right)}{\sin(\varphi_{2,n})}; \quad (22)$$

$$N_n=\sum_{i=1}^{2}\frac{\lambda_i}{a_i}C_i^2\left(\frac{\mu_n}{\sqrt{a_i}}R_i-\frac{1}{2}\left(\sin\left(2\left(\frac{\mu_n}{\sqrt{a_i}}R_i+\varphi_{i,n}\right)\right)-\sin(2\varphi_{i,n})\right)\right); \quad (23)$$

6 Discussion of research results of the nanomodified composite characteristics for the additive technology

Thus, the temperature dependence of heat capacity of the obtained material is studied (the maximum value is 5 kJ/kg°C) as well as a comparison of it with heat capacity of pure paraffin (8,2 kJ / kg°C) is carried out. The phase transition is followed by temperature changes of the volume. It is found out that synthesized composite has a smaller tendency of changing volume with an increase in temperature (2%), than pure (12%) and nanomodified paraffin (5%).

The study shows that it is possible to create materials - which are able to accumulate thermal energy – through the introduction of a component, which has high heat capacity, as for example nanomodified paraffin. Along with that, a use of nanomodified paraffin promotes a distribution of carbon nanotubes – "M" in a composite material such as polyethylene.

The mathematical modeling technique allows receiving analytical expressions which then can be applied in software of 3D printers.

7 Conclusions

1. The thermophysical and mechanical and physical properties of the obtained material are studied. It is found out that an extreme value of the temperature dependence of heat capacity changes its position on a graph.
2. The produced composite material can be employed for creating thermal storages due to its ability for phase transitioning with warmth of 5 kJ / kg°C and strength properties of 0,15 MPa.
3. The mathematical modeling technique is designed. This technique allows using analytical dependences for software of equipment, which realizes the additive principle of the creation of 3-dimensional models.

This paper was financially supported by the Ministry of Education and Science of the Russian Federation on the Program to improve the competitiveness of RUDN University among the world's leading research and education centers in the 2016-2020.

References

1. Barnatt CH. Nottingham, createspace independent publishing platform, **276** (2013)
2. Evans B. New york, apress, **306** (2012)
3. Lushin, E.N. Petersburg state polytechnical university journal. Physics and mathematics, **2**, 4 (2013)
4. N. Bhattacharjee, for a. Urrios, s. Kang, a. Folch, lab chip, **16** (2016)
5. A.K. AU, W. Huynh, L.F. Horowitz, A. Folch, Chem. Int. Ed. Engl, **55** (2016)
5. Coleman J.N., Khan U., Blau W.J., Gun'ko Y.K. Carbon. **44**, 9 (2006)
6. Zeng J.L., Cao Z., Yang D.W., Sun L.X., Zhang L., **101**, 1 (2009)
7. Fan, L., Khodadadi, J.M. A review. Renewable and sustainable energy reviews, **15** (2011)
8. Warzoha, R. J., Zhang, D., Feng, G., Fleischer, A. S. CARBON, **61** (2013)
9. Warzoha, R. J., Fleischer, A. S. Applied Energy, **154**, (2015)
10. Li, M., Guo, Q., Nutt, S. Solar energy, **146**, (2017)
11. Liu, X., Rao, Z. Thermochimica acta, **647**, (2017)
12. T.P. Dyachkova, Yu.A. Khan, N.V. Orlova, S.V. Kondrashov, Transactions TSTU. **22,** 232 (2016)
13. A. A. Aladinskiy, A. V. Rukhov, E. N. Tugolukov, T. P. Dyachkova, Transactions TSTU. **20**, 572 (2014)
14. Tugolukov E. N. the Monograph - Moscow. Mashinostroyeniye publishing house,. **100** (2004)

DOI: 10.1051/matecconf/201712901051

Technological supporting performance of machines and mechanisms

Oleg Burlachenko[1], *Yuri* Serdobintsev[1], and *Alexander* Lyashenko[1,*]

[1] Volgograd State Technical University, 400005 Volgograd, Russian Federation

Abstract. The problem of the machines and mechanisms working capacity technological support is considered in this article. The existing methodologies for ensuring operability were analyzed here. The authors revealed and justified the existing methodologies necessity improvement by application the mechanism updating phase into the lifecycle model. The model of the machines lifecycle, as well as its mathematical and technical-hardware support, was developed on the basis of the carried out research.

Nowadays ensuring the maximum possible machines operation is the primary task in the creation of any mechanical system, regardless of its complexity and role. The solution of this problem is to identify factors that have a significant influence on the studied mechanism operational properties. Only by managing each of these factors, it is possible to ensure the operability of any technical system.

Taking into consideration the fact that about 80% of machines functional losses are connected with the wear of their parts [1], it is necessary to create the conditions that allow the uninterrupted operation of each part to ensure the efficiency of the entire mechanism, it is especially important to pay attention to the most loaded tribocouples.

According to the collected data from various literature sources [2-5], one of the actively developing areas of increasing wear resistance are all kinds of influences on the machines and mechanisms particularly loaded parts surface layer. They distinguish a sufficiently large number of influence ways, such as surfacing, magnetic treatment, mechanical hardening, thermal method, electrophysical, combined, etc. In addition, the work is carried out in the field of lubrication and other promising methods of increasing wear resistance. The application of each of them has a certain influence on tribotechnical properties and, consequently, on the final operability of the entire mechanism. Therefore, it is required to develop the general methodology for selecting the most effective methods and to introduce the most appropriate one depending on the particular mechanism, the influences on it, and the environment where it is situated in.

A probabilistic approach is one of the options for solving the existing problem. It is used to predict the machine working state in a certain time period. This method is realized by constructing a lifecycle model of the most frequently failing conjugation in the form of a transient graph [6] from one phase (situation) to another. This allows to represent any tribocouple, as a stochastic system that changes its state during the lifecycle. Based on Solomentsev Yu.M., Serdobintsev Yu.P., and others [7-15] surveys, it is possible to assume that any mechanism can be an adaptive system, that is, adapting to a change in load, or to changed external influences. According to this methodology, we proposed the method of the most effective technological solution. Optimization in such a system in accordance with the obtained variables is performed by the optimization criterion. Precisely due to the accuracy of the choice of this criterion, the system will have the highest efficiency. In this connection, we proposed the criterion for choosing the most effective technological solution aimed at ensuring the mechanism operability:

$$K_{ij} = \min\left|\sum_{k=1}^{m} Z_{ijk} \times H_{ijk} \times P_{ijk}\right| \quad (1)$$

where, Kij is the resulted criterion estimating efficiency of j-th level i-th technology used to ensure the operability;
k is the number of the phase (situation) arising with the probability Pijk, thus causing the costs Zijk and the wear Hijk;
The value Zijk is the ratio of the cost of the machine or mechanism implementing to the resulting technical effect.

The wear is determined by two ways, depending on the geometric dimensions of the part, as well as its purpose and operating conditions. The first way is the so-called method of artificial bases.

According to this method, the holes with a known beforehand geometrical characteristics are cut out on the part's friction surface with a special tool, and their variation is monitored depending on wear. The thickness of the worn layer can be calculated according to the formula [12]:

$$H_{ijk} = h_1 - h_2 = \frac{1}{8}(l_1^2 - l_2^2)\frac{1}{r} \quad (2)$$

* Corresponding author: lyashenko2626@mail.ru

DOI: 10.1051/matecconf/201712901051

This method is used for parts and conjugations where it is possible to apply bases, without impairing the operational properties of the observed mechanism, for example on different shafts, pistons, etc. In other cases, the use of the labeled atoms method is possible. In this case, a radioactive substance is introduced into the part surface layer, and the wear is determined by measuring sensors, identifying the amount of radiation, detecting radioactive particles in any environment and measuring their quantity, calculating the wear of a particular part.

The mechanism optimization as an adaptive system according to this criterion, in comparison with existing ones, describes more precisely its lifecycle, takes into account the parts actual wear influence, as well as the costs and probability of failure in a given situation, on the technologies effectiveness used to maintain the working capacity.

To determine the tribocouples containment probability in a particular phase (situation) for a given resource, a matrix of transitions from one level to another was compiled. Using the lifecycle model in the form of a transition graph (Fig.1.) and the received matrix, we get:

- the containment probability of a part or a conjunction in the operation phase in the absence of failures:

$$P_{ij3,n} = \frac{\sum_{0}^{n} p'_{3,n}(3,n-1) + p'_{3,n}(5)}{1+\sum_{k=0}^{T-1} p'_k(k-1)} \quad (3)$$

where, T = 7 is the total number of phases (situations);

-the design phase:

$$P_{ij1} = \frac{p'_1(2) + p'_1(6) + p'_1(7)}{1+\sum_{k=0}^{T-1} p'_k(k-1)} \quad (4)$$

- the phase of production and promotion to the consumer:

$$P_{ij2} = \frac{p'_2(1)}{1+\sum_{k=0}^{T-1} p'_k(k-1)} \quad (5)$$

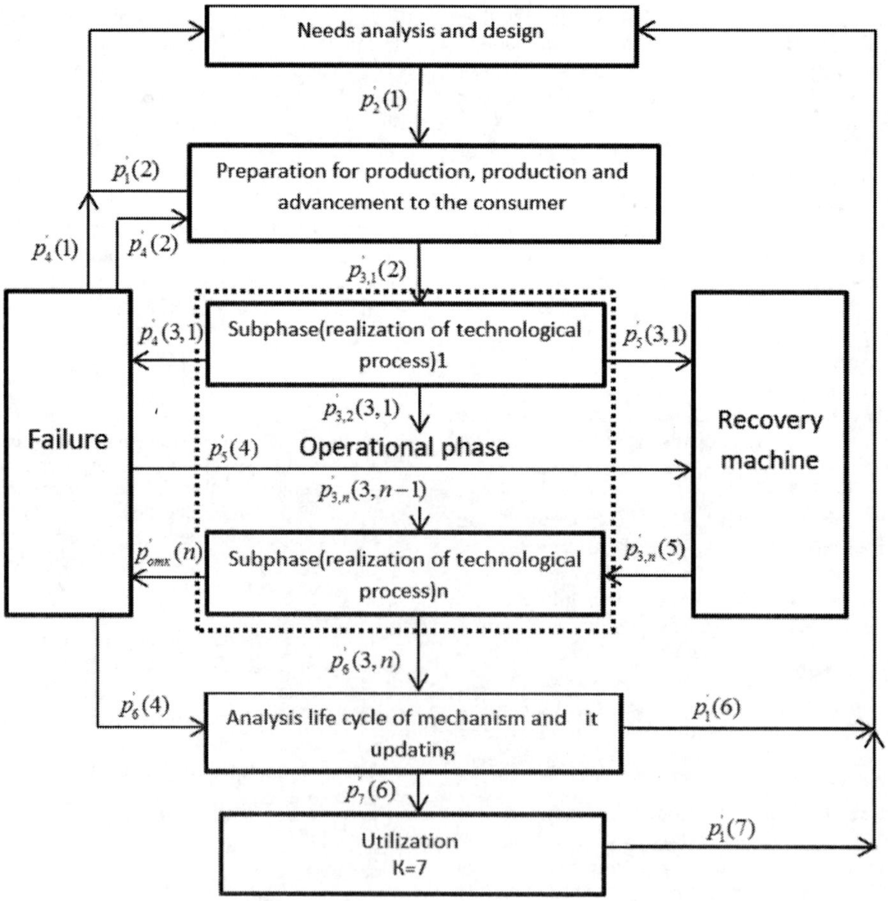

Fig.1. The model of the machine lifecycle due to the improvement phase in the form of a transient graph

-the failure phase:

$$P_{ij4} = \frac{p_4^{'}(3,n) + p_4^{'}(1) + p_4^{'}(2)}{1 + \sum_{k=0}^{T-1} p_k^{'}(k-1)} \quad (6)$$

-the recovery phase:

$$P_{ij5} = \frac{p_5^{'}(3,n) + p_5^{'}(4)}{1 + \sum_{k=0}^{T-1} p_k^{'}(k-1)} \quad (7)$$

-the improvement phase:

$$P_{ij6} = \frac{p_6^{'}(3,n) + p_6^{'}(4)}{1 + \sum_{k=0}^{T-1} p_k^{'}(k-1)} \quad (8)$$

-the phase of utilization:

$$P_{ij7} = \frac{p_7^{'}(6)}{1 + \sum_{k=0}^{T-1} p_k^{'}(k-1)} \quad (9)$$

When varying all the levels of controlled factors, various constraints, depending on the lifecycle phase, were taken into consideration, for example, by operating modes in the operation phase, by mechanical, physical, and technical processing, in the phases of improvement, recovery and production, etc.

As today many mechanical systems are created, each of them has various disadvantages. The example is the responsible tribocouples insufficient wear-resistant surface layer, in the case of its failure further full-scale operation of the entire mechanism is impossible. During this existing models machine failure, it was either restored by any of the known methods, or recycled as a mechanism with an inappropriate repair, and then the lifecycle repeated. The studied models [7-15] did not take into account such factor as the improvement of already created systems, whether parts' laser processing to increase the loaded parts surface layer wear resistance, or magnetic and other kinds of processing for the same purpose - all these methods are widely used in the machines and mechanisms production. And at the moment, when the bulk of scientific works is carried out to develop the already created mechanisms, we cannot neglect such stage in the machine lifecycle as the improvement phase. On its basis, we proposed the model where this factor is the next stage in the lifecycle after the mechanism failure. This allows to consider the failure not as a transition from the operational state to recycling, but rather to provide a failure as the way to improve any mechanical system, and subsequently ensure its higher performance, higher production indices, and in general to increase any machine operational activity, by all possible influences on failing out conjunctions.

In the proposed methodology, the concept of multi-loop control of the machines and mechanisms operational properties is considered as the basis, it describes each conjunction of the mechanism throughout the lifecycle. At the same time, the developed methodology allows to realize the functions of the adaptive optimal operability assurance search-free system. In other words, with the help of the described probabilistic methods, already in the initial phases of the mechanism creation, the proposed methodology makes it possible to choose the optimal solution to ensure operability by changing the tribotechnical properties in its parts surface layer. However, under conditions of an infinite number of influences, changing loads, temperature regimes, etc., throughout the operational activity, it is simply impossible to take into account all the influencing factors and their combinations in designing. Therefore, when a machine or mechanism passes through all the lifecycle phases, to solve the set out problem of the subsequent technology application to increase the wear resistance of the failed tribocouples surfaces, we need the information on all the influences that this mechanism experiences in a particular environment. Collected with the help of pre-installed, automated sensors, information will allow to correct the previously adopted technological influences and to implement adaptive control. We propose to combine the applicable sensors in a single diagnostic system with automatic data collection. This approach to monitor the mechanism parts actual state will reduce the number of failures, and subsequently, save the time that is usually spent on the machine recovery after the failure.

In the developed methodology, the structure of information support for choosing one of the alternative technological influences and their parameters for failing tribocouples corresponds to the algorithm of stage-by-stage access to resources of various levels in the process of ensuring the overall operability of the entire mechanism. (Fig.2.).

This block diagram [14] is suitable for optimized accepting of technological solutions in various mechanisms adaptive control.

To ensure selection of the most effective application of technological influences on the analyzed mechanism, all data, received from the sensors and other tracking devices, was saved in database. It has a hierarchical structure, and is based on characteristic equation [15] describing all the signs and characteristics of controlled mechanical system. It includes the values of contact interaction parameters, the calculated values, the values obtained from automatic sensors, in this case all sorts of situations under their serial numbers and valid values for required performance properties to ensure efficiency in accordance with degree of each component responsibility and its functionality entered in the database.

The implementation of the system for the selection effective technological solutions, and, consequently, to ensure operation of mechanical systems is carried out by our proposed scheme (Fig.3.).

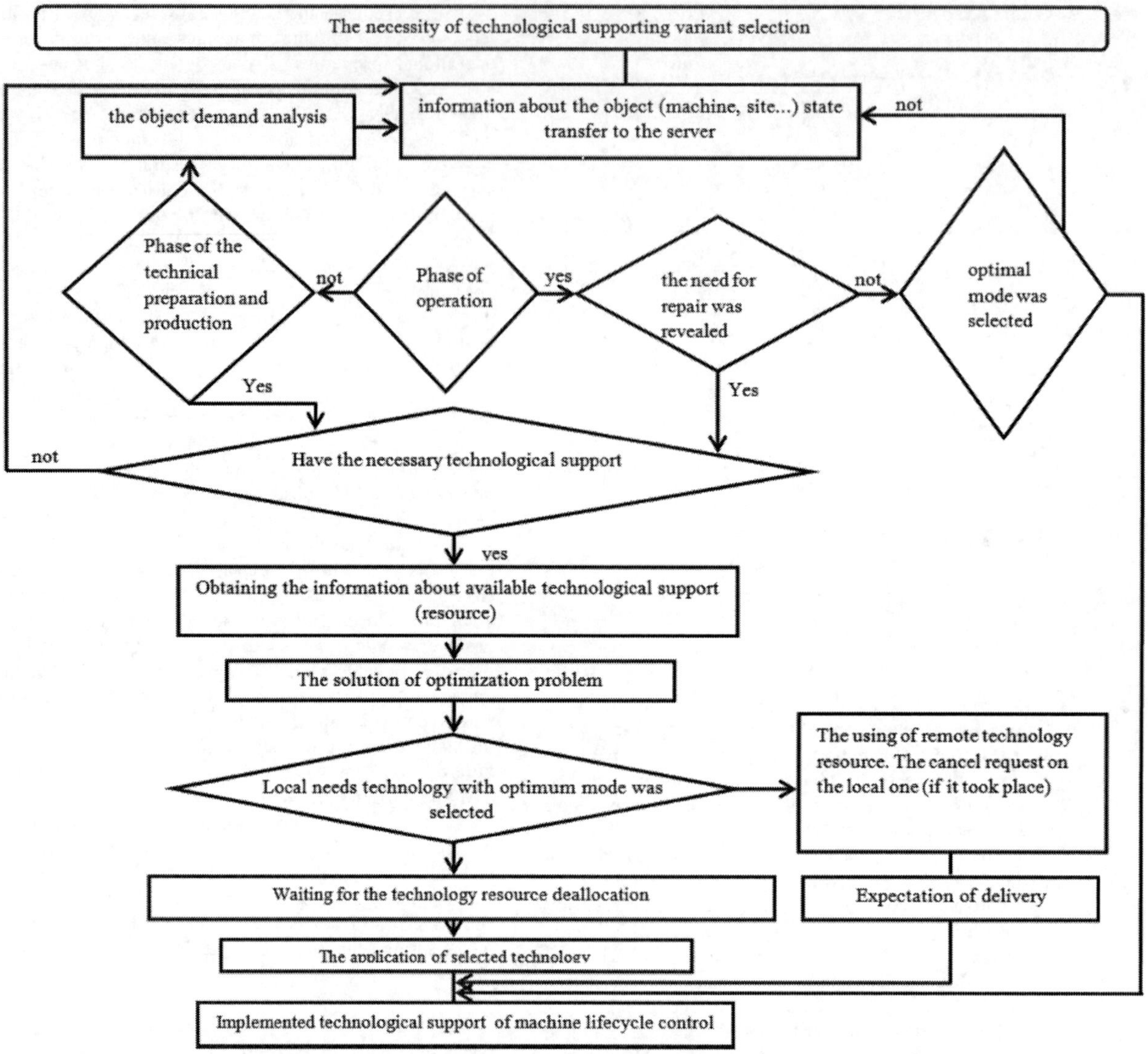

Fig.2. Algorithm of machine lifecycle control technological support

This scheme was developed on the basis of the operation [15] and improved by making new components for the machine performance maintaining.

We considered machine as the system consisting of many components and conjunctions that with certain probability can fail from time to time in the machine lifecycle. In this case, after failure, the perfection of mechanism is carried out by the means of one type strengthening surface processing use taking account factors, which leaded to the failure.

The proposed scheme consists of technical means aimed at diagnosis and monitoring of wear values in comparison with values from the database, and selection and implementation of specific technology to improve damaged tribocoupling, in accordance with received data. Mathematical integration of separate subsystems into a single hierarchical control system structure is the apparatus of queueing theory.

Thus, we developed the methodology of options and parameters for technological support evaluation, which allows to use the most effective technological solution according to the criterion, which includes technical and economic indicators. The improved lifecycle mechanism model including the process of improving and modification broken machine parts and components was put into the basis of proposed methodology.

In accordance with the proposed lifecycle machine model as adaptive systems, we chose the optimization criterion including factors that most broadly cover the lifecycle of any mechanism, which means better reflecting technical condition of system under study. This criterion allows to make right decisions on applications technologies improving wear resistance of mechanisms. Our proposed optimization criterion includes the ratio of each part deterioration, the cost of their creating, recovery and improvement, as well as probability of failure depending on lifecycle phase, where the considered mechanism is situated.

Additionally, we chose the algorithm of the machine lifecycle management technological support and designed the structural scheme of technical-hardware support of mechanical systems performance, where we tried to cover the process of machines and mechanisms work capacity supporting.

This approach, unlike existing, allows to apply the most effective methods of various tribocouples friction surfaces improving, to increase performance of mechanical system, based on the data obtained during the entire lifecycle of the studied mechanism. At the same time, taking into account the mechanism technical state constant monitoring, it allows to correct the decision taken earlier, depending on the changing operating conditions and the actual state of the machine components and parts.

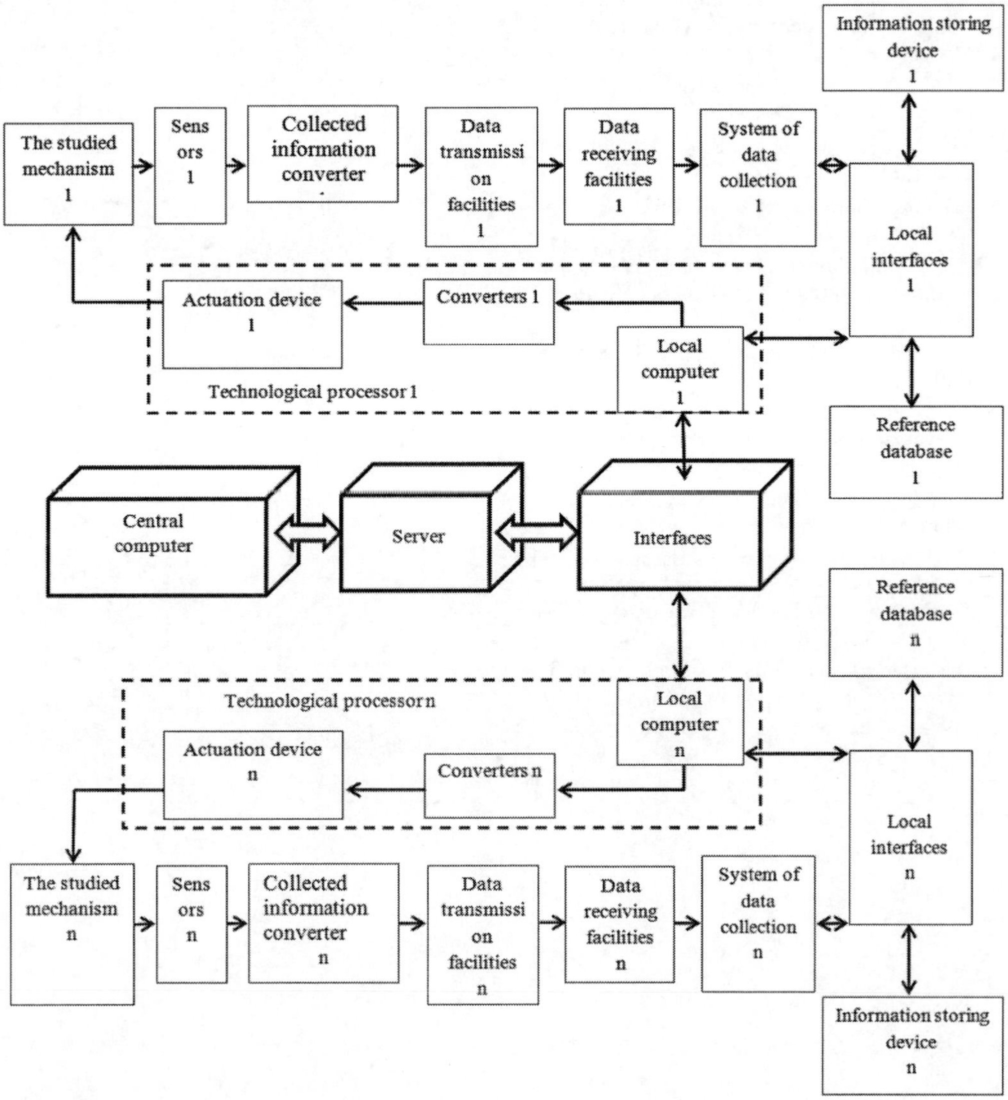

Fig.3. The generalized block diagram of technical-hardware support of mechanical systems performance

References

1. Zh.A. Popova, BSU Herald, **4**, 43-46 (2012).
2. I.V. Kragelskii, *Friction and Wear,* (Butterworths, London, 1965).
3. B. I. Kostetskii, I. G. Nosovskii, L. I. Bershadskiy, A. K. Karaulov, *Reliability and durability of the machines* (Techque, Kiev, 1975).
4. O.V. Burlachenko, M.V. Ivanov, Proc. Eng. **150,** 464-467 Elsev. Publ. (2016).
5. B. I. Kostetskii, *Friction, lubrication and wear in machines*, (Technique, Kiev, 1970).
6. V.E. Gmurman, *Probability theory and mathematical statistics*, **8** (Higher School, Moscow, 2002).
7. B.N. Orlov , A.I. Novichenko , N.B. Orlov , Environ. Engin., **4**, 88-90 (2015).
8. Yu. M. Solomencev, R.R. Zagidullin, E.B. Frolov, FRC CSC RAS, **4**, 77-87 (2010).
9. A.N. Maksimenko, Bull. of BRU, **2**, 31-40 (2012).
10. O.V. Burlachenko, *Technological support machine performance* (VSUAE, Volgograd, 2002).
11. A.V. Zazykin, S. V. Repin, N. K. Hovalyg, Bull. of Civ. Eng., **1**,85-94 (2009)
12. A. S. Pronikov, *Reliability of the machines*, (Mechanical engineering, Moscow, 1978).
13. V. I. Atopov, Y. P.Serdobintsev, O. K. Slavin, *Modeling of contact stresses*, (Mechanical engineering, Moscow, 1988).
14. Y.P. Serdobintsev, O.V. Burlachenko, A.G. Skhirtladze, *Improving the quality of functioning of technological equipment*, (TNT, Stary Oskol, 2010).
15. O.V. Burlachenko, *Affordable and comfortable housing to Russian citizens*, VSUASE, 65-67 (2009).

MATEC Web of Conferences **129**, 01052 (2017) DOI: 10.1051/matecconf/201712901052
ICMTMTE 2017

Equipment and technologies of air-plasma spraying of functional coatings

Viktor Kuzmin[1,3,*], *Igor* Gulyaev[1], *Dmitry* Sergachev[1], *Sergey* Vaschenko[1], *Elena* Kornienko[2], and *Alexander* Tokarev[3]

[1]Institute of Theoretical and Applied Mechanics SB RAS, Novosibirsk, Russia
[2]Novosibirsky State Technical University, Novosibirsk, Russia
[3]Siberian State University of Water Transport, Novosibirsk, Russia

Abstract. This article presents a short description of the DC plasma torch "PNK-50" structural features (ITAM SB RAS, Novosibirsk) used for spraying of functional coatings with powder materials as well as gives the results of thermophysical and technological studies of spraying regimes. We present preliminary results of the plasma torch supersonic modification development and the results of industrial approbation of the plasma torch and of multi-purposes functional coatings deposition processes.

1 Introduction

Plasma spraying is one of the effective methods of 0.1 mm and more thickness functional coatings applying. Practically any powder materials of 20…100 μm dimension (ceramic, metal, cermet, etc.) can be sprayed with the help of thermal plasma fluxes. Atmospheric plasma spraying by using air as a working gas (air-plasma spraying) deserves particular attention. This technology allows reducing the cost and payback period of equipment. Moreover, the plasma spraying allows realization of plasma jet temperature and dynamic characteristics which ensure the necessary conditions for the deposition of high-quality coatings of various categories.

The present article demonstrates the possibilities of modern air-plasma plant on the example of the plasma torch "PNK-50" and considers the approach to the development and optimization of deposition technological regimes using diagnostic equipment and examples of industrial approbation of technology.

2 The plasma torch for powder spraying

The plasma torch "PNK-50" is constructed according to a linear scheme with a sectioned interelectrode insert (IEI) [1]. It is designed for operation in turbulent (high-speed), transient and laminar (low-speed) regimes of plasma jets outflow. It allows high-efficiency deposition of both metallic and ceramic powders (Fig. 1, a). The gas-discharge chamber of the plasma torch is a channel widening from the cathode to the anode. It consists of the sections of IEI, which are electrically isolated from each other and from the electrodes (Fig. 1, b).

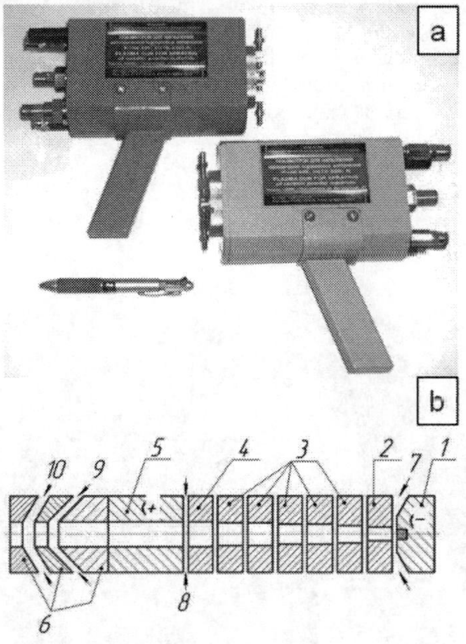

Fig. 1. Plasma torch «PNK-50»:
a – appearance of plasma torch version for manual operation;
b – plasma torch channel scheme:
1 - cathode; 2 – starting insert; 3 – IEI inserts; 4 – transition insert; 5 – anode; 6 – unit for annular injection of powder with gas-dynamic focusing; 7 – plasma-forming gas; 8 – shielding gas; 9 – powder with transporting gas; 10 – focusing gas.

The use of IEI with a different number of inserts (from 1 to 5) allows changing the average arc length and, consequently, the arc voltage and the power of the plasma torch for the implementation of different deposition regimes: "ceramics", "metal", "supersonic". In addition, the scheme of channel with IEI permits to repress unwanted pulsations of length and arc voltage.

* Corresponding author: vikuzmin57@mail.ru

The key element of the plasma torch construction is the original unit for annular injection of powder with gas-dynamic focusing (Fig. 1, b) [2]. This injection scheme ensures the material input in the axial part of the plasma flow, which is characterized by the higher values of speed and temperature. As a result, the efficiency of heating, acceleration of the particles and the deposition productivity are increased.

Air is used as the main working gas for plasma-forming, shielding, transporting and focusing. Small amount of combustible gas (propane-butane or methane) is added to the air which is used as protection (screen) of the plasma torch anode. If necessary, the combustible gas is added to the air which is used as a transporting and focusing gas. Air constitutes a high-enthalpy working gas being at the same time the most accessible and cheap one. The addition of combustible gas prevents the oxidation of the plasma torch anode working surface and the sprayed metallic powders.

3 Thermophysical studies and coatings deposition

The development and control of deposition regimes of coatings with specified functional properties requires a careful analysis of the sprayed material particles characteristics in the jet. Primarily, it is their speed and temperature at the deposition distance. In this work, we used the original system of express control. This system was developed by Yugra State University and ITAM SB RAS [3-5]. It allows scanning of a sprayed jet to measure velocity and temperature of the sprayed particles (Fig. 2).

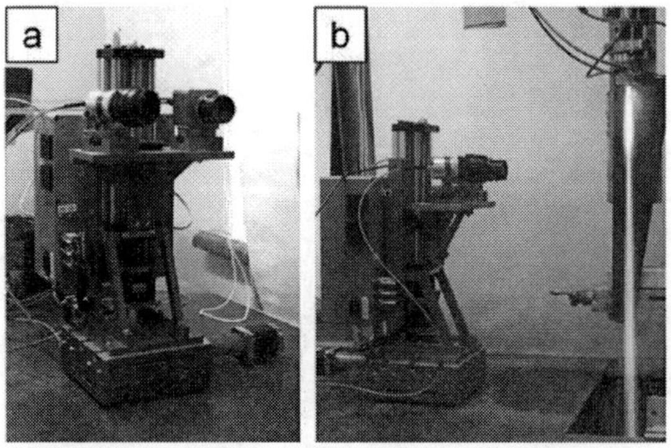

Fig. 2. Optical system for diagnostics of dispersed phase of spraying jet: a – appearance of diagnostics mounted on 3D-positioning system; b – analysis of self-fluxing NiCrBSiC powder spraying regime.

Fig. 3 presents the change graphs of the average values of the particles velocity and temperature along the jet length for different arc currents for spraying of self-fluxing powder (NiCrBSiC system) in order to demonstrate the spraying modes optimization technique. The data suggest that in order to achieve maximum velocity (270…290 m/s) and temperature (2400 K) of particles on this system, the deposition should be made at an arc current of at least 140 A and the spraying distance of maximum 200 mm.

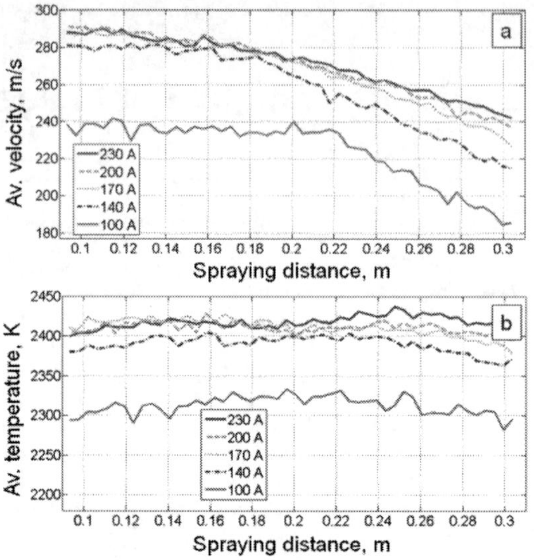

Fig. 3. Dependence of average particle velocity (a) and temperature (b) vs spraying distance for various electric arc currents. NiCrBSi powder 40…100 um. Plasma torch configuration «Metal».

The obtained data about the particles parameters in the sprayed jet were used to optimize the spraying regimes of the NiCrBSiC material. This material is used as wear-resistant, corrosion-resistant and heat-resistant coatings. Coatings spraying at the arc current of 140 A and at the distance of 170 mm are characterized by the absence of unmelted particles. Their porosity is of less than 1 % with more than 50 % of the pores having size smaller than 5 μm. The transmission electron microscopy method revealed that microcrystalline structure with grains average size of 80 nm was formed during deformation and quench of the material particles [6, 7]. This effect leads to the increase of the coatings hardness in comparison with the hardness of the original powder (more than 20 HRC).

4 Development of the supersonic plasma torch modification

The deposition of dense coatings from metal and carbide (WC/Co, WC/Ni) materials requires a high particle velocity (500...800 m/s and higher). For that, we calculated the geometry of the unit for annular injection of powder in the form of a supersonic Laval nozzle. The nozzle diameter in the critical section was 6 mm and the Mach number at the nozzle cut was 1.5. Integral characteristics of the plasma torch at supersonic regimes in correlation with the current strength of arc discharge are presented in Table 1.

The use of the optical system for diagnostics allows investigation of the spatial structure of a plasma jet, as well as measuring the velocity and temperature of a dispersed phase in the process. Fig. 4, a presents the jet during the deposition of Ni-Cr-Si-B-C wear-resistant material with 20...63 μm dimension.

Table 1. Properties of supersonic air plasma jet at various electric arc currents (I): arc voltage (U), thermal power (Q), static pressure (P), mean plasma temperature (T), mean plasma velocity (V).

I, A	U, В	Q, кВт	P, ата	T, K	V, м/с
110	285	22,1	0.79	2737	1484
140	278	26,5	0.85	3039	1564
180	268	32,5	0.92	3330	1642
220	262	38,1	0.96	3574	1737

The particles flux is seen to have a dense structure with angle divergence of not more than 5°. Fig. 4, b contains the image of the jet initial section (without powder) in which the system of compression rises (Mach's disks) of a slightly overexpanded supersonic flux is clearly seen. The blue color at the jet periphery corresponds to the protective envelope of propane-butane which is added to the transporting and focusing air to reduce the oxidation of the sprayed particles. The structure visualization of the supersonic jet obtained by mathematical processing of the monochrome image of the jet is shown in Fig. 4, c and d.

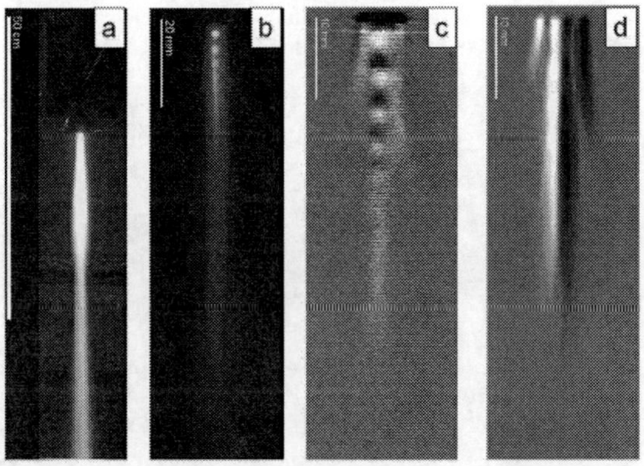

Fig. 4. Supersonic air-plasma jet: a – spraying process of NiCrBSiC powder, b – supersonic air plasma jet with shielding envelope of propane/butane, c – visualization of axial structure of jet – shock diamonds (Mach's discs), d – visualization of radial structure of jet – main central jet and shielding peripheral envelope.

The diagnostic of the sprayed jet showed that the use of a supersonic spraying regime allows significant increase in the average particle velocity at the spraying distance from 300 to 500 m/s. The stream of particles in the supersonic jet is more concentrated: the width of the jet is 9.6 mm compared to 14.7 mm in a subsonic jet. It is known that a concentrated sprayed jet provides more uniform values of the particles velocity and temperature in the sprayed spot.

This increase in the sprayed particles velocity in the supersonic regime allows achieving the values of the coatings microhardness of 693 HV that is 20…25 % higher in comparison with the subsonic regime (Table 2). The hardness increase can be explained by the formation of a substantially finer-grained structure of coatings in the case of supersonic spraying, as the closing time of the crystallization front of a liquid particle with a free surface at the impact against a substrate is inversely proportional to its velocity immediately before the impact. Accordingly, the higher is the velocity of the sprayed particles, the shorter is the time of their crystallization during the spreading on the substrate.

Table 2. Spraying modes oomparison of NiCrSiB powder with size 20…63 um

Spraying mode	Subsoni	Supersonic
Plasma torch electrical power, kW	48	52
Plasma jet characteristics		
Plasma velocity, m/s	1200	1650
Plasma temperature, K	3800	3200
Powders flow characteristics on spraying distance		
Spraying distance, mm	170	200
Powders velocity, m/s	307±62	504±68
Powders temperature, K	2845±145	2938±110
Jet diameter, mm	14.7	9.6
Coatings characteristics		
Overall porosity, %	3	5
Average size of pores, μm	13	10
Microhardness, HV	558±61	683±108

5 Industrial approbation of air-plasma spraying technology

The industrial plant of plasma spraying "Thermoplasma 50-01" was developed and manufactured on the basis of the plasma torch "PNK-50". It includes a power source, a control panel, a switching module with a start-up unit, two powder dozers and an unit of autonomous cooling of plasma torch (Fig. 5). The plant is equipped with two plasma torches. They can be used both in manual and mechanized versions. The channel of one of the plasma torches is assembled for the spraying of metallic powders, the channel of the other plasma torch was assembled for the spraying of ceramic powders. The switching of the power supply, cooling, gas supply, the unit for starting and feeding of the powder from one plasma torch to the other is done remotely from the control panel.

DOI: 10.1051/matecconf/201712901052

Fig. 5. Basic components of spraying facility «Thermoplasma 50-01».

Fig. 6 presents the photographs of a number of parts with various functional coatings deposited with the help of the plasma torch "PNK-50".

The high quality of the coatings obtained with the plasma torch "PNK-50" allowed to create a new technology for protecting of the details of the ships from hydroabrasive and cavitation wear. In this case a coatings flowing operation after the spraying is not required. The screw propeller of the small ship "RC-101D" with wear-resistant coatings looks outwardly practically not worn-out after the navigation in 2016 in hard conditions of the shallow waters of the Biya and Katun rivers (Fig. 7, a). This screw is suitable for exploitation in the next navigation period. The control screw propeller (without the protective coating) requires renewal during navigation (welding plates to worn edges and surfacing of the worn surfaces (Fig. 7, b).

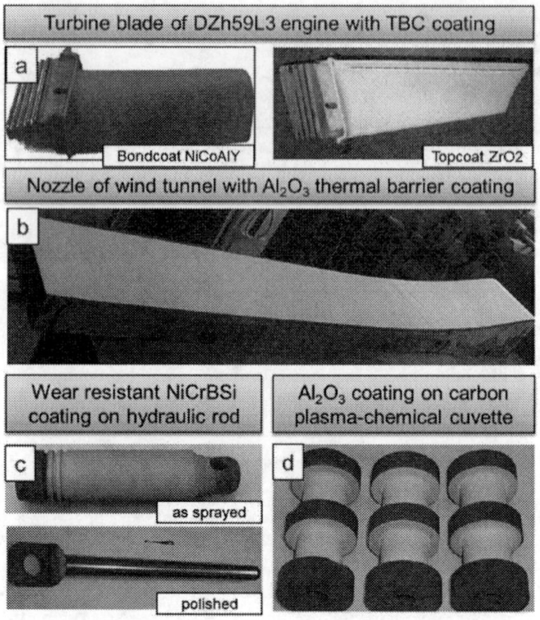

Fig. 6. Functional coatings produced by air-plasma spraying: a –DZh59L3 engine high pressure turbine blade with internal cooling restored with double-layer thermal barrier coating; b – nozzle element of wind tunnel with hydrogen heater. Coated surface 1300x300 mm, coating thickness 500 um; c – rod of hydraulic cylinder manipulating powered supports for mining facilities; d – carbon cuvettes for spectroscopic analysis of high-temperature decomposition in plasma-chemical reactor.

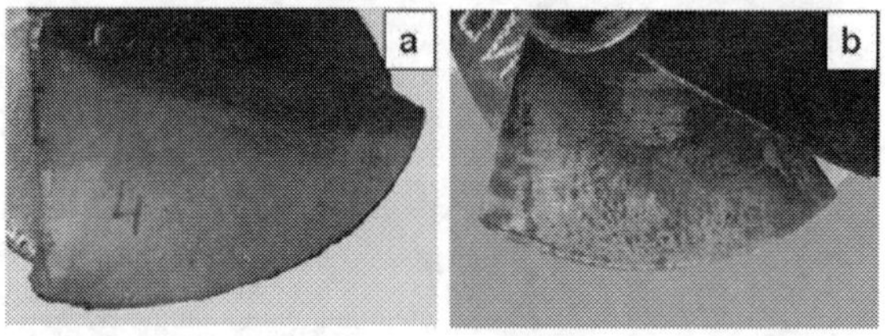

Fig. 7. Blades of project «KS-101D» boat propeller after navigation in 2016 season: a – blade with protective coating NiAl; b – blade without coating, required restoration during navigation period.

6 Conclusion

Air-plasma spraying of coatings is one of the most universal methods of gas-thermal spraying. It combines a wide range of the applied materials (from metals to ceramics and composites), high productivity and low cost of technology. Systematized studies of electric arc processes, interphase interaction in high-temperature dispersed streams, and coatings formation processes allowed developing of a new generation air plasmatron "PNK-50", as well as spraying technologies which have been introduced into production.

References

1. V.I. Kuzmin, A.A. Mikhalchenko, V.I.Nalivaiko, P.A. Chubakov, Russian Physics J. **50,** 85-88 (2007).
2. V.I. Kuzmin, A.A. Mikhal'chenko, O.B. Kovalev, E.V. Kartaev, N.A. Rudenskaya, Journal of Thermal Spray Technology. **21**,159-168 (2012)
3. A.V. Dolmatov, I. P. Gulyaev, V.I. Jordan, IOP Conference Series: Materials Science and Engineering **81**, 012041 (2015).
4. A.V. Dolmatov, I.P. Gulyaev, P.Yu. Gulyaev, V.I. Jordan, IOP Conf. Series: Materials Science and Engineering **110**, 012058 (2016).
5. 5. A.V. Dolmatov, I.P. Gulyaev, V.I. Kuzmin, E.A. Lyskov, K.A. Ermakov, Thermophysics and Aeromechanics **24**, 83-94 (2017).
6. E.E. Kornienko, E.J. Lapushkina, S.P. Vaschenko, I.P. Gulyaev, E.V. Kartaev, D.V. Sergachev, N. Kashapov, S. Sharifullin, I. Fayrushin, Journal of Physics: Conference Series **567,** 012010 (2014).
7. E. E. Kornienko, D. O. Mul', O. A. Rubtsova, S. P. Vaschenko, V. I. Kuzmin, I. P. Gulyaev, D. V. Sergachev. Thermophysics and Aeromechanics **23**, 919–927 (2016)

DOI: 10.1051/matecconf/201712901053

Manufacturing of thin-walled parts for machinery by selective laser melting

Vitaliy Bobyr[1,*], *Artem* Deev[1], *Anton* Zhukov[1],and *Pavel* Kuznetsov[1]

[1] NRC "Kurchatov Institute" – CRISM "Prometey", 191015, Saint-Petersburg, Russia

Abstract. The paper describes the technology of selective laser melting, as well as its capabilities in the manufacture of thin-wall honeycomb energy absorber (HEA). The effect of the technological parameters of the building process on the HEA walls' thickness is studied. Conformity analysis of the mass-dimensional characteristics of the finished composition with the predefined parameters of the 3D-CAD model is carried out. Dependencies of building parameterson the quality of the manufactured HEA are established, general recommendations for the practical use of technology in the creation of HEAare given.

1 Introduction

In recent years, one of the most promising ways of forming metal products with the required parameters is considered additive technology, in which the manufacturing of the product in the necessary spatial-geometric proportions occurs by sintering the powder layers with the help of a concentrated energy flux [1–2]. Of particular interest is the determination of the optimal technological parameters of the manufacturing process, which in turn will be decisive in determining the mechanical and operational properties of the product, manufacturability, material consumption and, ultimately, the cost of the product.

As is known, the honeycomb energy absorber (HEA) ensures the absorption of the impact energy, for example, in the case of its application in the landing spacecraft for soft landing, due to its deformation, taking into account the permissible overload. To accomplish this, the structure of the HEA must be quite complex and satisfy a set of mechanical properties.

The production of thin-wall HEA by the selective laser melting, in particular sintering of powder materials, is a promising direction for the manufacturing of complex shaped products for critical applications operating under intensive loads. Increasing capacity and speed of operation of modern technology require an improvement in the quality of energy absorbers to effectively quench energy during braking or in case of emergencies [3]. The using of laser technologies to create highly loaded parts and elements of technology will reduce the cost and time limits of production, as well as increase the reliability and durability of products, providing high physical and mechanical characteristics [4].

2 Materials and experimental methods

The raw material used is the powder of 321 asthenic stainless steel, manufactured by JSC "Polema" (Russia). The chemical composition of the powder, determined by X-ray fluorescent analysis methods on the Niton XL3t and LECO CS744 units, is shown in Table 1, from which it is seen that it complies with the ASTM A 240 standard.

Table 1. Chemical composition of used powders.

Basic alloying elements,% by weight	
C	0,12
Si	0,8
Mn	0,8
Cr	17,6
Ni	10,3
Mo	0,1
Ti	0,6

The particle size distribution of the powder measured in the Malvern Mastersizer 2000 using the Fraunhofer approximation's laser diffraction method is shown in Fig. 1. It can be seen that the 321 steel powder has a large fractional composition, but a small amount of fine particles is present. In general, the powder meets the requirements for powders for SLM technology.

*Corresponding author: vitalybobyr@yandex.ru

DOI: 10.1051/matecconf/201712901053

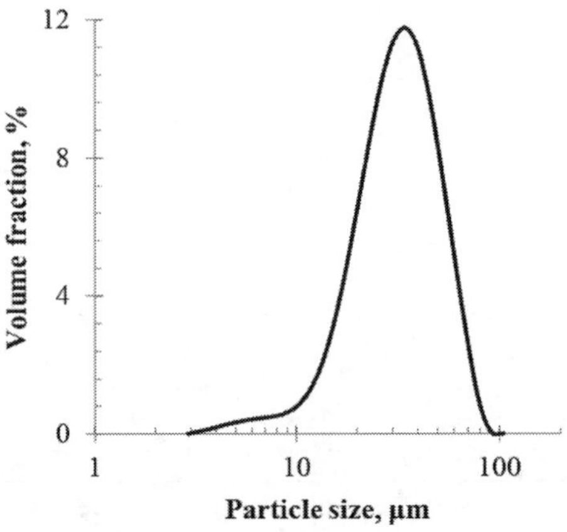

Fig. 1. The grain size distribution of 321 steel powders.

Selective laser melting of the powder was performed on the EOSINT M270 unit. The powder was melted under certain conditions: the 3-D model, developed in the AutoCAD environment, for the prototype of a HEA, is shown in Fig. 2.

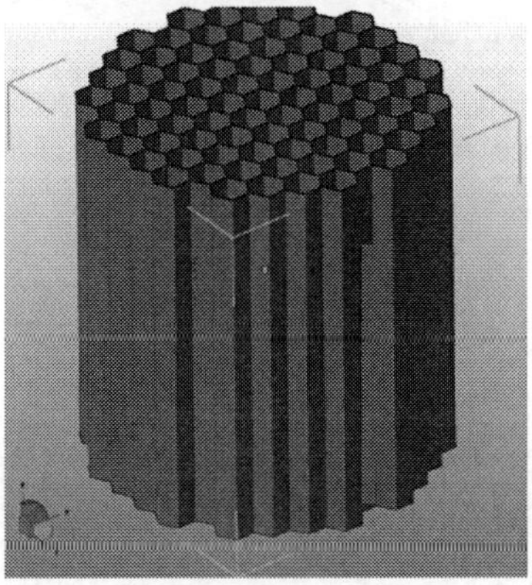

Fig. 2. Model of the HEAprototype.

The building of a detail by selective laser melting using a pre-designed model is carried out according to a certain algorithm presented on the Fig. 3 (for a single layer): DirectPart (a single detail layer):

- OuterSkin
- InnerSkin
- Counters

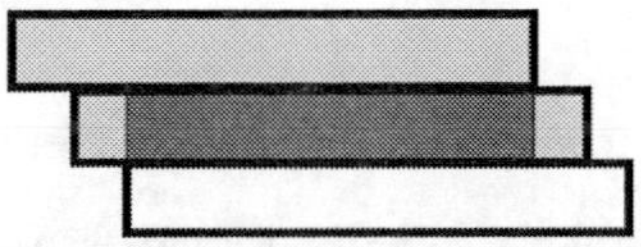

Fig. 3. Part of the detail with the overlay of three layers.

Proceeding from the fact that in the task it is necessary to build thin-walled honeycomb elements representing cylinders that are completely symmetric in the vertical direction (along the Z axis); Layers - the projections completely reflect each other without any displacements, and the wall thickness is 110 μm, 130 μm and 150 μm (Fig. 4).

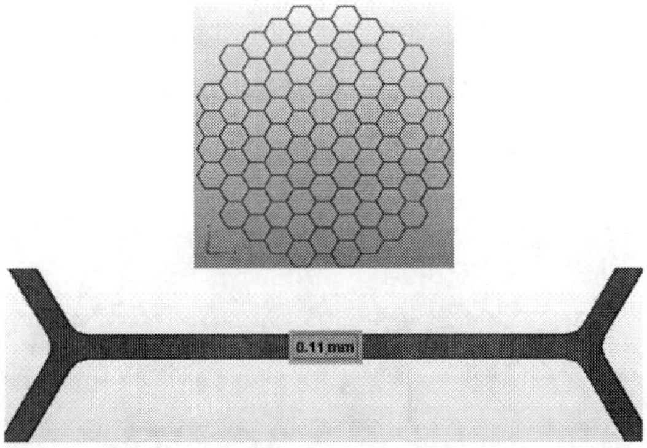

Fig. 4. Projection of a single layer of HEA in the horizontal plane XY, wall thickness 110 μm.

Accordingly, it was decided to build the HEA by the algorithm "in contours", i.e. excluding modes of additional melting of materialOuterSkinandInnerSkin (Fig. 5).

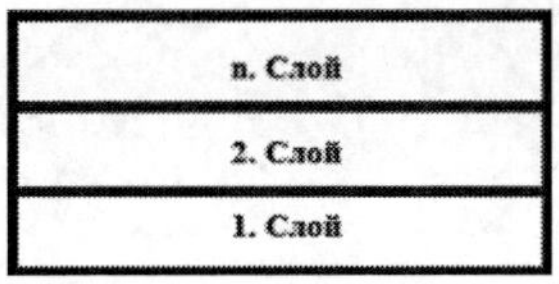

Fig. 5. Part of the detail with the overlay of three layers.

Thus, the required wall thicknesses of 110 μm, 130 μm, 150 μm are sufficient to build "in contours", which in turn will result in the lowest heat input and, correspondingly, in the production of items.

To improve the accuracy of the HEA manufacturing and complete elimination of rejects, it is necessary to select the material meltingmode in the contour algorithm (counters). The main parameters of this mode are laser radiation power P (W), scanning speed of the laser beam V (mm / s) and offset (μm). Nominal values prescribed in the process unit P = 120 W, V = 400 mm / s,offset = 0μm. Let us determine the coefficient of heat input C = P / VW · s / mm, equal to C = 0.3 W · s / mm. The offset parameter is responsible for shifting the laser beam relative to its center, with the positive offset shifting the beam toward the center (Fig. 6). Thus, for example, with offset = 10 μm, the centers of the laser beam will shift to each other and thus theoretically the wall thickness of the finished product will become less than the 3D model specified.

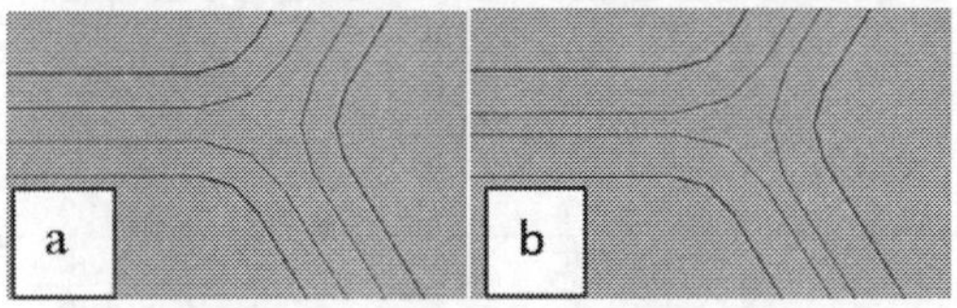

Fig. 6. Effect of the offset parameter on the path of the laser beam: offset = 0 (a), offset = 10 mm (b).

In Table 2, the samples building parameters with a selection of the coefficient of heat input C = 0.25; 12 C = 0.35, as a result of the change in power, as well as offset = 0; 0.002; 0.01 mm, are presented.

The construction scheme is shownon Fig. 7, while the honeycomb elements are located along the wall thicknesses from 150 µm to 110 µm in the direction of movement of the mechanical knife for leveling the powder.

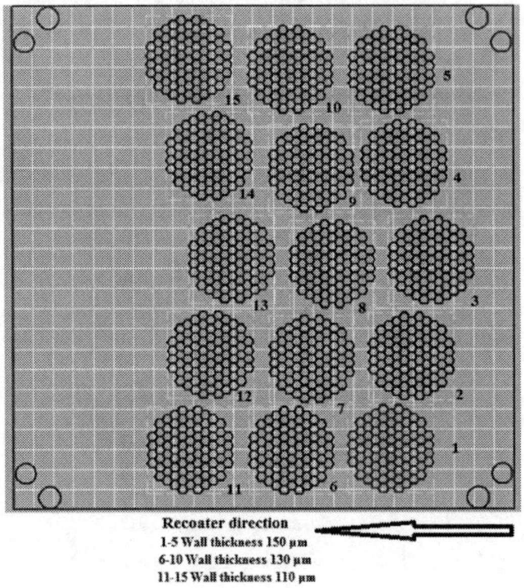

Fig. 7. Layout of honeycomb elements samples.

A large series of thicknesses (150 µm) is located in the first row with respect to the direction of motion of the laser blade. Since it is not uncommon for the blade to hit the surface of the product during its production, also taking into account rather small thicknesses, it was decided to arrange the blanks, thus, as shown in the figure above. In turn, this almost completely eliminates the external mechanical effect on honeycomb elements with a small thickness of the walls, which makes it possible to produce a HEA without surface defects and defects of the structure itself.

Table 2. The samples building parameters.

Sample №	Wall thickness, µm	Offset, µm	P, W
1	150	0	140
2	150	0	100
3	150	10	100
4	150	10	140
5	150	2	120
6	130	2	120
7	130	2	100
8	130	2	140
9	130	10	120
10	130	0	120
11	110	2	100
12	110	0	140
13	110	2	140
14	110	0	100
15	110	2	120

In the course of studying the results of HEA building on the first stage, it was suggested to make one more batch of products with the additional modes presented in Table 3, and also to build two gradient samples.

Table 3. Additional samples building modes.

Sample №	Wall thickness, µm	Offset, µm	P, W
1	150	0	80
2	150	10	80
3	130	0	80
4	130	10	80
5	110	0	80
6	110	0	90
7	110	0	85
8	110	0	95

The gradient sample is a customer's honeycomb element, divided into three parts in a horizontal plane, each of which is melted according to the specific mode indicated in Table 4.

Table 4. Gradient sample modes.

Sample №	Modes	Wall thickness, µm	Offset, µm	P, W
9*	part 1 (from the platform)	110	0	120
	part 2	110	0	100
	part 3	110	0	80
10*	part 1 (from the platform)	130	0	120
	part 2	130	0	100
	part 3	130	0	80

The obtained results of the building process for all the stages will make it possible to select those modes from the ones presented above for which the HEA made of the metal powders of stainless steels manufactured bytheselective laser melting will fully meet their specified 3D models both in geometry and in physical and mechanical properties.

3 Experimental results and discussion

Verification of the correspondence of the actual value of the wall thickness of thin-walled structures to the specified value:

With the electron microscope VEGA TESCAN, the images of the honeycomb structures shown in Fig. 8 (Platform No. 1) and 9 (Platform No. 2) were obtained, the purpose was to identify the possible presence of structural defects on the walls of the HEA.

The results of comparison of the theoretical and actual values of wall thicknesses of honeycomb structures are presented in Table 5.

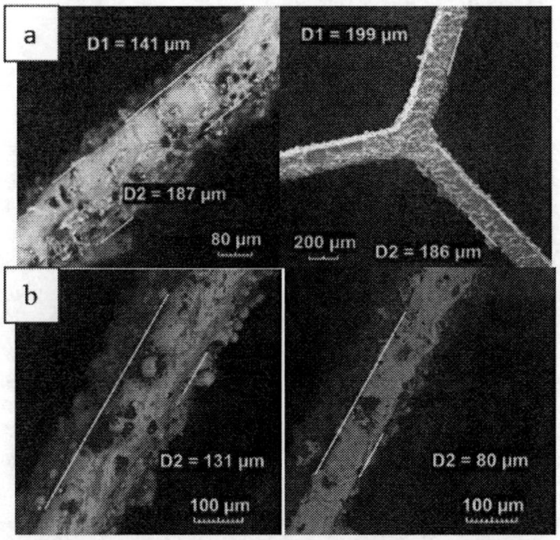

Fig. 8. Cellular structures from the platform 1 (a) and 2 (b).

As can be seen from the images shown in Fig. 8 (a) and (b), in general, the structure is slightly different in both cases. There are non-melted particles, but the structure continuity remains quite acceptable, there are no visible defects, such as microcracks caused by thermal stresses, and pores.

Table 5. Comparison of the theoretical and actual values of the wall thicknesses of HEA.

Sample №	Platform №1		Platform №2	
	Wall thickness, µm		Wall thickness, µm	
	Theor.	Fact.	Theor.	Fact.
1	150	145	150	129
2	150	139	150	100
3	150	116	130	102
4	150	159	130	89
5	150	151	110	80
6	130	128	110	85
7	130	117	110	82
8	130	132	110	90
9	130	129	110*	83*
10	130	127	130*	107*
11	110	92		
12	110	109		
13	110	124		
14	110	109		
15	110	108		

According to the presented data, it can be concluded that not always the actual values of the wall thicknesses of HEA coincide with the theoretically calculated ones. However, discrepancies can be leveled using the correction factor when creating a building model. The correction factor can be obtained by varying the power and speed parameters of the building.

The complex study of the influence of the technological modes on the HEA melting process showed the dependence of wall thicknessfrom the laser radiation power used. The values obtained can be explained by the presence of a small number of pores and other inclusions. Complex studies were carried out to determine the wall thicknesses, the mass of the HEA obtained, and also the static compression tests, since the main task of the HEA is to absorb the impact, the general data are given in Table 6.

Table 6. Characteristics of HEAbuild from 321 steel powder.

Sample №	Weight, g		Compression force, kN	
	Theor.	Fact.	Activation	Compression
1	37,3	39,8	56,8	45,2
2	37,3	34,4	46,2	33,3
3	32,3	29,1	32,8	23,9
4	32,3	36,0	44,1	35,8
5	36,3	36,1	48,5	36,4
6	31,4	31,3	36,0	28,0
7	31,4	28,2	30,3	23,5
8	31,4	35,7	43,9	34,6
9	27,4	29,9	33,8	24,0
10	32,4	32,3	41,0	29,8
11	26,5	23,9	21,6	16,7
12	27,5	34,3	39,6	30,9
13	26,5	35,1	39,2	30,2
14	27,5	24,8	23,5	18,7
15	26,5	28,3	30,9	22,4

The data showed in Table 6 relate to the process for the HEA manufacture with the melting of material between the contours. It should be noted that it was not possible to achieve full compliance with the given thicknesses, which could possibly be due to the multiparameter effect and the response of the material to thermal processes during laser melting. However, in a number of cases the difference between the theoretical and real HEA wall thicknesses is minimal and is of the order of 1-6 μm. In turn, the mass of the HEA is within +/- 5 g, which, in turn, is a fairly acceptable deviation.

Table 7. Characteristics of HEA build from 321 steel powder (construction in contours without shading).

Sample №	Weight, g		Compression force, kN	
	Theor.	Fact.	Activation	Compression
1	37,3	32,3	41,1	26,3
2	32,3	27,3	29,7	20,1
3	32,4	27,1	28,0	20,0
4	27,4	21,9	18,7	14,9
5	27,5	21,7	18,0	14,3
6	27,5	23,0	20,8	16,2
7	27,5	22,2	17,7	15,5
8	27,5	23,6	–	17,7
9*	27,5	25,0	16,7	14.6/18.3/23.9
10*	32,4	29,6	27,4	19.6/24.2/29.5

According to Table 7, the mass of the HEA is also within +/- 5 g. As for the HEA wall thickness values, in this case the construction was carried out at a laser radiation power of 80-95 W. It was assumed that in case of building only in circuits, it is necessary to use the minimum power. However, the variation in the values of the theoretical and actual thickness turned out to be larger than in the previous case. Below is a graph of the dependence of thickness on the technological modes, constructed on the basis of the above tables (Fig. 9).

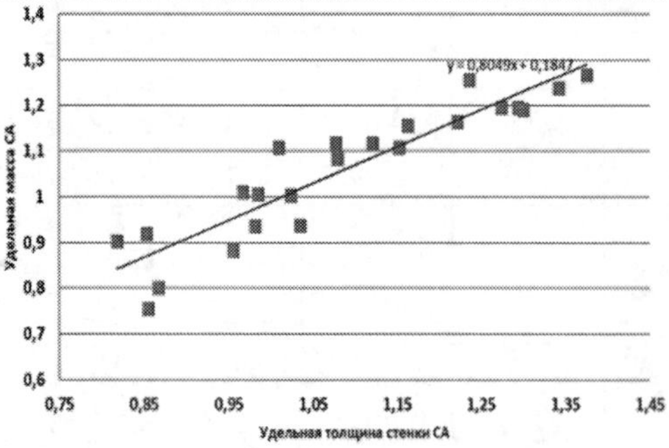

Fig. 9. Dependence of the specific mass on the HEA specific wall thickness.

This graph shows us the linear dependence of the specific mass on the specific wall thickness, thus confirming the absence or minimum number of pores and defects in the finished samples.

4 Conclusions

X-ray fluorescence analysis of the 321 metal powder showed that, on the whole, the elemental content of the alloying elements is within the permissible limits. The results of the granulometric analysis showed the particle size distribution of the powder material, where the largest volume percentage corresponds to a range of sizes on the order of 25-55 μm, which fully meets the requirements for the raw material used for the laser fusion EOSINT M270.

When working out the modes of metal powder melting, for the subsequent manufacture of HEA, taking into account the small wall thicknesses (minimum size of 110 μm), it was decided to build the part "in contours", thereby avoiding additional thermal effects, thereby preventing the possible formation of both surface (open pores), and internal (microcracks) defects. The obtained results for all stages of building will make it possible to select those modes in which build honeycomb elements will fully meet their specified 3D models both in geometry and in physical and mechanical properties, excluding unevennesses.

According to the data from the scanning electron microscope, it can be concluded that the actual values of the HEA walls thicknesses do not always coincide with the theoretically calculated values. However, discrepancies can be leveled using the correction factor when creating a building model. The correction factor can be obtained by varying the power and speed parameters of the building. Structural studies have shown that a small amount of non-melted particles are present, but the structure continuity remains quite acceptable, there are no visible defects such as microcracks caused by thermal stresses and pores.

In general, the studies carried out made it possible to draw several important conclusions: in general, to achieve a given (or actual) thickness, the laser sintering power should be equal in value to the HEA wall thickness. In this case, it is assumed that the deviation will be minimal, both in thickness and in mass, respectively. To prevent the effects of external mechanical stress on the honeycomb elements in the building process with different thickness, the arrangement of the preforms on the platform is desirable to implement in rows, from the maximum to the minimum thickness. The HEA building mode in contours without hatching (melting between contours) does not introduce any structural differences in comparison with the standard mode.

This work was supported by the grant of Russian Scientific Fund №15-19-00210.

References

1. B.K. Barakhtin., A.V. Voznyuk, A.A. Deev, A.S. Zhukov, *The structure – mechanical condition of*

addition material at hot plastic deform,Deformation and destruction of materials **2** (2017)

2. J. Sedlaka, D. Rican, *Study of Materials Produced by Powder Metallurgy Using Classical and Modern Additive Laser Technology*, Procedia Engineering **100,** 1232–1241 (2015)
3. V.F. Aleshin, *Landing devices of space vehicles (SV) on the basis of foams and honeycomb blocks* (Science and education: electronic scientific and technical edition of the Bauman Moscow State Technical University, Moscow, 2010)
4. A. Hussein, L. Hao, *Advanced lattice support structures for metal additive manufacturing*, Journal of Materials Processing Technology **213,** 1019–1026 (2013)
5. A.S. Oryshchenko, I.V. Gorynin, V.V. Bobyr, *Additive technologies based on composite powder materials*, Additive technologies in the Russian industry **22** (2015)

Mathematical models for estimating the degree of influence of major factors on performance and accuracy of coordinate measuring machines

Alexander F. Rezchikov[1], *Andrei* V. Kochetkov[2], and *Oleg* V. Zakharov[3,4,*]

[1]Institute of Precision Mechanics and Control of RAS, 24 Working st., Saratov, 410028, Russia
[2]Perm National Research Polytechnic University, 29 Komsomol pr., Perm, 614990, Russia
[3]Yuri Gagarin State Technical University, 77 Politecknicheskaya str., Saratov, 410054, Russia
[4]Samara National Research University, 34 Moscow Highway, Samara, 443086, Russia

Abstract. Increased productivity of Coordinate Measuring Machines, while providing given level of accuracy, is complex and actual problem of robotics and metrology. Therefore, this article provides a set of mathematical models that allow for prediction of values for performance and accuracy of advanced Coordinate Measuring Machines based on system dynamics method. The analysis of scientific literature and statistical data of exploitation CMMs Global Performance (DEA, Italy), CMM-750 (Lapik, Russia), FARO Arm 9 (FARO Technologies, USA) has identified 25 major factors of the process control and their mutual impact. Of the 16 basic factors are internal system parameters, 6 input actions and 3 external factors. As a result of the system analysis of the research object identifies the main characteristics of coordinate measuring machines that affect performance and measurement accuracy. We have done a general graph of causal relationships between modeled characteristics, on the basis of which is made up and numerically solved a system of nonlinear differential equations. In developing this system of equations is used in statistical data describing the cause-effect relationship between the internal model parameters and environmental factors. These solutions make it possible to practically interpret developed models and methods of system dynamics used in solving the problem.

1 Introduction

In developed countries, high-tech production, primarily the defense complex, use coordinate measuring machines (CMM). By the beginning of the XXI century in the world, operating more than 500,000 CMM. One third of this quantity is used in the United States, about the same in Germany, Great Britain and Japan. Active industrial development in China, South Korea, India, Brazil, is also accompanied by the rapid implementation of CMM. Complete CMM produced by only a few companies in the world.

The basis of most of the known of working capacity CMMs laid implementation of the Cartesian coordinate system. This feature has become a bottleneck in the way of a new generation of CMMs. Options traditional three-axis CMM when measuring complex spatial surfaces is not enough. An important breakthrough in this direction was the creation of the six-axis CMM Saratov group of experts, which included engineers, physicist's optics, electronics engineers, mathematicians and programmers. Six-axis CMM, using its main advantage - the six-axis control of the probe - are able to control the difficult to reach surface of products, to measure performance in a 2-5 times greater than traditional CMM.

For the current stage of development of precision engineering and instrumentation are relevant problem of increasing the precision, performance and reliability of the CMM. The solution of this problem is devoted to leading companies and scientists [1-17]. They generally, relate to the technological problems of building new types of measuring machines. At the moment, developers do not have the mathematical and information software, that allows on the stages of creation, operation and modernization, as well as in real time to analyze the mutual influence of the parameters of the CMM. These parameters are combined complex causal network and assess the impact on them of external factors and thereby optimize the design of perspective CMM is difficult.

This fact determines the relevance, practical and scientific significance as well as the economic feasibility to development of models and algorithms for improves the performance and accuracy of the six-axis CMMs.

2 Characteristics of the Research Object

Based on research findings required to develop software, which includes a complex of mathematical models.

* Corresponding author: zov17@mail.ru

These models allow to perform simulation the dynamics of changes in the values of the basic parameters of the CMM, affecting the accuracy and performance.

The structure of the software need to include:

- graph of causal relationships;
- system of nonlinear differential equations;
- regression models for the formal description of the causal relationships between the parameters of the models;
- algorithms to verify the adequacy of the developed models;
- information and logic scheme, that characterizes the processes of practical use of software developed in the structural sub-divisions of engineering enterprises.

In this article, we solve problems of paragraphs 1 and 2.

Scheme of six-axis CMM is shown in Fig. 1. All process operating control factors on CMM can be divided into three groups: internal parameters CMM input actions and external influences. The Interior parameters are set for design of CMM and operational factors. During the measurement of these parameters may be constant or vary. Input exposure determined by the parameters of the measured product, measurement tasks and mathematical software. External influences vary in function of time and related to the environment, are mostly perturbing character. These groups of factors affect both the performance and the accuracy of control.

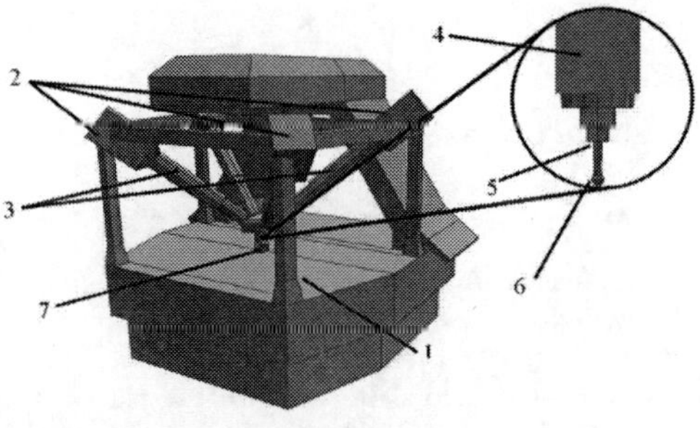

Fig. 1. Scheme of six-axis CMM: 1 – Table; 2 – Hinges; 3 – Telescopic Rod; 4 – Base of the Touch Sensor; 5 – Touch Sensor (probe); 6 – Spherical Tip of the Touch Sensor; 7 – Measured Product.

Performance X1 is the main parameter to be optimized CMM, which directly or indirectly affect all other parameters. By definition, the performance is an indicator of the efficiency of production, which characterizes output per unit of resources used, the factors of production. Therefore, performance of measurement will be determined by the following formula:

$$X_1 = \frac{V}{T \cdot k},$$

where V – volume of the measuring space, mm3; T – time of measurement, sec; k – coefficient of measurement complexity.

Accuracy X_2 is defined as the maximum error in the length measurement direction in parallel to the coordinate axes (ISO 10360-2):

$$X_2 = k_0 + k_1 L,$$

where k_0 – coefficient determined by the components of the measurement errors that are independent of measuring length (touch sensor error, electrical inertia, discontinuity and short-period errors coordinate displacements of nodes, errors associated with dynamic phenomena); k_1 – coefficient defined components, depending on the measured length L (long-period measurement error and coordinate movement of nodes, the impact of ambient temperature, deformation coordinate displacement devices).

Requirements for the simultaneous increase of productivity and accuracy are contradictory in many cases. Thus, as a criterion for the optimization of control process must be productivity, which the accuracy demands attributed to limitations problem.

Other important factors when measuring on CMM:

X_3 – mechanical design error, μm;
X_4 – structural rigidity, N/μm;
X_5 – error of the rotation hinges, rad;
X_6 – error linear displacement rods, μm;
X_7 – error fixing the moment of contact, μm;
X_8 – geometric accuracy of the spherical tip of the touch sensor, μm;
X_9 – error of laser interferometers, μm/m;
x_{10} – speed of linear motion along the axes, mm/s;
X_{11} – speed of angular motion along the axes, rad/s;
X_{12} – speed of rotation of the touch sensor; rad/s;
X_{13} – calibration time, sec;
X_{14} – application of the contactless scanning head, bool;
X_{15} – dimensions of the measured product, mm;
X_{16} – length for probe of the touch sensor, mm;
X_{17} – time for installation of the product in the process control, sec;
X_{18} – time of creation a mathematical model of the product, sec;
X_{19} – number of measurements to align the coordinate systems of the measured product and the CMM;
X_{20} – number of measurements at the control size;
X_{21} – number of measurements in the control form;
X_{22} – number of measurements at the control of deflection location for surface;
X_{23} – number of measurements for further analysis;
X_{24} – temperature fluctuation during measurement, degree/min;
X_{25} – deformation of the base and the base parts of CMM, μm.

3 Mathematical Models

Order to reflect of this interaction is built Count causal relationships shown in Fig. 2.

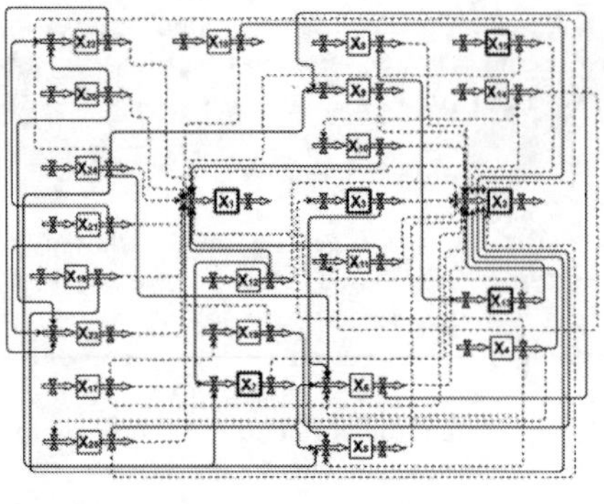

Fig. 2. Count causal relationships.

Based on the graph causal relationships to build a system of differential equations, each of which has the form:

$$\frac{dX_i}{dt} = X_i^+ - X_i^-; \quad F_i^+(\bullet) = S_i^+ \Pi_i^+; \quad F_i^-(\bullet) = S_i^- \Pi_i^-,$$

where X_i – simulated characteristic; S_i^+ and Π_i^+ – respectively the sum of external factors and the product of positively influencing variables; S_i^- and Π_i^- – adversely affecting the growth of variable X_i.

Thus, for the preparation of the differential equations of system dynamics is used relationship:

$$\frac{dX_i}{dt} = S_i^+ \Pi_i^+ - S_i^- \Pi_i^-.$$

The system of differential equations used for system simulation of the main characteristics of the CMM, has the following form:

$$\left\{\begin{aligned}
&\frac{dX_1(t)}{dt} = f_{9-1}(X_9(t)) f_{10-1}(X_{10}(t)) f_{11-1}(X_{11}(t)) C_3 - f_{12-1}(X_{12}(t)) f_{13-1}(X_{13}(t)) \times \\
&\times f_{14-1}(X_{14}(t)) f_{15-1}(X_{15}(t))(C_2 + C_4 + C_5 + C_6 + Y_1(t) + Y_2(t)); \\
&\frac{dX_2(t)}{dt} = f_{4-2}(X_4(t)) f_{13-2}(X_{13}(t))(C_4 + Y_1(t)) - f_{3-2}(X_3(t)) f_{5-2}(X_5(t)) \times \\
&\times f_{6-2}(X_6(t)) f_{7-2}(X_7(t)) f_{8-2}(X_8(t)) f_{9-2}(X_9(t)) f_{10-2}(X_{10}(t)) f_{11-2} \times \\
&\times (X_{11}(t)) f_{12-2}(X_{12}(t)) f_{16-2}(X_{16}(t))(C_1 + C_3 + Y_3(t)); \\
&\frac{dX_3(t)}{dt} = -f_{4-3}(X_4(t)); \\
&\frac{dX_4(t)}{dt} = -C_3; \\
&\frac{dX_5(t)}{dt} = f_{3-5}(X_3(t)) f_{16-5}(X_{16}(t)) Y_3(t) - f_{4-5}(X_4(t)); \\
&\frac{dX_6(t)}{dt} = f_{3-6}(X_3(t)) f_{16-6}(X_{16}(t)) Y_3(t) - f_{4-6}(X_4(t)); \\
&\frac{dX_7(t)}{dt} = f_{11-7}(X_{11}(t))(C_3 + Y_3(t)); \\
&\frac{dX_8(t)}{dt} = f_{6-8}(X_6(t)) Y_3(t); \\
&\frac{dX_9(t)}{dt} = -(C_2 + Y_1(t)); \\
&\frac{dX_{10}(t)}{dt} = -(C_2 + Y_1(t)); \\
&\frac{dX_{11}(t)}{dt} = Y_3(t) - Y_1(t); \\
&\frac{dX_{12}(t)}{dt} = Y_2(t); \\
&\frac{dX_{13}(t)}{dt} = f_{12-13}(X_{12}(t)) Y_2(t); \\
&\frac{dX_{14}(t)}{dt} = C_5 + C_6; \\
&\frac{dX_{15}(t)}{dt} = f_{14-15}(X_{14}(t))(C_5 + C_6); \\
&\frac{dX_{16}(t)}{dt} = f_{12-16}(X_{12}(t)) Y_3(t) - f_{4-16}(X_4(t)).
\end{aligned}\right.$$

d

For the numerical solution of the system is necessary to normalize the parameters and set the initial values. After normalizing the simulated variables and substituting the functions f_{i-j}, the system takes the form:

$$\left\{\begin{aligned}
&\frac{dX_1(t)}{dt} = (0{,}6X_9(t) + 0{,}15)(0{,}3X_{10}(t) - 0{,}15)(0{,}1X_{11}(t) + 0{,}15)C_3 + \\
&- \frac{X_{12}(t)}{0{,}9}(0{,}17X_{13}^3(t) + 0{,}05)(0{,}12X_{14}^4(t) + 0{,}07)(8X_{15}^3(t) + \\
&+ 0{,}04)(C_2 + C_4 + C_5 + C_6 + Y_1(t) + Y_2(t)); \\
&\frac{dX_2(t)}{dt} = (4X_4^4(t) + 0{,}07)(0{,}5X_{13}(t) + 9)(C_4 + Y_1(t)) - (-0{,}03X_3^2(t) + \\
&0{,}2X_3(t) + 0{,}1)(0{,}003X_6^2(t) + 0{,}035X_6(t) + 0{,}002)(0{,}95X_7(t) + 0{,}03\sin(t)) \times \\
&\times 0{,}75X_8(t) 0{,}75X_9(t) 0{,}7X_{10}(t) 0{,}68X_{11}^2(t)(0{,}0005X_{12}^2(t) + 0{,}003X_{12}(t) + \\
&+ 0{,}0004) 0{,}17X_{16}(t)(C_1 + C_3 + Y_3(t)); \\
&\frac{dX_3(t)}{dt} = -5 \times 10^{-10} X_4(t) - 0{,}07\sin(t); \\
&\frac{dX_4(t)}{dt} = -C_3; \\
&\frac{dX_5(t)}{dt} = (0{,}015X_3^2(t) + 0{,}004X_3(t))(0{,}003X_{16}^2(t) + 7 \times 10^{-8} X_{16}(t))Y_3(t) + \\
&- \frac{4 \times 10^{10}}{X_4(t)} - 0{,}007; \\
&\frac{dX_6(t)}{dt} = (0{,}78X_3(t) + 1{,}2)(0{,}003X_{16}^2(t) + 0{,}34X_{16}(t))Y_3(t) - \frac{4 \times 10^9}{X_4(t)} + 0{,}35; \\
&\frac{dX_7(t)}{dt} = (-3 \times 10^{-8} X_{11}^2(t) + 0{,}001X_{11}(t) + 0{,}5)(C_3 + Y_3(t)); \\
&\frac{dX_8(t)}{dt} = (0{,}02X_6^2(t) + 0{,}001X_6(t))Y_3(t); \\
&\frac{dX_9(t)}{dt} = -(C_2 + Y_1(t)); \\
&\frac{dX_{10}(t)}{dt} = -(C_2 + Y_1(t)); \\
&\frac{dX_{11}(t)}{dt} = Y_3(t) - Y_1(t); \\
&\frac{dX_{12}(t)}{dt} = Y_2(t); \\
&\frac{dX_{13}(t)}{dt} = (0{,}04X_{12}(t) + 9)Y_2(t); \\
&\frac{dX_{14}(t)}{dt} = C_5 + C_6; \\
&\frac{dX_{15}(t)}{dt} = (0{,}19X_{14}(t) + 25)(C_5 + C_6); \\
&\frac{dX_{16}(t)}{dt} = (3{,}5 \times 10^{-5} X_{12}^2(t) + 0{,}012X_{12}(t))Y_3(t) - \frac{30}{X_4(t)}.
\end{aligned}\right.$$

In developing this system of equations is used a statistical data describing the cause-effect relationship between the internal model parameters and environmental factors. It is shown the partial solutions for a model example at short time intervals in the form of cross sections performance and accuracy Hypersurfaces. These solutions make it possible to practically interpret developed models and methods of system dynamics used in solving the problem. It was found that a significant potential for growth performance while ensuring the specified accuracy of measurements is hidden in the optimization of the number, location and sequence of control points on difficult surfaces. Therefore, this direction is chosen as the most effective for further research. The results of these studies constitute the methodological basis for the development of new information and advising systems, which can be used to simulate and optimize the performance of existing and prospective Coordinate Measuring Machines.

4 Conclusion

 DOI: 10.1051/matecconf/201712901054

Numerically solving systems of nonlinear differential equations allows you to find the optimal value of the performance under the constraints on the accuracy. The practical implementation of this approach can be used when creating the concept and appearance of perspective Coordinate Measuring Machines, possessing the highest accuracy, reliability and performance at this stage of technological development.

The study was performed by a grant from the Russian Science Foundation (project №16-19-10204).

References

1. B.S. Brazhkin and V.S. Mirotvorskii, *Calculation of Curved Surfaces on Coordinate Measuring Machines*, Measurement Techniques **48,** pp. 657 (2003)
2. V.G. Domrachev, V.M. Grechishnikov and A.Zh. Chernyavskii, *Determination of the Oscillation Parameters of the Blades of Turbine-Driven Sets based on Nonlinear Approximation of the Signals of Primary Converters*, Measurement Techniques **56,** pp. 1242 (2014)
3. O.A. Yalovoy, O.V. Zakharov and A.V. Kochetkov. *The Centerless Measurement of Roundness with Optimal Adjustment*, IOP Conf. Series: Materials Science and Engineering **93**, pp. 012024 (2015)
4. O.V. Zakharov, I.N. Bobrovskij and A.V. Kochetkov, *Analysis of Methods for Estimation of Machine Workpiece Roundness*, Procedia Engineering **150**, pp. 963 (2016)
5. M. Sieniło, *Compensation of influence of elements eccentric positioning on the result of roundness deviation measurement of discontinuous sections by radial method*, Advances in Intelligent Systems and Computing **393,** pp. 361 (2016)
6. J. Caja, E. Gomez, P. Maresca and M. Berzal, *Development of a Calibration Model for Optical Measuring Machines*, Procedia Engineering **63,** pp. 225 (2013)
7. S.H. Mian and A. Al-Ahmari, *Enhance performance of inspection process on Coordinate Measuring Machine*, Measurement **47**, pp. 78 (2014)
8. F.A.M. Ferreira, J. de Vicente y Oliva and A.M.S. Perez, *Evaluation of the performance of coordinate measuring machines in the industry, using calibrated artefacts*, Procedia Engineering **63,** pp. 659 (2013)
9. R. D'Amato, J. Caja, P. Maresca and E. Gómez, *Use of coordinate measuring machine to measure angles by geometric characterization of perpendicular planes. Estimating uncertainty*, Measurement **47,** pp. 598 (2014)
10. A. Gąska, P. Gąska and M. Gruza, *Simulation Model for Correction and Modeling of Probe Head Errors in Five-Axis Coordinate Systems*, Appl. Sci. **6**, pp. 144 (2016)
11. R. Calvo, R. D'Amato, E. Gómez and R. Domingo, *Integration of Error Compensation of Coordinate Measuring Machines into Feature Measurement: Part II – Experimental Implementation*, Sensors, **16** pp. 1705 (2016)
12. Q. Huang, K. Wu, C. Wang, R. Li, K.-C. Fan and Y. Fei, *Development of an Abbe Error Free Micro Coordinate Measuring Machine*, Appl. Sci. **6,** pp. 97 (2016)
13. V.A. Pechenin, M.A. Bolotov, *Basing error in coordinate measurements of cylindrical gears*, Russian Engineering Research **36,** pp. 630-634 (2016)
14. V.A. Pechenin, M.A. Bolotov, N.V. Ruzanov, *Development of a method of ICP algorithm accuracy improvement during shaped profiles and surfaces control*, International Journal of Engineering and Technology **6**, pp. 2229-2235 (2014)
15. M.A. Badar, S. Raman and P. S. Pulat, *Experimental verification of manufacturing error pattern and its utilization in form tolerance sampling*, International Journal of Machine Tools and Manufacture **45**, pp. 63-73 (2005)
16. Małgorzata Poniatowska and Andrzej Werner, *Simulation tests of the method for determining a cad model of free-form surface deterministic deviations*, Metrol. Meas. Syst., **XIX**, pp. 151-158 (2012)
17. R. Ascione, G. Moroni, S. Petrò and D. Romano, *Adaptive inspection in coordinate metrology based on kriging models*, Precision Engineering **37,** pp. 44-60 (2013)

Study of the impact of treatment modes on hardness, deformability and microstructure of VT6 (Ti–6Al–4V) and VV751P (Ni-15Co-10Cr) alloy samples after selective laser sintering

Natalia V. Galkina[1,*], *Ekaterina A.* Nosova[1], and *Andrey V.* Balaykin[1]

[1]Samara National Research University, Russia, 34, Moskovskoye Shosse, 443086, Samara, Russia

Abstract. Selective laser sintering is an advanced method for obtaining sophisticated products and assembly permanent joints. This is particularly relevant for heat resistant alloys employed in aviation equipment. Heat treatment modes traditionally applied to the products are chosen in accordance with conditions of further product operation. In this paper there are given the results of experimental study of hardness, deformability and microstructure of samples after selective laser sintering of Ni-15Co-10Cr and Ti–6Al–4V alloy powders. It has been determined that Ni-15Co-10Cr alloy ageing increases the hardness and deformability of samples; these characteristics decrease if the ageing lasts for 9-19 hours. Annealing of Ti–6Al–4V alloy samples results in preserving original hardness. After complete annealing, the hardness of samples decreases from 32 ... 33HRC to 24 ... 26HRC. Microstructural studies showed that there are cracks between layers in the surface of Ti–6Al–4V alloy samples after sintering and not complete annealing. After full annealing, cracks' width and length decreased. Cracks in Ni-15Co-10Cr alloy samples' microstructure were not detected.

1 Introduction

Additive technologies have become popular in recent times. This is explained by the fact that final products can be derived by adding a layer of material [1]. Selective laser sintering is a metallic additive production in which parts are formed by melting of consecutive layers of powder particles using a laser beam [2 – 4]. Due to the fact that SLS enables to create functional parts of sophisticated forms, it is increasingly being used to manufacture parts in various areas of mechanical engineering, including aviation industry [5-7]. Heat resistant materials based on nickel and titanium alloys are well used in engine construction [8 - 11].

Due to the difference in the properties between sintered samples and those obtained by traditional methods, it is necessary to determine the reasons for these differences.

The purpose of this work was to indentify strength and plastic properties of sintered workpieces in comparison with traditionally used Ti–6Al–4V and Ni-15Co-10Cr alloy products, and to explain the result through a structure study with optical metallography methods.

2 Materials and Method

2x10x30 mm samples were grown out of Ni-15Co-10Cr and Ti–6Al–4V material with the method of selective sintering.

After growing, samples from Ni-15Co-10Cr material were quenched at a temperature of 1000°C for 1.5 hours, then at a temperature of 1180°C for 2.5-3 hours, then in the furnace for 50-60 min at a temperature of 1050°C; cooling together with the furnace to 1000 ° C takes place for 50 - 60 minutes; further cooling is in the air. After quenching, samples were aged at a temperature of 850°C for 3, 6, 9, 12 and 15 hours. Titanium alloy samples were annealed after sintering for one hour at a temperature of 750°C and 850°C.

After heat treatment was carried out, the samples were subject to bending test in a tool-stamp with an apex angle of 90° and a rounding radius of 10 mm [12].

In order to obtain a deformation index, the dimensions of bent samples were measured, particularly the height of curvilinear isosceles triangle's outer and inner sides and the length of the neutral layer. Deformability was evaluated according to the following formula:

$$d = \frac{H_{aver}}{l_H}, \quad (1)$$

where H_{aver} - is an average height of the curvilinear isosceles triangle; l_H - is a neutral layer length.

Investigation of the structure of deformability index shows that the greater is the value of this index, the higher is deformability.

* Corresponding author: galkina_nv@ssau.ru

 DOI: 10.1051/matecconf/201712901055

From the samples obtained, microsections were produced. The etching of titanium alloy samples was carried out by concentrated hydrofluoric acid HF for 5 seconds. Etching was made as follows: 180 ml of H2O, 180 ml of HCL, 120 ml of HNO3, 30 ml of HF. Etching duration is 1-2 minutes, washing carried out with flowing water, drying performed with filter paper. After etching, microstructure was studied with metallographic instrumental microscope METAM LV-32.

For each sample, the hardness of the side surface was measured according to Rockwell with a diamond point.

3 Results and Discussion

Dependence of hardness change on the duration of ageing is shown in Fig. 1 and Fig. 2.

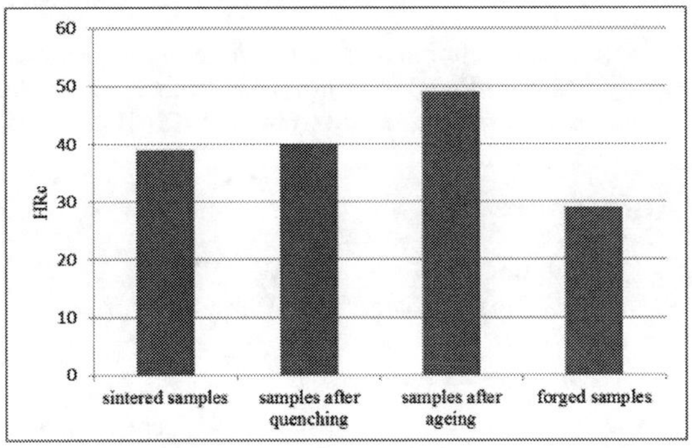

Fig. 1. Dependence of hardness change on the duration of ageing of Ni-15Co-10Cr alloy sample

Figure 1 shows that quenching and further ageing result in high levels of hardness. Compared to the values specified in Ni-15Co-10Cr compact alloy reference, the hardness should be 241-277HB, which corresponds to 24 ... 29HRC. These figures are significantly lower than those obtained after sintering, quenching and ageing.

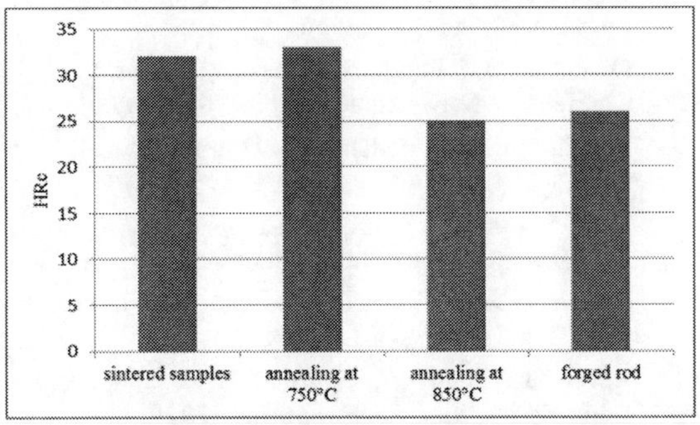

Fig. 2. Dependence of hardness change on annealing temperature for Ti–6Al–4V alloy sample

As you can see from Fig. 2, the hardness of Ti–6Al–4V alloy after not complete annealing at 750°C temperature for an hour remains rather high, but after full annealing at the temperature of 850°C, the hardness decreases from 32... 33 HRC to 24 ... 26HRC, which is about 30%. Ti–6Al–4V alloy rod after annealing, according to the standard, should have a hardness of 26..41 HRC. These figures show that the sintered samples are in the middle of specified limits, while the annealed samples' hardness corresponds to the lower boundary of the values specified in the standards.

Based on the measurements of the heights of curvilinear isosceles triangle, the deformability factor for each sample was calculated. As a result, there was determined the dependence of deformability factor on the time of ageing for Ni-15Co-10Cr material (Fig. 3) and the temperature of annealing for titanium alloy Ti–6Al–4V (Fig. 4).

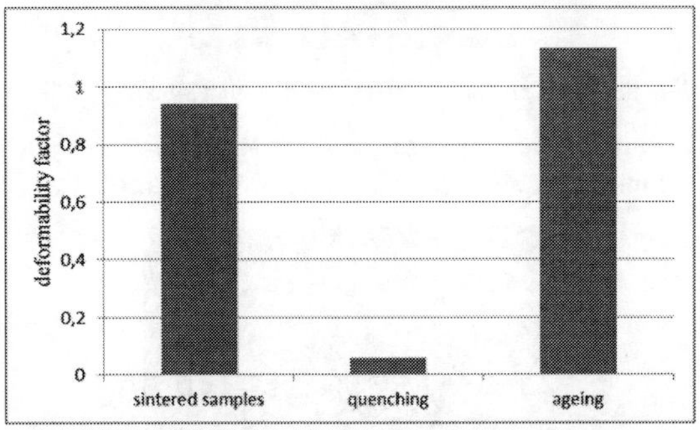

Fig. 3. Dependence of deformability factor on Ni-15Co-10Cr alloy treatment mode

The diagram in Fig. 3 shows that hardening and ageing increase deformability characteristics. For these geometric dimensions of curved samples, the relative elongation of the outer layer should be within 7.4 ... 8%. These values are lower than the elongation for Ni-15Co-10Cr alloy specified in standard, which amount to at least 12%.

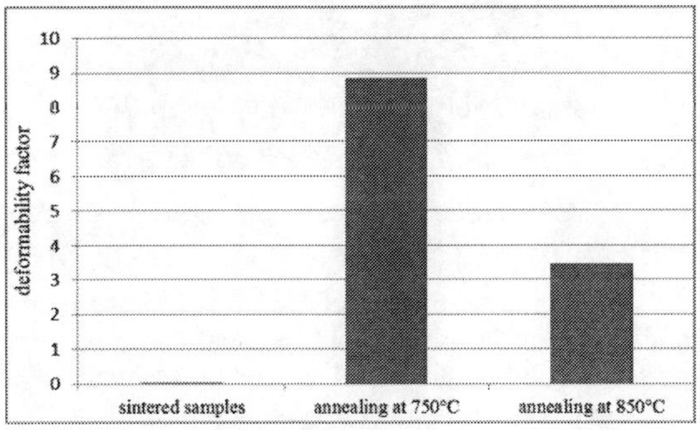

Fig. 4. Dependence of deformability factor on the temperature of annealing for Ti–6Al–4V alloy

The chart in Figure 4 shows that Ti–6Al–4V alloy deformability after annealing increases compared to the original sintered state. For such deformability characteristics, the degree of deformation of the outer layer of bent samples shall be 0.4 ... 1.25%. These values are an order of magnitude lower than the relative elongation specified in GOST for hot rolled rods from Ti–6Al–4V alloy, which shall be 8...12 percent.

 DOI: 10.1051/matecconf/201712901055

4 Microstructure

Microstructure of sintered samples is present on fig. 5-8.

Fig. 5. Microstructure of Ni-15Co-10Cr after sintering, 100

From Fig. 5, structure of Ni-15Co-10Cr alloy samples after sintering is uniform, free of cracks.

Fig. 6. Microstructure of sintered sample of Ti–6Al–4V alloy

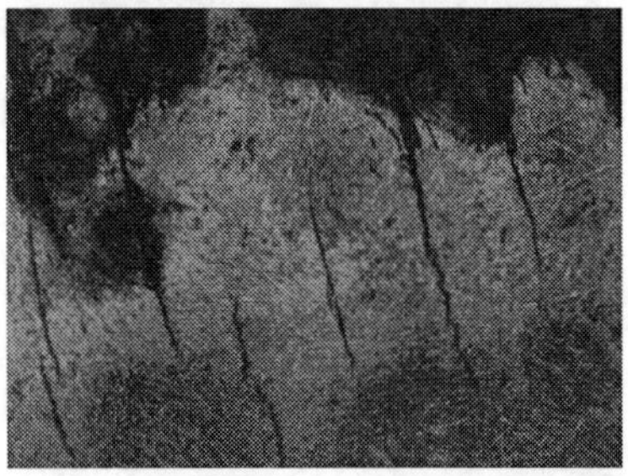

Fig. 7. Microstructure of sintered and annealed at 750°C sample of Ti–6Al–4V alloy

From Fig. 6-8, structure of Ti–6Al–4V alloy samples after sintering has cracks and unbaked particles. After incomplete annealing cracks stay in microstructure but after complete annealing they become less and shorter.

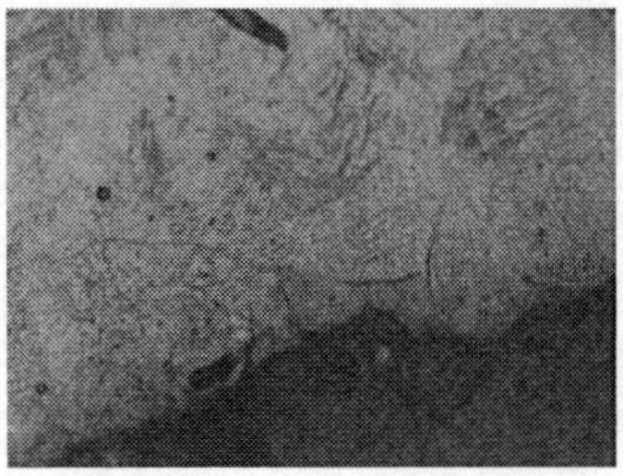

Fig. 8. Microstructure of sintered and annealed at 850°C sample of Ti–6Al–4V alloy

5 Conclusions

1. Sintered samplers obtained from powder of Ni-15Co-10Cr heat resistant alloy showed the decreased values of hardness, which is rises after quenching and aging. Comparison to forged rod, obtained samples had increased hardness due to over saluted structure of fast crystallized particles.
2. Samples sintered from Ti–6Al–4V alloy showed increased values of hardness in comparison to deformed standard rods. Annealing at 850°C lead to decreasing of hardness, approaching to forged state.
3. Deformability of sintered Ni-15Co-10Cr alloy samples goes down after annealing but grows up after aging.
4. Deformability of samples sintered form Ti–6Al–4V alloy has maximum values after incomplete annealing.
5. Decreased deformability after sintering in samples Ti–6Al–4V alloy due to interlayer cracks, which are absent in samples sintered from Ni-15Co-10Cr alloy

References

1. I. Gibson, D.W. Rosen, B. Stucker, *Additive Manufacturing Technologies* (Springer, 2010)
2. Z. Hu, H. Zhu, H. Zhang, X. Zeng, opt. Laser Technol, **87**, 17 (2017)
3. G. Casalino, S.L. Campanelli, N. Contuzzi, A.D. Ludovico, opt. Laser Technol, **65**, 151 (2015)
4. Q. Shi, D. Gu, M. Xia, S. Cao, T. Rong, Opt. Laser Technol., **84**, 9 (2016)
5. V.G Smelov., A.V Sotov, A.V. Agapovichev, IOP Conf. Series: Mat. Sc. Engin. **140**, 1 (2016)
6. A.V. Agapovichev, V.V. Kokareva, V.G. Smelov, A.V. Sotov, IOP Conf. Series: Mat. Sc. Engin. **156**, 1 (2016)
7. V.G. Smelov, A.V. Sotov, A.V. Agapovichev, Key Engin. Mat. **684**, 316 (2016)
8. D. Holec, D. Legut, L. Isaeva, P. Souvatzis, H. Clemens, S. Mayer Intermetallics, **61**, 85 (2015)
9. R. Pflumm, S. Friedle, M. Schütze, Intermetallics **56,** 1 (2015)
10. T. Shanmugasundaram, J. Guyon, J.P. Monchoux, A. Hazotte, E. Bouzy Intermetallics, **66,** 141 (2015)
11. X. Meng, W. Cai, F. Chen, L. Zhao Scr. Mater., **54,** 1599 (2006)
12. A.V. Balaykin, E.A. Nosova, N.V. Galkina, IOP Conf. Series: Mat. Sc. Engin. **142** (2016)

Direct formation of surface layer's quality parameters during combined treatment of two-phase stainless steel blanks

Eugeny Kiselev[1,*], and *Oleg* Blagovskiy[2]

[1]Ulyanovsk State Technical University, 432027 Ulyanovsk, Russia
[2]Ulyanovsk Mechanical Plant, 432008 Ulyanovsk, Russia

Abstract. The results of experimental studies of formation surface layer's quality parameters (residual stresses and phase composition) forming during combined turning and ultrasonic hardmetal burnishing of blanks made of 07X16H6 two-phase stainless steel are represented. Shows the effect of combined treatment regime on the value and depth of the arising compressive residual stresses and phase composition (volume of austenite). Also demonstrated that the changes in this parameters are distributed to a depth exteeding the allowance for subsequent processing.

1 Introduction

It is generally known that a significant impact on the performance of the critical parts (in particular, fatigue strength and corrosion resistance) have technological residual stresses (TRS) and phase composition (PC) of the material from which they are made. Using on the final operations of technological process (TP) of manufacturing parts processing methods such as grinding or high speed cutting, coupled with high temperatures in the contact zone of treatment dramatically increases the chance of phase transformations and the formation of the tensile stresses in surface layer, which significantly reduce fatigue and corrosion resistance of the finished product. This problem is particularly acute in the manufacture of parts of the two-phase stainless steels have low thermal conductivity and are prone to phase transformations, in particular, the reverse transition from the γ-iron to the α-iron.

Moreover, the greatest effect on corrosion resistance is the amount of an austenite phase (γ-Fe) having a face-centered cubic (FCC) lattice and has an increased chemical resistance and strength at high temperatures [1]. A typical representative of such material is AISI 301 steel (russian analog – 07X16H6), details of which operate in harsh environments and experience cyclic loading. The most widely to ensure a given level of residual stresses and phase composition of the surface layer of thermal methods used (natural or artificial aging, annealing for stress relaxation, partial or complete austenitization to form austenite phase by heating iron alloys to the transformation temperature or above) characterized by high energy and time consuming. Another effective way to accomplish this is to use the methods of surface plastic deformation (diamond or carbide burnishing, etc.).

As known, their limitation is the difficulty in the manufacture of thin-walled and low-rigidly parts. Moreover, using of operations of plastic deformation of the surface in the final stages of TP inevitably leads to the collapse of the austenite in the surface layer of workpieces and reducing the corrosion resistance of the treated surfaces of machine parts. Therefore, to provide the required mechanical properties of the surface layer is necessary to efficiently use the technological heredity (TH) phenomena, creating predetermined residual stresses and phase composition on operation preceding the final one.

Technological heredity is a transfer on the finished product in the manufacturing process errors, mechanical and physico-chemical properties of the initial blank or properties and errors that have emerged in that blank at individual manufacturing operations [2]. With the use of this phenomenon for a given level of technical requirements necessary to establish direct links between quality parameters formed on roughing and finish operations process, or to determine the influence of elements of treatment on the mechanical properties of the surface layer and has preset specifications of manufactured parts.

For example, consider a fragment of manufacturing process of parts made of the two-phase corrosion-resistant steel austenitic-martensitic class AISI 301, which includes the operation of combined turning and ultrasonic hardmetal burnishing (UHB) (equipped with a carbide cutter Korloy CCMT 060204-HPF NC8010; burnisher - plate Korloy CNMG 120408-HS NC9025), made on a lathe UT16P (УТ16П), and subsequent grinding operation (grinding machine 3M151V, grinding wheel 1-600x25x305 24AF8M7V35).

Samples in studies served rolls with a diameter of 60 mm and a length of 200 mm. Ultrasonic vibrations (frequency f = 22 kHz, amplitude A = 10-12 μm) was applied to the hardmetal burnisher from ultrasonic

* Corresponding author: kec.ulstu@mail.ru

generator UZG-641A (УЗГ-641A). Burnisher located opposite the turning tool with a lag of ≈ 5 mm.

Turning and grinding processing was carried out applying the tool to the zone of contact with the workpiece 3% aqueous solution semisynthetic coolant Vels-1 (flow rate: 10 l/min – turning; 40 l/min - grinding). Surface plastic deformation during combined treatment being fed into the nip with the hardmetal burnisher and blank with 100% concentrate coolant Vels-1 at a rate of 5 g per hour by applying it to the objects with a brush.

2 Results

As varied factors in determining the elements of the regime combined turning and burnishing provide minimum, maximum and average values of the residual stresses in the surface layer at a depth of grinding allowance Z_G considered cutting allowance t (0.5, 0.75 and 1 mm), the cutting speed V (47.1, 59.3 and 75.4 m/min), feed S (0.046, 0.054 and 0.064 mm/rev) and clamping force F (100, 200 and 300 N). Some results of experimental studies of residual stresses on the device "Siton-ARM" ("Ситон-АРМ") showed in Fig. 1 - 4.

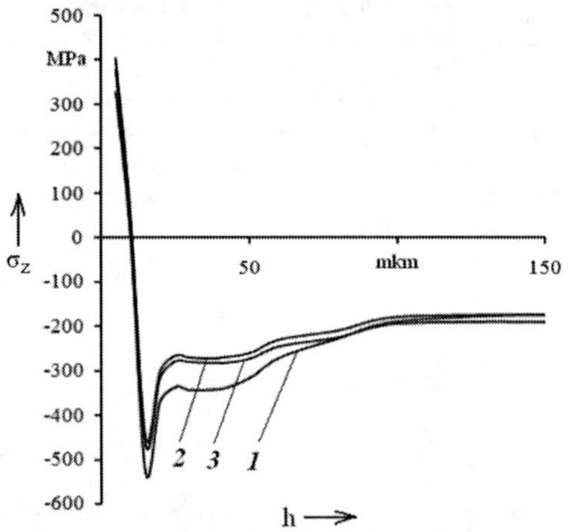

Fig. 1. Diagrams residual stresses after combined treatment by turning and ultrasonic hardmetal burnishing with various allowance: 1 – t = 0,5 mm; 2 – t = 0,75 mm; 3 – t = 1 mm; other parameters of processing – V = 59,3 m/min; S = 0,054 mm/rev; P = 200 N.

Shown in Fig. 1 - 4 show the stress distribution ambiguous effect regime change in all elements of the combined treatment on the stress but clamping force P. This is due, in our view, with the influence of various processing components of force and contact temperatures in areas of cutting and smoothing, as well as their phase transformations caused by the material of the surface layer.

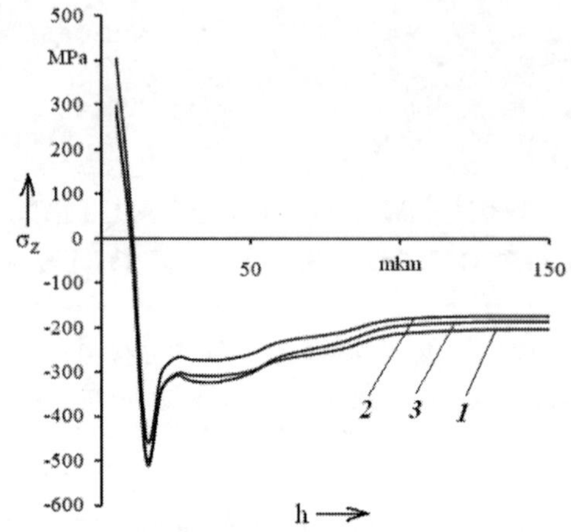

Fig. 2. Diagrams residual stresses after combined treatment by turning and ultrasonic hardmetal burnishing with various cutting speed: 1 – V = 47,1 m/min; 2 – V = 59,3 m/min; 3 – V = 75,4 m/min; other parameters of processing – t = 0,75 mm; S = 0,054 mm/rev; P = 200 N.

After analyzing the data, were selected three sets of elements for the combined treatment, providing maximum, medium and minimum residual stresses at a depth of subsequent grinding allowance (0.1 mm). Processed on these modes combined workpiece acted on the subsequent grinding. The diagrams illustrate the effect of technological heredity of residual stresses. If, after the combined treatment for a subsequent grinding depth were formed on the maximum value of the compressive residual stress, after removal of allowance on subsequent processing step, in forming the surface layer preform minimum tensile stress. The same dependence is observed for the remaining two pairs of diagrams.

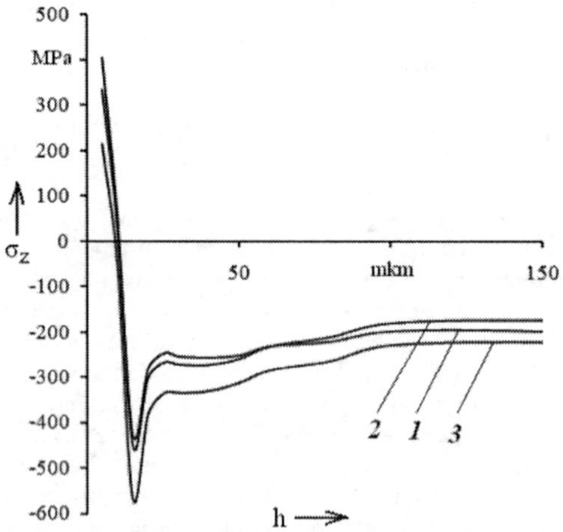

Fig. 3. Diagrams residual stresses after combined treatment by turning and ultrasonic hardmetal burnishing with various feed: *1* – S = 0,046 mm/rev; *2* – S = 0,054 mm/rev; *3* – S = 0,064 mm/rev; other parameters of processing – t = 0,75 mm; V_3 = 59,3 m/min; P = 200 N.

 DOI: 10.1051/matecconf/201712901056

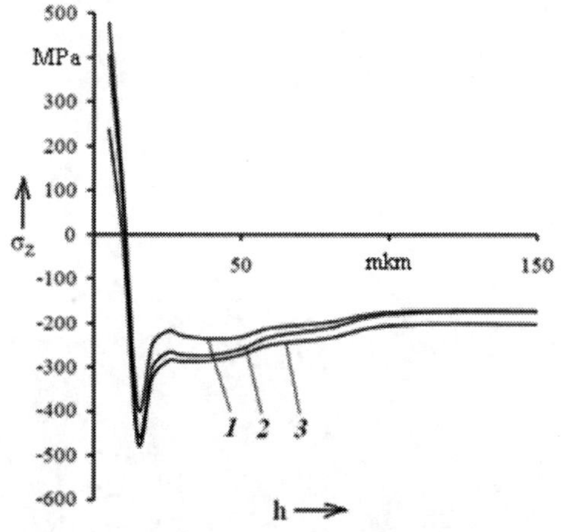

Fig. 4. Diagrams residual stresses after combined treatment by turning and ultrasonic hardmetal burnishing with various clamping force: *1* – $P = 100$ N; *2* – $P = 200$ N; *3* – $P = 300$ N; other parameters of processing – $t = 0{,}75$ mm; $V_3 = 59{,}3$ m/min; $S = 0{,}054$ mm/rev.

Experimental studies of technological inheritance of phase composition of the surface layer of the two-phase austenitic-martensitic corrosion-resistant steel AISI 301 produced by the X-ray measuring complex "RIKOR-7" ("РИКОР-7"), during which the intensity of the peaks monitored α-iron (martensite) and γ-iron (austenite) in grinded samples, which is then translated into the percentage by volume of austenite in surface layer. The results are shown in Fig. 5 - 6.

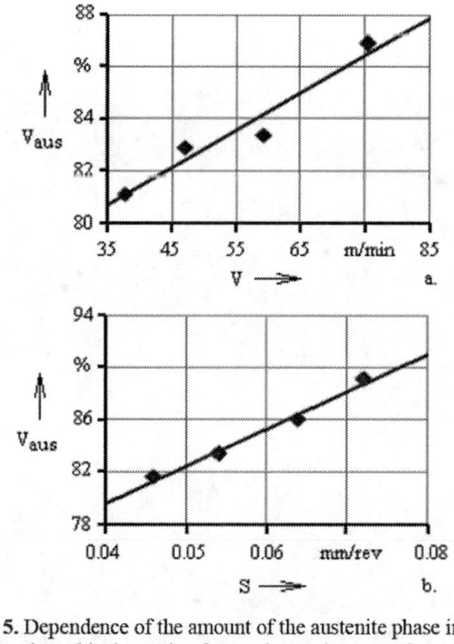

Fig. 5. Dependence of the amount of the austenite phase in the surface layer blanks made of corrosion-resistant steel AISI 301 of different regime elements of combined treatment turning and ultrasonic hardmetal burnishing.

Blanks were pretreated combined turning and burnishing with different regime elements (see Fig. 5), and then polished with infeed (see Fig. 6). Thus, the influence of determined elements for the pretreatment (allowance *t*, cutting speed *V*, feed *S* and clamping force *P*) on the percentage of austenite in grinded samples.

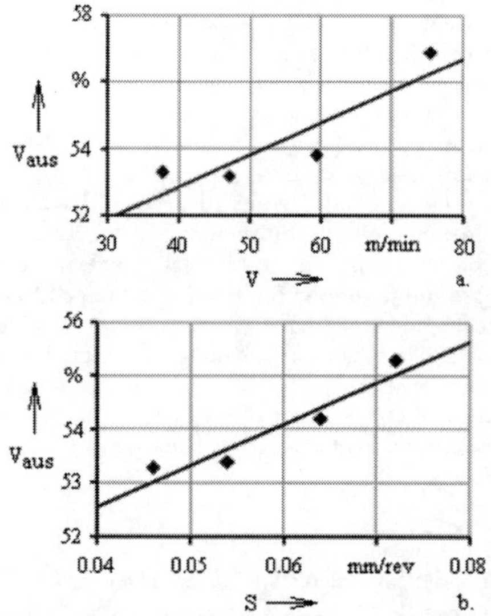

Fig. 6. Dependence of the amount of the austenite phase in the surface layer grinded blanks made of corrosion-resistant steel AISI 301 after combined treatment with the different regime elements.

After analyzing the experimental results, it can be argued that the cutting speed *V* and feed *S* have little effect on the amount inherited austenite (since changing his percentage is less than 5% of the surface layer). Growth of the allowance *t*, according experimental results, reduces the amount of austenite in grinded workpieces by almost 10%.

A similar pattern is observed with increasing of clamping force *P*, where the decline in the amount of austenite is 17%.

3 Summary

The results obtained are, in our opinion, as in the case of changing the residual stresses associated with the radial force components of the combined treatment, causing the austenite, and also the influence on the formation and decomposition of austenite grains in a superficial layer formed on the workpiece stage pretreatment residual stresses.

The results of the research, the following conclusions:

1. Using of combination treatment in the preliminary stages of the technological process allows to generate a sufficiently large in magnitude residual stresses at a depth greater than the amount of allowance for subsequent grinding. Taking into account that because of the large thermal stress in the process of grinding the surface layer formed blanks tensile residual stresses, the total value of stresses will be determined by their sum.

2. Significant impact force, which involves surgery combined treatment may have a negative effect on the phase composition of the surface layer of parts made of corrosion-resistant steel, causing it to decay of austenite. As the results of experimental studies, such as phase transformations extend deeper than allowance for subsequent treatment and due technological heredity phenomena, are transferred to the finished part.

3. Changing elements of the regime combined treatment can effectively provide a predetermined amount of residual stresses and phase composition of the surface layer of the items in accordance with the specifications of the finished product. However, the question of formation of residual stresses and phase composition during the preliminary combined treatment should be addressed comprehensively, as the choice of treatment regimen, providing a predetermined compressive stresses can lead to a substantial decrease in the austenite phase and, consequently, to reduce the corrosion resistance of surfaces of parts.

References

1. Y. Solntsev, *Metals and alloys. Handbook* (2003)
2. A. Dalski, *Technological ensuring of precision parts reliability* (1975)

The high-performance machining of the curved surfaces with the progressive cutting geometry of the end mill

Alexander Khramov[1], *Evgeniy* Kiselev[2,*], and *Evgeniy* Leksin[1]

[1]Haltec Ltd, 432071 Ulyanovsk, Russia
[2]Ulyanovsk State Technical University, 432027 Ulyanovsk, Russia

Abstract. There are some results of practical application of the solid carbide end mill developed in HALTEC Ltd company, using for high-performance machining of curved surfaces of workpieces, which have a complex spatial form. It's shown how to get high-performance machining of curved surfaces with a multidirectional curvature in heat-treatable and high strength Ti alloy Ti-6Al-2Sn-2Zr-2Mo-2Cr-0.15Si [Ti-6-22-22]. The real experience of the implementation of proposed end mill in the technological process of compressor blade machining is described.

1 Introduction

The machining of non-rigid workpieces, which have relatively small volume in large dimensions is a difficult task, which facing engineering technology at the moment. Due to the high ductility of technological system the high-performance machining of this workpieces usually impossible. Occurring in the surface layer of the treated parts after its removal from the machine residual stresses are the result of thermal and power intensity of the cutting, which is especially critical for non-rigid parts. Therefore, in practice, specialists are try to avoid high temperatures and cutting forces in the contact zone of the tool and the workpiece by reducing elements of the cutting conditions.

The most of non-rigid parts which used in the aircraft industry, have a complex spatial shape, including curved sections which are processed by milling on the multi-axis CNC machines. Their machining is carried out by end mills with spherical or toroidal end face, the choice each time determined by the capabilities of the equipment, the rigidity of the technological system, the geometric parameters of the specific milling, economic feasibility, etc. The main factor which is limiting the machining performance and, consequently, the cost of the final product is the small width of the cut, which affects the height of scallops machined surface. Obviously, for lowest roughness of the machined surface is important to have the more efficient radius of contact of the tool and workpiece, small step between two successive passages and minimum level of runout of tool.

But with a decrease of the step the machining time substantially increases. At the same time increase of the effective radius is currently achieved by using larger end mill diameter what is respectively increases the radius of spherical section or radius of the torus. It is known that the minimum roughness of the machined surface is achieved by the use of solid carbide end mills, but the use of such tools with a diameter bigger than 32 mm is not economically feasible due to their high cost.

Thus, all the cutting elements of the end mill (with ball or toroidal end face) – radius, end face and peripheral teeth may take part in cutting, although it should be noted that the machining of complex curved surfaces where depth of cut is relatively small, only radius of end mill is participates in cutting.

Thus, the development of the end mill, which combines the advantages of spherical and toroidal-shaped end and capable of leading high-performance machining of non-rigid parts is an urgent task.

The new end mill which combines two curved area – R_B and R_T and $R_B > R_{SHANK}$ and was developed in HALTEC Ltd (Ulyanovsk, Russia) is shown on Fig. 1 and 2. This end face design can significantly increase the effective diameter of the contact between end mill and workpiece, thereby significantly increasing the spacing between successive passes for high productivity.

A distinctive feature of the proposed end mill is the presence of the flat section W which is perpendicular to the axis of rotation (Wiper) and further cuts the scallops after consecutive passes. Sandvik Coromant (Sweden) was the first company who was using Wiper technology for machining for turning inserts T-max series. It's known that the use of inserts with Wiper geometry allows for doubling the feed while maintaining high-altitude surface roughness or halve these parameters while maintaining feed [1]. Later skills with turning inserts with Wiper geometry were extended to the milling inserts. Also tool companies invited to install the additional wiper insert on the cutter.

* Corresponding author: kec.ulstu@mail.ru

 DOI: 10.1051/matecconf/201712901057

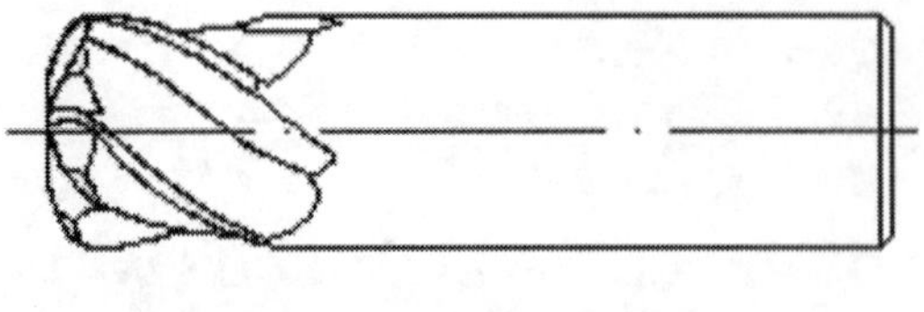

Fig. 1. General view of the proposed end mill.

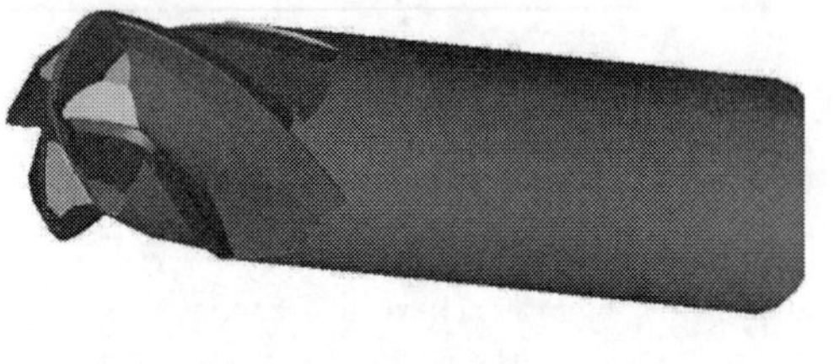

Fig. 2. 3D model of the proposed end mill in the window of WALTER Helitronic Tool Studio software.

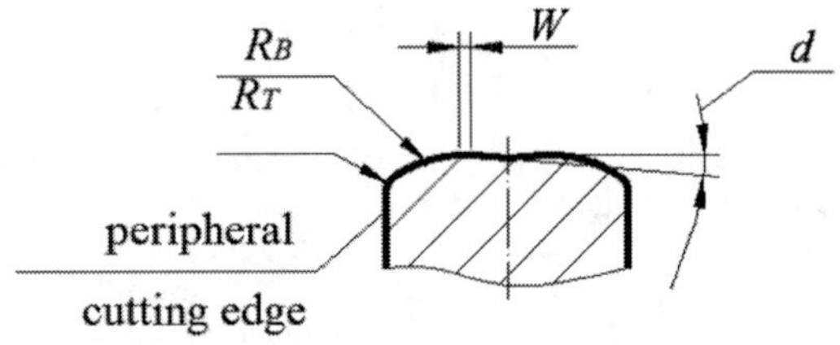

Fig. 3, a. Variant of end mill 1.

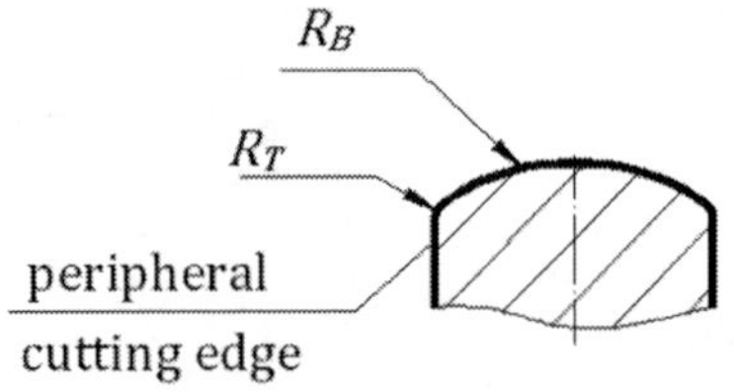

Fig. 3, b. Variant of end mill 2.

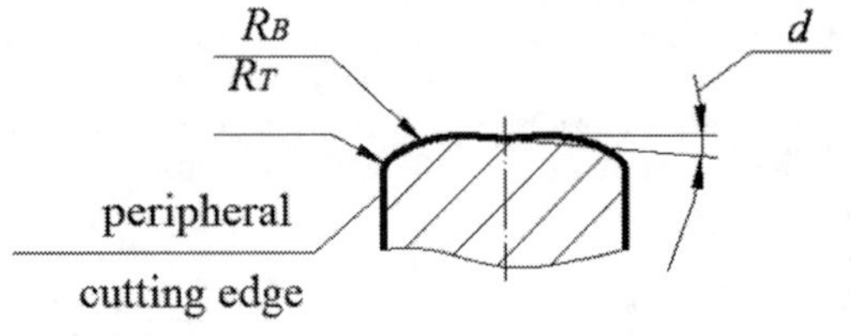

Fig. 3, c. Variant of end mill 3.

Wiper application section on the solid carbide tool is performed for the first time.

Furthermore, proposed end mill has a dish angle *d* to prevent friction of the end face teeth and machining surface.

Thus, in general case, the proposed mill is characterized by two radius sections R_B and R_T, the flat section *W* and dish angle *d* (Fig. 3a-c) [2].

2 Results

The basic ability to cut curvilinear surfaces of the workpiece on the example of milling titanium alloy Ti-6-22-22 is shown on Fig. 4 a) and b). Essence of the test is the milling of a curved convex and concave surface by the end face of tool. The workpiece was fixed in machine grip SCHUNK KSG125 and was machined with using of typical end mill (series MT4RXA, cutting diameter D = 20 mm, corner radius Rt = 3 mm, cutting length Lcut = 38 mm) and also with using of HALTEC proposed end mill (series MI4RXA, type 1, the cutting diameter D = 20 mm, corner radius Rtor=3 mm, ball section radius Rb = 15 mm, Wiper area = 0,1 mm, cutting length Lcut = 38 mm).

Practical recommendations how to use the proposed end mill for machining are shown in Table.

Table 1. Recommendations for use of the proposed end mill series MT4DXA

Type of machining	Variant of end mill from Fig. 3		
	1	2	3
Rough curved surfaces milling	0	+	+
Rough plane milling	+	0	+
Finish curved surfaces milling	–	+	+
Finish plane milling	+	–	–
Slot milling, pocket milling	+	–	0
Milling by the peripheral teeth	0	0	0

«+» – recommended; «0» – permissible; «–» – not recommended

Cutting conditions: depth of cut ap = 1.7 mm; width of cut ae = 7mm; cutting speed Vc = 40m/min; feed per tooth fz = 0,5 mm.

Typical end mill was broken after 6 minutes of machining (Fig 5, a). Proposed end mill kept possibility to work after 2 hours (Fig. 5, b)

In addition, the power intensity of the cutting by proposed end mill is about 40% lower than the typical end mill (Fig. 6).

Fig. 4, a. Test piece.

Fig. 4, b. Test piece after machining by proposed end mill.

The practical value of proposed end mill is the possibility of a significant increase of blade machining performance.

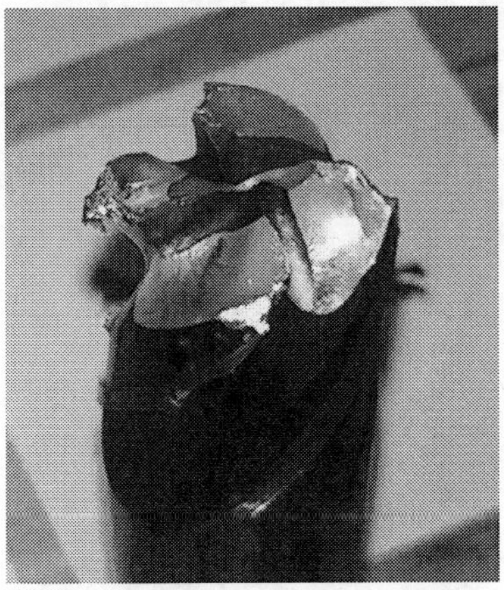

Fig. 5, a. Typical end mill series MT4RXA after 20 minutes machining of the test piece

The aim is to achieve maximum performance with satisfactory quality of machined surface and tool life for milling operation profile of the blade of compressor turbine engines. Used workpiece is stamping of titanium alloy TiAl6V with 2 mm allowance on each side.

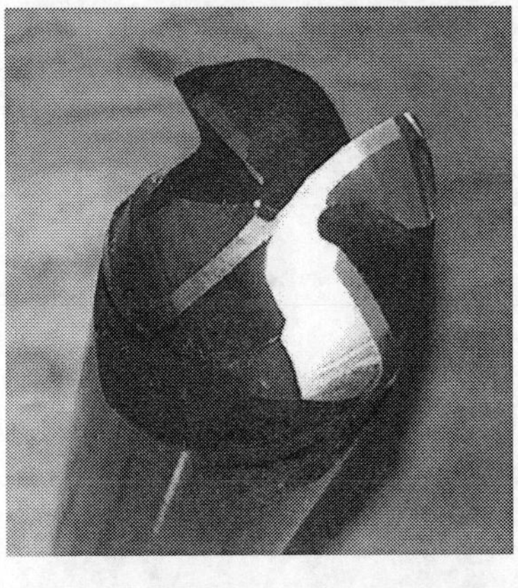

Fig. 5, b. Proposed end mill series MT4DXA after 120 minutes machining of the test piece.

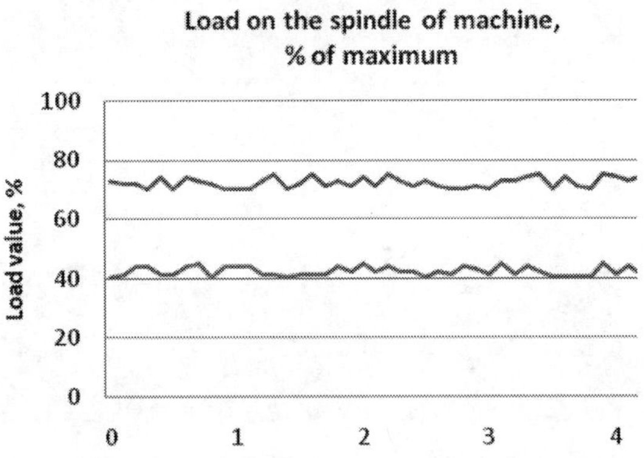

Fig. 6. Specific power intensity of the milling of the test piece. The time interval of 4 seconds is shown (steady cutting).

Workpiece clamping is carried out for the pre-treated surface of the blade lock in a specially designed device which is mounted on the faceplate, clamped in a self-centering three-jaw chuck of the MAZAK Integrex i-200 machining center. From the side of the shank the workpiece is pressed rear center with a force of 200 N (Fig. 7).

According to the accepted and proven technology, it is necessary to leave allowance for subsequent finish milling of 0.3 mm.

The modern technological process of machining of similar parts uses semi-finishing and finishing five-axis milling in one setup. According to the results of the used NC program the semi-finishing blade machining carried by toroidal end mill series MT4RXA, described earlier. Cutting conditions for reasonable tool life with satisfactory surface quality are: cutting speed Vc = 40m/min; feed per tooth fz = 0,06 mm; depth of cut ap = 2 mm; width of cut ae = 3 mm. Under such conditions, the machining time of semi-finishing blade profile operation was 27 minutes. Tool life until 0.3 mm flank wear is 108 minutes or four parts.

Exterior view of the workpiece after semi-finishing blade milling is shown in Fig. 8.

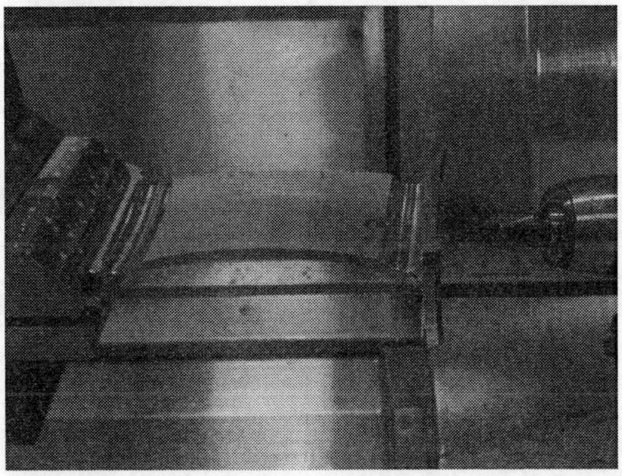

Fig. 7. Blade workpiece in the machine device.

DOI: 10.1051/matecconf/201712901057

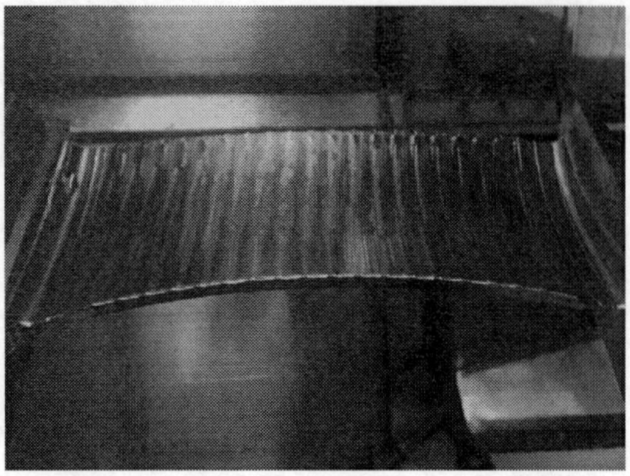

Fig. 8. Exterior view of the workpiece after semi-finishing blade milling.

Using of own proposed original end mill allowed to increase the feed more than 3 times - up to 0.2 mm / tooth with increasing the cutting width from 3 to 7 mm and reduce machining time to 5 minutes. It was noted more smooth machining, much less vibration and better surface quality by reducing the height of scallops. After 20 workpieces machining proposed end mill kept possibility to work: spalling and chipping on the cutting edge is not detected. So, further increase of the cutting feed is impractical due to low rigidity of the technological system.

3 Summary

Thus, the effectiveness of the proposed end mill in the machine parts production, in particular, for complex spatial form of difficult-to-cut materials, which often are used in high-tech sectors of the industry was experimentally validated

References

1. Sandvik Coromant Company. *http://www.sandvik.coromant.com/en-gb/knowledge/general_turning/selection-of-inserts/geometry-information/wiper-inserts.*

2. A.V. Khramov, E.N. Leksin, and I.V. Semdyakin, *RF Patent 145350* (2014)

Analysis of mechanism of carbide tool wear and control by wear process

Hoang Trung Pham [1,*], and *Julius. L.* Tchigirinsky [1]

[1] Volgograd State Technical University, 400005, 28 Lenin av., Volgograd, Russia

Abstract. The analysis of physic-mechanical and thermal physic properties of hard alloys depending on their chemical composition is conducted. The correlation of cutting properties and regularities of carbide tool wear with cutting conditions and thermal physic properties of tool material are disclosed. Significant influence on the tool wear of not only mechanical, but, in the first place, thermal physic properties of tool and structural materials is established by the researches of Russian scientists, because in the range of industrial used cutting speeds the cause of tool wear are diffusion processes. The directions of intensity decreasing of tool wear by determining rational processing conditions, the choice of tool materials and wear-resistant coating on tool surface are defined.

1 Analysis of the mechanism of carbide tool wear

Depending on the physical-mechanical properties of work and tool materials, the type of contact interaction, contact temperatures and contact stresses, the type of lubricant-cooling agents, their cooling effect, the possibility of lubricant-cooling agents penetration on the boundary surface of contact interaction and of chemical reactions passing there, as well as the possibility of diffusion processes passing through the boundary of work and tool materials, tool wear can have various physical nature and there may be various wear mechanisms.

The diversity of wear content and manifestations can be divided into two types: continuous dissolution at atomic level of worn material in the wearing; wear by removing discrete submicro- and micro-volumes from worn surface, having finite geometrical dimensions.

While cutting metals can take place following wear mechanisms: diffusion, adhesion-fatigue, oxidation, abrasive and wear as a result of cutting edge micro-spall [13]. The diffusion wear is most common of them, which can be described in the following way. Mutual diffusion of tool and work materials occurs while cutting in determination conditions of stable adhesion bond. During this process takes place, first, the carbides dissociation (mostly tungsten carbides) and subsequent diffusion of the elements (W, Ti, C) in the work material (direct diffusion dissolution); second, metal elements diffusion of the work material in the binding phase of hard alloy and its softening. The consequence is bond weakening, holding carbide grains or groups of carbides in the hard alloy, tear-out and ablation by their border volumes of work material. The result of both processes is the wear of tool contact surfaces.

According to the views of N. V. Talantov [14], the type of contact interaction in a certain way has an influence on the mechanism and intensity of tool wear, as well as forming main characteristics of the cutting process. It is known that while cutting carbon and low-alloy ferrite-pearlite steels with increase in cutting speed there is regular change of the following types of contact interaction: build-up, throbbing contact zone, and zone of relative stagnation, plastic and viscous contact. For high-alloy steels with the increase in cutting speed process is characterized by the regular change, mainly of the same types of contact interaction, as for ferrite-pearlite steels: interaction in cutting with cleavage chips formation, the build-up, the cutting with plastic yielding of contact layers of the work material, in which there are no phenomena of their full stagnation [10].

While analyzing the cutting process differences of high-alloy steels from carbon and low-alloy steels often compare their thermal conductivity values at a fixed temperature. The thermal conductivity decreases with the increase in contact zone temperature during cutting carbon and low-alloy steels, and for high-alloy – increases (Fig. 1) [10]. This is because while cutting carbon and low-alloy steels there is a positive feedback in temperature, i.e. increasing θ in the contact zone leads to a decreasing of heat flow into the chip due to reduction λ and consequently to a further increase in θ. Therefore, the transition with increase in cutting speed from built-up edge to the plastic yielding of contact layers is accomplished abruptly, in a narrow speeds range, and the appearance of this kind of interaction almost immediately is accompanied by the emergence of diffusion-viscous flow and crater wear. The opposite character of the dependence $\lambda(\theta)$ for austenitic and martensitic steels entails a negative feedback on the cutting temperature – with the increase in heat flow into the chip while temperature rising in the contact zone.

* Corresponding author: phamhoangtrung@mail.ru

This leads to a contrast weakening in temperature along the length of the contact and contributes to the process development of secondary strain hardening in the contact zone at a distance from the cutting edge.

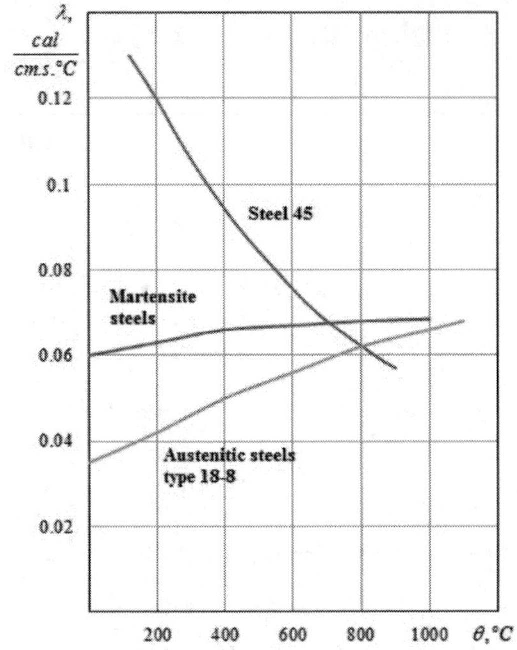

Fig. 1. The dependences of thermal conductivity coefficient of processing steels on temperature

There occurs the transition from built-up edge to a stagnant zone with an offset built-up while processing high-alloy steels with the increase in cutting speed. The main characteristic of this kind of interaction is the presence of fixed amounts of hardened metal at some distance from the cutting edge, calling the braking plastic yielding in the most part of the zone. The offset build-up is the result of secondary hardening of the contact layers of the work material.

The offset build-up disappears with further increasing of cutting speed and stagnant zone gives way to interaction with the plastic yielding of contact layers. However, the increasing of the heat flow into the chip with the increase in temperature prevents the deformation localization and the transition from plastic yielding of contact layers to viscous. Consequently the area of viscous contact (wear crater) is absent, and there is interaction with the plastic contact without viscous in a enough wide range of cutting speeds.

If when cutting ferrite-pearlite steels all kinds of interaction at the wear area on the back edge are offset relatively similar types on the front surface in the direction of high speeds, while processing high alloy steels this is not observed. For austenitic steel the transition to plastic yielding at the wear area is completed at lower cutting speeds than on the front surface. For martensitic steels the interaction with plastic contact is formed on the front surface and the wear area at close values of the cutting speed. This is due to the aforementioned differences of the thermal physic properties for these work material groups [10].

2 Control by wear process

Control by wear process of carbide tools can be implemented by determining the processing mode, the properties of tool and work material, and the features during processing and application on the working surface of the wear-resistant layer.

When establishing a stable adhesive bond between the work and tool materials along the actual tool surface, this is possible between these materials for the flow of two types of diffusions: heterogeneous and reactive. In the first case the solid solution is formed with the solvent lattice. In reactive diffusion at the boundary the chemical reaction can be to form intermetallic compounds or carbides. The formation condition of new phases during reactive diffusion is that substance participating in the process by its nature could form intermetallic compounds or carbides [13].

This can be observed when processing high alloy (titanium, cobalt) steels. The natural carbide coating is formed in the result of diffusion processes on the carbide tool surface under certain conditions. It consists of carbide-forming elements of steel and hard alloy elements. This process prevents the wear of hard alloy by the mechanism of the direct dissolution [10, 11].

In addition, there are series of processing technologies of working surfaces, giving them the additional hardening, the most effective of which is a method of applying to the cutting tool surface with special coatings of solid compounds.

By coating the working tool surface most effectively resists the typical types of tool wear. The most perspective innovative direction of the wear-resistant coatings modernization is the development of universal multilayer coatings, permitting in full degree to take into consideration of complex technological conditions of the cutting process.

The main reasons for the failure of tools with coatings are: the deformation of the wedge and the coatings diffusion wear. In this case in the capacity of matrixes under coating, we should choose a less plastic hard alloys. As coatings it is necessary to apply more wear-resistant compound [8]. Note that the application of coatings requires the absence of variable loads during processing.

At the level of the using additional technological operations when cutting hard-to-machine materials it is possible to apply advanced plastic deformation (APD). Machining with APD combines the advanced plastic deformation and subsequent cutting. By the beginning of the removal of the stock part work, which takes costs for cutter for plastic deformation already being performed by the device for the implementation of APD. This means that during machining with APD only part of work is implemented by implemented [8].

DOI: 10.1051/matecconf/201712901058

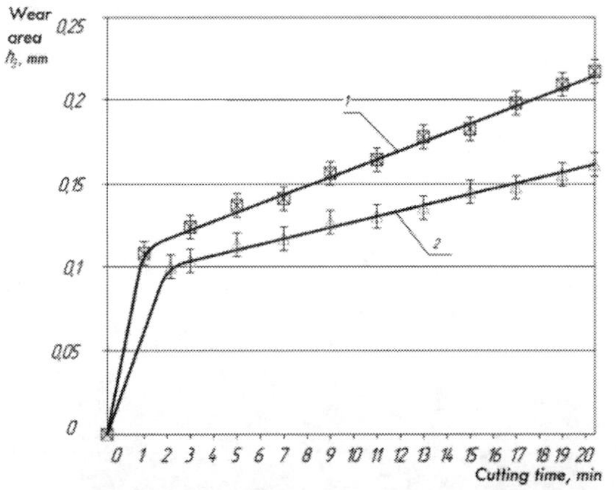

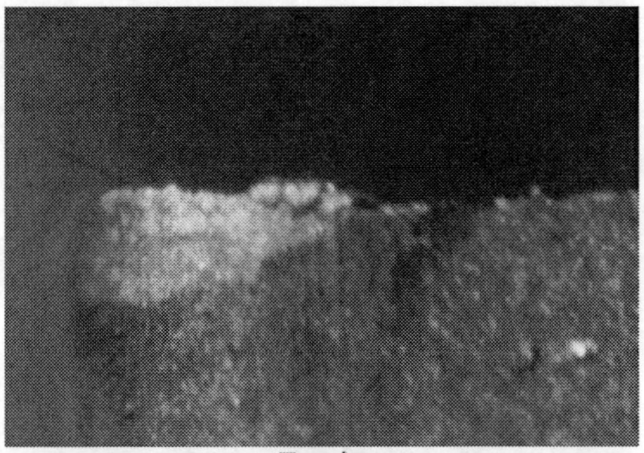

Turning

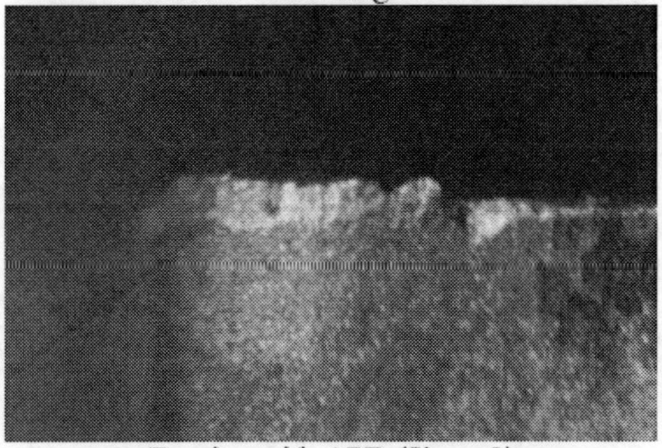

Turning with APD (K_{APD}=2)

Fig. 2. The dependence of the area size of the flank wear of the cutter from cutting time.
Steel 30CrMo4 – hard alloy WC 79-TiC 15-Co 6;
t_p=1 *mm*; v_p=90 *m/min*; s_o=0,256 *mm/rev*. (According to the source [8])

Scientists of Volgograd State Technical University (Norchenko P. A., Krainev D. V., Bondarev A. A., Ingemansson A. R.) have conducted a series of experiments to study the process of the cutting tool wear in the machining of structural carbon and alloyed steels and compared the results with traditional turning and turning with APD. It is discovered that while turning with APD a substantial reduction of tool wear of cutting blades is achieved on the main flank, thereby the processing productivity and the resource of the tool workability are increased in comparison with the traditional machining (Fig. 2) [7,8].

The application of advanced plastic deformation leads to a decreasing in intensity of the tool wear by reducing the cutting forces and the intensity of diffusion penetration of iron atoms to the structure of the hard alloy. The tool resistance increases from 25 to 100 %.

After cutting with advanced plastic deformation the surface roughness is reduced sharply (in many cases from 2 to 4 times), which permits to shorten the number of technological transitions, as well as to improve surface micro profile (in particular, the length of the reference line is increased).

The specific recommendations on the appointment of cutting modes with advanced plastic deformation were formulated. As fundamental criteria for the appointment of cutting modes has been selected: the minimum relative surface wear and minimum roughness of the machined surface [7].

References

1. *Classification of tool materials*: [Electronic resource]. – Access mode: http://studopedia.ru/4_17307_klassifikatsiya-instrumentalnih-materialov.html
2. *Hard alloys. The brands of hard alloys. The application of hard alloys*: [Electronic resource]. – Access mode : http://yaruse.ru/posts/show/id/949
3. N. C. Gerasimova, N. N. Gerasimov. *Tool materials*: *tutorial* (Mos. state technical university named N. E. Bauman. – M.: Kaluga, 2016)
4. N. N. Zubkov, *Tool materials for the manufacture of cutting tools* (Scientific edition of the Bauman MSTU named N. E. Bauman. – M.: Science and education, 2013)
5. Ju. S. Sharyn, T. I. Tishenina. *Handbook turner* (Environments Ural book publishing house: Sverdlovsk, 1985)
6. *The system of "Industrial catalogues"* / Industrial catalogs // Lab2U: [Electronic resource]. – Access mode: http://lab2u.ru.
7. P. A. Norchenko, *Increasing the efficiency of the cutting process of stainless steels of austenitic class with advanced plastic deformation*: *diss. cand. tech. sciences: 05.02.07.* (Volgograd, 2010)
8. A. A. Bondarev, *Study of the influence of advanced plastic deformation on the efficiency of the cutting process of structural steels*: *diss. cand. tech. sciences: 05.02.07.* (Volgograd, 2016)
9. Ju. M. Bikov, *Investigation of the regularities of carbide tool wear with wear-resistant coatings for the purpose of increasing its workability*: *diss. cand. tech. sciences: 05.03.01.* (Volgograd, 1983)
10. A. A. Lipatov, *Regularities of the cutting process high alloy steels and the ways to improve the workability of carbide tool*: *diss. cand. tech. sciences: 05.03.01.* (Volgograd, 1987)

DOI: 10.1051/matecconf/201712901058

11. Ju. L. Tchigirinsky, Pham Hoang Trung, A. A. Lipatov, *Features of Processing of Corrosion Resistant Steels*, MSF, Vol. **870**, pp. 598-602 (2016) – DOI 10.4028/www.scientific.net/MSF.870.598
12. N. V. Talantov, *Mechanism of carbide tool wear when machining steels*, Bulletin of mechanical engineering, N **7**, pp. 52-57 (1985)
13. N. V. Talantov, *Physical basis of the cutting process, the tool wear and the tool destruction* (M.: Mechanical Engineering,1992)
14. N. V. Talantov, *Physical basis of cutting process*, Physical processes in the cutting metals: collection of articles, Volgograd: VPI, 3–37 (1984)
15. N. V. Talantov, *Physical basis of the cutting process and the tool wear: tutorial.* (Volgograd: Volgograd truth, 1987)
16. A. S. Vereshchaka, I. P. Tretyakov. *Cutting tools with wear-resistant coatings* (M.: Mechanical Engineering, 1986)

 DOI: 10.1051/matecconf/201712901059

Power characteristics of the metal compounds formation process during the friction stir welding

Radmir Rzaev[1,*], *Alexander* Chularis[1], and *Larisa* Semyenova[1]

[1] Federal Budgetary Governmental Educational Institution of High Education "Astrakhan State University"

Abstract. An influence of the power characteristics on the formation process of the uniform metals compound during the welding with friction stirringis being examined in this article.A dependency between the machine-tool engine power input and the instrument tilt during the FSW for the aluminum alloy AD31, copper alloy M1, titanium alloy OT4-1 and steel St-3 low-alloyed has been explored. A question of the stabilization of power consumption process while the establishment of superplastic condition of welded metal during the FSW has also been reviewed. A dependency revealed between the power characteristics, the geometry of the formation, the rotation speeds, the longitudinal displacement of the tool and its dimensions for fixed values of the parameters during the FSW.

1 Introduction

More than fifty years has passed since the creation of a variety of the process of friction welding – the way of welding with friction stirring (FSW). However, a numerous publications devoted to various aspects of this process, both of the Russian and foreign researchers, are haven't been advanced very much in the issues of mechanisms of compound formation, and this is already more than two hundred different reports and a dozen of monographs. Although dozens of authors have already proposed various models of the FSW process: there are thermal, thermodynamic, kinetic, deformation, and, more recently, thermal deformation models based on physical principles of superplastic deformations with allowance for the large levels of plastic deformation in the evolution of plastic deformation modes developed by academician V.E. Panin and his co-workers [1-3,9].

The scarcity of information about the energy picture of the FSW process and the multifactor of dependencies of the properties of welded joints, the lack of correlation between the most significant parameters of the FSW process (tool rotation speed, tool tilt, speed of its translational motion along the welded edges – welding speed, axial force on the tool) – all of them have allowed to formulate the purpose of these operations: investigate the quantitative estimates of energy costs accompanying the compounds formation from various metals and OV during the welding on the modes that create and maintain a stable superplastic state in the zone of welded joint formation during the FSW.

2 Experimental

The research method consisted in obtaining the FSW compounds on the following coarse-grained sheet materials that could be transferred to the superplastic state. It's a deformed aluminum alloys AD31, technical copper M1, technical α-titanium VT1-0, (α + β)-doped titanium alloy OT4-1, ferrite-pearlite class type St-3 steels, steel #20, #09G2S perlite low-alloyed samples, an austenitic high-alloy type 18-10 – steel 12X18H10T, 2-3 mm-thick (see the Table 1)

Table 1. Characteristics of welded materials in FSW [4-6].

Mark of alloy	Size of grains at initial state	Melting temperature (T_{max}), °C	SP manifestation temperature (T_{sp}), °C (T_{max}/T_{sp})	Recrystallization temperature (T_{rc}), °C	The parameters of the manifestation process in the joint venture (elongation, the coefficient of velocity sensitivity)
AD31 (AD1)	NKK (approx. 0.35 micrometer)	565	270 (0,48)	266	Elongation to rupture 320%, 0,43
M1 copper	NKK (approx. 0.4 micrometer)	1083	270 (0,25)	250-300	-
VT1	NKK (up to 100 nanometers)	1820	700 (0,38)	600	m=0,4; n=2

* Corresponding author: radmir.82@mail.ru

OT4-1	MKK (approx. 1 micro meter)	1760	860 (0,49)	880	-
Steel 20	MKM (approx. 6-8 micro meters)	1520	730 (0,48)	760	Elongation to rupture 320%
18.Cr-10Ni-0.7Ti-0.6Mn-0.8Si-0.12C (12X18H10T)	-	1550	780 (0,51)	780	Max elongation 220%, 8×10-3

The welding was performed on a vertically milling machine equipped with a special tool made of sintered hard alloy. The tool had a Morse cone №1 on the mounting side in the drive head of the machine, this cone has smoothly passed into a cylinder, performing the function of a shoulder with a diameter of 14 mm. The shoulder's end surfacewas made in the form of an inverse cone. In the center of the tool, a pin was placed on its axis in the form of a truncated cone with diameters of 2/6 mm and a length along the generatrix equal to the thickness of the parts to be welded. Butt welded joints were welded at a speed of 900 rpm at the speed of longitudinal movement along the edges (welding speed 20 mm / min), the magnitude of the axial force on the tool, taking into account the properties of the welded metals 4-8 kgs.

3 Results and Discussion

When a rotating tool pin enters into the surface layers of welded materials, a friction of the pin and end of the shoulder occurs at the contact point in the first seconds of the process, a heat flux appears as a result of the dry-friction process of these parts of the tool with the weld metal. At the structural-microscale level, surface grains are involved in plastic deformation, which is realized due to the motion of sliding of lattice dislocations due to the presence of a strain gradient [11]. However, in large-crystal metals, most of the dislocations are blocked by the Cottrell and Suzuki atmospheres, as a result of which their start is difficult and for the propagation of plastic deformation sources of mobile dislocations are required, such as the large-angle grain boundaries and their triple junctions. The launch of the latter is carried out by means of a tool inserted into the metal to be welded, by means of which elastically deformable grains adjacent in the layer and in depth are involved in plastic deformation, providing expansion of the zone of plastic deformation to the deep layers of the metal located at the structural-mesoscopic scale level [9]. Owing to the deepening of the rotating and moving along the edges of the tool pin, several mesozones of plastic deformation localization are formed along the thickness of the welded plates and the expansion of mesozones to the entire cross section or thickness of the welded parts [9,11,12]. This stage characterizes the transition of local plastic deformation from the mesoscale to the macroscale level, that is, to the completion of the stage of macroplastic deformation. The transition to the structural-mesoscale level, when conglomerates and grain blocks are deformed together due to the launch of rotational mechanisms of plastic deformation and the transition to the mechanism of adhesion friction. As a result of these processes, the temperature increases in a welding zone, and the main powers of superplasticity – grain boundary diffusion [6, 7] and grain-boundary slip activated by it – begins to function [4-5, 11,12].

To gain the information about energy inputs at different stages of welded joint formation process, the value of the power consumed by the electric motor of the machine during the welded joints formation from metals, alloys at different angles of instrument tilt and the similar rates of rotating, the geometry the tool was oscillated.

Taking into account the foregoing, the curve describing the changes in energy consumption for the different stages of the compounds formation in FSW has three typical energy distribution regions (Fig. 1).

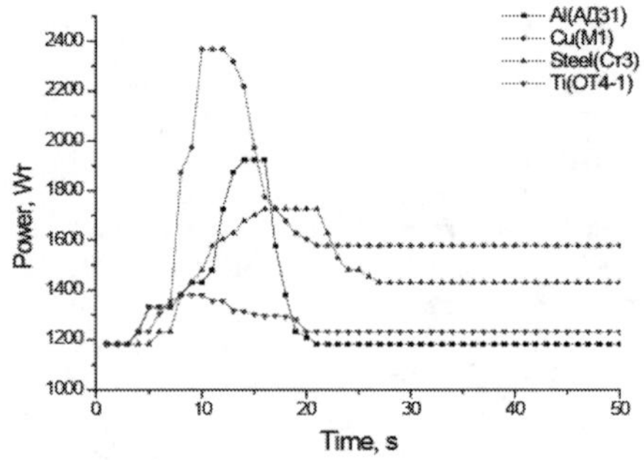

Fig. 1. Dependence between the power input during the FSW from time for AD31, M1, OT4-1, St-3 (tilt angle 1°)

The first one is the left, increasing, responsible for energy consumption during the logon period. Interacting parts and sources of "dry" friction at the micro- and first meso-scale levels of plastic deformation.

The second, medium, steeply dipping, demonstrates the onset of high levels at high temperatures, especially in the zone of pin, plastic deformation and transition to the superplasticity state and thermal stabilization of the welded joint. The third is the straight section of the curve parallel to the abscissa axis in the course of the welded FSW connections formation process in the established regime of superplasticity at the present macroscopic level (over the entire thickness of the welded parts), the level of plastical deformation with economical and stable FSW mode.

The described nature of the energy change during the welded FSW connections formation was established during the welding of AD31 (AD1) aluminum alloy, M1 copper alloy, welding the steels of different structural classes (St-3, steel 20, steel 12X18H10T), VT1-0 technical alloy and pseudo-α OT4-1 titanium alloy, Fig.2 (all metals are welded at the tool tilt at 3°).

DOI: 10.1051/matecconf/201712901059

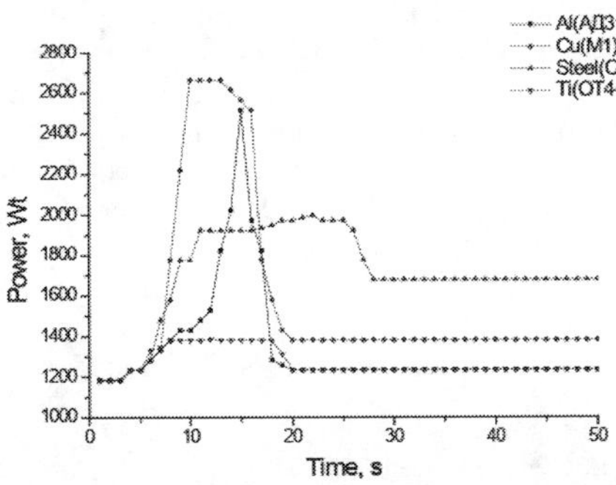

Fig. 2. Dependence between the power input during the FSW from time for AD31, M1, OT4-1, St-3 (tilt angle 3°)

Experimental metals and alloys used in the studies are representatives of a large range of structural ferrous and non-ferrous metals, differing in mechanical and thermophysical properties and, most significantly, in temperatures of transition to the state of superplastic flow, Table 1 [4-5,8-12].

As follows from the data analysis shown in Fig. 1 and 2, an important parameter of the FSW process is the tool axis angle tilt from the vertical by the "backward angle", since its efficiency determines the conversion of the frictional friction energy into thermal energy, the magnitude of the plastic flow of the solid metal (alloy) in the welding direction, the amount of plastificine under the action of a rotating pin of welded metal, which takes part in the welded seam geometry formation, including penetration through the thickness, and its scaling.

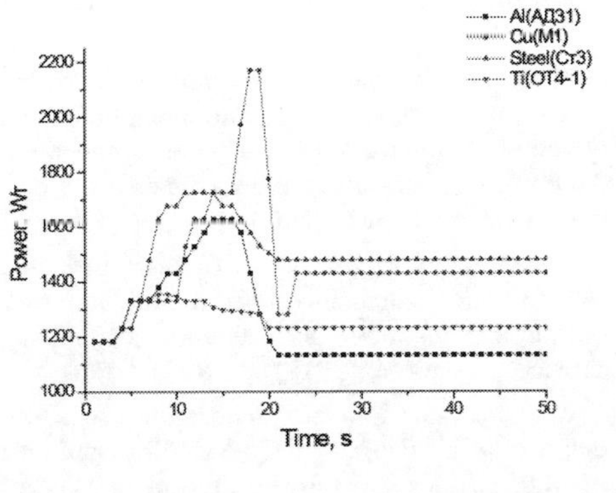

Fig. 3. Dependence between the power input during the FSW from time for AD31, M1, OT4-1, St-3 (tilt angle 5°).

In the practice of FSW, various researchers have used the tilt angles from one to five degrees. Measurements of energy consumption at the indicated angles of attack, Fig. 3 with identical values of the remaining parameters of the FSW process when aluminum, copper, steel and titanium were combined, showed that the regularity of the change in energy consumption differs in the initial stages of the FSW process in the form of durations and amplitudes of the fixing energy of the first and second stages until the process stability is achieved. In the third stage, the difference is manifested only in the absolute values of power costs. This can be probably explained by the kinetics of overcoming various mesoscale levels of plastic deformation on all experimental materials due to different values of the viscosity and stresses of the plastic flow of the welded materials that are established after reaching the superplastic transition temperature [4-5].

The analysis of the relative values of energy consumption in the FSW of aluminum, copper steels and titanium showed at first glance a paradoxical result with respect to titanium. Figs. 2 and 3, a reduction in the total energy consumption compared to other metals and alloys studied. But if we take into account the basic mechanisms responsible for the transition to the superplastic state, then it should be taken into account that, in contrast to other metals, where in addition to dynamic continuous recrystallization, anomalous and / or collective recrystallization, and in the transition to the mesoscale level of plastic deformation, and grain-boundary slip, processes (of small crystallographic shifts throughout the volume of the material) of a limited atom displacement by less than 20% of crystalline grid parameter, which can be observed during the twinning and polymorphous (β↔ α'(α")) transformations occurring in titanium alloys [10-12]. As a result of these transformations, the structure of the titanium alloy becomes more dispersed, which increases the superplasticity effect due to the activation of intergrain sliding [8,10-12].

At the same time, the increased energy consumption in the FSW of copper (Fig. 4) is probably due to significant kinematic viscosity, heat capacity, thermal conductivity, tendency to collective and even anomalous recrystallization, and increased values of the magnitude of plastic flow stress that are capable of removing material in the welding zone from the superplasticity regime [9,4-5].

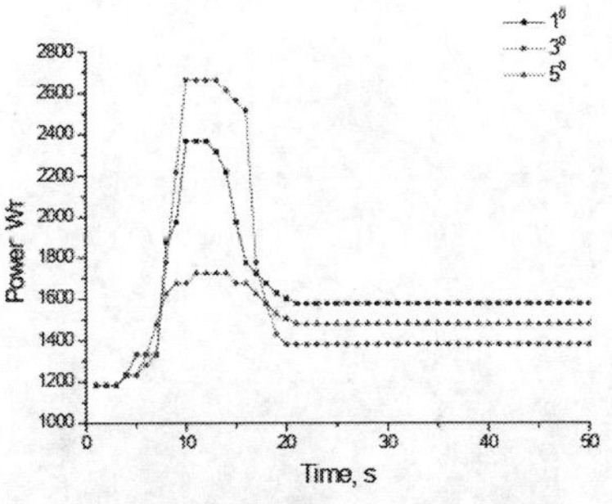

a)

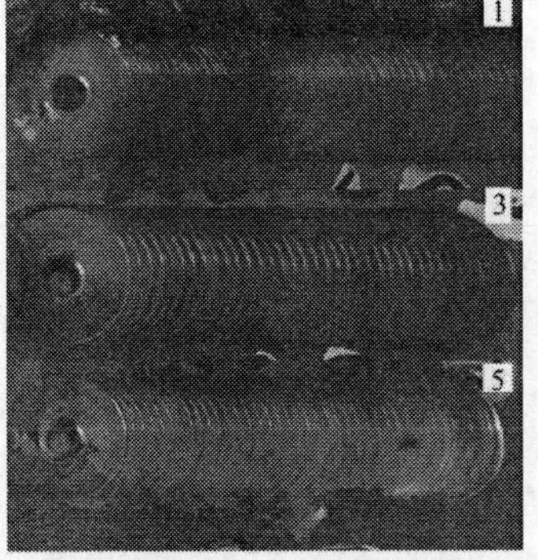

b)

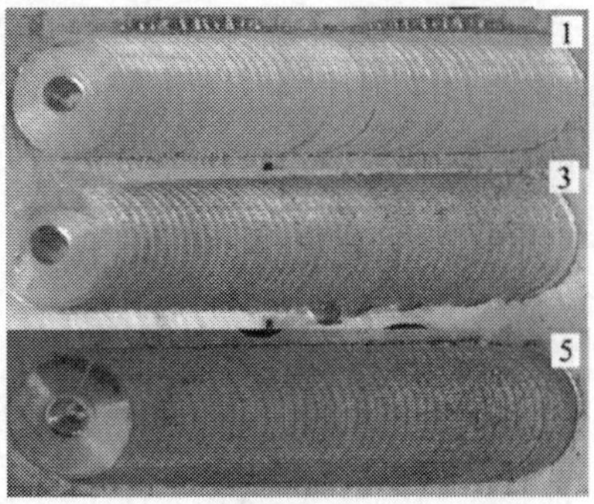

c)

Fig. 4. Dependence between machine-tool power consumption in the FSW from time (a) and the appearance of the welded samples of copper M1 (b) and AD31 (c) alloys for various tool angle tilts.

The geometry of the welded joints, the height of the reinforcement and the character of their scaling together with the state of the underside, the strength at break (on aluminum АД31- 200, on copper M1- 235, carbon steels - 380, titanium alloys- 420 MPa) are testifying about their satisfactory quality.

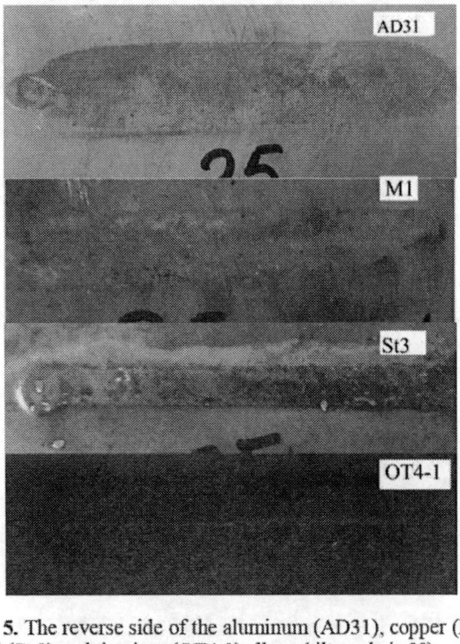

Fig. 5. The reverse side of the aluminum (AD31), copper (M1), steel (St.3) and titanium (OT4-1) alloys (tilt angle is 3°).

4 Conclusions

Thus, we can formulate the following conclusions:

1. For power consumption, the FSW cycle for all the studied metals has two regions: a starting cycle with a duration of up to 15-20 s. With a maximum power consumption, and steady, characterized by a constant value of power consumption, which are proportional to the stresses of the plastic flow of welded metals in the superplasticity regime.

2. With identical FSW conditions, the tendency of energy costs reduction is evident in connection with multiple mechanisms of plastic deformation manifestation and propagation during the weld formation and energy consumption increase under conditions of increased kinematic viscosity, heat capacity, thermal conductivity and plastic flow stresses.

3. Experimentally established a weakly proportional dependence of energy costs on the tool tilt angle and a rational tilt angle of three degrees is determined.

References

1. A.V. Panin, Scale levels of deformation in the surface layers of loaded solids and thin films, The dissertation author's abstract on competition of a scientific degree of the doctor of physical and mathematical sciences, Tomsk, (2009).
2. V.E. Panin, V.A. Likhachyov, Y.V. Grinyaev, *Structure levels of deformation of solids* (Nauka, Novosibirsk, 1985)
3. V.E. Panin, V.E. Egorushkin, A.V. Panin, The nature of localization of plastic deformation of solids, Physical Physics Magazine. **77**, 8 (2007)
4. O.A. Kaybishev, *Plasticity and superplasticity of alloys* (Metallurgiya, Moscow, 1975)
5. O.A. Kaybishev, *Superplasticity of industrial alloys* (Metallurgiya, Moscow, 1984)
6. Yu.R. Kolobov, R.Z. Valiev, G.P. Grabovetskaya et al., *Grain-boundary diffusion and properties of nanostructured materials* (Nauka, Novosibirsk, 2001)
7. G.P. Grabovetskaya, Grain-boundary diffusion and creep of submicrocrystalline metallic materials obtained by intensive plastic deformation methods, Author's abstract. dis. to the soot. uch. art. doct. fiz-mat. sciences, Tomsk, (2008)
8. S.S. Manokhin, Regularities of phase and structural transformations in an unalloyed titanite of mari VT1-0. Author's abstract. dis. to the soc. art. cand. tech. sciences, Belgorod, (2012)
9. T.V. Cherkasova, The dislocation structure of polycrystalline Cu-Al and Cu-Mn alloys and the ratio of the physics of plastic deformation. Author's abstract. dis. to the soc. art. cand. fiz-mat. sciences, Tomsk (2016)
10. A.A. Babarenko, I.V. Egiz, A.I. Khorev et al., Superplasticity of titanium alloys of different classes. Metallurgy and heat treatment of metals, **6** (1995)
11. E.F. Dudarev, G.P. Pochivalova, Yu. R. Kolobov, et al., Microplastic deformation of submicrocrystalline titanium at room and elevated temperatures, Izv. Higher Education, Physics, **55,** 7 (2012)
12. E.F. Dudarev, G.P. Pochivalova, Yu. R. Kolobov, et al., Effect of deep plastic deformation and subsequent annealing on true grain-boundary slip in coarse-grained and submicrocrystalline titanium, Physical mesomechanics, Special. Issue **7** (2004)

 DOI: 10.1051/matecconf/201712901060

Experimental implementation of direct laser alloying

Anton Vladimirovich Sotov[1], *Anton Vasilevich* Agapovichev[1,*] and *Vitaly Genadievich* Smelov[1]

[1]Samara University, 443086 Moskovskoye shosse 34, Russia

Abstract. A laser beam, as a source of energy, has been lately widely used in additive technologies for the manufacture of parts for the power engineering industry. One of the promising methods of additive manufacturing based on laser overlay welding of metal wires is the so-called Laser Engineering Net Shape. This work studies the structure and mechanical properties of the sample manufactured using a direct laser alloying unit designed for 07X18H10T stainless steel grade alloying. The allocation map of chemical elements has been examined in this work.

1 Introduction

The current trend for customization of products and reduction of time to market results in the increased demand from the industry for agile manufacturing technologies. These technologies should be robust and efficient from the standpoint of resources used and they should allow manufacturing quality products. The use of additive technologies enables us to ensure high level of production flexibility in relation to complex design features, while also enable us to directly manufacture parts using CAD models and without machining which results in saving time and resources [1,2].

Lately, in the aviation engine industry great emphasis has been placed on the technologies of additive manufacture using a metal base. This is due to the possibility of manufacturing reduced-weight (bionic) designs, which allows to increase weight-carrying capacity, reduce fuel consumption, and the level of atmospheric emission of pollutants [3-6].

Additive technologies (AT) are the technologies that allow for manufacture of an object using its digital model by way of layer-by-layer application of a material. These technologies may be grouped into two major groups depending on the method of material feeding: (1) Laser Metal Fusion (LMF) features layer-by-layer application of material and includes the technologies of selecting laser sintering (SLS), selecting laser melting (SLM), as well as electron-beam melting (EBM); (2) Laser Metal Deposition (LMD) directly feeds material to the treatment area, and this group includes the technologies of direct metal deposition (DMD), as well as the technology of Laser Engineered Net Shaping (LENS) [7,8].

LMF technology employs a layer-by-layer method of application of material (Figure 1). This method is based on three repeating stages. At the beginning, a thin layer of metal powder is placed on the platform using a special dosing device. At the second stage, a focused laser beam selectively fuses the powder layer, which is placed on the platform. And, at the third stage, the platform is lowered at the thickness of the layer, after which the cycle is repeated. Normally the thickness of a pad is anywhere from 30 μm to 50 μm therefore manufacture of whole parts may take alloying of thousands of pads.

During LMD process a fused pool is formed on the surface using a laser beam. At the same time, metal powder filler is fed into the fused pool. After hardening the filler forms single welded objects. Several welded objects located next to each other form layers or volumes.

The use of a powder pool in LMF technology enables to synthesize parts with complex geometric shapes. The disadvantages of LMF include its low process speed, relatively small size of the parts manufactured, and significant restrictions as to the use of various powder materials during the process of making parts. LMD technology, on the contrary, exercises higher speeds of generating parts with the possibility of changing powders during the process. However, LMD technology has significant limitations as to the complexity of geometric shapes of the parts manufactured.

LMD technology can be classified into two types depending on the material used: Powder-based (LENS-p) and wire-based (LENS-w) [10-12]. The use of wire materials for generation (growing) of parts has a number of advantages over the powder-based type, namely, higher speed of building of an object, lower cost of materials [13]. Such advantages allow the wire-based technologies to be competitive as compared to the powder-based ones. This work demonstrates an experimental test of the structural layout of a unit that employs LENS-w wire-based technology.

2 Materials and experimental methods

The samples have been manufactured using a laser technological installation which included a system of wire feeding, a system of moving the laser head, an

* Corresponding author: agapovichev5@mail.ru

 DOI: 10.1051/matecconf/201712901060

optical system of observation of the process of laser treatment. Fig. 2 shows the scheme of implementation of the process of making the samples using LENS-w method [14]. Samples for testing were manufactured on the equipment of CAM technology common use center (RFMEFI59314X0003).

In order to test the structural layout of the direct laser alloying unit they have manufactured samples made of 07X18H10T grade stainless steel wire.

Laser overlay of the samples has been performed on a platform in the medium of argon shielding gas. Once manufactured, the samples were separated from the platform using the method of electroerosive machining.

The study of microstructure of the samples has been conducted using Tescan Vega scanning-electron microscope. The study of chemical composition of the samples has been conducted using the energy-dispersive add-on unit of Tascan Vega scanning-electron microscope.

The study of mechanical properties of the samples has been conducted using INSTRON 8802 floor standing system for fatigue testing. Based on a series of preliminary experiments they have identified the modes for impulse laser irradiation used for the samples manufacture.

2.1. Description of the unit

The laser technological installation includes a solid-state laser with an irradiation wave length of 1.06 μm and with adjustable length of an irradiation impulse from 0.2 ms to 20 ms, the frequency of tracking the irradiation impulse from 1 Hz to 20 Hz, the diameter of a focused beam from 0.2 mm to 5 mm. The system of feeding the material allows for delivering a metal wire to the treatment area at the angle of 45 degrees relative to the laser irradiation. The diameter of the wire used is 0.4 mm. The system of movement steers the laser beam unit in relation to the surface of the sample synthesized in the direction of X, Y, and Z axes. The buildup platform remains steady throughout the entire process of generation of geometry. The optical observation system is a binocular with a shutter and lens. The cutoff filters which are built into the system ensure protection of the eyes of an operator from the impact of reflected and dissipated irradiation. The scheme of the experimental system LENS-w technology is presented in Fig. 1.

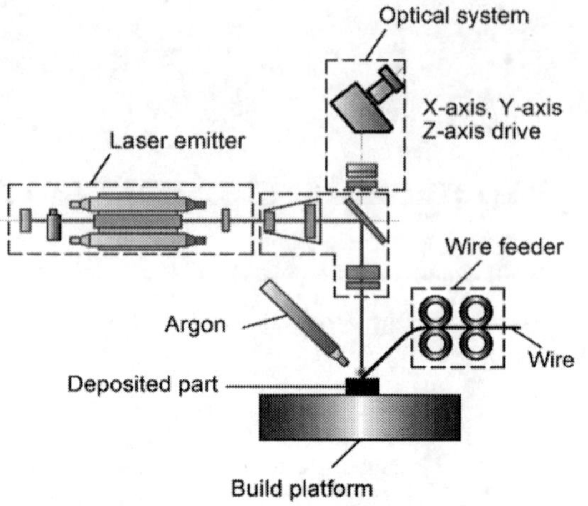

Fig. 1. The schematic of experiment LENS-w system.

The thickness of the build layer was 0.2 mm, which corresponds to half the diameter of the filler wire. The process of implementing LENS-w technology includes the following main steps: supplying protective gas to the treatment zone, melting of the wire by laser radiation, raising the laser head unit to the thickness of the build layer (Fig. 2).

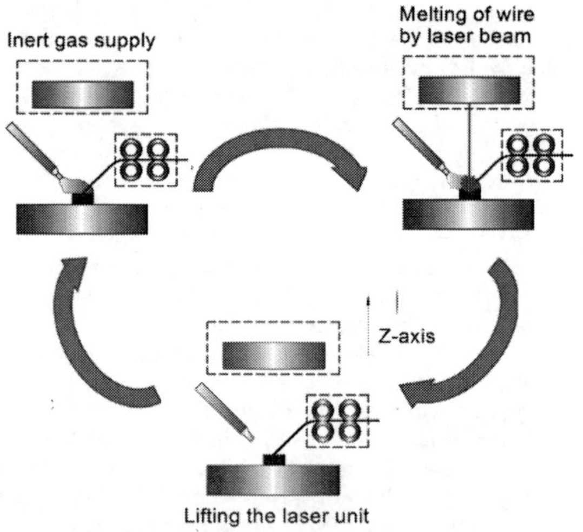

Fig. 2. The main steps of the LENS-w process.

3 Results

The results of mechanical pull tests of the samples manufactured using HTS 300 Mobile unit and made from 07X18H10T grade stainless steel wire are shown in Table 1 in comparison with reference properties of the material.

DOI: 10.1051/matecconf/201712901060

Table 1. Parameters of laser irradiation.

Sample #	Rupture strength, MPa	Tensile strain, ε_t %
1	487	15
2	594	24
Rod	590	20

Based on the tests results it is apparent that the mechanical properties of the samples made with HTS 300 Mobile unit and made from 07X18H10T grade stainless steel wire are comparable with the mechanical properties of the material supplied as rods.

In order to study microstructure of the samples they have performed fractographic investigations using the method of scanning-electron microscopy (Fig. 3).

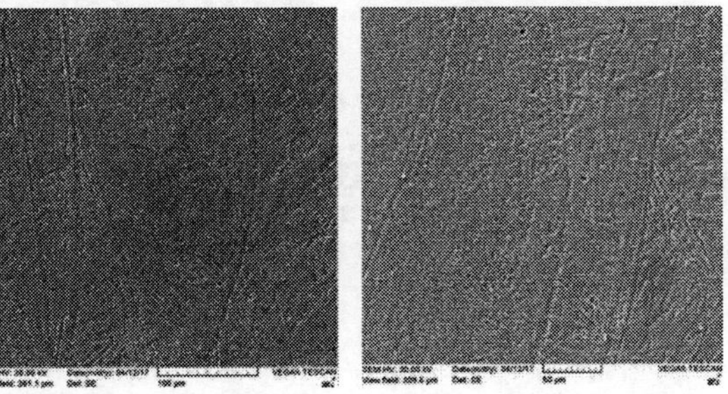

Fig. 3. Microstructure of the samples manufactured using HTS 300 Mobile unit and made of 07X18H10T grade stainless steel wire.

As shown in Fig. 3 the material's structure contains insignificant defects in the form of porosity with the diameter of less than 10 µm. There is no incomplete fusion and there are no cracks. Sandwich structure of the material is observed. The borders are clear and there is metallic linkage.

In order to study the borders between weld pads they have developed a map of allocation of the chemical elements in the microsection area (Fig. 4).

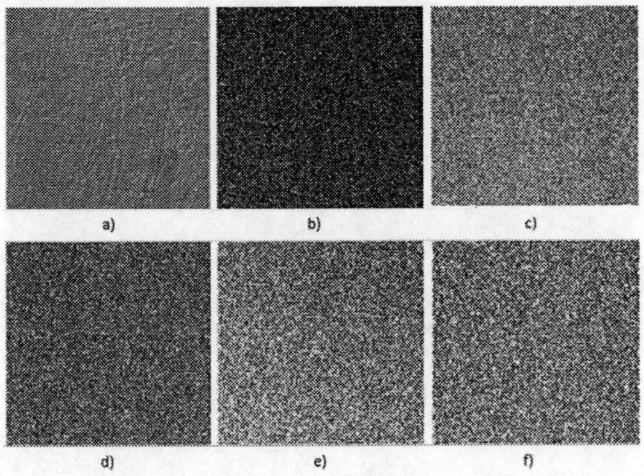

Fig. 4. Map of allocation of chemical elements
a) Electronic image of the microsection (x500); b) allocation of Si; c) allocation of Cr; d) allocation of Mn; e) allocation of Fe; f) allocation of Ni

As shown in Fig. 3 in the area between weld pads there was a clear cut borderline of allocation of chemical elements. This fact also suggests that the wire has not been sufficiently fused in the process of overlay welding.

4 Conclusion

The test the structural layout of the direct laser alloying unit has been conducted. Samples made of 20X13 grade stainless steel wire have been manufactured using 07X18H10T grade stainless steel. In order to manufacture the samples HTS 300 Mobile unit has been employed.

The tested installation scheme can be used for the manufacture of parts such as full-sized test plates for pull testing materials.

The samples thus manufactured have the shatter- and moldability indices which are comparable with those of reference properties of the material when supplied as rods.

The porosity with the diameter of up10 µm as identified on the surface of a polished section as a result of the metallographic tests is apparently inherited, and it was present in the source material – 07X18H10T grade stainless steel wire.

The analysis of microstructure of the samples showed the presence of a clear cut borderline between the weld pads which is also validated by the allocation maps of chemical elements.

Influence of the parameters of impulse laser irradiation on fusibility of planes needs to be further researched, as well as their influence on the mechanical properties of materials.

This work was supported -1098.2017.8) and by the Ministry of Education and Science of the Russian Federation (Grant № 9.1299.2017) and partly in the framework of the implementation of the Program "Research and development on priority directions of scientific-technological complex of Russia for 2014 – 2020".

References

1. T. Caffrey and T. Wohlers: Wohlers Report 2015: Additive Manufacturing and 3D Printing State of the Industry (2015)
2. A.V. Agapovichev, A.V. Balaykin and V.G. Smelov, Modern Applied Science. **9** (2015)
3. V.G. Smelov, A.V. Sotov and A.V. Agapovichev, IOP Conf. Series: Materials Science and Engineering, **140** (2016)
4. V.Sh. Sufiyarov, A.A. Popovich, E.V. Borisov and I.A. Polozov, Tsvetnye Metally = Non-ferrous metals, **1** (2015)
5. R. A. Vdovin and V.G. Smelov, J. of Engineering and Technology (IJET) **6**, 5 (2014)
6. A.V. Balaykin, A.I. Kondratev and N.G. Galkina, J. Key Engineering Materials **684** (2016)
7. G. N. Levy, R. Schindel, and J.-P. Kruth, CIRP Annals - Manufacturing Technology, **52**, 2 (2003)

8. Carl Zeiss Jena GmbH: Additive Manufacturing at ZEISS, (2015)
9. V.G. Smelov, A.V. Sotov, A.V. Agapovichev and T.M. Tomilina, IOP Conf. Series: Materials Science and Engineering **142** (2016)
10. E. Walach, B. Widrow, IEEE Transactions on Information Theory, **20**, 2 (1984)
11. A.B. Spierings, K.Wegener and G. Levy, Solid Freeform Fabrication Symposium SFF, Austin, TX, USA, (2012)
12. G. Kool and E. Amsterdam, The Future of Gas Turbine Technology - 5th International Conference, (2010)
13. K.S. Munir, Y. Li and C. Wen, Metallic Foam Bone: Processing, Modification and Characterization and Properties (2016)
14. V.G. Smelov, A.V. Sotov and A.V. Agapovichev, J. Key Engineering Materials **684** (2016)

Analytical study of modular cutting tools dynamic properties

Svetlana Lukina[1,*], and *Margarita* Krutyakova[1]

[1]Moscow polytechnic university, 38 B.Semyonovskaya st., Moscow 107023, Russian Federation

Abstract. The paper studies a comparative evaluation method for the modular cutting tools dynamic properties under the cutting forces loading with the help of the finite element numerical method. The method allows forecasting the modular cutting tool dynamic properties with identification of the confidence bounds for its exploitation in compliance with its intended use and reference operating conditions. The conducted modeling describes the modular cutting tool structure as a ranked set of structural components, such as a frame, a cassette, a cutting insert, etc., oriented towards a certain direction with some surfaces being contiguous and thus making contact interactions. The analytical model is represented by a multi-mass system in the form of elastic rods connected at elastic and damping joints. The research examined different tool structural component layout options, including those equipped/not equipped with a shim, having a radial/tangential (face or peripheral) location type of cutting components. The dynamic compliance values and vibrational modes at the natural frequencies were calculated for all structural component layout options. The calculation results showed an acceptable level of model convergence with the existing experimental data on the cutting tool condition express diagnostics.

1 Introduction

The relevant task of contemporary production is to enhance machining accuracy and quality while ensuring the established level of operational efficiency [1]. One of the key ways to ensure the necessary machining quality is the use of more progressive cutting tool structures.

The current state and development prospects for metal working is characterized by a broad use of modular cutting tools equipped with indexable cutting inserts which are mechanically clamped and made of hard alloys, ceramics and ultra-hard materials. The following cutting tools with mechanically clamped components are the most widely used ones: straight turning tools, side-facing tools, parting-off tools, boring tools; face and end mills and other tool with standardized principal dimensions [2-4].

A variety of machining situations alongside with underdeveloped calculation methods, unable to solve the problem of a rational tool structure choosing at the design stage, led to emergence of an extensive nomenclature of tools with the same intended uses.

At the moment, several thousands of such cutting tool structures with mechanically clamped cutting components are designed and applied; they differ by their overall dimensions and structural design, the ways cutting components are located and clamped [5,6].

The existing design methods allow for identification of cutting tool geometric and structural parameters while being unable to evaluate a variety of cutting insert locations and clamping ways [7].

Evaluation of the modular cutting tool dynamic properties is a relevant issue for design studies. The modular cutting tool dynamic properties manifest in cutting through the vibration process influencing the tool life and machining accuracy. The approaches for cutting tool dynamic properties forecast referred to in the available literature are based upon experimental express diagnostics and modeling of radially arranged cutting inserts without due regard to their clamping type [7,8].

The objective of our study is to develop the method for a comparative evaluation of cutting tool dynamic properties to identify the best design solutions intended for specific operating conditions.

To achieve the abovementioned objective, the paper sets forth the following research tasks:

- to form and provide an analytical description for the evaluation model of the modular cutting tool dynamic properties with due regard to a variety of cutting insert locations and clamping ways;
- to conduct analytical research of the influence of the modular cutting tool structural component design on the dynamic properties;
- to develop recommendations for application of the modular cutting tool structures with various ways of cutting component location and clamping.

2 Key provisions of the method for a comparative evaluation of the modular cutting tool dynamic properties

The developed method is underlain by the assumption that the general modular cutting tool structure defined by a system of design and geometric parameters may be represented as a ranked set of components (elements),

* Corresponding author: lukina_sv@mail.ru

such as a body, a cassette, a cutting insert, etc., oriented towards a certain direction [9,10]. Separate constituents of the tool structural component layout have some contiguous surfaces making contact interactions.

The peculiarities of the component contact surfaces, which have a real geometric contour defined by a set of parameters characterizing their position in relation to the tool coordinates system, dimension and form deviations, its micro relief states, presence of gap clearances (allowances) between these surfaces, are described by a subset of surface nodes $\{\Phi\}$. The subset $\{\Phi\}$ nodes are connected by imaginary lines or surfaces. As a result, the component surfaces are represented as a total of surface elements while their internal structure is represented in the form of a set of finite volume elements. These finite elements are interconnected and interact in the nodes determined by the tool structural component layout; interaction between the components (elements) is determined by the nature of the node connection (rigid, jointed, contact, etc.).

Each tool component is connected with a moving orthogonal coordinates system set in relation to some global system. The node positions are registered as per the coordinates system. The tool component movements and rotations with reference to the coordinates global system are considered as the movements and rotations of surface and finite volume elements connected with local coordinates systems.

To form a geometric element image, the nodes and their associated moving coordinates systems are assigned corresponding movements and rotations traced in relation to the global coordinates system in some certain time periods. To clamp the tool structural layout components in an assigned relative position, the authors introduce a set of rod elements having the stiffness parameters with stochastic or determined characteristics corresponding to the contiguous surface stiffness parameters. The number and the relative location of these rod elements is defined by the tool structural component layout [11,12].

The disturbing action system is explicitly described: the main vector {F} and the main load torque {M} imposed on the tool. The modular cutting tool elastic system is represented by a set of volume, surface and rod finite elements the motion equations of which are described by the Lagrange equations of the second kind and the d'Alembert's principle.

The modular cutting tool dynamic model is represented by the system of differential equations of the form:

$$[A]\{\ddot{U}\}+[Д]\{\dot{U}\}+[K]\{U\}=\{F\} \qquad (1)$$

where [A], [D] are inertia and damping matrices respectively; {U} is a column vector for movements of the finite element nodes; [K] is a global stiffness matrix for the modular cutting tool structure.

The method was implemented in the software by means of the finite element numerical method.

3 Implementation of the method for a comparative evaluation of the modular cutting tool dynamic properties

The developed method was implemented by the example of the design rationale for the modular turning tool and face mill structures. The authors considered four typical clamping and supporting options for a cutting insert: C, M, P, S at radial and tangential (peripheral or face) insert positioning.

Analytical research were conducted to check the capacity and to assess the validity of the developed models by the example of calculation of the structures for modular turning tools and face mills used for machining of medium-carbon steel work pieces 40X ($E=2\times10^5$ MPa, $\mu=0.3$). The taken cutting component (inserts) material was the steel grade T15K6 ($E=5.2\times10^5$ MPa, $\mu=0.3$). The clamping and cutting forces varied in the ranges of 50,...,200 daN and 100,...,1000 daN respectively.

Harmonic vibrations negatively influencing the tool wear rate and the machining quality represent the tool dynamic system inciting factor. The force vibration impact on the tool dynamic system is expressed through the correspondence of cutting forces along the coordinate axes ""Fig. 1"".

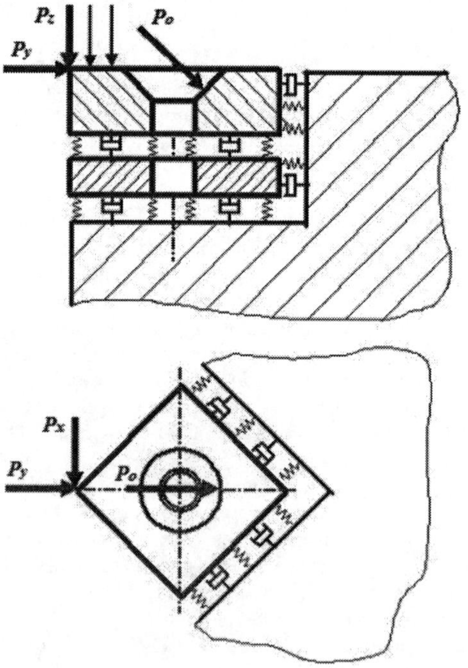

Fig. 1. The analytical model for evaluation of the dynamic properties of the modular face mills with radially located cutting inserts. The S-type clamping.

The calculations resulted in the natural frequencies of the modular cutting tool elastic system and the vibration modes at the modular cutting tool natural frequencies. The validity of the results was evaluated according to the correlation of calculated and experimental stiffness values in the reduced mass node points of the tool elastic system; the values of the first vibration natural frequencies; vibration modes at natural frequencies.

Comparison of dynamic properties for different insert positioning options (radial and tangential) demonstrated that, to ensure the minimum wear and to enhance the tool life, it is preferable to choose a radial positioning type (as dynamic compliance values belong to the next lower order in these directions) whereas a tangential positioning type is better for machining quality improvement. When cutting inserts are radially positioned in a low frequency area (up to 3,000 Hz) at the first natural frequency of the common elastic system, the modular cutting tool subsystem parameters have a kind of a compensating action which makes the dynamic compliance value 10% lower towards the normal in relation to the machined surface. Such effect is not observed when the insert is tangentially positioned.

The range of natural frequencies characterizing radially positioned modular mills is shifted towards a high frequency area. Tangentially positioned modular mills have the first mode frequency in the 24-27 kHz range (the range variation is given for various options of the insert clamping), its second mode is in the 42-57 kHz range while the radially positioned tool has the first and the second modes expressed in the ranges of 42-57 kHz and 69-71 kHz respectively. However, the dynamic compliance values towards the maximum wear direction (along the face surface) in the low-frequency area are 1.8 - 2 times lower comparing to the tangentially positioned tool. One can explain the "worse" dynamic properties towards the maximum wear direction of the tool with tangentially positioned inserts basing upon the analysis of modal compliance balances which reflect the contribution of specific components (joints, a cutting component body and a shim body, etc.) into the dynamic compliance value at this mode. Thus, the tool elastic system with tangentially positioned inserts has its joints "cutting insert - shim" and "shim - support of the mill frame seat" "working"; the bodies of the cutting component and the shim undergo elastic deformation at the second mode with a significant impact of the latter factor. In contrast, radially positioned cutting tools generally demonstrate the work of those junction points that are oriented towards coordinate directions in such a manner that the cutting edge top vibration level is lower towards the maximum wear direction. The dynamic compliance level towards the wear direction along the flank is higher for the radial tool, although the priority of this direction is significantly lower in terms of the vibration amplitude influence on the insert wear.

The numerical experiments for evaluation of the modular cutting tool dynamic properties showed that:

- in terms of a better dynamic stiffness towards the maximum tool wear direction along the face surface, the insert clamping options are ranged as follows: M, S, P, C;

- all other conditions being equal, the radial option for the cutting insert positioning is more dynamically stiff and provides for the higher tool wear resistance values comparing to the tangential one;

- cutting insert tangential positioning stipulates for the invariability of dynamic properties' values towards the normal as of the machined surface in terms of its clamping;

- cutting insert tangential positioning stipulates for the "work" of not only the joints, but also cutting and support component bodies through all the frequency range of the tool elastic system.

The obtained results allow one to forecast a better machining quality of tangentially positioned modular mills at a high sensitivity to the machining process stability assurance. Therefore, tangential positioning of the cutting inserts is mostly applicable to cutting tools used for finishing operations. To ensure better cutting tool wear resistance and life during roughing operations, it is recommended to position the inserts radially while meeting the requirements towards the frame seat supports, the accuracy of support location and insert clamping regularity. It is necessary to forecast a greater risk of failures for tangentially positioned inserts while in case of radial positioning, one shall foresee the breach of the frame seat due to dynamic vibrations in the modular cutting tool elastic system.

4 Conclusion

The research formal results allow forecasting of the modular cutting tool dynamic properties with identification of the confidence bound for their intended use and reference operating conditions, revelation of interrelation between the structural component layout factors and the machining accuracy output parameters at the technical production preparation stage.

This provides the basis for identification of the best solutions for various engineering requirements and limitations as well as for recommendations of the most efficient application of one or another structural component layout under specific operating conditions.

The developed method helps to significantly reduce the scope of conducted modelled and real experiments as it has the purpose of identification and prevention of the wrong cutting tool structure choice at the technical production preparation stage.

The proposed method will be developed as the evaluation of modular cutting tool dynamic properties at high-speed machining [13,14].

References

1. S. Zivkovic., L.Takic, N.Zivkovic, *The improvement of environmental performances by applying ISO 14001 standard – A case study,* Chemical Industry and Chemical Engineering Quarterly. vol. **19**, № 4, pp. 541–552.(2013) doi: 10.2298/CICEQ120513088Z
2. A.R. Maslov, *Konstruirovanie instrumental'noi osnastki* [Construction tooling] (Moscow, Yanus-K Publ., 2012) ISBN 978-5-8037-0557-4
3. V.A. Grechishnikov, S.N. Grigor'ev, S.V. Lukina, Ju.M. Solomencev, A.G. Shirtladze, V.I. Vlasov,

Processy i operacii formoobrazovanija i instrumental'naja tehnika (Moscow: Janus-K Publ. 280 p., 2006) ISBN 5-8037-0296-X.

4. A.A. Bekaev, Y.V. Maksimov, S.V. Lukina, *Predicting the surface quality in discontinuous cutting,* Russian Engineering Research. **№ 35 (10)**, pp.792 (2015)
5. S.V. Lukina, M.V. Krutyakova *Sovremennye problemy organizatsii i upravleniya instrumental'nym obespecheniem mashinostroitel'nykh proizvodstv* [Modern problems of organization and management of tool maintenance machine industry] (Moscow, MGTU «MAMI» Publ. 116 p., 2013).
6. S.V. Lukina, *Metodika optimizacii proizvodstvennoj dejatel'nosti promyshlennogo predprijatija na osnove kompleksa prognosticheskih reshenij v oblasti vysokotehnologichnyh proizvodstv* [The technique of optimizing of the production activity of industrial plant on the basis of a complex of predictive models of formation and selection of design innovative solutions in the field of hightech industries], Vestnik MSTU «STANKIN», **№1(32)**, pp. 125-129 (2015)
7. S.V. Lukina, *Povyshenie effektivnosti proektirovaniya sbornogo rezhushchego instrumenta na baze ustanovlennykh vzaimosvyazei konstruktorsko-tekhnologicheskikh i ekonomicheskikh reshenii. Diss. dokt. tekhn. nauk* [Improving the efficiency of the design of the cutting tool on the basis of established relationships design and technological and economic decisions. Dr. eng. sci. diss.] (Moscow, 1999)
8. A.S. Yanyushkin, D.V. Lobanov, D.A. Rychkov, *Automation of technological preparation of production at selection of cutting tools*, J.World of engineering and technology, **№ 6**. p.36-38. (2013)
9. S.V. Lukina, *Modelirovanie protsedur formirovaniya i vybora strukturnykh komponovok sbornykh rezhushchikh instrumentov na osnove setevykh graf-modelei* [Modeling procedures for formation and choice of structural component layout of modular cutting tools using network graph-models], Obrabotka metallov (tekhnologiya, oborudovanie, instrumenty) – Metal Working and Material Science, **№ 2**, pp. 28–30 (2009)
10. S.V. Lukina, *Avtomatizatsiya protsedur formirovaniya i vybora strukturnykh komponovok sbornykh rezhushchikh instrumentov na etape tekhnicheskoi podgotovki proizvodstva* [Automating procedures for formation and choice of structural component layout of modular cutting tools in step of technical preparation production], Vestnik Saratovskogo gosudarstvennogo tekhnicheskogo universiteta – Vestnik Saratov State Technical University, vol. **3**, **№ 1**, pp. 241–247 (2011)
11. S.V. Lukina, *Modeling of stresses-deformed state of inserted cutting tools during their production and operation,* Avtomatizatsiya i Sovremennye Tekhnologii, **№4**, pp.24 (2001)
12. V.A. Grechishnikov V.A., S.V. Lukina, A.I. Veselov, D.V. Makarov, *The forming processes modeling for geometric parameters of inserted cutting tools with regard to their assembly technology,* Avtomatizatsiya i Sovremennye Tekhnologii, **№4.** pp.32 (2001)
13. S.V. Lukina, *Otsenka konkurentosposobnosti sbornykh metallorezhushchikh instrumentov na etape ikh proektirovaniya i vybora* [Estimation of competitiveness of modular metal-cutting tools at a stage of their designing and a choice], Sborka v mashinostroenii, priborostroenii – Assembling in mechanical engineering and instrument-making, **№ 9.** pp. 3–8 (2011)
14. S.V. Lukina, *Formation of the system of the local indicators to assess the quality of the cutting tool at the stage of technical training of production,* Obrabotka metallov (tekhnologiya, oborudovanie, instrumenty) – Metal Working and Material Science, **№ 4(69)**, p. 43-50 (2015) doi: 10.17212/1994-6309-2015-4-43-50

MATEC Web of Conferences **129**, 01062 (2017) DOI: 10.1051/matecconf/201712901062
ICMTMTE 2017

Functional modelling for the automated control system of a threaded joints, such as threaded plug gauges

Mihail Shutikov [1,*], *Jurij* Solomencev[1], *Alexandr* Feofanov[1] and *Tat'jana* Grishina[1]

[1] Moscow State University of «STANKIN», 127055, 1, Vadkovsky per., Moscow, Russian Federation

Abstract. The article presents a methodology for modelling the control of a threaded plug gauge. There had been examined works of scientists who had studied the aspects of the average thread diameter control and studied the problems of providing and improving the quality of threaded joints. At a modern machine-building enterprise, threaded gauges are important in controlling threaded joints of parts. There had been built a functional model of the caliber life-cycle. The article shows possible applications of threaded gauges and examines the method of automating the average thread diameter control on a universal three-axis measuring video microscope. The application of this method is expedient in the conditions of modern production, since that usage reduces time expenses and requirements for the operators' qualification, as well as increases the reliability of the results of the average thread diameter control.

At the present stage, improving the quality of machines and mechanisms is one of the priorities in the engineering industry, along with providing a high technical level. The development of science and technology, changing needs of the society are caused by the necessity of creation of the equipment, increase productivity and facilitate the term [1, P. 448; 2, P. 5].

The solution of these problems determines the profitability and increases the safety of operation of machines and mechanisms. One of the most important aspects of the modern economy in the engineering industry is the reduction of material and time expenses for new products [3, P. 114; 4, P. 54].

This problem can be solved, including with the use of special measuring tools and equipment.

While engineering industries have particular difficulty controlling operations associated with the measurement of threaded surfaces, since for a reliable operation of the product necessary to ensure the quality of threaded connections. Threaded connections have extremely wide application and represent approximately 20% of the total number of compounds in modern products, and in some cases, their share reaches 70%. [5, P. 3].

However, due to the complexity of integrated control (which is especially important for internal threads) for both mass and small batch and individual production used control gauges, so you need to keep quality caliber at a level consistent with its functional purpose.

Integrated control of the thread parameters are mostly done by thread gauges which have to be maintained very precisely [6, P. 11].

Thus, it seems topical application of more effective techniques affecting factors reduce the time it takes to control the mean diameter of thread gauges and do not require highly skilled personnel.

The gauge is a control tool, reproducing the geometrical parameters of the elements of the product, determined by the given marginal linear or angular dimensions, and in contact with the element of the product on surfaces, lines or points (GOST 27284-87) [7, P. 3].

In the mass manufacture of products with threaded connections with the same size, are widely used tools of a rigid structure, such as a threaded plug gauges. Questions for the control of the mean diameter of the thread reflected in the works of domestic scientists: Z. G. Smilyanskogo, A. G. Skhirtladze, Y. M. Radkevich, V. I., Berkova, etc., as well as scientists and specialists of leading scientific and educational centers of the country. In recent years, much attention is paid to research of problems of ensuring and improving the quality of threaded connections on the basis of technological support and improve the quality of the working surfaces of the threads and their operational properties. In the solution of these problems is the great contribution made by such scientists as: A. N. Feofanov, A. V. Rybakov, A. N. Prokofiev, A. G. Suslov, A. V. Handozhko, etc.

However, the issues related to the improvement of thread control methods using automated computerized systems, allow to increase the reliability and effectiveness of the results of monitoring have not been sufficiently developed in the scientific and technical literature up to the present.

* Corresponding author: mi.shutikov93@yandex.ru

 DOI: 10.1051/matecconf/201712901062

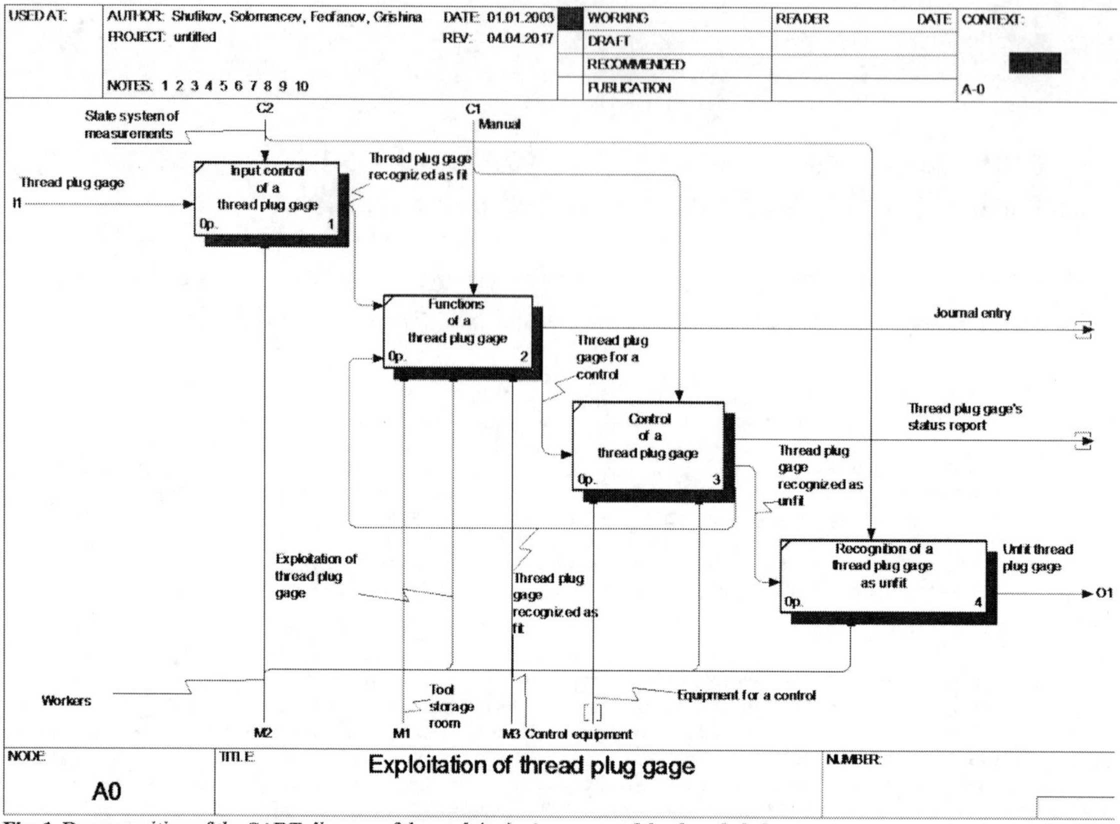

Fig. 1. Decomposition of the SADT-diagram of the exploitation's process of the threaded plug gages.

Assurance of the high quality of threaded calibers is a continuous process of high-quality production and controlling its quality. As follows from the diagrams SADT IDEF0 model (shown in Fig. 1) and its IDEF3 (Fig. 2), the operation of the gauges pass through certain stages of control.

We would like to note that in modern industries the need for their in-depth analysis and improvement is acute manifested. This is especially actual for integrated and automated enterprises. Under such conditions it is reasonable to use methods permitted to investigate the structure, characteristics and parameters of both the various production systems as a whole and their elements. The methodology of functional modeling could be accepted as one of such approaches.

The methodology of structural analysis is SADT (Structured Analysis and Design Technique) is a set of methods, rules and procedures for the construction of functional model of object of any subject area. This methodology is intended to reflect the functional structure of the object or process, that is, they produced the actions and relationships between these actions. In addition, it can be used to analyze the functions performed by the system, and to specify mechanisms through which they operate. And the application of this methodology to construct structural models in CAD could be very actual [5].

IDEF0 is used to create a functional model which is a structured image of the production system features and environment, or information and objects link those features [8, P. 62]. The IDEF0 model, or any model combination allows to formulate the "architecture" of the simulated system environment.

IDEF is a methodology, content architecture directly IDEF-diagrams, and performance improvement target.

As the decomposition of the IDEF0 model SADT-diagram shows one of the thread plug gauges life-cycle stages is the control its parameters. There are several methods to control the angle diameter in modern engineering, such as: two and three wires methods, projection one and the axial section method.

But all known methods in modern conditions do not provide the necessary economy of time resources and make strong requirements on skilled staff. It is necessary to improve the existing methods and enter the new ones. The management of the enterprise and the direct executors of the individual operations of the technological process in the field should be interested in improving the quality all the time [9, P. 285].

In this regard it is considered appropriate to develop a new method of monitoring the average diameter of thread plug gauges, which would be consistent with the requirements of mass production on the one hand, and also minimized the influence of human factor on the other side. This method is the measurement of average diameter of thread plug gauges on a universal three-axis video-microscope. Measuring microscope is designed for noncontact measurement of linear and angular

DOI: 10.1051/matecconf/201712901062

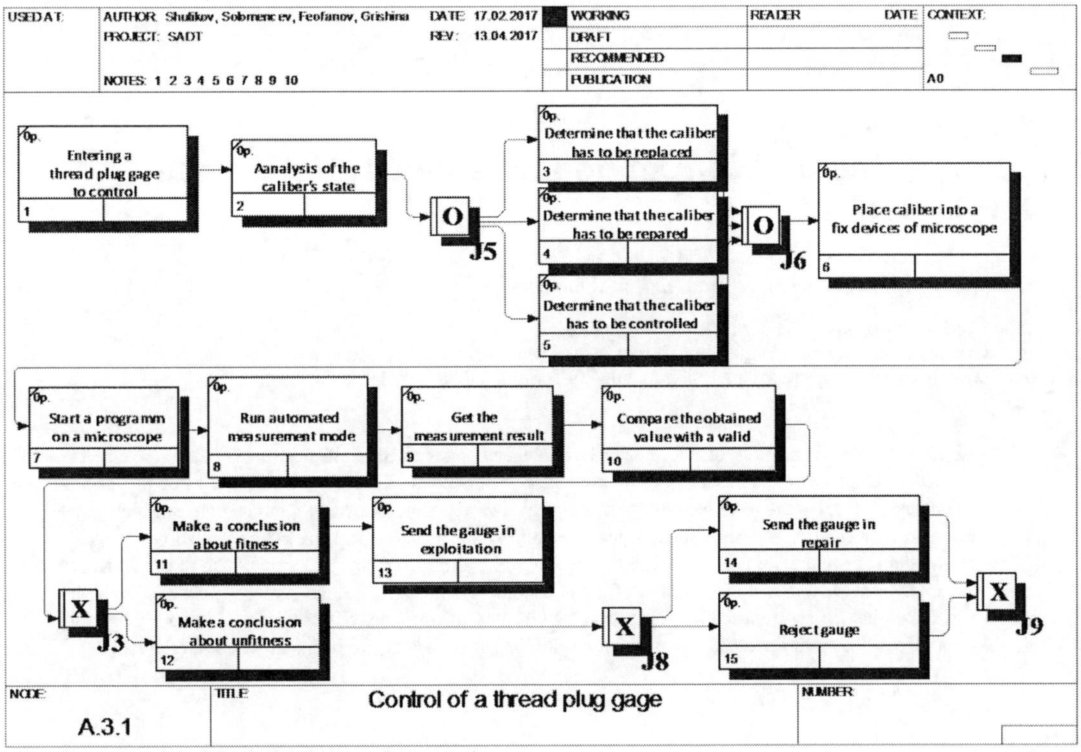

Fig.2. IDEF3 model of the threaded gauge-plug control.

dimensions of flat-surfaced parts, rotary bodies, threads and other precision instrumentation details.

The software "Saphir" had been used for the convenience work with microscope to automate the measurement method for the average diameter of thread plug gauges. To measure in auto mode, it's necessary first to measure the thread plug gauges in manual mode (the training mode), after the program store operations sequences and save them as file on PC that permits to measure thread plug gauges with the same data (diameter and leads) in auto mode. Figure 4 shows the process model for thread plug gauges monitoring using the IDEF3 diagram.

The introduction of the automated control method is justified in the modern realities of machine-building production, since process automation involves the acquisition of information flow for the collection and processing of data. As a result of the information statistics obtained, not only the time parameters are determined, but bottlenecks are also identified, in order to establish additional control or eliminate the process violation. And the method significantly reduces time resources for monitoring and the requirements for the personnel classification.

References

1. A. Feofanov, A. Skhirtladze, *The organization and execution of works on maintenance of industrial machinery: textbook for the stud. of secondary vocational schools* (Publishing center "Academy", Moscow, Russia, 2017)
2. A. Feofanov, M. Turapin, *Automated systems of identification and analysis of supply risks to improve the efficiency of engineering production* (Monography, 2016)
3. A. Feofanov, N. Sobrina, T. Grishina. *The basic methodology of substantiation of the technological characteristics of the product*, Vestnik MSTU "Stankin", **№4** (2015)
4. A. Feofanov, V. Korneeva, N. Negrimovskaja, *Calculation of the optimal number of experts and the amount of data to be detected based on the results of the preliminary survey*, Welding international, **№9** (2012)
5. A.Grigoriev, E-lectures. URL: http://masters.donntu.org/2010/fknt/bolotova/library/tez7.htm (2010).
6. A. Prokofiev, *Technological support and improvement the quality of screwed joints: dis. ... of PhD in Engineering: 05.02.08: defense of the thesis 01.07.08* (Bryansk: Bryansk state technical University, 392 p., 2008)
7. GOST 27284-87. Gauges. Terms and definitions. – state standard of the Russian Federation . Introduced 1988-01-01. 6 p. (2003)
8. RD IDEF 0 – 2000. Methodology of functional modeling IDEF0. Guidance document. Edition official. p. 62 (2000)
9. R. Bouncers, A. Feofanov, V. Korneeva, E. Nakhapetyan, *Qualimetry in mechanical engineering: the textbook*, p. 285 (2009)

MATEC Web of Conferences **129**, 01063 (2017)
ICMTMTE 2017 DOI: 10.1051/matecconf/201712901063

Solution of the problem of negative value of the linear technological size

Vasilii Masyagin[1,*], *Ruslan* Bazhenov[2], and *Andrei* Muholzoev[3]

[1]Omsk State Technical University, Mira 11, Omsk 644050, Russia
[2]Sholom-Aleichem Priamursky State University, Shirokaya 70a, Birobidzhan 679015, Russia
[3]National Research Tomsk Polytechnic University, Lenina 30, Tomsk 634050, Russia

Abstract. The calculation of the linear technological sizes is one of necessary design stages of technological processes of machining. A problem is that with performance of calculation of the linear technological sizes negative value of the technological size sometimes turns out. In the article consideedr an example of calculation, in which appears negative value of the technological size. The calculation is carried out on the technique with use of the chart of processing and the computer program which is carrying out automatic construction and the decision of the equations of dimension chains. As a result of the study, it was found that the appearance of a negative value when calculating the technological dimensions is not an error, this is due to the uncertainty of the positions of the technological size boundaries before calculation.

1 Introduction

The calculation of the technological sizes is one of necessary design stages of technological processes of machining. A theoretical basis of calculation is the theory of dimensional chains or tolerance charting, mathematical model is graph [1,2,3,4,5]. Nowadays given calculation is automated [6].

A problem is that with performance of calculation of the linear technological sizes negative value of the technological size sometimes turns out.

On the sense the technological size is always positive, so it is necessary to understand what is the cause of the negative value of the technological size, and what needs to be changed when obtaining a negative value of the technological size as a result of calculation.

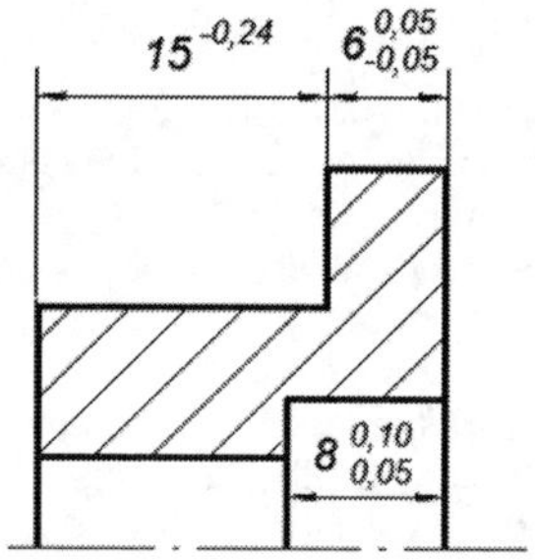

Fig. 1. The sketch of detail with linear design dimensions.

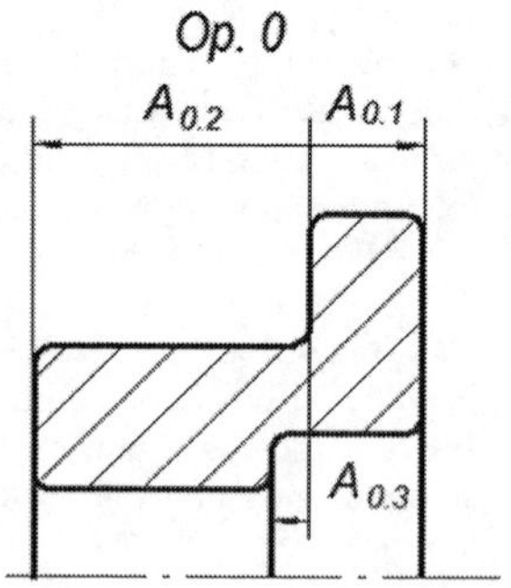

Fig. 2. The drawing of workpiece with linear technological dimensions.

2 The description of the research

Consider an example of calculation, in which there are negative value of the technological size. The calculation is carried out on the technique of Mordvinov [1] with use of the computer program which is carrying out automatic construction and the decision of the equations of dimension chains.

Let it is necessary to manufacture a detail (Fig. 1), having the linear design dimensions. The sketch of the workpiece and technological sketches with unknown linear technological dimensions are shown in Fig. 2 and Fig. 3.

The unknown technological dimensions are designated by a symbol A_{ij}, where the indexes designate the sequence number of the operation and the sequence number of the technological size.

[*] Corresponding author: masyagin@omgtu.ru

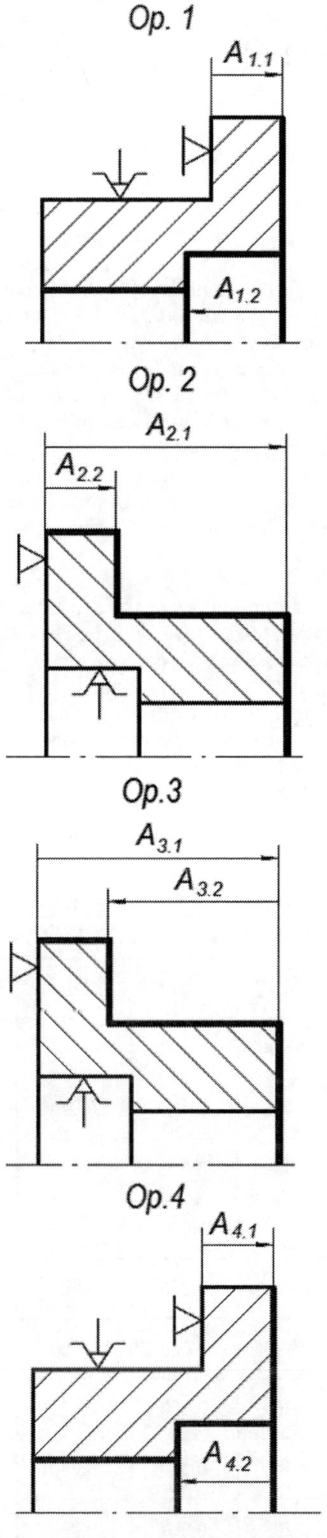

Fig. 3. The technological sketches with linear technological dimensions.

The technological dimensions are shown by one-sided arrows pointing towards the surfaces to be treated.

It should be noted that, according to the technological recommendations, all dimensions of the workpiece must be assigned from a processing base at the first operation (Fig. 3).

According to the sketches of the part, workpiece and technological sketches, the processing scheme is executed (Fig. 4). The formation of the processing scheme is reduced to the transfer of technological dimensions from the technological sketches and design dimensions from the sketch of the part to the contour of the workpiece.

The chart of processing of the linear technological sizes is shown in a Fig. 4. The chart of processing contains all the technological dimension chains with concluding links - the design sizes and the allowances, and component links - the technological sizes.

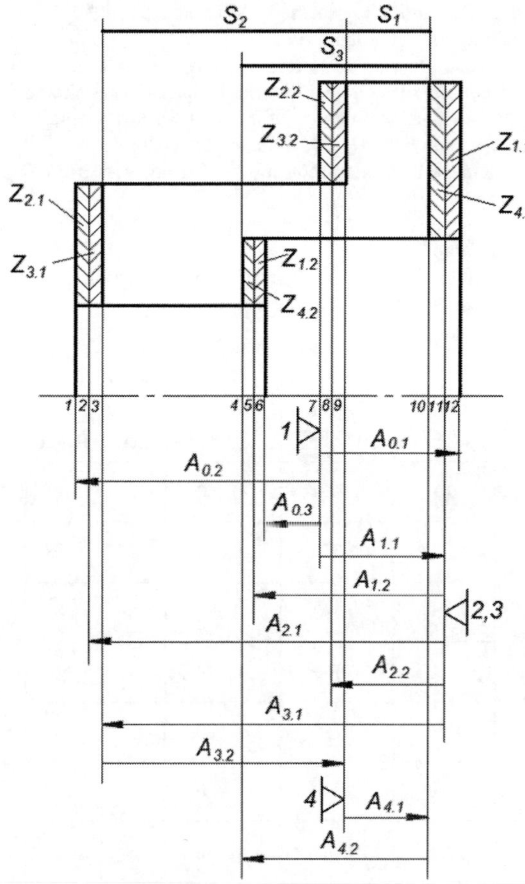

Fig. 4. The initial chart of processing.

The design sizes are designated by a symbol S_i with an index designating a sequence number of the design size.

The allowances are denoted by the symbol Z_{ij} with the index corresponding to the index of the technological size A_{ij}, according to which the allowance is removed.

The technological dimensions A_{ij} have the same designation as in Fig. 2 and Fig. 3.

The base surfaces are designated by a special symbol with the indication of the sequence number of operation.

All processable and processed surfaces are numbered from left to right in the growing order.

The values of deviations of the technological sizes and the values of the minimal allowances are determined by handbooks.

The constructed processing scheme must satisfy following conditions [1].

1. Only one arrow of technological size should indicate each surface, and no arrows should indicate the source base. In this example scheme satisfy the condition.

2. The number of technological dimensions, including the dimensions of the workpiece, should be equal to the sum of the quantities of design dimensions and allowances (excluding allowances, the removal of which changes the shape of the workpiece). In this example: 11 sizes of A_{ij}, 3 sizes S_i, and 8 allowances Z_{ij}. Therefore, 11 = 3 + 8. This satisfies the condition.

3. The number of technological dimensions, as well as the number of design dimensions and allowances should be one less than the number of surfaces. In this example, 12 surfaces and 11 sizes. This satisfies the condition.

The further calculation are carried out with use of the computer program.

The initial data for computations is given in Table 1 and Table 2.

Table 1 and Table 2 was formed according to the scheme presented in Figure 4.

Table 1 contains the following data.

Column 1 is a consecutive list of designations for all linear technological dimensions A_{ij}, starting from the dimensions of the workpiece and ending with the technological dimensions obtained at the finish operations.

Column 2 contains an upper tolerance deviation for the technological size with the corresponding sign.

Column 3 contains a lower tolerance deviation for the technological size with the corresponding sign.

Columns 4 and 5 contain, for each technological measurement, the numbers of surfaces (boundaries) that bound the size to the left and the right in accordance with the part processing scheme.

Table 2 contains the following data.

Column 1 contains a sequential list of designations for all of the design dimensions of the detail - S_i, then allowances for processing - Z_{ij}.

Column 2 contains the nominal dimensions of the detail according to the drawing are indicated (Fig. 1) and the minimum Z_{min} values of allowances, according to the handbooks.

Columns 3 and 4 contain the upper and lower deviations in the dimensions of Si, in accordance with the drawing of the part (see Fig. 1). For allowances, zero values are entered in both columns.

Columns 5 and 6 contains, for each design dimensions of the detail and for each allowance the numbers of the surfaces (boundaries) that bound the size or allowance to the left and to the right according to the part machining scheme are indicated.

Table 1. Technological dimensions.

Index of size	Upper deviation	Lower deviation	Boundary	
			Left	Right
0.1	0.450	-0.450	7	12
0.2	0.550	-0.550	1	7
0.3	0.300	-0.300	6	7
1.1	0.180	-0.180	7	11
1.2	0.180	-0.180	5	11
2.1	0.260	-0.260	2	11
2.2	0.180	-0.180	8	11
3.1	0.105	-0.105	3	11
3.2	0.090	-0.090	3	9
4.1	0.050	-0.050	9	10
4.2	0.025	-0.025	4	10

Table 2. Design sizes and allowances.

Index of size	Nomina l size	Upper deviatio n	Lower deviation	Boundary	
				Left	Righ t
1	6.000	0.050	-0.050	9	10
2	15.000	0.000	-0.240	3	9
3	8.000	0.100	0.050	4	10
1.1	0.490	0.000	0.000	11	12
1.2	0.490	0.000	0.000	5	6
2.1	0.490	0.000	0.000	1	2
2.2	0.490	0.000	0.000	7	8
3.1	0.200	0.000	0.000	2	3
3.2	0.200	0.000	0.000	8	9
4.1	0.200	0.000	0.000	10	11
4.2	0.200	0.000	0.000	4	5

DOI: 10.1051/matecconf/201712901063

Table 3. Calculated technological dimensions.

Index of size	Upper deviation	Lower deviation	Nominal size
0.1	0.450	-0.450	15.000
0.2	0.550	-0.550	10.000
0.3	0.300	-0.300	-2.150
1.1	0.180	-0.180	8.500
1.2	0.180	-0.180	7.500
2.1	0.260	-0.260	22.000
2.2	0.180	-0.180	7.500
3.1	0.105	-0.105	21.350
3.2	0.090	-0.090	14.900
4.1	0.050	-0.050	6.000
4.2	0.025	-0.025	8.075

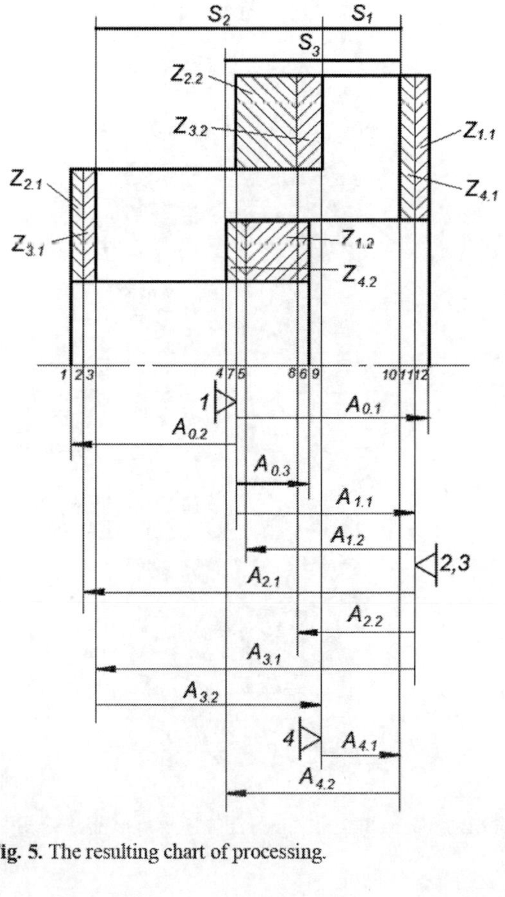

Fig. 5. The resulting chart of processing.

The results of computations is given in Table 3.

In Table 3, the column "Nominal size" contains the required values of technological parameters, which, with the corresponding upper and lower deviations, are inserted in the technological documentation.

As a result of the computations the negative value of the technological size for one of the sizes of workpiece $A_{0.3}$ turns out (-2.150).

Based on the calculation results, a new processing scheme was implemented (Fig. 4). In this processing scheme, the position of the boundaries is determined by the nominal values of the linear technological dimensions.

The actual position of the left and right boundaries of the size A0.3, as seen from Fig. 5, turned out to be the opposite to the position that was originally taken before the calculation, in accordance with the scheme in Fig. 4.

The given size A0.3 as a result of computations has negative value. Hence, the occurrence of negative value of the technological size is connected to change of real position of the left and right boundaries of the size in relation to initial.

From here follows, that the given size A0.3 should be designated in another way on the drawing of workpiece (Fig. 2), namely, as shown in the chart of processing (Fig. 5).

3 Conclusion

The result of calculating the process dimensions with obtaining a negative value is not an error, but is associated with the uncertainty of the positions of the size boundaries before calculation.

The actual positions of the size boundaries obtained as a result of the calculations may differ from the accepted positions of the boundaries before calculation.

When a negative value is obtained, it is necessary to change the positions of the boundaries of the corresponding technological size in the technological documentation.

References

1. B.S. Mordwinov, Feingerätetechnik, **24(5)**, 215-218, (1975)
2. K.A. Ngoi, C.C. Kai, Int. J. of Adv. Manuf. Technol., **8**, 175-181, (1993)
3. B.K.A. Ngoi, O.C. Teck, Int. J. of Adv. Manuf. Technol., **8**, 371–376, (1993)
4. G.A. Britton, G. Timm, Int. J. of Adv. Manuf. Technol., **20**, 448–453, (2002)
5. G. Thimm, G.A. Britton, S.C. Fok, Int. J. of Adv. Manuf. Technol., **24**, 261–271, (2004)
6. P. Ji, X. Jianbin, *Proceedings of the Institution of Mechanical Engineers*, **220**, 883-892, (2006)

An algorithm for controlling of cutting speed based on soft calculations

Maxim Bobyr[1,*], *Mariya* Luneva[1], and *Alexsey* Yakushev[1]

[1]South-West State University, Department of computer engineering, 305040 Kursk, Russian Federation

Abstract. The algorithm for controlling of cutting speed during machining of parts on equipment CNC is presented in the article. A program code has been developed for controlling of cutting speed on a three-axis milling machine CNC. The fuzzy-logical MISO system in which cutting speed depends on the rotation frequency of the cutting tool and the feed is presented.

1 Introduction

Machining cutting of products is a universal method of manufacturing of parts. Formation of high requirements to the final product requires an increase of the level of automation and intellectualization of production processes. If the automation of processes has now reached a high level through with use of various mechatronic complexes, then the applied intellectual systems are not sufficiently able to control complex systems [1, 2]. In this connection, the problem of increasing the level of intellectualization of control systems for equipment CNC arises.

2 Control system of a machine-tool with CNC

The experimental model of a 3-axis milling machine CNC is shown in Fig.1. The machine-tool includes: microcontroller Arduino Mega 2560, a control board of motors SH Ramps v.1.4, power supply 12 V, three drivers of stepper motor MP4988, three bipolar stepper motors, mini-drill with cutting tools.

The process of forming control signals to the actuating mechanisms of a machine-tool with CNC includes five steps:

Step 1. Software development in Arduino IDE for controlling the movement of a cutting tool along the axes of the machine-tool with CNC.

Step 2. Download of program code via serial COM-port from PC to microcontroller Arduino Mega 2560.

Step 3. Data transfer from the microcontroller Arduino Mega 2560 to the control board of motors SH Ramps.

Step 4. Data transfer from the SH Ramps to drivers of stepper motors MP4988.

Step 5. Data transfer from the drivers of stepper motors MP4988 with the subsequent transfer of impulses to bipolar stepper motors.

After that, the transfer of rotary motion from stepper motors using the shaft to the axes supports of machine-tool with CNC is carried out. The motion of supports of a machine-tool leads to the movement of the cutting tool along axes of the machine-tool.

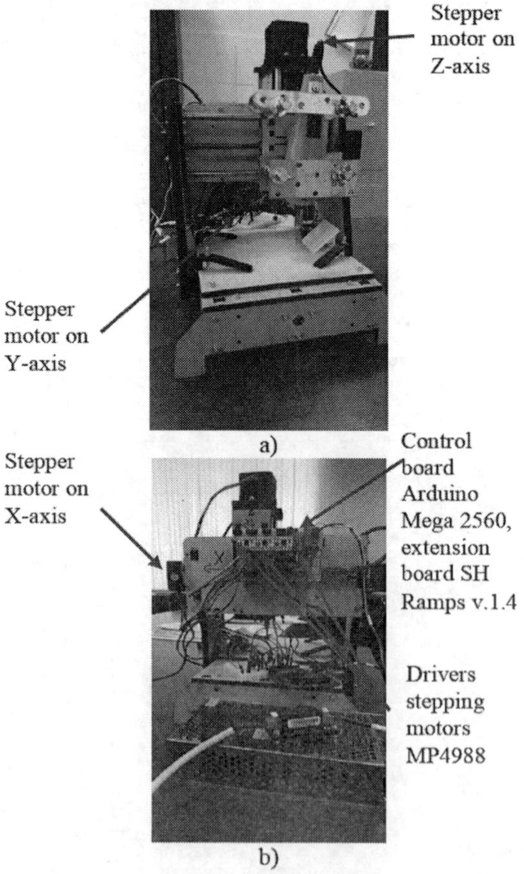

Fig.1. 3-axis milling machine with CNC: (a) front view; (b) rear view

* Corresponding author: fregat_mn@rambler.ru

The graphic model of the cutting process is shown in Fig.2.

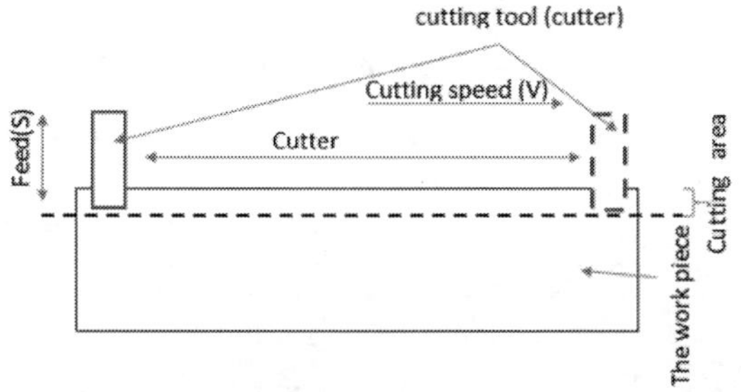

Fig.2. Process of cutting parts

3 Hardware-software model for control of the machine-tool

To connect MP4988 driver to the Arduino Mega microcontroller is used Table 1.

Table 1. Connecting the MP4988 driver pins to the Arduino Mega microcontroller.

Axis name	Pin	Pin number
X	DIR	55
	STR	54
	EN	38
Y	DIR	61
	STR	60
	EN	56
Z	DIR	48
	STR	46
	EN	62

Movement of the machine-tool supports is carried out along the shafts having thread of 2 mm. That is, in order the support of the machine-tool with CNC moved 2 mm, it is necessary to transfer 360 ° / 1.8 ° = 200 pulses through the STR channel.

The MP4988 driver works in several step modes: 1/2, 1/4, 1/8 and 1/16. Reducing the step size increases the accuracy of positioning machine-tool CNC supports. For example, in the 1/4 step mode, to move the support for 2 mm, it is necessary to transmit 800 pulses along the STR channel. The stepper motor is connected to the A1, A2, B1, B2 pins of the MP4988 driver.

To move supports along the axes of a three positioning machine-tool with CNC, the program code stored in the microcontroller is used. The program code for moving a support along the X-axis is shown need:

```
void setup() {
Serial.begin(9600);
pinMode (x_en, OUTPUT);
pinMode (x_step, OUTPUT);
pinMode (x_dir, OUTPUT);
digitalWrite(x_en, LOW); }
void x_ot_dvig(int imp, int skorost){
digitalWrite(x_dir, HIGH);
for (i = 0; i < imp; i++)  {
digitalWrite(x_step, HIGH);
delayMicroseconds(skorost);
digitalWrite(x_step, LOW);
delayMicroseconds(skorost);  }}
void x_k_dvig(int imp, int skorost){
digitalWrite(x_dir, LOW);
for (i = 0; i < imp; i++)  {
digitalWrite(x_step, HIGH);
delayMicroseconds(skorost);
digitalWrite(x_step, LOW);
delayMicroseconds(skorost);  }}
void flushReceive(){ while(Serial.available())
Serial.read();}
void loop() {
char val = Serial.read();
if (val == '1') {
x_ot_dvig(2000, 500); }
if (val == '2') {
x_k_dvig(2000, 500); } }
```

When using the program code, it was noticed that when changing the control parameters and then loading them into the Arduino Mega microcontroller, the program code written earlier in the Arduino Mega microcontroller is executed before using this code. As a consequence, the cutting tool does not move along the specified trajectory. To solve this problem, the program code need:

```
#define x_dir 55
#define x_step 54
#define x_en 38
int i;
void setup() {
pinMode (x_en, OUTPUT);
pinMode (x_step, OUTPUT);
pinMode (x_dir, OUTPUT);
digitalWrite(x_en, LOW); }
void loop() {
digitalWrite(x_dir, LOW);
for (i = 0; i < 4000; i++)  {
digitalWrite(x_step, HIGH);
delayMicroseconds(500);
digitalWrite(x_step, LOW);
delayMicroseconds(500);}
delay(2000);
digitalWrite(x_dir, HIGH);
for (i = 0; i < 4000; i++){
digitalWrite(x_step, HIGH);
delayMicroseconds(500);
digitalWrite(x_step, LOW);
delayMicroseconds(500);}
delay(2000);}
```

– to have separate functions for implementing movement of supports along each of axis;

– to use the function Serial.Available().This function allows to stop the execution of the program code until

DOI: 10.1051/matecconf/201712901064

the operator gives the start command from the computer keyboard.

4 Fuzzy MISO-system control of cutting speed

During the experiments it was found that depending on the change in the number of pulses (i), the amount of displacement of the supports along each axis changes. The time delay between transmission of pulses (t) affects the speed of passage of the cutting tool along the axis (v). Also, the speed of passage of the cutting tool along the axis depends from frequency of rotation of the spindle drill (n). The fuzzy MISO-system for control of cutting speed was created on the basis of this data [3, 4] (Fig. 3).

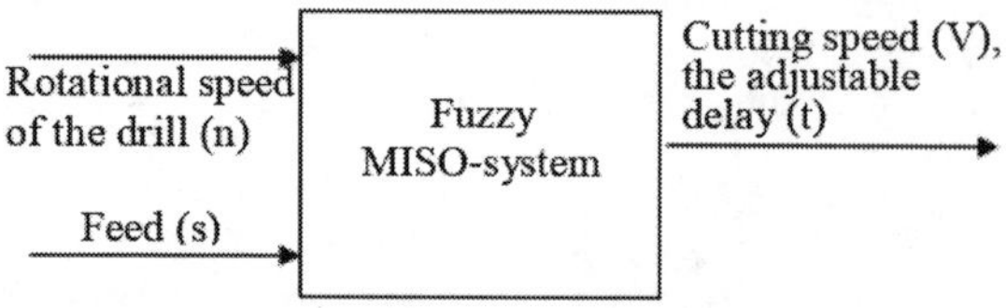

Fig.3. Fuzzy MISO-system control of cutting speed

The fuzzy system has two input variables:

$$n = \{n_1\}+\{n_2\}+\{n_3\}, \quad (1)$$

$$s= \{s_1\}+\{s_2\}+\{s_3\}, \quad (2)$$

and one output variable:

$$t = \{t_1\}+\{t_2\}+\{t_3\}+\{t_4\}+\{t_5\}, \quad (3)$$

where $n_1,\dots,n_3$ are terms of the input variable of fuzzy set n; $s_1,\dots,s_3$ are terms of the input variable of fuzzy set s; $t_1,\dots,t_5$ – terms of the output variable of fuzzy set t [5, 6].

Input and output variables are described by triangular membership functions (MF), the graphs of which are shown in Fig. 4, where μ (n), μ (s) and μ (t) are membership functions of the fuzzy sets n, s and t, respectively. [7, 8].

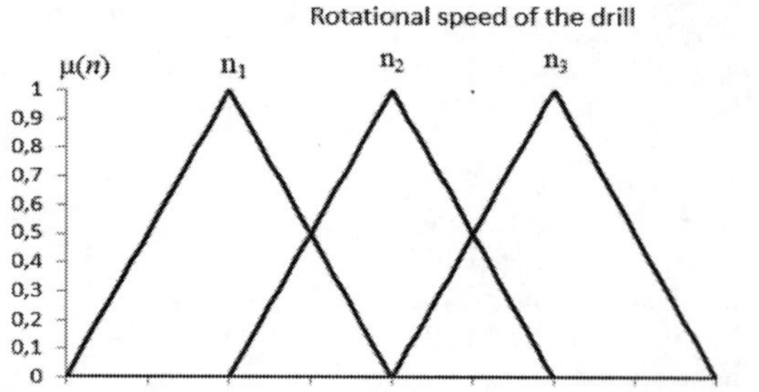

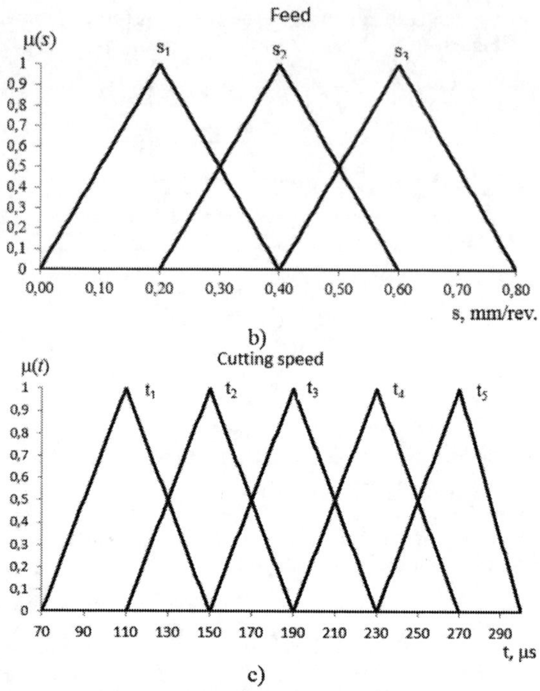

Fig.4. Graphics of triangular membership functions: *(a, b)* are input variables n and s, respectively; *(c)* is output variable t.

The MF is given by the following formula:

$$f(x\,;a\,,b\,,c)=\begin{cases}0, & x\le a\,;\\ \dfrac{x-a}{b-a}, & a\le x\le b\,;\\ \dfrac{c-x}{c-b}, & b\le x\le c\,;\\ 0, & c\le x\,.\end{cases} \quad (4)$$

The graphical view of the membership function is shown in Fig.5.

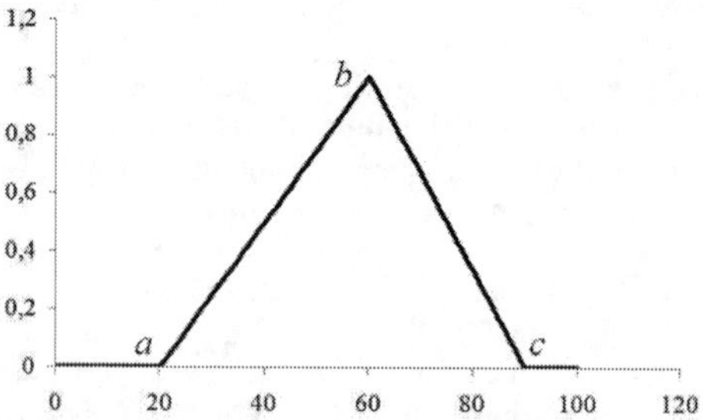

Fig.5. The membership function of triangular shape.

For fuzzification of the input variables were used the following data: the frequency of rotation of the drill spindle is 0 ÷ 8000 rpm. Feed of cutting tool is 0 ÷ 0.8 mm / rev. The delay time is limit of 70 ÷ 300 μs. If the time delay is less than 70 μs, the bipolar stepper motors do not rotate. The fuzzy knowledge base is given by nine fuzzy rules and is presented in Table 2 [9].

Table 2. Fuzzy knowledge base.

FR	If		That	FR	If		That	FR	If		That
FR_1	n_1	s_1	V_5	FR_4	n_2	s_1	V_4	FR_7	n_3	s_1	V_3
FR_2		s_2	V_4	FR_5		s_2	V_3	FR_8		s_2	V_2
FR_3		s_3	V_3	FR_6		s_3	V_2	FR_9		s_3	V_1

The defuzzification was carried out on the base of the method center of gravity [8, 9, 10]

$$V = \frac{\int_{min}^{max} v\mu'(t)dt}{\int_{min}^{max} \mu'(t)dt} \tag{5}$$

where min, max – limits of integration of fuzzy set; $\mu'(t)$ – degree of membership function truncated terms of output variable.

The structure of fuzzy inference with usage of soft arithmetic operations is presented in articles [11, 12 13]. The structure of algorithm for control of cutting speed is shown in Fig. 6.

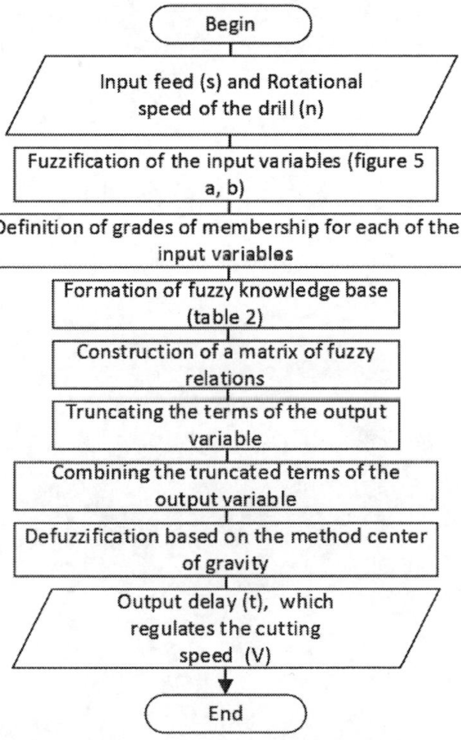

Fig.6. Fuzzy inference algorithm based on soft arithmetic operations

The resultant surfaces received during modeling of fuzzy inference with usage of soft and hard arithmetic operations are shown in Fig. 7.

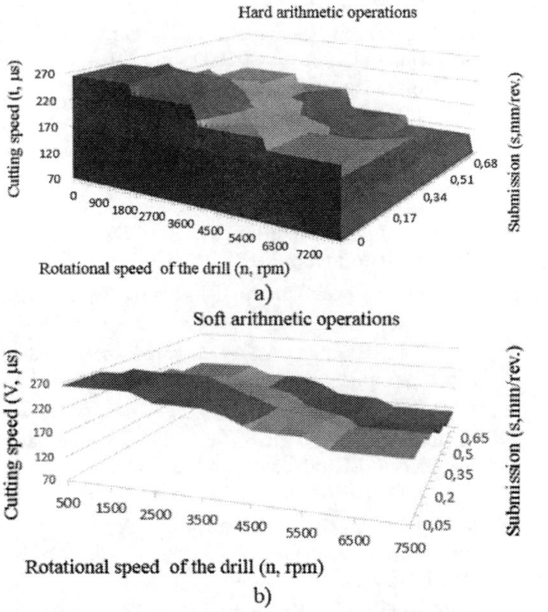

Fig.7. Fuzzy MISO-system with: (a) hard arithmetic operations; (b) soft arithmetic operations

Analysis of the graphs presented in Fig. 7 showed that during using in fuzzy inference of soft arithmetic operations and control of the cutting speed is performed more smoothly than with using of hard operations. Therefore, using soft arithmetic operations in fuzzy inference improves quality of the processed details. At the same time, the developed fuzzy MISO-system allows to extend the life of the cutting tools, since during the cutting, there is no overheating of the parts.

5 Conclusion

In this work the program code for controlling process of cutting on a three-axis milling machine with CNC is presented. The problem of deleting previous commands from the memory of the microcontroller, when loading the next command, was solved. Fuzzy MISO-system for controlling cutting speed, which depends on the rotation frequency of the spindle drill and the feed, is presented in the article. On the basis of the analysis of the resulting surfaces, it was concluded that, when using soft arithmetic operations, the accuracy in the machining of parts improves.

This work was supported by the grant of the President of the Russian Federation MD-707.2017.8 and the Goszadaniya: Agreement No. 2.3440.2017 / PCH.

References

1. Hui Liu, En Ming Miao, Xin Yuan Wei, Xin Dong Zhuang, International journal of machine tools and manufacture **113**, 35 (2016). DOI: 10.1016/j.ijmachtools.2016.11.001

2. Mehmet Emin Aktan, Nihat Akkus, Abdurrahman Yilmaz, Erhan Akdogan, MATEC Web of Conferences **45**, 05002 (2016). DOI: 0.1051/matecconf/20164505002
3. K. Yongjin, G.W. Fischer, T. Tzu-Liang, Journal of manufacturing systems **21,** 440 (2002). DOI: 10.1016/s0278-6125(02)80050-4
4. Earl-Juei Wang, Tai-Sheng Su, Ming-Hsien Chang, Journal of manufacturing systems **32**, 206(2013). DOI: 10.1016/j.jmsy.2012.10.003
5. A. Pegat, Nechetkoe modelirovanie i upravlenie 2013.
6. A. Gupta, H.Singh, A. Aggarwal, Expert Systems with Applications **38**, 6822(2011). DOI: 10.1016/j.eswa.2010.12.057
7. M.V. Bobyr, V.S. Titov, A.A. Nasser, Journal of Machinery Manufacture and Reliability **44**, 633 (2015). DOI: 10.3103/s1052618815070067
8. M.V. Bobyr, N.A. Milostnaya, S.A. Kulabuhov, Applied Soft Computing, (2017), (In Press) DOI: 10.1016/j.asoc.2017.05.040
9. M.V. Bobyr, S.A. Kulabuhov, N.A. Milostnaya, 2nd International Conference on Industrial Engineering, Applications and Manufacturing (ICIEAM), (2017), DOI: 10.1109/icieam.2016.7910970
10. M.V. Bobyr, A.S. Yakushev, N.A. Milostnaya, 2nd International Conference on Industrial Engineering, Applications and Manufacturing (ICIEAM), (2017), DOI: 10.1109/ICIEAM.2016.7910971
11. M. Liang, T. Yeap, A. Hermansyah, S. Rahmati, International Journal of Machine Tools and Manufacture **43**, 1497(2003). DOI: 10.1016/S0890-6955(03)00166-4
12. M.Bobyr, V.Titov, A.Belyaev, MATEC Web of Conferences **79**, 01052(2016). DOI: https://doi.org/
13. I.V. Chernova, S.A. Sumin, M.V. Bobyr, S.P. Seregin, Biomedical Engineering **49**, 263 (2016). DOI: https://doi.org/10.007/s10527-016-9545-y

The study of the process of chip segmentation in case of edge cutting machining of titanium alloy

Dmitry Sergeevich Ponyatov[1], *Vyacheslav Petrovich* Alekseev[1,*], and *Alexandr Isaakovich* Khaimovich[1]

[1] Samara University, 443086 Moskovskoye shosse 34, Russia

Abstract. This article presents the results of the studies of influence of the destruction criterion on the process of segmentation of chips in the process of treatment of VT6 titanium alloy. A series of experiments has been conducted in order to measure the cutting force, the geometry of actual chips, and the experiment's results have been analyzed and processed. Using DEFORM 2D software suite they have developed an imitation model of the milling process which is necessary for the analysis of the impact of cutting conditions on the process of chip formation.

1 Introduction

In the past several years due to the intensive growth of the technologies of metal processing using cutting it has become topical to study the processes of plastic deformation in the cutting edge area. Destruction of materials in the process of their cutting, i.e. separation of the surface layer from the main body of the workpiece processed is preceded by an intensive plastic deformation of the layer separated and the surface layer. The process of chip formation is accompanied by high effective velocities of deformation which are localized in the area of nominal shearing plane. The higher the level of localization of the focus of plastic deformation the less stressed is the area of chips outside of this focus. [1].

In the process of cutting the most intensive plastic deformation is observed in the small area near the cutting surface sized no more than 0.5 mm, while the velocities of deformation are around 10^6 exponential order and the temperature differential reaches 1,000°C. An adequate measurement of the actual situation with strain-stress state (SSS) and the temperature in the area with such small dimensions causes significant difficulties therefore the verification means for numerical imitational modeling are almost the only single source of information about the processes in the area of chip formation. Advanced numerical methods of calculation of the strain-stress state enable us to reduce the number of field tests in order to define the temperature and energy parameters of cutting depending on the modes and geometry of the relevant cutting tool [2].

In the course of the past several years the analysis using finite-element method (FEM) has become the main tool for modeling the cutting process [3-9]. T. Thepsonthi, T. Ozel used the FEM analysis in order to study the impact of a cutting tool with CBN coating for processing of Ti-6Al-4V titanium alloy. The modeling results have been additionally used for the evaluation of the tool's service life. M. Calamaz, D. Coupard, F. Girot have analyzed the impact of the behavior pattern of the material for KE modeling during chip formation when processing titanium alloys. The modified flow model by Johnson-Cooke accounts for the impact of deformation, velocity of deformation, and temperature on the flow stress, as well as introduces the effect of softening deformation. Using the finite-element analysis method Lorentzon, J, Järvstråt, N. Josefson, B.L. have studied the impact of various criteria of destruction on the process of chip formation, placing emphasis on the formation of segmented chips in the process of treatment of Inconel 718 alloy. It has been identified that transition from continuous chip formation to segmented formation had been stipulated by the heat factor and the destruction criterion.

In many studies [10-14] the adequacy of imitational modeling of chip formation in CAE system is determined using temperature- and energy-power parameters. In the process of a field test it is not always possible to adequately identify temperature- and energy-power parameters in the cutting area; therefore one of the indicators of evaluation of the adequacy of imitational modeling is morphology of the chips. The adequacy of imitational modeling using the chip morphology depends on the choice of the corresponding criterion of destruction which determines the moment of the material's exhaustion of plasticity in the course of deformation. As soon as the criterion reaches its peak value the material gets deformed.

The objective of this work is to increase the accuracy of numerical modeling of the cutting process using evaluation of morphology of chip formation in the process of treatment of VT-6 titanium alloy.

2 Experimental studies

In order to analyze the geometry of chips and energy-power parameters they have used the previous

* Corresponding author alexeev_v.p@mail.ru

DOI: 10.1051/matecconf/201712901065

experimental studies [15]. For the experiment they used the workpieces made of VT-6 titanium alloy with the dimensions L×H×D 40x24x6 (mm). The surface of each workpiece (sample) has been preliminarily treated to ensure that the initial roughness of all the samples is the same – Ra 1.25.

The experimental studies have been performed using a CNC machine ALZMETALLBAZ 15 CNC equipped with a Kistler dynamometric table which registered changes which comprised the cutting force Fx, Fy, and Fz in real time. The cutting tool was Ø 10 mm end milling cutter with 4 (Four) teeth.

In order to study the chip shape and its structure polished specimen of the chips have been prepared. In order to prepare micro polished specimen of the chips they used epoxide compound as filler. For etching of the samples they used the following solutions: HCl – 22; HNO3 – 15; HF – 41; H2O – 22. In order to study the microstructure they used METAM 1B-32 microscope. The cross sections of the chip segment with the main geometric features are shown in Fig. 1.

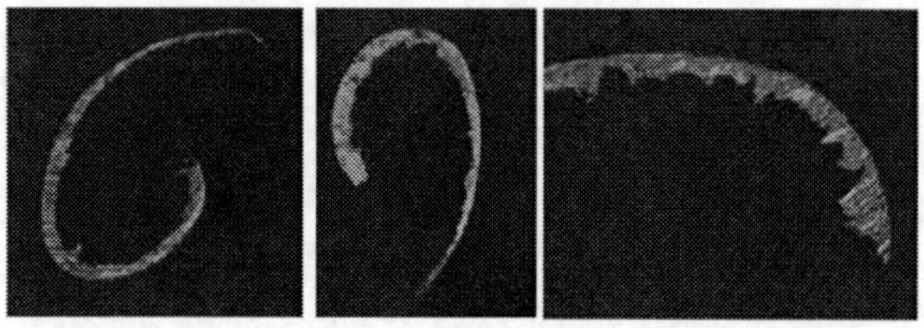

Fig. 1. Polished chips when processing VT-6 titanium alloy at the modes V = 60 m/min, t = 0.3 mm, S=0.05 mm/tooth (a); S=0.1 mm/tooth (b); S=0.2 mm/tooth (c).

In this work the study of strain-stress state in case of a lateral milling has employed the method of imitational finite-element modeling using DEFORM 2D software unit. The treated model is shown in Fig. 2.

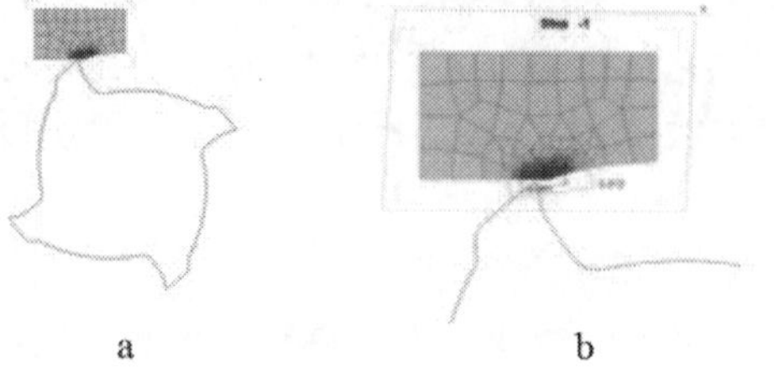

Fig. 2. Modeling of the milling process using DEFORM 2D: a – 2D model of the milling process; and b – areas of the grid condensation.

In the workpiece models they have made a cross section with the depth of 0.3 mm (Fig. 2a) which imitates the surface after operation of the previous tooth. This operation has been introduced with the purpose of reducing the time of calculation due to the optimal generation of the grid using the software suite. The tool is viewed by the model as an absolutely rigid body. They use the analog of VT-6 – Ti6Al4V titanium alloy. In order to limit loss of the volume it is necessary to condense the grid where a crack is expected to occur. The areas of the finite-element grid condensation: The internal area – cell size of 0.001 mm, average – 0.012 mm, external – 1 mm (Fig. 2b). The temperature of the tool and workpiece equals 20°C. The constant friction ratio (according to Siebel) of the tool and the workpiece is 0.12; and the constant friction ratio of chips and the workpiece is 0.5. Heat exchange with the environment is not taken into account.

3 Segmentation process modeling

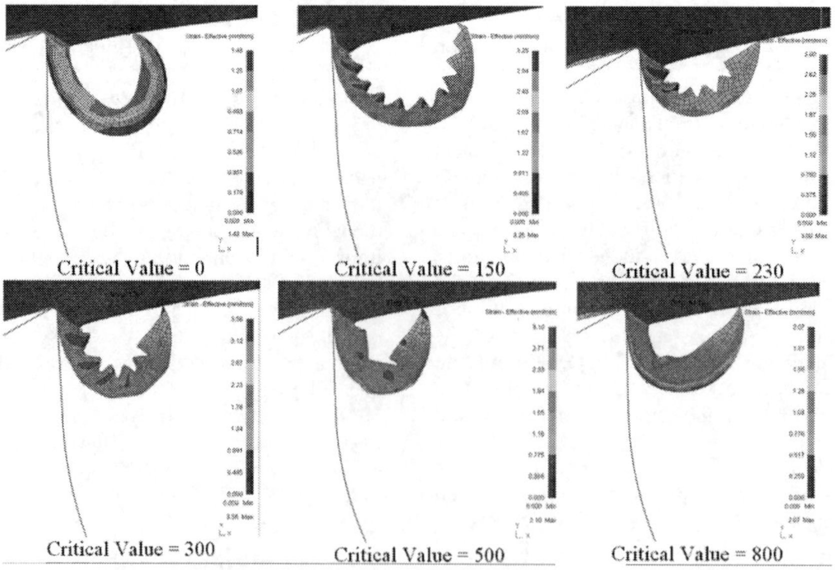

Fig. 3. Impact of the critical value of the destruction criterion on the chip morphology.

 DOI: 10.1051/matecconf/201712901065

One of the indicators of evaluation of the adequacy of imitational modeling is the chip morphology. Adequacy of imitational modeling using the chip morphology criterion depends on the choice of the corresponding destruction criterion which defines the moment of the material's exhaustion of its plasticity reserve in the process of deformation. Cockroft-Latham law has been chosen to describe destruction of the material. This criterion is empirical therefore the adequacy of this criterion depends on the limit value of plastic deformations at which the material begins to flow.

4 Discussion of the study results

In order to evaluate the adequacy of the models developed they have analyzed the impact of the critical value of destruction on the process of chip segmentation. In Fig. 3 they different types of chips are shown as achieved in the process of modeling for the following treatment modes: V=60 m/min, S=0.2 mm/rev., t=0.3 mm depending on the critical value of CL destruction. In the process of modeling the value of critical value of CL destruction has varied within the range from 0 to 800.

As shown in Fig. 3 when the critical value equals 0 chips have homogeneous features and as the criterion value increases the chips gain segmented shape. The period of formation of the segment increases all the way until the critical value which equals 200, and then it begins to drop. At the value of 500 the segments acquire incorrect shape and the chip thickness notably increases. When the value is above 800 the segment degenerated into two elongated parts with a sharp differential of thickness between them. Noteworthy, the frequency and type of generation of segmentation has a direct impact on the cutting force F_{pe3}.

The graphs of changes of the components of the cutting force overtime depending on the values of the critical value of the Cockroft-Latham criterion are shown in Fig. 4. As shown in Fig. 4 the cutting forces get reduced as the critical value increases until it reaches the value of 230, and then they observed a growth of forces and when the critical value reaches 800 the cutting forces reach their maximum values, and afterwards they practically do not fluctuate.

In order to evaluate the adequacy of the models developed they have analyzed and compared the results of the numerical and field tests of the chip morphology. Fig. 5 shows the images of a chip cross cut for the cutting modes: V=60 m/min, t=0.3 mm, and the three different loads: S=0.05; 0.1; 0.2 mm/rev., with the specified typical geometric parameters for the experiment and finite-element modeling. For finite-element models – based on the received geometrical parameters of the chips – they have selected the models which had been calculated using critical values of destruction criterion which equaled 300. As shown in

Fig. 5 the obvious sawtooth configuration of the cross cut of the chips is only typical for the load S=0.2 mm/rev., both for the numerical study and the experimental one. For the load S=0.1 mm/rev. evaluation of the segmentation using geometrical method is complicated as the height of dents of the segments and their depth are depths vary, on average, by 6 mkm. For the load S=0.05 mm/rev. the cross cut of the chips gains distorted shape from the standpoint of evaluation of the segmentation and in terms of its shape it is rather homogeneous and the only parameter which is used to typify it is the crosscut's average thickness. It should be noted that the result of simulation with a rather high degree of accuracy predicts the shape of chips for all three modes.

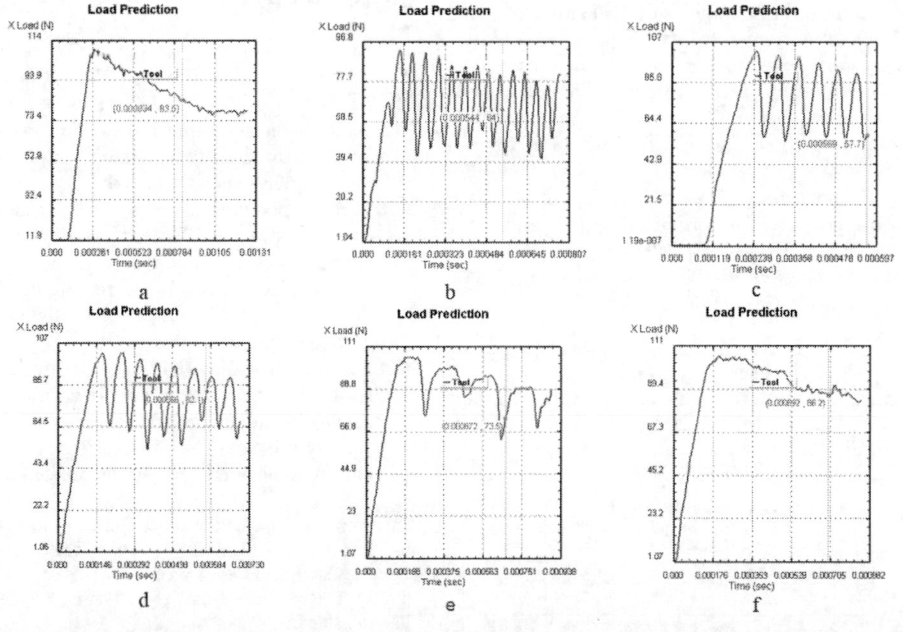

Fig. 4. Impact of the destruction on average cutting force at the critical values: a-0, b-150, c-230, d-300, e-500, e-800.

DOI: 10.1051/matecconf/201712901065

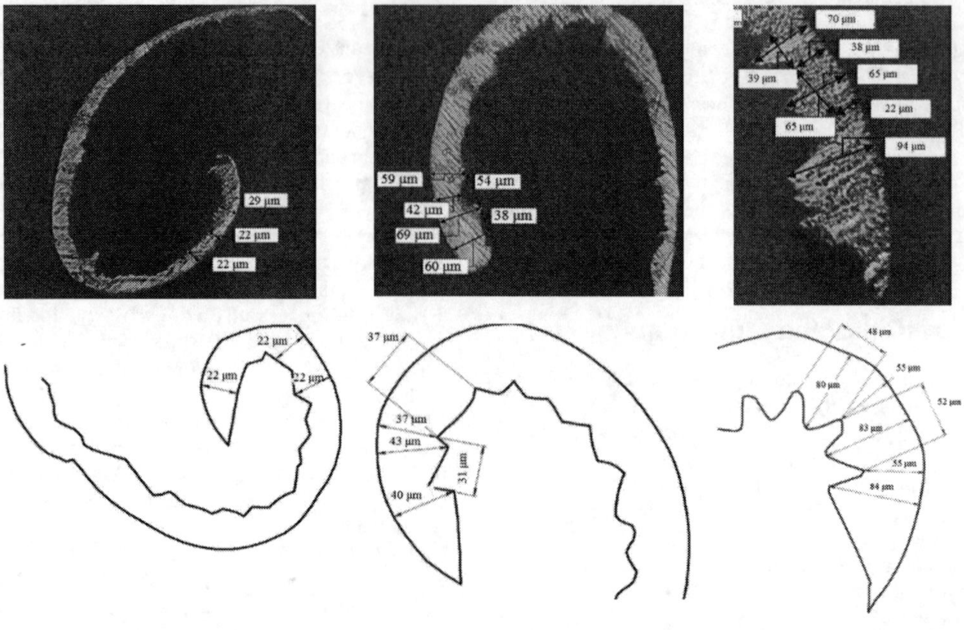

a b c

Fig. 5. Cross section of a chip cut for cutting modes, V=60 m/min, t=0.3 mm S=0.05 mm/rev. (a); 0.1 mm/rev. (b); 0.2 mm/rev. (c) for the experiment and finite-element modeling.

As shown in Fig. 5 the obvious sawtooth configuration of the cross cut of the chips is only typical for the load S=0.2 mm/rev., both for the numerical study and the experimental one. For the load S=0.1 mm/rev. evaluation of the segmentation using geometrical method is complicated as the height of dents of the segments and their depth are depths vary, on average, by 6 mkm. For the load S=0.05 mm/rev. the cross cut of the chips gains distorted shape from the standpoint of evaluation of the segmentation and in terms of its shape it is rather homogeneous and the only parameter which is used to typify it is the crosscut's average thickness. It should be noted that the result of simulation with a rather high degree of accuracy predicts the shape of chips for all three modes.

In Table 1 the cutting force values are shown as achieved in the process of experimental studies and the numerical modeling (CL=300) in the process of their treatment at the modes V=60-80 m/min, t=0.3 mm, and the three various loads: S=0.05; 0.1; 0.2.

Table 1. Cutting forces in the process of treatment of VT-6 titanium alloy

Cutting velocity, V, m/min	Load S, mm/tooth					
	0.05		0.1		0.2	
	Cutting force F, H					
	Modeling	Experiment	Modeling	Experiment	Modeling	Experiment
60	35.64	34.68	52.25	54.72	84.4	86.42
80	36.37	37.88	52.5	56.79	82.9	90.72

As apparent from Table 1 the experimental data of the cutting force are well harmonized with the data of numerical modeling with the ratio error of no more than 5%. For more adequate results it is required to further refine the data using rheological properties of the material during its treatment.

5 Conclusion

1) An imitation model for the cutting process has been developed using DEFORM 2D software suite;
2) Adequacy of the model developed based on the comparison of cutting forces and the shape of chips acquired in the process of the experiment and the modeling has been validated;
3) The 2D modeling data are well harmonized with the experimental data, while the error of forecasting the cutting force does not exceed the value of 5%. More a more adequate result it is required to further refine the data using rheological properties of the material in the process of its treatment.
4) The imitational model developed for lateral milling which includes experimentally verified values of the Cockroft-Latham destruction criteria is applicable for solving the tasks of optimization of the geometry of cutting tools and operating parameters of cutting.

This work has enjoyed the support of the Ministry of Education and Sciences of the Russian Federation under the framework of the « Research and Development of Priority Areas of Scientific and Technical Complex of the Russian Federation in 2014-2020 » Program.

DOI: 10.1051/matecconf/201712901065

References

1. N.I. Klushin, Metals Cutting Publishing House, (1958)
2. A.I. Khaimovich, Self-construction. Problems and prospects, **402** (2011)
3. M.A. Davies, Y. Chou, C.J. Evans, Natl. Inst. of Std. and Technology, (1996)
4. S. Atlati, B. Haddag, M. Nouari, M. Zenasni, International Journal of Machine Tools and Manufacture **51**, 9 (2011)
5. A.N. Afonin, Modeling of destruction of metals by plastic deformation in the DEFORM and LC-DYNA, **25** (2007)
6. D.V. Evdokimov, D.L. Skuratov, D.G. Fedorov, Russian Engineering Research **36**, 4 (2016).
7. M.G. Cockroft. D.J. Latham, J. Institute of Metals, **96** (1968)
8. A.N. Zhidyaev. M.A. Bolotov. I.L. Shitarev, Vestnik Rybinsk State Aviation Technological Academy P.A. Solovyov, (2010)
9. M.A. Bolotov, V.A. Pechenin, Journal of Materials and Environmental Science **7**, 9 (2016)
10. R.H Komanduri, I. Brown, Journal of Engineering for Industry, **103** (1981)
11. D.L. Skuratov, A.N. Zhidyaev, M.B. Sazonov, Res. J. of Appl. Sci. **9**, 11 (2016)
12. T. Thepsonthi. T. Ozel, Journal of Materials Processing Technology, **213** (2013)
13. T. Thepsonthi. T. Ozel, Journal of Materials Processing Technology, **221** (2015)
14. M. Galamaz. D. Coupard. F. Girot, International Journal of Machine Tools and Manufacture, **48** (2008)
15. V.P. Alekseev, A.I. Khaimovich, Key Engineering Materials, **684** (2015)

 DOI: 10.1051/matecconf/201712901066

Analysis of the structure of random packings of powder particles in laser additive technologies

Oleg Kovalev[1,2,*], *Andrey* Gusarov[2], and *Vitaly* Belyaev[1]

[1]Khristianovich Institute of Theoretical and Applied Mechanics SB RAS, Institutskaya str., 4/1, Novosibirsk, 630090, Russia
[2]Moscow State Technological university "STANKIN", Vadkovsky per., 3-A, Moscow, 127055, Russia

Abstract. Under consideration are the peculiarities of the computer simulation of random packings of the particles involved in the laser additive technologies of selective laser melting and sintering of powders. The results of definition of random packings of monodisperse solid spheres are presented. Numerical analysis and structural investigations of their morphology are carried out. The peculiar character of the theoretical and computer models used for the analysis of particle packings in real powder media is understood. The technique of ray tracing is proposed for the description of laser radiation absorption and scattering in the layer of the stainless steel powder with the structure of the loose packing of mono-size spherical particles.

1 Methods of random packings generation in a granular medium

The physics of laser beam and powder layer interaction includes such processes as radiation absorption and scattering, heat transfer, powder consolidation and formation of a remelted trace. These processes are important for the understanding of the additive laser technologies of selective laser melting (SLM) and selective laser sintering (SLS) of powders [1]. The properties of the fabricated parts depend on the internal structure of the particle layer, and also on the above mentioned processes. The problem of formation of dense powder packings via varying the shape and the size of the particles is today a complex problem focusing much attention [2 – 4]. The dependencies of the packing structure on the particle size, the density, the specific surface and the pore connectivity degree are the important characteristics.

Computer simulation is a helpful tool to study the structural peculiarities of the random packing, which can be used for the description of the laser interaction processes on the micro- and macro-scale levels.

The methods of generation of random packings of the granular material can be divided into the dynamic methods of expansion or compression of a randomly chosen free configuration of the particle system, and the methods of consequent filling by particles of a certain space in the force field as the packing is formed in a restricted region. Each method is normally adopted to the conditions of packing usage for the solution of a specific task [2 – 3]. In the case of the selective laser melting or sintering, the powder is deposited on a substrate in layers with a scraper blade or a roller, and any possibility of compaction is excluded. It is believed that the density and the structure of the powder layer deposited before the laser action approaches to the density and the structure of the loose random packing [1].

1.1 Method of generation of the random packing of spheres

This paper uses the previously developed technique of calculation of the packings of monodisperse spherical particles [5]. The algorithm of calculation of the particle center coordinates in the loose layer is based on the simulating consequence of individual sphere throwing. At the inelastic collision, their adhesion interaction is taken into account, which permits having the loose packings with different porosity via the variation of the physical properties and the size of the particles.

The mathematical problem is formulated in the terms of the discrete elements method. A cuboid-shaped container with the sizes of $L_x \times L_y \times L_z$ is considered in the Cartesian system of coordinates *Oxyz*. The rib sizes are preset as follows: $L_x = L_y = MD$, where M is the number of sphere diameters (10, 15, 20) arranged on the rib length. The container is filled via the consequent throw of N individual solid spheres of the same diameter $D = 2R$. Initial coordinates of sphere centers x_0, y_0, z_0 are specified within the $z_0 = L_z$ distance from the substrate, and x_0, y_0 are chosen with the aid of the random number generator inside the preset range by the axes *Ox*, *Oy*: $0 < x_0 < L_x$, $0 < y_0 < L_y$. The container side faces and the bottom restrict the motion of the thrown spheres.

The sphere motion is caused by the gravity force with due regard to the adhesion interaction between spheres which is described by the Van-der-Waals forces. The interaction at the sphere collision is deemed to be

* Corresponding author: kovalev@itam.nsc.ru

absolutely non-elastic. In the collision instant, the speed of the incident sphere is assumed to be zero, its farther motion continues along the lying sphere surface in the pure sliding mode, and the rolling sphere may separate and lose the contact, hence continuing its free fall. As two or more contacts appear, providing that the total force remains positive, the thrown sphere stops, and the coordinates of its center are registered. The appearing adhesion force F between the contacting solid spheres is directed along the line connecting their centers and results from inter-molecular interaction (from the Van-der-Walls forces) [6]. For two similar spheres with the radius R and distance between their surfaces d, the adhesion force F is written as [6]:

$$F(q) = -\frac{H_a}{12R}\frac{(1+q)}{(q^2+q+q)^2(q^2+q+q+1)^2}, \quad (1a)$$

For the particles contacting with walls:

$$F(q) = -\frac{H_a}{48R}\frac{1}{q^2(1+q)^2}, \quad (1b)$$

where $q = d/R$, H_a is the Hamaker constant. In the case of stainless steel, $H_a = 2.1\times10^{-20}$ J.

The details of the above described algorithm are presented in [5].

1.2 Calculation of container filling

Fig. 1 shows the Oxz projections of the filled containers for D = 0.5 and 5.0 μm. As the sphere diameter D decreases, the effect of the adhesion force on the packing structure becomes essential, container filling becomes more complicated, i.e. some opens appear between the walls and spheres, Fig. 1(a, b). In the presence of the adhesion and the gravity forces, the falling particles form the packing with an anisotropic structure.

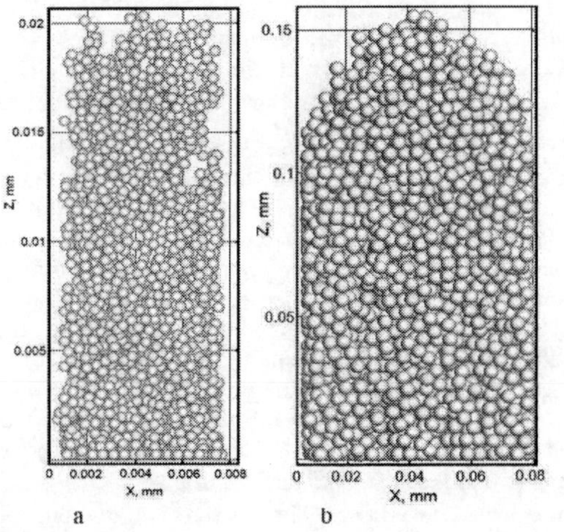

a b

Fig. 1. Projections of the containers in the Oxz plane, sphere diameters D, μm: 0.5(a); 5.0(b)

Owing to the adhesion forces which become more noticeable for the particles below 10 μm, the anisotropy appears to a greater extend over the container height.

2 Structural parameters of the packing

2.1 Theory

Such values as the porosity (or the density), the particle centers distribution, the pore specific surface and the tortuosity are the structural parameters of the packings which can be measured experimentally in packed granular media.

2.1.1 Porosity

Let the sphere of the radius R and center at the height z_i (Fig. 2(a)) cut out a circle of the area $\pi(R^2-(z_i-z)^2)$ on the horizontal plane z.

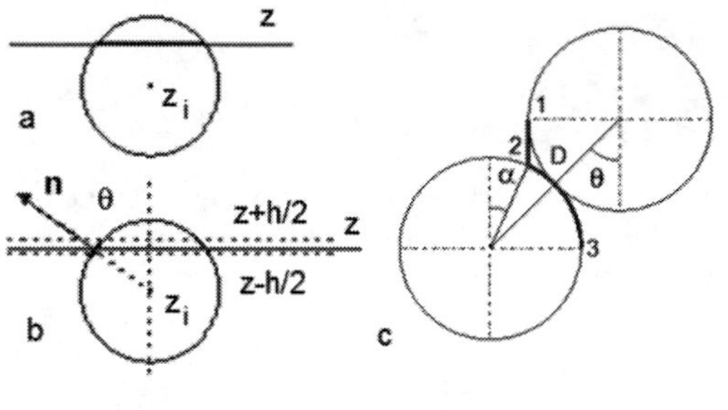

Fig. 2. Calculation of the porosity (a), the specific surface (b) and the tortuosity (c) for two adjoining spheres of the packed layer

If the area of the cross of the plane z and the container equals to $S(z) = const$, the value of the surface porosity ε_S or cross-section porosity at the z level equals to the fraction of this area free from spheres:

$$\varepsilon_S(z) = 1 - \frac{\pi}{S}\sum_i \max\left[0, R^2-(z_i-z)^2\right] \quad (2a)$$

To calculate the volume porosity ε_V, let us consider the volume $V(z) = Sz$ at the height z. Taking into account that the bottom area also equals to S, the volume porosity ε_V is:

$$\varepsilon_V(z) = 1 - \frac{4\pi R^3}{3S}\frac{1}{z}\sum_i \chi(z-z_i), \quad (2b)$$

where $\chi(x)$ is the Heaviside unit function, and the summing includes every sphere in the container.

Comparison of two ways (2a) and (2b) of calculating the packing porosity is of interest. According to [7], in the case of the homogeneous granular medium, the porosity value approximately equals to the value of the volume porosity of the system

DOI: 10.1051/matecconf/201712901066

$$\hat{\varepsilon}_S \approx \varepsilon_V. \quad (2c)$$

It is necessary to determine the validity range for this equation in our conditions.

2.1.2 Height distribution of sphere centers

The sphere area cut out by the layer of the thickness h, centered about the plane z, Fig. 2(b), is $2\pi Rh$, whereas the angle θ between the external normal **n** and vertical axis z is found in the equation: $\cos\theta = (z - z_i)/R$. Amount of sphere centers at the height z is calculated in accord with (3)

$$N_C(z) = \sum_i \frac{\max\left[0, (h/2)^2 - (z_i - z)^2\right]}{(h/2)^2 - (z_i - z)^2}, \quad (3)$$

where h is the height-wise subinterval. The formula (3) permits calculating the amount of the particles, which centers coordinates z_i get at the height z in the band $z_i \in (z - 0.5h, z + 0.5h)$.

2.1.3 The specific surface

The specific surface at the z level is found as the ratio of the sums of sphere surface areas inside this layer to its volume Sh:

$$A(z) = \frac{2\pi R}{S} \sum_i \frac{\max\left[0, R^2 - (z_i - z)^2\right]}{R^2 - (z_i - z)^2} \quad (4)$$

The sum in the right part (4) equals in fact to the amount of the particles which centers lie inside the layer restricted by the planes z-R and z+R.

Let the spatial distribution of the spheres be specified by the probability of sphere center finding in the volume unit $n(z)$. Then the sums in the equations (2a) and (4) can be approximately alternated by the integrals:

$$\varepsilon = 1 - \pi \int_{z-R}^{z+R} \left[R^2 - (z' - z)^2\right] n(z')\mathrm{d}z' \quad (5)$$

$$A = 2\pi R \int_{z-R}^{z+R} n(z')\mathrm{d}z' \quad (6)$$

If n does not depend on z, i.e. the packed layer is statistically homogeneous, integration gives the followings results: $\varepsilon = 1 - 4/3\,\pi n R^3$, $A = 4\pi n R^2$. Hence, n can be excluded:

$$A = \frac{3(1-\varepsilon)}{R} = \frac{6(1-\varepsilon)}{D} \quad (7)$$

As a result, we have the connection (7) between the porosity ε, specific surface, and mono-diameter D of the particle packing.

2.1.4 The pore tortuosity

The pore tortuosity characterizes the percolation properties of the packed structures and is found as the ratio of the path length of the tracing point around the porous structure labyrinth to the value of the shortest distance between the initial and final position of this point. Let us consider the passage of such a point though the packed layer pores by the shortest way under the action of the constant external force. Let the force be directed vertically downward. The path part 1-2-3 between two adjoining spheres with the diameter D is shown in Fig. 2(c). The geometry is specified by the inclination angle θ of the line connecting the sphere centers. The angle α is found in the equation: $\sin\alpha = 2\sin\theta - 1$. It follows from Fig. 2(c) that the height difference between the points 1 and 3 is $H = D\cos\theta$, whereas the path 1-2-3 is $S = H + 0.5D(0.5\pi - |\alpha| - \cos\alpha)$. The path 1-2-3 tortuosity for this angle θ is determined by the relation $S/H = 1 + \Omega - \Psi$, where $\Omega = (0.5\pi - |\alpha|)/(2\cos\theta)$, $\Psi = (\cos\alpha)/(2\cos\theta)$.

Then the packed layer tortuosity is found as:

$$\tau = 1 + \langle\Omega\rangle - \langle\Psi\rangle, \quad (8)$$

where $\langle ... \rangle$ is the averaging operator, which si defined as the averaging by the angle θ in the 2D case (2D) and by the solid angle in the 3D case (3D).

2.2 Calculation of the packing parameters

Fig. 3 (a, b) presents the varying surface ε_S and volume ε_V porosity over the container height z. Fluctuations ε_S and ε_V are most noticeable at low z, which are caused by the effect of the container lower boundary (bottom).

Due to the stochastic distribution of the sphere centers N_C, Fig. 4(a), the medium becomes more uniform at $z > 0.1$, and a certain fluctuation amplitude and frequency ε_S settles, Fig. 3(a). The monotonous variation of $\varepsilon_V(z)$, Fig. 3(b), is related with the fact that, according to (2b), the mean-volume porosity value is calculated in fact for each z. Based upon Fig. 3(a) and 3(b), the approximate equation (2c) is satisfied; this equation is valid for the homogeneously ordered granular structures, and is theoretically vindicated in the theory of granular and saturated porous media [7]. It is interesting to check the satisfiability of the ratio (2c) for the case of polydisperse packings.

Indeed, the distribution of the center coordinates $N_C(z)$, Fig. 4(a), in each container is uniform on average, except for the case D = 0.5 and 1.0 μm. At $z < 3h$, there are evident deviations. The reason is the container bottom effect, where smaller particles are concentrated denser. Fig. 4(b) presents the variation of the specific pore surface $A(z)$ over the container height. It is shown that for small particles with D = 0.5 and 1.0 μm, the packed layer has quite a developed specific pore surface.

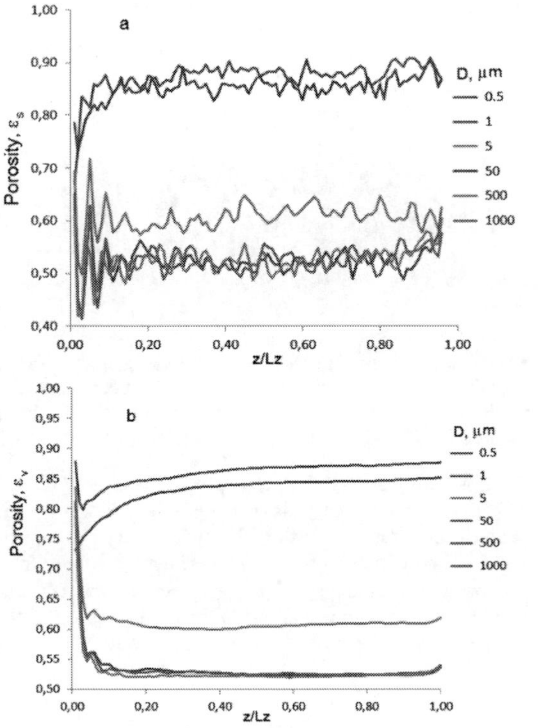

Fig. 3. Variation of the surface ε_S (a) and volume ε_V (b) porosity over the container height at the varying sphere mono-diameter D, μm: 0.5; 1.0; 5; 50; 500; 1000.

Fig. 5 demonstrates the 3D pore tortuosity over the height z (D = 5μm), where the trajectories of nine tracing points around the pore labyrinth are presented. Vertical sections of the trajectories are shown by solid lines, enveloping sections of sphere surfaces are shown with markers.

Table 1 presents the numerical values of the structural parameters of the studied packings.

Table 1. Packing structure characteristics

D, μm	ε_V	A, μm^{-1} (theory)	$\hat{A}$, μm^{-1} (calculation)	$\hat{\tau}_x$	$\hat{\tau}_z$
0.5	0.86	1.68	1.5	1.013	1.014
1	0.84	0.96	1.0	1.015	1.014
5	0.54	0.54	0.57	1.042	1.037
50	0.47	0.05	0.058	1.048	1.053
500	0.49	0.005	0.007	1.049	1.052

According to the formula (7) obtained for the case of the homogeneous distribution of solid spheres, the theoretical values of the specific surfaces A of mono-sphere packings, D from 0.5 to 500 μm were calculated by the values of the volume porosity ε_V and mono-diameter D.

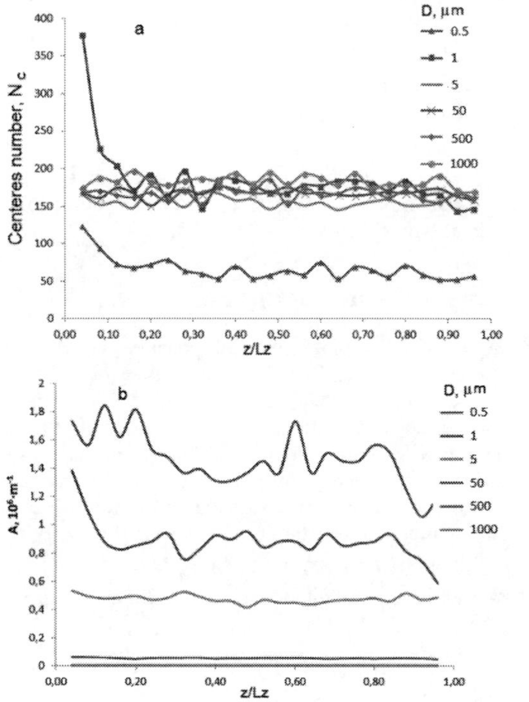

Fig. 4. Distributions of sphere center coordinates N_C (a) and specific pore surface A (b) over the container height at the varying mono-size of packing particles D, μm: 0.5; 1.0; 5; 50; 500; 1000.

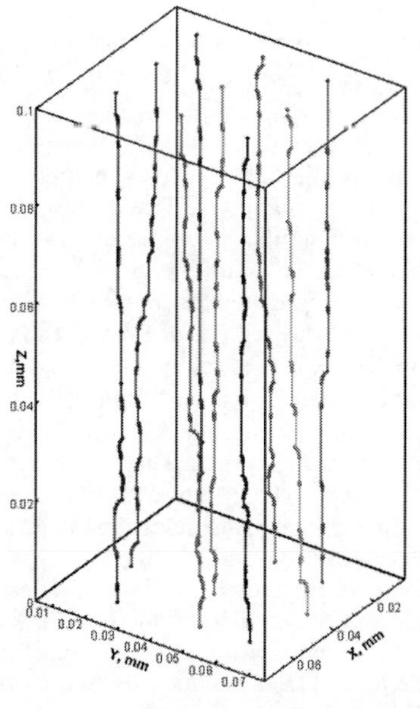

Fig. 5. Trajectories of the tracing points around porosity labyrinths calculated by the algorithm of Fig. 2(c)

DOI: 10.1051/matecconf/201712901066

The average values of the specific surfaces $\hat{A}$, according to the calculation presented in Fig. 4(b), are in good agreement with the theoretical values of A, calculated by the formula (7), see Table 1.

The characteristic (average) values of tortuosity $\hat{\tau}_x$, $\hat{\tau}_z$ of the studied packings in the directions Ox, Oz are presented here, respectively.

Based upon the comparison of $\hat{\tau}_x, \hat{\tau}_z$, it is evident that, the structure anisotropy does not appear so highly for the mono-size packings, from the viewpoint of the percolation properties.

3 Laser beam tracing method

The physics of the laser and powder particle layer interaction is such that the absorption of the laser radiation by the particle surface goes along with repeated re-reflection and scattering over the layer depth. The above constructed mono-size packing of particles were used to define the algorithm of laser beam tracing based upon the Gaussian beam presentation in the form of the finite-number discrete luminous elements [8].

Initial cross section of the Gaussian beam $\pi\omega_0^2$ with the radius ω_0 is divided into the finite number of discrete luminous elements n_{xy} (LE): $\pi\omega_0^2 = \sum_{k=1}^{n_{xy}} \Delta S_k$.

Each LE has its initial position with the coordinates x_k, y_k, z_k, and initial radiation energy $\Delta E_k = I(x_k, y_k, z_k)\Delta S_k$. First, the LE move by the trajectories parallel to the Oz axis. When interacting with the particle surface, the LE trajectory changes its direction in accordance with the geometrical optics law (the incidence angle equals to the reflection angle). Its energy ΔE_k decreases by the value of the energy absorbed by the particle surface in the incidence point. The numerical algorithm includes summing of the absorbed energy for each spherical particle from each LE, including the multiply reflected ones. The energy summed in such a way is averaged over the sphere surface. The absorbed laser energy is then used in the thermal balance of individual particles [9].

Fig. 6 presents the pattern of LE trajectory positions in the granular layer of steel spherical particles of the same radius (10 μm). The layer thickness is about 100 – 120 μm. It is seen that the incident beam does not pass throughout the whole layer thickness. The upper part of the layer (2 – 3 diameters) turns out to be heterogeneous at the construction due to the random particle throw.

The trajectories of multiply reflected LE penetrating into the layer depth are clearly seen. The temperature distribution over the particles in the layer is shown; it is calculated by the previously proposed model of discrete thermal elements [9]. Note that the thermal balance in this model regards only the absorbed energy of the laser radiation. Fig. 6 gives the calculation neglecting the contact thermal conductivity.

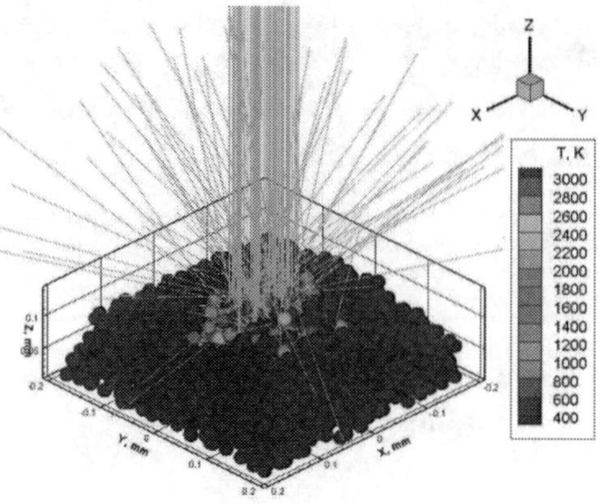

Fig. 6. Interaction of the laser beam and the randomly packed granular layer of spheres, 2,000 pieces. Particle material – steel, radiation wavelength 1.07 μm, power 100 W, beam radius 50 μm.

It is demonstrated that, to evaluate the laser radiation effect on powder materials, it is necessary to known the spatial distribution of the absorbed energy over the depth. The incident radiation may reflect repeatedly from the powder particle surface and penetrate inside. The proposed method consists of the detailed tracking of the LE path, with due regard to the interphase reflection and refraction.

4 Conclusions

The peculiarities of particle packing in powder systems have been considered. There are attractive Van-der-Waals' forces between small particles. As the particles get closer, these forces appear to a higher degree. Under the gravity conditions, as the packing of spherical particles forms, these forces influence the contact interaction and the spatial distribution of the particles. This factor considerably influences the internal structure and the physical properties of the packing, i.e. the packing density, the pore specific surface and the tortuosity, which is very important for the SLM and SLS technologies. Computer simulation of metal powder packing has been performed. Specific character of utilization of the computer models is detected for the analysis of the particle packings in real powder media.

Simulation by the ray tracing method has demonstrated that the penetration of the laser radiation into the powder layer is especially important for highly absorbing and reflecting metal powders. Correct combination of fine and large particle fractions permits having the packing with the properties favorable for the penetration and absorption of the laser radiation. It allows to improve the quality of the produced parts, reach denser sintering of powders and achieve good performance properties.

The authors gratefully appreciate the financial support from the Russian Scientific Fund (contract No. 15-19-00254, May 18, 2015).

References

1. I. Yadroitsev, *Selective laser melting: Direct manufacturing of 3D_objects by selective laser melting of metal powders* (Lambert Acad. Publ., 2010)
2. B.A. Klumov, Y. Jin, H.A. Makse, J. Phys. Chem. B. **118 (36)**, 10761–10766 (2014)
3. G. Parisi, F. Zamponi, Rev. Mod. Phys. **82** 789-845 (2011)
4. S. Torquato, F.H. Stillinger, Rev. Mod. Phys. **82,** 2633-2672 (2010)
5. O.B. Kovalev, I.O. Kovaleva, Journal of Applied Mechanics and Technical Physics **55,** 709-717 (2014)
6. H.C. Hamaker, Physica IV **10,** 1058-1072 (1937)
7. V.N. Nikolaevskij, K.S. Basniev, A.T. Gorbunov, G.A. Zotov, *Mechanics of saturated porous media* (Moscow: Nedra. 1970)
8. C.D. Boley, S.A. Khairallah, A.M. Rubenchik, Applied optics **54,** 2477-2482 (2015)
9. O.B. Kovalev, A.V. Gusarov, Int. J. Thermal Sciences **114,** 327–341 (2017)

MATEC Web of Conferences 129, 01067 (2017)
ICMTMTE 2017
DOI: 10.1051/matecconf/201712901067

The design of lightweight gas turbine engine parts using topology optimization

Aleksandra Dubrovskaya[1,*], *Konstantin* Dongauzer[1], and *Rustam* Faskhutdinov[1]

[1]Perm national research polytechnic university, Computation Mathematics and Mechanics Department 29, Komsomolsky prospect, Perm, 614990, Russian Federation

Abstract. This paper addresses a new approach to aircraft engine parts designing using the topological optimization algorithm for production by additive manufacturing. The full mathematical formulation which includes the problem statement within the elasticity theory and the SIMP optimization procedure is presented. The criterion particularities for evaluation of strength properties of the gas turbine engines parts according to the Aviation Regulations are specified. Optimization of the typical hinge bracket topology is performed, and the stress state of the part with the initial geometry and the optimized geometry is evaluated.

1 Introduction

Additive manufacturing is the newest production method which completely changes the approach to part designing. This method allows getting a complex structure with minimum post-process machining in one step literally. The structure complicity and shape intricacy of the part don't influence the production prime-cost, what matters is the weight of the part. Application of additive manufacturing allows a designer to eliminate limitations imposed by the strict rules of the technological capabilities in the production environment.

Even a slight weight reduction in an engine results in significant fuel savings and pollutant emissions reduction in the aerospace industry. That's why the industry's main objective is to design aircraft parts with lower weight while maintaining their prescribed functionality and life time. Rapid development of additive manufacturing allowed for recollection of a well-known, but up until now hardly applicable designing technology, that is topological optimization. Back then, manufacturing of optimized products was impractical and often impossible, as it is labour intensive and involves considerable investments in order to provide a complex geometry by conventional production methods. Fundamental nature of the topological optimization method performs the same idea, being the cornerstone of additive manufacturing, of getting the material exactly where it is needed. The methods of additive manufacturing and topological optimization united by the common concept are able to perform a leap-ahead in up-to-date domestic engine manufacturing.

The results of this work will be used in UEC-Aviadvigatel JSC when building up high-tech manufacturing of complex-shaped parts for aircraft and industrial gas turbine engines based on additive manufacturing from Russian metal powders.

2 Peculiarities in the evaluation of aircraft parts strength properties

Aircraft parts can undergo an entire range of combined-loaded conditions during operation dependent on the engine behaviour. It is a critical and knowledge-intensive task to define and take into consideration all the load profiles, and to bring all the parts to conformity with section 25 "Airworthiness Requirements for Carrier Aircrafts" and section 33 "Airworthiness Requirements of Aircrafts Engines" of the Aviation Regulations.

The following static strength requirements are applied to aircraft parts:

• The parts shall withstand the maximum operation loads with no dangerous fatigue deformations showing.

• The parts shall withstand design loads (that is the operation loads multiplied by safety factors) with no failure during three seconds.

The additional safety factor for the additive manufactured parts is not clearly specified in the Aviation Regulations. In this case, it was decided to take the safety factors equal to 1,5 (the factor used for castings) to ensure strength properties.

The fatigue resistance in the part is evaluated based on the following rule:

• The parts life shall be not less than 40000 cycles (80000 hours).

3 Theoretical basis of the topological optimization method

Generation of the optimal geometry design model for the part must be started from the definition domain identification. For this purpose, the procedure of geometry extension within the dimensions meeting the conditions of the part mounting into the assembly, and simplification of the initial topology of the item to be

* Corresponding author: dubrovskaya-as@avid.ru

designed are performed [1]. Static loading diagrams and loads variation rates are defined based on the operation force diagrams in different flight modes.

The topology optimization task is solved based on the calculated stress state of the part. As the aircraft parts must undergo no plastic strain during operation, it makes sense to chase the stress fields in 3D elastic formulation.

Let us consider a 3D elastic body with the volume V_0 and the boundary B fixed in the space along the boundary B_0 and subjected to the external loads complex $\hat{F}$. To find the stress tensor $\hat{\sigma}$, let us to write down the general form of the quasi-static moment equilibrium equation for this case :

$$-\nabla \cdot \hat{\sigma} = \nabla \hat{F}. \quad (1)$$

The physical relations of the elasticity theory can be defined as:

$$\hat{\sigma} = {}^4\hat{K} \cdot\cdot \hat{\varepsilon}^{el}, \quad (2)$$

where ${}^4\hat{K}$ is a fourth-rank elastic constants tensor, and $\hat{\varepsilon}^{el}$ is an elastic strain tensor.

Using the small-strain theory, let us write down the geometrical relations as:

$$\hat{\varepsilon} = \frac{1}{2}\left[\nabla u + (\nabla u)^T\right], \quad (3)$$

where u is a global displacement vector [2].

It is necessary to define the conditions of fixing u_0 at the boundary B_0, the boundary conditions for this task will then take on form of:

$$u\big|_{\Gamma_0} = u_0. \quad (4)$$

For the special case of the rigid fixing, $u_0 = 0$.

This task is solved by the finite elements method. Trigonal simplex elements with linear Lagrange approximation are used.

The SIMP (Solid Isotropic Materials with Penalization) method is used for topological optimization of a structure based on the specified stress fields. Its main idea is to get a tensor of the virtual specific binary densities in each element for effective material distribution. Initially, the specific density is prescribed as homogeneous and is equal to one in each element.

Let us assume that $\hat{x}$ is a tensor of the specific densities. The SIMP topological optimization method is based on prescription of the power laws between these densities and the elastic modulus. Such a dependency acts as a penalty function when related values of the power coefficient are more than one.

Let us assume that each element i has a correspondent E_{i0} – which is a Young's modulus of the structure material in this element, and its own distribution density x_i, which defines the relative Young's modulus $E_i(x_i)$ for it. Thus, it is correct that:

$$E_i(x_i) = E_{\min} + x_i^{\,p}\left(E_{i0} - E_{\min}\right),\ x_i \in [0,1] \quad (5)$$

where p is a weight coefficient of penalty functions, used to ensure a black-and-white solution, and $E_{\min}$ is some small value of the Young's modulus related to the empty domain and introduced to eliminate the confluence of the elastic constants tensor, [3].

The solution of the task resolves into achievement of minimum interchanges in the finite-element mesh points. The target function is structure compliance c. It is an integral characteristic which is inverse to stiffness and is a function of a specific density. The solution to minimize the prescribed target function is sought:

$$\min_{\hat{x}}\ c(\hat{x}) = u^T \cdot \hat{K} \cdot u = \sum_{i=1}^{N} E_i(x_i) u_i^T k_i u_i \quad (6)$$

where u_i and k_i are a displacement vector and an element of the elastic constants tensor for the element i respectively, N is the number of the elements in the optimization domain.

However, the weight of the optimized product acts as a restriction to the target function, i.e. the variation range of the virtual density is controlled via the final part material volume $V(\hat{x})$ and the prescribed volume faction f [4].

$$\frac{V(\hat{x})}{V_0} = f \quad (7)$$

Besides, the principle of virtual displacements acts as a restriction:

$${}^4\hat{K} \cdot u = \hat{F} \quad (8)$$

The result of the SIMP method is black-and-white density distribution. That is, the tensor $\hat{x}$ takes on not only specific values, but also has a binary form. Though, the disadvantage of the method is the problem of Checkerboard patterns in the received solution. The received densities distribution doesn't form a closed domain, but contains unlinked segments.

In order to ensure the solution consistency and prevent occurrence of Checkerboard patterns, some restrictions shall be imposed on the optimized geometry. The most favourable approach is to apply a filter to sensitivities, i.e. to define the elements with the maximum influence to the target function. The sensitivity filter changes the sensitivity $\partial c / \partial x_i$ as follows:

$$\frac{\widehat{\partial c}}{\partial x_i} = \frac{1}{\max(\gamma, x_i)\sum_{j=1}^{N} H_{ij}} \sum_{j=1}^{N} H_{ij} x_j \frac{\partial c}{\partial x_j} \quad (9)$$

where N is a set of the elements j, for which the distance $\Delta(i, j)$ between the center of the current element i is less than some critical radius of influence

DOI: 10.1051/matecconf/201712901067

r_{min}, γ is an artificially implemented parameter equal to a small positive number which is necessary to exclude a possibility of zero-divide; and H_{ij} is a weight coefficient, defined by:

$$\hat{H}_{ij} = \max(0, r_{min} - \Delta(i, j)) \qquad (10)$$

After the filter application, the original specific densities x_i, gained during optimization, are changed to:

$$\tilde{x}_i = \frac{1}{\sum_{j=1}^{N} H_{ij}} \sum_{j=1}^{N} H_{ij} x_j \qquad (11)$$

Therefore, the variables x_i serve for pre-optimization of the density in the domain of interest, while the filtered parameters $\tilde{x}_i$ correlate to the actual physical density of the optimized structure [5]. This notation is used in order to highlight the fact that application of the densities filter results in x_i losing its physical meaning. That's why the filtered binary specific densities $\tilde{x}_i$ will be the solution of the optimization task.

4 Application of the topological optimization method in aircraft parts designing

In order to test the topological optimization technique and to evaluate its applicability in UEC-Aviadvigatel JSC, a typical hinge bracket used in the up-to-date domestic aircraft engine (Figure 1) was chosen. There are eight pieces of such brackets in the engine; each weighs 392 grams.

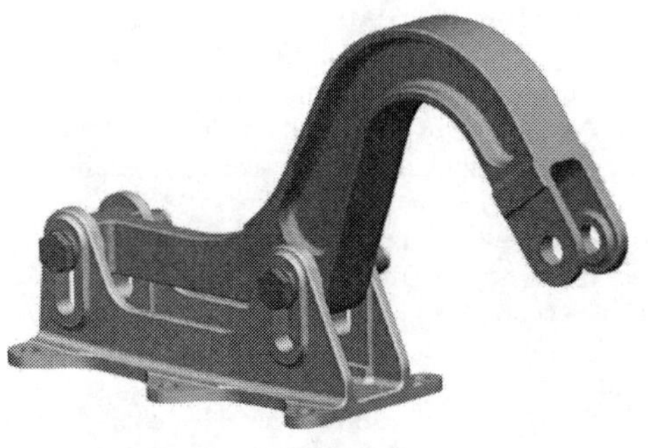

Fig. 1. An engineering model of the part chosen for designing with topological optimization method

In order to describe the stress state of the part, the whole nomenclature of aircraft engine main operation modes corresponding to the aircraft flight envelope is considered. There are 26 loads patterns identified for each of the 8 brackets mounted on the engine. The maximum loadings in each possible direction were considered to facilitate the task. The bracket fixing in bolt joints was prescribed as boundary conditions, with a possibility for tangential sliding being considered.

The simplification procedure in the initial topology of the part to be designed was performed to determine the definition domain of the part. The computational domain was extended within the prescribed dimensions which don't prevent the part from being installed into the joint connection.

The initial value of the optimized bracket weight chosen was 75% of the weight of the model with the initial geometry. This parameter can vary during optimization via changing of the critical volume fraction which ensures satisfaction of the stresses values limitation. Hereby, the main objective is to optimize the topology of the bracket of interest under working loading conditions and to reduce the weight of the structure by 25%.

The bracket topology was optimized in the specialized software allowing the particularities of the investigated problem formulation to be considered. It resulted in a 3D non-parametrized zone of the optimal material distribution in the definition domain (Figure 2).

Fig. 2. The material distribution zone for the optimal bracket geometry

According to these results, the CAD model of the bracket with the optimal geometry was designed for the prescribed task conditions (Figure 3). The weight of the optimized bracket was 294 grams. Thus, the bracket weight was reduced by 25%, which is 98 grams. Taking into account the number of this type brackets in the engine, the total gain in weight using the optimized bracket is 784 grams.

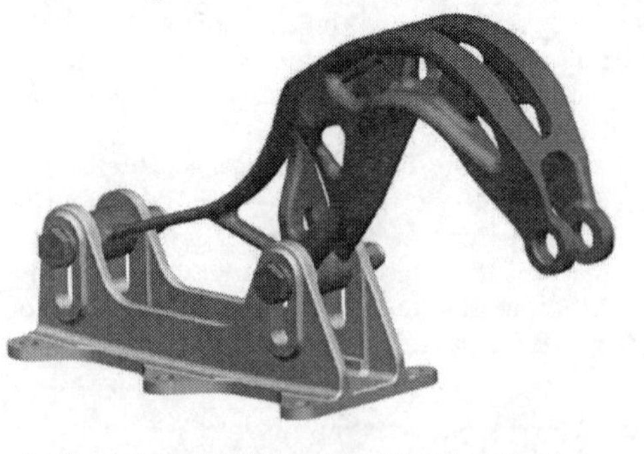

Fig. 3. The optimized CAD model of the hinge bracket

The topological optimization method was verified in the virtual space via software package Ansys Workbench.

DOI: 10.1051/matecconf/201712901067

The functional state of the bracket under investigation was simulated with the assistance of mathematical modelling at all the considered operational modes.

The maximum von Mises stress both in engineering, and in optimized model of the bracket was achieved under the same stress state pattern. Figure 4 shows the distribution of equivalent loadings under this load pattern in engineering and optimized bracket respectively.

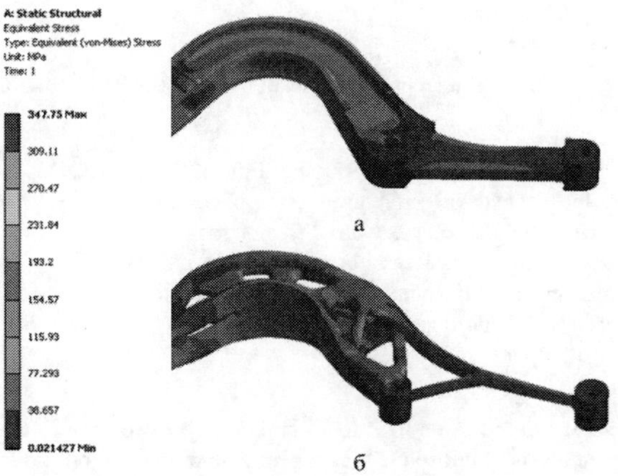

Fig. 4. The equivalent von Mises stresses in the maximum load pattern a) in the initial model, б) in the optimized model of the bracket

It may be noted that despite the 25% weight reduction, the optimized bracket is not only as good as the engineering model, but surpasses it in strength properties. The results of the optimization and verification calculations are given in Table 1.

Table 1. The Results of the Topological Optimization

Model	*m*, kg	*σ*, MPa	*P*, %	*δ*, %
Initial model	0.392	347	100	–
Optimized model	0.294	228	74.9	34.35

In Table 1, m is a weight of the bracket, σ is the maximum von Mises stress, P is a fraction of the initial bracket weight, δ is the level of the stress reduction in the bracket.

Conclusions

The topological optimization is an up-to-date effective method to create geometrical models and has a great application perspective in aircraft parts designing.

An optimized part is shape-intricate; such an approach to designing is often called a bionic design. It is technologically impossible to manufacture the similar geometry by conventional production methods. However, additive manufacturing methods allow for production of this bracket in one process step.

Currently, both bracket structures are ready for manufacturing by selective laser melting. The tooling compliant to the bracket operation loads is being developed for mechanical testing of the bracket.

This work was performed with financial support from Ministry of Education and Science of the Russian Federation (contract No. 02.G25.31.0168 dated 01.12.2015) within implementation of the RF Government Order No. 218 «On the measures of the State support for development of higher education institutions and organizations implementing integrated projects on high-tech manufacturing build-up».

References

1. A.A. Inozemtsev, A.S. Dubrovskaya, K.A. Dongauser, N.A. Trufanov, *Numerical Simulation of Cast Distortion in Gas Turbine Engine Components*, IOP conference series: materials science and engineering, №**84**, 012039 (2015) DOI: 10.1088/1757-899X/84/1/012039
2. R.N. Faskhutdinov, A.S. Dubrovskaya, K.A. Dongauzer, P.V. Maksimov and N.A. Trufanov, *Topology optimization of a gas-turbine engine part*, IOP conference series: materials science and engineering, №**177**, 012077 (2017) DOI: 10.1088/1757-899X/177/1/012077
3. D.J. Brackett, I.A. Ashcroft and R: Hague, *Topology optimization for additive manufacturing,* 22nd Annual International Solid Freeform Fabrication Symposium, Texas, USA, 348-362 (2011)
4. A.A. Krotkikh, P.V. Maksimovl, *The analysis of methods of refinement of the finite element model after topology optimization*, International research journal, No **01** Vol. **55**, pp. 91-94 (2017) DOI: 10.23670/IRJ.2017.55.071
5. E. Andreassen, A. Clausen, M. Schevenels, B. Lazarov, O. Sigmund, *Efficient topology optimization in MATLAB using 88 lines of code*, Structural and Multidisciplinary Optimization, 2010.:, DOI 10.1007/s00158-010-0594-7

Improvement of tribotechnological properties of oil-based cutting fluids by adding dry lubricants

Dilyaver Abdulgazis[1], *Ervin* Umerov[1,*] and *Umer* Abdulgazis[1]

[1]Crimean Engineering and Pedagogical University, 295015, Lane training, 8, Simferopol, Republic of Crimea, Russian Federation

Abstract. Having taken an oil component of cutting fluids used when machining difficult-to-machine materials, authors have searched for an additive which would have improved both tribotechnological properties and the cooling effect. Tribotechnological characteristics of widely used friction modifiers, such as graphite, molybdenum disulfide and serpentine, are analyzed. It has been show that the use of montmorillonite, a mineral that is widespread in nature, as an additive improves tribotechnological characteristics and scuffing load capacity of oil-based cutting fluids in general.

1 Problem statement

Basic properties required from oil-based cutting fluids are the improvement of hardness of a cutting instrument as well as the conditions of a working surface. Thanks to this, oil-based cutting fluids more or less determine cost-effectiveness and reliability of metal processing equipment. Oftentimes it is difficult to discriminate between the most important effects of oil-based cutting fluids. Edge-cutting machining of difficult-to-machine materials poses more significant requirements to both cooling and lubricating effect of oil-based cutting fluids [1].

At the same time, the technological tools currently used exhibit either the former or the latter effect, which prompts specialists in the field to search for an additive agent, which would have had the ability to improve tribotechnological properties and the cooling effect, taking oil-based cutting fluids as a basic material used during machining of difficult-to-machine materials. Therefore, an additive agent to an oil-based cutting fluid should exhibit endothermic and tribotechnological properties

2 Literature review

This double function necessary from an additive agent requires doing the search among non-traditional materials.

A promising trend in lubricant research has been the search for various kinds of fillers and additives, primarily solid and naturally occurring, which would have improved tribotechnological properties of liquid and grease lubricants. We believe that the search of endothermic properties is of equal importance.

Of all known endothermic processes which may be used to improve the cooling ability of oil-based cutting fluids, endothermic effect of crystalline hydrates was considered to be the most viable [2-5]. According to our analysis, addition of crystalline hydrates to oil-based cutting fluids does not improve their tribotechnological properties.

Since additives to oil lubricants, including engine oil, may have the same function that the additives to oil-based cutting fluids have, we will consider the improvements in this area as well. For instance, modern engine oils use additives which decrease friction coefficient; among those additives, graphite-based and molybdenum greases are the most popular ones.

Unique tribotechnological properties of graphite and molybdenum disulfide are attributed to a flat hexagonal structure that their crystal lattices possess. The distance between the closest atoms of graphite is 1.42 Å, while the distance between the closest planes is 3.4 Å [6]. Therefore, every crystal plane may be looked at as an isolated system of atoms. Atomic planes of graphite crystals are located parallel relative to one another.

Each of these planes consists of carbon atoms bonded covalently to each other to form regular hexagons. Adjacent layers are held together by weak van der Waals' forces, which are three to four times weaker than the covalent forces. This allows the layers to slide on each other [6].

Shear strength of graphite along crystal planes is incommensurably less than that of the graphite when covalent hexagonal bonds are broken. Additional lubricating effect of graphite is based on the fact that by sorbing between its crystal planes at normal temperature, water molecules relax interplanar bonds. Consequently, when the temperature is above 100°C thanks to evaporation effect, graphite greases exhibit reduced lubricating properties. In the absence of water, friction coefficient of surfaces divided by graphite layer may be as large as 0.3, while in the presence of sorbate water [7] it is approximately 0.05. This is the reason why graphite greases are used sparingly.

* Corresponding author: ervin777@yandex.ru

Boron nitride may form the same structure where atoms of boron and nitrogen alternate [6].

Molybdenum disulfide (MoS_2), which possesses similar hexagonal crystal lattice, is characterized by stable chemical bonds between molybdenum atoms, while sulfur atoms have stable bonds with molybdenum atoms form the structure which connects adjacent layers of molybdenum atoms to each other via weak van der Waals forces. Thus, relatively small shear strength is formed at the border between layers of sulfur atoms. Presence of water in this case exerts little influence, therefore, MoS_2 may be used at relatively high temperatures (up to 1000 °C). However, friction between its crystal lattices takes place in the absence of water.

Our assessment has also showed that, during thermal degradation in the cutting zone, molybdenum disulfide is capable of producing sulfur which may interact with cutting tool's material and consequently decrease its wear resistance and deteriorate the quality of mechanical and chemical structure of a working surface.

In this respect, metal carbides (Mo, V, W) may act more positively. At high temperature and pressure values in the cutting zone, metal carbides may penetrate into the lattices on the surfaces of a cutting tool and a workpiece, which in fact may improve strength of these lattices. Furthermore, it may improve wear resistance of a cutting tool and perform nanopatterning of workpiece's surface using alloying components.

Production of powdered molybdenum disulfide and stabilization of its ratio in oil lubricants is a time-consuming and complex process which is possible only at large production plants that have highly trained personnel and high-precision equipment at their disposal [7].

Another trend in the area of addition agents is [8] the application of natural fine-grain serpentine which may help to increase wear resistance of workpiece parts and decrease friction coefficient. Serpentines possess layered structure, are not soluble in oil products and do not react with them; plus, they are environmentally harmless. Paper [8] also underlines that natural serpentine contains a wide range of impurities, and in order for it to be used as an additive agent, it has to undergo enrichment (grinding, separation etc.), which drastically increases its production costs. Therefore, authors of [8] suggest replacing natural serpentine with the artificial one obtained using sol-gel method.

Action mechanism of serpentine as a friction modifier may be considered similar to that of the powdered molybdenum disulfide.

Considering the aforementioned limitations of dry lubricants like graphite, molybdenum disulfide and its alternatives, authors have strived to find nanomaterials with improved tribotechnological and (possibly) endothermic properties that may be used as an additive agent.

3 Overview

Bentonites, plate-shaped nanoclay minerals, have favorable physical and chemical properties. Among other well-known benefits that dry lubricants like molybdenum disulfide possess, it is also possible to provide fluid lubrication between layers of minerals via their hydro-wedging, which allows for improvement of tribotechnological properties.

The aforementioned has prompted researchers to conduct studies regarding viability of use of nanoclay mineral additives in oil-based cutting fluids, with the purpose of increasing hardness of a cutting tool and the quality of machining of difficult-to-machine materials. Application of nanoclay mineral additives instead of well-known friction modifiers (graphite, molybdenum disulfide, serpentine) helps to eliminate a range of problems:

1. Nanoclay minerals are widespread in nature; they may be found in Crimea, which gives them a huge advantage in the form of lower cost of production. They are obtained using the open-pit method and do not require highly qualified personnel and precision equipment.

2. Despite some similarities between the crystal lattices of molybdenum disulfide and bentonites, shearing of layers of the former happens in the absence of water, while in the case of nanoclay mineral additives it happens in the presence of hydrowedging, that is, in the presence of liquid friction between layers of the mineral which will eventually influence the tribotechnological and endothermic properties.

3. Necessity to use dispersants and stabilizing agents when adding molybdenum disulfide to engine oils thanks to their immeasurably large service life before replacement is eliminated if nanoclay mineral additives are added to oil-based cutting fluids.

One of the main components of bentonites is montmorillonite, which is a nanodisperse plate-shaped aluminosilicate consisting of blocks of nanominerals capable of swelling when hydrated. It creates hydrowedging areas between the layers. This phenomenon is what distinguishes tribotechnological properties of montmorillonite from those of the solid lubricants described above (Fig. 1).

Packages are the least structural component of bentonites in general and of montmorillonite in particular. The thickness of packages is 1.0 - 10 nm while linear dimensions are not more than 50 - 150 nm. The distance between packages at swelling (hydration) can increase up to 2.14 nm [4, 5].

 DOI: 10.1051/matecconf/201712901068

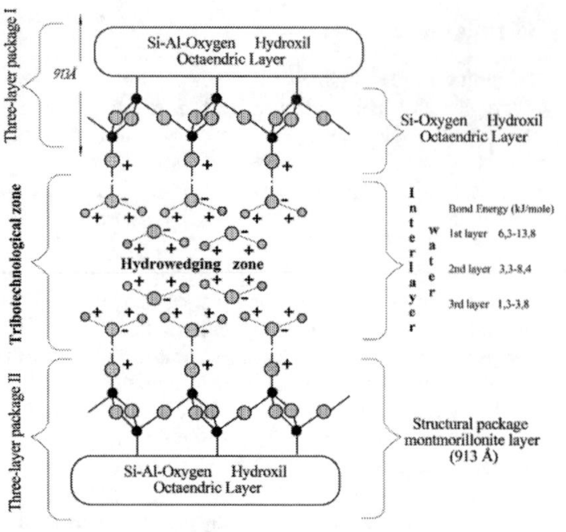

Fig. 1. The structural flow-chart of physical and chemical interaction of montmorillonite with water:
– OH hydroxyl; – Al aluminum; – Si silicon;
– O oxygen; – H hydrogen

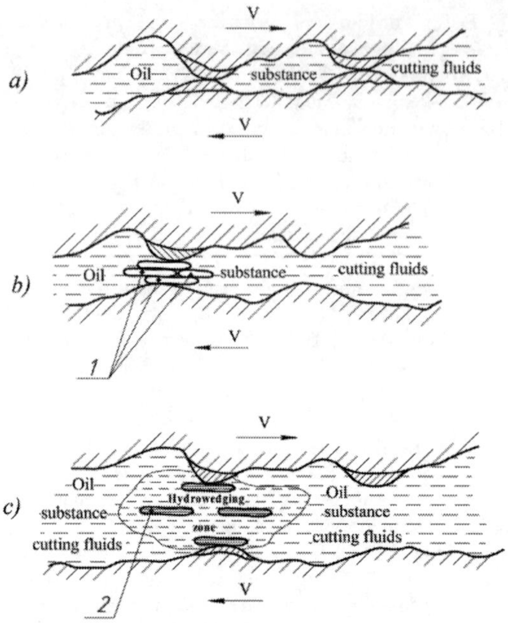

Fig. 2. Examples of implementation of junction and nanojunction coupling. Change of contact geometry:
a – of junction couplings (in the presence of dry junction);
b – in the presence of molybdenum disulfide as an additive to oil lubricants;
c – of nanojunction couplings, in the presence of montmorillonite as an additive to oil lubricants.
1 – crystal planes of either graphite or molybdenum disulfide;
2 – crystal blocks of montmorillonite;
▨ – area of high local deformation;
V – linear speed.

It shall be underlined that nanostructures have special properties intrinsic of nanolayers, nanocrystalls and nanoparticles connected to a so-called dimensional effect only. The property of the nanoclay mineral addition (NCMA) that is the most important to achieve the objective is the capacity of packages hydrowedging with superficially occluded water at hydration. As a result, it is possible to provide interpackage sliding due to occurring hydrolubrication thus improving tribotechnological properties of oil CLS in general (Figure 2).

It should be mentioned that recently modern motor oils have included additions modifying friction allowing for the decrease of the frictional factor between the rubbing engine parts. The most wide-known additions modifying friction include graphite and molybdenum disulphide. These unique properties are explained by the structure of their crystalline lattices. Binding between their layers is relatively weak similar to those of graphite and molybdenum disulphide and therefore it allows for the possible shifting at low voltage which determines decrease of the friction factor [6].

The use of NCMA for oil CLs eliminates a range of problems inherent in the use of molybdenum disulphide as an addition to motor oils:

1. Shift of molybdenum disulphide layers can be "dry" in NCMA; in case it is properly prepared in advance (hydrogenation), this shift can be accompanied by hydrolubrication, i.e. under fluid friction, which finally influences tribotechnical properties.

2. Nanoclay minerals are rather wide-spread natural mineral deposits. To prepare them for the procedure, one doesn't need to involve high-qualified staff and high-precision equipment.

3. There is also no need in dispersants and dispersion stabilizers which are required for molybdenum disulphide addition to motor oils.

Thermal action on the preliminarily wedged with superficially occluded water packages of nanominerals added to oil CLs in the cutting area leads to intense water evaporation providing extra endothermic reaction effect. Water vapours remain inside the system; when the temperature decreases, they return to its structure, and this is one of the properties characteristic of nanoclay minerals. As oil cannot interact with water, when the temperature lowers and the CL goes out from the cutting area, vapour is condensed and returns to the mineral structure recovering its cooling capacity.

The described process works similar to hydration-dehydration of crystalline hydrates described in the work [2]. Crystalline hydrates contain water in the defects of a crystalline lattice, and separating the first molecule requires bigger energy comparing to NCMA. Nanoclay minerals demonstrate the dehydration effect in a far more active way. The thermodynamic analysis [7] provides evidence that the endothermic effect of montmorillonite generally occurs in the temperature range of 50-200 °C with a maximum of 99 °C.

At heating of the adsorptive water layer of a nanoclay mineral particle up to the temperature 80-100 °C the weakest molecular binds are broken up first which triggers the first hydration step. Further, as the temperature increases up to 100-200 °C, medium strength molecular bind break up at the second turn, and when the temperature is more than 200 °C, the stronger molecular binds break up next, etc., thus creating a step-wise dehydration endothermic process.

Therefore, NCMA can be used as CLs endothermic component similar to crystalline hydrates. But as opposed to the latter, NCMA is more thermally and chemically stable which excludes a possibility of its adhering and burning on the surfaces of the cutting tool and machine at high temperatures.

The fact that NCMA can show an endothermic effect similar to crystalline hydrates is confirmed by bentonite thermogram given in the work [8]. It confirms that at 100–150 °C the first endothermic effect step takes place - i.e., elimination of adsorbing and interpackage water. The closing step takes place at the temperature of 500–650 °C with elimination of a hydroxyl group.

If necessary, temperature discontinuity between the first and closing steps of the endothermic reaction can be filled with similar effects provided by other mineral or crystalline hydrates.

The comparison of the temperature range of NCMA dehydration on the basis of montmorillonite with a working temperature of CLs in the cutting area that was measured experimentally [9,10] by means of cutter washing at various cutting modes with the flow of 0.15 l/min., shows that they are rather comparable and are equal to 76…221 °C. According to [11], CLs temperature, for example, at steel workpiece drilling with a drill of 50 size out of a rapid steel R6M3 can reach 270 °C.

Let us consider the opportunity of re-use or multiple use of oil CLs with NCMA. In this case a decisive meaning is attributed to the capacity of the considered addition of water re-absorption after the previous heating up to CLs working temperatures in the cutting area.

Basing upon the work [8], one may state that the previous heating mode can significantly change the following water absorption properties of NCMA. To confirm the above said, let us give a brief method and some results of the conducted experiment. Adding water to a NCMA sub-sample, we made it a paste-like, and after that the sub-sample was gradually dried at 200, 300, 400, 500, 600, 800 and 900 °C. After each drying stage and 10-minute temperature stabilization the sub-sample was re-hydrated.

Experiments demonstrated that NCMA heating up to the temperature 200 °C allows it to preserve its re-hydration properties by 61%. The most significant deterioration of NCMA re-hydration capacity is registered after its heating up to 500 °C. At the same time, NCMA re-hydrates, i.e. preserves the necessary properties. only by 37%. Such temperature threshold is rather satisfactory in terms of NCMA use in CLs as red hardness of rapid steel is 500 °C. At further heating up to 800–900 °C, NCMA virtually loses its re-hydration capacity, and thus, it becomes unsuitable for re-use as an addition to CLs.

Summing up the aforementioned, one can assume that NCMA overheating deteriorates its properties. The temperature limit can be considered 500 °C while admissible thermal optimum for endothermic effect of NCMA in CLs is 200 °C.

Results of experimental wheel tests aimed at determining the effectiveness of adding nanoclay mineral additives to oil-based cutting fluids are described in [14]. The analysis of these results as illustrated by drilling of various construction materials, has supported the idea that tribotechnological and endothermic properties of oil-based cutting fluids with nanoadditives improve. Significant increase in hardness of a drill thanks to the decrease of cutting thrust force and torque has been noted. Addition of nanoclay mineral additives to oil-based cutting fluids positively influences the formation of facings and increases the roughness of machined surfaces.

3 Conclusion

1. Well-known dry lubricants used as additives to oil lubricants to modify friction do not have heat-absorbing capacity, and the anti-friction and wear resistance properties of ones (graphite) are exhibited at relatively low temperatures (below 100 °C) when sorbed water is involved in friction of their crystal lattices, while friction in the crystal lattices of others (molybdenum disulfide, serpentine) takes place in the absence of water.

2. The most important NCMA property is the capacity of its packages to be hydrowedged with a superficially occluded water at hydrogenation. As a result, an opportunity of interpackage sliding is provided due to occurring hydro-lubrication, i.e. transfer from dry to fluid interpackage friction thus improving tribotechnological and antiscoring properties of oil CLs in general.

3. From the point of view of nanoclay minerals, montmorillonite, in particular, suitability as an addition to oil CLs to increase heat removal from the cutting area, they are characterized by a higher heating capacity of water phase of the boundary layer (hydro-wedging area) comparing to volumetric water. Moreover, they possess endothermic properties, which are important for oil-based cutting fluids.

4. When CLs with NCMA is exposed to low temperatures, the broken boundary adsorption layer is automatically recovered, i.e., re-hydrated releasing some energy in the environment that is equal to the energy obtained by means of contact with a heated tool surface. The latter will be convectionally distributed in CLs volume and removed from the system into the environment.

References

1. *Cutting fluids for metal cutting. Practical recommendation / Guide*, (M.: NIImash, 1979)
2. D. U. Abdulgazis, *Use of De-Hydration Endothermic Effect of Crystalline Hydrates to Increase a Heat-Removing Capacity of CLS on the basis of Vegetable Oils*, High technologies in mechanical engineering: Collected Works NTU "KhPI", Issue **2(11)**, P. 3-8 (2013)
3. D.U. Abdulgazis, E.R. Ablayev, *On the possibility of*

decreasing drilling temperatures by using cutting fluids with endothermic properties, Bulletin of Crimea Engineering and Pedagogical University, issue **20**, pp. 30–34 (2009)

4. D.U. Abdulgazis, *On methods of creating compositional vegetable oil- and crystalline hydrates-based cutting fluids,* Bulletin NTU KPI. Machinery manufacturing No. **46**, pp.109–110 (2005)
5. D.U. Abdulgazis, E.D. Umerov, *Bench and wheel tests of effectiveness of endothermic oil-based cutting fluids*, Bulletin of Crimea Engineering and Pedagogical University, issue **29**, pp. 62–67 (2011)
6. W.A. Harrison, *Electronic Structure and the Properties of Solids: The Physics of the Chemical Bond. – Vol.1* (translated into Russian) (M.: Mir, 1983)
7. *The Mechanism of Action of Friction Modifier* - Molybdenum Disulphide [Digital Resource]. Access mode: http://www.avsta.ru/files/flib/35.pdf.
8. V.P. Zarubin, *Development and research of tribotechnical properties of lubricants with powdered friction geomodifiers as fillers: abstract of Candidate of Sciences (PhD) thesis* (Ivanovo, 2007)
9. *Structural Froms and Mobility of Molecular Water in Montmorillonite* [Digital Resource]. Access mode: http://chem.kstu.ru/butlerov_comm/rus/home.htm/.
10. V.V. Goryushkin, *Technological Properties of Paleocene Bentonites of the Voronezh Anteclise and Opportunity of its Use*, Bulleting of Voronezh University. Geology, No. **1**, P.166-177 (2005)
11. V.G. Solonenko, *Enhancement of Cutting Tools Working Capacity* (Krasnodar, Rostov-on-Don: Kuban State Technological University, North Caucasian Department of the Academy of the Quality Problems in the Russian Federation, 1997)
12. V.G. Solonenko, M.G. Serikova, L.A. Solonenko, *Working Capacity of Auger Drills* (Krasnodar: Kuban State Technological University, 2004)
13. P.I. Yashcheritsin, Ye.E. Feldshtein, M.A. Kornevich, *Theory of Cutting: Manual* (Minsk: Novoe Znanie, 2006)
14. E.D. Umerov, D.U. Abdulgazis, U.A. Abdulgazis, *Increase of hardness of a cutting tool via using environmentally friendly additives to oil-based cutting fields with endothermic and tribotechnological properties*, Technical regulations in Eurasian Economic Space: collection of materials of All-Russian Conference with international participants. Yekaterinburg, pp.24-31 (2015)

Ensuring stable quality of ground parts by monitoring the process system parameters

Elmar Yagyaev[1,*], and *Boris* Shron[2]

[1] Crimean Engineering and Pedagogical University, Russian Federation
[2] Sevastopol State University, Sevastopol, Russian Federation

Abstract. The paper proposes a system for monitoring surface wheel-grinding by analyzing the dynamic alterations in output variables and the process system state parameters. It analyzes the intervals of the process system state parameter interrupts. State space analysis was employed to describe the behavior of the grinding process system. The dependencies obtained enable a complete description of how the system behaves at any interval of the main interrupt period. The analysis of such behavior is reduced to sequential calculation of its parameters at times t_1, t_2, ..., t_i at interrupt intervals T_1, T_2, ..., T_N. The proposed state space analysis method enables an independent evaluation of output perturbations. Besides, by analyzing the dependencies we had obtained, we concluded that for such evaluation, one would have to calculate not only the absolute values of output variables, but also their first and sometimes even second derivatives.

1 Problem statement

One current trend in mechanical engineering is that assembling joints are becoming considerable more accurate. This necessitates better quality and performance of finishing machining. Metal machines and complexes are equipped with highly efficient controllable drives, special information systems and other functional devices with improved performance. More and more automated equipment comes in use with every year. Analysis of the existing production facilities shows that automated equipment is only used to 70 or 80 percent of its capacity when it comes to cutting and grinding operations.

Let us analyze the reasons why automated equipment is used under-efficiently. When operating conventional machines, operators rely on their observation and operation experience to adjust the cutting parameters. One can machine a shaft in 30 or 40 seconds using a conventional machine, while it takes 100 to 120 seconds when using a CNC machine. This is due to the considerable alterations in the process system state that occur in machining. To ensure the specified quality as well as the stability of output variables, operations are configured for the worst system state.

As a rule, automated systems are coupled with the existing process engineering and control principles without reconsidering and improving the latter. The process is designed once and for all and is usually subject to no alterations over the entire production of a product whereas alterations occurring in the process system are ignored. Operations result in considerable alterations in the accuracy and stiffness properties of machines; the physical and chemical properties of the cooling lubricant may also considerable alter between replacements, and so do the environmental temperatures, the diameter of the grinding wheels, and the condition of their applied surface. Each workpiece has specific properties.

A relevant problem of material cutting theory consists in studying the improvements in performance achieved by monitoring the machining process, i.e. by analyzing the dynamic alterations in the output variables and the process system state parameters. Theoretical and experimental studies show that automated grinding operations can be performed 20 to 25 percent more efficiently; besides, such studies help stabilize the quality of final products.

Despite a significant amount of known research, automated grinding machines are not utilized to their full capacity. This is due to the great impact process system parameters, altering during the machining process, have on both performance and quality. When mounting a new workpiece on the machine, the system assumes a new initial state that is different from the preceding one. To ensure the required quality while keeping high performance in grinding operations, one has either to stabilize the state of the system or to control the process by means of the diagnosis data that factor in the alterations in the process system. When CNC-machining a workpiece, the second approach requires an evaluation of alterations in the process system parameters.

2 Literature review

We have analyzed papers on grinding operations and found out that the best theoretical descriptive models are proposed in [1–4]; if appropriately refined, they can factor in the multiplicity of factors affecting the output

* Corresponding author: elmar1875@gmail.com

variables and enable the monitoring of the dynamic alteration in the system parameters.

Modern machine control systems provide for in-process adjustments, which, however, require diagnostic data on the current state in the system and model parameters, as well as a link of the state parameters to the control parameters and the output variables.

The goal hereof is to substantiate the possibility of ensuring a stable quality of ground parts by monitoring the parameters of the process system.

Monitoring and diagnosis methods used in grinding are based either on evaluating the state parameters by using complex diagnosis equipment during a halt, or by analyzing the results of measuring the machined surface quality parameters and some of the process parameters (cutting forces, acoustic emissions, etc.). Both options are not suitable for automated production systems because they enable neither an unambiguous identification of the altered process system parameters nor any prediction of how the system could behave further.

3 Statement of the main material

The cutting process alters the parameters of both the process system and the object produced. Given this feature of cutting operations, represent the link of the output variables to the input ones and the state parameters as a function

$$Y_j(t) = F_j(X_j;\ Z_j(t);\ U_j(t);\ \Omega(t)), \qquad (1)$$

where $Y_j(t)$ *is* the vector of output variables in the *j*th cycle at time *t*; X_j*is* the vector of input variables; $Z_j(t)$ *is*is the vector of the process system state parameters in the *jth*cycle at time *t*; $U_j(t)$ *is*the corresponding vector of control effects; $\Omega(t)$is the vector of perturbation effects.

$$Y_j(t) = (r;\ \varepsilon;\ T;\ W;\ R_a;\ P_y;\ P_z),$$

where r is the radius of the workpiece; ε is the eccentricity; T is the depth of the defective layer; W is the waviness; R_a is the roughness of the surface; P_y, P_z are the cutting force components.

Dependency analysis shows that alterations in the process system state parameters vector brings about alterations not only in the numerical values of the output variables, but also in the function that represents the object parameter transformation law.

When developing a control program, one has to evaluate the deviations in the output variables and the current state of the PS. This is why one has to develop a real-time monitoring system based on analyzing the dynamic alterations in the output variables and the process system state parameters.

To solve this problem, the authors of [5, 6] decomposed the PS into the following subsystems: the machine, the fixtures, and tool, the workpiece, the cooling lubricant, and the cutting process.

By analyzing the decomposition of input and output variables plus the subsystem state parameters, one can identify the most significantly altered parameters in each subsystem to evaluate their effect on the output variables of clean-grinding.

Consider the specifics of how state parameters and output variables alter in a grinding operations; for such analysis, we use Yu. Tu's classification [7]. Intervals of the alterations in interrupt times T of the parameters of a process system consists of longer (higher-order) interrupt periods include several intervals of the sub-order. The intervals are structured in a hierarchy. For instance: when machining parts type shaft using a wheel-grinding machine, it was found that the time machine maintenance $T_{\text{ст}} = 7.5 \times 10^6$ s., time complete tool wear $T_{\text{ии}} = 6.9\times10^4$ s., time to tool dressing $T_{\text{пр}} = 2.4\times10^2$ s., time to process a single surface $T_{\text{пов}} = 20$ s., time to rotate the part $T_{\text{обд}} = 0.25$ s., time to rotate the wheel $T_{\text{обкр}} = 0.026$ s. These intervals are not constant and depend on the structural specifics of the system components, the control effects alterations law, and the state of the system at time t. Alterations in the subsystem parameters alter the output variables y by Δy.

After an interrupt in cutting, the parameters of the system or its subsystems are partially or fully restored (workpiece replacements, wheel dressing, tool or fixture replacement). As a result, the output variables y (material removal rate at the beginning of the next cycle alters by $\Delta(\Delta r)$, the roughness of the material at the beginning of the next cycle alters by ΔRa.

By generalizing the experience with gringing machines and the analyzsis above, we conclude that controlling the output variables y at time t does not enable the identification of the process system state parameters, the alterations of which cause deviations in the output variables. Such identification is enabled by controlling the alteration rate of $y\ (t)$ as the intervals of interrupt times T are considerably different for different subsystems [8].

Simulate the grinding state space.

We know the models for: a continuous system

$$\dot{z}(t) = A(t)z(t) + B(t)U(t); \qquad (2)$$

$$y(t) = D(t)z(t); \qquad (3)$$

a discrete system

$$z(t) = A(t)z(t-1) + B(t)U(t); \qquad (4)$$

$$y(t) = D(t)z(t); \qquad (5)$$

where $z(t)$ is the vector of the process system state parameters; $U(t)$ is the vector of control effects;

$A(t)$, $B(t)$, $D(t)$ are the matrices of the corresponding coefficients.

Kalman filters can be built to evaluate the process system state parameters based on the dependencies (2–5), whereas Kalman-Bucy filters can be used in case of significant errors in output variable and perturbation measurements [9]. When monitoring the system herein described, building such filters can be difficult as the vectors of control effects in (2) and (4) depend on the system state parameters; at any stage, they are determined by the optimum control method, e.g. when calculating the limit cycle for maximum performance. Besides, most dependencies that simulate the machining process are non-linear. When building a filter, matrices A and B are considered known whereas they have yet to be determined in diagnosing. Dependencies (2) and (4) do not contain system restoration data either. To factor in the restoration option for the process under analysis, rewrite the equations (4) and (5) as:

$$z_{i,j}(\tau) = z_{ij-1} + A_{ij}(j-1) + A_{zij}(\tau - \sum_{k=1}^{j-1} \tau_{ұk}), \tag{6}$$

$$y_i(t) = D_i(t) z_i(t); \tag{7}$$

$$\dot{y}_i(t) = D_i \dot{z}(t); \tag{8}$$

where A_{ij} is the operator that factors in the restoration of the ith parameter of the system upon completing the (j – 1)th cycle (i.e. before the jth cycle); A_{zij} is the operator that determines the alterations in the system parameter in the jth cycle; $\tau_{ұk}$ is the time to complete the kth cycle.

Besides, the system should be complemented with a procedure to find out whether the output variable alterations rate belongs to the alteration intervals $y_1(t)$, $y_2(t)$, … $y_m(t)$ of this or that interruption period:

$$\dot{y}(t) \in \begin{bmatrix} \dot{y}_1(t) \\ \dot{y}_2(t) \\ \dots \\ \dot{y}_m(t) \end{bmatrix} \tag{9}$$

By analyzing the dependencies (6) to (9), we can conclude that if the ratios in (9) allows one to identify the altered process system state parameter, we can find the value of this parameter based on the output variable value at any time.

Restoration and PS state parameter alteration operators in the equation (6) can crucially depend on the number of the current cycle. In grinding operations, wheel dressing (i.e. restoring its applied surface) can be carried out after machining a single workpiece or a batch of workpieces. Such restoration is not always complete. For instance, if we machine ten parts and then dress the cutter (i.e. make adjustments in the number of cutting edges, their distribution over the depth of the layer, or their protrusion above the bond), the wheel becomes 1 to 2 millimeters less in radius. This reduces the cutting speed by 0.2 to 0.5 percent. This may seem insignificant; however, we should take into account that before a wheel is worn out, it machines 1000 to 3000 parts and has its radius halved from 400 mm to 200 mm. This correspondingly halves the cutting speed.

We also have to evaluate the alterations in the uncontrollable process system state parameters that may occur when restoring its components and subsystems, i.e. during machine and fixture maintenance, tool replacement or dressing.

$$\Delta y_j = y_j - y_{j-1}; \tag{10}$$

$$\Delta z_j = D\Delta y_j. \tag{11}$$

State space analysis can be employed to describe the behavior of the grinding system. Consider how it could be applied to study the processes herein analyzed. To link the sequential states of the system in time, assume that the derivative of the state vector at time t depends on its current state rather than pre-development. This enables us to describe the process by a vector-matrix differential equation:

$$z'(t) = f[z(t), U(t), t], \tag{11}$$

where $z(t)$ is the vector of the process system state parameters; $U(t)$ *is* the vector of control effects; f is the vector function of state variables.

For an interval of τ, the differential equation (11) for a non-standard discrete system is written as:

$$\frac{dV(\lambda)}{d\lambda} = AV(\lambda), \tag{12}$$

where A is the extended system coefficient matrix;

$$\lambda = t - nT \text{and} 0 < \lambda \leq T; \tag{13}$$

$V(\lambda)$ is the column vector of the input variables m_i and the parameters z_j of the system

$$V(\lambda) = \begin{bmatrix} m(\lambda) \\ z(\lambda) \end{bmatrix}; \tag{14}$$

$$m(\lambda) = \begin{bmatrix} m_1(\lambda) \\ m_2(\lambda) \\ \vdots \\ m_n(\lambda) \end{bmatrix}; \tag{15}$$

$$z(\lambda) = \begin{bmatrix} z_1(\lambda) \\ z_2(\lambda) \\ \vdots \\ z_k(\lambda) \end{bmatrix}. \tag{16}$$

The equation (12) is solved as follows:

$$V(\lambda) = \Phi(\lambda)V(0^+), \tag{17}$$

where $\Phi(\lambda)$ is the transition matrix defined from (14) by replacing λ with 0^+; $V(0^+)$ is the initial conditions vector, written as:

$$\Phi(\lambda) = L^{-1}/sI - A/^{-1}, \tag{18}$$

whereL^{-1} is a inverse Laplace transformation; I is an identity matrix.

In terms of the variable t, the equation (17) is written as:

$$V(t) = \Phi(t-nT)V(nT^{+}). \quad (19)$$

By making substitutions in (19) $V(nT^{+}) = BV(nT)$, where$B$ is an initial conditions matrix, and introducing $H(t\text{-}nT) = \Phi(t-nT)B$, we finally obtain

$$V(t) = H(t-nT)V(nT). \quad (20)$$

Consider the sequential states of the system at $n = 0$; $n = 1$; ... $n = n-1$; direct and inverse Z-transformation results in the general system state equation solution:

$$V(nT) = Z^{-1}/I-z^{-1}H(T)/^{-1} \,.\, V(0). \quad (21)$$

By generalizing the equations (21) for a discrete system with a main interrupt period and a higher-frequency interrupt, we obtained the dependency [5]

$$V(t) = \Psi(t-nT)V(nT), \quad (22)$$

where $\Psi(t-nT)$ depends on the recurrent ratio

$$\Psi_i(t-nT) = H_i(t-nT-t_i)H_{i-1}(t_i - t_{i-1})\ldots H_1(t_2 - t_1)H_0(t_1), \quad (23)$$

where T is the duration of the main interrupt period; t_1, t_2, ..., t_i are the varying durations of higher-frequency interrupt intervals.

The dependency (22) enables a complete description of how the system behaves at any interval of the main period. It is convenient for modern computing as it involves a sequenced solution of the system in the space of the time t.

Consider the application of the ratio (22) to the process under analysis. If T_1, T_2, ..., T_N stands for the interrupt intervals in ascending order, a generalized dependency for the current time t can be written as a system:

$$\left.\begin{aligned} &V(t) = \psi_N(t - n_N T_N)V(nT_N) \\ &\psi_N(t - n_N T_N) = H_i(t - n_N T_N - t_i)H_i(t_i - t_{i-1}) \\ &\psi_{N-1}(t - n_N T_N - T_{N-1}) = \ldots \end{aligned}\right\} \quad (24)$$

4 Conclusion

Thus, to improve the efficiency of the monitoring system based by dynamic alterations in the output variables and the process system state parameters, we have to:

- classify the alterations in the process system state parameters;
- establish functional or statistical links of the output variables to the alterations in the process systems state parameters.

The proposed state space analysis method enables an independent evaluation of output perturbations. Besides, by analyzing the dependencies we had obtained, we concluded that for such evaluation, one would have to calculate not only the absolute values of output variables, but also their first and sometimes even second derivatives.

The analysis of such behavior is reduced to sequential calculation of its parameters at times t_1, t_2, ..., t_i at interrupt intervals T_1, T_2, ..., T_N. The solution can be simplified by means of parallel programming by viewing the multidimensional system as a total of n one-dimensional systems.

References

1. A.V. Korolev, Yu.K. Novoselov, *Theoretical and probabilistic fundamentals of abrasive processing. Part 1. State of the working surface of the abrasive tool* (Saratov: Izdatel'stvo Saratovskogo universiteta, 1987)
2. Yu.V. Petrakov, O.I. Drachev, *Teoriya avtomaticheskogo upravleniya tehnologicheskimi sistemami* (M.: Mashinostroenie, 2009)
3. Yu.K. Novoselov, *Dynamics of surface shaping in abrasive processing* (LAP LAMBERT Academic Publishing. – Saarbrucken, Deutschland, 2017)
4. A.A. Ignatev, M.V. Vinogradov, V.V. Gorbunov, *Monitoring stankov i protsessov shlifovaniya v podshipnikovom proizvodstve* (Saratov: SGTU, 2004)
5. Yu.K. Novoselov, E.E. Yagyaev, N.R. Kirienko, *Diagnostika operatsiy chistovogo shlifovaniya po dinamike izmeneniya vyihodnyih peremennyih i parametrov sostoyaniya tehnologicheskoy sistemyi*, Vestnik Sevastopolskogo natsionalnogo tehnicheskogo universiteta: sbornik nauchnyih trudov, V **107**. pp.170–173 (2010)
6. E.E. Yagyaev, L.B. Shron, *Diagnostika processa shlifovaniya po dinamike izmeneniya vyhodnyh peremennyh*, Technologiai Automatyzacja Montażu – zespołów, maszyniurządzeń. Nr **2(80)**, pp. 53–57 (2013)
7. Y. Tu, *Sovremennaya teoriya upravleniya* (M.: Mashinostroenie, 1971)
8. R. Dorf, R. Bishop, Sovremennye sistemy upravleniya [transl. by B.I. Kopylova] (M.: Laboratoriya bazovyh znanij, 2002)
9. S.M. Bratan, *Quality assurance and improving processing stability during the finishing and fine grinding*, Cutting & tool in technological system. International Scientific-Technical Collection, Edition **68**. pp. 34–39 (2005)

Role of laser radiation in activating anodic dissolution under electrochemical machining of metals and alloys

Kharis Rakhimyanov[1,*], *Nadezhda* Gaar[1], and *Aleksandr* Loktionov[1]

[1] Novosibirsk State Technical University, Faculty of Mechanical Engineering and Technologies, 630073 Prospekt K. Marksa 20, Novosibirsk, Russia

Abstract. The specific features of electrochemical dissolution of the 12X18H9T stainless steel, the OT-4 titanium alloy and the BK8 hard alloy in the sodium nitrate water solution exposed to 1.06 micrometer wavelength laser radiation were considered. It is found that depassivation of the anode surface is the main mechanism of laser activation in electrochemical dissolving of materials. It is established that the maximum efficiency of laser electrochemical machining is achieved at a pulse repetition frequency of 10 kHz laser radiation. It is connected with the photoactivation mechanism of electrolyte solution molecules, which increases their reaction capacity.

1 Introduction

Improving modern electrophysical technologies aims at increasing their efficiency in machining different metals and alloys. Introducing laser radiation in the machining zone is one of the perspective ways of intensifying the processes of anodic dissolving of current conducting materials. Different mechanisms of machining activation including those connected with laser radiation influence on the electrolyte are considered in investigations [1, 2]. Understanding the complex character of laser effect on the machining procedure providing different activation mechanisms, we tried to find the main mechanism and assess the influence of the laser radiation frequency on the process of anodic dissolving of the investigated materials.

2 Experimental Data

The 12X18H9T stainless steel, the OT-4 titanium alloy and the BK8 hard alloy were used as machined materials. The 10% sodium nitrate water solution electrolyte was chosen during the preliminary experiments [3]. Investigations on studying the anodic behavior of materials in laser radiation were conducted on the experimental installation, the design of which is described in [4]. It should be noted that this installation is unique and allows us to use laser radiation in a wide range of the wavelength (1.32; 1.06; 0.63; 0.53; 0.32 and 0.26 micrometers) by changing the pulse repetition frequency in the range from 1 Hz to 30 kHz.

The influence of laser radiation with a wavelength of 1.06 micrometer and the average power density of 1.05•106 W/m2 chosen according to the results of mathematical modelling presented in [5] was studied during the investigation. The condition that the used value of the average power density does not result in boiling the electrolyte was observed. The peculiarities of the electrochemical cell construction providing the introduction of laser radiation in the machining zone are presented in [6].

The perspective of using laser radiation for intensifying the electrochemical machining of investigated materials is shown in [7-9].

The investigation of the material anodic behavior was conducted using the potentiodynamic and potentiostatic methods by the IP-Pro M potentiostat-galvanostat.

3 Results of the Research and Discussion

Each material used in the investigation has limitations under electrochemical dissolving in the chosen electrolyte. It is confirmed by the data of potentiodynamic investigations characterizing the anodic behavior of the materials under electrochemical machining (ECM) in 10% sodium nitrate water solution (Fig. 1).

Small values of the current density determining the low speed of the anodic dissolving of the investigated materials can be explained by insufficient anion activity of the electrolyte used and the formation of passivating films of different chemical composition on the anode surface. In introducing the laser radiation in the electrochemical system it is possible to realize the following mechanisms of the activation process of anodic dissolving:

• Heat activation,
• Effect of light-hydraulic effect,
• Influence on passivating films,
• Photoactivation of electrochemical and chemical processes.

* Corresponding author: x.raximyanov@corp.nstu.ru

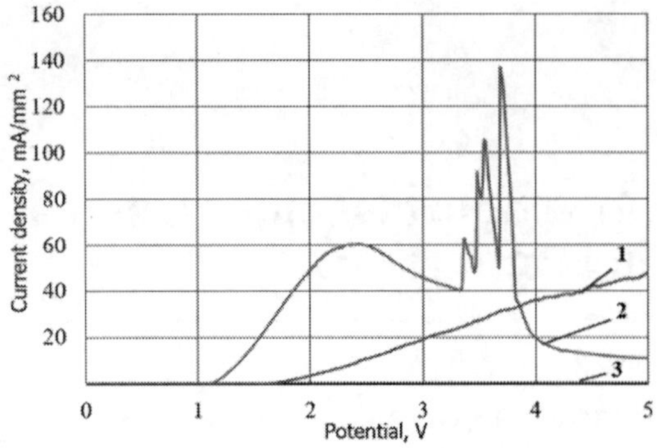

Fig. 1. Potentiodynamic curves during ECM in 10% NaNO3 water solution: 1 – 12X18H9T stainless steel; 2 – BK8 hard alloy; 3 – OT-4 titanium alloy.

The results of the laser electrochemical machining (LECM) of the investigated materials are presented as polarization curves in Fig. 2-4.

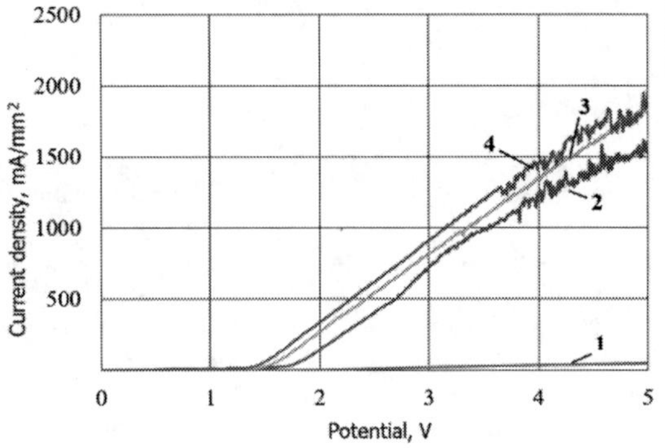

Fig. 2. Potentiodynamic curves for 12X18H9T stainless steel: 1 – during ECM; 2, 3, 4 –during LECM at pulse repetition frequencies: 2-2.5 kHz; 3–5 kHz; 4-10 kHz.

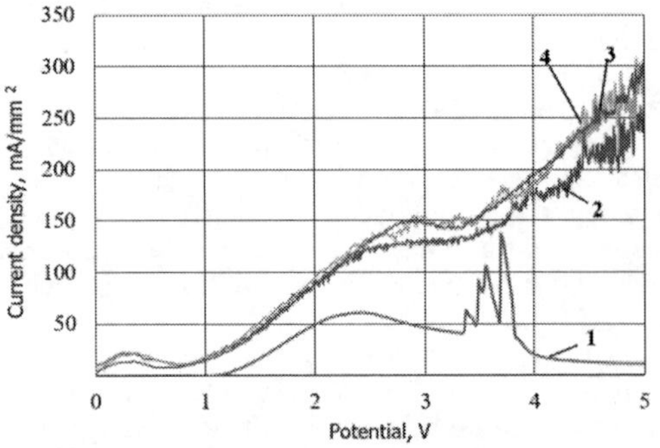

Fig. 3. Potentiodynamic curves for BK-8 hard alloy: 1 – during ECM; 2, 3, 4 – during LECM at pulse repetition frequencies: 2-2.5 kHz; 3-5 kHz; 4-10 kHz.

The analysis of the obtained results shows that introducing laser influence in the zone of electrochemical machining results in a significant activation of the anodic dissolving process, increasing the values of the current density by tens of times. It is more likely that all the mechanisms of activation mentioned above are involved to a different extent in

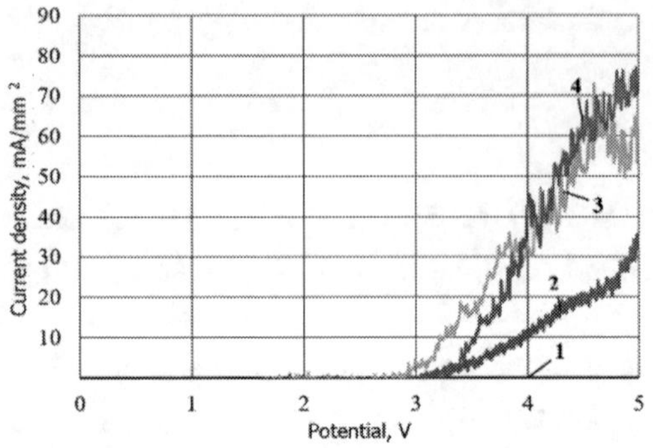

Fig. 4. Potentiodynamic curves for OT-4 titanium alloy: 1 – during ECM 2, 3, 4 – during LECM at pulse repetition frequencies: 2-2.5 kHz; 3–5 kHz; 4–10 kHz.

laser intensification of anodic dissolving. It is known that the main limitation in processes of electrochemical dissolving of metals and alloys is considered to be the appearance of passivating phenomena. Elimination of these limitations during LECM is confirmed by the data of polarization investigations using the potentiostatic method. The fall in the current density during the entire process (Fig. 5–7) does not happen for all the investigated materials during LECM, which shows the absence of passivating limitations in machining.

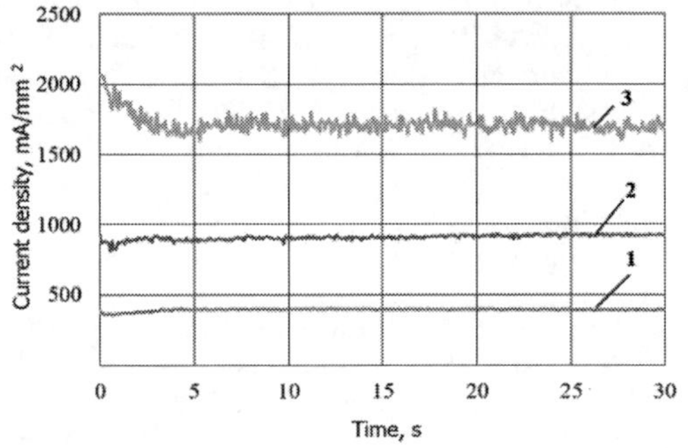

Fig. 5. Potentiostatic curves for 12X18H9T stainless steel during LECM with 10 kHz pulse repetition frequencies at potentials:1–2 V; 2–3 V; 3-5 V.

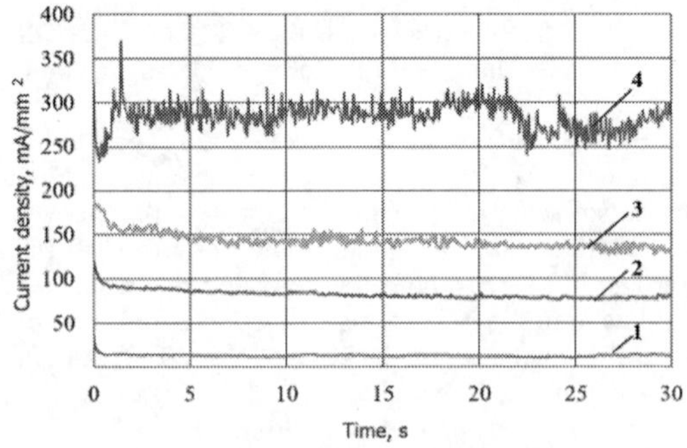

Fig. 6. Potentiostatic curves for BK8 hard alloy during LECM with 10 kHz pulse repetition frequencies at potentials: 1-1B; 2-2 V; 3-3V; 4-5 V.

 DOI: 10.1051/matecconf/201712901070

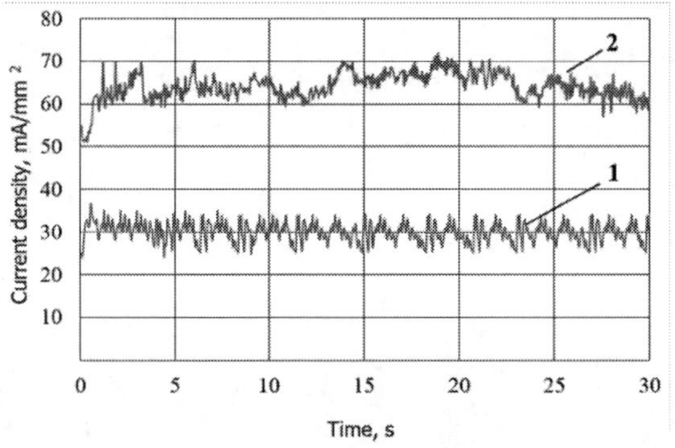

Fig. 7. Potentiostatic curves for OT-4 titanium alloy during LECM with 10 kHz pulse repetition frequencies at potentials: 1–4 V; 2–5 V.

Otherwise, the decrease in the current density during machining would be observed in forming different types of passivating films on the anode surface.

The analysis of the polarized investigation results presented in Fig. 2–4 shows that maximum efficiency in activation of anodic dissolving is achieved at a pulse repetition frequency of 10 kHz. The further frequency increase results in reducing the current density in the whole range of potentials and for all the investigated materials. The importance of pulse repetition frequency of laser radiation in intensifying the electrochemical machining procedure can be obtained by analyzing possible activation mechanisms. It is known that heat activation increases with the increase in temperature but heat effect decreases with increasing the laser radiation frequency under other equal conditions. Therefore, the intensification of the heat mechanism by increasing the frequency characteristic of radiation is impossible.

Intensification of anodic dissolving under laser influence on the ECM process takes place due to the action of light-hydraulic effect that eliminates passivation limitations. The appearance of acoustic shock waves in conducting laser radiation through liquid results in the deformation effect on the objects in the machining zone. The action of light-hydraulic effect under LECM results in destroying the films of different type on the anode surface. The increase in the pulse repetition frequency does not lead to strengthening the mentioned mechanism because the strength of the light-hydraulic effect increases in reducing the frequency characteristic if the average power density is constant. It is also unlikely that the increase in the pulse repetition frequency of laser radiation will provide the increase in the speed of anodic dissolving under the influence of radiation directly on passivating films. The maximum effect from the action of the mentioned activation mechanism in machining different materials would be observed at different frequencies, optimal for dissolving oxide films of a certain chemical composition.

Acceleration of electrochemical and chemical processes is also possible due to selective ionization of electrolyte solution molecules by stepwise photoactivation. It results in excitation of solution atoms and molecules, reducing the height of the energy barrier of the chemical reaction with their participation [10]. As a result, the reaction capacity of the solution anions as well as the velocity of anodic dissolving increases. Ionization selectivity is provided by choosing the frequency characteristic of laser radiation. In spite of the fact that materials of different types are used in investigations, the greatest value of the current density is observed at one value of pulse repetition frequency of laser radiation. This fact proves that the maximum photoactivation of 10% NaNO3 water solution molecules is achieved at a frequency of 10 kHz.

4 Conclusions

1. The results of the investigations show the efficiency of laser intensification of ECM for 12X18H9T stainless steel, BK8 hard alloy and OT-4 titanium alloy. The data of polarization investigations of the material anodic behavior prove the increase in the current density by tens of times in introducing laser radiation in the ECM process. The main factor in using all the mechanisms of the anodic dissolving activation process under LECM is the depassivation of the machined material surface in the zone of laser action.
2. Change in laser radiation frequency characteristic influences the efficiency of ECM intensification. It is shown that the maximum value of the current density for a group of machined materials in sodium nitrate electrolyte is achieved at a pulse repetition frequency of 10 kHz.

Thus, selecting the frequency of laser radiation, we can obtain maximum photoactivation of electrolyte solution molecules, which provides the increase in their reaction capacity.

References

1. N. P. Gaar, Kh. M. Rakhimyanov, *Puti intensifikacii jelektrohimicheskoj razmernoj obrabotki (Ways for intensification of electrochemical dimensional machining)*, Polzunov's almanac, Barnaul, I.I. Polzunov Altai State Technical University **4**, 191-192 (2008) (in Russian).
2. N. P. Gaar, Kh.M. Rakhimyanov, *Possible ways for intensification of dimensional electrochemical machining (DECM)*. IFOST. Novosibirsk- Tomsk, Russia, June 23-29, 2008, Proceedings of the third international forum on strategic technologies, Novosibirsk, Novosibirsk State Technical University 106-107 (2008)
3. N. P. Gaar, A. I. Zhuravlev, B. A. Krasilnikov, A. A. Loktionov, Kh. M. Rakhimyanov, *Method of selection electrolyte for laser-electrochemical treatment*, Obrabotka metallov (tekhnologiya, oborudovanie, instrumenty) **55**, 18-21 (2012) (in Russian)
4. N. P. Gaar, A. I. Zhuravlev, A. A. Loktionov, K. Kh. Rakhimyanov, Kh. M. Rakhimyanov, *Anodic behavior of titanium alloy OT-4 with electrochemical stripping in the conditions of laser influence,*

Obrabotka metallov (tekhnologiya, oborudovanie, instrumenty) **51**, 11–13 (2011) (in Russian)

5. N. P. Gaar, A. I. Zhuravlev, B. A. Krasilnikov, A. A. Loktionov, K. Kh. Rakhimyanov, Kh. M. Rakhimyanov, *Dimensional electrochemical machining of firm alloy BK8 in the conditions of laser influence*, Obrabotka metallov (tekhnologiya, oborudovanie, instrumenty) **50**, 3–7 (2011) (in Russian)
6. N.P. Gaar, *Anodnoe povedenie nerzhavejushhej stali 12H18N9T v uslovijah lazernoj aktivacii pro-ces-sa jelektrohimicheskoj razmernoj obrabotki (Anode behaviour of "12X18H9T" stainless steel in the conditions of laser activation of electrochemical dimensional machining process)*, Proceedings of the All-Russian Scientific Conference of Young Scientists "Science. Technology. Innovation", Dec. 3-5, **Part 2**, Novosibirsk, Novosibirsk State Technical University, 262-264 (2010) (in Russian)
7. N. P. Gaar, A. I. Zhuravlev, B. A. Krasilnikov, A. A. Loktionov, Kh. M. Rakhimyanov, *Installation for studying materials under of laser-electrochemical treatment combined radiation*, Obrabotka metallov (tekhnologiya, oborudovanie, instrumenty). **56**, 30-32 (2012) (in Russian)
8. N. P. Gaar, Kh. M. Rakhimyanov, *Matematicheskoe modelirovanie pri lazernoj termoaktivacii jel-ektrohimicheskoj razmernoj obrabotke (Mathematical simulaion during laser thermoactivation of electrochemical dimensional machining*), Modern technological systems in machine-building (MTSM-2006). Collection of repots abstracts of international scientific-technical conference, Barna-ul, I.I. Polzunov Altai State Technical University, 45-47 (2006) (in Russian)
9. N. P. Gaar, Kh. M. *Rakhimyanov, Jelektrohimicheskaja jachejka dlja issledovanija intensifikacii JeHRO lazernym izlucheniem (Electrochemical cell for investigation of DECM intensification by laser radiation)*, Proceedings of the first Russian Scientific-Technical Conference, Modern ways for machine-building and auto-transport development in Kuzbass, Oct. 24-25, Kemerovo, Kuzbass State Technical University, 251-254 (2007) (in Russian)
10. V.S. Letokhov, *Selektivnoe dejstvie lazernogo izluchenija na veshhestvo (Selective action of laser radiation on material)*, Uspekhi fizicheskikh nauk, Vol. **125**, issue 1. 57–94 (1978) (in Russian)

MATEC Web of Conferences **129**, 01071 (2017) DOI: 10.1051/matecconf/201712901071
ICMTMTE 2017

Twin-cutter form turning of commutator rings using cutters' progressive motion piezoelectric engines

Andrey Vozhzhov[1,*], *Eugene* Pashkov[1], *Vasiliy* Golovin[1], and *Pavel* Florya[1]

[1]FSAEO HO Sevastopol State University, 299053, 33 Universitetskaya str., Sevastopol, Russia

Abstract. The option of realization of the technological scheme of twin-cutter form turning via use of piezoelectric engines of progressive motion is hereby proposed. The cutting force vector constitutes the external perturbation effect. The scheme, design, transfer function of the cutter's piezoelectric engine of progressive motion is hereby presented.

Special attention is drawn to questions of improving the quality of the surface and the accuracy of particularly complex precision parts processing in the instrument-making industry. Various new materials, processing of which is associated with certain difficulties, constantly appear, the shape, surface accuracy and quality requirements to the part's surface become more complex, that's all demand the development of fundamentally new processing technologies, as the possibilities of traditional processing technologies in certain cases are limited and inefficient.

One of the most important elements that determine the reliability and durability of various control systems is the current-collecting unit - electrical contact point.

Due to the lack of reliable and durable contacts, it is often necessary to reduce the life time of expensive systems and products in general.

A special place among electrical contact points is taken by a low-current sliding contact points, that are widely used in the aircraft, space, ship control systems, in manufacturing automation equipment.

Low-current sliding contact points consist of a cylindrical commutator and brushes. Mechanically, a sufficiently large group of rings may be characterized as cylindrical with profile flute (Fig. 1). Manufacturing complexity of such parts is associated with both high requirements for accuracy and quality of the surface, and with their miniaturized dimensions.

The miniaturization of low-current sliding contact points structures is associated with the economy of noble metals used for their production, reduction of mass and size of the contact point units. Another reason for the miniaturization of the structure is the search for optimal conditions for contacting and operating parameters that increase reliability and durability. However, when the size of contact points decreases, their manufacture becomes more expensive, defect ratio increases, objectiveness of evaluation with regard to reliability and life cycle of contact points is significantly reduced.

Flute form turning operation is performed on special lathes machines with improved precision. Concurrently twin-cutter form turning is one of the effective technological solutions. Flute treatment by two simultaneously working cutters reduces the negative influence of cutting force deformation to minimum, thus leads to accuracy increase and reduces the height of the chatter marks on the processing surface.

Processing based on opposite placement of two cutters is the simplest solution. This will allow the process of dividing the feed between the cutters to be performed automatically at any numerical values of the feed amount since one cutter will always lag behind the other by half the revolution in the direction of rotation. Herewith the high processing capacity is ensured as well as sharpening and adjustment of the cutters are greatly simplified.

In practice, this method is implemented in two fundamentally different ways. Practical implementation of the method assumes a sequential arrangement of two cutters one behind the other, while the second cutter may operate in parallel with the first or may enter into operation at the final stage of first cutter processing. If both tools are statically adjusted to the same processing dimension, increased elastic release in the section of the first cutter will result not only the removal by the second cutter of the processed surface roughness left by the first cutter, but it will also cut off some of the left (due to the first cutter release) processing allowance.

The implementation of high-precision form turning of miniature parts, requires the creation of high-precision tracking drives of progressive motion, operating in the micro- and nanometer range with an accuracy from tenths to hundredth of a micrometer. Rotation microengines with mech-electric converters of rotary motion to progressive motion do not provide the required high accuracy and speed.

* Corresponding author: 0506773532@mail.ru

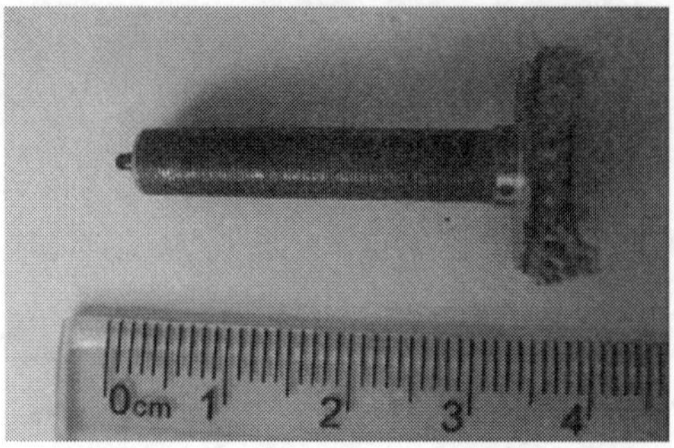

a)

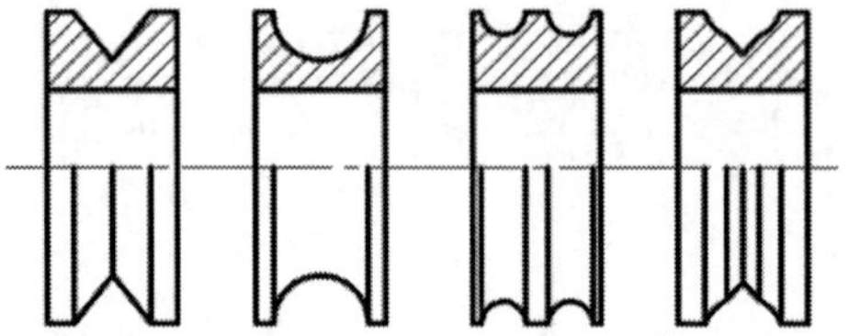

b)

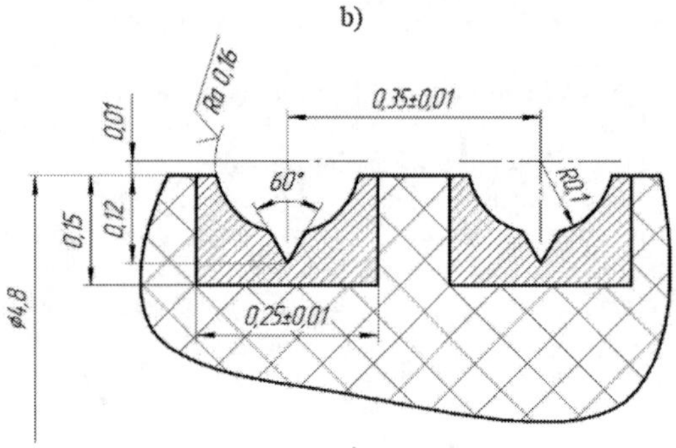

c)

Fig. 1. Collector (a), rings of low-current sliding contact points (b) and a typical flute profile for a brush (c).

The most effective for these purposes is the use of piezoelectric engines of progressive motion (PEPM), in which the conversion of electrical energy into mechanical energy is carried out due to the inverse piezoelectric effect [1].

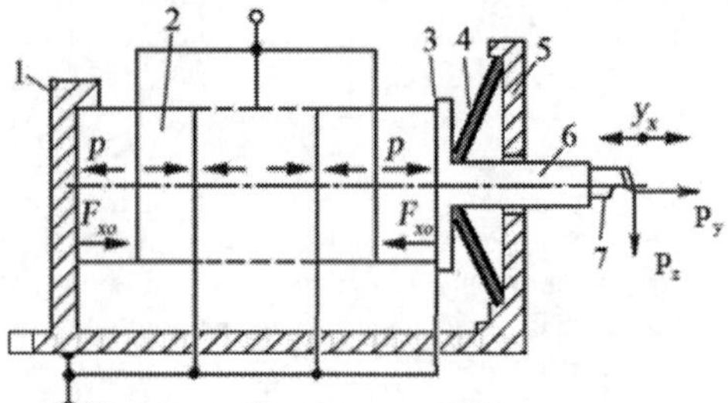

Fig. 2. Scheme of stack type PEPM for cutters

According to the constructive features, PEPMs are quite diverse. The simplest PEPMs, from the constructive aspect, are the tubular and stack constructions. Stacks are collected from individual piezoelements (PE) in the form of rods, disks, washers with a PE number in one stack, reaching 50 and over. The expediency of the PEPM stack construction is determined by the need to reduce the control voltage U_y, as $U_y = E_d \cdot a$, where E_d is the allowed electric field intensity of the piezoceramics; a - PE thickness [1]. The allowed electric field intensity E_d is limited at a certain level. For example, for a two-component piezoceramic (like PZT $Pb(Zr_{0.55}Ti_{0.45})O_3$) $E_{d.в} = 85$ *V/mm*; $f = 100$ *kHz*, and for a multicomponent ceramics (like PKR) $E_{d.в} \approx 60\ldots140$ V/mm, $f = 100$ *kHz*. Therefore in order to reduce U_y it is necessary to reduce the thickness of PE. At the same time, the stack is formed by gluing the unipolar surfaces of individual PE with the thicknesses $a \approx 1$ mm. A metal foil placed in the place of gluing is acting as an electrode. PE is connected mechanically in series, and electrically in parallel.

Fig. 2, [1] stipulates the scheme of stack type PEPM from disk shaped PE.

In order to increase the mechanical strength and stability PE engine have to be assembled by means of a prestressed elastic element 4 (for example, a rubber ring) installed between the flanges 3 and 5, that are pressed against each other by the force F_{xo} applied to the flange 1. The output element of the PEPM the cutter 6 is kinematically connected to the operating device SP. The mechanical tension T_{xo} in PE reaches $(1\ldots3)\cdot10^7$ N/m²; $|T_{xo}| = F_{xo}/A$; where A is the cross-sectional area of the PE.

An increase of the number of PEs in the stack increases the power of the PEPM and the cumulative value of movement of the output element – the cutter.

The static characteristic (Fig. 3) of a stack-type PEPM in the quasistatic linear conditions, with $\pm U_y$ can be approximately described by the expression [1]

$$\Delta y_x \approx k_x \cdot n \cdot d_{ij} U_y , \qquad (1)$$

where k_x – structural-technological ratio, taking into account the cycle and the operating mode of the engine; n – the number of PE in the stack; d_{ij} – piezoelectric unit; ij – index marks, determined by the PE polarization, affecting the deformation direction of PE.

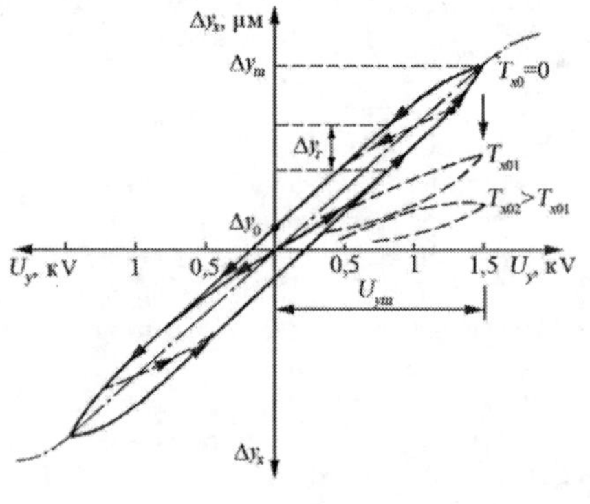

Fig. 3. Static characteristic of a stack-type PEPM

Minimum piezoelectric unit for piezomaterials of the barium titanate system at a temperature of no higher than 85° C is $d_{31} = 1{,}3 \cdot 10^{-8}$ m/V; for piezomaterials of the lead titanate system – $d_{31} = 0{,}2 \ldots 0{,}3 \cdot 10^{-8}$ m/V (at a temperature of 20° C) and $d_{31} = 0{,}4 \ldots 0{,}5 \cdot 10^{-8}$ m/V (at a temperature of 40° C) (the first index $i = 3$ refers to an electric field directed parallelly to the thickness, the second index, $j = 1$ refers to the deformation, perpendicular to the polarization direction) [5].

The sensitivity to the movement of the PEPM in quasi-static mode is determined by the ratio $\Delta y_x / U_y$, which is approximately 1,0 μm/kV for a single PE with $a = 1$ mm from modern piezomaterials (PZT) $(T_x = 0)$ [2]. Variation limits U_y in stack PEPM are $(1 \ldots 20) \cdot 10^2\ V$; the value Δy_x at $U_{y\max}$ reaches 50 μm or more.

When the control voltage is varied in the range $-U_{y\max} \le U_y \le +U_{y\max}$, the static characteristic $\Delta y_x = f(U_y)$ is nonlinear and has the form of hysteresis loops (fig. 3). In this case values k_x and $\Delta y_x / U_y$ are nonlinear variable parameters with maximum values in the initial (rectilinear) part of the characteristic, which may be approximately described by one of the known analytical expressions for hysteresis loops, for example, of the form (longitudinal polarization)

$$\begin{aligned} \Delta y_3\big|_{T_3=0} &\approx k_1 U_y + k_2 U_y^{\,3} + k_3 U_y^{\,5} - \\ &- k_4 U_{ym} \left(1 - U_y^{\,2} / U_{ym}^{\,2}\right)^{n/2} sign \frac{dU_y}{dt}, \end{aligned} \quad (2)$$

where U_{ym} – is the amplitude value U_y; $k_1 \ldots k_3$ – approximation ratios, determined from the three selected points of the main experimental curve ($k_1 \approx k_x n d_{33}$; $k_2 \approx 0{,}75 \ldots 0{,}88 y_m$; $k_3 \approx 0{,}95 \Delta y_m$); $k_4 = y_0 / U_{y_m}$; $n = 2,3,4,5 \ldots$ – value that depends on the shape of the experimental curve [3].

The transfer function of a PEPM as a harmonically linearized link will generally look:

$$\begin{aligned} W_{н} &= \Delta y_{x1} / U_y = q(U_{y_m}, \omega) + jq'(U_{y_m}, \omega) = \\ &= A_q(U_{y_m}, \omega) \exp\left(j\varphi_q \times (U_{y_m}; \omega)\right) \end{aligned}, \quad (3)$$

where $A_q(U_{y_m}, \omega)$ is the transfer ratio; $\varphi(U_{y_m}, \omega)$ – phase shift characterizing the nonlinear delay; q and q' are the ratios of harmonic linearization.

In the particular case of the control of a PEPM of a stack type with a constant voltage, its transfer function has the following approximate form [1]:

$$\begin{aligned} W_x(p) &= \frac{\Delta y_x(p)}{U_y(p)} \approx \frac{k}{T_1^2 T_2^2 p^2 + T_1^2 p^2 + T_2 p + 1} \approx \\ &\approx \frac{k}{T_0^2 p^2 + 2\xi T_0 p + 1} \end{aligned}, \quad (4)$$

where k, T – transmission ratio and time constants, respectively; ξ – ratio of relative attenuation.

As it appears from expression (4), the PEPM is a third- or second-order vibrational element, which is close to the conservative element for small values ξ.

In addition to the PEPM discussed above, other designs are known in which both the DC voltage control modes U_y and the mechanical oscillation modes of the PE at a frequency f_p are used. Mechanical vibrations of the PE are transformed into progressive motion of the moving part (cutter). If the frequency of oscillations of the PE $0 < f_p < 1kHz$, then the PEPM can be considered as a progressive mode one; when $1kHz < f_p < (1 \ldots 5)\ MHz$ there is practically uniform parallel motion of the moving part (MP).

Therewith, traction forces up to 20 N are provided, which is sufficient for realization of the technological scheme of twin-cutter form turning [4].

PEPM provide high dynamic properties in step-by-step, start-stop modes. The range of step change in the progressive motion PEPM can be from 10^{-9} up to 10^{-6} m and more, the speed of movement $V_c \approx 0{,}1 \ldots 0{,}15\ m/s$.

The equation of motion has the following form (5)

$$F_T - F_\Sigma = m \frac{dV}{dt}, \quad (5)$$

where m – the effective dynamic mass of the entire moving system; $\frac{dV}{dt}$ – linear acceleration.

Frequency variations, amplitude or phase control methods can be used to change the speed of the PEPM. The development of a control model for the cutting process on the operation of twin-cutter form turning and the implementation of the control over the process are the goal of further research.

References

1. R.G. Dzhagupov, *Piezoceramic elements in instrumentation and automation* (1986)
2. E.V. Pashkov, V.A. Kramar, A.A. Kabanov, *Tracking drives of industrial process equipment: teaching guide* (2013)
3. E.P. Popov, I.P. Paltov, *Approximate methods for investigating nonlinear automatic systems* (1960)
4. A.A. Vozhzhov, S.V. Borozdin, *Simulation of force interactions in the formation of flute surfaces of collector rings of low-current sliding contacts*, Izvestiya of the Tula State University. Engineering Sciences, **Issue 8. Part 1**, p. 187-193 (2016)
5. I.A. Gluzman, *Piezoceramics* (1972).

Numerical studies to determine spatial deviations of a workpiece that occur when machining on CNC machines

Roman Yu. Nekrasov[1,*], *Igor V.* Solovyov[1], *Alexander I.* Starikov[1], *Yulia A.* Tempel[1], and *Olga A.* Tempel[1]

[1] Industrial University of Tyumen, 38 Volodarskogo st., Tyumen, Russia, 625 000

Abstract. The article discusses the problem of increasing the accuracy of machining workpieces on machines with computer numerical control. Within the framework of the problem identified, the role of the finite element method for solving it is considered. Moreover, numerical simulation is primarily used for receiving estimates of operational and technological characteristics of the developed product, and in the present work we propose to use a method of the finite element analysis as a tool for ensuring the accuracy of shaping the surface of a workpiece in process systems for determining the deformation of the workpiece under the action of cutting forces and transformation of its geometric shape prior to direct machining based on identified deviations. That is, the possibility of applying the finite element method in the development of control programs for details such as bodies of revolution is considered.

1 Introduction

At present, due to sharp competition in the domestic and global markets, machine-building products are subject to strict requirements, provision of which is the main goal of enterprises of the industry.

High productivity and flexibility of production, and, as a result, reduced cost of finished products, allows one to provide machine tools with computer numerical control (CNC) which has wide application in modern machine building.

However, despite the fact that equipment equipped with CNC systems has high discreteness of movements, functions of adjusting the trajectory of the cutting tool, taking into account the wear of the cutting edge and other technical possibilities, the accuracy of processing does not always correspond to the established requirements. Therefore, increasing the accuracy of machining at this time is one of the most pressing issues in engineering technology and cutting theory, and one of the priority areas of machine-building production.

Therefore, many researchers and scientists in this field seek to solve this problem in various ways, but pursue the same goal - the output of a high probability of products that meet the established requirements, as well as preservation of material resources, energy, labor and time in the production of high-precision parts of modern production.

Computer correction of movements of the executive working bodies of the machine tool, development and investigation of intelligent control systems, automatic control of geometric accuracy and regulation of power parameters during processing are among the significant areas in the research of authors in the field of improving the accuracy and quality of processing. But, since a lot of factors influence the machining process, the urgent task for today is to replenish the already existing knowledge base in the area under consideration and increase the efficiency of modern machine-building production.

In addition, requirements for designing in machine building are constantly increasing, in connection with the introduction of new technological processes, modes, materials in the manufacture of parts with high requirements for dimensional accuracy and service life, so the use of numerical simulation is the most profitable.

One of the methods of numerical simulation which is widely applied in the field of engineering analysis is the finite element method (FEM). It serves as a more perfect instrument for making calculations when solving problems of ensuring the reliability of projects in the field of machine building.

In connection with the foregoing, the purpose of the study is: the use of numerical research methods and consideration of the possibility of their application in the development of control programs for processing parts such as bodies of revolution in order to improve the accuracy of their shaping.

2 Materials and Methods

To accomplish the research tasks, theoretical foundations and scientific positions formulated in the works of Russian scientists in the field of engineering technology, automation and control of technological processes, cutting metals were used. When performing the work, methods of calculus mathematics, mathematical modeling, matrix analysis and statistical processing of results of experiments were used. For the finite element simulation of the effect of cutting force components on

* Corresponding author: nekrasovrj@tyuiu.ru

DOI: 10.1051/matecconf/201712901072

the object of investigation, the software product SolidWorks Simulation was used.

3 Results and Discussion

Mathematical modeling and numerical studies of various phenomena and processes are becoming increasingly important. This is due to the ability to view complex objects in different environments, as well as the costs of conducting full-scale experiments.

When turning parts, the tool performs the function of shaping, that is, it must ensure that surfaces of specified sizes and shapes are obtained. The solution of such problems is largely determined by the kinematics of the cutting process. The kinematics of shaping by turning is considered as a combination of four components: shape of the resulting surface of the part; shape of the cutting tool producing surface; shaping movements of the cutter relative to the workpiece; position of the cutter and the workpiece relative to each other.

In order to provide the possibility of assigning the necessary technological conditions for the forming of a part within the specified tolerance limits, a number of the following tasks are realized: building a mathematical model for shaping the surface of a part with the help of an edge of a cutting tool, as well as carrying out numerical studies using the finite element method to determine spatial deviations of a workpiece that arise during machining on CNC machines under the action of cutting force components and the possibility of their compensation.

In the process of machining a workpiece, the machine is subjected to intensive impact from a whole range of operational loads of a random nature, as a result of which the trajectories of the forming elements change widely, which in turn leads to a change in the shape of the part and its accuracy parameters. And, as you know, increasing the accuracy of manufacturing parts increases the service life of machines and equipment. If accuracy of parts that make up the machine is inadequate, the quality of its functioning will be low and in the course of work significant dynamic loads can occur that lead to accelerated deterioration of the equipment and its destruction.

Many Russian authors [eg, 1,2,3,4] consider the influence of processing errors caused by cutting forces, thermal phenomena and wear of the cutting tool to be critical. They lead to a change in the shape and spatial position of both the machine executive organs and the workpiece.

Since cutting is a multifactor process and its complete analytical description is practically impossible at present [4,5], in this paper a narrow problem was solved aimed at modeling the transformation of the part model under the influence of elastic force deformations during the turning process and assessing their effect on the dimensional accuracy of the finished product. The calculation was made using a number of the above assumptions: calculations did not take into account thermal deformations that occur in the cutting area; operating conditions of the CNC machine tool, its life and accuracy were also not taken into account.

In connection with the foregoing, to determine specific values of spatial deviations of the real profile of the part from the nominal during processing under the action of elastic deformations and subsequently with their account to model the actual CAD model of the shaft, it is necessary to perform numerical simulation.

One of the methods of numerical simulation which is widely applied in the field of engineering analysis is the finite element method (FEM). It serves as a more perfect instrument for making calculations when solving problems of ensuring the reliability of projects in the field of machine building. Unlimited possibilities for realization of computer models on the basis of FEM led to the development of new methods of mathematical physics to solve problems of dynamics, stability, physical and geometric nonlinearity [6]. The FEM module is a powerful tool for mathematical solution of the issues of durability and overall design reliability.

Since in the total volume of machining the share of manufacturing body parts and details like body of rotation is about 55%, the shaft of steel grade Steel 40X (GOST 4543-71) served as the object of investigation. As an example, we consider a workpiece at turning, fixed in a lathe chuck without a rear center (Figure 1).

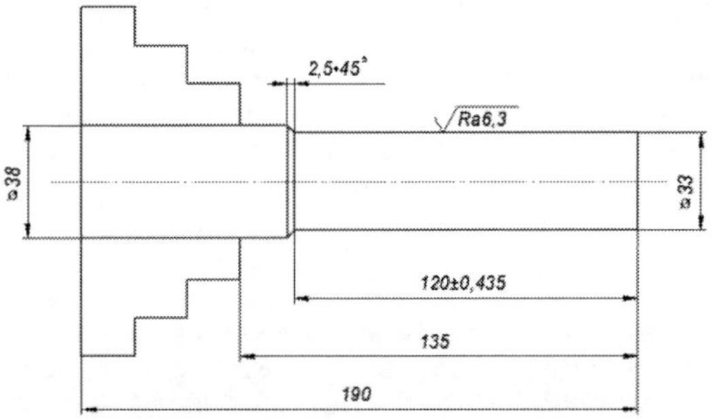

Fig. 1. Workpiece sketch.

For finite element analysis, calculating cutting force components that arise during the turning of a cylindrical surface of a piece with a diameter of 33 mm and a length of 120 mm with a single rough turning, with a cutting depth t = 2.5 mm, a cutting tool - a straight-turning tool T15K6 GOST 18878-73 with the size of the holder 16 × 25 and the following geometric elements of the tool: plan relief angle $\varphi°=45°$; rake angle $\gamma°=0°$; pitch angle of the main blade $\lambda°=0°$, a computational model is created with the physical-mechanical properties of the selected material. Properties of the selected material are as follows: elastic modulus $E=1.76\cdot10^{-5}$ MPa; Poisson's ratio $\nu=0.25$; density $\rho=7670$ kg/m^3. The set load is the components of the cutting force $P_{z,y,x}$ = 2798/1392/1392 N.

The deformed body is divided into finite elements. As a result of the breakdown, a grid is created from the boundaries of the elements. The intersections of these boundaries form nodes. Additional node points can be created at the boundaries of and inside the elements. An ensemble of all finite elements and nodes is the basis of

DOI: 10.1051/matecconf/201712901072

the finite element model of the deformed body. The discrete model should cover the area of the object under investigation fairly well.

The results of the finite element analysis show the change in the spatial shape of the object under study, the magnitude of the resulting displacements of which is shown in Figure 2, they are indicated in micrometers.

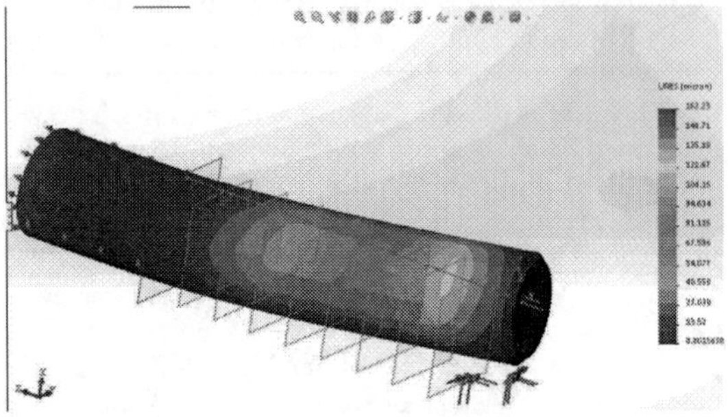

Fig. 2. Results of analysis of displacements of the finite element model.

As Figure 2 shows, after turning, the shape of the part is not cylindrical, but saddle-shaped as a result of deformation from the applied cutting forces.

As the designer provides the CNC programmer with the geometric model of the workpiece which is doomed in advance to the occurrence of errors related to the elastic force deformation of the process equipment, the solution to the problem lies in changing the geometric model of the workpiece by generating a control trajectory and creating a control program.

In connection with the above, based on the numerical simulation, the next step was the introduction of changes to the original geometric model recorded in the STEP format. To confirm the effectiveness of the proposed method for controlling the geometric accuracy of the workpiece on CNC machines, two batches of parts were tested, each of which consisted of five pieces processed by a straight-turning tool T15K6 GOST 18882-73 with a holder size of 16×25 on the turning center SMTCL CAK50135. The first batch was processed without taking into account the arising errors; the second batch of parts was processed according to the program with a modified trajectory of the tool tip displacement on the basis of the transformed CAD model.

To assess reliability indexes according to the quality parameters of the manufactured products, depending on the type of technical system and the evaluation purposes, experimental-statistical methods were used. When comparing the results of measurements and determining the reliability of the technological operation when processing parts without changing the control program and processed according to the proposed method of geometrical accuracy control, it can be said that the efficiency of the method is confirmed by experiments and analysis of data obtained in the experiment. Based on the analysis of distribution of a random variable, there is a constant shift in the mathematical expectation in a batch of parts processed by a control program with a transformable CAD model. In addition, standard deviation is reduced by 40.9%, and the range of variation - almost twofold.

4 Conclusion

In the modern scientific literature, much attention is paid to the quality and precision of metal working on machines with computer numerical control. In connection with this an urgent task was to identify the main trends and problems arising in this direction of the cognitive activity of the scientific community.

In the presented research, the following goal was achieved: increasing the dimensional accuracy of the shaping of parts such as bodies of revolution, the quality of their metalworking, the efficiency of using the production accuracy of technological equipment when turning on machines with computer numerical control.

To realize this goal, factors that affect the quality and accuracy of machining on CNC machines were analyzed. Based on this analysis, and also taking into account the fact that cutting is a rather complicated process, a narrow task was solved in the work to simulate transformations of the CAD model of the stepped shaft, considering error compensation that occurs during the action of the cutting force on the workpiece during turning.

An approach is proposed for determining elastic deformations of a workpiece under the influence of cutting forces and controlling the accuracy of machining on CNC machines using the finite element method.

Reliability of the obtained results is confirmed by numerical and physical experiments.

Reliability of the technological operation of machining of two batches of parts has been estimated using methods of statistical analysis. The first batch is parts manufactured according to the control program with the modified CAD model of the workpiece, and the second batch - without modification. Comparison of the results of assessment confirms the effectiveness of the proposed method for controlling the geometric accuracy of machining of shaft-like parts on CNC machines.

References

1. A.F. Lysenko, A.I. Izyumov, O.V. Goncharov, Bul. of Don State Tech. Univ., **14**, No.3(78) 96-102 [Electronic resource] URL: http:// Elibrary.ru (2014)
2. R.Yu. Nekrasov, A.I. Starikov, I.V. Soloviev, O.V. Bekareva, Tech. and Mat., **4**, 41-48, (2015)
3. I.A. Shchurov, I.S. Boldyrev, Bul. of South Ural State Univ., *Mechanical Engineering*, **12**, 143-149, (2012)
4. A.S. Kuznetsov, A.A. Drozdov, *Modern problems of science and education,* **2**, 165-170 [Electronic resource] URL: http:// Elibrary.ru (Penza, 2015)
5. I.V. Soloviev, Yu.A. Tempel, O.A. Tempel, *IPDM-2017* (St. Petersburg, 2017)
6. H.M. Muselemov, O.M. Ustarkhanov, I.A. Ferzaliev, *Science in Figures*, **1**, 14–16, (2016)

7. V.G. Shalamov, *Mathematical modeling in cutting metals* (Chelyabinsk, 2017)
8. A.A. Lasukov, V.V. Dureev, Mining Inf. and An. Bul., 2, 314-322 (2011)
9. A.S. Yanyushkin, S.O. Safonov, D.V. Lobanov, *Perfection of technological processes of machine-building productions* (Bratsk, 2006)
10. A.A. Lasukov, A.A. Mokhovikov, IFOST 2012 6357720, IOP Conf. Series: Mater. Sc. and Eng., 142 (2016)
11. U.S. Putilova, Yu.I. Nekrasov, A.A. Lasukov, Appl. Mech. and Mat., **682,** 192-195, (2014)

DOI: 10.1051/matecconf/201712901073

Innovative model of business process reengineering at machine building enterprises

Roman Yu. Nekrasov[1,*], *Yulia A.* Tempel[1], *Olga A.* Tempel[1], and *Boris V.* Barbyshev[1]

[1]Industrial University of Tyumen, 625000, Volodarskogo str., 38, Tyumen, Russian federation

Abstract. The paper provides consideration of business process reengineering viewed as a managerial innovation accepted by present day machine building enterprises, as well as ways to improve its procedure. A developed innovative model of reengineering measures is described and based on the process approach and other principles of company management.

1 Introduction

Machine building is one of the key industries in Russia. It is the branch which has provided approximately 15% GDP over the last four years. Machine building is also a resource base for scientific and technical progress. However, innovation activities at machine building enterprises are still at a quite low level. Only 8% of machine building products can be considered as conditionally innovative ones.

Therefore, upgrading the process base of the industry with up to date and prospective technologies, development of innovative products, improvement of machine building competitive ability on the whole, and conversion to the innovative model of economic development are of considerable importance in the scope of the programs aimed at modernization of Russian economy up to the year 2010.

Although production and process technologies have undergone significant transformations and IT technologies are used widely for managerial purposes, a lot of companies still rely on old and traditional principles of management as they usually did. As a consequence, innovative managerial decisions are to be introduced everywhere as innovations in management are far more inexpensive than those in technology. Furthermore, experts of the European Commission claimed that in the short run innovations in management would be of high importance for improvement of the competitive ability on the global market. The direct search for innovative products should be replaced by managerial innovations in these conditions; and that is a mainstream of strategic transformations in any company.

2 Organization management

The article presents the results of fundamental research on the role of administrative innovations in the field of machine-building production. In this study we were used to achieve a systemic approach that implements the review of the engineering enterprise as a complex hierarchical system, as well as the process approach to management. The dialectical method has allowed to carry out understanding of the specifics of modern organizations through innovative development and management.

The theoretical base of the research were scientific statements and conclusions set forth in the works of foreign scientists in the field of business process reengineering. Business process reengineering is one of innovative managerial technologies.

The work is becoming more urgent since nowadays the innovative development in the Russian industry faces some challenges, as stated before.

Moreover, reengineering, being one of the managerial innovations, is gaining in popularity because it alternates the way of business process implementation, making it more efficient, resulting, therefore, in a significant cut in expenditures, better manufacturing capability, improvement of quality, decrease in failures, losses or wastes, reduction of the employed staff (due to automation) and the increasing level of customer satisfaction.

At the same time its implementation is associated with diverse problems which might cause failures as demonstrated by the global experience and studies [1, 3, 5].

Everything mentioned above was relevant for selecting the research domain – business process reengineering.

The study aims to support the methodology of reengineering viewed as an innovation in process and systematic approaches.

Revolutionary processes of efficiency and performance improvement, automation and management optimization are based on reengineering business processes which makes it possible to cut production costs, increase efficiency of the applied machinery and equipment, improve quality of products and decrease the quantity of output that does not meet particular

* Corresponding author: nekrasovrj@tyuiu.ru

 DOI: 10.1051/matecconf/201712901073

requirements, advance customer satisfaction, and contribute to the competitive ability of the enterprise on the whole. Despite a number of advantages listed above the world practice demonstrates that this method involving a rigorous business rearrangement has some drawbacks as well.

Insufficient reengineering might be the reason for using inefficient methods in implementing algorithms of these measures. Although various models of reengineering implementation are available, a conclusion based on their analysis was made that they have a number of disadvantages, including the lack of the total control over each phase implementation, no interrelation between results of the first and final reengineering phases, the use of more general principles of this process, no concrete plan of works and appointment of employees responsible for their implementation, as well as no methods and possible means of reengineering on all its phases which would contribute to more efficient and resultant achievement of the specific goals and objectives. The shortcomings above can be eliminated fully or partially if reengineering is considered in terms of the process approach, i.e. implemented as a set of interdependent sub-processes (phases).

To reveal main trends, principles, methods and technologies of reengineering, and its implementation some algorithms were compared [1, 3, 4, 6], the conclusions below were made on the base of the obtained results.

If viewed as product life-cycle phases, a project of reengineering like any other project has its initial (pre-project) phase, phase of development, implementation and completion. All the studied methodologies of reengineering are similar in this aspect [1, 3, 4, 6].

The majority of the analyzed algorithms comprise four basic phases: preparation, analysis of the current situation in the company, reengineering implementation, introduction of a new business model; all of them are called differently though the content is similar.

As for the shortcomings of the available algorithms, these include no control of business process reengineering implementation and no strategic planning, as well as efficiency and performance of the taken measures can't be estimated. Furthermore, the authors have not addressed to all phases interrelation; the scopes of reengineering sub-processes are also not determined. Some measures are only listed within each phase, and technologies of their fulfillment are provided without describing methods and means of attaining goals. Agents and project participants are assigned though their particular responsibilities within the project of reengineering are not given. There is no constant improvement of the business processes. Moreover, some algorithms are not systematized; all their phases are not exactly defined and measures are not logical. The principles are also not available which measures of business process reengineering should be relied on.

Therefore, all the above drawbacks point to the fact that there is no proper algorithm available which can be used for the purpose of reengineering at any company; as a consequence it influences quality, efficiency and performance of business process reengineering, resulting sometimes in failures when implementing and significant financial and economic losses of the company.

The most of the present day companies are organized functionally, i.e. they have either a linear-functional or functional organization of management.

3 Results and discussion

The divisions in such organizations depend on their functions, but their relations with other divisions are not clear. The management considers subordination of subdivisions and the staff. This form of organizational structure fails to respond fast to strict requirements of the present day market.

First of all, it is caused by lacking relations between divisions of the company and its strategic goals, that is why, the form of company management in the scope of the process approach is regulated by the international standard ISO 9001. It makes urgent the use of process management and planning organizational transformations at the level of industrial complexes and corporations, implemented on the base of reengineering methods and principles. The organization here should be considered as an open system, as a complex of interrelated elements, such as personnel, structure, objectives and technologies which are oriented at fast responding to the challenges arising both in the company and out of it (external environment).

To sum up everything stated above it is necessary to systematize these measures, considering reengineering as a process which consists of several sub-processes (phases) aimed at reduction of possible failures when implementing.

To improve efficiency and performance of enterprises and corporations in various industries of Russian and global economy an algorithm of business process reengineering implementation is proposed which relies on theoretical fundamentals of business process reengineering and international standards ISO 9000.

The developed algorithm can be used in the following fields of economic activities: planning, development and testing of new products, purchases, logistics, marketing research, sales, management of relations with customers and suppliers, production processes etc.

The model of reengineering implementation is given in Figure 1.

This model comprises nine phases: preparation, development of the company model, selection and analysis of business processes, strategic planning, and the development of a new business (innovative idea), approval of the innovative idea, introduction of a new model, control, and continuous improvement. Moreover, this model takes consideration of conditions and principles of reengineering implementation, as well as the scheme of its phases with the definite action sequence.

DOI: 10.1051/matecconf/201712901073

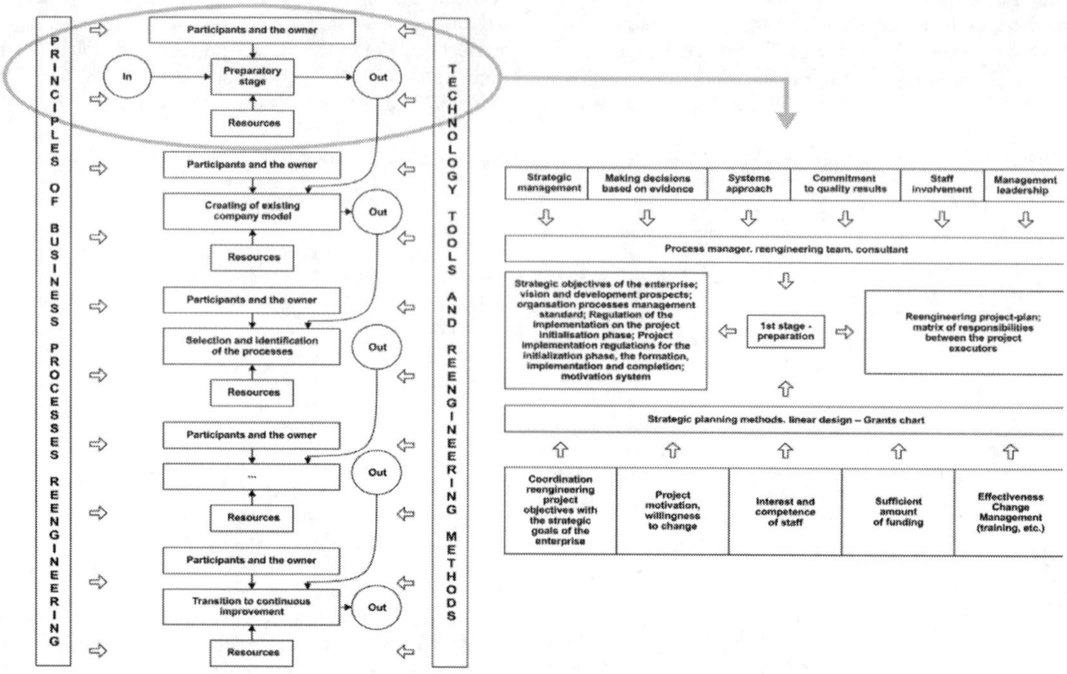

Fig. 1. The proposed innovative model of business project reengineering implementation

When developing the model under consideration a plan of works was framed for each phase as a Gantt chart. The schedule-plan can be specified and adjusted for different enterprises, depending on the type and kind of production, branch, economic activity and purposes of reengineering.

Furthermore, inputs and outputs were determined for each phase with respect to objectives and goals of the sub-process; that makes it possible to see the interdependence of the phases, since they are shown as sub-processes, where each subsequent output is an input for the other process (phase).

The initial input of reengineering implementation is a developed organization standard in the field of project management and regulations of project implementation in phases, it helps to determine the responsibilities of the project manager, heads of functional subdivisions, their rights and duties, this way a lot of conflicts might be avoided. Alongside with the standard in the field of project management a system of motivation is to be formed. The analysis of available motivation theories has revealed that Herzberg's two-factor motivation theory is the most efficient one, it allows encouraging of employees for actions in the scope of their professional activities by means of both motivating factors which are more important for their motivation and work conditions which hardly influence the satisfaction, but decline it if absent.

The developed model of business process reengineering also provides certain methods and means which further implementation of a particular phase; they include Gantt chart, SWOT-analysis and Porter's model to determine one's own competitive ability.

To implement each phase agents and participants are required: a leader – a top manager who approves the reengineering project and motivates; an organizational committee – a body consisting of managers who develop a general strategy of reengineering and monitor its implementation; a process manager – a person responsible for business processes and their reengineering; a team carrying out reengineering – a group of people who conduct diagnostics of the available process, its rearrangement and introduction of new rules; a consultant – a participant who is engaged in the development of reengineering methods and instruments.

Besides the above characteristics of the developed model the principles are of significant importance, which influence the key success factors of reengineering implementation. It is stated in international standards ISO 9000 that the principles of successful, efficient, and resultant reengineering comprise: strategic management; making fact-based decisions; process approach; systematic approach; orientation towards a consumer; personnel involvement, leadership of managers, and ambition for success.

4 Conclusions

The proposed model makes it possible to meet certain conditions of successful reengineering, such as coordination of project goals with the strategic purposes of the company, motivation of a new business model introduction, efficient and total control of changes, technological and methodical support, requirements of internal and external environment, sufficiency of provided resources for the project of business process

DOI: 10.1051/matecconf/201712901073

reengineering, as well as determination of all phases and interdependence of their implementation.

Furthermore, the approach under consideration helps to allocate basic resources, methods and means which are necessary for implementation of a phase, its principles and technologies.

The proposed process-oriented model of reengineering algorithm is an idea of strategic innovative management of company development and comprises systematic and process approaches which form a new notion of company management in the long-run perspective. This approach makes it possible to improve the interaction of subdivisions, increase efficiency and performance of the introduced measures due to logical consequence of actions, strict recourse and responsibility division at each phase.

The work has a certain practical value, i.e. development of the business process reengineering model which helps to meet definite requirements of successful reengineering and improve interaction of the subdivisions. Furthermore, it increases efficiency and performance of the introduced measures due to logical sequence of actions and strict division of recourses and responsibilities at each phase.

References

1. D.Z. Liu, M. Liu, P.S. Zhong, Advances in Materials Manufacturing Science and Technology; Jinan; China; 18 September 2004, Volume **471-472,** P. 770-774 (2004)
2. A.A. Datsyk, S.I. Grudina, S.G. Avdonina, A.I. Podgornaya, Topical issues of influence of "contraction reengineering" on structure of business management: Academy of Strategic Management Journal Volume **15**, Issue Special Issue **1**, P. 67-82 (2016)
3. I. Verboncu, A.A. Fucec, A.I. Serban, M. Rusu, Association Conference - Innovation Management and Education Excellence Vision 2020: From Regional Development Sustainability to Global Economic Growth, IBIMA, P. 753-764 (2016)
4. H. A Lal Bhaskar, International Journal of Productivity and Quality Management, Volume **19, Issue 1,** P. 98-115 (2016)
5. D.Grant, Business analysis techniques in business reengineering: Business Process Management Journal, Volume **22, Issue 1**, P. 75-88. (2016)

DOI: 10.1051/matecconf/201712901074

Role of adhesive interaction centers in control of chip formation and cutter wear

Vladimir.A. Kim[1], *Boris Ya.* Mokritskii[1], and *Anna V.* Morozova[2,*]

[1]Komsomolsk-on-Amur state technical University, 27, Lenin St., Komsomolsk-on-Amur, 681021, Russia
[2]Bryansk state technical University, 7, 50 let Oktyabrya bul., Bryansk, 241035, Russia

Abstract. Adhesive processes define the character of chips formation and a tool-wear rate. As a result of the investigation of chip inner surfaces and a cutting face of the tangential cutting plate of LNMX (LNUX) 301940 type (of different manufacturers including Sandvik coromant, Korloy, Kennenmetel, Pramet) after recovery turning a tread of railway thermo-strengthened wheel pairs it is established that there is a possibility of the process control in minimization of continuous chips formation based on the account of peculiarities in the adhesion interaction of the material under machining and the cutting one; a conditional chart of the distribution of bifurcation centers on deformation-energetic scales of tool material and chips is formed; there is developed a cutter plate of the own design minimizing a possibility of continuous chips formation.

1 Problem definition

A contact interaction at metal cutting is considered to be classified as a tribological process [19 - 24]. At that it is assumed that the loss of cutter efficiency occurs at the expense of structural degradation and destruction of its working surfaces [14 - 18]. An active role is played by adhesive mechanisms in these processes [10 - 13].

Adhesive bonds may become channels of diffusion conductance and contribute to the formation of atomic and chemical compounds resulting finally in adhesion forces increase [1].

Cutter wear may be presented as a multi-stage process of the formation and destruction of adhesion bonds between material to be worked and tool material [2].

The distributions of stress concentrators on working cutter surfaces defines the character of adhesion interaction with material to be worked and tool material wear [3].

The results of the fulfilled investigation of the morphology of chip inner surfaces and working surfaces of a cutter reveal the mechanism in the formation of adhesion bonds, a mutual transfer of material worked and tool material and cutter wear.

2 Research procedure

As an object of researches were chip inner surfaces and working areas of a front face of the tangential cutting plate of LNMX (LNUX) 301940 type made of TC8 alloy after recovery turning the tread surface in railway thermo-strengthened (wheel steel 2 with hardness of 350 HB according to SSR 398-2010) wheel pairs of freight cars.

Cutter plates of different manufacturers - leaders in the production of wheel-turning tools, in particular, are Sandvik coromant, Korloy, Kennenmetel, Pramet. For comparison there was used a plate of the own design the production is carried out at PJ-S Co. "Kirovgrad Plant of Hard Alloys". Figure 1 gives some idea of the plate design.

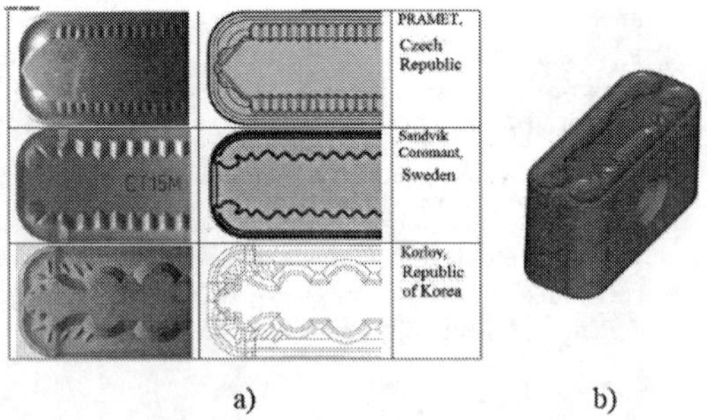

Fig. 1. Schematic representation (carried out with the assistance of V.V. Altukhova) of tangential cutting wheel-turning plates: a) geometry of a front face of cutting plates of different manufacturers; b) 3-D model of a cutting plate offered.

The machining of wheels used before in operation was carried out on the machine of NC UBB-112 type. The following parameters of a cutting mode were used: number of revolutions of a wheel pair was 20 – 25 rpm

* Corresponding author: niotiostu@gmail.com

(at the wheel of 950 mm diameter), a feed -20 – 40 mm/min, a cutting depth - 4 mm. The system of machine soft ware control was provided for the cutting power limitation. By virtue of this the application of a cutting depth exceeding 4 mm was impossible and machining according to the "up to a crust" method [4], that is, for the depth exceeding defects depth (of sliders, cracks) appeared at wheel operation was excluded. Owing to this the cutting is carried out on a defect and in case of necessity a finishing pass was performed.

The illustration of cutting modes used most often is shown in Figure 2.

In Figure 2 (and further in Fig. 4) there are marked typical areas of a wheel profile in a fright car: A – face area; B and C – inclines with different value of a slope; D – cylindrical area; E – flange fillet; F – flange straight-line portion; G – radius area of a flange tip.

The morphology of surfaces was analyzed by photos obtained with the aid of raster electronic SEM S-3400N microscope and Micro-200 optic metallograph microscope. The image processing was carried out by Image.Pro.Plus.5.1 special software allowing the calculation of a perimeter and an area of different micro-structure objects and the analysis of their state by the procedure of color segmentation [5].

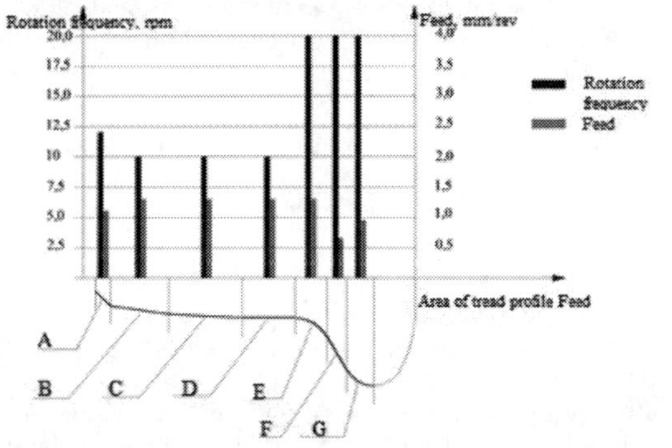

Fig. 2. The applied correlation of cutting mode elements at recovery wheel turning (formed with the participation of V.V. Altukhova during the fulfillment of the assignment on grant 9.251.2014/K of the Ministry of Education and Science of the Russian Federation).

3 Results of investigation and their analysis

During the investigation there was assessed a possibility of the realization of efficient (without continuous chips formation) chips formation and tool operational capacity. Tool working capacity was assessed on amount of wheels machined by one plate up to the achievement by it a critical (on roughness of wheel surface machined) value of wear (destruction). The difference in operational capacity does not have a fundamental character the number of wheels machined by one cutting edge of the plate varies within 5 – 8 work pieces. It is limited not only by the properties of tool material, but also by the presence of defects of operational origin on a wheel surface under machining, Figure 3.

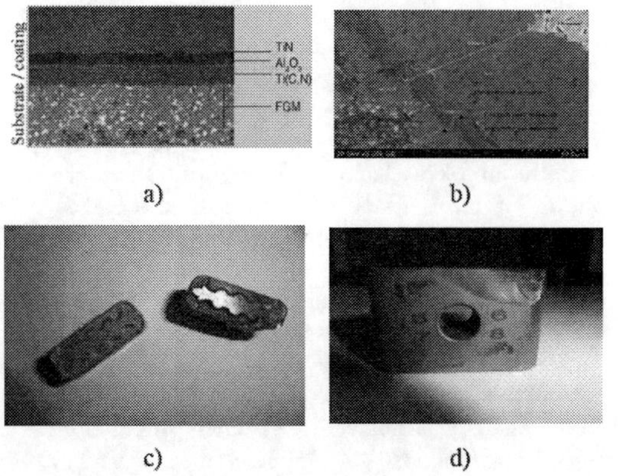

Fig. 3. The samples of common idea of tool material and cutting edges destruction

In Figure 3 there are shown samples: a) - structures and coatings on the plate of Pramet company (taken from a catalogue); b) – coatings developed by the authors; c) – initial state (left photo) and destruction of a cutting edge of Carloy plate; d) – destruction of a cutting edge of the plate manufactured by PJ-SC "Kirovgrad Plant of Hard Alloys".

The chips formation efficiency was assessed by minimum of continuous chips share at all stages (areas of a wheel profile), Figure 4. Continuous chips are bad not only from the point of view of cleaning and transportation, continuous chips may injure and damage machine parts, cutters and a wheel surface machined.

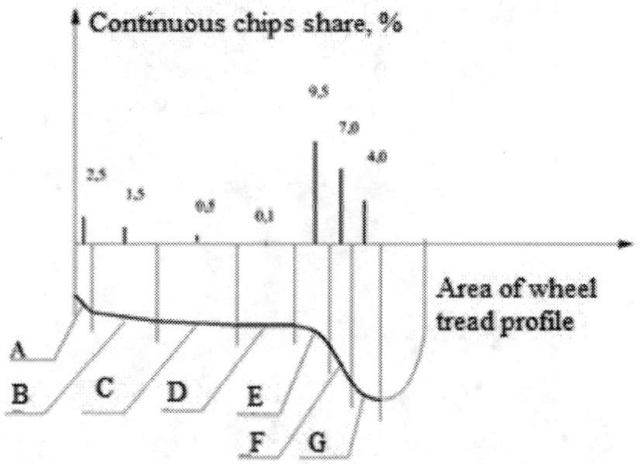

Fig. 4. Share (percentage of all wheel machining time) of the presence of continuous chips at machining different areas of the wheel profile by a cutting plate manufactured by Sandvik coromant Co. (formed with the participation of V.V. Altukhova at the assignment fulfillment according to the grant 9.251.2014/K of the Ministry of Education and Science of the Russian Federation).

Further is considered a possibility for the control of continuous chips formation minimization based on the account of peculiarities of the adhesive interaction of material under machining and cutting material. A general idea of chips formed is shown in Figure 5.

DOI: 10.1051/matecconf/201712901074

It is accepted that the aggregate of dissipative centers distributed in a certain volume of a solid should form a thermal source consisting of ultrashort pulses. Hence, a temperature field is an integral value of the results of thermal source function the gradient of which bears a smooth unbroken character. In Figure 6 there are shown photos of chips inner surfaces with the traces of the effect of thermal dissipative sources distributed at random on the surface as separate "isles". By annealing colors it is possible to define a level of temperatures achieved, a depth of location and a form of a primary source. The area of dark-yellow annealing colors points out to the temperature equaled to 230…2500C up to which there was heated steel under processing, a blue color corresponds to 3500C and higher, a grey background of annealing colors – 4000 and higher, but light-cherry color corresponds to the temperature of 720…9000C [6].

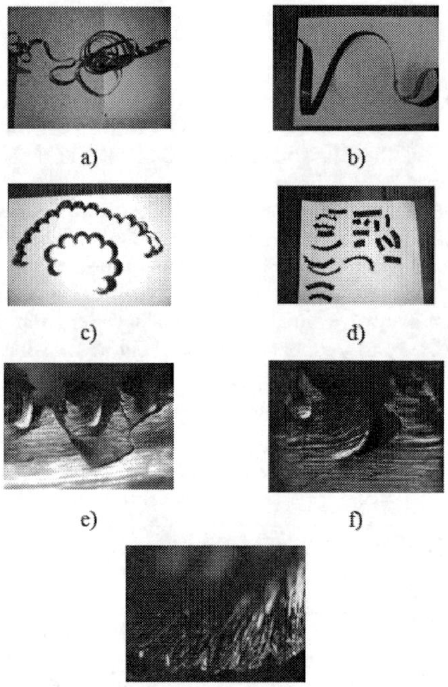

Fig. 5. Samples of entangled chips a) and creeping (b) continuous chips; chips (c) without deformation essential for division into segments; segmented (d) chips; samples of chips edges (e, f, g – obtained with the participation of V.V. Altukhova during the fulfillment of the assignment within the limits of the grant 9.251.2014/K of the Ministry of Education and Science of the Russian Federation) with unequal deformation.

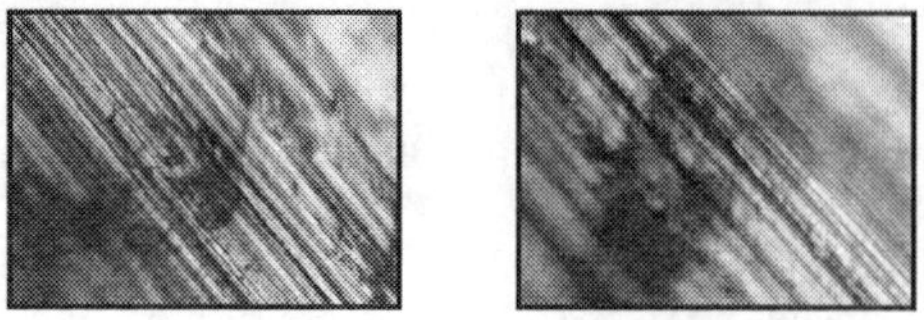

a) b)

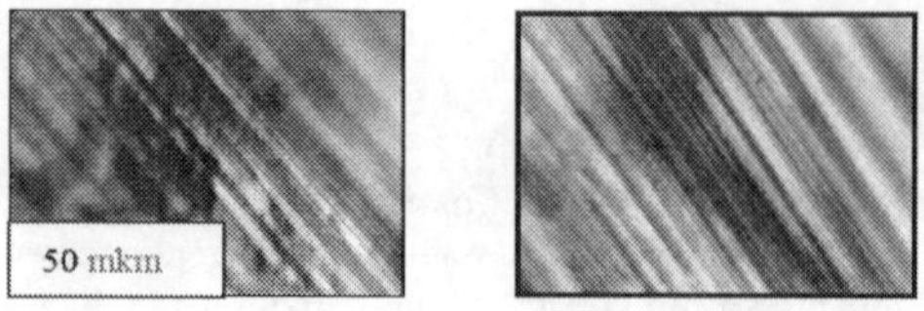

c) d)

Fig. 6. Traces of thermal sources with different intensity of heat generation (x400): a, b, c, d – digitized image of chips surfaces with a thermal image of thermal sources in chips the temperature of which is different by a color spectrum.

If a dissipative thermal source was located on the surface, then its central area had a light-cherry color and clearly distinguished limits at that the temperature of the source achieved 720 … 9000C. The temperature field under formation is characterized by a high gradient of *grad(T)* = (1,0…1,3).107 K/m.

The inner chip surface has mainly a blue color. The relief of the inner chip surface consisting of longitudinal grooves parallel to the direction of the motion of a layer cut does not distort the limits of areas with different annealing colors. It indicates that the basic thermal sources are located in the depth from the inner surface of chips. The density of such dissipative thermal sources in a layer cut off may achieve high values as contact layers of chips are subjected to the second plastic deformation and micro-scaled concentrators of stresses may move inwards at the expense of rotation phenomena in a plastic deformation. If we suppose that a temperature gradient takes the same value in different values, then the depth of thermal sources location makes 50…70 μm of the inner chip surface.

The period of time of unit dissipative source functioning makes thousandth fractions of a second and annealing colors development connected with the formation of oxide films is a more inertial process. In this connection a true temperature of the thermal pulse may achieve higher values than those which are defined by annealing colors.

In Figure 7 there is shown the inner chip surface on which traces of the adhesive interaction are absent. A thermal state of this surface corresponds to the image shown in Figure 6.d). It is seen that the inner layer is formed as a result of intensive plastic deformation of metal cut off which was at a high-plastic state. The existence of micro-cracks and spalling points to a high level of residual stresses spread evenly through the whole of the surface.

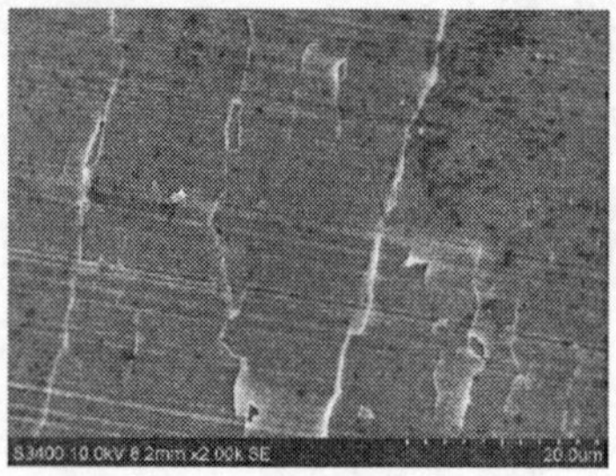

Fig. 7. The Inner surface of chips

For the formation of a strong adhesive bond it is necessary that a dissipative center on the inner side of chips or on a machined surface should be combined with a stress concentrator on the surface of a tool. An initiating thermal pulse arises in material to be machined since it is subjected to plastic deformation to a greater extent and the generation of dissipative centers or thermal pulse sources takes place easier. A concentrator of stresses on the surface of a cutter may be transformed into a source of a thermal pulse, but high durability and low plasticity of tool material and, as a consequence of this high energy of the activation of bifurcation centers make this process highly improbable. A concentrator of stresses in tool material can response as a thermal source only under the influence of relatively high outer force or thermal perturbations which can arise at the expense of chips formation dynamics during the cutting of hard-to-machine material and structurally heterogeneous one.

In Figure 8 are shown the inner surfaces of chips on which are marked the traces of formation and destruction of adhesive bonds and also a transfer of tool material. Some of these photos require an explanation.

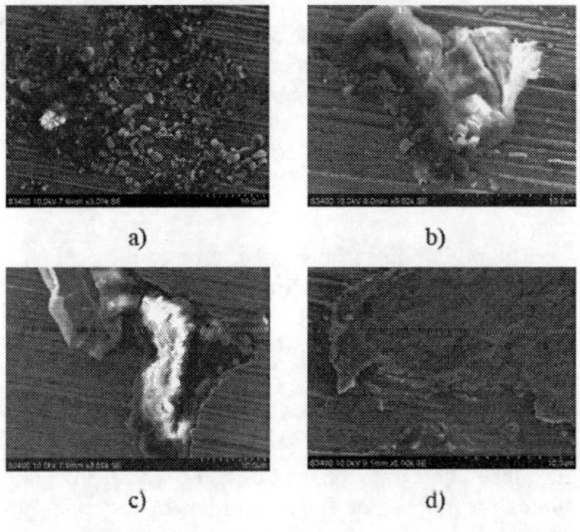

Fig. 8. Inner surface of chips with different options of adhesive bond destruction: a) – traces of adhesive bond break with a small transfer of tool material as separate grains of carbides; b) – adhesive bond break with the detachment of a buildup fragment; c) – buildup fragment of a cylindrical form; d) – "blurred" fragment of a buildup.

So, in Figure 8.b) there is defined a break of an adhesive bond with the detachment of relatively large buildup formed on the face surface of a hard-alloy plate which was subjected to an intensive plastic deformation. The formation of such a fragment is followed with the destruction of tool material which is manifested in the form of grains of tungsten carbide picked up upon a buildup detached.

Being a certain period of time in the area of a contact interaction of chips with the face surface of a cutting wedge such a fragment can obtain a cylindrical form at the expense of rotation dynamics (Fig. 8.c). The high hardness of buildup and presence of hard-alloy particles in it contributes to its introduction and charging into an inner chip surface. A longer existence of a particle in the contact process can result in its smearing on an inner chip surface (Fig. 8.d). In Figure 8.d it is shown that the layer formed at the expense of smearing has particles of tungsten carbides.

In Figure 9 are presented working areas of the cutting face of hard-alloy wheel-turning TC8 plate after turning wheel steel 2. In Figure 9.a) there is shown a structural degradation of a surface layer consisting in micro-cracks formation, in weakening and depletion of a cobalt bond between carbide grains. A basic reason of micro-cracks formation is cobalt diffusion and tungsten carbide oxidation with the formation of spinel compound on contact surfaces [5 - 8].

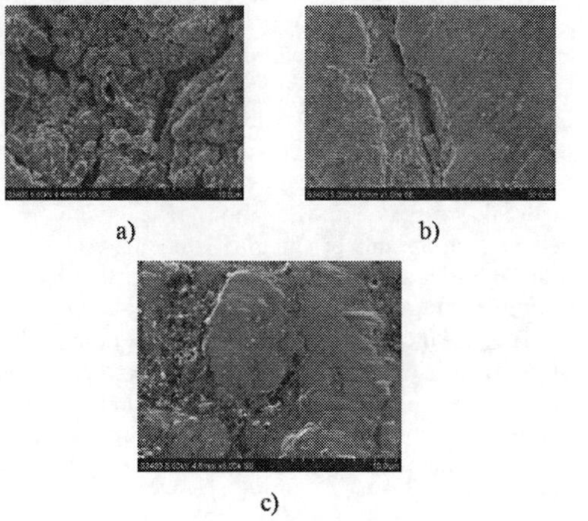

Fig. 9. Face surface of a hard-alloy TC8 plate: a) – micro-crack network; b) – crack with its filling by products of cutting products; c) – pickups of buildup elements.

As a result of this on contact surfaces of a hard-alloy cutting wedge there is formed a developed network of micro-cracks and pores which are filled with particles of material under machining taking into account that an inner chip layer is in the plastic state at a high temperature and pressure. In Figure 9.b) there is shown a micro-crack partially filled with products of the transfer of material worked from the inner surface of chips. As micro-cracks and pores are filled with fragments of material worked and their tamping there is formed a meso-scaled area for the formation on it a buildup or pickup. In this case the connection of material worked and a surface of a cutting wedge is carried out at the expense of an "anchor" effect. An adhesion mechanism remains a basic one in the matter of buildup formation. It embraces a larger area of a continuous contact and active centers of adhesive setting are dislocation outlets on card grains and a cobalt bond.

In Figure 9 c) there are shown buildups and pickups formed as a result of the transfer of material worked. It is evident that the transfer of material worked is followed with intensive plastic deformation and buildups themselves have a lamellar network. The formation of the primary layer of buildup and pickup have adhesive origin though a certain role can be played by "anchor" processes connected with the "flow-in" and filling different micro-irregularities and cavities in the material worked and their conservation at the expense of dry friction. The further growth of buildups is connected absolutely with the formation of metal ties between the fragments of material worked. A metallic or atomic bond is much more stronger than an adhesive one therefore the formation of cutter wear particles takes place if in the depth from the adhesive bond a weakened defect layer is formed.

A strong adhesive bond of the contact interaction of metal materials arises only at the imposition of areas with a high concentration of elastic stresses being in a non-equilibrium state and located on the inner chip surface and on a front face of the cutting wedge, or on the machined surface of a part and on a clearance face of a tool. A thermal pulse arises as a result of the release of elastic internal energy a certain part of which is spent for the adhesive bond formation at that friction processes between tool material and material under machining is transferred in the contact area between a buildup and material under machining. A basic mechanism of tool material wear is the formation of an adhesive bond between chips and products of the material machined transfer upon a cutting wedge and then the detachment of buildup fragments with tool material tear.

The strength of an adhesive bond and the probability of wear particle detachment at its destruction depend upon an energetic state and the overlapping degree of combining bifurcation areas of material under machining and tool material. Let us consider this process more thoroughly.

In Figure 10 there is shown a conventional chart of the distribution of bifurcation centers upon deformation – energetic scales of tool material (TM) and chips (C). A deformation-energetic scale is an aggregate of values in energy activation of different defects formation in the crystalline structure and types of deformation carries generated by them. Each bifurcation center represents a micro-local area consisting of defects of a crystalline structure and a field of elastic stresses round them. The energy of the bifurcation center activation depends upon a structural organization of defects of a crystalline structure which defines a type of the plastic deformation carrier or the structure of the primary source of destruction at the bifurcation algorithm realization.

On the deformation scale of the inner chips surface there is no the first point of bifurcation where occurs the counterparts change on the dislocation. It is connected with that the material machined joining a contact process with a tool is in such a stress-deformed state when its further deformation passes far from nonlinear elasticity.

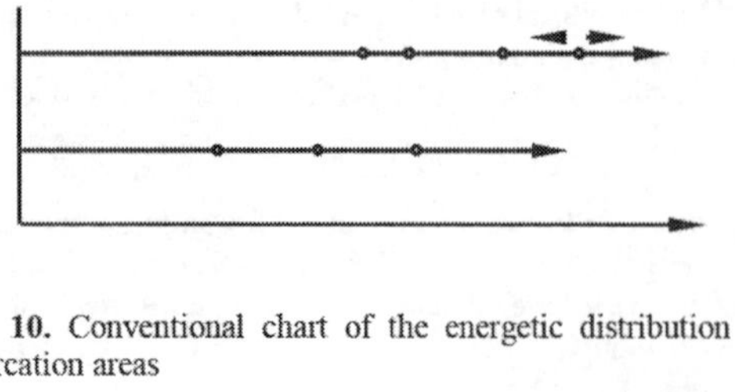

Fig. 10. Conventional chart of the energetic distribution of bifurcation areas

On the deformation scale of tool material there is no an area of rotary-translation whirlwinds. A high thermal stability (red-hardness), wear-resistance and strength of tool material do not allow fragmenting the structure at a local plastic deformation, that is, there is no capacity to the formation of active fragmented meso-structures. Therefore, a grain-boundary shift becomes the most power-consuming bearer of a plastic deformation in tool materials, and the fourth point of bifurcation functions only in the mode destruction, at that its position on the deformation scale bears a non-fixed character. It means that any grain-boundary shift in tool steel at any instant may become an action of discontinuity formation.

If we accept that the functioning of a dissipative process in the areas of a stress concentrator has thermo-activation origin and their energetic activity complies with the Gibbs distribution, then the probability of thermal pulse appearance can be expressed [14] through the equation:

$$p_j = exp\left(-\frac{E_j}{kT}\right), \tag{1}$$

where Ej – activation energy of a thermal pulse; k – Boltzmann ratio; T – temperature.

At that the value of a single thermal pulse makes:

$$q_i = E_j \cdot exp\left(-\frac{E_j}{kT}\right). \tag{2}$$

The probability of the integration of bifurcation areas of chips and a cutter is defined through:

$$p_a = exp\left(-\frac{E_{TMi}}{kT}\right)\cdot exp\left(-\frac{E_{Cj}}{kT}\right) = exp\left(-\frac{E_{TMi}+E_{Cj}}{kT}\right). \tag{3}$$

The same correlation characterizes the probability of the formation of adhesive bonds. Stronger adhesive bonds are formed at the integration of the most power-consuming points of dissipation and bifurcation, but the probability of such an integration is the least one. Hence, not all adhesive bonds produce wear particles at a break. The formation of strong adhesive bonds and the formation of wear particles in the areas of the dissipation of the concentrator of stresses generating a grain-boundary slip on the tool material surface is most likely.

4 Conclusions

The investigations carried out allow drawing the following conclusions:

1. The investigation of surfaces of chips and a cutter allow defining the functioning traces of adhesive processes revealing the comprehension of the essence in the formation, development and destruction of adhesive bonds, of mutual mass-transfer of material under machining and tool material and also cutter wear.

2. The analysis of the role of adhesive interaction centers allows developing measures for the management

DOI: 10.1051/matecconf/201712901074

of metal-cutting tool operational capacity and that of chips formation efficiency.

3. No manufacturer has developed a common cutting plate as the most efficient from the standpoint of working capacity and chip breaking. There is a considerable list of plates differed with the destination (for finishing, roughing, universal) plates and with the design of a front face. The simpler design of the front face is the narrower field of plate application. The most complicated designs of a front face widen the field of plate application, increase plate operational capacity and chip breaking efficiency. There is no plate design which could exclude completely the formation of continuous chips.

4. The topology of the plate front face of different designers is formed at the expense of the application of analogous design elements, topology variations are achieved through the combination of a limited quantity of fundamentally similar elements.

5. For the plates recommended for finishing is typical the existence of local and continuous projections and lugs located both in the areas of a radius connection of a major cutting edge and a minor cutting edge, and along a long side of the front face of the plate.

6. For the plates of a universal purpose is characteristic the existence of a circular cutting in the area of the radius connection of the major cutting edge and the minor cutting one. There are used also local projections located along a long side of the supporting surface and ended on the bottom of a circular groove or a flat hollow and also along one or several stepped projections located parallel to a short side of the front face at some distance from it.

7. The development of existing plates design by leading tool manufacturing companies is carried out according to the methodology of maximum durability support in a cutting edge at the creation of acceptable conditions of chips breaking.

But the research results of adhesive interaction mentioned above have shown that there are reasons for the application of a different methodology: to ensure an efficient chips breaking without plate operational capacity decrease.

More than this, the creation of an efficient chip forming surface can result in loads and temperatures transfer in a plate and, thereby, increase a period of its operational capacity.

References

1. B.V. Deryagin, N.A. Krotova, V.P. Smilga, *Solids Adhesion* (Moscow, Science, 1973)
2. V.A. Kim, F.Ya. Yakubov, *Self-organization in Processes of Cutter Strengthening, Friction and Wear* (Vladivostok, Dalnauka, 2001)
3. V.S. Ivanova, A.S. Balankin, I.Zh. Bunin, A.A. Oxogoev, *Synergetics and Fractals in Material Science* (Moscow, Science, 1994)
4. A.A. Vorobiyov, P.M. Terekhov, *Scientific Problems in Transport of Siberia and Far East* **2**, 121 (2016)
5. V.A. Kim, O.V. Bashkov, A.A. Popkova, *Proceedings of KnASTU*, **III-1 (23)**, 46 (2015)
6. T.N. Loladze, *Cutter Durability and Wear-Resistance* (Moscow, Mechanical Engineering, 1982)
7. A.D. Verkhoturov, F.Ya. Yakubov, V.A. Kim, L.A. Konevtsov, Ch.F. Yakubov, *Proceedings of KnASTU*, **III-1 (19)**, 65 (2014)
8. И.S. Kamenichny, *Short Reference-Book of Thermo-Technologist* (Moscow, Mashgiz, 1963)
9. Ch. Kittel, *Statistical Thermodynamics* (Moscow, Science, 1977)
10. M. G. Buonocore, Journal of dental research. **34(6)**, 849 (1955)
11. A. Zanchetta, P. Lefort, E. Gabbay, Journal of the European Ceramic Society **15(3)**, 233 (1995)
12. C. Huo, W. He, Z.Fan, Surface Technology **6**, 014 (2003)
13. S. Yuhui, Machine Development **5**, 013 (2004)
14. M. C. Shaw, *Metal cutting principles*, T. 2 (New York, Oxford university press, 2005)
15. E. Ceretti, M. Lucchi, T. Altan, Journal of Materials Processing Technology **95(1)**, 17 (1999)
16. Y. Altintas, *Manufacturing automation: metal cutting mechanics, machine tool vibrations, and CNC design* (Cambridge university press, 2012)
17. M. A. Davies, Y. Chou, C. J. Evans, CIRP Annals-Manufacturing Technology **45(1)**, 77 (1996)
18. E. O. Ezugwu, International Journal of Machine Tools and Manufacture **45(12)**, 1353 (2005)
19. I. M. Hutchings, P. Shipway, *Tribology: friction and wear of engineering materials* (Butterworth-Heinemann, 1992).
20. G. Stachowiak, A. W. Batchelor, *Engineering tribology* (Butterworth-Heinemann, 2013).
21. B. Bhushan, *Introduction to tribology* (John Wiley & Sons, 2013)
22. B. Bhushan, *Principles and applications of tribology* (John Wiley & Sons, 2013)
23. R. W. Carpick, M. Salmeron, Chemical Reviews **97(4)**, 1163 (1997)
24. D. F. Moore, International Series in Materials Science and Technology **14** (Elsevier, 2013)

DOI: 10.1051/matecconf/201712901075

High-temperature thermomechanical treatment of hollow track pins of track-type machines

Andrei Zasypkin[1,*], *Vyacheslav* Dement'yev[1], and *Mikhail* Sterkhov[1]

[1]FGBUN Institute of Mechanics, the Ural Branch of the Russian Academy of Sciences, Izhevsk, the Russian Federation

Abstract. Very often the track pins of heavy-loaded track-type machines have insufficient fatigue safety margin, which causes their frequent breakage. One of the most rational solutions of the problem is the use of the high-temperature thermomechanical treatment of hot-rolled tubular billets for track pins. The analysis of the investigation results and engineering developments shows the possibility of improving the performance characteristic of hollow items (shafts, axis pins, track pins for a caterpillar mover, torsion suspensions of an undercarriage, etc.) by combining the thermomechanical treatment strengthening, various methods for improving the quality of the surface of tubular billets and the application of developed technological schemes in the factories producing tubular billets and items with diameter of Ø 10 ÷ 60 mm and the parameters of safety margin, precision and roughness required for the durability of the items.

1 Introduction

The application of the tubular design of the track pins for a caterpillar chain in Russia is very limited. At present, hot-rolled tubular billets are used in track-type machines with an open hinge, where the cross-section area of the track pins provides the transverse strength with the safety margin of 2-3 in the absence of the limited weight requirements for the track pins (the track pins of industrial tractors T-330, T-500).

A significant increase in loading of modern track-type machines (TTM) connected with the requirement on the improvement of their technical features leads to high loading of the undercarriage parts. A track pin is an undercarriage part which fails very often and limits the life of the machine. The use of hollow track pins solves a number of problems connected with their performance characteristic; however, the use of the tubular design leads to the decrease in the transverse strength of the track pins. One of the ways for improving the transverse strength and performance characteristic of the hollow track pins operating in the condition of asymmetric constant-sign bending loading is when in the technological process of the track pins production, a strengthening high-temperature thermomechanical treatment (HTMT) with the deformation of a hot-rolled billet by spiral reduction (SR) with the simultaneous improvement of the quality of both the external and internal working surface is used [1]. The process of HTMT+SR is a complex of continuous actions on the tubular billet consisting of high-frequency heating to the temperatures higher than the austenization temperature, deformation in three tangentially placed idle rollers, and cooling in a sprayer device [2].

It is known that HTMT positively influences the mechanical characteristics of steels (static and cyclic strength, toughness and ductility, wear resistance, etc.). Unfortunately, the use of this high-efficiency treatment for improving the operating reliability of the machine parts is difficult due to its peculiar features; the anisotropy of properties and high hardness of a material make it difficult to process parts by cutting. The development and application of the HTMT technology is possible when the above peculiar features are taken into account at the stage of the designing of parts and development of the technology for their fabrication.

Earlier, all the investigations were mainly conducted for track pins with solid cross-section. The creation of HTMT makes it possible to use a hollow track pin design after a hot-rolled extra thick-walled billet has been obtained instead of a billet with the solid cross-section.

2 Materials and methods

The choice of a material is determined by the operational requirements to the parts. In the investigation of the purpose and the functional stress of the track pins for track-type machines, alloy steels 50G, 55HGR and 38HS for an open hinge operating in the conditions of abrasion wear and alloy steels 38HS and 30HGSN2A for a rubber-metal hinge undergoing high static and cyclic loading were used.

Based on the numerous investigations of the use of track pins from various steels conducted in VNIIStal [Research Institute of Steel] (Moscow) and the results of the investigations of the thermomechanical treatment (TMT) of track pins with the solid cross-section [1], steels 30HGSN2A and 38HS could be recommended for investigating TMT tubes for hollow track pins.

However, steel 38HS is not used for the tube production and the testing results [3] showed poor durability of the track pins from steel 38HS in the

* Corresponding author: oka592@rambler.ru

DOI: 10.1051/matecconf/201712901075

condition of constant-sign asymmetric bending loading; therefore, we have chosen steel 30HGSN2A for our studies and conducted the comparative studies of both steels.

Steel 30HGSN2A has good technological and service properties, a good combination of strength and ductility after quenching and low-temperature tempering (σ_B = 1700 – 1800 MPa; δ = 12%; KCU = 1.1 MJ/m^2), and high stability of austenite at overcooling.

The chemical composition of steels 30HGSN2A and 38HS used for fabricating track pins corresponds to GOST 8294-71 and GOST 7860-71, respectively.

As initial billets, hot-rolled tubes with diameters of Ø 50 and Ø 40 mm according to GOST 2590-76 were used for making track pins Ø 30 x15 mm and Ø 22 x 10 mm, respectively. The technological fabrication process is described in [1] in details. When the technology was adhered exactly, the hollow track pins made in accordance to the technology showed satisfactory results during bench tests and roadormance trials.

3 Results and discussion

In the process of the fabrication of the pilot batches of hollow track pins strengthened by HTMT, the quality level of the billets and items is studied. Based on the results of the studies, the dependences of the service properties and the quality level on the production process conditions are established. All the quality levels of the tubular goods can be divided into three groups.

The first group is the quality level geometrically characterizing a tube: the accuracy of the external diameter ($D_{external}$) and the thickness of the wall (W_T).

The second group of the quality levels characterizes the tube external condition, namely, the quality of the ends of the tubes, the quality of the outer and inner surfaces, and the denseness of the entire body of the tube (cracks, scale residue, etc, GOST 20847-75).

The third group of the quality levels is used for characterizing physicochemical properties of the tube: strength, ductility, chemical activity, etc.; they are mainly dependent on the tube material properties.

The accuracy evaluation is made for the tubes produced according to two manufacturing schemes of calibration with the use of deformation shown in Fig. 1.

The deformation on the triblet (Fig. 1*a*) is more labor intensive and it is used in the production of tubes with the internal diameter of Ø 10 mm and smaller than that. The tubes with a diameter larger than 10 mm are processed according to the second scheme (Fig. 1*b*).

The statistical estimation of the accuracy of the tube wall thickness and diameter is conducted according to the method of the Chelyabinskiy tube-rolling mill [3]. It should be noted that the final difference in the tube wall thickness strongly depends on the initial difference in the wall thickness of the billets after processing on the piercing mill. One of the ways for decreasing the difference in the tube wall thickness is the tube straightening by tube cross-roll straighteners of the type RVV 320×4 after sizing reduction (due to the decrease of the tube radial runout) before the turning machining of the external diameter.

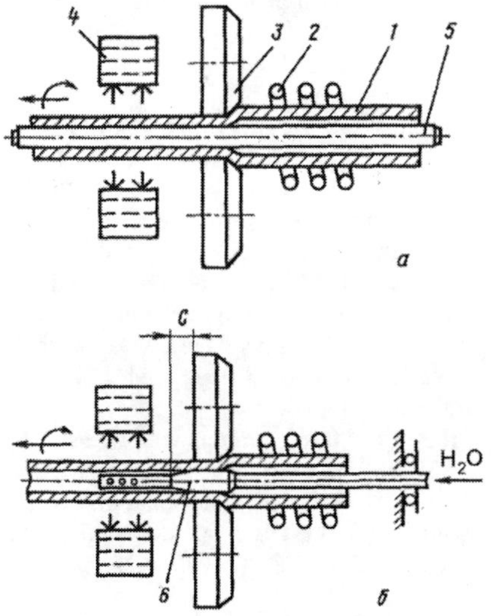

Fig. 1. The manufacturing schemes of the tube deformation on the triblet (a) and on the stub mandrel (b): 1 – a billet; 2 – an induction coil; 3 – deforming rollers; 4 – a sprayer; 5 – a triblet; 6 – a stub mandrel

The comparison of the two schemes of the calibration by deformation (Fig. 1) shows that at the processing of the tubes with $D_{external}$ = 22 mm and $d_{internal}$ = 10 mm according to the first scheme (Fig. 1*a*), the total deviation ω_Σ of the external diameter is smaller than for the same tubes processed according to the second scheme (Fig. 1*b*). However, the internal diameter obtained with the use of the second scheme is more exact.

The calibration of the rolled tubular products in the deformation mode provides the geometry fidelity at the level 11 of the quality class according to GOST 25347-82.

The investigation of the ***surface quality*** of the tubular billets shows that for obtaining a defect-free strengthened tubular billet it is necessary to include the process of removing a defective layer to the depth of 0.8…1.0 mm into the manufacturing process prior to the HTMT process or in the HTMT process and the follow-on operations of grinding and clearing with an abrasive tool to the depth of 0.7…0.8 mm.

The influence of the HTMT conditions on the ***steel mechanical properties*** has been studied in details in the papers by O.I. Shavrin, V.B. Dement'yev [1, 3] and other researchers. However, the influence of the technology of processing and the degree of the deformation at the straining of tubes on the steel mechanical properties is also of a certain interest [3].

Bench and performance testing

In the present paper, the results are presented of the investigations of hollow track pins from TMT rolled products made from steels 38HS and 30HGSN2A

DOI: 10.1051/matecconf/201712901075

(GOST 4543-78) with pressure shaping and without pressure shaping conducted according to the method developed in the Research Institute of Steel (Moscow). The loading conditions during bench testing are: σ_{max} = 800 MPa and σ_{min} = 200 MPa.

The testing results show that

- the durability of the track pins without pressure shaping made from steel 30HGSN2A exceeds the durability of the track pins from steel 38HS independent of the treatment conditions;
- the durability of the track pins without pressure shaping made from TMT tubular rolled products from steel 30HGSN2A is longer by a factor of four than that of the track pins from the same steel which has been quenched and tempered (319, 100 and 72, 400 cycles, respectively);
- the durability of the track pins with pressure shaping made from TMT rolled products from steel 30HGSN2A is ~4 times higher than that of the track pins with pressure shaping made from isothermally quenched steel 38HS (1, 073, 000 and 219, 400 cycles, respectively).

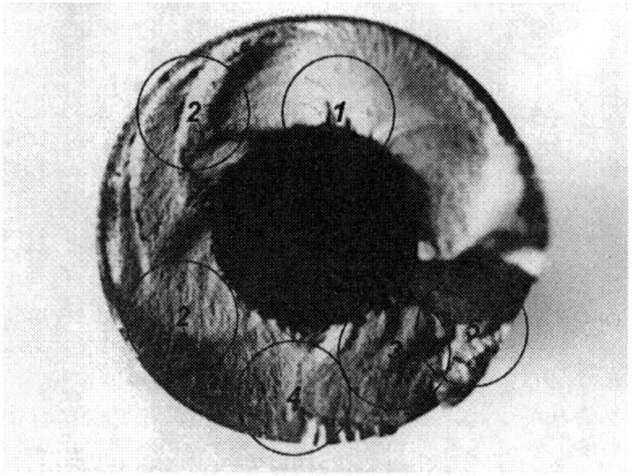

Fig. 2. A multinucleated crack on the internal surface of a hollow track pin undergone 305,000 loading cycles: 1 – region of the nucleation and development of the fatigue crack; 2 – region of fibrous fracture; 3 – region of cut; 4 – region of herringbone and scar fracture; 5 – region of brittle fracture.

Figure 2 shows the view of the fracture of a bench-tested hollow track pin.

Based on the extensive bench tests of the TMT hollow track pins ø22 × 10 mm, ø30 × 10 mm, and ø30 × 15 mm from steel 30HGSN2A, the pilot batches of track pins have been fabricated. The performance testing of the products with the track pins fabricated according to the developed technology shows that in their characteristics, the hollow track pins are not inferior and, in a number of cases, are superior to the track pins with the solid cross-section fabricated by the conventional technology [1].

One of the conditions for improving the durability of a hollow part is the complete elimination of defects of the internal surface and its strengthening by the methods of the surface plastic deformation in the combination with the HTMT process. Because of the residual stresses appearing during strengthening and the necessity of the fulfillment of the requirement to strength balance, the optimal range of the ratio of $d_{internal}$ / $D_{external}$ is within the limits of 0.4…0.6.

4 Conclusions

The technological process has been developed and tested for the fabrication of hollow track pins for track-type machines from an extra thick-walled hot-rolled tubular billet strengthened in the HTMT regime with better mechanical properties and longer durability in comparison with those of track pins fabricated with the use of conventional technology.

References

1. V.B. Dement'ev, O.I. Shavrin, L.N. Maslov, A.D. Zasypkin, *Kachestvo pal'tsev trakov – osnova nadezhnosti i dolgovechnosti gusenitsy* (Izhevsk: IPM UrO RAN, 2009) ISBN 978-5-89238-102-4
2. A.D. Zasypkin, *Uprochnenie trubchatykh zagotovok dlia izgotovleniia polykh pal'tsev trakov gusenichnykh mashin*, Traktory i sel'khozmashiny № 7, pp. 38-41 (2010)
3. O.I. Shavrin, V.B. Dement'ev, L.N. Maslov, A.D. Zasypkin, *Kachestvo poverkhnosti tsilindricheskikh izdelii s termomekhanicheskim uprochneniem* (Izhevsk: IPM UrO RAN, 2006) ISBN 5-89238-088-2

DOI: 10.1051/matecconf/201712901076

Influence of rigidity of the hardened parts on forming the shape accuracy during flat grinding

Yakov I. Soler[1], *Nguyen* Van Le[1,*], and *Mai* Dinh Si[1]

[1]Irkutsk Nation Research Technical University, Department of Technology and Equipment of Machinery Production, 664074, Irkutsk, Russian Federation

Abstract. At present time there is not information about the influence of rigidity of hardened steel parts on the parameters of flatness deviations: EFE_{max}, EFE_a, EFE_q, which are called correspondingly greatest, arithmetic average and quadratic average. To model the variable rigidity of workpieces, an arrangement with the frame structure was used. It allowed to simulate the variable compliance of parts with constant sizes in the direction of longitudinal feed of grinding machine. Searching for models of multiple dispersion analysis was carried out using the D-optimal plan of experiment. They predicted that in rough grinding all parameters of flatness deviations increase with the growth of the part rigidity. In the subsequent grinding stages, the increase of part rigidity is accompanied by the favorable decrease of the flatness deviations. The compliance of low-rigid parts in the finishing stage of grinding should be reduced using additional technological devices.

1 Introduction

Rigidity is one of the most important factors in manufacture of machines and plays a dominant role in formation of the accuracy and the quality of part surfaces [1, 2]. While manufacturing the compliant parts the following regularities are additionally taken into account: significant elastic and plastic deformations of the workpiece under influence of the cutting and inertia forces call the excessive static deflections; the low-rigidity parts are characterized by insufficient resistance to resist cutting forces, as a result of which the excessive forced oscillations (vibrations) are arisen; the workpiece distortion, generating the residual stress and decreasing their heat stability [2, 3].

Taking into account the above phenomena, there is a big complicacy while manufacturing the low-rigidity parts. In connection with the foregoing, the improvement of technological processes for processing the compliant parts is an important problem to increase the efficiency of manufacturing and the quality of engineering products [3].

Grinding process of hardened compliant parts was studied by Yang F., Zhang B. Khusainov A.Sh. and etc. There is not a consensus of opinion on the role of the rigidity of technological system (TS) while grinding the brittle materials, especially for provision of the dimension accuracy [4]. For example, Shore P. [5] considered that high rigidity has an important role for minimizing the defects of the ground surfaces. Jenkins H. and Kurfess T. [6] found that a high quality of the ground surfaces was received in a resonance when the dynamic rigidity of the TS is minimal. In paper [7], the experiments were carried out to determine influence of rigidity of the TS on the processing accuracy. It was found that increase of rigidity had no significant effect on the surface roundness and roughness. Zhang B. [8] confirmed that the residual stresses of the surface of the part from silicon nitride were decreased with increasing the TS rigidity. In papers [8, 4], the investigations were carried out to study influence of the TS rigidity on the forces, the cutting depth, the workpiece strength and the surface quality while grinding the ceramic parts. It has been established that the increase of the TS rigidity leads to increase the normal force and the cutting depth. There is a critical rigidity of the TS. The reduction of rigidity, which is lower than this value, can lead to a serious damage of the ground surface.

The problem of flat grinding of the thin-wall part was studied by Khusainova A.Sh. and Zubairova L.Kh. In [9] a technology of the defectless grinding was developed on the base of the decrease of the process heat-tensity. An investigation of forming the shape accuracy of thin plate during flat planetary face grinding was carried out in paper [10].

Unfortunately, until now there have been insufficient studies of processing the compliant parts, especially during grinding the flat parts. It is due to the complexity of simulation of their variable rigidity.

The purpose of this paper is to identify the influence of the rigidity on the formation of the form accuracy of the surfaces of the hardened steel parts using the statistical modeling the flatness deviations at difference stages of flat grinding.

* Corresponding author: nhatle007@gmail.com

2 Methodology of the experiment

In this study, the constructional steel 30KhGSA (σ_{UST}=1080 MPa, $\sigma_{0.2}$=835 MPa, δ=10 %) was used to investigate. It was subjected to heat-treat regimes: hardening and abatement. This steel are used in aircraft industry to manufacture responsible high-loaded parts of the flying vehicles, operating under alternating loads. The full-scale experiments were carried out on a planogrinding machine of model 3G71. The applied liquid lubricant is 5% emulsion Akvol-6 (TU 0258-024-00148845-98), which is supplied by sprinkling on the part in volume of 7–10 l/min. The objects of the study are specimens with sizes $B\times L\times H$ = 40×40×50 mm. Grinding was carried out along the plane $B\times L$. In this study a grinding wheel, which has the shape 01 and the size 250 × 20 × 76 mm and the characteristic 5NQ 46 I 6 V S3 (company *Norton*, the USA). An arrangement of variable compliance, having a ring-bridge structure, was used to regulate the rigidity. Its sketch is shown in Figure 1. An thick-walled fixed table from assembly-and-disassembly device (ADD) was fixed on this arrangement. Its deformation under the impact of the normal cutting force P_y can be neglected. Thus, the frame of the arrangement is deformed in the direction of the machine table under the impact of the tangential force P_z by the direction of the longitudinal feed s_1 (Figure 2).

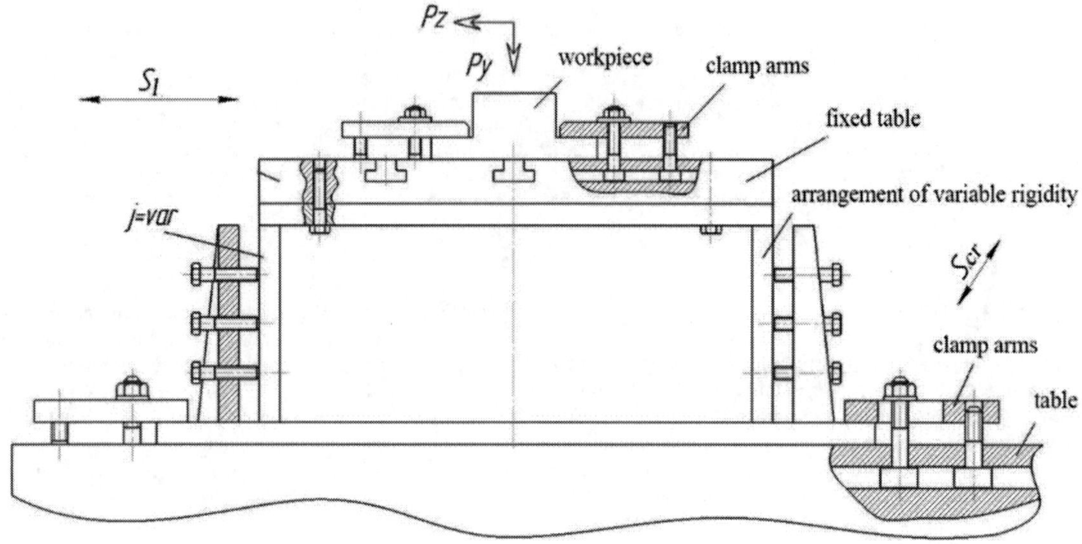

Fig. 1. Sketch of arrangement of variable rigidity.

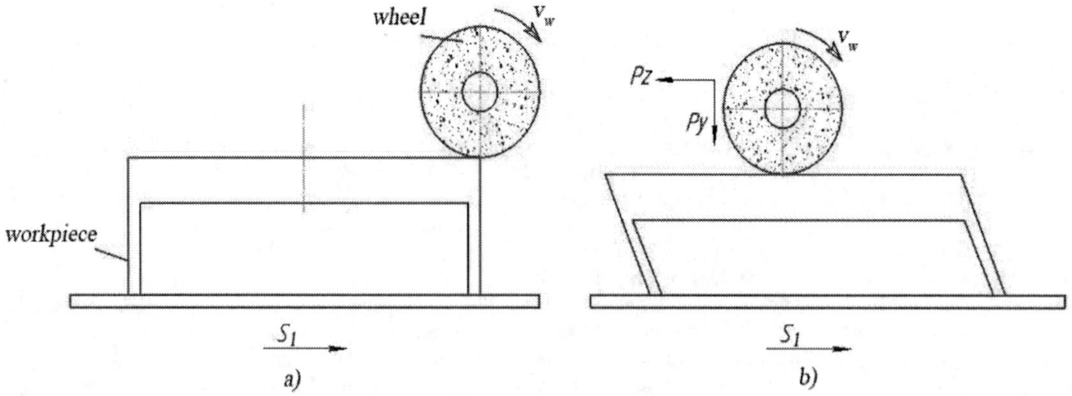

Fig. 2. Scheme of grinding the compliant parts: a) cutting of the wheel into the part; b) grinding process.

The variable rigidity of the arrangement and, therefore, of the part was provide by varying the height of the frame ribs and their pressing with bolts on a different distance from the setting base of the machine table

The deviation from the flatness of the ground surfaces were measured by a twisted-spring micrometer 2-IPM (TU 2-234-229-89). According to GOST 24642-81, they are represented by three indicators: EFE_{max} (main) and two additional ones – the arithmetic average EFE_a and the quadratic average EFE_q.

The workpieces were ground using D-optimal design of experiment in multiple dispersion analysis (MDA). Its matrix was presented in paper [11] under the condition specified in Table 1.

Table 1. Variation intervals and levels of natural and normalized factors

Factors	**Variation intervals**	**Factor levels**		
		lower (-1.0)	**base (0.0)**	**upper (+1.0)**
A **– longitudinal feed** s_l**, m/min**	5	5.0	10	15
B **– cross feed** s_c**, mm/double stroke**	4	2	6	10
C **– cutting depth** t**, mm**	0.0075	0.005	0.0125	0.020
D **– operational allowance** z**, mm**	0.1	0.1	0.2	0.3
E **– rigidity** j**, Н/мм**	5420	380	5800	11220

Initially, the MDA models were constructed using the last square method. The received models were tested for the adequacy in relation of signal-to-noise, as well as for multicollinearity and emission. The expediency of transforming the original model using the maximum likelihood method was determined by the Cox-Box criterion. Implementation of this approach is accomplished using a well-known statistical software Design-Expert. The procedure of finding models was described in detail in paper [11].

3 Research results and discussion

Based on the research and modeling results, the following models of the form accuracy indicators are obtained in Eq. (1) – Eq. (6). They were finally carried out using the maximum likelihood method:

the normalized form

$$\widehat{EFE}_{max} = exp(2.02 + 0.025\ A + 0.12\ B + 0.11\ C - 0.039\ D - 0.086\ E + 0.086\ A\ E - 0.062\ C\ D + \\ + 0.12\ C\ E + 0.10\ A^2 - 0.17\ E^2),\ \mu m; \quad (1)$$

$$\widehat{EFE}_a = (1.98 + 3.054 \times 10^{-3}\ A + 0.095\ B + 0.080\ C - 0.025\ D - 0.095\ E + 0.070\ A\ C + 0.10\ A\ E + \\ + 0.082\ C\ E + 0.079\ D\ E)^2,\ \mu m; \quad (2)$$

$$\widehat{EFE}_q = (2.12 + 0.028\ A + 0.11\ B + 0.10\ C - 0.026\ D - 0.100\ E + 0.066\ A\ C + 0.10\ A\ E - 0.049\ C\ D + \\ + 0.097\ C\ E + 0.061\ D\ E + 0.11\ A^2 - 0.13\ E^2)^2,\ \mu m; \quad (3)$$

the nature form

$$\widehat{EFE}_{max} = exp(2.16848 - 0.095456\ s_{пр} + 0.030782\ s_п + 14.56231\ t + 0.65020\ z - 1.58242 \times 10^{-5}\ j + \\ + 3.15766 \times 10^{-6}\ s_{пр}\ j - 82.89884\ t\ z + 2.88452 \times 10^{-3}\ t\ j + + 4.10247 \times 10^{-3}\ s_{пр}^2 - \\ -5.84124 \times 10^{-9}\ j^2),\ \mu m; \quad (4)$$

$$\widehat{EFE}_a = (2.61753 - 0.044938\ s_{пр} + 0.023637\ s_п - 19.54310\ t - 1.09861\ z - 1.10214 \times 10^{-4}\ j + 1.85587\ s_{пр}\ t + \\ +3.85356 \times 10^{-6}\ s_{пр}\ j + 2.00757 \times 10^{-3}\ t\ j + 1.45549 \times 10^{-4}\ z\ j)^2,\ \mu m; \quad (5)$$

$$\widehat{EFE}_q = (2.77559 - 0.12765\ s_{пр} + 0.026805\ s_п - 4.45867\ t - 0.088511\ z - 5.96140 \times 10^{-5}\ j + 1.75498\ s_{пр}\ t + \\ + 3.84351 \times 10^{-6}\ s_{пр}\ j - 65.98721\ t\ z + 2.38119 \times 10^{-3}\ t\ j + 1.12586 \times 10^{-4}\ z\ j + 4.45203 \times 10^{-3}\ s_{пр}^2 - \\ - 4.27128 \times 10^{-9}\ j^2)^2,\ \mu m. \quad (6)$$

The obtained models Eq. (1) – Eq. (6) allow to estimate the influence of each technological factor on the form accuracy of the ground parts. Models Eq. (1) – Eq. (3) additionally allow to predict the intensity and direction of the response variation using the coefficient and sign of mail factors. At that longitudinal feed (A), cross feed (B) and cutting depth (C) cause the growth of all researched accuracy indicators, and operating allowance (D) and rigidity (E) – their decrease. It can be seen from Eq. (1) – Eq. (3) that to reduce the flatness deviations it shout be decrease cross feed and cutting depth and increase the part rigidity. As will be shown below, the paired interactions of factors and the quadratic effects introduce additional influences on the studying phenomena.

Fig. 3 shows the influence of part rigidity on indicators of flatness deviations on different grinding stages. At that

DOI: 10.1051/matecconf/201712901076

the roughing, semi-finishing and finishing stages of grinding are carried out consecutively in the following modes: A=B=C=D=+1,0, A=B=C=D= 0,0, A=B=C=D=-1,0 (Table 1).

Fig. 3 shows that the part rigidity significantly influences on formation of the surface shape accuracy. When the part rigidity is varied from 380 N/mm to 11220 N/mm the parameters of the flatness deviations are changed to 1.4 times on the rough, to 1.3 times on semi-finishing and 2.2 times on finishing grinding.

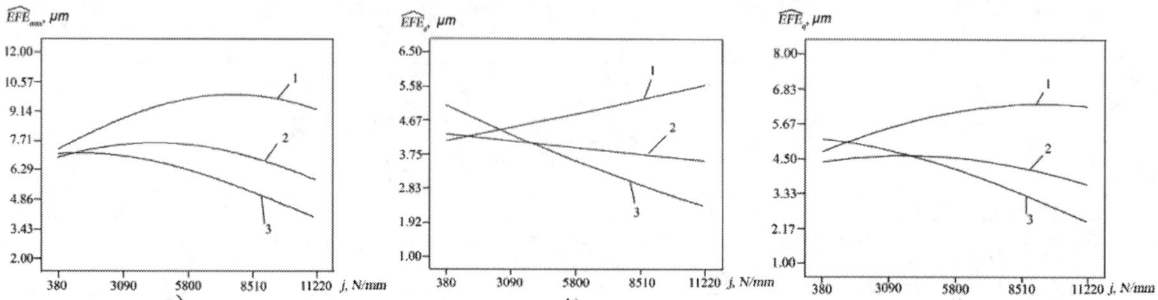

a) b) c)

Fig. 3. Influence of the part rigidity on the indicators of the flatness deviations: a – $\widehat{EFE}_{max}$; b – $\widehat{EFE}_a$, c – $\widehat{EFE}_q$, during stage grinding: 1 – rough; 2 – semi-finishing; 3 – finishing

In rough grinding stage (line 1), with increasing the part rigidity, the value of $\widehat{EFE}_{max}$ increases to maximum at j = 7629 N/mm, and with its further growth – decreases (line 1, Fig. 3a). A similar regularity was noted for the parameter $\widehat{EFE}_q$ (line 1, Fig. 3c) with displacement of extremum to the point j = 9980 N/mm. For the parameter $\widehat{EFE}_a$ the increase of rigidity leads to a monotonic increase in its value (line 1, Fig. 3b).

In semi-finish grinding stage, the observed regularities are preserved for $\widehat{EFE}_{max}$ and $\widehat{EFE}_q$ (line 2, Fig. 3,a,c). The greatest extremums of these lines $\widehat{EFE}_{max,q}=f(j)$ correspond to smaller rigidity of the parts. For parameter $\widehat{EFE}_a$, the increase of rigidity leads to a monotonic decrease in its value (line 2, Fig. 3b).

In finish grinding stage, the increase of part rigidity is accompanied by a decrease of all parameters of flatness deviations from the maximum to the minimum (line 3, Fig. 3).

The noted regularity of the influence of rigidity on the parameters of the form accuracy of ground part surface at different stages can be explained as follows. The rough grinding was carried out with the greatest intensive technological parameters that leads to an increase of the wheel impact when it cuts into the rigid workpiece and an decrease it form accuracy. Low-rigid parts j=380 N/mm have great damping properties, as materials with great internal friction. They quench this impact when cutting into the workpiece. It leads to increase the form accuracy while grinding the low-rigid parts at the rough stage. The obtained result in the rough grinding about the increase of the form accuracy of the low-rigid parts up to 1.4 time is a scientific novelty, because it is always recommended to increase its rigidity in the normative documents.

The semi-finish and especially finish grinding were carried out with a less intensive stock removal. By this reason, the wheel impact into the workpiece is decreased in comparison with the rough stage. The plastic deformations and vibrations, which are the feature of grinding low-rigid parts, increase their influence on the forming the flatness deviations. It is known that they were greater during processing the lower rigid parts than during grinding parts with high rigidity. It explains the decrease of parameters of flatness deviations while increasing the part rigidity.

The obtained results allowed to make the recommendations for increasing the form accuracy in finish grinding the low-rigid parts. One solution to solve this problem is the use of fixture devices as vacuum fixtures, adaptive fixtures using low melting temperature alloys or automated fixtures. Another alternative solution is the use of actuators that compensate online the deformation and vibration of the part. Finally, another solution is the study of the process by means of optimization of the cutting conditions and the study of the dynamic response of the workpiece during the machining. This is a predictive knowledge-based solution, flexible and universal, as it can be applied to every part, and far less expensive than others [11, 13, 14].

4 Conslusions

1. It was established that the rigidity of hardened steel parts had a significant influence on forming the shape accuracy of the ground surface.
2. It is revealed that in the rough grinding stage, the increase of the part rigidity leads to an increase of parameters of flatness deviations, and in the finishing stage – to a decrease of them.

3. For provision of the form accuracy of low-rigid parts during finishing stage of grinding, the additional measures should be used to increase their rigidity.

The authors gratefully acknowledge the support by Irkutsk National Research Technical University under Grant No. 02F/2016 and by The Ministry of Education and Science of the Russian Federation under Grant No. 9.6484.2017/BCh.

References

1. M. Ebrahimi, R. Whalley, Computers & Industrial Engineering, **38** (2000)
2. A. Świć, W. Taranenko, Eksploatacja i Niezawodnosc – Maintenance and Reliability, **14**, 3 (2012)
3. O.I. Drachev, *Control of technological heredity of parts with low rigidity* (Research-and-production cooperative "Association of Scientific Engineering Commercial Structures", 2011)
4. F. Yang, et al., Journal of Manufacturing Science and Engineering, **123** (2001)
5. P. Shore, British Ceramic Proceedings, 46 (1990)
6. H. Jenkins, T. Kurfess, Journal of Vibration and Control, 3 (1997)
7. B. Colding, Machinery, **76**, 7 (1970)
8. B. Zhang, Annals of the CIRP, **50**, 1 (2001)
9. A.Sh. Khusainov, *Increase in the efficiency of grinding operations of blanks of thin-walled parts by reducing the heat stress of the processing process* (Ulyanovsk State Tech. Un-t, Ulyanovsk, 1996)
10. L.Kh. Zubairova, *Technological support of the required accuracy of the shape of thin-walled parts with flat face planetary grinding* (Perm. Nat. reach. Polit. Un-t, Perm, 2014)
11. Nguyen Van Le, *Providing the required quality of surfaces during flat grinding of hardened steel parts with various rigidity* (IRNITU, Irkutsk, 2017)
12. Ya.I. Soler, Nguyen Van Le, D.Yu Kazimirov, Proceedings of higher educational institutions. Mechanical engineering, **4**, 673 (2016)
13. S. Ratchev, S. Liu, A. A. Becker, Journal of Materials Processing Technology, **162-163** (2005)
14. F.J. Campa, L.N. de Lacalle Lopez, G. Urbikain, D. Ruiz, Proceedings of the 2008 International Manufacturing Science and Engineering Conference, (2008)

DOI: 10.1051/matecconf/201712901077

The technological methods of surface layer modification in construction materials

Vladislav Smolentsev[1,*], and *Sergey* Safonov[1]

[1] Voronezh State Technical University, Voronezh city, 394026 Moskovskiy prospect, 14, Russia

Abstract. The paper deals with the choice of a method and methodology of planning the technological processes to ensure the required quality of the surface layer of parts, resting on the principle of utility proposed by authors, which is an overall indicator of efficient use of external and internal impacts upon operational characteristics of products with modified surface layers: 1) the modification of properties without allowance removal, with a change in surface layer characteristics; 2) the removal of the allowance portion until reaching the material layer, having the desired operational characteristics; 3) the coating of products with surface layers, made from materials with desired characteristics and having a structure in which the operational properties are obtained by imparting the preset characteristics (wear resistance, heat resistance, etc.) to parts (including those made from other materials), applying the useful properties of coatings deposited; 4) the deposition of coatings comprising the inclusions of conductive and dielectric materials (e.g. ceramics, abrasives, etc.). The paper discovers the ways of using the principle of utility to justify the effectiveness and the scope of preferred application of technological processes of surface layer modification. It presents the methods of automation in selecting the technical means for enhancing the operational properties of high-tech products.

1 Introduction

When developing new methods and technologies, using the principle of utility [1], one should take into account that, along with the possibility to significantly improve the performance parameters of products by modifying the surface layer, there are limitations, one of which is the need to ensure the minimum cost and complexity of product manufacture. Such parameter is attained by optimized combinations of processing methods, the informed choice of base and coating materials, or surface processing, the way of producing the surface layer with specific operational requirements for products, and the cost of labour and materials for product manufacture. The criteria of the principle of utility underlie the methodology of planning new technological processes, which, for every area of problem development, becomes a foundation for setting new scientific trends in mechanical engineering, particularly, vibrowave technologies [2] and combined processing methods [3, 4, 8].

In their works, the scientists of Voronezh have developed the program of scientifically informed choice and elaboration of new (mostly, at the level of inventions) methods of technological management of the surface layer quality to significantly improve the operational parameters of products. The program involves:

- the analysis of studies on the impact of materials upon the surface layer for improving the operational properties of products, with the study of well-known methods of surface layer modification and deposition of coatings using homogeneous and multicomponent materials. The result of this stage is a broad assessment of the useful effect of the surface layer upon attainment of higher operational parameters. The shortcomings of known methods and the level of their importance for final evaluation of the usefulness of technologies used have also been examined;

- the implementation of the principle of utility in manufacturing high-tech products [4, 5], involving the capacity analysis of enterprises of various forms of ownership, Russian and foreign scientists, universities, scientific and research institutes, design offices and the enterprises of small and medium-sized business. The result of this stage is technical-economical substantiation of the sequence of studies using the principle of utility;

- the rationale for the choice of overall numerical indicators of efficient use of the principle of utility for investigated methods of surface layer modification to attain higher performance indicators of products;

- the elaboration of the methodology for planning the combined technological processes [6, 7] on the basis of the principle of utility, combining several types of impacts which are most efficient for achieving the desired performance indicators of products [8]. Using the standard parts [9, 10], the analysis of expected results is conducted, suggesting the synthesis of compatible impacts among those really existing, being developed, or applied in industry.

The criteria for determining the usefulness of impacts upon the surface layer of products are shown in table 1.

* Corresponding author: enponomarenko@mail.ru

Table 1. The criterial evaluation of the vector of utility for various technological impacts

Types of impacts	Methods of impact	The vector of impacts upon performance characteristics of products (+ - rise, - - fall, 0 – no significant impact , oo – the impact under special conditions)					
		The recovery of initial perfor-mance parame-ters	Heat resistance	Wear resistance	Fatigue strength	Protec-tion from external impacts	Giving new operation properties
Without surface layer modification	Chemical and thermal *	0	0	+	+	+	oo
	Vibro-hardening	0	0	0	+	oo	oo
With removal of:							
- allowance	Mechanical** and electromechanical*** processing	+	0	oo	-	-	+
- coating	Magnetovibrational removal of thick coatings	+	0	oo	0	0	+
With buildup of:							
- a homogeneous layer	Electrospark coating****	0	+	+	-	+	+
	Laser	0	+	+	0	+	+
	Ion-plasma	0	0	+	0	0	+
- a layer with pellet inclusions	Laser	0	+	+	oo	oo	oo
	Ion-plasma	0	+	+	oo	+	oo
	Electrochemical coating*****	oo	+	+	-	0	oo

It presents the broad criterial assessment of the main effects, some of which can alter the vector of impact depending upon application conditions of the method. The table 1 does not show the generally known technological methods (e.g. cementation, nitriding, cyanidation), having fairly full coverage in reference books [6, 7]. Most technological methods and impacts, presented in the table, are at the level of inventions and are being considered for the first time, since their application in perspective products is quite preferable.

When growing a homogeneous layer (table 1), it is possible to obtain new characteristics of materials while preserving or recovering the useful properties, evaluated by the utility coefficient, due to chemical and thermal impacts during coating of materials which do not have a high degree of protection from external factors (particularly, corrosion in aggressive media). It is necessary for engines operating on aggressive fuels.

The inclusion of pellets in coatings (table 1) provides an opportunity to impart their special operational characteristics to the whole product, maintaining the useful properties of the base material. This can be exemplified by deposition of the cast-iron high-temperature coating onto the fusible aluminium alloy, having the lower specific mass but little protection from effects of aggressive media and low wear resistance.

A more detailed presentation of operational characteristics of products with modified surface layer properties is given in [5; 8].

The analysis of the impact of long-term studies of traditional and new technological processes upon the improvement of performance indicators of parts has shown the necessity for use of the available processing methods and their targeted integration into a single technological process, taking account of the operating conditions of parts in products [1, 2]. When modifying the surface layer, one can impart the appropriate combination of operational characteristics to objects, which is not feasible under individual technological impacts, or such processes prove inefficient. The example of such impacts is chemical and electrochemical modification of a material layer, allowing to obtain the parts with new properties, e.g. those using the small weight base made of light alloys with hard cast-iron coating, having high durability and wear resistance, also at elevated operating temperatures. The quantitative estimates of the efficiency of proposed technological methods for standard operations of high-tech product manufacture are listed below.

 DOI: 10.1051/matecconf/201712901077

2 The technological processes of surface layer modification

The analysis of the impact of long-term studies of traditional and new technological processes upon the improvement of performance indicators of parts has shown the necessity for use of the available processing methods and their targeted integration into a single technological process, taking account of the operating conditions of parts in products [1, 2].

When modifying the surface layer, one can impart the appropriate combination of operational characteristics to objects, which is not feasible under individual technological impacts, or such processes prove inefficient. The example of such impacts is chemical and electrochemical modification of a material layer, allowing to obtain the parts with new properties, e.g. those using the small weight base made of light alloys with hard cast-iron coating, having high durability and wear resistance, also at elevated operating temperatures. The quantitative estimates of the efficiency of proposed technological methods for standard operations of high-tech product manufacture are listed below.

The coatings under review meet the highest standards in terms of the principle of utility when using technological processes involving deposition, removal or modification of the surface layer without allowance removal. The purpose of using such technological processes is to impart new operational properties to products, in particular, the manufacture of a part with minimal weight, high anti-friction rates and wear resistance of friction units (i.e. the parts made of aluminium alloy with cast-iron coating). The examples of such parts are drive shafts (e.g. those of the type presented in fig.1). Here, there is a combination of aluminium alloys and steels with coatings ensuring the operation of a unit in strong oxidizers, which is only possible with surface protection of the material.

The greatest interest lies in the deposition of cast-iron coatings onto aluminium parts [8], allowing to obtain the properties of products, never achieved before, namely, minimization of mass along with high durability, protection from destruction and good anti-friction rates.

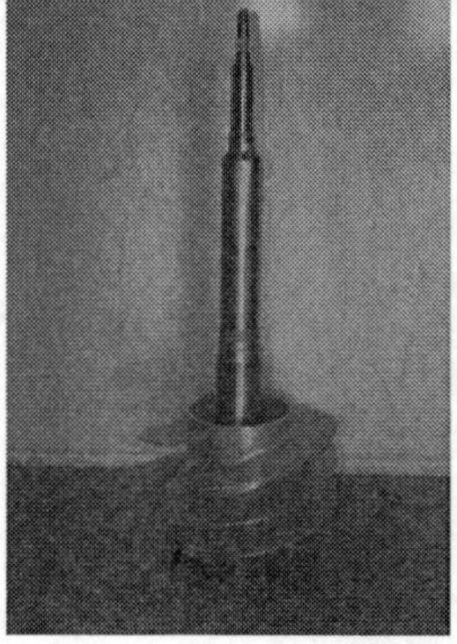

Fig. 1. A screw drive shaft for pumping aggressive media

The process of surface layer modification involving the allowance removal is used in bilateral processing of the working profile of aircraft jet engine blades (fig. 2).

Fig. 2. Working blades of an aircraft engine turbine

After removing the modified layer from blade airfoils (fig.2) and depositing special coatings, it became possible to significantly increase the operational properties of parts and an engine as a whole, which is shown in table 2.

In table 2 $\overrightarrow{K_1} = \frac{\Pi_p}{\Pi_o}$; $\overrightarrow{K_2} = \frac{\Pi_д}{\Pi_o}$, where Π_o – is the performance indicator of a product without modifying the surface layer; Π_p – the post-modification indicator; $\Pi_д$– the indicator after elaboration and application of combined methods of modification, developed by authors.

When using the technological equipment for electrochemical processing, some of its elements are used to supply the electric current to a workpiece and can intensively dissolve, which causes a rapid loss of installation bases and the necessity for repair or replacement of tooling. The complexity of solving the problem consists in the fact that most metal protective coatings are subjected to anodic dissolution, while the deposited dielectric materials do not provide the supply of technological current. A new method of protecting the technological tooling from dissolution has been developed (The Author's Certificate of the USSR № 310772. The method of protecting current-carrying elements made of titanium alloys during electrochemical processing. The Bulletin of Inventions, 1971, № 24). The evaluation indicators of efficiency coefficients are presented above.

Table 2. The improvement of operational properties of turbine blades

The main performance indicator	The method of impacting the surface blade layer	Rates of efficiency		The results of surface layer modification
		$\overrightarrow{K_1}$	$\overrightarrow{K_2}$	
The increase of heat resistance	The removal of a modified layer Special coating	1,5÷2,0	2,0÷3,0	The rise of the operating temperature in front of the turbine up to 200K
The increase of fatigue strength (lifetime)	Mechanical hardening	1,1÷1,2	1,2÷1,3	The increase of the product lifetime by 10-15%

Table 3. The assessment of efficiency of the developed methods of protection from technological tooling dissolution

Object of processing	The method of surface layer protection	Rates of efficiency		The results of surface layer modification
		$\overrightarrow{K_1}$	$\overrightarrow{K_2}$	
The tooling for processing the blades of aircraft engines	The combined processing in developed technological modes	1,2÷1,5	1,5÷2,0	The cost reduction of tooling by 15-20%
The tooling for electrochemical dimensional processing of helicopter blades	The surface layer transformations (the Author's Certificate № 310772)	2,0÷3,0	10÷30	The increase of tooling strength by 80-90 times; the reduction of cost by 18-20 times

3 Conclusion

The discussion on the efficiency of using the principle of utility for forecasting the possibilities of surface layer modification in metals with the purpose of improving the operational properties of durable product parts has shown the following:

- the criterial evaluation based on the principle of utility can underlie the importance and efficiency of using the known and newly developed methods of surface layer modification, as well as planning the processes of coating deposition and removal. The result of this stage is broad assessment of the useful impact of surface layer modification upon the attainment of higher performance parameters;
- the examples of implementing the principle of utility when creating high-tech products have been shown;
- the system of numerical indicators of the principle of utility has been developed for modifying the surface layer and increasing the performance parameters of products.

References

1. E.V. Smolentsev, V.G. Gritsyuk, B.I. Omigov, *The technological complexes for electrical processing methods* (VSTU, Voronezh, 2015)
2. S.V. Safonov, S.N. Grigoriev, V.P. Smolentsev, The Bulletin of Voronezh State Technical University, **2**, 19-26 (2015)
3. V.K. Sikka, S. Viswanathan, C.G. McKaamey, J. Struct. Intermetallics: Champion, 26-30 (1993)
4. A.M. Kadyrmetov, N.F. Kashapov, S.N. Sharifullin, A.I. Saifutdinov, S.A. Fadeev, IOP Conference Series: Materials Science and Engineering **134**, 1-7 (2016)
5. V.V. Ivanov, V.A. Lebedev, I.A. Pinahin, J. of Friction and Wear, **35**, 339-342 (2014)
6. V.I. Butenko, *Structure and properties of the surface layer of tribosystem components* (TTI JuFU, Taganrog, 2012)
7. E.V. Smolentsev, *Planning of electrical and combined processing methods* (Mashinostroenie, Moscow, 2005)
8. S.V. Safonov, V.P. Smolentsev, I.I. Koptev. J. Aircraft Engineering, **4**, 72-75 (2016)

DOI: 10.1051/matecconf/201712901078

Theoretical-probabilistic model of the rotary belt grinding process

Sergey Bratan[1,*], *Aleksandr* Kolesov[1], *Stanislav* Roshchupkin[1], and *Tatyana* Stadnik[1]

[1]Sevastopol State University,33, Universitetskaya St., Sevastopol, 299053, Russia

Abstract. At present, belt grinding is widely used in the leading branches of the national economy. The rotary belt grinding process has a complex stochastic nature, which leads to a dispersion of product quality parameters and does not allow to use all possibilities of this method, due to the lack of an adequate mathematical description of the process. The article is devoted to the development of the theoretical and probabilistic model of the rotary belt grinding process.

1 Introduction

The transition to market relations in the economy put forward the problems associated with the production of competitive products in engineering, car manufacturing, aircraft manufacturing and other industries. "The consumers need products that meet their needs and expectations" (ISO/CPM 2 9000: 2015). A high level of requirements for modern machines and devices has caused a number of problems associated with the creation of high-performance technological processes that ensure the manufacture of parts with specified quality parameters. The quality parameters determine the functional suitability of the parts and their performance characteristics. Such parameters are formed to the full at the final stages, including grinding and finishing processes. When manufacturing a wide range of parts for machines and devices, it is necessary, on the one hand, to maintain high demands about the quality of their geometric dimensions, roughness and the physical and mechanical state of the surface layer; on the other hand, to increase the productivity of their output according to market requirements. The task is not only to improve the accuracy of manufacturing parts in general, but also to ensure the stability of their parameters. The research in the field of creating technological processes for shaping the working surfaces of precision parts will improve the quality of parts processing, and hence the operational properties of the entire product.

One of the most progressive methods for product processing is grinding with an abrasive belt that moves at a high speed of up to 50 m/s [3].

An abrasive belt grinding is widely used in the aircraft-, car- and shipbuilding, in the bearing and woodworking industries, as well as in other branches of the national economy at all stages of the technological process of manufacturing parts: when roughing with the removal of large allowances, when finishing workpieces with low roughness, the required degree of accuracy and quality of the treated surface [4].

The abrasive belt grinding has a number of significant advantages in comparison with the abrasive wheel treatment. Unlike the abrasive wheels with rigid fastening of the cutting grains, the flexible belts make it possible to vary the contact area between the tool and the workpiece, which helps to control the magnitude and character of the contact pressure distribution, the number of cutting elements in the cutting zone in accordance with the physical and mechanical properties of the workpiece and the technical conditions of its processing [5].

In comparison with the abrasive wheel grinding, the belt grinding process is characterized by a higher productivity and a better surface quality of the processed part. The belt grinding provides the roughness *Ra* 0,63 ÷ 0,08.

In general, the belt grinding has the following advantages over the abrasive wheel treatment: high grinding performance; greater contact between the belt and the workpiece; a lower contact temperature during grinding; short preparatory-final time. The belt grinding equipment is simple in design and economical in operation. These circumstances make the advantages of abrasive belts undeniable in comparison with the abrasive wheel grinding in finishing stages when processing complex, dimensional and lengthy parts, rolled metal and drawing products [4, 5].

The disadvantages of the abrasive belt grinding: less dimensional accuracy and geometric shape during processing; difficulties in processing sharp ridges on parts; relatively low abrasive tape resistance and short service life due to inability to straighten (in most cases); belt extension during operation.

One of the most common methods of belt grinding is grinding with a free strand of the abrasive belt, where the abrasive belt comes into contact with a workpiece on the free strand, resulting in an increase of the contact area of

* Corresponding author: bratan@sevsu.ru

DOI: 10.1051/matecconf/201712901078

the belt with the workpiece due to a larger angle of coverage. This method is used for grinding shaped parts, internal and external surfaces, long pipes and slender parts. For processing axisymmetric nonrigid long parts when the rotation of the workpiece is impossible or difficult, the most optimal scheme, from the perspective of achieving surface roughness values $R_a = 0,63$ and below, is the rotary belt grinding method [3-5].

The processing method (Figure 1) is realized due to a combination of movements of the abrasive belt moving at a speed V_l, rotation of the faceplate 1 at a speed V_g, providing a circular feed. The longitudinal feed during one direction grinding is provided by moving the workpiece 5 along its axis, and it is absent from the plunge grinding pattern.

The abrasive tape 6 is located on three rollers – the leading roller 3, the spreading roller 4 and the driven roller 2.

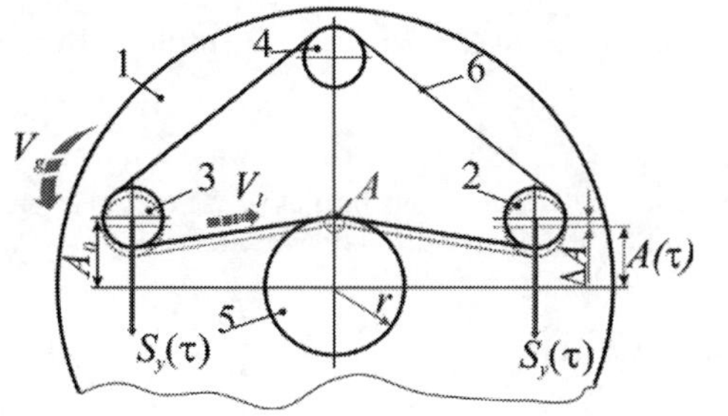

Fig. 1. Pattern of the rotary belt grinding process with two movable rollers. 1 – faceplate; 2 – driven roller; 3 – driving roller; 4 - spreading roller; 5 – workpiece; 6 - abrasive belt.

Largely, the rotary grinding makes it possible to provide the required technological parameters of the surfaces to be treated, however, the results are unstable and the shaping process is difficult to control. To stabilize the quality parameters, the technological modes are assigned based on the technologist's experience or the recommendations given in the reference materials. When predicting the state of a technological system (TS), the empirical models of the manufacturing procedure (MP) flow, obtained by the methods of multifactorial experiment planning, are used. The error of such models in some cases reaches 100 ... 300%, which does not allow to use them for calculating the cutting modes for rotary belt grinding. At the same time, the process of belt grinding has a complex stochastic nature, which leads to a dispersion of product quality parameters and does not allow to use all possibilities of this method, due to the lack of an adequate mathematical description of the process.

Ensuring the quality and efficiency of the rotary belt grinding operation is possible only on the basis of the development of stochastic models, taking into account the probabilistic nature of the process. A comprehensive solution to these problems is required, and the totality of these problems is an unresolved task to date.

Very often in the simulation, the operation is considered as a dynamic system [6]. One of the possible schemes for the decomposition of an operation as a dynamical system is considered in [7]. According to the functional features, operation is divided into subsystems of the machine, fixtures, tools, workpieces, lubricating and cooling manufacturing agents. Each of the subsystems has its own set of properties, state parameters, development history, the vector of input and output variables, and the vector of perturbing influences.

The central subsystem is the contact area of the workpiece and the tool. The area state parameters include size, shape, chip formation process, material removal rate, wear and abrasive material destruction. The dimensions of the area are in direct proportion with the state of the surface, the dimensions and the spatial arrangement of the abrasive belt and a workpiece. Due to changes in size, spatial location and state of the contacting surfaces over time, the contact area state parameters will change and, as a consequence, the output process parameters [6].

Therefore, the aim of this work is to establish the relationship between the processing modes with the current parameters of the contact area and the output quality parameters for the rotary belt grinding and to develop transient stochastic process models.

2 Development of the mathematical model of the rotary belt grinding process

To obtain a mathematical model that allows calculating the output quality parameters, let us consider the interaction process between the abrasive tool and the workpiece at the microlevel in rotary belt grinding (Fig. 2).

The probability of material removal taking into account the initial roughness is determined by the following formula [1]:

$$P(M) = 1 - e^{-a(y) - a(y,\tau)},$$

where $a(y)$ is the indicator determining the material removal probability at level y before the surface comes into contact with the workpiece and the belt; $a(y),\tau$ is the parameter determining the change in the material removal probability at level y in the contact area.

In the grinding process the material is removed in discrete portions, but the number of single sections passing through the cross-section is quite large (from several hundreds to several thousands per section length unit), which allows, when calculating the probability, to move from a discrete model to a continuous one.

DOI: 10.1051/matecconf/201712901078

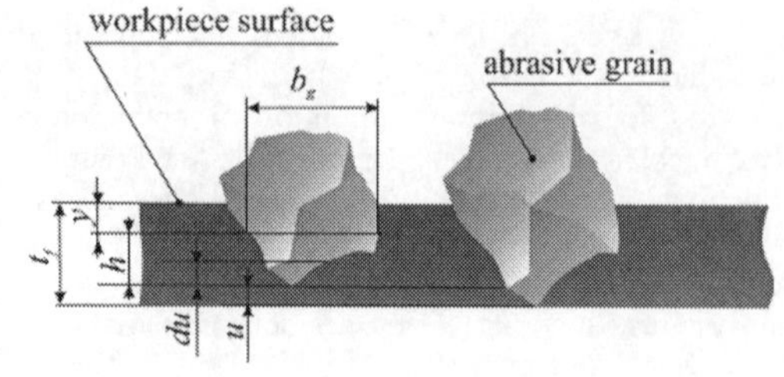

Fig. 2. Interaction pattern of an abrasive tool with a working surface at micro level.

When modeling the grain profile, in the simplest case, the rotation paraboloid determines the width of the vertex profile b_g at level h as follows:

$$b_g = 2\sqrt{2\rho_g h}\ ,$$

where ρ_g is the radius of rounding at the grain top.

The width of the vertex contour corresponding to the given level, in the simulation by a power law, is calculated from the following equation

$$b_g = C_b h^m = C_b[t_f - y - u]^m\ ,$$

where C_b and m are the coefficient and the exponent (determined experimentally).

With time $\Delta\tau$ a section is rotated at the angle $\Delta\varphi$ and through it a section with the arc length of $V_k\Delta\tau$ passes. Out of the total number of grains having passed through the section, the profile width b_g will have the grains whose vertices are located in the layer $1\Delta u V_k \Delta\tau$ of the belt. The number of such vertices is calculated according to their distribution density along the depth of the tool $f(u)$.

The number of grains having passed through section y within the time $\Delta\tau$:

$$\Delta\lambda = V_k n_g f(u)\Delta u \Delta\tau\ ,$$

where n_g is the number of grains per unit volume of the tool working layer; V_k is the circumferential speed of the tool; $f(u)$ is the distribution of the tool working grains in depth, which can be approximated by various dependencies, for example, by the power law [6]:

$$f(u) = \frac{(\chi + m - 0.5)}{H_u^{\chi+m-0.5}} u^{\chi-1}\ ,$$

where χ and m are parameters of the rotation paraboloid.

The change in the parameter $a(y,\tau)$ is determined by the increment in the sum of the transverse dimensions of the abrasive grain profiles:

$$\Delta a(y,\tau) = K_c n_g b_g f(u)\Delta u \Delta\tau\ ,$$

where K_c is the chip formation coefficient, taking into account the fact that not all the material is removed from inside a machining mark, and some of it is displaced and forms marks along the edges.

We obtain the integral equation for calculating the parameter $a(y,t)$ in the contact area

$$a(y,\tau) = V_k K_c n_g \int_{t_0}^{t} \int_{0}^{t_f - y} b_g f(u) du d\tau, \qquad (1)$$

After integration with respect to u and τ of dependence (1), we obtain:

$$a(t) = \frac{0{,}78\pi K_c \sqrt{2\rho_g} V_k n_g S_y^{\,2}(t-t_0)^3}{2H_u^{1,5}}\ .$$

Given that for the outer surface of the part $(y=0)$, $P(\overline{M}) = \beta_0$ when multiplying $P_{1,2}(\overline{M}) = P_1(\overline{M}) \cdot P_2(\overline{M})),$ we obtain the dependences for the probabilities of the material removal failure for the plunge grinding pattern:

$$1 - \beta_0 = exp\left\{ -a_0 - \frac{0{,}78\pi K_c \sqrt{2\rho_g} V_k n_g S_y^{\,2}(t-t_0)^3}{2H_u^{1,5}} \right\} \qquad (2)$$

Further, after taking the logarithm of the equation (2), we obtain:

$$\ln\frac{1}{1-\beta_0} + a_0 = -\frac{0{,}78\pi K_c \sqrt{2\rho_g} V_k n_g S_y^{\,2}(t-t_0)^3}{2H_u^{1,5}} \qquad (3)$$

The obtained formula (3) allows to predict not only the radial metal removal, but also the output quality parameters for the rotary belt grinding.

While going through the contact area surface of the part and the belt, the levels will shift to the center of the part. In this case, the material removal value is numerically equal to the level shift with the removal probability $P(V)=\beta_M$, and is calculated according to the following formula:

$$P(M) = 1 - exp(-\frac{\Theta_M}{S_y}[S_y(t-t_0)])^3 = 1 - P(\overline{M})\ , \qquad (4)$$

where $\Theta_M = \dfrac{0{,}78\pi K_c \sqrt{2\rho_g} V_k n_g}{2H_u^{1,5}}$.

When the grains continuously come into contact with the workpiece material, for particular values $\beta_M = 0{,}9973$, and $t_f = u$ (4) can be rewritten as follows:

DOI: 10.1051/matecconf/201712901078

$$ln0{,}0027 = -\frac{\Theta_M t_f^{\ k}}{S_y}. \quad (5)$$

Based on formula (5), we obtain an equation that allows one to calculate the value of the layer in which the surface roughness is distributed:

$$H = [-\frac{ln0{,}0027S_y}{\Theta_M}]^{\frac{1}{k}}.$$

The value of the layer in which the surface roughness is distributed in rotary belt grinding is numerically equal to the actual infeed depth and is directly proportional to the tool feed.

For a steady grinding process where a particular value of the coefficient $k = 3$, the value of the layer in which the roughness is distributed is calculated according to the function:

$$H = [\frac{47{,}31S_y H_u^{\ 1.5}}{\pi K_C n_g V_k \sqrt{2\rho_g}}]^{\frac{1}{3}}. \quad (6)$$

The temperature deformation value can be taken into account by the coefficient.

Function (6) makes it possible to obtain an equation for the material removal calculation

$$Q_\Sigma = t_f - [\frac{47{,}31S_y H_u^{\ 1.5}}{\pi K_C n_g V_k \sqrt{2\rho_g}}]^{\frac{1}{3}}.$$

Assuming that the profile of the ground surface can be a stationary random process with a normal ordinate distribution, we obtain an equation for estimating the values of the roughness altitude parameters.

The average arithmetic deviation of the profile R_a can be calculated (see Figure 3) taking into consideration the area of the protrusions F_1 and the area of the depressions F_2 measured from the profile midline.

As

$$dF_1 = t_p l dy_m \text{ and } dF_2 = (1-t_p)l dy_m,$$

so

$$R_a = \int_0^\infty t_p dy_m + \int_\infty^0 (1-t_p)dy_m,$$

where l is the length of the base section, mm; y_m is the distance from the points of the profile to its midline.

Calculation of R_a according to this function is possible when using numerical methods of integration.

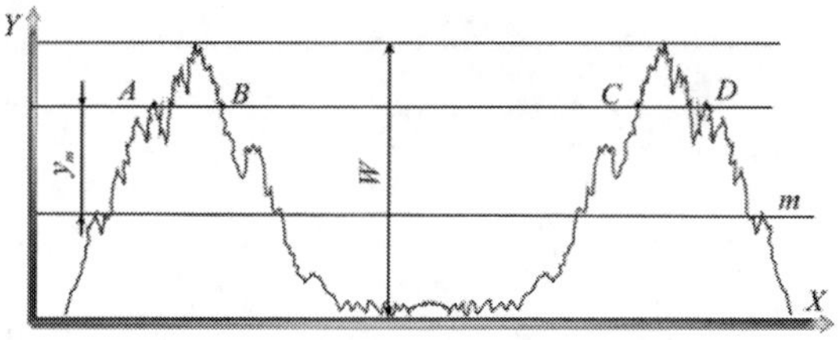

Fig. 3. Profilogram for calculating the distribution function of the profile ordinates of the treated surface.

The function for calculating R_a is simplified if we use the law of stationarity and normality of the random process, considered in this work, describing the profile ordinates.

For such a process, the mathematical expectation of the mean arithmetic deviation of the profile is proportional to the mean square deviation σ_{y_m} of the random ordinates y_m of the profile points [1].

$$M(R_a) = \sqrt{\frac{2}{\pi}K(0)} = \sqrt{\frac{2}{\pi}}\sigma_{ym}, \quad (7)$$

where $K(0)$ is the value of the correlation function for an interval $x_2 - x_1 = 0$ is equal to the variance of the random variable y_m.

To determine the root-mean-square deviation σ_{ym}, let us calculate the function $F_{Y_m}(Y_m)$ and the probability density $f_{Y_m}(Y_m)$ of distances Y_m from the midline to the points of the profile ground surface.

$$F_{ym}(y_m) = P(y < y_m) = \lim_{l \to \infty}\frac{1-\sum b_{mi}}{l} = P(M),$$

$$f(y_m) = \frac{3\Theta_M}{S_y}(W_m + y_m)^2\left\{exp\left[-\frac{\Theta_M(W_m + y_m)^3}{S_y}\right]\right\} \quad (8)$$

where W_m is the distance from the deepest depression to the profile midline.

For a normal process where $Y_m = 0$, $P(M) = 0{,}5$ [88]:

$$f(y_m) = \frac{1}{\sigma_y\sqrt{2\pi}}.$$

On basis of this, the second part of the equation (8), separated by curly braces, is equal to 0.5, and the average arithmetic deviation of the profile, taking into account the equation (7), is defined as:

$$R_a = \frac{2S_y}{\pi\Theta_M 3W_m^{\ 2}}.$$

Thus, when determining R_a one must first find the value W_m. In the general case, it is defined as the solution to the equation:

$$\Theta_M \sum_{i=o}^{n} (W_m - \sum_{i=1}^{i} \Delta Q_i)^{2.5} = \ln 2\,. \quad (9)$$

The solution of the equation (9) is possible only on the basis of numerical methods application, and this greatly complicates the calculation of the surface roughness parameters.

Let us note that formula (7) has a simple analytical form and only the calculation of W_m requires the use of numerical methods.

In [6], the function for approximate calculations of the profile arithmetic mean deviation is recommended:

$$W_m = \frac{H}{2}. \quad (10)$$

Then the function for R_a calculation takes the form of

$$R_a = \frac{2S_y}{3\pi\Theta_M \left(H/2\right)^2} \quad (11)$$

After plugging (6) into (11) we obtain:

$$R_a = \frac{S_y^{\frac{1}{3}}}{6\pi\Theta_M Z^2}, \quad (12)$$

where $Z = \left[\frac{28{,}895 H_u^{1.5}}{\pi K_C n_g V_k \sqrt{2\rho_g}}\right]^{\frac{1}{3}}$.

Equation (12) reflects the physical nature of the surface formation process and corresponds to the basic principles of the dimensional theory.

The complex parameter that combines the altitude and step parameters of the roughness is the relative reference length of the profile. By definition, this parameter coincides with the probability of metal removal failure. When $t_f = const$ the formula changes as follows:

$$t_{pj} = exp\left[-\frac{AK_C n_g V_k \sqrt{2\rho_g} S_y^{\,2} t^3}{8H_u^{1.5}}\right].$$

In addition to roughness, the depth of the defective layer is related to the surface quality parameters:

$$h_d = \alpha\xi(q - q_O)\left(\frac{h_q}{V_l}\right)^{\beta}, \quad (13)$$

where q_O is the defect-free level of the heat flux density; q is the heat flow; h_q is the half-width of the heat source; β is the coefficient determined by the nature of heat release (β = 0.8, ..., 0.9); α is the coefficient, depending on the thermal properties of the belt and the workpiece; ξ is the coefficient that takes into account the influence of the electric field on the depth of the defective layer.

The half-width of the heat source

$$h_q = 0{,}5\frac{\pi \cdot R_w \cdot \varepsilon}{180}, \quad (14)$$

where R_w is the workpiece radius; ε is the coverage angle of the workpiece with the abrasive belt.

The coverage angle of the workpiece with the abrasive belt (Figure 4):

$$\varepsilon = 2 \cdot \arccos\left[\frac{(R_1 + R_n)x_1 + y_1 \cdot \sqrt{y_1^{\,2} + x_1^{\,2} - (R_1 + R_w)^2}}{y_1^{\,2} + x_1^{\,2}}\right].$$

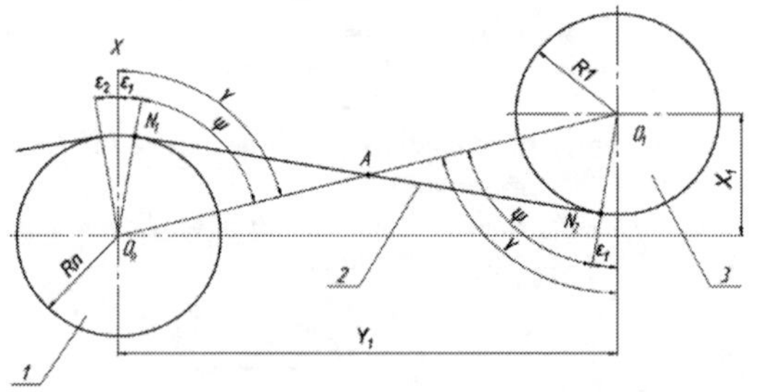

Fig. 4. Scheme for calculation of the coverage angle of the workpiece with the abrasive belt.

Plugging (14) into (13) we obtain a function for calculating the depth of the defect layer:

$$h_d = \alpha\xi(q - q_O)\left(\frac{0{,}5\pi \cdot R_w \cdot (\varepsilon/180)}{V_l}\right)^{\beta}.$$

3 Conclusions

A stochastic model of the rotary belt grinding process is developed. It takes into account the probabilistic nature of the process and allows to calculate the output quality parameters bearing in mind the changes in the system properties over time.

References

1. Yu. Novoselov, *Dynamics of surface formation in abrasive processing* (Publishing House: LAP LAMBERT AcademicPublishing, 2017).
2. S. Korchak, *Productivity of the grinding process of steel parts* (Car Manufacturing, Moscow, 1974)
3. S. Shilyaev, A. Ivanov, *Analysis of constructive and technological parameters of the rotary belt grinding device on basis of studies of the relative movement*

DOI: 10.1051/matecconf/201712901078

dynamics of an abrasive tape, Science and Education: scientific edition of the Bauman MSTU, **6** (2011)

4. A. Koryazhkin, D. Volkov, *Stochastic model of the abrasive belt wear for grinding the compressor blades of the gas turbine engine*, Vestnik RGATA, **2** (2011)
5. K. Chubukov, S. Shilyaev, *Development of a forming pattern for rotary belt grinding,* Topical issues of machine tool construction: materials of the All-Russian Scientific and Practical Conference. Izhevsk: Innova (2013)
6. S. Bratan, *Technological fundamentals of quality maintenance and stability improvement of high-performance finishing and fine grinding*, Thesis of Dr. of Sciences (Odessa, 2006)
7. S. Bratan, *Quality maintenance and stability improvement of finishing and fine grinding,* Cutting and tools in technological systems. Int. Scient.-Tech. Collection, **68**. (2005)
8. M. Jackson, *Machining with Abrasives* (Springer, New York, 2011)

DOI: 10.1051/matecconf/201712901079

Identification of removal parameters at combined grinding of conductive ceramic materials

Sergey Bratan[1,*], *Stanislav* Roshchupkin[1], *Alexandr* Kolesov[1], and *Boris* Bogutsky[1]

[1]Sevastopol State University,33, Universitetskaya St., Sevastopol, 299053, Russia

Abstract. The models of combined grinding for processing of conductive ceramic materials which allows to calculate the material removal and to optimize the machining process on the automated equipment were obtain. The models take into account that process has a stochastic nature and that material is removed by a combination of microcutting-chipping, brittle volumetric fracture and erosive material removal.

At the finishing of conductive brittle materials, for example, specialized ceramics and conductive ceramic coatings, combined grinding methods are used, in which the material is removed by a combination of microcutting-chipping, brittle volumetric fracture and erosive material removal.

Automation of such processes in modern conditions is possible in the presence of rational operation control cycles. Their development by empirical means does not allow to widely implement the processes of combined grinding due to the large volume of preliminary studies.

Creating a mathematical model of the machining process makes it possible to automate the design of the finish operation and optimize it according to the required criterion.

To obtain a mathematical model that allows to calculate the removal of material in the presence of microcutting-chipping, brittle volumetric fracture and erosive material removal, the process of interaction of the tool with the workpiece at the micro level is considered. Diamond grains are randomly distributed in the volume of the grinding wheel, so the process of their interaction with the material being processed is random [1,2].

The removal of material as a result of erosion and abrasive work is a random process, which can be characterized by a complete probability of material removal. For any point of the workpiece surface, one can consider a random event consisting in the joint failure of this point by erosion, microcutting-chipping and brittle volume destruction, then the total probability of removal is determined by:

$$P(\overline{M}) = P_1(\overline{M}) \cdot P_2(\overline{M}) \cdot P_3(\overline{M}), \qquad (1)$$

where $P_1(\overline{M})$ - is the probability that the material has not been removed due to micro-cutting; $P_2(\overline{M})$ - the probability that the material is not removed due to volumetric brittle fracture, $P_3(\overline{M})$ - the probability of erosion of the material. The probability of material failure at a time due to microcutting is determined by the dependence [2]

$$P_1(\overline{M}) = \exp\left[-a_0 - \int_{t_o}^{t} a_{1i}(y,\tau)d\tau \right], \qquad (2)$$

where y - distance from the outer surface of the workpiece to the considered level. The indicator a_1 for each pass is given by

$$\Delta a_1(y,\tau) = \kappa_c b_з(y,\tau)\Delta\lambda \,,$$

where k_c - chip formation ratio; $b_з(y)$ - grain width at the level y; $\lambda\Delta$ - the number of diamond grains that have passed through a unit cross section.

For conditions where brittle materials (various types of ceramics) are ground, k_c = 1. In the process of grinding, a certain number of grains perform a microcutting of the surface of the workpiece, and the other part of the grains immediately cleaves the material. Therefore, when calculating the indicator a_1 it is necessary to take into account only those grains that can cut the material, i.e. from the total number of grains it is necessary to subtract the grains that produce the spalling.

$$\Delta a_1(y,\tau) = b_з(y,\tau)\Delta\lambda(1 - P_{c\kappa}) \,, \qquad (3)$$

where P_{ck} - is the probability of volumetric brittle fracture of the workpiece material when interacting with the grain of the tool.

When the grain profile is approximated by a power law (in the simplest case, a paraboloid of revolution), we obtain

* Corresponding author: bratan@sevsu.ru

$$b_з(y,\tau) = C_b(t_ф - y - u)^m, \quad (4)$$

where C_b, m - are the coefficients of the grain shape; $t_ф$ the actual depth of microcutting; u - the distance from the conditional outer surface of the tool to the top of the grain.

Through a unit section 1·Δ u in time Δτ passes the Δλ grain of the tool

$$\Delta\lambda = V_к n_з f(u)dud\tau, \quad (5)$$

where $n_з$ - the number of grains per unit volume of the working layer of the tool; V_k - tool feed speed.

In turn, the distribution function of the number of working grains of the tool along the coordinate can be approximated by the dependence [2]

$$f(u) = \frac{\chi}{H_u^{\chi}} u^{\chi - 1},$$

where H_u the value of the layer of the working surface of the circle within which it is counted $n_з$.

After substituting (5) and (6) in (4) and integrating with respect to u and τ, it follows that (6):

$$a_1(y,\tau) = n_з V_к к_c \int_{t_0}^{t} \int_{0}^{t_ф - y} b_з f(u)(1 - P_{ск})dud\tau. \quad (6)$$

The probability of spalling P_{ck} is determined by the mode of grinding and the location of the grain in the volume of the tool. For its calculation can be used. the dependences of A. Korolyov [1], obtained by analyzing the process of dressing the abrasive tool and allowing to determine the probability of volumetric destruction of the grinding wheel grains. The process of dressing abrasive wheels in a certain sense is similar to the process of diamond grinding. However, these dependencies are complex for practical use. Therefore, it is desirable to replace them with approximations so that an analytic calculation of the integral (6) becomes possible. In this capacity, the dependence

$$P_{ск} = P_0\left[1 - \left(\frac{u}{t_ф}\right)^{\beta}\right]. \quad (7)$$

For the level $u = 0$ и $P_{ck} = P_0$, where P_0 it is determined by grinding modes and tool characteristics. For the level $u = t_ф$ (the outer surface of the workpiece) $P_{ck} = 0$. The coefficients P_0 and for specific grinding conditions can be calculated from the dependences [1]. for specific grinding conditions can be calculated from the dependences [1]. For this, it is sufficient to calculate only two values P_{ck} and u. Substituting (7) into (6), we obtain

$$a_1(y,\tau) = n_з V_к к_c \cdot \int_{t_0}^{t} \int_{0}^{t_ф - y} C_b(t_ф - y - u)^m \frac{\chi}{H_u^{\chi}} u^{\chi-1}\left[1 - P_0 + P_0\left(\frac{u}{t_ф}\right)^{\beta}\right] dud\tau \quad (8)$$

The integral (8) is calculated for $\tau = t_0$:

$$a_1(y,\tau) = n_з V_к к_c \cdot \int_{t_0}^{t} \int_{0}^{t_ф - y} \frac{C_b(t_ф - y - u)^m \chi u^{\chi-1}}{H_u^{\chi}} \cdot$$

$$\cdot(1 - P_0)dud\tau + +n_з V_к к_c \int_{t_0}^{t} \int_{0}^{t_ф - y} \frac{C_b(t_ф - y - u)\chi u^{\chi-1}}{H_u^{\chi}} \cdot \quad (9)$$

$$\cdot P_0\left(\frac{u}{t_ф}\right)^{\beta} dud\tau =$$

$$= \frac{C_b n_з V_к к_c \Gamma(m+1)\Gamma(\chi)(t - t_0)(t_ф - t_0)^{m+\chi}}{H_u^{\chi}\Gamma(m+\chi+1)(m+\chi+1)} \cdot$$

$$\cdot(1 - P_0) + C_b n_з V_к к_c \Gamma(m+1)\Gamma(\chi+\beta)P_0 \cdot$$

$$\cdot \frac{(t - t_0)(t_ф - t_0)^{m+\chi+\beta}}{t_ф^{\beta} H_u^{\chi}\Gamma(m+\chi+\beta)(m+\chi+\beta+1)}.$$

Taking into account the notation:

$$A = \frac{\Gamma(m+1)\Gamma(\chi)}{\Gamma(m+\chi+1)(m+\chi+1)}(1 - P_0);$$

$$B - \frac{\Gamma(m+1)\Gamma(\chi+\beta)}{\Gamma(m+\chi+\beta+1)(m+\chi+\beta+1)} P_0;$$

$C = \dfrac{C_b n_з V_к к_c}{H_u^{\chi}}$; (9) takes the form

$$a(y,\tau) = CA(t - t_0)(t_ф - y)^{m+\chi}(1 - P_0) +$$

$$+ \frac{CB}{t_ф^{\beta}}(t - t_0)(t_ф - y)^{m+\chi+\beta} P_0 = \quad (10)$$

$$= C(t - t_0)(t_ф - y)^{m+\chi}\left[A + \frac{B}{t_ф^{\beta}}(t_ф - y)^{\beta}\right].$$

After each time interval $\Delta\tau$, the value y changes by the amount of removal Δh

$$y(\tau_1) = y + \Delta h$$
$$y(\tau_2) = y + 2\Delta h$$
$$y(\tau_3) = y + 3\Delta h$$

....

Hence, taking (2) into account, we obtain (11)

$$P_1(\overline{M}) = \exp\left(\begin{array}{c} -\int\limits_{t_0}^{t} C(t-t_0)(t_\phi - y - \tau\Delta h)^{m+\chi} \cdot \\ \left(A + \frac{B}{t_\phi{}^\beta}(t_\phi - y - \tau\Delta h)^\beta \right) d\tau \end{array} \right), \quad (11)$$

where τ - is the current time.

After integrating over τ, expression (11) takes the form (12):

$$P_1(\overline{M}) = \exp\left(-G_1 \frac{(t_\phi - y - t\Delta h)^{m_1} - (t_\phi - y - t_0\Delta h)^{m_1}}{(t_\phi - y)m_1} + \right.$$
$$\left. + L_1 \frac{(t_\phi - y - t\Delta h)^{m_2} - (t_\phi - y - t_0\Delta h)^{m_2}}{(t_\phi - y)m_2} \right), \quad (12)$$

where $G_1 = CA; L_1 = CB,$ $m_1 = m + \chi + 1,$ $m_2 = m + \chi + \beta + 1.$

The probability of not removing material due to volumetric brittle fracture will be written as

$$P_2(\overline{M}) = \exp(-a_0 - \int\limits_{t_0}^{t} a_2(y,\tau) d\tau) .$$

For the case under consideration, the width of a single risk b_x, as a result of the interaction between the grain of the instrument and the ceramic part, is greater than its actual width $b_з$.

For the approximation b_x, we use the power-law dependence

$$b_x(y,\tau) = C_{bx}(t_\phi - y - u + \Delta h_x)^{m_x} ,$$

where Δh_x - is the value of the increment of material removal during the brittle fracture of the ceramic.

In this case, the exponent a_2 is equal to

$$\Delta a_2(y,\tau) = b_x(y,\tau)\Delta\lambda P_{ск} .$$

Transformations similar to the derivation of dependencies for $a_1(y)$, will lead to the dependence

$$P_2(\overline{M}) = \exp\left[-a_0 - \int\limits_{t_0}^{t} \begin{array}{c} c(t-t_0)(t_\phi + \Delta h_x - y - \tau\Delta h)^{m+\chi} \cdot \\ \cdot \left(A_2 + \frac{B}{t_\phi{}^\beta}(t_\phi + \Delta h_x - y - \tau\Delta h)^\beta \right) d\tau \end{array} \right] \quad (13)$$

Where $A_2 = \dfrac{\Gamma(m+1)\Gamma(\chi)}{\Gamma(m+\chi+1)(m+\chi+1)} P_0.$

After integrating over τ, expression (13) takes the form

$$P_2(\overline{M}) = \exp\left(-G_2 \frac{(t_\phi - y + \Delta h_x - t\Delta h)^{m_1}}{(t_\phi - y)m_1} - \right.$$
$$- \frac{(t_\phi - y + \Delta h_x - t_0\Delta h)^{m_1}}{(t_\phi - y)m_1} + \quad (14)$$
$$\left. + L_1 \frac{(t_\phi - y + \Delta h_x - t\Delta h)^{m_2} - (t_\phi - y + \Delta h_x - t_0\Delta h)^{m_2}}{t_\phi{}^\beta (t_\phi - y)m_2} \right),$$

where $G_2 = CA_2$.

To describe the probability of erosion failure, let us consider the processes taking place in the presence of erosion discharges in the processing zone.

Traces from erosion discharges can form on the surface of the billet (Fig. 1).

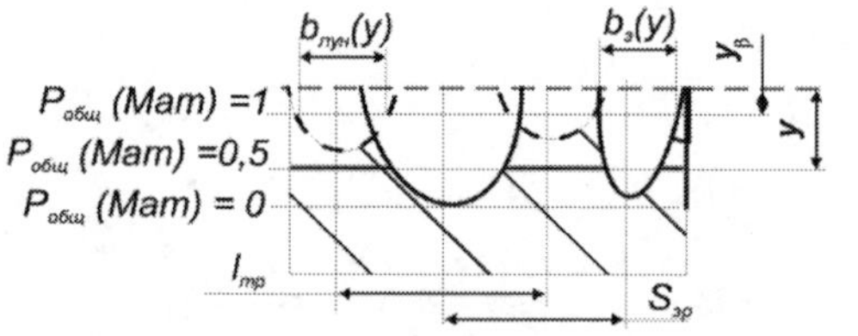

Fig.1. Calculation scheme

If two adjacent wells overlap, the material between them will be completely removed. This condition for any level y can be written:

$$0 < S_{эр} \le 0.5[b_{эр1}(y) + b_{эр2}(y)], \quad (15)$$

Where $S_{эр}$ - is the distance between two adjacent holes on the surface in the section under consideration, $b_{эр1}(y), b_{эр2}(y)$ - is the width of two neighboring lunar holes at the level y.

If the width of the holes at the level y is equal, then the material removal condition can be rewritten as follows: $0 < S_{эр} \le b_{эр}(y)$.

Obviously, the probability of erosive removal of the workpiece $P_3(y)$ at the level y will be equal to the probability that the random variable Sэр falls on the segment from 0 to $b_{эр}(y)$. If the density of the probability distribution of a continuous quantity $S_{эр}$ is denoted by *f(Sэр)*, then the probability of removing the material $P_3(y)$:

$$P_3(y) = P(0 < S_{эр} \le b_{эр}(y)) = \int\limits_{0}^{b_{эр}(y)} f(S_{эр}) dS_{эр} \quad (16)$$

Therefore, the appearance of holes on the surface of the workpiece with a distance Sэр after passing through a single section of the circle can be regarded as a stationary ordinary flow with a limited aftereffect. Since a large number of single tool surfaces pass through each section of the workpiece, the stream of events Sэр in the section under consideration is the sum of a large number of stationary, ordinary flows with a limited aftereffect.

From the theory of queuing it follows that the flow Sэр can be assumed arbitrarily close to the simplest one, and the distribution density

$$f(S_{эр}) = \lambda \cdot e^{-\lambda S_{эр}}$$

Then we get an expression for the probability of material removal:

$$P_3(y) = \int_0^{b_{эр}(y)} \lambda_{лун} \cdot e^{-\lambda_{лун} S_{эр}} dS_{эр} = 1 - e^{-\lambda b_{эр}(y)} = 1 - e^{-a_3(y)} \quad (17)$$

where $\lambda_{лун}$ - the average number of holes per unit length of the section of the workpiece.

Thus, the probability of erosion of the material at any of the considered levels y is determined by the width of the generating holes and their number per unit length of the cross section.

$$a_3(y) = \lambda_{лун} b_{эр}(y). \quad (18)$$

The shape of the hole is usually modeled by a spherical segment with a diameter at the surface - $d_л$ and depth - $h_л$ [3]. The surface on which the holes are formed is constantly shifted deep into the workpiece, because the workpiece material is removed. The width of the well bэр at the level $y - y_\beta$ from the surface of the occurrence of the discharge (Fig. 1) is determined from the geometric analysis of the spherical segment:

$$\left.\begin{aligned} & b_{эр}(y) = 0, && at\ y_\beta > y > y_\beta + h_л \\ & b_{эр}(y) = R\sqrt{h_л - y_{эр}}, && y_\beta \le y \le y_\beta + h_л \end{aligned}\right\}, \quad (19)$$

where R - is the index that determines the diameter of the well.

Since erosion discharges occur at the places where chipped particles of the material (shavings) are formed [3], and the share of random discharges in the total volume is negligible and can be neglected, the number of emerging holes depends directly on the number of cutting grains and on the structure of the tool.

To determine the probability of erosion destruction of the workpiece Δz when grinding the periphery of the circle, consider the section of the part in the contact zone.

During the time $\Delta\tau$ the section of the workpiece is displaced by a distance and through it passes a section of the circle with the length of the arc, $(V_к \pm V_и)\Delta\tau$, where $V_к$ - is the circumferential speed of the circle, $V_и$ - the speed of the workpiece. Of the total number of cutting elements on the surface of the tool, chips can form grains whose vertices are located in the layer of the circle in height $u \le t_ф - y_\beta$, where $t_ф$ - is the actual depth of cutting.

The number of chips $\Delta\lambda$ formed by the elementary layer of the tool is calculated from the grain distribution density *f3(u)* (5), the generated chips create a high local electric field strength and thereby cause the erosion discharge. However, studies show that not all chips can cause breakdown of the interelectrode gap. This is due to the fact that there is a pause between two digits at one point in space τ_n for the duration of the deionization of the gap [3]. The length of the necessary pause is determined by the medium filling the MEP.

Thus, the number of discharges occurring in the contact zone from the action of the elementary surface of the instrument, we determine:

$$\Delta N_{эр} = K_{эр}\Delta\lambda \quad (20)$$

where $K_{эр}$ - is a coefficient that takes into account the number of chips that do not produce a discharge.

Change the parameter $a_3(y,\tau)$ as the increment of the sum of the transverse dimensions of the holes (18) and calculate:

$$\begin{aligned} \Delta a_3(y,\tau) &= k_c b_{эр}(y)\Delta N_{эр} = \\ &= k_c b_{эр}(y) n_з f_з(u)\Delta u(V_к \pm V_и)\Delta\tau \end{aligned} \quad (21)$$

Replacing τ by $\frac{z}{V_u}$ calculate the value of the indicator $a_{эр}(y,\tau)$ at the point z of contact zone.

$$a_3(y,\tau) = Rk_c \frac{(V_к \pm V_и)}{V_и} n_з \cdot \int_0^{t_ф - y_\beta} \int_{-L}^{z} f_з(u)\sqrt{(h_л - y + y_\beta)}\,du\,dz \quad (22)$$

where L - half the length of the contact zone, z - the coordinate along the length of the contact zone.

Expression (22) represents a generalized dependence for determining the probability of erosive removal of the workpiece material.

Substituting values $h_з$ for the depth of the erosion hole hл and the distribution function of the grains with respect to the height *f3(u),* we will be able to calculate the total probability of removing the material $P_\Sigma(y,z)$ sequentially at all levels of the workpiece:

$$\begin{aligned} P_\Sigma(y,z) &= 1 - \overline{P}_1(y,z)\overline{P}_2(y,z)\overline{P}_3(y,z) = \\ &= 1 - \exp[-a_0(y,z) - a_1(y,z) - a_2(y,z) - a_3(y,z)] \end{aligned} \quad (23)$$

The common layer of the removed material y_β can be formally written:

$$y_\beta = \int_0^{t_ф} 1(P_\Sigma(y,z) - \beta_м)dy, \quad (24)$$

where $1(P_\Sigma(y,z) - \beta_м)$ - is the unit function $P_\Sigma(y,z)$, which when value $\beta_м$ takes a value of one; $\beta_м$ - the accepted confidence level is usually 0.95.

There was considered an approach to modeling of the combined grinding process allows to optimize the

DOI: 10.1051/matecconf/201712901079

processing and shorten the preparation time of effective processing cycles on automated equipment.

References

1. A.V. Korolyov, *Investigation of tool and workpiece shaping at abrasive cutting* (Saratov, 1975) (in Russ.)
2. Yu.K. Novoselov, *Dynamics of surface shaping in abrasive processing,* (LAP LAMBERT Academic Publishing. Saarbrucken, Deutschland, P.317 2017)
3. S. Bratan, *Technological bases of quality assurance and increase in stability of high-productive fine grinding, Doctoral thesis* (Odessa, 2006)
4. S. Bratan, *The probabilistic approach for modeling of electroerosion grinding,* Bulletin of SevGTU. Series – Automation and control, Sevastopol, SevGTU Publ., Vol.7, 1997, pp. 140-144. (in Russ.)
5. I.D. Marinescu, M. Hitchiner, E. Uhlmann et al, *Handbook of Machining with Grinding Wheels* (CRC Press, 2007)
6. L. Yang, C. Ren, X. Jin, *Experimental Study of ELID Grinding Based on the Active Control of Oxide Layer* J. Mater. Process. Technol. **210(13)**, 1748–1753 (2010)
7. K. Fathima,M. Rahman,A. Senthil Kumar, H. S. Lim, *Modeling of Ultra-Precision ELID Grinding*, J. Manuf. Sci. Eng **129(2)**, pp. 296-302 (2006)
8. M.Z. Rahim, S. Ding J. P.T. Mo, *Electrical discharge grinding of polycrystalline diamond – Effect of wheel rotation*, Machining Science and Technology 20(1), pp. 62-78 (2016)
9. S. K. Choudhury, V. K. Jain, M. Gupta, *Electrical discharge diamond grinding of high speed steel*, Machining Science and Technology **3(1)**, pp. 91-105 (1999)

 DOI: 10.1051/matecconf/201712901080

Forecasting the surface roughness of the workpiece in the round external grinding

Yurij Novoselov[1], *Vladimir* Bogutsky[1], *Leonid* Shron[1,*], and *Alexander* Kharchenko[1]

[1]Sevastopol State University,33, Universitetskaya St., Sevastopol, 299053, Russia

Abstract. The article gives an analysis of the influence of technological factors on the parameters of the surface roughness during the machining of workpieces with the abrasive tools, based on the use of the modern mathematical apparatus of the theory of probability. It is shown that when processing with abrasive tools, the formation of surface microrelief is directly related to the mechanism of material removal in the area of contact between the workpiece and the tool. The main parameters of the contact zone are its dimensions and parameters of the material removal processes, which are characterized by the probabilities of removal or non-removal of the material. The obtained dependences take into account the parameters of the state of the working surface of the tool and can be used to calculate the basic parameters of the surface roughness in round external grinding. The made estimation of the influence of technological factors on the roughness of the surface of the workpiece being processed allows for the selection of the most significant ones, the changes of which must be taken into account when designing grinding operations.

Microroughness of the surfaces of the parts have a significant effect on their performance properties. In this regard, the provision of roughness parameters is provided for the treatment of all surfaces. Criteria for roughness are very diverse. The most frequently used are the arithmetic mean deviation of the *Ra* profile, the largest height of the roughness of the *Rmax* profile, the height of the unevenness of the profile at ten *Rz* points, and the parameters of the reference surface curve.

When developing a machining operation, the problem arises of predicting the parameters of roughness, which provides a more reasonable choice of tools, cutting modes and other conditions for its implementation. This task is especially important in the design of operations with abrasive tools, which often result in technological processes [1, 2, 3]. The complexity of the problem lies in the fact that the process of interaction between the abrasive tool and the workpiece is stochastic and requires special approaches in the study. Nevertheless, at the present time the system principles of analysis based on the theory-probabilistic approach, system principles of analysis and possessing sufficiently high adequacy of the model, allowing to take into account at designing a large number of technological factors are developed [3, 4, 5, 6]. They are the basis for creating techniques, algorithms and programs and can be included in modern CAD systems.

The purpose of the article is to substantiate the method and sequence of solving the problem of predicting surface roughness parameters when machining parts with abrasive tools based on the use of modern mathematical models.

The formation of surface microrelief in abrasive processing is directly related to the patterns of material removal in the area of contact between the workpiece and the tool. The material in the zone is removed by single abrasive grains, i.e. The process of material removal is discrete. However, the number of unit slices is large enough that we can consider it as continuous and use the main provisions of statistical physics. For such a simulation, the concept of the probability of material removal is introduced, consisting in the probability of the event that the material at the point is removed, and the concept of the opposite event is the probability of not deleting the material [6, 7].

On the basis of a meaningful description it can be asserted that for the point *M*, Fig. 1, the section of the workpiece that is in the contact zone, the probability of material removal is higher than before the entrance of the section into the zone, and it increases with further passage through the section of the zone. When leaving the zone, the radius-vector of the surface area of the workpiece decreases with circular external grinding by the amount of radial material removal Δr, a roughness layer *H* remains on the surface, and their sum is equal to the depth of microcutting t_f. All three parameters are interrelated and determined under the known cutting conditions and the state of the working surface of the tool. The depth of microcutting is calculated on the basis of the solution of the balance equation of displacement in the technological system (TS), which for the process of round external grinding has the form [7]

* Corresponding author: shronlb@rambler.ru

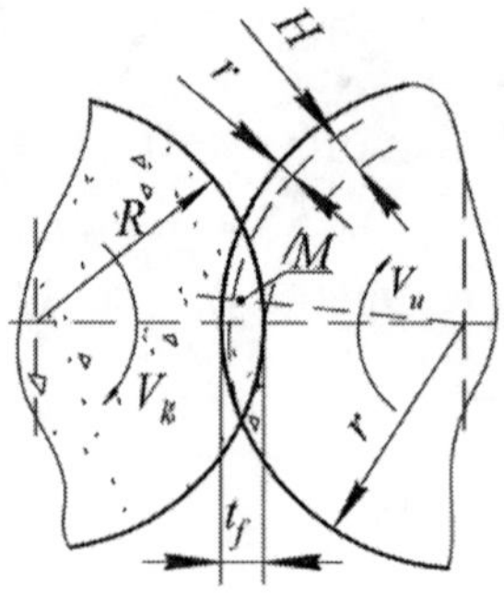

Fig. 1. Process scheme

$$\Delta A_{sj} = \Delta t_{fj} + \Delta R_j + \Delta r_{j-1} + \sum_{k=1}^{n} \Delta y_{yjk} + \sum_{k=1}^{m} \Delta y_{qjk}, \quad (1)$$

where ΔA_{sj}, Δy_{yjk}, Δy_{qjk}, - respectively, the changes in the center-to-center distance due to the presence of feed, elastic and temperature deformations during the time from the combination of the radius vector of the section of the workpiece surface with the plane passing through the center of rotation of the circle and the workpiece *j-1*-th and *j-th* contacts; ΔR_j - radial wear of the tool; Δr_{j-1} - radial material removal; Δt_{fj} - change of depth of microcutting at *j-th* contact; And - the sum of the increments of the elastic and temperature deformations of the TC elements. A plus sign is assigned if the increment decreases the depth of cut and the minus sign - if it increases.

Analysis of equation (1) shows that for forecasting the spatial position of the TS elements it is necessary to have information about the parameters of the quality of the workpiece, the initial profile of the tool, and analytical models determining the relationship of the depth of microcutting, radial material removal, circle wear, cutting forces, elastic and temperature deformations with technological Factors. Such models and examples of solving the balance equation by the iteration method are given in [7].

For the steady-state process of increment of elastic, temperature deformations and Δt_{fj} are equal to zero and equation (1) is simplified

$$\Delta A_{s\ j} = \Delta R_j + \Delta r_{j-1}. \quad (2)$$

For a known value of Δr, the cutting depth Δt_f is calculated [7]

$$t_f = 0{,}739\Delta r + \sqrt{0{,}546(\Delta r)^2 + \frac{13{,}66 V_u \Delta r}{K_s (V_k \pm V_u) n_g \sqrt{D_e \rho_g}}}; \quad (3)$$

where V_k,V_u - the speed of the circle and the speed of the workpiece; n_g is the number of abrasive grains per unit of the working surface of the circle; D_e - equivalent diameter, $D_e = Dd/(D+d)$; ρ_g is the radius of curvature at the top of the grain [12]; K_s is the chip formation coefficient, for typical grinding operations is given in [5, 8, 9].

To calculate the roughness parameters, consider the surface before and after the *j-th* contact (Fig. 2).

At a known depth of microcutting, the value of the surface roughness layer is calculated as

$$H = t_f - \Delta r. \quad (4)$$

We determine the position of the midline of the profile after contact, denoting the distance to it from the lower boundary of the roughness layer through W_m, and the coordinate of the profile point relative to the midline through y_m.

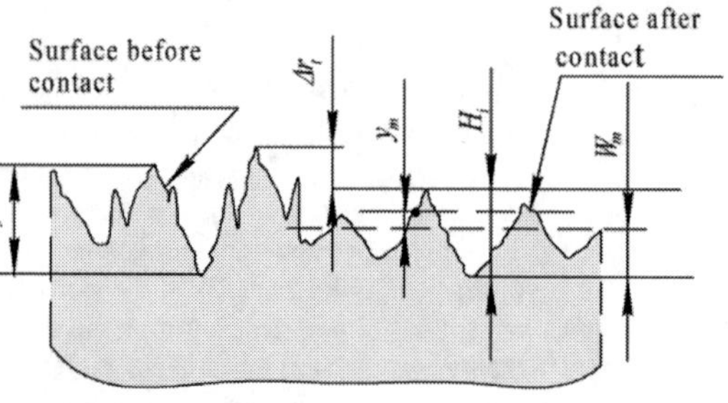

Fig. 2. Surface before and after contact

The average arithmetic deviation of the profile is then calculated

$$R_a = \frac{1}{L}\int_0^L |y_m| dx, \quad (5)$$

where L is the length of the base portion of the surface microrelief.

The integral in the dependence (5) determines the sum of the areas of the protrusions and valleys. They can also be determined by integration with respect to the variable y_m.

Since for the area of the protrusions $dF_1 = t_p L dy_m$ and the area of the depressions $dF_2 = (1-t_p) L dy_m$, Ldym, then

$$R_a = \int_0^{\infty} t_p dy_m + \int_{-\infty}^{0} (1 - t_p) dy_m \quad (6)$$

$$\text{or} \quad R_a = \int_0^{\infty} (1 - P(M)) dy_m + \int_{-\infty}^{0} P(M) dy_m, \quad (7)$$

where t_p is the relative length of the reference profile, the mathematical expectation of which is equal to the probability of not removing the material; $P(M)$ is the probability of material removal, calculated by [7].

The dependence for calculating Ra is greatly simplified if we use the position [10, 11] that the random process describing the ordinates of the profile is stationary and normal. For such a process, the mathematical expectation Ra is proportional to the mean square deviation σY_m of the random ordinates Y_m of the profile points

DOI: 10.1051/matecconf/201712901080

$$M[R_a]=\sqrt{\frac{2}{\pi}k(0)}=\sigma_{Ym}\sqrt{\frac{2}{\pi}}, \qquad (8)$$

where $k(0)$ is the value of the correlation function, for the interval $x_1=x_2$ is equal to the variance of the random variable Y_m.

To determine the root-mean-square deviation σ_{Ym}, we calculate the function $F_{Ym}(y_m)$ and the probability density $f_{Ym}(y_m)$ of the distances Y_m. For positive and negative values of y_m.

$$F_{Ym}(y_m)=P(Y_m<y_m)=\lim_{L\to\infty}\frac{L-\sum b_{mi}}{L}=P(M)$$

$$f_{Ym}(y_m)=\nu G\sum_{i=0}^{n}(W_m+y_m-i\Delta r)^{\nu-1}\times$$

$$\times\left\{exp\left[-G\sum_{i=0}^{n}(W_m+y_m-i\Delta r)^{\nu}\right]\right\} \qquad (9)$$

where n is determined from the condition [8]

$$W_m+y_m-n\Delta r\geq 0;\quad W_m+y_m-(n+1)\Delta r\leq 0;$$

$$G_k=\frac{\sqrt{\pi D_e}\,\Gamma(m+1)\Gamma(\chi)\chi K_s C_b(V_k\pm V_u)n_g}{\Gamma(m+\chi+3/2)V_u H_u^{\chi}}$$

$$\nu=m+\chi+\frac{1}{2}$$

where m and C_b are the parameters of the shape of the cutting edge of the abrasive grain [12]; χ - grain distribution parameter in the depth of the working surface of the tool; H_u - the depth of the working layer of the tool, within which the number of abrasive grains is calculated; $\Gamma(m+1)$; $\Gamma(\chi)$; $\Gamma(m+\chi+3/2)$ are the corresponding gamma functions.

Comparison of the probability densities of the ordinates of the profile, calculated from equation (9) and Gaussian law, confirmed the hypothesis that the random process - the ordinates of the polished profile surface - is close to normal. For such a process, $y_m=0$ и $P(M)=0,5$,

$$f_{Ym}(y_m)=\frac{1}{\sigma_{Ym}\sqrt{2\pi}}.$$

On the basis of this, the second part of equation (9), separated by curly braces, is 0.5, and the average arithmetic deviation of the profile, taking into account equation (8), is defined as

$$R_a=\frac{2}{\pi G\nu\sum_{i=0}^{n}(W_m-i\Delta r)^{\nu-1}} \qquad (10)$$

The distance W_m from the deepest depression to the midline of the profile is calculated from the condition

$$y_m=0,\ P(M)=0,5,\ G\sum_{i=0}^{n}(W_m-i\Delta r)^{\nu-1}-ln\,2=0.$$

At the value of the radial removal of the metal $\Delta r\geq W_m$

$$W_m=\left(\frac{ln\,2}{G}\right)^{1/\nu}.$$

As shown in [7], with the value of the radial metal removal $\Delta r\geq W_m$ – $W_m\approx 0,5(t_f - \Delta r)$ and at particular values $m=0,5$ (the shape parameter of the cutting edge of the abrasive grain [12] and $\chi=1,5$ (the grain distribution parameter for the depth of the working surface of the tool), *Ra* can be calculated as:

$$R_a=\frac{\sqrt{2}V_u H_u^{3/2}}{\pi^2 K_s(V_k\pm V_u)n_g\sqrt{D_e\rho_g}\sum_{i=0}^{n}(W_m-i\Delta r)^{3/2}}$$

at $\Delta r<W_m$; (11)

$$R_a=\frac{\sqrt{2}V_u H_u^{3/2}}{K_s^{0.4}(V_k\pm V_u)^{0.4}n_g^{0.4}D_e^{0.2}\rho_g^{0.2}}$$

$\Delta r\geq W_m$; . (12)

where H_u is the depth of the working layer of the tool, within which the number of abrasive grains is calculated.

Equation (12), in its structure and magnitude of the exponents for variable factors, is close to the power relations in the literature, but, unlike them, reflects the physical nature of the process of shaping and corresponds to the basic principles of the theory of dimensions.

The maximum height of the roughness of the *Rmax* profile and the height of the profile irregularities along the ten Rz points can be calculated from the layer value in which the surface roughness is distributed and the mathematical expectation of the distances Y_1, Y_2... Y_5 from the upper boundary of the layer to the five higher profile points and distances W_1, W_2... W_5 from the lower boundary of the layer to the five lowest points of the profile (Fig. 3).

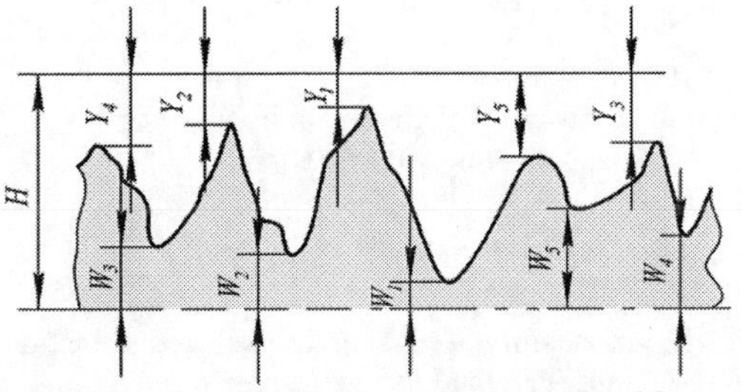

Fig. 3. Scheme for calculating the maximum height of the profile irregularities and the height of the irregularities by ten points

The distances considered are random and independent, on the basis of this [7, 13]

$$M[R_{max}] = H - M[Y_1] - M[W_1];$$

$$M[R_z] = H - \frac{1}{5}\sum_{i=1}^{5} M[Y_1] - \frac{1}{5}\sum_{i=1}^{5} M[W_1];$$

$$D[R_{max}] = D[Y_1] - D[W_1];$$

$$D[R_z] = \frac{1}{25}\sum_{i=1}^{5} D[Y_1] - \frac{1}{25}\sum_{i=1}^{5} D[W_1].$$

For a stationary process close to normal, we can assume that the distances from the upper boundary of the roughness layer to the most protruding vertices of the profile are distributed according to laws analogous to the distribution of distances from the depressions to the lower boundary of the layer. From this follows the approximate equality of mathematical expectations and variances Y_j and W_j..

If the density distribution of grain vertices along the depth of the tool is modeled by a power law, then the number of vertices of the grains passing beyond the last contact of the surface area with the circle is determined from equation [7]

$$\lambda = \frac{\tilde{A}(\chi)\chi(V_k \pm V_u)n_g\sqrt{\pi D_e}}{\tilde{A}(\chi + 3/2)V_u H_u^{\chi}} u^{\chi+1/2} = Qu^{\chi+1/2}$$

To calculate the mathematical expectations and variances, the corresponding distribution functions $F_W(w)$ are defined. Each of the deepest depressions of the microrelief can be viewed as a trace from a separate peak of the abrasive grain, so the probability of events $P(W_1<w)$, $P(W_2<w)$, ... $P(W_5<w)$, which consists in the fact that the distance from the lower boundary to the first, second, ..., fifth cavities is less than some fixed values of w, is determined by the probability of passing through the layer of section from $w = 0$ to W by at least one, two, ..., five vertices of grains . The probability of passing through a section of a section of length L of at least one vertex is determined by the consequence of the Poisson law

$$F_{W1}(w) = P(W_1 < w) = 1 - e^{-\lambda L}.$$

The probability of passing at least two vertices is less than the probability of passing at least one vertex by the probability of passing only one tops

$$F_{W2}(w) = P(W_2 < w) = P(W_1 < w) - \frac{\lambda L}{1!}e^{-\lambda L}.$$

The probability of passing through a cross section of at least three, four, and five grains

$$F_{W3}(w) = P(W_3 < w) = P(W_2 < w) - \frac{\lambda^2 L^2}{2!}e^{-\lambda L}$$

$$F_{W4}(w) = P(W_4 < w) = P(W_3 < w) - \frac{\lambda^3 L^3}{3!}e^{-\lambda L}$$

$$F_{W5}(w) = P(W_5 < w) = P(W_4 < w) - \frac{\lambda^4 L^4}{4!}e^{-\lambda L}$$

After a consistent determination of the distribution functions, probability densities, mathematical expectations, and dispersion of the distances W_i and Y_i [7], the equations for calculating $M[Rmax]$, $D[Rmax]$, $M[Rz]$, $D[Rz]$

$$M[R_{max}] = H - 2\Gamma\left(\frac{2\chi+3}{2\chi+1}\right)\left(\frac{Q}{L}\right)^{\frac{2}{2\chi+1}} \quad (13)$$

$$D[R_{max}] = 2\left[\Gamma\left(\frac{2\chi+5}{2\chi+1}\right) - \Gamma^2\left(\frac{2\chi+3}{2\chi+1}\right)\right]\left(\frac{Q}{L}\right)^{\frac{4}{2\chi+1}};(14)$$

$$M[R_z] = H - \frac{2}{5}\left(\frac{Q}{L}\right)^{\frac{2}{2\chi+1}}\sum_{i=1}^{5}\frac{1}{(i-1)!}\Gamma\left(\frac{2i\chi+i+2}{2\chi+1}\right);(15)$$

$$D[R_z] = \frac{2}{25}\left(\frac{Q}{L}\right)^{\frac{4}{2\chi+1}}\sum_{i=1}^{5}\frac{1}{(i-1)!}\left[\Gamma\left(\frac{2i\chi+i+4}{2\chi+1}\right) - \frac{1}{(i-1)!}\Gamma^2\left(\frac{2i\chi+i+2}{2\chi+1}\right)\right], (16)$$

where Q is denoted by $Q = \dfrac{\Gamma(\chi+1{,}5)V_u H_u^{\chi}}{\Gamma(\chi)\chi(V_k+V_u)n_g\sqrt{\pi D_e}}$.

The values of the parameter χ in calculating the probability of material removal, mathematical expectations and variance $Rmax$ and Rz can be different, since different parts of the real curve of the probability density of grain distribution over the depth of the tool $f(u)$ are modeled. For a particular value χ=1,5, equations (13) and (15) take the form

$$M[R_{max}] = H - 2\sqrt{\frac{2V_u t_f^{3/2}}{3n_g(V_k \pm V_u)L\sqrt{D_e}}}; \quad (17)$$

$$M[R_z] = H - 2{,}95\sqrt{\frac{V_u t_f^{3/2}}{n_g(V_k \pm V_u)L\sqrt{D_e}}}. \quad (18)$$

An example of an analytical calculation of the surface roughness parameters for the process of circular external grinding of samples of steel 9X2 with a diameter of 50 mm in circles 1 300x40x127 24A F60 L 4V is given in Table 1.

In the calculations, it is assumed: K_s=0,9; V_k=35 m/s; V_u=0,25 m/s; Transverse feedrate ΔA_s=3,33•10^{-9} m per one turn of the workpiece; n_g=5.2•10^6 1/m^2; ρ_g=21•10^{-6} m, the value of the base length L=0.8•10^{-6} m, ΔR=0.01r.

Table 1. Example of calculating the height parameters of the surface roughness

№ in order	Parameter	The equation
1	Δr=3,3·10-6 m/turn	(2)
2	$t_f=11,5\cdot10^{-6}$ m	(3)
3	$H=8,2\cdot10^{-6}$ m	(4)
4	$Ra=1,0\cdot10^{-6}$ m	(11)
5	$Rmax=7,3\cdot10^{-6}$ m	(17)
6	$Rz=6,5\cdot10^{-6}$ m	(18)

A comparison of the calculated and experimental values of the relative reference length of the profile for the grinding conditions considered in the example is given in [7].

Table 2 compares the calculated and experimental data obtained with oscillating grinding of blanks with longitudinal feed S_x from steel 9X2 with a diameter of 50 mm in circles of 1 300x40x127 24A F60 L 4V, HRC 55 ... 60. The repetition of experiments at each point of the experimental design is taken equal to nine

For the most part of the processing modes, the deviation of the calculated data from the experimental data does not exceed 20%, which makes it possible to recommend the considered sequence of predicting surface roughness parameters when designing grinding operations. The greatest deviation of the calculated data from the experimental ones is observed for mode 2. It is assumed in the calculations that the number of abrasive grains on the working surface does not depend on the regime while it is known that with decreasing speed and increasing feed, the probability of grain breaking and tearing out of the bundle increases Leads to a reduction in the number of cutting edges on the working surface of the tool.

Table 2. Experimental and calculated values of surface roughness during grinding with changing cutting conditions

№	V_k, m/s	V_u, m/s	S_x, mm/s	S_t, m/s	Ra, µm	
					[illegible]	[illegible]
1	35	0,66	16,6	0,533	1,17	1,2
2	20	0,33	16,6	0,533	1,57	3,0
3	20	0,66	8,3	0,533	1,54	1,6
4	20	0,66	16,6	0,266	1,08	1,2
5	35	0,33	8,3	0,533	1,20	1,0
6	35	0,33	16,6	0,266	0,85	0,6
7	35	0,66	8,3	0,266	0,83	0,7
8	20	0,33	8,3	0,266	1,10	0,8
9	27	0,50	12,5	0,400	1,10	1,0
10	27	0,50	12,5	0,400	1,10	0,9

Thus, it is recommended that the calculation of the surface roughness parameters during grinding be performed in the following sequence:

- at the first stage, the input variable operations are defined: the parameters of the state of the working surface of the tool, its geometric characteristics, the parameters of the workpiece, the grinding mode;
- in the second stage, the solution of the displacement balance equation, equations (1) and (2), calculate the radial material removal, the depth of micro cutting and the thickness of the surface roughness layer.
- in the third stage, select the dependence for calculating the roughness parameters from the dimensional analysis data and update the initial data if necessary;
- in the fourth stage, the required parameters for the surface roughness are calculated from the equations given; Give an assessment of their reliability;

If the parameters do not match, given by the designer, the selection of the cutting mode and the characteristics of the abrasive tool are corrected.

If necessary, the calculations are repeated with new values of the input variables.

References

1. S.M. Bratan, *Quality assurance and improving processing stability during the finishing and fine grinding*, Cutting & tool in technological system. International Scientific-Technical Collection. Edition **68,** P. 34–39 (2005)
2. Stephen Malkin, Changsheng Guo, GRINDING TECHNOLOGY. *Theory and Applications of Machining with Abrasives*, (Industrial press, New York,. – P.372. 2008)
3. Yu.K. Novoselov, D.A. Kainov, S.M. Bratan, *The construction of the model of a ground surface on the basis of a stochastic description of the properties of the tool,* Nauk. pr. Donec. nac. tehn. un-tu. Ser.: Mashinoobladnannja, Issue **7**. P.17-31 (2010)
4. Ioan D Marinescu, [et al.], *Tribology of abrasive machining processes*, (Published in the USA by Brent Beckley, William Andrew, Inc. Cover Art – P.764. 2004)
5. S.N. Korchak, *Productivity of process of grinding of steel parts* (M.: Mashinostroenie, P. 280 1974)
6. A.V. Korolev, Yu.K. Novoselov, *Theoretical and probabilistic fundamentals of abrasive processing.* (Part 1. State of the working surface of the abrasive tool, Saratov: Izdatel'stvo Saratovskogo universiteta, P.158. 1987)
7. Yu.K. Novoselov, *Dynamics of surface shaping in abrasive processing,* (LAP LAMBERT Academic Publishing. Saarbrucken, Deutschland, P.317 2017)
8. A.N. Reznikov, *Abrasive and diamond processing of materials*: *Handbook* (M.: Mashinostroenie, P.391, 1977)

9. Robert I. King, Robert S. Hahn, *Handbook of modern grinding technology* (Publ. Chapman and Hall, New York, P.369. 1986)
10. A.P.Husu, Yu.R.Vitenberg, V.A. Palkov, *Roughness of surfaces: the theoretical-probabilistic approach.*, (M.: Nauka, P.344. 1975)
11. Yu.R. Vitenberg, *Application of the correlation theory for the evaluation of polished surface*, Vestnik mashinostroenija, Issue **1**, P. 55–57 (1969)
12. L.V. Khudobin, *Analysis of the geometry of abrasive grains*, Mechanical engineering, electro instrumentation. Saratov: Privolzh. Book Publishing house. P. 6-20. (1966)
13. A.D. Wentzel, *Course of the theory of random processes* (M .: Nauka. Fizmatlit, P.420, 1996)

Parametric synthesis of assembly device for fitting rubber sealing ring into the inner groove of the basic item

Aleksey Tarakhovskiy[1,*], *Sergey* Sazonov[1], *Denis* Tokarev[1] and *Yuriy* Moroz[1]

[1]Sevastopol State University, 33, Universitetskaya St., Sevastopol, 299053, Russia

Abstract. On the basis of the block diagram of the assembly device technological system a group of morphological matrices of various hierarchy levels may be built. The analysis of the engineering system " the assembly device for assembly Seal of connections " is manufactured. The structure of the engineering system and e ë of units at different hierarchical levels is defined, the links between subsystems are detected. On the basis of the skeleton diagram of the assembly device the group of morphological matrixes, different levels of hierarchy is constructed. The synthesis of variants of structure of the assembly device is manufactured.

1 Introduction

Analysis of assembly device for assembling the sealing compound (elastic annular item – groove of the hole) shows that the assembly device possesses all the features of the engineering dynamic systems. Considering this problem of synthesis of optimal structures of assembly device can be regarded as a common task of the theory of synthesis of engineering systems, but it also has a number of features related to:

– the requirements to the assembly device;
– a physical feature of the assembly.

2 Synthesis of assembly device

Synthesis of assembly device may be configured basing on the use of structural and parametric optimization synthesis methods of engineering systems.

The synthesis problem solving sequence of the assembly device:

1. Formation of a plurality of the assembly device structures.
2. Assessment of structure options by extensional indexes.
3. Evaluation of options in accordance with stochastic processes.
4. Identification of "weak" places in the technological system.
5. Development and improvement of the "weak" places of the system.
6. Evaluation of the system effectiveness considering its verifications.
7. The implementation of the system.

In the first stage of solving the synthesis aim of the assembly device system the structure of the engineering system and its constituent subsystems should be determined. To do this, you must submit morphology (structure) of an engineering system, and then the connections between its elements (subsystems) and their parameters. After carrying out the study at one level (the level of the initial engineering system) analysis then in most cases is required to continue to lower hierarchical levels (at the levels of specific subsystems).

For this purpose, when forming a plurality of structures of different effects on the output indicators as the process system, and the assembly process are taken into consideration as attributes and characteristics of their technological elements. In this case it is advisable to apply a morphological analysis unit, representing the engineering system elements (or its functions) as morphological features.

However, this method is in its pure form can not be used, since the choice of efficient variations does not take into account the actual conditions of the assembly process of sealing compounds. Therefore, morphological analysis should be used as a unit on the first two stages of the synthesis to form a plurality of structures and assessment of options for extensional indexes.

On the basis of the block diagram of the assembly device technological system a group of morphological matrices of various hierarchy levels may be built. To do this in the technological system of the assembly device, some structural or functional characteristic for its morphological features should be highlighted. Each feature may characterize a structural unit of the system or its function or mode of operation of the system, i. e. parameters or characteristics of the system, from which the solution of the problem and the achievement of the main goal depend.

The list of various specific characteristics the expression of technical alternatives for each selected morphological feature is made. Features and their alternatives are presented in tabular form.

* Corresponding author: atarahovsky@gmail.com

Table 1. Morphological matrix of assembly device for assembling a sealing compound (elastic annular item – groove of the hole)

Subsystem - assembly mechanism (AM)	
subsystem feature	Feature characteristics
Deforming body - X1	assembly device X11 ... Basic item X1m
...	...
n. Restoring the deformed shape - Xn	n.1 Due to the elastic properties of the elastic ring item - X51 ... n.m Due to assembly device - Xnm

Similarly matrices for the other subsystems are constructed: the basic item (BI), a elastic ring item (ERI), the basic item of the supplying device (SBI) and feeder of elastic ring item (FERI).

The next step after filling the morphological matrices is a synthesis of options for the structure of the assembly device. To do this, from a variety of technological features of the subsystem - "assembly device" the most applicable ones for the given conditions should be selected.

Suppose we have M types of n-th feature of the subsystem, with m - the serial number of the given subsystem, then the variety of the subsystem conditions of "assembly device" can be expressed in the form of a matrix.

$$\|C6\| = \begin{Vmatrix} X_{11} & X_{21} & \dots & X_{N1} \\ X_{12} & X_{22} & \dots & X_{N2} \\ \dots & \dots & \dots & \dots \\ X_{1M_1} & X_{2M_2} & \dots & X_{NM_N} \end{Vmatrix} \tag{1}$$

By analogy, a variety of the subsystems conditions is expressed in the same way in the form of matrices: the basic item (BI), a elastic ring item (ERI), the basic item of the supplying device (SBI) and feeder of elastic ring item (FERI).

Subsystem features from morphological tables are chosen to build graphs for set of possible variants of "assembly device" which is equal to the number of paths in the graph. Figure 1 shows a graph of possible structural variants of the technological subsystem of "assembly device". Graphs of possible variations of other technological subsystems are constructed similarly.

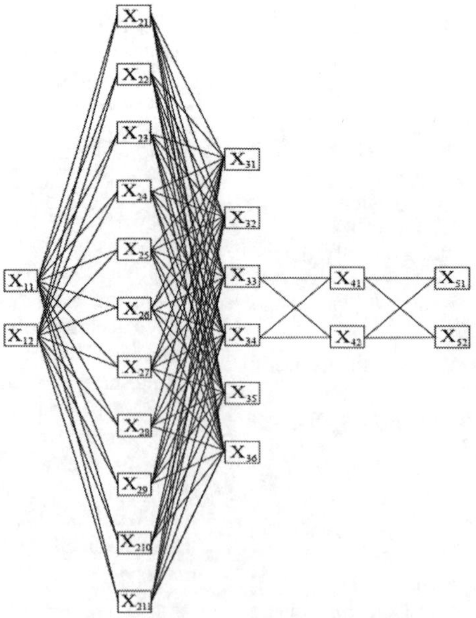

Fig. 1. Graph of possible structural variants of the technological subsystem of "assembly device".

When considering a large number of variants the solution synthesis should be carried out, starting from the basic elements of choices, including more and more significant features.

Let's carry out a set of technological subsystem variations of "elastic ring items." Made investigations have shown that the most frequent application is being made with sealing compounds with elastic ring items in the form of annular rubber rings of circular cross-section. They differ in design simplicity, a wide versatility, and durability, ease to manufacture, reliability in operation, and are used at wide temperature and pressure ranges. The most appropriate structure of subsystem "elastic ring item " is shown in Figure 2.

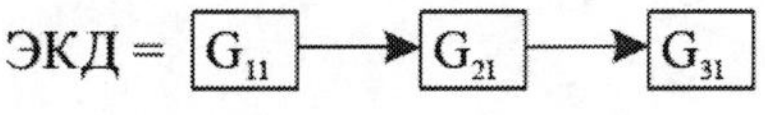

Fig. 2. structure of subsystem "elastic ring item "

where the G11 - kind of ERI (ring); G11 - diametrical sectional view of the ERI (simple round); G11 - number of installed elements (one).

Knowing the structure of the ERI a choice of technological subsystem variants of "assembly device" can be made. A patent search and analysis of the literature showed that when installing a seal ring is in contact with the tapered mandrel surface (basic item), this resulted in the sealing ring to be installed at a slant or in a twisted state (from arising a torque in the section). Tightening a ring in the groove increases tension on its surface, which leads to increased wear of the seal assembly and the system failure. The presence of levers and the punch in the device, gripping the sealing ring and feeding it into the hole of the basic item may cause jamming of the sealing ring between the levers (punch) and a tapered mandrel (basic item), and this will lead to cross-sectional deformation, which ultimately leads to a decrease in the quality of assembling products. Therefore, the next most suitable structural variant is as follows:

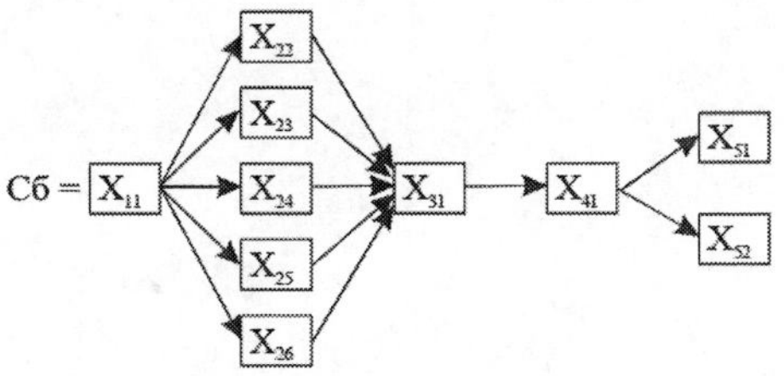

Fig. 3. The structural subsystem version "assembly mechanism"

Concerning given structure:

- the number of simultaneously mounted elements - 1;
- type of elastic deformation of the ring item
- form deformation ("Eight", "three-pointed star," "four-pointed star," "a multi star radiation", "projection into" "the two tabs inward");
- deforming body - levers;
- supply body - levers;
- restoration of the deformed shape of elastic ring item - either due to the elastic properties of the elastic parts, or due to the assembly of the device.

Similarly, there is variety of structural variants and other technical subsystems - assembly device for assembling of sealing compounds.

3 Conclusion

Thus, the morphological analysis reveals a few of the many options of rational structures. For complete information and an objective assessment of the preferred embodiment requires accounting of stochastic processes occurring in the system under real operating conditions. Conducted activities help to solve the problem at the level of synthesis of structurally combined optimization. Step parametric synthesis requires further theoretical and practical research of real structural assembly of devices that allow you to define "weak" places in the technological elements, and their improvement will lead to the eventual implementation of the most effective system.

References

1. V.M. Odrin, S.S. Kartalov, *Morphological analysis systems* (Kiev Naukova Dumka, 1977) (in Russian)
2. *Automation of search design,* Ed. by A.I. Polovinkin (M.: Radio and communication, 1981) (in Russian)
3. A.O. Kharchenko, *Development of flexible production modules based on the reliability of technological systems* (K.: knowledge Society, 1989) (in Russian)
4. T. Ritchey, *General Morphological Analysis: A general method for non-quantified modeling.* (1998)
5. T. Ritchey, *"Problem Structuring using Computer-Aided Morphological Analysis"*, Journal of the Operational Research Society (JORS), Vol. **57**, No. 7, (2006)
6. T. Ritchey, Wicked *Problems/Social Messes: Decision support Modelling with Morphological Analysis.* Berlin: Springer, (2011)
7. F. Zwicky, & A. Wilson (eds.), *New Methods of Thought and Procedure*: Contributions to the Symposium on Methodologies. (Berlin: Springer, 1967)
8. J.C. Jones, *Design Methods*, (Wiley, 1981)
9. Mark Sh. Levin, *Modular systems design and evaluation* (Springer, 2015)
10. J.C. Jones, *Design Methods* (Wiley, 1981).
11. Mark Sh. Levin, *Composite Systems Decisions*, (Springer, 2006). (in Russian)
12. Mark Sh. Levin, *Combinatorial Engineering of Decomposable Systems* (Kluwer, 1998).

DOI: 10.1051/matecconf/201712901081

13. Mark Sh. Levin, *Modular System Design and Evaluation* (Springer, 2015). (in Russian)
14. G. S. Altshuller. *To find the idea. Introduction to the theory of inventive problem solving*, (Petrozavodsk: Scandinavia, 2003). (in Russian)
15. T. Baker, *Attribution Standardization for Integrated Concurrent Engineering, M.S. Thesis* (Brigham Young University, 2005).
16. M. Balachandran, J.S. Gero. *A Knowledge-based approach to mathematical design modeling and optimization,* Engineering Optimization, vol. **12**, 91-115 (1987).
17. J.K Novoselov, A.Yu.Tarahovsky, *Justification requirements for mechanized and automated build process sealing connections with internal grooves,* Scientific notes of the Crimean Engineering and Pedagogical University, Number **27.** pp. 11-16, (2011) (in Russian)
18. V.V. Burenin, V.P. Ivaniv, *Estimation of durability of rubber O-rings and cuffs for rotating machines shafts mechanisms*, Chem. and Naftogaz. mechanical engineering, №**11**, pp 32 – 33, (2003). (in Russian)
19. ISO 11346: 2004. Rubber, vulcanized or thermoplastic - Estimation of life-time and maximum temperature of use. (in Russian)
20. A.Yu. Tarahovsky, *Methods deformation O-rings when installing them in the internal groove cylindrical surfaces*, assembly in mechanical engineering, instrument making, Number **11**. pp. 10-15. (2010) (in Russian)

 DOI: 10.1051/matecconf/201712901082

Investigation of thermal processes in abrasive pipe sampling

Elena Levchenko[1,*], and *Nikolay* Pokintelitsa[1]

[1]Sevastopol State University, Department of Mechanical Engineering, 299053 Sevastopol, Russia

Abstract. The paper presents the results of research on the process of interaction between the cutting circle and the cut pipe billet on the basis of heat-deforming analysis with the establishment of a relationship between processing conditions and cutting conditions with the design parameters of the abrasive tool.

1 Introduction

Abrasive cutting of materials, from the point of view of the physical processes occurring in the surface layers of the cut billets, is one of the highest-temperature and high-speed machining processes. The thermal process arising during the cut is characterized by temperatures between 200 – 800 °C, the instantaneous heating with velocities of hundreds of thousands of degrees per second. Following rapid heating, heat is removed from the surface layers deep into the pipe material at a cooling rate of approximately the same order as in heating.

The presence of heat exchange with the environment causes the redistribution of heat fluxes in the system "cutting circle - cut pipe-chip", and as a consequence, leads to a change in the basic kinetic parameters of the thermal process. The temperature deformations of parts along the length and diameter depend on the coefficient of thermal conductivity of their material [1, 3]. It is known that heat-resistant steel 12H18N9T is less heat-conducting than stainless steels 4H13, 3H13 and steel grades 08kp and 1ps. However, they have about 1,5 times the large coefficient of linear expansion and, as a consequence, higher temperatures appear in the cutting zone. Such circumstances determine the propensity of heat-resistant steels to high temperature deformations exceeding the deformation of stainless steels and structural steels of grades 08kp and 1ps. As is known, the quality parameters of the surface layer of the billet largely depend on the thermal factor. The physico-mechanical state of the surface layer of the cut billets is determined not only by the contact temperature, which can be measured experimentally, but by the entire space-time temperature field [1]. In particular, the gradients of temperatures and the rate of change of temperature are of paramount importance in the formation of the surface layer. All these factors have unique functional relations with the kinematic and dynamic parameters of the processing regime, the properties of the material, the characteristics of the circle, and so on. The quality control of the cut surface of the billet largely reduces to establishing these bonds and acting through them on the thermal regime of the cutting and the physical and mechanical properties of the surface layer. Therefore, the aim of this paper is to select and justify the method for determining the temperature in the treatment zone, taking into account the influence of the temperature factors accompanying the cutting process on the efficiency of the technological operation.

2 Main text

To determine the causes of the errors, a heat-deforming analysis of the process was carried out. For this purpose, the regularity of the temperature distribution in the billet has been established. It was assumed that the width of the groove of the workpiece increases only as a result of thermal expansion (Fig. 1).

In the process of work, the circle as a whole and its abrasive grains enter the metal of the workpiece. The cutting wheel is unevenly heated along the length of the working part.

Fig. 1. Scheme of cutting the workpiece.

Since the normal components of the cutting force in abrasive cutting reach large values, the most protruding grains and the binder holding them undergo an elastic deformation and the number of grains in contact with the metal surface increases substantially. Not all contacting grains will participate in the removal of metal. But if we evaluate the grains as heat sources that heat the metal, then the grains sliding along the surface will contribute no less than those that remove the shavings. Cutting grains located on the side surfaces of the circle absorb a generally smaller part of the energy, since it is also expended on the work of deformation and heating of the formed chips.

* Corresponding author: ealev1978@mail.ru

 DOI: 10.1051/matecconf/201712901082

At the initial moment of cutting due to intense heating, the material of the workpiece substantially extends toward the resulting groove. Then, as the tool moves in the feeding direction, the linear expansion proceeds more slowly. As a result, a section of the groove is formed, the profile of which differs significantly from the nominal.

The temperature deformations of the workpiece cause the shape of the groove to distort, which it would absorb after cooling (Fig. 2). The actual shape of the groove section is obtained taking into account the thermal deformations of the workpiece.

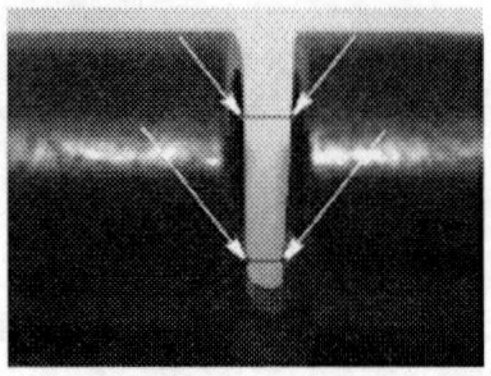

Fig. 2. Schematic of the change of the groove under Exposure to temperature deformations.

As shown by experimental studies [4], the process of mass cutting of abrasive grains during cutting, proceeding at very high speeds, creates a large number of high-temperature centers in the surface layer. The sources of heat arising during the cutting are the work of metal deformation and the work of external friction of abrasive grains on the metal surface. In contrast to cutting with a metal tool in abrasive cutting, the heat of cutting is mainly transferred to the mass of the metal of the part, the cutting wheel is practically not heat-conductive. Knowing the cutting temperatures and the thermal phenomena accompanying this process is necessary not only to obtain a qualitative surface layer, but also to establish the true picture of the process of removing the thinnest chips at high speeds.

It is known that the process of heat transfer in a stationary solid medium is carried out due to heat conductivity and is described by the equation [2]

$$\frac{d\Theta}{dt}=\frac{\lambda}{C\gamma}\left(\frac{\partial^2\Theta}{\partial x^2}+\frac{\partial^2\Theta}{\partial y^2}+\frac{\partial^2\Theta}{\partial z^2}\right). \quad (1)$$

In the presence of motion of a heat source, it is necessary to associate a coordinate system with a moving source. When the x-axis is aligned with the direction of motion of the heat source, the differential heat equation (1) is written in the form

$$\frac{d\Theta}{dt}=\frac{\lambda}{C\gamma}\left(\frac{\partial^2\Theta}{\partial x^2}+\frac{\partial^2\Theta}{\partial y^2}+\frac{\partial^2\Theta}{\partial z^2}\right)+V\frac{d\Theta}{dx}. \quad (2)$$

where V – is the velocity of the heat source, m/s; C – is the specific heat, J/(kg·K); γ – density, kg/m^3; λ – is the coefficient of thermal conductivity, W/(m·K).

To solve thermophysical problems in abrasive metal cutting processes, the source method (Green's method) is used, the essence of which is that if the law of temperature variation Θ (x, y, z, t) at point B (x, y, z) is known) Of the heat-conducting space from the heat source placed in the point A (x', y', z'), which allocated heat quantity Q at time t', then the temperature at point B (x, y, z) from the system of spatially distributed sources Is determined by superposition of temperature fields. The solution of the problem reduces to determining the function Θ (x, y, z, t) from the heat source Q (Green's function). In order for the solution Θ (x, y, z, t) to be the unique solution of the problem, it must simultaneously satisfy certain initial and boundary conditions.

The initial condition $\Theta = \Theta\ (x, y, z, t_0)$ determines the temperature distribution inside the investigated region at the initial instant of time.

The boundary conditions in the theory of thermal conductivity are divided into four types.

1. The boundary conditions of the first kind determine the temperature distribution at the boundary of the region at any instant of time, i.e. $\Theta\ (x_S, y_S, z_S, t) = \varphi\ (t, S)$.

2. A boundary condition of the second kind determines the density of the heat flux for each point of thermal space at any instant of time, i.e. $q(x_S, y_S, z_S, t)$.

3. A boundary condition of the third kind determines the heat exchange at the boundary of the region with the surrounding medium, which can be specified in the form of the Newton-Richman law:

$$q=\alpha\left(\Theta_O-\Theta_C\right);\ \alpha\left(\Theta_O-\Theta_C\right)+\lambda\left.\frac{\partial\Theta}{\partial n}\right|_S=0,$$

where Θ_O – is the surface temperature of the workpiece; Θ_C – ambient temperature; α – is the coefficient of heat exchange, depending on the properties of the material of the workpiece, medium and temperature.

4. The boundary conditions of the fourth kind express the equality of the heat fluxes at the interface between two heat-conducting media and are written in the form of the equation:

$$-\lambda_1\left.\frac{\partial\Theta_1}{\partial x}\right|_S=\lambda_2\left.\frac{\partial\Theta_2}{\partial x}\right|_S.$$

The distribution of thermal sources in space, the time of their action and motion are taken into account by integration.

The fundamental solution of the heat equation (2) is the source function with a continuously varying coordinate x' according to the law$(x'-Vt)$. Fig. 3 shows a thermal source moving in the XOY plane together with the moving coordinate axes [2].

At the initial instant of time, the source $\tau=t'$ occupies a position at a point $A(x'-V(t-t'),y')$, and at the final instant of time – at a point $A'(x',y')$. When moving the coordinate axes by a distance $V(t-t')$, the temperature at the point $B(x,y)$ is equal to the superposition from all positions of the moving heat source. When you move the coordinate axes, the square

 DOI: 10.1051/matecconf/201712901082

of the distance R^2 changes also in the final position of the coordinates axes (for $\tau = t$) at the point $A'(x', y')$:

$$R_K^2 = (x - x')^2 + (y - y')^2,$$

And in the initial position of the coordinate axes at the point $A(x' - V(t - t'), y')$:

$$R_u^2 = \left[x - x' + V(t - t')\right]^2 + (y - y')^2.$$

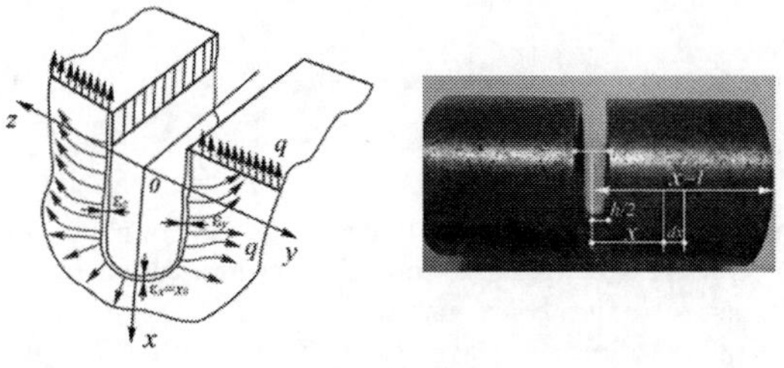

Fig. 3. Diagram of the movement of the heat source.

It is known that if in the n-dimensional space at the point A with the coordinates x'_1, x'_2, x'_3,..., x'_n at the time t', the amount of heat Q is released, then the body temperature at point B with the coordinates x_1, x_2 ,..., x_n at time $(t > t'_n)$) is described by the function [1]

$$\Theta(x_1, x_2, ..., x_n, x_1', x_2' ... x_n', t, t') = \\ = \frac{Q}{C\gamma\left[\sqrt{4\pi a(t - t')}\right]} \cdot \exp\left\{-\frac{R^2}{4a(t - t')}\right\}, \qquad (3)$$

where R^2 – is the square of the distance between points A and B,

where $R^2 = (x_1 - x_1')^2 + (x_2 - x_2')^2 + ... + (x_n - x_n')^2$;

a – is the coefficient of thermal diffusivity, m^2/s.

The function (3) is the solution of the fundamental heat equation (2). On the basis of this solution, we can write down the Green's function for various conditions of the problem.

As is known, the quality parameters of the surface layer of the billet largely depend on the thermal factor. If the cutting conditions and other processing conditions change in such a way that the amount of heat generated in the cutting zone increases, then the shape and size of the cutting groove should be expected to change.

On the basis of the foregoing, the issue of heat distribution between contacting bodies should be decided on the basis of the thermal contact between the two bodies: the cut billet and the abrasive wheel.

Consider the question of the proportion of heat going into the cutting circle. This task must be solved from the prediction of the discreteness of contact between the circle and the workpiece. If for a cut surface the rapid change of thermal impulses on the most active grains merges into a continuous background, then for the cutting circle the contact is always discrete, acting for a short time, depending on the speed of the circle. Advantageous heat dissipation on grains in comparison with a binder makes it possible to model each grain in the form of a half-space having a contact in a small area with another half-space moving relative to the first with the speed of supply of the lateral surfaces of the circle. The question of the distribution of heat between these half-spaces is solved on the basis of the equality of average contact temperatures. This condition [5 – 11] will lead to a formula for determining the input coefficient $\alpha_в$, which shows how much heat goes into the half-space (with respect to abrasive cutting is the coefficient $\alpha_в$ that shows how much heat goes into the workpiece):

$$\alpha_в = \frac{1}{1,25\dfrac{\lambda_K}{\lambda}\sqrt{\dfrac{a}{d_u \cdot V_K}} + 1}, \qquad (4)$$

where λ_K – coefficient of thermal conductivity of the cutting circle, W/(m·K); λ – coefficient of heat conductivity of the workpiece material, W/(m·K); a – coefficient of thermal diffusivity of the billet material, m^2/s; V_K – speed of the cutting wheel, m/s; d_3 – average diameter of the contact area of the cutting grain with the workpiece (wear surface on the grain), m.

This form of writing for the input coefficient is convenient for analysis, since the numerical value $\alpha_в$ can be judged from the size of the dimensionless complex in the denominator as compared to unity. With abrasive cutting at all practical regimes, the complex is considerably less than unity.

$$1,25\frac{\lambda_K}{\lambda}\sqrt{\frac{a}{d_u \cdot V_K}}. \qquad (5)$$

If we take into account that for most abrasive materials the heat conductivity is ten times less than the thermal conductivity of metals, the most approximate estimates show that the fraction of heat going into the circle can be neglected. In fact, if we set the values most typical for the cut for all the values entering into expression (5), for example, for steel 08kp:

$$\frac{\lambda_K}{\lambda} = 0,1;\ a = 8 \cdot 10^{-6} m^2/s;\ d_u \approx 0,1 \text{ mm};$$

$$V_K = 80 \text{ m/s},$$

We get

$$1,25 \cdot 0,1\sqrt{\frac{8 \cdot 10^{-6}}{0,1 \cdot 10^{-3} \cdot 80}} = 0,4 \cdot 10^{-2}.$$

Substituting this value into expression (4), we obtain that $\alpha_в \approx 0,99$, i.e., the circle receives no more than 1% of the total heat. Therefore, when calculating the contact temperature in the cut-off zone, the fraction of heat going into the cutting circle can be neglected.

DOI: 10.1051/matecconf/201712901082

The total density of the heat flux formed in the cutting zone

$$q = \frac{P_Z \cdot V_K}{F_K},$$

where P_z – is the tangential component of the total cutting force, N; V_K – speed of the cutting wheel, m/s; F_K – contact area of the circle and the cut workpiece, m^2.

Using the principle of superimposing temperature fields from elementary sources of heat with allowance for integration in space and time, it is possible to determine the temperature field of a source arbitrarily distributed in space and acting for any time interval according to the formula [3]:

$$\Theta(x) = \frac{q}{2C\gamma\sqrt{\pi a\tau}} \exp\left[-\frac{(x-h/2)^2}{4a\tau}\right],$$

where C – is the specific heat of the metal being cut, J/(кg·K); $h/2$ – the initial size of the groove half of the cut, mm; x – distance from the center line of the groove to the section in question, mm; τ – cutting time, s; γ – density of the material of the cut billet, kg/mm^3.

Fig. 4 shows a graph of temperature variation along the length of the workpiece when cutting a pipe billet from steel 08kp.

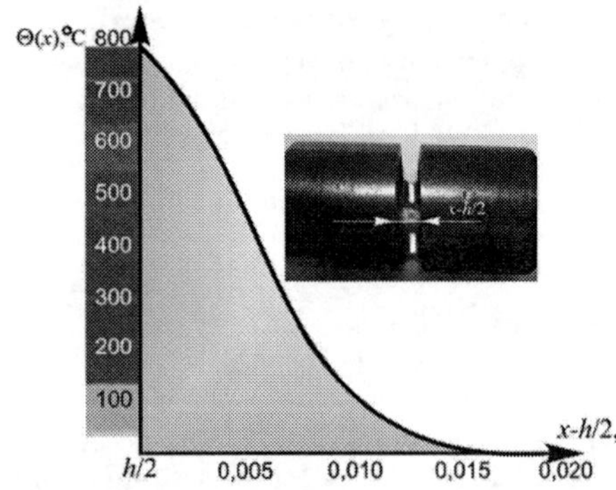

Fig. 4. Graph of temperature change along the length of the workpiece.
$d_3 = 42,4$ mm, workpiece material – steel 08kp, circle 400×4×32, 14AF30

It can be seen from Fig. 4 that the temperature varies along an extremal curve with a maximum of its values at a certain distance from the contact zone of the tool and the workpiece. This location of the maximum on the curve $\Theta = f(x)$ corresponds to the thermophysical nature of the processes occurring in the contact zone. As the source of heat generation moves, localization of the process of contact plastic deformation and, consequently, an increase in the intensity of heat release per unit volume of metal with the approach of the source of heat release and the contact surface. At a distance $x = C_1$, there is a transition from the plastic flow to the boundary friction and the change of the heat source from the volumetric to the source, in which heat release occurs in a thin boundary layer. Further, as the displacement along the contact increases, the intensity of heat release decreases. As a result, the following pattern is formed. Within the limits of plastic contact, the intensity of heat release in the zone of contact plastic deformations is higher than the intensity of heat sink into the tool. Therefore, there is an accumulation of heat and the temperature $\Theta(x)$ rises. At the contact site, the picture changes: the intensity of heat flow from the boundary friction zone to the tool is higher than the heat dissipation rate, so the temperature $\Theta(x)$ in this section decreases.

3 Conclusion

Analyzing the above mentioned, we can conclude that the process of abrasive cutting is accompanied by the release of a large amount of heat, which leads to the formation of a defect layer on the surface of the cut. It is established that the contact of the sides of the circle with the metal is due to the expansion of the material of the workpiece under the influence of heat generated by the cutting, the oscillation of the circle height and the non-uniformity of wear of its lateral surfaces.

The prospect of further research is to determine the kinematic parameters of the abrasive cutting process and their relationship to the geometry of the working surface of the cutting wheel.

References

1. V.A. Sipailov, *Thermal processes during grinding and surface quality control* (Mechanical Engineering, Moscow, 1978)
2. A.V. Yakimov, *Fundamentals of thermal phenomena in the grinding of machine parts* (OGPU, Odessa, 1997)
3. A.V. Yakimov, P.T. Slobodyanik, A.V. Usov, *Thermophysics of machining* (Lybid, Odessa, 2010)
4. Yu.K. Novoselov, *Dynamics of surface shaping in abrasive processing* (SevNTU, Sevastopol, 2012)
5. A.M. Tikhontsov, S.I. Chukhno, A.N. Korobochka, *Thermal processes in the machining of materials by cutting* (UMK VO, Kiev, 2012)
6. S.A. Vasin, *Cut materials: Thermomechanical approach to the system of interrelations in cutting* (MGTU, Moscow, 2001)
7. A.N. Reznikov, *Thermal processes in technological systems* (Mechanical Engineering, Moscow, 1990)
8. N.S. Penkin, A.N. Penkin, V.M. Serbin, *Fundamentals of tribology and tribotechnology* (Mashinostroenie, Moscow, 2008)
9. V.B. Strutinskiy, V.M, Drozdenko, *Dynamic processes in machine tools* (Osnova-Print, Kiev, 2010)

10. V.F. Terentyev, *Cyclic strength metallic materials* (Mashinostroenie, Moscow, 2001)
11. C. Duan, M. Wang, *Some metallurgical aspect of chips formed in high speed machining of high strength low alloy steel*, J. Sci. Scripta Materialia. **52**, 1001–1004 (2005)

DOI: 10.1051/matecconf/201712901083

Improvement of quality of details at round grinding in the conditions of a floating workshop

Sergey Bratan[1], *Ekaterina* Vladetskaya[1], and *Aleksander* Kharchenko[1,*]

[1]Sevastopol State University, Sevastopol, Russian Federation

Abstract. The results of laboratory studies of the grinding processing process are given taking into account the evaluation of the quality of the surfaces of the shafts with disturbing vibrational influences on the equipment of external forces that are the consequence of sea waves, as well as the neighboring operating equipment in a floating workshop. To determine the influence of external influences on the quality of processing during grinding, an experimental stand was used, including a cylindrical grinding machine with a complex of test equipment, as well as a compressor placed at a different distance from the machine, creating external vibration effects simulating external disturbances through the surface Floor and foundation on the machine bed, corresponding to the effects of the external environment and the operating neighboring equipment. The dependences of the frequency of mechanical oscillations (in the measuring zones-the base-bed and the machine table guides in the transverse and vertical directions) are obtained, the amplitude of the vibro-displacement in these zones from the conditions of the circular grinding machine and sea-wave levels. The changes in the error in the shape of the machined parts, the roughness of the machined surface, the waviness of the roll neck surfaces according to the basic and achieved variants are determined taking into account the use of new designs of effective vibration isolating supports and devices.

1 Relevance of the problem

Existence of such mobile mini-plants as floating workshops and road floating workshops with the production sites placed on decks including machining, having a wide range of the machine equipment, including grinders is characteristic of the industry of the seaside cities. Such workshops are capable to perform complex works on production and modernization of hi-tech products, knots and units in areas, remote from the main stationing sites. Feature of operation of the specified equipment is existence of the vibrations caused by impact on him of various sources reducing accuracy and raising a roughness of the processed surfaces during technological process [1].

Improvement of quality of processing of details in the conditions of floating workshops is directly connected with need of reduction of the compelled fluctuations of the machine transmitted through the surface of the deck from external sources including from influences of the external environment — disorders of a surface of the water. As a rule, on decks of floating workshops fixing of machines is provided with the base which doesn't provide sufficient vibration insulation owing to considerable external oscillatory influences, poorly studied so far. Their influence on quality of processing on the finishing operations complicated by oscillatory processes of different level wasn't investigated. In works [2, 3] modeling of interaction of the tool and the processed preparation in the conditions of floating workshops is executed. The dynamic system with difficult stationary and non-stationary vibration influences, and also shock influences from the external equipment and long sea nervousness through floating foundation and the surface of the deck is considered. On the basis of the systems concept the structure of operation is offered, basic provisions and methods of the analysis of process are formulated; input, output variables and parameters of a status of each of subsystems are defined.

To research of processes of grinding of necks of shaft for the purpose of determination of parameters of quality of the processed surfaces received at various levels of external influences under production conditions (fig. 1) and at the experimental stand (fig. 2) in laboratory works are devoted [4, 5].

On the basis of the morphological analysis and synthesis of structures of anti-vibration devices the morphological matrix is constructed, signs and characteristics of subsystems and elements are considered, communications between them are revealed. Options of structures of the anti-vibration device taking into account the essential signs and technical requirements defining properties of system are synthesized. Structural and layout optimization of anti-vibration devices is executed. Parametrical synthesis, theoretical and pilot studies of real designs of anti-vibration devices have led to creation of the new anti-vibration device (fig. 3, *a*) the machine of a floating

* Corresponding author: khao@list.ru

 DOI: 10.1051/matecconf/201712901083

workshop on the basis of a support (fig. 3, *b*) with rubber-metal and hydraulic dampers.

The purpose of this article is development of practical recommendations about improvement of quality of grinding of details in the conditions of floating workshops on the basis of results of tests of new anti-vibration devices in the conditions similar to conditions of a floating workshop.

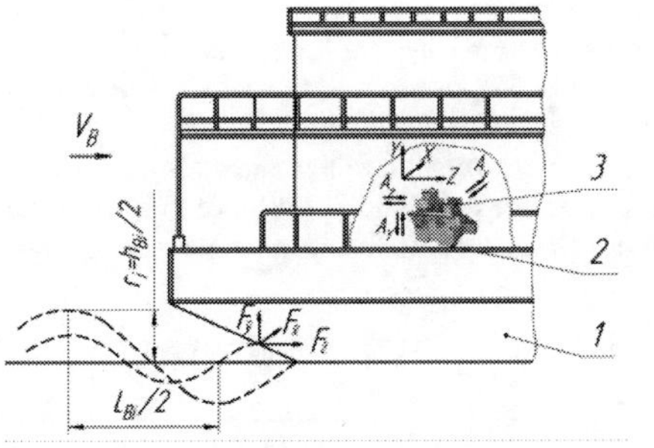

Fig. 1. Scheme of an arrangement the circular grinding machine on the site of a floating workshop: 1 – the floating basis, 2 – the surface of the deck, 3 – the machine 3B12

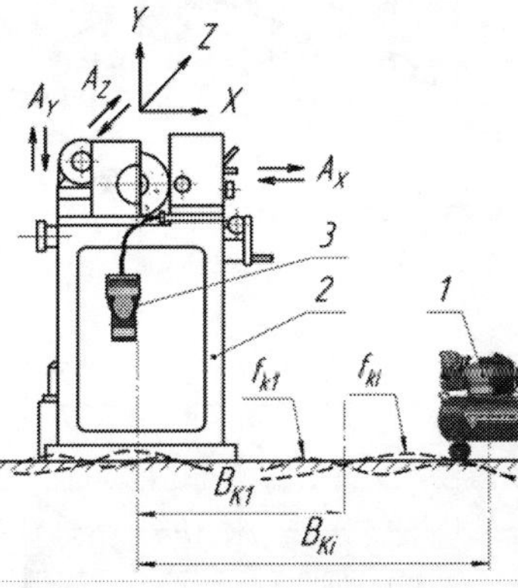

Fig. 2. Scheme of experimental the stand for imitation of the external vibration influences when grinding arising on a floating workshop: 1 – FINI Tiger 245 compressor; 2 – BUA 16 machine; 3 – «Vibrotest-MG4.01» vibrometer

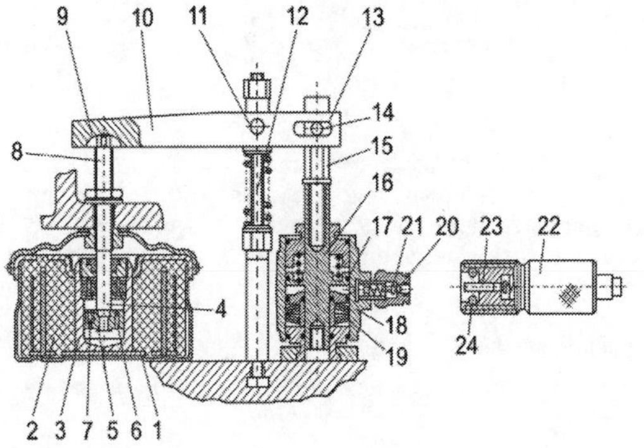

a

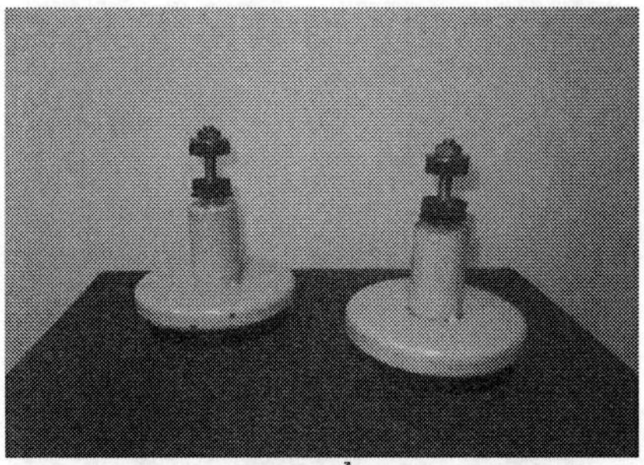

b

Fig. 3. A general view of a new design of the anti-vibration device(*a*) on the basis of a support (*b*) with rubber-metal and hydraulic dampers

2 Technique of carrying out researches

For definition of influence of external impacts on quality of processing when grinding used the laboratory equipment (fig. 2) as a part of the experimental stand which is turning on the circular grinding BUA 16 machine equipped with a control instrumentation complex. The compressor 1 (FINI Tiger 245) placed on different removal (*Bk1* ... *Bki*) from the machine 2, creating during the work on the different modes the external vibration impacts on technological system of the machine (*fk1* ... *fki*) imitating external indignations through a surface of a floor and the base on a machine bed. They correspond to impacts on the machine in a buoyant workshop of an external environment (sea disturbance) and the operating adjacent equipment, causing vibrations and vibration displacements with amplitudes of A_X, A_Y, A_Z respectively in the directions of its axes of coordinates. For determination of parameters of vibrations and amplitude of vibration displacements used the vibrometer 3 «Vibrotest-MG4.01». The carried-out stage of imitating modeling has allowed to define the main conditions under which carrying out grinding processing on the laboratory equipment with receiving the indicators of quality corresponding to these or those conditions of external influences (nervousness level, operation of the external equipment) on round grinding in a floating workshop [6] is possible. By the developed technique and under the set conditions carried out on the circular grinding machine processing of surfaces of necks of shaft ($d = 26,00^{+0,050}_{-0,025}$, *mm*) for kits of parts, imitating various levels of disorders and at the same time existence or lack of the working external equipment for each of parties.

For carrying out tests of new designs at the experimental stand (fig. 2) the machine was installed on anti-vibration support of a new design (fig.4), other processing and control and measuring equipment imitated in vitro different influences for sea nervousness and from work of the next objects.

Fig.4. The circular grinding BUA 16 machine on anti-vibration support of a new design (UA 36389 patent)

3 Main content of work

Results of comparative tests on indicators of vibration impacts on the circular grinding machine at installation on the base (basic option) or on anti-vibration support (the reached option) are received in vitro, corresponding to conditions of a floating workshop. Dependences of frequencies of mechanical oscillations (*fkx, fky, f'ox, foy*) from statuses of the circular grinding machine and level of disturbance $h_{3\%}$ are illustrated with diagrams on figs. 5. Lowering of frequency of mechanical oscillations (fig.5, *a*) *foy* in the lower part of a bed in the vertical direction (along a Y axis) on average from 1,4 times (in case of the rough sea and the inoperable machine) to 2,5 times in case of the rough sea and the operating machine. Vibromovements of *Sox* in the lower part of a bed in the cross direction (along axis X) decrease on average from 2,14 times (at a rough sea and the working machine) to 3,74 times at a rough sea and the idle machine. Vibromovements of *Soy* in the lower part of a bed in the longitudinal direction (along axis Y) decrease on average from 1,1 times (at weak nervousness and the idle machine) to 1,8 times at a rough sea and the working machine.

The analysis of indices in case of installation of the machine on anti-vibration support of new construction shows lowering of frequency of mechanical oscillations of *f' ox* in a zone of guides of the machine in the cross direction (along a X axis) on average from 1,8 times (in case of the considerable and strong disturbances of both the operating machine, and at the same time adjacent equipment and cooling) to 2,1 times in case of feeble disturbance (all three statuses of the machine) and the operating machine in case of the rough sea. Lowering of frequency of mechanical oscillations of *f 'oy* in a zone of guides of the machine in the vertical direction (along a Y axis) from 1,5 times (in case of the rough sea and the inoperable machine) to 1,8 times in case of feeble disturbance and the inoperable machine. Vibromovements (fig.5, *b*) *S'x* in a zone of guides of the machine in the cross direction (along axis X) decrease on average from 2,3 times (at strong nervousness and the working machine along with the next equipment and ventilation) to 3,8 times at weak nervousness and the idle machine. Vibromovements of *S'y* in a zone of guides of the machine in the longitudinal direction (along axis Y) decrease on average from 1,7 times (at strong nervousness and the working machine along with the next equipment and ventilation) to 2,9 times at a rough sea and the working machine.

When performing circular grinding processing of surfaces of necks of shaft results of quality check of the surfaces processed at the same time are received (table 1).

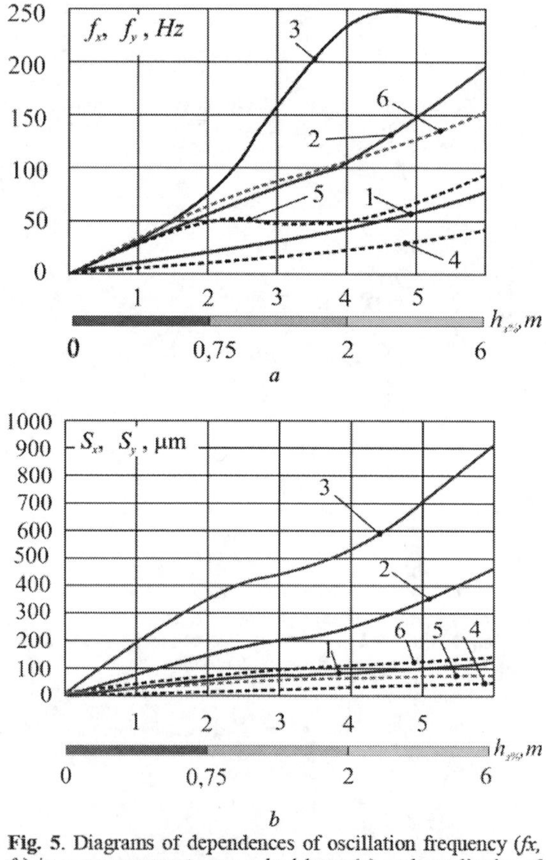

Fig. 5. Diagrams of dependences of oscillation frequency (*fx, fy*) in a measurement zone a bed base (*a*) and amplitudes of vibration displacement (*S'x, S'y*) on guides of the desktop of the machine (*b*) in case of different statuses of the circular grinding machine (1, 4 – it is disconnected; 2, 5 – works; 3, 6 – works along with the face grinding machine and cooling) from disturbance level

DOI: 10.1051/matecconf/201712901083

Table 1. Comparison of results of quality of surfaces of necks of the shaft processed by the basic and reached options

Beating of basic necks – Δ , *mkm* (basic / reached)			**Roughness of surface –** R_a , *mkm* (basic / reached)		
max	min	Dispersion σ^2, *mkm*^{2}	max	min	Dispersion σ^2, *mkm*^{2}
Weak waves					
0,025/0,016	0,010/0,011	6,2x10^{-6}/0,7x10^{-6}	0,63/0,50	0,41/0,32	1,3x10^{-3}/0,9x10^{-3}
0,048/0,029	0,009/0,008	4,2x10^{-5}/1,2x10^{-5}	0,80/0,63	0,51/0,42	2,3x10^{-3}/1,2x10^{-3}
Significant waves					
0,094/0,052	0,018/0,015	15,9x10-5/3,8x10^{-5}	1,00/0,82	0,46/0,54	8,1x10^{-3}/2,2x10^{-3}
0,123/0,079	0,008/0,017	3,7x10^{-6}/1,1x10^{-6}	1,25/1,02	0,50/0,48	15,6x10^{-3}/8,1x10^{-3}
High waves					
0,124/0,069	0,020/0,014	2,9x10^{-6}/8,4x10^{-5}	1,6/1,25	0,50/0,52	3,3x10^{-2}/1,5x10^{-2}
0,162/0,097	0,023/0,018	5,4x10^{-4}/1,7x10^{-4}	2,0/1,58	0,63/0,62	5,2x10^{-2}/2,6x10^{-2}
Waveform indices					

Weak waves			
W_{max}, *mkm*	W_z, *mkm*	S_w, *мкм*	L_w, *мкм*
5,6/5,1	4,8/4,2	192/169	960/845
7,2/5,8	5,3/3,9	265/211	1457,5/1170
Significant waves			
6,9/5,7	5,4/4,2	297/264	1782/1420
10,2/6,1	5,8/5,0	916/870	4580/4350
High waves			
15,4/7,8	9,9/6,1	574/526	3920/2530
23,8/10,2	10,6/7,4	1328/1170	6640/5850

The graphs $\Delta = f\ (h_{3\%})$ in fig. 6, a show that the difference in the deviation value of the shape of the roll neck surfaces (Δmax-Δmin) treated in the base version is greater than in the achieved variant with vibration isolating supports, respectively, by an average of 2.3 times when the circular grinding machine is operated and the external equipment is idle (from 1.9 for strong waves, to 3 for weak waves) and for 1.82 times for a surface grinding machine and ventilation (from 1.76 for strong waves to 1.86 for weak ones). Also, the graphs (fig. 6, *b*) $Ra = f\ (h_{3\%})$ of the roughness deviations for Ra max confirm the quality improvement with the use of new vibration-proof supports 1.25 times on average (both with external equipment and with a surface-grinding Machine and ventilation, and for all levels of excitement, the improvement in quality by roughness is approximately the same - from 1.22 to 1.28 times). Changes in the indices for Ra min are not significant at that, although also the achieved variant gives a decrease in roughness averagely from 1.03 times (with the operation of the circular grinding machine and non-working external equipment) to 1.08 times (when working simultaneously with the surface grinder and ventilation).

Dependence graphs $W = f\ (h_{3\%})$ of the waviness indices (fig.7,*a*) for *Wmax* and *Wz* of the roll neck surfaces from external influences and conditions of the circular grinding machine, as well as an example of the obtained profilogram in the basic version (*b*) and using new vibration-isolating supports (*c*) show that the decrease in the waviness of *Wmax* averages 1.39 times (when a cylindrical grinding machine is used and external equipment is not working) to 1.75 times (with a working circular grinder simultaneously with a surface grinding machine and ventilation). Similarly, in the parameter *Wz*, the decrease in the undulation with the use of new vibration proof supports averages 1.35 times, respectively, for processing in the absence of external equipment and 1.32 times with a working circular grinder simultaneously with a surface grinding machine and ventilation.

Comparison of the obtained profilogram with the basic version of the treatment with the cylindrical grinding machine installed on the foundation (fig.7,b) with the achieved variant with the use of new vibration isolating supports (c) clearly illustrates the advantages of the measures developed and methods for synthesizing rational variants of vibration isolation devices.

Thus, the vibration-isolating supports (fig. 3, *b*) of patent UA 36389 confirmed the achievement of the stated goal in the process of testing - improving the quality of grinding when working in a floating workshop. However, in the test specimens of the supports for use in the floating workshop, some fastening elements are necessary, since the technological equipment located on the decks of the floating workshops is exposed to the danger of displacement due to the detachment of the support base from the surface of the deck. To do this, they must be securely fixed using a lever system on the rack and equipped with a hydraulic cylinder with a spring battery. Such vibration-proofing devices (patent UA 51621) adapted to the conditions of floating workshops (patent UA 51621), which are created on the basis of laboratory-tested supports,

DOI: 10.1051/matecconf/201712901083

provide an improvement in the quality of grinding processing of parts by reducing the errors in their shape, reducing roughness and ripeness by reducing the levels of external and Internal vibrations, which are suppressed by vibration isolation supports and spring accumulators in hydraulic cylinders. In addition, it is possible to further modernize these devices in order to create automatic vibration protection and vibration isolation systems for grinding machines and other precision machines capable of modifying the damping properties of vibro-supports as a result of signals from vibration sensors during machining of parts in floating workshops.

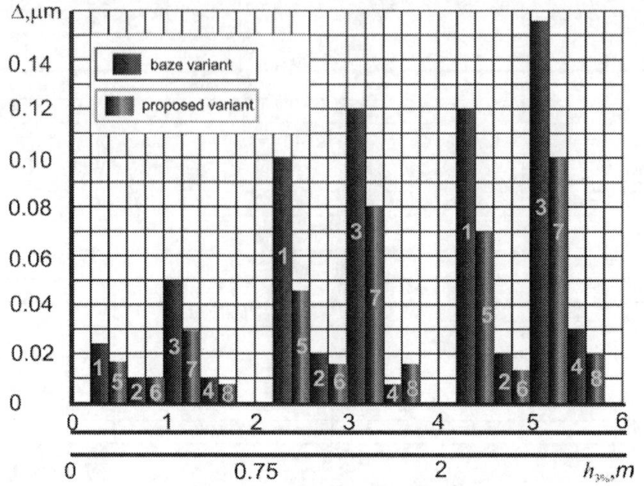

a

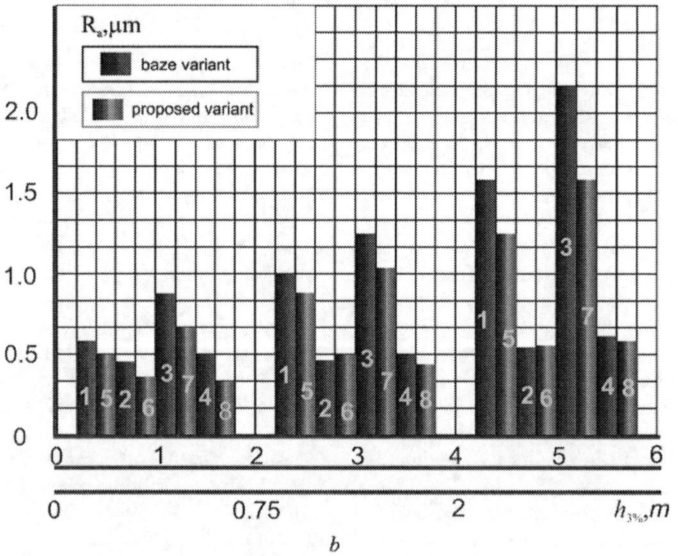

b

Fig. 6. Dependency curves of the deviations of the form $\Delta = f(h_{3\%})$ for Δmax (1, 3 - base, 5.7 - achieved variant) and Δmin (2, 4-base, 6, 8 - achieved variant) (*a*) , and also the roughness $Ra = f(h_{3\%})$ for *Ra max* (1, 3 - basic, 5, 7 - the achieved variant) and *Ra min* (2, 4 - base, 6, 8 - the achieved variant) (*b*) shafts from external influences and conditions of the circular grinding machine (1, 2, 5, 6 - works, 3, 4, 7, 8 - works simultaneously with a surface grinding machine and ventilation)

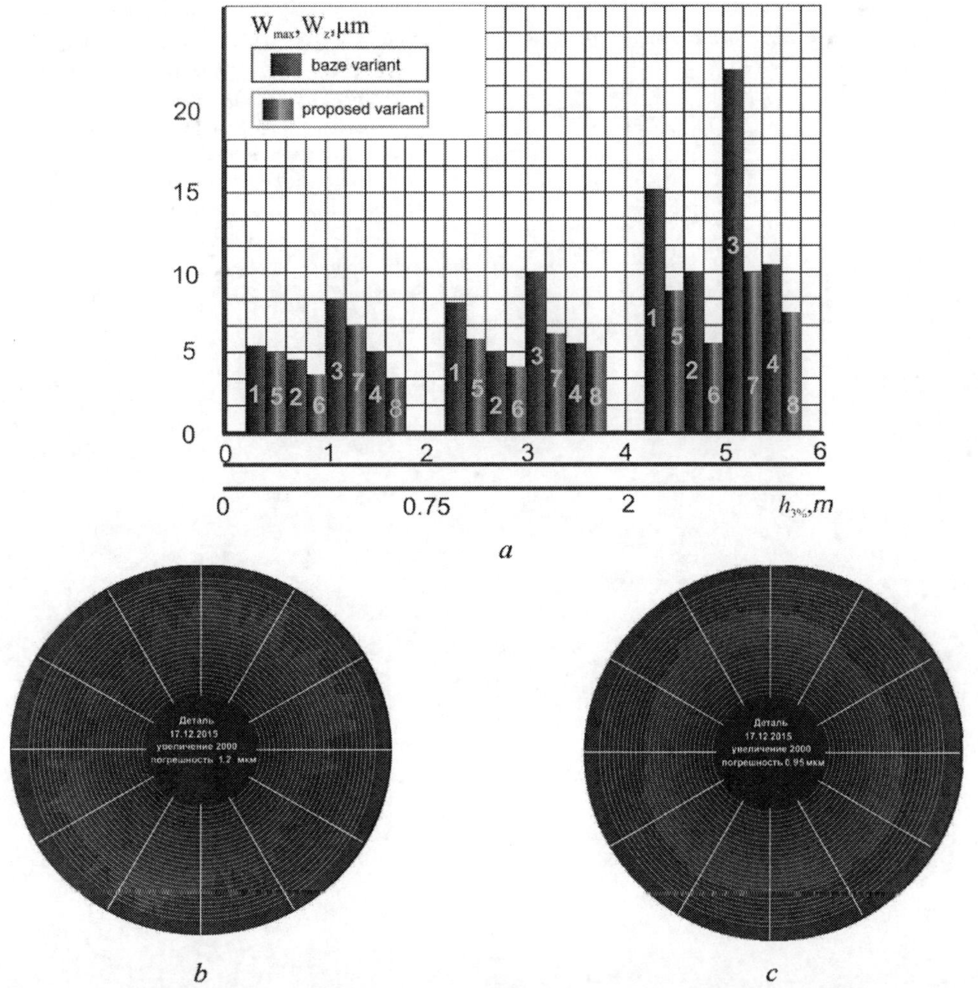

Fig. 7. Dependence graphs of the waviness indices $W = f(h_{3\%})$ for *Wmax* (1, 3 - base, 5.7 - achieved variant) and *Wz* (2, 4 - base, 6, 8 - achieved variant) (*a*) shafts from external influences and conditions of the cylindrical grinding machine (1, 2, 5, 6 - works; 3, 4, 7, 8 - works simultaneously with the surface grinding machine and ventilation), as well as an example of the roundogram obtained in the basic version (*b*) and using new vibration-proof supports (*c*)

Conclusions

This approach allowed us to experimentally investigate the influence on the quality of parts during grinding processing of sea waves affecting the floating base of the workshop, revealed the expediency of using a new vibration protection system of the machine for solving the problems of ensuring dynamic stabilization of the grinding process [1,2] and creating structures for effective vibration isolating supports and devices. The proposed constructions of vibration isolating supports and devices simplify the possibility of mounting and dismantling of technological equipment, contribute to improving the quality of machining of parts by reducing the errors of their shape by reducing external vibrational effects. The obtained results are the basis for further experimental studies of the parameters and characteristics of the technological system of a cylindrical grinding machine in the conditions of a floating workshop [6]. In addition to the above-mentioned measures in the conditions of floating workshops, the use of high-precision CNC machines equipped with adaptive control systems based on the development of new models of optimal systems with a stochastic process controller is possible.

References

1. E.A. Vladetskaya, S.M. Bratan, A.O. Kharchenko, D.O. Vladetsky, *Maintenance of quality of grinding processing by reduction of external disturbances in conditions of floating workshop*, Fundamental and Applied Problems of Engineering and Technology, № **6(314)**, P.88-103 (2015)
2. E.A.Vladetskaya, *Development of a shaping filter that simulates the dynamics of sea waves in a floating repair shop*, Bulletin of SevNTU, №.**150**, pp. 36-40 (2014)

3. A. Kharchenko, S. Bratan, E. Vladetskaya, *Increasing the accuracy of the round grinding process by efficient vibration isolation of the machin,* International Scientific Conference on engineering design and research of automotive vehicles and machines «SAKON'08», P. 47-58 (2008)
4. A. Kharchenko, E. Vladetskaya, V. Dolgin, S. Bratan, *Reaction of the dynamic system to an arbitrary signal by the example of a machine in the conditions of a floating workshop, Monografie* (Lublin, Polska: Politechnika Lubelska, 2015)
5. S.M. Bratan, E.A. Vladetskaya, *Investigation of the reliability of a vibration-isolating device of a grinding machine*, Science intensive technologies in mechanical engineering, № **9 (63),** p. 10-15 (2016)
6. E.A. Vladetskaya, *Modeling vibration exposure from external sources during grinding of details in laboratory conditions*, Jornal of modern technologies: Collection of scientific papers - Sevastopol: SevGU, № **4**, P.25-34 (2016)

DOI: 10.1051/matecconf/201712902001

Modern treatment technologies of carbon fibre for ensuring the high strength carbon fibre reinforced plastic production

Vladimir Nelyub[1], and *Galina* Malysheva[1,*]

[1]FGBOU VO Bauman Moscow State Technical University, 5, 2[nd]Baumanskaya Street, Moscow, 105005,Russian Federation

Abstract. The investigation results of physical and mechanical properties of carbon fibre and carbon fibre-reinforced plastics are presented. It is shown that the number of paramagnetic centres, which were determined by electron paramagnetic resonance method, were used as an index to define the chemical activity of the carbon fibre relative to the matrix. The number of paramagnetic centres in the carbon fibres and CFRP were analysed. The interlayers shear strength of CFRP based on epoxy resin was investigated.

Applications of carbon fibre-reinforced plastics structural products for various purposes are expanding every year due to their unique complex of strength and thermal properties [1, 2] and as well as the development of new moulding technologies [3-5].

It was shown in the paper [3] that oxygen-containing functional groups with the different chemical composition and stability are present on the surface of carbon fibres, which largely determine the magnitude of the adhesive interaction of the fibre-matrix pair.

Some of the functional groups are in ionized form, forming carboxylate ions. Depending on storage conditions, they are capable of hydrolysing according to various schemes, which leads to an increase (or decrease) in the reactivity of carbon fibres. It is quite difficult to quantitatively determine the concentration of oxygen-containing functional groups on the surface of carbon fibres, but it is possible to estimate experimentally the number of paramagnetic centres on the fibre and in CFRP, which also characterize their chemical activity.

The aim of this paper is to determine the number of paramagnetic centres on the fibre and in CFRP, depending on the carbon fibre processing technology.

1 Objects and methods

The objects of the research were domestic carbon fibres in the form of bundles and tape (VMN-4, LU-3, ELUR) and fibres with oxidized surface (LU-P и ELUR-P). Also, imported carbon fibres brands T-300, T-700, AS-4 and HTS-45 were used as research objects.

In the production of carbon fibre-reinforced plastics, an epoxy binder ENPB brand and vacuum infusion technology were used [6].

Interlayer shear strength (σ_{ab}) of six samples with cross section 6×6 mm^2 and length of 41 mm on the working base 30 mm was determined by short beam method according to ASTM D 2344.

The maxing rate of the active gripper of the testing machine was 0.5 mm/min. Investigations of electron paramagnetic resonance (EPR) were carried out by the radio spectrometer RE-1301. To determine the intensity of the EPR signal, the area under the absorption curve, which is proportional to the amount of PMC in the sample, was measured. Their absolute number (N_f) was estimated by comparing the spectrum intensity of the investigated sample and standard, which used diphenylpicrylhydrazyl.

The treatment of carbon fibres was carried out by two methods: thermal oxidation and plasma treatment. A modified Bulat-6 machine unit (at the E8 department of Bauman MSTU) was used for plasma processing. The ion source "AI-200" with an ion beam diameter of 200 mm was designed in Bulat machine, which allows to treat an entire spraying surface area. The surface of LU-P carbon tape was only treated by modified Bulat-6 machine.

2 Results and discussion

In this paper assume that the functions of paramagnetic centres have defects in isotropic matrix of carbon and graphite bundles and tapes in the fibrils, and localization of the electron density which occurred due to the formation of oxygen-containing functional groups. Thus, the activity of the carbon fibre relative to the polymer matrix can be characterized both by the number of paramagnetic centres and by the number of oxygen-containing functional groups. In this study, only the number of paramagnetic centres was measured experimentally. The results of the experimental

* Corresponding author: malyin@mail.ru

DOI: 10.1051/matecconf/201712902001

determination of the concentration of paramagnetic centres on the carbon fibre surface and in CFRP are given in Table. 1. The main disadvantage of this method is the large scatter of data, which is about 25%, which requires a large number of repeated measurements.

As follows from the obtained data, the number of paramagnetic centres in CFRP produced from the carbon fibre brands VMN-4 and LU-3, is almost twice higher than the same fibres in the initial state. At the same time, the number of paramagnetic centres in CFRP produced from the fibres brands ELUR, ELUR-P and LU-P is, on the contrary, significantly lower than the same fibres in the initial state.

Table1. The determining results of the number of paramagnetic centres

Carbon Fibre Brands	The number of paramagnetic centres, 10^{-17}, spin/cm³	
	Fibre, N_f	CFRP, N_{PTP}
VMN-4	1.4	2.7
LU-3	1.3	2.6
ELUR	3.4	1.7
ELUR-P	13.5	0.7
LU-P	7.9	1.1
T-300	3.5	3.7
T-700	7.1	1.2
AS-4	3.3	3.1
HTS-45	3.3	2.6

To explain this paradoxical phenomenon, the study assumes that paramagnetic centres are present on any carbon fibres, but they can be located on the surface (in this case, they are active and interact with the matrix), or they are located in the core of the fibre (in this case, they are not active and do not enter into interaction with the matrix). During the CFPR production, if there is interaction between the carbon fibre and epoxy matrix, then in this case must be inequality:$N_f > N_{КПК}$.

Thus, from the presented data in Table. 1, the fibres brands LU-3, VMN-4, T-300 are inert (they do not interact with the epoxy matrix), and the fibres Elur-P, LU-P, T-700, on the contrary, are active (there is a strong inter-component interaction between the fibre and the matrix).

Table 2 shows the values of the interlayers shear strength obtained from the carbon fibres, for which the number of paramagnetic centres was previously determined.

Table 2. Interlayer shear strength of CFRP

Carbon Fibre brands	Interlayer shear strength of CFRP, MPa
VMN-4	22
LU-3	31
ELUR	43
ELUR-P	78
LU-P	64
T-300	30
T-700	64
AS-4	46
HTS-45	34

All CFRP products were manufactured by using the same technology and same epoxy binder under the same conditions.

In this paper assumes that the more the number of paramagnetic centres on a carbon fibre and the less of them remain in CFRP, and the higher the adhesion interaction. The presented results confirmed this assumption, since the maximum value of the interlayer shear strength obtained by using carbon fibre brand Elur-P, which is characterized by the greatest difference between the number of paramagnetic centres in the fibre and carbon fibre reinforced plastics CFRP ($13.5-0.7= =12,8\times 10^{-17}$ spin / cm³)

In this study, a number of chemical activities of oxygen-containing functional groups located on the surface of a carbon fibre are compiled:

–COO >–COOH > –C=O > –OH.

As presented in the paper [3], the most chemically active functional groups are formed mainly on the defects of the quasi-amorphous matrix of carbon in areas saturated with nano-and micro-pores. To change the content of oxygen-containing functional groups on the surface of carbon fibre is possible as a result of oxidative treatment of carbon fibre. Oxidation of selected fibres was conducted under isothermal conditions in the temperature range 550-850°C. In addition to the oxidation method, the plasma treatment was also used by modified Bulat-6 machine. The following processing modes were used: voltage 300 V, discharge current 113 mA, working gas - argon.

The choice of optimal carbon fibre surface processing mode as a result of plasma treatment was carried out by the criterion of the CFRP interlayer shear strength. However, as a result of the conducted studies, it was not possible to find a treatment mode, in which it would be possible to increase the strength of CFRP. On the contrary, with this treatment, there were large mass losses and a significant decrease in

 DOI: 10.1051/matecconf/201712902001

the strengths of carbon fibre and CFRP. Probably, there was a drop in strength due to a significant damage of the structure of the carbon fibre. Thus, the conducted research did not allow to determine the optimal modes. However, work is underway in this direction, because this direction is very potential, since it is a simple and very effective way of changing the structure and chemical activity of carbon fibre.

Along with the use of plasma treatment, the study also carried out the oxidation on the surface of the carbon fibre by standard methods. The choice of the optimum mode of thermal oxidation was carried out according to the criteria of the highest strength and minimum mass loss. Mass loss was determined by thermo gravimetric analysis. The studies found that the mass loss at temperatures up to 650 °C does not exceed 1.5%. When the temperature is raised to 700 °C, it was a little over 4%, and the temperature increases to 750 °C,mass loss also increases to 7.5%.

Thus, as the studied results have established that the optimum oxidation temperature is 650 °C, the oxidation time at a given temperature must not exceed 10 min. Increasing in the duration of oxidation or temperature, there is an increase in the number of paramagnetic centres on the surface of the carbon fibre, but at the same time occur significant drop in the mass of the carbon fibre, which leads to a loss of its strength.

For the investigated carbon fibres, the number of paramagnetic centres on the fibre and in the CFRP were determined before and after oxidation. As a result of experimental studies, there was increase in the number of paramagnetic centres during the oxidation. For example, for LU-P fibre the number of paramagnetic centres increased 2.6 times. In CFRP, on the contrary, after oxidation, it decreased 1.4 times. Similar results were obtained for other carbon fibres brands.

Raman method allows a quantifying ratio of amorphous (active) and crystallite (passive) phases of the surface of carbon fiber. We carried out measurements at a diameter of the laser beam of 1 μm, laser wavelength of 473 nm, spectral resolution of 2 cm 1, delay time of 10 seconds. Area under spectrum curve (ID) with average frequency D characterized a scattering intensity of the radiation on ordering graphite-like fragments of surface of carbon fiber. Spectrum area (IG) at average frequency G characterized the intensity on unordered amorphous fragments. We calculated fraction of amorphous phase using equation:

$$f_S = \frac{I_G}{I_G + I_D}$$

Table 3 shows the values of the quantity of amorphous and crystallite phase of the surface of carbon fiber.

Table 3. Effect of oxidation on ratio of surface amorphous phase of carbon fiber

Brands of carbon fibers	Ratio of phase	
	Amorphous	**Crystallite**
LU before oxidation	0.38	0.62
LU after oxidation	0.53	0.47
ELUR before oxidation	0.43	0.57
ELUR after oxidation	0.65	0.35
AS-4 before oxidation	0.49	0.51
AS-4 after oxidation	0.6	0.4

Results of experimental determination of ratio of amorphous phase fS.. for investigated carbon fibers brands before and after their oxidation and CFRP strength at interlaminar shear are shown in Table 1. From the experimental data it follows that increase in 1.2 – 1.5 times of relative content of amorphous phase occurs at the oxidation. Also, there is a strength enhancement in 1.5 – 2.1 times of CFRPs at interlaminar shear at oxidation.

We found that a correlation between the values of strength at interlaminar shear τab and a ratio of amorphous phase fS is linear after analyzing data from Table 1. Thus the unique deterministic bond is presented and changing of amorphous carbon content on the surface of fiber increases CFRP strength at interlaminar shear.

Specific concentration of active centers per unit surface of carbon fiber is constant at oxidation. This allows us to assume that crystal-chemical structure of carbon fibers is of the same type, although they are made by different manufacturers from different raw materials using different technologies.

3 Conclusions

As a result of the conducted studies, the difference between the amount of paramagnetic centres on the surface of CFRP and in the initial fibre can be used as a quantitative characteristic of the inter-component interaction between the fibre and the epoxy matrix. It was shown that among all of the investigated carbon fibres brands, the most active fibre brands were Elur-P, LU-P, T-700.

The influence of the carbon fibres heat treatment modes on the magnitude of their mass loss and on the interlayer shear strength of CFRP was investigated. It was shown that the optimum value of the oxidation temperature was 650 °C for 10 min duration. By using the oxidation method, it was possible to increase the number of paramagnetic centres on the carbon fibre.

Effect of thermo-oxidation treatment of phase content of surface of carbon fibers was investigated using Raman spectroscopy. We showed that their adhesion to epoxy matrix is defined only by relative content of amorphous part of fiber surface.

DOI: 10.1051/matecconf/201712902001

References

1. G.V. Malysheva, Y.Y. Shimina, Fibre Chemistry, **46 (4)**, 237-240 (2014)
2. N.I. Baurova, Polymer Science - Series D, **2 (1)**, 54-57, (2009)
3. V.A. Nelyub, Fibre Chemistry, **47 (1)**, 40-42 (2015)
4. G.V. Malysheva, E.S. Akhmetova, Y.Y. Shimina, Polymer Science - Series D, **8 (1)**, 17-21 (2015)
5. V.A. Nelyub, A.S. Borodulin, L.P. Kobets, G.V. Malysheva, Polymer Science - Series D, 9 (3), 322-325 (2016)
6. N.I. Baurova, A.Y. Sergeev, Polymer Science - Series D. **7 (2)**, 140-144 (2014)

MATEC Web of Conferences 129, 02002 (2017) DOI: 10.1051/matecconf/201712902002
ICMTMTE 2017

Distinctive features of thermal treatment of potential tool steel 70X3G2FTR

Evgeniy Romashkov[1,*], *Svetlana* Krylova[1], and *Artur* Kuznetsov[1]

[1]Orenburg State University, 460018, Victory Avenue 13, Orenburg, Russian Federation.

Abstract. It was defined the influence of thermal treatment parameters on qualities and structure-phase composition of potential die steel for hot working 70X3G2FTR and made the simulation of temperature stressed condition of hammer die by thermal treatment.

1 Introduction

The problem of stamp tool performance incoordination despite numerous researches and longstanding searches doesn't't have ultimate solution by now and is physicometallurgical problem that has a big scientific-technical and practical value. The main parameter that determines the quality of stamp tool is its firmness that firstly depends on material and heat treatment conditions. That's why the appliance of new steel grades and also the development and improvement of their heat treatment conditions will provide complex approach to the solution of problem of quality and stamp steel performance incoordination [1-2].

2 Research procedure

The material for researches was developed on the base of multiply-factor experiment and regressive statistic analyses [3-5] experienced tool steel grade 70X3G2FTR (C – 0,67 %, Mn – 1,92 %, Si – 0,59 %, P – 0,022 %, S – 0,010 %, Cr – 2,92 %, Ni – 0,09 %, Cu – 0,08 %, B – 0,015 %, Al – 0,012 %, V – 0,062 %, Ti –0,030%). The researches were made on blocked by processing method samples. Deformation ratio by hammering was 30 %. The structure of hammered steel – laminar pearlite and carbides (structure firmness HRC 35-40). Critical points ($A_{c1н}$ = 760 °C and $A_{c1к}$ = 815 °C) were received by dilatometric method on complex Gleeble 3800.

Metallographic researches were made on optical microscopes Axio Observer, Axio 1 (ZEISS) zooming in from 100 to 1000 times and also on electronic focused beam microscope JEOLJSM-6460 LV. Transmission electron microscopy researches were done by the use of diffraction and dark-field analyses on electron microscope JEM 200-CX. The research of carbide base was accomplished by X-ray structural and chemical analyses of carbide fallouts. Carbides were extricated by samples anodic dissolution with 0,02 A/cm^2 conductance and as electrolyte was used solution 7,5 % KCl and 0,5 % of citric acid in water. Electrolyses time was 1,5- 3 h. From anode deposit containing all carbide phases were extricated special carbides. Chemical analyses of carbide deposits were made from two parallel isolated phases. The composition and type of carbide inclusions were defined by X-ray and X-ray diffraction methods. Carbide deposits X-ray patterns were received on diffractometer "DRON-2,0" (U = 20 kV, I = 12 mA) in Fe Kα-emission.

Mechanical and processing characteristics (hardness, small hardness, toughness) were defined by GOST 24622 – 92, GOST 595 – 88, GOST 19265 – 73.

Numeric evaluation of thermal fields and stress changing appearing n the process of thermal treatment was made in computer environment "ANSYS" with the use of finite-element method together with initial and boundary conditions.

3 Research results and their discussion

With the view of structure recrystallization, softening and residual strength level lowering, prevention of flake formation and providing of sufficient metal adaptability by mechanical treatment, hammered samples were incurred spheroidizing annealing. Taking into account these works for steel 70X3G2FTR was set the following annealing cycle: heat to the temperature 780 °C (holding 8 h.) with composite cooling: with the oven till 500 °C and then in the air.

Solubility test of addition elements in carbide fallouts proves that in the process of annealing in researched steel were extricated carbides of type Me_3C, MeC, Me_7C_3 и $Me_{23}C_6$. After annealing there is alloy cementite $(Fe,Cr)_3C$ in which about 28 % of iron atoms are displaced by chrome atoms, the content of manganese is not big and is about 3 %. In carbides with chrome dissolved up to 13 % Mn, 68 % Cr and to 1-2,5 % Ti, V.

* Corresponding author: evgeniyromashkov@yandex.ru

DOI: 10.1051/matecconf/201712902002

The research of steel fine texture after annealing showed that while annealing and following slow cooling happened partial $\alpha \rightarrow \gamma \rightarrow \alpha$ transformation. The base of steel structure 70X3G2FTR makes ferrite for which location density and non-equilibrium form of subgrain boarders are typical (figure 1a). Against the light sections of ferrite are observable chromium carbide entities of type $Me_{23}C_6$, Me_6C. Average size of evolution is about 0,1 micron by minimum distance between the species of (0,3-0,5) micron. On X-ray diffraction pattern there are reflexes ofHa α –phase of several orientations with zone axis (111), (113) and (311) and reflexes of chromium carbide $Cr_{23}C_6$; On figure 1b is shown ferrite space unit with the axle (111), arrow shows reflex (002) $Cr_{23}C_6$ with d-space of 5,311. Steel hardness after annealing is – HRC 20-22.

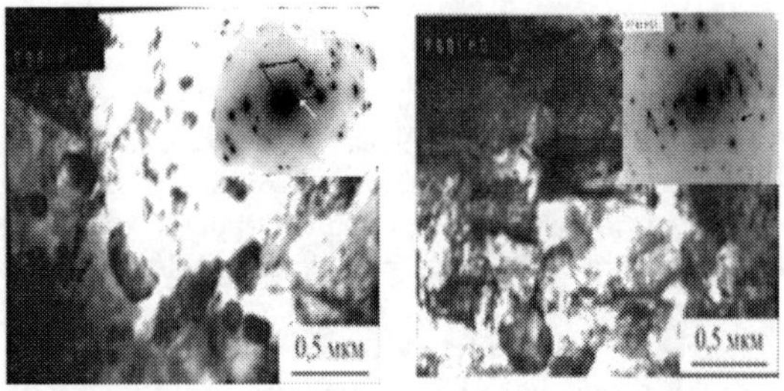

a) b)

Fig. 1. Steel structure 70X3G2FTR after spheroidizing annealing: a, b) bright-field image and X-ray diffraction pattern

In the capacity of hardening heat treatment was used quenching. The research of decay kinetics of undercooled austenite allowed to determine the heat treatment conditions: quenching was carried out by 1000 °C, the heating rate was 25 °C/h with oil cooling.

According to electron-microscopic researches in steel structure after quenching appears rack morphology martensite with high density of dislocation. The length of martensite racks is about 1,5 micron. Dark-field image of this structure section represents that martensite contains a lot of twins (figure 2 a). On corresponding X-ray diffraction pattern with matrix reflexes of α phase there are reflexes from twins located at a distance of 1/3 from the main ones. The structure contains also martensite like areas the origin of which is connected to resolving of austenite in intermediate mechanism. On figure 2b are presented the areas of bainitic structure; it's characteristic that the X-ray diffraction pattern received from these areas contains reflexes of retained austenite one of which is marked by an arrow. The hardness after quenching was HRC 57-62.

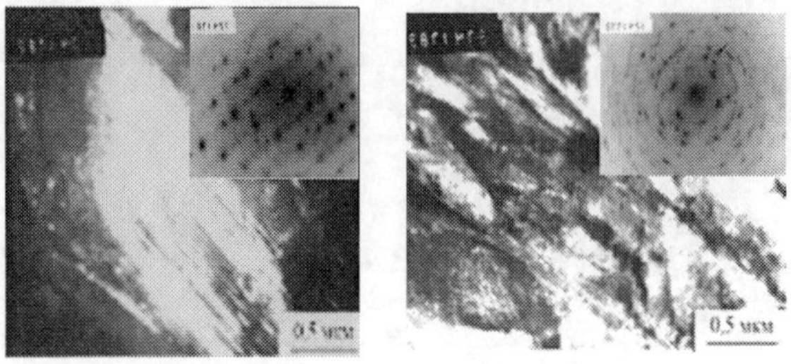

a) b)

Fig. 2. Steel structure 70X3G2FTR after quenching: a) dark-field image (martensite) and X-ray diffraction patternи; b) light-field image (bainite) and X-ray diffraction pattern

To research the processes of structure softening happening by die steel temper, preliminary hardened samples were subject to temper in interval of temperatures 300-700 °C with interval of 100 °C and duration of 1 to 10 hours.

The research results are presented as steel hardness dependence from temper parameters (figure 3).

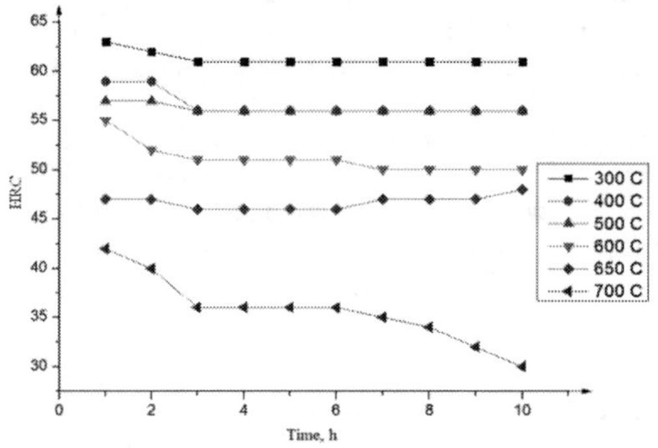

Fig. 3. The dependence of hardness of steel 70X3G2FTR from the temperature and temper duration

The presented dependence clearly demonstrates that by temper temperatures of 300-500 °C during 10 h. steel hardness doesn't change compared with initial hardened condition (HRC 57-62). The rise of temper temperature up to 500-600 °C through all the heat treatment conditions is recorded the hardness at level HRC 50-55 and that determines steel augmented stability to softening by temper in mentioned temperature range. This phenomena can be explained first by the presence in steel matrixes atoms of chrome, manganese, silicon that reduce diffusibility of carbons and slow cementite coalescence, secondly by pinning of dislocation with atmospheres from carbons and vanadium. This stabilization of hardness values by temper high temperatures can point at ageing of steel 70X3G2FTR.

The research of carbide phase forming by temper showed that carbides of type Me_3C and Me_7C_3 are phases of changeable composition in which are replaceable not only atoms of iron and chrome but also presented atoms of titanium and vanadium in different limits. For example by temper temperature changing 500 and 550 °C it's visible the growth in carbide fallout the quantity of alloy cementite but content of chrome and manganese increases slightly whereas by the temper of 600 °C there is a heavy increase of chrome by simultaneous decrease of Me_3C. It proves that alloy cementite $(Fe,Cr)_3C$ goes over special carbide Me_7C_3.

The researches results allowed to recommend the following heat treatment condition on hammer die for hot working from steel 70X3FTR:

1. Spheroidize annealing: heating to the temperature of 780°C with the holding of 8h, composite cooling: with oven up to 500 °C and then in the air;
2. Hardening: heating up to 1000 °C with the holding of 2-3 h, cooling in oil;
3. Temper: heating up to 500-600 °C with the holding of 5 h, cooling in the still air.

In this work by means of mathematical method approach in visual environment ANSYS were calculated temperature and stress conditions by heat treatment of hot working stamps from steel 70X3G2FTR. The basis of calculations is the finite elements method. Thermal fields analyses (figure 4) shows that at the stage of annealing the temperature distribution is more uniform. Temperature gradient in the section is not more than 50 °C (figure 4 a). The hardening makes the metal more stressed. The temperature difference increases and becomes about 100-120 °C (figure 4 b). However at the stage of temper there is equalizing strain and the temperature gradient becomes more equable in all section and is not more than 20-40 °C (figure 4 c).

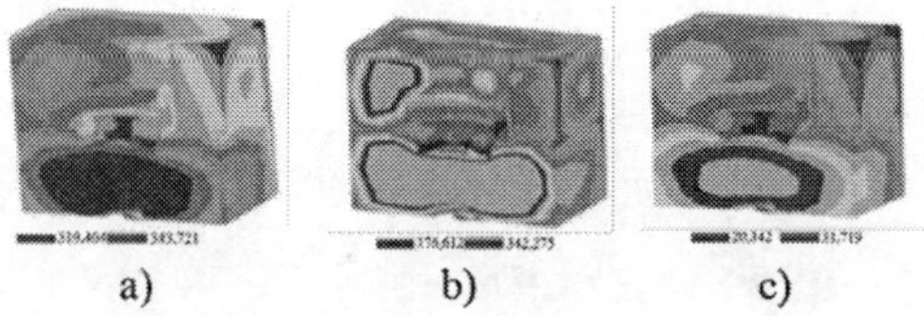

Fig. 4. Temperature fields distribution by steel cooling with temperature of annealing (a), hardening (b), temper (c)

Consequently, the offered heat treatment condition leads to receiving of dispersive ferrite cementite mixture with contained in it special carbides and provides balanced distribution of temperature fields in section of a large-size tool and the formation of low level of residual strength.

4 Conclusions

1. It was established that in the process of spheroidize annealing in steel are formed the carbides of following type: Me_3C, MeC, Me_7C_3 and $Me_{23}C_6$, $(Fe,Cr)_3C$, happens local transmutation $\alpha \rightarrow \gamma \rightarrow \alpha$ as the result of which the basis of steel structure makes ferrite with the high density of dislocation and non-equilibrium form of subgrain borders. Steel structure hardness after annealing is HRC 20-22.

2. It was established that in the process of hardening there is a formation of lath martensite structure with the hardness of HRC 57-62. It was shown that the main carbide phases dissolving by heating are the carbides on the base of Cr and alloy cementite $(Fe,Cr)_3C$. By that 0,5-0,6 % of carbide phase stays insoluble.

3. On the base of electron-microscopic and X-ray diffraction methods it was determined that by temper in temperature interval of 500-600 °C in the steel 70X3G2FTR arises dispersion strengthening by means of release of fine carbides of chrome, titan and vanadium of different types. Hardness after hardening and high tempering is HRC 50-55, stress limit is σ_B = 190 mPa, impact strength is KCU = 112 rJ/m^2.

4. By mathematical model approaches in software environment "ANSYS" it was defined that the offered heat treatment condition leads to uniform distribution of temperature fields and stresses in section of hammer stamp at all stages of thermal influence and also provides the best configuration of mechanical characteristics.

References

1. B.B.Vinokur, V.L. Pilyushenko, *Strength and Fragility of Structural Alloy Steel* (Kiev: Naukova Dumka, 1983)
2. B.B. Vinokur, V.L. Pilyushenko, *Hardening Capacity of Structural Steel* (Kiev: Naukova Dumka, 1970)
3. S.E. Krylova, V.I. Gryzunov, V.A. Lysov, *Application of Method of Prioritization for Optimal Durability Alloy Synthesis,* Materials of international scientific conference "Innovation Activity of Enterprises on Research, Processing and Receiving of Modern Constructional Materials and Alloys" (M.: Engineering, P. 370 -380, 2009)
4. S.E. Krylova (Svoykina), E.V. Bratkovsky, O.V. Samokhina, *On the Question of the Search of Advanced Technology of Impact Abrasive Steel Development*, Festschrift of scientific-theoretical department "Material Engineering and Metal Technology" (Orsk: Publishing house OSTU, P. 129 -138, 2004)
5. Patent № 2535148 Russian Federation, IPC C 22 C 38/38. *Instrumental Steel for Hot Working*, S. V. Kamantsev; applicant and patentee OJSC "Machine-Building Concern ORETO-YUMZ". -№ 2013100389/02. -Decl. 09.01.2013. -Publ. 10.12.2014

DOI: 10.1051/matecconf/201712902003

Investigation of the influence of hollow aluminosilicate microspheres on properties of elastomer fire- thermal protective coatings

Victor F. Kablov[1], *Oxana M.* Novopol'tseva[1], and *Vladimir G.* Kochetkov[1,*]

[1]Volzhsky Polytechnical Institute (branch) VSTU, 404121, 42a Engelsa Street, Volzhsky, Russia

Abstract. The paper considers the influence of the introduction of aluminosilicate microspheres on value of the Payne effect, rheometric, physico-mechanical thermophysical and fire-heat-shielding properties of the elastomer compositions used in the thermal protective coatings based on ethylene-propylene rubber. It was found that partial sorption of vulcanization accelerators and sulfur can occur on the surface of microspheres, followed by their desorption at the vulcanization temperature. This is confirmed by the appearance of a second peak on the vulcanization rate curve. The use of hollow aluminosilicate microspheres in the composition of elastomeric fire-heat-shielding coatings makes it possible to increase the efficiency of these compositions by reducing the thermal conductivity and density of the article while saving the optimum level of physico-mechanical properties. In accordance with the proposed mechanism for the interaction of the polymer matrix and microspheres, additional crosslinked structures are formed within the rubber matrix. An increase in the content of microspheres leads to an increase in the filler-filler interaction, which leads to a relative increase in thermal conductivity. However, the thermal conductivity of compositions containing relatively large amounts of the microspheres is still lower than that of the control. The introduction of microspheres slows down the heating of the sample during high-temperature tests, while reducing the loss of mass and the rate of linear combustion.

1 Introduction

Processes of filler-filler and elastomer-filler interactions are an important characteristic for the elastomer compositions working in dynamic conditions. Typically, filler-filler interactions or the formation of spatial grid is estimated by volume of Payne effect, i.e. reduction dynamic elastic modulus at low strain and increasing dynamic elastic modulus at high deformation. The value of Payne effect correlates with the dispersion degree of the active filler agglomerates and the quantity of maximum deformation per cycle – with material viscosity at the beginning of loading, and thereby a peak load (e.g., at the time of loading rollers).

When creating a fire and thermal protection materials increase the interaction of the filler-filler is not uniquely negative. At an acceptable reduction of physical and mechanical properties increase interaction filler-filler promotes the formation of additional spatial crosslinks exerting a reinforcing effect in the coke layer under conditions ablation erosion and breaking of the material the high velocity gas stream.

The purpose of the research is the investigation of the he influence of the introduction of aluminosilicate microspheres on value of the Payne effect, rheometric, physico-mechanical thermophysical and fire-heat-shielding properties of the elastomer compositions used in the thermal protective coatings based on ethylene-propylene rubber.

2 The experimental part

There were investigated the influence the introduction of aluminosilicate microspheres (MFS) on Payne effect in rubber based on ethylene-propylene caoutchouc containing 30 parts by weight filler (BS-120) and sulfuric curatives. The impact of the microspheres content on physical and mechanical characteristics of the elastomer composition and the Paine effect are presented in Table 1.

A study of the rheometric characteristics (MDR 3000 Professional, ASTMD2084-79) has shown that kinetic curves of vulcanization of samples containing small dosages of MSF (1-3 parts by weight) have flexures characteristic of two-stage vulcanization, which is confirmed by the appearance of a second peak on the vulcanization rate curve (Fig. 1).

This feature of the kinetics of vulcanization can be explained as a partial sorption of the sulfur and vulcanization accelerators by the surface of the microspheres, and followed their desorption at the vulcanization temperature, and by the flow of additional structuring processes (Fig. 2).

* Corresponding author: geminy-i@mail.ru

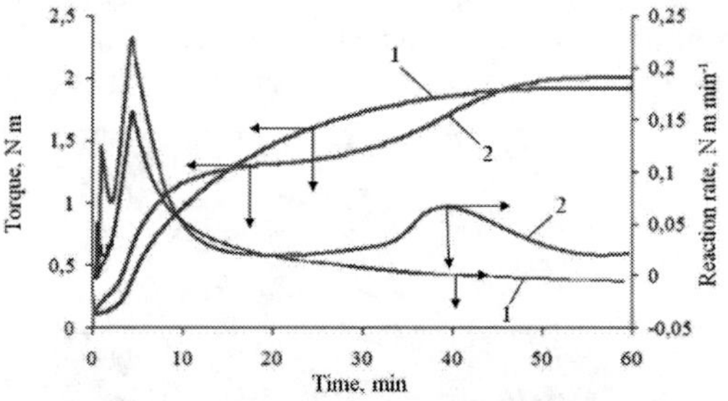

Fig. 1. Kinetic curves of vulcanization and reaction rate curves of the control mixture (1) and a mixture containing 1 wt. MFS (2)

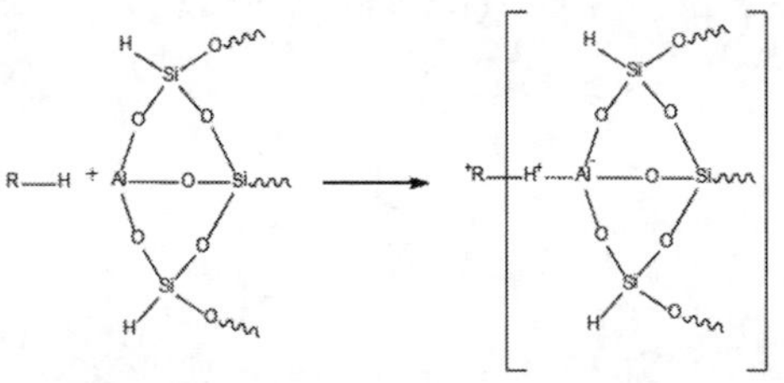

Fig. 2. The proposed scheme of interaction between the MFS and elastomer matrix

The microsphere surface is coated with the network of hydroxyl ions, which are responsible for the adsorption properties of aluminosilicates. Some authors, taking into account possible coordination bond at the Al atom in aluminosilicates, consider the aluminum atom with coordination number 3 as the active site. Upon interaction of macromolecules that form a carbocation with aluminosilicate, the coordination number of the aluminum atom becomes equal to 4 [7, 8] (fig. 2).The progress of such processes is also indicated by an increase in the Payne effect characterizing the degree of interaction of polymer-polymer, filler-filler and polymer-filler and decrease in swelling in toluene.

Fig. 3 shows the dependence of the elastic component of the shear modulus (ΔG ') of the strain amplitude for the test compositions.

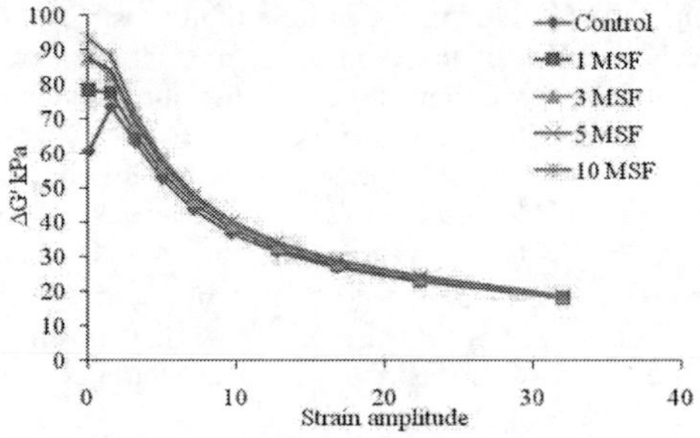

Fig. 3. The dependence of the elastic component of the shear modulus (ΔG') of the strain amplitude

Table 1. The impact of the microspheres content on Paine effect, rheological and physico-mechanical characteristics of the elastomer composition.

Parameter	MSP content, wt parts per 100 wt parts of rubber				
	control sample	1	3	5	10
Vulcanization characteristics of rubber stocks at 165°C					
M_{min} – minimal torque, N m	0,11	0,13	0,14	0,14	0,14
M_{max} – maximal torque, N m	1,92	2,13	2,19	2,01	1,97
τ_S – induction period, min	3,81	2,88	2,83	2,55	2,49
R_V – vulcanization rate, min^{-1}	4,29	4,33	3,71	3,92	4,01
ΔG' – Payne effect	55,37	59,44	64,22	65,29	69,74
Properties of vulcanizates (vulcanization at 165°C, 60 min)					
f_t – nominal tensile strength, MPa	10,3	12,9	12,1	10,9	9,4
ε_{rel} – relative elongation at break, %	643	650	530	500	480
θ_{res} – relative elongation set, %	20	20	20	16	16
ρ – density, kg m^{-3}	1060	1025	1014	1006	998
w – degree of swelling in toluene, %	264	243	219	184	175
Changes in vulcanizate properties after aging in air(125°C,72h)					
Δf_t – relative changes after aging, %	-35,1	-32,7	-33,5	-36,2	-36,9
$\Delta\varepsilon_{rel}$ – relative changes after aging, %	-40,5	-39,7	-40,2	-41,2	-42,0
Heat-protecting and thermal properties of vulcanizates					
$\tau_{T\to100\,oC}$ – warmed up to 100°C time, s	200	270	280	320	330
Δm – sample weight loss in the course of the test, %	28,78	17,87	17,15	15,38	16,25
$V_{l.c.}$ – linear combustion velocity, mm s^{-1}	24,12	23,01	21,42	20,87	20,08
C_m – heat capacity, W m^{-1} K^{-1}	1635	1641	1631	1621	1619
λ – conductivity, J kg^{-1} K^{-1}	0,23	0,19	0,19	0,20	0,21

 DOI: 10.1051/matecconf/201712902003

It follows from the above data, the increased content of MSF leads to increase Payne effect by 20-30 %, which indicates an increase in the filler-filler interactions in the rubber matrix. The constant value of the elastic component of the shear modulus for large deformation, demonstrates consistency in the contribution of the hydrodynamic effect module, polymer-filler interaction and "structure inside the rubber" [5].

The data in Table 1 show that the introduction of MSF allows increase conditional tensile strength by 20-40 %.

Thus, an increase in tensile strength conditional on introducing 1 wt part of microspheres may be due to the formation of additional adsorption bonds. The occurrence of such processes is also suggested by a decrease in the sample swelling.

However, upon further increase microsphere content conventional tensile strength is decreased that is apparently due to a decrease of rubber homogeneity.

To evaluate the flame and heat resistance of the samples, we determined the following parameters using the developed procedure: dependence of the temperature on the unheated sample surface on the time of the action of open fl ame from a plasma torch, the sample weight loss, and the linear combustion velocity.

In high-temperature heating, the temperature created on the sample surface was of the order of 2000 °C.

Introduction of microspheres decelerates the speed of the sample heating in high-temperature tests (Fig. 4).

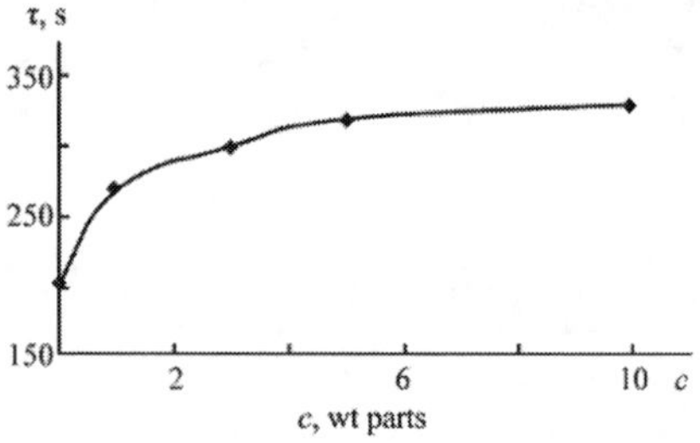

Fig. 4. Time τ in which the unheated sample surface is warmed up to 100 °C as a function of the microsphere content c.

Owing to the occurrence of additional cross-linking processes in the coking layer of samples containing microspheres, the time in which the unheated surface of the sample is warmed up to 100 °C increases (by 35–65 %), and the weight loss and linear combustion velocity decrease.

The forming coke structure is more uniform, and the pore diameter in it is smaller than in the control sample. The mean pore diameter in the prepyrolysis zone decreases from 350–420 to 110–120 μm (Figs. 5, 6).

Introduction of microspheres leads to a 10–20 % decrease in the thermal conductivity of the elastomer compound. An increase in the content of the microspheres can lead to the formation of "thermal conductivity bridges" due to enhancement of the fi ller–fi ller interaction (20–30 % increase in the Payne effect), which, in turn, leads to a relative increase in the thermal conductivity.

However, even at the MSP content increased to 10 wt parts, the thermal conductivity of the compound remains lower than that of the control sample.

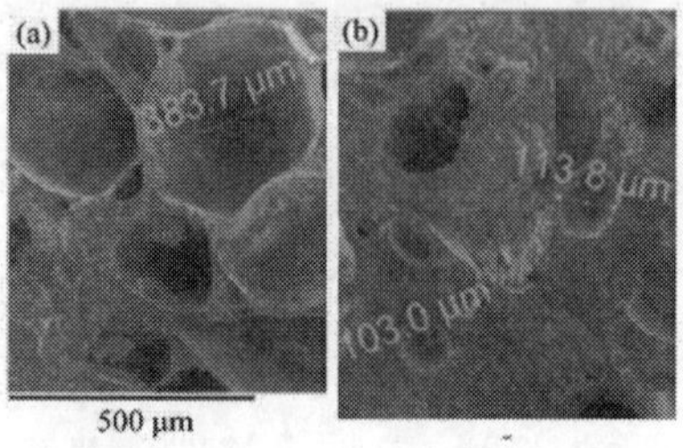

Fig. 5. Structure of the prepyrolysis layer of the samples after the test. Control sample (a) and sample containing 5 wt parts of MSF (b).

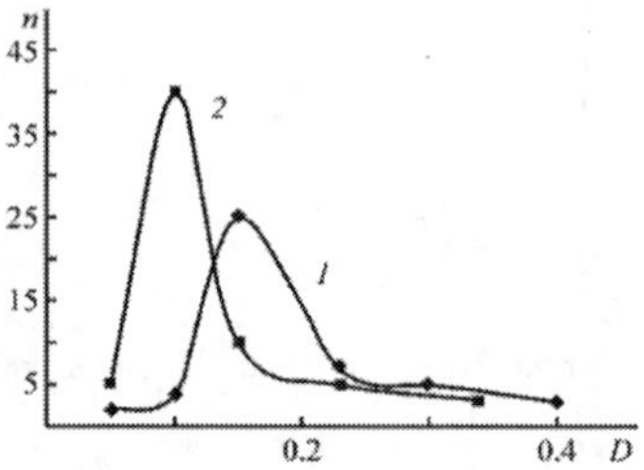

Fig. 6. Pore diameter distribution in the prepyrolysis layer of the samples. Relative amount of pores (n) and pore diameter (D). Control sample (1) and sample containing 5 wt parts of MSF (2).

3 Conclusions

The use of hollow aluminosilicate microspheres as components of elastomer flame-retardant and heatprotecting materials enhances the performance of these compounds owing to a decrease in the thermal conductivity and density of the item, with the optimum level of physicomechanical properties preserved. In accordance with the suggested mechanism of interaction of the polymer matrix and microspheres, additional crosslinked structures within the rubber matrix are formed. An increase in the content of microspheres leads to enhancement of the filler–filler interaction. Constant value of the elastic constituent of the shear modulus at high strains demonstrates the constancy of the contribution made to the modulus by the hydrodynamic effect, polymer–filer interaction, and "structure within the rubber." Hence, optimum choice of the rubber to microspheres ratio allows formation of additional cross-linked structures within the elastomer matrix, favoring manifestation of the reinforcing effect in the coke layer of the material under the conditions of erosion removal and detachment of the material with a high-velocity gas flow. Introduction of microspheres favors formation of a more regular dense finely porous coke structure upon high-temperature treatment, which, in turn, leads to an increase in the time in which the unheated surface of the sample is warmed up to 100 °C and to a decrease in the weight loss and linear combustion velocity.

DOI: 10.1051/matecconf/201712902003

The work was carried out with the financial support of the Ministry of Education and Science of the Russian Federation in the framework of the basic part of the state assignment for 2017-2019. (Project 4.7491.2017 / CU) on equipment purchased under the Strategic Development Program of Volgograd State Technical University for 2012-2016.

References

1. G.E. Zaikov, E.V. Kalugina, and K.Z. Gumargalieva *Fundamental regularities of thermal oxidation of heatresistant heterochain polymers*, Thermal Stability of Engineering Heterochain Thermoresistant Polymers, Utrecht: 280 p. (2004)
2. RF Patent 2552728, Publ. 2015.
3. RF Patent 2473751, Publ. 2013.
4. A.S. Inozemtsev, and E.V. Korolev, *Promyshl. Grazhd. Stroit.*, **10**, pp. 80-83 (2013)
5. RF Patent 2473504, Publ. 2013.
6. K.L. Kandyrin, and A.S. Sedov, *Vopr. Prakt. Tekhnol. Izgot. Shin*, **1 (50)**, pp. 93-100 (2010).
7. J.E. Huheey, *Inorganic Chemistry: Principles of Structure and Reactivity*, New York: Harper & Row, 3rd ed (1983)
8. M.M. Shul'ts, Soros. Obraz. Zh., **8**, pp. 25-38 (1997)

 DOI: 10.1051/matecconf/201712902004

Peculiarities of heavy electro-mechanical plating on medium carbon steel 30 and 38 ChN3MA

Olga V. Gorozhankina[1,*], *Yury R.* Kopylov[1], and *Alexander A.* Zhirkov[2]

[1]Voronezh State Technical University, 14, Moscow Avenue, Voronezh, 394026, Russia
[2]Bryansk State Technical University, 7, 50 let Oktyabrya bul., Bryansk, 241035, Russia

Abstract. One of the urgent problems in mechanical engineering is development of efficient inexpensive thick-layered coatings for repair worn-out large-sized parts. To solve these problems a comparatively new method of electro-plating with simultaneous mechanical impact of a tool upon coating under formation is promising enough. In this paper there are considered the peculiarities of plating through this method of thick-layer electro-mechanical iron plating on medium carbon steel 30 and 38 ChN3MA, and the composition, structure and properties of coatings obtained are analyzed. Coatings were applied upon crank main journals made of steel 38ChN3MA and upon models imitating crank main journals made of steel 30. Micro-hardness of the coating obtained, roughness, a chemical composition and a structure were analyzed. The results of the investigations have shown that a more significant role in hardness increase is played not by a chemical composition of medium carbon steel upon which it is applied, but a preparation of the surface and the technology of coating and also an electrolyte composition. The structure of electro-mechanical plating is less defective, contains a value of detrimental impurities considerably lower of allowable values and has a satisfactory adhesion strength with a substrate.

1 Problem definition

At present one of the urgent problems in mechanical engineering is development of efficient and inexpensive heavy coatings for repair of worn large-sized parts [9, 10, 11]. Such coatings must have an increased thickness, good durability of adhesion with a basis, sufficient hardness and wear-resistance. In world practice the reduction of worn-out large-sized parts is carried out by retailoring [12, 13, 14], sputtering [15, 16, 17], cladding [18, 19, 20] and so on. But all these methods suppose an intensive thermal impact upon a part under reduction that results in hard-removable buckling and shrinkage.

Therefore, more and more attention is paid to the process of galvano-mechanical metal deposition without temperature increase – flow-contact deposition of galvanoplatings with the mechanical activation of a cathode surface [5, 6, 7, 8]. It is well-known that the mechanical effect upon a part in the course of galvanic deposition allows increasing the allowable current density, a deposition rate, micro-hardness, coating thickness and also improving the adhesion with a basis and decreasing residual stresses.

2 Investigation procedure and equipment

The considered in this paper off-bath method of the restoration through an electro-galvanic dry topping in a flowing electrolyte with the mechanical effect by a smoothing tool in the course of deposition is realized at small temperatures (40 – 50 ^{0}C). It prevents buckling and shrinkage in parts and ensures the required thickness (1.5 – 1.8 mm), micro-hardness (600 - 650 HV) and low roughness (Ra 1.25 – 2.5 μm) stipulating for a small allowance for the further grinding of a layer restored. The method has small material and power expenses and a quick recoupment at the espense of the application of ecologically acceptable inexpensive electrolyte of iron chloride. The restored layer contains detrimental impurities considerably lesser of admissible values and has satisfactory strength of adhesion with the basis (200 – 250 MPa).

Taking into account that at the current moment the works on electro-galvanic mechanical coatings are precious few and the investigation of their composition and structure were practically not carried out and it seems to be expedient to consider this problem in relation to some medium carbon kinds of steel used for manufacturing large-sized parts which further are supposed to be reduced according to the given technology.

For the possibility of practical use of such coatings during the repair of worn large-sized parts (shafts) there was carried out an investigation on the application and properties of a electro-galvanic mechanical coating (EGMC) upon worn parts of a diesel engine crankshaft.

A crankshaft of diesel engine 1A-5D49 is produced of steel 38ChN3MA in Russia. Dimensions: length is 3951 mm; the diameter of main necks is ø 219 mm.

* Corresponding author: winter.07@mail.ru

DOI: 10.1051/matecconf/201712902004

Journals are nitrided, the depth of a nitrided layer is 0.5…0.7 mm. The requirements to journals reduced are the following: hardness of a reduced layer is to be not less than HV 460, at the thickness not less than 400 µm; the adhesion strength of the reduced layer with the basis not less than 100…150 MPa; roughness Ra is 0.16 µm.
Coatings are to be carried out on crank main journals cut out preliminarily from a basic part (Fig. 1). At that the journals friction wear at which the further operation is already impossible makes about 0.5…0.7 mm, that is, is equal to the depth of a nitrided layer. The residual margin of fatigue resistance is 55…70%. For plating there is used a specially developed device (Fig. 2).

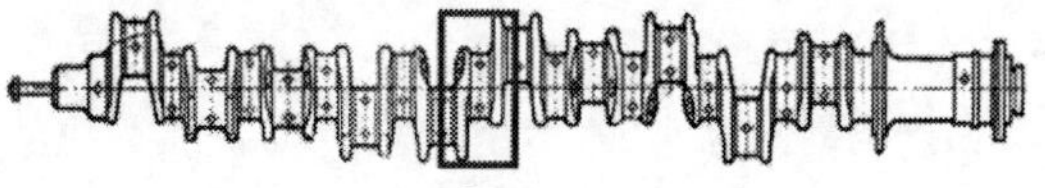

Fig. 1. Crankshaft of a diesel engine; 1 – crank main journal

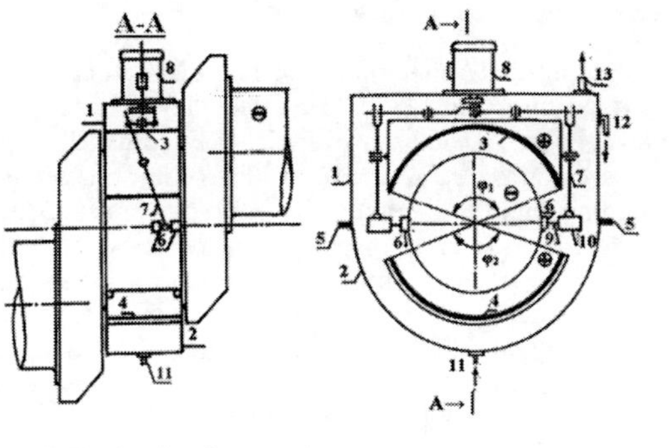

Fig. 2. Device for electro-galvanic mechanical plating: 1,2 – two halves of a split container; 3,4 – anode two parts; 5 – seal; 6 – tool for smoothing; 7,9 –levers; 8,10 – electrical drive; 11 – outlet for electrolyte feed; 12 – outlet for electrolyte discharge; 13 – outlet for removal of hydrogen formed.

The device contains an electro-galvanic cell as a split conatiner, anode consisting of two parts and a tool for smoothing a coating layer under formation. As anode there are used electrodes made of steel 30. As tools for the mechanical effect upon a layer under reduction the mineral ceramic plates are used. Flow electrolyte is fed into a cell. A part rotates, at current supply begins a process of iron sedimentation on a part surface and at the same time a surface of the part is smoothed by a mineral ceramic tool. At the same time the effort for tool pressure to the coating surface is normalized, a tool performs reciprocating movements.

A very significant moment is a preparation of a part surface to sedimentation, in particular, an electro-chemical pickling. As a rule, a pickling is carried out in a separate bath and during the transfer of the prepared part from the bath of pickling into the bath of sedimentation on the surface of the part there is appeared a thin juvenile layer of oxides which decreases considerably adhesion strength of a coating with main metal.

In our case pickling is carried out in the bath of dry topping without oxygen access in order to decrease or even exclude completely the formation of an oxide layer and increase the adhesive strength of a coating with main metal. After the surface preparation the electrolyte of pickling is pumped out from a cell, a bath is washed out and further an electrolyte of dry toppong is supplied in a cell in a flow mode (consumption 10 l/min) which is performed already up to the end of the process of electro-galvanic mechanical dry topping.

Sedimentation begins with the realization of the so-called accelerating mode:

- dc is supplied to the part under reduction in the course of 5 min at the least values of cathode current density $i_к = 500$ A/m^2;
- current density increases gradually up to operational values 2500 A/m^2 in the course of 10-15 min;
- after this for the obtaing of an electro-galvanic mechanical coating there is included a mechanical impact in the process which effects a surface of sedimentation by a polished electro-chemically neutral tool adjusted beforehand to the essential parameters. In the course of sedimentation a temperature, a concentration and acidity of electrolyte, current density are supported constantly.

The average value of the initial micro-hardness (before coating) of a shaft main journal was obtained according to 48 measurements and makes HV 325 units; an error from HV 18 up to HV 26. The average value of micro-hardness of the obtained EGM coating after polishing is obtained according to 120 measurements and makes HV 730 units; an error from HV 20 up to HV 54.

The thiskness of EGM coating for one side is 0.85 mm; the diameter of of a shaft journal with coating after polishing is ø 220.4 mm. The coating according to geometry and hardness corresponds to the technical specification.

Coating roughness Ra according to data of 10 measurements is 0.325 – 0.45 µm and does not exceed the requirements of the drawing Ra 0.63 µm.

These results are obtained at Voronezh Engineering Plant in the central measurement laboratory.

For convenient investigations with the aid of a metallographic microscope and for a more thorough study of the structure, EGM- and electrogalvanic (EG) coating is applied on the surface of cylindrical hollow samples made of steel 30. The dimensions of models imitating journals of crankshafts are the following: length 100 mm, external diameter ø 60 ± 0.2 mm, inner diameter ø 54 ± 0.2 mm (Fig. 3, a). The average hardness of a test surface of samples before dry topping HV 230-250.

a b

Fig. 3. Overview of the model of the part with an outer surface prepared for dry topping – a; metallographic sample for metallographic investigations cut out from a cylindrical model of the part - b

DOI: 10.1051/matecconf/201712902004

Electrolyte composition and sedimentation modes of EGM- and EG coatings are shown in Table 1.

Table 1. Electrolyte composition and sedimentation modes of EGM- and EG coatings

Sedimentation parameters	With mechanical effect (EGM)	Without mechanical effect (EG)
Electrolyte composition	$FeCl_2 \cdot 4H_2O$	$FeCl_2 \cdot 4H_2O$
Chloride iron concentration in electrolyte, C_{Fe}, kg/m^3	350	350
Electrolyte temperature, T, ^{0}K	333	333
Electrolyte acidity, pH	1,0	1,0
Cathode current density, $i_к$, A/м^2	2500	2500
Tool pressing force to coating surface, F, H	2	-
Frequency of tool reciprocating displacements, $\omega_н$, min^{-1}	40	-
Circumferential velocity of part rotation, V_0, m/sec	0,2	0,2
Electrolysis time support, t, h	6	6

3 Research results and their analysis

After having obtained a coating one cuts from the part some samples with dimensions about 22x22x3 mm and further they are under analysis (Fig. 3, b). In case of EG coating the surface for the sedimentation is prepared in a common way in a separate bath for pickling. A chemical composition of EGM coating and a basic metal are analyzed through Coulomb metering method with the aid of express-analyzers. At that the percentage of the content of carbon, hydrogen and sulfur in a surface layer (Table 2).

Table 2. Chemical composition of EGM coating and basic metal

Chemical composition of EGM coating and basic metal		
Chemical element	Content, %	
	Basic metal, steel 30	EGM coating
Carbon (C)	0.223	0.086
Hydrogen (H)	0.0042	0.0043
Sulfur (S)	0.0088	0.0021

According to the results of the investigation hydrogen content in EGM coating is small (0.004 – 0.005%) which corresponds approximately to hydrogen quantity (0.0042%) in basic material (steel 30) obtained in a metallurgical way. It may be explained by that the adatoms of hydrogen from the surface of the coating deposited are eliminated systematically and quickly enough by a smoothing tool that is why it is possible to suppose that such a low content of hydrogen in the coating obtained depends just upon this. As a rule in electro-plated coatings without a mechanical effect the hydrogen content is higher and makes 0.0098 – 0.01%, that increases brittleness.

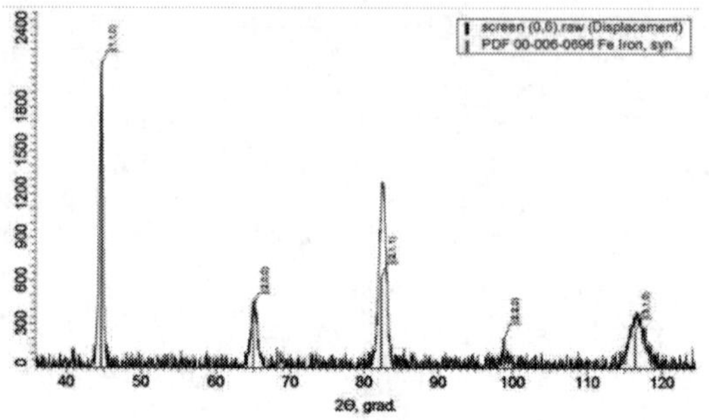

Fig. 4. Diffractogram of a heavy (1.8 mm for one side) EGM iron coating obtained on the diffrectometer of D2Phaser Bruker type with the use of Cu-radiation

The fulfilled analysis of X-ray diffractograms of EGM coating (Fig. 4) has revealed the existence of a solid solution on the basis of lettice α-Fe to which points out a certain displacement of diffraction maximums regarding their position on the diffractograms of pure iron plating. Such a displacement points out to the presence a small quantity of admixtures in an iron coating which is confirmed by the results of Coulomb metering and electronic analysis. The obtained diffractogram of EGM heavy iron coating is a common diffractogram of a polycrystal and represents a set of peaks on a smooth line of the background. According to data of the X-ray structural analysis the substance under investigation is single-phase.

In general case of electroplated coatings obtained in chloride, sulfated and other iron electrolytes there is observed an obvious lamination of the structure caused by layers formed occasionally on a cathode with a large content of hydroxides. On their surface there are formed layers of polycrystals with a metal structure and a considerably lesser content of hydroxides [1] (Fig. 5, b).

During the investigation of EGM heavy iron coatings on the optical microscope a lamination is not detected (Fig. 5, a). It is explained, in the author's openion, with that a smoothing tool affects occasionally and intensive upon coating crystals formed and contributes to electrolyte mixing at the cathode and also to the partial purification of a cathode layer from hydroxides which eliminates an interval in the formation of layers with hydroxide insertions. The application of the mechanical effect with a smoothing tool results in the formation of a more dense and even structure of EGM coating.

One of the significan parameters of EG and EGM coating is thiskness. A maximum obtained thickness of a coating and the dependence of its increase upon the time of a process for EGM dry topping and common EG dry topping without a mechanical effect are under investigation. The results of the investigation are shown in the graph (Fig. 6).

DOI: 10.1051/matecconf/201712902004

Fig. 5. Coating microstructure on a microsection: a – EGM coating; b - EG coating obtained at the same mode

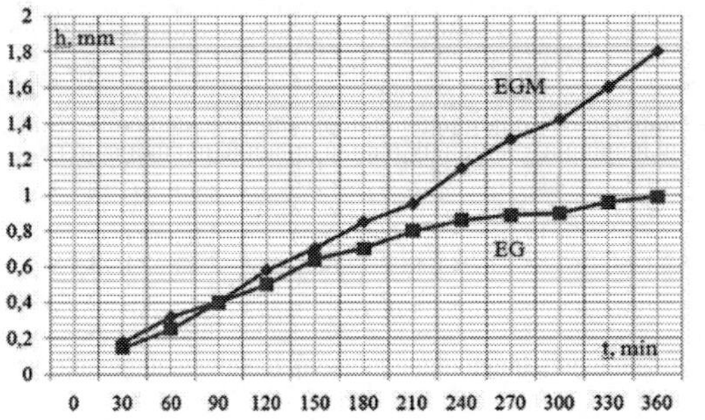

Fig. 6. Dependence of the thickness of EGM and EG coating upon the sedimentation time

The maximum achieved thickness of EGM coating during the 6 hours period of sedimentation makes 1.8 mm without a layer increase delay. In EG coating occurs an increase delay of the sedimentation thickness; a general thickness of a layer during the 6 hours period of sedimentation makes less than 1.4 mm. In such a way as compared with a common bath electro-galvanic dry topping an electro-galvanic mechanical dry topping allows obtaining a layer of the thickness large enough without time delay in the process efficiency.

The investigations of micro-hardness are carried out on a micro-hardness meter PMT-3. The micro-hardness of the main metal (steel 30) was 230 - 250 HV. The micro-hardness of EGM coating was 600 - 717 HV.

On the basis of data obtained there was formed a histogram (Fig. 7) showing the dependence of the micro-hardness of a layer obtained upon the steel grade of initial material.

The micro-hardness of EGM coating obtained through the plating on steel 38ChN3MA does not exceed substantially the micro-hardness of EGM coating obtained through the plating on steel 30. It could be drawn a conclusion that a more significant role is played in hardness increase in coatings obtained not by a chemical composition of medium carbon steel upon which a coating is applied, but a surface preparation and a plating technology and also an electrolyte composition. The adhesion strength (according to Ollard-Melkov method) of the coating with the basis, σ_{ad} at the surface preparation only by a standard pickling reaches values 38.5 – 57.7 MPa (for EG coating).

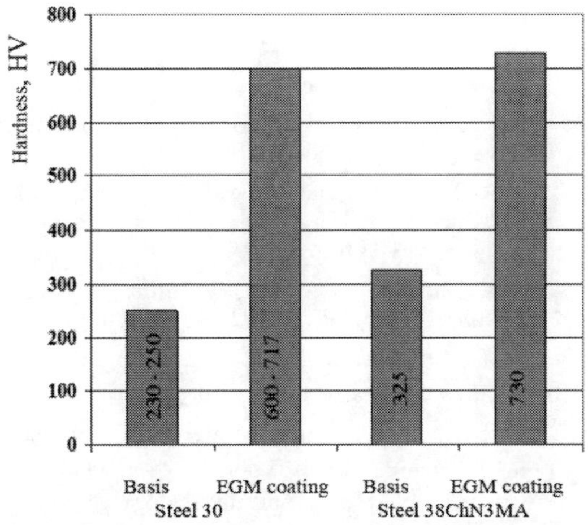

Fig. 7. Dependence of hardness of EGM coating upon a steel grade of initial material

Pickling in a dry topping bath prepared for surface sedimentation allows increasing values of the adhesion strength, σ_{ad} up to 201.2 – 247.4 MPa (for EGM coating), that is, approximately five times with regard to a common pickling in a sulfate bath. As a part is not transferred into another bath and it does not have contact with air, then on its surface there are no foreign matters and no oxide films are formed that ensures the best adhesion strength of a coating with the basis.

4 Conclusions

In accordance with data of Yu.N. Petrov, G.N. Zaidman et al. [2, 3] it was established that the peculiarity of the structure electrolytic iron is the utmost workhardening of this material and the 'frost' of its defects. Just this peculiarity forms the highest iron strength and determines its capacity to resist successfully friction and wear and also other motion deformations.

In our case the whole amount of EGM coating is work-hardened – strengthened in a layer-by-layer way by the impact of a smoothing tool and as a result of this there is formed a large density of dislocations. In such a way, a layer of electro-galvanic mechanical iron possesses a sufficient micro-hardness and wear-resistance which is confirmed by data shown in [4].

References

1. E.D. Pleshka, *Electric Methods of Surface Treatment* **44 (4)**, 17 (2008)

2. Yu.N. Petrov, G.V. Guriyanov, Zh.I. Bobanova, S.P. Sidelnikova, L.N. Andreeva, *Electrolytic Deposition of Iron* (Kishinyov, Shtiintsa, 1990)
3. I.M. Kovensky, V.V. Povetkin, Metal science of coatings (Moscow, *SP Intermet Engineering*, 1999)
4. Yu.R. Kopylov, A.V. Tolcheev, *Problems of Machines Quality and Their Competitive Ability* (Bryansk: BSTU, 2008).
5. M. Schlesinger, M. Paunovic, *Modern electroplating* (John Wiley & Sons, 2011)
6. P.C. Andricacos, IBM Journal of Research and Development, **42(5)**, 567 (1998)
7. M.R. Vaezi, S.K.Sadrnezhaad, L. Nikzad, Colloids and Surfaces A: Physicochemical and Engineering Aspects, **315(1)**, 176 (2008)
8. Q.X. Liu, S.Z. El Abedin, F. Endres, Surface and Coatings Technology, **201(3)**, 1352 (2006)
9. M. Berger, Surface Engineering, **18(3)**, 219 (2002)
10. Y. Zhang, Materials Letters, **69**, 86 (2012)
11. M. Berger, Surface Engineering, **18(3)**, 219 (2002)
12. R.L. Deuis, J.M. Yellup, C. Subramanian, Composites science and technology, **58(2)**, 299 (1998)
13. N. Murugan, R.S. Parmar, Journal of Materials Processing Technology, **41(4)**, 381 (1994)
14. H.J. Kim, B.H. Yoon, C.H. Lee, Wear, **249(10)**, 846 (2001)
15. H. Assadi, Acta Materialia, **51(15)**, 4379 (2003)
16. P. Fauchais, Journal of Physics D: Applied Physics, **37(9)**, R86 (2004)
17. Y. Yang, K.H. Kim, J.L. Ong, Biomaterials, **26(3)**, 327 (2005)
18. L. Shepeleva, Surface and Coatings Technology, **125(1)**, 45 (2000)
19. L. Sexton, Journal of Materials Processing Technology, **122(1)**, 63 (2002)
20. Y.P. Kathuria, Surface and Coatings Technology, **132(2)**, 262 (2000)

Dissipation mechanisms under irreversible deformation

Yury Alyushin[1,*], and *Sergey* Gorbatyuk[1]

[1] National University of Science and Technology "MISiS", 119991, Moscow, Leninsky prospect, 4, RF.

Abstract. The possibility is shown of the transition from reversible to irreversible deformation without additional energy from outside sources by changing the distribution of speeds and accelerations of the particles of the continuum. Two models of irreversible deformation are proposed. The first of which uses a different correlation between transverse and longitudinal deformations at each stage, while the second stage is accompanied by dissipation of energy, stored in the previous elastic stage. For the second model with ideal rigid-plastic continuum, the sequential formation of several families of planes of sliding was substantiated. After the formation of the first one a sliding starts in the second family, which consumes less energy. The consequence is the dissipation and reduction of loadings, the stand sliding on the first family. For the further development of strain it is required to increase loadings up to a critical value, corresponding to the beginning of the first stage. The both models are consistent with the experimentally observed mechanisms of irreversible deformation.

From among all the many different observed mechanisms of reversible and irreversible deformation slipping and twinning is usually considered as the main [1-3]. Selectivity of sliding plane is due to the line microflaws, called dislocations. Dislocation theory has proved very successful, including for explanation of bending and twisting of sliding planes.

However, by the different theoretic and really observed values of stress for developing dislocations, have moved from simple dislocaton models to complex ones by the early 1980s. The idea of double-level strategy for plasticity description, using autowaves of localized plastic flows, have come up. In work [3] relaxational nature of plastic flow with stress redistribution and reducing their total level through slide processes, was justified. Slide processes activate related spaces with appearance of new pockets thus ensures its autocatalytic nature, when the finish result returns a system to its initial state with subsequent repeating of similar cycles.

Similar mechanisms were found out in monocrystals [4], including a translation slip, when some layers of crystal's atoms slipping over other ones with movements equivalent in size to atom spacing. Deformation mechanisms are the same for any real material [5], but a standard physical analogue is still absent.

There is no a valid mathematical model and on the basis of solids mechanics. Moreover, in modern plasticity theory there is no a standard accounting practice of stress history from an initial to critical state. There are a number of contradictions, for example, about uniform stress and strain states in elementary units and possibility of tangent line ingredient`s jump at the surfaces of velocity discontinuity in a perfect plasticity model.

The objective of the work: using general accepted ratios, offer possible models of passing from reversible to irreversible deformation with conversion of stored energy from one species to another with subsequent its dissipation.

From the energy point of view [6-7] irreversible deformations can occur when the volume energy density, increasing under the influence of the elastic deformation, when the critical values can contribute to change the nature of the movement of the particles of the continuum and contributes to the running of the relaxation processes, in which energy is either not changed or reduced, and can return to its original steady value when a significant change in particle size taking into account the existing load.

Specific deformation power w is equal to the product tensors of the stress σ_{ij} and the strain rate s_{ij}

$$w = \sigma_{ij} s_{ij} = \sigma_{xx} x_{tx} + \sigma_{xy} y_{tx} + \sigma_{xz} z_{tx} + \\ + \sigma_{xy}(x_{ty} + y_{tx}) + \sigma_{yz}(y_{tz} + z_{ty}) + \sigma_{zx}(z_{tx} + x_{tz})$$

Here $s_{ij} = 0{,}5(\partial v_i / \partial x_j + \partial v_j / \partial x_i)$, $v_i = dx_i / dt$ – particle movement velocity components. Turning to the spherical tensor σ, s and diverter σ'_{ij} and s'_{ij}

$$\sigma_{ij} = \sigma + \sigma'_{ij}, \qquad s_{ij} = s + s'_{ij},$$

we obtain

$$\omega = \sigma_{ij} s_{ij} = \sigma div(\vec{v}) + \sigma'_{ij} s'_{ij}. \tag{1}$$

Where $\sigma = (\sigma_{xx} + \sigma_{yy} + \sigma_{zz})/3$ – the average stress. The first term describes the power spent on the volume change, the second - in the shape change. Both parts are invariant, the second term can be written in terms of the intensity of tangential stresses τ_e and shear strain rates s_i,

* Corresponding author: alyushin7@gmail.com

$$\omega' = \sigma'_{ij} s'_{ij} = \tau_e s_i ,$$

$$\tau_e^2 = (1/6)[(\sigma_{xx} - \sigma_{yy})^2 + (\sigma_{yy} - \sigma_{zz})^2 + \\ + (\sigma_{zz} - \sigma_{xx})^2 + 6(\sigma_{xy}^2 + \sigma_{yz}^2 + \sigma_{zx}^2)],$$

$$s_i^2 = 2[(s_{xx} - s_{yy})^2 + (s_{yy} - s_{zz})^2 + (s_{zz} - s_{xx})^2 + \\ + 3/2(s_{xy}^2 + s_{yz}^2 + s_{zx}^2)]/3. \quad (2)$$

We use the concept of volumetric energy density $u = \delta U / \delta V$ similar to the density of the material $\rho = \delta m / \delta V$. Operator δ correspond to infinitesimal values of functions, such as volumes, relating to fixed particle with an initial volume of δV_0 at a given moment of time $t > 0$. To increment δU function of an infinitely small time interval dt we obtain

$$d\delta U = d(u\delta V) = (u_t \delta V + u\delta V_t)dt = \\ = (u_t R + uR_t)\delta V_0 dt = [u_t + u\, div(\vec{v})]\delta V dt \quad (3)$$

The right side of the subscript t represents the substantial derivative of the function with respect to time. Comparing equations (1) and (3), and taking into account that the specific power is essentially a volumetric energy density increment rate, it can be argued that the average stress σ characterizes the volumetric energy density u (with the dimension of Pa), and the specific power to change shape $\omega_\phi = \tau_e s_i$ - the rate of change "volumetric energy density" u_t (with the dimension of Pa/s).

In view of this result, the criteria of the limiting value of stress intensity for elastic materials and energy used to change the shape, can be interpreted as the existence of limiting the rate of change (transfer) of energy in solids, as a criterion as the ratio τ_e / σ [8] as the existence of the maximum energy the rate of change on each of its particular level (taking into account the denominator).

Based on the above it can be assumed that the failure mechanism in solids is similar to the destruction of heat stroke: very high energy flow can not be transferred to an adjacent volumes and there is a "fusion" of the contact layer with a change in its physical properties.

The amount of plastic deformation processes, workpieces and their constituent particles is not changed. Then the first term on the right side of the equation (1) becomes equal to 0 and the deformation capacity of the product determines the intensities of the stress and strain rate $\omega = \tau_e s_i$. In accordance with the hypothesis of the constancy of the plastic constant $\tau_e = k = const$ can be argued that the deformation capacity mainly determines the intensity of shear strain rate s_i.

With any addiction $\tau_e = \tau_e(s_i)$ it can be argued that the development of deformation is possible without additional energy from outside sources by changing the divergence and rotor of vectors of speed $\vec{v}$ and acceleration $\vec{v}_t$, as [8]

$$s_i^2 = 2div(\vec{v}_t) + \\ + (rot(\vec{v}))^2 - 2(div(\vec{v}))_t - (2/3)(div(\vec{v}))^2 . \quad (4)$$

The equivalence of the above two forms of the equations for s_i^2 (2) and (4) it follows from the relations

$$f_{tx_i} = f_{x,t} + f_x x_{tx_i} + f_y y_{tx_i} + f_z z_{tx_i} ,$$

that follow from common rules of differentiation of implicit functions.

Moreover, the transition of reversible to irreversible deformation is possible with a homogeneous velocity field, also due to the localization of deformation in bands slip, when adjacent volumes can be considered rigid.

In particular, in a linear tension the energy is growing during elastic loading through volume and form changes. Volume recover to initial value is only possible with a redundant strain with size reduction across and increase the size in the direction of external force cotemporary. The stored energy shall be not less than critical, which is necessary for launching a mechanism of irreversible deformation. Process must be accompanied by reducing of energy potential of particles and discharge of excess part of energy as a heat.

Consider a possible two-staged mechanism of plastic flow by the example of linear tension in the direction of x axis. At the first stage $0 \le t \le t_s$ elastic strain occurs $0 \le \varepsilon_x \le \varepsilon_s$ with ratio of transverse to longitudinal deformations $|\varepsilon_y / \varepsilon_x| = \mu$ according to a motion equation (time parameter is deformation ε_x)

$$x = \alpha \exp(\varepsilon_x), \quad y = \beta \exp(-\mu\varepsilon_x), \\ z = \gamma \exp(-\mu\varepsilon_x), \quad (5)$$

where $\alpha = x_0$, $\beta = y_0$, $\gamma = z_0$ – are initial coordinates (Lagrange variables). As analysis is expected of both transverse and longitudinal deformations, motion equations were written with logarithmic measure of Hencky deformation $\varepsilon = \ln(l / l_0)$ unlike Cauchy measure $\varepsilon = (l - l_0)/l_0$, where l and l_0 – is a current and initial length. Derivatives $x_{i,p} \equiv \partial x_i / \partial \alpha_p$ from Euler variables $x_i \in (x, y, z)$ by Lagrange variables $\alpha_p \in (\alpha, \beta, \gamma)$

$$\partial x / \partial \alpha \equiv x_\alpha = \exp(\varepsilon_x), \\ \partial y / \partial \beta \equiv y_\beta = z_\gamma = \exp(-\mu\varepsilon_x) \quad (6)$$

determine a relative change of volume R, average length of bonds e, main energy invariant Γ_e^2, and standard deviation Γ of bonds' lengths of an infinitesimal parallelepiped from their average value e [9]

$$R = \delta V / \delta V_0 = | x_{i,p} | ,$$

DOI: 10.1051/matecconf/201712902005

$$e = (e_\alpha + e_\beta + e_\gamma)/3,$$

$$e_p^2 = x_p^2 + y_p^2 + z_p^2,$$

$$\Gamma_e^2 = x_{i,p}x_{i,p} = x_\alpha^2 + x_\beta^2 + x_\gamma^2 + y_\alpha^2 + ... + z_\gamma^2,$$

$$\Gamma^2 = \Gamma_e^2 - 3e^2. \quad (7)$$

Stored energy can lead to a spontaneous deformation $\Delta\varepsilon_x \geq 0$, which is different from deformation by ratio of transverse to longitudinal deformations at the first stage $|\Delta\varepsilon_y / \Delta\varepsilon_x| = \mu'$, and at the second stage $t_s \leq t \leq t_k$ provides volume recover of particles to initial value with a parallel growth of deformation in the direction of tension with increasing standard deviation of bonds` lengths of infinitesimal parallelepiped from their average value

$$x = \alpha \exp(\Delta\varepsilon_x), \qquad y = \beta \exp(-\mu'\Delta\varepsilon_x),$$

$$z = \gamma \exp(-\mu'\Delta\varepsilon_x). \quad (8)$$

To conserve on equations (8) unified with (5) system of time reckoning and therefore deformations, it is enough to use the principle of superposition [6], by replacing of Lagrange variables in equations (5) by the expressions for relevant Euler variables of system (8) if $\varepsilon_x = \varepsilon_s = const$

$$x = \alpha \exp(\varepsilon_s + \Delta\varepsilon_x), \; y = \beta \exp(-\mu\varepsilon_s - \mu'\Delta\varepsilon_x),$$

$$z = \gamma \exp(-\mu\varepsilon_s - \mu'\Delta\varepsilon_x). \quad (9)$$

At the beginning of the second stage using real values of derivatives (6) for invariant characteristics (7) we get (end values are given for $\mu = 0{,}25$) [10]

$$R = \exp[\varepsilon_s(1-2\mu)] = \exp(\varepsilon_s/2) = 0{,}5\varepsilon_s + \varepsilon_s^2/8,$$

$$3e = 3 + \varepsilon_s(1-2\mu) + 0{,}5\varepsilon_s^2(1+2\mu^2) =$$

$$= 3 + \varepsilon_s/2 + 9\varepsilon_s^2/16,$$

$$\Gamma_e^2 = 3 + 2\varepsilon_s(1-2\mu) + 2(1+2\mu)\varepsilon_s^2 = 3 + \varepsilon_s + 3\varepsilon_s^2,$$

$$\Gamma^2 = x_\alpha^2 + y_\beta^2 + z_\gamma^2 - 3e^2 = (2/3)[\varepsilon_s(1+\mu)]^2 =$$

$$= (50/48)\varepsilon_s^2. \quad (10)$$

At the second stage at value $\varepsilon_x = \varepsilon_s = const$ and growth $\Delta\varepsilon_x$ with the derivatives

$$x_\alpha = exp(\varepsilon_s + \Delta\varepsilon_x),$$

$$y_\beta = z_\gamma = exp(-\mu\varepsilon_s - \mu'\Delta\varepsilon_x), \quad (11)$$

invariant characteristics (7) are

$$R = \exp[\varepsilon_s(1-2\mu) + \Delta\varepsilon_x(1-2\mu')],$$

$$3e = 3 + \varepsilon_s(1-2\mu) + 0{,}5\varepsilon_s^2(1+2\mu^2) +$$

$$+ \Delta\varepsilon_s(1-2\mu') + 0{,}5\varepsilon_s^2[1+2(\mu')^2],$$

$$\Gamma_e^2 = 3 + 2\varepsilon_s(1-2\mu) + 2\Delta\varepsilon_x(1-2\mu') +$$

$$+ 2(\varepsilon_s + \Delta\varepsilon_x)^2 + 4(\mu\varepsilon_s + \mu'\Delta\varepsilon_x)^2,$$

$$\Gamma^2 = (2/3)[\varepsilon_s(1+\mu) + \Delta\varepsilon_x(1+\mu')]^2 \quad . \quad (12)$$

To different combinations of parameter values, appearing in the motion equation and characterizing continuum properties, will be consistent with different reduction value of volumes and quantity of output energy. Volume of particle recovers its initial value if the additional deformation satisfies the condition

$$\Delta\varepsilon_x = -\varepsilon_s(1-2\mu)/(1-2\mu').$$

In order that it remains positive (in this case the tension of specimen is continuing) shall be met condition $\mu' > 0{,}5$. Of course, a complete reduction of volume is not necessary and a range of possible change μ' may be too broad at the second stage. Reduction of volume also depends on a deformation ratio $\Delta\varepsilon_x / \varepsilon_s$. A way of full reduction of volume is possible, for example, if deformations remained equal at the first and second stages $\varepsilon_s = \Delta\varepsilon_x$, then $\mu' = 1 - \mu$, $e_\alpha = \exp(2\varepsilon_s)$, $e_\beta = e_\gamma = \exp(-\varepsilon_s)$. From (12) we get

$$R = 1, \quad e = 1 + \varepsilon_s^2, \; \Gamma_e^2 = 3 + 12\varepsilon_s^2, \; \Gamma^2 = 6\varepsilon_s^2.$$

As a result the volume recovers its initial value with significant changes of particle sizes, deviation Γ^2 increases almost in 6 times.

Using ratios $\tau_{pi} = \varphi x_{i,p}$ between stresses τ_{pi} and deformations of Lagrange $x_{i,p}$ [6], at the end of the first stage the total relative energy gain of particle is

$$E_{\partial e\phi} = \varphi\varepsilon_s(1+3\varepsilon_s).$$

According to equations (9) relieving at the second stage, when the volume recovers its initial value, reduces energy density of particles till $E_\phi = 6\varphi\varepsilon_s^2$, work of external forces almost completely transformed in heat and goes into environment [7-9]

$$E_{\partial ucc} = \varphi\varepsilon_s(1-3\varepsilon_s).$$

Given that the second stage must be accompanied by a release of heat then at this stage (isotropic continuum) the deformation ratio will be $\mu' = 1$, because deformations from increasing of temperature are the same in all directions $\Delta\varepsilon_y = \Delta\varepsilon_x$. In this case the full reduction of initial volume is possible if $\Delta\varepsilon_x = 0{,}5\varepsilon_s$ with dissipation of stored energy.

The described mechanism includes the changes of deflection behavior at the second stage, even allowing homogenous deformation by analogy with elastic area. This complicates understanding of dissipation mechanism, which is usually identified with Lueder's bands [1-4]. As point heat sources, such slidings stimulate dissipation of stored energy and periodic process execution of plastic flow in surrounding volumes.

To build the simplest energy model of localization of deformation zones, can use properties of an ideal plastic-rigid continuum and extremum principle [6], under which the power, required for deformation whatever kinematically-admissible field of velocities, is always higher than in a real process of deformation. It allows to find an upper bound of power or deformation loading and field of velocities under conditions close to actual practice.

For the plastic-rigid continuum kinematically-admissible is a field of velocities which met the condition of volume stability and boundary conditions for velocities [11]. The feature of the ideal plastic-rigid continuum is that a material may be both in solid and plastic states by intensity of tangential stress $\tau_e=k$. Under deformation of material with such properties inside object, can appear slide surfaces (fig. 1, 2), on both sides of which material is hard. These surfaces are similar to contacting surfaces of two, moving relative to each other, hard objects, but influencing tangential stress on them shall comply the passage condition of material plastic state and $\tau_{max}=k$.

For example, under tension of rectangular-sectioned specimen $h\times b$ suggests that sliding occurs on plane *ABCD* (fig. 1) with angles α and β. From condition of vector's complanarity *AB*, *AC* and *AD* [12] find an equation of plane of sliding *ABCD*

$$xtg\alpha + ytg\beta - z = 0$$

and its area if cross-section $f_0 = bh$

$$f_{12} = bh(1+tg^2\alpha+tg^2\beta)^{1/2}.$$

The volume of specimen remains immobile below the plane *ABCD*. In order to comply with boundary condition a displacement velocity of upper part relatively lower one $\vec{v}_{12} = u\vec{i} + v\vec{j} + w\vec{k}$ shall be parallel to the plane of sliding with ratio of components

$$w = v_0 = utg\alpha + vtg\beta$$

and velocity of point's relative movement on both sides of plane *ABCD*

$$v_{12} = \sqrt{u^2+v^2+(utg\alpha+vtg\beta)^2}\,.$$

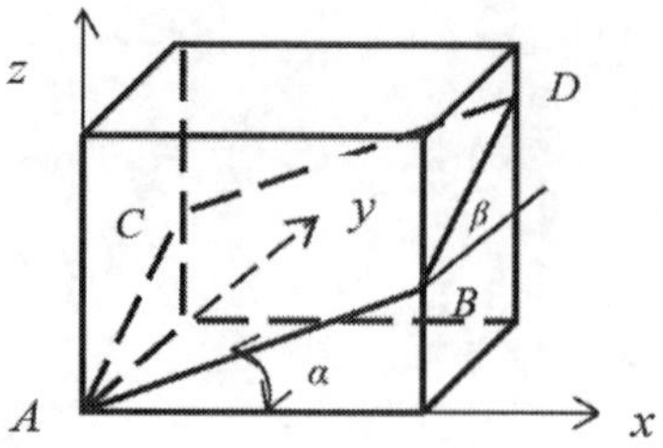

Fig. 1. Plane of sliding under linear tension.

Under the tangential stresses $\tau_{max}=k$ the capacity $W_{12} = kf_{12}v_{12}$ is required on shear, which should be from external reference source. Under specific loading on upper end q from condition of capacities` equality of external and internal forces

$$qbhv_0 = kf_{12}v_{12}$$

find

$$\frac{q}{k} = \frac{\sqrt{1+tg^2\alpha+tg^2\beta}}{tg\alpha+\psi tg\beta}\sqrt{1+\psi^2+(tg\alpha+\psi tg\beta)^2}\,, \quad (12)$$

where value $\psi=v/u$ gives the ratio of displacement velocities of the upper part in x and y direction relatively.

For all values ψ, α, β velocity field is kinematically-admissible. As these characteristics are independent, their values shall be defined from minimum condition of specific loadings.

$$\partial q/\partial\alpha = 0, \quad \partial q/\partial\beta = 0, \quad \partial q/\partial\psi = 0.$$

Subject to the conditions

$$\psi tg\alpha = tg\beta; \quad tg^2\alpha + tg^2\beta = 1, \quad (14)$$

the specific loadings (13) are $q = 2k$, including for the state of plane strain if $\alpha = 45^0$ and $\beta = 0$, $\psi = 0$. If properties are similar (isotropic continuum) in x and y directions then can be taken $\psi = 1$ for comparable sizes $b \approx h$ and from the second equation get $\alpha = \beta = 35°16'20''$. Exactly this result has been confirmed experimentally [9].

Loading diagram changes after the appearance of plane of sliding. Now the lower part is under tangential stress $\tau_{\max} = k$ on plane *ABCD,* which for simplicity of graphical representation will be characterized by a slope of its normal relative to vertical (a tension load) by angle α_1. While not excluding the appearance of new plane of sliding with angle α_2. In fig.2 physical plane and velocity hodograph shows a velocity of mobile zone 2 between 0 and 1 zones under conditions of equal of normal components of velocity on all planes of sliding. On plane f_{12} between 1 and 2 zones tangential stresses $\tau_{12} = \tau_{\max} = k$ generate a force $F_{12} = kbh/\cos\alpha_1$,

DOI: 10.1051/matecconf/201712902005

directed at angle α_1 to horizontal. The power is defined by scalar product of vectors of force and velocity with angle between them $(\alpha_1-\alpha_2)$

$$W_{12}=kv_{02}(b/\cos\alpha_1)\cos(\alpha_1-\alpha_2).$$

At the surface of new plane power consumption is

$$W_{02}=kv_{02}(b/\cos\alpha_2).$$

Deformation with the formation of secondary system of sliding is possible if $W_{12}\geq W_{02}$ or

$$\tau_{12}v_{02}(b/\cos\alpha_1)\cos(\alpha_1-\alpha_2)\geq\tau_{02}v_{02}(b/\cos\alpha_2).$$

If $\tau_{12}=\tau_{02}=k$ all angle values $0\leq\alpha_2\leq\alpha_1$ meet this condition, put in other words, the secondary bands shall be to initial ones from parallel to horizontal.

The most probable will be ones for which the required power are minimum. The location of such planes specifies a condition for minimum ratio

$$\theta=\frac{\cos(\alpha_1-\alpha_2)\cos\alpha_2}{\cos\alpha_1}.$$

From $d\theta/d\alpha_2=0$ get

$$\sin(\alpha_1-\alpha_2)\cos\alpha_2-\cos(\alpha_1-\alpha_2)\sin\alpha_2=0,$$

where $\sin(\alpha_1-2\alpha_2)=0$ or $\alpha_2=\alpha_1/2$.

Because the analyze made for arbitrary values of angles α_1 and α_2, this result may be expanded to any follow-up surfaces of slide.

Estimates suggest that sliding on planes of follow-up systems is possible, in this case at the previous family shear capacity exceeds the required up to 17,2% for the second system of sliding (with tilting angle 22,5^0), just for 4% for the third one(with tilting angle 11,25^0) and just for 1% for the forth one (with tilting angle 5,625^0).

Suchwise, every previous system of sliding may create the follow-up one, but their concurrent existence is impossible. Applied in zone 1 specific loading $q=2k$ is not enough to maintain sliding at the first two families of deformation bands. With using sliding velocities v_{02} and v_{12} on surfaces f_{02} and f_{12} with tilting angles α_2 and α_1 these specific loadings produce the power

$$W_q=qbh(v_{02}\sin\alpha_2+v_{12}\sin\alpha_1), \qquad (15)$$

while consumable power (if shear stress k) is

$$W_k=W_{02}+W_{12}=kbh(v_{02}/\cos\alpha_2+v_{12}/\cos\alpha_1). \qquad (16)$$

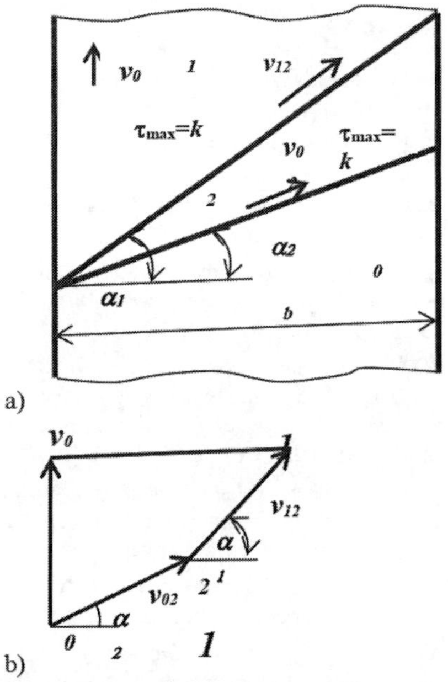

Fig.2. Scheme of pre- and post- deformation bands:
a) physical plane b) velocity hodograph.

Indicate velocity ratio $\psi=v_{02}/v_{12}$, and then a condition of capacity equality (15) and (16) can record as

$$qv_{12}(\psi\sin\alpha_2+\sin\alpha_1)=kv_{12}(\psi/\cos\alpha_2+1/\cos\alpha_1).$$

If $q=2k$ and $\alpha_1=45^0$ the equality for the first family

$$2(\psi\sin\alpha_2+\sin\alpha_1)=\psi/\cos\alpha_2+1/\cos\alpha_1$$

is only met if $\psi=0$ or $\alpha_2=\alpha_1$.

Suchwise, if achieved the specific loadings $q=2k$ a primary system of sliding appears with tilting angle $\alpha_1=45^0$. Below this plane in the volume of specimen shear on plane with tilting angle $\alpha_2=22{,}5^0$ becomes the preferred, because in the arisen conditions such deformation demands less power. It is not excluded the appearance of post- deformation bands, right up to horizontal ones. But external forces cannot support coexistence of two families, that`s why the sliding comes to an end on surface between 0 and 1 zones.

On the secondary sliding plane power reduction of deformation leads to load relaxation at 17,2% and appearance of "yield drop" on diagram "force-strain" which is observed experimentally [14]. Under static tension in condition of plane strain distraction of plated specimens under tension more often occurs at an angle not 45^0, but about 21^0 [13]. Consequently, slides are more significant in the second family than in the first one.

The considered model of the transition of reversible to irreversible deformations agree with the experimental data.

DOI: 10.1051/matecconf/201712902005

References

1. R. Honecomb, *Plastic deformation of metals* (M.: Mir, 1972)
2. S.M. Gorbatyuk, A.V. Kochanov, *Method and equipment for mechanically strengthening the surface of rolling-mill rolls*, Metallurgist, **56 (3-4)**, pp. 279-283 (2012)
3. L. B. Zuev, V. I. Danilov, S. A. Barannikova, *Physics of plastic flow microlocalization* (Novosibirsk: Nauka, 2008)
4. S.M. Gorbatyuk, V.M. Pavlov, A.N. Shapoval, M.S. Gorbatyuk, *Experimental use of rotary rolling mills to deform compacts of refractory metals*, Metallurgist, **42 (5-6)**, pp. 178-183 (1998)
5. E. Nikolaeva, *Shear mechanisms of plastic deformation of single crystals* (Perm: Publishing house Perm. GOS. tehn. University press, 2011)
6. Yu. Alyushin, *Energy principles of mechanics* (Lambert Academic Publishing, 2016)
7. Yu. Alyushin, *Energy model of reversible and irreversible deformations in the space of Lagrangian variables,* Advanced technology plastic deformation, M.: NITU MISIS, pp. 44-67 (2009)
8. Yu. Alyushin, M. Skripalenko, *Energy peculiarities and accelerations during reversible and irreversible deformations*, Journal of Machinery Manufacture and Reliability, **40 (2)**, pp. 154-160 (2011).
9. Yu. Alyushin, *The generalized model of reversible and irreversible deformation describing the processes in Lagrange form*, «Kuznechno-Shtampovochnoe Proizvodstvo», **6**, pp. 2-5 (1997).
10. Yu. Alyushin, *Mechanics of deformation processes in the space of Lagrangian variables* (M.: Mechanical Engineering, 1997)
11. A. Tomlenov, *Mechanics of metal forming* (M.: Mashgiz, 1963)
12. Yu. Alyushin, *Theory of metal forming* (Rostov-on-Don: RISHM, 1977)
13. A. Nadai *Flow and fracture of solids* (M.: Mir, 1969)
14. D. Bell *Experimental foundations of mechanics of deformable solids* (M.: Nauka, 1984)

MATEC Web of Conferences **129**, 02006 (2017)
ICMTMTE 2017
DOI: 10.1051/matecconf/201712902006

Calculation of equivalent magnetic permeability of ferromagnetic materials for modeling of three-dimensional eddy current fields

Sergey Karpushkin[1,*], *Aleksey* Glebov[1], and *Sergey* Karpov[1]

[1]Tambov State Technical University, 392000, Sovetskaya, 106, Tambov, Russia

Abstract. The paper presents a method of calculating a constant value of magnetic permeability of ferromagnetic materials, equivalent to the basic magnetization curve by the criterion of real power. The proposed method makes it possible to use linear differential equations of the electromagnetic field in the analysis of three-dimensional fields of eddy currents in ferromagnetic materials, thus significantly reducing the calculation time. We propose a method of calculating the power of inductors using two-dimensional axisymmetric models. The scope of the proposed methods is limited to the case of placement of the inductor inside the ferromagnetic body.

1 Introduction

The problem of calculating eddy current fields is of great practical importance for the analysis of electrical devices for different purposes. A special role in this equipment segment is given to induction heating devices, the functioning of which largely depends on the configuration of the electromagnetic field.

Eddy currents are induced most effectively in ferromagnetic materials, which are characterized by a non-linear relationship between induction and magnetic field strength. This dependence is expressed as the magnetization curve, and the stage of mathematical modeling it results in the formation of a system of nonlinear time-dependent vector differential equations in partial derivatives. The solution of such a system in a three-dimensional representation is quite complicated.

Assuming the linearity of magnetic properties of the material, the electromagnetic field equations are considerably simplified. The use of complex numbers makes it possible to get rid of differentiation with respect to time, and the model equations take a quasi-stationary form.

However, ferromagnetic materials with linear magnetic properties are not found in practice. In this regard, it is relevant to develop methods of calculating constant values of magnetic permeability, ensuring that the heating capacities in linear and non-linear calculations are the same. In other words, there is a need to replace the original magnetization curve with the equivalent value of magnetic permeability.

2 Method for calculating equivalent magnetic permeability

The electromagnetic field equations when using the concept of magnetic vector potential $\mathbf{A}$ (Wb/m) and electric scalar potential V (W) are of the form [1]:

$$\mathrm{rot}\left(\mu^{-1}\mathrm{rot}\mathbf{A}\right)-\mathrm{grad}\left(\mu^{-1}\mathrm{div}\mathbf{A}\right)+\gamma\frac{\partial\mathbf{A}}{\partial t}+\gamma\,\mathrm{grad}V=0\,; \quad (1)$$

$$\mathrm{div}\left(-\gamma\frac{\partial\mathbf{A}}{\partial t}-\gamma\,\mathrm{grad}V\right)=0\,, \quad (2)$$

where μ is absolute magnetic permeability, H/m; γ is specific electric conductivity, $\mathrm{ohm}^{-1}\mathrm{m}^{-1}$; t is time, s.

Equations (1)-(2) are written for an isotropic material in the existence of eddy currents and derived under assumptions about the absence of Maxwell's displacement current, as well as the absence of the phenomenon of magnetic hysteresis. Typically, in induction heating these phenomena can be disregarded.

To solve equations (1)-(2) a method of finite elements realized in various CAE-systems is the most effective. The authors of this study used ANSYS system of finite-element analysis. Modeling of an electromagnetic field described by equations (1)-(2) was done using SOLID97 finite elements (Fig. 1). Finite elements of tetrahedral shape are more versatile to split solids of arbitrary form.

Modeling of induction heating involves a step-be step analysis, using: 1) calculation of eddy currents with the use of SOLID97 finite elements; 2) thermal analysis using SOLID70 finite elements, the geometry of which is similar to those shown in Fig. 1. Thus, for electromagnetic and thermal analyses a unified finite element mesh is used; changing only the type of the finite elements when conducting the thermal analysis [2,3].

* Corresponding author: karp@mail.gaps.tstu.ru

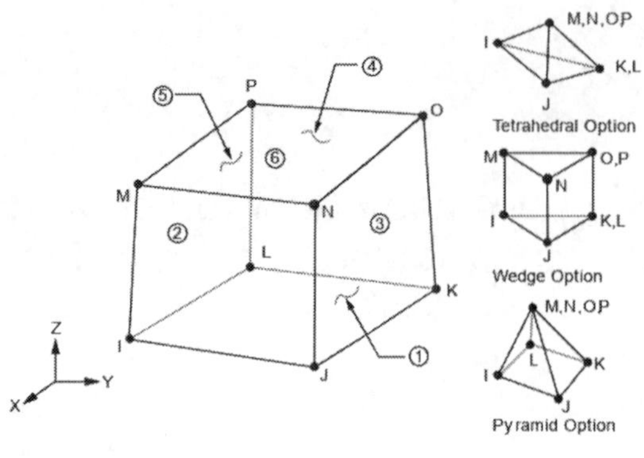

Fig. 1. Geometry of SOLID97 finite element of the ANSYS system

Solution of the three-dimensional nonlinear equations (1)-(2) using the finite element method (FEM) is costly in terms of computing time because of the low rate of convergence of iterations and time-dependent electromagnetic processes. At each time point in the nodes of the finite element mesh it is required to determine the four unknowns: electric scalar potential V and the three components of the magnetic vector potential **A**. If magnetic permeability is constant, then for sinusoidal supply voltage expressions (1)-(2) are simplified to linear equations in a quasi-stationary complex representation:

$$-\mu^{-1}\Delta\dot{\mathbf{A}} + j2\pi f\gamma\dot{\mathbf{A}} + \gamma \mathrm{grad}\dot{V} = 0\,; \quad (3)$$

$$\mathrm{div}\left(-j2\pi f\gamma\dot{\mathbf{A}} - \gamma \mathrm{grad}\dot{V}\right) = 0\,, \quad (4)$$

where j is an imaginary unit ($j^2 = -1$).

Time required to solve equations (3)-(4), as compared with that required to find solutions of the initial equations (1)-(2) can be reduced by more than one order of magnitude. However, the need for a constant value of magnetic permeability equivalent to the magnetization curve by average real power for one period prevents their direct application to calculate the eddy current fields in ferromagnetic materials.

This problem can be solved by a method based on the solution of two-dimensional nonlinear equations of electromagnetic field (in this case, the mathematical model is greatly simplified). Assuming perpendicularity of the external current to the XY plane, the magnetic vector potential has only one component A_z, and the scalar electric potential is not used [4]:

$$\mathrm{rot}\left(\mu^{-1}\mathrm{rot}A_z\right) + \gamma\frac{\partial A_z}{\partial t} = 0\,. \quad (5)$$

Thus, in every point of the calculation domain it is required to find only one unknown A_z (in a three-dimensional analysis the number of unknowns is four), i.e. it is not difficult to take into account nonlinear characteristics of ferromagnets in a two-dimensional problem. Two-dimensional models can be quite effectively in the cases, in which their application does not lead to distortion of the description of physical processes.

Linear equations (3)-(4) are simplified in a similar way in a two-dimensional formulation:

$$-\mu^{-1}\Delta\dot{A}_z + j2\pi f\gamma\dot{A}_z = 0\,. \quad (6)$$

The dependence of equivalent permeability on the characteristics of the inductor and its position was investigated by solving the equations (5) and (6) for the axisymmetric system, consisting of a ferromagnetic disc and an inductor (Fig. 2). The radius of the inductor location, the number of its winding turns, the current, the fill factor (the proportion of "pure" wire in the total volume of the slot), the position of the inductor relative to the outer surfaces of the disc ranged alternately. For each combination we conducted a "standard" non-linear calculation, followed by a series of linear calculations, in the course of which the equivalent value of magnetic permeability was determined.

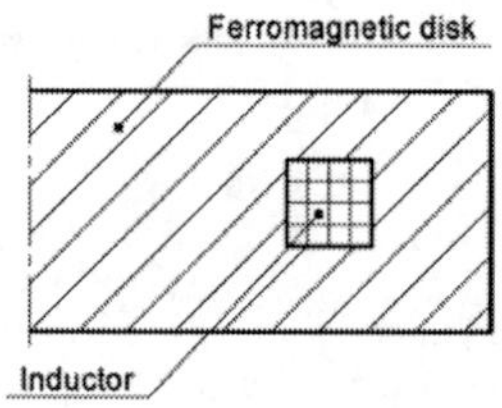

Fig. 2. Axisymmetric model with a disk and an inductor

The computational experiments resulted in the following findings:

- magnetic permeability is mostly affected by cross-sectional dimensions of the inductor and its magnetomotive force (MMF);
- the impact of the inductor location radius can be ignored if it exceeds the width of the inductor more than twice;
- the value of the distance from the inductor to the outer surfaces of the disc substantially affects the eddy currents if this distance is less than the doubled depth of penetration of electromagnetic waves in the body of the disc;
- the fill factor of the inductor slot with wire affects only the inductor heat generation (resistive heating);
- the location of the inductor inside the slot and the gap between the inductor and the slot walls do not affect the induced eddy currents due to shielding.

The findings allowed offering a method of determining the equivalent permeability, shown in Fig. 3. The solutions cycle continues as long as the values of real power the inductor in linear and nonlinear calculations (P_{lin} и P_{nlin} respectively) differ by more than permissible error value ε. The obtained value of magnetic permeability is used in three-dimensional calculation to solve the system of equations (3)-(4). The real power is calculated by the Joule-Lenz law. For a nonstationary model described by the nonlinear equation (5) the calculation of the average power in one period is done by the formula:

DOI: 10.1051/matecconf/201712902006

$$P_{nlin} = \frac{1}{T}\int_{t_{st}}^{t_{st}+T}\int_{\Omega}\frac{J_z^2}{\gamma}d\Omega dt\,, \qquad (7)$$

where T is the period of the current source fluctuations, s; Ω is the region of existence of eddy currents (the disk region); J_z is a z-component of the density vector of eddy currents J (in the two dimensional formulation the other components are equal to zero), A/m^2; t_{st} is conditional stabilization time of electromagnetic processes, s; when it finishes, changes in amplitude values of the current density J_z can be neglected.

Time t_{st} is determined at a stage of preliminary computational experiments and generally does not exceed $3T$.

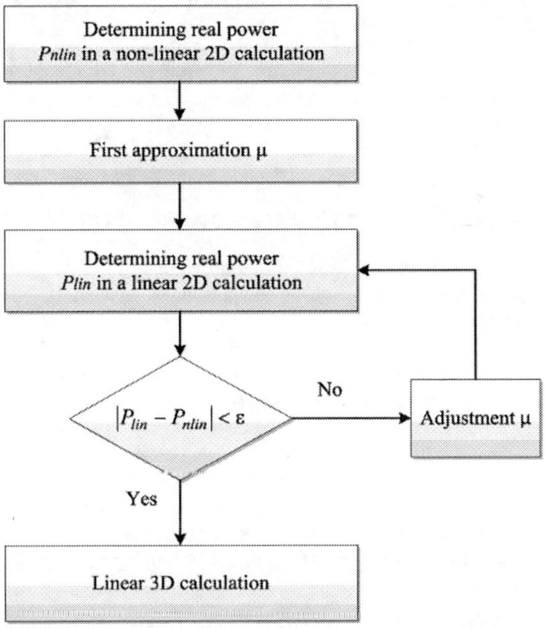

Fig. 3. A flowchart of determining the equivalent magnetic permeability

The eddy currents density in the two-dimensional formulation is calculated by the formula:

$$J_z = -\gamma\frac{\partial A_z}{\partial t}\,. \qquad (8)$$

When using the complex linear equation (6) the average power in one period will be

$$P_{lin} = \int_{\Omega}\frac{|J_z|^2}{2\gamma}d\Omega\,, \qquad (9)$$

where $|J_z| = \sqrt{\mathrm{Re}\,\dot{J}_z^2 + \mathrm{Im}\,\dot{J}_z^2}$ is the module of complex current density $\dot{J}_z$.

By using this method we obtained and approximated by power function the dependences of equivalent magnetic permeability of ferromagnetic structural steel and the MMF of the inductor sized 25×25 mm when connected to an ideal current generator (Fig. 4, curve 1) and an ideal voltage generator (Fig. 4 , curve 2). In the calculations, we used the magnetization curve described in [5].

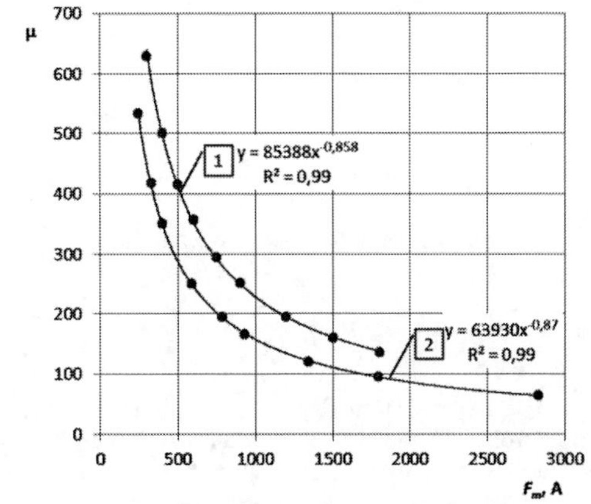

1 – when connected to a current generator,
2 – when connected to a voltage gnerator.

Fig. 4. Magnetic permeability dependence on the inductor MMF

The differences in curves *1* and *2* were caused by the transition to the linear description of electromagnetic processes: in fact, harmonic curves with nonlinear distortions characteristic of ferromagnetic materials were replaced with equivalent sinusoids. In this case the active power, which is released in the form of heat, is determined differently: the values of the power factor do not match. We note that curve *1* is theoretical, since the current generators in induction heating of industrial equipment are not used, i.e. in practical calculations curve *2* must be used.

A similar approach to solving three-dimensional nonlinear equations of an electromagnetic field of ferromagnetic parallelepiped with a regular mesh of finite elements was described in [6]. Unlike the methods discussed, magnetic permeability is determined for each wall-adjacent block of finite elements by using the corresponding number of one-dimensional calculations. However, a three-dimensional model is used in the iterative process. On the one hand, such an approach can provide a high accuracy because magnetic permeability is not averaged over the volume of the ferromagnetic material; in fact, the magnetic permeability field is determined. On the other hand, a large amount of computation for solving real problems is inevitable.

The proposed method seems to be a compromise: it provides acceptable accuracy of the solution, it does not require a large amount of computation, and it is easy to implement.

3 A method for calculating the inductors power

The proposed method of finding the equivalent magnetic permeability can be used in calculation of the inductors power. The studies of the impact of different factors on

 DOI: 10.1051/matecconf/201712902006

magnetic permeability show that any three-dimensional inductor can be accurately approximated by a two-dimensional model if the following two conditions are satisfied:

1) the inductor must be placed entirely within the heated body;

2) parts of the one or more inductors with multi-directional currents must be positioned from each other at a distance of at least 4-fold depth δ of the electromagnetic wave penetration in the body under consideration.

The value of δ is calculated by the formula:

$$\delta = \sqrt{\frac{T}{\pi\gamma\mu}}. \tag{10}$$

The wave penetration depth δ is arbitrary: it is the distance from the surface of the material with linear properties at which the magnetic field strength is attenuated by *e* times. The generated real power is proportional to the square of the magnetic field. Therefore, in a layer with a thickness of δ about 86.5% of the total energy is released from the surface.

If two of these conditions are satisfied, an axisymmetric two-dimensional analog of the original inductor is constructed. The principle of the transition to the two-dimensional model is based on the equality of the lengths of axial lines of the original inductor and the axisymmetric inductor, i.e., the radius of the latter is calculated by the formula:

$$R_{2D} = \frac{L_{3D}}{2\pi}, \tag{11}$$

where L_{3D} is the length of a three-dimensional inductor along the axial line, m.

Then, a two-dimensional axisymmetric model similar to that shown in Fig. 2 is constructed. In this case, a cross-sectional shape of the inductor is not modified.

The next step is the solution of the equation (5) and calculation of real power by the formula (7).

When calculating the power, this method allows taking into account non-linear characteristics of ferromagnetic materials and perform calculations of high-frequency inductors. For example, in accordance with (10) at 50 kHz the final dimensions of elements in the steel material should not exceed 0.04 mm near the magnetic field sources. In a three-dimensional formulation, a similar calculation can be done only by supercomputers.

The proposed method is also effective for a simplified calculation of temperature fields of induction heating devices, assuming a uniform heat generation in the volume of the slots for the inductors [7].

To test the proposed method, we used data of physical experiment on heating the induction plate sized 500×410 mm with four identical inductors used in vulcanization presses (for details, see [8]).

During the experiment, we recorded the power consumption readings for the plate, and its temperature at a place of control thermocouple location. The experimental results and calculated values of real power depending on the plate temperature are shown in Fig. 5. The average deviation of the calculated values from experimental data was 2.7%.

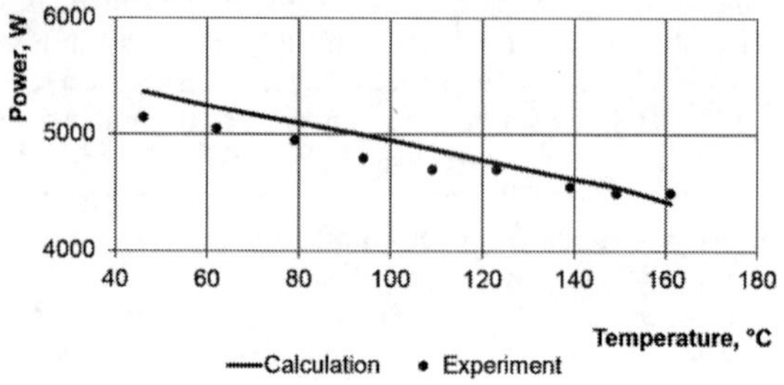

Fig. 5. Experimental data and calculated values compared

Since the calculation of the inductors power was made with application of the proposed method for determining the equivalent permeability, these results indirectly verify the accuracy of the magnetic permeability calculations.

4 Results and discussion

We described a method of determining the equivalent magnetic permeability of ferromagnetic materials. The proposed method is based on simplifying induction heating mathematical model equations to quasi-stationary linear equations in the complex representation. This simplification allows reducing the amount of computation by more than an order of magnitude.

Given the proposed method, we developed a method of calculating the inductors power, which can be effectively used in design of new inductors (including high-frequency inductors), as well as in a simplified calculation of temperature fields of induction heating devices.

This work was financially supported by Ministry of Education and Science of the Russian Federation within the basic part (project 8.7082.2017/8.9).

References

1. O. Bíró, K. Preis, IEEE Transactions on Magnetics. **25**(4), 3145-3159 (1989)
2. A.O. Glebov, S.V. Karpov, S.V. Karpushkin. Autom. Remote Control, **75**(6), 1120-1129 (2014)
3. S.M. Fu, Z.Y. Wu, S.Q. Cao, Advanced Materials Research, **753**, 1035-1039 (2013)
4. J.L. Coulomb, J.C. Sabonnadière *CAO en électrotechnique* (Hermes, 1985)
5. N. Takahashi, T. Nakata, H. Morishige, COMPEL, **14**(2/3), 57-75 (1995)
6. A. Canova, F. Dughiero, F. Fasolo, M. Forzan, F. Freschi, L. Giaccone, M. Repetto, IEEE Transactions on Magnetics, **45**(3), 1855-1858 (2009)

7. A.V. Ostroukh, A.O. Glebov, S.V. Karpov, S.V. Karpushkin, M.N. Krasnyanskiy. Am. J. Appl. Sci., **11**(6), 939-946 (2014)
8. S.V. Karpushkin, S.V. Karpov, A.O. Glebov. Transactions Tambov STU, **17**(1), 110-120 (2011)

MATEC Web of Conferences **129**, 02007 (2017) DOI: 10.1051/matecconf/201712902007
ICMTMTE 2017

Generating multi-component multi-layer coatings type mecraly using multi-cathode magnetron sputtering systems

Valery Kolesnik[1], *Yury* Rubanov[1], and *Yulia* Tokach[1,*]

[1]BSTU after V. G. Shukhov, Kostyukov str., 46, Belgorod, 308012, Russia

Abstract. The paper presents the results of experiments carried out to obtain multi-component coatings that have a broader range of properties compared to their single-component counterparts. We have studied the generation of coatings based on a Co (18…20%)-Cr (5…7%)-Al (0.3…0.4%)-Y (0.2…0.5) alloy. We have also studied the chemical composition of the coating, its thickness, and the transient zone condition. We present here the results of X-ray fluorescence and X-ray spectroscopy microanalysis of such coatings. It is shown that the microtopography of surfaces depends on the coating generation conditions; the set of chemical elements in each coating is similar for all specimen, their concentration depends on the deposition conditions, and the thickness distribution is constant; there is virtually no diffusion of coating and substrate materials. We have identified an insignificant percentage of substrate materials in the generated coating. The thicker the coating, the less substrate materials it contains.

1 Introduction

Scientific and technical literature contains a lot of information on how to generate multi-component coatings with specified characteristics and adjustable composition [1–9]. Such coatings are widely applied in air, rocket, and mechanical engineering. One of the main causes of gas-turbine engine compressor failure is the dust erosion of blade airfoils, which is particularly evident when aircraft and helicopters are operated at ground airfields or are flown low in excessively dusty air. On the other hand, improving the efficiency of ground and air turbines necessitates higher temperatures of input gases, which creates excessive temperature loads on the 'hot tract' parts, particularly on the blades. This is why the service life of engines directly depends on the service life of blades.

One of the ways to prolong the service life of engines it to protect compressor and turbine blades from erosive and thermal effects by applying protective coatings [10–14].

Vacuum sputtering is used to create quality precision coating. Widely used coating materials are MeCrAlY where the main component (Me) is cobalt, cobalt plus nickel, cobalt plus chromium, etc.

2 Methodology

To study how multi-layer coatings based on multi-component alloys can be generated, we experimentally sputtered a Co (18…20%)-Cr (5…7%)-Al (0.3…0.4%)-Y (0.2…0.5) alloy using a laboratory machine [15, 16].

Twelve cylindrical water-cooled targets were symmetrically located relative to the system axis and had a cathode potential. The material we sputtered was ten target cathodes made of a Co – (18…20%) Cr – (5…7%) Al – (0.3…0.4%) Y alloy (combinations varied). The electrical wiring circuit enabled us to measure the cut-off potential at target cathodes during the experiment. The substrate material we selected was different from the alloy components for easy X-ray fluorescence and X-ray spectroscopy microanalysis of coatings generated.

The sputtering parameters were as follows: 0.7 Pa working gas pressure, 2.2 kW discharge; the substrate holder had a positive constant potential $U_п$. The latter value was experimentally adjusted to maximize the deposition rate. Prior to the deposition of the coating, substrates had been subjected to ion cleaning.

Multi-layer coatings were generated in the following order:

- first, we sputtered a heat-resistant CoCrAlY alloy. Cut-off potential was applied to aluminum and nickel targets.
- the second layer of the coating was generated by sputtering all the targets located in a vacuum chamber. All the target cathodes had the same potential;
- the third layer was generated by sputtering the aluminum and nickel targets. Cut-off potential was applied to all the CoCrAlY cathodes;
- the outer layer was generated by sputtering an aluminum target. Nickel and alloy targets were under cut-off potential.

Given the physical characteristics of coating materials, the area of target cathodes, and how the current density changed thereon, we can pre-evaluate the generated coating composition.

On an isolated site, the thickness of single-component coating (i=1) was as follows:

$$\delta_i = K_{ui} \cdot K_{pi}(E_i) \cdot \frac{S_{Mi}}{S_{общ}} \cdot \frac{A_i \cdot m_a}{\rho_i \cdot t} \cdot \frac{j_{ki}}{e} \cdot t_{oc.i} \qquad (1)$$

---

[*] Corresponding author: tokach@bk.ru

DOI: 10.1051/matecconf/201712902007

where K_{ui} is the utilization factor of the ith sputtered material; S_{Mi} is the surface area of the ith sputtered component; $S_{общ}$ is the total area of the surface to which the coating is applied; K_{pi} (E_i) is the sputtering coefficient of the ith component that depends on the energy of plasma-generating gas ions bombarding the target cathode; A_i is the atomic weight of the ith component; ρ_i is the density of the ith sputtered material; j_{ki} is the current density for the ith target group; $t_{oc.i}$ is the time to sputter the ith component.

As we identified the thickness of the coating, we concluded that this value could be approximated as the total of thicknesses of each component if those deposited layer-by-layer. X-ray fluorescence analysis showed that the ratio of component-specific thicknesses is numerically equal to the percentage of these components in the coating (within the experimental margin of error).
Therefore, when generating multi-component coating, its thickness can be viewed as the total of all component-specific thicknesses:

$$\delta_\Sigma = \sum_{i=1}^{N} \delta_i \quad (2)$$

The mass flow density is inversely proportional to the square of the target-to-substrate distance [17]. As the substrate moves relative to the target cathodes, we can assume that distances from targets made of various materials to the substrate are roughly equal; the difference between such distances can be disregarded in further analysis.

The percentage of each component in the coating depended on the ratio of sputtered surfaces of target cathodes made of this component material to the total area of the receiving surface upon which the multi-component coating is deposited. The receiving surface in this case was the shell, in the plain of which the substrate moved. This fact is due to the radial discharge of plasma, a characteristic of the system under analysis [18].

Thus, by substituting (2) in (1) and factoring in the assumptions made, we obtain an expression to calculate the thickness of multi-component coatings:

$$\delta_\Sigma = \sum_{i=1}^{N} K_{ui} \cdot \frac{S_{Mi}}{S_{общ}} \cdot K_{pi}(E_i) \frac{A_i \cdot m_a}{\rho_i \cdot t} \cdot \frac{j_{ki}}{e} \cdot t_{oci} \quad (3)$$

This method is for calculating the percentage of components in multi-component coatings; it can be used for pre-evaluation of the coating thickness. Based on our calculations, the projected film thickness was 59.5 μm.

3 Studying the Multi-Layer Multi-Component Coatings

Coatings were studied using a JSMT 300 scanning electron microscope with a Phoenix accessory for X-ray spectroscopy (XRS) by EDAX, USA. The chemical composition of the coating surface was determined by X-ray fluorescence using an EX 6500 machine by BAIR, USA. XRF showed that the chemical elements sputtered were presented in the generated coating, see **Table** 1.

Table 1. XRF analysis results.

Element	Al	Cr	Fe	Co	Ni	Cu	Ti	Y
%	70.5	2.8	10.9	1.2	8.0	0.8	5.9	0.06

The generated coating was 59.00 μm thick, see Figure 1.

There was no clear boundary between the layers, which might be due to the following. When sputtering a target cathode group at the rest of the targets under cut-off potential, a portion of the sputtered material is deposited. When cut-off potential is switched, first the material deposited on that target cathode group is sputtered, whereafter the targets themselves are sputtered as well. This generates a transient zone between the layers.

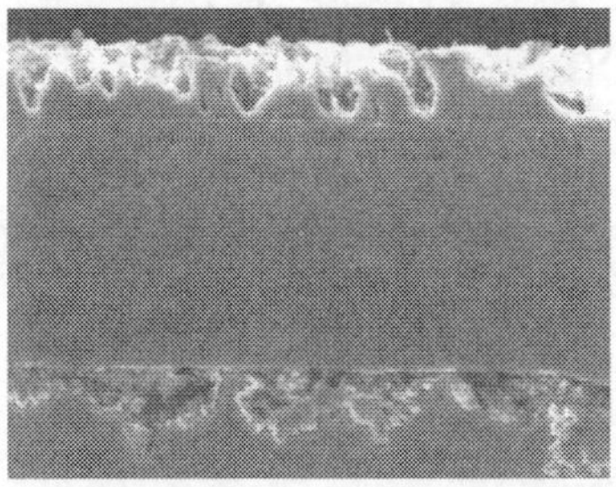

Fig. 1. Microsection of multi-layer multi-component coating.

The noted effect is a favorable factor for the formation of multilayer coatings. The system analyzed can generate multi-layer coatings with clear inter-layer boundaries. To do that, we have to experimentally find such cut-off potential that would prevent the deposition of sputtered materials on cut-off targets.

Based on X-ray spectroscopy microanalysis carried out in the spot-line scanning mode, we could see we generated a multi-layer coating with specified characteristics, see Figure 2, 3.

The percentage of chemical elements in the coating matched the order in which the main and the additional components had been modified, see Figure 2b.

In the initial sputtering of the CoCrAlY heat-resistant alloy, the percentage of components was preserved. The simultaneous sputtering of all target cathodes (heat-resistant, aluminum, and nickel ones) having the same potential results in a change in the coating component concentration. Applying cut-off potential to target cathodes made of CoCrAlY considerably reduces the concentration of cobalt and chromium while increasing the percentage of aluminum and nickel. When generating the outer layer, nickel and alloy target cathodes were under cut-off potential, while the aluminum target was sputtered, which resulted in a considerable change in the component composition of this layer. Experimental data were in line with the projected results. The theoretical and the experimental data did not diverge by more than 8–10%, which proves that multi-layer coatings can be generated with thickness-adjustable percentage of chemical elements.

Figure 2c shows how the concentration of impurities in the coating was distributed.

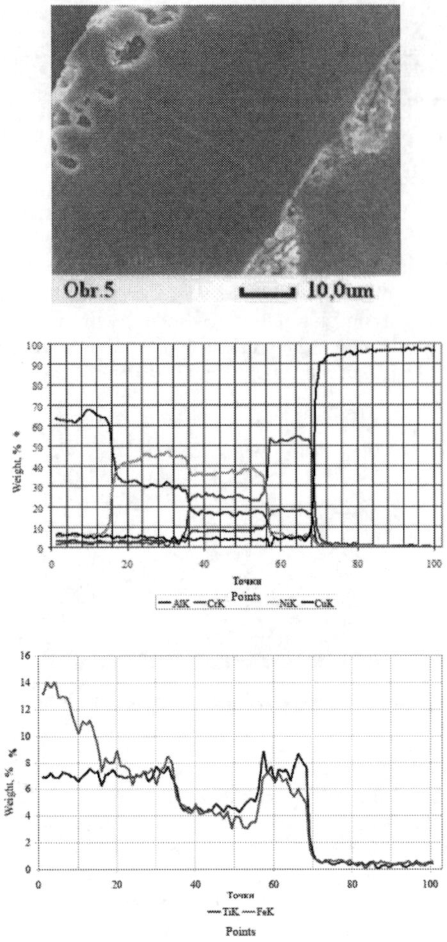

a) b) c)

Fig. 2. Results of spot-line XRS scanning:
a) the electron probe scanning line;
b, c) the distribution of chemical components along the scanning line.

The distribution of chemical components in the coating as determined in the mapping mode (see Figure 3) shows that we obtained multi-layer multi-component coating were the components were uniformly distributed in the layer thickness.

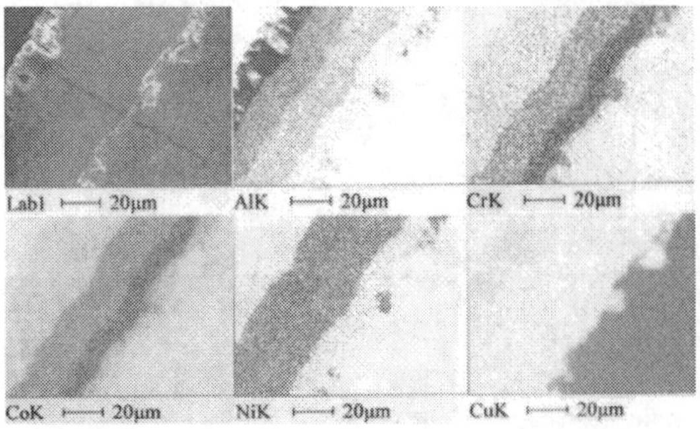

Fig. 3. Mapping-mode distribution of chemical elements

4 Conclusions

Studying the multi-component multi-layer coatings were generated revealed that:

- the coating contained the chemical elements sputtered;
- the thickness distribution of chemical elements was uniform, the boundary between the coating and the substrate was very clear, diffusion was virtually non-existent;
- the coating was dense;
- the system analyzed can generate multi-layer components.

References

1. A. A. Achimov, I. M. Tolmachyov, S. Yu. Udovichenko, Bulletin of Tyumen State University. **7**, 105 (2014)
2. S. A. Muboyadzhan, S. A. Budinovsky, V. V. Terekhova, Metal Studies and Thermal Treatment of Metals, **1**, 14 (2003)
3. A. N. Ramazanov, P. V. Beglikcheyev, Mechanical Engineering, Material Studies, and Thermal Treatment of Metals. **10**, 108 (2008)
4. Yu. P. Tarasenko, Physics and Chemistry of Metal Processing. **4**, 42 (2006)
5. A. G. Illarionov, A. A. Popov, *Technological and Operating Properties of Titanium Alloys* (Yekaterinburg, Ural University Publishing House, 2014)
6. V. M. Beresnyov, M. Yu. Kopeikina, S. A. Klimenko, Questions of Atomic Science and Technology. **1**, 152 (2008)
7. V. Beletsky, V. Kireyev, S. Knyazev, D. Chelapkin, Modern Electronics. **2**, 12 (2002)
8. Ye. V. Berlin, L. A. Seidman, *Ion Plasma Processes in Thin-Film Technology (*Tekhnosfera, 2006)
9. V. S. Mukhin, R. M. Kireyev, S. R. Schechtman, Bulletin of Ufa State Aviation Technical University. **12**, 212 (2011)
10. Patent 2490369 Russian Federation, V. Ye. Beletsky, Apparatus for Application of Multi-Layer Components to Devices (2013)
11. Patent 2379378 Russian Federation, D. D. Slivakov, V. S. Mitin, A Method for Ion Plasma Application of Multi-Component Film Coatings and a Machine for Such Application (2010)
12. Patent 2210620 Russian Federation, M. I. Guseva, M. V. Atamanov, Yu. V. Martynenko, Yu. V. Moskovskin, V. S. Mitin, A. V. Mitin, S. A. Shiryayev, A Method for Ion Plasma Application of Multi-Component Film Coatings, a Mosaic Target for Such Application, and a Method for Making Such Targets (2003)
13. US Patent 6488824, Sputtering apparatus and process for high rate coatings (2002)
14. US Patent 20130056348 A1, Vacuum coating apparatus and method for depositing nanocomposite coatings (2013)

15. V. V. Kolesnik, Yu. K. Rubanov, Yu. Ye. Tokach, Bulleting of V. G. Shukhov Bashkir State Technical University. **4**, 110 (2017)
16. Utility Model Patent 159075 Russian Federation, V. P. Kolesnik, V. V. Kolesnik, Yu. K. Rubanov, Ye. I. Yevtushenko, Multi-Component Multi-Layer Coating Apparatus (2016)
17. B. S. Danilin, *Use of Low-Temperature Plasma for Application of Thin Films* (Moscow, Energoatomizdat, 1989)
18. V. V. Kolesnik, V. G. Padalka, I. V. Lunev, Aerospace Engineering and Technology (Aviatsionno- kosmicheskaya tekhnika i tekhnologiya). **12**, 58 (1999)

MATEC Web of Conferences **129**, 02008 (2017) DOI: 10.1051/matecconf/201712902008
ICMTMTE 2017

Developing of porous composites on the basis of bi-polymers with aluminum particles and AlOOH nanoparticles for getting antifriction coatings

Natalia Antonova[1,*], *Ilya* Lisnichenko[1], and *Igor* Sibirka[1]

[1]Kamensk Institute (Affiliation) of SRSTU (NPI) named after M.I. Platov, 23 K. Marx Ave, Kamensk-Shakhtinsky, 347810, Russia

Abstract. Porous composites on the basis of bi-polymer of natrium-carboxymethylcellulose containing aluminum powder and nano-particles of aluminum oxohydroxide have been developed. With the help of electronic microscopy the sizes of the formed pores were defined: 600 – 1000 µm; the surface porosity of the coating was defined as well, it made 65%. Using methods of x-ray diffraction analysis and infrared spectroscopy it was stated that the phase composition of porous composites remains unchanged in the process of forming. The possibility to form antifriction coatings by filling porous matrixes with solid lubricants is shown on the example of molybdenum disulfide. When applying the coatings to steel surfaces the friction coefficient reduces by 60%. With the help of differential scanning calorimetry the limit of coatings thermo destruction was defined: 300 - 306° C.

1 Introduction

Highly porous materials with cell structure are demanded as filters, catalyst bearers, parts of light constructions, coatings with functional properties. Technologies of getting foamed metals on the basis of nickel, iron, copper, titanium, aluminum are developed. In production scales porous structured metals are produced by injecting inert gases into the melt, stimulating gases generating by using gas emitting reagents, doubling porous polymers structures with metals [1-4]. Polymer net structures, which do not contain metals, may be got easier as a result of systems self-organising [5, 6]. However, approaches, which help to form composite functional materials with cell structure on the basis of polymer matrixes with metal powders using controllable self-organisation, are not studied properly nowadays.

We got [7] porous film materials from water suspensions of natrium- carboxymethylcellulose bi-polymer (Na-CMC) with aluminum powder as a filling and nanoparticles of aluminum oxohydroxide (AlOOH) as microadditives in a relatively cheap and ecologically safe way. Changes in AlOOH quantity let us control the generated cells sizes within the diapasone of 100 – 1000 µm [8]. Biodegradable Na-CMC polymer is nowadays widely used in production and attracts researchers' attention as the basis for developing membranes and packaging films [9-11]. There are almost no works devoted to using porous organic films as antifriction materials. Such composites, produced with the help of using ecologically safe Na-CMC bi-polymer, were of great interest as porous matrixes suitable to develop materials which being combined with different solid lubricants let provide resistanse of the coating under the conditions of environment impact and contact strain. So arose the idea of the work presented: to use film pores for filling them with solid lubricants of different origin and dispersion and to form antifriction coatings to protect metal surfaces under friction conditions.

The powder of molybdenum disulfide (MoS_2) was chosen as solid lubricant. This lubricant is widely used and the results of the researches in the sphere of its antifriction properties are being published during the last 40 years. The friction coefficient of MoS_2 under vacuum conditions is within 0,005 – 0,05 limits [12], it increases when in the air and makes 0,15 – 0,3 due to the possibility of the lubricant to absorb moisture and oxygen [13-14]. To reduce adsorption the authors of the work [14] recommend increasing the density of coatings and decreasing free surface energy. Inserting MoS_2 powder into porous materials supposes becoming the lubricant in the pores more solid when the friction pair contacts dry and reducing the friction coefficient.

The aim of the work presented is to reveal the peculiarities of structure forming and antifriction properties of composite porous coatings developed on the basis of Na-CMC with Al and MoS_2 powders and designed to protect metal surfaces.

2 Materials and Methods

2.1 Materials

Porous films were got by mixing gel-like solution of purified Na-CMC polymer (with the concentration of 2,8%) and aluminum powder particles of spheric shape, the size of particles less than 40 µm. Distilled water was the solvent. Glycerin was used as plasticizer,

* Corresponding author: melnik1@rambler.ru

microadditives were nanoparticles of aluminum oxohydroxide – (AlOOH), which were synthesized in accordance with the technique [8]. Forming of samples was made in "wet" way on fluoroplastic wafers with the temperature of (55±1) °C. Films were dried until getting constant mass. The composition of the suspension used for films production is represented in table 1.

Table 1. The composition of the suspension for porous films production.

Components proportion, g.			
Water solution, Na-CMC	Glycerin	Al	AlOOH
100,00	3,00	2,50	0,20

In the process of work the possibility of using the formed porous films as buffer matrixes for developing antifriction materials was determined. It is known that purified watersoluble polymer Na-CMC, used as bond for aluminum powder, possesses adhesive properties. That is why the obtained porous coating after being moistened with water was spread onto samples surfaces made of 40Xsteel which had been degreased with acetone and afterwards hold inside the oven under the temperature of 25°C within 2 hours to provide the adhesion of the film to the metal surface. The lubricant was spread onto the surface of the film with open pores with a brush.

2.2 Techniques

Films morphology was studied with the help of electronically scanning microscopes Quanta 200 in the *Common Use Center* (*CUC*) *Nanotechnologies*, SRSTU (NPI) named after M.I. Platov, VEGA II LMU in the *CUC Mineral Raw Materials and Environment Condition Study Center*, SFU. Dispersion of aluminum powder and molibdenum disulphide powder was defined with the help of MicroTrac Bluewave analyser (*CUC Nanotechnologies*, SRSTU (NPI). X-ray diffraction analysis (XDA) of porous films was made with the help of diffractometer ARL X'TRA within the anglular spacing of 2θ from 5° to 80° with the interval of 0.05°, emission – $CuK_{\alpha 1}$, $K_{\alpha 2}$ (*CUC Nanotechnologies*, SRSTU (NPI). Infrared spectra (IR-spectra) of the films were registered by spectrophotometer FT-08, *Lumex*, within the spacing of 600÷4000 cm^{-1} using the technique of DFIR (disodered full inside reflection).

Antifriction properties of the materials were studied with the help of stand friction unit TMT-25 under the load up to 500 MPa and sliding speed 0,04 m/sec. The experiment lasted for 2 hours. The work environment was air. Termostability of obtained porous films was defined by the technique of differantial scanning calorimetry (DSC), with the help of METTLER TOLEDO calorimeter DSC-1 within the temperature spacing 30,00 - 600,00 °C with the sample heating speed of 20°C min^{-1} and air discharge of 50 ml/min^{-1} in a standard aluminum crucible.

3 Results and Discussions

3.1 Morphology and Structure of Porous Films

Figure 1a shows SEM-picture of composite film surface with cells, obtained on fluoroplastic wafer from Na-CMC suspension with Al and AlOOH powders. Typical pore fragment is shown in Figure 1b. Pores are of domed shape formed mainly by the polymer film with the thickness of 30 μm. Spaces between pores are filled with aluminum microparticles, the diaphragm thickness is about 300 μm. Pore bases are open; cell sizes reach 1000 – 600 μm. In the process of cells forming aluminum particles remained unchanged in thier size and shape. The thickness of porous film is about 100 μm. Total surface porosity of the film, defined with the help of Tescan software *Morphology*, made 65%.

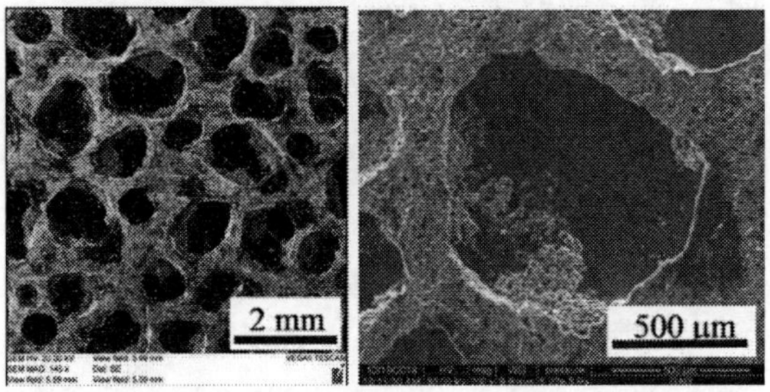

a) b)

Fig. 1. Surface of porous film with Al powder and AlOOH additives – (a), pore fragment – (b)

Results of dispersion analysis of Al and MoS_2 powders are given in histograms in Fig.2 where N means the total amount of enumerated particles, ΔN – the number of particles which got into the diameters spacing, d – particle size. Flat particles of molybdenum disulphide tend to agglutinate; Al and MoS_2 particles distribution according to their size is close. In both cases submicron fraction takes place, particles and agglomerates with the size of 10 – 20 μm prevail. Sizes of the cells, formed in the film, reach 1000 μm which supposes dense packaging of MoS_2 particles having smaller sizes into pore cells.

Diffraction pattern of the porous film is given in Fig 3. X-ray diffraction analysis of the formed porous composites showed the presence of metal aluminum in them (Al cubic cell, space group Fm-3m, elementary cell parameter *a*= 4.0494 Å, PDF card000-04-0787). Absence of AlOOH and aluminum oxides, taking place on the aluminum powder particles in the form of films, may be explained by their low content in the initial composition. Interplanar line spaces in the diffraction pattern of cell composite point at zero-valent condition of metal in the sample. Reflex in the 2θ=18-20 ° field, according to the data [13], is characteristic to cellulose crystallites. There were no structural changes of aluminum in the process of porous composite forming.

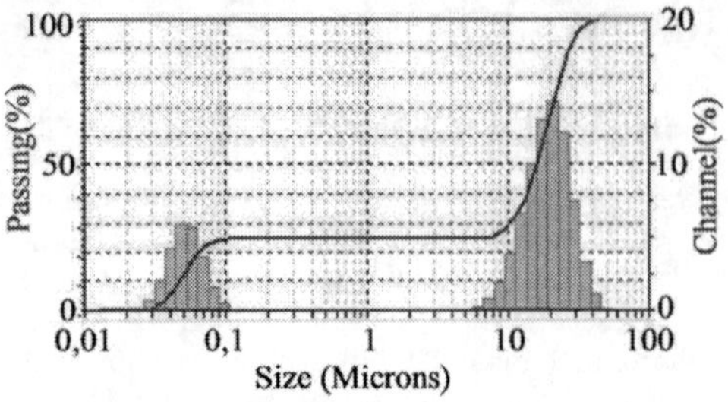

a)

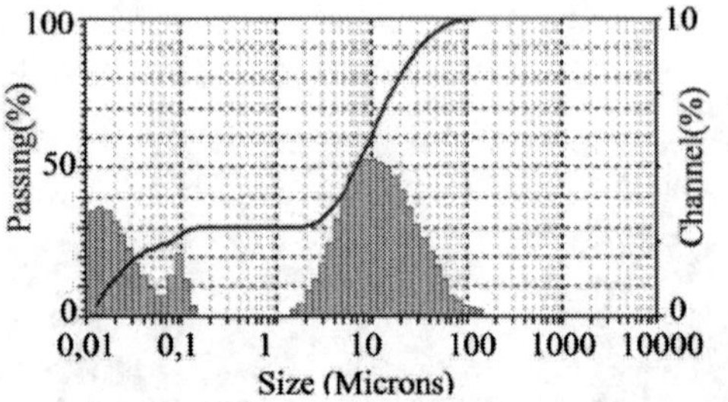

b)

Fig. 2. Histograms of particles distribution according to their size: Al powder – (a); MoS_2 powder – (b)

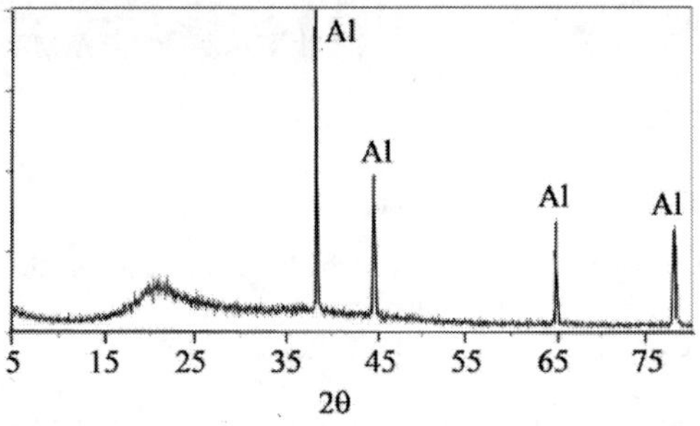

Fig. 3. Diffraction pattern of cell film: Na-CMC+Al

Diffraction pattern of the porous film is given in Fig 3. X-ray diffraction analysis of the formed porous composites showed the presence of metal aluminum in them (Al cubic cell, space group Fm-3m, elementary cell parameter a= 4.0494 Å, PDF card000-04-0787). Absence of AlOOH and aluminum oxides, taking place on the aluminum powder particles in the form of films, may be explained by their low content in the initial composition. Interplanar line spaces in the diffraction pattern of cell composite point at zero-valent condition of metal in the sample. Reflex in the 2θ=18-20 ° field, according to the data [13], is characteristic to cellulose crystallites. There were no structural changes of aluminum in the process of porous composite forming.

Changes in the organic component of porous composites in the process of forming them from solutions were analysed with the help of IR-spectroscopy. Fig. 4 shows IR-spectra of Na-CMC powder, film with empty pores and film with pores filled with molybdenum disulfide. Broad intensive stripe in 3600 – 3000 cm^{-1} field characterises frequencies of valent fluctuations of OH-groups, involved into intramolecular and intermolecular bonds. OH-groups' taking part in forming intermolecular hydrogen bonds shows itself in shifting of absorption stripe towards smaller frequencies [15]. For Na-CMC powder characteristic stripe of absorption makes 3550 cm^{-1}, for the film sample without lubricant – 3500 cm^{-1} and its intensity increases abruptly. Na-CMC macromolecules posess ionised groups which promote Na-CMC macroions repulsing each other. Staying by their conformational structure polyanionic Na-CMC macromolecules mutually repulsing form a space polymer net with distributed aluminum microparticles in the suspension. As a result hydrogen bonds between active hydroxyl groups of macromolecules and oxide shells of aluminum particles are formed. The set of stripes in 1350-700 cm^{-1} field is an individual characteristic for each substance [16]. For Na-CMC powder and porous composite films they are practically identical. Molybdenum disulfide, densely filling pores of the coating, forms tight solid film. IR-diffraction pattern is of little use for defining disulfides which is also proved by spectrum obtained for film with lubricant – absorption stripes almost do not exist in this case.

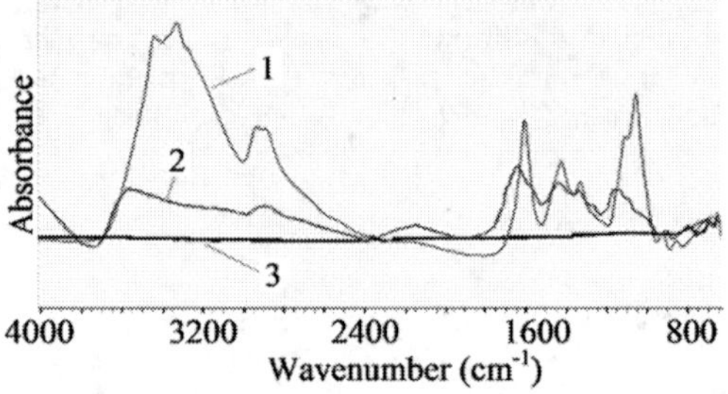

Fig. 4. IR-spectra of samples: Na – CMC powder – (1); Na-CMC+Al porous film – (2) ; Na-CMC+Al+MoS_2– (3)

Analyses of IR-spectra and diffraction pattern allow to conclude that initial components in films with generated cells retain their chemical composition unchanged.

3.2 Antifriction Properties of Porous Coatings with Molybdenum Disulfide

Friction pairs were tested dry. The counterbody was made of 47GT steel.

Two types of samples were studied:

- steel uncoated sample after being lathed;
- steel sample with porous coating containing MoS_2.

SEM–pictures of sample surfaces before tests are shown in Fig. 5A, 6A. Sample surfaces after tests on the friction unit are shown in Fig. 5b, 6b. The wear of the steel sample after the tests has mainly adhesive character. In the coated sample in the initial state the pores on the film surface are densely filled with the lubricant – MoS_2. Molybdenum disulfide has laminated structure which provides 'easy' glide of friction surfaces against each other, helping to lower friction coefficient and to increase wear resistance of friction-pair parts surfaces. In the process of friction pair work as the temperature gets higher, molybdenum disulfide powder is extruded from the pores and fills the surface of composite matrix on the basis of Na-CMC containing aluminum powder with lubricant film. After tests on the friction unit samples with composite coating obtain smooth shiny surface. Pores-cells with compacted lubricant are clearly seen on the sample surface, nevertheless there are no traces of solidification which, in turn, leads to wear resistance increase (Fig. 6b).

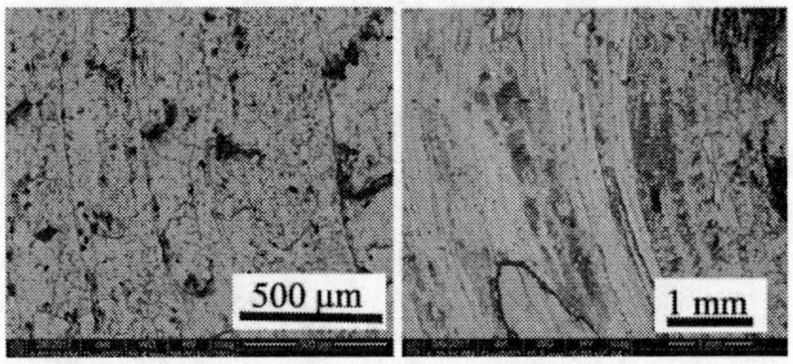

a) b)

Fig. 5. The steel surface: before – (a) and after – (b) friction tests

a) b)

Fig. 6. Surface coated with Na-CMC+Al+ MoS_2: before – (a) and after – (b) friction tests

After friction tests IR-diffraction pattern of coated steel sample showed the presence of polymer Na-CMC (3600 – 3000 cm^{-1} field, Fig. 7A). This fact witnesses of retaining of the coating polymer component on protected metal surface. The possibility to re-locate Na-CMC polymer onto the contacting the composite coating counterbody surface was defined. With this purpose IR-spectra of metal counterbody surface were compared before and after the tests (Fig. 7b). IR-spectrum of the initial steel counterbody surface does not have absorption stripe in 3000 – 2800 cm^{-1} field characteristic to OH-groups valence fluctuations, whereas after the tests it appears in counterbody surface spectrum. Consequently, the formation of friction re-location film takes place.

Changes of friction coeffitient C_F for studied samples, depending upon the applied load, is shown in Fig. 8a. For the sample, protected by porous coating with MoS_2 friction coefficient lowers dramatically – by 60% comparing with the friction coefficient for steel surface (Fig. 8a, curves 1 and 2). Its value does not practically change and makes 0,09. While contacting coated samples the counterbody temperature increases only by 15-16%, whereas while contacting steel surfaces it practically doubles (Fig. 8B). It is known that insertion of MoS_2 into antifriction material lets increase materials heat conduction by several times. Aluminum heat conduction is higher than that of steel by 4-5 times. That is why heat is better abducted by means of composite coating Linear wear of uncoated steel sample made 4,7 µm during the test time (2 hours), of coated sample – 12,3 µm. Consequently, after the test more than 80% of initial thickness of composite coating retains. In the process of friction pair work friction re-place of composite coating onto metal counterbody and formation of re-place film takes place.

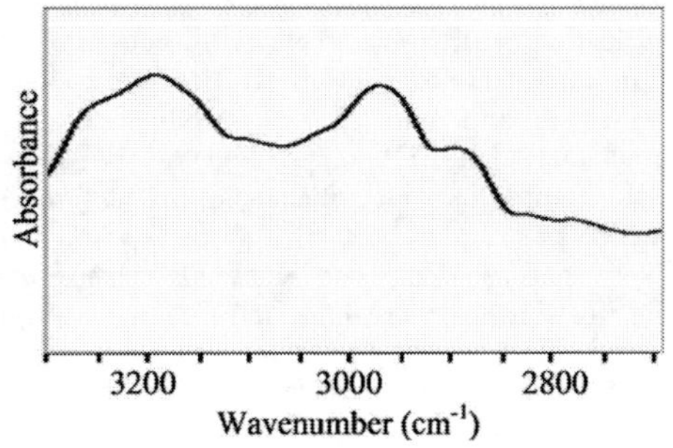

a)

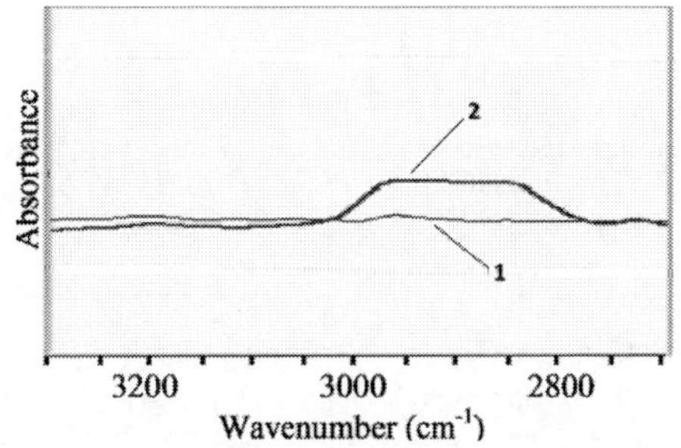

b)

Fig. 7. IR-spectra: (a) – of coated surface after friction tests; (b) – of counterbody surface before (1) and after (2) the tests

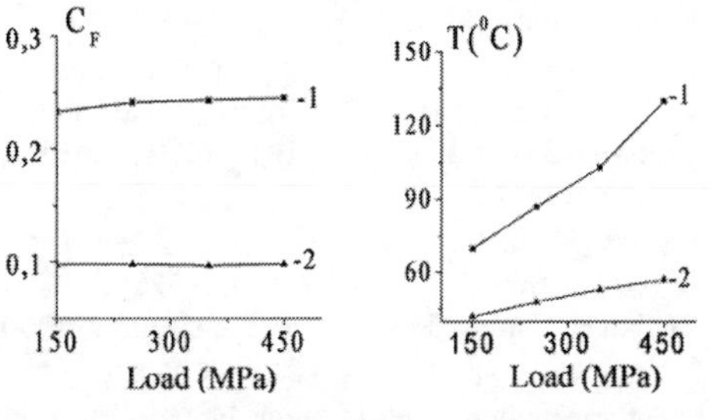

a) b)

Fig. 8. Influence of load, applied to samples 1-2, on: friction coefficient C_F – (a); counterbody temperature T – (b)

3.3 Scanning calorimetry of porous coatings with molybdenum disulfide

In the composite functional coatings under study thermally vulnerable component is biopolymer. Analysis, made by methods of scanning calorimetry showed that relatively small exopick takes place when the temperature is 306°C and corresponds destruction of cellulose components. However thermo destruction of the coating takes place when the temperature is 450°C which does not contradict data about thermal destruction of MoS_2 existing in literature sources.

4 Conclusion

The paper presented shows the possibility to form porous materials from natrium-carboxymethylcellulose suspensions with aluminum microparticles and aluminum oxohydride nanoparticles. In the process of solidification oriented open cells with the size of 1000 – 600 μm are generated in the films. Using methods of X-ray analysis and infrared diffraction pattern it was defined that chemical composition of the composite porous coating remains unchanged in the process of formation.

New porous composite materials perspective as buffer matrixes for creating antifriction coatings are developed. Filling pores with lubricant – molybdenum disulfide – allows to lower friction coefficient by 60% and works towards the increase of coating heat conduction and its wear resistance.

The limit for thermo destruction of antifriction coatings of 450°C is stated.

Authors express gratitude to Alexey Yatsenko, CUC Nanotechnology, SRSTU (NPI) named after M.I. Platov for help with fulfilling SEM-pictures of our materials.

References

1. V.N. Antsiferov, A.M. Makarov, A.A. Khanov, G.V. Bashkirtsev, Perspective Materials, **3**, 5 – 9 (2010)
2. E.V. Matygullina, Theoretial and Applied Problems of Forming Composite Materials with Controlled Micro and Macro Porosity on the Basis of Oxide Systems: author's abstract of technical doctoral dissertation. (Perm', 2011)
3. V.N. Antsiferov, The Way of Obtaining Highly Porous Cell Material. Patent № 2508962 the Russian Federation, № 201215000/02, 2014, Patent Bulletin № 7
4. M. F. Ashby, A.G. Evans, N.A. Fleck, L.J. Gibson, *Metal Foams: A Design Guide.* Hutchinson and H.N.G. Wadley (Butterworth-Heinemann, 2000)
5. D.V. Novikov, I.S. Kuryndin, V. Bukoˇsek, G.K. Yel'yashevitch, Solid Body Physics, **54**, № 11, 2176-2182 (2012)
6. D.V. Novikov, A.N. Kraovsky, Solid Body Physics, **54**, № 11, 2183-2188 (2012)
7. N.M. Antonova, O.S. Ovchinnikov, A.P. Babichev, Nanoengineering. **3,** 17-22 (2014)
8. N.M. Antonova, A. P. Babichev, V. Yu. Dorofeev, Protection of Metals and Physical Chemistry of Surfaces. **49,** № 7, 868-872 (2013)
9. Q. Chen, P. Yu, W. Huang, S. Yu, M. Liu, C. Gao, Journal of Membrane Science **492,** 312-321 (2015)
10. A.A. Oun, J. W. Rhim, Carbohydrate Polymers. **127,** 101-109 (2015)
11. Xiaoyan Lin, Ying Li, Zhe Chen, Chi Zhang, Xuegang Luo, Xinchen Du, Yuanhao Huang, Chemical Engineering Jornal **215-216,** 709-720 (2013).
12. A.V. Bondarev, Developing Nanocomposite, Wear-Resistant Coatings on the Basis of Titanum and Molibdenum Carbonitrides for Operating Within Temperature Interval of 25 – 700°C: Technical Candidate's Thesis. Moscow, 165 pp. (2014)
13. W. Lauwerens, J. Wang, J. Navratil, E. Wicers, Surface and Coatings Technology **131**, 16-21 (2000)
14. E.W. Roberts, W.B. Price, Proceedins of European Space Mechanisms and Tribology Symposium. **6** 73-78 (1996)
15. G.A. Petropavlovsky, *Gydrophilic Partially Substituted Cellulose Eteri and their Modification* by Chemical Cross-Linking. L.: Science, 298 pp. (1988)
16. L. Belami, *Infrared Spactra of Compound Molecules.* M.: Foreign Literature Publishing Office, 590 pp. (1963)

Simulation of boron and carbon fiber composite characteristics of the elasticity

Victor .P. Pavlov[1], *Vilina. M.* Kudoyarova[2,*], and *Alexander. A.* Philippov[1]

[1]Ufa State Aviation Technical University, Chair of Strength of Materials, st. K. Marcsa,12, Ufa, 450000, Russian Federation
[2]Ufa State Aviation Technical University, Chair of Aviation Thermal and Energy Engineering, st. K. Marcsa,12, Ufa, 450000, Russian Federation

Abstract. The use of hybrid fiber-reinforced polymer composites (FRP) opens up wide possibilities for creating structures that are most suitable for different operating conditions. The modelling of a multilayer FRP structure is the most important stage in the creation and manufacture of FRP products since the FRP structure determines the entire complex of elastic, strength and other properties. This paper proposes a method for the theoretical prediction of anisotropic elastic characteristics of material equivalent to a given FRP. A technique for calculating the stiffness coefficients of a homogeneous anisotropic material equivalent to a hybrid composite material having a magnesium matrix reinforced with boron and carbon fibers has been developed and tested in ANSYS. This paper has found that the results of the finite element analysis essentially depend on the dimension of the mesh, and it is necessary to use a mesh with a number of nodes exceeding $N = 1 \cdot 10^5$ to obtain sufficiently accurate calculated values of the elasticity coefficients.

1 Introduction

The FRP of two or more kinds of reinforcing fibers are widely used in modern technology-based products, e.g. boron and carbon fibers. For the strength and stiffness calculation, an equivalent homogeneous anisotropic material usually replaces the heterogeneous FRP, for which the mechanical properties will be determined [1-12].

The experimental simulation of properties FRP composites is very time-consuming [13-19], and a wide variety of structures FRP composites does not allow to study the characteristics for all possible variants. Therefore, computer simulation of the mechanical characteristics FRP materials for its various structures is very promising [8-12].

Two problems arise:

• What do we take as an equivalence for considering heterogeneous FRP composite and for homogeneous anisotropic material?

• How do we determine the effective elasticity characteristics of an equivalent anisotropic material?

It should be noted that there is no unequivocal solution of these problems, and each researcher can realize his own path.

This paper considers a technique for mathematical modeling of the elasticity characteristics of FRP having carbon and boron fibers, based on the FEM using ANSYS.

The proposed work solves the mathematical modeling problem of the elasticity characteristics FRP composite having carbon and boron fibers. The finite element method (FEM) is implemented by using the ANSYS.

2 Problem definition

The paper considers a FRP composite with a given mechanical characteristics (matrix and reinforcement) and orientation of the fibers.

This paper builds the theoretical prediction task of the anisotropic elastic material characteristics that are equivalent to the FRP composite. An equivalent material is considered to be obeying to the linear Hooke's law [20].

3 Formation a FRP Composite Representative Element

In order to predict the mechanical characteristics of the hybrid FRP, it is necessary to specify its structure and the mechanical characteristics of its components.

The real FRP structure has a significant indeterminacy which is associated with the fact that FRP consists of many fibers of small diameter, and it is almost impossible to dispose the fibers by a uniquely defined law. We can only orient the fiber's threads in space as a whole and even then not with absolute exactness. In addition, the fibers combined into the FRP thread will be located at different distances unknown for each individual thread.

It is possible to determine the average direction for the fiber bundle in the space and also determine the

* Corresponding author: Kudoyarova@inbox.ru

DOI: 10.1051/matecconf/201712902009

proportions of one or the other type fiber in the volume of hybrid FRP by studying the structure and composition of hybrid FRP experimentally.

Thus, we have assumed that we knew the directions of the reinforcing threads and the coefficients of their concentration in the considering volume of hybrid FRP. Based on this, we are applying some idealization and forming the representative element for FRP composite based on magnesium binder reinforced with carbon and boron fibers to calculate its effective elastic characteristics. It has a rectangular prism form with a_x, a_y, a_z side dimensions and contains a single boron fiber d_B = 100 mic diameter, fourteen carbon fibers d_C = 10 mic diameter. Boron fiber is oriented along to Y-axis, carbon fibers are directed along X-axis (Fig. 1).

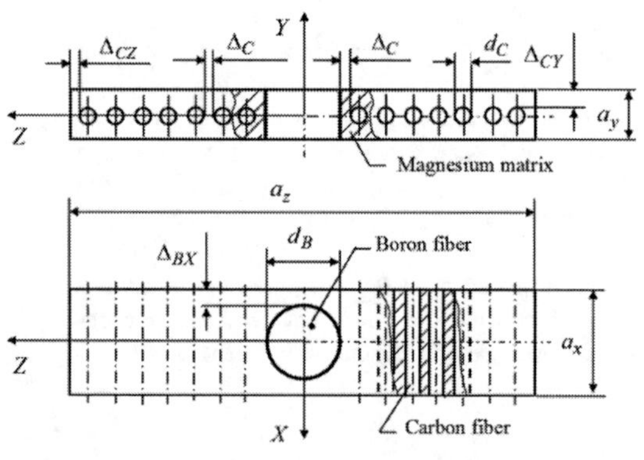

Fig. 1. Representative element for FRP composite

The distance between adjacent carbon fibers along the Y- axis and the minimum distance between boron and carbon fibers equals to Δ_C = 2 mic (fig. 1). The minimum distances from the carbon fibers to the side faces of the representative element that is perpendicular to the Y-axis and Z-axis are equal to Δ_{CY} = 1 mic and Δ_{CZ} = 1 mic, respectively, (fig. 1). The distance from the boron fiber to the side faces of the representative element that is perpendicular to the X-axis equals to Δ_{BX} = 5 mic (fig. 1).

In accordance with fig. 1 the representative element has the following sizes: $a_x = 1{,}1\cdot10^{-4}$ m, $a_y = 1{,}2\cdot10^{-5}$ m, $a_z = 2{,}7\cdot10^{-4}$ m.

We define for a representative element of the hybrid FRP for the indicated dimensions: total volume V, volume of the carbon component V_C, volume of the boron component V_B:

$$\begin{cases} V = a_x a_y a_z = 3{,}56\cdot10^{-13} \text{ m}, \\ V_C = 14\pi d_C^2 a_x / 4 = 1{,}21\cdot10^{-13} \text{ m}, \\ V_B = \pi d_B^2 a_y / 4 = 9{,}42\cdot10^{-14} \text{ m}. \end{cases} \quad (1)$$

On the basis of these expressions, we determine the volume coefficients of FRP reinforcement by carbon (Ψ_C) and boron (Ψ_B) fibers:

$$\begin{cases} \Psi_C = \dfrac{V_C}{V} = 0{,}339; \\ \Psi_B = \dfrac{\Psi_B}{\Psi} = 0{,}264. \end{cases} \quad (2)$$

The FRP reinforcement ratio for carbon fibers equal to $\Psi_C = 0{,}339$ and for boron fibers equal to $\Psi_B = 0{,}264$.

4 Modeling the elasticity characteristics of FRP composite by using FEM

This paper considers a homogeneous anisotropic material is accepted as the equivalent for the FRP with the relationship between stresses σ_x, σ_y, σ_z, τ_{xy}, τ_{yz}, τ_{xz} and deformations ε_x, ε_y, ε_z, γ_{xy}, γ_{yz}, γ_{xz} that is determined by six relations [20]:

$$\begin{cases} \sigma_x = E_{11}\varepsilon_x + E_{12}\varepsilon_y + E_{13}\varepsilon_z + E_{14}\gamma_{xy} + E_{15}\gamma_{yz} + E_{16}\gamma_{xz}, \\ \sigma_y = E_{21}\varepsilon_x + E_{22}\varepsilon_y + E_{23}\varepsilon_z + E_{24}\gamma_{xy} + E_{25}\gamma_{yz} + E_{26}\gamma_{xz}, \\ \sigma_z = E_{31}\varepsilon_x + E_{32}\varepsilon_y + E_{33}\varepsilon_z + E_{34}\gamma_{xy} + E_{35}\gamma_{yz} + E_{36}\gamma_{xz}, \\ \tau_{xy} = E_{41}\varepsilon_x + E_{42}\varepsilon_y + E_{43}\varepsilon_z + E_{44}\gamma_{xy} + E_{45}\gamma_{yz} + E_{46}\gamma_{xz}, \\ \tau_{yz} = E_{51}\varepsilon_x + E_{52}\varepsilon_y + E_{53}\varepsilon_z + E_{54}\gamma_{xy} + E_{55}\gamma_{yz} + E_{56}\gamma_{xz}, \\ \tau_{xz} = E_{61}\varepsilon_x + E_{62}\varepsilon_y + E_{63}\varepsilon_z + E_{64}\gamma_{xy} + E_{65}\gamma_{yz} + E_{66}\gamma_{xz}, \end{cases} \quad (3)$$

The stiffness coefficients of the equivalent material E_{ij}, i, j = $\overline{1{,}6}$ are determined on the basis of mathematical modeling of the deformation for a FRP representative element at six numerical experiments, in each of which only one of the strain components differs from zero.

In the first experiment $\varepsilon_x \neq 0$, $\varepsilon_y = \varepsilon_z = \gamma_{xy} = \gamma_{yz} = \gamma_{xz}$ =0. For this case the deformation of the representative element is shown in Fig. 2.

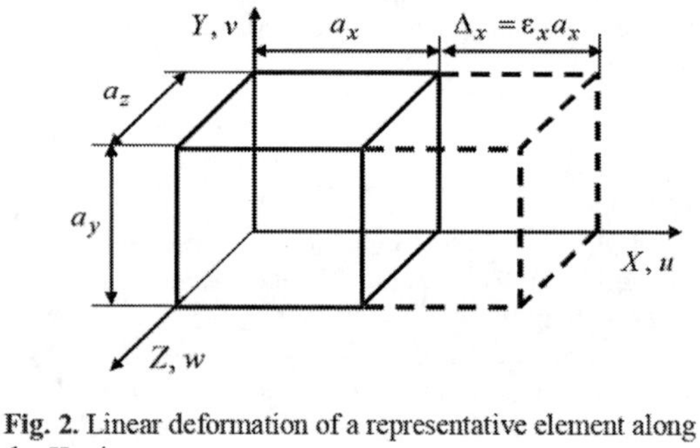

Fig. 2. Linear deformation of a representative element along the X-axis

For all three faces of the FRP representative element the averaged stresses $\overline{\sigma}_x^{(\varepsilon x)}, \overline{\sigma}_y^{(\varepsilon x)}, \overline{\sigma}_z^{(\varepsilon x)}, \overline{\tau}_{xy}^{(\varepsilon x)}, \overline{\tau}_{yz}^{(\varepsilon x)}, \overline{\tau}_{xz}^{(\varepsilon x)}$ at given deformation are determined in ANSYS. The expressions of stiffness coefficients: $E_{11} = \overline{\sigma}_x^{(\varepsilon x)} / \varepsilon_x$, $E_{21} = \overline{\sigma}_y^{(\varepsilon x)} / \varepsilon_x$, $E_{31} = \overline{\sigma}_z^{(\varepsilon x)} / \varepsilon_x$, $E_{41} = \overline{\tau}_{xy}^{(\varepsilon x)} / \varepsilon_x$, $E_{51} = \overline{\tau}_{yz}^{(\varepsilon x)} / \varepsilon_x$, $E_{61} = \overline{\tau}_{xz}^{(\varepsilon x)} / \varepsilon_x$ were obtained under substituting of averaged stresses in the first three equations of the system (1).

Similarly, the elasticity coefficients $E_{i2}, i = \overline{1{,}6}$ and

$E_{i3}, i=\overline{1,6}$ of an equivalent anisotropic homogeneous material are determined by specifying the deformation components different from zero $\varepsilon_y \neq 0$, and then $\varepsilon_z \neq 0$.

To determine the stiffness characteristics under shear the deformation of the representative element under shear strain $\gamma_{xy} \neq 0$, $\varepsilon_x = \varepsilon_y = \varepsilon_z = \gamma_{yz} = \gamma_{xz} = 0$ is considered (fig. 3), in which the averaged stresses $\overline{\sigma}_x^{(\varepsilon x)}, \overline{\sigma}_y^{(\varepsilon x)}, \overline{\sigma}_z^{(\varepsilon x)}$, $\overline{\tau}_{xy}^{(\varepsilon x)}, \overline{\tau}_{yz}^{(\varepsilon x)}, \overline{\tau}_{xz}^{(\varepsilon x)}$ are calculated, and then the corresponding elasticity coefficients $E_{i4}, i=\overline{1,6}$ are also calculated.

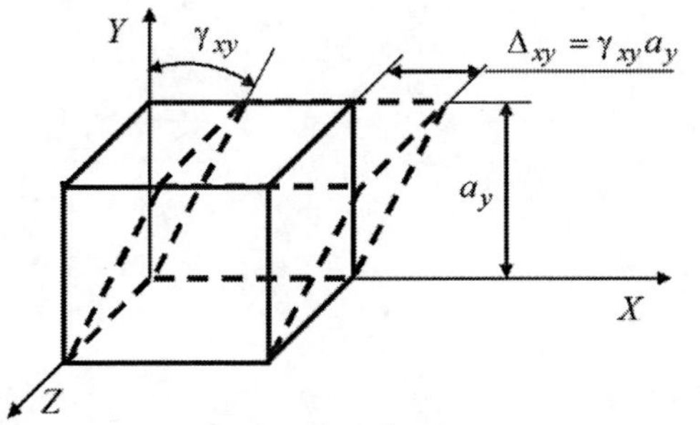

Fig. 3. Shear strain of the material of a representative element on the *XY*-plane

Then, the elasticity coefficients $E_{i5}, i=\overline{1,6}$ and $E_{i6}, i=\overline{1,6}$ are determined by specifying deformation components different from zero $\gamma_{yz} \neq 0$, $\gamma_{xz} \neq 0$.

5 Results

According to the proposed method, the FRP stiffness characteristics under consideration are calculated, assuming that the FRP reinforcing components and FRP matrix components are isotropic materials with modulus of elasticity *E* and Poisson's coefficient ν.

At the same time, we take into account that the physical-mechanical characteristics of carbon and boron fibers, which are given in the literature, have a very wide dispersion related to the specific technology of their production, and therefore we take for them the following values of mechanical characteristics averaged from the literature sources [1, 5]:

• The matrix made of the alloy ML-10 has $\rho_a = 1800$ kg/m³, $E_a = 0{,}44\cdot10^{11}$ Pa, $\nu_a = 0{,}28$, $\sigma_a^{us} = 245$ MPa,

• Carbon reinforcing fibers have: $\rho_C = 1800$ kg/m³, $E_C = 3\cdot10^{11}$ Pa, $\nu_C = 0{,}2$, $\sigma_C^{us} = 2000$ MPa,

• Boron reinforcing fibers have: $\rho_B = 2500$ kg/m³, $E_B = 4\cdot10^{11}$ Pa, $\nu_B = 0{,}25$, $\sigma_B^{us} = 3000$ MPa.

The following non-zero values of the stiffness coefficients for the FRP composite E_{ij} have been obtained as a calculation results:

$$\begin{cases} E_{1,1} = 1{,}41\cdot10^{11}, & E_{1,2} = 2{,}92\cdot10^{10}, & E_{1,3} = 4{,}60\cdot10^{10}, \\ E_{2,1} = 2{,}92\cdot10^{10}, & E_{2,2} = 5{,}90\cdot10^{10}, & E_{2,3} = 4{,}49\cdot10^{10}, \\ E_{3,1} = 4{,}60\cdot10^{10}, & E_{3,2} = 4{,}49\cdot10^{10}, & E_{3,3} = 1{,}81\cdot10^{11}, \\ E_{4,4} = 7{,}02\cdot10^{9}, & E_{5,5} = 7{,}45\cdot10^{9}, & E_{6,6} = 8{,}07\cdot10^{9}. \end{cases} \quad (4)$$

The remaining stiffness coefficients are equal to zero in view of the presence of FPR composite's plane symmetry.

6 Conclusion

1. The representative element structure of a hybrid FRP composite based on a magnesium matrix reinforced with boron and carbon fibers is proposed.
2. A technique for calculating the stiffness coefficients of a homogeneous anisotropic material equivalent to a hybrid composite material having a magnesium matrix reinforced with boron and carbon fibers has been developed and tested by using FEM in ANSYS.
3. It is found that the results of the finite element analysis essentially depend on the dimension of the mesh, and it is necessary to use a mesh with a number of nodes exceeding $N = 1\cdot10^5$ to obtain sufficiently accurate calculated values of the elasticity coefficients.

References

1. Yu. S. Pervushin, *Proektirovanie i prognozirovanie mekhanicheskih svojstv odnonapravlennogo sloya iz kompozicionnogo materiala (Designing and predicting the mechanical properties of a unidirectional composite material).* (Ufa: UGATU, 2002). (in Russ.)
2. I. G. Zhigun, V. A. Polyakov, *Svojstva prostranstvenno-armirovannyh plastikov (Properties of spatially reinforced plastics).* (Riga: Zinatne, 1978). (in Russ.)
3. V. V. Vasil'ev, *Mekhanika konstrukcij iz kompozicionnyh materialov (Mechanics of composite structures).* (M.: Mashinostroenie, 1988). (in Russ.)
4. *Kompozicionnye materialy: Spravochnik (Composite materials: Handbook)* (M.: Mashinostroenie, 1990). (in Russ.)
5. Yu. A. Mikhajlin, *Konstrukcionnye polimernye kompozicionnye materialy (Structural polymeric composite materials)* (SPb.: Nauchnye omsnovy i tekhnologii, 2008). (in Russ.)
6. S. A. Lur'e, Yusefi Shakhram, Mekhanika kompozicionnykh materialov i konstrukcij. vol. **3**, № **4**, pp. 76-92 (1996). (in Russ.)
7. *Spravochnik po kompozicionnym materialam (Handbook of Composite Materials).* (M.: Mashinostroenie, 1988). (in Russ.)
8. Yu. I. Dimitrienko, A. P. Sokolov, Vestnik MGTU im. N. E. Baumana. Ser. Estestvennye nauki, **№2**, pp. 57 – 67 (2008). (in Russ.)
9. Yu. I. Dimitrienko, A. P. Sokolov, Vestnik MGTU im. N. E. Baumana. Ser. Estestvennye nauki, **№2**. pp. 56–67 (2008). (in Russ.)
10. Yu. I. Dimitrienko, A. P. Sokolov, Informacionnye tekhnologii, № **8**. pp. 31 – 38, (2008). (in Russ.)
11. Ju.I. Dimitrienko, A.I. Kashkarov, *Konechno-*

DOI: 10.1051/matecconf/201712902009

jelementnyj metod dlja vychislenija jeffektivnyh harakteristik prostranstvenno-armirovannyh kompozitov, Vestnik MGTU im. N.Je. Baumana. Ser. Estestvennye nauki № **2**, pp. 95-108 (2002). (in Russ.)

12. Yu. I. Dimitrienko, A. P. Sokolov, A. I. Kashkarov, Aehrokosmicheskie tekhnologii. M.: MGTU im. N. E. Baumana, pp. 113 – 114 (2004). (in Russ.)
13. E. K. Ashkenazi, E. V. Ganov, *Anizotropiya konstrukcionnykh materialov: Spravochnik (Anisotropy of structural materials: Handbook).* (L.: Mashinostroenie, 1972). (in Russ.)
14. *Kompozicionnye materialy. T. 2. Mekhanika kompozicionnym materialov (Composite materials. vol. 2. Mechanics of composite materials).* (M.: Mir, 1978). (in Russ.)
15. *Kompozicionnye materialy. T. 7. Analiz i proektirovanie konstrukcij (Composite materials. vol. 7. Analysis and design of structures).* (M.: Mashinostroenie, 1978). (in Russ.)
16. Yu. M. Tarnopol'skij, I. G. Zhigun, V. A. Polyakov, *Prostranstvenno-armirovannye kompozicionnye materialy: Spravochnik (Spatially-reinforced composite materials: Handbook).* (M.: Mashinostroenie, 1987). (in Russ.)
17. Yu. V. Sokolkin, A. A. Tashkinov. *Mekhanika deformirovaniya i razrusheniya strukturno neodnorodnykh tel (Mechanics of deformation and destruction of structurally heterogeneous bodies).* (M.: Nauka, 1984). (in Russ.)
18. V. P. Pavlov, *Polzuchest' polimernykh kompozicionnykh materialov pri peremennykh povyshennykh temperaturakh. Eksperimental'nye issledovaniya i matematicheskoe modelirovanie (Creep of polymer composite materials at variable elevated temperatures. Experimental research and mathematical modeling).* (Ufa: Ufimsk. gos. aviac. tekhn. un-t, 2004). (in Russ.)
19. V. P. Pavlov, *Teplovaya deformaciya, prochnost' i termovyazkouprugost' stekloplastikov pri vysokoj peremennoj vo vremeni temperature v usloviyakh termodestrukcii. Eksperimental'nye issledovaniya i matematicheskoe modelirovanie (Thermal deformation, strength and thermal viscoelasticity of GRP at a high temperature-variable temperature under thermal destruction conditions. Experimental research and mathematical modeling).* (Ufa: Ufimsk. gos. aviac. tekhn. un-t, 2004). (in Russ.)
20. S. G. Lekhnickij, *Teoriya uprugosti anizotropnogo tela (Theory of elasticity of an anisotropic body).* (M.: Nauka, 1977). (in Russ.)

Mathematical model of the crystallizing blank`s thermal state at the horizontal continuous casting machine

Igor Yu. Kryukov[1], *Sergey M.* Gorbatyuk[2], and *Margarita G.* Naumova[2,*].

[1]Thyssen Krupp AG Representative office Russian Federation, Russia , 105120, Moscow, 2 Siromyatnichesky proezd, 1
[2] National University of Science and Technology MISIS, Russia, 119049, Moscow, Leninsky prospect, 4

Abstract. Present article is devoted to the development of the mathematical model, which describes thermal state and crystallization process of the rectangular cross-section blank while continious process of extraction from a horysontal continious casting machine (HCCM).The developed model took cue for the heat-transfer properties of non-iron metal teeming; its temperature on entry to the casting mold; cooling conditions of blank in the carbon molds in the presence of a copper water cooler. Besides, has been considered the asymmetry of heat interchange from blank`s head and drag at mold, coming out from fluid contraction and features of the horizontal casting mold. The developed mathematical model allows to determine alterations in crystallizing blank of the following factors with respect to time: temperature pattern of crystallizing blank under different technical working regimes of HCCM; boundaries of solid two-phase field and liquid two-phase filed; blank`s thickness variation under shrinkage of the ingot`s material

1 Introduction

The proposed mathematical model is meant for studying of the casting metal`s thermal states the at the horizontal continuous casting machine (HCCM) of fine silver alloy [1,2].

The casting mold of the investigated HCCM was made of graphite and is intended for casting of rectangular cross-section blanks [3, 4]. A feature of its design is a presence of a copper cooling unit (Fig. 1). The basic geometry of the horizontal graphite mold is shown in Fig 2 and of the copper cooler in Fig. 3 [5]. The graphite casting mold and the copper cooler are used for production of the rectangular section blanks 200x20 mm.

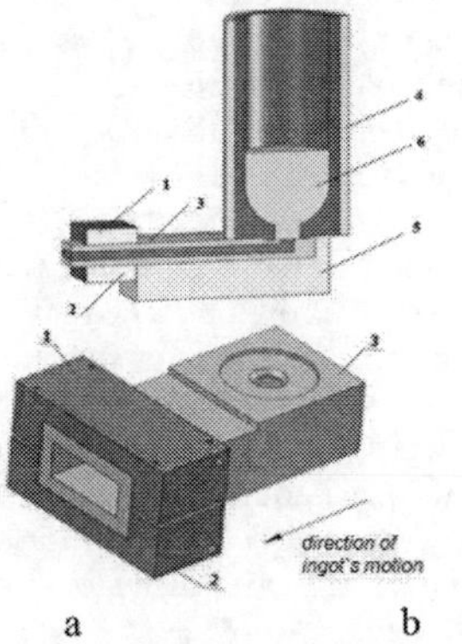

a b

Fig. 1 General arrangement drawing of the HCCM for production of the rectangular section blanks (a) and graphite mold with copper cooler (b): 1 – the top part of the cooler; 2-the bottom part of the cooler; 3- the top part of the graphite mold; 4-crucible; 5- heat-insulating material; 6- heat- insulator.

In the scientific works dedicated to the investigations of the ingot`s thermal state in the continuous-casting molds CCM [6, 7] pointed out, that in the liquid core of the cast slab the heat transfers as a result of combination of both heat conduction and convection. The reasons for this are natural convection of overheated melt, rabbling of the melt by force of supplied hot metal, electromagnetic effect and other exposures over the ingot`s liquid core. The motion of the molten metal in the unhardened part of the ingot significantly complicates the calculation of the ingot`s solidification, since towards the correct accounting of the heat- and mass transfer in the solution phase, should be taken into account not only the heat-transfer equation, but the system of equations of flows and continuity [6].

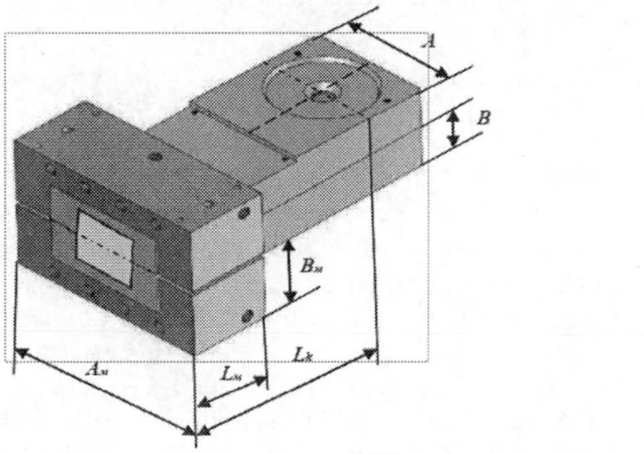

Fig. 2 Basic dimensions of the HCCM

* Corresponding author: qwerty-rita@yandex.ru

DOI: 10.1051/matecconf/201712902010

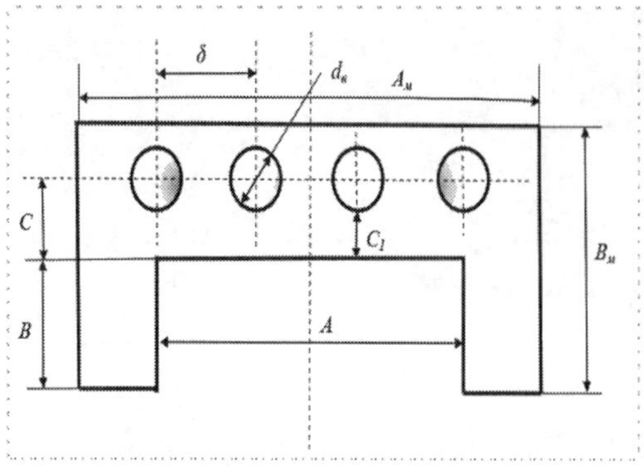

Fig. 3 Basic dimensions of the copper water cooler

The problem can be simplified by introducing an effective heat conductivity coefficient of the liquid phase λ_{ef}, whose value is extended several times in comparison to the molecular thermal conductivity coefficient λ_l.

$$\lambda_{ef} = \varepsilon_{con}\lambda_l \tag{1}$$

where ε_{con}- coefficient of convection, depending on the melt motion intensity . In view of low initial velocity of the metal stream flow rate in the HCCM , we possess the value ε_{con} =1÷2 [6]. Consequently, the effective heat conductivity of the liquid phase can outnumbers the conductivity of the stationary phase. The increasing of the effective heat conductivity in the hot metal while stirring, leads to a heat-transfer rise at the front of the phase transition, to the heating of the solid phase and to the associated heat-transfer rise at the ingot's surface.

When steady-state operating conditions the concerned horizontal graphite mold HCCM has in different parts the temperature from 100°C to 600°C. For this temperature range, for the thermal conductivity λ_t, has been obtained an equation of linear regression:

$$\lambda(t) = 127{,}28 - 0{,}11657t,\ \mathrm{W/(m\ K)}, \tag{2}$$

where λ_t- is the coefficient of conductivity of graphite in the direction perpendicular to extrusion.

With its application the mean integral value of conductivity in this temperature range is:

$$\lambda_{gr} = \bar{\lambda} = \frac{1}{600-100}\int_{100}^{600}\lambda(t)dt \approx 86{,}5\,\mathrm{W/(m\ K)} \tag{3}$$

It is well- known, that silver and its alloys are possible to cast at a temperature only 50 degrees above liquidus temperature. The casting speed, or the drawing speed generally depends on the blank's width, on the melt temperature on entering the casting mold and on properties of metal. The pacing factor, determining the casting speed is a thickness of a solid crust on exit from the casting mold. The water velocity in the canals generally depends on the width *a* and the thickness *b* of the blank and on the casting speed *v* (Table 1).

Table 1. Critical process parameters of ingot's draw-forming from HCCM

№	Name title	Parameter	Unit	Value
1	Degree of superheat of the melt on the exit from the crucible	ΔT_0	°C	50-100
2	Melt temperature on exit from the melting crucible	$T_0 = T_l + \Delta T_0$	°C	920-1020
3	Casting speed	v	sm/min	10-20
4	Cooling water temperature	T_w	°C	25
5	Water velocity in canals	W_w	м/s	1-4

2 Basic assumptions and limitations of the mathematical model

The scope of the blank's solidifying task is solved in the Cartesian coordinate system. For the origin of coordinates was taken a point of intersection of diagonals of rectangle, generated by the inner passage section of the casting mold, which is going through the crucible's axis. The orientation of the reference axes and the direction of the ingot's motion $\vec{v}$ is shown in Figure 4. It is apparent; that in general case the thermal process modeling area is a parallelepiped *ABCDA'B'C'D'*.

When developing of mathematical model of the ingot's thermal state were adopted the following assumptions and limitations:

1. We assume, that the temperature in the given point of the ingot in stationary casting regime is a time-independent quantity (in the Earth coordinate system), that is , we consider a quasipermanent thermal conditions;
2. We consider as the temperature significantly varies along the axes *x* and *z*. The thing is, that the thickness of the ingot *b* is much more less than its width *a*, the design of the mold and the cooler are such, that the cooling is more intensive on the side of faces *BB'C'C* and *AA'D'D* (Figures 3 and 4). The boundary conditions on the side face *ABCD* are same as on *A'B'C'D'*, that is, the symmetry of the temperature fields with respect to the plane *EFGH* is observed, providing $dT/dy=0$. To achieve the set modeling problem, for a sake of simplicity, we adobe the convention that $dT/dy \ll dT/dx$. The modeling work area of the ingot's thermal state is the rectangle *EFGH*.
3. The quasi-equilibrium model is used to describe the process of liquid metal solidification, with there is a belief that, there is no diffusional supercoooling in the mushy zone;
4. In the cooling zone (where copper cooler is used), on its border with the graphite mold, should be taken the average value of the thermal transmission coefficient, which , in its turn, depends on the cooler design and on parameters of quenching water ;

DOI: 10.1051/matecconf/201712902010

5. We assume that in the direction of axis Oz the heat is transferred only by convection. In witness of it, the Peclet number is the determining factor for the convection - heat conduction relationship:

$$Pe = \mathrm{Re} \cdot \mathrm{Pr} = \frac{W_z \Delta z}{a}, \tag{4}$$

Where Δz – typical length scale along the motion, where temperature gradient dT/dz is observed, this distance can be estimated as $L_k \approx 330$ mm=0,33 m; $W_z = v$ - the casting speed.

The minimum casting speed $W_z = v = 10$ sm/min =16,6 10^{-4}m/s, the thermal conductivity coefficient for solid phase in the neighborhood of Ts is a=1,22 10^{-4} m²/s. In these condition $Pe \approx 5$. In fact, for $Pe > 1$ the heat propagation through the heat conduction can be neglected. It means, that if we choose two different points 1 and 2, which have the same x, y coordinates and different z coordinate values ($z_2 > z_1$), then the temperature of the second point depends on the first point`s temperature, and the temperature of the first point doesn`t depend on the second point`s one for the selected direction of the ingot`s motion (Figure 4).

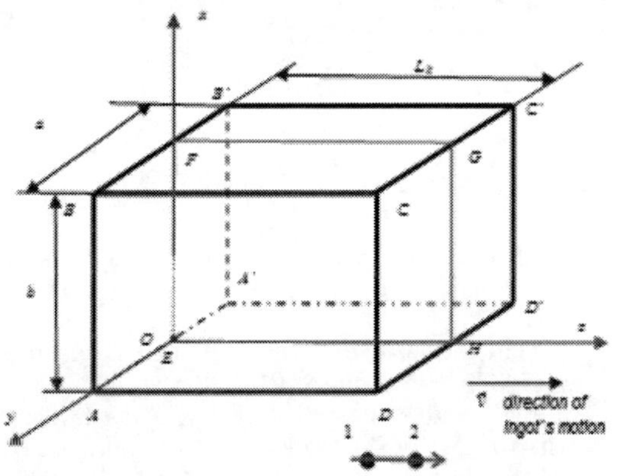

Fig. 4 Modeling area of ingot`s thermal state and the selected cartesian reference system.

3 Mathematical model of the ingot`s thermal state

The mathematical model of the ingot`s thermal state is based on the nonlinear equation of the heat conduction, which is complemented with initial and boundary conditions; at the same time the solidification heat is accounted according to the quasi-equilibrium model of the two-phase zone.

According to the first and the second suppositions, the temperature distribution in the solidifying ingot (in the frame reference associated with the Earth) can be described by the steady-state energy equation, taking into account the autothermal heat supply

$$c\rho v \frac{\partial T}{\partial z} = \frac{\partial}{\partial x}\left(\lambda \cdot \frac{\partial T}{\partial x}\right) + \frac{\partial}{\partial z}\left(\lambda \frac{\partial T}{\partial z}\right) + q_v(x, y, z, \tau) \tag{5}$$

Let us distinguish the control volume V_0 in the two-phase zone of the ingot and , in the same time, to introduct in the calculations the rate of the hard phase $\psi = V_{sol}/V_0$ - ratio of the hard phase volume to the total volume of the distinguished element of the two-phase zone (hard phase proportion).In the liquid phase parameter ψ is equal to zero $\psi|_{T=T_l} = 0$, for a solid part of the ingot it is equal to one $\psi|_{T=T_s} = 1$ and in the two phase zone it varies from zero to one; T_s and T_l - consequently the solidus and the liquidus temperatures of the blank`s material [6].

We will assume, that the heat source function q_v (x, y, z, τ) is directly-proportional to the local solidification rate :

$$q_v(x, y, z, \tau) = \rho L\, \partial\psi / \partial\tau \tag{6}$$

where heat of melting L acts as a proportionality coefficient.

We consider the thermal state of an object in the coordinate system, associated with an arbitrary cross section of the blank, which is in motion with the ingot. Taking into account the ratio $dz/dt = v$, we get:

$$c\rho \frac{\partial T}{\partial \tau} - \rho L \frac{\partial \psi}{\partial \tau} = \frac{\partial}{\partial x}\left(\lambda \frac{\partial T}{\partial x}\right) + \frac{\partial}{\partial z}\left(\lambda \frac{\partial T}{\partial z}\right) \tag{7}$$

In the general case the solidification speed $d\psi/dt$ in the Eq.7 is determined by conformity with the crystal growth in the supercooled melt. However, for the moment, the lows governing nucleation and crystal growth phenomena for most metals and alloys have not been entirely studied. In this regard, the solidification rate of the melt determines on the basis of the quasi-equilibrium hardening hypothesis [6]. According to this hypothesis, the equilibrium process is observed in case of rather slowly alloy`s crystallization (in other words if the diffusional transfer processes by impurity in between liquid and hard phases and intra- phase proceed completely). Then the relative value of the hard phase, which is in equilibrium with the liquid at a temperature T, can be determined from the alloy phase diagram by the lever rule.

In the simplest case, the cross section of the hard phase ψ is determined by the piecewise- continuous function, which diagram is shown in Figure 5.

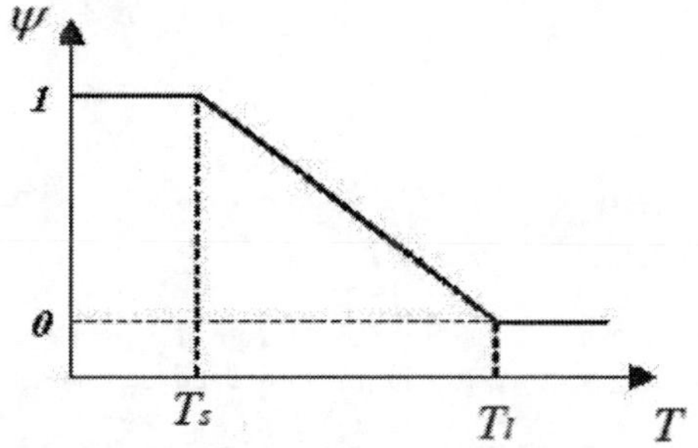

Fig.5 Function of the hard phase content ratio $\psi(T)$

$$\psi(T)=\begin{cases}0, \text{ if } T>T_l,\\ \dfrac{T_l-T}{T_l-T_s}, \text{ if } T_s\le T\le T_l,\\ 1, \text{ if } T<T_s,\end{cases} \quad (8)$$

In this case the crystallization rate of the alloy within the interval from T_l to T_s is a constant value and is equal to:

$$\frac{\partial\psi}{\partial T}=-\frac{1}{T_l-T_s} \quad (9)$$

Taking into account the assumptions of the possible neglect of the heat transfer due to the thermal conductivity in the Oz-direction, we may write the following equation:

$$\frac{\partial\psi}{\partial\tau}=\frac{\partial\psi}{\partial T}\frac{\partial T}{\partial\tau} \quad (10)$$

Applying Eq.10 to Eq. 7 , the last can be written as:

$$\frac{\partial T}{\partial\tau}\rho\left(c-L\frac{\partial\psi}{\partial T}\right)=\frac{\partial}{\partial x}\left(\lambda\frac{\partial T}{\partial x}\right)+\frac{\partial}{\partial y}\left(\lambda\frac{\partial T}{\partial y}\right)$$

or in greater detail:

$$\rho\cdot C_{ef}\frac{\partial T}{\partial\tau}=\frac{\partial\lambda}{\partial x}\frac{\partial T}{\partial x}+\lambda\frac{\partial^2T}{\partial x^2}, \quad (11)$$

where $\tau = 0 - L_k/v$; and C_{ef}, λ, ρ- effective heat thermal capacity, conduction and material density of the solidifying ingot.

The effective heat thermal capacity, conduction and material density of solidifying ingot are the piecewise continuous functions [7]:

$$C_{ef}(T)=\begin{cases}C_l, \text{ if } T>T_l;\\ C_T(T_s)\psi+C_l(T_l)(1-\psi)-L\dfrac{\partial\psi}{\partial\tau}, \text{ if } T_s\le T\le T_l;\\ C_T(T), \text{ if } T<T_s;\end{cases} \quad (12)$$

$$\lambda(T)=\begin{cases}\varepsilon_\kappa\lambda_l, \text{ if } T>T_l,\\ \lambda_\tau+\dfrac{\varepsilon_\kappa\lambda_l-\lambda_\tau}{T_l-T_s}(T-T_s), \text{ if } T_s\le T\le T_l,\\ \lambda_\tau, \text{ if } T<T_s,\end{cases} \quad (13)$$

$$\rho(T)=\begin{cases}\rho_l, \text{ if } T>T_l;\\ \rho_T\psi+\rho_l(1-\psi), \text{ if } T_s<T<T_l;\\ \rho_T(T), \text{ if } T<T_s.\end{cases} \quad (14)$$

To solve the boundary value problem, the Eq. 11 should be supplemented with the initial and the boundary conditions.

The initial condition defines the temperature field of the melt at the initial time on exit from the melting crucible (if $\tau=0$, $0<x<b$)

$$T(x,0)=T_0=T_L+\Delta T_0\cdot \quad (15)$$

For the sides *FG* and *EH* of the ingot (Figure 4), which are in contact with the working walls of the mold, the following boundary conditions should be set up.

To determine the boundary conditions for the heat-transfer from solidifying liquid was used the method of finite volumes. For the parallelepiped-shaped volume element $f\Delta x$, lying near the cooling wall surface (with an area f), the heat equilibrium conditions must be met. The capacity of the rejected heat by the mold q, at a metal temperature above T_s, is equal to the phase transition amount of heat (which is be govern by the effective heat capacity of the element of volume $C_{ef}\Delta x\rho\cdot dT/d\tau$) and to amount of heat transferred from the melt volume – $\lambda\delta T/\delta x$.

Accordingly, the boundary conditions are:

for $x=b$; $z=\tau\cdot v$,

$$\text{if } T>T_s,\ q_{up}(z)=C_{ef}\cdot\Delta x\cdot\rho\frac{dT}{d\tau}-\lambda\frac{\partial T}{\partial x}\Big|_{x=b};$$

$$\text{if } T\le T_s,\ q_{up}(z)=\lambda\frac{\partial T}{\partial x}\Big|_{x=b}; \quad (16)$$

for $x=0$; $z=\tau\cdot v$

$$\text{if } T>T_s,\ q_{dn}(z)=C_{ef}\cdot\Delta x\cdot\rho\frac{dT}{d\tau}-\lambda\frac{\partial T}{\partial x}\Big|_{x=0};$$

$$\text{if } T\le T_s,\ q_{dn}(z)=\lambda\frac{\partial T}{\partial x}\Big|_{x=0}, \quad (17)$$

where q_{up} (z), q_{dn} (z), - the heat flow rates towards casting mold interior from top and bottom surfaces of the ingot.

Thus, the obtained Eq. (11) - (17) are the mathematical model of ingot`s thermal state in the HCCM.

As a result of this scientific work was developed the mathematical model, which describes the thermal state and the crystallization process of the blank from a precious metal alloy, casting at the small capacity HCCM. Investigations, made on the base of this small capacity HCCM, allowed to determine the behavior in time of the ingot`s temperature field at different operating conditions of HCCM

References

1. V.E. Zinoviev, *Heat-transfer properties at high temperatures* (Moscow: Ref. ed. Metallurgy, 1989)
2. J. Bast, *Horizontal HCC-12000 unit for the continuous casting of semi finished products*, Metallurgist, Vol.**55**, № **1–2**. P.116–118 (2011)
3. S.M. Gorbatyuk, I. Yu. Kryukov, A.G. Radyuk, A.E. Titlyanov, *The casting mold of horizontal continuous casting machine of ingots from nonferrous metals and alloys*, Russian patent № 122052.2012.Bul. №32.

4. Yu. Bast, *Choosing of mold`s design by research temperature field in it,* Innovative fabricating technology: book of rep. of International scientific and technical conference (M.: Pub.h. MISiS, P. 537–541, 2011)
5. S.M. Gorbatyuk, I. Yu. Kryukov, A.G. Radyuk, A.E. Titlyanov, *Water-cooling device of casting mold of horizontal continuous casting machine of ingots from nonferrous metals and alloys*, Russian patent №120907.2012.Bul. №28.
6. Yu.A. Samoilovich, *Thermal process at continuous steel casting* (M.:Metallurgy, 1982)
7. A.I. Tsaplin, *Thermophysics in metallurgy*, Guide: pub. of Perm State Technical University, 2008)
8. S.M. Gorbatyuk, A.V. Kochanov, *Method and equipment for mechanically strengthening the surface of rolling-mill rolls*, Metallurgist, **56 (3-4)**, pp. 279-283 (2012)

MATEC Web of Conferences **129**, 02011 (2017) DOI: 10.1051/matecconf/201712902011
ICMTMTE 2017

Heat treatment of composite based on MAX-phases of the Ti-Al-C system

Sergey Galyshev[1,*], *Pavel* Bazhin[2], *Alexander* Stolin[2], *Fanil* Musin[1], and *Vasiliy* Astanin[1]

[1]Ufa State Aviation Technical University, 450000 Ufa, Russia
[2]Russian Academy of Sciences Institute of Structural Macrokinetics and Materials Science RAS, 142432 Chernogolovka, Russia

Abstract. This paper studies the effect of heat treatment performed at the MAX-phases formation temperature on the phase composition of the MAX/TiC-composite formed in the Ti–Al–C system in the conditions of free SHS-compression. The dependence between the titanium carbide (TiC) content and delay time before applying the load during free SHS-compression is defined. The experimental results show that the proposed heat treatment leads to carbide content reduction.

1 Introduction

In recent years MAX-phase materials have been increasingly used in industry due to their unique properties. The crystalline lattice of MAX-phases ($M_{n+1}AX_n$) is an alternation of intermetallic "MA" and carbide or nitride "MX" layers, where "M" is a transition metal; "A" is a IIIA or IVA group element, "X" is either C or N. This structure of MAX-phases determines a combination of useful metallic and ceramic properties in one material.

Due to the structure of MAX-phases, MAX-phase materials satisfy the highest requirements associated with high temperatures, corrosion environment and thermal cycling [1-4]. For the same reason, these materials can be used in various advanced industries, such as aircraft and aerospace engine building, electrochemical and electrical industries.

The properties of MAX-phase materials vary in a wide range depending on the phase composition. For example, TiC is exceptionally undesirable for high-temperature application of MAX-phase materials based on the Ti-Al-C system. This is connected with a large difference in the thermal expansion coefficients of the carbide and MAX phases, which results in high temperature embrittlement of the material due to the carbide phase. On the other hand, TiC significantly increases the wear resistance of material. It is important for application in areas that do not require high-temperature properties, for example as coatings for electrical contacts [5].

In case of the latter, one can speak of a dispersion hardened composite, in which the MAX phases are the matrix material, and the carbide precipitates function as a reinforcing component. The change in the carbide phase volume makes it possible to control the performance properties of the MAX/TiC-composite. Therefore, the search for this possibility is of indisputable practical interest.

To date, methods of powder metallurgy are the most widely used ones for producing materials based on MAX-phases [6]. These methods have some disadvantages such as large time and energy costs, many technological stages, etc. The techniques based on SHS-technology (self-propagating high-temperature synthesis) seem more promising for production of MAX-phase materials, as SHS allows fabricating refractory compounds with minimal energy consumption in the shortest time [7]. This means that the SHS-technology is more customer-oriented, which in its turn is an important economical feature of composite materials technology.

Phase formation in the Ti-Al-C system during SHS occurs at high rate and high temperature of the combustion reaction. This determines some difference from the process of equilibrium crystallization of these compounds. First, TiC forms as a result of reaction between Ti and C. At the same time, due to high reaction exothermicity of TiC formation, Al and remaining Ti melt and form the Ti-Al melt. Further, the Ti-Al melt interacts with the carbon surface. As a result the melt picks up carbon, and secondary TiC grains precipitate from the melt. MAX-phases Tix+1AlCx form from the melt, as well as during the interaction of the melt with TiC [9]. If the subsequent cooling occurs at rather high rates, the TiC cannot completely interact with the Ti-Al melt. The TiC remains in the volume of the synthesized material, and the MAX/TiC-composite forms.

The study of the equilibrium process of Ti-Al-C compound formation using the differential-thermal analysis during slow heating in the temperature range from 25 to 1400 °C revealed two stages [8]. At the first stage Ti_xAl_y type intermetallic compounds form by the reaction:

$$xTi + yAl \rightarrow Ti_xAl_y \quad (1)$$

* Corresponding author: galyshew@gmail.com

During the second stage, TiC forms, then it reacts with intermetallic compounds and aluminum melt to form MAX-phases $Ti_{x+1}AlC_x$ by the following reactions:

$$TiC + Ti_xAl_y \rightarrow Ti_{x+1}AlC_x; \quad (2)$$

$$TiC + Al \rightarrow Ti_{x+1}AlC_x. \quad (3)$$

Thus, one can assume that the TiC content in the Ti-Al-C-based composite can be controlled in two ways. First, if it is necessary to increase the TiC content, one should increase the cooling rate immediately after the synthesis. On the other hand, the cooling rate should be reduced if its fraction in the material is required to be decreased.

For example, it was shown in [9] that spontaneous air cooling of samples with the tool immediately after the free SHS-compression resulted in formation of composite with 20% TiC.

Another way to reduce the TiC content in the material is apparently heat treatment at a temperature close to the temperature of MAX-phase formation by reactions (2) and (3). However, this issue has not been studied in full and requires additional research.

The aim of this work is to determine the effect of high-temperature annealing on the TiC content in the MAX/TiC composite produced by free SHS-compression.

2 Materials and procedures

Powders of titanium (PTM), aluminium (ASD-4) and carbon black were used as initial materials. The powder mixture composition corresponded to the stoichiometric composition of the MAX phase of Ti2AlC. The initial cylindrical samples are 30 mm high and 30 mm in diameter, the relative density is 60%.

The free SHS-compression [10] is used to prepare samples of the MAX/TiC-composite material. This method consists in free compression of the synthesized material immediately after the synthesis (Fig. 1).

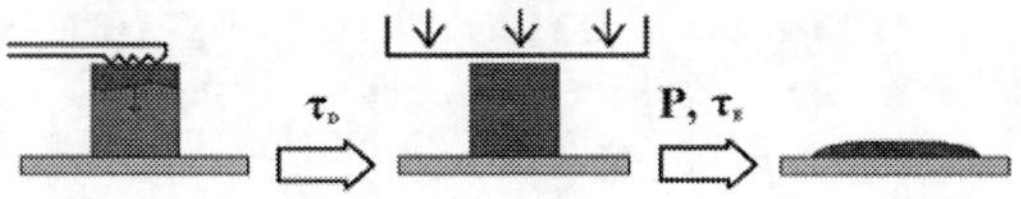

Fig. 1. Scheme of free SHS-compression

The time between the synthesis end and deformation start (delay time) is the main technological parameter, which determines the structure and properties of the produced material. The parameter varies from 0 to 10 seconds with a step of 2 seconds. The compression load is constant of 15 MPa.

Two series of samples are produced at different delay times. Then, one of the sample series is annealed at 1350°C in the air atmosphere for 3 hours and after cooled in water.

Both series of samples are cut, grinded and polished without etching. The microstructure before (PS - parent sample) and after annealing (HT_1350 - heat treatment 1350°C) is analyzed by means of a scanning electron microscope LEO-1450 with an energy-dispersive microanalyzer INCA Energy (EDS system). X-ray phase analysis is carried out on a DRON-3 (LOMO) equipment, using a copper anti-cathode X-ray tube at 40 kV of accelerating voltage and 30 mA of heating current. The microhardness of the samples is measured by the Vickers method, using a Duramin-2 equipment.

3 Results and discussion

The samples before annealing have a structure typical of materials based on MAX-phases of the Ti-Al-C system produced by the free SHS-compression method (Fig. 2). Light rounded inclusions of TiC particles in the composition with darker elongated grains of MAX phases are located in the intermetallic matrix of Ti-Al. The areas of black colour are pores.

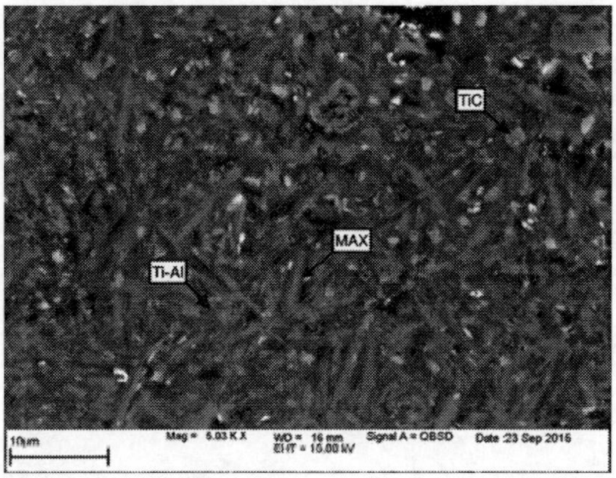

Fig. 2. Microstructure of samples obtained with a 4 seconds delay time in the initial state

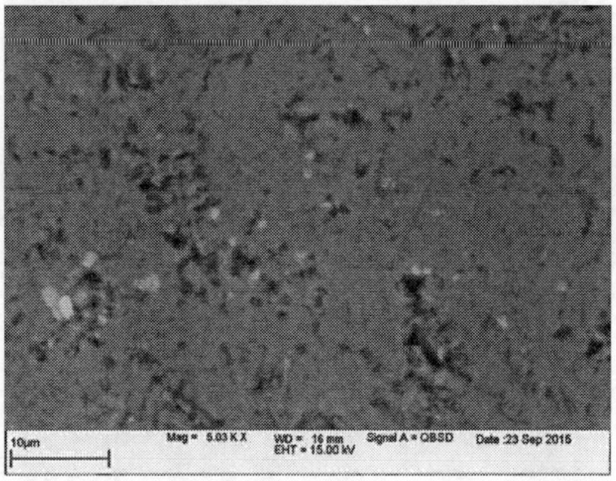

Fig. 3. Microstructure of samples obtained with a 4 seconds delay time after annealing

The metallographic analysis shows that the TiC content reduces from 27% to 11% when the delay time increases from 0 to 10 seconds (Fig. 4) Consequently, the phase formation process proceeds more completely with

DOI: 10.1051/matecconf/201712902011

a longer dwelling at a high temperature close to the synthesis temperature.

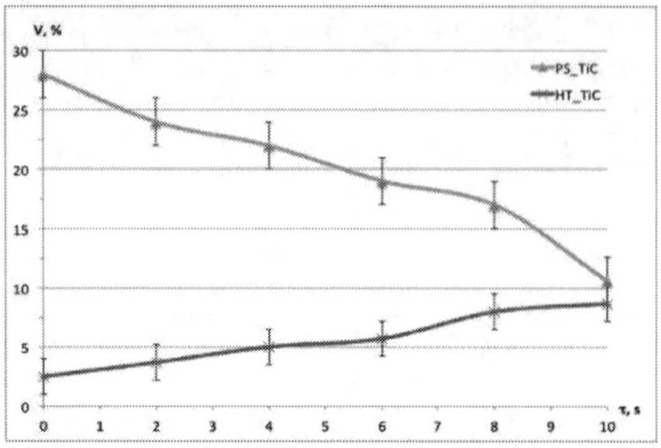

Fig. 4. Dependence of the TiC volume fraction on the delay time in the initial and annealed states

The TiC content significantly reduces when annealed at 1350°C (Fig. 4). The change in the TiC content obviously depends on the delay time, during which the material is produced. The content of TiC is maximally reduced in the samples produced without delay in time.

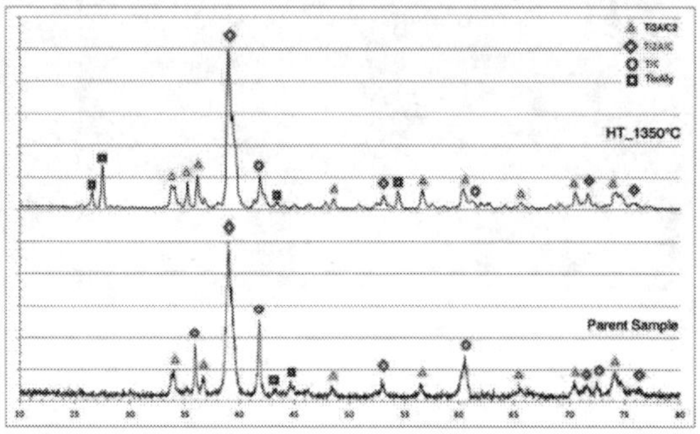

Fig. 5. X-ray diffraction spectrum of samples produced with a 6 seconds delay in the initial and annealed states

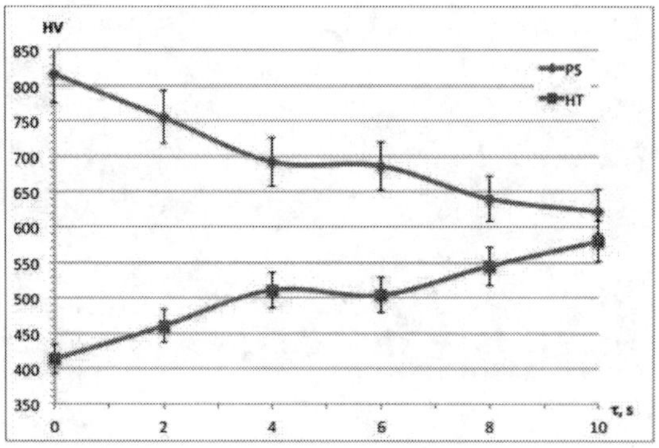

Fig. 6. Dependence of the composite microhardness on the delay time in the initial and annealed states

The TiC content reduction is confirmed by the X-ray spectra. The intensity of TiC peaks after annealing decreases (Fig. 5).

The obtained data correlate well with the microhardness measurements (Fig. 6). The material microhardness is determined by the content of the hardest phase - TiC (HV = 30 GPa). Samples with a high TiC content have a higher microhardness. On the other hand, the microhardness of the sample with the minimum carbide content produced is close to the microhardness of pure MAX-phases ($HV_{Ti3AlC2}$ = 4.9 GPa, HV_{Ti2AlC} = 4.5 GPa).

4 Conclusions

1. The dependence between the parameters of the MAX/TiC composite structure and the delay time prior to load application during the free SHS compression is established. The TiC content reduced from 27% to 11% when the delay time increased from 0 to 10 seconds.
2. The method to control the MAX/TiC-composite structure by high temperature annealing is shown. This heat treatment provides a decrease in the TiC content. The range of possible TiC content in the composite is from 3 to 27%.

This work has been performed with the financial support of the Ministry of Education and Science of the Russian Federation within the framework of the base part of the state task №11.6454.2017/BP.

References

1. R. Tomoshige, H. Tanaka, Arch. Metall. Mater. **59**, 1575 (2014)
2. T.A. Prikhna, S.N. Dub, A.V. Starostina, J. superhard mater. **34**, 102 (2012)
3. A.V. Starostina, T.A. Prikhna, M.V. Karpets, J. superhard mater. **33**, 307 (2011)
4. M.W. Barsoum, *MAX Phases: Properties of Machinable Carbides and Nitrides* (Wiley VCH, 2013)
5. A. Rahman, Z. Rahaman, AJMP **4**, 75 (2015)
6. M.W. Barsoum, T. El-Raghy, Am. Sci. **89**, 334 (2001)
7. A.G. Merzhanov, J. Mater. Chem. **14**, 1779 (2004)
8. E.A. Levashov, Yu.S. Pogozhev, D.V. Shtansky, M.I. Petrzhik, Izvestiya VUZ. PMFP (In Rissian) **3**, 91 (2008)
9. S.N. Galyshev, N.G. Zaripov, P. M. Bazhin, A. M. Stolin, PM (In Rissian) **11**, 63 (2015)
10. S.N. Galyshev, A.M. Stolin, D. Vrel, A. Hendaoui, P.M. Bazhin, A.E. Sytschev, Int. J. SHS **18**, 194 (2009)

On aging of high-strength stainless steel 08Cr15Ni5Cu2Ti

Tat'yana Makhneva[1,*], *Vyacheslav* Dement'yev[1], and *Sergei* Makarov[1]

[1]Institute of Mechanics, the Ural Branch of the Russian Academy of Sciences, T. Baramzinoi St.34. Izhevsk. Russia

Abstract. Based on our results and the results of the experimental studies previously conducted by other researchers, the analysis has been made of the influence of remelting methods on the character of the formation of the level of the impact strength KCU and its component KCV at aging of steel 08Cr15Ni5Cu2Ti prepared by two methods of remelting. The results of studying the influence of the remelting methods, heat treatment modes and impurity cleanness of steels on the level of KCU and the degree of its decrease at the temperatures of age hardening are presented. The factors causing the difference in the structure and phase composition in the steel formed by different remelting methods are established. The temperature range in which a significant decrease in KCU takes place has been determined. The hypothesis is offered about the role of interstitial impurities in the KCU decrease at aging.

1 Introduction

The desired set of properties of steels of this grade (high-strength Stainless) is attained by the selection of optimal conditions at quenching and aging in the temperature range of (400÷600)° C. The studied steel aging has been investigated in few works [1-3]; the matter was partially considered in monographs [4-9]. It is believed that the possible reasons for decreasing KCU at aging of chromium-nickel steels with the martensitic-austenitic structure are a small amount or absence of retained austenite, γ_{retain}, in the structure; coarse grain; the presence of embrittling phases on grain boundaries; the formation of segregations of harmful impurities and alloying elements, particularly, chromium, as the result of the separation process in the iron-chromium matrix [10], and their interaction with interstitial impurities [11-14].

The comparative study of the mechanical properties of steel 08Cr15Ni5Cu2Ti prepared by vacuum-arc and electroslag remelting (VAR and ESR) after quenching and aging shows that though the VAR-steel structure differs from that of ESR-steel in the grain size, larger tendency toward the grain growth at heating and small amount of γ_{retain}, the character of the formation of the mechanical properties is the same in both VAR- and ESR-steel [8]. However, the differece in structure, phase and chemical composition between remelting and even a smaller amount of gas impurities (Table 1) lead to the lower level of the VAR-steel mechanical properties. After aging for maximum strength, the impact strength appears to be the most unstable characteristics (inadequacy is 40% of the required level).

The investigation of the KCU formation in the studied steel prepared by the two remelting methods at aging and the determination of 'dangerous' temperature ranges with relation to the KCU decrease and factors influencing the position of the above temperature ranges is of interest.

2 Material and investigation techniques

The investigation was conducted on the industrial melts, the chemical compositions of which are given in Table 1. 1. For the investigation, samples from the steel prepared by the electroslag and vacuum-arc methods of remelting, ESR and VAR, were selected. The samples were heat-treated as follows: water quenching from the temperatures (950, 1000, 1050, 1200)°C. Aging was conducted in the temperature range of (300÷500)° C with the step of 25° C for 1 to 10 h. At quenching from 1000° C, the influence of the following preliminary conditions was studied: 1. Annealing 660° C, 6 h; 2. Annealing 660° C, 6 h × 2 times; 3. Annealing 520° C, 50 h; 4. Annealing 520° C, 50 h + 660° C, 6 h; 5. Annealing 750° C, 3 h + 520° C, 50 h + 660° C, 6h.
Before aging, the halves of the fractured samples after the impact strength test were checked for the presence of retained austenite ($\gamma_{retained}$) by the X-ray method. The impact strength (KCU) was determined according to GOSTs 9454-60 and 9455-60 (sample: type I); the crack

* Corresponding author: mah@udman.ru

Table 1. The chemical composition of steel 08Cr15Ni5Cu2Ti prepared by two methods of remelting

Remelt-ing methods	Content of element, %											
	C	Mn	S	P	Si	Cr	Ni	Cu	Ti	[O]	[N]	[H] cm³/100 g
ESR	0.07	0.59	0.007	0.017	0.45	14.05	5.33	2.03	0.03	0.0034	0.0032	3.8
VAR	0.05	0.48	0.006	0.020	0.43	14.22	5.64	2.16	0.06	0.0017	0.011	1.67

development work (KCV) was determined according to B.A. Drozdovsky method on non-standard samples with the section 11X10 mm, notch depth 2 mm and fatigue crack 1 mm [15]. The degree of the impact strength (ΔKCU) decrease in the ESR- and VAR-steel was investigated in different conditions of quenching, and in dependence on the isothermal holding time in the temperature range of annealing in the two-phase (α+γ)-region conducted prior to quenching, and aging.

3 Results and discussion

The investigation of the impact strength of the ESR- and VAR-steel quenched from 1000° C after aging in the studied temperature range shows that in the quenched condition both melts have high values of KCU, 2.6 and 1.6 MJ/m^2, respectively (Fig. 1, *a*). The impact strength decrease starts from the temperature of 350° C; in ESR-steel it is 1.4 MJ/m^2 and in VAR-steel – 0.6 MJ/m^2 (half of that of ESR-steel) with the minimum at the temperatures (425÷450)° C. Such sharp impact strength variation is accompanied by a significant change in the fracture character. After aging at (300÷375)°C, from being completely ductile, the fracture becomes half-brittle at (425÷450)° C (the fiber fraction decreases to 40-50% in the fracture). After aging at (475÷500)° C, the impact strength increases to the previous or even higher values at purely ductile fracture. Starting with 375° C, γ_{retain} becomes unstable and its amount decreases from 6.5% in the ductile condition to 4.5% in the brittle condition.

The impact strength testing of the samples after quenching, cold treatment and aging show the similar dependence but with a greater decrease in KCU, which is 0.2 MJ/m^2 and 0.1 MJ/m^2 for ESR-steel and VAR-steel, respectively (Fig. 1, *b*). Thus, the impact strength decrease in the studied steel is not connected with the γ_{retain} decomposition since in the ductile condition, the steel without austenite has the same KCU values as the samples without cold treatment. The decrease of the retained austenite amount contributes into a general decrease in KCU; however, it is not a determining factor.

The aging temperature range, in which the KCU decrease takes place, shifts to the side of lower temperatures with the increasing duration of holding from 1 h to 10 h for both ESR- and VAR-steel. At holdings for 3h, 7h, and 10 h, the minimal impact strength values correspond to 425° C, and at holding for 1 h - to 450° C.

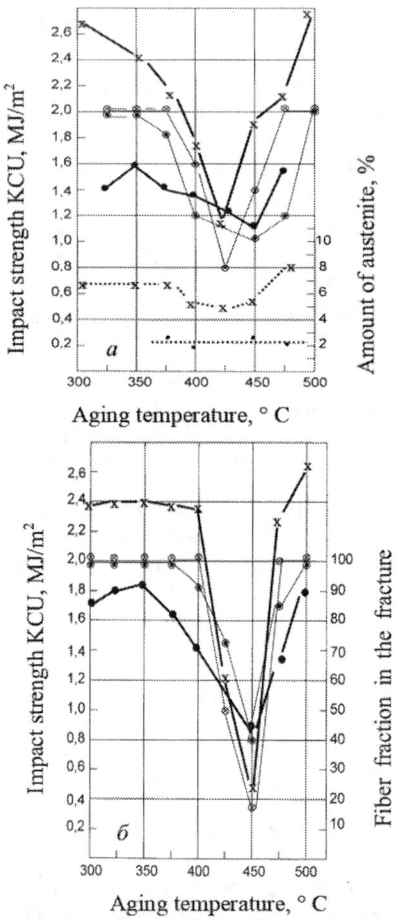

Fig. 1. The influence of the aging temperature on the impact strength (KCU), fiber fraction in the fracture and the amount of the retained austenite (γ) in steel 08Cr15Ni5Cu2Ti prepared by two remeltings methods.
a – after quenching from 1000° C;
b – *a* + cold-treatment at -70° C;
•- VAR, x – ESR – KCU and γ_{retain} (%);
• ©, ⊗ - ductile component (%) VAR and ESR, respectively. Aging for 3 h.

The degree of the impact strength decrease (ΔKCU) in the studied steels evaluated by the difference of KCU values in the ductile and brittle conditions (KCUduct - KCUbri) at holdings to 10 h in the 'dangerous' temperature range with relation to the KCU decrease is shown in Fig. 2. It can be seen that VAR-steel is much less tending to the loss of ductility than ESR-steel. The

ΔKCU value of VAR-steel increases steadily; in ESR-steel the processes leading to the impact strength decrease are taking place mainly for the first three hours (Fig. 2a, curve 2). At further holding to 10 h, the increase of ΔKCU is insignificant and it takes place due to the KCU decrease in the brittle condition. In addition, the change of the crack development work KCV at holding during aging also indicates the larger reliability of VAR-steel (Fig. 2b, curve 2). The results are confirmed by the shift of the transition temperature indicating the higher gas-and-interstitial-elements inclusion cleanness of VAR-steel [16]. However, in the VAR-melt with the high content of titanium required by OST (0.16%), the KCU decreases to the illegal values (Fig. 2c, curve 2).

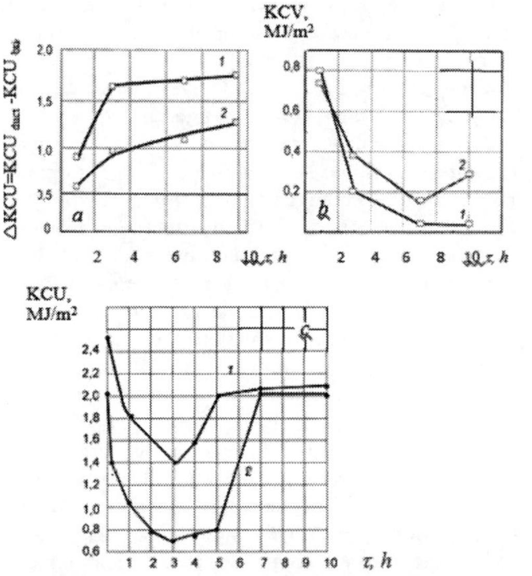

Fig. 2. The influence of the time of aging on the degree of the impact strength decrease (Δ KCU) and the work of the crack development (KCV) in the 'dangerous' temperature range. Steel 08Cr15Ni5Cu2Ti. Quenching temperature - 1000° C.
1 – ESR and 2 – VAR;
(*a*-2) and (*b*-2) - at Ti <0.06%; (*c*-2) - at Ti=0.16 %

On the electron micrograph of the sample with KCU = 0.28 MJ/m^2 (Fig. 3) it can be seen that the brittle fracture along the grain boundary has areas of ductile fracture (weakly elongated cups) situated along the retained austenite interlayers. Testing of cold-treated samples confirmed this fact. It can be suggested that the larger amount of austenite will make the KCU decrease smoother. The simultaneous presence of ductile and brittle components in the steel fracture indicates the competition of the two fracture mechanisms due to the strength balance of the boundaries and body of the grains strengthened by secondary phase precipitates. Unfortunately, the process of the ε-phase precipitation during aging has not been traced yet and it is impossible to link the temperature range of the ductility decrease with a certain degree of the dispersivity and distribution of the above phase.

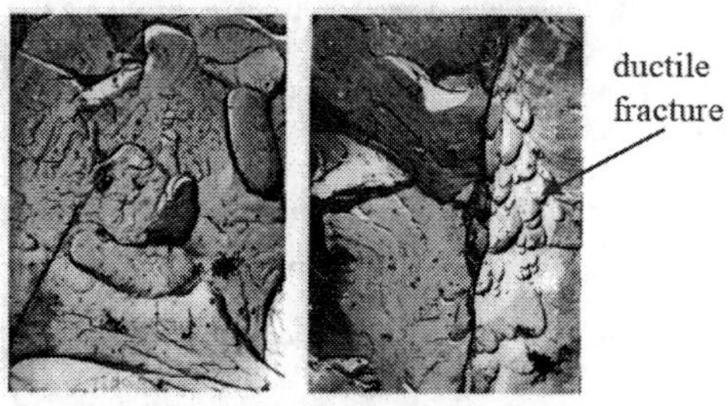

Magnification: 3000 × 1.5

Fig. 3. Fractographs of the sample of ESR-steel 08Cr15Ni5Cu2Ti (KCU = 0.28 MJ/m^2) after quenching from 1000° C and aging at 450° C, 2.5 h.

The degree and position of the KCU decrease interval are significantly influenced by the quenching temperature and the heat treatment conditions before quenching. If the studied steels quenching from 950-1050° C does not change the KCU-decrease temperature range, quenching from 1200° C makes the temperature range wider, especially for ESR-steel. The character of the influence of the quenching temperature and remelting method on ΔKCU is practically the same; both dependences have a minimum at 1000° C (Fig. 4). It means that at this temperature both remelts of the studied steel are less tending to the KCU decrease. In the range of 950° C - 1000° C the decrease of ΔKCU in ESR-steel is more intensive than that in VAR-steel. In addition, these dependences clearly show that independently of the austenization (quenching) temperature, at aging in the temperature range of (425÷450)° C, ESR-steel tends to the KCU decrease more significantly than VAR-steel at practically the same level of strength [8].

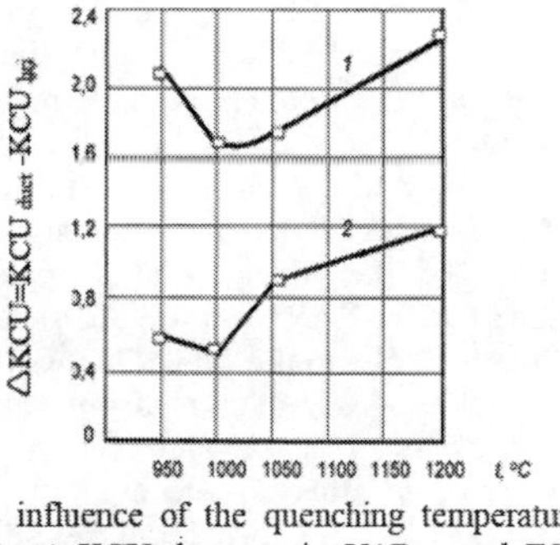

Fig. 4. The influence of the quenching temperature on the degree of the Δ KCU decrease in VAR - and ESR - steel 08Cr15Ni5Cu2Ti at aging for the strength maximum (425-450° C). 1 – ESR and 2 – VAR

The quenching temperature influences ΔKCU mainly via the grain size. At 1000-1050° C, the increase of ΔKCU in VAR-steel is more intensive than that in ESR-steel, and a noticeable growth of grain in VAR-steel starts at the temperature a little higher than 1000° C while in ESR-steel the grain growth starts at a much higher temperature (Fig. 5). However, the quenching temperature also influences the degree of ΔKCU with relation to obtaining the homogeneity of the solid solution at quenching. This is proved by the ΔKCU value which is higher after quenching from 950° C than that after quenching from 1000° C. At 950° C, the chromium (carbides, carbonitrides)-based complexes dissolve incompletely and even in the case of their complete dissolution, the homogeneity of the solid solution is not reached. It can be suggested that in the temperature range of (950÷1050)° C, the influence of two factors takes place, i.e. obtaining the more homogeneous solution leads to the KCU increase at aging in the temperature range of (425÷450)° C, and the grain size growth acts oppositely. At the temperatures higher than 1050° C, the grain growth makes a more significant contribution into ΔKCU.

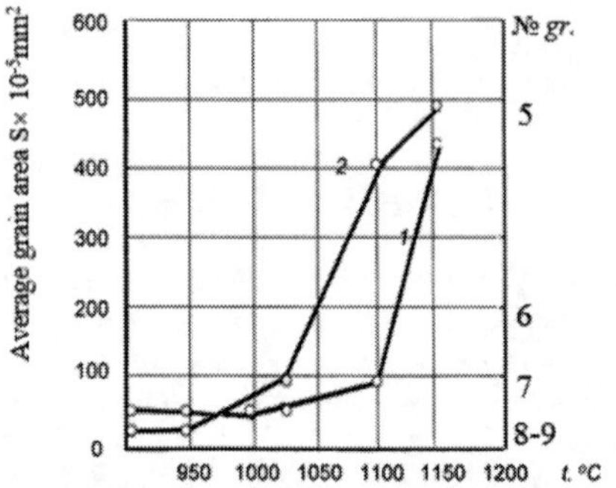

Fig. 5. The influence of the quenching temperature on the grain size in ESR-and VAR-steel 08Cr15Ni5Cu2Ti. 1- ESR, 2- VAR

In the heat treatment of the semifinished items from the studied steel, preliminary annealings in the two-phase (α+γ)-region are of special importance: one of them (750° C, 3 h + 520° C, 50 h) is conducted for removing diffusion-mobile hydrogen; the other (single or double annealing at 660° C, 6 h) stabilizes the mechanical properties. The investigation of the influence of annealings in the two-phase region conducted before quenching and followed by aging and of their combination show that annealings do not qualitatively change the character of the impact strength formation. The position of the 'dangerous' temperature range remains the same for VAR-steel (Fig. 6, *b*), and for ESR-steel it shifts by 25° C to the side of higher temperatures (Fig. 6, *a*). The stronger decrease of the ESR-steel impact strength is observed. However, in comparison with the samples after quenching, the annealings significantly increase the intensity of the KCU decrease after age-hardening: the most KCU decrease is at annealing (520° C, 50 h) at the VAR-steel aging – 0.7 MJ/m^2, and at the ESR-steel aging, annealing (750° C, 3 h + 520° C, 50 h + 660° C, 6 h) contributes into the KCU decrease - 0.8 mJ/m^2.

The question whether the influence of the processes taking place in the two-phase region still remains after heating for quenching is discussed in Ref. [13]. In the present work, the austenite condition obtained after double annealing and without annealing has been investigated by the method of internal friction and it is shown that the changes in the structure appearing at the stage of annealing in the (α+γ)-region remains after quenching due to the steadier and longer involvement of all the bulks of the metal into deformation in contrast to the avalanche-like beginning of the movement of dislocations in the metal without preliminary annealing.

At the same time, the influence of annealings on the tendency toward brittle fracture at decreasing temperature immediately after quenching is insignificant. The character of the serial curves is similar to that of the curves obtained after quenching from 1000° C without annealing: the curves of KCU = $f(T_{test})$ are flat for ESR- and VAR-steels without a distinct transition temperature, and the difference in the KCU level between the remelts remains [14]. Thus, the main contribution into the KCU decrease is made by the phase transformations at aging.

For understanding what causes the decrease of the impact strength and its component in the temperature range of (375÷450)° C and how the inclusion cleanness and chemical composition of the steel influence the impact strength, phase transformations at heating of hardened steel 08Cr15Ni5Cu2Ti have been analyzed. In the mentioned temperature range, in addition to the temper processes, precipitation of the disperse strengthening ε-phase rich in copper [5] and formation of concentration inhomogeneities in chromium [9,15], there are the processes of the redistribution of interstitial inclusions (carbon, nitrogen, oxygen) and harmful impurities as well. At heating temperatures (270÷320)° C, the setting of nitrogen, carbon and other atoms on dislocations with the formation of segregations takes place [13]. According to the results of the measurement of inner friction, the rise of the temperature to (380÷400)° C increases the coefficient of diffusion of the atoms and leads to the segregation dispersal. From the aging temperature of 320° C, the electrical resistance starts decreasing [2], and the impact strength also slightly decreases, and from the aging temperature of 350° C the impact strength decrease is more significant. Starting with 350° C the variation of the transition temperature is observed [14]: at the aging temperature of 375° C its shift is 40 deg. and at 400 °C – 60 deg.

Since the variation of the specific resistivity and the transition temperature shift indicate the separation processes characteristic of high-chromium ferrite, it is suggested that the separation is possible in the studied steel. The kinetic curve for the chromium regions formation in the studied temperature range of tempering and aging obtained by the NRG method [15] allows to conclude that at 450°C for 1 h, chromium-depleted and chromium-enriched regions are formed in the solid solution. In Ref. [16] it is stated that chromium enrichment of the regions is possible to 40%. At this temperature, the transition temperature shift in VAR- and ESR-steel is maximal [14]. However, there are no literature data that the separation significantly shifts the critical temperature of the transition of the steels of this grade to the brittle condition.

DOI: 10.1051/matecconf/201712902012

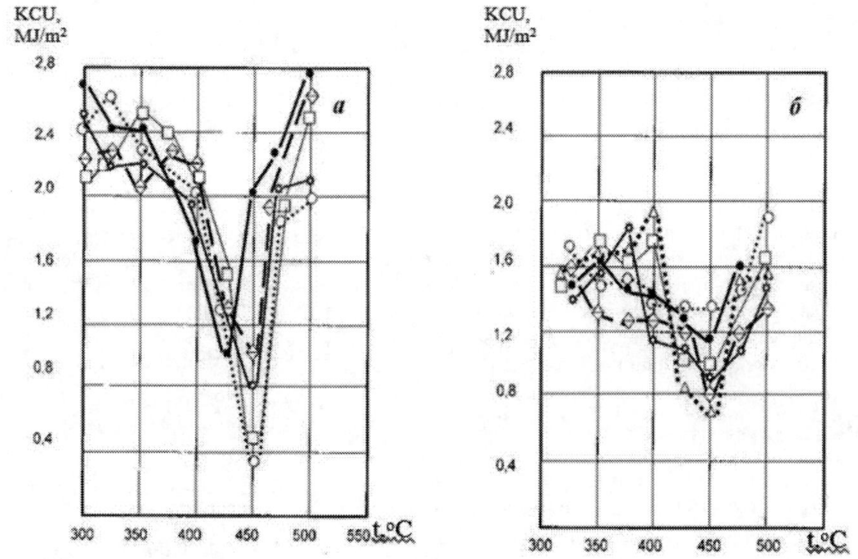

Fig. 6. The influence
and VAR (*b*) at differ

□ – 520° C, 50 h + 660° C, 6 h;
○ – 750° C, 3 h + 520° C, 50 h + 660° C, 6 h;
△ – 520° C, 50 h;
◇ – 660° C, 6 h. The time of aging is 3 h.

At the same time it is known [Problems of physics of strength, IFM Transactions ed. V.A. Arkharov] that interstitial impurities varying within 0.002 – 0.02% wt% can raise the transition temperature of pure chromium from (-100)° C to positive (above-zero) temperatures (~260)°C. The researchers in Ref. studying the influence of small amounts of impurities on the metal mechanical properties associate the transition of chromium from the ductile to brittle condition with the increase of the chromium resistance to plastic deformation and the decrease of σ_B under the influence of impurities, among which nitrogen produces the strongest influence on ductility. Thus, it can be suggested that the chromium-enriched regions can dissolve interstitial elements within the solubility limits at the studied temperature and contribute into the sharp manifestation of ΔKCU in the established intervals (Fig.1). The dimension of the appearing pure chromium regions is of importance; in the studied steel the dimension of theregions is ~5 nm according to X-ray and NGR data set.

The main hardening of the studied steel at aging takes place in the temperature range of (400÷450)° C according to the mechanism of dispersion hardening due to the precipitation of the disperse ε-phase based on copper and reaches (150÷200) MPa while the strength level in the hardened condition is up to (1150÷1200) MPa. By analogy with other steels and alloys, the corresponding decrease of the impact strength should be ~ (0.2÷0.4) MJ/m². In reality, the value of the KCU decrease does not correspond to the value of the martensite hardening due to the ε-phase precipitation. Thus, in addition to the influence of the ε-phase, the impact strength decrease happens due to another much stronger process. Possibly, this process is the formation of mixed regions consisting of chromium atoms and interstitial elements.

In addition, in all the experiments a very sharp transition from the low KCU values to the higher ones is noticed. It is sufficient to raise the aging temperature from 450 to 475° C for the impact strength to grow sharply to the values in the ductile condition. In this temperature range, the breakage of the coherency between the matrix and the strengthening ε-phase takes place. The increase of the temperature in the mentioned range at holding for 1 h cannot lead to the disappearance of the chromium-enriched and chromium-depleted regions and to the significant variation of the degree of the solid solution homogeneity. The ductility recovery most likely can be due to the decrease of the concentration of interstitial atoms in the chromium-enriched regions. At the dynamic equilibrium, the number of atoms getting to the chromium-enriched region is compensated by the number of atoms leaving the region.

Thus, the impact strength decrease in the studied temperature range can be divided into two stages. At the first stage (300÷400) °C the change of the properties can be explained by the martensitic matrix separation in chromium. At the second stage (400÷450)° C the loss of the matrix ductility takes place due to its hardening according to the dispersion hardening mechanism and further progress of the separation processes initiated by the precipitation of the disperse strengthening ε-phase. The chromium-enriched regions can be additionally 'embrittled' by interstitial elements in the amounts exceeding their equilibrium concentration.

The difference in the properties of the remelts of the studied steel is due to the difference in the short-range order structures [8] because of the non-equlibrium and inhomogeneity of the steel in the liquid state, gas-saturation, particularly, in interstitial elements, and high titanium content (Table 1).

The experiments show that in the aging process of steel 08Cr15Ni5Cu2Ti prepared by VAR- and ESR-methods, a significant impact strength decrease takes place. In practice, the KCU decrease remains unnoticeable due to the high value of the minimum KCU in a 'dangerous' temperature range. In the presence of retained austenite and fine grain, the impact strength level remains higher than the required one (0.8 MJ/m^2). In the case of coarse grain and small amount of austenite, the KCU decrease is very sharp as seen on the melt with high content of titanium (Fig.2, *curve* c).

4 Conclusions

1. The processes of the martensite decomposition in stainless steel 08Cr15Ni5Cu2Ti at heating in the tempering-and-aging temperature range of (270÷450) °C have been analyzed. In the mentioned range, in addition to the processes of tempering and precipitation of the disperse strengthening ε-phase, there are processes of the redistribution of interstitial elements (carbon, nitrogen, oxygen) and harmful impurities and the formation of the concentration inhomogeneities in chromium. Martensite of quenching decomposes into two solid solutions: the first is chromium-enriched and the second – chromium-depleted.

2. The temperature and time dependences of the variation of the impact strength KCU at aging of the studied ESR- and VAR-steel are established. In the aging temperature range of (375÷475)°C, the impact strength varies through the minimum at (425÷450) °C, the position of which depends on the remelting method, quenching temperature, modes of the preliminary treatment before quenching and time of holding in the 'dangerous' temperature range of aging (425÷450) °C.

- The influence of the remelting method on the character of the KCU formation is expressed as the degree of the decrease of $\Delta KCU = KCU_{duc} - KCU_{bri}$ in the 'dangerous' temperature range of aging. This value is much lower in VAR-steel, in which the aging processes influencing ΔKCU are developing much slower and do not have clearly expressed character than those in ESR-steel.
- The quenching temperature influences ΔKCU in both ESR- and VAR-steel due to the obtained homogeneity of the solid solution and the grain size.
- The existing modes of the preliminary annealings before quenching of ESR- and VAR-steels intensify the KCU decrease after aging, but at the same time, the value of the minimum of the impact strength in the 'dangerous' temperature range increases.
- The sharp impact strength decrease after aging is caused by the formation of mixed regions consisting of high-chromium martensite and interstitial atoms. Retained austenite plays the role of a background, on which the development of the processes leading to the KCU decrease takes place; therefore, there should be a sufficient amount of retained austenite in the steel structure.

3. It is shown that in the melt with a low content of titanium and gas impurities, despite the low level of γ_{retain} and KCU after quenching VAR-steel is more reliable after 1.5 h of aging for strength maximum.

The work is financially supported by the UrO RAN according to the program: Fundumental problems of mechanics and allied sciences in the study of multiscale processes in the nature and engineering (project number: 15-10-1-4)

References

1. L.V. Tarasenko, N.V. Uly'anova, V.P. Kucheryavy, Izv. Vuzov. Mashinostroyenie, 7 (1969)
2. M.N. Mikheev, M.M. Belenkova, R.N. Vitkalova et al, FMM, **47**, 6 (1979)
3. V.B. Spiridonov, V.S. Fridman, Yu.A. Rodionov et al, MiTOM, 10 (1974)
4. M.D. Perkas, V.M. Kardonsky, *High-strength maraging steels,* M.: Metallurgiya (1970)
5. Ya.M. Potak. *High-strength steels*, M.: Metallurgiya (1972)
6. V.A. Osminkin, dis. kand. tekh. nauk, Sverdlovsk (1979)
7. I.P. Konakova, dis. kand. tekh. nauk, Sverdlovsk (1988)
8. T.M. Makhneva, dic. doc. tekh. nauk, Izhevsk (2012)
9. Y.D. Solomon, L.M. Lehenson, Acta met, **26**, 3 (1978)
10. A. Cottrell, B. Bilby, Proc. Phys. Soc., A62, **49** (1949)
11. R. Smith, J. Rutherford, J. Met., **9**, 857 (1957)
12. R. B. Smith, Trans. Met. Soc. AIME, **218**, №1, 62 (1960)
13. V.M. Kondratov, Ye.S. Makhnev, V.A. Morozov, et al, Problems of strength, 2 (1977)
14. T.M. Makhneva, Metal Science and Heat Treatment (2012) - doi: 10.1007/S11041-012-9436-0
15. T.M. Makhneva, Ye.P. Yelsukov, E.B. Voronina, FMM, 5 (1991)
16. Sh.Sh. Bashkirov, N.G. Ivoylov, G.V. Kurbatov, et al, *Paramagnitny rezonans,* Kazan, KGU, 6 (1980)

 DOI: 10.1051/matecconf/201712902013

World market for nanomaterials: structure and trends

Elena Inshakova[1,*] and *Oleg* Inshakov[1]

[1]Volgograd State University, Department of International and Regional Economy, 400062 Volgograd, Russia

Abstract. Modern nanomaterials as well as the other advanced materials are aimed at scaled production application due to their unique physical and chemicals properties and the improved performance of final products, therefore at enforcing technological and economic progress of the countries. The article focuses on current state and key indicators of development of the world nanomaterials market, its key drivers and restraints, modern structure, emerging trends and prospects. Selected foreign and Russian databases, providing information on nanomaterials, which are applied in industrial production, and final products manufactured with the use of nanomaterials or nanotechnology are examined in the article. The authors come to the conclusion that information asymmetry, heterogeneity and even inconsistency of the information in the databases exist, thus making it difficult to obtain relevant business information both for manufacturers and consumers. These factors hinder development of nanoindustry in the Russian Federation, as well as national and regional markets for nano-enabled consumer products.

1 Dynamics and trends of the world nanomaterials market

The development and industrial application of innovative advanced materials is imperative for technological and economic progress of a country. Nanomaterials, with at least one external dimension in the size range of approximately 1 to 100 nanometers, due to their unique physical and chemicals properties resulting in improved performance and qualitative characteristics of final products and ample opportunities of applying in numerous end-user industries are widely used in energy storage, construction, electronics, aerospace, defense, production of sporting and other consumer goods, etc.

The modern drivers of the world nanomaterials market are as follows: increase in the market penetration of existing materials, decrease in the prices of nanomaterials, improving materials' properties, expending R&D activities related to the new materials, increase in public and private expenditure towards nanotechnology research, growing support of the government institutions, rapid new materials and applications' development [1-3], effective functioning the partnerships and strategic alliances of domestic and foreign companies and organizations, increasing number and collaboration among industry players including growing international research and production cooperation in nanotechnology industry.

The factors mentioned above have had a positive impact on the development of the world nanomaterials market and have resulted in the steady growth of the key market indicators. As to the conservative estimate by Mordor Intelligence, the global nanomaterials market was valued at about USD 4.1 billion in 2015 and expected to reach USD 11.3 billion by 2020, at a compound annual growth rate (CAGR) of over 22% during the forecast period 2017-2022 [4].

On the contrary, the new report by Allied Market Research optimistically projects the global nanomaterials market to reach more than USD 55.0 billion by 2022 from USD 14.7 billion in 2015, growing at a CAGR of 20.7% during this time period [5], while Deloitte Touche Tohmatsu Limited experts are much less optimistic declaring only 15.5% CAGR of global nanomaterials market in 2012-2019 [6, p. 4].

The pessimistic and optimistic views on the positive dynamics and prospects of the global value of nanomaterials have been presented at figures 1 [3, 4] and 2 [2, 5] below.

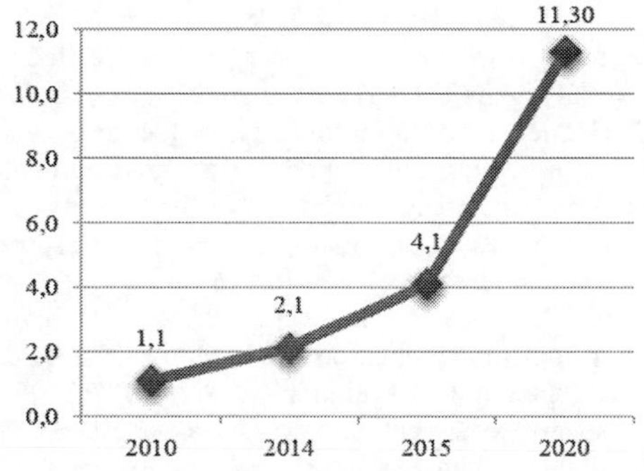

Fig. 1. Global value of nanomaterials, pessimistic view (USD billion).

* Corresponding author: inshakovae@mail.ru

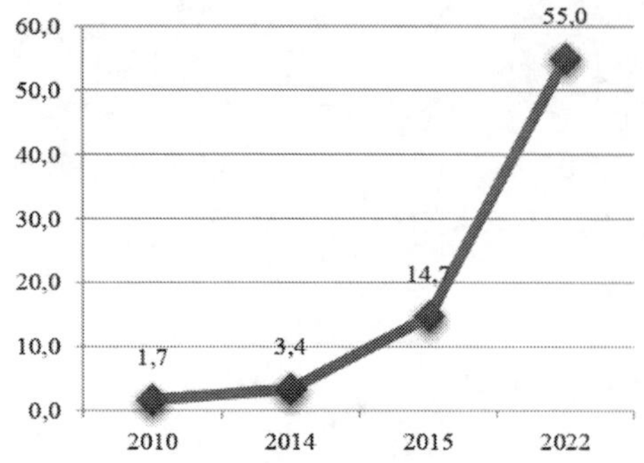

Fig. 2. Global value of nanomaterials, optimistic view (USD billion).

The European nanomaterials market generated revenue of more than USD 2.5 billion in 2015 and is expected to reach USD 9.1 billion by 2022 with a CAGR of 20.0% during 2016-2022 [7].

No matter what scenario of development of the global nanomaterials market has been taken into consideration, the fact is that it is expected to grow at a faster CAGR than the global nanotechnology market as a whole. The latter is projected by the RNCOS analysts to grow at a CAGR of around 17% during the forecasted period of 2017-2024 [8].

Major factors restraining development of the global nanomaterials market are: the growing concerns over their impact on human health and the environment (toxicity of nanomaterials themselves, presence of toxic solvents, release of hazardous intermediate compounds, toxicity of wastes resulted from nanomaterials' processing or manufacturing); stringent requirements in the frameworks of the government environmental regulation.

For example, the International Agency for Research on Cancer (IARC) states that oral exposure to titanium dioxide (known as E171 and presented in sweets, chocolate products, toothpaste, pharmaceutical products, etc.) becomes a factor of initiating and promoting early stages of colorectal carcinogenesis for humans.

The most important regulatory bodies related to nanomaterials in the US and the EU are:

– The United States Environmental Protection Agency (EPA), setting national standards of environmental protection and enforcing regulations in this sphere (The Toxic Substances Control Act, TSCA; Resource Conservation and Recovery Act, RCRA; etc.);

– The European Chemicals Agency (ECHA), facilitating the implementation of the obligations under the crucial regulations: Regulation (EC) No 1907/2006 of the European Parliament and of the Council on the Registration, Evaluation, Authorisation and Restriction of Chemicals (REACH); the Regulation (EC) No 1272/2008 on the classification, labelling and packaging of substances and mixtures (CLP Regulation); The Biocidal Product Regulation (BPR, Regulation (EU) 528/2012), and others.

In China, for example, it is the Ministry of Industry and Information Technology (MIIT), which has published new guidelines for industries that create new materials, such as graphene. The guidelines had been jointly developed by the MIIT, the National Development and Reform Commission, the Ministry of Science and Technology and the Ministry of Finance.

It should be mentioned the emerging trends in the global nanomaterials industry: growing demand for high strength and durable structural materials, new material development and product design, new and emerging market applications, falling nanomaterials' prices, enlarging the scale of the market due to the growing activities of the existing suppliers and the new entrants, as well as the increase in government support and R&D funding.

2 Structure of the world nanomaterials market by type of material, region and application

By material type, one can find several approaches to segmenting the global market for nanomaterials: a) carbon based, metal and non-metal oxides, metals, dendrimers, nanoclay, and nanocellulose [7]; b) nano-particles, nano-fibers, nano-tubes, nano-clays and nano-wires [4]; c) carbon black, carbon nanotubes, graphene, fullerene, nanofibers; silica fumes; clay; metal/alloys; ceramics [9]; and some others.

Allied Market Research experts team states that metal and non-metal oxide-based nanomaterials are the most widely applied in the industrial production with expected growth at a CAGR of 20.7% from 2016 to 2022 [5]. About 25% of the nanoproducts introduced into the markets are incorporated by titanium dioxide, silver, or silicon dioxide nanoparticles [14, p. 65] Titanium dioxide and silicon dioxide are mostly consumed metal and non-metal oxide-based nanomaterials. Consumption of silicon dioxide was 198 kilo tons in 2015 and is projected to reach 786 kilo ton by 2022, at a CAGR of 21.8%.

Nano silver due to its antibacterial and antimicrobial properties is among the most popular nanomaterials used in manufacturing consumer products by the companies in numerous industries, mostly electronics, IT, healthcare and beauty, textiles (20% of the world silver nanoparticles market).

Silver nanoparticles are considered by the Global Market Insights Inc. analysts to be the most commercialized nanoparticles, accounting for over 50% of the global nanomaterial consumer products in 2015 [10] with expectation of growth of the market at a CAGR of nearly 13% in 2016-2024 [11]. Prior areas of silver nanoparticles application are healthcare and life sciences, food and beverages packaging industry (improved packaging and active packaging), electronics and IT sector.

The global markets for nanoclay and nanocellulose also demonstrate stable growth of the indicators based on the enlarging spheres of their industrial application [5]. According to the Transparency Market Research (TMR) forecasts, the market revenue of nanoclay is expected to grow at a CAGR of 24.9%, and of nanocellulose market – 19% until at list 2020 [12, 13].

 DOI: 10.1051/matecconf/201712902013

The prospects of selected nanomaterials markets have been summarized in the table 1.

Table 1. Selected nanomaterials markets.

Selected nanomaterials market	Global market revenue by 2016, USD billion	Expected global market revenue by 2021, USD billion	Expected CAGR in 2016-2021, percent
Silver nanoparticles	1.1	3.0	13
Nanoclays	0,7	2,1	24,9
Nanocomposites	1.6	5.3	26.7
Quantum dots	0.61	3.4	41.3
Nanofibers	0.39	2.0	38.6
Advanced & nanoscale ceramic powders	14.6	22.3	8.9

Source: Silver Nanoparticles Market Size (2017); Global Silver Nanoparticles Market Trends (2017); Nanocellulose Market (2015); Nanoclay market (2016); Nanomaterials Market – Global Opportunity Analysis (2016).

Dynamics of the global nanomaterials demand by regions and sectors can be seen at figure 3 [2, 5].

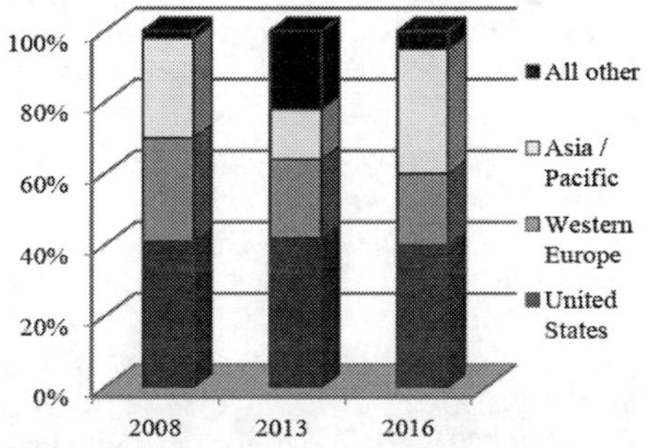

Fig. 3. Global demand for nanomaterials by regions (percentage).

In 2014, the US and Japan accounted for over a half of world demand for nanomaterials, while the EU countries and two high-income East-Asian countries, Taiwan and South Korea, represented an additional 34%.

The US led the global nanomaterials market in 2015 and is projected to keep this leading position until 2022 with the nanomaterials market revenue growing at a CAGR of 20.5%.

In 2016, North America and Asia-Pacific collectively accounted for 3/4 of the world demand for nanomaterials and are expected to maintain leadership throughout the examining period. At the same time, the highest CAGR of nanomaterials market revenues is projected in Asia-Pacific countries (21.4%), backed by North America and Europe.

Among the Asian countries the fastest growth in demand for nanomaterials is forecasted for China and India, while China is predicted to become the second largest market for nanomaterials following the U.S. with over 11% share of the global demand by 2022.

On the basis of applications, one should identify the following segments in the structure of the global nanomaterials market: electrical and electronics industry, construction, energy storage, healthcare, consumer goods, packaging, transportation, aerospace, water filtration, and others.

Nevertheless, the Allied Market Research analysts point out the industries with the highest opportunities for applying modern nanomaterials in production of the final goods: aerospace, sporting goods, automotive, energy storage, electronics and defence industries [7]. The electronics industry is expected to account for the highest share in the market (about 30%), while the aerospace sector - the fastest growing sector during the forecast period due to the increasing usage of nanometals, polymer nanocomposites and anti-corrosion coatings in the aircraft manufacturing.

3 Nanomaterials development and production: problems of representation in the databases

The development of applied research in the field of nanomaterials and nanotechnology can be represented by the positive dynamics of the patent activity of countries, that is by the growing number of patents issued by international and regional (see tables 2 and 3) registration systems.

The generalised information in this field one can get in the open access, for example, at the website of StatNano project, which has been launched with the support of Iran Nanotechnology Initiative Council (INIC).

In terms of the quantity of nanotechnology patents registered in the European Patent Office (EPO) in 2016, as to the StatNano database, Russia has been on the 29th place among 56 countries, while in the United States Patent and Trademark Office (USPTO) – has shared 21st place with Austria and Denmark among 59 countries under consideration [14, p. 36-37].

Table 2. Nanotechnology patents in EPO by countries (units).

Rank	Country	2012	2013	2014	2015	2016
1	USA	306	350	361	421	577
2	Germany	208	243	206	229	289
3	France	128	145	140	152	208
4	Japan	133	134	141	131	188
5	South Korea	49	55	73	52	105
6	UK	45	38	45	45	81
7	Switzerland	37	47	63	48	75
8	Netherlands	37	45	38	43	71
9	China	15	16	14	28	59
10	Italy	43	37	29	47	49
…	…	…	…	…	…	…
29	Russia	1	0	4	3	6
	World	1,133	1,255	1,289	1,415	2,006

Source: StatNano (2017).

Table 3. Nanotechnology patents in USPTO by countries (units).

Rank	Country	2012	2013	2014	2015	2016
1	USA	3,041	3,615	4,414	4,365	4,316
2	South Korea	430	501	667	839	914
3	Japan	526	587	889	902	819
4	Taiwan	407	425	551	500	514
5	China	225	270	357	393	416
6	Germany	205	248	496	307	301
7	France	140	176	347	242	210
8	Netherlands	103	115	155	156	136
9	UK	60	87	145	109	123
10	Canada	82	85	127	109	106
…	…	…	…	…	…	…
21	Russia	4	4	11	8	12
	World	5,342	6,354	8,681	8,588	8,484

Source: StatNano (2017).

The actual data on the produced nanomaterials, manufactures and products with the use of nanomaterials and nanotechnology are available at the websites of the Nanowerk Nanomaterial Database™, the StatNano Nanotechnology Products Database (NPD), the Netherlands NANO Supermarket, the DaNa2.0 (Data and Knowledge on Nanomaterials) in Germany, and some others abroad, as well as the Unified Computer Database on Nanomaterials and Nanotechnologies Used in the Russian Federation (website http://web.ion.ru/GM_1/GM.aspx), and the profiled registries of the Startbase – System of Supporting and Promoting Innovation (website https://www.startbase.ru/registries/), which are held by the Fund for Infrastructure and Educational Programs (FIEP), the member of RUSNANO group.

The Nanowerk Nano catalog contains 3883 nanomaterials, including 691 nanotubes, 2617 nanoparticles, 114 graphene items, 231 quantum dots, 99 fullerenes, 83 nanowires and 48 nanofibres from more than 150 suppliers worldwide. By the end of March 2017 the StatNano NPD has monitored 79 nanomaterials produced by 1188 manufacturers in 52 countries and applied by 13 industries in 642 product types.

Among 1574 items registered in the nanomaterials section of the Unified Computer Database on Nanomaterials and Nanotechnologies Used in the Russian Federation less than 7% are evident nanomaterials (including nanopowders, nanoparticles, nanocoatings, nanoclays, nanocomposites, etc.) while the rest of items are the final products produced with applying nanomaterials or nanotechnology. The similar problem is typical for the content of the nanomaterials section in the Startbase Nanotechnology Product Registry, for the DaNa2.0, and the NANO Supermarket.

Thus, the fact is that the majority of existing foreign and Russian databases under consideration represent asymmetry, heterogeneity and even inconsistency of the information, because they list both nanomaterials, which are applied in industrial production, and final products, which are manufactured with the use of nanomaterials or nanotechnology. This makes it difficult to obtain relevant business information for manufacturers and consumers, and, therefore, hinders consolidation of nanoindustry in the Russian Federation, as well as development of national and regional markets for nano-enabled consumer products.

This material is based upon work supported by the Russian Foundation for Humanities under project No. 16-02-00591 "State policy of the RF in the sphere of nanoindustry in an unfavorable external environment".

References

1. *Nanotechnology Market Outlook 2020* (RNCOS E-Services Pvt. Ltd., Noida, Uttar Pradesh, 2015)
2. *World Nanomaterials: Industry Study with Forecasts for 2016 & 2021. Study #2871* (The Freedonia Group, Cleveland, OH, 2012)
3. *Nanotechnology Update: U.S. Leads in Government Spending Amidst Increased Spending Across Asia* (Lux Research Inc., Boston, MA, 2015)
4. *Nanomaterials Market – Global Trends, Investment Analysis and Future Scope to 2022* (Mordor Intelligence, Hyderabad, Telangana, March 2017)
5. *Nanomaterials Market – Global Opportunity Analysis and Industry Forecast, 2014–2022* (Allied Market Research, September 2016)
6. D. Dickson, *Advanced Manufacturing in a Highly Connected World* (Deloitte Touche Tohmatsu Limited, 2015)
7. *Europe Nanomaterials Market by Type of Material, by End User – Opportunity Analysis and Industry Forecast, 2014–2022* (Allied Market Research, October 2016)
8. *Global Nanotechnology Market Outlook 2024* (RNCOS E-Services Pvt. Ltd., Noida, Uttar Pradesh, 2016)
9. T. Abraham, *Nanotechnology and Nano Materials: Types, Current/Emerging Applications and Global Markets* (Innovative Research and Products, Inc., 2011)
10. *Silver Nanoparticles Market Size by Application, Industry Analysis Report, Regional Outlook, Growth Potential, Price Trends, Competitive Market Share & Forecast, 2016–2024* (Global Market Insights Inc., January 2017)
11. *Global Silver Nanoparticles Market Trends 2017 Industry Report Forecast, 2016-2024* (Fractovia.org, Ocean View, Delaware, 2017)
12. *Nanocellulose Market – Global Industry Analysis, Forecast 2014–2020* (Transparency Market Research, 2015)

DOI: 10.1051/matecconf/201712902013

13. *Nanoclay market – Global Industry analysis, Size, Share, Growth, Trends and Forecast 2015-2023* (Transparency Market Research, 2016)
14. *StatNano 2016 – Status of Nano-science, Technology and Innovation* (Statnano Publications, 2017

DOI: 10.1051/matecconf/201712902014

Modelling of carbon nanotube functionalization processes

Tatyana Dyachkova[1,*], *Evgeny* Tugolukov[1], and *Elena* Burakova[1]

[1]Tambov State Technical University, 392000, Tambov, Russia

Abstract. In the present paper, kinetic regularities of oxidative functionalization of carbon nanotubes in nitric acid vapors are studied. The catalytic effect of transition metal oxide particles on the formation of carboxyl groups is shown. Based on the information about changes in the degree of functionalization and composition of gaseous reaction products, the effective values of the thermal effect of the implemented process are determined. A model of the temperature field of the reaction zone of the gas-phase oxidative functionalization of carbon nanotubes process is proposed, thereby making it possible to determine the structural and mode parameters of the equipment.

1 Introduction

Carbon nanotubes (CNTs) represent a whole class of materials that differ in length, diameter, number and shape of graphene layers, type of electrical conductivity, and volumetric morphology [1]. To date, methods allowing for obtaining the CNTs with specified parameters, including those on an industrial scale, have been developed [2]. Due to their unique mechanical and electrical properties, the CNTs often show high efficiency when used in composites with increased strength characteristics, thermal stability, resistance to hydrothermal aging [3-4], etc. However, when the CNTs with high free surface energy are introduced into polymers and solvents, they are prone to agglomeration phenomena, because of which the achieved effects on improving the properties of modifiable matrices are significantly reduced. Most often, to solve this kind of problem, the CNTs are subjected to various types of chemical treatment in order to form functional groups of different nature for improving dispersibility in polymers and solvents.

A number of papers show the advantages of functionalization by means of gas-phase treatment regarding the consumption of reagents, increase in the achieved indexes of the degree of functionalization, preservation of the structural integrity of surface graphene layers, volume morphology of materials and inclusions of transition metal particles in their composition [5]. The CNTs exposed to the gas-phase functionalization are used in composites with high electrical conductivity and ability to absorb electromagnetic radiation [6].

However, when scaling the processes of chemical treatment of the CNTs for mass production of their functionalized forms, it is necessary to develop conditions in working devices that ensure the production of materials of specified quality.

Considering the aforementioned, the aim of the present research was to experimentally study the kinetic regularities of the CNT gas-phase oxidative functionalization and mathematical modeling of the temperature field in the reaction zone of a device for implementing this process.

2 Materials and Methods

In experimental studies, "Taunit-M" cylindrical multi-walled CNTs ($d = 8 \div 15$ nm, $l \geq 2$ μm, $S \geq 300$ m^2 / g) synthesized at NanoTechCenter Ltd. (Tambov, Russia) via the CVD method over a Co/Mo catalyst were used (Fig. 1) . They were purified from the catalyst impurities using a hydrochloric acid solution, and then washed and dried.

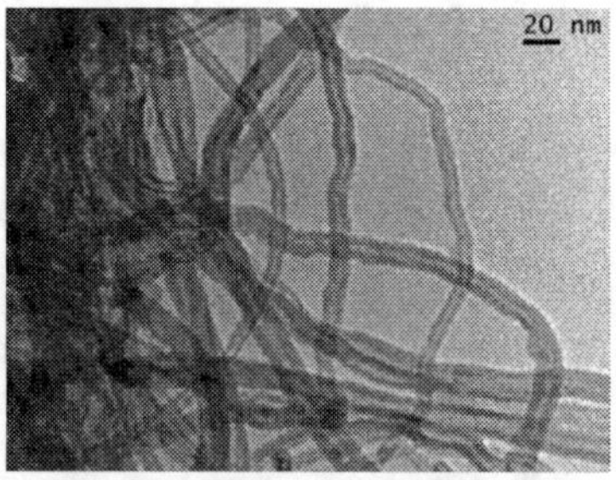

Fig. 1. TEM image of multiwalled CNTs "Taunit-M"

The oxidation in 65% nitric acid vapors was carried out at a temperature of 140 °C in a laboratory setup (Fig. 2), which allows for a dosed supply of an oxidizing reagent (1, 2) and an inert gas (3, 4, 5), evaporation of the oxidant , regulation and maintenance of temperature, sampling and measurement of gaseous product volume. The process was performed in a quartz glass tubular reactor consisting of two chambers, one of which was of the vaporizing type (6) for the oxidizing reagent, and the other was for introducing the CNTs (7). The reactor was placed in a tubular resistive electric furnace (8). To control the

* Corresponding author: dyachkova_tp@mail.ru

DOI: 10.1051/matecconf/201712902014

temperature, a chromel-alumel thermocouple (9) was used, whereas a microprocessor PID controller (10) was employed to fix the reagent supply. During the experiment, the volume was measured with a rotameter (11), and gaseous reaction products were sampled (12).lower case. You should leave 35 mm of space above the title and 6 mm after the title.

Contents of O_2, N_2, Ar, carbon oxides (CO, CO_2) and nitrogen oxides (N_2O, NO, NO_2) in the products were determined by a "Crystal - 2000M" gas chromatograph (GC). Two packed columns filled with zeolite CaA (l = 2 m, d = 3 mm) and a "Chromosorb 102" polymeric sorbent (l = 1 m, d = 3 mm) were used for separation. Concentrations of the components were quantified using a thermal conductivity detector (TCD). Helium was used as carrier gas.

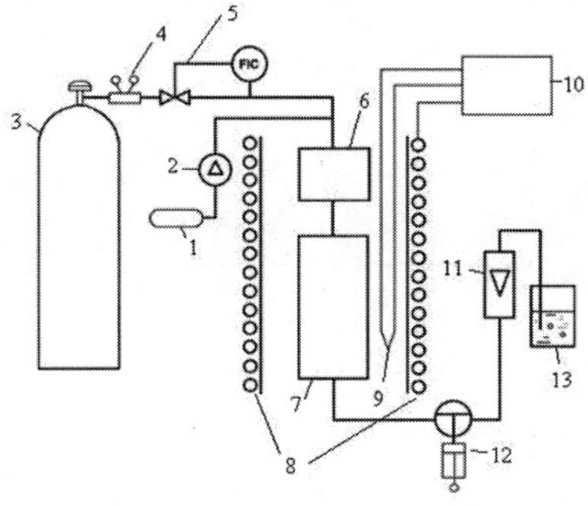

Fig. 2. A scheme of the laboratory setup to study the gas-phase functionalization of carbon nanotubes: 1 - reservoir with an oxidizing reagent; 2 - dispensing pump; 3 – inert gas-containing vessel; 4 - pressure regulator; 5- flow regulator; 6 – vaporizing-type chamber; 7 - reaction chamber; 8 - resistive furnace; 9 - thermocouple; 10- microprocessor temperature controller; 11 - rotameter; 12-gas sampler; 13 - water seal

The degree of the CNT functionalization using the carboxyl groups (D_f, mmol/g) was estimated according to the titrimetric method described elsewhere [7].

3 Experimental Results

The change in the degree of functionalization under the oxidation in the nitric acid vapors is shown in Fig. 3. During the process, the dependence of the D_f on the duration of the treatment of the previously purified CNTs reaches a certain limiting value corresponding to the saturation of their surface with the functional groups. The impurities (admixtures) of transition metal particles in the initial CNTs have a catalyzing effect on the formation of the functional groups. In the investigated interval of time, the D_f limiting value is not achieved.

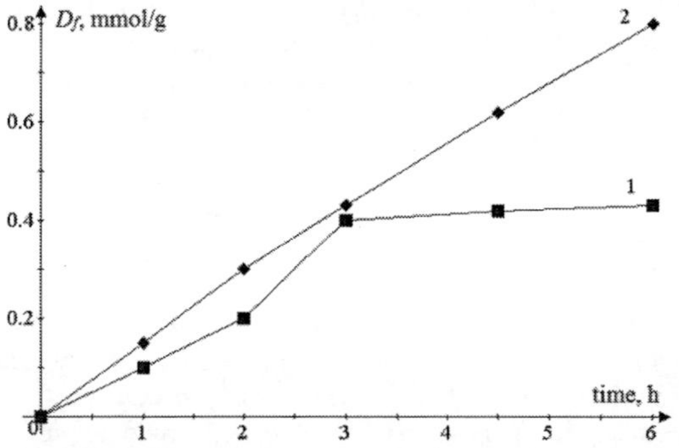

Fig. 3. The change in the degree of the functionalization with carboxyl groups during oxidation in nitric acid vapors at 140 °C for the CNTs previously purified from the metal oxide catalyst admixtures (1) and the initial CNTs (2)

The composition of the gaseous products of the CNT interaction with the nitric acid vapor after water vapor condensation includes oxides of nitrogen and carbon (Table 1).

Table 1. Volume(V) and composition of gaseous oxidation products of 1 g of the CNTs in nitric acid vapors at 140 °C according to gas chromatographic analysis data

CNT type	Time, h	V (norm. cond.), L	Component proportion in gas mixture, v. %			
			CO	CO_2	N_2O	NO_2
initial	1.5	0.78	12.2	5.3	0	82.5
	2.5	1.51	6.8	21.5	0.5	71.2
	3.5	2.01	6.8	17.7	0.6	74.9
	4.5	2.53	6.0	16.7	0.5	76.8
	6.0	3.17	5.1	13.6	0.9	80.4
purified	2.5	1.43	6.1	17.1	1.7	75.1
	3.5	1.95	5.7	15.8	1.2	77.3
	4.5	2.51	5.5	15.7	1.4	77.4
	6.0	3.11	4.8	12.8	0.9	81.5

Nitrogen dioxide formed by the reaction

$$4HNO_3 \rightarrow 4NO_2 + 2H_2O + O_2, \qquad (1)$$

is capable of oxidizing the carbon materials surfaces [8]. The gas mixture components can be arranged in the following order on the basis of the oxidative activity index: $NO_2 > N_2O \approx NO \approx O_2$.

It was shown in [9-10] that when nitrogen dioxide is reduced, nitric oxide (II) is predominantly formed. The absence of this component in the composition of gaseous products of CNT oxidation can be explained by the fact that NO is rapidly oxidized by oxygen formed by reaction (1). The absence of molecular nitrogen in products of the interaction of carbon with nitrogen oxides is indicated in [11]. In [12] it is noted that in addition to NO in the products of oxidation of carbon materials with nitrogen dioxide, N_2O is identified.

Thus, the information obtained on the composition of gaseous products of CNT oxidation in nitric acid vapor does not contradict the literature data. The chemistry of the oxidation of the CNT surface with the nitric acid thermal decomposition products can be described by a set of typical reactions, the most important of which are the processes of formation of oxygen-containing functional groups, processes of formation and transformation of nitrogen-containing surface formations, processes of elimination of oxygen-containing functional groups, processes of amorphous phase oxidation, and associated processes of interaction between reaction products. Water vapor present in the oxidizing mixture catalyzes the oxidation without changing the interaction mechanisms [13].

Nitric oxide (I) is identified in gaseous products ~2.5 h after the oxidizing reagent supply started in fairly small amounts (up to 1.7 v.%). The maximum N_2O concentration is observed in oxidation products purified from admixtures of the CNT metal oxide catalyst, the degree of functionalization of which using the oxygen-containing groups is lower than in the untreated CNTs subjected to the similar treatment. On this basis, it can be concluded that the catalyst particles, exerting an accelerating effect on the creation of functional groups, contribute to a decrease in the rates of the N_2O formation reactions.

The total volumes of gases released at equal intervals are determined by the rate of the oxidizing reagent supply. The predominant component of the analyzed mixtures of gaseous products is NO_2. Its concentration practically does not change during the process. It ranges from 71.2 to 81.5 v.%. The consistency of the NO_2 content is due to its excessive amount in the reaction mass and a low degree of conversion.

The obtained data were used to estimate the total thermal effect and heating rate of the reaction mass due to that effect (Table 2). For the oxidation of the CNTs containing the metal oxide catalyst admixtures, the total amount of the reaction heat released per unit time is 1.8 to 5 times higher than in the case of the oxidation of the purified CNTs. This fact lies in accordance with the assumption of the catalytic influence of the metal oxide inclusions on the process.

Table 2. Calculated values of the total amount of reaction heat (*qr*) and superheating of the reagents without considering thermal losses (*dt*) during the oxidation of 1 g of the CNTs in concentrated nitric acid vapors at 140 ° C

CNT type	Time, h	*qr*, J	*dt*, K
initial	1.5	15.07	3.01
	2.5	32.45	9.21
	3.5	26.37	7.52
	4.5	24.41	6.98
	6.0	19.54	4.08
purified	1.5	8.62	3.77
	2.5	6.45	6.20
	4.5	6.18	5.97
	6.0	4.96	3.51

4 Modeling of the Temperature Field of the Reaction Zone Inside the CNT Gas-Phase Oxidation Reactor

The purpose of thermal calculation is to determine the geometric characteristics of the reaction region (diameter and height) that provide an allowable volumetric temperature drop in the CNT bulk layer during the gas-phase functionalization.

The temperature field in the CNT bulk layer is modeled through solving the non-stationary heat conduction problem for a finite cylinder with a volumetric heat source.

Since the gas flow rate in the layer does not exceed 0.3 mm/s, the gas medium is assumed to be stationary, which provides an additional margin for the temperature difference.

The temperature of the medium from the lateral cylinder surface may be used as a reference point, thereby making it possible to implement homogeneous boundary conditions on this surface (Fig. 4).

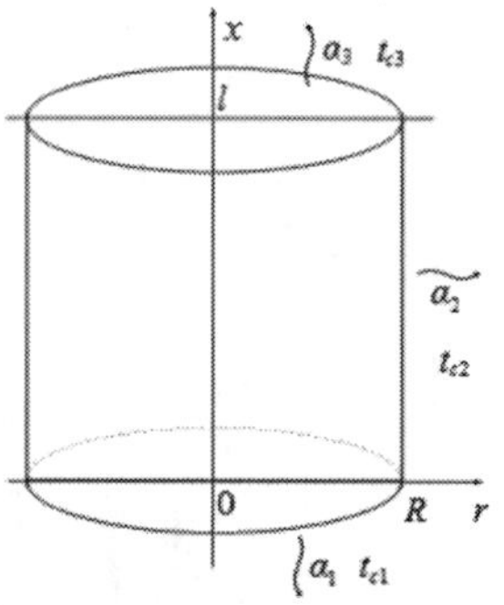

Fig. 4. A finite cylinder

The problem has the following form:

$$\frac{\partial t(x,r,\tau)}{\partial \tau} = a\left(\frac{\partial^2 t(x,r,\tau)}{\partial x^2} + \frac{\partial^2 t(x,r,\tau)}{\partial r^2} + \frac{1}{r}\frac{\partial t(x,r,\tau)}{\partial r}\right) + \frac{q_v}{c\rho}, \quad 0<x<l; \quad 0<r<R; \tag{2}$$

$$t(x,r,0)=f(x,r)-t_{c2}; \tag{3}$$

$$\frac{\partial t(x,0,\tau)}{\partial r} = 0; \tag{4}$$

$$\lambda\frac{\partial t(0,r,\tau)}{\partial x} - \alpha_1\left(t(0,r,\tau)-t_{c1}+t_{c2}\right)=0; \tag{5}$$

$$\lambda\frac{\partial t(l,r,\tau)}{\partial x} + \alpha_3\left(t(l,r,\tau)-t_{c3}+t_{c2}\right)=0; \tag{6}$$

$$\lambda\frac{\partial t(x,R,\tau)}{\partial r} + \alpha_2\, t(x,R,\tau)=0. \tag{7}$$

Here, $t(x, r, \tau)$ is the temperature field of a bounded cylinder as a function of the longitudinal coordinate x, the radial coordinate r, and the time τ; q_v is the specific thermal power of a volumetric heat source, J/m^3; λ, c, ρ - respectively, thermal conductivity, W/(m·K), heat capacity, J/(kg·K), density of the cylinder material, kg/m^3; $f(x, r)$ is the initial temperature distribution; α_1, α_2, α_3 -

heat transfer coefficients from the outer surfaces of the cylinder, W/(m^2·K); t_{c1}, t_{c2}, t_{c3} - ambient temperatures from the side of the corresponding surfaces; R, l - the radius and height of the cylinder, m, respectively.

The solution of problem (2) - (7), obtained by the method of finite integral transformations, has the following form:

$$t(x,r,\tau)=t_{c2}+\sum_{n=1}^{\infty}\sum_{m=1}^{\infty}\frac{1}{N_{mn}}\cdot\left(\frac{S+\frac{Q}{a}}{\mu_{mn}}+\left(F-\frac{S+\frac{Q}{a}}{\mu_{mn}}\right)exp\left(-\mu_{mn}^{2}a\tau\right)\right)\cdot sin(\nu_n x+\varphi_n)J_0(\gamma_m r), \quad (8)$$

where

$$F=\int_0^l\int_0^R\left(f(x,r)-t_{c2}\right)sin(\nu x+\varphi)rJ_0(\gamma r)\,dr\,dx, \quad (9)$$

$$S=\frac{R}{\gamma}J_1(\gamma R)\left(\frac{\alpha_3}{\lambda}(t_{c3}-t_{c2})\,sin(\nu l+\varphi)-\frac{\alpha_1}{\lambda}(t_{c1}-t_{c2})\,sin(\varphi)\right), \quad (10)$$

$$Q=\int_0^l\int_0^R\frac{q_v}{c\rho}sin(\nu x+\varphi)rJ_0(\gamma r)\,dr\,dx=\frac{q_v R}{c\rho\nu_n\gamma_m}\cdot\left(cos(\varphi_n)-cos(\nu_n l+\varphi_n)\right)J_1(\gamma_m R), \quad (11)$$

$$\varphi=arctg\left(\frac{\lambda\nu}{\alpha_1}\right), \quad (12)$$

ν - successive positive roots of the equation,

$$tg\,(\nu l+\varphi)=-\frac{\lambda\nu}{\alpha_3}, \quad (13)$$

γ – successive positive roots of the equation,

$$\frac{\lambda\gamma}{\alpha_2}J_1(\gamma R)-J_0(\gamma R)=0, \quad (14)$$

$J_1(r)$, $J_0(r)$ – Bessel functions,

$$\mu_{mn}^2=\gamma_m^2+\nu_n^2, \quad (15)$$

$$N_{mn}=\frac{R^2}{4}\left(J_1^2(\gamma_m R)+J_0^2(\gamma_m r)\right)\cdot\left(l+\frac{1}{\nu_n}\left(sin(\varphi_n)\,cos(\varphi_n)-sin(\nu_n x+\varphi_n)\,cos(\nu_n x+\varphi_n)\right)\right). \quad (16)$$

The solution to this problem was used to determine the design and mode parameters of the gas-phase functionalization equipment, at which the maximum temperature difference considering the volume of the CNT bulk layer during the process does not exceed 10 K.

The proposed approaches will give an opportunity to carry out modeling of processes occurring at the CNT oxidation using ozone, hydrogen peroxide vapors [14], and at secondary treatment of carboxylated CNTs using gaseous ammonia to form amide groups.

5 Conclusions

The particles of the transition metal oxide admixtures catalyze the formation of carboxyl groups on the CNT surface during the oxidation in nitric acid vapors and contribute to 1.8-5.0 increases in the total thermal effect. The technique of thermal calculation of equipment parameters for performing the process of CNT gas-phase oxidative functionalization is proposed. The data of the experimental studies were used to solve the equations of the mathematical model of the non-stationary temperature field in the working region inside the reactor where the process takes place.

The research was supported by the Russian Science Foundation (Grant No. 15-13-10038).

References

1. A.V. Melezhyk, A. V. Rukhov, E. N. Tugolukov, A.G. Tkachev, Nanosystems: Phys. Chem. Math. **4**, 247 (2013)
2. A. A. Aladinskiy, A. V. Rukhov, E. N. Tugolukov, T. P. Dyachkova, Transactions TSTU. **20**, 572 (2014)
3. C. Kingston, R. Zepp, A. Andrady, D. Boverhof et al., Carbon. **68**, 33 (2014)
4. Yu.I. Merkulova, S.V. Kondrashov, T.P. Dyachkova, P.S. Marakhovskii, G.Yu. Yurkov, Russ. J. Appl. Chem. **88**, 1848 (2015)
5. H.M. Kim, K. Kim, S.J. Lee, J. Joo et al., Curr. Appl. Phys. **4**, 577 (2004)
6. J. Yang, X. Li, C. Liu, G. Ma, Appl. Surf. Sci. **325**, 235 (2015)
7. H.P. Boehm, Carbon. **40**, 145 (2002)
8. B.R. Stanmore, V. Tschamber, J.-F. Brilhac, Fuel. **87**, 131 (2008)
9. A. Setiabudi, M. Makkee, J.A. Moulijn, Appl. Catal. B Envir. **50**, 185 (2004)
10. Z. Zhang, J.D. Atkinson, B. Jiang, M.J. Root, Z. Yan, Appl. Catal. B Envir. **148-149**, 573 (2014)
11. A.R. Chugtai, M.M. Atteya, J. Kim, B.K. Konowalchuk, D.M. Smith, Carbon. **36**, 1573 (1998)
12. U. Kirchner, V. Scheer, R. Vogt, J. Phys. Chem. A. **104**, 8908 (2000).
13. M. Jeguirim, V. Tschamber, J.F. Brilhac, P. Ehrburger, Fuel. **84**, 1949 (2005)
14. T.P. Dyachkova, Yu.A. Khan, N.V. Orlova, S.V. Kondrashov, Transactions TSTU. **22,** 232 (2016)

 DOI: 10.1051/matecconf/201712902015

Investigation of the thermophysical properties of the oxide layer formed by microarc oxidation on Al-Si alloy

Natalia Yu. Dudareva[1,*], *Pavel V.* Ivashin[2], and *Alexander B.* Kruglov[3]

[1]Ufa State Aviation Technical University, Internal Combustion Engine Department, 450000 Karl Marx st., 12, Ufa, Russian Federation
[2]Togliatti State University, Research department - 4, 445020 Belorusskaya st., 14, Togliatti, Russian Federation
[3]National Research Nuclear University «MEPhI», Department of Thermal Physics, 115409 Kashirskoe shosse, 31, Moscow, Russian Federation

Abstract. The results of the experimental study of the thermophysical properties of the coating formed by microarc oxidation (MAO) on an aluminum-silicon alloy AK12D are presented. The description of the research methodology, including the formation of the MAO-layer on the surface of laboratory samples and the study of their structure, composition and of thermophysical properties are given. The thermal conductivity of the coating was investigated by two methods: steady state and laser flash method. The tests were performed within the temperature range of 50-400° C. The obtained results showed that the coating have a rather low thermal conductivity (λ= 1.32…1.52 W/(m·K) at 100 °C). It is shown that a wealth of mullite (up to 60%) does not lower the thermal conductivity of the coating. It is assumed that the basic factors affecting the thermal conductivity of the oxide layer are small-scale porosity, crystallite sizes and amorphous phase.

1 Introduction

Microarc oxidation (MAO) technology is becoming increasingly popular in the industry nowadays [1-3]. This method of coating formation is very promising for details made from Al, Mg, Ti, Ta, Nb, Zr, Be and their alloys [4, 5]. MAO-coatings on aluminum alloys have a unique combination of the properties: high wear resistance, corrosion resistance and heat resistance [6-9].

Using of MAO-layers as thermal barrier coatings is greatly promising way of its application. The application of MAO-coatings to heated details of power equipment leads to reducing their heat stress enhances reliability and increases engine efficiency. This is very important for engine pistons, working at temperatures up to 400°C [10].

Application of MAO-coatings requires the knowledge of the basic thermophysical properties, such as specific heat capacity, thermal diffusivity and thermal conductivity. However, the thermophysical properties of MAO coatings are not completely investigated nowadays. Therefore, the reliable information about the thermophysical properties of MAO coating should be obtained experimentally.

2 Formulation of the problem

Investigation of the structure and thermophysical properties of the MAO-layers is a complex study. Therefore, the number of researches describing the thermophysical properties of MAO-layers is very insignificant nowadays.

There are a few papers, that describe the determination of the thermal conductivity of the MAO layer formed on a wrought aluminum alloy 6082 (with fraction Si of 0.7...1.3 %) in a silicate-alkaline electrolyte [11-13]. The thermal conductivity of the MAO-coating depends on its composition. It was revealed in [13] that the thermal conductivity of the mullite-rich coating decreases to λ=0.5±0.2 W/(m·K).

All the thermophysical data were obtained for the MAO-coatings formed on the wrought aluminum alloy 6082. However, in the industry, alloys with a high silicon content (Al-Si alloys) are most common. Therefore, the study of the thermophysical properties of MAO-layers formed on Al-Si alloys is an actual challenge. This research is aimed to this problem.

3 The Methodology of Research

3.1 Sample preparation

Samples of the eutectic aluminum-silicon alloy (Russian mark is AK12D; Si – 10…13%, Fe – 1.5%, Mn – 0.5%, Ti – 0.1%, Cu – 0.6%, Zr – 0.1%, Mg – 0.1%, Zn – 0.3%, Al – balance [14]) were used.

The MAO-coating was formed on a disk, with diameter of 68 mm and thickness of 10 mm. The MAO-process was performed in a silicate-alkaline electrolyte prepared with distilled water, Na_2SiO_3 (2.5 g/l) and

* Corresponding author: natalia_jd@mail.ru

KOH (2.5 g/l). The electrical parameters of the process were: the ratio of the anodic and cathodic current was kept constant Ia/Ic = 1, the average current density was j = 13±2 A/dm^2, the pulse frequency was 50 Hz. The process took 90 min. The electrolyte temperature did not exceed 36°C.

Coated disk was cut into samples to measure its thickness, porosity and to analyze the structure. The sample for steady state method of thermal conductivity measurement had sizes of 45 mm x 20 mm x 10 mm. The sample for laser flash method had diameter 9.9 mm and thickness of 1.9 mm.

Thermophysical properties of the alloy AK12D (the heat capacity – Cp_{sb} and the thermal conductivity - λ_{sb}) were investigated on samples with 9.9 mm diameter and 1.9 mm thick. The coefficient of linear thermal expansion (CLTE) – α_{sb} was studied on the sample with diameter of 6.0 mm and length of 25 mm with dilatometer DIL-402 at temperature range of 50 –400 °C.

MAO-layer was prepared as the powder for studies of heat capacity and phase composition.

3.2 Thickness, porosity and density

Scanning electronic microscope (SEM) "JEOL JSM 6390" was used for the analysis thickness and porosity of oxide layer. Fraction of pores was calculated by the superposition of a square grid on the image of cross section structure of the coating [15]. The SEM images obtained with backscattered electron mode.

Geometrical sizes and weight of the coated sample, coating thickness and density of substrate were used for the calculations of coating density. The density of aluminum alloy was measured with a special experiment. The samples were weighted using a Mettler Toledo XPE 26 microbalance, to ±2.5 μg precision. The geometrical sizes of the samples were measured using a LINKS digital measuring instrument with an accuracy of ±0.03 mm.

3.3 Phase composition

Phase composition was investigated by Rigaku Ultima IV X-ray diffractometer, using Cu-Kα radiation with 2Θ from 15 to 110°. MAO-layer prepared as a powder was used for XRD analysis as powders allow obtain more accurate results [16]. Quantitative phase analyzis was carried out with Rietveld method, using the Maud software.

3.4. Thermophysical properties

3.4.1 Heat capacity

Heat capacity of the MAO-layer within the temperature range of 50-400° C was measured using differential scanning calorimeter NETZSCH DSC 404F3 [17].

The measurements were made on the MAO-coating powder with a weight of 33.454 mg. A sapphire sample was used as the *Cp* standard. *Cp* measurement was carried out at 20 °C/min rate and isothermal hold at 400 °C during 3 min (the typical thermal program). Data were analyzed using the NETZSCH Proteus software.

3.4.2 Thermal conductivity

Steady state and laser flash methods both were used for thermal conductivity measurements. Coincidence of the results obtained with different methods increases reliability of the measurements.

Thermal diffusivity and conductivity of the coating and substrate both were determined with NETZSCH LFA 457 MicroFlash. This method determines thermal diffusivity with analysis of thermogram T(τ) of surface of the samples heated by laser flash W(τ) (Figure 1a). The shape of thermogram (Figure 1b) depends on thickness and thermal property of the sample layers.

Thermal diffusivity of the sample or the sample layer is calculated on the basis of the experimental thermogram. It is also possible to determine the heat capacity of the sample [18, 19]. The thermal diffusivity and thermal conductivity of the layers are calculated with Proteus LFA Analysis software, which contains methods for samples with one, two or three layers [19].

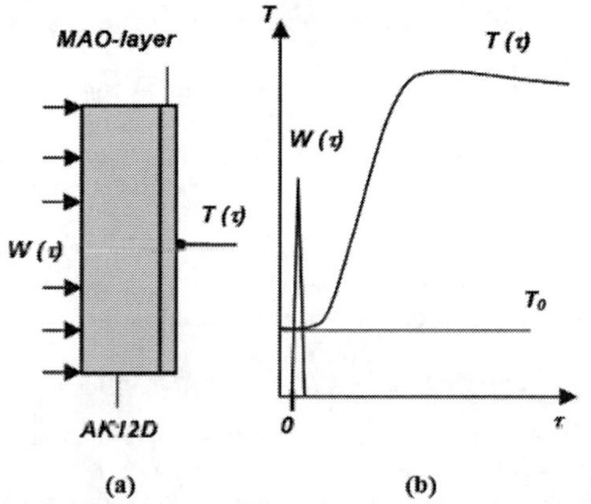

Fig. 1. Laser flash method: (a) pictorial schematic of the test, (b) thermogram.

The thermophysical properties of the AK12D alloy were determined on a sample without coating at the beginning of the experiment. The values of thermal diffusivity, heat capacity, CLTE and thermal conductivity were measured.

Thermal diffusivity both of coated and uncoated samples was measured. Thermal conductivity of MAO-coating was calculated with Proteus LFA Analysis software using thickness, density and Cp data of coating obtained in a special experiment. All the tests were performed within the temperature range of 50-400° C.

Multiple-repeated tests revealed the following experimental errors. For substrate: $\Delta a/a \leq 3$ %; $\Delta Cp/Cp \leq 5$ %; $\Delta\lambda/\lambda \leq 7$ %, for MAO-coating: $\Delta a/a \leq 5$ %; $\Delta\lambda/\lambda \leq 10$ %.

Thermal conductivity of MAO-coating was also measured with steady state method [20] within temperature range of 50-150°C. Scheme of experimental equipment is shown on Figure 2.

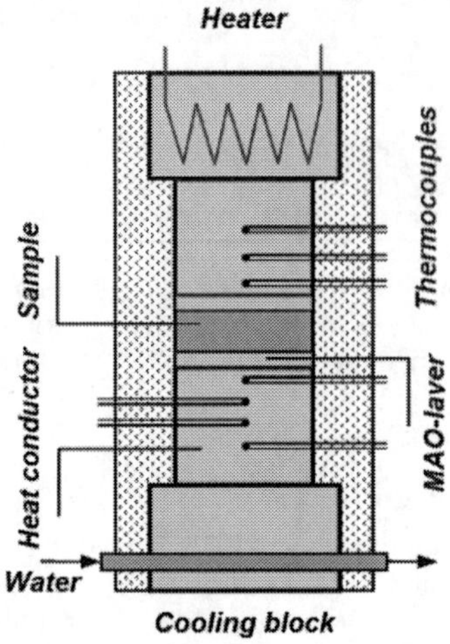

Fig. 2. Scheme of experimental equipment.

This equipment contains heater, cooler, two heat conductors, thermal insulator and four tie studs. Differential scheme of thermocouples connections was used to measurement of temperature gradient into heat conductors.

A thin layer of Arctic MX-2 thermal compound was applied on top and bottom of MAO-coated sample. Then the sample was put into measurement equipment, heater was warmed up with electrical power, cooler connected with circulated cooling water.

The signals of thermocouples were registered when thermal state of the sample and heat conductors became steady. Thermal conductivity calculated as:

$$\lambda_c = 2\delta_c\,[(\Delta T_{sm}/q) - (\delta_{sb}/\lambda_{sb}) - 2r_t]^{-1} \quad (1)$$

where λ_c – thermal conductivity of MAO-coating; δ_c – MAO-coating thickness; q – heat flux; ΔT_{sm} – the difference between temperatures of heater and cooler surfaces connected with the sample; λ_{sb} – thermal conductivity of substrate, δ_{sb} – substrate thickness; r_t – thermal resistance between sample surface and heat conductor surface.

Heat flux q and ΔT_{sm} were calculated by temperatures measured in heat conductors. λ_{sb} – was preliminary determined with the laser flash method. Thermal resistance r_t was determined in special experiments during repeatable measurements on samples made of copper and aluminum. Value obtained in this experiment is $r_t = (1.8 \pm 0.2)\cdot 10^{-5}$ (m^2·K)/W.

Calibration of steady state experimental equipment was made by repeatable tests on the sample made of stainless steel. Thermal conductivity of this steel is exactly known [21]. Uncertainties of thermal conductivity measurements with steady state method revealed in calibration experiment are less than 15% within temperature range of 50 – 150°C.

4 Experimental results

The transverse section of the sample, on which the thickness and porosity were measured, is shown in Figure 3.

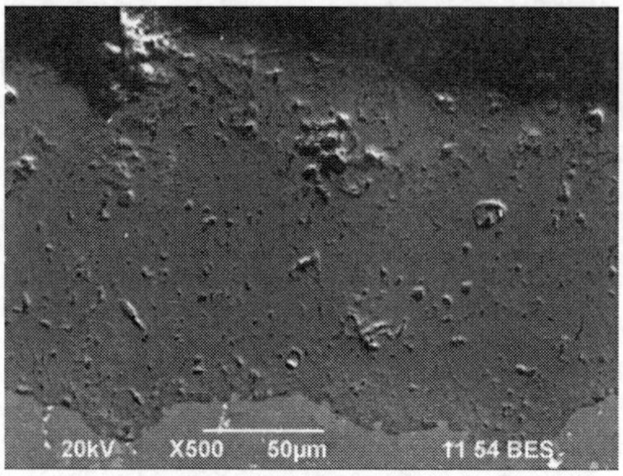

Fig. 3. Transverse section of the coating.

Studies of the macrostructure of the oxide layer showed that the average thickness of the MAO-coating is 154.1±3.7 µm. The through and closed pores are present in the structure of the coating. The average porosity is 13.1 ± 1.6%.

The density of aluminum alloy AK12D determined in this study is 2679±4 kg/m^3. The density of the MAO-coating is 2659±565 kg/m^3.

The following phases were identified by the X-ray diffraction analysis of the MAO layer: α-Al_2O_3, γ-Al_2O_3, $3Al_2O_3\cdot 2SiO_2$ (mullite), Al_2SiO_5 (sillimanite) [22, 23]. Phase composition of the MAO-coating is shown in Figure 4:

- α-Al_2O_3 (corundum) – 5.3±0.1%
- γ-Al_2O_3 – 34.8±1.8%
- $3Al_2O_3\cdot 2SiO_2$ (mullite) – 58.1±2.8%
- Al_2SiO_5 (sillimanite) – 1.8±0.3%.

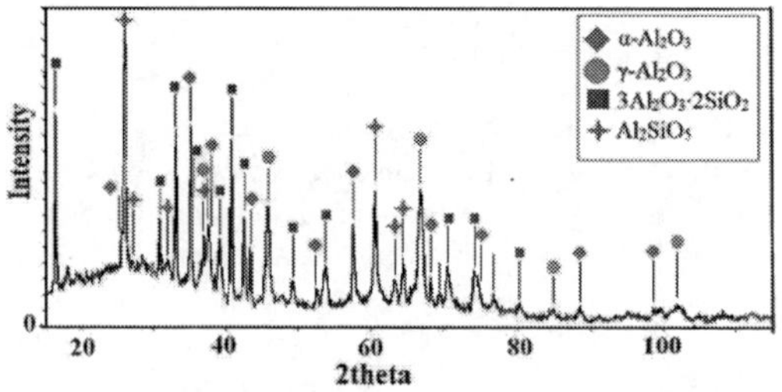

Fig. 4. Phase composition of MAO-coating.

The equations of the CLTE, heat capacity, thermal diffusivity and thermal conductivity of AK12D alloy were obtained as the result of thermophysical studies by the laser flash method:

$$\alpha_{sb} = (0.00841\cdot t + 20.45)\cdot 10^{-6} \quad (2)$$

$$Cp_{sb} = 808.4 + 0.4583\cdot t \quad (3)$$

$$a_{sb} = (-0.0302\cdot t + 74.05)\cdot 10^{-6} \quad (4)$$

$$\lambda_{sb} = 0{,}009 \cdot t + 161{,}6 \quad (5)$$

where α_{sb} – the coefficient of linear thermal expansion, K^{-1}; Cp_{sb} – the heat capacity, J/(kg·K); a_{sb} – thermal diffusivity, m^2/s, λ_{sb} – the thermal conductivity, W/(m·K), t – the temperature, °C.

The effect of temperature on the heat capacity of the MAO layer was found:

$$Cp_c = 830.82 + 1.0289 \cdot t - 0.0013 \cdot t^2 \quad (6)$$

Dependences of thermal diffusivity (a_c) and thermal conductivity (λ_c) of the MAO-coating on the temperature revealed by the laser flash method can be described as:

$$a_c = 1{,}198 \cdot t^{-0.1385} \cdot 10^{-6} \quad (7)$$

$$\lambda_c = 0{,}0002 \cdot t + 1{,}5132 \quad (8)$$

Thermal conductivity of MAO-coating obtained by the steady-state method:

$$\lambda_c' = 0{,}0017 \cdot t + 1{,}1503 \quad (9)$$

Thus, the thermal conductivity of the MAO-coating at 100°C obtained by the laser flash method is $\lambda_c = 1.52 \pm 0.34$ W/(m·K). The thermal conductivity determined by the steady-state method, is $\lambda_c' = 1.32 \pm 0.15$ W/(m·K) also at 100°C.

5 The discussion of the results

MAO-layer formed in this study contains a large fraction of mullite (about 60%). Based on the paper [13], where the mullite content was 50% and the porosity was about 10%, it was expected that the thermal conductivity of the MAO-layer would be ~ 0.5 W/(m·K). However, the independent thermophysical studies, carried out by two different methods, have shown that the coating has higher thermal conductivity: λ_c = 1.32…1.52 W/(m·K) at 100 °C. Thus, it can be concluded that the influence of mullite on thermal conductivity of the coating is not so significant as it expected.

It should be noted that the registered thermal conductivity of the oxide layer is lower than thermal conductivity of dense mullite (~ 10-15 W/(m·K)), dense sillimanite (~ 9.1 W/(m·K)), and dense aluminum oxide (~ 34.6 W/(m·K)) [24]. This fact can be explained if we suppose that the thermal conductivity of MAO-layer does not depend on the mullite volume, but depends on other factors: the crystallites size, the small-scale porosity and the fraction of the amorphous phase.

6 Conclusions

MAO-coating formed on the Al-Si alloy in the silicate-alkaline electrolyte has a high volume of mullite (~ 60%) and high level of porosity (~13%). The coating is characterized by large thickness of ~ 150 µm and a low density - 2659±565 kg/m³.

Thermal conductivity of a mullite-rich oxide coating is λ_c = 1.32 ... 1.52 W/(m·K) at 100°C. This value is significantly lower than thermal conductivity of the phases composing of the coating, but higher than in known studies, where the oxide layer also has a high volume of mullite.

The results of this work can be useful for thermal calculations when using the MAO-coating as thermal barrier for details of heat-stressed equipment.

This work has been partially performed with support of State Research Engineering Assignment (Project 11.3937.2017/PCh).

References

1. A.L. Yerokhin, X. Nie, A. Leyland, A. Matthews, S.J. Dowey, Surf. Coat. Technol, **122**, 73 (1999)
2. A.L. Yerokhin, L.O. Snizhko, N.L. Gurevina, A. Leyland, A. Pilkington, A. Matthews, Surf. Coat. Technol. **177 –178**, 779 (2004)
3. Y.J. Liu, J.Y. Xu,W. Lin, C. Gao, J.C. Zhang and X.H. Chen, Rev.Adv. Mater. Sci. **33**, 126 (2013)
4. R.O. Hussein, D.O. Northwood, X. Nie, Mater. Sci. and Applic. **5**, 124 (2014)
5. V.A. Koshuro, G.G. Nechaev, A.V. Lyasnikova, Techn. Phys, **10**, 1570 (2014).
6. A.L. Yerokhin, A.A. Voevodin, V.V. Lyubimov, J. Zabinski, M. Donley, Surf. Coat. Technol. **110**, 140 (1998)
7. J. Baxi, P. Kar, H. Liang, A. Polat, M. Usta, A.H. Ucisikc, Vacuum, **83**, 217 (2009)
8. P.B. Srinivasan, R. Zettler, C. Blawert, W. Dietzel, Material scharacterization, **60**, 389 (2009)
9. L.R. Krishna, K.R. Somaraju, G. Sundararajan, Surf. Coat. Technol. **163 –164**, 484 (2003)
10. N.Yu. Dudareva, O.P. Dombrovskiy, R.V. Kalschikov, I.A. Butusov, J. of Engineering Sci. and Technol. Rev. **6**, 10 (2015)
11. J.A. Curran, *Thermal and mechanical properties of plasma electrolytic oxide coatings* (Cambridge, 2006)
12. J.A. Curran, T.W. Clyne, Surf. Coat. Technol. **183**, 177 (2005)
13. J.A. Curran, H. Kalkancı, Yu. Magurova, T.W. Clyne, Surf. Coat. Technol. **201**, 8683 (2007)
14. Information on http://www.splav-kharkov.com.
15. S.A. Saltykov, *Stereometric metallography* (Moscow, Metallurgy, 1970) (in Russian)
16. R. Das, Md. E. Ali, Sh. B. Hamid, Rev. Adv. Mater. Sci. **38**, 95 (2014)
17. P.J. Haines, F.W. Wilburn, in: P.J. Haines (Ed.), *Thermal Methods of Analysis: Principles, Applications and Problems* (London, 1995)
18. W.J. Parker, R.J. Jenkins, C.P. Butler, G.L. Abbott. J. of Appl. Phys.. **32**, 1679 (1961)
19. Information on http://www.netzsch.com

20. J.H. Lienhard IV, J.H. Lienhard V, *A Heat Transfer Textbook* (Cambridge, Massachusetts, 2017)
21. *Tables of standard reference data.* Moskow (Standard Publishing House, 1994) (in Russian)
22. V. Dehnavi, X.Y. Liu, B.Li Luan, D.W. Shoesmith, S. Rohan, Surf. Coat. Technol. **251**, 106–114 (2014)
23. Z.C. Oter, A.E. Gulec, Y. Gencer, M. Tarakci, Acta phys. polonica, **4**, 1331 (2015)
24. J.F. Shackelford, R.H. Doremus, *Ceramic and Glass Materials. Structure, Properties and Processing* (Springer, 2008)

Influence of Al-Si alloy microstructure on the corrosion resistance of coatings formed by the microarc oxidation method

Natalia.Y. Dudareva[1,*], *Roman.V.* Kalschikov[1], and *Liliya.I.* Zaynullina[1]

[1]Ufa State Aviation Technical University, Ufa, Russian Federation

Abstract. The impact of the high-silicon aluminum alloy initial microstructure on the quality of the coating formed by microarc oxidation (MAO) has been studied. The MAO treatment is applied to AK12D samples in the initial coarse-grained state and after high pressure torsion. The following coating properties are studied: thickness, microhardness, porosity and corrosion resistance. It is established that the MAO layers properties depend on the base microstructure much. High pressure torsion applied to AK12D samples before MAO results in increase of the coating thickness by ~ 2 times. The microhardness of coatings reduces and their corrosion resistance degrades by ~ 10 times.

1 Introduction

The microarc oxidation (MAO) technique is one of the promising ways to form corrosion resistant coatings [1]. It enables expanding the application area of Al alloys in commercial production through formation of a modified surface layer of mainly high-strength oxides resistant to aggressive environment [2].

Al alloys find more application in aircraft and engine building due to their availability and simplicity in processing [3,4]. However, parts from Al alloys should possess not only definite strength properties but also high corrosion resistance for operation under such extreme conditions. Therefore, creation of the corrosion resistant surface on parts from Al alloys is a topical and complicated task.

The MAO essence is in migrating point microarc discharges taking place on the part surface under high voltage applied between the part in the electrolyte and the electrode [5]. As a result of these discharges the part surface layer transforms in the ceramic layer tightly bonded with the metal base. The quality and properties of the formed coating depend on different factors of the MAO: electrolyte composition, electrical regimes and processing duration [6-8]. The impact of these factors on the MAO layer properties has been studied well. One more important factor effecting the quality of the MAO layer is microstructure of the processed material.

It is known [9] that during the grain size change towards their reduction the volume characteristics of Al alloys such as strength, microhardness, etc change in a positive way. However, in modern scientific literature there is practically no information on quality and properties of MAO layers on Al alloys with ultrafine-grained (UFG) structure.

Thus, the aim of this work is the study of the initial microstructure effect on the structure and corrosion resistance of the oxide layer formed during MAO.

2 Experimental

2.1 Sample preparation

Eutectic silumin AK12D OST 192014-90 is chosen for the studies. The chemical composition of the material is 79–84 wt.% Al,13 wt.% Si,1.5 - 3 wt.% Cu, 0.2 wt.% Cr, 0.005 wt.% B, 0.7 wt.% Fe, 0.8-1.3 wt.% Mg, 0.3-0.6 wt.% Mn, 0.8-1.3 wt.% Ni, 11-, 0.02 wt.% Sn, 0.05- 0.2 wt.% Ti, 0.5 wt.% Zn.

In order to perform the studies on the samples with the base material structure, 3 samples with a diameter of 20 mm and 2 mm thick (group 1) are cut from a rod in the as-received state.

In order to produce samples with UFG structure 3 billets with a diameter of 20 mm and 2 mm thick are used. The UFG structure is formed via high pressure torsion (HPT). The HPT technique allows forming a nanostructure with non-equilibrium high-angle grain boundaries in the material during high pressure torsion [9]. The samples are processed by HPT via the following regime: 10 rotations with the velocity V = 1 rot/min, pressure P = 6 GPa, at a temperature of 20 °C. Prior to HPT the samples are water quenched with preliminary dwelling at T=515 °C for 1 hour. As a result 3 samples with UFG structure with a diameter of 20 mm and 1 mm thick are formed (group 2). Prior to the MAO treatment the samples with UFG structure are stored at -20 °C to avoid material ageing.

During MAO silicate and alkaline electrolyte on the basis of distilled water is used. The electrolyte composition is 1.5 g/l KOH and 3.5 g/l Na_2SiO_3. During the processing the ratio between the cathodic and anodic currents is constant Ia/Ik=1. The current density is 6.6…6.9 A/dm^2. The electrolyte temperature does not exceed 40 °C. The MAO process is performed till the complete end of sparking on the processed surface, and its duration is 180 minutes.

*Corresponding author: natalia_jd@mail.ru

DOI: 10.1051/matecconf/201712902016

2.2 Microstructure

The microstructure of the samples is studied with the help of an optical electron microscope "Olympus GX51" at a magnification of x500. The MAO layer studies, thickness measurement, viewing and imaging of pores are performed by a scanning electron microscope (SEM) « JEOLJSM-6490LV» at a magnification of x500.

The volume fraction of pores is determined by point application of the square mesh on the structure image.

The microhardness of the alloy and MAO-layer is determined on a microhardness measuring device StruersDuramin by the Vickers technique. A load of 100 grams is applied for 10 seconds. The microhardness of the coating is determined across the layer thickness.

2.3 Corrosion Resistance Studies

In order to test the corrosion resistance, aggressive solution of the following composition is used (chosen on the basis of recommendations listed in GOST 9.904-82):

- sodium chloride (NaCl) – 225 g/l;
- potassium nitride (KNO3) – 50 g/l;
- nitrogen acid (HNO3) – 5.5 g/l;
- base – distilled water.

The total volume of the prepared solution is 3 liters. The edges of the studied samples are covered with special varnish before immersion in the solution. The samples are put in the solution for 6 days (144 hours). After dwelling the samples are taken out of the solution, washed with water, dried and weighed. The corrosion rate is calculated via the formula:

$K = \Delta m/S*t$,

where K – the corrosion rate, [g/m^{2}h], Δm – change of the sample mass, g, S–the surface area (m^2), t–time of dwelling in the solution (h).

The mass of samples is measured with the help of the electronic scales Mettler Toledo XP26 before and after the tests. The measuring accuracy is up to 2.5 μg.

3 Results

The structure analysis of the samples of group 1 (AK12Д in the as-received state) demonstrates that the silicon phase is spread non-uniformly in the material structure. Areas enriched with silicon particles are observed. θ-phase ($CuAl_2$) precipitates with the size close to the size of silicon particles are also observed in the structure. The microstructure parameters of the alloy in the initial state are listed in Table 1 (group 1).

The microstructure of HPT-processed samples (group 2) differs considerably from the structure in the initial state (Fig. 1b). The silicon particles are distributed uniformly, the average grain size of the Al matrix reduces significantly. The microstructure parameters of the UFG alloy are listed in Table 1 (group 2).

Table 1. The microstructure parameters and microhardness of aluminum alloy depending on the group.

Parameters	Group	
	1	2
Average grain size of aluminum matrix	17±5μm	200±15nm
Average size of silicon particles,μm	3,2±0,7	4,3±0,7
Volume fraction of silicon particles, %	23±4	25±4
Volume fraction of θ-phase,%	3±1	3±1
Microhardness, HV	92±4	242±17

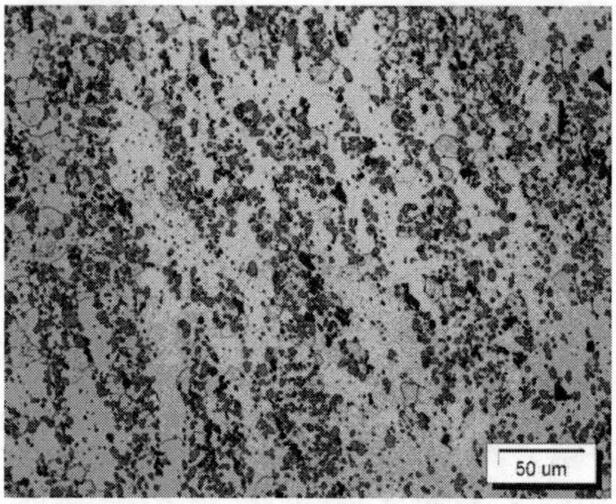

a

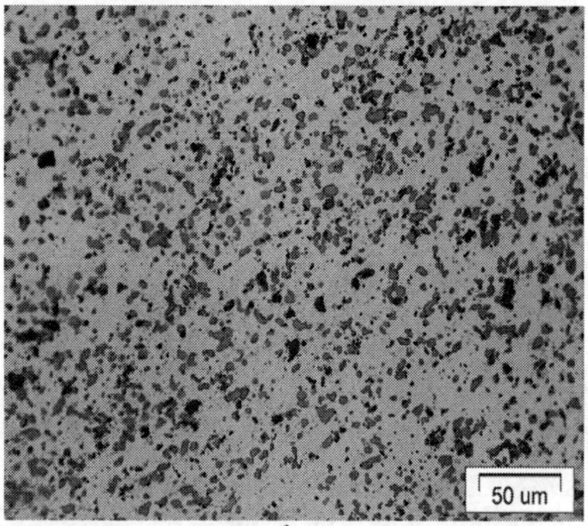

b

Fig. 1. Optical microstructure of samples: a – group 1; b – group 2.

As a result of the studies of the images of cross sections with the MAO layer, it is established that the thickness of the oxide layer is non-uniform in both groups. The average thickness value of the MAO layer on the samples of group 1 is h=41±15μm (Fig. 2a), and on the samples of group 2-h=88±18μm (Fig. 2b).

DOI: 10.1051/matecconf/201712902016

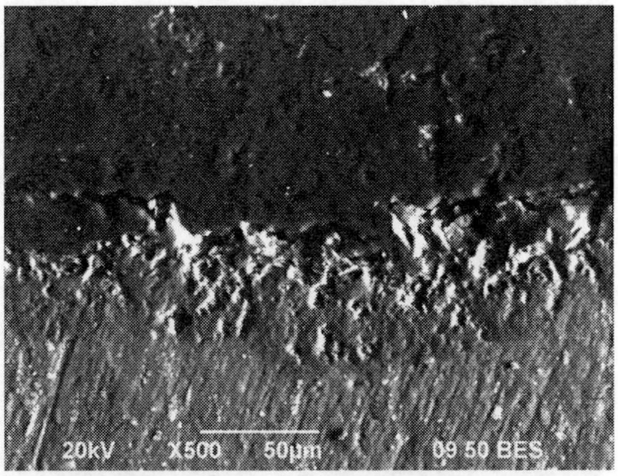

a

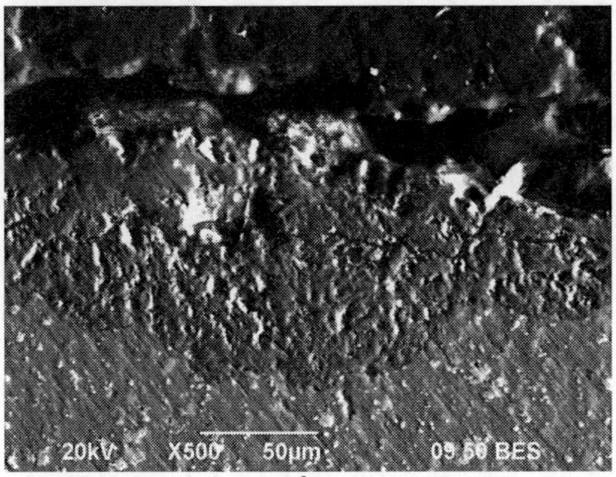

b

Fig. 2. Alloy AK12D (АК12Д) –coating system : a – group 1; b – group 2

The results of the porosity study of the MAO layers across the thickness are displayed in Fig. 3. Despite the difference in the distribution of pores and different thickness of oxide layers, the average porosity value of the coating is the same for both groups (Table 2).

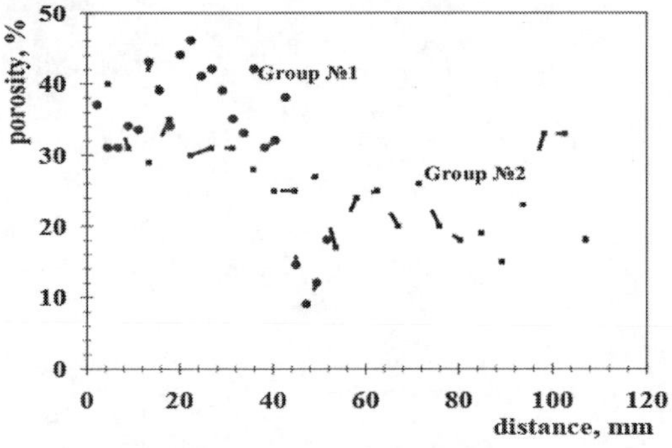

Fig. 3. Distribution of porosity in the MAO layer across the layer thickness on the samples of groups 1 and 2.

The results of microhardness measurements across the thickness of the MAO-layers are displayed in Fig. 4. The microhardness distribution is non-uniform through the MAO-layer thickness. The maximum value of the coating microhardness is observed in the samples of group 1 and is 900 HV. The samples of group 2 have the maximum microhardness value of the MAO-layer of 370 HV.

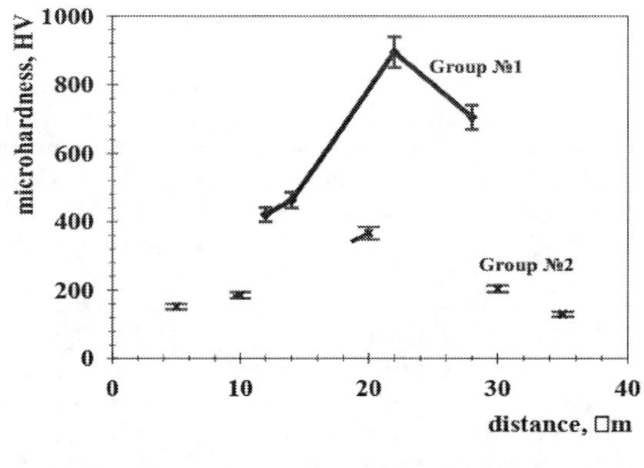

Fig. 4. Microhardness distribution in the MAO-layer across the layer thickness in the samples of groups 1 and 2.

The corrosion test results show that the corrosion rate in the samples of group 2 exceeds that in the samples of group 1~by 10 times (fig. 5a,b).

a

b

Fig. 5. Samples after corrosion resistance tests: a – sample of group 1, b – sample of group 2

The summarized results of the studies on the structure and corrosion resistance of the MAO-layers are listed in Table 2.

Table2. Properties of MAO layers.

Properties of the MAO layer	Group №	
	1	2
Thickness, t_{av}, μm	41±15	88±18
The average porisity, %	33±10	30±6
The average microhardness, HV	620±62	207±23
The average corrosion rate, K_{av}, g/(m^2·h)	0,229±0,001	2,678±0,006

4 Discussion

The results of the performed studies demonstrate strong impact of the initial microstructure of the alloy on the thickness, porosity, microhardness and corrosion resistance of MAO layers.

It is evident that the UFG structure of the material contributes to formation of MAO-layers with higher thickness but lower microhardness. Such a result can be explained by the fact that after HPT the structure of Al alloy is characterized by a small grain size of the Al matrix and developed dislocation structure. Such a structure contributes to deeper penetration of oxygen atoms/ions in the material of the substrate via intercrystallite defects during MAO, which results in increase of the coating thickness [10].

The microhardness reduction in the MAO layers after HPT can be explained by the fact that after HPT silicon particles are uniformly located in the base material of the sample. As a result during MAO interaction of Al and oxygen becomes difficult. In those parts where silicon comes out on the surface, coating is not formed or its quality is unsatisfactory - no very hard oxides are formed [11,12].

Usually the corrosion resistance directly depends on the coating porosity [13]. However, despite the same average porosity, the MAO layers formed on the samples with UFG structure differ considerably in lower corrosion resistance. By analyzing and comparing the images of polished coatings, one can see that the oxide layers on the UFG samples (Fig. 2b) are characterized by deep cracks passing across the coating. One may assume that these cracks lead to penetration of corrosion-aggressive environment into the sample material and cause active failure of it (Fig. 5b).

5 Conclusions

1. The structure of the sample material significantly impacts the thickness, porosity, microhardness and corrosion resistance of the MAO-layer.
2. The reduction of the structural elements sizes in Al alloy as a result of HPT contributes to decrease of the MAO-layer microhardness (~ by 3 times) and leads to increase of the MAO thickness (~ by 2 times).
3. The UFG structure of the substrate material contributes to formation of deeper cracks in the coating volume and reduction of the corrosion resistance of the MAO layer ~ by 10 times.

The reported study was partially supported by RFBR, research project №16-38-00827.

References

1. W. Xue, Applied Surface Science **253**, 6118 (2007)
2. H.X. Li, V.S. Rudnev, X.H. Zheng, T.P. Yarovaya, R.G. Song, J. of Alloys and Compounds **462**, 99 (2008)
3. R.H.U. Khan, A. Yerokhin, A. Matthews, X. Li, H. Dong, J. Surface and Coatings Technology **205**, 1679 (2010)
4. J.R. Davis, ASM International (1999)
5. A.L. Yerokhin, X. Nie, A. Leyland, A. Matthews, S.J. Dowey, J. Surface and Coatings Technology **122**, 73 (1999)
6. N.Y. Dudareva, Herald USATU **3**, 217 (2013) (in Russian)
7. G.M.M. Elhag, Ph.D. diss (MISA, Moscow, 2007) (in Russian)
8. P.M. Garinov, Ph.D. diss. (MISA, Moscow, 2009) (in Russian)
9. R.Z. Valiev, I.V. Alexandrov, *Bulk nanostructured metallic* materials, 397 (M: «Akademika», 2007)
10. N. Yu. Dudareva, Russian Aeronautics **51**, 321 (2008)
11. M. M. Krishtal, Advanced Materials Research **59**, 204 (2009)
12. M. M. Krishtal, Metal Science and Heat Treatment **46**, 377 (2004)
13. T. Wei, J. Journal of Alloys and Compounds **389**, 169 (2005)

Development of a model of a homogeneous continuous medium based on the material with defects in the form of hollows

Vadim Poroshin[1,*], *Andrey* Shlishevsky[2], and *Ksenia* Tsybulya[3]

[1]Aerospace Dept., Polytechnic Institute, South Ural State University (National Research University), Chelyabinsk, Russian Federation
[2]Aerospace Dept., South Ural State University (National Research University), Chelyabinsk, Russian Federation
[3]Social-humanitarian Institute, South Ural State University (National Research University), Chelyabinsk, Russian Federation

Abstract. In a body of the billet received by casting, inevitably there are defects accompanying this process: blisters, hollows and other discontinuities affecting the deformation and strength properties of the metal. The experience showed that direct calculation of the stress-strain state by the FEM of the body with a defective structure requires the use of supercomputer-level computing power. A medium with multiple randomly distributed discontinuities was schematized by a regular structure formed by a set of elements in the form of a regular tetrahedron with spherical hollows at its vertices. The proposed technique makes it possible to create the model of a continuous homogeneous medium, which is equivalent in its deformation properties to the original discontinuous material. Using this approach was got a power-law approximation of the stress-strain curve of the model medium. The last one represents the basic characteristic of the material, both under single-shot loading and (if we take into account the cyclic hardening) in analyzing the kinetics of the stress and strain fields' under repeatedly-variable loading. The discontinuity of the material was fixed using the Kolmogorov's deformation criterion of ductile fracture. This criterion was applied in evaluating the limit state of the valve chamber under operating conditions.

Nowadays the pipeline transportation carries out the relocation of more than a half of all volume of hydrocarbons. An integral part of any pipeline is a different type of pipe fitting, through which realize the distribution of flows, the regulation of the outturn or the complete overlapping of the main. A significant part of the main pipeline fitting are valves having the impressive sizes and considerable weight (about several tons). In this regard, shells of similar products are received, as a rule, by casting, during which various defects arise in the billet: metallic and non-metallic inclusions, discontinuities, blisters. This circumstance negatively affects durability of the finished product, so much more that the endurance of such hardwares can be up to 50 years. It is needless to recall that the fracture of even a single main valve entails not only serious economic losses, but it also can threaten an ecological catastrophe.

Estimating the strength reliability of such constructions is usually based on two criteria: strength under a single-shot static loading (in particular, for control tests after manufacturing), and durability under repeatedly-variable loading under operating conditions. In both cases, it is necessary to calculate the stress-deformed state of the object, taking into account the inelastic behavior of the material under single and repeatedly-variable loading in combination with the corresponding criteria for achieving the limit state. Such an analysis you inevitably have to perform numerically, for example, by the finite element method (FEM) with the help of recognized computing power. At the same time, experience shows that direct calculation of the stress-strain state of the body with randomly distributed defects on its volume, even under single-shot, and, even more, under repeatedly-variable, loading, is very laborious process and requires the use of supercomputer-level computing power.

In connection with the foregoing, we propose the technique for determining the parameters of the model of a continuous homogeneous medium, which is equivalent in its deformation properties to the original material with multiple randomly distributed discontinuities.

Such an approach is not absolutely new, however in known for us works it was mainly used for ceramic — classical brittle materials [1-3].

With regard to plastic materials, the problem arises in obtaining a stress-strain curve taking into account material inelasticity. To resolve it, the initial medium with defects was schematized with a regular structure formed by a set of elements in the form of a regular tetrahedron with spherical voids at its vertices (Fig. 1).

At the first stage we studied the effect of the parameters (the distance h between the defects and eccentricity e — displacements with respect to the medial surface of the valve chamber) of the structure single element (Fig.2) at a safety factor of the limit stress and the limit plastic strain. The diameter of the defect was assumed to be constant, equal to the maximum permissible for the most stressed zone of the shell in

* Corresponding author: poroshin2816@mail.ru

accordance with the regulatory documentation of manufacturer.

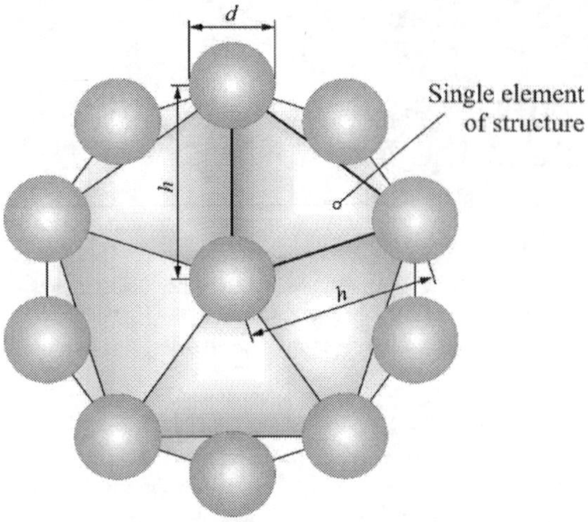

Fig. 1. The regular structure and its element — a tetrahedron.

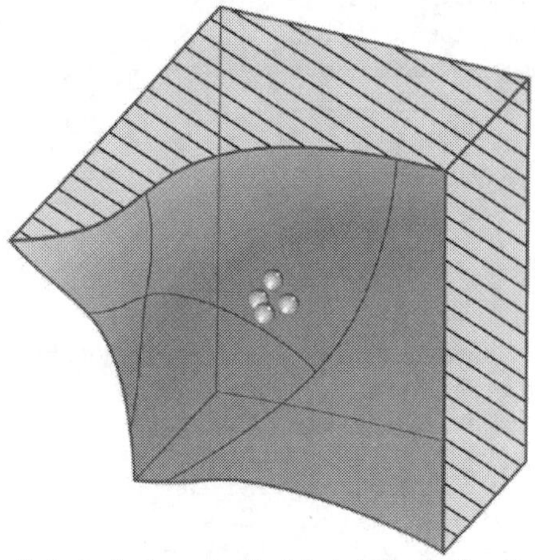

Fig. 2. A single element with defects in the shell wall.

Numerical analysis was performed by the finite element method using the ANSYS Workbench software. To increase the accuracy of calculation and reduce its labor intensity we used the submodeling procedure — allocation of the zone which is needs to be analyzed and setting boundary conditions for this fragment by transferring obtained by calculation of the common «global» model of forces and displacements on the submodeling nodes. Such a technique is correct provided that the features of the allocated zone, in particular the stress concentrators, do not introduce a significant perturbation into the stress-strain state at the boundaries of the selected fragment. This method allows significantly reduce the total number of finite elements (FE) and use a more dense, refined mesh.

Thus, one of the most loaded regions of the valve chamber (the root region of the rib) was identified, where due to the relatively complex geometric shape there are high probability of existence of casting defects (Fig. 3).

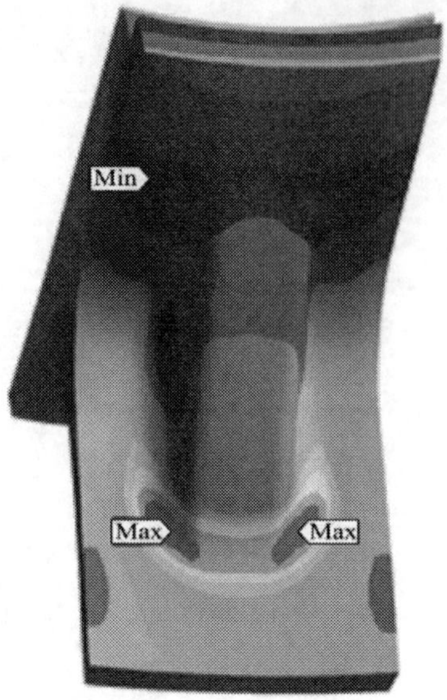

Fig. 3. The root part of the valve chamber ribs with the different size regions of equivalent von Mises stress under the working pressure (σ_{eq}^{max} = 255 MPa; inelastic calculation).

The use of a mesh of an irregular thickening hexahedral configuration made it possible to describe the stress and strain fields near spherical concentrators quite satisfactorily. The correctness of the FE object representation was confirmed by comparison with known solutions for linearly elastic material [4]. Fig. 4 shows the change of the theoretical concentration coefficient k_t with increasing distance between two spherical defects. As they are retired, the coefficient k_t tends to a value of 2.05, which is characteristic of a single concentrator; The error of the numerical solution does not exceed 2.5%.

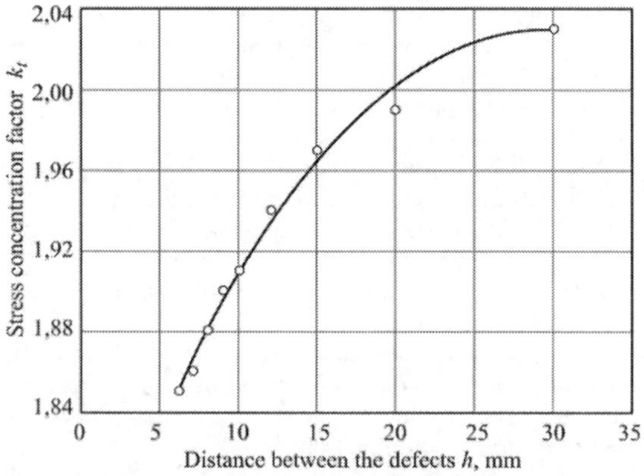

Fig. 4. The change of the theoretical concentration coefficient with increasing distance between two spherical defects.

In the same way, the finite element mesh was selected for four spherical concentrators located at the vertices of the tetrahedron. It is noted that when $h \approx d$, the greatest stresses occur at the places of tangency of discontinuities, in this case the concentration coefficient is equal to 5.57.

Taking into account that the value of the strain in the concentration region is usually significant, the strain-stress curve of the steel was approximated by a power-law dependence by Ramberg-Osgood:

$$\bar{\sigma} = K\,p^{m}, \tag{1}$$

where $\bar{\sigma} = \sigma(1+\varepsilon) = \sigma\exp(e)$ – true stress; σ, ε – conditional stress and conditional (sometimes called as engineering) strain; $e = \ln(1+\varepsilon)$, $p = \ln(1+\varepsilon_p)$ – total and plastic logarithmic strains; K, m – constants of the material of the conditions under consideration.

During the study, the distance between the hollows was increased in 0.5 mm increments until the mutual influence of the stress fields around each concentrator became negligible. The variation of the eccentricity e of the defect location with respect to the median surface of the shell wall we made in both directions (towards the more loaded region — the internal surface of the wall — the value e was taken as positive, toward the less loaded region — the external wall surface — negative). With each combination of the parameters h and e we counted the values of the actual safety factor of the limit stress and strain,

$$k_p = \frac{p_{\text{int}}^{f}}{p_{\text{int}}}, \qquad k_\sigma = \frac{\sigma_{\text{int}}^{f}}{\sigma_{\text{int}}}, \tag{2}$$

corresponding to the current stress-strain state; $p_{\text{int}}^{f} = p_{\text{int}}^{f}\left(\frac{\sigma_0}{\sigma_{\text{int}}}\right)$ – limit (corresponding to the discontinuity) logarithmic plastic strain;

$$\sigma_0 = \frac{\sigma_1 + \sigma_2 + \sigma_3}{3} \tag{3}$$

– mean stress;

$$\sigma_{\text{int}} = \frac{1}{\sqrt{2}}\sqrt{(\sigma_1-\sigma_2)^2 + (\sigma_2-\sigma_3)^2 + (\sigma_3-\sigma_1)^2} \tag{4}$$

– stress intensity;

$$p_{\text{int}} = \frac{\sqrt{2}}{3}\sqrt{(p_1-p_2)^2 + (p_2-p_3)^2 + (p_3-p_1)^2} \tag{5}$$

– logarithmic plastic strain intensity; σ_i – principal stresses; p_i – principal logarithmic plastic strains.

We determined the limit value p_{int}^{f} with the aid of the deformation criterion of V. L. Kolmogorov [5], which represents the condition of ductile fracture under a complex stress state of general form under the conditions of a single-shot static load. This criterion can be written in two forms:

a) in terms of strain:

$$\frac{p_{\text{int}}^{f}}{p_f} = a\exp\left(b\frac{\sigma_0}{\sigma_{\text{int}}}\right); \tag{6}$$

here $p_f = \ln\left(\frac{1}{1-\psi}\right)$ – ductile resource under uniaxial tension (ψ – reduction of the area at fracture); a, b – constants of the material of the conditions under consideration, which are determined in experiments on tension, compression, or pure shear;

b) taking into account the relationship between stress intensity and strain one by means of a generalized (general) stress-strain curve, in terms of stresses:

$$\frac{\sigma_{\text{int}}^{f}}{\bar{\sigma}_f} = \left[a\exp\left(b\frac{\sigma_0}{\sigma_{\text{int}}}\right)\right]^{m}; \tag{7}$$

σ_{int}^{f} – stress intensity at the moment of fracture; $\bar{\sigma}_f$ – ultimate tensile strength (actual stress at fracture).

Fig. 5 shows the field of the logarithmic plastic strain intensity in the boundary layer of the group defect, which corresponds to the internal pressure in the hollow of the valve under operating conditions. The maximum value of $p_{\text{int}}^{\max} = 0{,}32\,\%$, while the intensity of the logarithmic plastic strain at the moment of fracture corresponding to this stress state $p_{\text{int}}^{f} = 63.7\,\%$; the actual safety factor for the limit strain $k_p \cong 200$ – it would seem that the value is more than sufficient, but in this case the limit stress safety factor k_σ is only 1.38!

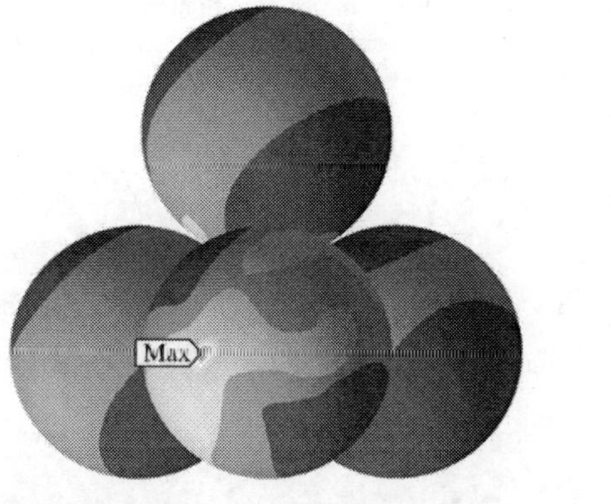

Fig. 5. Distribution of the intensity of logarithmic plastic strain in the boundary layer of the group defect under the working pressure (distance $h \approx 6$ mm; eccentricity $e = 0$; $p_{\text{int}}^{\max} = 0{,}32$ %).

Such an essentially nonlinear connection between the parameters «stress intensity» ~ «strain intensity» is determined by a slight hardening of the steel ($m = 0.06$) in the process of inelastic deformation.

It is worth noting that the given value of the safety factor k_σ is calculated by the limit stress σ_{int}^{f}. If to determine with respect to the ultimate tensile stress σ_B then this value will decrease to $k_\sigma = 1.12$. In the case of estimating by the yield point $\sigma_{0.2}$, as is customary in engineering practice, the safety factor is less than unity – $k_\sigma = 0.96$, as it was expected in the presence of stress concentration.

 DOI: 10.1051/matecconf/201712902017

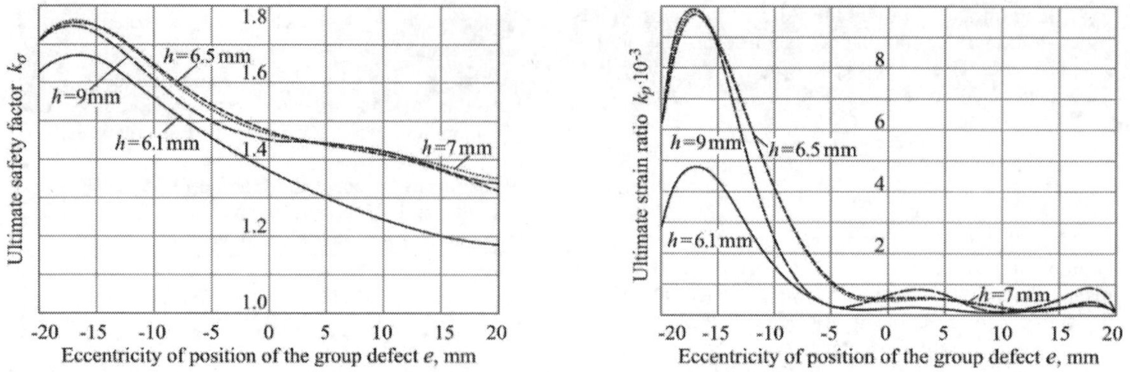

Fig. 6. Dependences of the safety factors by the limit strain and stress in connection with the distance between the discontinuities and the eccentricity of the arrangement of the group defect.

The correlation dependences of the safety factors by the limit strain and stress in connection with the distance between the discontinuities and the eccentricity of the arrangement of the group defect are illustrated by graphs (Fig. 6).

The evaluation of the durability of products under repeatedly-variable loading conditions by the criterion of fatigue fracture is made on the basis of a model of a continuous homogeneous elastoplastic medium, which makes it possible to use verified by engineering practice criterion relationships and the acceptable labor intensity methods of calculating. In this connection, the problem arises of modeling a medium with the indicated properties equivalent in its deformation characteristics to the original material with multiple randomly distributed discontinuities.

The equivalence criteria are the two conditions called by the authors «an integral stress»

$$\sigma_\Sigma = \frac{1}{V_0}\int_V \sigma_{\text{int}}\, dV \tag{8}$$

and «an integral deformability»

$$\varepsilon_\Sigma = \frac{1}{V_0}\int_V |e_i|\, dV \tag{9}$$

or

$$p_\Sigma = \frac{1}{V_0}\int_V |p_i|\, dV \tag{10}$$

– in the case of calculations in the parameters «true stress» ~ «logarithmic plastic strain»; index «i», as before, means belonging to principal stresses and strains.

Using this approach, was solved the problem of identifying the model – obtaining a stress-stain curve depending on the porosity of the material – based on the results of at least four (in the case of approximating the diagram by the Ramberg-Osgood power-low function) numerical experiments (Fig.7). Two of them are produced under the assumption of linearly elastic behavior of the material, the rest of them are elastoplastic. By the degree of porosity is meant the ratio of the volume of spherical hollows to the total volume of the substructure.

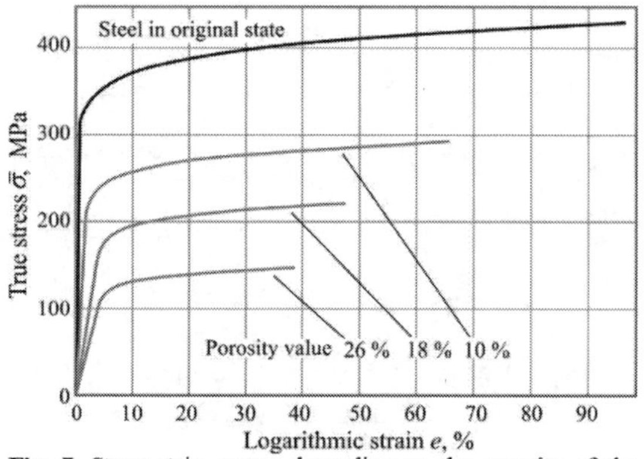

Fig. 7. Stress-stain curves depending on the porosity of the material.

In our opinion, further development of this work is in the comprehensive experimental verification of the proposed approach and improvement of criteria that establish the equivalence (in the said meaning) of a continuous model and real porous medium on this basis. In addition, it is proposed to develop a method of reckoning the stress concentration in the neighborhood of an actual defect, but absent in the model medium.

References

1. J.D. Clayton, R.H. Kraft, R.B. Leavy, Int. Journal of Solids and Structures, **49**, pp. 2686-2702 (2012)
2. A.I. Slutsker, A.B Sinani, S.S. Ordanyan, V.I. Betekhtin, A.A. Kozhushko, A.G. Kadomtsev, Phys. of the Solid State, **50**, no.8, pp. 1450-1457 (2008)
3. A.V. Ignatova, O.A. Kudryavtsev, S.B Sapozhnikov, PNRPU Mechanics Bulletin **4**, pp. 121–137 (2015)
4. R.E. Peterson, *Stress concentration factors* (John Wiley and Sons, Inc., 1974)
5. V.L. Kolmogrov, *Mechanics of metal forming (in Russion)* (Publishing house of Ural state technical University, 2001)

DOI: 10.1051/matecconf/201712902018

Carbon fiber reinforced magnesium alloy in a Ti-6Al-4V shell

Vasily Astanin[1,*], *Daniya* Nurullina[1], *Adel* Zaripova[1], *Pavel* Soloviev[1], and *Fanil* Musin[1]

[1]Ufa State Aviation Technical University, 450008 Karl Marx str. 12, Ufa, Russia

Abstract. Continuous carbon fiber reinforced magnesium alloy pieces in SMC Ti-6Al-4V shell have been fabricated using pressure infiltration. Similar temperatures (~700 °C) for superplastic formation of the shell and melting of the alloy allow this to be done in one step. The quality of infiltration of the molten alloys is found to be proportional to load. A limiting parameter in increasing the infiltration pressure is the strength of the welded bonds. Structure, fracture parameters and mechanical properties are discussed.

1 Introduction

Lightweight composite materials have attractive properties and, thus, are attractive for manufacturing of critical components and parts of aviation and space technology, power and transport devices. High specific strength and specific Young modulus give advantages, and metal matrix composites (MMC) also have good electrical and thermal conductivity.

Magnesium and aluminium alloys are the most promising ones to be applied as matrix material. According to many researchers, magnesium alloys have an advantage over aluminium, because they have a lower density [1]. Despite the fact that it is traditionally believed that magnesium alloys have low corrosion and erosion resistance, which makes it difficult to use MMC based on them, some sources show better corrosion resistance of magnesium alloys than aluminium ones [2, 3].

Low density of components and high strength of carbon fibers provide MMC with significant advantages over traditional titanium alloys [3, 4]. Magnesium alloys have one more feature - the low Young modulus, which is a drawback when using non-reinforced magnesium alloys, but being as part of MMC, magnesium alloys, therefore, have a damping ability and help to reduce the level of vibrations and noise that occur during operation of products.

Thus, MMCs based on magnesium alloys reinforced with carbon fibers are very promising for such products as aircraft engine compressor blades [1, 5].

2 Experimental

2.1 Materials and methods

The properties of magnesium impose requirements to its processing method. Both magnesium and its alloys have excellent fluidity, low melting point (~ 650 ° C), which makes them an excellent choice for casting. However, active interaction with oxygen requires protection by inert gases during processing.

The choice of a particular magnesium alloy to produce a composite is due to the following factors: its strength, cast properties and good adhesion to carbon fiber. The first two factors are fairly obvious and are determined by the doping system, and the latter should be considered in more detail. Liquid magnesium alloy itself does not wet carbon fibers either at atmospheric pressure or in low vacuum, and does not form a chemical bond with it. One can overcome this problem through use of magnesium alloy with moderate (8-10%) aluminum content. Aluminum, in contrast to magnesium, reacts actively with carbon fiber to form needle particles of Al_4C_3 carbides, which both provide fiber and matrix coupling. It is noted that higher aluminum content in the alloy is undesirable, since it leads to the growth of carbides to the size critical for impact properties and degradation of the fiber [6]. Thus, alloys of the AZ91 family and their common Russian counterpart ML5 are suitable for composite formation. AZ91magnesium alloy is used as the matrix material in the present research.

Table 1. Chemical composition of common Mg alloys [7,8]

Alloy	Al	Zn	Mn	other
AZ91D	8.3…9.7	0.35…1.0	0.15…0.5	~0.10
ML5	7.5…9.0	0.2…0.8	0.15…0.5	<0.5

When working under difficult conditions, protective edges and casings are used to protect composite materials from external impact. The protective shell can be made in the form of coating of a finished product, or it can be formed from the technological process in the production process.

The original method used in this work consists in fabrication of composite material directly in the non-removable titanium shell, which is thin and strong. This makes possible accurate reproduction of the shell

*Corresponding author: v.astanin@gmail.com

DOI: 10.1051/matecconf/201712902018

complex shape in the manufacturing process, so as not to increase substantially the product mass. These requirements are satisfied by thin sheets of Ti-6Al-4V alloy with submicrocrystalline (SMC) structure. The strength of the Ti-6Al-4V alloy in this state reaches 1350 MPa, and the superplastic deformation temperature, at which the product molding is facilitated, is reduced from 900 °C to 600...700 °C [9]. This temperature range is in good agreement with the melting point of magnesium alloy AZ91, the liquidus point of which is 650 °C.

The technology of composite products in shells assumes the production of ultrafine-grained sheets; tight welding of the shell ensuring the weld seam strength at the level of the base material strength; precise molding, which does not require further machining, except polishing; complete infiltration of the composite, which requires vacuum and sufficient pressure of the melt; compatibility of components by the criteria of equal strength, thermal expansion and bond strength at the interfaces [10, 11, 12].

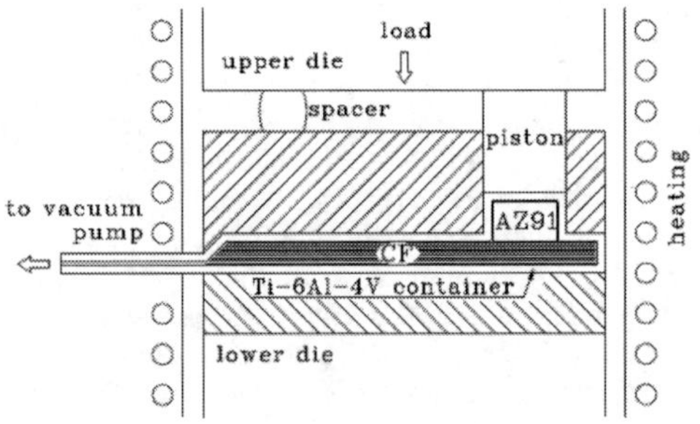

Fig. 1. Set-up of infiltration method.

The first problem is successfully solved by isothermal rolling under the conditions of low-temperature superplasticity [13]. The solutions of the others are shown in this paper, the main idea of which is to conduct all technological operations at the same temperature and in one cycle, which is possible due to the nanostructured state of the titanium alloy.

The initial rod of Ti-6Al-4V alloy was subjected to all-round forging with the decreasing temperature [13]. The produced billets were rolled on an isothermal rolling mill and then subjected to creep annealing. As a result, a structure was formed consisting of equiaxed fragments with an average size of 160 nm.

Table 2. Mechanical properties of the SMC Ti-6Al-4V alloy at elevated temperatures in the rolling direction.

T, °C	σ_{max}, MPa	δ, %
550	310	300
600	110	430
700	45	720

The produced material exhibits superplasticity in the temperature range 550-700 °C. The mechanical properties of such material at elevated temperatures are given in Table 1.

Carbon fibers UKN-5000 with no coating with a tensile strength of 3000 MPa were used.

2.2 Production of composite material in a container

Fig. 2.. Microstructure of AZ91 - UKN-5000 composite in case of insufficient (a), in case of sufficient pressure (b).

The container samples are subjected to superplastic molding to form cavities for the carbon fiber prepreg and the magnesium alloy billet. During assembly, parts of the container are hermetically sealed around the perimeter by contact welding, then by argon-arc welding. The air is evacuated through the nozzle of the container placed in the mold, and then the container is heated in an electric furnace installed on the electromechanical press. A load was applied after heating at a temperature higher than the melting point of magnesium alloy. The vacuum was maintained as low as 10^{-2} bar throughout the cycle until the load was removed.

The described method of composite material fabrication is protected by a patent [14].

2.3 Microstructure and properties of the composite material

Infiltration under pressure resulted in the titanium shell bonded with the magnesium matrix, which, in turn, penetrated the inter-fiber space. No pores and

DOI: 10.1051/matecconf/201712902018

detachments were found at the interface of the matrix-titanium alloy. The latter could appear as a result of thermal stresses during the cooling of the packet due to the difference in the thermal expansion of titanium and magnesium.

Apparently, in the process of cooling under pressure, thermal stresses can relax due to deformation of the magnesium alloy.

At low pressure magnesium melts practically do not wet carbon fibers [6, 11], which prevents successful infiltration and leads to the formation of inter-fiber voids. The fibers are assembled into bundles within which they come into contact, and the matrix remains on the periphery of the bundles (Fig. 2, a). Such a structure does not contribute to fabricated full value composite. Overcoming the forces of surface tension was achieved by selecting the temperature and pressure (Fig. 2, b).

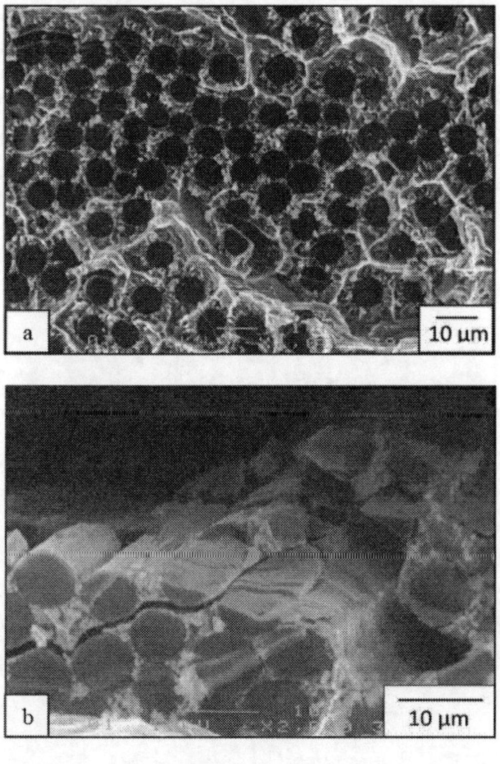

Fig. 3. Microstructure of the fracture of the composite at the sufficient infiltration pressure, combined brittle (a) and bundle (b) fracture.

It is clearly seen at a high magnification (Fig. 3, a) that the matrix completely surrounds the fibers, filling the gaps less than 1 µm. The fracture surface is homogeneous, there is no detachment of the matrix from the fibers. The destruction of the matrix itself is viscous. The magnesium alloy has a microcrystalline columnar structure around the fibers, which is indicative of the epitaxial nature of the crystallization. The latter circumstance in combination with the fracture character indicates to a strong bond between the fibers and the matrix.

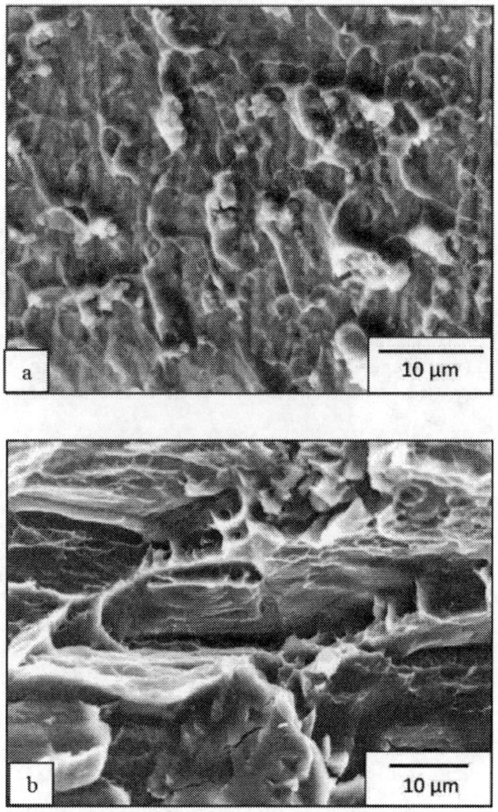

Fig. 4. The surfaces of the shell detachment from the composite matrix - the shell side (a), the magnesium alloy side (b).

The destruction nature has a complex branched relief, indicating to a high fracture energy, the fibers fail along the single surface with the matrix (Fig. 3, b). In some places there are secondary cracks that split the fiber, but the fiber does not detach from the matrix, this indicates a good strength of the matrix interaction with the fiber.

Tensile tests along the direction of fibers laying show that the destruction of the composite and the shell does not occur simultaneously. The composite is destroyed first, while the shell continues to deform plastically for some time.

The evaluation of the mechanical properties shows that at a 30% lower mass, the composite material has strength close to that of the monolithic Ti-6Al-4V alloy of about 1100 MPa.

3 Conclusions

Fiber reinforced lightweight materials with magnesium alloy matrices are promising to be used in lightweight applications under high dynamic and static operating loads. But it is necessary to overcome several obstacles to promote the use of magnesium matrix composites. Among them are manufacturing processes, the influence of reinforcing components, the influence of fiber on the microstructure and mechanical properties. The main conclusions made as a result of earlier work can be summarized as follows.

The fracture behavior of the composite material samples indicates good binding of the fiber and matrix, and at the same time there is no excessive embrittlement of the composite.

The good adhesion of the magnesium alloy to the titanium shell also contributes to the overall strength and stiffness of the product.

As a result of the work carried out, it can be concluded that the infiltration of a composite in the vacuum envelope from SMC titanium alloy by means of a press yields satisfactory results.

At the same time, economic feasibility and the most attractive specific mechanical properties can be achieved only with significant optimization of the technological process.

This method can be used to develop a technology to produce parts from magnesium-carbon composite in a protective shell.

The work was carried out using the equipment of the «Nanotech» Engineering center. http://nanotech.ugatu.ac.ru/

References

1. P. Cortes, W. Cantwell. Comp. B **37**, (2006)
2. AeroMag. (2006). Aeronautical application of wrought magnesium. A proposal for the 6th European framework program. FP6-2003-Aero-1
3. R. Supplit, Th. Koch. U. Schubert. Corros. Sci, (2007).
4. A. Novikov. Composite materials in GTE parts. Aviation industry, **11**, (1987) *in Russian*
5. A. Ivakh, A. Ryzhov, B. Galimkhanov. Bulletin of the USATU, 2, (2008) *in Russian*
6. W. Lacom, H.P. Degischer, P. Schulz. Key Eng. Mater. **679** (1997)
7. ASTM B94-13. (2013). Standard Specification for Magnesium-Alloy Die Castings
8. GOST 2856-79. Magnesium Casting Alloys. Grades. *in Russian*
9. G. Salishchev. Mechanical properties of titanium alloy VT6 with submicrocrystalline structure. Metals, **6**, (1999) *in Russian*
10. A. Metcalfe. Interfaces in Metal Matrix Composites **Vol. 1**, (1974)
11. V. Vasiliev. Composite materials. Reference book. M .: Mashinostroenie. (1990) *in Russian*
12. V. Astanin. Microstructural design of advanced materials and composites based on them. Ufa: USATU. (2000) *in Russian*
13. V. Astanin. Aviation and technological systems: Interuniversity Collection of scientific papers, (2004) *in Russian*
14. V. Astanin, V. Astanin, Patent No. 118571. RU. (2011). *in Russian*

DOI: 10.1051/matecconf/201712902019

Nanostructured thin film coatings with different strengthening effects

Yury Panfilov[1,*]

[1]MSTU named after Bauman, Electronic Technology in Machinery Chair, 105005 Moscow 2-d Baumanskay St., 5, Russia

Abstract. A number of articles on strengthening thin film coatings were analyzed and a lot of unusual strengthening effects, such as super high hardness and plasticity simultaneously, ultra low friction coefficient, high wear-resistance, curve rigidity increasing of drills with small diameter, associated with process formation of nanostructured coatings by the different thin film deposition methods were detected. Vacuum coater with RF magnetron sputtering system and ion-beam source and arc evaporator for nanostructured thin film coating manufacture are represented. Diamond Like Carbon and MoS_2 thin film coatings, Ti, Al, Nb, Cr, nitride, carbide, and carbo-nitride thin film materials are described as strengthening coatings.

Analysis of publications on functional and strengthening coatings published at special journals, e.g. "Strengthening technology and coatings", demonstrated that nanostructured thin film coatings are one of the main trend to increase of machine parts and of cutting and deformation tools durability, and decrease of mechanisms friction coefficient, as well. Choice of thin film deposition method and working has the grate influence to the strengthening effect and functional characteristics of the coatings.

The main methods of strengthening coatings manufacture are PVD (Physical Vapor Deposition): ion-plasma sputtering and ion-beam deposition and arc evaporation with the reactive gas inlet into vacuum chamber. Vacuum coater with three types of sources – RF magnetron sputtering system, ion-beam source, arc evaporator – was elaborated for nanostructured thin film coating manufacture (Fig. 1).

The main of the technological and design versions are followings: substrate surface activation before thin film deposition, ion-beam assisted deposition, thin film deposition under angle between film-forming particle flow and substrate surface, multi-component and multilayer thin film deposition, as well.

Wear-resistant and antifriction strengthening coatings are manufactured by means of PVD methods. Thin film coatings structure and the main characteristics of the coatings are determined by thin film deposition process.

Gas pressure and substrate temperature and thin film rate of growth are the main parameters of thin film deposition process.

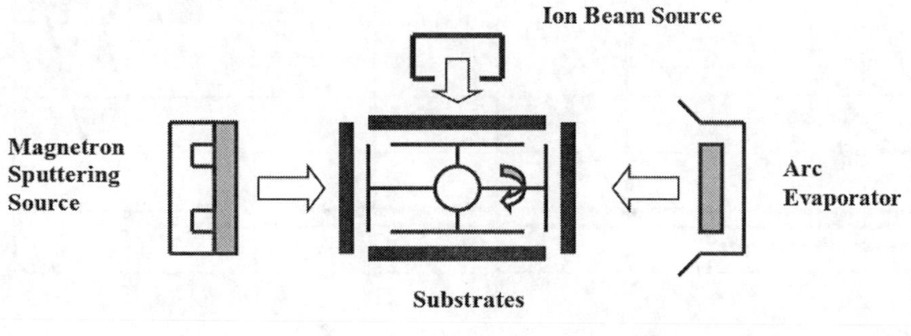

Fig. 1. Thin film deposition methods: magnetron sputtering, arc deposition, ion beam deposition and assistance.

An important application of thin film deposition in machine building is the strengthening of cutting and deformation tools by applying wear-resistant coatings. Using a composition with an optimum number of layers in the correct configuration ensures a high coating efficiency. The mono- or multilayer coatings are composed of nitride, carbide, and carbo-nitride compounds based on complex systems of high-melting-point metals (e.g., Ti, Nb, and Cr).

* Corresponding author: panfilov@bmstu.ru

DOI: 10.1051/matecconf/201712902019

Thin film material and process variety (molecules flow and energy, substrate temperature) are the main advantages of PVD methods of a cutting and deformation tools wear resistance increasing. The compositions of titanium with silicon, boron and other materials provide an opportunity to nanostructure thin film wear-resistant coating manufacturing. Ion-plasma magnetron sputtering of multicomponent targets produced by means of the Self-dispersion High-temperature Synthesis (SHS) [1] is the method of this coating preparation.

High hardness (up to 70 GPa) and adhesion and low friction coefficient of TiBSiN, TiBCrN, TiAlCN and other coatings deposited by means of the SHS targets sputtering are thanks to thin film grains minimum dimensions of 1 – 4 nm [2].

Cutting tool wear resistance tests were provided by means of stainless steel samples drilling: 4 mm drills without wear-resistant coating and with thin film strengthening coatings, such as TiN and TiBSiN and TiBCrN, were used in the test. The operating gas pressure of 1 – 2 Pa, magnetron target voltage of 500 V, ion current density of 800 A/m^2, reactive gas (nitrogen) concentration of 10 – 22 % were the main thin film deposition parameters. Thickness of the coating was 400 – 700 nm. The coatings were deposited via vacuum coater shown at Fig. 1.

Nanostructured strengthening coatings permit to increase of cutting tool durability and to increase of cutting process, as well. Cutting process increasing permits to decrease of pay for time of an expensive equipment. Dependence holes drilling time t on the drilled holes number n when an equal loading to drill is shown at Fig. 2.

The TiBSiN thin film coating increased durability of the drills by about 3 – 4 times in comparison with drills with TiN wear-resistant coating when stainless steel drilling. It was mainly due to decreasing of friction coefficient – a difference between chips (Fig. 3) was the result of the difference between the friction coefficients of the other wear-resistant coatings. The difference between chips was unexpectedly effect because chip of the drill with TiBCrN coating has small number of curves – the drill "went in steel as in butter".

Nanostructured thin film deposition by means of separate plasma of arc charging with carbon target provides of Diamond Like Carbon (DLC) coating manufacture with the next characteristics: minimum surface defects, high hardness (38 – 42 GPa) and adhesion, friction coefficient less than 0.2 at all friction way. DLC coating friction characteristics were tested at special experimental equipment.

DLC coating hardness and wear resistance were much bigger than indenter material (tempering ball) hardness and wear-resistance. So, there was wear of indenter, but not coating!

Unexpected strengthening effect was discovered after DLC coating deposition on drills with diameter of 50 – 300 μm for print circuit drilling – DLC thin film provides both of wear resistance increasing and curve rigidity increasing of drills [3]. Curve rigidity increasing of drills permits to increase preciseness of hall centers location after print circuit drilling.

Experimental research shown that curve rigidity of drills 0.105 mm diameter with DLC coating became up to 20% tougher in comparison with the drills without coating. Small diameter drill has very small rigidity so relational input strengthening coating to rigidity increasing may be very considerable (Fig. 4).

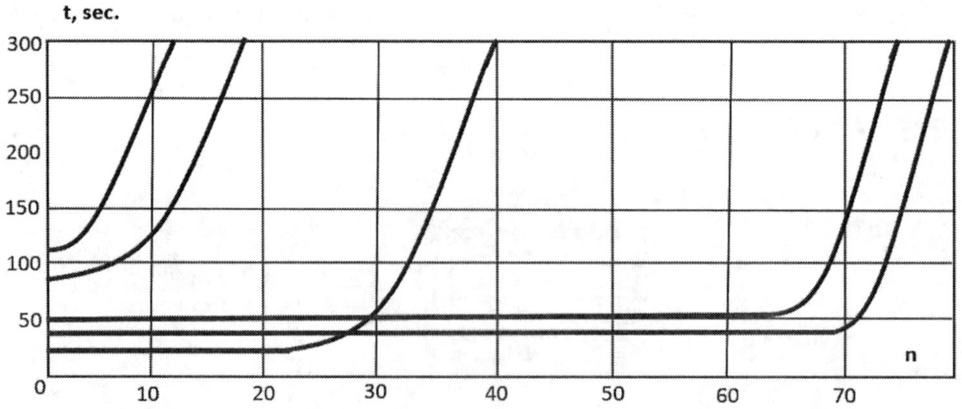

Fig. 2. Dependence holes drilling time t on the drilled holes number n when an equal loading to drill: 1 – without coating and with coatings: 2 – TiN, 3 – TiBCrN, 4 – TiBN+20%Si, 5 –TiBSiN.

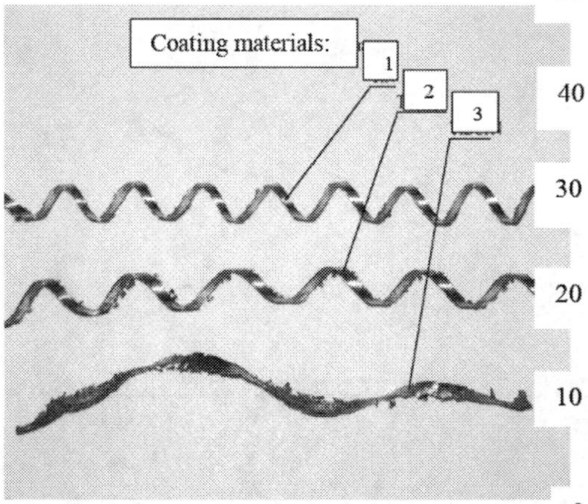

Fig. 3. Photo of drilling chips in dependence of coating materials: 1 – TiN, 2 – TiBSiN, 3 – TiBCrN.

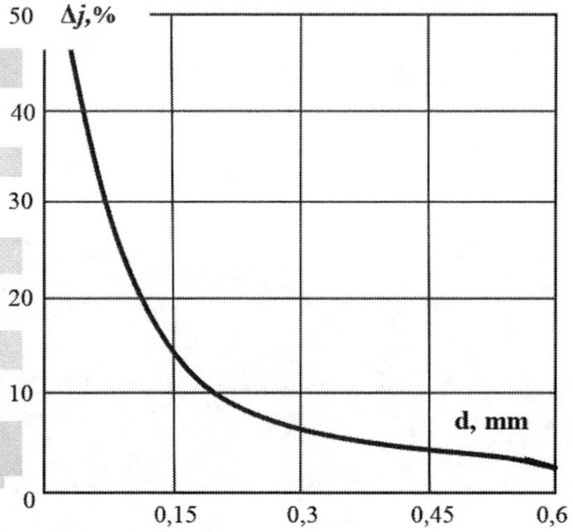

Fig. 4. Relational increment of curve rigidity Δj of different diameter *d* drills with DLC coating.

Experimental test results of print curcuit drilling by 0.3 mm drills with DLC coating demonstrated that root-mean-quare error of drilled halls centers was decreased on 28.9% or 6.5 μm in comparison with drilling by an initial drills (Fig. 5).

DLC strengthening coatings can be deposited by PVD and CVD (Chemical Vapor Deposition) methods. By means of vacuum coater (see Fig. 1), PVD methods only that is magnetron sputtering and arc evaporation of carbon targets were realized.

The best strengthening characteristics were obtained via arc evaporation with plasma flow separation because carbon ions only with high energy run into the substrate. So, the optimum correlation between deposited molecules mass and energy took place.

Ultra low friction effect was obtained by using antifriction thin film coating of MoS_2D_x compound (D is dope element) at mormal conditions. Magnetron sputtering of two targets of MoS_2 and Ga was used for manufacturing of this coating with sliding coefficient of 0.001 – 0.0001 [4, 5].

Model which explained of essence of antifriction action of MoS_2D_x coating with ultra low friction effect in normal condition at air was elaborated – dope element decreased of sliding friction between layers of MoS_2. Dependance sliding coefficient of MoS_2D_x coating deposited by RF magnetron sputtering on testing time is shown at Fig. 6. The tests were carried out by disk-sphere scheme at air on normal conditions.

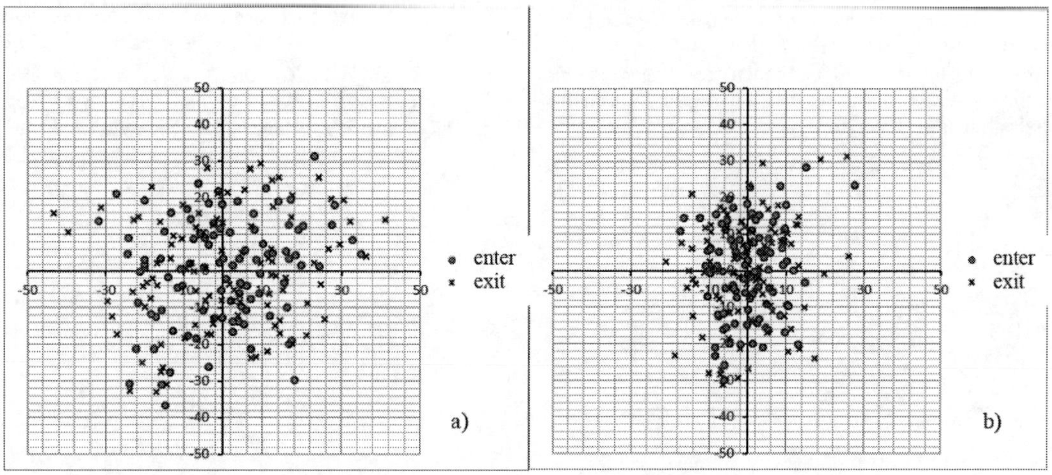

Fig. 5. Deviation of drilled halls centers (μm) from nominal location for 0.3 mm drill diameter: a) drills without coating, b) drills with DLC coating.

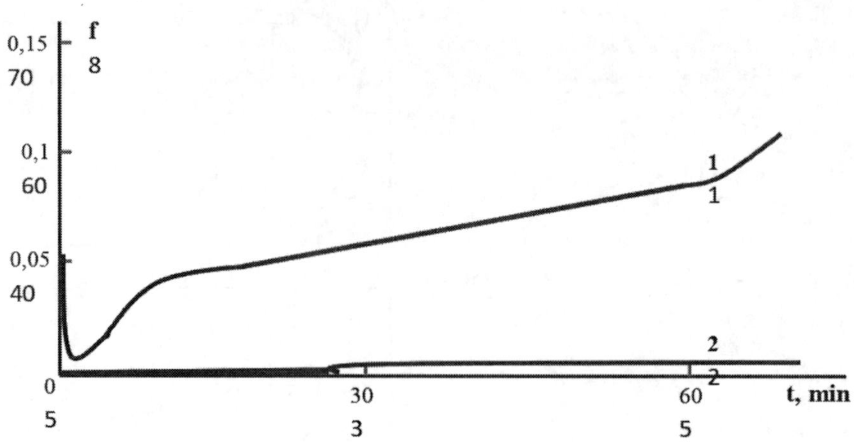

Fig. 6. Dependance of sliding coefficient *f* on testing time *t* of two coatings: 1 – MoS_2, 2 – MoS_2D_x.

Thin film coatings with ultra low friction coefficient can be deposited by means of magnetron sputtering method at vacuum coater represented at Fig.1. The antifriction thin film coating of MoS_2D_x compound was deposited via two magnetron sources. Besides this, such coating can be manufactured by magnetron source with combined target which consists of two materials: MoS_2 and dope element D.

Conclusion

Method and process of thin film deposition determine coating material energy and mass flow, thin film coating structure and functional chracteristics, as well. Besides this, unusial strengthening effects, such as super high hardness and plasticity simultaneously, ultra low friction coefficient, high wear-resistance, curve rigidity increasing of drills with small diameter, as well, are depended from coating material energy and mass transfer from source to substrate.

Vacuum technological equipment for strengthening thin film coatings deposition was developed. The coater includes vacuum dry pumps, ion-beam and ion-plasma sources of substrate surface cleaning and activation before thin film deposition, DC and RF magnetron sputtering systems, arc evaporator without drop phase and with plasma flow separation.

References

1. Y. Karabasov and others, New materials (MISIS, 2002)
2. E. Levashov, D. Shtansky, Y. Panfilov, *Multi-component nanostructure thin film wear-resistant coatings,* 7th International Conference on Nanostructured Materials **35**, (2004)
3. Y. Litvak, Y. Panfilov, *Strengthening coatings application for increasing of micro-size drilling precision of print circuit,* Strengthening technology and coatings **1**, 5 (2016)
4. M. Nozhenkov, *About weak interactions of particles in solid state materials,* Nanoengineering **3**, 7 (2012)
5. M. Nozhenkov, *Cristal structure and properties of the coatings made by vacuum ion-plasma methods,* Strengthening technology and coatings **9**, 5 (2012)

Structural investigation of 70Cu/30Fe based cast alloy obtained by combined use of centrifugal casting-SHS process and forging

Vitalii V. Sanin[1,*], *Michael R.* Filonov[1], *Vladimir I.* Yukhvid[2], and *Yurii A.* Anikin[1]

[1]National University of Science and Technology MISiS, Leninsky pr. 4, Moscow, 119049 Russia
[2]Institute of Structural Macrokinetics and Materials Science, Russian Academy of Sciences, Chernogolovka, Moscow, 142432 Russia

Abstract. This research is to test a combined chemical-technological chain of material production. The tests are made on a 70Cu/30Fe system. The technology consists of: 1- SHS of a cast material of a specific composition and 2- Mechanical and heat treatment, including: continuous swaging, where the workpiece is heated to 850^0 C and then drawn cold. In this study first used SHS-metallurgy method for producing, in the molten pseudoalloy Cu-Fe system (Cu- 70%, Fe- 30 wt.%). It was found that the high temperatures melt the SHS provides increased solubility of Cu in Fe. Then, in the crystallization, by decreasing the melt temperature and reduce the solubility limit of copper is released in the form of small dispersed particles. The observed structure is typical only for SHS alloy.

1 Introduction

There exists a well-known broad class of binary and multi-component metallic materials with limited solubility when liquid or solid. In some cases, such materials have unique characteristics.

Alloys with limited solubility (LS) or pseudoalloys (PA) have long been impossible to obtain with traditional methods of metallurgy. When trying to produce such alloys, metallurgists face a number of challenges, which are mostly down to a considerable difference in specific masses and melting points as well as a strong tendency to stratification in the liquid or in the solid state in a wide range of temperatures and concentrations.

According to the literature, Cu-Fe alloys are becoming of even greater interest [5,7,8,9]. Alloys of the Cu-Fe system with a Cu-based matrix (i.e. an increased concentration of Cu) are promising for creating high performance hard magnetic materials. [5,7] The system has an interesting structural peculiarity which is that its component phases actually do not interact, which enables selective effects on such structural components to cause targeted changes in the properties of the material. After necessary dispersion [5] of ferromagnetic Fe particles is achieved, and their form becomes isotropic, one can expect high magnetic energy values as it is known [5,7] that extended ferromagnetic particles of critical size do not prevent magnetic reversal in a non-ferromagnetic matrix.

This is why developing low-cost and simple technology to produce such alloys and materials based thereon and to adjust their physico-mechanical properties is still a relevant problem when it comes to the production of such materials.

This research is to test a combined chemical-technological chain of material production; the tests are made on a 70Cu/30Fe system. The technology consists of:

(I)- SHS of a cast material of a specific composition,

(II)- Mechanical and heat treatment, including: continuous swaging, where the workpiece is heated to 850^{0}C and then drawn cold.

2 Research Methodology

2.1 Synthesis of a cast alloy by means of SHS

To synthesize the alloy, we used SHS [3,4,10]. This method is about using chemical energy produced in the reaction of highly exothermic thermite alloys (combustion), making this a very energy-efficient method for production of cast materials. Synthesis does not take long (dozens of seconds), and the top surface of ingots is protected from oxidation with an oxide melt (Al_2O_3), which enables synthesis in atmospheric conditions.

To synthesize a cast Cu-Fe alloy with 70% Cu and 30% Fe, we used powder Cu and Fe oxides as well as Al for reduction. Source powder particles did not exceed 100 µm in size. Exothermic compositions were prepared as follows: components were dosed, then mixed in a planetary mixer for 15 minutes, then charge was put into graphite 40- or 80-mm forms.

The general chemical synthesis reaction can be written as follows:

$$Cu_2O + Fe_2O_3 + Al \rightarrow [Cu\text{-}Fe] + Al_2O_3 + Q \quad (1)$$

* Corresponding author: sanin@misis.ru

 DOI: 10.1051/matecconf/201712902020

where the weight of components was adjusted based on thermodynamic calculations and actual analysis of alloy composition.

To intensify the gravitational separation of the alloy and aluminum oxide as well as to induce convective mixing of the melt components, the synthesis was done with a centrifugal SHS machine [5] with 2 to 50a/g overloads, where a is centrifugal acceleration, g is gravitational acceleration.

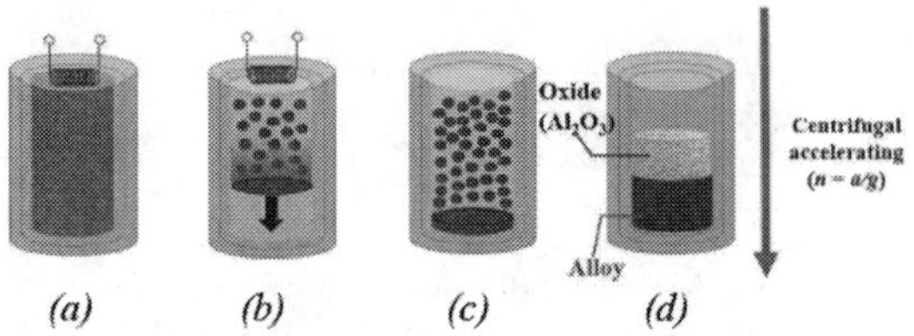

Fig. 1. (a) preparing the mixture, (b) SHS combustion, (c) gravitational separation of phases, (d) crystallization and cooling of the melt.

Figure 1 presents the main stages of the SHS method, i.e. component dosage, mix preparation, and putting the powder mixture into graphite forms. Then the form is placed on the centrifugal rotor; upon the necessary RPM is reached, the mixture is ignited, and combustion synthesis begins. The combustion temperature exceeds the melting point of both source and final reaction components, which enables one to get cast products.

Due to the mutual insolubility and significant difference in specific weights, gravity causes reciprocal separation of the metallic (target) and the oxide (Al_2O_3) phase. As a result, an ingot of the alloy is formed at the bottom of the form, whereas oxide solution is generated in its upper part. High gravity generated in the centrifugal machine inhibits the dispersion of combustion products during the synthesis, intensifies the separation of metallic (alloy) phase and the oxide (corundum) phase, and homogenizes the composition of the alloy, which is important for producing the PA and form a highly dispersive structure of the final product.

2.2 Continuous swaging (forging)

Continuous swaging was done using a C41-250 pneumatic hammer. This model has the following characteristics: falling parts weigh 250 kg, maximum impact energy equals 5.6 kJ, 140 hits per minute-1; the drive is Y180L-4-B3. Continuous swaging consists in consecutively hitting and moving the workpiece, whereby only a part of the workpiece is located between the dies as it sustains the impact. After each compression, the workpiece is moved by a distance that less than the length of the die. To make forgings quickly, we used rolling-impression dies, as they prevent axial fractures thanks to no intensive lateral flow of the metal.

2.3 Drawing

Drawing was done using an MB1 drawing machine. The machine consists of the main drawing unit, a bath, a wire spool, a wire-pointing machine, and a mechanism to handle finished products. When passing the drawing hole, rods become thinner and longer while unchanged in volume. Drawing was done using a multiple-drawing machine at a rate of 4.5 m/s with a die step of 0.2 mm.

2.4 Microstructure Research

The microstructure of the SHS-made cast alloy was studied using an Ultra-55 based Zeiss Ultra plus auto-emission ultra-high resolution scanning electron microscope. This microscope is able to perform point analysis, analysis along an arbitrary line, analysis over an arbitrary area, analysis of an array of points; it is also capable of mapping the distribution of elements over the surface of the sample as well as to identify phases. Magnification is 12 to 1,000,000. Acceleration voltage is 0.02V to 30kV. Proble current is 4 pA to 20 nA.

The chemical composition of the melt was tested using an OBLF emission spectrometer with CCD detectors.

This structure of the SHS alloy was studied by transmission electron microscopy (TEM) using a JEM-2100 machine made by JEOL, Japan.

3 Experiment Results and Discussion

We first conducted a series of experiments to SHS-synthesize a 70Cu/30fe alloy and optimized the synthesis condition; we researched how overloads could affect the parameters of the process. The first stage is about using chemical energy produced in the reaction of highly exothermic thermite alloys (combustion), making this a very energy-efficient method for production of cast materials. Synthesis does not take long (dozens of seconds), and the top surface of ingots is protected from oxidation with an oxide melt (Al_2O_3), which enables synthesis in atmospheric conditions. To ensure the uniform distribution of iron particles in the copper matrix, experiments were done using a centrifugal SHS machine [10] where centrifugal forces cause overloads within a range of 2 to 50a/g, where a is the centrifugal acceleration, g is the gravitational acceleration. The main process parameters as a function of a/g are presented in Figure 2. Experimental data revealed a considerable increase in the combustion rate from 4 to 9 cm/s depending on the SHS a/g value. The most intensive increase in rate (U) (more than 20 times) was observed within overloads of 2 to 20 a/g.

Microanalysis of the 70Cu/30Fe SHS alloy (Figure 3a) and the distribution of elements in structural components (Figure 3) confirmed the presence of dispersive precipitations varying in level, both in the Cu matrix and in the dispersed Fe particles.

DOI: 10.1051/matecconf/201712902020

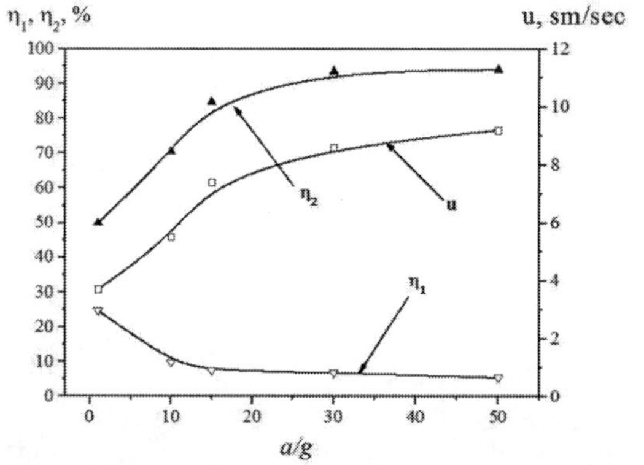

Fig.2. Effect of overloads (a/g) on the combustion rate (u), mixture dispersion (η_1.), and the extent to which the metal phase transformed into an ingot (η_2.)

Note that microanalysis of local alloy areas always identified the presence of both components in the area analyzed. Given the low mutual solubility of the alloy components, this result is very interesting.

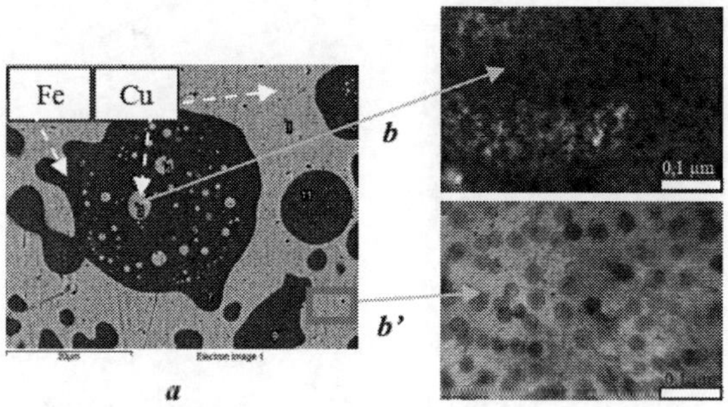

Fig. 3. EDS microanalysis of 70Cu/30Fe SHS alloy (a); microphotograph of the SHS alloy structure.

SHS alloy microstructure was studied at a greater magnification (Figure 3b) to find out that the SHS-made alloy has a multi-level hierarchial structure. Analysis of these structures led to a conclusion that a tri-level structure had been formed. The first level (Figure 3a) is characterized by uniform distribution of 10-30 µm Fe particles across the ingot. The second elvel is formed by uniformly distributed 2–0.5 µm Cu particles (Figure 3a) in the precipitated Fe particles of the 1st structural level. Greater-magnification TEM revealed a third level formed by uniformly distributed sub-micron 20–50 nm Fe particles (Figure 3b) in the copper particles of the 2nd level. We have to note that this multi-level microstructure was observed not only in the Fe particles; the first-level copper matrix contained dispersive precipitations of Fe particles (Figure 3b'). This is perhaps tied to the specifics of the SHS method. SHS mixture combustion temperature exceeds 2500 K. This ensures increased solubility of Cu in Fe. At the crystallization stage, as the temperature of the melt and the solubility limits are lowered, copper is precipitated in the form of small dispersed particles.

The next technological stage is mechanical and heat treatment to achieve necessary dispersion and ensure isotropy of ferromagnetic Fe particles in the Cu matrix. SHS is first heated in a muffle furnace to 850^0; then it is continuously swaged. At this stage, SHS workpieces are transformed into 10-mm rods, while the quality of the final material is improved. For the microstructure of a longitudinal cut of our sample, see Figure 4b. Then we wanted to retrieve longitudinal structural components from the 8-mm cylindrical rod cleansed of scales; for that purpose, we would draw it. The long workpiece was subjected to plastic deformation by drawing it through a hole in the die; the diameter of the hole was less than the cross-section of the workpiece. Thus we obtained samples that were reduced from 8 to 4.5 mm (deformation degree ε=67%, see Figure 4c) and from 8 to 3 mm (ε=86%, see Figure 4). Figure 4d presents the microstructure of the samples

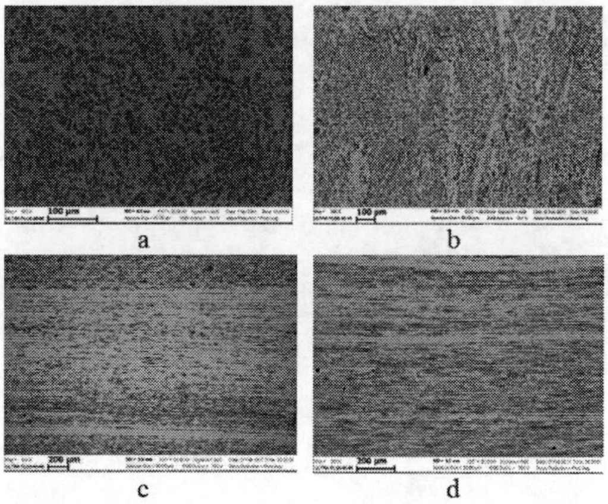

Fig.4. Microstructure of SHS samples (a) after continuous swaging (b) and being drawn to ⌀4.5 mm (c) or ⌀4.5 mm (d).

4 Conclusion

The paper is the first to present, test, and discuss a combined method to make limitedly-soluble alloys. With evidence from a 70Cu/30Fe alloy, we demonstrate that this SHS alloy features multi-level distribution of mutually unmixable components, whereby a pseudoalloy with uniformly distributed structural components is produced.

Based on the microstructure and chemical analysis of alloys made by SHS and mechanical and heat treatment, one can say that combustion-produced alloys have a homogeneous structure with uniform distribution of all structural components in the sample, which is of great practical interest for making isotropic and anisotropic hard-magnetic materials with great magnetic energy.

SHS-produced Cu–Fe alloys can be readily used to regulate their structure and to fabricate bulk materials with longitudinally oriented structural constituents.

Data analysis led to conclusion that these materials and combined SHS could be promising for making and generating dimensional nano-structure materials. In the future, studying these alloys to learn how to control their structure and properties could be of interest.

References

1. B.G. Livshic, Ya. L. Lineckiy, V.S. Kraposhin. Ph. *Properties of metals* (Moscow: Metallurgiya, 1980)
2. Yu.S. Avraamov, A.D. Shlyapin, *Alloys with limited solubility in the liquid phase* (Moscow: Intercontact Nauka, 1980)
3. V.V.Sanin, M.R. Filonov, V.I. Yukhvid, Yu.A. Aninkin., D.M. Ikornikov, Rus. J. Non-Fer Met. (to be published)
4. Merzanov A.G. *Combustion processes and synthesis of materials* (Chernogolovka: Isman, 1998)
5. G. Dublon, F. Habbal, J.L. Bell, J. Appl. Phys., **53**, 11 (1982)
6. V.N. Sanin, D.M. Ikornikov, D.E. Andreev, V.I. Yukhvid, J. Non-Fer Met. **3** (2013)
7. W. Yufei, G. Haiyan, H.Yanfeng, D. Yongbing, W. Jun, S.Baode, J.Alloys and Compd. **69**, 1 (2017)
8. L. Sheng, L.Feng, Y.Wei, Transactions of Non-Fer. Met. society of China, **27**, 1 (2017)
9. M. Ghannami, C. Gómez-Polo, G. Rivero, A. Hernando, Europh. Let., **26**, 9 (1994)
10. V.N. Sanin, V.I. Yukhvid, A.G.Merzhanov, Int. J. Self-Prop. High-Temp. Synth, **11**, 1 (2002)

Multi-cycle rolled aluminum alloy 3103 sandwiches: mechanical properties and stamp ability

Ekaterina Nosova[1,*], *Yaroslav* Erisov[2], and *Fedor* Grechnikov[2]

[1]Samara University, Metal Technology and Aviation Material Science Department, 443086, Moskovskoye shosse, 34, Samara, Russia
[2]Samara University, Metal Pressure Working, 443086, Moskovskoye shosse, 34, Samara, Russia

Abstract. Constructional part producing by sheet stamping of multilayer composites requires the stamping ability data. The aim of a work is to estimate mechanical properties, stamping ratio and anisotropy indexes of 2, 4, 8 and 12 layer sandwiches produced from aluminium alloy AA3103. The pieces were received by the cold rolling. Interoperation annealing was at 500°C for 1 hour. Charts of tensile strength, yield stress, elongation depending on layer thickness were composed. It was found that cold strain hardening does not disappear after annealing if the foil's thickness become 0.4 mm and less. Microstructure analysis has shown a good contact between layers for all samples and thicker outer layers.

1 Introduction

Multilayer rolling is one of popular processes for the fine structured materials producing including nano scale materials [1]. Initial alloys have to be very ductile to allow a high reducing ratio for a diffusion welding of separate layers [2, 3, 4]. For this reason joining of sheets is performed in hot state or additionally assisted by electricity, impulse magnetic etc. [5, 6]. But heating can lead to recrystallization and disappeared borders between single foils [7].

Absent borders may decrease fatigue strength of a final construction part [8]. Aluminium alloys have a high ductility in cold and heated condition but low strength. Aluminium alloy AA3103 is used for welded tanks, petroleum and oil transport tubes, radiators and another constructional parts and links. Using a multilayer structure these parts may have a higher fatigue especially during cyclic load in comparison to compact material [9].

Presence of a strength oxide film on the sheet surface defenses alloys from corrosion and can be used as reinforcement bodies in aluminium matrix in the layer composites at cold rolling joining [10]. The thinner layers of produced composite lead to more bulk part of the oxide phase therefore the strength is higher and ductility is less. This composition of properties makes a difficulty at sheet stamping of layer composite sandwiches [11, 12]. However, at present there are ways of stamp ability increasing due to rational texture set in aluminium alloys, which are published by authors of this work. The aim of present work is to produce of multilayer sheets from aluminium alloy AA3103 and to research their mechanical properties and stamp ability.

2 Experimental procedure

The commercial aluminium alloy AA3103 sheet with 1 mm thickness was used as initial blanks for producing multilayer specimens. At a first step pieces were cut to 250×70×1 mm3 and annealed at 500°C for 1 hour in muffle furnace. Then one plain side of the samples was scratched for roughness with a metallic brush at the angle ≈45° to a rolling axis and alcoholic treated for a fat removing. Pare of the samples was gathered with a faced scratched side inside and front edge of sandwich was joined to prevent a displacement by adhesive making a line dots with a distance nearly 10 mm between them. After that, sandwich was rolled with a thickness reducing degree 55-60%. Further samples were annealed at 500°C for 1 hour for softening. Procedure of metallic brush scratching, joining and cold rolling was repeated. Thereby the 2, 4, 8 and 12 layer samples were made. A thickness of sandwich and single layer after each pass of cold rolling are presented at table 1.

Table 1. Thickness of sandwiches and single layer after cold rolling

Number of a pass	Initial thickness, mm	Final common thickness, mm	Number of layers	Expected average single layer thickness, mm
1	2	0.8…0.9	2	0.4…0.45
2	1.6….1.8	0.72…0.96	4	0.24…0.36
3	1.92….2.88	0.86…1.15	8	0.11…0.14
4	2.88…4.32	1.29…1.73	12	0.11…0.14

After rolling and annealing samples for uniaxial tensile test were cut from the panels in along and across direction to the rolling axis. Data of tensile strength, yield stress, elongation were received and charts of their

* Corresponding author: eanosova@mail.ru

 DOI: 10.1051/matecconf/201712902021

dependence from layer thickness were made. Also, a stamping ratio $n=\sigma_{0.2}/\sigma_b$ and anisotropy indexes (%) were calculated by formulae:

$$A_{\sigma b} = 100 * \frac{\sigma_{b0} - \sigma_{b90}}{0.5(\sigma_{b0} - \sigma_{b90})} \quad (1)$$

$$A_{\sigma 0.2} = 100 * \frac{\sigma_{0.20} - \sigma_{0.290}}{0.5(\sigma_{0.20} - \sigma_{0.290})} \quad (2)$$

$$A_{\delta} = 100 * \frac{\delta_0 - \delta_{90}}{0.5(\delta_0 - \delta_{90})} \quad (3),$$

here σ_{b0}, σ_{b90} – tensile strength (MPa) of along (0) and across (90) samples, $\sigma_{0.20}$, $\sigma_{0.290}$- yield stress (MPa) of along and across samples, δ_0, δ_{90}- elongation (%) of along and across samples.

Microstructural analysis was provided to illustrate contact state between layers and to evaluate the thickness of layers. For this purpose a polished sample was cut along the rolling axis. Etching was during 2 minutes at 20°C in solute consisted of HNO_3 – 180 ml, HF – 30 ml, H_2O – 180 ml, HCI – 180 ml. After etching sample was washed under water flow and dried with a soft paper. Research of microstructure was made by optical microscopy using AXIOVERT microscope.

3 Results

3.1 Mechanical properties

Dependence of tensile strength, yield stress and elongation from a single layer thickness are presented in fig.1-3.

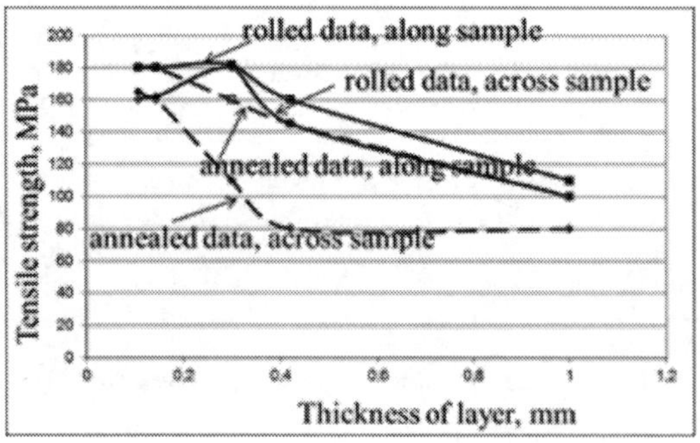

Fig. 1. Charts of the tensile strength depending on the layer' thickness

From fig.1, the along samples' tensile strength grows with the thickness reduce at the cold rolled state and achieves the highest value at the thickness of nearly 0.3 mm. Further reducing of a thickness does not lead to increasing of the tensile strength. Annealing of along samples results to a tensile strength decrease on nearly 10-12 percent for all thickness values, but at thickness of 0.12-0.14 mm there is no a difference between tensile strength data of annealed and cold rolled state. Across samples do not show the tensile strength growth until thickness reduces from 1 to 0.4 mm. Then tensile strength significantly goes up at both (rolled and annealed) state and takes the greatest values, equal to along samples at cold rolled state, at the thickness of single layer 0.3 mm. When thickness achieves 0.12-0.14 mm the across samples' thickness have the uniform values and annealing does not lead to decreasing of the tensile strength.

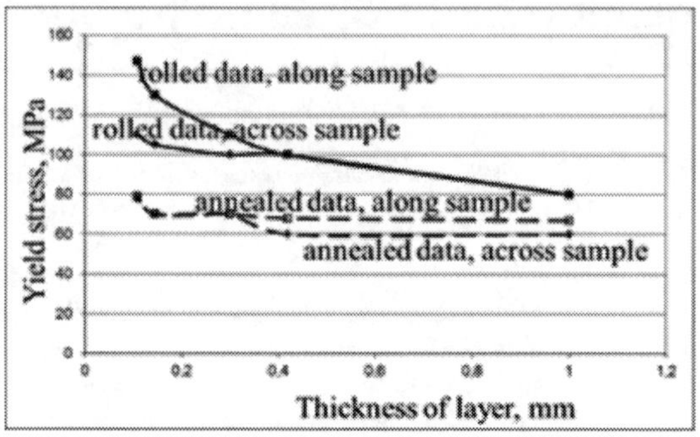

Fig. 2. Charts of the yield stress depending on the layer' thickness

Fig.2 shows yield stress grows if the layer thickness goes down. If the thickness of single layer at cold rolled state changes from 1 to 0.4 mm values of across and along samples at rolled state are equal. Then the thickness becomes less, the difference between values of yield stress becomes more obviously. Difference between yield stress values for along and across samples at annealed state is not very large as for cold rolled state, but difference between cold rolled and annealed state becomes more significantly for along samples with the single layer thickness reduces.

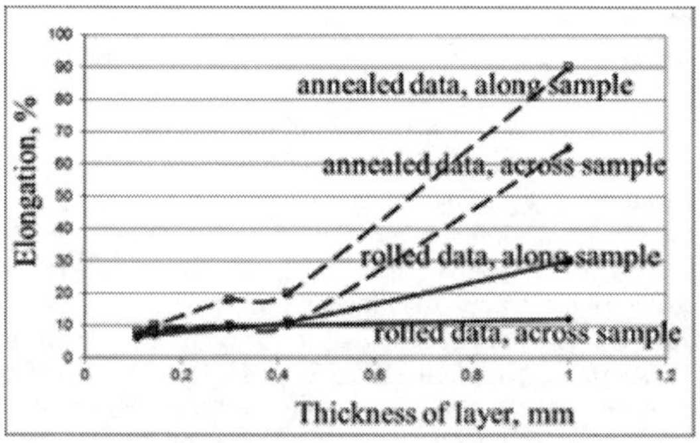

Fig. 3. Charts of the elongation depending on the layer' thickness

Elongation of samples reduces with the single layer thickness goes down (fig.3): the less single layer thickness gives the less elongation. Therefore ductility reduces while foil becomes thinner. The along samples seem to be the most ductile at annealed state. The rest type of samples shows elongation not higher than 10% if layer' thickness is 0.12-0.4 mm.

DOI: 10.1051/matecconf/201712902021

3.2 Stamping ratio

Stamping ratio $n=\sigma_{0.2}/\sigma_b$ of along samples at rolled state decreases if the layer' thickness become thinner from 1 to 0.3 mm, but for thinner layers (0.12…0.3 mm) this indexes rises (fig.4).

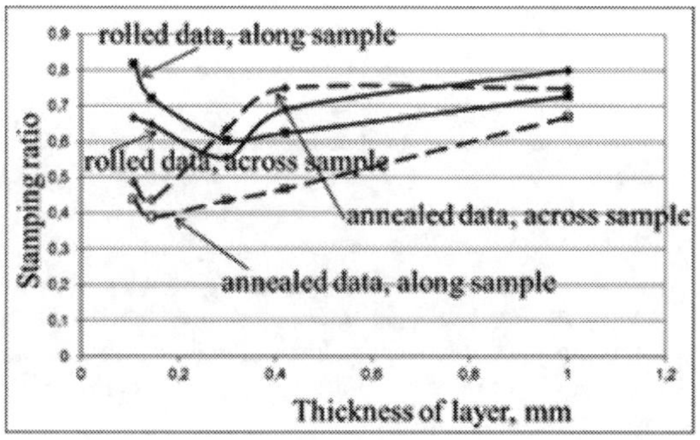

Fig. 4. Charts of the stamping ratio depending on the layer' thickness

For across samples the stamping ratio goes slightly up if the layer thickness moves from 1 to 0.4 mm, then stamping ratio drops until the thickness achieves nearly 0.3 mm and after that goes up again. Therefore the best stamp ability of samples becomes at the layer' thickness nearly 0.3 mm at the cold rolled state. Annealed along samples show continuously going down of stamping ratio with layer' thickness reduces. Across samples does not change the stamping ratio in the layer' thickness interval of 0.4…1.0 mm. The less than 0.4 mm down to 0.12 mm layer' thickness leads to the less stamping ratio. Therefore the thinner foils give the better stamp ability.

3.3 Anisotropy indexes

Anisotropy indexes (fig. 5) shows that tensile strength difference between along and across samples at rolled state keeps continuous values if layer thickness reduces from 1 to 0.4 mm.

In the interval of layer thickness there is a gap of anisotropy and at layer thickness 0.12-0.14 mm values of the strength anisotropy come back to values 10-11% as for thickness 0.4-1.0 mm.

For annealed state the strength anisotropy index rises with layer' thickness reduces from 1 to 0.4 mm. The thinner foils have the less anisotropic tensile strength. A behaviour of the yield stress at cold rolled state keeps low constant values for foil's thickness of 0.4-1.0 mm. Then yield stress anisotropy rises with foil's thickness reduces.

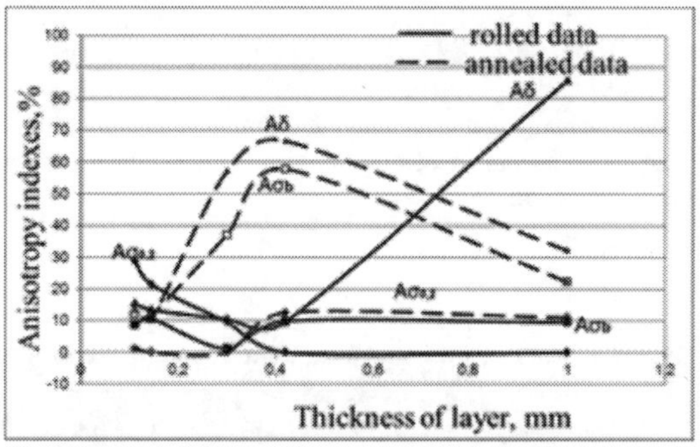

Fig. 5. Charts of the anisotropy indexes depending on the layer' thickness

At annealed state the foiled sandwiches are more anisotropic but stay continuous at the same thickness 0.4-1.0 mm. Thinner foils become less anisotropic. The elongation at cold rolled state decreases the anisotropy until thickness achieves 0.4 mm from 1.0 mm and after that slightly goes up for the thinner foils. At annealed state elongation shows increasing of anisotropy with reducing of foil's thickness until 0.4 mm. Thinner foils trends to have more isotropic elongation. Therefore the foil's thickness seems to be a critical value for mechanical properties' anisotropy.

3.4 Microstructure of sandwiches

Fig.6 shows a micro structure of two-layer (a), four-layer (b) and twelve-layer (c) samples.

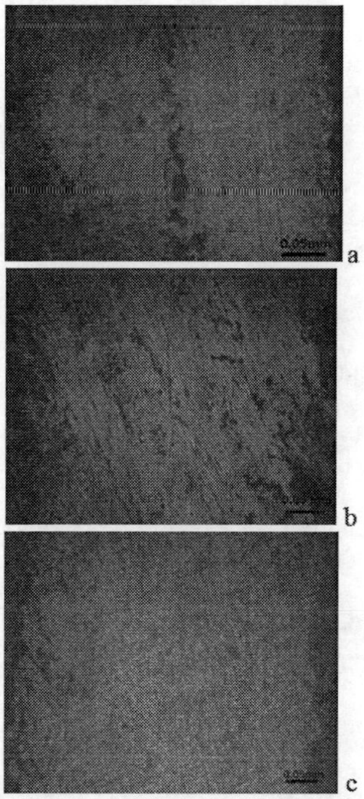

Fig. 6. Microstructure of sandwiches: (a) two-layer, (b) four-layer, (c) twelve-layer

It can be seen that there is some delamination in two layer sandwich and there is no delamination of layers in four and twelve layer sandwiches. Thickness of layers is homogeneous and equal but the outer layers are thicker than inner layers. It can be seen more obviously for four layer sandwich (fig.6. b).

4 Discussion

Received data shows that common reducing ratio 110-120% leading to layer thickness 0.4 mm after second pass of rolling does not allow to eliminate the cold-hardening during annealing. Thin foils with a thickness 0.12-0.14 mm have a high tensile strength. Moreover across and along samples do not significantly differ at the rolled and annealed condition.

Very close values of a yield stress for across and along samples at cold state keep a similar behaviour at annealed condition. Loss of yield stress after annealing take place for all reducing ratio and layer thickness.

Elongation decreases with the foil's thickness reduces at rolled and annealed condition excepting along samples. Although elongation decreases and stays very low for the layer thickness 0.12-0.4 mm, its values twice higher than values for another states. Despite a low ductility for thin foils sandwiches, stamping ratio appears very low for 8- and 12-layer samples with a foil's thickness 0.12-0.14 mm due to yield stress loss and keeping high values of the tensile strength. Results of research [10] shows the tensile strength of the reference cold rolled pure aluminium foil having thickness 14mkm is nearly 35MPa and common elongation 2% and 52 MPa and 3.5% respectively for tensile strength and elongation of the foil at heat treated (annealed at 425°C during 7 min) state.

Low values of a stamping ratio shows a better stamp ability that is why the most thin foil's sandwiches demonstrates good stamp ability and less elongation anisotropy.

Anisotropy indexes shows more anisotropic behaviour of mechanical properties at annealed state in comparising to the cold rolled state. The anisotropic annealed behaviour can be explained by more ductility after annealing due to opportunity of dislocation sliding. But at the foil's thickness is less than 0.4 mm, ductility goes down because distance of a dislocation developing become short and not enough for sliding. This model is described by Costache et al. in [13]. Borders of layers which are observed on fig.6 become a barrier for further dislocation move. Microstructure of welded layers of aluminium alloy AA6111-T4 in [14] shows a fine grain structure on the interfacial surface. This area seems to be as a border with a higher strength which limits the plastic deformation. Therefore opportunity for cold-hardening reduces due to a dislocation density growth and ductility becomes limited, as proved by Lederer., Kunčick et al. [9, 15].

5 Conclusions

1. Reducing of the foil's thickness until 0.4 mm at packed cold rolling 0f AA3003 sheets leads to strength growth. Moreover annealing at 500°C for 1 hour does not allow softening for sandwiches which single layers have the thickness less than 0.4 mm.

2. Despite low ductility of thin-foil's sandwiches their stamping ratio takes the lowest values and good stamp ability for samples which foils have the thickness 0.12-0.14mm.

3. Anisotropy indexes at annealed state have a trend to decrease if the foil's thickness achieves values 0.4 mm and less. The cold rolled sandwiches keep a constant anisotropy of strength properties if the foil's thickness is in the interval 0.4…1.0 mm. The thinner foils give a higher anisotropy of the yield stress.

Obtained results can be used for further reducing of layers thickness and multiplying of them in composite, studying of the grain structure and the distribution of reinforcing particles such as ceramics and intermetalisc. Also is in plan to obtain multilayer composites from rich alloyed aluminium alloys.

References

1. M.Lipińska, L.Olejnik, A.Pietras, A.Rosochowski, P.Bazarnik, J.Goliński, T.Brynk, M.Lewandowska, Mater. Des., **88** (2015)
2. R.Kebriaei, I.N.Vladimirov, S.Reese, J. Mater. Process. Technol.., **214** (2014)
3. U. Dilthey, *Encyclopedia of Materials: Science and Technology* (Elsevier, Amsterdam, New York, 2001)
4. X.Liu, X.-H. Liu, M. Song, X.-K. Sun, L.-Z. Liu, Trans. Nonferr. Met., **26** (2016)
5. M.-K.Ng, L.Li, Zh.Fan, R.X. Gao, E.F. Smith III, K.F.Ehmann, J.Cao, CIRP Ann.-Manuf. Tecchn., **64** (2015)
6. L.Kunčická, R.Kocich, P.Král, M.Pohludka, M.Marek, Mater. Lett., **180** (2016)
7. M.-N.Avettand-Fènoël, G.Racineux, L.Debeugny, R.Taillard, Mater. Des., **98** (2016)
8. A.Diehl, U. Engel, M.Geiger: 4M 2006 - Second Intern. Conf. on Multi-Material Micro Manufacture (2006)
9. M. Lederer, V. Gröger, G. Khatibi, B. Weiss, Mater. Sci. Eng., A **527** (2010)
10. J. Singha, A. Chauhan, Ceram. Int., **42** (2016)
11. Y.Birol: Eng. Fail. Anal., **28** (2013)
12. S. Cottrino, P. Viviès, D. Fabrègue, E. Maire, Acta Mater., **81** (2014)
13. E.M. Costache, N. Nanu, B. Chirita, G. Brabie, Int. J Mech. Sci., **69** (2013)
14. F. Haddadi, D. Tsivoulas, Mater. Charact., **118** (2016)
15. L. Kunčick, R. Kocicha, P. Kral, M. Pohludka, M. Marek, Mater. Lett., **180** (2016)

Structure and wear resistance of Ti-TiC-TiB layers obtained by non-vacuum electron beam cladding

Olga Lenivtseva[1,*], *Lyubov* Chuchkova[1], and *Dina* Krivezhenko[1]

[1] Department of Materials Science in Mechanical Engineering, Novosibirsk State Technical University, 630073, K. Marks av., 20, Novosibirsk, Russia

Abstract. In this study structure and tribotechnical properties of cp-titanium after non-vacuum electron beam cladding of powder mixture containing boron carbide were investigated. Structural investigations were carried out using optical and scanning electron microscopy and X-ray analysis. The thickness of cladded layers was 1.3…2.5 mm. The beam moving speed was not influence the phase composition of coatings. The main phases of the surface layers were α- titanium (α′-titanium), titanium carbide TiC and titanium boride TiB. It was determined that the average value of microhardness of the samples formed with the speed 25 mm/s was three times higher in comparison with substrate metal. A decrease in the beam moving speed led to an increase in the heat input per unit area and was accompanied by a reduction in the coating microhardness. To evaluate the wear resistance, friction test of the obtained materials against fixed abrasive particles was performed. The maximum relative wear resistance was exhibited by the coatings after cladding of 20 wt. % boron carbide and 30 wt. % titanium with the beam moving speed of 25 mm/s. Their wear resistance was 1.53 times higher as compared to cp-titanium.

1 Introduction

Titanium is a unique metal. At the present time, titanium alloys are widely used in aircraft industry, shipbuilding, mechanical engineering, automobile production, rocket engineering, oil and gas industry and medicine. Due to its physical and mechanical properties, titanium and its alloys are universal structural materials [1, 2]. High corrosion resistance, strength, low weight and durability provide titanium products with undeniable advantages over the similar products made from other metal alloys.

However, titanium alloys have several significant disadvantages. These are its low tribotechnical properties. Thus, the rate of abrasive wear of titanium alloys is 7 times higher than the rate of wear of carbon steel 1080 [3]. In contact with other structural materials, titanium alloys tend to severe abrasive and adhesive wear [3-5].

Recently, a lot of studies have been devoted to improving the tribotechnical properties of titanium alloys by the formation of composite coatings on their surface (TMC – titanium matrix composites) using laser and electron beam technologies [6-14]. Particles of carbides, borides, and silicides of metals act as hardening components in these materials. In addition, the base metal acts as a binder for ceramic particles, providing a good metallurgical bond.

The first researches that dedicated to the applying of lasers for surface hardening of materials appeared in 1970 [15, 16]. In these studies great attention was paid to the processes of melting and crystallization of the materials, which have a significant effect on the structural-phase state and properties of the layers. The preparation of reinforcing particles in the cladded layers is possible with several approaches: the introduction of hardening particles directly into the melting metal [8-11] or the simultaneous melting of the surface layer of substrate material and the powder mixtures. When applying the first approach, the particles do not have enough time for dissolving in the base material and retain their original shape and size.

Feeding of the powders into the molten bath occurs in a stream of inert gas or is dosed through a hopper. When applying the second approach, the morphology of the particles slightly affects the initial composition of the powder mixtures, due to their complete dissolving in the melt bath during the treatment [17-21]. Both approaches increase in the stability and wear resistance of the titanium surface layers.

In this study, the method of non-vacuum electron beam cladding of boron carbide and titanium powder mixture on titanium plates (VT1-0 alloy) was used to synthesize hardening titanium carbide and titanium boride particles. This technology is characterized by high performance and allows forming the coatings with a thickness of 1 to 3 mm [14, 22-28]. The microstructure, phase composition and tribotechnical tests of the coatings were investigated.

2 Materials and methods

Cp-titanium plates with dimensions of 100x50x12 mm were used as substrates. A powder mixture, that was

* Corresponding author: lenivtseva_olga@mail.ru

DOI: 10.1051/matecconf/201712902022

used to synthesize hardening particles in the surface layers of titanium, consisted of 20 wt. % of boron carbide (average particle size 20 μm) and 30 wt. titanium (average particle size 100 μm). To provide protection of the melt from air oxygen 40 wt. % of CaF_2 and 10 wt. % of LiF were added to the powder mixture. The electron beam treatment of the samples was carried out an ELV-6 electron accelerator produced by the Budker Institute of Nuclear Physics (Novosibirsk). The scheme of the electron accelerator is represented in [29, 30]. Table 1 shows the technological parameters of the process.

Table 1. Technological processing parameters.

Electron energy	E=1.4 MeV
Beam current	I=28 mA
Distance from the outlet window to the workpiece	L=90 mm
Beam size (the Gaussian diameter)	d=12 mm
Scanning frequency	f=50 Hz
Scanning width	l=50 mm
Beam moving speed	V_1=15, V_2=20 and V_3=25 mm/s

Metallographic analysis was carried out using a Carl Zeiss Axio Observer Z1m optical microscope and a Carl Zeiss EVO 50 XVP scanning electron microscope. Evaluation of the hardening particles sizes was carried out on micrographs taken from non-etched polished samples using the ImageJ program.

The phase analysis of the coatings was performed on θ-θ ARL X'TRA diffractometer using Cu Kα radiation. Diffraction patterns were registered in a stepwise mode with a step 0.05 °. The dwell time was 12 s. During the X-ray diffraction study a pinhole collimator was used. Identification of phases was performed using PDF-4 database.

The Wolpert Group 402 MVD tester was used to measure the distribution of microhardness over the depth of the cladded layers. Measurements were performed on cross-sections in the direction from the surface of the cladded layer towards the substrate. The indenter load was 0.98 N.

The evaluation of the wear resistance level of the obtained materials was carried out under friction test against fixed abrasive particles in accordance with the Russian standard GOST 17367-71 (the closest analogue is the ASTM G132-96 standard) [31]. The testing scheme is presented in [30]. The wear resistance was estimated based on the ratio of the linear dimension loss of the standard sample to the tested sample. Titanium alloy VT1-0 was used as a standard sample.

3 Results

During this study, several microstructural zones of the obtained materials were revealed: the zone of the melted material (cladded layer), the heat affected zone (HAZ), and the zone with the unaltered titanium structure. Cladding of boron carbide and titanium carbide powders at the beam moving speed V_1=15 mm/s leads to the formation of the layers with a thickness of 2.5 mm (Fig. 1 a). An increase in beam moving speed is accompanied by a decrease in the coatings thickness from 1.7 to 1.3 mm at the speed of V_2=20 and V_3=25 mm/s, respectively (Fig. 1 b, c). This effect is due to the change of the heat input per unit area. According to [28], taking into account the reflection of electrons from the surface, the input energy per unit area is 2.89, 3.61 and 4.81 kJ/cm^2 at the speed of 25, 20 and 15 mm/s, respectively.

It should be noted that all layers have high quality. However, non-dissolved particles of boron carbide with the size ~ 15 μm were determined in the structure of the coatings. Mainly they are located near the heat affected zone. Reduction of the beam moving speed to 15 mm/s is accompanied by almost complete dissolution of boron carbide particles. In addition, cracks were revealed in the cladded layer of sample 2 (V_2 = 20 mm/s) (Fig. 1 b).

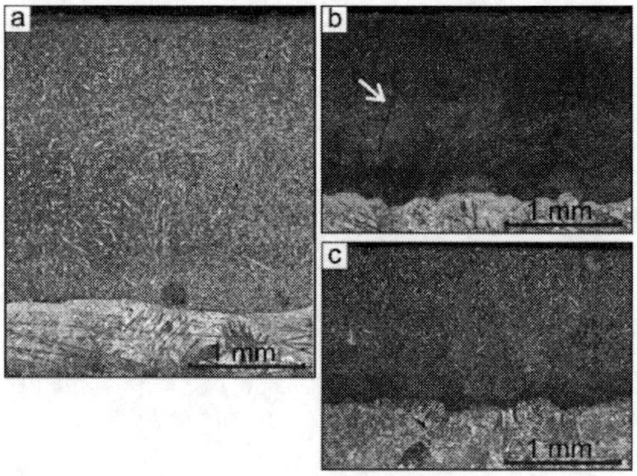

Fig. 1. The cross-section of the layers obtained by non-vacuum electron beam cladding of boron carbide and titanium at a beam moving speed V_1=15 (a), V_2=20 (b), and V_3=25 mm/s (c).

Figure 2 shows a typical X-ray diffraction pattern taken from the surface of the cladded layer. An analysis of the obtained data showed that a change of the beam moving speed was not affected the phase composition of cladded layers. According to the obtained data three phases were detected in the cladded layers: α-titanium (α′- titanium) with a hexagonal lattice, titanium carbide with a cubic lattice, and titanium monoboride with an orthorhombic lattice. No reflexes of boron carbide phase on the X-ray diffraction patterns were found. The displacement of the titanium carbide peaks towards large angles indicates a change in the lattice parameters. Considering the wide range of homogeneity of this phase, it can be assumed that nonstoichiometric titanium carbide observed in these coatings has TiCx formula, where x <1.

The final structure of the cladded layers depends on the cooling conditions of the material during crystallization. Considering different speeds of heating and cooling of zones located from the surface, a gradient structure formed during electron beam cladding. The greatest supercoiling degree was observed at the boundary «molten bath – base metal». Consequently, the layers obtained at higher speed (V_3=25 mm/s) had the higher supercoiling degree, the size of structural components is lower (Fig. 1). The average size of titanium boride crystals was equaled to 65...70 μm. The maximum length of TiB was about 260 μm. Decreasing the beam moving speed to V_1=15 mm/s is accompanied by an increase in boride crystals dimensions to 510 μm.

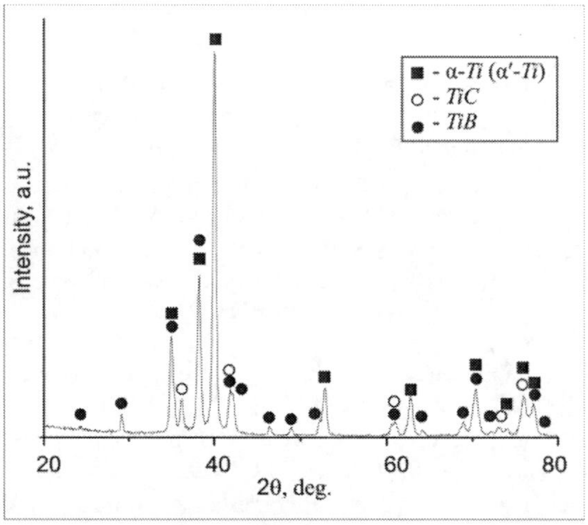

Fig. 2. Diffraction patterns of the cladded layers.

The mechanism of titanium boride crystals growing is determined by the type of their crystal lattice. X-ray diffraction analysis showed that titanium monoboride has an orthorhombic lattice (structural type FeB). Titanium boride mainly grows in the longitudinal direction and in the plane of the cross-section it has the shape of hollow hexagonal prisms cut at different angles or fine needle-like crystals (Fig. 3). Figure 3a shows that due to a lack of boron atoms and a high cooling rate, the shape of titanium boride crystals was not regular. An interesting feature is the growing of titanium carbide dendrites on the faces of boride crystals (Fig. 3b).

Changing of microhardness over the depth of the cladded layers indicates the formation of gradient structure after electron beam cladding (Fig. 4). The average microhardness level rises from 452 to 503 HV with an increase in beam moving speed from 15 to 25 mm/s, respectively. The peaks of the microhardness are due to indenting the hard particles (TiB and TiC), which hardness is much higher than the hardness of the substrate.

In the tests against fixed abrasive particles, the wear resistance of cp-titanium was chosen as a reference value. Analysis of tribological test results indicates that the significant differences in the relative wear resistance of the investigated layers were not observed. In the conditions of friction test against fixed abrasive particles the cladded layers demonstrated higher wear resistance (1.42…1.56) relative to the cp-titanium samples.

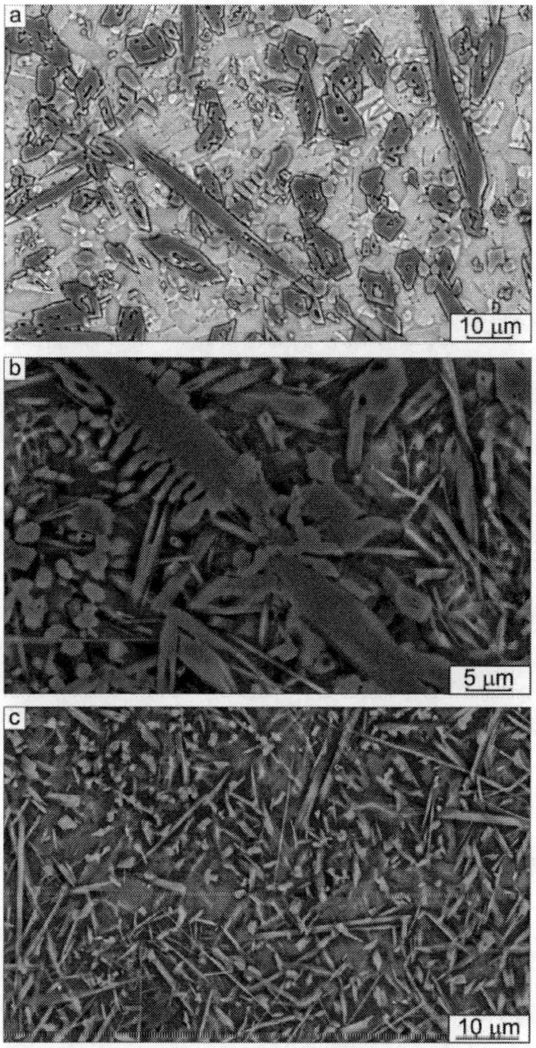

Fig. 3. The structure of cladded layers obtained by scanning electron microscopy.

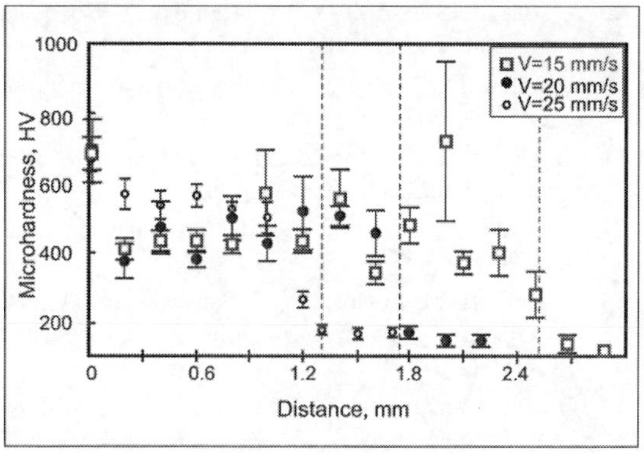

Fig. 4. Distribution of microhardness over the depth of cladded layers.

4 Conclusions

Non-vacuum electron beam cladding of powder mixtures containing boron carbide on cp-titanium plates provides the formation of qualitative hardening layers with a thickness of 1.3...2.5 mm. Changing the beam moving speed affects not only the thickness of the layers, but also the size of the hardening particles (TiC and TiB) and the properties of obtained coatings.

Cladding with a beam moving speed of 25 mm/s allows increasing the average microhardness level up to 503 HV, which is about 3 times higher than the hardness of the substrate. Surface hardening of cp-titanium with carbides and borides of titanium improves resistance of materials under abrasive wear conditions. Relative wear resistance of the obtained layers under an impact of fixed abrasive particles is 1.56 times higher compared to cp-titanium.

The reported study was funded by RFBR, according to the research project No. 16-33-60066 mol_a_dk.

References

1. G. Lütjering, J. Williams, *Titanium* (Springer Berlin Heidelberg, Berlin, 2007)
2. C. Leyens, M. Peters, *Titanium and Titanium Alloys: Fundamentals and Applications* (Wiley-VCH Verlag GmbH & Co. KGaA, Weinheim, 2005)
3. H. Dong, *Surface Engineering of Light Alloys* (Woodhead Publishing, 2010)
4. P.D. Miller, J.W. Holladay, Wear, **2**, 133-140 (1958)
5. E. Rabinowicz, Metal Progress, **65**, 107-110 (1954)
6. R. Yazdi, S.F. Kashani-Bozorg, Mater. Chem. Phys. **152**, 147-157 (2015)
7. W.F. Wang, L.S. Jin, J.G. Yang, F.J. Sun, Surf. Coat. Technol. **236**, 45-51 (2013)
8. V. Amigó, J.J. Candel, P. Franconetti, Materials Science Forum, 299-304 (2012)
9. J.J. Candel, V. Amigó, J.A. Ramos, D. Busquets, Surf. Coat. Technol. **204**, 3161-3166 (2010)
10. Y. Chen, D. Liu, F. Li, L. Li, Surf. Coat. Technol. **202**, 4780-4787 (2008)
11. V. Ocelík, D. Matthews, J.T.M. De Hosson, Surf. Coat. Technol. **197**, 303-315 (2005)
12. F. Weng, C. Chen, H. Yu, Mater Design, **58**, 412-425 (2014)
13. O. Lenivtseva, E. Golovin, V. Samoylenko, D. Mul, D. Golovin, Advanced Materials Research, **1040**, 784-789 (2014)
14. O.G. Lenivtseva, N.S. Belousova, E.A. Lozhkina, T.A. Zimoglyadova, V.V. Samoylenko, L.V. Chuchkova, IOP Conference Series: Materials Science and Engineering, **156**, Art. 0120212016 (2016)
15. J.D. Ayers, R.T. Schaffer, Metal, **33**, 19-23 (1981)
16. C.W. Draper, C.A. Ewing, J. Mater. Sci. **19**, 3815-3825 (1984)
17. K.-m. Zhang, J.-x. Zou, J. Li, Z.-s. Yu, H.-p. Wang, Transactions of Nonferrous Metals Society of China, **20**, 2192-2197 (2010)
18. X. Zeng, T. Yamaguchi, K. Nishio, Optics and Laser Technology, **80**, 84-91 (2016)
19. Y. Lin, J. Yao, Y. Lei, H. Fu, L. Wang, Optics and Lasers in Engineering, **86**, 216-227 (2016)
20. Y. Lin, Y. Lei, X. Li, X. Zhi, H. Fu, Optics and Lasers in Engineering, **82**, 48-55 (2016)
21. L.L. Bai, J. Li, J.L. Chen, R. Song, J.Z. Shao, C.C. Qu, Optics and Laser Technology, **76**, 33-45 (2016)
22. O.G. Lenivtseva, I.A. Bataev, M.G. Golkovskii, A.A. Bataev, V.V. Samoilenko, N.V. Plotnikova, Appl. Surf. Sci. **355**, 320-326 (2015)
23. O.G. Lenivtseva, D.V. Lazurenko, V.V. Samoylenko, Applied Mechanics and Materials, **682**, 14-20 (2014)
24. O. Lenivtseva, E. Golovin, V. Samoylenko, D. Mul, D. Golovin, **1040**, 784-789 (2014)
25. D.O. Mul, E.A. Drobyaz, I.K. Chakin, V.V. Samoylenko, V.S. Lozhkin, R.A. Dostovalov, Obrabotka metallov (tekhnologiya, oborudovanie, instrumenty), **60**, 115-120 (2013)
26. J. Cheol Oh, D.-K. Choo, S. Lee, Surf. Coat. Technol. **127**, 76-85 (2000)
27. K. Euh, J. Lee, S. Lee, Metall. Mater. Trans. A. **32**, 2499-2508 (2001)
28. J.C. Oh, S. Lee, M.G. Golkovski, Metall. Mater. Trans. A. **32**, 2995-3005 (2001)
29. I.A. Bataev, A.A. Bataev, M.G. Golkovsky, A.Y. Teplykh, V.G. Burov, S.V. Veselov, Surf. Coat. Tech. **207**, 245-253 (2012)
30. I.A. Bataev, A.A. Bataev, M.G. Golkovski, D.S. Krivizhenko, A.A. Losinskaya, O.G. Lenivtseva, Appl. Surf. Sci. **284**, 472-481 (2013)
31. *Standard Test Method for Measuring Abrasion Using the Dry Sand/Rubber Wheel Apparatus*, (ASTM International, 2010)

DOI: 10.1051/matecconf/201712902023

Multilayered Nb-Al composite manufactured by explosive welding

Iuliia Maliutina[1,2,*], *Daria* Lazurenko[1], and *Maxim* Esikov[1,2]

[1]Department of Materials Science in Mechanical Engineering, Novosibirsk State Technical University, K. Marks 20, 630073 Novosibirsk, Russia
[2]Laboratory of Physics of High Energy Density, Lavrentyev Institute of Hydrodynamics, SB RAS, 15 Lavrentyev pr., 630090 Novosibirsk, Russia

Abstract. In this study, the multilayered composite materials consisting of niobium and aluminum foils were produced by explosive welding. The conducted microstructural studies with the use of scanning electron microscopy (SEM) have shown that the boundaries of welded seams have predominantly a wave structure with a partial local melting of interacting materials. At the niobium/aluminum boundary, in the mixing zones, micro volumes with a non-uniform chemical composition were presented. Mechanical tests have shown that the strength properties, as well as the toughness of composite materials, are at an intermediate level relative to the initial materials.

1 Introduction

For many years researchers are interested in multilayer materials containing intermetallic phases, so-called metallic-intermetallic laminate (MIL) composites. To obtain materials of this type, various methods, including methods of deposition, cladding, casting, accumulation roll bonding have been used to produce multilayer compositions such as Ti-Al-Nb, Ni-Al, Nb-Fe and others [1-5]. Such materials have enhanced mechanical properties, including strength, toughness, fatigue characteristics, etc., which makes them interesting objects for research.

Among these methods, explosive welding followed by heat treatment is the most effective method of MIL composites manufacturing. It is well known that the welding process consists in accelerating the flyer plate because of the explosive located on it and its collision at very high speeds with the fixed plate. As a result of the high pressure and temperature originated at the contact point of the plates, high degrees of deformation are occurred, which ensure the formation of high-strength joints. The method makes possible to combine incompatible materials, and also to combine materials that differ in their thickness without significantly changing of their properties [6]. Thus, the bimetallic composite of dissimilar materials can be easily obtained by explosive welding [7-10]. This method also has been successfully used to obtain MIL composites consisting of alternating 10 titanium plates and 11 aluminum plates [11], as well as a seven-layer composite consisting of alternating plates of nickel and aluminum [2]. The materials involved in the formation of the above materials, namely the Ti-Al and Ni-Al, are the most interest, because of the intermetallic compounds formed during their interaction are promising candidates for the aircraft industry.

Less attention is paid to the Nb-Al alloys, which have obvious advantages over Ti-Al and Ni-Al alloys. For example, intermetallic compounds of niobium and aluminum have higher melting point (1680-2060 °C depending on the composition), and their density is less or comparable to the density of the studied intermetallic compounds (4.54-7.28 g/cm^3 depending on the composition) [12]. Therefore, composites containing intermetallic compounds based on niobium and aluminum are of great scientific and practical interest.
The paper presents the results of structural studies and mechanical properties of multilayered compositions consisting of thin foils of niobium and aluminum obtained by explosive welding.

2 Experimental procedure

Multilayered compositions were obtained by explosive welding technology. The outer layers were thin plates of VT14 titanium alloys ($\alpha + \beta$ alloy) of 0.6 mm thick and OT4 (near α-alloy) of 0.3 mm thick (Fig. 1). Ammonite 6ZhV explosive of 15 mm thick was placed onto the VT14 titanium plate, used as a flyer plate. Titanium plates located outside the composite were used as a protection for niobium during heating, and also for testing the possibility of welding niobium with various titanium alloys. The A5 aluminum foils of 0.25 mm thick and Nb-1 niobium foils of 0.2 mm thick were alternately arranged between them. The chemical composition of the initial materials is given in Table 1. The distance between the foils and plates was about 1 mm. Thus, a multilayered composition consisting of 7 thin plates with a total thickness of 2 mm was obtained.

* Corresponding author: iuliiamaliutina@gmail.com

DOI: 10.1051/matecconf/201712902023

Table 1. The chemical composition of initial materials.

Materials	Chemical composition, wt. %						
	Ti	Al	Mo	V	Mn	Fe	Nb
VT14	Bal.	4.5	3.4	1.2	-	-	-
OT4	Bal.	4.6	-	-	1.1	-	-
A5	-	Bal.	-	-	-	0.25	-
Nb-1	-	-	-	-	-	-	100

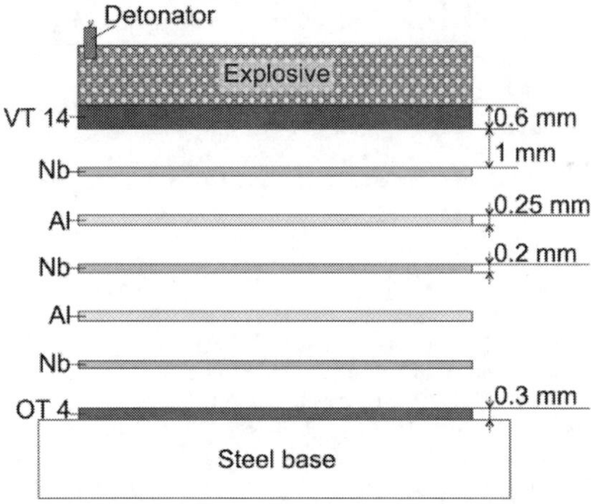

Fig. 1. The scheme of explosive welding of a seven-layer composite consisting of VT14 and OT4 titanium alloys, as well as alternating foils of niobium and aluminum.

Samples for microstructural studies were cut from the central part of the composite corresponding to the longitudinal direction of the shock wave propagation. Specimens were prepared by grinding and polishing using a fine dispersion of silicon oxide with a particle size of 0.05 μm. For chemical etching of titanium alloys, Kroll's was used. Microstructural studies were performed on a scanning electron microscope in a mode of secondary and back-scattered electrons. The chemical composition of reaction products formed in the zones of plates impact was analyzed by energy-dispersive microanalyzer.

The Vickers microhardness of the analyzed composites was determined on the polished transverse specimens in the direction from the VT14 plate to the OT4 plate through all the layers. The mechanical properties of the laminate composites were evaluated in the static tensile and impact tests.

3 Results and discussions

A detailed study of the boundaries formed during explosive welding between different materials is shown in Fig. 2 and 3. The boundaries between VT14 and OT4 titanium, as well as niobium, have a typical wave structure with local mixing areas. Near the welded seam in the titanium plates a thin layer of 5 μm, consisting of deformed grains, is found. This behavior indicates on a severe plastic deformation in the zone of a contact point.

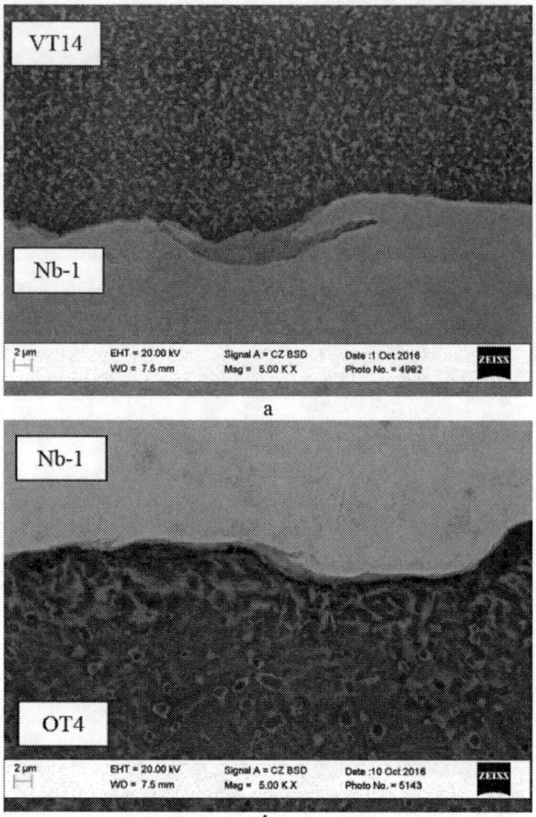

a

b

Fig. 2. SEM image of the VT14/ Nb and Nb/OT4 boundaries joint.

At the niobium/ aluminum boundary, there are many sections corresponding to the zones of partial melting and mixing of the two materials. The SEM image shows a nonuniform distribution of aluminum and niobium in various parts of the mixing region, due to high crystallization rates of the alloys and the nonequilibrium state of the explosive welding process.

a

DOI: 10.1051/matecconf/201712902023

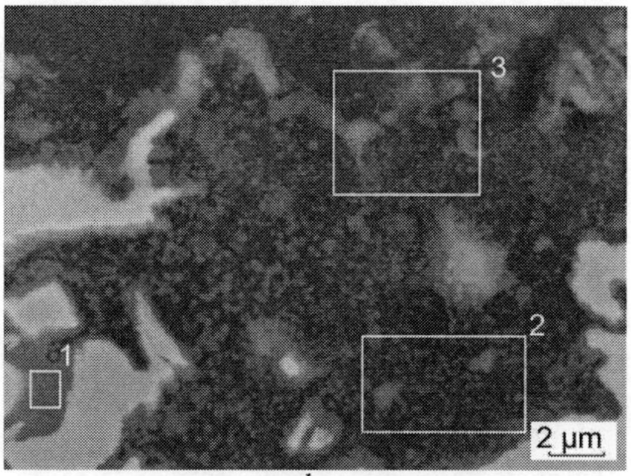

b

Fig. 3. SEM image of the boundary of the Nb/Al joint (a) and the mixing zone of these materials (b).

The results of the energy dispersive analysis confirmed the different elemental composition (Table 2). The concentration of niobium varied from 3 to 26 at. %.

Table 2. The content of niobium and aluminum in different parts of the mixing zone (at. %).

Analyzed area	Nb	Al
1	3.16	96.84
2	14.9	85.10
3	26.1	73.9

The microhardness of the composites was measured in the direction from the plate of the VT14 titanium alloy to the OT4 titanium plate through all the layers. The microhardness results of the composite after explosive welding are shown in figure 4. The average microhardness of the initial materials was 260 HV for VT14, 33 HV for Al, 118 HV for Nb and 232 HV for OT4. In the mixing zones of niobium and aluminum, a significant increase in the microhardness was not observed, due to the nonuniform distribution of the high-strength phase and the high aluminum content in these sections.

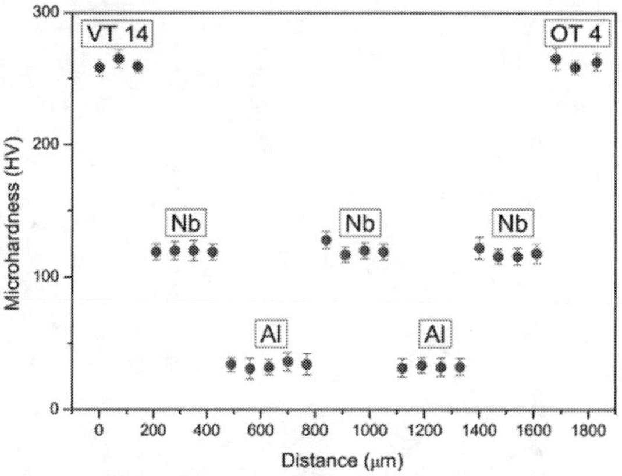

Fig. 4. Microhardness distribution in different composite layers.

The results of the mechanical properties of the composites obtained in comparison with the materials contained the composites are shown in Table 3. According to this data, the strength properties, as well as the impact strength are at an intermediate level with respect to the initial materials.

Table 3. Mechanical properties of the composite material and the initial materials.

Analyzed materials	Ultimate tensile strength, MPa	Tensile yield strength, MPa	Specific elongation, %	Impact strength, J/sm^2
Composite	534	335	-	42,5
Nb-1	425	355	23	37
A5	60	-	20-28	100-120
VT14	1050	895	8-10	45-50
OT4	635-685	550-660	10-20	35-40

4 Conclusions

Thus, it was established that explosive welding allows the formation of multilayer compositions consisting of thin niobium and aluminum foils. In accordance with microscopic analysis, it was shown that the welded seams formed during the process have a wave structure with the formation of local mixing areas. In the mixing zone of niobium and aluminum by scanning electron microscopy was detected that the composition is nonuniform so that the content of niobium varies from 3 to 26 at. %. The results of mechanical tests showed that the strength properties, as well as the impact strength, are at an intermediate level with respect to the initial materials.

The obtained results indicate the possibility of chemical compounds formation at the welding interface of niobium and aluminum foils. For the subsequent growth of the formed intermetallic compounds, it is necessary to select the heat treatment mode of the explosively welded samples, and also to study the kinetics of the phase transformations that occur during heating.

The work was carried out in the frame of project No 11.7662.2017/BCh

References

1. S. Simões, A.S. Ramos, F. Viana, O. Emadinia, M.T. Vieira, M.F. Vieira, J. Mater. Eng. Perform. **25,** 4394 (2016)
2. I.A. Bataev, T.S. Ogneva, A.A. Bataev, V.I. Mali, M.A. Esikov, D.V. Lazurenko, Y. Guo, A.M. Jorge Junior, Mater. Design **88** ,1082 (2015)
3. P. Qu, L. Zhou, V.L. Acoff, Mater. Charact. **107,** 367 (2015)
4. P. Qu, L. Zhou, H. Xu, V.L. Acoff, Metall. Mater. Trans. A **45,** 6217 (2014)

5. I. A. Bataev, K. Hokamoto, H. Keno, A.A. Bataev, I.A. Balagansky, A.V. Vinogradov, Met. Mater. Int. **21,** 713 (2015)
6. Iu.N. Maliutina, N.V. Stepanova, A.G. Cherkov, L.V. Chuchkova, Obrab. Met. **69,** 61 (2015)
7. F. Grignon, D. Benson, K.S. Vecchio, M.A. Meyers, Int. J. Impact Eng. **30,** 1333 (2004)
8. P. Manikandan, K. Hokamoto, A.A. Deribas, K. Raghukandan, R. Tomoshige, Mater. T. JIM **47,** 2049 (2006)
9. Iu.N. Maliutina, V.I. Mali, I.A. Bataev, A.A. Bataev, M.A. Esikov, A.I. Smirnov, K.A. Skorokhod, The Scientific World Journal **2013,** 1 (2013)
10. H. Paul, L. Lityńska-Dobrzyńska, M. Prażmowski, Metall. Mater. Trans. A **44,** 3836 (2013)
11. I.A. Bataev, A.A. Bataev, V.I. Mali, D.V. Pavliukova, Materials and Design **35,** 225 (2012)
12. E.P. Barth, J.K. Tien, S. Ucjo, S. Kambara, Mater. Sci. Eng. A - Struct **A153,** 398 (1992)

In-situ study of structural evolution of tribological materials using synchrotron radiation

Vladimir Burov[1,*], *Ivan* Bataev[1], and *Alexander* Smirnov[1]

[1] Novosibirsk State Technical University, Materials Science Department, 630073, 20 Karl Marx Ave., Novosibirsk , Russia

Abstract. In this paper, we propose an approach for estimation of structural changes at the surface of metallic materials during running-in stage of friction. The application of synchrotron microbeam for in-situ observation of structural evolution in thin layers subjected to friction load is substantiated. A design of experimental apparatus intended for in-situ studies of friction processes using synchrotron radiation is described. This apparatus maintains constant position of the sample surface with respect to the incident beam. The ring shaped experimental sample with a thickness of 1 mm is pressed with constant pressure against a pin shaped indenter and rotated with constant speed. The pressure on the sample is determined by the mass of the load placed in the loading basket. The surface of the sample that comes out of indenter is probed by synchrotron beam. The proposed experimental scheme allows receiving X-ray diffraction patterns even under condition of severe plastic deformation of the surface. The preliminary results on the synchrotron assisted study of the dry friction of a Cu sample are described.

1 Introduction

One of the main reasons for the quality loss of engineering products used in movable joints is the initial stage of wear that leads both to a change in the dimensions and roughness and to a change in the structural state of the surface layers. X-ray diffraction study of the worn surfaces provides a valuable insight on the irreversible processes of structural evolution and formation of thin surface layers with a gradient structure [1, 2]. Knowledge of sequence of rapid structural changes in the surface layers during the initial stages of sliding friction is important to predict the reliability and lifetime of the movable joints of mechanisms. The process of structural evolution during running-in stage (from initial state to the steady irreversible state) can be investigated using synchrotron x-ray diffraction [3].

2 The objectives of the study

Synchrotron diffractometry stations, which have a wide range of applications for study of rapid processes, differ in intensity of radiation, beam diameter and provide variety of universal devices that enable recording the diffraction patterns by 1d and 2d detectors.

It is known, that the process of dry friction is accompanied by the structural changes of the material surface layers [1-4]. This process leads to formation of the nanostructured layer, which substantially determines lifetime of parts of the movable joints. The application of the synchrotron radiation for tribological studies allows understanding the mechanisms of wear of materials with different chemical and phase compositions at different loads and friction speeds. In ref. [5] one can find a description of an experiment, when synchrotron radiation was used to probe the surface of D16 (Al-Cu-Mg) alloy during friction against high speed steel indenter. In ref. [6,7] an application of synchrotron radiation for in-situ study of steel SUJ2 – sapphire friction pair is described. Both these experiments provided interesting insight on mechanisms of structural evolution during the friction.

It should be noted, that when the synchrotron beam is focused to 1-5 μm spot the precise positioning of the friction surface with respect to the incident beam becomes extremely important since the friction leads to plastic deformation and wear. Thus, the position of the beam on the sample surface may change during the friction test. This is especially important in grazing-angle geometries or similar diffraction schemes.

The design of apparatus used by Kajita et. al. [6,7] provided high precision positioning and allowed using a microscope to get the images of the friction surfaces during experiments. This was provided by usage of transparent sapphire counterpart. However, one can rarely meet sapphire in real friction pairs. In our study, we developed a friction apparatus that allows application of different materials of both a sample and a counterbody (indenter). Fig. 1 represents the arrangement of the surface layer being investigated with respect to the X-ray beam. The unit is equipped with a control system that allows setting the rotation speed of the sample within the range of 0...5 s^{-1} and any number of rotation cycles, as well as the remote control (Fig. 2). In addition, a computer with control interface can be connected via a remote control.

* Corresponding author: v.burov@corp.nstu.ru

 DOI: 10.1051/matecconf/201712902024

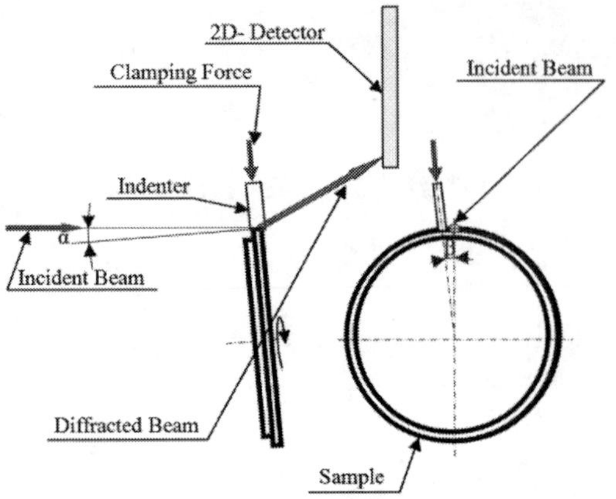

Fig. 1. The arrangement of a rotating sample and an indenter with respect to a sliding X-ray beam

This setup is designed to be used at microfocus beamlines of European Synchrotron Radiation Facility (ESRF) with diameter of the incident beam in the range 1…3 μm. The apparatus is designed in such way, that the surface of indenter is kept at constant position during the tests. The sample is pressed against indenter with the required load. The sample has a shape of a disc with 1-mm width rim on the outer curved or circular surfaces.

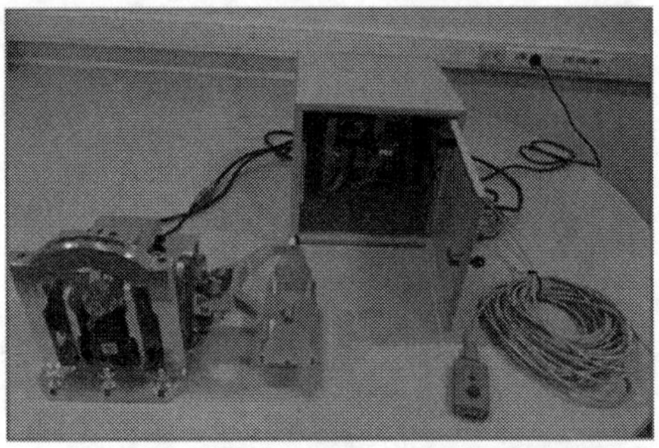

a

b

Fig. 2. Experimental apparatus designed for in-situ synchrotron assisted studies of rapid structural evolution that occurs at the surface layers during dry friction. (a) Overall view of the apparatus together with the control unit. (b) The apparatus placed at the beamline ID13 at ESRF.

The diameter of the sample should be in the range 60-70 mm. The working area of the indenter has dimensions of 1 x 5 mm (Fig. 3). In the initial moment of time the sample is placed in such way that its contact area with the indenter is 1 mm^2 (the possible increase of the contact surface due to the plastic deformation of the sample is neglected). Due to the fact, that the sample is pressed against the indenter and the pressure is kept constant during experiment, the relative arrangement of the incident beam and the sample stays the same during experimental time providing that indenter doesn't wear during the whole time of the experiment.

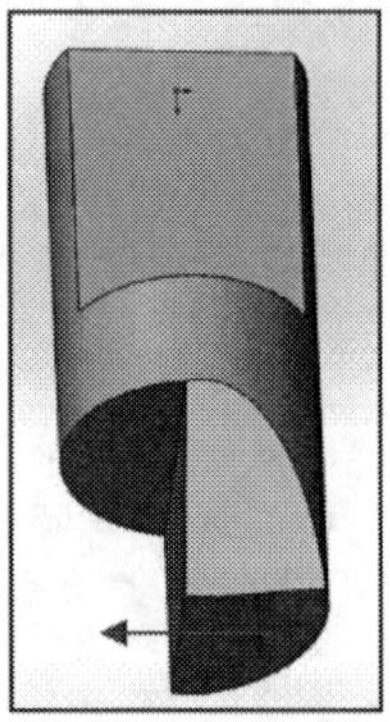

Fig. 3. The geometry of the indenter. The arrow shows the direction of the sample rotation.

The friction load is provided by test weights that are placed in the loading basket. The basket is attached to the body of apparatus via a lever. The body of tester moves inside the frame using low friction linear guides. In order to avoid shadowing of diffraction pattern by indenter the probing position is shifted from the friction position 2-8° in direction opposite to the sample rotation direction. This leads to 2-20 ms temporal delay between the friction moment and the probing moment.

The experimental apparatus allows using 2D X-Ray detector Eiger 4M at ESRF beamline 13. This provides frequency of 750 Hz of data acquisition with exposition time less than 2.6 ms.

3 Results of preliminary experiments

The friction of a copper sample against a high-speed steel indenter was studied using the apparatus to test new experimental technique. The copper discs were annealed in a furnace at 850 °C to minimize initial strains and provide a nondeformed surface. After the heat treatment, the sample was chemically polished to remove the oxide layer. The load on the sample was 40 MPa and the rotation speed was 60 rpm. A total number of the sample revolutions during the time of the experiment was 5500. The exposition time of detector was equal to the period of one revolution. Thus, first X-ray diffraction pattern contained information from all points of the sample subjected to friction during one revolution. The second frame contained information from all points of the sample that experienced friction load twice. And so on. The angle between the incident beam and the sample surface was 5°.

DOI: 10.1051/matecconf/201712902024

The examples of diffraction patterns received during the experiment are shown in Fig. 3. One may notice, that the diffraction patterns, received in static conditions (Fig. 4a and 4c) have sharp diffraction spots, which correspond to a polycrystalline material with limited amount of rather large individual grains. However, the diffraction rings after the experiment became more uniform and less spotty indicating that plastic deformation of the surface layer occurred leading to grain refinement. The diffraction rings, received in dynamic conditions were rather uniform (Fig. 3b). Since the exposition time was quite long a lot of individual grains passed under the beam and no sharp spots could be observed. One may also notice that only half of the rings are seen on the figure. The lower parts of the rings are absent due to the reflection geometry of diffraction, when lower part of the diffraction cones are absorbed by the sample.

Fig. 5 shows integrated diffraction patterns in coordinates "azimuthal angle - 2θ". The diffraction rings can be indexed as 111, 200, 220 and 311 as it is typically observed on diffraction patterns of Cu.

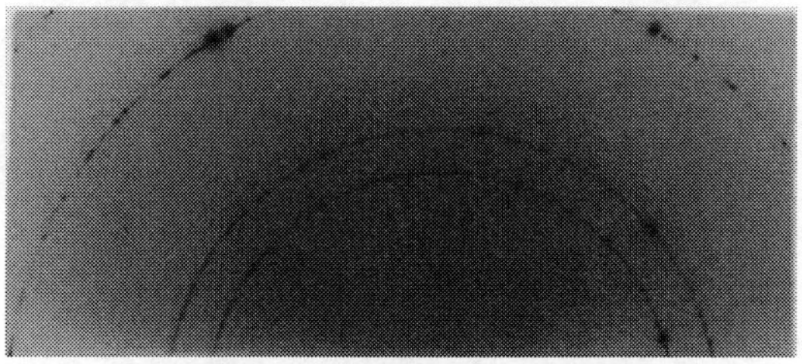
a

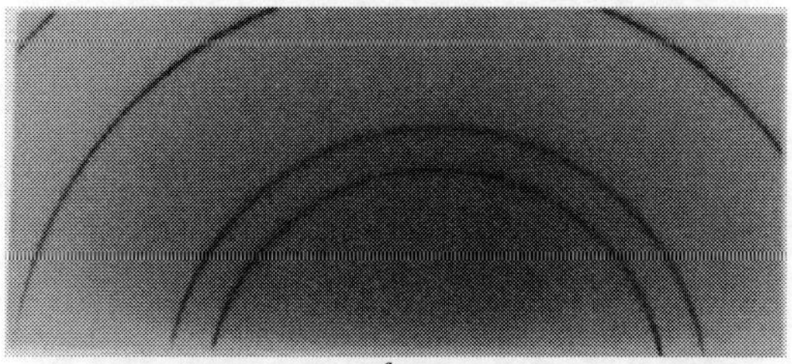
b

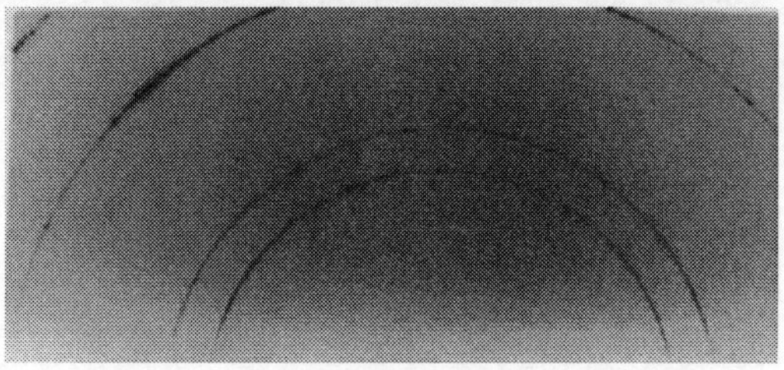
c

Fig. 4. X-ray diffraction patterns obtained from the sample surface in (a) static condition before the experiment, (b) in dynamic condition during the experiment and (c) in static condition after the experiment.

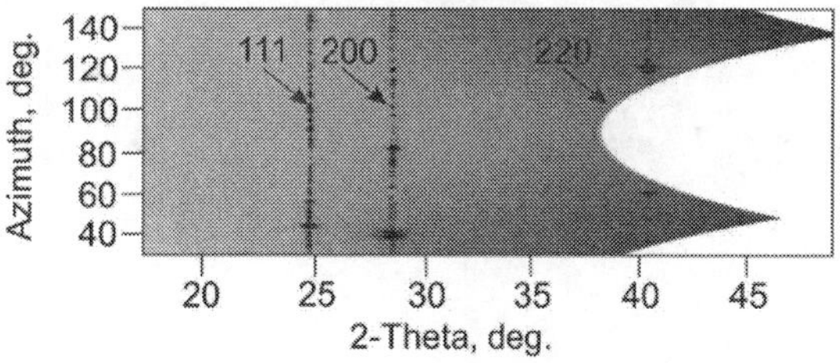

a

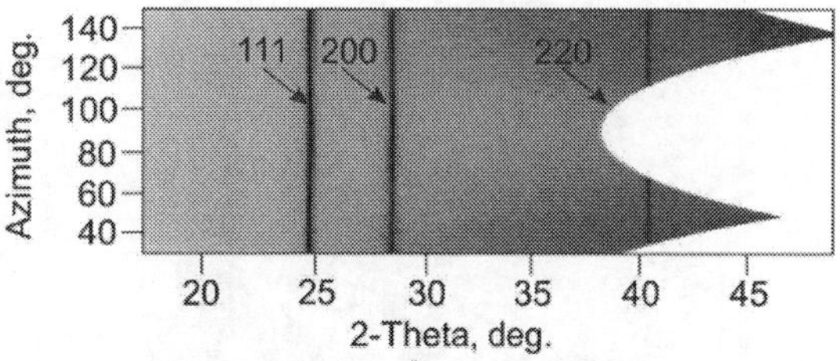

b

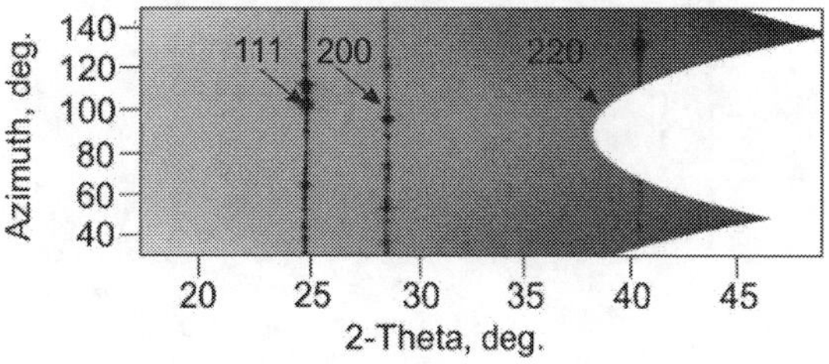

c

Fig. 5. Diffraction patterns of Cu sample (a) before the beginning of the experiment (b) during the experiment and (c) after the end of experiment.

Fig. 6 shows variation of the 111 peak intensity with the time of the experiment. One may observe that the intensity has noticeably decreased indicating possible plastic deformation and structural transformation at the surface layer. One may also observe broadening of 3 peaks of Cu (Fig. 7) indicating possible grain refinement and accumulation of defects and corresponding microstrains.

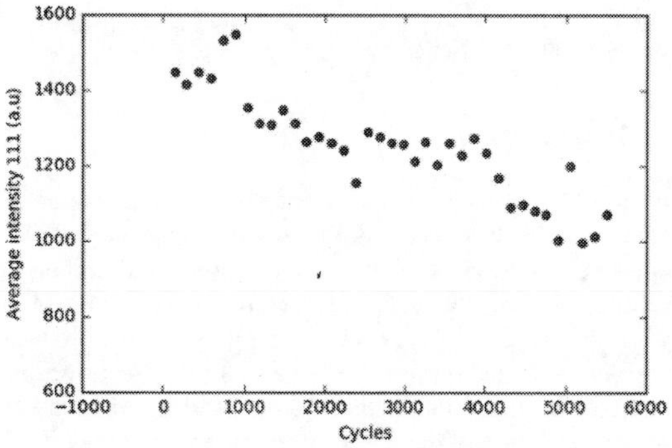

Fig. 6. Average intensity of 111 peak of Cu during friction against steel indenter.

DOI: 10.1051/matecconf/201712902024

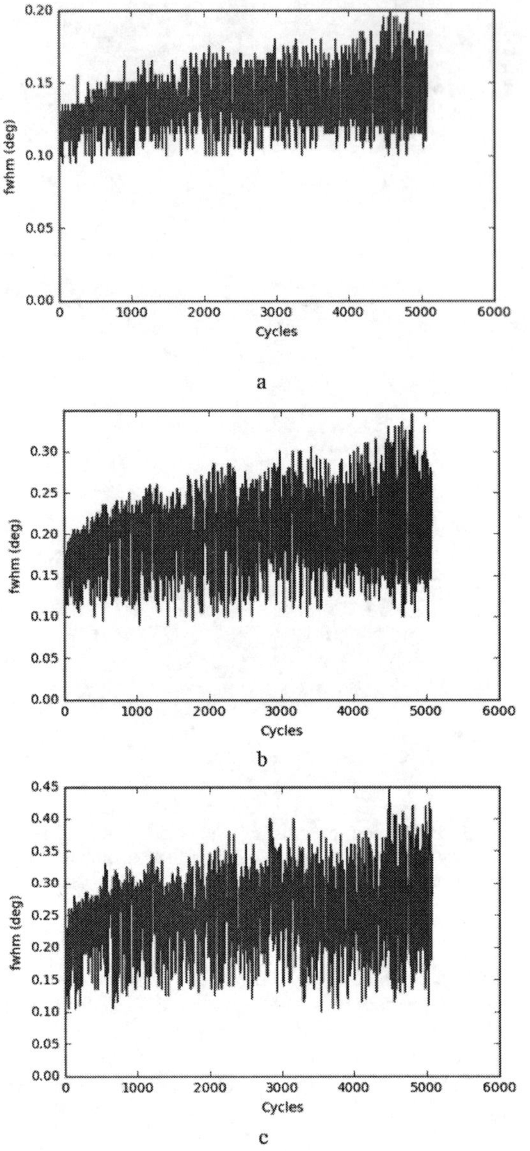

Fig. 7. Broadening of (a) 111, (b) 200 and (c) 220 peaks of Cu versus number of friction cycles.

4 Conclusion

A novel approach to investigate the friction processes using synchrotron assisted diffraction was proposed. This approach allows probing the sample and receiving information about surface structure of materials under friction "on the fly". Using the copper-steel friction pair we showed that during the friction there are noticeable changes in the peaks intensities and widths which is due to accumulation of defects and structure refinement. In the nearest future, the algorithm will be developed that will allow retrieving the information about most important structural characteristics of materials (coherent scattering region size, microstrains, microstresses, dislocation density, etc).

This research was carried out using financial support of the Ministry of Education and Science of Russian Federation (program "Investigations and developments in the priority areas of the scientific and technological complex of Russia in 2014-2020"), agreement № 14.586.21.0026, unique identifier of agreement RFMEFI58616X0026.

References

1. A.V. Makarov , P.A. Skorynina, A.L. Osintseva, A.S. Yurovskikh, R.A. Savrai. Obrabotka metallov (tehnologiya, oborudovanie, instrumenty), **69**, 80-92 (2015)
2. N.N. Soboleva, A.V. Makarov, I.Yu. Malygina. Obrabotka metallov (tehnologiya, oborudovanie, instrumenty), **4**, 79-85 (2013)
3. V.P. Kuznetsov, A.V. Makarov, S.G. Psakhie, R.A. Savrai, I.Yu. Malygina, N.A. Davydova. Physical mesomechanics, **17**, 250-264 (2014).
4. A.V. Kolubaev, O.V. Sizova, E.A. Kolubaev. Tribologie und Schmierungstechnik, **55**, 45-48 (2008)
5. A.A. Bataev, I.A. Bataev, V.G. Burov, S.V. Burov. Russian Physics Journal, **59**, 314-316 (2016)
6. S. Kajita, K. Yagi, T. Izumi, J. Koyamachi, M. Tohyama, K. Saito, J. Sugimura. Tribol. Lett, **57**, 11 (2015)
7. S. Kajita, T. Izumi, J. Koyamachi, M. Tohyama, K. Saito, J. Sugimura. Tribol. Lett, **61**, 16 (2016)

Experimental research steel grade «Steel 45» to abrasion wear, depending on the exit angle macrostructure fibers on the contact surface

Maksim Gennadievich Allenov[1,*], *Dmitrii Gennadievich* Allenov[2], and *Oleg Aleksandrovich* Belokurov[1]

[1]BMSTU, Engineering Technology Department, 105005 Baumanskaya 2-ya str. 5 Moscow, Russia
[2] RUDN University, Engineering academy, 117198 Miklukho-Maklaya str. 6 Moscow, Russia

Abstract. This paper describes the experimental research of the influence of the angle of exit of microstructure fibers to the contact surface wear abrasion steel grade «steel 45». The experimental dependence of abrasion wear depending on output angle of macrostructure fibers to the contact surface was obtained (considered angles: 0^0, 15^0, 30^0, 45^0, 60^0, 75^0, 90^0). A method for obtaining samples with a specified directional fiber structure was development. Two experiments were carried out: on a friction machine of translational type and on a friction rotational type machine. The article shows the difference in wear with different types of friction: rotation motion simulates wear in bearings, translational motion simulates the movement of the piston in the cylinder or the operation of the guides.The effect of the exit angle of the macrostructure fibers on wear is determined. An analysis of the experimental data was carried out and the angles of the yield of fibers on the contact surface were found at which maximum and minimum wear was attained.

1 Introduction

The fibrous composition of the metal macrostructure is obtained from the ingots dendritic structure contaminated with nonmetallic inclusions, during casting and subsequent processing by pressure. Fibers macrostructures are stretched in the direction of the most intense flow of the metal. This macro-structure cannot be eliminated by thermal treatment. The distribution of macrostructure fibers is detected by etching on the macrostructure, after which they are visible to the naked eye.

In the metal with the fibrous composition of the macrostructure, anisotropy of the mechanical properties is observed depending on the direction of the fibers. Similarly, the anisotropy of mechanical properties affects wear on the abraded surfaces as a result of contact with other parts [1]. This effect is observed in the work of friction pairs, for example, in an internal combustion engine (shafts, root and conrod necks, etc.). On the faculty of "Technology of pressure treatment" (MT-6) MSTU. N.E. Bauman was studied the creation of a directed fibrous structure of the macrostructure in forgings of critical details [2, 3]. A study of abrasion wear as a function of the yield angle of the macrostructure fibers to the contact surface was carried out [4]. However, no study of abrasion wear with modern types of oils including antiwear and antifriction additives was carried out. In addition, we investigated the effect only at 3 different exit angles of the fiber on the working surface (0^0, 45^0, 90^0), which does not allow to determine the increase in the operability of the obtained parts in the process of designing the forging process, depending on the yield angle of the macrostructure fibers on working surfaces. When stamping forgings, most often, the exit angle of the macrostructure fibers on the working surface is in a certain range, and does not have a definite value.

Therefore, an experimental study of the effect of the exit angle of macrostructure fibers with a small step on abrasive wear is an urgent task in the development of technological pro cesses of hot forging. Earlier, in the article [5], the influence of the exit angle fiber in the translational motion of friction particles was considered.

2 Research methodology

In this paper, we studied the effect of the exit angle of the macrostructure fibers on the contact surface in step of 15 degrees on the abrasive wear on the abrasion of steel 45. A special friction machine of translational and rotational type was used to study abrasive wear (Fig. 1, Fig. 2).

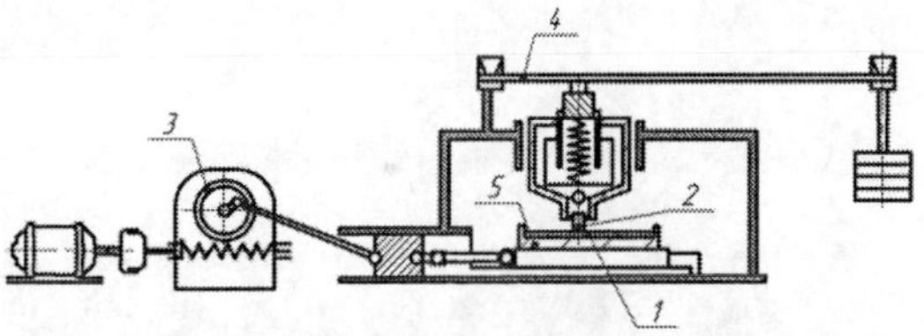

Fig. 1. Friction machine of the translational type: 1 – lower sample; 2 – upper sample (pin); 3 – crank mechanism; 4 – level device; 5 – bath for oil.

* Corresponding author: okdell@mail.ru

The operating principle of the friction machine of the translational type (Fig. 1): the upper sample 2 is pressed by the lever device 4 to the self-adjusting slide bar on which the lower sample 1 is located. The slide moves reciprocally from the crank mechanism 3. The length of the samples can vary from 30 to 70 mm; the maximum number of double strokes per minute is 47. The load varies in the range from 100 to 3000 N. A self-adjusting sample that provides running-in both initial and after intermediate weighing, facilitates work of the machine. Tests can be carried out in a variety of liquid fluid, since the sample can be placed in a bath 5.

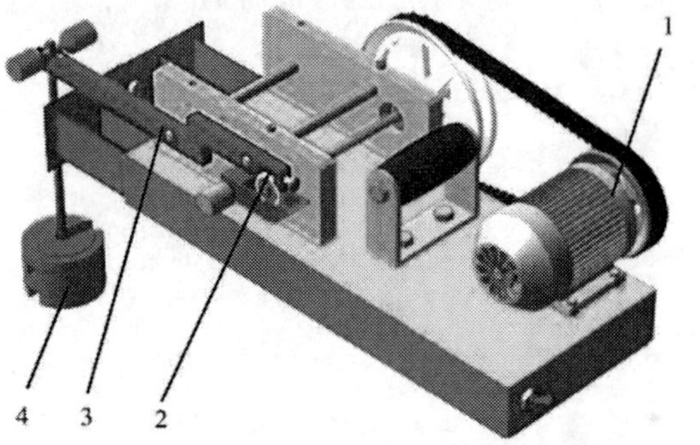

Fig. 2. Friction machine of the rotation type: 1 – electric motor; 2 – friction unit with bath for oil; 3 – level device; 4 – loads.

The principle of operation of the rotational type of friction machine (Fig. 2) is as follows: The electric motor drives 1 the disc in the friction unit 2 via a belt driving. The upper sample is pressed by the lever device 3 to the rotating disk by means of loads 4. There is also an oil bath in the friction unit. Rotational speed is 400 rpm.

The choice of these machines is due to the simplicity of the form of the samples for testing. The tests were carried out in a motor oil environment with antifriction and antiwear additives, motor oil of class 10w40 (API SG / CD standard) was used. The test scheme is shown in Fig. 3.

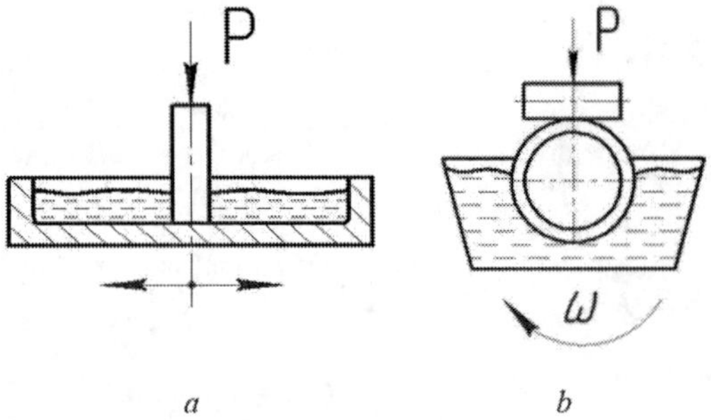

Fig. 3. Test scheme: a – translational type; b - rotation type.

As the samples were chosen square plates with dimensions 20x20x8.5 mm (Fig. 4, a), cut out on a milling machine made of steel rolled products with a section of 30x30 mm. Scheme for cutting samples with a given exit angle of macrostructure fibers in Fig. 4, b. Samples would be cut at different angles: 0^0, 15^0, 30^0, 45^0, 60^0, 75^0, 90^0. Sample material steel 45 was chosen as the material for samples. The choice of steel 45 is due to the fact that it is one of the materials for the production of crankshafts obtained by hot stamping technologies. The choice of a rolled rod for making samples allows one to determine the fiber direction of the macrostructure in the samples. The material of the pin, which causes abrasive wear, is also steel 45. The roughness of the samples and the pin is Ra 1.25 (grinding).

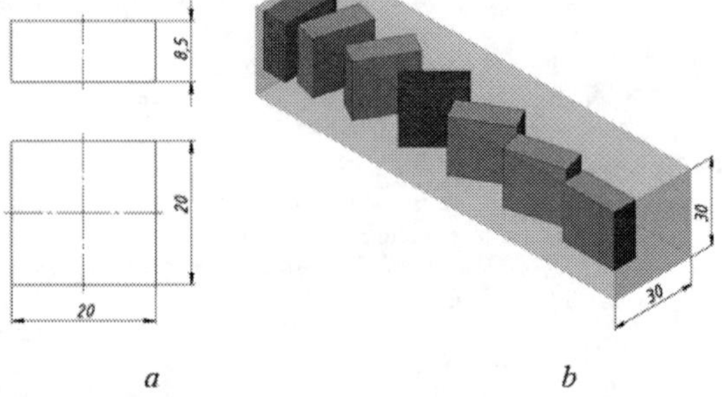

Fig. 4. Sample and methods for obtaining it: a – sample size; b – scheme for cutting samples.

3 Experiments

3.1 Carrying out the experiment on a machine of translational type

To fix the sample on the friction machine of the translational type, a special device (Fig. 5) was prepared in the plate form with a square cut for the sample. A 28 x 96 transition plate is necessary to reduce the size of the sample. Otherwise, the samples with a different exit angle of the macrostructure fibers would have to be cut out of hot-rolled steel with a section of 100x30 mm.

Fig. 5. Special device to fix the sample.

The samples were placed on a test machine (Fig. 6), the test mode and the load were set. The following test parameters were chosen: load - 15 loads (2100 N), test time - 3 hours, number of double strokes per minute - 47. Motor oil for testing was applied only at the initial time,

DOI: 10.1051/matecconf/201712902025

simulating critical wear. Wear was evaluated by measuring the mass of the sample before and after the experiment.

Fig. 6. Experiment on friction machine.

Mass measurements of the sample were carried out on an analytical balance Cubis MSA225S-100DI. The graduation mark of weights - 0,01 mg, accuracy class according to GOST R 53228-2008 - I. To reduce the measurement error, each weighing was performed 3 times.

3.2 Carrying out the experiment on a machine of rotation type

Fig. 7. Fixing the sample in the friction unit: 1 – sample; 2 – disk; 3 – bath for oil.

Samples were installed on a test machine (Fig. 7), the test mode and the load were set. The following test parameters were selected: load - 4 and 5 loads (560-700 N), test time - 6 minutes with 5 loads, 11 minutes with 4 loads. The test oil was poured into the bath to a level. Wear was evaluated similarly to the previous experiment.

4 The results of the experiments

The results of the experiments are tabulated (Table 1, 2, 3) and a plot of Δm (difference in mass before and after abrasion) is plotted on the exit angle of the macrostructure fibers on the sample surface (Fig. 8,9).

Table 1. Results of the experiment 1. Abrasive wear on a machine of translational type. 15 loads, 180 minutes.

№	Exit angle of fibers, ⁰	Mass before abrasion, g	Mass after abrasion, g	Δm,mg
1	0	22,200120	22,197902	2,218
2	15	22,133115	22,130817	2,298
3	30	21,691490	21,687977	3,513
4	45	21,745010	21,739950	5,060
5	60	20,498097	20,491811	6,2867
6	75	20,796477	20,798874	7,6067
7	90	20,386467	20,378361	8,1067

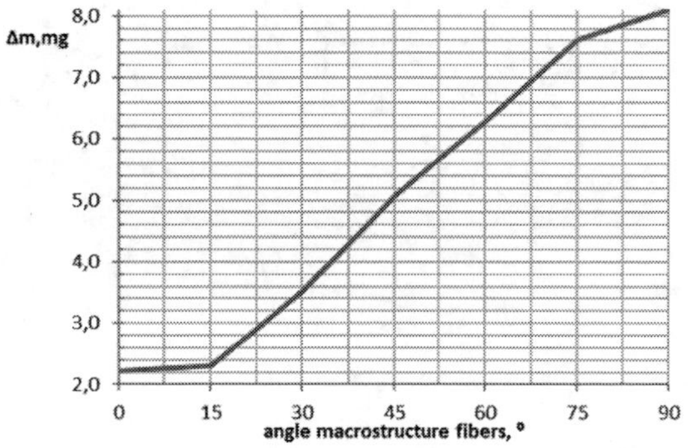

Fig. 8. Graph of dependence of Δm on the yield angle of macrostructure fibers for the experiment on a machine of translational type.

Table 2. Results of the experiment 2. Abrasive wear on a machine of rotation type. 4 loads, 11 minutes.

№	Exit angle of fibers, ⁰	Mass before abrasion, g	Mass after abrasion, g	Δm,mg
1	0	22,1646410	22,16068667	3,9533
2	15	22,1569223	22,15326521	3,7667
3	30	22,15692009	22,15233333	4,5867
4	45	22,15233333	22,14766667	4,6667
5	60	20,2024401	20,19751333	4,9267
6	75	20,4988112	20,4910902	7,7200
7	90	20,7945721	20,78317333	11,3967

Table 3. Results of the experiment 3. Abrasive wear on a machine of rotation type. 5 loads, 6 minutes.

№	Exit angle of fibers, ⁰	Mass before abrasion, g	Mass after abrasion, g	Δm,mg
1	0	20,5518901	20,54634667	5,5433
2	15	22,1276500	22,12146667	6,1833
3	30	22,1777113	22,1708607	6,8500
4	45	18,7643001	18,757120	7,1800
5	60	21,1643612	21,15666333	7,6967
6	75	20,7880202	20,77521333	12,8067
7	90	20,3783611	20,36505333	13,3067

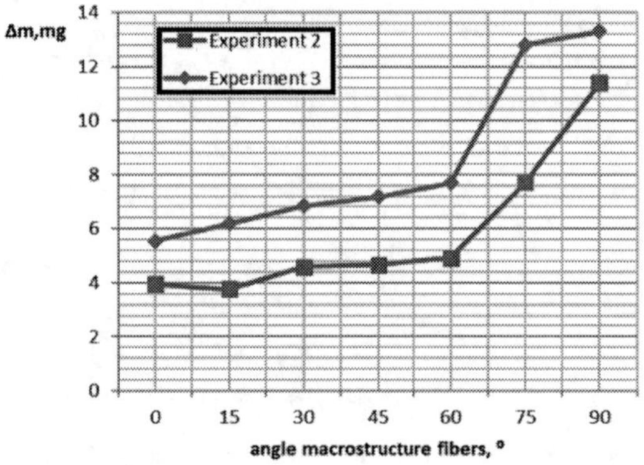

Fig. 9. Graph of dependence of Δm on the yield angle of macrostructure fibers for the experiment on a machine of rotation type.

According to the graphs presented, it can be seen that wear increases with an increase in the exit angle of the macrostructure fibers.

We should note that the on graph of the experiment 1 (Fig. 8) we observe 2 smoothed regions: in the corners area 0^0 ... 15^0 and in the corners 75^0 ... 90^0. In the range of angles from 0^0 to 15^0 there is an insignificant increase in wear of 27.4% ... 28.3% of the maximum wear (8.106 mg). Therefore, when stamping forgings with an exit angle of the macrostructure fibers to contact surfaces in the range 0^0 ... 15^0, we obtain a minimal wear on these surfaces with progressive wear, which increases the working life of the part. Nevertheless, the nature of the graphs indicates that the wear process for the translational and rotational motion is different. In the forward motion, the dependence of wear on the fiber exit angle increases continuously (Fig. 8), with rotational motion (Figure 9) - the graphs have an explicit inflection point at an angle of 60^0, after which wear rises sharply. Also, the graphs of experiments 2 and 3 (Fig. 9) do not contain smoothed regions. This does not contradict the assumption that the range of angles from 0^0 to 15^0 is the most preferable and there is minimal wear.

5 Conclusions

1. As a result of the conducted research, it was revealed that the exit angle of the macrostructure fibers to the contact surface significantly affects the wear process of the working surfaces of the parts.
2. The process of wear differs for translational and rotational motion.
3. The maximum wear occurs at the most unfavorable angle of 90 °; the least wear is at an angle of 0 °. The wear at an angle of 90^0 is 3.65 times greater than at an angle of 0^0 for translational motion. The maximum wear is 2.4 times the minimum for a rotational movement.
4. In the range of angles from 0^0 to 15^0, there is an insignificant increase in wear. After reaching the angle of 60^0, a significant increase in wear values occurs depending on the fiber exit angle. Proceeding from this range from 60^0 to 90 is undesirable for ingots obtained by hot stamping.
5. For forward motion at the top of the graph after reaching 75^0, the rate of increase in wear values decreases. At an angle of 75 °, wear is 93.8% of the maximum value.
6. Therefore, it is recommended to design processes of hot stamping so that the exit angle of the macrostructure of the fibers to the contact surface ranged from 0^0 to 15^0.

This scientific research is supported by the Foundation for the Promotion of Small Forms of Enterprises in the Scientific and Technical Sphere under the program "UMNIK" on the theme "Development of the technological process of cross-wedge rolling of the blank for hot stamping of the crankshaft in order to improve the mechanical properties" under contract No. 11294GU / 2016 of 04/07/2017.

References

1. M.P. Kozochkin, A.N. Porvatov, J. of Frict. and wear **5**, 389 (2014)
2. E.I. Semenov, O.A. Belokurov, V.Ju. Lavrinenko, Blank prod. in mech. Engineering **5**, 28 (2005)
3. M.G. Allenov, O.A. Belokurov, Eng. J.: Sc. and Inn., **10**, http://engjournal.ru/articles/1422/1422.pdf (2015)
4. E.I. Semenov, O.A. Belokurov, V.Ju. Lavrinenko, Mech. of def. body and met. working with pressure, **1**, 55 (2001)
5. M.G. Allenov, O.A. Belokurov, Bull. of MSTU Stankin, **4**, 67 (2016)

Investigation of structure and properties of linear acetylenic carbon coatings for medical polymeric materials

Valentin Morozov[1,*], *Alexey* Zhdanov[1], and *Leonid* Belyaev[1]

[1] Department of Mechanical Engineering, Vladimir State University named after Alexander and Nikolay Stoletovs, Vladimir, Russia

Abstract. Results of researches of structure and properties of the nanostructured carbon coatings deposited on a polyurethane substrate are presented in article. Possibility of application of the scanning probe microscope methods and sclerometer method for quality control of the nanostructured carbon coatings is shown. The results of researches showing that the nanostructured carbon coatings deposited on polyurethane have high degree of adhesion to a substrate, reduce porosity of surface and increase its durability and wear resistance, increase elasticity and coefficient of restitution at deformation are given.

1 Introduction

In developing of biomedical materials one of the most important the problem is a biological compatibility which is caused by influence of the biological environment on a material and influence of a material on surrounding fabrics. Materials suitable for these purposes are the polymers steady against influence of biological environments, and also not undergoing metabolic destruction in an organism.

At the present time absolutely biocompatible polymeric material doesn't exist. The main groups of polymers which are using in medicine aren't deprived in this regard a set of shortcomings. Thus, need of search of such way of protection of biomedical materials which would keep their positive properties for a long time is obvious, and at the same time would provide their sterility and biological inertness, определяемою with their functional purpose.

Biological compatibility of carbon with sp-hybridization of atoms is a priori predicted by Rouf R. et Williams D. in 1975 [1]. Besides, the oriented layers of sp-carbon are capable to change lattice parameter, i.e. are labile, pass ions and molecules of liquids, working as the molecular filter. The last property is explained by van der Waals' interaction of chains among themselves that does possible their drawing apart at interaction with molecules and ions [2-3]. Other major property of coatings based on linear acetylenic carbon (LAC) is uniqueness of their mechanical properties: adhesions; durability and elasticity [4-6].

Actively being applied in medicine polyurethane is among polymers. One of scopes of polyurethane is their use at production of the artificial ventricles of heart (AVH) for left ventricle assists systems (LVAS) and artificial heart (AH) devices.

Primarily, the choice of polyurethane is due to its physical and mechanical properties: elasticity or rigidity, mechanical strength, transparency. Besides that, the following factors are taken into account: manufacturability of production and low labor intensity of processing, the possibility of sterilization of the final product made from the selected material.

However purely technical polyurethane – biologically active material, and it can't be used for products of medical appointment. The polyurethane used in medicine, has own polymeric structure and contains high percent of the additives providing decrease in their toxicity and demanded mechanical properties. Therefore it is necessary to use such methods of preparation of a material which would keep demanded mechanical characteristics and provided biocompatible properties of a material.

One of solutions of this problem is modification of surface layer of the main material. For modification of surfaces of polymers, including AVH it is used or texturing of [7] surfaces, or depositing various on a chemical composition and the nature of coatings [8, 9]. Thus, one of the perspective directions on creation of biomedical materials with high mechanical properties and biocompatible indicators is modifying of surfaces of polymers due to formation of single-layer, multilayered, simple or difficult chemical composition coatings.

In this work results of researches of physico-mechanical and adhesive properties of the carbon coatings deposited on a surface of polyurethane for its further application as a material for production of AVH are described.

2 Materials and Methods

Carbon coating thickness ~ 500 Å, deposited on polyurethane VITUR PF TM-0333-95 by ion plasma deposition method. VITUR PF TM-0333-95 was selected as material, which have a positive toxicological certificate issued by FGBI "All-Russian Scientific and Research Organization for Medical Engineering" of Federal Service on sur-

* Corresponding author: blv_vlsu@mail.ru

 DOI: 10.1051/matecconf/201712902026

veillance in healthcare" (Moscow, Russia). The basic physico-mechanical properties of these materials are shown in the table 1.

Table 1. The basic physico-mechanical properties of VITUR PF TM-0333-95

Ultimate tensile strength, MPa	Elongation at break, %	Shore hardness, method A	Minimal tear resistance, N/mm	Stress at 100 % strain, MPa
30	380	95	120	5

The structure of deposited LAC film is shown in figure 1.

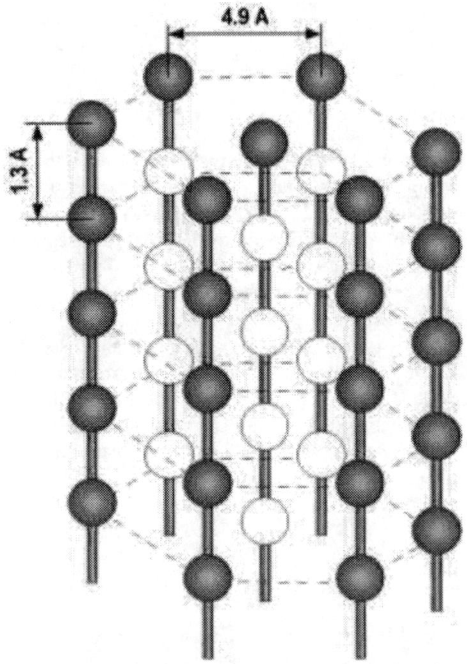

Fig. 1. The model image of deposited LAC film structure

The structural analysis and measurement of physico-mechanical properties was carried out on a scanning electron microscope (SEM) "Quanta 200 3D", atomic-force microscope (AFM) "NTEGRA Aura", scanning nanohardness tester "Nanoskan-3D". The main difference of scanning nanohardness tester of the "Nanoskan" series from classical scanning probe microscopes (SPM) is application of a piezo-ceramic probe sensor with high bending stiffness of the cantilever. Using of the resonance oscillation mode allows exercising control of contact between probe and surface in two parameters: change of amplitude and oscillation of a probe. The resonant operating mode of a probe provides high stability of amplitude and frequency of oscillation and guarantees soft contact between probe and material surface during scanning process. On the basis of the scanning nanohardness tester "Nanoskan-3D" for measurements of mechanical properties of a surface the method of a measuring indentation according to recommendations of the international standard ISO 14577, and also a sklerometer method is realized. The sklerometer method has a several advantages than cave-in methods at measurement of hardness of films at a nanoscale [10]. Scratching with variable loading gives the chance to determine at the same time some parameters of a film within one measuring procedure: area of elastic interaction, threshold loading at which plastic deformation, conditions film peeling begins. Comparative researches of physico-mechanical characteristics of an initial surface of polyurethane and sample with the put covering it was made on the scanning nanohardness tester "NanoScan-3D". The Berkovich tip of indenter was applied to mechanical tests.

3 Results

Micrographs of polyurethane surfaces received by means of SEM are presented in fig. 2. Application of this method of research allows assuming only that result of deposition on polymer surface of the thin film having non-continuous which quantity decreases with increase in thickness of a coating is formed.

Obviously, that this technique of research of such coatings (films) is low-informative and demands application of other methods of structural researches, for example – scanning probe microscopy (SPM).

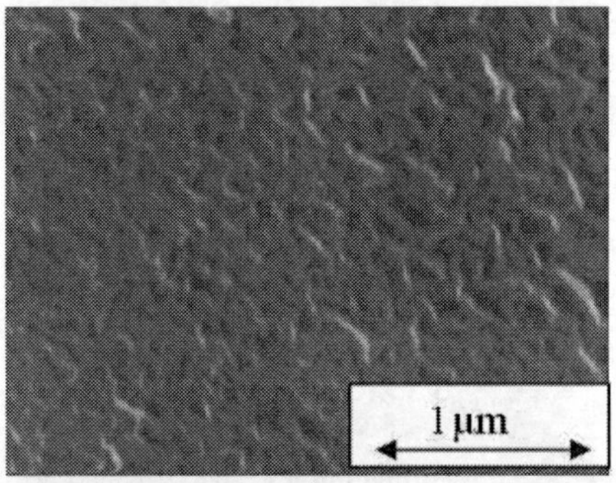

a)

b)

Fig. 2. Micrographs (SEM) of polyurethane surface: a) without film; b) with LAC deposited film

AFM-image of a surface samples with film and without coating are presented in fig. 3. The presented images

DOI: 10.1051/matecconf/201712902026

have distinctions, confirming coatings about existence on a polyurethane surface. The flat and volume image of a sample with a carbon film is submitted in fig. of 3 c, d.

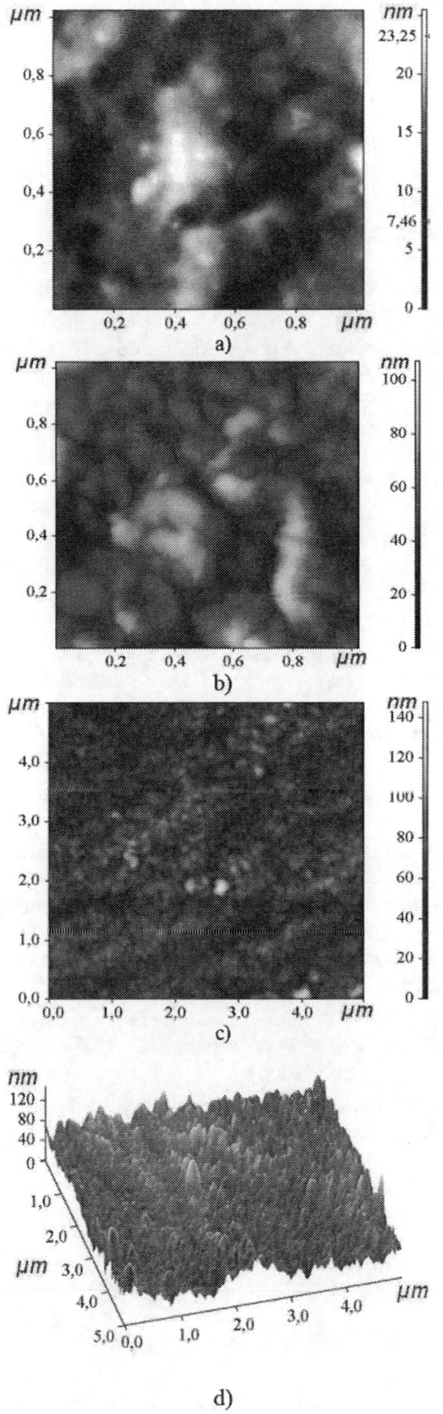

Fig. 3. AFM-images of polyurethane surface: a) without film; b-d) with LAC deposited film

The analysis of the surface relief of sample allowed determining main parameters of microroughnesses of surface with the deposited coating: maximum roughness depth (Rmax) – 147.843 nm; arithmetic average of the absolute values of the roughness profile (Ra) – 11.931 nm; average maximum height of the profile (Rz) – 125.573 nanometers.

The method of a measuring indentation obtained data on values of hardness, the elasticity module, and coefficient of elastic restoration. Range of the maximum loadings at an indentation was from 100 µN to 5 mN. Feature of studied materials is the high coefficient of restitution at the indentation, reaching ~ 95%. This effect is manifested in the fact that the load curve on the «load-indentation» chart almost coincides with the discharge curve (fig. 4).

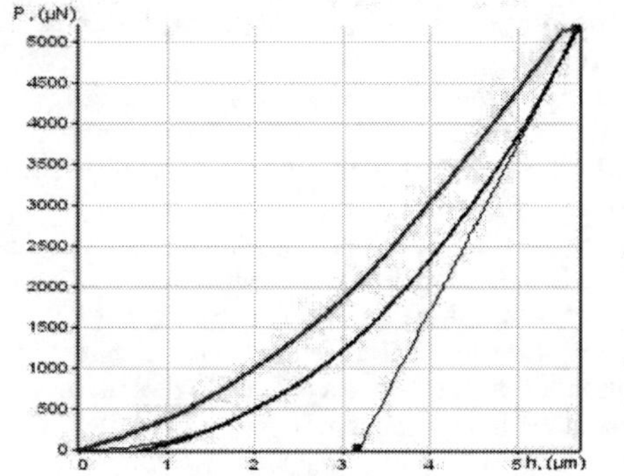

Fig. 4. Correlation between load P and depth of indentation h for coated polyurethane substrate

On the presented diagram of dependence of hardness from depth of indentation of an indenter (fig. 5 a) values of hardness of samples and without it practically coincide with a coating within casual variability of data this results from the fact that the loading curve almost coincides from curve unloading.

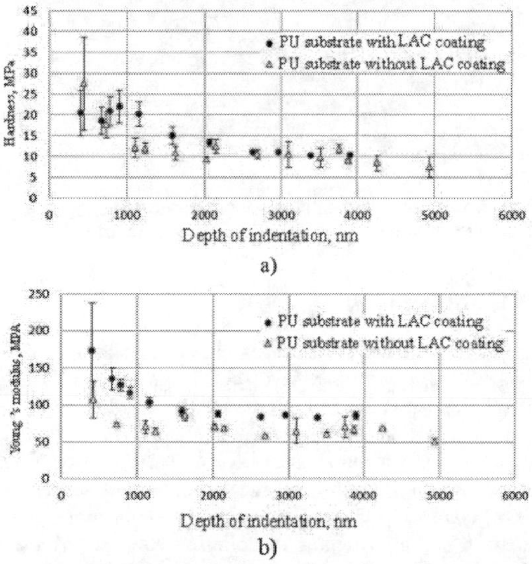

Fig. 5. Mechanical properties of research samples (the load range: 0,1 to 5 mN): a) hardness; b) Young's modulus

At the same time, the values of the module of elasticity calculated on the same experimental data (fig. of 5 b), show considerable (1.5-2 times) module excess on a surface with a coating.

The presented data (fig. 6) show that essential distinction of mechanical properties is observed with a depth of indentation less than 1 μm. This size is connected with a ratio of depth of an indentation and film thickness: with a depth of indentation of the indenter, a film comparable to thickness measured value generally is defined by properties of a substrate. At the same time, the measured module of elasticity of a surface with a film remains to higher, than at an initial material at depths of indentation of the indenter, many times exceeding film thickness. It testifies that the coating keeps elastic properties and doesn't destroy at large relative deformations.

The sklerometer method was applied to research of durability and adhesive properties of coating with variable loading. Scratches about 75 μm length were put on surfaces of samples, load changed from 0 to 5 mN. The scratch test of polyurethane without coating leads to formation of scratches with indistinctly outlined edges up to 500 nm in depth. Similar experiment on a sample with a covering (fig. 6) doesn't lead to essential deformation of a surface, except for flows at the end of scratch with the maximum loading 5 mN, connected, apparently, with partial transfer of a material on a surface. Thus film flaking from a substrate that speaks about good adhesive properties of a covering isn't observed.

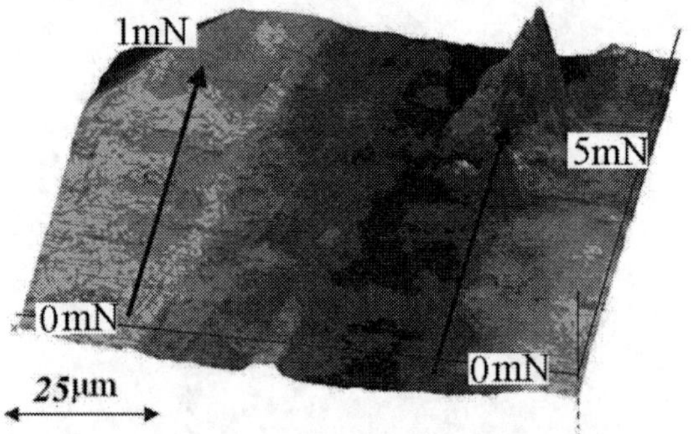

Fig. 6. AFM-image of surface sample with LAC after variable load scratch test: 0 to 1 mN (left) and 0 to 5 mN (right)

4 Conclusions

Results of the conducted researches of a relief and physico-mechanical properties of coverings on the basis of LAC deposited on polyurethane, the SPM methods, a nanoindentation and sklerometer method show that deposited coatings reduce porosity of a surface, increase its durability and wear resistance, increase elasticity and coefficient of elastic restoration at deformation. The coating has high degree of adhesion to a substrate and doesn't exfoliate at mechanical influence. It is shown that the SPM methods, a measuring indentation and sklerometer method can be used for quality control of carbon coatings on products from polyurethane.

References

1. D.Williams, R. Rouf, *Implants in surgery* (in Russian), Medicine, Moscow, (1978).
2. A. F. Aleksandrov, M.B. Guseva, Yu. V. Korneeva, et al., *The results and perspectives of biocompatible forms linear acetylenic carbon application in medicine* (in Russian), Integral., **5** (61) 27-31, (2011).
3. V.G. Babaev, N.D. Novikov, M.B. Guseva, *GCK – carbon – unique nanophase of carbon* (in Russian), Nanotechnologies: development, application., **1** 53-68 (2010).
4. D.A. Malinin, Yu. G. Aleksandov, N.D. Novikov, *Use linear acetylenic carbon coating for the purpose of protection of polymeric artificial limbs of an eardrum against destruction and immunological reactions*, Nonmaterial, Nanotechnologies - medicine., **10**, 64-69 (2007).
5. N.V. Dobrova, *Research of vascular artificial limbs from carbon contain fiber Vitlan*. Topical issues of the organization, prevention and surgical treatment of diseases of the main vessels (in Russian), Medicine, Moscow, (1991).
6. D.A. Malinin, Yu. G. Aleksandov, A.M Al-Naser, *Way of preservation of the implants applied in myringoplasty*, Modern problems of physiology and hearing pathology **1**, 103 – 104 (2011).
7. P. B. Kwant, *Implantable electromechanical displacement blood pumps: systematic design and validation methods*, (2007) Available at: (http://darwin.bth.rwthaachen.de/opus3/volltexte/2008/2380/pdf/Kwant_Paul.pdf) .
8. J. Sarna, R. Kustosz, R. Major, J. Lackner, B. Major, *Polish Artificial Heart – new coatings, technology, diagnostics*, Bull. Pol. Ac.: Tech. **58** (2), 329–335 (2010).
9. J. Lavee, Y. Paz, *Mechanical Alternatives to the Human Heart: Paracorporeal Assist Systems*, IMAJ, 4, 125 – 130, (2002).
10. A.S. Eseinov, K.V. Gogolinsky, V.N. Reshetov, *Measurement of mechanical properties of superfirm diamondlike carbon coverings* (in Russian), News of higher education institutions. Chemistry and chemical technology, 7 51-54 (2011).

Effect of drying methods of Al_2O_3-GO powder mixture on the properties and microstructure of sintered composites obtained by spark plasma sintering

Pavel Fokin[1,*], *Pavel* Peretyagin[1], and *Anton* Smirnov[1]

[1]Moscow State University of Technology "STANKIN", Vadkovskij per. 1, Moscow, 101472, Russian Federation

Abstract. Well-dispersed alumina ceramic-graphene oxide (GO) powder mixtures were fabricated using a colloidal processing route and three drying techniques were applied: in vacuum, spray and freeze dryers. Dense composites were obtained via spark plasma sintering, the technique reducing GO to graphene in situ during the sintering process. The mechanical properties of the sintered composites were investigated. The composites obtained by spray drying have shown the fracture toughness of 4.5 ± 0.4 MPa·m$^{1/2}$, and hardness of 18.4 ± 0.8 GPa with an extremely high dispersion of graphene within the ceramic matrix.

1 Introduction

The design of ceramic composite materials comprising two or more phases where at least one shows dimensions in the nanometre range has gained increased attention during the past few years.

Alumina (Al_2O_3) is the most widely used oxide ceramic material. Its applications are widespread, and include spark plugs, tap washers, pump seals, electronic substrates, grinding media, abrasion resistant tiles, cutting tools, bioceramics, body armour, laboratory ware and wear parts for the textile and paper industries [1]. However, the greatest drawback of ceramics is their low fracture toughness.

Graphene seems as a suitable filler for fracture toughness improvement in ceramic composites. The addition of graphene as a second phase with its unique combination of thermal [2], electrical [3] and mechanical properties [4] improves the performance of the bulk composite.

Graphene oxide (GO) is graphite that has been oxidized to intersperse the carbon layers with oxygen molecules, and then reduced, to separate the carbon layers completely into individual or few layer graphene. In addition, GO has been considered widely as a prominent precursor and a starting material for the synthesis of this processable material. GO is of great interest due to its low cost, easy access, and widespread ability to convert to graphene. Scalability is also a much desired feature.

Therefore, in order to improve mechanical properties, especially fracture toughness, GO was added to the corresponding ceramic matrix [5].

Spark Plasma Sintering (SPS) was chosen as rapid consolidation technique [6-9] that allowed reducing the sintering temperature and avoiding undesirable grain growth of alumina.

Moreover, the characteristics of powder mixtures were influenced by the different drying methods and, consequently, sintered samples properties were depend on them as well. However, the aspects of drying methods have been poorly studied yet.

The purpose of the present study was to obtain dense Al_2O_3-GO ceramic composites with enhanced mechanical properties and to describe the influence of different drying methods on the quality of a powder mixtures and, consequently, on sintered materials.

2 Materials and characterization

2.1 Raw materials

Commercially available α-alumina nanosized powder (Plasmotherm, Moscow, Russia) with a chemical purity of 99.5% and average particle size of 180 nm was used as a matrix material. Graphene oxide was synthesized by using a modified Hummer's method through oxidation of commercial graphite powder with a chemical purity of 99.9% [10].

2.2 Powder processing and sintering

Suspension of Al_2O_3-0.5 vol.% GO was prepared using a colloidal method [5]. Al_2O_3 powder were added to 100 mL of water at a pH 10, previously fixed by adding NH_4OH. The Al_2O_3 powder was dispersed by stirring for 30 min. Then, the appropriate amount of the graphene oxide suspension was dropwise to the ceramic powder

* Corresponding author: pashafokin94@mail.ru

DOI: 10.1051/matecconf/201712902027

dispersions. The mixed powder suspensions were stirred for 1 h.

Slurry of powder was prepared using distilled water as liquid media via dry ball mill (ML-1, Kaluga, Russia), containing alumina balls, in polyethylene containers at 150 rpm during 24 h.

After homogenization mixture was dried using three different equipments: Mini Spray Dryer B-290 (BÜCHI Labortechnik AG, Flawil, Switzerland) spray dryer equipped with a nozzle, in a vacuum drying oven VO 200 (Memmert GmbH Co. KG, Schwabach, Germany) and FreeZone 2.5 freeze-drying system (LabConco, Kansas, MO, USA).

The spray dryer was composed of a cylindrical chamber with an 800 mm internal diameter and 564 mm height, followed by a conical chamber of 564 mm high with a 60° angle. The powder was separated off the stream of hot air in a cyclone and collected directly into a glass pot. Operational conditions of the spray dryer were set as follows: feed flow rate: 20 mL/min; air flow rate: 130 m^3/h; inlet air temperature 200 ± 2°C; outlet air temperature, varies with inlet temperature.

Meanwhle, in freeze-drying system the suspension was dried at the collector temperature of -50 ± 2°C. Furthermore, the shell temperature and the chamber pressure were kept at +23 ± 2◦C and 0.02 ± 0.01 mbar, respectively, during the entire process.

These methods make it is possible to obtain a "ready-to-sinter powder" without sieving.

On the other hand, after drying in vacuum oven at the 90 ± 1°C the resulting powder was ground in an agate mortar and subsequently passed through a 75 µm sieve in order to avoid agglomeration.

Powder densification was performed by SPS KCE-FCT-H-HP-D25-SD (FCT Systeme GmbH, Rauenstein, Germany) at a maximum temperature of 1000-1400°C (increments of 100°C), reached under vacuum at a heating rate of 100°C/min, and an applied pressure of 80 MPa. The sintered specimens had diameters of 20 mm and thicknesses of 4 mm. For comparison purposes Al_2O_3 powders without GO were SPSed following the same sintering cycle.

2.3 Microstructural characterization

The true density of powder was measured using a gas pycnometer AccuPyc II 1340 (Micromeritics, Norcross, USA). The device operation principle involves sample placing into a calibrated chamber filled with helium. To clean the sample surface from water vapors and impurities, one need to start from several helium blowdowns, and after that the analysis itself can be conducted. The gas molecules penetrate the tiniest sample pores. In other words, the sample pushes out the gas volume equal to the volume of the true solid phase only. In case the sample weight was preliminary entered, the device will calculate its density.

Scanning electron microscopy (SEM) characterization was carried out on polished down to 1 µm surfaces by VEGA 3 LMH (SEM Tescan, Brno, Czech Republic).

The density of the sintered samples (ρ) was measured in distilled water using Archimedes' principle and was compared with the theoretical value, calculated according to the rule of mixtures. Raman spectra of as-prepared powders and sintered samples were collected to identify the phase composition. The Raman setup is composed of a laser (DXR^{TM}2 Raman Microscope, Thermo Fisher Scientific, MA, USA) with a wavelength of 532 nm and a laser power of 2.0 mW. The laser beam was focused through an optical microscope's 50× objective lens to a spot size of 50 µm on the studied area (from different spots, at an interval of 200 nm). The accumulation time for each Raman spectrum was about 10 s.

Vickers hardness, *Hv*, was measured on polished surfaces using a Vickers diamond indenter (QNess A10 Microhardness Tester, Salzburg, Austria), applying a load of 1 kg and an indentation time of 10 s. The hardness results were averaged over 10 indentations per specimen.

Fracture toughness (K_{Ic}) was determined from Vickers indentations obtained with a load of 10 kg for 10 seconds. The indentation sizes and crack lengths were measured using SEM. The method and formulas for calculating *Hv* and K_{Ic} have been reported elsewhere [11].

3 Results and discussion

3.1 Microstructure and Raman characterization

The morphology of Al_2O_3-GO obtained of the scanning electron microscope shown in Fig.1. The powder particles after spray drying (Fig.1A) are spherical and have a narrow size distribution (d_{50} ≈ 20 µm). The microstructure of the powder obtained by freeze-drying (Fig.1B), consists of separate fragments of the layered structure in the form of flakes with an average grain diameter (d_{50} ≈ 400 µm). Particles, dried in a vacuum oven (Fig.1C), have a variety of shapes with an average grain size (d_{50} ≈ 200 µm).

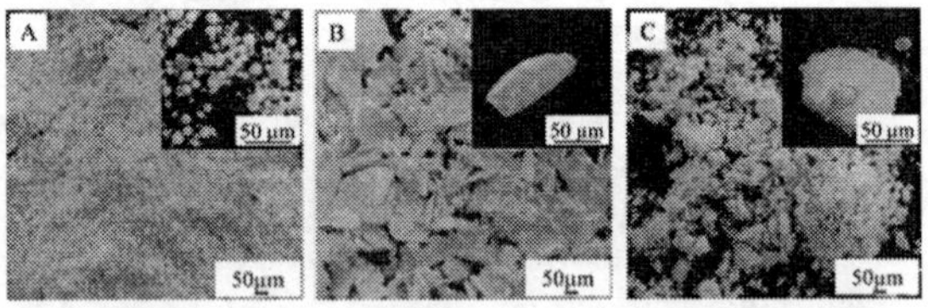

Fig. 1. SEM images of Al_2O_3-GO powder mixtures obtained by spray (a) freeze (b) and vacuum (b) drying.

Raman spectroscopy is an effective method of chemical analysis for studying the composition and structure of substances, allowing to see the thermal recovery of graphene oxide. Fig. 2 a-c shows the Raman spectra that correspond to the mixtures before sintering by SPS; they reveal that all compositions consist of alumina and typical GO peaks. The broad G peak and negligible 2D peak are characteristic of sp^1, sp^2 and sp^3 hybridized carbon–carbon bonds in graphene. The D bands at ~1350 cm^{-1}confirm the lattice distortions. On the other hand, for the specimens sintered at the temperature of 1400°C (Fig.2 d-f), the intensity of the D peaks decreases while G

DOI: 10.1051/matecconf/201712902027

peaks increases. This indicates partial heat recovery of graphene oxide, which occurs at a temperature of 1400°C.

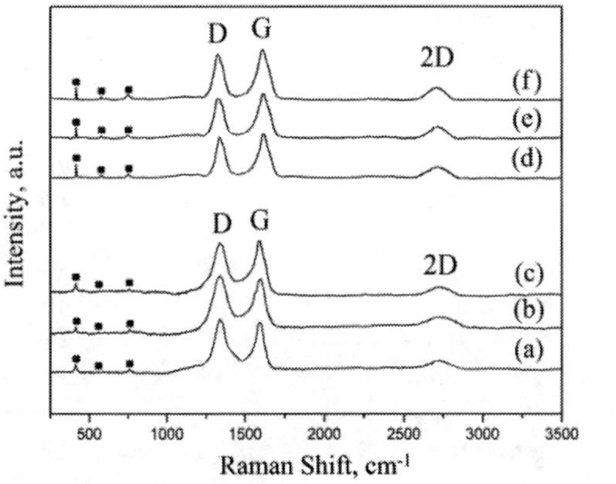

Fig. 2. Raman spectra of as-prepared (a-c) and sintered (d-f) at 1400°C powders Al_2O_3-GO, prepared spray, vacuum and freeze drying, respectively. "▪" label denotes alumina peaks. "D", "G" and "2D" mark GO peaks before and graphene peaks after sintering, respectively.

Figure 3 demonstrates the fracture surface of spray dried and sintered Al_2O_3-GO (Fig. 3B) composite and Al_2O_3 (Fig. 3A) ceramic as well. The average grain size was found to be 3 ± 0.38 μm and 0.8 ± 0.33 μm for Al_2O_3 and Al_2O_3-GO samples, respectively.

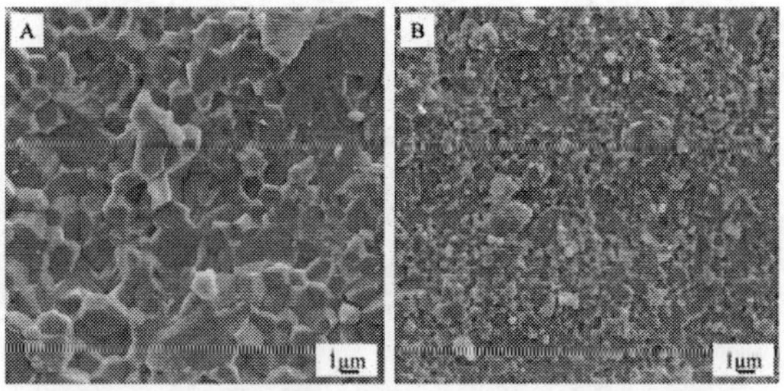

Fig. 3. SEM images showing the representative morphology of Al_2O_3 (A) ceramic and spray dried Al_2O_3-GO (B) ceramic composite fracture surface.

3.2 Mechanical properties

The density of all sintered samples increased with increasing sintering temperature and reach its maximum value (99 – 99.8 % th.) at 1400°C regardless of drying methods applied.

The influence of sintering temperature on hardness and fracture toughness of sintered composites presented on Fig.4 and Fig.5, respectively. Spray dried and sintered at a maximum (1400°C) temperature samples showed the highest value of hardness (≈ 18.4 ± 0.8 GPa). It is explained by the fact that this drying process much faster than other methods and allows to avoid agglomeration of GO in ceramic matrix that often acts as a weaker source for decrease a material's mechanical properties.

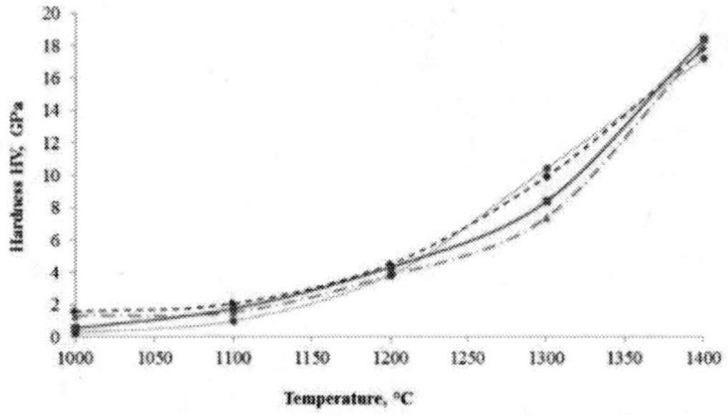

Fig. 4. Influence of sintering temperature on hardness of sintered composites obtained by vacuum (- - -), spray (——), freeze (— · —) drying, and monolithic ceramic Al_2O_3 (···).

Figure 5 shows that the measured fracture toughness of all samples revealed a similar trend, i.e. increased to a maximum value with increasing sintering temperature. The spray dried and sintered Al_2O_3-GO composites exhibited the highest fracture toughness of 4.5 ± 0.4 MPa·m$^{1/2}$ as compared to 3.6 ± 0.4 MPa·m$^{1/2}$ measured for the monolithic ceramic. This value confirms that graphene is well dispersed in the ceramic matrix as well.

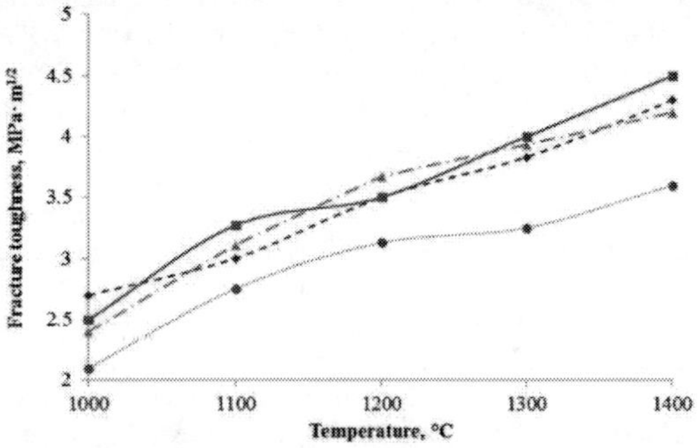

Fig. 5. Influence of sintering temperature on fracture toughness of produced composites obtained by vacuum (- - -), spray (——), freeze (— · —) drying, and monolithic ceramic Al_2O_3 (···).

Fig. 6 shows the Vickers indentation cracks induced on the surfaces of Al_2O_3 ceramic and spray dried Al_2O_3-GO composite. Fig. 6A, corresponding to the Al_2O_3 ceramic, shows a crack path that is mainly comprised of transgranular fractures which, consequently, implies a lower fracture toughness than that of the Al_2O_3-GO spray drying composite. Graphene-reinforced ceramic matrix composites (Fig. 6B) show singly and doubly deflected crack. This effect may be due to the nanodispersed graphene nature. Graphene consists of nanoflakes with a high aspect ratio and could leads to deflection of a crack, additional energy dissipation, and hence, to the enhanced fracture toughness of composites. In addtition, the fracture surface of Al_2O_3-GO composites show a more intricate surface (Fig. 3B) due to a more intergranular fracture and, as a consequence, a higher fracture energy.

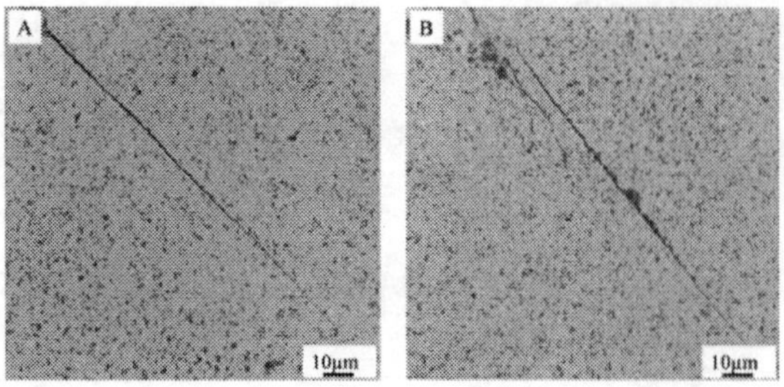

Fig.6. Vickers indentation cracks induced on the surfaces of: Al_2O_3 (A) and Al_2O_3-0.5 vol.% GO (B) samples sintered at 1400°C.

4 Conclusion

Al_2O_3-0.5 vol.% graphene oxide composites have been successfully fabricated by combining a colloidal processing route and Spark Plasma Sintering. Results showed that the addition of low graphene oxide contents to an Al_2O_3 ceramic matrix simultaneously improved fracture toughness and hardness. This mechanical upgrade takes place when small quantities of graphene are homogeneously dispersed in the matrix. The composites containing 0.5 % GO obtained after spray drying method possess the highest fracture toughness (4.5 ± 0.4 MPa·m$^{1/2}$) and hardness (18.4 ± 0.8 GPa) among all studied materials due to obtained a homogeneous and agglomerate-free dispersion of GO inside the alumina matrix. Compared to the Al_2O_3 samples, the samples with graphene oxide in a percentage ratio increased the rate of fracture toughness and hardness by 28% and 6%, respectively. Further scientific research in this area are needed for understanding of impact of graphene/ceramic interfaces on the reinforcing mechanism in composites. The uniform dispersion of graphene and composite's structure retention and control which affect the microstructural evolution during processing might also represent the prospective goal for further research.

Authors would like to thank The Ministry of the Russian Federation supported this work in the frame of Governmental Regulation of the Russian Federation No. 220, 9 April 2010 by contract 14.B25.31.0012, 26 June 2013.

References

1. I.W.M. Brown, W.R. Owers, Curr. Appl. Phys. **4**, 171 (2004)
2. A.A. Balandin, S. Ghosh, W.Z. Bao, I. Calizo, D. Teweldebrhan, F. Miao, C.N. Lau, Nano Lett. **3**, 902 (2008)
3. T. Stauber, N. M. R. Peres, F. Guinea, Phys. Rev. B **76**, 205423 (2007)
4. C. Lee, X. D. Wei, J. W. Kysar, J. Hone, Science **321**, 385 (2008)
5. S. Grigoriev, P. Peretyagin, A. Smirnov, W. Solís, L. A. Díaz, A. Fernández, R. Torrecillas, J. Eur. Ceram. Soc. **37**, 2473 (2017)
6. C. F. Gutiérrez-González, M. Suarez, S. Pozhidaev, S. Rivera, P. Peretyagin, W. Solís, L. A. Díaz, A. Fernandez, R. Torrecillas, J. Eur. Ceram. Soc. **36**, 2149 (2016)
7. N. W. Solís, P. Peretyagin, R. Torrecillas, A. Fernández, J. L. Menéndez, C. Mallada, L. A. Díaz, J. S. Moya, J Electroceram. **38**,119 (2017)
8. S.S. Pozhidaev, A.E. Seleznev, N.W. Solis Pinargote, P. Yu. Peretyagin, Mechanics & Industry **16**, 710 (2015)
9. C.F. Gutierrez-Gonzalez, N.W. Solis Pinargote, S. Agouram, P. Y. Peretyagin, S. Lopez-Esteban, R. Torrecillas, Mechanics & Industry **16**, 703 (2015)
10. W. S. Hummers Jr., R. E. Offeman, J. Am. Chem. Soc. **80**,1339 (1958)
11. A. Smirnov, H. D. Kurland, J. Grabow, F. A. Müller, J. F. Bartolomé, J. Eur. Ceram. Soc. **35**, 2685 (2015)

Comparative studies on mechanical properties of WC-Co composites sintered by SPS and conventional techniques

Yuri Pristinskiy[1,*], *Nikita* Peretyagin[1] , and *Nestor Washington* Solis Pinargote[1]

[1]Moscow State University of Technology "STANKIN", Vadkovskij per. 1, Moscow, 101472, Russian Federation

Abstract. Spark plasma sintering (SPS) is an extremely fast solidification technique for compounds that are difficult to sinter within the material group metals, ceramics, or composites thereof, SPS uses a uniaxial pressure and a very rapid heating cycle to consolidate these materials. With SPS the main benefit is the ability to control the WC grain size due to the short sintering times at high temperature. Additionally, its allows to avoid negative reactions between WC and cobalt and to minimize the formation of undesirable phases in sintered composites. The WC-6wt.% Co cermet prepared by SPS processing achieves the enhanced mechanical properties with the hardness of 18.3 GPa and the fracture toughness of 15.5 MPa·m$^{1/2}$ in comparison to standard reference tungsten carbide/cobalt material.

1 Introduction

Today, the tungsten carbide (WC) and cobalt (Co) alloy is a widely spread material for tool making [1]. It is used extensively in machining, mining, and making drilling tools [2,3]. Thanks to the presence of cemented carbides, such tools are harder and more heat-resistant. However, they are less strong and are prone to brittle fractures [4]. Increasing the amount of WC fraction and decreasing the growth of its grain during sintering can considerably improve the hardness and toughness; at the same time, in order to increase fracture resistance, the content of cobalt have to be increased [5,6]. Thanks to their good hardness and fracture toughness balance, carbides with 5 to 10 wt. % Co are the most promising materials for making cutting inserts [6]. To improve mechanical characteristics and performance of carbide tools, new processing methods of sintering should be used.

As of today, carbide cutting inserts are conventionally made by such methods as hot pressing [7] and hot isostatic pressing [8]. However, these methods have a number of drawbacks like low sintering rate, high equipment costs, low reproducibility of the output, etc. From this perspective, spark plasma sintering seems a very promising method [9,10].

Spark plasma sintering (SPS) or pulsed electric current sintering (PECS) is a sintering technique utilizing uniaxial force and a pulsed (on-off) direct electrical current (DC) under low atmospheric pressure to perform high speed consolidation of the powder. This direct way of heating allows the application of very high heating and cooling rates, enhancing densification over grain growth promoting diffusion mechanisms allowing maintaining the intrinsic properties of nanopowders in their fully dense products [11,12]. The heating rate during the SPS process depends on the geometry of the container/sample ensemble, its thermal and electrical properties, and on the electric power supplier. Heating rates as high as 1000 °C/min can be achieved. As a consequence, the processing time typically takes some minutes depending on the material, dimensions of the piece, configuration, and equipment capacity [13].

In addition, high pressure and heating rate allow decreasing sintering temperature lower than eutectic temperature (≈ 1320°C) for WC-6wt.% Co system. Therefore, it permits to avoid the appearance of η-phase and prevents the growth of WC grains.

The purpose of the present study was to investigate the dependence of the basic SPS processing parameters, such as temperature, pressure, and time on the mechanical properties of WC-6wt.% Co composites.

2 Materials and characterization

2.1 Raw materials

Commercially available submicron powder WC-6 wt.% Co (Plasmotherm, Moscow, Russia). with an average tungsten carbide particles size of 1.5μm and average cobalt particle is 40 nm was used as an initial material.

2.2 Powder processing and sintering

Powder densification was performed by SPS KCE-FCT-H-HP-D25-SD (FCT Systeme GmbH, Rauenstein, Germany) at a maximum temperature of 1000-1500°C (increments of 50°C), reached under vacuum at a heating rate of 100°C/min, and an applied pressure of 30 MPa. The final temperature and pressure were maintained for 5 min. The sintered specimens had diameters of 20 mm and thicknesses of 5 mm. Taking into account that these parameters were chosen to study only the effect of

* Corresponding author: y.pristinskiy@gmail.com

sintering temperature on the mechanical properties of the samples.

It is known that applied pressure and dwelling time affect the mechanical properties of samples as well [14]. Therefore, for subsequent studies dwelling time were increased up to 10 min and pressure was varied in the range from 30 to 80 MPa.

The commercially available WC-6wt.% Co cermet fabricated by hot press sintering at 1400°C was used as the reference material.

2.3 XRD characterization

XRD analyses (Empyrean diffractometer, PANalytical, Almelo, Netherlands, Cu-Kα radiation, wavelength 1.5405981 Å, accelerating voltage 60 kV, beam current 30 mA) of the raw powder as well as of the sintered samples were conducted in a step scanning mode at diffraction angles 2θ ranging from 25° to 100° (step size 0.05°).

2.4 Microstructural and mechanical characterization

The true density of the raw powder was measured using a helium pycnometer AccuPyc II 1340 (Micromeritics, Norcross, USA). The device operation principle involves sample placing into a calibrated chamber filled with helium. To clean the sample surface from water vapors and impurities, one need to start from several helium blowdowns, and after that the analysis itself can be conducted. The gas molecules penetrate the tiniest sample pores. In other words, the sample pushes out the gas volume equal to the volume of the true solid phase only. In case the sample weight was preliminary entered, the device will calculate its density.

Scanning electron microscopy characterization was carried out on polished down to 1 μm surfaces by VEGA 3 LMH (SEM Tescan, Brno, Czech Republic).

The density of the sintered samples (ρ) was measured in distilled water using Archimedes' principle and was compared with the theoretical value, calculated according to the rule of mixtures.

In order to quantify the WC average grain size, the sintered and polished samples were thermally etched in vacuum for 20 min at a temperature of 1200°C, and the linear intercept method (LIM) was used [15].

Vickers hardness, *Hv*, was measured on polished surfaces using a Vickers diamond indenter (QNess A10 Microhardness Tester, Salzburg, Austria), applying a load of 1 kg and an indentation time of 10 s. The hardness results were averaged over 10 indentations per specimen.

Fracture toughness (K_{1c}) was determined from Vickers indentations obtained with a load of 10 kg for 10 seconds. The sizes of the corresponding indentations and crack lengths were measured using SEM. The *HV* and K_{1c} of refence materials were measured using the same equipment and under the same testing conditions. The method and formulas for calculating *Hv* and K_{1c} have been reported elsewhere [16].

3 Results and discussion

3.1 XRD analyses

X-ray diffraction was carried out for the starting powder and for the sintered sample (1300°C / 10 min / 80 MPa) (Fig.1). XRD analysis revealed the presence of tungsten carbide (■) and cobalt (♦) phases, without any additional peaks appearing.

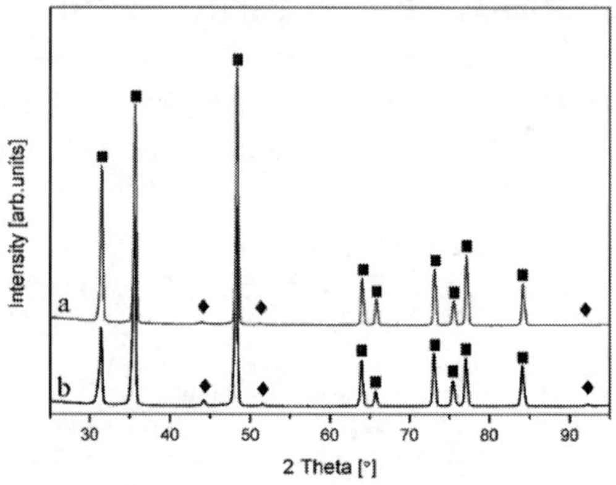

Fig. 1. XRD patterns of raw powder (a) and SPS sintered (1300°C/10min/80MPa) sample (b).

3.2 Microstructure characterization

Typical thermally etched microstructure of the SPS sintered WC-6% Co cermet is shown in Fig. 2. The grains have fairly equiaxed shapes. No residual porosity could be detected in microstructure. The average grain size of WC was about 0.7 ± 0.5 μm.

Fig. 2. SEM images of polished and thermally etched sections of SPS sintered (1300°C/10min/80MPa) WC-6% Co cermet. Light grey phase is carbide particles.

3.3 Mechanical properties

The influence of sintering temperature on density, hardness and fracture toughness of sintered composites presented on Fig.3. All samples sintered at 1300°C and higher temperature were reached the maximum densification (99.0–99.5% th.). High sintering temperature leads to an increase in relative density and decrease in porosities due to the formation of a liquid phase during sintering.

 DOI: 10.1051/matecconf/201712902028

Samples sintered at 1300 °C showed the fracture toughness of 10 ± 0.4 MPa·m$^{1/2}$, and hardness of 17.3 ± 0.8 GPa, that higher than values obtained for the reference material.

However, it is worth mentioning that further increases in temperature do not result in significant changes in hardness while fracture toughness becomes considerably lower (Fig. 3). It is explained by the fact that at temperature higher that eutectic intermediate phases (Co_3W_3C, W_2C or WO_3) can occur and play a negative role in fracture toughness.

Nevertheless, it needs to be noticed, that the mechanical performance of the test specimens was studied only dependent on the differentiating sintering temperatures, at the constant sintering pressure and dwelling time.

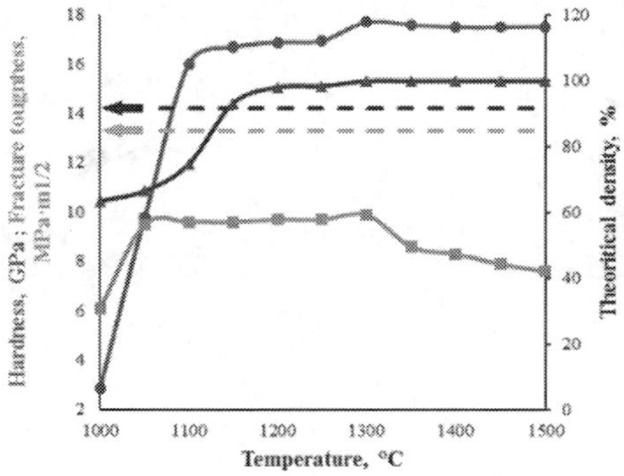

Fig. 3. Influence of SPS temperature on hardness (——), density (——), and fracture toughness (——) of sintered cermet. Fracture toughness (---) and hardness (---) of reference material is presented as well.

In order to investigate the effect of consolidation pressure and dwelling time on the mechanical properties of the cermet at the optimal temperature, dwelling time were increased up to 10 min and pressure was varied in the range from 30 to 80 MPa. Fig. 4 shows the effect of applied pressure on the *HV* and K_{Ic} of WC-6% Co composites.

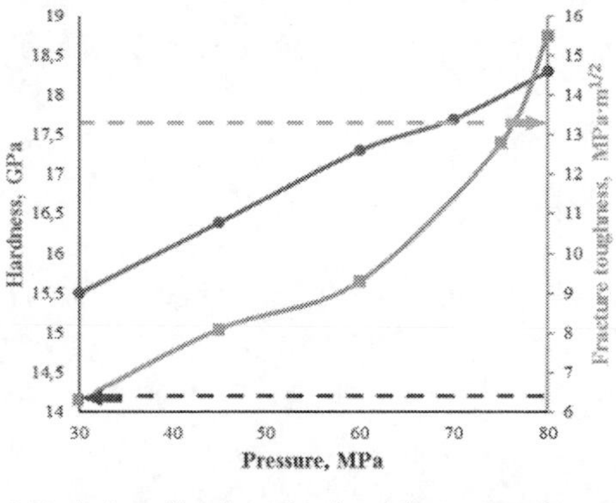

Fig. 4. Evolution of hardness (——) and fracture toughness (——) of WC-6% Co composites sintered at 1300°C with dwelling time 10 minutes under differentiating pressures. Fracture toughness (---) and hardness (---) of reference material is presented as well.

This data shows that the mechanical behavior of these composites improves with higher pressure / dwelling time and reaching a maximum hardness of 18.3 ± 0.6 GPa and a maximum fracture toughness of 15.5 ± 0.3 MPa·m$^{1/2}$ at 80 MPa.

This could be explained by the fact when pressure is applied to the compacted powder, it is transferred through the contacts between the particles and with increasing consolidation processes, both the area and the number of contacts between the powder particles increase. The pressure is manifested in "mechanical" effects, leading to the mixing of particles, the destruction of agglomerates, etc. In addition, the pressure enhances the driving force associated with the surface tension, which in turn leads to the effects of flow and creep, increases diffusion from the grain boundaries. Thus, the applying of pressure during sintering makes it possible to achieve high density at the same temperatures and, accordingly, improve the mechanical properties of the sintered material.

Fig. 5 shows the Vickers indentation cracks induced on the surfaces of sintered sample. The crack path goes along the cobalt binder rather than tungsten carbide grains and not propagate rapidly due to cobalt plasticity. Thus, homogeneous distribution of metal binder in SPSed cermet is crucial not only for better WC grain contiguity, but also can prevent brittle fracture and cracks propagation and, consequently, increase fracture toughness.

Fig.5. Vickers indentation cracks induced on the surfaces of SPSed cermet. Dashed area denotes where close-up (left upper corner) was taken.

In addition, these high mechanical values have been confirmed from the XRD data, that not identify any additional η- and θ- phases that could have negative effects on the mechanical properties (Fig.1b).

4 Conclusion

WC-6% Co cermets have been successfully fabricated by Spark Plasma Sintering. Results showed that all free basic sintering parameters (temperature, pressure, time) have a significant effect on the mechanical performance of sintered materials. High pressure (80 MPa) can decrease down to 1300°C sintering temperature and

DOI: 10.1051/matecconf/201712902028

simultaneously enhance fracture toughness (15.5 MPa·m$^{1/2}$) and hardness (18.3 GPa). Compared to the refence material, the SPSed WC-6% Co samples in a percentage ratio increased the rate of fracture toughness and hardness by 17% and 29%, respectively. The results presented in this study are promising, but further work is needed to optimize the sintering process to produce the material of superior characteristics.

Authors would like to thank The Ministry of Education of the Russian Federation supported this work by contract No. 14.577.21.0199, a unique identifier of contract RFMEFI57715X0199.

References

1. P. Siwak, D. Garbiec, Trans. Nonferrous Met. Soc. China **26,** 2641 (2016)
2. G.-Q. Shao, X.-L. Duan, J.-R. Xie, et al., Rev. Adv. Mater. Sci. **5 (4),** 281 (2003)
3. E. Lassner, W.-D. Schubert, *Kluwer Academic Publishers* (Plenum Press, London, 2000).
4. M. Eriksson , M. Radwan , Z. Shen, Int. J. Refrac. Met. Hard Mater. **36,** 31 (2013)
5. K. Hwan-cheol, S. In-jin, Y. Jin-kook, D. Jung-mann, Int. J. Refrac. Met. Hard Mater. **25**, 46 (2007)
6. S. Wei, S. Ye-Xi, Y. Hai-Lin, Z. Xian-Qi, R. Jian-Ming, Construction and Building Materials, **91**, 133 (2015)
7. C.-Chang, S. Lan, Tang Hua, Q. Xuan-Hui. Int. J. Refrac. Met. Hard Mater, **25**, 53 (2007)
8. W. Chong-Bin, S. Xiao-Yan, Z. Shi-Xian, Z. Li, Liu Wen-Bin, Int. J. Refrac. Met. Hard Mater. **28**, 567 (2010)
9. C.F. Gutiérrez-González, M. Suarez, S. Pozhidaev, S. Rivera, P. Peretyagin, W. Solís, L.A. Díaz, A. Fernandez , R. Torrecillas, J. Eur. Ceram. Soc. **36,** 2149 (2016)
10. S. S. Pozhidaev, A. E. Seleznev, N. W. Solis Pinargote and P. Yu. Peretyagin, Mechanics & Industry **16**, 710 (2015)
11. C. F. Gutierrez-Gonzalez, N. W. Solis Pinargote, S. Agouram, P. Y. Peretyagin, S. Lopez-Esteban and R. Torrecillas, Mechanics & Industry **16**, 703 (2015)
12. N. W. Solís, P. Peretyagin, R. Torrecillas, A. Fernández, J. L. Menéndez, C. Mallada, L. A. Díaz, J. S. Moya, Electroceram, **38**,119 (2017)
13. D. I. Yushin, V. Smirnov, N. Solis Pinargote, P. Yu. Peretyagin, V. A. Kuznetsov and R. Torrecillas, Rus. Ing. Res. **36**, 410 (2016)
14. R. Rumman , Z. Xie , S.-J. Hong , R. Ghomashchi, Materials and Design, **68,** 221 (2015)
15. J.C. Wurst, J.A. Nelson, Ceram. Soc. **55(2)**, 109 (1972)
16. A. Smirnov, H. D. Kurland, J. Grabow, F. A. Müller, J. F. Bartolomé, J. Eur. Ceram. So. **35**, 2685 (2015)

AUTHOR INDEX

AUTHOR INDEX

AUTHOR INDEX

AUTHOR INDEX

AUTHOR INDEX

International Conference on Modern Trends in Manufacturing Technologies and Equipment (ICMTMTE 2017)

MATEC Web of Conferences Volume 129 (2017)

Sevastopol, Russia
11 - 15 September 2017

Part 2 of 2

Editors:

Sergey Bratan
Sergey Borbatyuk
Sergey Leonov
Stanislav Roshchupkin

ISBN: 978-1-5108-5189-4

Printed from e-media with permission by:

Curran Associates, Inc.
57 Morehouse Lane
Red Hook, NY 12571

Some format issues inherent in the e-media version may also appear in this print version.

Printed by Curran Associates, Inc. (2017)

For additional information, please contact EDP Sciences – Web of Conferences
at the address below.

EDP Sciences – Web of Conferences
17, Avenue du Hoggar
Parc d'Activité de Courtabœuf
BP 112
F-91944 Les Ulis Cedex A
France

Phone: +33 (0) 1 69 18 75 75
Fax: +33 (0) 1 69 28 84 91

contact-edps@webofconferences.org

Additional copies of this publication are available from:

Curran Associates, Inc.
57 Morehouse Lane
Red Hook, NY 12571 USA
Phone: 845-758-0400
Fax: 845-758-2633
Email: curran@proceedings.com
Web: www.proceedings.com

TABLE OF CONTENTS

PART 1

PART 2

 DOI: 10.1051/matecconf/201712902029

Study of characteristics of steel specimens after selective laser melting of powder materials MS1

Anton Zhukov[1,*], *Boris* Barakhtin[1], and *Artem* Deev[1]

[1] NRC "Kurchatov Institute" – CRISM "Prometey", 191015, Saint-Petersburg, Russia

Abstract. At present, there is a great interest in the technologies of laser melting of metallic powders for the any shape items production. Over the past decade, the process of selective laser melting (SLM) of various powders based on alloys and steels was mastered. However, not so long ago, many studies were aimed at SLM stainless steels, including MS1 steel. The effect of laser radiation power, scanning speed, pitch, building direction on the microstructure and mechanical properties of the final product is studied. The influence of the parameters of the initial material, such as the chemical composition, the shape and size of the particles, and their distribution is indicated. Since the SLM process is sensitive to building parameters it is necessary to obtain more information about the mechanical and structural properties of standard samples in order to better understand the capabilities of this technology.

1 Introduction

In recent decades, selective laser melting of stainless steel powders has received much attention [1-4]. Data have become widely in demand in various industries, and the manufacture of products from powder materials is quite a promising direction.

As the object of investigation, maraging steel MS1 was chosen in this work. A special interest in this brand is that, in contrast to the austenitic steels, which are quite common for use in the SLM process, alpha-gamma-alpha phase transition takes place. Numerous investigations have revealed a number of features of the structure and properties of powdered martensitic steels [5-7]. It is established that there is an anisotropy of the mechanical properties with respect to the building parameters. This is due to the peculiarities of the grain structure – the grains are more elongated in the building direction than across the building direction [8].

In this connection, it seems important to try to find out whether it is possible, by changing the parameters of melting, to achieve a lower anisotropy.

The aim of the work was to study the structure, mechanical properties and the nature of the fracture surface of standard samples made of maraging steel powders MS1 by the method of selective laser melting (SLM) under different laser parameters. The purpose of the study is to determine the influence of the building parameters on the anisotropy of the mechanical properties of grown samples.

2 Materials and experimental methods

The raw material used is the powder of maraging stainless steel MS1, manufactured by EOS GMBH (Germany). The chemical composition of the powder, determined by X-ray fluorescent analysis methods on the Niton XL3t and LECO CS744 units, is shown in Table 1, from which it is seen that it complies with the ASTM A 240 standard.

Table 1. Chemical composition of used powders.

Basic alloying elements,% by weight	
C	1,0
Si	0,1
Mn	0,1
Cr	0,28
Ni	18,1
Mo	5,3
Co	9,3
Ti	1,0

The particle size distribution of the powder measured in the Malvern Mastersizer 2000 using the Fraunhofer approximation's laser diffraction method is shown in Figure 1. It can be seen that the MS1 steel powder has a large fractional composition, but a small amount of fine particles is present. In general, the powder meets the requirements for powders for SLM technology.

* Corresponding author: jouan2@gmail.com

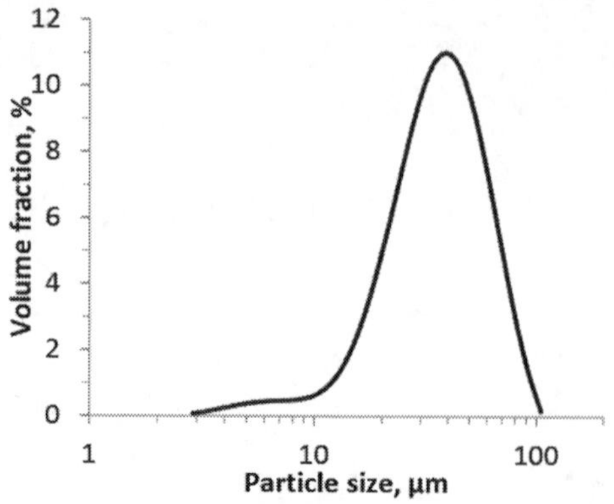

Fig. 1. The grain size distribution of MS1 steel powders.

Selective laser melting of powders was performed on the EOSINT M270 equipment. The powders were melted at 190 W and scanning speeds of 600, 650, 700, 750, 800 mm / s. The thickness of the melted powder's layer was 40 µm. For analyzing the effect of speed and power, a specific power value (in other words, the heat input) calculated as the Heat input = Laser power / Scan velocity [W · s / mm] was used.

3D models for mechanical testing for tension, according ASTM E8, and impact strength, according ASTM D 6110, were designed to evaluate the structure and mechanical properties. It should be noted that the U-shaped notch of horizontal samples was implemented in the model. In addition, due to a large amount of samples, their numbering was done during the modeling. An example of manufactured samples is shown on the Figure 2.

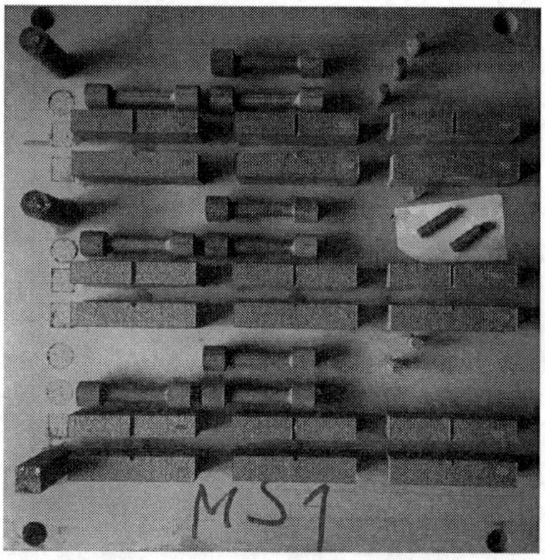

Fig. 2. Appearance of the built specimens on a platform.

Mechanical properties were determined at room temperature on a universal Zwick / Roell Z250 test machine and Metrocom-300 drop hammer, respectively. The hardness of the samples was determined using a Zwick / Roell ZHU 750 top.

The density of the samples was determined by the method of hydrostatic weighing on the analytical scales Mettler Toledo Excellence Plus XP.

Metallographic studies were performed on the optical microscope Axio Observer A1m with Clemex Vision PE computer program. Metallographic samples were made from impact strength samples. The etching of the surface to reveal the structure was carried out in a 10% aqueous solution of oxalic acid.

3 Experimental results and discussion

The measured values of the density and porosity of the samples are shown in Figure 3, from which it can be seen that both characteristics are close to the characteristics of the MS1 steel samples obtained by traditional metallurgical technologies. The values obtained can be explained by the presence of a small number of pores and other inclusions.

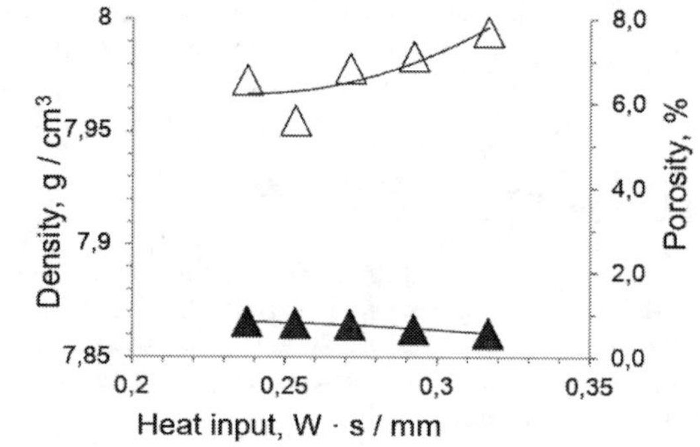

Fig. 3. Dependences of the density (open symbols) and porosity (painted symbols) of MS1 steel samples vs heat input.

It is known [9] that the strength properties of steels with "aging" martensitic, in which alpha-gamma-alpha phase transition takes place, increase with heat treatment after quenching. In our case, the material obtained by melting was subjected to annealing at a temperature of 490 °C for 1 hour.

Dynamic bending tests with fracture toughness assessment (Figure 4A) and failure performance (Figure 4B) showed that in the metal in the state "immediately after melting" (without annealing) the crack resistance is 20 ± 5 and 25 ± 5, respectively. The revealed differences are caused by the anisotropy of the macrostructure of the metal formed under different modes of sample building.

It was also found that with increasing energy deposition, the fracture toughness values do not decrease, but the effect of anisotropy practically disappears. The revealed features indicate an improvement in the powder's melting quality. However, it can be assumed that the observed tendency will be affected by the thickness of the deposited layer and the dispersion of the powder, since the quality of the melting is determined not only by the energy input, but also by the heat exchange between the particles.

 DOI: 10.1051/matecconf/201712902029

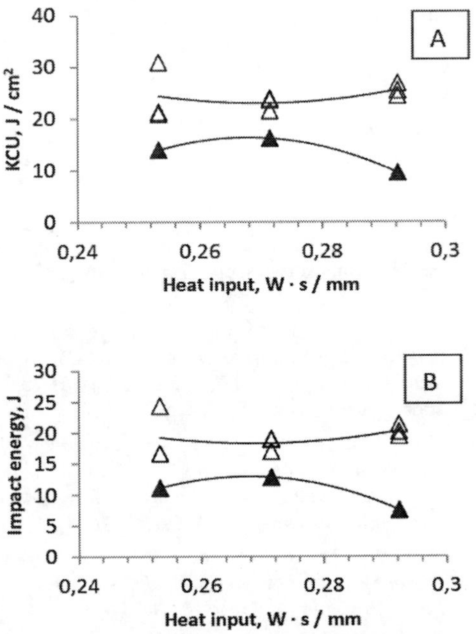

Fig. 4. Impact strength / KCU (A) and impact energy (B) of the test samples without annealing (open symbols) and after quenching (painted symbols).

Measurement of the hardness value on the faces of standard samples showed that this characteristic does not depend on the location of the preforms on the platform and varies exponentially with increasing heat input, reaching a maximum at a heat input of ~ 0.27 (Figure 5). The maximum hardness is about 350 HB, which is slightly higher than the standard values for this steel at 325 HB. After the annealing, the nature of the hardness dependence on the energy deposition did not change and the hardness numbers amounted to about 500 HB.

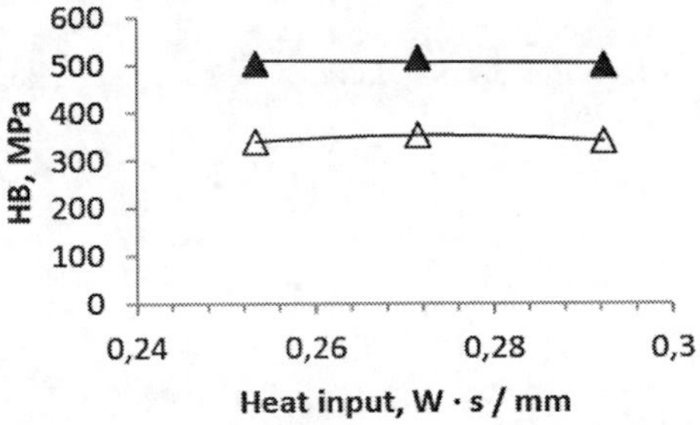

Fig. 5. Hardness of the test samples without annealing (open symbols) and after quenching (painted symbols).

As expected, after quenching, the strength of the metal increased, and the resistance to crack development decreased. In this case, the values of the viscosity were determined with a large spread for the samples created at a heat input of ~ 0.27. This fact allows us to assume that at a heat input of ~ 0.27 in the metal structure, the alloying elements (Mo, Co, Ti) are distributed so that dispersoids of different morphology and thermodynamic stability are formed during annealing, as it is known [10] that in maraging steels the hardening particles are Ni_nTi_m in the temperature range 300 – 400 °C and $Fe_p(Mo\ Co)_q$ at temperatures of 550 – 650 °C. In complete agreement with the dynamic bend test data, the results of uniaxial stretching of the samples are found. It is established that the "traditional" complex of mechanical parameters ($\sigma_{0.2}$, UTS, δ) reaches a maximum with minimum variance (spread) after melting with heat input ~ 0.27 (Figure 6).

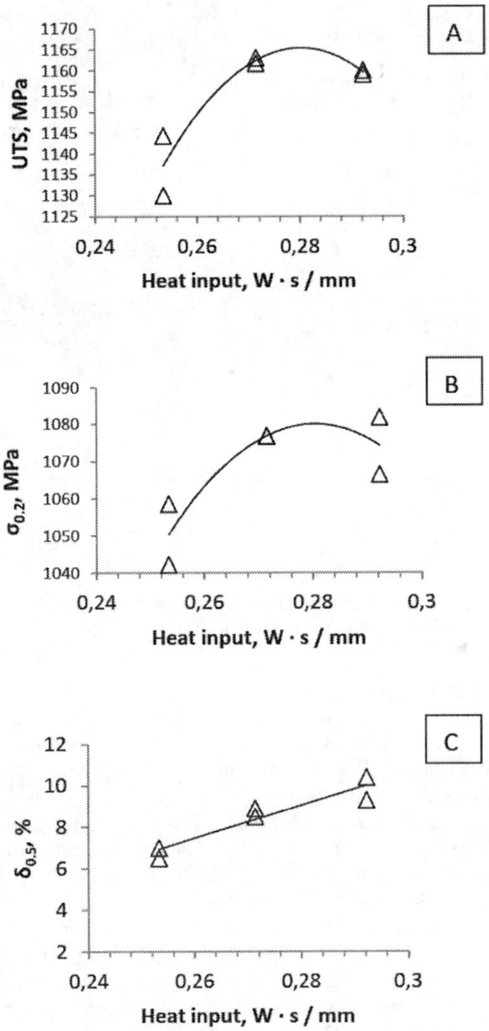

Fig. 6. Tensile strength (A), yield strength (B) и elongation (C) of the test samples.

Metallographic study revealed areas of heterogeneous distribution of chemical elements of the basic composition.

The increase of a heat input in the melt zone is accompanied by a decrease of the size of liquation, which indicates an improvement in the quality of melting. This fact can be explained by the dispersion of the powder used. After annealing on the sections, traces of dispersoids of intermetallic compounds that promote hardening of the metal are discernible (Figure 7).

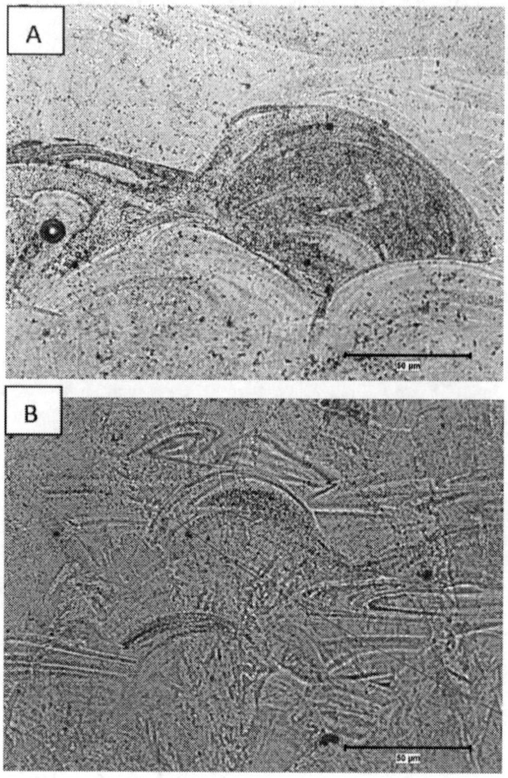

Fig. 7. Structure of selective laser melted samples without annealing (A) and after quenching (B).

4 Conclusions

1. As the heat input in the melting zone increases, the quality of alloy melting improves, and the effect of anisotropy decreases.
2. According to the data of dynamic bending and uniaxial stretching of samples, the complex of mechanical properties of the resulting melted alloy depends on the chemical composition of the powder material.
3. The mechanical characteristics of the steel obtained by selective laser melting of the powder composition MS1 are subject to the influence of subsequent heat treatment (annealing, aging).
4. In the technology of selective laser melting, the properties of the resulting melted alloy depend not only on the heat input, but also on the dispersion of the powder used.

This work was supported by the grant of Russian Scientific Fund №15-19-00210. Experimental studies were performed on the equipment of the collective use center "Composition, Structure and Properties of Structural Materials" NRC "Kurchatov Institute" – CRISM "Prometey", Saint-Petersburg, Russia.

References

1. Giedeon N. Levy *The role and future of the laser technology in the additive manufacturing environment*, Physics Procedia **5**, 65–80 (2010)
2. A. Hussein, L. Hao, *Advanced lattice support structures for metal additive manufacturing*, Journal of Materials Processing Technology **213**, 1019–1026 (2013)
3. Di Wang, Y. Yang, *Study on the designing rules and processability of porous structure based on selective laser melting (SLM)*, Journal of Materials Processing Technology **213,** 1734–1742 (2013)
4. V.P. Perevertov, I.K. Andronchev, M.M. Abulkasimov, *Technologies of processing materials by concentrated energy flow*, Reliability and quality of complex systems, **3 / 11,** 69–79 (2015)
5. P. Hanzl, M. Zetek *The Influence of Processing Parameters on the Mechanical Properties of SLM Parts*, Procedia Engineering **100,** 1405–1413 (2015)
6. P. Li *Constitutive and failure behaviour in selective laser melted stainless steel for microlattice structures*, Materials Science & Engineering **A 622,** 114–120 (2015)
7. J. Sedlaka, D. Rican *Study of Materials Produced by Powder Metallurgy Using Classical and Modern Additive Laser Technology*, Procedia Engineering **100,** 1232–1241 (2015)
8. K. Saeidi, X. Gao, Y. Zhong, Z.J. Shen *Hardened austenite steel with columnar sub-grain structure formed by laser melting*, Materials Science & Engineering **A 625**, 221–229 (2015)
9. M.L. Bernstein, A.G. Rakhshtadt *Metallurgy and heat treatment of steel* Reference book. In 3 volumes. (Mashinostroenye, Moscow, 1983)
10. L.V. Tarasenko, M.V. Unchikova. *Mechanical properties of corrosion-resistant steel after double aging* Engineering Journal: Science and Innovation **3** (2013)

DOI: 10.1051/matecconf/201712902030

Formation of a state with high strength and electrical conductivity in the Cu-0.5%Cr-0.2%Zr alloy

Elena Sarkeeva [1,4,*], *Marina* Abramova [1,4], and *Wei* Wei[2,3,4,*]

[1] Ufa State Aviation Technical University, 12 K. Marx, 450000 Ufa, Russia
[2] School of Materials Science and Engineering, Changzhou University, 1 Gehu Road, Changzhou 213164, P.R .China
[3] Jiangsu Key Laboratory of Materials Surface Science and Technology, Changzhou University, Changzhou 213164, P.R. China
[4]Joint Laboratory of Functional Nanostructured Metals, Changzhou University, Changzhou 213164, P.R. China

Abstract. The effect of severe plastic deformation by high pressure torsion (HPT) and subsequent aging on the formation of a high-strength state with sufficient electrical conductivity in the Cu-0.5%Cr-0.2% Zr (wt.%) alloy is investigated. It is shown that as a result of HPT and subsequent aging a high-strength state with a tensile strength of ≈ 900 MPa and a conductivity of 70% IACS can be formed.

1 Introduction

Low-alloy dispersion-hardened copper alloys of the Cu-Cr-Zr system have a high electrical and thermal conductivity [1, 2]. The introduction of a small addition of zirconium leads not only to an increase in strength properties, but also to a significant improvement in ductility at elevated temperatures. Alloys of this system are traditionally used in applications where a combination of high mechanical strength, heat resistance and electrical conductivity for manufacturing of contact wires, spot welding electrodes, fusion reactors, *etc.* is required.

The main factors affecting both the strength and the ductility of these alloys are the grain size, dislocation structure, and concentration of alloying elements in the solid solution, as well as concentration, distribution and size of dispersed particles of the second phases both in the body and at the grain boundaries.

Previous studies have shown that hardening with conventional aging is limited due to the low solubility of Cr and Zr in the copper matrix [3]. There are also many studies on the character, composition and distribution of particles formed during the aging of these alloys [4-6]. These studies have shown that particles of the composition $CrCu_2$, Cu_3Zr, Cu_5Zr are precipitated.

The conventional industrial scheme for processing copper alloys of the Cu-Cr-Zr system includes 3 stages: 1 stage - preliminary heat treatment to obtain a supersaturated solid solution (SS), stage 2 - cold deformation to increase strength due to cold work, stage 3 - post-deformation heat treatment, during which the SS decays and dispersed-hardening particles of secondary phases precipitate. As a result, in the alloys of the Cu-Cr-Zr system strength values of the order of 400-500 MPa are attained at an electrical conductivity of 75-80% of the conductivity of pure copper (IASC). But even higher strength is required by manufactures.

Recent studies have shown that severe plastic deformation (SPD) can significantly increase the strength of materials due to the formation of ultrafine-grained (UFG) structure [7,8]. Previous studies of the strength and electrical conductivity of Cu-Cr-Zr alloys subjected to SPD and subsequent aging showed that a combination of high strengths of about 688 MPa with a conductivity of 59% IACS, or 599 MPa at 82% IACS can be achieved [9-13]. Thus, the combination of SPD and subsequent aging seems to be a promising direction for the simultaneous formation of high strength and electrical conductivity in the alloys of the Cu-Cr-Zr system. It worse to note, however, that as a result of SPD performed by equal channel angular pressing (ECAP) a microstructure with an average grain size of 250-300 nm is formed in Cu-Cr-Zr alloys. The degree of plastic deformation achieved by such a treatment reaches ~ 9.2.

It is known that SPD performed by HPT allows achieving of a much larger grain refinement as a result of strain degrees equal to ~ 157. In [14], Cu-Cr-Zr alloys subjected to SPD by HPT (the number of turns N = 5) with subsequent aging resulted in a tensile strength of about 750 MPa with a conductivity of 59% IACS.

Thus, it can be seen that as a result of HPT with subsequent aging it is possible to substantially increase the strength properties while maintaining a sufficiently high level of electrical conductivity. In this paper, we investigated the effect of the strain degree which was 2 times greater than in [14] on the strength and electro conductivity of the Cu-0.5Cr-0.2Zr alloy.

2 Experimental

The copper alloy with chromium and zirconium Cu-0.5% Cr-0.2% Zr (wt.%) was chosen as the material for the studies. In order to obtain a supersaturated solid solution, the as-received rods with a diameter of 20 mm were

* Corresponding author: benjamin.wwei@163.com, sarkeeva.e@inbox.ru

subjected to high-temperature heat treatment at 1000 °C for 0.5 hours, followed by quenching in a 5% NaCl solution. As a result, the initial state (IS) was formed.

Furthermore, the rods were cut into discs 2 mm in thickness and subjected to HPT at room temperature. The disks were deformed under an applied pressure of 6 GPa. The number of revolutions N = 10, which corresponded to the true degree of ~ 157. HPT was carried out in a "hole" with a depth of 0.2 mm. As a result of the HPT, samples of Ø20 mm and a thickness of h = 2 mm were obtained.

Aging was carried out in air in an oven SUOL 45 at 450 °C for 1 hour.

To study the microstructure, an optical microscope (OM) OLIMPUS GX51 and a transmission electron microscope (TEM) JEOL 2100 were used.

The microhardness of the test samples was measured on a Duramin instrument under a load of 100 g and with a holding time of 10 seconds for the load.

Uniaxial tensile tests were carried out on an Instron 8801 small specimen testing machine at room temperature at a speed of 10^{-3} s^{-1}. Samples with dimensions 10×4×4 mm were cut perpendicular to the radial section of the disks. Three samples were tested for each condition.

Conductivity studies were carried out at a temperature T=23 °C by an eddy current method on flat samples. The accuracy is ±2%.

All investigations of microstructure and properties were carried out for regions located at a distance equal to half the radius of the disks.

3 Results and discussion

Fig. 1 shows the structure of the Cu-0.5%Cr-0.2%Cr alloy in the initial state. In general, the structure is characterized by a sufficiently high degree of homogeneity. The average grain size is 140±8 μm. At a given quenching temperature, the dissolution of the particles does not take place completely, a fine-dispersed phase is observed.

As already shown in previous studies, the composition of these precipitates corresponds to $CrCu_2$ [4, 5]. In addition, as shown in [6], the particles of Cu_3Zr and Cu_5Zr are also represented in the microstructure. The particles are arranged non-uniformly both in volume and in grain boundaries. The largest of them reach a size of 20 μm.

Electron microscopic images of the state of the alloy after HPT and HPT and the subsequent aging are shown in Fig. 2, and 3. As a result of the HPT, a UFG structure was formed. The boundaries of the structure elements are mostly blurred. The electronogram for this state is ring-shaped with well-defined individual reflexes evenly distributed around the ring (Fig. 2).

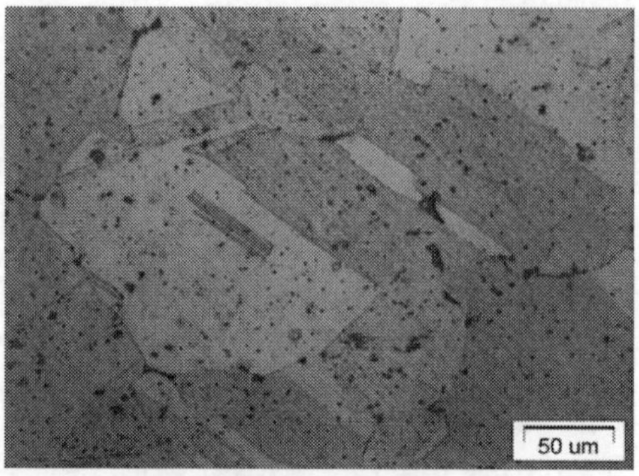

Fig. 1. The optical metallography of the IS state.

Fig. 2. TEM of the HPT state.

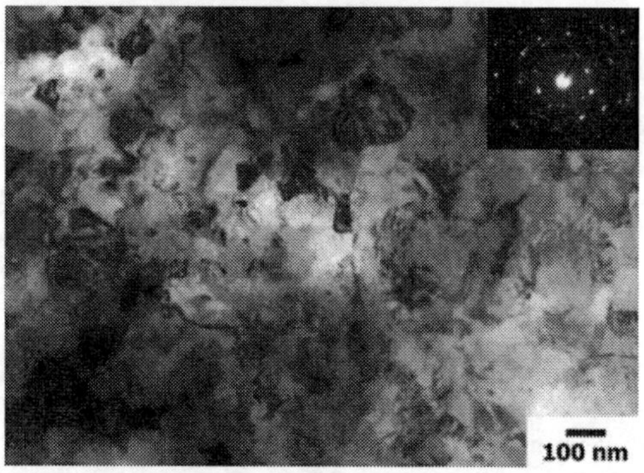

Fig. 3. TEM of the HPT + aging state.

Analysis of the dark field electron microscopic images for the HPT shows that the average size of the formed microstructure elements is 135±40 nm (Fig. 2). After HPT, no large particles are practically observed. Dispersed particles of about 5 nm in size are clearly visible.

As a result of aging, a microstructure, consisting of grains with clearly delineated boundaries, is formed (Fig. 3). The banded contrast along them indicates their perfection. The improvement of the microstructure is the result of relaxation processes, which are accompanied by the growth of grains. However, at the same time grains with blurred boundaries are preserved. The heterogeneity inside some of them indicates the preservation of the stressed state. The size of individual grains reaches 340 nm, and the average grain size is 140±50 nm.

In this case, part of the grains have a size of 20-40 nm. Particles of the second phase are visible in the microstructure. The shape of these particles is different. There are particles of round, oval, a regular hexahedron, a rhombus shapes, and also triangular particles. In the structure there are particles of different sizes. The size of round and oval particles is about 5 nm, the size of particles with other shapes is 20 nm. The average particle size is 10 nm. The particles are arranged non-uniformly both in volume and in grain boundaries.

Investigation of mechanical properties showed that the HPT leads to an increase in the microhardness values up to 2925 ± 190 MPa as compared to the microhardness of initial state equal to 845±20 MPa, i.e. 3.5 times (Table 1, Fig. 4).

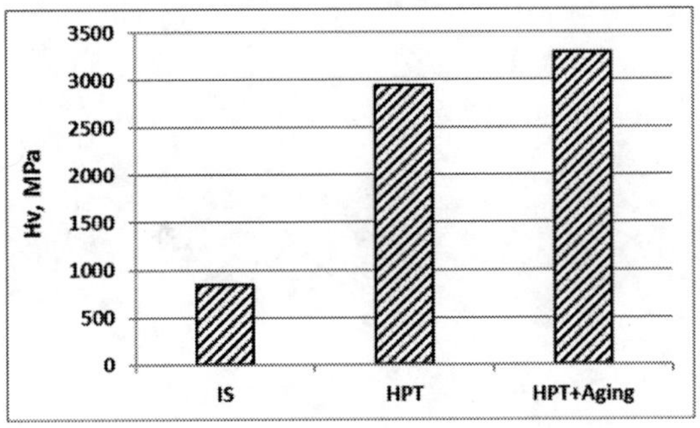

Fig. 4. The microhardness of Cu-Cr-Zr alloy after different treatments.

Fig. 5 shows the engineering strain-strain curves in coordinates σ_u-ε, obtained for samples in different states. In the initial state, the ultimate strength of the sample is 226±5 MPa. After HPT the strength of the Cu-Cr-Zr alloy is increased up to 820±5 MPa. Subsequent aging leads to an increase in strength up to 900±15 MPa.

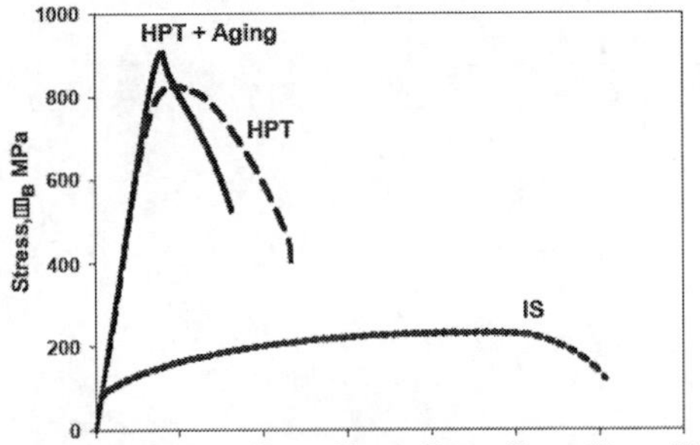

Fig. 5. The dependence of the flow stress upon strain ε, $\dot{\varepsilon}=10^{-3}$ s^{-1}, T=293K.

The results of mechanical tensile tests are well correlated with the results of microhardness measurements in Table 1.

Table 1. Mechanical characteristics of different states.

State	Hv, MPa	σ_u, MPa	$\sigma_{0.2}$, MPa	ε, %	IACS, %
IS	845±25	226±5	84±1	62±1	31
HPT	2925±150	820±5	730±20	22±1	33
HPT+ Aging	3290±110	900±15	868±40	16±1	70

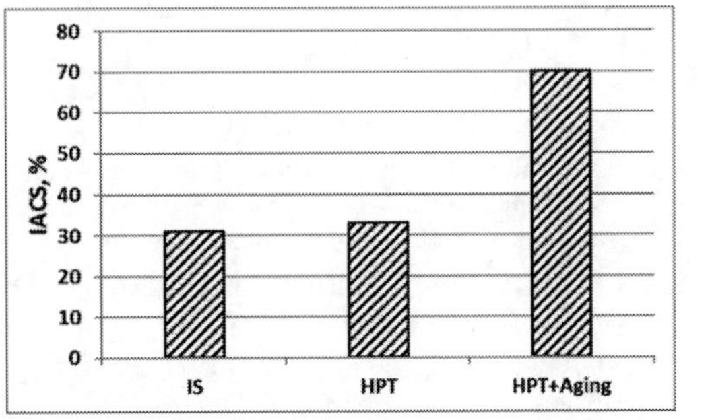

Fig. 6. The electrical conductivity after various treatments.

As can be seen, aging leads to an additional increase in microhardness from 2925±150 to 3290±110 MPa, as well as to the restoration of electrical conductivity to 70% of IACS in Fig. 6. Hardening is due to the precipitation of dispersed particles during aging, which inhibits the growth of grains during aging and compensate for softening due to a certain growth of grains. The value of the electrical conductivity as a result of aging increases from 33% to 70% of IACS (Fig. 6).

4 Conclusions

Thus, as a result of the HPT, a high-strength UFG state with a grain size of 140 nm was formed in the Cu-Cr-Zr alloy. The aging leads to an additional increase in strength up to 900 MPa. In this case, the electrical conductivity is restored to 70% of the IACS. The increase both in strength and electrical conductivity indicates the occurrence of deformation-stimulated decomposition processes in a supersaturated solid solution.

The work is executed by the financial support of the Ministry of education and science within the framework of the project No. 16.1969.2017/PCh and the Science and Technology Bureau of Jiangsu Province, P.R. China under grant No. BY2016029-19.

References

1. H. Fuxiang, M. Jusheng, N. Honglong, G. Zhiting, L. Chao, G. Shumei, Y. Xuetao, W. Tao, L. Hong, L. Huafen, Scr. Mater. **48** (2003)
2. W.X. Qi, J.P. Tu, F. Liu, Y.Z. Yang, N.Y. Wang, H.M. Lu, X.B. Zhang, S.Y. Guo, M. S. Liu, Mater. Sci. Eng. A. **343** (2003)
3. N. Bochvar, Cr-Cu-Zr (Chromium-Copper-Zirconium). In G. Effenberg, S. Ilyenko, editors. Non-Ferrous Metal Systems. Part 2. Landolt-

DOI: 10.1051/matecconf/201712902030

Börnstein - Group IV Physical Chemistry, Springer Berlin Heidelberg; V **11**C2 (2007)

4. W.X. Qi, J.P. Tu, F. Liu, Y.Z. Yang, N.Y. Wang, H.M. Lu, X.B. Zhang, S.Y. Guo, M. S. Liu, Mater. Sci. Eng. A. 343 (2003)
5. J.P. Tu, W.X. Qi, Y.Z. Yang, F.Liu, J.T. Zhang, G.Y. Gan, N.Y. Wang, X.B. Zhang, M.S. Liu, Wear **249** (2002)
6. N.Y. Tang, D.M.R. Taplin, G.L. Dunlop, Mater. Sci. Tech. **1** (1985)
7. R.Z. Valiev, R.K. Islamgaliev, I.V. Alexandrov, Progress in Mater Sci. **45** (2000)
8. R.Z. Valiev, I.V. Alexandrov, *Bulk nanostructured metal materials: preparation, structure and properties* (Akademkniga, Moscow, 2000)
9. R.K. Islamgaliev, K.M. Nesterov, Y. Champion, R.Z. Valiev, Mater. Sci. Eng. **A. 63** (2014)
10. W. Wei, K.X. Wei, F. Wang, Q.B. Du, I.V. Alexandrov, J. Hu, Mater. Sci. Eng. A. **528** (2011)
11. G. Purcek, H. Yanar, O. Saray, I. Karaman , H.J. Wear **311** (2014)
12. Vinogradov A., Ishida N., Kitagawa K., Kopylov V., Acta Mater. **53** (2005)
13. Vinogradov A., Patlan V, Suzuki Y., Kitagawa K., Kopylov V., Acta Mater. **50** (2002)
14. D.V. Shangina, N.R. Bochvar, M.V. Gorshenkov, H. Yanar, G. Purcek, S.V., Mater. Sci. Eng. A. **650** (2016)

 DOI: 10.1051/matecconf/201712902031

Effect of cold rolling on microstructure and mechanical properties of Cu-0,5%Cr-0,2%Zr subjected to ECAP with various NUNBERS of passes

Marina Abramova [1,4,*], *Elena* Sarkeeva [1,4], and *Wei* Wei [2,3,4,*]

[1] Ufa State Aviation Technical University, 12 K. Marx, 450000 Ufa, Russia
[2] School of Materials Science and Engineering, Changzhou University, 1 Gehu Road, Changzhou 213164, P.R. China
[3] Jiangsu Key Laboratory of Materials Surface Science and Technology, Changzhou University, Changzhou 213164, P.R. China
[4]Joint Laboratory of Functional Nanostructured Metals, Changzhou University, Changzhou 213164, P.R. China

Abstract. The paper studies the effect of the combination of ECAP and cold rolling on the structure, mechanical properties and coefficient of friction of low-alloyed bronzes. It is shown that the use of ECAP together with rolling leads to the formation of a UFG structure and an increase in strength characteristics, while the coefficient of friction is increases. Subsequent heat treatment results in additional hardening by the separation of fine particles, which also reduces the coefficient of friction.

1 Introduction

Low-alloy bronzes of the Cu-Cr-Zr system are widely used to create electrically conductive products from which enhanced strength tests are required [1, 2]. One of the priority areas of application of these bronzes is the manufacture of contact wires. The contact wire must have high strength, good electrical conductivity, and be wear-resistant.

Traditionally, scheme is used to obtain high strength in copper alloys, this treatment usually includes 3 stages. The first stage is high-temperature heat treatment followed by quenching into water to obtain a supersaturated solid solution (SS) of alloying elements in a copper matrix. The second stage is cold deformation, which leads to an increase in the strength of the matrix due to hard work. The third stage is aging, when the solid solution decomposes and, accordingly, the additional dispersion hardening of the alloy. Numerous studies have demonstrated the possibility of a significant improvement in physicomechanical properties through using of methods of severe plastic deformation (SPD). Previous studies have shown that if at the 2nd stage of deformation the traditional methods of deformation are replaced, for example by rolling, drawing, by any SPD method, a combination of high strengths of about 688 MPa can be achieved in the alloys with a conductivity of 59% IACS, or 599 MPa at 82% IACS [9-13]. If the combination of ECAP operations with subsequent cold rolling is applied at stage 2, it is possible to obtain more optimal functional properties.

The purpose of this work was to study the effect of the SPD degree in the equal-angle angular pressing (ECAP) in combination with subsequent cold rolling on the tribological characteristics of the Cu-0.5Cr-0.2Zr alloy.

2 Experimental methods

The copper alloy with chromium and zirconium Cu-0.5% Cr-0.2% Zr (weight) was chosen as the material for the studies. The material was studied in coarse-grained (CG) and UFG states. To obtain a supersaturated solid solution, a high-temperature heat treatment was carried out at 1000 ° C for 0.5 hours, followed by quenching in a 5% NaCl solution. (the initial state). For the samples, 12 × 12 × 80 mm in the form of rectangular parallelepipeds were used. ECAP was carried out at room temperature at a speed of 0.2 mm / s with an internal tool angle ψ = 110 ° and an external angle φ = 0 ° for ECAP along route A, without rotating the rods between successive passes. The number of passes was 1, 2, 4, 8. The degree of deformation for each pass was 0.8. Rolling was carried out at room temperature with a reduction of 55%.

Aging was carried out in air in an oven SUOL 45 at 450 ° C for 1 hour.

All investigations were carried out in the longitudinal section of the samples.

The OLIMPUS GX51 optical microscope (OM) and the transmission electron microscope (TEM) -JEOL 2100 were used to study the microstructure. Foil for TEM studies was prepared by electrolytic thinning in a solution of 30% orthophosphoric acid and 70% ethyl alcohol.

The microhardness of the test samples was measured on a Duramin instrument under a load of 100 g and with a holding time of 10 seconds for a load. Conductivity studies were carried out at a temperature T = 23 ° C by an eddy current method on flat samples measuring 12 × 12 × 1 mm. Surfaces of the samples before the test were subjected to a two-step polishing starting with mechanical polishing followed by final polishing on a diamond suspension.

* Corresponding author: benjamin.wwei@163.com, abramovamm@yandex.ru

 DOI: 10.1051/matecconf/201712902031

Tribological tests were carried out on the Nanovea (finger-plane) installation when the indenter was made of 95X18 steel was moved around the circumference at a speed of 2400 mm / min under dry friction conditions for 40 minutes. The load on the indenter was 2N. To investigate the nature of the tracks, the JEOL JSM-6490LV raster microscope was used.

3 Results and discussion

Fig. 1 shows the microstructure of Cu-0.5Cr-0.2Zr alloy in initial state. The microstructure of the alloy in the initial state consists of equiaxed grains with an average size of about 100 ± 5 µm and contains inclusions of a finely dispersed strengthening phase.

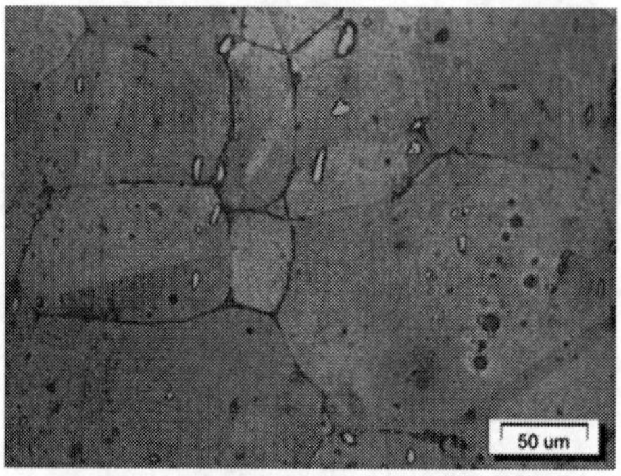

Fig. 1. Optical metallography of Cu-Cr-Zr alloy in the initial state.

The particles reach sizes up to 12 µm. As already shown in previous studies, the composition of these precipitates corresponds to $CrCu_2$ [2, 6].

After 1-8 passages of the ECAP with subsequent rolling, the structural elements that are stretched along the deformation direction are significantly crushed. The particles line up in the direction of deformation as shown in Fig. 2a-f.

a

b

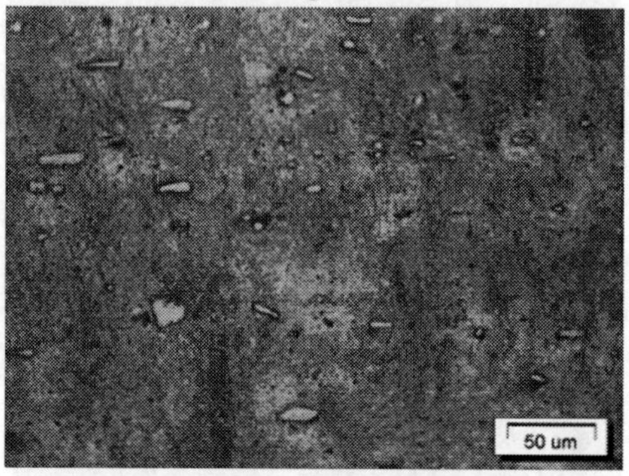

c

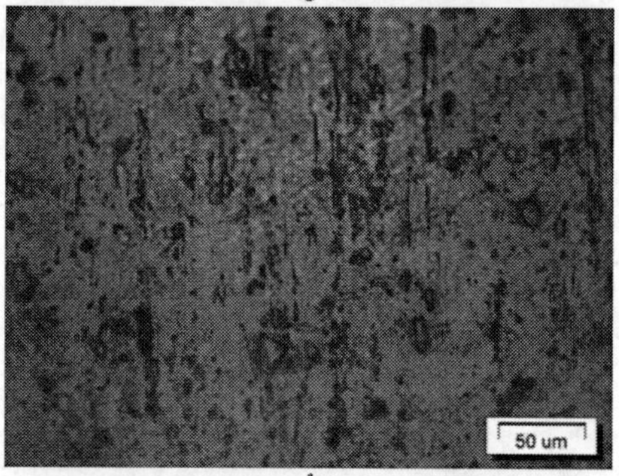

d

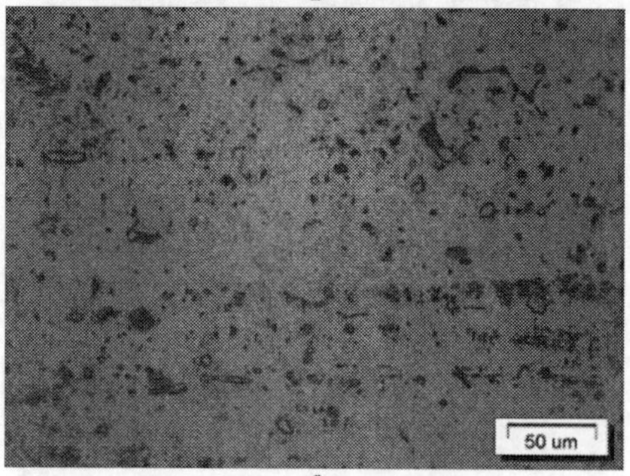

f

Fig. 2. Optical metallography of the Cu-Cr-Zr alloy after the ECAP with rolling: a - initial state + CR, b - 1 pass ECAP + CR, в - 2 passages ECAP + CR, г - 4 passages ECAP + CR, д - 8 passes ECAP + CR.

DOI: 10.1051/matecconf/201712902031

Analysis of the fine structure after ECAP in combination with cold rolling indicates a strong grinding with the formation of a grain-subgrain structure (Fig. 3a-f). A high density of dislocations is located in the microstructure, which are located both inside the grains and at grain boundaries. A complex diffraction contrast on the light-field image indicates the presence in the grains of large internal stresses that have arisen due to the application of large shear strains. It should be noted that with an increase in the degree of deformation, there is no significant increase in the defect density. Small particles are visible in the places of accumulation of dislocations. Probably, these particles are an obstacle to the movement of dislocations, which leads to an additional hardening of the material. Electronograms are typical for ultrafine-grained structures.

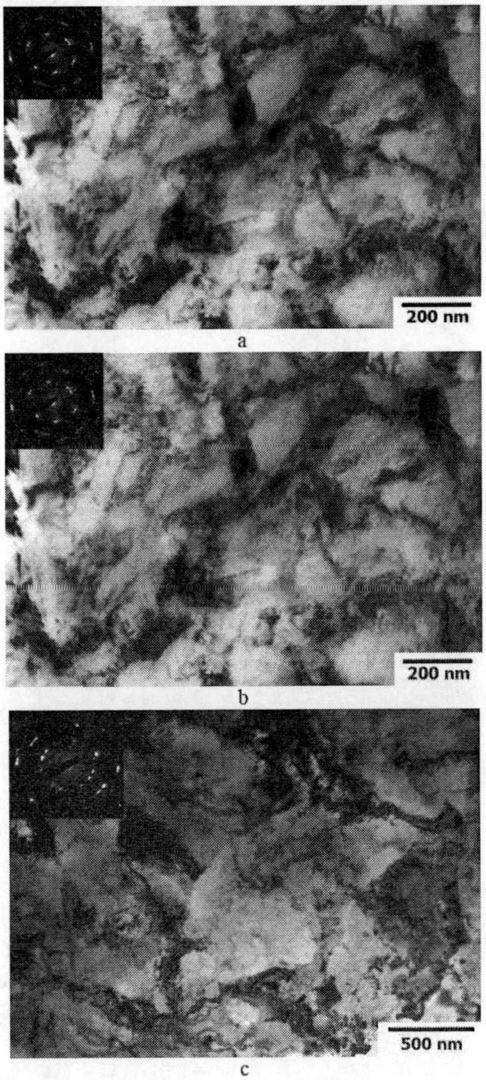

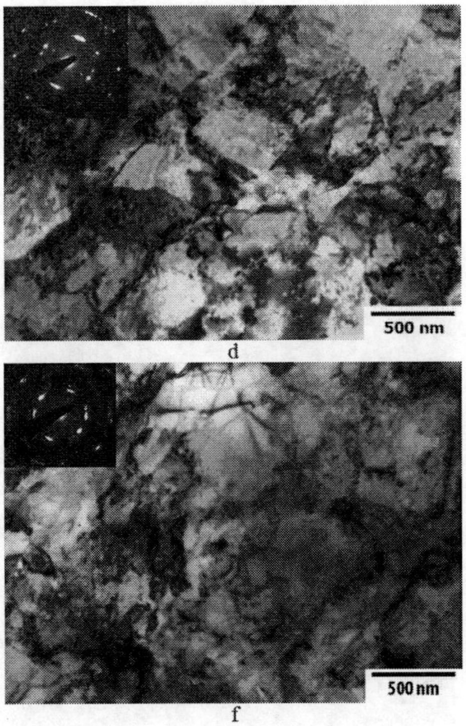

Fig. 3. TEM of Cu-Cr-Zr alloy after rolling: a - initial state + CR, b - 1 pass of ECAP + CR, c - 2 passages of ECAP + CR, d - 4 passages of ECAP + CR, f - 8 passages of ECAP + CR.

Measurements of the microhardness showed that with increasing strain, the microhardness increased from 950 MPa to 1980 MPa after 8 ECAP passes, which is 1.7 times higher than for the CG state (Table 1). The most intensive deformation hardening takes place after the rolling of the initial state. After the 1st, 2nd pass of the ECAP, microhardness does not change. Further, the microhardness increases monotonically to a maximum value of 1980 MPa. An annealing was carried out at a temperature of 450 ° C for 1 hour to determine the behavior of the samples after deformation. The states with the highest microhardness were chosen for postdeformation treatment. As a result of aging microhardness increased to 2600 MPa (Table 1), this is due to the release of dispersed particles of secondary phases. Which are clearly visible in photographs of fine structure after 8ECR and subsequent aging presented in Fig. 4. They are located non-uniformly both in the body and along the grain boundaries. Also fragments of grains are clean from defects, since annealing causes not only the release of particles, but also the return of the structure.

According to the results of tribological tests, the coefficient of friction in the initial state is 0.6 (Table 1). The subsequent increase in the degree of deformation leads to an increase in the coefficient of friction. The greatest coefficient of friction 0.9 was obtained after 4 and 8 passes of ECAP with subsequent rolling. This may be due to the brittleness of the UFG of copper and the development of cracks in the near-surface layer.

Apparently in our case it also fails to maintain a sufficient level of plasticity.

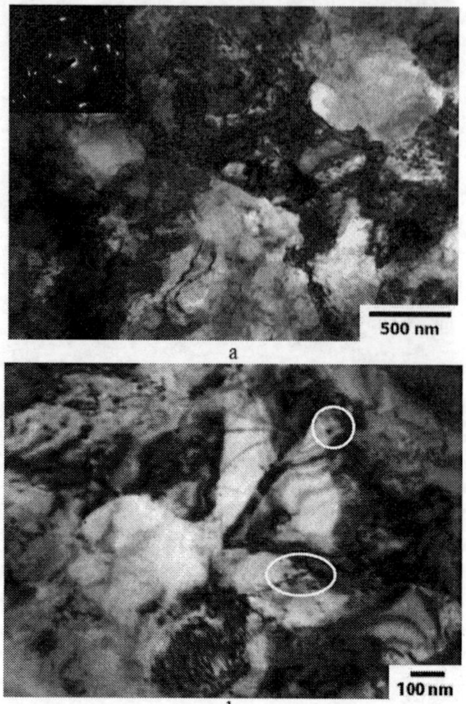

Fig. 4. TEM of the Cu-Cr-Zr alloy and rolling. The clear-field image: a) after 8 ECAP passes, typical shear bands, b) the observed particles are separated by circles.

Table 1. The mechanical properties of the alloy Cu-Cr-Z.

States	Hv, MPa	σ_B, MPa	Friction coefficient	Electro conductivity (IACS %)
SS+CR	1650±20	520	0.6	31
1 ECR	1650±30	590	0.8	28
2 ECR	1650±40	585	0.8	28
4 ECR	1700±30	630	0.9	27
8 ECR	2000±30	660	0.9	28
8 ECR+ Aging	2600	-	0.1	69

However, subsequent aging leads to a sharp decrease in the coefficient of friction to a value of 0.1. Perhaps this is due to the fact that fine particles were formed during annealing, which led to a further increase in the strength of the material. Thus, in the process of wear, the primary larger, stronger particles are dyed from the surface of the material and then abrasive wear continues. In the case of an increase in strength due to the release of fine particles in the matrix, the extrusion of coarse particles is observed to a lesser extent, which leads to a decrease in the coefficient of friction.

Thus, as a result of ECAP in combination with rolling and subsequent aging, it is possible to reduce the coefficient of friction to 0.1. At the same time, one of the most important operational parameters - the electrical conductivity is restored to 69% IACS (Table 1)

4 Conclusions

The effect of intense plastic deformation of ECAP in combination with rolling on the microstructure, mechanical properties and wear resistance of Cu-0.5Cr-0.2Zr alloy was investigated. It was revealed that as a result of ECAP the microhardness increased to 2600MPa and the coefficient of friction decreased to 0.1. In this case, the electrical conductivity reaches 69% of IACS.

The work is executed by the financial support of the Ministry of education and science within the framework of the project No. 16.1969.2017/PCh and the Science and Technology Bureau of Jiangsu Province, P.R. China under grant No.BY2016029-19.

References

1. H. Fuxiang, M. Jusheng, N. Honglong, G. Zhiting, L. Chao, G. Shumei, Y. Xuetao, W. Tao, L. Hong, L. Huafen, Scr. Mater. **48** (2003)
2. W.X. Qi, J.P. Tu, F. Liu, Y.Z. Yang, N.Y. Wang, H.M. Lu, X.B. Zhang, S.Y. Guo, M. S. Liu, Mater. Sci. Eng. A. **343** (2003)
3. R.Z. Valiev, R.K. Islamgaliev, I.V. Alexandrov, Progress in Mater Sci **45** (2000)
4. R.Z. Valiev, I.V. Alexandrov, Bulk nanostructured metal materials: preparation, structure and properties (Akademkniga, Moscow, 2000)
5. G. Purcek, H. Yanar, O. Saray, I. Karaman, H.J., Wear **311**, 149. (2014)
6. J.P. Tu, W.X. Qi, Y.Z. Yang, F.Liu, J.T. Zhang, G.Y. Gan, N.Y. Wang, X.B. Zhang, M.S. Liu, Wear **249** (2002)
7. A.P. Zhilyaev, I. Shakhova, A. Belyakov, R. Kaibyshev, T.G. Langdon, Mater Sci. **49** (2014)
8. R.Z. Valiev, A.P. Zhilyaev, T.G. Langdon, Bulk Nanostructured Materials: Fundamentals and Applications (John Wiley & Sons, Hoboken, New Jersey, 2014)
9. R.K. Islamgaliev, K.M. Nesterov, Y. Champion, R.Z. Valiev, Mater. Sci. Eng. A. **63** (2014)
10. W. Wei, K.X. Wei, F. Wang, Q.B. Du, I.V. Alexandrov, J. Hu, Mater. Sci. Eng. A. **528** (2011)
11. Vinogradov A., Ishida N., Kitagawa K., Kopylov V., Acta Mater. **53** (2005)
12. Vinogradov A., Patlan V, Suzuki Y., Kitagawa K., Kopylov V., Acta Mater. **50** (2002)
13. D.V. Shangina, N.R. Bochvar, M.V. Gorshenkov, H. Yanar, G. Purcek, S.V., Mater. Sci. Eng. A. **650** (2016)

Effect of the stacking fault energy on the mechanical properties of pure Cu and Cu-Al alloys subjected to severe plastic deformation

Liliya Zaynullina[1,4], *Igor* Alexandrov[1,4,*], and *Wei* Wei[2,3,4,*]

[1]Ufa State Aviation Technical University, Ufa, Russian Federation
[2]School of Materials Science and Engineering, Changzhou University, Changzhou 213164, PR China
[3]Jiangsu Key Laboratory of Materials Surface Science and Technology, Changzhou University, Changzhou 213164, PR China
[4]Joint Laboratory of Functional Nanostructured Metals, Changzhou University, Changzhou 213164, P. R. China

Abstract. The effect of stacking fault energy (SFE) on the mechanical properties of pure Cu and alloys of Cu-2.2%Al and Cu-4.5%Al subjected to severe plastic deformation (SPD) was investigated. SPD was performed by equal channel angular pressing (ECAP) at room and cryogenic temperatures. It is established that the increase in the weight concentration of Al in the Cu matrix (a reduction of SFE) and decreasing the ECAP temperature leads to an increase of the strength characteristics. The observed tendency is caused by increasing of the role of deformation twinning.

1 Introduction

The stacking fault energy (SFE) and temperature are important parameters determining the mechanisms and peculiarities of the microstructure refinement during the deformation of metallic materials [1]. In general, lowering of SFE in the result of the alloying metal materials, and reducing the deformation temperature, promotes the processes of deformation twinning. While deformation twinning and dislocation slip are two main competing mechanisms of plastic deformation in metals and alloys under relatively low temperatures [2].

It is known that in face-centered cubic (FCC) metals and alloys the possibility of twinning is higher in the materials with lower SFE values [3 - 5]. Twinning often observed in FCC materials, such as Cu-Zn alloys [4] with a low SFE, while in alloys with high SFE, such as Al twins are rare [6].

In recent years, among researchers a growing interest in twinning in metals at the nanoscale is observed [7,8]. The main reason for this interest is the experimental evidence that the strength of the materials increases with increasing density of twin boundaries. Twin boundaries affect the strength as well as the conventional grain boundaries, even at nanoscale thicknesses of twins. In addition, the grain refinement to the nano regime may occur by fragmentation of nanoscale twins [9, 10]. It was shown that the dynamic plastic deformation at low temperatures in alloys with a low SFE promotes the formation of high density of nanoscale twins [11, 12].

Recent studies have demonstrated the fact that the methods of severe plastic deformation (SPD) are promising methods of forming ultrafine-grained (UFG) and nanostructured (NS) states, including metallic materials with the various SFE [13, 14]. The contribution of the previously listed mechanisms in the deformation of these materials varies with their microstructure refinement. It also depends on the SFE and deformation temperature.

In [15] it was shown that in the Cu-Al alloys subjected to SPD by high pressure torsion, implemented at room temperature (RT), SFE plays an important role in the deformation process, since the strength of the alloy increases with decreasing SFE, and uniform elongation is also increased, with the exception of the alloys containing the lowest SFE (16 wt.% Al), where the uniform elongation is reduced. It was concluded that there is an optimum SFE (5 wt.% Al) to achieve the maximum uniform elongation in the Cu-Al alloys after HPT processing.

The analysis of these experimental results by the method of numerical modeling showed that the nonmonotonic change of ductility with SFE of the considered materials is due to the fact that in alloys with 5 wt.% Al work hardening is the main mechanism of hardening in tension and it increases due to the accumulation of dislocations in the inner areas of grains at the twin boundaries [16]. At the same time, the increase of Al concentration up to 16 wt.% retards the annihilation processes. As a result, the possibility of further accumulation of dislocations inside small grains is limited, which leads to the decrease of the degree of homogeneous deformation. This significantly increased concentration of deformation vacancies contributes to the destruction of the sample.

Studies of mechanical properties of Cu-Al alloys subjected to ECAP and further cold rolling at the liquid nitrogen temperature showed an increase in strength and ductility due to increasing contribution of twinning with decreasing SFE [17, 18].

* Corresponding author: igorvalexandrov@yandex.ru, benjamin.wwei@163.com

DOI: 10.1051/matecconf/201712902032

The aim of the present research is the analysis of the influence of SFE on the microhardness and mechanical properties of pure Cu and Cu-Al alloys with 2.2 and 4.5 wt.% Al subjected to ECAP at room temperature and the liquid nitrogen temperature.

2 Experimental

The materials used in this work are Cu and Cu-Al alloys with 2.2 and 4.5 wt.%. Al. The SFEs of these alloys are approximately 78 mJ/m^2, 25 mJ/m^2, 12 mJ/m^2, respectively. The square rods with a size of 12.0 mm × 12.0 mm × 80 mm were annealed in vacuum at 1073 K for 2 h, slowly cooled to 773 K and then quenched in water with the aim to avoid the possible short-range ordering (SRO) in Cu-Al alloys which could lead to inhomogeneous plastic deformation during a metal forming process [19]. These states are marked as SS states. ECAP was conducted in air at room and liquid nitrogen cryogenic temperature with a pressing speed of 0.25 mm×s^{-1} using a die with the inner and outer channel intersection angles $\varphi = 120°$ and $\psi = 0°$ correspondingly. All samples were pressed by one ECAP pass with the equivalent strain 0.7. Some samples before ECAP were immersed in liquid nitrogen for more than 5 min until no bubbling. The ECAP die was cooled by liquid nitrogen during plastic deformation [18].

Microhardness measurements were conducted on Struers Duramin hardness testing machine by Vickers method under the applied load of 100 grams for 10 seconds. The measurements were performed at points located at a distance of 25 μm from each other. The results of measurements were averaged for at least 10 points.

Mechanical tensile tests were carried out on a tensile testing machine Instron 8801 at room temperature with a strain rate of 0.24 mm/min. For tests flat samples cut along the direction of deformation with an initial working part 4 mm were used.

The optical micrographs were obtained from the ECAP flow plane characterization. Thin foils to prepare samples for transmission electron microscope (TEM) investigations were cut from the mentioned plane. Viewing and photography of samples were carried out on optical microscope Olympus GX51 at magnifications ×100. Thin foils were examined using a TEM JEM-2100 in dark and light fields at an accelerating voltage of 160-200 kV. Final thinning of the foils was performed by the method of ink-jet electrolytic polishing in a special unit Tenupol-5 using the electrolyte of the following composition: 33% nitric acid and 67% methanol at -25°C with a current of 180-200 mA.

3 Results and analysis

3.1 Microhardness and tensile properties

Fig. 1 contains a chart representing the results of the microhardness measurement of for pure Cu and Cu-Al alloys in the studied conditions. As it follows from the analysis of the diagram, there are two obvious patterns of change in microhardness. The first trend consists in the increase of microhardness with decreasing SFE (increasing Al content in the investigated Cu-Al alloys). This trend is typical for all similar conditions of the investigated materials, including SS states, and the states after ECAP at RT and ECAP at LNT. The second trend is the increase of microhardness with decreasing temperature ECAP for all relevant states of the investigated materials.

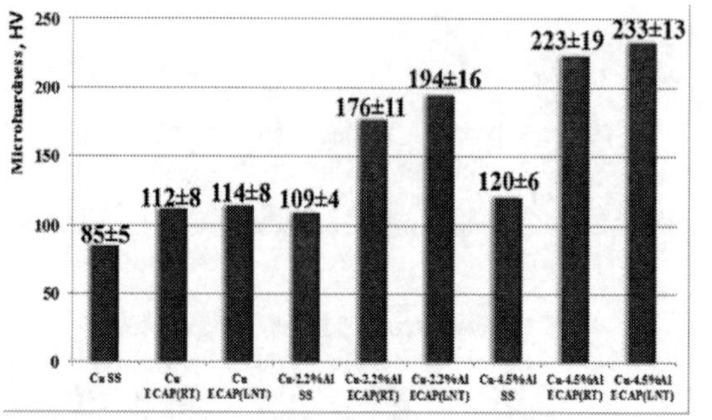

Fig. 1. Microhardness diagram for Cu, Cu-2.2%Al, Cu-4.5%Al in SS states and states after ECAP at RT and LNT.

In the SS condition, the microhardness of samples of pure copper is 85±5 Hv, alloys Cu-2.2%Al and Cu-4.5%Al - 109±4 Hv, 120±6 Hv, respectively. After conducting ECAP there is a significant increase in the hardness of the material relative to the hardness of the SS state. Microhardness of Cu after ECAP at room temperature is 112±8 Hv, and after ECAP at cryogenic temperature - 114±8 Hv. The microhardness of the alloys Cu-2.2%Al and Cu-4.5%Al after ECAP at room temperature is 176±11 Hv and 223±19 Hv, respectively, while after ECAP at cryogenic temperature of 194±16 Hv and 233±13 Hv respectively.

The results of experimental studies of the engineering tensile curves of samples of the investigated states are presented in Fig. 2.

The values of tensile strength, calculated on the basis of the engineering tensile curves for SS states are equal to 130 MPa for pure Cu, 155 MPa for Cu-2.2%Al alloy and 176 MPa for Cu-4.5%Al alloy, respectively. After ECAP at RT tensile strength is 150 MPa for pure Cu, 282 MPa for Cu-2.2%Al alloy and 478 MPa for Cu-4.5%Al alloy. After ECAP at LNT tensile strength equal to 276 MPa for pure Cu, 302 MPa for Cu-2.2%Al alloy and 533 MPa for Cu-4.5%Al alloy. As can be seen, both the decrease in SFE and decrease of ECAP temperature lead to higher tensile strength of the investigated materials. Thus, there is a full correlation of the detected tendencies related to tensile strength with the tendencies related to microhardness for the studied states.

The obtained values of the elongation for SS states are 70% for pure Cu, 32% for Cu-2.2%Al alloy and 44% for Cu-4.5%Al alloy, respectively. After ECAP at RT elongation is equal to 39% for pure Cu, 15% for Cu-2.2%Al alloy and 11% for Cu-4.5%Al alloy. After ECAP at LNT elongation is 14% for pure Cu, 19% for Cu-2.2%Al alloy and 8% for Cu-4.5%Al alloy.

DOI: 10.1051/matecconf/201712902032

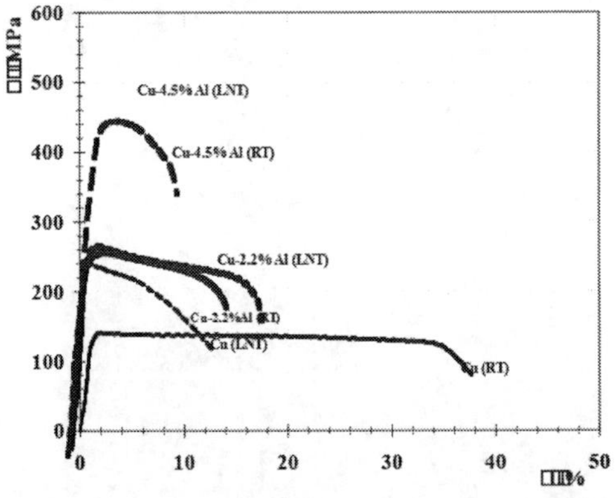

Fig. 2. Tensile engineering stress-strain curves of Cu, Cu-2.2%Al, Cu-4.5%Al after ECAP at RT and ECAP at LNT.

Thus, according to the obtained results it can be concluded that for the selected alloys with low and high SFE after ECAP at LNT there is an increase in strength, however the ductility decreases. However, for the Cu-2.2 wt.% Al alloy with an intermediate value of SFE after ECAP at LNT there is an increase in strength and elongation. In [20] the optimum SFE at which there is increased ductility of the alloy Cu-Al is reported. This can be explained by low dislocation density revealed in the observed fine structure (Fig. 4 a). As is known [21], the elongation can be improved in materials with lower density of dislocations, since there is more space for further accumulation of dislocations. Based on these results one can assume that a low dislocation density in the microstructure of the Cu-2.2%Al alloy led to further accumulation of dislocations during tensile testing, which led to hardening and increased ductility.

3.2 Microstructure characterization

Optical microstructure of Cu and Cu-Al alloys with 2.2, 4.5%Al show significant changes as a result of ECAP at room and cryogenic temperatures (Fig. 3 a - f).

After conducting ECAP at room temperature in the Cu, Cu-2.2%Al, Cu-4.5%Al alloys the boundaries of the former grains are seen. Shear bands are observed as well. The direction of the shear bands depends on grain orientation (Fig. 3 a - c). The shear occurs in three directions.

When the temperature of the ECAP is decreased, the shear bands become thinner, the distance between them decreases (Fig. 3 b, d). With increasing Al content, namely, in the alloy Cu-4.5%Al, there steps on shear bands are observed (Fig. 3 d).

TEM images indicate the presence of deformation twins and dislocations in Cu-2.2%Al, Cu-4.5%Al alloys after ECAP at cryotemperature (Fig. 4 a, b). The number of deformation twins increases with decreasing the processing temperature.

Thus, one can conclude that deformation twinning plays an increasingly crucial role in grain refinement and microstructural evolution during SPD [22 - 24]. Since the thickness of twins decreases with decreasing SFE, the twin can easily be converted into nanoscale grains with uniform distribution through the complex interaction of twins, dislocations and shear bands [25].

In addition to the development of the increased microstructural homogeneity, the average grain size is also decreasing with decreasing SFE and lower processing temperature due to the limited dynamic recovery and change of the dominant refinement mechanism from the dislocation one to twin fragmentation mechanism [26].

4 Conclusions

(1) Two obvious tendencies in the microhardness evolution depending on the SFE for the analogous states of pure Cu and Cu-Al alloys are observed. One is that the increase of the microhardness with decreasing SFE (increasing of the Al content) in the investigated Cu-Al. The other is the increase of the microhardness with decreasing ECAP temperature for all the investigated states of the mentioned materials.

(2) The decrease of SFE and ECAP temperature promotes the increase of the tensile strength of the investigated metallic materials.

(3) The microstructural investigations revealed the increased twinning activities with lowering of SFE and ECAP temperature.

(4) For a Cu-2.2%Al lowering the ECAP temperature helped to improve both strength and ductility. It seems that the reason for that is the balance achieved between the dislocation and twinning deformation mechanisms.

The work is executed by the financial support of the Ministry of education and science within the framework of the project No. 16.1969.2017/PCh and the Science and Technology Bureau of Jiangsu Province, P.R. China under grant No. BY2016029-19.

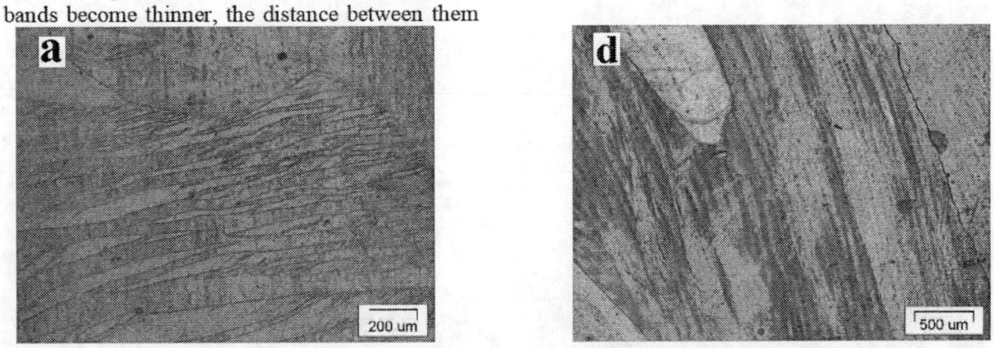

DOI: 10.1051/matecconf/201712902032

Fig. 3. Optical microscopy: a - Cu, b - Cu-2.2%Al, c - Cu-4.5%Al after ECAP at RT and d - Cu, e – Cu-2.2%Al, f - Cu-4.5%Al after cryoECAP.

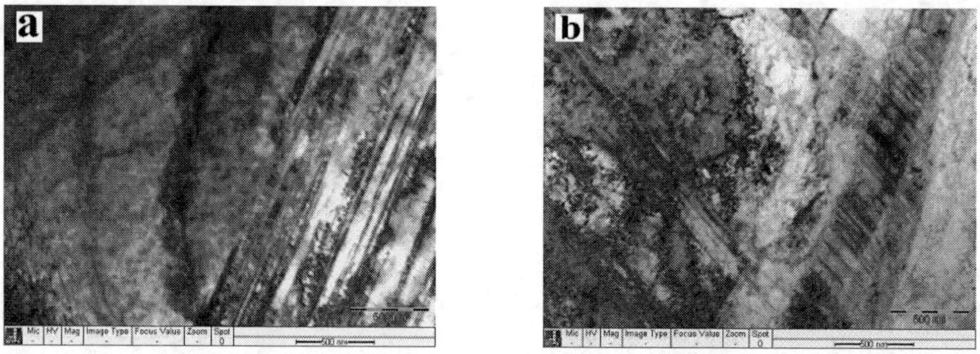

Fig. 4. TEM images: a-Cu-2.2%Al, b-Cu-4.5%Al after cryoECAP.

References

1. S.V. Subramanya, J. Wang, W.W. Jian, A. Kauffmann, H. Conrad, J. Freudenberger, Y.T. Zhu, Materials Science and Engineering A **527**, 7624 (2010)
2. J.W. Christian, S. Mahajan, Prog. Mater. Sci. **39**, 1 (1995)
3. T.H. Blewitt, P.R. Coltman, J.K. Redman, J. Appl. Phys. **28**, 651 (1957)
4. E.E. Danaf, S.R. Kalidindi, R.D. Doherty, Metall. Mater. Trans. A **30A**, 1223 (1999)
5. F. Greulich, L.E. Murr, Mater. Sci. Eng. **39**, 81 (1979)
6. G.T. Gray III, Acta Metall. **36**, 1745 (1988)
7. Y. Zhang, N.R. Tao, K. Lu, Acta Mater. **56**, 2429 (2008)
8. Y.H. Zhao, J.F. Bingert, X.Z. Liao, B.Z. Cui, K. Han, A.V. Sergueeva, A.K. Mukherjee, R.Z. Valiev, T.G. Langdon, Y.T. Zhu, Adv. Mater. **18**, 2949 (2006)
9. Y. Zhang, N.R. Tao, K. Lu, Acta Mater. **56**, 2429 (2008)
10. Y.S. Li, N.R. Tao, K. Lu, Acta Mater. **56**, 230 (2008)

11. W.S. Zhao, N.R. Tao, J.Y. Guo, Q.H. Lu, K. Lu, Scripta Materialia **53**, 745(2005)
12. Z.S. You, L. Lu, K. Lu, Scripta Materialia **62**, 415 (2010)
13. R.Z. Valiev, R.K. Islamgaliev, I.V. Alexandrov, Progress in Materials Science **45**, 103 (2000)
14. R.Z. Valiev, T.G. Langdon, Prog. Mater. Sci. **51**, 881 (2006)
15. X.H. An, Q.Y. Lin, S.D. Wu, Z.F. Zhang, R.B. Figueiredo, N. Gao, T.G. Langdon, Scripta Materialia **64**, 954 (2011)
16. I.V. Alexandrov, R.G. Chembarisova, L.I. Zainullina, K.X. Wei, W. Wei, J. Hu, DOI: 10.1557/jmr.2016.451
17. X. San, X. Liang, L. Cheng, L. Shen, X. Zhu, Trans. Nonferrous Met. Soc. China **22**, 819 (2012)
18. W. Wei, S.L. Wang, K.X. Wei, I.V. Alexandrov, Q.B. Du, J. Hu, Journal of Alloys and Compounds **678**, 506 (2016)
19. S. Matsuo, L.M. Clarebrough, Acta Metall. **11**, 1195 (1963)
20. S.D. Wu, X.H. An, W.Z. Han, S. Qu, Z.F. Zhang, Acta Mater. Sin **46**, 257 (2010)
21. P.Xue, B.L. Xiao, Z.Y. Ma, Scr.Mat. **68**, 751 (2013)
22. X.N. An, Q.Y. Lin, Z.F. Zhang, R.B. Figueiredo, N. Gao, Philos Mag. **91**, 3307 (2011)
23. Y.H. Zhao, X.Z. Liao, Y.T. Zhu, Z. Horita, T.G. Langdon, Mater Sci Eng A **463**, 22 (2007)
24. X.H. An, Q.Y. Lin, S.D. Wu, Z.F. Zhang, Mater Sci Eng A **527**, 4510 (2010)
25. S. Qu, X.H. An, H.J. Yang, C.X. Huang, G. Yang, Q.S. Zang, Z.G. Wang, S.D. Wu, Z.F. Zhang, Acta Mater **57**, 1586 (2009)
26. X.H. An, S.D. Wu, Z.G. Wang, Z.F. Zhang, Acta Materialia **74**, 200 (2014)

 DOI: 10.1051/matecconf/201712902033

Formation and properties of composite nanostructured PEO-coatings on metals and alloys

Dmitry V. Mashtalyar[1,2,*], *Igor M.* Imshinetsky[1] , and *Sergey L.* Sinebryukhov[1]

[1]Institute of Chemistry Far Eastern Branch of Russian Academy of Sciences, 690022 Vladivostok, 159 100-letiya Vladivostoka Pr., Russia
[2]Far Eastern Federal University, 690950 Vladivostok, 8 Sukhanova St., Russia

Abstract. Results of investigation of the incorporation of zirconia and titanium nitride nanoparticles into the coatings formed on magnesium alloy by plasma electrolytic oxidation are presented. Comprehensive research of electrochemical and mechanical properties of obtained coatings was carried out. It was established that the polarization resistance of the samples with a coating containing zirconia nanoparticles is in two fold higher than for the sample with base PEO-coating. One of the important reasons for improving the protective properties of coatings formed in electrolytes containing nanoparticles consists in enhanced morphological characteristics, in particular, the porosity decrease and increase of thickness and resistivity of porousless sublayer in comparison with base PEO-layer. Incorporation of zirconia and titanium nitride particles into the coating increases the mechanical performances. The coating containing nanoparticles have greater hardness and are more wear resistant in comparison with the coatings formed in the electrolyte without nanoparticle.

1 Introduction

Magnesium and its alloys have been used widely in various fields of engineering, such as aircraft, electronic, automobile owing to the combination of low density, high strength to weight ratio, higher damping capacity and high machinability. However, high corrosion activity and low wear resistance of magnesium alloys significantly have limited their applications. A proper surface treatment may improve wear and corrosion resistance of the substrate. There are some surface treatment techniques established for protection of magnesium and its alloys. These techniques include chemical conversion plating, anodizing, gas-phase deposition, electrodeposition, organic coatings and sol-gel technique.

Plasma electrolytic oxidation (PEO), also called micro-arc oxidation (MAO), is an economic, efficient, and environmentally friendly technology capable of producing a stable oxide coating on the surface of light metals (Al, Mg, Ti, Zr, Ta, etc.) or alloys [1, 2]. The oxide coatings have controllable morphology and composition, excellent adhesion to substrate, good electrical and thermal properties, high microhardness, wearproof and corrosion resistance. The PEO formation is induced by micro-discharges on the metal surface and accompanied by gas evolution. Various processes, including chemical, electrochemical, thermodynamic, and plasma-chemical reactions occur at the discharge sites, due to increased local temperature (10^3 to 10^4 K) and pressure ($\approx 10^2$ MPa). These processes are responsible for modifying the structure, composition, and morphology of obtained oxide coatings. Coatings formed by the PEO process usually contain crystalline and amorphous phases with constituent species originating from both the metal and electrolyte. Recent developments in this area are focused on the addition of nanoparticles into the electrolyte, aiming at in-situ incorporation them into PEO-coatings [3–15]. The corrosion resistance was effectively improved by the addition of TiO_2 nanoparticles in the electrolyte [5]. An increased coating strength has been shown for PEO-coatings formed in electrolytes with powders of Al_2O_3, TiO_2, Cr_2O_3, ZrO_2, TiC, B_4C, BN [9]. Suspensions of Al_2O_3 [16] have been employed to produce coatings with enhanced thermomechanical properties, reduced surface porosity and increased wear-resistance.

The main aim of the present study is to investigate the influence of the incorporation of nanoparticles zirconia (ZrO_2) and titanium nitride (TiN) into the coatings via PEO process in the phosphate-based electrolyte in the relationship with their structural and corrosion performances and mechanical properties.

2 Methods

Rectangular plates with size of 15 mm × 20 mm × 2 mm made of the magnesium alloy MA8 (1.5 wt. % Mn; 0.15 wt. % Ce; Mg – balance) were used as samples. Prior to oxidation, for the surface standardization, the samples underwent mechanical treatment by grinding paper of

* Corresponding author: madiva@inbox.ru

different grain sizes (600, 800, and 1200), washed with distilled water, and degreased by ethanol.

Two different kind of nanomaterials were used as additive, namely zirconia with particle size of up to 100 nm and titanium nitride with particle size of 20 nm. Three different electrolytes consisting of 5.0 g/l NaF, 15.0 g/l Na_2SiO_3 termed as bath A, which was then modified by adding 3 g/l ZrO_2 termed as bath B or 3 g/l TiN termed as bath C. Since at the plasma electrolytic oxidation the coating growth proceeds, as a rule, at the anodic polarization of treated sample, to ensure the maximal embedding of particles into the coating, an anionic surfactant was used as a stabilizer of the disperse system. The process of coatings formation was carried out using a plasma electrolytic oxidation apparatus. A conventional reversive thyristor rectifier was used as a power supply. The frequency of polarizing pulses was 300 Hz; the duty cycle was 50 %. All the samples were treated in the two-stage bipolar PEO-mode. At the first stage, the anodic component was set galvanostatically at a current density of 0.5 A/cm^2; the cathodic phase was set potentiostatically at -30 V. The duration of the first PEO-stage was 200 s: here, the anodic voltage attained 300 V. During the second stage, within 600 s the anodic component was changed potentiodynamically from 300 to 200 V, whereas the cathodic one – from -30 to -10 V.

The element composition of surface layers was determined using the energy dispersive X-ray fluorescence analysis (EDXRF) on an EDX-800HS spectrometer (Shimadzu, Japan).

Electrochemical properties of the obtained layers have been investigated with the use a VersaSTAT MC electrochemical system (Princeton Applied Research, USA). The studies have been carried out in the three-electrode cell at room temperature in a 3 % NaCl solution. A platinum mesh has been used as a counter electrode. The saturated calomel electrode (SCE) was used as a reference electrode. The surface area of the sample exposed to the electrolyte was equal to 1 cm^2.

Studies of mechanical properties, in particular, determination of microhardness and elasticity modulus of the coatings material were carried out using a DUH-W201 Dynamic Ultramicrohardness Tester (Shimadzu, Japan). Measurements of the universal microhardness were carried out on a cross-section using a Vickers indenter at a load of 50 mN.

Adhesion properties of surface layers were studied by scratch test on a Revetest (CSM Instruments, Switzerland). Experiments were carried out at a preset trace length of 5 mm chosen empirically and a gradual load increase from 1 to 20 N at a rate of 9.5 N/min. For scratch testing, a Rockwell diamond indenter with an angle of 120° and a tip radius of 200 μm was used. The study was carried out by measuring the friction force and the friction coefficient, recording the acoustic emission, and fixing critical loads, at which characteristic destruction traces emerged. For each type of coating, the following parameters were determined: the load at which the first cracking occurs, L_{C1}; the load at which the extent of fracture events increases, L_{C2}; loads, at which coating scratching down to metal occurred, L_{C3}.

Tribological tests were performed using a Tribometer (CSM Instruments, Switzerland) in accordance with the 'ball-on-plate' scheme. A ball of a diameter of 10 mm made of alumina was used as a counterbody. All the studies were carried out under dry friction and ambient conditions (25 °C and relative humidity of 50 %). A normal load was equal to 10 N. The linear rotation speed was equal to 50 mm/s, the track diameter was equal to 10 mm. The coatings wear, I, was calculated using the normalization of the sample volume loss during tests ΔV_{samp} on the values of run (N) and applied load (P): $I = \Delta V_{samp}/(NP)$. The sample volume loss is as follows: $\Delta V samp = S \cdot l$, where l is the trace length and S is the area of cross section of the wear trace.

3 Results and discussions

According to the data obtained using scanning electron microscopy (Fig. 1), the morphology of surface layers obtained in baths A, B and C differs substantially. The surface of the coating containing zirconia nanoparticles (Fig. 1(b)) is smoother in comparison with the convolute surface of the coating obtained in the base electrolyte without nanoparticles (Fig. 1(a)) and in the titanium nitride-containing electrolyte (Fig. 1(c)). The porosity of coating fabricated in the base electrolyte is equal to 5.0 %. This parameter for PEO-layers formed in baths B and C was equal to 2.2 and 4.1 %, respectively. The porosity of the coatings containing nanoparticles is lower due to partial sealing of the pores. In case of using nanosized particles of titanium nitride, the surface possesses the highest roughness parameter (Ra = 2.0 μm) as compared to the base PEO-coating (Ra = 1.23 μm) and PEO-layer with ZrO_2 nanoparticles (Ra = 1.21 μm).

The element composition of coatings was determined by EDXRF analysis (Table 1). The incorporation of nanoparticles in the coating is nearly equal for both zirconia and titanium nitride. The concentration of zirconium and titanium attained 1.7 and 1.8 wt.% for coatings formed respectively in bath B and C. It was found that most of the nanosized particles are concentrated in the surface layer of PEO-coating.

To evaluate the corrosion resistance of the obtained coatings, their electrochemical properties were investigated. Analysis of polarization curves (Fig. 2) enable one to conclude on a substantial positive effect of nanosized zirconia and titanium nitride in the composition of the forming electrolyte on anticorrosion properties of the fabricated coatings. Corrosion currents density (I_c) of the samples with coatings formed in in baths B are significantly lower in comparison with the sample with base PEO-layer (Table 2). One of the important reasons for improving the protective properties of coatings formed in baths B is the enhancement of morphological characteristics, in particular blocking the micropores and cracks, giving rise to a denser coating, which acts as an effective barrier to suppress corrosion. Corrosion current density for coatings containing TiN nanoparticles increased in 1.5 fold in comparison with coatings obtained without the nanopowder (Table 2). The deterioration of the protective properties of the coatings containing TiN nanoparticles is mainly explained by high conductivity of titanium nitride. Besides, the coatings obtained using nanoparticles contain chemical compounds (ZrO_2, TiN) having much higher chemical stability as compared to that of the main components of the base PEO-layer.

Table 1. EDXRF results of the sample treated by PEO in baths A, B and C.

Sample	Mg (wt.%)	O (wt.%)	Si (wt.%)	Zr (wt.%)	Ti (wt.%)	Na (wt.%)	Mn (wt.%)	Others (wt.%)
Bath A	37.8	38.4	18.5	-	-	1.2	3.0	1.1
Bath B	36.6	36.6	19.1	1.7	-	1.9	3.5	0.6
Bath C	40.1	38.0	15.3	-	1.8	2.6	1.7	0.5

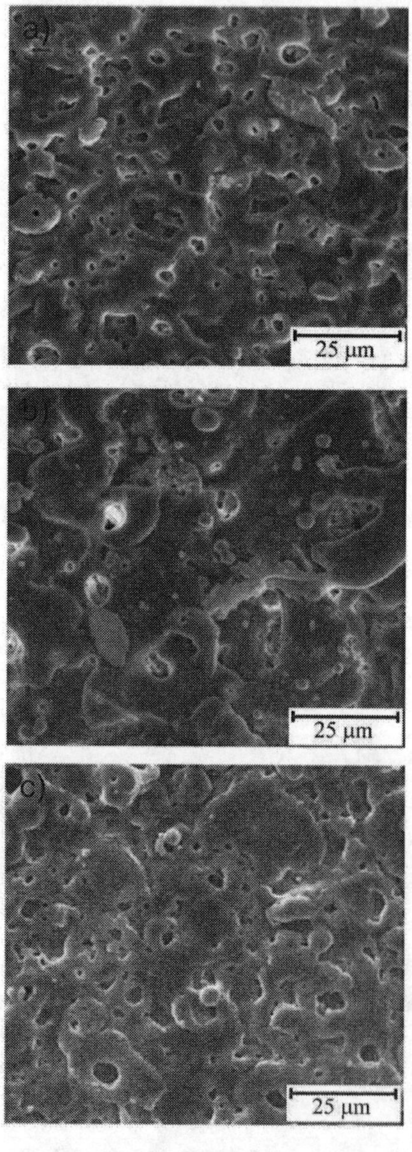

Fig. 1. SEM images of the surface of PEO coatings formed in the different electrolytes: (a) bath A, (b) bath B and (c) bath C.

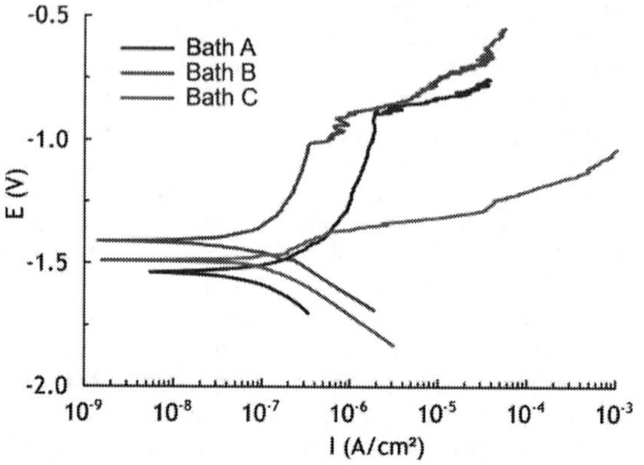

Fig. 2. Potentiodynamic polarization curves of the samples treated by PEO using baths A, B, and C

Table 2. Corrosion performances of MA8 magnesium alloy samples.

Sample	R_p ($\Omega\cdot cm^2$)	I_c (A/cm^2)	E_c (V vs. SCE)
Bath A	2.8×10^5	1.2×10^{-7}	-1.45
Bath B	6.1×10^5	5.3×10^{-8}	-1.42
Bath C	1.7×10^5	$1.8\cdot10^{-7}$	-1.47

Evaluation of the mechanical properties of the fabricated coatings was carried out using the method of microhardness testing. Obtained results (Table 3) testified, that the layers containing nanoparticles are harder in comparison with the coatings formed in bath A ($H_\mu = 2.1 \pm 0.3$ GPa). The coating containing TiN nanoparticles possesses the highest microhardness (4.2 ± 0.5) GPa. ZrO_2 nanoparticles provide somewhat lower microhardness value, (3.7 ± 0.8) GPa. Such increase is, probably, related to the composition and properties of the embedded nanoparticles and the surface layer morphology.

Table 3. Microhardness and Young's modulus for the magnesium alloy MA8.

Sample	Microhardness (GPa)	Young's modulus (GPa)
Bath A	2.1 ± 0.3	60 ± 5
Bath B	3.7 ± 0,8	97 ± 12
Bath C	4.2 ± 0.5	98 ± 10

Adhesion of coating to substrate was evaluated using the scratch test. To determine the coating material

 DOI: 10.1051/matecconf/201712902033

elastoplastic properties, the repeated registration of the indenter trace profile was carried out after unload. The value of the load corresponding to the first cracks (L_{C1}) for all types of the studied coatings is equal to 5.0 ± 0.3 N. The load, at which the coating is worn down to metal (L_{C3}), varies from 14.3 to 20.4 N for the coatings under study (Table 4).

Table 4. Mechanical characteristics of coatings determined by scratch test.

Sample	L_{C1} (N)	L_{C2} (N)	L_{C3} (N)
Bath A	5.0 ± 0.3	8.4 ± 0.5	13.8 ± 0.2
Bath B		9.4 ± 0.7	14.3 ± 0.7
Bath C		10.5 ± 0.9	20.4 ± 1.9

Tribological tests were performed to evaluate wearproof of samples under study. Fig. 3 shows the dependencies of the friction coefficient μ on the number of cycles. Tests were carried out until the coating wear down to metal. In general, the trend to positive effect of embedded nanomaterials into the coatings composition on their mechanical properties is preserved (Fig. 3, Table 5). Wear of coatings containing nanoparticles is lower than for the samples without additives. Due to higher microhardness and lower porosity the coatings with ZrO_2 and TiN nanoparticles withstand a 1.4–1.9 fold more number of load cycles, as compared to the base PEO-layer. One should mention that the slope angle of the curve of friction coefficient for the coatings obtained in electrolytes with nanoparticles is smaller than that for the base PEO coating. This fact is related to differences in the morphology and chemical composition of the coatings under study.

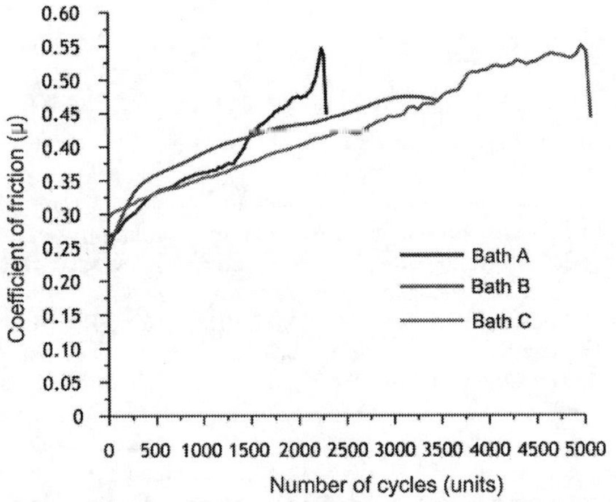

Fig. 3. Dependence of the friction coefficient of the corundum ball/sample on number of cycles for samples treated by PEO using baths A, B, and C.

Table 5. Wear of coatings.

Sample	Wear rate (mm³/(N·m))	Cycles
Bath A	$(4.1 \pm 0.4)\times10^{-5}$	2560
Bath B	$(2.9 \pm 0.2)\times10^{-5}$	3540
Bath C	$(1.9 \pm 0.1)\times10^{-5}$	4865

4 Conclusion

The incorporation of ZrO_2 and TiN nanoparticles into the coating via PEO of magnesium alloy MA8 was successfully achieved. As a result, the corrosion protection and mechanical properties of the coatings were improved considerably. Embedded nanoparticles modify the morphological structure, for instance, the porosity is decreased. The insertion of nanoparticles into the coating composition increases the microhardness and wear resistance in almost 2 fold as compared to the PEO-coating without nanoparticles.

This work was supported by the Russian Science Foundation (№ 14-33-00009).

References

1. Q. Li, J. Liang, Q. Wang, *Modern Surface Engineering Treatments,* 75–99 (2013)
2. F. Walsh, C. Low, R. Wood, K. Stevens, J. Archer, A. Poeton, A. Ryder, Trans. Inst. Met. Finish. **87,** 122-135 (2009)
3. X. Lu, M. Mohedano, C. Blawert, E. Matykina, R. Arrabal, K. Kainer, M. Zheludkevich, Surf. Coat. Technol. **307,** 1165-1182 (2016)
4. X. Li, B. Luan, Mater. Lett. **86,** 88–91 (2012)
5. Y. Song, X. Sun, Y. Liu, Mater. Corros. **63,** 813–818 (2012)
6. R. Bhuyan, T. Kumar, D. Goswami, A. James, A. Perumal, D. Pamu, Mater. Sci. Eng., B **178,** 471–476 (2013)
7. S.V. Gnedenkov, S.L. Sinebryukhov, D.V. Mashtalyar, I.M. Imshinetskiy, A.V. Samokhin, Yu.V. Tsvetkov, J. Nanomater. **2015,** 1-12 (2015)
8. D.V. Mashtalyar, S.V. Gnedenkov, S.L. Sinebryukhov, I.M. Imshinetskiy, A.V. Puz', J. Mater. Sci. Technol. **33,** 461-468 (2017)
9. V. Malyshev, K. Zorin, Appl. Surf. Sci. **254,** 1511–1516 (2007)
10. Y. Wang, D. Wei, J. Yu, S. Di, J. Mater. Sci. Technol. **30,** 984-990 (2014)
11. M. Daroonparvar, M.A.M. Yajid, N.M. Yusof, H.R. Bakhsheshi-Rad J. Alloys Compd. **688,** 841-857 (2016)
12. S.V. Gnedenkov, S.L. Sinebryukhov, D.V. Mashtalyar, I.M. Imshinetskiy, A.V. Samokhin, Yu.V. Tsvetkov, Vacuum. **120,** 107-114 (2015)
13. C.-J. Hu, M.-H. Hsieh, Surf. Coat. Technol., **258,** 275-283 (2014)
14. S.V. Gnedenkov, S.L. Sinebryukhov, I.A. Tkachenko, D.V. Mashtalyar, A.Y. Ustinov, A.V. Samokhin, Y.V. Tsvetkov, Inorg. Mater. Appl. Res., **3,** 151–156 (2012)
15. M. Shokouhfar, S.R. Allahkaram, Surf. Coat. Technol. **291,** 396-405 (2016)
16. Y. Khokhryakov, P. Butyagin, A. Mamaev, J. Mater. Sci. Lett. **40,** 3007-3008 (2005)

Multifunctional polymer-containing coatings on magnesium alloys

Dmitry V. Mashtalyar[1,2,*], *Konstantine V.* Nadaraia[1,2], and *Sergey V.* Gnedenkov[1]

[1]Institute of Chemistry Far Eastern Branch of Russian Academy of Sciences, 690022 Vladivostok, 159 100-letiya Vladivostoka Pr., Russia
[2]Far Eastern Federal University, 690950 Vladivostok, 8 Sukhanova St., Russia

Abstract. The results of surface modification of magnesium alloys by plasma electrolytic oxidation (PEO) and post-treatment in suspension of tetrafluoroethylene (TFE) telomeric solution or superdispersed polytetrafluoroethylene (SPTFE) dispersion have been presented. Electrochemical, tribological properties and wetting ability of obtained protective composite coatings were investigated. The fabricated coatings decrease both the corrosion current density (1.7×10^{-9}–5.4×10^{-11} A/cm^2) and wear (9.8×10^{-6}–7.6×10^{-7} mm^3/(N m)) by orders of magnitude as compared to bare magnesium alloy (5.3×10^{-5} A/cm^2 and 3.8×10^{-3} mm^3/(N m)) and base PEO-coating (2.4×10^{-7} A/cm^2 and 1.7×10^{-3} mm^3/(N m)). Moreover, coatings obtained by means of PEO and polymer dispersion dipping possess hydrophobic and superhydrophobic properties. Such multifunctional coatings have high corrosion resistance and good friction coefficient under dry wear conditions that extends the application field of magnesium alloys.

1 Introduction

Plasma electrolytic oxidation (PEO) is an advanced surface modification process, which leads to the formation of similar to ceramic and well-adhered coatings on the valve metals such as aluminium, magnesium, titanium, and a wide range of their alloys [1–4]. There are many advantages of PEO-coatings, such as a high hardness, corrosion and wear resistances and biocompatibility. PEO-layers provide better protection than conventional anodic oxide films, which makes PEO suitable for many industrial applications. In spite of many advantages, the PEO-coatings can have structural defects such as cracks and pores, which could dramatically reduce the corrosion resistance of even relatively thick coatings. There are several attempts to reduce or seal such porosity. The post-treatments are often applied in order to seal the pores and improve the protection effect in practical industrial applications where the PEO-layer is used as a base for composite coating formation [5–7].

In recent years, there have been numerous reports in the literature about PEO-coatings on Mg alloy with post-treatments. The post-treatments of PEO-based coatings can usually exhibit significantly better corrosion protection ability than the single PEO-coatings, suggesting better application prospects.

PEO allows forming on the magnesium alloy protective layers, which have a convolute, rough surface. Such structure can be served as a basis for application at the PEO-surface polymeric materials in order to provide additional protection for this product [8,9]. In this regard, polymers due to their chemical stability represent a particular interest. This paper describes the results of investigation of protective properties of multifunctional composite coatings (CC) on magnesium alloys with various polymer.

2 Methods

Rectangular plates of a size of 15 mm × 20 mm × 2 mm made of the magnesium alloy MA8 (1.5 wt. % Mn; 0.15 wt. % Ce; Mg – balance) were used as samples. Prior to oxidation, for the surface standardization, the samples underwent mechanical treatment by grinding paper of different grain sizes (600, 800, and 1200), washed with distilled water, and degreased with ethanol. Silicate-fluoride solution was used as electrolyte for PEO [3, 6-8]. The process of coating formation was realized using the tyristor rectifier as a power supply. Coatings were formed in bipolar PEO-mode. In the first step the anodic current density was specified of 0.61 A·cm^{-2} (voltage increases up to 270 V), cathodic component was kept potentiostatically at –40 V. In the second step the voltage was gradually changed down to 200 V in the anodic part and up to –10 V in the cathodic one. The duration of the first and second phases of the PEO was 200 s and 600 s, respectively.

Two different solutions for the composite coating formation were used. TFE telomeric solution in acetone and aqueous suspension of SPTFE (trademark Forum®) were used as a polymer-containing component for formation of composite coatings. The polymer was deposited by dipping of the sample with PEO-coating in

* Corresponding author: madiva@inbox.ru

 DOI: 10.1051/matecconf/201712902034

one of two used solutions. Time of the immersion in suspensions was not more than 10-15 s. After application, the samples were air dried for 30 minutes at 25 °C. Next step was heat treatment at 250 °C for coatings, which were obtained by application of telomeric solution (the number of applications was 3) (hereinafter CC#1) and at 315 °C for coatings, which were obtained by application of SPTFE (the number of applications was 4) (hereinafter CC#2).

Electrochemical properties of coatings were investigated using electrochemical system VersaSTAT MC (Princeton Applied Research, USA). Measurements were carried out in a three-electrode cell at room temperature in a 3 % NaCl solution. Platinum mesh was used as a counter electrode. Calomel electrode was used as a reference electrode. The exposed area of the sample was equal to 1 cm^2. Potentiodynamic measurements were carried out at sweep rate of 5 mV s^{-1} from $E_c - 0.15$ V up to $E_c + 0.5$ V. During the impedance measurements, sinusoidal signal with amplitude of 10 mV was used. A spectrum was recorded at the open circuit potential (E_c) in the frequency range from 0.01 Hz to 0.1 MHz.

The wettability of the obtained composite coating was evaluated by sessile drop method on Drop Shape Analyser DSA100 (KRÜSS, Germany). Solution of 3 % NaCl in distilled water was used for contact angle measurements.

Tribological tests were performed using a Tribometer (CSM Instruments, Switzerland) in accordance with the 'ball-on-plate' scheme. A ball of a diameter of 10 mm made of alumina was used as a counterbody. All the studies were carried out under dry friction and ambient conditions (25 °C and relative humidity of 50 %). A normal load was equal to 10 N. The linear rotation speed was equal to 50 mm/s, the track diameter – to 10 mm.

The surface morphology of the obtained composite coating was studied with use a Scanning Electron Microscope (SEM) EVO 40 (Carl Zeiss, Germany).

3 Results and discussions

Data analysis of SEM-images identified the influence of applying polymer and subsequent heat treatment on the surface morphology of the formed layers (Fig. 1). At surface treatment of the base PEO-coating (Fig. 1a) with the telomeric solution the number of pores and defects on the surface decrease and the surface becomes more uniform (Fig. 1b). In the case of applying SPTFE (Fig. 1c) a uniform film was formed with no visible defects.

The potentiodynamic polarization curves and calculated corrosion parameters are presented in Fig. 2 and Table, respectively. These results enable one to conclude that application of polymer at the surface of the base PEO-coating has a positive effect on the protective properties of the both formed composite layers.

In the case of applying at the surface treatment the TFE telomeric solution the corrosion current density (I_c) reduces, and the polarization resistance (R_p) increases by 4 orders of magnitude as compared to the uncoated material and by 2 orders of magnitude as compared to the PEO-coating (see Table).

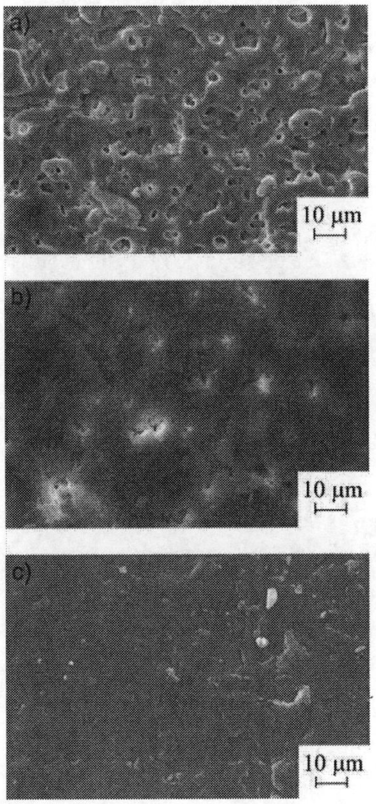

Fig. 1. SEM-image of PEO-coating (a) and composite coatings obtained using the TFE telomeric solution (b) and suspension of SPTFE (c)

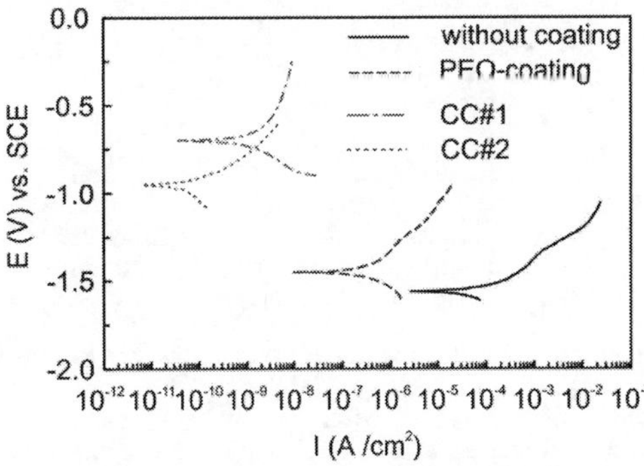

Fig. 2. Polarisation curves obtained in 3% NaCl solution for the studied samples without coating (solid line), with PEO-coating (dashed line), with composite coating obtained with triple application of TFE solution (CC#1, dashed-dotted line) and with composite coating obtained with fourfold application of SPTFE (CC#2, dotted line).

Formed coatings, which were obtained using SPTFE suspension, have even higher protective properties. For such composite coating I_c reduces and R_p increases by 6 and 4 orders of magnitude as compared to the uncoated alloy and PEO-coating, respectively (see Table).

DOI: 10.1051/matecconf/201712902034

Moreover, for both types of composite layers the value of E_c increases at least 0.5 V as compared to one for the PEO-coating. The conclusion can be done that CC#1 has lower values of R_p, and a higher I_c as compared to CC#2 (see Table). Apparently, this is due to the greater uniformity of the film obtained using a suspension SPTFE (Fig. 1c) as compared to the film obtained using telomeric solution (Fig. 1b).

Analysis of electrochemical impedance spectroscopy data revealed a significant difference in the morphology of the composite coatings depending on type of polymer. Observed transformation of the peak position and amplitude of the phase angle (Fig. 3b) indicats about the changes in the surface state: porosity, homogeneous of composite layers.

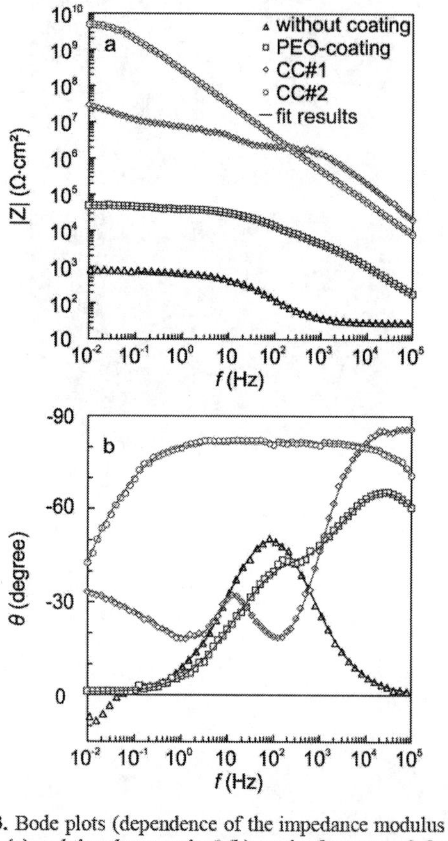

Fig. 3. Bode plots (dependence of the impedance modulus $|Z|_{f\to 0\,\mathrm{Hz}}$ (a) and the phase angle θ (b) on the frequency f) for the studied samples without coating (square symbols), with PEO-coating (triangle symbols), with composite coating obtained with triple application of TFE solution (CC#1, diamond symbols) and with composite coating obtained with fourfold application of SPTFE (CC#2, circle symbols). Impedance spectra contain experimental data (scatter plot marked by symbols) and theoretical fitting curves (lines), which simulate the experimental results by means of equivalent electrical circuits.

In order to describe the charge transfer mechanism at the coating/electrolyte interface, three kinds of equivalent electrical circuits (EEC) were used. For the uncoated sample the one-*R-CPE*-circuit was applied (Fig. 4a). For samples with coatings two kinds of EEC were used: two *R-CPE*-chain circuit (base PEO-coating) (Fig. 4b) and three *R-CPE*-chain circuit (composite coatings) (Fig. 4c).

The using three-*R-CPE*-chain circuit associated with the appearance of the third time constant on the spectra (Fig. 3b). This is a consequence of the sealing air by polymeric plug in the pores of PEO-layer. It should be noted that the dependence of the phase angle θ on the frequency f for CC#2 (Fig. 3b) is capacitive, which suggests a high uniformity of the obtained coating. According to fitting of experimental impedance spectra by a presented EES it was established that TFE- and SPTFE-treatment of the PEO-coating results in the increase of the thickness of the inner porousless sublayer and the geometric thickness of the coating as well.

The results of experiments enable one to conclude that multistep treatment of the PEO coated sample with telomeric solution increases the impedance modulus $|Z|_{f\to 0\,\mathrm{Hz}}$ up to 10^7 $\Omega\cdot\mathrm{cm}^2$, and in the case of the treating with suspension of SPTFE up to 10^9 $\Omega\cdot\mathrm{cm}^2$, that is almost by 3 and 5 orders of magnitude higher than one for the PEO-coating (Fig. 3a, Table). Apparently, all abovementioned reasons of improving the characteristics of composite coating associated with the sealing the pores and defects, which are in the base PEO-coating.

Results of wettability investigations for samples revealed the high hydrophobic properties of CCs (see Table). The highest values of the contact angle (171°) were obtained for the CC#1. Contact angle values of greater than 150° indicate that this sample possesses the superhydrophobic properties. In accordance with the results of previous studies a superhydrophobicity is caused by multimodal roughness of surface. Apparently, in the case of using of the TFE telomeric solution such structure is formed as a result of the partial destruction of the polymer compounds and subsequent aggregation during heat treatment.

As a result of data analysis about changing the friction coefficient during test (Fig. 5) there are at least 2 stages of the wear for composite polymer containing coatings. The first stage is an abrasion of polymer film. In this stage the friction coefficient μ varies from 0.10 to 0.15 (Fig. 5) in dependence on the forming method of the composite layer. The second stage (abrasion of the PEO-layer) is characterised by a sharp increase of the friction coefficient (Fig. 5).

At full abrasion of the coating in the contact zone, the friction coefficient attains of 0.55 and more. Analysis of tribological data revealed a positive impact of incorporation of polymer particles into the PEO-coating to improve the wear resistance of composite layers (Tab.). Thus, the applying of TFE telomeric solution reduces wear of the CC#1 by 3 orders of magnitude as compared to the base PEO-coating. For composite coating, obtained using SPTFE suspension the wear was decreased by 4 orders of magnitude in comparison with the base PEO-layer. Such increasing of the wear resistance is associated with the formation of the polymer-containing layer in the outer part of the PEO-coating, which has a low friction coefficient and acts as a dry lubricant.

 DOI: 10.1051/matecconf/201712902034

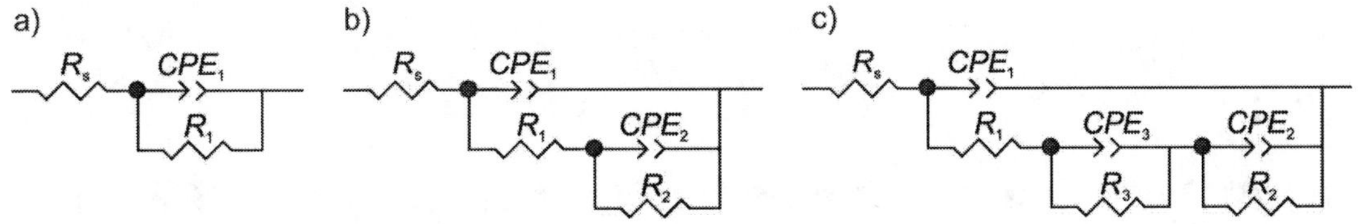

Fig. 4. The equivalent electrical circuits used for the fit of the experimental impedance spectra: one-*R–CPE*-chain EEC (a); two-*R–CPE*-chain EEC (b); three-*R–CPE*-chain EEC (c).

Table 1. Corrosion, tribological properties, and wettability of investigated samples

Type of coating	R_p ($\Omega \cdot cm^2$)	E_C (V vs. SCE)	I_C (A/cm^2)	$\|Z\|_{f \to 0\ Hz}$ ($\Omega\ cm^2$)	CA (°)	Wear rate ($mm^3/(N \cdot m)$)
Without coating	$(0.5 \pm 0.1) \times 10^3$	-1.61 ± 0.07	$(5.3 \pm 0.2) \times 10^{-5}$	$(0.8 \pm 0.03) \times 10^3$	38 ± 1	$(3.8 \pm 0.2) \times 10^{-3}$
Base PEO	$(2,8 \pm 0.3) \times 10^5$	-1.48 ± 0.06	$(2.4 \pm 0.07) \times 10^{-7}$	$(4.3 \pm 0.1) \times 10^4$	49 ± 2	$(1.7 \pm 0.09) \times 10^{-3}$
CC#1	$(3.2 \pm 0.5) \times 10^7$	-0.70 ± 0.02	$(1.7 \pm 0.05) \times 10^{-9}$	$(2.9 \pm 0.09) \times 10^7$	171 ± 2	$(9.8 \pm 0.5) \times 10^{-6}$
CC#2	$(4.9 \pm 0.4) \times 10^8$	-0.95 ± 0.05	$(5.4 \pm 0.2) \times 10^{-11}$	$(1.9 \pm 0.06) \times 10^9$	140 ± 2	$(7.6 \pm 0.4) \times 10^{-7}$

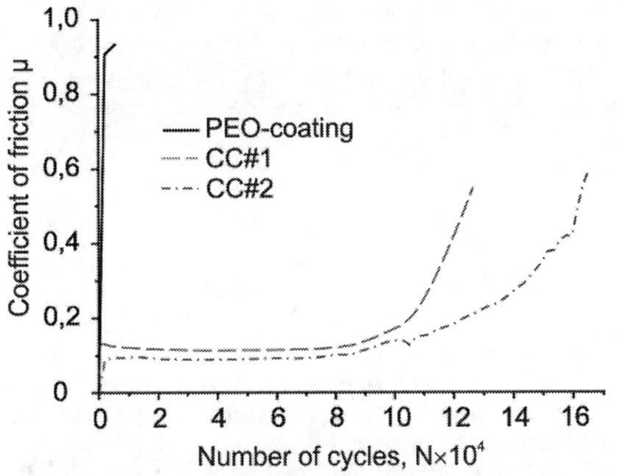

Fig. 5. Dependence of the friction coefficient on the number of cycles for samples made of magnesium alloy MA8 with PEO-coating (solid line), with composite coating obtained with triple application of TFE solution (CC#1, dotted line) and with composite coating obtained with fourfold application of SPTFE (CC#2, dashed-dotted line).

4 Conclusion

New way of the protective composite layers formation using PEO method and subsequent treatment of PEO-coating with fluoropolymer materials was developed. The obtained polymer-containing coatings decrease significantly the corrosion current density and the wear, as compared to unprotected alloy. Values of the contact angle attain up to 171°.

According to electrochemical impedance spectroscopy data and calculated parameters of the elements of equivalent electrical circuits, the SPTFE- and TFE-treatment of the PEO-coating results in the increase of the thickness of the inner porousles sublayer and the composite coating in whole.

This work was supported by the Russian Science Foundation (№ 14-33-00009).

References

1. S.L. Sinebryukhov, A.S. Gnedenkov, O.A. Khrisanfova, S.V. Gnedenkov, Surf. Eng. **25** 565-569 (2009)
2. C.S. Dunleavy, I.O. Golosnoy, J.A. Curran, T.W. Clyne, Surf. Coat. Technol. **203** 3410-3419 (2009)
3. J. Liang, P.B. Srinivasan, C. Blawert, W. Dietzel, Corros. Sci. **52** 540-547 (2010)
4. H. Duan, K. Du, C. Yan, F. Wang, Electrochim. Acta **51** 2898-2908 (2006)
5. Z. Yan, M. Wu, H. Qin, Y. He, X. Yu, L. Chen, Surf. Coat. Technol. **209** 331-336 (2017)
6. M. Mohedano, E. Matykina, R. Arrabal, B. Mingo, A. Pardo, Appl. Surf. Sci. **346** 57-67 (2015)
7. Z. Wang, L. Wu, Y. Qi, W. Cai, Z. Jiang, Surf. Coat. Technol. **204** 3315-3318 (2010)
8. S.V. Gnedenkov, S.L. Sinebryukhov, D.V. Mashtalyar, V.M. Buznik, A.M. Emel'yanenko, L.B. Boinovich, Prot. Met. Phys. Chem. Surf. **47** 93-101 (2011)
9. S.V. Gnedenkov, S.L. Sinebryukhov, D.V. Mashtalyar, K.V. Nadaraia, A.S. Gnedenkov, V.M. Bouznik, Corros. Sci. **111** 175-185 (2016)

 DOI: 10.1051/matecconf/201712902035

Simulation of crystalline pattern formation by the MPFC method

Ilya Starodumov[1,*], *Vladimir* Ankudinov[2], and *Peter* Galenko[1,3]

[1]Ural Federal University, Laboratory of Multi-Scale Mathematical Modeling, 620002, Ekaterinburg, Russia
[2]Udmurt State University, Universitetskaya str. 1, building 4, 426034, Izhevsk, Russia
[3]Friedrich-Schiller-Universität Jena, Physikalisch-Astronomische Fakultät, D-07743 Jena, Germany

Abstract. The Phase Field Crystal model in hyperbolic formulation (modified PFC or MPFC), is investigated as one of the most promising techniques for modeling the formation of crystal patterns. MPFC is a convenient and fundamentally based description linking nano- and meso-scale processes in the evolution of crystal structures. The presented model is a powerful tool for mathematical modeling of the various operations in manufacturing. Among them is the definition of process conditions for the production of metal castings with predetermined properties, the prediction of defects in the crystal structure during casting, the evaluation of quality of special coatings, and others. Our paper presents the structure diagram which was calculated for the one-mode MPFC model and compared to the results of numerical simulation for the fast phase transitions. The diagram is verified by the numerical simulation and also strongly correlates to the previously calculated diagrams. The computations have been performed using software based on the effective parallel computational algorithm.

1 Introduction

The phase field crystal (PFC) model [1] has been proposed to incorporate the physics naturally embedded on nano-length scales and on diffusive time scales. The PFC-model uses the free energy functional of the Swift-Hohenberg (SH) form that describes pattern formation in Rayleigh-Benard convection [2]. The PFC-model may also be expressed as a conserved version of the Swift-Hohenberg equation and as an efficient method for simulating liquid-solid transitions [3,4], colloidal solidification [5], dislocation motion and plasticity [6,7], glass formation [8], epitaxial growth [9,10], grain boundary premelting [11], surface reconstructions [12], and grain boundary energies [13].

In the present paper, numerical calculations based on the Phase Field Crystal (PFC) model are presented which may lead to the solution of many practically important problems in metallurgical industry [14]. The determination of the conditions of the technological process for metal castings with predetermined properties, prediction of crystal structure defects during casting, evaluation of the quality of special coatings, and many others are among them can be start to solve beginning from micro-level analysis using PFC-method [14, 15].

1.1 Free energy of system and MPFC equation

The modified phase field crystal model (MPFC) describes an atomic density field $\phi(x,t)$, which is conserved variable, and expressed by the sixth order in space and second order in time equation [14-17]:

$$\tau\frac{\partial^2\phi}{\partial t^2}+\frac{\partial\phi}{\partial t}=\nabla^2\mu\,. \tag{1}$$

Here t is the time, τ is the relaxation time of the atomic flux to its stationary state, and μ represents the chemical potential. The chemical potential may be obtained from functional of free energy:

$$F(\phi,\nabla\phi,\nabla^2\phi)=\int_\Omega\left(f(\phi)-\phi\nabla^2\phi+\frac{1}{2}(\nabla^2\phi)^2\right)d\Omega \tag{2}$$

associated to the computational domain Ω. The Eq. (1) includes the chemical potential μ as a variational derivative of the free energy functional F:

$$\mu(\phi)=\frac{\delta F(\phi)}{\delta\phi}=\frac{\partial f(\phi)}{\partial\phi}+2\nabla^2\phi+\nabla^4\phi. \tag{3}$$

The function of the homogeneous reference energy density f takes on as:

$$f(\phi)=\frac{(1-\varepsilon)\phi^2}{2}+\alpha\frac{\phi^3}{3}+\frac{\phi^4}{4}, \tag{4}$$

where ε is the dimensionless undercooling $\varepsilon=(T_c-T)/T_c$ with T_c – critical temperature of phase transition, T – temperature of the system, and the term α – the coefficient which controls the metastability of transition. For all our calculations, we state $\alpha=0$ for the unstable phase transitions modeling with zero potential barrier of the phase transformation. Introduction of the inertial term with τ in Eq. (1) changes the class of equation from

* Corresponding author: ilya.starodumov@urfu.ru

 DOI: 10.1051/matecconf/201712902035

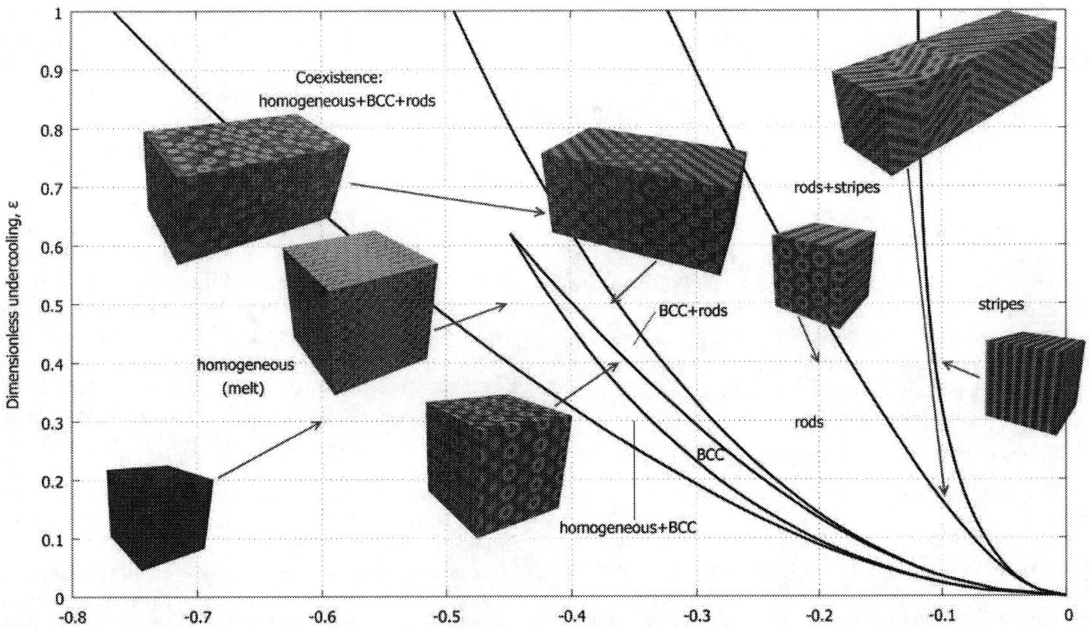

Fig. 1. Computed structures in comparison with the structural diagram "dimensionless undercooling ε – relative atomic density ϕ". calculated in the present work. The existence and coexistence regions remarked right on the diagram : homogeneous Eq. (5), body centered cubic BCC Eq. (6), rods Eq. (7), stripes Eq. (8). The top-left region for $\varepsilon>0.6$, $\phi<0.4$ represents region of coexistence of three structures, where the regions of coexistence of BCC+rods, homogeneous+BCC and homogeneous+rods intersect and match each other exactly. As it can be found on the depicted results of direct computation this region exposed to form irregular structures. For the structures depicted in the diagram we have employed the following parameters (for the illustrations from left to right): $\varepsilon=0.3$, $\phi=-0.6$; $\varepsilon=0.65$, $\phi=-0.43$; $\varepsilon=0.5$, $\phi=-0.45$; $\varepsilon=0.4$, $\phi=-0.36$; $\varepsilon=0.5$, $\phi=-0.369$; $\varepsilon=0.4$, $\phi=-0.2$; $\varepsilon=0.17$, $\phi=-0.1$; $\varepsilon=0.4$, $\phi=-0.1$.

parabolic to hyperbolic and plays significant role in the analysis of fast phase transformations [16, 17].

2 The diagram of structures and analytical solutions

The structure diagrams are calculated using the solution of the Maxwell area rule, using the chemical potentials and free energy functionals for each structure. The structure selection is based on the thermodynamic rule of the minimal energy for a virtually existed structure. The structure diagram is constructed in coordinates "dimensionless undercooling – relative atomic density" with the amplitude expansion of the free energy (2) for a given crystalline phase [18-20].

2.1 Free energy of crystalline states

Using the summation of reciprocal lattice vectors [14] we can make the amplitude expansion and get the dimensionless density profiles indicating the symmetry and properties of a given phase (homogeneous, BCC, rods, stripes). For all structures we choose as a reference the homogeneous phase relative density ϕ:

$$\phi_h = \phi, \tag{5}$$

$$\phi_{BCC} = \phi + 4\eta[\cos(qx)\cos(qy) + \cos(qx)\cos(qz) + \cos(qy)\cos(qz)], \tag{6}$$

$$\phi_{rods} = \phi + 4\eta\left[\frac{1}{2}\cos(qy) - \cos(\frac{\sqrt{3}}{2}qx)\cos(\frac{1}{2}qy)\right], \tag{7}$$

$$\phi_{stripes} = \phi + 4\eta\cos(qx). \tag{8}$$

We substitute the amplitude representations of each structure Eqs. (5-8) to the free energy Eq. (2) and get a dependence from the complex atomic density vector amplitude module η, reciprocal lattice vector q and dimensionless atomic density ϕ. Following the minimization procedure, an equilibrium wave vector $q=q_{eq}$ and minimize the amplitude η are obtained numerically [18, 19]. Finally, we substitute this roots to the free energy (2) and obtain the form of the free energy for each structure.

2.2 Construction of structure diagram

For the determined above free energies, we construct the diagram of the coexistence of 3D structures in terms of the expressions F_h (for homogeneous phase), F_{BCC}, $F_{stripes}$, F_{rods} dependent from driving force ε and dimensionless atomic density ϕ. With this aim, we write the Maxwell area rule [13, 14, 18-21] and solve the

DOI: 10.1051/matecconf/201712902035

equations for the entire range of control parameter values $0 \leq \varepsilon \leq 1$. The example of the system of equations showed below representing the conditions for the coexistence of the homogeneous and BCC phases:

$$\mu_{BCC}\ (\phi_{BCC},\ \varepsilon) = \mu_h\ (\phi_h,\ \varepsilon), \tag{9}$$

$$F_{BCC}\ (\phi_{BCC},\ \varepsilon)\ -\ F_H\ (\phi_h,\ \varepsilon) = \mu_H\ (\phi_h,\ \varepsilon)\ (\ \phi_{BCC}\ -\ \phi_h\ ). \tag{10}$$

Fig. 1 shows the domains of existence and coexistence of homogeneous (liquid or metastable melt) phase, the crystalline BCC, rods and stripes structures. We calculate the coexistence separately for the each pair of the structures and then select the existence regions for each structure by its minimal free energy as a criteria for selection for a given region of parameters.

3 Numerical calculations of structure formations by the MPFC method

The numerical formulation of the MPFC equation is based on the isogeometric analysis (IGA) [22, 23] for the spatial discretization. Such method allows us to perform the C^2-continuous functions for the discretization of the sixth-order differential equation in primal form. The integration in time is executed by the generalized-α method which was suggested by Chung and Hulbert [24]. Thus an implicit finite element method (FEM) for solving the problem of modified PFC was performed. Within this procedure, a finite-element regular hexagonal mesh is used for simulations. The details of the algorithm and its software implementation are given in [25] where, in particular, was shown that the numerical solution of the MPFC problem is consistent to the diagram obtained by Jaatinen and Ala-Nissila in [26]. The computational efficiency of such algorithm was shown in [27,28]. In [29,30], the analysis of solution's accuracy and physical stability was presented using this algorithm.

The present numerical simulations have been provided using the special software [25] based on the PETIGA and PETSc libraries [31]. For the three dimensional computational domain the spatial mesh has been generated by C^2-continuous quadratic elements. We set an initial distribution with a small spatial gradient of the atomic density. Such an initial distribution has no influence on the modeled crystal pattern in our simulations. The boundary of the computational domain has as periodic conditions.

4 Results and discussions

The initial state of the atomic density has been characterized by the average value of ϕ. Then, the system is quenched with a strength given by the driving force parameter ε. These two values define a position in the phase diagram (see Fig. 1). The solid lines in Fig. 1 represent the structure diagram calculated for three dimensional structures given by the static PFC-equations with free energy (2) and chemical potential (3) in one-mode approximation. As it can be seen in the Fig. 1, the present modeling gives the following set of structures: (1) stripes, (2) mixture of stripes and rods, (3) rods, (4) mixture of BCC and rods, (5) BCC structure, (6) mixture of BCC and liquid, (7) multiple patterns highly sensitive to values of undercooling and atomic density, (5) liquid. All of these modeled structures are perfectly consistent with the regions of the predicted structure diagram. Also is important to note that our diagram correlates with the diagram Jaatinen et al. [26] that also was verified by the presented method [25]. The coincidence of the classical PFC model predictions and numerical dynamical calculations shows the viability of the presently developed algorithm for the numerical solution of the modified three dimensional PFC-equation. The most complicated region of the diagram is the coexistence of three structures (the top-left region for $\varepsilon>0.6$, $\phi<0.4$, where homogeneous-BCC-rods regions of coexistence intersect) in which dynamical MPFC system has a tendency to evaluate to irregularly mixed structures. These mixed structures may be obtained in modeling very often that confirms the idea about evolution of the system through a series of the metastable states.

The modeling of the real systems could face problems with the determination of the crucial parameters, which have to be overcomed by measuring the structure factor of the liquid and crystal state. The necessary data my be provided by the X-Ray diffraction measurements which gives us the reference parameters of the liquid state (density), equilibrium wave number and structure factor for both phases. The perspective system for the calculation of the nature of transition may be the pure Iron with the Homogeneous–BCC–FCC–BCC phase transformations [18-21]. Moreover there is an attempt to use the molecular dynamics data for such calculations [13, 21].

The evolution away the mixed and defected structures can be provided by introducing the noise term. The noise is affected on the structure with the "shake-up" effect [30, 32]. Indeed, locally metastable and structurally mixed states have to be correlated in time and in space, due to the overlapping of the fluctuating atomic configurations [33,34]. In this case, a colored noise should be added to the MPFC equation. Previously, in the analysis of spinodal decomposition, it was notable shown that the consistently introduced colored noise influenced the pattern formation in a great extent [34].

5 Conclusions

Structure diagrams were calculated for the one-mode PFC model and compared with the numerical simulation of the full MPFC equation for the fast phase transitions. The computations were performed using parallel algorithms based on isogeometric analysis, which is known extension of the finite element method. The evolution of three dimensional crystal structures to their steady equilibrium state was simulated for various atomic densities and temperatures using the MPFC-equation and compared with the stationary solutions for regions of structural coexistence. The latter is obtained from amplitude expansion of free energies of the PFC-equations. The obtained diagram verified by the direct simulation and also it is strongly correlate to the diagram Jaatinen et. al. [26].

The proposed method makes possibilities to extend the description of the fast dynamical phase transitions by the calculation for metastable structural boundaries. The MPFC formalism allows us to model the bunch of the real liquid–crystalline and liquid–smectic phase transitions even for two-dimensional systems [15,18,19]. The extension of the method with the two-mode approximation creates the way to model the complicated three dimensional crystalline structures (such as crystals with FCC- and HCP-lattices). Further studies on the simulation of crystalline patterns by the MPFC method will allow developing an effective software tool for research and develop of materials in industry, including manufacturing and processing technologies.

This work was supported financially by the Russian Foundation for Basic Research project No. 16-38-00839 mol_a, author V. A. acknowledges support from "UMNIK" program of the Innovation Support Foundation of Russian Federation, grant No. 0020468. I. S. acknowledges support from the Ministry of Education and Science of the Russian Federation (project no. 1.9527.2017).

References

1. K. R. Elder, M. Grant, Modeling elastic and plastic deformations in nonequilibrium processing using phase field crystals, Phys.Rev. E **70**, 051605 (2004)
2. M.C. Cross, P.C. Hohenberg, Rev. Mod. Phys. **65**, 851 (1993)
3. G. Tegze, L. Granasy, G. I. Toth, F. Podmaniczky, A. Jaatinen, T. Ala-Nissila, Phys. Rev. Lett. **103**, 035702 (2009)
4. G.I. Toth, G. Tegze, T. Pustai, G. Toth, L. Granasy, J. Phys.: Condens. Matter **22**, 364101 (2010)
5. S. van Teeffelen, R. Backofen, A. Voigt, H. Lowen, Phys. Rev. E **79**, 051404 (2009)
6. J. Berry, M. Grant, K.R. Elder, Phys. Rev. E **73**, 031609 (2006)
7. P.Y. Chan, G. Tsekenis, J. Dantzig, K.A. Dahmen, N. Goldenfeld, Phys. Rev. Lett. **105**, 015502 (2010)
8. J. Berry, K.R. Elder, M. Grant, Phys. Rev. E **77**, 061506 (2008)
9. K.R. Elder, M. Grant, Phys. Rev. E **70**, 051605 (2004)
10. Z.-F. Huang, K.R. Elder, Phys. Rev. B **81**, 165421 (2010)
11. J. Mellenthin, A. Karma, M. Plapp, Phys. Rev. B **78**, 184110 (2008)
12. C.V. Achim, M. Karttunen, K.R. Elder, E. Granato, T. Ala-Nissila, S.C. Ying, Phys. Rev. E **74**, 021104 (2006)
13. A. Jaatinen, C.V. Achim, K.R. Elder, et al., Phys. Rev. E **80**, 031602 (2009)
14. N. Provatas, K. Elder, *Phase-Field Methods in Materials Science and Engineering*, 312 (Wiley, 2010)
15. P. K. Galenko, H. Gomez, N. V. Kropotin, et al., Phys. Rev. E **88**, 013310 (2013).
16. P. K. Galenko, D. A. Danilov, V. G. Lebedev, Phys. Rev. E **79**, 051110 (2009)
17. D. Jou, P. K. Galenko, Phys. Rev. E **88**, 042151 (2013)
18. V. Ankudinov, P. K. Galenko, IOP Conf. Ser.: Mat. Sci.-Eng. **192**, 012019 (2017)
19. V. E. Ankudinov, P. K. Galenko, N. V. Kropotin, et al. J. Exp. Theor. Phys. **122**, 298 (2016)
20. E. Asadi, M. A. Zaeem, Comput. Mat. Sci. **105**, 110 (2015)
21. K.-A. Wu, A. Karma, Phys. Rev. B **76**, 184107 (2007).
22. J. Cottrell, T. Hughes, Y. Bazilevs, *Isogeometric Analysis: Toward Integration of CAD and FEA* (Wiley, 2009)
23. T. Hughes, J. Cottrell, Y. Bazilevs, Comput. Methods Appl. Mech. Eng. **194**, 4135 (2005)
24. J. Chung, G. Hulbert, J. Appl. Mech. **60**, 371 (1993)
25. J. Bueno, I. Starodumov, H. Gomez, P. Galenko, D. Alexandrov, Comput. Mat. Sci. **111**, 310 (2016)
26. A. Jaatinen, T. Ala-Nissila, J. Phys.: Condensed Matter **22**, 205402 (2010)
27. I. Starodumov, E. Pavlyuk, L. Klyue, M. Kovalenko, A. Medyankin, Ceur-WS Proceedings **1513**, 109 (2015)
28. I. Starodumov, E. Pavlyuk, S. Abramov, L. Klyuev, P. Galenko, D. Alexandrov Bulletin of Udmurt University. Mathematics, Mechanics **26**, 445 (2016)
29. I. Starodumov, N. Kropotin, AIP Conference Proceedings **1759**, 020136 (2016)
30. I. Starodumov, P. Galenko, D. Alexandrov, N. Kropotin, IOP Conference Series: Materials Science and Engineering **192**, 012008 (2017)
31. S. Balay and W.D. Gropp and L.C. McInnes and B.F. Smith, *Efficient Management of Parallelism in Object Oriented Numerical Software Libraries, Modern Software Tools in Scientific Computing,* 163 (Birkhauser Press, 1997)
32. I. Starodumov, P. Galenko, N. Kropotin, D. Alexandrov AIP Conference Proceedings, Submitted (2017)
33. D. Jou, P. K. Galenko, Physica A **336**, 149 (2006)
34. P. Galenko, D. Kharchenko, I. Lysenko, Physica A **389**, 3443 (2010)

MATEC Web of Conferences **129**, 02036 (2017)
ICMTMTE 2017
DOI: 10.1051/matecconf/201712902036

Perspectives and principles of formulation of legal basis of diagnostic quality control of materials and products of nanostructure composites

Gennady K. Baryshev[1,*], *Igor* A. Tutnov[1,2], and *Aleksandr* P. Biryukov[1]

[1]National Research Nuclear University MEPhI (Moscow Engineering Physics Institute), Russian Federation
[2]National Research Center "Kurchatov Institute", Moscow, Russian Federation

Abstract. In the paper we discuss the problems of perspective of formulation of legal basis of diagnostics quality control of materials and products of nanostructured composites for various application, including the perspective markets of the National Technology Initiative. We formulate the basic principles including main statements of the typical manufacturing method of diagnostics and quality control of materials and products of nanostructured composites.

1 Introduction

At present the technological development of Russian Federation is coincided with the realization of the National Technoly Initiative.

The National Technology Initiative (NTI) is a programme aimed at shaping fundamentally new markets and creating conditions for the global technology leadership of Russia by 2035. The NTI includes integrated solutions to define the key technologies, necessary changes in norms and rules, efficient financial and personnel development measures, involvement and encouragement mechanisms for experts with the necessary skills. The choice of technologies is based on major global trends, with network technologies being the priority. Such technologies are people-focused, as people are the end consumers.

The basis for the current systematic work is the NTI matrix which comprises new markets, new technologies, institutes and infrastructure/resources. The NTI areas include a group of the markets of the future:

- EnergyNet (distributed power generation from personal power to smart grid and smart city);
- FoodNet (system of personal production and food and water delivery);
- SafeNet (new personal security systems);
- HealthNet (personal medicine);
- AeroNet (distributed systems of unmanned aerial vehicles);
- MariNet (distributed systems of unmanned maritime transport);
- AutoNet (distributed network of unmanned management of road vehicles);
- FinNet (decentralized financial systems and currencies);
- NeuroNet (distributed artificial elements of consciousness and mentality).

The NTI roadmaps through 2018 shape the ecosystem of future markets. It is important to mention that not all NTI roadmaps have been approved at this moment. As for the implementation of approved roadmaps (NeuroNet and others), this has recently started.

Another important problem being solved is the rapid import substitution of special products of technical purpose. Improvement of systems of life cycle management and planning of new technologies for perspective markets should rely on verified and approved physical basis of products manufacturing, including products for specific application, designed and manufactured on the base of nanostructured composite materials [1,2].

At the moment the so called "smart materials" are of certain interest – these are the materials which operating properties change depending on the operating conditions. For example, this can be a significant increasing (or decreasing) of electrical resistance of the surface by gas sorption from the environment. The operation of semi-conductor gas sensors is based on this principle. Another example is the change of transparency of layers by changing of the oxidation state of metals of anode or cathode electro chromic cover, that can be applied for design and development various indicator devices.

Application of "smart materials" – i.e. materials that controllably change their operating properties as a reaction to the change of the environment – in nuclear and other hi-tech industries allows to effectively solve a number of engineering and technical tasks, increase reliability, durability other important technological parameters of products and technical systems. At the moment application of materials with the memory of form effect, sensors of non-destructive testing on the

* Corresponding author: GKBaryshev@mephi.ru

DOI: 10.1051/matecconf/201712902036

base of materials with the application of piezoelectric and magnetostrictive effects is developing [3]. Composite materials for aviation and space industries and markets are no less perspective thematic directions on smart materials.

The range of methods and ways of application of materials and products with the given service properties, including materials bases on nanostructured composites, is respectively wide at the moment [4-6]. At the same time, many of the technologies are still at the stage of research and development, prototypes and industrial samples. The task of formulation of legal technological basis of manufacturing of new products, for the purposes of technological substitution and application at the perspective markets of the National Technological Initiative, including providing the process of quality control with the development of technological maps of such processes for the complete life cycle of items, products and systems, is connected not only with the research of technologies of development of new materials, but also with the development and application of typical manufacturing methods of research and control of service properties of the new class of materials and products during their fabrication.

2 The general method of research of electrophysical properties of nanostructured composites

We have developed the general method of research of electrophysical service properties of nanostructured composites.

The development of the method was relied on commonly known and applied methods of measurement of electrophysical properties of conductors [7-9], regarding the following main principles:

1. Principle of verification of the single information field of numerical results of research of electrophysical service properties of nanostructured composites.

This principle provides complete and general enough picture of the set of step processes for real realization of any general method of original methodology of research of electrophysical service properties of nanostructured composite materials or products. This principle in our case defies the mostly close by the criteria of sufficiency of the expected final result physical principles of manufacturing processes in part of application of perspective methods of in-situ research and quality control of composite materials and products. Together with this the principle formalizes the set of information measuring procedures and processes of the applied method of research of electrical resistivity, thermal electromotive force, contact potential difference and other electrophysical properties, defies the sufficient completeness and adequateness of these processes and procedures. The same principle sets the barriers of limitations for research of electric conductivity of nanostructured composites, which should mostly be concentrated on the problems of provision of legal basis of provision of metrological unity of any research process for any quality control of materials and products of that class and provision of the guarantees of high level of accurateness and logical authenticity of the newly obtained information. At the same time, the fundamental principles of the state system of provision of testing measurement unity [10] are applied without change of correcting them.

2. The second principles, in our case of formulation of adequate methods of research of electrophysical service properties of nanostructured composites, is the principle of realization of systematic approach for performing the procedures of virtual measurement and analysis of the information obtained via the procedures. The principle is adopted for the purpose of later imitation modelling of service properties of prototypes and samples of nanostructured composite materials and products and evaluation of the results of such modeling in the form of information data flows on electric resistivity and other parameters, first of all for direct current.

The principle allows to increase the objectiveness of the estimations of the adopted accuracy and authenticity of measured data on phenomenological and numerical levels of information flows of service information of samples of nanostructured composites for research.

3. The third principle of importance for provision of diagnostic measuring procedures for measurement of thermal electromotive force and contact potential difference of nanostructured composite samples is formulated as follows. The principle of priority for object-independent methods of measurement of thermal electromotive force and contact potential difference for nanostructured composite wires and other samples.

The measuring methods for thermal electromotive force and contact potential difference, developed on the given principle, in application to samples of nanostructured composites, would allow to abstract from the featured of quantitate containment and heterogenic allocation of chemical elements through the total of the volume of the samples for measurement

The application of this principle would allow to take into account only sustainable effects of influence of impurity inclusions on the parameters of thermal electromotive force and contact potential difference of the composite materials and products.

The main step procedures of the developed general method are:

1. Planning the measurement procedure and preparation of the initial data for research of service properties of the samples of nanostructured composites.
2. Development of the documented program of performing of diagnostic research of electrophysical properties of composite samples.
3. Typical instrumental research of electrophysical service properties of nanostructured composites.
4. Estimation of reliability of the results of research of electrophysical service properties of nanostructured composites.
5. Demonstration, validation and presentation of final data results of service properties of the tested samples.

The developed general method takes into account the contemporary image of the features of technological processes of manufacturing of nanostructured composite

DOI: 10.1051/matecconf/201712902036

materials and products, degrading of service properties and other perspective items on the duration of their product life cycle, legal requirements for safety, reliability and effectiveness, bases on modernized diagnostic procedures of research of service properties of wires and other composite items, contemporary technical systems and information measuring systems, applies the perfect instruments of commonly used methods of experiment planning, including methods for estimation of errors and reliability at obtaining data of service characteristics of materials and products on the base of nanostructured composites [11].

3 Development of physical principles of the technology of manufacturing of materials and items of nanostructured composites

It is important to obtain the methodology and methods of integral engineering and financial analysis of risks of development and application of new measuring and quality control technologies considering the whole process of manufacturing of materials and items of nanostructured composites.

When new measuring systems for quality control of nanostructured composites are designed, it is important to mark the objectives not only for analysis and recognition of engineering problems, but also to prove recommendations to solve this problems, based on rational financial and other costs. The search of optimal solutions at the process of design and implementation of new samples of devices and measuring systems for quality control of materials and products of nanostructured composites, as well as problem definition of new directions of innovation development for economic effectiveness and practical usefulness of diagnostic control of such materials for perspective markets of the National Technology Initiative, require integral and system vision of all possible risks of investment projects [12].

First of all it cares the development of serial production of measuring devices and systems of quality diagnostics for nanostructured composites. For nuclear industry, at the moment, criteria of effectiveness, approved by international practice and recommended by IAEA are applied for estimation of practicability of serial production of measurement techniques for nuclear power plant production [13] – NPV, IRR, DPP – on development of new devices and information measuring systems.

Nevertheless, these economic criteria do not take into account technical risks of investment projects concerning reliability of application of devices and measuring systems in industry. Thereafter the problem of development of scientific categorical apparatus of design of perspective samples of measuring systems with vector-type account integral analysis of engineering and financial risks, mostly for the stages of their development, protoryping and industrial operation. To solve this problem in application to products and materials of nanostructured composites for nuclear industry it is important to specify the basic processes of monitoring of safety of nuclear fuel elements. The next step is to choose the technology of diagnostic control and monitoring of nanostructured materials quality, suggest an innovative and perspective technical device to realize this technology, estimate the practical importance and perspectives of development of a new device or measuring system.

It is also of importance to know possible methods of monitoring and management of engineering and financial risks; to perform the system analysis of main factors of direct and indirect influence on conditions of production of new samples of nanostructured composites.

We in our case understand the following when describing the physical principles of the technology of manufacturing of materials and items of nanostructured composites:

- provision of the main processes of manufacturing preparation in the aspect of development of manufacturing regimes for various items on the basis of research and testing of their electrophysical properties;
- development of technical conditions of provision of inter-operation and final quality control of manufactured products after measurement and testing of their electrophysical properties;
- development of technological maps of technological operations during the process of manufacturing of materials and items of nanostructured composites;
- development of conditions for quality control of the technological process of manufacturing of materials and items of nanostructured composites;
- development of specific instruments, including measuring systems for non-stopping manufacturing quality monitoring of any items of nanostructured composites;
- including in the process the whole cycle of the main process and optional procedures for development and application of technologies of manufacturing of materials and items of nanostructured composites;
- provision the guarantees of safety of the manufacturing technology and final products of nanostructured composites;
- provision of industrial safety and ecological acceptability of manufacturing process of materials and items of nanostructured composites.

Application of the described general method of research of electrophysical properties of nanostructured composites would allow one to provide certification of technical items of materials of such class in industry. Moreover, adopting the service properties of engineering materials of this class to the obligatory safety requirements of contemporary hi-tech industries and perspective markets of the National Technology Initiative require the development of a typical method, including the development of prototypes of information measuring systems of non-destructive testing and technical diagnostics, approbation of the developed general method in the real industrial conditions of environmental influence on materials and items of nanostructured composites. These specific conditions are mostly limiting the application of the developed method in the conditions of real manufacturing process in the nearest future.

Still, the developed main principles and steps of the method would allow to make a conclusion about the effective development of principles of the technology of manufacturing of materials and items of nanostructured composites. The described general method may become the basis of the system of quality control management and the system of metrological provision of the future manufacturing and obtaining products with the given quality, as well as the legal basis of diagnostic quality control of materials and products of nanostructure composites

4 Conclusion

Formulation of the legal basis of diagnostic quality control of materials and products of nanostructure composites should be based mostly on scientific principles that are the core of the verified and approbated typical industrial methods. We have developed the general method of research of electrophysical properties of nanostructured composites that realizes this principles, has a feature of sufficient scientific novelty, and has all the characteristics of correspondence with the system of provision of unity of measurements and quality control of high technology products.

The results of application of the general method of research of electrophysical properties of nanostructured composites in the context of formulation of physical principles of new technologies of manufacturing of innovation products would allow to gain the new level of development of the system of quality management of such products, provide decreasing the terms of performance in increasing the effectiveness of manufacturing processes of development of such products, as well as increase the controllability and effectiveness of functioning of companies of high tech industries and perspective markets of the future for all the stages of the life cycle process of materials and products of nanostructured composites.

References

1. Strategy of the National Security of Russian Federation till 2020, approved by the Decree of the President of Russian Federation of May 12th 2009 #537
2. Decree of the Government of Russian Federation of April 15th 2014 #328 "About the Approving of the State Program of Russian Federation "Development of the Industry and Increasing of Its Competitiveness"
3. V.M. Baranov, A.M. Karasevich. G.A. Sarychev, *Testing and quality control of materials and engineering elements* (Higher school, Moscow, 2004)
4. V.I. Pantsyrnyy, V.I. Surin, A.P. Biryukov, G.K. Baryshev, E.P. Varyatchenko. Non-ferrous Metals **2** (2014)
5. A. S. Kondrateva, M. Mishin, A. Shakhmin, M. Baryshnikova and S. E. Alexandrov. Phisica Status Solidi **12**, 7 (2015)
6. A.I. Voronin, G.K. Baryshev, Yu. V. Bozhko, A.A. Usenko, V. Yu. Zueva, K.I. Litvinova, I.V. Petrova, M.A. Seredina, V.V. Hovailo, Bulletin of the Lebedev Physics Institute **42**, 7 (2015)
7. Blatt, Shreder, Foils, Grage, *Thermoelectric power of metals* (Metallurgia, Moscow, 1980)
8. Anatoly Zharin, *Contact Potential Difference Techniques as Probing Tools in Tribology and Surface Mapping* (NanoScience and Technology, Springer, New York, 2010)
9. Green, *Surface properties of solid state bodies* (Mir, Moscow, 1972)
10. RMG 29-99. Russian State System of Provision of Unity of Measurements.
11. Hicks, *Basic principles of experiment planning* (Moscow, 2013)
12. G.K. Baryshev, I.A. Tutnov. Yu. V. Bozhko, Proceedings of 2016 13th International Scientific-Technical Conference on Actual Problems of Electronic Instrument Engineering (APEIE) (2016)
13. Berens W. Hawranek, Peter M., *Manual for the preparation of industrial feasibility studies* (UNIDO, Vienna, 1991)

Development of methodical instruments of obtaining, research and control of electrophysical service properties of nanostructure composites

Gennady K. Baryshev[1,*], *Anastasia* S. Kondratyeva[2], and *Aleksandr* V. Berestov[1]

[1]National Research Nuclear University MEPhI (Moscow Engineering Physics Institute), Russian Federation
[2]National Research University St. Petersburg State Polytechnical University, Russian Federation

Abstract. The article highlights the main problems of development of methodical instruments of obtaining, research and control of electrophysical service properties of nanostructure composites. Some of the results obtained at the research performed by scientists from National Research Nuclear University MEPhI (Moscow Engineering Physics Institute) and National Research University St. Petersburg State Polytechnical University are presented. The results are discussed in the focus of development of a new general method of control of electrophysical service properties of materials and products of nanostructured composites during the manufacturing process.

1 Introduction

The national problem of rapid development of Russian technologies and manufacturing of new competitive products for the purpose of import substitution is actual at the moment [1,2]. These new products could be manufactured of nanostructured composites in various areas – nuclear and space industries, perspective markets of the National Technology Initiative, etc.

Nickel oxide, for example, is related to the class nanostructured composites, applied as a high-sensible and selective oxide sensor. Such sensors could find their application in systems of monitoring of the environment, medical diagnostic devices, etc.

Semiconductor Heusler alloys appeared very promising thermoelectric materials as well, with better service properties – physical quantities thermal emf coefficient, conductivity, and thermal conductivity, than traditional thermoelectric materials [3].

Another class of nanostructured materials that is of significant interest are electro technical materials, such as composite wires for power lines, operating in severe climatic conditions and other applications [4].

The set of methods of manufacturing of materials and products with the given operating conditions, including products of nanostructured composites, is rapidly growing at the moment. At the same time, most of these methods are at the research and testing stage, and samples and prototypes of nanostructured composites need general industrial methods and methodical instruments of obtaining, research and control of their service properties, including electrophysical service properties, to be developed. In this paper we describe some of the results obtained by our research group in this area on such materials of nanostructured composites as layers of ferrous-group metals and composite powders.

2 Development of methodical instruments of obtaining layers of ferrous-group metals

First, we will provide some information on the perspective technology and instruments of obtaining layers of ferrous-group metals related to the class of nanostructured composited with wide perspectives of application.

Thin layers of ferrous-group metals have a set of properties such as electrochromism, sensor activity, magnetic properties. One of the most attractive methods of obtaining these layers is the process of chemical sedimentation from gas phase that provides formation of high-quality layers on samples of complex form. To provide this the technology of chemical sedimentation of thin layers of ferrous-group metals from gas phase is developed, physical and chemical principles of the process are investigated, as well as practical properties of layers.

It is essential that for the last decades most innovations realized in high-technology economics sector and industry are based on application of engineering technologies and new materials. Thereafter, the so called "smart materials" - a unique class of materials which properties change under external factor in a specific way we expect them to change - are in most of interest thereafter.

Russian Federation is now in the active phase of the process of industrial modernization and development of digital economics. Major part of industrial equipment

* Corresponding author: GKBaryshev@mephi.ru

 DOI: 10.1051/matecconf/201712902037

has been in operation since the Soviet Union times. Still, what the industry need first of all are analytical devices and systems. Thus we may state that a market of chemical semiconductor sensors is a perspective economic niche in Russia.

The most known worldwide oxide semiconductor sensor producers are City Technology, Alphasense, Applied Sensor and Figaro Inc. Russian companies hold a very little percent of the market, being mostly at the stage of research and development. The years of 2020-2025 may be the time of total implementation of MEMC-technology sensors, with Russian companies being outside this technological trend, if they stay where they are, with the current level of research and development [5]. The growth of non-military purpose sensors market would be increasing rapidly in the nearest decade.

Most perspective materials to be applied in gas sensors are ferrous-group metals oxides. The surface of material is modified in the interaction with the environment. The application of semiconductor oxide layers as a high-sensitive and selective sensors is based on this effect [6,7]. Sensors of this kind can detect hydrogen, nitrogen oxide, carbon monoxide, ammonia, organic materials vapors such as acetone, formaline, methanol. They are used in systems of continuous environmental monitoring, fire alarms, medical diagnostics devices, catalysts for organic synthesis of biomass pyrolysis in bioenergetics [8].

Besides, thin layers of ferrous-group metals have a feature of electrochromism [7] – an ability to change color at the process of electron extraction, that can be used, for example, in energy-saving monitors. On the base of these materials chemically compatible capacitor electrodes for oxide isolators and conductive transparent electrodes for optical devices (light- and photo-diodes, computer monitors etc.) can be produced [6,7,9].

Two most commonly used methods to obtain layers of ferrous-group metals are magnetron sputtering in oxide atmosphere and sol-gel synthesis. The main limitation of the first one is inability to control the morphology of obtained layers, for the second one this is a large amount of admixtures in the product and inability to control the material structure in large scales. This way we consider a perspective method for obtaining thin layers of different materials the chemical sedimentation from gas phase that has a number of advantages [9]: high growth velocities, simplicity and low price of equipment, availability to control the product features by the parameters of the process (temperature, partial pressure of reagents, resident time). One of the key advantages of the chemical sedimentation from gas phase technology is easiness of technology scaling to mass-production.

We will now describe the technology more thoroughly. To solve a specific task we need a layer of material with given properties. Research of physical and chemical laws of chemical sedimentation from gas phase would allow us to develop certain engineering and industrial technologies [10].

First, one needs to study the process of gas-phase transforming in the Cp2Me-O2-Ar and Cp2Me-O2-O3-Ar systems, this would allow to:

1. Formulate a model of chemical reactions and processes of obtaining the layers of ferrous-group metals in the systems;
2. Estimate how sedimentation temperature influences the process;
3. Estimate the oxidant influence in the system.

It is also important to study the influence of main technological parameters of the Cp2Me-O2-Ar and Cp2Me-O2-O3-Ar systems on sedimentation velocity, this would allow one to estimate the time of the process limiting stage. This part includes studying of how the following technological parameters influence the layer sedimentation velocity:

1. Active zone temperature;
2. Active zone pressure;
3. General velocity of gas flow;
4. Supporting plate type (including crystallographic orientation);
5. Partial pressure of reagents.

It is a feature of such systems for chemical sedimentation to pass both in kinetic and diffusion modes, depending on the specific parameter values, with the kinetic mode transforming into the diffusion one when temperature increases.

The complex study of the layers obtained (chemical composition, morphology, crystallographic structure) would allow to get most complete data about their properties. Such data obtained would be the physical basis for the development of industrial technology.

3 Developing methodical instruments of research and control of electrophysical properties of composite powders at cold pressing

Another significant problem is developing the necessary methodical instruments of control of electrophysical properties of composite powders at cold pressing.

According with the practice of obtaining cold-pressed powders in various matrixes the heterogeneity of the density at the height of the backfill is considerable if ratio of the diameter to height of the matrix is less than 1:10. This manufacturing research problem could be solved by the development of specific methodical instruments: information measuring system for monitoring and control of the changes of porosity during the pressing process. Special attention to the development of mobile sensors should be paid when designing and developing controls of information-measuring and information-analytical systems for that purpose.

Automation and effective management of powder processing requires designing of positioning system of sensors, that monitoring filling level, and obtaining data of formed pressed powder. Data collection should be

DOI: 10.1051/matecconf/201712902037

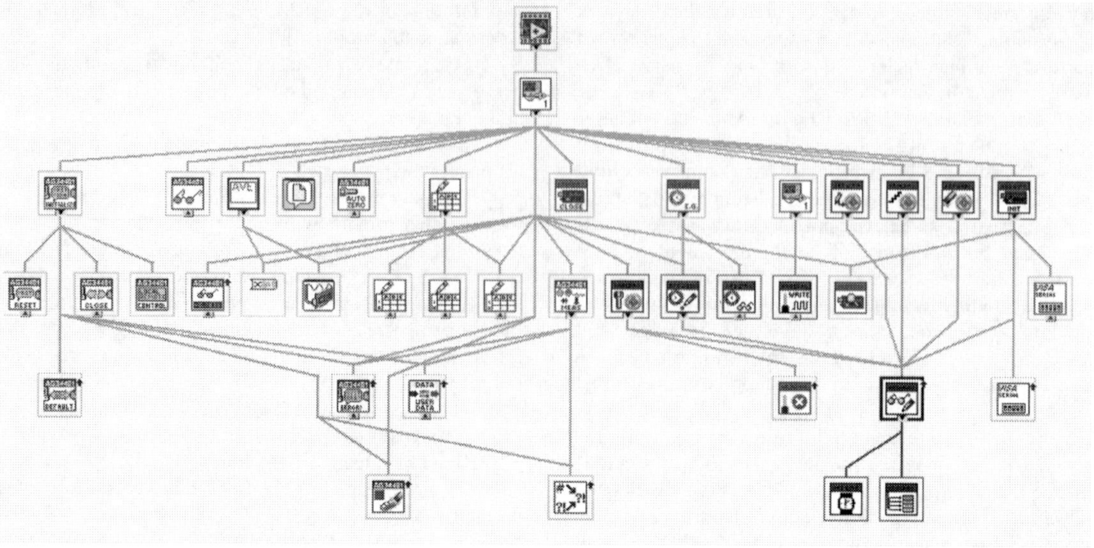

Fig. 1. The hierarchy of the virtual instrument

carried out by programmable logic elements included into information-analytical system [11].

Software development for solving task was implemented by NI LabVIEW 8.6.

A sensor-positioning system based on method of electrical resistance measuring was developed. The main sensors installed on the mobile carriage are developed to provide automation of the measurement and control of porosity during the process of cold pressing. Depending on the height of the used backfill measuring circuit may include one or more sensors.

Kinematic scheme of positioning mechanisms, as a rule, has two main parts: the operating mechanism (or sensor) and its actuator. The actuator`s carriage serves as base frame for sensor by moving along linear guideway. To achieve the necessary dynamic quality of the positioning mechanism and also to meet the requirements of interchangeability the carriage should satisfy the following requirements [12]:

• ease of movement with no backlash and clearances;
• high precision movement;
• low weight;
• high rigidity;
• convenience for installation and alignment

These requirements can be met by choosing the appropriate engineering materials of the carriage, its geometric shapes, precision linear guides with ball bearings, design of their fabrication, providing a preload to the base side of the guide. Switching on the actuator will be blocked in case of lack of tightness of the measuring mandrel or opened top cover.

Because the height of the filling powder decreases at the process of pressing, it is necessary to use a system, which includes the height level backfill sensor (or density sensor), to work jointly with the positioning system of the carriage.

The dimensional difference for the carriage positioning must not exceed 10-20 mcm. The chosen positioning scheme is similar to that of applied for computer floppy discs. We use the stepping engine as a driver to move the carriage with the aid of an actuator screw. Such a system does not have high dimanic parameters but is still characterized by relative simplicity and high level of reliability. The controller of powder level height is integrated into the carriage control system. To control positioning the computer program has setting commands. For example, the "set the sensor" command is applied to move the carriage with the sensor to the given position and connect it with the measuring device. Search commands are applied for control sensors as well as for database operating. The latest procedure is necessary for selecting a sample group of data relative to the same powder height level, comparing averaged readings for different powder level height or comparing data for different powders etc. The recording commands group includes commands performing both data recording as well as recording the control information (personal address, numerical regions, keys). For example, for recording the data obtained into a file the "record data" command is applied. This command is to go after the search commands (searching for the identifier or key equality). One may perform the command in case the data array to be written fits the assigned memory volume. The operating commands include such commands as "check input-output", "confirm condition" and a number of others.

By method of electrical resistivity we can estimate the porosity of powder depending on the applied pressure. Measurement of the powder electrical resistance was carried out using a high-performance digital multimeter Agilent 34401A according to two methods: wire and potentiometric circuits. According to user manual multimeter Agilent 34401A could provide measurement of the resistance by using two-wire and

 DOI: 10.1051/matecconf/201712902037

four-wire circuits. In both cases electric current flows from high-potential nest HI through the test sample.

We also have developed the information measuring system for registration diagnostic signals, which includes the protocol of data exchange between PC, digital multimeter and virtual device (Fig 1.)., that displays on the monitor screen current and the integral values of the resistance (Fig. 2).

Further improvement of measurement tools and methods by using the achievements of modern information technologies will allow achieving a greater sensitivity for measuring thickness of research and control of electrophysical properties of composite powders at cold pressing.

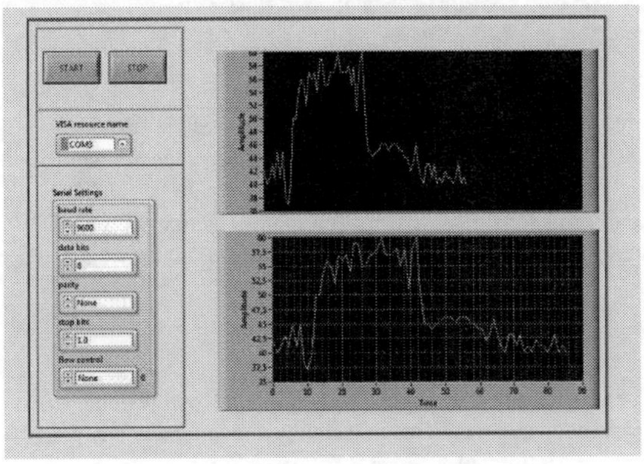

Fig. 2. The outer panel of the virtual instrument

References

1. Strategy of the National Security of Russian Federation till 2020, approved by the Decree of the President of Russian Federation of May 12th 2009 #537
2. Decree of the Government of Russian Federation of April 15th 2014 #328 "About the Approving of the State Program of Russian Federation "Development of the Industry and Increasing of Its Competitiveness"
3. A.I. Voronin, G.K. Baryshev, Yu. V. Bozhko, A.A. Usenko, V. Yu. Zueva, K.I. Litvinova, I.V. Petrova, M.A. Seredina, V.V. Hovailo, Bulletin of the Lebedev Physics Institute **42**, 7 (2015)
4. A. V. Putilov, A. K. Shikov, V. I. Pantsyrny, A. E. Vorobieva, V. A. Drobyshev, Tsvetnye Metally (Non-ferrous Metals) **3** (2008)
5. V.V. Zheltova, Saint-Petersburg National Research University of Information Technologies, Mechanics and Optics (2013)
6. Jin-Kyu Kang, Shi-Woo Rhee, Thin Solid Films **391** (2001)
7. B. Subramanian, M. Mohammed Ibrahim. J. Mat. Sci: Mater. Electron **20** (2009)
8. Jianfen Li, Rong Yan, Bo Xiao, David Tee Liang, Dong Ho Lee, J. Am. Chem. Soc. (2008)
9. Pavel Moravec, Jiří Smolík, Helmi Keskinen, Jyrki M. Mäkelä, Snejana Bakardjieva, Valeri V. Levdansky, Mat. Sci. and App. **2** (2011)
10. Gennady K. Baryshev, Anastasia S. Kondratyeva, Proceedings of 2016 International Conference on Control, Instrumentation, Communication and Computational Technologies (ICCICCT) (2016)
11. Aleksandr P. Biryukov, Gennady K. Baryshev, Vitaliy I. Surin, Yuri Yu. Vostrenkov, Proceedings of 2016 13th International Scientific-Technical Conference on Actual Problems of Electronic Instrument Engineering (APEIE) (2016)
12. G.K. Baryshev, E.A. Olevsky, V.I. Surin, A.I. Maksimkin, Information technologies in design and manufacturing **2** (2013)

Transformation of the microstructure and properties of ultrafine-grained TiNi alloys during the processing by ECAP-conform via the isothermal regime

Anna Churakova[1,2,3*], *Dmitry* Gunderov[1,2,3], and *Georgy* Raab[1]

[1]Ufa State Aviation Technical University, 12 K. Marx str, Ufa 450008 Russia
[2]Institute of Molecule and Crystal Physics, Ufa Research Center RAS, 151 pr. Oktyabrya, Ufa 450075 Russia
[3]Saint Petersburg State University, 7/9 Universitetskaya nab., St. Petersburg, 199034 Russia

Abstract. This paper examines a technique for producing a TiNi alloy with an ultrafine-grained structure and enhanced mechanical characteristics. It is demonstrated that the use of the ECAP-Conform technique enables producing TiNi alloy samples with an ultimate tensile strength of up to 1300 MPa and a rather high level of ductility. Studies show that in this alloy, there forms a banded deformation microstructure with a structural element size of up to 300 nm, and during subsequent annealings there forms a grain structure with a grain size of about 200 nm.

1 Introduction

Alloys based on titanium nickelide are well-known functional materials with the shape-memory effect (SME), conditioned by thermoelastic martensitic transformations. These alloys stand out among materials having functional characteristics, enhanced strength and ductility, and a good set of performance properties: a long life, corrosion resistance, biocompatibility, etc. [1-2]. For many cases of application, especially in critical or small-sized items, the level of mechanical and functional properties that TiNi alloys possess in their normal coarse-grained state is insufficient. Since physico-mechanical properties are structurally sensitive ones, methods of deformation-and-thermal treatment are traditionally used for their enhancement.

New possibilities in the control of the physico-mechanical properties of metals and alloys are opened by the formation of an ultrafine-grained (UFG) structure by severe plastic deformation (SPD) processing [3,4]. A promising direction in the enhancement of service properties of TiNi alloys is the formation of a nanostructured state in them through SPD processing [5-12]. At present, well-developed in physical materials science is the fabrication of bulk amorphous and nanocrystalline materials using such SPD techniques as high-pressure torsion (HPT) and equal-channel angular pressing (ECAP) [3,5]. Both of these techniques are based on shear deformation and enable effecting large strains without fracture of the billets [3,5]. However, the HPT technique enables producing only small-sized samples for research purposes. For the manufacture of many items, in particular, dental implants, nanostructured TiNi is required, in the form of rod-shaped semi-products with a diameter of 4-5 mm. Fabrication of large-sized TiNi alloy billets with a refined structure and enhanced properties is possible through the use of ECAP processing.

Nevertheless, the conventional ECAP technique has significant disadvantages that limit its use in industry: a relatively small productivity, the ability to produce only relatively short cylindrical samples, as well as the need to remove the edge parts of billets that have undergone repeated working during the deformation processing, which reduces the metal utilization factor. Free from these disadvantages is the technique of equal-channel angular pressing combined with Conform (ECAP-C) [13-16]. ECAP-C solves the above-mentioned problems, typical of conventional ECAP: it enables producing by deformation immediately a long-length rod (semi-product); this technique has a high metal utilization factor and a high productivity [16]. The first experiments on the ECAP-C processing of TiNi alloys were already conducted earlier [17]. However, the ECAP-C facility available at that time made it possible to perform isothermal pressing at temperatures not higher than 200°C. As a result of 2-4 cycles of ECAP-C processing at 200°C with intermediate annealings, a subgrain nanostructure was produced in the samples [17]. Still, when the number of cycles was increased to more than 4, the TiNi samples fractured [17]. As of today, our research team has created an ECAP-C facility with a working temperature of up to 500°C. The present work studies the effect of ECAP-C processing at an elevated temperature on TiNi, using this new facility, with a view to increase the number of processing cycles, produce in TiNi a UFG state with refined grains and enhanced mechanical characteristics.

* Corresponding author: churakovaa_a@mail.ru

DOI: 10.1051/matecconf/201712902038

2 Materials and research methods

2.1 Material under study

As the object of study, the $Ti_{49.1}Ni_{50.9}$ alloy with a large Ni content with respect to the stoichiometry with the temperature A_f=10°C was selected. At room temperature, the major phases in this alloy are austenite with a B2 crystalline lattice and the Ti_2Ni_3 phase, enriched with Ni. In order to form a solid solution based on TiNi, quenching of the alloy was conducted from the homogeneity region (from 800°C) in water. During the subsequent heating of the quenched alloy $Ti_{49.1}Ni_{50.9}$, decomposition of the solid solution took place, with a successive precipitation of the excess phases Ti_3Ni_4, Ti_2Ni_3, $TiNi_3$.

2.2 Research methods

The schematic diagram of the ECAP-C facility is presented in Fig. 1.

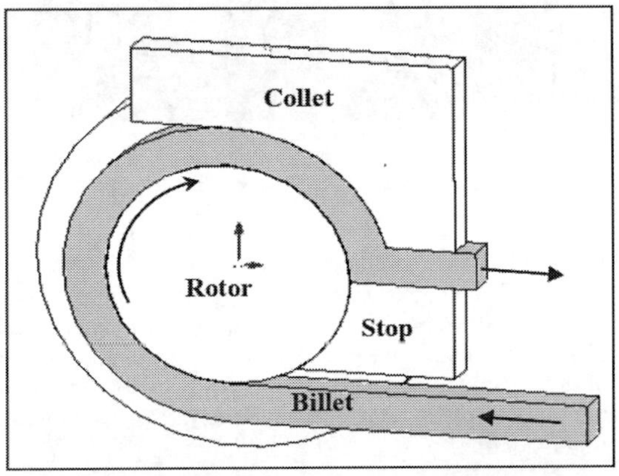

Fig. 1. Schematic diagram of the experimental facility for ECAP-Conform processing.

The facility created at Ufa State Aviation Technical University enables processing long-length rods of various metals with a diameter of up to 12 mm, both at room temperature and at elevated temperatures. In the present study, a variant of this facility was used that enabled the pressing of rods with an initial diameter of 10 mm. The following regime of ECAP-C processing was applied: two cycles at a temperature of 500°C (the condition «ECAP-C n=2»), then the same billet was subjected to two cycles of ECAP-C processing at a temperature of 400°C (the condition «ECAP-C n=4»), and finally two cycles at a temperature of 450°C (the condition «ECAP-C n=6»), the angle of channels intersection was 130°. The temperatures were varied, starting with 500°C, in order to evaluate the possibilities of ECAP-C processing of a $Ti_{49.1}Ni_{50.9}$ alloy billet without fracture. Then the temperature was decreased to 400°C (a temperature comparable to the temperature of conventional ECAP processing) and increased to 450°C, on the basis of evaluation of the billet's integrity. After every two cycles, a sample was cut off from the billet for further investigation.

Microstructural study was performed using an OLYMPUS GX5 optical microscope. Microhardness was tested using a Micromet-5101 microhardness tester under a load of 1 N. Tensile mechanical testing was performed on a tensile testing machine, designed at the Institute of Physics of Advanced Materials, Ufa State Technical University, with an upper load limit of 200 kg and a measurement accuracy of 5%. The tests were conducted at room temperature, at a strain rate of $10^{-3}s^{-1}$, using flat micro-samples with a gauge section of 0.25×1.0 mm and a gauge length of 3 mm. Structural studies by transmission electron microscopy (TEM) were performed using a JEOL - 2100 transmission microscope with an accelerating voltage of 200 kV.

3 Results and discussion

3.1 Microstructural studies by metallography and transmission electron microscopy

According to optical metallography (OM), in the initial state the $Ti_{49.1}Ni_{50.9}$ alloy had an austenitic structure with a grain size of about 50-60 μm (Fig. 2, a). In the samples processed by ECAP-C (Fig. 2, c), with the help of OM one can see old boundaries of individual grains, and at the same time, regions in the boundaries of the initial grains have a strongly refined structure. Observation by OM of old grain boundaries and regions with different degrees of deformity can be accounted for by the fact that the crystallographic orientations of neighboring grains are directed at different angles with respect to the deformation axis, so the internal structure of grains undergoes deformation in a different manner. However, this issue requires further study.

The resolution of the OM method is not sufficient for showing the microstructure of samples after ECAP-C at the micro-scale.

a

DOI: 10.1051/matecconf/201712902038

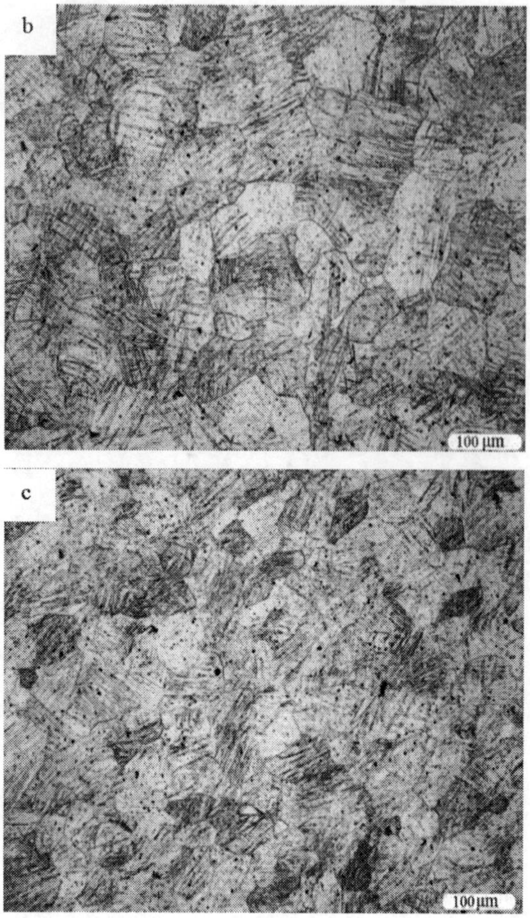

Fig. 2. Microstructure of the $Ti_{49.1}Ni_{50.9}$ alloy: a) initial state; b) ECAP-C, n=4, T=400°C; c) ECAP-C, n=6, T=450°C.

TEM studies demonstrate that in the $Ti_{49.1}Ni_{50.9}$ alloy subjected to ECAP-C there forms a banded microstructure with a high dislocation density and a developed fragmented structure inside the bands. The sizes of the deformation bands and fragments are presented in table 1. In the cross section of the rod after processing by ECAP-C with the number of passes n=2, regions are observed, having a banded deformation structure (Fig. 3) with a band width of about 500 nm and a length of several microns. Inside the bands, a high dislocation density is revealed. The electron diffraction pattern, showing a strong azimuthal and radial blurring of reflections, also confirms the formation in the alloy of a refined structure with a developed dislocation substructure and a high level of internal stresses due to work hardening.

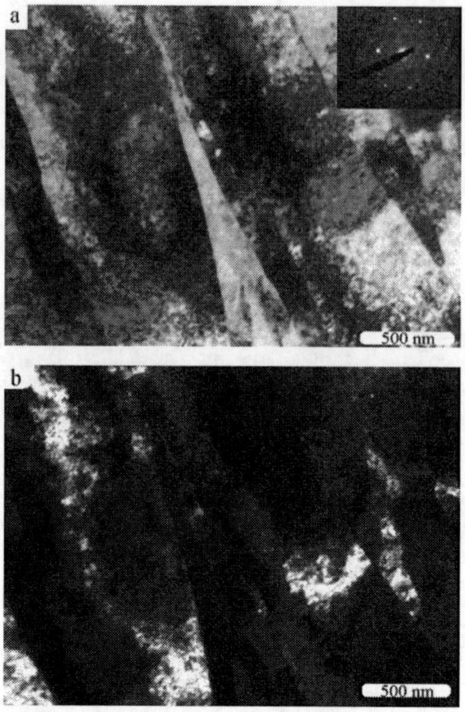

Fig. 3. Microstructure of the $Ti_{49.1}Ni_{50.9}$ alloy subjected to ECAP-C (n=2, T=500°C): bright-field image (a); dark-field image (b).

As the strain increases to 4 passes of ECAP-C, the structure becomes even more fragmented (Fig. 4). In the longitudinal section there form subgrains having rough dislocation walls, the dislocation density in the interior of grains/subgrains grows, which leads to an increase in internal stresses and, consequently, to progressing fragmentation.

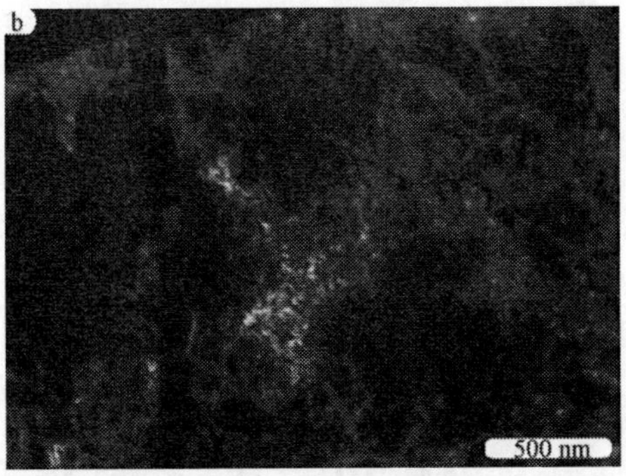

Fig. 4. Microstructure of the $Ti_{49.1}Ni_{50.9}$ alloy subjected to ECAP-C (n=4, T=400°C): bright-field image (a); dark-field image (b).

As the strain, further increases to 6 passes of ECAP-C, the size of fragments inside the bands decreases to 150 nm. In some regions, twins are observed, induced by deformation or martensitic transformation (Fig. 5).

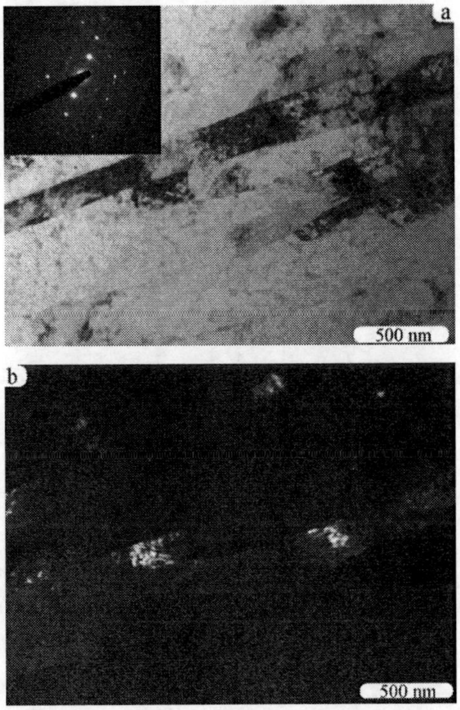

Fig. 5. Microstructure of the $Ti_{49.1}Ni_{50.9}$ alloy subjected to ECAP-C (n=6, T=450°C): bright-field image (a); dark-field image (b).

Subsequent annealing at 500°C after ECAP-Conform processing with the number of passes n=6 leads to the formation of a structure with a grain size of ~ 200 nm (Fig. 6).

Note should be made that in the condition «ECAP-C n=6» after annealing at 500°C, the grain/subgrain boundaries are revealed much more distinctly (Fig. 6) than in the conditions prior to annealing (Fig. 4-5). This can be attributed to the redistribution of dislocations, accumulated during ECAP-C processing, to the boundaries of grains and subgrains, and the transformation of loose and blurred boundaries into thinner and clearer boundaries.

Fig. 6. Microstructure of the $Ti_{49.1}Ni_{50.9}$ alloy after ECAP-C (n=6, T=450°C) + annealing T=500°C.

Table 1. The results of the estimation of bands and fragments.

Condition	The width of the fragments	The length of the fragments	The width of the bands
ECAP-C n=2	450 ÷ 580	450 ÷ 650	340 ÷ 500
ECAP-C n=4	300 ÷ 500	380 ÷ 590	140 ÷ 200
ECAP-C n=6	210 ÷ 320	290 ÷ 450	210 ÷ 380
ECAP C n=6 + annealing 500°C	130 ÷ 230	180 ÷ 340	250 ÷ 400

3.2 Mechanical properties of $Ti_{49.1}Ni_{50.9}$ alloy

The results of mechanical tests show that ECAP-C processing leads to an intensive growth of strength (σ_B) and yield strength (σ_y), as compared to the initial as-quenched condition (table 2).

Table 2. Mechanical properties of the TiNi alloy.

Condition	σ_B, MPa	$\sigma_{0.2}$, MPa	σ_m, MPa	δ, %
Initial	963	565	300	21
ECAP-C n=2	1060	835	210	27
ECAP-C n=4	1150	930	360	24
ECAP-C n=6	1320	1205	343	24
ECAP-C n=6 T=450°C+annealing 500°C	1120	970	290	27

Already after 2 passes the strength characteristics increase to 1060 MPa. After ECAP-C processing with the number of passes n=6, the ultimate tensile strength grows to 1320 MPa, the elongation being 24 %. The dislocation yield strength grown from an initial value of

565 MPa to 835 MPa at n=2 and to 1205 MPa at n=6. It should be noted that the relative increment, resulting from ECAP-C processing, in yield strength is visibly higher than the one in ultimate tensile strength (110% and 40%, respectively).

The significant increase in dislocation yield strength as a result of ECAP-C processing makes it possible to expect enhanced characteristics of the shape memory effect – reactive stresses and reversible strain, as was earlier observed in TiNi alloys subjected to conventional ECAP [6].

The results of microhardness testing after ECAP-C processing with the number of passes n=2-6 show that as strain increases and, consequently, as the structure becomes more refined, microhardness growth occurs (table 3), which is in agreement with the data from the mechanical tests.

Table 3. Results of microhardness testing.

Condition	Microhardness (HV), MPa
CG	2585±110
ECAP-C n=2	3100±85
ECAP-C n=4	3450±170
ECAP-C n=6	3740±150
ECAP-C, n=6 + annealing 500°C	3515±110

4 Conclusions

1. Using ECAP-Conform processing with the number of passes n=2, 4, 6, integral rods of the $Ti_{49.1}Ni_{50.9}$ alloy have been produced. TEM studies demonstrate that in the alloy there forms a banded deformation structure with a structural element size of up to 150 nm, after ECAP-C processing with n=6;
2. Annealing at a temperature 500°C of the alloy in the condition after ECAP-C processing with n=6 has led to the redistribution of the dislocations, accumulated during ECAP-C processing, to the boundaries of grains and subgrains, and to the formation of a structure with a grain/subgrain size of ~ 200 nm;
3. The performed mechanical tests show that as a result of increasing number of ECAP-C cycles, the ultimate tensile strength grows from an initial value of 963 MPa to 1320 MPa for ECAP-C processing with n=6, while the dislocation yield strength increases from an initial value of 565 MPa to 1205 MPa. Meanwhile, ductility is retained at an acceptable level of 25%.

This work was supported by grant No. 6.37.204.2016 from Saint Petersburg State University. This work was also supported by RFBR grant No. 16-38-00242.

References

1. V. Brailovski, S. Prokoshkin, P. Terriault, F. Trochu, *Shape Memory Alloys: Fundamentals, Modeling and Applications* (Ecole de Technologie Superieure (ETS), Universite du Quebec, Montreal, Canada, 2003)
2. K. Otsuka, X. Ren, Prog. Mater. Sci. **50,** (2005)
3. R.Z. Valiev, Y. Estrin, Z. Horita, T.G. Langdon, M.J. Zehetbauer, Y.T. Zhu, JOM **58** (2006)
4. R.Z.Valiev, A.P. Zhilyaev, T.G. Langdon, *Bulk Nanostructured Materials: Fundamentals and Applications* (John Wiley & Sons, Inc., Hoboken, New Jersey, 2014)
5. A.V. Sergueeva, C. Song, R.Z. Valiev, A.K. Mukherjee, Mater. Sci. Eng. A **339** (2003)
6. I.Yu. Khmelevskaya, S.D. Prokoshkin, I.B. Trubitsyna, M.N. Belousov, S.V. Dobatkin, E.V. Tatyanin, A.V. Korotitskiy, V. Brailovski, V.V. Stolyarov, E.A. Prokofiev, Mater. Sci. Eng. A **481–482** (2008)
7. V.G. Pushin, R.Z. Valiev, Y.T. Zhu, D.V. Gunderov, T.E. Kuntsevich, N.I. Kourov, A.N. Uksusnicov, L.I. Yurchenko, Mater. Trans. **47** (2006)
8. V.G. Pushin, V.V. Stolyarov, R.Z. Valiev, T.C. Lowe, Y.T. Zhu, Mater. Sci. Eng. A **410–411** (2005)
10. R.Z. Valiev, D.V. Gunderov, A.V. Lukyanov, V.G. Pushin. J. Mater. Sci. **47** (2012)
11. D. Gunderov, A. Lukyanov, E. Prokofiev, A. Churakova, V. Pushin, S. Prokoshkin, V. Stolyarov, R. Valiev. Mater. Sci. Forum. **738-739** (2013)
12. E.A. Prokofiev, J.A. Burow, E.J. Payton, R. Zarnetta, J. Frenzel, D.V. Gunderov, R.Z. Valiev, G. Eggeler, Adv. Mater. Eng. **12** (2010)
13. G.I. Raab, E.P. Soshnikova, R.Z. Valiev, Mater. Sci. Eng. A **387 – 389** (2004)
14. G.J. Raab, R.Z. Valiev, T.C. Lowe and Y.T. Zhu, Mater. Sci. and Eng. A **382** (2004)
15. G.I. Raab, F.F. Safin, T.C. Lowe, Y.T. Zhu, and R.Z. Valiev, Ultrafine Grained Materials IV. TMS Ultrafine Grained Materials IV (2006)
16. G.I. Raab, R.Z. Valiev, D.V. Gunderov, T. Lowe, A. Misra, Yu. Zhu Long-length, Mater. Sci. Forum, **584-586** (2008)
17. G.I. Raab, A.V. Polyakov, D.V. Gunderov, R.Z. Valiev, Russian Metallurgy (Metally), **5** (2009)

DOI: 10.1051/matecconf/201712902039

Strength and fatigue of an ultrafine-grained Al-Cu-Mg alloy

Elvira Khafizova[1,]*, *Rinat* Islamgaliev[1], and *Vil* Sitdikov[1]

[1]Ufa State Aviation Technical University, 12 K.Marx St., 450008 Ufa, Russia

Abstract. The dependence of strength and fatigue on microstructure of the Al-Cu-Mg alloy has been investigated. Various microstructures of the alloy were produced: the one with a coarse-grained (CG) structure after T6 heat treatment; the one with a homogeneous ultrafine-grained (UFG) structure and the one with a bimodal (mixed) structure, both processed by equal-channel angular pressing (ECAP). The mean grain size and morphology of precipitates were studied by transmission electron microscopy. The ultimate tensile strength and the fatigue endurance limit were determined using the tensile and fatigue tests of standard specimens. It is established that the formation of a homogeneous UFG structure and of a bimodal (mixed) structure alloy contributes to a significant increase in microhardness by 16% and 60%, and an increase of the ultimate tensile strength by 20 and 52%, respectively, as compared to the samples subjected to T6 heat treatment. Fatigue tests show that the alloy with a bimodal (mixed) structure has the highest fatigue endurance limit, 45% higher than in the sample subjected to T6 heat treatment. In contrast, the formation of a homogeneous UFG structure enables increasing the fatigue endurance limit by 15% only.

1 Introduction

Aluminum alloys of the Al-Cu-Mg system are widely used as structural materials. Today, an important task is to enhance their strength and fatigue in order to expand their application areas in the aircraft industry and increase the service life of parts.

One of the ways to solve this task is the formation of an ultrafine-grained structure by severe plastic deformation (SPD) [1-4].

It is well known that grain refinement in various metals and alloys is accompanied by a significant enhancement in strength and a decrease in ductility [5-8]. Meanwhile, the structural features and mechanical properties of UFG aluminum alloys essentially depend on the regimes of severe plastic deformation and a material's chemical composition, determining the contributions of solid-solution strengthening and precipitation hardening. Besides, UFG materials exhibit enhanced grain-boundary diffusion, which in the conditions of SPD may lead to a decrease in the aging temperature and time, required for the formation of precipitates [8, 9-11]. In their turn, the phase composition, volume fraction and size of precipitates may have an essential effect on the strength and fatigue of UFG aluminum alloys. Therefore, defining the UFG structure parameters that lead to the enhancement in the strength and fatigue properties of aluminum alloys is an important task.

As demonstrated by experimental research, an increase in the strength of aluminum alloys through grain refinement does not always lead to an increase in fatigue [12-15]. At present, two approaches are proposed for raising the fatigue of UFG materials. The first method is additional annealing after SPD processing, leading to a reduction in the initial dislocation density and internal microstresses, as well as to the relaxation of non-equilibrium grain boundaries. However, the use of this approach may result in the coarsening of dispersed particles and, consequently, to the softening of UFG aluminum alloys. The second approach ethod is the formation of a bimodal structure, where micron-sized grains are responsible for the ductility, while nano-sized grains are responsible for the strength [16]. The second approads has not been sufficiently studied so far. As a consequence of the above, the structural features determining the fatigue of UFG aluminum alloys are still a subject of study.

In this connection, the aim of the present work is to establish the relationship between the structural features and fatigue of an aluminum alloy of the Al-Cu-Mg system, and to study the effect of cyclic loads on a material's structure, using the Al-Cu-Mg alloy as an example.

2 Material and Research Methods

As the materials for the study, the aluminum alloy of the Al-Cu-Mg system was selected (Russion designation AK4-1). The alloy's chemical composition is presented in table 1.

Table 1. Chemical composition of the Al-Cu-Mg alloy, wt. %

Fe	Si	Mn	Ti	Al	Cu	Ni	Mg	Zn
0.98	0.05	0.04	0.02	93.8	2.33	1.04	1.65	0.02

The alloy was investigated in three states. «State 1» was produced by T6 heat treatment, where a coarse-grained structure with a mean grain size of 100 μm was

* Corresponding author: ela.90@mail.ru

 DOI: 10.1051/matecconf/201712902039

formed. Produced using: heating to a temperature of 530°C, holding at this temperature for one hour, water quenching, aging at a temperature of 190 °C for 7 hours and air cooling. Using equal-channel angular pressing (ECAP), two other types of structures were produced: «State 2» - a homogeneous UFG structure (ECAP angle 120°, route Bc, diameter 40mm, number of passes 6, 160°C) and «State 3» - a bimodal (mixed) structure (ECAP angle 90°, route Bc, diameter 20mm, number of passes 2, 160°C) [17,18]. Their structural models are presented in fig. 1.

In all the states, there are present Al_9FeNi intermetallic particles which are not shown in the models, since their sizes and volume fractions remain unchanged. Also present are the dispersed particles of two phases, Θ (Al_2Cu) and S (Al_2CuMg), which promote strengthening and thermal stability in the AK4-1 alloy. In «State 1» the S phase has a rod-like shape, the Θ phase has a globular shape, and their sizes reach 2 μm. The structural feature of «State 2» is that the UFG structure has grains with sizes of 1.5 μm and a width of 500 nm in the longitudinal section and 480 nm in the transverse section, and there are present dispersed particles of the phases Θ', Θ and S, having sizes of 20, 60 and 80 nm, respectively. The feature of «State 3» is a bimodal (mixed) structure, where in the longitudinal and transverse sections a UFG structure is formed, with grain sizes of 250 nm, and bands with a width of 200-300 nm, and there are also present dispersed particles of the phases Θ', Θ and S, having sizes of 10, 40 and 60 nm, respectively.

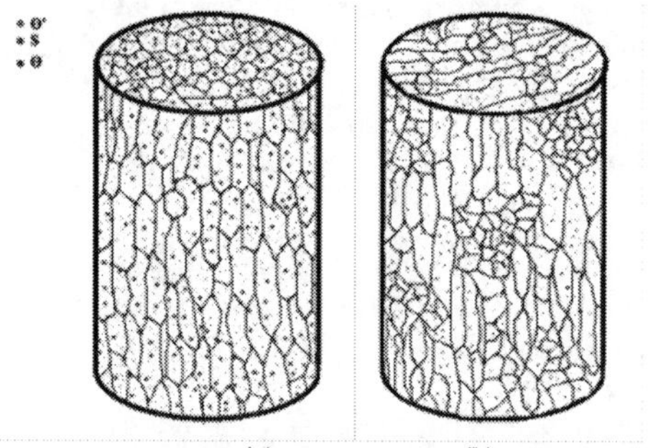

(a) (b)

Fig. 1. Structural models of the samples after different types of treatment: (a) «State 1»; (b) «State 2»;

Microhardness was tested by the Vickers method, using a Buehler «Micromet 5101» device with a pyramidal diamond indenter under a load of 1 N and with a time under load of 10 seconds.

The strength under tension was tested, using cylidrical specimens with a gauge length of 15 mm and a diameter of 3 mm. The tests were conducted on an Instron tensile-testing machine at 20°C and 175°C with a crosshead speed of 1 mm/min ($\sim 1 \cdot 10^{-3} s^{-1}$).

The fatigue tests were performed using two methods. In the first case, the specimens were studied in testing machines – high-frequency pulsators Amsler 100 HFP 5100 and Amsler 300 HFP 5100 (Zwick/Roell) under the axial loading of a specimen at a temperature of 20°C, the stress ratio was $R_\sigma = 0.1$ with 10^7 cycles to failure. In the second case, the fatigue tests were conducted as cantilever-bending tests with rotation at a frequency of 50 Hz using stress ratio $R_\sigma = -1$ and 10^7 cycles to failure.

In all the states, the specimens, subjected to high-cycle fatigue testing, were studied in terms of the effect of cyclic loads on the material's structure in the vicinity of the fracture zone (see Fig. 2) by transmission electron microscopy and X-ray diffraction analysis.

Structural studies were performed using transmission electron microscopy on a JEOL - 2100 microscope with an accelerating voltage of 200 kV, equipped with an INCAx-sight attachment that enables elemental energy-dispersive analysis.

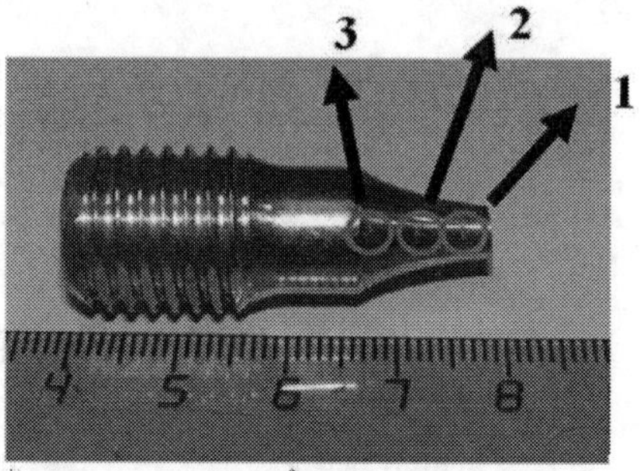

Fig. 2. Specimen of the Al-Cu-Mg alloy after fatigue tests: 1, 2, 3 - the areas under study

X-ray diffraction of the samples was performed using a Rigaku Ultima IV diffractometer in Cu-Kα radiation with a wave length of λ=1.54418 Å.

3 Results and Discussion

Formation of «State 2» and «State 3» in the alloy has contributed to a considerable increase in microhardness by 16% and 60%, respectively, as compared to «State 1». Ultimate tensile strength has grown by 20 and 52%, respectively, and at the same time, elongation to failure has decreased two-fold. Tensile tests of the UFG alloy at a temperature of 175°C demonstrate that UTS has decreased only slightly, which confirms the thermal stability of the mechanical properties of the alloy in the UFG state up to a temperature of 175°C (Table 2).

The enhanced strength of the UFG samples is the result of a strong grain structure refinement and precipitation hardening, since the reduction in the elongation to failure can be attributed to the presence of internal stresses around the particles and in the vicinity of grain boundaries, impeding the development, generation and movement of dislocations. In the UFG samples, there is also observed a short stade of strain hardening. As it is well known, at the stage of strain hardening there occurs an increase in dislocation density. In the UFG material the distance between grain boundaries is much larger, therefore dislocations reach the opposite grain boundary faster and become annihilated. This can explain the short stage of strain hardening in the UFG samples and, consequently, the small uniform elongation .

Table 2. Mechanical properties of the Al-Cu-Mg alloy

State	t, °C	HV, MPa	σ_B, MPa	$\sigma_{0.2}$ MPa	δ, %
«1»	20°C	1220±8	370±3	320±5	16±2
	175°C	-	360±4	306±5	18±3
«2»	20°C	1408± 37	460±8	420±5	7±1
	175°C	-	430±5	394±4	8±1.5
«3»	20°C	1980±20	566±10	520±6	8±1
	175°C	-	530±4	495±5	7±2

The large difference between «State 2» and «State 3» can be attributed to the difference in dislocation density, sizes of precipitates and structural elements ((sub)grains) [17,18]. Apparently, due to the long exposure at temperature of aging and large strains in «State 2» there is observed a noticeable increase in the sizes of the second phases, accompanied by a loss of coherence. Due to the strong distortions of the crystal lattice and the activation of diffision processes, the material, exposed to the temperature of aging for a long period of time, loses a part of the strengthening effect of these particles.

Fatigue tests of the «tension-compression» type under an asymmetrical cycle R=0.1 demonstrate that in the alloy the formation of a homogeneous UFG structure («State 2») has led to an increase in the fatigue endurance limit by 15% as compared to «State 1»(Fig. 3).

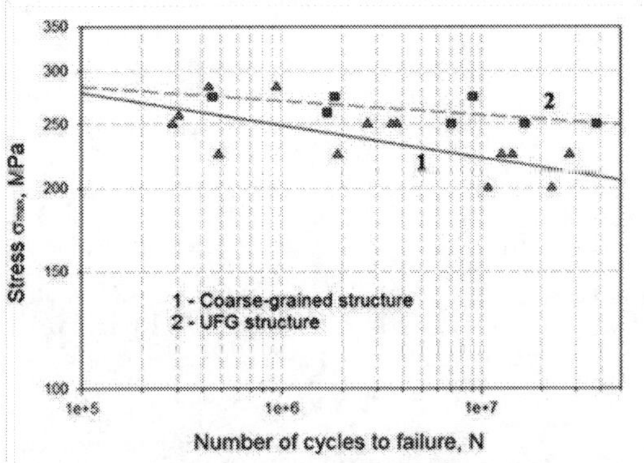

Fig. 3. The stress amplitude as a function of the number of cycles for the Al-Cu-Mg alloy in « State 1» (the lower curve) and « State 2» (the upper curve).

Fig. 4 shows the stress-cycle curves for the alloy in «State 1» and «State 3». Fatigue tests reveal that after ECAP processing the fatigue endurance limit is 250 MPa, whereas in the state after T6 it was 170 MPa. Thus, it can be concluded that ECAP processing enables increasing the fatigue endurance limit of the alloy approximately by 47% as compared to the T6 heat treatment.

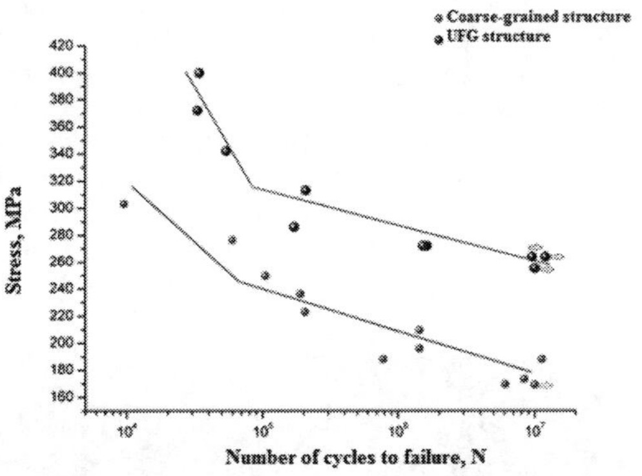

Fig. 4. The stress amplitude as a function of the number of cycles for the Al-Cu-Mg alloy in «State 1» (the lower curve) and «State 3» (the upper curve).

Note should be made that for standard specimens, the resistance to high-cycle fatigue under an asymmetrical cycle (R=0.1), as compared to a symmetrical cycle (R= -1), has a higher value of σ_{max}.

Apparently, the increase in fatigue strength in the ECAP-processed samples was achieved due to a strong grain refinement, as well as due to the presence of particles ensuring precipitation hardening.

Study of the effect of cyclic loads on the material's structure in the vicinity of the fracture zone by X-ray diffraction reveals that the closer is a zone to the fracture surface, the more microdistortions of the crystal lattice are observed ($<\varepsilon^2>^{1/2}$), the higher is the dislocation density ($\rho_\perp$), and, correspondingly, the smaller is the coherent scattering domain size. Also observed is a decrease in the lattice parameter (*a*), which is related to the precipitation and coarsening of dispersed particles in the fracture region for UFG and CG samples (Table 3), which has also been confirmed by TEM.

Table 3. The results of X-ray diffraction analysis of samples in «State 1» and «State 2»

	Zone	*a*, Å	$<\varepsilon^2>^{1/2}$, %	$\rho_\perp$, 10^{14} m^{-2}
State1	1	4.0546±0.0001	0.072±0.002	0.64
	2	4.0561±0.0001	0.053±0.004	0.46
	3	4.0584±0.0001	0.047±0.001	0.36
CG before fatigue		4.0587±0.0001	0.045±0.001	0.32
State2	1	4.0538±0.0001	0.191±0.003	3.67
	2	4.0540±0.0001	0.185±0.002	3.36
	3	4.0541±0.0001	0.173±0.002	3.12
UFG before fatigue		4.0542±0.0001	0.167±0.002	2.95

In the structure of the UFG samples, subjected to fatigue tests, dispersed particles of three phases have been found. Comparative studies of all the three regions reveal that the particles of the phases Θ and S, located closer to the fracture surface, have grown in size to 230 nm (Fig. 5a). In addition, their volume fraction has increased. The closer to the fracture zone, the higher is the dislocation density in the grain interiors. There are observed areas where particles of the Θ' phase have a high density. In the second zone, the volume fraction of the particles and their

 DOI: 10.1051/matecconf/201712902039

size are smaller in comparison with the first zone (Fig. 5b). In the third zone, there is observed a structure corresponding to a regular structure for this state prior to fatigue testing (Fig. 5c).

In the samples after T6 heat treatment, irrespective of the type of loading, similar changes have also occurred. There is also observed a high dislocation density in the vicinity of the fracture zone, particles of the Θ phase have appeared at grain boundaries, particles of the S phase are found in the structure as well.

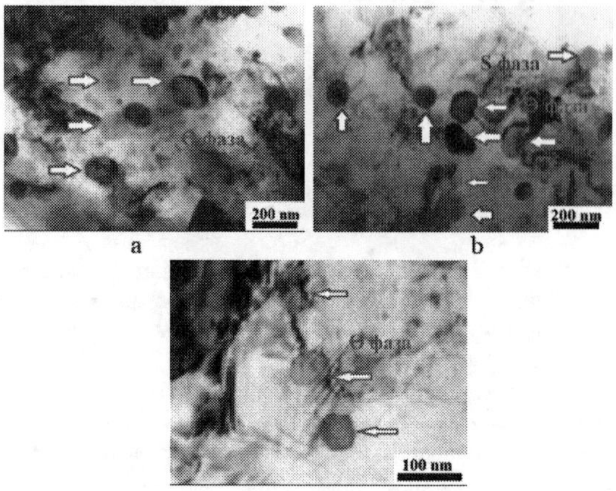

Fig. 5. Typical images of the structure after the fatigue tests for the UFG specimen in "State 2": a – zone 1, b – zone 2, c – zone 3.

On the basis of these data, it is possible to make an assumption that during the fatigue tests there occurs activation of diffusion processes in the sample's fracture zone, and the particles coarsen as a result of dynamic aging.

4 Conclusions

Thus, this study has demonstrated that the Al-Cu-Mg alloy with a bimodal (mixed) structure is capable of exhibiting the values of UTS and endurance limit that are 1.5 times higher than the respective values of the sample subjected to the T6 heat treatment, while a homogeneous UFG structure leads to a smaller growth in the material's UTS and fatigue endurance limit.

For this alloy, of great importance is the processing temperature and the time of processing at this temperature, since during the deformation there occurs activation of diffusion processes, which reduces the time of precipitation of dispersed particles playing an important role in the strengthening.

It has been established that during fatigue testing in all the states there is observed activation of diffusion processes, leading to further precipitation and coarsening of particles due to the heating of the sample in the fracture zone.

This work was conducted under RFBR project No. 16-38-00507 performed by young scientists («My first grant»).

References

1. R. Z. Valiev, R. K. Islamgaliev, I. V. Alexandrov, Progress in Materials Science **45**, 103-189 (2000) DOI:10.1016/S0079-6425(99)00007-9
2. R. Z. Valiev, T.G. Langdon, Progress in Materials Science **51**, 881-981 (2006) DOI:10.1016/j.pmatsci.2006.02.003
3. A.P. Zhilyaev, T.G. Langdon, Progress in Materials Science **53**, 893–979 (2008)
4. R.Z. Valiev, A.P. Zhilyaev, T.G. Langdon, *Bulk Nanostructured Materials: Fundamentals and Applications* (John Wiley & Sons Inc., 2014)
5. R. K. Islamgaliev, M. A. Nikitina, A. F. Kamalov, Materials Science Forum, Vols. **667-669**, 331-336 (2011)
DOI: 10.4028/www.scientific.net/MSF.667-669.331
6. G. Nurislamova, X. Sauvage, M. Murashkin, R. Islamgaliev, R. Valiev, Phil. Mag. Lett. **88 (6)**, 459–466 (2008)
DOI:10.1080/09500830802186938
7. O. Kulyasova, R.K. Islamgaliev, B. Mingler, M. Zehetbauer, Mat. Sci. Eng. A. **503**, 176–180 (2009) DOI: org/10.1016/j.msea.2008.03.057
8. E. Bobruk, I Sabirov, V Kazykhanov, R Valiev, M Murashkin, IOP Conf. Series: Materials Science and Engineering **63**, 012116 (2014)
DOI:10.1088/1757-899X/63/1/012116
9. W. Huang, Z. Liua, M. Lin, X. Zhou, L. Zhao, A. Ning, S. Zeng, Materials Science and Engineering A **546**, 26 – 33
(2012).DOI:org/10.1016/j.msea.2012.03.010
10. E.V. Bobruk, M.Yu. Murashkin, V.U. Kazykhanov and R.Z. Valiev, Rev. Adv. Mater. Sci **31**, 14 – 34 (2012)
11. I.G. Brodova, I.G. Shirinkina, A.N. Petrova, O.V. Antonova, V.P. Pilyugin, The Physics of Metals and Metallography **111(6)**, p. 630–638 (2011) DOI:10.1134/S0031918X11050036
12. Y. Estrin, A. Vinogradov, International Journal of Fatigue **32**, 898–907 (2010)
DOI:org/10.1016/j.ijfatigue.2009.06. 022
13. V. Patlan, A. Vinogradov, K. Higashi, K. Kitagawa, Mater Sci Eng A300, 171-182 (2001).
14. A. Washikita, K. Kitagawa, V.I. Kopylov, A. Vinogradov, Ultrafine Grained Materials **II**, TMS 341-350 (2002)
15. J. May, M. Dinkel, D. Amberger, H.W. Höppel, M. Göken, Metal Mater Trans A **38**, 1941-1945 (2007) DOI: 10.1007/s11661-007-9110-0
16. Y. Estrin, A. Vinogradov, Acta Materialia **61(3)**, 782–817 (2013) DOI:org/10.1016/j.actamat.2012.10.038
17. E. Khafizova, R. Islamgaliev, V. Sitdikov, Deformation and fracture of materials **6**, 25-29 (2015) (in Russian)
18. E. Khafizova, R. Islamgaliev, G. Klevtsov, E. Merson. Materials Physics and Mechanics **24(3)**, 232-241 (2015)

DOI: 10.1051/matecconf/201712902040

Structure determination and composite wing spar stress-strain state estimation

Pavel Solovyev[1], *Andrey* Gomzin[1,*], *Yuriy* Pervushin[2], *Fanil* Musin[1] , and *Sergey* Galyshew[1]

[1]FSBEE USATU, Department of Materials Science and Physics of Metals, 450008, Ufa, Russia
[2]FSBEE USATU, Department of Materials Resistance, 450008, Ufa, Russia

Abstract. The paper presents the results of studies on the stress-strain state of composite wing spars with different lay-up structures. The object of research is a U-shaped cross-section composite wing spar. On the basis of the analysis of the stress-strain state, the most favorable structure of a wing spar from the viewpoint of minimizing the risk of stress state and providing the highest stiffness. Recommendations are given on lay-up structure stacking (compared with typical structures) for more effective perception of external loads by a wing spar.

1 Introduction

Currently there is a clear tendency in the world practice of aircraft building to increase the share of composite materials in airframes construction. This tendency is due to the fact that composite materials have the best set of physical and mechanical properties, especially high specific strength and stiffness characteristics, high long-term strength, low sensitivity to stress concentrators. Extension of application areas of composites is conditioned by their introduction in design of airframe load-bearing elements.

Use of composite materials in load-bearing airplane elements (Fig. 1.b) allows making a fuselage and wings (Fig. 1.a) not only light and strong, but also to decrease assembly time by reducing the number of assembly units, while increasing reliability of a design.

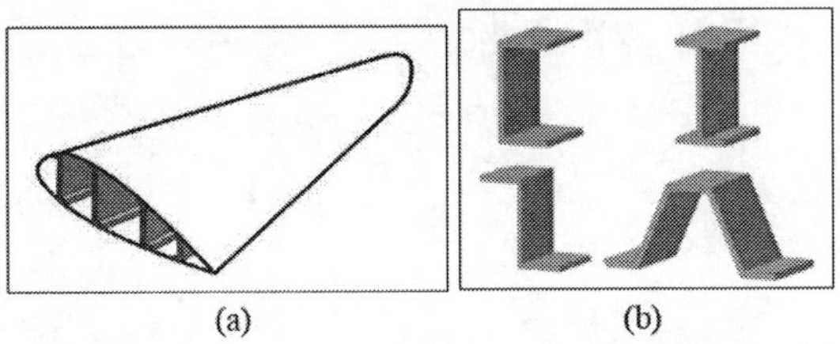

Fig. 1. Cross-section of an aircraft wing with a longitudinal set of wing spars (a) and several profiles types of airframe load-bearing elements (b).

A significant feature of composite materials is the ability to vary a composite structure to obtain the highest properties in those directions in which a product is exposed to maximum loads. However, calculation of composite materials products is a complicated and laborious problem.

The solution of a similar problem could be significantly simplified, if there were reference databases on the variants of composite materials structures, depending on the purpose of a load-bearing element and loads acting on it.

In practice, usually the following typical structures are used in composite structures: unidirectional (Fig. 2.a); bi-directional (Fig. 2.b); three-layer structure (Fig. 2.c); quasi-isotropic structure (Fig. 2.d).

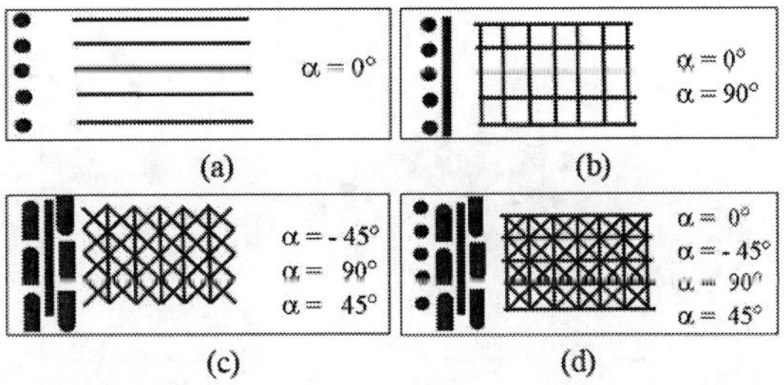

Fig. 2. Typical lay-up structures for products from composite materials.

For example, in certain world scientific sources [1] the opinion prevails that it is most convenient to design laminated composites using layered cells. These cells consist of several layers oriented at no more than four different angles and repeated the required number of times to reach a needed thickness of composite. However, this approach does not allow designers to obtain maximum efficiency of those composites.

So, the major disadvantage of such typical structures is that they do not take into account individual loading features of specific elements.

Therefore, typical structures are rarely used for multi-load products, as a rule, without making any changes.

All of the above-mentioned makes it urgent to determine typical and individual structures of major load-bearing elements of an aircraft (spars, stringers, ribs,

* Corresponding author: andre-gomzi@yandex.ru

frames), that provide favorable perception of an external load field.

2 Research goals and objectives

The purpose of this research is to establish the patterns for choosing basic typical composite structures for the design and manufacture of aircraft fuselage load-bearing elements, depending on their stress-strain state. To achieve this goal, the following tasks are solved in this study:

1. Analysis of the stress-strain state of a spar with a U-shaped section;
2. Choice of a typical layer stacking structure of a spar made from composite material, providing a favorable stress state and deformation behavior;
3. Practical recommendations on the choice of rational structures for aircraft load-bearing elements.

3 Computing experiment

A spar with a U-shaped cross-section is chosen to perform this experiment. The real force factors acting on the spar element are shown in Fig. 3.

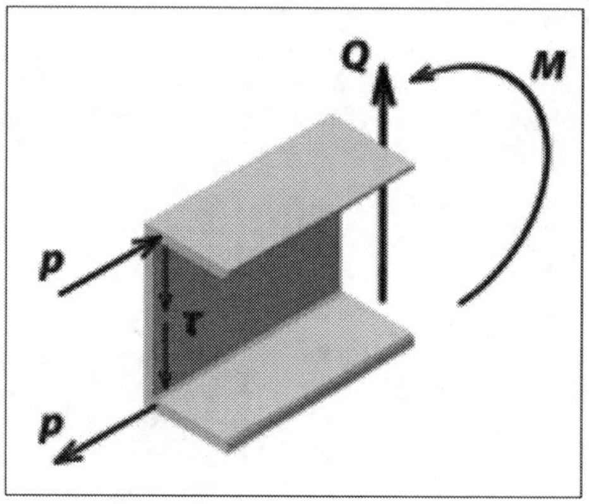

Fig. 3. Spar loading scheme.

The spar flanges operate in the tension/compression mode induced by the bending moment, and the spar web takes up shearing forces.

3.1 Initial data

The modeled spar material in this research is carbon fiber reinforced plastic based on a carbon ribbon and a polyimide matrix.

Carbon ribbon: density ρ = 1.7 g/cm³; tensile strength $\bar{\sigma}^{+}_{в1}$= 2750 MPa; elastic modulus along fibers $E_{в1}$ = 210 GPa; elastic modulus across fibers $E_{в2}$ = 8 MPa, Poisson ratio $\nu_{в12}$ = 0,15.

Polyimide matrix: density ρ = 1.2 g/cm³; tensile strength $\bar{\sigma}^{+}_{м}$= 114 MPa; elastic modulus $E_{м}$ = 4.2 GPa; Poisson ratio $\nu_{м}$ = 0.3.

The physical and mechanical characteristics of the composite unidirectional layer are calculated by the method considered in [2, 3] and are presented below.

The unidirectional composite elastic characteristics at the standard temperature t = 20°C: E_{11} = 127.7 GPa; E_{22} = 6.2 GPa; E_{33} = 6.2 GPa; G_{12} = 4.3 GPa; G_{13} = 4.3 GPa; G_{23} = 1.56 GPa; ν_{12} = 0.226; ρ = 1.5 g/cm³.

The unidirectional composite strength characteristics at the standard temperature t =20 °C: $\bar{\sigma}_{11p}$= 1.4 GPa; $\bar{\sigma}_{11c}$= 0.95 GPa; $\bar{\sigma}_{22p}$= 93.1 MPa; $\bar{\sigma}_{22c}$= 93.1 MPa; $\bar{\tau}_{12}$= 52.4 MPa.

The geometrical parameters of a spar: height h=100 mm, width b=40 mm, length l=1 m.

The initial layers stacking structure of the composite spar is $(0°)_n$. The direction of the main coordinate system is along the longitudinal axis of the spar. The number of layers is n=13. The thickness of one layer is h_c=2 mm.

The boundary conditions: fixed support on one of the tips, pressure p = 10 kPa, uniformly distributed over the lower flange of the spar. These boundary conditions are chosen on the basis of the analysis of force factors acting on the real wing spars of an aircraft.

3.2 Simulation of loading of the unidirectional structure spar

The composite spar is simulated in the ANSYS software using the ANSYS Composite Prepost module. The number of model elements is about 26,000.

Figure 4 displays the spar deformation caused by the above-mentioned force factors. The figure shows that under the bending loads, the spar not only bends, but also twists.

Figure 5 shows the distribution of the safety factor for the composite spar.

The RF safety factor is determined in accordance with the selected strength criterion – in this case the Mises-Hill criterion, which is widely used to assess the risk of the stressed state in the layers of the composite material during the ply strength analysis [4,5] – using the formula:

$$RF = \left(IRF\right)^{-1} = \left(\sqrt{H_{cr}}\right)^{-1} = \left(\sqrt{\left(\frac{\sigma_1}{X}\right)^2 - \frac{\sigma_1\sigma_2}{XY} + \left(\frac{\sigma_2}{Y}\right)^2 + \left(\frac{\tau_{12}}{S}\right)^2}\right)^{-1} \quad (1)$$

where X, Y и S – the ultimate stresses in the conditions of uniaxial loading of the unidirectional layer in the direction parallel and perpendicular to the fibers and under the shear in the plane of the layer, respectively, σ_1, σ_2, τ_{12} – the acting stresses in the direction of the principal axes of elasticity of the unidirectional layer.

The graph demonstrates that the minimum value of the safety factor is determined in the most loaded element and is equal to RF_{min} = 11.06. In general, the spar strength is ensured for the given loads.

DOI: 10.1051/matecconf/201712902040

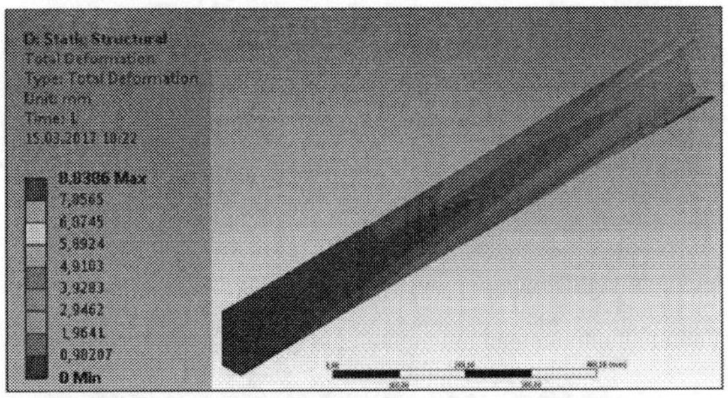

(a)

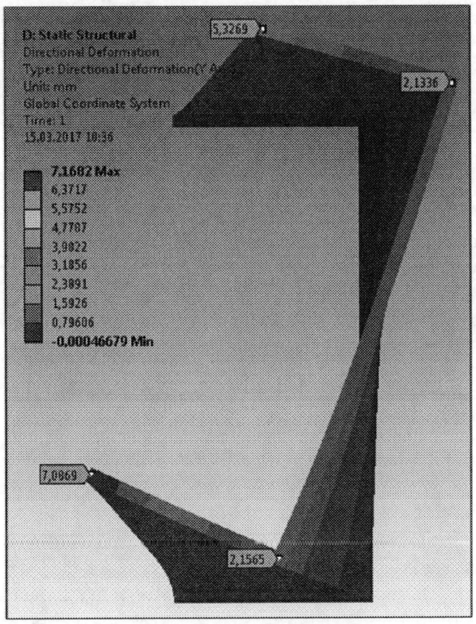

(b)

Fig. 4. Deformation of the spar under the action of specified loads

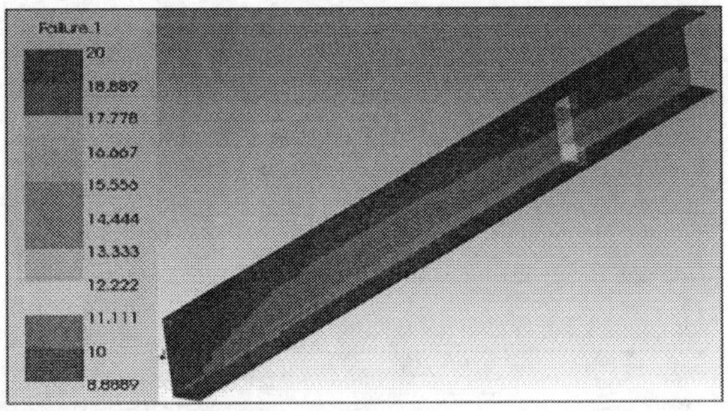

Fig. 5. Distribution of the safety factor for the composite spar

By analyzing the distribution of the reserve factors, one can conclude that the most loaded are the outer layers with numbers №1 and №13.

Figures 6-11 show the stresses in two outer layers of the spar oriented at an angle 0°. The green arrows in the figures show the directions of fiber orientation.

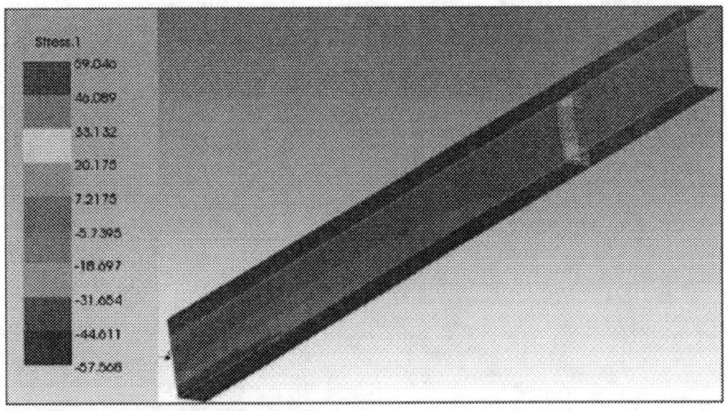

Fig. 6. Distribution of normal stresses (along fibers) σ_1 in the spar layer №1 oriented at an angle 0° (MPa).

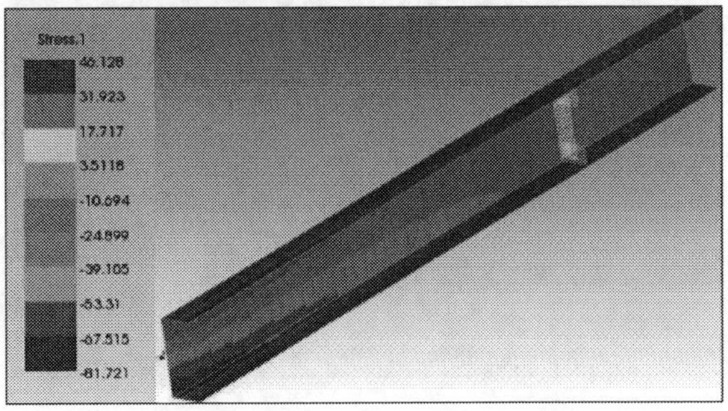

Fig. 7. Distribution of normal stresses (along fibers) σ_1 in the spar layer № 13 oriented at an angle 0° (MPa).

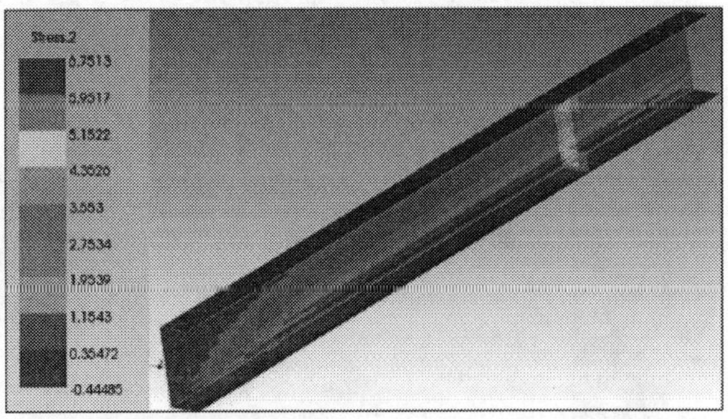

Fig. 8. Distribution of normal stresses (across fibers) σ_2 in the spar layer №1 oriented at an angle 0° (MPa).

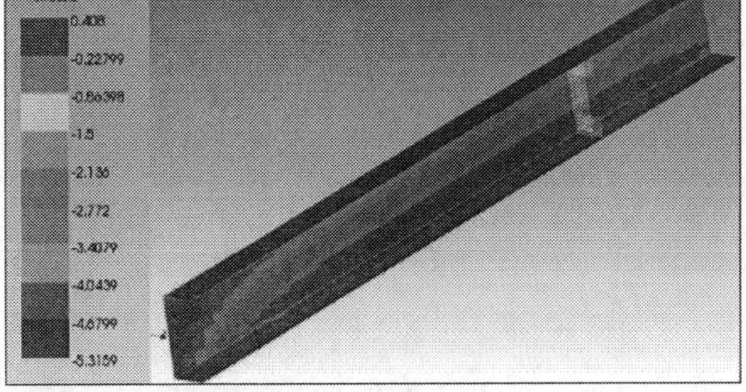

Fig. 9. Distribution of normal stresses (across fibers) σ_2 in the spar layer №13 oriented at an angle 0° (MPa).

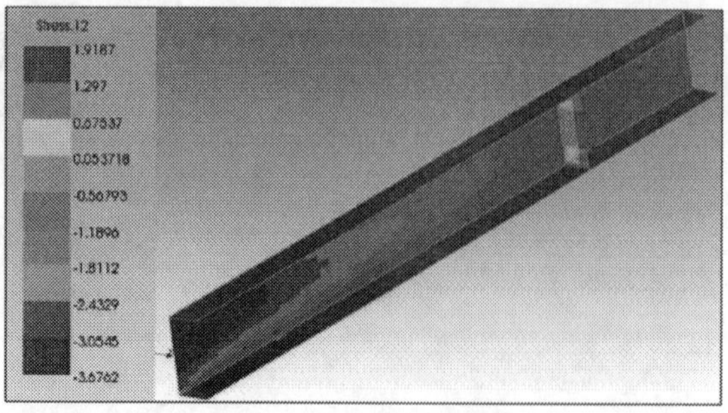

Fig. 10. Distribution of shear stresses in ply plane τ_{12} in the spar layer №1 oriented at an angle 0° (MPa).

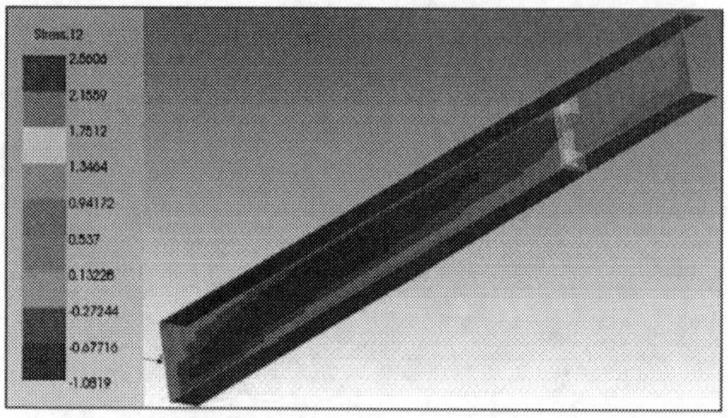

Fig. 11. Distribution of shear stresses in ply plane τ_{12} in the spar layer №13 oriented at an angle 0° (MPa).

Fig. 12 displays the distribution of stresses and safety factors in the most loaded element over all layers and the distribution of the elastic properties of the layered element as a function of direction.

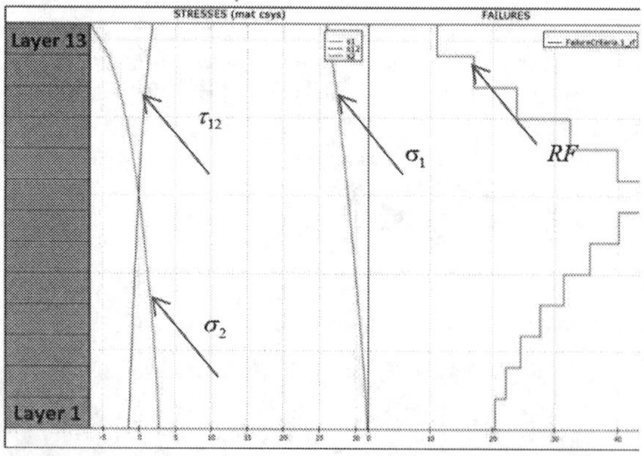

Fig. 12. Distribution of stresses, safety factors and elastic properties in the most loaded element.

In order to determine which of the stresses acting in the unidirectional layer are the most riskful, it is necessary to find the maximum of the following expressions entering into the Hill criterion:

$$\max\left(\frac{\sigma_1}{X};\frac{\sigma_2}{Y};\frac{\tau_{12}}{S}\right) \tag{2}$$

if the first fraction is maximum, then the most riskful are the normal stresses along the fibers, if the second one is maximum – then the riskful stresses act across the fibers, if the third one is maximum – then riskful stresses are tangential ones in the layer plane.

In the studied case:

For layer №1

$$\frac{\sigma_1}{X}=\frac{31.7}{1400}=2.3\%; \qquad \frac{\sigma_2}{Y}=\frac{2.75}{93.1}=3\%;$$

$$\frac{\tau_{12}}{S}=\frac{1.42}{52.4}=2.7\%;$$

For layer №13 (3)

$$\frac{\sigma_1}{X}=\frac{26.7}{1400}=1.9\%; \qquad \frac{\sigma_2}{Y}=\frac{-4.26}{-93.1}=4.5\%;$$

$$\frac{\tau_{12}}{S}=\frac{1.32}{52.4}=2.5\%;$$

The analysis of the graphs and formulas (2) indicates that the prevailing stresses in the most stressed layers are the stresses across the fibers σ_2. In this regard, for a more favorable perception of the external loads field, it is recommended to add reinforcement in this direction.

3.3 Modeling of spar loading with typical structures

Other typical structures of the spar are considered, and their influence on its stressed state and deformation behavior are estimated.

The results of the finite element modeling of the considered structures are given below:

1. $(0°)_{13}$: deflection U_Y = 2.16 mm; twist angle φ = 7.08°; $RF_{lay=1}$ = 20.5; $RF_{lay=13}$ = 11.06; $\sigma_{1\ lay=1}$ = 31.7 MPa; $\sigma_{1\ lay=13}$ = 26.7 MPa; $\sigma_{2\ lay=1}$ = 2.8 MPa; $\sigma_{2\ lay=13}$ = - 4.3 MPa; $\tau_{12\ lay=1}$ = 1.42 MPa; $\tau_{12\ lay=13}$ = 1.32 MPa;
2. $(0°/90°)_3/0°/(90/0)_3$: deflection U_Y = 3.53 mm; twist angle φ = 9.19°; $RF_{lay=13}$ = 12.77; $RF_{lay=12}$ = 20; $\sigma_{1\ lay=13}$ = 0.2 MPa; $\sigma_{1\ lay=12}$ = -14.5 MPa; $\sigma_{2\ lay=13}$ = -1.7 MPa; $\sigma_{2\ lay=12}$ = - 0.2 MPa; $\tau_{12\ lay=13}$ = 3.2 MPa; $\tau_{12\ lay=12}$ = -2.3 MPa;
3. $(0°/45°/-45°)_2/0°/(-45°/45°/0°)_2$: deflection U_Y = 2.98 mm; twist angle φ = 7.14°; $RF_{lay=12}$ = 10.88; $RF_{lay=11}$ = 11.3; $\sigma_{1\ lay=12}$ = 10.4 MPa; $\sigma_{1\ lay=11}$ = -13.6 MPa; $\sigma_{2\ lay=12}$ = -1.1 MPa; $\sigma_{2\ lay=11}$ = 0.5 MPa; $\tau_{12\ lay=12}$ = -4.6 MPa; $\tau_{12\ lay=11}$ = 4.2 MPa;
4. $(0°/45°/90°/-45°)_3$ / 0°: deflection U_Y = 3.34 mm; twist angle φ = 7.50°; $RF_{lay=1}$ = 11.66; $RF_{lay=5}$ = 12.1; $\sigma_{1\ lay=1}$ = -73.2 MPa; $\sigma_{1\ lay=5}$ = -70.8 MPa; $\sigma_{2\ lay=1}$ = 0.2 MPa; $\sigma_{2\ lay=5}$ = -0.3 MPa; $\tau_{12\ lay=1}$ = -0.47 MPa; $\tau_{12\ lay=5}$ = -0.2 MPa.

It is clear from the above-written that the most favorable structure for the spar from the viewpoint of ensuring the minimal deformations (both deflections and twisting of the free tip) is a unidirectional structure with the parameters: $((0°)_n$, φ = 7.08°). From the standpoint of the stressed state risk, the most preferable structure is the mutually

 DOI: 10.1051/matecconf/201712902040

orthogonally reinforced structure $(0°/90°)_3/0°/(90/0)_3$ (RF_{min}=12.77). In general, under the given loads, any structure satisfies the composite spar strength condition. The permissible safety factor for composite products [RF] = 2.0 ÷ 3.0.

4 Conclusion

During the finite element modeling of the U-shaped composite wing spar on the basis of the carbon ribbon and the polyimide matrix, the main regularities of the spar stacking structure effect on its stressed state and elastic deformation behavior have been revealed.

The most favorable typical structure from the viewpoint of the load-bearing element rigidity is a unidirectional structure $(0°)_n$. However, from the standpoint of the stress state risk, the preferred structure is the one with $(0°/90°)_m/0°/(0°/90°)_m$. In general, the strength of the spar under the specified loads is ensured.

For more favorable perception the external loads field while maintaining high stiffness of the spar, it is recommended to establish 20-30% of layers reinforced at an angle of 90° or close to it in the unidirectional structure.

References

1. R.M. Jones, *Mechanics of Composite Materials* (Taylor&Francis, 1999)
2. V.S. Zhernakov, Y. S. Pervushin, P. V. Solovyev, Vestnik UGATU **19 №3 (69)**, 139-149 (2015)
3. Y.S. Pervushin, V.S. Zhernakov, *Fundamentals of mechanics, design and manufacturing technology of items from layered composite materials: teaching guide* (USATU, Ufa, 2008)
4. C. Herakovich, *Non-elastic properties of composite materials* (Mechanics, Moscow, 1978)
5. L. Brautman, R. Crock, *Composite materials: Mechanics of composite materials* (Mashinostroenie, Moscow, 1978)

MATEC Web of Conferences **129**, 02041 (2017) DOI: 10.1051/matecconf/201712902041
ICMTMTE 2017

The use of fluorinated CNTs as epoxy resin modifiers

Alexander N. Blohin[1,*], *Tatiana P.* Dyachkova[1] , and *Artem K.* Sukhorukov[1]

[1]Tambov State Technical University, Sovetskaya street, 106, Tambov, 392000, Russia

Abstract. The effect of fluorination conditions for carbon nanotubes (CNTs) on their chemical composition and structure was studied by IR-Fourier, X-ray photoelectron and Raman spectroscopy, X-ray diffraction analysis, thermogravimetry and electron microscopy. The CNT treatment using fluorine does not cause noticeable damages to side walls and a pronounced increase in the degree of defectiveness, but contributes to an increase of up to 2.26 times in the specific surface area. From the viewpoint of a reinforcing effect, when introduced into a polymer matrix based on a Bisphenol-F-type epoxy resin, the fluorinated CNTs are noticeably more effective than the initial ones. At the fluorinated CNT concentration of 0.1 wt.% in the composite, the tensile strength, Young's modulus at stretching, flexural strength and Young's modulus at bending increase by 50, 74, 60 and 66 %, respectively, whereas the thermal stability and electrical conductivity of the composite practically do not change.

1 Introduction

Recently, carbon nanomaterials, especially single-walled and multi-walled carbon nanotubes (CNTs), have been widely used as fillers to improve the properties of polymer composites [1]. In many cases, the tendency of CNTs to agglomerate, low polarity of their surface, and low adhesion to polymer matrices do not allow for achieving the necessary efficiency indicators [2-12]. To enhance the affinity of the fillers to polymers and solvents, various methods of plasma, chemical and mechanical processing are used. One of the most promising ways of functionalizing the CNT surface is direct fluorination using treatment with fluorine gas at elevated temperature [13-17].

In this regard, the purpose of the present work was to study the effect of fluorination conditions on the chemical composition and structural characteristics of CNTs and the strength characteristics of composites based on a Bisphenol-F-type resin modified with initial and fluorinated CNTs.

2 Experimental

After fluorinating the CNTs, a broad diffuse band appears at 1200 cm-1 in the IR spectra (Fig. 1) due to the presence of covalent C-F bonds [1-2], and the intensity of bands in the range of 1720-1550 cm-1 increases, which is apparently caused by the availability of carboxyl groups formed during the fluorination due to the technological impurity of oxygen in fluorine, as reported elsewhere for similar processes of polymer fluorination [1-2].

Another reason may be production of surface defects with increasing process temperature. The nature of the IR spectra changes shows an increase in the number of C-F bonds formed with increasing the duration and temperature of the CNT treatment using fluorine. It was noted that the areas of the bands at 1200 cm-1 for the CNT samples subjected to fluorination for 2 h at 150 and 250 ° C are approximately the same. However, at a higher temperature, more "defective" oxygen-containing groups are formed on the CNT surface.

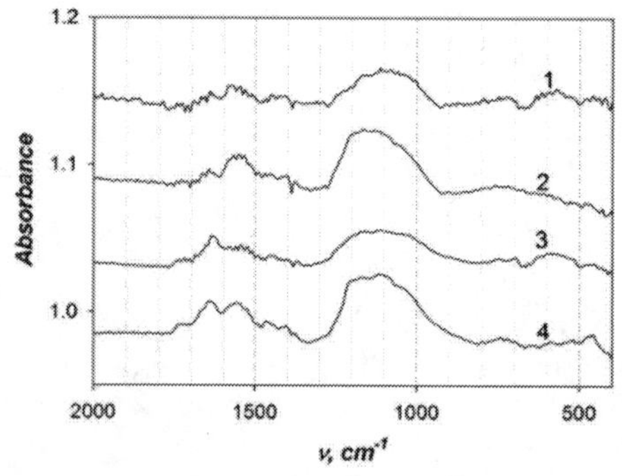

Fig. 1. IR spectra of the initial (1) and fluorinated (2-4) CNTs. Fluorination was carried out by treating the CNTs with undiluted fluorine at a fluorine pressure of 0.7-0.8 atm. Fluorination temperature (°C) and treatment time: 150 ° C and 2 h (2), 150 ° C and 10 min (3), and 250 ° C and 2 h (4).

According to the X-ray photoelectron spectroscopy data, the initial samples include 2 % of atmospheric oxygen. After the direct fluorination at a pressure of 0.7 atm and a temperature of 150 oC for 2 h, the ratio of the atomic fractions of carbon, oxygen and fluorine is 78: 1.3: 20.7. Consequently, the atomic ratio of F/C was 0.265. In the case of formation of chemical bonds between carbon and fluorine, the type of the C1s-spectrum of carbon

[*] Corresponding author: cha-cha@rambler.ru

 DOI: 10.1051/matecconf/201712902041

changes as the fluorine concentration increases (Fig. 2). In that spectrum, several absorption bands can be distinguished at the formation energy of 284.5 (1), 286.2 (2), 288.1 (3), and 289.1 (4) eV. Peak 1 corresponds to a signal from carbon atoms coordinated only by carbon.

Comparing the reference [3-4] data with the data obtained herein, the remaining peaks may be interpreted as follows (Fig. 2):

-peak 4 corresponds to the carbon, which has one bond with fluorine (C-F), the other three bonds are carbon-carbon (C-C*), where C * is carbon having one bond with the fluorine. This indicates the transformation of the sp2-hybridization of the atoms of the initial CNTs into the sp3-hybridized state, since all four bonds of the carbon atom are implemented;

-peak 3 corresponds to the carbon atom, in which one bond is occupied by the fluorine, the other two - by the carbon C * having one bond with the fluorine;

-peak2 corresponds to the carbon atom not directly bound to the fluorine, but having three or four C-C * bonds. If one assumes the presence of four bonds, then the appearance of the fourth bond is possible when disclosing the bond of the sp2-hybridized atom and the formation of the sp3-hybridized bond with the neighboring atom.

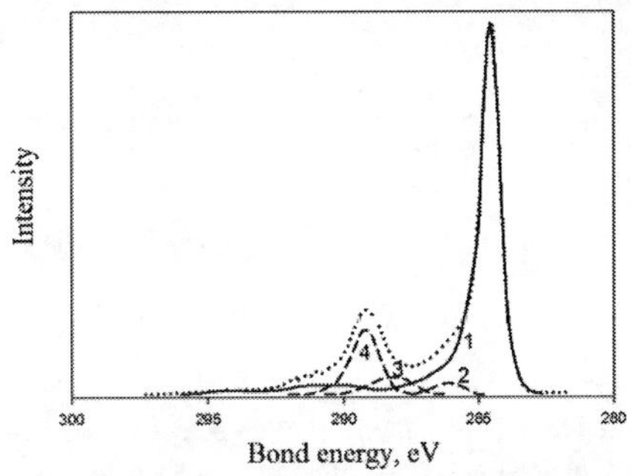

Fig. 2. X-ray photoelectron spectra obtained for the initial CNTs (continuous line) and the CNTs fluorinated at 150 oC CNT (1, dotted line). 2, 3 and 4 are the lines distinguished during the fitting for the fluorinated CNTs. The total spectrum of the fitting is not shown in the figure, since it practically coincides with the spectrum experimentally measured within the line thickness.

The formation of the "fluorine-carbon" bonds has the following peculiarity. They are formed by specific "islets", meaning that the carbon atom is bound to the fluorine atom surrounded by the similar atoms bound to F. The localization of such "islets" is due to the presence of defective sites on the surface of the initial CNTs.

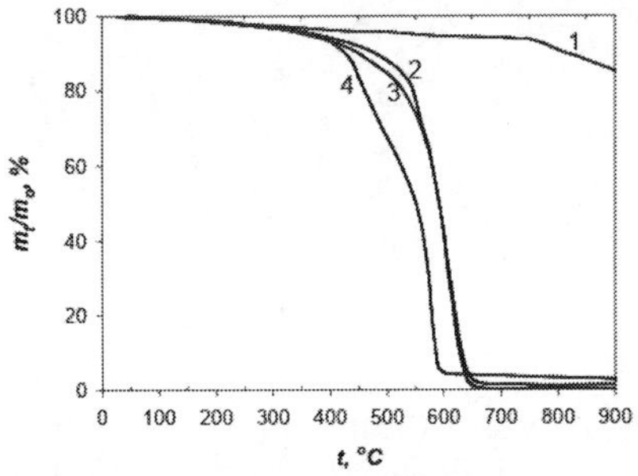

Fig. 3. Thermogravimetric data obtained in argon medium for the initial (1) and fluorinated (2-4) CNTs. Fluorination performed by treatment using undiluted F2 at 0.7-0.8 atm. Temperature and processing time: 150 ° C and 2 h (2), 150 °C and 10 min (3), and 250 °C and 2 h (4).

According to the thermo gravimetric data (Fig. 3), the thermal stability of the initial and fluorinated CNTs at temperatures up to 300-350 °C does not practically differ, and at higher temperatures the functionalized CNTs are less stable, which is associated with the dissociation of the C-F bonds. The CNTs treated with the fluorine at 250 °C (curve 4 in Fig. 3) undergo degradation at a temperature lower than for the CNTs processed at 150 ° C (curves 2-3 in Fig. 3), which is obviously associated with a higher degree of defectiveness.

The effect of the fluorination on the structure and degree of defectiveness was investigated by Raman spectroscopy, transmission electron microscopy (TEM), and X-ray diffraction analysis. The Raman spectra (Fig. 4) and X-ray diffraction patterns of the initial and fluorinated CNTs differ a little from each other, thereby indirectly confirming the previously expressed opinion that after the fluorination, the formation of "carbon-fluorine" bonds occurs at sites of primary defects of the CNT side walls, and no new defects take place.

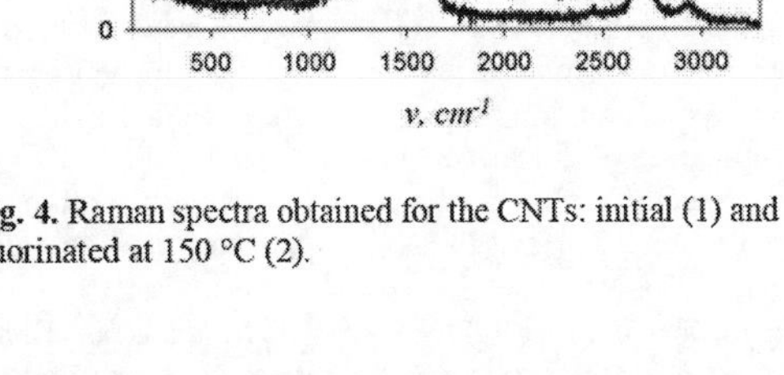

Fig. 4. Raman spectra obtained for the CNTs: initial (1) and fluorinated at 150 °C (2).

 DOI: 10.1051/matecconf/201712902041

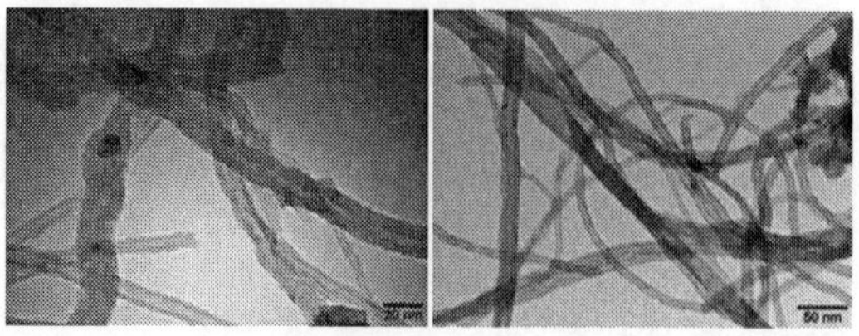

Fig. 5. TEM images of the CNTs fluorinated at 150 °C.

During fluorination, the CNT specific surface area increases (Table 1). The greatest increase, 2.26-fold, was recorded for the sample treated at 150 °C. Increasing the fluorination temperature to 250 ° C or introducing the oxygen into the fluorinating mixture gives a lesser effect. The increase in the specific surface area during the fluorination can positively affect the strengthening effect of the CNTs when incorporating them into the epoxy composition.

Thus, the samples fluorinated at 150 °C have the same degree of fluorination as the CNTs fluorinated at 250 °C, but have a higher specific surface area and fewer defects. Based on the data obtained, it can be assumed that the reinforcing effect of the CNTs fluorinated at 150 °C will be maximum, and further on, to strengthen epoxy composites, these CNTs were used.

Table 1. The fluorination effect on the specific surface area of the "Taunit-M" CNTs. The pressure of the fluorinating mixture is 0.7 atm, and the fluorination time is 2 h.

Sample	Fluorination mixture	Fluorination temperature, °C	Specific surface area, m^2/g
Initial CNTs	-	-	105
Fluorinated CNTs	F_2	250	166
Fluorinated CNT	$80\%F_2+20\%O_2$	150	180
Fluorinated CNT	F_2	150	238

Both the initial and fluorinated CNTs were firstly dispersed directly in the epoxy resin, due to shear flows in a thin gap between rotating rolls, and then sonicated in a thin layer

The dispersion composition of the obtained materials was studied on a Nicomp 308ZLS instrument by the dynamic light scattering method, which makes it possible to estimate the effective particle size of the disperse phase, which would be a true size if these particles were spherical.

Since both the CNTs and the agglomerates formed by them are not spherical, the results of the measurements allow for evaluating only relative changes in the disperse composition of the suspensions analyzed.

As seen from the analysis of the particle effective size distribution diagram (Fig. 6), the suspension containing the initial CNTs is polydisperse. The average effective size of the smallest agglomerates was found to be 418 nm. The second peak shows an effective particle size of about 7 μm, and the number of large agglomerates slightly dominates over the number of smaller ones.

The suspension containing the fluorinated CNTs is monodisperse, and the average effective agglomerate size was found to be about 900 nm, and larger agglomerates cannot be observed. Thus, the CNT fluorination contributes to a better dispersion of these in the epoxy binder.

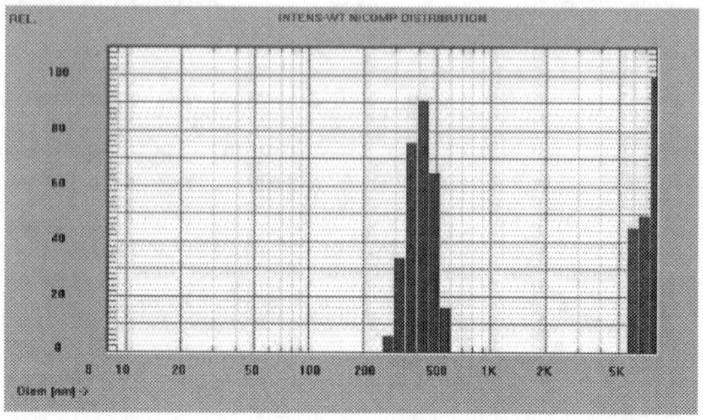

a

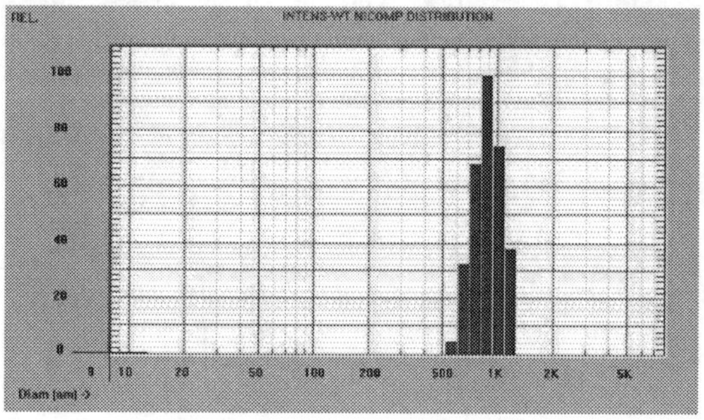

b

Fig. 6. Effective size distribution of the initial (a) and fluorinated (b) CNTs in the BFE-170 epoxy monomer dissolved in isopropyl alcohol. CNT concentration - 0.1 wt.%.

The use of the CNTs leads to a significant increase in the mechanical characteristics of the epoxy composites (Bisphenol-F-type resin and L-19 polyaminoamide-based hardener) on the basis thereof (Table 2, Figs. 7-8).

The fluorinated CNTs provide better effect compared with the initial ones. The greatest effect is achieved with 0.1 wt.% of the fluorinated CNTs in the composite: the tensile strength increases by 50 %, the Young's modulus at stretching increases by 74%, the flexural strength increases by 60%, and the Young's modulus at bending increases by 66%. As can be seen from Figs. 7-8, the results obtained after studying the effect on the tensile and flexural strength and the corresponding moduli (in percent), are superior to the literature data obtained using unmodified single- and multi-walled CNTs, amino-functionalized, oxidized and fluorinated CNTs, grafted epoxy groups, and nano-polyaniline [5,13-18].

Only in one of the above mentioned papers [16], the same increase in the tensile strength as in the experiments described herein, is reported; however, the amount of multi-walled amino-functionalized nanotubes introduced into the epoxy composition is 1%. In our experiments, the

DOI: 10.1051/matecconf/201712902041

same effect was achieved at the fluorinated CNT concentration of 0.1 %, i.e., 10 times less than in [16].

Table 2. Mechanical properties of the epoxy composites

Concentration of CNTs introduced into the composite, wt.%	Tensile strength, MPa	Young's modulus, MPa	Breakig elongation,%	Flexural strength, MPa	Young's modulus, MPa
0 (unfilled composite)	18.2±1.09	849±51	4.3±0.22	32±1.60	933±47
0.01 % of initial CNTs	20±1.1	1077±44	4.2±0.2	42±1.7	1100±55
0.1 % of initial CNTs	24±1.2	1137±46	4.9±0.2	45±1.8	1251±50
0.5 % of initial CNTs	23±1.1	1245±50	4±0.16	43±1.7	1290±52
0.01 % of fluorinated CNTs	22±0.88	1352±54	4±0.15	45±1.75	1187±48
0.1% of fluorinated CNTs	27±1.05	1479±56	3.5±0.14	51±2.0	1550±59
0.5% of fluorinated CNTs	25±1.0	1588±65	3.2±0.13	48±1.8	1611±61

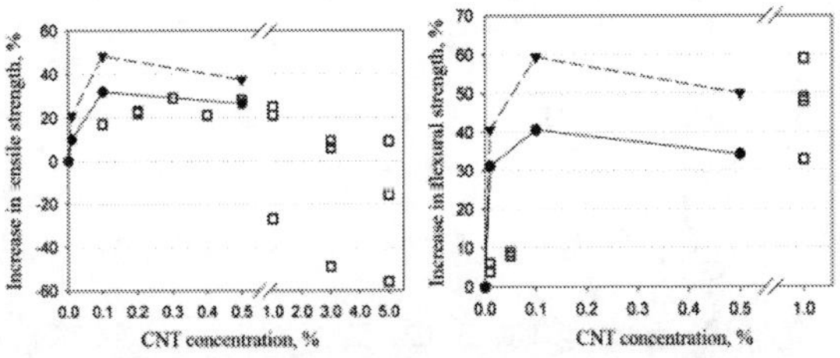

Fig. 7. Dependences of the increase in the tensile and flexural strength on the concentration of the initial (circles, continuous curve) and fluorinated (triangles, dashed curve) CNTs (wt.%). Unfilled squares refer to literature data on hardening using various nanomaterials [5,13-17].

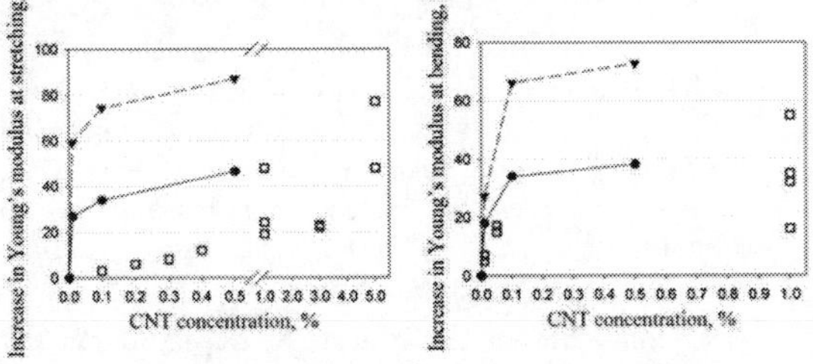

Fig. 8. Dependences of the increase in the Young's modulus at stretching and bending on the concentration of the initial (circles, continuous curve) and fluorinated (triangles, dashed curve) CNTs (wt.%). Unfilled squares refer to literature data [5,13-17].

According to the TGA data, the unfilled composites based on the Bisphenol-F-type resin begin to collapse at about 300 °C (1, Fig. 9). The introduction of the initial CNTs results in a significant decrease in the thermal stability of the composite (2, Fig. 9). On the contrary, the availability of the fluorinated CNTs in the composite practically does not lead to a deterioration of the thermal stability of the composite (3, Fig. 9) at temperatures up to 300 °C, very slightly degrades the thermal stability in the range of 300-370 °C, and slightly improves it in the range of 380-430 °C.

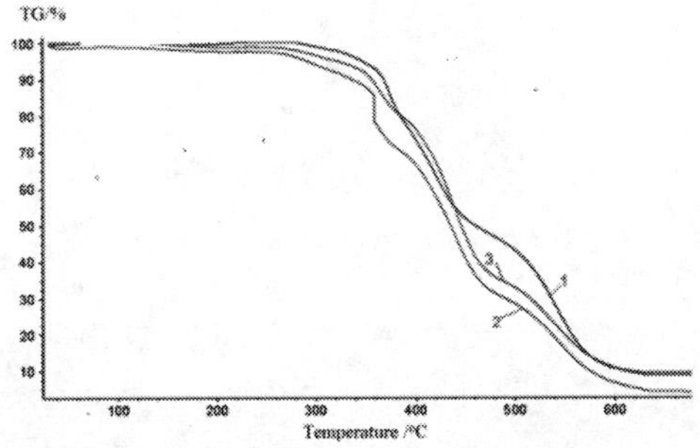

Fig. 9. TGA data obtained in air medium for the epoxy composites: (1), filled with 0.1 wt.% of the initial CNTs (2), and filled with 0.1 wt.% of the fluorinated CNTs (3).

The results on the measurement of the electrical resistivity of the composites are given in Table. 3. In the percentage region of up to 0.5 wt. %, the CNTs cause to a decrease in the resistivity of the epoxy bisphenol matrix from 274 to 6 MOhm.m. The introduction of the fluorinated CNTs does not lead to an improvement in the electrical conductivity of the epoxy bisphenol matrix, and the composite remains insulator.

Table 3. The effect of the CNT additive on the resistivity of the epoxy composite.

	Initial CNTs			Fluorinated CNTs		
Concentration of CNTs introduced into the epoxy matrix	0.01%	0.1%	0.5%	0.01%	0.1%	0.5%
Resistivity	274 MOhm m	19 MOhm m	6 MOhm m	1.4 GOhm m	1.4 GOhm m	1.4 GOhm m

The analysis of the SEM images presented in Figs. 10-12 shows that the rupture surfaces of all the composites have a structure close to each other, independent of the presence or absence of the initial and fluorinated CNTs.

For the filled composites, the CNTs were not found on the rupture surface, which may be explained by fairly good adhesion of both the initial and fluorinated CNTs to the epoxy matrix. This statement lies in agreement with the fact that these CNTs harden the polymer composition at the concentration of 0.1 % (Fig. 7). However, it can be emphasized once again that the fluorinated CNTs are more effective for hardening the polymer composition than the initial ones.

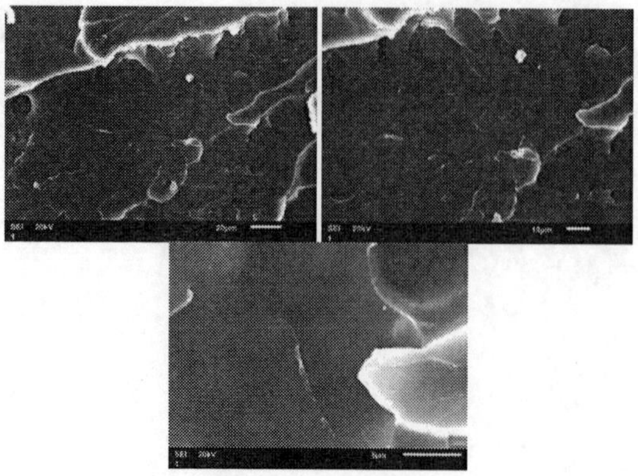

Fig. 10. SEM images of rupture surfaces of the unfilled composite.

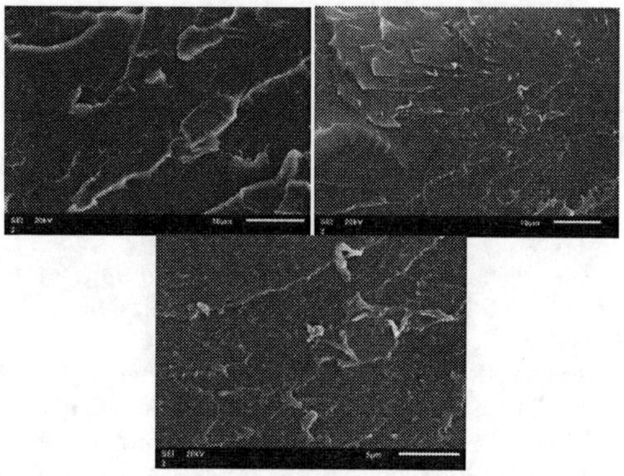

Fig. 11. SEM images of rupture surfaces of the composite containing 0.1 wt. % of the initial CNTs.

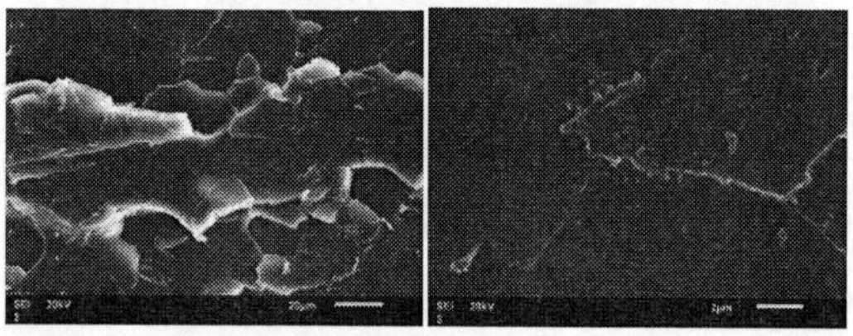

Fig. 12. SEM images of rupture surfaces of the composite containing 0.1 wt. % of the fluorinated CNTs.

3 Conclusion

Both the initial and fluorinated CNTs can be effectively used for hardening epoxy resin composites when their concentration in those materials is about 0.1 wt.%. The tensile and flexural strength and Young's modulus are significantly improved. The fluorinated CNTs are more effective as fillers compared with the initial ones.

This was confirmed by mechanical tests and electron microscopy analyses. Owing to the increase in the CNT specific surface area during the fluorination and formation of polar functional groups, the adhesion between the matrix (epoxy resin) and the filler is increased, thereby contributing to the improvement in the mechanical properties of the composites.

In this regard, the CNTs treated with fluorine at 150 °C are the most efficient. An increase in the fluorination temperature reduces the strengthening effect of the CNTs, which may be due to damages of their surface

The achieved tensile strength and flexural strength of the epoxy composites based on the "Bisphenol-F"-type epoxy resin reinforced with the fluorinated CNTs significantly exceeds the available literature data for similar materials when the CNT concentration is less than 5 wt.%. The fluorinated CNTs used as fillers did not degrade the thermal stability of the composites and did not affect the dielectric properties.

The research was funded by the Russian Science Foundation under Grant No. 15-13-10038.

References

1. Y.-S. Lee. T.H. Cho. B.K. Lee. J.S. Rho. K.H. An. Y.H. Lee, Surface properties of fluorinated single-walled carbon nanotubes,J. Fluorine Chem, **120** (2003)
2. G. Socrates, *Infrared Characteristic Group Frequences/3rd edition, Chichester* (N.Y.; Brisbane; Weinheim, Toronto, Brisbane, Singapore: John Wiley & Sons, 2001)
3. J.-M. Lee, S.J. Kim, J.W. Kim, P.H. Kang, Y.C. Nho, Y.-S. Lee, *A high resolution XPS study of sidewall functionalized MWCNTs by fluorination*, J. Industr. Engng. Chem., **15** (2009)
4. I. Crassous, H. Groult, F. Lantelme, D. Devilliers, A. Tressaud, **C.** Labrugere, M. Dubois et al., *Study of the fluorinationof carbon anode in molten KF-2HF by XPS and NMR investigations*, J. Fluorine Chem., **130** (2009).
5. X. Zhang, Q. He, H. Gu, H.A. Colorado, S. Wei, Z. Guo, *Flame-etardant Electrical Conductive Nanopolymers Based on Bisphenol F Epoxy Resin Reinforced with Nano Polyanilines*, ACS Appl. Mater. Interfaces., V. **5**, P. 898–910 (2013)
6. M. Moniruzzaman, F. Du, N. Romero, K.I. Winey, *Increased flexural modulus and strength in SWNT/epoxy composites by a new fabrication method*, Polymer **47** (2006)
7. Zhengtao Yang, K. McElrath, J. Bahr, N.A. D'Souza, *Effect of matrix glass transition on reinforcement efficiency of epoxy-matrix composites with single walled carbon nanotubes, multi-walled carbon nanotubes, carbon nanofibers and graphite*, Composites: Part B. **43** (2012)
8. R. Chaos-Moran, A. Salazar, A. Urena, *Mechanical Analysis of Carbon Nanofiber/Epoxy Resin Composites*, Polymer Comp. **32** (2011)
9. Y. Zhou, F. Pervin, S. Jeelani et al., *Improvement in mechanical properties of carbon fabric-epoxy composite using carbon nanofibers*, J. Mater. Processing Technol. **198** (2008)
10. Schulte. Influence of different carbon nanotubes on the mechanical properties of epoxy matrix

compos*ites - A comparative study*, Comp. Sci. Techn. **65** (2005)

11. L.Q. Liu, H.D. Wagner, *Rubbery and glassy epoxy resins reinforced with carbon nanotubes*, Comp. Sci. Techn. **65** (2005)
12. G.S. Baronin, D.V. Pugachev, D.E. Kobzev, P.V. Kombarova, *Investigation of strength properties and impact viscosity of ceramic polymeric composites based on abs-copolymer and uhmwpe, liquid and solid-phase extrusion*, Bulletin of TSTU. **17** (2011)
13. R.G. de Villoria, A. Miravete, J. Cuartero, A. Chiminelli, N. Tolosana, *Mechanical properties of SWNT/epoxy composites using two different curing cycles*, Composites Part B: Engng. **37** (2006)
14. Z. Wanga, H.A. Colorado, Z.-H. Guo, H. Kim, C.-L. Park, H. T. Hahna, S.-G. Lee, K.-H. Lee, Y.-Q. Shang, *Effective Functionalization of Carbon Nanotubes for Bisphenol F Epoxy Matrix Composites*, Materials Research. **15** (2012)
15. D.H. Wang, S. Sihn, A.K. Roy, J.-B. Baek, L.-S. Tan, *Nanocomposites based on vapor-grown carbon nanofibers and an epoxy: Functionalization, preparation and characterization*, Europen Polymer J. **46** (2010)
16. Y. Zhou, F. Pervin, L. Lewis, S. Jeelani, *Experimental study on the thermal and mechanical properties of multiwalled carbon annotube-reinforced epoxy*, Materials Sci. Engng. **452** (2007)
17. M. Theodore, M. Hosur, J. Thomas, S. Jeelani, *Characterization of epon 862 reinforced withfunctionalized,* MCNT's// http://www.iccmcentral.org/Proceedings/ICCM16proceedings/ contents/pdf/MonA /MoAA2-01sp_theodorem222779p.pdf
18. I.V. Anosova, M.O. Telegin, *Development of technology of modification of cnt series "Taunit" by polyaniline. Problems of technogenic security and sustainable development: Sat. sci. art. young scientists, graduate students and students*, TSTU, **4** (2013)

DOI: 10.1051/matecconf/201712902042

Theoretical and experimental investigations of the formation mechanisms of residual deformations of fibrous layered structure composites

Vitaly N. Paimushin[1,2], *Sergey A.* Kholmogorov[1], and *Ildar B.* Badriev[2,*]

[1]Kazan Tupolev National Research Technical University, 420111, 10 Karl Marx Str., Kazan, Russia
[2]Kazan Federal University, 420008, 18 Kremlyovskaya Str., Kazan, Russia

Abstract. On the example of a unidirectional fibrous composite made of unidirectional carbon fibre composite and cold-hardening epoxy XT-118, tests of specimens with a stacking scheme $[+45^0/-45^0]_{2s}$, where s is the number of monolayers, were carried out for cyclic loading by tension and compression. When the samples are stretched, both the fibers and the epoxy of composite are under shear stresses and stretching in the transverse direction by normal stresses, and under compression, under shear stresses and compression by normal stresses across the fibers. It is shown that in each loading cycle there is a consolidation of the material associated with the rearrangement of its microstructure, an increase in the secant modulus of elasticity and a decrease in the magnitude of the deformation increment at each cycle. The mechanism of accumulation of residual deformations under multiple loading of test samples was qualitatively studied and their level formed in the composite after each loading cycle was quantitatively found.

1 Statement of the problem

In the products of space technology for special purposes, nowadays, structural elements from fibrous composite materials (CM), in particular, carbon plastic, are widely used. Such structural elements in the process of operation under multiple (multi-cycle) loading (in particular, thermal loading) should preserve the stability of geometric sizes and shapes. It was found that when ground-based testing of such products and during their operation, distortions of the measured parameters are always fixed, apparently due to the formation of residual deformations in the design. All the formation mechanisms of such residual deformations in fibrous composites in the process of exploitation of the structural elements made from them have not yet been fully revealed, which raises an urgent problem in the world science in the field of mechanics and machine building. We note that a numerical study of the geometrically and physically nonlinear problem of the three-point bending of test samples from layered fibrous composites having a rectangular cross-section was carried out in [1]. The problems of the bending of sandwich plates with a transversally soft core are discussed in [2-6].

2 Experiments on cyclic deformation

An extensive scientific literature [7–12], etc is devoted to the study of nonlinear deformation under cyclic loading of unidirectional fibrous composites. In particular, in work [12] such a study is carried out for a unidirectional composite on the basis of glass fibers under shearing conditions and repeated cyclic loading of specimens with a stacking scheme $[+45^0/-45^0]_{2s}$ (s is the number of monolayers). On the basis of the obtained experimental data, the secant shear modulus G_{12} is calculated and a conclusion is made about its fall after loading the sample with ten cycles. The authors explain the fall of the shear modulus by the degradation of the composite under conditions of shear deformations forming in the composite.

The carried out studies on the shear of test samples from fibrous composites with a polymer matrix show that the dependences between the tangential stresses and the corresponding shear deformations are essentially nonlinear, under conditions of loading and subsequent unloading, differ significantly among themselves, with residual deformations always being fixed on them. In particular, such studies of nonlinear deformation of a unidirectional fibrous composite on the basis of unidirectional carbon fibre composite and XT-118 epoxy under shearing conditions and repeated loading by tensile and compressive testing of specimnes with a stacking scheme $[+45^0/-45^0]_{2s}$ were carried out. For tensile tests, specimens $[+45^0/-45^0]_{2s}$ (s=2) with an average thickness of h=0.56 mm, a width of b=24.60 mm, and a working part length of l=110 mm were used. For compression tests, the specimens were manufactured with an average thickness of h=4.126 mm, a width of b=24.91 mm and a length of the working part of l=25 mm. The tests were carried out two weeks after the manufacture of the test samples using the technology

* Corresponding author: ildar.badriev1@mail.ru

described in [13] and eight months after their manufacture, when the polymerization of the binder composite could be considered complete.

When the stack of layers with the above laying scheme is stretched, both the fibres and the epoxy in the orthotropic axes are stretched in the direction of the fibres and the shearing by tangential stresses, and under compression, under compression and shearing by shearing stresses. The experiment was carried out under conditions of the same maximum normal stresses σ_x in each "load-unload" cycle. This type of loading is sufficiently accurately equivalent to the work of a composite in a real construction, where the stresses should not exceed a certain specified (operational) value. At the previously found value of the ultimate (failure) stress σ_x^B in the case of tensile tests, the specimens were brought to stresses $\sigma_x^{\max}$=0, the magnitude of which was $\sigma_x^{\max}$=65 MPa, by repeated loading. Measuring of axial deformations was carried out with the help of a contact extensometer with a base of measurement 50 mm. The deformation diagrams obtained under conditions of tension with the cycles "load-unloading" with the speed of the crosshead movement of 0.5 mm/min are shown in Fig. 1 for samples [+45^0/-45^0]$_{2s}$, s=2, that were kept eight months after their manufacture, and for the same specimens held for two weeks after manufacturing, are shown in Fig. 2.

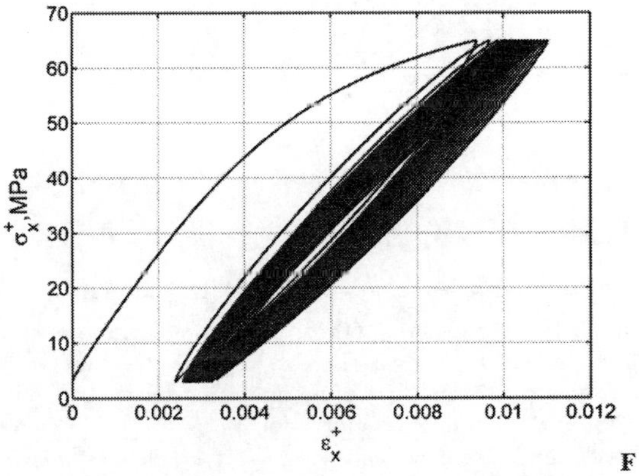

F

Fig. 1. Cyclic stretching diagram ($\sigma_x^{\max}$=65 MPa, exposure for eight months).

Comparing the results, it can be seen that holding the specimens at room temperature for eight months led to an increase in their rigidity by almost half. The loading and unloading branches do not coincide, forming a loop of hysteresis, and the axial strain ε_x on each subsequent cycle with the same σ_x in the cycle increases by a value of $\Delta\varepsilon_x^{(k)}$ (Fig. 1), where k is the cycle number. As the number of cycles $\Delta\varepsilon_x^{(k)}$ increases, this indicates the process of hardening the fibrous composite at each loading cycle. Such a process is called the process of adaptability of a composite material. It should be pointed out that for a given $\sigma_x^{\max} < \sigma_x^B$ there is a limit number of cycles k=N at which $\Delta\varepsilon_x^{(N)} \to \Delta\varepsilon_x$. Magnitude of $\Delta\varepsilon_x$ doesn't depend of cycle's number, but depends of loading rate.

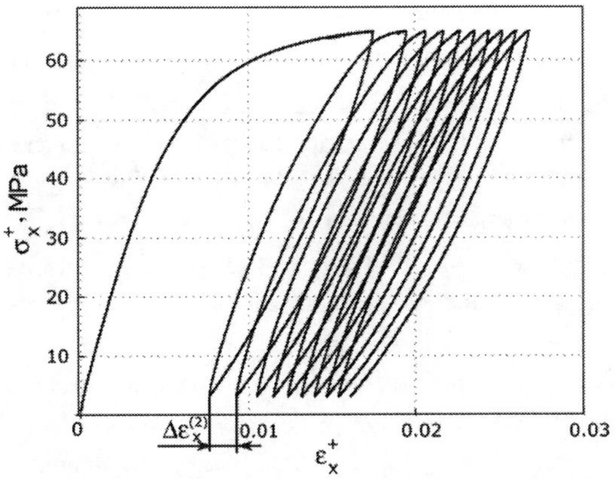

Fig. 2. Cyclic stretching diagram ($\sigma_x^{\max}$=65 MPa, exposure for two months).

The deformation diagram $\sigma_x^+ = \sigma_x^+(\varepsilon_x)$ for the first two loading-unloading cycles for clarity on modified scales is shown in Fig. 3 with the limit points A_1, A_2 corresponding to $\sigma_x^{\max}$=65 MPa, and B_1, B_2 corresponding to $\sigma_x^{\max}$=4 MPa. These diagrams have the form of straight lines passing through the points $0A_1$, A_1B_1, B_1A_2, A_2B_2.

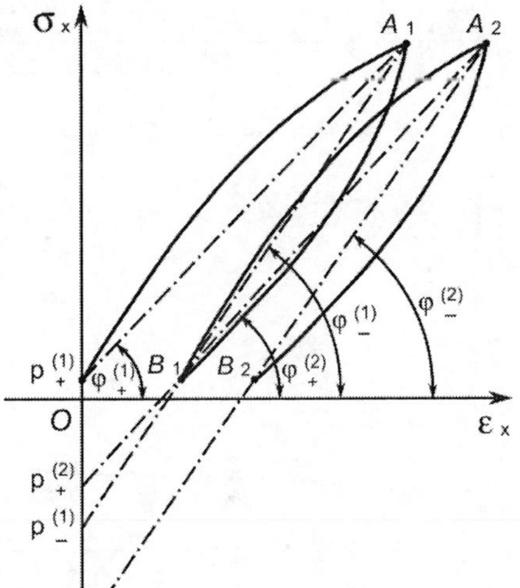

Fig. 3. Cyclic deformation scheme ($\sigma_x^+ = \sigma_x^+(\varepsilon_x)$).

The angles between these lines and the axis $0\varepsilon_x$ are also shown, and the table 1 shows the values

 DOI: 10.1051/matecconf/201712902042

$E_+^{(i)} = \text{tg}\varphi_+^{(i)}$, $E_-^{(i)} = \text{tg}\varphi_-^{(i)}$ representing the secant modulus of elasticity at the i-th load cycle (with the index "+") and unloading (with the "-" index). In addition, the values $p_+^{(i)}$, $p_-^{(i)}$ indicating coordinates of intersection points of the straight lines A_iB_i, B_iA_{i+1} with the axis $0\sigma_x^+$ are given. The changes of values $\Delta\varepsilon_x^{(i)}$ on the first ten loading cycles can be traced through the table, and the dependence $\Delta\varepsilon_x^{(i)} = \Delta\varepsilon_x^{(i)}(i)$ is shown in Fig. 4. Analyzing the above results, it can be seen that even on the tenth cycle with the highest degree of accuracy the equalities $E_+^{(10)} \approx E_-^{(10)} \approx 8.31$ GPa, $\Delta\varepsilon_x^{(10)} \approx 0$, $p_+^{(10)} \approx p_-^{(10)} \approx -18.5$ MPa are fulfilled. Results similar to those described above were also obtained at $\sigma_x^{\max}$=55 MPa and $\sigma_x^{\max}$=45 MPa. It has been established that as the reduction decreases of $\sigma_x^{\max}$=65 MPa, a smoother decrease in values of $\Delta\varepsilon_x^{(i)}$ is observed, as well as a slight increase of the parameters $E_+^{(i)}$, $E_-^{(i)}$, $p_-^{(i)}$.

Table 1. The results of the tests for the multicyclic stretching.

i	$E_+^{(i)}$, GPa	$E_-^{(i)}$, GPa	$\Delta\varepsilon_x^{(i)} \cdot 10^3$	$p_+^{(i)}$, MPa	$p_-^{(i)}$, MPa
1	6.59	8.9	2.39	3.45	-18.14
2	8.48	8.84	0.29	-17.12	-20.64
3	8.59	8.69	0.09	-19.98	-20.91
4	8.6	8.63	0.01	-20.67	-20.95
5	8.56	8.51	-0.03	-20.75	-20.32
6	8,.9	8.47	-0.01	-20.26	-20.08
7	8.47	8.44	-0.004	-20.08	-19.72
8	8.45	8.4	-0.004	-19.74	-19.27
9	8.42	8.34	-0.005	-19.31	-18.6
10	8.31	8.31	-0.001	-18.52	-18.53

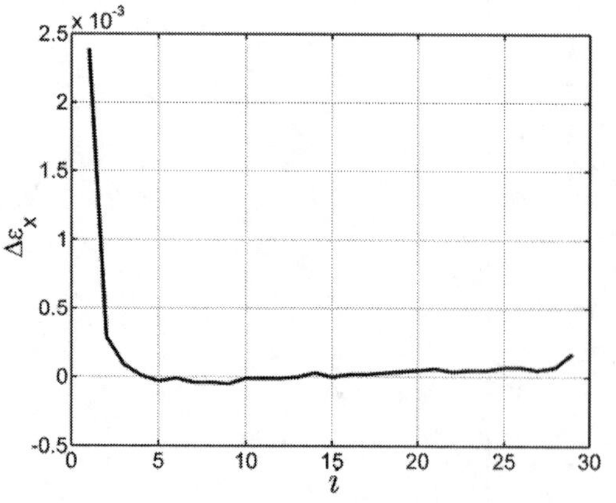

Fig. 4. Deformation increments at each loading cycle ($\sigma_x^{\max}$=65 MPa)

Figure 5 shows the deformation diagram for compression tests of specimens held for two weeks after their manufacture. The speed of traversing the machine in the experiment was 0.5 mm / min. It should be noted that the axial strain $\varepsilon_x^- \approx 0.02$ at the first loading cycle corresponds to a higher value of the normal stresses $\sigma_x^{\max}$=-80 MPa, which is 19% higher than $\sigma_x^{\max}$=65 MPa for the same $\varepsilon_x^- \approx 0.02$. The main regularities of the curve $\sigma_x^- = \sigma_x^-(\varepsilon_x^-)$ (Fig. 2) remained similar to the curve $\sigma_x^+ = \sigma_x^+(\varepsilon_x^+)$ (Fig. 5).

In order to determine the tangential shear modulus as a function $G_{12}^{(k)} = G_{12}^{(k)}(\gamma_{12})$, where γ_{12} is shearing strain, and to study its variation in each loading cycle of the tested specimens of the diagrams $\sigma_x^+ = \sigma_x^+(\varepsilon_x^+)$ and $\sigma_x^- = \sigma_x^-(\varepsilon_x^-)$ under tension (Fig. 2) and compression (Fig. 5), they were processed in accordance with the technique proposed in [14].

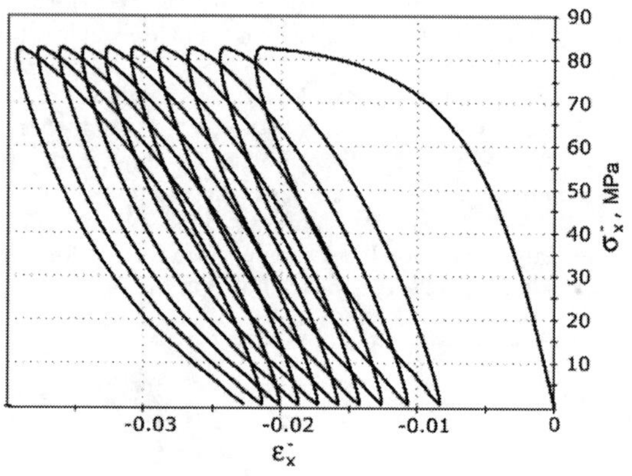

Fig. 5. Cyclic compression diagram ($\sigma_x^{\max}$=-80 MPa, exposure for eight months).

The dependencies $G_{12}^{(k)} = G_{12}^{(k)}(\gamma_{12})$ for the first and the tenth loading cycles (both in tension and compression) are shown in Fig. 6. In this figure, the line –□-• corresponds to the first cycle under tension, –▲–□□ to the first cycle under compression, – –•– – □to the tenth cycle under tension and – –▲– – to□ the tenth cycle upon compression.

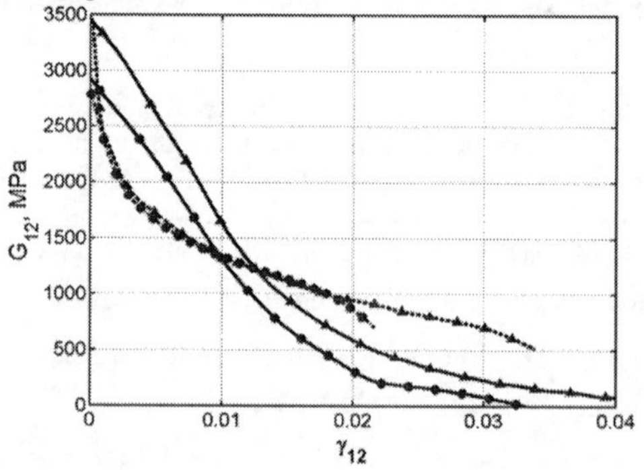

Fig. 6. Dependencies $G_{12}^{(k)} = G_{12}^{(k)}(\gamma_{12})$ for the first and the tenth loading cycles.

It can be seen that the curve $G_{12}^{(k)-} = G_{12}^{(k)-}(\gamma_{12})$ lies above the curve $G_{12}^{(k)+} = G_{12}^{(k)+}(\gamma_{12})$ over the entire range. The value of $G_{12}^{(1)-}$ at $\gamma_{12} = 0$ is 3400 MPa, and value of $G_{12}^{(1)+}$ □ is 2750M Pa. This difference between the two curves remains practically constant throughout the entire range of variation γ_{12}. The curves $G_{12}^{(10)-}$ and $G_{12}^{(10)+}$ practically coincide on all the range of γ_{12} presented in the tests, and on the sector $0 < \gamma_{12} < 0.01$ they are below the curves $G_{12}^{(1)-}$ and $G_{12}^{(1)+}$, and in the section $\gamma_{12} > 0.01$ □ they are above the curves $G_{12}^{(1)-}$ and $G_{12}^{(1)+}$. Curves $G_{12}^{(10)-}$, $G_{12}^{(10)+}$, $G_{12}^{(1)-}$ and $G_{12}^{(1)+}$ intersect in a neighborhood $\gamma_{12} \approx 0.01$. The diagrams for tension and compression for the first and subsequent cycles are non-linear over the entire range of variation. The diagrams for tension and compression for the first and subsequent cycles are non-linear the whole range of change γ_{12}.

3 Conclusion

In contrast to the effects known for loading multilayer composites in a direction transverse to the fibers and describing their models (developed by V.V. Vasilyev and then by A.N. Alfutov and co-authors), under repeated loading and unloading by tangential stresses on each cycle, a change of deformation diagrams is observed. The effect has been established experimentally in that there is such a limiting amplitude value of the stresses formed during testing, to which the parameters of the deformation diagrams stabilize, when exceeded, the hysteresis loop widens up to failure. The loss of structural strength is caused either by the destruction of elements in a sufficiently large area, or by the development of cracks. In fibrous composites, their failure is caused, in particular, by loss of stability or reinforcing fibers at the micro-level, or a fiber bundle (monolayer) at the meso-level, or the entire structural element at the macro-level, often accompanied by stratification of the composite structure. This effect, as a rule, is visually observed in experiments both in the compression of test samples and in their three-point bending. These types of tests are currently regulated by both Russian and international standards. The process of destruction caused by the described effects of the first two species, in the mechanics of composite materials and structures made of them, is practically not taken into account in modeling and analysis of their strength. Therefore, the actual problem is the creation of refined models of deformation and destruction of multilayer composites, taking into account the possibility of realizing non classical buckling of fibers at the micro-level (in scales of the cross-sectional dimension of the fiber), a fiber bundle at the meso-level (in the scale of one monolayer), taking into account the interaction of fibers and a fiber bundle with their surrounding matrix, as well as shear and flexural-shear buckling at the macro-level (in the scale of the thickness of structural elements in the form of a plates and shells and cross-sectional dimensions of structural elements in the form of rods), theoretical modeling of tests of specimens from fibrous composites for single and multiple stretching, compression, stretching-compression and three-point bending with appropriate experimental studies, as well as significant refinement of fiber composite testing standards with experimental confirmation of the results of theoretical studies. All these studies are necessary to show that in the process of deformation of structural elements from laminated fibrous composites in monolayers in which a shear stressed-deformed state is formed, structural changes occur in the composite both under loading and unloading of the structure, leading to the formation of Residual deformations. It is also necessary to identify and theoretically describe the mechanism for the formation of such structural changes.

The results of the research are obtained within the framework of the state task of the Ministry of Education and Science of Russia No. 9.5762.2017, project No. 9.1395.2017 and due to the grant of the Russian Scientific Foundation (project No. 16-11-10299).

References

1. I.B. Badriev, M.V. Makarov, V.N. Paimushin, *Mathematical Simulation of Nonlinear Problem of Three-point Composite Sample Bending Test, Procedia Engineering*, **150**, 1056–1062 (2016)
2. I.B. Badriev, M.V. Makarov, V.N. Paimushin, *Numerical Investigation of Physically Nonlinear Problem of Sandwich Plate Bending, Procedia Engineering*, **150**, 1050–1055 (2016)
3. I.B. Badriev, M.V. Makarov, V.N. Paimushin, *Solvability of a physically and geometrically nonlinear problem of the theory of sandwich plates with transversal-soft core*, Russian Mathematics, **59**, 10, 57–60 (2015)
4. I.B. Badriev, M.V. Makarov, V.N. Paimushin, *Contact statement of mechanical problems of reinforced on a contour sandwich plates with transversally-soft core*, Russian Mathematics, **61**, 1, 69–75 (2017)
5. I.B. Badriev, G.Z. Garipova, M.V. Makarov, V.N. Paimushin, R.F. Khabibullin, *Solving physically nonlinear equilibrium problems for sandwich plates with a transversally soft core*, Lobachevskii Journal of Mathematics, **36**, 4, 474–481 (2015)
6. I.B. Badriev, M.V. Makarov, V.N. Paimushin, *On the interaction of composite plate having a vibration-absorbing covering with incident acoustic wave*, , Russian Mathematics, **59**, 3, 66–71 (2015)
7. S. Ogihara, S. Kobayashi, K.L. Reifsnider, *Characterization of nonlinear behavior of carbon/epoxy unidirectional and angleply* laminates,

Advanced Composite Materials: The Official Journal of the Japan Society of Composite Materials, **11**, 3, 239–254 (2003)

8. T.A. Bogetti, C.P.R. Hoppel, V.M. Harik, J.F. Newill, B.P. Burns, *Predicting the nonlinear response and progressive failure of composite laminates*, Composites Science and Technology, **64**, 3-4, 329–342 (2004)
9. M. Kashtalyan, C. Soutis, *Stiffnes degradation in cross-ply laminates damaged by transverse cracking and splitting*, Composites: Part A, **31**, 335–351 (2000)
10. A.M. Dumansky, L.P. Tairova, *The prediction of viscoelastic properties of layered composites on example of cross ply carbon reinforced plastic*, World Congress on Engineering, **II,** London, UK 2-4 July, 2007, 1346–1351 (2007)
11. W.Van Paepegem, I. De Baere, J. Degrieck, *Modelling the nonlinear shear stress-strain response of glass fiber-reinforced composites. Part I: Experimental results*, Composite Science and Technology, **66**, 1455-1464 (2006).
12. A.M. Dumansky, L.P. Tairova, I. Gorlach, M.A. Alimov, *A design-experiment study of nonlinear properties of coal-plastic*, Journal of Machinery Manufacture and Reliability, **40**, 5, 483-488 (2011).
13. R.A. Kayumov, S.A. Lukankin, V.N. Paimushin, S.A. Kholmogorov, *Identification of mechanical properties of fiber-reinforced composites,* Uchenye Zapiski Kazanskogo Universiteta. Seriya Fiziko-Matematicheskie Nauki, **157**, 4, 112–132 (2015).
14. V.N. Paimushin, D.V. Tarlakovskii, S.A. Kholmogorov, *On non-classical buckling mode and failure of composite laminated specimens under the three-point bending*, Uchenye Zapiski Kazanskogo Universiteta. Seriya Fiziko-Matematicheskie Nauki, **158**, 3, 350–375 (2015).

MATEC Web of Conferences **129**, 02043 (2017) DOI: 10.1051/matecconf/201712902043
ICMTMTE 2017

Determination of parts tolerances in friction knots of grey and nodular cast iron with thermal diffusion platings vanadium and chrome

Alexander A. Veselovsky[1], *Roman R.* Dema[2], *Olga B.* Kalugina[2,*], and *Sergey P.* Nefedyev[3]

[1] Head of research center JSC KONAR, Yenisei street 8, 454000, Chelyabinsk ,Russia
[2] Federal state Budgetary Institution of Higher Education "Nosov Magnitogorsk State Technical University" (FSBIHE "NMSTU"), Lenin Avenue, 38, Magnitogorsk city, 455000, Russia.
[3]Limited Liability Company «Factory of lubricants» OKTAMIX», Leningrad street 49, office 14, 455000, Magnitogorsk, Russia

Abstract. In the study, experiments with a vanadium coating of samples were conducted, to establish the limits of the increase in the landing dimensions and the designation of limiting deviations in dimensions and the designation of limiting deviations in the dimensions of the manufactured parts. The maximum deviations relative to dimensions of cast iron parts with thermal diffusion platings vanadium and chrome in friction knots of machines and mechanisms were explored. The plate thickness and the linear sizes increment of gray and nodular cast iron hardened products were determined. Were obtained uniform in thickness carbid platings as a result of the experiments on producing of thermal diffusion vanadium and chrome plating.

1 Introduction

Thermal diffusion platings allow to create the highly rigid and wear-resistant platings on parts surface in friction knots by carbide-stabilizing elements. However the diffusion of carbide-stabilizing elements into cast iron presents certain difficulties

Carbon presence on the cast iron surface conduces bonding in carbides of adsorbed diffusants considerably complicating further diffusion deep into a product[1].

Theoretically the diffusion of vanadium and chrome into the cast iron is possible only during the initial period of saturation until their content will not reach the stoichiometric composition for carbide forming.

However considering border-line conditions of first genus it is possible to deduce that the diffusion of these elements deep into the material is inconspicuous.

Therefore the creation of thermal diffusion platings of preset thickness is provided by the counter diffusion of carbon from the deep layers of material to the zone of reactive diffusion[2].

The objective of this work is to determine magnification limits of coupling sizes and the purpose of the maximum deviations relative to dimensions and the the purpose of the maximum deviations relative to dimensions of manufacturing parts with due regard for further thermal diffusion vanadizing and cromodizing by contact process using powder mixture[3].

2 Equipment, materials and the procedure of experiments

The pieces with inside diameter of 20 mm and outer diameter of 40mm from gray iron and nodular cast iron were used for experiment.

Vanadizing was developed using ferrovanadium (60% on mass) inert filler aluminum oxide(40% on mass), and activator NH4Cl (4% on mass, extra 100%), cromodizing was developed using ferrochrome (50% on mass), aluminum oxide (50% on mass) and activator (4% on mass).

The saturation temperature was 1000-1020^{0}C, time of hardening was 5 hours for vanadizing process and 10 hours for chromodizing process. Processes of thermal diffusion saturation was developed at the plant(reworked retort furnace) which scheme is presented in figure 1.

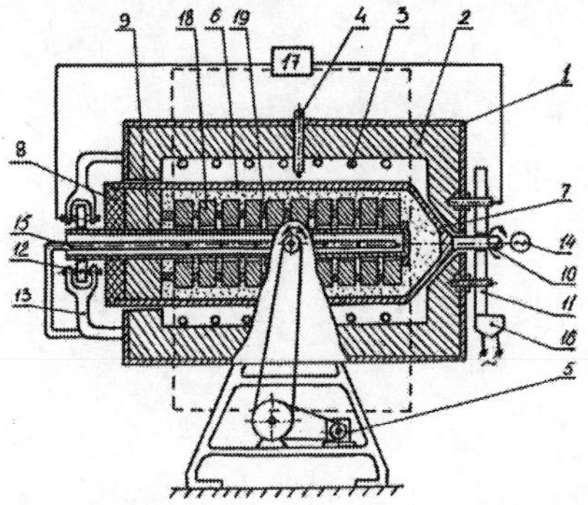

Fig. 1. Structural drawing of the plant for the thermal diffusion saturation.

The plant operation takes place in the following way. The heating chamber 1 with cylinder muffler 2, electric heating unit 3 and heat meter 4 is brought in a vertical

*Corresponding author: kalugina.olga@bk.ru

DOI: 10.1051/matecconf/201712902043

position with a help of the swivelling device 5 (is marked by dashed lines in drawing) when the retort is placed front cover down(with the removed back cover 8).Ring parts 18 are done on closed bottom tube (from gray and nodular iron) closely to each other or through the distance pieces 19 which help to protect some elements of parts from future plating on the one side and to provide the equability of plating on another side. The retort 6 is filled up with working mixture(furnace-charge).

Then the retort 6 is leakproof locked by the back cover 8 with parts are done on close bottom tube. The heating chamber 1 is brought to horizontal position by swivelling device 5. The shaft 10 is fixed by front rollers 11, the closed bottom tube 9 is fixed by back rollers isolated from the heating chamber 1 by electrical isolator 13.

The shaft 10 of the retort 6 is brought to rotation by driving mechanism 14 with rotation frequency 3-10 rpm. The External heat of the retort 6 is performed from muffler 2 by electric heater 3 supplied from variable transformer (is not presented in the drawing). The temperature control of treatment schedule is provided by heat meters 4 installed on the muffler 2, heating chamber 1 and additional heat meters 15 placed is the closed bottom tube 9 and fixed on the heating chamber 1. This temperature control provides uniform heating by length and cut of retort 6 that guaranties the improving quality of parts processing. Under the external heat of the retort 6 rotating on front 11 and back rollers 12, the thermochemical treatment processing of parts 18 goes on.

Additionally to the rotation agitation, with oscillation superposition (low-frequency or ultrasonic) from the resonator 16 to the retort 6 through the front rollers 11 or their axes transformer (are not presented in the drawing) in it create the vibrofluidized bed. Vibroliquefaction is performed with vertical vibration frequency of the retort 6 within the diapason 15-25 Hz and the amplitude 1-3mm. The vibrofluidized bed enables to forcibly lead active substance to the saturation surface increasing the speed of saturation and intensifying diffusion surface alloyage. This provides a high level of gas permeability of the furnace charge that enhances the uniformity of the plating.

3 Results and discussion

As a result of the experiment on the creation of thermal diffusion platings on gray and nodular iron, it was obtained vanadium platings (figure 2) which microhardness came to 26 000 -28 000 mPa.

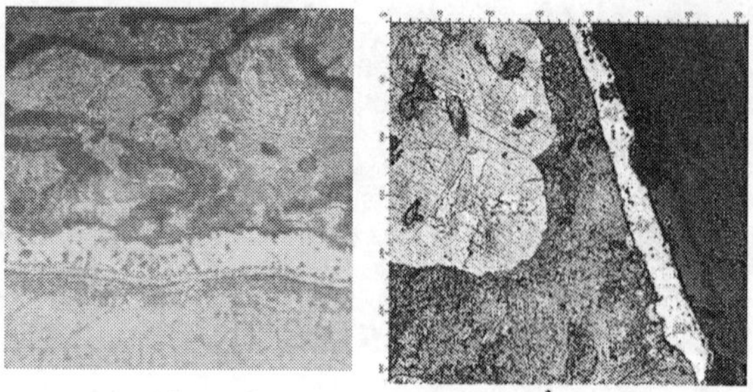

a b

Fig. 2. Microstructure of vanadium plating (x500): on gray (a) and nodular(b) iron.

Microstructure of platings obtained after chromodizing of pieces is presented on Figure 3.

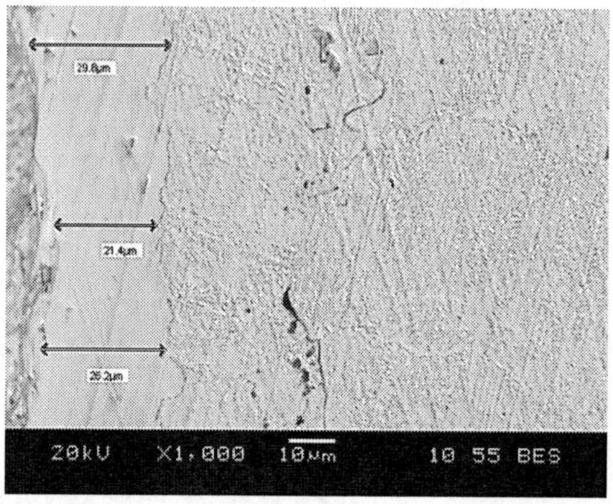

Fig. 3. Microstructure of chromium plating on nodular iron(x1000)

Formation mechanism of thermal diffusion vanadium plating can be explained by scheme presented on figure 4.

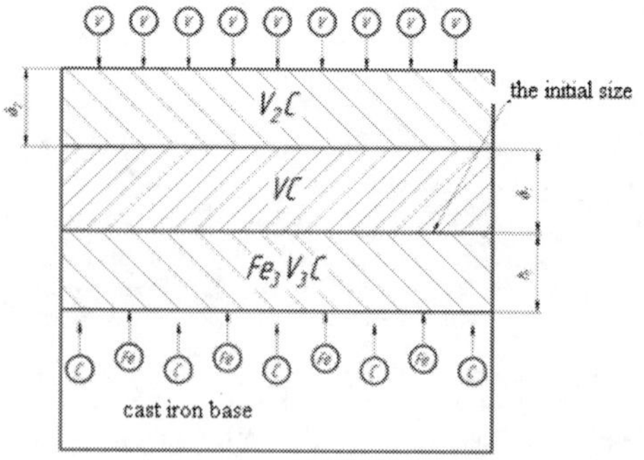

Fig. 4. Scheme of formation of diffusion vanadium plating.

High concentration of vanadium on the iron part surface, low concentration of carbon in this area on one side and lower free energy potential on the other side create favourable conditions for vanadium carbide formation V_2C.

Vanadium carbide VC has higher free energy potential than Vanadium carbide V_2C [5].

The formation of carbide phase slows down the

 DOI: 10.1051/matecconf/201712902043

vanadium diffusion from plating deep into a part, because its atomic mobility of chemical compound is low and insufficient for bond opening and vanadium transferring deep into a part.

The diffusion of ferrum into iron into the area of reactive diffusion slows down for the same reason. As a result the diffusion carbide h_2 –layer is formed in the outer zone of a part and the external edge of layer shifts to the side of the physical growth by a quantity $\delta 1$.

As a result of the carbon counter diffusion, V_2C vanadium carbide transforms into VC vanadium carbide:

$$V_2C + C = 2VC \quad (1)$$

The slowdown in the process of the rejection of a saturation element from outer zone deep into a part will lead to increasing of its concentration on the surface of a part. As a result of the carbon counter diffusion and deposition of a hardening element on a part surface, $h_1$carbide layer is formed, and its external edge shifts by a quantity δ_2[4].Shifting of the external edge of the diffusion layer by a quantity δ_2 will depend on the diffusion depth of the hardening chemical element in the material of h_2 part, parameters difference of ferrum lattice and formed chemical compounds – carbids.

The major change of δ_2 layer sizes takes place as a consequence of active vanadium or chrome formation on the part surface and depends on mass-transfer coefficient of saturated chemical compound from reaction space to a part surface, and also owing to carbon counter diffusion of iron to a diffusion layer and formation of a carbide phase.

Thus, the increment of the linear sizes of a lpart can be determined by a formula:

$$\Delta d = \delta_1 + \delta_2 \quad (2)$$

The layer is situated under the plating is a cementite alloyed by vanadium Fe_3V_3C [6]. Carbon concentration reaches stoichiometry for cementite creation. Vanadium increasing liability of iron to a chilling effect helps formation of the chilling effect in the zone bording to a plating and decomposes in it partially.

The sequence of the phase formation under cromodizing is presented of figure 5. As a result of this the increasing of tolerances of coupling size is observed.

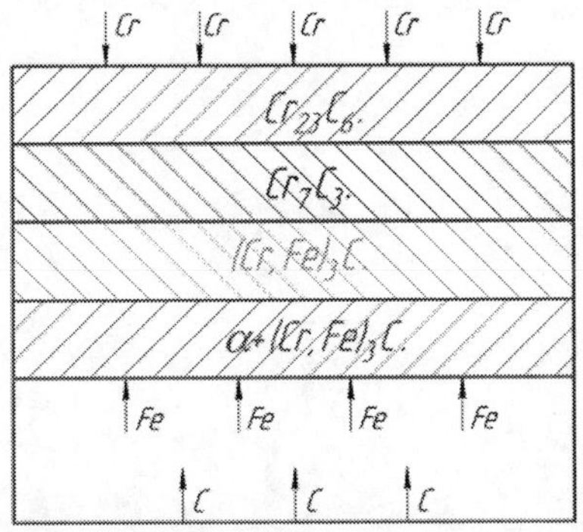

Fig. 5. Phases which formed under thermal diffusion saturation of gray and nodular iron with chrome.

Time-temperature curve was done to determine the best time and temperature of vanadium process and chromodizing process of gray and nodular iron. They are presented on figures 6,7.

As we can see on figures 6, the hardened layer thickness increase permanently on all time line. However for vanadium platings, the increasing of hardening time over 6 hours provides spall fracture and peeling of plating. For chrome platings ,the increasing of hardening time over 10 hours provides strong slowing of the plating formation speed and inconspicuous increasing of it.

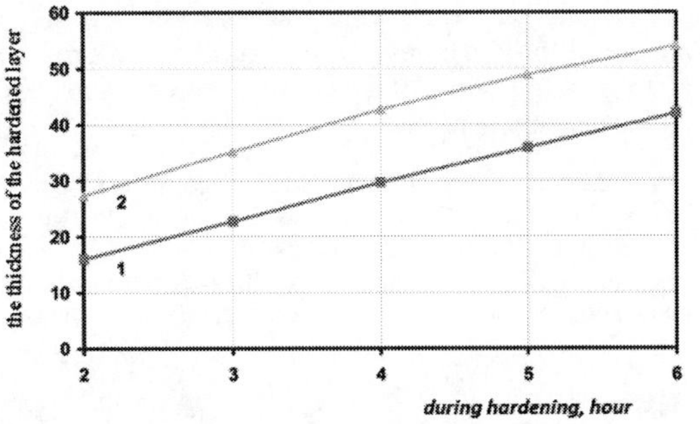

Fig. 6. Dependece of thickness of hardened vanadium plating δ on saturation time (1000^{0}C): 1- gray iron ($y=10,474x^{0,7531}$) 2- nodular iron ($y= 18,163x^{0,649}$)

To indicate the variance of the coupling sizes of a part as a result of thermal diffusion and determination tolerances to the basic parts sizes, experiments on identification of increments of the linear sizes were done. The variance is indicated by size comparison detected by micrometer before and after thermal diffusion saturation (figure 7)

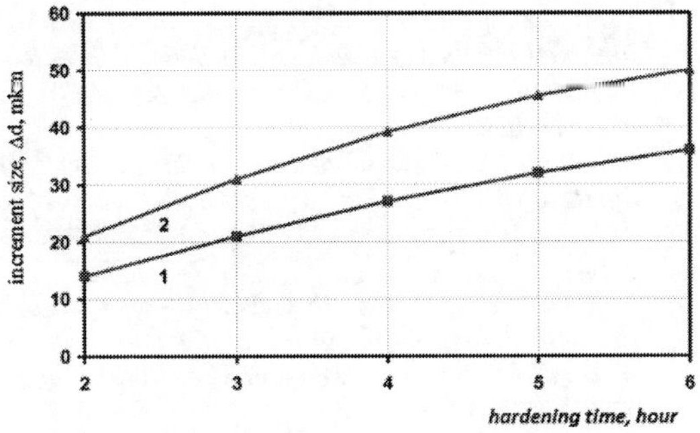

Fig. 7. Dependence of increments of the linear sizes (on a side) Δd on hardening time (1000^{0}C) under vanadizing (a) and chromodizing (б): 1 – gray iron, 2-nodular iron.

Comparing obtained results with the results on figure 8, we can conclude that increments of the linear sizes of hardened iron pieces under vanadizing and chromodizing approximately equals the thickness of produced platings.

DOI: 10.1051/matecconf/201712902043

4 Conclusions

As a result of the experiments on producing of thermal diffusion vanadium and chrome plating, uniform in thickness carbid platings were made. Their microhardness reaches 26 000- 28 000 mPa under vanadizing and 13000-16000 mPa under chromodizing. Besides that, the increment of exterior sizes of hardening parts is observed after hardening relative to ingoing size (before hardening). The mechanism of this hardening is proved theoretically. In this regard at purpose of maximum deviations of tolerances of the sizes of iron parts which are exposed to thermal diffusion saturation by vanadium and chrome, it is necessary to consider an increment of the sizes by the quantity of produced plating: 40-60 micron for vanadium process and 45-60 for chromodizing process.

The authors are thankfully acknowledge the financial support by the Ministry of Education and Science of the Russian Federation for the project № 11.2054.2017/Project task within the framework of the state task for 2017-2019 (Number for publication: 11.2054.2017 / 4.6) to carry out this investigation.

References

1. An-Hui Lu, Guang-Ping Hao, Qiang Sun, Xiang-Qian Zhang, Wen-Cui Li, *Chemical Synthesis of Carbon Materials With Intriguing Nanostructure and Morphology*, Macromolecular Chemistry and Physics,. **213**, 10-11. pp. 1107–1131, (2012)
2. William M. Steen, *Laser Surface Treatment*, Chapter of Laser Material Processing. Springer London, pp 172-219, (1991)
3. S.P. Nefediev, R.R. Dema, N.Sh. Tyuyaterakov, A.N. Morozov, M.V. Kharchenkov, M.A. Belotserkovskiy, M.A. Levantsevich, *Chromomargane coating fusion area structure obtained by plasma powder weld deposit*, Problems of ferrous metallurgy and materials science **1**. pp. 68-72. (2016)
4. V.K.Ilyin, *Restoration and hardening of parts in agricultural machinery by methods of thermal diffusion saturation* (Dokt, Diss. Kazan, 290 p., 2004)
5. A.A.Veselovsky, I.M. Yachikov, *Increasing of durability of parts from gray iron for telescopic and hydraulic cylinders*, Monography (Magnitogorsk, 118p., 2009)
6. A.A.Veselovsky, A.N.Zavalishchin, *Research of the microhardness of the aluminized layers on the gray iron obtained by thermal diffusion hardening*, The hardening technologies and platings **5**. p18-21. (2007)

Investigation of the structure of wear resistant coatings obtained by plasma powder surfacing

Alexey N. Shapovalov[1], *Olga.B.* Kalugina[2,]*, and *Roman. R.* Dema[2]

[1] National University of Science and Technology «MISiS», Leninskiy prospekt, 4, 119049, Moscow, Russia
[2] Nosov Magnitogorsk State Technical University, Lenin Avenue, 38, Magnitogorsk city, 455000, Russia

Abstract. The paper studied the structure of the deposited coating, formed on 45 steel with a plasma-powder surfacing wear-resistant white cast iron. The effect of surfacing conditions and other technological influences on the weld pool during surfacing - current modulation, rapid cooling of the weld beads by blowing air and the rapid cooling of the flowing water of the substrate. Tests have been conducted to determine the plasma-powder coating modes of high alloyed wear-resistant coatings which provide in the middle of the metal deposit the structure and properties that ensure maximum wear resistance. Recommendations of using of cooling systems modes has given to obtain an even distribution of microhardness over the fusion zone .

1 Introduction

Consolidating surfacing of the details of mining machines and metallurgical equipment involves certain technological difficulties. First of all it is connected with excessive tendency of weld metal products to crack formation due to high tension appearing in highly alloyed weld metal and in the welding area. The main reason of the tension in the welding area is the absence of discreteness of the chemical and structural-phase compositions. To avoid cracking plasma-powder surfacing is made at higher values of current and, hence, at maximum melting of metal base and its penetration in the weld metal [1]. It helps to reduce the gradient of the chemical composition in the welding area and as a consequence to reduce the cracking risk. But it leads to strong adulteration of the weld metal with the metal base and finally to the reduction of the performance attributes of the coating structure and increased consumption of expensive surfacing materials. In majority of cases high current and strong metal base melting allows to eliminate the risk of cracking, however the weld metal doesn't meet the requirements to the hardness and wear resistance.

In comparison with the other ways of arc surfacing plasma-powder surfacing of wear resistant coating by twin-arc plasmatron allows to eliminate the chance of cracking the weld metal. It becomes possible due to the effective spending of the plasma arc power on the formation of the molten bath at minimal melting of the metal base thus providing with the preset composition of the weld metal in the first layer of the coating [2, 3]. At that in the area of weld metal and metal base fusion there appears a durable austenitic interlayer separating base metal from the weld metal [4]. The forming of the austenitic interlayer in the welding area leads to the formation of layered composite coating [5] which in its turn contributes to the increase of the structure discreteness and mechanical properties of the welding area. However, even at optimum modes of welding minimal adulteration of the weld metal with the metal base when plasma-powder surfacing is inevitable [6-8]. Due to the welding in the narrow intermediate area there appears a new transitional alloy with an averaged composition in accordance with the share of the main and the welding materials. At that there are strong concentration gradients of alloying elements, appearing not in the whole area of welding but within the close limits of adjacent diffuse layer and the heat-affected area which can result in cracking as well.

Structural-phase state and performance attributes of intermediate alloys in the welding area will depend on their chemical composition, which can be approximately determined by analysing the corresponding charts [7]. In its turn the chemical composition of the intermediate alloys in the welding area depends on the compositions of the main and weld metal and the level of welding, i.e. mode of plasma-powder surfacing.

In this connection it turns out to be the issue of the day to determine the modes of plasma-powder surfacing with high-alloyed wear-resistant coatings which would ensure maximum wear resistance in the intermediate part of the weld metal, durability and operability in the welding area.

2 Experimental

The work investigates the structure of coating of 315Х19Ф3 type on steel 45 when plasma-powder surfacing at different modes by applying additional

* Corresponding author: kalugina.olga@bk.ru

technological impacts - current modulation. The work also gives evaluation of the current modulation influence on the formed structure as well as phase composition of the weld metal and the welding area.

Plasma-powder surfacing was made by twin arc plasmatron at maximal amperage of the subsidiary arc 40 A. The additive material was the powder, containing 19% of chromium, vanadium, 3% (of 315Х19Ф3 type), specially developed for wear-resistant plasma-powder surfacing details of mining machines and metallurgical equipment [8]. The surfacing was made on steel 45 patterns according to the modes, indicated in the table1. To adjust the thermal cycle of surfacing and the depth of melting we applied impulse current (current modulation) as well as additional cooling the base by water or blowing.

Table 1. Modes of surfacing

Mode of surfacing	Weld metal type	Impulse amperage, A	Pause amperage, A	Modulation period, c	Additional effects
1	315Х19Ф3	120	-	-	-
2		180	-	-	-
3		180	-	-	Cooling the base by water
4		120	60	1	-
5		180	80	0,5	-
6		180	-	-	Blowing

To get the maximal thickness of coating when surfacing powder expenditure was 6 kg per hour.

3 Results

According to [6] when raising the amperage from 120 A to 180 A, ceteris paribus, the share of the base metal in the coating increases from 3 % to12 %. Under these conditions the welding area becomes the most sensitive to the adulteration with metal base. The microstructure of the welding area is the most sensitive to the modes of surfacing and additional technological impacts and as a consequence becomes an indicator of the share of the metal base in the coating.

When surfacing the powder similar to 315Х19Ф3 at mode 1 with amperage 120A there appears a structure of eutectic type in the middle part of the weld metal. The analysis of the difractogram, taken from the middle part of the weld metal, showed 5 lines, related to the matrix latitude (ranet centered cubic lattice) with a parameter a = 3,6081Å, which is a bit greater than the period of pure γ-iron aγ= 3,571 Å [9-11]. The coincidence of the comparative intensity of the matrix lines and the table values of the line intensity for γ-iron allows to conclude that the matrix of the weld metal is an austenite [10]. Morphological peculiarities also allow to determine it as alloyed γ-iron.

In the volume of the austenite matrix carbide phase is equally distributed. There are particles of a pencil form distinctive for M_7C_3 as well as fan-shaped eutectic on the base of carbide M_7C_3 with microsolidity 9400 mega Pascals. The aggregate quantity of carbide phase is 24, 5 % with an averaged distance between carbides equal to 3,37 micrometers (Fig 1).

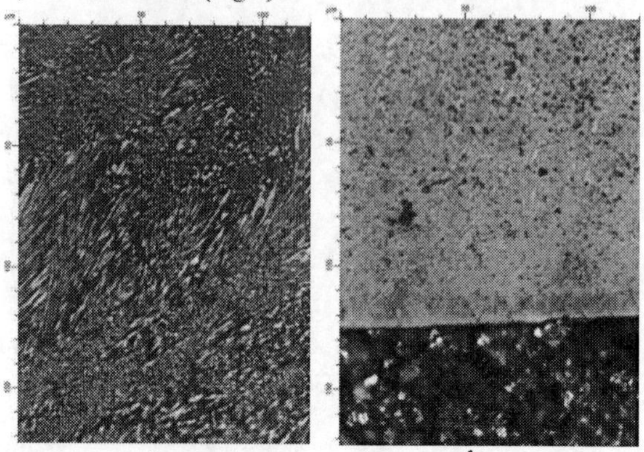

a b

Fig 1. Microstructure of coating welded at amperage 120 A: a – middle part of the weld metal, x500; b – welding area, x500.

The data of metallographic analysis indicate to the presence of light ball-shaped particles in the welding area and the lower part of the middle area of the coating (Fig 2 a). The particles, having similar structure, can be found in the metal base along the welding line (Fig 2 b). The microthickness of the particles in the weld metal is 13000 mega Pascals, but in the metal base it is 12200 mega Pascals. The data of the X-ray structural analysis, microthickness measuring and metallographic analysis allow to come to a conclusion that light ball-shaped particles in the welding area and in the middle part of the middle area are cementite.

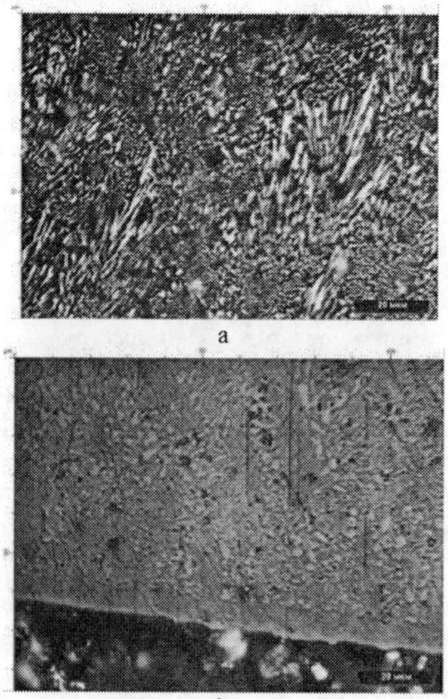

a

b

Fig 2. Cementite particles in the coating 315Х19Ф3 welded at amperage 120 A: a – in the lower part of the middle area, x500; b – in metal base, x200.

We can suppose that when welding the metal base by arc the molten bath is made by the particles of the additive powder as well as ferrite and perlite grains along the welding line. The cementite laminae of perlite have no time to dissolve in the liquid phase of the molten bath, but coagulate and resurface to a certain altitude, staying in the weld metal. The absence of the metallographically distinguished coagulated particles of cementite in the middle area of the weld metal confirms the supposition that cementite gets to the coating from the metal base as in the process of mixing in the middle part of the coating the cementite released from melted perlite grains dissolves and becomes invisible for metallographic analysis. The presence of cementite causes a negative effect on the wear resistance and performance attributes of the coating, that is why it's necessary to provide such welding modes, in which adulteration with metal base will be minimal.

The powder surfacing of 315Х19Ф3 type in the mode 2 at amperage 180 A leads to the formation of the under eutectic structure of coating (Fig 3) instead of the eutectic structure which can appear when surfacing at amperage of 120 A. It can be explained both by a smaller transition coefficient of such alloying elements [8] as carbon and chromium and by a considerable overheating of the molten bath and lower speed of its crystallization. The mentioned processes result in the crystallization of coating, which moves by thalweg in the direction of under eutectic structures [12].

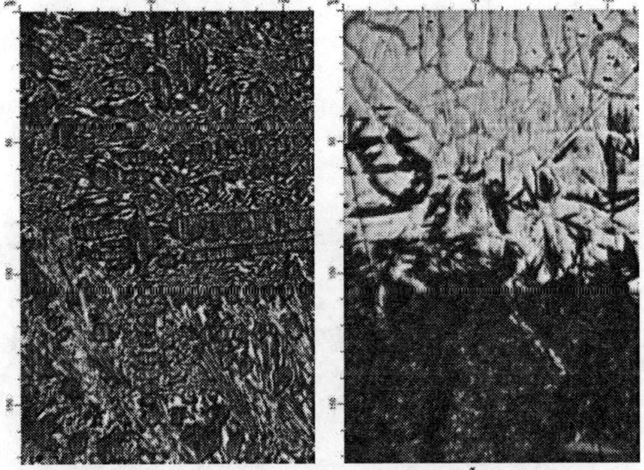

a b

Fig 3. Microstructure of the coating welded at amperage 180 A: a – the middle area of the weld metal, x500; b – welding area, x500.

Raising arc pressure when surfacing at amperage 180 A leads to strong metal base melting and its penetration to the weld metal. As a result in the welding area on the side of the weld metal at 2..4 micrometers distance from the metal base there appears a martensite layer, being on the lines and in the center of crystallites having a dendritic structure. The presence of martensite in the welding area predetermines the tendency of such coating to cracking. In the central part of the weld metal a structure of under eutectic type forms. Austenite dendrites are arranged chaotically, there is no general direction of growth (expansion). It proves a high overheating rate of the central part of the molten bath and low speed of its crystallization. In the interdendritic space there is austenite carbide eutectic of lamellar shape built on the base of chrome carbide similar to M_7C_3.

Water cooling of the metal base leads to an accelerated cooling of the coating itself. That is why after surfacing in mode 3 in the welding area there forms an austenite layer 10 …15 micrometers thick form which dendrites of γ-solid solution with ill-defined axes of the second order grow in the direction of the weld metal. The interdendritic space of the weld metal is filled up by highly dispersed triple eutectic on the base of carbides Cr_7C_3, VC and γ-solid. The microstructure of the coating is shown at Fig 4.

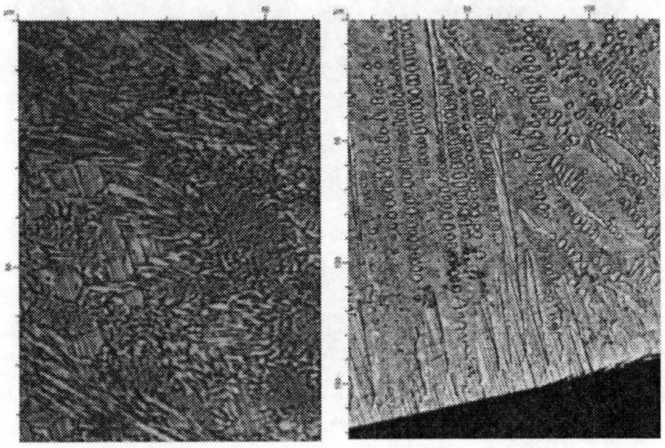

a b

Fig 4. The microstructure of the coating made at the amperage 180 A with metal base water cooling: a – middle part of the weld metal, x1000, b – welding area, x500.

The surfacing by a modulated current allows to reduce the overheating rate of the molten bath due to the recurrent deceleration of the welding current from the impulse current to the pause current. At the moment of current impulse metal base melting is taking place and a molten bath is forming (point 2), at the moment of a pause cooling of the molten bath and its size diminishing is happening (point 2). At the moment of plasmatron moving regarding the weld part metal crystallization at lesser degree of heating is being caused [13-15].

The thermal cycle of plasma surfacing can be regulated by changing the parameters of current modulation, i.e. impulse and pause amperage as well as impulse and pause time. As a result of current modulation we see an increase of heat sink of the molten bath, that takes place in the direction perpendicular to the metal base and the environment. The microstructure gets a purposeful character of crystallization, the size of the crystals diminishes and thus the tendency to crystal cracking decreases [16-19].

Due to surfacing in mode 4 at current modulation 120\60 A and frequency 1 Hz there forms a structure of the welding area (Fig 6) similar to the structure of the welding area at the amperage 120 A without current modulation. In the middle part of the weld roll there appears a structure, consisting of dendrites γ-solid solution and austenite and chrome carbide eutectic on the base of chrome carbide of Me_7C_3 type. Dendrites of austenite have a purposeful from of growth with

DOI: 10.1051/matecconf/201712902044

developed axes of the first order , which can become 100 to 120 micrometers in length and axes of the second order, being 50 …70 micrometers in length.

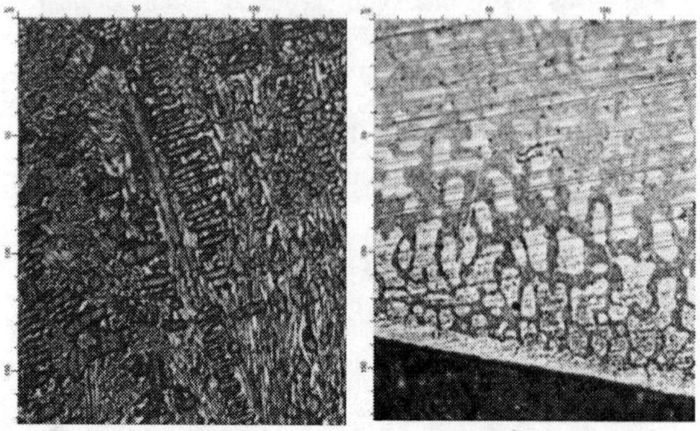

a b

Fig 5. Microstructure of the coating made with current modulation 120\60 A: a – middle part of the weld metal, x 500; b – welding area, x500

At that the main axes of dendrites are arranged in the direction perpendicular to the metal base and in the higher layers of the roll in the direction perpendicular to the surface. The greater length of the axes of the second order in comparison with the surfacing in mode 1 is explained by a strict geometrical orientation of the redundant phase of the γ-solid solution, correlation of the first and the second orders and the absence of interferences in the way of dendritic crystals growing resulting in their considerable size [20-22].

When surfacing with current modulation 180\80 A and frequency 2 Hz (mode 5) the pressure of the arc on the metal base is maximal among all investigated modes of surfacing. As a result, along with a considerable metal melting one can observe great metal mixing in the welding area and pressing the martensite layer much further from the welding boundary (Fig 6). Along with it the amount of martensite in the welding area decreases to a great extent and the welding area itself becomes extensive and homogeneous in the chemical composition. In the middle of the weld roll there forms a structure consisting of dendrites of γ-solid solution having a constant orientation regarding each other and the metal base. The main axes of austenite dendrites reach 200…250 micrometers in length and 15 … 20 micrometers in width. The axes of the second order also reach a great length of 100..150 micrometers and width of 15-20 micrometers correspondingly.

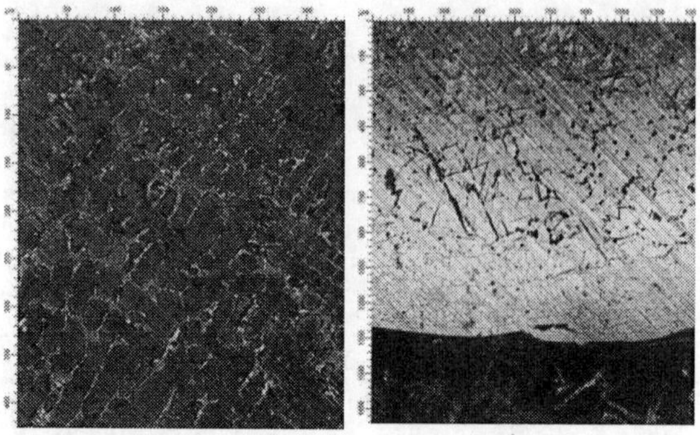

a b

Fig 6. Microstructure of coating made with current modulation of 180/80 A: a – middle part of the weld metal, ×500; b – welding area, ×50.

In the interdendritic space there is eutectic of lamellar construction $\gamma + Me_7C_3$ with an average interlamellar distance of 4.7 micrometers и microsolidity 4250 mega Pascals, as well as rough carbide laminae of irregular shape, being found mostly on the boundary of eutectic and dendrtic areas.

When surfacing powder of 315Х19Ф3 type in mode 6 (general current 180 A, blowing 0,2 mega Pascals), a coating structure of eutectic type is formed (Fig 7). By metallographic analysis one can reveal insignificant quantity of initial carbides M_7C_3, 3,2 % (Fig 7 b) with an average size in diameter 16,4 micrometers. This little quantity of chrome carbide of M_7C_3 type doesn't provoke changes in the eutectic structure in whole. Austenite in its entirety is in eutectic colonies of laminar and rosette structure. Microsolidity of eutectic as well as laminar and rosette structure is similar and comprises 9800 mega Pascals.

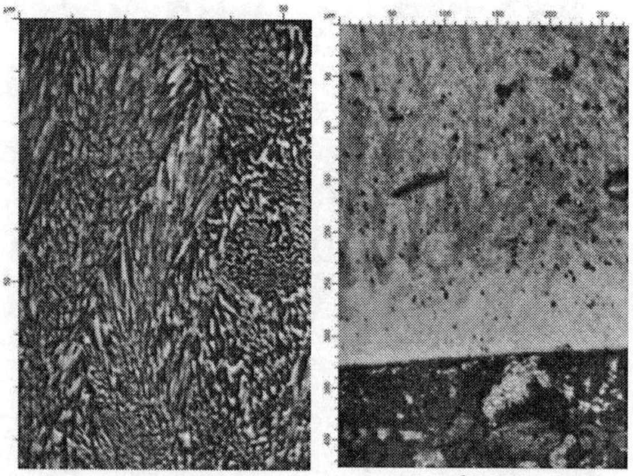

a b

Fig 7. Microstructure of the coating made at amperage 180 A with blowing 0,2 mega Pascals: a – morphology of eutectic, ×1000; b – welding area, ×200.

The plasma powder surfacing with cooling the coating by air causes considerable approaching of carbide particles in eutectic till 1,2 micrometers, and colonies orienting perpendicular to the metal base which predetermines high microsolidity of such eutectic and wear-resistance of the coating in whole.

4 Discussion

The above stated analysis allows to conclude that introduction of additional technological impacts on the molten bath when plasma-powder surfacing allows to change the course of crystallization of the coating thus changing the type of the formed structure, as due to the introduction of current modulation a structure of eutectic type starts to form. The structure becomes of an eutectic type when being cooled by an air flow under pressure 0,2 mega Pascals.

At that current modulation and accelerated air cooling result in the increase of dispersion of the forming structure and approaching of carbides in eutectic. Intensively weld rolls are cooled by blowing (surfacing mode 6). This mode is also characterized by a considerable overheating of the molten bath above the temperature of liquidus alloy which provides complete dissolution of chrome and vanadium in the melt and formation of dispersed eutectic structure.

Surfacing at amperage 180 A with the introduction of forced cooling of weld metal allows to form a welding area with a gradual transition of properties from the metal base to the weld metal due to the formation of crystallized metal base of high alloyed austenite and absence of martensite in the welding area.

The introduction of current modulation and blowing allows to get an even distribution of microsolidity over welding area due to the formation of crystallized layer consisting of austenite without martensite crystals.

5 Conclusions

Surfacing on the current 180 A with the introduction of forced cooling weld metal, fusion zone allows you to create a smooth transition from the properties of the substrate to the weld metal due to the formation of high-alloy austenitic crystal layer, and to prevent the formation of martensite in the weld zone.

The introduction of modulation current and air purge allows to obtain a uniform distribution of Micro fusion zone by forming a crystal layer composed of austenite, martensite without the crystals.

More effectively increases the dispersibility structure blowing air cooling roller, as compared with running water cooled substrate.

The authors are thankfully acknowledge the financial support by the Ministry of Education and Science of the Russian Federation for the project № 11.2054.2017/Project task within the framework of the state task for 2017-2019 (Number for publication: 11.2054.2017 / 4.6) to carry out this investigation.

References

1. A.N. Emelyushin, S.P.Nefedyev. "Investigation of the structure and im pact-abrasive wear resistance of coatings of the FE-C-CR-MN-SI system, additionally alloyed with nitrogen", Welding International. **27**, 2. pp. 150-153. (2013).
2. S.P.Nefedyev, R.R Dema., S.A.Nefedyeva, A.V.Yaroslavtcev, "Microstructure of cast iron after plasma bleaching", Journal of Chemical Technology and Metallurgy.**50**, 2. pp. 213-216. (2015)
3. V.A.Korotkov, "Improvement in arc quenching", Heavy engineering.**6**. pp. 34-37. (2004).
4. V.A.Korotkov, "Wear resistance of plasma-hardened materials", Journal of Friction and Wear.. **32**, 1. pp. 17-22. (2011)
5. S.N.Berdnikov, A.E.Pozin, A.A.Podosyan, K.N.Vdovin , "New narrow-wall design for the mold of continuous slab-casting machines", Steel in Translation. **41**, 5. pp. 410-412. (2011).
6. Karin Graf, Ana Sofia Clímaco Monteiro D'Oliveira, "Microstructural stability and wear performance of a Ni based alloy PTA coating", Coatings Technology. **106**. pp. 156-161. (2002)
7. T. Naguy, J. Kolek and T Anderl, "Engineers Evaluate Cold Spray Coating Processes", J. Therm. Spray. Technol., **14**, 3, pp. 300-301.(2005)
8. N.Ben-Oved, O. Kesler. "A New Technique for the Rapid Manufacturing of Direct Oxidation Anodes for SOFCs", Adv. Mater. Res. **15–17**. pp. 287–292. (2007)
9. O.Kesler, "Plasma Spray Processing of Solid Oxide Fuel Cells", Materials Science Forum. **439–543**, pp. 1385–1390. (2006).
10. F.Gitzhofer, M.Boulos, J.Heberlein, R.Henne, T.Ishigaki, T.Yoshida, "Integrated Fabrication Processes for Solid-Oxide Fuel Cells Using Thermal Plasma Spray technology", MRS Bulletin. **25**. pp. 38–42. (2000)
11. D. Mariotti, P. Maguire, C.M. O.Mahony, and J. A.McLaughlin, "Analysis of excitation processes and electron temperature changes from spectral data in a dc micro plasma discharge", Plasma Sources Sci. Technol., **13**, 4, p. 576–581, Nov. (2004).
12. J. G. Eden and S.-J. Park, «Microcavity plasma devices and arrays: A new realm of plasma physics and photonic applications,» Plasma Phys. Control. Fusion, **47**, 12, p. 83–92, Dec. (2005).
13. José Roberto Tavares Branco, Robert Gansert, Sanjay Sampath Christopher C. Berndt, Herbert Herman, , "Solid Particle Erosion of Plasma Sprayed Ceramic", Coatings Materials Research, **7**, 1, pp. 147-153.(2004)
14. H. T.Fan, and Wu, S. M., 1995, "Case studies on modeling manufacturing processes using artificial neural networks". Journal of Engineering for Industry, **117**, p. 412-417.
15. K.N.Vdovin, D.A.Gorlenko, A.N.Zavalishchin, "Influence of industrial tempering on the composition of complex cast iron". Steel in Translation. **43**, 5. pp. 288-290. (2013)
16. K.N Vdovin., D.A.Gorlenko, A.N.Zavalishchin, "Carbide transformations in tempering of complexly

alloyed white cast iron", Metal Science and Heat Treatment. **57,** 5. (2015)

17. S.V.Anakhov, Y.A Pykin, "Comparative analysis of gas-dynamic factors at the noise radiation of plasmatrons". Welding production. **1**, pp. 31-36. (2005)
18. S.V.Anakhov, "Effect og active gas mixtures on the efficiency of plasma surface hardening", Welding International. **27**, 1. pp. 77-82. (2013)
19. S.V.Anakhov, Y.A Pykin., S.A Shakurov., "Structural-functional principles of design in electric plasma technologies", Welding International. **28**, 7. pp. 579-582. (2014).
20. A.N. Yemelyushin, A.B.Sychkov, V.P.Manin, M.A.Sheksheyev, "Investigation of the structure and mechanical properties of welded joints in steels of the K56 strength grade in different welding conditions", Welding International. **28**, 1. pp. 70-74. (2014)
21. A. Nady, H. Bonnefoy, V. Klosek, M. H. Mathon, A. Lodini and A. Baczmanski, "Finite element analysis and neutron diffraction evaluation of residual stresses in satellite coating produced by PTA process", JCPDS-International Centre for Diffraction Data. pp. 454-461. (2009)
22. A.Zikin, S. Ilo, P.Kulu, I.Hussainova, C.Katsich, E. Badisch, "Plasma tra nsferred arc hardfacing of recycled hardmetal reinforced nickel-matrix surface composites", Mat. Sci. (Medžiagotyra), **18**, pp. 12–17. (2012)

MATEC Web of Conferences **129**, 02045 (2017) DOI: 10.1051/matecconf/201712902045
ICMTMTE 2017

The evolution of the welded joints structure at low cycle fatigue

Ekaterina Poyarkova[1,*], and *Iskander* Kuzeev[2]

[1]Orenburg State University, Orenburg, Russia
[2]Ufa State Petroleum Technological University, Ufa, Russia

Abstract. The effect of cyclic damage on the structure evolution and the change in the micro stresses of the crystal cell in characteristic zones of dissimilar welded joints of steels strength classes X60 and X70 is estimated. It is shown, that with increase in the number of loading cycles in less strong metal of the welded joint, relaxation of internal stresses, realized by splitting individual ambits of ferrite grain or by the initiation of micro cracks, is observed. The results of fractographic analysis of a less reliable weld zone at different levels of samples fractures of accumulated damage are presented.

1 Introduction

Trends in the development of the modern oil and gas industry are due to the impressive volume of requirements growth for the survivability of structures as full-fledged constructions. Their individual components from the diverse positions of the need to ensure long-term and trouble-free operation. The presence of welded joints in such structures is inevitable.

Since the process of development and accumulation of damages both of structures materials in general, and of their welded joints in particular, is localized; the resource of the entire construction, according to [1], is determined by the "resource of their critical zones". Such localization is caused by the presence of stress concentrators and a combination of complex interactions of various competing factors, among which it is possible to isolate selectively the evolution of the structure at different scale levels, the relaxation of stresses and the intensity of manifestations in the weakened places of the construction associated with their geometric features.

The issues of increasing the safety of operation of critical engineering facilities by preventing fragile destruction of their elements are becoming more urgent and acute every year. This is true for oil and gas pipelines that have heterogeneous butt joints. One of the reasons for their destruction is internal structural and mechanical changes occurring in the metal.

The purpose of this work was to assess the effect of cyclic damage on the trajectory of the evolution of structural heterogeneity of dissimilar welded joints of low-carbon steels for welded structures of 09G2FB and 17G1SU.

2 Materials and methods of research

To analyze the structural changes in dissimilar welded joints in the accumulation of cyclic damages, low-carbon steels for welded structures were chosen as the most commonly used in the pipeline systems of the oil and gas industry of the Russian Federation, grades 09G2FB (strength class X70) and 17G1SU (strength class X60). The chemical composition of these steels is regulated by normative documents for the material and is given in Tables 1 and 2.

Table 1. The chemical composition of structural steel 09G2FB for TS 13657842-150-2006, mas. %

C	<0,12	**Cu**	<0,3
Si	0,17-0,37	**P**	<0,035
Mn	1,4-1,8	**S**	<0,035
Cr	<0,3	**A**	<0,008
Nb	0,7-0,9	**V**	0,8-0,9

Table 2. Chemical composition of structural steel 17G1SU according to GOST 5520-79, mas. %

C	<0,17	**Cu**	<0,3
Si	0,4-0,6	**P**	<0,035
Mn	1,15-1,6	**S**	<0,04
Cr	<0,3	**A**	<0,008
Nb	<0,3	**V**	<0,08

The reliability of obtained results was ensured by the use of welded samples with a thickness correlates with real sizes of devices, produced according to GOST 25502-79. Welded samples were obtained by manual electric arc welding (direct current with a current strength of 80-100 A, an electrode of the UONI-13/55 brand with a diameter of 4 mm) of two plates of sheet products from the above-mentioned steels of increased strength. The direction of the samples cutting (along the sheet) was selected from the loading conditions of the products and the technology for material obtaining.

Durability of dissimilar welded joints from steels in the low-cycle region was determined under cyclic elastoplastic loading [2, 3]. As a test machine, the facility of IR 5113-100 was used. The number of loading cycles was set at 0.2; 0.4; 0.6 and 0.8 from the

* Corresponding author: yarko.e@mail.ru

DOI: 10.1051/matecconf/201712902045

destructive (N_i/N_p). The characteristic zones of the investigated welded joints were the zones of the main metals forming part of the welded joint, the central zone of the weld metal and the zone of thermal influence on the side of both metals.

Metallographic analysis of dissimilar welded joints 09G2FB-17G1SU at different levels of metal accumulation of cyclic damage was performed on optical microscopes Neophot-2, AxioObserver.Dlm and using a scanning electron microscope JEOL JSM-6460LV with a console of energy-dispersive analysis INCA-energy in reflected and secondary electrons at accelerating voltage 25 кВ. Grain evaluation was carried out by comparison with scales in accordance with GOST 5639-82.

Glagolev (point) and Rosival (linear) methods performed the quantitative evaluation of the steels phase composition in welded joints. The ratio of ferrite and perlite in various zones of welded joints was determined with the help of the image analysis program by the automatic evaluation method "SIAMS 700".

Defects in the crystal structure in various zones of welded joints were studied by X-ray diffraction on DRON-3 and DRON-UM1 diffractometers in iron radiation at a voltage of 30 kV and a current value of 20 mA in the range of angles 2θ from 18 to 90° in steps Δ(2θ) = 0.02° and an exposure of 10 s at a point. The micro deformations of the cells are determined. The dislocation density values in the characteristic zones of the investigated samples of dissimilar welded joints are calculated. The results were processed using a program for the analysis and visualization of scientific and statistical data SigmaPlot.

3 The results of the study and their discussion

3.1 Metallographic analysis of a heterogeneous welded joint of steel 09G2FB-17G1SU

As is known, the steels 09G2FB and 17G1SU of the investigated welded joint refer to low-carbon, low-alloyed, with ferritic-perlite structure welded without restrictions. In the heat-affected zone of these steels, martensitic decomposition is almost completely absent. Welding of such steels does not cause any difficulties.

The initial structural state of the investigated welded joints was studied by scanning electron microscopy; the results are shown in Fig. 1. The structures of the welded joint throughout the section are ferrite-pearlite (F + P) mixtures whose morphologies depend on the area of the welded joint. In the microstructure of the base metal 9G2FB a segregation banding that coincides with the direction of sheet is observed. Large grains of sorbitol-like perlite (P) and sections of the ferritic (F) grid, which have an equiaxed and obvious dendritic structure, represent the weld metal. The cross-section of the grains undergoes a significant change with distance from the smelting boundary, which is due to the growth of the crystals from the smelting line towards the heat removal in the direction of displacement of the heat source. The microstructure of the base metal of grade 17G1SU also has a structure - F + P, obvious segregation banding is observed. The grains have a flattened elongated shape due to technological features of the sheet method of production.

A) B) C)

Fig.1. Microstructures of characteristic zones of heterogeneous welded joint: A) the base metal 09G2FB; B) the metal of the weld center; C) the base metal of 17G1SU

In the area of the transition zone from the base metal 09G2FB to the seam, the microstructure is identified as a ferrite-carbide mixture (Fig. 2, a). Microstructure of the transition zone from the seam to the base metal 17G1SU– F+C (ferrite-cementite) mixture (Fig. 2, b).

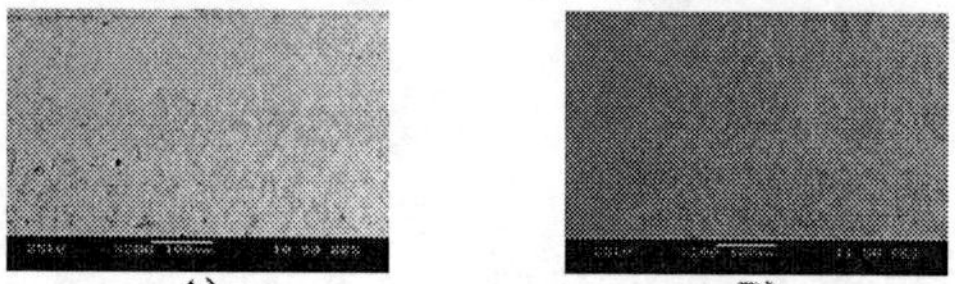

A) B)

Fig. 2. Microstructures of zones of thermal influence of a heterogeneous welded joint: A) HAZ from the steel side 09G2FB; B) HAZ from the steel side 17G1SU

The ratio of ferrite and perlite in the characteristic zones of the welded joints, established with the help of the SIAMS 700-image analysis program, made it possible to note a marked reduction in the percentage of the pearlitic component at the center of the seam relative to the base metal of the stronger material (steel grade X70 of 09G2FB). The growth of thermal influence of both steels was observed in these zones. It was found that the micro hardness of ferrite varies insignificantly across all sections of the welded joint. The micro hardness of perlite grows in the areas of the welded joint with increased percentage of it.

3.2 Analysis of structural changes of dissimilar welded joints under cyclic damage

Fig. 3 shows the evolutionary changes in the steel microstructure over the welded joint cross section with the accumulation of cyclic damage at the mesoscale structural level.

In the process of cyclic loading, the structures of all characteristic zones of a heterogeneous weld joint undergo some changes. At the initial stage of damage accumulation (level N_i/N_p=0.2), significant structural changes in the HAZ of both metals are observed. The degeneration of the ferrite's Widmanstatten figures begins because of the formation of smaller and equiaxial subgrains, an increase in the dispersity of the distribution of ferrite and perlite occurs in the main metal 17G1SU and 09G2FB.

DOI: 10.1051/matecconf/201712902045

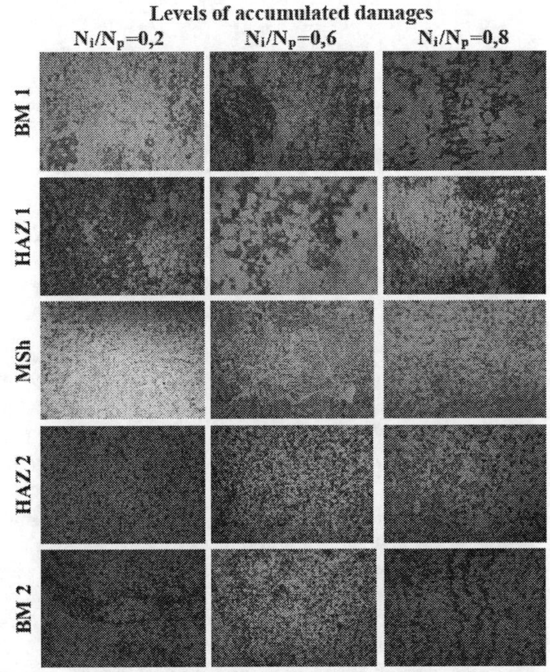

Fig. 3. Change in the microstructure of steel along the welded joint cross-section with accumulation of cyclic damage, 200^x:
BM 1 – the basic metal of steel 09G2FB;
HAZ 1 – zone of thermal impact from steel 09G2FB;
MSh – metal of the central zone of a joint;
HAZ 2 – zone of thermal impact from the side of steel 17G1SU;
BM 2 – the basic metal of the steel 17G1SU

With cyclic damages of 0.6 N_i/N_p, degradation processes affect the dendritic structure of the joint, crushing and leading to equiaxity of its grain. In the remaining zones of the connection grain enlargement is observed. In the main metal, 09G2FB, the discharge bands, which are a sign of irreversible structural changes, are poorly visible.

At a level of 0.8 N_i/N_p in the main metal of 17G1SU, the discharge bands are much brighter, signaling the formation of a pre-destruction zone.

In metallographic studies, it was found that because of cyclic loading in the structure of the welded joint, its complete rearrangement took place. First of all, this refers to the morphology of the carbide phase. Due to the production process high-strength strips steel in the initial state had a structure in which the lines of cement particles of the globular shape were clearly identified. They were called in [4] jets of a peculiar plastic flow, and "convenient structural labels" decorating these lines. The directivity of such lines in the structure of the base metals of the compound in the initial state indicates a correspondence between the natures of the plastic deformation of the material to the laminar (-layered) flow regime. In this case, the laminated nature of the plastic flow of material with increasing degree of cyclic damage could be identified only in the surface volumes of the metal.

In the process of cycling in the near-surface layers of the basic metal regions, structural features were also observed. They looked like vortex patterns in the form of closed and open rings and ellipses, which were sufficiently detectable in this structure by the method of optical metallography. The maximum geometric dimension of such a pattern in the cross section did not exceed 25 μm, and the surface layers themselves had a thickness of 20 to 60 μm.

Earlier in [4-9], the cause of the formation of vortex patterns in the structure was established. It was caused by the appearance at a certain time of loading of unintended torque in the surface layers of the material being deformed. This is a direct proof of the presence in these layers of explicit signs of a rotational vortex flow, which is associated with special test conditions in which the surface layers are located.

According to [9], plastic deformation arises immediately in several regions of boundaries. Both the cessation of the development of localized flow certain regions and the beginning of the other regions propagation (perpendicular to the loading axis) in the form of bands of different sizes (shape of ellipse) are possible in the early stages of loading.

At the levels of accumulated damages in the range from 0.2 N_i/N_p to 0.6 N_i/N_p, vortex motions in the base metals of the dissimilar weld joint are found. They are created by the propagation of plastic fronts, which move in the region of the action of maximum stresses. However, the direction of motion changes periodically because of stress relaxation in the regions of plastic deformation. After the process of plastic fronts joining coming from opposite surfaces, the development of the second and subsequent layers of plastic deformation, localized in zigzag strips begins. The analogy was found for the localization of plastic deformation inside the strip with the formation of a system of conjugate bands of a smaller scale. It was found that not only individual fragments confined by localization bands of a smaller scale but also their conglomerates are involved in the vortex motion. In the process of cyclic damages accumulation in the microstructure of welded joints different zones, a significant distortion of the crystal takes place. In addition, defects of a crystalline structure arise. At the initial stages of damages accumulation, stress relaxation in ferritic grains is carried out by dislocation sliding and subsequent grain-boundary slip. With an increase in the number of cycles, the relaxation of internal stresses is realized by splitting the individual boundaries of the ferrite grains, or by the initiation of microcracks, mainly in a less strong metal.

Fig. 4 shows that with an increase in the number of loading cycles in a less strong metal of a dissimilar welded joint (base metal - steel 17G1SU), microstress accumulation takes place. It determines the formation in the microstructure of this zone and in the region of the «fault strips». The process of deformation in this case can be attributed to nanocrystalline, because it occurs by the displacement of individual grains of the layers.

On the contrary, in the basic metal of stronger steel (steel 09G2FB), transformation (or recoil) of microstresses is observed. This metal behaves similarly

DOI: 10.1051/matecconf/201712902045

to the metal of thermal influence zone located next to it. Separate areas of such a heterogeneous weld joint are under the influence of internal stress relaxation mechanisms. It was noted that at the pre-destruction stage (0.8 N_i/N_p) the microstress growth manifests itself in all characteristic regions of the welded joint.

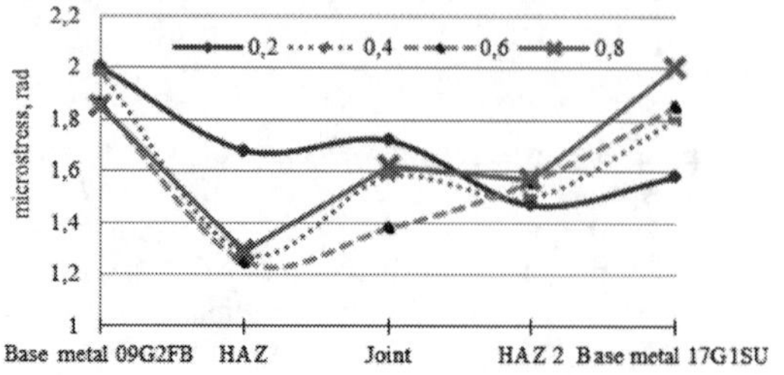

Fig. 4. Changes in the values of microstress over the section of the welded joint from the level of accumulated cyclic damages

With an increase in the number of loading cycles, the value of microdeformation hardly increases in the metal of the thermal influence zone of the less strong metal (steel 17G1SU), which is part of the welded joint, due to accumulation of microstresses. At the same time, the base metal from the same steel is in a more stressed state than the other bond zones, which causes characteristic damages in this zone. In the joint, the microdeformations tend to decrease only to the level of accumulated cyclic damages of 0.6 N_i/N_p. At the stage of prefracture of the welded joint in this opposite area, microdeformations growth is noted.

The averaged values of the dislocation density in the characteristic regions of the dissimilar welded joint were determined by calculation methods. The dynamics of the change in the process of low-cycle fatigue is shown in Fig. 5. It is obvious that an increase in the level of accumulated damages leads to significant changes in the dislocation structure of materials in the characteristic zones of welded joints.

On graphical dependencies, in almost all regions of the welded joint, a decrease in the density of dislocations is observed, except of the zone of the base less strong metal, where the growth of microstresses was previously observed. The maximum density of dislocations is established at the pre-destruction stage in steel grade 17G1SU, which was caused by a decrease in the distance between dislocations and an increase in the average value of interdislocation interaction forces.

At a level of accumulated damage of $0.8N_i/N_p$ in the structure of the HAZ metal of less stable steel, an "ensemble of strongly interacting dislocations" [10], with its inherent collective effects, is localized. Their role in the process of metal accumulation of damage in the system "defects - dislocations" becomes the main one.

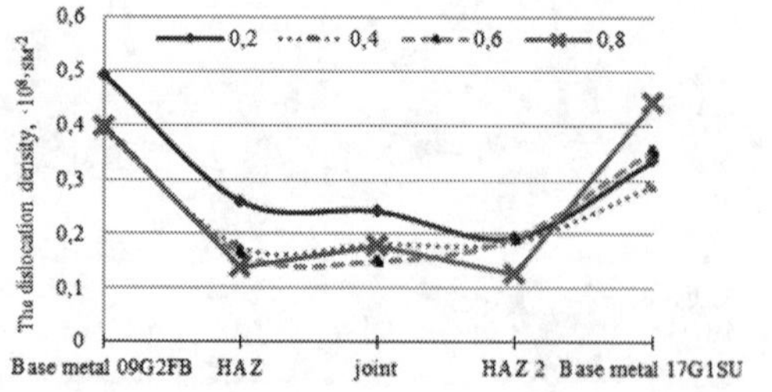

Fig. 5. Distribution of dislocation density along the welded joint cross-section at different levels of cyclic damage

Earlier, in [11,12], the dependence of the energy of dislocation substructures on the degree of ordering of dislocation location was established. The ordering of the dislocation location in the characteristic zones of the welded joint with the accumulation of fatigue damage causes reduction of the long-range field of substructures by elastic dislocation fields Because of the chemical and structural-mechanical inhomogeneity throughout the volume of the welds. In this case, the total energy of the dislocation system decreases.

It is found that the formation of ordered dislocation formations inevitably leads to a decrease in the energy of the "new" structures in comparison with the energy of the "old" structures even with equal degrees of deformation or dislocation densities.

Along with changes in the structural and mechanical state of welded joints under fatigue damage conditions, a change in the mechanism of their destruction from viscous transcrystalline to brittle fracture occurs. Embrittlement presents an increased danger, since the destruction of a welded structure in this case can occur at stresses less than the yield strength, in the form of spontaneous propagation of a crack under the action of the stored elastic energy accumulated by the structure. The changes in the nature of welded joints fracture during the cyclic loading is shown in Fig. 6 in the form of microfractograms of the samples fractures surfaces of the most dangerous section of the welded joint. These are the boundaries of the metal between the zone of thermal influence of the less strong weld metal (steel grade 17G1SU) and the base metal (at multiplicity of 200^x and 2000^x). Analysis of fracture surfaces was carried out in two regions: the central part of the fracture (1/2 of the sample thickness) and the surface surface of the sample disruption surface (not more than 1/8 of the sample thickness from the surface).

The nature of the fracture of the dangerous zone of the welded joint obtained at room temperature not subjected to cyclic loads was identified as viscous. Cyclic loading of the sample to the level of accumulated damage N_i/N_p=0.2 leads to the appearance of bundles clearly visible in the investigated region with an increase of 200x. They have the form of transverse grooves (dips), the formation of which is associated with the presence of elongated nonmetallic inclusions in the type of sulphides [13]. A small fraction of the surface of the fracture is occupied by large pits, in which particles of the sulfide type and of oxides of various sizes also found. The central part of the fracture surface is also

DOI: 10.1051/matecconf/201712902045

covered with pits, but their dimensions are much smaller than in the near-surface zone. In the pits, sometimes the presence of particles of a nonmetallic nature can be observed.

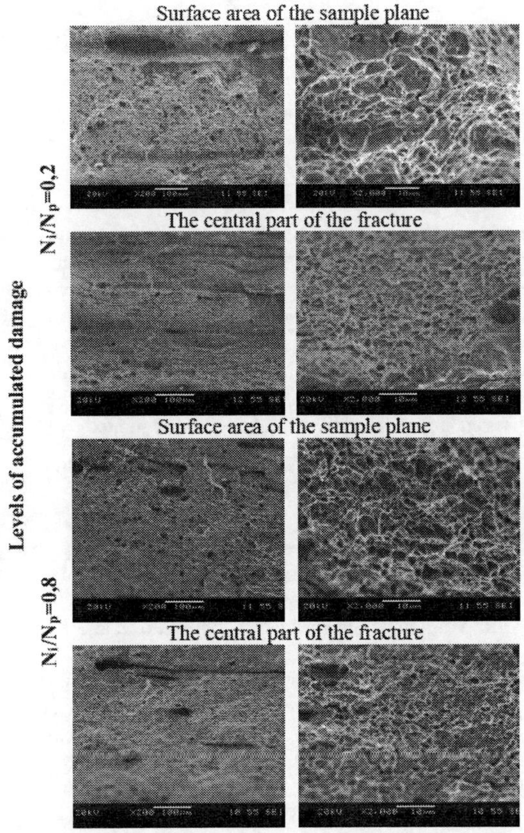

Fig. 6. Microfractogram of the surfaces of specimen fractures in the region of a heterogeneous weld joint at the boundary of the heat-affected zone and the base metal of the less stable 17G1SU steel at different levels of accumulated damage

With an increase in the number of loading cycles from 0.2 N_i/N_p to 0.8 N_i/N_p, the microstructure of the fracture surface becomes more smooth, the longitudinal grooves (dips) are less deep. The surface of destruction now has a quasi-brittle character, which is clearly expressed with a large increase. The fracture of the central zone contains the areas of separation and fatigue grooves. All illustrations confirm that cyclical fracture is inherent in the combination of fatigue grooves and intergranular fracture.

4 Conclusions

Complex metallographic analysis including X-ray diffraction analysis revealed structural changes in the metal of dissimilar welded joints under cyclic damage, distribution of microhardness, degree of distortion of the crystal lattice as the duration of cyclic loading increased. In the process of cyclic loading, structures reminiscent of vortex patterns in the form of closed and open rings and ellipses at the level of optical metallography capabilities are revealed in the near-surface layers of the main metal regions. With an increase in the number of loading cycles at the microscale level, microstresses accumulate in a less strong metal, which leads to the formation of discharge bands in the region adjoined to it.

The dislocation density distribution across the cross-section of a dissimilar welded joint is obtained, which, qualitatively, does not change with increasing number of cycles before failure, which is an important result for understanding the scale hierarchy of the emerging energy dissipation processes. In the field of low-cycle fatigue, meso- and macro-scale levels are involved, when rotations play the main role, and processes do not associated with dislocations. An in-depth analysis of the revealed regularities is made and recommendations are given. In order to describe the plastic deformation in the case of cyclic damage to welded joints, it is necessary to take into account the misorientation of structures based on the disclination theory. In fact, it is justified that exactly the rotations of the crystal cell determine the processes of intensive accumulation of damages in the region of low-cycle fatigue.

References

1. V.A.Skudnov, M.K. Chegurov, Tehnolog. Met. **3**, 6-8 (2008)
2. I.R. Kuzeev, E.V Poyarkova, Neftegaz. Delo, **9**, 80-86 (2011)
3. E.V. Poyarkova, I.R. Kuzeev, Vest. Tambov. Uni., **18,** 1676-1678 (2013)
4. V.A. Bataev, NGTU, 404 (2002)
5. A.A. Bival'kevich, A.A. Bataev, V.G. Burov, V.A. Bataev, S.V. Hlebnikov, Polzunov. vest., **2**, 44-46 (2005)
6. V.E. Panin, V.V. Kibitkin, V.S. Pleshanov, Ju.I. Pochivalov, N.A. Lebedeva, Fiz. Mezomeh., **7**, 385-388 (2004)
7. A.N. Savkin, V.P. Bagmutov, V.I. Vodop'janov, Izvest Vuz. Mash., **12**, 6-14 (2007)
8. A.A. Shanjavskij, A.L. Nikitin, T. Palin-Luc, C. Bathias, Fiz. Mezomeh., **17**, 59-68 (2014)
9. V.A. Romanova, R.R. Balohonov, Tjazh. Mash., **9**, 27-32 (2010)
10. V.V. Rybin, *Metallurgija*, (Moscow, 1986)
11. Je.V. Kozlov, V.A. Starenchenko, N.A. Koneva, Metally., **6**, 152-161 (1993)
12. N.A.Koneva, Sorosov. Obrazov. Zhur., **6,** 99-107 (1996)
13. E.V. Poyarkova, I.R. Kuzeev, Neftegaz. Delo, **6,** 239-244 (2008)

MATEC Web of Conferences **129**, 02046 (2017) DOI: 10.1051/matecconf/201712902046
ICMTMTE 2017

Non-destructive test of steel structures by conical indentation

Alexey Beskopylny[1,*], *Andrey* Veremeenko[1], *Elena* Kadomtseva[1], and *Natalia* Beskopylnaia[1]

[1]Don State Technical University, sq. Gagarin, 1, Rostov-on-Don, Russia

Abstract. The problem of mechanical properties measurement of steel structures by non-destructive method of impact cone indentation is considered. The method is based on analytical solution and experimental data of a problem of impact indentation by cone into elastic-plastic half-space. Non-destructive dynamic indentation method is one of more efficient for mechanical properties assessment because of compact instrument that makes it possible of measuring at any point of structure. The scheme of impact indentation was considered and the problem of elastic-plastic impact is solved. The device with a computer way of information processing was created and tested for determining the characteristics of the existing structures. The method was adopted and tested on bridge structures, butt welded joints, civil engineering structures and others.

1 Introduction

Organization of all types of non-destructive quality control is associated with the measurement of mechanical properties of structural elements and machine parts. However, currently, these properties can be determined only indirectly through the manufacture and special samples testing.

During manufacture structural elements are treated by complex types thermal, mechanical, chemical or another power, after which the material properties change. The most important mechanical properties are yield strength, tensile strength, elongation, Charpy impact strength and hardness. It needs to know these characteristics at all stages of the life cycle structure [1,2]. Resent researches [3,4,5] show that characteristics of conical indentation allow us evaluate creep parameters and fracture toughness for some materials.

The shape of the indenter plays an important role in mechanical characteristics determining and specifies the algorithm test [6,7]. In practice widly known indenters in the form of a sphere [8,9], cone [10,11], pyramid, flat punch [12]. A large amount of testing of materials in industry poses the problem of the real structure testing instead of laboratory standart method. The destruction of most machine parts and structural elements is dynamic in nature or is a consequence of the dynamic application of the load. The complexity of the processes occurring during rapid load application resulted in a large number of methods for impact testing of materials in industry.

Quite an attractive scheme of testing the hardness when the indenter approaches the measured surface with an initial velocity V_0 , the measure of the ratio of the speeds of approach and rebound. In this scheme used a number of indenters: spherical, conical, pyramidal. Studies of German companies, producing devices for measuring the hardness of materials, it is shown that such a scheme of hardness test is the most stable.

However, such methods require a high degree of preparation of the surface to be measured and can lead to indentors failure.

In this article, another test scheme is considered in which the hammer flies up to in the indenter rested on the surface to be measured. Dynamic measurement methods, in our opinion, are preferred because they allow to develop compact and convenient devices that could be used at any point of real structures.

2 Problem definition

It is known that the strain velocity during the test can significantly change the mechanical properties of the material. Therefore, in our case, it is advisable to use a range of speeds to 3 m/s. In that velocity range wave processes and inertial forces do not have a significant effect on the material reaction and can be neglected. So quasistatic models of stroke is applicable.

The shape of the indenter plays important role in the tests. For example during indentation of spherical indenters their prints are not geometrically similar for wide range of material properties. Conical and pyramidal indenters do not have these disadvantages because they give geometrically similar prints in the entire range of indentation depths.

We consider dynamic scheme (**Figure 1**) in which the striker impacts the initially immovable conical indenter with angle γ. The speed of the indenter in this case changes from zero to some maximum value V_{max}, then falls to zero and changes sign during the rebound. This scheme is softer in comparison with the flying indenter and in this case the strain velocity will have less

[*] Corresponding author: besk-an@yandex.ru

effect on the change in mechanical properties, which can be effectively used in measurements.

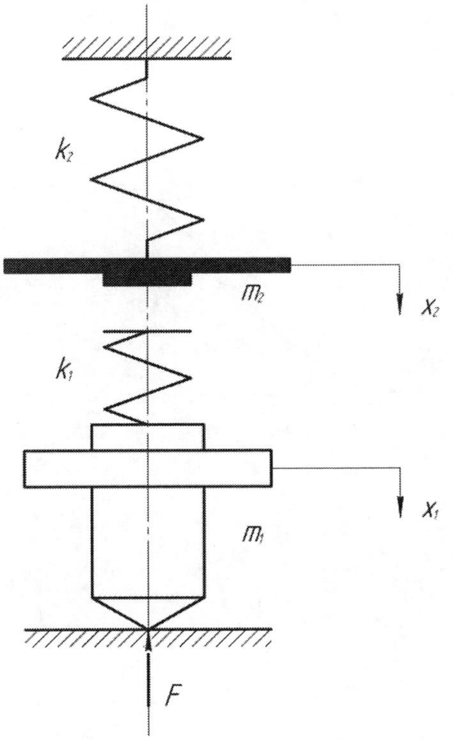

Fig. 1. Mechanical model of impact conical indentation.

The motion of the elements of the system is described by equations

$$
\begin{aligned}
m_1\ddot{x}_1 &= k_1\left(x_2 - x_1\right) - F\left(t\right) \\
m_2 x_2 &= k_2\left(x_{st} - x_2\right) - k_1\left(x_2 - x_1\right)
\end{aligned} \qquad (1)
$$

at $x_1 = 0, x_2 = 0, \dot{x}_1 = 0, \dot{x}_2 = V_0, x_{st} = const$.

The problem of indentation of an elastic cone in elastic-plastic half-space by force F that changes with time (Figure 2) is considered. We consider the cylindrical coordinates system; z-axis is directed to coincide with the axis of the cone and facing down (inside the half-space). In the contact area $\Omega_1(\alpha)$ of cone with a half-space the friction is neglected, outside the area of contact $\Omega_2(\alpha)$ stresses σ_{zr} and σ_{zz} equate to zero. Plastic deformation begins under the Tresca conditions, the material at the same time characterizes the plastic constant k=0,5σ_Y, where σ_Y is yield strength, and indenter material undergoes only elastic deformation.

The displacement u of any point of interacting solids is the sum of the elastic u_e and plastic u_p components:

$$u = u_e + u_p \qquad (2)$$

The distribution of pressure in the contact zone has the form elastic solutions for any values of the force F

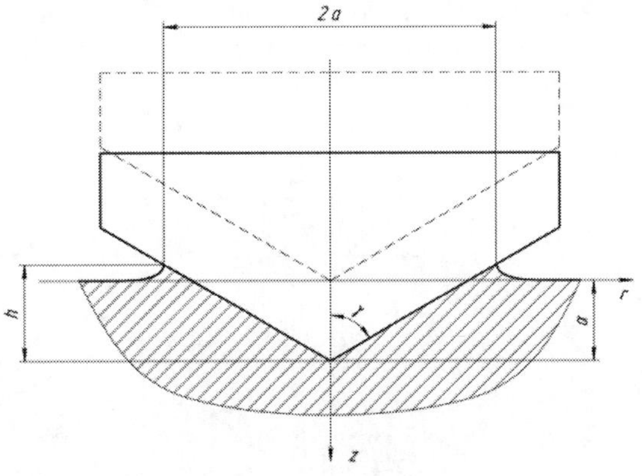

Fig. 2. The model of conical indentation.

$$q\left(F,r\right) = -q_0 \ln\left(a/r - \sqrt{\left(a/r\right)^2 - 1}\right), \qquad (3)$$

where $q_0 = P/\left(\pi a^2\right)$ is the average pressure, a - the contact area radius.

The increment of plastic deformations occurs when the conditions are performed

$$q_0 = \lambda k; \quad dF/dt > 0, \qquad (4)$$

where λ = 5.7, $2k = \sigma_Y$, σ_Y is material yield strength.

The average contact pressure does not exceed

$$q_0 \le \lambda k$$

Full indenter displacement

$$\alpha = \alpha_e + \alpha_p = h_p\left(1 - \delta^*\right) + \alpha_e, \qquad (5)$$

where h_p - depth of the residual print, taking into account the height of the inflow,

δ^* - coefficient determined at experiment.

So the dependence of indentation force *F(α) is*

$$\alpha = C F^{1/2}, \qquad (6)$$

$$C = ctg\left(\gamma\right)\left(1-\delta^*\right)\left(\lambda k\pi\right)^{-1/2} + \left(1 + \left(2\delta^* - 2\right)/\pi\right)\left(\lambda k\pi\right)^{1/2} E^{-1}.$$

Substituting the expression for *F(α)* (6) in equation (1) we get the dependence of the depth *S(t)*, the velocity *V(t)* and acceleration *W(t)* and contact force *F(t)* during impact. The resulting equation contains material yield strength σ_Y that can be determined during test. Other mechanical properties, such as tensile strength, elongation and impact strength, are closely related to the three parameters of the indentation: the depth of indentation *S*=α, speed *V* and acceleration *W*.

3 Results and discussion

The theoretical and experimental studies made it possible to design of an instrument [13] for determining the mechanical properties of steels (Figure 3), which consist of handle 1, spring 2, striker 3, body 4, indenter holder 5, sensor 6 and 9, body part 7, damper 8.

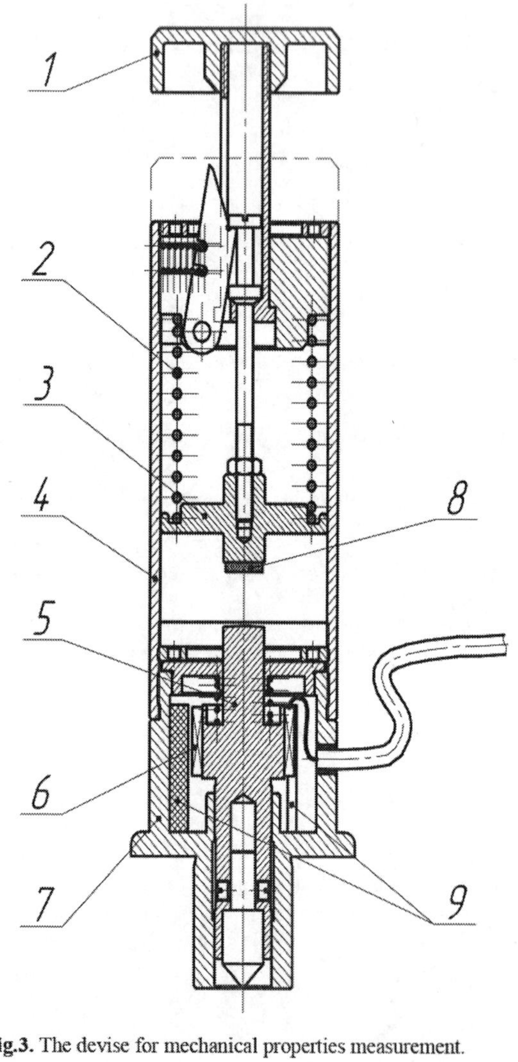

Fig.3. The devise for mechanical properties measurement.

During measurement striker 3 hits over the indenter holder 5 and sensor 6 registers signal of velocity *V(t)*. The dependences *S(t)* and *W(t)* are obtained by integrating and differentiating.

Figure 4 shows dependences of the depth *S(t)*, velocity *V(t)*, and acceleration *W(t)* during impact indentation of a cone 90° for two kinds of steel 100 HB and 370 HB. The registration of the maximum and minimum values on these graphs makes it possible to determine the number of mechanical characteristics of steel.

The comparison of the obtained models with the experimental data was carried out on various structural steels in a wide range of mechanical characteristics. The proposed method makes it possible to estimate the yield strength with an error of 7%, the tensile strength with an error of 5%, the elongation of 10%, and the toughness of 15%.

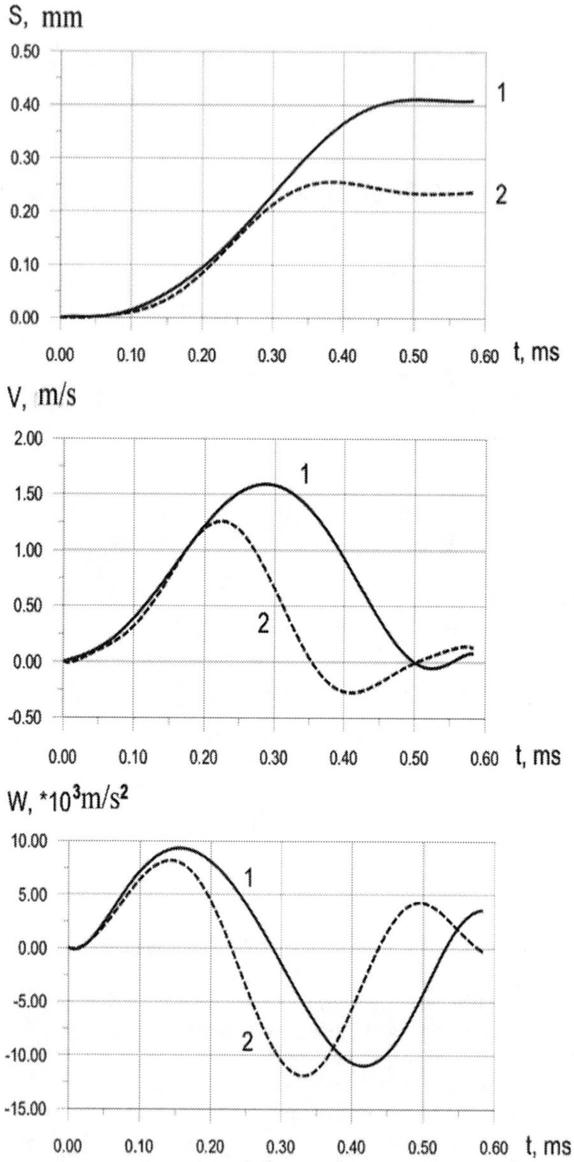

Fig.4. Dependences of the depth *S(t)*, velocity *V(t)*, and acceleration *W(t)* during impact indentation of a cone 90 °. 1 – steel 100 HB, 2 – steel 370 HB.

Subsequent computer processing of the signal is as follows. The original function of the indenter velocity *V(t)* is defined on [0, T] into n nodal points. The function is approximated by a cubic spline on each segment $t \in [t_i, t_{i+1}]$

$$V(t) = y_{i+1}\tau + (1-\tau) y_i +$$
$$+h\left[\left(M_i - \Delta y_i^*\right)\tau(1-\tau)^2 - \left(M_{i+1} - \Delta y_i^*\right)\tau^2(1-\tau)\right] \quad (7)$$
$$\tau = (t - t_i)/h; \quad \Delta y_i^* = (y_{i+1} - y_i)/h; \; h = t_{i+1} - t_i$$

Quadrature integration of the spline functions leads to simple expressions for displacement *S(t)*

$$S(t) = \frac{h}{2}\sum_{i=0}^{n-1}\left(y_{i+1} + y_i + \frac{h}{6}(M_i - M_{i+1})\right) \quad (8)$$

To construct the acceleration curve W(t) the original function of the velocity V(t) was approximated by a polynomial of the 9th degree. Differentiation of this polynomial gives a function of acceleration *W(t)* at a given period of time $t \in [0,T]$

4 Conclusion

Existing methods for determining the mechanical characteristics of steels do not allow measurements directly on the structure in operating conditions with a high degree of accuracy. Developing a method and a series of portable instruments that allow controlling mechanical properties by non-destructive test can solve such a problem.

On the basis of analytical, design and metrological developments, new portable impact devices for determining the mechanical properties of steels have been created and their accuracy has been estimated.

The impact energy of the striker is 0.16 J, which corresponds to the work done by the force of indentation on the Rockwell instrument for a material with a hardness of 45 HRC. For the hit device scheme shown in Fig. 1, the energy of impact is determined by the formula

$$E = \frac{k_2}{2} z_0^2 \left[\sin\left(\arccos\frac{z}{z_0}\right)\right]^2$$

where z_0 is the preliminary compression of the spring, m;

z - coordinate of the holder of the indenter relative to the equilibrium position of the spring, m;

k_2 - rigidity of the main trigger spring, N / m; k_2 = 500 N / m.

The model of the device with a computer way of information processing was tested in determining the characteristics of the existing structures [14], butt welded joints [15] and others.

References

1. D.M. Belenkyi, A.N. Beskopylnyi, N.L. Vernesi, News of higher educational institutions. Construction **1**, 99-102 (2003)
2. A.N. Beskopylny, I.G. Kadomtsev, N.I. Beskopylnaya, Theoretical foundations of civil engineering, **83-84** (2005)
3. A.S. Chepurnenko, V.I. Andreev, B.M. Yazyev, A.N. Beskopylny, MATEC Web of Conferences **67**, 06059 (2016)
4. S. Chatterjee, Sanjay Panwar, K. Madhusoodanan, A. Rama Rao, Nuclear Engineering and Design **305**, 9-17 (2016)
5. J. Dean, J. Campbell, G. Aldrich-Smith, T.W. Clyne, Acta Materialia **80**, 56-66 (2014)
6. A.R.H. Midawi, C.H.M. Simha, M.A. Gesing, A.P. Gerlich, International Journal of Solids and Structures **104–105**, 81-91 (2017)
7. A.N. Beskopylnyi, A.A. Lyapin, Recent trend in Science and Technology management **23-29**, 36-45 (2016)
8. J. Dean, T.W. Clyne, Mechanics of Materials **105**, 112-122 (2017)
9. A.S. Useinov, K.S. Kravchuk, A.A. Rusakov, I.V. Krasnogorov, A.P. Kuznetsov, T.V. Kazieva, Physics Procedia **72**, 194-198 (2015)
10. I. Mohagheghian, W.J. Stronge, G.J. McShane, European Journal of Mechanics **61**, 134-150 (2017)
11. I.I. Argatov, International Journal of Solids and Structures **48**, 3444-3452 (2011)
12. Wei-Min Chen, Yang-Tse Cheng, Min Li, Materials Science and Engineering **527**, 5613-5618 (2010)
13. D.M. Belenkii, A.N. Beskopylnyi, N.N. Beskopylnyi, E.K. Polibin, B.A. Pesenko Method for determining mechanical characteristics and the device for its implementation. Patent No. 2079831 (1997)
14. D.M. Belenkii, A.N. Beskopylnyi, N.L. Vernezi News of Higher Education Institution. Construction. **1**, 99-102 (2003)
15. D.M. Belenkii, A.N. Beskopylnyi, N.L. Vernezi, L.G. Shamraev Welding International, №**11** **(8)**, pp. 642-645 (1997)

Quality assessment of structural steel in mining equipment manufacture

Aleksandr Korotkov[1,*], and *Lidia* Korotkova[1]

[1]Kuzbass State Technical University, Department metal-cutting, machines and tools, 650000, Kemerovo, Russian Federation

Abstract. The relevance of this work is to form a methodical, scientific-based approach and the application of this experience in solving problems of quality control and improvement, as well as to identify ways of eliminating defects and small lifetime of the materials, used in critical parts and units of mining equipment. The aim of scientific research of the article is the study of discrepancy of structural steels quality problem, used for the manufacture of parts and components of mining equipment, established by standards. The goals are also, development and practical application of quality control methods in delivery condition and at all main stages of parts manufacture in accordance with the current state standards. As a result of steel quality complex control in delivery condition, it was possible to identify: the deviation of the chemical composition of steels and alloys, macrodefects, microdefects, deviation in hardness. Practical recommendations to eliminate and reduce and prevent the defects in parts and units used in mining engineering were developed. The paper presents practical experience in evaluation the quality of construction materials, which may be useful in design, manufacture and operation of mining industry and related manufacture industries.

1 Introduction

High demands are maintained to the reliability of mining equipment parts and one of the main issues of this demand is that the quality of the materials used must be established by standards. Technological control of steels and alloys in mining industry enterprises is an obligatory element of management system of goods quality, the task of the system is to prevent the release of defective goods.

In recent years, the problem of relevance of metals quality correspondence to the established standards increased. Its solution is possible through the creation of laboratories with modern metallographic equipment, training of qualified staff and the development of necessary methodical support to control materials quality.

The KuzSTU department "Metal-cutting machines and tools" studied some aspects of this problem on the example of steels and alloys quality assessment used in enterprises of Kuzbass coal mining, and the ways of its solution were suggested. The correlation of intense mining equipment parts wear with the quality of metal was studied; and recommendations for repair, recovery and work life extension of machine parts for such goods were developed.

2 Materials and Methods

In the course of the research the method of structural steels quality control used in mining engineering was developed and implement. The method is based on the current state standards, and its application allows to solve the problem of provision of the required characteristics for the parts of mining equipment.

The method of steel quality control on the basic parameters of the microstructure is shown in Table 1.

Steels quality control should include the following basic steps:

- structural steel quality control in delivery condition for compliance with standards of the following parameters: visual and measuring control, chemical analysis, mechanical and technological tests, metallographic tests;

-analysis of thermal processing at all major technological stages of production, the development of recommendations for the initial heat treatment to eliminate the defects of the microstructure, the adjustment of hardening heat treatment in order to ensure the necessary properties to complex parts and goods;

- recommendations' formulation for rejecting defected metal parcels on the basis of their quality control results in delivery condition.

After visual and measuring control of steels their chemical analysis is carried out, since it is the main characteristic.

In particular, the full range of physical, mechanical and technological properties of steels depends on the content of chemical elements and chemical impurities. Macroanalysis is also an important step in steels quality control, it allows to assess the quality of the metal and its previous processing, affecting the property of metal under heat treatment during subsequent processing. Microanalysis reveals structural defects as in delivery

* Corresponding author: korotkov.a.n@mail.ru

 DOI: 10.1051/matecconf/201712902047

condition and also at the stages of initial and hardening heat treatment. Indicators of metals mechanical properties are measured according to the strength, hardness, relevant elongation, contraction, toughness and etc.

Table 1. The main parameters of the microstructure and control conditions of structural steels quality

Controlled index	Sample condition (HT); Conditions of sample preparation	Microscope extansion; Index assessment method
1	2	3
Decarbonized layer	Heat treatment; Polishing in diametral plane *a**; etching operation by 2–4% alcoholic solutionn HNO_3	100ˣ and more; Measurement by ocular ruler on two samples on GOST 9012, allowed 2 %×*d*, not more
Pearlite microstructure	Heat rolled, annealed Polishing in diametral plane *a**; etching operation 2–4% alcoholic solution HNO_3	500ˣ (450–600); Comparison with the standard scale on GOST 9454, allowed grade 4, not more
Carbide network	Heat treatment; Polishing in diametral plane *a**; etching operation 2–4% alcoholic solution HNO_3	100ˣ and more; Comparison with the standard scale on GOST 9454, allowed grade 3, not more
Carbide segregation	Heat treatment; Polishing in diametral plane *a**; etching operation 2–4% alcoholic solution HNO_3	90–110ˣ; Comparison with the standard scale, allowed grade 1–2, not more
Banded orientation	Non annealed state; Polishing in diametral plane *в***; etching operation 24% alcoholic solution HNO_3	90–110ˣ; Comparison with the standard scale on GOST 1763, allowed grade 2–3, nor more
Microporocity	Heat treatment; Polishing in diametral plane *a**	90–110ˣ; Comparison with the standard scale, allowed grade 2, not more
Nonmetalic inclinations	Heat treatment Polishing in diametral plane *a**	90–110ˣ; Comparison of the most contaminated place with the scales 1, 2, 3 of standard on GOST 8233, allowed grade 2–3, not more
Grain size	Heat rolled, annealed; Polishing in diametral plane; etching operation to blackening	100ˣ 200ˣ 400ˣ; Method according to grain size. The grades on the scale of the standard on GOST 5640, allowed grade 6–8, not lower

3 Results and Discussion

As the examples of the application of the developed steels quality control method the practical experience of such developments for the Kuzbass enterprises is shown.

In the course of development of production of hydraulic roof supports the problem of low strength cylinder was discovered. To eliminate it, the complex of the following activities was done:

- structural steel used for the manufacture of hydraulic cylinders was explored;
- deviations from the standard indicators were determined;
- hardening heat treatment to provide the required mechanical properties of goods was designed;
- recommendations were developed of initial heat treatment in order to remove defects and to provide the desired technological properties of steel in delivery condition.

Chemical analysis allowed identifying the supplied steels for compliance with steel grades. It was found that these heat rolled steel pipes are made from steels 45, 30ChGSA and 35Ch [1, 2]. According to the basic elements the chemical composition of all steels was agreed with standards GOST 4543 [3], but unacceptable overstated sulfur content was found in these steels.

Metallurgical studies (Table 1) include an assessment of contamination by nonmetallic inclusions, grain size determination, banded orientation, microstructure research (on GOST 8233, GOST 5639, GOST 1778) [4].

 DOI: 10.1051/matecconf/201712902047

Microstructure researches have shown the difference in heat treatment of steel in delivery condition: pipes from steel 45 are normalized heat rolled ones with ferrite pearlitic structure. Pipes from 35ChGSA steel and steel 35Ch are an improved heat rolled ones with the structure of granular sorbite (on GOST 4543, GOST 10243) [5, 6, 7].

In some parcels of pipes from steel 45 defective overheated structures were found with signs of widmanstatt with unacceptable grain size corresponding to the 5th grade. In the pipes from steel 35Ch the defect of structures was found – it is the inclusion of graphite in the structure of sorbite. In all parcels the defects were found in the form of non-metallic inclusions with admissible limit values (on GOST 1778) [8, 9].

Thus, metallographic studies indicate the low quality of the metal in delivery condition.

In the results of the complex control the company was recommended:

-to use air-hardening as initial heat treatment, and to use refining as strengthening treatment – that will provide the desired mechanical complex of properties to hydraulic cylinders (on GOST 9012, GOST 9454, GOST 1497);

-to organize systematic complex quality control of steels in the enterprise;

-to correct thermal treatment, taking into account monitoring results.

Another example of the developed method application relates to the production of a drilling tool. In particular, the problems associated with low-strength of holders of worm cutting tools from steel 35ChGSA were studied [10, 11, 12, 13].

The researches were carried out at the main stages of the production process: after forging with various variants of initial heat treatment and after strengthening heat treatment.

Quality control of steel was carried out in accordance with GOST 4543. The presence of non-metallic inclusions, pores, flakes, forging cracks was investigated. The microstructure was studied at all major stages of the heat treatment process. The analysis was conducted using a metallographic studies and micro-hardness measurements. The fracture character of cutting tools was determined by fractures type, received in the result of the destruction of a tool during operation under actual production conditions.

Blanking operation consisted of forging and subsequent heat treatment. The working part of the cutting tool holder was subjected to forging, which was initially heated by high frequency currents (HFC). The temperature for the beginning and end of the forging was not controlled. After hot treatment by pressure two variants of initial heat treatment were used: I variant - annealing (at 700 ^{0}C during 6 hours); II variant - refining (heat treatment at 880 ^{0}C in oil followed by blazing-off at 600 ^{0}C for 1.5 hours, cooling - air).

In microstructure of the forged sample a number of unacceptable defects was identified: large grain (3-4 grades); widmanstatten structure; ferrite network; the presence of flakes, non-metallic inclusions above the admittible limits [4, 5, 6, 7].

The results of microstructure studies and measurement of microhardness of testify to the heterogeneity of microstructure and hence to the irregularity of properties along the axis of the cutting tool holder. Here there are three distinct zones. The first zone is located directly in the cutter head (up to the depth of 10 mm), it has a characteristic structure of overheated steel in the form of widmanstatten structure. The second zone is transitional and extremely heterogeneous (with the depth of 10 to 25 mm); it includes large size grain ferrite pearlitic structure with ferrite network along grain boundary with inclusions of sorbitic pearlite. The third zone is located in the tail of the cutting tool (at a distance of 25 mm), refers to a non-deformable section of the sample and has a fine grain ferrite pearlite structure of normalize steel. The heterogeneity of the microstructure in the forged sample is confirmed by microhardness measurements.

Thus, on the basis of technological inheritance studies, based on the structure researches, microhardness measurements, analysis of fractures [12,13], the adjustment has been made in heat treatment of worm of the cutting tool from steel 35ChGSA:

-in the capacity of initial heat treatment, it is recommended to apply a complete annealing (840-860 ^{0}C), that allowed to obtain homogeneous structure along the whole length of the cutter, to remove defects in the structure (large grain, structural heterogeneity), acquired at the stage of forging;

- in the capacity of strengthening heat treatment it is recommended to apply blazing-off (250 ^{0}C) after heat treatment (880 ^{0}C) with rapid cooling in oil, instead of the average, which improves the toughness of the steel (1-1.2 mJ / m3);

-the enterprise was recommended to increase the requirements to the supplier because of the low quality of supplied steel, associated with unacceptably large number of non-metallic inclusions and the presence of harmful flakes in the microstructure.

The parts wear of a high-pressure pump (70-90 atmospheres) was also investigated (the pump is used by coal mining enterprises), due to high wear and tear of its parts, and rapid failure of the whole pump mechanism. The object of the study was the leading pump shaft and gears, front infill plates and a plain bearing line. As a result of metallographic tests, it was estimated that the shaft is made of low carbon steel containing 0.178% of carbon, and it is alloyed with 1.058% manganese and 0.940% chromium. The analogue of such steel is a national steel grade 18ChG GOST 4543 [3].

Thermal processing of a shaft provides the hardness 57-59 HRC. The microstructure of the shaft over the entire section consists of a low-carbon martensite. The structure of landing part of the shaft is heterogeneous: the shaft core is composed of ferrite pearlitic normalized structure, and in the zone of contact with the gear the hardness is 51-53 HRC, which corresponds to the martensite structure [4].

The gear is made of high carbon steel type Ch, ShCh15. It is heat treated for maximum hardness of 60-64 HRC. The microstructure consists of martensite and derived carbide inclusions which are distributed

DOI: 10.1051/matecconf/201712902047

uniformly and have a dispersed structure with a size of 1-2 microns. The layer of ferrite fine grained steel is rolled on the outline of the gear surface. The layer thickness is 0.08-0.10 mm.

Studies of front infill plates of a pump showed that they are made of carbon-free ferrite steel and their surface is coated with a layer of powders of cast tin bronze. The powder on the infill plate is sprayed [14, 15]. Analogs of these bronzes can be done from national tin bronze of Br05S25 Br08S12 grades according to GOST 613-79.

As a result of the research and determined chemical composition of the pump parts, it becomes possible to manufacture them from national materials mentioned above.

4 Conclusion

As a result of this research, the following conclusions can be formulated:

1. There is a problem with low-quality steel goods in delivery condition (metal deviations according to chemical composition, the macroscopic defects presence, the defects presence in the microstructure, mechanical properties discrepancy), which is projected to enterprises of mining engineering.

2. A method of steel quality control in delivery condition and during manufacturing process of parts was developed, contributing to the improvement of the quality, and therefore provision of the specified performance characteristics of the various engineering goods under the manufacture of mining equipment.

3. Due to the metals quality control (in particular, the establishment of their chemical composition and heat

treatment features), it is possible to repair and restore the parts and goods of foreign production from national metals and alloys.

4. In general, it should be mentioned that the metal quality control problem is becoming increasingly important and to solve it the factory laboratories should be developed, or centralized laboratories should be created (e.g., on the basis of technical institutes) which are capable to solve such alloy sheet during rolling. Structural materials: properties, microstructure and processing.

References

1. Y.P. Solntsev. *Metals and alloys. Reference book.* (ANO NPO Professional, St. Peterburg, 2003)
2. A.S. Zubchenko. *Database of steels and alloys.* (Mechanical engineering, Moscow, 2001)
3. *Handbook of steels.* (Sendvikens Trykeri AB, Sweden, 2013)
4. H. Hiorns. *Metallography* (An Introduction to the Study of the Structure of Metals, 2009)
5. G. Zlateva. *Microstructure of Metals and Alloys: An Atlas of Transmission Electron Microscopy Images.*

(Taylor&Francis Group, 2008)

6. B. Bramfitt. *Metallographer's Guide: Practices and Procedures for Irons and Steels.* (ASM International, 2002)
7. E. Berghof-Hasselbächer. *Atlas of Microstructures.* (Main Germany, 2013)
8. M. Shtremel, MiTOM. **8**, (1980)
9. M. Gergiev, MiTOM, **2**, (1974).
10. L. Korotkova, MiTOM, **3**, (1988)
11. G. Muchin, MiTOM, **4**, (1982)
12. A. Amosov, Bulletin of universities. Ferrous metallurgy, **1**, (2004)
13. S.-J. Kim, *Structurual materials: properties microstructure and prosessing.* (2014)
14. A. Korotkov, Applied Mechanics and Materials, **682** (2014)
15. M. Tuttle Structural *Analysis of Polymeric Composite Materials.* (Hardcover Book, 2007)

DOI: 10.1051/matecconf/201712902048

Metamaterials based on wedge-shaped electrodynamic structures

Vladimir Mitrokhin[1], *Georgy* Kuvyrkin[1], and *Dmitry* Ryzhenko[1*]

[1]Bauman Moscow State Technical University, 105005 Moscow, Russia

Abstract. The paper studies a possibility of simulation of artificial composite media with negative values of the real part of the equivalent dielectric (magnetic) permittivity, by the use of segments of hollow composite waveguides with cylindrical guided waves in evanescent mode. Reactive evanescent fields of wedge-shaped waveguide eigenmodes are formed in the evanescent region before the critical section of the waveguide which separates the quasistatic field region from the distributing field of the evanescent waveguide mode. The possibility of simulation is determined by the equivalence of dispersion equation of the eigenmode propagation constant and the dispersion equation for the electric (magnetic) permittivity of plasma-like medium if cut-off frequency and electric (magnetic) plasma frequency of the medium are equal.

1 Introduction

In recent years, developers of microwave devices and antennas have been attracted by metamaterials and their use in microwave engineering [1-9]. Different combinations of infinitesimal electric and magnetic dipoles (wire structures, posts, bars, frames of various configurations, rings, etc.) are used for implementation of metamaterials. Often, waveguide sections operating in cutoff mode (evanescent mode) are used in combination with the above mentioned dipoles. Artificial composite medium proposed in [11] was made by placing a regular array of metallic posts between two parallel metallic plates. Further simplification of such model was achieved in [12] by using only one waveguide due to its dispersion properties.

2 Metamaterials based on wedge-shaped electrodynamic structures

In hollow metallic waveguides, E_{mn} and H_{mn} eigenmodes satisfy dispersion relation [10]:

$$\Gamma = k\sqrt{\mu\varepsilon}\ , \qquad (1)$$

where Γ is eigenmode propagation constant; $k = \omega\sqrt{\varepsilon_0\mu_0} = 2\pi/\lambda$ - phase constant of empty space with ε_o and μ_o parameters – dielectric permittivity and magnetic permeability of vacuum; ω - the angular frequency; λ - the wavelength in vacuum; μ - the magnetic permeability (for hollow waveguide μ= 1); ε_{eff} - the effective dielectric constant given by

$$\varepsilon_{\text{eff}} = 1 - f_{\text{cr}}^2 / f^2\ , \qquad (2)$$

where $f=\omega/2\pi$ is the operating frequency; f_{cr} - the cutoff frequency of the considered type of eigenmode, e.g., H_{mn}.

Considering that the dispersion relation (1) with account of (2) is identical of that of a lossless plasma medium model with plasma frequency $f_{pe} = f_{cr}$ [10], a hollow metallic waveguide can be regarded as one-dimensional plasma with respect to EM wave propagation along the axial direction of the waveguide. Moreover, the equivalence between waveguide field spread below cutoff frequency of the eigenmode and EM wave propagation in artificial plasma in the form of evanescent field is theoretically established in [13] and experimentally confirmed in [14]. The equivalence exists between waveguide and electric plasma for evanescent H_{mn} modes of rectangular and circular waveguides. The discussed equivalence shows that medium with negative dielectric permittivity can be simulated by means of H_{mn} modes at frequencies below cutoff, while medium with negative magnetic permeability can be simulated by means of E_{mn} modes at frequencies below the corresponding cutoff.

With such approach double-negative medium is implemented by placing a periodic array of split ring resonators inside a rectangular or circular waveguides operating at frequencies below corresponding cutoff frequencies of H_{10} and H_{11} modes [3, 4, 6]. The periodic array of resonators is in fact artificial anisotropic magnetic plasma with permeability tensor [13]:

$$\vec{\mu} = \begin{pmatrix} \mu_{\text{pl}} & 0 & 0 \\ 0 & 1 & 0 \\ 0 & 0 & 1 \end{pmatrix}, \qquad (3)$$

$$\mu_{\text{pl}} = 1 - \frac{f_{\text{pm}}^2 - f_0^2}{f^2 - f_0^2 - i\gamma_{\text{m}} f / 2\pi}, \qquad (4)$$

* Corresponding author: dsr@bmstu.ru

DOI: 10.1051/matecconf/201712902048

where f_{pm} is magnetic plasma frequency (when Lorenz model of the considered medium is used); f_o is resonant frequency; γ_m - magnetic plasma losses.

In such double-negative medium, the EM wave propagates with propagation constant

$$\Gamma = k\sqrt{(1-f_{cr}^2/f^2)(1-f_{pm}^2/f^2)}\,, \qquad (5)$$

In this structure, H_{mn} modes can propagate at frequencies at which both brackets in the radical expression have the same sign. Backward wave is formed at frequencies lower than both f_{cr} and f_{pm}. Forward waves propagate at all frequencies that are higher than both f_{cr} and f_{pm}. Between the latter, there is a stopband. When $f_{cr} = f_{pm}$. all types of waves become propagating.

For E_{mn} modes of the rectangular waveguide, dispersion relation can be expressed as

$$\Gamma = k\sqrt{\varepsilon\mu_{\text{eff}}}\,, \qquad (6)$$

where μ_{eff} is effective relative magnetic permeability defined by

$$\mu_{\text{eff}} = 1 - f_{cr}^2/f^2\,, \qquad (7)$$

where f_{cr} is the cutoff frequency of the considered E_{mn} mode; for hollow waveguide, $\varepsilon=1$. Considering that the dispersion relation (3) with account of (4) is identical of that of a lossless plasma medium model with plasma frequency $f_{pm} = f_{cr}$ [14], a hollow metallic waveguide can be regarded as one-dimensional plasma with respect to EM wave propagation along the axial direction of the waveguide. In [14], the authors propose to simulate artificial medium with negative magnetic permeability by using a hollow square waveguide operating at frequencies below cutoff frequency of the lowest E_{11} mode of the E_{mn} eigenmode family.

Double-negative metamaterial medium is implemented by placing a periodic array of wires inside the square waveguide operating at frequencies below the cutoff frequency of the E_{11} mode [13, 14]. The array structure is in fact artificial anisotropic magnetic plasma with dielectric permittivity tensor

$$\ddot{\varepsilon} = \begin{pmatrix} \varepsilon_{\text{pl}} & 0 & 0 \\ 0 & \varepsilon_{\text{pl}} & 0 \\ 0 & 0 & 1 \end{pmatrix}, \qquad (8)$$

$$\varepsilon_{\text{pl}} = 1 - \frac{f_{pe}^2 - f_0^2}{f^2 - f_0^2 - i\gamma_e f/2\pi}\,, \qquad (9)$$

where f_{pe} is electric plasma frequency (when Lorenz model of the considered medium is used [6]); f_o is resonant frequency; γ_e - electric plasma losses.

In such double-negative medium, the EM wave propagates with propagation constant

$$\Gamma = k\sqrt{(1-f_{cr}^2/f^2)(1-f_{pe}^2/f^2)}\,. \qquad (10)$$

The E_{mn} modes propagate at frequencies at which round parenthesis in the radical expression have the same sign. Backward wave is formed at frequencies lower than both f_{cr} and f_{pe}. Forward wave propagates at frequencies that are higher than both of the above.

Regions with negative dielectric permittivity and magnetic permeability in nonhomogenious waveguides with spherical and cylindrical guided waves are implemented by means of evanescent reactive field of the corresponding eigenmodes before critical section of the waveguide separating the quasistatic field region from the distributing field of the evanescent waveguide mode [1,10,19].

In wedge-shaped waveguides with cylindrical guided waves (in ρ, φ, y coordinates) (Figure 1), Maxwell equations can be solved in the form of superposition of "electric' and "magnetic" waves, with their components determined via U and V potential functions [10]. Terms "electric" («E_{mo}») and "magnetic" («H_{mo}») are quoted. It means that such splitting does not comply with the accepted waveguide classification, i.e. y coordinate does not correspond to propagation direction. The waves are cylindrical and propagate in the direction of the ρ^0 radial coordinate. The U and V functions satisfy Helmholtz equations, with particular solutions given by

$$\left.\begin{matrix} U \\ V \end{matrix}\right\} = \left\{\begin{matrix} A \\ B \end{matrix}\right. Z_{p_m}(k\rho) \begin{matrix} \sin \\ \cos \end{matrix}\left(p_m\varphi + \frac{m\pi}{2}\right), \qquad (11)$$

where A and B are constant coefficients determined by excitation conditions; $Z_m(k\rho)$ is the radial dependence of the field satisfying boundedness condition at $k\rho \to 0$, radiation condition at $k\rho \to \infty$, and being continuous at the critical section [15]:

$$k\rho = (k\rho)_{\text{cr}} = \sqrt{p_m^2 - 0{,}25}\,, \qquad (12)$$

where $p_m = m\pi/(2\gamma)$ are eigenvalues determined from boundary conditions; m – whole numbers; 2γ - wedge angle.

Considering the above conditions, radial dependence should be written as:

$$Z_{p_m}(k\rho) = \begin{cases} H_{p_m}^{(2)}[(k\rho)_{\text{cr}}]J_{p_m}(k\rho), & k\rho \le (k\rho)_{\text{cr}}, \\ J_{p_m}[(k\rho)_{\text{cr}}]H_{p_m}^{(2)}(k\rho), & k\rho \ge (k\rho)_{\text{cr}}. \end{cases} \qquad (13)$$

where $Jp_m(k\rho)$ are cylindrical Bessel functions of the first kind and p_m order, $Hp_m^{(2)}$ are cylindrical Hankel functions of the second kind (exp(iωt) time dependence of the field).

Transverse (relative to ρ^0 direction) $\mathbf{E}_\perp$, $\mathbf{H}_\perp$ field components constitute superposition of transverse EM field components for «E_{mo}» and «H_{mo}» waves:

$$\mathbf{E}_\perp = \sum_m \mathbf{E}_{\perp e}^m + \sum_m \mathbf{E}_{\perp m}^m\,, \qquad (14)$$

$$\mathbf{H}_\perp = \sum_m \mathbf{H}_{\perp e}^m + \sum_m \mathbf{H}_{\perp m}^m\,, \qquad (15)$$

where

 DOI: 10.1051/matecconf/201712902048

$$\mathbf{E}^{m}_{\perp e} = \frac{1}{\rho}\frac{\partial^2 U}{\partial y \partial \phi}\phi^0 + \left(\frac{\partial^2 U}{\partial y^2} + k^2 U\right)y^0, \quad (16)$$

$$\mathbf{H}^{m}_{\perp e} = -i\omega\varepsilon_a \frac{\partial U}{\partial \rho}\phi^0, \quad (17)$$

$$\mathbf{E}^{m}_{\perp m} = i\omega\mu_a \frac{\partial V}{\partial \rho}\phi^0, \quad (18)$$

$$\mathbf{H}^{m}_{\perp m} = \frac{1}{\rho}\frac{\partial^2 V}{\partial y \partial \phi}\phi^0 + \left(\frac{\partial^2 V}{\partial y^2} + k^2 V\right)\mathbf{y}^0, \quad (19)$$

Here, φ^0, y^0 are unit vectors along the φ and y coordinates.

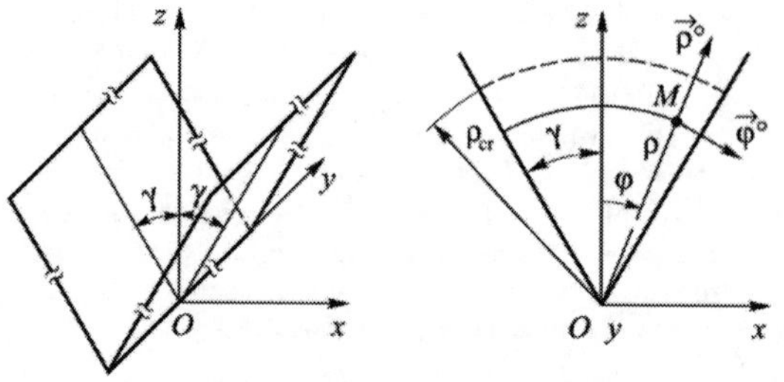

a b

Fig. 1. Wedge-shaped waveguide: *a*) general view; b) in Cartesian and cylindrical coordinate systems.

The Γ propagation "constant" of the eigenmodes is determined by means of logarithmic derivative of radial dependence [16]:

$$\Gamma = -ikZ'_{p_m}(k\rho)/Z_{p_m}(k\rho), \quad (19)$$

where prime mark means full derivative of the argument. Since Γ depends on ρ, The word "constant" is quoted.

Using asymptotic representation of Hankel function at $k\rho \to \infty$, it is possible to show that at narrow wedge angles

$$\Gamma \approx k\sqrt{1-(p_m^2-0,25)^2/(k\rho)^2} = k\sqrt{1-f_{cr}^2/f^2}, \quad (20)$$

where

$$f_{cr} = c\sqrt{p_m^2-0,25}/(2\pi\rho_{cr}), \quad (21)$$

coinciding with (1) – (7) for regular waveguides. It allows to use the above methods for regular waveguides in this case too.

In the first example, let us consider a «E_{10}» mode, for which EM field components are given by:

$$E_y = k^2 U = Ak^2 Z_{p_1}(k\rho)\sin(p_1\phi + \frac{\pi}{2}), \quad (22)$$

$$H_\phi = -i\omega\varepsilon_a \frac{\partial U}{\partial \rho} - Ai\omega\varepsilon_a Z'_{p_1}(k\rho)\sin\left(p_1\phi + \frac{\pi}{2}\right), \quad (23)$$

Impedance for this type of wave is

$$Z_{«E_{10}»} = E_y / H_\phi = Z_0 k / \Gamma, \quad (24)$$

where

$$Z_0 = \sqrt{\mu_a / \varepsilon_a}, \quad (25)$$

Z_0 is free space characteristic impedance with μ_a and ε_a parameters.

Comparing (20) and (1) at μ=1, we get

$$-iZ'_{p_m}(k\rho)/Z_{p_m}(k\rho) = \sqrt{\varepsilon_{eff}}, \quad (26)$$

From which

$$\varepsilon_{eff} = -\left[Z'_{p_m}(k\rho)/Z_{p_m}(k\rho)\right]^2, \quad (27)$$

If $k\rho \le (k\rho)_{cr}$,

$$\varepsilon_{eff} = -\left[J'_{p_m}(k\rho)/J_{p_m}(k\rho)\right]^2, \quad (28)$$

If $k\rho \ge (k\rho)_{cr}$, expressions (7) and (1), given μ=1, with account of relation (13) for propagation of the considered wave type after critical section, give

$$-\frac{2}{\pi(k\rho)}\left[J^2_{p_m}(k\rho) + N^2_{p_m}(k\rho)\right]^{-1} = \sqrt{\varepsilon_{eff}}, \quad (29)$$

where $Np_m(k\rho)$ are Neumann functions.

The relation gives for $k\rho \ge (k\rho)_{cr}$

$$\varepsilon_{eff} = \frac{4}{\pi^2(k\rho)^2}\left[J^2_{p_m}(k\rho) + N^2_{p_m}(k\rho)\right]^{-2}. \quad (30)$$

Consider the effective dielectric permittivity for «E_{10}» mode of the wedge-shaped waveguide with 2γ=π/2 wedge angle. In this case,

$$p_m = p_1 = \pi/(\pi/2) = 2, \quad (31)$$

then

$$(k\rho)_{cr} = \sqrt{p_1^2 - 0,25} \approx 2, \quad (32)$$

Note that $(k\rho)$ variable in expression (28) and (30) can be expressed as:

$$\varepsilon_{eff} = \begin{cases} -\left\{\left[\frac{f_{cr}}{f} - J_0\left(\frac{2f}{f_{cr}}\right)\Big/ J_1\left(\frac{2f}{f_{cr}}\right)\right]^{-1} - \frac{f_{cr}}{f}\right\}^2, & f < f_{cr} \\ \left\{(\pi f/f_{cr})\left[J_2^2(2f/f_{cr}) + N_2^2(2f/f_{cr})\right]\right\}^{-2}, & f > f_{cr} \end{cases} \quad (33)$$

The dependence is shown in the diagram in Figure 2. As we see, evanescent region of the wedge-shaped waveguide has negative dielectric permittivity. Double-negative medium can be implemented by placing an array of split-ring resonators inside the wedge-shaped region according to the structure of the «E_{10}» (22) EM field, i.e. resonator lattice plane should be perpendicular to the magnetic field vector lines.

In the second example, let us consider «H_{10}» mode, for which EM field components are given by:

DOI: 10.1051/matecconf/201712902048

$$E_\phi = i\omega\mu_a \frac{\partial V}{\partial \rho} = Bi\omega\mu_a k Z'_{p_1}(k\rho)\cos\left(p_1\phi+\frac{\pi}{2}\right), \quad (34)$$

$$H_y = k^2 V = Bk^2 Z_{p_1}(k\rho)\cos(p_1\phi+\frac{\pi}{2}), \quad (35)$$

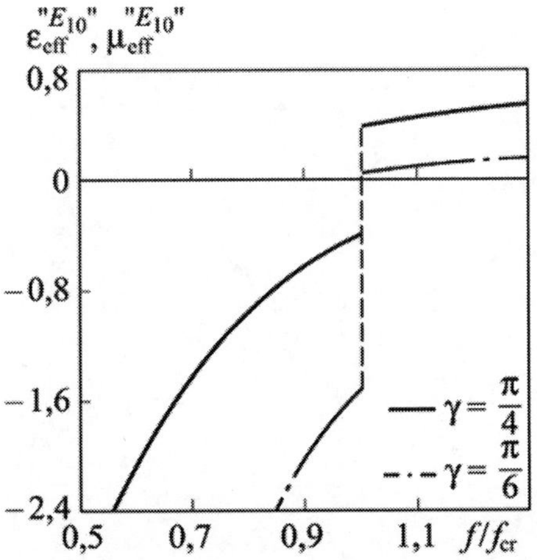

Fig. 2. Effective dielectric permittivity $\varepsilon_{eff}^{"E_{10}"}$ and magnetic permeability $\mu_{eff}^{"H_{10}"}$ of the wedge-shaped region at «E_{10}» and «H_{10}» waves, correspondingly, for different wedge angles.

Impedance for this type of wave is

$$Z_{"H_{10}"} = E_\phi / H_y = -Z_0\Gamma / k, \quad (36)$$

From expressions (6) and (19) at ε=1, we get

$$-iZ'_{p_m}(k\rho)/Z_{p_m}(k\rho) = \sqrt{\mu_{eff}}, \quad (37)$$

From which

$$\mu_{eff} = -\left[Z'_{p_m}(k\rho)/Z_{p_m}(k\rho)\right]^2, \quad (38)$$

For $k\rho \leq (k\rho)_{cr}$ μ_{eff} is determined by formula (28), with ε_{eff} substituted by μ_{eff}.

If $k\rho \geq (k\rho)_{cr}$, then, by comparing expressions (6) and (19), given ε=1 with account of relation (13) for propagation of the considered wave type after critical section, we get expression (30) for μ_{eff} if ε_{eff} is substituted by μ_{eff}.

The effective dielectric permittivity for «H_{10}» mode of the wedge-shaped waveguide with 2γ=π/2 wedge angle proves to be equal to (33) if ε_{eff} is substituted by μ_{eff}. This dependence is also shown in the diagram in Fig. 2. As we see, evanescent region of the wedge-shaped waveguide has negative magnetic permeability. Double-negative medium can be implemented by placing a periodic array of thin wires inside the wedge-shaped region according to the structure of the «H_{10}» (34) EM field, i.e. wires direction should coincide with the direction of the electric field.

3 Conclusion

The research that has been carried out allows to make conclusions as follows:

- a region with negative dielectric permittivity can be implemented based on hollow wedge-shaped waveguide at frequencies below the cutoff frequency of the «E_{10}» mode.
- a region with negative magnetic permeability can be implemented based on hollow wedge-shaped waveguide at frequencies below the cutoff frequency of the «H_{10}» mode.

The research work was supported by Ministry of Education and Science of the Russian Federation under state task No 8.4868.2017/8.9.

References

1. V.N. Mitrokhin, D.S. Ryzhenko and N.Yu. Fadeeva, Antennas **12(199)**, 4 (2013)
2. Y. Dong and T. Iston Microwave magazine **5** 39 (2012)
3. D.S. Ryzhenko, V.N. Mitrokhin, V.A. Tyagunov, *Microwave Bandpass Filter* (Utility Model Patent RU 113077U, 2012)
4. D.S. Ryzhenko, V.N. Mitrokhin, V.A. Tyagunov, *Microwave Bandstop Filter* (Patent for Invention RF 2459321, 2012)
5. A.K. Sarychev and V.M. Shalaev, *Electrodynamics of metamaterials* (Moscow: Scientific world, 2011)
6. V.N. Mitrokhin, D.S. Ryzhenko and V.A. Tyagunov, Physics of Wave Processes and Radiotechnical Systems 14(3), 43 (2011)
7. V. Slyusar The First Line **34** 44 (2010)
8. V.N. Mitrokhin and D.S. Ryzhenko Vestnik SONIIR 4(26) 70 (2009)
9. Yu.V. Gulyaev, A.N. Lagar'kov and S.A. Nikitov, Herald of the Russian Academy of Sciences **78(5)**, 438 (2008)
10. N.S. Golubeva and V.N. Mitrokhin, *Fundamentals of SHF Radio Electronics: Study Guide* (Moscow: The Bauman University Publishing House, 2008)
11. D.R. Smith, W.J. Padilla, D.C. Vier, S.C. Nemat-Nasser and S. Schultz, Phys. Rev. Lett. **84(18)**, 4184 (2001)
12. R. Marques, J. Martel, F. Mesa and F. Medina, Phys. Rev. Lett. **89(18)**, 183901 (2002)
13. C. Camacho-Penalos, J. Esteban, T.M. Martin-Ginerero and E. Marques-Segyra, *27th ESA Antenna Technology Workshop* (Santiago de Compostela, Spain, March 9-11) 461-468 (2004)

14. J. Esteban, C. Camacho-Penalos, J.E. Page, T.M. Martin-Ginerero and E. Marques-Segyra, Transaction оп Microwave Theory and Techniques **53(4)**, 1506 (2005)
15. V.N. Mitrokhin, Radiotekhnika **4**, 43 (1999)
16. V.N. Mitrokhin, Radiotekhnika **3,** 62 (1985)
17. V.N. Mitrokhin, Radiotekhnika **7,** 31 (1989)
18. V.N. Mitrokhin, Herald of the BMSTU. Instrument Engineering **1,** 53 (1990)
19. V.N. Mitrokhin, Herald of the BMSTU. Instrument Engineering **1,** 66 (1991)
20. V.N. Mitrokhin, Radiotekhnika **8,** 64 (2002)

The application of statistical methods to control the quality of the contact pairs of electrical connectors

Konstantin V. Podmasteryev [1], *Vladimir V.* Markov[1], and *Anna V.* Morozova[2,*]

[1] Oryol State University name's of I.S. Tyrgenev, Department «Device-building, Mertology and Sertification», 302026 Orel 95 Komsomolskaya Street, Russia
[2] Bryansk State Technical University, Department Innovative Development of University, 241035 Bryansk 7, 50 let Oktyabrya bul., Russia

Abstract. The development of electronic technology in the direction of improving its performance and speed, accompanied by the decrease of its dimensions and mass. This trend requires a new generation of electrical connectors differing from the basic models of a large number of contacts and density of placement on the insulator, having low contact resistance and high insulation resistance, high reliability and long life. The article describes the way of practical application of statistical methods for quality control of contact pairs of electrical connectors. The objects of study of effectiveness of statistical methods of quality management selected two typical details contact pairs received by turning on automatic lathes: pin and socket. The article considers the process of collecting statistical information about the technological process automatic turning of parts of the contact pairs. Shows the results of processing statistical information about the quality of the details of the contact pairs. The use of statistical methods for quality control of contact pairs of electrical connectors allows to correctly configure technological equipment according to the results of making a test batch of parts. Recommendations are made to improve the quality of the contact pairs of connectors. Compiled by technique of quality management of contact pairs according to the results of the statistical analysis of technological processes of their manufacture.

1 Introduction

The quality standard of products is determined by the perfection of technological processes of its manufacture [5, 6, 8, 9, 23]. The technological process of production is accompanied by a manifestation of many factors [11, 14, 18, 19, 22] that vary randomly [10, 12, 16, 20, 21]. Therefore, to evaluate the degree of perfection of technological processes it is necessary to use statistical methods of information analysis [7, 13, 15, 17, 20]. Statistical methods allow us to estimate the basic properties of the process: willingness, precision, and stability. Typical examples of batch production are contact pairs of electrical connectors. They are manufactured in batch sizes from hundreds to tens of thousands of pieces. For the manufacture of parts of contact pairs are used by the automatic equipment. For example, parts of the "body of rotation" are manufactured by turning on lathes. Therefore, technological processes for the manufacture of contact pairs must obey the known laws of distribution of probabilities. The purpose of this paper is a summary of the methodology of application of statistical methods for quality control of contact pairs of electrical connectors and a demonstration of the results of experimental studies of the effectiveness of its application.

2 Description of research objects

The objects of study of effectiveness of statistical methods of quality management selected two typical details contact pairs received by turning on automatic lathes: pin and socket. Blanks for pin and socket contact pairs is brass rod, profile "round".

The manufacturing process of the contact pairs of rectangular electrical connectors consists of a complex of technological operations:

1) the complex of technological operations for blanking production, the purpose of which is to obtain workpieces of the type "Strip" of metal sheet materials or workpieces of the type "Rod" from the hire of round or hexagonal (depending on the design features of the pins and sockets contact pairs);

2) the complex of technological operations of primary production, the purpose of which is the shaping of parts of type "Pin" and "Socket" (two main details of the contact pairs); for the production of parts of type "Socket" is the basic method of forming is cold stamping, production of parts of type "Pin" can be used as cold stamping and lathe machine operation. Note that the processes of formation of pins and sockets are the main contributors to the formation of the electrical parameters and mechanical characteristics of the contact pairs collected in the connector;

* Corresponding author: niotiostu@gmail.com

3) the complex of technological operations of applying a conductive coating on the surfaces of the contact pair by the method of electroplating. As coating materials are used metals with a low specific electrical resistance (silver, copper, gold, Nickel) and alloys with special electrical properties (alloy, "tin-bismuth" and others);

4) the complex of technological operations for the preparation of contact pairs for Assembly in the connector. The essence of this stage consists in removal of stamped pins and sockets with a common bandwidth. The peculiarity of the implementation of technological processes of preparation of contact pairs for Assembly lies in the subjective nature of the results due to the large number of manual operations performed by a forager connectors;

5) the complex of technological operations of the connector Assembly: Assembly of contact pairs in the insulator and secure them. These operations are performed using special jigs, so the results of the operation may be subject to the objective laws of distribution of probabilities.

To provide confidence in the required level of product quality, all manufacturing operations documented in the individual work instructions; data on progress are regularly recorded.

In the individual job description provides criteria for satisfactory completion of specific work and its compliance with quality standards. Standards of quality of performance of the work issued in the form of rules, posters, pictures or samples. Workplaces include products showing the worker the desired quality of performance of its work.

A mandatory element of the quality management system at this stage are inspection and testing. To reduce production losses from biased decisions seek to control quality of products at the points of greatest sensitivity to changes.

The manufacturing process of a pin contact pair involves turning processing of outer cylindrical surfaces, and cutting off the finished part from the bar. The object of measurement selected outer diameter of the finished pin, equal to 3.7 mm with a unilateral tolerance of body parts equal to 0.16 mm.

The process of manufacture of socket contact pairs involves turning processing of inner and outer cylindrical surfaces, milling of flats and cutting of the finished part. The object of measurement selected inside diameter of the finished nest, equal to 8.14 mm with a unilateral tolerance of body parts, is equal to 0.06 mm.

Turning parts contact pairs were produced on automatic lathe type "NEXTURN". This lathe machine has a numerical control system and resolution of the positioning of the parts to 0.001 mm.

As means for measuring dimensions of parts of the contact pairs of connectors were used:

1) to measure the outside diameter of the pin is smooth micrometer with measuring range 0...25 mm division value 0.001 mm;

2) to measure the internal diameter of the socket shadow projector Nikon with measuring range 0...300 mm division value 0.001 mm.

The control was subjected to two batches of the finished parts of the contact pairs. For each type of the details of the parties selected a sample size of 50 pieces. Selection of products for the size measurement was carried out continuously in the process of their manufacture on automatic lathe.

3 The methodic of experiment

To improve the objectivity of the process of collecting and processing statistical information on the quality of the parts and contact pairs atonement process equipment developed a methodology which includes several steps.

1 Analysis of working drawings of parts. According to the analysis of working drawings of the pin and socket of a selected size, which will be the basis of the statistical analysis of the disposition of equipment.

2 Preparation of measuring instruments. Measuring tools (micrometer smooth and the shadow projector) directly before the measurements were calibrated to determine the measurement error. The measurement error was determined by repeated measurements of the block size limit measures of the lengths corresponding to the size of parts of the contact pairs. For micrometer error made ±0.004 mm. For the shadow of the projector, the error amounted to ±0.005 mm.

3 The preparation of the equipment. Before the experiment lathe machine is configured for turning the set of parts and made a test piece. After making a trial of parts of lathe machine began to manufacture serial parts.

4 The conduct of the experiment. Measure the dimensions of the pin and socket were carried out at the operator's lathe machine. The finished part is removed from the hopper lathe machine, cleansed of lubricant and subjected to measurement. The measurement results recorded in the Protocol. Thus on each object of the studies are shown the measurement results of 50 pieces.

5 Processing of measurement results. Statistical processing of results of measurements of the dimensions of the parts conducted in the following order.

Determined the value of the configuration center, lathe machine, as mid-tolerance part size:

$$me = d_{min} + \frac{Td}{2}, \qquad (1)$$

where me – the median of the tolerance field size, mm; d_{min} – smallest allowable dimension value, mm; Td – dimension tolerance, mm.

Estimated arithmetic mean value of dimensions:

$$\bar{d} = \frac{\sum_{i=1}^{n} d_i}{n}, \qquad (2)$$

where d_i – the value of the diameter of the i-th part, mm; n – number of parts, amount of detail ($i = 1 \ldots n$).

The estimated standard deviation of the sizes:

DOI: 10.1051/matecconf/201712903001

$$S_d = \sqrt{\frac{\sum_{i=1}^{n}(d_i - \bar{d})}{n-1}} \quad (3)$$

Field size tolerance for each part is broken into intervals of width 0.01 mm. Is the width of the intervals selected on the basis of the error measurement means, taking into account the safety factor $k = 2$.

To determine the frequency of hit of values of dimensions of parts in each interval (to percent):

$$p = \frac{n_j}{N} \cdot 100\% \ , \quad (4)$$

where n_j – the number of values that fall in the j-th interval; N – the total number of dimension values (N = 50).

The calculated ratio between half the dimension tolerance of parts and standard deviation. In this relation, taking into account the hypothesis about the normal distribution of the dimensions of parts, using the table of standardized normal distribution defined by the probability of getting the size of the parts in the boundaries of tolerance and probability of presence in a batch of defective parts.

6 Create of histograms. The experimental data are processed using the formulas (1) to (4), histograms of the sizes of parts of the contact pairs. Coordinate histograms are: x-axis – dimensions of parts are within tolerance; and the y – axis the frequency of contact values of the sizes of items in each interval (to percent).

The histograms indicated: arithmetic mean and standard deviation of the size of the part; the border and the median (middle) of the tolerance field.

7 Analysis of mood and precision equipment. According to the histogram, the coordinates of the average size and the middle of the tolerance field determined by the correct setting of technological equipment.

Against half of the tolerance field of the part size to the standard deviation of the size determined by the accuracy of the technological process, as the probability of the presence in the party of defective products.

8 Formation of proposals to improve the quality of the contact pairs of electrical connectors. The results of the analysis of statistical data proposed action to improve the quality of the products.

As an example, in figures 1 and 2 show histograms of the distribution of sizes of pin and socket contact pairs of the electrical connector.

4 Discussion of results

Consider the results of processing statistical information about the quality of the parts of contact pairs of electrical connectors and attunement process equipment.

Figure 1 shows a histogram of the distribution of the dimensions of the pin contact pairs. According to statistical data, plotted on a histogram, it is possible to draw the following conclusions.

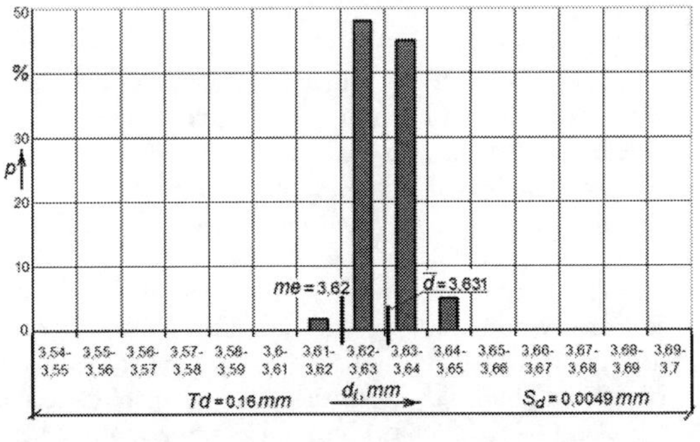

Fig. 1. A histogram of the distribution of the dimensions of the pin contact pair.

1 The shape of the histogram matches the normal distribution of probabilities. Statistical processing of results of measurements confirms this. Therefore, the manufacturing process automatic turning pin is stable.

2 The arithmetic average of the size of a pin does not coincide with the middle of the tolerance field size. It is offset 0.011 mm to the upper limit. Therefore, there is an increased probability of making the pins larger than the upper limit.

3 To improve the quality of the pins we recommend that you configure automatic lathe the size of 3.61 mm and not in the middle of the tolerance field (me = 3.62 mm). After this you need to make a trial batch of parts in an amount not less than 20 pieces to test the coincidence of the middle of the tolerance field and the mean value of the size of a pin.

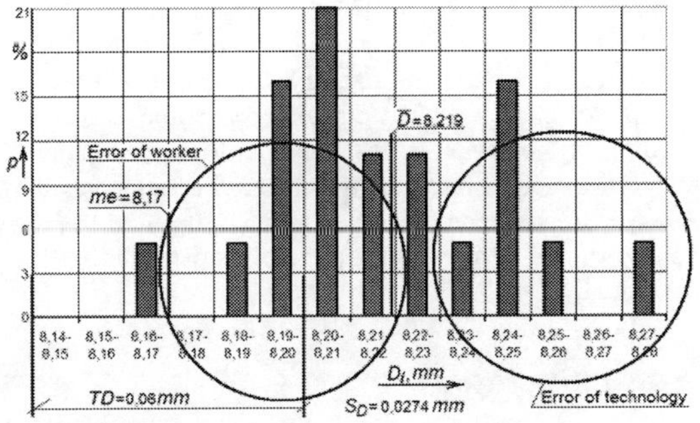

Fig. 2. Histogram of sizes of socket contact pairs.

4 Attitude is half of the size tolerance of the pin to the arithmetic mean is 16.36. In the table of the standardized normal distribution this corresponds to the probability of occurrence of defective products for size with a tolerance Td = 0.16 mm, equal to 0.002 %. Utmost precision, this lathe machine is the size tolerance Td_{MIN} = 0.02 mm; with this tolerance, the likely share of the marriage reaches 5%. This means that this automatic lathe parts with tolerances less than 0.02 mm can be obtained only in the manufacture and subsequent sorting.

For this scheme, will analyze the technological process of manufacture of socket contact pairs. A histogram of the distribution of the dimensions of the slot shown in figure 2. According to statistical data, plotted on a histogram, it is possible to draw the following conclusions.

1 The shape of the histogram corresponds to a combination of two normal probability distribution, mathematical expectation which are offset from each other by about 0.04 mm. This means that in the formation of the size of the pins involved two independent but stable random processes.

2 The arithmetic mean size of the nest did not significantly coincide with the middle of the tolerance field size. It is shifted by 0.05 mm to the upper limit, and this shift is almost equal to the dimension tolerance of the socket ($Td = 0.06$ mm). From the histogram it is seen that this leads to appearance of defective parts that are larger than the upper size limit. The results of the calculation of the probability of occurrence of defective parts is set, it is equal to 19.71 %, and this kind of marriage is unrecoverable, as the study size of socket contact pairs internal.

3 To improve the quality of nests is necessary, first of all, to establish not one but two causes of defective parts. The first possible reason is incorrect configuration of lathe machine: it is set not to the middle of the size tolerance (me = 8.17 mm), while the upper size limit (D_{MAX} = 8.2 mm). The reason marriage provides the "burst" frequency of contact sizes in intervals in the range of 8.19 to 8.21 mm. the Second reason is probably the deformation of a thin wall socket when it is boring lathe cutter. This reason gives the "burst" frequency of contact sizes in intervals in the range from 8.24 to 8.25 mm.

Given the two possible causes of defective parts, to improve the quality of nests is recommended:

1) to configure automatic lathe to the middle of the tolerance field size – the size of 8.17 mm;

2) to change the technological route, automatic turning of the socket so that the operation for boring the inside diameter were carried out to the outer grinding. This will ensure a thick wall when boring the inner diameter of the socket and, consequently, less deformation.

5 Conclusions

The use of statistical methods for quality control of contact pairs of electrical connectors allows to correctly configuring technological equipment according to the results of making a test batch of parts. Then the probability of occurrence of defective products at the stage of serial production of parts can be significantly reduced.

The authors consider it their pleasant duty to express my gratitude to the workers and engineering-technical workers of JSC "Karachevsky factory "Electrodetail" for useful participation, greater technical assistance in performing experiments, moral support and valuable advice.

References

1. V.E. Gmurman, *Theory of probabilities and mathematical statistics* (Higher School, Moscow, 1999)
2. V.V. Efimov, *Statistical methods in the control quality of production* (KnoRus, Moscow, 2013)
3. G.I. Utkin, Fundamental and Applied Problems of Equipment and Technology **6**, 114 (2009)
4. S.F. Korndorf, Fundamental and Applied Problems of Equipment and Technology **1**, 78 (2010)
5. S. R. Bowling, European Journal of Operational Research **159(3)**, 636 (2004)
6. C. H. Chen, M. B. C. Khoo, Computers & Industrial Engineering **57(3)**, 1080 (2009)
7. P. Winkel, N. F. Zhang, Scandinavian Journal of Clinical and Laboratory Investigation **64(3)**, 195 (2004)
8. S. Bandini, F. Sartori, Artificial Intelligence for Engineering Design, Analysis and Manufacturing, **24(01)**, 17 (2010)
9. J. Sun, International Journal of production economics **114 (1)**, 149 (2008)
10. P. Schäffler, W. Rajner, Advanced Engineering Materials **6(6)**, 452 (2004)
11. Y. H. Chun, IIE Transactions **41 (9)**, 831 (2009)
12. C. H. Chen, Journal of Quality and Technical **17(2)**, 99 (2010)
13. Y. Zhang, Y. Teng, Y. Zhang, Chemical Engineering Science **65(6)**, 2153 (2010)
14. E. Hull, K. Jackson, J. Dick, Requirements engineering (Springer Science & Business Media, 2010)
15. L. Huber, Validation of analytical methods (Agilent Technologies. Germany, 2010)
16. D. E. Seborg, Process dynamics and control (John Wiley & Sons, 2010)
17. D. C. Montgomery, Statistical quality control, **7** (2009)
18. D. Reay, C. Ramshaw, A. Harvey, Process Intensification: Engineering for efficiency, sustainability and flexibility (Butterworth-Heinemann, 2013)
19. J. R. Evans, W. M. Lindsay, Managing for quality and performance excellence (Cengage Learning, 2013)
20. L. D. Miles, Techniques of value analysis and engineering (Miles Value Foundation, 2015)
21. M. A. Razzhivina, B. A. Yakimovich, A. I. Korshunov, Pollack Periodica 10(2), 17 (2015)
22. S. N. Grigoriev, Measurement Techniques **11**, 1311 (2013)
23. R. Shankar, M. Sundararajan, Indian Journal of Science and Technology, **8(15)** (2015)

MATEC Web of Conferences **129**, 03002 (2017)
ICMTMTE 2017
DOI: 10.1051/matecconf/201712903002

Identifying the objects in the structure of an e-model by means of identified formal parameters in the design and engineering environment

Mariya Lavrentyieva[1], *and Aleksey* Govorkov[1,*]

[1] Irkutsk National Research Technical University, 664074, Irkutsk, the Russian Federation

Abstract. This research concerns the issues of automating the design and engineering in aircraft-making. It proposes a method for identifying the structural objects in the e-model of a product by means of formalized parameters. The objects of a part and their interlinkages are described in the form of a structural-geometric model. We therefore analyse the procedure of using this model, whereby production rules as well as set theory are applied. We cite a few basic rules of how to identify the links between individual parts, which may be used to assess the manufacturability of such design, to develop the manufacturing and assembly routes, as well as to design the tooling.

1 Introduction

The diversity of technical literature and the recommendations featured in the mechanical engineering design analysis [2] contain the basic information about the structure, the composition, and the characteristics of individual parameters used in the analysis. As of today, when designing a product in CAD/CAM/CAE/PDM systems, it is optimal to analyze the structure of a product in early production, i.e. during design and engineering of aircraft. Thus, to reduce the time it takes to design the documentation and start making the product, the designer and the technologist are tasked to:

- choose modern designs and an optimal option for manufacturing and layout;
- optimally choose the design of the product depending on its intended function;
- use standard, library-provided structural components when modeling a product, say, in NX.

There are no methods and tools for the formal analysis of design composition and structure as such, which prevents automated solution of design and engineering problems. This increases the likelihood of suboptimal solutions, makes the production more time-consuming and labor-intensive.

To solve various problems related to automating the engineering stage of pre-production, we have to design a structural-geometric model of the product.

By using contour recognition method, we can select the geometric primitives that limit the surfaces forming the contour of each structural component

To make recognition more reliable, we used mathematical logic to develop a set of rules that classify the structural components of a part, to find their spatial orientation, their surfaces that face the aerodynamic outlines, their dimensions, etc. In early stages of working with a part, more rules have to be generated because linkages are harder to identify. In terms of design and geometry, a part is uniform and indivisible, its components being non-existent individually; however, in terms of technology, each structural component is individual and independently formable.

The data set obtained enables an engineer to analyze the part as a whole and to identify the nature of its linkages to other parts within an assembly unit, i.e. to find the pairing surfaces and select boundary linkages by means of production rules.
First-level rules are generated in stages [2].

2 First level rules

2.1 Assigning the code depending on the type of lines

There are straight lines, curves, circles, and arcs. To create a structural combination, we assign each type of those lines a code, see Fig.1., where: *a* stands for an arc; *b* stands for a line; *c* stands for a circle; *d* stands for a curve.

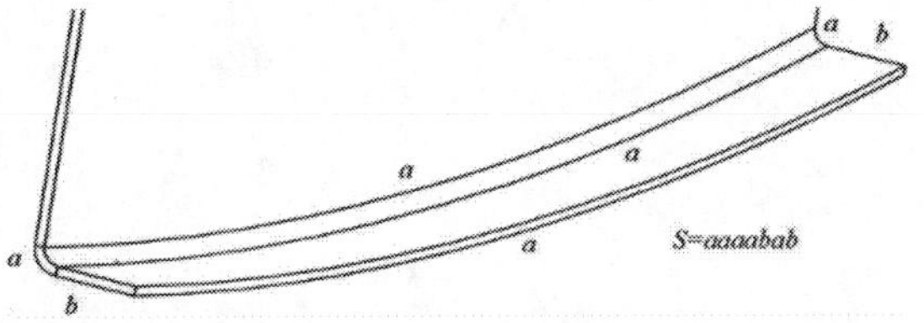

Fig. 1. Components of the code of a board as a structural component

* Corresponding author: govorkov_as@istu.edu

2.2 Find the boundary and pairing (crossing) points of the lines.

Find points *gi* that are common for various lines *li* (at least two points, start point *gis* and end point *gic*); find for which structural components *ai* those points coincide. Each point has coordinates *x*, *y*, *z*, i.e. where the coordinates of points belonging to one surface boundary line overlap the coordinates of the points of the boundary line of another surfaces, those surfaces intersect:

$$gis = gjs \rightarrow \{xis, yis, zis\} = \{xjs, yjs, zjs\};$$
$$gic = gjc \rightarrow \{xic, yic, zic\} = \{xjc, yjc, zjc\};$$

where *gis* and *gjs* are the starting points of the lines; *gic* and *gjc* are the end points.

$$gis, gic \in li; gjs, gjc \in lj;$$

where *li* and *lj* are lines;

$$lj \in pi; lj \in pj;$$

where *pi* and *pj* are surfaces;

$$lj = li \Rightarrow pi \cap pj; pi \subseteq ai, pj \subseteq aj \Rightarrow ai \cap aj.$$

Fig. 2. A common line with start and end points

2.3 Identify the necessary projections tp

To identify some structural components unambiguously, we have to analyze its contour in various projections, i.e. for some products, the set of lines may actually consist of multiple sets. This means that if upon the retrieval of primitives, the code of a component coincides with the code of another component while the structural characteristics do not, we need to use additional sets of primitives taken from other projects.

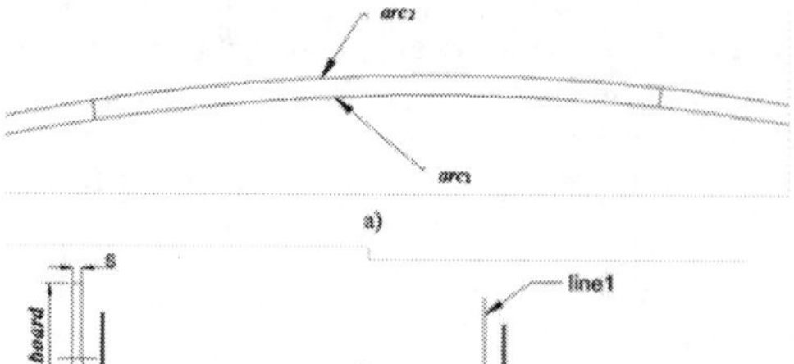

a) longitudinal surface boundary contour of a board; b) transverse surface boundary contour of a board; c) elementary contour dissected into the constituent line types (thickness not factored in).

Fig. 3. Set of board parameters

2.4 Find the dimensions of lines *tp*.

Depending on the type of lines, we find the numeric parameters: length l, diameter *D*, radius *R*, and the number of spline segments. Each component contains a set of points, and as mentioned above, each point has its own coordinates, i.e. the length is measurable as a distance between the coordinates of its end points:

$$g_{is} = \begin{Bmatrix} x_{is} \\ y_{is} \\ z_{is} \end{Bmatrix}, g_{ic} = \begin{Bmatrix} x_{ic} \\ y_{ic} \\ z_{ic} \end{Bmatrix};$$

$$dt_p = \sqrt{(dx)^2 + (dy)^2 + (dz)^2},$$

where *gis* is the starting point of the line; *gic* is the end point; *tp* is the inter-point distance.

Transform this expression by multiplying it and dividing it by *dt*:

$$dt_p = \sqrt{\frac{(dx)^2 + (dy^2) + (dz)^2}{(dt^2)}} dt;$$

Then divide each term in the numerator by the denominator and present the result as

$$dt_p = \sqrt{\left(\frac{dx}{dt}\right)^2 + \left(\frac{dy}{dt}\right)^2 + \left(\frac{dz}{dt}\right)^2} dt \Rightarrow$$

$$dt_p = \sqrt{(x')^2 + (y')^2 + (z')^2} dt,$$

where x', y' and z' are the derivatives of functions $x(g)$, $y(g)$ and $z(g)$ with variable g.
Then

$$t_p = \int_{g_{is}}^{g_{ic}} \sqrt{(x')^2 + (y')^2 + (z')^2} dt.$$

2.5 Determine the basic structural component

The basic structural component a_0 is the reference point for locating the structural components a_0 and is characterized by superior overall dimensions *S*, the maximum number of linkages t_c with other structural components, the lack of linkages to hanging vertices; it is also the platform for other structural components in the part:

$$S_0 > S_i,$$

where is *S0* the area of the basic structural component; S_i is the area of a structural component,

$$t_c(a_0) > t_c(a_i)$$

where $t_c(a_0)$ is the number of linkages of the basic structural component; $t_c(a_i)$ is the number of linkages of a structural component.

DOI: 10.1051/matecconf/201712903002

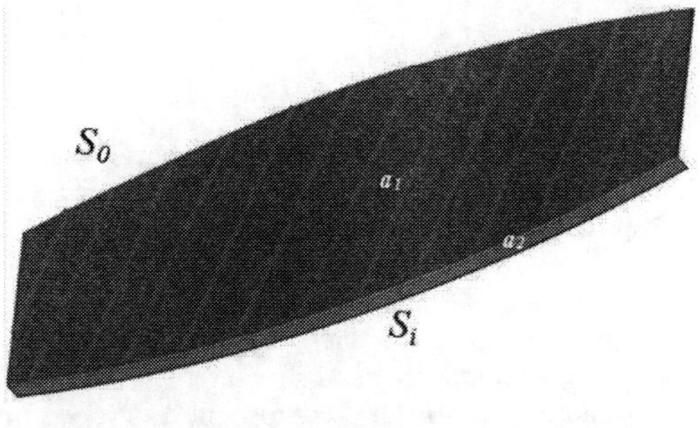

Fig. 4. Determining the basic structural component

2.6 Orientation of structural components relative to the basic one

As we have found the basic structural component a_0, the orientation of other structural components a_i shall be determined relative to the basic one made for the zero point of reference.

$$A_i=\{x_1, x_2,..., x_j\};$$
$$X_i=\{g_1, g_2, ..., g_i\};$$
$$G_i=\{x, y, z\}.$$

The displacement vector $\overline{k_\iota}$ of the structural component ai relative to a0 is calculable using the formula:

$$\overline{k_\iota} = \{g_0 - g_i\} = \begin{Bmatrix} x_0 - x_i \\ y_0 - y_i \\ z_0 - x_i \end{Bmatrix},$$

where g_0 is a point of the basic structural component; g_i is a point of the structural component to compare.

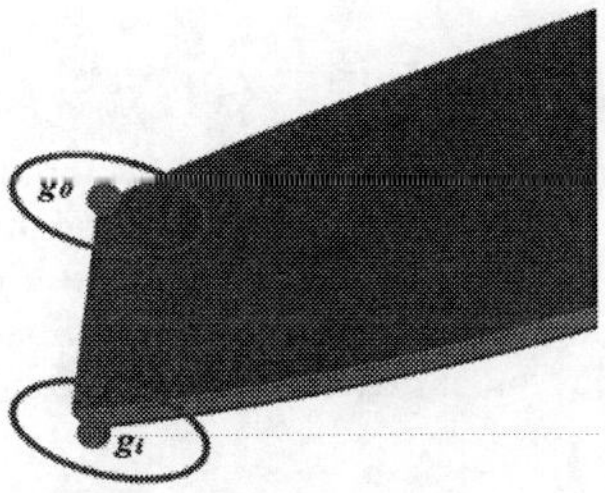

Fig. 5. Points of the base and compared features

2.7 Facing the aerodynamic outline

The outline is the surface of the airframe that is streamlines by airflow in flight. The theoretical outline is one that is specified in the theoretical drawings or the mathematical model of the surface. For convenience, denote the outline as *w*. The outline surface is formed by the surfaces of the structural components belonging to the parts that form the actual outline as a result of assembly.

Whether this or that structural component a_i belongs to the outline w in the theoretical e-model (TEM) of the product can be found out using the formula:

$$a_i \in w_i fp(a_i) \cap p(w).$$

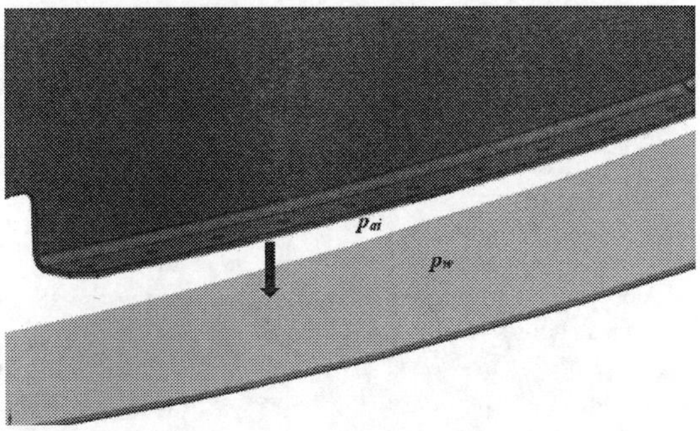

Fig. 6. Outline surface and board surface, paired

As mentioned above, we use the contour recognition method that underlies this set of rules. These rules are a minimum for generating the second-level rules because it enables us to identify the types of the structural components of a part, their location in relation to each other, by linking the primitives, the number of components of the same type, and the total number of the structural components in a part.

For a broader picture of the structure of a part, its structural components are classified to optimize the product of modeling by reference to the identified specifics of construction [1].

Thus, the identified design and engineering parameters match a specific design component, the curvilinear board in our case. This identification fits each and every structural component of the part, which further enables the classification of the part itself.

Part classifier is designed to select a specific group or family based on whether the part has or lacks specific properties.

Considering the second-level operations with the product, recognition as such is not used in this set of rules because the use of the solution to the first problem combined with production rules enables the analysis of parts in the assembly context. In this case, we only need to formulate a few basic rules to identify the linkages between individual parts.

Second-level rules consist of the following stages [1, 10].

3 Second-level rules

3.1 Determining the basic part

Which part shall be the basic part (a rigid part or a low-rigidity part) determines to a great extent the sequence and the method of assembly; this is the basis, upon which the basic component identification criteria are determined.

As with the basic structural component, the basic part has the largest number of linkages t_c and as a rule, has superior overall dimensions S_{be}.

$$S_{be} > S_e$$

where $S_{бэ}$ is the area of the basic component;
$S_э$ is the area of a component,

DOI: 10.1051/matecconf/201712903002

$$t_c(m_0) > t_c(m_i)$$

where $t_c(m_0)$ is the number of linkages of the basic part;
$t_c(m_i)$ is the number of linkages of a part.

3.2 Finding all the pairing surfaces of various components

We have to find all the pairing surfaces for each component to make the final decision regarding the type of a structural component, which has been identified when solving the first-level problem. To find out whether the surfaces are paired, at least three points of each surface should share their coordinates.

$$g_{i1} = g_{j1} \rightarrow \{x_{is}, y_{is}, z_{is}\} = \{x_{js}, y_{js}, z_{js}\};$$
$$g_{i2} = g_{j2} \rightarrow \{x_{ic}, y_{ic}, z_{ic}\} = \{x_{jc}, y_{jc}, z_{jc}\};$$
$$g_{i3} = g_{j3} \rightarrow \{x_{ic}, y_{ic}, z_{ic}\} = \{x_{jc}, y_{jc}, z_{jc}\};$$

where g_{is} and g_{js} are the starting points of the lines;
g_{ic} and g_{jc} are the end points.

$$g_{i1}, g_{i2}, g_{i3} \in p_i;$$
$$g_{j1}, g_{j2}\ g_{j3} \in p_j;$$

where pi and pj are surfaces;

$$p_i \cap p_j.$$

3.3 Identification of components

After we have finally determined the composition of structural components and the pairings of their surfaces, we can identify the type of part this component belongs to.

$$m_i = m_1 \rightarrow m_i = \{a_{1-1}, a_{1-2}, a_{1-i}\},$$

where m_i is any part; m_1 is the part to identify.

3.4 Spatial orientation of components relative to the basic one

As all parts have been identified, and their pairing surfaces established, we can find out their coordinate position relative to each other and get an idea of the general contour and dimensions of the product as a whole. Unlike structural components, each part has a single point that can be used to find the spatial position thereof. The point is the center of mass point $g_ц$, and the position is found using the displacement matrix M_{per} relative to the coordinate system used in the assembler file.

$$M_{per} = M_r \times M_v,$$

where M_{per} is the displacement matrix; M_r is the rotation matrix; M_v is the transition matrix.

$$M_{per} = \begin{vmatrix} xx_0 & xy_0 & xz_0 \\ yx_0 & yy_0 & yz_0 \\ zx_0 & zy_0 & zz_0 \end{vmatrix} \times \begin{vmatrix} x_v \\ y_v \\ z_v \end{vmatrix},$$

$$\begin{vmatrix} x_v \\ y_v \\ z_v \end{vmatrix} = \begin{vmatrix} x_0+x_i \\ y_0+y_i \\ z_0+x_i \end{vmatrix}.$$

Solving the second-level problem helps get an idea of the product in general and of the structure and composition of each of its parts. In terms of methodology, there is no familiar semantic binding to a particular denomination; we merely assign a type of part and therefore obtain a product consisting of n parts of determined composition and structure.

To automate structural component data processing, it makes sense to use all-Russian technological process and transition classifiers.

The sets of components in such parts differ, which means we have to model and manufacture them in different ways.

Determination of the design and engineering parameters of a part as well as classifying it as a certain type have a considerable impact on its subsequent basing. We introduce several rules.

4 Third-level rules

4.1 Determining the composition of structural components ai

According to study cases, typical parts have certain structural components unique to this type of part:

$$a_{i1}, a_{i2}, a_{i3} \in m_i;$$

$$a_{j1}, a_{j2}, a_{j3} \in m_j,$$

where m_i is an ith type part; m_j is a jth type part.

4.2 Determining the semi-finished product

When finding out the structural and engineering properties of a part, the choice of the semi-finished product is very important; by default, profile semi-products are considered more rigid than sheet semi-products; a wall made from a plate is more rigid than one made from a sheet.

4.3 Finding the constant thickness s of a part

Beside the physical and mechanical properties of semi-finished products, some of which are added to the e-model when designing it, the thickness of material is of importance, too; as a rule, sheet-made parts have a constant thickness, while plate-made ones may have significant variations in thickness.

4.4 Finding the center of mass point $g_ц$

The center of mass point has its own coordinates in each individual part. It is a reference point for finding the location and pairings of parts in an assembly unit. The center of mass coordinates are the closest for two parts of similar dimensions and a complete pairing on one surface:

$$g_{ij} = \{x_{ij}, y_{ij}, z_{ij}\},$$

where $\{x_{ij}, y_{ij}, z_{ij}\}$ are the coordinates of the center of mass..

Thus, a classification enables clear grouping of parts, which enables using them to create basic models, analyze their typical formulation, and identify the most optimal option.

As noted above, all parts under analysis that form the structure of nodes and units in an aircraft can be rationally represented as combinations of typical structural components. The choice of their composition, design options, the analysis of their compatibility enable a formalization of the recognition process for each specific part and the entire assembly unit.

We therefore obtain a classified product not bound to semantics, with a complete inventory of parameters selected on the basis of the rules developed [10]. Rules for selecting the relevant parameters are in its turn formulated depending on the ultimate purpose; they enable further transfer of data for the desired output.

The research herein presented is supported financially by the Fund for the Promotion of Small Scientific and Technical Enterprises as a part of the project 'Developing a System for Automated Analysis of Aircraft Engineering Products in Preproduction', the UMNIK 16–10 contest.

References

1. S.A. Aksenov, E.N. Chumachenko, A.V. Kolesnikov, S.A. Osipov, *Determination Of Optimal Conditions For Gas Forming Of Aluminum Sheets,* Procedia Engineering **11** th., P. 1017 1022 (2014)
2. S.A. Aksenov, I.Y. Zakhariev, A.V. Kolesnikov, S.A. Osipov, *Characterization of superplastic materials by results of free bulging tests,* Materials Science Forum. Vol. **838–839**, P. 552–556 (2016)
3. P.E. Chimitov, *Generating the image of a product for designing the airframe assembly process.* Reshetnyov's, Readings: Proceedings of the XII International Scientific Conference in the Memory of the Designer General of Rocket Space Systems, Academician M. F. Reshetnyov, P.183–184 (2008)
4. V.E. Gozbenko, S.K. Kargapoltsev, D.N. Kornilov, N.V. Minaev, A.I. Karlina, *Definition of the main coordinates of the car with two-level spring suspension,* International Journal of Applied Engineering Research. Vol. **11**. Issue 20, P. 10367–10373 (2016)
5. V.A. Grechishnikov, R.M. Khusainov, D.R. Akhkiyamov, S.Y. Yurasov, O.I. Yurasova, *Identifying the primary rigidity axes in the elastic system of a metal-cutting machine,* Russian Engineering Research Vol. **36**, No. 8, pp. 673–67 (2016)
6. G.K. Irzaev, *Expert methods of industrial products manufacturability management* (Moscow: Infra-Inzheneria, 2010)
7. G.K. Irzaev, *Research and modeling data streams of design and engineering adjustments in the development and medium-batch manufacturing of products,* Organizator proizvodstva Vol. **52.** Issue 1, P. 131–135 (2012)
8. G.K. Irzaev, *Expert methods for controlling the manufacturability of industrial products: a monograph* (Moscow: Infra-Inzheneriya, 2010)
9. A.A. Kabanov, *Object model for the analysis of aircraft products with account of inheritance for the assessment and control of the efficiency of manufacturing systems in discrete mechanical engineering at the stage of their organizational and engineering design or modernization,* Kazan: Scientific and Technical Bulletin of the Volga Region, p. 161–168 (2014)
10. R.M. Khusainov, I.F. Sharafutdinov, *Methods of assessing the dynamic stability of the cutting process using Unigraphics NX,* IOP Conf. Series: Materials Science and Engineering **134,** 012042 (2016)
11. P.M. Krastyaninov, R.M. Khusainov, *Selection of equipment for machining processing of parts using NX and Teamcenter programs,* IOP Conf. Series: Materials Science and Engineering **134,** 012041 (2016)
12. A.A. Mitin, *Use of production rules to display the integrity limits in graphical CAM editors,* Proceedings of the Oryol State Technical University. Series: Information Systems and Technology. Oryol: Oryol State Technical University, P.173–177 (2008)
13. K.A. Odnocurtcev, *Laboratory equipment for automated control of assembly tooling components,* Bulletin of Irkutsk State Technical University. Issue **12 (59)**, P. 41–47 (2011).
14. A.K. Shmakov, V.V. Mironenko, K.K. Kirishina, A. S. Stanislavchik, V.V. Kotov, *Effect of the average velocity of the free part of the semifinished product on the process of pneumothermal forming in the superplastic regime,* Metallurgist. –Vol. **57,** Issue 1–2, P. 8–12 (2013)

MATEC Web of Conferences **129**, 03003 (2017) DOI: 10.1051/matecconf/201712903003
ICMTMTE 2017

Using a discrete product model to determine the design element junctures

Mariya Lavrentyieva[1], and *Aleksey* Govorkov[1,*]

[1] Irkutsk National Research Technical University, 664074, Irkutsk, Russian Federation

Abstract. This paper describes a discrete model of a standard airframe part. The model is intended for determining the junctures between the design elements of a node (unit) in the pre-assembly procedures. We herein propose an algorithm selecting base surfaces for basing the product being assembled; selections are made on the basis of the 3D model of such product.

1 Introduction

Making pre-production processes (PPP) more efficient has recently become one of the most relevant questions. The current trends of scientific and technical progress enable significant enhancements in many areas of mechanical engineering, including aircraft building; however, assembling still involves intensive manual labor, slowing down the development of this industry as a whole.

The current state of the art allows for using various software packages to make pre-production processes in engineering more efficient.

2 Relevance and scientific importance

In pre-production, the main objective is basing the part, i.e. ensuring it is positioned correctly. After a part is based, it must be locked. To ensure it does not move in the chosen coordinate system, you must impose six bilateral geometric relationships, for which they would need a set of reference anchor points. If one or more of them is missing, the part will have one or more degrees of freedom. This means that, in the direction of the missing anchor points, the position of the part will be determined by its juncture to another part.

3 Problem Statement

Two or more adjoining parts are referred to as joint parts. The surfaces in such a juncture are referred to as joint surfaces.

Each part is generated in its own coordinate system; when assembled in a single file, their coordinates are adjusted relative to the working coordinate system (WCS) of the assembler file, which means that before one can talk about the junctures, they have to know the minimum number of anchor points and the position of each part in the assembly, as these factors are crucial for basing.

4 Theory

Let us use a beam calculation model that holds for those structures wherein the size along one axis significantly exceeds their cross-section, whereby there is no abrupt change in the rigidity and the anchoring elements, as shown in figure 1.

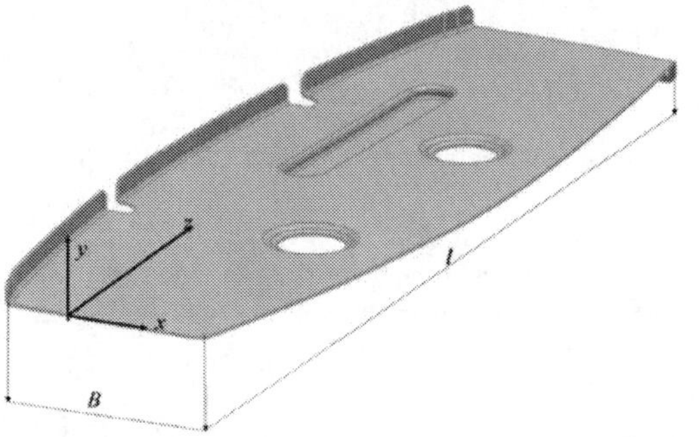

Fig. 1. Rib wall

To determine the deflection of the part, let us select a stripe along the generatrix, where there is a uniformly distributed load q along the length and a pressing force at the anchoring points, as shown in Figure 2.

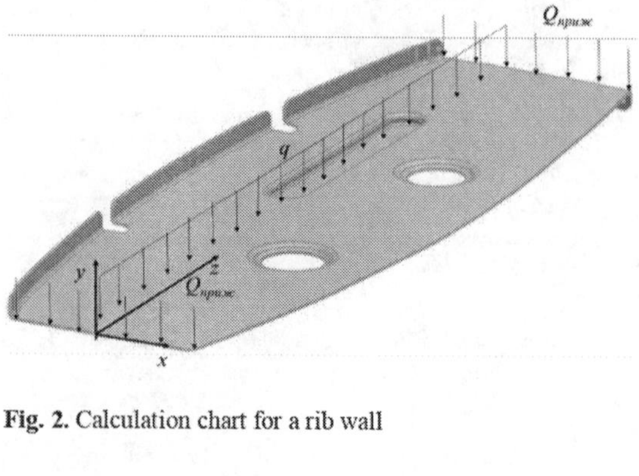

Fig. 2. Calculation chart for a rib wall

* Corresponding author: govorkov_as@istu.edu

Then, the maximum deflection in the middle is calculated using the formula (1), while the maximum distance between the anchor points is calculated using the formula (3).

$$w_{\max} = \frac{ql^4}{8EJ_x} + (\delta_{бфу} - \delta_{дет}) = \frac{\rho Fgl^4}{8EJ_x} + (\delta_{бфу} - \delta_{дет}), \quad (1)$$

where **E** is the first genus modulus of elasticity for the material of the part, MPa; **Jx** is the moment of inertia of the section relative to the axis x–x, m4; **EJx** is the rigidity of the profile, N·m²; **ρ** is the density of the material of the profile, kg/m³; **g** is the free-fall acceleration, m/s²; **F** is the area of the profile cross-section, m²; **δдет** is the part manufacturing error, mm; **δбфу** is the manufacturing error of base anchoring devices of assembly fixtures, mm.

Using the formulas (3.2)

$$w_{\max} \le w_{доп}, \quad (2)$$

where w_{max} is the maximum deflection, mm; $w_{доп}$ is permissible deflection, mm.
and (3.3.), we obtain:

$$l \le \sqrt[4]{\frac{8(w_{доп} - \delta_{бфу} + \delta_{дет})EJ_x}{\rho Fg}} \quad (3)$$

Substituting all the values in the formula (3), we obtain the maximum distance between the anchor points lmax for all the parts made from beam-type profiles.

If the part anchoring is vertical, the deformation caused by the pressing force and the deformation caused by the own weight of the part are less than the maximum deflection wmax. For all of the cases where the preliminary value of the distance between the anchor points lmax, thereby ensuring the rigidity of the parts in assembly.

To simplify the search for the location by the parts coordinate measurement point, let us calculate the point of the center of mass of the part gц. The coordinates of the center of mass are always specified relative to the working coordinate system (WCS), which for the assembly units of aeronautical engineering parts shall coincide with the principal axes of the aircraft.

Using the movement matrix Mt relative to the working coordinate system of the assembler file.

$$M_t = M_r \times M_v,$$

Where **Mt** is the movement matrix; **Mr** is the rotation matrix; **Mv** is the transition matrix.

$$M_{trans} = \begin{vmatrix} xx_0 & xy_0 & xz_0 \\ yx_0 & yy_0 & yz_0 \\ zx_0 & zy_0 & zz_0 \end{vmatrix} \times \begin{vmatrix} x_v \\ y_v \\ z_v \end{vmatrix} = \begin{vmatrix} x_0+x_i \\ y_0+y_i \\ z_0+x_i \end{vmatrix}.$$

Accordingly, the coordinate placement of the other parts of the assembly unit shall be calculated by the center of mass relative to the working coordinate system of the assembly.

5 Practical significance and experimental results

The position of each part is now known, and we can start determining the junctures of the parts of the assembly unit. In essence, the principle is still valid that was used to identify the inter-element linkages to determine the structure of the part.

For instance, for parts 1 and 2 that are sets M1 and M2 consisting of i surfaces pi, each surface contains at least three points, but in fact it has many more points; therefore, we get a set Mi={p1, p2,…pj} consisting of subsets Pi={g1, g2,…gj}, where gj is a point belonging to the surface.

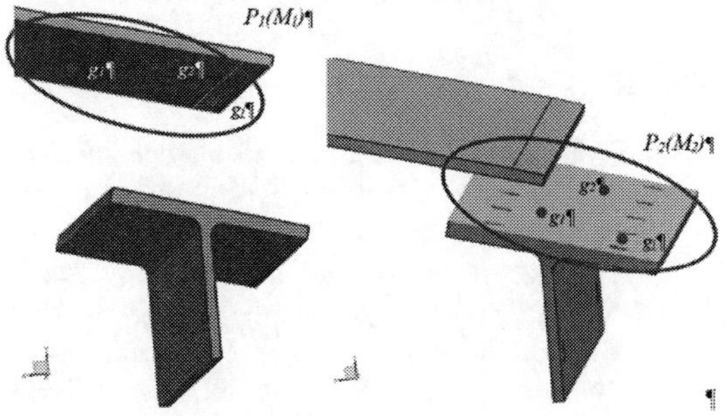

Fig. 3. Juncture surfaces of the parts in an assembly unit

In this case, the subset of points {g1, g2,…gj}∈ p1 (M1) and {g1, g2,…gj}∈ p2 (M2), i.e. the subset of points gj is coincident for surfaces of different parts p1 (M1) ∩ p2 (M2), which in its turn means these surfaces are juncture surfaces, and parts 1 and 2 are joint through these surfaces p1 ⊆ M1, a p2 ⊆ M2⇒ M1∩M2.

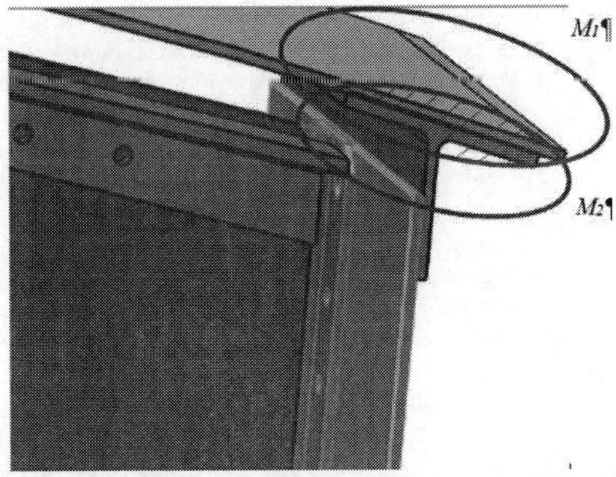

Fig. 4. Juncture of parts in an assembly unit

To determine that the subset {g1, g2,…gj} is the common subset for surfaces pi of different parts, we compare the coordinates x, y, z for all gj, and if the coordinates coincide, the points coincide, too.

As a result, we come to know the position, the surfaces, and the junctures for every part in the assembly unit; taking into account the first analysis, we know all the lines, all the existing surfaces, and even the points. Each of those primitives is a base (assembly base, technology base, or design base); however, it does not make sense to use all of them; we have to single out the bases that are relevant to the problem to solve.

5 Conclusion

Using a discrete product model to determine the design element junctures allows for determining:
• the structure of the product;
• the design and technological parameters of the product;
• the set of the bases of the product.

It also makes it possible to obtain data for identifying the relevant bases to later solve the problems related to designing the technological tooling for product assembly.

References

1. Y.D. Amirov, *Design for manufacturability*, Biblioteka konstruktora (M.: Mashinostroenie, 1990)
2. A.A. Cherepashkov, N.V. Nosov, *Computer technology, modeling and automated systems in mechanical engineering* (Infolio. Volgograd, 2009)
3. A.A. Cheslavskaya, V.V. Mironenko, A.V. Kolesnikov, N.V. Maksimenko, V.V. Kotov, *Choosing an efficient method for forming parts by means of an engineering analysis performed with the use of a CAE system*, Metallurgist, №**58**. Issue **11–12**, pp. 1051–1059 (2015)
4. P. Emad, N Abouel, A. K. Kamrani, *A new methodology. for extracting manufacturing features from CAD system*, Computers & Industrial Engineering **51,** pp. 389–415 (2006)
5. R. Duda, *Image recognition and scene analysis,* Richard O. Duda, Peter E. Hart (M.: Mir, 1976)
6. A. S. Govorkov, *Technique of designing of the product of aviation technics with maintenance of the set criteria of adaptability to manufacture*, Journal of International Scientific Publications: Materials, Methods & Technologies, Vol. **5**. Issue **3**, pp. 156–161 (2011)
7. A.S. Govorkov, A.S. Zhilyaev, *The estimation technique of the airframe design for manufacturability,* IOP Conference Series: Materials Science and Engineering, Vol. **124**, Issue **1**, pp. 01 (2014)
8. V.E. Gozbenko, S.K. Kargapolcev, V.V. Kondratiev, A.I. Karlina, N.V. Minaev, *Vertical dynamics of the vehicle taking into account roughness gauge,* Proceedings of the XV International Academic Congress "Fundamental and Applied Studies in the Modern World", V. **XV.** "Oxford University Press", pp. 373–383 (2016)
9. V.E. Gozbenko, S.K. Kargapoltsev, D.N. Kornilov, N.V. Minaev, A.I. Karlina, *Simulation of the vibration of the carriage asymmetric parameters in MathCAD,* International Journal of Applied Engineering Research (IJAER), V. **11**, Number **23**, pp. 11132–11136 (2016)
10. V.A. Grechishnikov, R.M. Khusainov, D.R. Akhkiyamov, S.Y. Yurasov, O.I. Yurasova, *Identifying the primary rigidity axes in the elastic system of a metal-cutting machine,* Russian Engineering Research, Vol. **36**, No. **8**, pp. 673–676 (2016)
11. R.M. Khusainov, I.F. Sharafutdinov, *Methods of assessing the dynamic stability of the cutting process using Unigraphics NX.* IOP Conf. Series: Materials Science and Engineering **134**, 012042 (2016)
12. I.M. Kolganov, *The ways of design for manufacturability increasing on airplane industry*, **CHast' 1**: Uchebnoe posobie (Ul'yanovsk: UlGTU, 2003)
13. P.M. Krastyaninov, R.M. Khusainov, *Selection of equipment for machining processing of parts using NX and TEAMCENTER programs,* IOP Conf. Series: Materials Science and Engineering **134,** 012041 (2016).
14. Y.I. Karlina, A.S. Govorkov, *Design and Technology Parameters of the Nomenclature of Products Made When Automating Pre-Production Processes; Choosing the Basic CAD System of the Enterprise for Creating a Digital Product Mockup*, Advanced Technologies. System Analysis. Modeling, Issue **3 (51)**, pp. 49–55, (2016).
15. A.S. Govorkov, A.S, ZHilyaev, *Practical application of the "DFM analysis system" during the process control products aircraft equipment,* Trudy MAI, Issue **74**, pp. 21 (2014).
16. R. Akhatov, A. Govorkov, A. Zhilyaev, *Software solution designing of "The analysis system of workability of industrial product" during the production startup of aeronautical products*, International Journal of Applied Engineering Research, Vol. **10**, Issue **21**, pp. 42560–42562, (2015).
17. A.K. Shmakov, V.V. Mironenko, K.K. Kirishina, A.S. Stanislavchik, V.V. Kotov, *Effect of the Average velocity of the free part of the semifinished product on the process of pneumothermal forming in the superplastic regime,* Metallurgist, Vol. **57**, Issue **1–2**, pp. 8–12 (2013).
18. P. Subrahmanyam, M. Wozny, *An overview of automatic feature recognition techniques for computer-aided process planning.* Computers in industry, Vol. **26**, pp. 1–21 (1995)
19. J. Tou, *Pattern recognition princples* (Moscow: Mir, 1978).

A method for determination of mass-exchange and capacitive characteristics of absorbers for life support systems

Pavel V. Balabanov[1,*], *Guzel R.* Kamaletdinova[1], and *Maria G.* Muravyeva[1]

[1]Tambov State Technical University, 392000 Tambov, Sovetskaya 106, Russia

Abstract. We describe a method for determining mass-exchange and capacitance characteristics of chemisorbents of carbon dioxide. Through a cartridge equipped with a chemosorbent, a gas-air mixture is blown continuously from a sealed container of a small volume into which the impurity of carbon dioxide is continuously introduced. The gas-air mixture, which has been purified from the impurity, is again introduced into the tank. The concentration of carbon dioxide in a sealed volume is monitored by means of the gas analyzer; the effective mass transfer coefficient and the maximum capacity of the chemisorbent for carbon dioxide are calculated from the obtained data.

The development of methodology for certification of parameters of regeneration processes for the maximum capacity of chemisorbents, mass exchange coefficients for carbon dioxide and oxygen for substances used in life support systems is an urgent task [1-5]. Previously, theoretical aspects of the dynamics of chemisorption by regenerative substances were developed in [1], mathematical models for the functioning of collective and individual protective equipment, on the basis of which the parameters of air regeneration processes were determined in [2, 3]. Methods and equipment for the automated investigation of the characteristics of chemisorbents were proposed in [4, 5].

It is noteworthy that parameter identification was carried out on the basis of rather complex mathematical models, containing a system of equations in partial derivatives using numerical calculation methods, which required the creation of special software [6-10].

At the same time, the practice of certification of devices for air regeneration has shown that there is a need for an automated method for determining the mass exchange and capacitance characteristics of chemisorbents (absorbers) that would be universal for substances and cartridges of any configuration.

The purpose of this paper is to develop such a method.

We consider a life-support system with low volume V to 100 dm^3, for which the exchange multiplicity Q/V (Q is the volume of ventilation) is so large that the influence of the volume V on the concentration of carbon dioxide can be neglected.

The equation for the dynamics of chemisorption CO_2 can be written in the form

$$\frac{\partial \varphi}{\partial \tau}+\frac{\partial C}{\partial \tau}=-w\frac{\partial C}{\partial x}+D\frac{\partial^2 C}{\partial x^2}, \qquad (1)$$

where φ, C are current absorption and concentration of dioxide in the gas phase, respectively, x is the current coordinate counted from the inlet of the gas flow to the chemisorbent layer, D is coefficient of longitudinal diffusion, w is linear velocity, τ is time.

We assume that the cartridge is filled with chemosorbent, the process is irreversible and its velocity is determined by a universal equation of the form [1, 2, 5, 11]

$$\partial\varphi/\partial\tau=\beta C(1-\varphi/\varphi_0), \qquad (2)$$

where φ_0 is end capacity for carbon dioxide; β is mass transfer coefficient.

An approximate allowance for the irregular transfer of the absorbed component in the flow within the layer can be carried out by the method proposed in [12]. In this case, the effect of a longitudinal transfer may be taken into account by an effective mass transfer coefficient

$$\beta_{ef}\approx k\beta,$$

where

$$k\in[k_1;k_2],\ k_1=\frac{1}{1+y},\ k_2=\frac{2}{1+\sqrt{1+4y}},\ y=D\beta/w^2.$$

The initial conditions are assumed to be zero

$$C(x,0)=\varphi(x,0)=0. \qquad (3)$$

To determine the boundary conditions, we assume that the concentration C_0 existing at the initial time is defined as the quotient obtained by load division for impurities q by the total volume of ventilation

* Corresponding author: pav-balabanov@yandex.ru

$$C_0 = q/Q. \quad (4)$$

The concentration of carbon dioxide at the inlet to the cartridge is an unknown function of time and is composed of a constant concentration C_0 due to the presence of a source of impurity release and concentration at the outlet of the cartridge $C_0\Psi(\tau)$. We assume that the flow of carbon dioxide in the volume starts with the time $\tau=L/w$, where L is sorbent bed length. At the inlet to the cartridge there exists concentration $C_0+C_0\Psi(\tau)$. The amount of carbon dioxide released in time $d\tau$ from the cartridge is equal to $SwC_0\Psi(\tau)d\tau$, and the amount of impurity (carbon dioxide) supplied to the cartridge (carbon dioxide) is $SwC_0(1+\Psi(\tau))d\tau$.

Usually the device is used to the maximum permissible concentrations, for example, of carbon dioxide. If the initial concentration CO_2 is 0.5% (in this case q=30 dm^3/h, Q= 6 m^3/h), the end concentration CO_2 is 3.0%, then for the volume of 100 dm^3 accumulation of CO_2 must not exceed 2.5 dm^3. The latter means that the device (cartridge) must absorb practically the entire volume of carbon dioxide fed into the life support system during the time of its actual operation.

Thus, the amount absorbed in the cartridge in a time $d\tau$

$$d\tau S\int_0^L \frac{\partial(\varphi+C)}{\partial\tau}dx = -d\tau S\int_0^L w\frac{\partial C}{\partial x}dx \approx SwC_0 d\tau \quad (5)$$

Hence

$$C_0 \approx -\int_0^L \frac{\partial C}{\partial x}dx. \quad (6)$$

Given the dimensionless variables

$$u=\frac{C}{C_0},\ \xi=\frac{\beta_{ef}x}{w},\ t=\frac{\beta_{ef}C_0}{\varphi_0}(\tau-\frac{x}{w}),$$

from equation (6) we obtain

$$1 \approx -\int_0^{\xi_L}\frac{\partial u}{\partial\xi}d\xi. \quad (7)$$

We note that under the usual operating conditions of the cartridge the condition

$$-w\int_0^{\xi_L}\frac{\partial C}{\partial\xi}d\xi >> \int_0^{\xi_L}\frac{\partial C}{\partial t}d\xi, \quad (8)$$

is satisfied.

We use the expression for the distribution of the dimensionless concentration of the absorbed substances in the mobile phase inside the device obtained by solving the system of equations (1), (2) with the initial condition $\varphi(x, 0)=const$ and boundary condition for concentration $C(0,\tau)=C_0\Psi(\tau)$. The solution of this problem has the form

$$u=\frac{\Psi(t)}{e^{-\int_0^t\Psi(t)dt}\left[e^{\xi}-1\right]+1}. \quad (9)$$

Differentiating the last expression with respect to ξ and substituting the result in (7), we obtain

$$1=\int_0^{\xi}\frac{\Psi(t)\exp(\int_0^t\Psi(t)dt)\exp\xi}{\left[\exp(\int_0^t\Psi(t)dt)+e^{\xi}-1\right]^2}d\xi. \quad (10)$$

After integrating expression (10) with respect to ξ and making simple transformations, we obtain

$$\exp(\int_0^t\Psi(t)dt)=[\Psi(t)-1][\exp\xi_L-1]. \quad (11)$$

The logarithm and differentiation of the resulting expression gives an equation for finding $\Psi(\tau)$

$$\Psi^2(t)=\Psi'(t)+\Psi(t), \quad (12)$$

which by substituting $\Psi(\tau)=1/z$ is reduced to the linear equation having a solution

$$\Psi(t)=\frac{\exp(-t)}{\exp(-t)+Const}. \quad (13)$$

where $const=\exp(-\xi_L)$ is found by substituting the last expression into equation (11). Consequently,

$$\Psi(t)=\frac{\exp(t-\xi_L)}{1-\exp(t-\xi_L)}. \quad (14)$$

Expression (14) is applicable to $t<\xi_L$. Passing to dimensional variables, after simple transformations we obtain equation

$$\tau=\frac{\varphi_0}{wC_0}L+\frac{\varphi_0}{\beta_{ef}C_0}\ln\frac{C}{C_0+C}, \quad (15)$$

that is the basis of the method for determining the mass-exchange and capacitive characteristics of sorbents.

For a cartridge with a cylindrical layer and a radial flow direction: $W=Q/(2\pi h)$, $L=(R_1^2-R_2^2)/2$, where h is the height of a cylindrical layer, R_1, R_2 are large and small radii of the cylindrical layer, respectively.

Proposed method can be represented as a sequence of the following operations.

The absorbent cartridge 3 is filled with the chemosorbent under the study, then the cartridge inlet is connected to the outlet of the flow agitator (Fig. 1). The outlet of the cartridge is connected with the flange fitting of the tank1 having the volume V. Then the fan and the agitator for blowing gas-air mixture from the tank 1 through the cartridge with a constant flow Q and temporary flow q are turned on. From cylinder 2 carbon dioxide is fed into the tank, simulating the load on the device. The gas-air mixture purified from carbon dioxide is returned to the volume V, where it is mixed with CO_2 newly supplied by the fan. During the test, the concentration C of carbon dioxide in the tank 1, the consumption of gas-air mixture flown through the cartridge and the consumption of CO_2 from the cylinder are recorded. Based on the experimental data obtained, the parameters α_1 and α_2 are found

$$\tau = \alpha_1 \ln \frac{C}{C_0 + C} + \alpha_2, \tag{16}$$

where τ is current time, C_0, C are initial and current concentrations of CO_2 in the tank 1.

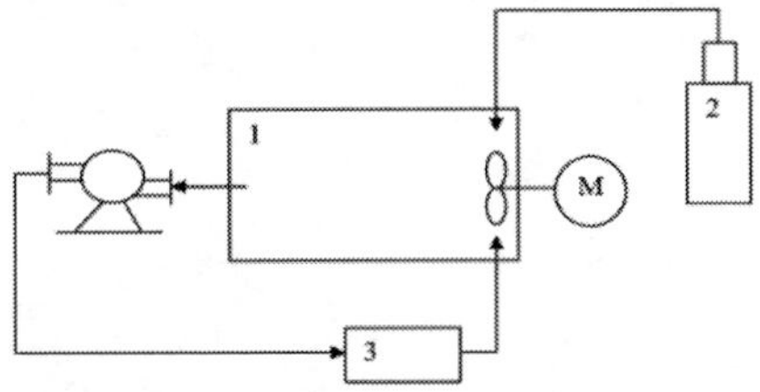

Fig. 1. Installation diagram for testing

The effective mass transfer coefficient and the limiting capacity for carbon dioxide under the experimental conditions are found from the expressions

$$\beta_{ef} = \frac{\varphi_0}{\alpha_1 C_0}, \quad \varphi_0 = \frac{\alpha_2 w C_0}{L}. \tag{17}$$

As an example, we give the results of tests of absorbers based on $LiOH$ in the model of a cartridge of cylindrical shape with a diameter of 117 mm. In the experiments, we used chemisorbents in the form of cylindrical blocks with channels. The length of the layer, depending on the thickness of the test blocks, varied in the range 195 ... 210 mm. On average, the multiplicity of transfer Q/V was equal to 18 h^{-1}. The supply of carbon dioxide was varied in such a way that q/Q was within the range of 0.0107…0.0124. The tests were carried out at room temperature of the medium and the relative humidity of gas-air mixture of 80 ± 10 %. Fig. 2 shows the curves of changes in the concentration of carbon dioxide in time; the results were obtained in the two experiments with the absorber blocks. In the first experiment (curve 1) the mass of each block was 1.12 times larger than that in the second one (curve 2). Fig. 3 shows the experimental curves plotted in semilogarithmic coordinates.

The results of finding mass transfer characteristics of the investigated absorbers are given in the Table 1.

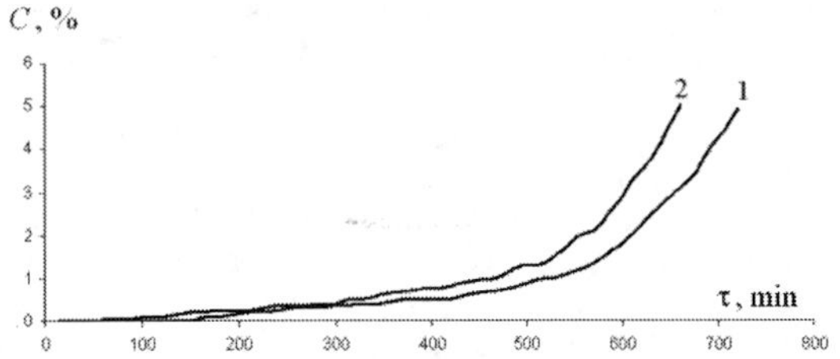

Fig. 2. Time variation in concentration

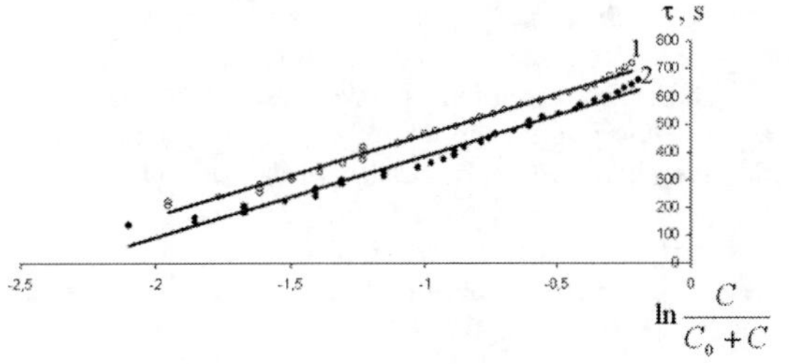

Fig. 3. Experimental curves in semilogarithmic coordinates

Table 1. The results of the determination of the mass transfer characteristics of the investigated absorbers

№ test	Process characteristics			Mass transfer characteristics	
	Q/V, h^{-1}	q/Q	L, m	β_{ef}, 1/s	φ_0, m^3/m^3
1	18	0,0124	210	1,23	260
2	18	0,0107	195	1,18	224

The root-mean-square errors in the determination of mass transfer characteristics calculated by the formulas

$$\left(\frac{\Delta\varphi_0}{\varphi_0}\right)_{sc} = \sqrt{\left(\frac{\Delta\alpha_2}{\alpha_2}\right)^2 + \left(\frac{\Delta w}{w}\right)^2 + \left(\frac{\Delta C_0}{C_0}\right)^2 + \left(\frac{\Delta L}{L}\right)^2},$$

$$\left(\frac{\Delta\beta_{ef}}{\beta_{ef}}\right)_{sc} = \sqrt{\left(\frac{\Delta\alpha_1}{\alpha_1}\right)^2 + \left(\frac{\Delta\varphi_0}{\varphi_0}\right)^2 + \left(\frac{\Delta C_0}{C_0}\right)^2},$$

did not exceed 7% for φ_0 and 9% for β_{ef}. In the calculations, we took into account that the errors in measuring the consumption of carbon dioxide q did not exceed 1%, of the gas-air mixture consumption Q did not exceed 2.5%, the layer length and the block diameters did not exceed 2%. Parameters α_1 and α_2 were determined by approximating the dependences shown in Fig. 3 as the equation of a straight line. In order to increase the accuracy, we used only the experimental data for which condition C>0,2 was satisfied because at lower concentrations the consumption of carbon dioxide and gas-air mixture can vary, and the errors in the measurement of concentrations can occur. Errors in finding α_1 and α_2 did not exceed 5%. The root-mean-square errors of C_0, calculated by (4) did not exceed 3%, the velocity of gas-air mixture flow calculated by the known consumption Q, did not exceed 3%.

We note that the obtained values β and φ_0 for lithium absorber blocks were different from those given in [2]. Thus, in particular, the limiting capacity for CO_2, equal to 216…260 m^3/m^3 was close to the stoichiometric under normal conditions (temperature 0 and pressure 760 mm Hg). This is due to the difference in test conditions (mainly for concentrations CO_2 at the inlet to the layer). With a decrease in the supply of carbon dioxide to the volume, the values β and φ_0 will approach the data given in [2], but the total duration of the experiment will increase substantially.

Thus, we proposed an "express" method, which allows determining the chemisorption characteristics of the absorbers relatively quickly and in a wide range of impurity concentrations. The method can be used in the study of the quality of carbon dioxide absorbers, as well as in the certification of absorption devices for life support systems.

Financial state support from Russian Scientific Foundation (Project15-19-10028 "The development of the theoretical foundations for creating the research base for human respiratory protection systems from the chemical and biological dangerous nature factors") on the base of Tambov State Technical University is gratefully acknowledged.

References

1. S.V. Mishchenko, P.V. Balabanov, A.A. Krimshtejn, S.V. Ponomarev, Trans. TSTU, **16**, 870 (2010)
2. A.A. Krimshtein, S.V. Plotnikova, V.I. Konovalov, B.V. Putin, J. Appl. Chem. USSR, **65**, 2031 (1992)
3. A.A. Krimshtein, S.V. Plotnikova, V.I. Konovalov, B.V. Putin, Rus. J. Appl. Chem., **66**, 1356 (1993)
4. P.V. Balabanov, A.A. Krimshtejn, S.V. Ponomarev, Ind. Labor. Diagn. Mater., **77**, 34 (2011)
5. S.B. Putin, S.V. Gudkov, E.N. Tugolukov, N.P. Kozlova, S.Yu. Alekseev, A.Yu. Khromov, Trans. TSTU, **14**, 90 (2008)
6. J. Li, L. Jin, S. Wang, Z. Zhang, Y. Xu, Q. Li, Adv. Mater. Res., **726**, 363 (2013) DOI:10.4028/www.scientific.net/AMR.726-731.363
7. J.H. Kim, Y.K. Park, S.K. Jeong, Korean J. Chem. Eng., **27**, 320 (2010) DOI: 10.1007/s11814-009-0291-7
8. S. Choi, J.H. Drese, C.W. Jones, ChemSusChem, **2**, 796 (2009) DOI: 10.1002/cssc.200900036
9. S. Wang, S. Yan, X. Ma, J. Gong, Energy Environ. Sci., **4**, 3805 (2011) DOI: 10.1039/c1ee01116b
10. C.-H. Yu, C.-H. Huang, C.-S. Tan, Aerosol Air Qual. Res., **12**, 745 (2012) DOI: 10.4209/aaqr.2012.05.0132
11. A.M. Kudryavtsev, A.V. Kudryavtseva, A.A. Krimshtejn, S.V. Plotnikova, Trans. TSTU, **1**, 59 (1995)
12. V.L. Kolin, G.I. Bagrov, G.A. Musakin, V.P. Semenov, in *Kinetics and dynamics of physical adsorption*, 98 (Nauka, Moscow, 1973)

MATEC Web of Conferences **129**, 03005 (2017) DOI: 10.1051/matecconf/201712903005
ICMTMTE 2017

The evaluation of the correlation between entropy and negentropy in the structure of a technical system

Aleksander Dulesov[1], *Denis* Karandeev[1,*], and *Tatyana* Krasnova[1]

[1]Katanov Khakass State University, 655017 Lenina ave. 92, Abakan, Russia

Abstract. The role of information uncertainty in the structure of a technical system is defined. The application possibility of the R. Hartley's and C.E. Shannon's approaches in determination of entropy and negentropy of structural content of the system are selected and justified. The connection between entropy and negentropy in the case of system states changes is justified. The determination example of the quantitative values of information that characterizes connection association between entropy and negentropy is done.

1 Introduction

According to cybernetics, a technical system must not only maintain its properties (for example, to ensure a high reliability level [1]), but also to evolve. Therefore, information also develops after the system. The change of structural connection of the system in time is reflected through information. And the pre-processing of this information allows to monitor the system condition ("health"). The processed information allows a control subject to develop acceptable solutions that are aimed to maintain the state according the required level. Thus, timely and operational information is required to stabilize the system through reconstruction of broken structural connections between elements. The development and stability of the control system depends on information awareness of the system, connections with the environment, modern instruments of control and experience of system interaction.

In the operation of mechanisms and machines, information value consists in the measure determination of its uncertainty or information entropy in the system. The determination of its value allows to combine information uncertainty when it is important to consider the influence of random factors of different nature. At the same time, information contents can be represented through entropy of both quantitative and qualitative forms [2]. Entropy participates in the whole cycle of the system operation: collection of statistical information about system states → information processing and its analysis → obtaining entropy → searching of removal uncertainty → determining control resources → control and results estimation of removal uncertainty.

It is obvious, optimally using the resources in the operation process, we both increase quality and reliability of system operation [3] and temporarily reduce the uncertainty by introducing it to the known information.

2 The R. Hartley's and C.E. Shannon's connections of measures with the entropy of a technical system

Let's consider the possibilities application of information measure. We have a system structure (for example, as a graph), where the connections (edges) – elements of the system and the nodes (peaks) – concatenations of elements. Parameters are only for graph edges: the states of elements, the probability of events occurrence changing the element state. It is necessary to determine information value in order to estimate the system state and to pass on variations of the uncertainty removal in the system.

R. Hartley offered a combinatorial determination of an information amount using the logarithmic function following the principle of additivity and having connected information concept to implementation of the choice from a set of opportunities [4]. His approach to information retrieval can be regarded as the choice of one system state from a finite preassigned set of N equally probable states of its elements:

$$I = \log_2 N, \qquad (1)$$

where I – the information amount in bits, $N=2^I$ – the number of equally probable states or events. If $N = 2$ (a choice from two opportunities), then $I = 1$ bit.

In spite of the fact that the logarithm base in expression (1) can be various, we have assumed the base is 2. It is explained by the following reason: in this example, only two states (for example, operable and non-operable state, enabled and disabled, "yes" and "no") are considered. Further, these states can be designated as "1" and "0".

The Hartley's binary logarithm (1) allows to determine entropy of the set of equally probable states. It

* Corresponding author: den_dr_house_1991@mail.ru

DOI: 10.1051/matecconf/201712903005

is can be removed through the information amount, compensating for any states (for example, unwanted).

As the combinatorial approach to the determination of the information amount allows the various combinations of the system state or its element, then the entropy H can characterize the hypothesis about a possible arrangement of the states. In this case, it can pass on to the states probabilities. Hartley's formula is written so:

$$I = \log_2 N = \log_2 (1/p) = -\log_2 p, \quad (2)$$

and all states of N have the equally probable result: $p = 1/N$, $N = 1/p$.

R. Hartley's approach is based on the combinatorics theory and simple clear intuitive assumptions. When we look at the formula (2), we can tell: all the states have equal probability and we can find information in bits, thereby determining the number of the elements that generating this information. For example, when $N = 8$, then $I = 3$. It means that there are 3 elements, each of which generates 2 equally probable opposite states and in total 8 states is generated.

According to the additivity properties: all the independent elements (sources), which realize couples of the states simultaneously, generate the same information as the system itself. Thus, the uncertainty of the system is equal to the uncertainties amount of its elements. However, if we don't have details about the system state as source of a message, then there is an uncertainty about which of the element states or set of the states reflected on the state of the entire system. If we have information about the states that are generated by the selected element among the possible states, we can say that the existence of this information will reduce the information uncertainty about the system state.

The quantitative information measure meets the condition of the monotone increasing and increase of the choice possibility, i.e. the system states amount. In addition, as there is the additivity property then states total amount: $N = N_1 + N_0$, and the required function with equal probabilities elements states must meet the condition: $f(N_1 \cdot N_0) = f(N_1) + f(N_0)$. Here N_1 and N_0 – the number of the opposite states. These quantitative values indicate that we have divided the states on the qualitative character (for example, the number of the operable N_1 and non-operable N_0 element states).

If we consider the system as analysis then according to (2) we can write the equation by $N1 = N0$:

$$\log_2 N_1 = \log_2 N_0, \quad (3)$$

where private information about the occurrence of the operable states in the left formula part, and non-operable in the right formula part.

The equality (3) testifies about equilibrium of opposite states, whereas the formula (2) testifies about the maximum of the information I. For example, Hartley's model gives an indication of the equilibrium gas in a closed system when for all the molecules the equality of the states probabilities is caused by thermodynamic equilibrium. The equilibrium according to (3) is unsuitable for the majority of technical systems, because there equilibrium "is displaced" in the area that is ensuring the robustness of system structure functioning. For such systems the ratio of the states should not be in favor of the entropy (information uncertainty), because $N_1 >> N_0$. In this case, Hartley's approach is of little use due to the inequality of the probabilities opposite states.

Further, we are going to pass on to C.E. Shannon's approach in the determination of information amount. Shannon has applied the probabilistic approach to the information measurement (the information as the removal uncertainty, which reduces entropy) [5, 6]. In a technical system, the elements states follow each other occurring as events and have their own duration in time. In the operation process, registration of the states allows to determine the probability of the occurrence of events based on accumulated statistical data. The connection between the states probability and information amount, which obtained in case of an event approach, expressed by the Shannon's formula:

$$I = -\sum_{i=1}^{N} p_i \log_2 p_i, \text{ by } \sum_{i=1}^{N} p_i = 1, \quad (4)$$

where I – the information amount; N – the number of possible events; p_i – the probability of i-th event.

In Shannon's formula (4), we think, he reflected the possibility of the information differentiation on a qualitative character [2]:

$$I = -\left(\sum_{i=1}^{N_1} p_i \log_2 p_i + \sum_{j=1}^{N_0} q_j \log_2 q_j\right), \quad (5)$$

$$\text{by } \sum_{i=1}^{N_1} p_i + \sum_{j=1}^{N_0} q_j = 1,$$

where p_i and q_j – the probability of the opposite states (for example, p_i – the probability of an operable element state, q_j – the probability of a non-operable element state).

When we consider the states of a technical system based on the results of the processing statistical data, it becomes obvious that $p_i >> q_j$. Therefore, the value of the left term in the formula (5) exceeds the value of the right term. Note that: equality of these terms will be reached when $p_1 = q_1 = \ldots = p_i = q_j = \ldots$, that coincides with the relation (3) and indicates the equilibrium of the opposite states or the presence of chaos (disorder) in the system. However, as noted earlier, this equilibrium (lack of the order) is unacceptable in case of the operation of technical systems. In order to preserve and maintain a high scientific and technical level, it is necessary to involve additional resources in the system, for example, timely maintenance of equipment, scheduled preventive maintenance, etc. Then the number of the type "1" states (that measured by the number of negentropy) will greatly exceed the number of the type "0" states (that measured by the number of entropy). In case of information measurement on a state system, when it is necessary to maintain the order, we must neutralize a part of an entropy (to remove the information uncertainty) through the introduction into it of the negentropy [7]. In our opinion, the connection of these information components confirms the principle of the information mutual exclusion.

3 Detailing of connection the "entropy-negentropy"

Considering the system structure and its states, however, the question of the connection between entropy and negentropy requires a detailed study. In information theory [8], both information components are determined through a "probability of the state ", because the Nature laws and human behavior are subordinated to the probabilistic dominance. The growth of the "0" type states in relation to "1" type and, as a result, to entropy is tells us about the aging process of the object (destruction). In turn, the negentropy as order measure is aimed at removal uncertainty and increasing of the "1" type state. The exchange between information components provides the principle of the mutual exclusion. At the same time, it should be appreciated that exchange between information components is carried out within the full information about the closed system structure. Therefore, the increasing of one of components reduces the opposite component with the same value. These information components are changed to self-law and their absolute values are practically independent of each other.

Further, according to our example of structure states, we are going to consider the correlation between entropy and negentropy. For simplicity, from the standpoint of reliability [9], we are going to consider only the presence of the form "1" - operable states and "0" - non-operable states [10-13].

The regulation sets the rules on timely preventive maintenance of equipment (elements) to maintain a high reliability level in the system operation [14]. All repair and interrepair periods are specified in time based on the operating experience of similar facilities, or empirical way for newly constructed facilities. The process of systematic purposeful actions on the part of the control subject are connected with the determination of negentropy amount, which can be determined by the Shannon's formula:

$$I = -\sum_{i=1}^{N_1} p(A_i^{(j)}) \log p(A_i^{(j)}), \quad (6)$$

by $\sum_{i=1}^{N_1} p(A_i^{(j)}) = 1$,

where $A_i^{(j)}$ – an event $i = \overline{1, N_1}$ that has probability p_i or the proportion of the time t_i / T, $T = \sum t_i$ in the interval T, the reflecting signs (planned repair; interrepair period) of the element j. Additionally, we note the following: formula (6) is applicable to a structure element, because it's a planned outage should not significantly affect the operability of the entire system.

The nature impact on the system though isn't considered aggressive, however, introduces some chaos, which indicates the occurrence of the states "0" and the presence of the entropy. The entropy "will replace" negentropy, because the information about the element won't be changed.

Information about the state of the entire system can be determined by the expression:

$$I_\Sigma = I + H, \quad (7)$$

where I and H – respectively, negentropy and entropy.

The expression (7) reflects information about the states, which connected with the implementation of specified functions and opposite states that violate these functions. When $I = H$ – utter chaos or a balance between the opposite states. The task of a control subject: to bring the system through an orderly influence in the area where the condition $I >> H$ is satisfied, when the order is considerably exceeds chaos. Moving from chaos to order the negentropy increases, i.e. the anti-entropic process is carried out. L. Brillouin has shown that the quantity of the negentropy is exactly equal to the decreasing in entropy [15]. Thereby, the control subject tries to maintain equilibrium at the required level.

Look at the example. Let the system consists of n=3 of elements, where each can be fitted in one of two opposite states with equal probabilities. All elements are independent. It is necessary to determine the information components about system state.

Full or total information amount about the possible system states can be determined by the expression (4). Since the statement of the problem is associated to existence of the equally probable events, then quantity of the all equally possible states of the system - $N=2^n=2^3=8$ and states variety of forms "1" and "0" for each of 3 elements: 1,1,1; 1,1,0; 1,0,1; 1,0,0; 0,1,1; 0,1,0; 0,0,1; 0,0,0. The occurrence probability each of 8 states is equal to 1/8. Then we will receive required information value according to the expression:

$$I_\Sigma = \sum_{i=1}^{N} I_i = \sum_{i=1}^{N} \frac{1}{N} \log_2 \frac{1}{N} = \sum_{i=1}^{8} \frac{1}{8} \log_2 \frac{1}{8} = 3, \quad (8)$$

where I_i =3/8 – the number of the private information about the state of the i system.

The following follows from a statement of the problem (8): the maximum of information amount (in bits) is determined and we have $I = H = 1.5$ (or utter chaos) for the variety of 8 system states. This ratio doesn't suit us. For example, we are able to provide a high order in the system due to the control actions. Then the elements states will be mostly of only one type, for example: 1,1,1; 1,1,0; 1,0,1; 1,1,0; 0,1,1; 0,1,1; 0,0,1; 0,0,0. These 8 system states are placed in random order and have the condition: one state of type "0", which inherent in the element, doesn't result in state "0" of the entire system. There are two qualitatively different types in this combination of the system states: 6 states correspond to the qualitative type "1" and 2 states correspond to the qualitative type "0". Let's determine the information components about system states:

$$I = \frac{N_1}{N}\log_2\frac{1}{N} = \frac{6}{8}\log_2\frac{1}{8} = 2.25; \quad (9)$$

$$H = \frac{N_0}{N}\log_2\frac{1}{N} = \frac{2}{8}\log_2\frac{1}{8} = 0.75. \quad (10)$$

In the expressions (9) and (10), the probability of generating of the private information amount about the system state of the type "1" is equal 6/8 and type "0" is equal to 2/8. The property (7) is confirmed according to this example.

4 Conclusion

The negentropy characterizing with systematic actions of the control subject removes uncertainty (entropy) and it is the quantitative information measure. The removing of the uncertainty is expressed through the changes of conditions that imposed on the system and its entropy. The preserving of system structure stability is a fight against entropy, which reflects aging process in technical systems.

In technical systems, the information attachment from the control subject allows to accumulate the ability to probability differentiation of the negative reactions on the external environment influence through the equation (4). At the same time, the system structure is changed due to the using of material resources, i.e. the entire complex of signs. The necessaty to increase the negentropy – is the necessaty to compensate the entropy as much as the entropy can be decreased itself. As a result, the control subject tends to that the probability of one type "has won", for example a type "1". However, the technical system can't reach absolutely deterministic state, because the Nature looks out for the existence of the system and elements lifecycle to make them more perfect and replace.

We can determine the entropy and negentropy amounts based on C.E. Shannon's formulas for discrete systems due to the processing of statistical data about the system states and executing the differentiation of the probability about the changing control conditions. The problem solution of the removal uncertainty will allow to plan actions to reduce entropy based on the negentropy principle of L. Brillouin's information about the structural system changes.

References

1. A. S. Dulesov, D. Yu. Karandeev, *Construction of optimal structure of technical system using "branch and bound" by criteria of profitability and reliability [in Russian]*, Journal Safety & reliability of Power Industry **33**, pp. 56-59 (2016)
2. A. S. Dulesov, N. V. Dulesova and D. Yu. Karandeev, *Delimitation indicator of level of reliability of technical system on the qualitative character: entropy approach [in Russian]* Fundamental research **2**(4), pp. 477-481 (2016)
3. I.A. Ushakov, *Solving of optimal redundancy problem by means of a generalized generating function*, Journal of Information Processing and Cybernetics archive **24**(4), pp. 219 – 222 (1988)
4. R. V. L. Hartley, *Transmission of Information*, Bell System Technical Journal 7(3), pp. 535–563 (1928)
5. C. E. Shannon, Mathematical *Theory of Communication*, Bell System Tech. J. **27**(1), pp. 379-423 (1948)
6. C. E. Shannon, *Communication Theory of Secrecy Systems*, Bell System Tech. J. **28**, pp. 656-715 (1949)
7. E. H. Liiv, *Infodinamika. Infodynamics. Generalized entropy and negentropy. [in Russian]* (Tallinn:AO Juhisjelu, 1998)
8. Thomas M. Cover, A. Thomas Joy, *Elements of Information Theory* (New Jersey: Wiley and Sons, second edition, 2006)
9. Reda Farag, Haldar Achintya, *A novel reliability evaluation method for large engineering systems*, Ain Shams Engineering Journal, pp. 1-13 (2016)
10. A. S. Dulesov, D. Y. Karandeev and N. N. Kondrat, *How to define amount of information for the simplest structure of system [in Russian]*, Vestnik of Katanov Khakass State University) **15**, pp. 7-9 (2016)
11. A. S. Dulesov, D. Y. Karandeev and N. N. Kondrat, *Definition of amount of information entropy in structure of the technical system by method of the minimum ways [in Russian]*, Modern high technologies **2**(3), pp. 425-429, (2016)
12. A. S. Dulesov, D. Y. Karandeev and N. N. Kondrat, *Definition of amount of information entropy in structure of the technical system by method of the minimum sections [in Russian]*, Fundamental research **2**(3), pp. 425-429 (2016)
13. A. S. Dulesov, D. Y. Karandeev and N. N. Kondrat, *Possibilities of application of expression of Shannon when determining amount of information [in Russian]*, Vestnik of Katanov Khakass State University **15**, 5-7 (2016)
14. Charles E. Ebeling, *An Introduction to Reliability and Maintainability Engineering* (McGraw-Hill Companies, Inc., Boston, 1997)
15. L. Brillouin, *Science and Information Theory*, (Academic Press, 1956)

DOI: 10.1051/matecconf/201712903006

Robust stabilization of the multilinked object

Yuliya Lezhnina[1,*], *Galina* Ternovaya[2], and *Elena* Evsina[1]

[1]Astrakhan State University of Architecture and Civil Engineering, CAD&M department, 414056, str. Tatischeva, 18, Astrakhan, Russia
[2]Astrakhan State Technical University, Department of higher and apply mathematics, 414056, str. Tatischeva, 16, Astrakhan, Russia

Abstract. We consider the problem of robust control for a multilink object in the presence of the signal and parametrical uncertainty when the measurement of the derivatives of input and output signals of the local subsystems is not available. Proposed and justified the use of auxiliary circuits for each of the subsystems that eliminates the use of the vector of regression during the generate of the control signal, where by the reduced order closed-loop control system. The advantage of the proposed robust control law is that it remain unchanged when exposed to unknown disturbances, and in their absence. In the first case, the system is dissipative, while in the second case, it one is asymptotically stable.

1 Introduction

Problem of control of uncertain objects is one of the classical problems of control theory. Complex multiply linked objects take a special place among object of control, the control of which is complicated by the linkages between local subsystems. The large number of the interconnected subsystems influencing the each other complicates the traditional control problems, and the requirement of the decentralization comes to the forc [1, 2]. The utilization of the decentralized algorithms corresponds to the nature of large interconnected systems, because it expects the distribution of the system components in space. And in addition to this, the decentralized management structure allows getting more qualitative and trusted control system, because it brings the governing body to the object and simplifies the system structure considerably [3, 4, 5].

In this paper, we consider that class of multilinked objects, which mathematical models can be brought to a canonical form [6, 7]. To assess the perturbations introduced auxiliary circuit and the observer derivative. Compensation of the perturbations produced by the control law with opposite sign to the assessment of disturbances. The control law is constructed in such way as to offset the impact of local interconnections of subsystems [8, 9, 10, 11]. There are developed a lot of methods of construct of adaptive and robust systems under parametric uncertainty of the object in the modern control theory. There are developed method extended errors, the method bypass, the algorithms of high order. At the core of these methods is the use of the filter or shunt device. A control signal is generated as the scalar product of vector of adjustable parameters and vector of regression. However, the task of creating a simple to implement and low-dimensional algorithms is still relevant. In this paper, we consider the stabilization problem for a dynamic object.

2 Statement of the problem

Let us examine the interconnected system with its dynamic processes which are described by the equations in the local subsystems so:

$$Q_i(P)y_i(t) = R_i(P)\left(u_i(t) + f_i(y_i,t) + \sum_{i=1, i\neq j}^{k} y_{sij}(t) \right), \quad (1)$$

$$\begin{aligned} s_{ij} &= A_{sij}s_{ij} + B_{sij}y_j, \quad y_{sij} = L_{ij}s_{ij}, \\ i &\neq j, \quad i = \overline{1,k}. \end{aligned} \quad (2)$$

where $Q_i(P), R_i(P)$ - are linear differentiation operators whose elements depend on the vector of unknown parameters $\xi \in \Xi$; Ξ – is a known aggregate of possible values of the vector ξ; $\deg Q_i = n_i$, $\deg R_i = m_i$; $f_i(y_i,t)$ – is an unknown bounded function; $y_i(t)$ – is a scalar controlled variable of i-th subsystem which is accessible for measurement; $u_i(t)$ – is a scalar control action; $s_{ij} \in \Re^{n_{ij}}$, $y_{sij} \in \Re$; $A_{sij} \in \Re^{n_j \times n_j}, B_{sij} \in \Re^{n_j}$ – are unknown numerical matrixes; $L_{ij} = [1,0,\ldots,0]$ – is a row matrix of the appropriate orders. In this case, the equation of the control object is presented in canonical form.

The equations (1) describe the dynamic processes in the local subsystems, and (2) in the cross coupling. Their transfer functions look like this as in

$$W_{ij}(\lambda) = L_{ij}(\lambda I_{n_j} - A_{sij})^{-1}B_{sij} = R_{sij}(\lambda)/Q_{sij}(\lambda) \quad (3)$$

where $i \neq j$, the orders of polynomials $R_{sij}(\lambda), Q_{sij}(\lambda)$ are respectively equal to the following m_{ij}, n_{ij}; I_{n_j} – is an identity matrix of order.

[*] Corresponding author: lejninau@gmail.com

 DOI: 10.1051/matecconf/201712903006

Hypothesis.

A1. Set Ξ are known.

A2. Polynomials $R_i(\lambda)$ are Hurwitz.

A3. The orders of Polynomials are known , the relative degree a object is known too $n_i - m_i > 1$.

A4. The perturbation actions are the bounded quantity, that is $|f_i(y_i,t)| \leq C_i, \quad C_i > 0$.

A5. The matrix A_{sij}, B_{sij} is unknown, but they are Polynomials $Q_{sij}(\lambda)$ Hurwitz.

3 Robust stabilization of the multilinked dynamic object

We first solve the problem of the asymptotic robast stabilization when there are no external disturbances in the object (1), i.e. $f_i(y_i,t) = 0$. We have to do is to design the control system for which one condition will meet

$$\lim_{t\to\infty} |y_i(t)| = 0, \tag{4}$$

but at the same time, use of the measurable values of the other subsystems isn't allow in the local control subsystems. Decentralized robust control is implemented only if there are used only the current information about the system for construction of local control units for each subsystem. Have the equation of the control object (1), written in canonical form as

$$Q_i(P)y_i(t) = R_i(P)\left(u_i(t) + \sum_{j=1, i\neq j}^{k} y_{sij}(t)\right)$$

where $Q_i(P), R_i(P)$ – is a linear differentiation operator with a constant rates. In the generation of control law will take estimate of the output signal with opposite sign. So it form is

$$u_i(t) = -\theta_i T_i(P) y_i(t) \tag{5}$$

where a number $\theta_i > 0$ and a linear differentiation operator $T_i(P)$ of order $n_i - m_i - 1$ are chosen, according to the Hurwitz of polynomial

$$Q_{0i}(\lambda) = Q_i(\lambda) + \theta_i R_i(\lambda) T_i(\lambda) \tag{6}$$

the function $\bar{y}_i(t)$ is an estimate of a signal $y_i(t)$ which is realized by the observer [12]

$$\dot{\bar{x}}_i = F_{0i}\bar{x}_i + H_i(y_i - \bar{y}_i), \quad \bar{y}_i = L_{0i}\bar{x}_i. \tag{7}$$

Here $\bar{x}_i \in \Re^{n_i - m_i}$; $L_{0i} = [1,0,\ldots,0]$; $\chi_i > 0$ – is the little number, $F_{0i} = \begin{bmatrix} 0 & I_{n_i-m_i-1} \\ 0 & 0 \end{bmatrix}$, $H_i^T = \left[\frac{h_{1i}}{\chi_i}, \ldots, \frac{h_{n_i-m_i-1}}{\chi_i^{n_i-m_i-1}}\right]$;; the vector H_i is chosen by special method so that the matrix $F_i = F_{0i} + H_i L_{0i}$ was the Hurwitz, where $\bar{H}_i^T = \left[-h_{1i}, \ldots, -h_{n_i-m_i-1}\right]$; $I_{n_i-m_i-1}$ – is an identity matrix of order $(n_i - m_i - 1) \times (n_i - m_i - 1)$. It is obvious, that the control law can be technically realizable, as it contains the known or measurable values. At the time, the solution of the problem gives the following statement.

The Statement 1 If the hypotheses A.1 – A.5 are implemented, then the control law (5) with the observer (7) implements of the target condition (5).

Proof of Proposition 1. Inputting the control (5) to the equation of object (1) with (6), we'll get the last input-output model in the matrix vectorial form

$$\dot{x}_i = A_{0i}x_i + B_i(\bar{y}_i - y_i) + B_i \sum_{j=1, i\neq j}^{k} y_{sij}, \quad y_i = L_i x_i, \tag{8}$$

where $B_i \in \Re^{n_i}$ – is a vector whose components are the coefficients $\theta_i R_i(P) T_i(P)$; A_{0i} – is a Hurwitz matrix in the form of a characteristic polynomial of the Frobenius $Q_{0i}(\lambda)$; $B_i^T = [0,\ldots,0,b_0,\ldots,b_m]$. Following the [13], we introduce into consideration nullity vector η_i as

$$\dot{\eta}_i = \frac{1}{\chi_i} F_i \eta_i + B_{1i}\dot{x}_{21i} + D_i^{-1} B_{1i} L_{3i} \eta_i \tag{9}$$

where $D_i = diag\{\chi_i^{n_i-m_i-1}, \ldots, \chi_i, 1\}$, x_{1i} – is a component part of the state vector, namely $x_i^T = [x_{1i}^T, x_{2i}^T]$.

We introduce the augmented state vector $z_i = col(x_i, \eta_i)$ and write the equation (8), (9) in the composite matrix vectorial form

$$\begin{aligned} \dot{z}_i &= B_{1pi}\dot{x}_{21i} + D_i B_{3pi} L_{3i} \eta_i + \\ &+ \frac{1}{\chi_i} A_{pi} z_i + B_{2pi} \sum_{j=1}^{k} y_{sij} \\ y_{pi} &= L_{pi} z_i, \end{aligned} \tag{10}$$

and at the same time matrixes A_{pi} will be Hurwitz.

Let us take the Lyapunov function in the form of

$$V = \sum_{i=1}^{k} \left\{ z_i^T(t) P_i z_i(t) + \sum_{j=1}^{k} s_{ij}^T P_{sij} s_{ij} \right\} \tag{11}$$

in which P_i, P_{sij} – are the positive definite symmetric matrixes satisfy the matrix relation:

$$\begin{aligned} & P_i A_{pi} + A_{pi}^T P_i = -Q_{pi}, \\ & Q_{pi} = Q_i + (\rho_{1i} + \rho_{2i}) I_{2n_i - m_i - 1}, \\ & P_{sij} A_{sij} + A_{sij}^T P_{sij} = -Q_{pij}, \\ & Q_{pij} = Q_{sij} + \rho_{sij} I_{n_{ij}}. \end{aligned} \tag{12}$$

Because the matrixes A_{pi} and A_{sij} are Hurwitz, there are matrixes P_i and P_{sij}, that satisfy the matrix relation (12). Let us calculate the total derivative of the function (12) on the trajectories of the system (10), (2), with (12). Then the derivative of the Lyapunov function will look like

 DOI: 10.1051/matecconf/201712903006

$$\dot{V} \le \sum_{i=1}^{k} \left\{ -\alpha_1 V - \frac{\rho_{1i}}{2\chi_i} \|x_i\|^2 \right\} \le -\alpha_1 V,$$

the synthesized system is stable, because the target condition is satisfied (4).

4 Robust stabilization of the multilinked dynamic object under bounded disturbances

However, in practice it is difficult to create the situation for control of technical objects in ideal conditions when it is not exposed to external disturbances. The case where external disturbances are bounded is the simplest. Next, let us solve the problem of a decentralised stabilization when the external disturbances are present in the object (1), i.e. $f_i(y_i,t) \ne 0$. We had to design a control system for which next condition will be satisfied

$$\lim_{t\to\infty} y_i(t) < \delta, \tag{13}$$

In accordance with the approach presented in [13, 14] let us ask one local control law in the form of (5) where a number $\theta_i > 0$ and the linear differentiation operator $T_i(P)$ of order $n_i - m_i - 1$ are chosen from the reasons of Hurwitz polynomial (6), and the function $\bar{y}_i(t)$ is an estimate of output $y_i(t)$. Then the equation (1) will look like

$$Q_{0i}(P)y_i(t) = R_i(P)\left(\theta_i T_i(P)(y_i(t) - \bar{y}_i(t)) + f_i(y_i,t) + \sum_{j=1, i\ne j}^{k} y_{sij}(t) \right) \tag{14}$$

Implementing the control law (5) requires getting the estimate $\bar{y}_i(t)$ and its $n_i - m_i - 1$ derivatives, for which we will use the observer (7). It is obvious, that the control law can be technically realizable, as it contains the known or measurable values.

The Statement 2. If the hypotheses A.1 – A.5 are implemented, then the control law (5) with the observer (7) implements the limited nature of system trajectory (14).

Observation. But it should be noted, that choosing the number θ_i of greater value, and the value χ of smaller, we can achieve the target condition (14).

Proof of Proposition 2. Let us convert the equation (14) to the vectorial matrix form

$$\dot{x}_i(t) = A_{0i}x_i(t) + B_i\left(\tau_i^T(x_{1i}(t) - \bar{x}_i(t)) + f_i(y_i,t) + \sum_{j=1, i\ne j}^{k} y_{sij}(t) \right), \tag{15}$$
$$y_i(t) = L_i x_i(t),$$

where $x_i \in \Re^{n_i}$; $y_i \in \Re$; A_{0i} – is a Hurwitz matrix in Frobenius form with characteristic polynomial $Q_{0i}(\lambda)$; $\tau_i \in \Re^{n_i - m_i}$ – is a vector whose components are the coefficients $\theta_i T_i(P)$; $B_i^T = [0,...,0,b_{0i},...,b_{mi}]$; $L_i = [1,0,...,0]$; x_{1i} – is a component part of the state vector, namely $x_i^T = [x_{1i}^T, x_{2i}^T]$:

$$\dot{x}_{1i}(t) = F_{0i}x_{1i}(t) + B_{0i}\left(\frac{1}{b_{0i}}x_{21i} + \tau_i^T(x_{1i}(t) - \bar{x}_i(t)) + f_i(y_i,t) + \sum_{j=1, i\ne j}^{k} y_{sij}(t) \right),$$
$$\dot{x}_{2i}(t) = F_{1i}x_{2i}(t) + B_{1i}c_{1i}^T x_i(t) + B_{1i}\left(\tau_i^T(x_{1i}(t) - \bar{x}_i(t)) + f_i(y_i,t) + \sum_{j=1, i\ne j}^{k} y_{sij}(t) \right), \tag{16}$$
$$y_i(t) = L_{0i}x_{1i}(t),$$

where the matrix F_{1i} has the same structure as F_{0i}; $B_{0i}^T = [0,...,0,b_{0i}]$; $x_{1i} \in \Re^{n_i - m_i}$; $x_{2i} \in \Re^{m_i}$, the vector components c_{1i} are the coefficients $Q_{0i}(\lambda)$ with opposite signs; $B_{1i}^T = [b_{1i},...,b_{mi}]$; x_{21i} – is the first component of vector x_{2i}; $L_{0i} = [1,0,...,0]$. Following the [13], we introduce the vector (9). Then from (7) and (16) we have

$$\dot{\eta}_i(t) = \frac{1}{\chi}F_i\eta_i(t) + B_{0i}\left(\tau_i^T D_i\eta_i(t) + \frac{1}{b_{0i}}x_{21i} + f_i(y_i,t) + \sum_{j=1, i\ne j}^{k} y_{sij}(t) \right) \tag{17}$$

Structure of matrix $F_{0i}, B_{1i}, L_{0i}, D_i, H_i$ is such that the equalities are satisfied

$$D_i^{-1}(F_{0i} - H_i L_i)D_i = \frac{1}{\chi}F_i, \qquad D_i^{-1}B_{0i} = B_{0i},$$

Let us introduce block diagonal matrix

$F = diag\{F_1,...,F_k\}$; $B_0 = diag\{B_{01},...,B_{0k}\}$,
$D = diag\{D_1,...,D_k\}$; $B_0 = diag\{b_{01}^{-1},...,b_{0k}^{-1}\}$;
$A_0 = diag\{A_{01},...,A_{0k}\}$; $B = diag\{B_1,...,B_k\}$,
and the vectors $\eta = col(\eta_1,...,\eta_k)$; $\tau = col(\tau_1,...,\tau_k)$,
$x = col(x_1,...,x_k)$; $x_{21} = col(x_{211},...,x_{21k})$;
$f = col(f_1,...,f_k)$;
$y_{si} = col(y_{si1},...,y_{sik})$; $y_s = col(y_{s1},...,y_{sk})$,

Then the equation (15), (17) in the composite form will look like

$$\dot{\eta}(t) = \frac{1}{\chi}F\eta(t) + B_0(\tau^T D\eta + B_0 x_{21} + f(t) + a_s y_s), \tag{18}$$
$$\dot{x}(t) = A_0 x(t) + B(\tau^T D\eta + f(t) + a_s y_s(t)),$$

where $\eta(t) \in \Re^{n-m}$; $x(t) \in \Re^n$; a_s – is a block matrix with the blocks I_k, I_k – is a identity matrix of order $k \times k$:

DOI: 10.1051/matecconf/201712903006

$$a_s = \begin{bmatrix} I_k & 0 & \cdots & 0 \\ 0 & I_k & \cdots & 0 \\ \vdots & \vdots & \ddots & \vdots \\ 0 & 0 & \cdots & I_k \end{bmatrix}.$$

Let us take the Lyapunov function

$$V_1 = \eta^T(t)P\eta(t) + \sum_{i=1}^{k}\sum_{j=1,j\neq i}^{k} s_{ij}^T M_{sij} s_{ij}, \qquad (19)$$

where P, M_{sij} – are the positive definite symmetric matrixes, at the same time $P = diag\{P_1,\dots,P_{n-m}\}$. To take account the fact that the matrixes are block diagonal, we will produce that positive definite symmetric matrixes P_i must satisfy the equations for every subsystem

$$P_i F_i + F_i^T P_i = -Q_i - \rho_i I_{n_i - m_i},$$

$$M_{sij} A_{sij} + A_{sij}^T M_{sij} = -Q_{sij} - \rho_{sij} I_{n_j},$$

where $\rho_i > 0; \rho_{sij} > 0; \quad Q_i = Q_i^T > 0; Q_{sij} = Q_{sij}^T > 0.$ Since the matrixes F_i and A_{sij} are Hurwitz we see that there are the matrixes P_i and M_{sij}, with the ratio (19).

Let us take the Lyapunov function

$$V_2 = x^T(t)Hx(t), \qquad (20)$$

where H – are the positive definite symmetric matrixes, at the same time $H = diag\{H_1,\dots,H_n\}$. To take account the fact that the matrix H is a block diagonal, we will produce that positive definite symmetric matrixes H_i must satisfy the equations for every subsystem

$$H_i A_{0i} + A_{0i}^T H_i = -Q_{1i} - \rho_{1i} I_{n_i},$$

where $\rho_{1i} > 0;\ Q_{1i} = Q_{1i}^T > 0.$ Since the matrixes A_{0i} are Hurwitz we see that there are the matrixes H_i which satisfy these matrix relation. To take account the estimated derivative of the functions (19), used (20) and the last inequality, we will calculate the total derivative of the Lyapunov function $V=V_1+V_2$, and we'll get the inequality

$$\dot{V}(x_i(t),\eta(t)) \le -\beta_3 V + \beta_4,$$

where $\quad \beta_4 = \sum_{i=1}^{k} \beta_{2i} \sup_t C_i^2,$

$$\beta_3 = \min\left\{0.25\alpha_i; \frac{1}{\chi}\frac{\lambda_{\min}(Q_i)}{\lambda_{\max}(P_i)}; \frac{\lambda_{\min}(Q_{1sij})}{\lambda_{\max}(M_{sij})}\right\},$$ from which the dissipative system follows

$V \le \dfrac{\beta_4}{\beta_3}$ and the estimate $\lim_{t\to\infty} |x_i| \le \dfrac{\beta_4}{\beta_3}$.

This value can be reduced by choosing a little number χ which leads to an increase of the $1/\chi$, and θ_i in the (5) greater. Unfortunately, we cannot get a more accurate estimate.

5 Example

Consider the multilinked control plant, dynamic processes which are described by the equations

$$(P^3 + a_1P^2 + a_2P + a_3)y_1(t) =$$
$$= (b_1P + b_2)u_1(t) + s_1y_2(t)$$
$$(P^3 + a_4P^2 + a_5P + a_6)y_2(t) =$$
$$= (b_3P + b_4)u_2(t) + s_2y_1(t)$$

Aggregate of possible values of the vector $\Xi = \{a_i, s_i, b, l = 1{,}6 : a_i \in [-8;5], b \in [1;8]\}$.

The equation of the auxiliary circuit has the form $(P^2 + 2P + 1)\bar{y}_i = u_i,\ i = 1{,}2.$

Let's used observer

$$\dot{\zeta}_i = \begin{bmatrix} 0 & 1 \\ 0 & 0 \end{bmatrix}\zeta_i + \begin{bmatrix} 3/\mu \\ 3/\mu^2 \end{bmatrix}(\vartheta_i(t) - \bar{\vartheta}_i), \qquad i = 1{,}2.$$
$$\bar{\vartheta}_i = \begin{bmatrix} 1 & 0 \end{bmatrix}\zeta_i$$

The control law is in the form of

$$u_i(t) = -\frac{\alpha}{k_m}(\zeta_{1i}(t) + 2\zeta_{1i}(t) + \zeta_{2i}(t)),\ i = 1{,}2.$$

General block diagram of the control object shown in fig. 1. Here we have the multilinked control object consisting of *K* local subsystems. Control signal and the output signals from all the other *K*-1 local subsystems are inputs of each local subsystem. The local control unit generates a control signal, using an estimate of the state vector of the object that is obtained with the observer. Thus is implemented a decentralized control.

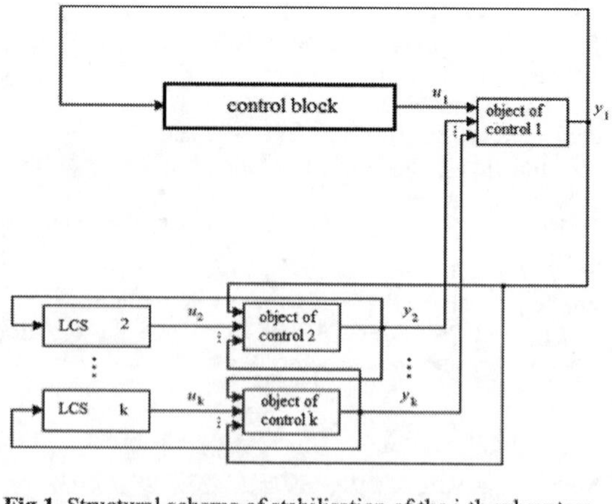

Fig.1. Structural scheme of stabilisation of the i-th subsystem

This object is synthesized with a decentralized robust controller for stabilization of each of the local subsystems, the structure of which is shown in Fig.2. The control signal is generated with a opposite sign to

DOI: 10.1051/matecconf/201712903006

the assessment of variables of the internal state. Compensation of external and structural disturbances is possible because introducing the auxiliary circuit. The disadvantage of the proposed control law is that the absolute magnitude of the coefficients K and α is impossible to find analytically. Their value is chosen at the modeling stage.

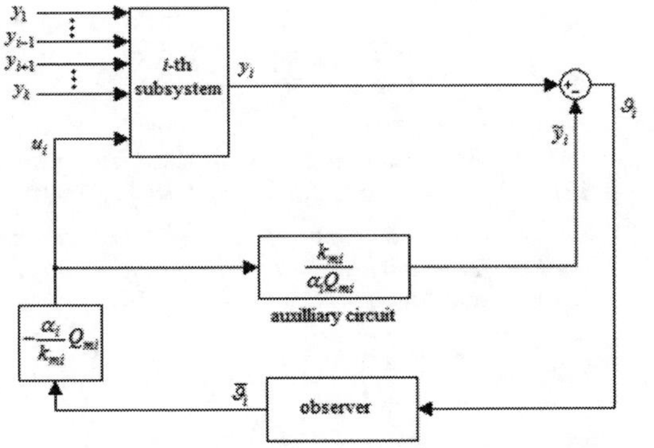

Fig.2. Structural scheme of robast control of the i-th subsystem

Fig. 3 shows the simulation results of the stabilization system of the unstable object in the presence of external disturbances

$$f_1(t) \neq 0, \quad f_2 \neq 0,$$
$$y_1(0) = \dot{y}_1(0) = \ddot{y}_1(0) = 2,$$
$$y_2(0) = \dot{y}_2(0) = \ddot{y}_2(0) = -1$$

There are follows external disturbances in the object

$$f_1(t) = 1 + \sin 0{,}1t + \sin 10t,$$
$$f_2(t) = 1 + 2\sin 0{,}3t + \sin 10t.$$

Consider the control object, dynamic processes in which case are described by the equation

$$(P^3 + a_1P^2 + a_2P + a_3)y_1(t) = b_1(u_1(t) + f_1(y,t) + y_{s12}(t)),$$
$$y_{s12} = 3/(p+2)y_2,$$
$$(P^3 + a_4P^2 + a_5P + a_6)y_1(t) = b_2(u_2(t) + f_2(y,t) + y_{s21}(t)),$$
$$y_{s21} = 5/p^2 + p + 1y_1.$$

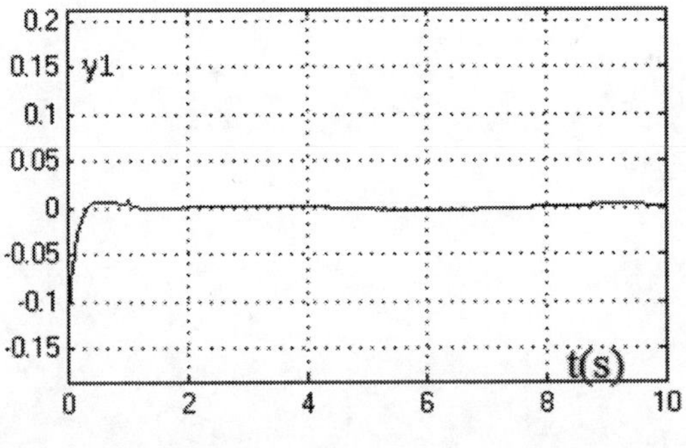

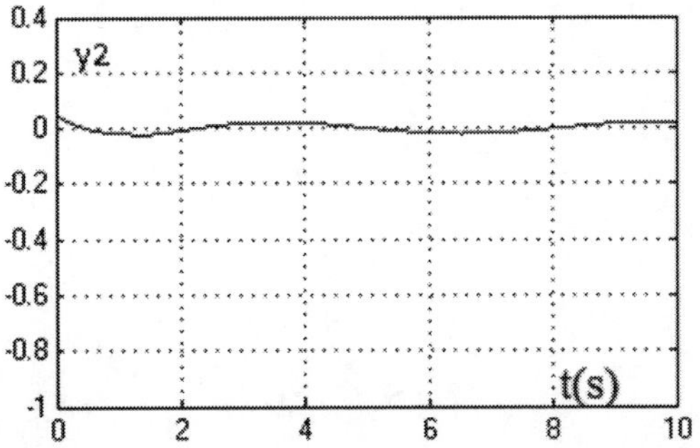

Fig.3. The trajectories of the system's outputs in robust stabilization

Aggregate of possible values of the vector $\Xi = \{a_l, s_l, b, l = 1{,}6 : a_l \in [-5;5], b \in [1;5]\}$.

External disturbances are present as $f_1(t) = \sin 2t, f_2(t) = 2\sin 3t$.

We define the control law in the form (5), where $T_i(P) = \tau_{1i}P^2 + \tau_{2i}P + \tau_{3i}$.

Get $\theta_i = 20, \tau_{1i} = 16$, $\tau_{2i} = 8, \tau_{3i} = 2$.

In this case, the polynomials $Q_{0i}(\lambda), (i = 1{,}2)$ are Hurwitz. To obtain the observer (7) will take the value $\chi = 0.01$, and vector $H_i^T = [-3;-3;-1]$, $i = 1{,}2$.

Fig. 4 shows a model diagram of the first subsystem of the object (1) dissipative stabilization.

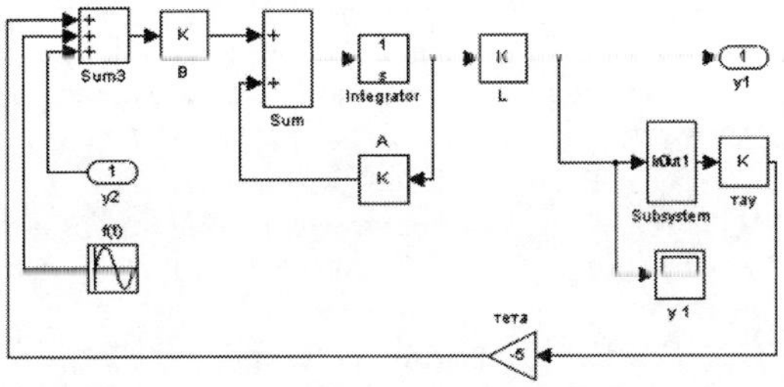

Fig.4. Structural scheme of robast control of the i-th subsystem.

Fig. 5 shows the simulation results of dissipative system of stabilization with perturbations $f_1(t) = \sin 2t, f_2(t) = 2\sin 3t$.

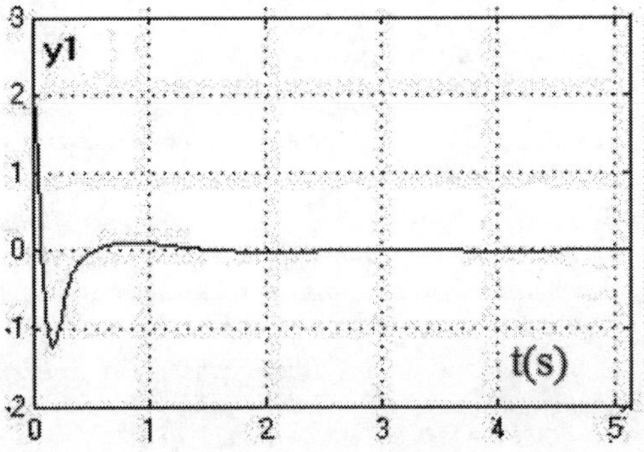

DOI: 10.1051/matecconf/201712903006

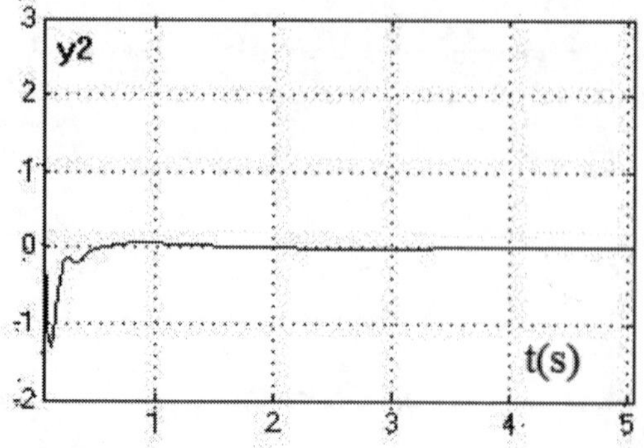

Fig. 5. The trajectories of the system's outputs in robust stabilization.

5 Discusion

The advantage of the proposed robust control law (5) is that it remains the same in the presence of unknown perturbation actions and in their absence. Only in the first case the system will be dissipative, while the second case it will be asymptotically robust.

References

1. Y.A. Lezhnina, G.N. Ternovaya, E.R. Galyauv, I.Y. Kvyatkovskaya, *Decentralized robust control over robot manipulator*, Applied Mechanics and Materials, **437** (2013)
2. I.B. Furtat, J.V. Chugina, A.L. Fradkov, *Robust control of multi-machine power systems caused by perturbation of mechanical input power and variable unknown communication time-delay*, IFAC-PapersOnLine **49(14)** (2016)
3. A. M. Tsykunov, *Robust control algorithms with compensation of bounded perturbations*, Automation and Remote Control, **7** (2007)
4. Yu. Lezhnina, I. Petrova, G. Ternovay, *Path Control for a Manipulator While Tracking a Given Trajectory*, Applied Mechanics and Materials, pp. 752-753 (2015)
5. Yu. Lezhnina, G. Ternovaya, V. Zaripova, *Robust Adaptive Control of the Dynamic Multilinked Object: Control of Robot Manipulator*, Progress in Systems Engineering Proceedings of the Twenty-Third International Conference on Systems Engineering (ICSEng 2014, Las Vegas, Nevada, August 19, 2014), (2014)
6. A. M. Tsykunov, *Robust output control of linear dynamic objects*, Mekhatronika, avtomatizatsiya, upravlenie **8** (2008)
7. E. A. Parsheva, Yu. A. Lezhnina, *Robust decentralized control with disturbance compensation with nonlinear multivariable objects*, Mekhatronica, avtomatizatsiya, upravlenie **8** (2011)
8. E. A.Parsheva, *Robust Decentralized Control with Scalar Output of Multivariable Plants with Uncertain Structures*, 6th IFAC Symposium on Robust Control Design, **6(1)** (2009)
9. V.O. Nikiforov, *Robust high-order tuner of simplified structure*, Automatica, **8(35)** (1999)
10. A. A. Bobtsov, *Robust control algorithm of uncertain object without measuring derivatives of adjusted variable*, Autom. Telemekh **8** (2003)
11. I. B. Furtat, A. L.Fradkov, A. M. Tsykunov, *Robust Synchronization of Linear Networks with Compensation of Disturbances*, Proceedings of 18th IFAC World Congress **18(1)**
12. A. N. Atassi, H. H. Khalil, *A separation principle for the stabilization of a class of nonlinear systems*, IEEE Trans. on Automatic Control **9(44)** (1999)
13. G. Tao, P. A. Ioannou, *Model reference adaptive control for plants with unknown relative degree*, IEEE Trans. Automat. Control **6(38)** (1993)
14. V. A. Brusin, *A class of singular distributed adaptive systems*, Autom. Telemekh. **4** (1995)

Numerical simulation of the flow of a gas-liquid medium in a circular channel at cooling a high-temperature metal cylinder with a variable cross-section

Sergey Makarov[1,*], *Vyacheslav* Dement'yev[1,*], and *Tatyana* Makhneva[1,*]

[1] Institute of Mechanics of Ural Branch of the Russian Academy of Sciences, 34, T. Baramzinoy St., Izhevsk, 426067, Russia

Abstract. The numerical simulation results of the flow of a gas-liquid medium over the high-temperature metal cylinder surface in a circular channel are presented. The flow over the solid metal cylinder with a variable cross-section by a gas-liquid medium is considered. Based on the mathematical model taking into account the cooling-media-flow axial symmetry relative to the cylinder longitudinal axis, the results of the numerical simulation of the process of solving the two-dimensional nonstationary problem of the coupled heat exchange of the gas-liquid medium flow and metal cylinder are obtained. The control volume approach is used for solving the differential equation system. The flow field parameters are calculated using the algorithm SIMPLE. For the iterative solution of linear algebraic equations, Gauss-Seidel method with under-relaxation is used. The numerical calculation results are obtained for the flow hydrodynamic parameters of the gas-liquid medium at cooling the high-temperature metal cylinder by taking into account vaporization. The values of temperatures, flow velocities, and vapour distribution in the computational domain are determined. The variation intensity of the metal cylinder surface temperature depending on the gas-liquid flow velocity is analyzed.

1 Introduction

In metallurgy and engineering industry in the production of items from metals and alloys, the creation of rational conditions for cooling processed high-temperature metal bodies is an important problem in the procedures of strengthening of machine components. The selection of the heat-transfer properties of cooling media and methods and modes of their supply during cooling depend on the required cooling rate which is determined by the intensity of the heat transfer between a body and a medium. The formation of the required conditions of heat removal during the heat treatment of items from metals and alloys is paid much attention to. The works in this direction were conducted by N.V. Zimin, G.F. Golovin, D.B. Budrin, M.V. Kondratov, G.N. Kheyfets, A.C. Kadinova, Yu.G. Eysmond, V.G. Labeysh, K.Z. Shepelyakovskiy et al. The investigation results allowed to create a general pattern for the process of the cooling of heated metal workpieces at thermal treatment. However, practically in all the works, particular solutions and general conclusions are given only in a certain required range of cooling operation conditions for the standard set of properties of the workpiece material and cooling media. In this connection it is still important to conduct research studies in this direction.

The choice of the mode for cooling high-temperature metal workpieces mainly depends on the distribution of hydrodynamic and heat-transfer parameters of the liquid cooling flow.

The process of cooling of high-temperature metal bodies by a gas-liquid media flow is investigated under the leadership of A.M. Lipanov, Academician of RAN in the Institute of Mechanics of the Ural Branch of RAN. In Ref. [1], the numerical solution of the problem of cooling of a high-temperature solid metal cylinder by the air-water flow is presented, and the cooling liquid flows are considered quasi-stationary. In Ref. [2], the results of the mathematical simulation of cooling axisymmetric metal workpieces from structural alloyed steel by quasi-stationary flows of cooling liquid are given. The case of the flow over the high-temperature cylinder surface by a nonstationary one-dimensional water flow in the longitudinal axis direction is considered in Ref. [3]. In Ref. [4], the mathematical model of cooling a hollow cylindrical metal workpiece by longitudinal quasi-stationary water flows is considered. The algorithm of the numerical solution of the problem and the results of the numerical parametric calculations are given. The determination of the temperatures of the cooled cylinder was made by taking into account the time of the relaxation of thermal stresses and internal heat sources caused by polymorphic transformations in the material. The calculation results for the parameters of the heat exchange between the cylinder and medium flow depending on the geometry, thermophysical properties and the process time were obtained on the basis of the

* Corresponding author: ssmak15@mail.ru

criterial dependences for the determination of the heat-transfer conditions.

In Ref. [5], the mathematical model of the conjugate problem of the convective heat exchange at cooling a heated metal cylinder by the cooling-medium laminar flow is considered. Cooling occurs without liquid boiling. In Ref. [6], the mathematical model is offered for the coupled heat-exchange in the heterogeneous system ‘solid body – gas-liquid medium’ taking into account vaporization in the cooling longitudinal liquid flow by the surface of the high-temperature metal cylinder. The numerical algorithm was used for studying the cooling of the metal cylindrical workpiece from structural steel in Ref. [7].

The goal of the present work is to obtain the results of the numerical simulation of a gas-liquid flow in a circular channel at cooling a high-temperature metal cylinder with a variable cross-section and to investigate the variation of the temperature of the metal cylinder surface during cooling depending on the value of the gas-liquid flow velocity.

2 Problem statement

Let us consider the flow over a solid cylinder with a variable cross-section by a gas-liquid medium. The metal cylinder consists of two areas: the first of radius r_{m0} and length L_0 and the second of radius r_m and length of $L - L_0$. In the longitudinal direction of the horizontal axis x, the cylinder is flowed by the liquid flow with the initial velocity u_0. The thickness of the moving liquid is determined by the distance of the annular gap: on the area L_0 – from r_{m0} to r_l, and on the area $L - L_0$ – from r_m to r_l. The thickness of the external ring is determined by the distance from r_l to r_{m1}. Calculation points 1 and 2 are on the inner and outer boundary of the cylinder vertical surface, respectively, at $x = L_0$. The physical diagram of the computational domain is shown in Fig. 1.

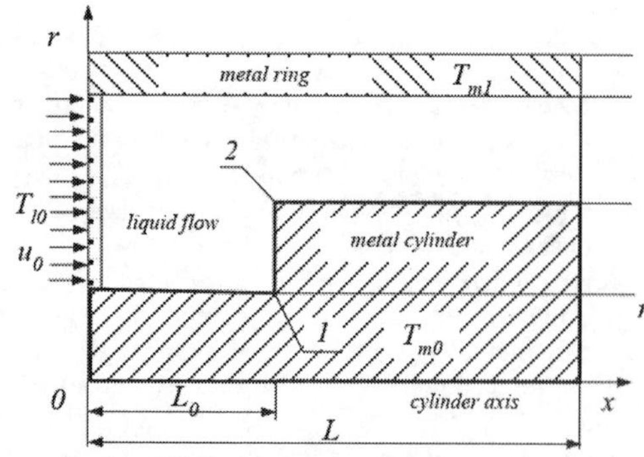

Fig.1. Physical diagram of the flow over the cylinderwith the variable cross-section

In the description of the process the following is assumed:

the liquid flow is considered axisymmetric relative to the longitudinal axis of the cylinder; the gravity forces action can be neglected;

the liquid flow is considered laminar;

the calculation parameters in the direction r vary continuously, and on the boundaries between the metal and liquid, the matching conditions in heat parameters are fulfilled;

by the cylinder surface, there is vaporization in the liquid flow;

the thermophysical parameters of the gas-liquid medium are determined by the relations describing the properties on the saturation line;

it is considered that the vapor density is much smaller than the liquid density and is described by the perfect gas equation;

the liquid vaporization is assumed to be equilibrium; the vapor pressure is equal to that of the liquid.

3 Solution procedure

The mathematical model of the coupled heat transfer of the two-dimensional non-stationary flow of the gas-liquid medium and the metal cylinder taking into account the axial symmetry of the cooling medium flow relative to the longitudinal axis of the cylinder is considered in Refs. [6, 7].

The system of equations describing the gas-liquid medium flow is as follows:

$$\rho\frac{\partial u}{\partial t}+\rho u\frac{\partial u}{\partial x}+\rho v\frac{\partial u}{\partial r}= =-\frac{\partial p}{\partial x}+\frac{\partial}{\partial x}\mu\frac{\partial u}{\partial x}+\frac{1}{r}\frac{\partial}{\partial r}r\mu\frac{\partial u}{\partial r} \quad (1)$$

$$\rho\frac{\partial v}{\partial t}+\rho u\frac{\partial v}{\partial x}+\rho v\frac{\partial v}{\partial r}= =-\frac{\partial p}{\partial r}+\frac{\partial}{\partial x}\mu\frac{\partial v}{\partial x}+\frac{1}{r}\frac{\partial}{\partial r}r\mu\frac{\partial v}{\partial r}-\mu\frac{v}{r^2} \quad (2)$$

$$\frac{\partial \rho}{\partial t}+\frac{\partial(\rho u)}{\partial x}+\frac{1}{r}\frac{\partial(r\rho v)}{\partial r}=0 \quad (3)$$

$$\rho c\frac{\partial T}{\partial t}+\rho c u\frac{\partial T}{\partial x}+\rho c v\frac{\partial T}{\partial r}= =\frac{\partial}{\partial x}\lambda\frac{\partial T}{\partial x}+\frac{1}{r}\frac{\partial}{\partial r}r\lambda\frac{\partial T}{\partial r}-\dot{m}_V Q_V \quad (4)$$

$$\rho\frac{\partial Y_V}{\partial t}+\rho u\frac{\partial Y_V}{\partial x}+\rho v\frac{\partial Y_V}{\partial r}=\dot{m}_V \quad (5)$$

The specific mass velocity of the vaporization is found from the heat balance equation:

$\dot{m}_V=\left(\rho c\Delta T^*\right)/Q_V$ where the reduced heat flux is determined by the following relation:

$$\Delta T^*=\begin{cases}0, & \textit{если}\ T(t+\Delta t)<T_s\\ \left[T(t+\Delta t)-T_n(t)\right]/\Delta t, & \textit{если}\ T(t+\Delta t)>T_s\end{cases}$$

$T(t)=max(T(t);\ T_s)$, Δt – time increment,

T_s – saturation temperature.

The energy equation for the metal cylinder and the ring has the form:

$$\rho_m c_m \frac{\partial T_m}{\partial t} = \frac{\partial}{\partial x}\lambda_m \frac{\partial T_m}{\partial x} + \frac{1}{r}\frac{\partial}{\partial r} r\lambda_m \frac{\partial T_m}{\partial r} \qquad (6)$$

The effective thermophysical parameters of the gas-liquid medium are determined as follows:

$$\Gamma = \Gamma_1 Y_v + (1 - Y_v)\Gamma_2, \quad \text{where} \quad \Gamma = \{\rho, \mu, \lambda\},$$
$$\Gamma_1 = \{\rho_v, \mu_v, \lambda_v\}, \quad \Gamma_2 = \{\rho_l, \mu_l, \lambda_l\};$$
$$c = (c_v \rho_v Y_v + c_l \rho_l (1 - Y_v))/\rho.$$

At the calculation of the field of the gas-liquid medium flow, the values $\{\rho, \mu, \lambda\}$ are interpolated on the bounds of the control volume. The boundary conditions are taken as those in Refs. [6 - 8]. In the equations, the following symbols are used: c – specific heat capacity, *J/(kg·grad)*; ρ - density, *kg/m³*; λ - heat conduction, *W/(m·grad)*; μ – dynamic viscosity, *Pa·s*; p –pressure, *Pa*; T – temperature, *K*; t – time, *s*; r – radial coordinate, *m*; x - longitudinal coordinate, *m*; L – computational domain length, *m*; u – velocity component along x; v – velocity component along r, *m/s*; Y – volume concentration, Q – specific heat of vaporization, *J/kg*. Indices: m – metal, l –liquid, v – vapor, 0 – initial value, s – saturation parameters.

The system of differential equations (1) – (6) is solved by the control volume approach. The parameters of the flow field (1)-(3) are calculated using the algorithm SIMPLE [9]. For accelerating the iteration process convergence, the sequential bypass of the points of the computational grid in four directions is used [6]. For the calculations, the grid having a convergent profile on the boundaries 'metal cylinder-liquid' and 'liquid-metal ring' and the sizes of the boundary cells $10^{-5}\times10^{-5}$ m is used at $r = r_m$ and $r = r_l$ for metal and liquid. As moving from the boundaries of the joint surfaces, the size of the grid cell increases according to the linear law with the coefficient 1.2. The computation step over the longitudinal coordinate x is constant, 10^{-4} m.

4 Results of the numerical calculations

For the numerical calculation, let us take $r_{m0} = 0.005$, $r_m = 0.01$ m, $r_l = 0.02$ m, $r_{ml} = 0.03$ m, $L_0 = 0.01$ m, $L = 0.11$ m, $T_{m0} = 800°C$, $T_{l0} = 20°C$. The initial temperature of the metal ring is $T_{ml} = 20°C$. The cylinder and ring are made out of steel 30 HGSN. The cooling liquid is water. The thermophysical parameters are taken according to Refs. [10, 11]. The initial velocity of the liquid flow is $u_0 = 0.1$ *m/s*. The computation time is $t = 1$s. It is believed that at the initial time, in the circular channel the liquid medium has the initial velocity equal to the velocity of the flow.

Figures 2 and 4 show the values of the temperatures, flow velocity along the longitudinal coordinate, distribution of the vapor concentration in the computational domain at the gas-liquid flow over the high-temperature metal cylinder with the variable cross-section. The parameter values are given in the coordinates of the numbers of the grid cells in the longitudinal and radial directions.

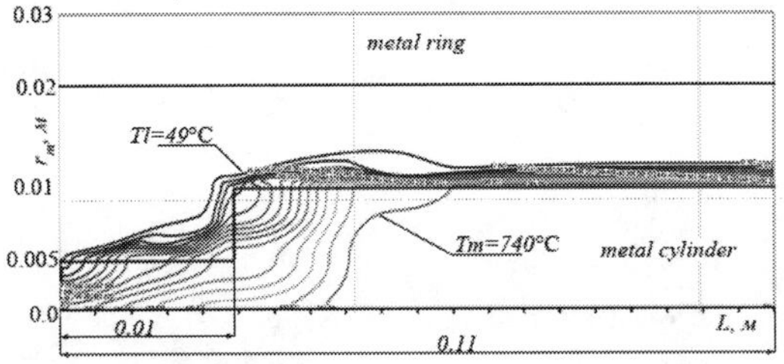

Fig.2. Temperature field

As can be seen from the conducted calculations, the temperature decrease of the cylinder from the boundaries of the gas-liquid medium contact in the direction of the axis takes place irregularly. Mostly the near-boundary regions situated perpendicularly to the liquid flow direction are cooled. The area L-L_0 of the surface of the cylinder of radius r_m is cooled almost in two times more intensive than the other surface areas (Fig. 2). It is explained by the irregular distribution of the gas-liquid flow velocity and formation of the regions with lower velocities (Fig. 3). The decrease of the liquid flow velocity in the near-boundary regions leads to a significant overheating of the liquid which results in the intensive vaporization (Fig. 4) and increase of the of the metal cylinder surface temperature.

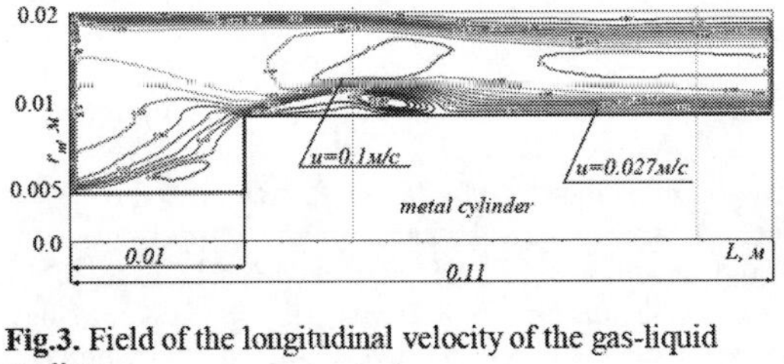

Fig.3. Field of the longitudinal velocity of the gas-liquid medium *u*

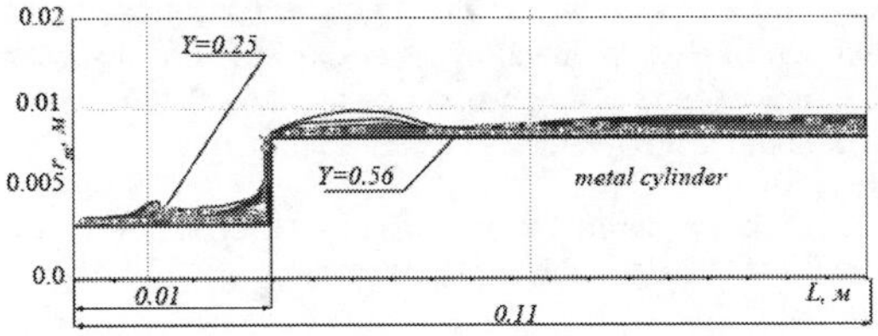

Fig.4. Field of the vapor concentration *Y*

Let us evaluate the influence of the gas-liquid medium flow velocity on the intensity of the metal cylinder surface cooling. The calculation will be conducted for two values of the initial velocity of the liquid flow *0.1 m/s* and 1 m/s. In Fig. 5, the variations of the temperature of the metal cylinder surface depending on the velocity of the gas-liquid medium flow are presented.

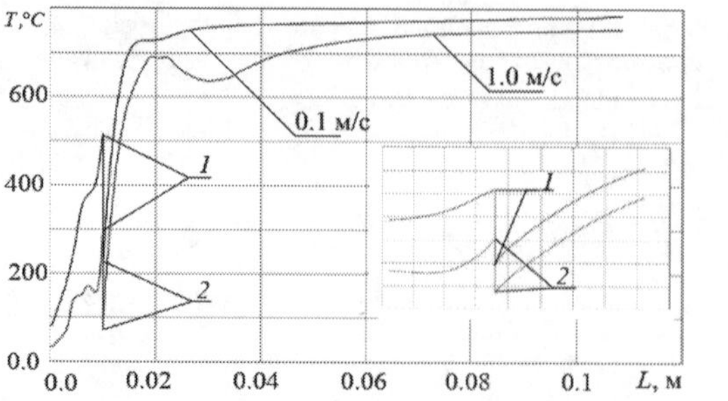

Fig.5. Metal cylinder surface temperature

At $x = L_0$, the surface temperature varies in the step-wise manner. The temperature at point 1 is higher than that at point 2 (see Fig. 1). For the flow velocity 0.1 m/s, at $x = L_0$, the range of the temperature variation between points 1 and 2 over the vertical cylinder surface is (515 – 192)°C. At the flow velocity 1 m/s, it is (305-73)° C, respectively. It can be seen that with increasing velocity of the gas-liquid flow the intensity of cooling is increasing, and the surface of the cylinder with a larger radius is cooled more intensively. The results of the calculation show that for the equal time of calculation when the initial velocity of the cooling water flow increases tenfold, the temperature of the cylinder surface decreases on average by 50° C more than in the other case. The calculated values of the cylinder temperature confirm the regularities obtained at the conduction of natural experiments [12].

5 Conclusion

Using the mathematical model of the gas-liquid medium flow moving over the heated surface of the high-temperature metal cylinder with the variable cross-section, the numerical values of the temperature, gas-liquid medium flow velocity and distribution of the vapor concentration in the computational domain have been obtained. The presented results of the numerical calculations have been obtained by taking into account the vaporization in the liquid. The variation of the temperature of the metal cylinder surface with respect to the time of the cooling process depending on the value of the medium flow velocity has been studied.

The work is sponsored by the Combined program of the Ural Branch of the Russian Academy of Sciences № 15-10-1-4.

References

1. A.M. Lipanov, S.S. Makarov, Mashinostroenie i inzhenernoe obrazovanie. 1 (2014)
2. S.S. Makarov, K.E. Chekmyshev, S.N. Khramov, E.V. Makarova, **3**, 63 (2014)
3. S.S. Makarov, K.E. Chekmyshev, E.V. Makarova. Nauchno-tekhnicheskie vedomosti SPbGPU, **4**, 207 (2014)
4. A.M. Lipanov, S.S. Makarov. Khimicheskaya fizika i mezoskopiya, **16**, 4 (2014)
5. S.S. Makarov, A.I. Karpov, E.V. Makarova, Khimicheskaya fizika i mezoskopiya. **18**, 1 (2016)
6. A.M. Lipanov, S.S. Makarov, A.I. Karpov, E.V. Makarova, Teplofizika i aeromekhanika. **24**, 1 (2017) doi: 10.1134/S0869864317010061
7. S.S. Makarov, V.B. Dement'yev, E.V. Makarova, Procedia Engineering **150** (2016) doi: 10.1016/j.proeng. 2016.06.734
8. S.S. Makarov, Scientific and Technical Journal of Information Technologies, Mechanics and Optics **17**, 2 (2017) doi: 10.17586/2226-1494-2017-17-2-324-331.
9. S. Patankar, *Numerical Heat Transfer and Fluid Flow* (Press Roman, 1980)
10. M.P. Vukalovich. *Thermophysical properties of water and vapor* (M. Mashinostroenie, 1967).
11. V.G. Sorokin, M.A. Gervas'yev, *Steels and alloys. Steel grade guide* (M.: Intermet Inzhiniring, 2001)
12. V.G. Labeysh. *Liquid cooling of high-temperature metal* (L.: Izd-vo Leningr. un-ta, 1983)

MATEC Web of Conferences **129**, 03008 (2017) DOI: 10.1051/matecconf/201712903008
ICMTMTE 2017

The study of the possibility of using innovative technologies for saving engineering 3d objects

Ekaterina Sokolova[1,*]

[1]NCIMM, Information Technologies Department, 362021 Vladikavkaz, Russia

Abstract. The prospect of using 3D printers in machine engineering is economically feasible since these devices significantly speed up the process of developing new products, substantially reduce the risks of design errors, reduce the cost of layout, and now their prices make them affordable for most world enterprises. The article studies the possibility of using innovative technologies for saving engineering 3D objects.

1 3D technologies in machine engineering

Car industry widely uses fast prototyping as an integral part of the design process. Obtaining prototypes by means of traditional methods (mechanical processing, casting) takes several weeks or even months and is a complicated, longtime and expensive design stage; contracts signing for the high-quality prototypes production is costly and may lead to a delay in the product release. The fast design cycle in the automated manufacturing process requires the use of up-to-date solutions based on fast prototyping, which allows to get items of almost any shape with different material properties quickly and inexpensively.

The prospect of using 3D printers is economically feasible since these devices significantly speed up the process of developing new products, substantially reduce the risks of design errors and the cost of layout, and now their prices make them affordable for most Russian enterprises. 3D printers with the best accuracy, quality, performance of prototyping, are successfully used in engineering. A 3D printer creates a prototype based on a 3D computer model created in any of the 3D CAD programs.

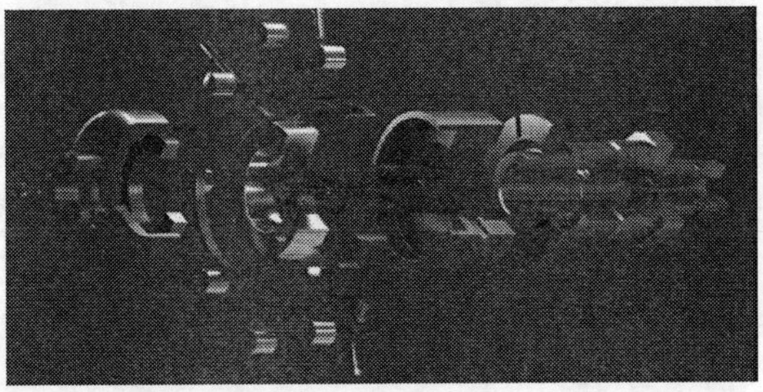

Fig. 1. 3D computer model

There are a lot of ways of prototyping usage, for example, you can print a transparent part of an engine casing or a gearbox and check the oil circulation and the operation of the whole system.

Fig. 2. 3D component parts

Also, there is the possibility of making models, which allows you to skip the stage of template production.

After scanning a cloud of points is obtained, which will need pre-processing, aligning, closing holes. At the output 3DS, ASE and STL file are obtained (other possible formats are dae, fbx, ma, txt, obj, ply, wrl, x3d, zpr).

Fig. 3. An example of a 3D model of a machine engineering object.

Then the scanned part and the pre-prepared point cloud file are subject to the part's testing and verifying. This requires the 3D file of the model. Part verifying is

* Corresponding author: katya_sea@mail.ru

actually the process of creating a discrepancy card, i.e. a 3D colored diagram or several sectional views where one line shows the ideal part profile and the other shows the profile of the actual (scanned) part. Based on the above, the main problem in the process of saving three-dimensional models will be the algorithm that uses the selected format. In the case of using lossy algorithms, artefacts cannot be ruled out, which is unacceptable in the case of prototyping.

To solve the stated problem we propose the mathematical model that takes into account special features of 3D models.

2 Mathematical model development

It is required to develop a mathematical model for processing a 3D model, which will take into account the features of the 3D printing process.

To do that it is necessary to develop a compression method that will turn the pixels of the three-dimensional object into a bit set of points that will be shorter than the original digital representation.

Given the known information about the structure of images obtained as a result of 3D scanning, it was decided to use:

- system and information-and-communication approaches, various computer graphics methods, 3D modelling as well as the theory of sets. The program architecture was developed with the help of component-oriented methods and object-oriented programming.

The formation of a 3D model occurs by converting images from several cameras into a set of spatial points, which can be represented as a multidimensional array.

A specific feature here is the use of effective implementation of the known methods of group encoding to reduce the amount of the stored file without loss of quality, a new method of pixel transformation (bit set of values) that allows to foresee and prevent the appearance of artefacts when saving a file.

To solve the problem that arises due to the existence of the principle of constructing a multidimensional Cartesian space and to order points in a multidimensional space, the correspondence can be written in the following form:

$$P^n = P_1^1 \times P_1^1 \times \cdots \times P_n^1 \qquad (1)$$

where for every coordinate $P^n \in P^1$ there is a certain order.

Therefore, in order for all points to be ordered, it is necessary to introduce additional conditions and limitations (mathematical formulas, algorithms, etc.). Depending on the chosen limitation, the algorithm of points enumeration will be chosen; besides the larger the rage, the less informative it will be.

The task of ordering the indicated points is carried out with the following condition:

- points adjacent to each other should not fall outside the range.

If you depict the entire object as consisting of lines, you can consider them as curves filling the space.
A formula can be determined for each of those curves:

$$I(k) = \sum_{j=1}^{n} j q_j (k) \qquad (2)$$

For each k-th element of the sequence, which we take as l(k), only one q function can be other than zero.

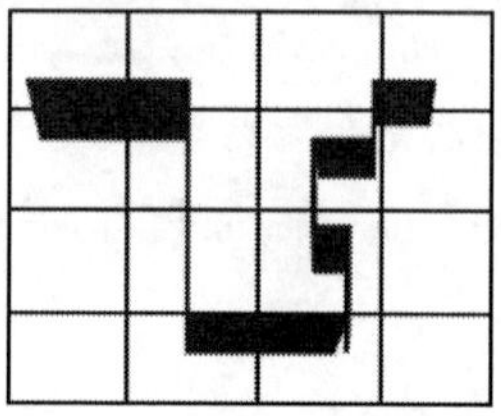

Fig. 4. An example of an object consisting of lines

Let's assume that x_0, y_0, z_0 is the relative displacement of the object if we take the global coordinate system as the basis, then:

$$x_0 = \min x_i \, , y_0 = \min y_i \, , z_0 = \min z_i$$

The lengths of the sides of the parallelepiped Sx, Sy, Sz are defined as follows:

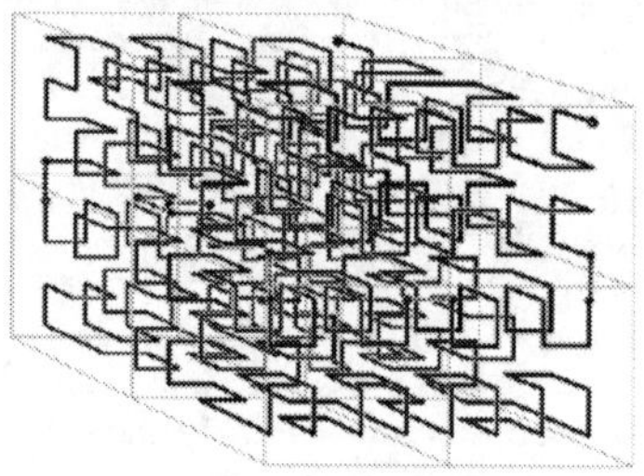

Fig. 5. Representation of an object in the form of lines in space

It is necessary to calculate the difference between the displacements and find the maximum value modulo.

Let's base on a three-dimensional coordinate system, which can be represented as a parallelepiped. Consequently, various transformations of the point cloud will be carried out in the selected coordinate system which includes the point cloud. The above procedure is adapted to the given task.

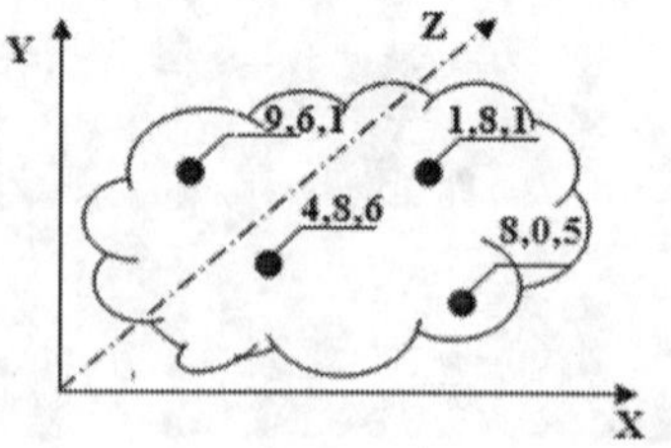

Fig. 6. Coordinates of a 3D object.

As you can see from figure 6, if you take any point in the point cloud, you can represent it in the Cartesian

 DOI: 10.1051/matecconf/201712903008

coordinate system in case you choose relative positioning at the origin of the coordinates, taking into account the hardware specifics of the scanner.

Based on peculiarities of the practical implementation of the three-dimensional scanning process, it is possible to convert the initial set point cloud. With the newly introduced links between points, we can get a polygonal model. Let us denote the cloud of points as *D*, and the set of faces as *V*; at that each face can be represented as a triangle with given vertices.

The model constructed above makes it possible to describe a set of points that do not have a specific structure, therefore there is excessive information on the object. Also, in case of changing the initial data, the model must be corrected using full enumeration of all the elements of the model.

Let us imagine the developed mathematical model, where *D* is the point cloud, *G* is the curve in which the order of points is described, and *R(G)* is the function describing the principle of reference to points *G*. Below is its mathematical description:

Let us introduce the following definitions:

O – 3D object;
M – memory in bites;
T – translation table;
$T_b(i)$ – bit code position;
B - block consisting of a three-bit union of positions;
x, y, z - 3-bit code header in a block;
T_D – displacement table;
i, j, k – codes, additional bits, distance in the displacement table.

For each three-dimensional object, you need to specify a set of its vertices and their connections to each other.

Each pixel is represented by a bit code (Figure 3 shows a numerical interpretation of code values).

Then each of the codes must be placed in the byte memory while observing the condition: each following bit code is greater than the previous one.

The position of the bit code is constantly checked when stored in memory.

Then the obtained sequence is stored in the block, and at the same time the information is analyzed with the help of the construction of the bit tree and the displacement table based on it.

All of the above can be combined in the form of a mathematical model:

$$\begin{cases} \forall O: \\ T \in M(b)T, where \\ \forall\, i \in (1,m): Tb_{i-1} < Tb_i \\ B = b_x \cup b_y \cup b_z \\ \forall\, Ts(i,j,k) \equiv Tb \\ I(k) = \sum_{j=1}^{n} jq_j\,(k) \\ x_0 = \min x_i\,, y_0 = \min y_i\,, z_0 = \min z_i \end{cases} \quad (3)$$

To simplify the automation of the encoding process, we create the nodes of the binary tree by the example of a three-dimensional object based on 8 vertices $ABCDA_1B_1C_1D_1$.

Each vertex has the following coordinates:

A - 9,6,1
B - 1,8,1
C- 4,8,6
D- 8,0,5
A_1- 0,7,7
B_1- 7,7,7
C_1- 4,4,0
D_1- 0,1,0

Let's make a gradation by the number of occurrences:

'1' -3, '4'-3, '5'-1, '6'-2, '7'-5, '8'-3, '9'-1, '0' -5

Then let's order them in ascending order:

9,5,6,8,4,1,7,0.

Let's start building a binary tree:

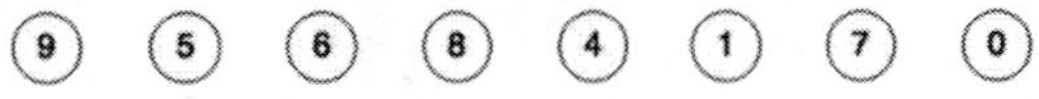

Fig. 7. Sorting vertices by number of mentioning.

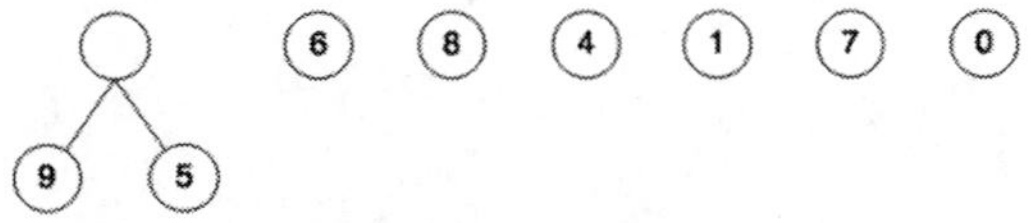

Fig. 8. Building the 1st layer of the tree.

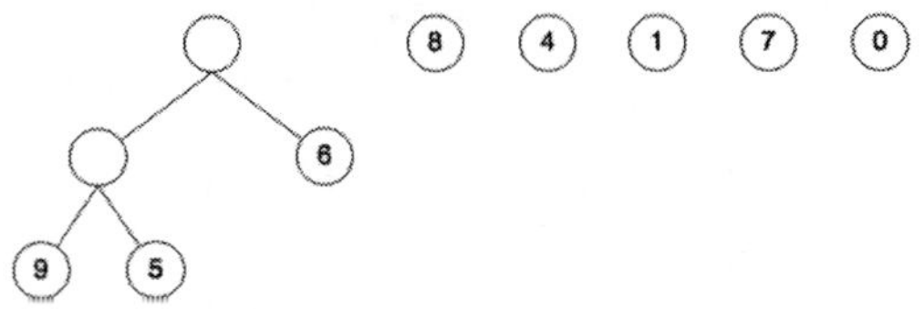

Fig. 9. Building the 2nd layer of the tree.

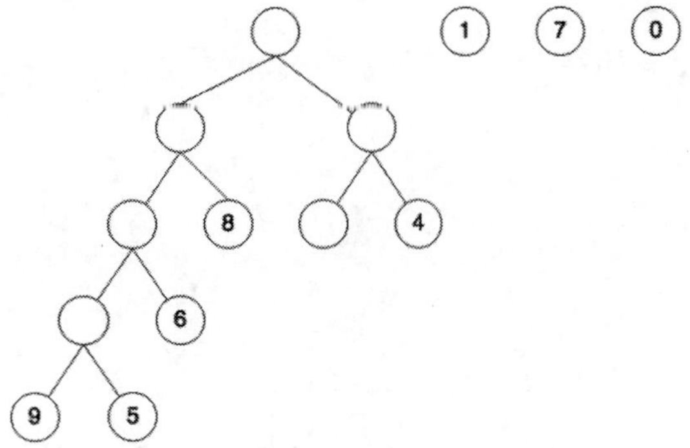

Fig. 10. Building the 3rd layer of the tree

DOI: 10.1051/matecconf/201712903008

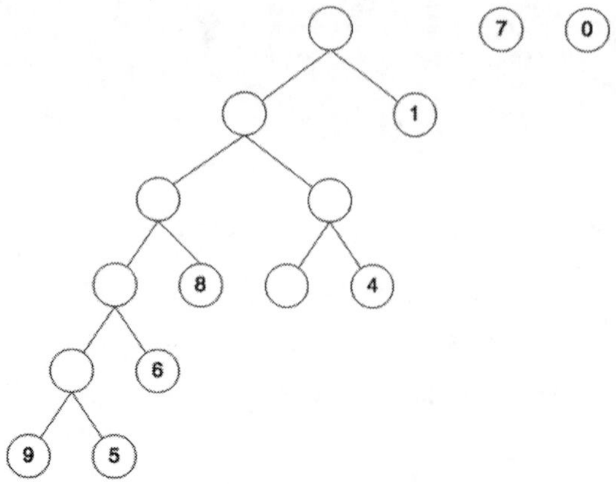

Fig. 11. Building the 4th layer of the tree.

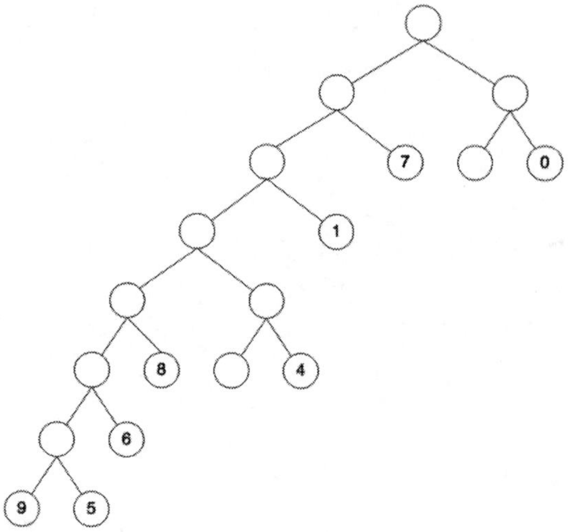

Fig. 12. Building the 5th layer of the tree.

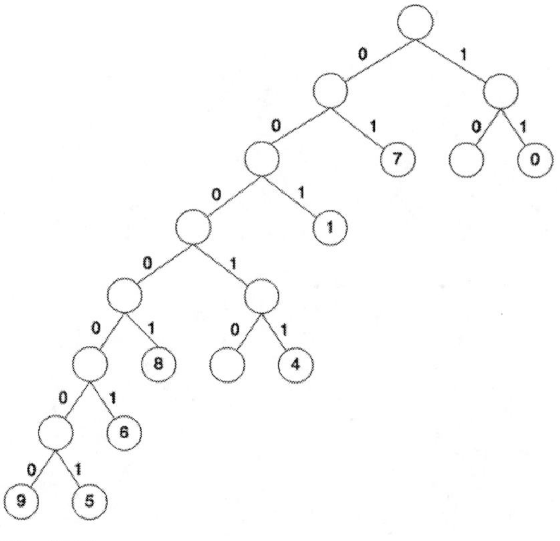

Fig. 13. Binary tree

3 Experiments

Quantitative characteristics of the product:
- the mathematical model will allow to process 3D images (*.obj, *.3ds, *.ase);
- the operating time of the mathematical model depends on the volume of processed images (directly proportional to the size of the downloaded file).

We conducted a number of experiments on the basis of which we constructed a diagram showing the dependence of the operation time of the mathematical model on the file size (Figure 14).

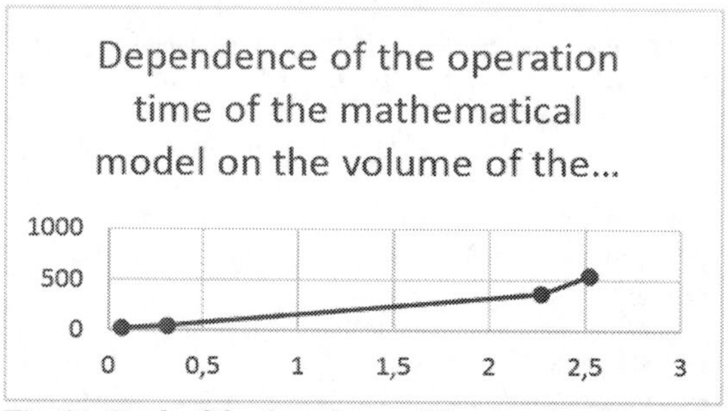

Fig. 14. Graph of the dependence of the operation time of the mathematical model on the volume of the processed images

The created lossless compression method for 3D models images for engineering will allow to obtain images without the appearance of artefacts with the probability of 0.87.

The appearance of artefacts directly depends on the number of restored points of the object. Figure 15 shows histograms representing the percentage of restored object points and the likelihood of artefacts.

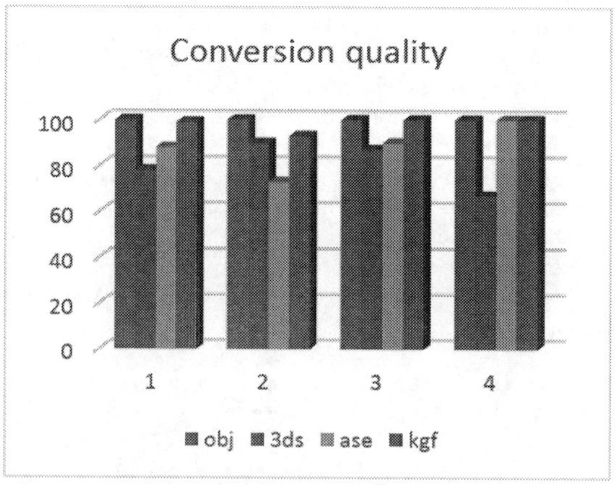

Fig. 15. Likelihood of artefacts

4 Conclusion

As a result of experimental studies, the possibility of using innovative technology for 3D engineering objects storing was confirmed. A fast design cycle in automated manufacturing process requires the use of up-to-date solutions based on fast prototyping, which allows to get

DOI: 10.1051/matecconf/201712903008

products of almost any shape with different material properties quickly and inexpensively. Taking into account the proposed mathematical model it also became possible to reduce the size of the stored models and avoid the appearance of artefacts.

References

1. Zhi-liang Zhu, Wei Zhang, Kwok-wo Wong and Hai Yu, Inf. Sci. **6** 1171-86 (2011)
2. Wei Zhang, Kwok-wo Wong, Hai Yu and Zhi-liang Zhu Commun, Nonlinear Sci. Numer. Simul. **3** 584-600 (2013)
3. Xingyuan Wang, Lin Teng and Xue Qin, Signal Process **4** 1101-08 (2012)
4. Yicong Zhou, Weijia Cao, and Philip Chen, Signal Process **8** 197-207 (2014)
5. A.A. Sokolov, A.S. Miroshnikov, E.A. Sokolova, Gornyi Zhurnal **12** 83-86 (2016)
6. M.A. Murillo-Escobar, C Cruz-Hernández, F Abundiz-Pérez, R.M. López-Gutiérrez and O.R. Acosta Del Campo, Signal Process **109** 119-131 (2015)
7. A.M. Kumaritov, E.A. Sokolova, A.A. Sokolov, Gornyi Zhurnal **2** 94-96 (2016)
8. Yingqian Zhang and Xingyuan Wang, Inf. Sci. **20** 329-351 (2014)
9. Wei Zhang, Kwok-wo Wong, Hai Yu and Zhi-liang Zhu Commun, Nonlinear Sci. Numer. Simul. **8** 2066-80 (2013)

Analysis of the time reserve influence on the technological cell productivity

Yuriy E. Obzherin[1,*], *Stanislav M.* Sidorov[1], and *Sergey N.* Fedorenko[2]

[1] Sevastopol State University, Higher mathematics Department, Sevastopol, Russia
[2] Sevastopol State University, Department of Instrument Systems and Automation of Technological Processes, Sevastopol, Russia

Abstract. Time redundancy, widely used in technical systems, is one of the methods to improve the reliability and efficiency of technical systems. In this paper, the effect of instantly replenishing the time reserve on the productivity of a technical system is considered. Formulas of productivity of a technological cell in cases of random, constant and exponential time reserve are received, numerical results are given.

1 Introduction

Time redundancy [1-10, 16] is a method to improve the reliability and efficiency of technical systems, in which the system in the process of operation is given the opportunity to spend some additional time, called a reserve, to restore characteristics. The time reserve can be used for switching the structural reserve, detecting and eliminating system failures, repetition of work impaired by failures, waiting for the load to be operational. The presence of a time reserve is important in ergonomic systems [18] to correct possible operator errors.

A time reserve can be created by increasing the time allocated to the system for the task and called operational time. It occurs when you create a performance margin of the entire system or its individual devices, and without increasing the operational time. The presence of a time reserve is characteristic for technical systems that have a functional inertia [3, 9, 11]. The sources of the time reserve are intermediate stores, warehouses, reserve power sources.

One of the features of the time redundancy, which determines its dignity, is that improving the performance indicators of the system is often not related to the increase in the amount of equipment and the use of additional funds, but is based on the introduction or use of the time reserve available in the system itself.

For systems with time redundancy, a malfunction is not necessarily accompanied by a system failure, since there is the possibility to restore serviceability for a reserve time.

In most works, the influence of the time reserve on the reliability of the system is analyzed. In this paper, a semi-Markov model of a technological cell (TC) is constructed, taking into account the processing time of a unit of production and the presence of an instantly replenished time reserve. Using the constructed model, formulas for the productivity of the TC are found and the influence of the time reserve on productivity is analyzed.

2 Description of the system

We consider the functioning of the system S, consisting of a TC that processes products and has a random instantly replenished time reserve. At the initial time, the TC begins to process the products.

TC processing time for unit of production is a random variable (RV) α_1 with the distribution function (DF) $F_1(t)=P(\alpha_1 \le t)$ and distribution density (DD) $f_1(t)=\lambda_1 e^{-\lambda_1 t}$. TC time to failure is RV α_2 with DF $F_2(t)=P(\alpha_2 \le t)$ and DD $f_2(t)=\lambda_2 e^{-\lambda_2 t}$. When the TC failure has occurred, the restoration of TC is beginning, TC restoration time is RV β with DF $G(t)=P(\beta \le t)$ and DD $g(t)=\mu e^{-\mu t}$. Time reserve is RV δ with DF $R(t)=P(\delta \le t)$ and DD $r(t)$.

The failure of the system S occurs at the time when the recovery time of the TC becomes more than the time reserve and continues until the recovery of the TC. It is assumed that by the time of recovery of the TC, the time reserve is completely restored (instantly replenished time reserve).

In the event of a failure of the TC, the processing of the unit of production continues at the expense of the time reserve until the system fails. Then, after the recovery of the TC, the processing of a new unit of production begins. It is assumed that RV $\alpha 1$, $\alpha 2$, β, δ are independent and have finite mathematical expectations.

To describe the system S operation, Markov renewal process (MRP) $\{\xi_n, \theta_n; n \ge 0\}$ and corresponding semi-Markov process (SMP) $\xi(t)$ with discrete-continuous phase state space [8, 12, 13, 14, 15] are used.

* Corresponding author: objsev@mail.ru

DOI: 10.1051/matecconf/201712903009

Introduce the following set E of semi-Markov system states:

$$E = \{1, 10, 11, 20, 21, 1\overline{0}x, 1\overline{1}x, \omega\}.$$

The meaning of state codes is the following:
1 – TC is able to work, the processing of unit production begins;
10 – TC is able to work, the processing of a unit production is finished (instant state);
11 – TC is able to work, the processing of a new unit of production begins;
20 – TC restoration has beginning, the processing of a unit production is continuing by the time reserve, system is able to work;
21 – TC restoration has finished, the processing of a unit production is continuing, time reserve is instantly replenished;
$1\overline{0}x$ - TC restoration is continuing the processing of a unit production has finished by the time reserve, system is able to work (instant state); x is time till the end of TC restoration;
$1\overline{1}x$ - TC restoration is continuing the processing of a new unit of production begins by the time reserve, system is able to work; x is time till the end of TC restoration;
ω - TC restoration is continuing, time reserve has finished, the processing of a unit production stopped, system failure.

2.1 The system of integral equations for stationary distribution definition

For finding the stationary characteristics of the system, it is important to find stationary distributions of embedded Markov chain (EMC) $\{\xi_n; n > 0\}$.

The general equation for the stationary distribution $\rho(B)$ is as follows:

$$\rho(B) = \int_E \rho(dx) P(x, B), \tag{1}$$

where $P(x,B)$ – transition probabilities of EMC.

Denote by $\rho(1), \rho(\omega), \rho(10), \rho(11), \rho(20), \rho(21)$ the values of EMC $\{\xi_n; n > 0\}$ stationary distributions in the states $1, \omega, 10, 11, 20, 21$ and assume the existence of stationary densities $\rho(1\overline{0}x), \rho(1\overline{1}x)$ for the states $1\overline{0}x, 1\overline{1}x$ respectively.

Using formula (1) and the transition probabilities of the EMC, we obtain that the system of integral equations for the EMC $\{\xi_n; n > 0\}$ stationary distributions looks like the following:

$$\begin{cases}
\rho(1) = \rho(\omega) = \rho_0, \\
\rho(10) = \rho_0 \dfrac{\lambda_1}{\lambda_1 + \lambda_2} + \rho(11)\dfrac{\lambda_1}{\lambda_1 + \lambda_2} + \rho(21)\dfrac{\lambda_1}{\lambda_1 + \lambda_2}, \\
\rho(11) = \rho(10), \\
\rho(20) = \rho_0 \dfrac{\lambda_2}{\lambda_1 + \lambda_2} + \rho(11)\dfrac{\lambda_2}{\lambda_1 + \lambda_2} + \rho(21)\dfrac{\lambda_2}{\lambda_1 + \lambda_2}, \\
\rho(21) = \rho(20)\displaystyle\int_0^\infty \mu e^{-\mu y} e^{-\lambda_1 y} \overline{R}(y) dy + \int_0^\infty \rho(1\overline{0}y) dy \int_0^y \mu e^{-(\mu+\lambda_1)z} dz, \\
\rho(1\overline{0}x) = \displaystyle\int_x^\infty \rho(1\overline{0}y) \lambda_1 e^{-(\mu+\lambda_1)(y-x)} dy + \rho(20)\int_0^\infty \lambda_1 e^{-(\mu+\lambda_1)y} r(x+y) dy, \\
\rho(1\overline{1}x) = \rho(1\overline{0}x), \\
\rho(\omega) = \rho(20)\displaystyle\int_0^\infty e^{-(\mu+\lambda_1)y} r(y) dy + \int_0^\infty \rho(1\overline{0}x) e^{-(\mu+\lambda_1)x} dx, \\
\rho(1) + \rho(\omega) + \rho(11) + \rho(10) + \rho(20) + \\
+\rho(21) + \displaystyle\int_0^\infty \rho(1\overline{0}x) dx + \int_0^\infty \rho(1\overline{1}x) dx = 1.
\end{cases} \tag{2}$$

The last equation of the system (2) is a normalization requirement.

2.2 The case of a general time reserve

Let the time reserve has general distribution R(t)=P(δ≤t). By means of the method of successive approximations [12], the system of equations (2) is provided to have the following solutions:

$$\begin{cases}
\rho(1) = \rho(\omega) = \rho_0, \\
\rho(10) = \rho_0 \dfrac{\lambda_1}{\lambda_2} \dfrac{\lambda_1}{\lambda_1 - \mu\int_0^\infty \overline{R}(t) h_{\overline{g}}(t) dt}, \\
\rho(11) = \rho(10), \\
\rho(20) = \rho_0 \dfrac{\lambda_1}{\lambda_1 - \mu\int_0^\infty \overline{R}(t) h_{\overline{g}}(t) dt}, \\
\rho(21) = \rho_0 \dfrac{\mu\int_0^\infty \overline{R}(t) h_{\overline{g}}(t) dt}{\lambda_1 - \mu\int_0^\infty \overline{R}(t) h_{\overline{g}}(t) dt}, \\
\rho(1\overline{0}x) = \rho_0 \dfrac{\lambda_1}{\lambda_1 - \mu\int_0^\infty \overline{R}(t) h_{\overline{g}}(t) dt} \displaystyle\int_0^\infty r(x+t) h_{\overline{g}}(t) dt, \\
\rho(1\overline{1}x) = \rho(1\overline{0}x), \\
\rho(\omega) = \rho_0 \dfrac{\lambda_1 \left[\int_0^\infty e^{-(\mu+\lambda_1)t} r(t) dt + \int_0^\infty e^{-(\mu+\lambda_1)x} dx \int_0^\infty r(x+t) h_{\overline{g}}(t) dt \right]}{\lambda_1 - \mu\int_0^\infty \overline{R}(t) h_{\overline{g}}(t) dt},
\end{cases} \tag{3}$$

where

DOI: 10.1051/matecconf/201712903009

$\bar{R}(t)=1-R(t)$, $\tilde{g}(t)=\lambda_1 e^{-(\lambda_1+\mu)t}$,

$h_{\tilde{g}}(t)=\sum_{n=1}^{\infty}\tilde{g}^{*(n)}(t)$, $\tilde{g}^{*(n)}(t)$ - nth fold convolution of function $\tilde{g}(t)$. For function $h_{\tilde{g}}(t)$ has an equation

$$h_{\tilde{g}}(t)=\tilde{g}(t)+(\tilde{g}*h_{\tilde{g}})(t),$$

which can be used to find it.

The value of constant ρ_0 is obtained from the normalization requirement and is not included in the final formulas of TC productivity.

For finding the productivity of the technological cell (P_{TC}), we use the formula obtained in [17]:

$$P_{TC}=\lim_{t\to\infty}\frac{N_A(t)}{t}=\frac{\int_A \rho(dx)}{\int_E m(x)\rho(dx)}, \quad (4)$$

where

$\rho(dx)$ is the EMC stationary distribution, which is defined by formula (3),

m(x) are average values of semi-Markov process sojourn times in states,

$N_A(t)$ – the number of visits of a semi-Markov process in a subset of states

$$A=\{10x,1\bar{0}x\}.$$

When a semi-Markov process visits the states of this subset, the processing of a unit production ends.

The sojourn times in states of the system (2) are determined as follows:

$$\begin{aligned}&\theta(1)=\theta(21)=\theta(11)=\lambda_1\wedge\lambda_2,\\&\theta(10)=\theta(1\bar{0}x)=0,\\&\theta(1\bar{1}x)=\lambda_1\wedge\beta\wedge x,\theta(\omega)=\beta,\end{aligned} \quad (5)$$

where $\wedge$ is the minimum sign.

Therefore, system average sojourn times m(x) in states have the form:

$$\begin{aligned}&m(1)=m(21)=\mathrm{m}(11)=\frac{1}{\lambda_1+\lambda_2},\mathrm{m}(10)=\mathrm{m}(1\bar{0}x)=0,\\&m(1\bar{1}x)=\int_0^x\bar{F}_1(t)\bar{G}(t)dt,m(\omega)=M\beta.\end{aligned} \quad (6)$$

In this case, using the stationary distribution EMC (3) and system average sojourn times in states (6), the TC productivity is calculated by the formula:

$$P_{TC}=\frac{\mu}{\mu+\lambda_2}(\lambda_1+\lambda_2\int_0^\infty \bar{R}(t)h_{\tilde{g}}(t)dt). \quad (7)$$

Using formula (7), we find the TC productivity formulas for particular cases of the time reserve distribution laws.

2.3 The case of exponential time reserve

Consider the case, when the time reserve is random and has an exponential distribution with DF $\mathrm{R}(t)=1-e^{-\tau t}$.

In this case the solution of system (2) has the form:

$$\begin{cases}\rho(1)=\rho(\omega)=\rho_0,\\ \rho(10)=\rho_0\dfrac{\lambda_1(\mu+\tau)}{\lambda_2\tau},\\ \rho(11)=\rho(10),\\ \rho(20)=\rho_0\dfrac{\mu+\tau}{\tau},\\ \rho(21)=\rho_0\dfrac{\mu}{\tau},\\ \rho(1\bar{0})=\rho_0\dfrac{\lambda_1}{\tau},\\ \rho(1\bar{1})=\rho(1\bar{0}).\end{cases} \quad (8)$$

Using the stationary distribution and formula (4), we obtain the formula for the productivity of the TC:

$$P_{TC}=\frac{\lambda_1\mu(\lambda_2+\mu+\tau)}{(\lambda_2+\mu)(\mu+\tau)}. \quad (9)$$

2.4 The case of a costant time reserve

Consider the case, where the time reserve is constant, namely $\delta=\tau=const$. In this case $\mathrm{R}(t)=1(t-\tau)$.

The solution of system (2) can be written as follows:

$$\begin{cases}\rho_0=\rho(\omega)=\rho(1),\\ \rho(10)=\rho(11)=\rho_0\dfrac{\lambda_1}{\lambda_2}e^{\mu\tau},\\ \rho(20)=\rho_0 e^{\mu\tau},\\ \rho(21)=\rho_0(e^{\mu\tau}-1),\\ \rho(1\bar{0})=\rho(1\bar{1})=\rho_0\dfrac{\lambda_1}{\mu}(e^{\mu\tau}-1).\end{cases} \quad (10)$$

Substituting the obtained stationary distributions in (4), we have the following formula:

$$P_{TC}=\frac{\lambda_1}{\mu+\lambda_2}(\mu+\lambda_2(1-\mathrm{e}^{-\mu\tau})). \quad (11)$$

Let us analyze, for example, the effect of the time reserve on the TC productivity, using the formula (11).

Example.

DOI: 10.1051/matecconf/201712903009

Let $M\alpha_1 = \frac{1}{12}\text{h}$, $M\alpha_2 = \frac{1}{2}\text{h}$, $M\beta = \frac{1}{3}h$, that is $\lambda_1 = 12(\frac{1}{h})$, $\lambda_2 = 2(\frac{1}{h})$, $\mu = 3(\frac{1}{h})$, and the time reserve varies from 0 to 2 hours in 0.2 increments.

Figure 1 shows the influence of the time reserve on the productivity of the technological cell. In this case, the productivity of the TC without a time reserve is 7.2 units/hour. It follows from formulas (9) and (11), that the time reserve significantly affects on the productivity of the TC.

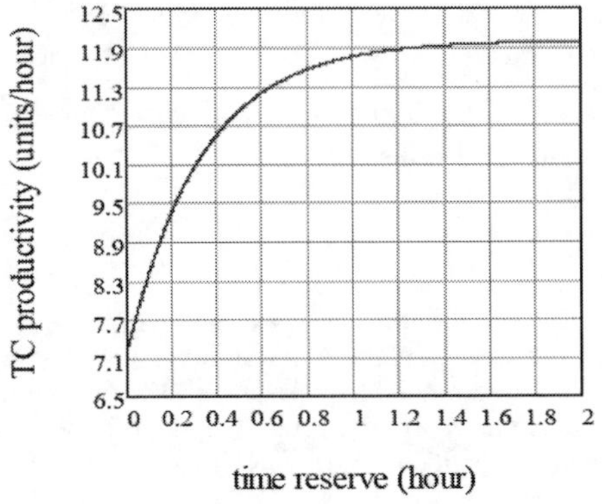

Fig. 1. The influence of the time reserve on the productivity of the TC.

3 Conclusions

In this paper, a semi-Markov model of the technological cell, taking into account the processing time of a unit of production and the availability of an instantly replenished time reserve, is constructed. To construct the model, we used the theory of semi-Markov processes with a common phase space of states. A system of equations is constructed for the stationary distribution of the embedded Markov chain of a semi-Markov process and its solution is found by the method of successive approximations.

When the productivity of a technological cell is found, cases of exponential, general type and constant time reserve are considered. The analysis of the influence of the time reserve on the technological cell productivity in the case of a constant time reserve was carried out. It is shown that the presence of an instantly replenished time reserve significantly affects on the productivity of the technological cell.

We note the connection between the model considered and the random walk theory.

The results of this work can be used to analyze the influence of the time reserve on the productivity of technical systems for various purposes. The applied technique can be used in the construction of multicomponent systems with different types of time reserve.

To model multicomponent systems with a time reserve, it is proposed to use phase-coherence algorithms [13], which are an effective tool for simplifying the analysis of complex systems.

The work has supported by a grant RFBF № 15-01-05840

References

1. G. N. Cherkesov, *Reliability of Technical Systems with Time Redundancy* (Sovietskoe Radio, Moscow, 1974) (in Russian)
2. B.P. Kredentser, *Prediction of Reliability of Systems with Time Redundancy.* (Naukova Dumka, Kiev, 1978) (in Russian)
3. Yu.N. Rudenko, I.A. Ushakov, *Reliability of Energy Systems* (Nauka, Novosibirsk, 1989) (in Russian)
4. Yu.E. Obzherin, A.I. Peschansky, Cybern. Syst. Anal., Vol. **30**, No. 6. (1994)
5. Yu.E. Obzherin, A.I. Peschansky, Cybern. Syst. Anal., Vol. **40**, No. 5. (2004)
6. Yu. E. Obzherin, A.V. Skatkov, J. Math. Sci., Vol. **57**, No. 5. (2010)
7. I.A. Ushakov, *Probabilistic Reliability Models* (Wiley, 2012)
8. Y.E. Obzherin, E.G. Boyko, *Semi-Markov Models: Control of Restorable Systems with Latent Failures* (Elsevier Academic Press, London, 2015)
9. V.Y. Kopp, Yu. E. Obzherin, A.I. Peschanskiy, *Stochastic models of automized system with time reservation* (Publishing of Sevastopol State Technical University, Sevastopol, 2000) (in Russian)
10. G.N. Cherkesov, *Reliability of hardware-software complexes*, (Peter, St. Petersburg, 2005) (in Russian)
11. Lisnianski, A. and G. Levitin, H. Ben-Haim, Reliability Engineering & System Safety, Vol. **67**, Issue 2 (2010)
12. A.B. Vasileva, N.A. Tihonov, *Integral equations* (Fizmatlit, Moscow, 2003) (in Russian)
13. V.S. Koroluk, A.F. Turbin, *Markovian restoration processes in the problems of system reliability* (Naukova Dumka, Kiev, 1982) (in Russian)
14. J. Jansen, N. Limnios (Eds.), *Semi-Markov Models and Applications* (Kluwer Academic Publishers, The Netherlands, 1999)
15. N. Limnios, G. Oprisan, *Semi-Markov Processes and Reliability* (Springer Science+Business Media, New York, 2001)
16. G.N. Cherkesov, Izv. AS USSR, Ser. Tehn. kibern., № **6**, p.73–78 (1986)
17. V.M. Shurenkov, *Ergodic Markovian processes* (Nauka, Moscow, 1989) (in Russian)
18. E.A. Leontev, *Reliability of economic information systems* (Izd-vo Tamb. gos. tehn. un-ta, Tambov, 2002) (in Russian)

DOI: 10.1051/matecconf/201712903010

Automated decision-making system based on genetic algorithm in managing maintenance process

Aleksey Nikishenko[1,*], *Elena* Boyko[2], and *Yuriy* Obzherin[2]

[1] KB KA Ltd., 299003 Sevastopol st. Revolution 1905, 12,, Russia
[2] Sevastopol state University, Department of higher mathematics, 299053 Sevastopol st. Universitetskay, 33, Russia

Abstract. An automated decision-making system in managing the maintenance process was developed. The problem of vector optimization was set. A genetic algorithm with domination sorting and elite selection was used to find a multitude close to the Pareto set in multi-criteria optimization.

1 Introduction

The choice of strategy and frequency of monitoring, as well as the maintenance method refers to solutions that have a serious long-term impact. Therefore, strategic decisions of this kind deserve careful analysis, including mathematical modeling, and, most importantly, in the face of constantly changing factors (the quality of components, the cost of modules and materials, the size of the expected profit) require the creation of an automated decision-making system (ADMS). The structure of the ADMS was developed taking into account modern requirements for such systems and analysis of existing standard structures. Its peculiarity is division into mutually independent and interacting parts. The "decision making" is a specific function of the ADMS, which distinguishes it from automated computer (information) systems. An analysis of various literature sources [1-3] showed that genetic algorithms (GA) show good applicability for solving optimization problems.

2 Structure of ADMS in managing maintenance process

ADMS in managing maintenance process is a multi-level system designed to organize and manage the maintenance process of the production system (PS), which is based on the assessment of the controlled object state during operation (the technical condition monitoring - TCM). ADMS has to take into account problematic production situations, as well as material and financial costs.

It should be noted that the maintenance process includes:

- the TCM process of the PS or its components;
- the maintenance process itself.

In general, the ADMS consists of three blocks. The first block is a block for collecting, processing and storing of information. It is the most voluminous and labor-intensive, in terms of organization. This block is not directly related to the decision-making process, but it provides the necessary base for making these or that decisions. The second block is an information system whose function is to make the prepared information foreseeable and suitable for processing. The third block is the decision support system (DSS). This block works directly with the solutions and allows to choose a specific solution based on the embedded algorithm or in the process of interaction with the decision maker (DM).

The "decision making" function is designed to find the area of optimal solutions for multi-criteria optimization or to choose the optimal solution to the problem situation associated with the malfunction of the equipment. To implement this function, the ADMS includes a set of calculation and decision-making models obtained using the theory of semi-Markov processes with a common phase space of states [4] and GA, as well as a database creation and maintenance system that provides them with information.

The general concept of building the ADMS is based on a hierarchical principle that provides the distribution of management functions between the interrelated structural levels: information, strategic, tactical and executive. The hierarchical structure of the ADMS in managing maintenance process of the PS is shown in Figure 1. The hierarchical organization of the ADMS ensures collection of data about the controlled object at the information level, choice of distribution functions and methods of monitoring at the strategic level, formation of quality criteria for PS functioning and their quantitative assessments at the tactical level, as well as management of PS maintenance process in the process of PS functioning at the executive level. Moreover, an operator controls its functioning.

The input of the system receives information from the DM and from the database. At the output of such system it is possible to obtain the following information: the efficient control frequency for the selected mode, the efficient tactics of maintenance, the technical and

* Corresponding author: sklibur@rambler.ru

economic performance indicators of the PS for the selected mode.

ADMS is implemented in the form of a software system, the structural diagram of which is presented in Figure 2. It contains five main modules and includes a large number of algorithms combined into computational models that allow solving various technical problems. DSS modules, as the most complex in the organization, will be considered in more detail.

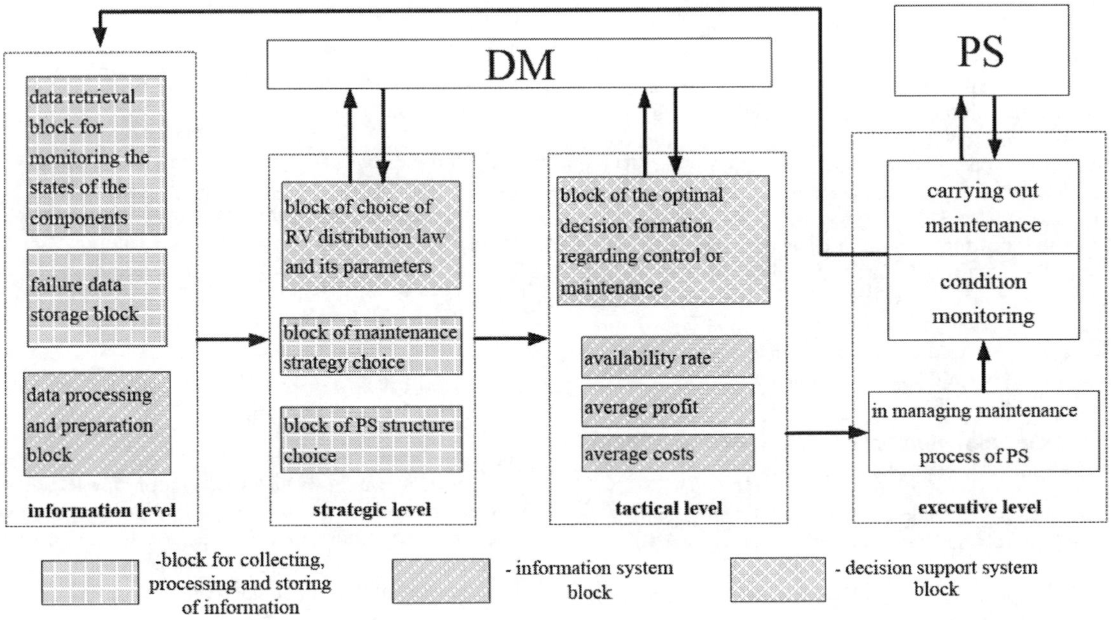

Fig. 1. Structural diagram of ADMS in managing maintenance process.

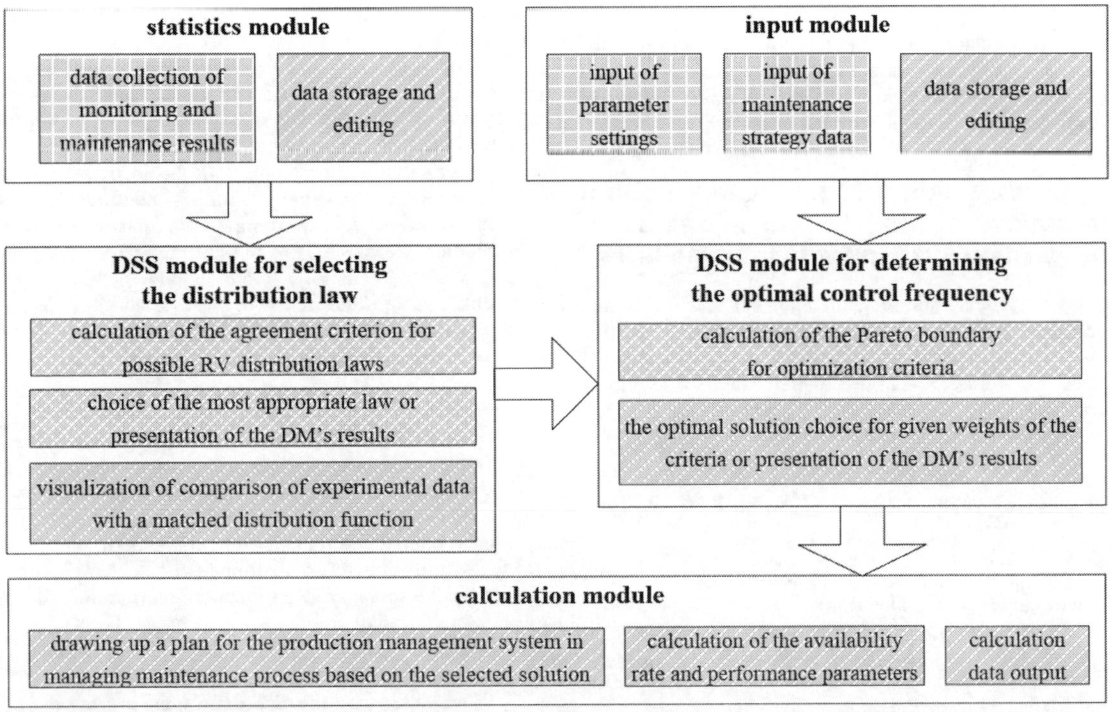

Fig. 2. Structural diagram of ADMS in managing maintenance process.

2.1 DSS module for choice of the random variable distribution law and/or its parameters

The DSS module makes it possible to determine the law of random variables (RV) and/or its parameters based on the results of monitoring the presence of latent failures (LF) [5].

At the input of the DSS block, the vector of experimental values of the RV is fed by the choice of the distribution law. Before the processing, the vector is sorted and filtered (foreign objects, for example, symbols are deleted). Verification of the hypothesis about the supposed law of RV distribution is carried out in accordance with Pearson's agreement criterion (criterion χ^2) [6], [7].

The following laws are used as the supposed laws of the failure times distribution in the module by default [8]: exponential distribution; Weibull-Gnedenko distribution; gamma distribution; normal distribution; lognormal distribution; uniform distribution. This list can be supplemented with user functions.

2.2 DSS module for selection of the optimal management process

The decision to choose a tactic of PS management is based on solving the optimization task of selecting the most effective frequency for PS monitoring.

The operation efficiency is the degree to which the actual result of the operation corresponds to the required one [9]. In practical research of effectiveness, the problem of assessing effectiveness and the problem of choosing a rational mode of action (choice of tactics) are usually singled out. The assessment of effectiveness involves formation of the purpose of the operation, the choice of the effectiveness criterion, which quantitatively expresses the degree to which the actual result corresponds to the required one.

3 Optimization of automated control frequency for multi-component production systems with latent failures

For the system under consideration, the reliability and economic parameters of the wave soldering production unit used to describe PS with TCM and maintenance in scientific works [10-12] are implemented as optimality criteria:

1) Stationary availability:

$$\phi_1(\overline{\tau}) = K_\Gamma(\overline{\tau}) \to \max_{\overline{\tau}\in\mathbf{T}}; \quad (1)$$

2) Average profit per unit of calendar time:

$$\phi_2(\overline{\tau}) = S(\overline{\tau}) \to \max_{\overline{\tau}\in\mathbf{T}}; \quad (2)$$

3) Average costs per unit of time of the efficient system functioning (with a minus sign):

$$\phi_3(\overline{\tau}) = -C(\overline{\tau}) \to \max_{\overline{\tau}\in\mathbf{T}}. \quad (3)$$

Here $\overline{\tau} = (\tau_1,...,\tau_N)$ — is the vector of the variable parameters of the problem, the admissible values region of the vector $\overline{\tau}$ forms a multitude $\mathbf{T}$.

We set the three-criteria optimization problem [1] for the management process of component-wise control: to find the values of the variable parameters τ_i, $i=\overline{1,3}$, which, in the limit, ensure the fulfillment of all conditions (1) — (3). The vector optimality criterion $(\phi_1(\overline{\tau}),\phi_2(\overline{\tau}),\phi_3(\overline{\tau}))$ is defined on the multitude $\mathbf{T}$, and the DM strives to maximize each value of the particular optimality criteria $\phi_i(\overline{\tau})$, $i=\overline{1,3}$:

$$\max_{\overline{\tau}\in\mathbf{T}}(\phi_1(\overline{\tau}),\phi_2(\overline{\tau}),\phi_3(\overline{\tau})) \quad (4)$$

Usually, the optimality criteria are contradictory, and they reach their optimal values at various points in the multitude of admissible values of the variable parameters vector $\mathbf{T}$. Therefore, a preliminary solution to the problem is a multitude of compromise solutions, called the Pareto set. This multitude includes solutions that cannot be improved immediately by all the criteria.

3.1 Multi-criteria optimization using genetic algorithm for finding solutions

For the system under study, there are three system performance criteria, which depend on three control parameters (on N parameters for the N-component system). In the optimization process, it is necessary to determine the control parameters values that will ensure the efficiency of the system operation.

This task has the following peculiarities:

- multi-criteria;
- a large number of alternative solutions;
- the objective function can have many local extremums in view of consideration of the general form of the RV distribution law.

It is obvious that a simple enumeration of possibilities can take too much time. Thus, there is a problem of finding within the acceptable time the optimal or close to the optimal solution.

The solving of problems with the above-mentioned features makes it possible to simplify the usage of GAs, which are used to determine the frequency of TCM installation blocks close to the optimum from the point of view of the vector objective function (4).

GAs have the following advantages (+) and disadvantages (-) [2]:

+ no information about the objective function is required;
+ gaps existing on the response surface have little effect on the optimization efficiency;
+ relative resistance to falling into local optima;
+ suitability for solving a wide range of large-scale tasks;
- it is problematic to find an exact global optimum;
- difficult to apply to isolated functions;
- ineffective to apply to optimize a function that takes a long time to calculate;
- it is not possible to find the optimal coding of parameters for all tasks.

When describing the GA, definitions borrowed from genetics are used.

 DOI: 10.1051/matecconf/201712903010

A population is a finite multitude of individuals. The individuals belonging to the population in the algorithm are determined by a genotype with the multitude of parameters of the problem coded in it. A genotype is a set of chromosomes of a given individual. A chromosome is an ordered sequence of genes. A gene is an atomic element of a genotype, in particular, a chromosome. Locus is a multitude of gene positions. An adaptability function (fitness function, assessment function) is a function that determines the degree of adjustment of a given individual in the population.

Let us examine the use of GA to solve the problem of optimizing the process of component-wise TK of latent failures of a multicomponent PS.

Let us consider an *N*-component system with a component-wise TCM. The individual in this case will be the control frequency management plan. The vector of times $\overline{\tau}$ determining the control frequency of each of the *n* components is a characteristic of the individual — a genotype. The genotype consists of *N* chromosomes. The *I* chromosome determines the control frequency of the *i* component. As a fitness function to assess the suitability of the individual the objective three-criteria function is used (4).

Each chromosome can be represented as a binary code, where the number of bits k_i is determined by the number of possible values of waiting times for monitoring τ_i. It is obvious, that the genotype is also represented as a binary code with a bit depth $k = \sum_{i \in \overline{1,N}} k_i$.

In the genome k_1 of the lower orders is formed the first chromosome, in the next k_2 of the orders, the second chromosome, etc. The structure of an arbitrary genome is shown in Figure 3.

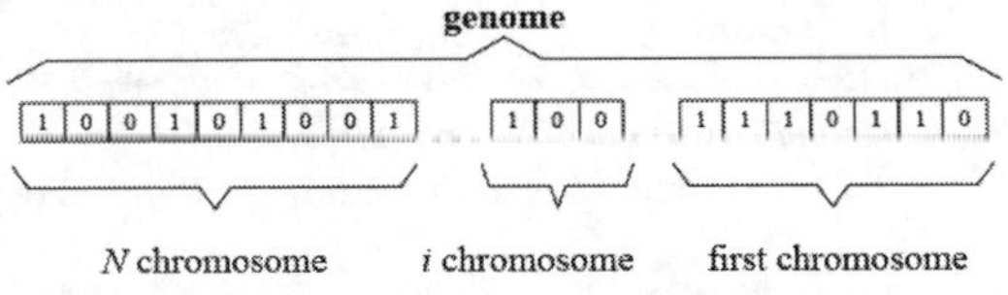

Fig. 3. The genome structure.

Taking into account the specifics of the problem, the GA with domination sorting and elite selection was chosen (NSGA-II) [13, 14].

Let us describe the work of the chosen GA step by step, taking into account the selected genetic operators (crossover operators, mutations, population generations) and their parameters:

1. After the start of the algorithm, a population of R individuals is created. The initial population is randomly generated using a uniform distribution.
2. The selection is carried out with the help of the pair tournament selection method.
3. A cross-breeding operation is performed for the selected individuals (parents). Creation of the descendants occurs using a homogeneous crossover. In it, the cross-breeding is performed on the basis of a binary vector, the length of which equals the length of the genome in the locus. Each position of the vector determines the mutual replacement of the locus of the parental individuals. As a result of cross-breeding, the descendants R/2 are created.
4. The missing part of the individuals (R/2) up to twice the population size (2*R) is formed due to a mutation of already obtained individuals.
5. The individuals are assessed on the level of adaptability. R of the best adapted individuals are included in the new generation, the others are eliminated.
6. The steps 2 - 5 are repeated, if the exit condition has not been reached.

The block diagram of the used GA is shown in Figure 4. The algorithm stopping condition is invariance of the best individuals for 50 generations or the achievement of the critical size of the generation's number (500). A high proportion of the mutation of individuals and the applied exit criterion make it possible to reduce the probability of finding a local optimum of the system. At the same time, the use of NSGA-II and the selection tournament allows to accelerate the process of global optimum allocation.

At the end of the algorithm, among the resulting R individuals, unique ones are selected that make up a multitude that is close to the Pareto set for the set optimization problem.

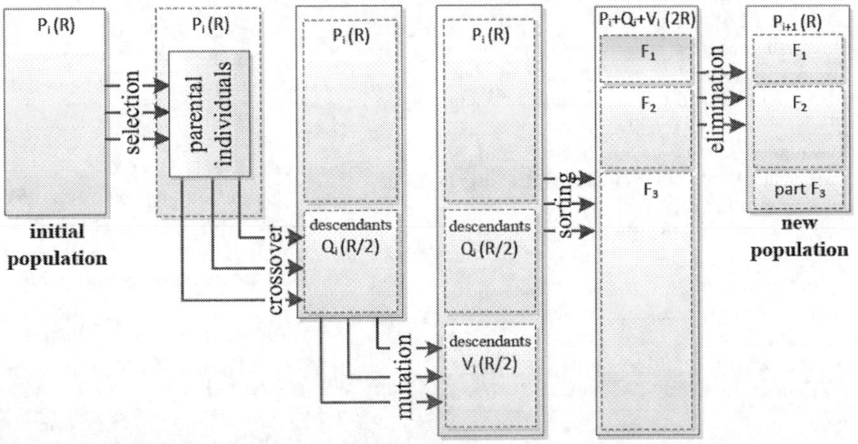

Fig. 4. Block diagram of the used GA.

4 Example of ADMS usage

Let us consider the possibilities of ADMS when solving the problem of optimizing the monitoring frequency of the flux density, the solder quality and the temperature parameters of the system at the WS330-SF Essemtec wave soldering unit [15].

Three TCM are distinguished in the operating PS: flux density monitoring, solder quality monitoring and temperature parameters monitoring. The flux density is monitored every 1.5 hours; in accordance with monitoring results the flux is replaced and a subsequent correction of its density takes place. The solder quality is checked irregularly, on average once every 8 months. The temperature parameters are regulated without carrying out the TCM, if solder defects have been detected. At a wave soldering unit, the soldering of products is carried out in one shift on average for 3 hours per day. In this mode of operation, the percentage of brazed solder joints is 10.2% of the total number of soldered joints after the installation.

In these conditions, ADMS allows to automate the decision making process; and the decisions are made on the basis of quantitative assessments of the PS technical and economic performance.

In order to solve the problem of "Optimization of component-wise monitoring frequency", the RV parameters determined by the DSS block were used as the initial statistical data to determine the distribution law on the basis of passive production experiment data [16] and obtained as a result of an expert evaluation for the preheating block.

For each monitoring, the following is defined: the minimum monitoring time period, the minimum time step, that the performer can provide. Based on these data, the code length for each of the chromosomes and for the genome, which unambiguously describes the chosen tactics of conducting the component-wise monitoring, is determined.

For the case under consideration, within the constraints imposed, ~ 8.3 million (223) of different decisions are possible with respect to the monitoring frequency. The calculation of three system efficiency criteria for one monitoring tactic takes an average of 0.21 second on a PC (2.66 GHz, 4 GB of RAM). To compare, a full calculation of an area close to the Pareto border using the GA (5 populations of 100 individuals), takes 90 seconds on an identical PC. As a result of using the GA, 32 different solutions were obtained.

The objective function values at the Pareto boundary are achieved with the following values of the input parameters (monitoring frequency):

- the monitoring frequency of the fluxing agent varies within the limits of 46 - 50 minutes;
- the monitoring of temperature parameters in view of its high speed and low cost should be carried out continuously, or daily;
- the optimal frequency of the solder quality monitoring varies from continuous to up to 13 days in between monitoring.

At the same time, depending on the chosen strategy, the reduction in the percentage of brazed joints down to 6.3-3.2% is predicted.

5 Conclusion

The ADMS structure was developed and described. The ADMS building concept is based on a hierarchical principle. Two modules of the decision support system were developed: the DSS module for choosing the RV distribution law and/or its parameters, the DSS module for the selection of the TCM optimal frequency. The problem of vector and scalar optimization was posed. To find a multitude close to the Pareto set for multi-criteria optimization of the TCM frequency of the PS with LF, the GA with sorting by domination and elite selection was chosen. To reduce the vector objective function to a scalar one, one of many known methods can be used, for example, the weighting coefficients method.

The research was carried out under the financial support of RFBR within the scope of the scientific project No 15-01-05840.

References

1. A.V. Lotov, I.I. Pospelova, *Multi-criteria Decision-making Tasks* (2008)
2. T.V. Panchenko, *Genetic Algorithms* (2007)
3. V.V. Frolov, Eastern European Journal of Advanced Technology **3/9**, 60-65 (2012)
4. V. S. Koroluk, *Stochastic System Models* (1993)
5. Yu.E. Obzherin, E.G. Boyko *Semi-Markov Models. Control of Restorable Systems with Latent Failures.* (2015)
6. V.I. Denisov, *Applied Statistics. The Rules for Verifying Agreement of the Experimental Distribution with the Theoretical One*, **1** (1998)
7. B.Yu. Lemeshko, Reliability and Quality Control **11**, 3-17 (1997)
8. F. Baykhelt, P. Franken, *Reliability and Maintenance. Mathematical Approach* (1988)
9. V.S. Avduyevskiy, *Reliability and Efficiency in Engineering*, **1** (1986)
10. Yu. Obzherin, A. Nikishenko, European International Journal of Science and Technology, **Vol. 2 No. 6**, 143-152 (2013)
11. A.I. Peschanskiy, A.I. Kovalenko, Automat. and telemech. **12**, 112–126 (2016)
12. Ye.G. Boyko, Automation of Processes and Control: Newsletter SevSTU **108**, 52–56 (2010)
13. K. Deb, S. Agrawal, A. Pratap, T. NSGA-II. IEEE Transactions on Evolutionary Computation **6**, 182-197 (2002)
14. Deb, Kalyanmoy, *Multi-Objective Optimization Using Evolutionary Algorithms* (2001)
15. WS 330-SF. *Wave Soldering System. Installation and User Manual* — Essemtec, (2007)
16. A.N. Nikishenko, Automation of Processes and Control: Newsletter SevSTU **146**, 68–76 (2014)

Analytical evaluation of the roller bearing service life

Aleksey Antsupov[1], *Viktor* Antsupov[1], and *Mikhail* Slobodianskii[1,*]

[1]Department of Design engineering and operation of metallurgical machines and equipment, Nosov Magnitogorsk State Technical University, 455000 Magnitogorsk city Lenin Street, 38 , Chelyabinsk Region, Russian Federation

Abstract. An analytical method has been developed for the design evaluation of the service life of roller bearings by the criterion of contact endurance of elements that does not require carrying out the experimental studies. It is based on the kinetic approach to the description of the solids destruction process and the theory of parametric reliability of machine parts. In contradistinction to the standard empirical dependence, which is currently used, the method allows one to take into account the effect of the material properties, and basic design and operational parameters on the service life. The results of the theoretical studies of the service life of roller bearing No. 212 (DIN 5412 T1) are presented for the quantitative assessment of this effect. They showed that the temperature-force loading conditions or material properties that were not taken into account might result in the prediction error of 1.5 or more times.

1 Setting research objectives

At present, a standard empirical dependence formulated based on the statistical analysis of bench test results is used in order to estimate the rated life L_{10} (in hours) of roller bearings (RB) [1, 2]:

$$L_{10} = \left(\frac{C}{P}\right)^{10/3} \cdot \frac{10^6}{60 \cdot n}, \quad (1)$$

where C is a dynamic radial or axial load capacity of bearings, which is given in the standard [3] for each bearing size; P is a dynamic equivalent radial or axial load, which is determined by the dependences [1]; n is a constant rotation frequency of one of the bearing rings relative to another in rpm [2].

The standard [1] does not regulate the conditions of bench tests; it is only noted that a service life for bearings made of widely used material, being of a good quality, with a good quality of manufacture, and operating under normal operation conditions is determined by condition (1). It is obvious, that changes in the operation conditions comparing with the bench test conditions are not reflected in dependence (1). Additional bench tests and the introduction of new empirical coefficients into calculated dependence (1) are necessary to record such changes. For instance, in accordance with [1, 4], modification factors are introduced into equation (1) taking into account the effect on the design life of various reliability requirements, lubrication conditions, lubricant contamination, a fatigue load limit, oil film thickness, an axial load component and others. However, the introduction of additional coefficients reduces the accuracy of service life prediction and significantly increases material and time costs in stages of the bearing support development.

Thus, empirical dependence (1) does not describe physical regularities of the process of RB material structure damageability and does not allow one to predict the time of their failure with sufficient reliability because of surface contact fatigue failure. The issue on the theoretical evaluation of the RB design life without carrying out the experimental research remains open in contemporary literature despite the significant success of the machine parametric reliability theory [5, 6] and the kinetic concept of structural materials destruction [7, 8].

Therefore, the purpose of current research is the development of an analytical method for the design evaluation of the roller bearing service life by the criterion of the surface endurance of elements, as well as the carrying out the theoretical studies on the assessment of the effect of temperature-force operating conditions on their durability.

2 The predicting method of the RB service life by the criterion of the fixed outer ring endurance

The design life of the roller bearing under study is determined by the moment of occurrence of spalling of one of the rings or rolling bodies (rollers). To reduce the article length, we represent a method for predicting the average RB service life with a rotating inner ring by the contact strength criterion of a fixed outer ring. In addition, we consider temperature-force loading conditions stationary. The following groups of parameters are the initial data for the calculation.

The first group. The geometric and kinematic characteristics of the RB, the operation conditions: d ,

* Corresponding author: m.slobodianskii@gmail.com

D, B are internal and external diameters of the RB, its width; D_W, l, z are roller diameter and length, their number; $n_в$ is the rotation frequency of the inner ring; $F_r = const$ is the radial load; $T_0 = const$ is the ambient temperature; $T_{CM} = const$ is the steady operating temperature of the lubricant.

The second group. The physical and mechanical characteristics of the material of the outer ring of the RB: E and K are the modulus of linear and bulk elasticity; G is the modulus of elasticity in shear; μ is Poisson's ratio; ρ is the density; V_{am} is the atomic volume; HV and HB are Vickers and Brinell hardness; σ_B is the strength limit; σ_T is the yield point; σ_r is the endurance limit.

The third group. The thermophysical characteristics of the material of the ring: ΔH_{TB} is the enthalpy of melting of the material in the solid state; c is the specific heat capacity of the material; α_0 is the coefficient of linear thermal expansion.

The fourth group. The physical constants: k is the Boltzmann's constant; h is the Planck's constant.

In accordance with the thermodynamic concept of predicting the reliability of machine parts [6, 9, 10], the RB mean life under stationary load conditions can be determined by the equation:

$$t_{50} = (u_{e*} - u_{e0}) / \dot{u}_e, \tag{2}$$

where u_{e0} is the latent energy density of the defects in the structure of the ring material in the initial state:

$$u_{e0} = \frac{\left((0{,}067 \cdot HV)^{1{,}2} \cdot k_\sigma\right)^2}{6 \cdot G}; \tag{3}$$

k_σ is a complex structural parameter:

$$k_\sigma = \left(6{,}47 \cdot 10^{-6} \cdot HV + 0{,}12 \cdot 10^{-2}\right)^{-1};$$

u_{e*} is the critical density of the latent energy of the material of the most loaded volume of the outer ring. The equation for its calculation is shown below.

The rate of damageability $\dot{u}_e$ of the structure material in condition (2) is determined from the kinetic equation of the form [9]:

$$\dot{u}_e = k_{cmp} \cdot \frac{2 \cdot k \cdot T_f}{h} \cdot U(\sigma_0, T) \times \\ \times \exp\left(-\frac{U(\sigma_0, T) \cdot V_{am}}{k \cdot T_f}\right) \cdot sh\left(\frac{A_f^0 \cdot V_{am}}{2 \cdot k \cdot T_f}\right). \tag{4}$$

We define the parameters entering into equation (4) in the sequence of their influence on each other [9, 10]:

The average temperature of self-heating of the deformable volume of the outer ring:

$$T_f = T_{CM} + \frac{T_* - T_{CM}}{44 \cdot \nu_0}, \tag{5}$$

where T_* is the local temperature in the contact zone:

$$T_* = \frac{T_{CM} \cdot (3 \cdot \sigma_T)^2}{(3 \cdot \sigma_T)^2 - \sigma_{max}^2}; \tag{6}$$

σ_{max} is the maximum stresses acting on the outer ring surface in the most loaded contact zone with the roller and determined from the reference dependencies [4] for each type of the RB;

$\nu_0 = T_0/870$ is the coefficient of uneven distribution of the internal energy density over the volume of the material.

The coefficient of resistance of the structure of the deformable volume of material in agreement with Le Chatelier's:

$$k_{cmp} = \frac{2 \cdot (u_{e*} - u_{e0})}{3 \cdot u_{e*} - u_{e0}}, \tag{7}$$

where u_{e*} is the critical density of the hidden energy:

$$u_{e*} = \Delta H_{TB} - c \cdot \rho \cdot T_f. \tag{8}$$

The activation energy of the damage process of the material structure at temperature T_f and stress σ_0 (taking into account the specific work of volume change A_V^0):

$$U(\sigma_0, T) = U(p_{T,0}) - \Delta U_T - A_V^0, \tag{9}$$

where $U(p_{T0})$ is the initial activation energy taking into account the thermal pressure at T_0 and $\sigma_0 = 0$ determined from the empirical dependence of V.V. Fedorov [9, 10];

ΔU_T is a change in the activation energy from self-heating temperature $T_f, °C$:

$$\Delta U_T = 1{,}5 \cdot \alpha_0 \cdot K \cdot T_f; \tag{10}$$

A_V^0 is the specific energy of volume change under the conditions of the equivalent static load:

$$A_V^0 = \frac{M_r^2 \cdot \varphi_\sigma^2 \cdot \sigma_0^2}{2 \cdot K}; \tag{11}$$

M_r is the coefficient of equivalence of the cyclic and static stresses:

$$M_r = \left[\frac{\sigma_T \cdot (65 + 0{,}46 \cdot HV)}{\sigma_r^2}\right]^{0{,}5}; \tag{12}$$

DOI: 10.1051/matecconf/201712903011

φ_σ is the coefficient of overvoltage on the interatomic bonds:

$$\varphi_\sigma = k_\sigma \cdot v_0^{0,5}; \tag{13}$$

σ_0 is the average hydrostatic stress at the most dangerous point of the loaded volume under the working surface of the ring at the critical depth [9, 11]:

$$\sigma_0 = \frac{\sigma_1 + \sigma_2 + \sigma_3}{3}; \tag{14}$$

σ_1; σ_2; σ_3 are the main stresses at the dangerous point: $\sigma_1 = -0{,}18 \cdot \sigma_{max}$; $\sigma_2 = -0{,}976 \cdot \sigma_{max}$; $\sigma_3 = -0{,}78 \cdot \sigma_{max}$.

Specific energy of forming under conditions of the equivalent static load:

$$A_f^0 == \frac{M_r^2 \cdot \varphi_\sigma^2 \cdot \sigma_i^2}{6 \cdot G}, \tag{15}$$

where σ_i is the intensity of normal stresses at the most dangerous point under the working surface of the ring [9, 11]:

$$\sigma_i = \sqrt{\frac{(\sigma_1 - \sigma_2)^2 + (\sigma_2 - \sigma_3)^2 + (\sigma_1 - \sigma_3)^2}{2}}. \tag{16}$$

After the average design life t_{50} had been found by condition (2), a ninety percentile life t_{90} was estimated under the assumption of a normal distribution of the time to failure t_* as a random variable (quantile $u_{90} = -1{,}282$) with the value of the coefficient of variation $v_t = 0{,}125$:

$$t_{90} = t_{50} + u_{90} \cdot v_t \cdot t_{50} \approx 0{,}84 \cdot t_{50}. \tag{17}$$

An algorithm and the program "Durability of the RB", which makes it possible to estimate t_{50} and t_{90} under given stationary load conditions, were created based on the proposed method (2) – (17).

The theoretical studies have been carried out with the use of this program to assess the effect of temperature-force operating conditions on the roller bearing life.

3 Analysis of computer experiment results

We carried out the theoretical studies for single-row roller bearing no. 212 (DIN 5412 T1) with short cylindrical rollers. In the example given in the study [4], its nominal life calculated from the standard empirical dependence (1) with the radial load $F_r = 5600\,H$ and the rotational speed of the inner ring $n_в = 800\,rpm$ was $L_{10} = 41757\,h$.

The initial groups of bearing parameters were matched such that the theoretical life t_{90} at $F_r = 5600\,H$ and $n_в = 800\,rpm$ was approximately equal to the empirical one $L_{10} = 41757\,h$ to ensure that the results of studies on the life time of RB no. 212 on empirical relationship (1) and theoretical method (2) – (17) were comparable. RB no. 212 manufactured, for instance, from SHX15 with hardness $HB \approx 5820\,MPa$ and yield strength $\sigma_T \approx 2000\,MPa$ corresponds to this condition. In this case, the stable operating temperature of the lubricant should be equal to $T_{см} \approx 80°C$. Then, its theoretical design life is $t_{90} = 41743h$ and is approximately equal to $L_{10} = 41757\,h$. The graphical interpretation of these solutions is shown in Figure 1 in the direction of the a-b pointers.

This figure also shows the change in the empirical life L_{10} (curve ISO 281:2007) and the theoretical life t_{90} (curves 1-9) of RB no. 212 when changing the operating conditions: $F_r = 1000 \div 8000\,H$ and $T_{см} = 20 \div 100°C$.

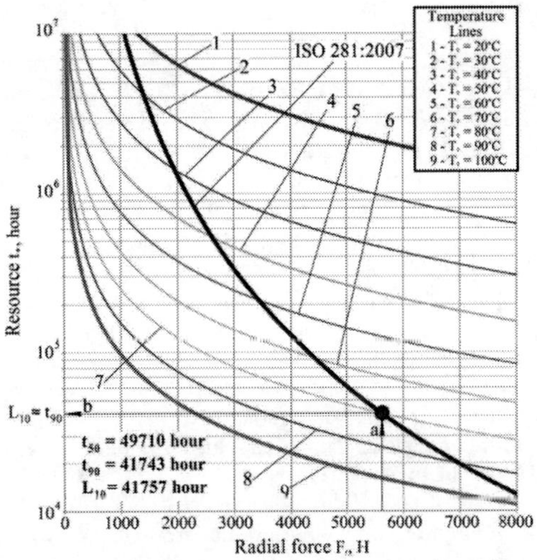

Fig. 1. The effect of temperature-power parameters on the service life of bearing no. 212 (DIN 5412 T1).

The results show the following:

- the operating temperature is the most important parameter, which the empirical dependence does not explicitly take into account (1). The figure shows that, in the example considered, a change in the temperature in the range $T_{см} = 20 \div 100°C$ at constant load $F_r = 5600\,H$ can result in a change in the life time by 2 orders of magnitude;
- the change in the radial load in the specified range results in a change in the life by 3 orders of magnitude (from ~ $10^7\,hours$ to ~ $10^4\,hours$). At the same time, the corresponding change in the theoretical life is only one order (from ~ $10^4\,hours$ to ~ $10^3\,hours$). A more abrupt decline in the empirical curve is evidently because dependence (1), simultaneously with the increase of the load F_r, implicitly takes into account the natural

 DOI: 10.1051/matecconf/201712903011

increase in the operating temperature of the bearing lubricant in bench tests. The nomogram shows that the operating temperature should increase from $T_{CM} \approx 20°C$ to $T_{CM} \approx 100°C$. The theoretical curve is created for one constant value $T_{CM} \approx 80°C$.

It is obvious that the proposed method allows one to investigate the effect of other initial parameters on the value of the project life of the RB under study without carrying out additional experiments. For instance, calculations show that a change in the hardness of bearing steels in the range $HB \approx 5820 \div 6700 MPa$ [4] under other equal conditions of the example under consideration results in the change in the life time of no less than 1.5-1.7 times. The empirical curve also does not consider this.

The possibilities described favourably distinguish the proposed analytical approach from the unambiguous experimental dependence for estimation of the roller bearing service life by the dynamic load capacity.

4 Conclusion

1. It is noted that now at the design stage of bearing supports their life is determined by the standard empirical dependence that was obtained by the results of bench tests. The change in operating conditions requires additional experimental studies that significantly increases the cost and lengthens the development stages.
2. An analytical technique for predicting the life of roller bearings by the energy criterion of contact endurance is formulated based on the modern kinetic approach to the destruction of solids and the thermodynamic theory of strength of structural materials. It allows one to take into account the effect of the properties of RB materials, the main design, and operational parameters on the durability without additional experiments.
3. The theoretical studies carried out showed that taking into account changes in the properties of materials or temperature-force operating conditions can result in a significant change in the service life no less than 1.5 and more times.

References

1. ISO 281:2007 *Roller bearings. Dynamic load rating and basic rated life*
2. M.N. Ivanov, V.A. Finogenov, *Machine parts: Textbook for machine-building specialties of high schools* (Higher Education, Moscow, 2008)
3. GOST 3395-89 *Roller bearings. Types and constructive performances*
4. L.Ya. Perel, A.A. Filatov, *Rolling bearings: Calculation, design and maintenance of supports: Handbook* (Mechanical Engineering, Moscow, 1992)
5. A.S. Pronikov, *Machine Parametric Reliability* (N.E. Bauman MSTU Publ., Moscow, 2002)
6. A.V. (Jr.) Antsupov, A.V. Antsupov, V.P. Antsupov *Theory and practice of ensuring the reliability of machine parts by the criteria of kinetic strength and wear resistance of materials* (Publishing house NMSTU, Magnitogorsk, 2015)
7. V.R. Regel, A.I. Slutsker, E.E. *Tomashevskyi, Kinetic character of strength of solid bodies* (Chief editorial board of physical and mathematical literature of the publishing center «Science», 1974)
8. G.P. Cherepanov, Strength problems **11**, pp.3-8. (1989)
9. A.V. Antsupov, A.V. Antsupov(Jr.), V.P. Antsupov Procedia Engineering **150**, p.726-733 (2016).
10. V.V. Fedorov, *Basics of ergodynamics and synergy of deformable bodies* (Izd HPE "KSTU", Kaliningrad, 2014)
11. K.V. (Ed.) Frolov, *Modern tribology: Results and perspectives* (Publishing House of LCI, Moscow, 2008)

DOI: 10.1051/matecconf/201712903012

An approach to building a cross-platform system for the collection and processing of diagnostic information about working technological equipment for industrial enterprises

Petr A. Nikishechkin[1,*], *Ilya* A. Kovalev[1], and *Anatolii* N. Nikich[1]

[1]MSTU STANKIN, 127055 Vadkovskiy per. 3a, Moscow, Russia

Abstract. The article is devoted to the trends of building a modern smart manufacturing and solving information transmission problems from technological equipment to the higher levels of enterprise management. The theoretical and practical aspects of developing solutions for collecting the diagnostic information about the work of technological equipment, the construction of its computational kernel, and results of its tests are presented.

Data acquisition from the working area where particular technical operations are carried out is a major and important problem for a modern high-tech manufacturing [1,2]. Through the use of obtained data not only different diagnostic data can be traced (errors and warnings of CNC systems, temperature, vibration while cutting work, tool wear and etc.), but also the information about machine capacity and work of its controller (for example machine hours). Modern data transfer and information processing about work of the production equipment to higher levels of an enterprise is on a direct basis of economical indicators not only of a single section, workshop, but also of the whole enterprise, as it allows to react immediately to emergency situations and to increase productivity by process flow optimization [3-5].

Modern technologies represent clusters for various processing operations. Often one controller can maintain up to 5 machines where it is necessary to monitor technology process. Herewith there can be more than 20 controllers on the line, operated process units – more than 100 (at large machine-building enterprises – «Uralvagonzavod», «KEMZ», «AvtoVAZ» - more than 1000).

Currently flourishes the tendency to enterprise development in accordance with the conception «Industry 4.0» and construction of digital production with high degree correlation between all levels of the enterprise. Large complex of production machinery from various manufacturers with different data exchange protocols often complicates operating monitoring processes and entangles the process of technical data transmission to higher levels of the enterprise (figure 1).

The paper suggests to work out the system the main task of which is classified accumulation and heterogeneous information processing into the single data-base [6] and its transmission to the higher enterprise levels.

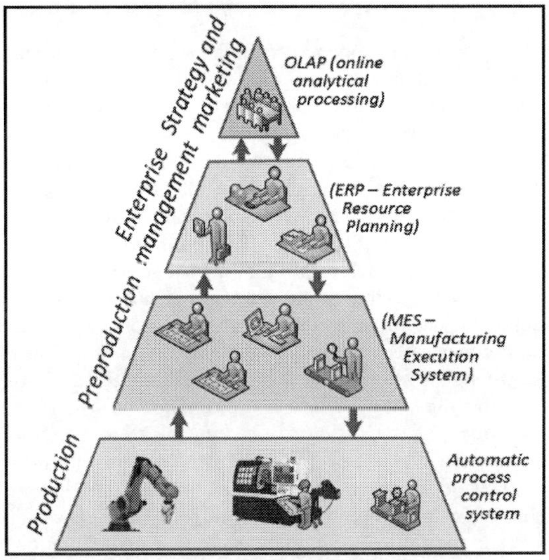

Fig. 1. Levels of modern digital production arrangement.

In the furtherance of this goal it is suggested to develop cross-platform software-based controller able to carry out compilation of data from different processing equipment. The conducted analysis showed that there are no analogues of the developed product in the domestic market. The concurrence with foreign solutions is avoided because of the another price tier and also due to the fact that most of the foreign workings represent proprietary software. The general structure of proposed solution is showed in figure 2 and can be both an autonomous solution that transmits various diagnostic data to the net and integrated to the control system of the higher level, for example, CNC system.

* Corresponding author: pnikishechkin@gmail.com

DOI: 10.1051/matecconf/201712903012

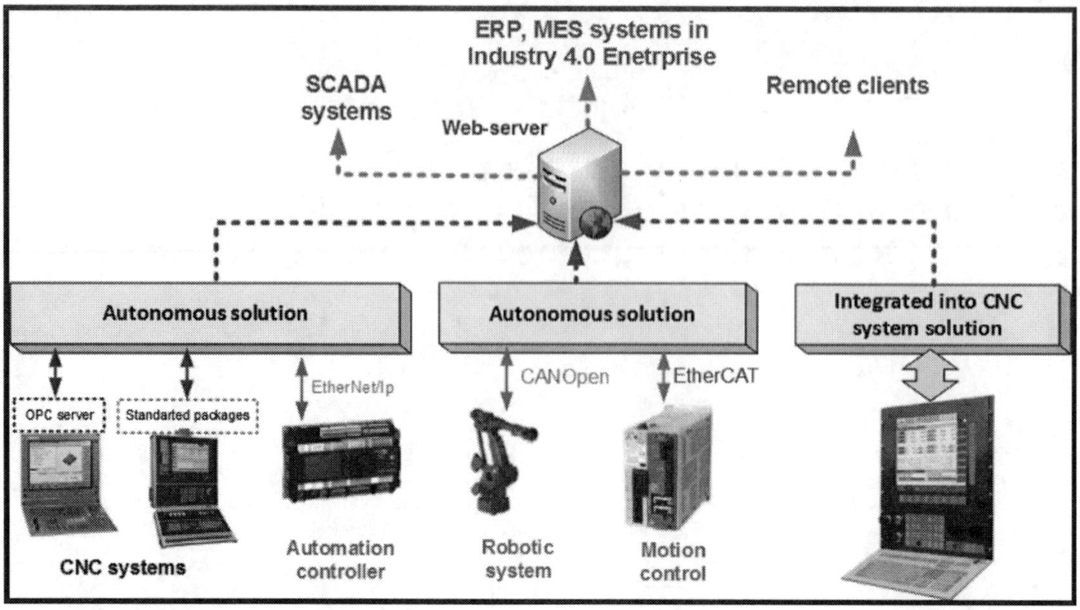

Fig. 2. Levels of modern digital production arrangement.

The main implementation of the suggested solution is at the application layer and it is a programmed computation module that allows to connect to the equipment of various types using best known interconnect protocols (EtherCAT, SERCOS, CANBus) and to aggregate the data to the integrated Webserver and then the information will be transmitted to the higher production levels (SCADA, MES, remote client users).

The main implementation of the suggested solution is at the application layer and it is a programmed computation module that allows to connect to the equipment of various types using best known interconnect protocols (EtherCAT, SERCOS, CANBus) and to aggregate the data to the integrated Webserver and then the information will be transmitted to the higher production levels (SCADA, MES, remote client users).

Developed the organization of the main modules of the suggested solution. The central part shows the main computation module implementation – computational kernel that works in real-time mode and carries out data acquisition from the equipment of various types. This module interacts with terminal clients and also with automated equipment.

The main task of the basic computational module (kernel) is data acquisition and information processing in real-time mode from the equipment of various types through different fieldbuses. Flexibility, generality of the solution and the possibility of using it with a set of control equipment from different manufacturers is possible due to the abstract level implementation (execution units, interconnect protocols and etc.), that is often observed in domestic enterprises, even within the same workshop.

It is suggested to implement the basic computational module in accordance with the modular approach: a terminal module of the top level of the user space, a mathematical module where basic calculations occur, a cross-platform module, a driver module that operates in the address space of the selected OS. These modules are implemented in C ++ using object-oriented programming methods. Currently, the solution is being adapted to the work with NCT201 CNC systems (NCT, Hungary), AxiOMAControl (MSTU "Stankin", Russia), as well as with servocontrollers IntDrive (BoschRexroth) and DS (NCT, Hungary) and obtaining diagnostic data from them.

In addition to working with standard equipment to get the diagnostic information on the operation of the technological unit it is supposed a possibility of different sensor types connection (vibration, thermo), the interaction with them can be carried out with remote input/output devices (using analog inputs). Thus, it is possible to expand the area of the diagnosed equipment and to receive more extensive and reliable information about the operation of technological units and, in general, of the workshop/enterprise.

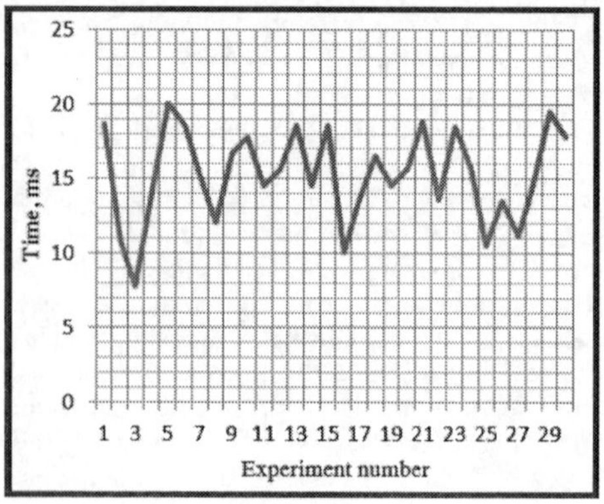

Fig. 3. Scanning tests of the net device elements

During tests it was found out that the amount of used program elements tied to the structure of processing facilities can increase sufficiently (the implementation of connection of five milling machines worked out in about 2000 components in the integrated environment). Scanning such amount of elements (figure 3 there are 100 elements, figure 4 - 1500 elements) took up about 2 seconds.

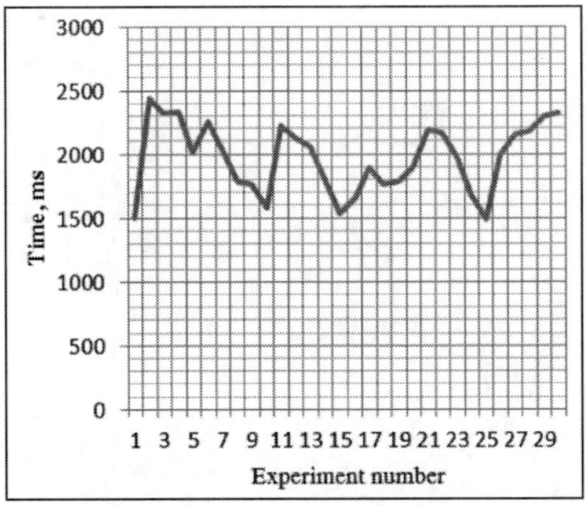

Fig. 4. Scanning tests of the net device elements

Taking into consideration the fact that there can be more than 15 production units in one data line, the total number of polled processing facilities projected to the control program can amount to more than 10 thousand.

It was decided to use the method of asynchronous messages with maximum data burst filling instead of the method of polling type request (one request-one answer) to increase the high speed response. In case of the new approach for introduced example the average time of getting data from all nodes fall by 50 ms, also to get more process pieces this time can be maintained within 100 ms, by means of asynchronous messages only from those devices that changed its state [13].

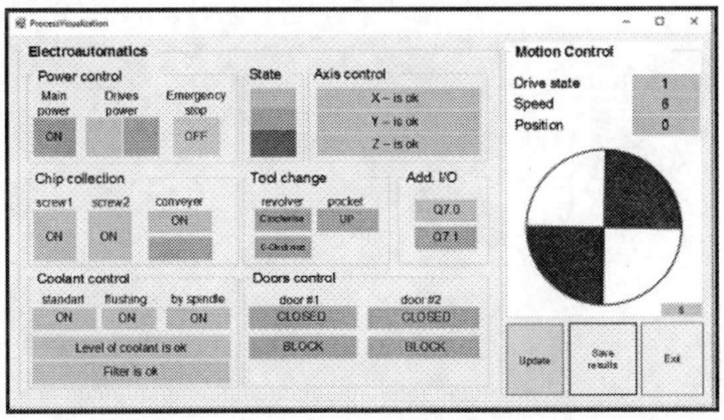

Fig. 5 Example of window process visualisation

The figure 5 shows the implementation of process visualization for a milling machine. The proposed approach allows to display what is happening in the subsystem of electroautomatics, working on fieldbus EtherCAT, as well as take readings from engines working on the Sercos fieldbus

Along with that one of the important characteristic while computational kernel development of the suggested solution is its cross-platform both at the system and hardware level. It allows to use the various base of its building-up and integrate it to different working systems. At the system level the work fulfillment in various operation systems is realized due to the use of cross-platform pools, mutexes, timers and function wrapping and devices of shared memory: Windows-family, OS based on Linux, also with Windows system extension – system RTX of IntervalZero company [14-15].

Cross-platform at the system level also allows to perform module porting to different hardware platforms – it is realized the possibility of the main module work on the processors of various type: x86 (Intel, AMD) or ARM including domestic ones (Elbrus 4C, Baykal-T, M). It helps to use module both on personal computers and on single-board computers, and to draw out single compact devices for automation of technological processes with the possibility of remote control and access to data that is an important condition in smart manufacturing.

Consequently, in case of this approach the following tasks can be solved: heterogeneous hardware unification into the consolidated information area, monitoring of the equipment operation status, data acquisition about machine and operator capacity, rapid response to emergency situations, forecasting of possible contingency, the possibility of transferring information about production to the upper levels of the enterprise (matching for development concept Industry 4.0), and also downtown time minimization [12,14-16].

Practical implementation and use of the suggested solution at production enterprises will provide an opportunity to get a new scheme for data acquisition and information processing from the machine line and other manufacturing equipment with the use of domestic hardware and software components. Implementing of cross-platform approach in the solution will make it possible to decrease, and in the future to avoid dependence on foreign elements, and thus to provide technological and information security of domestic enterprises.

References

1. G.M. Martinov., L.I. Martinova, *Development of the base computing platform of CNC for construction of specialized control systems*, Vestnik MSTU "Stankin", **№1** (24), P. 92-97. (2014)
2. G.M. Martinov, A.B. Lyubimov, A.I. Bondarenko, A.E. Sorokoumov, I.A. Kovalev, *An approach to building a multiprotocol CNC system*, Automation and Remote Control, **Vol.76**, No. *1*, P. 172-178 (2015)
3. G.M. Martinov, A.S. Grigoryev, P.A. Nikishechkin. *Real-time diagnosis and forecasting algorithms of the tool wear in the CNC systems*. Advances in Swarm and Computational Intelligence. **Vol.9142**. P. 115-126 (2015)

4. S.N. Grigoriev., G.M. Martinov, *An ARM-based Multi-channel CNC Solution for Multi-tasking Turning and Milling Machines,* Procedia CIRP. **Vol.46**, P. 525–528 (2016)
5. S.N. Grigoriev., G.M. Martinov, *The Control Platform for Decomposition and Synthesis of Specialized CNC Systems*, Procedia CIRP, **Vol.41**, P. 858-863 (2016)
6. S.N. Grigorev, V.A. Dolgov, A.V. Krasnov, A.A. Kabanov, N.S. Andreev, *A Method of Technologic Audit of Technical Re-Equipment Projects in Aircraft Production Enterprises*, Izv.Vuz. Av. Tekhnika, **Vol.58**, no. *2*, P. 103–108 (2015)
7. G.M. Martinov, A.I. Obuhov, L.I. Martinova, A.S. Grigoriev, *An Approach to Building a Specialized CNC System for Laser Engraving Machining.* Procedia CIRP, **Vol. 41**, P. 998-1003 (2016)
8. G.M. Martinov, R.A. Nezhmetdinov, S.V. Sokolov, *The Principles of Constructing a Toolchain for Monitoring and Setup Mechatronic Equipment Parameters Based on Integration of Specialized Software Components into Control System Structure,* Mekhatronika Avtomatizatsiya, Upravlenie, **No. 7**, P. 45–50 (2012)
9. L.I. Martinova, S.V. Sokolov, P.A. Nikishechkin, *Tools for Monitoring and Parameter Visualization in Computer Control Systems of Industrial Robots*, Advances in Swarm and Computational Intelligence. 6th International Conference, **Part II**, P.200-207. (2015)
10. G.M. Martinov, R.A. Nezhmetdinov, P.A. Nikishechkin, *Development of visualization means and debugging of control programs for magnetics integrated into CNC system*, Vestnik MSTU "Stankin", **№4** (23), P. 134-138 (2012)
11. G.M. Martinov, P.A. Nikishechkin, A.S. Grigoryev, N.Y. Chervonnova, *Organization of interaction of the main components of the CNC AxiOMA Control for integration into its new technologies and solutions,* Automation in the industry, **№5**, P.10-15 (2015)
12. R.A. Nezhmetdinov, P.A. Nikishechkin, R.L. Pushkov, S.V. Evstafieva, *Practical aspects of the software-based controller to control electroautomatics vertical milling machines Quaser MV184,* Automation in the industry, **№5**, P. 15-18. (2016)
13. N.V. Kozak, R.A. Abdullaev, I.A. Kovalev, N.Yu. Chervonnova, *Realization of the logical task of CNC and task of industrial safety on the basis of external computing SoftPLC modules*, Automation in the industry, **№5**, P. 28-30 (2016)
14. G.M. Martinov, N.V. Kozak, *Construction of a specialized CNC system for a five-axis planer-milling machining center*, STIN, **№8**, P.2-6 (2015)
15. I.A Kovalev, S.V. Sokolov, L.I. Martinova, S.V. Rybnikov, *Construction of a specialized CNC system for thread grinding machines*, Automation in the industry, **№5**, P.38-40 (2015)
16. G.M. Martinov, N.V. Kozak, R.A. Abdullaev, I.A. Kovalev, *Construction of a specialized distributed control system for a precision machining center VMG 50*, Automation in the industry, **№6**, P. 16-20 (2014)

Pareto optimization using the method of ant colony

Olga *Chengar*[1,*], *Elena* Savkova[2], *Elena* Vladimirova[1], and *Nikolay* Sapozhnikov[3]

[1]Sevastopol State University, Department of Information Technologies and Computer Systems, 299053, University Str., Sevastopol, Russian Federation
[2]Donetsk National Technical University, Department of Automated Control Systems, 83084, Artyoma Str., Donetsk, Ukraine
[3]Federal state owned military educational institution of the Higher professional Education "the black sea higher navy order of the red star academy by P.S.Nakhimov", 299028, Dybenko Str., Sevastopol, Russian Federation

Abstract. The article deals with the development of an effective mathematical and algorithmic apparatus for solving multicriteria problems of optimizing scheduling in engineering using the Pareto principle. The proposed algorithm uses the instrumental capabilities of the ant colony method, which makes it possible to obtain the solution closest to the "preferred" point of the criteria space that is given by the user. The variant of forming the objective function as a weighted sum of separate criteria is considered.

1 Introduction

The real optimization problem is usually solved for several criteria. A number of scientific works and publications have been devoted to this issue [1-5]. But the problem of constructing the Pareto set approximation (and thus of the Pareto front) is not completely solved. This problem lies in the fact that the standard methods of multicriteria optimization do not allow determining the whole set of effective solutions, especially in the scheduling tasks for machine-building production. To use even the simplest of such methods with high accuracy Pareto set approximation and / or high computational complexity of target functions, generally unacceptably high computational resources are required. Therefore, at present, alternative methods are developing intensively.

Usually, these algorithms are based on evolutionary algorithms and, more often, on genetic algorithms [6]. Then the corresponding Pareto-approximation methods are called evolutionary, and in the area under consideration they are the most promising developments. The basic principle of these methods is not the use of evolutionary algorithms, but the rules for the formation of fitness functions that ensure the movement of individuals in the population, ultimately, in the direction of the Pareto set [6,7]. The evolution of these individuals can proceed according to laws different from the laws used in evolutionary algorithms, for example, according to the laws motion of the ant population in the ant algorithm.

Thus, the task of multicriteria optimization, based on the so-called "population" Pareto-approximation methods, becomes urgent. The aim of the work is the development of an effective mathematical and algorithmic apparatus for solving multicriteria optimization problems on the basis of the ant colony method, which can easily be adapted to given conditions with additional constraints in the problem.

2 The research task formulation

Multicriteria optimization is based on finding a solution that is the best at the same time for several functions [8,9]. To this end, special methods have been developed for to differ significantly from the standard technique, which has been focused on the optimization of one function. While using the Pareto optimization approach for multicriteria problems, there are two types of search in target space [10]:
1) constant search direction (in the case of fixed weights in the target function);
2) multiple search direction.
In general, the problem of multicriteria optimization can be formulated as follows (1):

$$\max(\min)\ \{z_1=f_1(x),\ z_2=f_2(x),\dots,\ z_q=f_q(x)\} \\ g_i(x)\le 0;\ \ x\ge 0;\ \ i=1,2,\dots,m \tag{1}$$

where $f_q(x)$ is a function-criterion; q is number of criteria; $g_i(x)$ is task limitation; m is the number of constraints.

Here, the solution search space is defined (2):

$$S=\{x\in R^n \mid g_i(x)\le 0,\ \ i=1,2,\dots,m,\ x\ge 0\} \tag{2}$$

where $x\in R^n$ is vector of variable parameters.

In the case of multicriteria optimization, graphical interpretation in the space of criteria is often used Z (3).

$$Z=\{z\in R^q \mid z_1=f_1(x), z_2=f_2(x),\dots,z_q=f_q(x),\ x\in S\} \tag{3}$$

where $z\in R^q$ is values vector of q target functions.

Thus, Z is the set of images in S (fig. 1 and fig. 2).

* Corresponding author: OVChengar@sevsu.ru

DOI: 10.1051/matecconf/201712903013

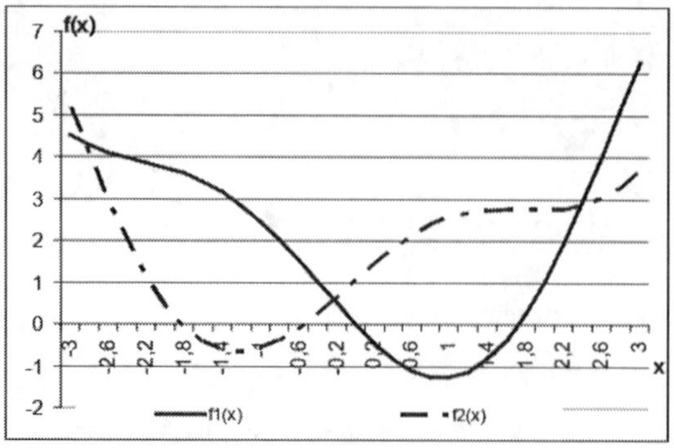

Fig.1. Target functions in the solution space S.

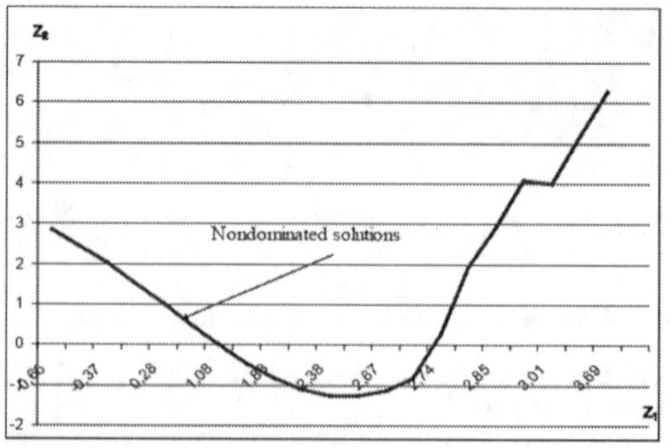

Fig.2. Target functions in the space of criteria Z.

If the solution is not dominated by any other solution, then it is called non-dominant or optimal in the sense of Pareto. It should be noted that multicriteria tasks are fundamentally different from single-criteria ones [10]. In the latter case, it is necessary to find a solution that is better than all the others. In the case of multicriteria optimization, there is not necessarily a solution that is the best with respect to all the criteria separately due to possible conflicts.

The decision can be better with respect to one criterion and worse with respect to other criteria [7, 10]. According to Pareto's dominance principle, the potential solutions of the multicriteria problem are divided into dominated and non-dominated solutions [10].

A solution X is said to be dominant if there exists a solution Y that is not worse than X by all criteria, that is, for all optimizable functions $f_i\ (i=1,\dots,q)$: $f_i(x) \le f_i(y)$ for all $1 \le i \le k$ while maximizing the function f_i and $f_i(x) \ge f_i(y)$ for all $1 \le i \le k$ while minimizing the function f_i. Algorithms, which are based on ant colonies do not impose any requirements on the kind of target functions and constraints and allow to obtain a lot of potential solutions in various directions of global search. Therefore, in case of multicriteria optimization, the search for more than one path of ants' population movement along the graph vertexs is performed, and multiple paths for each ants population optimal in the sense of Pareto are searched. Then the general structure of the multicriteria approach based on the method ant's colonies can be represented by the following algorithm (fig. 3).

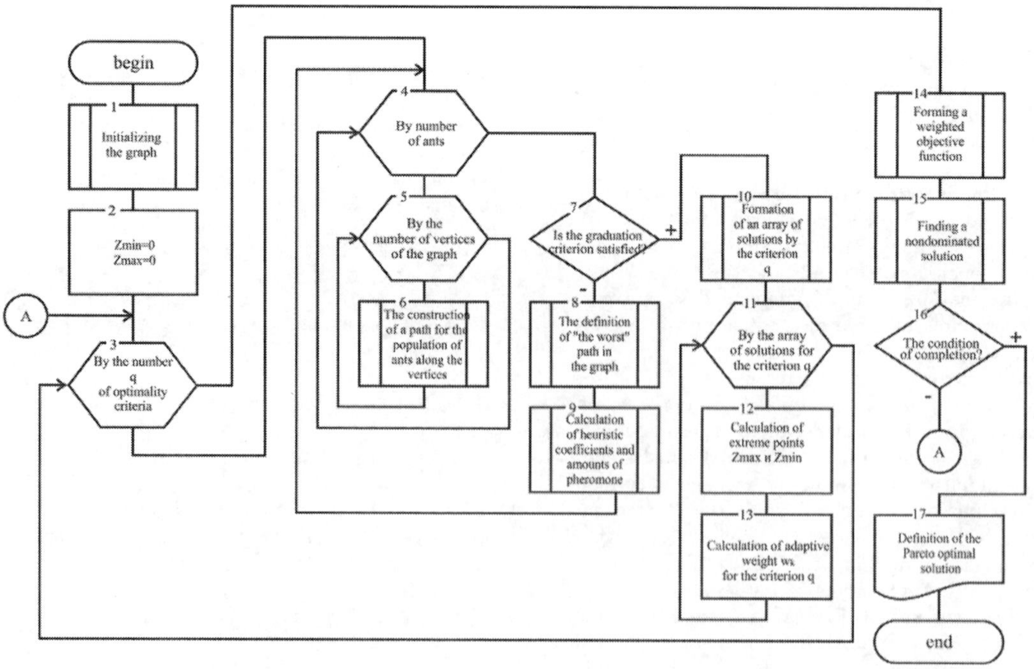

Fig.3. Generalized algorithm for solving a multicriteria problem based on the method ant's colonies.

In fact, the method ant's colonies refers to the methods of meta-strategy, therefore, in order to solve a specific problem, in addition to the basic parameters the ant algorithm, it is necessary to select and justify the basic criteria for optimality.

Justification of this choice or that criterion of efficiency is responsible and far from always obvious task. The difficulty is in that the various optimality criteria are often contradictory, optimization by one criterion leads to a deterioration in quality by another criterion. There are several ways to solve this problem and the main one is the reduction of the multicriteria problem to the one-criterion one, for example, using the integrated criterion, as the sum of particular criteria with some expertly assigned coefficients (the so-called additive convolution method). However, such an approach also requires a serious justification, first of all, because of the subjective arbitrariness in choosing the structure of the integrated criterion and the appointment of coefficients.

Since the multicriteria optimization is a natural development of ordinary numerical or combinatorial optimization, many developed methods have been extended to this more general case [7]. While using the ant algorithm for multicriteria optimization, the central question is the construction of the objective function [11,12]. Recently, with the rapid development of evolutionary and population optimization methods, several approaches have been developed [13,14], one of which is the weighted sum method, the essence of which is to determine the weight of each criterion included in the objective function [15].

This approach is one of the most popular and natural development of classical optimization methods, in which the "general" target function is constructed from separate target functions in the form of a weighted sum (4).

$$F(x)=\sum_{i=1}^{k} w_i f_i(x) \qquad (4)$$

where weights $w_i \in [0,1]$ and $\sum_{i=1}^{k} w_i = 1$

Here, each target function $f_i(x)$ is assigned its own weight w_i and the problem reduces to the scalar case. Different values the weights w_i lead to different solutions in the sense of Pareto. The scalar value of the new target function is then calculated by summing the weighted values q of the optimality criteria. For parallel search for multiple solutions, weights are not fixed, which makes it possible for the ant algorithm to expand the front in all directions.

In the weighted sum method, weight determination is performed either randomly, which makes it difficult to find a Pareto optimal solution, or with the use of an adaptive weighted approach.

In the presented algorithm, at each iteration for a certain optimization criterion, a lot of decisions are formed based on the functioning the ant's algorithm. Further, for the solutions being investigated, the maximum and minimum extremal points in the space of given criteria are determined (5):

$$z^{+}=\{z_1^{max}, z_2^{max}, \ldots, z_q^{max}\}$$
$$z^{-}=\{z_1^{min}, z_2^{min}, \ldots, z_q^{min}\} \qquad (5)$$

For each criterion, the maximum and minimum values are determined as follows (6):

$$z_k^{max} = \max\{f_k(x) \mid x \in P\},\ k=1,2,\ldots,q$$
$$z_k^{min} = \min\{f_k(x) \mid x \in P\},\ k=1,2,\ldots,q \qquad (6)$$

where z_k^{min} and z_k^{max} are minimum and maximum value for the k-th target according to the specified criterion; q is number of optimality criteria; P is set of solutions by a given criterion.

As a result, we obtain a hyperplane defined by two extremal points, and which contains all the current solutions. These two extremal points are updated at each iteration. The adaptive weight of the k-th goal is determined by the relation (7).

$$w_k = \frac{1}{z_k^{max} - z_k^{min}},\ k=1,2,\ldots,q \qquad (7)$$

For each goal, based on the selected performance criteria, average weighted weighting factors of significance are established, which are normalized within the group (8).

$$w_k^{норм.} = \frac{w_k}{\sum_{i=1}^{q} w_i},\ k=1,2,\ldots,q \qquad (8)$$

Then, at each iteration, the weighted target function is determined according to the following expression (9).

$$z(x)=\sum_{k=1}^{q} w_k^{норм.}(f_k(x)-z_k^{min}) \qquad (9)$$

The schematically proposed algorithm can be represented in the form a graph (fig. 4) reflecting the sequence of constructing an adaptive moving line in the space of optimality criteria Z in order to search for Pareto-optimal solutions.

Initially, the algorithm assumes zeroing of all values Z^{min} and Z^{max}. Since extreme points are updated at each iteration, the weights are updated accordingly. Expression (9) represents a hyperplane defined by the following extremal points in the set of current solutions (10).

As shown in fig. 4, an adaptive moving line is determined by extremal points (z_1^{max}, z_2^{min}) and (z_1^{min}, z_2^{max}). Thus, a rectangle defined by extremal points (z_1^{max}, z_2^{min}) and (z_1^{min}, z_2^{max}), is the minimum rectangle that contains all the current solutions. All

 DOI: 10.1051/matecconf/201712903013

explored Pareto solutions lie in the space Z and during the process of finding a solution for each population of ants, the hyperplane successively approaches a positive (or negative) ideal point. So this method allows you to adjust the weights of the objective function and directs the search for solutions in the right direction.

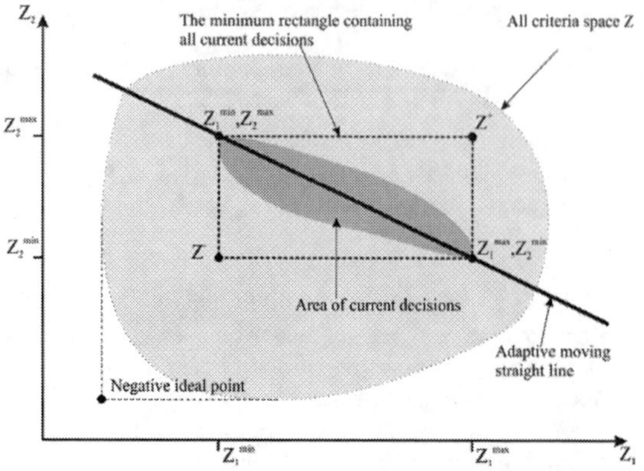

Fig.4. Adaptive weights and adaptive hyperplane in the space of criteria Z.

Since each the optimality criteria can have its own solution accuracy, which is achieved with different numbers of iterations (n_t), for determining n_t Experimental researches of all single-criterial problems were carried out and the number of iterations was determined, at which accuracy would be achieved corresponding to the technological conditions. Therefore, as a condition of stopping while solving a multicriteria problem, the maximum value from the total number of iterations is accepted.

Thus, the multicriteria optimization method, which was described above, allows one to search through the entire solution space for each ants population and choose those paths of motion that will be optimal in the sense of Pareto.

3 Conclusions

The variety of methods for constructing Pareto optimization points to the relevance of the problem were considered in this article. The development of this method in the direction of evolutionary and population optimization methods seems perspective.

The computational complexity of the particular optimality criteria is not the same in different parts the region of variable parameters and for tasks of schedule optimization the large dimension. Many modern Pareto optimization methods do not take this condition into account. In this regard, the development of adaptive and self-adaptive Pareto-optimization methods based on such promising optimization algorithms as the method of ant colonies is relevant.

The algorithm, which was proposed in this article for solving a multicriteria problem using the instrumental capabilities of the ant colony method makes it possible to construct an approximation not of the entire Pareto front, but of its part closest to the user-defined "preferred" point in the criteria space. This approach is realized by forming an integrated criterion, for example the sum of private criteria with some coefficients, where the "general" target function is constructed from separate target functions in the form of a weighted sum.

The properties inherent in ant algorithms contribute to the their effective application in solving multicriteria problems because ant algorithms are based on the use a multitude of potential solutions in the global search in various directions and do not impose any requirements on the form of target functions and constraints.

Thus, the proposed approach of the Pareto dominance concept, based on the ant algorithm, is of considerable interest.

References

1. E. Zitzler, K. Deb, L. Thiele, Evolutionary Computation Vol. **8(2)**, P. 173-195. (2000).
2. K. Deb, *Multi-objective optimization using evolutionary algorithms* (UK: Wiley. 2001)
3. S.Mostaghim, J. Teich. In: Swarm Intelligence Symposium, P. 26– 33 (2003)
4. D.Moor, D. Mukhlisullina, Science and education: electronic scientific and technical publication Vol. **4**, P. 123-129 (2010)
5. V.Guliashki, H.Toshev, Ch.Korsemov, Problems of Engineering Cybernetics and Robotics Vol. **60**, P. 42 – 54 (2009)
6. V. Podinovsky, V. Nogin, *Pareto-optimal solutions of multicriteria problems* (RU: Moscow. 2007)
7. V. Nogin, *Decision-making in a multicriteria environment: a quantitative approach.* (RU: Moscow. 2005)
8. D.Ashlock, *Evolutionary Computation for Modeling and Optimization* (Springer. 2006)
9. I. Sobole, R. Statnikov, *The choice of optimal parameters in problems with many criteria* (RU: Moscow. 2006)
10. O. Chengar, Donbass-2020: Perspectives of development through the eyes of young scientists. P. 339-343 (2010)
11. J. Ryu, S. Kim, Proceedings of the 2009 Winter Simulation Conference (WSC), P. 623-633 (2009)
12. O. Chengar, Problems of Information Technology. Bulletin of Kherson National Technical University, Vol. **1(46)**. P. 212-217 (2013)
13. O. Chengar E. Savkova. Journal of East Ukrainian National University. **13 (167)**, P. 239-245 (2011).
14. Y. Skobtsov, A. Sekirin, S. Zemlyanskaya, O. Chengar, V. Skobtsov, S.Potryasaev, Advances in Intelligent Systems and Computing, vol.**466**, P.453-462 (2016)
15. Y. Skobtsov, O. Chengar, V. Skobtsov, A.Pavlov, Advances in Intelligent Systems and Computing, vol.**573**, P.456-465 (2017)

Mathematical modelling of electrode systems of electrostatic filtration processes

Leonid Ivanovich Koltunov[1,*], and *Alexander Stepanovich* Kizhuk[1]

[1]Department of Engineering Cybernetics, Belgorod State Technological University n.a. V.G. Shukhov, Kostyukov Str., 46, 308012, Belgorod, Russia

Abstract. The up-to-date air filtration and ionization systems and units for premises, buildings and constructions have been analyzed. The peculiarities of mathematical modeling with the results of numerical and experimental research of electrostatic systems and units, distributed and autonomous, are presented. The peculiarities of discrete models of objects under research in the operator form are given. The computing algorithms, the numerical modeling software environment structure and the functioning peculiarities of automatic control local systems of electrostatic objects as part of automated dispatch control system have been developed.

1 Introduction

The modern electrostatic systems and units, dealing with electrostatic purification, ionization and ozonation of air, are used at rocket launching systems, submarines, in hospitals, health resorts and other facilities; they are also embedded in car fascia panels, used in refrigerators, LC-displays, air humidifying devices etc. It should be noted that these electrostatic systems provide the relatively high quality of air at the lowest energy consumption, unlike ventilation and air conditioning systems, in which the issues of energy saving and providing the high quality of air usually contradict each other. In this regard the development of electrostatic systems in conditions of energy saving is a topical problem [1].

The analysis of the modern development of electrostatic systems has shown that it develops in several ways. The development of the first way is connected with the name of professor A.L. Chizhevsky. He has designed a facility for generating negative air ions by means of creating high electrical voltages on the «needle-shaped» electrodes. The main peculiarity of these facilities is the application of negative electrodes (e.g., ionizer of «Aeroion-25» type). There also appear air ionizers to provide the balance of positive and negative air ions (e.g., the «Yantar» ionizer).

The second way, dealing with two-stage electrostatic precipitators for ventilation and air conditioning systems, was developed in the former USSR in 60-ies. But the main boost for the development of these electric precipitators was given by the company «Plymoth» (Sweden) in the sphere of purifying air from welding, oil and other finely-dispersed aerosols in the industrial conditions. Electric precipitators of the similar type (LLC «Elstat», Moscow) are used also for the purification of the intake air in ventilation systems in buildings of various purposes. The main peculiarity of these facilities is the application of two chambers: ionization chamber for the particles charging and precipitation chamber for catching them.

The third way appeared at the turn of the century almost simultaneously in a few developed countries, in a form of ionizing air cleaner for various spheres of application: for personal use (e.g., model XJ-888 of the company AirComfort, Italy); for transport (e.g., model CP-3 of the company Chung Pung, South Korea); for residential buildings and offices (e.g., models CV-N63CX of the company Sharp, Japan and XJ 3500 of the company AirComfort, Italy); for industrial enterprises (model Eagle 5000 of the American company «EcoQuest International»); for embedding into ventilation and air conditioning systems (model DuctwoRx of the company «EcoQuest International») etc. The peculiarity of these facilities consists in creating the air environment with the required balance of fast positive and negative air ions in the premises of buildings and constructions, and the conditions for the air self-purification in a certain area of premises as well as its purification by filtering through this unit. The development of such appliances is essential at the present stage, as it deals with the protection of environment and the human health [2].

2 Analysis of existing methods and models

The analysis of the existing methods and models of electrostatic units research has demonstrated that analytical methods are not fit for determining the electrical potentials distribution field in the interelectrode systems of these appliances. At present, there are still no analytical expressions for the modern electric precipitator systems, taking into account their

* Corresponding author: koltunov_l_i@mail.ru

appearance in the 60-ies of the last century, and as for the rapidly developing new trend of autonomous filtration systems with the simultaneous air ionization, it is developing almost empirically. Studying these units on the basis of computational experiments and numerical methods is the only way of obtaining relatively accurate results. Among the main methods of solving boundary value problems we should single out the finite difference method, which has a number of applications for the numerical study of electrostatic units with account of their structural features [3] (for example, flat polarizing electrodes with corona wire electrodes).

The purpose of this work is developing the basis for determining the characteristic regularities of electrostatic units for air filtration and ionization, and improving their efficiency with account of a wide range of structural features on the basis of integrated research with the use of up-to-date mathematical modeling and computational experiment technologies.

There is a lot of standard software for solving boundary value problems with differential equations in partial derivatives on the basis of using various numerical methods (MatLab, ELCUT, FemLab, ROSTWERK etc.). The usage of standard software is limited considerably for solving the boundary value problems, dealing with studying electrostatic units for air filtration and ionization, but there is the «IMPULS» modeling program, registered in the Russian Agency for software and database legal protection and aimed at calculating breakdown paths in electro-hydraulic units' interelectrode systems. On its base, with account of modernization, the electrostatic units can be studies with the subsequent creation of the software environment for the numerical modeling of such units.

The analysis of filtration processes' features in electrostatic units has demonstrated that they have both ionization areas for the particles charging, and electrostatic precipitation areas, where the unipolar corona can be also used. In some types of air cleaners the corona appears simultaneously on corona-forming electrodes in different parts of these units and at various high-voltage potentials on them. On the basis of analyzing the structures of electrostatic units for air filtration and ionization the basic standard layouts for their modeling have been designed.

For example, one of the modeling diagram for multichamber air cleaners with account of symmetry is shown in Fig. 1, (G_i – researched area boundaries $B_3(x, y)$; K_1 and G_6 – corona-forming electrodes, the dimensional ratio is determined with the type of various electrode systems of electric precipitators).

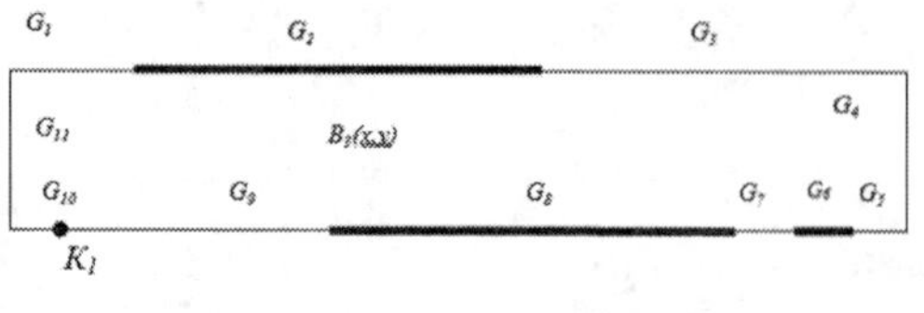

Fig. 1. The modeling diagram for multichamber air cleaners.

By means of studying the electrostatic units' electric fields a generic mathematical model is selected and substantiated, for which the boundary value problems with mixed-type boundary conditions, taking into account the basic limitations of the researched objects and the design features of these units, are formulated. The basis of the model is the approach, at which there is no need to determine the nonlinear dependence between the current density and the electric field intensity in the interelectrode systems of these objects [4]. The problem of calculating the electric field of electrostatic units' interelectrode space consists in determining the electric potential field φ and the electrical function of the flux χ, on basis of which the required parameters are calculated (electrostatic field intensity, effective ionization area etc.) with the subsequent determining of the range of appropriate rations of these objects' parameters.

For the boundary value problems of electrostatic units the conditions of finite difference method adaptation and transition to the finite-difference approximation of elliptical equations and boundary conditions are fulfilled. At this the peculiarities of the research objects' discrete models in the operator form are presented.

1. For example, for a boundary value problem (Fig. 1) with account of the second-order elliptical equation and the mixed-type boundary conditions the initial equations in operator form to determine the electric potential field φ in node (i, j) of the discrete area $B_{g3}(x, y)$ with the boundaries Gi are as follows.

$$\Omega_h\varphi^{(h)} \equiv \begin{cases} \Omega_{xx}\varphi_{mn} + \Omega_{yy}\varphi_{mn} = 0, (x_m, y_n) \in B_{g3}(x,y), \\ \varphi_{mn} = c_1, (x_m, y_n) \in K_1, \\ \varphi_{mn} = c_2, (x_m, y_n) \in G_8, \\ \varphi_{mn} = c_3, (x_m, y_n) \in G_6, \\ \varphi_{mn} = 0, (x_m, y_n) \in G_2, \\ \Omega_n\varphi_{mn} = 0, (x_m, y_n) \in G_1, G_3, G_4, G_{11}, \\ \Omega_n\varphi_{mn} = 0, (x_m, y_n) \in G_5, G_7, G_9, G_{10}. \end{cases} \quad (1)$$

Here $\Omega_{xx}=\partial^2/\partial x^2$; $\Omega_{yy}=\partial^2/\partial y^2$; $\Omega_n=\partial/\partial n$. The constants c_i are determined by the values of high-voltage potentials on electrodes K_1, G_6, G_8 with account of standard industrial prototypes of these objects.

2. The problem of determining the electrical function field of the flux χ for an electrostatic unit (Fig. 1) is the following. The function χ in node (i, j) of the discrete area $B_{g3}(x, y)$ with the boundaries G_i is determined with account of the second-order elliptical equation and the mixed-type boundary conditions, which in operator form are as follows:

$$\Omega_h\chi^{(h)} \equiv \begin{cases} \Omega_{xx}\chi_{mn} + \Omega_{yy}\chi_{mn} = 0, (x_m, y_n) \in B_{g3}(x,y), \\ \Omega_n\chi_{mn} = 0, (x_m, y_n) \in K_1, G_2, G_6, G_8, \\ \chi_{mn} = c_1, (x_m, y_n) \in G_1, G_{10}, G_{11}, \\ \chi_{mn} = c_2, (x_m, y_n) \in G_9, \\ \chi_{mn} = c_3, (x_m, y_n) \in G_7, \\ \chi_{mn} = 0, (x_m, y_n) \in G_3, G_4, G_5. \end{cases} \quad (2)$$

 DOI: 10.1051/matecconf/201712903014

The mathematical models are intended for studying the characteristic regularities of electrostatic units for air filtration and ionization processes, and for determining the peculiarities of electrostatic precipitators and autonomous ionizing air cleaners with the purpose of improving their efficiency. The developed base allows researching the existing design solutions based on empirical approach at designing these appliances. So, the basis for the integrated research of electrostatic units with the use of up-to-date mathematical modeling technologies is presented [5].

The calculation data of the electrostatic units' interelectrode systems and of the experimental research are presented on a per-unit basis with account of dimension and similarity laws in relation to the objects' basic parameters. So, without the alteration of ionization chamber parameters the maximum value of the effective ionization area (F^*_e) depends on the location of corona-forming electrode and the gap between the electrostatic precipitator's chambers. At reducing the gap δ from 30 mm to the minimum value the effective value F*e increases from the value $F^*_{e(min)}$ to the maximum $F^*_{e(max)}$ within the range from 1 to $m_{11} = 1,43$ ($m_{11} = F^*_{e(max)}/F^*_{e(min)}$), whereas the maximum value $F^*_{e(max)}$ increases by 1,25 times. With account of the alteration of the electrode length (L^*_1) in electrostatic precipitators' ionization chamber the maximum values of the effective ionization area increases by 2,5 times (Fig. 2). Through the research one of the important regularities of electrostatic units was revealed; it was determined that the increase of effective ionization area F*e of electrostatic precipitators is connected, firstly, with the ionization chamber's fringe effects, which move closer at its deminishing, which results in the considerable increase of F*e (by several times), and secondly, with the chamber's fields superposition.

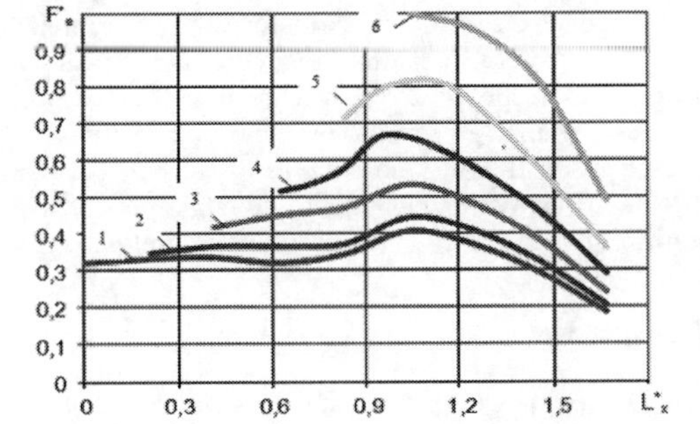

Fig. 2. Functional dependences $F^*_e = f(L^*_x)$ with account of the alteration L^*_1 (δ = const):1 – $L^*_1 = 1,625$; 2 – $L^*_1 = 1,458$; 3 – $L^*_1 = 1,25$; 4 – $L^*_1 = 1,042$; 5 – $L^*_1 = 0,833$; 6 – $L^*_1 = 0,625$.

To take into account the important features of electrostatic units the computer modeling methods have been developed, which allows determining their rational parameters with the possibility of creating the maximum effective ionization area and with account of air-flow rate in the interelectrode systems of these appliances. With account of the tendency of the consumed energy resources' continuous cost increase, the quasioptimal dimensions of these appliances allow improving their efficiency considerably.

At the designing stage of the single-chamber ionizing air cleaners' interelectrode systems it's necessary to take into account the possibility of improving their efficiency by means of selecting the appropriate ratios of their chambers' dimensions and with account of the corona-forming electrode location. For example, for ionizing air cleaners with the air consumption in the form of «ionic wind» at the chamber size alteration from $L^*_1 = 1,625$ to $L^*_1 = 0,625$ the maximum value $F^*_{e(max)}$ increases by approximately 3 times (Fig. 3).

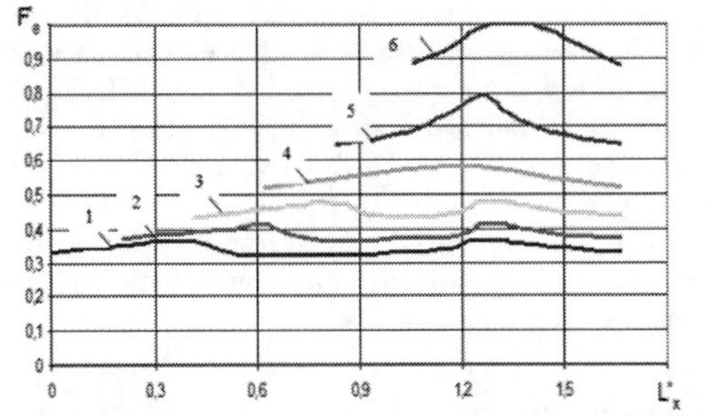

Fig. 3. Functional dependences $F^*_e = f(L^*_x)$ with account of the alteration L^*_1:1 – $L^*_1 = 1,625$; 2 – $L^*_1 = 1,458$; 3 – $L^*_1 = 1,25$; 4 – $L^*_1 = 1,042$; 5 – $L^*_1 = 0,833$; 6 – $L^*_1 = 0,625$.

The computational experiments with multichamber ionizing air cleaners (for example, «Super-Plus-Turbo» of the first modification) have shown that selecting a needle electrode in the backward semisphere of this unit has allowed obtaining better results in comparison with other types of electrodes. The example case of calculation data of a plane-parallel electrostatic field in multichamber ionizing air cleaners is shown in Fig. 4.

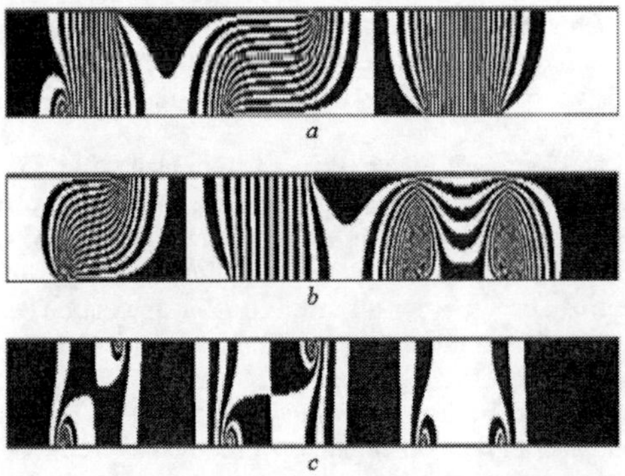

Fig. 4. The standard field calculation data for multichamber ionizing air cleaners: *a* – distribution of the electrostatic field's equipotential surfaces with the given pitch $\Delta\varphi$;*b* - distribution of force surfaces of the field intensity flux with the given pitch $\Delta\chi$;*c* - distribution of electrostatic field's equal intensities surfaces with the given pitch ΔE.

As follows from analyzing the existing models of single-chamber ionizing air cleaners, the corona-forming electrode is placed with regard to grounded plate electrodes. As the calculation data show the design of these appliances is not too optimal and efficient, in contrast with the two-stage electrostatic precipitators, for example, EF, EFO, EF/SP, EFO/SP (of companies «PlymoVent» (Sweden) and «SovPlim», St.Petersburg), or electric precipitators of EFVA type (LLC «Elstat», Moscow), which are developing in the right direction. It was found that the development of multichamber ionizing air cleaners' models (e.g., «Super-Plus-Turbo» of the 2nd modification) on the basis of empirical approach resulted in significant deterioration of ionizing air cleaners' properties as compared to the initial model due to installing saw-shaped corona-forming electrodes in the forward semisphere instead of wire electrodes and excluding the needle electrode, as the field intensity in the backward semisphere of the unit decreased by several times as compared to the initial model.

It was established that the mathematical model of a two-stage electrostatic precipitator, as a control object, can be presented as an aperiodic link, while the time constant of the transition process, dealing with the alteration of positive air ions concentration (N_+^*), is within the scope of tens of seconds.

3 Practical research

The xperimental research of multichamber ionizing air cleaners with account of the alteration of the type of corona-forming electrodes and their location has shown that that the relative specific concentration of negative air ions (N_-^*)can change by 4 times (air ions meters of 2 types - «Sapfir-3k» and «MAS-01» were used). These results qualitatively coincide with the numerical computation data of the electrostatic field intensity alteration of multichamber ionizing air cleaners for various types of electrodes. It was established that with the increase of distance from air ions sources (ionizing air cleaners), firstly, the average relative specific concentration of both positive and negative air ions at the exit of these units is decreased, and secondly, the average relative velocity of the «ionic wind» is also decreased.

The comparative analysis of the calculation data, of the known and obtained experimental data for the objects under research and other types of electrostatic units has shown that the made assumptions for the calculation of electrostatic units for air filtration and ionization on the basis of second-order elliptical equations with the constant coefficients with account of mixed-type boundary conditions are substantiated. The qualitative and quantitative coincidences of mathematical modeling results of the objects under research with the experimental data and the findings of other authors confirm the validity of the obtained results.

The peculiarities of the general algorithm of numerical study for electrostatic units for air filtration and ionization are presented. Taking into account that in practice it's not always possible to implement the optimality criterion for electrostatic units, so the task of quasioptimal units structure synthesis consists in providing the maximum effective ionization area F*e(m) with account of alteration of both the electrodes locations and distances between them, the ratios of chambers' sizes and gaps between them, and the potentials on the electrodes of the researched object at the specified ratios between them. In this regard the problem of determining rational parameters of electrostatic units is also considered, which can be established with the use of the dependences, obtained in the course of work. At this, according to the developed computer modeling methodology, namely, with account of its second stage, the acceptable rational parameters of ionization chamber are established.

As a result of the carried-out research there were presented the peculiarities of structures and circuit designs of automated ventilation systems on base of energy-efficient equipment with the use of two-stage electrostatic precipitator and autonomous ionizing air cleaners with the possibility of ionizing air in the premises of buildings and constructions, with the automated ventilation system being part of automated dispatch control system (ADCS) of the distributed electric power systems of buildings.

4 Conclusions

It was established that the mathematical model of a two-stage electrostatic precipitator, as a control object, can be presented as an aperiodic link, while the time constant of the transition process, dealing with the alteration of positive air ions concentration (N_+^*), is within the scope of tens of seconds [6].

As a result of the carried-out research the peculiarities of structures and circuit designs of automated ventilation systems on base of energy-efficient equipment with the use of two-stage electrostatic precipitator and autonomous ionizing air cleaners are presented.

From analyzing the numerical modeling results of two-stage electrostatic precipitators it was established that the effective ionization area (F_e^*) can be increased within the scope of 25% by changing the location of corona-forming electrode and the gap between the chambers, and with the additional account of fringe effects of ionization chamber the maximum effective ionization area ($F^*_{e(max)}$) increases by 2,5 times.

References

1. O. Seppanen, *The energy-efficient ventilation systems for providing the high-quality indoor climate*, V., h., air cond., h. sup. and con. th. ph. (AVOK). **5**, 23-28 (2000)
2. S.A. Chistovich, V.K. Averyanov, Y.Y. Tempel, S.I. Bykov, *Automated heating and heat supply systems* (Stroyizdat, Leningrad, 1987)
3. V.M. Paskonov, V.I. Polezhaev, L.A. Chudov, *Numerical modeling of heat and mass exchange processes* (Nauka, 1984)
4. I.P. Vereschagin, V.E.Bobrikov, *Parameter selection at the electric field computation by the*

equivalent charge method, **1**, 38-44 (Electricity, 1988)

5. Y.A. Gol`tsov, A.S. Kizhuk, V.G.Rubanov, Int. J. Pharm. Techn. **12,** (2016)
6. A.N. Potapenko, A.V. Belousov, L.I. Koltunov, E.A. Potapenko, News of H. Ed. Inst. Problems of power industry, **5-6**, 69-84 (2006)

DOI: 10.1051/matecconf/201712903015

Analysis of quality characteristics in the multicriteria synthesis problem of the production schedule

Olga Chengar[1,*], *Elena* Mashchenko[1], *Darya* Zamorenova[2], and *Andrey* Phedosov[3]

[1]Sevastopol State University, Department of Information Technologies and Computer Systems, 299053, University Str., Sevastopol, Russian Federation
[2]Sevastopol State University, Department of Instrument Systems and Automation of Technological Processes, 299053, University Str., Sevastopol, Russian Federation
[3]Federal state owned military educational institution of the Higher professional Education “the black sea higher navy order of the red star academy by P.S.Nakhimov”, 299028, Dybenko Str., Sevastopol, Russian Federation

Abstract. The article considers the possibility of applying the Pareto optimization method in combination with the ant algorithm for solving the task of scheduling the loading of process equipment in a flexible production system. Quality characteristics of the obtained production schedule of the machining section are analyzed depending on the choice of various combinations of optimization criteria. As a result of the experiments, carried out in the test case, it was found that with increasing number of criteria, the number of iterations required for a sufficiently accurate description of the Pareto boundary increases, which entails an increase in the algorithm running time, but the quality characteristics do not decrease significantly.

1 Introduction

There is currently considerable experience in addressing the challenges of building a discrete type production schedule, which is represented in the works of both domestic and foreign experts [1- 5]. However, with the increase in the degree of automation of industrial enterprises, there is a growing need for increasingly powerful intelligent systems for short-term production planning, especially for in-house planning to compile a detailed schedule for loading technological equipment in a real production environment. Note that the use of known classical methods that explicitly or implicitly assume various simplifications in the formulation of the planning problem often does not yield acceptable results. Effective means for solving such problems is to use the methods of «Natural computing », which allow to receive sub-optimal solutions to problems of real work situations in a short time [1, 3]. Ant algorithms (AA) are the prominent representative of «Natural computing » and they are based on the behavior of an ant colony as a population of potential solutions and adapted to solve combinatorial optimization problems, first of all, search for different paths on graphs [2, 4.6]. Thus, it can be argued that AA have shown themselves to solve combinatorial optimization problems, one of which is the task of scheduling production equipment for flexible production systems (FPS).

Evaluating the effectiveness of the resulting production schedules can be made by one of the criteria:

- minimizing the violation as the deadlines for orders;
- maximizing the average load factor of flexible manufacturing modules (FMM);
- minimization of the cycle time manufacturing parts;
- minimizing the time changeovers;
- uniform loading equipment;
- minimization of not completed production;
- minimization of the number of using FMM.

However, it does not create the best in a real production environment, but rational plan of production. In connection with this approach is valid, providing the best using of not one efficiency criterion, and using the groups of criteria [7]. Priorities can change quickly based on current production needs. So, in practice, in a real production environment need to create a schedule of works, based simultaneously on a number of criteria, and thus the resulting timetable should be sub-optimal with respect to the selected criteria simultaneously. So the task of multi-criteria optimization becomes urgent in the production schedule, based on the method of ant colonies, the difficulty of working with a prospective nature of optimality conditions for a variety of criteria.

2 Pareto optimization on the basis of the "direction" of ant algorithms with adaptive weights

In [7,8], it is proposed the Pareto method, which optimizes the production schedules based on the "directional" ant algorithm "to the account several factors affecting the production planning. To ensure a high efficiency of the production area it is proposed to form a schedule FPS download, which is the result of the decision multicriteria based on Pareto dominance concept.

When the multi-criteria optimization is searched more than one way of artificial populations of ants on the

* Corresponding author: OVChengar@sevsu.ru

tops of the graph, and a variety of options for ways to each population, are optimal in the sense of Pareto. Therefore, the central issue is to form the target function. The analysis [9.10] showed the efficiency of the method of the weighted sum, wherein the weight is determined for each criterion, entering into the objective function. This approach is one of the most popular and natural development of classical optimization methods, where the "general" objective function is constructed from separate objective functions in the form of a weighted sum:

$$F(x) = \sum_{i=1}^{k} w_i f_i(x), \tag{1}$$

where weights are

$$w_i \in [0,1] \text{ and } \sum_{i=1}^{k} w_i = 1 \tag{2}$$

The criteria were selected proceeding from the fact that optimization goals did not contradict each other, optimization was carried out in one direction (minimization or maximization), and also based on the requirements of the real production situation. Based on the above, three efficiency criteria were selected, the combinations of which are supposed to be experimentally investigated to find the effective solution of the optimal problem in the sense of Pareto:

– minimization of violation of deadlines for orders;
– minimization of the duration of the cycle of manufacturing parts;
– minimizing the time for equipment changes.

There are different opportunities and the various staging and tasks to optimize the production schedule for this search. In all productions, it is required among all possible ways of the graphic and analytic model to find a solution in which the resulting schedules will be the best.

There are combinations of possibility proposed performance criteria for further experimental studies:

I) minimize the violation of the deadlines for the manufacture of the order with the minimum duration of the cycle of manufacturing parts;

$$\begin{aligned} &Z_1 = \min f_i(x) = Tf_i(x) - Tpl_i(x), (i=1,\ldots,n) \\ &Z_2 = \min f_i(x) = \sum_{i=1}^{n}\left(\sum_{j=1}^{m} To_{ij}(x) + \sum_{j=1}^{m} Tnp_{ij}(x)\right), (i=1,\ldots,n; j=1,\ldots,m) \end{aligned} \tag{3}$$

II) minimize the violation of the deadlines for the manufacture of the order while minimizing the adjustment times of the equipment;

$$\begin{aligned} &Z_1 = \min f_i(x) = Tf_i(x) - Tpl_i(x), (i=1,\ldots,n) \\ &Z_2 = \min f_i(x) = \sum_{i=1}^{n}\sum_{j=1}^{m} Tn_{ij}(x), (i=1,\ldots,n; j=1,\ldots,m) \end{aligned} \tag{4}$$

III) minimize the length of the manufacturing of parts while minimizing the adjustment times of equipment;

$$\begin{aligned} &Z_1 = \min f_i(x) = \sum_{i=1}^{n}\left(\sum_{j=1}^{m} To_{ij}(x) + \sum_{j=1}^{m} Tnp_{ij}(x)\right), (i=1,\ldots,n; j=1,\ldots,m) \\ &Z_2 = \min f_i(x) = \sum_{i=1}^{n}\sum_{j=1}^{m} Tn_{ij}(x), (i=1,\ldots,n; j=1,\ldots,m) \end{aligned} \tag{5}$$

Next in the series we will examine these options for multi-criteria optimization for load planning task m in FMM machining manufacturing site for production of n batches of parts x, simultaneously on two performance criteria proposed in [7, 8].

It was experimentally found that during the production program for a month, it is consisting of 3 parts in a total amount of about 200 pcs., The maximum number of iterations necessary to obtain a Pareto suboptimal solutions of $n_t = 100$.

In Table 1, the experimentally obtained values of the parameters of the "directed" ant algorithm for the multicriterial optimization problem for the selected two targets Z1 and Z2 are shown simultaneously for the above selection of efficiency criteria.

The convergence of the algorithm depends largely on the following control parameters [3,6]:

– the maximum number of iterations n t;
– value of the initial concentration of pheromone τ 0 at the transition artificial ants vertex graphic and analytic models corresponding to manufacturing operations;
– value of the initial concentration of pheromone in the transition τ 0 "elite" artificial ants vertex graphic and analytic models corresponding storage operations;
– the value of the positive constant α, which determines the effect of pheromone concentration;
– value of the positive constant β, which determines the effect of heuristic information.

Table 1. The control parameters of ant algorithm for finding Pareto-optimal solutions based on two criteria

Options \ Criteria		I	II	III
Denomination	Range of	Value	Value	Value
τ_0 (for all ants)	(0; 1)	0,090	0,150	0,095
τ_0 (for "elite" ants)	(0; 1)	0,380	0,310	0,425
α	(0,3; 0,5)	0,39	0,34	0,41
β	(0,5; 0,8)	0,67	0,73	0,71

Figure 1 is a graphical implementation of the results obtained by the first variant by selection criteria. For greater clarity, provided the task is displayed in the space of criteria in order of approximation of solutions to the Pareto-optimal.

The points are marked by the principle for all the points dominated the points obtained during the algorithm. Straight noted successive approximation line to the adaptive non-dominated solutions from the amount of artificial ant populations. The point of the conditional minimum, that is the solution of the problem under consideration, is the lower point of intersection of the admissible region with the adaptive line, built on a given number of iterations.

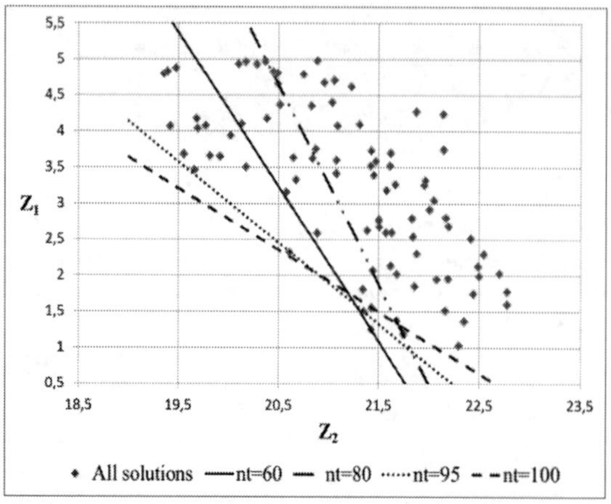

Fig.1. Pareto-optimization of the two-criteria problem: minimization of the violation of the deadlines for manufacturing the order with the minimum duration of the manufacturing cycle of parts.

Similar actions were also made for the second and the third variants of selection criteria and results are summarized in Table 2.

Table 2. Results of the solution of test tasks

Variant of the criterion choice	min duration manufacturing parts, h.	min time adjustment of FMM, h.	min violation of deadlines, h.
I	20,4	-	1,25
II	-	5,04	1,45
III	19,1	6,75	-

The results of the solution of the proposed test tasks by the two-criteria "directional" ant algorithm make it possible to conclude that the modification of the method of ant colonies and the concept of Pareto dominance have been carried out-the points are located almost uniformly and over the whole set of criteria, the adaptive straight line moves successively to the unmodified points, thus approaching the Pareto-Optimal solution.

Similar studies were carried out for a problem with three criteria. According to their results, we can conclude that with increasing amounts of a growing number of criteria the population of artificial ants are increasing too, and it is needed for sufficiently accurate description of the Pareto frontier, which entails an increase in the operation time of the algorithm. In addition, the effectiveness of the solutions obtained by individual criteria is noticeably reduced, which is far from always justified in the real production environment.

3 Analysis of the characteristics of the quality of production schedules

For experimental studies and validation of the developed models for capacity planning process equipment as an object was selected organizational workflow process area of machining parts, such as bodies of rotation, with buildings in PJC "Azovobschemash" (Mariupol). This production site is designed for machining couplings, washers, disks, nipples of locks, rings with a diameter of up to 400 mm and a length of up to 500 mm (60-70 thousand pieces / year with two shift operation), it has a medium-series production type. On this site, for each product an average of 3 to 5 operations, depending on its type, and up to 12 items can be processed. There is a block diagram of a portion of machining in Fig.2. It is composed of six cargo-lifting equipment, transport module and an automated warehouse.

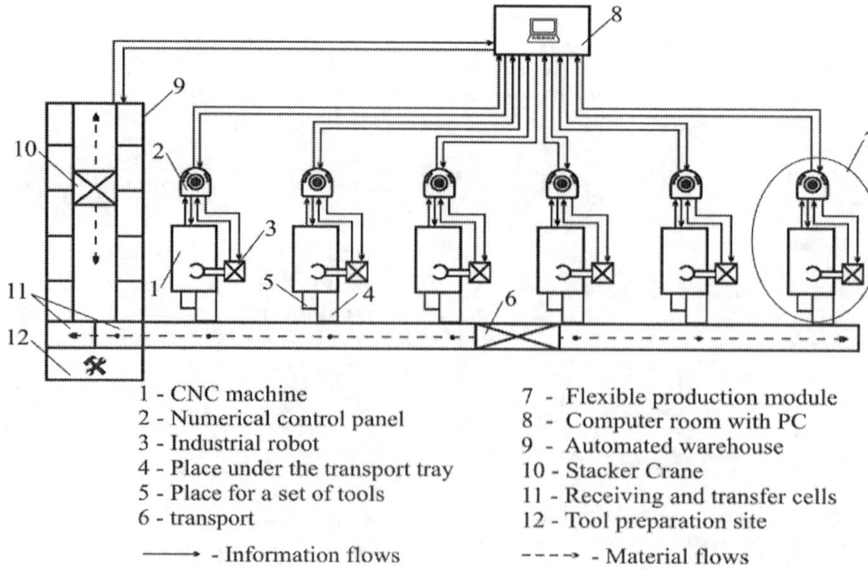

Fig.2. Structural diagram of the production section of machining parts such as bodies of rotation

In the process of functioning of the FPS site machining with time, there is a need to change the nomenclature of manufactured products, production program, etc. Clarification of all transformations of the production site allows to ensure high efficiency of work in the changed operating conditions. With reference to the selected object of research, a series of computational experiments was carried out. To assess the adequacy of the model in production were selected information on the work performed in a few month s in order to compare with the results obtained in E experimentally on the basis of the developed model. At the same time estimated and camping by the following factors:

– observance of works and technological operations in the model of cause-effect relationships;
– obtaining planned indicators of the schedule (loading, downtime and equipment changeover, compliance with the deadlines).

Since the production site under consideration is built on the basis of CNC machines of the same technological purpose, the analysis confirmed the relative uniformity of loading and downtime of the FMM. Losses on readjustment are small enough and make up about 3.5%, and downtime is about 25% of the total simulation time.

The main reasons for downtime are:

– absence of blanks (7%);
– lack of tools (5%);
– waiting for service by the transport module (10%);
– equipment breakdown (2%);
– other reasons (1%).

The above mentioned data show that the automated transport can not cope with the performance of applications coming from FMM. Thus, the main possibility of "sealing" the schedule is to reduce the downtime of the process equipment for the first three reasons mentioned above. Also, analysis of indicators of the functioning of the automated transport it showed that its load factor is close to 1, significantly exceeds the load lifting equipment, as The FPS is serviced by one transport, which moves the blanks and tools in two directions.

Therefore, the transport system has always been the bottleneck of the FPS and the issue of how to unload the transport is acute, thereby increasing the loading factor of the FMM.

With the application of the developed methods has been set up under the proposed model of FPS parameters described above. The evaluation of the results showed the observance of cause-effect relationships in all experiments and a significant improvement in the key parameters of the production schedule.

To analyze the production schedule obtained, the data on the performance of the production program, obtained after the implementation of the proposed model, are summarized in Table 3, which is built on the diagram (Fig.3), clearly showing the results of the experiment.

In the course of the experiment, multicriteria problems with the use of a "directed" ant algorithm with adaptive weights are also considered.

Table 3. Average values of the parameters of the production program before and after modeling

Average value \ Type of equipment	Before simulation		After simulation	
	FMM	Transport	FMM	Transport
The share of adjustment	0,040	0,162	0,028	0,085
The share of idle time	0,283	0,030	0,186	0,010
Coefficient of loading	0,690	1,000	0,801	0,940

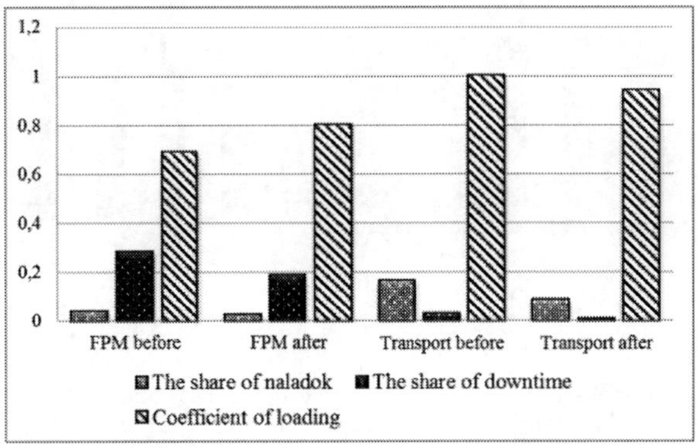

Fig.3. The diagram of the averaged values of parameters of loading, idle times and adjustments of the process equipment before and after modeling

The analysis of the results was carried out in comparison with the solutions obtained in solving the single-objective problem. The following options for choosing the optimality criteria were considered:

I. Maximizing the average load factor of process equipment

II. Minimization of the violation of the deadlines for the manufacture of an order with the minimum duration of the cycle of manufacturing parts

III. Minimization of the violation of the deadlines for the manufacture of the order, while minimizing the adjustment times of equipment

IV. Minimization of the violation of the deadlines for the manufacture of an order with the minimum duration of manufacturing of parts and the time of adjustment of equipment

Studies basic quality characteristics obtained by solving the optimization problem of the production schedule, have shown that the introduction of additional performance criteria, the task of choosing the optimal solution much more complicated, but the effectiveness of methods of an existing member is not reduced. So in Fig. 4 the graphs are showing the difference between the values of the basic obtained quality characteristics with one-criterion and multiobjective optimization.

DOI: 10.1051/matecconf/201712903015

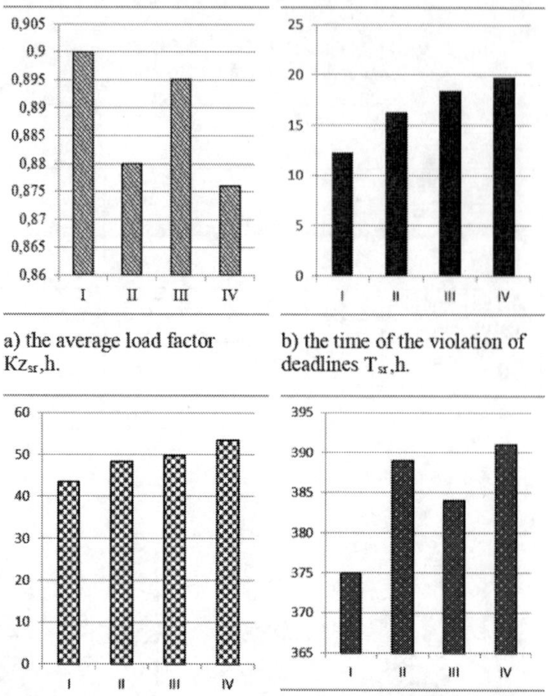

a) the average load factor Kz_{sr},h.

b) the time of the violation of deadlines T_{sr},h.

c) the equipment set-up time Tn,h.

d) the duration of production cycle Tc,h.

Fig. 4. The values of the quality characteristics of the production schedule, depending on the choices of performance criteria

An experimental study of the Pareto optimization method for the production schedule of FPS based on a "directed" ant algorithm with adaptive weights allows us to assert that, according to various criteria of effectiveness, the discrepancy between the results of one-criterion and two-criteria problems is in the range of 1 to 3%, and the one-criterion and three-criteria problems are within 6 ÷ 8,5%, which is acceptable. Also made computer experiments have shown that the effectiveness of the directional ant algorithm increases with the dimension of the problem.

References

1. M. Dorigo Swarm Intelligence, *Ant Algorithms and Ant Colony Optimization*, Reader for CEU Summer University Course «Complex System». P. 1–34. (2001)
2. Y. Skobtsov, D. Speransky, *Evolutionary computation* (The National Open University "INTUIT", 2015)
3. O. Chengar E. Savkova, Journal of East Ukrainian National University **13 (167)**, P. 239-245 (2011)
4. O. Chengar, Journal of SFU. Technical Science, **4 (141)**, P. 185-193 (2013)
5. Y. Skobtsov, A. Sekirin, S. Zemlyanskaya, O. Chengar, V. Skobtsov, S.Potryasaev, Advances in Intelligent Systems and Computing, vol.**466**, P.453-462 (2016)
6. O. Chengar, Bulletin of Kherson National Technical University, **1 (46),** P. 212-217 (2013)
7. Y. Skobtsov, O. Chengar, V. Skobtsov, A.Pavlov. Advances in Intelligent Systems and Computing, vol.**573**, P.456-465 (2017)
8. O. Chengar, Problems of Information Technology. Bulletin of Kherson National Technical University, **1 (15)**, P. 211-216 (2014)
9. E. Zitzler, K. Deb, L. Thiele, Comparison of Multiobjective Evolutionary Algorithms, Vol. **8(2)**, P. 173-195 (2000)
10. K. Deb, *Multi-objective optimization using evolutionary algorithms* (UK: Wiley. 2001)

MATEC Web of Conferences **129**, 03016 (2017) DOI: 10.1051/matecconf/201712903016
ICMTMTE 2017

Component simulation in problems of calculated model formation of automatic machine mechanisms

Igor Telegin[1,*], *Alexander* Kozlov[1] and *Alexander* Zhirkov[2]

[1]Lipetsk State Technical University, 398600, 30, Moskovskaya St., Lipetsk, Russia
[2]Bryansk State Technical University, 241035, 7, Bulvar 50-letiya Oktyabrya, Bryansk, Russia

Abstract. The paper deals with the problems of the component simulation method application in the problems of the automation of the mechanical system model formation with the further possibility of their CAD-realization. The purpose of the investigations mentioned consists in the automation of the CAD-model formation of high-speed mechanisms in automatic machines and in the analysis of dynamic processes occurred in their units taking into account their elasto-inertial properties, power dissipation, gaps in kinematic pairs, friction forces, design and technological loads. As an example in the paper there are considered a formalization of stages in the computer model formation of the cutting mechanism in cold stamping automatic machine AV1818 and methods of for the computation of their parameters on the basis of its solid-state model.

One of the basic problems in the field of production and modernization of automatic machines, in particular, multi-position cold pressing automatic machines (CAM) or, a more general name of press machines is the increase of reliability, productivity and durability.

The analysis of reasons in automatic machine failures shows that in the overwhelming majority of cases in the ground are dynamic (oscillating) processes the intensity of which increases sharply with the increase of speed modes in equipment operation [1 – 9]. When operating the pressing machines with the productivity from fifty up to some hundred products per minute the most substantial influence upon such parameters as reliability and durability is carried out by dynamic processes developing in auxiliary systems: of a blank transfer between technological positions, cutting and others to the accurate positioning of their executive units are made particularly strict requirements [10 – 13].

Current methods of the dynamic phenomena account at designing mechanisms consist in the substitution of the latter by some abstract designs – models. But, in order that the model reflected really dynamic processes in a real mechanism it is necessary that the structure itself and a mathematical description should be developed theoretically correct and substantiated and quantitative characteristics of its parameters should be defined with the sufficient degree of accuracy.

The investigations of dynamic processes in mechanisms on the basis of models allowed taking into account elasto-inertial characteristics of their units give a possibility already at the first stage of designing to determine not only loads in these units and their kinematic parameters, but to carry out numerical experiments for the determination a maximum productivity and to determine efficient speed modes for a mechanism operation in particular and an automatic machine as a whole [13 – 15].

The investigation of the mechanic system dynamics is:

• the formation of the successive approximation file: real mechanism – dynamic model – mathematical model – simulator numerical realization – computer model;

• the development of the method for a quantitative and qualitative definition of the level of dynamic processes occurred in the mechanism and affecting its reliability and working capacity under the influence of outer factors (operating speed, technological forces) and changes of inner parameters (kinematic pair wear, structure loads, adjustment parameters).

The described above procedure of a dynamic mechanism analysis is known [1, 4, 11]. But, in spite of the constant improvement connected with the mass (complexity) number increase, with the account of gaps in kinematic connections, with the specification of elastic tie characteristics, with dissipation parameters, friction forces, external loads it did not gain a wide dissemination. Basic reasons of this were as follows:

• low degree of accuracy and a large labor intensity in the computation of model parameters, in the first place, elastic-inertial properties;

• large labor intensity of works connected with the development of a dynamic model of a specific mechanism, its mathematic description and a program of realization;

• impossibility of the assessment of authenticity of results obtained without experimental investigation fulfillment [13 – 15].

* Corresponding author: vv.telegin@yandex.ru

DOI: 10.1051/matecconf/201712903016

The purpose in the development of the procedure of dynamic processes analysis in mechanisms consists in the decrease of labor intensity in dynamic investigations of mechanisms and increase of their credibility. At the heart of it there are methods of component modeling and object-oriented techniques [2, 7, 16], and also the methods of solid-state modeling [14, 15].

The mechanism can be considered as a totality of some elements connected between each other by different ties. By means of these ties the mechanism elements interact with each other. With the term ‘element’ we will identify a separately taken part of the mechanism, its certain fragment or, vice versa, some parts simultaneously. This definition of an element corresponds to the ‘object’ concept – one of basic in the theory of a component modeling of complex dynamic systems and object-oriented techniques [2, 13 – 15, 16].

For the visual presentation of the information on mechanisms there are used diagrams (schematic circuits, kinematics, and others) 2D and 3D drawings and also dynamic models. To the point all these mechanism presentations (not only the latest ones) are at the same time their models of different complexity and purpose. Evidently, the object can contain both a single part or even its element and some parts. Any processes in such an object can be modeled in any way and, in accordance with the principles of encapsulation and polymorphism, at the given stage their nature and mathematical-, program-, or physical description are not under consideration.

Such an approach is of interest both at the solution of problems connected with the development of dynamic and mathematical models of mechanisms and, what is particularly significant, their program realizations for the further computer processing [2, 13, 16].

Let us consider a stepwise process in the formation of the object presentation of a cutting mechanism in cold pressing automatic machine AB1818 and its mathematical description. At each stage of this process is carried out an addition of a new unit to the mechanism, to its object presentation – new object, and to the mathematical description – response of the unit joined.

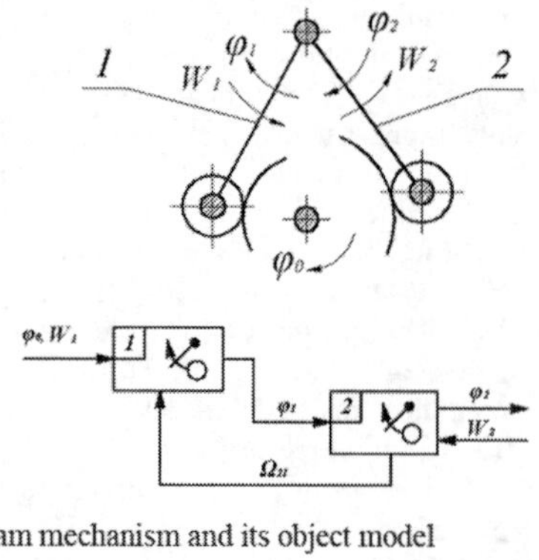

Fig. 1. Cam mechanism and its object model

The formation of mathematical models of the equations of motion (1) – (9) is shown in Figures 1, 2 and 3 and carried out in the assumption that each object is modeled by a single-mass oscillating system described by a differential equation of the second order. Certainly, a real model and its mathematical description may be of any complexity.

Mathematical model (Fig. 1):

$$\begin{cases} J_1\ddot{\varphi}_1 = -c_1(\varphi_1 - U_1) - b_1(\dot{\varphi}_1 - \dot{U}_1) + W_1 + \Omega_{21} \\ J_2\ddot{\varphi}_2 = -c_3(\varphi_2 - \varphi_1) - b_3(\dot{\varphi}_2 - \dot{\varphi}_1) + W_2 \end{cases} \quad (1)$$

Where external load:

$$W_1 = -M_{mp1} - M_{zam}, \qquad W_2 = -M_{mp2} + M_{zam} - c_2(\varphi_2 - U_2) - b_2(\dot{\varphi}_2 - \dot{U}_2). \quad (2)$$

Disturbance:

$$\Omega_{21} = c_3(\varphi_2 - \varphi_1) + b_3(\dot{\varphi}_2 - \dot{\varphi}_1). \quad (3)$$

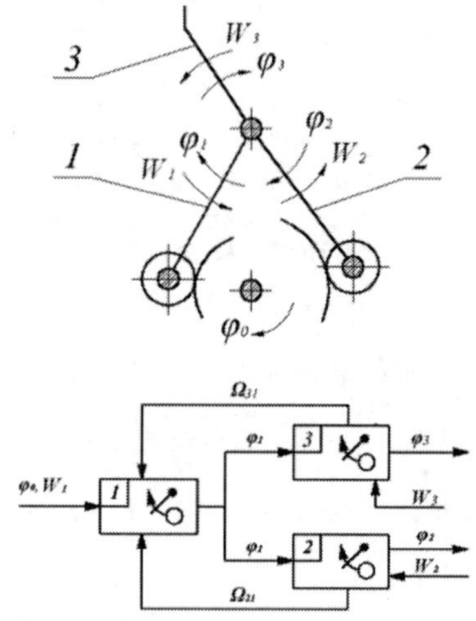

Fig. 2. Cutting mechanism without a knife rod and its object model

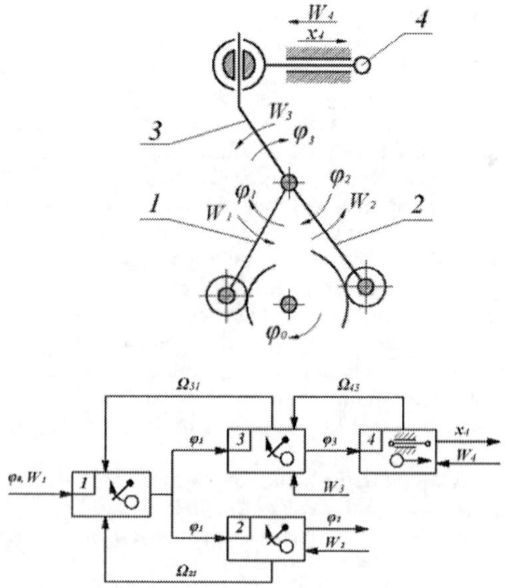

Fig. 3. Mechanism of cutting and its object model

Mathematical model (Fig. 2):

$$\begin{cases} J_1\ddot{\varphi}_1 = -c_1(\varphi_1 - U_1) - b_1(\dot{\varphi}_1 - \dot{U}_1) + W_1 + \Omega_{21} + \Omega_{31} \\ J_2\ddot{\varphi}_2 = -c_{12}(\varphi_2 - \varphi_1) - b_{12}(\dot{\varphi}_2 - \dot{\varphi}_1) + W_2 \\ J_3\ddot{\varphi}_3 = -c_3(\varphi_3 - \varphi_1) - b_3(\dot{\varphi}_3 - \dot{\varphi}_1) + W_3 \end{cases} \quad (4)$$

Where external load:

$$W_3 = -M_{mp3}\,. \quad (5)$$

Disturbance:

$$\Omega_{31} = (c_3(\varphi_3 - U_3) + b_3(\dot{\varphi}_3 - \dot{U}_3))U_3'\,. \quad (6)$$

Mathematical model (Fig. 3):

$$\begin{cases} J_1\ddot{\varphi}_1 = -c_1(\varphi_1 - U_1) - b_1(\dot{\varphi}_1 - \dot{U}_1) + W_1 + \Omega_{21} + \Omega_{31} \\ J_2\ddot{\varphi}_2 = -c_{12}(\varphi_2 - \varphi_1) - b_{12}(\dot{\varphi}_2 - \dot{\varphi}_1) + W_2 \\ J_3\ddot{\varphi}_3 = -c_3(\varphi_3 - \varphi_1) - b_3(\dot{\varphi}_3 - \dot{\varphi}_1) + W_3 + \Omega_{43} \\ m_4\ddot{x}_4 = -c_4(x_4 - U_4) - b_4(\dot{x}_4 - \dot{U}_4) + W_4 \end{cases} \quad (7)$$

External load:

$$W_4 = -P_{mp4} - P_{mex}\,, \quad (8)$$

Disturbance:

$$\Omega_{43} = (c_4(x_4 - U_4) + b_4(\dot{x}_4 - \dot{U}_4))U_4'\,. \quad (9)$$

In Figures 1 – 3 through positions 1 and 2 there are designated levers of a forward and reverse stroke of the cutting mechanism (Fig. 4), 3 – lever connecting levers 1 and 2 with knife rod 4. φ_0 – rotation angle of cams rotating at constant speed φ_1, φ_2, φ_3, x_4 – rotation angles of levers and a linear displacement of a knife rod, W1 … W4 – external loads (2, 5, 8) affecting units of the mechanism (friction forces, technological force, spring pressing force (M_{zam})) with rigidity, c_{12} connecting levers 1 and 2.

In equations (1), (4), (7) J_1, J_2, J_3 and m4 – moments of inertia and a mass of corresponding units, c_1 … c_4 – rigidity, b_1 … b_4 – parameters of dissipation, U_1 … U_4 – functions of the position of mechanism units. Impacts upon a current object of other objects (3), (6), (9) connected with it are designated as Ω_{21}, Ω_{31}, Ω_{43}.

Fig. 4. Cutting mechanism KhVA AV1918 (3D-model).

It should be particularly emphasized that a set of objects presenting a mechanism (object presentation) and a mechanism itself (Fig. 4) are adequate each other. The substitution of reality with a model is carried out only at the level of an object. In such a way, the accuracy of dynamic investigation results is defined with 'goodness' of those objects used in these investigations.

And there is one more very significant circumstance. A possibility to create own user's objects out of already available 'reliable' objects having no care regarding the matter of adequacy [13].

The solution of problems connected with the formation of object-oriented models, with the computation of its parameters and the fulfillment of dynamic investigations is automated. The complex of corresponding programs is combined into a united system – ***dam*** (dynamic analysis of mechanisms). The present application gives wide potentialities connected with the control of input parameters of the mechanism, a computation process, the results of computations and their visualization [13 – 15].

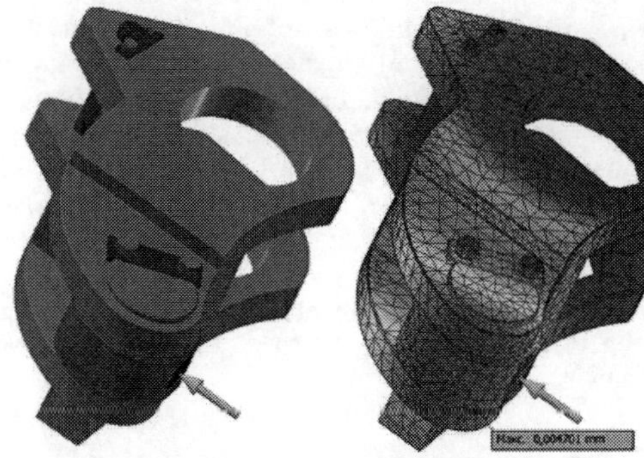

Fig. 5. Rigidity computation for the direct motion lever

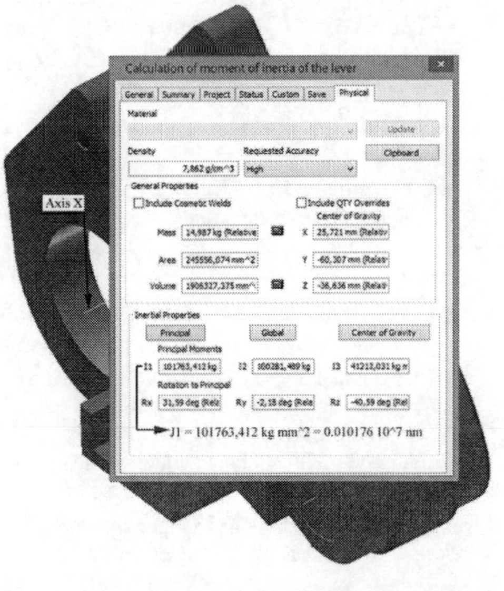

Fig. 6. Computation of the inertia moment for the direct motion lever

Current methods of 3D-modeling including modules of the finite-measuring analysis allow calculating with sufficiently high accuracy elastic-inertial characteristics (rigidity, inertia moments and mass) constituting mechanism models [14, 15]. To the number of programs allowing the solution of similar problems belongs Autodesk Inventor Professional in the environment of which there is created a solid-state model of the mechanism shown in Figure 4.

In Figure 5 there is shown a procedure for the computation of rigidity for the lever of direct motion. The deformable elements of the lever are loaded with force of 1000 N, a radius of reference point rotation is equal to 200 mm (160 – lever + 40 – roller). The reference point displacement under the influence of moment of force equal to 200 N·m – 0.004701 mm or $0.23505 \cdot 10^{-4}$ rad. Lever rigidity – $0.859 \cdot 10^{7}$ N·m/rad.

In Figure 6 there is shown a procedure for the definition of unit inertia characteristics, in this case, a moment of inertia regarding the axis of coinciding with its axis of rotation.

The combined application of the solid-state modeling systems with the module of a finite element method (FEM) and the system of a dynamic analysis developed on the basis of the material of this paper allows carrying out dynamic investigations simultaneously with static computations of strength and works connected with the effectiveness increase of mechanisms under development.

The results were obtained within of the state assignment of the Ministry of Education and Science of Russia (project No. 11.9505.2017/8.9).

References

1. I.I. Wolfson, *Oscillations of Machines with Mechanisms of Cyclic Action* (1990)
2. J. Rumbaugh, *Object-Oriented Modeling and Design* (1991)
3. N.I. Levitsky, *Machine and Mechanism Theory* (1979)
4. V.L. Veits, M.Z. Kolovsky, A.E. Kochura, *Dynamics of controlled machine aggregates* (1984)
5. *Vibrations in Engineering: Reference Book*, **3**, (1980)
6. I.A. Krivosheev, D.A. Akhmetzyanov, *Modeling dynamic processes in a complex system* (2003)
7. V.M. Ponyatsky, E.N. Davydov, V.G. Fedorischeva, A.V. Osadchiy, IV Kislinsky, *General and applied mechanics,* Bulletin of the Nizhny Novgorod University. N.I. Lobachevsky, **4 (2)**, 283–284, (2011)
8. A.M. Kozlov, E.V. Kiryushchenko, *Bulletin of Ufa State Aircraft Technical University,* **4,** 117-123, (2012)
9. A.M. Kozlov, A.A. Kozlov, *Russian Engineering Research*, **29**(7), 743-746 (2009).
10. *Forging and stamping: Reference Book*, **3**, (1987)
11. Yu.A. Miropolsky, *Cold Volumetric Stamping on Automatic Machines* (2001)
12. *Vibrations in Engineering: Reference Book*, **5**, (1981)
13. V.V. Telegin, *Proceeding of Samara Scientific Center of the RAS*, **12**, **4(3)**, 623–628 (2010)
14. V.V. Telegin, *Proceedings of Samara Scientific Center of the RAS,* **14**, **4(5)**, 1306-1309, (2012)
15. V.V. Telegin, *Proceedings of TulSU. Serie. Mechanics of Deformable Solid and Metal Forming*, **1**, 197–206, (2004)
16. E.S. Benkovsky, Yu.B. Kolesov, Yu.B. Senichenkov, *Dynamic System Practical Modeling* (2002)

DOI: 10.1051/matecconf/201712903017

Numerical solution of the model problem of CCRF-discharge at atmospheric pressure

Viktor Zheltukhin[1], and *Violetta* Chebakova[2,*]

[1]Kazan National Research Technological University, 68 Karl Marx Street, 420015, Kazan, Russian Federation
[2]Kazan Federal University, 18 Kremlin street, 420008, Kazan, Russian Federation

Abstract. This work describes a 1D mathematical model of capacitive coupled RF discharge between symmetrical electrodes in argon at atmospheric pressure in a local approximation. Electrons, atomic and molecular ions, metastable atoms and argon dimmers as well as ground-state atoms are considered. A simplified diagram of argon excited states when two metastable and two resonance states are replaced with the uniform level. Such diagram is frequently used to simulate argon plasma due to efficient mixing of these layers at electron impacts. Velocity factors of electron impact processes were calculated using Boltzmann equation with a glance to electron-electron collisions. This work describes numerical algorithm of mathematical model implementation, which is based on finite-dimensional approximation of the problem using difference schemes together with iteration process. The software was developed to implement iterative processes using MatLab. Characteristics of atmospheric pressure capacitive coupled RF discharge at interelectrod distance 20 mm are calculated.

1 Introduction

Low-temperature plasma is widely used for various spheres of science and technologies. Plasma is created by various types of gas discharges. An important role is assigned to high-frequency discharges, and particularly, capacitive coupled radio-frequency (CCRF) discharges. Argon is often used as a plasma gas [1-2].

Experimental and calculation methods, which complement each other, are used to tie internal and external parameters of the discharge and allow solving a lot of problems of physics and chemistry of low-temperature plasmas.

In our earlier works we found that, as a rule, all investigations, which are dedicated to simulation of CCRF-discharge in argon, considered that plasma consists of electrons, atomic ions, metastable atoms and ground state atoms. Dependencies of molecular and atomic ions densities in CCRF argon plasma at atmospheric pressure on gas temperature is researched in the article [3].

It is found that molecular ions prevail at the temperatures near to 500 K, and density of atomic ions increases at the temperatures equal to 1500 K and above. This fact is the reason for including molecular ions and dimmers to kinetic diagram of argon in this study. Moreover, earlier the authors tested different types of boundary conditions as for balance equations of charged particles and for balance equation of gas temperature.[4,5] Self-consistent 1D model of CCRF-discharge in argon at atmospheric pressure is offered because of this investigation.

The authors have developed approximate algorithm of numeric implementation of the model; this algorithm is based on finite-dimensional approximation together with using of iterative method. It should be noted that different iteration methods intended to solve non-linear problems including partial derivatives, were offered earlier (see, for example, [6-10] at al.). But the problem studied herein has a range of peculiar features, and particularly, availability of different time scales of change of the main characteristics of the steady state of CCRF-discharge at low pressure. Moreover, gradients of charged particles density and electric field intensity, electron temperature in near electrode layers at the boundary of computational domain is the specific feature of the problem too.

2 Setting the problem

Here we study CCRF-discharge between two parallel plate electrodes, one of which is grounded and another is connected to CCRF-generator, and distance between electrodes is less then size of the electrodes. Under such conditions electric field is close to potential one and discharge is uniform along electrodes thereby we may use 1D model, which allows considering kinetics of plasma-chemical processes in diffusive-drift performance.

Assessment of time and distance, required for electrons to lose energy obtained from field, shows that to simulate CCRF-discharge at atmospheric pressure, we can use local approximation, under which the parameters of plasma electronic component (diffusion, mobility as well as moderate energy, velocity of plasma and chemical reactions, etc.) depend on local value of the

* Corresponding author: vchebakova@mail.ru

reduced electrical field E/N, where E is electric field intensity, N is density of ground states atoms [11].

The following features of CCRF-discharge is considered in the model: presence of domains without quasi-neutrality, fast time change of applied voltage, as well as availability of processes involving metastable atoms, molecular ions and dimmers. Efficient mix of four lowest electronically excited states allows replacing them with a unified layer [12, 13]. Let us assume that b is the distance between electrodes, the grounded electrode locates at the point x = 0 and the charged one locates at x=b (axis Ox is directed perpendicular to the surface of the electrodes).

Processes, occurring inside of CCRF-discharge at atmospheric pressure, can be described by the following initial-boundary value problems and Cauchy problem.

2.1 Convection-diffusion equation for atomic ions

$$dn_+/dt+dG_+/dx=R_1n_eN+R_2n_m^2+R_3n_mn_e-R_4n_en_+$$
$$-R_5n_e^2n_++R_{12}n_en_{2+}-R_{11}n_+N^2 \quad (1)$$
$$+R_{20}Nn_{2+},0<x<b,\ t>0,$$

where n_e, n_+, n_{2+} are densities of electrons, atomic and molecular positive ions, respectively, n_m is density of metastable atoms, $G_+=-D_+dn_+/dx+\mu_+n_+E$ is density of atomic ion flow, μ_+ is mobility of atomic ions, D_+ is diffusion coefficient of atomic ions, $E=-d\varphi/dx$ is intensity of electric field, φ is electric field potential.

Hereinafter R_i,i=1…20, are velocity factors of plasma-chemical reactions [14-17].

Boundary conditions at x=0 shall be chosen as:

$$G_+(0,t)=(-n_+(8kT_a/\pi/m_+)^{1/2}/4+\mu_+n_+E)|_{(0+,t)},\ E<0, \quad (2)$$

and

$$G_+(0,t)=(-n_+(8kT_a/\pi/m_+)^{1/2}/4)|_{(0+,t)},\ E>=0, \quad (3)$$

at $x=b$ – shall be chosen as:

$$G_+(b,t)=(n_+(8kT_a/\pi/m_+)^{1/2}/4+\mu_+n_+E)|_{(b-,t)},\ E>=0, \quad (4)$$

and

$$G_+(b,t)=(n_+(8kT_a/\pi/m_+)^{1/2}/4)|_{(b-,t)},\ E<0, \quad (5)$$

where m_+ is atomic ion mass, k is a Boltzmann constant. We assume that temperatures of atoms, atomic ions and excited atoms are equal T_a.

2.2 Convection-diffusion equation for ionized dimers

$$dn_{2+}/dt+dG_{2+}/dx=R_{10}n_m^2+R_{11}n_+N^2-R_{12}n_en_{2+}-R_{13}n_en_{2+}$$
$$+R_{16}n_{2*}^2+R_{17}n_mn_{2*}-R_{20}Nn_{2+},\ 0<x<b,\ t>0, \quad (6)$$

where n_{2*} is density of argon dimers, μ_{2+} is mobility of molecular ions, D_{2+} is diffusion coefficient of molecular ions, $G_{2+}=\mu_{2+}n_{2+}E-D_{2+}dn_{2+}/dx$ is density of molecular ion flow.

By the same way as above, let us set the following boundary conditions:

Boundary conditions at x=0 shall be chosen as:

$$G_{2+}(0,t)=(-n_{2+}(8kT_a/\pi/m_{2+})^{1/2}/4+\mu_{2+}n_{2+}E)|_{(0+,t)},\ E<0, \quad (7)$$

and

$$G_{2+}(0,t)=(-n_{2+}(8kT_a/\pi/m_{2+})^{1/2}/4)|_{(0+,t)},\ E>=0, \quad (8)$$

at $x=b$ – shall be chosen as:

$$G_{2+}(b,t)=(n_{2+}(8kT_a/\pi/m_{2+})^{1/2}/4+\mu_{2+}n_{2+}E)|_{(b-,t)},\ E>=0, \quad (9)$$

and

$$G_{2+}(b,t)=(n_{2+}(8kT_a/\pi/m_{2+})^{1/2}/4)|_{(b-,t)},\ E<0. \quad (10)$$

Here m_{2+} is mass of atomic ion.

2.3 Convection-diffusion equation for electrons

$$dn_e/dt+dG_e/dx=R_1n_eN+R_2n_m^2+R_3n_mn_e-R_4n_en_+-R_5n_e^2n_+$$
$$+R_{10}n_m^2-R_{13}n_en_{2+}+R_{16}n_m^2+R_{17}n_mn_{2*},$$
$$0<x<b,\ t>0, \quad (11)$$

where $G_e=\mu_en_eE-D_edn_e/dx$ is density of electron flow, μ_e is velocity of electrons, D_e is coefficient of electron diffusion.

Boundary conditions at x=0 shall be chosen as:

$$G_e(0,t)=(-n_e(8kT_e/\pi/m_e)^{1/2}/4-\mu_en_eE)|_{(0+,t)},\ E>=0, \quad (12)$$

and

$$G_e=(-n_e(8kT_e/\pi/m_e)^{1/2}/4-\Upsilon E(\mu_+n_++\mu_{2+}n_{2+}))|_{(0+,t)},\ E<0 \quad (13)$$

at $x=b$ – shall be chosen as:

$$G_e=(n_e(8kT_e/\pi/m_e)^{1/2}/4-\mu_en_eE)|_{(b-,t)},\ E<0, \quad (14)$$

and

$$G_e=(n_e(8kT_e/\pi/m_e)^{1/2}/4-\Upsilon E(\mu_+n_++\mu_{2+}n_{2+}))|_{(b-,t)},E>=0 \quad (15)$$

where Υ is secondary electron emission factor, T_e is temperature of electrons, m_e is mass of an electron.

We assume that at normal conditions gas contains ~10^4 sm^{-3} of charged particles. Let us assume this value as the initial condition for complete density of positive charged particles and electrons.

2.4 Poisson's equation for electric potential

$$-d^2\varphi/dx^2=q_e(n_++n_{2+}-n_e)/\varepsilon_0, \quad (16)$$

including boundary conditions $\varphi=V_a\sin(\omega t)$, on charged electrode at $x=b$ and φ=0, on grounded electrode at x=0. Here q_e is charge of an electrod, ε_0 is electric constant, V_a is voltage peak-to-peak amplitude on charged electrode, ω is radian frequency of electromagnetic field.

 DOI: 10.1051/matecconf/201712903017

2.5 Balance equation for metastable density

$$dn_m/dt\text{-}dG_m/dx=R_6Nn_e+R_{18}n_en_{2*}\text{-}R_2n_m^2\text{-}R_3n_mn_e\text{-}R_7n_m$$

$$\text{-}R_8Nn_m\text{-}R_9n_mn_e\text{-}R_{10}n_m^2\text{-}R_{15}N^2n_m\text{-}R_{17}n_mn_{2*}\text{-}R_{19}n_mN^2 \quad (17)$$

where $G_m=D_mdn_m/dx$ is density of flow of metastable atoms, D_m is diffusion factor of metastable argon atoms.

Boundary conditions for this equation are:

$$G_m(0,t)=(-n_m(8kT_a/\pi/m_a)^{1/2}/4)|_{(0+,t)}, \quad (18)$$

$$G_m(b,t)=(-n_m(8kT_a/\pi/m_a)^{1/2}/4)|_{(b-,t)}, \quad (19)$$

Let us assume that electrode surface is catalytic, when metastable atoms decay completely on it, then $n_m(x,0)=0$.

2.6 Kinetic equation for argon dimer density

$$dn_{2*}/dt=R_{19}n_mN^2+R_{15}N^2n_m\text{-}R_{14}n_{2*}\text{-}R_{16}n_{2*}^2$$

$$\text{-}R_{17}n_mn_{2*}\text{-}R_{18}n_en_{2*}+R_{19}n_mN^2. \quad (20)$$

Let us take zero initial conditions as well as for metastable atoms: $n_{2*}(x,0)=0$.

2.7 Kinetic equation for neutral atoms

$$dN/dt=\text{-}R_1n_eN+R_2n_m^2+R_4n_en_++R_5n_e^2n_+\text{-}R_6Nn_e+R_7n_m$$

$$+2R_{13}n_en_{2+}+R_{14}n_{2*}\text{-}R_{15}N^2n_m+2R_{16}n_{2*}+R_{17}n_mn_{2*}$$

$$+R_{19}n_mN^2+R_8Nn_m+R_9n_mn_e\text{-}R_{11}n_+N^2+R_{12}n_en_{2+} \quad (21)$$

initial conditions based on ideal gas equation:

$$P/(kT_a(x,0))=N(x,0) \quad (22)$$

2.8 Kinetic equation for neutral atoms

Since fluctuations of atomic temperature in scope of average values taken within the period of field change are negligibly small, the balance equation for atomic temperature can be considered as average temperature within the period of variation of electric field, identified with frequency generated by a high-frequency generator. Moreover, we consider that coefficient of energy transfer from ions to atoms equals to 1.

$$d(-\lambda_adT_a/dx)/dx=[j_iE]+[QNn_e], \quad (23)$$

where $j_i=e(G_++G_{2+})$ is ion current, λ_a is heat transfer ration of atom-ion gas, Q is energy obtained by the heavy particles because of elastic collisions with electrons.

Let us assume that electrode is cooled with water. Therefore, the boundary conditions for this equation are the following:

$$-\lambda_adT_a/dx=-\chi(T_a-T_w),\ x=0, \quad (24)$$

$$-\lambda_adT_a/dx=-\chi(T_a-T_w),\ x=b, \quad (25)$$

where χ is complete heat transfer factor, T_w is the temperature of water cooling electrode.

3 Simulation of the discharge

Numerical algorithm of simulating of CCRF-discharge at atmospheric pressure is the same as the ones used for simulation of the same problem at low pressure which is described in the paper [18,19]. This algorithm is based on the method of transferring of non-linearity in equation coefficients to the lower time layer. Non-linear square terms in the right hand of the equations are linearized due to Newton scheme. Solution of equations for gas temperature was realized once per T period using Jacobi's iteration process.

Calculating of charged particle flow is needed for determine of conductivity current. Density of averaged ion flow is also included into the first part of equation for atomic and ion temperature. The problem is that numerical differentiation results significant inaccuracy in calculation of flow density due to high gradients of solution and coefficients of balance equations. So, we used Scharfetter-Gummel's approach to calculate flow density [20]. We determine charged particles densities by using implicit difference scheme, then we determine charged particle flow per the algorithm. This calculations sequence allows us rejecting a restriction on a scheme time step, which is associated with Courant condition.

4 Calculation data

Results of simulation of CCRF discharge at atmospheric pressure, interelectrode distance 2 mm, and frequency f=13.56 MHz, came near results of researches [21, 22] and gave good fit with them. We perform simulation of the problem at atmospheric pressure, interelectrode distance 20 mm, and frequency f=13.56 MHz (Fig.1)

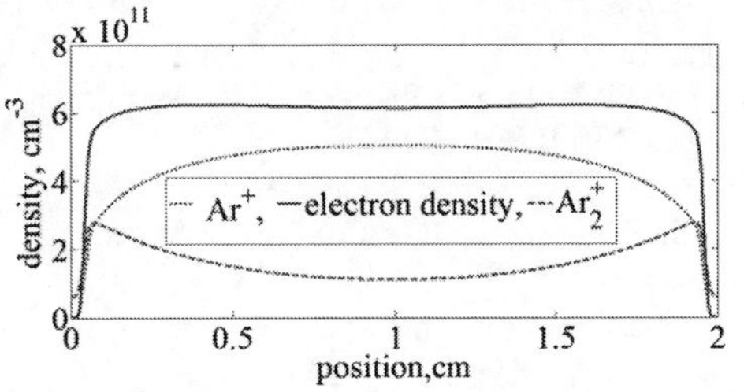

Fig. 1. Distribution of averaged density of charged particles at interelectrode distance of 20 mm.

It is found that maximum of gas temperature about 2000 K is achieved in the middle of the discharge, and the minimum about 930 K is achieved near the electrodes. Difference between maximum and minimum of gas temperature becomes significantly affecting on charged particles distribution.

The density of charged particle distributions in case of interelectrode distance d=20 mm becomes more observable then in case of d=2 mm.

The most significant changes occur in distribution of $Ar2^+$. The density in the middle of the charge is almost constant in case of d=2 mm, but local maximum emerges close to the near-electrode layer when interelectrode distance is in case of d=20 mm, and local minimum is reached in the middle of charge area.

Ratio of minimum to maximum density of molecular positive ion of argon value is reached about 0.43, in other words, density of molecular ions in the middle of discharge gap drops twice. However, quasi-neutrality of plasma in the ambipolar area survives.

Ratio of dimmers to excited atoms also changes in case of d=2 mm, the maximum is equal approximately 0.78, but in case of d=20 mm, maximum value is about 0.38. Note that dimmer density can be considered permanent in case of d=2 mm, but in case of interelectrode distance is 20 mm, the drop of density occurs at the same point, where gas temperature is reached maximum.

The work is performed according to the Russian Government Program of Competitive Growth of Kazan Federal University and work was supported by the Russian Foundation for Basic Research projects nos. 16-11-10299.

References

1. A. Fridman, *Plasma Chemistry* (Camb. Univ. Press, 2008)
2. P. Chebert, N. Braithwaite, *Physics of Radio-Frequency Plasmas* (Camb. Univ. Press, 2011)
3. E. C Martinez, Y. Kabouzi, K. Makasheva, M. Moisan, Phys. Rev. E **70**, 066405(2004)
4. V.Ju. Chebakova, V.S. Zheltukhin, V.T. Dubrovin, Appl. Math. Sci. **10,** 21, 1013-1022 (2016)
5. V.Ju. Chebakova, A.F. Gaisin, V.S. Zheltukhin, IOP Conf. Ser.: Mat. Sci. and Eng., **158,** 012024 (2016)
6. I.B. Badriev, M.V. Makarov, V.N. Paimuhin, IOP Conf. Ser.: Mater. Sci. and Eng. **158,** 012011 (2016)
7. I.B. Badriev, V.V. Banderov, E.E. Lavrentyeva, O.V. Pankratova, IOP Conf. Ser.: Mater. Sci. and Eng., **158**, 012012 (2016)
8. I.B. Badriev, V.V. Banderov, O.V. Pankratova, A.I. Shangaraeva, IOP Conf. Ser.: Mater. Sci. and Eng., **158**, 012013 (2016)
9. I.B. Badriev, M.V. Makarov, V.N. Paimushin, Proc. Eng., **150**, 1050-1055, (2016)
10. I.B. Badriev, M.V. Makarov, V.N. Paimushin, Proc. Eng., **150**, 1056-1062 (2016)
11. A.A. Kudryavtsev, A.S. Smirnov, L.D. Tsendin, *Physics of Glow Discharge*, (Publ. House "Lan", Saint Petersburg, 2010)
12. C.M. Ferreira, J. Loureiro, A. Ricard, J. Appl. Phys. **57,** 82, 82-90 (1985)
13. N.A. Diatko, Yu.Z. Ionokh, A.B. Meshanov, A.P. Napartovich, Physics Reports **31, 10** 939-953 (2005)
14. Xi-Ming Zhu, Yi-Kang Pu, J. Phys. D: Appl. Phys. **43** 015204 (2010)
15. I.L. Epstein, M. Gavrilović, S. Jovircević, N. Konjević, Yu.A. Lebedev, A.V. Tatarinov, Eur. Phys. J. D., **68**,334 (2014)
16. E. Karoulina, Yu.A. Lebedev, J. Phys. D: Appl. Phys., **25** 401-412 (1992)
17. G.J.M. Hagelaar, L.C. Pitchford, Plasma Sources Sci. Techn., **14**, 722–733 (2005)
18. V.T. Dubrovin, V.Ju. Chebakova, V.S. Zheltukhin, Proc. Eng., **150**, 1041–1045 (2016)
19. I.B. Badriev, V.Yu. Chebakova, V.S. Zheltukhin, IOP Conf. Series: J. Phys.: Conf. Ser., **789**, 0120042 (2017)
20. A.A. Kulikovsky. A. More, J. Comp. Phys., **119**, 149–155 (1995)
21. N.P. Balcon, A. Aanesland, G.J.M. Hagelaar, R. Boswell, J.P. Boeuf, 28th ICPIG, 957–960 (2007)
22. R.M. Askhatov, V.Yu. Chebakova,V.S. Zheltukhin, IOP Conf. Ser.: Mat. Sci. Eng., **158**, 012009 (2016)

 DOI: 10.1051/matecconf/201712903018

Computer-aided design system for GTE blades electrochemical machining technology

Maksim Nekhoroshev[1,*], *Nikolay* Pronichev[1], and *Gennadiy* Smirnov[1]

[1]Samara University, 34 Moskovskoye Shosse, Samara, 443086, Russia

Abstract. The paper presents the results of computer-aided design of GTE blade electrochemical machining. Developed a method to use ANSYS to calculate electric fields in the electrochemical cello under ECM. We created an electronic reference base and refined the dissolution model applicably to PECM. In addition, we developed a method for profiling tool electrodes DC ECM and PECM. The calculation module uses parametric relationships with Siemens NX by means of Excel files to produce output TE designs as a 3D computer models. The subsystem developed contains an ECM shaping model, has a modular structure, and is implemented as a wizard.

1 Introduction

When creating a generic methodology for computer-aided design of ECS operations for geometrically complex parts, where computational technology is applied end-to-end, one needs to address a set of design problems, e.g. to design a methodology for calculating the profile of a tool electrode using CAE systems.

Papers [1,2] give a detailed description of methods for computer-aided design of DC ECS operations. As current industries widely employ impulse ECS for geometrically complex parts being made of hard-to-machine materials, the process necessitates integrated research for computer-aided design of such operations with the use of CAD/CAM/CAE/PLM systems. PECS-machined surfaces dissolve in a non-stationary environment. All electrode processes occur within the flow of current and re-occur periodically. During pauses, surfaces stop dissolving while the gradients of ion concentrations in the near-electrode layers decrease; polarization-caused potential steps alter. When modeling the process of electrochemical dissolution, theoretical description of its kinetics is impossible. For practical use, we developed a theoretical-experimental model for calculating PECS shaping; the model should enable sufficiently accurate virtual profiling of tool electrodes (TE).

When developing the model, we assumed the following:

• electrodes were profiled at specified surface points, the coordinates of which were changed sequentially;

• the entire machining process was broken down into short intervals equal to the duration of the pulse package (for better accuracy, cycles can be as short as a single pulse);

• to calculate the removal per cycle for each point, the process flow conditions were assumed to be quasi-stationary, while the interelectrode gap (IEG) was assumed to be constant;

• for the assumed duration of machining cycles, for each point the potential steps were determined from the experiment as the averaged value φ_{eqv}.

• a standard experiment to determine φ_{eqv} was carried out for each machine whereby power supply pulse parameters as well as experiment conditions were known.

• the dependency $j=f(a)$ was read, where j is the current density, a is the interelectrode gap;

• all ECS parameters were consistent with the previously accepted values and remained constant; the gap changed discreetly within the specified limits;

• time to record each current value was minimized to ensure the condition $a=const$ would hold for each point.

2 Materials and methods

When developing methods for automated design of 3D EC shaping operations, we carried out an informational and analytical review of approaches to similar problems.

In paper [3], the authors propose a model that was developed when solving the related problems above, whereby some parameters were either assumed to be constant, or the range within which they could change would be pre-specified based on other parameters and preliminary experimental research of samples, machined correspondingly. The authors modeled the shaping process for the specified tool electrode profile as well as copying the TE profile using a reference blade.

Paper [4] is also about solving the problem of TE profiling based on modeling the processing occurring in IEG. For better accuracy of the model, the surface machined was discretized to improve the accuracy of computer modeling. They have also developed a special shell to simplify the construction of free surfaces.

* Corresponding author: maxnogood@ssau.ru

DOI: 10.1051/matecconf/201712903018

Paper [5] dwells upon special physical and mathematical models used to develop a computer-aided ECS process design system module. The module enables the simulation of processes occurring in an electrochemical IEG.

Paper [6] covers the issue of optimizing the electrochemical machining of hard-to-machine materials. Its authors have developed a mathematical algorithm and obtained mode parameters regression equations. They believe that using this algorithm reduces the time to find the optimal parameters, which is crucial for cost reductions in ECS operations design.

In paper [7], they used 3D ECS modeling with volumetric distribution of parameters in IEG being factored in; however, the calculation process itself would become significantly longer.

In papers on similar topics [8,9,10,11], they solved the problems of modeling the ECS process in geometrically complex IEG, an inter-blade gap of a mono-wheel, to describe the complex electrode trajectory. Simulation-obtained data were used in CAD/CAM systems for CNC machining. In the process of modeling, some parameters were assumed to be constant and would be determined by preliminary experiments. The solution optimized the electrolyte flow rate as well as the shape of the tool electrode.

In papers [12,13], the researchers determined the optimal blade airfoil machining parameters to minimize the labor intensity and maximize the cost-effectiveness of the process. This principle can be used to optimize the ECS process.

In particular, results of researching titan alloys [14,15] were used to define the parameters of machined materials.

Summarizing the above, it is safe to say that ECS modeling, TE profiling, and optimization of machining parameters are topical problems that many authors are striving to solve while using different approaches.

The method we propose herein is based on mathematical models as well as analyzing the spatial evolution of arbitrary surfaces and the physico-chemical processes in IEG and near-electrode layers.

3 Creating a method for computer-aided ECS of GTE blades

In accordance with the standard experiment methodology designed, we carried out research of PECS conditions for NiCr alloys. These materials are widely used for making GTE blades. Fig. 1 presents the results for EI-961 steel and EP-718 alloy. Experiment results were processed using the methods presented in [1,2] and automatically recorded in the Teamcenter database, see Fig. 2.

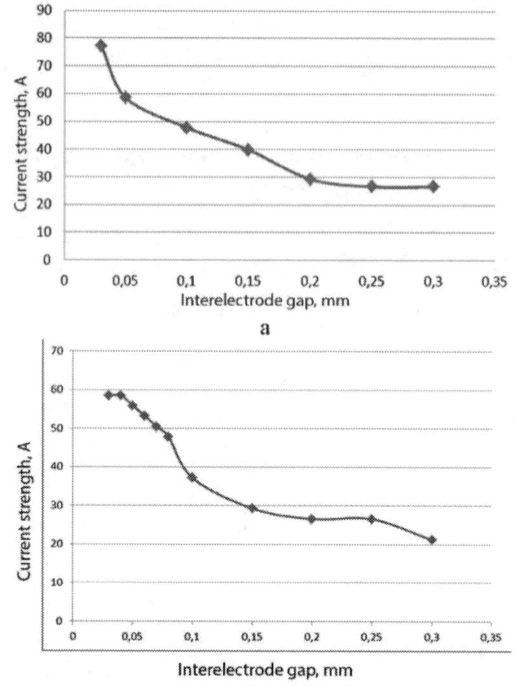

Fig. 1. Dependency of amplitude current on the inter-electrode gap for alloys: a. EI-961, b. EP-718.

To solve the problems of computer-aided design of GTE blade electro-chemical machining technology, we:

1. Developed a method to use ANSYS to calculate electric fields in the electrochemical cell under ECS. The calculations made confirmed the results obtained by analyzing resistance paper and showed that the computer model was consistent with the law of potential (current) distribution in the electrochemical cell. Based on this research, we created a database with material and electrolyte characteristics, dependencies of current output, and polarization curves for DC ECS. These data were presented in a tabular format for subsequent iterative calculations in ANSYS [1,2].
2. We also experimented to make readings of pulse current characteristics in the entire range of working gaps for various materials and electrolytes so as to create databases for computer-aided design. We also created an electronic reference base and refined the dissolution model applicably to PECS.
3. We developed a method for profiling tool electrodes for DC ECS and PECS. This methodology is based on calculating the TE profile when reverse-copied using the CAD model developed. It is essentially about iterative adjustments of TE shape. For ECS TE design, we chose a GTE blade airfoil as a sample [1,2].
4. To facilitate the use of this method for factory technologists, we developed an application that simplified running the algorithm and reduced the modeling process to a wizard that had simplified

DOI: 10.1051/matecconf/201712903018

standard settings. The module profiles tool electrodes by reverse copying for DC ECS and dimensional pulse ECS of GTE compressor blade airfoils.

5. The calculation module uses parametric relationships with Siemens NX by means of Excel files to produce the output TE designs as 3D computer models. For an example, we created a model of tool electrodes for GTE blades.

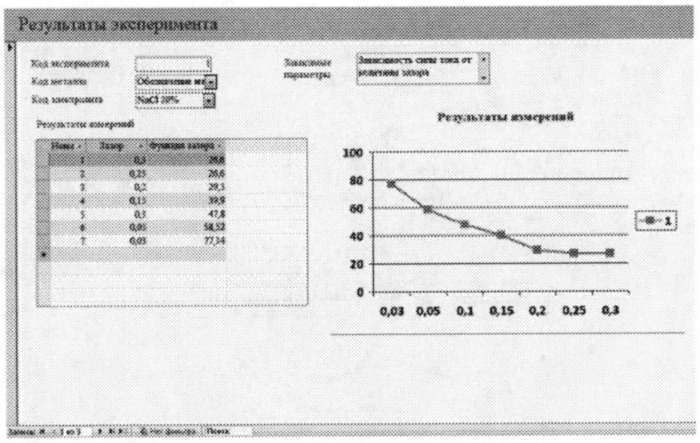

Fig. 2. Adding the experimental data to the database.

To improve the efficiency of using this output, we had to create a Siemens NX information parameters model for CNC shaping of tool electrodes designed based on computer modeling of GTE blade ECS. This model can be implemented in the environment of the calculation module developed earlier that is used to profile tool electrodes for pulse electrochemical machining of GTE blades, whereby the mathematical model developed is applied.

Let us analyze this computer-aided ECS design method as applicable to ECS of GTE blade airfoils, as it is such geometrically complex surfaces that reveal all problems that may arise in design. Below are the detailed steps of how the computer-aided design system was created.

Blade airfoil surfaces models and workpieces to machine were generated in the integrated CAE system. Current density was calculated iteratively in ANSYS with machining parameters being factored in. Based on the distribution of potentials and currents, the profile of the workpiece surfaces machined would change. Calculations were iterated until the first point of the dissolution boundary was on the lower blade airfoil tolerance limit.

The computer-aided ECS design subsystem solves both its primary problem (TE profiling) and the reverse problem of evaluating the geometric parameters of blade airfoils machined with TE of specified profile. When developing such a subsystem, it is important to implement wizard-like modeling, as the entire design process includes a set of interrelated procedures [1,2], which enables 'virtual' electrochemical machining of blades. Essentially, implementing a design wizard means that machining of test parts to refine the process is replaced with 'machining' their geometric models in calculational experiments.

Shaping analysis is crucial for process design because at the modeling stage it enables one to evaluate the possibility of making appropriate blades where allowance values are known, with the accuracy of blade geometry being factored in. This is an important issue for designing a blade ECS technology for low-allowance workpieces. Localization of machining depends on the properties of electrolyte, machined material, and machining parameters. When using ECS, the structure of minimum allowance changes as it has to include an additional component with the real localization characteristic being factored in. The properties of the Machined Material - Electrolyte system were determined experimentally and added to our Teamcenter database, where they are highlighted in the Experiment table.

Computer-aided TE profiling is the most complex and demanding problem. To solve it accurately means to face significant challenges due to the necessity to factor in the changes in dissolution rates at various points on the movable boundary. On this basis, for automating the design of EC shaping operations we used an approach where the reverse EC shaping problem was down to solving a chain of direct problems, i.e. we used the method of sequential iterations. Based on their solutions, we adjusted the working surface of TE as well as the machining parameters.

The output of the computer-aided EC shaping design subsystem is geometric models of the workpiece and the profile part of tool electrodes, machining parameters, and other technical data. These data were transmitted to the integrated CAD/CAM system to design a TE-making technology, in particular to develop CNC programs to obtain the shaping surfaces of TE.

When creating the computer-aided ECS design subsystem, we developed information models of DC ECS and PECS shaping [1].

These two methods have a common basis; the distribution of potentials and currents in IEG was simulated using ANSYS. However, the assumptions we made and the boundary condition determination principles we adopted to solve this engineering problem were significantly different from what the authors of [1,2] describe.

For the dissolution of surfaces of DC-machined workpieces, the dependency of potential on the current density was assumed as the boundary condition and presented in the form of polarization curves for the specified part material and electrolyte. These data were chosen automatically from the Teamcenter databases.

When analyzing EC shaping operations, processes with considerably different time and space scales would be calculated using different discretization intervals for time and space.

As the evolution (and thus the IEG configuration) of machined surfaces occurs at a considerably lower rate than the rate at which hydrodynamic and diffusion processes achieve steady state, we assumed that EC shaping processes within the IEG occur in a quasi-stationary mode.

DOI: 10.1051/matecconf/201712903018

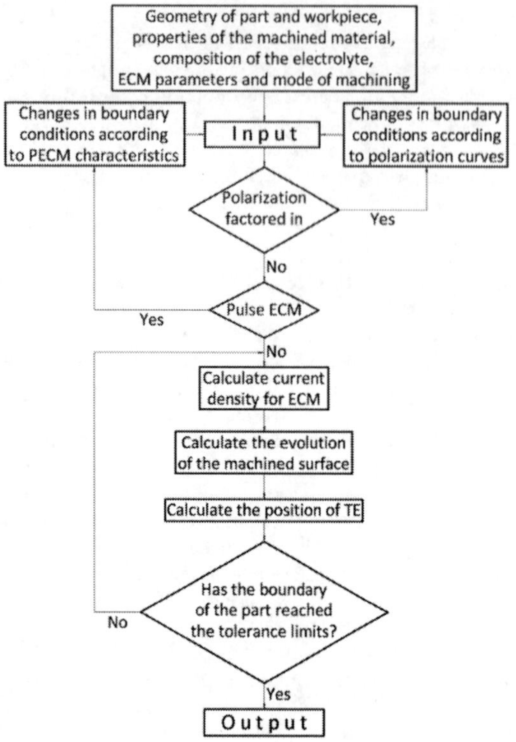

Fig. 3. EC shaping analysis chart

In line therewith, we implemented the following EC shaping analysis chart, see Fig. 3:

• calculate the current density in the IEG volume based on the ANSYS solution;

• interpret the values and directions of the current in each finite element mesh node into an etching boundary offset;

• determine the altered workpiece shape.

After the new position of the machined surface was determined, we checked whether the machining of the workpiece was complete. If machining was not complete, the calculation would be reiterated in the above sequence for the new position of the machined surface. Otherwise, calculations would stop and their results would be recorded in a file for further processing and visualization.

For pulse EC shaping, calculations are done similarly, but the anode and cathode potential values are picked from a database formed on the basis of experimental current readings.

To synthesize EC shaping operations, one has to determine the working surface of TE and the machining parameters, including the choice of EC shaping process (DC or pulse) to ensure that the specified surface would be produced with appropriate accuracy and maximized performance. Fig. 4 presents an EC shaping synthesis (design) chart.

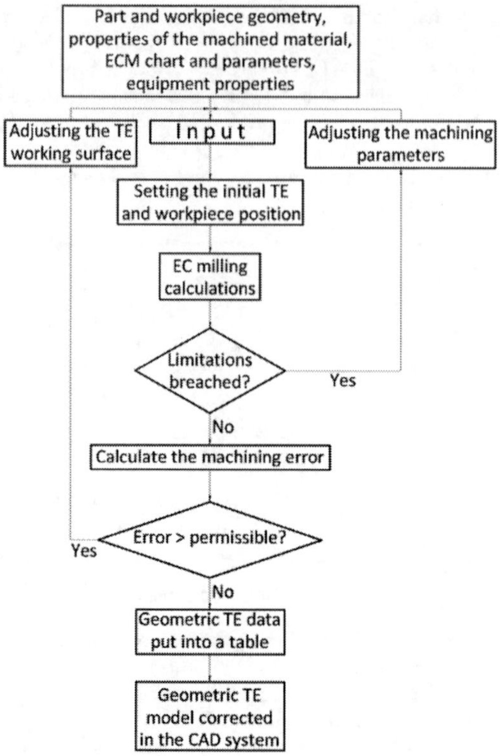

Fig. 4. EC shaping design chart

4 Conclusion

Therefore, this research summarizes all specific methods earlier developed and enables computer-aided design of GTE blade electrochemical machining, whether DC or pulse.

For more accurate profiling, we obtained experimental dependencies of current on the gap value for PECS operations performed with a pilot machine. Based on these studies, we calculated the values of φ_{eqv} which were then used in EC shaping process models as boundary conditions. This approach factors in the mutual impact of parameters on the evolution of PECS-dissolved surfaces.

References

1. M.V. Nekhoroshev, N.D. Pronichev and G.V. Smirnov, *Open Mechan Engin. J.* **8** (2014)
2. M.V. Nekhoroshev, N.D. Pronichev and G.V. Smirnov, *Proceedings of the Samara Scientific Center, Russian Academy of Sciences*, V. *15*, 6(4) (2013)
3. F. Klocke, M. Zeis, S. Harst, A. Klink, D. Veselovac and M. Baumgärtner, *Procedia CIRP*, V. **8** (2013)
4. T. Paczkowskia and J. Zdrojewski, *Procedia CIRP* **42** (2016)

5. J. Kozak, Lecture Notes in Electrical Engineering V. **170** (2013)
6. R. Mukherjee and Chakraborty, *International J. of Advan. Manuf. Tech.*, V. **64**, I. 5-8 (2013)
7. M. Purcar, A. Dorochenko, L. Bortels, J. Deconinck and B. Van den Bossche, *J. of Mater. Proces. Techn.*, V. **203**, I. 1-3 (2008)
8. J. Zhao, F. Wang, J. Xu and Y. Liu, *Hangkong Xuebao/Acta Aeronautica et Astronautica Sinica*, V. **34**, I. 12 (2013)
9. F. Wang, J. Xu and J. Zhao, *Hangkong Xuebao/Acta Aeronautica et Astronautica Sinica*, V. **31**, I. 12 (2010)
10. D. Zhu, D. Zhu and Z. Xu, *Inter. J. of Adv. Manuf. Tech.*, V. **62**, I. 1-4 (2012)
11. R.Wu, D. Zhang and J. Sun, *Res. J. of Applied Sciences, Engin. and Tech.*, V. **3**, I. 9 (2011)
12. E.M. Korovin, A.N. Lunev and V.V. Tsareva, *Russian Aeronautics*, V. **55**, I. 1 (2012)
13. A.N. Lunev, L.T. Moiseeva, and M.V. Solomina, *Russian Aeronautics*, V. **50**, I. 2 (2007)
14. A.I. Khaimovich and A.V.Balaykin, *ARPN J of Engin and Applied scien.*, **№9(10)** (2014)
15. A.V. Balaykin, E.A. Nosova and N.V. Galkina, *IOP Conf. Series: Mater. Sc. and Engin.*, V. **156**, I 1 (2016)

MATEC Web of Conferences 129, 03019 (2017) DOI: 10.1051/matecconf/201712903019
ICMTMTE 2017

Prediction of electronic nanodevices technical status and reliability based on analysis of their performance parameters kinetics under the influence of external factors

Mstislav Makeev[1,*], *Sergey* Meshkov[1], and *Vladimir* Sinyakin[1]

[1]Bauman Moscow State Technical University, 105005 Moscow, Russia

Abstract. The problems of technical diagnostics of electronic nanoobjects and devices based on them are considered. Algorithms for constructing the operability and reliability prediction fields of electronic nanoobjects and devices based on them in the specified operation conditions are given.

1 Introduction

One of the promising areas for the space industry development is the creation of a small spacecraft system, which will be entrusted with the solution of a wide range of tasks, both purely commercial and related to national security and the competitiveness of the country. Nanodevices that make it possible to give the equipment new qualitative characteristics and to increase the efficiency of performing the functions assigned to them are used for spacecraft hardware creation. The priority task of such nanodevices creation is the use of domestic materials, technologies and element base.

Strict requirements are imposed to the space-based nanodevice reliability. They are dictated by hard operating conditions on the one hand and the increasing spacecraft active shelf life on the other. The most reliable and complete reliability index of devices are usually obtained from the operation results. However, obtaining such information takes a lot of time. Traditional test methods in many cases also do not allow to confirm a given level of object reliability due to the presence of hidden defects. One of the promising directions in the development of methods for assessing the reliability and quality of nanodevices for space purposes is to predict their future status.

2 Problem formulation

The problem of the technical system reliability increasing with reliability prediction development on specified exploitation period in specified conditions is solved by technical diagnostics (TD), the scientific base of which are outlined particularly in [1, 2]. One of the main problems of TD is the identification of the object status under conditions of limited information. Object status identification is based on identification algorithms, which are usually considered as classification problems.

Identification algorithms in technical diagnostics are based on the diagnostic models, which establish relation between object status and its representation in the diagnostic signals dimension. An important part of identification problem is decision-making rules. Enumerated elements of the TD structure need to be developed regarding to the specific object of diagnostics.

The purpose of this project is the development of the major elements of TD structure regarding to electronic nanocomponents, which characteristics determine the indices of devices based on it.

An example of such promising electronic nanoscale components are resonant tunneling diodes (RTDs) based on multilayered nanoscale AlGaAs nanoheterostructures. RTD operation principle is based on the physical effect of resonance tunneling of electrons in the nanocomponent heterostructure, which layer thickness is in the range from units to dozens of nanometers. The prospects of RTD for modern radio electronics is caused by the following circumstances:

- RTD operates in the range of temperatures and other external factors necessary for technical applications.
- Technologies and facilities for RTD production are validated in the framework of micro- and nanoelectronics technologies.
- The limit frequency for RTD spreads up to units of THz [3-6], which makes the RTD a promising device for SHF, EHF, and THF electronics.
- It is possible to control the shape of the current-voltage characteristic (CVC) and create a diode with the optimum CVC shape for the specific type of nonlinear conversion by changing the parameters of the heterostructure layers (thickness, chemical composition).

It was shown in [7-10] that the possibility of optimizing the CVC shape, as well as the presence of a falling part on it, makes it possible to create a number of nonlinear radio signals converters with improved characteristics based on RTD. The set of possible

* Corresponding author: mc.stiv@gmail.com

DOI: 10.1051/matecconf/201712903019

nonlinear transformations using RTD is very wide: generation of radio signals, frequency modulation, mixing of radio signals, amplitude detection, rectification, frequency labels grid generation and etc.

The following tasks are solved in the work with regard to electronic nanodevices:

- development of decision-making rules. The rules are meant for decision making on finding the parameters of a nanodevices in the field of operability.
- development of a methodology for constructing a reliability prediction of the nanocomponents and a devices based on it.

Herewith, two hypotheses are accepted:

- Reliability of nanocomponent and device based on it are based on gradual failure model [11].
- Degradation of device parameters is determined by nanocomponent degradation. Nanocomponent degradation leads to the changes of its CVC shape, resulting in device parameters (indices) changes. The failure of device and nanocomponent occurs, when at least one of the device indices falls outside of set limits range (operability field).

3 Decision-making rules

The rules of decision-making are meant for decision making of nanocomponent parameter location in the operability field. Fig. 1 shows in simplified form fields of operability ***O***, correct operation ***CO*** and catastrophic failure ***CF*** in the context of RTD for radio signals mixers.

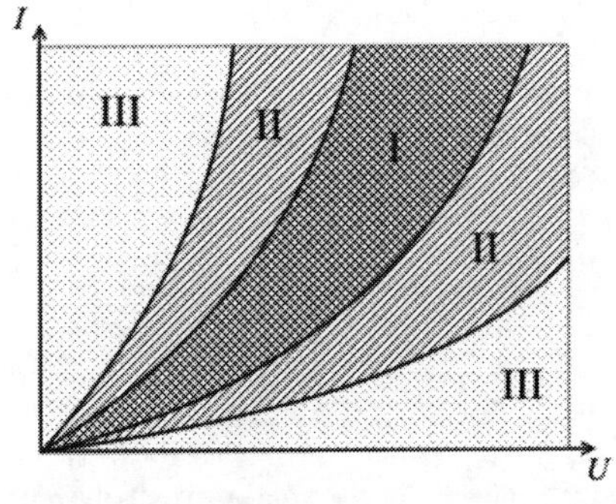

I - operability field ***O***, II - correct operation field ***CO***, III - catastrophic failure ***CF*** field

Fig. 1. Fields of nanocomponent operability ***O***, correct operation ***CO*** and catastrophic failure ***CF*** in the context of RTD for radio signals mixers.

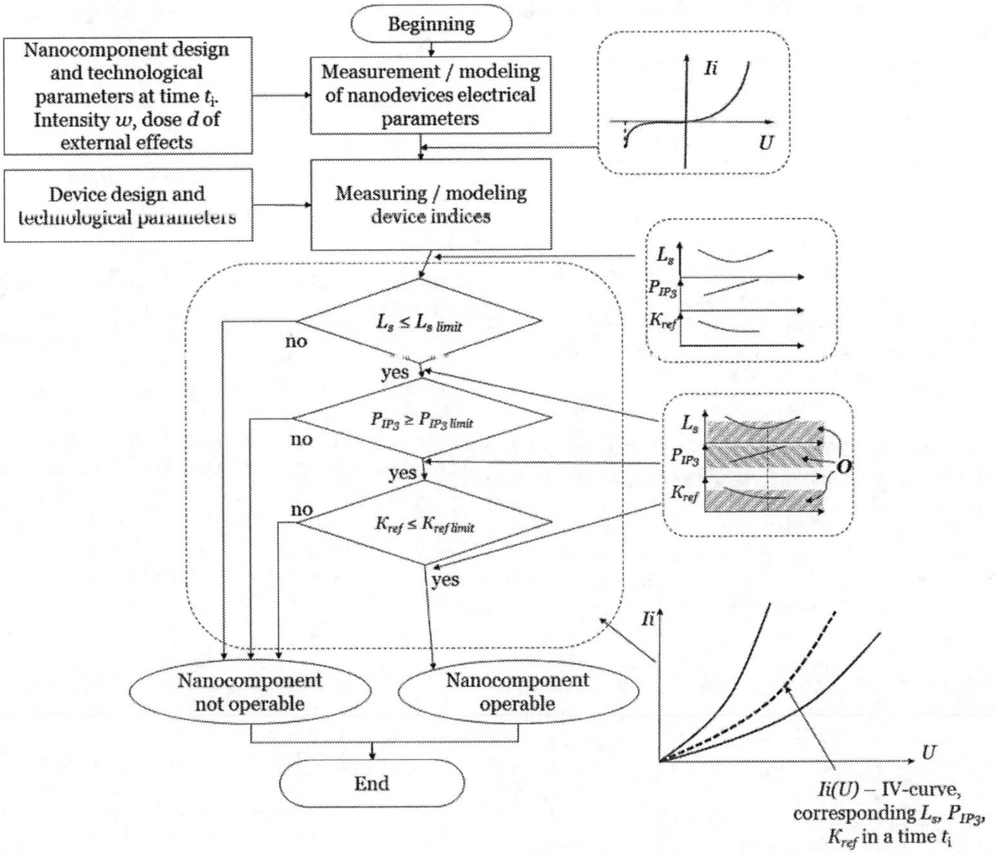

Fig. 2. Decision-making algorithm.

The assignment of nanocomponent parameters to a particular category depends on the characteristics of the device based on it.

Decision-making algorithm of relating nanocomponent parameters to a certain category is shown in Fig.2. If we consider a radio signals mixer as the device the indices of which is determined by a nanocomponent, then we consider the conversion loss L_s, the dynamic range P_{IP3} and the input signal reflection coefficient K_{ref} as the indices. We will assume that the mixer operability field is determined by the conditions:

$$L_s \leq L_{s\ limit}, \quad (1)$$

$$P_{IP3} \geq P_{IP3\ limit}, \quad (2)$$

$$K_{ref} \leq K_{ref\ limit}. \quad (3)$$

Special feature of decision-making of the nanocomponent parameters classification involves operability field $\boldsymbol{O}$ transformation from device indices space into nanocomponent parameters plane, so that one point in device indices space $\{L_{si}, P_{IP3i}, K_{refi}\}$ corresponds to one curve $I_i(U_i)$.

At the first stage, the nanocomponent electrical parameters, which are transferred to a software module that calculates the device's indices, are measured or calculated. The result of the device simulation is its indices (1)-(3). In general, the program module responsible for calculating the device indices can be replaced by a measuring module, but this replacement involves the need for prototyping the device.

In the following algorithm module calculated device indices are compared with the limitations (1-3). If all indices (1-3) are fallen within corresponding limitations, it is concluded that the device and nanocomponent parameters are laid in operability field $\boldsymbol{O}$. If any of device parameters is out of the limitations framework, it is concluded that device and nanocomponent parametric failure is detected, in other words, the nanocomponent parameters is related to ***CO*** or ***CF*** field.

4 Reliability-prediction technique of nanocomponent and device based on it

The technique of reliability prediction of a nanocomponent and a device based on it includes a decision-making algorithm. The block diagram of the algorithm of the reliability prediction of a nanocomponent and a device based on it is shown in Fig. 3. The technique is based on the sequential modeling of cycles of destabilizing effects (artificial aging) on nanocomponent, the dose of which gradually accumulates in each cycle and causes to a change in the parameters of a nanocomponent and a device based on it. The intensity w, the dose d, the duration t and the number of stress cycles are chosen based on operating conditions, a priori information on the sensitivity of nanocomponent parameters to destabilizing effects, the required accuracy of the predictive reliability and modeling laboriousness.

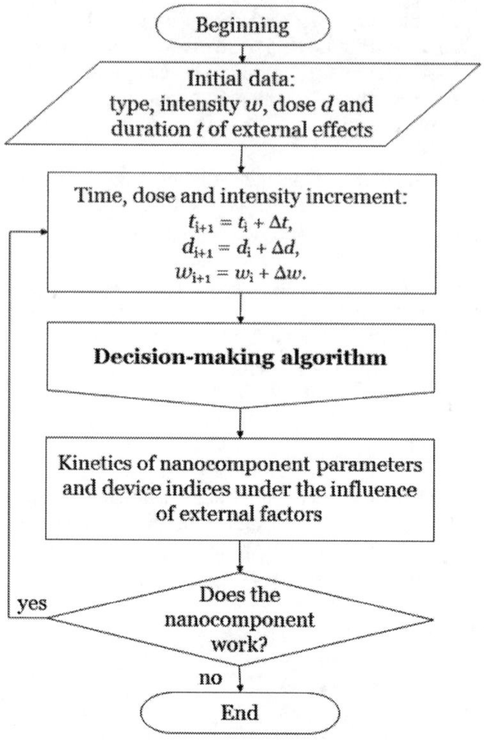

Fig. 3. The block diagram of the algorithm of the reliability prediction of a nanocomponent and a device based on it.

The technique of reliability prediction assumes conducting nanocomponent accelerated aging, during which destabilizing effects accelerate the degradation processes. The type, intensity and duration of effects are selected in such a way that the acceleration occurs only for a failure mechanism, which is typical for normal operating conditions of the device [12]. In practice for the semiconductor devices degradation rate increasing an elevated temperature is often used. The number of stress cycles is chosen at least from 4 to 5 so that the nanocomponent parameters kinetics can be estimated. The minimum intensity and duration of a stress cycle are chosen, so that the consequent change of the nanocomponent parameters were not less than by an order of magnitude larger than the error of the measuring instrument. The maximum intensity and duration of a stress cycle are limited by the risk of nanocomponent catastrophic failure. Thus, nanocomponent parameters comes to the "decision-making rules" block input after one cycle of destabilizing effects, at the output - the classification of a nanocomponent as failed, or workable.

The exit from the cycle "destabilizing effect - decision-making rule" is carried out immediately after the device and nanocomponent failure at the time t_i is fixed. As a result kinetics of the nanocomponent parameters $I(w,d,t)$ and device indices under the effect of external factors are revealed. On this basis estimation of the nanocomponent time to failure under given conditions is carry out.

The results of the RTD parameter kinetic studies under the influence of the elevated temperature are shown in [13-16].

The given algorithm makes it possible to obtain a point deterministic estimation of the time to failure of a nanocomponent and a device based on it. In addition, algorithm will allow constructing the MTBF statistical distribution and determining the necessary reliability characteristics of a production batches by adding to the algorithm technological errors of a nanocomponent and the device based on it, as well as the operational effects randomness.

5 Conclusions

The nanocomponent operability field is determined by the operability field of the device based on it, its parameters and application conditions. The nanocomponent time to failure is determined by its operability field and application conditions.

Consideration of the proposed technique and algorithms with respect to the mixing RTD does not limit their generality. The considered technique and algorithms can be applied to the construction of the operability field and time to failure forecast under given conditions of any nanocomponent of any device whose indices depend on the nanocomponent parameters.

The considered technique and algorithms can be used in production conditions during testing and planning of technological operations for monitoring the nanocomponent parameters.

The research work was supported by Ministry of Education and Science of the Russian Federation under state task № 16.1663.2017/4.6.

References

1. I.A. Birger, *Tehnicheskaja diagnostika* (Mashinostroenie, Moscow, 1978)
2. H. Czichos (ed.), *Handbook of technical diagnostics: fundamentals and application to structures and systems* (Springer Science & Business Media, 2013)
3. H. Mizuta, T. Tanoue, *High-speed and functional applications of resonant tunnelling diodes. In The Physics and Applications of Resonant Tunnelling Diodes* (Cambridge university press, 2006)
4. S. Suzuki, M. Asada, A. Teranishi et al., Appl. Phys. Lett. **97** (24) 242102 (2010)
5. H. Kanaya et al., Applied Physics Express **5** (12) 124101 (2012).
6. T. Maekawa et al., Applied Physics Express **9** (2) 024101 (2016).
7. Y.A. Ivanov, S.A. Meshkov, I.A. Fedorenko et al, J. Commun. Technol. Electron **55** 921-927 (2010)
8. Y.A. Ivanov et al., Microwave & Telecommunication Technology (CriMiCo), 2014 24th International Crimean Conference, IEEE. 1063-1064 (2014)
9. A.G. Gudkov, Y.A. Ivanov, S.A. Meshkov et al, Biomedical Engineering **49** 98-101 (2015)
10. V.Y. Sinyakin, M.O. Makeev, S.A. Meshkov, J. Phys.: Conf. Ser. **741** 012160 (2016)
11. A.S. Pronikov, *Parametric reliability of machines* (Bauman University Publishing House, Moscow, 2002)
12. O.P. Gludkin, *Metody i ustrojstva ispytanij RES i EVS* (Vysshaja shkola, Moscow, 1991)
13. M.O. Makeev, Y.A. Ivanov, S.A. Meshkov, Journal of Physics: Conference Series **584** 012014 (2015)
14. M.O. Makeev, S.A. Meshkov, Y.A. Ivanov, Key Engineering Materials **724** 48-52 (2017)
15. M.O. Makeev, Y.A. Ivanov, S.A. Meshkov et al, 2015 5th International Workshop on Computer Science and Engineering: Information Processing and Control Engineering, WCSE 2015-IPCE. 260-265 (2015)
16. M. O. Makeev, S. A. Meshkov, AIP Conference Proceedings **1858** 020001 (2017)

Research of the combined extremum-seeking control system dynamics on mathematical and computer models

Dmitriy Alexandrovich Bushuev[1,*], *Vasiliy Grigorievich* Rubanov[1], and *Dmitriy Vitalyevich* Titov[2]

[1]BSTU n.a. V.G. Shukhov, Department of Engineering Cybernetics, 308012, Belgorod, Kostyukova Str, 46, Russia
[2]SWSU, Department of Computer Science, 305040, Kursk, 50 let Oktyabrya Str., 94, Russia

Abstract. The article introduces the process of studying the dynamics of the combined extremum-seeking control system, which is used for grinding-mixing unit automatic balancing. The structure of the proposed system is shown and the algorithm of its operation is described. On the basis of the harmonic balance and Goldfarb's methods, the values of quality factors of the extremum-seeking control system on the first cycle were obtained in the absence of static map drift and the stability of the system was established. The co-simulation model of the combined extremum-seeking control system implemented in the MATLAB / Simulink software is presented and described. MSC.Adams model of the grinding-mixing unit, which is a plant in the co-simulation model of the system, is presented. The results of co-simulation of the extremum-seeking control system dynamics are presented and analysed.

1 Introduction

Extremum-seeking control (ESC) systems are designed to ensure the minimum or maximum of the plant cost function in real time of its operation.
ESC systems, depending on the seeking algorithm, are divided into the following [1]:
- Systems using the definition of derivative of the cost function;
- Stepper systems that respond to the sign and the value of the discrete increment of the output signal;
- Systems using determination of the direction of motion to the extremum by the phase shift between the input perturbation signal and the output signal;
- System using storing of the extremum, reacting to the difference between the optimum reached at the previous instants of time and the current value of the output and its derivatives.

Extremum-seeking control has found wide application in various sectors: in the automotive industry (anti-lock braking systems [2-3], internal combustion engines [4], electromechanical valves [5], cruise control systems [6], in aircraft construction (formation flight systems [7-8]), in robotics (autonomous robotics and mobile robots [9-11], manipulator positioning systems [12]), in the optimal adjustment of PI controllers [13], neural network and fuzzy controllers [14-15], and etc. [16]. In the Belgorod State Technological University named after V.G. Shukhov the combined ESC system for automatic balancing of centrifugal grinding-mixing (GMU) units has been proposed [17]. This system automatically changes the counterweights position during the work of the unit, in order to establish and maintain a minimum of supporting bearings vibration. To reduce the losses for "search and yaw" inherent in ESC, the program control channel is included in this system which, starting from the second loading cycle of the grinding material, realize the movement of the counterweights along the drift trajectory found on the first cycle. In [18], the process of obtaining an approximate mathematical model of GMU is described and the results of the combined ESC dynamic simulation in the MATLAB/Simulink software are shown. There is a linear drift model for the plant parameters, in accordance with alternately loading the maximum amount of material into each of the grinding chambers.

However, often in practice, a single or batch dosing mode is used. To evaluate the performance of the proposed combined ESC method for GMU automatic balancing in the above dosing modes it is a necessity to perform dynamics co-simulation of the mechanical virtual plant, designed in the MSC.Adams software, and the control system, implemented in MATLAB/Simulink. This makes it possible to approach to the conditions for the operation of the automatic GMU vibration suppression system in real modes, and also to take into account the possible errors in the mathematical model that arose as a result of approximation. To compare the results of computer simulation with analytical calculations, as well as to assess the stability of the extremum-seeking (ES) control loop, a system analysis using approximated mathematical model and the harmonic balance method is performed in this paper.

1.1 System description

Figure 1 shows the functional diagram of the proposed system.

* Corresponding author: dmbushuev@gmail.com

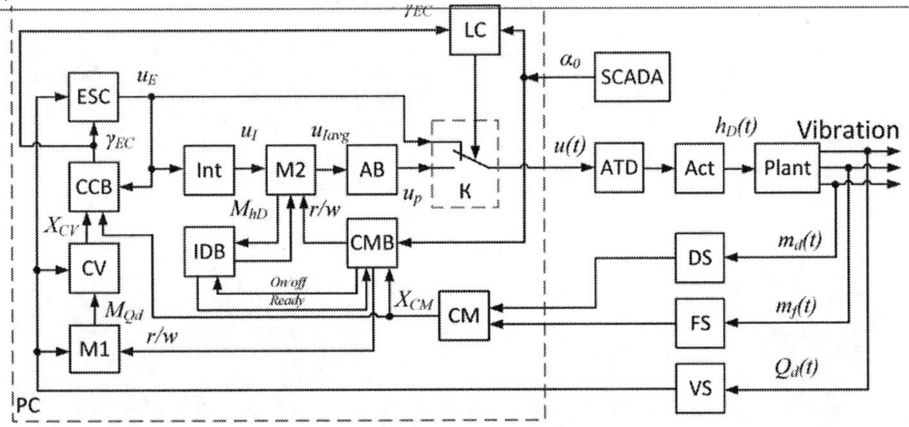

Fig. 1. Function block diagram of the system.

As an ESC algorithm in the ESC block uses a method of storing extremum with reverse zone defined by the necessary noise immunity. It is implemented movement of the counterweights with a constant speed by amplifying and transducing device ATD and actuator Act. The static map of the ES controller is shown in Figure 2.

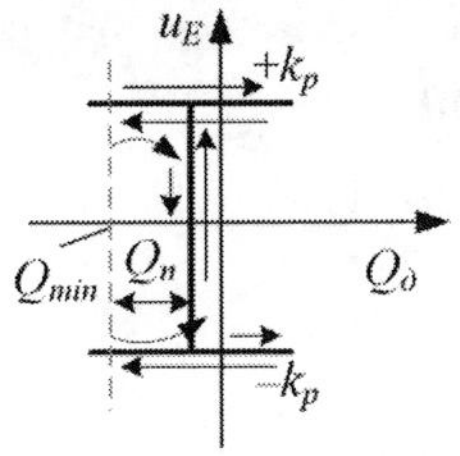

Fig. 2. The static map of the ES controller.

The combined control system, according to the algorithm in the programmable real-time controller PC, on the first cycle of material feeding which is determined by the appropriate loading signal α_0 from the corresponding SCADA system, turn on the ES controller in the mode of constant extremum-seeking. At the same time, integrated using the integrator Int control signals from the ESC U_I are stored in the memory block M2 and the corresponding values of the plant output Q_d are recorded in the memory block M1. The termination of the grinding is determined in the mass comparator CM when the mass m_f of the feed material is equal to the mass of discharged m_d, measured by the feed sensor FS and the discharge sensor DS, respectively. The X_{CM} signal from the CM block via the memory control block CMB turns on drift identification block IDB that determines trajectory of drift by averaging (for example, using the moving average method with weighted local quadratic regression) and the resulting drift path is placed in the memory block M2 in a matrix form.

On subsequent grinding cycles, after the getting of the loading signal α_0 and the measuring of the mass of the feed material, the memory control block CMB and the LC logic circuit (which is switching the key K) sets the program control on the drift path obtained at the first cycle, converting it to the control action u_p using the approximation block AB. In the case of an identical drift, program control system maintains the minimum of the plant output without seeking signals and yaw losses. At the same time, the current vibration level determined by the vibration sensor VS is constantly compared with the same value stored in memory M1. If the difference exceeds by the setting deadband, determined from the condition of the permissible deviation error, the vibration comparator CV sends X_{CV} signal to the controller control block CCB, which starts the refinement seeking algorithm consisting in turning on the ESC in three movements with the last movement, equal to the halftime of the previous, thereby reducing the error of tracking the extremum accumulated as a result of program control. A more detailed algorithm of the system operation is described in [18].

2 Research of the combined ESC system dynamics on the basis of a mathematical model

Approximated dynamic mathematical model of three-chamber grinding-mixing unit oscillations with an actuator in the form of differential equations is given by [17]:

$$\begin{cases} \dfrac{d}{dt}\left(T_a \dfrac{dh_D(t)}{dt} + h_D(t)\right) = k_a U(t), \\ Q(h_D, t) = \begin{cases} k_1(t)\cdot\left(h_D - h_{D0}(t)\right)^2 + Q_0(t),\ h_D \le h_{D0}, \\ k_2(t)\cdot\left(h_D - h_{D0}(t)\right)^2 + Q_0(t),\ h_D > h_{D0}, \end{cases} \\ T_{out}\dfrac{dQ_d}{dt} + Q_d = Q(t - \tau_{Dout}), \end{cases} \quad (1)$$

where T_a – actuator time constant, h_D – counterweights position, k_a – transfer factor of the actuator, U – control action, $Q=f(h_d,t)$ – the static map of the plant, k_1, k_2 – the parabola coefficient, h_{D0}, Q_0 – coordinate of the unimodal static map minimum, Q_d – the output of the plant dynamic part, T_{out} – the time constant of the plant dynamic part, τ_{Dout} – delay of the plant dynamic part.

DOI: 10.1051/matecconf/201712903020

For the GMU virtual prototype, the numerical values of the quantities in the system of equations (1) are shown in Table 1.

Table 1. Quantities included in the plant description

Parameter	Value
T_a, s	0.1
k_{act}, cm/(s·V)	0.6
k_1, μm/cm	17.8
k_2, μm/cm	13.2
h_{D0}, cm	9.418
Q_0, μm	253.9
T_{out}, s	0.01
τ_{Dout}, s	0.147

On the first cycle of material feeding, the control action u (t) is formed only by the ES controller, which works in the seeking mode. For such a nonlinear system, the structure and filter hypothesis is fulfilled (since the actuator and the plant dynamic part exhibit the properties of a low-pass filter), then for the dynamics analysis it is possible to use the method of harmonic balance. Quality factors of the ESC system: the amplitude a_1 and ω is the frequency of self-oscillations in the absence of drift (i.e. $Q_0 = h_{D0} = 0$) can be found by using the Goldfarb's method.

According to the Nyquist criterion, the appearance of a self-oscillation mode in a closed harmonic linearized system occurs when the following condition is satisfied

$$W_e(j\omega) = -\frac{1}{J_2(a_1)}, \quad (2)$$

where $W_e(j\omega)$ is the equivalent complex frequency response of the plant with an actuator equal to

$$W_e(j\omega) = \frac{2k_p(k_1+k_2)}{\pi}\cdot\left(W_a\left(j\frac{\omega}{2}\right)\right)^2 W_D(j\omega)e^{-j\frac{\pi}{2}}, \quad (3)$$

and $J2(a_1)$ is the equivalent complex transfer factor of the nonlinear ES controller, which, as a result of the harmonic linearization of the static map (see Fig. 2), can be defined as

$$J_2(a_1) = \frac{4k_p}{\pi a_1} e^{j\left(\pi - \arcsin\frac{Q_n - a_1}{a_1}\right)}, \quad (4)$$

where Q_n is the dead space of ES controller, accepted for this plant (equal to 10 μm).

To solve equation (2) with respect to a_1 and ω, amplitude-phase characteristics of $W_e(j\omega)$ and -1/ $J2(a_1)$ according to Goldfarb's method are constructed (Fig. 3). As a result, it was found that the amplitude of the oscillations $a_1 = 5.3$ μm and the frequency ω=1.84 rad/s. The found periodic solution is stable, since with increasing amplitude a_1 from 0 to *inf* the curve -1/ $J2(a_1)$ crosses the complex frequency characteristic of the linear part $W_e(j\omega)$ in the direction from the inside outward. In the case of the static map drift the differential equations (1) are non-autonomous, which makes it difficult to use analytical methods to estimate the dynamics of the ESC. Therefore, to analyze the dynamics of the ESC system with the drift of a static map, computer simulation is used.

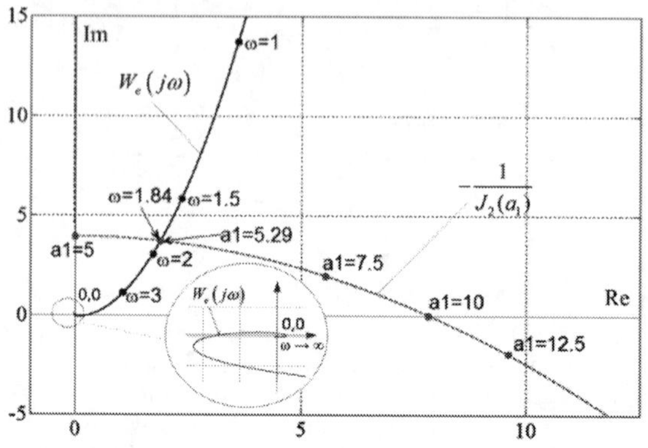

Fig. 3. Determination of the amplitude and frequency of self-oscillations of the ESC by the Goldfarb's method

3 Co-simulation model of the combined ESC system

A dynamic model of the combined ESC system with the integration of a virtual prototype of GMU with automatic balancing created in the MSC.Adams software is shown in Fig. 4.

The functions of the memory block M2 and drift identification block IDB are performed by the "Avg" subroutine, which performs the averaging using weighted local quadratic regression. The "PSGoal" block, in accordance with the drift trajectory obtained on the first cycle, sets the value of the speed that is fed to the integrator "Integrator" of the actuator. The ES controller is executed using two subprograms in the form of S-functions: "ESCLC" defines a learning search by the method of storing an extremum on the first feeding cycle, "ESCWC" performs the functions of the ES controller on the remaining cycles and the M1 memory block in the training mode. To control of the different modes turning on, the sub-programs Subsystem1 and Subsystem are used. Relay element K is implemented using the "Relay" block.

To create a kinematic connection between the eccentric and additional shaft, a gear module from Adams/Machinery is used and a "Bearing" is used in the "Compliant" mode to set the bearings, which allows approximately obtaining the oscillations of the GMU bearing supports. The radial stiffness of the bearings is set equal to $k = 1.23 \times 10^4$ N/cm, and the damping factor c=10 N·s/cm. In order to change the grinding material loading Δm, during the GMU operation, GFORCE forces are added to each CM_marker of the grinding chambers. The format of such functions is FX=Varval(deltam)* *ACCX(CM_marker), where ACCX is the function that determines the acceleration along the x-axis, for

DOI: 10.1051/matecconf/201712903020

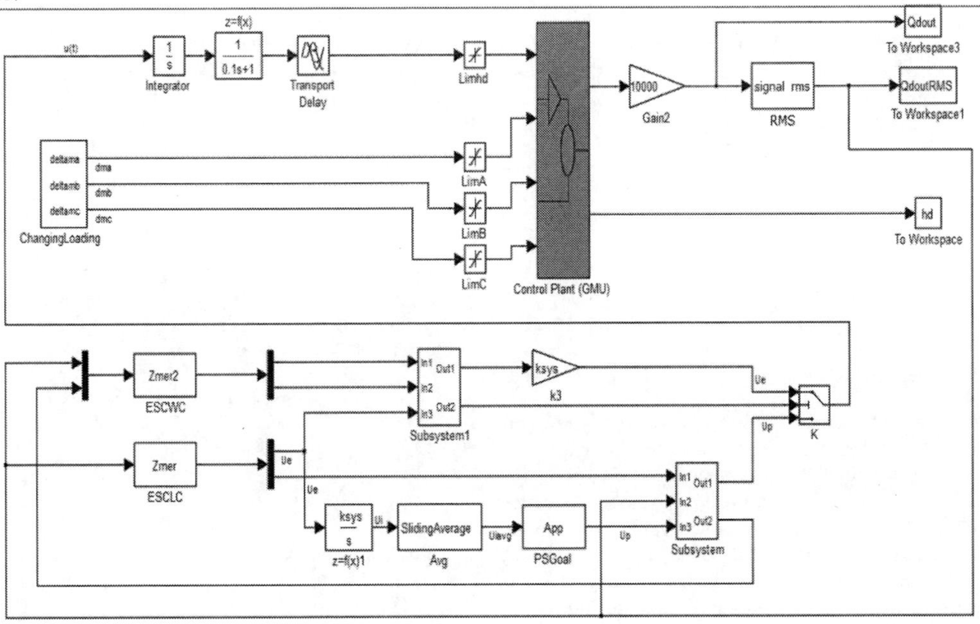

Fig. 4. Co-simulation model of the combined ESC system.

example, to change the loading of the grinding chamber A, the function FX takes the following form: Varval(deltama)*ACCX(A.cm)-Varval(deltama)*g.

The model of the GMU in the MSC.Adams software is presented in Fig. 5.

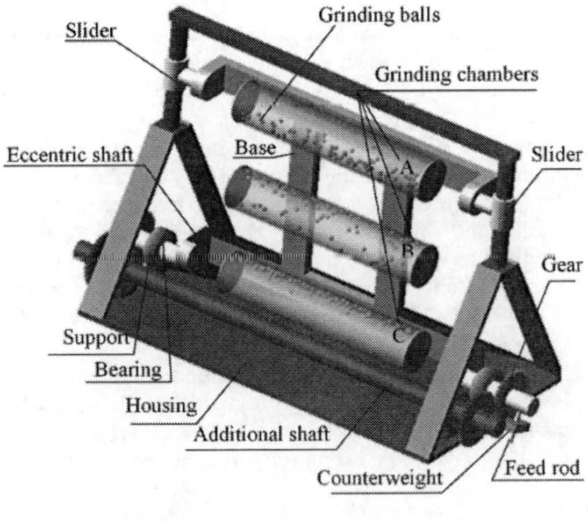

Fig. 5. Adams model of the GMU.

4 Simulation results and discussion

In the virtual GMU prototype, the load factor of grinding bodies in the grinding chambers is determined by the $K_L = 25\%$, which corresponds to a mass of 13.8 kg. Two variants of the periodic operation of the unit with automatic balancing were simulated:

• Single dosing of the maximum amount of grinding material into the upper grinding chamber;

• Batch dosing of grinding material into the upper grinding chamber.

For the first variant, the "ChangeLoading" subroutine specifies the change in the loading of grinding chambers, shown in Fig.6. In this case, the maximum amount of material in the lower chamber is not more than 9 kg.

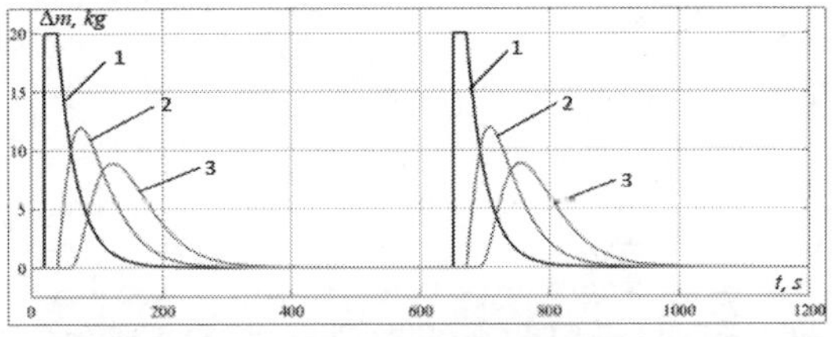

Fig. 6. Graphs of changing the loading of the GMU grinding chambers with a single dosing of the maximum amount of grinding material: 1 - chamber A (Δm_A), 2 - chamber B (Δm_B), 3 - chamber C (Δm_C).

As a result of the combined ESC system dynamic simulation, the following results were obtained (Figures 7-8).

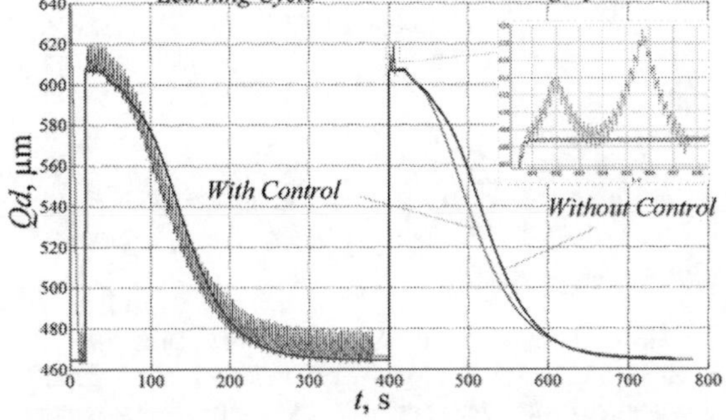

Fig. 7. RMS of the virtual prototype vibration displacement in a single dosing mode.

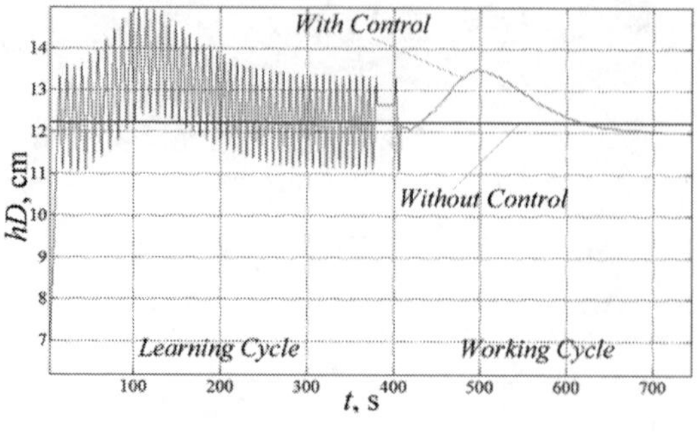

Fig. 8. Counterweights position.

At the initial moment of the GMU operation, the position of the counterweights is h_D=6.5 cm and the RMS of the vibration displacement Q_D = 640 μm. The work of the ES controller on the first cycle of operation allowed to search for the minimum value of vibration (464 μm), thereby reducing the level of oscillations by 28%, after which a yaw is observed near this value. As can be seen, the values of the amplitude and frequency of the output plant oscillations in the absence of the drift coincide with the results obtained from the analysis by the harmonic balance method of the system with the plant given by the mathematical model. On the second grinding cycle, the ES controller was turned on once when the set tracking error threshold was exceeded at the beginning after the material was loaded, thus, yawing was practically absent. The maximum decrease in the level of RMS vibration displacement compared to the level corresponding to the fixed position of the counterweights (h_D=12.2 cm) on the training cycle is 20.5 μm (3.7%) and in the mode of the program control - 18.5 μm (3.4 %).

For a comparative evaluation of the vibration suppression efficiency, the integral index I_{ef}, is represented in the form

$$I_{ef} = \frac{1}{t_2 - t_1}\int_{t_1}^{t_2}\left(Q_{wc}(t) - Q_c(t)\right)dt, \qquad (5)$$

where t_1, t_2 - initial and final moments of the efficiency calculation, $Q_{wc}(t)$, $Q_c(t)$ are the values of the control plant output without a control and with a regulator, respectively.

The value of the integral index in the interval from t_1 =20 s to t_2 = 780 s (i.e., from the beginning of the grinding process to the end of the second cycle) is equal to I_{ef} = 1.65.

In the batch mode, the operation of the combined system was investigated by grinding 42 kg of material, dosed at 14 kg per portion, dispensed at a given time interval.

The "ChangeLoading" subroutine specifies the following change in grinding load (Fig. 9). The maximum amount of material in the bottom grinding chamber is not more than 12.7 kg.

The co-simulation results of the combined ESC system of the GMU automatic balancing with the load factor of grinding bodies K_L=25% are shown in Fig. 10.

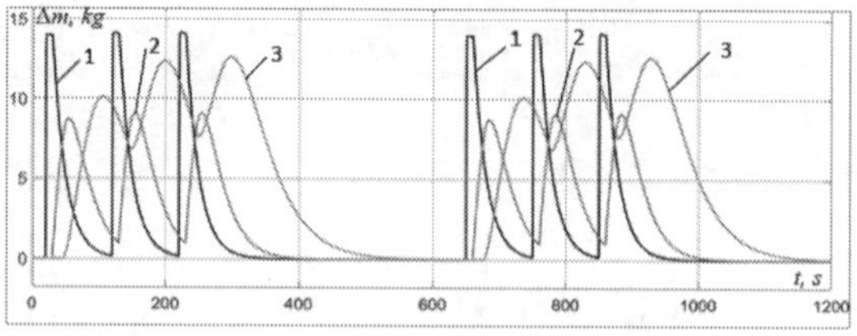

Fig. 9. Graphs of changing the loading of the GMU grinding chambers with a batch dosing of the maximum amount of grinding material: 1 - chamber A (Δm_A), 2 - chamber B (Δm_B), 3 - chamber C (Δm_C).

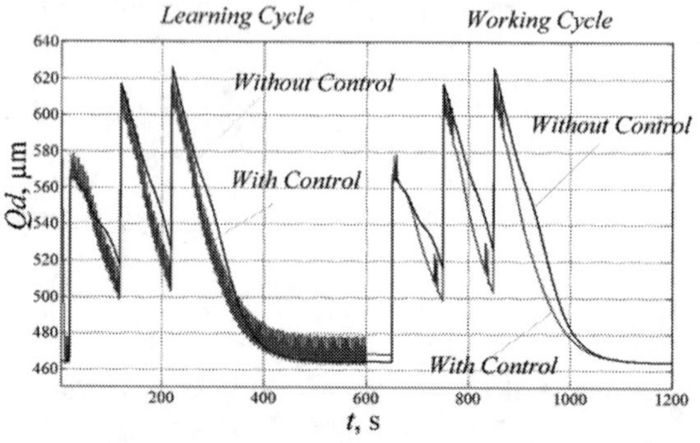

Fig. 10. RMS of the virtual prototype vibration displacement in a batch dosing mode.

Using of the combined ESC system on the training cycle ensured a maximum reduction in the level of vibration displacement RMS by 31 μm (5.6%), but there were large yaw losses. In a while program control ensured yawing only at the moments corresponding to a strong drift of the static map, and this was also facilitated by the low value of the deadband (5 μm), above which the ES controller was turned on for three seeking oscillations. The maximum decrease in the level of oscillations in the working cycle was 30 μm (5.5%). The integral efficiency index over the interval from t_1 = 20 s to t_2 = 1200 s (on two grinding cycles) exceeded the value of E_{ef} = 7.42.

Increasing the load factor of grinding bodies K_L to a level of 30% (16.6 kg) leads to a shift of the extremum position, as a result of which the balance error with the aid of a fixed located counterweights increases (Fig. 11).

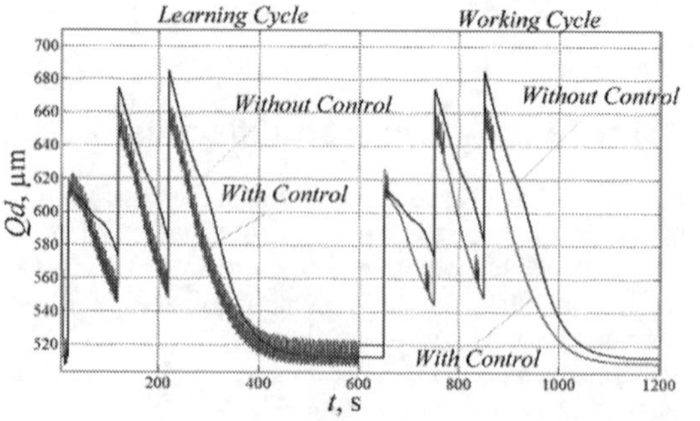

Fig. 11. RMS of the virtual prototype vibration displacement in a batch dosing mode with K_L=30 %.

DOI: 10.1051/matecconf/201712903020

Using the developed system, it was possible to significantly reduce the average level of oscillations (the efficiency index I_{ef} is more than 15.1 in the interval from t_1 = 20 s to t_2 = 1200 c.) The maximum decrease in the vibration displacement RMS level was 44 μm (7.5%).

For the comparative efficiency evaluation of the combined ESC system on the training (when combined ESC system operated as a classical ESC system with an extremum storing) and the working cycles, in comparison with the fixed counterweight, indexes I_{ef} were also determined at the corresponding time intervals. The results of a comparative analysis of the combined ESC system efficiency are presented in Table 2.

Table 2. Comparative analysis of the combined ESC system efficiency

Dosing mode	Efficiency index, I_{ef}			
	Training Cycle	**Working Cycle**	**Both Cycles**	**The increase in efficiency due to the program control**
Single (20 kg, K_L=25 %)	-0.3	3.6	1.65	more than 100 %
Batch (42 kg, K_L=25 %)	5.6	9.2	7.42	39 %
Batch (42 kg, K_L=30 %)	13.7	16.53	15.1	17 %

5 Conclusions

The application of the program control algorithm in the combined system is much more effective than using an ESC system with an extremum storing due to the presence of search and yaw losses in the last one. Based on the results of co-simulation, it is also evident that the effectiveness of reducing the level of undesired oscillations increases with increasing degree of filling of the bottom grinding chamber with a material or grinding bodies.

Using a continuous loading mode with a constant feed rate for grinding a limited amount of material (for example, the same 42 kg) can cause a significant reduction in the accuracy of balancing by the fixed counterweights at the start and in the end of grinding process, when not all the chambers are full. Therefore, using of the developed system is also justified in the continuous loading mode. The proposed control principle can be applied to the extremum-seeking control of low-inertia plants with unknown unimodal static maps with cyclic drift.

Research is carried out with the financial support of The Ministry of Education and Science of the Russian Federation within the Public contract project 2.1396.2017/PCh.

References

1. E.S. Rybchikova, M.Yu. Rybchikov, A.I. Sunargulova, R.V. Tankov, V.Yu Perevalov, *Mechatronics., Automation., Control*, **16**, 5 (2015)
2. S. Drakunov, ¨U. ¨Ozg¨uner, P. Dix, B. Ashra, *IEEE Trans. on Control Sys. Tech.*, **3**, pp. 79–85 (1995)
3. C. Zhang and R. Ord´or˜nez, *IEEE Trans. on Autom. Control*, **52**, 3 (2007)
4. N. J. Killingsworth, S. M. Aceves, D. L. Flowers, F. Espinosa-Loza, M. Krstic, *IEEE Trans. on Control Sys. Tech.*, **17**, 6 (2009)
5. K. S. Peterson, A. G. Stefanopoulou, *Automatica*, **40**, pp. 1063–1069 (2004)
6. A. Rahnama, M. Xia, S. Wang, P.J Antsaklis, *Proc. of the American Control Conf.*, art. no. 7526629, pp. 6109-6114 (2016)
7. P. Binetti, K.B. Ariyur, M. Krstić, F. Bernelli *Proc. of the American Control Conf.*, **4**, pp. 2848-2853 (2002)
8. B. Zuo, Y. A. Hu, *Proceedings of Fifth World Congress on Intelligent Control and Automation*, pp. 3302–3305, (2004)
9. J. Cochran, M. Krstic, *IEEE Trans. on Automatic Control*, **54**, 4, (2009)
10. C. G. Mayhew, R. G. Sanfelice, A. R. Teel, *Proc. of the 2007 American control conf.*, pp. 1185–1190 (2007)
11. M. S. Stankovic, D. M. Stlipanovic, *Proc. of the 48th IEEE Conf. on Decision and Control*, pp. 4541–4546 (2009)
12. V. Koropouli, A. Gusrialdi, S. Hirche, D. Lee, *Control Engineering Practice*, **52**, pp. 1-14 (2016)
13. M. Guay, D.A. Dochain, *Automatica*, **77**, pp. 61-67 (2017)
14. L. Gurvich *Proc. of the IEEE Convention of Electrical and Electronics Engineers*, pp. 18-21 (2004)
15. Y. A. Hu, B. Zuo, *Int. J. of Inf. Tech.*, **11**, 6, (2005)
16. Y. Tan, W.H. Moase, C. Manzie, D. Nesic, I.M.Y. Mareels, *Proc. of 29th Chinese control conf.*, pp 14–26 (2010)
17. D.A. Bushuev, V.G Rubanov, D.V. Velichko, *ARPN Journal of Engineering and Applied Sciences*, **11**, pp. 12288-12294 (2016)
18. V.G. Rubanov, D.A. Bushuev *Patent RU* No. 2600688 (2016)

MATEC Web of Conferences **129**, 03021 (2017)
ICMTMTE 2017
DOI: 10.1051/matecconf/201712903021

Experimental test of the optimal rotation for the finite stiffness cantilever rod

Aleksandr I. Bokhonsky[1,*], *Mikhail M.* Maistrishin[1], and *Aleksandr I.* Ryzhkov[1]

[1]Sevastopol State University, Sevastopol, Russian Federation

Abstract. One of the simplest types of skew-symmetric optimal controls for the motion of an object as an absolutely rigid and deformable body was theoretically researched and experimentally verified. An elastic system with one degree of freedom was observed as an example, and it was shown, that for an acceptable minimum possible time an absolute rest could be achieved at the end of the motion. The time of motion was found from the system of transcendental equations, which are moment relations in relative motion (relative displacement and relative velocity at the end of translational motion are equal to zero). Experimental test was completed for the optimal translational rotation of the elastic cantilever rod with a distributed mass and the absolute rest was achieved after optimal turning through a predetermined angle (in the shortest possible time) around the axis passing through the end of the rod. Results of the research can find application in: optimal control of the motion of finite stiffness objects in automated production, creation of manipulators of a minimal mass, in the case of mounting non-rigid large-sized structures in outer space, in a number of other areas of modern technologies.

1 Introduction

Papers [1-7] and others are devoted to the theory of oscillations of elastic systems with a finite and infinite number degrees of freedom. The methods of optimal control of oscillations of systems with lumped and distributed parameters are shown in [5-7], where method of moments was used in the search of a control for deformable systems' oscillations.

In [8-10] results of searching of a wide class of skew-symmetric optimal controls for the translational motion of elastically deformable systems with a finite and infinite number degrees of freedom are presented, and examples of technical applications' characteristic are given. Papers [10-14] are devoted to the problems of optimal translational motion using the algorithm of the complete inverse problem of the variations calculus ("reversion calculus").

A generalization of the algorithm for the solutions of the complete inverse problem of the variations calculus (from a function to a functional) is the formulation of the reversion principle of optimality [11]. According to it, during minimal possible time (which is determined from the moment relations in the relative motion of the elastic system's period) restored (by a constructed motion) functional-criteria assumes a stationary value.

The experimental verification of the optimal translational parallel motion of an elastic system with one degree of freedom was performed for the first time [14] using FESTO laboratory equipment in Sevastopol State University (Sevastopol, Russia).

The purpose of the research is to use the previously developed technique of experimental research of the translational parallel motion of an elastic object to test the optimal translational rotary motion of a cantilever elastic rod with a distributed mass.

2 Description of the experimental setup

In Fig. 1 is a simplified block diagram of an experimental setup implementing optimal rotation for a rod about a fixed axis. The operating principle of the experimental setup's control system is based on implementation of the control law with the outputting of a command for the servo motor controller generating voltage of a certain frequency.

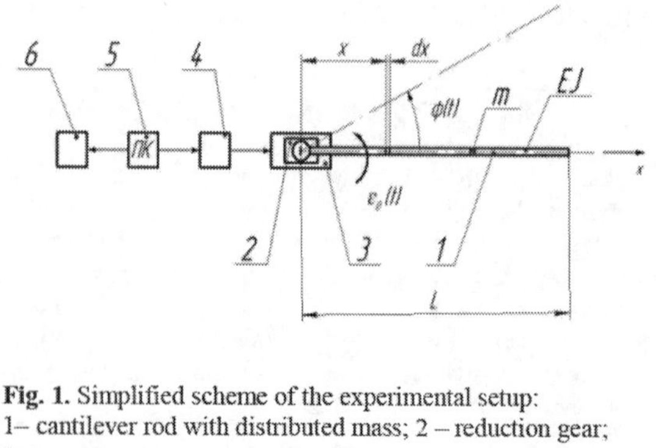

Fig. 1. Simplified scheme of the experimental setup: 1– cantilever rod with distributed mass; 2 – reduction gear; 3 – servomotor; 4 – servo controller; 5 – PC; 6 – Arduino Mega 2560.

* Corresponding author: bohon.alex@mail.ru

3 Solution for the equations of the frame relative motion

Similarly to the optimal translational motion described in [10], the acceleration of the translational rotary motion is taken the following form:

$$\varepsilon_e(t) = a \cdot \sin(pt)\,, \tag{1}$$

where $a, p = const$. The boundary conditions are given:

$$\varphi_e(0) = 0,\ \ \dot{\varphi}_e(0) = 0,\ \ \varphi_e(T) = \varphi_*,\ \ \dot{\varphi}_e(T) = 0, \tag{2}$$

where T is a total time of a portable rotary motion, φ_* is a maximum angle of rotation of the rod in the end of motion. After integrating (1), taking into account the boundary conditions (2), we obtain:

$$\begin{gathered}\varphi_e(t) = \frac{a}{p}\cdot(t - \frac{1}{p}\sin pt),\\ \varepsilon(t) = a\cdot\sin(pt), \qquad \omega_e(t) = \frac{a}{p}\cdot(1-\cos pt)\end{gathered} \tag{3}$$

where $a = \dfrac{\varphi_* p^2}{2\pi}$. When the rod had been rotated by an angle φ_* during a time T, the translational rest would be achieved.

In the relative motion in the interval $T \ge t \ge 0$ flexural vibrations of the elastic rod could be observed. They are caused by the optimal translational rotary motion and could be described (without taking into account resistance to movement) by the following equation:

$$EJ\frac{\partial^4 W(x,t)}{\partial x^4} + m\frac{\partial^2 W(x,t)}{\partial t^2} = -f(x,t), \tag{4}$$

where EJ is a rod's flexural stiffness (E – is Young modulus; J – is an axial moment of inertia of rod's cross-section); m – is a distributed mass (kg/m); $f(x,t) = \varepsilon_e(t)\cdot x\cdot m$, where $\varepsilon_e(t) = \dfrac{\varphi_* p^2}{2\pi}\sin(pt)$ is a angular acceleration.

It is assumed that in this experiment the appearance of the elastic rod' first form (mode) of bending vibrations is most likely. According to the Fourier method [2, 3], taking into account the first mode of forced bending vibrations, following is accepted:

$$W_1(x,t) = W_1(x)\cdot\sin(pt)\,. \tag{5}$$

After substitution (5) in (4) and transformations, the fourth-order differential equation for the first mode of oscillation takes the following form:

$$\frac{d^4 W_1(x)}{dx^4} - k^4 W_1(x) = -\frac{\varphi_* p^2 m}{2\pi\cdot EJ}, \tag{6}$$

where $k^4 = \dfrac{mp^2}{EJ}$.

The solution of the inhomogeneous equation (6) is found using the functions of A.N. Krylov. The general solution of the inhomogeneous equation (6) could be written:

$$\begin{gathered}W_1(x) = A\cdot s(x) + B\cdot T(x) +\\ C\cdot U(x) + D\cdot v(x) + \frac{\varphi_* p^2 m}{2\pi\cdot k^4 EJ},\end{gathered} \tag{7}$$

where

$$s(x) = \frac{\cosh(kx)+\cos(kx)}{2};\ T(x) = \frac{\sinh(kx)+\sin(kx)}{2};$$
$$U(x) = \frac{\cosh(kx)-\cos(kx)}{2};\ V(x) = \frac{\sinh(kx)-\sin(kx)}{2}.$$

Taking into account the boundary conditions for the rod with a anchorage and the free end:

$$\begin{gathered}W_1(0) = 0, \quad W_1'(0) = 0,\\ W''(L) = 0, \quad W'''(L) = 0.\end{gathered} \tag{8}$$

After determining the constants A, B, C, D solution of equation (7) takes the following form:

$$\begin{aligned}W = &-\frac{1}{4}\frac{1}{k^5(1+\cosh(kL)\cdot\cos(kL))\pi EI}\cdot\\ &\cdot(fp^2m(-2kx\cdot\cosh(kL)\cdot\cos(kL)+\sin(kx)\cdot\\ &\cdot\cosh(kL)\cdot\cos(kL)+\sinh(kx)\cdot\cosh(kL)\cdot\\ &\cdot\cos(kL)-\cos(kL)\cdot\sinh(kL)\cdot\cosh(kx)+\\ &+\cos(kL)\cdot\sinh(kL)\cdot\cos(kx)+\cosh(kL)\cdot\\ &\cdot\sin(kL)\cdot\cosh(kx)-\cosh(kL)\cdot\sin(kL)\cdot\\ &\cdot\cos(kx)-\sinh(kL)\cdot\sin(kL)\cdot\sinh(kx)+\\ &+\sinh(kL)\cdot\sin(kL)\cdot\sin(kx)-2kx+\sinh(kx)\\ W = &-0.001262134811\cdot\sinh(1.570796327x)-\\ &-0.001262134811\cdot\sin(1.570796327x)+\\ &+0.001376145229\cdot\cosh(1.570796327x)-\\ &-0.001376145229\cdot\cos(1.570796327x)+\\ &+0.003965113454x,\end{aligned} \tag{9}$$

where $k = \dfrac{\pi}{L}$.

For the numerical analysis with subsequent experimental verification, following initial data are accepted: specific weight of material is $\gamma = 7800$ kg/m³; Young modulus is $E = 2{,}1\cdot 10^{11}$ Pa; rod's diameter is $d = 0{,}55\cdot 10^{-2}$ m; rod's length is $L = 1$ m.

Calculated values are:

$$m = \gamma\cdot\frac{\pi d^2}{4} = 0{,}1853\ \text{kg/m},\ J = \frac{\pi d^4}{64} = 0{,}1853\,M^4.$$

DOI: 10.1051/matecconf/201712903021

According to [2], the frequency of the first tone of the oscillations is equal to $\omega_1 = \frac{\lambda_1^2}{L^2}\sqrt{\frac{EJ}{m}} = 25.0.8\ s^{-1}$, where for the first mode $\lambda_1 = 1{,}875$ was accepted.

Period of oscillations' first tone is $T_1 = \frac{2\pi}{\omega_1} \approx 0.25$ s; further accepted: $p = \frac{\omega_1}{2} = 12.54 s^{-1}$; so in this case the total motion time is $T = \frac{2\pi}{p} = 0.5$ s.

Graph of the function $W_1(x,t) = W_1(x)\cdot\sin(pt)$ when $T \geq t \geq 0$ is shown in fig.2.

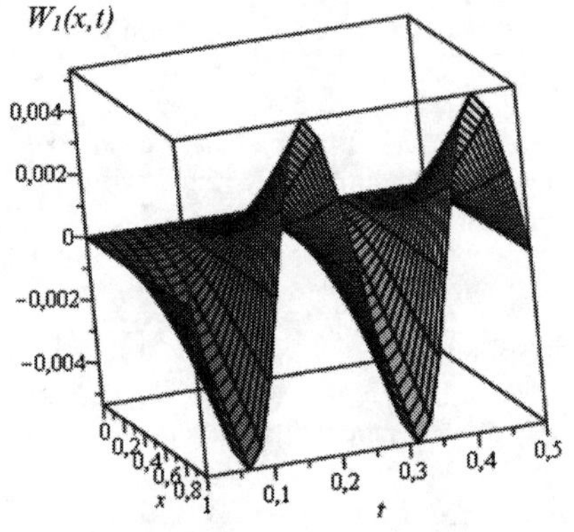

Fig. 2. Oscillations of the rod during relative motion

As it comes from the fig.2, at the moment $t = T$ comes relative rest, and, taking into account translational rest (fig.3), absolute rest will be achieved.

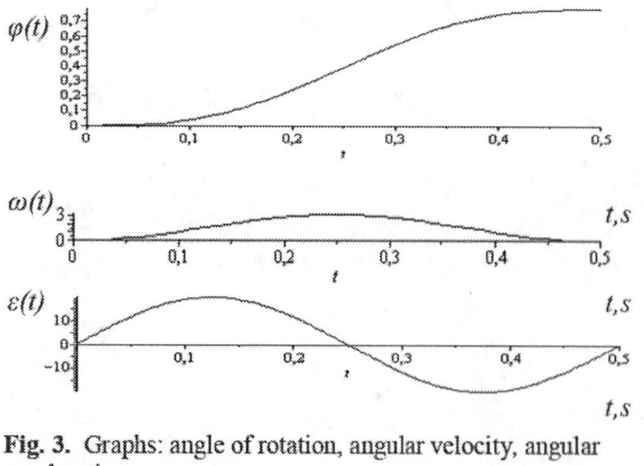

Fig. 3. Graphs: angle of rotation, angular velocity, angular acceleration

4 The experimental procedure

1. During the experimental research following equipment has been used: controller Simatic S7-300 CPU-313C; specialized controller SEC-AC-305 of the servomotor; rod with length L=1m and diameter d=0,0055m; accelerometer MPU 6050, attached to the end of rod and connected to Arduino Mega 2560.
2. According to the program downloaded in the controller S7-300, the output voltage was calculated and applied to the input of the servo controller SEC-AC-305 and converted into the corresponding angular velocity of the output shaft of the servomotor.
3. Results of the experimental observations of relative oscillations were approximated by analytical function using MathCAD (Genfit function).
4. The maximum experimental relative displacements of the end of the rod are comparing with the theoretical ones, and the relative error is estimating.

5 Conclusion

The results of experimental studies confirm the practical possibility to carry out, with sufficient accuracy, skew-symmetric control laws of translational rotary motion of elastic objects with distributed parameters. Optimal motion controls for the objects with finite stiffness could be applied in a various field of technologies.

References

1. V.A. Svetlitsky, *Collection of problems on the theory of oscillations* (Vysshaya shkola, Moscow, 1973).
2. V.L. Biderman, *Theory of mechanical oscillations* (Vysshaya shkola, Moscow, 1980).
3. I.M. Babakov, *Theory of oscillations* (Nauka, Moscow, 1968).
4. Y.G. Panovko, *Introduction to the theory of mechanical oscillations* (Nauka, Moscow, 1980).
5. I.A. Karnovsky, *Methods for optimal control of oscillations of deformable systems* (Vishcha shkola, Kiev, 1982).
6. A.G. Butkovsky, *Methods for managing systems with distributed parameters* (Nauka, Moscow, 1975).
7. N.N. Krasovskii, *Theory of motion control* (Nauka, Moscow, 1968).
8. A.I. Bokhonsky, S.Y. Zolkiewski, *Modeling and analysis of elastic systems in motion* (Wydawnictwo Politechniki, Gliwice, 2011).
9. A.I. Bokhonsky, N.I. Varminskaya, *Variational and reversion calculus in mechanics* (Publishing house of SevNTU, Sevastopol, 2012).
10. A.I. Bokhonsky, N.I. Varminskaya, M.I. Mozolevsky, *Optimum control of the translational motion of deformed objects: theory and technical applications* (Publishing house of SevNTU, Sevastopol 2007).
11. A.I. Bokhonsky, *Reversible Optimality Principle* (Vuzovsky uchebnik: INFRA, Moscow, 2016).
12. A.I. Bokhonsky, *Actual problems of variations calculus* (Palmarium Academic Publishing, Saarbrücken, 2013).
13. A.I. Bokhonsky, A. Buchacz, M. Placzek, A.Wrobel, *Modeling and investigation of discrete-continuous*

DOI: 10.1051/matecconf/201712903021

vibrating mechanic systems with damping (Wydawnictwo Politechniki, Gliwice, 2011).

14. A.I. Bokhonsky, M.M. Maistrishin, E.O. Balakanov, I.V. Gusakov, *Experimental verification of optimal control of the translational motion of a deformable object,* Optimization of production processes: collection of scientific papers, **12,** pp.66-72, (2010).

 DOI: 10.1051/matecconf/201712903022

Modeling of monitoring processes of structurally heterogeneous technological objects

Aleksandr Skatkov[1], *Victoria* Shevchenko[1], *Dmitry* Voronin[1,*], and *Dmitriy* Moiseev[2]

[1] Sevastopol State University, IT&CS department, 299053 Sevastopol, Russian Federation
[2] ChVVMU, Sevastopol, Russian Federation

Abstract. The tasks of technological information capture, taken from industrial enterprise monitoring systems, and its further transfer to the level of management and decision-making are considered in the article. The problems arising at the stage of integration of isolated automation systems into a single information space of the enterprise have been analyzed. A model of the monitoring system of structurally heterogeneous technological objects has been proposed. It allows to estimate events' probabilities of the following types: successful information processing, computing equipment downtime.

1 Introduction

The modern automated control system of an industrial enterprise, as a rule, has an open architecture, including the following levels: object level (Field Management); level of data capture and dispatching (Process Management); enterprise resource management application level (Business Management) [1, 2].

The lower level (object level) contains different technological devices – sensors, controllers, electric drives, actuators – that are used for the implementation of regulatory and management actions on the primary controlled objects. Information flows got from lower-level systems should be available for all top-level applications and users. They can be used, for example, for analysis and control operations. The next level of management comprise SCADA (Supervisory Control And Data Acquisition) and DCS (Distributed Control Systems) systems. This level provides the following processes: 1) secondary processing of data obtained from primary monitoring systems; 2) organization of information storage and management of technological processes in production.

The production management systems are situated at the third level of the hierarchy. The main functions of this level are the following:

1) automated control systems integration;

2) creation of a common information space for production and technical data;

3) real-time processing and presentation of data got from the lower levels of automation. There are the following classes of information-control systems: MRP (Material Requirements Planning), MRP II (Production Resource Planning), ERP (Enterprise Resource Planning) и ERP II (Resource Management and Enterprise Relationships) [3, 4].

The actual task of industrial enterprises automation contains a single information space creation that will be able to perform the following functions:

- objective and rapid assessment of the current situation in the manufacture process;
- rapid optimal decision-making in enterprise management;
- elimination of information and organizational barriers between management and technological levels.

Technical, technological and organizational problems that arise at the stage of integration of heterogeneous systems, which control technological objects, are considered in this paper. A four-level model of data processing has been proposed. It allows to estimate the information processing quality for the entire CAM lifecycle, which starts from information capture (got from sensors and monitoring devices) and ends with distributed reporting information storage.

2 Features analysis of the of automated enterprise systems integration into a single information space

Integration of the existing enterprise automated systems into a single information and management economic infrastructure usually is carried out in the two directions: «horizontal» and «vertical».

The goal of horizontal integration is to ensure the real-time capture of data got from all systems that carry out control and management functions of technological process. They are the following: heterogeneous and territorially distributed automated process control systems (APCS), telemechanic systems, automated workplaces of manual information input, etc. The complexity of horizontal integration lies in the territorial

* Corresponding author: dima@voronins.com

DOI: 10.1051/matecconf/201712903022

distribution of existing APCS, variety of their data formats and software-technological tools.

Vertical integration provides automation of data exchange between the level of APCS and enterprise management level. To realize vertical integration, it's necessary to implement a reliable and efficient system providing processing and transmitting of data flows between APCS and automated enterprise management complexes. The following aspects should be mentioned: 1) APCS process data in real time; 2) automated enterprise management complexes use aggregated (synthetic) information for a certain period.

Analysis of modern information management systems architectures [5] allows to single out a generalized information exchange scheme that takes into account the horizontal and vertical integration of automated systems (fig. 1).

Thus, the data sources can be divided into three types: technological, laboratory and management. Technological are got from SCADA, DCS, PLC (programmable logic controller); laboratory were received from LIMS (Laboratory Information Management System), management – from the control room workstations. The main components of the information exchange scheme are data processing systems (DPS), which in turn are determined by a combination of program modules, data sets and technical means that ensure their storage and processing. Data is transmitted to real-time databases and DPS using specialized interfaces. Then it is accumulated in the specialized archives of long-term storage. The calculation of technological, technical and economic indicators is performed based on the collected input information and values from archival sources in DPS.

The storage duration of archives has two terms: short and long. By the degree of data retrieval the system files can be divided into the following types:

1) Realtime databases archives, which store information about the parameters of the production system obtained from monitoring;

2) Production time databases, which store information for technological purposes;

3) Databases and datamart designed to provide rapid information transfer to external systems.

Data archives also contain normative and reference information, supporting the decision-making process of operational dispatch and staff control. For these purposes the system includes specialized knowledge bases, analytical and simulation models complexes. To support the system structure management a complex of engineering administration is used. Special client applications allow to view the archived data using the required views or forms (graphics, tables, etc.). Monitoring systems used for distributed technological objects, as a rule, are not allocated to a separately functioning information system. Thus, they are realized

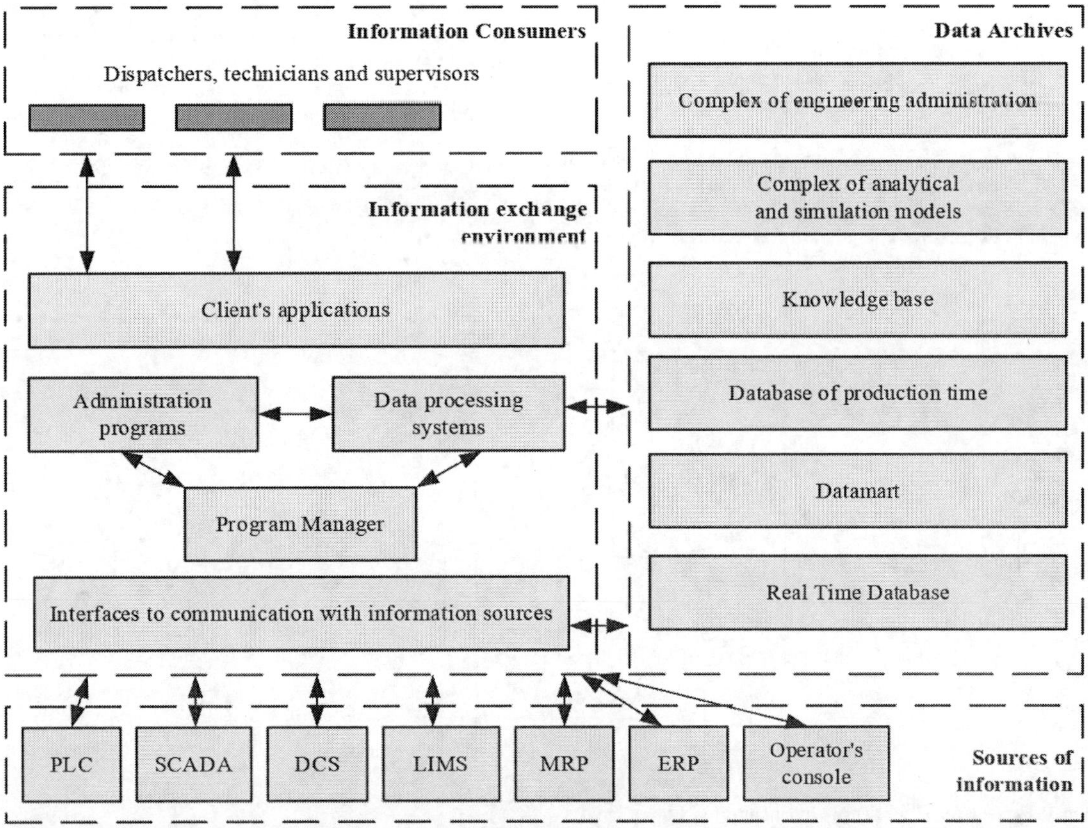

Fig.1. The scheme of information exchange in automated enterprise systems

 DOI: 10.1051/matecconf/201712903022

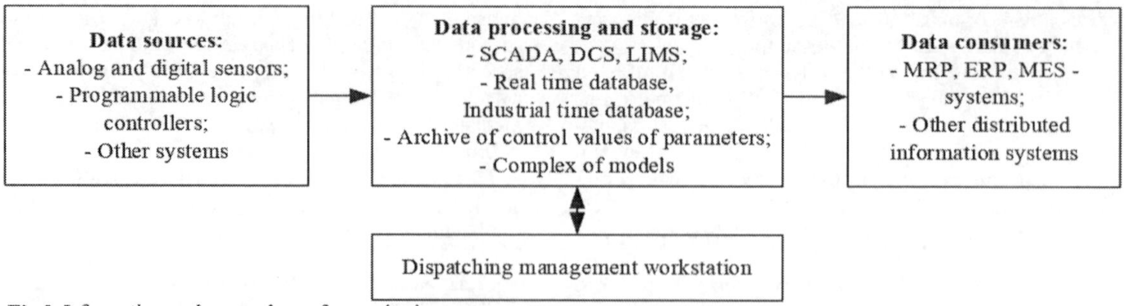

Fig.2. Information exchange scheme for monitoring systems

as a network of interacting information subsystems. This is due to the complexity of the structural and functional organization of the control object, large volume of controlled and processed parameters, territorial distribution, etc. Information exchange scheme for monitoring systems is represented on the fig. 2. Nontriviality of the problem of vertical integration – based on data got from monitoring information systems, APCS and CAM – is caused by the following factors:

- There is a large amount of primary data, and, as a result, there are problems with productive access to them by communication channels and storage systems;
- Complexity of integration processes administration and so on;
- Redundancy of primary data;
- The need to support a large number of industrial protocols for data access;
- Various manufacturers of APCS;
- The need to ensure information security in the exchange of data at the interface between the office and technological networks.

It should be noted that vertical unidirectional integration is required in the overwhelming majority of cases where monitoring systems are the sources and external information systems are the receivers of data.

3 Conceptual model of the monitoring system for structurally-heterogeneous technological objects

The main subsystems of distributed monitoring system are the following: subsystem of collection, accumulation, storage, retrieval of information dissemination [6] and data analysis subsystem [7]. Cloud computing technologies are used in modern industrial automation systems in order to save computing resources and to improve the management of information exchange and integration. A generalized scheme of the monitoring system is shown in fig. 3, a detailed description of data processing and storage networks based on cloud technologies is given in [8].

Regardless of the configuration of the computing infrastructure, when performing labor-intensive functional tasks in the cloud computing environment, the process of collecting and processing data, as a rule, includes four main stages [9]:

1) Data capture and data transfer from primary monitoring systems for multidimensional objects (sensors).

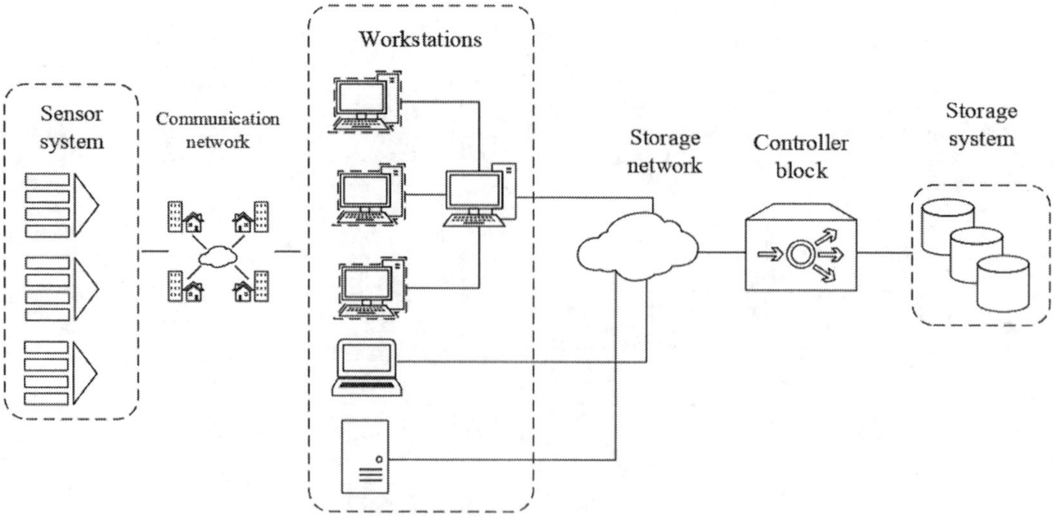

Fig.3. The scheme of the distributed monitoring system

DOI: 10.1051/matecconf/201712903022

2) Primary data processing (filtering, classification and clustering processes).

3) Entering user data about the research process.

4) Archiving the information received during the working period.

Next, the formal model of the monitoring system for structurally-inhomogeneous objects is considered. It has been constructed using the approaches of the theory of queuing systems and graph representation of states and transitions of the system. The process of monitoring distributed production facilities can be represented by a four-phase queuing system with a finite number of states. The mentioned stages of data processing correspond to these phases (fig. 4).

In the first phase of processing, it is assumed that the number of sensors in the monitoring system is predetermined and data processing cycle – consistent, i.e. the information from the next sensor is processed only after processing of the previous data. In this case, the part of the computing infrastructure responsible for the current stage of data processing can be represented as a single-channel processing device.

The second phase of processing corresponds to the stage when the transmitted primary monitoring data is filtered and combined into clusters according to the classification characteristics indicated by users of the monitoring system (corresponds to the block of workstations in fig.3). At this stage, data belonging to the same cluster is combined into packets and is labeled for further processing. The model for the second phase of processing can be described using a multi-channel queuing system with limited queue. The issues of data clustering and description of the characteristic features of individual data packets lie outside the scope of the considered problem.

The third phase corresponds to the stage where data is consistently processed and supplemented by users. Status of data describe the class of its archival storage ("working array / online archive / long-term storage"). At this stage, the only channel of data processing is the workstation server.

The fourth phase corresponds to the stage of data archiving. To model the storage network, this phase can be represented as a multi-channel queuing system with a multidimensional incoming stream and a limited queue. A multidimensional input stream consists of three parts: the flow of writing data to a working array; operational archive; long-term storage archive.

The limitation of monitoring system consideration in conditions of stationary leads to serious systemic errors in decision-making. The reasons for the stationary disturbance can be the following: failures in the technological process of data processing, loss of information and retransmission of data in transmission networks and data storage networks, possible technical problems with the cloud infrastructure equipment or software, the human factor, etc. Therefore, more adequate is to present the monitoring process of a structurally heterogeneous object in the form of a multi-phase non-stationary queuing system with variable values of the data entry and processing intensities. The development of the analytical model was based on the provisions works [10]. They have been extended taking into account the transition from a single-phase to a four-phase model and the specific features of distributed computing for production processes automation.

4 Analytical model of the monitoring system

The first processing phase is considered. It contains capturing and transfer operations with data got from the multidimensional object monitoring sensors. Consistently the system input receives N processing requests from the N system sensors. In order to simplify the analytical calculations let's suppose, that the distribution of time intervals between the moments of incoming requests are described by an exponential law with intensities $\{\lambda_1^{(k)}, \ldots, \lambda_i^{(k)}, \ldots, \lambda_N^{(k)}\}$ depending on the request's number, where k – is the processing phase number (k = 1,4), i – is the request's number. The law of processing times distribution is exponential (with intensities also depending on the request's number – $\{\mu_1^{(k)}, \ldots, \mu_i^{(k)}, \ldots, \mu_N^{(k)}\}$ respectively).

The state of the system at each time is characterized by the number of requests, denoted as $i=\{0, N\}$, and the number of processed requests, denoted as $j=\{0, N\text{-}i\}$. $P_{k,i,j}(t)$ denotes the probability of the event comprising the system's staying in the (k, i, j)-th state. The described system can be represented as a Markov chain with a discrete set of states and a continuous time.

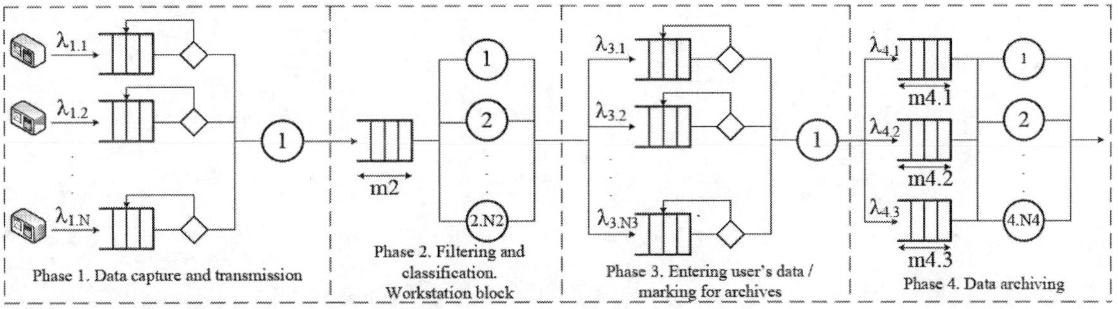

Fig.4. The scheme of the model of the data storage system used for objects monitoring

The system status number is coded as: <*Processing phase number* > – is a number from 1 to 4, corresponding to the number of the data processing phase in the distributed transmission and storage monitoring network; < *Line number*> – is a number of unprocessed requests in the system; < *Column number* > – is a number of requests that have been already processed in the system.

Transition from the state (k, i, j) into the state $(k, i+1, j)$ indicates that the $(k, i+j+1)$-th request has entered the system. Transition from the (k, i, j)-th state into the $(k, i-1, j+1)$-th state indicates that the $(j+1)$-th request has been processed. For example, at the beginning of the simulation, the system is in the (1, 0, 0) state. Then the system input receives a processing request – the transition to the (1, 1, 0)-th state accrues. Furthermore, if the system receives another request, it will come to the (1, 2, 0)-th state. If the single request have been successfully processed (data transfer to workstations in the primary processing system, see fig. 4) the system come in (1, 0, 1)-th state.

Total number of states N_c is calculated by the formula (1) [11].

$$N_c = \sum_{k=1}^{k=N+1} k \quad (1)$$

A diagram of the transitions between system states for the first processing phase is shown on the fig.5.

To calculate the values of the probabilities of finding a system in each of the states, it is necessary to solve Kolmogorov's system of differential equations relatively to $P_{k,i,j}(t)$, where i,j – th equation in general form can be represented as (2).

$$\begin{aligned}\frac{dP_{k,i,j}}{dt} &= \delta(i)(P_{k,i-1,j}\lambda_{k,i+j} - P_{k,i,j}\mu_{k,j+1}) + \\ &+\delta(j)P_{k,i+1,j-1}\mu_{k,j} - \\ &-\delta(N-i-j)(P_{k,i,j}\lambda_{k,i+j+1})\end{aligned}, \quad (2)$$

where $\delta(x)$ – Heaviside function.

To determine the values of the probabilities of finding the system in states (k,i,j) the system of differential equations (2) should be decided on $P_{k,i,j}(t)$ under the following initial conditions: $P_{k,i,j}(0)=1$, if $i+j=0$ and $P_{k,i,j}(0)=0$ in other cases. The system of differential equations (2) can be solved in various ways, for example, using the Runge-Kutta method, which has an advantage of the simplicity of software implementation. The probability of i requests being in the system at each point in time can be represented as (3).

$$P_{k,i}(t) = \sum_{j=0}^{N-1} P_{k,i,j}(t) \quad (3)$$

The (4)-th equation represents the expected number of requests in the system. The (5)-th equation describes the probability of j requests proceeding at each point of time. The probability of proceeding of not less than q requests at each point of time can be found using the (6)-th equation.

$$N_i(t) = \sum_{i=1}^{N} iP_{k,i}(t) \quad (4)$$

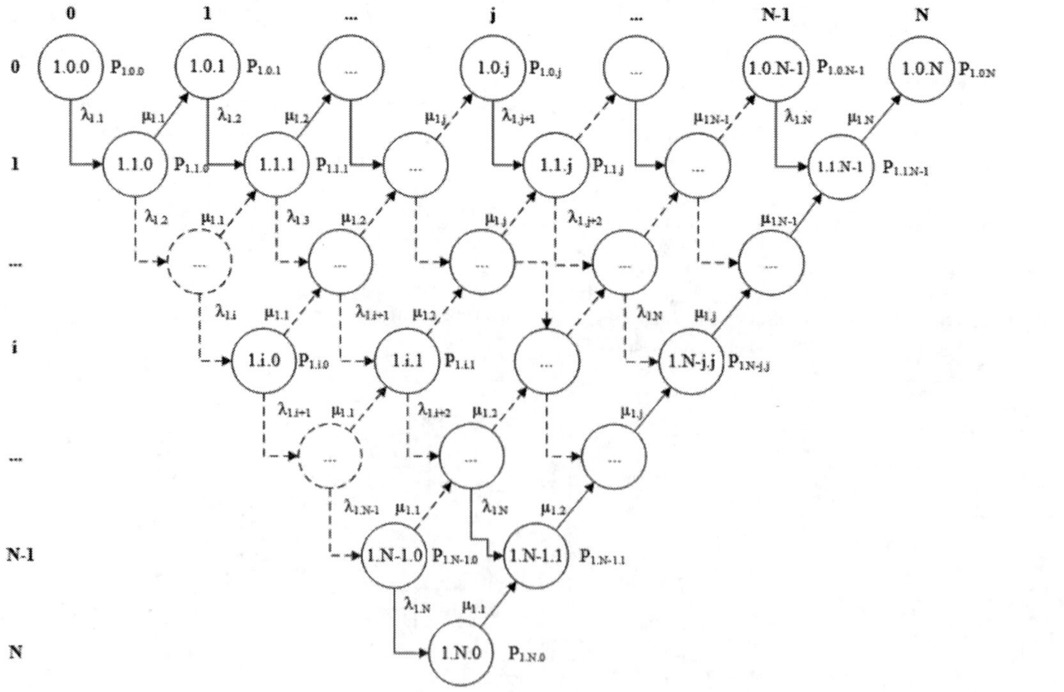

Fig. 5. Diagram of transitions between system states for the first phase of data processing in the monitoring system

$$P_{k,j}(t) = \sum_{i=0}^{N-j} P_{k,i,j}(t) \qquad (5)$$

$$P_{k,q}(t) = \sum_{i=q}^{N-j} P_{k,j}(t) \qquad (6)$$

Sources of non-stationarity of the system are the intensity of the arrival and proceeding time of requests, depending its number. An algorithm for calculating the probabilities of a similar type of model is given in works [10, 11].

The first data processing phase was considered to illustrate to usage of the proposed model. The following parameters were used: 1) arrival rates (requests per second): λ_1=0.085; λ_2=0.08; λ_3=0.075; λ_4=0.07; λ_5=0.065; λ_6=0.06; λ_7=0.055; λ_8=0.05; λ_9=0.045; λ_{10}=0.04; 2) intensiveness of request processing: μ_1=0.18; μ_2=0.14; μ_3=0.1; μ_4=0.6; μ_5=0.2; μ_6=0.1; μ_7=0.08; μ_8=0.07; μ_9=0.06; μ_{10}=0.05.

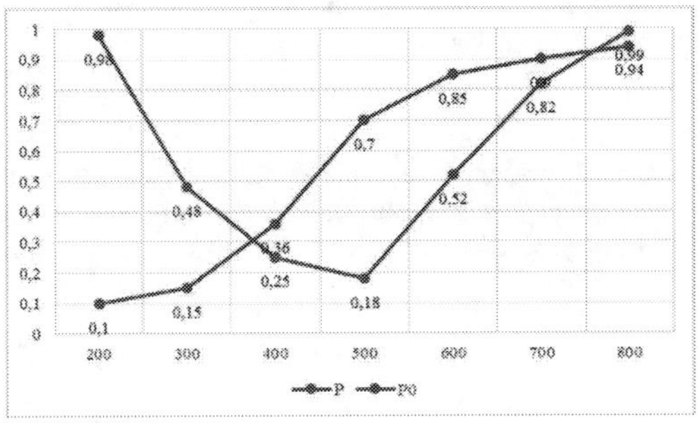

Fig. 6. The results of the calculation of the model of the first data processing phase in the monitoring system

On the fig. 6 the P curve shows the relationship between the probability of requests processing and the simulation time. Curve P0 shows dynamics of change in the probability of monitoring system's downtime.

5 Conclusion

The analytical model of the monitoring system, which has been proposed in this paper, can be used as an element of analytical and simulation models complex used in a decision-support system managing the integration of automated enterprise systems into a single information space. Its usage makes it possible to estimate such parameters as: probability of processing data streams by types; to determine the maximum package amount of requests that can be guaranteed processed in the system. These parameters are essential in assessing the quality of information processing for distributed heterogeneous monitoring objects.

Prospects for further research. Later on, based on the conceptual model of a four-phase queuing system, which has been proposed in this article, it is planned to develop a complex of simulation models for estimating the average processing time for requests packets by types, taking into account the dynamics of value and aging of monitoring information. It is also planned to conduct model experiments to fill the knowledge base of the decision support system for managing the integration of automated enterprise systems into a single information space.

This work was supported by the Russian Foundation for Basic Research (project № 15-29-07936).

References

1. A. A. Ambarcumjan, A. S. Hadeev, *Analysis of the functionality of maintenance management systems and equipment repair,* Problemy upravlenija **6,** pp. 2-12, (2005).
2. A. V. Vicentij, E. V. Harionovskij, *Technologies of data access in modern SCADA-systems,* Trudy Kol'skogo nauchnogo centra RAN **7,** pp. 203-210, (2011).
3. A.A Musaev, *Integration of automated control systems for large industrial enterprises: principles, problems, solutions,* Avtomatizacija v promyshlennosti **10,** pp. 40-45, (2003).
4. Chyong Din Tjau, *Interaction of open systems of industrial automation state and problems,* Informacionno-upravljajushhie sistemy **2-3,** pp. 52-57, (2003).
5. D.P. Volodin, A.M. Marasanov, T.A. Tihonova, *Architecture of distributed enterprise information system,* Matematicheskoe modelirovanie **8,** pp. 31-36, (2002).
6. I.V. Hmel'nickaja, *Methodological approaches to monitoring large-scale socio-economic systems,* Jekonomika i upravlenie **1(33),** pp. 100-103, (2012).
7. S.P. Kovaljov, *Aspect-oriented approach to the design of monitoring systems for large-scale objects,* Problemy informatiki **3(4),** pp. 5-18, (2009).
8. *Information Storage and Management: Storing, Managing, and Protecting Digital Information in Classic, Virtualized, and Cloud Environments* (Piter, Saint-Petersburg, 2016) 544.
9. A. Skatkov, E. Mashchenko, V. Shevchenko, D. Voronin, Actors *Interactions Research in Cloud Computing Environments Using System Dynamics Methodology,* Proceedings of the 18th FRUCT & ISPIT Conference, Saint-Petersburg, Russia. FRUCT Oy, Finland, pp – 612-619 (2016).
10. V.P. Bubnov, V.I. Safonov, *Development of dynamic models of non-stationary service systems* (Lan', Saint-Petersburg, 1999) 64.
11. V.P. Bubnov, A. S. Eremin, S. A. Sergeev. *Features of software implementation of the numerical and analytical method for calculating models of non-stationary service systems,* Trudy SPIIRAN **1(38),** pp. 218-232, (2015).

Numerical modeling of mechanical behavior of clinch connections at breaking out and shearing

Dmitri V. Berezhnoi[1,*], *Reza* Shamim[1], and *Irina S.* Balafendieva[1]

[1]Kazan (Volga Region) Federal University, 18 Kremlyovskaya Street, Kazan, 420008, Russian Federation

Abstract. This article describes an approach to constructing the defining relationships between increment of true stresses and true deformations, with considering the contact interaction of elastoplastic deformed bodies among each other. Within the framework of finite element method, solving these problems in case of "breaking out" and "shearing" in the clinch joint, the stress fields in the zone of the clinch connection are defined, and recommendations are given for realizing the process of their creation.

1 Introduction

The article assigned to investigate a contact interaction structure in elastic-plastic condition. Here had brought monographs [1-6] and articles [7-22] as classical examples, it should be noted that in all these mentioned works occurs large deformations. In these problems, one of the most apprehensions to build calculation contrivance is the decomposition of total deformations or their velocities into the elastic and plastic components. As the one of the suggested methods is the additive decomposition either the velocity of deformation tensor, or the multiplicative expansion of the strain gradient tensor. However, current work is colourful considerably the problem was solved both for the axisymmetric case and for the three-dimensional state. We called three deformations of elongations and three components of shear deformations, which had shown further as true strain. The units of appropriate areas after had distorted ascribe with the true stresses. The method, which used incompletely provided in these works [23-29].

2 Statement of the problem

We take the components of true deformations as relative elongations of linear elements (summation over i)

$$\varepsilon_{ii} = \varepsilon_i = (1+2e_{ii})^{0.5} - 1 \qquad (1)$$

and sinuses of the shear angles are

$$\varepsilon_{ij} = \sin\gamma_{ij} = 2e_{ii}\,(1+2e_{ii})^{-0.5}\,(1+2e_{jj})^{-0.5}, \qquad (2)$$

where

$$e_{ij} = (u_{i,j} + u_{j,i} + u_{k,j}\,u_{k,i}) / 2, \qquad (3)$$

and u_i – components of displacement vector. True stresses according to V.V. Novozhilov [30] are the components of stress vectors in expansion in orthonormal primary basis vectors in the deformed state of the body σ_{ij}, which through them can express the generalized (according to V.V. Novozhilov) stresses τ_i in next equation

$$\tau_{ij} = (S_i\,\sigma_{ij}) / (S_{i0}\,(1+\varepsilon_j)). \qquad (4)$$

We can transform these relations to the form, which had written on the next line

$$\tau_{11} = \sigma_{11}\,(1+\varepsilon_2)\,(1+\varepsilon_3)\cos\gamma_{23} / (1+\varepsilon_1), \qquad (5)$$

$$\tau_{12} = \sigma_{12}\,(1+\varepsilon_3)\cos\gamma_{23} = \sigma_{21}\,(1+\varepsilon_3)\cos\gamma_{13} = \tau_{21}, \qquad (6)$$

where S_i and S_{i0} respectively are unit of deformed area and unit of the undeformed body (indices 1, 2, 3 are interchanged cyclically), although physical relationships are written in the form between the true stresses and strains [24].

To calculate elastoplastic deformation, the Huber-Mises plasticity criterion was chosen in the form of

$$\sigma_i - H(\varkappa) = 0, \qquad (7)$$

where $\varkappa$ – is the scalar hardening parameter, and

$$\sigma_i = (3\,s_{kl}\,s_{kl} / 2)^{0.5} \qquad (8)$$

it is the intensity of the true stresses, also components of the deviator of the true stress defined as s_{kl}. The relationship between increment of the components of tensor of true stress σ_{ij} and true strain tensor ε_{ij} at the stage of elastic-plastic deformation is adopted in form of the Prandtl- Reuss equations for hardening the material. For implementation of the contact interaction, the method, which used had presented in [24]. The problem was solved both for the axisymmetric case and for the three-dimensional state.

* Corresponding author: berezhnoi.dmitri@mail.ru

DOI: 10.1051/matecconf/201712903023

3 Analytical model and numerical results

3.1 Modling of the calculated area

In this work, we performed a numerical simulation of the mechanical behavior for clinch joints at "breakout" and "shear" modes. The process of creating clinch connection, as well as the task of "breakout" in clinch connection are realized within the framework of the axisymmetric case, in the task of "shearing" the clinch connection based on three-dimensional analytical model. The geometry of the computational area, sketchily shown in Figure 1. The forming limit curve of steel sheet obtained as result of the experiment and shown in Fig. 2. The material from which the matrix and the punch are made is considered absolutely rigid. The coefficient of friction between the steel sheets, as well as between the die and punch, is assumed equal $\alpha_{fr=0.1}$.

The calculation carried out, for the case when applied, clamping force $P = 9\ t$. For leading in Figure 1, should know the dimensions, which presented on next line $r1 = 0.013\ m$, $r2 = 0.033\ m$, $r3 = 0.07\ m$, $t1 = 0.006\ m$, $r1 = 0.004\ m$.

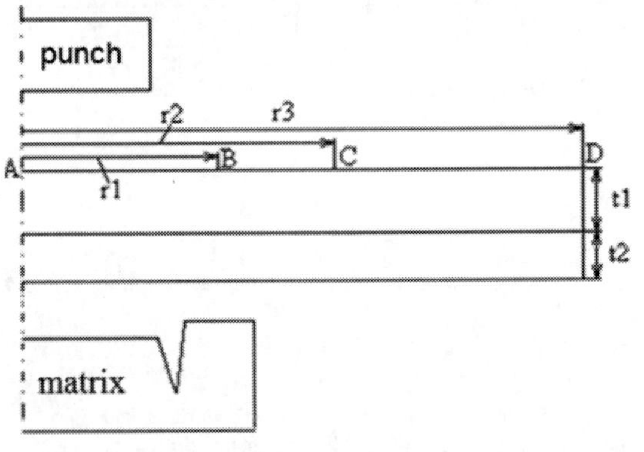

Fig. 1. Scheme of clinch connection zone.

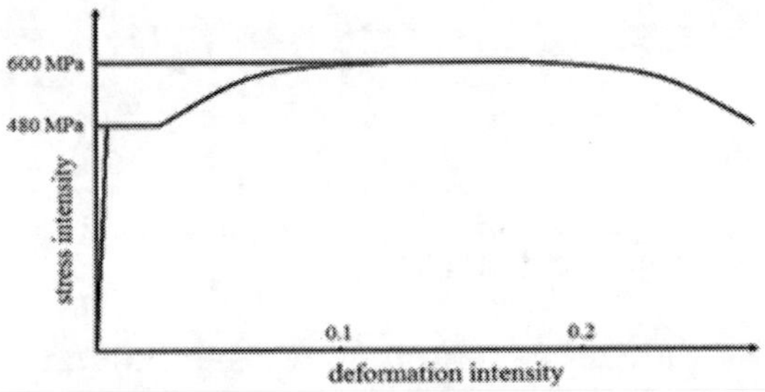

Fig. 2. The curve deformation of steel sheet material in clinch connection.

3.2 The task of creating a clinch connection

The task of creating a clinch-connection was solved in several stages. At the first stage, the clamping condition was reproduced in one-step. At the second stage, the punch was loaded kinematically. The loading is realized in 125 steps, loading step is $\Delta H = 0.0001$ m. At the third stage of loading, the punch and matrix are removed. The punch is removed in 125 steps (its kinematic loading in the opposite direction is carried out), loading step is $\Delta H = -0.0001$ m. At the fourth stage, the tension of the clamp is removed for 1 step. At the last fifth stage, the matrix is removed for 73 steps (from it the kinematic loading along the axis of the clinch joint has carried out), loading step is $\Delta H = 0.0001$ m. The thickness of the neck in this case is t=0.00089 m, and the value of the internal groove is $f = 0.00049$ m. In figure, 3 illustrated the dependence of the force in duration of its punching progress into the clinch joint made steel sheets. The maximum force in the punch measures up values $P = 357\ kN$. Figures 4 and 5 show, respectively, the intensity of the stress according to Mises and tangential stresses along the cross section of the steel sheets.

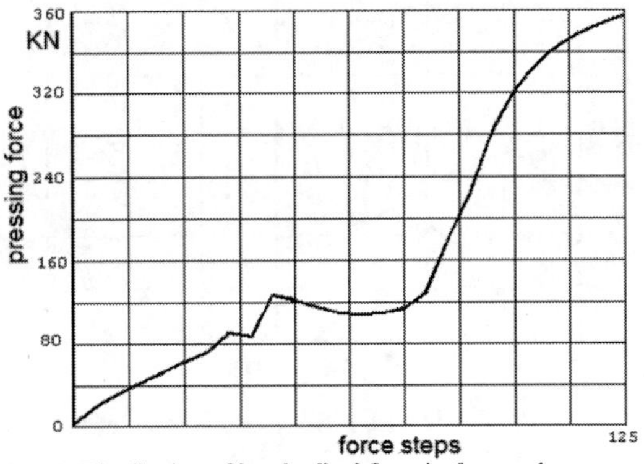

Fig. 3. Distribution of longitudinal force in the punch according to loading steps.

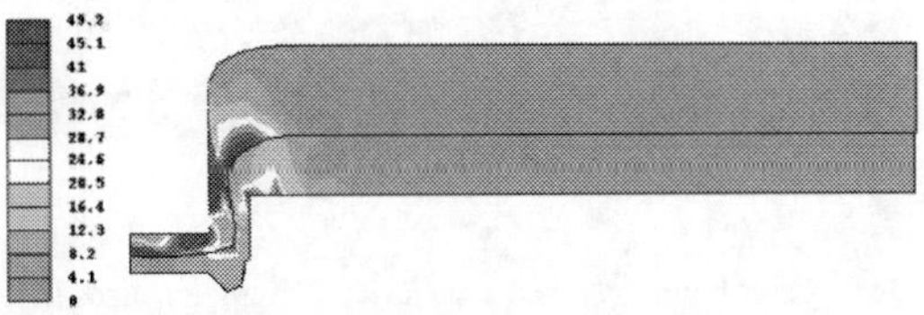

Fig. 4. Distribution of stress intensity according to Mises in the zone of the clinch joint after removal of the punch and matrix.

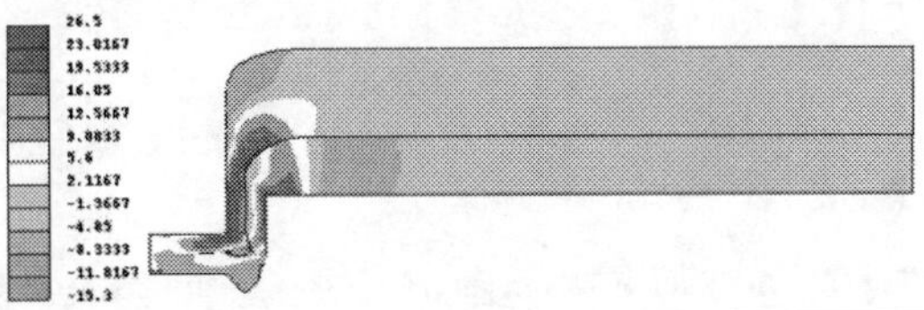

Fig. 5. Distribution of tangential stresses in the zone of the clinch joint after removal of the punch and the matrix

3.3 The task of "breaking out" in the clinch joint

The problem of "breaking out" in clinch joints in an axisymmetric state with considering the geometry of the clinch connection and the residual stresses on it, solved

DOI: 10.1051/matecconf/201712903023

earlier. The lower sheet of the clinch joint is fixed, the upper sheet pulled out from it (the kinematic loading is applied on at upper surface at the top sheet). The step of kinematic loading is equal to $\Delta H = 0.0001$ m. "Breaking out" duration occurs in 66 steps. Figure 6 illustrates the dependence of the value of the breaking out force in the process of separation of steel sheets from the clinch joint. Maximum pressure effort is $P_v = 96\ kN$. Figures 7 and 8 show, respectively, the intensity of the stress according to Mises and tangential stresses along the cross section of the steel sheets.

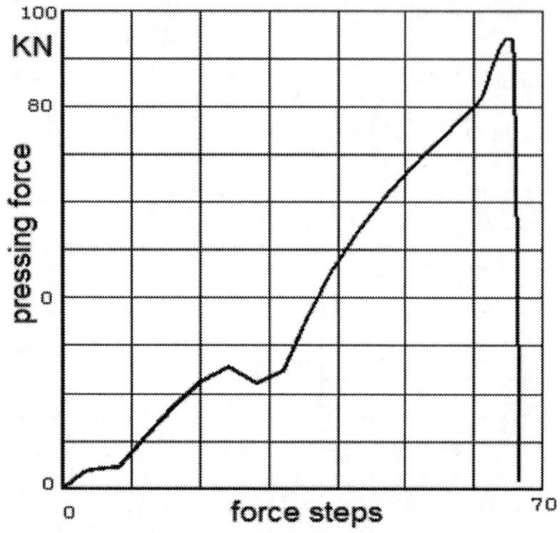

Fig. 6. Distribution of force in the "breakout" case in clinch joint by loading steps.

Fig. 7. Distribution of stress intensity according to Mises in the zone of the clinch joint after "breaking out" on the top sheet.

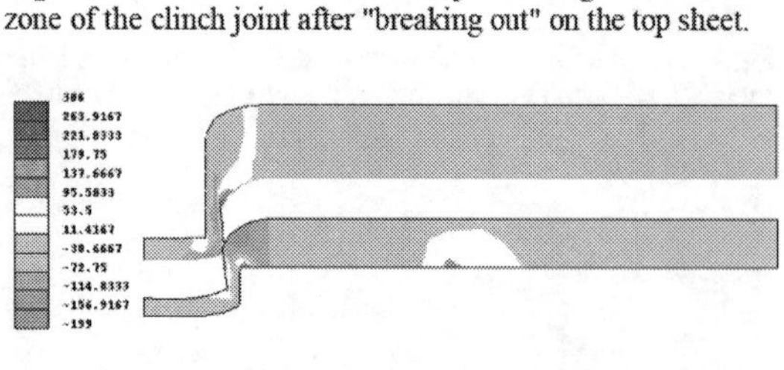

Fig. 8. Distribution of tangential stresses in the zone of the clinch joint after "breaking out" on the top sheet.

3.4 The task of “shearing “in the clinch joint

The task of “shearing “in the clinch joint is solved in a three-dimensional state, also solved with considering the geometry of the clinch connection and the residual stresses on it. The bottom sheet of the clinch joint is fixed, the upper one is shearing (the kinematic loading at outline points of the top sheet is specified). The step of kinematic loading is equal to $\Delta H = 0.0001$ m. The shear in this case happens in 9 steps, after which the computational process diverges. This moment can be considered as disconnecting point on the clinch joint.

Figure 5 illustrates the dependence of the amount of shear force during the separation of steel sheets of the clinch joint. Maximum shear force is $P_s = 69\ kN$. Figures 10 and 11 show, respectively, the intensity of the stress according to Mises and tangential stresses along the cross section of the steel sheets. By way of illustration, the stresses are given for the central cross section of the clinch joint (through this cross section passes a line connecting the point of application of shearing force and the clamping point).

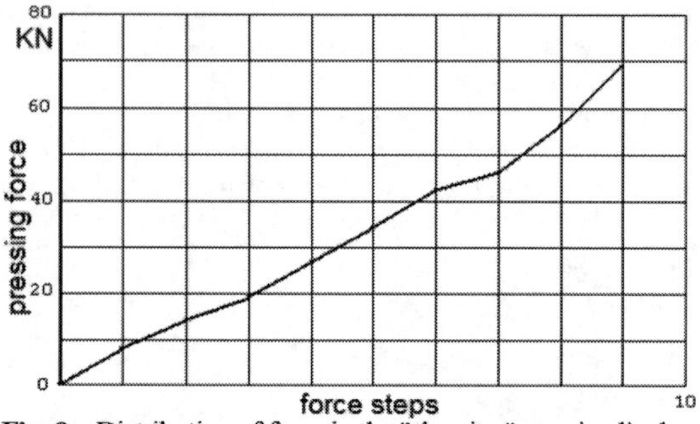

Fig. 9. . Distribution of force in the "shearing" case in clinch joint by loading steps.

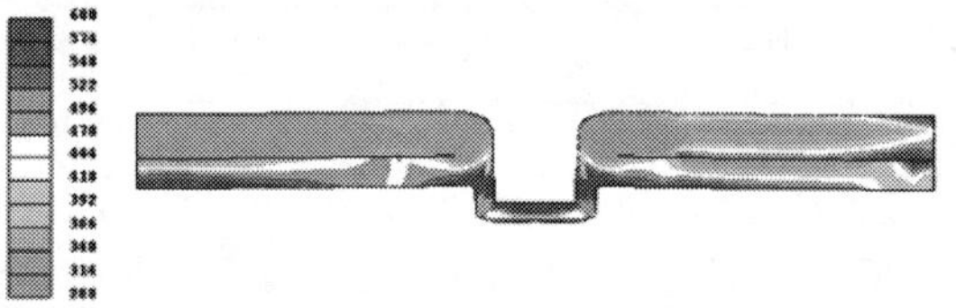

Fig. 10. Distribution of stress intensity according to Mises in the zone of the clinch joint after "shearing" on the top sheet.

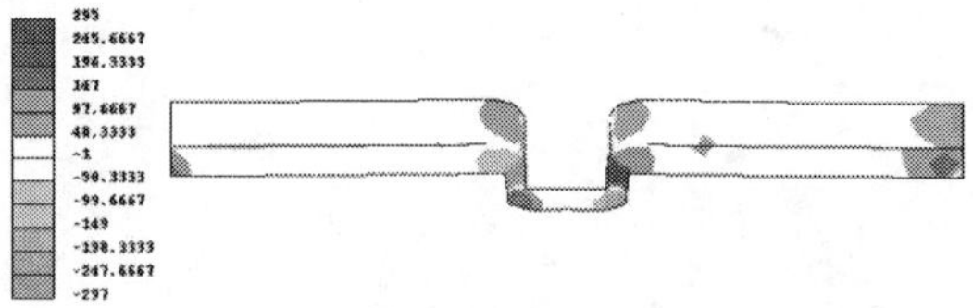

Fig. 11. Distribution of tangential stresses in the zone of the clinch joint after "shearing" on the top sheet.

4 Analysis of results and conclusions

For implement the creation of clinch connections, it is also not necessary to model the entire working part of the matrix: it is sufficient to know the geometry of the boundary of the upper part of the matrix. A special relation should be given for modelling the angular regions. The effect of the step size of kinematic loading on the convergence of the finite-element solution for various stages of the solution in this problem was investigated. It clarified that the value of kinematic

DOI: 10.1051/matecconf/201712903023

loading step directly depends on the degree of sampling in the calculated area at the selected finite-element partition, the step size should not exceed from characteristic size of the height of the final element in the zone of the punch implementing. The further reduction, leads only to increase in the time of solving the problem. The effect of kinematic consolidating on the lower surface of the lower steel plate was investigated. The computational experiment showed that the availability of the kinematic boundary conditions on this part (absence of vertical displacements), provides the clamping force which is applied to the top sheet of the clinch connection, practically does not influence the solution of the problem. The effect of the pressing force amount on top sheet of the clinch joint was studied. It is noted that the clamping force in 3 *t*, 5 *t*, 7 *t* is not sufficient to ensure that the steel sheets do not disperse during the creation of the clinch joint, and by force of 14 *t* it creates additional stress concentration zones and significantly changes the geometry of the lower steel sheet of the clinch joint.

The reported study was supported by RFBR and Government of the Republic of Tatarstan research projects No. 15-41-02555, 15-41-02557.

References

1. D. Kolarov, A. Baltov, N. Boncheva, *Mechanics of plastic mediums* (Mir, Moscow, 1979), 302 (In Russian).
2. A.S. Sakharov, V.N. Kislooky, V.V. Kirichevski, N. Altenbach, U. Gabbert, Y. Dankert, Kh. Kepler, Z. Kochyk, *The finite element method in the mechanics of solids* (Higher School, Kiev, 1982), 480 (In Russian).
3. P.V. Trusov, A.I. Shveikin, *The theory of constitutive relations*, (Perm State Technical University, Perm, 2008), 243 (In Russian).
4. S.N. Korobeynikov, *Nonlinear deformation of solids* (Novosibirsk, 2000), 262 (In Russian).
5. V.I. Levitas, *The large elastic-plastic deformation of materials under high pressure* (Naukova Dumka, Kiev, 1987), 232 (In Russian).
6. A.A. Pozdeev, P.V. Trusov, Y.I. Nyashin, *The large elastic-plastic deformation theory, algorithm, application* (Nauka, Moscow, 1986), 232 (In Russian).
7. A.I. Abdrakhmanova, L.U. Sultanov, Materials physics and mechanics, **26 (1)**, 30–32 (2016)
8. A. Golovanov, L. Sultanov, Applied mechanics, **41 (6)**, 614–620 (2005)
9. R. Davydov, L. Sultanov, (DEStech publications inc., USA, 2013), 64–67.
10. L.U. Sultanov, Procedia earth and planetary science **15**, 119–124 (2015)
11. R.L. Davydov, L.U. Sultanov, Journal of engineering physics and thermophysics, **88 (5)**, 1280–1288 (2015)
12. L.U. Sultanov, Lobachevskii journal of mathematics **37 (6)**, 784–790 (2016)
13. A.I. Abdrakhmanova, I.R. Gariffulin, R.L. Davydov, L.U. Sultanov, L.R. Fakhrutdinov, Applied mathematical sciences, **9 (118)**, 5907–5914 (2015)
14. I.B. Badriev, V.V. Banderov, G.Z. Garipova, M.V. Makarov, R.R. Shagidullin, Applied Mathematical Sciences **9 (82)**, 4095–4102 (2015)
15. I.B. Badriev, V.V Banderov, M.V. Makarov, V.N. Paimushin, Applied mathematical sciences, **9 (78)**, 3887–3895 (2015)
16. I.B. Badriev, M.V. Makarov, V.N. Paimushin, Russian mathematics **3 (59)**, 66–71 (2015)
17. I.B. Badriev, V.V Banderov, O.A. Zadvornov, Applied mechanics and materials), **392**, 188–190 (2013)
18. I.B. Badriev, M.V. Makarov, V.N. Paimushin, Procedia engineering **150**, 1056–1062 (2016)
19. O.A. Sachenkov, V.I. Mitryaikin, T.A. Zaitseva, Yu.G. Konoplev, Applied Mathematical Sciences **8 (159)**, 7889–7897 (2014)
20. O. Sachenkov, L. Kharislamova, N. Shamsutdinova, E. Kirillova, Yu. Konoplev, IOP Conference series, Materials science and engineering **98**, 012015 (2015)
21. R.R. Galiullin, O.A. Sachenkov, R.F. Khasanov, P.S. Andreev, International journal of applied engineering research **10 (24)**, 44855–44860 (2015)
22. F.A. Shigapova, R.F. Mustakimova, G.T. Saleeva, O.A. Sachenkov, International journal of applied engineering research **10 (24)**, 44711–44714 (2015)
23. D.V. Berezhnoi, V.N. Paimushin, V.I. Shalashilin, Mechanics of solids **6 (44)**, 837–851 (2009)
24. D.V. Berezhnoi, V.N. Paimushin, Journal of Applied Mathematics and Mechanics **75 (4)**, 447–462 (2011)
25. M.K. Sagdatullin, D.V. Berezhnoi, Applied Mathematical Sciences **8 (35)**, 1731–1738 (2014)
26. D.V. Berezhnoi, A.A. Sachenkov, M.K. Sagdatullin, Applied Mathematical Sciences **8 (127)**, 6341–6348 (2014)
27. D.V. Berezhnoi, A.A. Sachenkov, M.K. Sagdatullin, Applied Mathematical Sciences **8 (143)**, 7107–7115 (2014)
28. D.V. Berezhnoi, I.S. Balafendieva, A.A. Sachenkov, L.R. Sekaeva, IOP Conference Series, Materials Science and Engineering **158**, 012018 (2016)
29. M.R. Shamim, D.V. Berezhnoi, IOP Conference Series, Materials Science and Engineering **158**, 012083 (2016)
30. V.V. Novozhilov, *Fundamentals of nonlinear elasticity* (Gostechizdat, Moscow-Leningrad, 1948), 211 (In Russian).

Interrelation of the concept of "uncertainty" in the information theory and theory of measurements

Vadim Kopp[1], *Alexey* Balakin[1,*], *Natalya* Balakina[1], and *Vladimir* Preys[2]

[1]"Sevastopol State University", Sevastopol, Russian Federation
[2]"Tula State University", Tula, Russian Federation

Abstract. The problem of determining the optimal number of multiple measurements based on the type of the error distribution density of the measuring means is considered. The laws of distribution of a random variable are obtained, they ensure the extreme values of dispersion at specified entropy. It is indicated that the given law corresponds to the known compositional one which ensures maximum entropy under the restrictions on variation limits of the random variable and at specified dispersion. In particular, under certain conditions, it corresponds to an abridged normal, uniform and bimodal laws. It is shown that two dispersion values correspond to a single entropy value. This effect is provided due to the fact that the random variable is concentrated on a finite interval. The theorem on the interrelation of entropy and dispersion of a random variable is proved which allows us to reconcile the concept of uncertainty used in the information theory with the same concept used in the up-to-date standards of technical measurements. It is shown that an abridged normal distribution law providing maximum of entropy at specified dispersion, at the same time provides the minimum of dispersion at specified entropy, and the bimodal law provides the maximum of entropy at the maximum of dispersion. The conclusions are based on the solution of two variational problems with isoperimetric constraints. The results of modeling allowing to evaluate the correctness of the conclusions are presented.

The issues of quality improvement of products in the machine-tool industry are inextricably linked to the increase of the measurement accuracy [1, 2].

In many cases, the toughening of the requirements to the product quality lies in the implementation of more stringent requirements to the tolerances on controlled parameters which in its turn leads to an increase in the requirements for the accuracy of parameters' and product characteristics' measurement. Thus, one of the main tasks facing the metrological services of enterprises is to ensure the required accuracy of measurements performed at various stages of production. Besides, in conditions of acute market competition the requirements imposed on the quality of manufactured products also increase, which in its turn leads to the need of measurements accuracy improvement [3].

Having analyzed the literature we can conclude that improving the accuracy of measurements contributes to the development of both individual branches of the national economy and of thre world scientific and technological progress as a whole.

The methods used today in order to improve the accuracy of measurements are given in the interstate standardization recommendations CIS 64-2003 "SSEUM. Efficiency ensuring in the management of technological processes. Means and methods for measurements accuracy increase" [4].

One of the methods for increasing the accuracy of measurements is to use multiple measurements. This method allows to increase the accuracy of measurement without acquisition of new more accurate means of measurement. However, in this case, the question of the need to determine the minimum number of measurements necessary to ensure a given accuracy arises. For this purpose as indicated in [5, 6] it was proposed to determine the number of measurements based on the error distribution density of the measuring instrument on the basis of the expression:

$$n = \frac{\sigma^2}{\varepsilon^2},$$

where ε^2 - are the required fidelity values determined from the distribution density and the confidence level, σ^2 - is the dispersion of the measuring instrument.

Thus, the aim of the research is to determine the form of the error distribution density of the measuring instrument and to derive on its basis the relation connecting the dispersion with the differential entropy.

In most cases it is assumed that the law of error distribution of the instrument is normal. In some cases this is permissible and is widely used, and in some cases this introduces significant errors. Indeed, the error of the

* Corresponding author: AlBalakin@sevsu.ru

 DOI: 10.1051/matecconf/201712903024

device can not change from minus infinity to plus infinity. We are to consider the fact that the limits of the change in the random variable which is the error of the device are specified in the certification. In addition, it is proposed to determine the distribution density from the maximum entropy [7, 8, 9] of the random variable (for the worst case) by using the principle of Jain's maximum entropy [10]. The density of the distribution is defined as the solution of the variational problem [5]: to determine the density $p(x)$ that delivers the extremum to the functional

$$\max_{p(x)} \to \left\{ H_{\partial u\phi}(x) = -\int_a^b p(x)\cdot \log p(x)\ dx \right\},$$

under the following restrictions:

$$\int_a^b p(x)dx = 1; \qquad (1)$$

$$\int_a^b x^2 \cdot p(x)dx = \varepsilon^2 . \qquad (2)$$

The solution of the problem has the form:

$$p(x) = e^{\mu_1 - 1} \cdot e^{\mu_2 \cdot x^2} \ , \ a \le x \le b \ ,$$

where μ_1, μ_2 - are the Lagrange multipliers. They are determined from conditions (1), (2).

The solution of the problem shows that the maximum of entropy is achieved with the known compositional distribution law [5], in accordance with which at $\sigma^2 > \frac{(b-a)^2}{12}$ it is two-module, at $\sigma^2 = \frac{(b-a)^2}{12}$ – uniform, at $\sigma^2 < \frac{(b-a)^2}{12}$ – abridged normal. These cases are determined by the sign and the value of the coefficient λ_2. Here a and b are the ranges of variation of the random variable.

As is known, the maximum possible value of entropy $H_{\partial \max}$ is attained at $\sigma^2 = \frac{(b-a)^2}{12}$ and equals to $H_{\partial \max} = \log(b-a)$.

Experimental studies were carried out on two types of measuring instruments: with an inductive converter and with a raster photoelectric converter that confirmed that the law of distribution of their errors is abridged normal [11]. This case is provided when $\mu_2 < 0$.

In itself, the use of entropy as a measure of uncertainty is not entirely convenient and illustrative. As a measure of uncertainty the standard deviation σ of the measurement result [12, 13, 14] is used more often, uniquely related to the dispersion. Then the question arises: how do the entropy and dispersion of a random variable concentrated on a finite interval, and in this case distributed according to the compositional law, correlate.

In determining the entropy natural logarithms were used. As further studies showed the same value of entropy corresponds to two different values of the dispersion, which is clearly demonstrated on the graph (Fig. 1) constructed when $a = -1$, $b = 1$, $\sigma^2 \in [0{,}06;0{,}7]$.

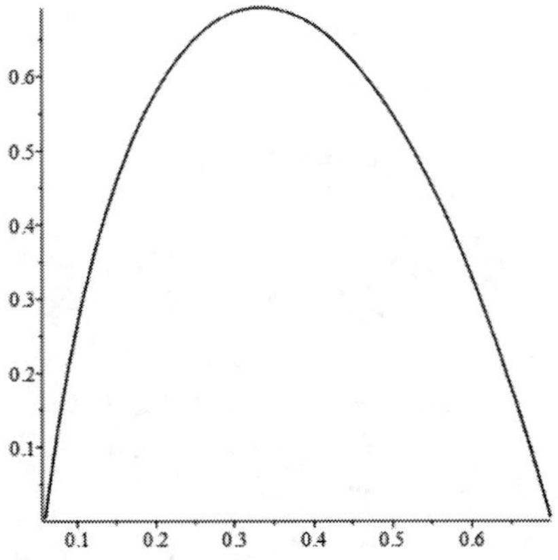

Fig. 1. Dependence of entropy on the value of dispersion

Is it true that this law provides the minimum and maximum dispersion T a given entropy. To this end, two inverse variational problems were being solved: the determination of the types of distribution densities ensuring the minimum and maximum dispersion of a random variable concentrated on a finite interval at a given entropy.

Both problems were solved for the case of a centered random variable.

Problem 1 which makes it possible to determine the form of the distribution density that ensures the minimum of the dispersion $D(x)$ for a given differential entropy H_{∂}, is formulated as follows.

To determine the density $p(x)$ that delivers an extremum to the functional

$$\min_{p(x)} \to \left\{ D(x) = -\int_a^b x^2 p(x)\cdot dx \right\}, \qquad (3)$$

under the restrictions:

$$-\int_a^b p(x)\cdot \ln p(x)\ dx = H_{\partial}, \qquad (4)$$

$$\int_a^b p(x)dx = 1 . \qquad (5)$$

Thus, in this case there is a variational problem for a conditional extremum with isoperimetric constraints.

The Euler equation for the functional (3) and constraints (4), (5) takes the form:

$$F_p^{*} - \frac{d}{dt} F_{p'}^{*} = 0 .$$

 DOI: 10.1051/matecconf/201712903024

Where $F^{*'}_{p}$, $F^{*'}_{p'}$ – are the derivatives of the Lagrange function with respect to $p(x)$ and with respect to $p'(x)$.

In this expression:

$F^* = F_0 + \sum_{j=1}^{n} \lambda_i F_i$; λ_i - are the constants; $n = 2$, and the functions F_0, F_1, F_2 have the form: $F_0 = -x^2 p(x)$; $F_1 = -p(x) \cdot \ln p(x)$; $F_2 = p(x)$.

Given that $\frac{d}{dt} F^{*'}_{p'} = 0$, we have:

$$F^{*'}_{p} = \left[F_0 + \left(\sum_{j=1}^{n} \lambda_i F_i \right) \right]' p = 0$$

Substituting this expression into the Euler equation and taking into account the expressions for F_0, F_1, F_2 we will finally obtain:

$$-x^2 - \lambda_1 [\ln p(x) + 1] + \lambda_2 = 0 .$$

This expression determines the form of the distribution density, which ensures the minimum of the dispersion $D(x)$:

$$p(x) = e^{-\frac{x^2}{\lambda_1}} e^{\frac{\lambda_2 - \lambda_1}{\lambda_1}} . \quad (6)$$

Substituting (6) into the constraints (4), (5) and numerically solving the obtained system of equations, we find the values of the coefficients λ_1, λ_2.

Having changed the variables and taking into account that

$$\lambda_1 = 2\sigma^2 ; \; e^{\frac{\lambda_2 - \lambda_1}{\lambda_1}} = M \text{, and } \lambda_1 > 0 ,$$

we obtain expressions for the density and distribution function in the following form:

$$p(x) = M e^{-\frac{x^2}{2\sigma^2}} ; \quad (7)$$

$$F(x) = M \int_a^x e^{-\frac{x^2}{2\sigma^2}} dx .$$

Then, in accordance with the Euler-Poisson integral and taking into account the truncation of the normal law, we have:

$$M = \frac{C}{\sigma\sqrt{2\pi}} , \quad (8)$$

where

$$C = \frac{\sigma\sqrt{2\pi}}{\int_a^b e^{-\frac{x^2}{2\sigma^2}} dx} .$$

Substituting (8) into (7), we obtain:

$$p(x) = \frac{C}{\sigma\sqrt{2\pi}} e^{-\frac{x^2}{2\sigma^2}} , \; F(x) = \frac{C}{\sigma\sqrt{2\pi}} \int_a^x e^{-\frac{x^2}{2\sigma^2}} dx .$$

Thus, the resulting law is abridged normal.

Consequently, the minimal dispersion at a given entropy, as well as the maximum entropy at a given dispersion, is provided by an abridged normal law.

Thus, it is proved that the law providing the given entropy and having a minimum dispersion, provided that $D(x) < \frac{(b-a)^2}{12}$ is an abridged normal law. Here $D(x)$ - is the dispersion of the obtained law.

Let us define the functional relationship between dispersion and entropy. We substitute expression (6) into the constraints (4), (5):

$$-\int_{-\beta}^{\beta} e^{\frac{-x^2}{\lambda_1}} e^{\frac{\lambda_2 - \lambda_1}{\lambda_1}} \cdot \ln e^{\frac{-x^2}{\lambda_1}} e^{\frac{\lambda_2 - \lambda_1}{\lambda_1}} dx = H_\partial , \quad (9)$$

$$\int_{-\beta}^{\beta} e^{\frac{-x^2}{\lambda_1}} e^{\frac{\lambda_2 - \lambda_1}{\lambda_1}} dx = 1 . \quad (10)$$

We submit expression (9) in the form:

$$-\int_{-\beta}^{\beta} e^{-\frac{x^2}{\lambda_1}} e^{\frac{\lambda_2 - \lambda_1}{\lambda_1}} (\ln e^{\frac{-x^2}{\lambda_1}} + \ln e^{\frac{\lambda_2 - \lambda_1}{\lambda_1}}) dx = H_\partial$$

Having transformed this expression, taking into account (10), we obtain

$$D(x) = (H_\partial - 1)\lambda_1 + \lambda_2 . \quad (11)$$

Expression (11) uniquely characterizes the relationship between the minimal dispersion $D(x)$ and the given differential entropy H_∂ under the condition $D(x) < \frac{(b-a)^2}{12}$.

Problem 2, which makes it possible to determine the form of the distribution density that provides the maximum of the dispersion $D(x)$ for a given differential entropy H_∂, differs from the previous one only in the form of the functional:

$$\max_{p(x)} \to \left\{ D(x) = \int_a^b x^2 p(x) \cdot dx \right\} .$$

Similarly solving this problem we get:

$$p(x) = e^{\frac{x^2}{\lambda_1}} e^{\frac{\lambda_2-\lambda_1}{\lambda_1}} . \qquad (12)$$

Provided that $\lambda_1 > 0$ the law obtained corresponds to the one bimodal.

The results of the simulation are shown in Fig. 2.

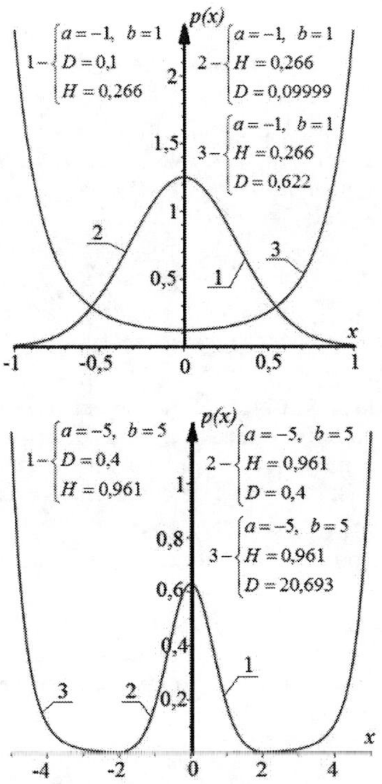

1 - abridged normal
2 - law ensuring minimum dispersion
3 - law ensuring maximum of dispersion

Fig. 2. Results of modeling

As for the problem of determining the dispersion minimum, we define a functional connection between the maximum dispersion $D(x)$ and the given differential entropy H_∂.

Let us substitute the expression for the distribution density (12) into the constraints (4), (5):

$$-\int_{-\beta}^{\beta} e^{\frac{x^2}{\lambda_1}} e^{\frac{\lambda_2-\lambda_1}{\lambda_1}} \cdot \ln e^{\frac{x^2}{\lambda_1}} e^{\frac{\lambda_2-\lambda_1}{\lambda_1}} dx = H_\partial , \qquad (13)$$

$$\int_{-\beta}^{\beta} e^{\frac{x^2}{\lambda_1}} e^{\frac{\lambda_2-\lambda_1}{\lambda_1}} dx = 1 . \qquad (14)$$

We represent expression (13) in the form:

$$-\int_{-\beta}^{\beta} e^{\frac{x^2}{\lambda_1}} e^{\frac{\lambda_2-\lambda_1}{\lambda_1}} (\ln e^{\frac{x^2}{\lambda_1}} + \ln e^{\frac{\lambda_2-\lambda_1}{\lambda_1}}) dx = H_\partial$$

Converting this expression to (14)

We obtain a functional relationship between the maximum dispersion and the given entropy H_∂.

$$-D(x) = (1 - H_\partial)\lambda_1 - \lambda_2 . \qquad (15)$$

The expression (15) uniquely characterizes the relationship between the maximum dispersion $D(x)$ and the given differential entropy H_∂ under the condition $D(x) > \frac{(b-a)^2}{12}$.

The results obtained can be formulated as a theorem.

Theorem on the relation of a dispersion to the differential entropy of a continuous centered random variable concentrated on a finite interval. If the continuous centered random variable is concentrated on a finite interval bounded by the limiting values a, b, then, under the condition $D(x) < \frac{(b-a)^2}{12}$ the minimum dispersion $D(x)$ and the differential entropy H_∂ of the indicated random variable are related by a relation $D(x) = (H_\partial - 1)\lambda_1 + \lambda_2$, where the constants λ_1 and λ_2 are defined as solutions of the equations system

$$-\int_{a}^{b} e^{\frac{x^2}{\lambda_1}} e^{\frac{\lambda_2-\lambda_1}{\lambda_1}} \cdot \ln e^{\frac{x^2}{\lambda_1}} e^{\frac{\lambda_2-\lambda_1}{\lambda_1}} dx = H_\partial ;$$

$$\int_{a}^{b} e^{\frac{x^2}{\lambda_1}} e^{\frac{\lambda_2-\lambda_1}{\lambda_1}} dx = 1$$

and when the condition $D(x) > \frac{(b-a)^2}{12}$ is satisfied, the differential entropy H_∂ and the maximum dispersion $D(x)$ of the indicated random variable are related by the correlation $-D(x) = (H_\partial - 1)\lambda_1 + \lambda_2$, where the constants λ_1 and λ_2 are defined as the solutions of the equations systems:

$$-\int_{a}^{b} e^{\frac{x^2}{\lambda_1}} e^{\frac{\lambda_2-\lambda_1}{\lambda_1}} \cdot \ln e^{\frac{x^2}{\lambda_1}} e^{\frac{\lambda_2-\lambda_1}{\lambda_1}} dx = H_\partial ;$$

$$\int_{a}^{b} e^{\frac{x^2}{\lambda_1}} e^{\frac{\lambda_2-\lambda_1}{\lambda_1}} dx = 1 .$$

Consequence. The extremal values of the dispersion for a given differential entropy of a continuous centered random variable concentrated on a finite interval bounded by the limiting values a, b, are uniquely provided by the compositional law of the form:

$$p(x) = e^{\mu_1 - 1} \cdot e^{\mu_2 \cdot x^2}, \quad a \le x \le b ,$$

where the constants μ_1 and μ_2 are defined as solutions of the equations system

DOI: 10.1051/matecconf/201712903024

$$\int_a^b e^{\mu_1 - 1} \cdot e^{\mu_2 \cdot x^2} dx = 1;$$

$$\int_a^b x^2 \cdot e^{\mu_1 - 1} \cdot e^{\mu_2 \cdot x^2} dx = D_3,$$

where D_3 is the specified dispersion value.

On the basis of the studies carried out it can be concluded that the abridged normal distribution law ensures a maximum of entropy for a given dispersion and at the same time a minimum of dispersion at a given entropy, and the bimodal law provides the maximum of entropy at the maximum of the dispersion.

Let us note once again that the concept of entropy which is a measure of uncertainty in the information theory, in this case, differs with the concept of uncertainty of the measurement accuracy used as a certainty criterion and given in [15, 16], and the correlations (12) and (17) relate these two concepts with unique functional dependencies.

Another important conclusion that can be drawn from the studies is that the distribution density obtained from the maximum of the entropy of a random variable concentrated on a finite interval for a given dispersion uniquely corresponds to the distribution densities of the laws that ensure the minimum and maximum of the dispersion for the same entropy.

As further research there will be the analysis of the applicability of the obtained distribution law in various areas of mechanical assembly production in the machine-tool industry.

References

1. L.N. Gavrilov, N.A. Borodachev, R.M. Abrashitov. *Accuracy of production in mechanical engineering and instrument making* (Mechanical Engineering, Moscow, 1973)
2. N.G. Bruevich., E.A. Pravotorova, V.I. Sergeev. *Fundamentals of the theory of the accuracy of mechanisms*: (Nauka, Moscow, 1988)
3. V.P. Bulatova (red.), I.G. Friedlander (red.), *Fundamental problems of the theory of accuracy*. (Science, St. Petersburg , 2001)
4. RMG 64-2003 "GSI. Ensuring efficiency in the management of technological processes. Methods and ways to improve the accuracy of measurements »
5. V.Ya. Kopp, A.A. Skidan, A.I. Balakin, O.V. Filipovich. Proceedings of the Odessa Polytechnic University. *Analysis of differential entropy in technical measurements in machine-tool construction,* **1(27)**. PP. 214 – 219 (2007)
6. V.Ya. Kopp, A.A. Skidan, A.I. Balakin, O.V. Filipovich. Zb. Sciences. The praise of SNYAETA. *Optimum estimation of the number of measurements of instrument systems,* **1(21)**. PP.231 – 238 (2007)
7. V.I. Dmitriev. *Applied information theory* (Higher education. Shk. Moscow, 1989).
8. P.V. Novitsky. *Fundamentals of the information theory of measuring devices*. (Energia, Leningrad, 1968).
9. P.V. Novitsky, I.A. Zograf *Estimation of errors in measurement results* (Energoatomizdat, Leningrad 1991).
10. E.T. Jane. Proceedings of the Institute of Electrical and Electronics Engineers. *On the logical justification of the maximum entropy methods*, **9(70)**. PP. 33 – 51 (1982).
11. V.Ya. Kopp, A.A. Skidan, A.P. Vasyutenko. Bulletin of SevNTU. Process Automation and Control. *Optimum estimation of the number of measurements of instrument systems,* **(36)**. PP. 75 – 80 (2002).
12. GOST 8.207-76 State system for ensuring the uniformity of measurements. Direct measurements with multiple observations. Methods for processing the results of observations. Basic provisions.
13. GOST R ISO 5725-2002. Accuracy (correctness and precision) of measurement methods and results. Part 2.
14. RMG 43-2001 State system for ensuring the uniformity of measurements. Application of the "Guide to the expression of measurement uncertainty".
15. T. Barashkova, R. Laaneots. *Uncertainty of the result of measurements* (Tallinn Technical University, Tallinn, 2003).
16. I.P. Zakharov, V.D. Kukush. *Theory of uncertainty in measurements* (Konsum, Kharkov, 2002).

Effectiveness analysis of the kulback's information measure in the technological objects monitoring

Alexander Skatkov[1], *Alexey* Brjuhoveckij[1,*], and *Dmitriy* Moiseev[2]

[1]Federal Public Autonomous Educational Institution of the Higher Education «Sevastopol State University», Russian Federation, 299053, Sevastopol
[2]Federal state owned military educational institution of the Higher professional Education "the black sea higher navy order of the red star academy by P.S.Nakhimov" Russian Federation, 299028, Sevastopol

Annotation. The technological objects' state monitoring is focused on the following purposes: to ensure the production processes effective functioning and to improve the quality and competitiveness of the products. The aim of the proposed scientific approach is the improvement of the known statistical methods used for production processes monitoring. It allows to minimize the first and the second kind errors when evaluating the technological objects' states. The usage of Kulback's information measure allows to compress large amounts of data without losing their validity. This approach mostly satisfies the requirements for the implementation of known methods of technological objects' state monitoring. These requirements are the following: processing of large amounts of current information for setting the parameters of the monitoring and control system; real time decision-making using the monitoring results on the production process' state; timely detection of possible precritical conditions of the technological process, which can be associated with great economic damage or even catastrophic consequences.

1 Introduction

In modern conditions, special attention is paid to production processes' state monitoring and products' quality increasing. In addition, the scope of application of so-called critical systems and infrastructures is significantly expanding [1]. Their criticality is manifested in the fact that even small changes in parameters or structures can lead to catastrophic failures, accidents or even catastrophes. Therefore, functioning of such facilities requires the continuous operation of parametric statistical control. The main purpose of such applications lies in systematic analysis and interpretation of the processes occurring in a heterogeneous production environment. The organization of technological objects' state monitoring is associated with the following conflict situation. On the one hand – the provision of real-time mode dictates the need to use small-volume samples; on the other hand – there is a need to ensure the required reliability level of the hypotheses acceptance based on the monitoring results. Thus, it is necessary to make a compromise decision in the conditions of a priori information deficit and initial statistical uncertainty, which should guarantee risks minimization for technological objects functioning under performance requirements on speed and reliability of situation recognition procedures. Such solutions should be sought by compressing large volumes of processed data without losing their validity.

A widely used approach *MSPC (Multivariate Statistical Process Control)* is a natural development of the approach *SPC (Statistical Process Control)*. According to the work [2], MSPC is a multifaceted activity associated with multifactor monitoring, modeling and process management, which is aimed at stabilizing the functioning of the enterprise, increasing its efficiency and ultimately producing high-quality products or providing quality services. One of the reasons why these control methods provide necessary but insufficient measures to obtain the final product of the proper quality is that it is necessity to control not only the quality of the final product but also the production process itself, that consists of a sequence of various operations of a complex structure.

The known solutions for operational monitoring of technological processes lies in the oil and gas industry [3]. Models and classification of monitoring systems from the viewpoint of queuing theory are considered. In the work [4] a system of monitoring of distributed technological objects is proposed. Intelligent monitoring systems for large-scale distributed technological objects are considered. A distinctive feature of such systems is the need to determine the actual global state of the system, assumed to be unchanged for the planning horizon.

* Corresponding author: bryukhovetskiy@sevsu.ru

 DOI: 10.1051/matecconf/201712903025

Knowledge of the actual global state of the system can be used to solve various complex optimization problems. Its representation in the system is called the information model of the state.

In [5], the author presents probabilistic statistical methods of decision theory, explaining how to detect spontaneously arising effects using observations of the realization of a random process. In conditions, including production processes, most of the problems of data analysis are related to the study of stochastic systems in which the detection of significant but rare information situations is often of decisive importance. Due to the mentioned circumstances, the known methods of production processes state monitoring cannot be a perfect operational tool, since they do not provide the required quality of control over the state of production processes. In particular, with their help it is difficult to trace the moments of changes in individual properties of the technological objects states (such as accuracy, stability, quality indicators and others). It is advisable to search for such solutions within the framework of the statistical theory of hypothesis acceptance [6]. Since, as it is well known, the possibilities of an analytical solution of such problems are very limited, it is suggested to use simulation methods as an alternative. One of the main tasks of technological objects monitoring is the synthesis of measurement metrics that can be used for the assessment of the information situation based on small samples of data. In the work [7] it is proposed to use Kulbak's distance as an information measure.

Despite a large number of practical systems used to monitor technological objects, there is an urgent need for their continuous improvement, as the complexity and scale of production systems are constantly increasing, and, consequently, the quality requirements of the products are increasing too. The main advantages of the proposed method for assessing the state of a technological facility on the basis of Kulbak's information measure are the following: sensitivity to changes in technological objects' state, low computational complexity, adaptability to external influences. Let's call the estimation of the inhomogeneity of technological object's state as J-effect, and let's call the device for its detection as J-detector. Thus, the J-effect describes instability in the production process that can lead to equipment operation violation.

2 The proposed decision

Software of the proposed research complex contains the following structural components:

- a subsystem for generating random samples (with a priori preset parameters close to real samples of complex systems' parameters);
- a subsystem of complex management;
- a subsystem of statistical data processing;
- a subsystem of H_0 hypothesis acceptance (about J-effect absence) or the acceptance of competing hypothesis H_1, confirming the J-effect presence. The experiments were carried out with various combinations of input parameters, including the following:
- setting the law for the distribution of input data;
- Sample size;
- Number of intervals;
- Confidence level;
- Collection of statistical data for different time intervals;
- Estimation of the dynamics of the change in the information state of technological objects;
- Decision-making on the presence of J-effect based on a given criterion.

There is a need to compare two samples of the monitoring results to determine the significance of its qualitative change. The set of observations is a set of measurements of technological objects' states –
$X=\{X_1, X_2,\dots, X_N\}$, where $X_i=\{x_1, x_2,\dots, x_K\}$, N - number of samples; K - volume of each sample; x_i - measured value of technological object. The question is whether the differences between X_i and X_j can be considered significant, or the differences between them should be attributed to the random dispersion of the values of the test feature.
Kulback's information measure (Kulback's distance) between X_i and X_j can be calculated according to the rule (1):

$$D(X_i||X_j) = \sum_{l=1}^{K} pl \log\frac{pl}{ql}, \quad (1)$$

where p_l and q_l are respectively the probabilities of the events when values from samples X_i and X_j belong to the l- interval.

Kulback's distance for distribution of X_i relatively to X_j can be calculated the following way:

- $D(X_i||X_j) \leq Q$ – J-effect absence,
- $D(X_i||X_j) > Q$ – J-effect presence,

where Q – distance limit (depending on the criticality of the process object).
Thus, the null hypothesis H_0 is accepted when

$$D(X_i||X_j) \leq Q$$

J-effect absence – the state is stable, otherwise the hypothesis H_1 is accepted – there is a qualitative change in the state.

The problem of modeling the J-effect detection can be solved using the following algorithmic scheme:
1) N samples of X_i of a given volume K are generated from a population having a predetermined distribution law;
2) Two samples are selected (X_i and X_j) to determine the Kulback's distance;

3) Confidence level is given depending on the criticality of the technological object;

4) The H_0 hypothesis is formulated;

5) The calculated criterion value Q is determined (at a given confidence level to test the null hypothesis);

6) Critical areas of values of the criterion for testing the null hypothesis are determined – areas of stable / unstable J-effect detection;

7) For the specified number of times the algorithmic points (2 – 6) are repeated for other pairs of samples X_i and X_j. Then a conclusion is made about an assessment of their differences.

On the fig. 1 the main moduls of the software system are presented.

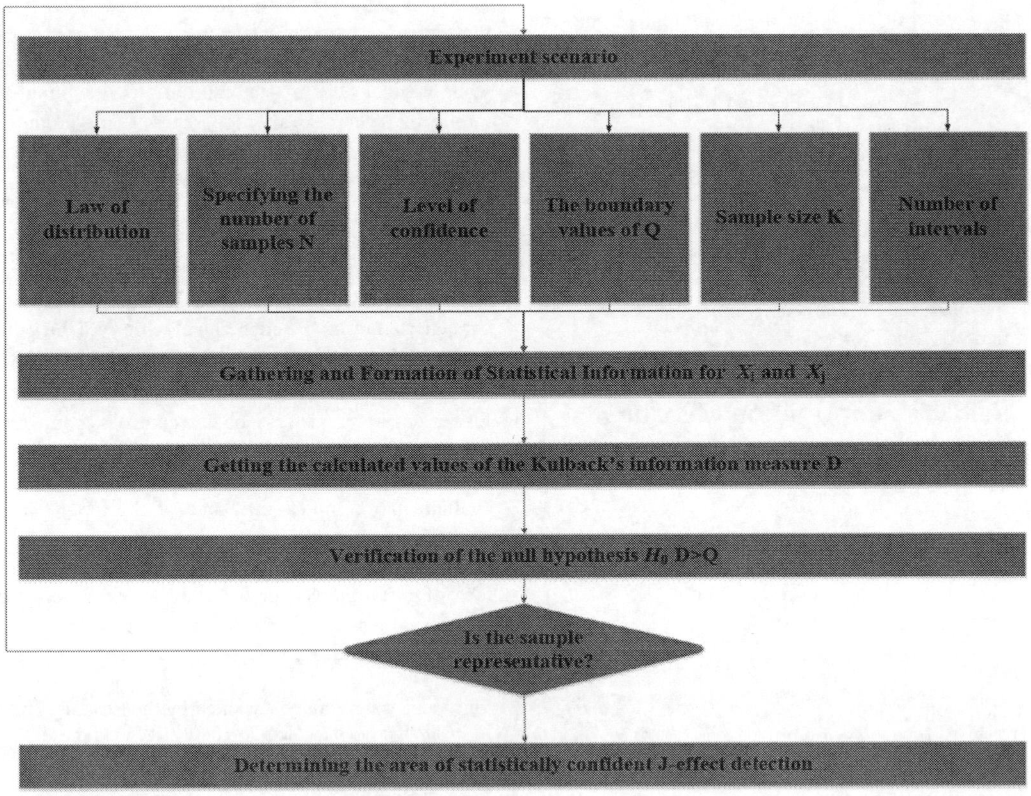

Fig. 1. Block diagram of the J-effect detection software system

The graphical programming environment *Simulink* from the package *MATLAB* have been used to illustrate the process of technological objects' states changing detection.

The process of J-effect detection based on using of Kulback's information measure can be considered on the following example. The task is to control the λ parameter that is distributed according to the normal law with mathematical expectation $M(\lambda)$ and the standard deviation $\sigma(\lambda)$.

The experiments have been carried out to answer the following question: what effect can the parameters K, N and $\sigma(\lambda)$ render on the changes in the states of technological objects and on the definition of stable recognition areas of J-effect. The following settings has been used:

– Sample sizes varied in the ranges: 50 <= K <=5000;

– Number of samples $X=\{X_1, X_2, \ldots, X_N\}$, $N = 4, 7, 10$;

– Expected value $M(\lambda) = 1$;

– $\sigma(\lambda)$ varied in the range [0,05 ; 0,35] in increments of $\Delta\sigma(\lambda)$=0,05.

In order to ensure the universality of the measured values processing and independence from the controlled metrics the software system normalizes the values of the samples X_j according to the rule (2).

$$x_{i\,norm} = (x_i - M_x) / M_x, \quad (2)$$

3 Results of experiments

The results of the experiment are presented on the fig. 2. The following settings were used:

Sample size K = 5000, 500, 50;
Number of samples N = 7;
$M(\lambda) = 1$;
$\sigma(\lambda)$ = [0,05; 0,1; 0,15; 0,2; 0,25; 0,3; 0,35].

A series of experiments was performed to assess J-effect. The values of the following samples were compared:

$$X_1 - X_2, X_1 - X_3, \ldots, X_1 - X_7 .$$

Thus, the following patterns in J-effect detection were established as a result of simulated experiments.

1) Stable J-effect detection for the technological objects with the given input parameters is observed at the sample size K>500 and $\sigma(\lambda) \geq 0{,}15$.

2) When using the sample size $K \leq 80$ J-effect is unstable in the range or $0{,}15 \leq \sigma(\lambda) \leq 0{,}3$.

3) J-effect stably is not observed when K<50 over the entire range of variation $\sigma(\lambda)$.

4) During the experiments it was found that the higher the value $\Delta\sigma(\lambda)$ is, the more effectively the quality status is monitored technological objects. Thus, the maximum J-effect is observed when comparing the results of experiments obtained for samples $X_1 - X_7$.

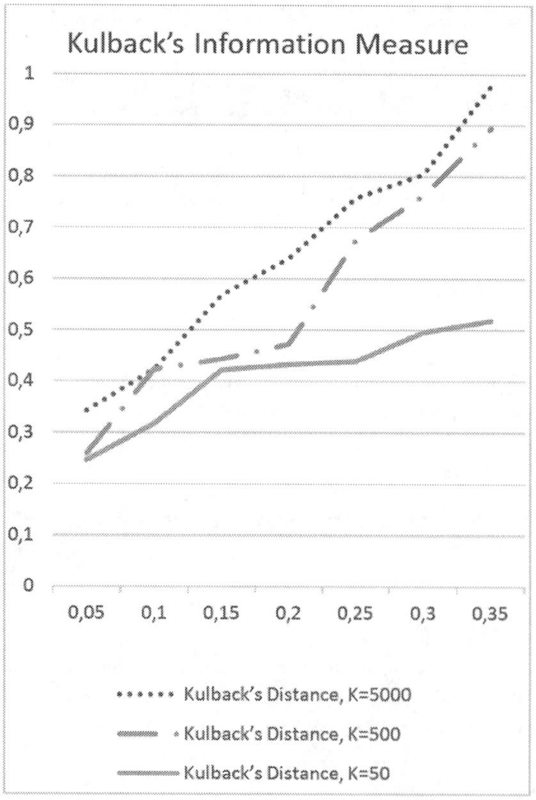

Fig. 2. The results of the experiments on the J-effect presence assessment

The obtained results shows that the application of the model on the basis of Kulbak's information measure allows to provide statistical stability of the area of confident / uncertain J-effect detection when monitoring the qualitative changes in the state of technological objects. The general recommendation for assessing detection is as follows. Depending on the purpose of the monitored system (is it critical or not), the decision-maker is authorized to set the optimal values of confidence level – p and the corresponding value of distance limit Q . There are two goals: Ensure a high reliability of the values of the characteristics of the monitored process facilities, to minimize the production losses. The results obtained in the conducted experiments with models are used as a basis for information technology for monitoring the technological objects' state and are used as input to the intellectual decision support system. This system is to ensure the minimal losses that can be caused by:

– untimely decision-making (cost of information loss),
– state control costs.

Each type of loss is compared with the coefficient – α and β respectfully: $\alpha + \beta = 1$. This system contains evaluation matrix, the elements of which are the values of losses, and inputs - threshold values of Kulbak's distance– Q_m , m – Number of gradations of Q; f_t – frequency of measurement of process object control parameters and values $\sigma(\lambda)$.

In the plans for further research, it is proposed to evaluate the complex influence of the input parameter values on statistical sensitivity, statistical stability, areas of confident / uncertain detection under other distribution laws of the parameters under study, for a different number of intervals, and to build a decision support system for detecting the J-effect.

This work was partially supported by the Russian Foundation for Basic Research (project № 15-29-07936).

References

1. A. Skatkov, Information technology for critical infrastructures, 306, (2012)
2. I. Kablashova, Theory and methods of ensuring the quality of production processes, 173, (2005)
3. E Ohotnikov, Bulletin of Tyumen State University, 7, 80, (2006)
4. S. Andryushkevich, Bulletin of the Tomsk Polytechnic University, **317**, 35, (2010)
5. A. Shiryaev, Probabilistic-statistical methods in the theory of decision-making, 144, (2014)
6. S. Ayvazyan, Statistics and the Foundations of Econometrics, 1000, (1998)
7. S. Kulback, Information theory and statistics, 408, (1967)

MATEC Web of Conferences **129**, 03026 (2017) DOI: 10.1051/matecconf/201712903026
ICMTMTE 2017

Recursive estimation of the parts production process quality indicator

Oleg Filipovich[1,*], *Vadim* Kopp[1], *Nataliya* Voloshina[1], and *Alexander* Bokhonsky[1]

[1]Sevastopol State University, Sevastopol, Russian Federation

Abstract. Consideration is given to a mathematical representation for manufacturing of batch parts on a metal-cutting machine tool. Linear dimensions of machined parts are assumed to be the major quality indicator, deviation from these dimensions is determined by size setting of machine tool and ensemble of random factors. It is allowed to have absolutely precise pre-setting of machine tool, effects from setup level offsetting due to deformation in process equipment on the specified indicator are disregarded. Consideration is given to factors which affect the tool wear, with two definitions of tool wear being provided. Reasons for development of random error in processing, dependence of measurement results on error as well as distribution laws and some parameters of random values are provided. To evaluate deviation of size setting value in each cycle, it is proposed to apply a recursive algorithm in description of investigated dynamic discrete process in the space state. Kalman filter equations are used in description of process model by means of first-order difference equations. The algorithm of recursive estimation is implemented in the mathematical software Maple. Simulation results which prove effectiveness of algorithm application to investigate the given dynamic system are provided. Variants of algorithm application and opportunities of further research are proposed.

Consideration is given to machining of part batch N made with a cutting machine tool. A cycle is understood as completion of total operational sequence for each part made on a single piece of production machinery.

Mathematical description of part fabrication process seen from the aspect of achieving a required quality indicator is presented in detail in [1]. The linear size at some fixed section is selected as a major criterion of machining accuracy while errors of part shape are neglected. Achievement of required indicator is associated with a complex fabrication process under conditions of dynamically non-deterministic environment. One of possible objectives here is to obtain a forecast for deviation value of dimensional parameter with the maximum feasible precision. The purpose of this study is to apply known algorithms to procure recursive estimates for the mentioned indicator.

In real-life conditions any manufacturing process is realized as affected by multiple random disturbances which depend on the strength and oscillations of workpiece allowance, deformations during machining, value of cutting tool wear etc. Taking this fact into account, the quality indicator value (deviation of linear size from some nominal value) can be written as:

$$x_k = \overline{x_k} + \xi_k , \qquad (1)$$

where $\overline{x_k}$ – value of machine tool size setting in k-n cycle, which represents the total of determined trend and random functional component of processing error; ξ_k – random departure of part dimensions from the setting level.

Drift of setting level in k-n cycle of manufacturing process can be expressed in the following form:

$$\overline{x_k} - x_0 + \mu_k + l_k , \qquad (2)$$

where

x_0 – initial offset in the setting level;

μ_k – offset in the setting level due to deformation of manufacturing machinery;

l_k – offset in the setting level due to wear of cutting tool.

To simplify further explication let us assume that the setup was absolutely accurate to match zero level, i.e. $x_0 = 0$.

There are thermal and mechanical deformations of manufacturing system to be distinguished. Thermal deformations depend on the design of separate machine tool assemblies, amount of coolant and degree of its heating, extension of dimension chains, cutting tool characteristics and bluntness degree, feed, cutting speed and depth, material and workpiece size, ambient temperature, friction forces etc. Mechanical deformations depend on stiffness of joints, cutting depth, feed rate, on size, hardness and quality of a workpiece, quality and size of a cutting tool, cutting speed, degree of

* Corresponding author: ophisl@yandex.ru

DOI: 10.1051/matecconf/201712903026

heating in cutting tool assemblies, lubricant etc. When considering the case of piece machining on metal-cutting machine tools with disturbances related mostly to severe wear of cutting tool, we can neglect drifting of setting level due to deformation of manufacturing machinery, i.e. $\mu_k = 0$.

Value of tool wear in each cycle of process depends on such factors as machining allowance value or material hardness, which are of random nature. Therefore, on the whole, wear is a random parameter whose value depends neither on values in preceding cycles nor on total wear. Let us consider two cases.

Case 1. If dispersion of this value is not significant, then the trend can be regarded as uniform and represented as linear function

$$l_k = \alpha k, \; (k = \overline{1, N}), \tag{3}$$

where

α – cutting tool wear rate which falls within one cycle.

In this case the value l_k is a systematic component of error and a determined value.

Case 2. If dispersion of wear value in each cycle is significant, and the wear factor is prevalent in (1), then it can be regarded as a stochastic process with independent increments. The value l_k in this case can be represented in the form of [2]

$$l_k = \alpha k + \sum_{i=1}^{k} \gamma_i, \; (k = \overline{1, N}), \tag{4}$$

where

γ_i – independent increment in i-n cycle (stochastic variable distributed in accordance with the Gaussian law with zero mathematical expectation and dispersion σ_γ^2).

There are random errors developed in each manufacturing process cycle during machining of parts which are known as “instantaneous” processing errors or purely random errors ξ_k. Their sources lie in non-homogeneity of work pieces in their machining allowances and hardness, in non-homogeneity of cutting tool structure and in random changes of cutting modes. Salient feature of sequence $\{\xi_k\}$ formed actually by random errors of manufacturing process is mutual independence of values from each other. Research demonstrates that the random value ξ_k is distributed – independent of k – with zero mathematical expectation, as a rule, while its dispersion retains constancy in sufficiently stable and debugged manufacturing process for a long period of time [2, 3, 4].

By substituting (2) in (1), we obtain – with due account for made assumptions

$$x_k = l_k + \xi_k. \tag{5}$$

To determine the parameter x_k, it is measured after completion of each cycle by passive control system. Results of z_k measurements under certain assumptions can be expressed by equation

$$z_k = x_k + \delta_k, \tag{6}$$

where

δ_k – measurement error.

On condition that systematic errors are missing or compensated let us determine the distribution law for the random component of measurement error. It was demonstrated in work [5] that the maximum differential entropy of continuous random variable constrained by finite limits provides for the composition law, which, depending on the dispersion value, manifests itself as truncated normal, uniform or bimodal. Let us assume that at certain dispersion restrictions the measurement error δ_k has the truncated normal distribution whose density function at zero mathematical expectation equals

$$f(\delta) = \frac{1}{C\sigma_\delta \sqrt{2\pi}} e^{-\frac{\delta^2}{2\sigma_\delta^2}}, \; \delta \in [-a, a], \tag{7}$$

where

$[-a, a]$ – limits of random value variation;

σ_δ, C – law parameters,

$$C = \int_{-a}^{a} f(\delta) d\delta. \tag{8}$$

To solve the task of target indicator assessment, let us apply the recursive algorithm by describing the dynamic process under study in the state space form. Recursive algorithms of estimation (Kalman filter) became widespread when solving a number of identification, evaluation, filtration and control tasks [6…22]. Let us assume that the process model is described by first-order difference equations. In our case there is a given discrete scalar system

$$\begin{aligned} x_{k+1} &= x_k + l_k + \xi_k \\ z_{k+1} &= x_{k+1} + \delta_{k+1} \end{aligned} \tag{9}$$

where

x_{k+1} and x_k – values of quality indicator at $(k+1)$-n and k-n cycles of manufacturing process respectively;

l_k – a known value that determines evolution of system;

ξ_k and δ_k – stationary random values whose distribution laws and moment characteristics were described earlier.

When determining the value l_k, let us assume that the value of tool wear has linear dependence on k (case 1).

Let us assume that all random values are independent and the system is stable, fully controllable and observable. The Kalman filter equation for such system will be of the following form

DOI: 10.1051/matecconf/201712903026

$$\hat{x}_{k+1} = \hat{x}_k + K_{k+1}\left(z_{k+1} - \hat{x}_k\right), \qquad (10)$$

$$K_{k+1} = \frac{V_{\varepsilon k} + \sigma_\xi^2}{V_{\varepsilon k} + \sigma_\xi^2 + \sigma_\delta^2} \qquad (11)$$

on initial conditions $\hat{x}_0 = 0$, $\sigma_{\xi 0}^2 = 0$,

where

$\hat{x}_{k+1}, \hat{x}_k$ – estimates of values x_{k+1} and x_k;

K_{k+1} – Kalman gain whose optimal value minimizes the RMS estimation error

$$V_{\varepsilon k+1} = M[\varepsilon_{k+1}^2] = \frac{\sigma_\delta^2\left(V_{\varepsilon k} + \sigma_\xi^2\right)}{V_{\varepsilon k} + \sigma_\xi^2 + \sigma_\delta^2}, \qquad (12)$$

where

M – symbol of mathematical expectation;

ε_{k+1} – estimation error,

$$\varepsilon_{k+1} = x_{k+1} - \hat{x}_{k+1}. \qquad (13)$$

The algorithm (10…12) is implemented in the Maple system with the following given data:

1. $N = 20$ parts,
2. $\alpha = 3 \cdot 10^{-6}$ m/cycle,
3. $\sigma_\delta = 2 \cdot 10^{-6}$ m,
4. $a = 2\sigma_\delta = 4 \cdot 10^{-6}$ m,
5. $\sigma_\xi = 1 \cdot 10^{-6}$ m.

Results from incremental determining of values x_k, z_k и $\hat{x}_k$ are summarized in Table 1.

Table 1. Simulation results

No. of cycle, k	x_k, 10^{-6} m	z_k, 10^{-6} m	$\hat{x}_k$, 10^{-6} m
1.	0,0	1,2	–
2.	2,6	3,4	3,8
3.	5,0	5,4	6,2
4.	7,9	7,1	8,3
5.	11,4	9,9	10,7
6.	14,3	13,6	13,7
7.	17,2	14,4	15,8
8.	20,9	19,4	19,0
9.	24,0	24,9	23,2
10.	26,4	26,2	26,2
11.	29,5	27,8	28,6
12.	33,6	31,6	31,6
13.	35,9	34,1	34,4
14.	38,0	39,9	38,4
15.	40,6	39,1	40,5
16.	43,0	45,2	44,2
17.	45,1	46,6	46,9
18.	47,2	47,6	49,0
19.	50,2	49,9	51,2
20.	53,1	53,4	53,9

Of special importance is comparison of value ε_k and difference

$$\Delta_k = x_k - z_k \qquad (14)$$

at every cycle. The plot that illustrates their change is displayed in Fig. 1.

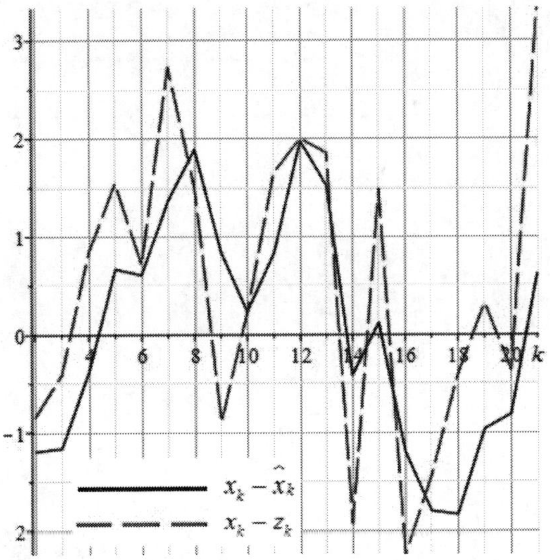

Fig. 1. Results from determining ε_k and Δ_k values depending on the number of cycle k

As it can be seen from the results presented in Table 1 and in Fig.1, the mean relative value ε_k does not exceed 2%, which is a sufficiently good estimation in the framework of task to be solved. At the same time, the mean relative difference between x_k and z_k is approximately 1.5 times higher than this value, thus confirming the effectiveness of estimation. The estimation algorithm for this task is not divergent, the gain becomes stable fairly quickly (Fig.2).

The values of the function $V_\varepsilon(k)$ are displayed in Fig. 3. It can be seen that after 10 cycles this function goes to a certain limit.

The results obtained can be used to improve operation efficiency of data management and control systems installed on metal working machines, which provide for their independent setup and adjustment during operation. It is also proposed to apply this algorithm further on to evaluate probability indicators for batches composed of two various products and assembled jointly with the use of group interchangeability technique (selective assembly).

 DOI: 10.1051/matecconf/201712903026

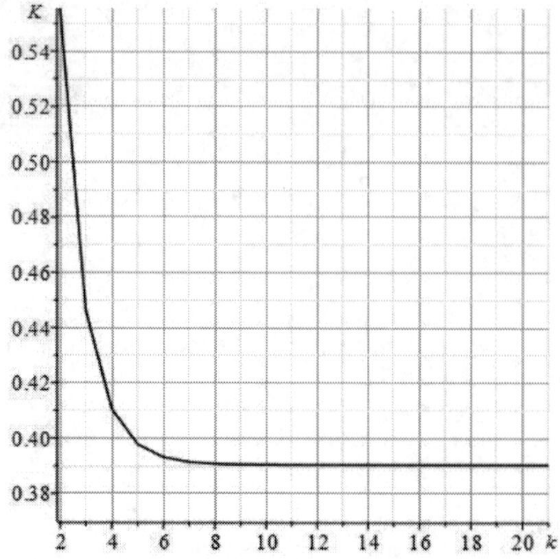

Fig. 2. Change in coefficient K_k depending on the number of cycle k

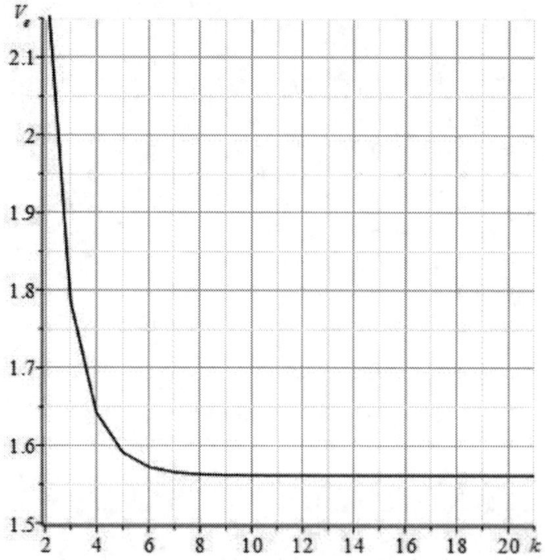

Fig. 3. $V_e(k)$ values

Prospects of further studies lie in development of estimation algorithm with due account for available control actions and more complicated relationships in the wear of cutting tools used on various machine tools.

References

1. Z.Sh. Geyler, *Self-tuning active control systems* (Machine Building, Moscow, 1978)
2. A.N. Gavrilov (red.), *Precision of production in mechanical engineering and instrument making* (Machine Building, Moscow, 1973)
3. S.S. Volosov, *Fundamentals of precision dimensional active control* (Machine Building, Moscow, 1969)
4. M.S. Nevelson, *Automatic control of metalworking accuracy* (Machine Building, Leningrad, 1973)
5. V.Ya. Kopp, A.A. Skidan, A.I. Balakin, O.V. Filipovich, *Analysis of differential entropy in engineering measurements in mechanical engineering and instrument making*, Proc. of the Odessa Polytechnic University, **1(27)**, PP. 214 – 219 (2007)
6. R.E. Kalman, *A New Approach to Linear Filtering and Prediction,* Transaction ASME Journal of Basic Engineering, **86**, PP. 35–45 (1960).
7. A.V. Balakrishnan, *Kalman Filtering Theory.* (Optimization Software, Inc., 1984)
8. J.C. Spall, *Introduction to Stochastic Search and Optimization: Estimation, Simulation, and Control* (Hoboken, New Jersey: Wiley, 2003).
9. G. Welch, G. Bishop, *An Introduction to the Kalman Filter* (Los Angeles, 2001).
10. B.D.O. Anderson, J.B. Moore, *Optimal Filtering* (New York: Dover, 2005).
11. Yu.E. Voskoboinikov, *Recursive estimation of the state vector of the dynamic system* (Novosibirsk: NSTU, 2014)
12. Yu.E. Voskoboinikov, *The divergence criterion and the algorithm for adapting the recurrent algorithm for estimating the state vector,* Scientific bulletin of the NSTU, **3(60)**, PP. 7-22 (2015)
13. R.G. Brown, P.Y.C. Hwang, *Introduction to Random Signals and Applied Kalman Filtering, Second Edition* (John Wiley & Sons, Inc., 1992)
14. A. Gelb, *Applied Optimal Estimation* (MIT Press, Cambridge, MA., 1974)
15. Mohinder S. Grewal, P. Andrews Angus, *Kalman Filtering Theory and Practice* (Upper Saddle River, NJ USA, Prentice Hall, 1993)
16. O.L.R. Jacobs, *Introduction to Control Theory, 2nd Edition* (Oxford University Press, 1993)
17. Richard Lewis, *Optimal Estimation with an Introduction to Stochastic Control Theory* (John Wiley & Sons, Inc., 1986).
18. Peter S. Maybeck, *Stochastic Models, Estimation, and Control* (Academic Press, Inc., 1979)
19. J.C Bertein, R. Ceschi, *Processus stochastiques et filtrage de Kalman* (Hermes, 1998)
20. C.K. Chui, G. Chen, *Kalman Filtering* (Springer-Verlag, 1987)
21. Mourad Barkat, *Signal Detection and Estimation* (Artech House Inc, 2005)
22. I.N. Sinitsyn, *Kalman and Pugachev Filters* (Moscow, 2006)

MATEC Web of Conferences **129**, 03027 (2017)
ICMTMTE 2017
DOI: 10.1051/matecconf/201712903027

Test method of distribution functions of the semi-markovian system dwell times in states

Mikhail Zamoryonov[1,*], *Darya* Zamoryonova[1], and *Vladimir* Preis[2]

[1]Sevastopol State University, Russian Federation, 299053, Sevastopol
[2]Tula State University, Russian Federation, 300012, Tula

Abstract. The article presents and proves the theorem on the stay times in the trajectory states taking the repeated returns into account. A method for determining the distribution functions of the stay times of the semi-Markov system in the states, called the method of trajectories, is introduced. An example of semi-Markovian modeling systems proving the correctness of the proposed method is examined.

1 Introduction

In many cases the hierarchical approach is employed for modelling of the complex semi-Markovian (SM) systems for their model development. In this case it is necessary to abut separate elements of the system with each other, for what it is necessary to know the distribution functions (DF) of the operation times of these elements.

In the case of SM systems the integral equations of Markov renewal are solved (MRE). The article proposes new method of SM systems modelling and its application to the study of functioning of the production element (PE). The model allows to assess the reliability impact of PE on its performance under the following hypothesis: in case of PE failure the production service is interrupted, and after the performance restoration of the element the production service continues in consideration for the time of the interrupted service (non-devaluate failures) [1, 2].

The proposed method is based on the theorem presented below.

2 Problem formulation

Consider the SM system with the common phase space of states M. Select the two subsets M_+ and M_- in the phase space of states M of the SM processes, such that $M_+ \cup M_- = M$. In the future, we will only talk about the subset M_+, as all the above will be true for the subset M_-. The times of single stay in the states $S_i \in M_+$ are random variables (RV) α_i having mathematical expectations m_i, functions $F_i(t)$ and density $f_i(t)$ of distribution, with the images in the complex domain $F_i(s)$ and $f_i(s)$ respectively. RV θ_Σ - stay time period in the subset M_+ having mathematical expectation T_+. The times of the multiple stays of the system in the state $S_i \in M_+$ due to the repeated entering into them during the θ_Σ period are the RV θ_i having mathematical expectations m_i^θ, functions $F_i^\theta(t)$ and density $f_i^\theta(t)$ of distribution, with images in the complex domain $F_i^\theta(s)$ and $f_i^\theta(s)$ respectively. The distribution of embedded Markov chain (EMC) corresponds to the described SM process, and is characterized by specific frequency ρ_i of entering each of the states $S_i \in M$ and P_{ij} transition probabilities from states $S_i \in M$ to states $S_j \in M$. Not all the states $S_i \in M_+$ can have the direct transition from M_+ to M_-, that is why it is advisable to select the subset $E \subset M_+$ of states $S_e \in E$ of the set M_+ from which the direct transition into the set M_- is possible.

Let us introduce the definitions.

Trajectory - the set of states in which the system is to stay in order to go out of the subset M_+ into M_- (or vice versa), and $M_+ \cup M_- = M$, where M - is the whole phase space of states. The trajectory of the system stay in the subset M_+ starts in the state in which there is direct transition from the subset M_-. Such trajectory is ended up by the state, from which there is direct transition to the subset M_-. If such final state has two or more different transitions into various states of the subset M_-, than the two or more different trajectories are formed.

A particular trajectory is any part of the particular trajectory containing the state from which there is direct transition to the subset M_- will be referred to as particular trajectory of renewal.

The stationary distribution of EMC and transition probabilities are defined for each trajectory, assuming

* Corresponding author: zamoryonoff@gmail.com

that the states of the trajectory form the set $M_+^k \subset M$, and all the other states are combined in the subset M_-^k such that $M_+^k \cup M_-^k = M$, where k – is the number of trajectory. And also, the probabilities P_k^T of the k-th trajectory on the basis of the transition probabilities of the system, and $\sum_k P_k^T = 1$.

3 Problem Solution

The theorem on the times stay in the states of trajectory in view of the repeated returns.

If the discrete ergodic SM system with the known stationary distribution for the selected k-th trajectory with the states $S_i \in M_+^k$ the functions $F_i(t)$ and density $f_i(t)$ of distribution are known, the function $F_i^k(s)$ and the density $f_i^k(s)$ of the time distribution θ_i of the system stay in the state S_i of this trajectory, taking into account repeated enterings it in the images domain are:

$$F_i^k(s) = \frac{F_i(s)}{c_i^k - \left(c_i^k - 1\right) f_i(s)}, \qquad (1)$$

$$f_i^k(s) = \frac{f_i(s)}{c_i^k - \left(c_i^k - 1\right) f_i(s)}$$

and in the original domain, they look like:

$$F_i^k(t) = \frac{1}{c_i^k} F_i(t) + \frac{1}{c_i^k} \sum_{m=1}^{\infty} \left(1 - \frac{1}{c_i^k}\right)^m F_i^{(*)m}(t), \quad (2)$$

$$f_i^k(t) = \frac{1}{c_i^k} \cdot f_i(t) + \frac{1}{c_i^k} \sum_{m=1}^{\infty} (1 - \frac{1}{c_i^k})^m f_i^{(*)m+1}(t),$$

where $c_i^k = \dfrac{\rho_i}{\rho_e P_{ej}}$ - is coefficient of increasing the stay time of the system in the states $S_i \in M_+$ by re-entering it;

ρ_i - is fixed rate of the system stay in the state S_i for the selected trajectory;

ρ_e - is fixed rate of the system stay in the state S_e (a final state of the current trajectory from which there is direct transition to the subset M_-) for the selected trajectory;

P_{ej} - is the probability of transition from state S_e to S_j belonging to the subset M_-.

Theorem proof.

Stay times of the system in the states ($S_i \in M_+^k$) increase by the system re-entering with some probability in these states during the stay of the entire system in the subset M_+^k, and there is nothing else due to what they can increase which is obvious. But this means that the number of the system entering follows the geometrical law of distribution with the P_i probability that the system will come out of this state, and the probability that it will remain is $1 - P_i$. The graph that describes the behavior of the system in given state at certain distribution function $F_i(t)$ of the stay time in this state S_i has the form shown in **Fig.** 1.

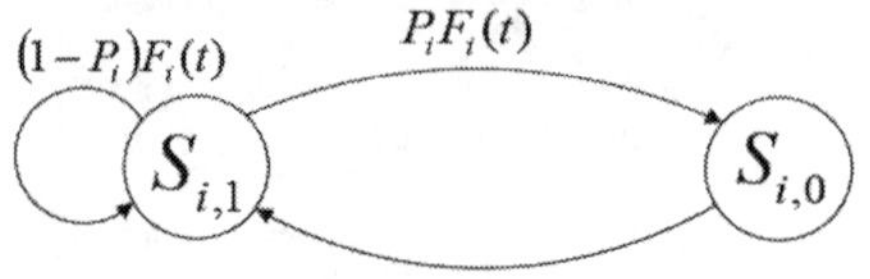

Fig. 1. Graph of states corresponding to the geometric distribution.

The legend are the following: $S_{i,0}$ - the instantaneous state corresponds to the system coming out of state S_i; $S_{i,1}$ - the studied state S_i, the total stay time in which it is being determined; P_i - the probability of transition of the system from $S_{i,1}$ to $S_{i,0}$; $(1-P_i)$ - the probability of the system re-entering the state $S_{i,1}$; $F_i(t)$ - is the DF of RV α_i having mathematical expectation m_i.

According to this graph let us make Markov renewal equation:

$$F_i^k(t) = (1 - P_i) \int_0^t f_i(t-s) F_i^o(s) ds + P_i F_i(t) \cdot \qquad (3)$$

Iterating the resulting equation (3), we have [3]:

$$F_i^k(t) = P_i F_i(t) + P_i \sum_{m=1}^{\infty} (1 - P_i)^m f_i^{(*)m}(t) * F_i(t). \quad (4)$$

The density of distribution $f_i^k(t)$ is as follows:

$$f_i^k(t) = P_i \cdot f_i(t) + P_i \sum_{m=1}^{\infty} (1 - P_i)^m f_i^{(*)m+1}(t)$$

The unknown that is to be determined here is P_i.

It is known that the geometric distribution law leads to $m\theta_i = \dfrac{m_i}{P_i}$ and this leads to the following:

$$P_i = \frac{m_i}{m\theta_i}. \qquad (5)$$

Let us find $m\theta_i$.

According to the known theorem [1] T_+ can be accurately determined. For the discrete states the formula is:

$$T_+ = \frac{\sum\limits_{i\in M_+} m_i \rho_i}{\sum\limits_{e\in E\subset M_+} \sum\limits_{j\in M_-} P_{ej}\rho_i} = \sum_{e\in M_+} m_i \frac{\rho_i}{\sum\limits_{e\in E\subset M_+} \sum\limits_{j\in M_-} P_{ej}\rho_e}.$$

Then $m\theta_i^k$:

$$m\theta_i^k = m_i \cdot \frac{\rho_i}{\sum\limits_{e\in E\subset M_+} \sum\limits_{j\in M_-} P_{ej}\rho_e}. \quad (6)$$

Let us introduce the coefficient c_i^k of the increase of the system stay time in the states S_i equal to

$$c_i^k = \frac{\rho_i}{\sum\limits_{e\in E\subset M_+} \sum\limits_{j\in M_-} P_{ej}\rho_e}.$$

Since there is only one way out for the current trajectory of the state S_e - the final state of the trajectory from which there is direct transition to the state S_j of the subset M_- , the coefficient c_i^k of time increase spent by the system in the states S_i would be:

$$c_i^k = \frac{\rho_i}{\rho_e P_{ej}}.$$

Then:

$$m\theta_i^k = m_i \cdot c_i^k,$$

and the probability P_i is determined by the condition of the increase m_i up to the value $m\theta_i$ that is of equality $m\theta_i^k = m_i \cdot c_i^k$, which implies that the sought probability P_i on the basis of (5) equals to:

$$P_i = \frac{m_i}{m\theta_i^k} = \frac{1}{c_i^k}. \quad (7)$$

Substituting (7) into (4) we obtain (2).

Applying the Laplace transformation to the formula (4), we obtain

$$F_i^k(s) = P_i F(s) + P_i \sum_{m=1}^{\infty} (1-P_i)^m f_i^m(s) F(s) =$$

$$= P_i F_i(s) \sum_{m=0}^{\infty} (1-P_i)^m f_i^m(s) \quad (8)$$

The resulting expression is an infinitely decreasing geometric progression. Let us take the limit of the expression $\sum\limits_{m=0}^{\infty}(1-P_i)^m f^m(s)$:

$$\lim_{n\to\infty} \sum_{m=0}^{n} (1-P_i)^m f_i^m(s) = \frac{1}{1-(1-P_i)f_i(s)}.$$

Then the expression (8) will look like:

$$F_i^k(s) = \frac{P_i F_i(s)}{1-(1-P_i)f_i(s)}. \quad (9)$$

Accordingly, the image of DD $f_i^k(t)$ will look like:

$$f_i^k(s) = \frac{sP_i F_i(s)}{1-(1-P_i)f_i(s)} = \frac{P_i f_i(s)}{1-(1-P_i)f_i(s)}. \quad (10)$$

Substituting (7) into (9) and (10) we will obtain

$$F_i^k(s) = \frac{\frac{1}{c_i^k} F_i(s)}{1-\left(1-\frac{1}{c_i^k}\right) f_i(s)} = \frac{F_i(s)}{c_i^k - \left(c_i^k - 1\right) f_i(s)},$$

$$f_i^k(s) = \frac{\frac{1}{c_i^k} f_i(s)}{1-\left(1-\frac{1}{c_i^k}\right) f_i(s)} = \frac{f_i(s)}{c_i^k - \left(c_i^k - 1\right) f_i(s)}.$$

The theorem is proved.

Corollary 1.

The time θ_Σ^T of the system stay in the trajectory is the sum of all the RV θ_i $\theta_\Sigma^T = \sum\limits_1^n \theta_i$, which means that the DF F_Σ^T of time θ_Σ^T - of the system stay in the trajectory is sequence of convolution of all DF F_i^θ :

$$F_\Sigma^T = F_0^\theta * F_1^\theta * F_2^\theta * \ldots * F_i^\theta * \ldots * F_n^\theta,$$

where n – is the number of states included in the trajectory.

Corollary 2.

Based on the formula of total probability the expression for the distribution function θ_Σ of the system stay time in the subset M_+ is:

$$F_\Sigma = P_1^T \cdot F_1^T + P_2^T \cdot F_2^T + \ldots + P_i^T \cdot F_i^T + \ldots + P_n^T \cdot F_n^T.$$

Corollary 3.

The stay times in the states of the entire system are determined by the following expression:

$$F_i^\Sigma = P_1^T \cdot F_i^1 + P_2^T \cdot F_i^2 + \ldots + P_j^T \cdot F_i^j + \ldots + P_n^T \cdot F_i^n$$

Corollary 4.

DF of the stay times in the particular trajectories of renewal correspond to EMC for the distribution function of system stay times in the subset M_+ with given initial state. Thus, the exact solution of the Markov renewal equation is obtained for the discrete SM system.

Note that in general, the algorithms of phase consolidation (for example, the one proposed in [2]) used for the transition from the systems with continuous set of states to the systems with discrete states are rough, that is why it is incorrect to speak about the exact solution of the Markov renewal equation in the continuous phase space.

The Method Of Trajectories

Step One. The transition from system with continuous states to system with discrete states $S_i \in M_+$. For that the DF F_i of the system stay times in the new discrete states, P_{ij} transition probabilities of these states to other states (transitional probabilities), the specific frequencies ρ_i of entering the states (stationary distribution of EMC) and fixed probabilities of stay in the states (the stationary distribution of the SM process) are determined. The procedure is performed with the help of known methods of SM systems modelling.

Step Two. Selection of all possible trajectories of transition from the subset M_+ to the subset M_-. Moreover, each state of the system is included in one or several trajectories at once.

Step Three. The stationary distribution of EMC and transition probabilities for each trajectory are determined.

Step Four. Based on the foregoing theorem, replaced by the stay times in the states α_i are replaced by θ_i - for them the densities and distribution function of system stay times in the states $S_i \in M_+$ are determined, in view of the repeated returns in accordance with the presented theorem.

Step Five. In accordance with the theorem of total probability, the probabilities P_k^T of each of the trajectories realization are determined on the basis of transition probabilities of the Markov chain.

Step Six. We find the distribution function of the stay time in M_+, regardless of the initial state, which is defined as the weighted sum (the mixture) of DF of each trajectory emerging from the state S_i. The coefficients of the mixture are found on the fifth step of the P_k^T probability of trajectories.

Let us consider specific example implementation of the proposed method of modelling.

We will describe the operation of the element [1, 2]. The service time per unit of output PE – RV α_1 with the DF $F_1(t) = P\{\alpha_1 \le t\}$. Uptime PE - RV α_2 with risk factors with DF $F_2(t) = P\{\alpha_2 \le t\}$, the renewal time of PE – RV β_2 with DF $G_2(t) = P\{\beta_2 \le t\}$. RV $\alpha_1, \alpha_2, \beta_2$ assumed to be independent having finite mathematical expectations and variances; DF $F_1(t), F_2(t), G_2(t)$ has densities $f_1(t), f_2(t), g_2(t)$. In case of PE failure the production service is interrupted, and after the performance restoration of the element the production service continues in consideration for the time of the interrupted service.

It is necessary to define the DF $F_\varepsilon(t)$ RV ε - of the cycle time service of PE production unit in view of its failures, the mean and variance of said RV, as well as the performance of PE.

To describe the operation of the system we use the Markov renewal process (MRP) and the corresponding SM process (SMP) with the states:

10x - PE is operational, the service of the next production unit has begun; the remaining time before failure of PE, as well;

11x - momentary state corresponding to the end of the maintenance unit of production; the remaining time before failure of PE, equals ;

20th - was restored and continued efficiency PE interrupted service production unit; the time remaining before the end of the interrupted service as well;

21x - has failed PE, service unit of production is interrupted; the time remaining before the end of the interrupted service as well.

The transition graph of the system - in **Fig.** 2.

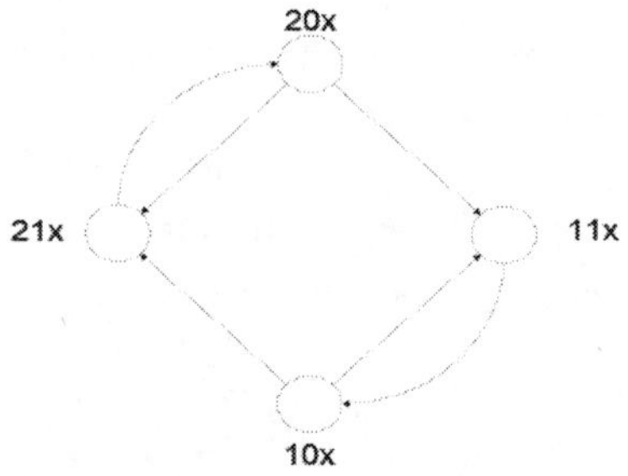

Fig. 2. The states graph of the system with continuous states.

The phase space of states is as follows:

M = {10x, 11x, 20x, 21x}.

We describe the densities and EMC transition probabilities:

$$p_{10x}^{11y} = f_1(x-y), 0 < y < x; \; p_{10x}^{21y} = f_1(x+y), y > 0;$$

$$p_{20x}^{21y} = f_2(x-y), 0 < y < x; \; p_{20x}^{11y} = f_2(x+y), y > 0;$$

$$P_{11x}^{10x} = 1; \; P_{21x}^{20x} = 1.$$

We denote $\rho_{10}(x), \rho_{11}(x), \rho_{20}(x), \rho_{21}(x)$ of the density of the stationary distribution for the states of 10x, 11x, 20x, 21x, respectively. The system of integral equations for the densities of stationary distributions is:

$$\begin{cases} \rho_{11}(x) = \int\limits_x^\infty f_1(y-x)\rho_{10}(y)dy + \int\limits_0^\infty f_2(x+y)\rho_{20}(y)dy; \\ \rho_{21}(x) = \int\limits_x^\infty f_2(y-x)\rho_{20}(y)dy + \int\limits_0^\infty f_1(x+y)\rho_{10}(y)dy; \\ \rho_{10}(x) = \rho_{11}(x); \rho_{20}(x) = \rho_{21}(x); \\ \int\limits_0^\infty (\rho_{10}(x) + \rho_{11}(x) + \rho_{20}(x) + \rho_{21}(x))dx = 1. \end{cases}$$

In [1] it is shown that the solution of this system is determined by formulas

$$\rho_{10}(x)=\rho_{11}(x)=\rho_0\overline{F_2}(x);\ \rho_{20}(x)=\rho_{21}(x)=\rho_0\overline{F_1}(x),$$

where the constant ρ_0 is explicitly from the normalization state.

Stay times in the states

$$\upsilon_{10x}=x\wedge\alpha_1;\ \upsilon_{11x}=0;$$
$$\upsilon_{21y}=\beta_2;\ \upsilon_{20y}=y\wedge\alpha_2.$$

DF of the stay time in the states

$$\overline{F}_{10x}(t)=\bar{1}_x(t)\cdot\overline{F}_1(t);\ F_{21y}(t)=G_2(t);$$
$$\overline{F}_{20y}(t)=\bar{1}_y(t)\cdot\overline{F}_2(t).$$

As mentioned above, let us complete the transition to an equivalent system with discrete states. The graph of such system is similar to the graph in **Fig.** 3. As shown in **Fig. 3**, the graph states have no continuous components, that is, the system is discrete. To go to discrete system phase consolidation algorithm (PCA) is used. It is proposed in [2].

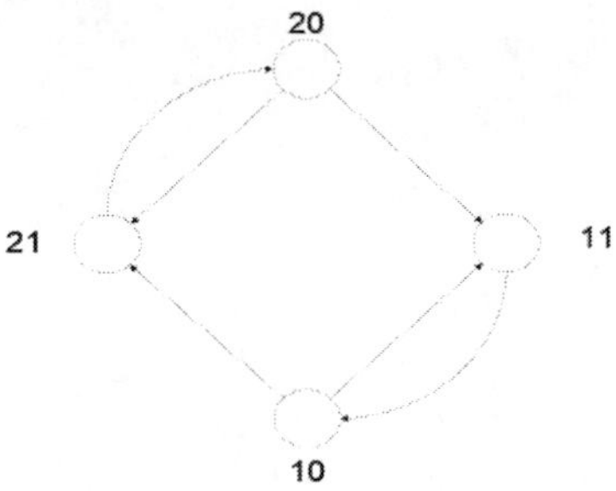

Fig. 3. The graph of the system states with discrete states.

It is necessary to define the transition probabilities, the stationary distribution of EMC and the stay times of distribution function in the discrete states 10 and 20 of the formulas [2]:

$$\hat{P}_{k,r}=\frac{\int_{E_k}^{\infty}\rho(dx)P(x,E_r)}{\rho(E_k)} \qquad (11)$$

$$\hat{F}_k(t)=\frac{\int_{E_k}\rho(dx)F_x(t)}{\rho(E_k)} \qquad (12)$$

For the state 20:

$$\rho_{21}=\rho_{20}=\rho_0\int_0^{\infty}\overline{F}_1(x)dx=\rho_0 m_{\alpha_1}$$

According to the formula (11) we find the transition probabilities:

$$P_{20}^{21}=\frac{\int_0^{\infty}\overline{F}_1(x)F_2(x)dx}{m_{\alpha_1}};\ P_{20}^{11}=\frac{\int_0^{\infty}\overline{F}_1(x)\overline{F}_2(x)dx}{m_{\alpha_1}}.$$

We find the DF system for stay time in the discrete state 20 using (12):

$$\hat{F}_{20}(t)=\frac{\rho_0\int_0^{\infty}\overline{F}_1(x)dx\left[1-\bar{1}_x(t)\cdot\overline{F}_2(t)\right]dx}{\rho_0 m_{\alpha_1}}.$$

Let us transform the numerator:

$$\int_0^{\infty}\overline{F}_1(x)dx\left[1-\bar{1}_x(t)\cdot\overline{F}_2(t)\right]dx=m_{\alpha_1}F_2(t)+\overline{F}_2(t)\int_0^{t}\overline{F}_1(x)dx.$$

Here

$$\hat{F}_{20}(t)=\frac{m_{\alpha_1}F_2(t)+\overline{F}_2(t)\int_0^{t}\overline{F}_1(x)dx}{m_{\alpha_1}}.$$

For the state 10:

$$\rho_{11}=\rho_{10}=\rho_0\int_0^{\infty}\overline{F}_2(x)dx=\rho_0 m_{\alpha_2}.$$

According to the formula (11) we find the transition probabilities:

$$P_{20}^{11}=\frac{\int_0^{\infty}F_1(x)\overline{F}_2(x)dx}{m_{\alpha_2}};\ P_{20}^{21}=\frac{\int_0^{\infty}\overline{F}_1(x)\overline{F}_2(x)dx}{m_{\alpha_2}}.$$

We find the DF system for stay time in the discrete state 10 using (12):

$$\hat{F}_{10}(t)=\frac{\rho_0\int_0^{\infty}\overline{F}_2(x)dx\left[1-\bar{1}_x(t)\cdot\overline{F}_1(t)\right]dx}{\rho_0 m_{\alpha_2}}.$$

Let us transform the numerator:

$$\int_0^{\infty}\overline{F}_2(x)dx\left[1-\bar{1}_x(t)\cdot\overline{F}_1(t)\right]dx=m_{\alpha_2}F_1(t)+\overline{F}_1(t)\int_0^{t}\overline{F}_2(x)dx.$$

Here

$$\hat{F}_{10}(t)=\frac{m_{\alpha_2}F_1(t)+\overline{F}_1(t)\int_0^{t}\overline{F}_2(x)dx}{m_{\alpha_2}}$$

Let us define the stay time of distribution function of the system in the states 21 and 20 of the system based on the repeated enterings according to formula (1):

DOI: 10.1051/matecconf/201712903027

$$F_{20}^{2}(s)=\frac{\hat{F}_{20}(s)}{c_{20}-(c_{20}-1)\hat{f}_{20}(s)},$$

where $c_{20}^{2}=\frac{\rho_{20}}{\rho_{20}\cdot P_{20}^{11}}=\frac{1}{P_{20}^{11}}$.

$$F_{21}^{2}(s)=\frac{\hat{F}_{21}(s)}{c_{21}-(c_{21}-1)\hat{f}_{21}(s)},$$

where $c_{21}^{2}=\frac{\rho_{21}}{\rho_{21}\cdot P_{20}^{11}}=\frac{1}{P_{20}^{11}}$.

There are two subsets:

$$M_{+}=\{10,21,20\}$$
$$M_{-}=\{11\}$$

We define trajectory to the subset of system (Fig. 4):

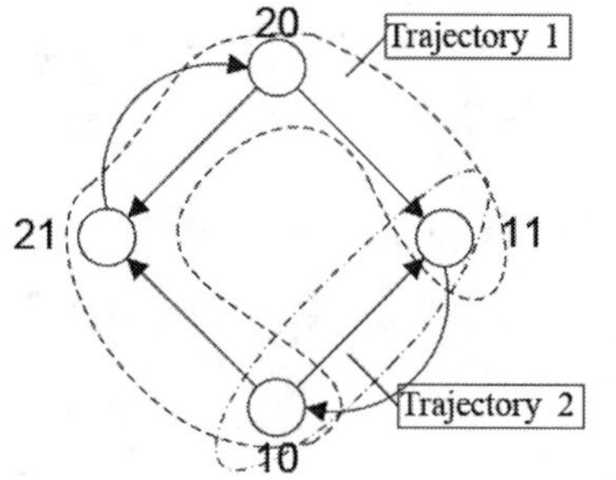

Fig.4. The trajectories of system subsets

$T_1=\{S_{10}S_{11}\}$, **4.** $T_2=\{S_{10}S_{21}S_{10}S_{11}\}$.

With serial convolution we determine the stay time of distribution function of each trajectory.

$F_1^T(t)=F_{10}(t)$, $F_2^T(t)=F_{10}(t)*F_{21}^2(t)*F_{20}^2(t)$,

where * - sign of the convolution operation.

Let us find the probability of each of trajectories

$$P_1^T=P_{10}^{11},\ P_2^T=P_{10}^{21}.$$

Hence, it becomes possible to determine the stay time of distribution function $F_{\Sigma}^{\theta}(t)$ of the system in the subset, regardless of the initial state as a weighted sum (a mixture) of the stay times of distribution function of the subset of each trajectory with coefficients equal to the probability of those trajectories:

$$F_{\Sigma}^{\theta}(t)=F_1^T(t)\cdot P_1^T+F_2^T(t)\cdot P_2^T.$$

Solutions for $F_{обсл}(t)$ DF cycle time maintenance PE because of its failure, and also for the mathematical expectation and variance obtained by known method using the Markov renewal equation, given in [4]:

$$F_{обсл}(t)=1-\frac{1}{M\alpha_2}\Bigg[M\alpha_2\overline{F}_1(t)+\int_0^t\overline{F_2}(x)dx\int_x^t f_1(s)\overline{G_2}(t-s)ds+$$
$$+\sum_{n=1}^{\infty}\int_0^t\overline{F_2}(x)dx\int_x^t f_1(s)F_2^{\bullet(n)}(s-x)ds\int_s^t g_2^{\bullet(n)}(y-s)\overline{G_2}(t-y)dy\Bigg].$$

$$M\theta=\frac{M\alpha_1(M\alpha_2+M\beta_2)}{M\alpha_2}$$

$$D\theta=(1+\frac{2M\beta_2}{M\alpha_2})D\alpha_1+\frac{M\alpha_1}{M\alpha_2}(D\beta_2+$$
$$+(M\ \beta_2)^2(1-\frac{M\alpha_1}{M\alpha_2}))+\frac{2(M\ \beta_2)^2}{M\alpha_1}\int_0^{\infty}\overline{F}_1(s)H_2(s)ds.$$

Results comparing the two methods are presented in Fig. 5.

The initial data for the modelling are:

DF, and distributed by the generalized Erlang law of second order with the parameters $\upsilon_1,\upsilon_2;\lambda_1,\lambda_2;\mu_1,\mu_2$ respectively. Moreover,

$$f_1(t)=\frac{\upsilon_1\upsilon_2(e^{-\upsilon_1 t}-e^{-\upsilon_2 t})}{\upsilon_2-\upsilon_1},$$

where υ1 = 5,505 (h-1); υ2 = 54,495 (h-1).

$$f_2(t)=\frac{\lambda_1\lambda_2(e^{-\lambda_1 t}-e^{-\lambda_2 t})}{\lambda_2-\lambda_1},$$

where λ1 = 0,0275 (h-1); λ2 = 0,272 (h-1);

$$g_2(t)=\frac{\mu_1\mu_2(e^{-\mu_1 t}-e^{-\mu_2 t})}{\mu_2-\mu_1},$$

where μ1 = 0,667 (h-1); μ2 = 2,000 (h-1).

Fig. 5 shows the modelling results.

Fig. 5a shows that the results of the proposed modelling of the exact method and of the approximate one almost coincide. It is possible to see the difference in the curves only with 4,000-times magnification in **Fig.** 5b.

Mathematical expectation and variance RV -time cycle service of PE production unit in view of its failures is defined in two ways:

- known method using the Markov renewal equation: $m_{обсл}=0.209999999h$; $D_{обсл}=0.069076$;
- the proposed method: $m_{\Sigma}^{\theta}=0.210000000h$; $D_{\Sigma}^{\theta}=0.069423$.

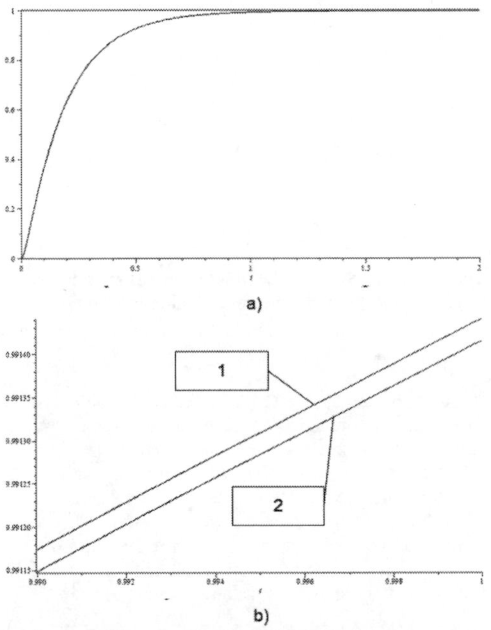

Fig. 5. Comparison of the modelling results with the help of the classic approximate method of successive approximations, and of the precise method proposed: 1 – with approximate method; 2 - the solution offered with the precise method.

4 Conclusions

The results of modelling prove the correctness of the proposed method. The obtained exact solution is of great interest for it allows to determine accurately the DF of time between the events in the stream of the resulting superposition of the renewal stream and alternating renewal stream. In the future it is expected to approve the proposed method for the modelling of other systems.

The research was partially supported by the Russian Foundation for Basic Research (research project No. 15-01-05840 a.).

References

1. V.S. Korolyuk, A.F. Turbin, *Markov renewal processes in reliability problems of systems*, Naukova Dumka (1982)
2. V.S. Korolyuk, A.F. Turbin, *Stochastic models of systems*, Naukova Dumka (1989)
3. V.Y. Kopp, M.V. Zamoryonov, U.E. Obzherin, O.V. Filipovich, *Methods for semi-Markov systems modelling*, Adaptive Systems of automatic control: Interdepartmental sci-tech collection. №**1 (26)**. pp. 208-221 (2015)
4. V.Y. Kopp, Y.E Obzherin, A.I. Peschanskiy, *Automated lines modelling* (Sevastopol University SevNTU, 2006)

Semi-markov model of processing requests to the cloud storage

Mikhail Zamoryonov [1,*], *Victoria* Shevchenko [1], *Elena* Vladimirova [1], and *Nikolay* Sapozhnikov [2]

[1]Sevastopol State University, Department of Information Technologies and Computer Systems, Sevastopol, Russian Federation
[2]Federal state owned military educational institution of the Higher professional Education "the black sea higher navy order of the red star academy by P.S.Nakhimov" Russian Federation, 299028, Sevastopol

Abstract. This paper presents a semi-Markov model as an important part of the computer-aided manufacturing and a modular system of cloud storage, which affects the functioning of the whole process. The residence times of the system in states and the probability of system transitions are determined. There is a stationary distribution of the embedded Markov chain. The residence times of the system in states with allowance for repeated hits are determined by the theorem on the distribution functions of the residence times of the system in states with allowance for repeated hits. When using the trajectory method, the distribution function of the time for the complete processing of the read request by such a system is determined. A comparison of the expectation time complete processing the read request obtained in this study and obtained by the known formula for determining from the literature the expectation residence time in the subset of the system states.

1 Introduction

Computer-aided manufacturing today is a complex-structured multi-purpose hardware and software systems operating on a wide range of problems with the use of advanced information technology and advanced software and hardware tools, to effectively carry out operational planning and management of the whole enterprise as a whole. At the same trends in the development of such systems is largely due to the transfer of software geographically distributed industrial companies by means of Internet communications in the virtual space. The introduction of the concept of virtual enterprise (extended enterprises) allows to provide technological support for remoting professionals and individual units of mechanical assembly production at the stages of design, production, marketing and other stages life cycle of products.

One of the most promising information technologies in this direction is the technology of cloud computing. Cloud computing, according to [1], provide a model for convenient network access on demand to the total pool of configurable in computing resources (networks, servers, storage, applications and services) that can quickly identify and provide with minimal management effort or minimal intervention on the part of the supplier of cloud services [2]. The introduction of cloud computing technology has some significant advantages [3-6]: the enterprise by eliminating the need to invest substantial sums to machine tool making, purchase of expensive software (ERP, PDM, CAD / CAM / CAE-systems); Economy of computing resources; Possibility of mobility of workplaces; Scalability, which increases the efficiency of computing resources through the use of virtualization technologies; saving on maintenance of the IT department, these functions are moving to outsourced cloud providers.

One common use for this cloud for industrial automation is the placement of the cloud of part or all SCADA-system (Supervisory Control and Data Acquisition). Cloud technologies are used by SCADA applications in two main ways: 1) the SCADA system is installed locally, directly connected to the control network, and also sends information to the cloud where it can be stored and distributed from there. 2) the SCADA system is fully deployed in the cloud and remotely connected to the control network. The data received from these systems in a network cloud storage for later use MES-systems (Manufacturing Execution System) and other industrial automation systems. It should be noted that the volumes of transmitted, processed and stored information are very significant and can be attributed to the Big Data. In this connection, the problems associated with determining the performance of these systems become particularly topical. We also note that mathematical models allow not only to determine productivity, but also to indicate ways for increasing it.

The primary characteristic processing systems and data storage is waiting for the response to the request or the response time. Given that the response time is a random variable that depends on a number of factors different for each particular implementation of the system, the definition of this time is not an easy task. As part of further studies, the inventors have focused on one of the technologies of data storage systems – block storage [7] as the most widely used for industrial automation systems.

*Corresponding author: zamoryonoff@gmail.com

The purpose of this paper is to determine the distribution function (FF) of the response time of a block cloud storage system to a data read request.

Consider a block system of cloud storage of general data. This system works as follows. Modular storage system (NAS) provides computer systems access to a block-level storage volumes. In this environment, the file system is created on the basis of computer systems, and access to the data is carried out through the network at the block level. A block storage system consists of one or more controllers and a storage system. In its turn, the controller storage system consists of three key components: a client interface (external ports and controllers), cache memory and the server part. The structure of the block-type storage intelligent system is shown in Fig. 1, a detailed description is given in [7].

The Input/Output request is transferred from the computer system via an external port and processed using the server's cache memory. Read request can be served directly from the cache, if it found the necessary data to the user. In modern storage systems, the client interface, cache memory and server part are usually integrated into a single panel (which is called the storage processor or storage controller).

The client interface provides data exchange between the storage system and the computer system. It consists of two components: external ports and external controllers. Typically, the client interface has redundant controllers for high availability, and each controller contains several ports that allow multiple computer systems to be connected to an intelligent data storage system.

Fig. 2 shows the read operation using the cache memory subject to the availability (a) and unavailability of (b) in the data cache.

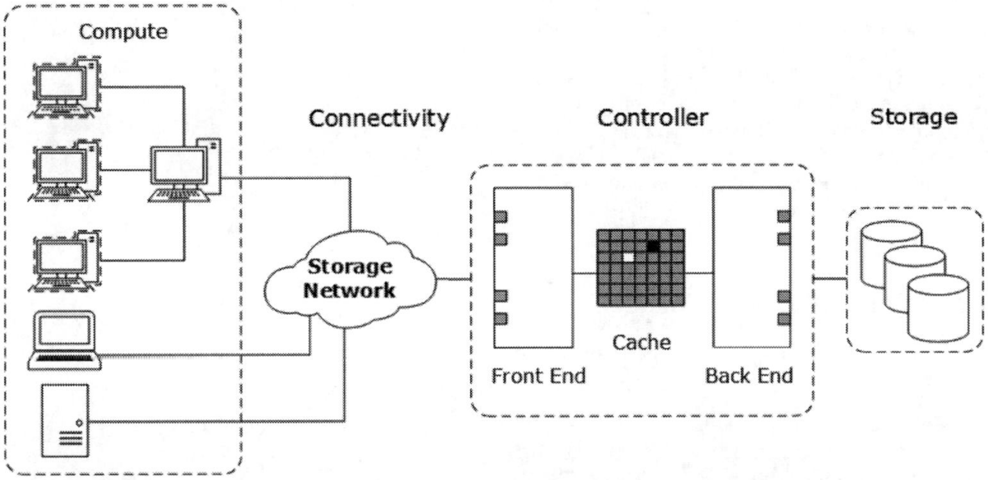

Fig. 1. Components intelligent data storage system.

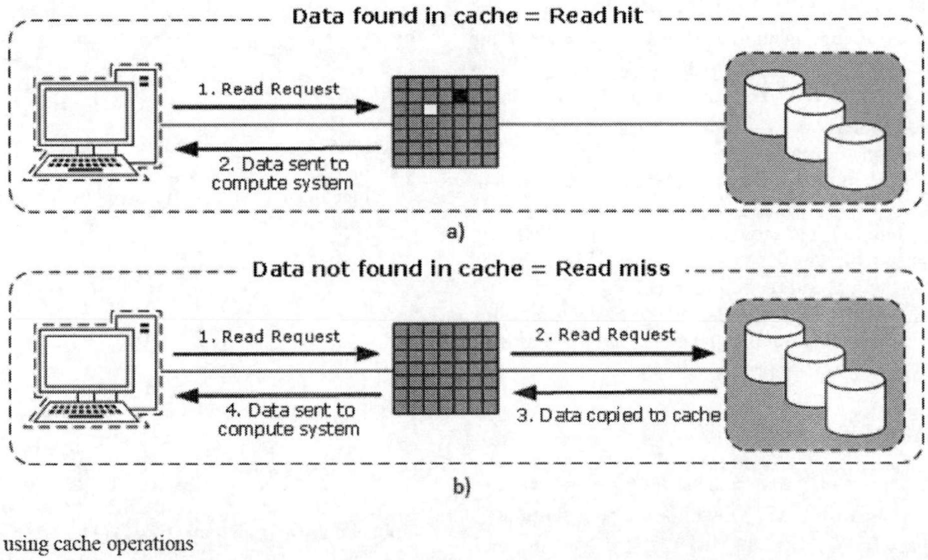

Fig. 2. Read using cache operations

When the computer system sends a read request, the storage controller reads the RAM label to determine the availability of the necessary data in the cache. If the requested data is found in the cache, this is called a hit in the read cache or hit, while reading, and the data is sent directly to the computer system without performing operations in the internal storage. This provides for the computer system a short response time (about milliseconds). If the requested data is not found in the cache, then this is called a cache miss. In this situation, the data must be read from the storage system. The server part gains access to the appropriate storage device and retrieves the requested data. The data is placed in the cache memory, and then sent to the computer system using the client interface. Cache misses increase the response time for Input/Output operations.

The performance of reading is measured by the hit rate of read operations or the frequency of hits, which is usually expressed as a percentage. This ratio is the ratio of the number of read hits to the total number of read requests. The higher the hit rate of read operations is, the higher the performance of these operations is.

It should be noted that the response time also depends on the reliability of data transmission networks.

To simulate storage block type is proposed to use trajectories method given in [8]. This method allows one to obtain an exact solution of the Markov recovery equations for a system with a discrete phase space of states.

2 Formulation of the problem

We pose the problem in the following way: Distribution functions (DF) $F_1(t)$ and $G_1(t)$ of the random variables (RV) α_1 and γ_1,which are sometimes processing read requests from the user to the cache memory and from the cache memory to the storage data server (SDS), respectively, and DF $F_1(t)$ and $G_1(t)$ of random variables α_2 and γ_2 – time between failures in the channels of communication between the user and the memory modules and between the cache and SDS. In addition, DF $R(t)$ RV β, being the time search data in the cache memory and the probability p of finding the desired information in the memory are known. Also the probability of finding the data in the cache memory of the system p is know P.

When building the model it is assumed that when a failure occurs in the data transmission channel of the user performing a read request is restarted.

In the current study, it is necessary to determine the time distribution function for the complete processing of the read request by the specified system.

3 Building a model

The functioning of the system is described by a semi-Markov process (SMP) $\xi(t)$ with a discrete-continuous phase space of states [9, 10, 11, 12, 13]. We introduce the following set E of Semi-Markov states of the system:

$$E = \{S_0, S_1, S_2, S_3\}$$

Let us decipher the meaning of the state codes:
S0 - the request is executed (instantaneous state);
S1 - a request from the user is received (or when the query has failed in data transmission channel) the re-execution request is beginning;
S2 - searches for the data in the cache memory modules;
S3 - the data in the cache memory is not found, a request is made to SDS (or during a query to SDS, the failure occurred in the data channel), the process is beginning with the start request.

Count system states is shown in Fig. 3.

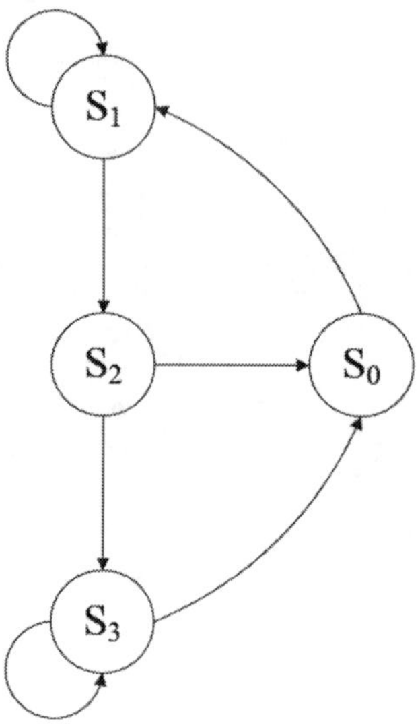

Fig. 3. System graph.

Times of stay θ_0, θ_1, θ_2,θ_3 of the system states *S0, S1, S2* and *S3* are equal to:

$$\theta_0 = 0;\ \theta_1 = (\alpha_1 \wedge \alpha_2);\ \theta_2 = \beta;\ \theta_3 = (\gamma_1 \wedge \gamma_2),$$

where $\wedge$ is a sign denoting a minimum of random variables.

Then the DF of the times of the system stay in the states are equal:

$$F_{S0}(t) = 0;$$

$$\overline{F}_{S1}(t) = \overline{F}_1(t) \cdot \overline{F}_2(t);$$

$$F_{S2}(t) = R(t);$$

$$\overline{F}_{S3}(t) = \overline{G}_1(t) \cdot \overline{G}_2(t).$$

We describe the probabilities of transitions of the embedded Markov chain (ECM) (1).

 DOI: 10.1051/matecconf/201712903028

$$\begin{cases} P_0^1 = 1; \\ P_1^1\{\alpha_1 > \alpha_2\} = \int_0^\infty \overline{F}_1(t) f_2(t)dt; \\ P_1^2\{\alpha_1 < \alpha_2\} = \int_0^\infty f_1(t)\overline{F}_2(t)dt; \\ P_2^0 = p; \\ P_2^3 = 1 - p; \\ P_3^3 = \int_0^\infty \overline{G}_1(t) g_2(t)dt; \\ p_3^0 = \int_0^\infty g_1(t)\overline{G}_2(t)dt. \end{cases} \quad (1)$$

Using system (1), we write the system of equations for determining the stationary distribution of the embedded Markov chain $\rho(x)$ [14, 15, 16]:

$$\begin{cases} \rho_0 = \rho_2 \cdot P_2^0 + \rho_3 \cdot P_3^0 \\ \rho_1 = \rho_0 \cdot P_0^1 + \rho_1 \cdot P_1^1 \\ \rho_2 = \rho_1 \cdot P_1^2 \\ \rho_3 = \rho_2 \cdot P_2^3 + \rho_3 \cdot P_3^3 \end{cases} \quad (2)$$

Normalization condition:

$$\rho_0 + \rho_1 + \rho_2 + \rho_3 = 1. \quad (3)$$

Solving the system of equations (2) using condition (3), we obtain:

$$\begin{cases} \rho_0 = \dfrac{P_1^2 \cdot P_3^0}{2 \cdot P_1^2 \cdot P_3^0 + P_3^0 + P_1^2 \cdot P_2^3}; \\ \rho_1 = \dfrac{P_3^0}{2 \cdot P_1^2 \cdot P_3^0 + P_3^0 + P_1^2 \cdot P_2^3}; \\ \rho_2 = \dfrac{P_1^2 \cdot P_3^0}{2 \cdot P_1^2 \cdot P_3^0 + P_3^0 + P_1^2 \cdot P_2^3}; \\ \rho_3 = \dfrac{P_1^2 \cdot P_2^3}{2 \cdot P_1^2 \cdot P_3^0 + P_3^0 + P_1^2 \cdot P_2^3}. \end{cases}$$

We represent the set of all states in the form: $E = E_+ \cup E_-$, $E_+ = \{S_1, S_2, S_3\}$ $E_- = \{S_0\}$. A subset of states E_+ describes the process of servicing the read request.

To determine the DF complete request processing time for reading system it is used the method of trajectories [8]. For this we use the theorem on the DF time the system spends in the states considering re-entering [8].

Define the trajectories of the system's output into a subset $T_1 = \{S_1 S_2\}$, $T_2 = \{S_1 S_2 S_3\}$.

We define the residence time distribution function of the system in the states 21 and 20 and with the DF repeated hits by the formulas (1):

$$F_{S1}^{\theta}(s) = \frac{F_{S1}(s)}{c_1 - (c_1 - 1) f_{S1}(s)},$$

$$F_{S3}^{\theta}(s) = \frac{F_{S3}(s)}{c_3 - (c_3 - 1) f_{S3}(s)}.$$

where $c_1 = \dfrac{\rho_1}{\rho_0 \cdot P_0^1} = \dfrac{1}{P_1^2}$; $c_3 = \dfrac{\rho_3}{\rho_0 \cdot P_0^1} = \dfrac{1}{P_3^0}$.

With sequential convolution DF time of system stay is defined in each path:

$$F_1^T(t) = F_{S1}^{\theta}(t) * F_{S2}(t);$$

$$F_2^T(t) = F_{S1}^{\theta}(t) * F_{S2}(t) * F_{S3}^{\theta}(t);$$

where * is the sign of the convolution operation.

Let us find the probability of realizing each of the trajectories:

$$P_1^T = p; \qquad P_2^T = 1 - p,$$

moreover, $P_1^T + P_2^T = 1$.

Hence it becomes possible to determine the DF $F_{\Sigma}^{\theta}(t)$ of the time of the system stay in the subset E_+ regardless of the initial state, as a weighted sum of (a mixture of) the DF system residence times in the subset each trajectory with coefficients equal to the probability of implementing these trajectories

$$F_{\Sigma}^{\theta}(t) = F_1^T(t) \cdot P_1^T + F_2^T(t) \cdot P_2^T. \quad (4)$$

The results of simulation by the formula (4) shown in Fig. 4.

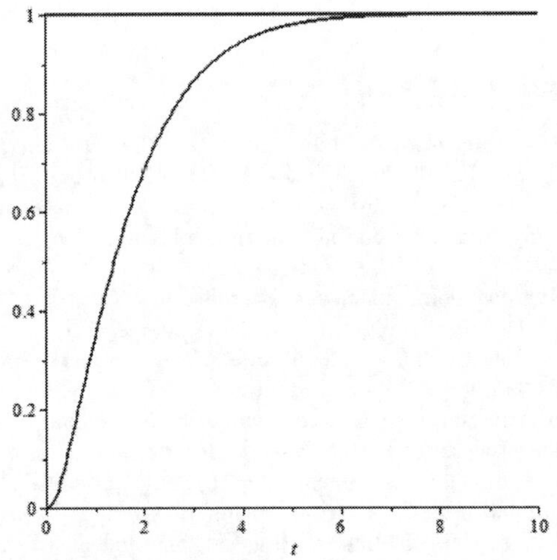

Fig. 4. Type of DF of full read request processing time.

The initial data for modeling are DF $F_1(t)$, $F_2(t)$, $G_1(t)$, $G_2(t)$, $R(t)$, which are corresponded to the generalized second-order Erlang law with parameters λ_1, λ_2; μ_1, μ_2; υ_1, υ_2; ν_1, ν_2; τ_1, τ_2, respectively. And:

$$f_1(t)=\frac{\lambda_1\lambda_2(e^{-\lambda_1 t}-e^{-\lambda_2 t})}{\lambda_2-\lambda_1},\ f_2(t)=\frac{\mu_1\mu_2(e^{-\mu_1 t}-e^{-\mu_2 t})}{\mu_2-\mu_1},$$

$$g_1(t)=\frac{\upsilon_1\upsilon_2(e^{-\upsilon_1 t}-e^{-\upsilon_2 t})}{\upsilon_2-\upsilon_1},\ g_2(t)=\frac{\nu_1\nu_2(e^{-\nu_1 t}-e^{-\nu_2 t})}{\nu_2-\nu_1},$$

$$r(t)=\frac{\tau_1\tau_2(e^{-\tau_1 t}-e^{-\tau_2 t})}{\tau_2-\tau_1},$$

where λ_1=1,101 (s^{-1}); λ_1=10,899 (s^{-1});

μ_1=0,0688 (s^{-1}); μ_2=0,681 (s^{-1});

υ_1=1,101 (s^{-1}); υ_2=10,899 (s^{-1});

ν_1=0,0688 (s^{-1}); ν_2=0,681 (s^{-1});

τ_1=11,01 (s^{-1}); τ_2=108,99 (s^{-1}).

And also the probability of finding data in the cache memory of the system p=0,4.

Let us compare the values of the mathematical expectation of the function (4) obtained by us and the mathematical expectation determined by the expression [14]:

$$T_+=\frac{\sum_{i\in M_+} m_i\rho_i}{\sum_{e\in E\subset M_+}\sum_{j\in M_-} P_{ej}\rho_i}. \quad (5)$$

The mathematical expectation of the distribution function obtained by us is 1.7037037 sec.

While determined by formula (5) - 1.7037037 sec.

4 Conclusions

A comparison of the values of the mathematical expectation of the time for the complete processing of the read request by the data storage system determined using the distribution function obtained by us and obtained using the formula given in [16] showed a complete coincidence of the results.

The results allow to determine the response time for reading by the cloud storage system request threads cloud storage, which allows to evaluate the impact of the information system performance in the adaptability of the process and, if necessary, to make a decision to increase the speed of the system. The resulting simulation of the distribution function allows the use of this model in the construction of models of more complex hierarchically subordinate information and production systems.

This work was supported by grant s Russian Foundation for Basic Research № 15-01-05840, № 15-29-07936.

References

1. NIST Special Publication 500-307 Cloud Computing Service Metrics Description, NIST Cloud Computing Reference Architecture and Taxonomy Working Group. The Web: http://www.nist.gov/itl/cloud / upload the / RATAX-Cloud Service Metrics Description-DRAFT-20141111.pdf
2. A.V. Skatkov, V.I. Shevchenko, *Extension of the reference model of the cloud computing environment in the concept of large-scale scientific research*, Proceedings of ISP RAS. № **6**, pp. 285-306 (2015)
3. F. Doelitzscher, T. Ruebsamen, T. Karbe, M. Knahl and C. Reich, *Sun Behind Clouds - On Automatic Cloud Security Audits and a Cloud Audit Policy Language*, International Journal on Advances in Networks and Services, № **6**, (1&2) (2013)
4. S. Singha, S.D. Satav, *A Survey Paper on Cloud Storage Auditing with Key Exposure Resistance*, International Journal of Science and Research (IJSR), Vol. **4**, pp. 1821-1823. (2015)
5. B. Wang, B. Li and H. Li. *Public Auditing for Shared Data with Eff i cient User Revocation in the Cloud*, Web: iqua.ece.toronto.edu/papers/bwang-infocom13.pdf
6. M. Hogan, F. Liu, A. Sokol and J. Tong, *NIST Cloud Computing Standards Roadmap*, Computer Security Division Information Technology Laboratory Gaithersburg, MD 20899-8930 (2011)
7. From data storage to information management. Peter. P.544 (2016)
8. M. V. Zamoryonov, V. Ya. Kopp, Yu. E. Obzherin, D.V. Zamoryonova, *Approbation of the method of trajectories on the example of modeling the process of functioning of the production element with devaluating failures*, Proceedings of Tula State University. Technical science № **8**, pp. 57-71 (2015)
9. Yu. E. Obzherin, E.G. Boyko, *Semi-Markov Models. Control of Restorable Systems with Latent Failures* (Elsevier, Academic Press, 2015)
10. A.I. Peschansky, *Semi-Markov Models of One-Server Loss Queues with Recurrent Input* (LAP LAMPERT Academic Publishing, 2013)
11. F. Bayhelt, P Franken, *Reliability and maintenance. Mathematical approach*, Radio and Communication, (1988)
12. V.S. Korolyuk, *Superposition of Markov recovery processes*, Cybernetics №4, pp. 121 – 124 (1981)
13. V.S. Korolyuk, *Semi-Markov processes and their applications* (Science, Dumka, 1976)
14. V.S. Korolyuk, A.F. Turbin, *Markov recovery processes in reliability problems of systems* (Naukova Dumka, 1982)
15. V.S. Korolyuk, *Stochastic models of systems* (Naukova Dumka, 1989)
16. K. Rainshke, I.A. Ushakov, *Evaluation of the reliability of systems using graphs*, Radio and Communication (1988)

MATEC Web of Conferences **129**, 03029 (2017)
ICMTMTE 2017

DOI: 10.1051/matecconf/201712903029

Feedback linearized trajectory-tracking control of a mobile robot

Aleksey Kabanov[1,2,*]

[1]Federal Research Centre "Computer Science and Control" of Russian Academy of Sciences, Moscow, Russian Federation
[2]Sevastopol State University, 299053 Sevastopol, Russian Federation

Abstract. This paper is devoted to the designing of a trajectory-tracking control system for a unicycle-type mobile robot. Synthesis of the trajectory control law is based on the feedback linearization method and a canonical similarity transformation of nonlinear affine system in state-dependent coefficient form. The result of experimental test of the trajectory control system for mobile robot Rover5 is presented.

1 Introduction

The problem of mobile robot (MR) control is relevant over the years because of the wide range of theoretical problems and practical applications associated with it. There is a great number of works in the field of motion control of MR that are published in recent decades. Their reviews can be founded in [1-4].

Trajectory-tracking control problem is a typical control problem for MR. It is mainly concerned with the design of control laws that force a MR to reach and follow a time parameterized reference trajectory [5]. Thus, as a rule, simple non-linear models are operated, which characterize only the kinematic relationship between the MR's motion parameters and parameters that are predetermined by a reference trajectory.

At this point it is possible to mark out two basic approaches to design the tra-jectory-tracking system. The first approach is based on linear control design methods and tangent linearization of system model about the reference trajecto-ry. At that, a pole placement method or linear optimal control method are usually used to adjust controller parameters [6-8]. Another approaches is based on non-linear control design methods, such as feedback linearization (FL), Lyapunov techniques [9-11].

A lot of publications are devoted to the trajectory-tracking control problem for MR with nonholonomic constraints by means of FL (static and dynamic). In [9] applicability of static FL for setpoint regulation problems and for trajectory-tracking problems is considered. In [11] the trajectory-tracking control problem based on dynamic FL is considered. It is shown that dynamic FL is an efficient design tool leading to a solution simultaneously valid for both trajectory tracking and setpoint regulation problems. The main drawback of such solution is aligned with the raise of the system order and the order of the dynamic feedback.

In this paper, we propose another approach for static FL control design for tra-jectory-tracking, based on representing the original nonlinear system into a state-dependent coefficient form and applying the canonical similarity transformation, that allow getting the system to canonical form. This similarity transformation allow accomplishing linearization of a system without determining of a virtual system output.

The remaining sections of the paper proceed as follows: Section 2 describes the trajectory- tracking control problem statement; Section 3 is devoted to the problem of designing of trajectory control system for MR based on FL method; Section 4 describes the structure of the trajectory control system, results of ex-perimental tests for MR Rover5; some conclusions are shown in Section 5.

2 Formulation of a Trajectory-Tracking Control problem for MR

The simplest model of a nonholonomic MR is the unicycle. Let us consider the kinematic model of a unicycle-type MR motion in a horizontal plane (Fig.1a) [1-4]:

$$\begin{aligned} \dot{X} &= V\cos\varphi = \frac{1}{2}\cdot\cos\varphi(V_1+V_2), \\ \dot{Y} &= V\sin\varphi = \frac{1}{2}\cdot\sin\varphi(V_1+V_2), \\ \dot{\varphi} &= \omega = \frac{1}{2d}\cdot(V_2-V_1), \end{aligned} \tag{1}$$

where X, Y are coordinates of MR posture in the earth coordinate system; V_1, V_2, V are linear velocities of left and right wheels and velocity of MR; ω is a MR angular velocity; φ is an angle between the vector V and the axis OX.

We suppose that there is a given trajectory of MR

$$p_{ref}(t) = (X_{ref}(t), Y_{ref}(t), \varphi_{ref}(t)).$$

* Corresponding author: kabanovaleksey@gmail.com

 DOI: 10.1051/matecconf/201712903029

MR trajectory-tracking errors (Fig. 1b) is defined as follows [1-4]

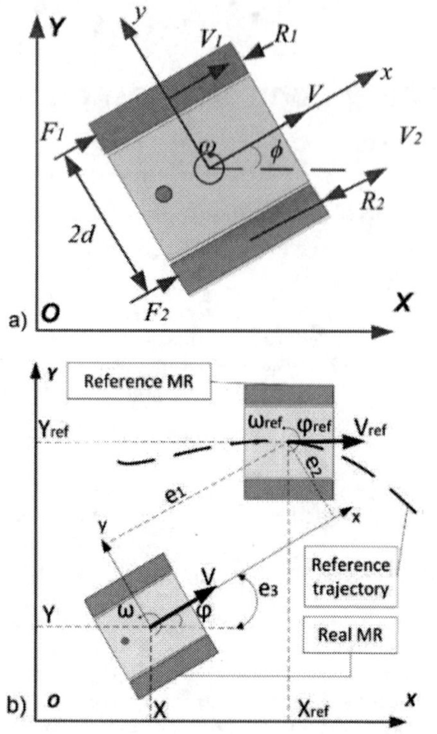

Fig. 1. a The MR on a plane. **b** Statement of the trajectory-tracking control problem for MR.

$$\begin{aligned} e_x &= X_{ref} - X, \\ e_y &= Y_{ref} - Y, \\ e_\varphi &= \varphi_{ref} - \varphi, \\ e_1 &= e_x \cos\varphi + e_y \sin\varphi, \\ e_2 &= e_y \cos\varphi - e_x \sin\varphi, \\ e_3 &= e_\varphi. \end{aligned} \tag{2}$$

Differentiating (2), we obtain the dynamic model of MR trajectory following errors:

$$\begin{aligned} &\dot{e}_1 = \omega_{ref} \cdot e_2 + u_1, \ \dot{e}_2 = -\omega_{ref} \cdot e_1 + V_{ref} \cdot \sin e_3, \\ &\dot{e}_3 = u_2, \ u_1 = V_{ref} \cdot \cos e_3 - V, \ u_2 = \omega_{ref} - \omega. \end{aligned} \tag{3}$$

The problem is to find control inputs u_1 and u_2, such that system (3) be an asymptotically stable, i.e. $e_i \to 0, i = 1,2,3$ for $t \to \infty$.

3 Control law design

3.1. Feedback Linearization of MIMO nonlinear system

Nonlinear MIMO system in state-dependent coefficient form is considered:

$$\dot{x}(t) = A(x)x + B(x)u, \tag{4}$$

where $x \in R^n$ is a state vector; $u \in R^r$ is a control vector; matrix $A(x) - (n \times n)$ and $B(x) - (n \times r)$ are continuous differentiable matrices and have bounded derivatives. Let us suppose that the system (4) is controllable.

The feedback linearization problem of the system (4) consists in searching a nonsingular similarity transformation $z = T(x)x$ and a control input $u = u_{FL}$ by means of which the system (4) can be transformed into a linear canonical form.

The procedure of constructing the similarity transformation is given in detail in [13]. Here is an example of the system conversion for the case $u \in R^2$. Applying the similarity transformation $z = T(x)x$ to the system (4) we get

$$\begin{gathered} \dot{z} = A_C(z)z + B_C(z)u, \\ A_C = TAT^{-1} + \dot{T}T^{-1}, \ B_C = TB, \end{gathered} \tag{5}$$

$$A_C = \left(\begin{array}{c|c} J_{r_0} & E_{r_0 r_1} \\ \hline E_{r_1 r_0} & J_{r_1} \end{array} \right), B_C = \left(\begin{array}{c|c} e_{p_1 p_1} & B_{12} \\ \hline 0 & e_{p_2 p_2} \end{array} \right),$$

$$J_k = \left(\begin{array}{c|c} 0 & I \\ \hline a_{k1} & a_{k2} \dots a_{kk} \end{array} \right), E_{kl} = \left(\frac{0}{a_{k1} \dots a_{kl}} \right),$$

$$B_{12} = \left(\frac{0}{b_{12}} \right), \ I = diag(1,...,1);.$$

$$\begin{gathered} J_k - (k \times k), \ E_{kl} - (k \times l), B_{12} - (r_0 \times 1), \\ I - (k-1, k-1), k = r_0, r_1; \ l = r_0, r_1; \ m = 2. \end{gathered}$$

Matrix B_C can be presented in terms of

$$\begin{gathered} B_C = \overline{B}C, \ C - (2 \times 2), \\ \overline{B} = \begin{pmatrix} e_{p_1} & 0 \\ 0 & e_{p_2} \end{pmatrix}, C = \begin{pmatrix} 1 & b_{12} \\ 0 & 1 \end{pmatrix}. \end{gathered}$$

Since C is nonsingular, there is a new equivalent input $\overline{u} = Cu$. Feedback linearizing control for system (4) can be defined in the next form

$$\begin{gathered} u = u_{FL} + \upsilon = C^{-1}(K_{FL}z + \upsilon), \\ K_{FL} = -\begin{pmatrix} k_{11} & k_{12} & \dots & k_{1n} \\ k_{21} & k_{22} & \dots & k_{2n} \end{pmatrix}. \end{gathered}$$

Instead of elements k_{ij}, it is necessary to take corresponding elements of J_k, E_{kl}.

3.2 Design of Trajectory-Tracking Control Law

Trajectory-tracking control loop is designed on the basis of FL method (4), (5) for tracking error dynamic system (3), which can be written in a state-dependent coefficient form:

DOI: 10.1051/matecconf/201712903029

$$\dot{e} = A(e)e + Bu,$$
$$e = (e_1 \quad e_2 \quad e_3)^T, \; u_K = (u_1 \quad u_2)^T,$$
$$A(e) = \begin{pmatrix} 0 & \omega_{ref} & 0 \\ -\omega_{ref} & 0 & a_{23} \\ 0 & 0 & 0 \end{pmatrix}, \; B = \begin{pmatrix} 1 & 0 \\ 0 & 0 \\ 0 & 1 \end{pmatrix}, \qquad (6)$$
$$a_{23} = V_{ref} e_3^{-1} \sin e_3.$$

The similarity transformation for system (6) under condition of controllability of pair of matrices $(A(e), B)$ (holds for all V_{ref}, ω_{ref}, except $V_{ref} = 0, \omega_{ref} = 0$) is

$$\tilde{e} = T(e)e, \; T(e) = \begin{pmatrix} 0 & -\omega_{ref}^{-1} & 0 \\ 1 & \dot{\omega}_{ref}\omega_{ref}^{-2} & a_{23}\omega_{ref}^{-1} \\ 0 & 0 & 1 \end{pmatrix}. \qquad (7)$$

Under transformation (7), the system (6) takes the form

$$\dot{\tilde{e}} = \tilde{A}(e)\tilde{e} + \overline{B}\overline{u}, \; u = C^{-1}\overline{u},$$
$$\tilde{A}(e) = \begin{pmatrix} 0 & 1 & 0 \\ \tilde{a}_{21} & \tilde{a}_{22} & \tilde{a}_{23} \\ 0 & 0 & 0 \end{pmatrix}, \overline{B} = \begin{pmatrix} 0 & 0 \\ 1 & 0 \\ 0 & 1 \end{pmatrix}, C = \begin{pmatrix} 1 & \tilde{b}_{22} \\ 0 & 1 \end{pmatrix},$$
$$\tilde{a}_{21} == -\frac{\omega_{ref}\ddot{\omega}_{ref} - \dot{\omega}_{ref}^2 + \omega_{ref}^4}{\omega_{ref}^2},$$
$$\tilde{a}_{22} = -\frac{\dot{\omega}_{ref}}{\omega_{ref}}, \; \tilde{b}_{22} = \frac{V_{ref}\sin e_3}{\omega_{ref} e_3},$$
$$\tilde{a}_{23} = \frac{\sin(e_3)(\dot{\omega}_{ref}V_{ref} - \dot{V}_{ref}\omega_{ref})}{e_3\omega_{ref}^2}.$$

The resulting control law is

$$u = C^{-1}(K_{FL} + G_{ST})T(e)e,$$
$$K_{FL} = -\begin{pmatrix} \tilde{a}_{21} & \tilde{a}_{22} & \tilde{a}_{23} \\ 0 & 0 & 0 \end{pmatrix}, \qquad (8)$$
$$G_{ST} = -\begin{pmatrix} g_{11} & g_{12} & 0 \\ 0 & 0 & g_{23} \end{pmatrix}, g_{11}, g_{12}, g_{23} > 0.$$

Numerical values of g_{11}, g_{12}, g_{23} may be calculated by a pole placement method. When $V_{ref} = const$, $\omega_{ref} = const \neq 0$ we have a special case:

$$T(e) = \begin{pmatrix} 0 & -\omega_{ref}^{-1} & 0 \\ 1 & 0 & a_{23}\omega_{ref}^{-1} \\ 0 & 0 & 1 \end{pmatrix},$$
$$u_2 = -g_{23}e_3, \; u_1 = \omega_{ref}e_2 - g_{12}e_1 + \qquad (9)$$
$$+\frac{g_{11}e_2 + g_{12}V_{ref}\sin(e_3) - g_{23}V_{ref}\sin(e_3)}{\omega_{ref}}.$$

The control law (9) is not applicable when $\omega_{ref} = 0$. So on the basis of this law it is impossible to realize a tracking along a straightforward trajectory. To overcome this disadvantage, it is possible to change the similarity transformation:

$$\tilde{e} = T(e)e, \; T(e) = \begin{pmatrix} 0 & -1 & 0 \\ \omega_{ref} & 0 & a_{23} \\ 0 & 0 & 1 \end{pmatrix},$$

which leads to transformed system

$$\dot{\tilde{e}} = \tilde{A}(e)\tilde{e} + \tilde{B}C^{-1}\overline{u},$$
$$\tilde{a}_{21} == -\omega_{ref}^2, \; \tilde{b}_{22} = -a_{23}.,$$
$$\tilde{A}(e) = \begin{pmatrix} 0 & 1 & 0 \\ \tilde{a}_{21} & 0 & 0 \\ 0 & 0 & 0 \end{pmatrix}, \tilde{B} = \begin{pmatrix} 0 & 0 \\ \omega_{ref} & \tilde{b}_{22} \\ 0 & 1 \end{pmatrix},$$
$$C^{-1} = \begin{pmatrix} \omega_{ref} & -\tilde{b}_{22} \\ 0 & 1 \end{pmatrix}.$$

In this case, the resulting control law has the form (8) where $\tilde{a}_{22} = \tilde{a}_{23} = 0$. It can be written in a scalar form

$$u_1 = \omega_{ref}(e_2 + g_{11}e_2 + g_{12}V_{ref}\sin(e_3)) -$$
$$- g_{12}e_1\omega_{ref}^2 - g_{23}V_{ref}\sin(e_3), \; u_2 = -g_{23}e_3. \qquad (10)$$

Finally, for the required values of MR linear and angular velocities, we have:

$$V_{req} = V_{ref}\cos(e_3) - u_1, \; \omega_{req} = \omega_{ref} - u_2.$$

4 The experimental results

Object of the study is a MR Rover5 by DAGU [8]. Power section of robot includes two electric drives based on the DC Motors. As a control unit Arduino UNO controller is used. To control the motors Rover5 Explorer PCB board is used. It is equipped with two FET H-bridges rated at 4A. Wireless communication interface is implemented through the usb-programmable wireless module Wixel.

Some parameters of the Rover5: track length $l = 0{,}245$, m; driving wheel's radius $r_w = 0{,}03$, m; robot's mass $m = 0{,}83$, kg; distance between tracks $d = 0{,}19$, m. For measuring the motor's speed MR Rover5 is equipped with two quadrature encoders. Encoder's resolution is 1000 pulses for 3 turns of the driving wheel.

An eight-shaped reference trajectory was taken as a test trajectory. At the initial time the robot is at the point with coordinates (0,25; -0,25), the trajectory starts to move from the point (0,0). The frequency of harmonic signals $X_{ref}(t)$ and $Y_{ref}(t)$ is set so that the trajectory $p_{ref}(t)$ forms an "eight" for 40 seconds.

The experimental results on a real Rover5 robot are shown in Figure 2.

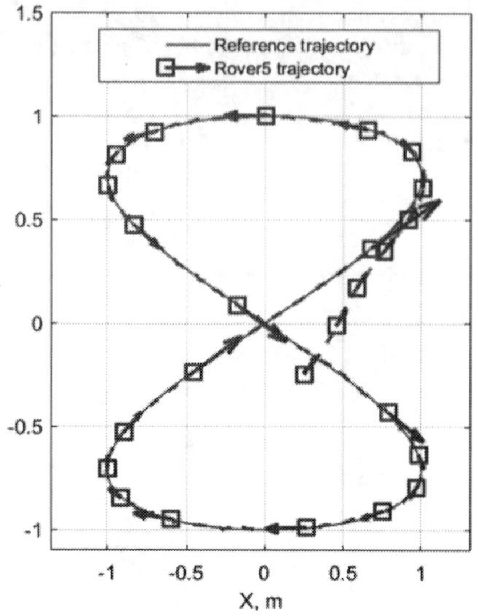

Fig. 2. The experimental results: reference and real trajectory of MR Rover5.

Reference and real Rover5's trajectories are shown on the fig.2. The required V_{req} and real V linear velocities, required ω_{req} and real ω robot's angular velocities are shown on the fig.3.

The results show us that the system has an acceptable quality. Robot enters the reference trajectory for 5 seconds, then deviations are $|e_X| \leq 0.015$, m, $|e_Y| \leq 0.01$, m, $|e_\varphi| \leq 0.05$, rad, the relative tracking error doesn't exceed 2% after the robot reached reference trajectory.

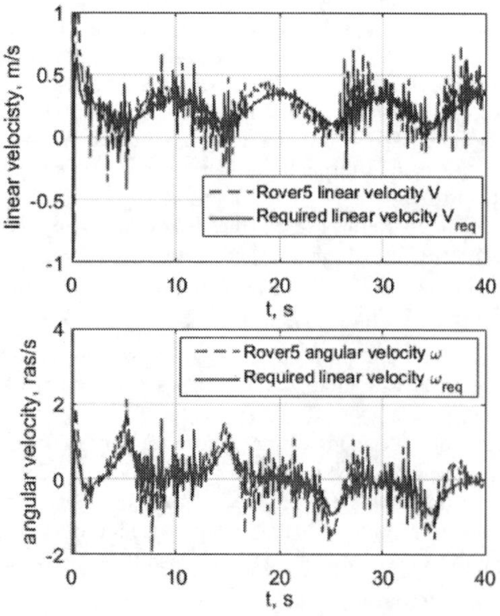

Fig. 3. The experimental results: reference and real trajectory of MR Rover5.

5 Conclusions

The problem of MR trajectory-tracking control was considered. The design of trajectory control loop is based on FL problem for tracking error dynamic. In general case, the control law is a tracking error feedback with time-varying coefficients. Non-stationary nature of the problem is determined by the time-varying values of reference linear and angular velocities of the robot. In such a case when these velocities are constant, the problem becomes stationary and the feedback's coefficients also become time-invariant.

Experimental verification of developed trajectory-tracking control system was performed on mobile robot Rover5. The results of experiment confirm the acceptable quality of designed control system.

This work was supported by the Russian Science Foundation (project No. 17-11-01220)

References

1. B. Siciliano, O. Khatib, *Springer handbook of robotics* (Springer, Berlin Heidelberg, 2016)
2. B. Siciliano, L. Sciavicco, L. Villani, G. Oriolo, *Robotics: modelling, planning and control.* (Springer, London, 2009)
3. K. Kozlowski, *Robot motion and control: recent developments* (Springer, London, 2006)
4. G. Cook, *Mobile robots: navigation, control and remote sensing* (Wiley-IEEE Press, 2011)
5. A.P. Aguiar, J.P. Hespanha, IEEE Trans. on AC, **52**(8) (2007)
6. G. Klančar, D. Matko, S. Blazič, Proc. of the 13th Mediterranean conf, on control and automation, Limassol, Cyprus, June 27-29, (2005)
7. C.D. Charalambous, A. Lambis, X. Li, Proc. of 16th Mediterranean conf. on control and automation. Ajaccio, France, 946-951 (2008)
8. A.A. Kabanov, WSEAS Trans. on Sys. Cont., **9** (2014)
9. B. D'Andréa-Novel, G. Campion, G. Bastin, Int. J. of Robotics Research. **14(6)**, 543-559 (1995)
10. W.E. Dixon, D.M. Dawson, E. Zergeroglu, Nonlinear control of wheeled mobile robots. Springer, London (2001)
11. G. Oriolo, A.D. Luca, M. Vendittelli, IEEE Trans. on Cont. Sys. Tech., **10**(6) (2002)
12. I.E. Zuber, J. App. Math. and Stoch. Anal. **11(4)** (1998)
13. A.A. Kabanov, Int. Siberian Conf. on Cont. and Com. (SIBCON). Proc. National Research University Higher School of Economics. Russia, Moscow, May 12-14 (2016)

Negative valuation of undesirable outputs in evaluating the efficiency of production systems

Chiang Kao[1,*]

[1] Department of Industrial and Information Management, National Cheng Kung University, Tainan, Taiwan, Republic of China

Abstract. This paper extends a data envelopment analysis (DEA) model that considers only desirable outputs in measuring efficiencies to incorporate undesirable outputs by requiring the associated multipliers to be negative. It is shown that the proposed model is the ratio-measure counterpart of the difference-measure model existed in the literature. It has the technology of variable returns to scale, and the modification made to it means that it also has the property of null-jointness, satisfies the axiom of weak disposability, and has positive shadow prices, considered as pitfalls in modeling undesirable outputs. In addition to the economic interpretation, the ratio-measure of efficiency of the proposed model better shows the relative performance of a unit than the difference-measure does, and its simpler form also provides a clearer geometric interpretation of efficiency measurement with undesirable outputs.

1 Introduction

An output is considered as undesirable in the production process if more of it being produced from the same amount of input is not wanted. Typical examples of undesirable outputs are waste water and pollutants generated in producing the main, desirable products. As environmental issues are becoming greater concern, the number of studies on undesirable outputs has grown. How to appropriately measure the efficiency of a production unit that produces both desirable and undesirable outputs using the data envelopment analysis (DEA) technique has also attracted numerous studies. Leleu [1] reviewed many articles on the modeling of undesirable outputs, highlighted some problems that can arise, and proposed a model that avoids these. This model also provides an economic interpretation of the production process.

In this note a DEA model extended from the BCC model of Banker et al. [2] that incorporates undesirable outputs by assigning negative values to the associated multipliers is proposed. This model is shown to be a ratio-measure counterpart of the difference-measure model of Leleu [1], maintaining the strengths of this while in a simpler form. Moreover, the ratio-measure of efficiency used in this work, in the range of zero and one, provides a better idea of the relative performance of a production unit than the difference-measure, in the range of zero and an unknown upper bound, does.

In the next section we will introduce the extended BCC model, in ratio form, and discuss its relationship with Leleu's model. Then, in Section 3, the properties of this model will be explored from the envelopment form. Finally, some conclusions will be drawn in Section 4 from the discussion of this model.

2 Ratio model

Banker et al. [2] proposed a DEA model, usually referred to as the BCC model, to measure the relative efficiency of a set of n production units that apply multiple inputs X_{ij}, i=1,…, m, to produce multiple outputs Y_{rj}, r=1,…, m, under the technology of variable returns to scale. All the outputs are assumed to be desirable. The output model, in ratio form, is as follows:

$$\begin{aligned} \frac{1}{E_0} = \min. \ & \frac{\sum_{i=1}^{m} v_i X_{i0} + v_0}{\sum_{r=1}^{s} u_r Y_{r0}} \\ \text{s.t.} \ & \frac{\sum_{i=1}^{m} v_i X_{ij} + v_0}{\sum_{r=1}^{s} u_r Y_{rj}} \geq 1, j=1,\ldots,n \\ & u_r,\ v_i \geq 0,\ r=1,\ldots,s,\ i=1,\ldots,m \\ & v_0 \text{ unrestricted in sign} \end{aligned} \quad (1)$$

where the multipliers u_r and v_i can be considered as the values of the corresponding outputs and inputs, respectively, as perceived by the decision maker, and E_0 is the efficiency of the unit being evaluated.

Suppose this model is extended to include undesirable outputs B_{kj}, k=1,…, t. Since these outputs are undesirable, the values perceived by the decision maker must be negative. We use $-w_k$, where $w_k \geq 0$, to represent the multiplier of the undesirable output B_k. The extended BCC model, with undesirable outputs, can thus be formulated as follows:

$$\frac{1}{E_0} = \min. \ \frac{\sum_{i=1}^{m} v_i X_{i0} + v_0}{\sum_{r=1}^{s} u_r Y_{r0} + \sum_{k=1}^{t} (-w_k) B_{k0}} \quad (2)$$

* Corresponding author: ckao@mail.ncku.edu.tw

$$\text{s.t.} \quad \frac{\sum_{i=1}^{m} v_i X_{ij} + v_0}{\sum_{r=1}^{s} u_r Y_{rj} + \sum_{k=1}^{t} (-w_k) B_{kj}} \geq 1, j=1,\ldots,n$$

$$u_r, v_i, w_k \geq 0, r=1,\ldots,s, i=1,\ldots,m, k=1,\ldots,t$$

$$v_0 \text{ unrestricted in sign}$$

This model obviously has multiple solutions, and if $(\boldsymbol{u}, \boldsymbol{v}, \boldsymbol{w})$ is a solution, then so is $(c\boldsymbol{u}, c\boldsymbol{v}, c\boldsymbol{w})$ for $c>0$. Due to the nature of multiple solutions, it is conventional to set the denominator of the objective function to one as a normalization constraint, and to minimize the numerator alone. This is equivalent to applying the variable substitution technique of Charnes and Cooper [3] to transform a fractional program into a linear one. The resulting linear program, usually called the multiplier form, is:

$$1/E_0 = \min. \sum_{i=1}^{m} v_i X_{i0} + v_0 \quad (3.0)$$

$$\text{s.t.} \quad [\sum_{r=1}^{s} u_r Y_{rj} + \sum_{k=1}^{t} (-w_k) B_{kj}] - (\sum_{i=1}^{m} v_i X_{ij} + v_0) \leq 0,$$

$$j=1,\ldots,n \quad (3.1)$$

$$\sum_{r=1}^{s} u_r Y_{r0} + \sum_{k=1}^{t} (-w_k) B_{k0} = 1 \quad (3.2)$$

$$u_r, v_i, w_k \geq 0, r=1,\ldots,s, i=1,\ldots,m,$$

$$k=1,\ldots,t \quad (3.3)$$

$$v_0 \text{ unrestricted in sign} \quad (3.4)$$

At optimality, $[\sum_{r=1}^{s} u_r Y_{r0} + \sum_{k=1}^{t} (-w_k) B_{k0}] - (\sum_{i=1}^{m} v_i X_{i0} + v_0) = 0$ is the production frontier corresponding to the unit being evaluated, provided the solution is unique. The terms in square brackets in constraint set (3.1) are the observed aggregate output, and the terms in parentheses are the aggregate output expected to be produced from the inputs of this unit, and their ratio is the output efficiency of this unit shown in the ratio model (2). Constraint set (3.1) states that when the multipliers selected by this unit to calculate its efficiency are applied to calculate the efficiency of any unit, they must be less than or equal to one to ensure that all units lie below the frontier of $[\sum_{r=1}^{s} u_r Y_{r0} + \sum_{k=1}^{t} (-w_k) B_{k0}] - (\sum_{i=1}^{m} v_i X_{i0} + v_0) = 0$.

Following the notation of the extended BCC model, the model proposed by Leleu [1], in what he called dual form, can be formulated as:

$$\min. \varphi \quad (4.0)$$

$$\text{s.t.} \quad (\sum_{r=1}^{s} u_r Y_{rj} - \sum_{k=1}^{t} w_k B_{kj} - \sum_{i=1}^{m} v_i X_{ij}) - (\sum_{r=1}^{s} u_r Y_{r0} - \sum_{k=1}^{t} w_k B_{k0} - \sum_{i=1}^{m} v_i X_{i0}) \leq \varphi, j=1,\ldots,n \quad (4.1)$$

$$\sum_{r=1}^{s} u_r g_r^Y + \sum_{k=1}^{t} w_k g_k^B = 1 \quad (4.2)$$

$$\sum_{r=1}^{s} u_r Y_{r0} - \sum_{k=1}^{t} w_k B_{k0} + \varphi \geq 0 \quad (4.3)$$

$$u_r, v_i, w_k \geq 0, r=1,\ldots,s, i=1,\ldots,m, k=1,\ldots,t \quad (4.4)$$

where $\boldsymbol{g}=(\boldsymbol{g}^Y, \boldsymbol{g}^B)$ is a direction vector and v_i, u_r, w_k are the shadow prices of the ith input, rth desirable output and kth undesirable output, respectively. Leleu [1] provided an economic interpretation of this model. Constraint set (4.1) states that the difference between the profit of any Unit j and the unit being evaluated, which is the profit inefficiency, must be less than or equal to φ, and the objective function (4.0) is to find the minimum profit inefficiency φ of the focal unit. Constraint (4.2) is the conventional normalization constraint used to solve the problem of homogeneity of degree one of costs, revenues, and profits in dual prices. Constraint (4.3) requires the efficient revenue from the desirable and undesirable outputs to be positive. Finally, constraint set (4.4) restricts all the shadow prices to be nonnegative.

Denote $\varphi = (\sum_{i=1}^{m} v_i X_{i0} + v_0) - [\sum_{r=1}^{s} u_r Y_{r0} + \sum_{k=1}^{t} (-w_k) B_{k0}]$ as the difference between the expected aggregate output and the observed aggregate output, which is the output inefficiency in a difference form, and is the objective function of Leleu's model (4.0). This is in contrast to the ratio form of $1/E_0 = (\sum_{i=1}^{m} v_i X_{i0} + v_0) / [\sum_{r=1}^{s} u_r Y_{r0} + \sum_{k=1}^{t} (-w_k) B_{k0}]$ of the extended BCC model. If $\varphi = (\sum_{i=1}^{m} v_i X_{i0} + v_0) - [\sum_{r=1}^{s} u_r Y_{r0} + \sum_{k=1}^{t} (-w_k) B_{k0}]$ is substituted into constraints (4.1) and (4.3), then $[\sum_{r=1}^{s} u_r Y_{rj} + \sum_{k=1}^{t} (-w_k) B_{kj}] - (\sum_{i=1}^{m} v_i X_{ij} + v_0) \leq 0$ and $\sum_{i=1}^{m} v_i X_{i0} + v_0 \geq 0$ are obtained, respectively. The former is constraint (3.1) of the extended BCC model. For the latter, since we have $(\sum_{i=1}^{m} v_i X_{i0} + v_0) \geq [\sum_{r=1}^{s} u_r Y_{r0} + \sum_{k=1}^{t} (-w_k) B_{k0}] = 1$ in the extended BCC model, the condition of $\sum_{i=1}^{m} v_i X_{i0} + v_0 \geq 0$ is obvious. When the direction vector of $\boldsymbol{g}=(\boldsymbol{g}^Y, \boldsymbol{g}^B)$ is replaced with $\boldsymbol{g} = (\boldsymbol{Y}_0, -\boldsymbol{B}_0)$, constraint (4.2) becomes $\sum_{r=1}^{s} u_r Y_{r0} + \sum_{k=1}^{t} w_k (-B_{k0}) = 1$, which is the same as constraint (3.2). Finally, constraint set (4.4) is the same as constraint set (3.3), and constraint (3.4) is specific to the extended BCC model, which has no effect on Leleu's model (4). In sum, the extended BCC model is equivalent to Leleu's model, except the former expresses inefficiency in a ratio form, while the latter expresses it in a difference form.

With this relationship, the economic interpretation of Leleu's model also applies to the extended BCC model. The major difference between these two models is that the extended BCC model expresses inefficiency in a ratio form, with one as the lower bound, and whose inverse is conveniently converted to an efficiency measure, while Leleu's model expresses inefficiency in a difference form, with a lower bound of zero and an unknown upper bound, which makes a transformation into an efficiency measure difficult. One situation common to these two models is that they usually have multiple solutions, and it is difficult to identify which ones in this case are the real values perceived by the decision maker or the shadow prices of the factors.

3 Envelopment model

The extended BCC model (3) is slightly simpler than Leleu's model (4). Its dual, which is usually called the envelopment form, is also simpler than that of Leleu's model. The envelopment form of Leleu's model is:

$$
\begin{aligned}
&\max.\ \delta \\
&\text{s.t.}\ \sum_{j=1}^{n}\lambda_j(Y_{rj}-Y_{r0}) \ge \sigma Y_{r0}+\delta g_r^Y, && r=1,\ldots,s \\
&\sum_{j=1}^{n}\lambda_j(B_{kj}-B_{k0}) \le \sigma B_{k0}+\delta g_k^B, && k=1,\ldots,t \\
&\sum_{j=1}^{n}\lambda_j(X_{ij}-X_{i0}) \le 0, && i=1,\ldots,m \\
&\sum_{j=1}^{n}\lambda_j+\sigma=1 \\
&\sigma,\ \lambda_j \ge 0, && j=1,\ldots,n
\end{aligned}
\tag{5}
$$

which is difficult to provide geometric interpretations of, and the technology of variable returns to scale is also difficult to recognize. In contrast, the envelopment form of the extended BCC model, formulated as follows, is better in this aspect:

$$
\begin{aligned}
&1/E_0=\max.\ \eta \\
&\text{s.t.}\ \sum_{j=1}^{n}\lambda_j Y_{rj} \ge \eta Y_{r0}, && r=1,\ldots,s \\
&\sum_{j=1}^{n}\lambda_j B_{kj} \le \eta B_{k0}, && k=1,\ldots,t \\
&\sum_{j=1}^{n}\lambda_j X_{ij} \le X_{i0}, && i=1,\ldots,m \\
&\sum_{j=1}^{n}\lambda_j =1 \\
&\lambda_j \ge 0, && j=1,\ldots,n
\end{aligned}
\tag{6}
$$

The constraint $\sum_{j=1}^{n}\lambda_j=1$ clearly indicates the technology of variable returns to scale.

For geometric interpretations, this model is similar to the conventional BCC model that treats the undesirable outputs as inputs, except that the undesirable outputs of the unit being evaluated, B_{k0}, are also attached with the distance parameter η. This is the key feature of this model. Without attaching η to B_{k0}, the model implies that B_k is strongly disposable, and that it can be disposed of freely, which is unreasonable for an undesirable output. In contrast, there are several merits when η is attached to both the desirable and undesirable outputs. First, the model satisfies the condition of weak disposability for outputs. Second, the null-jointness condition that "if B_k=0 then Y_r=0" is satisfied. Third, the model becomes a directional distance one by expressing the right-hand sides of the output constraints, ηY_{r0} and ηB_{k0}, as $Y_{r0}+(\eta-1)Y_{r0}$ and $B_{k0}-(\eta-1)(-B_{k0})$, respectively, where the direction is $(\boldsymbol{g}^Y, \boldsymbol{g}^B)=(\boldsymbol{Y}_0,-\boldsymbol{B}_0)$. Therefore, once the desirable outputs are increased, the undesirable outputs must also be increased, and at the same rate.

To illustrate different ideas of efficiency measurement with undesirable outputs, consider five production units, *A*, *B*, *C*, *D*, and *E*, depicted in Figure 1, using one unit of input *X* to produce different combinations of the desirable output *Y* and undesirable output *B*. The production frontier constructed under the assumption of strong disposability of outputs is *TET'*, that under weak disposability is *OCET'*, and that by treating the undesirable output as an input is *SACES'*. The frontier constructed from the extended BCC model is *OCES'*, which is similar to that of weak disposability, except that it extends eastward horizontally beyond Unit *E*, instead of downward vertically. There are three inefficient units, *A*, *B*, and *D*, under this technology. In measuring the efficiency, for example of Unit *B*, the undesirable output-desirable input exchange approach uses Unit *C* as the target to result in the efficiency of Y_B/Y_C. However, since the undesirable output is produced with the desirable output, as the desirable output of Unit *B* increases from the level of Y_B to Y_C, the undesirable output will increase at the same rate to reach point $\hat{B}$, where $Y_{\hat{B}}=Y_C$. One cannot dispose of the undesirable output of $\hat{B}$ to that of Unit *C*, and cannot use Unit *C* as the target to measure efficiency. Unit *B* in this case must increase both the desirable and undesirable outputs further to reach point B^* on the frontier, and use it as the target to obtain the inefficiency of $\eta=OB^*/OB$. The idea of Leleu is to use a direction $\boldsymbol{g}$ to reach a point B° on the frontier from *B*, and use the absolute difference between *B* and B° as the inefficiency measure φ.

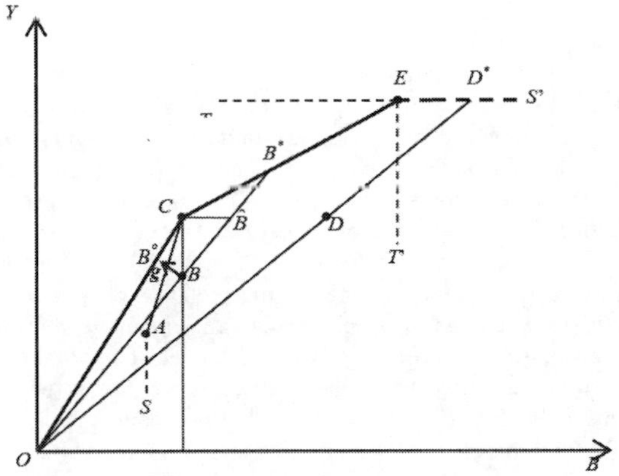

Fig. 1. Frontiers and efficiency measures under different assumptions.

Similar to other BCC-type models, the extended BCC model also faces difficulties in determining appropriate efficiency measures for weakly efficient units, for example D^*, lying on the horizontal frontier line segment *ES'*. A chain effect of this situation is the inappropriate efficiency measure of those units calculated by using the weakly efficient units as the target, for example Unit *D*. The amendment for the BCC-type models is to impose a small non-Archimedean number ε as the lower bound for the multipliers [4]. When u_r, v_i, $w_k \ge 0$ in constraint set (3.3) of the multiplier form of the extended BCC model are replaced with u_r, v_i, $w_k \geq \varepsilon$, the corresponding envelopment model then becomes:

$$1/E_0 = \max.\ \eta + \varepsilon\left(\sum_{r=1}^{s} s_r^+ + \sum_{k=1}^{t} t_k^- + \sum_{i=1}^{m} s_i^-\right)$$

$$\begin{aligned} \text{s.t.}\ & \sum_{j=1}^{n} \lambda_j Y_{rj} - s_r^+ = \eta Y_{r0}, \quad r=1,\ldots,s \\ & \sum_{j=1}^{n} \lambda_j B_{kj} + t_k^- = \eta B_{k0}, \quad k=1,\ldots,t \\ & \sum_{j=1}^{n} \lambda_j X_{ij} + s_i^- = X_{i0}, \quad i=1,\ldots,m \\ & \sum_{j=1}^{n} \lambda_j = 1 \\ & \lambda_j \geq 0, \quad j=1,\ldots,n \end{aligned} \tag{7}$$

The value for the lower bound ε must be appropriately determined to be able to produce proper efficiency measures for those units on the frontier segment *ES*', and those using them as the target to measure efficiency.

4 Conclusion

This paper proposes an extended BCC model to measure the efficiency of production units with undesirable outputs by requiring the associated multipliers to be negative. It is shown that this model is equivalent to the one proposed by Leleu [1], except the inefficiency is expressed in a ratio form, rather than a difference form. Moreover, this model does not have the weaknesses noted by Leleu with regard to modeling undesirable outputs. Specifically, the model has a correct form of variable returns to scale technology, it satisfies the condition of null-jointness, the undesirable outputs are weakly disposable, and the shadow prices of the undesirable outputs are positive. Due to its relationship with Leleu's model, the extended BCC model also shares the economic interpretation of Leleu's model. Its simpler form also makes the geometric interpretation clearer.

The extended BCC model has an advantage that is not possessed by Leleu's model. That is, it produces a relative measure of how far a unit is from the target in the range of (0, 1), while Leleu's model gives an absolute distance measure, which cannot be converted to a scale of (0, 1) to show a unit's relative performance as compared to other units.

Finally, the idea proposed in Leleu [1] of requiring all units to have the same shadow price for the undesirable output, which is similar to the idea of a common-weight in Kao and Hung [5], can also be applied to the extended BCC model by minimizing the total inefficiency ratios of all units. However, the objective function in this case is nonlinear, and the solution process will be made easier if it can be transformed into a linear one.

References

1. H. Leleu, *Shadow pricing of undesirable outputs in nonparametric analysis*. European J. Operational Research **231**, 474-480 (2013)
2. R.D. Banker, A. Charnes, W.W. Cooper, *Some models for estimating technical and scale efficiencies in data envelopment analysis,* Management Science **30**, 1078-1092 (1984)
3. A. Charnes, W.W. Cooper, *Programming with linear fractional,* Naval Research Logistics Quarterly **9**, 181-186 (1962)
4. A. Charnes, W.W. Cooper, *The non-Archimedean CCR ratio for efficiency analysis: A rejoinder to Boyd and Färe,* European J. Operational Research **15**, 333-334 (1984)
5. C. Kao, H.T. Hung, *Data envelopment analysis with common weights: The compromise solution approach,* J. Operational Research Society **56**, 1196-1203 (2005)

DOI: 10.1051/matecconf/201712904002

Implementation of statistical process control methods as a way to reduce production costs and improve product quality

Sergey Egorov[1,*], *Alexey* Kapitanov[1], and *Dmitriy* Loktev [2]

[1]Moscow state university of technology "STANKIN", 127055 Moscow Vadkovskiy per., 1, Russia
[2]Technopolice Group, 117218 Moscow Dm. Ulyanova street, 42/1, Russia

Abstract. The article analyzes various situations of building a quality assurance system at an enterprise. The article includes the analysis of influence of various factors on the decision-making on the principles of building a quality assurance system, the conclusion about the need to build the quality assurance system aimed not at preventing the assembly of parts with deviations but on creating a process producing parts without deviations. The stages of implementation of the statistical process control system are considered. Moreover, the article analyzes the characteristics of measuring methods and tools that can be included into the quality assurance system, and examines the various process states allowing making a conclusion about the possibility of their controllability. The example of a comprehensive solution of the statistical process control system is given.

1 Introduction

During the lunch of any product, the quality requirements (as opposed to the production process requirements) are determined at the product design stage and cannot be changed at the manufacturing stage. A designer determines the main parameters of a finished product. These requirements are almost always determined by the market conditions - a consumer, on the one hand, and competitors, on the other hand. In other words, a product should completely meet the consumer's needs and at the same time be not worse and, most importantly, cheaper, than a competitor's product. Tightening of market competition results in constant tightening of the product requirements. After determining the requirements to a finished product, the designer determines the requirements to the parameters of individual parts. These requirements are converted into tolerance zones for certain sizes. By that, the designer proceeds from the premise that the technological process necessarily implements the tolerance zones assigned to it, and a product will have the planned parameters. The tolerance zone assigned by the designer is determined only by possible changes in the product parameters and very seldom takes into account possible difficulties in manufacturing of parts.

2 Problem statement - ideal and realistic approach to technological processes

After the designing stage, it is necessary to develop the technological process for individual parts. Here the first difficulties appear. The production engineer planning the technological process tries to select equipment and process parameters based on the design requirements.

Fine tolerances (determined in accordance with the market requirements) may not meet the equipment capabilities or may require the use of modern and more accurate tools. In any case, there is almost always a need to purchase new equipment and tools. But the production engineer is always sure that these investment needs will not be met. This can be partly explained with the lack of current assets (because a new product is only put on production and so far, it does not bring any money, hence the financing of manufacture of this new product should come from the funds earned on other projects or received from investors). Another reason may be the restriction concerning the product cost. As it has been already mentioned above, there are always market restrictions concerning the cost of a finished product. Taking into the account the project planned profitability, we determine the possible level of product cost and, as a consequence, the ultimate level of costs. In many cases, the calculation shows that implementation of the requirements specified by the designer without serious investments and, accordingly, without exceeding the planned level of costs is impossible. And in this case, the production engineer has to strike a compromise with the designer or to solve the problem using the methods not described in this article [1,5].

Therefore, the production engineer was able to develop the technology meeting the necessary product requirements. The process of implementation of this technology starts. The parts are manufactured, inspected and shipped to a consumer as part of the product. New problems can appear at this stage again. It may be found

* Corresponding author: egorovsergey@yandex.ru

out that the parts do not correspond to the established tolerances. It may be found out at the stage of technical control and then the necessary measures are taken within the production. The worst situation is to receive a customer's complaint. This means that the inconsistency with the tolerances has been identified in the product in general and has resulted in the mismatch of the parameters of a finished product with the agreed specification.

In any of these "negative" scenarios, there is the analysis stage, the implementation of appropriate changes in technology, but the result is often similar to the one discussed above. The most unpleasant thing is that the inconsistency with the tolerances of individual parts or the inconsistency of the entire product with the agreed specification is not found out always, but at random times. The management natural reaction is toughening of the technical control requirements.

And here the manufacturer stands on the proven path - to ensure the product quality due to 100% verification of manufactured parts. This method of quality assurance has surely the right to exist, but it is fundamentally different from the methods used by the global leading companies for last two or three decades.

3 Problem analysis

Quality assurance due to 100% control is a traditional method, which is still used at many industrial enterprises. When using this method (Fig. 1), the problem of quality assurance is solved quite simply - it is necessary to separate good parts from bad ones. Moreover, it does not matter where (on what operation) and for what reason such defective product was manufactured. It is only important to detect it in time.
Good parts are assembled and shipped to customers, but bad parts are further analyzed to determine whether it is possible to make good parts from them (recoverable defect) or not (irrecoverable defect).

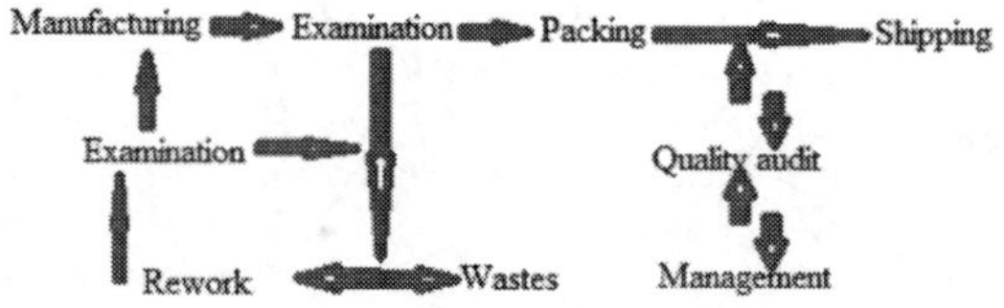

Fig. 1. Quality assurance due to 100% control.

It is quite obvious that the more parts are manufactured and the more complex these parts are (the more features should be controlled), the more controllers are required to check these parts. It is also obvious that the detected defects should be recovered - such parts should be completed (if it is a recoverable defect) or new parts should be manufactured instead of products with irrecoverable defects. The need to recover defects results, on the one hand, in increase in the production plan, and, on the other hand, in increase in production costs. The increase in the production plan cannot be eternal. If there are reserves of equipment performance, they are used to correct recoverable and irrecoverable defects. If the performance of processes is not enough, they start to assemble parts that are on the boundary between good and defective ones. If these parts are not enough, the production refers to the designer to get his/her permission to manufacture products with deviations. The designer has to expand the tolerances for the manufactured parts, and the previously defective products start to be assembled, but it helps to solve the performance problem. In some time, the quality of parts can improve, and the old design tolerances can be returned. Then the situation may become worse and everything may happen again.

The application of 100% control methods significantly increases the production costs. As it has been already shown above, if the problem of only determining defective products is solved, but not of preventing of its receipt, then the costs are substantially increased due to the recovery of defects. However, in addition to this cost item, the production cost of a part increases significantly due to the expenses for measuring. If a part (for example, an engine block) has about 1,000 different parameters (features) to be controlled, and 300 such blocks are manufactured per day, the quality assurance group should measure and make decisions on the compliance of 300,000 parameters per day. Such number of measurements requires an appropriate number of personnel and appropriate number of measuring tools that significantly increases the production costs and reduces the possible profit of this enterprise.

The necessity of total inspection of products is caused by the fact that the technological process, as a rule, does not always produce defective products. The situation, when the process produces only the defective products won't be considered here. This is related either to the incorrect tolerances, or to the complete unfitness of the process for manufacture of such products. Such situations are quite rare. In other cases, the process produces good products for some time, then defective ones, then again good ones, then the products, which are on the border between good and defective, and so on. The quality assurance system makes different decisions at different times, either accepting products, or directing them to be completed, or referring to the designer to get his/her permission.

With using this traditional approach, the management of an enterprise has to constantly make decisions related to the need to respond to the lack of suitable parts and incur additional costs, which are sometimes significant, to control products and correct defects.
The modern quality assurance system is aimed not at preventing the assembly of parts with deviations, but at creating a process that would produce parts without any deviations. In fact, if a process produces suitable parts at certain intervals, then it starts producing parts with deviations, then again suitable ones, and such transition occurs repeatedly, it means that there are some causes of that. It can be concluded that if these causes are found and eliminated, then the process will always produce only suitable products [2,3,4].

Such modern approach is fundamentally different from the abovementioned traditional quality assurance approach due to 100% control of products. It also requires new attitude of enterprise management and production to quality assurance issues. To implement this approach, it is necessary to collect data on product parameters for a certain period of time, analyze them, find sources of changes and eliminate them. Then it is necessary to maintain the process monitoring the product parameters from time to time. If the process remains stable, you can be sure that only suitable parts will be assembled. The solution of the product quality issue will result in significant profit increase due to reduction of costs for 100% technical control, completion of recoverable defects and production of additional parts instead of irrecoverable defects. At the same time, the implementation of this new method will require certain costs, which are sometimes significant, which are quickly compensated due to increased profit.

It is possible to implement the analysis of existing problems affecting the quality of products, using the mathematical statistics methods. In this case they speak about the use of statistical process control. The concept "statistical process control" is defined as the use of statistical methods to achieve the quality of products or services by protecting the process from undesirable changes.

Statistical process control is related to mathematical statistics.

4 Implementation of the statistical process control system

Any process in nature is characterized by variability. It is affected by a huge number of different factors that change the process itself and its result. Considering cutting, we can find dozens of factors affecting the process. For example, hardness of a workpiece is not a constant value and even the hardness of workpieces from the best suppliers varies in the range of 10 ... 30 HB units. Even positional repeatability of the best machine is no better than 1...5 µm. And so on. It means that changing of these factors will inevitably lead to the changes in parameters of the processed parts.

However, all these factors affect the process to a very small extent and, in most cases; it is very difficult to single out the impact of one of these factors. Such causes of process changing are called ordinary. The process remains stable due to the fact that there are many ordinary causes and each of them has its small effect, which is often compensated by other ordinary causes. From a statistical point of view, a stable process has constant distribution (Fig. 2 on the left).

If the process becomes instable, it is always associated with the appearance of specific causes or with a change in the set of ordinary causes. The causes of process changing that can be detected and identified are called specific. Any specific cause may lead to process instability. Then from a statistical point of view the process has inconstant distribution (Fig. 2 on the right).

In this case, the task of statistical process control is detection and elimination of specific causes of changing of product parameters, leading to the appearance of a product that does not meet the specified parameters.

The implementation of statistical process control system is carried out in several stages.

Statistical process control is based on analysis and interpretation of the initial data obtained from various sources. If the product geometrical parameters are evaluated, the initial data is obtained from various measuring tools. Besides, discrete features with two states (for example, presence or absence of a defect) may also be used.

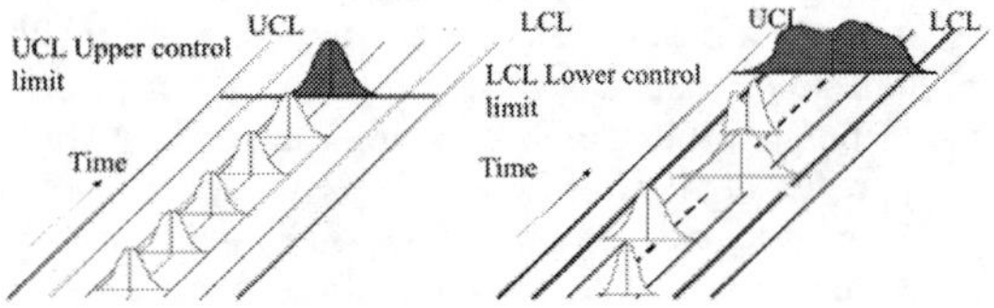

Fig. 2. Stable and instable process.

Therefore, at first it is necessary to determine a set of features for control of suitability of a part. Such features are dimensions and parameters of a part with certain tolerances. Depending on the role of controlled features in functioning of a finished product, they are divided into critical, very significant, significant and insignificant. Depending on the type of a feature, it is possible to determine different requirements to the stability of its implementation.

Various features (quality parameters) appear at different stages of the technological process. Therefore, at the second stage of the implementation of the statistical process control system, it is necessary to determine where, how and when the selected parameters will be measured. "Where" means both technological process operation and the organization of a measuring point. "How" implies the use of necessary measuring tools (more on this later) and the implementation of the measurement process itself. "When" in this case refers to determining a sample volume of parts for measuring and the measurement time interval of a given sample.

When choosing a measuring point and measuring tools, it is necessary to take into account the need to ensure reliability of the data send to the statistical analysis system. Data reliability is related to two main factors - unambiguousness of the obtained data and independence of data transmission from subjective factors. The first factor is associated with the use of certain measuring tools. To date, the use of digital measuring tools is an operational requirement to the implementation of the statistical process control system.

Indeed, it is difficult to talk about data reliability when reading information from a vernier of a conventional measuring tool because each operator sees slightly different values, and it calls into question unambiguousness of the obtained data. The second factor is also associated with the use of digital measuring tools and requires a direct connection of a measuring tool with the system for fixing the transmitted values. Human intervention in this process (an operator takes readings from a measuring tool and then manually enters them

DOI: 10.1051/matecconf/201712904002

into the system for fixing values) results in the risk of deliberate or accidental data distortion [6,7].

Concluding the brief description of this stage of implementation of the process control system, we note that for further analysis of process controllability and for reflecting current statistics, the transmitted data should be accompanied by the full information on a product being measured and process parameters. Only in this case it is possible to conduct an effective analysis of the arising specific causes.

After determining parameters and methods for their measuring, it is required to make sure that the used measuring tools allow objectively assessing the quality of the selected parameters. In other words, a measuring tool should be suitable for measuring this value. This is not a matter of physical fitness (for example, inability to measure roughness parameters using the tools for measuring linear dimensions), but it is a matter of correspondence between the maximum error and the uncertainty of measurements of necessary accuracy associated with the tolerance zone of the measured parameter. There are various norms determining the ratio between the tolerance zone of the measured parameter and the maximum error of measuring tools.

This is always the first step in determining the suitability of a measuring tool. If the maximum error of a measuring tool exceeds 30% (GOST requirement) of the tolerance zone of the measured parameter, then such measuring tool will be considered not suitable in almost all cases. In addition, measuring resolution must not exceed 5% of the tolerance zone of the measured parameter. If these two requirements are met, then the measurement system suitability is checked under realistic measurement conditions with the participation of real controllers. Various methods are practically used to assess the suitability of measuring tools. They determine the suitability indices Cg and Cgk, G R&R parameters (convergence and reproducibility) as well as some other parameters of suitability of measuring tools.

The whole complex of assessing measuring tools is standardized and included into the field of statistical methods. The European automotive industry uses VDA 5 standard, the American automotive industry uses the set of standards called MSA (Measuring System Analysis). The ISO standard regulating the suitability of measuring tools has recently appeared - ISO 22514-7:2012 [8,9].

After the parameters under evaluation are determined, the suitability of appropriate measuring tools is checked, they proceed to the next stage of the implementation of process control system - the check of equipment suitability for manufacturing of parts of certain precision.

To do this, at first, one part is processed on the used equipment. Based on the measurement results, appropriate corrections are made to obtain parameters in the middle of the given tolerance zone. Then five parts are processed successively, the selected parameters are fully measured and the spread of the obtained results is estimated. At this level, it is possible to make necessary corrections. For example, if the processing of five parts reveals a trend of resizing due to tool deterioration, it is possible to include appropriate software or hardware to eliminate this trend. And then fifty parts are processed on the machine. Based on the results of this processing, short-term reproducibility indices can be determined. Since the sample is small, international standards set quite high requirements for this parameter. If these requirements are met, then we can say that this equipment is suitable for the implementation of the required tolerance zones of the processed product.

Thus, after verification of suitability of the used equipment and measuring tools, we can proceed directly to the implementation of process control.

We are only interested in two questions - whether the process is stable (controllable) and, if so, how stable is this controllable state. In other words, we are interested whether the process can produce suitable products after appropriate adjustment and how reliable it is. The process indices - indices of reproducibility and process suitability - are used to assess the ability of the process to remain controllable.

If the process is statistically controllable, the process indices can be used to predict the further process behavior and, as a consequence, the possible share of defective products. Depending on this information, it is possible to make decisions on periodicity of product control. However, process indices work effectively only if the process is statistically controllable, i.e. there are no specific causes of process variability. This fact should be confirmed.

Control charts are used to assess process stability and to identify the presence of specific causes. The theory and practice of using control charts were developed in the middle of 20th century and has been constantly improving since that moment. Two control charts are made (Fig. 3) - mean values and standard deviations (or swings) - to control the process. The first chart allows assessing stability of the mean value position of the observed parameter. The second chart shows dispersion (spread) of the values of the considered parameter. Control charts should be used only in pairs, otherwise there is a risk of making a false decision.

At the beginning of using control charts, all necessary calculations were made manually. Although control charts are based on the serious theory of mathematical statistics, the calculations are quite simple. With the development of computer technology, the calculation of control charts was fully automated.

When creating each control chart, first the values of upper and lower control limits are calculated based on the data of the measured value. Then, the current state of the process is assessed to sample the data, and it is recorded as a point on the control chart. The main sign of process instability is location of one of the points beyond the control limits (see Fig. 3, marked in red). There are other criteria for assessing process controllability in accordance with control charts (certain trends of the location of points).

If a point is located beyond the control limits, the responsible personnel decide on the need to adjust the process. They identify a specific cause, which has affected the process stability, and take measures to eliminate the damage from this specific cause. After that, the process is further controlled until the control charts show full controllability of the process.

Control charts allow qualitatively assessing the presence of specific causes and effectiveness of the measures taken to eliminate them. If a control chart shows that the process is statistically controlled (no process control criterion is violated), then it is possible to assess how stable the process is.

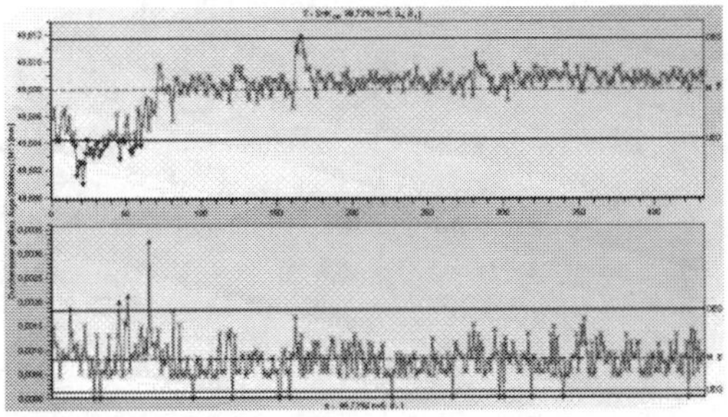

Fig. 3. Example of control charts.

Special indices as the indices of reproducibility and process suitability are used to quantify the process stability. The reproducibility index C_p shows the ratio of permissible process spread (i.e., width of the tolerance zone) to actual process spread (Fig. 4). The value $C_p = 1$ corresponds to the process reproducibility limit (it corresponds to the control limits on the control charts). In the absolute majority of cases, the minimum value of process reproducibility is more than 1 (usually 1.33). In this case, the actual process spread should not exceed 75% of the tolerance zone, and the process guarantees to obtain no more than 84 defective parts per million of manufactured parts.

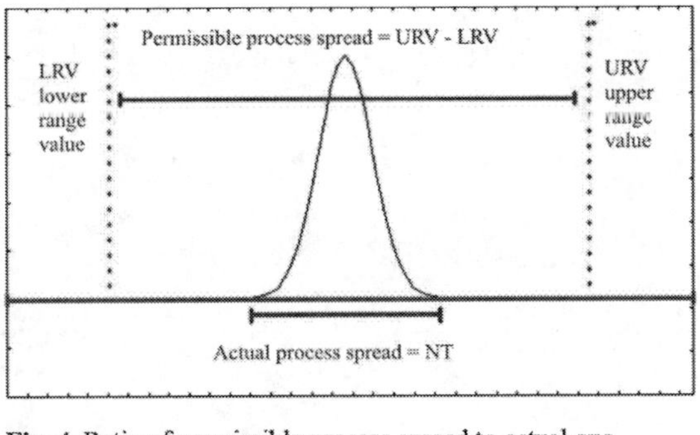

Fig. 4. Ratio of permissible process spread to actual one.

The reproducibility index C_p reflects the process potential, but it does not take into account the location of actual spread limits relative to the nominal value of the considered value. The reproducibility index C_{pp} is used to consider the process centering. This index assesses the distance between the mean process value and the nearest tolerance zone boundary. The higher this index is, the better the process is centered [10,11]. Absolutely centered processes have similar C_p and $C_{pк}$.
In certain cases, the index of process suitability P_p and $P_{pк}$ are used instead of the reproducibility index C_p and $C_{pк}$. The difference between these groups of indices is the method of assessing actual process spread. In general, all four indices are united with a common name of the process capability values.

Thus, the combination of control chart analysis and calculation of process capability values makes it possible to determine qualitatively and quantitatively whether the process is in a statistically controllable state. If not, there is a certain algorithm of actions and certain tools to bring the process to a controlled state. Without going into details, we can mention such tools as a cause and effect diagram (Ishikawa chart/fishbone graph) and a Pareto chart.

5 Comprehensive solution - conclusions and perspectives

One may feel that the implementation of statistical process control requires significant costs, both mental and material. In fact, it is not so. With the help of qualified consultants, the statistical process control system may be implemented in a short period of time with optimal investments.

To do this, you need to meet only two basic conditions - to equip the production with modern measuring tools, which reliably allow obtaining and transmitting process data, and to implement the control system itself. It can be easily done with the help of specialized software. One of the most famous brands in this field is Q-DAS (Germany), which offers a full set of software for the implementation of statistical process control and is a world leader in this direction.

The concept of statistical process control software product based on modern measuring tools is called CAMERA and implemented at six levels (Fig. 5).

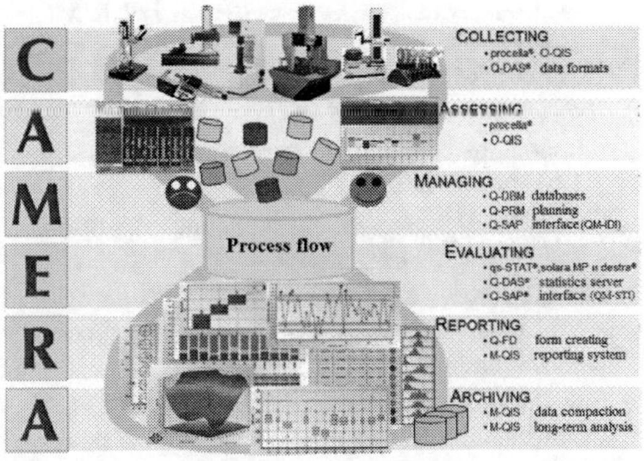

Fig. 5. Concept of CAMERA software product.

The first level includes the collection of data using various measuring tools suitable for measuring the considered parameters and creating process parameter values in sufficient volume and with sufficient reliability. Specialized software products (procella My.SPC and O-QIS) are used to connect the measuring tools with the systems for assessing process parameters and to obtain data in the required format. It should be noted that the data formats created by Q-DAS are supported by almost all suppliers of measuring systems.

The second level includes primary assessment of the obtained data using the same software products. The assessment results in getting the process parameters - process flow, histograms, control charts, indices of reproducibility and suitability, etc. These assessment results can be presented in various forms depending on their recipient [12-15].

At the third level the data are transmitted to the central database.

The fourth level includes a deeper analysis of the obtained data, if necessary. Solara.MP software implements the analysis of suitability of measuring tools. The software product qs-STAT is designed to obtain almost any statistical process evaluations, and destra software allows optimizing the studied process using statistical methods (for example, regression and variational analysis).

The fifth level includes reporting on the conducted evaluations. Reporting forms and content can be changed depending on their recipient.

Finally, the sixth level includes archiving of the received data for its further storage and long-term analysis.

The research is performed with financial support of the Russian Ministry of Education and Science (project № 11.1817.2017/4.6).

References

1. D. Loktev, S. Egorov, A. Kapitanov, V. Mitrofanov, and T. Egorova, Russian Engineering Research Vol. **36**, No. 4, pp. 312–317 (2016)
2. Sergey B. Egorov, Alexey V. Kapitanov, Vladimir G. Mitrofanov, Leonid E. Shvartsburg, Natalia A. Ivanova, and Sergey A. Ryabov, IEJME — Mathematics Education Vol. **11**, No. 7, pp. 2213-2225 (2016)
3. S. Egorov, A. Kapitanov, D. Loktev, T. Egorova International journal of experimental education № **10**-1, pp. 127-129 (2016)
4. D. Loktev, S. Egorov, A. Kapitanov, V. Mitrofanov, T. Egorova, STIN **№ 9,** pp. 13-19 (2015)
5. S. Egorov, A. Kapitanov, T. Egorova, V. Mitrofanov, International journal of experimental education № **12**-1, pp. 128-129 (2015)
6. S. Egorov, A. Kapitanov, T. Egorova, D. Loktev, V. Mitrofanov, Innovation's **№08** (202), pp. 85-90 (2015)
7. D. Loktev, S. Egorov, A. Kapitanov, A. Loktev, V. Mitrofanov, T. Egorova. *Materials of the VI International Scientific Conference*, Vol. **II**, p. 276 (2015)
8. S. Egorov, D. Loktev, A. Kapitanov, V. Mitrofanov, T. Egorova, Eurasian Union of scientists **№5**, pp. 51-54 (2015)
9. S. Egorov, A. Kapitanov, D. Loktev, A. Loktev, V. Mitrofanov, Vestnik MSTU «Stankin» **№4**(35), pp. 83-88 (2015)
10. S. Egorov, A. Kapitanov, V. Mitrofanov, T. Egorova, WCSE (2016).
11. Sergey B. Egorov, Alexey V. Kapitanov, Vladimir G. Mitrofanov, Leonid E. Shvartsburg, Natalia A. Ivanova and Sergey A. Ryabov, International journal of environmental & science education Vol. **11**, No. 11, pp. 4065-4078 (2016)
12. S. Egorov, A. Kapitanov, D. Loktev, Materials Science Forum,Vol. **870**, pp. 397-403, (2016)
13. S. Egorov, A. Kapitanov, D. Loktev, II International scientific and technical conference "Industrial Engineering" (2016).
14. D. Loktev, S. Egorov, A. Kapitanov, V. Mitrofanov, T. Egorova, Innovation's **08**(214), p. 113-119 (2016)
15. D. Loktev, S. Egorov, A. Kapitanov, V. Mitrofanov, T. Egorova, Problems of mechanical engineering and automation **№3,** p. 110-119 (2015)
16. S. Egorov, Fundamental research **№6**, vol. 5, pp. 920-927 (2014)
17. S. Egorov, Fundamental research **№8**, vol. 1, pp. 26-31 (2014)

DOI: 10.1051/matecconf/201712904003

Metasystem approach to increase the load factor FMS

Alexandr Pishchukhin [1,*]

[1]Orenburg state university, Orenburg, Russia

Abstract. The work is devoted to increasing the efficiency of using flexible manufacture systems (FMS) by increasing the load factor. Unlike the well-known multi-agent approach to production management, we are investigating the addition of a metasystem level allowing flexible combination of the decentralized control method with centralization. The metasystem approach allows to reduce management to flexible switching of functioning technologies on the basis of the replacement rule, while describing FMS as a metasystem consisting of classification of the system analyst J. Klir from structured data systems. Production events are simulated in virtual space with the help of a multi-agent approach. At the same time, the number of requests from the waiting queue is chosen such that it becomes possible to optimize the loading of the FMS in the simplest case by the full-scan method, if the queue is not available, then the initiative of choice is provided to communications of agents-technologies and agents-equipments. Experimental studies were carried out on the FMS model including six pieces of equipment divided into three groups with a stream of six applications, each of which has a technological route of six operations each.. The experimental results showed the consistency of the developed approach by increasing the load factor of the FMS.

1 Introduction

Universal flexible manufacture system (FMS) has in its arsenal a significant set of production equipment. And although in its operation they try to apply the method of group technology [1], there is equipment that does not participate in the production process, which is bad for the load factor (on the ratio of the employment time to the total running time of the FMS). This problem is solved by compiling an optimal schedule [2] with strict centralization of its implementation, but with a sufficiently large load, many of its variants appear and it is difficult to select an optimization method that ensures sufficient speed. On the other hand, with increasing complexity, you can move to decentralized management and use the simulation method of modeling the production environment [3-5], such as agent-oriented programming. However, this eliminates the possibility of optimization where it is possible. Let us examine the question of the combination of centralization and decentralization through timely switchovers. A metasystem approach to management, developed from the beginning of the year 2000, can serve as a theoretical basis for solving switching issues and their timeliness [6].

2 Theory

In accordance with J. Klir's systemology [7], there are two ways of integrating systems: creating a structured system and creating a metasystem (See Figure 1). Each system is described by a set of variables, the changes of which are considered on a set of parameters. In structured systems, integration is performed over sets of variables under the assumption that they all have the same parametric set. Thus, the elements of structured systems are systems with different sets of variables, but with the same parametric sets. In metasystems, on the contrary, systems are integrated over parametric sets, regardless of whether these systems have one set of variables or not.

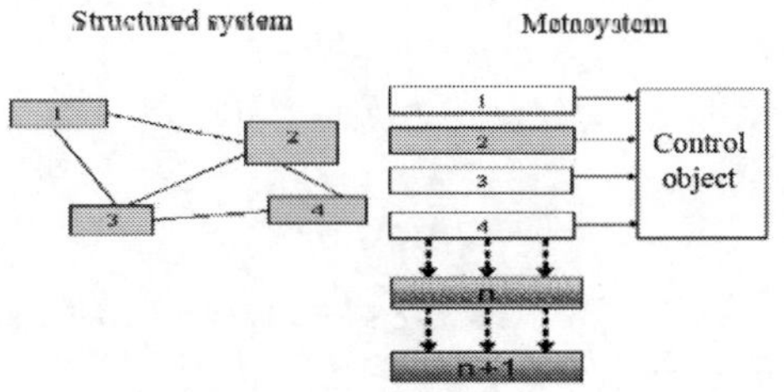

Fig. 1. System Integration Scheme.

As a consequence of this definition, the difference between the two ways of integrating systems, also in that the structured system reacts to external influences by changing the magnitudes of the variables in the relationships between the elements, the metasystem (which acts as a slider) can, in addition, replace the functioning system, that is reacts more drastically. In addition, in a structured system, all the elements function simultaneously, so if any one element fails, the entire system ceases to function. In a metasystem, a functioning system is selected by the replacement rule and, upon failure, can be replaced by another capable of functioning.

* Corresponding author: pishchukhin55@mail.ru

DOI: 10.1051/matecconf/201712904003

For these reasons, the metasystem has a higher survivability a priori. By survivability, it should be understood as a longer working capacity due to better adaptability [8].

The metasystem is defined as a triple:

$$\mathbf{MS} = (\mathbf{W}, S, r) \tag{1}$$

where **W** is a parametric set; S - can be a set of any systems, in this case it is a set of structured systems whose parametric sets are subsets of **W**; r is a replacement procedure realizing a certain function of the form

$$r: \mathbf{W} \rightarrow S \tag{2}$$

This formula defines the replacement function.

Metasystems are divided into two large classes [6]: sequential and parallel actions, so the replacement function for the second class selects a group of simultaneously functioning systems. For the most effective implementation of the replacement of one functioning system by another, it is necessary to set and solve six metasystem tasks [9]: identification of subbands for the efficient functioning of systems; assessment and improvement of the required level of system availability for use; development of a switching strategy for individual or groups of simultaneously functioning systems; identifying and ensuring the compatibility, coherent interaction of systems; optimal redistribution of limited system-wide resources; optimal synthesis of the metasystem.

3 Data and methods

Let us now look at the FMS and present it in the form of a metasystem that receives the flow of service requests, as was done in [10], consisting of structured systems, which in turn consist of data systems

$$\mathbf{MSD} = (\mathbf{W}, SD, r) \tag{3}$$

With the rule of replacement of the technological process

$$r: \mathbf{W} \rightarrow SD. \tag{4}$$

A lot of parameters in this case **W** will be represented by an assortment of manufactured products. For example, for a machining FMS, this can be: product housings, shafts, gears, levers, bushings and so on. Then the set SD represents the set of technological processes for manufacturing the above assortment, since the structured systems themselves are integrated as follows from the above theory, we will consider them to be components of structured systems-data systems. Each of these data systems includes a description of the operations of the manufacturing process (cutting modes, types of equipment, tools, devices). Thus, each of the technological processes includes a set of interrelated operations (structure). For example, $SD_i = \{({}^xV, {}^xD) | x \in N_q\}$, where ${}^xV$ is the set of binding variables for the data system x, ${}^xD$ is the set of data systems, q is the number of elements in the structured system, and N is the number of data systems in the j- th element of a structured system.

In the case of FMS, the structured system will simulate a technology that includes associated data systems for the mastering equipment, the product design with surface treatment methods, and technological processing modes. Metasystem accordingly will be at the top level of management, in its charge will be all resources of FMS in the form of technological, transport and storage equipment

In addition, the solution of the six above-mentioned tasks of the metasystem approach will allow to structure the optimization of the production process in the FMS. When solving the first problem, it is important to remember that the metasystem arises there and then, when and where the range of the problem being solved is so great that it is not blocked by the functioning of one system, or this system does not function effectively in some parts of the general range. Therefore, the problem arises of identifying the boundaries separating the subranges of the effective functioning of systems. For example, in a FMS one and the same surface can be processed by different technologies. These subbands can be identified using a multidimensional feature space.

The second problem is related to the fact that there is a period of time in which unused systems stand idle and, therefore, lose their readiness for use. For technologies included in the FMS, the loss of readiness is associated with equipment malfunction, lack of blanks, raw materials, materials used for this technology. Readiness can be assessed by the probability that the system will function at the time of switching to it. It is possible to control this probability only by creating favorable conditions for the appearance of the desired event, or eliminating the reasons for its non-appearance.

The third task of the metasystem approach is prognostic and is linked to the FMS with the production schedule [11,12].

The fourth task involves some interconnection of systems, for example at the structural level, when combinations of some systems in operation give a greater effect than their other combinations. For example, the functioning of a technology for manufacturing a product on the market and the technology using the resulting waste, or the functioning of two related technologies with a dramatically changing demand for each type of products manufactured with their help - the "emissions" of the market can be skillfully smoothed, flexibly managing the release volumes.

The fifth task arises only when using the group technology method in FMS, while the need for their coordination, optimization of resources directed to each system, to achieve the most effective metasystem, comes to the fore. At the same time, system-wide resources are directed to a greater extent in the system that most increases the metasystem effect.

Finally, the sixth task is connected with the optimization of the set of systems in the metasystem. As a criterion, one can choose the specific effect coefficient, that is, the ratio of the effect of including this technology in the metasystem portfolio to the costs associated with its acquisition and development, and then systems with the

DOI: 10.1051/matecconf/201712904003

largest specific effect coefficient are included in the portfolio first. This task, primarily related to optimizing the set of equipment in the FMS

When planning the loading of FMS, which is represented as a metasystem, it is necessary to select the replacement rule in such a way that the load factor of the equipment is the maximum. And in this case, a certain generalized load factor is required because it must take into account the loading of many pieces of equipment. It can be taken in the form:

$$K_z = \frac{\sum_1^N k_i}{N \cdot T}, \qquad (5)$$

where N is the number of pieces of equipment, T is the total running time of the FMS, k_i is the load factor of the i-th equipment unit.

4 Methods or model

In this paper, we use an agent-based model for planning a loading strategy for equipment, as having several advantages [13-15]:

- a continuous stream of events at the entrance allows the system to automatically react to changes in the state of orders and resources in real time;
- every order and every resource (in this case equipment) of the enterprise receives its program agent in the virtual space, which is triggered by the events that occur, to improve its current performance;
- Agents of each order and resource build their own, but connected in a common network of schedules, in such a network the schedule can contain hundreds of thousands of interrelated transactions;
- the basis for the approach to planning is not a complete or partial combinatorial search of options, but the identification and resolution of conflicts through negotiation of agents and the achievement of compromises - just as people do in real life,
- the created plan is launched for execution, during which the metasystem monitors the execution and initiates re-planning in case of discrepancies between the plan and the fact.

Multi-agent systems are a promising direction of simulation modeling, considered as independent from around 1990, when Job Shochem [16] formulated the main concepts of this new paradigm for solving complex problems.

However, this approach means fully decentralized management, in which the global minimum of managerial costs for the result can be missed. It is proposed to apply a multi-agent (agent-oriented) approach, in which there are active agent-technologies in the virtual space, whose purpose is to continuously assess the level of their readiness for launch and a metasystem that monitors the continuity of the production process and replaces the functioning technology as the production resources are freed. The metasystem in the regular case is guided by a previously developed technology switching strategy. Agents-technologies also prepare and launch their technology in accordance with this strategy, as well as in the case of an order from the metasystem.

The implementation of the described model was carried out with the help of an appropriate computer program. For simplicity, only standard switching for six technologies was simulated in it. At the same time, the usual multi-agent management of FMS, in which the preparation and launch of technology into production began in a decentralized manner, was compared with multi-agent, in which the metasystem first developed an optimal strategy in the virtual space and combined it in management with the production environment.

Each agent-technology monitors the freed machines and for operations that are not implemented in their technological map. If such conditions are found, it raises the "flag". The metasystem follows the equipment and does not allow downtime of any machine. As soon as the operation on any machine ends, a technology agent with a raised "flag" is searched. So the metasystem functions in the normal mode when the flow of orders arrives. If orders are known in advance, it tries to simulate in the virtual space communication agents-equipment and technology agents. With a small number of orders, as in the case considered in this paper, with six machines and six technologies with six operations, each has 720 variants of permutations, so the task can be optimized by the method of full search.

5 Results

The results are shown in Figure 2, where different colors depict functioning technologies and their switching from one equipment to another. On the vertical axis are six types of equipment that implement these technologies - the absence of color on a horizontal line means simple equipment. The horizontal axis represents time. As you can see, the figure shows an ordinary example of a production schedule with a load factor of 0.83. Preliminary optimization of the production process in the virtual space by the method of full enumeration allows you to bring the load factor to 0.917.

6 Conclusions or discussion and implication

Thus, the metasystem approach allows us to concentrate the researchers' attention on optimizing the process at the strategic level, freeing up routine operational management, giving freedom to decentralized management using active agents in the virtual space at the tactical level. The mechanism of metasystem

DOI: 10.1051/matecconf/201712904003

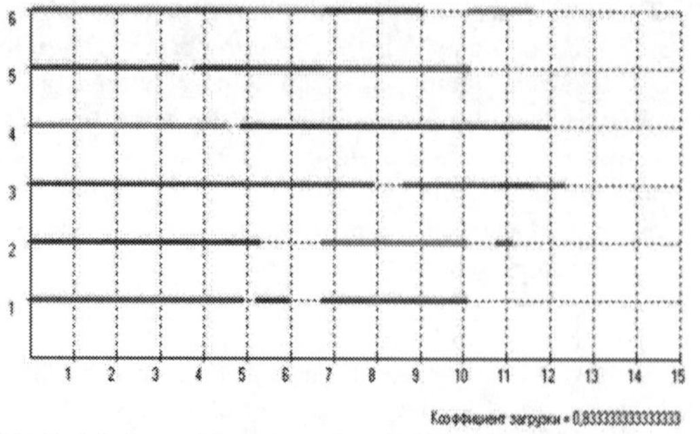

Vertical 1-6 - equipment number, horizontally – time

Fig. 2. Scheme of technology switching by equipment groups.

switchings, while providing apparatus for describing FMS as metasystems, consisting of structured systems that describe the technologies that appear in orders. The obtained experimental results allow to draw a conclusion about the operability and advantage of the agent-oriented method, combined with the metasystem approach to managing the FMS with respect to increasing its load factor

Further studies may be related to the detailed development of techniques for assessing the availability of technology for operation and management in emergency situations, which will increase the accuracy of the proposed method and its introduction into the practice of machine-building enterprises.

References

1. S.P. Mitrofanov, *Group technology of machine-building production* (Mechanical Engineering, Leningrad, 1983)
2. V. Vinod, R. Sridharan, Scheduling a dynamic job shop production system with sequence-dependent setups: An experimental study Robotics and computer-integrated manufacturing **24,** 3, pp. 435-449 (2008)
3. Mohammad M. Hamasha, Azmi Alazzam, Sa'd Hamasha,Multimachine, *Multimachine Flexible Manufacturing Cell Analysis Using a Markov Chain-Based Approach,* IEEE transactions on components packaging and manufacturing technology, **5** 3, pp. 439-446 (2015)
4. M. Savsar, Majid Aldaihani, *A Stochastic Model for Analysis of Manufacturing Modules,* Applied mathematics & infor-mation sciences **6,** 3, pp. 587-600 (2012)
5. Yue Wu, *A stochastic model for production loading in a global apparel manufacturing company under uncertain-ty*, Production planning & control, **22,** 3, pp. 269-2812, (2011)
6. S.V. Mironov, A.M. Pishchukhin, *Metasystem approach in management* (Publishing house: Orenburg State. Univ., Orenburg, 2005)
7. J. Klir, *Systemology. Automation of the solution of system problems* (Radio and Communication, Moscow, 1990)
8. A.M. Pishukhin, *Increasing the survivability of FMS based on metasystem replacements*, Izvestiya Samara Scientific Center of the Russian Academy of Sciences, **18,** 4-1, pp. 134-138. (2016)
9. S.V. Mironov, A.M. Pischukhin, *Metasystem approach to the examination of hazardous production facilities,* Successes of modern natural science **6**, p. 81-82, (2005)
10. A.M. Pishchukhin, T.A. Pishchukhina, *The Control Simulation of the Enterprise on the Basis Metasystem Approach,* Universal Journal of Control and Automation **1,** 4, p. 98-102 (2013)
11. Takahiro Kono, Keiji Ogawa, Toshiki Hirogaki, *Autonomous distributed agv system based on taxi transportation strategy: effect of multiple-load agvs on conveyance efficiency,* Proceedings of the asme international manufacturing science and engineering conference **2,** , pp. 553-560 (2010)
12. Mohammed Ali, Subhash Wadhwa, *The effect of routing flexibility on a flexible system of integrated manufacturing,* International journal of production research **48,**19, pp. 5691-5709 (2010)
13. Yu. A. Ivashkin, *Multiagent modeling in the Simplex3 simulation system* (Laboratory of Knowledge, Moscow, 2016)
14. G.F. Akhmedyanova, O.S. Eroshenko, A.M. Pishukhin, *An agent-oriented approach to the modeling of the learning process,* Fundamental Research, 11 (part 3) (2013)
15. A.M Pishchukhin, G.F. Akhmedyanova, *Methodological support of agent-oriented modeling of the learning process,* The University Complex as a Regional Center for Education, Science and Culture Materials of the All-Russian Scientific and Methodological Conference, p. 3071-3074. (2014)
16. Y. Shoham, *Agent Oriented Programming,* Journal of Artificial Intelligence 60 (1), pp. 51-92 (1993)

The development of discrete-address manufacturing planning method

Yuri Polyanskov[1], *Artem* Lototsky[2], and *Aleksandr* Madanov[2,*]

[1]ULSU, Competence Center "Aviation Technologies and Aviation Mobility", 432017 Ulyanovsk City, L'va Tolstogo st, h.42, Russian Federation
[2]ULSU, Scientific Research Centre of CALS-technologies, 432017 Ulyanovsk City, L'va Tolstogo st, h.42, Russian Federation

Abstract. The article describes the main stages of building the discrete-address manufacturing planning method. The developed discrete-address method ensures manufacturing of part sets "Just-in-Time" for high-variety manufacturing, using the vector planning in the space of the states of parts in the set.

In the context of limited demand for products and underutilization of capacity, the priority consists in "improving the efficiency of machine builders by cutting manufacturing and nonmanufacturing costs".

This statement applies to Russian aircraft builders designed for batch production. The machine builders' development strategy options in the current environment have been examined in detail in [9].

According to the customer-oriented market classification by Gartner Group [6], this type of manufacturing cannot be characterized as job manufacturing, as it is based on a pre-engineered process. It does not fully fit the Make-To-Order category either, since most commonly, orders would contain special requirements providing for variability of the products and relevant changes in the existing package of the engineering design documentation. This type of manufacturing differs from Assembling-To-Order, as due to the shortage of current assets, the enterprises lack current reserves. Therefore, the production cycle envisages the manufacturing of virtually the entire set of parts and sub-assemblies "from scratch". The part replicability of this type of manufacturing depends on the product specification and may vary from a single part up to relatively large batches.

Cutting the output of parts at the stage of manufacturing finished products in the context of the company's unchanged spatial and management structure alters the manufacturing typology. The operation consolidation coefficient K_{oc} that pursuant to GOST 3.1121-84 USTD is determined in accordance with formula (1) increases due to the decrease in the denominator.

$$K_{oc} = \frac{\sum_i K_{opi}}{K_{ws}} \quad (1)$$

where K_{opi} is the number of operations performed at the i-th work station within a month; K_{ws} is the number of work stations.

Depending on this indicator, if $20 \leq K_{oc} < 40$, the manufacturing type in question is small-batch manufacturing, if $K_{oc} \geq 40$, then – one-off manufacturing [7].

The need to supply finished piece-products by the deadlines under the contracts sets high requirements to the key components of the operations management system: day-to-day manufacturing planning and dispatching. Along with the manufacturing quality, the tasks of managing the parts and sub-assemblies material flow (that increases in size as it approaches the main assembly shop) take center stage.

This type of manufacturing represents a multi-level pyramid with the assembly of the item to be supplied at the top and with the manufacturing of individual parts at the bottom. The intermediate levels comprise the assembly of sub-assemblies that increase in size, transform in units, sub-systems and systems of the finished product as they move upwards. Meanwhile, as the material increases in size, the number of components goes down. All the discrete components of the said flow have a designated purpose in accordance with the specifications and are to be supplied to the next hierarchical level "Just-In-Time" manufacturing system in order to cut the production costs.

The balance of the production load in this context is ensured either through the key orders or through cooperation. All the material flows and the components thereof in this order and work system need to be identified.

Therefore, individual parts and sub-assemblies made to their orders have a specific purpose, "Just-In-Time" delivery address. Their manufacturing cycles are to be integrated into the overall order cycle so that they were synchronized in time with the works at the higher hierarchical levels of the assembly.

This type of discrete-address manufacturing is significantly different from conventional discrete Make-To-Stock manufacturing. Moreover, the supply of the same parts Made-To-Stock as spare parts for the same

* Corresponding author: madanov.alexandr@yandex.ru

DOI: 10.1051/matecconf/201712904004

finished product does not represent discrete-address manufacturing, as it has no specific address or dates of application, even though in case of repair works, if there is a demand for specific parts unavailable in the spare-part kit, such parts will be manufactured under the discrete-address cycle.

It is commonly assumed that the issues of current calendar planning (CCP) of job machine building are poorly formalizable and represent the most challenging domain of manufacturing planning [1, 7, 8]. The methods and quality of CCP for specific manufacturing depend on the planning unit, on the criteria used to estimate the manufacturing launch time of the said unit and on the arrangements for the method of its flow along the manufacturing route [1, 7].

The basic unit of discrete manufacturing flow is the part. Let's examine the manufacturing process of an abstract part D_j, where D stands for its name, and j – for its address.

Assume that the said part is provided with the complete package of engineering design documentation, including the route sheet. Let's assume that the manufacturing route of part D_j comprises a finite set m of process operations.

Note that the only nonrenewable and unrecoverable manufacturing resource is time, and therefore, the part manufacturing process by definition is a dynamic process evolving in time. Time is the independent variable, the argument of any manufacturing process that may quite naturally be taken as the X-axis of the manufacturing chart.

Note that not all scholars treat time as a manufacturing resource. Thus, in some recent studies of manufacturing planning, e.g. in [2], the list of resources includes "capital, manpower, materials (raw supplies), power, tools, machine-tools, information, knowledge". The other sources on the day-to-day management of flexible manufacturing systems [3] view resources as "operation material aids (machine-tools, tools, appliances, lubricating fluids, etc.), power and communications". Time is viewed as the duration of the planning period representing the time limit for accomplishing the end". In the textbook on machine building engineering and day-to-day management [7], the resources are conceptualized as tangible components of manufacturing – the means and objects of manufacturing, while time "serves as the measure of labor input per unit and as the time period required to manufacture it. In the first instance, time defines the performance level, while in the second – the rate of converting resources into products".

If we treat time as a nonrenewable resource, we can determine its value proceeding from a unidimensional time model of the manufacturing process to a two-dimensional, cost-time model of the process enabling us to examine manufacturing costs as a dynamic process, as a function of time.

Let's illustrate the above considerations by developing an ideal discrete-address manufacturing model. The salience of developing an ideal model has been emphasized in [7], as the ideal manufacturing process "provides for the minimization of space and time consumption for the conversion of source materials into the necessary products. Therefore, the principles of managing the ideal process turn into the principles (rules) of saving time and space at the engineering stage".

Generally, the duration of the manufacturing cycle for the parts of one-off (small-batch) production T_{mc} may be calculated using formula (2) [7]:

$$T_{mc} = \sum_{i=1}^{m} \frac{t_{pci}}{K_{opi}} + \sum_{i=1}^{m-1} \frac{t_{ioi}}{K_{opi}} + \sum_{i=1}^{K_{op}} t_{ni}, \quad (2)$$

where: t_{pci} – piece-calculation duration of the *i*-th operation; K_{opi} – *i*-th operation rate, hereinafter assumed to be equal to 1 ($K_{opi} = 1$); t_{ioi} – the duration of the interoperation period between (i – 1) and the i-th operations composed of the interoperation waiting time and the interoperation queue time; t_{ni} – the natural holding time at the *i*-th operation.

The manufacturing process of part D (we will omit index j at this point, as only one part is examined so far) may be treated as a sequential flow of process operations in time i = {0, 1, …, m} converting the state of the part from the blank part (D_0) to the finished part (D_m). Therefore, throughout the manufacturing process, part D, undergoing changes, consecutively transits k states, k = {0, 1, …, m}, resulting from the i-th process operation immediately preceding each consecutive process operation. Assume that the state of the part during the i-th operation remains unchanged and corresponds to state D_{k-1}.

Each i-th operation coincides with the time period for the relevant operation (τ_i) specified in the manufacturing route. In case of ideal process, $\tau_i = t_{pci}$. times τ_i vary for different operations. Taking the manufacturing launch time of the part for the benchmark, i.e. assuming $t_0 = 0$, the transition of the part into state k at discrete time t_k will be defined by the sum of preceding operations times τ_i, i.e.

$$t_\kappa = \sum_{i=0}^{k} \tau_i = t_{k-1} + \tau_i, \text{ where } k = 0, 1, \ldots m. \quad (3)$$

Time $t_D = t_m = \sum_{i=0}^{m} \tau_i$ – is the full manufacturing time of the part. Note that this is the minimum time for manufacturing the part strictly subject to the manufacturing route. The issue concerning the potential optimization of the manufacturing route is not examined here. The relocation time of the part, e.g. relocation of the part between the shops, should it be significant, may also be included in the manufacturing route as the relevant i-th operation changing the location of the part, which we will equally treat as state transition from state D_{i-1} to D_i. Therefore, the time axis t breaks down into m unequal intervals as depicted in Fig. 1.

Let's turn to the Y-axis, since the part state transition function that we will refer to as the manufacturing function (not to be confused with the production function in microeconomics). As over the course of manufacturing, part D undergoes multiple states {D_0, D_1, …, D_k, …D_m} from the blank part D_0 to the finished part D_m, the percentage of completion of the part along the manufacturing route changes from 0% (blank part D_0) to 100% (finished part D_m). Therefore, we can introduce the availability function g_k that, corresponding to the part states D_k, will naturally reflect the manufacturing process. The completion function $g_k(t)$ introduced this way has

DOI: 10.1051/matecconf/201712904004

certain weaknesses, specifically, it fails to provide us with the Y-axis creation algorithm, i.e. with the procedure for the quantitative breakdown of the section [0%, 100%] into m intervals corresponding to the percentages of part completion between two adjacent states D_{k-1} and D_k.

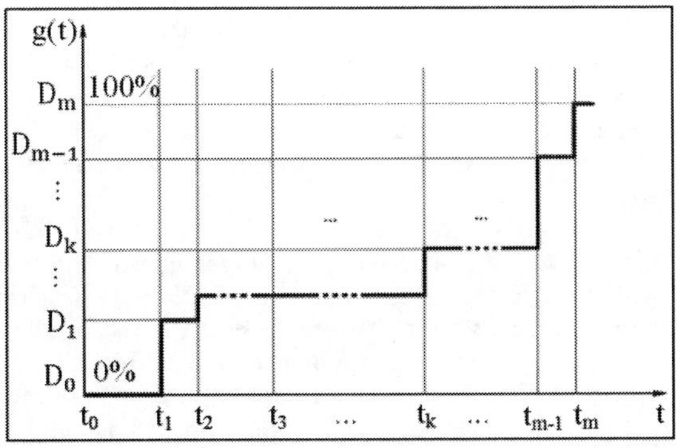

Fig. 1. State transitions of the part during the manufacturing process.

To fill in these gaps, let's use the cost approach. As the part is being manufactured, the enterprise invests its resources that raise the cost of the part from the cost of the blank part D_0 to the cost of the finished part D_m. These costs may be expressed through the cost of a conventional standard hour ψ_i of the i-th process operation measured as Rub/h. Generally, process operations may be performed at a variety of manufacturing areas in a variety of shops (e.g. post-machining application of galvanic coating). Moreover, these may comprise the intershop logistics operations or natural holding processes included in the route sheet. Therefore, in a general case, variation in the standardized resource intensity of operations should be considered that is defined through the cost of the i-th conventional standard hour ψ_i = var.

Therefore, apart from the dataset for the time of manufacturing operations $\{\tau_i, i = 0, 1, \ldots, m\}$, the task of developing the manufacturing model requires data on the cost of the relevant conventional standard hour, i.e. a dataset for the standardized cost of the relevant operations $\{\psi_i, i = 0, 1, \ldots, m\}$.

The cost of the i-th manufacturing operation vi is calculated according to formula (4)

$$v_i = \psi_i \cdot \tau_i \tag{4}$$

The current manufacturing cost of the part in the k-th state is determined as

$$p_\kappa = \sum_{i=0}^{k} v_i = \sum_{i=0}^{k} (\psi_i \cdot \tau_i) \tag{5}$$

Therefore, the source data for developing the model of part D manufacturing process is matrix $TD = \left\{ \begin{matrix} \tau_i, \\ \psi_i, \end{matrix} i = 0, 1, \ldots m \right\}$. Use equations (3) and (5) to obtain the part manufacturing matrix from matrix TD:

$$PD = \left\{ \begin{matrix} t_k, \\ p_k(t_k), \end{matrix} k = 0, 1, \ldots m \right\}. \tag{6}$$

In matrix PD, the time function $p_k(t_k)$, is the step function of time with discontinuity of the first kind.

Matching each state of the part D_k to the relevant dot on the plane (p, t) $D_k \leftrightarrow (p_k, t_k)$ and connecting these dots in the form of a piecewise broken line w(t) as provided on Fig. 2, we get the part manufacturing trajectory in monetary terms. Therefore, function $p_k(t_k)$ has the meaning of the part manufacturing function.

Putting a straight line through the dots (p_0, t_0) and (p_m, t_m), corresponding to the initial and the final (finished) state of the part will produce the linear model of part manufacturing:

$$v(t) = \psi_D \cdot t, \tag{7}$$

where the proportionality constant $\psi_D = p_m/t_m$ has the meaning of the cost of the conventional average standard hour for manufacturing part D.

It is obvious that if $\psi_i = \psi$ = const, then formula (7) implies that $p_m = \psi \cdot t_m$ and $\psi_D = \psi$. This means that the part manufacturing trajectory will be defined through a linear function – rectangle diagonal (Fig. 2). In case of the large number of operations with the variable cost of the conventional standard hour ψ_i = var, a more accurate approximation of v(t) may require some other methods, e.g. least-square polynomial approximation.

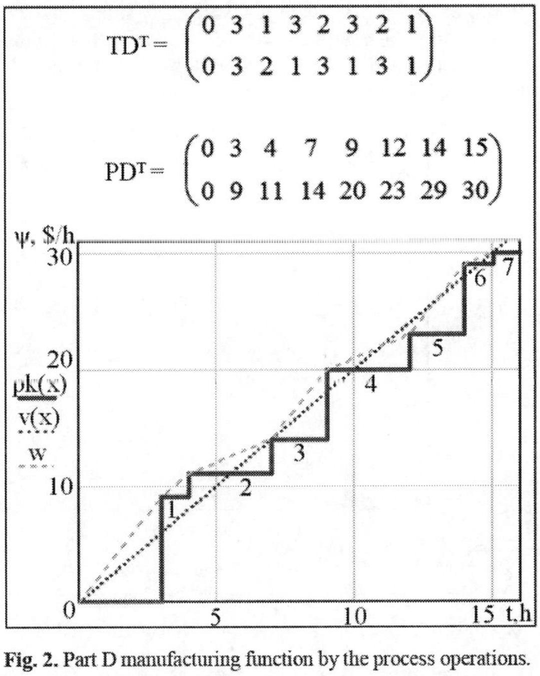

Fig. 2. Part D manufacturing function by the process operations.

Fig. 2 illustrates this through the example of part D manufacturing function based on the transposed process matrix TD, where the first line depicts the times of process operations τ_i in standard hours, and the second – the cost of the relevant standard hours ψ_i, in abstract relative units.

Fig. 2 illustrates the part D manufacturing function by the following process operations: 1 – rough machining, 2 – grinding, 3 – transportation, 4 – chrome plating, 5 – transportation, 6 – drilling, 7 – marking.

If we normalize axis p by p_m, i.e. divide p_k/p_m, we get the normalized cost scale for manufacturing part P_k that can be expressed in percentage. But previously, we have expressed the completion of part g_k in percentage. Therefore, the transformation $g_k = P_k = p_k/p_m$ has a conceptual meaning assigning the meaning of the relative part manufacturing cost to the part completion concept. Formally, g_k and P_k are two different variables, but within the model under consideration, they are isomorphic. Therefore, let's assume that the part relative completion functions G(t) and functions $g_k(t)$ and $P_k(t)$ are identical $G(t) \equiv g_k(t) \equiv P_k(t)$.

Similarly, we can move on to the dimensionless normalized scale of the part manufacturing time $T_k = t_k/t_m$.

Fig. 3 depicts the manufacturing function of part D above on the normalized plane.

Therefore, we get a normalized linear model V(T) of part manufacturing, where the part manufacturing optimum trajectory is the unit square diagonal. Meanwhile, function V(T) represents a linear approximation of the completion function G(T).

As a matter of convenience, the part manufacturing process may be considered equisignificant either in the rectangle (p_m,t_m) on the plane (p,t), or in the rectangle $(1,t_m)$ on the plane (G,t) or in the square (1,1) on the normalized plane (G,T). In the latter case (similarly to the other two), the area beneath line V(T) or, to be more precise, under the step function G(T) characterizes the resource intensity of part manufacturing.

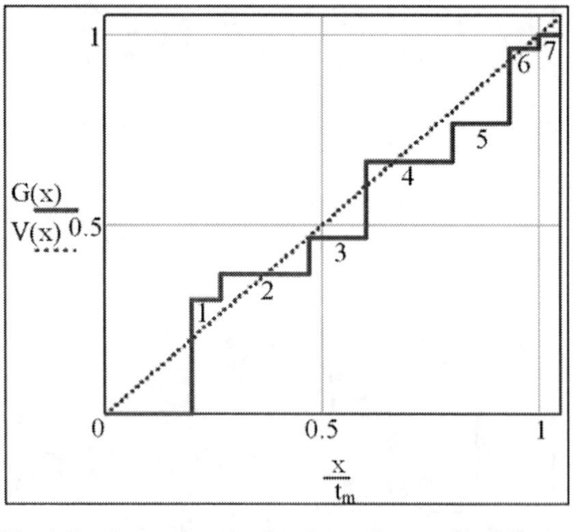

Fig. 3. Part D manufacturing function on the normalized plane.

Deviations from or irregularities of the manufacturing route on the normalized plane (G,T) may theoretically look like a deviation of the actual manufacturing time of part T_f from the unit value: $\Delta T = T_f - 1$. The negative value of ΔT (faster than provided in the manufacturing route) either points at the error in the route (we will rule out this possibility by sticking to the previous assumption regarding the completeness and accuracy of the documentation package), or at the irregularities in the process parameters (speed of cutting, duration of chrome plating, etc.). I.e. ΔT should be treated as an indirect sign of the worsening quality of the part. The positive value of ΔT, i.e. delay in manufacturing the part by time $\Delta t = \Delta T \cdot t_m$, apart from direct consequences, also indicates the increase in the actual manufacturing cost of the part by $\Delta p = \psi_D \cdot \Delta t$.

Let's proceed to the consideration of the planning unit through a set of parts for sub-assembly K. Take the simplest case, where the set consists of two parts K = {D1, D2}, each of them characterized by its own process matrix: TD1 and TD2 respectively. In a general case, these matrices differ from each other, both in terms of content and quantity of process operations $m1 \neq m2$, $\tau 1_i \neq \tau 2_i$, and in terms of respective standard hours cost $\psi 1_i \neq \psi 2_i$.

Creation of the relevant X-axis G1(t) and G2(t) presents no difficulty and is virtually similar to the case with a single part. This is due to the fact that in physical terms, parts D1 and D2 are two physically unrelated items and therefore, the relevant X-axes are orthogonal and turned 90° away from each other. Accordingly, the normalized planes (G1,T) and (G2,T) are orthogonal and cross along X-axis T. These planes may be brought into coincidence by turning axis T 90°.

The task of creating time axis T may be challenging to a certain extent, if the part manufacturing times are unequal $t_{m1} \neq t_{m2}$, i.e. in the majority of real situations. This is due to the fact that the normalization of the common real time axis t by heterogeneous values $1/t_{m1}$ and $1/t_{m2}$ results in the emergence of heterogeneous times T1 and T2 in the manufacturing model K, which is unacceptable for application. Therefore, the time axis should be normalized by the larger manufacturing time value. Suppose, that $t_{m1} > t_{m2}$, then axis t should be normalized by t_{m1} value. In this case, the square (G2,T) shall become smaller along the X-axis and the relevant width thereof $T_{m2} = t_{m2}/t_{m1} < 1$ changes by $\Delta T_m = (t_{m1} - t_{m2})/t_{m1}$.

The manufacturing launch time of the part t_{laun} plays an important part in the manufacturing planning practice. The optimum launch time t_{laun}^{opt} for a single part is determined so as to minimize the part resource intensity – the area beneath line V(T), and therefore, $t_{laun}^{opt} = - t_m$, where the benchmark $t_0 = 0$ is the launch of the assembly or successive hierarchical level manufacturing. Note that this conclusion is strictly in conformity with the Just-In-Time manufacturing strategy.

The same resource intensity minimization principle is universal and applicable to a set of parts. Within a set of parts, the part with the largest t_m stands out, the so called core part [7], t_{m1} in the example above. This time determines the so called critical path of the set manufacturing. Manufacturing of the other parts is launched with a time lag against the critical part. The optimum value of this time lag is equal to $t_{laun} = t_{m1} - t_{m2} > 0$. The same time is the acceptable time lag for the launch of part D2. Bringing the relative planes (G1,T) and (G2,T) in coincidence for the case above subject to $t_{m2} = 0{,}5t_{m1}$ is depicted on Fig. 4. The launch times for parts D1 and D2 are the intersection points of X-axis by lines V1(T) and V2(T), i.e. $T1_{laun} = -1$, $T2_{laun} = -0.5$, at axis T scale equal to $1/t_{m1}$.

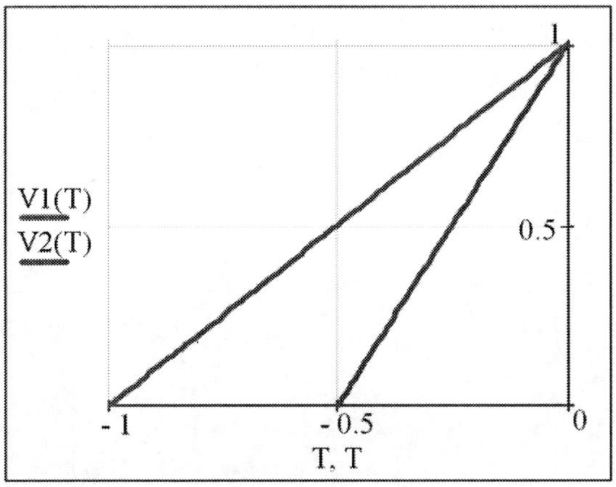

Fig. 4. Bringing relative planes (G1,T) and (G2,T) in coincidence for a two-part set, $t_{m2} = 0.5t_{m1}$

Let's consider the practicality of applying the linear manufacturing model for the set as whole using the example of our model. For illustrative purposes, assume that $\psi_{D1} = \psi_{D2}$. Then the manufacturing cost of the set will be equal to:

$$p_K = p_{m1} + p_{m2} = \psi_{D1} \cdot (t_{m1} + t_{m2}) = 1.5 \cdot p_{m1} \quad (8)$$

The relative cost of the parts in the set will account for $P1 = p_K/p_{m1} = 0.667$, $P2 = p_K/p_{m2} = 0.333$. The resource intensity of set S_{Set} linear model is equal to the area beneath the unit square diagonal $S_{Set} = 0.5$. The resource intensity of the by-part manufacturing model for the set will equal the sum of the areas of the relevant resource intensity triangles for part D1 $S_{D1} = 0.333$ and for part D2 $S_{D2} = 0.083$, ultimately – $S_{Sp} = 0.416$. As a result, the linear model produces an inflated resource intensity value (20.2% above the by-part model). The estimation above is depicted on Fig. 5. Therefore, the decision on the practicality of the linear model for set manufacturing should be made individually for each case with consideration of the fact that the resulting value will invariably be inflated.

As modern computing machinery does not impose tight constraints on the computational complexity of manufacturing planning procedures typical for manual methods, it appears practicable to apply the by-part planning model for the manufacturing of sets in the form of parts state matrix composed of the parts state vectors $K_n = \{D_1, D_2, \ldots, D_j, \ldots D_n;\ j = 1, \ldots, n\}$. For practical purposes, it is practicable to present matrix K_n in the form of matrix Kg_n of parts completion vectors g_{jk} superimposed on the common discrete axis of state transition time.

An example of such matrix made for a three-part set with $m_1 = 4$, $m_2 = 3$ and $m_3 = 4$ transiting through the states at different times is provided on Fig. 6.

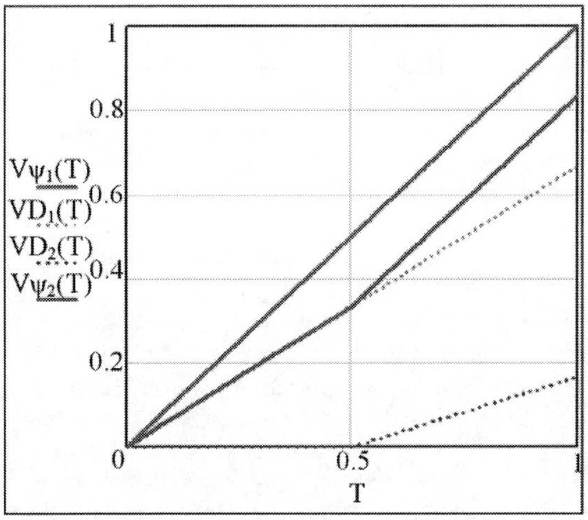

Fig. 5. Linear vs by-part model of set manufacturing.

The number of rows in the matrix equals n+1 = 4, corresponding to the total number of parts plus the common time scale T_k. The number of columns is equal to the sum of all states, i.e. $m_k = 1 + \sum_j^n m_j = 12$. If the times when multiple parts transit from one state to another coincide, dimensionality m_k is automatically reduced by the number of such coincidences. Our matrix has 2 coincidences, as all the parts transit to the end state simultaneously at the last discrete time, which is in conformity with the strategy of the set resource intensity minimization. Hence, the actual dimensionality of matrix $m_{ka} = 12 - 2 = 10$. Columns T4 and T5 may be excluded from the matrix, as the part does not change states at these times (rows are provide for illustrative purposes).

$$Kg = \begin{pmatrix} T_0 & T_1 & T_2 & T_3 & T_4 & T_5 & T_6 & T_7 & T_8 & T_9 & T_{10} & T_{11} \\ g_{1,0} & g_{1,1} & g_{1,2} & g_{1,2} & g_{1,2} & g_{1,2} & g_{1,3} & g_{1,3} & g_{1,3} & g_{1,3} & g_{1,3} & g_{1,4} \\ g_{2,0} & g_{2,0} & g_{2,0} & g_{2,1} & g_{2,1} & g_{2,1} & g_{2,1} & g_{2,1} & g_{2,1} & g_{2,2} & g_{2,2} & g_{2,3} \\ g_{3,0} & g_{3,0} & g_{3,0} & g_{3,0} & g_{3,0} & g_{3,0} & g_{3,0} & g_{3,1} & g_{3,2} & g_{3,2} & g_{3,3} & g_{3,4} \end{pmatrix}$$

Fig. 6. Vector plan matrix

Therefore, the developed discrete-address method ensures manufacturing of part sets "Just-in-Time" for high-variety manufacturing, using the vector planning in the space of the states of parts in the set.

As for the choice between the "pushing" and the "pulling" strategy for the model under consideration, it may be argued that they have different benchmarks: launch time for the pushing strategy and delivery time for the pulling strategy. Meanwhile, the pulling strategy is to operate in the context of "negative time", which is impossible in the absence of current reserves.

The work is performed according to the Russian Federation Program of the Ministry of Education and Science (№2.1816.2017/PCh).

DOI: 10.1051/matecconf/201712904004

References

1. M.I. Bukhalkov, *Management Planning* – **2nd ed.**, 400 (INFRA-M, 2003)
2. E. Westcamper, M. Decker, L. Endowby, A.I. Grabchenko, V.L. Dobroskok, *Intro into Industrial Engineering*, 376 (NTU KPI, 2008)
3. V.F. Gornev, V.V. Emelyanov, M.V. Ovsyannikov, *FMS Day-to-Day Management*, 256 (Mashinostroenie, 1990)
4. R.R. Zagidullin, *Machine-Building Management Using MES, APS, ERP Systems*, 372 (TNT, 2011)
5. K.S. Kulga, I.A. Krivosheev, *Models and Methods for the Creation of an Integrated IS to Automate Technical Training and Manage Aircraft and Machine Building*, 377 (Mashinostroenie, 2011)
6. M. Polyakova, *Discrete Manufacturing in Russia*, **No. 02**, (2008)
7. N.S. Sachko, *Machine Building Engineering and Day-to-Day Management*, 676 (Novoe Znanie, 2005)
8. E.B Frolov., R.R. Zagidullin, *MES Systems. Top-Down, Inside-Out. The Criteria We Pick* (ERPNEWS, 2007)
9. V.N. Tsuglevich, *Development Strategy of Russian Machine Builders in the Current Context*, **44** (2003)

Modeling of reconfigurable manufacturing systems for determining their performance

Alexey Kapitanov[1,*], and *Vladimir* Mitrofanov[1]

[1] Moscow State Technological University «Stankin», Vadkovskiy lane, 1, Moscow, 127055, Russian Federation

Abstract. A method of determining flexible manufacturing system performance (throughput capacity) based on identifying elementary controlled events that occur during FMS operation and controlling this flow based on Shannon approach to evaluation of the quantity of information required for control is proposed. The behavior of the main FMS elements in the manufacturing process is suggested to be described using three types of random flows of square pulses – state, availability and propagation flows. The proposed mathematical model enables to choose the optimum relationship between FMS power margins, parts and tools supply and throughput capacity.

1 Introduction

Modeling is the most important tool while creating flexible manufacturing systems (FMS). The main types of models, applied to FMS shall be analytical and simulation ones (Table 1) [1-12].

The main goal of FMS control is the exact realization of the determined production plan. Where the qualitative indicators are represented by the maximum efficiency, implementability (coordination with production capacities), and information throughput capacity of control system required in the process of plan targets' implementation [13-18].

The last question for FMS control systems determines the conditions in which computer-aided control system (CACS) is put in the process of plan implementation. It may seem that the effect achieved with increase in the degree of optimality of the planning algorithm is completely offset by additional costs of computer-aided operational control system.

The determining of FMS performance (throughput capacity) is particularly important at its design stage. This requires the development of a model that adequately reflects the process and uses statistical data, available for relatively easy measurement on the operating FMS, as the source [19-24].

Regular elementary controlled events with different characteristics can be identified for FMS in mechanical engineering and instrument engineering, which are characterized by discreteness of products, consisting of multiple parts and assemblies, manufacturing facility, consisting of a number of machines with selected modes for moving and machining of products (various parts grouping for individual and mass manufacture).

Startup of batch manufacture and adjustment of the process of its passage through the facility during its machining refer to such events in FMS. The volume of data to be received and processed by CACS for FMS when controlling an elementary event is usually quite steady and depends on the number of alternatives considered in the decision-making process. Their number is determined by the batch type, number of parts in a batch, time interval between sequential startups and the nature of startup and release (scheduled, ahead of schedule). Shannon approach to evaluation of the quantity of information required for high quality control of elementary events' flow that occur in the process of FMS operation can be used with such choice of FMS elementary controlled events [25-29].

The elementary event control function involves synchronization of operation of the main elements of manufacturing process in accordance with the defined plan.

2 Main part

Behavior of the main FMS elements in the production process can be described using three types of random flows of square pulses, known as the flows of state, availability and propagation.

State flows $P(t)$ are generated from impulses with an amplitude of 1, random pulse duration and pauses between them. They characterize the performance of equipment, attachments, apparatus, and control software etc. over time, and the duration of their recovery, and are determined by the laws of pulses and pauses duration distribution.

Availability flows $A(t)$ are generated by pulses with variable amplitude and duration, characterizing the and

* Corresponding author: av.kapitanov@stankin.ru

DOI: 10.1051/matecconf/201712904005

Table 1. Main FMS Models

Model Type	FMS	Models Characteristics	Use of models
Models of FMS facilities			
	analytical	Models of transport devices Models of processing centers Models of diagnostic measurement systems based on differential equations	During the stage of development for calculation of parameters During the operation stages for calculation of regimes
	simulation	Models of vehicles, robotic manipulators, and machining systems response to random influences	During the stage of development for evaluation of the elements' parameters in case of application of an particular FMS
Models of FMS processes			
	analytical	Models of state flows synchronization of various FMS elements, including conveyor (e.g. based on the model of coincidence of random pulse flows)	Calculation of FMS throughput capacity for various sets of parts
	simulation	Models of FMS comprehensive performance evaluation under various laws of elements' state distribution	Calculation of FMS operation modes

resources by units of equipment, they are determined by the laws of amplitudes' distribution and pulses duration.

Pulse propagation flows *B(t)* characterize the flow of elementary controlled events passing through a particular equipment unit during machining. They are represented in two ways: as pulses with zero duration and variable amplitude, or as pulses with single amplitude and variable duration.

In both cases pulses have random repetition interval and variable parameters — either amplitude or duration, proportional to the number of parts that have been machined on this equipment unit.

Propagation flows are determined by the amplitudes' distribution laws, duration of pulses and pauses between them.

Synchronous behavior of production process elements is achieved by controlling, which obviously corresponds to time coincidence of random flows' pulse parameters characterizing elements that determine the possibility of implementation of this part operation on this equipment unit. The result is the coincidence flow that corresponds to the propagation flow of machined products through an equipment unit. Application of n-coincidence operation

$$n(t) = \bigwedge_{i=1}^{n} P_i(t) \tag{1}$$

is preceded by a number of additional operations: addition of the availability flow *A (t)* to the propagation flow *B (t)*

$$S_k\{A(t); B(t)\} = S_{k-1}\{A(t); B(t)\} + B(t_k) \tag{2}$$

where $t_k < t < t_{k+1}$, $S_0\{A(t); B(t)\} = A(t)$ the separation of the propagation flow from the availability flow pool (reserves) for different parts operations time

$$f(t) = \min\{A(t) \vee B(t)\} \tag{3}$$

conversion of the propagation flow into the state flow

$$P(t) = \sum_{k=1}^{\infty}\left\{\overset{o}{1}(t - t_k) - \overset{o}{1}[t - t_k - B(t_k)]\right\}; \tag{4}$$

conversion of the state flow into the propagation flow

$$B(t) = a(t)\int_{t_k-\tau_k}^{t_k} dt$$

$$a(t) = \begin{cases} 0\ npu\ t \neq t_k; \\ 1\ npu\ t = t_k, \end{cases} \tag{5}$$

where t_k, τ_k - are the random moments of state flow pulse occurrence and their duration; separation (subtraction) of the propagation flow from the availability flow $R_k\{A(t); B(t)\} = R_{k-1}\{A(t); B(t)\} - B(t_k)$ for $t_k < t < t_{k+1}$, $R_0\{A(t); B(t)\} = A(t)$;

$$B(t_k) \le R_{k-1}\{A(t); B(t)\} \tag{6}$$

The specified operations enable description of movement of the products' material flow (deliveries, rejects) and precede the time coincidence of single amplitude pulses *n(t)*. Further, the object of discrete manufacture is interpreted as discrete communication channel transmitting the propagation flow under noise conditions determined by parameters of state flows and *n*-coincidence operation. The actual and the required throughput capacity of the elementary facility (machine) control system can be determined based on the theorems of information theory, by calculating the entropy per elementary event of the source - $H(x)$ for all input flows, receiver — $H(y)$ for all output propagation flows (coincidence flow) and based on n-coincidence operation - conditional entropy $H_x(y)$. The actually provided throughput capacity per λ equipment unit with j_λ parts machined on it, and the startup intensity of production batches $N_{j\lambda}/T$

$$C_\lambda = \max \frac{N_{j\lambda}}{T}\left[H(y)_{j\lambda} - H_x(y)\right] \quad (7)$$

With simultaneously used k equipment units the throughput capacity of FMS will be:

$$C' = \sum_{\lambda=1}^{k} C_\lambda \quad (8)$$

Additionally required throughput capacity from CACS for FMS, attributed to one controlled event, can be determined in accordance with Shannon's 10th Theorem. Here, CACS for FMS is considered as corrector of the channel, which is FMS.

For the entire FMS with simultaneous use of k machines the additionally required CACS for FMS throughput capacity

$$\Delta C' = \sum_{\lambda=1}^{k} \max\left\{\frac{N\left[H(x)_{j\lambda} + H_x(y)_{j\lambda} - H(y)_{j\lambda}\right]}{T}\right\} \quad (9)$$

at the same time the batch startup and release control require particular time regulated actions ensuring complete processing of information, decision-making and managerial activities.

3 Conclusion

Extension of FMS throughput capacity by means of CACS enables the implementation of a larger number of elementary controlled events at an average, while ensuring the required statistical parameters of the products propagation flow at FMS output.

The estimate obtained from (9) represents the lower limit of CACS for FMS throughput capacity extension in order to ensure the admissible parameters of $H(y)_{j\lambda}$ for all λ and $j - C_{\min} = (C' + \Delta C')$. The upper limit of throughput capacity extension C_{max} is determined from the ratio, connecting FMS resources involved in manufacture during control automation E_τ with difference between the source entropy value of the output flows $H_x(y) = H_0$ and the value achieved by extending CACS for FMS throughput capacity. This dependence can be determined as

$$E_\tau = E_{\max}\left[1 \;\; a_1 e^{k_1[H_0 \; H_\tau]}\right] \quad (10)$$

and is similar to the dependence obtained by Academician V.A. Trapeznikov for complex manufacturing systems, at $H_{T_1} \leq H(y)_{add}$ C_{max} is determined from the expression

$$\left[\frac{d E_T}{d H_T}\right]_{at\, H_{T_i} \to \min} \geq \frac{d\left[F(c) \mid H_{T_i}\right]}{d H_\tau} \quad (11)$$

where $F(c)$ determines the costs of CACS for FMS depending on its throughput capacity. In this case FMS entropy is $-H_{T_i}$. The following relationship should be observed for efficiently designed system with optimum throughput capacity C_{opt}:

$$C_{\min|at\, L_i \to var} = H_i(x)\left[1 \;\; a\, e^{kL_i}\right] \quad (12)$$

In addition to FMS control automation, the increment of ΔC value can be ensured by increasing the mean amplitude value of availability flows (reserves, manufacturing capacity, and FMS equipment). ΔC value depends on the distribution of total resources between individual flows L_i. When L_i is changed for one of the participants in n-coincidence of elementary flows operation

$$\Delta C_{|at\, L_i \to var} = H_i(x)\left[1 \;\; a\, e^{kL_i}\right] \quad (13)$$

where $H(x)$ – if the input flow entropy; a and k – are constants.

The use of (13) enables choosing the optimum relationship between FMS power margins, supply of components and tools and CACS for FMS throughput capacity.

The reported study was funded by RFBR, according to the research project No. 16-31-60079 mol_a_dk

References

1. H. ElMaraghy, International Journal of Flexible Manufacturing Systems Vol. **17**(4), pp. 261-276 (2006)
2. G. Schuh, V.Stich Springer (2012)
3. F. Karl, G. Reinhart, M. Zaeh. 45th CIRP CMS, pp. 608-613 (2012)
4. D. Telgen, L. Moergestel, D. Ceglarek, Robotics and Computer-Integrated Manufacturing. Vol. **43**, pp. 30–38 (2017)
5. V. Bryuhanov, Vestnik MSTU Stankin Vol. **2**, pp. 56-62 (2010)
6. V. Voronenko, E. Mikhailov, Y. Sokolov, Vestnik MSTU Stankin Vol. **3** (34), pp. 29-33 (2015)
7. V. Mitrofanov, A. Kapitanov, A. Popov, ONIKS, (2013)
8. V. Chekanin, A. Chekanin, Information technology and computer systems. Vol. **2**, pp. 37-45 (2014)
9. K. Mittal, P. Jain An. Procedia Engineering Vol. **69**, pp. 1125-1129 (2014)
10. A. Azab, H. ElMaraghy, P. Nyhuis, J. Pachow-Frauenhofer, M. Schmidt, CIRP Journal of Manufacturing Science and Technology Vol. **6**, pp. 110–119 (2013)
11. M. Bruccoleri, G. Lo Nigro, G. Perrone, P. Renna,

S. Noto La Diega, Annals of the CIRP Vol. **54,** pp. 433-436 (2005)
12. M. Abbasi, M. Houshmand, International Journal of Advanced Manufacturing Technology Vol. **54**, pp. 373–392 (2011)
13. A. Tsarev, Problems of engineering and automation. Vol. **3**, pp. 33-41 (2014)
14. B. Cherpakov, A. Feofanov, STIN Vol. **6,** pp.3. (2004)
15. V. Merchants, A. Feofanov, Industrial and manufacturing engineering. Vol. 2, pp. 66-67 (2010)
16. T. Grishina, V. Mitrofanov, A. Feofanov, Vestnik of Belgorod state technological University n. a. V. G. Shukhov Vol. **3**, pp. 158-160 (2014)
17. A. Feofanov, Y. Milkina, Economics and management in mechanical engineering Vol. **1**, pp. 5-7 (2011)
18. A. Musatov, A. Apple, V. Kulakov, The scientific review Vol. **5,** pp. 253-262 (2014).
19. R. Adgamov, S. Dmitriev, T. Karimov, A. Khayrullin, Scientific-technical Vestnik of the Volga region Vol. **6,** pp. 72-75 (2012).
20. B. Gorshkov, I. Samartsev, Vector of science of Togliatti state University. Vol. **3,** pp. 53-56 (2014)
21. L. Shvartsburg, E. Butrimova, N. Drozdov, Safety Vol. **2,** pp. 27-30 (2012)
22. L. Shvartsburg, Y. Zvenigorodsky, N. Bukeikhanov, Vestnik MSTU Stankin Vol. **2,** pp. 14-17 (2011)
23. A. Kutin, M. Turkin, Vestnik MSTU Stankin Vol. **4,** pp. 41-47 (2010)
24. A. Kutin, M. Turkin, Vestnik MSTU Stankin Vol. **3,** pp. 36-40 (2011)
25. A. Ananiev, Vestnik MSTU Stankin Vol. **2,** pp. 65-68 (2010)
26. A. Shabarov, Vestnik MSTU Stankin. Vol. **3,** pp. 136-143 (2011).
27. V. Mitrofanov, T. Grishina, A. Feofanov, Industrial and manufacturing engineering Vol. **8,** pp. 43-45 (2015)
28. A. Feofanov, O. Feofanov, I. Junín, Vestnik MSTU Stankin Vol. **4,** pp. 95-99 (2009)
29. A. Kapitanov, Procedia Engineering Vol. **150,** pp. 832 – 836 (2016)

MATEC Web of Conferences **129**, 04006 (2017) DOI: 10.1051/matecconf/201712904006
ICMTMTE 2017

A new paradigm of industrial system optimization based on the conception "Industry 4.0"

Olga Maksimchuk[1], and *Tatyana* Pershina[1,*]

[1] The department of management and urban development and construction, Volgograd State Technical University, Volgograd, Russia

Abstract. Globalization, urbanization, digitalization, climate change create new trends in improving of the competitiveness of the national economy by acting industrial system optimization in the format of the "Industry 4.0" concept, aimed at integrating of physical operations and related processes of production systems into a single information space. However, this can not be done outside the solution of the problem of increasing the competitiveness of cities - the concentration of production systems. Energy efficiency ensuring is the main feature of production system optimizing of the new generation "Industry 4.0". The aim of the research is to reveal the potential of a new paradigm for production systems optimizing based on the "Industry 4.0" concept in the regions of Russia. The article shows historical experience and features of production system optimization based on the concept of "Industry 4.0" in the regions of Russia; practical aspects of introduction of digital technologies in production systems on the basis of "smart city" concepts aimed at resource and energy saving; the rationale of the advantages of the original local automated system for energy efficiency managing at industrial enterprises which have a high potential for introducing and stimulating of energy saving effects in the optimization of production systems.

1 Introduction

Nowadays, we can observe five global megatrends in the global community: climate change, demographic changes (population aging), globalization, urbanization and digitalization. About 54% of the world's population lives in cities, and by 2030 this index will increase to 66-70%. At the same time, most of the energy resources and polluting substances are formed and consumed in the urban environment.

The search for answers to the question of how to develop and improve competitiveness in modern cities, in order to reduce energy consumption in the conditions of increasing urbanization and at the same time ensure environmentally friendly comfortable living of the population, is implemented by researchers in all corners of the globe [1-7]. And increasingly, as a solution to the above-mentioned problems, it is proposed to introduce elements of an "smart" city, an "smart" region.

2 «Smart» city and «Industry 4.0»

Currently, the implementation of the concept of "smart" city is taking place during the transition to the fourth industrial revolution, called "Industry 4.0".

The first revolution began with the invention of the steam engine by James Watt and created the primary industrialization in Europe throughout the 18th and 19th centuries.

The second revolution occurred at the beginning of the 20th century with Henry Ford's invention of the assembly line, due to this, it was possible not only to create a mass market, but also to make the car available.

The third revolution began in the 1960s, when a computer was invented, and later - industrial robots (Fig. 1).

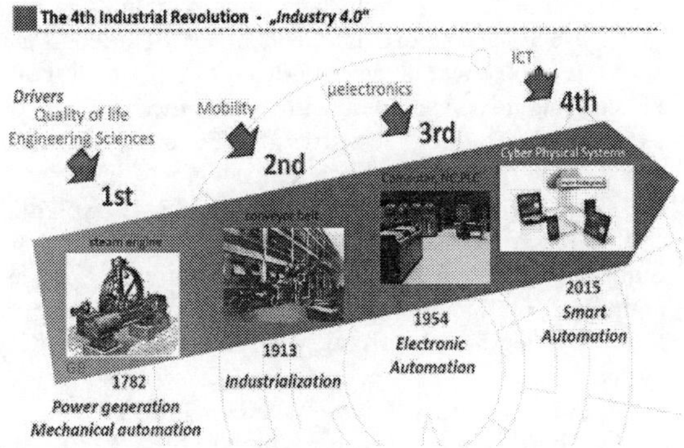

Fig. 1. Types of industrial revolutions.

Analysts at McKinsey & Company suggest "Concept 4.0" as the new business processes that are being introduced in the industry in the 21st century. At the same time, the first concept - is "thrifty" production of the 1970s, the second - the phenomenon of outsourcing production to developing countries in the 1990s, the third - the wide automation of zeroes [8].

The industrial concept "Industry 4.0" is a global, complex, multi-level organizational and technical system based on the integration of physical operations and

* Corresponding author: tatpersh@yandex.ru

 DOI: 10.1051/matecconf/201712904006

related processes into a single information space consisting of 6 subsystems:

1. PLM (Product Lifecycle Management).

Complexes of object-oriented application software, also known as PLM-systems, are intended for structuring of the automation of physical and information process control throughout the entire product life cycle.

2. Big Data .

Big Data is a set of data of traditional and digital sources inside and outside the company. The data is collected everywhere: from sensors, company employees, counterparty companies, customers, service centers, suppliers, etc., after that, the data is structured, analyzed and used. The main task of enterprises when working with Big Data, to interpret the data for further use best way.

3. Smart Factory .

The basis of the Smart Factory concept is the connection of individual stages (operations) of the production process, from the stage of product design and planning of the production resources use to executive mechanisms. The foundation of "Smart Factory" is the concept of "Digital Manufacturing". Digital Manufacturing is an organizational and technical system of adequate production simulation, based on advanced simulators based on CAD. Today, there are several production virtualization systems, such as: DELMIA Digital Manufacturing (Dassault Systèmes); Tecnomatix (Siemens PLM Software); Factory Design Suite (Autodesk); Visual Components and others.

4. Cyber-physical systems.

Cyberphysical system is an organizational and technical concept of information flow management, integration of computing resources into physical production processes.

5. Internet of Things (IoT).

The Internet of things is a simple definition of a grandiose concept. IoT is the concept of a computer network of physical objects (things) equipped with built-in technologies to interact with each other or with the external environment, considering the organization of such networks as a phenomenon which is able to restructure economic and social processes, excluding from the part of actions and operations the need for human participation. IoT connects things that are connected to the Internet with each other and ensures their collaboration, managed by cloud computing systems.

6. Interoperability (functional compability).

Interoperability - the most important factor of the industrial concept Industry 4.0. Without functional compability, it is impossible to create an efficient integrated industrial system "Industry 4.0". High efficiency is achieved by the rational system management of automation of physical production operations and related processes integrated into a single information space.

The Industrial Internet of Things (IIoT) is a key part of the "Industry 4.0" concept and means the interconnected functioning of information and operational technologies in production. This is not just a trend, it is a real way to solve specific tasks, from operational safety to financial indexes. The industrial Internet of things is a B2B direction ("business for business"). This includes M2M solutions (machine-machine interaction), big data, cloud, robotics, etc.

Abroad, one of the directions of the strategic development of the economy is the digitization of all spheres, and mainly of industrial production. Therefore, some countries have come close to the ideal model of Industry 4.0, that is, to digitized and automated production, which, in real time without human participation, is managed by intelligent systems. These technologies are being implemented at enterprises of such global corporations as: General Electric, BASF, Boliden, Honda, Bosch, British Petroleum, Daimler, Goldcorp, Intel, Konecranes, Nestle, Coca-Cola, Osram, Potash Corporation, Pirelli, SKF, SEAT, Siemens, Akzo Nobel, The Anglo Platinum Group and others.

3 The features of the implementation of the "Industry 4.0" concept in Russia

Currently, Russia has a target re-equipment program (2015-2020), which implies a comprehensive modernization of state-owned enterprises, namely, the introduction of innovative technologies, the purchase of new equipment. In connection with the entry into the new phase of industrial development, questions are raised about the effectiveness of new equipment use, the recoupment of the costs and the profit increase of the enterprise are studied. Today, many Russian enterprises are modernizing, which means that there is a chance to equip production with intellectual technologies and make them technological leaders, that will make Russian-made products more competitive in the world market.

In Russia, the technologies of the Internet of things have not acquired a mass character, and the way to Industry 4.0 will not be easy, but examples of implementation in various industries have already existed (Table. 1).

Table 1. Examples of IoT implementation in Russia.

Direction	Company	Tasks
System integrators and IT services	Technocore, Sibintec, Aiteko, Revolta Engineering	- selection of optimal platforms - development of new solutions - implementation of technologies at enterprises
Information security systems	Kaspersky Lab	- security in IoT solutions
IT services and Internet platforms for transport	Transport TV, StarLine, Yandex + KAMAZ	- production of unmanned vehicles - the provision of information and media services on public transport - the organization of a safety system for motor transport - deployment of smart transport infrastructure

IT solutions in energy industry and utilities	Prospective linear technologies, Strizh Telematics	- delivery of measuring equipment and various solutions for the energy and utilities sector
IT solutions for agriculture	Neoflex, Revolta Engenneering	- supply of intelligent systems for optimizing and improving of the efficiency of agriculture
IT systems and devices for the Smart Home	GS Group, Aqua-care	- development of solutions for Smart Home
Electronics	Mikron, Angstrem, T8, T-Platforms	- creation of platform hardware solutions, networks and infrastructure for IoT - supply of microelectronic components and sensors, computers, network and telecommunication equipment
Link	MTS, MegaFon, Beeline, Tele2, Rostelecom	- stimulating the introduction of IoT technologies - development of M2M services

Source: made by authors

Digital technologies in production allow to create a single production cycle, participation in which will take only machines and mechanisms. The transition to digital production is a five-level pyramid.

Business level - EPR

Production level - MES

Technological level - CAM / CAE, PDM

Level of Design-CAD

Monitoring system – MDC

The business level is the top level - it's EPR. On the basis of the EPR system, purchase and production orders are made. The production level is the MES system, which help to create the equipment loading drawing. Technological level - study in CAM / CAE, PDM systems. Level of design - drawings and 3D model are designed in the CAD system. The monitoring system is the last base level on which everything else is built. MDC-systems increase the efficiency of production without significant investment, while solving other problems. These systems operate on the basis of monitoring protocol programs on CNC machines. The information received from the protocol programs is sent to the server, where it subsequently generates reports for individual services and managers. The monitoring system is the starting point of "Industry 4.0". It does not automate production fully, but with the data obtained through monitoring, a single information space of the enterprise is created. Abroad, monitoring is an indispensable object of production. In Russia, the monitoring of production is carried out by several systems: AIS « Dispatcher », developments from «Signum », «Solver », X-Tensive, etc. AIS «Dispatcher » has more than 50 implementations in large plants. Thus, data to analyze the effectiveness of MDC-systems and the first results exist, it remains to reach the mass level.

4 Local automated energy saving management system at industrial enterprises

At present, in the Russian Federation, one of the most important strategic purposes set by the President of the Russian Federation is the reduction by 2020 of the energy intensity of the domestic economy by 40% [9]. On the way to achieve this aim, all organizations (institutions) of business, government, society should be energy efficient. Energy audit as one of the main instruments of the movement towards energy efficiency does not allow to organize permanent monitoring and analysis of energy consumption, it is not free from errors, subjectivism of assessments and corruption schemes in implementation in connection with which alternative methods and tools of energy efficiency management at the local level are required [10] .

The analysis showed that, unfortunately, none of the existing software products do not fully match to requirements that are presented to them. For example, workplace SRO "Energypasport" allows only to fill out the energy passport and does not have any additional functions for auto-settlements, conversion etc. A more successful software product is the «E-Pass » workplace, which is designed to automate the work of energy auditors in conducting and finalizing the results of the energy survey, but it is not suitable for implementing all energy management functions at the enterprise.

Foreign software products such as AMR systems (Automatic Meter Reading), PLC (Power Line Communication), ENSI (Energy Saving International) and their localized Russian versions are also on the Russian market [11-13]. But, work with these products is only available to trained energy auditors, and there are no tools for local energy efficiency management, that is, directly at the level of organizations (institutions) and business enterprises, authorities, society.

To solve this problem, within the concept "Industry 4.0" (a subsystem of the Smart Factory), scientists from the Volgograd State Technical University developed a local automated system for managing energy efficiency at industrial enterprises [14-16].

It is based on the Firebird DBMS, which, firstly, is free for distribution and use, and secondly, is ideally suited for a combination of factors: ease of maintenance / free / speed. DBMS Firebird offers concurrency, high efficiency and powerful language support for stored procedures and triggers. It is a commercially independent project of C and C ++ programmers, technical advisors and supporters of the formation and strengthening of multi-platform relational database management systems, based on the source code released by Inprise Corporation (now known as Borland Software). Now we can say with certainty that FireBird is a developed industrial software that guarantees transactional data integrity when using it with multiple users that conforms to ACID rules:

 DOI: 10.1051/matecconf/201712904006

1.Atomicity — atomicness of transactions;
2.Consistency — data integrity;
3.Isolation — user access control;
4.Durability — long-term storage of data [17].

The energy efficiency management system allows analyzing the modes of operation of complex aggregates and systems, predicting their technical condition depending on specific operating conditions, planning the necessary procedures to reduce production risks, increasing the energy efficiency of fixed assets use, reducing energy production costs in terms of energy efficiency based on accumulated indicators, telemetric information.

The energy management system allows:

1) to automate the process of energy efficiency management in the enterprise;

2) to optimize the cost of energy audit, as compared with the involvement of an external energy auditor, by preventing the likelihood of technical and design errors in the filling of energy passports / declarations;

3) to reduce time spent on data processing and preparation of Energy Saving Programs

4) to fill in energy declarations by experts of budgetary organizations;

5) to raise the qualification level of employees in the process of working with the system.

Using the software allows to make the process of registration of the survey results / filling out the energy declaration more convenient and simple, which in turn allows to reduce costs and increase the efficiency of enterprises of various forms of ownership.

As a result of approbation of the proposed automated energy efficiency management system at the enterprise of the industry, the Program of energy saving measures was developed and successfully implemented.

Table 2. The final Program of energy-saving measures of the enterprise JSC "Volzhsky Abrasive Plant".

Measures	Annual economic effect, thous. rub.	Payback period, years	Cost of measures, thous. rub.
Quickly-paid measures			
Introduction of energy management in the enterprise	525,180	0,75	395,120
Replacement of fluorescent lamps with lamps of the same standard size of less power	5,547	1	3
Translation of boiler houses from liquid fuel to gas and use of energy-saving boilers	716,4	0,3	360
Medium-cost measures			
Installation of a variable-frequency drive at a compressor station	1125,378	2,2	2475,83
Installation of heat flow regulators	1350	1,07	1450
Long-term measures			
Thermal insulation of hot water pipelines	73,5	3,7	270
Reconstruction of autoclaves	944,784	3,45	3262
Creation of the infrared heating system	163,8	4	678,881
Introduction of the automated system of commercial electricity metering	439,59	3,25	1430
TOTAL	5344,179		10324,8

The final Program of energy-saving measures is designed for a long-term period. It is planned to receive profit already for the first year in the amount of 5344.179 thousand rubles with a total cost of 10 324.832 thousand rubles. In the future, it is also planned to receive profit in this amount depending on the payback period of energy-saving measures.

It should be noted that some measures not only save energy costs, but also have a positive impact on the environmental situation, which, if possible, should be reflected in the final indicators.

At the enterprise JSC "Volzhsky Abrasive Plant" according to the forecast before the implementation of the project of the energy efficiency increase and implementation of the Program of energy-saving measures, energy costs in the cost structure by the end of 2017 will amount to 129107 thousand rubles, and according to the forecast after the Program will amount 122522,543 thousand rub. Thus, the reduction of energy costs in the structure of the cost of production and services of JSC "Volzhsky Abrasive Plant" due to the introduction of the proposed measures will amount 5.1%.

The real dynamics of electricity prices, real disposable income of the population may turn out to be somewhat different due to the inevitable error (uncertainty) of the used initial data and some unaccounted factors (including options for reforming the electric power industry and housing and communal services).

5 The analysis of factors hindering the implementation of the concept «Industry 4.0» in domestic enterprises

As shown by the analysis, the introduction of the concept of "Industry 4.0" at industrial enterprises in Russia is very late. This process is inhibited by a number of factors:

1. Economic factors:
- lack of own funds;
- high cost of innovation.

2. Production factors:
- low innovative potential of the company;
- lack of qualified personnel;

- the susceptibility of the company to innovation.

An essential advantage of resolving production factors is that technologies of "Industry 4.0" can be implemented in stages. Individual intellectual elements allow to modernize existing production assets, gradually bringing them closer to the standards of the future. Starting with individual units, it is possible to gradually transform the whole industrial enterprise. To implement the program for the introduction of industrial Internet and the transition to "Industry 4.0", it is necessary to implement retraining of the managerial personnel and senior management. Because for the effective management of enterprises operating on the basis of "Industry 4.0", absolutely new management skills are required in comparison with the management of traditional enterprises. This approach minimizes the amount of time which is necessary for the enterprise to transit to "Industry 4.0".

In addition to the re-qualification of the company's managers, new knowledge and skills will be required from the engineering department. For this, the training programs that are being used now will have to be radically changed, putting training in programming and teaching IT technologies at the equal level with the profile training. For effective work at the enterprise working on the basis of "Industry 4.0", knowledge of programming languages will be required in order to intervene in the workflow of digital production in a timely manner. These requirements will give impetus to the educational sphere, which will entail the training and retraining of existing workforce, on the basis of newly opened centers of excellence. But in addition to the positive moment in the concept of "Industry 4.0", there is also a negative point, that full automation of the technological process in production will entail a massive reduction of workplaces.

It is clear that to implement the "Industry 4.0", the company's own funds are not enough. To do this, it is necessary to resort to borrowed funds. Sberbank provides a loan "Financing of investment projects", aimed at the reconstruction, modernization, existing production. Also an effective tool for financing new production projects of Russian enterprises is the Production Development Fund, as well as the Fund for the Development of Internet Initiatives.

6 Conclusion

The received scientific results have a practical focus and can be applied in the process of developing and implementing of energy conservation policies aimed at the formation and effective implementation of the energy saving potential of industrial enterprises and companies.

References

1. C.A. Jenne, R.K. Cattell, Energy Economics, **5**, 114 (1983)
2. M. Marques, R. Neves-Silva, Journal of Cleaner Production, **88**, 115 (2015)
3. Y. Kim, A. Aravkin, H. Fei, A. Zondervan, and M. Wolf, IBM Journal of Research and Development, **1** (2016)
4. V. Blass, C.J. Corbett, M.A. Delmas and S. Muthulingam, Energy, **65**, 560 (2014)
5. V.U. Baldin, N.I. Danilov, G.I. Khudyakova, WIT Transactions on Ecology and the Environment, **2**, 1208 (2014)
6. G.J. Tsekouras, M.A. Tsaroucha, C.D. Tsirekis, A.D. Salis, E.N. Dialynas and N.D. Hatziargyriou, International Journal of Electrical Power and Energy Systems, **6**, 1222 (2011)
7. S. Rajabi, S. Behairy, Procedia Engineering, 1088 (2016)
8. S. Bouton, D. Cis, L. Mendonca and etc. Annual report McKinsey & Company, (2013)
9. M. Belyaev, O. Maksimchuk, O. and T. Pershina, *Management of energy consumption in the enterprises of housing and communal economy*. Volgograd: VolgGASU.144 (2009)
10. S. Gheorghe, N. Golovanov, H. Albert, C. Stanescu and D. Ilisiu, Proceedings of International Conference on Harmonics and Quality of Power, ICHQP, 430 (2014)
11. A.G. Van Engelen, J.S. Collins, Proceedings of the Annual Hawaii International Conference on System Sciences (2010)
12. A. Al-Mofleh, S. Taib, H. Al-Gulman, Proceeding of the 5th International Symposium on Mechatronics and its Applications (2008)
13. R. Hackl, E. Andersson, S. Harvey, *Chemical Engineering Transactions*, **21**, 301 (2010)
14. O. Maksimchuk, T. Pershina, G. Golikova, N. Borisova, N. and S. Ivashova, *The concept of energy saving management in housing and communal services: a systematic approach.* Volgograd: Crouton.284 (2015).
15. T. A. Pershina, *The system development of the energy expenses management in the enterprises of housing and communal services*. Volgograd.174 (2006)
16. T. A. Pershina, Journal of Volgograd State University, **7**, 133 (2008)
17. H. Borry, FireBird. *Guide Database Developer*. St. Petersburg : BHV-Petersburg.104 (2006)

MATEC Web of Conferences **129**, 04007 (2017) DOI: 10.1051/matecconf/201712904007
ICMTMTE 2017

Functional management of the economic activities of industrial systems

Elena Matveeva[1,*], and *Svetlana* Simagina[2]

[1]Povolzhskiy State University of Telecommunication and Information, Department of Economic and Information Systems, 443010 Samara, Russia
[2]Samara University, Department of Mathematical Methods in the Economy, 443086 Samara, Russia

Abstract. This article refers to the methods of functional management of Industrial systems. In detail a management is examined at intershop level. An effective management is provided by the decision of operative economic and operational business tasks answering the decouplig of productive tasks to the workshops. Their permission is provided by realization of four base functions of management planning, control, adjusting, account. It is shown that qualificatory is a defining feature of planning. The methods of estimation and optimization of the batch, methods of estimation and remapping of the machine loading and workplace capacity, methods of simulation and optimization of the procedures are presented. For the exposure of rejections of productive process from the level set by planning, the tasks of account, control, analysis and adjusting are examined, that allows to warn possible deviations of actual motion of production from the expected model.

1 Introduction

The market demand rapid change and the integration to the World market result in the necessity to produce competitive products – both technically and cost-effectively. This provides the prospect of the development of industrial systems. In market conditions the most effective industrial system should be adaptable to the permanent irregularity of the environment. This leads to the wide range of the product mix.

Convergence to meet the market requirements result in the expansive growth of the quantity of the short-run industrial systems which are managed by specific and complicated way, especially when we talk about major economical activities management.

The efficiency of the whole industrial system depends in tote on the business processes of all the phases and levels of major production management. The economical activities management is the process of specific decision-making on the ground of the stated objective, data collecting and processing, major business processes monitoring, and staff activities coordination [1].

2 Main exposition

The production business object management system is generally divided between the levels of inter-shop and inner-shop planning.

The production program for the shops subject to their specialization is developed at the inter-shop level. Relatively to the assembling the outrunning of the part launchings depends on the part manufacturing cycles, they can be estimated both for the different phases of production and for each part peculiarly [2].

The inner-shop management provides the solving of the operational problems in accordance with the decomposition of the shop production goals. The solving of those problems is provided by means of realization of the four basic management functions

$$\sum_{i=1}^{n}\sum_{j=1}^{n}\left(P_{ij}+K_{ij}+Z_{ij}+Ч_{ij}\right), \quad (1)$$

where P – planning, K – monitoring, Z – regulation, Ч – record keeping, i – processed parts, j – procedures [4, 7].

The planning is the key function in the developed inner-shop management system, where the goal (planning period output) and the goal achievement mode (calendar timetable of the parts moving through workplaces) are specified. The decomposition of the operational problems, informatization and cybernation of basic business processes permit to overstep from process specialization to the higher level – specialization of the object locked sections (OLS). The OLS input data is the detailed plan generated as a result of solving of the operational planning problems at the inter-shop level. This plan assigns the OLS management goal [3].

For the realization of the OLS management functions the following tasks are stated and solved:

1. At the phase of planning. The optimal (providing balanced machine loading) batch sizes are estimated. Basing on that the machine loadings by type of procedure and by each workplace are estimated

*Corresponding author: helen_matveeva@mail.ru

according to the accepted specialization. The simulation of the parts processing is made;

2. At the phase of monitoring. The work shift goal achievement and the parts moving through workplaces are monitored. The interoperation stock volumes are reviewed;

3. At the phase of Regulation. Based on the actual accomplishment of the workplace shift goal and the deviation from the calendar timetable the work shift goals are revised for the next step of regulation;

4. At the phase of Record keeping. The actual accomplishment of the stated plans is identified. The collected data are used for the correction of the next planning period business process management [4, 5].

3 Estimation and optimization of the batch sizes

Estimation of the optimal batch sizes – n_{onm} (estimated monthly) is a double critical task with minimizing of the combined losses from the disequilibrium between the required and the actual supply of the parts for the assembling shops $F(n,p_1)$, and between the costs determined by the laboriousness of the parts processing – $T(n,p_2)$, wherein p_1- the planned batch size $n_{nл}$, p_2 – the batch size minimum limit n_{np}.

The particular efficiency indicators $F(n,p_1)$ and $T(n,p_2)$ determine the minimized function of function

$$\hat{J} = \min_{n=n_{onm}} (F(n, p_1) + T(n, p_2)), \tag{2}$$

where $\hat{J}$ – optimality criteria, and dictate the contradictory requirements to the batch size value.

Losses $F(n,p_1)$ from the disequilibrium (ε) between the required (specified) $n_з$ and the actual supply $n_ф$ $\varepsilon = n_з - n_ф$ – directly depends on the batch size value - $\frac{dF(n, p_1)}{dn} > 0$. And, the smaller batch sizes should be formed based on the minimization of the losses $F(n)$. The laboriousness of the processing $T(n,p_2)$, contrary to the losses $F(n,p_1)$, is in inverse relation to the batch sizes – $\frac{dT(n, p_2)}{dn} < 0$. When n is low the costs of the machine tooling and the continuous correction of the plans increase significantly. And, the bigger batch sizes should be formed based on the minimization of the costs $T(n,p_2)$.

It is stated that:

1. The losses from the disequilibrium $F(n,n_{nл})$ increase drastically when the planned supply is poorer than the required one $n_{nл} < n$;

2. Significant increase of the laboriousness arise from $n_{np} > n$, as a result of the sharp increase of the machine tooling costs (amount of time and other resources) [1, 4].

Formalizing the developed regularities depending parametrically on $n_{nл}$ and n_{np} , correspondingly, by the degree correlations for the sensitivity functions

$$\left.\frac{dF(n,n_{nл})}{dn}\right|_{n_{nл}<n} >> \left.\frac{dF(n,n_{nл})}{dn}\right|_{n_{nл}>n} \tag{3}$$

$$\left.\frac{dT(n,n_{np})}{dn}\right|_{n<n_{np}} << \left.\frac{dT(n,n_{np})}{dn}\right|_{n>n_{np}} \tag{4}$$

and, based on the analysis above, it is determined that the sensitivity of the laboriousness of the parts processing is higher than the losses from the production disequilibrium

$$\left|\frac{dT(n,n_{np})}{dn}\right| > \frac{dF(n,n_{nл})}{dn} \tag{5}$$

The asymptotic solution for the parameterized task of optimization is found as follows $n_{onm} = sup\{n_{nл}, n_{np}\}$ (Fig. 1)

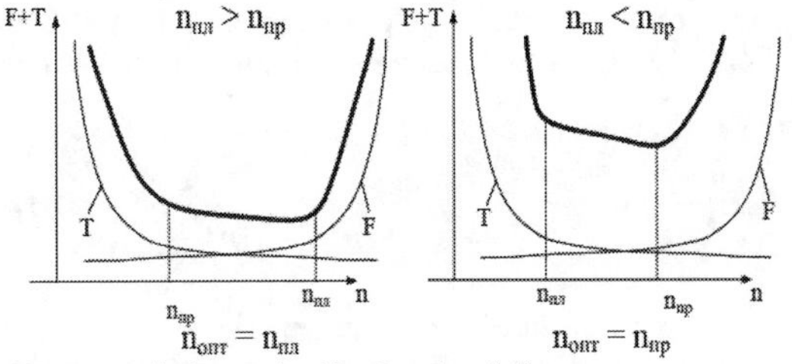

Fig. 1. Batch size optimality functional diagrams

Based on the solved task the batch size optimization is made for each part on each procedure reducing the machine tooling cases number to the minimum, subject to the reasonable workplace capacities and the assembling shop demands for the planned period. The comparison between the time for the machine tooling and the part processing period planned length is showed at Figure 2 for the five types of procedures, for the implemented method of the economical activities management, and for the developed optimization algorithm.

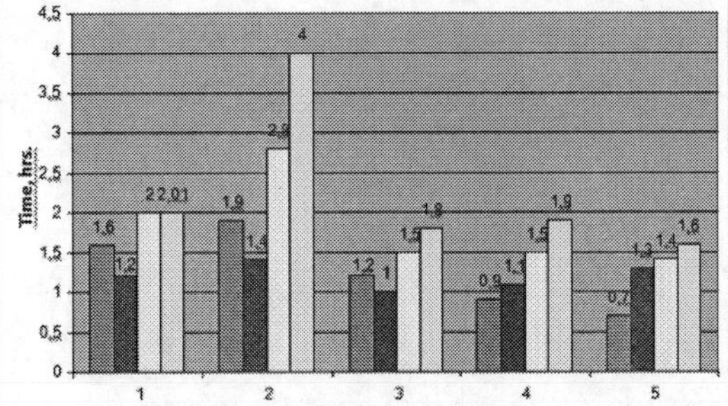

Fig. 2. Comparison of the machine tooling time and the batch processing length diagram

Based on the implementation of the developed algorithm the batches were formed with the processing times exceeding the machine tooling time by 20,1% at

DOI: 10.1051/matecconf/201712904007

the first procedure and increasing up to by 56,1% through the fifth procedure. The reduced batch processing laboriousness went 37% down [1].

4 Estimation and remapping of the machine loading and workplaces capacity

The task of the optimization of the machine loading ($ЗО_{ij}$) is stated as minimization of the production plan tightness (P_{ij}) and maximization of the workplaces (PM_{ij}) when the production mix is numerous, and the production volume is unstable by months: $\left\{\begin{array}{l}(P_{ij})\to\min\\(PM_{ij})\to\max\end{array}\right\}\mapsto ЗО_{ij}$ [1, 5].

The solution of the laboriousness and machine loading factors estimation tasks is made by each type of work for the required production terms – year, month, week, etc. – for the whole enterprise ($BP_{ij}^{З}$), for the shops ($BP_{ij}^{Ц}$), for the production sections ($BP_{ij}^{ПЗУ}$), for the workplaces (BP_{ij}^{PM}). $ЗО_{ij} = \sum_{k}^{12} BP_{ij}$.

$$BP_{ij} = \left\{BP_{ij}^{З}; BP_{ij}^{Ц}; BP_{ij}^{ПЗУ}; BP_{ij}^{PM}\right\}, \tag{6}$$

where k – planned period (mon.).

Analysis of the real enterprises operation indicates the substantial irregularity of the machine loading and workplaces capacity.

The task of the minimization of the irregularity of the machine loading by the type of work and by the workplaces is formalized by the optimality functional

$$\hat{J} = \min_{TO} \sum_{i}\sum_{j} \left|K_{ij}(TO_a) - K_{ij}^{норм}\right|, \tag{7}$$

where K_{ij} – machine loading factors i – type of work, j – workplace; $K_{ij}^{норм}$ – loading standard; TO_a – $a=1,2,3\ldots$ – type of work technological procedures set.

Factors K_{ij} form the quadratic matrix with the size $n \times m$

$$K = \begin{bmatrix} K_{11} & K_{12} & \ldots & K_{1m} \\ K_{21} & K_{22} & \ldots & K_{2m} \\ \ldots & \ldots & \ldots & \ldots \\ K_{n1} & K_{n2} & \ldots & K_{nm} \end{bmatrix}, \tag{8}$$

where n – machine types quantity; m – workplaces number.

The exact solution of the optimization task corresponds the significant complication. For the solution of the task the approximate successive task decomposition is implemented.

For the first step, the machine loading factors are taken as variables for n types of machines corresponding to the average workplaces capacity.

$$K_i = \frac{1}{m}\sum_{j=1}^{m} K_{ij}, \quad i = 1,2\ldots n \tag{9}$$

For the second step, K_i – are estimated. For their determination the optimization task with functional is solved

$$\hat{J} = \min_{TO} \sum_{i=1}^{n} \left|K_i(TO_a) - K_i^{норм}\right| \tag{10}$$

For the third step, the similar task of the minimization of the irregularity of the workplaces with functional is solved for the estimated machine loadings

$$\hat{J} = \min_{TO} \sum_{j=1}^{m} \left|K_j(TO_a) - K_j^{P}\right| \tag{11}$$

The analysis and remapping of the loading is made in several steps

The comparison of the OLS machine loading diagrams is showed at Figure 3 by the types of work corresponding to the existing and the developed planning methods.

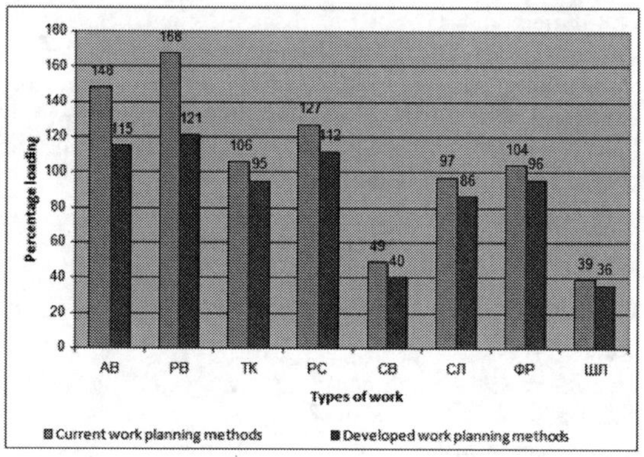

Fig. 3. Machine loading diagrams by the types of work for the different planning methods (AB – automatic machine, PB – capstan lathe, TK – turning lathe, PC – boring lathe, CB – drilling lathe, СЛ – locksmith's, ФР – milling lathe, ШЛ – grinding lathe)

The implementation of the developed work planning method permits to reduce the irregularity of the machine loading by 1,4 times from 33,5% to 24,4%, to eliminate the overload of two technological procedures, and to reduce the overload of the rest of the works by 3,2 times [1, 4].

The comparison of similar actual and estimated capacity diagrams for the different workplaces is showed at Figure 4.

The implementation of the developed planning method reduces the irregularity of the capacity by 2,3 times down to 16%, reduces the average overload of workplaces by 2,5 times from 49,3% to 20%, and reduces the maximum overload by 3 times [1, 4].

The developed methods of reasonable loading by types of work and by workplaces could be implemented for the different calendar and planning periods.

DOI: 10.1051/matecconf/201712904007

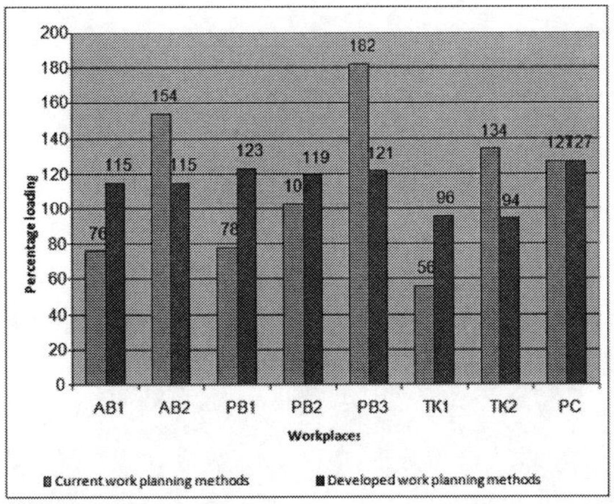

Fig. 4. Workplaces capacity diagram for the different planning methods

5 Simulation of the production procedure

The task of the improvement of planning stipulates the simulation of the parts movement through the workplaces of the OLS with the optimization of the processing order providing the in-time arrival to the assembling and the reasonable machine loading.

The minimization of the processing cycle duration ($T_ц$) is taken as optimization criteria.

$$T_{цj} = \sum_{j=1}^{m} a_j + t_k \cdot n \;\rightarrow \min \qquad (12)$$

where $t_к$ – all parts processing time at the last workplace (min); a_j – drift of the beginning of the last procedure relative to the first one.

The solution of the task is possible with the implementation of the F.I. Paramonov's method based on the simulation of the phases of the parts processing business process. The basement of the forming of part processing simulator is the time of the outrunning of the part processing launchings at the interfacing workplaces. The optimization of the parts processing order is made by the minimum sum of differences between the processing times at the connected workplaces [6].

The solution is the production business process calendar timetable developed based on the equilibrium between the program goals and sizes of the batches launched into production at different shops, and on the coordination of the production terms at the interfacing business process phases [4, 6].

The fragment of the simulation of the parts А, Б, В, Г, Д production process is showed at Figure 5.

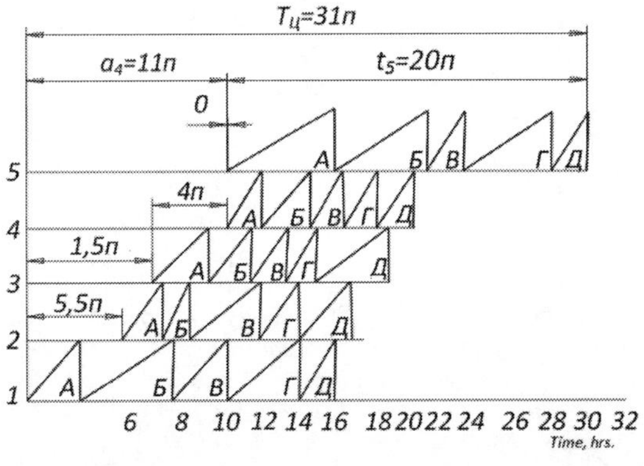

Fig. 5. The fragment of the simulation of the part А, Б, В, Г, Д batches processing.

The time movement of the processed parts is showed for the coupled workplaces 1, 2, 3, 4, 5. The diagram determines the time of the outrunning of the work start at the first workplace of each couple. The time depends on the correlation between the processing times at the coupled workplaces.

The planned documents for the production section and for the workplaces are made as a result of the simulation. They are the production mix assignment, parts processing calendar time schedule by workplaces, machine loading schedule for the month.

6 Record keeping, monitoring, analysis and regulation of the production

For in-time detection of the production process deviations the tasks of record keeping ($Ч_i$), monitoring (K_i), analysis (A_i) and regulation (Z_i) are being solved.

The variety of the working shift goals at each production section and workplace, the variety of the batches, of the procedures, etc. are taken into account for the solution of those tasks.

The production regulation task is the decision-making on the operational elimination of the production process deviations. The following complex of the tasks is being solved at the phase of the regulation Z_i={ $a_{зij}$; r_{ij}; n_{ij}; Δn; $K_{зi}$; $T_{доп}$; T_Σ}: analysis of the dayshift assignment actual achievement for the past step of the regulation ($a_{зij}$), decision-making for the necessity of the including of the unaccomplished items to the next step of the regulation (r_{ij}), Launched batches and additional batches laboriousness estimation (n_{ij}), determination of the times of launching (Δn) next step machine loading estimation ($K_{зi}$), – analysis and correction of the machine loading ($T_{доп}$), working out the dayshift assignments (T_Σ) for the next step of the regulation [7].

The tasks of the record keeping are the collecting of the daily information upon parts availability at the interoperation stock. Based on that the achievement of the dayshift and production mix assignments is analyzed, and the necessity of the correction of the planned

DOI: 10.1051/matecconf/201712904007

assignments for the next step of the regulation is determined.

7 Conclusion

The organization of the works at the OLS should be addressed to the prevention of the possible deviations of the actual production from the estimated simulator, to the maximum possible maintenance of the launching-output times order for each batch by the procedures. This is provided by the following conditions:

1. Strict regulation of the functional links and circulation of documents between the departments of the shop and whole factory;
2. Workplaces planned-prophylactic service system;
3. Specially developed methods of the monetary and moral incentives for the achievement of the dayshift and month assignments [4, 8].

The problem of the improvement of the day-to-day management system should be solved integrally with the implementation of the modern means and methods of the business system, what management efficiency allows to promote by economic activity of the productive systems.

References

1. A.C. Ivanova, V.V. Pirogov, C.N. Polotovskiy and E.A. Matveeva, *Integrated manufacturing and control systems for machine-building enterprises* (Samara, OOO AC-Gard Publ., 2011)
2. A.N. Troshin, *Automatic manufacture control system for machine-building enterprise* (Moscow, Statistika Publ., 1978)
3. J.V. Adayev, *Short-term team production planning* (Penza, Penz. Politekhn. Inst. Publ., 1983)
4. E.A. Matveeva, *Best management practice in job lot production*, Computer and information technologies bulletin, no. **2**, pp. 29-37 (2007)
5. E. Matveeva, S. Simagina, *Manufacturing process Optimization at Enterprises*, Key engineering materials, vol. **684**, pp. 409-413 (2016)
6. F.I. Paramonov, *Manufacturing processes simulation* (Moscow, Mashinostroyeniye Publ., 1984)
7. E.A. Matveeva, *Development of generic integrated enterprise resource planning systems for various branches of industry*, Intelligence, Innovation, Investments. Proc. Int. Forum "Russia as emerging society: economy, culture, management, pp. 30-35. (2011)
8. E.A. Matveeva, *Implementation and economic performance indices model of integrated enterprise resource planning systems*, Samara State Universuty of Economics bulletin, no. **6**(80), pp. 44-50 (2011)

MATEC Web of Conferences **129**, 04008 (2017) DOI: 10.1051/matecconf/201712904008
ICMTMTE 2017

Identification of the technical product characteristics of process scoping studies

Alexander Feofanov[1,*], *Natal'ja* Kim[2], *Evgenij* Frolov[1], and *Tat'jana* Grishina[1]

[1]Moscow State University of «STANKIN», 127055, 1, Vadkovsky per., Moscow, Russian Federation
[2] OOO "OMS Management", 143441,69 km MKAD str.,15, Krasnogorsk district, Moscow region, Russian Federation

Abstract. The analysis of methods for converting customer requirements into optimal technical characteristics of the product was made, a conclusion was made about the applicability of the methodology for the deployment of quality functions. Deficiencies in methodology have been identified and a decision has been made about the need to apply the principles of the method of analyzing hierarchies and expert assessments within the methodology of the deployment of quality functions. Presented organizational-methodical provision of the process for identifying the technical product characteristics for the purpose of optimization of process scoping studies. The technique of identification of technical characteristics of a product is offered. Expert groups were identified in the context of the processes stages of identifying technical characteristics of the product, principles for the formation of priority matrices were recognized. Reviewed the application of techniques based decision support system with three-level data store.

In the conditions of a dynamically changing market and in order to support the state program "Development of Industry and Enhancing Its Competitiveness" in the development and application of advanced industrial technologies, in particular information technologies, most enterprises of the machine-building complex are aimed at developing and expanding the sales market. The goal can be achieved by minimizing the error at the design stage of the product and its pre-project research by identifying those technical characteristics of the product whose target indicators affect the demand of consumers.

Based on the findings, it was decided to optimize the stage of pre-project research by developing a system of models and a technique for identifying technical characteristics.

The most reliable method for converting customer requirements into optimal product specifications is the methodology for the deployment of quality functions [1,2]. The idea of deploying quality functions appeared in Japan in the late 1960s. The first book, "Deploying Quality Functions," was written by Prof. Shigeru Mizuno and Yohji Akao and published in Japan in 1978 [3].

The disadvantage of the method is an unregulated procedure for making interim and final decisions [4,5].

In order to regulate the decision-making processes, the hierarchy analysis method (MAI) was analyzed. MAI is a mathematical tool for a systematic approach to complex decision-making problems. The method was developed by the American mathematician Thomas Saati, his first works about the method were published in American scientific journals since 1979 [6].

The disadvantage of the method is the probability of mismatched estimates in the case of an incorrectly formed expert group [7,8].

The analysis of modern mechanisms for the formation of expert groups made it possible to determine the mechanism for evaluating potential experts with the aim of forming expert groups. Based on the conducted research, applying the principles of MAI, curly RFK matrix can be represented as a graph hierarchy structure (Fig. 1) [9].

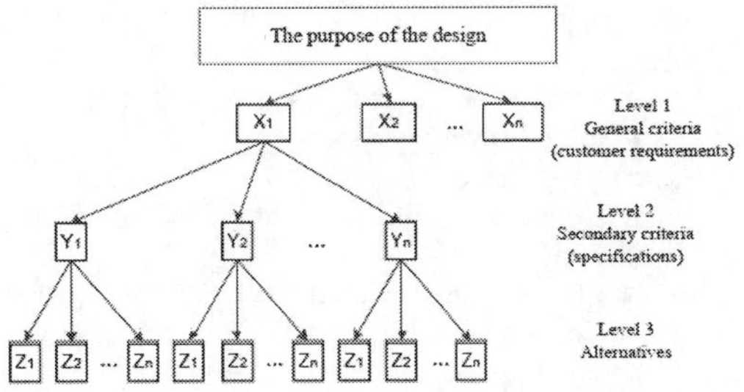

Fig. 1. Interpretation of the figured matrix of the RQF methodology in the form of a graph.

The picture mentions one branch of the graph, showing the relationship of customer X1 requirements with technical specifications and alternatives in achieving the design goals.

The system of models for identification of the product`s technical characteristics at the stage of its design is based on the methodology of the RFK taking into account the principles of the MIA and methods of expert assessments [10].

* Corresponding author: feofanov.fan1@yandex.ru

DOI: 10.1051/matecconf/201712904008

Based on the proposed system of models, a technique for identifying product specifications (MITCI) was developed. To carry out subjective paired comparisons, based on the hierarchy analysis method, for evaluation experts are guided by a scale of relative importance (Table 1) [6].

Table 1. Scale of relative importance

Degree of preference	Definition	Comments
0	Incomparable	The expert finds it difficult to compare
1	Equal importance	Equal contribution of the two types of activities in goal
3	A moderate superiority of one over the other	Experience and judgment give the effortless superiority of one activity over another
5	A significant or strong superiority	Experience and judgment give strong superiority of one type of activity over another
7	A significant advantage	One type of activity is given so strong superiority that it becomes practically significant
9	Very strong superiority	The evidence of the superiority of one type of activity over another is confirmed most strongly
2, 4, 6, 8	Intermediate solutions between two adjacent judgments	Apply in compromise case

A scale of relative importance makes it possible to display a number of qualitative results of comparisons to quantitative indicators. Experts form priority matrices, which they fill with respect to the main goal for each customer requirement, product technical specification and alternative (enterprise competitors) [11].

The opinions of several experts are taken into account of the following ways:

- geometric mean;
- Calculation taking into account the qualification of the expert.

In order to take into account the opinions of several experts having equal weight, it is used the geometric mean calculated by the following formula

$$a_{ij}^{\text{агр}} = \sqrt[n]{a_{ij}^{(1)} \times a_{ij}^{(2)} \times \ldots \times a_{ij}^{(n)}} \quad (1)$$

$a_{ij}^{(k)}$ – evaluation of the element belonging to the i-th row and the j-th column of the matrix of paired comparisons of the k-th expert.

This criterion is applicable in view of the following statement: if two experts with an equilibrium opinion designate the estimates of *a* and *1 / a* when comparing objects, which in calculating the aggregated estimate gives unity and indicates the equivalence of the compared objects. In the event that the opinions of experts are unequal, the experts' judgment is averaged taking into account their qualifications. To determine the weighting coefficients of experts, use the hierarchical structure of the criteria presented in (Fig. 2).

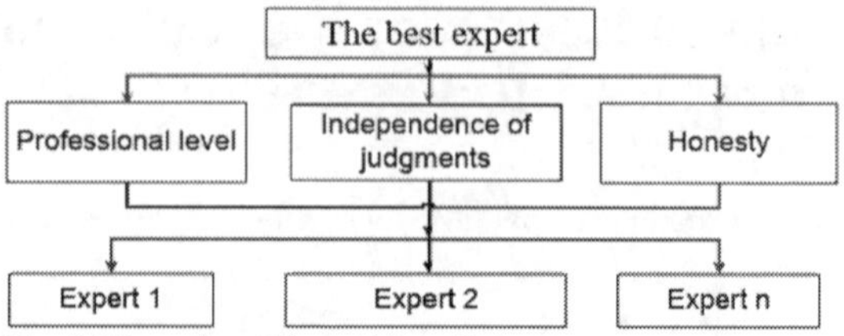

Fig. 2. Hierarchy of the ranking of experts

The calculation of the aggregated estimates in the case involving n experts, having different significance, is calculated as follows:

$$a_{ij}^{\text{агр}} = \sqrt[n]{a_{ij}^{(\alpha_1)} \times a_{ij}^{(\alpha 2)} \times \ldots \times a_{ij}^{(\alpha n)}} \quad (2)$$

$a_{ij}^{(\alpha_k)}$ - the object evaluation, conducted by the k-th expert with weight coefficients α_k.

Herewith $\sum_{k=1}^{n} \alpha_k = 1$

That is, the equality is valid:

$$a_{ij}^{(\alpha_1)} = \alpha_1 a_{ij}^{(1)} \quad (3)$$

Based on the analysis of scientific works on the problem of forming expert groups and assessing the opinions of experts, it can be concluded that the criterion for assessing the competence of experts is not always applicable, because the effectiveness of the expert depends more on the interest in the results of work, in good faith attitude to work than from the values of the competence factors of experts.

Within MITHI it is recommended to distinguish the components of the expert group in accordance with the stages MITHI (Table 2) [12].

Table 2. The expert group in the framework of MITHI

№	Position of the staff members of the expert group	Process
1	Analyst, marketer	Definition of the list of requirements of the customer
2	Engineer-designer, marketer, analyst	Ranking of customer requirements
3	Engineer-designer,	Definition of technical characteristics, assignment of local priorities
4	Engineer-designer, economist, analyst	Assessment of organizational complexity
5	Representatives of top management	The final decision on the appointment of technical characteristics

Consider the application of MITHI based on a Decision Support System (DSS) with a three-tier data

DOI: 10.1051/matecconf/201712904008

repository. In general, the architecture of an automated identification system specifications (ASID) consists of three levels: data Sources, "data Warehouse" and "data showcase ".

The architecture of ASITH at the stage of pre-project research is shown in (Fig. 3).

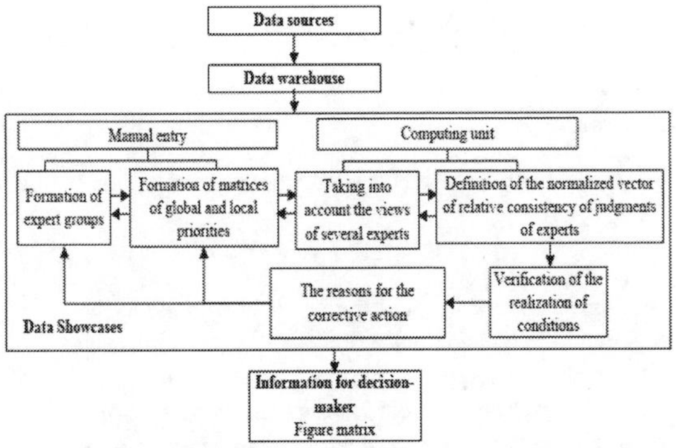

Fig. 3. ASITH architecture at the stage of pre-project research

Block "Data Sources" contains the following information: the external and the internal regulations of the enterprise, structured by types of activities; archival documentation describing the old models of products; database of expert evaluations; results of monitoring of the external environment.

The "Data warehouse" block includes product information structured according to the blocks corresponding to the stages of its life cycle.

The "Data Showcases" block consists of the following modules: the formation of expert groups, the formation of expert solutions, the definition of local and global priorities, the definition of a normal vector and the relative consistency of expert judgments, the causes of inconsistencies and corrective actions.

The "Information for the decision-maker (LPR)" block includes a graphic interpretation of the MITHI that is presented to senior management with the aim of deciding on prioritization of product specifications and the possibility of improving their targets.

"Computing unit", which is providing block, consists of an automated module that includes the following blocks: the system for forming matrices of local and global priorities, the system of determining the index of consistency of experts ' evaluations, the formation of graphical interpretation of MITHI, determining the method of selecting exert groups.

A feature of the proposed technique for identifying technical characteristics of the product is the regulated order of identification of technical characteristics of the product and the adoption of expert decisions [13].

Application of the principles of automation of the pre-project study phase allows to reduce the risk of expert error and the probability of error "human factor" by using a wide range of statistical estimates and contributes to reducing the human, time and financial resources required for the process of identifying the technical characteristics of the product [14,15].

References

1. E. Taver, *Introduction to quality management. Tutorial,* 368 p. (2012).
2. O. Gludkin, N. Gorbunov, A Gurov, Iu Zorin, *Total quality managment: textbook for high schools*, 600 p. (2001)
3. K. Isikava. *Japanese quality management methods*. 215 p. (1988).
4. T.Grishina, *Probabilistic argumentation and decision-making in management of the automated production*, Mechatronics, automation, management, **№1**, pp. 48-52 (2012)
5. T.Grishina, *The validity of decision-making in management of the automated production*, Vestnik MGTU «Stankin», **№ 3 (21)**, pp.70-75 (2012)
6. T. Saati, *Decision making with dependence and feedback* /Translation from English. Scientific edition board A. Andreichikov, O. Andreichikova, 360 p. (2011)
7. R. Khvastunov, A.Feofanov, V. Korneeva, E. Nakhapetian, *Qualimetry in mechanical engineering*, 285 p. (2009)
8. V. Korneeva, N. Negrimovskaia, A. Feofanov, *The calculation of optimal number of experts and volume data identified by results of preliminary survey*, Tekhnologiia mashinostroeniia, **№ 12,** pp. 35-39, (2012).
9. N. Shokhrina, A. Feofanov, *Automation of the design stage of production through the application of the methodology of the development of the quality function*, Tekhnologiia mashinostroeniia, **№5 (155)**, pp. 89-92. (2015)
10. N. Shokhrina, A. Feofanov, *Automated system of determination the priority technical specifications on the designing stage*, RIA «Standarty i kachestvo», **№3 (945)**, P. 108 (2016)
11. A. Feofanov, *Methods of selection of experts to expert groups*, Tekhnologiia mashinostroeniia, **№10**, pp.58-67 (2008)
12. A. Feofanov, Y. Eleneva, T. Grishina, A. Skhirtladze, N. Negrimovskaia, *Participation in organization of production activities of structural divisions of the enterprises of mechanical engineering, textbook for high schools*,144 p. (2014)
13. N. Shokhrina., A. Feofanov, T.Grishina, *The basic of the methodology of the substantiation of technical specifications*, Vestnik MGTU «Stankin», **№ 4 (35)**, pp. 113-116 (2015)
14. V. Mitrofanov, T. Grishina, A. Feofanov, *Management of automated technological systems and simulation efficiency when making decisions*, Tekhnologiia mashinostroeniia, **№8 (158)**, pp. 43-46 (2015)
15. A. Skhirtladze, V. Voronov, V. Boriskin, *Automation of production processes in mechanical engineering: textbook*, 600 p. (2013)

MATEC Web of Conferences **129**, 05001 (2017) DOI: 10.1051/matecconf/201712905001
ICMTMTE 2017

Pneumatic formwork used in strengthening of structural elements during reconstruction of buildings and structures

Susanna Abramyan[1,*], *Vladimir* Polyakov[1], and *Oganes* Oganesyan[1]

[1]Institute of Architecture and Civil Engineering of Volgograd State Technical University, 1 Akademicheskaya str., Volgograd, 400074, Russia

Abstract. The paper discusses the importance of building reconstruction and the need for improving the available formwork systems used for strengthening of structural elements during reconstruction. Citing the results of the analysis of the Russian and foreign scientific publications, the authors emphasize the necessity of replacing costly wooden formworks with high-technology reusable all-purpose and mobile formwork systems. Considering that such requirements are met by rubber-fabric formworks, the authors have developed a pneumatic formwork and showed by way of experiment that the labor intensity of installation and dismantling of such formworks is much lower as compared with conventional wooden formworks, which makes this research original. The paper further describes the design of the suggested formwork and the installation and dismantling techniques. The versatility of the formwork is confirmed by its ability to easily adapt to any structural elements (pier foundations, columns, beams etc.) ensured by the flexibility of rubber-fabric materials. This system can be used for structural elements having various geometric shapes (round, rectangle, trapezoidal) and dimensions. Their transportability is warranted not only by the dimensions and flexibility, but also by a small weight. The use of pneumatic formworks contributes to resolving the problem of conservation such valuable natural material as wood.

1 Introduction

The increasing scopes of reconstruction of buildings and structures all around the world necessitate the invention of fundamentally new approaches to the performance of construction and installation works. This shift is facilitated not only by the use of new high-strength materials, such as composite reinforcement, fiber glass, microcement, composite concrete etc. [1-6], but also the availability of high-technology devices, appliances and fittings, which also defined the problematics considered in this paper. The analysis of the studies carried out by national and foreign scientists [7-11] shows that current reconstruction projects involve replacement of old wooden ceilings with reinforced concrete slabs and making superstructures after preliminarily enhancing the bearing capacity of underlying structural elements (foundations, columns, beams etc.). Thereby, the strengthening of columns alone avails of a plethora of conventional methods, such as: use of reinforced concrete or metallic collars; reinforced concrete collars with confinement reinforcement; vertical bracing; side relieving elements; reinforced concrete jacket; stressed clamps; welding steel angles on main reinforcement; removable relieving housing; reinforced concrete extensions etc. Some of the above methods are also used for strengthening of other structural elements. The choice of a particular method always depends on the condition of a respective structural element and the specifics of the technology used for the performance of the works.

However, with the advent of the recent innovations, the method of reinforced concrete extension has proved to be the most versatile as it may in fact be used for all of the above listed structures. This involves the use of a formwork system or extension of a structure by concrete spraying [12-14].

Whereas the construction using cast-in-situ structural elements relies on various formwork systems, reconstruction of buildings and structures uses mostly small-panel wooden forms, where labor inputs in installation and dismantling dramatically exceed the labor intensity of the main process: reinforcing and (where necessary) concreting. Also, as the experience shows, customized formworks are often have to be used, which is explained by non-identical damages of load-bearing structural elements.

The use of timber formwork is in itself associated with numerous inconveniences: low turnaround capacity, considerable consumption of timber, imperfect assembly and inferior quality of the surface of structures after concreting.

The analysis of certain works [15-19] shows that conventional wooden formworks are recently finding increasing competition on the part of permanent formwork systems made of corrugated sheets, polystyrene foam or chipboard, and large-scale reconstruction projects dealing with thin-wall shells of

* Corresponding author: susannagrant@mail.ru

vault or dome structures opt for pneumatic formwork. These are pneumatic formworks that prove the most sophisticated and highly reusable.

Therefore, the relevance of this research is underlain by the need to introduce new formwork systems for reconstruction and overhaul of buildings and structures, because highly reusable formwork systems used for construction of new buildings are not always suitable for repair and refurbishment purposes.

2 Main text

2.1 Goals and tasks

Assuming that reconstruction is required to be based on a technology-savvy work process, which can be achieved through a reduction in its labor intensity, the goal of this research is to identify the methods for reducing labor inputs associated with the installation and dismantling of formworks during strengthening of certain structural elements using extension technologies, and for ensuring a high turnaround capacity of formwork systems.

The main tasks in the accomplishment of these goals were as follows: engineering of a pneumatic formwork for strengthening certain load-bearing structural elements (individual foundations, columns, beams, belts) of buildings and structures using extension technologies; evaluating the labor intensity of the installation and dismantling of such formworks using experimental methods.

2.2 Pneumatic formwork for strengthening of structural elements during reconstruction of buildings and structures

2.2.1 Design of suggested pneumatic formwork

A pneumatic formwork is a formwork consisting of a shape-forming air-supported flexible membrane.

Let us consider the design features of the suggested pneumatic formwork (fig. 1).

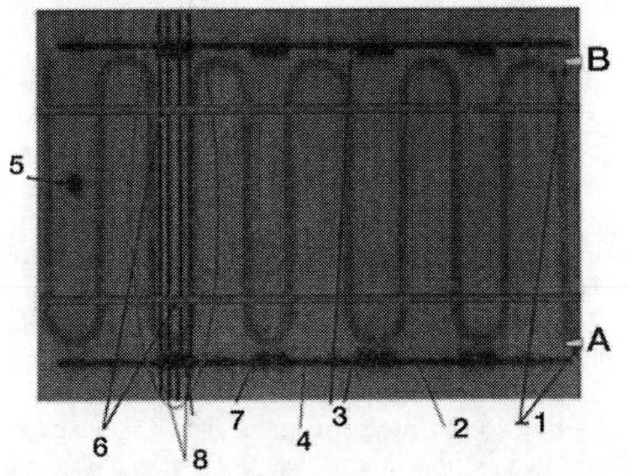

Fig. 1. Design of suggested formwork: 1- air supply valves; 2- internal zigzag channel; 3- lashing straps with locks; 4- belt; 5- injector; 6- additional lashing straps; 7- tubular holes in the belt; 8 - rigid bars

It consists of two layers of rubber-impregnated materials, which are normally used in the manufacturing of air bags, rigidly attached to each other along their edges (other special-purpose or reinforced vinyl textiles are also suitable for this purpose) [20, 21]. The formwork is provided with two valves (1) for air supply. Through the valve *A*, the air is supplied into the internal zigzag channel (2) of the formwork in to ensure it has the necessary rigidness before being finally installed in the design position.

The zigzag channel is formed by stitching two layers of a rubber-impregnated material, with a distance between the stitches of 10-15 mm. In order to facilitate the preliminary fastening of the formwork in the design position, the formwork is tightened with a hook and loop fastener with a width of 100 mm. The air is supplied through the valve *Б* only after the formwork is finally installed in the proper place and attached with lashing straps (3) having ratchet type locks. Lashing straps are fastened with the use of the belt (4). The concrete mixture is supplied through the injector (5).

To provide additional rigidness of the formwork (and consequently to appropriately shape the structure to be strengthened and achieving the necessary geometric dimensions), the formwork is provided with additional lashing straps (6) made of nylon. Before attaching the additional belts, rigid bars (8) are placed in the tubular holes (7) of the belt along the entire short side of the formwork.

If bare main reinforcement is found when strengthening structures, such reinforcement is provided with weld-on additional reinforcement using reinforcement stubs.

To preserve the protective layer and avoid damaging the formwork material, the additional reinforcement is fitted with plastic stops.

2.2.2 Pneumatic formwork installation procedure by the example of strengthening a reinforced concrete column without additional reinforcement

The scope of works associated with the operation of the suggested pneumatic formwork includes: ensuring the initial stability of the formwork before installation by supplying air through the valve *A* (fig. 2); installation of the formwork, i.e. wrapping the column with the pneumatic formwork in such a way as to coordinate the injector with the damaged part of the column (fig. 3); preliminary fastening of the formwork with lashing straps (fig. 4), supplying air into the shell of the formwork through the valve *Б*, final attachment (locking the lashing straps; making the formwork rigid with the use of the rigid bars, attaching the additional belts), supplying concrete through the injector and concreting the column. Before the installation, the surface of the pneumatic formwork which will contact the concrete is covered with emulsion grease. The formwork is dismantled after the concrete reaches at least 70 % design strength. After that, the formwork is cleaned and assembled (fig. 5), i.e. wrapped in a roll for storage or transportation to another destination.

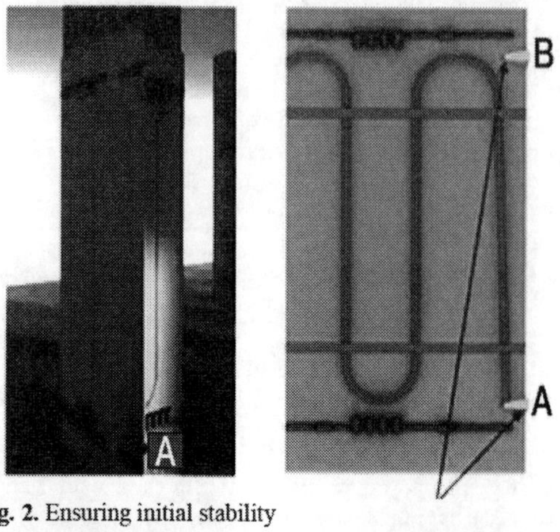

Fig. 2. Ensuring initial stability

Fig. 3. Wrapping the column with the pneumatic formwork.

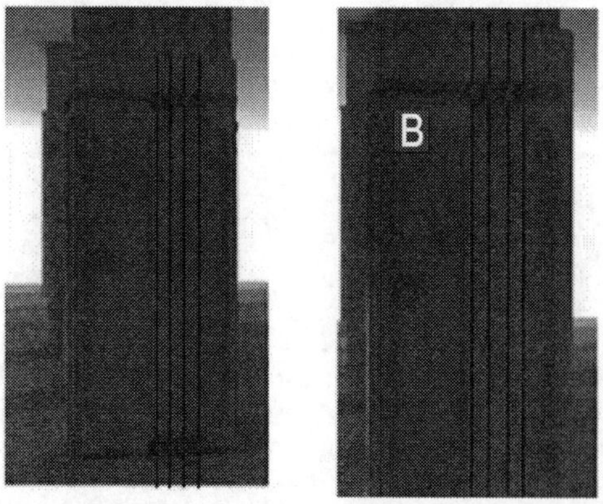

Fig. 4. Preliminary fastening of the formwork with lashing straps, supplying air info the shell of the formwork through the valve B, final locking of the lashing straps.

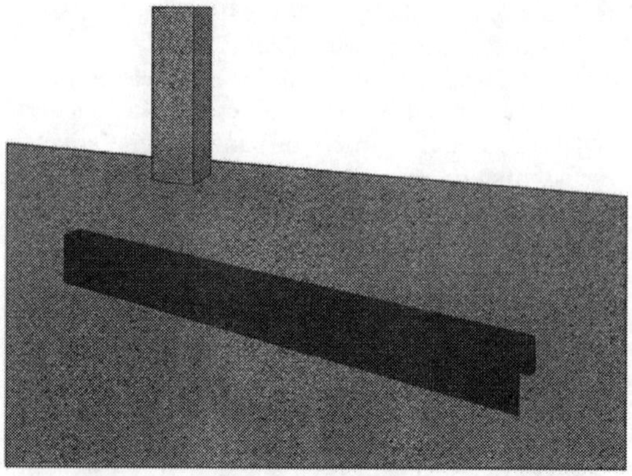

Fig. 5. The assembled pneumatic formwork

3 Experimental part

A model of the pneumatic formwork was built using the available materials (an inflated laminated vinyl mattress with dimensions 2x1.5 meters) in order to determine the labor intensity of its installation and dismantling. The only difference of the suggested structure consisted in that the internal zigzag channels in the formwork were replaced with reinforced 9mm silicon tubes (fig. 6).

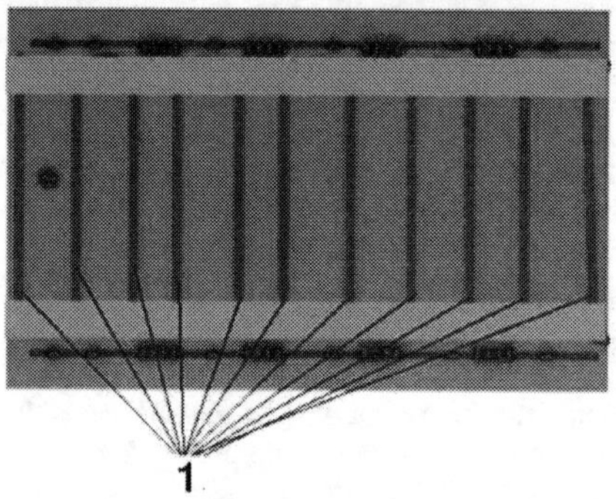

Fig. 6. Internal strengthening of experimental pneumatic formwork with silicon tubes (1).

Such tubes are used in medicine and food industry. The work was performed as follows: marking the place for installation of the formwork; placing the formwork preliminarily fixed with a hook and loop fastener, supplying the air by electric pump; fastening with lashing straps; placing rigid bars in the tubular holes of the belt, fastening with additional lashing straps.

The pneumatic formwork was installed on a column with a cross section 400 x 400 mm by two workers. The thickness of the formwork after air inflation was 3 cm. According to the time study, the labor intensity during the formwork installation was 0.28 man-hours, and the labor intensity during the dismantling was 0.19 man-hours. The comparison of the labor intensity of installation and dismantling with regard to pneumatic and wooden formworks is visualized in fig. 7.

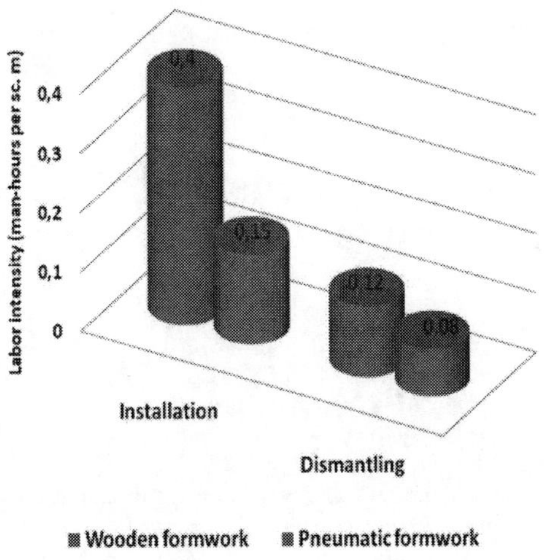

Fig. 7. Labor intensity of installation and dismantling of pneumatic and wooden formworks, man-hours/m^2

The labor intensity of the installation and dismantling of wooden formwork was determined according to the effective regulatory documents.

The experimental studies have shown that the formwork is required to have air supply restrictors. They will allow unifying the use of formworks for structures (such as column, piers etc.) varying in the perimeter shape and configuration.

4 Summary

The research has shown that the use of innovations for improving labor productivity and saving materials and natural resources is one of the key focuses of repair and construction works during reconstruction of buildings and structures.

The theoretical and experimental studies conducted by the authors in writing this paper enabled them to make the following conclusion: the use of pneumatic formworks is namely one of the areas offering potential for reduction of costs while ensuring the appropriate quality of works.

This paper is original in that it suggests, for the first time, to use pneumatic formworks for strengthening of pier foundations, columns and beams as part of reconstruction of buildings and structures.

The research is practically relevant in that it has identified the key advantages of pneumatic formworks for strengthening structural elements, as it is installed and dismantled with lesser labor inputs as compared with wooden formworks, since wooden formworks often lose their stability during dismantling, with prevents their further use. Pneumatic formwork is easy to dismantle and can be fast cleaned of concrete sags. It fits all purposes and can be used for columns of any cross sections varying in the configuration and dimensions. It is transportable because it can be rolled and moved to another place with the least effort. And, what is no less important from the environmental point of view, the use of pneumatic formworks allows saving such valuable natural material as wood.

References

1. D. Lau, QW. Qiu, A. Zhou, CL Chow, Construction and Building Materials, **126**, 573 (2016)
2. HY Shi, WQ Liu, H. Fang, Y. Bai, D. Hui, Composites Part B-Engineering, **108**, 364 (2017)
3. WT Gan, LK GAO, SL Xiao, WB Zhang, XX Zhan, J. Li, Journal of Materials Science, **52**, 3321 (2017)
4. S.G. Abramyan, R.Kh. Ishmametov, Procedia Engineering, **150**, 2133 (2016)
5. M.N.Ibragimov, V.V.Sjomkin, A.V.Shaposhnikov, Academia. Architecture and construction, **4**, 114 (2016) (In Russian)
6. A. A. Shilin, V. A. Pshenichnyj, D. V. Kartuzov, Prospects for Use of Composite Reinforcements, 181 (2007) (In Russian)
7. V. Pukhkal, V. Murgul, M. Garifullin, Procedia Engineering, **117**, 624 (2015)
8. T. Wang, X. Zhong, XF Xu, Fundamental Research in Structural Engineering: Retrospective and Prospective, **1**, 607 (2016)
9. P. Podnebesov, V Teryanik, Procedia Engineering, **150**, 1733 (2016)
10. ZH Bian, SJ Fu, J. Li, Modeling and Computation in Engineering II, 197 (2013)
11. GQ Bi, Applied Mechanics and Materials, **438-439,** 711 (2013)
12. V. E. Rusanov, Bulletine of SibADI, **5 (27)**, 65 (2012) (In Russian)
13. V Babichenko, V Daneliuk, S. Kiriliuk, O. Poddubnyi, Modern Industrial and Civil Construction, **9 (no 3)**, 131 (2013)
14. I. Emelyanova, Bulletine of HNADU, **72**, 107 (2016) (In Russian)
15. R. S. Fedjuk, Bulletine of RUDN, Series: Enig Surveys, **4**, 105 (2013) (In Russian)
16. I. A. Rumyantseva, Bulletine of АГТУ, Series: Maritime Equipment and Technologies, **3**, 112 (2015) (In Russian)
17. E.L. Ayrumyan, I.A. Rumyantseva, PGS **4**, 25 (2007) (In Russian)
18. A.N., Malakhova, A.S. Balakshin, Bulletine of MGSU, **3**, 81 (2013) (In Russian)
19. G.M. Bad'in, S.A. Sychev, Modern Building Construction and Reconstruction Technologies, 288 (2013) (In Russian)
20. D. Veenendaal, M. West, P. Block, Structural Concrete, **12 (3)**, 164 (2011)
21. V.N. Mel'kumov, A.N. Tkachenko, D.A. Kazakov, Scientific Herald of the VGASU, **3**, 7 (2015)

The research of contamination regularities of historical buildings and architectural monuments by methods of computer modeling

Andrey A. Kuzmichev[1,*], *Valeriy N.* Azarov[1], and *Alexander V.* Kuzmichev[1]

[1]Institute of Architecture and Civil Engineering of Volgograd State Technical University (IACE of VSTU), The Department of Life Safety in Civil Engineering and Municipal Economy, Akademicheskaya str., 1, 400074, Volgograd, Russia

Abstract. Due to the active step of urbanization and rapid development of industry the external appearance of buildings and architectural monuments of urban environment from visual ecology position requires special attention. Dust deposition by polluted atmospheric air is one of the key aspects of degradation of the facades of buildings. With the help of modern computer modeling methods it is possible to evaluate the impact of polluted atmospheric air on the external facades of the buildings in order to save them.

1 Introduction

Nowadays due to the active step of urbanization and rapid development of industry, the appearance of historical buildings and architectural and cultural monuments of modern cities requires special attention. Dust deposition by polluted air is one of the key aspects of the deterioration of the facades of buildings. Owing to the modern computer modeling methods, it is possible to evaluate the impact of dusty ambient air on the external facades of buildings in order to save them, because it is a demonstration of architectural, structural and design features of various temporary periods.

2 Current state of the problem of contaminated air influence on the appearance of historical buildings and architectural monuments

Air pollution is associated with natural processes, such as cosmic dust, volcanic activity, dust forms as a result of destruction of rocks etc., also with anthropogenic processes, associated with the development of industry, agriculture and household economy, transport activity etc. [1,10].

The World Health Organization acknowledged the negative impact of polluted atmospheric air on the health of inhabitants of cities, and the most dangerous pollutants for human, in addition to nitrogen oxides, sulfur dioxide, carbon monoxide and ground-level ozone, are particulate matters – PM_{10} and $PM_{2,5}$. Their inclusion in the list of especially dangerous pollutants gave a powerful impetus to the development of a large number of studies on the treatment of these particles in different conditions and for different types of economic activities in order to reduce their content in the atmospheric air.

Based on studies conducted in Italy, the effect of air pollution on cultural heritage of the country was found. It turned out, that about 30% of destructions and damages to monuments and works of art took place because of air pollution [2]. Joint effect of aggressive solid, liquid and gaseous contaminants in the air combined with high humidity of ambient air reduce the endurance term of the different structural pieces and in general all building and facilities [3]. On the external facades of buildings black crusts are formed and dust is deposited due to impact of gases and particles, leading to undesirable aesthetic effects and also compromising the integrity of the materials [4].

3 Visual ecology as a new scientific direction

In the study of contamination regularities of historical buildings and architectural monuments it is necessary to consider the following aspects in total in order to achieve the best results in this field: physical aspect (it considers degradation of the exterior facades of buildings by means of dust, acid deposition, etc.), as well as visual aspect, covering visual perception of the environment [1].

Visual ecology is a science which studies and explains origins and mechanisms from the standpoint of modern psychophysiological science and gives theoretical basis for development of ecological principles of creation of material objects, which would respond the standards of a comfortable visual perception of the human environment [5]. The essence of the comfortable visual environment concept can be explained by the theory of «saccade automation» position in the example of the natural environment – a forest [6]. The term «saccade», introduced in the usage by Russian scientist V.A. Filin, represents quick eye movements aimed to change an observed object. It emphasizes the need of

* Corresponding author: andrew_9207@mail.ru

 DOI: 10.1051/matecconf/201712905002

constantly updating the attentional objects [7]. In a forest landscape the number of elements for an eye analysis is uncountable, including visual dominants (single tree, open views, water, etc.). Upon the infinite transition from one element to another element it is capable for our eyes to have a rest [6].

4 The experimental part

4.1 Experimental Installation Description and the Methods of Research

In order to reduce the level of aggressiveness of urban visual environment by maintaining the external appearance of historical buildings and monuments of architecture and culture, we conduct the researches allowing to study in total physical and visual aspects of contamination of building constructions. Data were obtained and processed for the period of 2014 - 2016. It was solved to create two sampling sites: №1 - in Central district of Volgograd (Russia) in close proximity to the main highway of the city; №2 - in Krasnoslobodsky district (Volgograd region, Russia) away from busy roads (habitat). In order to study visual aspect of contamination of building constructions, the sampling sites were painted in the colours of buildings of the Stalin Era architecture (Fig. 1). Therefore, owing to the visual analysis of the sampling sites we can make conclusions on the extent of degradation of the external facades (dust, discoloration etc.) [5, 8, 9].

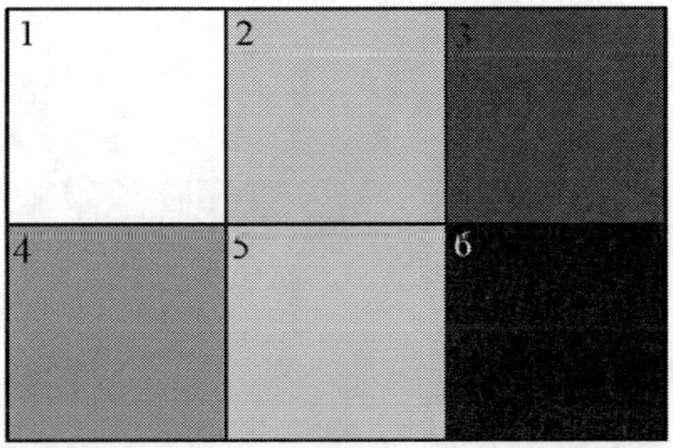

Fig. 1. Model of the sampling site: 1-white, 2-beige, 3-vinous, 4-grey, 5-yellow, 6-black.

The microscopic analysis of the dust samples were produced by using a scanning electron microscope Versa 3D Dual Beam. The chemical (elemental) composition of the samples were studied by scanning transmission electron microscopy (STEM). The dispersion analysis of the samples were produced by microscopic studies of micrographs using the computer software Image J (Fig. 2) [24]. According to the results, integral and differential curves, histograms of numerical distribution of particle's diameter to the probabilistic-logarithmic coordinate system are modeled, where $D(d_p)$,% - a share of a total mass of particles, which consists of particles with a diameter less than d [14, 22, 23].

Further in the computer software STATISTICA [25] the values of Pearson's criterion and Kolmogorov's criterion were obtained. The significant level of 5% was adopted (Fig. 3). The observing values of the criterions should be less than the theoretical values for submission to the logarithmic-normal distribution [16, 17, 18].

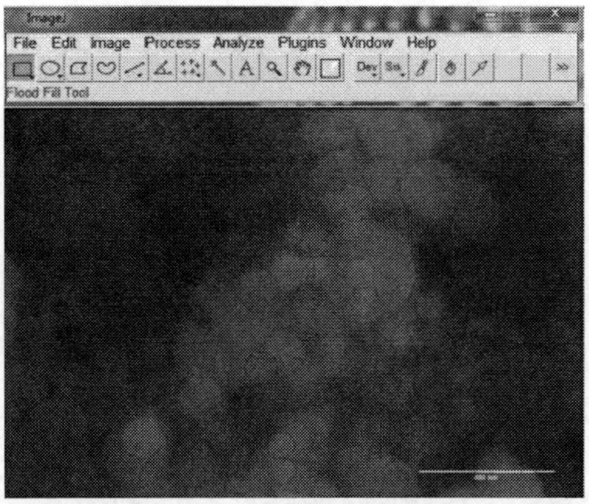

Fig. 2. Micrograph of the dust sample in the computer software Image J.

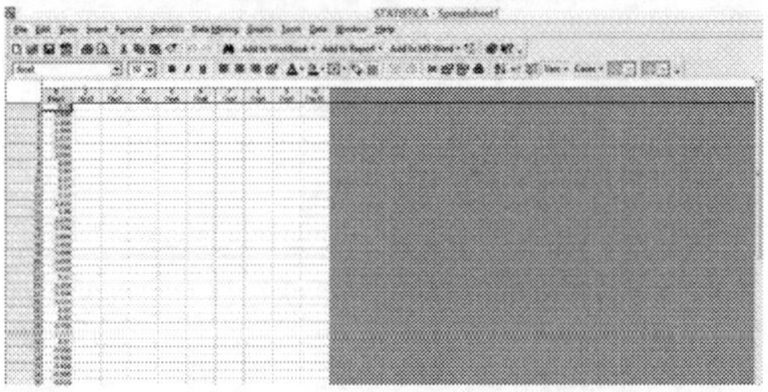

Fig. 3. The computer software STATISTICA.

4.2 Processing of the results

The results of the experimental studies of the dust samples were processed in selected urban environment and habitat for the period of 2014-2016. The results of the samples taken in the autumn 2015 on the site № 1 (Fig. 4) and in the autumn 2016 (Fig.5) were compared.

In the Fig. 4 the histogram of numerical distribution of the dust particles, selected in the autumn 2015 on the sampling site №1, is submitted. The amount of particles was 100 units. The size of the measured particles is in range from 0.79 μm to 57.232 μm. The predominant sizes of dust particles are 2.5-5 μm. The observed values of the criterions are less than the theoretical values (Table 1).

In the Fig. 5 the histogram of numerical distribution of the dust particles, selected in the autumn 2016 on the sampling site №2 is submitted. The amount of particles was 100 units. The size of the measured particles is ranged from 1.249 μm to 25.178 μm. The predominant sizes of dust particles are 8-10 μm. The observed values of the criterions are less than the theoretical values (Table 2).

DOI: 10.1051/matecconf/201712905002

The highest concentration of fine particles is observed during the summer period in urban environment (Fig. 6). The size of the measured particles is in range from 0.559 μm to 25.983 μm. The predominant sizes of dust particles are 2-4 μm. The observed values of the criterions are less than the theoretical values (Table 3).

Based on the results the distributions are close to the logarithmic-normal. At the same time the size of particles in the habitat is lager than in the urban environment [19, 20, 21].

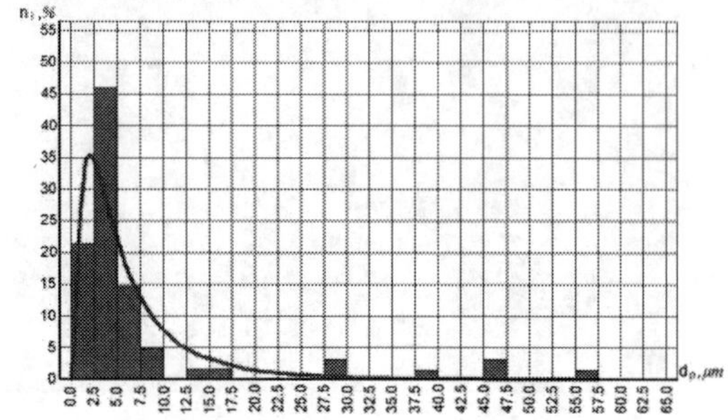

Fig. 4. The histogram of numerical distribution of dust particles on the sampling site №1 (autumn 2015).

Table 1. Values of the criterions according to the data on the site №1 (autumn 2015).

Pearson's Criterion χ2		Kolmogorov's Criterion	
Observed Value	Theoretical value	Observed Value	Theoretical value
13.00845	67.328	0.11476	0.13403

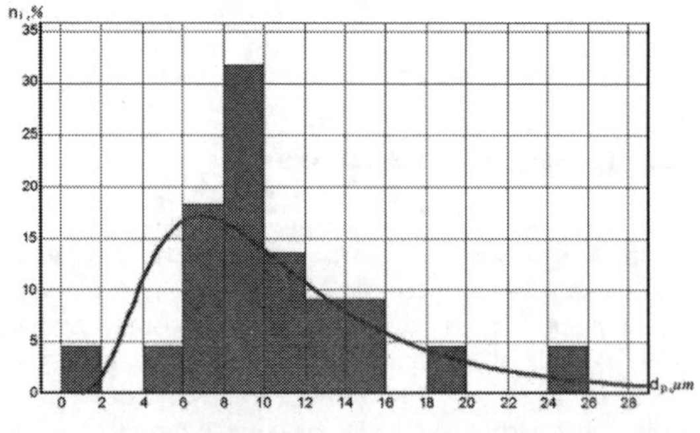

Fig. 5. The histogram of numerical distribution of dust particles on the sampling site №2 (autumn 2016).

Table 2. Values of the criterions according to the data on the site №2 (autumn 2016).

Pearson's Criterion χ2		Kolmogorov's Criterion	
Observed Value	Theoretical value	Observed Value	Theoretical value
11.31524	67.328	0.11763	0.13403

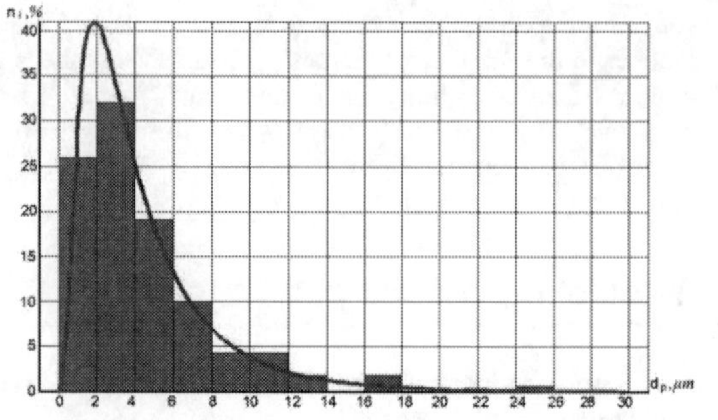

Fig. 6. The histogram of numerical distribution of dust particles on the sampling site №1 (summer 2016).

Table 3. Values of the criterions according to the data on the site №1 (autumn 2016).

Pearson's Criterion χ2		Kolmogorov's Criterion	
Observed Value	Theoretical value	Observed Value	Theoretical value
1.69971	67.328	0.01773	0.13403

For an assessment of regularities of distribution of particulate composition of the dust samples, integral curves of mass distribution on equivalent diameters of particles in probabilistic-logarithmic grid were considered, where D (d_p),% - a share of a total mass of particles, which consists of particles with a diameter less than d [11, 12, 13, 15].

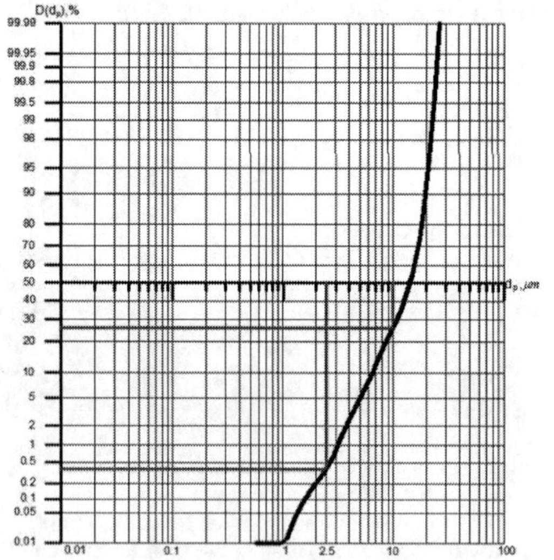

Fig. 7. Integral curve of mass distribution of dust particles on diameters on the sampling site №1 (summer 2016)

The integral curve (Fig. 7) shows, that Kolmogorov's law is observed only on certain areas, it can be represented by a two-unit spline. In general one can speak about a truncated logarithmic-normal distribution.

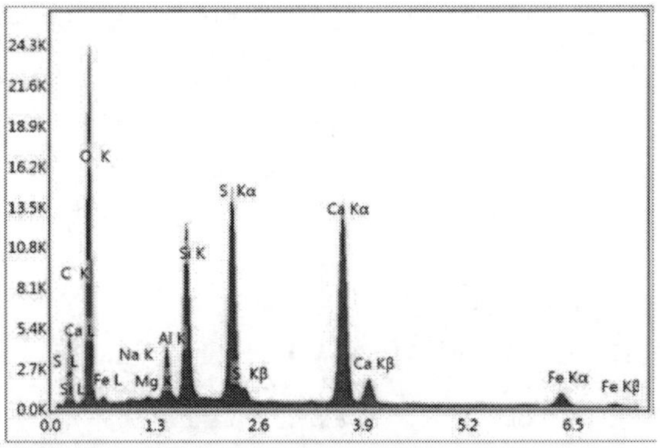

Fig. 8. Chemical (elemental) composition on the sampling site №1 (autumn 2016)

The analysis of chemical (elemental) composition of the samples, taken in the autumn 2016 on the site № 1 (urban environment) and on the site 2 № (habitat) shows, that the main components of the samples are the following substances: O, Si, Ca, S, Ti, was less significant content of components, such as Al, K, Na, Mg, Fe. There is similar elemental composition on both sampling sites. Concentration on the site № 1 is higher because it is located in the urban environment, except for the concentration of Si, which is higher on the site № 2 (habitat).

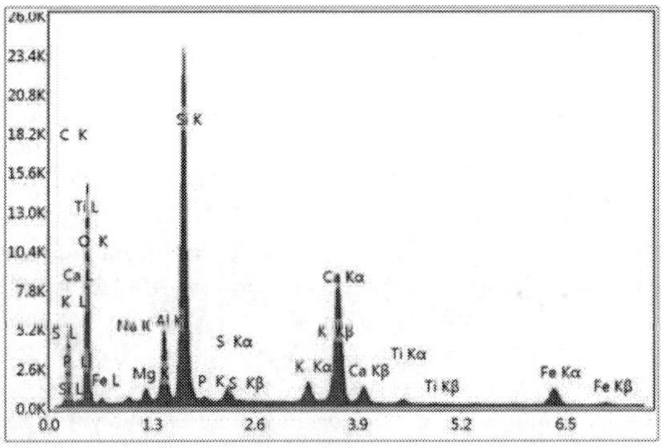

Fig. 9. Chemical (elemental) composition on the sampling site №2 (autumn 2016)

Besides there is a new type of particles-iberulites on the dust sample taken on the site №1 in the autumn 2016 (Fig. 10). This is a type of spherical aerosol particle with a vortex, made of a complex mineral assemblage that also contains biological remains. The most likely source areas based on inherent mineral components are the Sahara and Sahel, while other minerals are the results of atmospheric neoformation. The appearance of iberulites is a discontinuous phenomenon, occurring mainly during summer period and in dry periods corresponding to the highest total suspended particle contents. Thus, a correlation between amounts of dust and iberulites can be established; indeed, often their presence coincides with red dust rains and periods with no heavy rains [4].

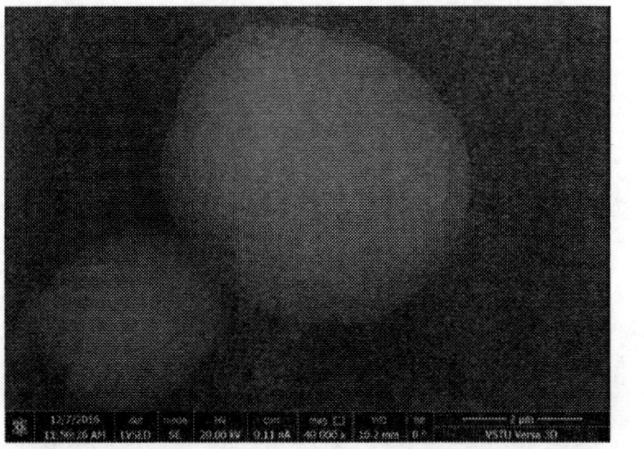

Fig. 10. Micrograph of the dust sample on the site № 1 (autumn 2016)

5 Conclusion

The need to preserve architectural heritage of historical and cultural value is getting more significant. At the same time it is very important to develop environment dynamically to sustain life activities in general. As modern life high rate does not allow an individual permanently or most of the time to be in a natural environment, the primary goal of visual ecology is to adapt urban environment to a comfortable perception by human. Due to the implementation of methods described in the article for the safety of building constructions and architectural monuments can be achieved the best results in this field of town planning [1].

References

1. A.A. Kuzmichev, V.F. Loboyko, *Impact of polluted air on the appearance of buildings and architectural monuments in the area of town planning*, Procedia Engineering **150**, pp. 2095-210 (2016)
2. V.N. Azarov, A.A. Kuzmichev, *Pollution of building constructions and architectural manuments as one of the aspects of visual ecology*, Sotsiologiya Goroda [Sociology of City], no.**2**, pp. 76-86 (2015)
3. V. Kucera, S. Fitz, *Direct and indirect air pollution effects on materials including cultural monuments.* Water, air and soil pollution, Vol.**51**, №**1**, pp.153-154 (1995)
4. V. Kontozova-Deutsch, C. Cardell, M. Urosevic et al., *Characterization of indoor and outdoor atmospheric pollutants impacting architectural monuments: the case of San Jerónimo Monastery (Granada, Spain)*. Environ Earth Sci. **63**:, pp.1433-1445 (2011)
5. V.N. Azarov, A.A. Kuzmichev, *The totality of physical and visual aspects when studying contamination of building structures and monuments of architecture*, Sotsiologiya Goroda [Sociology of City], no.**3**, pp. 28-42 (2016)
6. A.V. Gorodkov, S.I. Saltanova, *Ekologiya vizual'noi sredy : uchebnoe posobie dlya vuzov. [Ecology of visual environment: work-book for higher education*

institutions]. 2nd edition (Saint-Petersburg. Lan' Publ., 2013)

7. A.A. Kuzmichev, *On the problems of visual ecology of the urban environment,* Problemy okhrany proizvodstvennoy i okruzhayushey sredy: sb. materialov i nauch. trudov inzhenerov-ekologov [Problems of protection of working space and the environment. Set of proceedings and scientific works of the environmental engineers]. Volgograd, PrinTerra-Design Publ., iss. **6**. 103 p. (2015)
8. A.A. Kuzmichev, V.N. Azarov, *The study of impact of air pollution on external appearance and perception of building constructions and architectural monuments*, Biosfernaya sovmestimost: chelovek, region, tekhnologii [Biosphere compability: human, region, technologies], **1(13),** pp. 86-96 (2016)
9. Nazim Y. Orudjev, Mikhail B. Lempert, Ivan Osaulenko, Natalia A. Salnikova, Andrew A. Kuzmichev, Alla G. Kravets, *Computer — Based visual analysis of ecology influence on human mental health,* 2016 7th International Conference on Information, Intelligence, Systems & Applications (IISA), IEEE Conference Publications, pp. 1-6 (2016)
10. V.N. Azarov, E.V. Gorshkov, R.M. Sarkisov, *Construction sector of economy and atmospheric air in industrial cities*, Sotsiologiya Goroda [Sociology of City], no.**4**, pp. 71-78 (2012)
11. Azarov V. N., Tertishnikov I. V., Marinin N. A. [Rating of PM10 and PM2.5 as Social Standards of Life Quality in Areas of Building Industry Enterprises Location]. Zhilishchnoe stroitel'stvo [Housing Construction], 2012, no. 3, pp. 20—23
12. V.N. Azarov, I.V. Tertishnikov, Ye.A. Kalyuzhina, N.A. Marinin, *About concentration estimation of fine dust (PM10 u PM2,5) in air,* Vestnik Volgogradskogo gosudarstvennogo arhitekturno-stroitel'nogo universiteta. Seriya: Stroitel'stvo i arhitektura [Bulletin of Volgograd State University of Architecture and Civil Engineering. Series: Civil Engineering and Architecture], iss. **25(44)**, pp. 402—1061 (2011)
13. V.N. Azarov, N.A. Marinin, D.V. Jogoleva, *About a Concentration Estimation Disperse a Dust (PM10 and PM2,5) In Atmosphere of Cities,* Izvestiya Yugo-Zapadnogo gosudarstvennogo universiteta [Proceedings of the South-West state university], **5(38),** part 2, pp. 144—149 (2011)
14. V.N. Azarov, V. Yu. Yurk"yan, N.M. Sergina, A.V. Kovaleva, *Methodology of the microscopic analysis of dust disperse composition with the use of a personal computer (PC)*, Zakonodatel'naya i prikladnaya metrologiya [Legal and applied metrology], no. **1**, pp. 46—48 (2004)
15. V.N. Azarov, N.A. Marinin, I.V. Burba, *Estimation of PM[10] and PM [2,5] dust concentration with the help of the bisection method.* Ekologicheskaya geologiya: teoriya, praktika i regional'nye problemy: materialy vtoroi mezhdunarodnoi nauch.- praktich. konf., g. Voronezh, 4—6 okt. 2011 g. [Ecological geology: theory, practice and regional problems. Proc. of 2 Int. sci.-pract. conf., Voronezh, 4—6 October, 2011]. Voronezh, "KOMPIR" Center of documentation, pp. 285—287 (2011)
16. V. Azarov, N. Barikaeva, *Monitoring of Fine Particulate Air Pollution as a Factor in Urban Planning Decisions*, Procedia Engineering **150,** pp. 2001–2007 (2016)
17. V. Azarov, M. Trokhimchyk, O. Sidelnikova, *Research of Dust Content in the Earthworks Working Area*, Procedia Engineering **150,** pp. 2008–2012 (2016)
18. V.N. Azarov, D. P. Borovkov, A. M. Redhwan. *Application of swirling flows in aspiration systems,* International Review of Mechanical Engineering, vol. **8**, iss. **4**, pp. 750-753 (2014)
19. V.N. Azarov, D.V. Lukanin, D.P. Borovkov, A.M. Redhwan, *Experimental study of secondary swirling flow influence on flows structure at separation chamber inlet of dust collector with counter swirling flows*, International Review of Mechanical Engineering, vol. **8**, iss. **5**, pp. 851-856 (2014)
20. V.N. Azarov, I. N. Logachev, K. I. Logachev et al., *Methods of reducing the power requirements of ventilation systems Part 4. The oretical prerequisites with swirling air flows.* Refractories and industrial Ceramies, vol. **55**, iss. **4,** pp. 365-370 (2014)
21. V.N. Azarov, N. V. Menzelitseva et al., *Main Trends of Conditions Normalizing at Cement Manufacturing Plants*, International Review of Civil Engineering, vol. **6**, no. **6**, pp. 145-150 (2015)
22. Natalia A. Salnikova, Boris A. Lempert, and Mikhail B. Lempert, *Integration of Methods to Quantify the Quality of Medical Care in the Automated Processing Systems of Medical and Economic Information.* In: Communications in Computer and Information Science, vol. **535**, CIT&DS 2015, Volgograd, Russia, pp. 307-319 (2015)
23. Valery A. Kamaev, Ilya P. Mikhnev, Natalia A.Salnikova, *Natural Radionuclides as a Source of Background Irradiation Affecting People Inside Buildings*, Procedia Engineering, vol. **150**, pp. 1663-1672 (2016)
24. V.N. Azarov, N.M. Sergina, *Metodika mikroskopicheskogo analiza dispersnogo sostava pyli s primeneniem personal'nogo komp'yutera (PK).* Dep. v VINITI. № 1332- 80002. 15.07.2002 [*Methodology of the microscopic analysis of dust disperse composition with the use of a personal computer (PC).* Depp. in VINITI. no. 1332-80002. 15 July, 2002].
25. Bureeva N.N. Mnogomerniy statisticheskiy analiz s ispolzovaniem PPP "STATISTICA" [Multivariate statistical analysis using computer software "STATISTICA"]. Nizhny Novgorod, 2007, 112p.

Automation of production of concrete mix

Egor Popello [1,*], and *Victoria* Gurieva[1]

[1]Orenburg State University, Orenburg, Russia

Abstract. Computer-aided design of concrete leads to a reduction in terms of production, the exclusion of product deficiencies, improve the quality of manufactured products. This approach allows to produce the concrete to exact physical and mechanical characteristics, which makes the design more reliable and economical. The software package will allow: to reduce the settlement time in the design of concrete mixture, to improve the efficiency of the staff of the laboratory building, to improve the quality of the concrete mix due to higher accuracy of calculations, to apply a flexible approach to the design of concrete mixture in question of introducing new chemical additives and their characteristics.

1 Introduction

Modern construction is unthinkable without concrete mixtures. They are used in the construction of many man-made structures beginning from building foundations finishing with bridges of high complexity. The minimum possible time and quality of execution are the main criteria that apply to the production of concrete [1-8]. Time for execution order for concrete production depends on many factors such as the level of the process, qualification of personnel and automation of equipment. Apart from the characteristics of the raw materials, the quality of the product depends on all the above factors. Today the need for a concrete with different physical and mechanical properties increase. At the same time, it imposed more stringent requirements on the quality of concrete. In one factory the task of increasing the volume of the concrete issue is solved in the following ways: increasing in production areas, parallelization and extension of technology; performance improvement of existing processes and service of their equipment. But the task of raising the quality of the concrete is the main criterion for the entire production [9-12]. In one factory performance improvement concrete mixing due to the automation of measuring the parameters of the initial components, the automation of the calculation of recipes, raw material supply and preparation of the mixture, dispensing the finished product and its quality indicators.

The technological process of preparation of the concrete mixture with the calculation of the composition and laboratory tests are about 30 days [13-15]. In the works of Doctor of Science prof. Malinina LA, prof. Mikhailov VV, Mironov SA, Frenkel IM, Bazhenov YM and Gorchakov GI noted that the selection of the composition of the concrete mix is one of the most important operations of technological process of production reinforced concrete structures. As a result, the relationship between design components concrete mixture is determined that ensures the strength of concrete in the structure, the mobility of the concrete mix and the concrete efficiency, estimated from the minimum flow rate of the cement. In the adjustment process carried out preliminary tests in order to obtain accurate depending on the properties of concrete, concrete mix of water-cement ratio and other factors. Carrying out these tests, the processing of the results carried out by mathematical methods factorial experiment planning [16]. Thus, it is necessary to automate the design of concrete.

There is a number of difficulties in the process of automation of the design composition of concrete mix. The main difficulty is that the characteristics of the components of the concrete mix and the concrete, which correspond to a satisfactory quality, are in certain ranges of values. The result is that during the installation relationships between parameters concrete, concrete quality prediction have a problem, a promising solution that consists in the use of artificial intelligence in the process.

The object of research is the technological process of manufacturing of concrete for heavy concrete.

The subject of research is the process of computer-aided design composition of concrete mixtures.

The aim of this work is reduction the complexity, time designing of concrete mixture and decreasing of defect in the design process for a given quality of concrete.

Based on the goal, the following tasks are:

- To analyze the planning of automation system for the design of the experiment the composition of the concrete mix (CM) and to analyze the system of selection of the composition of concrete mixtures based on use of the composition of chemical additives.

- To develop a simulation model of an automated system (AS) CM on the basis of neural networks (NN).

In the process of solving the problems were used the theory of neural networks, statistical methods, experimental design and factorial analysis. Experimental studies were carried out on the basis of production of LLC "Strojgarant" (Orenburg). Mathematical models of implementation and optimization procedures performed on a computer using a software application. Development

* Corresponding author: peplos@inbox.ru

 DOI: 10.1051/matecconf/201712905003

of the AS concrete mixture and planning trials using neural networks was to develop an algorithm and program of the automated designing composition of concrete mixtures.

The general scheme of the process of designing the composition of the concrete mixture is shown in Fig. 1.

The diagram shows the input (left) and output (right) variables. On function entry: P1 - characteristics of the aggregates, P2 - desired characteristics of the concrete mix and of concrete, P3 - characteristic chemical additives. At the exit - mass performance component of ready-mixed concrete, where C is cement, S is sand, CrS is crushed stone, W is water, ChA is chemical additives.

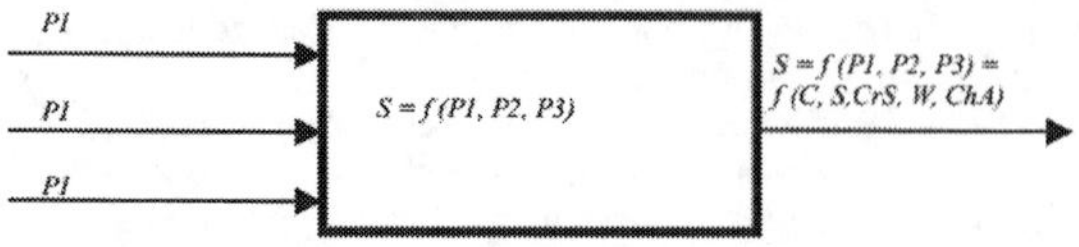

Fig. 1. Scheme of design process of concrete mixture.

Based on neural network, our schemes and techniques as described above, the calculation algorithm settlement and experimental method for composition of heavy concrete aggregates has been developed (Fig. 2).

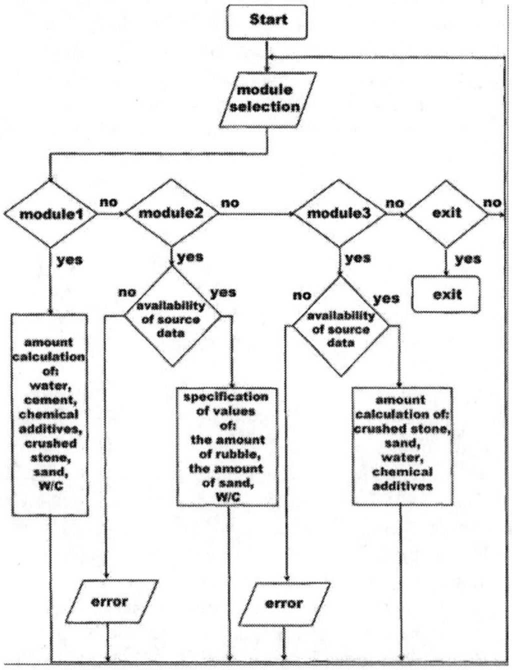

Fig. 2. The structure of the quantitative composition calculation algorithm, the CM.

Three modules stand out in the algorithm. The first module is a theoretical design, the second module is a composition correction, the third module is the preparation of the composition to the transfer of production. Since it is necessary to provide an adjustment of the results and the flexibility of the system, it is used modularity algorithm. The algorithm is shown in Fig. 2. At the entrance of module 1: data aggregates (P1), requirements for the design result (P2) and the data on additives (P3) Approximate value of the required water-cement ratio (W / C) is calculated on the basis of the above data. The amount of water for mixing the concrete mixture needed to obtain the desired workability is determined by the maximum size of crushed stone (M_cr, mm) and the required mobility (OK, cm), or stiffness (ST, c). Consumption of cement is determined by the water-cement ratio and the amount of water. Next, the calculation is based on the fact that the sum of the absolute volumes that make up the material is equal to 1 $м^3$ tightly laid concrete mix, then the absolute volume of the cement paste and the absolute volume of the mixture of aggregates, the mass of each of the aggregates (sand and gravel) and chemical additives (or additives) of 1 $м^3$ determined. Some of the data which were obtained as a result of the module 1, we apply the algorithm module 2 in the adjustment process W/C by changing the amount of water for mixing with the mixture. Adjustment of concrete that's just theoretically designed in Module 1, along with the changes that are obtained in module 2, we provide the module 3 in order to transfer it into production. This is done by making changes in the quantitative composition of the aggregates with regard to their humidity (W,%) and conversion to a dry weight of additives which are liquid structure. As a result, the module 3 provides a ready structure of production, which is subjected to laboratory testing for the correct design and modification.

Output function of the final composition of the concrete mix, which is represented in Fig. 1, according to the algorithm and regulatory requirements [13-15], is calculated by the following formula:

$$S = f\ (P_1, P_2, P_3) \tag{1}$$

each of the input functions P_i determined as:

$$P_1 = f(R_C, \mathrm{M}_{CR}^{CS}, p_{SC}, p_S, W_S, W_{CS}, \frac{S}{CS}) \tag{2}$$

$$P_2 = f\ (R_{\text{C.requir.}} \frac{W}{C} \text{OK}), \tag{3}$$

$$P_3 = f\ (\text{К}_A, p_A), \tag{4}$$

R_C - cement Activity (kg / cm²) ; M_{CR}^{CS} - size gravel unit ; p_{CS} и p_S - density of crushed stone and sand, respectively ; W_S и W_{CS} - sand and gravel humidity (%) ; S/SC - the required ratio between sand and gravel by weight ; W/C - the desired ratio of water to cement weight ; OK - the desired value of the mobility of the concrete mix after it was made (cm) ; $R_{\text{C.requir.}}$ – the value of the strength of concrete at 28 days after the production of concrete mix (kg/cm^2) ; К_A - amount of dry matter additive volume (% of the solution) ; p_A- additives density.

Function parameters P_1 и P_3 are measured by instruments. The required ratio of sand and gravel, water cement relation and mobility (stiffness) of the mixture are taken as parameters to a function P_2. Further discussed in more detail the method of designing a heavy concrete structure with the help of computational and experimental method.

 DOI: 10.1051/matecconf/201712905003

$$W/C = f\left(R_C, R_{C.requir.}\right), \quad (5)$$

The amount of water for mixing 1 m^3 of mixture is defined as:

- if the mixture is mobile:

$$W = f\ (M, M_{CR}^{CS}), \quad (6)$$

- if the mixture is stiff:

$$W = f\ (R, M_{CR}^{CS}), \quad (7)$$

whereP is the mobility (OK) of the mixture, cm;

ST is the rigidity of the mixture, c.

Then we calculated the amount of cement for 1 $м^3$:

$$C = f\ (W/C, W), \quad (8)$$

Calculate the absolute volume of cement paste (l):

$$V_T = f\ (C, A, W, p_{CS}, p_S), \quad (9)$$

The absolute volume of the mixture of aggregates (sand and gravel) (l):

$$V_P = f\ (V_T\ ) \quad (10)$$

Then determines the content (mass) of aggregates (sand and gravel):

$$P = f\ (V_P, p_P) \quad (11)$$

Including rubble:

$$CS = f\ (P, r) \quad (12)$$

where r - the specified weight ratio of sand and gravel, which is determined by the formula:

$$r = f(S, CS) = S/CS \quad (13)$$

The amount of sand is determined as the difference between the mass of aggregates (P) and a mass of rubble (CS):

$$S = f(P, CS) \quad (14)$$

Then recalculate the amount of aggregates takes place in accordance with their humidity, properties and composition of the chemical additives is transferred to the production [17-18].

Currently, a significant number of automated systems offered on the market of software products, they are designed to automate a variety of activities, processes, companies, organizations. A review of existing analogues has been taken to review the program developed by the All-Russian Scientific Research Institute of Physico and Radio Engineering Measurements. The program has a number of advantages such as ease of use, reliability and accuracy of the calculations performed, the high speed of computing.

But in the process of program testing the following weaknesses have been identified in the actual production conditions: firstly, the possibility of introducing the complex of chemical additives in the process of designing the composition of the concrete mix is not considered, secondly, there is no possibility of the water-cement ratio mixture correction; third, it does not counted the humidity of concrete components (sand, gravel), fourth, denied access to the software code. The lack of capability data, the program can not be used effectively in production. the cause of the development of its own software was the inability to integrate functions into the overall project due to the closure of the code. Therefore, the creation of the application is a necessary condition for improving the efficiency of production. A result of analyzing analog disadvantages the standards and recommendations of the experts, the author has developed a software application (SA) for the design the composition and parameters of "MCM" concrete mixture ("Multiple Concrete mixture»). The purpose of the program is the reduction of time and increase the efficiency of the design process and, as a consequence, increase the quality of concrete in its industrial production in the construction industry enterprises.

The system should operate on IBM compatible personal computers. You need a processor with a clock speed of 1 GHz or higher RAM 512 MB of free space on at least 20 MB of hard disk, keyboard, mouse. Just requires the following software: the operating system is not lower than Microsoft Windows 7, Microsoft .NET Framework 4.0, SQLServer [19-24].

Each of the three modules of SA calls user interface which represents the end user dialog. This entry is required for further calculations. The appearance of the user interface is shown in Fig. 3.

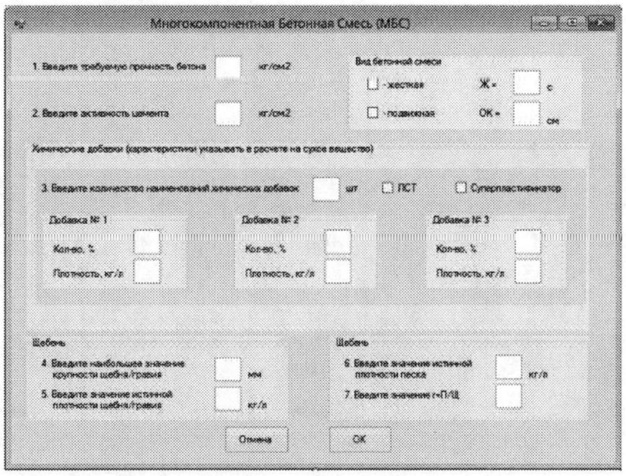

Fig. 3. The main program window.

2 Insights

Thus, the efficiency of the developed program is as follows:

- time reduction calculations for designing concrete mixture;
- increase employee productivity construction laboratories;
- improving the quality of the concrete mix due to higher accuracy of calculations;
- a more flexible approach to the process of designing the composition of the concrete mix in the issue of the

introduction of new chemical additives and their characteristics.

References

1. V.G. Mikulski, *Construction Materials* (Textbook.- M.: ACB, 2004)
2. K.N. Popov, M. V. Caddo, *Construction materials and products* (Tutorial. M.: High school, 2002)
3. S.S. Kaprielov, A. V. Sheynfeld, V. G. Batrakov Concrete and reinforced concrete, *Complex concrete modifier brand MB-01*, **5**, 38, (1997)
4. V.I. Kalashnikov, Building materials, *Scientific Terminology of the new generation of concretes*, **3**, 103, (2011)
5. S.S. Gordon, Mechanization construction, *Production of cold-resistant concrete structures*, **7**, 22, (2008)
6. J.A. Karasev, *Improving the efficiency of cement dispersion systems with water in a metastable state* (Author. dissertation of the candidate of technical sciences, 2008)
7. L.I. Dvorkin, O.L. Dvorkin, *Fundamentals of Concrete* (OOO "Story-Concrete", 200)
8. W. Ken, *Concrete mix design, quality control and specification* (Academic Press, MA, 2010)
9. V.A. Nevsky, *Construction Materials: textbook / under the general editorship* (D: Phoenix, 2007)
10. V.S. Izotov, Y. Sokolova, *Chemical additives for concrete modifications* (monograph M.: Kazan State University of Architecture and Civil Engineering "Paleotypes" Publisher, 2006)
11. I.L. Chulkova, Herald SibADI, *Automated calculation of heavy concrete composition and prediction of its properties,* **6**, 28, (2012)
12. L.I. Kastornov, Additives to concrete and mortar (Phoenix, 2005)
13. SP 130.13330.2011, *Manufacture of precast concrete products and* (Gosstroy SSSR Moscow: USSR State Committee TSITP, 1988)
14. SNIP 82-02-95, *Federal (typical) Element of cement consumption in the manufacture of concrete and concrete products and structures* (Gosstroy Russia.- M.: FSUE CPP, 1996)
15. GOST 310.1-76 - GOST 310.3-76, GOST 310.4-81, GOST 310.5-80, GOST 310.6-85, *Cements. Test methods* (M. Publisher standards, 1985)
16. Y.M. Bazhenov, *Methods for determining the composition of different types of concrete*, (Tutorial Manual for high schools. M. Stroyizdat, 1975)
17. *Guidance on the use of chemical additives in concrete* (NIIZhB Gosstroy SSSR. M: Stroyizdat, 1980)
18. *Concrete and concrete products* (Materials for concrete production. M.: Publishing Standards, 1980)
19. E.C. Mamdani, C. Assilian, *An Experiment in Linguistic Synthesys with a Fuzzy Logic Controller*, Intl. Man-machine Studies **7**, 22, (1975)
20. B. Steinauer, T. Kathmann, G. Mayer, T. Becher, Heft, *Einsatzkriterien für Betonschutzwände, Berichte der Bundesanstalt für Straßenwesen*, **5**, 112 (2004)
21. O. Yagashita, O. Itoh, M. Sugeno, Ed. Amsterdam, *Application of fuzzy reasoning to the water purification process, in Industrial Applications of Fuzzy Control*,19, (1985)
22. G.G. Kulikov, T.V. Breykin, V.Y. Arkov, *Intelligent information systems* (Textbook. Benefit / Ufimsk. state. aviation. tehn., 1999)
23. P.A. Gudkov, *Systems of artificial intelligence* (Method. instructions, 2007)
24. O. Herman, J.O. Herman, *Artificial Intelligence* (textbook, Belarusian National Technical University, 2013)

DOI: 10.1051/matecconf/201712905004

Simulation and estimation of parameters in reconfigurable multipoint forming processes of plates in the creep mode

Konstantin Bormotin[1,*], *Sergey* Belykh[1] , and *Vin* Aung[1]

[1] Komsomolsk-na-Amure State Technical University, 681013, Komsomolsk-on-Amur, Russia

Abstract. Mathematical models and numerical methods for the optimal solution of inverse forming problems monolithic panels with contact stamp devices have been developed. Using the minimization criterion for damage in functionals of inverse problems, the optimal laws of deformation in creep are found. The algorithms implemented in MSC.Marc allow to model and evaluate the parameters of the technological processes for the manufacture of parts, in particular reconfigurable multi-point form-creation.

1 Introduction

Lately in the aircraft industry, large monolithic and monolithic prefabricated panels are used more and more as parts, which allow increasing strength and stiffness of structures with simultaneous reduction of masses and improvement of aerodynamic performances. Monolithic panels utilize reinforced membranes fabricated in conjunction with the stiffening ribs of one plate or sheet [1]. Setting the aerodynamic shape of such monolithic panels is performed by shaping processes undertaken in plastic deformation modes, both under normal and increased temperatures.

The development of metal forming processes is basically either through the creation of highly efficient technologies based on optimization of regimes, or the development of new technological processes. Establishment of the most effective technological process of deformation leads to the problem of choosing optimal technological parameters. In any case, the most important criteria necessary for assessing the effectiveness of technological processes and their optimal management of technology parameters are determined for the solution, the objective function characterizing the process and being a function of various factors is introduced. In a number of works, a set of problems was solved to optimize the parameters of the force loading and deformation of thin sheet billets under pulsed methods of shaping, in particular, the multiparametric problem of optimal shape formation was posed and solved [2].

The process of deformation in the creep regime allows for economically and efficiently forming large parts. The use of shaping such a regime avoids many manual operations by mechanical molding.

The deformation-strength properties of the creep of structural aluminum alloys depend significantly on the state of supply, the thickness of the slab, and the direction of cutting the blanks from the plate (slab) for making samples and the sign of the applied load during the tests [3]. It is experimentally shown in works [4-10] that forming in the creep regime and at stresses not exceeding the elastic strength of material, provides for higher residual strength resource and less damage of components compared to the method of conventional cold forging. The results confirm the prospects of creep using in the manufacturing processes of critical structures parts. This is especially important for modern civil and transport aircraft, since the airworthiness standards are continually being increased and are currently based on residual strength estimates and damage tolerance of the basic design elements [11, 12].

Such processes, in addition, will allow controlling the level of material damage, coordinated with technological limitations, due to the optimal choice of the strain path in time.

Development of technical decisions in ways of shaping does not guarantee high accuracy of final geometrical shape of the part. For example, lately shaping of products from sheets and panels is considered as a promising solution, using configurable core punch (die), which allows changing the shape of workpiece. Accuracy of the part shape obtained by pressure processing technologies of materials under specified process parameters depends on accuracy of calculated and manufactured tooling shape (stamps, punch, rod configurable punch) that determines proactive panel shape. Proactive panel shape shall provide specified residual curvature of the panel after its release from the power tooling. In this regard, inverse problem rises: to determine the shape of tooling, which creates such proactive curvature of the panel that provides specified residual shape of the panel after unloading. In addition, in order to minimize damage to parts and deviations from specified shape under existing limitations in production, formulation of process parameters optimal control problems and development of methods of their solution are required.

* Corresponding author: cvmi@knastu.ru

DOI: 10.1051/matecconf/201712905004

It's shown in works [13, 14] that panel formation in the creep regime is limited by deformation and shows a large value of material spring-back. To account the spring-back and to obtain the required curvature of the panel it is proposed to use multi-point forming (MPF) in the creep regime [15].

The transition of forming matrixes and punches from rigid to flexible structures, enabling changes to their shape under the required geometry have opened broad possibilities for the application of reconfigurable tooling system in the variety of technological processes. The forming surface of the punch as well as the matrix formed by two coaxially arranged rods systems, each placed in an individual position by the numerical control, allows for the adaptation of the tooling system for manufacturing parts of various configurations. The mechanisms of rods positioning can be mechanical, hydraulic or combined. There are also differences in the configuration of the rods, their tips and the quality of formed surface of the punch (the matrix).

Most of the research exploring the possibility of using the reconfigurable rod devices is aimed to develop the technology of work-piece manufacturing from thin sheet material. In works [16-19] design features of rod systems are considered and the main approaches to definition of the loads acting on each rod element are proposed. Results of searches for the most rational form of rod element are also presented.

Results of complex research for the prediction of defect occurrence of multi-point forming of sheet material by utilizing finite-element analyses (FEM) are shown in works [20-22]. The obtained data allowed for the optimization of the control system for the punch and matrix rods and gave the opportunity to manufacture the parts using double curvature from the sheet material with account of the effects of material spring-back.

2 Formulation and method for solving inverse optimal forming problems

Let $V \subset R^3$ be a bounded domain with a sufficiently regular boundary *S*. The contact surface of rigid bodies (stamps) with deformable body is designated through S_c ($S_c \subset S$). Denote by $u = (u_1, u_2, u_3)$, $\tilde{u} = (\tilde{u}_1, \tilde{u}_2, \tilde{u}_3)$, $\bar{u} = (\bar{u}_1, \bar{u}_2, \bar{u}_3)$ - the vectors of current, residual displacements of deformable body and the vectors displacements contact surface of rigid bodies; $u, \tilde{u} \in [W_2^1(Q)]^3$, $Q = V \times \{0 \le t \le T\}$, $\bar{u} \in [W_2^1(Q_c)]^3$, $Q_c = S_c \times \{0 \le t \le T\}$. The symbol $(\cdot,\cdot)_S$ denotes the inner product in $L_2(S)$: $(u,v)_S = \int_S \sum_{i=1}^{3} u_i v_i dS$. The corresponding norm is given by $\| u \|_S = \sqrt{(u,u)_S} = \left(\int_S \sum_{i=1}^{3} {u_i}^2 dS \right)^{1/2}$.

The inverse problem of a kinematic forming by contact rigid stamps can be represented in the form of a quasi-static variational principle with the functional [23]:

$$J(\dot{\bar{u}}, \dot{u}, \dot{\tilde{u}}) = \frac{1}{2\varepsilon_1} \| \dot{\bar{u}} - \dot{\bar{u}}^* \|_{S_c}{}^2 + W_c + a(\dot{u}, \dot{u}) +$$

$$+ a(\dot{\tilde{u}}, \dot{\tilde{u}}) + \frac{1}{2\varepsilon_2} \| \dot{\tilde{u}} - \dot{\tilde{u}}^* \|_S{}^2, \ \varepsilon_1 > 0, \ \varepsilon_1 \to 0, \quad (1)$$

$$\varepsilon_2 > 0, \ \varepsilon_2 \to 0,$$

where $\dot{\tilde{u}}^*$, $\dot{\bar{u}}^*$ is a given residual displacement rate and current contact displacement rate in moment time *t*; $t \in [0,T]$ – time deformation under load; W_c - the quasistatic contact potentials received by imposing of contact conditions on the equations of bodies motion by method of multipliers of Lagrange or by method of penal functions [24] and differentiation on *t*; potential form are given by:

$$a(\dot{u}, \dot{v}) = \int_V \frac{\partial W(\dot{\varepsilon}_{ij}(\dot{u}))}{\partial \dot{\varepsilon}_{ij}(\dot{u})} \dot{\varepsilon}_{ij}(\dot{v}) dV,$$

$$a(\dot{\tilde{u}}, \dot{\tilde{v}}) = \int_V \left(\frac{\partial W(\dot{\tilde{\varepsilon}}_{ij}(\dot{\tilde{u}}))}{\partial \dot{\tilde{\varepsilon}}_{ij}(\dot{\tilde{u}})} \right) \dot{\tilde{\varepsilon}}_{ij}(\dot{\tilde{v}}) dV,$$

$$W(\dot{\varepsilon}_{ij}) = \frac{1}{2} c_{ijkl} \dot{\varepsilon}_{ij} \dot{\varepsilon}_{kl} - c_{ijkl} \dot{\varepsilon}_{ij} \dot{\varepsilon}_{kl}^c,$$

$W(\dot{\tilde{\varepsilon}}_{ij}) = \frac{1}{2} c_{ijkl} \dot{\tilde{\varepsilon}}_{ij} \dot{\tilde{\varepsilon}}_{kl} - c_{ijkl} \dot{\tilde{\varepsilon}}_{ij} \dot{\varepsilon}_{kl}^c$, c_{ijkl} are the components of the symmetric elastic constant tensor; $\dot{\varepsilon}_{kl}^c$ are the creep strain rates; $i, j, k, l = 1, 2, 3$; $\dot{\varepsilon}_{ij} = \frac{1}{2}(\dot{u}_{i,j} + \dot{u}_{j,i})$, $\dot{\tilde{\varepsilon}}_{ij} = \frac{1}{2}(\dot{\tilde{u}}_{i,j} + \dot{\tilde{u}}_{j,i})$.

By consideration of the inverse problem with functional in the form of (1) it is supposed that in the considered time *t* there is a contact of rigid stamp with deformable body, otherwise these contact potentials shouldn't enter functional.

The problem of optimal deformation is formulated as follows: it is required to find a way to deform an element of the medium for a given time *T*, so that at the time *t*=*T*, the specified creep strains $\dot{\varepsilon}_{kl}^{c*}$ are obtained and the damage parameter Ω is minimal.

Minimizing the dissipation power for each time point will give the minimum value of the dissipated work $A = \int_0^T \sigma_{ij} \dot{\varepsilon}_{ij}^c dt$ and accordingly the damage parameter [25, 26].

Using the weighting coefficients method, the problem of optimal deformation by the functional [25, 26]

$$J(\dot{\bar{u}}, \dot{u}, \dot{\tilde{u}}) = \frac{1}{2\varepsilon_1} \| \dot{\bar{u}} - \dot{\bar{u}}^* \|_{S_c}{}^2 + W_c + a(\dot{u}, \dot{u}) + \quad (2)$$

$$+ c_1 \{ a(\dot{\tilde{u}}, \dot{\tilde{u}}) + \frac{1}{2\varepsilon_2} \| \dot{\tilde{u}} - \dot{\tilde{u}}^* \|_S{}^2 \} + c_2 \int_V \sigma_{ij} \dot{\varepsilon}_{ij}^c dV,$$

where $c_1 > 0$, $c_2 > 0$ - weighting coefficients.

Thus, in the case of small deflections, the optimal deformation is linear $u_z = \frac{t}{T} u_z^*$, in the case of large deflections with the assumptions made above, optimal deformation is carried out by the nonlinear law $u_z = \sqrt{\frac{t}{T} u_z^*}$ [25].

In the inverse problem of uniaxial stretching of a rod in the creep theory with minimal damage, the dependence of the rational loading path on the weight coefficients and time in the following form [26]:

$$p = \left[\frac{EB(1+c_1)}{c_2} + \left(p^{*(1-n)} - \frac{EB(1+c_1)}{c_2} \right) e^{c_2(1-n)t} \right]^{\frac{1}{1-n}},$$

where $p = p(t)$ - tensile forces applied to the rod, *E, B, n* – material constants и p^* - initial load. A similar expression is constructed for displacements.

It is proposed in [27] to determine the optimal loading path by formulas

$$u_z(t, x_1, x_2) = \frac{f(t)}{f(T)} u_z^*(x_1, x_2), \quad (3)$$

$$f(t) = \left[\frac{EB(1+c_1)}{c_2} + \left(\left[\frac{1}{B} \right]^{\frac{1}{n}} - \frac{EB(1+c_1)}{c_2} \right) e^{c_2(1-n)t} \right]^{\frac{n}{1-n}},$$

where the values of the weight coefficients are found by the method of local variations.

Taking into account discretization of the variation (1), (2), the finite element equations for the solution of a contact problem are formed [24, 28]

$$ {}^{t+\Delta t}\mathbf{K}^{(i-1)} \Delta \mathbf{U}^{(i)} = {}^{t+\Delta t}\mathbf{R}^{(i-1)}, \quad {}^{t+\Delta t}\tilde{\mathbf{K}}^{(i-1)} \Delta \hat{\mathbf{U}}^{(i)} = {}^{t+\Delta t}\hat{\mathbf{R}}^{(i-1)}$$

where ${}^{t+\Delta t}\mathbf{K}^{(i-1)}$, ${}^{t+\Delta t}\tilde{\mathbf{K}}^{(i-1)}$ - are the tangent stiffness matrices (in matrices ${}^{t+\Delta t}\mathbf{K}^{(i-1)}$ are already included the additional elements which are formed from contact restrictions), ${}^{t+\Delta t}\mathbf{R}^{(i-1)}$, ${}^{t+\Delta t}\hat{\mathbf{R}}^{(i-1)}$ - the vector of internal and external forces. The top indexes $t + \Delta t$ of quantity indicate time for which it is calculated. The top indexes $(i-1)$ indicate number of iteration at correction of the solution by Newton-Rafson's method. The solution of the following step is found on a formula ${}^{t+\Delta t}\mathbf{U} = {}^{t}\mathbf{U} + \Delta\mathbf{U}$.

As a result of the solution of the first problem (2) of the given movements we receive the deformed model with distribution of stress and strain. The second problem on the basis of data on initial stress and strain determines unloading movements. After that it is possible to find residual nodal movements $\tilde{\mathbf{U}} = \mathbf{U} + \hat{\mathbf{U}}$.

Several procedures have been developed to treat contact problems including the use of Perturbed or Augmented Lagrangian methods, penalty methods, [24, 28] and direct constraints. MSC.Marc divides contact problems into two domains [29]: the first is when a deformable body makes contact with a rigid surface and the second is when a deformable body makes contact with another deformable body or itself. A target node on the deformable body has no constraint while contact does not occur. Once contact is detected, the degrees of freedom are transformed to a local system and a constraint is imposed such that $\Delta u_{normal} = \mathbf{v} \cdot n$, where **v** is the prescribed velocity of the rigid surface.

The iterative method for solving the inverse forming problem in displacement of a contact surface is offered

$$\bar{u}_i^{k+1} = \bar{u}_i^k + \alpha^k (\tilde{u}_i^* - \tilde{u}_i^k) + \beta^k (\bar{u}_i^k - \tilde{u}_i^k) \square \quad (4)$$

where $0 < \alpha^k < 2$, $\beta^k \to 0$ at $k \to \infty$, $i = 1, 2, 3$.

3 Numerical results for the inverse problems

The problem of movement's determination of the cores of reconfigurable tool creating such anticipatory curvature of the panel which provides the set residual form of the panel after unloading is considered. For modeling the process of forming the work-piece from plate with a thickness of 45 mm, finite-element model (FEM) in the CAE system MSC.Marc has been developed (Fig.1). The perfectly rigid body with a tip diameter of 50 mm has been utilized for the rod model. Distances between the centres of one tip of a core with another are made by 250 mm. Dimensions of work-piece are 1750x750x45 mm. The characteristics of the AK4-1T material (aluminum alloy) were used in the computations. The material is isotropic and has identical elasticity characteristics under tension and compression, namely, the Young's modulus E = 7000 kg/mm^2 and Poisson's ratio ν = 0.4. Steady state creep under both compression and tension in the experiments is described by Norton's law with different coefficients B for each of these types of strain:

- compression: $B_1 = 0.25 \cdot 10^{-14} (kg/mm^2)^{-n1} (h)^{-1}$, $n_1 = 8$;
- tension: $B_2 = 0.5 \cdot 10^{-14} (kg/mm^2)^{-n2} (h)^{-1}$, $n_2 = 8$.

Fig. 1. Finite element model of work-piece and surface with the set curvature

Conditions of contact are modeling slipping without friction.

According to manufacturing process the necessary detail with the set curvature obtain after chipping and milling of previously formed work-piece.

As initial displacement of contact bodies $\overline{u}_i^0$ (4) distances from rigid bodies to a surface with required curvature taking into account work-piece thickness are accepted. The residual displacements in an iterative method are determined in nodes of a deformable body nearest to projections of central points of contact bodies to the surfaces of work-piece.

The iterative method is realized in MSC.Marc system by means of a set of user procedures [30].

The numerical model with the set boundary conditions and definitions of projections of contact bodies is implemented by means of user programs in MSC.Patran.

As a result of calculation the anticipatory configuration of work-piece is found (Fig. 2).

In Fig. 3 the work-piece configuration after unloading in comparison with the set surface geometry is shown.

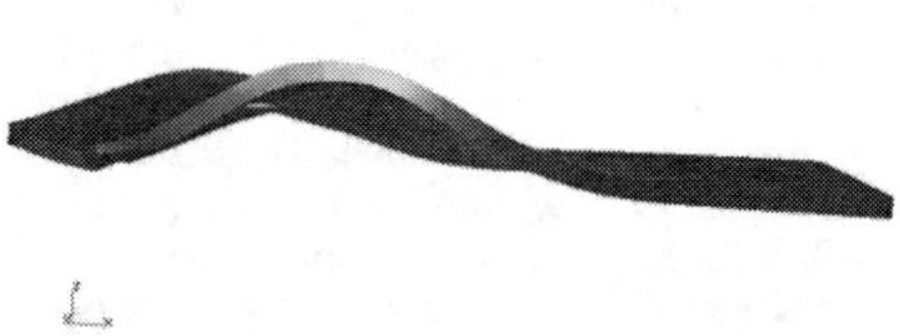

Fig. 2. Anticipatory configuration of work-piece in comparison with the set surface

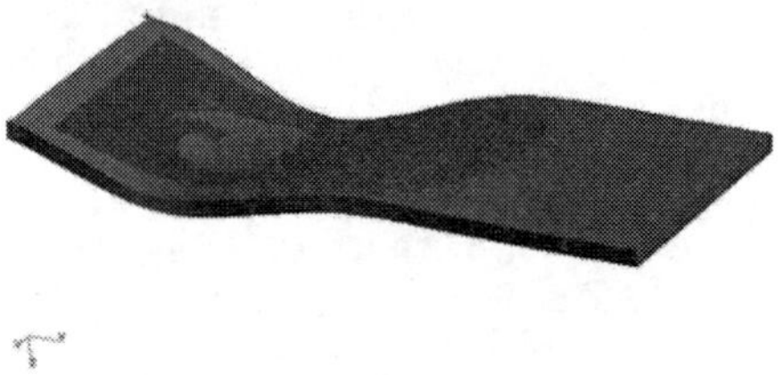

Fig. 3. Residual configration of work-piece in comparison with the set surface

For the iteration method (4), the best convergence coefficients were determined: $\alpha^k = 2-(1+k)^{-1}$, $\beta^k = 0.1(1+k)^{-5}$. As a result of solving the inverse problems of shaping with different laws of plates deformation, it is possible to determine the optimal path for stresses and dissipated work (Table 1).

Table 1. Maximum values of stress intensity and dissipated work under different deformation laws.

	σ_i, kg/mm²	A, kg/mm²	e_k, mm
$u_z = \frac{t}{T} u_z^*$	36.4	22	28.6
$u_z = \sqrt{\frac{t}{T}} u_z^*$	33.3	20.9	27.8

The convergence of the iterative method in the mean square norm ($e_k = \left(\sum_{S_c} \left(\tilde{u}_z^k - \tilde{u}_z^0\right)^2\right)^{1/2}$, S_c - where is the lower contact surface of the plate and k is the iteration number, $\tilde{u}_z^k$, $\tilde{u}_z^0$ - node parameters describing the residual displacement.

A rational loading path can also be found in the form (3).

4 Conclusions

The formulation of inverse quasistatic problems of creep theory in the form of optimal control makes it possible, with the help of the developed algorithm, to find numerical rational solutions for more complex details that are free from representations of the ideal plate and shell, in particular in the formation of wing panels. The developed algorithm is proposed to determine the anticipatory shape (snap-in of a reconfigurable bar device) with the law of change during the deformation process, ensuring minimum damage to the part.

Algorithms developed can be used in industrial applications, such as the forming of aircraft wing panels. It allows simulating and efficiently evaluating the parameters of the parts manufacturing technological processes.

This research was financially supported by the Russian Foundation for Basic Research (16-31-60038 mol_a_dk).

References

1. V.A.Portnoy, A.V. Portnoy, Vestnik KGTU im. A.N. Tupoleva, **2**, 9-13 (2006) [in Russian]
2. V.K. Boresevich, M.V. Zagerniak, V.V. Forge-stamp manufacture, **2**, 38–41 (2009) [in Russian]
3. B.V. Gorev, I.Zh. Masanov, Technology of mechanical engineering, **7**, 13-20 (2009) [in Russian]
4. O.V. Sosnin, B.V. Gorev, I.V. Lyubashevska, Journal of applied mechanics and technical physics, **38**, № 2, 293–297 (1997)
5. B.V. Gorev, V.A. Panamarev, V.N. Peretyat'ko, Steel in Translation, **41**, №6, 461–463 (2011)
6. B.V. Gorev, V.A. Panamarev, and V.N. Peretyat'ko, Steel in Translation, **41**, № 4, 275–277 (2011)
7. O.V. Sosnin, A.F. Nikitenko, B.V. Gorev, Journal of Applied Mechanics and Technical Physics, **51**, № 4, 608–614 (2010)
8. I.A. Banshchikova, B.V. Gorev, I.Y. Tsvelodub, Journal of Applied Mechanics and Technical Physics, **48**, № 5, 751–754 (2007)
9. I.A. Banshchikova, B.V. Gorev, I.V. Sukhorukov, Journal of Applied Mechanics and Technical Physics, **43**, № 3, 448–456 (2002)
10. Y. C. Lin, Y.-Q. Jiang, X.-M. Chen, D.-X. Wen, and H.-M. Zhou, Materials Science and Engineering A, **588**, 347–356 (2013)

11. B.D. Annin, A.I. Oleinikov, K.S. Bormotin, Journal of Applied Mechanics and Technical Physics, **51**, № 4, 579–589 (2010).
12. F.C. Ribeiro, E.P. Marinho, D.J. Inforzato, P.R. Costa, G.F. Batalha, Journal of Achievements in Materials and Manufacturing Engineering, **43**/1, 353-361 (2010).
13. R. Arabi Jeshvaghani, M. Emami, H.R. Shahverdi, and S. M. M. Hadavi, Materials Science and Engineering A, **528**, № 29-30, 8795–8799 (2011).
14. P. R. C. Junior, C. De Moura Neto, and D. A. Wade, Materials Research, **17**, № 3, 603–611 (2014).
15. R. Arabi Jeshvaghani, H. Zohdi, H.R. Shahverdi, M. Bozorg, S.M.M. Hadavi, Materials characterization, **73**, pp. 8–15 (2012).
16. D. F. Walczyk, J. Lakshmikanthan, and D. R. Kirk, Journal of Manufacturing Systems, **17**, № 4, 287–296 (1998).
17. D. F. Walczyk and D. E. Hardt, Journal of Manufacturing Systems, **17**, № 6, 436–454 (1998).
18. E. Haas, R. C. Schwarz, and J. M. Papazian, Journal of Manufacturing Processes, **4**, № 1, 77–85 (2002).
19. D. Simon, L. Kern, J. Wagner, and G. Reinhart, *Procedia CIRP*, **17**, 853–858 (2014).
20. S.Z. Su, M.Z. Li, C.G. Liu, C.Q. Ji,R. Setchi, J. Larkiola, I. Panteleev, I. Stead, R. Lopez, Key Engineering Materials, **504-506**, 839-844 (2012).
21. J.J. Chen, M.Z. Li, W. Liu, C.T.Wang, The International Journal of Advanced Manufacturing Technology, **25**, 935–939 (2005).
22. S. Y. Hwang, J. H. Lee, Y. S. Yang, and M. J. Yoo, CAD Computer Aided Design, **42**, № 11, 1001–1012 (2010).
23. K.S. Bormotin, Computational mathematics and mathematical physics, **53**, №12, 1908-1915 (2013).
24. P. Wriggers, Computational Contact Mechanics (2006).
25. Bormotin K.S., Oleinikov A.I., Journal of Applied Mechanics and Technical Physics, **53**, №5, 751-760 (2012) .
26. Bormotin K.S., Journal of Applied and Industrial Mathematics, **6**, №4, 421-430 (2012).
27. Bormotin K.S., Computational methods and programming, **15**, 222-228 (2014) [in Russian].
28. K.-J. Bathe, Finite element procedures (1982).
29. Marc 2016, Vol A: Theory and User Information, MSC.Software Corporation http://www.mscsoftware.com/product/marc.
30. Bormotin K.S., Belykh S.V. and Vin Aung, Computational methods and programming, **17**, 258-267 (2016) [in Russian].

Analysis of the providing environmental safety supervision in construction and reconstruction of facilities in the urban territories

Vadim Bespalov[1] and *Ekaterina* Kotlyarova[2,*]

[1]Don State Technical University, Department of Environmental Engineering, 344022 Rostov-on-Don, Russia
[2]Don State Technical University, Department of Urban Planning and Building Design, 344022 Rostov-on-Don, Russia

Abstract. Economic development of the urban territories supposes preservation of industrial enterprises and industrial complexes in the infrastructure of populated places which connected with the decrease of the ecological safety level of the urban territories. The solution of the problem of ecological safety supervision in this areas is connected with the providing the optimized balance between economic development and comfortable environment conditions. On the first stage of solving this problem the authors carried out factor evaluation of the negative influence level caused by relevant sources on the environment based on complex consideration of three groups of factors: nature-oriented, social and economic.

Modern economic urban territories development inseparably connected from the one hand with production sphere development which means preserving industrial enterprises and industrial complexes in the infrastructure of the populated areas and from the other hand it is connected with the maximum level provision of environmental safety in the urban territories focused on ecological norm. We shouldn't forget about the postulates of steady ecological and rational development of the urban areas where the particular place is taken by the problem of environmental safety provision of the built areas including built-up, recreational and industrial areas [1-4].

It is necessary to point out the fact that over the recent years urban environment formation, including its architectural peculiarities, is connected both with designing stages and further building of the new urban elements and sites in built-up, industrial and recreational zones and also with the existing sites reconstruction. These include separate industrial enterprises and industrial complexes which are significant by their land area included into the urban areas structure. In spite of the cost-effectiveness of preserving industrial sites in the total urban areas structure, they are losing their function of city-forming centers every year as we could see several decades ago. Particularly striking examples are the large settlements in the second half of the nineteenth century in England, France, Germany, and others. At that time such functions of industrial enterprises led to the fact that many medium-sized towns and large cities transformed into the places of the concentration of different industrial enterprises of various industries. They in total negatively affected the environment, polluting the air of surrounding areas, soil and water sites, significantly lowering the environmental safety level and the level of comfortable living in urban areas. This situation caused the necessity of reducing environmental stress in urban areas by the means of large industrial sites and industrial complexes dispersal. They were moved outside built-up and recreational zones to the external borders of urban areas.

Beginning approximately from 1960s industrial zones formation in urban areas and placing into urbanized zones large industrial sites and then creating industrial clusters began to be carried out by other principles. More and more careful considering of ecological factors and negative effect level measurement on the environment of such sites in regional schemes of the territorial planning of populated areas began to be held on a regular basis. With this we can confidently claim that was the beginning of ecosystem orientation in the city planning in our country. However a newly-formed or reconstructed industrial zone somehow should be placed in maximum possible accessibility for the staff and next to the main highways for delivering raw materials, marketing and other. In this regard it is necessary to consider that with the urbanization development which now could be expressed not only in particular settlement expansion but also in area's strengthening by the means of urban agglomerations forming, the situation could happen when earlier free from industrial sites built-up and recreational areas could again come close to already replaced industrial sites and separately formed industrial zones.

Providing an efficient balance between economic development and comfortable condition of the environment is from our point of view the key problem of environment safety supervision in such territories. On the first stage of solving this problem is necessary to

* Corresponding author: ekkot.arch@gmail.com

DOI: 10.1051/matecconf/201712905005

complete factor evaluation of the negative influence level of relevant sources on the environment.

Long-term observation analyses related to the impact of different sources to the negative influence on the environment allow to conclude that the industrial enterprises and industrial complexes mentioned above take the second place after motor transport. However these questions of sustainable transport infrastructure development currently draw the attention of the large amount of native and foreign scientists, and also professionals in designing and building fields, as the most urgent problem. For its solving "road maps" are created in many Russian and foreign cities But collaborative environmentally safe running of the industrial sites with other elements of urban areas hasn't been examined up to the end.

In present conditions of cities' development along with the industrial enterprises as substantial sources of the negative effect on the environment large shopping centers (malls, retail parks etc.) appeared. As among premises of such complexes apart from directly trading spaces there are more over decades all around the world a large number of feeding points, enormous outdoor and indoor parking spaces, cinemas, bowling, ice skating rinks and many other shopping and entertainment and supporting infrastructure objects, aimed at an endless attracting of the customers.

For negative influence evaluation of the industrial enterprises and modern shopping and entertainment malls on environment of urban territories they use different scientific approaches which are characterized by a certain level of monopolarity. We have offered the approach based on a complex recording of three basic groups of factors: nature-oriented, social and economic [5].

For the nature-oriented group we offer to consider [6]:

- Peculiarities of industrial enterprises and modern shopping and entertainment malls location as the objects of man-made impact on the environment. They can be determined by using the electronic topographic map of the area with coordinates, enterprises location and relevant to each sources of negative influence: pollutants emission (hard, liquid spray and gaseous), discharge of sewage waters, noise, vibration, electromagnetical, radioactive and terminal radiation,
- Characteristics of the water-supply resources, water disposal systems, electric power system providing exploitation of these enterprises, and also minimum distance between the borders the enterprises and the nearest residential areas and buildings,
- Characteristics of each source of negative influence on the environment of each considered enterprise, including kinds of pollutants ended in the environment, their mass emissions, the efficiency of clearing of the relevant engineering and ecological systems of the enterprises,
- The condition of each component of the environment (air basin, water resources, soil) of the considered urban territory including factors defining severe weather conditions for the area.

Factor evaluation of the environmental safety of the industrial enterprises according to nature-oriented factors should be completed in accordance to the algorithm given below.

On the electronic map of the examined area it is necessary to highlight the borders of the industrial zone, location of the main sources of negative influence on the environment, borders of buffer zone, waste collection site and others. It's desirable to update the map with the available monitoring or given information about physicochemical and quantitative characteristics of all pollutant kinds (declared, rated and real).

For the air basin evaluation it's necessary:

– To define the main sources of air pollution,
– Classified the sources in accordance to the amount of emissions and priority of carrying out air protection measures,
– To elicit the air pollution level,
– To define the level of maximum one-time and average daily pollutant concentration from each elicited source of pollutant emission.

Finding over limits zones of standard measures of relevant substances concentration on the bases of using the following basic methods:

– Instrumental, based on using automatic econometers, constantly measuring pollutant concentration in emissions. More often this method is used for controlling the main and the most spreading pollutants,
– Instrumental and laboratory, when previously exit gas sample collection is taken in the pollutant source, then their analys is carried out in a lab environment,
– Calculation, which is based on mass pollutant emission in accordance with the data of feed composition and fuel in processing method etc.,
– Indicative, concluding the using of the indicative elements which measure their color depending on pollutant concentration in the worked out gas sample,
– Controlling, which is the defining the real pollutant level from enterprise emissions in reference points beyond its borders. After that the real pollutant level must be compared with reference value. The comparison is carried out in accordance with wind streamline for the particular area. This method is implemented controlling the large amount of emission sources both controlled and uncontrolled.

For the evaluation of water resources condition in the considered area it's necessary:

– To define sanitary and hygiene characteristics of water objects providing exploitation of the considered enterprises,
– Waterworks conditions in the considered area, including water-economic balance,
– To define for each considered enterprise the main water pollution sources (waste water disposal).

Water objects observation points is firstly organized on basins and drains which have primary domestic value and also subjected to large pollution by industrial and domestic sewage. Background observation points are organized on not apparently polluted by water waste water objects and drains.

In evaluation of soil condition we can include:

- The defining soil pollution level and sum characteristics of chemical elements data concentrated in soil,
- The defining of natural and recreation resources in an area (the presence of parks, gardens, mini-parks, boulevards, internal landscape gardening),
- The calculation of nature and recreation resources (which are available) rational using level,
- The defining of functional value of the examined land,
- The defining of characteristics of the soil loss level.

At that it's necessary to take into account that nature and recreational resources have three functions: pollutant assimilation, filling the air basin with oxygen, satisfaction of aesthetic needs of people. All these functions can be defined as the components of a single main function of the environment – the function of providing the life.

Characteristics consideration of the land functional value is necessary to know the borders of the negative influence from the enterprises on the environment and to limit this influence on the bases of using either technological or special engineering and ecological measures. In work [7] methodical bases of the defining the severity of landscaped urban areas functional value are given. Nature and recreational resources of the area include recreational landscapes, hydro and mineral resources, bioclimate. Favorable environment condition is the necessary factor for natural recreation resources availability.

For natural and recreational resources evaluation different methods are applied. At that the most spreading and appropriate to the common recreational analysis of the area is the evaluation of the level of landscape usefulness, bioclimatic and ecological characteristics for recreational using.

For the physical factors (vibration, noise, magneto-electric, ionisable and heat radiation) evaluation influencing on the environment it's necessary:

- To define the sources of negative influence on the environment and find out their intensity,
- To define discomfort zones within which the tolerated level of the physical factors negative influence is exceeded.

In the group of social factors we offer to include [8]:

- The number of population in the area,
- Density of population,
- Children and adult population state of health,
- Children and adult number of death,
- Infrastructure of urban area and its planning,
- The living level of comfort in the examined area,
- Conveniences of urban territory,
- Availability and the list of social objects in the examined area.

Listed social factors can be used for the living environment quality evaluation in the considered area [9,10], and their violation leads to global health and social threats for the population [11,12].

It's known [11,12] that population health is the main indicator of social wellbeing and proper economic functioning of the society. The raised level of ecological danger in the considered area firstly affects the health condition of people living on the territory which is affected by negative influence of the territory. The results of numerous researches proved that with the influence of negative environment the quantity of diseases raises sharply especially among children, also there is the raise of lethal level.

Under the conditions of health decline of the population which is explained by stressed conditions of which living conditions and environment condition demands towards the organization of ecological safety of the urban territories are toughened.

Man-made influence on the population living and working in the examined area leads to the raise of pathology and others factors of human health: from common physical and psychical development to peculiarities of age progress and death rate and also all forms of diseases. This idea is the main form of monitoring the health of the population in social and hygiene system which is necessary to consider as the basic instrument for the health evaluation of the population and the revealing of environmental factors defining its forming. Important integral index recommended by WHO as the characteristic of the health condition of the population and its living standards of one or another region is the life expectancy.

The main difficulty is in defining the level of influence of each factor on the health condition of the population.

In a group of economic factors we offer to include:

- Taxes, payoffs, contributions and other cash flows into municipal treasury,
- Liabilities of investment nature aimed at the social sphere development and economic development of the city and consequently the region in all,
- Building and ground rent,
- Economic damage caused by limit-exceeding environmental pollution.

From one hand economic factors define the profit and the efficiency of the industrial objects, large shopping and entertainment malls, power stations and others located in the examined area and allow to analyze the financial opportunity of the area but from the other hand reflect the economic loss caused by environmental pollution.

Thus in the result of carried out analyses we can define the main factors of economic, social and ecological nature defining the level of ecological security of the industrial enterprises and modern shopping and entertainment malls in the urban territories. This will allow us on the next stage of the research to develop and describe relevant criteria of the ecological safety of industrial areas and separate enterprises in the urban territories.

Industrial zones as a part of urbanized territories are one of the main resources of the city development. However it would be mistakenly to think that the main task of them – the maximum launch of these sites for further profiting. Each industrial complex should have its own model of the development, based on its location, demands and considering the interests of the residents of nearby areas and future consumers. Aspects of creating comfortable and attractive urban environment also have considerable importance. In addition, it is necessary to conduct a detailed analyses of building influence or reconstruction objects of this type on the further running of the city in the whole and also, which is importantly, with necessary preserving and enhancing environmental safety of urban territories.

References

1. A.G. Lazarev, E.V. Kotlyarova, *Problems of sustainable development of industrial zones in the structure of residential areas of large cities*, Izvestiya RGSU. № **16**, P. 13-18 (2012)
2. Frederick R. Adler, Colby J. Tanner, *Urban Ecosystems: Ecological Principles for the Built Environment* (Cambridge University Press, 2013)
3. Richard T. T. Forman, *Urban Ecology: Science of Cities* (Cambridge University Press, 2014)
4. S.G. Sheina, A.A. Khamavova, T*echnique for the Russian federation regional territories assessment used to create industrial parks network*, Procedia Engineering, Vol. **150**, P. 1960-1965 (2016)
5. V.I. Bespalov, E.V. Kotlyarova, *The main principles of methodic perfection of the social-ecological-economic evaluation of the industrial zones environment in large industrial cities*, Inzhenernyy vestnik Dona, vol. **18, № 4,** p. 137-144 (2011)
6. V.I. Bespalov, E.V. Kotlyarova, *Mathematical description of ecological, economic and social criteria of the complex evaluation of the industrial areas*, TERRA ECONOMICUS, vol. **9**, № **4-3**, p.121-123 (2011)
7. V.L. Bogdanov, V.V. Dmitriev, A.I. Chistobaev, *Ecological monitoring of Saint-Petersburg environment condition*, Izvestiya rgo, vol.**132**, № **6**, p. 13-20 (2000)
8. E.V. Kotlyarova, *Analyses of social-ecological-economic condition of the industrial territories of Rostov-on-Don, Biznes. Obrazovanie. Pravo,* Vestnik Volgagradskogo institute biznesa, № **1**, p. 104-107 (2012)
9. V.I. Taburkin, M.V. Doronina, *Social ecology as the theoretical basis of the natural resource use efficiency*, Vestnik Gosudarstvennogo agrarnogo universiteta Severnogo Zauralia, № **2 (29)**, p. 84-90 (2015)
10. U.A. Rahmanin, *Updating human ecology problems and environment hygiene and the ways of their solution*, Gigiena i sanitaria, № **5**, p. 4-8 (2012)
11. S.V. Kuzmin, V.B. Gurvich, O.V. Dikonskaya, S.V. Yarushin, S.V. Romanov, A.S. Kornilkov, S*ocial and hygiene monitoring – integrated system of evaluation and risk supervision for population health at the regional level*, Gigiena i sanitaria, № **1**, p. 30-32 (2013)
12. T.M. Ushmaeva, *Damage index of the population health in the management of developing the society*, Economica prirodopolzovaniya, № **2**, p. 40-63 (1995)

Creation on the basis of the first theory of strength model equal stressed cylinder exposed to power and temperature loads

Vladimir Andreev[1,*], and *Andrey* Leontyev[1]

[1] National Research Moscow State University of Civil Engineering, 26 Yaroslavskoye Shosse, Moscow, 129337, Russia

Abstract. The paper discusses the axially symmetric plane problem theory of elasticity for inhomogeneous thick-walled cylinder under the action of external pressure and stationary temperature field. The essence of the inverse problem is to determine such dependence of the modulus of elasticity on the radius at which the stress state of the cylinder will be given. It should be noted that an equal stressed cylinder does not coincide with an equal strength cylinder [1] since when the modulus of elasticity of a material changes by modifying its composition, as a rule, its strength properties also change.

1 Introduction

We assume that in the cylinder is carried out flat strain state. Consider the axisymmetric problem for a radially inhomogeneous cylinder under the action power and temperature loads.

Temperature field caused by the heating of the inner surface of the shell and is stationary. Power load is an external uniform pressure, which creates a prestressed state of the shell.

2 The resolving equation

In [2], a resolving equation is given for the stress σ_r for an axisymmetric problem (plane deformed state) for a radially inhomogeneous body:

$$\sigma_r'' + \varphi(r)\sigma_r' + \psi(r) = f(r), \quad (1)$$

where for the case $\nu = const$

$$\varphi(r) = \frac{3}{r} - \frac{E'}{E}; \; \psi(r) = -\frac{1-2\nu}{r(1-\nu)} \cdot \frac{E'}{E};$$
$$f(r) = -\frac{E(1+\nu)\varepsilon_f'}{r(1-\nu^2)} \quad (2)$$

Here, $E = E(r)$ and ε_f - forced (in this case temperature) strain, which is calculated by the formula:

$$\varepsilon_f = \alpha_T T(r), \quad (3)$$

where $\alpha_T = const$ – coefficient of linear temperature expansion.

Substituting (2) and (3) in (1) we obtain the equation:

$$\sigma_r'' + \left(\frac{3}{r} - \frac{E'}{E}\right)\sigma_r' - \frac{1-2\nu}{r(1-\nu)} \cdot \frac{E'}{E}\sigma_r = -\frac{E(1+\nu)\alpha_T T'}{r(1-\nu^2)}. \quad (4)$$

The stationary temperature field in the cylinder, on the inner boundary of which the temperature T_0 is maintained and on the outer surface $T = 0$, is described by the formula:

$$T(r) = T_0 \frac{\ln(b/r)}{\ln(b/a)}. \quad (5)$$

Here a and b – respectively inner and outer radii of the cylinder, $T_0 = 100° \text{C}$.

To determine the constants of integration of equation (4) using the boundary conditions

$$r = a, \quad \sigma_r = 0; \quad r = b, \quad \sigma_r = -p_b = 50\,\text{MPa}. \quad (6)$$

3 Direct problem for a homogeneous cylinder

For a homogeneous material, equation (4) is transformed to the form

$$\sigma_r'' + \frac{3}{r}\sigma_r' = -\frac{E(1+\nu)\alpha_T T'}{r(1-\nu^2)}. \quad (7)$$

The solution of equation (7) taking into account (5) and (6) has the form:

$$\sigma_r = \frac{199,95}{2r^2} - 99,975 + 36,05\ln(r/a). \quad (8)$$

From the equation of equilibrium

* Corresponding author: asv@mgsu.ru

$$\frac{d\sigma_r}{dr}+\frac{\sigma_r+\sigma_\theta}{r}=0 \quad (9)$$

we can determine the stresses σ_θ:

$$\sigma_\theta = r\sigma_r' + \sigma_r . \quad (10)$$

From the third Duhamel-Neumann relation, setting $\varepsilon_z = 0$, we can obtain an expression for the stresses σ_z:

$$\sigma_z = \nu(\sigma_r + \sigma_\theta) - \alpha_T ET . \quad (11)$$

Fig. 1 shows the calculation results obtained with the following initial data: $\nu = 0.2$; $a = 1$ m; $b = 2$ m; E_0=2·10^{4}MPa; $\alpha_T = 1\cdot 10^{-5} 1/°C$.

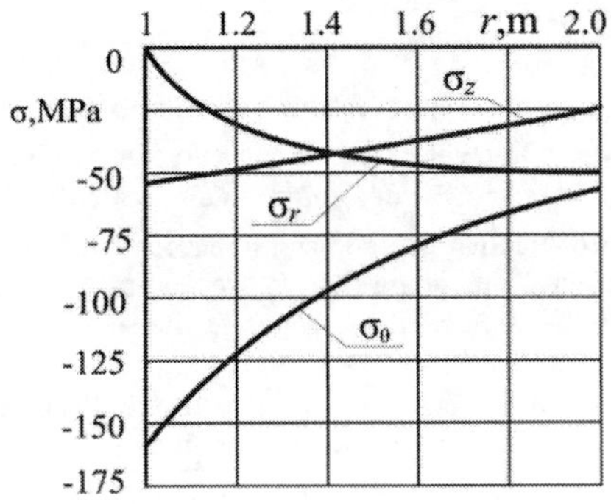

Fig. 1. Stresses in homogeneous cylinder

We note that the solution of the problem was obtained under the assumption that the cylinder is sufficiently long and a flat deformed state is realized in it.

However, in practice, the structures have a finite length and, in the presence of free ends, the stresses σ_z at the ends will be zero.

As shown in [3], from the condition $\int\limits_F \sigma_z dA = 0$ one can calculate the average stress $\sigma_z^* = -\int\limits_F \sigma_z dA / A$, which should be added to the calculated stresses. Then in the middle part of the cylinder there will be a stress state corresponding to the finite cylinder.

4 Equal stressed cylinder. Inverse problem

The idea of the method of creating a model of equal-stress structures based on the results of numerous calculations inhomogeneous bodies in which it is shown that if in a certain area of the body the modulus of elasticity is less than in a homogeneous material, the stresses in this area are also reduced, and vice versa. Below is the solution of the inverse problem in which wanted the function *E* (*r*) decreasing near the inner surface of the cylinder which reduces the stresses in this area. In this case, based on the theory of strength of maximum normal stresses, a stress state is determined that satisfies the condition $\sigma_{max} = \sigma_\theta = const$, which corresponds to the model of an equally stressed cylinder [4, 5]. Note that since in this problem all the stresses are negative, then the maximum is considered the largest in absolute value of stresses.

Substituting $\sigma_\theta = \sigma_0 = const$ in equilibrium equation (8) we find

$$\sigma_r' = \frac{\sigma_0 - \sigma_r}{r} . \quad (12)$$

The solution of this equation is

$$\sigma_r = \frac{K}{r} + \sigma_0 . \quad (13)$$

From the boundary conditions (6) one can determine the constants *K* and σ_0: *K*=100MPa·m, $\sigma_0 = \sigma_\theta = -$100MPa.

Substituting function (13) in resolving equation (4) after some transformation we find first-order differential equation for determining the function *E* (*r*):

$$E' - \frac{K}{r\left[A(1-k) - kr\sigma_0\right]}E - \frac{B}{A(1-k) - kr\sigma_0}E^2 = 0 \quad (14)$$

where

$$B(r) = -\frac{\alpha_T T' \cdot r}{1-\nu}, \quad k = \frac{1-2\nu}{1-\nu} .$$

Equation (14) is the Bernoulli equation [4], solving it we obtain the desired dependence *E*(*r*):

$$E = \frac{1}{\dfrac{B}{a^m}\dfrac{(ar+b)^n}{r^n}G + \dfrac{(ar+b)^n}{r^n}D} . \quad (15)$$

Here

$$G(r) = \ln(ar+b) + \frac{nb}{ar+b} - \frac{nb^2}{2(ar+b)^2} + \frac{nb^3}{3(ar+b)^3}$$

$$-\frac{b^n}{n(ar+b)^n}; \quad a = -k\sigma_0, \; m = n+1, \; b = K(1-k).$$

The "reduction of the elastic modulus near the inner contour" mentioned above should be understood not with respect to a homogeneous material, but with respect to the value of *E* near the outer contour of the cylinder. Therefore, the constant *D* can be determined from two different boundary conditions for *E*:

1 – *r*=*a*, *E*=*E*$_0$=20000 МПа;

2 – *r*=*b*, *E*= *E*$_0$=20000 МПа.

In Fig.2. the graphs of the dependence *E* (*r*) corresponding to the two indicated variants are presented.

DOI: 10.1051/matecconf/201712905006

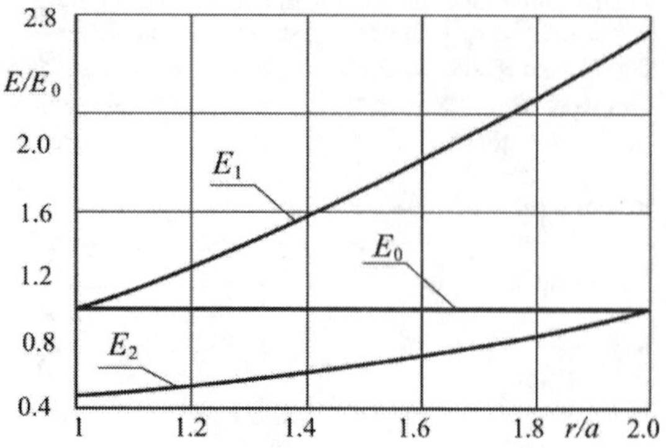

Fig. 2. Dependencies of the elastic modulus in the inhomogeneous (E_1, E_2) and homogeneous (E_0) cylinders

To verify the obtained results there was obtained the solution of the direct problem, by direct substitution of the dependence (14) into (4). Fig. 3 shows the stresses diagrams σ_r and σ_θ. As was to be expected, the results of stress calculations were the same for both variants. It should be noted that in an equal stressed inhomogeneous cylinder, the stress $\sigma_{\theta\max}^{inhom} = -100\,\text{MPa}$, while in a homogeneous cylinder $\sigma_{\theta\max}^{hom} = -160\,\text{MPa}$. Thus, by introducing a coefficient $\beta = \sigma_{\theta\max}^{hom} / \sigma_{\theta\max}^{inhom}$, it is possible to determine how many times the greatest stresses in a homogeneous cylinder are greater than in a similar inhomogeneous one. In this case this coefficient is equal to $\beta = 1.6$.

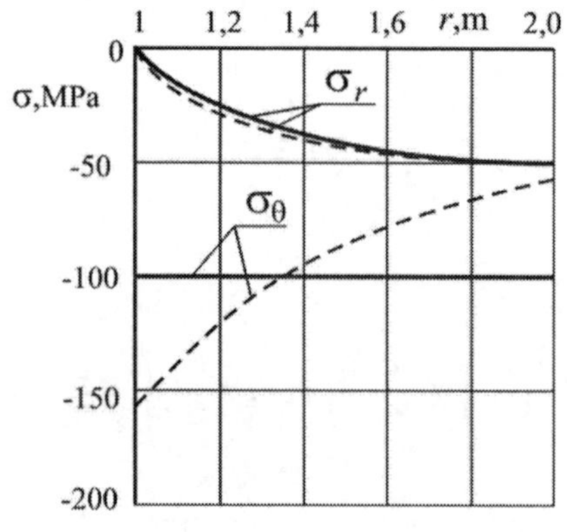

Fig. 3. The stresses in the inhomogeneous (—) and homogeneous (- - -) cylinders

Especially it is necessary to stop on stresses σ_z, the diagrams of which are shown in Fig. 4. At first sight, it is surprising difference in stresses σ_z for two dependencies of *E (r)* shown in Fig. 2 since the stresses

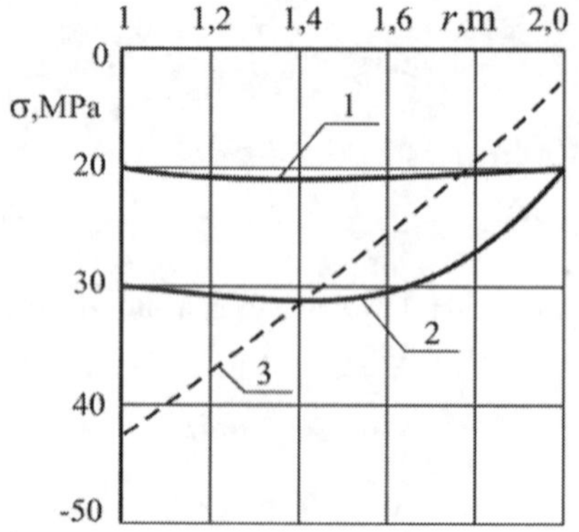

Fig. 4. Stresses σ_z in a thick-walled cylinder:
1, 2 — inhomogeneous cylinder;
3 — homogeneous cylinder

σ_r and σ_θ for these two variants of calculation practically coincides. This fact is explained as follows. Stresses σ_z for a flat deformed state are calculated from the third Duhamel-Neumann relation (11) which contains the "temperature" term, depending on *E*. If we consider the stresses at the edges of the segment [*a, b*], then in this example *T*(*a*) = 100°C, and *T*(*b*) = 0. Therefore, at the point $r = b$ due to the absence of a "temperature" term, the stresses coincide, and at the point $r = a$ differ by an amount

$$\Delta\sigma_z = \alpha_T \left[E_1(a) - E_2(a)\right] \cdot T(a)\,. \tag{16}$$

Taking into account that $E_1(a) = E_0 = 2 \cdot 10^4$ MPa, and $E_2(a) = 0,45 \cdot E_0 = 0,9 \cdot 10^4$ MPa, we obtain $\Delta\sigma_z$ (a) = 10 MPa, which coincides with the result shown in Fig. 4.

5 Conclusions

Models equal stressed cylinders on the basis of other theories strength (maximum shear stress, energy and others) were published in [5-7].

In conclusion, we note that an equal stressed cylinder does not coincide with an equally strength cylinder [3], since the modulus of elasticity of a material changes by modifying its composition, as a rule, its strength properties also change.

There are some materials in which, by modifying the composition, it can be achieved that the modulus of elasticity will vary substantially, and the strength is negligible. Such a material is polymer concrete [8] for which the corresponding mechanical characteristics are given in Table 1.

Table 1. The physical and mechanical properties of polymer concrete filled with silica flour.

The degree of filling silica flour	The modulus of elasticity E_b, 10^{-4}MPa	The ultimate compressive strength, MPa
-	3.10	142
50	4.50	146
100	7.10	160
200	10.5	148
300	13.7	132
400	16.7	115

Construction of models of equally strength concrete and reinforced concrete thick-walled cylindrical and spherical shells under the influence of only power loads is devoted to the thesis [9].

The author and his co-workers made attempts to create a thick-walled cylinder with a continuous inhomogeneity of deformation characteristics based on a polymer filled with quartz flour of different factions [10]. During curing, the mold filled with the composite rotated and the filler particles moved towards the outer wall of the mold. Unfortunately, the result was a non-continuously inhomogeneous cylinder, and a two-layer cylinder whose outer layer was stiffer than the inner layer.

In the above [9], is also considered a model of cylindrical shells, consisting of several layers of different hardness. It is shown that in such models, even with a small number of layers, the efficiency coefficient can reach 1.67 - 1.87 (for a three-layer - five-layer cylinder).

This work was financially supported by the Ministry of Russian Education (state task #7.1524.2017/PCh)

References

1. S.P. Timoshenko, J. Goodier, *Theory of Elasticity* (Moscow, Nauka, 1979)
2. V.I. Andreev, *Mechanics of inhomogeneous bodies* (Moscow, Publishing House of the ASV, 2002)
3. V.I. Andreev, I.A. Potekhin, *Equal stressed and equal strength structures. Modeling and creation,* Building materials, equipment, technologies of the XXI century **6** (2009)
4. E. Kamke, *Handbook of ordinary differential equations* (Moscow, Nauka, 1976)
5. V.I. Andreev, I.A. Potekhin, *Construction of a model of an equally stressed cylinder on the basis of the second and fourth strength theory*, Proceedings of the 16th Russian-Polish-Slovak seminar "Theoretical Foundations of Construction", Zilina, Slovakia (2007)
6. V.I. Andreev, I.A. Potekhin, *Iterative method for constructing a model of an even-strength cylinder,* Building Mechanics of Engineering Constructions and Structures, **1** (2008)
7. V.I. Andreev, A.S. Chepurnenko, B.M. Yazyev, *Model of Equal-stressed Cylinder based on the Mohr Failure Criterion,* Advanced Materials Research. **887-888** (2014)
8. V.V. Paturoev, *Polymer concrete* (Moscow, Stroyizdat, 1987)
9. V.I. Andreev, I.A. Potekhin, *Optimization of the strength of thick-walled shells* (Moscow, MGSU, 2011)
10. B.M. Yazyev, V.I. Andreev, R.A. Turusov, *Some problems and methods of mechanics of macro-inhomogeneous viscoelastic media* (Rostov-on-Don, 2009)

MATEC Web of Conferences **129**, 05007 (2017)
ICMTMTE 2017
DOI: 10.1051/matecconf/201712905007

Forecasting ofuseful life of the structures of a production building during operation

Valeriya Pshenichkina[1,*], *Kseniya* Suhina[1], and *Artem Zhukov*[1]

[1]Volgograd State Technical University, 400081 Academicheskaya 1, Russia

Abstract. The article presents a probabilistic concept of practical evaluation of the individual resource of structures of industrial buildings and structures in operation. The results of calculation of a monolithic reinforced concrete overpass for safety and durability are given. The quantitative values of the resource are obtained for a given security of operability.

1 Introduction

At present, there is a certain contradiction in the calculation of structures between the sufficiently high level on which the forces and movements of the system are determined by the methods of structural mechanics, the theory of elasticity, plasticity, and the approximate evaluation of the reliability of buildings and structures in two groups of limiting states only at the beginning of their operation.

The main drawback of the method of limiting states is that it does not take into account the factor of time and wear. These factors are necessary for the establishment of reasonable overhaul periods, assessment of the state of the structures for the purposes of reconstruction, safety assessment and the forecast of the remaining service life. In the deterministic formulation and without taking into account the time factor, it is impossible to solve the main task of calculating structures and bases - to provide the required level of reliability throughout the life of the operation.

Quantitative methods for assessing the safety and resource of building objects, which are based on a probabilistic concept, the application of probability models and methods of reliability theory, are not yet widely used in practical studies. This is due to the complexity of carrying out probabilistic calculations that require a large amount of statistical data, the undeveloped engineering techniques. At the same time, the tasks of estimating the residual resource, with sufficient accuracy for practice, can be solved in the correlation approximation, which does not require a significant increase in the volume of information.

2 Probabilistic concept of durability evaluation

The rectangu Resource - this is an indicator of longevity, characterizing the reserve of possible operating time of the object. The estimation of the resource of a building is carried out on the basis of theoretical design models taking into account statistical data on loads, impacts, materials and operating conditions of similar buildings or structures. The forecasting of the resource is an integral part of the theory of structural reliability and is of a probabilistic nature, which is due to the lack of complete information about the object and the spread of longevity indicators. Unlike the design stage, resource forecasting at the operational stage is carried out for a particular building or structure, therefore the main resource concept in operation is the individual residual resource or remaining service life-the operating time from this point in time to the limit state. Since in service conditions, between-repair periods are assigned individually according to the technical state of the facility, the concept of an individual resource is used up to the nearest medium or major repair.

Forecasting an individual resource includes a wide range of tasks. At the first stage, an assessment is made of the technical condition of the structure of the facility, the detection of defects and damage by methods of nondestructive testing. The second stage determines the actual load and the actual operating conditions of the structures under study. At the final stage, reconciliation of the actual resource with reliability indicators at the design stage; Prediction of the optimal residual life on the basis of the data obtained, as well as an assessment of the likelihood of failure of structures. In this case, the predicted resource must be coordinated with the designated service life of the building.

Currently, the assessment of the loss of performance of an object, especially for local injuries, is mainly based on a deterministic approach to its output parameters. For load-bearing structures of buildings, the output parameters that determine their operability are the bearing capacity and rigidity, by the variation of which it is possible to judge the course of the process of wear and tear. Assessment of the degree of damage on the output parameters is used in the technical examination of building structures of buildings and structures, as they

* Corresponding author: vap_hm@list.ru

 DOI: 10.1051/matecconf/201712905007

are influenced by many factors that are difficult to describe.Such factors include: corrosion of steel and reinforced concrete structures, rotting of wooden structures, wetting of structures and associated unfreezing of concrete and masonry, oiling of concrete, fatigue failure of steel structures, alteration of ground conditions for foundations, cracks in reinforced concrete and stone structures, elevated Deformation of structures - deflections, death, loss of stability, fluctuations, damage from fire during fires, as well as errors in the design and erection Constructions (human factor) [1]. Diagnostic signs of defects of all types of building structures - ferro-concrete, stone, metal, wooden, and also their bases and foundations, - allow to estimate approximately the category of their technical condition. In this case, the serviceable condition meets the requirements of the design standards, the rules of production and acceptance of construction works. Inoperative condition corresponds to the unsuitable design for operation. Types of characteristic damage and structural defects, as a rule, are reduced to special tables, in which for each defect index, the maximum values are determined in the serviceable state and the minimum values for an inoperable state [2,3,4].

Deterministic evaluation of damages by output parameters approximately establishes the level of reliability of the object at the time of its technical examination, but does not allow to fully determine the causes of the change in output parameters and determine the numerical characteristics of the aging process that occurs in the materials of structures. Such an assessment does not provide an opportunity to obtain a reliable forecast of the residual resource of the building as a whole and its structures. Therefore, the calculation of the residual life of building objects should be carried out using the methods of reliability theory.

Knowledge of the laws of the change in the output parameters of the system makes it possible to calculate the average time of safe operation of the structure, and taking into account the stochastic nature of the process - to obtain quantitative values of the risk of failure of the design when the output parameter exceeds the maximum permissible value. Considering the aging processes as random, it is convenient to reduce them to more simple regularities, highlighting the stochastic component in the form of a random variable or a stationary random function. The apparatus of the theory of random functions can be applied as a differential function expressing the speed of the process $v(t)$, and to an integral function describing the change in the degree of damage $u(t)$. But, since it is the speed of the damage process that is determined by the physics of the phenomenon, it is more convenient to use it in computational modeling. Then, the yield parameters $X(t)$ are considered as a function of the wear rate $X(t) = \phi[v(t)]$ [5].

The law of distribution of service life (operating time) to failure is formed depending on the nature of the course of the wear process in the structures, the level of their initial reliability and safety requirements. However, in practical calculations, from a theoretical distribution laws of random variables known in probability theory, a law is chosen that reflects the statistical regularities of the processes with a high degree of reliability, and can be used to solve the problems of estimating the resource. The basis for choosing the distribution law and its parameters are experimental studies, operational data on the operation of similar objects, theoretical premises. So, with wear and other gradual failures, as a rule, apply a normal law. When external factors that lead to failures, irrespective of its state and duration of previous work, act on the design, that is, when sudden failures occur, they can be described by exponential or uniform distributions [5,6].

It should be noted that direct application of the laws of the distribution of service life can not be recognized as the main for building structures. First, the law is chosen formally, it does not reflect the processes of the formation of a refusal. Secondly, to obtain experimental data that allows us to judge the law of distribution, it takes a long time, since the final stage of damage is fixed-failure.

In [1], devoted to the reliability of machines and structures, V.V. Bolotin considered in general terms all known models of failures, mainly for machine-building structures, which at present it is not possible to use when predicting the resource of building structures during operation due to the lack of sufficient information on the regularities of resource development of various types of structures.

At the same time, the use of instruments and equipment for non-destructive methods for controlling the material of structures (ultrasonic and magnetic equipment, strain gauges, mechanical measuring instruments) allows the processes of accumulation of damages and aging, a decrease in strength and deformation parameters, generalized by means of such a measure of damage as the rate of decrease of the strength parameter . This measure of damage is not directly related to the characteristics of the physical aging processes of the material, but it allows the development of simple and affordable methods for estimating the resource at all stages of the design.

Let's consider a practical method of estimating and forecasting a resource by the example of reinforced concrete structures of an exploited industrial facility. The practice of the operation of structures indicates that the resistance of the elements begins to decrease only after some random time interval $T1$. The value of $T1$ is determined experimentally. Then the random lifetime of the construction is $T_{res} = T1 + T_\gamma$, where T_γ is the operating time of the structure to the limiting state.

To illustrate the calculation of the durability of the structure, long-term experimental and estimated estimates of the performance of a monolithic reinforced concrete overpass for rubber mixers at the Voltyrsky tire plant in Volzhsky have been used. The trestle 144 m long is divided into four blocks by temperature-sedimentary seams. On the platform there are 14 rubber mixers, with the work of which there are dynamic loads. The trestle structure is a frame spatial system with an upper structure in the form of a ribbed plate 6.5 m wide, 0.4 m thick and a base plate 6.5 m wide and 0.6 m thick. Span of the frame 4.25 m, column spacing 6 , 0 m. Under the project, concrete M-200 (class B15), valves of class A-III (A400). The flyover has been in operation since 1964. In 1980, design mixers were replaced with new, more productive ones. In 1980, a survey was carried out of horizontal and vertical oscillations in a number of sections of the top plate of the overpass and the strength of concrete. It is established that in some zones of the upper plate the amplitudes of the horizontal oscillations in the stationary mode exceed the maximum permissible value $A_g = 0.1$ mm for machines with rotating parts at $n = 750$ rpm. Mass testing of concrete overpass designs with non-destructive methods showed that there was a decrease in strength, the average statistical value of which was $\overline{R}_b = 0.130 \cdot 105$ kN / m2. For the beginning of the resource life was adopted in 1980, and the value of $T1$ in 1996 was 16 years.

Spatial calculation of the trestle for static and dynamic effects allowed to establish the stress-strain state of all structural elements. For an example of calculation for durability, a bolt of one of the intermediate frames of the flyover was considered. Safety of the bolt element at the beginning of operation. The calculated values of R_S, R_b, b, h_0, M were accepted by uncorrelated random variables with the following numerical characteristics: mathematical expectations $\overline{R}_b = 0.145 \cdot 105$ kN / m2; $\overline{R}_S = 4.785 \cdot 105$ kN / m2; $\overline{M} = 316$ kN · m; $\overline{h}_0 = 0.66$ m; $\overline{b} = 1.96$ m; Standards $\sigma_{R_b} = 1957$ kN / m2; $\sigma_{R_S} = 4.785 \cdot 104$ kN / m2; $\sigma_M = 47.4$ kN · m; $\sigma_{h_0} = 0.02$ m; $\sigma_b = 0.059$ m.

The height of the compressed section of the element $x = \dfrac{A_S \overline{R}_S}{\overline{R}_b \overline{b}} = 0{,}03$ m.

Bearing capacity of section $\overline{\Phi} = \overline{R}_b \overline{b} x(\overline{h}_0 - 0.5x) = 549.34$ kN·m. The mathematical expectation of the function of working capacity [7] $\overline{S} = \overline{\Phi} - \overline{M} = 233.34$ kN·m. The standard of the health function σS is determined from expression

$$\sigma_S^2 = \sum_1^n \left(\frac{\partial S}{\partial x_i} \right)^2 \sigma_{x_i}^2 , \qquad (1)$$

where the function S is represented in the form

$$\overline{S} = \overline{R}_S A_S \left(\overline{h}_0 - 0{,}5 \frac{\overline{R}_S A_S}{\overline{R}_b \overline{b}} \right) - M ;$$

$$\frac{\partial S}{\partial R_S} = A_S \overline{h}_0 - \frac{\overline{R}_S A_S^2}{\overline{R}_b \overline{b}} ; \quad \frac{\partial S}{\partial R_b} = \frac{A_S^2 \overline{R}_S^2}{2 \overline{b} \overline{R}_b^2} ;$$

$$\frac{\partial S}{\partial h_0} = A_S \overline{R}_S ; \quad \frac{\partial S}{\partial b} = \frac{A_S^2 \overline{R}_S^2}{2 \overline{R}_b \overline{b}^2} ; \quad \frac{\partial S}{\partial M} = -1{,}0 .$$

Denoting the expressions for the partial derivatives by the symbols a, b, c, d, e, we obtain the formula for the standard of the health function

$$\sigma_S = \sqrt{(a \cdot \sigma_{R_S})^2 + (b \cdot \sigma_{Rb})^2 + (c \cdot \sigma_{h_0})^2 + (d \cdot \sigma_b)^2 + (e \cdot \sigma_M)^2} = 73.58$$

Coefficient of variation of the health function

$$k_S = \frac{\sigma_S}{\overline{S}} = 0{,}315$$

Reliability index [8] $\beta = \dfrac{\overline{\Phi} - \overline{F}}{\sigma_S} = \dfrac{\overline{S}}{\sigma_S} = 3{,}17$. The value of the function of the normalized normal distribution for the obtained reliability index $\Phi(\beta) = 0.49924$, then the reliability of the crossbar section will be $P(S) = 0.5 + 0.49924 = 0.99924$, respectively, the risk is $P_f = 0.00076$.

The security of the construction is higher than the safety limit $\beta = 3.17 > 3.0$, which characterizes its operability during the whole operation period of the structure.

Evaluation of durability [9,10,11]. Indicators durability are gamma T_γ resource (time between the limit state) and technical resource $T_{res} = T1 + T_\gamma$. The evaluation of the longevity of elements is carried out for a given security of their operability

$$P[T \geq T_{res}] = P[T \geq T_\gamma] = P(S_r > 0) = \Phi(\beta_S), \quad (2)$$

where r is the final calculated cross section of the process $S(t)$.

The construction of the design for the limit state has been established since 1980, when $T1 = 16$ years, and the strength of the concrete was $\overline{R}_b = 0.130 \cdot 105$ kN / m^2. The bearing capacity of the section $\overline{M} = 316$ kN · m is $\overline{\Phi}_н = \overline{R}_b \cdot \overline{b} \cdot x \cdot (\overline{h}_0 - 0{,}5x) = 547.98$ kN · m. Functionality function $\overline{S}_н = \overline{\Phi}_н - \overline{M} = 231.98$ kN · m. The standard of the health function $\sigma_{S_н} = 73.382$ kN · m. Reliability index $\beta = 3.16$, $\Phi(\beta) = 0.4992$, $P(S) = 0.9992$.

According to the 1996 data, it was found that: $\overline{R}_b = 0.105 \cdot 105$ kN / m^2; $\chi = 0.041$ m; $\overline{\Phi}_K = 544.48$ kN · m; $\overline{M}_K = 350$ kN · m; $\overline{S}_K = 194.48$ kN · m; $\sigma_S = 72.89$ kN · m; $\beta = 2{,}67$.

The wear rate of the section of the element in the time interval $t_u = 1996 - 1980$ year $\overline{V}_R = \dfrac{\overline{S}_н - \overline{S}_к}{t_u} = 2{,}34$ kN · m / year. Function of working capacity of the element with allowance for wear $S(t) = S_r - t \cdot V_r$, where $S_r = S_H$.

The reliability index is found from formula

$$\beta_S = \frac{\overline{S}_r - t\overline{V}_R}{\sqrt{\sigma_r^2 + t^2\sigma_{vR}^2}}. \quad (3)$$

We transform this expression

$$\beta_S \sqrt{\sigma_r^2 + t^2\sigma_{vR}^2} = \overline{S}_r - t\overline{V}_R.$$

$$\beta_S^2\left(\sigma_r^2 + t^2\sigma_{vR}^2\right) = \overline{\overline{S}}_r^2 - 2\overline{S}_r t\overline{V}_R + t^2\overline{V}_R^2.$$

We denote: $A = \overline{V}_R^2 - \beta_S^2\sigma_{vR}^2$; $B = -2S_r\overline{V}_R$; $C = S_r^2 - \beta_S^2\sigma_S^2$, then the resource T_γ for a given security will be

$$T = \frac{-B - \sqrt{B^2 - 4AC}}{2A}. \quad (4)$$

Define the resource for three cases of system security: $P(S) = 0.95$, $\beta_S = 1.64$; $P(S) = 0.98$, $\beta_S = 2.06$; $P(S) = 0.99$, $\beta_S = 2.34$. The results of the calculation are presented in the table 1.

Table. 1. The results of the resource calculation

$P(S)$	β_S	A	B	C	T_γ (year)	T_{res} (year)
0,95	1,64	5,16	-1087	3,93·104	46,3	62,3
0,98	2,06	4,97	-1087	3,10·104	33,6	49,6
0,99	2,34	4,82	-1087	2,43·104	25,2	41,2

Thus, after 5-6 years, counting from 2000, it is necessary to conduct a regular survey of trestle structures and repair and restoration works.

4 Conclusions

The above probabilistic methodology allows us to predict the onset of the limiting state in time using only two statistical moments of design parameters: mathematical expectation and variance (standard).
Numerical values of the individual residual life of the operated structures of the production building are obtained by the criterion of the probability of failure-free operation P (S). This is the rationale for planning the operating mode and preventive measures, leading to an increase in the average life of the structures to be repaired, and the establishment of an optimal service life.

References

1. V. Bolotin, Mechanical Engineering, **165,** 312 p. (1984)
2. A. Dobromyslov, Reference manual., 72 p. (2004)
3. Recommendations for assessing the reliability of building structures by external features. 43 p. (2001)
4. BCH 53-86 (p), Rules for assessing the physical wear of residential buildings. (1986)
5. A. Pronikov, Mechanical Engineering, 592p. (1978)
6. G. Poryvay, Manual for high schools, 384p. (1983)
7. A. Rzhanitsyn, Theory of calculating building structures for reliability, 240 p. (1978)
8. V. Raiser, Scientific publication, 384 p. (2010)
9. V. Pshenichkina, K. Sukhina, State. Architect-builds.Un-t., **35**, 85 p. (2014)
10. V. Pshenichkina, K. Sukhina, Construction mechanics of engineering structures and structures **1**(2015).
11. V. Pshenichkina and other, International Review of Civil Engineering 7, pp 158. (2016)

 DOI: 10.1051/matecconf/201712905008

Calculation of a three-layer plate by the finite element method taking into account the creep of the filler

Anton Chepurnenko[1,*], *Andrei* Savchenko[1], and *Svetlana* Yazyeva[1]

[1]Don State Technical University, 344022 Sotsialisticheskaya 162, Russia

Abstract. In the article the derivation of the resolving equations for the calculation of a three-layer plate taking into account the creep of the middle layer by the finite element method is given. Rectangular finite elements are used. The problem reduces to a system of linear algebraic equations. An example of calculating a three-layer plate hinged on the contour and loaded with a uniformly distributed load is considered. A comparison of the results with the solution based on the finite difference method is presented.

1 Introduction

Three-layer structures are widely used in many industries, including aircraft construction, shipbuilding, civil and industrial engineering, etc. Such structures, as a rule, consist of two outer layers with high mechanical characteristics (steel, aluminium, fiberglass) and a lightweight filler located between them. Porous polymers (foams) are widely used as a filler. For polymers in addition to elastic properties, a pronounced rheology is characteristic. The calculation of three-layered structures taking the creep into account is considered in [1-3]. Resolving equations for a triangular finite element of a three-layer plate and shell are given in [1-2]. In the present paper rectangular finite elements with higher accuracy will be considered.

2 Derivation of resolving equations

The rectangular finite element of the three-layer plate is shown in Fig. 1-2. Each node of this element has 5 degrees of freedom: displacements in the plane of the upper layer u_i^- and v_i^-, displacements in the plane of the lower layer u_i^+ and v_i^+, as well as deflection w_i.

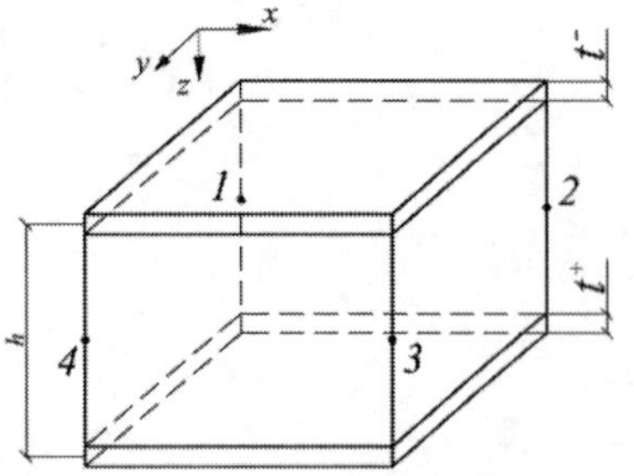

Fig. 1. Rectangular finite element of a three-layer plate: general view.

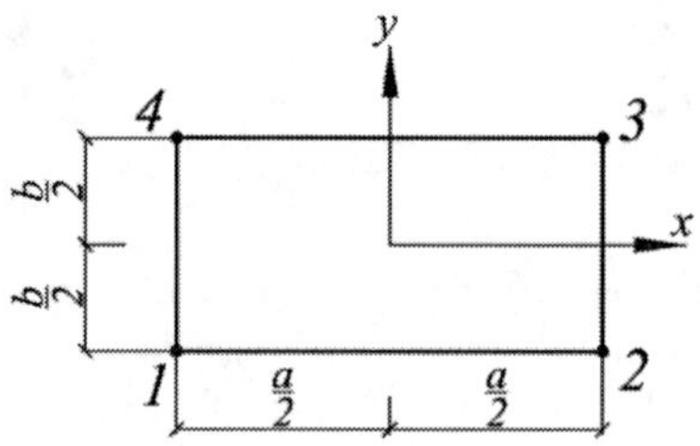

Fig. 2. Rectangular finite element in the plane.

For the displacement field within the element, we take the following approximation:

$$u^{+(-)} = N_1 u_1^{+(-)} + N_2 u_2^{+(-)} + N_3 u_3^{+(-)} + N_4 u_4^{+(-)}$$
$$v^{+(-)} = N_1 v_1^{+(-)} + N_2 v_2^{+(-)} + N_3 v_3^{+(-)} + N_4 v_4^{+(-)} \quad (1)$$
$$w = N_1 w_1 + N_2 w_2 + N_3 w_3 + N_4 w_4,$$

where N_1, N_2, N_3, N_4 – form functions.

$$N_1 = \frac{1}{ab}\left(\frac{a}{2}-x\right)\left(\frac{b}{2}-y\right);\quad N_2 = \frac{1}{ab}\left(\frac{a}{2}+x\right)\left(\frac{b}{2}-y\right);$$
$$N_3 = \frac{1}{ab}\left(\frac{a}{2}+x\right)\left(\frac{b}{2}+y\right);\quad N_4 = \frac{1}{ab}\left(\frac{a}{2}-x\right)\left(\frac{b}{2}+y\right), \quad (2)$$

where a, b – dimensions of the finite element.

The coordinates x and y in formulas (2) are measured from the centre of the finite element.

The deformation vector of a finite element is written as:

$$\{\varepsilon\} = \left\{\varepsilon_x^+ \quad \varepsilon_y^+ \quad \gamma_{xy}^+ \quad \varepsilon_x^- \quad \varepsilon_y^- \quad \gamma_{xy}^- \quad \gamma_{zx}^m \quad \gamma_{yx}^m\right\}^T,$$

where ε_x^+, ε_y^+, γ_{xy}^+ – deformations of the lower layer, ε_x^-, ε_y^-, γ_{xy}^- – deformations of the upper layer, γ_{zx}^m, γ_{yx}^m – deformations of the filler.

* Corresponding author: anton_chepurnenk@mail.ru

In the technical theory of three-layer plates, the relationship between displacements and deformations has the form:

$$\varepsilon_x^{+(-)} = \frac{\partial u^{+(-)}}{\partial x};$$
$$\varepsilon_y^{+(-)} = \frac{\partial v^{+(-)}}{\partial y};$$
$$\gamma_{xy}^{+(-)} = \frac{\partial u^{+(-)}}{\partial y} + \frac{\partial v^{+(-)}}{\partial x}; \qquad (3)$$
$$\gamma_{zx}^{m} = \frac{u^+ - u^-}{h} + \frac{\partial w}{\partial x};$$
$$\gamma_{zy}^{m} = \frac{v^+ - v^-}{h} + \frac{\partial w}{\partial y}.$$

Substituting (1) in (3), we obtain the following relationship between the node displacements and deformations in the matrix form:

$$\{\varepsilon\} = [B]\{U\},$$

where $\{U\}$ — vector of nodal displacements.

$$\{U\} = \begin{Bmatrix} \{\rho_1\} \\ \{\rho_2\} \\ \{\rho_3\} \\ \{\rho_4\} \end{Bmatrix}, \quad \{\rho_i\} = \{u_i^+ \quad v_i^+ \quad u_i^- \quad v_i^- \quad w_i\}^T.$$

Matrix $[B]$ has a size of 8x20. Non-zero elements of the matrix $[B]$ are defined as follows:

$$B_{1,1} = B_{3,2} = B_{4,3} = B_{6,4} = B_{7,5} = \frac{\partial N_1}{\partial x}; \; B_{1,6} = B_{3,7} =$$
$$= B_{4,8} = B_{6,9} = B_{7,10} = \frac{\partial N_2}{\partial x}; \; B_{1,11} = B_{3,12} = B_{4,13} =$$
$$= B_{6,14} = B_{7,15} = \frac{\partial N_3}{\partial x}; \; B_{1,16} = B_{3,17} = B_{4,18} = B_{6,19} =$$
$$= B_{7,20} = \frac{\partial N_4}{\partial x}; \; B_{2,2} = B_{3,1} = B_{5,4} = B_{6,3} = B_{8,5} = \frac{\partial N_1}{\partial y};$$
$$B_{2,7} = B_{3,6} = B_{5,9} = B_{6,8} = B_{8,10} = \frac{\partial N_2}{\partial y}; \; B_{2,12} =$$
$$B_{3,11} = B_{5,14} = B_{6,13} = B_{8,15} = \frac{\partial N_3}{\partial y}; \; B_{2,17} = B_{3,16} =$$
$$= B_{5,19} = B_{6,18} = B_{8,20} = \frac{\partial N_4}{\partial y}; \; B_{7,1} = B_{8,2} = -B_{7,3} =$$
$$= -B_{8,4} = \frac{N_1}{h}; \; B_{7,6} = B_{8,7} = -B_{7,8} = -B_{8,9} = \frac{N_2}{h};$$
$$B_{7,11} = B_{8,12} = -B_{7,13} = -B_{8,14} = \frac{N_3}{h}; \; B_{7,16} = B_{8,17} =$$
$$= -B_{7,18} = -B_{8,19} = \frac{N_4}{h}.$$

The partial derivatives of the form functions are written as:

$$\frac{\partial N_1}{\partial x} = -\frac{1}{ab}\left(\frac{b}{2} - y\right); \quad \frac{\partial N_2}{\partial x} = \frac{1}{ab}\left(\frac{b}{2} - y\right);$$
$$\frac{\partial N_3}{\partial x} = \frac{1}{ab}\left(\frac{b}{2} + y\right); \quad \frac{\partial N_4}{\partial x} = -\frac{1}{ab}\left(\frac{b}{2} + y\right);$$
$$\frac{\partial N_1}{\partial y} = -\frac{1}{ab}\left(\frac{a}{2} - x\right); \quad \frac{\partial N_2}{\partial y} = -\frac{1}{ab}\left(\frac{a}{2} + x\right);$$
$$\frac{\partial N_3}{\partial y} = \frac{1}{ab}\left(\frac{a}{2} + x\right); \quad \frac{\partial N_4}{\partial y} = \frac{1}{ab}\left(\frac{a}{2} - x\right).$$

The resolving equations will be obtained using the variational principle of Lagrange. The potential energy of deformation of a three-layer plate considering creep is defined as follows:

$$W = \frac{1}{2}\int_A t^+(\sigma_x^+\varepsilon_x^+ + \sigma_y^+\varepsilon_y^+ + \tau_{xy}^+\gamma_{xy}^+) +$$
$$+ t^-(\sigma_x^-\varepsilon_x^- + \sigma_y^-\varepsilon_y^- + \tau_{xy}^-\gamma_{xy}^-) + \qquad (4)$$
$$+ h\left[\tau_{zx}^m(\gamma_{zx}^m - \gamma_{zx}^{m*}) + \tau_{zy}^m(\gamma_{zy}^m - \gamma_{zy}^{m*})\right]dA,$$

where γ_{zx}^{m*}, γ_{zy}^{m*} – creep deformations of the filler, A – area of the finite element.

Expression (4) can be rewritten as:

$$W = \frac{1}{2}\int_A \{N\}^T(\{\varepsilon\} - \{\varepsilon^*\})dA = \frac{1}{2}\int_A \{N\}^T([B]\{U\} - \{\varepsilon^*\})dA,$$

where $\{\varepsilon^*\} = \{0 \;\; 0 \;\; 0 \;\; 0 \;\; 0 \;\; 0 \;\; \gamma_{zx}^{m*} \;\; \gamma_{zy}^{m*}\}^T$ creep strain vector,

$\{N\}^T = \{N_x^+ \;\; N_y^+ \;\; N_{xy}^+ \;\; N_x^- \;\; N_y^- \;\; N_{xy}^- \;\; Q_{zx} \;\; Q_{zy}\}$ — vector of internal forces, $N_x^{+(-)} = \sigma_x^{+(-)} \cdot t^{+(-)}$, $N_y^{+(-)} = \sigma_y^{+(-)} \cdot t^{+(-)}$, $N_{xy}^{+(-)} = \tau_{xy}^{+(-)} \cdot t^{+(-)}$, $Q_{zx} = \tau_{zx}^m \cdot h$, $Q_{zy} = \tau_{zy}^m \cdot h$.

The relationship between deformations and internal forces has the form:

$$\{N\} = [D](\{\varepsilon\} - \{\varepsilon^*\}) = [D]([B]\{U\} - \{\varepsilon^*\}) \qquad (5)$$

where $[D]$ — block matrix of elastic constants.

$$[D] = \begin{bmatrix} [D^+] & & \\ & [D^-] & \\ & & [D^m] \end{bmatrix},$$

$$[D^{+(-)}] = \frac{Et^{+(-)}}{1-\nu^2}\begin{bmatrix} 1 & \nu & 0 \\ \nu & 1 & 0 \\ 0 & 0 & (1-\nu)/2 \end{bmatrix}, \quad [D^m] = G_m h\begin{bmatrix} 1 & 0 \\ 0 & 1 \end{bmatrix},$$

G_m — shear modulus of the filler.

 DOI: 10.1051/matecconf/201712905008

Substituting (6) in the expression for the potential energy, we get:

$$W=\frac{1}{2}(\{U\}^T\int_A[B]^T[D][B]dA\{U\}-\{U\}^T\int_A[B]^T[D]dA\{\varepsilon^*\}-$$
$$-\{\varepsilon^*\}^T\int_A[D][B]dA\{U\}+\int_A\{\varepsilon^*\}^T[D]\{\varepsilon^*\}dA)=$$
$$=\frac{1}{2}\{U\}^T\int_A[B]^T[D][B]dA\{U\}-\{U\}^T\int_A[B]^T[D]dA\{\varepsilon^*\}+$$
$$+\frac{1}{2}\int_A\{\varepsilon^*\}^T[D]\{\varepsilon^*\}dA.$$

The total energy represents the difference between the potential energy of deformation and the potential of external forces:

$$T=W-A,$$

where $A=\{U\}^T\{F\}$, $\{F\}$ — vector of external nodal forces.

Differentiating the total energy with respect to the vector of nodal displacements, we obtain:

$$\frac{\partial T}{\partial\{U\}}=[K]\{U\}-\{F\}-\{F^*\}=0,$$

where $[K]=\int_A[B]^T[D][B]dA$ — stiffness matrix, $\{F^*\}=\int_A[B]^T[D]dA\{\varepsilon^*\}$ — the contribution of creep strains to the right-hand side of the system of linear algebraic FEM equations.

Exact expressions for the matrix $[K]$ and vector $\{F^*\}$ coefficients were obtained using functions for the symbol variables of the Matlab package. The vector $\{F^*\}$ has the form:

$$\{F^*\}=\begin{Bmatrix}\{F_1\}\\ -\frac{G_m h}{2}\left(a\gamma_{zy}^{m*}+b\gamma_{zx}^{m*}\right)\\ \{F_1\}\\ \frac{G_m h}{2}\left(-a\gamma_{zy}^{m*}+b\gamma_{zx}^{m*}\right)\\ \{F_1\}\\ \frac{G_m h}{2}\left(a\gamma_{zy}^{m*}+b\gamma_{zx}^{m*}\right)\\ \{F_1\}\\ \frac{G_m h}{2}\left(a\gamma_{zy}^{m*}-b\gamma_{zx}^{m*}\right)\end{Bmatrix},$$

where $\{F_1\}=\frac{abG_m}{4}\begin{Bmatrix}\gamma_{zx}^{m*}\\ \gamma_{zy}^{m*}\\ -\gamma_{zx}^{m*}\\ -\gamma_{zy}^{m*}\end{Bmatrix}.$

The stiffness matrix is not presented here due to its bulkiness.

3 Results and discussion

A three-layer rectangular hinged plate (Fig. 3) was calculated with the following initial data: plate thickness h = 8 cm, modulus of elasticity of sheaths $E = 2\cdot10^5$ MPa, Poisson's coefficient of sheaths ν = 0.3, sheath thickness $t^+=t^-=$ 1.5 mm, shear modulus of the filler $G_m=2.5$ MPa, dimensions of plate $a=b=3$ m. The uniformly distributed over the area load $q=2$ kPa acts on the plate. As the law of creep, the equation of the linear theory of heredity was used:

$$G_m\gamma_i=\tau_i+\int_{-\infty}^{t}\tau_i K(t-\tau)d\tau,\quad i=(xz,yz). \tag{6}$$

The core of creep was assumed to be exponential:

$$K(t-\tau)=C_m e^{-\alpha_m(t-\tau)};\quad C_m=\alpha_m=0{,}077\,\frac{1}{\text{hour}}.$$

With an exponential kernel, the creep law (6) is easily represented in a differential form:

$$G_m\frac{\partial\gamma_i}{\partial t}+G_m\alpha_m\gamma_i=\frac{\partial\tau_i}{\partial t}+(\alpha_m+C_m)\tau_i. \tag{7}$$

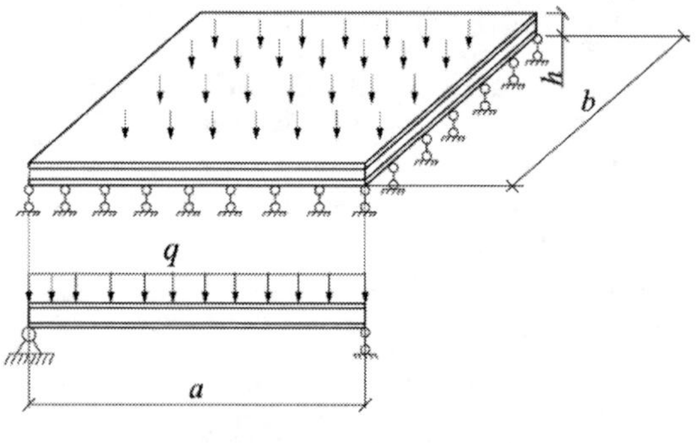

Fig. 3. Settlement scheme of the plate

Expressions (6) and (7) contain the values of the total shear deformation of the filler, which represent the sum of elastic deformations and creep strains:

$$\gamma_i=\frac{\tau_i}{G_m}+\gamma_i^*. \tag{8}$$

Using (8), we can express from (7) the growth rates of creep strains:

$$\frac{\partial\gamma_i^*}{\partial t}=\frac{C_m}{G_m}\tau_i-\alpha_m\gamma_i^*.$$

The calculation was carried out by the step method, the creep strains at the time $t+\Delta t$ were determined as follows:

$$\gamma_i^*(t+\Delta t)=\gamma_i^*(t)+\frac{\partial\gamma_i^*(t)}{\partial t}\Delta t.$$

 DOI: 10.1051/matecconf/201712905008

This method is also used in [4-9]. The following boundary conditions were assumed:

$$x = 0, x = a: \quad w = 0, \quad v^+ = v^- = 0;$$

$$y = 0, y = b: \quad w = 0, \quad u^+ = u^- = 0.$$

Fig. 4 shows the curve of growth of the deflection in the center of the plate. The solid line corresponds to the solution obtained by the authors using the finite element method, the dashed line is the solution by the finite difference method according to the method described in [10]. When $t = 0$ the results are the same and at $t \to \infty$ the difference is 1.83%.

Stresses in the skins and filler do not change during creep. The stress σ_x^+ and τ_{xy}^+ distribution in the lower skin is shown in Fig. 5 - 6. The stresses in the upper skin under the boundary conditions (10) in absolute value coincide with the stresses in the lower skin. The distribution of tangential stresses in the filler is shown in Fig. 7.

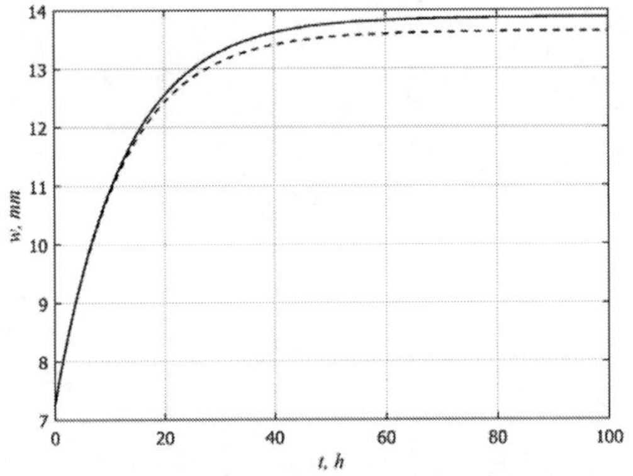

Fig. 4. Graph of growth of deflection in the center of the plate: solid line – finite element method, dashed line – finite differences method.

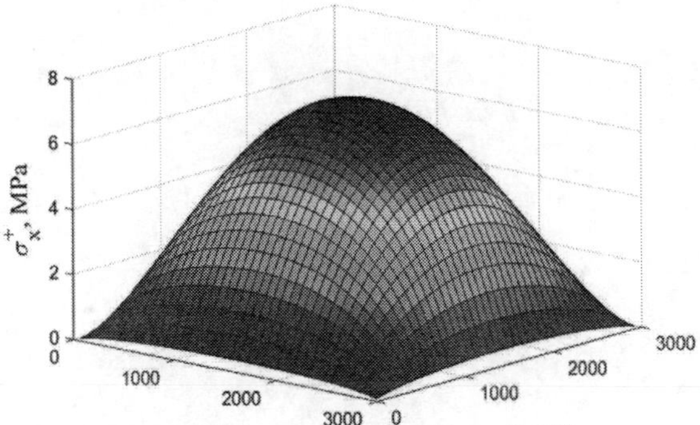

Fig. 5. Distribution of normal stresses in the lower skin.

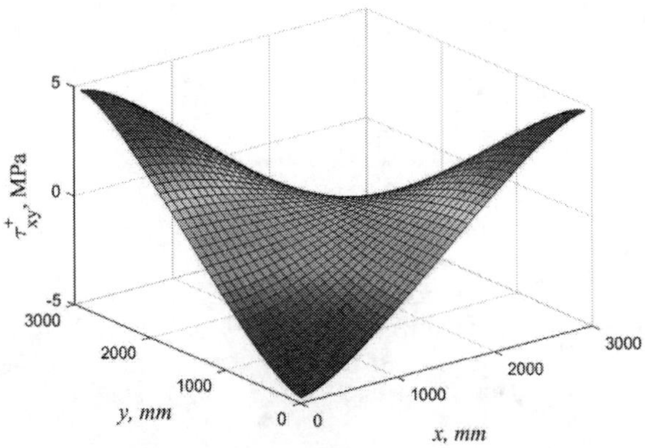

Fig. 6. The distribution of tangential stresses in the bottom skin.

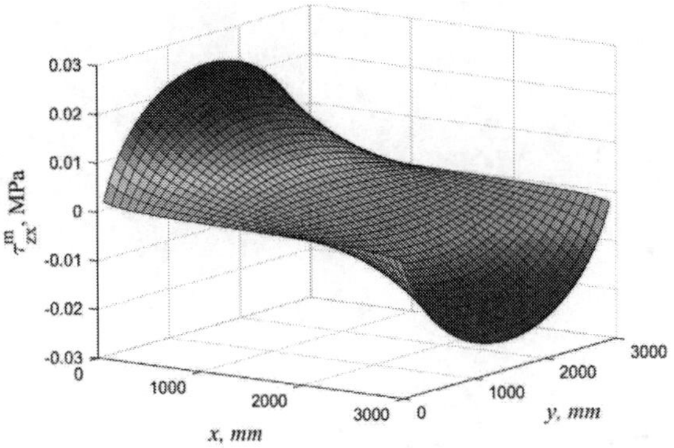

Fig. 7. Distribution of tangential stresses in the filler.

4 Conclusions

The equations obtained are applicable for arbitrary creep laws of the filler, including nonlinear ones. The correctness of the equations and the reliability of the results are confirmed by a comparison with the solution on the basis of the finite difference method. It is established that under linear creep law the stresses in the shells and filler do not change during the creep process.

References

1. A. Chepurnenko, L. Mailyan, B. Jazyev, Procedia Engineering, **165**, pp. 990 – 994 (2016)
2. V. Andreev, B. Yaziev, A. Chepurnenko, S. Litvinov, Vestnik MGSU, **7**, pp. 17-24 (2015)
3. B. Yaziev, A. Chepurnenko, S. Litvinov, S. Yaziev, Bulletin of the Dagestan State Technical University, **2**, pp.47-55 (2014)
4. S. Litvinov, L. Trush, S. Yazyev, Procedia Engineering, **150,** pp. 1686–1693 (2016)
5. S. Litvinov, E. Klimenko, I. Kulinich, S. Yazyeva, International Polymer Science and Technology, **2**, pp. 23-25 (2015)

6. S. Litvinov, L. Trush, A. Dudnik, Engineering Bulletin of the Don, **2**, 2016. http://www.ivdon.ru/en/magazine/archive/n2y2016/3560
7. V. Andreev, A. Chepurnenko, B. Yazyev, Advanced Materials Research, **1004-1005**, pp. 257-260 (2014)
8. V. Andreev, E. Barmenkova, Applied Mechanics and Materials, **204-208**, pp. 3596-3599 (2012)
9. V. Andreev, A. Avershyev, WIT Transactions on the Built Environment. Fluid Structure Interaction VII , pp. 123-132 (2013)
10. V. Andreev, B. Yazyev, A. Chepurnenko, Advanced Materials Research, **900**, pp. 707-710 (2014)

Creep calculation for a three-layer beam with a lightweight filler

Viacheslav Chepurnenko[1,*], *Batyr* Yazyev[1], and *Xuanzhen* Song[1]

[1]Don State Technical University, 344022 Sotsialisticheskaya 162, Russia

Abstract. In the article the technique of calculation of a three-layer beam with a lightweight filler taking into account the creep of the middle layer is given. The problem reduces to a second-order differential equation, which is solved numerically by the method of finite differences. An example of a calculation is presented for a hinged at the ends beam under the action of a uniformly distributed load. The linear Maxwell-Thompson equation is used as the creep law. Solution was performed in software package Matlab.

1 Introduction

Products containing polymers are more prone to creep than many other building materials. Since the middle layer of three-layer panels is often made of foams, this property must be taken into account in the design. The consideration of creep is associated with the solution of complex systems of integro-differential equations [1-7], which significantly hinders the use of the apparatus of creep theory in the practice of engineering calculations. In this paper, a comparatively simple method for calculating a three-layer beam considering creep is given, which can be used in engineering practice.

2 Formulation of the problem

We consider a hinged-supported at the ends beam, loaded with a uniformly distributed load q (Fig. 1).

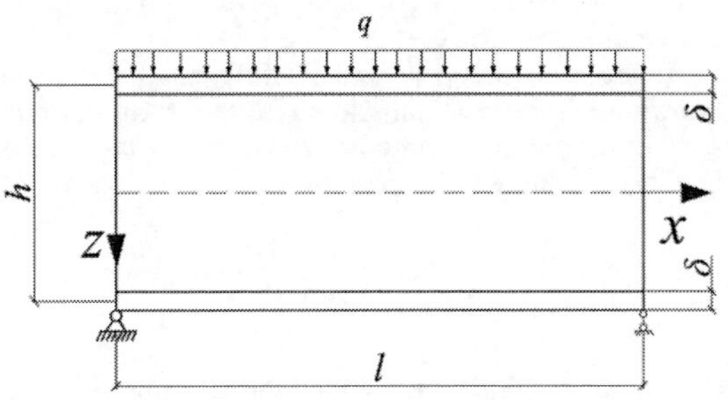

Fig. 1. Calculation scheme.

In deriving the equations, the following hypotheses are accepted:

1. The bending moment is completely perceived by the outer layers.

Proceeding from this hypothesis, the bending moment is related to the stresses in the upper and lower sheathing as follows:

$$M(x) = \sigma^{+} b\delta \frac{h}{2} - \sigma^{-} b\delta \frac{h}{2}, \qquad (1)$$

where b – beam width, h – beam height, δ – thickness of the sheathings, σ^{+} – stress in the bottom sheathing, σ^{-} – stress in the upper sheathing.

Expression (1) can be rewritten as:

$$\frac{M(x)}{I} = \frac{\sigma^{+} - \sigma^{-}}{h}, \qquad (2)$$

where $I = b\delta \frac{h^2}{2}$ – moment of inertia of the sheathings.

2. The displacements along the thickness of the filler are distributed according to a linear law:

$$u_m(z) = \frac{u^{+} + u^{-}}{2} + \frac{u^{+} - u^{-}}{h} z, \qquad (3)$$

where u^{+} and u^{-} – respectively displacements of the lower and upper skin, u_m – displacement of the middle layer.

3. The transverse force is completely perceived by the filler, the filler works only on shear, and the tangential stresses along its thickness are constant. Proceeding from this hypothesis, the transverse force is defined as follows:

$$Q = \tau_m bh = G_m \gamma_m^{el} bh, \qquad (4)$$

where τ_m – shear stress in the filler, G_m – shear modulus of the filler, γ_m^{el} – elastic shear deformation of the filler, representing the difference between total shear deformation and creep deformation:

$$\gamma_m^{el} = \gamma_m - \gamma_m^{*}. \qquad (5)$$

---

[*] Corresponding author: anton_chepurnenk@mail.ru

3 The derivation of the resolving equations

The stresses in the sheathings are related to the displacements as follows:

$$\sigma^{+}=E\frac{\partial u^{+}}{\partial x};\quad \sigma^{-}=E\frac{\partial u^{-}}{\partial x},\tag{6}$$

where E — modulus of elasticity of sheathings.

Substituting (5) into (2), we obtain:

$$\frac{M}{EI}=\frac{1}{h}\frac{\partial}{\partial x}\left(u^{+}-u^{-}\right)\tag{7}$$

The displacement and deformation of the filler are related by the Cauchy equation:

$$\gamma_m=\frac{\partial u_m}{\partial z}+\frac{\partial w}{\partial x}=\frac{u^{+}-u^{-}}{h}+\frac{\partial w}{\partial x}.\tag{8}$$

We substitute (8) in (5), and then (5) in (4):

$$Q=G_m bh\left(\frac{u^{+}-u^{-}}{h}+\frac{\partial w}{\partial x}-\gamma_m^{*}\right).\tag{9}$$

According to Zhuravskii's theorem, the transverse force is related to the intensity of the uniformly distributed load as follows:

$$\frac{\partial Q}{\partial x}=-q.$$

Differentiating (9) with respect to x, we obtain:

$$G_m bh\left(\frac{1}{h}\frac{\partial}{\partial x}\left(u^{+}-u^{-}\right)+\frac{\partial^2 w}{\partial x^2}-\frac{\partial\gamma_m^{*}}{\partial x}\right)=-q.\tag{10}$$

Substituting (7) into (10), we obtain the basic resolving equation:

$$\frac{\partial^2 w}{\partial x^2}=-\frac{M(x)}{EI}-\frac{q}{G_m bh}+\frac{\partial\gamma_m^{*}}{\partial x}.\tag{11}$$

4 Method of solving the problem

Equation (11) is solved numerically by the method of finite differences. The bending moment and shear force for considered beam are determined as follows:

$$M(x)=\frac{qx}{2}(l-x);\quad Q(x)=\frac{ql}{2}-qx.$$

The boundary conditions have the form:

$$w_{x=0}=w_{x=l}=0.$$

The time interval is divided into n_t steps Δt. At the first stage we perform the solution of the elastic problem ($\gamma_m^{*}=0$). A uniform grid in x with step Δx is introduced. The second derivative of the deflection in the node with the number i using central differences is written in the form:

$$\left.\frac{\partial^2 w}{\partial x^2}\right|_i=\frac{w_{i+1}-2w_i+w_{i-1}}{\Delta x^2}.$$

As a result, the problem reduces to a system of linear algebraic equations:

$$[A]\cdot\{X\}=\{B\},$$

where $[A]$ – matrix of coefficients, $\{X\}=\{w_1\ w_2\ \ldots\ w_{n+1}\}^T$ – vector of unknown nodal displacements, $\{B\}=\{0\ f_2\ f_3\ \ldots\ f_n\ 0\}^T$ – vector of free terms.

The coefficients f_i are defined as follows:

$$f_i=-\frac{M_i}{EI}-\frac{q}{G_m bh}+\left.\frac{\partial\gamma_m^{*}}{\partial x}\right|_i.$$

Matrix $[A]$ has the form:

$$[A]=\begin{bmatrix}1&0&0&0&\ldots&0&0&0\\ a_2&b_2&c_2&0&\ldots&0&0&0\\ 0&a_3&b_3&c_3&\ldots&0&0&0\\ \ldots&\ldots&\ldots&\ldots&\ldots&\ldots&\ldots&\ldots\\ 0&0&0&0&\ldots&a_n&b_n&c_n\\ 0&0&0&0&\ldots&0&0&1\end{bmatrix},$$

where $a_i=c_i=\frac{1}{\Delta x^2}$, $b_i=-\frac{2}{\Delta x^2}$.

It follows from formula (4) that for a statically determinate beam, the tangential stresses do not change during the creep process.

If the creep law is given in differential form, then knowing the creep strain and stress at the current time, we can determine the rate of growth of the creep strain and its magnitude at the time $t+\Delta t$:

$$\gamma_m^{*}(t+\Delta t)=\gamma_m^{*}(t)+\frac{\partial\gamma_m^{*}}{\partial t}\Delta t.$$

Numerical differentiation of the function γ_m^{*} with respect to x is performed by the formula:

$$\left.\frac{\partial\gamma_m^{*}}{\partial x}\right|_i=\frac{\gamma_{m,i+1}^{*}-\gamma_{m,i-1}^{*}}{2\Delta x}.\tag{12}$$

The formula (12) is applicable only for internal grid nodes. For the first and last node, it is not necessary to calculate the derivative $\frac{\partial\gamma_m^{*}}{\partial x}$.

DOI: 10.1051/matecconf/201712905009

5 Results and discussion

The test task was solved with the following initial data: $l = 1.5\ \text{m}$, b = 0.1 m, h = 0.06 m, $q = 0.82$ kN/m, $G_m = 25\ \text{MPa}$, $E = 0.71 \cdot 10^5$ MPa, δ = 1 mm.

We used the deformation law, which is described by the Maxwell-Thompson equation [8]:

$$\kappa \frac{\partial \gamma_m}{\partial t} + H\gamma_m = n \frac{\partial \tau_m}{\partial t} + \tau_m, \tag{13}$$

where $\kappa = nG_m = 56\ \text{MPa} \cdot \text{day}$ – coefficient of viscosity of the filler, n – time of relaxation, $H = 15\ \text{MPa}$ – long shear modulus of the filler.

The total deformation is represented as the sum of elastic deformation and creep deformation:

$$\gamma_m = \frac{\tau_m}{G_m} + \gamma_m^*. \tag{14}$$

Substituting (14) into (13), we obtain an expression for the growth rate of the creep strain:

$$\frac{\partial \gamma_m^*}{\partial t} = \frac{1}{\kappa}\left[(1 - \frac{H}{G_m})\tau_m - H\gamma_m^*\right]. \tag{15}$$

The resulting curve of the deflection growth in the middle of the span is shown in Fig. 2.

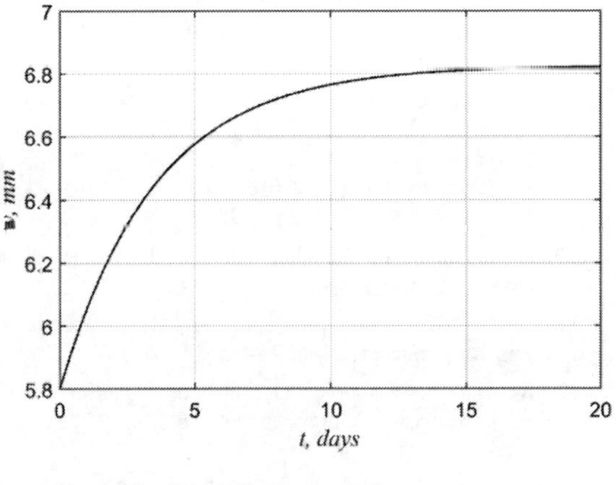

Fig. 2. The graph of the deflection growth

We note that under the law of creep (15), the problem can be solved analytically. The law of creep (15) can be rewritten in the form:

$$\frac{\partial \gamma_m^*}{\partial t} = k_1 - k_2\gamma_m^*, \tag{16}$$

where $k_1 = \dfrac{\left(1 - \dfrac{H}{G_m}\right)\tau_m}{nG_m}$, $k_2 = \dfrac{H}{nG_m}$.

We divide the variables in Eq. (16):

$$\frac{\partial \gamma_m^*}{k_1 - k_2\gamma_m^*} = \partial t. \tag{17}$$

Integrating the expression (17), we obtain:

$$-\frac{1}{k_2}\ln\left|k_1 - k_2\gamma_m^*\right| = t + \ln C(x),$$

where $C(x)$ – arbitrary function of integration which is determined from condition:

$$\gamma_m^*\Big|_{t=0} = 0 \rightarrow -\frac{1}{k_2}\ln\left|k_1\right| = \ln C(x) \rightarrow C(x) = k_1^{-1/k_2}.$$

The dependence of the creep strain on time takes the form:

$$\gamma_m^*(t) = \frac{k_1}{k_2}\left(1 - e^{-k_2 t}\right) = \tau_m\left(\frac{1}{H} - \frac{1}{G_m}\right)\left(1 - e^{-\frac{H}{nG_m}t}\right). \tag{18}$$

Differentiating expression (18) with respect to x, we obtain:

$$\frac{\partial \gamma_m^*}{\partial x} = \frac{\partial \tau_m}{\partial x}\left(\frac{1}{H} - \frac{1}{G_m}\right)\left(1 - e^{-\frac{H}{nG_m}t}\right) =$$

$$= \frac{1}{bh}\frac{\partial Q}{\partial x}\left(\frac{1}{H} - \frac{1}{G_m}\right)\left(1 - e^{-\frac{H}{nG_m}t}\right) =$$

$$= -\frac{q}{bh}\left(\frac{1}{H} - \frac{1}{G_m}\right)\left(1 - e^{-\frac{H}{nG_m}t}\right).$$

Equation (11) for considered beam takes the form:

$$\frac{\partial^2 w}{\partial x^2} = -\frac{1}{EI}\frac{qx}{2}(l - x) - \frac{q}{G(t)A}, \tag{19}$$

where $\dfrac{1}{G(t)} = \dfrac{1}{G_m} + \left(\dfrac{1}{H} - \dfrac{1}{G_m}\right)\left(1 - e^{-\frac{H}{nG_m}t}\right)$, $A = bh$.

Integrating equation (19) with respect to x, we obtain:

$$\frac{\partial w}{\partial x} = -\frac{1}{EI}\left(\frac{qx^2 l}{4} - \frac{qx^3}{6}\right) - \frac{q}{G(t)A}x + C_1(t).$$

$$w(x) = -\frac{1}{EI}\left(\frac{qx^3 l}{12} - \frac{qx^4}{24}\right) - \frac{q}{G(t)A}\frac{x^2}{2} + C_1(t)x + C_2(t).$$

Functions $C_1(t)$ and $C_2(t)$ we determine from the boundary conditions:

$$w\big|_{x=0} = 0 \rightarrow C_2(t) = 0;$$

$$w\big|_{x=l} = 0: \quad -\frac{1}{EI}\left(\frac{ql^4}{12} - \frac{ql^4}{24}\right) - \frac{q}{G(t)A}\frac{l^2}{2} + C_1(t)l = 0 \rightarrow$$

$$\rightarrow C_1(t) = \frac{ql}{2G(t)A} + \frac{ql^3}{24EI}.$$

The expression for the deflection in the middle of the span takes the form:

$$w\big|_{x=\frac{l}{2}} = \frac{5ql^4}{384EI} + \frac{ql^2}{8G(t)A}. \quad (20)$$

At $t = 0$, the maximum deflection is defined as

$$w_{\max}\big|_{t=0} = \frac{5ql^4}{384EI} + \frac{ql^2}{8G_m A}. \quad (21)$$

Furthermore, at $t \rightarrow \infty$ substitution in the formula (20) gives:

$$w_{\max}\big|_{t\rightarrow\infty} = \frac{5ql^4}{384EI} + \frac{ql^2}{8HA}. \quad (22)$$

Substituting the initial data into the formulas (21) and (22), we obtain at $t = 0$ $w_{\max} = 5.77$ mm and at $t \rightarrow \infty$ $w_{\max} = 6.79$ mm. The results obtained numerically almost coincide with the analytical solution.

6 Summary

When solving a test problem, one of the simplest creep laws was considered, which is valid for linearly viscoelastic material. However, Eq. (11) allows to specify an arbitrary creep law, including a nonlinear one. For polymers, the nonlinear Maxwell-Gurevich equation [9] agrees well with the experimental data. The method for determining the relaxation constants included in this equation is given in [9-10].

It is necessary to note that an analogous problem was solved in [11].The Voigt law was used, which has the form:

$$\tau_m = \kappa \frac{\partial \gamma_m}{\partial t} + H\gamma_m. \quad (23)$$

The authors of the paper [11] have reduced the problem to a system of two differential equations, which is not very convenient.

If the stress τ_m is constant, then the derivative $\frac{\partial \tau_m}{\partial t}$ vanishes and the results obtained on the basis of the Maxwell-Thompson law must coincide with the results based on the Voigt law.

Comparison of our results with work [11] is not possible, since in this paper the authors do not give all the initial data necessary for calculating.

An experimental study of the creep of foam filler and three-layer beams is carried out in [12-13].However, the authors use Findley's power law, in which the time is contained in an explicit form. Such a law cannot be used for complex loading regimes.

Thus, the question of approximating the creep curves of the filler on the basis of the Maxwell-Gurevich equation remains valid.

References

1. S. Litvinov, L. Trush, S. Yazyev, Procedia Engineering, **150**, pp. 1686–1693 (2016)
2. S. Litvinov, E. Klimenko, I. Kulinich, S. Yazyeva, International Polymer Science and Technology, **2**, pp. 23-25 (2015).
3. V. Andreev, A. Avershyev, WIT Transactions on the Built Environment. Fluid Structure Interaction VII , pp. 123-132 (2013).
4. V. Andreev, E. Barmenkova, Applied Mechanics and Materials, **204-208**, pp. 3596-3599 (2012)
5. V. Andreev, A. Chepurnenko, B. Yazyev, Advanced Materials Research, **1004-1005**, pp. 257-260 (2014)
6. S. Litvinov, L. Trush, A. Dudnik, Engineering Bulletin of the Don, **2**, 2016. URL : http://www.ivdon.ru/en/magazine/archive/n2y2016/3560.
7. V. Andreev, B. Yazyev, A. Chepurnenko, Advanced Materials Research, **900**, pp. 707-710 (2014)
8. A. Chepurnenko, L. Mailyan, B. Jazyev, Procedia Engineering, **165**, pp. 990 – 994 (2016)
9. A. Chepurnenko, V. Andreev, A. Beskopylny, B. Jazyev, MATEC Web of Conferences, 06059 (2016). URL: http://www.matec-conferences.org/articles/matecconf/pdf/2016/30/matecconf_smae2016_06059.pdf
10. A. Dudnik, A. Chepurnenko, S. Litvinov, Plastic masses, **1-2**, pp. 30-33 (2016)
11. A. Gavrilov, Yu. Osetinsky, Light construction of building coverings, pp. 76-82 (1976).
12. M. Garrido, J. Correia, F. Branco, T. Keller, Journal of Composite Materials, **48(18)**, pp. 2237–2249 (2014).
13. M. Sa et al. Composite Structures, **93**, pp. 2409-2418 (2011)

MATEC Web of Conferences **129**, 05010 (2017) DOI: 10.1051/matecconf/201712905010
ICMTMTE 2017

Energy method in solving the problems of stability for a viscoelastic polymer rods

Serdar Yazyev[1,*], *Maria* Kozelskaya[1], *Grigory* Strelnikov[1], and *Stepan* Litvinov[1]

[1] Don State Technical University, 344022 Sotcialisticheskaya 162, Rostov-on-Don, Russia

Abstract. With the energy method in the form of Ritz-Timoshenko solved the stability problem for a polymer rod under axial compression for the clamping-free edge option. The proposed form of loss of stability is chosen as a sum of functions with indeterminate coefficients. The shape functions for various fastenings and the represented fastening of the rod in particular are considered. The result is obtained numerically using the MatLab complex. A study was made of the long critical loads. It is shown that if the compressive force does not exceed the long-term critical one, then stability loss does not occur.

1 Introduction

Analysis of the load-carrying capacity of the structure, in addition to strength calculation, should include the stability issues of the entire system and its individual elements. This is important for construction, mechanical engineering, since many structural elements work on compression and can lose stability at stresses much less than destructive [1-4]

To reduce weight, cost and labor, and also to improve the quality of building structures and their elements, it is necessary to develop and introduce new types of light and lightweight structures. A significant role in the development of effective designs is performed by improving the calculation methods taking into account the actual work of the material [9-12].

One of the factors that have a significant effect on the stress-strain state and the deformability of the structures is creep.

The aim of the article is to solve the stability problems of rods, taking into account initial imperfections, various variants of fixing, and arbitrary creep laws.

2 Formulation of the problem

According to the Ritz-Timoshenko method, the assumed form of loss of stability is chosen as a sum of functions with indeterminate coefficients $a_i\ (i=1,...n)$:

$$v(x)=\sum_{i=1}^{n} a_i f_i \tag{1}$$

As f_i we take functions from x, satisfying the kinematic boundary conditions of the problem, i.e. those that relate to the deflection and rotation angles, regardless of a_i.

The resolving equations for determining the unknown coefficients of the series are obtained from the condition of the minimum of the total potential energy of the Э system:

$$\frac{\partial Э}{\partial a_1}=0,\ \frac{\partial Э}{\partial a_2}=0,\ ...,\frac{\partial Э}{\partial a_n}=0 \tag{2}$$

The total potential energy of the rod is written in the form:

$$Э=\Pi-A, \tag{3}$$

Where Π – potential energy of deformation,
$A=F\Delta$ – work of external forces.
The potential strain energy has the form:

$$\Pi=\frac{1}{2}\int_V \sigma\varepsilon^{el}\,dV=\frac{1}{2}\int_V E(\varepsilon^{el})^2\,dV. \tag{4}$$

Here ε^{el} is meant an elastic deformation that represents the difference between total deformation and creep strain:

$$\varepsilon^{el}=\varepsilon_0-\frac{d^2v}{dx^2}y-\varepsilon^{*} \tag{5}$$

3 The derivation of the resolving equations

Substituting (5) in (4), we obtain:

* Corresponding author: ps62@yandex.ru

$$\Pi = \frac{1}{2}\int_0^l E(x)dx\int_A \left\{ \begin{matrix} \varepsilon_0^2 + \left(\frac{d^2v}{dx^2}\right)^2 y^2 + \left(\varepsilon^*\right)^2 + \\ +2\varepsilon^* \frac{d^2v}{dx^2} y - 2\varepsilon_0\varepsilon^* \\ -2\varepsilon_0 \frac{d^2v}{dx^2} y \end{matrix} \right\} dA =$$

$$= \frac{1}{2}\int_0^l E(x) \left\{ \begin{matrix} I(x)\left(\frac{d^2v}{dx^2}\right)^2 + 2\frac{d^2v}{dx^2}\int_A \varepsilon^* y dA + \\ +\int_A (\varepsilon^*)^2 dA - 2\varepsilon_0 \int_A \varepsilon^* dA + \varepsilon_0^2 A \end{matrix} \right\} dx.$$

Approximation of the ends of the elementary fragment dx of the rod (see Figure 1) during buckling can be determined by formula:

$$\Delta(dx) = dx \cdot \left(\cos\left(\frac{dv_0}{dx}\right) - \cos\left(\frac{dv}{dx} + \frac{dv_0}{dx}\right) \right) =$$

$$-dx \cdot 2\sin\frac{\frac{dv_0}{dx} + \left(\frac{dv}{dx} + \frac{dv_0}{dx}\right)}{2} \cdot \sin\frac{\frac{dv_0}{dx} - \left(\frac{dv}{dx} + \frac{dv_0}{dx}\right)}{2} =$$

$$= 2 \cdot \frac{\left(\left(2 \cdot \frac{dv_0}{dx} + \frac{dv}{dx}\right) \cdot \left(\frac{dv}{dx}\right)\right)}{4} = \left(\frac{1}{2}\left(\frac{dv}{dx}\right)^2 + \frac{dv}{dx}\frac{dv_0}{dx}\right) dx.$$

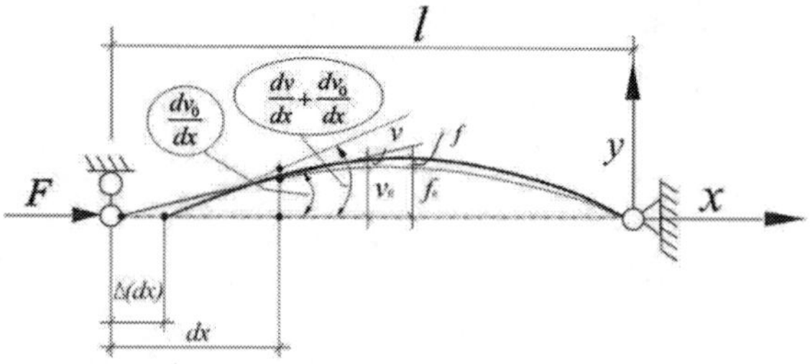

Fig. 1. To the determination of the approach of the ends of the rod during buckling.

Then the convergence of the ends of the whole rod will be written in the form:

$$\Delta = \frac{1}{2}\int_0^l \left\{ \left(\frac{dv}{dx}\right)^2 + 2\frac{dv}{dx}\frac{dv_0}{dx} \right\} dx$$

The work of external forces is defined as follows:

$$A = F\Delta = \frac{1}{2} F \int_0^l \left\{ \left(\frac{dv}{dx}\right)^2 + 2\frac{dv}{dx}\frac{dv_0}{dx} \right\} dx.$$

The total potential energy of the system is written in the form:

$$Э = \frac{1}{2}\int_0^l E(x) \left\{ \begin{matrix} I(x)\left(\frac{d^2v}{dx^2}\right)^2 + 2\frac{d^2v}{dx^2}\int_A \varepsilon^* y dA \\ +\int_A (\varepsilon^*)^2 dA - 2\varepsilon_0 \int_A \varepsilon^* dA + \varepsilon_0^2 A \end{matrix} \right\} dx -$$

$$-\frac{1}{2} F \int_0^l \left\{ \left(\frac{dv}{dx}\right)^2 + 2\frac{dv}{dx}\frac{dv_0}{dx} \right\} dx. \qquad (6)$$

After substituting (1) in (6) and then minimizing it with unknown coefficients a_i we obtain:

$$\frac{\partial Э}{\partial a_j} = \sum_{i=1}^n a_i \left\{ \begin{matrix} \int_0^l E(x) I(x) \frac{d^2 f_i}{dx^2}\frac{d^2 f_j}{dx^2} dx \\ -F\int_0^l \frac{df_i}{dx}\frac{df_j}{dx} dx \end{matrix} \right\} +$$

$$+\int_0^l E(x)\frac{d^2 f_j}{dx^2}\left\{ \int_A \varepsilon^* y dA \right\} dx -$$

$$-F\int_0^l \frac{df_j}{dx}\frac{dv_0}{\partial x} dx = 0, \; j = 1 \ldots n$$

Or in the matrix form:

$$([K] - F[B])\{X\} = \{C\}, \qquad (7)$$

Where $X = \begin{Bmatrix} a_1 \\ \ldots \\ a_n \end{Bmatrix}$, $\{C\} = \begin{Bmatrix} c_1 \\ \ldots \\ c_n \end{Bmatrix}$, $K = \begin{bmatrix} k_{11}\; k_{12} \ldots k_{1n} \\ \ldots\ldots\ldots\ldots \\ k_{n1}\; k_{n2} \ldots k_{nn} \end{bmatrix}$,

$$B = \begin{bmatrix} b_{11}\; b_{12} \ldots b_{1n} \\ \ldots\ldots\ldots\ldots \\ b_{n1}\; b_{n2} \ldots b_{nn} \end{bmatrix}.$$

$$k_{ij} = \int_0^l E(x) I(x) \frac{d^2 f_i}{dx^2}\frac{d^2 f_j}{dx^2} dx;$$

$$b_{ij} = \int_0^l \frac{df_i}{dx}\frac{df_j}{dx} dx;$$

$$c_j = -\int_0^l E(x)\frac{d^2 f_j}{dx^2}\left\{ \int_A \varepsilon^* y dA \right\} dx + F\int_0^l \frac{df_j}{dx}\frac{dv_0}{dx} dx.$$

Instant critical force can be determined from the condition that the determinant of the system (7):

$$\det([K] - F[B]) = 0. \qquad (8)$$

Thus, the problem of determining the critical force reduces to a generalized vector equation. In the case where the matrix $[B]$ is not degenerate, system (7) can be multiplied by $[B]^{-1}$:

$$\left([B]^{-1}[K] - F[E]\right)\{X\} = [B]^{-1}\{C\}, \qquad (9)$$

DOI: 10.1051/matecconf/201712905010

where $[E]$ – unit matrix.

The first critical force will correspond to the minimum of the eigenvalues of the matrix $[B]^{-1}[K]$.

4 Method of solving the problem

Consider the function of the form $f_i(x)$ when fixing the rod embodiment "jamming - free edge" (see Figure 2). The initial deflection is defined as follows:

$$v_0(x) = f_0\left(1-\cos\left(\frac{\pi x}{2l}\right)\right). \quad (10)$$

The function takes the form of:

$$f_i(x) = 1-\cos\left(\frac{\pi x i}{2l}\right). \quad (11)$$

The derivatives of (10) and (11) are written in the form:

$$\frac{dv_0}{dx} = \frac{\pi f_0}{2l}\sin\frac{\pi x}{2l}; \quad (12)$$

$$\frac{df_i}{dx} = \frac{\pi i}{2l}\sin\frac{\pi x i}{2l};$$
$$\frac{d^2 f_i}{dx^2} = \frac{\pi^2 i^2}{4l^2}\cos\frac{\pi x i}{2l}. \quad (13)$$

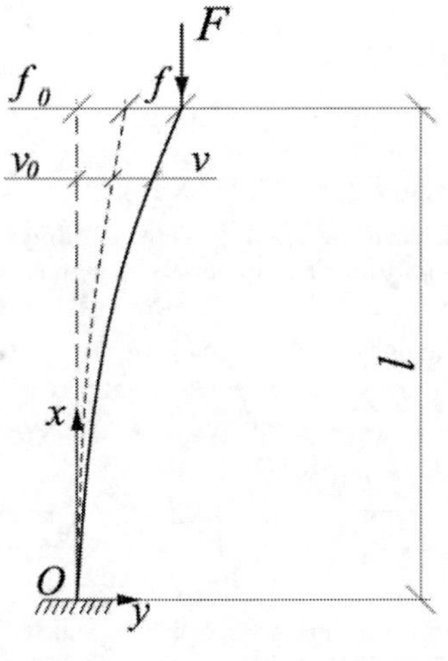

Fig. 2. The option of embodiment a "jamming-free edge."

The integrals in the formulas of the coefficients of the system of equations are written in the form:

$$\int_0^l \frac{df_i}{dx}\frac{df_j}{dx}dx = \left(\frac{\pi}{2l}\right)^2 ij\int_0^l \sin\left(\frac{\pi x i}{2l}\right)\cdot \sin\left(\frac{\pi x j}{2l}\right)dx = \begin{cases} 0, при\ i\neq j; \\ \dfrac{\pi^2 ij}{8l}, при\ i=j. \end{cases} \quad (14)$$

$$\int_0^l \frac{df_j}{dx}\frac{dv_0}{dx}dx = \frac{\pi^2 j f_0}{4l^2}\int_0^l \sin\left(\frac{\pi x j}{2l}\right)\cdot \sin\left(\frac{\pi x}{2l}\right)dx = \begin{cases} 0, при\ j\neq 1; \\ \dfrac{\pi^2 j f_0}{8l}, при\ j=1. \end{cases} \quad (15)$$

In the case of constant stiffness of the rod k_{ij} the coefficients take the form:

$$k_{ij} = EI\int_0^l \frac{d^2 f_i}{dx^2}\frac{d^2 f_j}{dx^2}dx = EI\left(\frac{\pi}{2l}\right)^4 i^2 j^2 \cdot \int_0^l \cos\left(\frac{\pi x i}{2l}\right)\cos\left(\frac{\pi x j}{2l}\right)dx = \quad (16)$$

$$= \begin{cases} 0,\ при\ i\neq j; \\ \dfrac{EI\pi^4 i^2 j^2}{32 l^3}, при\ i=j. \end{cases}$$

5 Results and discussion

We consider a rod of rectangular constant cross-section from epoxy resin EDT-10. Rod sizes: $l = 157мм$, $b = 15мм$, $h = 8мм$, $f_0 = 0{,}16мм$, The rod is acted upon by force $F = 0.165Кн$. This problem was solved by M. Y. Kozelskaya in [2], but by a different method. She received the critical time value - $t_{кр} = 1ч43мин$.

The Euler force for the problem under consideration: $F_э = \dfrac{\pi^2 EI}{4l^2} = 0.189Кн$. Prolonged critical force: $F_д = \dfrac{H}{E}F_э = 0.097Кн.$

Figure 3 shows a graph of deflection height. Critical time $t_{кр} = 1ч41мин$. The rod loses stability over 130 hours. For a criterion of stability loss was taken condition: $f \rangle \dfrac{l}{10}$.

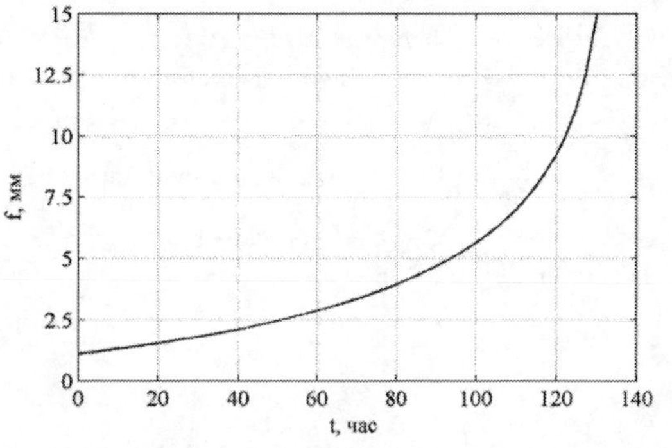

Fig. 3. The growth of the deflection at F=0.165 кН

This problem was solved both by the finite difference method and by the Ritz-Timoshenko method.

 DOI: 10.1051/matecconf/201712905010

Comparison of the value of the deflection at different times is shown in table 1.

Table 1. Comparison of deflections of the rod, obtained by the Ritz-Timoshenko method and using the finite difference method.

t, hour		20	40	60	80	100	120	130
f, mm	*Ritz-Timoshenko method*	1.5425	2.1131	2.8719	3.9430	5.6411	9.2379	14.930
	Finite difference method	1.5439	2.1151	2.8747	3.9472	5.6480	9.2533	14.989

The change in deflection as a function of x and t is shown in Fig. 4.

When solving the Ritz-Timoshenko method, the number of terms in the series n assumed to be equal to 4. The number of intervals for x and y is equal to 50.

From the table presented, it can be seen that the results are the same, which indicates their reliability, as well as the reliability of the calculation methodology developed by the authors.

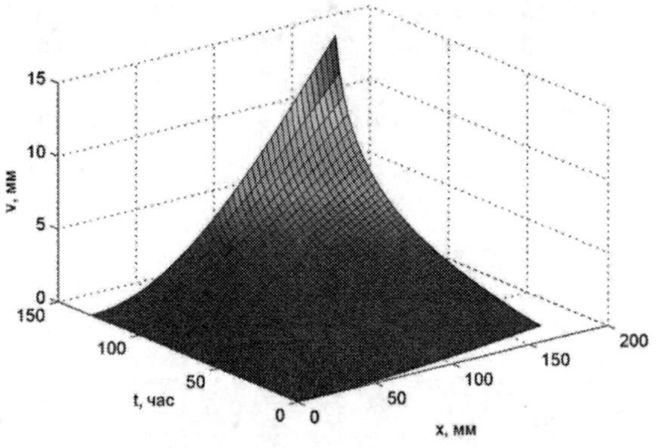

Fig. 4. The growth of the deflection at F=0.165 кН

The test problem was also solved for $F = 80H$, which is less than $F_{дл}$. T the theoretical value of the deflection at the end of the creep process: $f_\infty = \frac{0.16}{97.6/80-1} = 0.727$ мм. The growth curve of the deflection when $F = 80H$ is shown in Figure 5. At $t = 15000ч$, $f = 0,726$.

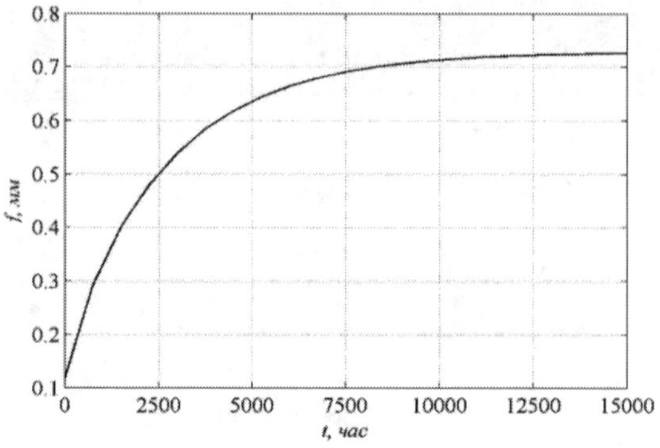

Fig. 5. The growth of deflection at F=80 H.

6 Conclusions

The resolving equations are obtained and a method based on the finite difference method is developed, the Ritz-Timoshenko method for calculating the stability of creep rods of variable stiffness with allowance for initial imperfections, various fixing variants, and arbitrary creep laws.

For compressed polymer rods that obey the nonlinear Maxwell-Gurevich equation, a technique has been developed for determining long-term critical loads. It is established that if the compressive does not exceed a prolonged critical one, then stability loss does not occur.

References

1. B. Yazyev, A. Chepurnenko, S. Litvinov, S. Yazyev. Bulletin of the Dagestan State Technical University, **2**, pp.47-55 (2014)
2. M. Kozelskaya. Longitudinal bending of rods from mesh and linear polymers with nonlinear creep: dis. Cand. Tech. Sciences. Rostov-on-Don, 2013. 131 p.
3. A. Chepurnenko, L. Mailyan, B. Jazyev. Procedia Engineering, **165**, pp. 990 – 994 (2016)
4. A. Chepurnenko, V. Andreev, B. Yazyev. Vestnik MGSU, **1,** pp. 101-108 (2013)
5. S. Litvinov, L. Trush, S. Yazyev. Procedia Engineering, **150,** pp. 1686–1693 (2016)
6. S. Litvinov, E. Klimenko, I. Kulinich, S. Yazyeva. International Polymer Science and Technology, **2**, pp. 23-25 (2015).
7. V. Andreev, A. Chepurnenko, B. Yazyev. Advanced Materials Research, **1004-1005**, pp. 257-260 (2014)
8. V. Andreev, A. Avershyev. WIT Transactions on the Built Environment. Fluid Structure Interaction VII , pp. 123-132 (2013).
9. V. Andreev, B. Yazyev, A. Chepurnenko. Advanced Materials Research, **900**, pp. 707-710 (2014)
10. E.S. Klimenko, E.H. Amineva, S.V. Litvinov, S.B. Yazyev, I.I. Kulinich, *Stability of Compressed Heterogeneous Rods with Account for the Physical Nonlinearity of the Material*. Rostov-on-Don,

DOI: 10.1051/matecconf/201712905010

Rostov State University of Civil Engineering Publ., **4**, pp. 77 (2012)

11. A.S Vol'mir, *Stability of Deformable Systems* (Moscow, Nauka Publ., 1975)
12. P.A. Belous, *Stability of a Polymer Rod Exposed to Creeping with Regard for Its Initial Curvature,* Works of Odessa Polytechnic Institute. **2**, pp. 43—46 (2001)

Recipe-technological aspects of improving the properties of non-autoclaved aerated concrete

Anatoliy Shuisky[1,*], *Sergey* Stelmakh[1], *Evgeniy* Shcherban[1], and *Elena* Torlina[1]

[1]Construction and Architecture Academy at Don State Technical University, Rostov-on-Don, Russian Federation

Abstract. The paper dwells upon non-autoclaved aerated concrete technology issues relating to the effect of various composition and technological factors on the aerated-concrete non-autoclave products. It presents the results of experiments carried out to study the effect of three factors: the order in which components are added to the AC mixture; the efficiency of adding a complex sodium additive to the mixture; the effect of adding the structuring additive, the 'off-cut'. Comparative analysis of physico-mechanical test results revealed that adding the structuring additive increases the compression strength of aerated concrete while also making it denser.

1 Introduction

As of today, aerated concrete is popular in construction as an efficient heat insulation. Producing high-quality aerated concrete products is very relevant.

Aerated concrete is either a cement or lime mortar, classified as lightweight concrete, in which air-voids are entrapped in the mortar matrix by means of a suitable aerating agent. Broadly speaking aerated concrete falls into the group of cellular concrete (microporite being the other). The prominent advantage of aerated concrete is its lightweight, which economizes the design of supporting structures including the foundation and walls of lower floors. It provides a high degree of thermal insulation and considerable savings in material due to the porous structure. By appropriate method of production, aerated concrete with a wide range of densities can be obtained thereby offering flexibility in manufacturing products for specific applications (structural, partition and insulation grades). There have been several investigations on the properties of aerated concrete in the past.

Although aerated concrete was initially envisaged as an insulation material, there has been renewed interest on its structural characteristics in view of its lighter weight, savings in material and potential for large-scale utilization of wastes like pulverized fuel ash. Hence, it was felt essential to compile and review the available literature on aerated concrete. This review aims to classify the studies on aerated concrete related to its material structure and properties. Gas-forming chemicals are mixed into lime or cement mortar during the liquid or plastic stage, resulting in a mass of increased volume and when the gas escapes, leaves a porous structure. Aluminum powder, hydrogen peroxide/bleaching powder and calcium carbide liberate hydrogen, oxygen and acetylene, respectively. Among these, aluminum powder is the most commonly used aerating agent. Efficiency of aluminum powder process is influenced by its fineness, purity and alkalinity of cement, along with the means taken to prevent the escape of gas before hardening of mortar. In the case of Portland cements with low alkalinity, addition of sodium hydroxide or lime supplement the alkali required.

Aerated concrete can be non-autoclaved or autoclaved based on the method of curing. The compressive strength, drying shrinkage, absorption properties etc. directly depend on the method and duration of curing. The strength development is rather slow for moist-cured products. Autoclaving initiates reaction between lime and silica/alumina bearing ingredients.

Properties of concrete such as strength, permeability, diffusivity, shrinkage and creep are intimately related to its porosity and pore size distribution. Thus, the characterization of the pore structure is extremely important, and more so in the case of aerated concrete where the porosity may be as high as 80%. Porosity and pore size distribution of aerated concrete varies considerably with the composition and method of curing [1].

Russian and foreign scientists have developed theories on the aerated concrete structuring processes as well as the basics of producing high-quality materials that have the necessary physico-mechanical properties and sufficient service life.

The important problem here consists in developing and organizing the production of pasty gassing agents for construction material industries. To this end, we have to identify the most efficient of theexisting gassing agents based on such criteria as reactionary capacities, technological versatility, and long-term storage under various conditions [2].

* Corresponding author: rgsu_tvvb@mail.ru

DOI: 10.1051/matecconf/201712905011

The optimal aerated concrete structure is very and uniformly porous; cells are optimally distributed by size, while the defects of interporous partitions are minor to none.

Gassing kinetics induced by the agent as well as the changes in the rheological properties of the interporous material are crucial for the formation of the pore structure in aerated concrete [3, 4, 5, 6, 7, 8].

We paid particular attention to the kinetics of the rheological properties of the interporous materials as observed during the swelling of cellular-concrete mixtures [9, 10, 11, 12, 3, 13].

The interporous partitions in such an AC mixture is a complex kinetic system where the changes in rheological properties mostly depend on the structuring of the cement. Both Russian and non-Russian scientists have studied the structuring of concretes and cements: A.A. Baikov, Yu.M. Butt, P.P. Budnikov, R.L. Berger, S. Brunauer, A.V. Volzhensky, J. Verbeck, H. Le Chatelier, F. Lawrence, Yu.S. Malinin, V. Mikhaelis, O.P. Mchedlov-Petrosyan, V.B. Ratinov, I.N. Kruglitsky, L.A. Pashchenko, P.L. Rebinder, M.M. Sychev, V.V. Timashev, et al. When the cement is water-mixed (gauged), it triggers continuous and complex physico-chemical transformations that occur in parallel with the dissolution, hydration, topochemical reactions, coagulations, crystallizations, etc. By measuring the intensity of these processes, we can adjust the rheological properties of the mixture at the various stages of cement curing, thus controlling the structure and properties of the cured material.

Swelling occurs when three intensively evolving processes—structuring of the cement and the interporous partitions, heat release, and gassing induced by the gassing agent—overlap. Slight variations in technological parameters result in drastic changes in the swelling kinetics, making the properties of concrete less stable. This assumption prejudges the search for solutions to eliminate these disadvantages.

Based on the achievements of physico-chemical mechanics, a branch founded by P.A. Rebinder, O.P. Mchedlov-Petrosyan worded the principles of controllable structuring induced by hydration curing; the most important principles are the quality of the original mixture component and the conditions under which it is prepared (homogenization, the time and intensity of mechanical impact).

Mixtures used in the production of aerated concrete should have low critical shear stress point while also having sufficiently high viscosity. This assumption is referred to when deciding how to control the aerated concrete structuring process [14, 12, 15, 16]. One of the methods used to induce controllable changes in the rheological properties is to apply vibration and impact during the swelling [9, 17, 16, 18, 12, 3, 15]. Vibration with varying frequency and amplitude can change and stabilizing the swelling kinetics [3, 19]. However, vibration makes the entire process more energy-intensive while also producing considerable noise.

Another efficient method used to adjust the rheological properties of the mixture being swollen consists in adding surfactants and superplasticizers that lower the critical shear stress point [14, 11, 20, 13, 21] and slow down the structuring process.

There are other ways to control the changes in aerated concrete structuring, including (i) use of raw materials based on coarse-ground sands [22, 23, 24, 25, 21], (ii) dispersive reinforcing by means of fibrous additives [26], (iii) modifying the structure by means of polymer additives [27, 28], etc.

In-depth analysis of how technological factors affect the properties of the AC mixture and the operating performance of such concrete, and the subsequent use of analytical data to control the structuring process is enabled by identifying the set of patterns in the physico-chemical processing occurring in micro- and macrostructuring of aerated concrete; identifying the specifics of destructive processes occurring in aerated concrete at all the structuring phases; determining the criteria to evaluate the quality of the AC structure at swelling, with the structuring process being subsequently optimized based on such criteria; developing reliable methods and instruments to control and study the cellular-concrete mixture swelling phenomena.

2 Experimental program and research results

The Cementing Substance, Conrete, and Construction Ceramics Department of the Construction and Architecture Academy at Don State Technical University has carried out a series of experiments to study the presence and effect of various factors.

One of such factors is the order in which components are added to the mixture. In our studies, we assumed the basic component addition order that is used at an enterprise located in Rostov Oblast.

For this basic technology, water is dosed at t=45°C to the tank of the AX mixer; after the mixer is turned on, add sand, dry soda and sodium sulphate additives, and then cement. The components are mixed for four minutes upon gauging. Then the mixture is supplemented with aluminum suspension and mixed for another 20 seconds.

The experimental order was as follows: dose water to the mixer tank at t=45°C and switch on the mixer engine; then add soda and sodium sulfate and prepare the additive solution for 15 seconds; then add cement and mix the suspension for 30 seconds; then dose sand and mix for four minutes upon gauging; then doze aluminum suspension to the mixture and mix it for another 20 seconds.

We've made samples of the AC1 (factory-used) and AC3 (experimental) mixtures.

Those were compared on the following bases: average density, compression strength, and structural quality factor.

Figure 1 presents the results of AC sample tests taken after seven days of normal curing.

The results of AC1 and AC3 tests led us to the conclusion that the compression strength had increased 1.5 times while the structural quality factor had been

DOI: 10.1051/matecconf/201712905011

almost doubled while the density increase was insignificant.

The results encouraged us to use the proposed component addition and AC mixture preparation procedure in further experiments.

The next factor affecting the quality of non-autoclaved concrete is the addition of the so-called offcut, the porous mixture pieces that are cut off after swelling is over and the necessary elastic strength is achieved.

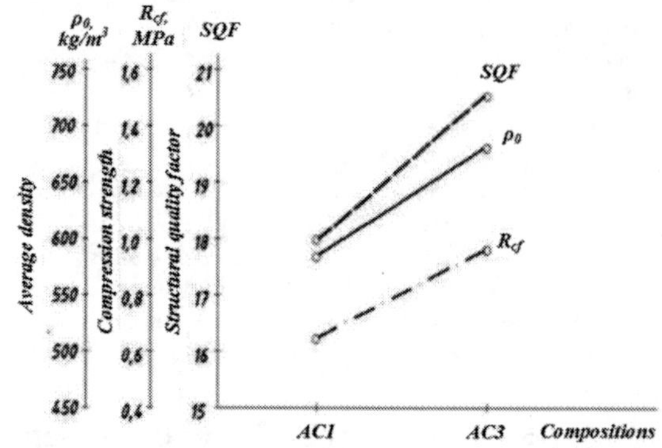

Fig. 1. Test results

When adding the off-cut to the AC mixture being prepared, hydrated cement grains should become the hydrate neoplasm crystallization centers to boost the AC structuring and curing process. In an experiment we carried out, sodium sulfate used at factories as the curing accelerator was replaced with the off-cut.

For the experiment, we made the control AC1 mixture; the off-cuts produced therefrom was added to the similarly prepared AC2 mixture instead of sodium sulphate. The off-cut consumption was 30% of the mass of cement.

Figure 2 presents the results of the experiment.

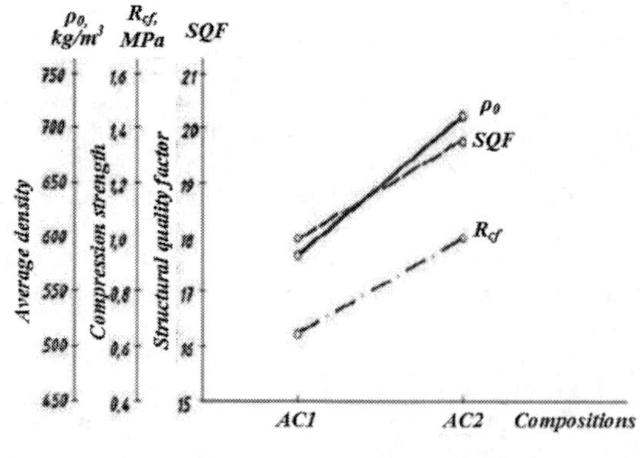

Fig. 2. Test results

Analysis of these data revealed that adding the off-cut to the mixture had produced stronger and denser aerated concrete. The structural quality factor of AC1 was 9% lower than that of the experimental AC2 mixture.

The experiment proved that the off-cut was an efficient additive. This enables both AC waste recycling and reduction of production costs as sodium sulphate is no longer needed.

To evaluate how the off-cut amount in the mixture affects the quality of aerated concrete, we experimented with varying amounts. We produced the control (factory) AC1 mixture and took off-cuts from there to add it to AC2, AC3, AC4, and AC5 mixtures where its mass was respectively 30, 10, 20, 25 percent of the cement mass.

Figure 3 presents the results of the experiment.

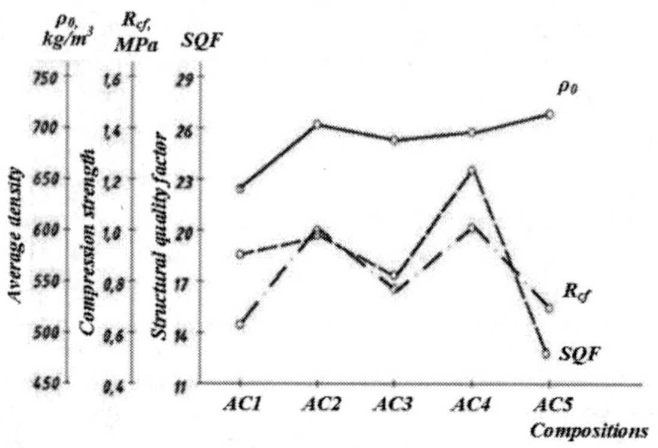

Fig. 3. Test results

Analysis of these data revealed that adding the off-cut in the amount equal to 20% of the cement mass produced stronger aerated concrete. The structural quality factor of AC4 was 20.4% higher than of AC2, 31% higher than that of AC1.

At the next stage, we decided to find out how efficient is complex sodium-sulphate and off-cut additive when it comes to improving the quality of non-autoclaved aerated concrete.

We experimented to find out the optimal amount of sodium sulphate in the mixture combined with the previously optimized amount of the off-cut. The compositions we studied were as follows: AC1 was the control (factory) mixture with sodium sulphate at 1.54% of the mass of cement; AC4 was AC1 where sodium sulphate was replaced with the off-cut at 20% of the cement mass; A1 was 20% off-cut and 1.23% sodium cement; A2 was 20% off-cut and 2.4% sodium cement; A3 was 20% off-cut and 3.7% sodium cement.

Figure 4 presents the results of the experiment.

 DOI: 10.1051/matecconf/201712905011

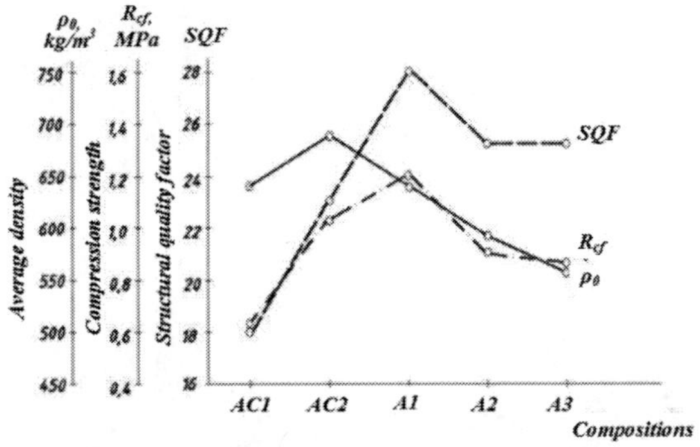

Fig. 4. Test results

Analysis of these data revealed that adding sodium sulphate in the amount of 1.23% of the cement mass plus the optimum dosage of the off-cut (20%) produced the strongest aerated concrete with the highest structural quality factor, i.e. the best sample was A1.

3 Conclusions

Thus, based on the literature review and experimental studies, we made a conclusion on how the analyzed composition and technology aspects affect the properties of non-autoclaved AC products.

Adding the off-cut to the mixture had produced stronger and denser aerated concrete. The structural quality factor of AC1 was 9% lower than that of the experimental AC2 mixture.

Adding the off-cut in the amount equal to 20% of the cement mass produced stronger aerated concrete. The structural quality factor of AC4 was 20.4% higher than of AC2, 31% higher than that of AC1.

Adding sodium sulphate in the amount of 1.23% of the cement mass plus the optimum dosage of the off-cut (20%) produced the strongest aerated concrete with the highest structural quality factor, i.e. the best sample was A1.

References

1. N. Narayanan, K. Ramamyrthy, *Structure and properties of aerated concrete: a review*, Cement & Concrete Composites, **22**, 321-329 (2000)
2. A.I. Shujskij, *Optimizacija processov strukturoobrazovanija i povyshenie kachestva gazobetonnyh izdelij, dis. ... kand. teh. nauk* (Rostov-na-Donu, 1983)
3. G.Ja. Kunnos, *Vibracionnaja tehnologija betona* (Izd. lit. Po stroitel'stvu, 1967)
4. A.P. Merkin, *Nauchnye i prakticheskie osnovy uluchshenija struktury i svojstv porizovannyh betonov, dis. ... dokt. teh. nauk*, (MISI, M, 13-46, 130-139, 1973)
5. A.Ja. Pylaev, A.I. Minas, E.S. Savin, *Razrabotka matematicheskoj modeli dlja opredelenija optimal'nyh uslovij vspuchivanija gazobetona*, Tez. dokl. II Vsesojuznogo simpoziuma, Reologija betonnyh smesej i ee tehnologicheskie zadachi, Riga, 172-173 (1976)
6. G.P. Saharov, P.V. Kornienko, *Obrazovanie optimal'noj struktury jacheistogo betona*, Stroitel'nye materialy, **10**, 30-33 (1973)
7. A.B. Solodovnik, *Kontinual'naja model' vspuchivanija zhidkosti s puzyr'kami, dis. ... kand. teh. nauk* (Riga, 1969)
8. A.B. Solodovnik, *Model' izmenenija poristosti vspuchivajushhejsja dvuhfaznoj sfery, Issledovanija po mehanike stroitel'nyh materialov i konstrukcij*, (t. U, Zinatne,1970)
9. K.Je. Gorjajnov, B.O. Atrachev, T.N. Nazarova, *Sovershenstvovanie vibroformovanija krupnorazmernyh massivov iz jacheistogo betona*, Stroitel'nye materialy, **8** (1978)
10. K. Je. Gorjajnov, S.K. Gorjajnova, *Tehnologija teploizoljacionnyh materialov i izdelij*, M, Strojizdat, 213-252 (1982)
11. K.Je. Gorjajnov and others, *Vlijanie dobavok na process formirovanija struktury i povyshenie kojefficienta kachestva gazosilikata*, VNIISTROM, Avtoklavnye silikatnye materialy i konstrukcii. Vjazhushhie materialy, **22 (50)**, 16-23, (1972)
12. S.A. Krzheminskij, B.B. Kryzhanovskij, S.G. Danilova, *Vlijanie tehnologicheskih faktorov na svojstva gazosilikatnyh materialov*, Sb. tr. ROSNIIMS, **15** (1959)
13. A.P. Merkin, *Napravlennoe izmenenie reologicheskih svojstv jacheistobetonnyh smesej na vseh stadijah proizvodstva polufunkcional'nymi PAV*, Reologija betonnyh smesej i ee tehnologicheskie zadachi, Tez. dokl. II Vsesojuznogo simpoziuma, Riga, 117-118 (1976)
14. Ju.P. Gorlov, A.P. Merkin, A.A. Ustenko, *Tehnologija teploizoljacionnyh materialov* (M, Strojizdat, 84-92, 1980)
15. A.Ja. Pylaev, *Issledovanie processa vspuchivanija i svojstv gazosilikata, dis. ... kand. tehn. nauk* (Rostov-na-Donu, 1977)
16. D.G. Zemcov, S.A. Krzheminskij, G.Ja. Kunnos and others, *Jacheistye betony na osnove smesej povyshennoj vjazkosti*, VNIISTROM, **20**, 14-34 (1971)
17. A.V. Dobrovol'skij, N.P. Sazhnev, *Proizvodstvo izdelij iz jacheistyh betonov s primeneniem udarnoj ploshhadki,* Promyshlennost' avtoklavnyh materialov i mestnyh vjazhushhih, Ref. inf. VNIIJeSM, **11**, 8 (1979)
18. G.V. Krasnova, *Proizvodstvo i primenenie v stroitel'stve jacheistyh betonov*, Promyshlennost' avtoklavnyh materialov i mestnyh vjazhushhih, Ref. inf. VNIIJeSM, **11**, 8, (1981)
19. K.K. Jeskusson and others, *Issledovanie tehnologii proizvodstva jacheistogo betona s uluchshennymi stroitel'no-tehnicheskimi svojstvami*,

NIPIsilikatobeton, Proizvodstvo i primenenie silikatnyh betonov, Tallin, 40-49 (1979)

20. V.I. Kalashnikov, Ju.S. Kuznecov, N.I. Makridin. *Ispol'zovanie himicheskih dobavok s uchetom tehnologicheskih osobennostej proizvodstva gazobetona*, Tez. i dokl. IX Respublikanskoj konferencii Dolgovechnost' konstrukcij iz avtoklavnyh betonov, Part 1, Tallin, NII stroitel'stva Gosstroja SSSR, 151-155 (1981)
21. T.A. Uhova, N.G. Krivickaja, *Opyt primenenija kompleksnyh dobavok na osnove superplastifikatorov pri proizvodstve jacheistyh betonov*, Tez. dokl. IV Respublikanskoj konferencii Dolgovechnost' konstrukcij iz avtoklavnyh betonov, Part 1, Tallin, NII Stroitel'stva Gosstroja JeSSR, 146-148 (1981)
22. A.P. Merkin, M.I. Zejfman, *Osnovnye napravlenija jekonomii toplivno-jenergeticheskih resursov v tehnologii jacheistyh betonov*, Povyshenie jeffektivnosti proizvodstva i primenenie industrial'nyh izdelij iz jacheistogo betona v narodnom hozjajstve, Tez. dokl., Nikolave, 6-8 (1980)
23. A.P. Merkin, Ju.P. Gorlov, M.I. Zejfman, *Povyshenie treshhinostojkosti jacheistogo betona za schet formirovanija racional'noj struktury silikatnogo kamnja*, Tez. dokl. III Respublikanskoj konferencii, Tallin, 57-61 (1978)
24. A.P. Merkin, M.I. Zejfman, I.B. Udachkin and others, *Snizhenie jenergoemkosti proizvodstva i povyshenie kachestva jacheistobetonnyh panelej pri ispol'zovanii peska kompozicionnogo sostava*, Stroitel'nye materialy, **3**, 4-5 (1981)
25. G.P. Saharov, K.I. Popov, S.S. Bataev, *Povyshenie stojkosti jacheistogo betona na grubomolotom peske pri kratkovremennom i dlitel'nom nagruzhenii*, Tez. dokl. III Respublikanskoj konferencii Dolgovechnost' konstrukcij iz avtoklavnyh betonov, Tallin, NII stroitel'stva Gosstroja JeSSR, 61-65 (1978)
26. E.M. Chernyshov, A.G. Baranov, A.M. Krohin, *Povyshenie kachestva jacheistyh betonov putem uluchshenija ih struktury*, Beton i zhelezobeton, **1** (1977)
27. I.B. Udachkin, Ju.V. Gontar', *Jeffektivnye sposoby povyshenija vodozashhitnyh svojstv jacheistogo betona*, Promyshlennost' avtoklav. mater. i mestnyh vjazhushhih. obzor. inf. VNIIJeSM, **8**, 25-30 (1980)
28. Z.V. Shuljupina, N.A. Kozlova, P.R. Taube, *Itogi dlitel'nogo ispol'zovanija PAV obrabotki aljuminievoj pudry,* Materialy III konferencii po jacheistym betonam, Privolzhskoe knizhnoe iz-vo, Saratov-Penza (1966)

DOI: 10.1051/matecconf/201712905012

Effect of caustic soda on the intensity of gassing in the production of non-autoclaved aerated concrete

Alexander Halyushev[1,*], *Mikhail* Holodnyak[1], and *Muhuma* Nazhuyev[1]

[1]Construction and Architecture Academy of Don State Technical University, Rostov-on-Don, Russian Federation

Abstract. This article discusses the effect of caustic soda on the intensity of gassing in the production of non-autoclaved aerated concrete. It has been found out that adding 0.8 and 1.0 percent of caustic soda increases the amount of released hydrogen by 20–40%. This is due to caustic soda increasing the alkalinity of the medium, which intensifies gassing. Exceeding the optimum amounts of caustic soda rapidly increases the elastic strength and worsens the 'swelling' of the AC mixture. Physico-mechanical tests of aerated concrete samples showed a considerable improvement in properties compared to the control composition.

1 Introduction

Aerated concrete refers to concrete having an excessive amounts of air voids. These air bubbles are created to reduce the density of the concrete and provide good thermal insulation. Aerated concrete was first produced in Sweden in the 1930s and can also be referred to as cellular, foamed or gas concrete [1]. It was originally designed for use as building blocks, but today, the applications have been expanded to larger units, such as floors and walls that have a low weight. This provides great advantages in the ease of handling as the constructions can be easily assembled by hand with little power. Aerated concrete is produced similarly to pre-cast concrete from very fine aggregate, cement and water. The term ''concrete'' is inappropriate since the material contains no coarse aggregate particles. After mixing and placing, the slurry is usually steam-cured in an autoclave to improve its properties. Manufacturing of aerated concrete is very efficient and economical, as it takes little manpower to produce [2]. The air pores in aerated concrete are usually 0.1 to 1 mm in diameter. They can be formed by a few different methods but the most common is by the addition of aluminium powder which is added to the mixing ingredients at about 0.2% to 0.5% by dry weight of cement. The porous structure of aerated concrete is generated by adding water-mixed solid components of the mixture and a gas-forming additive to the suspension; the most common additive is highly dispersed aluminum powder. Metallic aluminum and calcium oxide hydrogen that is released upon the hydrolysis of cementing clinker minerals reach; the reaction releases molecular hydrogen as shown in Eq. (1) [3, 4, 5]:

$$Al + 3Ca(OH)_2 + nH_2O \rightarrow 3CaO + Al_2O_3 + nH_2O + 3H_2 \uparrow \quad (1)$$

As hydrogen is releases, the mixture swells due to the coincidence of maximum gassing period and the viscoplastic condition of the mixture [6].

The optimal aerated concrete structure is very and uniformly porous; cells are optimally distributed by size, while the defects of interporous partitions are minor to none.

Gassing kinetics induced by the agent as well as the changes in the rheological properties of the interporous material are crucial for the formation of the pore structure in aerated concrete [7, 8, 9, 10, 11, 12].

At this moment, the mixture is deformed freely under the effect of the released gas; however, the latter does not leave the mixture. The subsequent curing of non-autoclaved aerated concrete affects the final strength of the products which depends on the high "cement-water" chemical potential. This is why non-autoclaved aerated concrete requires highly active cementing substances [13]; compared to autoclaved concrete [12, 14, 15], it uses considerably more Portland cement. On the other hand, increased cement consumption in non-autoclaved cellular concrete causes a considerably greater shrinkage of concrete in operation; such shrinkage may reach 2–3 mm/m [12, 14].

Cross-linking is one of the key processes in cellular concretes. Mixtures used in the production of aerated concrete should have low critical shear stress point while also having sufficiently high viscosity. This assumption is referred to when deciding how to control the aerated concrete structuring process [16, 17, 18, 19]. Controlling this process is crucial for making aerated concrete with the properties one needs [20, 21]. The specific feature of new technologies is the effective control of the cross-linking process [22]. In this regard, researchers are interested in intensifying the hydration and improving the structural quality of non-autoclaved concrete cementing substance to ensure sufficient strength at reduced density.

* Corresponding author: khaljushev@mail.ru

This opened the door for replacing sand and lime (partly or wholly) by industrial waste by-products [23–26]. The advantage of this method is that it can be used to recycle materials which possess a very low reactivity at room temperature. However, the utilization of by-products in autoclaved building materials is not only controlled by the suitability of these materials for this purpose but also by the local economy and the competitive position of other building materials within the area. Technically and economically, it may be possible to partly replace sand or lime with by-products if this is accompanied with improving the end products or reducing the production cost through reduction of the autoclaving time or temperature.

To form optimally structured cement stone, one has to note the predominance of coagulation structures at the first stages of cross-linking. This is crucial for substantiating the optimal parameters of the joint effects of physico-chemical and mechanical factors during the mass transition of phases at the onset of the cross-linking process [27]. The possibility of accelerating the condensation structuring of non-autoclaved cellular concretes eliminates the need for autoclaving while keeping the qualitative parameters on par [28]. It is vital to increase the hydration of cement and the crystallization of hydration products by adding curing accelerators [18].

There are other ways to control the changes in aerated concrete structuring, including (i) use of raw materials based on coarse-ground sands [29, 30, 31, 32, 33], (ii) dispersive reinforcing by means of fibrous additives [34], (iii) modifying the structure by means of polymer additives [35, 36], etc.

2 Experimental programme

This paper is to study the joint effect of caustic soda and sodium sulphate on the intensity of gassing in the production of non-autoclaved aerated concrete.

The mixture was mixed for 2 minutes in a planetary agitator. The AC mixture so made was shaped into 10-cm cubes which were then held in heat treatment chambers at 70°C in the τ (2+6+3) mode. At the same time, a portion of the mixture was loaded in a PGV-2 instrument to determine the gassing kinetics, see Figure 1. Table 1 presents the compositions of AC mixtures at a design density of D600.

Table 1. Compositions of AC mixtures

Name of the material	Consumption per 1 m³, kg		
	CN№1	CN№2	CN№3
Cement	312	312	312
Sand	255	255	255
Water	232	232	232
Aluminum powder	0.578	0.578	0.578
Caustic soda	-	2.43	3.12
Sodium sulfate	-	-	4.8

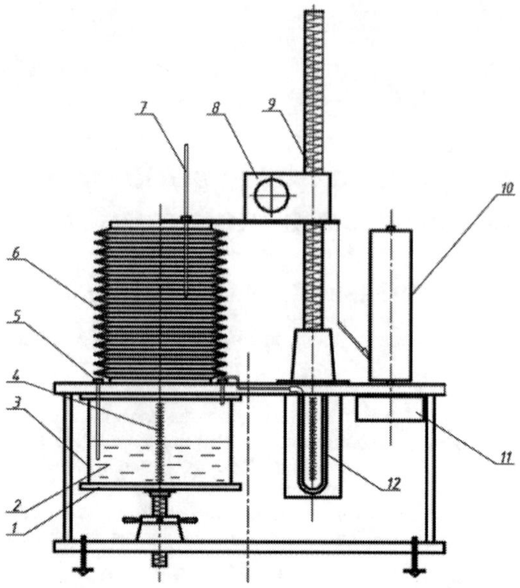

Fig. 1. PGV-2 gassing instrument diagram (front view): 1: the lift table; 2: AC mixtures; 3: 3-liter glass reaction container; 4: scale; 5: thermometer; 6: volumetric flask; 7: thermometer; 8: lifting beam; 9: lifting bar; 10: cylinder for attaching the recorder paper; 11: reversible engine; 12: U-shaped gauge to control the gassing;

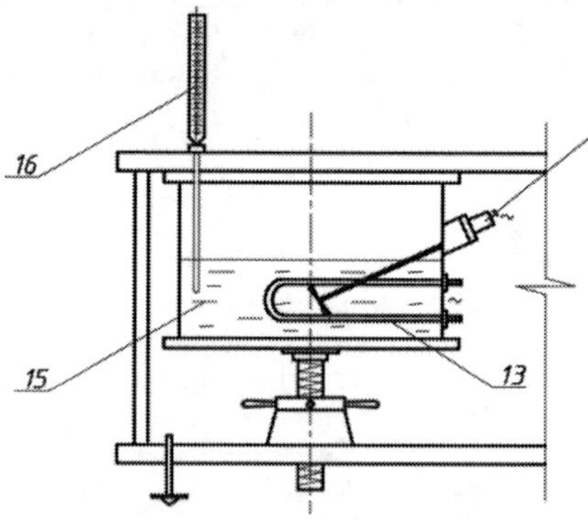

Fig. 2. PGV-2 gassing instrument diagram (side view): 13: heating element; 14: mixer; 15: supersaturated alkali solution; 16: thermometer.

The instrument we used consisted of the reaction container 3 with the scale 4 to determine the level of AC mixture fill and swelling. The measure its temperature, we used the electronic thermometer 5. The lifting table 1 presses the reaction contained against the upper base of the instrument. A rubber gasket seals the internal volume of the instrument. The volumetric flask 6 shaped as a think corrugated pipe is attached to the upper base of the instrument. Its upper part is closed with a lid that has a built-in mercury thermometer to measure the temperature of the gas medium above the surface of the mixture in the instrument.

Volume is measured with the lifting beam 8. Meters are plotted onto millimeter paper. The paper is rolled onto the drum of the recorder 10, and the latter is

DOI: 10.1051/matecconf/201712905012

powered by the reversible engine 11. The internal pressure is controlled by a U-shaped gauge.

The gassing kinetics parameters are determined as follows: AC mixture is put into the volumetric flask mounted on the instrument. The electronic thermometer is put into the mixture and sealed. Sometime later, the following parameters are recorded: the temperature of the mixture $t_{B(i)}$, the temperature of the gas medium above $t_{2(i)}$, the volume of the variable volumetric flask $V_{\Pi(i)}$, the volume of the mixture in the flask $V_{B(i)}$. As the gassing reaction ends, calculate the gassing kinetics parameters: the swelling coefficient $K_{B(i)}$, the gas diffusion coefficient $D_{(i)}$, the total gassing volume $V^{H_2}_{B(i)}$, and the temperature change $t_{B(i)}$; that is done using the following formulas:

$$K_{B(i)} = \frac{V^{H_2}_{(i)}}{V^{H_2}_{(i)}}; \quad (1)$$

$$V^{H_2}_{B(i)} = \frac{293.15 \times P_0 \beta_B(i) \times (V_B(i) - V_B(1))}{11325 \times (273.15 + t_B(i))}; \quad (2)$$

$$V^{H_2}_{D(i)} = 293.1ж5 P_0 \beta_{2(i)} \times \frac{V_{0(i)} - V_{B(i)}}{273.15 + t_{2(i)}} - \frac{V_{0(i)} - V_{B(i)}}{273.15 + t_{2(i)}}; \quad (3)$$

$$V^{H_2}_{(i)} = V^{H_2}_{B(i)} + V^{H_2}_{D(i)}; \quad (4)$$

$$D_{(i)} = \frac{V^{H_2}_{D(i)}}{V^{H_2}_{(i)}}; \quad (5)$$

where $V_{0(i)}$ is the total internal volume of the instrument, cm^3; $V^{H_2}_{B(i)}$is the volume of hydrogen in the AC mixture, cm^3; $V^{H_2}_{D(i)}$ is the volume of hydrogen diffused from the mixture, cm^3; $V^{H_2}_{(i)}$is the total volume of hydrogen generated during the swelling, cm^3; P_0 is the atmospheric pressure, Pa; $\beta_{2(i)} = 1 - \alpha_0 \times g_t \times (273 + t_{2(i)})$ is the relative gas content in the medium above the mixture; $\beta_{B(i)}$is the relative content of gases in the mixture; $\alpha_0 = 4.555$ cm^3/g. grad.; g_t is the absolute density of the gas medium at 100% relative humidity, g/cm^3.

Figure 2, 3 presents the results of calculating the gassing kinetics parameters.

It has been found out that adding 0.8 and 1.0 percent of caustic soda increases the amount of released hydrogen by 20–40%. This is due to caustic soda increasing the alkalinity of the medium, which intensifies gassing as per (2):

$$2NaOH + 4Al + 4H_2O = 2NaO \cdot Al_2O_3 \cdot H_2O + 3H_2 \uparrow \quad (6)$$

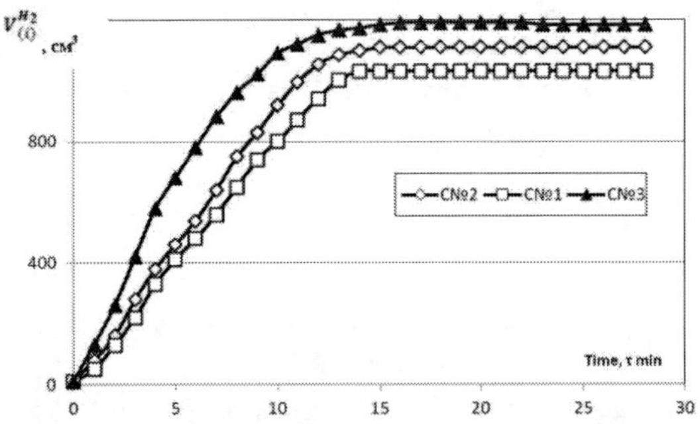

Fig. 3. How the volume of released gas ($V^{H_2}_{(i)}$, cm^3) in the AC mixture depends on time (τ, min).

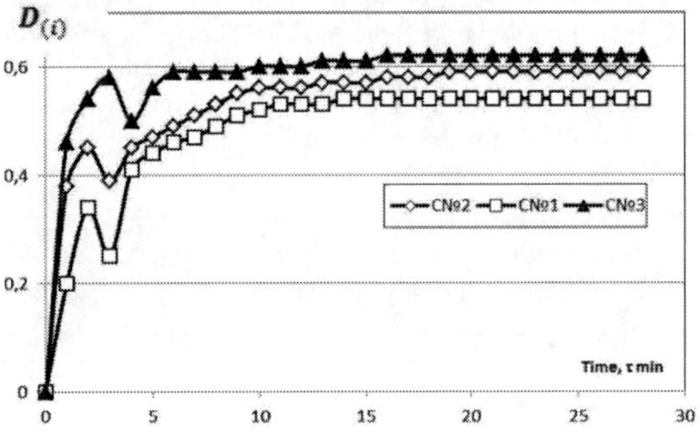

Fig. 4. How the gas diffusion coefficient $D_{(i)}$ in the AC mixture depends on time (τ, min).

Exceeding the optimum amounts of caustic soda rapidly increases the elastic strength and worsens the swelling of the AC mixture. Physico-mechanical tests of aerated concrete plus caustic soda samples showed a considerable improvement in properties compared to the control composition see Table 2.

Table 2. Physico-mechanical tests of AC samples: results

Indicator	Composition code		
	CN№1	**CN№2**	**CN№3**
Average dry density, kg/m^3	638	621	606
Compressive strength, MPa	1.8	2.2	2.6

Therefore, we have experimentally confirmed that removing caustic soda from the AC mixture lowers its alkalinity. As a result, gassing slows down. This makes aerated concrete denser, causes surface deficiencies and results in a greater failure rate.

References

1. A.M. Neville, *Properties of Concrete* (Pitman Books, London, 1981)
2. RILEM, *Autoclaved Aerated Concrete* (RILEM Technical Committees 78-MCA and 51-ALC, E & FN Spon, London, 1993)
3. *PFA Utilization, Central Electricity Generating Board* (Gibbons, Wolverhampton, 1972)

4. I.V. Udachkin, Ju.V Goncharov. *Jeffektivnye sposoby povyshenija vodozashhitnyh svojstv jacheistogo betona*, Strojizdat, 22-26, (1980)
5. L.I. Dvorkin, O.l. Dvorkin, *Osnovy betonovedenija* (S.Piterburg, "Stroj-beton", 2006)
6. M.Ja. Krivickij, N.I. Levin, V.V. Makarichev, *Jacheistye betony (tehnologija, svojstva i konstrukcii* (Strojizdat, 1972)
7. G.Ja. Kunnos, *Vibracionnaja tehnologija betona*, Izd. lit. Po stroitel'stvu (1967)
8. A.P. Merkin, *Nauchnye i prakticheskie osnovy uluchshenija struktury i svojstv porizovannyh betonov, dis. ... dokt. teh. nauk* (MISI, M, 13-46, 130-139, 1973)
9. A.Ja. Pylaev, A.I. Minas, E.S. Savin, *Razrabotka matematicheskoj modeli dlja opredelenija optimal'nyh uslovij vspuchivanija gazobetona,* Tez. dokl. II Vsesojuznogo simpoziuma, Reologija betonnyh smesej i ee tehnologicheskie zadachi, Riga, 172-173 (1976)
10. G.P. Saharov, P.V. Kornienko, *Obrazovanie optimal'noj struktury jacheistogo betona*, Stroitel'nye materialy, **10**, 30-33 (1973)
11. A.B. Solodovnik, *Kontinual'naja model' vspuchivanija zhidkosti s puzyr'kami, dis. ... kand. teh. nauk* (Riga, 1969)
12. A.B. Solodovnik, *Model' izmenenija poristosti vspuchivajushhejsja dvuhfaznoj sfery, Issledovanija po mehanike stroitel'nyh materialov i konstrukcij, t. U* (Zinatne, 1970)
13. G.P. Saharov, E.P. Skorikov, *Neavtoklavnyj jenergojeffektivnyj porobeton estestvennogo tverdenija*, Izvestija VUZov, **7**, 48-54, (2005)
14. V.A Lotov., N.A. Mitina, *Vlijanie dobavok na formirovanie mezhporovoj peregorodki v gazobetone neavtoklavnogo tverdenija*, Stroitel'nye materialy, **1**, 2-6 (2003)
15. V.A. Lotov, N.A. Mitina, *Osobennosti tehnologicheskih processov proizvodstva gazobetona*, Stroitel'nyen materialy, **4**, 21-22 (2000)
16. Ju.P. Gorlov, A.P. Merkin, A.A. Ustenko, *Tehnologija teploizoljacionnyh materialov* (M, Strojizdat, 84-92, 1980)
17. S.A. Krzheminskij, B.B. Kryzhanovskij, S.G. Danilova, *Vlijanie tehnologicheskih faktorov na svojstva gazosilikatnyh materialov*, Sb. tr. ROSNIIMS, **15** (1959)
18. A.Ja. Pylaev, *Issledovanie processa vspuchivanija i svojstv gazosilikata, dis. ... kand. tehn. nauk* (Rostov-na-Donu, 1977)
19. D.G. Zemcov, S.A. Krzheminskij, G.Ja. Kunnos and others, *Jacheistye betony na osnove smesej povyshennoj vjazkosti,* VNIISTROM, **20**, 14-34 (1971)
20. D.I. Shtakel'berg, V.Je. Mironov, *Termodinamicheskij analiz vspuchivanija i doavtoklavnogo tverdenija gazobetona,* Tehnologicheskaja mehanika betona, Riga, 105-113 (1979)
21. R. Cabrillac , B. Fiorio ., A. Beaucour , H. Dumontet , S. Ortola, *Experimental study of the mechanical anisotropy of aerated concretes and of the adjustment parameters of the introduced porosity*, Construction and Building Materials. Vol. **20**, No **5**., P. 286-295 (2006)
22. Ju.M. Bazhenov, *Novomu veku - novye betona*, Stroitel'nye materialy, oborudovanie, tehnologii XXI veka, **2**, 10-11 (2000)
23. H. Andre, E. Urs, M. Thomas, *Fly ash from cellulose industry as secondary raw material in autoclaved aerated concrete,* Cem. Concr. Res. **29 (3)** 297– 302 (1999)
24. D. Briesemann, *Structural elements and masonry of autoclaved aerated concrete*, Preprints-RILEM International Symposium on Autoclaved Aerated Concrete. Organ at Swiss Fed. Inst. of Technol., Sponsored by: RILEM, Paris, pp. 241–249 (1982)
25. K.C. Verma, *Scenario of utilization of fly ash in India*, Irrig. Power J. **51 (2),** 69– 74 (1994)
26. S.A.S. El-Hemaly, A.S. Taha, H. El-Didamony, *Influence of slag substitution on some properties of sand–lime aerated concrete*, J. Mater. Sci. **21 (4),** 1293– 1296 (1986)
27. N.B. Ur'ev, I.S. Dubinin, *Kolloidnye cementnye rastvory* (Strojizdat., 1980)
28. G.P. Saharov *Jacheistyj beton: novyj jetap razvitija*, Tehnologii betonov, **6**, 12-13 (2006)
29. T.A. Uhova, N.G. Krivickaja, *Opyt primenenija kompleksnyh dobavok na osnove superplastifikatorov pri proizvodstve jacheistyh betonov*, Tez. dokl. IV Respublikanskoj konferencii Dolgovechnost' konstrukcij iz avtoklavnyh betonov, Part 1, Tallin, NII Stroitel'stva Gosstroja JeSSR, 146-148 (1981)
30. A.P. Merkin, M.I. Zejfman, *Osnovnye napravlenija jekonomii toplivno-jenergeticheskih resursov v tehnologii jacheistyh betonov*, Povyshenie jeffektivnosti proizvodstva i primenenie industrial'nyh izdelij iz jacheistogo betona v narodnom hozjajstve, Tez. dokl., Nikolave, 6-8 (1980)
31. A.P. Merkin, Ju.P. Gorlov, M.I. Zejfman, *Povyshenie treshhinostojkosti jacheistogo betona za schet formirovanija racional'noj struktury silikatnogo kamnja*, Tez. dokl. III Respublikanskoj konferencii, Tallin, 57-61 (1978)
32. A.P. Merkin, M.I. Zejfman, I.B. Udachkin and others, *Snizhenie jenergoemkosti proizvodstva i povyshenie kachestva jacheistobetonnyh panelej pri ispol'zovanii peska kompozicionnogo sostava*, Stroitel'nye materialy, **3**, 4-5 (1981)
33. G.P. Saharov, K.I. Popov, S.S. Bataev, *Povyshenie stojkosti jacheistogo betona na grubomolotom peske pri kratkovremennom i dlitel'nom nagruzhenii*, Tez. dokl. III Respublikanskoj konferencii

Dolgovechnost' konstrukcij iz avtoklavnyh betonov, Tallin, NII stroitel'stva Gosstroja JeSSR, 61-65 (1978)

34. E.M. Chernyshov, A.G. Baranov, A.M. Krohin, *Povyshenie kachestva jacheistyh betonov putem uluchshenija ih struktury*, Beton i zhelezobeton, **1** (1977)
35. I.B. Udachkin, Ju.V. Gontar', J*effektivnye sposoby povyshenija vodozashhitnyh svojstv jacheistogo betona*, Promyshlennost' avtoklav. mater. i mestnyh vjazhushhih. obzor. inf. VNIIJeSM, **8**, 25-30 (1980)
36. Z.V. Shuljupina, N.A. Kozlova, P.R. Taube, *Itogi dlitel'nogo ispol'zovanija PAV obrabotki aljuminievoj pudry*, Materialy III konferencii po jacheistym betonam, Privolzhskoe knizhnoe iz-vo, Saratov-Penza (1966)

 DOI: 10.1051/matecconf/201712905013

Thin issues products of processing waste heaps as raw materials for ceramic wall products

Vladimir Kotlyar[1,*], and *Khungianos* Yavruyan[1]

[1] Architectural and Construction Academy Don State Technical University, 344022, Sotsialisticheskaya str. 162, Rostov-on-Don, Russia

Abstract. Shown high perspectivity of production of wall ceramics products on the basis of fine-grained gob pile processed products with a high content of coal constituent. The characteristics of these materials, which are actually a ready-made ceramic fusion mixture, as well as the characteristics of the products obtained depending on the roasting temperature, are given. The established relationship between the various properties of the products obtained, the compressive strength of which is from 10 to 19 MPa, at a low density. Recommendations on the main technological parameters of production are given. It is emphasized that low prime cost of products is caused: excluding the costs of raw materials associated with the development and maintenance of fields; practical absence of costs for mass preparation; drying of products due to heat extraction from the furnace; lack of fuel costs necessary for roasting. It is pointed out that in the production of wall ceramics products based on fine-grained waste products of waste heaps, additional sources of income appear that are associated with a reduction in capital and current costs for the maintenance of dumps of FGPPP and the use of excess heat in various directions.

1 Introduction

Nowadays more and more attention for producing wall ceramics is attributed to different technogenic raw materials of coal variety. This is determined by several reasons, major of them are:

- the necessity to produce products in modern economic conditions at minimum net cost, that is achieved by reducing expenses on roasting and processing of raw material;
- a growing demand for products of high efficacy with middle density class 0,7-0,8 (less that 700-800 kg/m^3) and heat conduction coefficient of bond in dry condition until 0,20 Wt/ (m · °C). First of all, it concerns cellular porous ceramic stones from 2,1 to 14.9 NF in size:
- changed economic conditions which made technogenic raw material more available for "brick makers" due to a range of reasons;
- developed roaster furnaces for the raw material with high content of fuel component and with an opportunity to remove excess heat.

The usage of coal industry wastes as a raw material for producing wall ceramics has been known for a long time. In the Soviet Union this area was thoroughly investigated in 70-80's of the past century [1-3]. There were major efforts put into laboratorial-technological researches. However, due to some reasons, this did not lead to practical appliance. Coal industry enterprises were not much interested in it – the main task was coal. Adjusting to brick factories had no economic sense for them. Independent gob pile development appeared to be quite problematic and costly for brick factories.

The situation has changed dramatically in the recent years. Coal enriching plants now consider coal enrichment by-products as goods and by all means are trying to adjust to a definite consumer in terms of characteristics and price [4-8]. Non-burnt gab pile is especially actively developed. The thing is that the coal content in them is from 10 to 25%. And the development of gab pile for coal extraction is today more economically profitable then underground coal mining. Only in the territory of Eastern Donbass there are about 500 mine heaps and enriching factories with mass about 600-700 mln. tons. In spite of the development, the heaps do not get smaller in volume, because during the process of extracting and enriching for 1 ton of clean coal it is necessary to take about 3 tons of by-products. Moreover, in the recent years the technologies of extracting and enriching coal have significantly advanced, as far as it is possible to regulate by-products properties according to coal constituent, grain composition, moist and so on.

2 Experiment

We have carried out researches on the possibility of receiving wall ceramics on the basis of fine-grained gob pile processed products with high content of coal constituent. On extracting coal during gob pile processing there are some kinds of materials formed. They are coarse-fractioned materials with the size of grains from 2 to 70 mm. They almost do not contain coal, they are separated into fractions (2–5; 5–10; 10–20; 20–40; 40–70 mm) and are used in construction industry. And fine-fractioned materials with the size of grains from 0 to 2 mm which do not have generally accepted names. They are divided into two types according to

* Corresponding author: diatomit_kvd@mail.ru

 DOI: 10.1051/matecconf/201712905013

their fractional composition : fraction 0,5-2 mm, notionally referred to as prefilt, and 0-0,5 mm fraction, named by us as – Fine-grained gob pile processed products (FGPPP).

In greater degree due to its properties, fine-grained gob pile processed products appears to attract greater interest as a basic raw material for producing wall ceramics, because the prefilt is a leaner for clay masses. Fine-grained gob pile processed products can carry different amount of coal constituent, but in modern technologies it is mostly 20-25%. Our goal was to determine the possibility of receiving high efficacy wall ceramics on basis of fine-grained gob pile processed products. It is traditionally acknowledged that for producing brick of high quality, losses at roasting for coal wastes should not exceed 20% [1, 3]. However, there is a lack of experimental data on this topic and they are quite controversial.

The average chemical composition of fine-grained gob pile processed products of Eastern Donbass is presented in table 1. In contrast to clay raw material, fine-grained gob pile processed products possess high heat losses indices during roasting – 25-29% due to the coal constituent. If not taking into account this constituent, heat losses on roasting would change in the interval 5-8%. The content of SiO_2 and Al_2O_3 without coal constituent is the same as in semi-acid clay raw material. The peculiarity is a high iron oxide and potassium oxide content that corresponds to mineral composition.

Table 1. Average chemical composition of fine-grained gob pile processed products of Eastern Donbass (%, in mass).

P.o.i.	SiO_2	Al_2O_3	Fe_2O_3 gen.	CaO	MgO	SO_3 gen.	K_2O	Na_2O	TiO_2
25-29	34-38	13-17	7-12	1-4	0,5-2	0,4-1,5	3-4	0,5-2	0,5-1

The prevailing component of fine-grained gob pile processed products of Eastern Donbass is stone-like clay rocks of different lithification degree. Their main constituents are – chlorite, hydrous micas, micas, kaolinite, quartz, field spars, oxides and iron hydroxide (fig. 1).

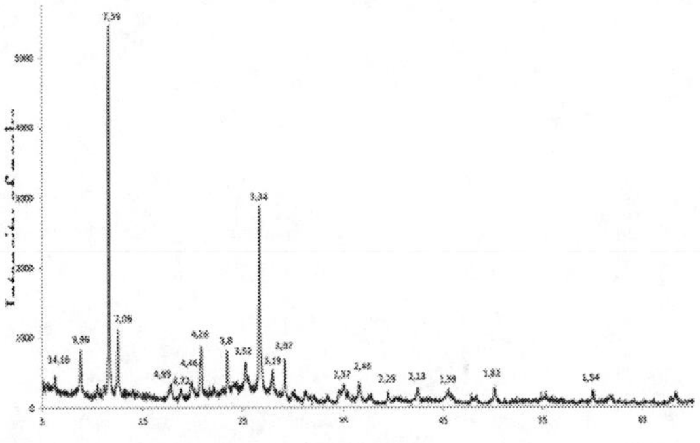

Fig. 1. X-ray pattern of typical fine-grained gob pile processed products of Eastern Donbass

The peaks of angles 9 and 18° indicate presence of hydrous micas and micas. The peaks of angles 6, 12 and 19° refer to the presence of chlorite and kaolinite. The peaks 3,34; 4,26; 2,58 Å clarify the existence of quartz. A slight peak of 3,19 Å signifies the presence of field spars. The presence of montmorillonite in coalenriching products is still a question, but some authors point out its existence in coalenriching products.

The ash content of proper coal grains is 5–8 %, the volatile content is 3-4%. This is an important aspect for selecting optimal roasting regime of fine-grained gob pile processed products. Coal devolatilization happens at 300–350 °C, it is proved by exo-effect at 321 °C (fig. 2). Antracite flaming occurs at 630 °C, what is clearly evident by exo-effect and mass loss. However, on FGPPP based brick roasting in industrial furnaces this happens at the temperature of 700–750 °C.

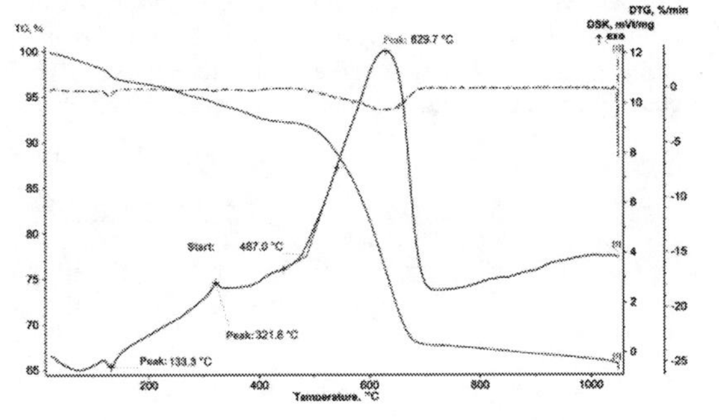

Fig. 2. Heat pattern of fine-grained gob pile processed products of Eastern Donbass

From the economic point of view the cost of FGPPP heat calory is 15-20 times lower in comparison with gas or pure coal. For example, the price for fine coal is 4000-5000 rubles per a ton, for power-generating coal is 3000-4000 rubles, for gas – 8000-1000 rubles per a ton, and for FGPPP is 100-200 rubles per a ton.

In fact, fine-grained gob pile processed products are prepared fusion mixure for producing wall ceramics. The main mass-preparing operation is making the mass in a mixing machine normally moisted for forming, mixing moisture content is 18–20 %. After the enriching process, the FGPPP moisture should be 10-14%. The plasticity is 9-12 units. Air shrinkage in comparison with clay is not really high, it is 3,5-4,5%. The average granulometric composition was measured with the help of a lazer diffraction particle size analyzer «NanoTec 2» (Germany) with water dispergation of the material and ultraonic machining (capacity 70 Wt, frequency 36 Hz). The average granulometric composition is presented in table 2.

Table 2. Results of granulometric FGPPP composition measurements

Fractions contents, mm, %					
0,5–0,25	0,25–0,05	0,05–0,01	0,01–0,005	0,005–0,001	< 0,001
4–6	6–10	21–25	23–27	28-32	7–11

DOI: 10.1051/matecconf/201712905013

Molding properties of FGPPP based mass are quite satisfactory for extrusional (plastic) type of products molding, as well as for a compressional one. However, a thorough choice of technological parameters of products molding according to the type of molding and types of products is necessary.

FGPPP with high coal constituent – more than 20%, are of special interest. According to the coal constituenr content (organic carbon) there are several types of rocks : low-carbon rocks (C_{org} = 0–8 %), intermediate-carbon (C_{org} = 8–20 %) and high-carbon (C_{org} more 20%). It is acknowledged that if the carbon content is more that 20%, the strength of products will be quite low [1,2].

3 Results and discussion

The carried out researches gave us interesting results. On figure 3 there are interrelations of strength limit on compression, on bending and average density for FGPPP based plastic molding samples with a coal constituent of 21,8%.

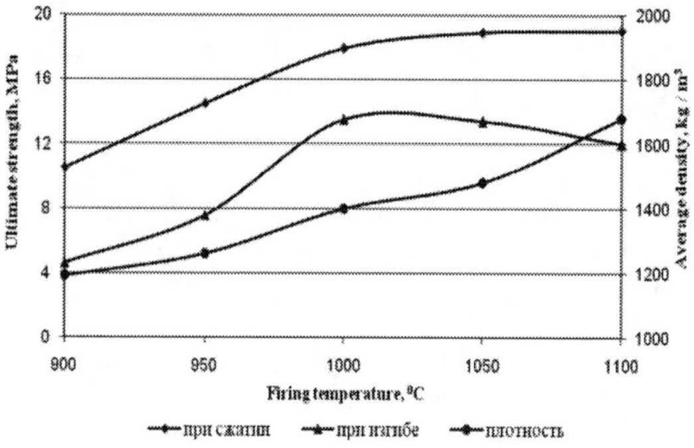

Fig. 3. Interrelation of strength limit on compression, on bending and average density with roasting temperature

As it is evidenced, the received products possess quite acceptable for wall ceramics characteristics, which are significantly influenced by roasting temperature. So, the average density of solid-bodied products at the temperature of 900 °C is 1190 кг/m3 and as the roasting temperature rises till 1050 °C it increases to 1480 кг/m^3. The strength limit on compression at the roasting temperature 900 °C is 10,5 MPa, and as the temperature rises till 1050 °C it increases to 18,9 MPa. The strength lomit on bending also increases within the roasting temperature from 4,6 to 13,5 MPa. These indices are quite acceptable for load-carrying constructions. For achieving the average products density of 750 kg/m^3 at the roasting temperature 900 °C the porosity should be 37%, at the roasting temperature 1050 °C - 50%. Consequently, the strength limit on compression also decreases proportionally and it will be 5–10 MPa, that is also acceptable for highly-efficient ceramic stones.

Products based on fine-grained gob pile processed products have high porosity, that is 55% at the roasting temperature of 900 °C and 45 % at the roasting temperature of 1050°C. This significzntly decreases heat conductivity of products. High porosity does not affect negatively frost resistance of products, as far as a brick carries enough amount of reserve pores of the size more than 200 mkm (fig. 4).

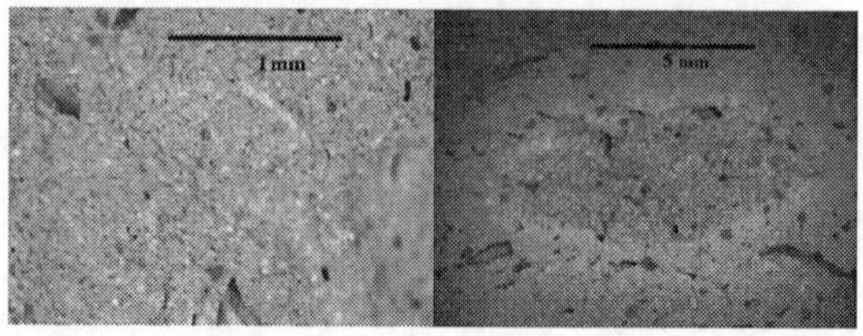

a) б)

Fig. 4. FGPPP based ceramic brick roasted at 900 °C (a) and 950 °C (b)

Special attention in the wall ceramics production should be attributed to roasting process. Products should be roasted at a regime with isometric timing and oxidizing environment at the temperature of coal contituent burning. Taking into account that the fuel content in FGPPP significantly exceeds the amount necessary for roasting, the furnace construction should provide an opportunity of extracting excess heat and gradual highly-controlled airing in the period of coal burning and carbon residue.

Under roasting temperatures of 1050 °C and higher, there are signs of overburning starting to appear, due to high iron oxide constituent that tranfers out of a trivalent form into a bivalent one (fig. 5).

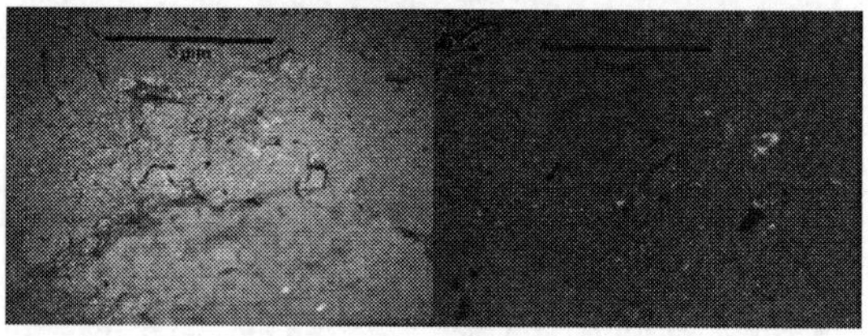

Fig. 5. FGPPP based ceramic brick roasted at 1050 °C

Appearing at this process FeO is a strong fluxing agent, significantly increasing roasting process. This is illustrated by a rapid growth of fire shrinkage beginning at temperatures 1030–1050 °C (fig. 6).

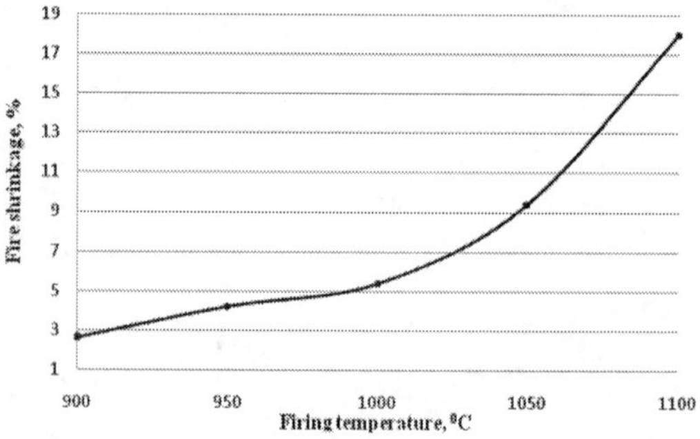

Fig. 6. Fire shrinkage of FGPPP based ceramic brick

DOI: 10.1051/matecconf/201712905013

4 Conclusion

The carried out researches have shown that basing on fine-grained gob pile processed products with high coal onstituent (more that 20%) it is quite possible to produce lightweight wall ceramics with good physical-mechanical characteristics. In addition to that, the net coat of such products is minimal due to :
- excluding expenses on raw materials connected to the development and maintanence of fields ;
- practical absence of mass preparation expenses ;
- products drying by extracting excess heat out of the furnace ;
- absence of roasting fuel expenses.

According to preliminary calculations, the net cost of conventional FGPPP based brick is 2,5-3,0 rubles. Moreover, FGPPP based wall ceramics production gives additional sources of income connected to the reduction of capital and operating expenditures for FGPPP heaps maintanence and the use of excess heat in quite various directions.

References

1. V.N. Burmistrov, D.A. Varshavskaya, V.T. Novinskaya and others, VNIIESM, 44 p, (1976)
2. M.G. Lundina, VNIIESM, 96 p., (1974)
3. A.Z. Zolotarskii, E.Sh. Sheiman, Manufacture of ceramic bricks, Moscow: VSh. 264 p., (1989)
4. A.Yu. Stolboushkin, G.I. Storozhenko, Stroitel'nye materialy, **4**, pp. 43–46, (2011)
5. V.D. Kotlyar, A.V. Ustinov, V.Yu. Kovalev, Yu.V. Teriohina, A.V. Kotlyar, Stroitel'nye materialy, **4**, pp. 44–48 (2013)
6. V.D. Kotlyar, A.V. Ustinov, Naukovedenie, **3 (16)**, (2013) Identification number of the article in the journal: 44TRGSU313. http://naukovedenie.ru/index.php?p=issue-3-13-RGSU (date of access 21.02.2017)
7. V.D. Kotlyar, A.V. Kozlov, A.V. Kotlyar, U.V. Teriohina, Vestnik MGSU, **10**, pp. 95–105, (2014)
8. B.K. Kara-sal, V.I. Kotel'nikov, T.V. Sapelkina, Estestvennye i tekhnicheskie nauki, **2**. pp. 160–163, (2015)

DOI: 10.1051/matecconf/201712905014

Statistical modeling for risk assessment at sudden failures of construction equipment

Tatyana Rogovenko[1], and *Marina* Zaitseva[1,*]

[1]Don State Technical University, Rostov-on-Don, Russia

Abstract. This paper applies the statistical modeling method to provide a quantitative assessment of risks emerging at sudden failures of construction equipment. The risk mathematical model takes into consideration the distribution parameters of operation, downtime and repair hours determined for a general population of the failed part finite volume. The proposed modeling algorithm is applied to assess risks at sudden dipper stick failures. The modeling results allow one to evaluate the tolerable period for the equipment use under the set risk levels.

1 Introduction

The relevance of research conduct in risk assessment for construction equipment is growing along with the enhancement of its security assessment methods [1-10].

Sudden failures represent one of the risk sources during construction equipment operation. In general, construction equipment is repairable, and in this connection, there is a vital necessity to assess risks with the account of the operation (running) influence and time as well as the recovery time. The multi-parameter Weibull model flexibility enables one to use it for approximation of the experimental data obtained in relation to recovery and running intervals including the small data selections [11]. The transition from the Weibull distribution parameters assessment for small data selections towards the general population finite volume (GPFV) parameters may be performed by means of the well-proven and successfully applied FISP method [4, 12].

2 Statement of Problem

The quantitative risk assessment for the construction equipment t operation time can be calculated as a product of an undesirable event occurrence, i.e., a sudden failure, $P(t)$, and the consequences of this event in the form of damage $D(t)$ in compliance with the standard [1, 13]:

$$R(t) = P(t)D(t).$$

Hence, new tasks arise: assessment of the operation time, downtime and repair time distribution parameters as well as determination of the damage caused by such failure for each of the time intervals. Solution of the mentioned problems requires application of the probability and statistics methods to assess distribution parameters for a data selection with a subsequent transition to the GPFV of the parts the failure of which is considered to be an undesirable event [4,12].

The quantitative risk assessment mathematical model at the sudden failure can be represented in a certain way. Let us introduce the following notation: n – failure number, m – time interval number identified during the risk assessment process (in what follows m=3), C_{Ri}– repair costs after the i failure, t_{ij} – length of the j time interval after the i failure (i = 0, 1,..., n; j = 1, 2,..., m) (Fig. 1).

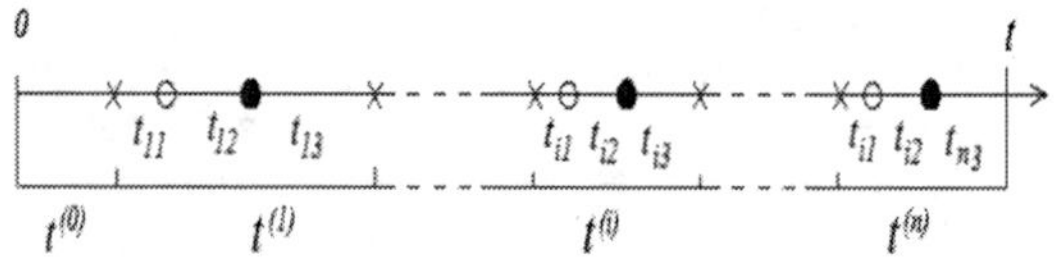

Fig. 1. Intervals tij on the time scale

Each of the t_{ij} time intervals has its own probability distribution $F_j(t_{ij})$ and the damage function $D_j(t_{ij})$.

The operation time to assess risks is define using the formula

$$t = t^{(0)} + \sum_{i=1}^{n}\sum_{j=1}^{m} t_{ij}.$$

The repair costs C_{Ri} can be determined in compliance with the procedure described in [14].

Obviously, for the notation introduced: $t_{01}= t^{(0)}$ – running time until the first failure, $t_{02} = 0$, $t_{03} = 0$, $C_{R0} = 0$.

The risks from sudden failures n occurrence for the time t including the m time intervals may be calculated applying the following formula:

---

* Corresponding author: marincha1@rambler.ru

$$R(t)=\sum_{i=1}^{n}\left(C_{Ri}+\sum_{j=1}^{m}F_j(t_{ij})D_j(t_{ij})\right).$$

3 Distribution parameters assessment

The three-parameter Weibull distribution was chosen as the probability distribution for each of the t_{ij} time intervals (operation time, downtime and repair time):

$$F(t)=1-exp\left(-\left(\frac{t-c_S}{a_S}\right)^{b_S}\right),$$

where a_S, b_S, c_S are parameters of the scale, form and location respectively, determined by the maximum likelihood method for the small volume data selection n_S = 5-15.

The FISP method can be helpful for the transition from the distribution parameters assessment over the a_S, b_S, c_S small data selections towards the assessment of a_G, b_G, c_G for the failed parts GPFV [12].

The FISP method is essentially about sequential simulation of finite intermediate subpopulations (FISP) $N_T = 10^2$ - 10^4 using small samples of size n_S = 5 - 15, which is followed by modeling the finite general population (FGP) $N_G = 10^4$ - 10^6 using the correlative dependencies obtained in [15].

For such simulation, FISPs are determined by the number of additional values of random variable m_{ad} based on the distribution law and depending on the extent of order statistics $R_{n+1} = x_{i+1} - x_i$ as well as their distribution law:

$$m_{ad}=\sum_{i=0}^{n_S}(n_{ad_i}+\Delta_i),$$

where the components are calculated using the formula:

$$n_{ad_0}=[\frac{F^{*}\cdot(N_T-n_S-2)}{(x_1-A)\cdot\sum_{j=1}^{n_S-1}\left(\left(F(R_{jj+1})+F^{*}+F^{**}\right)/R_{jj+1}\right)}],$$

$$n_{ad_i}=[\frac{F(R_{ii+1})\cdot(N_T-n_S-2)}{R_{ii+1}\cdot\sum_{j=1}^{n_S-1}\left(\left(F(R_{jj+1})+F^{*}+F^{**}\right)/R_{jj+1}\right)}]$$

for i=1, …, n_S-1 ,

$$n_{ad_{nS}}=[\frac{F^{**}(N_T-n_S-2)}{(B-x_{n_S})\cdot(\sum_{j=1}^{n_S-1}\left(\left(F(R_{jj+1})+F^{*}+F^{**}\right)/R_{jj+1}\right)}]$$

$$F^{*}=\min_{1\le i\le n_S-1}\{\frac{F(R_{ii+1})(x_1-A)}{R_{ii+1}}\},$$

$$F^{**}=\max_{1\le i\le n_S-1}\{\frac{F(R_{ii+1})(B-x_{n_S})}{R_{ii+1}}\}.$$

where $F(R_{ii+1})$ - distribution of order statistics:

$$F(R_{ii+1})=\frac{n_S!}{(i-1)!(n_S-i+1)!}\times$$

$$\int_{0}^{R_{ii+1}}\int_{A}^{c_S-R_{TS}}P^{i-1}(x)p(x)p(x+R_{ii+1})\times$$

$$\times[1-P(x+R_{ii+1})]^{n_S-i+1}dxdR,$$

$P(x)$, $p(x)$are the probability function and density, A and B are the limits of the simulated range.

The value of Δ_i is determined by successive approximations so that $N_T = n_s+m_{ad}+2$. If the order statistics are equal, the measurement error of the random variable is assumed as the extent value for simulating the additional values.

When debugging the FISP method, the homogeneity of the source population and the FISP was evaluated using the Wilcoxon test; the Smirnov-Kolmogorov test was used to verify the hypothesis that the distribution functions of the general populations the samples were taken from coincide in the entire domain. It was evaluated the share of FISPs N_W=69-75%. that are Wilcoxon-homogeneous to the initial population for a significance of α=0.01; the share of FISPs N_K=75-89%. that share the general population with the source population (based on the Smirnov-Kolmogorov test for a significance of α=0.01; 0.001.

4 Risk modeling

The statistical risk $R(t)$ modeling is intended for determining of the construction equipment operation critical time t^* after which the risks exceed the set R_p level.

In compliance with the preliminarily established laws for distribution of the operation time, downtime and repair time $F_j(t_{ij})$ (j = 1, 2,…, m), risk modeling shall be repeated until the subsequent $R^*(t)$ value exceeds the pre-set level. The calculations may be repeated multiple times to determine the confidence bounds for the t^* critical time assessment. Figure 2 shows the aggregate algorithm for risk modeling.

5 Application

The FISP simulation and quantitative risk assessment algorithm was implemented as a part of the MRM software suit developed by the Motor Vehicles, Construction and Road Machines Department, Don State Technical University, to study the problems of machine reliability and risk analysis.

The algorithm is applied to assess risks at sudden fatigue failures of the crane-type excavator EK-14 dipper stick.

The distribution parameters $F_j(t_{ij})$ basic data after the FISP method [15] application for risk modeling are given in the Table 1 below.

DOI: 10.1051/matecconf/201712905014

The $D_j(t_{ij})$ damage functions are taken as being linearly dependent on time with the coefficients equal to the hourly cost of the excavator performance.

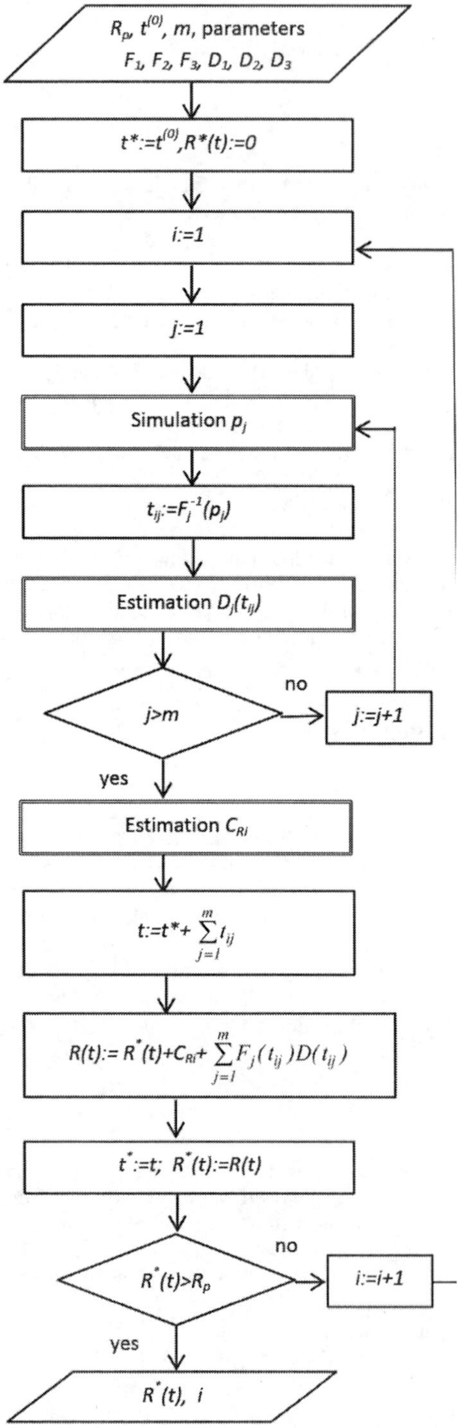

Fig. 2. Risk modeling algorithm

Figures 3-5 present the time interval distribution functions for selections and GPFV.

Figure 6 demonstrates one of the options of the *R(t)* function modeling by means of accrual.

The multiply repeated *R(t)* modeling allowed for determination of the variability interval for the operation time t^* critical values under different R_p risk levels reduced to the C_E. excavator cost. The results are given in the Table 2.

Table 1. The initial data for modelling

t_{ij}	a_G	b_G	c_G
t_{i1}	1850	2.1	16550
t_{i2}	10	2	1.5
t_{i3}	116	1.35	5.7

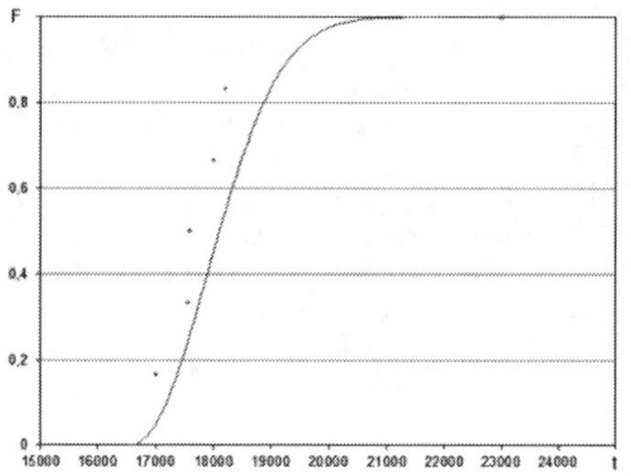

Fig. 3. Dipper stick operation time distribution

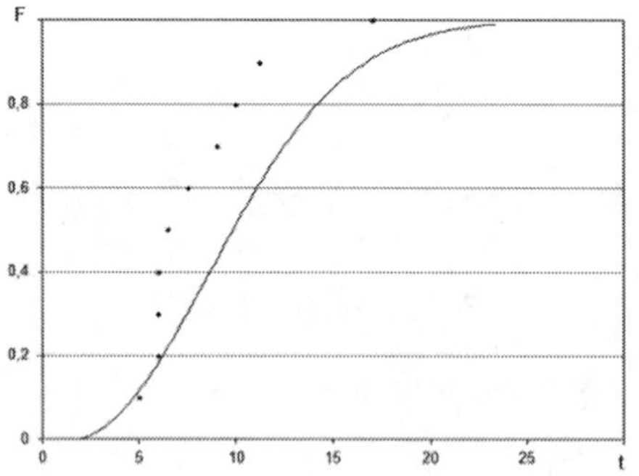

Fig. 4. Downtime distribution after a dipper stick sudden failure

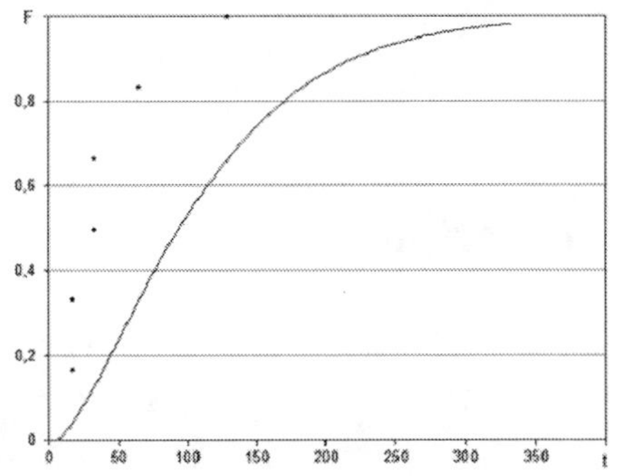

Fig. 5. Dipper stick repair time distribution

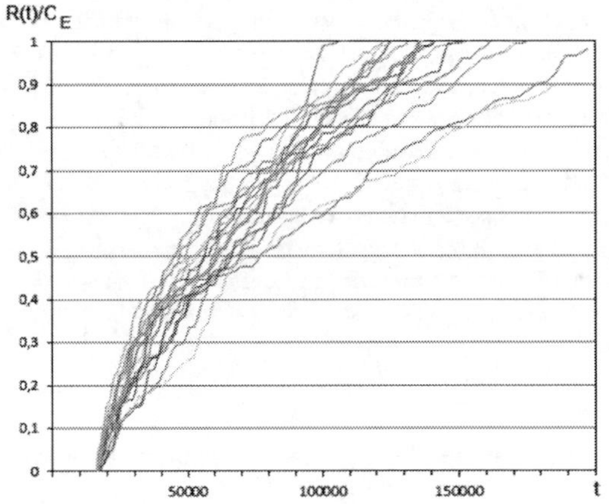

Fig. 6. Accumulated risks related to excavator cost

Table 2. Limits of the t^*critical values

R_p/C_E	t^*_{min}	the average value t^*	t^*_{max}
0.1	18355	21053	24222
0.2	21819	27760	36764
0.5	44762	59976	77495

6 Conclusions

Therefore, the proposed risk quantitative assessment mathematical model takes into account distribution parameters for the operation time, downtime and repair time for the failed parts GPFV at sudden failures of construction equipment. Risk statistical modeling allows taking a decision in terms of the further equipment use practicability and develop measures to mitigate sudden failure risks.

References

1. *Handbook of risk theory*. S. Roeser, R. Hillerbrand, P. Sandin,M. Peterson(Springer, 2012)
2. S.N. Goncharenko, E.V. Dement'eva. Min. Inf. & Anal. Bul. (GIAB), **12**, 95-104 (2010)
3. V.E. Kas'yanov, T.N. Rogovenko, Rus. Engin. Res., **6** (1999)
4. V. E. Kas'yanov, T.N. Rogovenko, M M Zaitseva, Bul. of Rost. St. Trans. Un., **1**, 16-20 (2010)
5. B. Y. Renaulta, J. N. Agumba, MATEC Web Conf. (IBCC 2016) 66, 00008 (2016)
6. R. Stasiak-Betlejewsk, A. Czajkowska, MATEC Web Conf.(CoSME'16) 94, 04011 (2017)
7. V. E. Kas'yanov, L.P. Schulkin,ScienceRev., **10-2**, 437-440 (2014)
8. V. E. Kas'yanov, L.P. Schulkin, ScienceRev., **10-3**, 671-674 (2014).
9. V. E. Kas'yanov, L.P. Schulkin, ScienceRev., **11-3**, 782-784 (2014)
10. D.Z. Evseev, S.N. Fil,ScienceRev., **11-2**, 482-484 V. E. Kas'yanov, T.N. Rogovenko, M M Zaitseva, Bul. of the Don St. Tech. Un., **1**, 91-94 (2010).
11. H. Rinne, *The Weibull Distribution* (2008)
12. M. M. Zaitseva, *Obespecheniye zadannogo ustalostnogo pesursa detaley odnokovshovogo excavatora s ispolzovaniem malyh viborok* (2010)
13. ISO/IEC 31010:2009 *Risk management — Risk assessment techniques* (IDT)
14. T.N. Rogovenko, M M Zaitseva, Eng. Jour. Don, **4**, (2016)
15. V. E. Kas'yanov, T.N. Rogovenko, I. V. Topilin, Rus. Eng. Res., **3**, 9-11 (2003)

 DOI: 10.1051/matecconf/201712905015

Technical-economic justification of the choice of heat source autonomous heating schemes

Sergei Tikhomirov[1,*], *Gennadiy* Golubov[1], and *Pavel* Syichev[1]

[1]Don State Technical University, 344022 Sotsialisticheskaya 162, Russia

Abstract. One of the priority tasks of improving the fuel efficiency in the Russian Federation is the development of district heating systems. Such a policy spelled out in the Federal law "On heat supply" dated 27.07.2010 No. 190-FZ, the number of bylaws and regulations of the government. However, there is strong growth of low-rise, primarily residential construction in the country. Today major Russian cities are surrounded by settlements and gardening, where the laying of heat networks economically not justified. In the current environment are higher requirements for energy efficiency schemes for Autonomous heat supply heat sources to meet the seasonal and year-round loads, and also search of alternative ways of using the heat systems with heat pumps and solar systems. The paper discusses options for using in the circuits Autonomous heating such heat sources, like traditional and condensing hot water boilers. In this experiment you performed the study on the impact of modes of heating systems to customers in fuel efficiency. This is very important for technical applications results efficient operation of the systems of Autonomous heat supply.

1 Introduction

Optimization of heat supply systems was significant enough issue in the market of energy efficient sources of heat. Therefore, the wide application among all other find systems of Autonomous heating, which actively introduced in the last decade.

Currently, more acutely raises the question of energy efficiency. On this basis, the most urgent item of a larger program of optimization and modernization of the heat supply is the comparative characteristics of different systems of independent heat supply, the nominal power of which will show the advantages of one over the other on the basic commercial and operational indicators [1].

The aim of this work was to reduce energy consumption in the production of heat.

For solving the optimization problem of the scheme of Autonomous heat supply to reduce the operating cost was conducted research in a pilot plant consisting of a heating apparatus, low-temperature and conventional heating systems of the consumer, metering systems gas flow and the resulting thermal energy [2].

2 Objectives of the research

Efficient operation schemes of Autonomous heat supply requires fine tuning for each element. The main methods of fuel economy today are working to improve the efficiency of the heat source, which often serve as gas boilers [3]. If for a major boiler performance cards are prepared for all possible situations, the influence of heating domestic heating devices on their efficiency is considered to be insignificant [4].

The task of the pilot study was to test the influence of the thermal performance of traditional and condensing domestic boilers in the two typical modes by the value of their efficiency.

The object of the research was the current model schemes of heating system. The structure of the scheme included:

- a consumer of heat with variable load up to 30 kW consisting of a tank with cold tap water heat exchanger for heat recovery, fan coil unit, traditional high temperature heating system with radiators and floor heating circuit;
- the storage tank capacity of 500 liters to compensate for the daily non-uniformity of the heat load;
- piping system;
- a source of heat to cover the heat load of consumer;
- control of measured parameters [5], including flow meters and temperature sensors straight and return main boiler circuit, the heat accumulator and pump user groups. In addition to measuring consumption of heat was determined by the cost of fuel gas meters that are located at each boiler. The efficiency of boilers were determined as direct and reverse balance. For this purpose we used regular gas analyzer Testo 310.

3 Methods of research and results

To achieve this goal, we used the following methodology for conducting the experiments:

* Corresponding author: sergtihomirov@yandex.ru

Before any work was performed visualization of all devices involved in the execution of skilled works. Checked the operation of the instrumentation system of metering the flow of gas and water piping, tightness of valves and faucets to ensure reliable and uninterrupted operation of the whole system.

Readings of gas consumption measured by meter stamps of Grand-3,2 (Gmax/Gmin=3,2/0,04 m³/h), and heat consumption by heat energy meter type M-Cal MC1,5 (Gmax/Gmin=3,0 m³/h 30 l/h). The values recorded in the measurement Protocol.

In the works include each boiler, when it asked desired mode of operation of the heating system.

Before starting the boiler open the valves DN 15 on omitting, taps on the forward and return lines, piping, faucets hot and cold water boiler with the heated environment, the faucet to the consumer.

For both boilers typical modes of operation, in which the research was conducted was the temperature in the coolant supply/return lines 40/30 ⁰C and 80/60 ⁰C. After achieving the required temperature at the outlet of the boiler in the measurement Protocol were made following testimonies: the temperature at the inlet and the outlet of the boiler, inlet and outlet from the consumer, the consumption of water before and after the boiler, power boiler and consumers. Every time after time was filmed readings of metering gas flow and heat meter.

In the technical theory of three-layer plates, the relationship between displacements and deformations has the form:

$$Q_i^d \cdot G = Q_{hs} + Q_{loss} \quad (1)$$

$$N \cdot \tau = \eta \cdot Q_i^d \cdot G \quad (2)$$

where Q_d^i – is the low heat of combustion, kJ/m³;

G – natural gas consumption, m³/hour;

Q_{hs} – heat, useful transferred to the heating system, kJ;

Q_{loss} – heat loss, kJ;

η – the efficiency of the heating apparatus;

N – thermal capacity, kW;

τ – the time experience, seconds.

Table 1. Test results of heating of the condensation boiler according to the temperature regime 40/30 ⁰C.

№	G, m³/hour	t₁, ⁰C	t₂, ⁰C	N, kW	η
1	4,394	39,7	34,7	5,834	109,5
2	4,441	39,9	34,9	5,889	107,8
3	4,618	39,5	35,8	4,279	105,6
4	4,689	39,5	35,9	4,215	104,7
5	4,83	39,3	36	3,997	104,2

Table 2. Test results of the heating boiler convection under temperature regime 40/30 ⁰C.

№	G, m³/hour	t₁, ⁰C	t₂, ⁰C	N, kW	η
1	41,443	40	30,9	11,2	84,2
2	41,684	41,3	34,5	8,4	86,5
3	41,77	41,55	34,6	8,55	83,7
4	41,871	41,9	35,1	8,4	82,4
5	41,96	42,3	35,7	8,3	81,9

Table 3. Test results of heating of the condensation boiler according to the temperature regime 80/60 ⁰C.

№	G, m³/hour	t₁, ⁰C	t₂, ⁰C	N, kW	η
1	11,601	72,6	67,8	5,409	83,17
2	11,676	67,7	66,3	1,64	80,43
3	11,714	70,3	65,6	5,32	90,04
4	11,766	69,5	64,7	5,343	85,28
5	11,821	68,8	64,1	5,338	91,67

Table 4. Test results of the heating boiler convection under temperature regime 80/60 ⁰C.

№	G, m³/hour	t₁, ⁰C	t₂, ⁰C	N, kW	η
1	57,823	80,4	65,9	16,1	80,9
2	58,01	80,5	65,6	16,6	83,3
3	58,185	80,6	68	15,66	85
4	58,41	80,6	67,2	16,8	91,8
5	58,77	81,1	63,7	21,6	90

Testing domestic heating boilers, combination systems of Autonomous heating system of the building is made in accordance with the requirements of normative documents for direct measurement, using calibrated instruments and the error estimate.

 DOI: 10.1051/matecconf/201712905015

The pilot study revealed the dependence of the efficiency of household heating devices by the heat output at various modes of their work, the optimum conditions of a system of Autonomous heat supply in General, confirmed ways to reduce operating costs [6,7].

4 Evaluation of economic efficiency of the scheme Autonomous heating

The economic viability of the various schemes of Autonomous heat supply is determined on the basis of the comparative economic efficiency of capital investments necessary to implement such actions, i.e. co-deliver costs and the results obtained in these costs [8]. Cost is the most appropriate solution, which reduced the cost of Π, rub/year which are the sum of operating costs C and capital investments K, reduced to the same dimension in accordance with the regulatory coefficient of comparative efficiency of capital investments E_n:

$$\Pi_i = C_i + E_n \cdot K_i \rightarrow \min \quad (3)$$

where i – the sequence number of variant design decisions.

The amount of the required capital investment K, rub, determined by the formula:

$$K = K_{rs} + K_{sm} + K_{pi} \quad (4)$$

where K_{rs} - cost design and estimate work, rub;

K_{sm} - the cost of construction works (including commissioning works), rub;

K_{pi} - the cost of a square of flooring that will host equipment systems, rub/m^2.

Given that the overhaul of the plant is performed not every year, you need the appropriate costs P_{ki}, rub/year to distribute by year using the following formula:

$$P_{ki} = \frac{(C_{ki} + n_{yeari})}{n_{res}} \quad (5)$$

where C_{ki} - the cost of one major repair, rub;

n_{yeari} - annual fund operating time of the equipment or component of the device, h/year;

n_{res} - resource time to the first major repair hours, the value of which for equipment operating under normal conditions.

Annual costs for maintenance and overhaul maintenance of equipment systems of autonomous heating system P_{mi}, rub/year, with the impending inflation is determined by the formula:

$$P_{mi} = \sum_{T=1}^{Tcc} (n_{cr} \cdot C_{cr} \cdot (\frac{n_{gas}}{n_{res}}) + C_{sm} + n_{ce} \cdot C_{ce} \cdot n_i \cdot C_i \cdot \frac{n_{gas}}{n_{res}}) \cdot \frac{(1+\gamma_{sn})^T}{T_{cs}} \quad (6)$$

where n_{cr} - number of current repairs in the period between two overhauls the previous and following him;

C_{cr} - the cost of one repair in the current year, the start of the heating system, rub;

C_{sm} - annual costs for system maintenance for the first year of its validity, rub;

n_{ce} - number of cleanings of the equipment per year;

C_{ce} - cost of one cleaning the equipment in the same year of the system, rub/hour;

n_i - number of inspections of the installation in the period between two overhauls;

C_i - the cost of one examination in the same year, rub/insp.

Energy costs E consumed by the pumps of systems of heating, ventilation, air conditioning and automation of these systems are determined by formula, not taking into account the coming inflation of energy prices:

$$E = (\sum_{n=1}^{n``} W_e^n \cdot n_h^n \cdot n_{day}^n + \sum_{q=1}^{q`} W_e^q \cdot n_h^q \cdot n_{day}^q) \cdot C_e \quad (7)$$

where W_e^n, W_e^q - estimated costs of electricity, respectively, by the pumps of the heating system and automation equipment, kW;

n_{day}^n, n_{day}^q - duration of the operation of this equipment, hours/day;

n_h^n, n_h^q - the same day/year.

$n``$, $q`$ - the number of pumps and automation panels

C_e - electricity at two betting rate, (rub/kW·h).

The cost of gas, rub/year, occurring prior to the installation of equipment, can be subject to price increase for the overspent part determined by the formula:

$$\sum T_{ni`} = T_{ni} + T_{ni}^{\partial on} + 2\Delta T_{n`} \quad (8)$$

where 2-additional fee for excessive part, in shares from $\Delta T_{n`}$;

$\Delta T_{n`}$ - payment for the specified part of the gas, rub/year, determined in accordance with the meter readings.

The work was carried out economic assessment and comparison of schemes autonomous heating with different heat sources (conventional and condensing boilers).

Table 5. Comparative characteristics of the condensing and traditional hot water boilers in the traditional scheme of an autonomous heating without solar collector and heat pump.

The type of boiler	**Convection**	**Condensation**
The cost of design work, rub	30000	30000
The cost of construction and installation, rub	100000	100000
The cost of a square of flooring, rub	40000	40000
The cost of the equipment, rub	73000	91000
Capital investments, rub	**243000**	**261000**
The cost of gas, rub/year	31332	26733
Electricity costs, rub/year	3275	3275
The cost of overhaul and repair, rub/year	4301	5001
Operating cost, rub/year	**38908**	**35009**
Operating costs and depreciation of equipment, rub/year	**51058**	**48059**

Table 6. Comparative characteristics of the condensing and traditional hot water boilers in the combined scheme of an autonomous heating with boiler and solar collector.

The type of boiler	**Convection**	**Condensation**
The cost of design, rub	30000	30000
The cost of construction and installation, rub	400000	400000
The cost of a square of flooring, rub	70000	70000
The cost of the equipment, rub	273000	291000
Capital investments, rub	**773000**	**791000**
The cost of gas, rub/year	27723	23654
Electricity costs, rub/year	9407	9407
The cost of overhaul and repair, rub/year	4301	5002
Operating cost, rub/year	**41431**	**38063**
Operating costs and depreciation of equipment, rub/year	**80081**	**77613**

Table 7. Comparative characteristics of the condensing and traditional hot water boilers in the combined scheme of an autonomous heating with boiler, solar collector and heat pump.

The type of boiler	**Convection**	**Condensation**
The cost of design, rub	30000	30000
The cost of construction and installation, rub	400000	400000
The cost of a square of flooring, rub	100000	100000
The cost of the equipment, rub	723000	741000
Capital investments, rub	**1253000**	**1271000**
The cost of gas, rub/year	9986	8520
Electricity costs, rub/year	30323	30323
The cost of overhaul and repair, rub/year	14496	15197
Operating cost, rub/year	**54805**	**54040**
Operating costs and depreciation of equipment, rub/year	**117455**	**117590**

5 Conclusions

The study of developments in the field of condensing and traditional heating technology of Autonomous systems of heating, allowed to systematize the data and to justify the direction and objectives of the study.

The developed model feasibility study selection of the heat source Autonomous heating schemes, allowing to analyze the features of reducing operational costs in the annual cycle depending on the set-up and thermal-hydraulic regimes of its work.

The factors to assess the impact, modal and thermo-physical parameters on the determination of the efficiency of the combined scheme of an Autonomous heating with multiple heat sources.

Designed experimental stand, methodology and testing program condensing and traditional boilers, which allows to determine the dependence of the efficiency of the heat source from the thermal-hydraulic and climatic factors.

A study of the impact on efficiency of the heat generator combined circuit Autonomous heating, operating conditions and climatic factors.

References

1. P.A. Khavanov, ABOK, **1,** 34 (2004)
2. *S.A.*Tikhomirov, O.K. Mazurova, *P. A.* Syichev, Construction and Architecture – 2015, **1,** 191-192 (2015)
3. P.A. Khavanov, Vestnik MGSU, **7,** 429-435 (2011)
4. S.G. Sheina, S.A. Tikhomirov, E.N. Minenko, International Journal of Applied Engineering Research, **10 (12),** 31389-31402 (2015)
5. *S.A.*Tikhomirov, V.G. Fedorovsky, P.V. Stasenko, Construction – 2015, **1,** 229 (2015)
6. O.E. Rakhnov, I.Y. Saklakov, A.D. Potapov, Vestnik MGSU, **11,** 177-187 (2013)
7. P.A. Khavanov, E.B. Solovyova, Scientific review, **4,** 128-130 (2013)
8. A.I. Eremkin, T.I. Koroleva, G.V. Danilin, V.V. Byzev, A.G. Averkin, *The economy of energy conservation in heating, ventilation and air-conditioning,* (2008)

Simulation of the process and system of power plants exhaust gases purification during construction and re-construction of housing and utilities infrastructure of urban areas

Ekaterina Lysova[1,*], *Oksana* Paramonova[1], and *Oksana* Gurova[1]

[1] Department of Environmental Engineering, Federal State-Funded Educational Institution of Higher Education Don State Technical University, Rostov-on-Don, Russian Federation

Abstract. Urban territories shall provide for favorable life activity of people alongside with preservation of ecological balance, however, they are peculiar city-planning "focuses" on which ecological problems are maximally concentrated. Air basin of urban territories is the most dynamic and most vulnerable of all the components of environment exposed to maximal anthropogenic load. The solution of the actual problem of increase of ecological safety of urban territories on basis of imitational simulation of the process and system of purification of exhaust gases of power plants of housing and utilities infrastructure with the help of management of parameters substantially influencing the ecological efficiency and energy intensity value of the purification process is represented in the article. Management of parameters is possible on the stage of designing of system for purification of exhaust gases of power plants, which substantially optimizes the choice of ways and means for decrease of air media of urban territories.

1 Introduction

Under the modern conditions of development of urban territories the problem of pollution of air basin is rather pressing, because the air basin is subject to maximum anthropogenic load and is the most dynamic of all components of the environment [1-3]. Production of heat and electrical energy by power plants is accompanied with substantial environment pollution [4]. However, power plants are an important object of housing and utilities infrastructure of urban media and a necessary element of the system of life support of urban territories. They cannot be located outside the urban territories which is related with minimization of losses, both material and energetic [5].

Main contribution into the pollution is made by exhaust gases of power plants with which polluting substances come into the air basin. Structure of exhaust gases and amount of discharged polluting substances depends on the type of burnt fuel [6].

2 Relevance, scientific importance of the issue with brief literature review

Relevance of the article is explained by constant increase of air basin pollution, which is a problem of the majority of big cities and is related to the necessity of provision of ecological safety of urban territories, creation of favorable conditions for life of the population, and environmental protection.

Notwithstanding the sustainable trend of decrease of the volume and change of the structure of exhausts of polluting agents into the air basin during fuel burning, power plants are still taking one of the leading positions by contribution to atmospheric air pollution.

Works of such authors, as Belik S.E., Bespalov V.I., Boguslavskiy E.I., Valdberg A. Yu., Gibbs V., Gluzberg V. E., Greene X., Zhuravlev V. P., Zhuravlyov V. P., Ishchuk I. G., Kouzov P. A., Lane V., Livchak I.F., Logachev I. N., Minko V. A., Mishner Y, Petryanov-Sokolov I. V., Rehbinder P. A., Rekun V. V., Rusanov A.A, Saranchuk V. I., Segal I. Ya., Sidorenko V. F., Sou S., Spurny To., Sutugin A. G., Uzhov V. N., Fuchs N. A., Shtokman E. A., Strauss W. and many other authors are dedicated to research and evaluation of negative impact of power plants on the environmental components, studying of characteristics of polluting aerosol, including as a disperse system, its behavior under the conditions of external impact, examination of series of tasks related to purification, as well as improvement of technique of air purification [7-12].

However, it is important to make a right choice of constructive elements and working characteristics of the process and system of exhaust gases purification on the designing stage. It will provide for maximum ecological efficiency and energy economic feasibility of the process.

Thus, relevance of research aimed at simulation of the process and system of purification of exhaust gases of power plants of housing and utilities infrastructure and management of parameters influencing efficiency and energy intensity of the purification process does not cause any doubts.

* Corresponding author: katerina.lysova0803@gmail.com

DOI: 10.1051/matecconf/201712905016

3 Task setting

To choose optimal process and system of exhaust gases purification of power plants, including on the designing stage, it is necessary to provide for possibility of quality process modelling. Solution of this task supposes construction of a physical model of decrease of pollution of air media for the conditions of operation of power plants and obtaining of parametric dependencies of efficiency and energy intensity value as the resulting parameters of purification process.

4 Theoretical part

Based on earlier carried out physical simulation of the process of pollution of air media [13] we have constructed a physical model of the process of decrease of air media pollution [14], which illustrates consecutive deliberate impact on polluting aerosol with external disperse systems on each stage of pollution process.

Decrease of concentration of polluting agents **(PA)** contained in exhaust gases of power plants could be achieved by realization of two main cycles of the process [14]:

1 – decrease of pollution of technological raw materials (different types of organic fuel) or technological equipment (boiler units) with the aim of prevention of exhaust of polluting agents particles from the volume of technological raw materials.

2 – decrease of pollution of air media with additional implementation on the operating or designed manufacturing and technological equipment (boiler unit) of special engineering and technical events (systems) with the aim of destruction of aerosol polluting agents at the account of division of disperse phase and dispersing medium.

First cycle of the process for power plants is more rational [15-17] and could include recirculation of exhaust gases, feeding of excess air amount into the burning zone, fuel desulfurization, additional of chemical substances, moisture sprinkling or steam into the furnace, preliminary fuel heating, etc. However, its use is economically unjustified, and main obstruction is manufacturing and technological requirements which contain limitations on parameters of characteristics of additional disperse systems.

With the purpose of decrease of air basin pollution of urban territories by power plants the second cycle of the process is used. During organic fuel burning there is no need in organization of the stage of pollutants collection, since they from the source of generation come into duct works. Thus, in our case, the second cycle of the process of decrease of air medium pollution includes implementation of the process of purification and dispersion (fig. 1).

Interconnection of parameters of external influences and gas aerosol is reflected in the physical model of the process of decrease of air medium pollution (fig. 1) through changes in the properties parameters (PP), energy parameters (W_S) and persistence (U) of interacting systems.

In the process of organic fuel burning at the power plants aerosol is created which lowers its persistence and transforms into the «Initial–I» («I–I») disperse system. Its persistence increases, since in the process of burning energy is constantly fed into this system.

In the process of purification disperse system «I–I» interacts with preliminary prepared «Additional–II.1» («A–II.1») disperse system. As the result of this interaction two disperse systems are created:

1 – «Residual–II.2» («R–II.2») disperse system, which is characterized by increased content of PA particles and increased persistence. «R–II.2» disperse system goes into the utilization system.

2 – «Intermediate–II.1» («In–II.1») disperse system – polluting aerosol with certain residual concentration of polluting agents, directed either into the atmosphere (with adherence to maximum allowable emission), or into the zone of forced dispersion. At that «In–II.1» disperse system is characterized by increased value of persistence as an aerosol.

In the process of forced dispersion the disperse system «In–II.1» is purposefully influence by «Additional–II.3» («A–II.3») disperse system, preliminary prepared by parameters. As the result two disperse systems are created in aerosol state:

1 – «Residual–II.3» («R–II.3») disperse system represented by maximum number of polluting agents particles. After interaction with «A–II.3» disperse system it lowers its persistence as a polluting aerosol and is trying to transfer into the state of polluting agents material with increase, and then stabilization of its persistence.

2 – «Residual–II.4» («R–II.4») disperse system represented by minimal amount of polluting agents, it stays in the aerosol state, its persistence in time slightly grows, and then stabilizes. This system forms residual concentration of polluting agents in air basin of the reviewed territory.

As the result of consecutive impact on «Initial», «Intermediate» and «Residual» disperse systems with external additional disperse systems in accordance with laws of mass and energy preservation, transformation of interacting systems happens. This transformation consists of the following: newly created disperse systems are distinguished with properties parameters, energy parameters and persistence from the interacting ones.

From the position of the theory of persistence of disperse systems the essence of the process of decrease of air medium pollution lies in destruction of a disperse system – polluting agents being in different aggregate states. Mathematically the essence of the process of decrease of air medium pollution taking into consideration probability of implementation of each of its stages as dependent events in our cases is expressed by the formula [14, 18, 19]:

 DOI: 10.1051/matecconf/201712905016

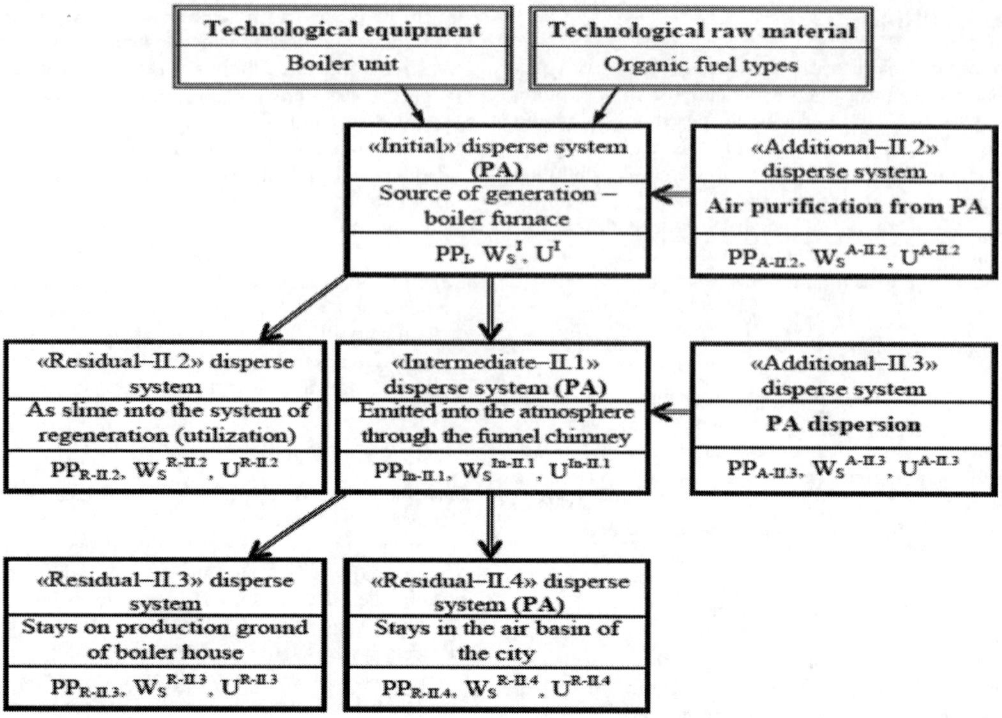

Fig. 1. Flow chart of the physical model of the process if decrease of air medium pollution for power plants

$$P_{DAMP} = 1 - (1 - P_{PP}) \cdot (1 - P_{DP}), \quad (1)$$

where P_{PP} – probable implementation of the purification process; P_{DP} – probability of implementation of dispersion process.

The created physical model of air medium pollution allows to suggest and simulate the system of purification in which on components of the exhaust gases shall be consecutively and purposefully influenced by corresponding external impacts.

In accordance with the suggested classification scheme of systems of control of polluting agents [20] for air purification from polluting agents contained in exhaust gases of power plants the following main methods are used [21]:

- hydrodynamic method (use of fluid as external impact managing the persistence of gas and solid PA contained in exhaust gases);
- aerodynamic method (use of gas (air) flows);
- electromagnetic method (use of electromagnetic field);
- thermophysical method (use of thermal field).

Out of known and traditionally applied methods for purification of exhaust gases of power plants we have chosen hydrodynamic method as one of the most widely spread and being rather highly ecologically efficient. Implementation of hydrodynamic method of purification of exhaust gases from gaseous matter is carried out by the following methods (and technical devices):

- use of steam, fog (steam chambers, for generators);
- use of foam (foam generators);
- by barbotage through a liquid layer (various types of barboters);
- by surface contact with a static layer of liquid;
- by irrigation with dispersed liquid (low-pressure, high-pressure, pneumohydroirrigation, irrigated gas ducts);
- by combination of listed methods (scrubbers, wet cyclones) under this method.

Determination of optimal working parameters of the process of hydrodynamic purification of exhaust gases, improvement of the process, simulation and management of the process are based on analysis of parametric dependencies of its resulting parameters – ecological efficiency and energy intensity value [22]. These criteria, in our opinion, are complementary and characterize the process of decrease of air basin most fully.

We understand the term «ecological efficiency» as the degree of decrease of concentration of polluting agents in the air basin at the account of implementation of the process of purification related to the concentration of polluting agents going into the air basin without purification.

To determine efficiency of purification of exhaust gases with hydroirrigation we have used probabilistic approach taking into consideration consecutive implementation of four stages of the process of interaction of the system «gas-fluid» both between each other and with the environment. At that the efficiency of this or that process is understood as the probability of its implementation.

The process of hydrodynamic purification includes 4 stages:

1 – approximation of gas molecule with a drop to the distance when intermolecular forces of molecule interaction of gas and fluid start acting, at that, in our opinion, the underlying are turbulent and diffusion forces;

2 – transfer of gas molecule into the area on the boundary of phases division where, in our opinion, the main influence is carried out by absorption and molecular forces.

3 – penetration of the molecule into the volume of fluid where, in our opinion, it is necessary to take into account diffusion forces and chemical reactions between molecules of gas and fluid;

4 – settling of drops with molecules diluted in them from gas-air stream due to gravitation forces.

The term «energy efficiency» we understand as the share of energy used for achievement of the aim of the process of decrease of air medium pollution (useful energy) relative to the energy used for implementation of the process in general.

We understand useful energy as the energy used for interaction of gas stream with dispersed fluid:

- kinetic energy of fluid drops, into the boundary layer of which the gas molecules have diffused;
- kinetic energy of gas aerosol;
- energy of absorption interaction of gas and fluid molecules;
- energy of chemical interaction of gas molecules and irrigating fluid.

We consider as consumed the energy for irrigation:

- energy parameter of draft agitator characterizing energy consumed for organization of air flow with certain characteristics;
- energy parameter of pumping station characterizing energy consumed for organization of liquid flow with certain characteristics.

5 Practical importance, suggestions and results of implementation, results of experimental research

To simulate the process and system of purification of exhaust gases of power plants of housing and utility sector it is necessary in each certain case of their practical application to determine corresponding resulting parameters, which include, first of all, ecological efficiency and energy intensity value. Examination of physical essence of the process of hydrodynamic purification of exhaust gases allowed to carry out mathematical description of these resulting parameters.

Parametrical dependence of ecological efficiency of the process of gas purification with hydroirrigation is written as:

$$E_{ef}^{hir} = \left[1 - \left(1 - \frac{2}{3} \cdot \frac{1{,}8B}{\mu_a \cdot v_d \cdot D_g \cdot d_d^2}\right)^{2/3} x\right.$$
$$x\left(1 - 0{,}032 \cdot q \cdot \sqrt{\frac{d_d \cdot \rho_l \cdot \sigma_{l-g}}{Q_l \cdot D_g \cdot \mu_a \cdot v_d}}\right)\Bigg] x$$
$$x\left(1 - \frac{\rho_l \cdot d_d^2 \cdot v_d \cdot e^{\alpha} \cdot E_a \cdot \chi}{\frac{\rho_l \cdot d_d^2 \cdot v_d}{36\mu_a \cdot d_d} + 10^{An_1}}\right) \quad (2)$$

where B – the constant of intermolecular interaction, J·m²; μ_a – dynamic air viscosity, Pa·s; v_d – mean drop speed, m/s; D_g – the coefficient of gas diffusion, m³/s; d_d – the mean surface-bulk drop diameter, m; q – the electrical charge of liquid aerosol, Kl/kg; ρ_l – the density of irrigating liquid, kg/m³; σ_{l-g} – the superficial tension of fluid on the border «liquid-gas», N/m; Q_l – the irrigating liquid consumption, m³/c; α – the coefficient of absorption for gases and water; E_a – the activation energy, kJ/mole; χ – the absorption acceleration in the liquid phase during chemical reaction in it; A – the coefficient of fullness of ductwork closure with the flame body of dispersed fluid; n_1 – the coefficient characterizing the speed mode of fluid aerosol distribution in the active purification zone.

Parametric dependence of energy intensive value of the process of hydrodynamic purification is written as:

$$E_{hir}^{E} = \frac{E_{ef}^{hir} \cdot v_d \cdot h_a \cdot B_1}{9{,}8H_{н} + \frac{2800 \cdot (H_{нт} - H_{п})^{1{,}5} \cdot d_a^2}{\mu\pi \cdot d_n^2 \sqrt{H_{н} \cdot g}}} x$$
$$x\left(0{,}5\rho_l + \frac{10^{-26}}{g\mu_a \cdot d_d} + \frac{6e^{\alpha} \cdot \sigma_{l-g} \cdot (1 - cos\chi)}{v_d^2}\right) \quad (3)$$

where h_a – the length of active zone of irrigation flame, m; B_1 – the coefficient taking into account direction and value of the speed of stream of polluting agents in the active zone of the purification process; $H_н$ – the pressure of fluid created by the pump directly in front of the irrigator, Pa; $H_{нт}$ – the full pressure of air in the sucking in cross-section of draught activator (ventilator), Pa; $H_п$ – loss of pressure in the network of ducts from the sucking in cross-section of draught activator (ventilator) to the initial border of active zone of the purification process, Pa; d_a – the equivalent diameter of cross-section of active zone of the purification process, m; μ – the coefficient of consumption of nozzle opening of irrigator; d_n – the diameter of nozzle opening of irrigator, m.

The parametric dependencies of ecological efficiency and energy intensity value obtained by us allow carrying out of simulation modelling with further improvement of the process of hydrodynamic purification of exhaust gases of power plants. Simulation is carried out on basis of management of functionally independent parameters, changes of which lead to simultaneous increase of value of ecological efficiency and energy intensity value.

Of substantial importance is the fact that management of parameters is possible on the stage of designing of the system of purification of exhaust gases of power plants. It substantially optimizes the choice of methods and means for decreasing of pollution of air medium of urban territories.

6 Conclusions

The solution of the actual problem of increase of ecological safety of urban territories on basis of imitational simulation of the process and system of purification of exhaust gases of power plants of housing and utilities infrastructure with the help of management of parameters substantially influencing the ecological efficiency and energy intensity value of the purification process is represented in the article.

To achieve the set goal we have:

1) built the physical model of the process of decrease of air medium pollution for the conditions of operation of power plants showing the interconnection between the main physical objects participating in this process and allowing to offer the structure of the purification system in which the pollution agents are consecutively and purposefully influenced with external impacts with the purpose of decrease of air medium pollution at the account of decrease of their stability;

2) obtained parametrical dependencies of efficiency and energy intensity value allowing to determine ways for improvement of the process of purification of exhaust gases of boiler houses with the purpose of maximally economic provision of regulatory pollution of urban territories on basis of the analysis and management of functionally independent parameters of the process, changing of which leads to simultaneous increase of values of efficiency and energy intensity value.

Practical implementation of hydrodynamic purification of exhaust gases of power plants carried out by us on basis of modelling has shown that it is impossible to offer unique solution of the purification system, everything depends on original characteristics of the facility – production and technological parameters, characteristics of technological raw materials, parameters of properties of polluting agents of exhaust gases, air medium characteristics, technological parameters of irrigation implementation, physical and chemical characteristics of irrigating fluid, parameters of exhaust source (funnel chimney).

The authors express their sincere gratitude to the staff of the Department of Environmental Engineering, Federal State-Funded Educational Institution of Higher Education Don State Technical University for advice and recommendations kindly given while writing of this article.

References

1. I. Ilchenko, News of Higher Educational Institutions. The North Caucasus region. Series: Natural sciences, **6**, 79-83 (2009)
2. V. Denisov, A. Kurbatova, I. Denisova, *Ecology of the city* (Moscow, MarT, 2008)
3. Yu. Sedunov, S. Avdyushin, E. Borisenkov, O. Volkovitsky, N. Petrov, R. Reytenbakh, et al., *Atmosphere. Reference book (reference data, models)* (Leningrad, Gidrometeoizdat, 1991)
4. S. Paliwal, H. Chandra, A. Tripathi, IJMET, **4**, 2-37 (2013)
5. E. Lysova, Modern situation and prospects of technical sciences development, 79-81 (2015).
6. V. Bespalov, E. Lysova, Collection of scientific papers SWorld, **2**, 19-24 (2015)
7. P. Rayst, *Aerosols – an introduction to the theory* (Moscow, Mir, 1987)
8. H. Greene, W. Lane, *Aerosols – dust, fumes and mists* (Leningrad, Himiya, 1972)
9. S. Bretschneider, *The Properties of Gases and Liquids*, (Leningrad, Himiya, 1966)
10. P. Kouzov, L. Skryabina, *Methods for determination of physical and chemical properties of industrial dusts* (Leningrad, Himiya, 198).
11. N. Fuchs, *The mechanics of aerosols* (Moscow, Academy of Sciences Publishing House, 1955)
12. N. Strakhova, L. Ovchinnikova, A. Pleskachev, *Technical solutions on protection of air environment against the polluting substances* (Rostov-on-Don, Rostov State Civil Engineering University, 2002)
13. V. Bespalov, E. Lysova, O. Paramonova, L. Ganicheva, T. Pirozhnikova, A. Mishchenko, Collection of scientific papers SWorld, **3**, 3-10 (2014)
14. V. Bespalov, E. Lysova, Scientific Review, **7**, 914-917 (2014)
15. V. Kotler, S. Belikov, *Industrial heating plants: combustion of fuels and protection of the atmosphere* (St Petersburg, Energotech, 2001)
16. P. Roslyakov, V. Zinina, Power systems, **12**, 60-62 (1991)
17. I. Segal, Power systems, **3**, 5-8 (1989)
18. E. Lysova, Yu. Eletskaya, Construction – 2003, 31-33 (2003)
19. E. Lysova, L. Lisutina, Current trends in science and education, **1**, 88-89 (2013)
20. V. Bespalov, S. Meshcheryakov, O. Gurova, *Assessment of processes and calculation of apparatuses of environmental protection* (Rostov-on-Don, Rostov State Civil Engineering University, 2007)
21. V. Bespalov, N. Samarskaya, O. Paramonova, E. Lysova, Scientific Review, **9**, 921-924 (2014)
22. V. Bespalov, E. Lysova, Yu. Eletskaya, Life safety. Labor and environmental protection, **7**, 8-9 (2003)

 DOI: 10.1051/matecconf/201712905017

About influence of some superplasticizers on hydration and the structure of hardened cement paste

Yulia Koryanova[1,*] and *Grigory* Nesvetaev[1]

[1]DSTU, Department technology of construction production, 344000 Rostov-on-Don, sq. Gagarin, 1, Russia

Abstract. Current construction can not be imagined without the use of high-quality mortars and concretes obtained by using high-tech and workable mixtures with lower water content. Obtaining such mixtures in current conditions is impossible without the use of superplasticizers. The use of superplasticizers in concrete technology requires an answer to the question of the influence of superplasticizers on the deformation-strength properties of cement stone. There is a well-known dependence "composition-technology-structure-properties", from which it follows that the strength and deformation properties of cement stone directly depend on the hydration of cement stone in the early stages and structure formation. The influence of some types of superplasticizers on the hydration and structure of cement stone, namely, total, open and conditionally-closed porosity, total contraction, autogenous shrinkage, contraction porosity and hydration heat is considered in the article.

1 Introduction

Current construction can not be imagined without the use of high-quality mortars and concretes obtained by using high-tech and workable mixtures with lower water content. The production of such mixtures under current conditions is impossible without the use of superplasticizers (SP). Application of the SP in concrete technology requires an answer to the question of the influence of SP on the strength and deformation properties of cement stone. There is a well-known dependence "composition-technology-structure-properties", from which it follows that the deformation-strength properties of cement stone directly depend on the hydration of cement stone in the early stages and structure formation. In this connection, the investigation of the influence of the SP on hydration and the structure of cement stone is an important scientific problem [1-3].

In the literature there are very contradictory data on the influence of SP on such indicators as strength and deformation properties of cement stone. Numerous domestic and foreign studies have shown that the effectiveness of different SP depends on a number of parameters such as the type of cement used, cement grinding thinness, the type of aggregate, and the type of modifying additives used [4-8]. However, in the first place, it is necessary to take into account the influence of the SP on hydration and early structure formation, as the factors that predetermine the physic-mechanical properties of the cement stone.

The purpose of this work is to study the influence of some types of SP on hydration and the structure of cement stone. To achieve this goal, it is necessary to solve the following tasks: to study the influence of some types of SP on the hydration process, by studying the hydration heat and autogenous shrinkage in the early stages of hardening cement stone, and also to investigate the influence of the SP on early structure formation by studying the formation of porosity and own free deformations of the cement stone.

The influence of the SP was studied using Glenium ACE 430 (hereafter 430); Glenium 323 MIX (hereafter 323); Glenium 115 (hereafter 115); Melment F10 (hereafter M F10); Structuro (hereafter St); SikaViscocret (hereafter Si); Glenium 51 (hereafter 51); Rheobuild 181K (hereafter 181); Polyplast SP-1VP (hereafter SP1); Glenium 30 (hereafter 30) and Melflux 2641 (hereafter 2641) for total (Pt), open (Po) and conditionally-closed porosity (Pcc), total contraction (TC), autogenous shrinkage (CS), contractional porosity (CPr) and hydration heat of cement stone made of cement paste (CP) with a value of W / C = 0.27 using portland cement (PC) made by plants "Proletary" (3 batches) (hereafter, Pr1, Pr2 and Pr3), "October" (hereafter O), "Verkhnebakansky" (hereafter V), "Pervomaisky" (hereafter Pe), "Volskcement" (hereafter Vo) and CEM I 42.5 made in Europe (hereafter T).

2 Influence SP on porosity

As is known, porosity of cement stone predetermines all its physic-mechanical properties and, consequently, affects the properties of concrete. According to [9-12], the SP can influence the process of hydration of cement stone and, as a consequence, its porosity. According to the data of [12], the total porosity of cement stone in the presence of SP can vary from +5,5 to -4,7%, which causes a change in the strength of cement stone from -25 to +22%. Depending on the porosity of the cement stone,

* Corresponding author: koryanova.yi@mail.ru

its modulus of elasticity, creep parameters, and shrinkage change [10, 11].

The dosage of the SP was selected from the condition for obtaining the magnitude of the limiting shear stress of a cement slurry with a value of W/C = 0.27 not more than 10 Pa, which, according to [10], makes it possible to obtain self-compacting concrete mixtures. The results of the studies are presented in Table 1.

Table 1. Porosity of cement stone.

№	Cement	SP	Porosity of cement stone		
			P_t	P_o	P_{cc}
1	Pr1	Not	0,285	0,262	0,023
2		430	0,271 (95)	0,265 (101)	0,006 (25)
3		181	0,319 (112)	0,302 (115)	0,017 (71)
4	O	Not	0,3	0,274	0,026
5		430	0,286 (95)	0,283 (103)	0,003 (11)
6		181	0,328 (110)	0,304 (111)	0,024 (93)
7		323	0,306 (102)	0,294 (107)	0,012 (46)
8	V	Not	0,292	0,273	0,019
9		430	0,295 (101)	0,285 (104)	0,011 (59)
10		323	0,298 (102)	0,281 (103)	0,019 (100)
11		115	0,295 (101)	0,281 (103)	0,013 (68)
12	Pe	Not	0,316	0,289	0,027
13		430	0,306 (97)	0,293 (101)	0,013 (47)
14		323	0,314 (99)	0,289 (100)	0,024 (90)
15		115	0,345 (109)	0,319 (110)	0,026 (94)
16		181	0,343 (108)	0,323 (117)	0,02 (73)
17	Pr2	Not	0,296	0,285	0,011
18		30	0,283 (96)	0,276 (97)	0,006 (62)
19		2641	0,294 (99)	0,286 (100)	0,008 (82)
20	Pr2 + NTF	Not	0,308 (104)	0,297 (104)	0,011 (108)
21		30	0,314 (106)	0,3 (105)	0,013 (126)
22		2641	0,311 (105)	0,296 (104)	0,014 (137)

As follows from the presented in Table 1 data, in the presence of a SP both the total and the open and, especially, the conditionally-closed porosity of the cement stone can change. The increase in open porosity with the introduction of the SP occurs to a greater extent with respect to the increase in the total porosity, and therefore the conditionally-closed porosity decreases.

If we adopt the following scale for assessing the compatibility of a SP with particular cement:

- the SP does not increase the total and open porosity of the cement stone relative to the standard - "excellent";
- SP increases either the total or open porosity of the cement stone relative to the standart by no more than 1.03 times - "good";
- SP increases either the total or open porosity of the cement stone relative to the standard more than 1.03 times - "satisfactory",

then the compatibility evaluation of the SP and cement studied will look like this:

Table 2. Compatible SP with cement.

PC	Assessment of compatibility of cements with SP		
	«excellent»	«good»	«satisfactory»
Pr	30; 2641	430	181
V		323; 115	430
O		430	181; 323
Pe	323	430	181; 115

As is known, in the provision of frost resistance of concrete an important role is played by the ratio of conditionally-closed and open porosity. According to [12], frost resistance of concrete can be determined by the formula

$$F = 307 \cdot K_{frs}^{0.57} \quad (1)$$

where as the criterion of frost resistance is used the ratio of conditionally-closed (Pcc) and open (Po) porosity:

$$K_{frs} = \frac{Pcc}{0{,}09 \cdot P_o} \quad (2)$$

In Table 3 shows the values of the criterion of frost resistance of cement stone in accordance with formula (2).

Table 3. Calculated values of the criterion of frost resistance of cement stone.

SP	The values of the K_{frs} criterion for cements (in parentheses - in% of the standard)			
	Pr	O	V	Pe
Not	$0{,}985^1$ / $0{,}411^2$	1,057	0,787	1,051
430	$0{,}239\ (24)^1$	0,117 (11)	0,327 (42)	0,485 (46)
181	$0{,}608\ (62)^1$	0,887 (84)		0,689 (66)
323		0,45 (43)	0,594 (76)	0,708 (67)
115			0,486 (62)	0,894 (85)
30	$0{,}261\ (64)^2$			
2641	$0{,}335\ (82)^2$			
$Not^{2,3}$	0,425 (103)			
$30^{2,3}$	0,49 (119)			
$2641^{2,3}$	0,543 (132)			

Notes: 1 - for the first batch; 2 - for the second batch; 3 - with NTF

As follows from the presented in Table 1 and 3, when the SP is introduced into the CP, the frost resistance criterion worsens (formula (2)), mainly due to a decrease in the value of the conditionally-closed porosity. This

 DOI: 10.1051/matecconf/201712905017

fact can negatively affect the frost resistance of cement stone. In studies [10], a decrease in the frost resistance of cement stone was recorded with a constant value of W/C in the presence of some SP. In connection with this, when designing compositions of concrete with increased requirements for frost resistance, it is necessary to provide the required value of conditionally-closed porosity through, for example by using of air-entraining agent (AEA). It is not ruled out that in this case it is necessary to increase the dosage of the AEA to compensate for the missing value of the conditionally-closed porosity in the cement stone from the SP.

3 Influence SP on contraction

As is known, with PC hydration, the volume of product of reaction is less than the total volume of substances entering the reaction (cement + water). This phenomenon is called total contraction [13]. The total reduction in the volume of the hardening system «cement + water = cement paste» (TC) consists of a reduction in the external geometric volume of the hardening CP, i.e. CS or autogenous shrinkage, and uniformly distributed practically spherical cavities with a diameter of several tens of microns, so-called CPr, formed in the test volume. Thus, TC = CS + CPr [1]. The measurement of contraction is usually carried out by the Le Chatelier method, while the value of TC is measured on a non-isolated from the environment (liquid) CP, and the value of CS is measured on an isolated from the environment (liquid) CP. There are two main periods of development of contraction:

- before formation in the structure of the crystalline framework, when the CP is in the plastic state, and the values of TC and CS are equal;
- after the formation of a crystalline framework in the structure of the hardening CP, when the difference between the values of the TC and the CS is noted, and the formation of the CPr begins (this period in [11] is defined as the period of deformation development in the postplastic stage, and the value of the CS in this period, from the appearance of the crystalline framework, according to [11], has an influence on the early cracking of cement stone and concrete).

The development of contraction is directly related to the hydration of cement [13], therefore, the change in the hydration kinetics, in particular, in the presence of chemical additives, incl. SP can, due to a change in the duration of 1 and 2 periods of the process of contraction, both the magnitude of the CS, which determines the early cracking, and the magnitude of the CPr. In [10, 11], the changes in CPr and CS in the presence of SP and (or) some additives affecting the kinetics of hydration of PC were recorded.

In Table 4 presents the results of studying the influence of some SP produced by BASF on the process of contraction and the magnitude of CS and CPr. From the presented in Table 4 data is obvious:

- the kinetics of the process of contraction is influenced by both the type and dosage of the SP;
- in the presence of SP the value of TC at the age of 3 - 4 days slightly changes both in the direction of increase and in the direction of decrease, which indicates a some influence of the SP on the hydration process;
- the main result when using SP is an increase the value of the CS, which can be considered as a negative factor, in connection with which it can be asserted that a study of the influence of the SP on the CS value will make it possible to select a rational combination "PC + SP", which reduces the risk of early cracking due to development CS;
- there is a decrease in the value of CPr in the presence of a SP.

Table 4. Data on PC contraction in the presence of a SP.

Cement	SP	The values of the quantities			
		τ_1[1], h	TC, ml/ 100 g C[4]	CS, mm/m (share of TC)[5]	CPr, ml/100 g C (% volume of CP)
V	-	less 6	2,2[2]	3,0[2] (0,35)	1,43[2] (2,44)
	G323 0,47%	3	2,58	8,4 (0,64)	0,92 (1,57)
	G323 0,55%	12 - 24	2,42	3,4 (0,64)	0,86 (1,47)
	G115 0,35%	12	2,38	4,2 (0,41)	1,4 (2,39)
	G115 0,4%	24	2,43	4,3 (0,63)	0,89 (1,52)
	G430 0,37%	6 - 12	2,49	1,15 (0,76)	0,59 (1,0)
	G430 0,45%	36	2,4	0,95 (0,8)	0,47 (0,8)
Pe	-	12	2,15	1,89 (0,44)	1,2 (2,0)
	G323 0,47%	36	2,04	1,99 (0,68)	0,65 (1,1)
	G323 0,55%	12	1,82	3,1 (0,35)	1,17 (2,0)
	G115 0,35%	12	1,86	3,32 (0,44)	1,04 (1,77)
	G115, 0,4%	12	1,98	2,18 (0,28)	1,44 (2,44)
	G430, 0,37%	36	2,09	1,45 (0,78)	0,47 (0,8)
	G430, 0,45%	24	2,11	1,8 (0,61)	0,83 (1,4)
O	-	12	2,04	2,86 (0,5)	1,01 (1,7)
	G323, 0,47%	12	2,0	2,35 (0,3)	1,4 (2,38)
Pr1	G430, 0,37%	12	2,46	7,14 (0,59)	1,0 (1,7)
Pr2	-	5	3,0[3]	2,9[3] (0,32)	2,04[3] (3,48)
	G30, 0,32%	1	2,94	6,2 (0,44)	1,65 (2,81)
	M 2641, 0,36%	1 - 3	3,1	5,15 (0,36)	1,97 (3,36)

Notes: 1 - the duration of the first period of development of the process of contraction; 2, 3 - at the age of 96 and 72 hours respectively; G-series "Glenium"; M-series "Melflux"; 4 - C = cement; 5 - share of total CS from TC

Since, as noted above, the SP has an insignificant influence on the value of TC at 3 to 4 days of age, a significant change in the values of CPr and CS in the presence of SP is associated with the kinetics of the process of contraction (hydration) in the early period. In particular, on the first day of hydration there is a lag in varying degrees of the TC value of Verkhnebakansky cement with all SP from the sample without SP, which indicates some inhibition of the hydration process in the compositions with the SP in this period.

The change in the kinetics of the formation of a crystalline frame in the early period influences the change in the ratio between CPr and CS, i.e. an increase in time τ_o (duration of period 1) leads to a decrease in the CPr and an increase in the CS.

Thus, with the introduction of a SP into the composition of the concrete mixture, due to the possible influence of the SP on the hydration process of the PC and, as a consequence, on the kinetics of the contraction process, it is possible to change the duration of the period prior to the onset of intensive formation of the crystalline frame in the structure of the hardening cement stone, CPr and CS - with the increase of this period, i.e. when the hydration process slows down in the early period, the CPr increases and the CS decreases. Since CS has a significant influence on the cracking of cement stone (concrete) in the early period of structure formation [10, 11], and CPr has an influence on the frost resistance of concrete [12], and the ratio of CPr and CS of cement stone can vary widely [14], the question of the influence of a SP on these quantities requires special study.

4 Influence SP on hydration heat

According to [15-17], three main periods of cement hydration - induction, accelerated and delayed - are distinguished on thermokinetic dependence. In our opinion, the following three sections are of primary interest in the integral curve of heat of hydration of the PC in the early period of hydration: 1 - due to the heat of wetting and hydration of alkalis, free lime, partial hydration of C_3A and the formation of primary ettringite; 2 - the so-called induction period, during which the hydration heat is practically not fixed; 3 - period of intense hydration heat (accelerated) due to hydration of the basic clinker minerals. Then comes the period of slowing down the rate of hydration.

The SP can have a significant influence on the hydration heat parameters of cement [18], and for practical applications it is very important to select the most compatible "cement-SP" pair, since the SP can influence all the properties of the cement stone [9]. For each of these areas, certain quantitative indicators can be used, which depend on the type of cement, the presence of additives, and temperature conditions, i.e. can characterize the features of the hydration process. For the first stage, for example, the hydration heat quantity Q1 is important, since, as is known, the heat of wetting for powders depends on their dispersion. It is clear that in the amount of hydration heat Q1 in the first section there are "noises" - the heat of hydration of alkalis and free lime, partial hydration of C_3A and the formation of the primary ettringite. But if we consider the influence of the SP on the hydration heat in the first stage for particular cement, we can indirectly assess the influence of the SP on the dispersion of cement by changing the hydration heat Q1. As is well known, the heat of wetting for powders depends on their specific surface and varies, according to some data, in a very wide range, from 10^{-5} to 10^{-3} cal/cm^2, which, when recalculated on a PC, roughly corresponds to a value from 0.12 to 20 kJ/kg. It is clear that for PC this range will be much narrower.

According to the data of [15], hydration heat in the initial period with hydration of C_3S is 2.3 to 4.9 kJ/kg, depending on the dispersity of the binder, and the duration of this stage is from 20 to 38 minutes. In our studies (Table 5), the values for PC without SP, respectively Q1 from 2.15 to 6.6 kJ/kg, and the duration of the first period from 5 to 50 minutes.

Table 5. Influence of SP on the value of Q1 and the duration of the first stage of hydration.

SP	Hydration heat in the first stage kJ/kg / Duration of the first stage, min							
	Specific surface of cements, cm^2/g							
	2930	3270 – 3600						4140
	Cements							
	Pr1	Pr2	Pr3	V	O	Pe	Vo	T
Not	2,6 / 5	4,9 / 7	2,2 / 20	5,9 / 30	2,2 / 15	2,2/ 25	4,6 / 12	6,6 / 50
30	3,0/ 5							
2641	5,2/ 5							
SP1		11,6 / 10					8,7 / 10	6,4 / 26
M F10		9,9 / 19					10,4 / 19	6,8 / 7
St		10,2 / 22					8,8 / 19	13,8 / 44
Si		13,8 / 240					12,4 / 25	9,9 / 160
51		12,3 / 23					9,9/ 17	8,4 / 20
323				4,5 / 60	4,4 / 20	4,9 / 20		
430			2,4 / 10	5,8 / 60	1,4 / 175	1,8 / 180		
181			8,2/ 60		2,3/ 45	7,6/ 60		
115				2,6 / 70		2,6 / 20		

From the presented in Table 5 data, it is obvious that in cements without SP, the duration of the first period naturally increases with the decreasing size of cement particles, and the amount of hydration heat during this period depends not only on cement grinding thinness, which is also quite natural, since hydration heat in this

 DOI: 10.1051/matecconf/201712905017

period, as already noted, is related with the presence of free lime, etc.

With the introduction of the SP, the increase in the hydration heat in the first period is noted mainly, which can be estimated as the disperger influence of the SP (some exception is V + 115; O + 430; Pe + 430; T + SP1). It is also noted, as a rule, the extension of the first period. This may be due to both the growth of the actual specific surface due to the dispersion of cement and the retarding influence of the SP on the hydration of free lime and other compounds in the early period. It can be assumed that the more the hydration heat in the first period occurs with the introduction of the SP and the less the elongation of the first period, the more effective the SP with this cement is.

Estimating the duration of the induction period allows us to determine the influence of the SP on slowing down the hydration process in the early period, i.e. classify the SP on the influence on the setting time (SP with a retarding influence or not). According to [15], the duration of the induction period during hydration of C_3S is from 1.1 to 4 hours, depending on the dispersity. In our studies, the duration of the induction period for PC without SP was 1.8 to 3.1 hours (Table 6).

Table 6. Influence of SP on the duration of the induction period.

SP	The duration of the induction period, h							
	Specific surface of cements, cm²/g							
	2930	3270 – 3600						41 40
	Cements							
	Pr1	Pr2	Pr 3	V	O	Pe	Vo	T
Not	2,5	1,8	3,1	2,2	3,0	3,0	2,5	2, 3
30	5,5							
2641	6,5							
SP-1		2,3					5,6	3, 5
M F10		7,0					5,3	3, 9
St		6,6					22,5	22 ,5
Si		54,4					57	44 ,5
51		19,2					17,9	3, 15
323				8,3	3,0	3,5		
430			9,1	5	10,5	5,2		
181			13		14,4	3,0		
115				5,8				

From the data presented in Table 6, it is obvious that when the SP is administered, as a rule, the duration of the induction period increases and in some cases, very significantly. It is clear that increasing the duration of the induction period to 44 and even 57 hours for practical use in most cases is undesirable. Some increase in the duration of the induction period can positively affect the increase in the stability of the concrete mix. Therefore, the influence of the SP on the change in the duration of the induction period can also be used as a criterion for assessing the compatibility of PC and SP.

The hydration heat rate in the third period can characterize the influence of the SP on the hydration process. According to [15], the duration of the third period of hydration because of C_3S hydration is 6.3 to 14.2 hours, depending on the size of cement particles, and the rate of hydration heat at this stage is from 8.2 to 19.8 kJ/kg · h. The total hydration heat at this stage is 44.8 to 90.9 kJ/kg. According to our data, the rate of hydration in the third period for PC without SP was from 2 to 25 kJ/kg · h, and the total hydration heat in the third period was from 26 to 94 kJ/kg (Table 7).

Table 7. The magnitude of hydration heat and its rate in the third period.

SP	The amount of heat released, kJ /Average hydration heat rate in the third period, kJ/kg · h							
	Specific surface of cements, cm²/g							
	2930	3270 – 3600						4140
	Cements							
	Pr1	Pr2	Pr3	V	O	Pe	Vo	T
Not	69 / 12	94 / 14	81 / 15	55 / 7,5	26 / 2	45 / 5	62 / 11	87 / 25
30	68 / 13							
264 1	68 / 15							
SP1		82 / 10					67 / 6	77 / 24
M F10		95 / 9					53 / 7	80 / 19
St		62 / 1					36 / 1	56 / 9
Si		42 / 3					26 / 0,5	46 / 7
51		59 / 7					40 / 4	79 / 7
323				87 / 9	52 / 1	34 / 1,5		
430			86 / 9	59 / 7	35 / 2	45 / 3		
181			80 / 6		41 / 0,6	30 / 0,4		
115				86 / 7		37 / 2		

The hydration heat rate in the third period can characterize the influence of the SP on the hydration process:

- since the hydration heat during hydration of various minerals is different, the change in rate can be a consequence of selective adsorption of the SP;
- it is obvious that the hydration heat will depend on the area of contact "cement-water", in connection with which the dispergation of cement particles in the presence of a SP can cause increasing of hydration heat.

From the presented in Table 7 data, it is obvious that the SP can have a significant influence on both the hydration heat in the third period and the rate of hydration. Such information can be very useful when selecting a «cement-SP» pair, for example, for concreting massive structures that require the minimization of the hydration heat of concrete. It is clear that minimum change the total hydration heat and the rate of hydration when using SP characterized small influence of the SP on the hydration process. As a confirmation, we can consider the relationship between the rate of hydration heat in the third period and the tensile strength of cement stone at the age of 28 days.

Thus, the analysis of the SP influence on the change the parameters of heat of hydratation of the PC gives information for a preliminary assessment of the "compatibility" of a particular PC with a particular SP, including taking into account the specific application (temperature conditions, requirements for the conservation of concrete mix and limiting hydration heat). To fully assess compatibility, it is necessary to identify the influence of SP on porosity, strength, and deformation properties of cement stone [18-19].

5 Conclusions

An analysis of the influence of the SP on the changes in the parameters of hydration and structure formation in the early periods provides information for a preliminary assessment of the "compatibility" of a particular PC with a particular SP, incl. taking into account the specific application (temperature conditions, requirements for the conservation of concrete mix and limiting hydration heat). For mix proportions of concrete with increased requirements for frost resistance and crack resistance, it is necessary to provide the required amount of conditionally-closed porosity, CPr, CS and their ratio. To fully assess the compatibility and values of the parameters of conditionally-closed porosity, CPr, CS and their ratio, it is necessary to determine the influence of the SP on the porosity, strength, and deformation properties of the cement stone.

References

1. G.V. Nesvetaev, Building materials **10,** 23-25 (2006)
2. Yu.Yu. Galkin, S.A. Udodov, Sat. sci. works on the results of the international. scientific-practical. conf. "Innovations in the field of technical sciences", 32-36 (2017)
3. M.S. Saidumov, M.Sh. Salamanova, T.S.A. Murtazaeva, M.S.M. Hubayev, Innovations and investments **8**, 163-166 (2015)
4. G.V. Nesvetaev, I.V. Korchagin, Yu.I. Potapova, Scientific review **7-3**, 837-841 (2014)
5. Yu.M. Bazhenov, V.R. Falikman, Materials I vseros. conf. of concrete and reinforced concrete, 91-101(2001)
6. G.V. Nesvetaev, I.V. Korchagin, Yu.I. Potapova, Scientific review **7-3,** 842-846(2014)
7. V.G. Batrakov, Materials I vseros. conf. of concrete and reinforced concrete, 184-187 (2001)
8. G.V. Nesvetaev, I.V. Korchagin, Yu.I. Potapova, Scientific review **8-3,** 907-913 (2014)
9. V.I. Kalashnikov, V.S. Demyanova, A.A. Borisov, Izvestiya vuzov. Construction **1,** 39-42(1999)
10. G.V. Nesvetaev, A.N. Davidyuk, Building materials, **1,** 44-46 (2010)
11. G.V. Nesvetayev, G.S. Kardumyan, Concrete and reinforced concrete **1,** 12 -15 (2013)
12. J. Bensted, World cement, 27-32 (1991)
13. V.V. Nekrasov, Izvestiya of the Academy of sciences of the USSR **6,** 592 – 610 (1945)
14. J. Bernal, J. Jeffery, H.F.W. Taylor, Mag. of concr. res. **11, 47** (1952)
15. A.V. Usherov-Marshak, V.Kh. Prokhorov, N.V. Smirnova, et al., Cement **2,** 7-8 (1980)
16. A.V. Usherov-Marshak, A.M. Urzhenko, VI International congress on the chemistry of cement, 235-237 (1974)
17. M. Fukuhara, S. Goto, K. Asaga, M. Daimon, R. Kondo, Cem. concr. res. **11,** 407-414 (1981)
18. G.V. Nesvetaev, Building materials **3,** 24 – 28 (2008)
19. V. Ramachandran, Proc. 3 intern. congr. on polymers in concrete **V. II** (1981)

DOI: 10.1051/matecconf/201712905018

Industrial robots application in the construction of buildings and structures

Gennady Verzhbovskiy[1,*] and *Alexey* Reshetnikov[1]

[1]Don State Technical University, Gagarin square 1, Rostov-on-Don, 344000, Russia

Abstract. Proposals on the use of modernized industrial robots in the construction of low-rise buildings are formulated. The necessary parameters of such a mechanism are established. The time necessary for building the walls of a two-story house is determined. Features of the robots use on the construction site are described.

1 Introduction

As a rule, sets of products for the erection of buildings and structures, produced within one enterprise, are characterized by constant dimensions and high quality, that turns the construction process into a rigorous set of simple operations for the sequenced lifting, transfer and installation in a specific place and in a certain order of the structure details. In industry, similar actions have long been performed by robotic manipulators. When they are used in production, there are a number of positive points:

- in the process of performing monotonous operations, the "human factor" does not appear;
- they can work in environment dangerous for human life and health;
- they can work around the clock, without days off and holidays;
- the accuracy of part positioning in the assembly is much higher than that of a person;
- the percentage of defects is minimized.

Robots in construction are no longer fiction. Back in 1984, the USSR issued the state standard "Manipulators for construction and installation works. General technical requirements" [1]. However, in Russia they are used in construction very rarely. First, the robotization of building is constrained by the fact that each project is unique and requires the use of various high-tech solutions. Secondly, the use of robots, for example, in the creation of concrete frameworks, is associated with the need to use expensive construction mixtures with hardening high speed, formless and nonarmature construction technologies, otherwise the efficiency of using complex equipment will be lost. Third, the cost of robots is still high.

Experts believe that in future robots will be actively used in low-rise construction [2]. Their application allows to reduce time and money costs by several times, and also provides high accuracy of construction. Several years ago, an attempt to create a construction robot designed for the mechanization of masonry was undertaken by scientists from the Robotics and Microelectronics Laboratory of the Russian Academy of Sciences. The main reason for the unpopularity of such technologies in Russia is that the cost of housing does not depend on the cost of labor, so, it is more profitable to use labor of migrants, who receive less wages than professional workers.

Abroad, the process of robotics is very intense. Thus, Fabio Gramasio and Matthias Kohler, professors at the Institute of Architectural Technology ETH Zurich in Switzerland, were among the first to use robots in the field of architecture and design [3]. Since 2006, they have investigated various production technologies and the application of a wide range of different materials to create buildings by robots. Their studio Gramazio & Kohler produces complex designs, which are then used to build unusual structures in Zurich, London, Barcelona, New York, and other places (Figure 1).

Fig. 1. The works of the studio Gramazio & Kohler

* Corresponding author: vergen2005@yandex.ru

The peculiarity of these structures is that they are assembled from separate elements, installed in the design position with a high degree of accuracy by a special algorithm.

Scientists in Japan, the United Kingdom and the United States are simultaneously developing 3D building printers that "print" buildings from monolithic reinforced concrete. There is information about similar studies in Spain - the project "Stone Spray". All this proves that the use of robots in construction is a promising direction in the development of the industry.

2 Problem formulation

Accuracy and constancy of the factory-made products sizes and their predetermined location in the building structure make it possible to automate the process of erecting the building. To solve the task, it is necessary to determine first the required parameters of the robot manipulator.

Industrial robots (IR) are characterized by a set of basic data such as load capacity, number of degrees of freedom, work area, mobility, positioning accuracy and speed of movement [4]. Load-carrying capacity is the greatest total effort developed by the operating bodies of the IR. If the weight of the element exceeds 160kg, the robots are considered as heavy ones.

The number of degrees of freedom is the sum of all manipulator brush possible movements without taking into account the gripping and clamping. In general, six degrees of freedom are necessary to move and orient bodies in space, although two or three degrees are sufficient for certain operations. As it turns out from the following discussion, when erecting buildings, it is irrational to limit the movement of the manipulator.

The working area is defined by the space in which the gripper can operate. It is characterized either by the volume serviced by the manipulator or by the magnitude of the greatest displacement along each coordinate and around. Since we are considering low-rise buildings, we agree to limit the working area to 25x20 meters in plan and 15 meters in height.

Mobility is the ability of a robot to perform local, regional and global movements. Local ones called the movements of the manipulator, necessary for the final installation of the product in the design position. They are rotations and minor linear movements of the working member. Regional ones are movements of the grip in different areas of the working space, determined by the size of the arm links, and also commensurate with the size of the equipment to be serviced. Global one is a movement of the manipulator at distances exceeding both the dimensions of the industrial robot itself and the equipment size. Due to the fact that building materials are usually stored near the object under construction, the IR must be able to perform all of these types of movements. In the literature such mechanisms are called nonstationary.

Positioning accuracy is the greatest deviation of the gripping device from a given position at a certain fixed point with multiply repetition of the cycle. The most common industrial robots with positioning accuracy of about 0,2-0,3mm are applied, and such accuracy is quite sufficient for construction.

Most modern manipulators have a linear arm speed up to 1,5m/s, and angular velocity up to 180grad/s. Taking in account the construction of the facility is conducted in the open air and in various weather conditions, it is irrational to use the maximum values, so it is suggested to limit the speed to 1m/s and 90grad/sec. In this case, the robot will be able to transfer a separate element along the entire working area in just 25 seconds, so it is a good indicator.

3 Results and its discussion

As a result, we can conclude that for the construction of low-rise buildings and structures, a heavy non-stationary robot with six degrees of freedom of the manipulator, positioning accuracy of 0,2-0,3mm, average speeds and adjustable working area with maximum dimensions of 25x20x15m (LxWxH) is needed. To facilitate the installation of extended elements, such as joists, it is convenient to use a technique with two grippers, so another condition is the presence of two independent working bodies. Finally, the various operations carried out on the construction site require the use of replaceable attachments.

A serious proof of the expediency use robots can be the determination of the time spent on the wall blocks installation. It is assumed that the building is erected in a "dry" way and interlocking and interchamber fixators are used to fix the blocks [5].

Suppose that we need to install 3 blocks at the same time at a height of 6 meters to the wall located in the 15 meters from the place of their storage. Let's consider that the height of a stack is 1 meter, and the robot, whose actuator is above the installation site at an altitude of 6,5 meters, can perform motions and rotations only sequentially. It is more convenient to calculate the required time in a tabular form.

Table 1. Calculation of the time required for the robot to install three wall blocks (beginning)

	Contents of the operation	Duration of the operation, s
1	Moving the manipulator to the storage location	15
2	Positioning of the gripper above the blocks	5
3	Descent from a height of 6,5 meters and blocks capture	10
4	Lifting of the gripper at 1 meter, its rotation and placement above the interlace clamps	8
5	Release and lock of latches	3
6	Lifting of the gripper at 1 meter, its rotation and placement above the interlock clamps	8
7	Release and lock of latches	3

 DOI: 10.1051/matecconf/201712905018

Table 1. Calculation of the time required for the robot to install three wall blocks (end of table)

8	Lifting the manipulator to a height of 6,5 meters	8
9	Rotate the gripper to the position required to install the blocks	3
10	Moving the manipulator to the installation site	15
11	Positioning the gripper	5
12	0,5 meter descent and units installation	3
13	Lifting the manipulator at 1 meter and turning it to 90° to insert the interlace clamps	3
14	1 meter descent and fixing installation	3
15	Lifting of the manipulator by one meter and its rotation to 90° for insertion of interlock clamps	3
16	1 meter descent and fixing installation	3
17	Lifting the gripper to 0,5 meters and turning it to 180° for subsequent operations	3
	Total	101

When determining the time required for performing the operations in Table 6.4, one second was added to each change the way the manipulator moves. Taking in account possible unforeseen circumstances, let's increase the time required for installing the units up to two minutes. With the parallel operation of the two grippers, it takes about twenty seconds to fix one element to the wall. During 23 hours the robot will be able to install 4100 wall blocks. Approximate calculations show that the walls of a two-storey building in this way can be erected in one day instead of six, required by a team of builders. The indicated term can be even less if in the robot base design there is a site for installing a stack with mounted elements.

Let us turn to the consideration of applied questions connected with the use of such equipment on the construction site. First of all, it is a possible robot design. It is proposed to be performed as a reach truck. Reach truck is a type of a high-lift stacker with a counterweight and a retractable mast that is one of the high-tech types of warehouse equipment (figure 2).

Fig. 2. Reach track with counterweight and telescopic lift

The robot on the specified mechanism basis should move along the building under construction on special guides, and instead of the forks intended for carrying cargoes, it should have the manipulator installed on it. Two similar mechanisms are installed on both sides of the object and controlled by one computer with a specialized program. The approximate location of the robot on the construction site is shown in Figure 3.

The robotic construction process has its own characteristics and is somewhat different from the standard human one. So, for example, when moving an object into nature, there is no need to secure all the centering axes. It is quite simple to define three any characteristic points of a building that do not lie on one straight line. It is the most convenient to indicate the position of the corners of the structure.

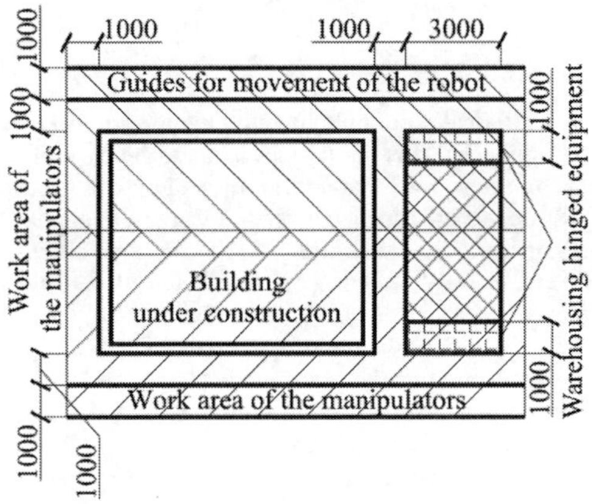

Fig. 3. Possible robots location on the construction site

After removal of the object, installation of guides for the robot movement is carried out. In principle, these can be flat slabs laid on a planned base. For example, they can be the composite road surfaces "MDP-MOBISTEK" shown in Figure 4 [6].

Fig. 4. Prefabricated composite road surface "MDP-Mobystek"

The result of this operation should be a surface without kinks at the joints. Global misalignments of guides, associated with the relief of the site, are completely acceptable. Prior to construction, the mounted on the rails robots produce a high-altitude

DOI: 10.1051/matecconf/201712905018

survey of the base profile and form the necessary corrections to the assembled parts coordinates.

The next stage is the organization of storage areas for materials and attachments. Builders determine the location of inventory pallets with removable devices for manipulators relative to the guides and prepare a place from which the building elements will be taken automatically. The supply of materials to this zone is made by workers. The most convenient is the stacks with products installation on special carts, which are placed on the site in the spaces appropriate for them. They can be allocated with special restrictors fixed in the storage area. Precise positioning simplifies the robots work and minimizes the possible errors. The pallets stacking to the carts must be carried out outside the working area of the robots. The order of parts placement in the storage is determined by the assembly instruction.

4 Conclusions

It is theoretically possible to automate the whole process of building, but it is hardly advisable at the current level of technology development. So, for example, a robot can easily cope with drilling wells or digging trenches for foundations, but backfilling is more convenient by workers. Separate actions on the device of engineering networks in a building are easier for man to carry out. We can specify some other operations, where the use of robots is economically unprofitable. Thus, we are talking about the joint work of robots and man, as it is organized in a number of industrial productions.

Round-the-clock robotic construction assumes three-shift work of the personnel, therefore the standard brigade of builders is unlikely to be reduced, but a significant reduction in the construction time will reduce the cost of providing workers and the costs associated with the operation of construction equipment.

The proposed innovative technology makes it possible to organize an analogue of industrial production on the construction site, which will allow creating objects with high quality and in a short time.

References

1. *Manipulators for building. General technical requirements*, GOST 26055-84 (Moscow, Publishing house of standards, 1984)
2. V.V. Nemchenko, *Building robots - a new class of cars*, Workshop. Modern construction **3**, 25 (2013)
3. *Architects using robots to create amazing and beautiful structures*, Tzoom-news portal. Informational-educational project, Available: http://www.tzoom.com.ua/arxitektory-ispolzuya-robotov-sozdayut-udivitelnye-i-krasivye-stroeniya/. Last accessed 2.12.2013 (2013)
4. A.G. Bulgakov and others, *Automation and robotization of building processes and productions* (Moscow, Russian Engineering Academy, 2006)
5. G.B. Verzhbovskiy, *Low-rise prefabricated buildings and structures made of composite materials* (Rostov-on-Don, LLC Publisher Bara, 2015)
6. *Mobile road surface MDP-Mobystek*, Available: http://gazpromss.ru/production/road_reinforcing/84/. Last accessed 13.10.2013 (2013)

DOI: 10.1051/matecconf/201712905019

Stability of direct circuits integrating the equations of motion in the simulation of the dynamics of destruction

Leonid Panasjuk[1], and *Galina* Kravchenko[1,*]

[1]Don State Technical University, pl. Gagarina, 1, Rostov-on-Don, 344010, Russia

Abstract. In this paper, it is discussed the approximation of the nonlinear equations of motion of mechanical systems used on different operators achieved level displacements and displacement increments. For the numerical solution of the problem, we used the scheme of direct integration of the equations of motion, which in the case of a linear problem reduces to the known Newmark scheme. The stability parameter of the integration scheme is determined. Analytical calculations are confirmed on the test meeting of the model task.

1 Introduction

The main task of structural mechanics is the development of such methods of analysis of structures for strength, rigidity and stability which with sufficient economic efficiency would ensure the safety of structures during the entire period of their life. Today, the increased requirements are applied to the methods of calculations - the analysis of structures should provide a comprehensive outlook on their performance at all stages of the life cycle. To solve a similar problem, it is necessary to use refined calculation models in order to adequately represent all the features of the system response to external perturbations in a numerical experiment. Thus, developing of effective methods of modeling of the dynamic response of non-linear problems, including problems of fracture mechanics, is an urgent task of modern structural mechanics.

The classification of non-linear problems of mechanics today considers three factors that determine the nonlinearity of the resolving equations. The first type - physical nonlinearity. Here the refined nonlinear relations between deformations and stresses, or between their increments, are considered. Rejection of the Navier hypothesis of small displacements leads to a refinement of geometric constraints and calculation in geometrically nonlinear statement. Classification of nonlinear problems of K. Novozhilov is complemented by the concept of constructive nonlinearity (at first time this kind of nonlinearity was presented by S. Timoshenko), which takes into account the change in the construction scheme at certain stages of its existence. The important aspect - fracture mechanics - can be considered as a special case of physically and structurally nonlinear creations.

A sufficiently comprehensive consideration of the features of geometry of the calculated structures, their mutual influence, physical and geometric dependencies could only be possible today when using numerical calculation methods. Solving of dynamic problems in a nonlinear design leads to the fact that in the method of principal coordinates it is necessary to measure many times the frequency and shape of free oscillations. Therefore, because of the high complexity of the process of solving the spectral problem, even in a truncated formulation, the complexity of the principal coordinate method increases catastrophically.

Recently, in order to find dynamic response, the direct integration schemes are being used, in which a series of problems is being solved sequentially at small time phases, where the resolving equations are algebraic. In this case, at one phase in time, the complexity of the solution corresponds to the statics problem. Solving more complex and time-consuming spectral problem here is not necessary.

In problems of the linear dynamics of the rigidity of the elements of the systems, the equations of equilibrium of a small element and the continuity equation do not change for the entire time interval where the solving of a problem is taking place. However, in nonlinear dynamics problems, the parameters of the basic equations change on each time layer. First the parameters of the linearized (within a small step on the time) equations of motion must be defined. These parameters seem to be "frozen" and are considered constant within the temporal layer. After determining the components of the stress-strain state of the system at the end of the time phase, the parameters of the equations of motion corresponding to the next phase are being determined. That is, within one time phase, a quasilinear problem with current parameters corresponding to the stress-strain state at the beginning of the time phase is considered. The simple loading hypothesis, which is used in many versions of the theory of plasticity, is being implemented within one time phase, but, in general, the process of complex loading of the system is applied.

* Corresponding author: galina.907@mail.ru

2 Theoretical part

One of the options of the equations of motion for an ensemble of finite elements linearized in a time phase has the form (1):

$$M\ddot{q}(t)+C\dot{q}(t)+K_{c,n}q(t)=P(t) \qquad (1)$$

and onwards M – is a mass matrix, C – a deformation matrix, K, n – a "secant" rigidity matrix for phase n of the time for the ensemble of finite elements, q (t) – a vector-function of nodal displacements, P (t) - the load vector function given to the nodes of the system. Matrix K, n is formed during the first time phase in accordance with the parameters of the stress-strain state of the system achieved at a time t = t n . It is assumed that in the time segment [t n, t n +1] the overall rigidity of the system does not change. Thus, at the time phase the potential energy of deformation is approximated by the quadratic function (2), and therefore the equations of motion are represented in the form (1):

$$\Pi_2=0.5(q(t))^T R(q(t))=0.5(q(t))^T K_c q_n\, q(t)= \qquad (2)$$
$$=0.5(q(t))^T K_{c,n}\, q(t)$$

The second form of the equations of motion obtained on the basis of the decomposition of functional of the total potential energy in the range achieved at time t n of the level of stress-deformed state into a Taylor line with retention of terms up to second order and is presented as following (3):

$$M\ddot{q}(t)+C\dot{q}(t)+K_{\kappa,n}\Delta q(t)+K_{c,n}q_n=P(t) \qquad (3)$$
$$\Delta q(t)=q(t)-q_n,\ t\in[t_n,t_{n+1}]$$

For the approach used in (3), the linearization is designed in advance in the form of variation, so the overall algorithm of the method of finite elements forming "secant" K, n and "tangent" rigid matrices of the system is sustained (4):

$$K_{\kappa,n}=\int_{(V)}\Phi^T H_n\Phi\, dV, \qquad (4)$$

$$K_{c,n}q_n=\int_{(V)}\Phi^T\sigma_n\, dV,\ H_{n,ijkl}=\frac{\partial^2 U}{\partial\varepsilon_{ij}\,\partial\varepsilon_{kl}}$$

The disadvantage of (1) in comparison with (3) is a lower quality of approximation of the system reaction at a time phase. Indeed, it is straightforward showing that in (1) the augmentation in the reaction of the system is determined by the "secant" rigidity, which is not utterly correct:

$$M\ddot{q}(t)+C\dot{q}(t)+K_{c,n}q(t)=M\ddot{q}(t)+C\dot{q}(t)+$$
$$+K_{c,n}\Delta q(t)+K_{c,n}q_n$$

However, according to the formula, within the time phase (1), it corresponds to the equations of motion of the linear system. Therefore, for any numerical integration (1) any sustainable method can be used, such as Newmark, Tett-Wilson, and others.

It should be noted that if at the dawn of the development of direct integration methods the question of improving the accuracy of approximation at the time phase was considered to be dominant, later the question of numerical stability of methods became critical. Moreover, many methods that use the best approximation of the displacement vector function have an instability area, and their application is limited. A conditionally stable method of central differences can serve as an example.

Thence, we consider absolutely stable method for integrating the nonlinear equations of motion (3), here presented in the form (5). Earlier it was demonstrated that equations (5) can be handled as an extension of one of the best Newmark's methods for nonlinear problems.

$$\begin{cases}\left(2M+\Delta t\beta C+0.5\Delta t^2\beta^2 K_{K,n}\right)q_{n+1} \\ s_{n+1}=-s_n+2(q_{n+1}-q_n),\ s=\Delta t\,\dot{q}\end{cases} \qquad (5)$$

In (5), the parameter β is assigned due to the absolute stability conditions scheme. That is, in contrast to the classical form of the Newmark method, a directed search for the stability parameter is performed in (5). Previously, with no major justification, it was recommended that parameter β is defined as a form a ratio of squares of circular pitch frequencies in the oscillations of linear problems with the "secant" and "tangent" rigidity (6):

$$M\ddot{q}(t)+K_C q=0\Rightarrow\omega_C, \qquad (6)$$
$$M\ddot{q}(t)+K_K q=0\Rightarrow\omega_K,$$
$$\beta=\frac{(\omega_C)^2}{(\omega_K)^2}=\frac{q^T K_C q}{q^T K_K q}$$

However, this result (the ratio of the squares of frequencies) was obtained for the oscillator, and the propagation of the formula for a system with several degrees of freedom is unreasonable.

This paper shows that for a system with several degrees of freedom of the stability the parameter β should be defined according to (7)

$$\beta=\max\left(1, ejg\left(\left[K_{K,n}\right]^{-1}\left[K_{C,n}\right]\right)\right) \qquad (7)$$

In this paper, it is demonstrated that such an assignment of the stability parameter according to (7) corresponds to the Neumann absolute stability criterion: the spectral radius of the integration scheme does not exceed unity.

Indeed, the stability test is carried out by considering the free oscillations of a nonlinear system in the absence of damping. The system (5) in this case can be written in the form (8):

$$\begin{cases} R\,q_{n+1} = \left(R - \frac{1}{\beta} K_{C,n}\right) q_n + \left(R - 0.5 K_{C,n}\right) s_n, \\ R = \frac{2}{\beta \Delta t^2} M + 0.5 \beta K_{K,n} \\ s_{n+1} = -s_n + 2(q_{n+1} - q_n),\ s = \Delta t\, \dot{q} \end{cases} \tag{8}$$

The matrix $\boldsymbol{R}$ in (8) is symmetric positive definite. We perform the splitting (8) with respect to the eigenvectors of the matrix $\boldsymbol{R^{-1}K_{c,n}}$. Let μ_i be the eigenvalues of the matrix $\boldsymbol{R^{-1}K_{c,n}}$, we obtain for the ith splitting form (9):

$$\begin{Vmatrix} \varsigma_{i,n+1} \\ \xi_{i,n+1} \end{Vmatrix} = \begin{Vmatrix} 1 - \frac{\mu_i}{\beta} & 1 - 0.5\mu_i \\ -\frac{2\mu_i}{\beta} & 1 - \mu_i \end{Vmatrix} \bullet \begin{Vmatrix} \varsigma_{i,n} \\ \xi_i \end{Vmatrix} \tag{9}$$

For each eigenvalue μi, the representation (10) holds for the eigenvectors $\boldsymbol{Z_i}$ of the matrix $\boldsymbol{R^{-1}K_{c,n}}$:

$$\mu_i = \frac{z_i^T K_{C,n} z_i}{z_i^T K_{K,n} z_i} = \frac{z_i^T K_{C,n} z_i}{\frac{2}{\beta \Delta t^2} z_i^T M z_i + 0.5\beta z_i^T K_{K,n} z_i} \tag{10}$$

Due to the positive definiteness of the quadratic form $\frac{2}{\beta \Delta t^2} z_i^T M z_i$, (10) implies (11):

$$\mu_i = \frac{z_i^T K_{C,n} z_i}{\frac{2}{\beta \Delta t^2} z_i^T M z_i + 0.5\beta z_i^T K_{K,n} z_i} \le \frac{2}{\beta} \frac{z_i^T K_{C,n} z_i}{z_i^T K_{K,n} z_i} \tag{11}$$

For the purpose of the assignment of the stability parameter β, according to (7), the Rayleigh inequality holds $\frac{z_i^T K_{C,n} z_i}{z_i^T K_{K,n} z_i} \le \beta$, which implies the restriction (12) of the eigenvalues of the matrix $\boldsymbol{R^{-1}K_{c,n}}$:

$$\mu_i \le \frac{2}{\beta} \frac{z_i^T K_{C,n} z_i}{z_i^T K_{K,n} z_i} \le 2 \tag{12}$$

Obviously, that when the restriction (12) is satisfied, the spectral radius of the matrix operator in (9) does not exceed unity, i.e. The Neumann absolute stability condition for the scheme of direct integration of the nonlinear equations of motion (5) is satisfied.

Numerically, for a system with a large number of degrees of freedom, the realization of formula (7) can be quickly carried out by the Lyusternik iterative method. Note that the matrix $[K_{k,n}]^{-1}K_{c,n}$ is not symmetric, it takes a large amount of memory and the laboriousness of its construction is great. In order not to form the matrix $[K_{k,n}]^{-1}K_{c,n}$, the following modification is suggested within the limits of one iteration of the Lyusternik algorithm (13):

$$\begin{cases} \bar{\rho}_{i+1} = K_{c,n}\, \rho_i, \\ K_{k,n}\, \hat{\rho}_{i+1} = \bar{\rho}_{i+1}, \\ \rho_{i+1} = \frac{1}{\left\| \hat{\rho}_{i+1} \right\|} \hat{\rho}_{i+1} \end{cases} \tag{13}$$

3 Experimental part

We consider model illustrative examples for nonlinear systems with several degrees of freedom. The stability of the solutions is shown according to (7) and the instability according to (6).

Example 1: Consider an illustrative example of free oscillations of a flat bracket made of a physically nonlinear material (Fig. 1). The number of dynamic degrees of freedom equal 2. In a state of equilibrium when working in rest mode, the static force P, from which in a state of rest in a pair of voltages: σ1 = 1.463σT, σ1 = 1.837σT. The value of the "tangent" module for both rods 0.1E, the "secant" module is 0.26E and 0.2E for the first and second rods, respectively. Filling the "tangent" and "secant" stiffness matrices:

$$K_K = \begin{Vmatrix} 378 & 96 \\ 96 & 72 \end{Vmatrix},\ K_C = \begin{Vmatrix} 906 & 192 \\ 192 & 144 \end{Vmatrix},\ K_K^{-1} \bullet K_C = \begin{Vmatrix} -2.6 & 0 \\ -0.8 & 2 \end{Vmatrix}$$

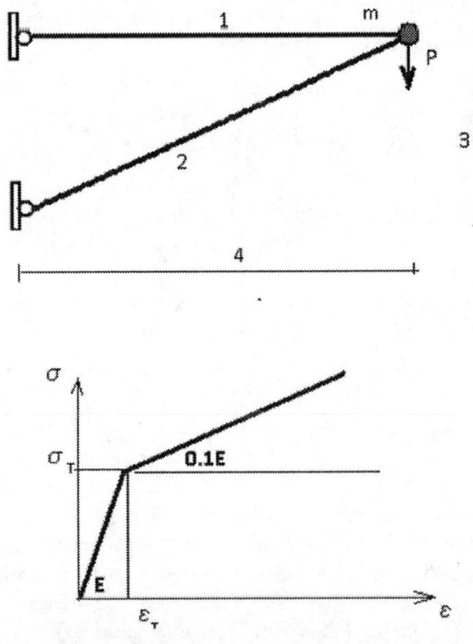

Fig. 1. Example of physically nonlinear frame

Table 1 shows the values of the stability parameter, assigned according to different methods.

Table 1. Values of the stability parameter.

$ejg([K_{K,n}]^{-1}[K_{C,n}])$	$\frac{\omega_C^2}{\omega_K^2}$	$\frac{q^T K_{C,n} q}{q^T K_{K,n} q}$
2.60	2.216	2.138

Table 2 shows the values of the spectral radius of the resolving operator for different values of the time step.

Table 2. Values of the spectral radius.

Δt	$ejg([K_{K,n}]^{-1}[K_{C,n}])$	$\frac{\omega_C^2}{\omega_K^2}$	$\frac{q^T K_{C,n} q}{q^T K_{K,n} q}$
0.001	0.9999	1.0001	1.004
0.01	0.9981	1.0057	1.0082
0.1	0.9728	1.0293	1.0345
0.5	0.98432	1.0837	1.0923
1.0	0.9926	1.0999	1.1214
1000	0.9999	1.1013	1.1524

As can be seen, only the stability parameter, assigned according to the proposed procedure, ensures the fulfillment of the Neumann criterion over the entire range of the step of integration.
Example 2: Oscillations of a system with nonlinear restoring force.
We consider the numerical solution of the known model equation:

$$\ddot{X} + X + X^3 = 0 \quad (14)$$

For (14), the relation between the amplitude and the oscillation period is known:

$$\frac{T_A}{T_0} = \frac{2}{\pi}\frac{1}{\sqrt{1+A^2}} F\left(k, \frac{\pi}{2}\right), \quad T_0 = 2\pi \, (A = 0) \quad (15)$$

In (15), A is the amplitude, T_A is the oscillation period, and *F()* is the normal elliptic Legendre integral of the first kind.
Approximation of the equation of motion (3) for this example has the form (16):

$$\ddot{x}(t) + K_{K,n}\Delta x(t) + K_{C,n} x_n = 0,$$
$$K_{K,n} = 1 + 3(x_n)^2, \; K_{C,n} = 1 + (x_n)^2 \quad (16)$$

A series of test calculations is performed for different amplitudes of free oscillations and a different step of integration over time. Table 3 shows the selection of the main results, in each cell two values are given - the absolute value and the relative error (in percent).

Table 3. Results on free oscillations of a nonlinear system.

Амплитуда X0	Δt=2π/n, n	T δT(%)	A δA(%)	Точное T
0.5	10	6,0729 -5,40	0,4928 1,43	5,7617
	25	5,8218 -1,04	0,4995 0,11	
	50	5,7822 -0,36	0,4999 0,01	
	100	5,7722 -0,18	0,5000 0,00	
	200	5,7697 -0,14	0,5000 0,00	
1.0	10	5,3920 -12,91	0,9480 5,20	4,7752
	25	4,8711 -2,01	0,9957 0,43	
	50	4,7944 4,7746	0,9994 0,06	
	100	4,7746 -0,01	0,9999 0,01	
	200	4,7749 -0,002	1,0000 0,00	

4 Conclusions

The circuit has both a relatively amplitude errors and phase error. The phase error exceeds the value of the error in the amplitudes. It is expressed in an increase in the period of free oscillations in the approximate solution. Amplitude accuracy reduces the calculated amplitude relatively accurate. Both errors decrease rapidly with a decrease in the integration step. With the improved stability parameter of the computational process, we obtain a stable numerical solution. Parameter assignment algorithms used above stability is not guaranteed stability of the iterative process.

References

1. S. H. Razavi, A. Abolmaali, M. Ghassemieh, *A weighted residual parabolic acceleration time integration method for problems in structural dynamics*, Computational Methods in Applied Mathematics, **7 (3),** 227–238 (2007)
2. A. Chopra, *Dynamics of Structures: Theory and Applications to Earthquake Engineering, 3rd ed.* (Prentice-Hall, Upper Saddle River, New Jersey, 2007)
3. T. Belytschko, W.K. Liu, B. Moran, *Nonlinear Finite Elements for Continua and Structures, 3rd ed.* (John Wiley & Sons, Chichester, UK, 2000)
4. K.J. Bathe, *Finite Element Procedures* (Prentice-Hall, Englewood Cliffs, New Jersey, 2003)

Innovative methods of monitoring for deformations of man-made loaded areas

Svetlana Sheina[1,*], and *Lidia* Girya[1]

[1]Don State Technical University, Gagarin square 1, Rostov-on-Don, 344000, Russia

Abstract. This article presents the technique of RS data use for monitoring of man-made loaded environment deformations in municipalities, explores the methods of geological hazards forecasting. Based on the results of the research conducted in the territory of Rostov-on-Don, areas of geological hazards expressed by ground surface subsidence have been detected.

1 Relevance

At the present time there are explicit hazardous man-made processes in big cities, such as subsiding deformations of loess soil, ground surface subsidence, ground water rise, soil slips, karst processes and others. With the development of existing and new territories the current monitoring of hazardous processes should be improved for their sustainability. [1-5]

The main issue in information gathering on the parameters of such processes is that they are unevenly distributed geographically and over time. The lack of data is a result of high economic, labour costs and technical difficulties of their obtaining. That makes it difficult to forecast the hazardous geological processes and phenomena, which impedes making efficient decisions on ground protection. Modern innovative technologiesof Earth remote sensing (ERS) could help in the matter. They enable non-contacting studying of immense territories to a high precision,assessing of process dynamics for an extended period of time, detecting of zones of uneven ground subsidence that can arise from existence of hazardous geological processes and phenomena.

2 State of knowledge of the matter and research task Interferometric survey

Using the method of phase shift in reflection response for remote radar sensing was first suggested by D. Richman in 1972. [2] But its practical use was hindered by inadequate technical capabilities. The first survey results under the proposed methodology were obtained only in 1986-1989 by R. M. Goldstein and H. A. Zebker [3], following which specified demands on accuracy of space vehicle equipment were formulated.

Persistent Scatterer Interferometry was developed in 1999 in Milan Polytechnic University. The analysis of multipass chains of interferometric radar surveys (30 and more surveys provide the maximum accuracy of estimation of vertical surface deformations (2-4 mm). It is implemented in consistent execution of several stages [3].

At the first stage, radar scan of surface is carried out by two radars simultaneously, and interferometric chain of radar images is obtained. At the second stage, the images are processed. The entire algorithm of interferometric processing is implemented in the software system SARscape (the developer is SARMAP, Switzerland), incorporated in the programme ENVI (Exelis VIS, USA) [7].

As a result of calculations, average annual shift velocity, graph of vertical shifts for the period of 30 surveys for any fixed point can be determined. The data obtained are used for mapping of ground surface shifts, high-precision measurements of building deformations.

At the present time the application of such methods is growing rapidly, and their opportunities are steadily increasing.

Today there are series of consecutive images for the territory of Russia, radar images which can be processed by interferometric techniques with obtaining maps of shifts and deformations of buildings and ground surface for the period of 20 years and more. Since 2002 the satellite ENVISAT with onboard radar ASAR was launched. Its surveys are performed using the regime that allows to combine the results with the products from satellites ERS-1 и 2, which provides the creation of digital elevation models with 20-metre spatial resolution. In 1995 Canada launched the satellite Radarsat-1, able to conduct shooting with spatial resolution from 100 to 7 m, and broad angle and frame area variation. Next satellite Ralarsat-2 (2007) had 1-100-metre spatial resolution and better aspect ratio. With 1-metre resolution the image area is 18x8 km, and with 7 m - 150x170 km, 3 m – 50x50 km.

Authors' suggested approach to the monitoring of deformations of industrial and civil constructions is based on the application of results on ground surface remote sensing, carried out with multipass radar

* Corresponding author: rgsu-gsh@mail.ru

 DOI: 10.1051/matecconf/201712905020

interferometric survey technique [3]. It consists in recording of radar signal reflected from the ground surface at very short wavelengths by space vehicles. Measurement results processing on several dozens of images allows to determine the position of satellite in space with high precision, to measure the distance to reflective surfaces, and to control the shifts of monitored objects in millimetric range. Multiple surveys of one and the same area from recurrent satellite orbits allow to record the shifts over the years.

3 The using of monitoring system for the deformations of man-made loaded territories

Since 2003 the scientists of RSSU at the Department of Urban building and economy were studying the monitoring system of geotechnical conditions. As source material the maps of K. A. Merkulova, the leading expert of «RostovDonTisiz» (on geotechnical conditions, ground water level, rate of ground water level rise), were taken and digitized in ArcGIS. Software product for the analysis of geotechnical surveys results was developed [2].

Classification of geological risk by rate of ground water level rise based on the analysis of hazardous geotechnical processes was developed, presented in Table 1. The map of geological risk, based on it, was charted in Rostov-on-Don using the software package ArcView, presented in Figure 1.

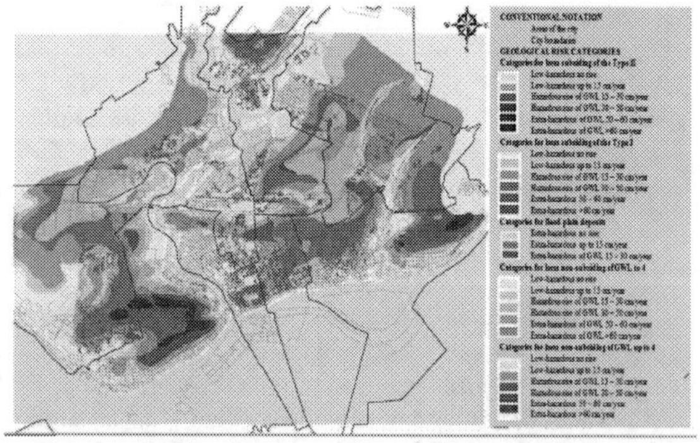

Fig. 1. Electronic map of Rostov-on-Don with geological risk zones.

In order to enhance the monitoring system the remotely sensed data from radar satellites COSMO-SkyMed-1-4 (e-GEOS, Italy) for the period of May 2011 to November 2013(40 images) (Figure 2) was procured by the scientists at the Department of Urban building and economy of DSTU. Spatial resolution – 3 m, image area – 40x40 km, length of wave – 3.1 cm. Image processing is performed with EXELIS VIS:ENVI и SARscape Interferometric Stacking software products. Figure 3 shows the data of interferometric survey, presented in the form of point objects, holding information about vertical shifts of ground surface.

As a result of ERS data analysis conducted in Rostov-on-Don, Figure 3 shows the areas with hazardous shifts of ground surface, which in their turn, cause the deformation of buildings and constructions, bridges, roadbed, railway lines and others.

Table 1. Classification of geological risk by rate of ground water level rise.

Group of geological engineering conditions	Geological engineering conditions	Ground water depth level	Rate of GWL rise, cm/year	Risk categories
1	Loess subsiding of the Type II	>8(10)	No rise	Low-hazardous
			up to 15	
			15 – 30	Hazardous
			30 – 50	
			50 – 60	Extra-hazardous
			>60	
2	Loess subsiding of the Type I	3-8 m	No rise	Low-hazardous
			up to 15	
			15 – 30	Hazardous
			30 – 50	
			50 – 60	Extra-hazardous
			>60	
3	Flood-plain deposits	0.5-3 m	No rise	Extra-hazardous
			up to 15	
			15 – 30	
4	Loess non-subsiding	2.5-4 m	No rise	Low-hazardous
			up to 15	
			15 – 30	Hazardous
			30 – 50	
			50 – 60	Extra-hazardous
			>60	
5	Loess non-subsiding	4.1-10 m	No rise	Low-hazardous

Fig. 2. Plot coverage Rostov-on-Don with satellite data COSMO-SkyMed-1-4 (e-Geos. Italy) for the period of May 2011 to November 2013.

DOI: 10.1051/matecconf/201712905020

Fig. 3. Focuses of vertical shifts, Rostov-on-Don.

The comparison of electronic maps with the focuses of vertical shifts based on ERS data, and electronic map of Rostov-on-Don with geological risk data allowed to define the most hazardous zones of ground surface subsiding in the areas of hazardous geological processes (Figure 4).

Fig. 4. Plot coverage Rostov-on-Don with satellite data COSMO-SkyMed-1-4 (e-Geos. Italy) for the period of May 2011 to November 2013.

In particular, the electronic map with zones of geological risk indicates the possibility of hazardous geological processes of the second group on subsiding of the second type, as can be seen from remotely sensed data.

4 Abstracts

An advantage of using the interferometric survey over the geodetic observations is that the large area rather than single object is covered, and the acquisition of data on vertical deformations of surface is less labour-intensive.

Comparison of remotely sensed data with data from monitoring of hazardous geological processes using geotechnical results gives a complete and reliable picture on technological hazards that allows, in its turn, to make adequate managerial decisions aimed at reduction of natural or man-made hazards in the development of built-up areas.

References

1. A.V. Gridnevskiy, Inzhenernyy Vestnik Dona, **3**, (2013)
2. A.V. Gridnevskiy, R.B. Matveiko, Scientific review, **11-3**, 678-681 (2014)
3. A.V. Gridnevskiy, AA. Hamavova, Scientific review, **11**, 83-386 (2014)
4. S.G. Sheina, L.V. Girya, Inzhenernyy Vestnik Dona, **3**, (2012)
5. L.V Girya, V.V Belash., S. V. Horenkov, K. S. Petrov, Inzhenernyy Vestnik Dona, **4**, (2013)
6. Russian Standard VSN 53-86
7. Software by EXELIS VIS:ENVI and SARscape Interferometric Stacking.
8. P.O. Skobelev, E.V. Simonova, A.A. Zhilyaev, V.S. Travin. Procedia Computer Science, **103**, 396-402 (2017)
9. S.M. Gandhi, B.C. Sarkar. Chapter 4, Remote Sensing Techniques, Essentials of Mineral Exploration and Evaluation, 81-95 (2016)

Nanomodified constructional fiber-reinforced concrete

Vladimir Perfilov[1,*], and *Viktoriya* Gabova[1]

[1]Volgograd State Technical University, Oil and Gas Constructions Department, Avenue V.I. Lenin, 28, 400005, Volgograd, Russia

Abstract. Nanotechnology of receipt of the high-strength fiber-reinforced concrete which don't have analogs, using nanocarbon additives, the modified gel on a structure nanolevel, applied knitting is developed for improvement of physicomechanical properties on micro and nanolevel that allowed to create concrete with exceptional properties.

1 Introduction

The researches conducted for the last years convincingly show that fiber reinforcing improves mechanical characteristics of concrete, contributes resistance of concrete to impact of aggressive environment, allows to reduce thickness of designs and significantly to reduce or to completely exclude an armature expense. Thus, favorable conditions for decrease of material capacity and labor input of construction are created. However so far the use of fiber-reinforced concrete is practically absent because of high cost of steel fiber and lack of special technology of its stacking. One of the main reasons of insufficient application of fiber-reinforced concrete is lack of understanding by industrial organizations of great advantages of their application, and also lack of science-based regulating documentation. For the solution of this and other scientific problems connected with rational selecting of components of fiber-reinforced concrete the offered project is directed.

Fiber-reinforced concrete, being heterogeneous material, is characterized by microstructure of the cement gel, and also macrostructure caused by interaction knitting and fillers. For increase of durability and crack resistance of concrete at the macrolevel use of basalt fiber fibers blocking development of cracks is perspective [1-12].

Proving prospects of application as the disperse-reinforced component of basalt fiber fiber-reinforced concrete is necessary to pay special attention to studying of physicomechanical characteristics of basalt fiber-reinforced concrete. Having studied the principles of structurization of this type of concrete is possible to improve, influencing at different levels, structure of cement and sand matrix.

2 Main part

The disperse-reinforced concrete and solution represents the cement and basalt composite having the increased durability of compressed cross-sections bent and also the stretched elements of designs.

Basalt fiber contributes at stage of mixing of components to formation rheological of the homogeneous mixture possessing high rates of plasticity, and also which is characterized by its homogeneity. Considering process of structure formation at the microscopic level between the cement stone and basalt fiber in the contact zone "basalt fiber - cement matrix", there are new growths which arise owing to interaction of hydroxide of calcium of portland cement and oxide of silicon of basalt fiber. It is known that basalt fiber is characterized by the high degree of adhesion to the cement matrix. For this reason the field of engagement of fiber and matrix rather close, providing full contact of surfaces, excepting cracks and gaps. At the time of hardening, received basalt fiber-reinforced concrete, has three-dimensional the reinforced microstructure of cement composite with more evenly redistributed efforts arising in concrete blocking formation and development of cracks.

Thus, the basalt fiber reduces shrinkage and cracking at the microlevel, increasing ability of concrete matrix to deformation, excepting destruction during the shrinkage, contributing conservation of internal energy of concrete. During hardening of concrete and its shrinkage, the basalt fiber as if connects edges of cracks, reducing thereby risk of fracture.

Besides disperse reinforcement the modifying components also exert different impact on microscopic level. Practice shows that change of properties for purpose of improvement of quality of concrete composites is impossible without directional regulation of his formation. It is reached by introduction to mix of chemical additives.

Formation of microstructure of cement stone on which depend the main physicomechanical properties of concrete begins with the supramolecular level of dispersion of particles of size $3–5\cdot10^{-9}$ m (3–5 nanometers). The calcium hydrosilicates which are formed as a result of interaction of the cement knitting with water during an initial stage of hydration at the supramolecular level have size of elementary germs of new crystalline phase to 3 nanometers. The defects of

* Corresponding author: vladimirperfilov@mail.ru

structure in the form of dislocations having the similar sizes also begin to be formed at this stage of formation of cement stone.

Now researchers with mistrust treat of use of nanocarbon additives in solutions and concrete. It, apparently, is connected with the fact that it is difficult to present how miserable dosages of nanoadditives can be subsequently the centers of crystallization of new growths of cement stone.

The following can serve as an explanation of positive influence of nanoadditives on physicomechanical properties of solutions and concrete.

It is known that when using plasticizers in amounts of nanoadditives, commensurable with dosage, there is their full dissolution in mixing water which contributes to adsorption of molecules of additive on surface of components of mixture with formation of the thinnest layer increasing wettability of solution. As a result, the water-cement relation by increasing the mobility of mixture decreases and conditions for homogeneous distribution of all components of solutions and concrete improve. It, in turn, contributes formation of dense homogeneous structure of the hardened concrete with improved physicomechanical characteristics. However the plasticizing water-soluble additives directly don't participate in interaction with hydrosilicates of new growths of cement stone. Their role consists only in improvement (acceleration) of conditions of formation of microcrystalline structure of cement stone.

It is possible to strengthen gel structure of cement stone on nanolevel by use of the multilayered nano-hallow carbon tubes allowing to significantly change physicomechanical parameters. It is connected with the fact that nanotubes are directly involved in processes of crystallization of cement stone due to adhesion and directed microstructural formation of hydrosilicates. Nanotubes on which surface, including internal cavities, there is adhesion and subsequent interaction of products of cement hydration have hundreds of times greater durability in comparison with cement hydrosilicates. As a result the strongest particles of carbon nanoadditives become bases (centers) of crystallization of new growths of cement stone that leads to microhardening of structure and significant improvement of physicomechanical properties of the hardened composite materials. Nanotubes are material with a huge number of loosely coupled valent electrons, besides, that the atomic framework has a high stability. For this reason nanoadditives, being on limit of the section of phases in the polarized condition, it adgezirutsya adgesiruta and directed forms crystalline hydrates. Formation of the fibrillar microstructure allowing to increase physicomechanical properties of a bazalt fiber-reinforced concrete on micro and a nanolevel is result of nanomodification [13-28].

For a greater hardening of structure appropriate use of various high-strength fiber fibers on which surface products of hydration of cement stone together with nanoadditives will also adhesivity is reasonable.

Nanotechnology of receipt of the high-strength fiber-reinforced concrete which don't have analogs, using nanocarbon additives, the modified gel on a structure nanolevel, applied knitting is developed for improvement of physicomechanical properties on micro and nanolevel that allowed to create concrete with exceptional properties [14,15].

Nanotubes represent hollow cylindrical tubes from one or several atoms of carbon. The multilayered nanotubes of "Taunit" received in LLC "Nanotechcenter" of the Tambov state technical university are of the greatest interest.

For more complete dispergating of water-insoluble nanocoal family of additives application of the known low-resourceful mixing installations is insufficient. Now the most effective are ultrasonic devices with a frequency from 20 to 26 kHz, for example, ultrasonic dispergators of the USD brand or ultrasonic bathtubs. Application of the above-stated installations allows the most complete distribution of particles of nanocarbon additives on all amount of water of mixing water. The advantage of such technology is also long stability of the received suspension, up to two weeks.

Thus, mixing water activated by nanocarbon additives by means of ultrasonic hashing possesses a high reactionary capability for the most complete interaction with minerals of cement and receipt of dense and strong microstructure of solutions and concretes.

In researches nanostructuring of cement stone was made by means of introduction in solution of carbon nanotubes "Taunit" representing cylindrical tubes from one or several atoms of carbon. Outer diameter of nanotube is 15 – 40 nanometers; internal diameter is 3 – 8 nanometers, and length – to several microns. The main advantages of nanotubes is their small dosage in concrete constituting thousand and ten-thousand shares of percent from weight knitting at durability to several GPa that in hundreds times more of durability of hydrosilicates.

Thus, result of purposeful influence on regulation of processes of structurization on macro, micro and nanolevels is receipt of the constructional fiber-reinforced concrete possessing improved strength characteristics, crack resistance, frost resistance and other considerable higher operational properties.

Task of scientific research is the increase of efficiency of process of receipt of the modified fiber-reinforced concrete mixture directed to increase on its durability at compression and stretching in case of bend due to use of the components hardening the structure of fiber-reinforced concrete on micro and nanolevels.

The purpose of this stage of researches was studying of influence of "Taunit" nanoadditive on properties of fiber-reinforced concrete mixture and improved fiber-reinforced concrete. As the modifying additives used carbon "Taunit" nanomodified received in LLC "Nanotechcenter" of the Tambov state technical university.

For the purpose of studying of influence of additives preliminary experiments were made on the basis of cement and sand mixture of structure cement:sand =1:2 prepared in a spherical cup within 5 minutes for avoidance of " false setting" of portland cement.

As initial materials for production of samples of 40x40x160 mm the portland cement of production of CJSC "Oskolcement" of the PTs M500 D0-N brand, sand of CJSC "Orlovsky sand pit" of the Volgograd region, basalt microprofile of production of the LLC NPO "Vulkan company" in number of 1,4 kg/m^3 was used. Nanoadditive was entered in percentage terms from the mass of cement together with mixing water.

Complexity of production of solution with "Taunit" additive is connected with the fact that carbon nanotubes are insoluble in water. Therefore preparation of solution of "Taunit" nanoadditive was performed by means of ultrasonic dispergator of UZD1-0.1/22 with a frequency of 22-26 kHz within 30 seconds. Formation of samples was made on the laboratory vibroplatform.

Determination of durability of control samples of solution was carried out at the age of 28 days according to GOST 10180 with use of the hydraulic testing machine of the "TECNOTEST company" of the KD 150-R15 brand with availability of the second channel with the device C 362/FN. Therefore received indicators on durability on compression and on stretching in case of bend.

Results of testing, the hardened samples from equally mobile of solution mixes are provided in the table.

The analysis of results of testing showed that introduction in mix of "Taunit" nanoadditive promotes increase in durability at compression at 56,44% and at stretching in case of bend for 74,1% in comparison with samples without additive in case of identical mobility of solution mix (see structure No. 5).

Table 1. Influence of "Taunit" additive on properties of fine-grained fiber-reinforced concrete

№ structure	Amount of "Taunit" additive of the mass of cement	Spread of cone, mm	Durability, MPa	
			bend	compression
1	0	107	2,7	42,86
2	0,001	110	2,93	58,37
3	0,005	111	4,2	60,17
4	0,01	108	4,3	62,11
5	0,05	108	4,7	67,05
6	0,1	110	3,3	66,35

3 Conclusions

By results of scientific research the computer program for matching of structure of fiber-reinforced concrete with nanoadditives is developed [29].

The new documentation including prescription structures of fiber-reinforced concrete using original designs of fibers, the nanodimensional complex additives allowing to modify gel (on nanolevel) a microstructure of the applied knitting is developed. Recommendations about manufacturing techniques of products on the basis of developed structures of fiber-reinforced concrete are made. For practical application of new structures of fiber-reinforced concrete at the entities of the construction industry, and also in monolithic construction Technical conditions were developed [30]. They extend to the constructional heavy structural fiber-reinforced concrete with reinforcer from fiber "Mixarm" having the following subjects to application: monolithic construction, coverings of industrial floors, roads and airfields, tunnel constructions, bank vaults, pipelines, railway cross ties, piles, sidewalk coverings, wall panels, panels of overlapping and covering, tile and many other construction products and designs.

References

1. V.A. Perfilov, *Fine-grained fiber-reinforced concrete,* Monograph. Ministry of science and education of the Russian Federation. Volgograd state university of architecture and civil engineering, Volgograd: VSUACE, 126 p. (2015)
2. V.A. Perfilov, D.V. Oreshkin, M.O. Zubova, D.L. Neizvestnyj, I.G. Lukina, A.N. Chebotayev, *Ballasting of underwater pipelines using a basalt fiber and boring slime,* Construction of oil and gas wells by land and by sea, **No. 11.** Pp. 40-41 (2012)
3. D.V. Oreshkin, A.N. Chebotaev, V.A. Perfilov, *Disposal of drilling sludge in the production of bulding materials,* Procedia Engineering, **T.11.** Pp. 607-611 (2015)
4. V.A. Perfilov, D.V. Oreshkin, D.Y. Zemlyanushnov, *Concrete strength and crack resistance control*, Procedia Engineering 2. Сер. "2nd International Conference on Industrial Engineering, ICIE 2016", Pp. 1474-1478 (2016)
5. V.A. Perfilov, D.V. Oreshkin, V.S. Semenov, *Environmentaliy safe mortar and grouting solutions with hollow glass microspheres,* Procedia Engineering 2. "2nd International Conference on Industrial Engineering, ICIE 2016", Pp. 1479-1484 (2016)
6. V.A. Perfilov, A.V. Kotlyarevskaya, U.V. Kanavets, *Influence of microreinforcing fibres and plasticizing additives on properties of fine-grained concrete,* Bulletin of Volgograd State University of Architecture and Civil Engineering. Series: Civil Engineering and Architecture, **Iss. 44 (63), P. 2**. Pp. 111-118 (2016)
7. V.A. Perfilov, M.O. Zubova, *The influence of superplasticizers on the property is fibrous concrete,* Internet-Vestnik VolgGASU, **No. 1(37),** paper 11 (2015).
8. V.A. Perfilov, M.O. Zubova, *Fine basalt fibre concrete,* Bulletin of Volgograd State University of Architecture and Civil Engineering. Series: Civil Engineering and Architecture, **Iss. 38(57),** pp. 85-93 (2014)
9. V.A. Perfilov, M.O. Zubova, *Basalt fiber fiber as main component of dispersed fibrous reinforcement concrete,* Bulletin of the Donbass national academy of construction and architecture, **No. 3 (101),** Pp. 146-148 (2013)

10. V.A. Perfilov, M.O. Zubova, D.L. Neizvestnyj, *Fiber-reinforced concrete with basalt fillers and superplasticizers,* Reliability and durability of construction materials, designs and foundations of the bases. Materials of the 6th International science-technical conference. Volgograd, VSUACE, Pp. 248-250 (2011)
11. V.A. Perfilov, M.O. Zubova, D.L. Neizvestnyj, *Application of basalt fibers and the modifying additive for increase in strength characteristics of heavy concrete,* News of higher educational institutions. Construction, **No.12,** Pp. 46-49 (2011)
12. V.A. Perfilov, A.V. Atkina, O.A. Kusmartseva, *Application of the components modifying the microreinforcing for increase in durability of cellular materials,* News of higher educational institutions. Construction, **No. 9,** Pp. 11-14 (2010)
13. V.A. Perfilov, A.V. Atkina, O.A. Kusmartseva, U.V. Kanavets. *Fiber-reinforced concrete with macroreinforces and nanoadditives,* Social and economic and technological problems of development of a construction complex of the region. Science. Practice. Education. Materials III of the All-Russian scientific and technical conference, Pp. 236-238 (2009)
14. V.A. Perfilov, U,V, Alatortseva, A.A. Tyurin, *Fiber-reinforced concrete of the accelerated curing.* News of higher education institutions. Construction. Novosibirsk, **No.**1, Pp. 48-51 (2009)
15. V.A. Perfilov, U.V. Alatortseva, M.I. Dmitruk, I.L. Zhoga, *Application of the modyfying nanoadditives for increase in durability of fiber-reinforced concrete,* News of higher education institutions, Construction. Novosibirsk, **No. 8,** Pp. 17-19 (2009)
16. B.V. Gusev, *Application of nanotechnologies and nanomaterials in construction: opinions of experts,* Construction: new technologies – the new equipment, **No. 1,** Pp. 17-28 (2012)
17. V.R. Falikman, *Nanomaterialy and nanotechnologies in modern concrete.* Industrial and civil engineering, **No. 1,** Pp. 31-34 (2013)
18. M.E. Yudovich, A.N. Ponomarev, *Nanomodification of plasticizers, regulation of their properties and strength characteristics of cast concrete..* Stroyprofil, **No. 6,** Pp. 49-51 (2007)
19. U.V. Alatortseva, *Fiber-reinforced concrete based on nanostructural knitting..* Actual problems in construction and architecture. Education. Science. Proceedings of the 66th All-Russian scientific-technical conference on the results of the research University in 2008. Samara: SSUABCE, Pp. 209-210 (2009)
20. M. Vauchsky, *Nanomodification of concrete - the absolute reality.* Construction: new technologies, new equipment, **No. 2,** Pp. 47-52 (2009)
21. A.I. Vakhmistrov, V.I. Morozov, Yu.V. Pukharenko, A.N. Dmitriev, U.H. Magdeev, *Effective fiber-reinforced materials and products for construction,* Industrial and civil engineering, **No. 10,** Pp. 43-44 (2007)
22. B.V. Gusev, *Development of nanoscience and nanotechnologies.* Industrial and civil engineering, **No. 4,** Pp. 45-46 (2007)
23. A.S. Korolev. *Fine-grained concrete with nanoadditives of synthetic zeolite,* Concrete and reinforced concrete, **No. 6,** Pp. 13-15 (2008)
24. V.A. Perfilov, U.V. Alatortseva, *Constructional fiber-reinforced concrete with the use of modifying nanoadditives,* Materials of the Second International forum on nanotechnologies on October 6-8, 2009 - Moscow: Rusnanotech Group, Pp. 374-375 (2009)
25. V.A. Perfilov, U.V. Alatortseva, A.V. Atkina, O.A. Kusmantseva, *The nanomodified structure of the knitting fiber-reinforced concrete,* Nanotechnologies and nanomaterials: current state and prospects of development in conditions of Volgograd region: materials of the all-Russian scientific-technical conference, Volgograd: Volgograd state university, Pp. 373-380 (2009)
26. V.A. Perfilov, U.V. Alatortseva, A.V. Atkina, *Fiber-reinforeced concretes with the use of modifying nanoadditives,* Materials of the International week of construction materials devoted to the 65 anniversary of formation of construction and technological faculty of MGSU. - Moscow: MGSU, Pp. 105-110 (2009)
27. Yu.V. Pukharenko, *Modern state and prospects of application fullerenic of nanostructures in cement compositions,* Siberian industrialist **No.3,** Pp.30-31 (2008)
28. A.A. Smolikov, *Concrete reinforced by nanofibers.* Concrete and reinforced concrete, **No. 4,** Pp. 8-9 (2009)
29. V.A. Perfilov, S.P. Mityaev, *Calculation of fiber-reinforced concrete with nanoadditive.* from 29.04.2009. The Certificate on the state registration of the computer program. No. 2009612195
30. V.A. Perfilov, U.V. Alatortseva, *Fiber-reinforced concrete mixture with hardener from fiber "Mixarm" and nanoadditives",* Technical conditions of TC of 574510-002-02068077-2008.

Protection system of hydraulic drive of road construction machinery

Nikolay Fomenko[1], *Oleg* Burlachenko[1], and *Maxim* Ivanov[1,*]

[1]Volgograd State Technical University, Department of Construction Production Technology, Avenue V.I. Lenin, 28, 400005, Volgograd, Russia

Abstract. Tendencies of development of ways of protection of a hydraulic drive of vehicles from unauthorized emission of working liquid at destruction of high pressure holes are analyzed. The analysis of work of traction vehicles, detects the reasons at which there is a destruction of high pressure hoses of hydraulic system and the emission in the atmosphere of working liquid that violate environmental safety. Technical solution is proposed for elimination of negative consequences of destruction of high pressure hoses, which author's rights are protected by the patent. Its essence is, that the locking device is further supplied with a rod with the axial through channel, which will allow a plunger in case of a gap of head pressure hoses reliably to block radial holes in a rod, to provide that high reliability of the overlap of supply of working liquid in head pressure hoses in case of their destruction and effective protection of hydraulic drive against unauthorized emission of working liquid.

1 Introduction

Researches show that at damage of high pressure hoses of hydraulic machines [1-11] is released into the atmosphere the significant amount of working liquid, equal to nearly a volume of a hydraulic tank. It leads to deterioration in an ecological situation. Despite the taken measures for restriction of pressure in high pressure hoses and decreasing upon them cyclic loadings, the problem of search of technical solutions, caused by the improvement of the design elements of hydraulic system, for example, the flexible of high pressure hoses [12], or new methods and the principles of protection of hydraulic drive [13-23] remains urgent.

Operational properties of a prototype [13], as the closest to an effective objective, have a low efficiency of operation of the locking device of protection system of hydraulic drive. It is caused by the fact, that in the time of the message of an input cavity through radial holes of a plunger and the channel drain union with the line of discharge, pressure in the input cavity decreases and because of oscillatory process of a plunger and prevents the overlap of the channel drain union. The arising fluctuations of the locking regulating element – of plunger are followed by intensive blows of the valve about a saddle and fluctuations of pressure in system that is a cause of wear and losses of tightness of the valve. At the same time, time of overlapping of the channel of the output fitting, reduces speed of the locking device.

To the disadvantage of this protection system can be attributed not reliable overlapping of a stream of working liquid by conical valve, contacting to a saddle, which requires for close contact of pairing considerable spring force, that causes stiff operation of the valve locking devices, wear of the valve, decrease in the operating reliability of the device and environmental safety of the use of hydraulic drive machines. The disadvantage of this protection system is also what closed not compressible the volume of working liquid, formed between the plunger at the end of his stroke and an emphasis, prevents reliable the overlap of the channel of the output fitting and provokes leak of working liquid in the damaged hydraulic line at operation of the locking device, leads to growth of losses of working liquid.

Besides, the disadvantage of this protection system of hydraulic drive is that, is not regulated the threshold of operation of the locking device, determined by differential pressure between input and output cavities. It causes to low universality locking device for installation in protection system of hydraulic drive of various machines.

2 Main part

System is intended for protecting hydraulic drives of machines from unauthorized discharge of working fluid from hydraulic system. System includes hydraulic tank, pump connected by pressure hydraulic line with hydraulic drive through distributor, and locking device including housing with inlet and outlet cavities, inlet, outlet and drain unions with channels and stop. Locking device is equipped with a rod with axial through channel, one end rigidly fixed in the stop. In the axial through channel of the rod there are radial holes, thread and adjustment screw for adjusting actuation threshold of the locking device. Other end of rod is located in the axial

* Corresponding author: clevermax18@mail.ru

channel of plunger with possibility of piston displacement along the rod. System is equipped with check valve installed on the pressure hydraulic line between the pump and a locking device, two shutoff valves and hydraulic lock, the control cavity of which is connected with drain union with the housing channel and through one shutoff valve is connected to the drain hydraulic line, input cavity of hydraulic lock is connected to pressure hydraulic line between the pump and check valve and an outlet cavity of hydraulic lock is connected to drain hydraulic line. Input cavity of the locking device housing is connected to drain hydraulic line through the inlet connection pipe with channel and other shutoff valve. Effect: higher operational efficiency, response time and universality of the locking device.

The essence of the proposed technical solution consists is that the adjusting screw, installed in the axial channel of a rod from an emphasis, allows to change the section of the radial holes, through passage reporting input and output cavities through the axial channel of a rod and, as a result, by throttling of a stream of working liquid to regulate differential pressure, between output and input cavities, the defining threshold of operation of the locking device. It allows to carry out control of a threshold of operation of the locking device excluding false operation, connected with pressure decline in an output cavity of the locking device, or when draining of working liquid in the neutral position of the control valve spool, or idle operating mode of hydraulic drive. Settings of a threshold of operation of the locking device provide high universality of the locking device at installation in systems of the protection of hydraulic drive of various machines.

The proposed protection system is equipped with the hydraulic lock and two shutoff valves, which allow to provide constancy pressure in the input cavity at destruction head hoses of power hydraulic drive, thus draining the working liquid from the input cavity does not occur, since the channel drain union locking device is connected with the deaf administering of a cavity of the hydraulic lock and on the drain hydraulic line closed shutoff valve. Discharge in a hydrotank of the working liquid pumped is carried through the opening hydraulic lock the operating signal in the form of liquid pressure in an input cavity of the locking device appearing when moving a plunger to extreme right situation at destruction of head pressure hoses. At the same time the check valve and the shutoff valve lock working liquid on an inlet of the locking device and exclude an overflowing it from an input cavity to pressure and drain hydraulic lines at the begun discharge of the working liquid pumped from the pressure hydraulic line through input and output cavities of the hydraulic lock in a hydrotank.

Thus, the input cavity will be blocked on an inlet by the check valve, at the exit the plunger, blocking radial holes on a rod, and the channel drain union is reported from the deaf administering cavity. Therefore, in an input cavity of the locking device will be supported by the plunger with a spring the constant pressure of the liquid, which has appeared in the closed volume of the input cavity. Constant pressure in the input closed cavity of the case excludes fluctuations of a plunger, allows to hold constantly blocked radial holes of a rod and opened the hydraulic lock, which provides the draining of liquid in a hydrotank at a rupture of head pressure hoses. It significantly increases speed of the locking device and effectiveness of the protection system of hydraulic drive.

The protection system of hydraulic drive works as follows (fig. 1). In the operating state systems of protection shutoff valves 17 and 26 are closed. In a static state protection system of hydraulic drive, that is in the neutral position spools of the distributor 7, the working liquid is supplied by pump 2 through the check valve 5, the input channel of the fitting 10 in an input cavity A of the housing 9 and further comes on channel C on the inside of the rod 19 and his radial holes 22 to an output cavity of B, and through the output channel of the fitting 11 to the entrance distributor 7. In this position of the spools of the distributor 7 is drained of working liquid to the drain hydraulic line 8 through his overflow valves. Pressure of working liquid in cavities of A and B at the same time almost identical. The plunger 15 is supported a spring 14 and is located in the extreme left position, at the same time radial holes 22 of rods 19 are open and report among themselves through channel C of a rod 19 cavities to A and B locking devices 6.

When moving the spools of the distributor 7 the position of the rising, the working liquid is supplied by pump 2 through the check valve 5, the input channel of the fitting 10 in an input cavity A of the housing 9 and further comes on channel C of the rod 19 and his radial holes 22 in the output cavity B, and through the output channel of fitting 11 is fed to the entrance distributor 7, which feeds the working pressure of hydraulic drive 3. The liquid stream at the same time creates differential pressure between cavities of A and B locking devices, the backwater from which on the plunger 15 is compensated by a spring 14. The edges of radial holes 22 turned to an end face of a burt 20, spaced from the right end face of a plunger 15 at sufficient distance of e, at which in the regular working condition of a hydraulic system there is no their closing with a cylindrical surface of axial channel 16 of a plunger, and there is also no opening of the channel drain union 12 of the left end face of a plunger, as the edge of the channel drain union is located at the same distance of e from an end face of the input fitting 10.

At a rupture of head pressure hoses of supply of hydraulic drive pressure in an output cavity of B significantly decreases and, because of the increased stream of working liquid from an output cavity of B, there is considerable differential pressure in cavities of A and B.

By effort from differential pressure in cavities of A and B case 9 the plunger 15, overcoming resistance of a spring 14, moves to the right against the stop 13, closing the cylindrical surface of axial channel 16 radial holes 22 with the overlap equal to the distance S from edges of radial holes to the end face of the burt 20 of the rod 19. At the same time supply of working liquid from an input cavity of A in output cavity of B stops. Considerable overlap of radial holes 22 of surface of axial channel 16 of plunger 15 excludes infiltration of working liquid

DOI: 10.1051/matecconf/201712906001

from an input cavity of A in output cavity B case 9 through a gap between a rod 19 and axial channel 16 of a plunger because of big hydraulic resistance in a gap.

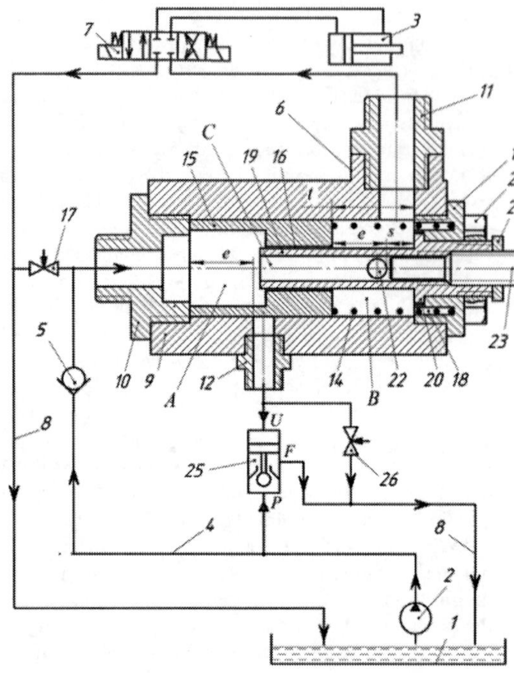

Fig.1. Protection system of hydraulic drive road construction machinery

In an emergency condition of a hydraulic system at a rupture of hoses of power hydraulic drive the plunger 15 under the influence of differential pressure moves in extreme right position on the distance of his full speed t and its left end face opens of the channel drain union 12 by which at the closed valve 26 pressure of working liquid as the hydraulic operating signal, is transmitted to the deaf control cavity U hydraulic lock 25. At the same time the hydraulic lock opens and through his input P and output F is ignored working liquid from the discharge hydraulic line 4 of the pump 2 to drain into the drain hydraulic line 8 and pressure in the pressure hydraulic line from the pump to the check valve 6 considerably decreases. The pressure drop in the pressure hydraulic line can cause the return outflow of working liquid from cavity A to the pump 2 under action on a plunger 15 the compressed springs 14 when the plunger at destruction of head pressure hoses settles down in extreme right position and rests against an stop 13. However, the check valve 5 excludes the return outflow of liquid to the pump from a cavity A case 9 of the locking device 6.

In an emergency condition of a hydraulic system at a rupture of hoses of power the cavity A locking devices is closed, as the valve 17 her communications with the drain hydraulic line 8 is closed, the holes 22, reporting cavity A with an output cavity of B are blocked by a surface of axial channel 16 of a plunger 15, the check valve 5 doesn't pass working liquid from cavity A to the pump 2, and the drain union 12 with the channel is reported from the deaf administering cavity U hydraulic lock 25. In such condition of the locking device 6 pressure in a cavity of A will be constant and the plunger 15 at an emergency rupture of head pressure hoses will remain not movably in the position of full overlapping of radial holes 22 the surface of axial channel 16 of a plunger.

Constant pressure in the input closed cavity A case 10 of the locking device 6 excludes fluctuations of a plunger 15, allows to hold constantly blocked radial holes 22 and opened the hydraulic lock 25 which provides discharge of liquid in a hydrotank 1 at a rupture of head pressure hoses.

For restoration of an initial condition of the protection system after elimination of a emergency condition of the protection system of hydraulic drive, at the switched-off pump 2 open shutoff valves 17 and 26. At the same time pressure in an input cavity A housing 9 of the locking device 6 and the control cavity U hydraulic lock 25 decreases, under the influence of a spring the 14 plunger 15 moves to the left to an initial state. The working liquid from a cavity of A is forced out by a plunger 15 through the valve 17, and from the control cavity U hydraulic lock 25 through the valve 26 to the drain hydraulic line 8 and further in a hydrotank 1.

The threshold of operation of the locking device determined by differential pressure between his output B and input A cavities is regulated by the throttling of a stream of working liquid by means of the adjusting screw 23 changing the flow section of the holes 22. The screw 23 performs exact tuning of a threshold of operation of the locking device 6, his sensitivity for differential pressure between output A and input B cavities that allow to achieve its maximum speed, exclude its false operations and provide high universality of application him in protection system of hydraulic drive of various machines.

3 Conclusions

Thus, the proposed technical solution allows to increase the efficiency of protection system of hydraulic drive against unauthorized emission of working liquid from hydraulic system, to provide speed of the locking device, operational reliability and an environmental safety of use of hydraulic drive of working bodies of machines.

References

1. S.A. Astapchik, V.S. Golubev, A.G. Maklakov, *Laser technology in the mechanical engineering and metals industries.* (Minsk: Belorus. science, 2008)
2. V.N. Fomenko, *Development of systems of protection of hydraulic actuators of mechanisms of a hinge plate of traction and special transport cars.* (Diss. Candidate of Technical Sciences. Volgograd, 2000)

3. N.G. Dyurgerov, *Technological methods of recovery and improve the wearing capacity of machine parts: Tutorial, Part.* (Donetsk State Technical University, Rostov-on-Don, 2006)
4. A. Harnoy, H. Rachoor, *Modeling of dynamic friction in lubricated line contacts for precise motion control.* Tribology Trans. 601 - 612 (1995)
5. A. Harnoy, B. Friedlan, *Dynamic friction model of lubricated surfaces for precise motion control.* Tribology conf. New Orleans. 24 - 27 (1993)
6. N.A. Kovalev, *Applied mechanics.* (M.: Higher school, 2000)
7. I.M. Fedorchenko, I.N. Frantsevich, I.D. Radomyselsky, *Powder metallurgy. Materials, technology, properties, fields of application: reference book. (*Kiev: Naukova thought, 1985)
8. D.N. Garkunov, *Tribotechnology: design, manufacture and operation of the machine.* (Publishing House of the ICCA, Moscow, 2002)
9. F. Findik, *Investigation of explosive welding parameters and their effects on microhardness and shear strength.* 659–664 (2003)
10. Yu.G. Dorofeev [etc.], *Industrial technology of hot pressing of powder products.* (M.: Metallurgy, 1990)
11. V.B. Ilitsky, V.V. Mikityansky, L.M. Serdyuk, *Machine-tool accessories, Design and Technology software operating properties.* (Engineering, Moscow, 1989)
12. N.A. Fomenko, V.I. Bogdanov, V.N. Fomenko, *The high pressure pipeline.* Patent RF, no. RU 25111926 C 2 F 15 B 20/00, F 16 L 11/20
13. N.A. Fomenko, V.I. Perel'miter, V.N. Fomenko, *System of Hydraulic Control Protection.* Patent RF, no. RU 15763 U 17 F 15 B 21/00
14. V.N. Fomenko, V.I. Perel'miter, N.A. Fomenko, V.P. Shevchuk, *Hydraulic System.* Patent RF, no. RU 15764 U 17 F 15 B 21/00
15. N.A. Fomenko, S.V. Dubinskiy, G.I. Golobuta, G.P. Lyshko, *System of Hydraulic Control Protection.* Patent RF, no. SU 1813937 A 1 F 15 B 20/00
16. I.P. Ksenevich, V.A. Nasirov, *Device of emergency protection of the vehicle.* Application for the invention No. 4820466/06 (048212) of 01.03.91
17. V.I. Perel'miter, *Hydraulic system.* Patent RF, no. SU 1822471 A 3 F 15 B 20/00
18. A.M. Dalsky, B.M. Bazrov, A.S. Vasiliev, *Technological heredity in engineering production.* (Publishing house of the Moscow Aviation Institute, Moscow, 2000)
19. V. A. Altman [etc.], *Influence of an alloying on physico-mechanical interaction and frictional characteristics of powder materials with solid lubricants,* Powder metallurgy, **2,** 67-69 (1980)
20. N.A. Fomenko, V.I. Bogdanov, S.V. Alexikov, V.N. Fomenko, S.A. Bogdanov. *System of protection of hydraulic actuator.* Patent RF, no. RU 2571240 C 1 F 15 B 20/00
21. Yu. K. Bobkov, V.P. Shevchuk, V.G. Chernyshev. *Method of protection of a hydraulic system.* Patent RF, no. SU 1550255 A 1 F 15 B 20/00
22. V.I. Bogdanov, N.A. Fomenko, O.V. Burlachenko, S.V. Alexikov, S.A. Bogdanov, V.N. Fomenko. *System of protection of hydraulic actuator.* Patent RF, no. RU 2579531 C 1 F 15 B 20/00
23. Yu.P. Serdobintsev, O.V. Burlachenko, A.G. Skhirtladze. *Improving of the quality of functioning of technological equipment.* (A monograph, Staryi Oskol, TNT Publ., 2010)

Prediction of leakage in the fixed mechanical seal

Anatoly A.Asheichik[1,*], and *Mohammad Reza* Bahrami[1]

[1]Peter the Great St.-Petersburg polytechnic university, St.-Petersburg, Russian Federation

Abstract. The questions of influence of the shape of contact surfaces on leakages through rubber seals in fixed connection of subassemblies are considered in the article. It is known from practice of operation of seals of various designs that the shape of contact surfaces and consequently also the shape of diagram of stresses in a contact zone considerably influences on value of leaks Linking leakage magnitude and distribution of contact stresses enables, firstly, more precisely calculate the amount of leakage for existing seals, and, secondly, to optimize the shape of the seals in their design in each case. As the result of experimental studies on the introduction of the rubber gasket ring fixed indenters different profiles found that by optimizing the shape of the indenter magnitude of leakage can be reduced by 10 times.

1 Introduction

Leakage through a circular fixed mechanical seal can be defined by the formula [1]

$$Q = \frac{\pi}{6\ln R_1/R_2}\frac{\Delta p}{\eta}\psi_0 R_z^3 \exp(-\frac{3\sigma_n}{kE}) \qquad (1)$$

where Q is leakage through the seal R_1 and R_2 - the outer and inner radii of gaskets; Δp is pressure drop; η is the viscosity of the sealing medium ψ_0 is the coefficient of micro-roughnesses form; R_Z is the sealing surface roughness parameter; σ_n is the normal stress on the contact point; E is the seal modulus; k is the coefficient taking into account the specific features of the physical model 0.05 to 0.3.

Formula (1) derived from the Navier-Stokes equations for laminar flow of gas without considering the centrifugal forces. From the analysis of the above expression for leak detection, follows that they are largely dependent on the magnitude of the stresses in the zone of contact of the seal with the sealing surface. The magnitude of this stress is known [1- 4], which depends on the preload pressure p_{ko} and in some designs of seals depends on the drop pressure of sealed medium Δp. Usually in the formula (1) the average value of the normal stress is using [5-7].

From experimental practices of various designs of seals [8-10] is known that the shape of the contact surfaces and, consequently, the shape and stress distribution in the contact area significantly affects the amount of leakage. In literature, however, practically there is no information about the effect of contact stress distribution along the width of the contact area on leakage. Thus establishing a relation between leak size and distribution of contact stresses will allow firstly to better calculate the amount of leakage for existing seals and secondly to optimize the shape of the seals in their design in each case.

2 Apparatus for investigations and test method

Scheme of apparatus for studying leakage through mechanical seal is shown in Fig. 1. The appearance of the elements of the structure with photographs of the ring indenter and their schemes are shown in Fig. 2.

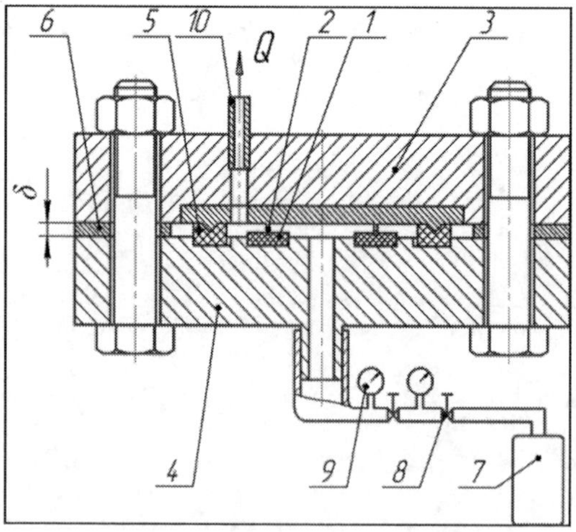

Fig. 1. The apparatus for leakage studying through mechanical seal.

The object of the study is a fixed mechanical ring seal 1. In tests, the rubber O-ring clamped between two steel flanges 3 and 4. In the inner cavity seal supplied pressurized air, 0 – 1.5 *MPa*. Air was supplied from a gas cylinder 7, where its pressure was 15 *MPa* through the gas reducer 8. Accuracy of adjustment and

* Corresponding author: aseichik52@mail.ru

measurement of overpressure was 0.01 *MPa*. leakage measured from the outside of the gasket (10 fitting).

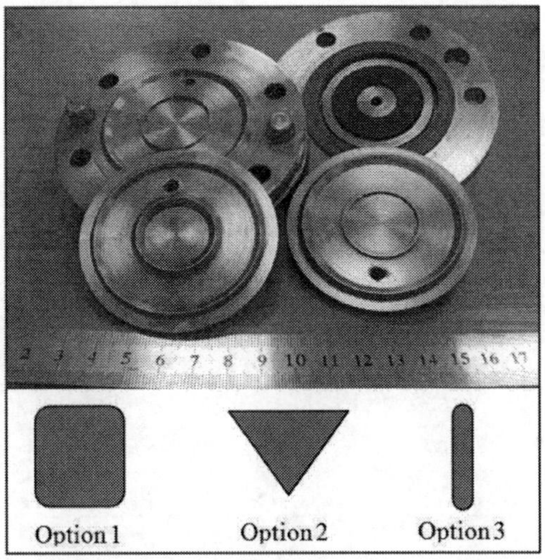

Fig. 2. Appearance of the elements of the structure and the scheme of indenters variants.

The dimension of the rubber O-ring with the rectangular section 1 is as follows: outer diameter of 36 *mm*, inner diameter of 16 *mm* and height of 4 *mm*. It was placed in an annular groove on the steel flange 3 with the depth of 2.6 *mm* and the diameter is equal to the diameter of the gasket and glued. In the opposite flange laid removable ring indenters 2 of different shapes (3 options, see Fig. 2).

In the first embodiment, it was a rectangular protrusion 1.8 *mm* wide with radii of 0.3 *mm* at the edges (width of the flat portion 1.2 *mm*). In the second embodiment, the protrusion section is an equilateral triangle with angles at the apex of 90 ° and the radius at the tip of 0.5 *mm*. In the third embodiment, it was rectangular protrusion 0.4 *mm* wide with the radius of 0.2 *mm* at the top. The roughness surface of protrusion corresponded to $R_a = 1.25\ \mu m$. Static modulus of rubber was $E = 3.2\ MPa$. Average diameter of the protrusions in all cases was 25.8 *mm*, and the height was 1.8 *mm*. The preload pressure p_{ko} was created by six bolting joints which are spaced evenly over flanges. Movement was regulated and measured by instrumental measurement tiles 6 (see Fig. 1) with the accuracy of 0.01 *mm*.

During the test, the rubber O-ring was placed in the groove on the steel flange. By means of bolting and measuring plates 6 was set in the rubber specified minimum implementation δ of the annular protrusion – indenter, made from the first embodiment. Then, through the gas reducer in the internal cavity of the seal was applied the specified minimum air pressure and the leakage was measured. After this, the air pressure was increased and leakage measurement was repeated. After measuring the leakage over the entire range of pressures, the implementation value δ has been increased to the following amount, and all measurements were repeated.

3 Results of experimental studies

As the result of the experimental studies were obtained the dependence of leakage from excessive pressure at different values of the implementation of the protrusion in the rubber. These dependencies for the first variant form of the indenter are presented in Figures 3-5.

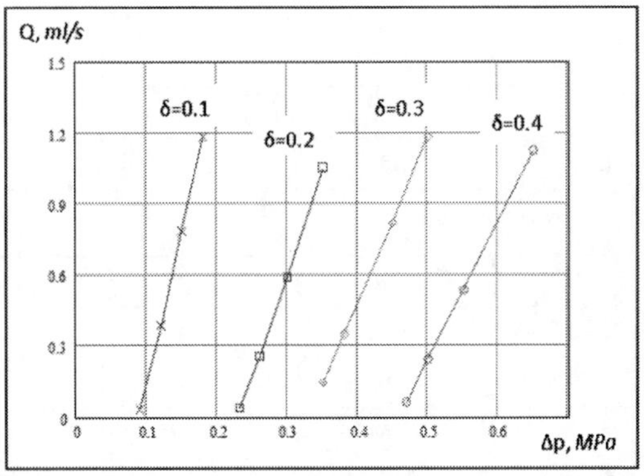

Fig. 3. Dependence of leaks from the pressure drop at different implementations δ in the rubber indenter made of embodiment 1.

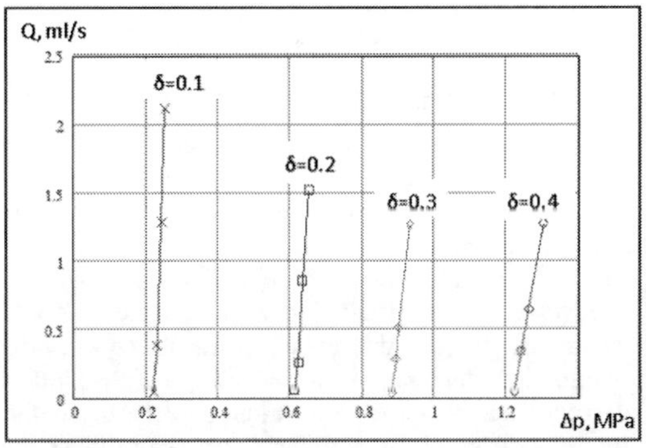

Fig. 4. Dependence of leaks on the pressure drop for various insertions of δ into the rubber.

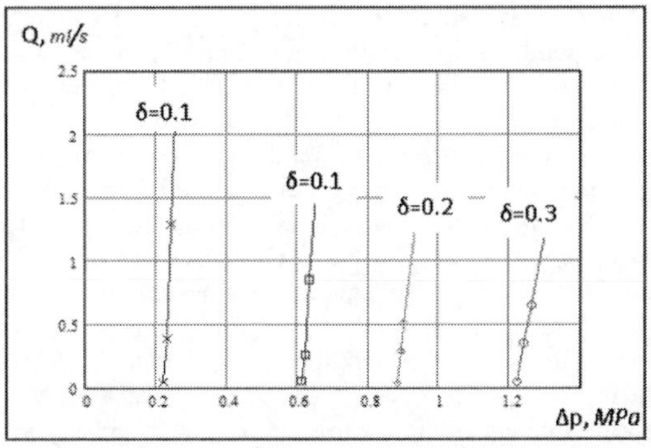

Fig.5. Dependence of leakages on the pressure drop for various insertions of δ into the indenter rubber manufactured according to the variant 3.

DOI: 10.1051/matecconf/201712906002

Table 1. Results of geometric similarity coefficient calculation.

δ, *mm*	R_1, *mm*	R_2, *mm*	R_1/R_2	$\ln R_1/R_2$	$\left(\ln\frac{R_1}{R_2}\right)_1 \Big/ \left(\ln\frac{R_1}{R_2}\right)_j$
1st embodiment					
0.1	13.72	12.08	1.1358	0.12646	1
0.2	13.77	12.03	1.1446	0.13506	1
0.3	13.80	12.00	1.1500	0.13977	1
0.4	13.80	12.00	1.1500	0.13977	1
2nd embodiment					
0.1	13.20	12.60	1.0476	0.04650	2.72
0.2	13.30	12.50	1.0640	0.06204	2.17
0.3	13.50	12.30	1.0976	0.09313	1.50
0.4	13.60	12.20	1.1148	0.10868	1.29
3rd embodiment					
0.1	13.06	12.74	1.0251	0.02479	5.10
0.2	13.10	12.70	1.0315	0.03101	4.37
0.3	13.10	12.70	1.0315	0.03101	4.50
0.4	13.10	12.70	1.0315	0.03101	4.50

Analysis of the results showed that for all variants forms of indenters and for all values of the implementations of the indenter in the rubber observed linear dependence of the leakage of the sealed medium pressure drop that corresponds to the formula (1).

However, the obtained dependencies does not allow to compare the results of experiments with various embodiments indenters since radiuses R_1 and R_2 (macrogeometry), used in formula (1), in these three structures differ sharply. Values of these radii and the ratio $(\ln R_1/R_2)_1/(\ln R_1/R_2)_j$ which may be called the geometric similarity ratio, in comparable embodiments with different implementations of the indenter in the rubber δ are given in table 1. The index j, here corresponds to the embodiment which is compared with the first embodiment, that is selected as the basic embodiment.

The use of geometric similarity coefficient allows lead dependencies obtained experimentally for different indenters to the basic embodiment (embodiment 1). This alignment makes it possible to eliminate the influence of all factors on the leakage except concentration of the stress created by the indenter in embodiments 2 and 3. From the leakage calculations regarding the similarity factor we can say leak in the second embodiment is approximately 2 times less than in the first (basic) embodiment and in the third embodiment, the difference reaches 10-times.

Analysis of the expression (1) allows determining the excess of the maximum stress σ_{max} over medium stresses σ_m. This can be accomplished by the use of the results of the two series of experiments. In the first series for the base embodiment (first) , $\sigma_{max} = \sigma_m$. In the second series, σ_{max} should significantly exceed the average stress, achieved for the second and third embodiments form indenters. Transforming the formula (1), we obtain

$$\sigma_{\max} - \sigma_m = \frac{1}{3}\cdot \mathrm{k\,E}\cdot ln\left\{\frac{\mathrm{Q_1}}{\mathrm{Q_j}}\cdot\frac{\Delta p_j}{\Delta p_1}\cdot\frac{\left(\ln\frac{R_1}{R_2}\right)_1}{\left(\ln\frac{R_1}{R_2}\right)_j}\right\}, \quad (2)$$

where the value

$$B = \frac{Q_1}{Q_j}\cdot\frac{\Delta p_j}{\Delta p_1}\cdot(\ln\frac{R_1}{R_2})_1 \Big/ (\ln\frac{R_1}{R_2})_j \,,$$

can be called the general similarity coefficient.

Value σ_m can be defined as

$$\sigma_m = E/\varepsilon, \quad (3)$$

where ε is the relative deformation which in this experiment can be defined as

$$\varepsilon = \delta/h, \quad (4)$$

where h is the thickness of the rubber gasket $h = 4$ *mm*.

Then, using the experimental values of the pressure drops and the related leak from Fig. 2 and of similar

Table 2. The calculation results of the experimental ratio σ_{max}/σ_m.

δ *mm*	Q_1 *ml/s*	Q_j *ml/s*	Δp_1 *MPa*	Δp_j *MPa*	*B*	ε	σ_m *MPa*	σ_{max} *MPa*	σ_{max}/σ_m
comparing the second embodiment with the base (first)									
0.1	1.280	0.640	0.24	0.18	4.08	0.025	0.08	0.155	1.93
0.2	1.260	0.540	0.93	0.44	2.40	0.050	0.16	0.207	1.29
0.3	1.270	0.724	1.30	0.88	1.78	0.075	0.24	0.271	1.13
comparing the third embodiment with the base (first)									
0.1	1.280	0.780	0.24	0.15	5.10	0.025	0.08	0.168	2.10
0.2	1.260	0.588	0.93	0.30	4.37	0.050	0.16	0.218	1.36
0.3	1.270	0.820	1.30	0.45	4.50	0.075	0.24	0.287	1.20

dependencies for indenters for embodiments 2 and 3, and taking $k = 0.05$, we obtain experimental ratio σ_{max}/σ_m . All results of these calculations are summarized in Table 2.

4 Conclusions

As the result of experimental studies on the implementation of circular indenters of various profiles in the rubber gasket fixed found that by optimizing the shape of the indenter, magnitude of leakage can be reduced by 10 times. Establishing relation between the leak and the size of the contact stress distribution allows more precisely calculate the amount of leakage for existing seals, as well as to optimize the shape of the seals in their design.

References

1. A.I. Golubev, L.A. Kondakov, *seals and sealing machines: Directory* (Mechanical engineering, Moscow, Russia, 1986)
2. S.O. Lazarev, A.A. Asheichik, V.L.Polonsky, Investigation of the influence of stress distribution in the contact area on the leakage through the rubber seal in a fixed joint (Publishing House of STU, St. Petersburg, Russia, 311-312, 2006)
3. A.A. Asheichik, S.G. Chulkin, *Experimental Mechanics* (Polytechnical University Press, St. Petersburg, Russia, 2008)
4. A.A. Asheichik, V.L. Polonsky, *Experimental study of elastomers and polymers for the oil industry*, (Polytechnical University Press, St. Petersburg, Russia, 2015)
5. D.A. Denischenko, *Determination of the activation energy of elastomers by the experimental-theoretical method* (Week of Science SPbPU: Part 1, St. Petersburg, Russia 65-67, 2014)
6. A.A. Kukina, *Prediction of the change in the strength properties of rubber in its thermal aging* (Week of Science of SPbPU: Part 1, St. Petersburg, Russia 63-65, 2014)
7. S.O. Lazarev, V.L. Polonsky, *Computational mechanics,* (Publishing house of Polytechnic. University, St. Petersburg, Russia, 2007)
8. G.R. Khachatryan, B.V.Ratner, *Investigation of the antifriction properties of metal-filled rubbers*, Caoutchouc and Rubber, №**6**, 33-35 (1981)
9. G.M. Bartenev, Structure and relaxation properties of elastomers, *Chemistry*, Moscow (1979)
10. N.P. Kirillov, N.P. Burenin, V.I. Polyansky and others, *Structures of mechanical seals for rotating shafts of centrifugal pumps of rocket refueling systems*, Herald of MADI, №**1**, 18-22 (2012)

Optimization of geometry of the sealing cuff of the piston of volumetric oil and gas hydraulic aggregates

Oleg Dushko[1], *Vladimir* Perfilov[1,*], and *Vladimir* Yaroshik[1]

[1]Volgograd State Technical University, Oil and Gas Constructions Department, Avenue V.I. Lenin, 28, 400005, Volgograd, Russia

Abstract. The use of modern sealing elements of volumetric hydroaggregates in the oil and gas industry is hampered by lack of scientifically based constructive decisions. The implementation of this complex and urgent challenge is possible due to optimization of geometry of the elastic seal (cuff) by criterion of uniform distribution of contact tensions on all its length that it will promote uniform wear of the seal. It is possible to manage this process change of initial geometry of the cuff.

1 Introduction

Increase the reliability and durability of volumetric hydroaggregates is particularly acute especially now. First of all it concerns the equipment working in the oil and gas industry, where separate components and aggregates are exposed to simultaneous influence of high contact tensions and temperatures, containing aggressive abrasive environments, that substantially reduces a resource and reliability of work not only separate components and aggregates, but also in whole oil-extracting complexes [1,2].

From operating experience of machines and mechanisms it is known that reliability of any technical system directly depends on reliability of separate components and details constituting this system. Due to the above, one of the main directions of development of the domestic industry is to increase the wear resistance and durability of machines and equipment, repeated decrease in repair work is. Particularly relevant this question in oil-extracting and petrochemical industries in relation to the sealing elements and friction pairs of volumetric hydroaggregates[3-6].

In the process of pumping of the oil products, drilling mud containing solid-phase particles there is an abrasive wear of surfaces of friction pairs, which that conducts to a leakage of liquid and reduce a resource of work of the equipment [6-8]. Working capacity, reliability, fire and environmental safety of production objects significantly depends on perfection, reliability and resource of work of heavy-duty friction pairs, and condensed elements of different function. On sea oil and gas complexes, and also on machine-building, chemical, petrochemical, some overworking and other enterprises volumetric hydroaggregates and automated pneumatic test facilities, which basis is constituted by driving pumps of direct reciprocating action gain distribution. Volumetric aggregates with mechanical and pneumatic actuator are effective, first of all, when using in objects with increased fire and environmental danger [9-14].

2 Main part

The weakest spot, of volumetric hydroaggregates, especially in case of 50– 150 MPa, consolidations which often use liquids with limited greasing properties are. For example, drilling of oil and gas wells is impossible without system of circulation of drilling mud and its basis – drilling pumps. Mud pumps – volumetric hydraulic machine – are characterized by high capacity and high hydraulic pressure.

On installations of OAO "Volgograd plant of drilling equipment" capacity of pumps reaches 600 kW, and the pressure – 25 MPa; on OAO "Uralmash" – respectively 950 kW and 40 MPa (unique installations for superdeep drilling are not considered). Cylinder-piston group of pumps works in hard conditions - in the presence of solid particles (abrasive) in drilling fluid, which temperature reaches 100 °C. Such conditions predetermine limited service life of pump seals, which when drilling of deep wells (3000 meters or more) sometimes are constituted by only 40 hours. It forces to contain as a part of the drilling rig at least two pumps (one in a hot allowance). In the same time conditions the equipment for forcing of cementing, clay and other trade and squeezing liquids in oil and gas wells in process of their drilling or repair of UNB-125x63 (analogue of CA-320) works.

Thus, increase of resource of cuff to a resource of other basic elements of system will allow to create reliable and economical hydroaggregate that, finally, will lead to increase in overall performance of all processing equipment in general [15-19]. The solution of this complex problem is possible due to optimize of geometry of an elastic seal (cuff) by criterion of uniform distribution of contact tensions on all its length that will promote uniform wear of the seal [20,21]. It is possible

* Corresponding author: vladimirperfilov@mail.ru

DOI: 10.1051/matecconf/201712906003

to manage this process change of initial geometry of the cuff.

The effective analytical method of determination of tension and deformations in a cuff allowing to receive the required solution of a task of optimization for real time was developed for search of the optimal solution. The essence of this method consists in determination of the intense-deformed condition of the cuff and contact tensions in cuff in case of its interaction with the cylinder for the set any initial geometry of the cuff and pressure of working liquid in the cylinder.

For calculation of intense-deformed condition of an elastic seal of the piston working course when on volumetric hydromachine the piston 1 under influence of force F creates working pressure p repumped liquid in the cylinder 4 (Fig. 1). Sealant 2, prevulcanized to the piston, interferes with penetration of working liquid 3 into a gap at the internal surface of the cylinder 4. At the same time sealant is affected by hydrostatic pressure p of liquids and friction forces on the contact surface "sealant – cylinder".

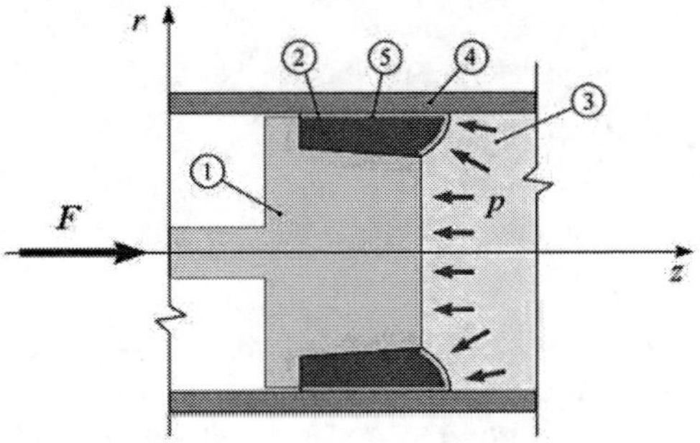

Fig. 1. Working piston stroke: 1 – metal core; 2 – elastic sealant; 3 – working liquid; 4 – cylinder; 5 – modified layer

For creation of preliminary consolidation, a forward part of a sealant (a bell-shaped part) is executed with a diameter big the internal diameter of the cylinder plug. It is necessary to emphasize that the seal material (rubber mixture) is uniform and isotropic in all amount, except area of contact with the cylinder. Here the surface layer 5, for increase in a resource, can be modified – in rather thin surface layer module of elasticity significantly higher than the module of elasticity of the main amount of material of the cuff [22-26].

We consider that for the solution of an objective there is possible an application of a mathematical apparatus and methods of the linear theory of elasticity. It follows from the fact that, first, the sealant is in tension (Fig. 1), axisymmetric concerning an axis z, and tensions and deformations changes only along two coordinate axes z and r, and secondly, deformations of sealant are small ($\varepsilon<10\%$) owing to what communication between tensions and deformations can be described by a linear dependence (Hooke's law).

During the carried-out calculation, the intense-deformed condition of a cuff and contact tension "cuff – cylinder" for the set any initial geometry of a cuff and working pressure of liquid is determined that forms the basis of the solution of a task of optimization of geometry of a sealant.

In Fig.2 system of geometry of the cuff and loading for working piston stroke in system of coordinates z-r is shown, and in Fig. 3 settlement scheme of geometry of the cuff in not deformed condition is shown.

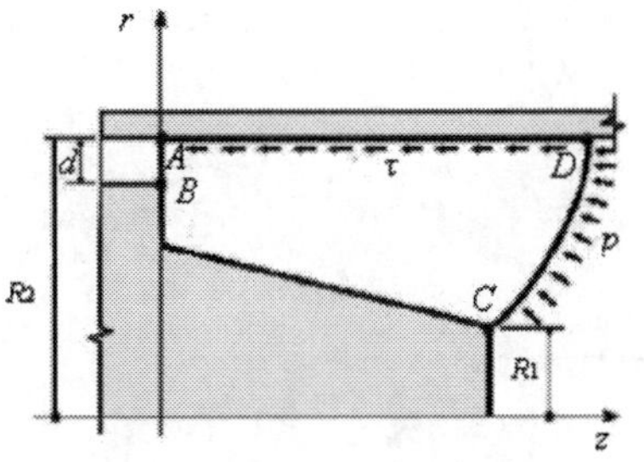

Fig. 2. Cuff geometry for the working course

For definition of the intense-deformed condition of a sealing cuff the known method of function of tension and designations, standard for the theory of elasticity, and the rule of signs for tension and movements is used. However, for the solution of a task the function of tension has been entered. Let's note that boundary conditions on each contour are met approximately: their exact compliance is carried out only in separate points of a contour which number can be any. At the same time increase in these points leads to increase accuracy of solution of a task (fig. 3).

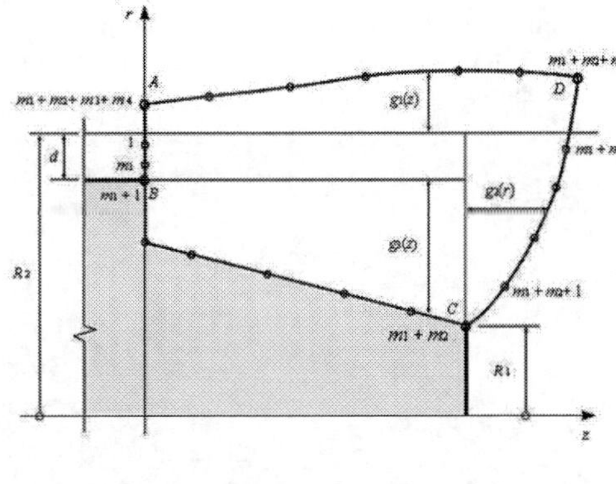

Fig. 3. Initial geometry of the cuff in a free state

The received decision precisely satisfies to all equations of the theory of elasticity at the confidant, as much as exact, performance of boundary conditions, and comes down to the solution of system of the linear algebraic equations. The order of system of equations is equal to the doubled number of points of contour for satisfaction to boundary conditions.

Contact tangent tension is provided by power series through unknown coefficients which are defined from balance conditions.

Function of purpose defines compliance of the decision top optimum and is accepted in the form of weighted combination of the decisions determining the sum of squares of deviations of contact tensions from an average at various stages of attrition of the surface. At the same time function of purpose gives chance to define an optimal configuration of the cuff for the most uniform distribution of contact tensions in work process and deleting of material.

In Fig.4 initial optimal configuration of the cuff (curve 3 – in full-scale, curve 2 – in increased scale) and the corresponding distribution of contact tensions (curve 1) without cuff material attrition in work process is shown.

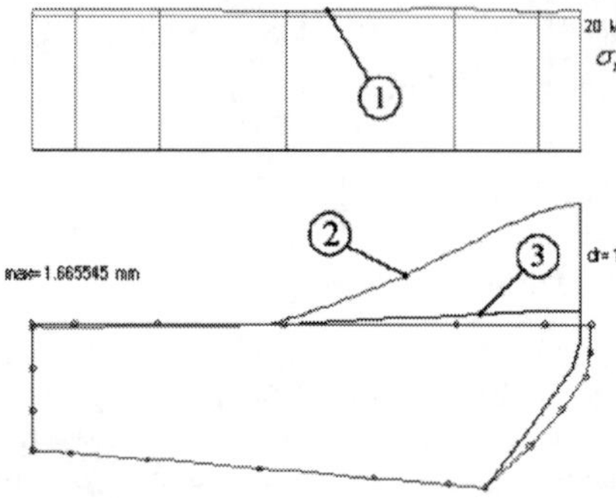

Fig. 4. Optimal configuration of cuff without considering erasure: 1 – distribution of contact tensions; 2 – optimal geometry of cuff in increased scale; 3 - optimal geometry of cuff in full-scale.

In Fig.5 initial optimal configuration of the cuff for receiving the maximum possible uniform distribution of contact tensions in work process at her deleting on depth of 0,5 mm (curve 3 – in full-scale, curve 4 – in increased scale) and the corresponding distribution of contact tensions (curve 1 - beginning of work, curve 2 – erased surface).

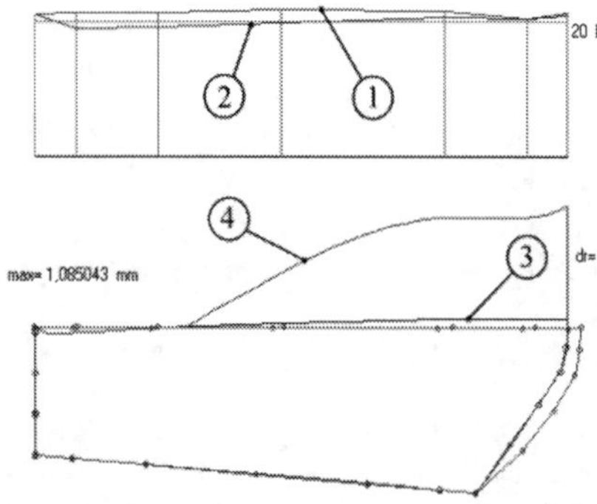

Fig. 5. Optimal configuration of cuff subject to erasure: 1 – distribution of contact tensions at beginning of work; 2 – distribution of contact tensions after attrition; 3 – optimal geometry of cuff in full-scale; 4 - optimal geometry of cuff in increased scale.

3 Conclusions

Thus, it is possible to manage process of wear of an elastic sealant by change of its initial geometry. Calculation of initial optimum, from the point of view of uniform wear, surface geometry of seal can be performed using a mathematical apparatus and methods of linear theory of elasticity. [27, 28].

Modification of sealants more than by 5,6 times raises a resource of pistons of mud pumps, and universal consolidations increase efficiency of hydropneumatic devices of reciprocating action with high and ultrahigh pressure, including several times a resource of pistons of mud pumps (when testing on the stand) [29,30]. Reliability of elaborated provisions, conclusions and recommendations is confirmed with practice of designing, the solution of analytical tasks on the computer, testing at stands with imitation of severe conditions of operation, production and approbation of samples as a part of the installations working in actual practice.

References

1. O.V. Dushko, V.I. Pindak, A.D. Polikarpov, A.V. Shcherbin, *Possibilities of increase the durability of oil fields pump units and decrease their consumption*, Geology, drilling and development of oil fields of Prikaspiya and the Caspian sea, **Iss. 61,** Pp. 182-191 (2003)
2. O.V. Dushko, *Tribology of seals of oil and gas volume hydrounits (*Volgograd: publishing house of VolgGASU, 2005)
3. H. Yuan, J. Guo, Z. Su, Y. Sun, J. Qi, *Research on reliability of domestic drilling mud pump piston-cylinder linerfriction pair,* Zhongguo Jixie Gongcheng, **Vol. 24, No. 11.** Pp. 1425-1430 (2013)
4. E.F. Romanenko, V.I. Kolmykov, *Analysis of geological exploration mud pumps component wear and damage,* Chemical and Petroleum Engineering, **Vol. 49, No. 11-12,** Pp. 811-815 (2014)
5. E.K. Isakaev, V.B. Mordynskii, A.S. Tyuftyaev, Y. M. Avilkin, V. A. Katarzhis. *Increasing the service life of rapidly wearing components of mud pumps,* Welding International, **T.26, No.1,** Pp.40-44 (2012)
6. A.A. Asheichik, V.L. Polonsky, *Experimental study of elastomers and polymers for oil industry* (St. Petersburg, 2015)
7. O.V. Dushko, V.I. Pindak, *Loading and wear resistance of rubber piston drilling and oil fields pumps,* Construction of oil and gas wells on land and at sea, **No. 5,** Pp. 23-25 (2014)
8. V. Spitas, C. Spitas, G.A. Papadopoulos, T. Costopoulos, *Experimental investigation of load sharing in multiple gear tooth contact using the stress-optical method of caustics,* Strain, **T. 47. No.** SUPPL. 1, Pp. 227-233 (2011)
9. N.R. Tailor, V.V. Kuzin, *Model of tensely deformed condition of the surface layer of ceramic if abrazivostruynoy abrasive jet processing,* Vestnik MSTU Stankin, **No. 1 (36),** Pp. 59-63 (2016)
10. V.D. Atanasov, I.I. Dement'ev, S.A. Nemykin, B.I. Poletaev, *Mathematical models of intense-deformed*

ostoyaniye elastic elements of aircraft structures, Bulletin of the Russian Academy of rocket and artillery Sciences, **No. 2,** Pp. 71-75 (2015)

11. V.V. Glagolev, V. L. Glagolev, A. A. Markin, *Stress-strain state of elastoplastic bodies with crack.* Acta Mechanica Solida Sinica, **T. 28. No. 4,** Pp. 375-383 (2015)
12. S.V. Reznik, P.V. Prosuntsov, A.V. Azarov, *Modelling of the temperature and stressed-strained states of the mirror a reflector of space antenna,* Journal of Engineering Physics and Thermophysics, **T.88. No. 6,** Pp. 978-983 (2015)
13. V.V. Murav'ev, L.V. Volkova, A.V. Platunov, V.A. Kulikov, *An electromagnetic-acoustic method for studying stress-strain states of rails,* Russian Journal of Non-destructive Testing, **Vol. 52, No. 7,** Pp. 370-376 (2016)
14. I.P. Miroshnichenko, V.P. Sizov, *Definition of stress-strain state of layered anisotropic elliptic construction under action of impulse load,* Journal: Advanced Materials Studies and Applications. Ser. "Nanotechnology Science and Technology", Pp. 353-390 (2015)
15. Yu.K. Mashkov, Zn. Shepherd, M. Y. Baibaratskaya, O. A. Mamaev, *Polymeric composites materials in tribology (*M., 2004)
16. G.M. Sorokin, *Tribology of steels and alloys* (M.: Nedra, 2000)
17. D.N. Garkunov, *Tribology* (M.: publishing house of ICCA, 2001)
18. V.M. Khokhlov, *Design of wear-resistant friction units of sliding* (VIMAGE, Bryansk, 2000)
19. G.M. Sorokin, *Problems of metallurgy and tribology in petroleum engineering.*. Izv. Universities, The oil and gas, **No. 4,** Pp. 21-24 (2003)
20. R.N. Suleymanov, A.S. Galeev, G.I. Bikbulatov, *Overall performance of pumping aggregates* (Ufim. state petroleum University, Ufa T., 2004)
21. A.K. Shekunov, V.A. Vasilchenko, *Regulating hydrodevices for hydraulic systems of oil-field machines and an oil and gas complex petroleum engineering,* Oil and gas mechanical engineering, **No. 7,** Pp. 74-75 (2004)
22. A.V. Chichinadze, E.M. Berliner, E.D. Brown, M. etc., *Friction, wear and lubrication,* Tribology and tribotechnics (M., 2003)
23. V.N. Vinogradov, G.M. Sorokin, M.G. Kolokol'chikov, *Abrasive wear* (M.: Mechanical engineering, 1990)
24. P.I. Malenko, *Acoustic spectroscopy of the processes of structure formation on the rubbing surfaces* (Tul'skiy state university, Tula, 2002)
25. V.N. Prokop'ev, N.A. Usol'tcev, E.A. Zadorozhnaya, *Fundamentals of tribology: text of lectures* (South. - Urals. state. un-t. Chelyabinsk, 2001)
26. L. A. Timofeeva, L. V. Proskurin, V. N. Ostapchuk, S. S. Timofeeva, *Management of conditions of antifrictional and frictional friction units,* Heavy mechanical engineering, **No.** 3, Pp. 27-28 (2002)
27. S.V. Fedorov, *Bases of triboengineering and physico-chemical prerequisites of the theory of compatibility* (Kaliningr. state. tekh. un-t, Kaliningrad, 2003)
28. L.V. Krasnichenko, *Tribotechnical properties of friction pairs reverse,* Vestn. Donskoy state. tekh. University, **T.3, No.3**. Pp. 281-292 (2003)
29. V.G. Novikov, A.V. Zharov, Tribology, *Friction of metals and elastomers: [studies benefit]* (Yarosl. state. tekh. university, dep. internal combustion engines; Yaroslavl, 2004)
30. V.A. Vlasov, *Research of influence of high normal pressures on frictional properties of elastomers,* Machines and technology of conversion of rubbers, polymers and rubber mixes. – Yaroslavl, Pp. 10-12. (1981)

DOI: 10.1051/matecconf/201712906004

Methods and means for the in-house training of mining machine operators

Vladimir Velikanov[1,*], *Natalja* Dyorina[1], and *Azat* Abdrakhmanov[1]

[1]Nosov Magnitogorsk State Technical University, Magnitogorsk, Russia

Abstract. This study investigates the quality issue of the in-house training process for mining machine operators. The authors prove the urgency of the designated problem. The changes in modern society, as well as the development of science and technology have a direct impact on the vocational education system. This paper describes the main aspects of the in-house training process of mining machine operators; define the essence, structure, contents, and main directions of its revitalization. The following solutions are proposed in order to improve the quality of the in-house training process: to use the original method based on a rating system of the operator knowledge evaluation, active and interactive forms of using modern training technologies. The authors conducted testing techniques in mining enterprises with the aim of confirming the adequacy of the suggested approaches. The results are given in the work. It was proposed that the methods and tools integration has a positive impact on professional training system.

1 Introduction

Raising the population's welfare is known to depend critically upon the economy growth ensure, which is inextricably connected with its human potential. The "Doctrine of the Russian Federation National Security" states a high level of education quality as one of the most important elements of the country national security [1]. This fact implies overcoming the Russian Federation scientific, technical, and technological dependence on external sources, market expansion, and faster growth of high-tech products, the financial and material resources concentration in priority areas of science and technology.

To date, the production sphere enterprises have a substantial innovative potential, but they cannot implement it due to insufficient quantity, quality, and sometimes the lack of specialists having the necessary knowledge and mastering the latest technology that meet the needs of the age.

The mining industry is known to be one of the most promising and strategically important sectors of the Russian economy. In accordance with the "Energy Strategy of Russia for the period till 2030», mining should reach a fundamentally new level of its development [2]. It has been demonstrated that it is necessary to solve a number of issues in order to achieve these goals. They include weak development of the Russian mining machinery and growing industry dependence on imported technology and equipment, physical and moral deterioration of production assets, which in turn determines the less satisfactory profitability of the mining industry and productivity, great losses in the raw material processing and extraction. The other problems can be described as high-energy intensity of the production, as the lack of new formation specialists, who are able to quick and adequate development of the technique and technology, by providing growth of new advanced competencies. The work has emphasized that eliminating the imbalance between supply and demand of labor to solve the professional and qualifying tasks can be added to the list of problems.

About 490-500 units mining shovels and hydraulic excavators are in operation on the large iron ore quarries of Russia and the CIS, and 90% of them are the cars made by domestic producers (PO "Uralmash", "IS CARTEX", PA "Izhora Works", and others.). Furthermore, 80% of the total fleets of excavators are the shovels with a bucket of 5-10 m3, released back in the 80s of the last century. It was found that the share of mining equipment of foreign production in 2012 were less than 5% (in amount) [3, 4].

Excavator fleet status indices for the coal industry are given in Table. 1 [5].

Table 1. Excavator fleet status for the coal industry in the whole

№ п/п	Excavator model	Amount, item	
		on stock	in work
1.	Domestic brand excavators	1254	1113
2.	in percent, at hand	100	88,8
3.	Foreign brand excavators	233	205
4.	in percent, at hand	100	88,0
5.	total = domestic brand excavators+ foreign brand excavators	1487	1318
6.	in percent, at hand	100	89
7.	domestic brand excavators in percent as percentage of total	15,7	15,6

* Corresponding author: rishik_00@mail.ru

In 2016, it is planned to increase the total volume of iron ore, as well as an increase in coal production, which implies an increase in the annual volume of overburden operations.

In light of this, it can be assumed that a significant increase of quarries and open pits mined rock throughput can be achieved by the technical re-equipment of the parks with modern samples of domestic excavation equipment of the new generation of CMMS (crawler mounted mining shovel)-12K, CMMS -18P, CMMS -32R, as well as those made by foreign manufacturers Hitachi (Japan), Komatsu (Japan), Kobelko (Japan), Caterpillar (the USA) etc. Therefore, the open cast mining needs human resources, relevant for the qualification requirements of modern technologies of mining and equipment in order to achieve high productivity.

The need for a change of approaches in the field of the recruitment policy has been reflected in the "Development Strategy of Labor Training System and of Applied Qualifications Formation in the Russian Federation for the period till 2020" (hereinafter - the Strategy), which defines the main sustainable directions of the state policy in the field of skilled workers (employees) and mid-level professionals training in Russia [6].

The Strategy declares that a significant trend in the development of organizations and enterprises personnel potential is the creation of their own centers and personnel training programs. In common, this situation corresponds to the global trend of increasing the role of employees' in-house training at the premises of corporate training centers.

Furthermore, it has been suggested that the staff in-house development is the most important factor of the enterprise long-term success. The process of the management system organization in the field of training for mining companies involves the development and implementation of corresponding recruitment policy at each enterprise. Its main areas must become the following ones: to identify the needs of the enterprise in the labor force both in its quantity and in quality; to introduce the definite measures for improvement in professional skills use and increase; to monitor the quality of training; to develop the further in-house training path, taking into account the current labor market and the specifics of the mining industry.

The training center of an enterprise is created in the structure of its staffing services and is a new type of an intra-school. Its main tasks include the organization of skilled workers training with a focus on specifics and the main directions of the production process development, as well as taking into consideration the existing level of the student training.

However, some unresolved problems remain: the low prestige of the training system for the labor and the applied qualifications; the system imbalance and lack of the vocational education system reaction to changes in the labor market; lack of the qualified personnel, caused by the general demographic situation in the Russian Federation; lack of regulatory and methodical documents defining the professional skills, experience and professional education requirements.

The presence of the unresolved problems made it possible to determine the aim of the work: the control system organization in the field of labor training for the mining industry should be oriented to increasing the in-house training efficiency of mining machinery and transport-technological complexes operators. The aim decomposition is illustrated in Figure 1.

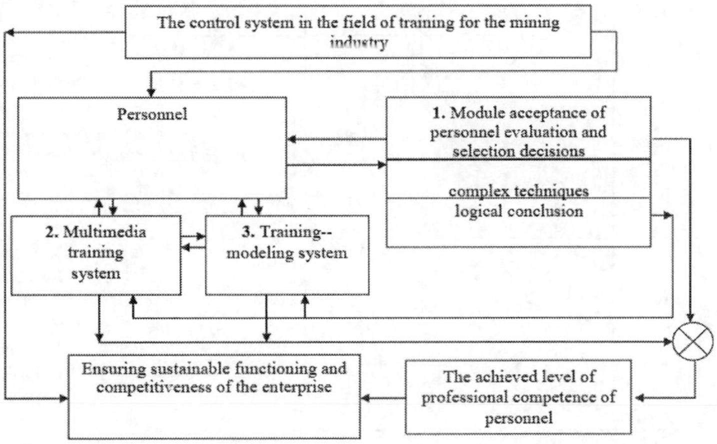

Fig. 1. The aim decomposition.

It is necessary to solve the following tasks in order to achieve the aim of this study:
1. To carry out the set-theoretic analysis of the training systems structures for identifying the specifics of mining machinery and transport-technological complexes operators training.
2. To develop the functional structure of the information-analytical control system in the field of the personnel training, tools for its implementation and the operation evaluation.
3. To develop a quantifying evaluation method of important indicators showing the readiness level of

mining machinery and transport-technological complexes operators, where the indicators affect the professional activity efficiency.

4. To design and implement a multi-functional training facility for mining machine and transport-technological complexes operators training.

5. To use the developed system of mining machinery and transport-technological complexes operators training by trial and to evaluate the system effectiveness.

The investigation underlines that the Russian Federation training technologies are intensively evolving in the aerospace, military, aviation, chemical industry, in the preparation of the operational personnel for electric power facilities, as well as simulators for technological equipment are developed in construction and mining.

The traditional teaching method for fueling machine operators is known to be the exercises on real hardware. However, when being trained on real fueling machines, difficulties may arise in the implementation of such important principles of learning as phasing and continuity; there is no objective control and self-control action of a student's correctness. It must be emphasized that high-performance equipment is diverted from the production process and it wears out in the course of training, further power consumption happens [7].

It should be noted that the term "refueling machine" is quite applicable to the considered cyclical mining excavators. For an operational cycle an excavator performs the following operations: dredging (cutting and filling the bucket); deducing the bucket with soil from the bottom; moving the bucket filled with soil to the place of unloading; unloading soil from the bucket to the dump or into a vehicle; bucket movement (swing) to the bottom; lowering the bucket to prepare for the next step of digging [7]. Additionally, it is demonstrated in Figure 2 that in accordance with the classifications transport and loading machines belong to the same group - the moving and transport-technological systems.

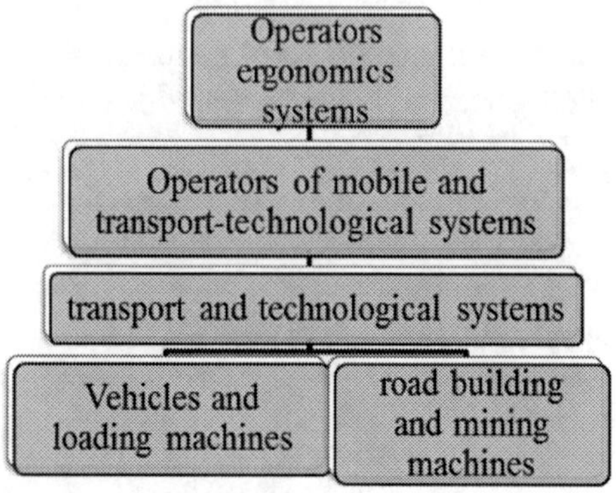

Fig. 2. Classification of the ergonomic systems operators.

The scientific literature presents a technique to assess the effectiveness of the simulators use in the personnel training (Table. 2)

Table 2. Evaluation of training facilities

Application Area	Author	Dependence
engineering and manufacturing systems	R.A. Fayzrakhmanov, I.S. Polevshchikov [7]	The researchers calculate comprehensive assessment of the level of operators professional skills formation, based on the use of fuzzy sets theory
electrical power engineering	S.I. Magid, I.SH. Zagretdinov, L.P. Muzyka [8]	$P_{ob} = P_{cy} P_{int} P_{inf} P_{mat} P_{erg} P_{ps}$, P_{cy} —adequacy of objectives and circumstances, P_{int}- interior adequacy, P_{inf}- information adequacy, P_{mat} - the adequacy of mathematical modeling, P_{erg} - ergonomic adequacy, P_{ps} - psychological adequacy
aviation industry	L.A. Proshkina [9]	$K = K_0 + (aK_1 + bK_2 + cK_3)$, K – efficiency ratio of the simulator, K0 – coefficient, taking into account the initial level of training, K1, K2, K3 – coefficients, taking into account, respectively, knowledge and skills, a, b, c – weights

Changes in the structure and composition of the excavator fleet, the introduction of modern technologies of mining operations and the increase in mining operations are determined by the need to update the content and quality of the mining machinery and transport-technological complexes operators training.

Furthermore, for the successful operation, mining enterprises should not only have high technical equipment and qualified staff in order to make effective use of its production facilities.

The authors have examined the issue of results appraisal of in-house training based on the developed technique with the aim of specifying further studies [10-14].

DOI: 10.1051/matecconf/201712906004

2 Methodology

The range of professional activity tasks were defined owing to the analysis of the operators labor activities (for example, drivers of career dredges), to the study of the technological regulations and instructions requirements for mining machinery and transport-technological complexes, owing to accounting expert opinions. The authors suggested assessing the quality of training and simulator training of operators on the generalized criteria - the activity efficiency coefficient (k_{ef}). The procedure includes three main components: an analysis of the components that make up the structure of the production activities of operators; evaluation of each group, and the indicators that determine the production efficiency; forming the preliminary report and analyzing the results.

The procedure involves the following steps:

1. The procedure of classification groups and the distribution of key indicators in the group that determine the effectiveness of the operator's activity. The first group includes the kind of activity defining "professional skills" of an excavator operator - Pi, the second one includes "theoretical knowledge».
2. Carrying out the importance estimation of groups and indicators within the group.
3. Rationing the key indicators.
4. Expert information collection and processing - monitoring cards processing.
5. Effectiveness coefficient determination of the mining machinery and transport and technological complexes operators.
6. The results analysis and the preliminary report formation.

In accordance with the obtained efficiency coefficient, a decision is made by the team leader (the decision maker) (Table 3).

Table 3. Rules of adopting decisions on the bases of $k_{\text{эф}}$.

Values $k_{\text{эф}}$	Operator Readiness level	Ergatic system state
$0{,}75 < k_{\text{эф}} < 0{,}9$	High	mining machine control at a high level (HL) - high performance (HP)
$0{,}55 < k_{\text{эф}} < 0{,}75$	Average	Mining machine control at a sufficient level (SL) - high performance (HP)
$0{,}35 < k_{\text{эф}} < 0{,}55$	Lowered	Mining machine control at a satisfactory level (SL) - Low performance (LP)
$k_{\text{эф}} < 0{,}35$	Low	Mining machine control at a critical level (CL) - Low performance (LP)

3 Results

Methodology testing was carried out at the mining enterprises of the Southern Urals. The total amount of sampling for the experimental verification was obtained in the survey n = 400 excavator drivers.

The monitoring results illustrated that promising ones are the operators with experience of 5-10 years, from the point of view of the performance of their production functions, as the received and earned value of theoretical knowledge, backed by achieved skills, allow to realize the potential of machinery and to ensure performance in compliance with specifications (Fig. 3).

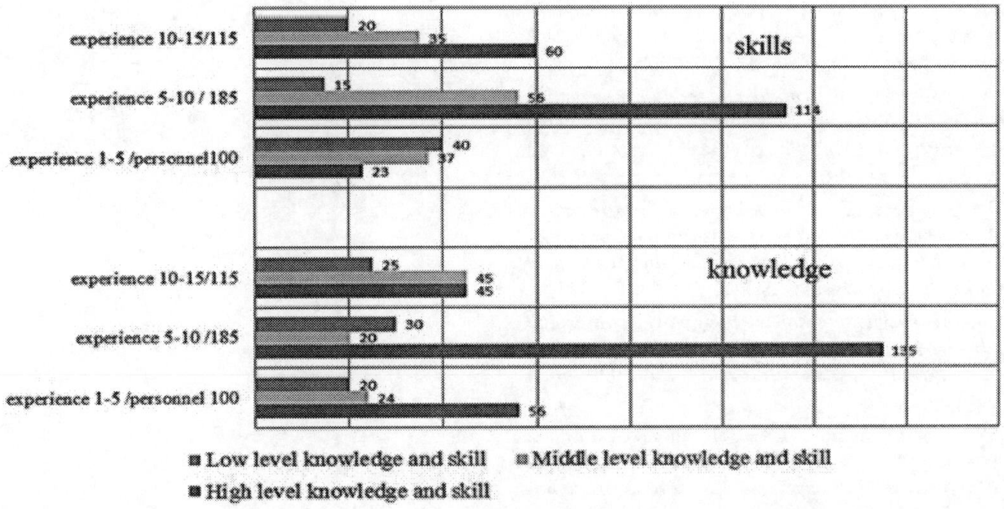

Fig.3. Final results of the evaluations for excavator drivers distribution into groups that determine the operation efficiency.

4 Conclusion

1. It was found that the change in the excavator park structure, modernization and improvement of technical and operational characteristics of the machines, the introduction of modern mining operations technologies lead to a shortage of the professional staff, and non-compliance of their competence level, of the modern mining requirements.
2. Our study developed fundamentals of complex estimation methodology for training quality and proposed a list of key indicators that determine the efficiency of the operators, which differ from the previously known possibility of comprehensive and continuous monitoring of the control system components in the field of training within the general concept of the mining enterprise sustainable functioning.
3. A pilot testing methodologies was carried out for a comprehensive assessment of the training quality of mining machinery and transport-technological complexes operators on the basis of the integral index (k_{ef}), comprehensively taking into account the professional skills, qualifications and work performance.

References

1. *Doktrina natsional'noy bezopasnosti RF [Homeland security doctrine of the Russian Federation]*, Moscow.
2. *Energeticheskaya strategiya Rossii na period do 2013 goda [Energy Strategy of Russia for the period till 2030]*, Moscow.
3. S.P. Reshetnyak, A.V. Samolazov, N.I. Paladeyeva *Burovoye i gorno transportnoye oborudovaniye zhelezorudnykh kar'yerov Rossii i stran SNG*. Gornaya promyshlennost'. **5**, 18 (2009).
4. K.YU Anistratov, *Razrabotka metoda formirovaniya struktury kompleksnoy mekhanizatsii gornykh rabot na kar'yerakh*, Abstract of thesis. Apatity, **43** (2013)
5. L.I. Andreyeva, *Metodologiya formirovaniya tekhnicheskogo servisa gorno-transportnogo oborudovaniya na ugledobyvayushchem predpriyatii*, (Doctoral thesis, Ekaterinburg, 2004)
6. *Strategii razvitiya sistemy podgotovki rabochikh kadrov i formirovaniya prikladnykh kvalifikatsiy v Rossiyskoy Federatsii na period do 2020 goda* (Moscow, 2015)
7. R.A. Fayzrakhmanov, I.S. Polevshchikov, *Analiz metodov i sredstv avtomatizatsii protsessa obucheniya operatorov proizvodstvenno-tekhnologicheskikh sistem (na primere operatorov peregruzochnykh mashin*, Sovremennyye problemy nauki i obrazovaniya **5**, (2013) URL: science-education.ru/111-10494.
8. I.SH. Zagretdinov, S.I. Magid, O.B. Danilov et al. *Realizatsiya novykh tekhnologiy dlya trenazherov i obuchayushchikh sistem Kharanorskoy GRES*, Energosberezheniye i vodopodgotovka. 2, (2004)
9. L.A. Proshkina, *Razrabotka kontseptual'noy skhemy obespecheniya konkurentnosposobnogo funktsionirovaniya predpriyatiy trenazherostroyeniya*, Nadezhnost' i kachestvo 2008. Trudy mezhdunarodnogo simpoziuma 457 (2008)
10. V.S. Velikanov, *Testovyye metodiki i trenazhernyye sredstva v sisteme povysheniya professional'nogo masterstva operatorov gornykh mashin*, Gornyy zhurnal **9**, 131 (2012)
11. V.S. Velikanov, E.A. Ilina, N.V Dyorina, *Obosnovaniye konstruktivnykh i skhemnykh resheniy upravleniya ergonomichnost'yu kar'yernykh ekskavatorov*, Procedia Engineering. **150**, 1215 (2016)
12. V.S Velikanov, *Metodika otsenki professional'noy kompetentnosti operatorov gorno-transportnykh mashin*, Gornaya promyshlennost' **2**, 114 (2012)
13. V.S. Velikanov, A.A. Shabanov, *Otsenka professional'noy kompetentnosti operatorov gornotransportnykh mashin v usloviyakh nechetkoy informatsii*, Perspektivy razvitiya gornotransportnogo oborudovaniya: Sbornik statey. Otdel'nyy vypusk Gornogo informatsinno-analiticheskogo byulletenya, Moscow: Publishing House "Mountain Book". **2**, 117 (2012)
14. N.A. Osintsev, *Safety management of production in the workplace with the application of the apparatus of the theory of fuzzy sets*, Vestnik of NMSTU **4**, 83 (2008)

The system of monitoring of epilepsy in a sleep

Alexey Gorbunov[1*], *Vladimir* Podolskiy[1], and *Vladimir* Egorov[1]

[1]Tambov state technical university, 392000, 106, Sovetskaya str,, Tambov, Russia

Abstract. The slow development of the pathological process of neurodegeneration in the brain structures through appointment of appropriate medications can significantly reduce the risks of disability and improve the patient's quality of life. Currently, it is in many cases possible to carry out therapeutic treatment. For best results, the early diagnosis and instrumental monitoring of the treatment strategy, that is, the monitoring system. System analysis of information about motor activity of the patient during his sleep is likely to monitor epilepsy, including remote, and to predict the patient's condition.

1 Relevance

Brain disease - stroke and dementia, schizophrenia and depression, epilepsy and diseases movements and many others - are not only global economic burden to society, but also to a substantial decline in the quality of life of millions of people worldwide. Improvement of diagnostics, treatment and rehabilitation of diseases of the brain this is a problem of high priority in the medical world and its solution requires joint efforts of specialists not only in medicine, but also many others, including in engineering and IT.

Epilepsy is a chronic noncommunicable disorder of the brain that affects people of all ages. Globally, about 50 million people have epilepsy, one of the most common neurological diseases on a global scale. Most of the people suffering from this disease does not receive treatment they need, while at present, the epilepsy responds to treatment about 70 % of cases. Almost 80% people with epilepsy live in low and middle income. Treatment of epilepsy in most cases is based on a daily intake of inexpensive drugs cost about $5-10 per year. After 2 to 5 years of successful treatment, the medication can be discontinued in approximately 70% of children and 60% adults with no subsequent relapses. In addition, for successful treatment the most important is the ability to set the correct diagnosis at the earliest stages of the disease, as well as the correct therapeutic approach, because timely diagnosis has a significant influence on the development of the pathological process of neurodegeneration of brain structures through the appointment of appropriate drugs, it is significantly slowing down [1].

2 Methods of diagnosis of epilepsy

2.1 Electroencephalography (EEG) and Video Electroencephalography (VEEG)

Currently, the most common and most available instrumental method of diagnostics of brain disorders, including epilepsy is electroencephalography (EEG), which is held in the majority of cases only on an outpatient basis in all cases, only medical personnel with appropriate qualifications. This is due to the necessity of special training of the patient for the procedure, the EEG and the specifics of carrying out the procedure during which held the test sample and which can last several hours. According to the who a shortage of health personnel with relevant expertise is one of the main reasons hindering the successful fight against this disease [1]. In addition, prejudices, entrenched behind this disorder because it causes social problems that can hinder the circulation of people for treatment and the identification they have this disorder. In the end, this leads to the fact that up to three quarters of people with epilepsy may not receive the treatment they need, which can significantly improve their quality of life. Given the above, we can conclude that for a successful anti-epilepsy essential monitoring of this disease accessible to all social groups with the aim of early diagnostics and for observation of dynamics of treatment, with the aim not only of choosing the right tactics of treatment, but also minimize the side effects of antiepileptic drugs due to their prolonged use.

It is acknowledged that currently the most reliable method of instrumental diagnosis of epilepsy is the so-called night VEEG monitoring. This method is carried out during sleep of the patient and also the procedure provides for synchronized EEG with her the video recording of motor activity of the patient. The main disadvantages of this method does not allow its massive use is its high cost (at least $400 per session) and use only in a clinical setting. In addition, typically the session the night VEEG is conducted during one night in hospital and it can make a significant interference in the measurement result is known "first night effect" in which the neural network of the left hemisphere of the brain remains in a state of increased activity, as evidenced by the researchers of Brown University [2].

* Corresponding author: alexey.gorbunov@mail.ru

DOI: 10.1051/matecconf/201712906005

Thus, from the point of view of Metrology, the process of instrumental diagnosis of epilepsy is reduced to the analysis of nonstationary process with a single implementation. But the acceptable accuracy of the analysis of nonstationary process can be achieved only under certain conditions, the chief of which, as shown in the work Julius S.Bendat and Allan G.Piersol [3], is the presence of an increasing ensemble of realizations of the process, i.e. as many as possible of the research conducted under identical conditions . Therefore, to improve the reliability of the study, it is necessary to spend some times under the same conditions, which is currently impossible.

2.2 Studies based on the analysis of motor activity during sleep

Research conducted at the University of Minnesota [4], showed that the bumps, knocks and sudden movements during sleep may signal onset of Parkinson's disease. The researchers found that nearly half of the people who performed in sleep the sudden movement, for 10 years had neurological disorders.

Researchers from the University of Toronto revealed a relationship between increased motor activity during sleep and the early disorders of the brain. In their opinion, if a person is dramatically spreads in a dream hands, moving her legs – in a word, performs a variety of involuntary movements of extremities, such behavior should be alerted and be the cause of going to the doctor-the neurologist, since the spread of pathology from the area of the brain that controls the physiology of sleep, for others its structure, it could trigger diseases such as Alzheimer's or Parkinson's disease [5].

We have made the assumption that the increased motor activity of the limbs of the patient at the time of his sleep can also be caused by epilepsy and thus, the analysis of this activity can be the basis, at least, for a more in-depth studies in this direction.

3 The need for instrument monitoring of epilepsy

As can be seen from the presented material and by direct statements in the works now for a successful response required epilepsy monitoring system, or even a long instrumental monitoring which should be inexpensive in production and available during the operation, which can be used by the patient without medical personnel in the home that best suits obtaining ensemble of realizations under similar conditions. The results of research for their computer analysis can be provided by the patient, not only personally, but using IT, because the problem of lack of proper medical treatment for epilepsy is particularly relevant to residents of small settlements, especially in rural areas. So, according to WHO, the risk of premature death in people with epilepsy in rural areas is three times higher than urban. In addition, the need for a prolonged instrumental monitoring is due not only control the tactics of therapeutic treatment, but also the need to minimize negative side-effects of drugs caused by the long course of their admission, especially in women of childbearing age.

4 System analysis of paroxysmal accelerometry for monitoring of epilepsy in a sleep

In accordance with the above, we have developed a system for the monitoring of epilepsy during sleep of a patient consisting of measuring device and motor activity recording and software for analysis of the information received.

The sensitive element of the device for measuring and recording motor activity is a triaxial accelerometer sensor (analog or digital) with a limit of measurement of ±2g (for example, ADXL327 or ADXL346 company Analog Divices) measurement signal which is processed by the microcontroller (e.g. Atmel ATmega328) and then recorded in a file on the microSD card. The device is powered from the element type is AAA, through step-up device to 3.3 V. This choice of components and method of registration makes the device size slightly larger than, for example, a fitness tracker, but it is made to minimize the cost and operation of the device, and a method of registering data does not require the patient to any additional electronic gadgets and skills with him that makes our system accessible to all social classes of society. The patient is required only to fix the device on your wrist or ankle and turn it on before bed and off after sleep. Then the information from the card is transmitted to the specialist for computer analysis using application software, and the patient will be able to do this not only personally, but through any available means of telecommunication, such as Internet or mobile connection. This architecture of the system should significantly improve the quality of life of patients from small towns, as the lack of specialists in epileptology is particularly relevant for such locations, and in this case they will be able to obtain the necessary specialist advice without the need for personal visits to medical facilities, which may be at a considerable distance [6,7].

The device received the Russia patent for useful model RU 168584 U1, 11.01.2016 [8].

Currently, with the help of our system comparative studies in 10 healthy individuals (control group) and 10 patients with an established diagnosis of "epilepsy".

For processing of the obtained results we took into account the main provisions for the analysis of nonstationary process with a single implementation, described in Chapter 12 [3]. For this one session night monitoring were divided into equal temporary files, which allowed us to obtain multiple implementations under the same conditions, and this allows for conventional statistical analysis, since short time intervals of the studied process can be considered quasi-stationary.

In the analysis of data about physical activity of patients in these groups during sleep, were identified following her graduation, numerically equal to the following:

$$A = \frac{\sum_{i=1}^{N} n_i}{N} \cdot 100\% \quad (1)$$

$$N = \left\lfloor \frac{T}{t} \right\rfloor \quad (2)$$

$$\frac{U(t)_{x_{\max}}}{U(t)_{x_{\min}}} - 1 = A_x \quad (3)$$

$$\frac{U(t)_{y_{\max}}}{U(t)_{y_{\min}}} - 1 = A_y \quad (4)$$

$$\frac{U(t)_{z_{\max}}}{U(t)_{z_{\min}}} - 1 = A_z \quad (5)$$

$$n_i = \begin{cases} 1, (A_x > 0.15) \vee (A_y > 0.15) \vee (A_z > 0.15) \\ 0, (A_x \le 0.15) \wedge (A_y \le 0.15) \wedge (A_z \le 0.15) \end{cases} \quad (6)$$

A – coefficient of motor activity; T – one implementation; t – duration of interval partitioning; U – the measured value of motor activity; x,y,z – spatial coordinate axis; *max, min* – maximum and minimum value; Ax, Ay, Az – the relative peak amplitude peak-to-peak.

If the value is t=60 [s] and the condition (6), value (1) in healthy persons does not exceed 9% , while 80% of individuals from this group it does not exceed 7%, and persons from the first group it from 9% to 12% in individuals receiving therapeutic help and above 12% the rest. In our opinion, this provides a basis for a conditional division of motor activity when monitoring night's sleep on the following areas: to7% - "green", from 7 to 9 % - "yellow", from 9 to 12% orange, from 12% - "red" and could be an indicator for the assessment of the patient by computer analysis of recorded data.

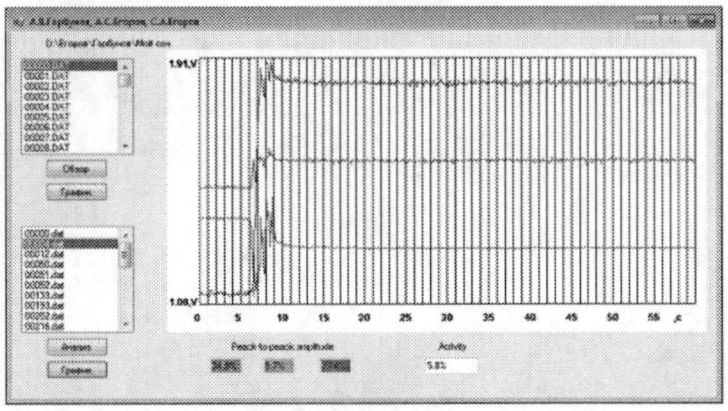

Fig.1. A patient in the "green" zone.

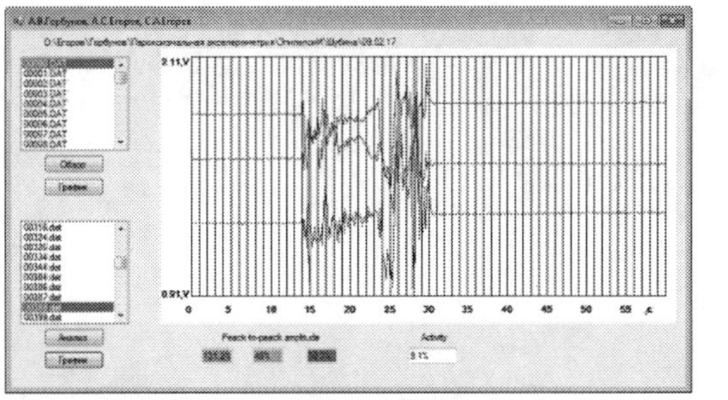

Fig.2. A patient in the "orange" zone.

It should be noted that the shape and duration obtained data are characteristic differences (Fig1, Fig.2).

5 Summary

We believe that our proposed system of analysis of paroxysmal accelerometry for monitoring epilepsy in the dream is promising for an effective anti-epileptic as as a complement to functional analysis, which will assist in the detection of the disease, and in the process of instrumental monitoring of treatment, especially in the regime of telemedicine. In our opinion, the division of motor activity during the monitoring of a night's sleep on the area to 7%, 7 to 9 %, 9 to 12%, more than 12% can be an indicator for the assessment of the patient by computer analysis of recorded data.

References

1. The website of the World Health Organization //URL: http://www.who.int/mediacentre/factsheets/fs999/en //(2017)
2. The website of the Brown University, News from Brown, Asleep somewhere new, one brain hemisphere keeps watch// URL: https://news.brown.edu//2016/04/sleep (2016)
3. Julius S. Bendat, Allan G. Piersol, *Random Data: Analysis and Measurement Procedures,* (1986 by John Wiley & Sons, Inc.)
4. The website of the University of Minnesota, Medical Bulletin. The pursuit of Parkinson's// https://www.med.umn.edu/news-events/medical-bulletin/pursuit-parkinson%E2%80%99s (2017)
5. University of Toronto. "Potential cause of severe sleep disorder discovered, implications for Parkinson's disease." ScienceDaily. ScienceDaily, 15 June 2011. URL:http://www.sciencedaily.com/releases/2011/06/110615103228.htm
6. S.V. Frolov, S.N. Makoveev, S.V. Semenova, S.G. Farea , *Present-Day Tendencies in the Development of Medical Information Systems,* Transactions of the TSTU **V.16**, (2010)
7. S.V. Frolov, M.A. Lyadov, O.A. Ostapenko, A.Yu. Potlov, I.A. Komarova, *Automated Information*

DOI: 10.1051/matecconf/201712906005

System to Monitor Schoolchildren's Health, Development, Implementation Experience, Results, Transactions of the TSTU **V.19**, (2013)

8. A.V. Gorbunov, A.S. Egorov, S.A. Egorov, Patent RU 168584 U1, 11.01.2016

MATEC Web of Conferences **129**, 06006 (2017)
ICMTMTE 2017
DOI: 10.1051/matecconf/201712906006

Evaluation of residual stresses in lock valve elements of petrochemical productions

Elena Nikolaeva[1,*] and *Artem* Mashukov[1]

[1]Irkutsk National Research Technical University, 83, Lermontov Str., Irkutsk, 664074, Russia

Abstract. Lock valves of the pipeline is an important and responsible product. It is designed to shut off the flow of liquid or gaseous fluid. During transportation ambient temperature reaches 400°C, and the pressure - over 30 MPa. Leakage of the fluid through the details and components of lock valves is not permitted. The batch of high-pressure valves was withdrawn after some time of its usage at the petrochemical enterprise. Premature wear of input branch pipe saddles was dis-covered. The cause of destruction of the input branch pipe saddles was erosion be-cause of abrasive particles, which are present in the working fluid. Locking node of the high-pressure armature is highly loaded element. The welding depositions are widely used to increase the strength of the valve locking node. It is established that under recommended heat treatment of the valve an ordered microstructure is formed and residual stress in the welding deposition zone is reduced. The article shows the relationship of residual stress with microstructure of the material. The magnitude of residual stresses in high-pressure valve locking node - in the brunch pipe saddle and the adjacent zones - was determined by x-ray method.

1 Introduction

"Stem-branch pipe" contact patches are the most destructible. Build-up welding is one of the main technological methods of increasing the reliability and service life of valve, and it is used in various industries [1, 2]. Sealing surface of valves is hard-faced by alloys which are tearing, corrosion and erosion resistant in aggressive environments.

Building-up should provide not only good corrosion resistance but also abrasive wear resistance in the technological environments of petrochemical production through the useful life. Therefore, in the manufacture of sealing nodes of valves the corrosion resistant materials with high hardness are preferred [3].

2 Materials and methods

To strengthen the branch pipe saddle of high pressure valve DN25 [4], made of steel Fe-09% C-2% Mn-1% Si, the hard-facing with flux-cored wire (fcw) EnDOtec DO*04 was carried out. Flax-cored wire EnDOtec DO*04 has advantages over other filler materials: it possesses high hardness and corrosion resistance. Also, the hard-faced layer must be resistant to thermal shock and temperature fluctuations, be quenchable and workable by cutting and not show a tendency to crack [5]. It is expected the hard-facing has high resistance the abrasive wear and "metal-metal" type friction, thermal wear hardness and deformation resistance [6], tempering resistance up to 650 °C, scale resistance – up to 900°C.

Figure 1 shows the section of valves, which were rein-forced by hard-facing.

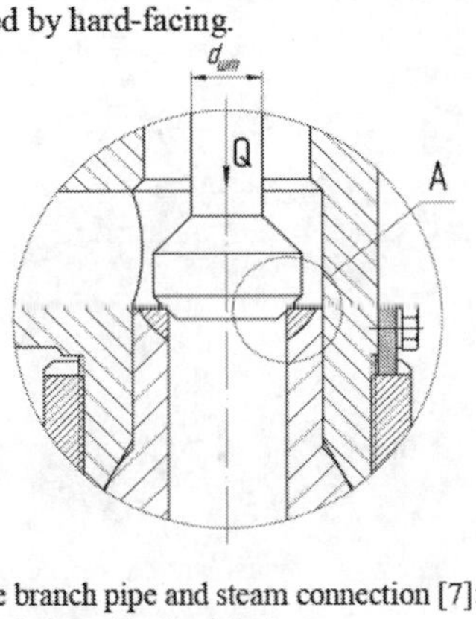

Fig. 1. The branch pipe and steam connection [7]: "A" – section is reinforced by hard-facing.

Building-up was performed in three layers. Mechanical processing with hard-alloy tools was carried out after build-up welding. The authors of the publications [8, 9] in-vestigated the processes of metal cutting with different methods.

To improve the properties of building-up and branch pipe material the heat treat-ment was carried out: oil quenching from 1040°C and at 612 °C. After quenching and tempering the branch pipe was cooled in oil. The quenching and tempering holding time is 1 hour.

The content of chemical elements in the weld-deposited metal was determined by spectrophotometer UNIKO 1201, the contents of phosphorus and sulfur – by analyzer METABAK CS-30 (see table. 1).

* Corresponding author: nicolaeva-ep@yandex.ru

DOI: 10.1051/matecconf/201712906006

Table 1. Weight content of element, %.

C	Si	Mn	Co	Cr
0,13	0,23	0,26	9,81	8,69
Mo	**W**	**S**	**P**	**Fe**
2,0	0,08	0,008	0,011	Balance

Branch pipes were investigated directly after mechanic processing, after build-up welding, after build-dup welding and subsequent heat treatment. Branch pipes were cut on the machine Discotom-10. Samples for microanalysis were prepared using grinding-and-polishing system Tegramin-25. The microstructure was studied on an optical light microscope with magnification from 50 up to 1000 times. 5% solution of nitric acid in ethyl alcohol was used to detect the microstructure of the base metal; for building-up - water 10% solution of nitric and hydrofluoric acid mixture was used. Microhardness was measured.

Residual stresses were measured on the inner surface of the branch pipe. X-ray diffractometer Xstress 3000 G3/G3R (figure 2) was used to measure residual stresses.

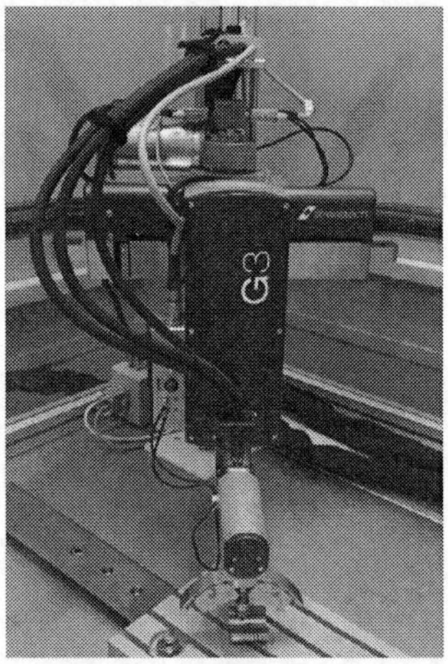

Fig. 2. Measurement of residual stresses on the inside of the branch pipe: the goniometer in position «omega».

Residual stresses in the building-up and the adjoining areas were determined by x-ray sin2 ψ-method [10, 11]. The optical scheme of the x-ray photography with "omega" goniometer configuration was used to record the position of the diffraction maximum. X-ray photography options were Cr-Kα radiation. We used two detectors – A and B. The angle of diffraction, 2θ, is 156.4°. Method for automated stress analysis is "Cross correlation" method; the Poisson's ratio is 0.3; elastic modulus is 210000 MPa.

Also the Barkhausen noise was measured. Application of the method is deter-mined by correlation between the amplitude of Barkhausen noise (mp) and micro-structure, hardness and residual stresses [12, 13].

Digital Barkhausen noise analyzer Rollscan-300 was used. The authors successfully applied this device to control sur-face hardening by running rollers and shot-shock treatment [14, 15]. The amplitude of Barkhausen noise was determined on the same sections of the branch pipe that the re-sidual stresses. Measurement parameters of Barkhausen noise are voltage of 4 V and frequency of 80 kHz. Fig. 3 shows the recorder Rollscan 300 and the position of the sensor in the measurement.

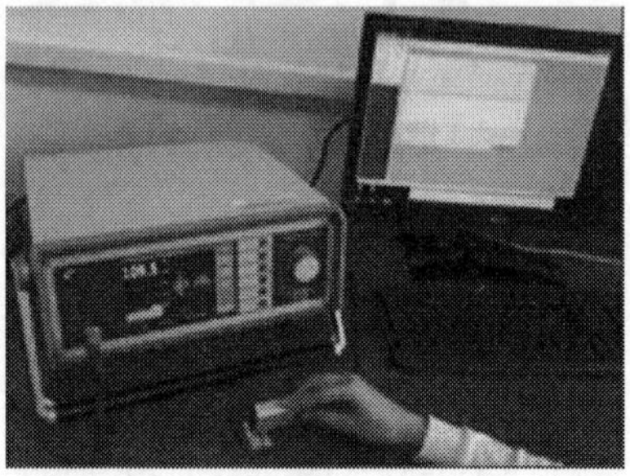

Fig. 3. Measurement of Barkhausen noise amplitude using the analyzer-Rollscan-300.

3 Results and discussion

The measurement results are shown in tables 2 and 3. In the build-up layer significant compressive stresses form, which are corresponded to minimal amplitude values of Barkhausen noise. Under the influence of heat treatment residual stresses increase. The lower level of compressive stress, the higher Barkhausen noise amplitude is.

Table 2. Amplitude of the Barkhausen noise in a branch pipe.

The section are taken measurements	Barkhausen noise amplitude	
	the branch pipe with a building-up before heat treatment	the branch pipe with a building-up after heat treatment
Building-up	86	128
Heat-affected zone (HAZ)	158	271
Base metal	283	451

It should be noted that the results of the noise measurements are influenced by the position of the sensor: along or across article. Therefore, to compare results in the same type of articles is necessary to adhere to the chosen schemes and the measurement parameters.

Residual stresses of the branch pipe after machining have been -398 MPa.

Table 3. The residual stresses in a branch pipe.

The section are taken measurements	Residual stresses, MPa	
	the branch pipe with a building-up before heat treatment	the branch pipe with a building-up after heat treatment
Building-up	-542	-252
Heat-affected zone (HAZ)	-387	-333
Base metal	-384	-343

The microstructure of building-up before heat treatment consists of dendrites, oriented in opposite of heat removal direction (Fig. 4, a).

High-alloy martensite (Fig. 4, b) is formed in the building-up after heat treatment.

Microstructure of build-up layers is very different. It takes place because of thermal effect on the build-up layer when each subsequent layer is over-laid on it.

Before thermal treatment the micro-hardness spread is observed in the build-up metal. Each new layer of building-up has higher micro-hardness than the previous layer: 533…525 HV; 495... 502 HV; 488…494 HV - in the last top layer, second layer and first layer, respectively.

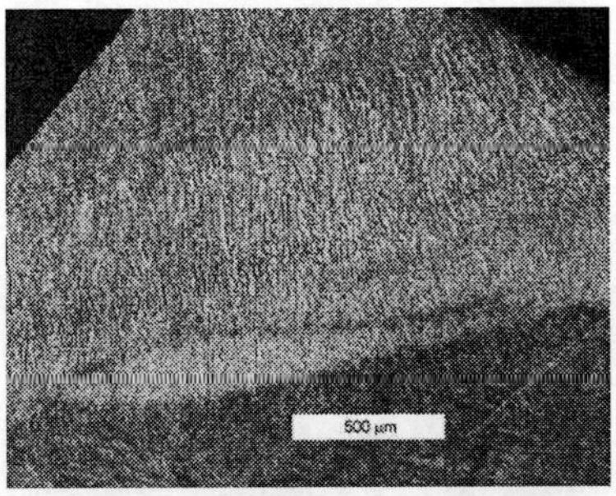

a

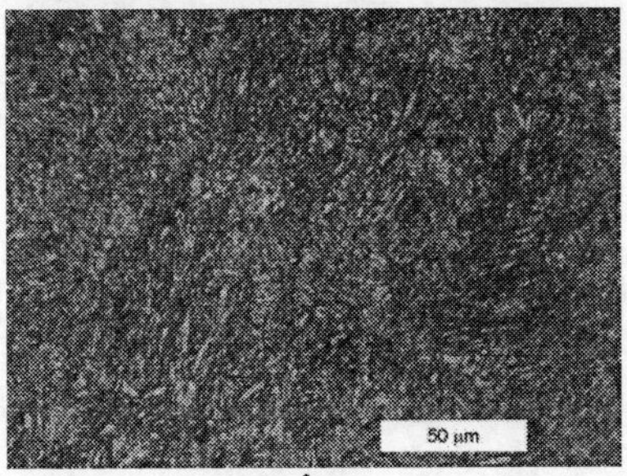

b

Fig. 4. The microstructure of building-up: a – before the heat treatment, dendrites, b – after heat treatment, high alloy martensite.

On the segment, which is adjoining to the building-up, there is a crude ferrite-pearlite structure (widmanstatten) with micro-hardness of 167...172 HV.

Throughout the micro-section there is a clear interface between the build-up and the base metal (Fig. 5, a). After heat treatment, the boundary between the base and build-up metal is “blurred” (Fig. 5, b), value and scatter of micro-hardness decrease, widmanstetter re-mains, but the grains become small, and the microstructure is generally more homo-geneous.

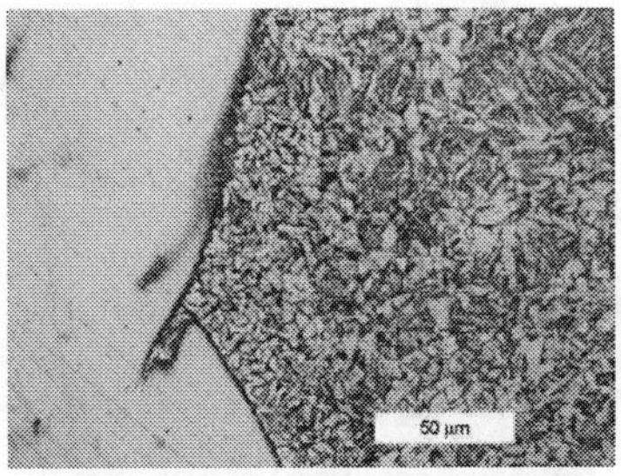

a

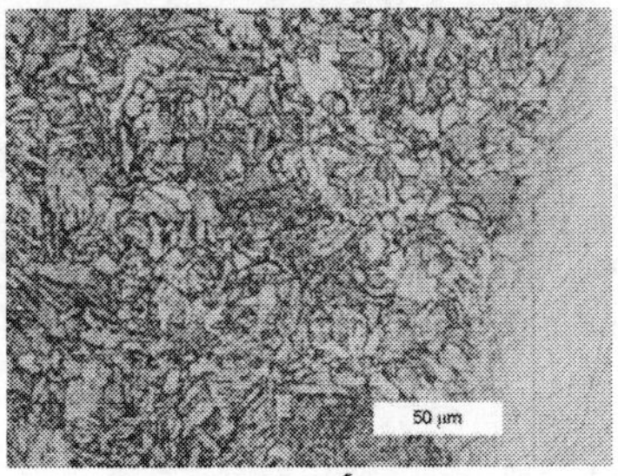

b

Fig. 5. The zone of transition from base metal to build-up metal: a – before heat treatment; b – after heat treatment.

The micro-hardness of the base metal after heat treatment slightly reduces to 161...163 HV. The micro-hardness of building-up is: the first layer - 351...352 HV, the second layer - 383…385 HV; the last top layer, – 424...435 HV.

4 Conclusion

The highest value of residual stresses is defined on the inner surface of the branch pipe saddle in the contact zone of the "branch pipe-stem". The conducted research al-low to reveal a high level of compressive residual stress in building-up of branch pipe saddles of high pressure valves – on the order of -541,5 MPA.

It was confirmed a favorable effect of heat treatment, which improves the homo-geneity of the building-up microstructure and the adjoining layers of valve material.

Homogeneous microstructure provides reduce of residual stresses level by 50%. It has a beneficial effect on the strength of built-up layer at the areas of "branch pipe-stem" contact and ensures the reliability of the pressure-tight joint.

The study confirmed the high degree of safety, which ensures reliability of struc-ture enhanced by building-up within a specified period of operation. Compressive stresses prevent the development of cracks near possible stress concentrators.

Barkhausen noise method can be used for nondestructive examination (NDE) of build-up ferromagnetic materials, including building-up of branch pipes of high pres-sure valves.

Based on the conducted studies the serial production of the angular high-pressure lock valves DN 25 was set.

Authors are grateful to the Director General of OJSC «IrkutskNIIhimmash» A. Kuznetsov, head of the laboratory of the tightness department A. Gorbunov, employees of O. Semenova, A. Lantin for supporting this research.

References

1. S. Zaydes, A. Mashukov, A. Tatarinov, Vestnik IrGTU, **5 (112)**, 37 (2016)
2. S. Gaisin, S. Zaides, Vestnik IrGTU, **6 (89)**, 45 (2014)
3. G. Gridin, Mater. of Sci. and tech. conf., Irkutsk, **IX**, 49 (2011)
4. G. Gridin, Chemical and Petroleum Engineering, **10**, 26 (2011)
5. E. Eremin, A. Losev, Welding International, **3**, 11 (2015)
6. Standard RF 053-2008
7. V. Pogodin, *Shut-off valves for high operating parameters. Research and design* (BrGU, Bratsk, 2016)
8. K. Ahmadi, A. Savilov, Int. J. of Machine Tools and Manufacture, **89**, 208 (2015)
9. A. Nikolaev, IOP Conf. Ser.: Mater. Sci. Eng., **177**, 012080 (2017)
10. V. Hauk, *Structural and residual stress analysis by nondestructive methods: evaluation, application, assessment* (Elsevier, Amsterdam, 1997)
11. L. Suominen, M. Khurshid, J. Parantainen, Procedia Eng. 5th Fatigue Design Conf. **66**, 181 (2013)
12. E. Nikolaeva, D. Vlasov, IOP Conf. Ser.: Mater. Sci. Eng., **177**, 012113 (2017)
13. E. Nikolaeva, D. Nikulin, MTSAM, **2 (50)**, 73 (2016)
14. E. Nikolaeva, E. Gridasova, V., Izv. SSC RAS, **17 (2-1)**, 125 (2015)
15. E Nikolaeva, Materials Science Forum, **870**, 500 (2016)
16. E. Nikolaeva, Izv. SSC RAS, **15 (6-2),** 428 (2013)
17. A. Karamyshev, A. Nikolaev, ISTU Bulletin of Youth, **3**, 8 (2016)

 DOI: 10.1051/matecconf/201712906007

The parameter's optimization of the meteorological rocket's correcting engine to reduce the zone of the used blocks' falling

Nicholas Mitiukov[1,*], *Timothey* Kasatkin[2], and *Guzalia* Samigullina[3]

[1]Moscow Economic Institute, 109390 Moscow, Artyukhinoy st. 6/1, Russian Federation
[2]Kalashnikov Izhevsk State Technical University, Graduate School, 426069 Izhevsk, Studencheskaya St. 7, Russian Federation
[3]East-Europe Institute, 426008 Izhevsk, Pushkinskaya St. 268, Russian Federation

Abstract. On the example of the meteorological rocket MMR-06, the optimum parameters of the trajectory correction engine to reduce the zone of the used blocks' falling was selected. It is shown that the optimum mass and optimal maneuvering time are achieved for engines with the medium-thrust. Low-thrust engines, despite the minimum weight, provide a long maneuver time and because of their large number, the total mass increases. The large-thrust engines are very heavy and capable of causing dangerous lateral overloads.

1 Introduction

The Meteorological rockets are widely used for meteorological and geophysical researches. They have the greatest value in the study of the upper layers of the atmosphere, inaccessible for ordinary balloon atmospheric sounding [7]. The effectiveness of the flight task of a meteorological rocket is determined by the time of its presence on the altitude range [5, 8]. And in this case, the issues of public safet come into conflict with the flight task [2].

Meteorological rockets, as a rule, are triggered on a ballistic trajectory with large initial angles of pitch (about 85-90°), and are not corrected in flight. In this case, even when reached the upper layers of the stratosphere, the size of the exclusion zone, into which the blocks or fragments of the rocket can fall, do not exceed a dozen kilometers. However, in the study of certain high-altitude echelons this is not always justified. Thus, the optimal program for researching the upper layers of the stratosphere, in order to reach the desired level and stay there for a maximum time, the rocket must to have an angle of pitch of about 5-10° at the entrance to the lower boundary of the echelon. For such parameters, the range of the rocket, and hence the size of the exclusion zone, should be about 200 km and above. But the launch of meteorological rockets is usually the most necessary for a densely populated European part of Russia and therefore the size of the exclusion belt for the possible dropping of rocket's blocks and parts is very limited [2].

The trajectory correction engines may to be one of the ways to overcome this crisis. In this case, it is possible to launch the rocket with pitch angles of 85-90°, and when entering the required altitude level, it can to be corrected to 5-10° (Fig. 1).

The task of this research was based on optimization of parameters of the trajectory correction engine for an uncontrolled meteorological rocket to reduce the size of the exclusion zone.

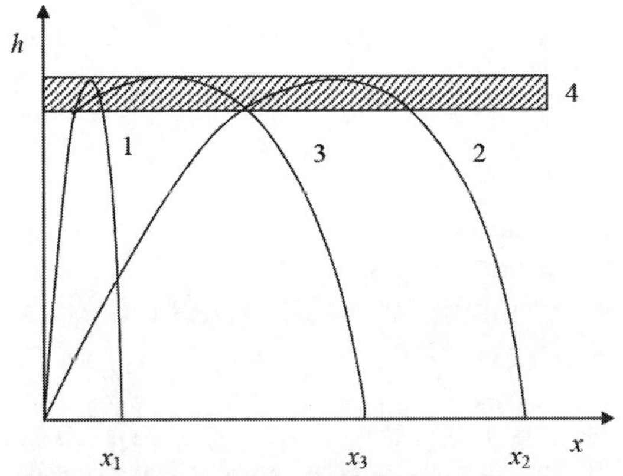

Fig. 1. The trajectories of the meteorological rockets (h – height, x – range): 1 – traditional rocket (range x_1); 2 – rocket for investigation of a certain altitude level (range x_2); 3 – rocket with a trajectory's correction (range x_3); 4 – examined altitude level. Note: x is a maximum range of a missile's fragments fall.

2 Research methodology

The methods of optimization theory were used in this work. There was a virtual design of the engine for the correction of the trajectory of the MMP-06 meteorological rocket with variation of initial parameters and determination of the mass and ballistic parameters [4]. The objective function was the condition of minimum mass for a given maneuvering time (changing the pitch angle from 85° to 10°).

* Corresponding author: nico02@mail.ru

3 Virtual engine design

The main dimensions of the engine are shown in Fig. 2. Usually unguided rockets are stabilized on the trajectory by rotation. The rotation period of the rocket determined the operation time of the engine (size $d_1 - d_2$). If the operating time exceeds a quarter of a revolution, there were unwanted lateral impulses. Therefore, each engine starts to work when there was 45° before the shooting plane and ends when it is turned more the 45° after the shooting plane. The outer diameter of the solid propellant charge (d_1) and its length were determined from the condition of the arrangement. Thus, two variable parameters that determined the thrust and appearance of the engine were obtained: the pressure in the combustion chamber, which affects the diameter of the critical section of the nozzle ($d_{кр}$) and the initial combustion surface, which affects the dimensions of the solid propellant charge (and the length L_3).

Since the effect of the appearance of the correction engine on the mass-dimensional characteristics is not unambiguous, the virtual design of the engines was carried out with a variation of the chamber pressure and the surface of combustion area.

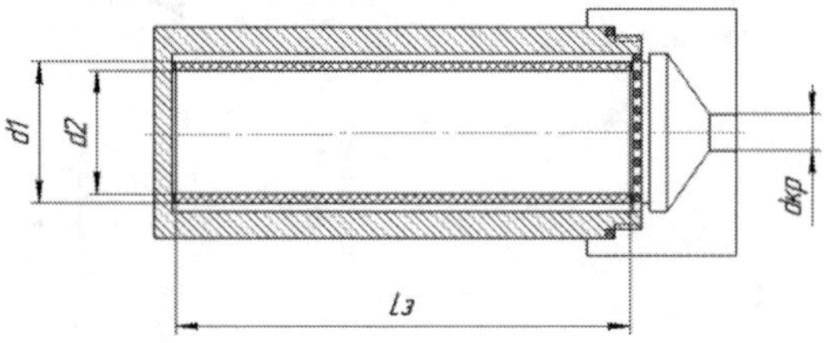

Fig. 2. Calculation scheme of the correcting engine.

4 Determination of the required number of the engines

Since the impulse from one engine was too small for the complete correction of the trajectory, it was necessary to determine the number of pulses equal to the number of engines.

The engine starts its operation at the time t_0, when the heeling angle is $\varphi = -45°$ before the shooting plane. The time to enter to the nominal parameters is comparatively small to the operation time of the engine, therefore the cyclogram of the thrust P can be considered rectangular, since the turn-off time is also too small, the engine stops operating at the time $t_к$, when the heeling angle is $\varphi = 45°$. However, for the correction pulse, the thrust it is not important, but the projection of thrust on the plane of firing is more important (Fig. 3):

$$P_b = P \sin \varphi. \tag{1}$$

To determine the required number of pulses (and the engines), there was produced the numerically integrated of the equation of rotation of the rocket relative to its center of mass:

$$\frac{d^2\theta}{dt^2} = \frac{1}{\Im}\left(P_b \ell_{т} - R\ell_{a}\right), \tag{2}$$

where θ – is the pitch angle; $\Im$ – is the moment of inertia of the rocket relative to its center of mass; $\ell_{т}$ – the thrust arm of the correcting engine (the distance between the point of application of thrust and the center of mass); R – aerodynamic lifting force; ℓ_a – the arm of the lifting force (the distance between the center of mass and the center of pressure).

The cyclogram of correcting force for one variant of the engine's design is shown in Fig. 4. Since the correcting engines are uniformly distributed around the perimeter of the shell, one engine is being run after the next, which is at this moment in the most suitable position, and it is switched on. At the altitude where the trajectory is corrected, the density of air is too small, and the value of the lifting force is not large [1], [3]. Therefore, it can perform the task of stabilizing the missile, and it is necessary first to give pulses to increase the angular velocity, and then to reduce, so that at the end of the cycle of work the angular velocity is zero (Fig. 5). In this case, the law of pitching is obtained as it is shown in Fig. 6 [6].

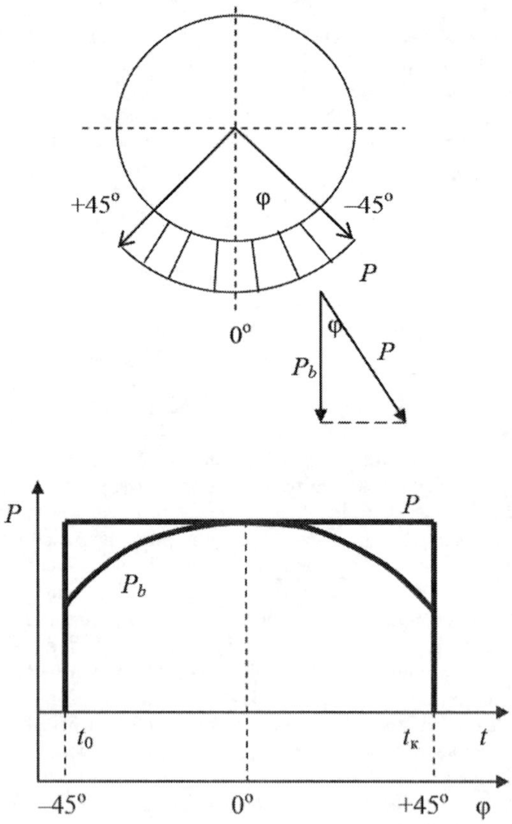

Fig. 3. Working cyclogram of the one correction engine.

DOI: 10.1051/matecconf/201712906007

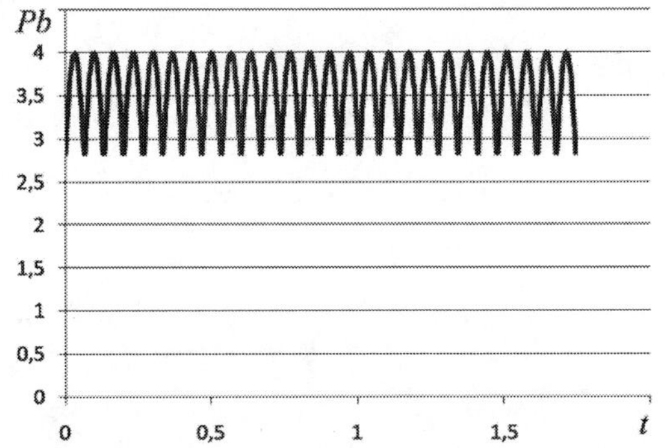

Fig. 4. The dependence of the thrust P_b (H) on the time t (c) of one of the engine variants.

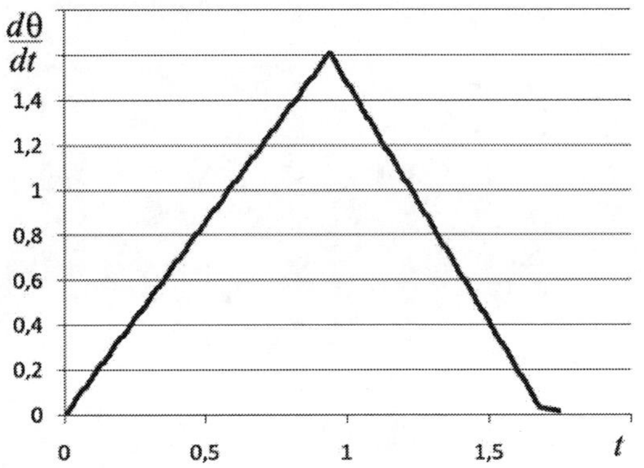

Fig. 5. Dependence of the angular velocity $d\theta/dt$ (rad/s) on the time t (c) of one of the engine variants.

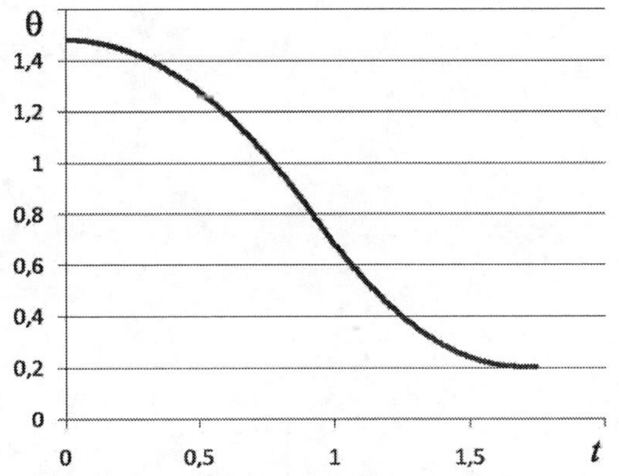

Fig. 6. Dependence of the pitch angle θ (rad) on the time t (c) of one of the engine variants.

5 Solving the problem of optimizing the appearance of the engine

Most clearly, the calculations results are presented in the form of a three-dimensional surface. In Fig. 7 the surface of the engine mass from the variable parameters p and S is shown. It can be seen that the chamber pressure has relatively little effect on the mass. On the other hand, the effect of the combustion area S is not unique. The quantity of the large-thrust engines is not much, but they are heavy themselves, which is displayed on the plane of Fig. 7 as a sharp increase in mass for large areas. On the other hand, the low-thrust engines are light, but a great number of these engines is necessary for correction, which also leads to an increase in the total mass in the region of small burning surfaces S. And thus, medium-thrust engines in the "failure" of the surface in the region $S = 3.847$ cm^2.

Fig. 8 shows the correction time surface t from the variable parameters p and S. In the topside view, this surface is shown in Fig. 9. It can to divided into two critical areas. Region I is the region of low-thrust engines. At a certain maximum value of t, the system passes into a critical state, when the momentum is insufficient to overcome the aerodynamic moment. Area II is the area of powerful engines. At some minimum critical value of t, dangerous lateral accelerations start acting on the rocket, which can lead to destruction. The isoline of the optimal turn times is between them. In this case it was taken equal to 1 second. As a result, the optimal geometry of the engine is obtained in the lower part of the plane, at the intersection of the lines 1 s and 3.847 cm^2.

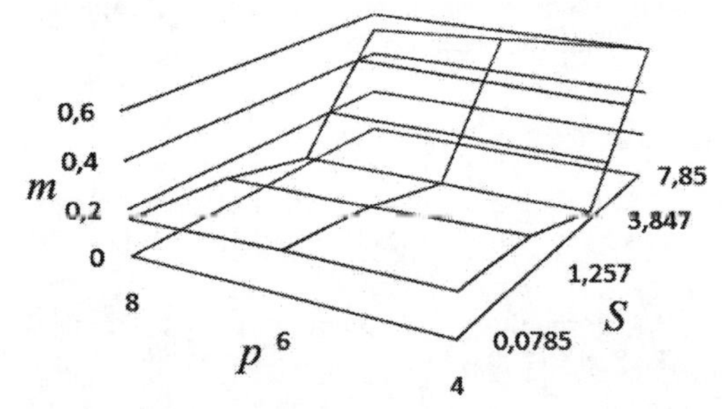

Fig. 7. Mass of correction engine m from the chamber pressure p and the burning area S.

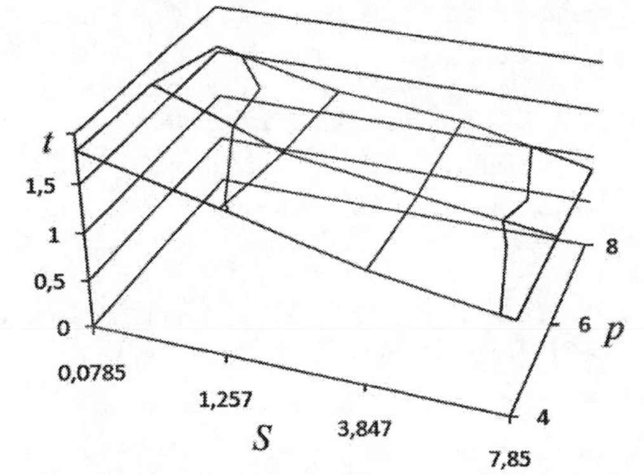

Fig. 8. Time of trajectory correction t from the chamber pressure p and combustion area S.

DOI: 10.1051/matecconf/201712906007

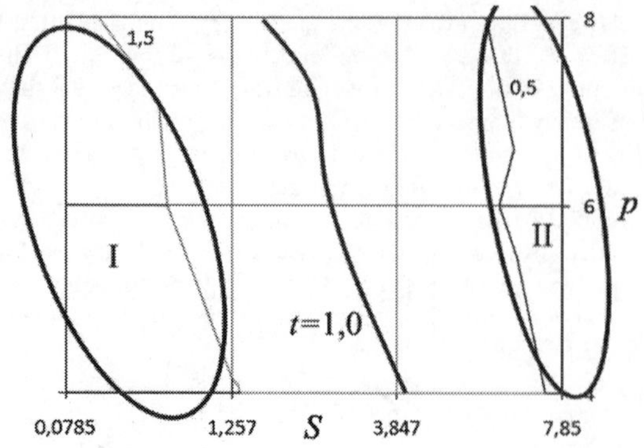

Fig. 9. Time of trajectory correction *t* as isolines.

5 Conclusions

On the example of the MMR-06 meteorological missile, the optimal parameters of the trajectory correction engine were chosen to reduce the size of the falling zone of the used blocks. It is shown that the optimum mass and optimal maneuvering time are achieved for medium-thrust engines. Low-thrust engines, despite the minimum weight, provide a long maneuver time and because of their large number, the total mass increases. The engines of large traction themselves are heavy and capable of causing dangerous lateral overloads.

References

1. T.A. Kasatkin, European Journal of Technology and Design, **4** (2014). DOI: 10.13187/ejtd.2014.6.173
2. E.G. Klimova, Yu.N. Morokov, G.S. Rivin, A.I. Borodulin, Atmospheric and Oceanic Optics T.18, **5-6** (2005)
3. N.V. Mitiukov, Y.V. Ganzy, E.L. Busygina, European Journal of Technology and Design, **2** (2014). DOI: 10.13187/ejtd.2014.4.81
4. N.W. Mitiukov, S.N. Khramov, *Gas-jet pulse systems* (Izhevsk, 2000)
5. G. Morar, E.L. Busygina, N.W. Mitiukow *Wschodnie partnerstwo – 2015* (Przemyśl, 2015). DOI: 10.17686/rusnauka_2015-200362
6. Yu.P. Nagovitsina, Pervy shag v nauku, **1** (2017)
7. V.A. Yushkov, V.V. Fedorov, Yu.N. Gvozdev, V.N. Bespalov, V.M. Kuznetsov, Mir Izmerenii, **9** (2011)
8. G. Morar, New University, Ser. Technical Science, **5-6** (2015) DOI: 10.15350/2221-9552.2015.5-6

DOI: 10.1051/matecconf/201712906008

Fracture strength for metal structure of a crane

Vladimir Mironov[1,2], *Olga* Lukashuk[2,*], and *Dmitriy* Ogorelkov[2]

[1]Institute of Engineering Science, RAS (Ural Branch), Yekaterinburg, Russia
[2]Ural Federal University, Yekaterinburg, Russia

Abstract. Anthropogenic catastrophes, as a rule, are caused directly by severe violations of operating rules set for technical systems and devices. One accident is considered in the article, where the rotating section of a tower crane fell down. A conjecture based on calculations is proposed as to what could have possibly caused the fracture of a bolted joint between the slewing unit and mast shell.

1 Introduction

The problem of fracture strength of metal structures is topical in many fields of technology. And especially so in the engineering of lifting and carrying machines, where, along with the demand for lower metal intensity and power inputs, excessive requirements are set for product safety [1-4]. Therefore, it is reasonable to supplement typical calculations made for nominal loads with those for bearing capacity of a structure under extreme conditions when fractured elements start to appear.

Excessive loads could be caused by a fall of lifted load, weather cataclysms, severe violations of operating rules, etc. As a rule, such loads are dynamic by their nature and often lead to fracture of a structure, material losses and human deaths. The goal of auxiliary calculations in such a case is to determine ultimate loads and check the strength of the structure.

As a rule, the emergence of anthropogenic hazards is related to various discrepancies appearing due to failure to comply with different requirements. The article analyzes the situation which happened with a tower crane POTAIN MD 265 B J12 operating at a regular angular velocity and without a load. The analysis of operating conditions and the pattern of crane fracture has revealed several major discrepancies: fatigue cracks in the shell under the rotating section (Fig. 1, a), a full-circle trail where the rotating section were touching the base (Fig. 1, b), dilapidated teeth of the slewing ring (Fig. 1, c). One could not have missed any discrepancy from that list during day-by-day visual examination. Jamming of the slewing mechanism could be caused by any factor mentioned above.

According to the ISO 9001 standard [1], when a person in charge of crane operation learns about any discrepancy, he must initiate the correction procedure to determine the causes of the violation and eliminate them.

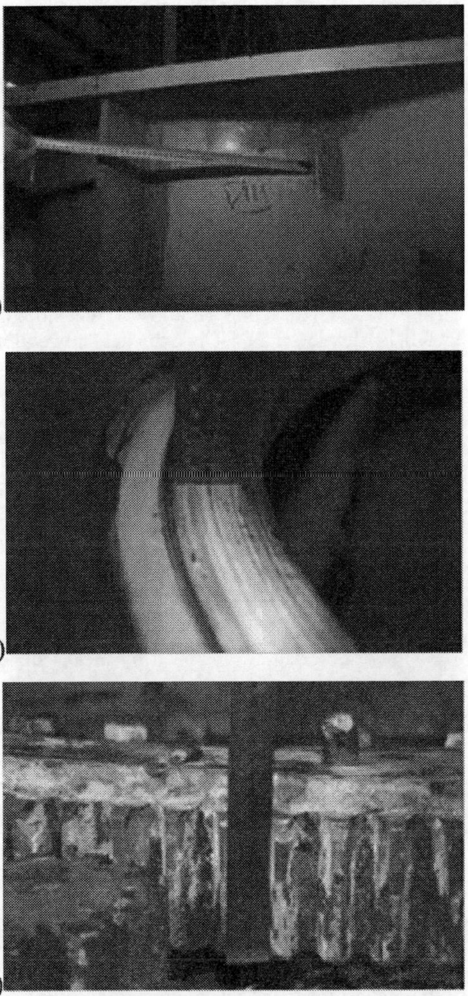
a) b) c)

Fig. 1. a) welded cracks, b) full-circle trail, c) slewing ring.

---

[*] Corresponding author: oldim96@mail.ru

DOI: 10.1051/matecconf/201712906008

The shell body shows signs of welded cracks, thus indicating the correction which was made. But there is no guarantee of its helping to maintain normal conditions for the operation of balls in the slewing unit. The second and third discrepancies could have been caused by the repair work, which was introduced by means welding open fatigue cracks, and led to wedging of the slewing mechanism. Other factors might have played a role in jamming, like, loose bolts by which the fixed part of the slewing unit was fastened to the shell or incomplete set of those.

It is supposed that bolted flange joint of the rotating section was fractured due to wedging of the slewing unit. Notwithstanding the real reasons behind jamming of the slewing mechanism, it is necessary to find out whether or not it could be the cause of such an accident. The goal of the work is to determine dynamic loads when wedging of a slewing mechanism occurs and check the strength of bolts similar to the ones fractured during the accident. To check the working hypothesis on what caused the rotating section of the crane to fall, the following tasks were solved:

- dynamic models were built, and peak forces determined, for the motions of the jib and counterweight (plus its frame), all considered as reduced localized masses;
- severing reduced forces, bending and torsion moments applied to bolting joints were evaluated;
- a group bolted joint was evaluated, with equivalent stresses determined and pie chart of their distribution among bolts of the joint plotted;
- finite element analysis for the counterweight frame was carried out to check the rigidity of the structure and establish whether or not it is expedient to reinforce the bolted joint unit;
- the conclusion was made on what could have possibly caused the rotating section to fall, and one proposed variant of engineering changes was considered.

2 Jib and counterweight vibrations after slewing unit wedging

Fig. 2 shows schematics of the crane rotating section. A lifting jib of $L_J = 65\,m$ in length and a counter-jib of $L_C = 18\,m$ in length are fastened by means of flapping hinges to a column of $h = 2{,}4\,m$ in height and supported by guy lines. The accident occurred when the crane was rotating at an angular velocity of $\omega = 0{,}08378\ rad/s$ and with no load on its hook. A 12,83-ton mass of the lifting jib is reduced to its end, and $Q_{JR} = 31{,}5\ kN$. A 9,83-ton mass of the counter jib is reduced to the load, and, with a 18-ton mass of the counterweight taken into account, $Q_C = 200{,}7\ kN$. A circle diameter d of the bolts which attach the base of the rotating section is equal to 1494 *mm*. Elastic forces F_J and F_C, which are to be determined, act in a horizontal plane and are caused by vibrations of the jib and counterweight which follow wedging of the slewing unit.

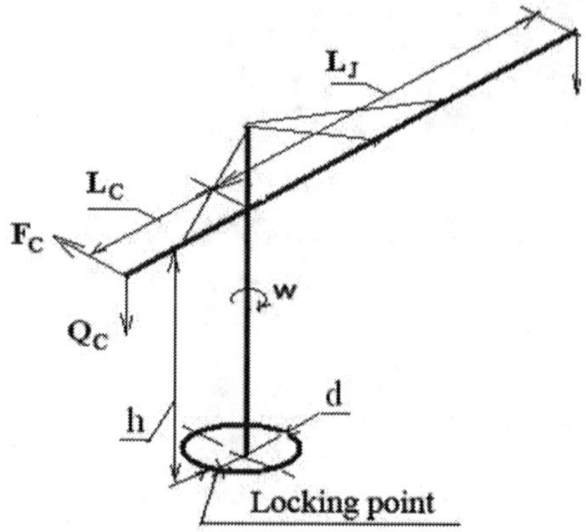

Fig. 2. Schematics of the crane rotating section.

Considering the trajectory curvature as negligible, let us select the scheme of translational motion for localized jib and counterweight masses: $m_J = 3{,}2075\ tn$ and $m_C = 20{,}4587\ tn$.

Using D'Alembert's (kinetostatics) principle, we establish the equation of motion in the form of equilibrium condition for a mass m and with inertia forces taken into account. It would be defined as following:

$$m\ddot{x} + cx = 0. \qquad (1)$$

Having defined an angular frequency as $p = \sqrt{c/m}$, we can reduce equation (1) to the canonical form of

$$\ddot{x} + p^2 x = 0.$$

The solution of a homogeneous differential equation with constant coefficients takes the form of [5]

$$x = C_1 \cdot \sin(pt) + C_2 \cdot \cos(pt),$$

where C_1, C_2 are integration constants determined on starting conditions. For $t = 0$: $x_0 = 0$, $\dot{x}_0 = \omega R$, where R is the jib or counter-jib length.

The jib and counterweight have a similar pattern of motion, and specific parameters are determined by an initial speed and angular frequency

$$x_J = \frac{\dot{x}_{0J}}{p_J}\sin(p_J t),\ x_C = \frac{\dot{x}_{oC}}{p_C}\sin(p_C t). \qquad (2)$$

The initial speed at the moment of wedging equals the linear motion speed of the jib end $\dot{x}_{0J}$ and counterweight $\dot{x}_{0C}$:

$$\dot{x}_{0J} = \omega \cdot L_J = 5{,}446\ m/s,\ \dot{x}_{0C} = \omega \cdot L_C = 1{,}508\ m/s,$$

where ω is the linear speed of the slewing mechanism.

In the first version of calculations, the rigidities of the jib and the counter-jib in a horizontal plane were determined by the formula (well-known from the

DOI: 10.1051/matecconf/201712906008

strength of materials) for a girder restrained at one of its ends: $c = 3EJ / L^3$.

The corresponding rigidities were

$$c_J = 3{,}0696\ kN/m;\ c_C = 1381{,}6847\ kN/m,$$

while angular frequencies – $p_J = 0{,}9783\ s^{-1}$ and $p_C = 8{,}218\ s^{-1}$, and periods of jib and counterweight vibrations – $T_J = 6{,}42\ s$ and $T_C = 0{,}76\ s$. Then, using the second formula in (2), we find the peak travel of the counterweight for $t_* = T_C / 4 = 0{,}19\ s$, and, after that, the maximal force

$$F_C = c_C \cdot x(t_*) = 253{,}5\ kN.$$

At that moment, the maximal force applied to the jib $F_C(t_*) = 3{,}16\ kN$.

Then, force factors reduced to the plane in which severing of the bolts occurred were determined:

- bending moment in the vertical plane directed to the jib

$$M_1 = Q_C L_C - Q_{JR} L_J = 1565{,}1\ kNm;$$

- bending moment in the perpendicular plane

$$M_2 = (F_C - F_J)h = 1051{,}4\ kNm;$$

- tension moment in the horizontal plane

$$M_T = F_C L_C + F_J L_J = 4357{,}6\ kNm;$$

- cross-sectional severing force

$$Q_{sev} = F_C - F_J = 250{,}34\ kN.$$

3 Calculation of bolted flange joint between slewing unit base and jib

Assume that wedging of the mechanism occurs on the side of the counterweight close to bolt 24 (Fig. 3).

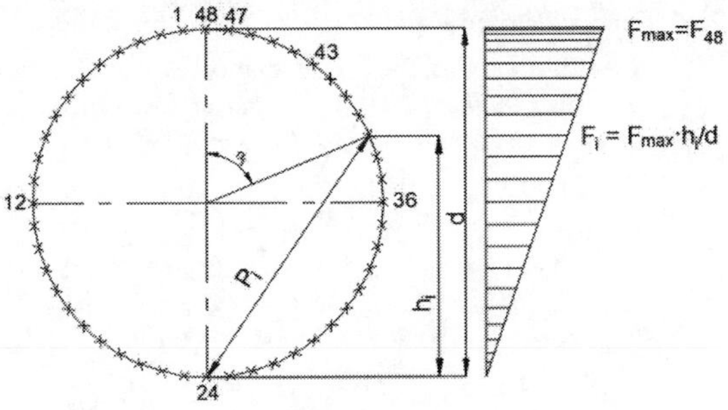

Fig. 3. Diagram of bolt locations and tensile strain distribution.

The direction 24-48 is a trail of the plane of the moment M_1 action, and the direction 12-36 is the same for the moment M_2. Suppose, as a common case, that a tensile strain in bolts F_i is in direct proportion to a distance h_i. Then, on the condition of equality between the moments of external forces and the total moment of forces in bolts, tensile strains are determined for all the bolts in the group joint.

Next, normal stresses at the cross-section near the thread bottom along which the bolts broke during the accident were found. It was established that maximal calculated stress is in bolt 43: $\sigma_{43}(M_1, M_2) = 341 MPa$. Since the joint had been loosened, tightening stresses and friction forces are neglected.

Then, it is necessary to allow for the concentration of stresses in the thread where fracture occurred (Fig. 4).

Fig. 4. Fractured bolt.

Bolts of 10.9 strength grade are made from high-strength steel with the yield point (yield stress) at $\sigma_Y = 900\ MPa$ and the breaking point (ultimate stress) at $\sigma_U = 1040\ MPa$. Concentration factor in the thread for such steels is estimated by different sources to be in the range $k \in [3{,}8 - 5{,}4]$, then, for k = 4, the stress calculated for the most loaded bolt $\sigma_{C43} = k\sigma_{43}(M_1, M_2) = 1364 MPa$.

Such stress noticeably exceeds the ultimate stress for the bolt material and is somewhat higher than the dynamic yield stress [6, 7] $\sigma_{YD} = 1{,}3\sigma_U = 1350 Mpa$. But, to make a fair assessment of the bolt joint strength, it is necessary to allow for tangential severing stresses at the calculated cross-section along the groove bottom. The joint effect of normal and tangential stresses was evaluated on the fourth theory of strength

$$\sigma_e = \sqrt{\sigma_C^2 + 3\tau^2}.$$

Fig. 5 shows a circular diagram for equivalent stresses in the bolted joint for the maximal F_C. The stresses calculated for bolts 10÷32 of the black zone (Fig. 5) exceed static yield stress of the material, and, for the zone of bolts 4÷38, they are even higher than the dynamic yield stress. The bolt which turned out to be loaded the most was bolt 44. Guaranteed fracture of these bolts would lead to fracture of the whole joint during the first semi-cycle of counterweight vibrations.

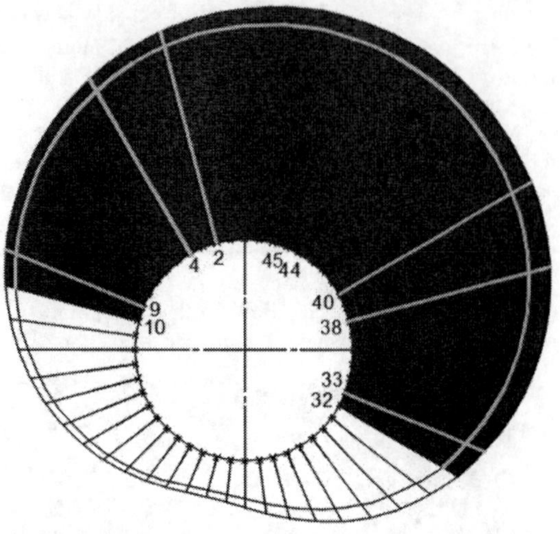

Fig. 5. Circular diagram of equivalent stresses.

Next, one needs to consider the necessity of strengthening the flange joint or developing a new design for the section which joins the slewing unit base and mast. To achieve these goals, an elemental analysis of the counterweight frame was carried out in the software environment of SolidWorks for the peak force F_C. The software allowed to build a solid model, and then to study it in the SolidWorks Simulation add-on using its finite element analysis tool [8].

4 Finite element calculation of counterweight frame

The counterweight frame material S235 is analogous to a domestic steel St3sp5 with the yield stress of 235 *MPa*. The distribution pattern for calculated normal stresses in the counterweight frame in relation to the action of the peak force F_C is shown in Fig. 6. The maximal normal bending stress at the location of frame fastening is 283 *MPa*. Therefore, strengthening the flange joint could lead to the limit yielding state of the counterweight frame in the case of slewing unit wedging.

Fig. 6. Distribution of stresses on inertia force.

The calculated frame rigidity $c_C = 903,4087\,kN/m$ is somewhat lower than that obtained using formulas from the strength of materials. Accordingly, calculated equivalent stresses in bolts (bounded by the internal curve in Fig. 5) turned out to be lower, as well. This correction did not make any significant difference in terms of the conclusions on the cause of the accident when wedging of the slewing unit occurred. The zones of yield (bolts 9-33) and fracture (bolts 2-40) became narrower, and bolt 45 turned out to be loaded the most.

5 Conclusion

Locking of the slewing init of an unloaded crane which operates at its working angular velocity is an unacceptable catastrophic event. The forces of gravity and inertia of jib and counterweight masses cause, in such a bolted flange joint as the one considered here, such stresses which considerably exceed ultimate breaking loads for bolts of 10.9 strength grade.

Since practically all possible ways of strengthening the bolts have been tried out, new design solutions are needed to mount the slewing unit. For example, a ring belt to resist the severing force or spot welding of a flange joint might be introduced. But, it should be remembered that an increase in strength of a flange joint could lead to fracture of the counterweight metallic structure.

The only sure way of securing the work capacity of a crane is to adhere closely to the operating rules and eliminate emergency locking of the slewing unit.

References

1. BS EN ISO 2001:2008. Quality Management Systems – Requirements. 4th. Ed. (Management Centre, Brussels, 2008)
2. S.A. Kazak, *Bridge Crane Dynamics* (Mashinostroenie, Moscow, 1968)
3. D.V. MacCollum, *Crane hazards and their prevention* (ASSE, USA, 2005)
4. H. Ouyang, Mech. Syst. Signal Pr., **25**, 2039 (2011)
5. V.I. Smirnov, *Course on Higher Mathematics* (Nauka, Moscow **2**, 1974)
6. H.A. Rahmatulin, Ju.A. Dem'janov, *Strength under Intensive Short-term Loads* (Izd. fiziko-matematicheskoj literatury, Moscow, 1961)
7. I.G. Emel'yanov, V.I. Mironov, A.V. Kuznetsov, AIP Conf. Proc., **1785**, 040014 (2016)
8. A.A. Aljamovskij, A.A. Sobachkin, E.V. Odincova, etc., *Computer-aided Design in SolidWorks for Engineering Purposes* (BHV-Peterburg, Saint Petersburg, 2005, 800 p.)

MATEC Web of Conferences **129**, 06009 (2017)
ICMTMTE 2017 DOI: 10.1051/matecconf/201712906009

Influence of the track gradient on the contact temperature at the wheel-rail zone for open-pit locomotives

Ambartsum Keropyan[1], *Alla* Gerasimova[1,*], and *Kirill* Goloshapov[1].

[1]National University of Science and Technology "MISIS", street the Lenin prospect,4,Moscow,119049, Russia

Abstract. For open-pit locomotives, application of the existing equations for calculation of the temperature for main-line railway transport is inappropriate, since it doesn`t take into account the specific conditions of the open-pit mining, such as increased track gradients (up to 60‰, in another words 60m per 1 km). Therefore we proposed to calculate the temperature in the contact zone at the wheel-rail interface according to the derived equation, which takes into account the track gradient of the open pit`s rail truck. The calculation results, which were made according developed computational technique, show satisfactory repeatability with the experimental results. As the result of theoretical and experimental investigations have been discovered, that the range of the ruling gradient of the transport incline affects on the temperature changes in the wheel-rail contact zone and characterized by temperature factor of the rail track gradient k_{tgr}.The results of calculations, made according developed method, show satisfactory repeatability with the experimental results. The approximation error of the mining actual operating conditions factor K_{mc} in the range from 25‰ to 45‰ is less than 5%.

1 Introduction

One of the criteria for energy processes estimation, ongoing in the contact zone at the wheel-rail system while torque transmission, is a temperature, which also is indicative of stress-related characteristics changes of materials of interacting solids, as consequence, it characterizes their serviceability and durability. Recommended equation for calculation of the temperature in the contact zone at the wheel-rail interface for general-purpose railways [1,2] after some manipulations can be represented as:

$$\theta = \frac{0{,}0576K_{new/worn} \cdot f \cdot U_{rs} \cdot \sqrt{a_1 \cdot V_m}}{\lambda (2a)^{0{,}5} \cdot A}, \qquad (1)$$

where θ - temperature in the contact zone at the wheel-rail system, °C;

$K_{(new/warn)}$- contact conditions index, accordingly for the new rail profile (new) and for the warn-in rail profile (worn), J

$K_{new} = \frac{4}{3}\pi a^2 b \sigma_{max}$ - for the new rail profile;

$K_{worn} = \pi a^2 m \sigma_{max}$ - for the worn-in rail profile;

f - coefficient of friction;

U_{rs} - relative slip ratio, %,

$U_{rs} = (V_{slip}/V_m) \cdot 100$,

a_1 – thermal diffusivity, m²/s, for steel $a_1 = 11{,}9 \cdot 10^{-6}$,

λ - thermal conductivity, W/(m*°C), for steel λ=44,0

a- half-width (major semiaxle) of the contact zone, m

b- minor semiaxle of the contact zone, m

m - contact zone width (for worn-in rail profile), m

A - contour area of the contact, m²

V_m - locomotive`s speed of motion, m/s

V_{slip} - profile sliding velocity of the locomotive`s wheel tread, m/s

σ_{max} - maximum bearing pressure at the contact zone, MPa

Purpose of the work is to investigate the influence of the ruling gradient of the industrial railway on the temperature in the contact zone at the wheel-rail interface.

2 Objects and methods of research

For open-pit locomotives, the application of Eq. (1) is inappropriate as it does not take into account the open-pit`s special conditions- increased track grades, up to 60‰. That is why we propose to calculate the temperature in the contact zone at the wheel-rail interface according the further formula:

$$\theta = \frac{0{,}0576K_{new/worn} \cdot K_{mc} \cdot U_{rs} \cdot \sqrt{a_1 \cdot V_m}}{\lambda (2a)^{0{,}5} \cdot A}, \qquad (2)$$

where K_{mc} - factor, which takes into account mining actual operating conditions.

The nominal dimensions of the contact area can be calculated by Hertz-Belyayev equation [3,4], however, while calculating were used more accurate data, obtained experimentally during working period [2].

The mining actual operating conditions factor K_{mc} can be defined as:

$$K_{mc} = (1 + k_{tgr} \cdot i) \qquad (3)$$

* Corresponding author: allochka@rambler.ru

where k_{tgr} - temperature coefficient, which depends on the track gradient,
i- inclination of the track gradient, ‰
Taking into account Eq. (3), we deduce:

$$\theta = \frac{0{,}0576K_{new/worn}\cdot f\cdot(1+k_{tgr}\cdot i)U_{rs}\cdot\sqrt{a_1\cdot V_m}}{\lambda(2a)^{0,5}\cdot A} \quad (4)$$

Investigations, presented in the scientific work [2], were made under the conditions of general-purpose railway tracks, for 80% of them maximum incline of track gradient is not more than 9‰ [5]. That is why, for track gradients less than 10‰ we apply the temperature factor of the rail track gradient $k_{tgr}=0$. For such operating conditions $K_{mc}=1$ and the temperature can be calculated using the following Eq. (1).
Dependence of coefficient k_{tgr} on the process variables of wheel-rail interaction can be obtained from the Eq. (4)

$$k_{tgr} = \frac{\theta\cdot\lambda\cdot(2a)^{0,5}\cdot A - U_{rs}\cdot 0{,}0576\cdot K\cdot f\cdot\sqrt{a_1\cdot V_m}}{0{,}0576\cdot U_{rs}\cdot K\cdot i\cdot f\cdot\sqrt{a_1\cdot V_m}} \quad (5)$$

Experimental researches into the determination of k_{tgr} values were carried out on the tunnel locomotive "SCHÖMA–350" (Germany), on the base of JSK"Mosmetrostroy" company, in conditions of under construction subway stations. Were measured temperatures and slip velocities in the contact area at the wheel-rail system for different motion speeds and with different rail track gradients Fig.1. The temperature measurements were carried out by infrared temperature measuring instrument "testo 845" (thermovisor), sliding – by the procedure, contained in the engineering manual [6].

Temperature and sliding velocity measuring device for taking measurements in the contact zone at the wheel-rail system shown in figure 1.

a)

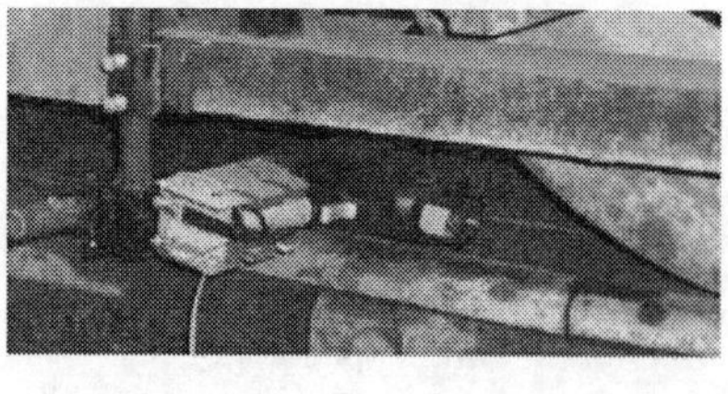

b)

Fig.1. Temperature and sliding velocity measuring device for taking measurements in the contact zone at the wheel-rail system: a) general view; b) unit of an infrared thermometer "testo 845".

The measured results are shown in Table 1.

Table 1. Experimental values of sliding velocity and temperature in the contact zone at the wheel-rail system for tunnel locomotive "SCHÖMA–350"

Input data		Measured data (average values)	
V_m,m/s(km/h)	i, ‰	U_{rs}, %	θ, °C
1,39 (5,0)	5,0	3,4	82,0
1,39 (5,0)	25,0	7,8	195,0
1,39 (5,0)	45,0	12,3	330,0
2,78 (10)	5,0	5,6	191,0
2,78 (10)	25,0	13,2	465,0
2,78 (10)	45,0	17,5	650,0

Values of the factors, needed for the formula evaluation (4), are given in Table 2.

Table 2. Values of the factors for calculations of the temperature in the contact zone at the wheel-rail system

P,k N	Kind of the rail profile	$a\cdot10^{-3}$, m	$A\cdot10^{-4}$, m²	K, J	a_1, m²/s	λ, W/(м·°C)
175	Worn-in	3,30	2,65	605	$11{,}9\cdot10^{-6}$	44,0

Equation (4) allows to take into account the influence quantity of several factors on the temperature θ in the contact zone at the wheel-rail system, such as contour area of contact A, axle loading P, resultant sliding velocity U_{rs}, friction coefficient f, locomotive speed V_m and value of rail track gradient i. The results of the experimental data treatment are shown in Table 3.
From the physical point of view, $K_{(new/worn)}$ is an amount of energy, which releases in the contact zone at the wheel-rail system, where friction coefficient is f=1 and resultant sliding velocity is U_{rs}=100%. On the assumption, that coefficient index $K_{(new/worn)}$ can be used in an impact analysis of different parameters (which are determining conditions of wheel-rail interengagement; on the amount of energy, released in-between coworking surfaces while traction mode motion [2]

Hereafter: $K_{(new/worn)}=K_{worn}=K$.

Table 3. Calculated values of coefficients k_{tgr} and K_{mc} according to experimental data

Input data		Experimental data		Data processing result	
V_m m/s(km/h)	i,‰	U_{rs}, %	θ, °C	$k_{tgr}\cdot10^{-3}$	$K_{mc}=1+k_y\cdot i$
1,39 (5,0)	5,0	3,4	82,0	≈0(0,05)	1,0
1,39 (5,0)	25,0	7,8	195,0	1,5	1,0375
1,39 (5,0)	45,0	12,3	330,0	2,5	1,1125
2,78 (10)	5,0	5,6	191,0	≈0(0,1)	1,0
2,78 (10)	25,0	13,2	465,0	1,33	1,0333
2,78 (10)	45,0	17,5	650,0	2,13	1,0959

DOI: 10.1051/matecconf/201712906009

In the work [2] are given values of K depending on axle loading P, on size of the contact contour area A and on the level of worn-in for conjugated surfaces at the wheel-rail system. It is also shown, that increase of the axle load is accompanied by the increase of coefficient K and, in the same time, by increasing of released energy at the contact zone. However, with increasing of the contact area, the specific value of released at the wheel-rail contact zone energy is substantially reducing. Accordingly, it may be affirmed, that the worn-in rail profiles (i.e. profiles, which ensure conformal contact) has notably less amount of released heat in the contact zone than the new.

By analyzing Equation (4) it is very easy to note, that with decreasing of the contact conditions coefficient K; of resultant sliding velocity U_{rs}; of the friction coefficient f and with increasing of the contacting area of two associated surfaces A, the temperature in the contact zone θ should decrease, and will be increasing with increase of the track gradient.

Received (with account for data, shown in Table 3) dependence of the temperature coefficient k_{tgr} on the track gradient is shown in Figure 2.

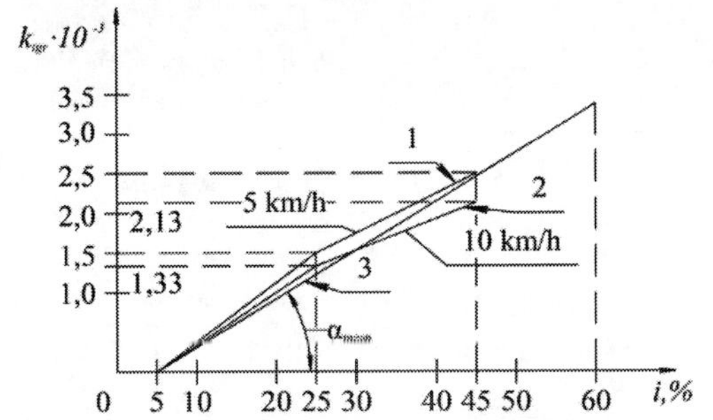

Fig.2. Change of the temperature coefficient k_{tgr} with inclination of the track gradient : 1,2- experimental curves; 3- approximating curve.

According to Figure 2 can be obtained approximation function k_{tgr}:

$$K_{tgr} = (i-5)\cdot tg\alpha_{mean}, \quad (6)$$

where α_{mean} –the mean value of gradient angle of approximation curve of the temperature coefficient k_{tgr}, determined by least square method.
Substituting this value in Eq.(3) we obtain:

$$K_{mc} = [1+(i-5)\cdot tg\alpha_{mean}\cdot i], \quad (7)$$

On rearrangement, this result can be written as:

$$K_{mc} = 1 + i^2\cdot tg\alpha_{mean} - 5\cdot i\cdot tg\alpha_{mean} \quad (8)$$

Taking into account the results, obtained after processing of the experimental dataset, we can define $tg\alpha_{mean}$:

$$tg\alpha_{mean}=6{,}3\cdot 10^{-5};\quad \alpha_{mean}=0{,}0036°. \quad (9)$$

Substituting values of Eq.(9) in Eq.(8) we obtain:

$$K_{mc} = 1 + 6{,}5\cdot 10^{-5}\cdot i^2 - 3{,}15\cdot 10^{-4}\cdot i \quad (10)$$

Figure 3 shows dependence diagrams of coefficient K_{mc} from the rate of rail track gradient.

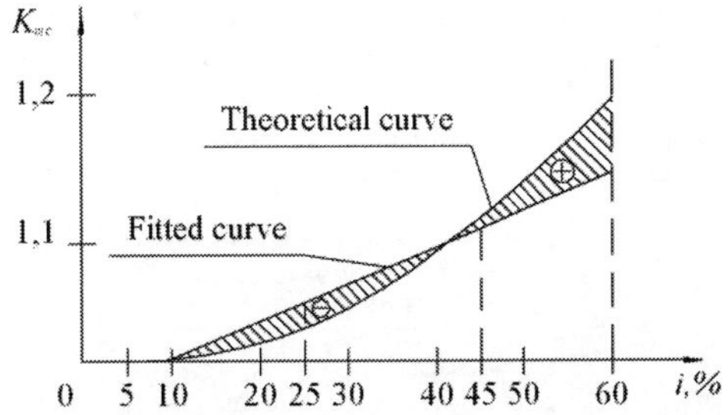

Fig.3. Dependences of theoretical and curve-fitted constants of mining conditions influence K_{mc} on inclination of the rail track gradient.

3 Results and discussions

Analysis of experimental results (presented in Table 3), at bound temperature of 300°C [7], shows, that the difference under a substitution of a cuspidal function with a linear one is a $\Delta K_{mc} = 0{,}05$, while the error is not more than 15°C or 15%.

In the Table 4 shown values of theoretical (calculated) and curve-fitted constants K_{mc} and given error of approximation Δ.

Table 4. Values of theoretical and curve-fitted constants K_{mc}.

i, ‰	5	10	25	45	60
$K_{mc.th.}$	1,0	1,00315	1,0315	1,1134	1,2079
$K_{mc.app.}$	-	1,0	1,048	1,106	1,15
Δ, %		- 0,315	+ 1,6	- 0,66	- 4,79

In the scientific researches [8, 9] have been shown, that under wheel-rail contracting accompanied with high thrust loads (typical for open-pit locomotives) deformations of surface roughness have plastic feature. In this respect, the nominal contact area of interacting surfaces can be defined by Hertz-Belyayev equations [4] and is taken to be equal to the size of the contour contact area [10].

In consideration of the results of complexes of science works, held by several researches [1,11,12], furthermore, taking into account studies of open-pit`s railways contamination , given in science works [13,14] and based on criteria that the temperature in the rail-wheel contact must be less then 300°C, which was determined in the works [7,15]. From Eq. 2 can be determined the assumed value of the allowable slip velocity U_{rs} of open-pit locomotives wheels.

$$U_{rs} = \frac{\theta \cdot \lambda \cdot (2a)^{0,5} \cdot A}{0,0576K \cdot K_{mc} \cdot f \cdot \sqrt{a_1 \cdot V_m}} \quad (11)$$

The Fig.4 shows dependences of allowable range of the wheelslip velocity on the rail track gradient at the temperature of 300°C, at 15km/h and under the axleload of 270kN for the new and worn-in locomotive's wheel profiles, plotted according to the Eq. (11)

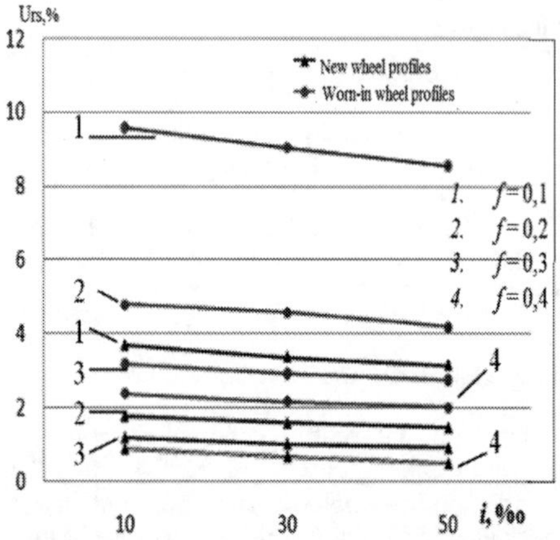

Fig.4. Charts of the allowable range of the wheelslip velocity for new and worn-in wheel profiles at the temperature of 300°C, at 15km/h and under axleload of 270kN for the friction coefficient range from 0,1 to 0,4.

4 Conclusions

1. Has been determined that the ruling track gradient value of the transport incline affects on the temperature at the wheel-rail contact zone and is characterized by temperature factor of rail track gradient k_{tgr}.
2. Value of the temperature factor of the rail track gradient k_{tgr} over the range 0 to 10‰ is k_{tgr}=0, and calculation of the temperature at the wheel-rail system for the open-pit mining conditions should be done according equations recommended for general-purpose railways.
3. With angle increase of the track gradient, the temperature at the wheel –rail system grows up.
4. The results of calculations by the suggested method, show reasonably repeatable results with the experimental. The approximation error of the mining actual operating conditions factor K_{mc} over the range 25‰ to 45‰ is less then ±5%.

The work was carried out with financial support from the Ministry of Education and Science of the Russian Federation in the framework of increase Competitiveness Program of NUST "MISIS", implemented by a governmental decree dated 16th of March 2013, № 211.

References

1. A.V. Chichinadze, *Calculation and investigation of external friction during braking* (Science, Moscow, 1967)
2. K.P. Flyachinskiy, *Effect of conditions of interaction of wheels and rail by frictional processes in the contact zone* (Ph.D. thesis, Moscow, 1993)
3. S.V. Amelin, Proceedings of LIIZHT, **210** (1963)
4. G.S. Pisarenko, A.P. Yakovlev, V.V. Matveyev, *Handbook on strength of materials* (Naukova dumka, Kiev, 1975)
5. V.D. Kuz`mich, V.S. Rudnew, S.Y. Frenkel, *The theory of locomotive traction* (Marshrut Pub, Moscow, 2005)
6. A. M. Keropyan, Yu. E. Babichev. J. Mining equipment and electromechanics, **2**, 24 (2015)
7. A. M. Keropyan, K.I. Shakhova. J. Mining equipment and electromechanics, **9**, 30 (2013)
8. C. M. Gorbatyuk, et al., J. Metallurgist, **42**, 178 (1998)
9. A. A. Gerasimova, A. G. Radyuk, L. M. Glukhov, J. Metallurgist, **58**, 397 (2014)
10. N. B. Demkin, *Contacting of rough surfaces* (Science, Moscow, 1970)
11. M.I. Ruditsyn, *Handbook on strength of materials* (Higher education, Minsk, 1970)
12. Yu. M. Luzhnov, *Nanotribology of wheel-rail adhesioin. Reality and opportunities* (Intekst, Moscow, 2009)
13. N.M. Belyaev, *Strength of materials* (State Publishing House of technical and theoretical literature, Moscow, 1976)
14. A. M. Keropyan, J. Friction and Wear, **1**, 78 (2016)
15. A. M. Keropyan, S.M. Gorbatyuk, ICIE, **150**, 406 (2016)

DOI: 10.1051/matecconf/201712906010

Impact of the liquid level in the jet-film contact devices on the heat-and-mass transfer process

Oksana S. Dmitrieva[1], *Andrey* V. Dmitriev[1,*], *Ilnur* N. Madyshev[2], and *Leonid* V. Kruglov[1]

[1]Kazan State Power Engineering University, 420066, Kazan, Krasnoselskaya str., 51, Russia
[2]Kazan National Research Technological University, 420015, Kazan, Karl Marx str., 68, Russia

Abstract. For the purpose of increase in efficiency and performance of the heat-and-mass transfer processes, the design of jet-film contact device has been developed. The advantage of this device is a low hydraulic resistance, high mass-transfer efficiency with relatively low energy demands, as well as soiling resistance, efficient operation at low liquid loads. In this article the authors present the research results of heat-transfer from the flowing down liquid film to the gas flow inside of the proposed device. When calculating the Nusselt number, the authors used a correction coefficient, taking into account the impact of the liquid level in the device.

1 Introduction

Currently, the energy efficiency and conservation are a top five strategic courses of the priority technological development of Russia's economy. According to the International Finance Corporation and the World Bank, the energy consumption in Russia may be reduced in industrial infrastructure by more than twofold [1]. The strategic objective is the reduction of the energy intensity of the gross domestic product by not less than 2.3 times by 2030 [2]. In this regard, the relevant objective is to increase the energy conservation (saving) and energy efficiency of the industrial units of petrochemical plants. Different mass-transfer processes and apparatuses are applied in chemical, petrochemical and refining branches of industry. Basically, the efficiency of the behavior of such processes is determined by the perfection of the used contact devices, at that material and energy costs are strongly dependent on their design.

Nowadays, apparatuses with tray-type and packing contact devices are widely used in the industry, but have several disadvantages. In the tray-type column apparatuses it is necessary to provide separating zones, which significantly increase the vessel volume and, as a consequence, steel intensity. In the packing columns we can observe uneven distribution of liquid and gas phases in the cross section of the working area, which significantly reduces the efficiency of mass-transfer processes.

Heat-and-mass transfer equipment is the most promising for carrying out the processes of rectification under vacuum, absorption of gases at low pressures, as it has a low hydraulic resistance and high mass-transfer efficiency. Some research papers of the authors are devoted to study of the liquid flows in the mass-transfer apparatuses [3-6]. Many designs are inefficient from the point of view of achieving a high intensification of the processes at minimum energy demands. Therefore, it is necessary to develop contact devices, which can provide the most efficient and intense hydrodynamic modes with minimal hydraulic resistance.

2 Description of the device and its operation

The aim of this work is to develop effective apparatuses, based on the use of jet-film contact devices in relation to the processes of rectification, absorption, desorption and evaporation.

The authors have developed the jet-film contact device [7], consisting of parallel square drain cups with vertical walls (Fig. 1).

Fig. 1. One stage of the jet-film contact device: *1* – drain cup; *2* – perforated flow baffle plates; *3* – recurved leaves.

* Corresponding author: ieremiada@gmail.com

DOI: 10.1051/matecconf/201712906010

Vertical walls, with slots for installation of appropriate drain cups, serve as supports. There are holes in the bottoms for intensification of the mass-transfer process at high values of the width of the drain cup (b > 60 mm). When liquid disperses through the holes, the jets disintegrate with drops formation. Thus, there are circular segments in the bottom of the device, which are curved to the bottom and required for the liquid distribution over the surface of the baffle plates. Each stage of the contact device is formed by a row of drain cups in a checkerboard order. Moreover, the downstream stage has an offset of drain cups, forming a vertical checkerboard order of them. The liquid is distributed with the formation of a stable film flow, when the liquid jets move over the surface of the baffle plates. Then, the formed film collides with the liquid surface, inside of the drain cups, breaks itself up and engages the gas volume inside. There is an impingement of drops, falling from the top plate, and a sharp decrease of the kinetic (motional) energy, which, in the ideal case, goes into the energy of formation of a bubble surface. For this reason, developed and constantly updated additional contact surface of phases is formed, which is determined by the presence of relatively small gas bubbles in the liquid layer and drops, departing from it.

In addition, the presence of holes in the baffle plates leads to the alignment of the profile of concentrations in the flowing down liquid film and its more uniform distribution over the surface, and to the decrease of the steel intensity of the proposed design.

The process of heat-and-mass transfer on the contact devices and in the apparatus as a whole is determined by the hydrodynamic environment. The study of hydrodynamic environment will help to understand the causes of qualitative changes of efficiency, to define the scope and specifics of operation, to identify further areas for the development engineering of the design of the device.

3 Studies of heat-transfer from the flowing liquid film to the gas flow

During the studies of heat-transfer from the flowing film of liquid to the gas flow, it was found that reduction of the liquid level in the drain cup causes the reduction of heat-transfer intensity up to 30%. Consequently, for obtaining satisfactory results of calculation of the heat-transfer from a flat baffle plate, we need to add a fudge (correction) factor that takes into account the impact of the liquid level in the drain cup.

The research was conducted on the example of heat exchange between air and a flat baffle plate. The air temperature was 20°C, the temperature of the baffle plates was 40°C. The width of the drain cup was b = 60 mm, and the height was h_1 = 30 mm. The liquid level in the drain cup h_2 was varied from 15 to 30 mm. The research objective was to determine the average heat-transfer coefficient from the flat baffle plate to the air flow. The velocity of the gases was varied in such a way that Reynolds number was varied in the range from 726 to $7.8 \cdot 10^4$.

For the analytic definition of average coefficients of momentum, heat and mass transfer, they widely use models of boundary layer (by Prandtl, Karman, Landau – Levich, Taylor, Clauser, etc.). The quantitative correlation between friction and heat-transfer can be found using Chilton-Colburn analogy and its development [8]. Taking into account the damping of turbulent fluctuations in the viscous sublayer, the most promising is the model of diffusion boundary layer by Landau – Levich and other authors [9].

Criteria equation of the convection heat-transfer in the gas phase, in case of liquid flow over the flat baffle plate at turbulent flow regime, can be recorded as follows [10]:

$$\mathrm{Nu}_G = 0.037\,\mathrm{Re}_G^{0.8}\,\mathrm{Pr}^{0.43}, \qquad (1)$$

where $\mathrm{Nu}_G = \alpha b/\lambda$ – Nusselt number in the gas phase; α – heat-transfer coefficient, W/(m^2·K); b – width of the contact element, m; λ – thermal conductivity coefficient, W/(m·K); Pr – Prandtl number.

Reynolds number in the gas phase was determined by the equation $\mathrm{Re}_G = bW_G/\nu_G$, where W_G – actual gas velocity, m/s; ν_G – coefficient of kinematic gas viscosity, m^2/s.

4 Research findings

Research Findings The results of numerical studies are found overestimated (Fig. 2). The studies established that there are at least two modes of the heat-transfer, while the start of the second mode corresponds to the value of the Reynolds number $\mathrm{Re}_G = 10^4$.

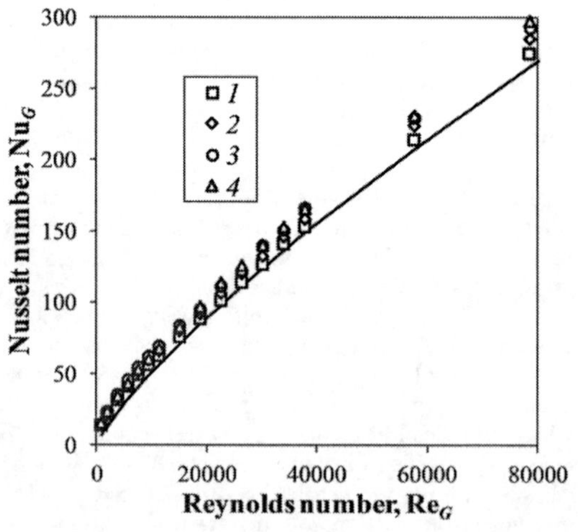

Fig. 2. The dependence of Nusselt number in the gas phase on Reynolds number at different liquid levels in the drain cup of the jet-film contact device h_2/h_1 = 0.5 (*1*); 0.666 (*2*); 0.83(*3*); 1.0 (*4*); solid line – calculation according to the (Eq. 1), points – experiment, liquid – water.

To clarify the calculation of Nusselt number, taking into account the liquid level in the drain cup Nuh, we added the correction coefficient χ_h = $\mathrm{Nu}_h/\mathrm{Nu}_G$, depending on the ratio of h_2/h_1, where h_1 – height of the

DOI: 10.1051/matecconf/201712906010

vertical walls of the drain cups, m; h_2 – height of the liquid layer in the contact element, m. Using the calculated data in Fig. 3 as the base, we may record with high accuracy:

$$\chi_h = 1.16(h_2/h_1)^{0.16}. \quad (2)$$

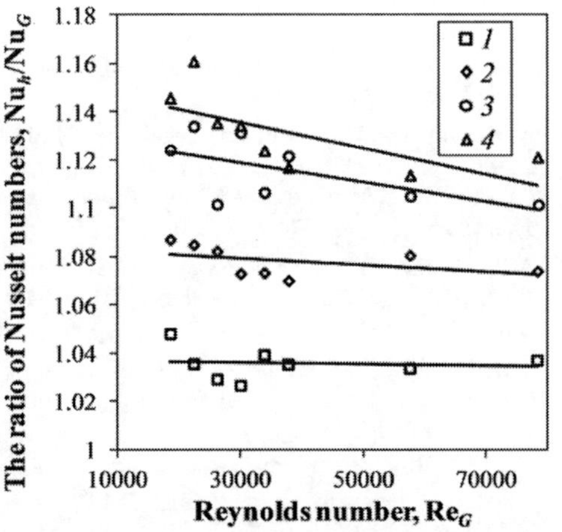

Fig. 3. The dependence of the ratio of Nusselt number, taking into account the liquid level in the drain cup, to Nusselt number in case of flowing around the flat plate, on Reynolds number at different ratio of h_2/h_1 = 0.5 (*1*); 0.666 (*2*); 0.83(*3*); 1.0 (*4*); solid line – experience-based data fitting, points – experiment.

Then the average heat-transfer coefficient, when the flow is around the flat plate, can be determined from the modified criteria equation:

$$Nu_h = 0.037\chi_h \mathrm{Re}_G^{0.8} \mathrm{Pr}^{0.43}, \quad (3)$$

where χ_h – coefficient, considering the impact of the liquid level in the drain cup.

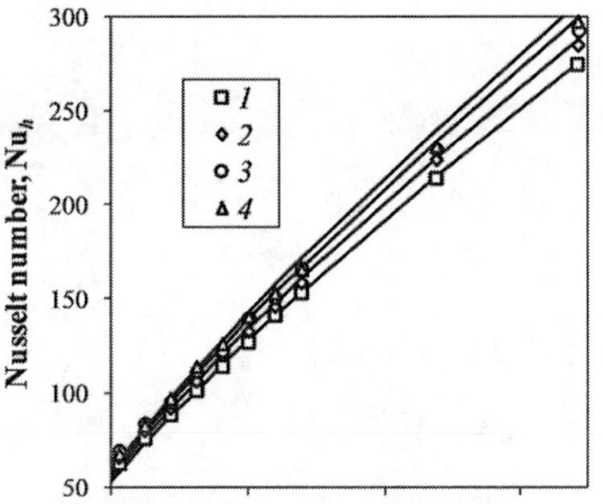

Fig. 4. The change of Nusselt number, obtained by the (Eq. 3), Reynolds number and different fill level of the drain cup h_2/h_1 = 0.5 (*1*); 0.666 (*2*); 0.83 (*3*); 1.0 (*4*); solid line – calculation according to the (Eq. 3), points – results of numerical simulation in Ansys Fluent.

Correlation of the data, obtained by the (Eq. 3) with the results of numerical simulation is shown in Fig. 4. The maximum values of heat transfer coefficients are observed when drain cups are completely filled. This is due to the lack of air flow expansion and as a result, a greater flow rate of air relative to the walls. The values of heat transfer coefficient increase by more than 30%. The maximum thermal flow is observed at higher Reynolds numbers.

We have simulated the heat-transfer process from a flat baffle plate to the air. As shown in Fig. 5, relative error of calculation, according to the modified criteria equation, that takes into account the impact of the liquid level in the drain cup, does not exceed ±9%, in the range of the Reynolds numbers from 10^4 up to $7.8 \cdot 10^4$. This is quite acceptable for engineering calculations.

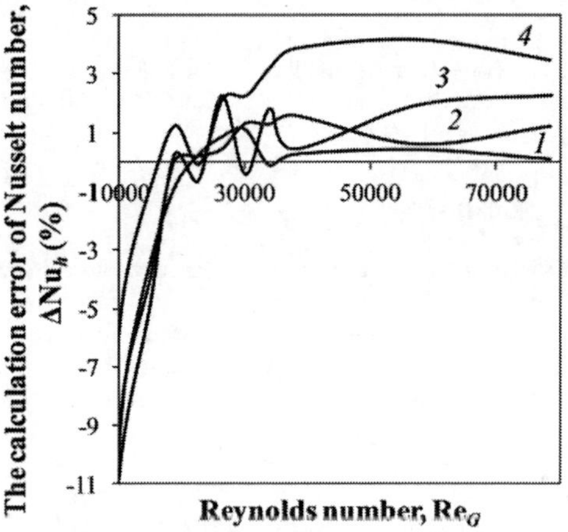

Fig. 5. The calculation error by the (Eq. 3) at different ratio h_2/h_1 = 0.5 (*1*); 0.666 (*2*); 0.83(*3*); 1.0 (*4*); the heat-transfer process from a flat baffle plate to the air.

To further verify the adequacy of calculations by the (Eq. 3), we have simulated the heat-transfer process from a flat baffle plate to the water. Underestimated values of Nusselt number, especially at high values of Reynolds number, were here observed. When using the fudge (correction) factor, the calculation accuracy was significantly increased (Fig. 6).

Calculations show that the proposed design can create the optimal conditions for a wide variety of heat-mass transfer processes while providing a sufficiently high efficiency and relatively low energy costs for the organization processes. Developed delivery devices use energy of the liquid, flowing downwards, which reduces the flow resistance for the gas phase. In addition, by reducing the separation space, we can also reduce the size of units with the installed heat-mass transfer stream-bubble contact devices.

Thus, the proposed contact device should be designed so that the liquid level in the drain cups must be maximal. This is especially true for devices operating at sufficiently high gas velocities.

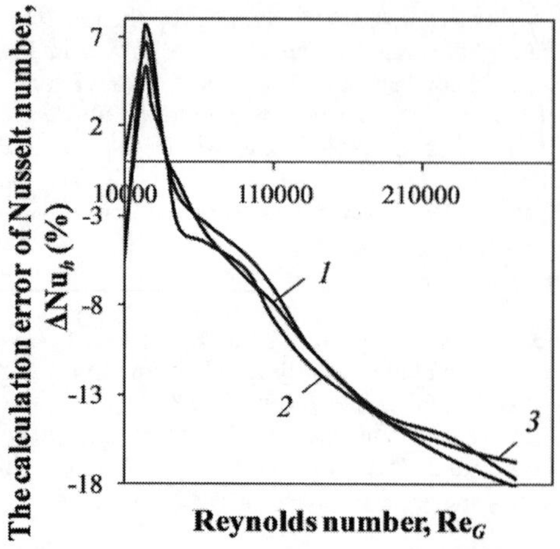

Fig. 6. The calculation error of Nusselt heat transfer number by the (Eq. 3) at different liquid levels in the drain cup of the jet-film contact device h_2/h_1 = 0.5 (*1*); 0.666 (*2*); 0.83(*3*); the heat-transfer process from a flat baffle plate to the water.

5 Conclusion

On the basis of the conducted research we can draw the following conclusions:

1) organization of the original interaction between gas and liquid allows to intensify the mass-transfer processes both in liquid and in gas phases with a relatively simple design of the device;

2) contact of the phases is based on using the energy of the flowing down liquid stream, eliminating the necessity of redistribution of the liquid over the cross section of the apparatus;

3) modified criteria equation for the calculation of the average heat transfer coefficient from the flowing down liquid film to the gas flow, takes into account the impact of the liquid level in the drain cup;

4) jet-film contact devices will allow to create energy-efficient apparatuses with a wide range of stable operation for rectification, absorption of gases.

The reported study was funded by RFBR, according to the research project No. 16-38-60081 mol_a_dk.

References

1. Energetic Efficiency in Russia. Burried Reserve. An account prepared by experts of world bank, International finance corporation and Center of effective use of energy, 2008. http://www.cenef.ru/file/FINAL_EE_report_rus.pdf.
2. Energetic Strategy of Russia for the Period to 2020, Approved by Russian Federation Government Order No. 1715-r on November 13, 2009.
3. G.Q. Wang, X.G. Yuan, and K. Yu, *Review of Mass-Transfer Correlations for Packed Columns*, Ind. & Eng. Chem. Res., **44**, 23, 8715 (2005).
4. A. Senol, U. Dramur, *Performance test and design considerations of a column packed with a new ceramic packing*, Chim. Acta turc, **23**, 2, 145-155 (1995).
5. A.V. Dmitriev, O.S. Makusheva, K.V. Dmitrieva, and A.N. Nikolaev, *Contact mass exchanger to increase output of active tower units*, Chem. and Petrol. Eng., **47**, 5-6, 319-323 (2011).
6. J. Maćkowiak, *Model for the prediction of liquid phase mass transfer of random packed columns for gas-liquid systems*, Chem. Eng. Res. and Des., **89**, 8, 1308-1320 (2011).
7. A.V. Dmitriev, O.S. Dmitrieva, and I. N. Madyshev, *Determination of the Mass-Transfer Coefficient in Liquid Phase in a Stream-Bubble Contact Device*, Therm. Eng., **63**, 9, 674-677 (2016).
8. A.P. Sinha, De Parameswar, *Mass Transfer: Principles and Operations* (PHI Learning Private Limited, New Delhi, 2012).
9. L.D. Landau, E.M. Lifshitz, *Fluid Mechanics*. Vol. 6 of Course of Theoretical Physics (Butterworth-Heinemann, UK, 1987).
10. F.F. Tsvetkov, B.A. Grigorev, *Teplomassoobmen* [Heat and mass transfer] (MPEI Publ., Moscow, 2011).

DOI: 10.1051/matecconf/201712906011

Compressible material flow in cylindrical channel with variable cross section

Elena Pryanishnikova[1,*], *Nadezhda* Belyaeva[1], and *Alexandr* Stolin[2]

[1]Department of Mathematical Modelling and Cybernetics, Syktyvkar State University named after Pitirim Sorokin, Russia
[2] Institute of Structural Macrokinetics and Material Science of Russian Academy of Science, Russia

Abstract. In the mathematical model of the flow of compressible material the effect of friction and the slip velocity of the material at the side boundary surfaces are considered. The dependence of the slip velocity on the average velocity at the entrance of the channel is built.

1 Introduction

Mathematical modelling of the compressible frictional flow of material is important theoretical [1, 2, 3] and practical task [4, 5]. Corresponding models are applying to describe and to control technological processes of composite material formation. For example, solid-phase plunger extrusion is a process of one-sided compression under press and extrusion. A material is moved from cylindrical chamber in the calibre across conical domain.

2 Problem statement

The process of the compressible material flow in a cylindrical channel with variable cross section under the effect of one-sided pressure **P=P(t)** applied to the end surface (fig. 1) is considered. The movement of the material occurs in three areas: cylindrical channel of radius R_1 (*I*), transition zone (*II*), the channel of radius R_2 (*III*). A friction force $\mathbf{F}_{fr}$ between the channel wall and the material counteracts the flow.

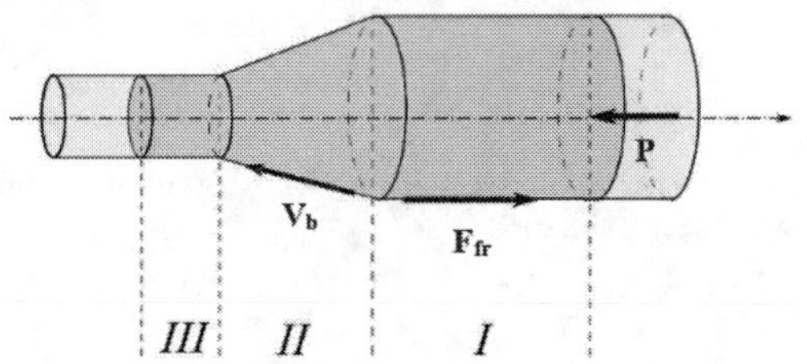

Fig. 1. The model of material flow.

In general, the movement of the material is described by the system consisting the equations for the density determined by the ratio $\rho_1 \cdot \rho$, where ρ_1 is density of the incompressible base of the material, $\rho = \rho(r, \varphi, z; t)$ is relative density, the flow velocity and the stress tensor $\Pi = (\sigma_{ij}(r, \varphi, z; t))$, $i, j = r, \varphi, z$:

$$\frac{\partial(\rho_1\rho)}{\partial t} + div(\rho_1\rho\mathbf{V}) = 0, \quad (1)$$

$$\rho_1\rho\left(\mathbf{F} - \frac{d\mathbf{V}}{dt}\right) + div\Pi = 0, \quad (2)$$

$$\Pi = \left[-p + \left(\xi - \frac{2}{3}\mu\right)div(\mathbf{V})\right]I + 2\mu\Phi, \quad (3)$$

here (1) is the continuity equation, (2) is the Navier-Stokes equation [6], **F** are mass forces acting on the material. The Newton's law of viscosity (3) is accepted as the differential equation of the state, where μ, ξ are the variable dynamic and volume viscosity depended on density, I is unit tensor, Φ is strain rate tensor.

2.1 Statement of the problem in cylindrical coordinate system

A cylindrical coordinate system r, φ, z is used to describe the problem. The axis of the channel symmetry is assumed as the axis z which positive direction is opposite to the direction of the material movement. At the time t the right-hand boundary of zone *I* is at the position $z = z_1(t)$, the boundary of zone *III* is denoted as $z = z_3(t)$. Transition zone *II* has a conical shape with a taper angle α, its boundaries are constant and denoted as $z = 0$ and $z = z_2$. The radius of the cross section of the transition zone is defined by the formula $R(z) = z\tan\alpha + R_2$.

Thus the movement of the material occurs in a region with variable boundaries.

* Corresponding author: pryanysh@inbox.ru

DOI: 10.1051/matecconf/201712906011

Due to the symmetry of the channel, the functions do not depend on the angular position and the tangential velocity v_φ, shear stresses $\sigma_{r\varphi}$ and $\sigma_{z\varphi}$ are equal to zero.

In view of the Reynolds number smallness: viscosity of the incompressible base of the material is large, the equation of motion (2) is replaced by the conditions of equilibrium. It is assumed that external forces **F** do not exist, pressure p is equal to zero. The motion in zone *III* is supposed to be unobstructed.

Under these assumptions the equations (1)-(3) in the cylindrical coordinate system with the appropriate initial and boundary conditions in the area *I* can be written as follows

$$\frac{\partial\rho}{\partial t}+\frac{1}{r}\left(\frac{\partial}{\partial r}(r\rho r_r)\right)+\frac{\partial}{\partial z}(\rho v_z)=0, \tag{4}$$

$$\frac{1}{r}\frac{\partial}{\partial r}(r\sigma_{rr})+\frac{\partial\sigma_{rz}}{\partial z}-\frac{\sigma_{\varphi\varphi}}{r}=0, \tag{5}$$

$$\frac{1}{r}\frac{\partial}{\partial r}(r\sigma_{rz})+\frac{\partial\sigma_{zz}}{\partial z}=0, \tag{6}$$

$$\sigma_{rr}=\left(\xi+\frac{4}{3}\mu\right)\frac{\partial v_r}{\partial r}+\left(\xi-\frac{2}{3}\mu\right)\left(\frac{v_r}{r}+\frac{\partial v_z}{\partial z}\right), \tag{7}$$

$$\sigma_{\varphi\varphi}=\left(\xi+\frac{4}{3}\mu\right)\frac{v_r}{r}+\left(\xi-\frac{2}{3}\mu\right)\left(\frac{\partial v_r}{\partial r}+\frac{\partial v_z}{\partial z}\right), \tag{8}$$

$$\sigma_{zz}=\left(\xi+\frac{4}{3}\mu\right)\frac{\partial v_z}{\partial z}+\left(\xi-\frac{2}{3}\mu\right)\cdot\frac{1}{r}\frac{\partial}{\partial r}(rv_r), \tag{9}$$

$$\sigma_{rz}=2\mu\left(\frac{\partial v_r}{\partial z}+\frac{\partial v_z}{\partial r}\right), \tag{10}$$

initial and boundary conditions

$$\rho|_{t=0}=\rho_0, \tag{11}$$

$$v_i|_{t=0}=0,\ i=r,z,\quad \sigma_{ii}|_{t=0}=0,\ i=r,\varphi,z,\quad \sigma_{rz}|_{t=0}=0, \tag{12}$$

$$\sigma_{zz}|_{z=z_1(t)}=P(t), \tag{13}$$

$$v_z|_{r=R_1}=v_b,\quad v_r|_{r=R_1}=0, \tag{14}$$

$$\sigma_{rz}|_{r=R_1}=\tau_{fr},\quad \sigma_{rr}|_{r=R_1}=0,\quad \sigma_{\varphi\varphi}|_{r=R_1}=0, \tag{15}$$

here $\mu=\mu_0\rho^s$, $\xi=\mu\rho/(1-\rho)$. At the initial time the distribution of density, velocity, deformation tensor (12) are set. The relation (13) defines the force in the zone *I*, (14) and (15) define intensity of friction τ_{fr} and the slip velocity v_b at the wall. In particular, in case of sticking $v_b=0$.

Let us write the projections of the relations (1)-(3) in region *II*:

$$\frac{\partial\rho}{\partial t}+\frac{1}{r}\left(\frac{\partial}{\partial r}(r\rho r_r)\right)+\frac{\partial}{\partial z}(\rho v_z)=0, \tag{16}$$

$$\frac{1}{r}\frac{\partial}{\partial r}(r\sigma_{rr})+\frac{\partial\sigma_{rz}}{\partial z}-\frac{\sigma_{\varphi\varphi}}{r}=0, \tag{17}$$

$$\frac{1}{r}\frac{\partial}{\partial r}(r\sigma_{rz})+\frac{\partial\sigma_{zz}}{\partial z}=0, \tag{18}$$

$$\sigma_{rr}=\left(\xi+\frac{4}{3}\mu\right)\frac{\partial v_r}{\partial r}+\left(\xi-\frac{2}{3}\mu\right)\left(\frac{v_r}{r}+\frac{\partial v_z}{\partial z}\right), \tag{19}$$

$$\sigma_{\varphi\varphi}=\left(\xi+\frac{4}{3}\mu\right)\frac{v_r}{r}+\left(\xi-\frac{2}{3}\mu\right)\left(\frac{\partial v_r}{\partial r}+\frac{\partial v_z}{\partial z}\right), \tag{20}$$

$$\sigma_{zz}=\left(\xi+\frac{4}{3}\mu\right)\frac{\partial v_z}{\partial z}+\left(\xi-\frac{2}{3}\mu\right)\cdot\frac{1}{r}\frac{\partial}{\partial r}(rv_r), \tag{21}$$

$$\sigma_{rz}=2\mu\left(\frac{\partial v_r}{\partial z}+\frac{\partial v_z}{\partial r}\right). \tag{22}$$

The boundary conditions in the transition zone are written as follows:

$$v_z|_{z=0}=-\frac{k_1|\sigma_{zz}(0,t)|^m}{\rho_1\rho(0,t)},\quad \left.\frac{\partial v_r}{\partial z}\right|_{z=0}=0,\quad \left.\frac{\partial\sigma_{rz}}{\partial z}\right|_{z=0}=0, \tag{23}$$

$$\sigma_{rz}|_{r=R(z)}=\tau_{fr}\cos\alpha,\quad \sigma_{rr}|_{r=R(z)}=\tau_{fr}\sin\alpha, \tag{24}$$

$$v_z|_{r=R(z)}=v_b\cos\alpha,\quad v_r|_{r=R(z)}=v_b\sin\alpha. \tag{25}$$

The first relation in (23) sets the velocity at the boundary $z=0$ of zones *II* and *III*.

3 Solution of the problem

To avoid the difficulties associated with the consideration of two-dimensional model (4)-(25) and to analyze the results, the averaging method [2, 7], is applied to the system. Average value of function $f(r)$ is defined as

$$\overline{f(r)}=\frac{2}{R^2}\int_0^R rf(r)dr.$$

Let us write the averaged system in region *I*, using previous designations:

$$\frac{\partial\rho}{\partial t}+\rho\frac{\partial v_z}{\partial z}+v_z\frac{\partial\rho}{\partial z}=0, \tag{26}$$

$$\frac{\partial\sigma_{rz}}{\partial z}-\frac{3}{2R_1}\sigma_{\varphi\varphi}=0, \tag{27}$$

$$\frac{2}{R_1}\tau_{fr}+\frac{\partial\sigma_{zz}}{\partial z}=0, \tag{28}$$

 DOI: 10.1051/matecconf/201712906011

$$\sigma_{rr} = -\left(\frac{3}{2}\xi + 5\mu\right)\frac{v_r}{R_1} + \left(\xi - \frac{2}{3}\mu\right)\frac{\partial v_z}{\partial z}, \quad (29)$$

$$\sigma_{\varphi\varphi} = \left(-\frac{3}{2}\xi + 4\mu\right)\frac{v_r}{R_1} + \left(\xi - \frac{2}{3}\mu\right)\frac{\partial v_z}{\partial z}, \quad (30)$$

$$\sigma_{zz} = \left(\xi + \frac{4}{3}\mu\right)\frac{\partial v_z}{\partial z}, \quad (31)$$

$$\sigma_{rz} = 2\mu\left(\frac{\partial v_r}{\partial z} + \frac{3}{R_1}(v_b - v_z)\right), \quad (32)$$

$$\rho\big|_{t=0} = \rho_0,\ v_i\big|_{t=0} = 0,\ i = r, z, \quad (33)$$

$$\sigma_{ii}\big|_{t=0} = 0,\ i = r, \varphi, z,\ \ \sigma_{rz}\big|_{t=0} = 0, \quad (34)$$

$$\sigma_{zz}\big|_{z=z_1(t)} = P(t). \quad (35)$$

System (16)-(25) in the transition zone takes the form:

$$\frac{\partial\rho}{\partial t} + \frac{2}{R(z)}\rho v_b \sin\alpha + \rho\frac{\partial v_z}{\partial z} + v_z\frac{\partial\rho}{\partial z} = 0, \quad (36)$$

$$\frac{2}{R(z)}\tau_{fr}\sin\alpha + \frac{\partial\sigma_{rz}}{\partial z} - \frac{3}{2R_1}\sigma_{\varphi\varphi} = 0, \quad (37)$$

$$\frac{2}{R(z)}\tau_{fr}\cos\alpha + \frac{\partial\sigma_{zz}}{\partial z} = 0, \quad (38)$$

$$\sigma_{rr} = \left(\xi + \frac{4}{3}\mu\right)\frac{3}{R(z)}(v_b\sin\alpha - v_r) + \left(\xi - \frac{2}{3}\mu\right)\left(\frac{3v_r}{2R(z)} + \frac{\partial v_z}{\partial z}\right), \quad (39)$$

$$\sigma_{\varphi\varphi} = \left(\xi + \frac{4}{3}\mu\right)\frac{3v_r}{2R(z)} + \left(\xi - \frac{2}{3}\mu\right)\left(\frac{3}{R(z)}(v_b\sin\alpha - v_r) + \frac{\partial v_z}{\partial z}\right), \quad (40)$$

$$\sigma_{zz} = \left(\xi + \frac{4}{3}\mu\right)\frac{\partial v_z}{\partial z} + \left(\xi - \frac{2}{3}\mu\right)\frac{2}{R(z)}v_b\sin\alpha, \quad (41)$$

$$\sigma_{rz} = 2\mu\left(\frac{\partial v_r}{\partial z} + \frac{3}{R(z)}(v_b\cos\alpha - v_z)\right), \quad (42)$$

$$v_z\big|_{z=0} = -\frac{k_1|\sigma_{zz}(0,t)|^m}{\rho_1\rho(0,t)},\ \ \frac{\partial v_r}{\partial z}\bigg|_{z=0} = 0,\ \ \frac{\partial\sigma_{rz}}{\partial z}\bigg|_{z=0} = 0. \quad (43)$$

3.1. Determination parameters of the problem

Assuming that the friction at wall does not depend on the axial coordinate $\partial\tau_{fr}/\partial z = 0$, equation (28) can be integrated over z:

$$\int_z^{z_1(t)}\frac{2}{R_1}\tau_{fr}ds + \int_z^{z_1(t)}\frac{\partial\sigma_{zz}}{\partial s}ds = 0,$$

$$\frac{2}{R_1}\tau_{fr}(z_1(t) - z) + \sigma_{zz}(z_1(t)) - \sigma_{zz}(z) = 0.$$

Using (35) we can express $\sigma_{zz}(z)$:

$$\sigma_{zz}(z) = P(t) + \frac{2}{R_1}\tau_{fr}(z_1(t) - z).$$

Integrating the equation (38) we can define an axial component of the stress tensor in zone *II*

$$\sigma_{zz}(z) = \sigma_{zz}(z_2) + \frac{2\tau_{fr}\cos\alpha}{\tan\alpha}\ln\left(\frac{R_1}{R(z)}\right).$$

We get the gradient of velocity from (31)

$$\frac{\partial v_z}{\partial z} = \frac{\sigma_{zz}}{\xi + \frac{4}{3}\mu}$$

and integrate

$$\int_{z_2}^{z}\frac{\partial v_z}{\partial s}ds = \int_{z_2}^{z}\frac{\sigma_{zz}}{\xi + \frac{4}{3}\mu}ds,$$

as result

$$v_z(z) - v_z(z_2) = \int_{z_2}^{z}\frac{\sigma_{zz}}{\xi + \frac{4}{3}\mu}ds. \quad (44)$$

To calculate the integral of the right hand part of (44) we use the method of trapezoids. Let us denote it as $J(z)$, then we obtain the recurrent formula for calculating the integral:

$$J(z) = \int_{z_2}^{z}\frac{\sigma_{zz}}{\xi + \frac{4}{3}\mu}ds = J(z - \Delta z) + \int_{z-\Delta z}^{z}\frac{\sigma_{zz}}{\xi + \frac{4}{3}\mu}ds =$$

$$J(z - \Delta z) + \frac{\Delta z}{2}\left[\frac{\sigma_{zz}(z)}{\xi(z) + \frac{4}{3}\mu(z)} + \frac{\sigma_{zz}(z - \Delta z)}{\xi(z - \Delta z) + \frac{4}{3}\mu(z - \Delta z)}\right],$$

$$J(z_2) = 0.$$

The equations (26), (27), (29), (30), (32), (36), (37), (39)-(42) are solved numerically by using the method of differences scheme.

4 Numeric results

It is assumed that the initial distribution of the material density is given by the function

DOI: 10.1051/matecconf/201712906011

$$\rho_0(z)=\rho_0+(\rho_m-\rho_0)z,$$

the friction intensity is defined as

$$\tau_{fr}=k_{fr}P(t),$$

where k_{fr} is the coefficient of the friction. The slip velocity is assumed [5] to be equal to

$$v_b=\beta\tau_{fr}^c,$$

here β and c are parameters that depend on the material properties. Analysis of the numerical experiment results shows that the process is possible at such values of the coefficient β that $|v_b|\le|v_z|$.

The results of numeric experiments in case of $\beta=0.1$ ($v_b=-2.2\cdot10^{-5}\ m/s$) are presented below.

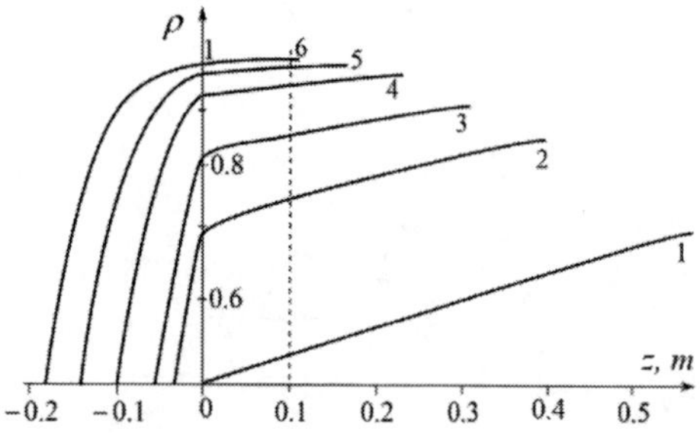

Fig. 2. Distribution of the relative density $\rho=\rho(z;t)$; full time 503.51 s; full length of billet 0.203 m; t(s): 1(0 s); 2(59 s); 3(114.3 s); 4(224.19 s); 5(279.12 s); 6(443.9 s); 7(498.83 s)

Fig.2 shows the space-time distribution of the relative density. The curves to the right of the axis ρ correspond to the distribution density of the material in zones *I* and *II* (designated by dashed border). It follows from the nature of curves that material is compacted with time, thus the porosity of the material $\Pi=1-\rho$ decreases. This fact, for example, indicates the properties of composite products.

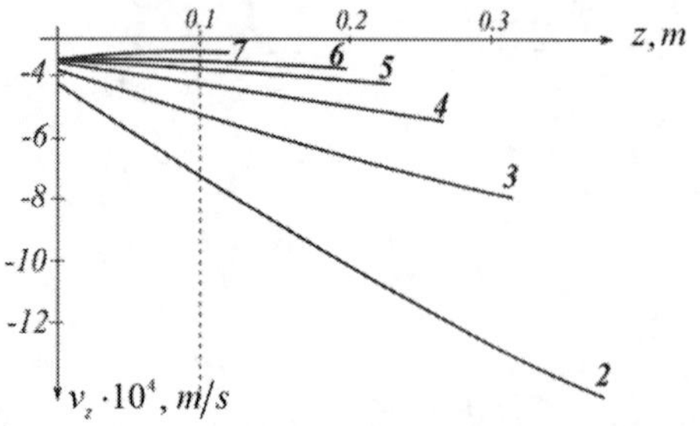

Fig. 3. Distribution of the axial component of velocity $v_z=v_z(z;t)$

Velocity, similarly to the stress, decreases from the boundary surface to the hole (fig. 3), the distribution is almost linear. The material becomes more compacted and the velocity values decrease. This effect is accorded with the results obtained in [1, 3, 5]. Due to the changes in the geometric shape and the material density in the region *II* the flow of the material is slowed down substantially the velocity is equal to the slip velocity at the walls.

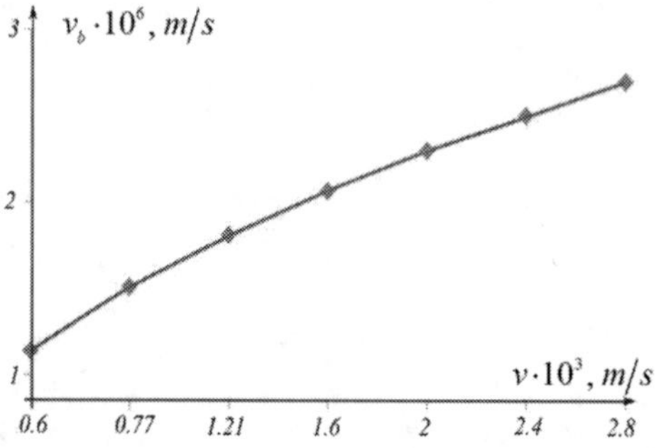

Fig. 4. The dependence of the slip velocity from the average velocity at the inlet of the channel

To determine the dependence of the velocity slip on the walls form the average speed at the entrance to the channel the extensive numerical experiment was conducted, obtained points are represented in fig. 4. This dependence can be approximated by the third degree polynomial

$$v_b=0.0222v^3-0.4345v^2+4.8075v+7.0143,$$

which coincides with the results obtained, for example, in [8]. This polynomial can be used to specify the functional dependence of the slip velocity from the velocity on the boundary layer of the material.

5 Conclusions

In the research the mathematical model of the composite material flow in cylindrical channel with variable cross section is presented. The form of the boundary conditions is suggested, the numerical experiment was carried out. The dependence estimate of the slip velocity on the average velocity of the material at the entrance is obtained.

References

1. N.A. Belyaeva, E.A. Pryanishnikova, Vestnik of Syktyvkar University. Ser. 1, **15**, 31-44 (2012)
2. E.A. Pryanishnikova, N.A. Belyaeva, Vestnik of Syktyvkar University. Ser. 1, **16**, 20-27 (2012)
3. N.K. Evstigneev, A.G. Knyazeva, PNRPU Mechanics Bulletin, **1**, 59-71 (2010)
4. A. Perrot, D. Rangeard, Y. Melinge, F. Micaelli, P. Estelle, C. Lanos, Rheologica Acta, **51/8**, 743-754 (2012)
5. M. Gupta, ANTEC, 1191-1196 (2011)
6. J. M. McKelvey, *Polymer Processing* (Khimiya, Moscow, 1965)

DOI: 10.1051/matecconf/201712906011

7. S.I. Khudyaev, *Threshold phenomena in non-linear equations* (Moscow, FIZMATLIT, 2003)
8. I.I. Ivitsky, Technology audit and production reserves, **5/3 (19)**, 8-11 (2014)

DOI: 10.1051/matecconf/201712906012

Development of criteria for a magnetic hydrocyclones

Boris Avdeyev[1,*]

[1]Kerch State Maritime Technological University, 298309 Kerch, Russia

Abstract. New criteria of a magnetic hydrocyclone - average value, magnetic field density and magnetic energy in a hydrocyclone are introduced in the article. Due to simplicity of a design and Ease of production and service, magnetic hydrocyclones have found broad application in virtually all sectors of industry, but more conspicuously advantages of their use reveal in purification systems of cooling lubricant while metal working. The dependence between field distribution in the working chamber and its average value hasn't been mathematically stated still. Complexity of criteria development and the determination of average value of the field is that the field strength changes depending on radius and height of the device. It is impossible to calculate the efficiency of cleaning from ferromagnetic impurity without knowledge of average value of magnetic field in the working chamber of a hydrocyclone therefore this problem is topical. The assumptions made in this paper are listed. Formulae are given, and numerical calculation of criteria of a hydrocyclone with radial magnetic field is shown by an example. Average value of the field, field density and magnetic energy in the working chamber of the device of inertial type has been defined on the basis of the dependence between field strength on the one hand both the radius and height of the field on the other hand which was obtained earlier.

1 A problem statement

Cooling lubricant which withdraw heat from workpiece, grease a surface, and delete residue are applied in the majority cases while metalworking [1-3]. Mechanical clearing of any technical liquid is carried out by one or various combination of four well-known methods: filtrational method, inertial method, gravitational method and separation (under the action of external fields). Devices of inertial type have found broad application in many industries and agriculture due to low cost, ease of manufacturing and maintenance, high efficiency, reliability and productivity [4-7].

Electric field is added to inertial force for the purpose of increase in efficiency of extraction of mechanical impurities [8]. A magnetic hydrocyclones which have begun to be used effectively since the fourth decade of XXth century can be referred to such devices. Devices vary not only in the geometrical dimensions, but also in an arrangement of sources of magnetic field, field parameters and other one (fig. 1) that demonstrates vigorous inventive work at combination of a magnetic and inertial method of clearing. Hydrocyclones with radial magnetic field, presented in fig. 2 a found the greatest use.

2 Analysis of the last achievements and publications

Magnetic hydrocyclones are fit for use if fast clearing of large amount of the cooling lubricant is required. However, the field of application isn't limited to it: magnetic hydrocyclones can successfully be applied to extraction of ferromagnetic particles from the discharge liquor, to receive iron powders and iron-ore concentration, etc.

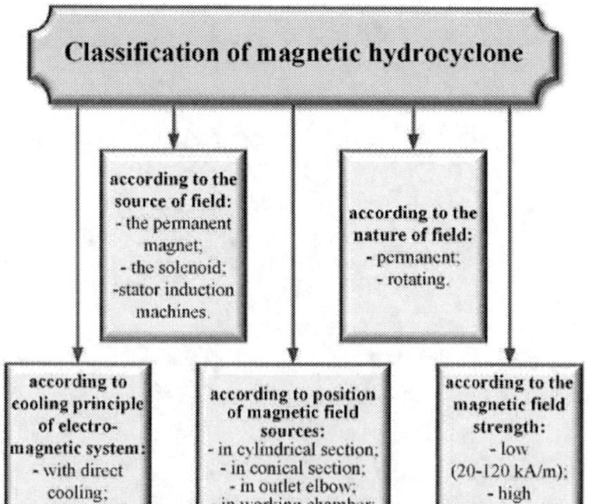

Fig. 1. Classification of magnetic hydrocyclones.

The most important parameters of devices with the superimposed magnetic field for clearing of viscous media from magnetic impurities are values of magnetic field in the working chamber and its configuration,

* Corresponding author: dirigeant@mail.ru

DOI: 10.1051/matecconf/201712906012

necessary for calculation of forces of coagulation and the magnetic force affecting on a particle or a floccule in each point of the working chamber [9].

3 Objectives and tasks of work

The objective of this article is criteria development for evaluation of magnetic hydrocyclones on the basis of an evaluation of magnetic field in the working chamber of the device. The dependence between field distribution in the working chamber and its average value hasn't been mathematically stated still. The problem is in what in value of tension of the field changes depending on radius and height of the device. The criteria for evaluation of efficiency of the field given by J. Chen in [9] aren't acceptable in a consequence of need to know field strength in each point of space both complexity, and on occasion and impossibility of calculations of these criteria.

4 The basic material of research

The magnetic hydrocyclone (MHC) was developed as an outgrowth of a conventional hydrocyclone for the purpose of providing additional external magnetic force in addition to gravitational and centrifugal forces which cause grading and separation. Adding an external magnetic force, it is possible to increase the efficiently.

A vortex motion is being created in the separation space due to the tangential feed of the fluid being purified (fig 1 a). The liquid rotating at the outer wall of the apparatus goes down and in the center of apparatus goes upward. The down flow to the outer wall of the hydrocyclone is more important than gravity.

The particles are repelled to the wall of magnetic hydrocyclone under the action of centrifugal force, and they are transported to the bunker by the downward flow near the wall. The magnetic field is produced by a coil supplied by DC voltage, which is located between the main body and the vortex finder.

The particles coagulate with each other at the input of the apparatus. The magnetic field causes particles to attract each other, so the centrifugal force effecting the particles becomes stronger and operating efficiency rises.

The method of calculation of efficiency of a magnetic hydrocyclone is given in [8]. There is an average value of field strength among the parameters affecting efficiency of the device. Due to the fact that the field in the working chamber is inhomogeneous (fig. 2 b), and is changed depending on radius and height of a hydrocyclone, it is advantageous to find such method of evaluation of its average value which adequately gives a true picture.

Magnetic field distribution was investigated in work [10]. The equations describing the change of magnetic field in the working chamber on radius and height of a hydrocyclone are given below:

$$\begin{cases} H(R,Z) = H_1'(Z) + \left(H_0'(Z) - H_1'(Z)\right) \cdot e^{-\frac{(R-R_0)}{R_1-R_0} \cdot n}; \\ H_1'(Z) = H_2 + (H_1 - H_2) \cdot e^{-\frac{Z \cdot n}{Z_1}}; \\ H_0'(Z) = H_3 + (H_0 - H_3) \cdot e^{-\frac{Z \cdot n}{Z_1}}, \end{cases} \quad (1)$$

where H_0-H_3 – magnetic field strengths, which are measured in the reference points (fig. 2), A/m; R_0 – inner radius of hydrocyclone, m; Z_1 – the height of the cylinder part of hydrocyclone, m; n – empirical coefficient, dimensionless quantity.

Since the field is radial, it is possible to neglect coordinate φ [11,12]. Actually the field won't be absolutely symmetric because the magnetic circuit is absent at the point of connection to the body of an inlet elbow and the field value will reach the maximum values at edges of a magnetic circuit. However, the field curvature has no significant effect on the general separation process therefore this phenomenon can be neglected [13-15].

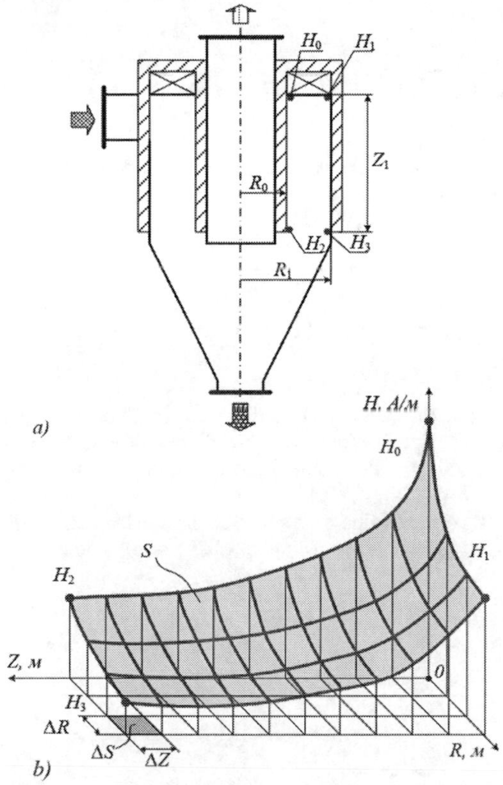

Fig. 2. Magnetic hydrocyclone (a) and changes in magnetic field strength in the working chamber of hydrocyclone (b)

Average value of fields at radial consideration can be calculated:

$$\langle H_S \rangle = \frac{1}{S} \sum_i \sum_j H(R_i, X_j) \cdot \Delta R_i \cdot \Delta Z_j, \quad (2)$$

where $H(R_i,Z_j)$ – value of magnetic field strength in infinitesimal area ΔS_i.

$$S=(R_1-R_0)\cdot Z_1 . \quad (3)$$

The product of H_i and S_i in geometric sense represents value under the plane S`.

Let's express (2) in the integrated form

$$\langle H_S\rangle=\frac{1}{S}\iint_{S`} H(R,Z)\mathrm{d}S` . \quad (4)$$

Expression (1) is cumbersome therefore let's use the simplified distribution formula of magnetic field which is measured both radially and throughout a height. A dependency with a high degree of precision represents real magnetic field in the working chamber of a hydrocyclone:

$$H(R,Z)=\left[H_3+(H_0-H_3)\cdot e^{-\frac{Z\cdot n}{Z_1}}\right]\cdot\left(\frac{R_0}{R}\right)^N . \quad (5)$$

Let's convert double integral (4) to iterated one, we get:

$$\langle H_S\rangle=\frac{1}{S}\int_0^{Z_1}\mathrm{d}Z\int_{R_0}^{R_1}\left[H_3+(H_0-H_3)\cdot e^{-\frac{Z\cdot n}{Z_1}}\right]\cdot\left(\frac{R_0}{R}\right)^N \mathrm{d}R . \quad (6)$$

Lowering calculations, $\langle H_S\rangle$ will take the following form:

$$\langle H_S\rangle=\frac{1}{(R_1-R_0)\cdot Z_1}\cdot\frac{R_0^N}{N-1}\left(\frac{1}{R_0^{N-1}}-\frac{1}{R_1^{N-1}}\right)\times \left[H_3\cdot Z_1+\frac{Z_1\cdot(H_3-H_0)}{n}\left(1-e^{-n}\right)\right] . \quad (7)$$

For the case which is described in [model], average value of field strength is equal to 6,887 kA/m at the following values: H_0=25 kA/m; H_3=5 kA/m; R_0=0,04 m; D=0,12 m; Z_1=0,4; N=2,5; n=3.

Average value of the field in the working chamber in cylindrical coordinate system can be written down:

$$\langle H_V\rangle=\frac{1}{V}\sum_i\sum_j\sum_k H(R_j,\varphi_i,Z_k)\cdot R\cdot\Delta R_i\cdot\Delta\varphi_j\cdot\Delta Z_k , \quad (8)$$

where

$$\Delta V=R\cdot\Delta R_i\cdot\Delta\varphi_j\cdot\Delta Z_k \quad (9)$$

- volume element in cylindrical coordinate system (fig. 3).

Integrated form of equation will take a form:

$$\langle H_V\rangle=\frac{1}{V}\iiint_V H(\varphi,R,Z)\mathrm{d}V . \quad (10)$$

Let's convert triple integral (9), we get:

$$\langle H_V\rangle=\frac{1}{V}\int_{R_0}^{R_1}R\mathrm{d}R\int_0^{2\pi}\mathrm{d}\varphi\int_0^{Z_1}\left[H_3+(H_0-H_3)e^{-\frac{Z\cdot n}{Z_1}}\right]\left(\frac{R_0}{R}\right)^N \mathrm{d}Z . \quad (11)$$

$\langle H_V\rangle$ will take a form:

$$\langle H_V\rangle=\frac{1}{\pi\cdot(R_1^2-R_0^2)\cdot Z_1}\cdot\frac{2\pi\cdot R_0^N}{N-2}\left(\frac{1}{R_0^{N-2}}-\frac{1}{R_1^{N-2}}\right)\times \left[H_3\cdot Z_1+\frac{Z_1\cdot(H_3-H_0)}{n}\left(1-e^{-n}\right)\right] , \quad (12)$$

where

$$V=\pi\cdot(R_1^2-R_0^2)\cdot Z_1 \quad (13)$$

- volume of working chamber of hydrocyclone, m³.

Average value of field strength $\langle H_V\rangle$ is equal to 6,045 kA/m at the same values, at which $\langle H_S\rangle$ was calculated. Relative error $\langle H_S\rangle$ with the respect to $\langle H_V\rangle$ is equal to 4,2%, it is quite allowable.

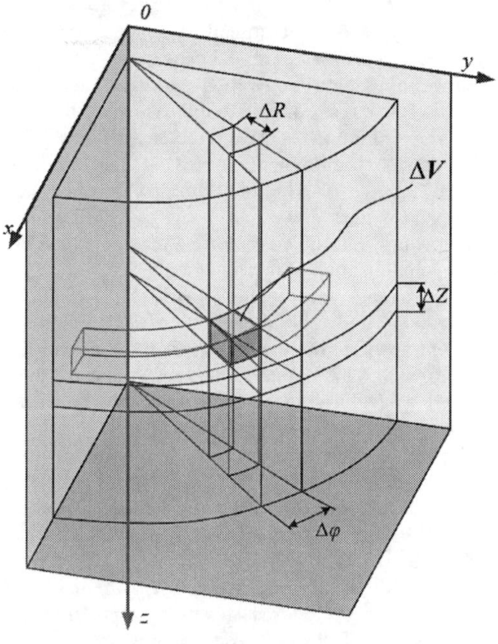

Fig. 3. Concerning calculating volume element ΔV

The area density of magnetic field is calculated:

$$\rho_S=\frac{\langle H_S\rangle}{S} . \quad (14)$$

The volume density of a magnetic field:

$$\rho_V=\frac{\langle H_V\rangle}{V} . \quad (15)$$

Magnetic energy at volume V is defined as follows:

$$W_M = \frac{1}{2}\int_V H \cdot B dV \,, \qquad (16)$$

where B – magnetic induction, T.
Provided that

$$B = \mu_0 \cdot H \,, \qquad (17)$$

magnetic energy of field write down as:

$$W_M = \frac{1}{2}\mu_0 \int_V H^2 dV \,, \qquad (18)$$

where μ_0- magnetic constant, H/m.
Let's substitute (5) and (9) into (18), we get:

$$W_M = \frac{\pi \cdot R_0^{2N}}{N-1}\left(\frac{1}{R_0^{2N-2}} - \frac{1}{R_1^{2N-2}}\right) \times \\ \times \left[H_3^2 \cdot Z_1 + \frac{Z_1}{2n}(H_3 - H_0)(4H_3 + 1)\left(1 - e^{-2n}\right)\right] . \qquad (19)$$

Magnetic energy for magnetic hydrocyclone is equal 71,28 kJ to at the same values at which $\langle H_S \rangle$ was calculated.

5 Conclusion

Magnetic field of a hydrocyclone can be characterized by three values: average value of the field in the working chamber; volume density of magnetic field; magnetic energy. Average value of the field in the working chamber is necessary value for calculation of efficiency curve. Use of these criteria will allow to evaluate effect of magnetic field on motion of particle in working chambers of magnetic hydrocyclones.

References

1. Popov, A. Dugin, Int. J. Adv. Manuf. Technol. **76** (2015)
2. S. Chernyi, IOP Conf. Ser.: Mater. Sci. Eng. **124** (2017)
3. V.P. Astakhov, S. Joksch, *Metalworking Fluids (MWFs) for Cutting and Grinding* (Woodhead Publishing, 2012)
4. V. Budnik, S. Chernyi, Procedia Eng. **150** (2016)
5. V.A. Zhukov, E.P. Masyutkin, B.A. Avdeyev, IOP Conf. Ser.: Mater. Sci. Eng. **177** (2017)
6. L.-Y. Chu, W.-M. Chen, X.-Z. Lee, Sep. Purif. Technol. **26** (2002)
7. R.A. Augusto, J.R. Martins, Min. Proc. and Extr. Met. Rev. **22** (2001)
8. G. Chen, *Design and analysis of magnetic hydrocyclone* (M. Eng. Thesis, McGill University, Montreal, 1989)
9. B.A. Avdeyev *Improving the efficiency of the engine oil cleaning in ship diesel engines through the use of magnetic hydrocyclones* (Zebra, Ulyanovsk, 2016)
10. V.I. Prosvirnin, S.P. Golikov, B.A. Avdeyev, Bulletin of the Kherson National Technical University **46** (2013)
11. R.J. Freeman, N.A. Rowson, T.J. Veasey, I.R. Harris, IEEE Transactions on Magnetics **30** D (1994)
12. N. Nezami, B. Hajighorbani, International conference on sustainable development, strategies and challenges With a focus on Agriculture, Natural Resources, Environment and Tourism (2015)
13. J.R. Portela, J. Sánchez-Oneto, M. Belén García-Jarana, J. Martinez de la Ossa, Chem. Eng. Trans. **17** (2009)
14. R.H.M.G. Nabben, R.P.J. Duursma, A.A. Kamperman, J.L. Lagerberg, Ironmaking Steelmaking **25** **(5)** (1998)
15. A. Nyrkov, A. Shnurenko, S. Sokolov, S. Chernyi, & V. Korotkov, Procedia Engineering **178** (2017).

Eigenvibrations of a bar with load

Anton A. Samsonov[1], *Sergey* I. Solov'ev[1,*], and *Pavel* S. Solov'ev[1]

[1]Kazan Federal University, 420008 Kazan 18 Kremlevskaya Street, Russian Federation

Abstract. The differential eigenvalue problem describing eigenvibrations of an elastic bar with load is investigated. The problem has an increasing sequence of positive simple eigenvalues with limit point at infinity. To the sequence of eigenvalues, there corresponds a complete orthonormal system of eigenfunctions. We formulate limit differential eigenvalue problems and prove the convergence of the eigenvalues and eigenfunctions of the initial problem to the corresponding eigenvalues and eigenfunctions of the limit problems as load mass tending to infinity. The original differential eigenvalue problem is approximated by the finite difference method on a uniform grid. Error estimates for approximate eigenvalues and eigenfunctions are established. Investigations of this paper can be generalized for the cases of more complicated and important problems on eigenvibrations of beams, plates and shells with attached loads.

1 Introduction

Let us formulate the differential eigenvalue problem governing longitudinal eigenvibrations of the bar–load system. Assume that the elastic bar axis occupies in the equilibrium position the segment $[0,l]$ on the Ox axis. Denote by $\rho(x)$ the volume mass density, $E(x)$ the elasticity Young modulus, $S(x)$ the square of transversal cut of the bar. Suppose that the end $x=0$ of the bar is rigidly fixed while the end $x=l$ is free, at the point $x=l$ of the bar a load of mass M is rigidly joined. Then the longitudinal deflection $w(x,t)$ of the bar at a point x at time t satisfies the following equations

$$\frac{\partial}{\partial x}\left(p(x)\frac{\partial}{\partial x}w(x,t)\right)=r(x)\frac{\partial^2}{\partial t^2}w(x,t),\ x\in(0,l), \quad (1)$$

$$w(0,t)=0,\ -p(l)\frac{\partial}{\partial x}w(l,t)=M\frac{\partial^2}{\partial t^2}w(l,t), \quad (2)$$

where $t>0$, $p(x)=E(x)S(x)$, $r(x)=\rho(x)S(x)$, $x\in[0,l]$.

The eigenvibrations of the bar–load system are characterized by the function $w(x,t)$ of the following form $w(x,t)=u(x)v(x)$, $x\in[0,l]$, where $v(t)=a_0\cos\sqrt{\lambda}t+b_0\sin\sqrt{\lambda}t$, $t>0$; a_0, b_0, λ are constants. Then equations (1) and (2) lead to the following differential eigenvalue problem of second order: find numbers λ and nontrivial functions $u(x)$, $x\in[0,l]$, such that

$$-(p(x)u'(x))'=\lambda r(x)u(x),\ \ x\in(0,l), \quad (3)$$

$$u(0)=0,\ \ p(l)u'(l)=\lambda Mu(l). \quad (4)$$

Problem (3), (4) has an increasing sequence of positive simple eigenvalues with limit point at infinity. To the sequence of eigenvalues, there corresponds a complete orthonormal system of eigenfunctions. In the present paper, limit properties as $M\to\infty$ of eigenvalues and eigenfunctions of parameter dependent eigenvalue problem (3), (4) with parameter M are studied.

Parameter eigenvalue problems are applied for the investigation and solution of eigenvalue problems with nonlinear dependence on the spectral parameter. Nonlinear eigenvalue problems arise in various applications [1-3]. Numerical methods for solving matrix eigenvalue problems with nonlinear dependence on the spectral parameter were constructed and investigated in the papers [4-13]. Mesh methods for solving differential nonlinear eigenvalue problems were studied in the papers [14-16]. The theoretical basis for the study of nonlinear spectral problems is results obtained for linear eigenvalues problems [17-23]. In the papers [24-30], numerical methods for solving applied nonlinear boundary value problems and variational inequalities have been studied.

2 Variational statement of the problem

Let $\mathbb{R}$ be the real line, $\Omega=(0,l)$, $\bar{\Omega}=[0,l]$, $\Lambda=(0,\infty)$. As usual, we denote by $L_2(\Omega)$ and $W_2^1(\Omega)$ the real Lebesgue space and the real Sobolev space, respectively, with norms $|\cdot|_0$ and $\|\cdot\|_1$, where

$$|v|_0=\left(\int_0^l (v(x))^2\mathrm{d}x\right)^{1/2}, \quad (5)$$

* Corresponding author: sergei.solovyev@kpfu.ru

$$\|v\|_1=(|v|_0^2+|v|_1^2)^{1/2},\ |v|_1=|v'|_0. \quad (6)$$

Set $H=L_2(\Omega)$, $V=\{v:\ v\in W_2^1(\Omega),\ v(0)=0\}$. For a continuous function $v(x)$, $x\in\bar{\Omega}$, we denote

$$|v|_{0,\infty}=\max_{x\in\bar{\Omega}}|v(x)|. \quad (7)$$

Note that the space V is compactly embedded into the space H, any function from V is continuous on $\bar{\Omega}$. The semi-norm $|\cdot|_1$ is a norm over the space V, which is equivalent to the norm $\|\cdot\|_1$. There exist constants c_0 and c_1 such that $|v|_{0,\infty}\le c_0|v|_1$, $|v|_0\le c_1|v|_1$, for any $v\in V$.

Define sufficiently smooth functions $p(x)$, $r(x)$, $x\in\bar{\Omega}$, for which there exist positive constants p_i, r_i, $i=1,2$, such that $p_1\le p(x)\le p_2$, $r_1\le r(x)\le r_2$. Introduce a number $\mu\in\Lambda$ and bilinear forms $a:V\times V\to\mathbb{R}$, $b:V\times V\to\mathbb{R}$, $c:V\times V\to\mathbb{R}$, by the formulas

$$a(u,v)=\int_0^l p(x)u'v'dx,\quad b(u,v)=\int_0^l r(x)uvdx, \quad (8)$$

$c(u,v)=u(l)v(l)$, for any $u,v\in V$.

For $\mu=M$, the differential problem (3), (4) is equivalent to the following variational eigenvalue problem: find $\lambda\in\mathbb{R}$, $u\in V\setminus\{0\}$, such that

$$a(u,v)=\lambda(b(u,v)+\mu c(u,v))\quad \forall v\in V. \quad (9)$$

The number $\lambda=\varphi(\mu)$ that satisfies (9) is called an eigenvalue, and the function $u=u^\mu$ is called an eigenfunction of problem (9) corresponding to λ. The set $U(\lambda)$ that consists of the eigenfunctions corresponding to the eigenvalue λ and the zero function is a closed subspace in V, which is called the eigensubspace corresponding to the eigenvalue λ. The dimension of this subspace is called the multiplicity of the eigenvalue λ. If the dimension of an eigensubspace is equal to unity, then the corresponding eigenvalue is said to be simple. The pair (λ,u) with λ and u satisfying (9) is called an eigensolution or eigenpair of problem (9).

3 Parameter eigenvalue problem

Let us formulate the following existence results for problem (9). Set

$$R(v)=R_\mu(v)=\frac{a(v,v)}{b(v,v)+\mu c(v,v)} \quad (10)$$

for any $v\in V\setminus\{0\}$, $\mu\in\Lambda$. If W is a subspace of the space V, then we denote

$$W_a^\perp=\{v:\ v\in V,\ a(v,w)=0\ \forall w\in W\}. \quad (11)$$

Denote by $S_k(W)$ the set of all k-dimensional subspaces of the space W for $k\ge1$. The set $S_0(W)$ consists of $\{0\}$ alone. Put $S_k=S_k(V)$ for $k\ge0$.

Problem (9) has an increasing sequence of positive simple eigenvalues with limit point at infinity $\lambda_k=\varphi_k(\mu)$, $k=0,1,\ldots$, $0<\lambda_0<\lambda_1<\ldots<\lambda_k<\ldots$, $\lambda_k\to\infty$ as $k\to\infty$. These eigenvalues correspond to a complete orthonormal system of eigenfunctions $u_k=u_k^\mu$, $k=0,1,\ldots$, which can be chosen according to one of the conditions:

a) $a(u_i,u_j)=\delta_{ij}$, $b(u_i,u_j)+\mu c(u_i,u_j)=\delta_{ij}/\lambda_i$, $i,j=0,1,\ldots$;

b) $a(u_i,u_j)=\lambda_i\delta_{ij}$, $b(u_i,u_j)+\mu c(u_i,u_j)=\delta_{ij}$, $i,j=0,1,\ldots$

Functions u_k, $k=0,1,\ldots$, form a complete system in the space V. The following relations hold: $u_k(l)\ne0$, $k=0,1,\ldots$ The following variational properties are valid:

$$\lambda_{k-1}=\min_{v\in(E_{k-1})_a^\perp\setminus\{0\}}R(v)=\max_{v\in E_k\setminus\{0\}}R(v), \quad (12)$$

$$\lambda_{k-1}=\max_{W\in S_{k-1}}\min_{v\in W_a^\perp\setminus\{0\}}R(v)=\min_{W\in S_k}\max_{v\in W\setminus\{0\}}R(v), \quad (13)$$

where $E_k=E_k^\mu=\mathrm{span}\{u_0,u_1,\ldots,u_{k-1}\}$, $k=1,2,\ldots$, $E_0=E_0^\mu=\{0\}$, $(E_0)_a^\perp=V$.

The following inequalities hold:

$$\alpha_1|v|_1^2\le a(v,v)\le\alpha_2|v|_1^2, \quad (14)$$

$$\beta_1|v|_0^2\le b(v,v)\le\beta_2|v|_0^2, \quad (15)$$

$0\le c(v,v)\le\gamma_2|v|_1^2$, for any $v\in V$, $\alpha_i=p_i$, $\beta_i=r_i$, $i=1,2$, $\gamma_2=c_0^2$.

The functions $\varphi_k(\mu)$, $\mu\in\Lambda$, $k=0,1,\ldots$, are continuous and decreasing.

4 Limit properties of parameter problem

Define the subspaces $V_0=\{v:\ v\in W_2^1(\Omega),\ v(0)=0,\ v(l)=0\}$ and $V_1=(V_0)_a^\perp$ of the space V. Introduce the following linear eigenvalue problems.

Find $\lambda^{(0)}\in\mathbb{R}$, $u\in V_0\setminus\{0\}$, such that

$$a(u,v)=\lambda^{(0)}b(u,v)\quad\forall v\in V_0. \quad (16)$$

Find $\lambda^{(1)}\in\mathbb{R}$, $u\in V_1\setminus\{0\}$, such that

$$a(u,v)=\lambda^{(1)}c(u,v)\quad\forall v\in V_1. \quad (17)$$

Problem (16) has an increasing sequence of positive simple eigenvalues with limit point at infinity $\lambda_k^{(0)}$,

$k=1,2,\ldots$, $0<\lambda_1^{(0)}<\lambda_2^{(0)}<\ldots<\lambda_k^{(0)}<\ldots$, $\lambda_k^{(0)}\to\infty$ as $k\to\infty$. The corresponding eigenfunctions $u_k^{(0)}$, $k=1,2,\ldots$, form a complete orthonormal system in the space V_0.

Since $\dim V_1=\operatorname{codim}V_0=1$, the eigenvalue problem (17) has one eigenvalue $\lambda^{(1)}$ which is positive and simple. There exists a unique normalized eigenfunction $u^{(1)}$ such that $a(u^{(1)},u^{(1)})=1$, $u^{(1)}(l)>0$, corresponding to the eigenvalue $\lambda^{(1)}$. Moreover, the functions $u_0^{(0)}=u^{(1)}$, $u_k^{(0)}$, $k=1,2,\ldots$, form a complete orthonormal system in V such that $a(u_i^{(0)},u_j^{(0)})=\delta_{ij}$, $i,j=0,1,\ldots$ The eigenvalue and the eigenfunction of the eigenvalue problem (17) are defined by

$$\lambda^{(1)}=\left(\int_0^l p(y)^{-1}\mathrm{d}y\right)^{-1}, \tag{18}$$

$$u^{(1)}(x)=\int_0^x p(y)^{-1}\mathrm{d}y\left(\int_0^l p(y)^{-1}\mathrm{d}y\right)^{-1/2},\quad x\in[0,l]. \tag{19}$$

Theorem 1. The convergence $\varphi_0(\mu)\to 0$ as $\mu\to\infty$ holds, and the following asymptotic representation is valid $\varphi_0(\mu)=\lambda^{(1)}\mu^{-1}+o(\mu^{-1})$ as $\mu\to\infty$. Moreover, the convergence $u_0^\mu\to u^{(1)}$ in V as $\mu\to\infty$ holds, where $a(u_0^\mu,u_0^\mu)=1$, $u_0^\mu(l)>0$, $\mu\in\Lambda$, $a(u^{(1)},u^{(1)})=1$, $u^{(1)}(l)>0$.

Theorem 2. The convergence $\varphi_k(\mu)\to\lambda_k^{(0)}$, $u_k^\mu\to u_k^{(0)}$ in V, $u_k^\mu(l)=O(\mu^{-1})$, as $\mu\to\infty$ holds, where $a(u_k^\mu,u_k^\mu)=1$, $u_k^\mu(l)>0$, $\mu\in\Lambda$, $a(u_k^{(0)},u_k^{(0)})=1$, $a(u_k^{(0)},u_k^\mu)>0$, $k=1,2,\ldots$

Theorems 1 and 2 are proved with using results from [15, 16].

5 Finite difference approximation of the problem

Let us partition the interval $[0,l]$ by equidistant points $x_i=ih$, $i=0,1,\ldots,N$, into elements $e_i=[x_{i-1},x_i]$, $i=1,2,\ldots,N$, $h=l/N$, and let V_h denote the subspace of the space V consisting of continuous functions v^h linear on each element e_i, $i=1,2,\ldots,N$. For $u^h,v^h\in V_h$, we introduce approximate bilinear forms

$$a_h(u^h,v^h)=\\=\sum_{i=1}^N hp(x_i-h/2)(u^h(x_i-h/2))'(v^h(x_i-h/2))', \tag{20}$$

$$b_h(u^h,v^h)=\\=\sum_{i=1}^N h(r(x_{i-1})u^h(x_{i-1})v^h(x_{i-1})+\\+r(x_i)u^h(x_i)v^h(x_i))/2. \tag{21}$$

We approximate the original eigenvalue problem (9) by a finite-dimensional eigenvalue problem: find $\lambda^h\in\mathbb{R}$, $u^h\in V_h\setminus\{0\}$, such that

$$a_h(u^h,v^h)=\lambda^h(b_h(u^h,v^h)+\mu c(u^h,v^h))\ \forall v^h\in V_h. \tag{22}$$

Problem (22) has an increasing finite sequence of positive simple eigenvalues λ_k^h, $k=0,1,\ldots,N-1$, $0<\lambda_0^h<\lambda_1^h<\ldots<\lambda_{N-1}^h$. These eigenvalues correspond to an orthonormal system of eigenfunctions u_k^h, $k=0,1,\ldots,N-1$, which can be chosen according to one of the conditions:

a[h]) $a_h(u_i^h,u_j^h)=\delta_{ij}$, $b_h(u_i^h,u_j^h)+\mu c(u_i^h,u_j^h)=\delta_{ij}/\lambda_i^h$, $i,j=0,1,\ldots,N-1$;

b[h]) $a_h(u_i^h,u_j^h)=\lambda_i^h\delta_{ij}$, $b_h(u_i^h,u_j^h)+\mu c(u_i^h,u_j^h)=\delta_{ij}$, $i,j=0,1,\ldots,N-1$.

Functions u_k^h, $k=0,1,\ldots,N-1$, form a complete orthonormal system in the space V_h.

Denote $y_i=u^h(x_i)$, $i=0,1,\ldots,N$, $\tilde{p}_i=p(x_i-h/2)$, $r_i=r(x_i)$, $y_{x,i}=(y_{i+1}-y_i)/h$, $y_{\bar{x},i}=(y_i-y_{i-1})/h$. Then problem (22) can be written in the following finite difference form

$$-(\tilde{p}y_{\bar{x}})_{x,i}=\lambda^h r_i y_i,\quad i=1,2,\ldots,N-1, \tag{23}$$

$$y_0=0,\quad \tilde{p}_N y_{\bar{x},N}=\lambda^h h\left(\frac{1}{2}r_N+\frac{\mu}{h}\right)y_N. \tag{24}$$

By c we denote various positive constants independent of h.

Theorem 3. Suppose that conditions b) and b[h]) are valid. Then the estimates $|\lambda_k^h-\lambda_k|\le ch^2$, $|u_k^h-u_k|_1\le ch$, hold for sufficiently small h, $a(u_k^h,u_k)>0$.

Proof. For any v^h, $w^h\in V_h$ we have the following estimates

$$|a_h(v^h,w^h)-a(v^h,w^h)|\le ch^2|v^h|_1|w^h|_1, \tag{25}$$

$$|b_h(v^h,w^h)-b(v^h,w^h)|\le ch^2|v^h|_1|w^h|_1. \tag{26}$$

Now desired error estimates are proved with using results from [19-23].

To illustrate theoretical results of Theorems 1 and 2, we have solved the eigenvalue problem (9) for $p(x)=1$, $r(x)=1$, $l=1$, using the finite difference scheme (23), (24) for $N=100$, $\mu\in[0,2]$. Fig. 1 shows the graphs of the functions $\lambda_k=\varphi_k(\mu)$, $\mu\in[0,2]$, $k=0,1,\ldots,5$, and the eigenvalues $\lambda_k^{(0)} = (k\pi)^2$, $k = 1,2,\ldots,5$, of the limit eigenvalue problem (16). We see that the experimental results are consistent with the theoretical results in Theorems 1 and 2; namely, $\varphi_0(\mu) \to 0$, $\varphi_k(\mu) \to \lambda_k^{(0)}$, as $\mu \to \infty$, $k = 1,2,\ldots,5$. Note that investigations of the present paper can be generalized for the cases of more complicated and important problems on eigenvibrations of beams, plates and shells with attached loads.

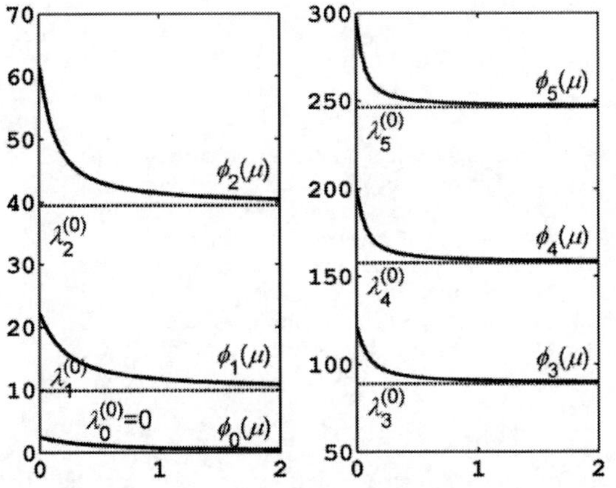

Fig.1. Limit properties of the eigenvalues $\lambda_k = \varphi_k(\mu)$, $\mu \in [0,2]$, $k = 0,1,\ldots,5$.

This work was supported by Russian Science Foundation, project no. 16-11-10299.

References

1. A.V. Goolin, S.V. Kartyshov, Surv. Math. Ind. **3**, 29-48 (1993)
2. T. Betcke, N.J. Higham, V. Mehrmann, C. Schröder, F. Tisseur, ACM Trans. Math. Software **39**, 7 (2013)
3. V.A. Kozlov, V.G. Maz'ya, J. Rossmann, *Spectral Problems Associated with Corner Singularities of Solutions to Elliptic Equations* (AMS, Providence, 2001)
4. R.Z. Dautov, A.D. Lyashko, S.I. Solov'ev, Russ. J. Numer. Anal. Math. Modelling **9**, 417-427 (1994)
5. A. Ruhe, SIAM J. Numer. Anal. **10**, 674-689 (1973)
6. V. Mehrmann, H. Voss, GAMM-Mit. **27**, 1029-1051 (2004)
7. F. Tisseur, K. Meerbergen, SIAM Rev. **43**, 235-286 (2001)
8. D. Kressner, Numer. Math. **114**, 355-372 (2009)
9. X. Huang, Z. Bai, Y. Su, J. Comput. Math. **28**, 218-234 (2010)
10. H. Schwetlick, K. Schreiber, Linear Algebra Appl. **436**, 3991-4016 (2012)
11. W.-J. Beyn, Linear Algebra Appl. **436**, 3839-3863 (2012)
12. A. Leblanc, A. Lavie, Eng. Anal. Bound. Elem. **37**, 162-166 (2013)
13. X. Qian, L. Wang, Y. Song, J. Comput. Appl. Math. **290**, 268-277 (2015)
14. A.V. Kregzhde, Differ. Uravn. **17**, 1280-1284 (1981)
15. S.I. Solov'ev, Differ. Equations **50**, 947-954 (2014)
16. S.I. Solov'ev, Differ. Equations **51**, 934-947 (2015)
17. S.I. Solov'ev, Differ. Equations **30**, 1138-1146 (1994)
18. S.I. Solov'ev, Differ. Equations **38**, 752-753 (2002)
19. S.I. Solov'ev, Differ. Equations **46**, 1030-1041 (2010)
20. S.I. Solov'ev, Differ. Equations **47**, 1188-1196 (2011)
21. S.I. Solov'ev, Differ. Equations **48**, 1028-1041 (2012)
22. S.I. Solov'ev, Differ. Equations **49**, 908-916 (2013)
23. S.I. Solov'ev, Appl. Numer. Math. **93**, 206-214 (2015)
24. I.B. Badriev, G.Z. Garipova, M.V. Makarov, V.N. Paymushin, Res. J. Appl. Sciences **10**, 428-435 (2015)
25. I.B. Badriev, L.A. Nechaeva, PNRPU Mech. Bull. 3, 37-65 (2013)
26. I.B. Badriev, M.V. Makarov, V.N. Paimushin, Proc. Engin. **150**, 1050-1055 (2016)
27. I.B. Badriev, G.Z. Garipova, M.V. Makarov, V.N. Paimushin, R. F. Khabibullin, Lobachevskii J. Math. **36**, 474-481 (2015)
28. I.B. Badriev, M.V. Makarov, V.N. Paimushin, Russ. Math. **59**, 57-60 (2015)
29. I.B. Badriev, M.V. Makarov, V.N. Paimushin, Proc. Engin. **150**, 1056-1062 (2016)
30. I.B. Badriev, V.V. Banderov, V.L. Gnedenkova, N.V. Kalacheva, A.I. Korablev, R.R. Tagirov, Appl. Math. Sciences **9**, 5697-5705 (2015)

Finite difference approximation of electron balance problem in the stationary high-frequency induction discharges

Sergey I. Solov'ev[1,*], *Pavel* S. Solov'ev[1], and *Violetta* Yu. Chebakova[1]

[1]Kazan Federal University, 18 Kremlevskaya Street, Kazan 420008, Russian Federation

Abstract. The problem of finding the minimal eigenvalue corresponding to a positive eigenfunction of the nonlinear eigenvalue problem for the ordinary differential equation with coefficients depending on a spectral parameter is investigated. This problem arises in modeling the plasma of radio-frequency discharge at reduced pressures. The original differential eigenvalue problem is approximated by the finite difference method on a uniform grid. A sufficient condition for the existence of a minimal eigenvalue corresponding to a positive eigenfunction of the finite difference nonlinear eigenvalue problem is established. Error estimates for the approximate eigenvalue and the corresponding approximate positive eigenfunction are proved. Investigations of this paper generalize well known results for eigenvalue problems with linear dependence on the spectral parameter.

1 Introduction

In the present paper, we study the nonlinear eigenvalue problem of finding the minimal eigenvalue $\lambda \in \Lambda$, $\Lambda = [0,\infty)$, corresponding to a positive eigenfunction $u(x)$, $x \in \Omega$, $\Omega = (0,\pi)$, $\bar{\Omega} = [0,\pi]$, satisfying the following equations

$$-(p(\lambda s(x))u'(x))' = r(\lambda s(x))u(x), \quad x \in \Omega, \tag{1}$$

$$u(0) = u(\pi) = 0. \tag{2}$$

We assume that $p(\mu)$, $r(\mu)$, $\mu \in \Lambda$, and $s(x)$, $x \in \bar{\Omega}$, are smooth positive functions. We also assume that the function $p(\mu)$, $\mu \in \Lambda$, is nondecreasing and bounded and the function $r(\mu)$, $\mu \in \Lambda$, is nondecreasing and unbounded. The nonlinear eigenvalue problem (1), (2) is approximated by the finite difference method on a uniform grid. A sufficient condition for the existence of a minimal eigenvalue corresponding to a positive eigenfunction of the finite difference nonlinear eigenvalue problem is established. Error estimates for the approximate eigenvalue and the corresponding approximate positive eigenfunction are proved.

Nonlinear eigenvalue problems of the form (1), (2) arise in modeling the plasma of radio-frequency discharge at reduced pressures. An inductive coupled radio-frequency discharge has found broad applications in diverse technological plasma processes, such as processing textiles and leather-fur half-finished products, metals, hydrogen accumulation by silicon powders, synthesis of oxygen-free ceramic materials, and obtaining carbide and boride materials for nuclear and processing industry [1]. A more effective and qualitative choice of constructive solutions in designing inductive coupled radio-frequency devices requires mathematical models, because some technological characteristics of the plasma cannot be measured.

Nonlinear eigenvalue problems also arise in various fields of science and technology [2-4]. Numerical methods for solving matrix eigenvalue problems with nonlinear dependence on the parameter were constructed and investigated in the papers [5-13]. Mesh methods for solving differential nonlinear eigenvalue problems were studied in [14-16]. The theoretical basis for the study of nonlinear eigenvalue problems is results obtained for linear eigenvalues problems [17-23]. In the papers [24-30], numerical methods for solving applied nonlinear boundary value problems and variational inequalities have been studied.

2 Variational statement of the problem

Let $H = L_2(\Omega)$ denote the real Lebesgue space with norm $|\cdot|_0$ and inner product $(\cdot,\cdot)_0$ defined by

$$|v|_0 = \left(\int_0^\pi (v(x))^2 \mathrm{d}x \right)^{1/2}, \tag{3}$$

$$(u,v)_0 = \int_0^\pi u(x)v(x)\mathrm{d}x \;\; \forall u,v \in H. \tag{4}$$

By $V = \{v : v, v' \in H, v(0) = v(\pi) = 0\}$ we denote the real Sobolev space with norm $|\cdot|_1$ and inner product $(\cdot,\cdot)_1$ defined by

* Corresponding author: sergei.solovyev@kpfu.ru

$$|v|_1=\left(\int_0^\pi (v'(x))^2 dx\right)^{1/2}, \quad (5)$$

$$(u,v)_1=\int_0^\pi u'(x)v'(x)dx \ \forall u,v\in V. \quad (6)$$

For $\mu\in\Lambda$, we introduce the bilinear forms defined by the formulas

$$a(\mu,u,v)=\int_0^\pi p(\mu s(x))u'v'dx, \quad (7)$$

$$b(\mu,u,v)=\int_0^\pi r(\mu s(x))uvdx, \quad (8)$$

for any $u,v\in V$, and the Rayleigh ratio

$$R(\mu,v)=\frac{a(\mu,v,v)}{b(\mu,v,v)} \quad \forall v\in V\setminus\{0\}. \quad (9)$$

Set $K=\{v:\ v\in V,\ v(x)>0,\ x\in\Omega\}$.

The differential eigenvalue problem (1), (2) is equivalent to the following variational eigenvalue problem: find the least $\lambda\in\Lambda$, $u\in K$, $b(\lambda,u,u)=1$, such that

$$a(\lambda,u,v)=b(\lambda,u,v) \quad \forall v\in V. \quad (10)$$

For fixed $\mu\in\Lambda$, we introduce the following linear parametric variational eigenvalue problem: find the least $\gamma(\mu)\in\mathbb{R}$, $u\in K$, $b(\mu,u,u)=1$, such that

$$a(\mu,u,v)=\gamma(\mu)b(\mu,u,v) \quad \forall v\in V. \quad (11)$$

The following variational property is valid

$$\gamma(\mu)=\min_{v\in V\setminus\{0\}} R(\mu,v). \quad (12)$$

Let $\gamma_i(\mu)$, $u_i(\mu)=u_i(\mu,x)$, $x\in\bar{\Omega}$, $\mu\in\Lambda$, $i=1,2,...$, be eigenvalues and eigenfunctions satisfying equation (3),

$$\gamma_1(\mu)<\gamma_2(\mu)<...<\gamma_i(\mu)<..., \quad (13)$$

$$a(\mu,u_i(\mu),u_j(\mu))=\gamma_i(\mu)\delta_{ij}, \quad (14)$$

$$b(\mu,u_i(\mu),u_j(\mu))=\delta_{ij}, \quad (15)$$

$i,j=1,2,...$, $u_1(\mu)\in K$, $\gamma_1(\mu)=\gamma(\mu)$, $\gamma_i(\mu)\to\infty$ as $i\to\infty$, functions $u_i(\mu)$, $\mu\in\Lambda$, $i=1,2,...$, form a complete system in the space V. The minimal eigenvalue λ of problem (10) is the least root of the equation $\gamma(\mu)=1$, $\mu\in\Lambda$.

Theorem 1. If $p(0)>r(0)$, then there exists a minimal simple eigenvalue of problem (10) corresponding to a positive eigenfunction.

The result of Theorem 1 is proved by analogy with [15, 16].

3 Finite difference approximation of the problem

Let us partition the interval $[0,\pi]$ by equidistant points $x_i=ih$, $i=0,1,\dots,N$, into elements $e_i=[x_{i-1},x_i]$, $i=1,2,\dots,N$, $h=\pi/N$, and let V_h denote the subspace of the space V consisting of continuous functions v^h linear on each element e_i, $i=1,2,\dots,N$. For $\mu\in\Lambda$, u^h, $v^h\in V_h$, we introduce approximate bilinear forms

$$a_h(\mu,u^h,v^h)= =\sum_{i=1}^N hp(\mu s(x_i-h/2))(u^h(x_i-h/2))'(v^h(x_i-h/2))', \quad (16)$$

$$b_h(\mu,u^h,v^h)= =\sum_{i=1}^N h(r(\mu s(x_{i-1}))u^h(x_{i-1})v^h(x_{i-1})+ +r(\mu s(x_i))u^h(x_i)v^h(x_i))/2. \quad (17)$$

Set $K_h=\{v^h:\ v^h\in V_h,\ v^h(x)>0,\ x\in\Omega\}$.

We approximate the original eigenvalue problem (10) by a finite-dimensional eigenvalue problem: find the least $\lambda^h\in\Lambda$, $u^h\in K_h$, $b(\lambda^h,u^h,u^h)=1$, such that

$$a_h(\lambda^h,u^h,v^h)=b_h(\lambda^h,u^h,v^h) \quad \forall v^h\in V_h. \quad (18)$$

For fixed $\mu\in\Lambda$, we introduce the following linear parametric variational eigenvalue problem: find the least $\gamma^h(\mu)\in\mathbb{R}$, $u^h\in K_h$, $b_h(\mu,u^h,u^h)=1$, such that

$$a_h(\mu,u^h,v^h)=\gamma^h(\mu)b_h(\mu,u^h,v^h) \quad \forall v^h\in V_h. \quad (19)$$

Denote

$$R_h(\mu,v)=\frac{a_h(\mu,v^h,v^h)}{b_h(\mu,v^h,v^h)} \quad \forall v^h\in V_h\setminus\{0\}. \quad (20)$$

Then the following variational property is valid

$$\gamma^h(\mu)=\min_{v^h\in V_h\setminus\{0\}} R_h(\mu,v^h). \quad (21)$$

Let $\gamma_i^h(\mu)$, $u_i^h(\mu)=u_i^h(\mu,x)$, $x\in\bar{\Omega}$, $\mu\in\Lambda$, $i=1,2,...,N-1$, be eigenvalues and eigenfunctions satisfying equation (19),

$$\gamma_1^h(\mu)<\gamma_2^h(\mu)<...<\gamma_{N-1}^h(\mu), \quad (22)$$

$$a_h(\mu,u_i^h(\mu),u_j^h(\mu))=\gamma_i^h(\mu)\delta_{ij}, \quad (23)$$

$$b_h(\mu,u_i^h(\mu),u_j^h(\mu))=\delta_{ij}, \quad (24)$$

$i,j=1,2,\dots,N-1$, $u_1^h(\mu)\in K_h$, $\gamma_1^h(\mu)=\gamma^h(\mu)$, functions $u_i^h(\mu)$, $\mu\in\Lambda$, $i=1,2,\dots,N-1$, form a complete system in the space V_h. The minimal eigenvalue λ^h of problem (18) is the least root of the equation $\gamma^h(\mu)=1$, $\mu\in\Lambda$. Put $\sigma_h=(\sin(h/2)/(h/2))^2$.

Theorem 2. If $p(0)\sigma_h>r(0)$, then there exists a minimal simple eigenvalue of problem (18) corresponding to a positive eigenfunction.

Proof. Using the variational property of the minimal eigenvalue of problem (19), we derive

$$\gamma^h(0)=\min_{v^h\in V_h\setminus\{0\}}R_h(0,v^h)= \tag{25}$$

$$=\min_{v^h\in V_h\setminus\{0\}}\frac{\int_0^\pi p(0)((v^h(x))')^2\,dx}{\int_0^\pi r(0)(v^h(x))^2\,dx}= \tag{26}$$

$$=\frac{p(0)}{r(0)}\min_{v^h\in V_h\setminus\{0\}}\frac{\int_0^\pi ((v^h(x))')^2\,dx}{\int_0^\pi (v^h(x))^2\,dx}=\frac{p(0)}{r(0)}\sigma_h>1. \tag{27}$$

Denote

$$s_1=\min_{x\in\Omega}s(x),\quad p_1=\sup_{\mu\in\Lambda}p(\mu). \tag{28}$$

Then we obtain

$$\gamma^h(\mu)=\min_{v^h\in V_h\setminus\{0\}}R_h(\mu,v^h)= \tag{29}$$

$$=\min_{v^h\in V_h\setminus\{0\}}\frac{\int_0^\pi p(\mu)((v^h(x))')^2\,dx}{\int_0^\pi r(\mu)(v^h(x))^2\,dx}\le \tag{30}$$

$$\le\frac{p_1}{r(\mu s_1)}\min_{v^h\in V_h\setminus\{0\}}\frac{\int_0^\pi ((v^h(x))')^2\,dx}{\int_0^\pi (v^h(x))^2\,dx}= \tag{31}$$

$$\le\frac{p_1}{r(\mu s_1)}\sigma_h\to 0 \tag{32}$$

as $\mu\to\infty$, since $r(\mu)\to\infty$, as $\mu\to\infty$.

Thus, the continuous function $\gamma^h(\mu)$, $\mu\in\Lambda$, has the properties $\gamma^h(0)>1$ and $\gamma^h(\mu)\to 0$ as $\mu\to\infty$. This implies the existence of a minimal root of equation

$$\gamma^h(\mu)=1,\quad \mu\in\Lambda, \tag{33}$$

which defines the minimal eigenvalue λ^h of problem (18). The minimal eigenvalue λ^h is simple and corresponds to a positive eigenfunction, because $\gamma^h(\mu)$ with $\mu\in\Lambda$ is the simple eigenvalue of the parametric problem (19) corresponding to a positive eigenfunction. This completes the proof of the theorem.

Denote $y_i=u^h(x_i)$, $i=0,1,\dots,N$,

$$\tilde p_i(\lambda^h)=p(\lambda^h s(x_i-h/2)), \tag{34}$$

$$r_i(\lambda^h)=r(\lambda^h s(x_i)), \tag{35}$$

$y_{x,i}=(y_{i+1}-y_i)/h$, $y_{\bar x,i}=(y_i-y_{i-1})/h$. Then problem (18) can be written in the finite difference form

$$-(\tilde p(\lambda^h)y_{\bar x})_{x,i}=\lambda^h r_i(\lambda^h)y_i,\ i=1,2,\dots,N-1, \tag{36}$$

$$y_0=y_N=0. \tag{37}$$

By c we denote various positive constants independent of h. Put

$$a'(\mu,u,v)=\int_0^\pi p'(\mu s(x))s(x)u'v'\,dx, \tag{38}$$

$$b'(\mu,u,v)=\int_0^\pi r'(\mu s(x))s(x)uv\,dx, \tag{39}$$

for $\mu\in\Lambda$, $u,v\in V$. Then the function $\gamma^h(\mu)$, $\mu\in\Lambda$, is continuous and the following formula is valid

$$\gamma'(\mu)=a'(\mu,v,v)-\gamma(\mu)b'(\mu,v,v) \tag{40}$$

for $\mu\in\Lambda$, $v=u_1(\mu)$.

Theorem 3. Suppose $\gamma'(\lambda)\neq 0$, λ, u, and λ^h, u^h, are eigenvalues and eigenfunctions of the eigenvalue problems (10) and (18). Then the estimates

$$|\lambda^h-\lambda|\le ch^2,\quad |u^h-u|_1\le ch, \tag{41}$$

hold for sufficiently small h.

Proof. For $\mu\in\Lambda$, v^h, $w^h\in V_h$ we have the estimates

$$|a_h(\mu,v^h,w^h)-a(\mu,v^h,w^h)|\le ch^2|v^h|_1|w^h|_1, \tag{42}$$

$$|b_h(\mu,v^h,w^h)-b(\mu,v^h,w^h)|\le ch^2|v^h|_1|w^h|_1. \tag{43}$$

Then the error estimates of Theorem 3 are proved with using results from [19-23]. This completes the proof of the theorem.

To illustrate the theoretical results of Theorems 1 and 2, we have solved the eigenvalue problem (10) for $s(x)=\sqrt{\pi-x}+1$, $x\in\bar\Omega$,

$$p(\mu)=\begin{cases}-\mu^3+\mu^2+\mu+2, & \mu\in[0,1],\\ 3, & \mu\in(1,\infty),\end{cases} \tag{44}$$

$$r(\mu)=\begin{cases}1, & \mu\in[0,1],\\ (\mu-1)^2/2+1, & \mu\in(1,\infty),\end{cases} \tag{45}$$

using the finite difference scheme (36), (37) for $N = 100$, $\mu \in [0,3]$. In this case, $p(0) = 2$, $r(0) = 1$, $\sigma_h = 0.99992$; therefore, the conditions of Theorems 1 and 2 are valid. Fig. 1 shows the graph of the function $\gamma(\mu)$ of the parametric eigenvalue problem (11) and the minimal simple eigenvalue $\lambda = 1.3248$ of the nonlinear eigenvalue problem (10). We see that the experimental results are consistent with the theoretical results in Theorems 1 and 2. Note that investigations of the present paper generalize well known results for eigenvalue problems with linear dependence on the spectral parameter.

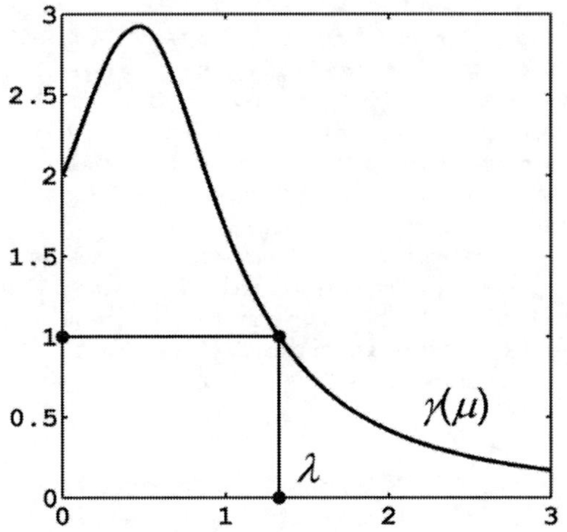

Fig. 1. The minimal eigenvalue λ.

This work was supported by Russian Foundation for Basic Research (Project nos. 16-01-00301, 16-31-00378).

References

1. I.Sh. Abdullin, V.S. Zheltukhin, N.F. Kashapov, *Radio-Frequency Plasma-Jet Processing of Materials at Reduced Pressures: Theory and Practice of Applications* (Izd. Kazan. Univ., Kazan, 2000) [in Russian]
2. A.V. Goolin, S.V. Kartyshov, Surv. Math. Ind. **3**, 29-48 (1993)
3. T. Betcke, N.J. Higham, V. Mehrmann, C. Schröder, F. Tisseur, ACM Trans. Math. Software **39**, 7 (2013)
4. V.A. Kozlov, V.G. Maz'ya, J. Rossmann, *Spectral Problems Associated with Corner Singularities of Solutions to Elliptic Equations* (AMS, Providence, 2001)
5. R.Z. Dautov, A.D. Lyashko, S.I. Solov'ev, Russ. J. Numer. Anal. Math. Modelling **9**, 417-427 (1994)
6. A. Ruhe, SIAM J. Numer. Anal. **10**, 674-689 (1973)
7. V. Mehrmann, H. Voss, GAMM-Mit. **27**, 1029-1051 (2004)
8. F. Tisseur, K. Meerbergen, SIAM Rev. **43**, 235-286 (2001)
9. D. Kressner, Numer. Math. **114**, 355-372 (2009)
10. X. Huang, Z. Bai, Y. Su, J. Comput. Math. **28**, 218-234 (2010)
11. H. Schwetlick, K. Schreiber, Linear Algebra Appl. **436**, 3991-4016 (2012)
12. W.-J. Beyn, Linear Algebra Appl. **436**, 3839-3863 (2012)
13. X. Qian, L. Wang, Y. Song, J. Comput. Appl. Math. **290**, 268-277 (2015)
14. A.V. Kregzhde, Differ. Uravn. **17**, 1280-1284 (1981)
15. S.I. Solov'ev, Differ. Equations **50**, 947-954 (2014)
16. S.I. Solov'ev, Differ. Equations **51**, 934-947 (2015)
17. S.I. Solov'ev, Differ. Equations **30**, 1138-1146 (1994)
18. S.I. Solov'ev, Differ. Equations **38**, 752-753 (2002)
19. S.I. Solov'ev, Differ. Equations **46**, 1030-1041 (2010)
20. S.I. Solov'ev, Differ. Equations **47**, 1188-1196 (2011)
21. S.I. Solov'ev, Differ. Equations **48**, 1028-1041 (2012)
22. S.I. Solov'ev, Differ. Equations **49**, 908-916 (2013)
23. S.I. Solov'ev, Appl. Numer. Math. **93**, 206-214 (2015)
24. I.B. Badriev, G.Z. Garipova, M.V. Makarov, V.N. Paymushin, Res. J. Appl. Sciences **10**, 428-435 (2015)
25. I.B. Badriev, L.A. Nechaeva, PNRPU Mech. Bull. 3, 37-65 (2013)
26. I.B. Badriev, M.V. Makarov, V.N. Paimushin, Proc. Engin. **150**, 1050-1055 (2016)
27. I.B. Badriev, G.Z. Garipova, M.V. Makarov, V.N. Paimushin, R. F. Khabibullin, Lobachevskii J. Math. **36**, 474-481 (2015)
28. I.B. Badriev, M.V. Makarov, V.N. Paimushin, Russ. Math. **59**, 57-60 (2015)
29. I.B. Badriev, M.V. Makarov, V.N. Paimushin, Proc. Engin. **150**, 1056-1062 (2016)
30. I.B. Badriev, V.V. Banderov, V.L. Gnedenkova, N.V. Kalacheva, A.I. Korablev, R.R. Tagirov, Appl. Math. Sciences **9**, 5697-5705 (2015)

MATEC Web of Conferences **129**, 06015 (2017) DOI: 10.1051/matecconf/201712906015
ICMTMTE 2017

Role of asphalt paver productivity in direction development of road mechanical engineering

Marina Karpushko[1,*], and *Igor* Bartolomei[1]

[1]Perm National Research Polytechnic University, 614000, Komsomolsky prospekt, 29, Perm, Russia

Abstract. Road construction machinery is one of the largest and stable sectors of national economy. With the help of road-building machinery, construction, maintenance and repair of highways are carried out. Modern system of road and building machines is a complex of high-performance machines and mechanisms, of large and small capacity and productivity. Expansion of paved roads network of both federal and local values, use of resource-saving technologies, increasing of pace and quality of work, ensuring of reliability and durability of highways have a major impact on the development of road construction machinery. The level of foreign road and construction machinery puts forward complex requirements for compliance with world safety and environmental standards, increasing equipment mobility, expanding areas of effective use and, finally, increasing its capacity and productivity. Article contains approach to the calculation of the operational productivity of asphalt pavers that takes into account technical characteristics and conditions of works production and contributes to the reliability of decisions on organization and management of road maintenance works on urban roads.

Roads are of strategic importance for any country, and for Russian Federation especially because of the vast territory which it occupies. Their quality, pace of construction, possibility of using fundamentally new technologies, structures and materials, as well as the cost, are largely determined by the availability of modern highway engineering machinery. In this regard, Federal Road Agency Rosavtodor, guided by the Transport Strategy of Russian Federation, develops interdepartmental programs for the development of road machinery.

The main focus is on import substitution, by creating conditions for the production of machinery and equipment on the territory of Russian Federation, rising of technical level and reliability of existing machines, creating fundamentally new models of road construction machinery and equipment. As a result, in recent decades, plants have mastered the production of new models of highway engineering machinery.

Technology of highways construction provides the use of high-performance mechanized links of road machinery. For upper layers of asphalt concrete pavements arrangement, a set consisting of an asphalt paver and a link of road rollers is used. In many aspects the quality and service life of the pavement depends from this set of machines.

Determining factors in the intensity and quality of repair works of hot mix asphalt concrete pavements are: production of the mixture at the asphalt concrete plant, delivery of the mixture to an object by motor transport, preparation of the required existing coating operations, laying and compaction processes of mixture (Figure 1).

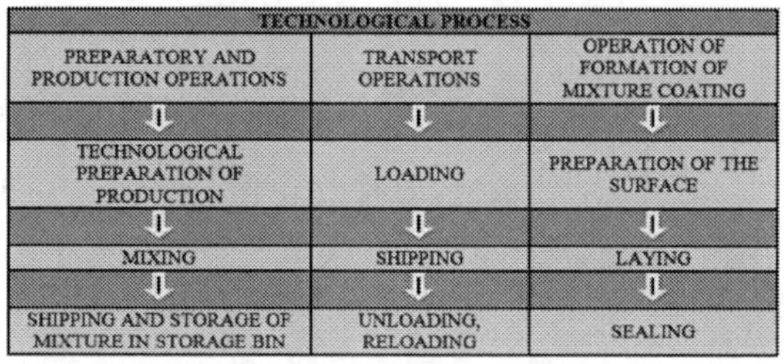

Fig. 1. Schematic of the process and operations performed during the repair of asphalt concrete pavement.

The assessment of the selected operations performed on the basis of simulation modeling and collection of statistical data of road-building enterprises work showed that under conditions of sufficient motor transport the intensity of work is mainly determined by the work of the laying link (coefficient of influence 0,565):

$$I = 0{,}271 \cdot P_{\text{п}}^{0{,}385} W_{\text{т}}^{0{,}051} P_{\text{у}}^{0{,}565} \quad (1)$$

where I - intensity of works; $P_{\text{п}}$ - performance of asphalt concrete plant; $W_{\text{т}}$ - performance of the transport unit; $P_{\text{у}}$ - performance of a laying link; 0,271 - coefficient reflecting influence on production of work's intensity of unaccounted factors; 0,385 / 0,051 / 0,565 - coefficients of influence of technological factors on the intensity of the implementation of road renovations.

The correlation coefficient – 0,98, standard deviation 0,107. Due to the high performance of the asphalt

* Corresponding author: mkarpushko@gmail.com

DOI: 10.1051/matecconf/201712906015

concrete plants their impact is 1,4 times less (coefficient of influence 0,385). Influence of motor transport insignificant (0,051).

Nowadays in development of construction and repair of asphalt concrete pavements technology there is tendency to use high-productive asphalt pavers. Pavers of American production are characterized by high productivity, which is provided by a high operating speed and large paving width [1, 2]. Asphalt pavers of European production: «Demag», «Vogele», «ABG», «Baukema», «Dynapac», etc., in addition to increased productivity, have the ability to pre-compact the mixture to standard coefficient of consolidation with active working elements - tamper and vibratory plates (Figure 2).

These companies produce both tracked and wheeled pavers with different paving width: 0,80-16,0 m (tracked), 1,5-9,1 m (wheel); and capacity up to 1500 tons / hour (Vogele Super 2500), up to 1900 tons / hour (crawler asphalt paver Vogele Super 1900-2). Hydraulic spreader or hinged widener can increase the width of the working elements of the paver, with maximum thickness of the layer of 110-500 mm at an operating speed of 1,08 - 110 m / min. [3].

Fig. 2. Mix laying by Vogele Super 1803-1 asphalt paver on Universitetsky Ave. in Volgograd, Russia.

Tracked and wheeled pavers, which are capable to provide a preliminary compaction of asphalt-concrete mixes, are produced by Russian companies [1]. Track models: DS-195, DS-179, DS-189; wheeled: DS-155, DS-173, SD-404b, DS-191, AK-7, DS-181 (Figure 3).

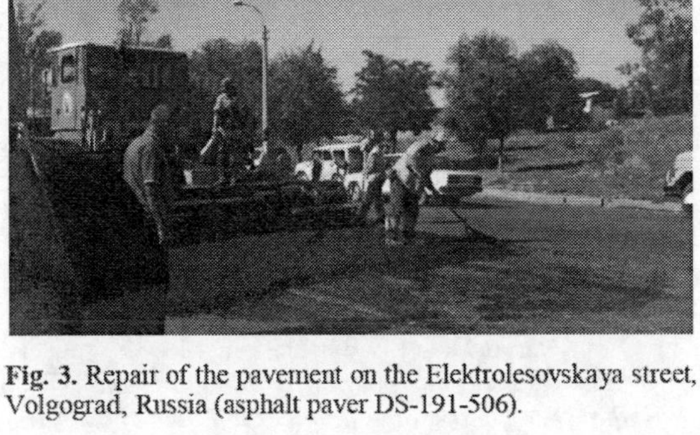

Fig. 3. Repair of the pavement on the Elektrolesovskaya street, Volgograd, Russia (asphalt paver DS-191-506).

If laying is made by strips of different width and at the same time one or several asphalt pavers are used, different methods of work's organization are applied accordingly. For example, if laying is made on width of 1,0-12,0 m - laying is made by one asphalt paver. With a laying width of 9 m and more two and more asphalt pavers are used. If the technology requires the work of several asphalt pavers, then subject distance between them is 10-15 m (maximum 30 m) and they are put with a ledge.

Knowing the cooling time of asphalt mixture at the permissible temperature intervals (τy) and operating speed of the paver (Va/y), you can determine the length of the strip to be laid for roller operations (Sy):

$$S_y = V_{a/y} \cdot \tau_y \qquad (2)$$

In order to make the process of paving continuous and in the right temperature range, it is recommended to take the length of laying strips equal to the length of the clamping, and the length of the strip to be compacted by the rollers should be more than 10-15 m, in order to ensure their acceleration and braking, as well as to satisfy the following condition [4, 5]:

$$S_y = [S] \qquad (3)$$

where [*S*] - recommended minimum length of the seal area, m; *Sy* - length of the strip, to be laid by paver for the same time, m.

The efficiency of paver unit is determined by the productivity of asphalt paver, as it distributes the mixture on divisions, provides all required parameters of layer and ultimately provides front work operation for roller link. Modern models of asphalt pavers of European and American manufacturers have a maximum passport capacity 500-700 tons / hour. It is calculated from the conditions of the maximum possible thickness and width of the packing, operating speed and uniform movement of the stacker.

In production conditions, the coincidence of all the parameters is rare, so in practice the productivity of machines is 1,5-2 times or 250-350 tons / hour lower, in some cases does not exceed 100-150 tons per hour, in rare cases it reaches 150-200 tons / hour, which is associated with a low supply of a mixture from asphalt concrete plant. Therefore, this performance is taken into account while units recruiting. But there is a reserve of up to 1,5-2 times in increasing the intensity of work, which confirms a comparison of the possible and actual productivity of pavers.

Productivity of paver is distributed according to the normal law or β-distribution (Figure 4). The coefficient of variation varies in the range 0,17-0,56.

Calculations of the productivity of pavers according to formulas overestimate the shift production of machines to 20-30% [7]. Assigning performance based on time norms, is only possible for individual domestic pavers.

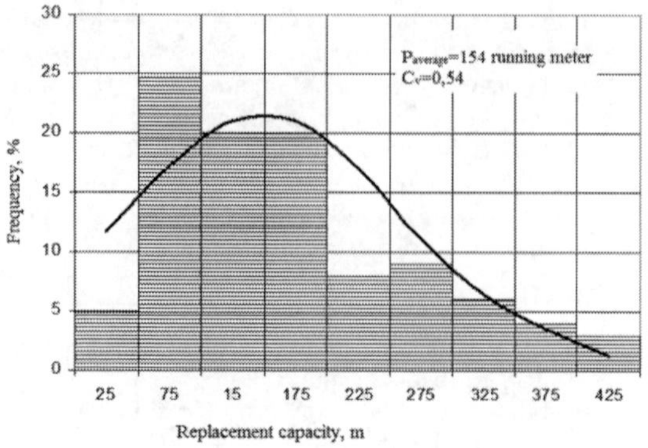

Fig. 4. Distribution of replacement capacity of DS-1 paver (according to [6]).

On the basis of the regression dependences of the maximum permissible thickness of the mixture laying in the coating h, machine shifting from width H and working speed of the paver Vук, it was proposed to perform the justification of the performance of new modern pavers. Statistics were compiled for 98 asphalt pavers [3], which allowed to propose the following dependencies:

For crawler pavers:

$$P_y = \frac{0{,}0019 \cdot H^{0{,}359} \cdot h^{2{,}102} \cdot V_{\text{ук}}^{0{,}112}}{K_o \cdot K_п}, \text{ t / h} \qquad (4)$$

For wheeled pavers:

$$P_y = \frac{0{,}0007 \cdot H^{1{,}039} \cdot h^{2{,}022} \cdot V_{\text{ук}}^{0{,}299}}{K_o \cdot K_п}, \text{ t / h} \qquad (5)$$

where K_O - coefficient that takes into account the decrease in pavers productivity in the dark, as the road reconstruction works in urban conditions is preferably carried out at night time (Table 1); $K_п$ - coefficient of influence on labor productivity in warm season of weather conditions (Table 2).

Table 1. Values of coefficient K_o, depending on work place lighting conditions [8].

Lighting conditions		K_o
Natural lighting	Normal, daytime	1,0
	Bright sunny days in southern regions	1,1-1,2
	Twilight, moonlit night	1,4-1,6
Artificial lighting	Normal, using fixtures	1,1-1,2
	With spotlights	1,3-1,4

Table 2. Values of operating conditions coefficient $K_п$ for warm period of the year [8].

Weather		$K_п$
Air temperature, °C	Precipitation	
0… +25	Dry weather	1,0
+10… +25	Precipitation > 5 mm/day	1,2-1,3
0… +10	Precipitation > 5 mm/day	1,4-1,5
Above +25	Dry weather	1,4-1,6

When the weather is dry and the ambient temperature is from 0 to + 25 °C maximum productivity is reached, which decreases during rain and wind conditions, as well as at the exit out of comfortable temperature range limits. The correlation coefficient – 0,80-0,85, standard deviation 0,42.

All of the above indicates the complexity of fulfilling the specified parameters of the asphalt-concrete paving. Today, without the use of high technologies, it is impossible to obtain high-quality road surface. The proposed approach to the calculation of the operational productivity of asphalt pavers takes into account technical characteristics and conditions of works production and contributes to the reliability of decisions on organization and management of road maintenance works on urban roads.

References

1. V.K. Kuznetsov, *Maintenance and current repair of city roads*. Moscow, Alfa-Print, P. 151 (1998)
2. M.P. Kostelyov, *Modern methods and means of pavement patching*, Road engineering, № **5**, pp. 56-59 (2001)
3. Catalogue of road construction equipment, St. Petersburg, (2000-2008)
4. A.V. Zubkov, *Technique of technological processes development of hot asphalt mixtures pavement's construction*, Herald TSTU, vol. **13**, № **1**, P. 52 (2007)
5. A.V. Zubkov, *Influence of factors on duration of hot asphalt mixtures laying during highways construction,* Russia Roads of XXI century, № **4**, pp. 42-45 (2006)
6. V.M. Mogilevich, *Winter organization and technology of road construction*, Moscow, P. 276 (1971)
7. A.Z. Apartsev, *Flow charts and rules of laying and compaction of hot asphalt mixes during highways coatings application*, St.-Petersburg; P. 51 (1966)
8. M. Eddous, *Methods of decision-making*, Moscow: Audit: UNITY, P. 590 (1997)
9. M.O. Karpushko, E.N. Karpushko, I.L. Bartolomei, *World experience of operational management in construction,* Globalization and its Socio-Economic Consequences 16th International Scientific Conference Proceedings. ZU - University of Zilina, 5th -6th October 2016, Rajecke Teplice, Slovak Republic, pp. 823-830 (2016)
10. E. Popkova, S. Meshkova, E. Karpunina, E. Karpushko, M. Karpushko, *Developing countries as new growth poles of post-crisis global economy*, Contemporary Economics, vol. **10**, № **2**, pp. 175-186 (2016)
11. E.N. Karpushko, I.L Bartolomei, M.O. Karpushko, *Study of using the possibility of textile sand piles at the base of the automobile road folded by saline soils*, Procedia Engineering. vol. **150**, pp. 2287-2292 (2016)

MATEC Web of Conferences **129**, 06016 (2017) DOI: 10.1051/matecconf/201712906016
ICMTMTE 2017

Mechanics of unloading of a rough surfaces pre-loaded joint

Peter Ogar[1], *Denis* Gorokhov[1,*], and *Elena* Ugryumova[1]

[1] Bratsk State University, 665709 Makarenko 40, Bratsk, Russia

Abstract. In this paper we study the problem of the contact characteristics changing on a decrease of the load applied to the preloaded joint of roughness surfaces. The penetration of a rigid sphere (indenter) into the elastic hardenable half-space is considered originally. The elastic crater restoring by unloading is also considered. In elastic-plastic material's describing, Hollomon's pawer law is used. To describe a contact of a rigid rough surface with an elastic plastic half-space, the discrete model of a rough surface is used. Microasperities are represented as a set of identical spherical segments, the height distribution of which corresponds to the bearing profile curve of the surface. To describe the bearing profile curve, incomplete beta function ratio is used. The relations of relative contact areas η and η_e and gap densities at the joint Λ_e and Λ_{ce} on dimensionless loading F_q and F_{qe} at loading and unloading for different values of ε_y and n are given. The obtained results are of practical importance for the sealing ability prediction of fixed sealing joints at the design stage, in particular for tightness supply of flange couplings and high pressure vessels seals.

1 Introduction

Many operational properties of joints of machine parts and devices, including tightness, are defined by contact interaction of rough surfaces [1]. In most cases, by contacting metal rough surfaces, the contact is elastic-plastic. In [2, 3] for the description of an elastic-plastic contact, the authors use the kinetic sphere indentation's load-displacement diagram. As it follows from [1], the sealing ability of a sealing joint is defined by the relative contact area and the density of gaps in a joint. From this point of view, the loading of sealing joints at their assembly by the increased load and maintenance of their tightness by lower load to providing the minimum weight and size characteristics is of practical interest. Especially it is actual at the flange joints and high pressure vessels design.

To solve this problem it is necessary to know how the contact characteristics (the relative contact area and the density of gaps) change during unloading preloaded fixed joint.

2 Contact of a rigid sphere with the elastic-plastic half-space

Let us consider the indentation of the rigid sphere (indenter) into the elastic-plastic half-space and the elastic crater restoring. For the description of an elastic-plastic hardening material we use the Hollomon's power law

$$S=\begin{cases}\varepsilon E, & \varepsilon \le \varepsilon_y;\\ \sigma_y^{1-n}E^n\varepsilon^n, & \varepsilon > \varepsilon_y,\end{cases} \quad (1)$$

where n is the strain hardening exponent, $\varepsilon_y = \sigma_y / E$, σ_y is the yield strength, E is the elastic modulus.

By using a spherical indenter, the degree of deformation changes at all stage of loading and the primary result of such test is the kinetic indentation's load-displacement diagram [4] (Fig. 1).

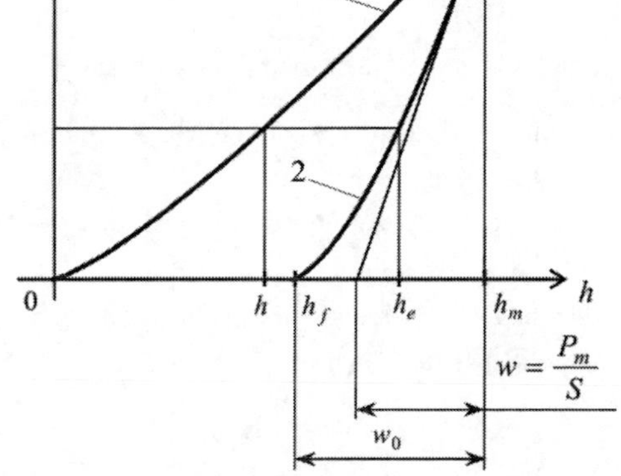

Fig. 1. The typical kinetic indentation's load-displacement diagram: 1 corresponds to loading, 2 corresponds to unloading

According to [3], the load-displacement branch can be written as

---
[*] Corresponding author: gorokhov@brstu.ru

$$P = C_1 h^{\alpha}, \tag{2}$$

and the unload-displacement branch can be written as

$$P_e = P_m \left(\frac{h_e - h_f}{h_m - h_f} \right)^{\gamma} \tag{3}$$

where C_1, C_2 are constants; α and γ are exponents.

The initial unloading stiffness is

$$S = \left. \frac{dP_e}{dh_e} \right|_{h_e = h_m} = \frac{P_m \cdot \gamma}{h_m - h_f} = \frac{P_m \cdot \gamma}{w_0}. \tag{4}$$

From Fig. 1 follows that

$$w = \frac{P_m}{S} = \frac{w_0}{\gamma}, \quad \gamma = \frac{w_0}{w}. \tag{5}$$

In [3] parameter γ is determined by means of a calculation

$$\gamma = \frac{3 - 2c^2 \bar{h}_m}{2 - c^2 \bar{h}_m}, \tag{6}$$

where $\bar{h}_m = h_m / R$, R is the sphere radius, $c^2 = h_c / h$, h_c is the depth at which the contact of an indenter and a material under load P_m takes place.

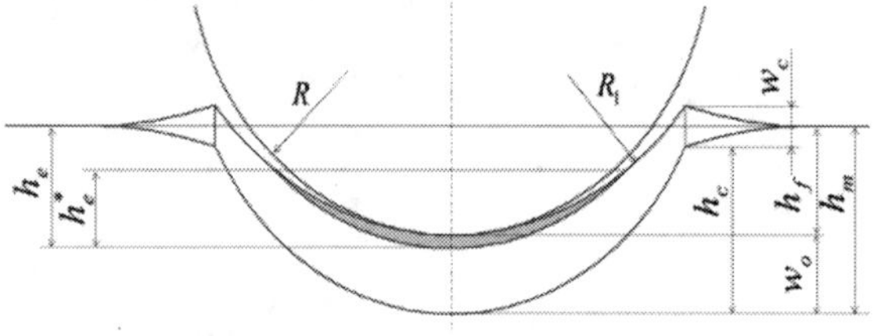

Fig. 2. Schematic illustration of a crater under unloading of a spherical indenter

The problem of determination of the contact area at elastic restoring was considered in [5], however the author did not take into account effects of "sink-in/pile-up". Under loading to the contact depth h_e^* (Fig. 2), the subsequent elastic restoring will be equal to $h_e - h_f$, in a similar under loading to the contact depth h_c, the elastic restoring will be equal to $h_m - h_f$. Let's make an assumption that the ratio takes place

$$\frac{h_e^*}{h_e - h_f} = \frac{h_c}{h_m - h_f},$$

then we have

$$h_e^* = h_c \frac{h_e - h_f}{h_m - h_f}, \qquad h_f \le h_e \le h_m. \tag{7}$$

At $h_e = h_f$ we have $h_e^* = 0$, at $h_e = h_m$ we have $h_e^* = h_c$, i.e. boundary conditions are carried out.

The contact area at crater restoring is

$$A_e = \pi a_e^2 = 2\pi R h_e^*. \tag{8}$$

3 Contact of a rigid rough surface with the elastic-plastic half-space

3.1 The discrete model of a rough surface

Let us use the discrete model of a rough surface, in which microasperities are presented by identical spherical segments with the distribution of segments' peaks on height corresponds to the bearing profile curve of the real surface [1]. To describe the distribution of the reference curve is used incomplete beta function

$$\eta(\varepsilon) = \frac{B_{\varepsilon}(p,q)}{B(p,q)}, \tag{9}$$

where $B_{\varepsilon}(p,q)$ and $B(p,q)$ are incomplete and complete beta functions;

$$p = \left(\frac{R_p}{R_q}\right)^2 \left(\frac{R_{max} - R_p}{R_{max}}\right) - \frac{R_p}{R_{max}}, \quad q = p\left(\frac{R_{max}}{R_p} - 1\right). \tag{10}$$

where R_p, R_q, R_{max} are height roughness parameters according to ISO 4281/1–1997.

In this case, the density of the asperities distribution on height function is

$$\varphi_n'(u) = \frac{u^{p-2}(1-u)^{q-2}\left[(p-1)(1-u) - (q-1)u\right]}{\varepsilon_s^{p-1}(1-\varepsilon_s)^{q-1}}, \tag{11}$$

where ε_s is determined from the condition $\varphi_n(\varepsilon_s) = 1$ [1].

Geometrical parameters of a spherical segment are: the height is ωR_{max}, where $\omega = 1 - \varepsilon_s$; a_c is the radius of the base of the spherical segment.

$$R = \frac{a_c^2}{2\omega R_{max}}, \tag{12}$$

where $R >> R_{max}$.

At elastic contact, the relation between the relative size of indentation of the single asperity and the relative load is defined according to the Hertz theory.

$$\frac{P_{ei}}{E^* R^2} = \frac{4}{3}\left(\frac{h_i}{R}\right)^{\frac{3}{2}}, \tag{13}$$

where E^* is the reduced (contact) elastic modulus.

For elastic-plastic contact we use the equation from [6]

DOI: 10.1051/matecconf/201712906016

$$\frac{P_{epi}}{E^* R^2} = e^{-B}\left(\frac{h_i}{R}\right)^A, \quad (14)$$

where $A = A(\varepsilon_y, n)$ and $B = B(\varepsilon_y, n)$.

3.2 The relative contact area

This problem has been considered in detail by [7]. When using the Eq. (14) for asperities of a rough surface, it is necessary to take into account that

$$h_i = (\varepsilon - u)R_{\max}, \quad (15)$$

$$\frac{h_i}{R} = \frac{(\varepsilon - u)\cdot 2\omega R_{\max}^2}{a_c^2} = \left(\frac{\varepsilon - u}{2\omega}\right)\cdot\left(\frac{2\omega R_{\max}}{a_c}\right)^2, \quad (16)$$

where ε is the relative approach a rough surface and the half-space; u is the initial distance to the peak of i-th asperity.

When the indentation of a rigid rough surface is equal ε, the total load P is described by equation

$$P = \int_{\varepsilon-\varepsilon_e}^{\varepsilon} P_{ei} dn_r + \int_0^{\varepsilon-\varepsilon_e} P_{epi} dn_r, \quad (17)$$

where ε_e is the relative boundary an elastic contact, dn_r is the number of peaks in a layer du,

$$dn_r = n_c \varphi_n'(u)du, \quad n_c = \frac{A_c}{\pi a_c^2}. \quad (18)$$

Let us consider a case, when $\varepsilon_e \to 0$.

Substituting Eq. (18) in Eq. (17), we have

$$\frac{P}{A_c} = q_c = \int_0^{\varepsilon} q_{cepi}\cdot\varphi_n'(u)du, \quad (19)$$

where $q_{cepi} = \frac{P_{epi}}{\pi a_c^2}$.

By denoting

$$\frac{q_c a_c}{\omega R_{\max} E^*} = F_q \quad (20)$$

and by taking into account Eq. (14), (16), (19), we will finally obtain

$$F_q(\varepsilon) = \frac{2^{2(A-1)} e^{-B}}{\pi}\left(\frac{\omega R_{\max}}{a_c}\right)^{2A-3}\cdot\int_0^{\varepsilon}\left(\frac{\varepsilon - u}{2\omega}\right)^A \varphi_n'(u)du. \quad (21)$$

For the real contact area, similar to Eq. (17), we have

$$A_r = \int_0^{\varepsilon} A_{repi} dn_r. \quad (22)$$

Determining A_{ri}, we take into account that

$$h_{ci} = c_i^2 \cdot h_i. \quad (23)$$

According to [8], we have

$$c_i^2 = \frac{h_{ci}}{h_i} = M^{\frac{2}{N}}\left(2\frac{h_i}{R}\right)^{\frac{2}{N}-1}, \quad (24)$$

where $M = M(\varepsilon_y, n)$ and $N = N(\varepsilon_y, n)$.

For the relative contact area $\eta = A_r/A_c$, we will finally obtain

$$\eta(\varepsilon) = (2M)^{\frac{2}{N}}\left(\frac{2\omega R_{\max}}{a_c}\right)^{2\left(\frac{2}{N}-1\right)} \times \int_0^{\varepsilon}\left(\frac{\varepsilon - u}{2\omega}\right)^{\frac{2}{N}} \varphi_n'(u)du. \quad (25)$$

Let us consider the unloading process. For a single spherical asperity according to Eq. (3), we have

$$P_{ei} = P_{epi}\left(\frac{h_{ei} - h_{fi}}{h_i - h_{fi}}\right)^{\gamma_i}. \quad (26)$$

For approximate calculations it is possible to accept $\gamma_i \approx 1{,}5$.

By taking into account that

$$h_i - h_{fi} = w_{0i}, \quad (27)$$

Eq. (3) can be written as

$$P_{ei} = P_{epi}\left(\frac{h_i - \Delta h - h_{fi}}{h_i - h_{fi}}\right)^{\gamma} = P_{epi}\left(\frac{w_{0i} - \Delta h}{w_{0i}}\right)^{\gamma}. \quad (28)$$

Similar to the equation Eq. (16) for $\Delta h/R$, we have

$$\frac{\Delta h}{R} = \frac{\Delta_\varepsilon}{2\omega}\left(\frac{2\omega R_{\max}}{a_c}\right)^2. \quad (29)$$

It should be noted that under unloading of a rigid rough surface, the size Δh is identical for all asperities.

Using data from [2], we determine w_{0i} as

$$\bar{w}_{0i} = \frac{w_{0i}}{R} = \frac{e^{B}\bar{h}^{A-\frac{1}{N}} K_{\alpha 0}}{\pi(2M)^{\frac{1}{N}}}, \quad (30)$$

where $K_{\alpha 0} = 2^{2\alpha-1}\alpha\,\mathrm{B}(\alpha,\alpha)$, $\alpha = A(\varepsilon_y, n)$.

By taking into account Eq. (16), we have

$$\bar{w}_{0i} = \frac{K_{\alpha 0} e^{-B}}{\pi(2M)^{\frac{1}{N}}}\left(\frac{\varepsilon - u}{2\omega}\right)^{A-\frac{1}{N}}\left(\frac{2\omega R_{\max}}{a_c}\right)^{2\left(A-\frac{1}{N}\right)}. \quad (31)$$

Then for unloading process, similar to Eq. (21), we will obtain

DOI: 10.1051/matecconf/201712906016

$$F_{qe}\left(\varepsilon_y, n, \xi, \Delta_\varepsilon\right) = \frac{2^{2(\alpha-1)} e^{-B}}{\pi}\left(\frac{\omega R_{\max}}{a_c}\right)^{2A-3} \times \\ \times \int_0^\varepsilon \left(\frac{\xi - u}{2\omega}\right)\left(H\left(\varepsilon_y, n, \xi, u, \Delta_\varepsilon\right)\right)^m \varphi'_n(u)du, \tag{32}$$

where

$$H\left(\varepsilon_y, n, \xi, u, \Delta_\varepsilon\right) = \begin{cases} \dfrac{w_{0i}\left(\varepsilon_y, n, \xi, u\right) - \Delta_\varepsilon}{w_{0i}\left(\varepsilon_y, n, \varepsilon, u\right)}, \quad w_0 > \Delta_\varepsilon; \\ 0, w_0 \le \Delta_\varepsilon. \end{cases} \tag{33}$$

Using Eq. (7) and Eq. (8), similar to Eq. (25), will have

$$\eta_e\left(\varepsilon_y, n, \xi, \Delta_e\right) = (2M)^{\frac{2}{N}}\left(\frac{2\omega R_{\max}}{a}\right)^{2\left(\frac{2}{N}-1\right)} \times \\ \times \int_0^\varepsilon \left(\frac{\xi - u}{2\omega}\right)^{\frac{2}{N}} H\left(\varepsilon_y, n, \xi, u, \Delta_\varepsilon\right)\varphi'_n(u)du, \tag{34}$$

where $H\left(\varepsilon_y, n, \xi, u, \Delta_\varepsilon\right)$ is defined by Eq. (33).

It should be noted that

$$\left(\Delta_\varepsilon\right)_{\max} = \frac{w_{0i}\left(\varepsilon_y, n, \xi, 0\right)}{2\omega}\left(\frac{a_c}{R_{\max}}\right)^2. \tag{35}$$

The fig. 3 shows the relations of loading $F_q(\varepsilon)$ and unloading $F_{qe}(\xi-\Delta\varepsilon)$ of a contact of a rigid rough surface with the elastic-plastic half-space for different values ε_y and n calculated on Eqs. (21) and (32).

The fig. 4 shows the relations the relative contact area η and η_e versus the dimensionless force elastic-geometric parameter F_q and F_{qe} under loading and unloading, respectively.

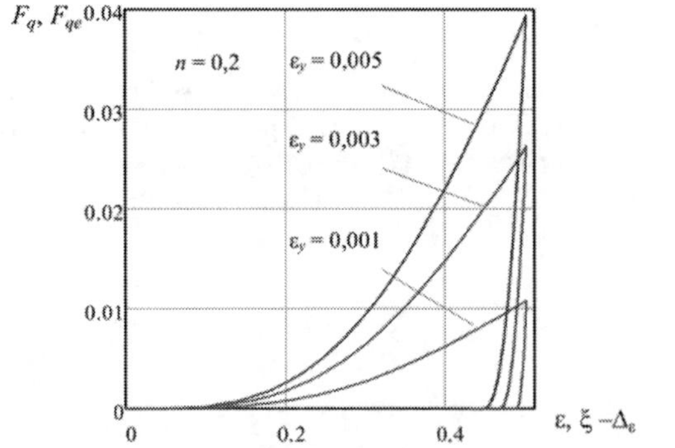

Fig. 3. The loading and unloading diagram of a contact of a rigid rough surface with the elastic-plastic half-space

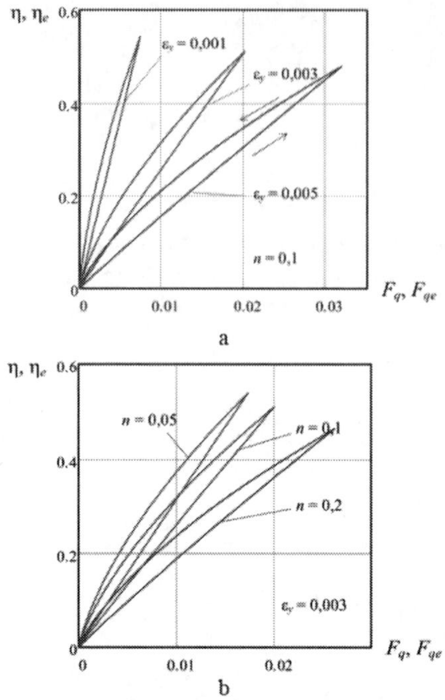

Fig. 4. The relative contact area versus the dimensionless force elastic-geometric parameter $\eta(F_q)$ and $\eta_e(F_{qe})$ for different values ε_y and n

3.3 The density of gaps in the joint

During indentation of rough surface on the value $\varepsilon R_{\max}$, contact of every single asperity is followed by "pile-up/sink-in" effects, i.e. by plastic extrusion of the half-space's material and its elastic squeezing [9]. For gaps volume is followed that

$$V_c = A_c R_{\max}\left(1 - K_p\right) - A_c \varepsilon R_{\max} + V_e - V_p. \tag{36}$$

where V_e is the total volume increase due to the elastic material squeezing for all the contacting asperities, V_e is the total volume decrease due to the plastic material extrusion for all the contacting asperities.

Correspondingly, the joint's gaps density is

$$\Lambda_c = \frac{V_3}{A_c R_{\max}} = 1 - K_p - \varepsilon + \Lambda_e - \Lambda_p, \tag{37}$$

where $\Lambda_e = V_e/\left(A_c R_{\max}\right)$, $\Lambda_p = V_p/\left(A_c R_{\max}\right)$, K_p is the profile fill factor

$$K_p = \int_0^1 \eta(\varepsilon)d\varepsilon. \tag{38}$$

For a symmetric profile (when $\alpha = \beta$) $K_p = 0{,}5$.

The total volume due to the elastic squeezing of all asperities is

DOI: 10.1051/matecconf/201712906016

$$V_e(\varepsilon_y,n,\varepsilon)=\int_0^{\varepsilon} V_{ei}(\varepsilon_y,n,\varepsilon,u)n_c\cdot\varphi'(u)du, \quad (39)$$

where

$$V_{ei}=2\pi\int_{a_r}^{a_{\varepsilon}} ru_{ze}(r)dr, \quad (40)$$

u_{ze} is the elastic displacement of surface points out of the contact area [10].

From the eq. for spherical segment volume with height h_f and unloaded crater radius ρ_i, the volume of plastically displaced material, attributable to a single hole, is

$$V_{pi}(\varepsilon_y,n,\varepsilon,u)=\pi\rho h_{fi}^2-\frac{\pi}{3}h_{fi}^2=\pi h_{fi}^2\left(\rho_i-\frac{h_{fi}}{3}\right) \quad (41)$$

and the total displaced volume is

$$V_p(\varepsilon_y,n,\varepsilon)=\int_0^{\varepsilon} V_{pi}(\varepsilon_y,n,\varepsilon,u)n_c\varphi_n'(u)du. \quad (42)$$

Substituting Eq. (42) and Eq. (39) in Eq. (36), we determine the the density of gaps in the joint during rigid rough surface indentation into the elastoplastic hardening half-space as

$$\Lambda_c(\varepsilon_y,n,\varepsilon)=1-K_p+\Lambda_{ee}(\varepsilon_y,n,\varepsilon)-\Lambda_p(\varepsilon_y,n,\varepsilon) \quad (43)$$

Carrying out similar arguments for the unloading process, we obtain

$$V_{ee}(\varepsilon_y,n,\xi,\Delta_\varepsilon)=\int_0^{\xi} V_{eei}(\varepsilon_y,n,\xi,u,\Delta_\varepsilon)n_c\cdot\varphi_n'(u)du. \quad (44)$$

The density of gaps in the joint during unloading is

$$\begin{aligned}\Lambda_{ce}(\varepsilon_y,n,\xi,\Delta_\varepsilon)=1-K_p-(\xi-\Delta_\varepsilon)+\\+\Lambda_{ee}(\varepsilon_y,n,\xi,\Delta_\varepsilon)-\Lambda_p(\varepsilon_y,n,\xi)\end{aligned} \quad (45)$$

The fig. 5 shows the densities of gaps Λ_c and Λ_{ce} versus the dimensionless force elastic-geometric parameter F_q and F_{qe} under loading and unloading for different values ε_y and n.

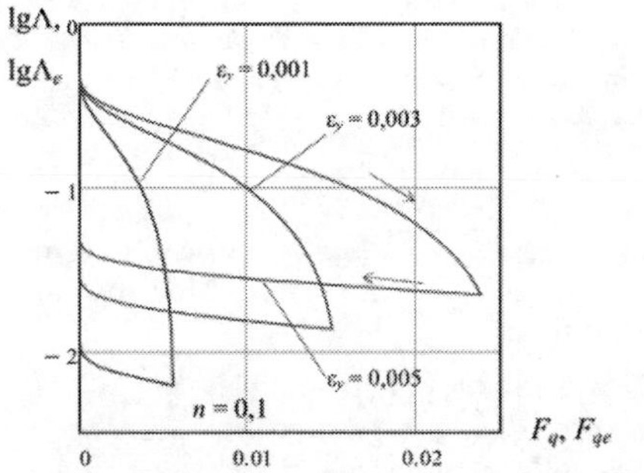

a

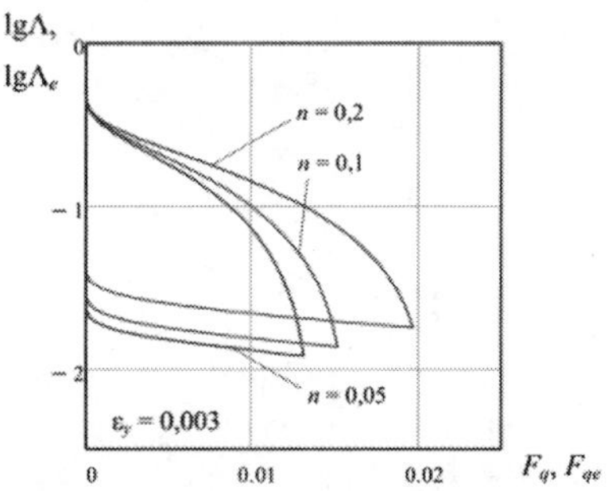

b

Fig. 5. The densities of gaps versus the dimensionless force elastic-geometric parameter $\Lambda_c(F_q)$ and $\Lambda_{ce}(F_{qe})$ or different values ε_y and n

4 Application of the obtained results to practical calculations

For example, the sealing ability of the joint of an elastoplastic contact is considered [11]. The intensity of the volume rate (per unit length along the seal perimeter) of ideally compressible gas through a sealing joint with a uniform distribution of contact pressure q_c is

$$Q=C_u\cdot C_q, \quad (46)$$

where C_u is the dimensionless permeability functional that characterizes the sealing ability of the joint

$$C_u(\varepsilon_y,n,\varepsilon)=\frac{(\Lambda(\varepsilon_y,n,\varepsilon))^3\cdot\nu_n(\varepsilon_y,n,\varepsilon)}{4(1-\eta(\varepsilon_y,n,\varepsilon))^2}, \quad (47)$$

$$C_q=\frac{R_{\max}^3(p_1^2-p_0^2)}{2l\mu}, \quad (48)$$

where p_1 is the environment pressure, p is the atmosphere pressure, μ is the dynamic viscosity of the environment, l is the width of the sealing area, ν_n is the rate of effective micro-channels.

At pressures above 16 MPa at a leakage calculations, it should to use the model of a real gas [12].

The rate of effective micro-channels [11] is

$$\nu_n(\varepsilon_y,n,\varepsilon)=\begin{cases}(1-x^*), & x^*<\frac{1}{3};\\ 3^{n-1}(1-x^*)^n, & x^*\geq\frac{1}{3};\end{cases} \quad (49)$$

where n is the number of rows of microasperities on the sealing zone width.

$$x^*(\varepsilon_y,n,\varepsilon)=\int_0^{\varepsilon} x_i^*(\varepsilon_y,n,\varepsilon,u)\varphi'(u)du, \quad (50)$$

where

$$x_i^*\left(\varepsilon_y,n,\varepsilon,u\right)=\begin{cases}0, & \eta_i\left(\varepsilon_y,n,\varepsilon,u\right)<\eta_i^*;\\ 1, & \eta_i\left(\varepsilon_y,n,\varepsilon,u\right)\geq\eta_i^*;\end{cases} \quad (51)$$

$\eta_i^* = 0{,}5...0{,}55$ is the critical value of η_i.

Using Eqs. (25), (43) and (49) – (51), it is possible to determine the dimensionless permeability functional during loading of the sealing joint. For unloading, it should to use Eqs. (36) and (45) and also to take into account that for determining of $\nu_{ne}\left(\varepsilon_y,n,\xi,\Delta_\varepsilon\right)$

$$\eta_{ei}\left(\varepsilon_y,n,\xi,u,\Delta_\varepsilon\right)=\eta_i\left(\varepsilon_y,n,\xi,u\right)\cdot H\left(\varepsilon_y,n,\xi,u,\Delta_\varepsilon\right) \quad (52)$$

The fig. 6 shows the densities of gaps C_u and C_{ue} versus the dimensionless force elastic-geometric parameter F_q and F_{qe} under loading and unloading for different values ε_y and n.

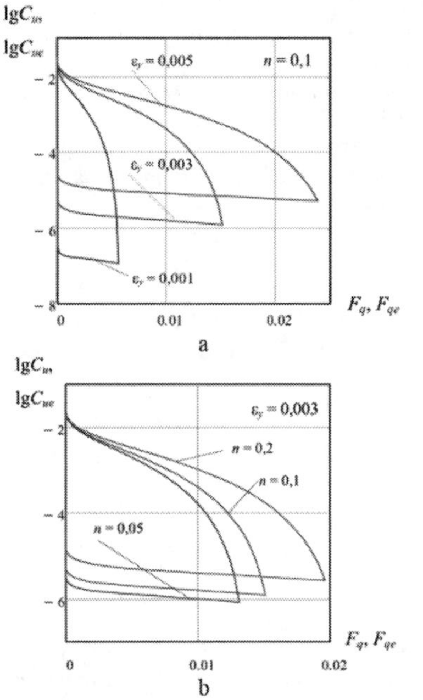

Fig. 6. dimensionless permeability functional versus the dimensionless force elastic-geometric parameter $C_u(F_q)$ and $C_{ue}(F_{qe})$ for different values ε_y and n

5 Conclusions

1. Using a discrete model of a rough surface, the authors obtained dependences for determining the relative contact area and the density of the gaps in the joint during indentation of a rigid rough surface into the elastic-plastic half-space and the subsequent unloading.

2. This makes it possible to predict the sealing ability of the sealing joint at various operating conditions, leading to unloading pre-loaded sealing joint.

3. When load fully reduced, the functional permeability functional and hence the amount of leakage are reduced by about half order. This effect can be used to reduce the weight and size characteristics of the flanged joints and gaps of high-pressure vessels by loading it by an increased effort and maintaining required tightness by an lower effort.

4. The role of the material hardening parameters ε_y and n is shown. More efficient in terms of providing a predetermined tightness by an lower effort is ε_y.

References

1. N.B. Demkin, E.V. Ryzhov, *Surface quality and contact of machine parts,* 244 (Mashinostroyeniye, Moscow, 1981)
2. A.A. Lankov, *Probability of elastic and plastic deformations while compressing metal rough surfaces*, Treniye i smazka v mashinakh i mekhanizmakh **3**, 3-5. (2009)
3. P.M. Ogar, V.A. Tarasov, *Kinetic indentation application to determine contact characteristics of sphere and elastoplastic half-space*, Adv. Mater. Res. **664**, 625-631 (2013)
4. S.I. Bulychev, V.P. Alekhin,. *Testing of materials by continuous depression of an indenter,* 224 (Mashinostroyeniye, Moscow, 1990)
5. A.P. Kovalev, *Basic laws of the indentation of a spherical indenter and evaluation of physical and mechanical properties of the surface layer of machine elements,* Uprochnyayushchiye tekhnologii i pokrytiya **1,** 36-41. (2007)
6. J.-M. Collin, G. Mauvoisin, P. Pilvin, Materials characterization by instrumented indentation using two different approaches, Materials and Desing, **31,** 636-640 (2010)
7. P.M. Ogar, D.B. Gorokhov. *Influence of materials hardenability parameters on the machine parts characteristics after unloading*, Key Engineering Materials **723**, 369-375 (2017)
8. X. Hernot, O. Bartier, Y. Bekouche, R. El Abdi, G. Mauvoisin, *Influence of penetration depth and mechanical properties on contact radius determination for spherical indentation,* Int. Journal of Solids and Structures, **43**, 4136-4153 (2006)
9. P.M. Ogar, V.A. Tarasov, A.V. Turchenko, *Contact of a rigid rough surface with elastic-plastic half space*, Systems. Methods. Technologies, **13**, 17-22 (2012)
10. P.M. Ogar, V.A. Tarasov, *Influence of axisymmetric form of load on stressedly-deformed state of elastoplastic half-space*, Systems. Methods. Technologies **5**, 14-20 (2010)
11. Yu.N. Alpatov, V.A. Tarasov, A.V. Turchenko, *Impact of the hardenable material characteristics on the joints sealing property,* Sovremennye tekhnologii. Sistemny analiz. Modelirovaniye, **2**, 83-88 (2012)
12. D.B. Gorokhov, A.A. Sobolevsky, *Leakage variation of gas environment through the joint of releasable connections with temperature increasing,* Sovremennye tekhnologii. Sistemny analiz. Modelirovaniye, **3**, 70-76 (2016)

The elastic-plastic contact of a single asperity of a rough surface

Peter Ogar[1,*], *Denis* Gorokhov[1,*], and *Sergey* Belokobylsky[1]

[1] Bratsk State University, 665709 Makarenko 40, Bratsk, Russia

Abstract. A penetration of spherical asperity into the elastic-plastic hardening half-space is described. The elastic-plastic material properties correspond to Hollomon's power law. In this case the empirical Meyer law relating a spherical indentation load with an indentation diameter *d* is used. Initially, the Meyer law is not related to the mechanical characteristics of the test material. The study used the relations between the strain hardening exponent *n* and the Meyer law constant obtained by S.I. Bulychev. The effects relating to elastic punching and plastic displacement of material are taked into account. It is shown that there is no need to define Meyer law constants. Expressions relating the value of the relative load magnitude to the relative indenter penetration magnitude are presented. The scope of application of the proposed equations is defined. A comparison of the obtained results with the experimental data and published data of the finite element analysis is given.

1 Introduction

Widespread use in tribology finds a discrete model of roughness, in which the asperities are presented as a set of bodies of regular geometric shapes, for which solutions of contact problems are available [1, 2]. In this case, the asperity model in the form of a spherical segment is considered to be optimal.

Problems of spherical asperity elastic-plastic penetration are not sufficiently studied and the solutions suggested require clarification and improvement [3]. One of important problems to be taken into account is material hardening. The authors' approach to solving this problem is given in [4-6 et al.]. The method consists in the use of the kinetic indentation load-displacement diagram and the similarity method of deformation characteristics. In this case, the notion "plastic hardness" is used as characteristics of material resistance to contact plastic strain. Plastic hardness has the form

$$HD = K_h(\varepsilon_y, n)\cdot\sigma_y,$$

where σ_y is the yield strength, $K_h(\varepsilon_y,n)$ is a parameter defined by the double indentation technique [4] using the results of the finite element analysis [7], ε_y and n are the characteristics of Hollomon's elastic-plastic material.

In a number of works [8, 9] the empirical Meyer law linking the spherical indentation load and an indenter diameter was used to allow for material hardening in solving the tribomechanics problems. In [9] an influence of some physical and mechanical properties of real materials on the features of contact elastic-plastic deformation formation is emphasized. However, the limitation of the method is that it does not explicitly take into account elastic-plastic characteristics of the hardened material.

The aim of this work is to use the above mentioned hardened material characteristics in the Meyer law application.

2 Problem solution

In describing elastic-plastic characteristics of the hardened material Hollovon's power law is widely used. According to this law the relation between the true stress S and the strain ε under uniaxial tension or compression is described by equations

$$S = \begin{cases} \varepsilon E, & \varepsilon \le \varepsilon_y; \\ K\varepsilon^n, & \varepsilon \ge \varepsilon_y; \end{cases} \quad (1)$$

where E is an elastic modulus, n is a strain-hardening exponent.

The constant K is determined from the equality condition for σ at ε_y. Then the second expression in Eq. (1) can be written as

$$\frac{S}{\sigma_y} = \left(\frac{E\varepsilon}{\sigma_y}\right)^n = \left(\frac{\varepsilon}{\varepsilon_y}\right)^n, \ \varepsilon \ge \varepsilon_y. \quad (2)$$

where $\sigma_y \approx S_y$, σ_y - yield strength, $\varepsilon_y = \sigma_y/E$.

Taking into accord that the limiting uniform strain $\varepsilon_u = n$, the strain hardening exponent can be defined from following Eq. as [10]

$$n\ln n - n(1+\ln\varepsilon_y) - \ln\frac{\sigma_u}{\sigma_y} = 0, \quad (3)$$

* Corresponding authorgorokhov@brstu.ru

 DOI: 10.1051/matecconf/201712906017

where σ_u is the tensile strength.

E. Meyer was the first to describe a material behavior in the elastic-plastic domain. He related the load P to the indentation diameter d as

$$P = Ad^m. \quad (4)$$

The empirical Meyer law is often written as

$$\frac{4P}{\pi d^2} = A^*\left(\frac{d}{D}\right)^{m-2}. \quad (5)$$

where m, A, A^* are constants. A^* has a dimension of strength.

The equation in the left part is a mean contact area pressure referred to as the Meyer hardness

$$\frac{4P}{\pi d^2} = \frac{P}{\pi a^2} = p_m = HM, \quad (6)$$

where a is the radius of the contact area.

Using the maximum Meyer hardness concept we have

$$HM_{\max} = HM\big|_{d=D},$$

from equations (5) and (6) it follows that

$$p_m = HM_{\max}\left(\frac{d}{D}\right)^{m-2} = HM_{\max}\left(\frac{a}{R}\right)^{m-2}. \quad (7)$$

The maximum Meyer hardness relates to the Brinell hardness as [11]

$$HM_{\max} = 2HB/k_m, \quad (8)$$

$$k_m = m^{\frac{m}{2}}(m-1)^{1-m}(m-2)^{\frac{m-2}{2}}. \quad (9)$$

From [12] it follows that

$$\sigma_u = k_\sigma \cdot HB, \quad (10)$$

where $k_\sigma = 0.333$ for carbon and pearlitic steel, for other materials the values of k_σ are given in [13, 14]. As it shown in [14], in practice well justifies the dependence

$$k_\sigma = 0.333 \cdot (\delta_u/\psi)^{0.25}$$

where δ_u is the uniform elongation, ψ is the deformation in a print of a spherical indenter.

According to the data obtained by S.I. Bulychev [15], the limiting uniform strain ε_u corresponding to σ_u is equal to

$$\varepsilon_p = n = 0.961(m-2) \quad (11)$$

From here we have

$$m = 2 + 1.041n. \quad (12)$$

From Eq. (10) with regard to Eqs. (2) and (11) it follows that

$$HB = \frac{\sigma_y}{k_\sigma}\left(\frac{n}{e}\right)^n \cdot \varepsilon_y^{-n}. \quad (13)$$

where e is base of natural logarithm.

For strain (load) $P = \pi a^2 p_m$ with regard to Eqs. (7), (8) and (13), we have

$$P = 2\pi\frac{\sigma_y a^2}{k_\sigma \cdot k_m}\left(\frac{n}{e}\right)^n \varepsilon_y^{-n}\left(\frac{a}{R}\right)^{m-2},$$

$$\frac{P}{E^* R^2} = \frac{2\pi}{k_\sigma \cdot k_m}\left(\frac{n}{e}\right)^n \varepsilon_y^{1-n}\left(\frac{a}{R}\right)^m. \quad (14)$$

With regard to (12) we have

$$\frac{P}{E^* R^2} = \frac{2\pi}{k_\sigma \cdot k_n}\left(\frac{n}{e}\right)^n \varepsilon_y^{1-n}\left(\frac{a}{R}\right)^{2+1.041n}. \quad (15)$$

where E^* is reduced elastic modulus,

$$k_n = \frac{(2+1.041n)^{1+0.5205n}}{(1+1.041n)^{1+1.041n}}(1.041n)^{0.5205n}.$$

As $a = \sqrt{2Rh_c} = \sqrt{2Rc^2h}$, where h_c is the depth at which the sphere contact with a half-space takes place, h is the depth of indentation from the initial surface level, we have

$$\frac{P}{E^* R^2} = \frac{2\pi}{k_\sigma \cdot k_n}\left(\frac{n}{e}\right)^n \varepsilon_y^{1-n}\left(\frac{2c^2h}{R}\right)^{1+0.5025n}. \quad (16)$$

The parameter $c^2 = h_c/h$ can be defined based on the data obtained in [16]

$$c^2 = c^2(n) = 1.4e^{-0.97n}, \quad (17)$$

Equation (17) is in good agreement with the results of FE simulations [17].

A lower limit is $\bar{a}_y = a_y/R$ corresponding to ε_y. According to [15] under spherical indentation, the strain is defined by the relation

$$\varepsilon = \alpha\left(\frac{d}{D}\right)^\beta = \alpha\left(\frac{a}{R}\right)^\beta, \quad (18)$$

where

$$\alpha = 0.15 + 1.85(m-2)^{1.4},\ \beta = 1.18\left(1+(m-2)^{1.6}\right).$$

Or with regard to Eq. (12) it follows that

$$\alpha = 0.15 + 1.85(1.041n)^{1.4},\ \beta = 1.18\left(1+(1.041n)^{1.6}\right).$$

 DOI: 10.1051/matecconf/201712906017

Then we have

$$\overline{a}_y = \left(\frac{\varepsilon_y}{\alpha}\right)^{\frac{1}{\beta}}, \qquad (19)$$

and the corresponding load is defined from Eq. (15).

An upper limit is defined from the condition $h_c = c^2 h^* = R$.

With regard to Eq. (12) we have

$$\frac{h^*}{R} = 1.4e^{-0.97n}. \qquad (20)$$

The corresponding load is defined from Eq. (16).

3 Comparison with experimental data and published data of the finite element analysis

For estimation of the obtained results, we will compare it with the experimental study given in [18]. In this paper for calculating parameters of elastic-plastic contact, the authors used the generalized deformation curve and method of variable elasticity parameters. The experimental study were to determine the contact radius curvatur during indentation hardened steel ball (HRC 63…64) with a radius of curvature 2.5 mm into material samples specimens in the table. Mechanical properties of the material of specimens defined by results of tension.

To the analytical dependences for each material from the Eq. (3), the values of hardening exponent n were determined. The values of ε_y were determined by the formula

$$\varepsilon_y = \frac{\sigma_y}{E^*} = \sigma_y\left(\frac{1-\nu_i^2}{E_i} + \frac{1-\nu^2}{E}\right), \qquad (21)$$

where ν_i, E_i are Poisson's ratio and the elastic modulus of the indenter.

Table 1. Mechanical properties of materials [18]

No	Materials	σ_y, MPa	σ_u, MPa	E, GPa	ν
1	Armko-iron	256	410	210	0.3
2	Steel 45	480	725	204	0.26
3	Steel 30XГСА	667	942	215	0.3
4	Steel 30XГСА	1207	1344	215	0.3
5	Copper M2	69	196	132	0.35
6	Duraluminium	265	392	72	0.3
7	Titanium BT6	687	883	118	0.32

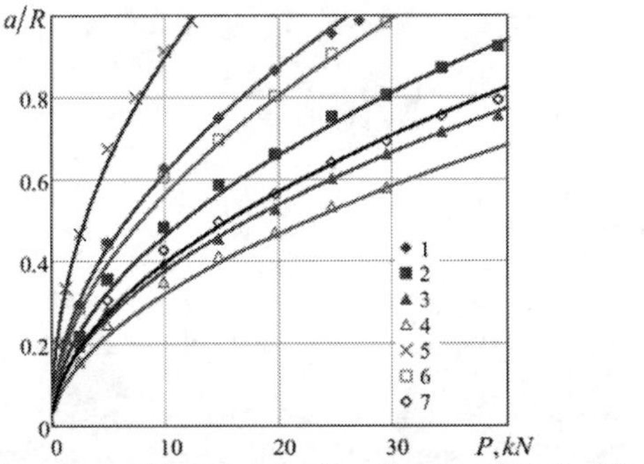

Fig. 1. Dependence of the relative radius of contact of the load P.

The fig. 1 show the experimental data [18] are presented by points, the coordinates were have been "digitized" in the processing of the results. The lines denote the corresponding dependence, calculated by the Eq. (15).

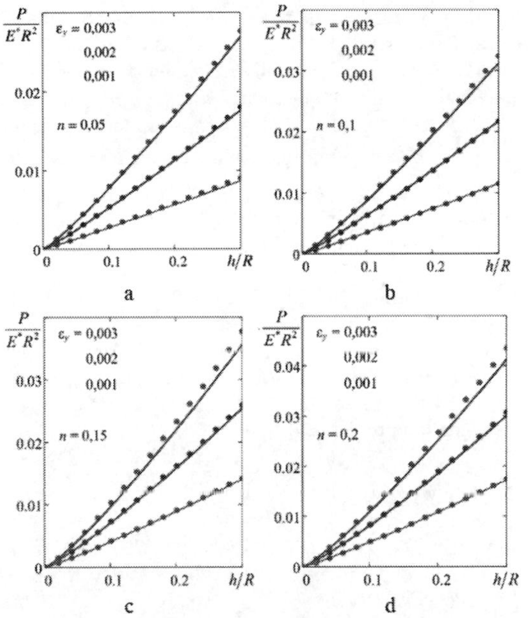

Fig. 2. Dependences $\overline{P}-\overline{h}$: for $\varepsilon_y = 0{,}001$; $0{,}002$; $0{,}003$; *a*) $n = 0{,}05$, *b*) $n = 0{,}1$, *c*) $n = 0{,}15$, *d*) $n = 0{,}2$; solid lines - according to expression (18); dashed - according to the expression (16).

Let us also compare the results with the data of the finite element analysis [19].

$$\frac{P}{E^* R^2} = e^{-B(\varepsilon,n)}\left(\frac{h}{R}\right)^{A(\varepsilon,n)}, \qquad (22)$$

where

$$A(\varepsilon_y, n) = \frac{\left[\begin{array}{l}0{,}93 + 290{,}65\varepsilon_y - 23408{,}85\varepsilon_y^2 + \\ + 546404.43\varepsilon_y^3 + 3{,}11n + 0{,}77n^2\end{array}\right]}{\left[\begin{array}{l}1 + 194.49\varepsilon_y - 17101{,}85\varepsilon_y^2 + \\ + 407006{,}05\varepsilon_y^3 + 2{,}1684n\end{array}\right]},$$

$$B(\varepsilon_y, n) = \frac{\begin{bmatrix} 5{,}30 + 81{,}12\varepsilon_y - 21284{,}04\varepsilon_y^2 + \\ + 381357.70\varepsilon_y^3 - 8{,}34n + 1{,}81n^2 \end{bmatrix}}{\begin{bmatrix} 1 + 437{,}16\varepsilon_y - 9993{,}45\varepsilon_y^2 + \\ + 0{,}34{,}05n - 3{,}43n^2 \end{bmatrix}}.$$

The fig. 2 show the dependences calculated by Eq. (16) and Eq. (22). As it follows from Fig. 1 and Fig. 2, there is a good agreement between the analytical dependences, the experimental data and the results of the finite element analysis.

Thus, the proposed approach suggests an alternative to a more complex method for describing elastoplastic penetration of a sphere on the basis of the kinetic indentation diagram [5], which was used in solving problems of elastoplastic contacting of rough surfaces.

4 Conclusions

1. Using the empirical Meyer law initially not linked to the physical and mechanical properties of materials, we obtained Eqs. (15) and (16), which allows *defining* a contact area radius and spherical indentation depth into an elastic-plastic hardening half-space.
2. Characteristics of elastic-plastic material such as $\varepsilon_y = S_y / E^*$ and strain hardening exponent n are taken into account. It is shown that there is no need of determine the Meyer law constants m and A^*.
3. The scope of the obtained equations application is defined.
4. A comparison of the results obtained by the authors with the published results of spherical indentation into elastic-plastic half-space obtained by the experimental data allows recommending the given relations for the solution to problems of surface plastic deformation problems.

References

1. N.B. Demkin, Eh.V. Ryzhov, *Surface quality and contact of machine parts,* 244 (Mashinostroyeniye, Moscow, 1981)
2. I.G. Goryacheva, N.M. Dobychin, *Contact problems in tribology,* 256 (Mashinostroyeniye, Moscow, 1988)
3. M.M. Matlin, S.L. Lebsky, A.I. Mozgunova, *Laws of elastoplastic contact problems in surface plastic hardening,* 218 (Mashinostroyeniye-1, Moscow, 2007)
4. P.M. Ogar, V.A. Tarasov, A.V. Turchenko, *Effect of the hardenable material characteristics on a spherical asperity elastoplastic indentation,* Systems. Methods. Technologies **11**, 29-34 (2011)
5. P.M. Ogar, V.A. Tarasov, *Kinetic Indentation Application to Determine Contact Characteristics of Sphere and Elastoplastic Half-Space,* Adv. Mater. Res. **664**, 625-631 (2013)
6. P.M. Ogar, V.A. Tarasov, D.B. Gorokhov, A.V. Turchenko, *Contact mechanics of rough surfaces,* 282 (Brstu, Bratsk, 2016)
7. H. Lee, J.H. Lee, G.M. Pharr, *A numerical approach to spherical indentation techniques for material property evaluation,* J. Mech. Phys. Solids **53**, 2037-2069 (2005)
8. N.K. Myshkin, M.I. Petrokovets, *Friction, lubrication, wear. Physical basis and engineering applications of tribology,* 368 (FIZMATLIT, Moscow, 2007)
9. A.N. Bolotov, V.V. Meshkov, O.V. Sutyagin, M.V. Vasil'yev, *The influence of hardening on the characteristics of the surface microasperities elastoplastic contact,* Izvestiya of the Samara Scientific Center of the Russian Academy of Sciences, **15**, 313-315 (2013)
10. P.M. Ogar, D.B. Gorokhov, *Parameters for elastic-plastic body to calculate contact characteristics under the sphere indentation,* Systems. Methods. Technologies, **29**, 28-32 (2016)
11. P.M.Ogar, V.A. Tarasov, D.B. Gorokhov, A.V. Turchenko, *The account of hardening material characteristics in applying of Meyer's empirical law,* Systems. Methods. Technologies, **20**, 50-54 (2013)
12. M.P. Markovets, *Determination of mechanical properties of metals hardness,* 191 (Mashinostroenie, Moscow, 1979)
13. M. Gaško, G. Rosenberg, *Correlation between hardness and tensile properties in ultra-high strength dual phase steels – short communication,* Materials engineering - Materiálové inžinierstvo, **18**, 155-159 (2011)
14. S.I. Bulychev, V.P. Alekhin, *Testing of materials by continuous depression of an indenter,* 224 (Mashinostroenie, Moscow, 1990)
15. S.I. Bulychev, *Hardness and hysteresis at the yield strength,* Deformation and fracture of materials, **1**, 41-45 (2011)
16. S. Kucharski, Z. Mroz, *Indentation of plastic hardening parameters of metals from spherical indentation tests,* Materials Science and Engineering A. **318**, 65-76 (2001)
17. X. Hernot, O. Bartier, Y. Bekouche, R. El Abdi, G. Mauvoisin, *Influence of penetration depth and mechanical properties on contact radius determination for spherical indentation,* International Journal of Solids and Structures, **43**, 4136-4153 (2006)
18. M. Matlin, E. Kazankina, V. Kazankin, *Mechanics of unitial contact,* Mechanika, **76**, 20-23 (2009)
19. J.-M. Collin, G. Mauvoisin, P. Pilvin, *Materials characterization by instrumented indentation using two different approaches,* Materials and Desing, **31**, 636-640 (2010)

 DOI: 10.1051/matecconf/201712906018

Analysis of vibroprotection characteristics of pneumatic relaxation seat suspension with capability of vibration energy recuperation

Mikhail Lyashenko[1], *Pavel* Potapov[1], and *Azamat* Iskaliev[1,*]

[1]Volgograd State Technical University (VSTU), department «Transport vehicles and engines», 400005, Volgograd, Lenin Ave 28, Russia

Abstract. This paper proposes mechanism and control algorithm for pneumatic relaxation system of suspension with vibration energy recuperation applied to standard vehicle operator seat ("Sibeko" company). Mathematical model of the seat pneumatic relaxation suspension with two additional air volumes was created. Pneumatic motor – recuperator activated by means of air flow from the one additional volume to another is installed in air piping between additional volumes. Computational research was made in Matlab/Simulink. Amplitude-frequency characteristics of transmission coefficient for standard and proposed suspensions were plotted for preliminary evaluation of vibration protection properties of seat suspension. Performed comparative analysis of amplitude-frequency characteristics shows that noticeable improvement of vibration protection properties of pneumatic relaxation suspension system with vibration energy recuperation in comparison with standard system both in region of resonance disturbances and in above-resonance region. Main ways for further improvement of vibration protection properties of proposed system were marked out.

At the modern stage of development of ground vehicles high values of operational characteristics (for example, specific power and operational speeds) led to the problem of improvement of the functional characteristics of suspension systems. These systems must provide necessary riding comfort during performing of technological operations and effective protection of operator from unhealthy influence of the low-frequency vibrations and noise.

At the present time, operator's seat suspensions with nonlinear characteristics are believed to be the most effective. For example, suspensions with the pneumatic elastic element are widely used in modern tractors [6,7]. Standard unregulated shock absorber installed in these suspensions provides necessary damping of vibrations in the resonance zone but makes worse protection from vibrations in the above- resonance zone [2]. There are ways with various efficiency for solving of this problem, for example, using of active suspensions [3,8]. But the results of the performed analysis show that the most preferable one is the self-regulated suspension system with the semi-active control that doesn't require external power sources and operate effective in the whole frequency diapason of impacts. Also, availability of function of recuperation of absorbed vibration energy clearly increases efficiency factor and efficiency of the suspension in comparison with other suspensions that have standard unregulated dampers which dissipate vibration energy. This paper presents the proposed controllable pneumatic relaxation suspension of seat with the feature of recuperation of vibration energy. This suspension was created on base of the driver seat made by "Sibeko" company.

The research was aimed at determination of ways for improvement of protection from vibration for the one-support suspension system by means of using of the pneumatic relaxation suspension with availability of vibration energy recuperation. Object of research was widely using in automotive technique "Sibeko" seat [5] that has pneumatic suspension with automatic regulation of static position.

In results of the stand experiments static elastic-damping characteristic of the standard seat suspension was obtained [1]. Analysis of the characteristic shows that significant area of hysteresis curve indicates a presence in the suspension mechanism (mainly, in hydraulic shock absorber) of dry friction forces with high values. These forces negatively affects on seat protection from vibrations with high-frequency excitations [2]. In accordance with the proposed concept of control of suspension stiffness, energy absorption and its further recuperation must takes place on areas at the end of compression and rebound strokes. At these stages relative velocity of suspension mass changes sign and potential energy reaches extreme values. The authors propose scheme and the method of control of the one-support suspension which allow implementation of the relaxation suspension that meet requirements of proposed concept (Fig. 1).

*Corresponding author: asamat-iskaliev@mail.ru

 DOI: 10.1051/matecconf/201712906018

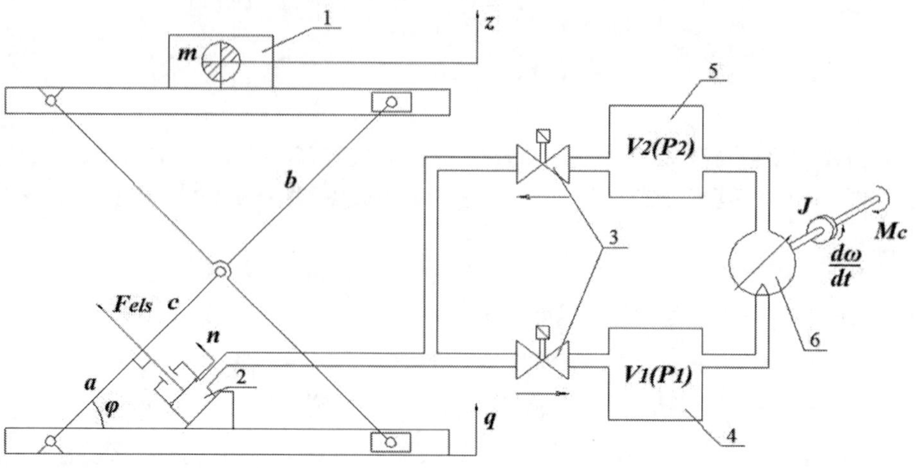

Fig. 1. Design scheme of the pneumatic suspension with two additional volumes: 1 – sprung body with mass m; 2 – displacement of elastic element; 3 – controlled electric valves; 4,5 – additional volumes; 6 – pneumatic motor

Absorption of vibration energy is performed by means of operation of valves 3. Opening of valves 3 is performed in the next manner: the first valve allows flow of a gas mass from the volume 2 to the first volume 4 at the end of compression stroke and the second valve – from the volume 5 to the volume 2 at the end of rebound stroke. Pneumatic motor (recuperator of energy) installed between additional volumes performs effective work by means of variable difference of air pressure in mentioned volumes.

The mathematical model was created in accordance with the main statements of theoretical mechanic and gas dynamic.

Equation of dynamic of the sprung body with mass m:

$$m \cdot \ddot{z} = F_{els} \cdot \cos\varphi - F_n - F_f, \tag{1}$$

where $\ddot{z}$ – current value of acceleration of sprung body, *m/sec²*; F_{els} – elastic force, *N*; φ – current value of the tilt angle between lever of directing mechanism and base, *rad*; $F_n = m \cdot g$ ($g = 9{,}81\,\mathrm{m/sec^2}$) – gravity force, *N*; F_f – friction force, *N*.

Elastic force:

$$F_{els} = (P_p - P_a) \cdot S_p, \tag{2}$$

where P_p – current value of air pressure in working chamber, *Pa*; P_a – atmosphere pressure, *Pa*; S_p – area of piston in the elastic element, *m²*.

Current value of the tilt angle between lever of directing mechanism and base:

$$\varphi = \arcsin\left(\frac{z - q}{(a + b + c)} + \sin\varphi_0\right), \tag{3}$$

where z, q – current values of displacements of base and sprung body, *m*; a, b, c – components of length of lever of direction mechanism, *m*; φ_0 – initial value of the tilt angle between lever of directing mechanism and base, *rad*.

Friction force:

$$F_f = T_0 \cdot sign(\dot{z} - \dot{q}), \tag{4}$$

where T_0, *N*, (value was defined in experiment [1]); $\dot{z}, \dot{q}$ – current velocities of sprung body and base, *m/sec*.

Kinematic harmonic impact on base:

$$q = A_0 \cdot \sin(\omega \cdot t), \tag{5}$$

where $A_0 = 0{,}03\ m$ – amplitude of excitation; ω – frequency of excitation, *sec⁻¹*; t – time, *sec*.

Speed of change of air pressure in working chamber:

$$\dot{P}_p = \frac{\chi \cdot P_p \cdot \dot{m}_{2p}}{m_p} - \frac{\chi \cdot P_p \cdot \dot{m}_{p1}}{m_p} - \frac{\chi \cdot P_p \cdot S_p \cdot \dot{n}}{V_p}, \tag{6}$$

where χ – polytrophic index; m_p – current mass of air in working chamber, *kg*; $\dot{m}_{2p}, \dot{m}_{p1}$ – current air flow rate from the second additional volume to working chamber and from working chamber to the first additional volume respectively, *kg/sec*; $\dot{n}$ – current velocity of piston rod in working chamber of elastic element, *m/sec*; V_p – current volume of air in working chamber, *m³*.

Current volume of air in working chamber:

$$V_p = V_{p0} + S_p \cdot n, \tag{7}$$

where V_{p0} – initial static volume of air in working chamber, *m³*; n – current coordinate of piston rod position in working chamber of elastic element, *m*.

Current coordinate of piston rod position in working chamber of elastic element:

 DOI: 10.1051/matecconf/201712906018

$$n = a \cdot (tg\varphi - tg\varphi_0). \quad (8)$$

Speed of change of air pressure in the first additional volume:

$$\dot{P}_1 = \frac{\chi \cdot P_1 \cdot \dot{m}_{p1}}{m_1} - \frac{\chi \cdot P_1 \cdot \dot{m}_{12}}{m_1}, \quad (9)$$

where P_1 – current air pressure in the first additional volume, *Pa*; m_1 – current air mass in the first additional volume, *kg*; $\dot{m}_{12}$ – current air flow rate from the first additional volume, *kg/sec*.

Speed of change of air pressure in the second additional value:

$$\dot{P}_2 = \frac{\chi \cdot P_2 \cdot \dot{m}_{12}}{m_2} - \frac{\chi \cdot P_2 \cdot \dot{m}_{2p}}{m_2}, \quad (10)$$

where P_2 – current air pressure in the second additional value, *Pa*; m_2 – current air mass in the second additional volume, *kg*.

Air pressure in system at static position of sprung body:

$$P_{p0} = P_{10} = P_{20} = \frac{F_g}{\cos\varphi_0 \cdot S_p} + P_a, \quad (11)$$

where P_{p0}, P_{10}, P_{20} – air pressure in working chamber and in additional volumes at static position of sprung body, *Pa*.

Air density in system at static position of sprung body:

$$\rho_{p0} = \rho_{10} = \rho_{20} = \rho_a \cdot (\frac{P_{p0}}{P_a}), \quad (12)$$

where $\rho_{p0}, \rho_{10}, \rho_{20}$ – air density in working chamber and in additional volumes at static position of sprung body, *kg/m³*; ρ_a – air density at atmosphere pressure, *kg/m³*.

Current air mass in working chamber:

$$m_p = \rho_{p0} \cdot V_p \cdot (\frac{P_p}{P_{p0}})^{\frac{1}{\chi}}. \quad (13)$$

Current air mass in the first additional volume:

$$m_1 = \rho_{10} \cdot V_1 \cdot (\frac{P_1}{P_{10}})^{\frac{1}{\chi}}, \quad (14)$$

where V_1 – value of the first additional volume, m^3.

Current air mass in the second additional volume:

$$m_2 = \rho_{20} \cdot V_2 \cdot (\frac{P_2}{P_{20}})^{\frac{1}{\chi}}, \quad (15)$$

where V_2 – value of the second additional volume, m^3.

Equation of rotational dynamic of shaft of pneumatic motor:

$$\frac{d\omega}{dt} = \frac{(P_1 - P_2) \cdot F_\psi \cdot r - M_c}{J}, \quad (16)$$

where $\frac{d\omega}{dt}$ – current angular acceleration of pneumatic motor shaft, sec^{-2}; F_ψ – area of influence of air positive pressure, m^2; r – length of arm of influence, m; M_c – resistance torque of pneumatic motor drive, $N{*}m$; J – inertia moment of pneumatic motor shaft, $kg{*}m^2$.

Resistance torque of pneumatic motor drive:

$$M_c = M_{c0} + \sigma_0 \cdot \omega^2, \quad (17)$$

where M_{c0} – initial resistance torque ("rest"), $N{*}m$; σ_0 – correction coefficient; ω – angular speed of pneumatic motor shaft, sec^{-1}.

Flow section of air line:

$$F = \frac{\pi \cdot d^2}{4}, \quad (18)$$

where d – diameter of air line cross-section, m.

Flow of working body from one volume to another could be assumed adiabatic if flow sections are sufficiently small.

Current air flow rate from working chamber to the first additional volume:

$$\frac{P_1}{P_p} > 0{,}528$$

$$\dot{m}_{p1} = \mu \cdot F \cdot \sqrt{\frac{2 \cdot k}{k-1} \cdot \frac{P_p \cdot m_p}{V_p} \cdot \left[(\frac{P_1}{P_p})^{\frac{2}{k}} - (\frac{P_1}{P_p})^{\frac{k+1}{k}} \right]}; \quad (19)$$

when:

$$\frac{P_1}{P_p} \leq 0{,}528$$

$$\dot{m}_{p1} = \mu \cdot F \cdot \sqrt{\frac{2 \cdot k}{k-1} \cdot \frac{P_p \cdot m_p}{V_p} \cdot \left[(\frac{2}{k+1})^{\frac{2}{k-1}} - (\frac{2}{k+1})^{\frac{k+1}{k-1}} \right]}, \quad (20)$$

where μ – coefficient of air flow through controlled valve; k – adiabatic index.

Current air flow rate from the first additional volume to the second additional volume:

when

$$(P_1 - P_2) \cdot F_\psi \cdot r - M_c > 0,$$

$$\frac{P_2}{P_1} > 0{,}528$$

$$\dot{m}_{12} = \mu^* \cdot F_\psi \cdot \sqrt{\frac{2\cdot k}{k-1}\cdot\frac{P_1\cdot m_1}{V_1}\cdot\left[\left(\frac{P_2}{P_1}\right)^{\frac{2}{k}} - \left(\frac{P_2}{P_1}\right)^{\frac{k+1}{k}}\right]}; \quad (21)$$

when

$$(P_1 - P_2)\cdot F_\psi \cdot r - M_c > 0,$$

$$\frac{P_2}{P_1} \le 0{,}528$$

$$\dot{m}_{12} = \mu^* \cdot F_\psi \cdot \sqrt{\frac{2\cdot k}{k-1}\cdot\frac{P_1\cdot m_1}{V_1}\cdot\left[\left(\frac{2}{k+1}\right)^{\frac{2}{k-1}} - \left(\frac{2}{k+1}\right)^{\frac{k+1}{k-1}}\right]}, \quad (22)$$

where μ^* – coefficient of air flow in cavity of pneumatic motor drive.

Current air flow rate from the second additional volume to working chamber:

when:

$$\frac{P_p}{P_2} > 0{,}528$$

$$\dot{m}_{2p} = \mu \cdot F \cdot \sqrt{\frac{2\cdot k}{k-1}\cdot\frac{P_2\cdot m_2}{V_2}\cdot\left[\left(\frac{P_p}{P_2}\right)^{\frac{2}{k}} - \left(\frac{P_p}{P_2}\right)^{\frac{k+1}{k}}\right]}; \quad (23)$$

when

$$\frac{P_p}{P_2} \le 0{,}528$$

$$\dot{m}_{2p} = \mu \cdot F \cdot \sqrt{\frac{2\cdot k}{k-1}\cdot\frac{P_2\cdot m_2}{V_2}\cdot\left[\left(\frac{2}{k+1}\right)^{\frac{2}{k-1}} - \left(\frac{2}{k+1}\right)^{\frac{k+1}{k-1}}\right]}. \quad (24)$$

Proposed algorithm of valves working is presented in table 1.

Table 1. Control algorithm of valves working

Compression stroke		
$\dot{m}_{p1} > 0;\ \dot{m}_{2p} = 0$	$\dot{z} - \dot{q} < 0$	$0 > z > -A_0$
$\dot{m}_{p1} = 0;\ \dot{m}_{2p} = 0$		$z \le -A_0$
$\dot{m}_{p1} > 0;\ \dot{m}_{2p} = 0$	$\dot{z} - \dot{q} = 0$	$z < 0$
$\dot{m}_{p1} = 0;\ \dot{m}_{2p} > 0$	$\dot{z} - \dot{q} > 0$	$z < 0$
Rebound stroke		
$\dot{m}_{p1} = 0;\ \dot{m}_{2p} > 0$	$\dot{z} - \dot{q} > 0$	$0 < z < A_0$
$\dot{m}_{p1} = 0;\ \dot{m}_{2p} = 0$		$z \ge A_0$
$\dot{m}_{p1} = 0;\ \dot{m}_{2p} > 0$	$\dot{z} - \dot{q} = 0$	$z > 0$
$\dot{m}_{p1} > 0;\ \dot{m}_{2p} = 0$	$\dot{z} - \dot{q} < 0$	$z > 0$

To evaluate potential characteristics of protection from vibration of the proposed suspension, it is necessary to determinate its work for absorption of vibration energy. Given that, $z - q = -\lambda$, where λ - suspension deformation, in accordance with the elastic-damping characteristic (fig. 2) obtained with the algorithm (table 1) it is possible to define work for absorption of vibration energy and recuperator efficiency coefficient.

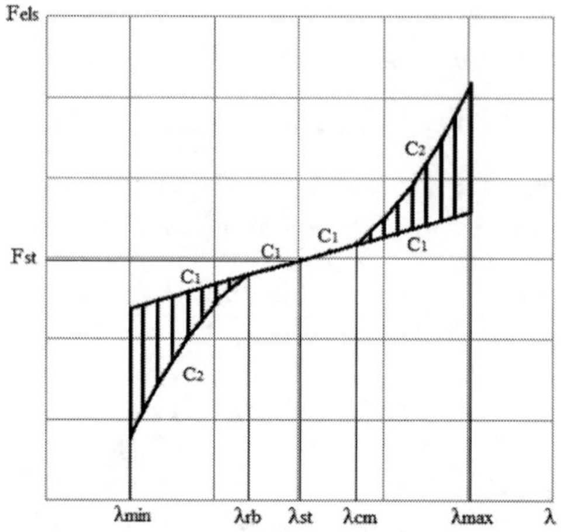

Fig. 2. Theoretical elastic-damping characteristic

Work for absorption done by the suspension during one period of vibrations (fig. 2):

$$L_{abs(t=T)} = \int_{\lambda_{cm}}^{\lambda_{max}} (c_2 - c_1)\cdot\lambda\cdot d\lambda + \int_{\lambda_{min}}^{\lambda_{rb}} ((c_1 + c_2)\cdot\lambda + c_1\cdot(\lambda_{min} - \lambda_{rb}))\cdot d\lambda, \quad (25)$$

where c_1, c_2 – suspension stiffness values at opening and closing of the controlled valves accordingly; $\lambda_{min}, \lambda_{max}$ – suspension deformations at the ends of rebound and compression strokes; $\lambda_{rb}, \lambda_{cm}$ – suspension deformations corresponds to the moment of closing of the controlled valves; λ_{st} – values of suspension static deformation.

Power of the recuperator:

$$\int_0^{N_{rec}} dN_{rec} = \int_0^{\omega} M_c \cdot d\omega = M_{c0}\cdot\int_0^{\omega} d\omega + \sigma_0 \cdot \int_0^{\omega} \omega^2 d\omega,$$
$$N_{rec} = M_{c0}\cdot\omega + \sigma_0\cdot\frac{\omega^3}{3}. \quad (26)$$

Efficience work done by recuperator during one period of vibrations:

$$\int_0^{L_{rec(t=T)}} dL_{rec} = \int_0^{T} N_{rec} dt. \quad (27)$$

Efficiency coefficient of recuperator:

$$\eta_{rec} = \frac{L_{rec(t=T)}}{L_{abs(t=T)}}. \quad (28)$$

Presented mathematical model was created in Simulink Matlab. Equations were solved by means of

DOI: 10.1051/matecconf/201712906018

Dormand-Price method with variable integrator step. Example of a model part for one differential equation (9) in Simulink is presented in fig. 3.

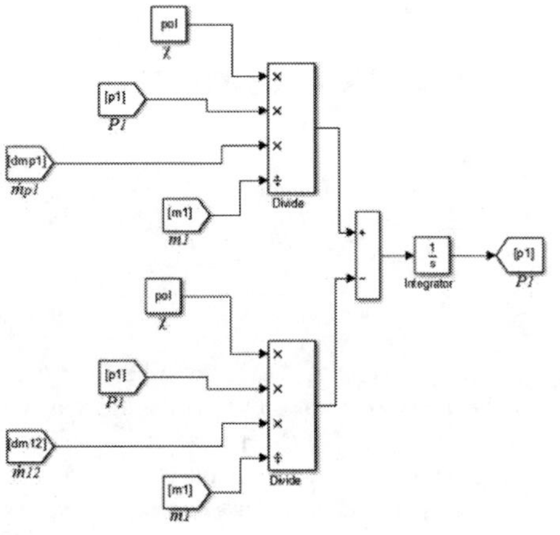

Fig. 3. Example of a model part for one differential equation (9) in Simulink

Graphs of amplitude-frequency characteristics of transfer coefficients (fig. 4) for standard and proposed suspensions (fig. 2) were plotted to initially evaluate characteristics of protection from vibrations.

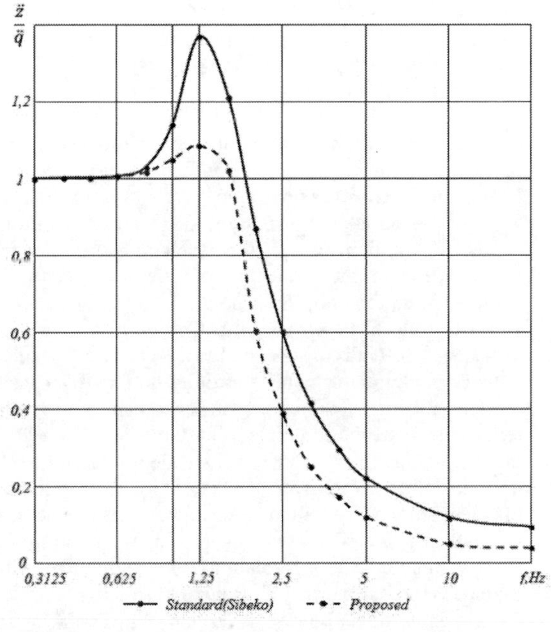

Fig. 4. Amplitude-frequency characteristics of transfer coefficients

Results analysis shows that the proposed suspension system has significantly better protection from vibrations both in resonance zone, and in high-frequency zone in comparison with the standard "Sibeko" suspension. Values of transfer coefficient of the proposed suspension system in resonance zone of impacts are 1.19 times less and in over-resonance zone 2 times in comparison with transfer coefficient of the standard suspension.

Results of presented research show that the proposed mechanism and control algorithm for the pneumatic relaxation system with vibration energy recuperation could be efficiency used for both one-support, and multi-support suspension systems of ground vehicles. Further improvement of proposed suspension systems would be performed by means of design search of possibilities for increasing of value of absorbed vibration energy. It allows significantly decrease hydraulic damping or even fully reject standard hydraulic shock absorber in prospective.

References

1. A. Iskaliev, M. Ljashenko, Proceedings of IV international scientific and practical conference "Scientific and technical progress: topical and prospective directions of future",181-183 (2016)
2. V. Novikov, I. Ryabov, K.V. Chernyshov, *Vibroprotection characteristics of automotive suspensions* (RPK Politekhnik, Volgograd, 2009)
3. O. Polivaev, A. Yushin, O. Kostikov, Tractors and agricultural machinery, **9**, 47-49, 2008
4. A. Pozdeev, Truck, **6**, 2-6, 2011
5. http://sibeko-russia.ru/catalog/8/27/1253.html.
6. http://www.claas.ru/produkte/traktoren.
7. http://www.deutz-fahr.com/ru-RU/products/tractors.
8. A. Yushin, *Improvement of using efficiency of tractor-transport aggregate by means of modernization of suspension of tractor seat: abstract of thesis for candidate of technical science degree* (Voronezh, 2007)

MATEC Web of Conferences 129, 06019 (2017) DOI: 10.1051/matecconf/201712906019
ICMTMTE 2017

Simulator as a tool of training to modern equipment management

Gulnara Ahmedyanova[1,*]

[1]Orenburg state university, Orenburg, Russia

Abstract. In the work, the learning process with the use of a simulator was studied. In this case, both the design and algorithmic content of the simulator, as well as the trainee, must pass their part of the path, only in this case the result of learning is maximized. Theoretically, it is shown that the effectiveness of simulator training is primarily a function of the cognitive-operational and professional-personal aspects of the trainee's competence. The experiment confirmed that, despite the differences above the indicated qualities, the result can be estimated as the sum of their estimates.

1 Introduction

Modern technology is characterized by the tendency of continuous complication and increase in the speed of the processes occurring in it [1]. This, in turn, requires highly skilled professionals to manage it. The latter can not be achieved without sufficient practice for real management in various situations.

Of course, it is best to gain this practical experience on real equipment, but this is not always possible and often unsafe. In such conditions, the quasi-professional activity of the student is necessary, as the most appropriate tool for developing the necessary skills. Quasi-professional activity is understood here as maximally close to reality, so the best way to implement it is training on the simulator. An example is: training an astronaut to dock with an international space station, training a pilot, training a driver, training a surgeon, etc..

2 Theory

Trainers in the modern sense have appeared, when there was a necessity of mass preparation of experts for work or on the same type of equipment, or with similar working actions, and also for military needs. Recently, in connection with the rapid computerization of the world community, with the creation of sophisticated technology, the exploitation of which is associated with a risk to the lives of not only one person, but also of mankind as a whole, a whole industry has emerged-simulator technologies. Simulator technologies are complex compound, simulation and modeling systems, computer programs and physical models, special techniques designed to prepare the person for making quality and quick decisions. In modern simulators and in training and education programs based on them, the principles of development of practical skills with simultaneous theoretical preparation are laid [2].

Simulator technologies have emerged and developed most in those areas where errors in training on real objects can lead to extreme consequences, and their elimination - to greater financial costs[3-6].

The simulator in the general case is a software and hardware complex. The dynamic response of the simulator should be as close as possible to the response of systems and components of the real object, for which virtual reality is widely used in simulators [7-12].

On the other hand, the simulator is an important representative of laboratory equipment. Unlike any other laboratory equipment, it is possible to learn exactly professional activity, rather than develop research skills. Under the simulator is usually understood as a training device for training, that is, for developing skills for the management of any machine or any mechanism. In another interpretation, the simulator is a mechanical, software, electric or combined training device that imitates artificially various loads or circumstances (situation). Simulator can be training (imitation) or sports.

To ensure its functions in accordance with the requirements of the principle of system accordance [13], the simulator must have an extended interface with trainees. When creating an interface, it is necessary to take into account the psychological characteristics of a person, for example, his sharp reaction to red, while the green color calms him, the information in the right side of the field of view draws attention faster and so on.

A part of the path must be trainee. When working on a simulator, it requires, apart from specialized knowledge: concentration, attentiveness, quick response, resourcefulness. Therefore, before working on the simulator, one should study a certain theoretical course.

Figure 1 shows a generalized scheme of the trainee's work on the simulator. Both the main components and the direct links between them are shown.

* Corresponding author: ahmedyanova@bk.ru

 DOI: 10.1051/matecconf/201712906019

The indicator of the effectiveness of training on the simulator in the general case can be represented in the

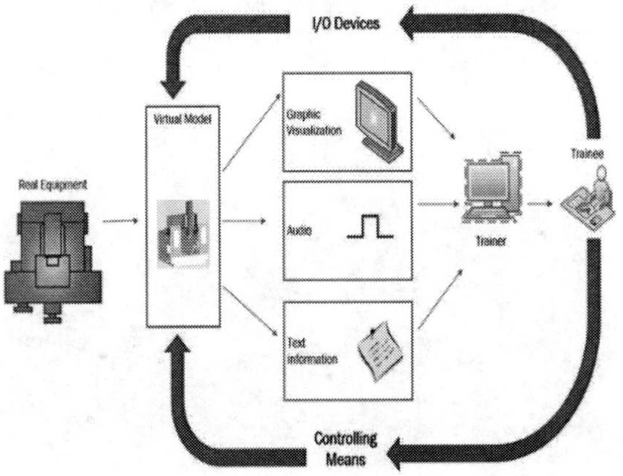

Fig.1. The scheme of the virtual simulator

form of some dependence on the totality of parameters of the simulator and the qualities of the trainees type [14]:

$$F = F(N, M, S, P, V), \tag{1}$$

where N is the set of typical situations for working out typical skills necessary for equipment management in standard situations;

M is the set of abnormal situations modeled on the simulator, (with the possibility of changing both their number and the intensity of the component faults of different classes, the types and parameters of the laws of distribution of time intervals between the time of receipt of various failures, etc.);

S - set of ergonomic factors;

P - the totality of theoretical knowledge about the managed equipment (cognitive-operational aspect of the trainee's competence [15]);

V - set of psychological qualities of the trainee, influencing the speed and correctness of making managerial decisions (professional-personal aspect of the trainee's competence [15]);

The first three sets N, M, S reflect the impact on the results of the learning process of the technical indicators of the simulator itself, while the last two sets reflect the influence of the personality traits of the trainee. More clearly in the linear expansion, the structure of the efficiency index can be expressed by the formula:

$$Q = \frac{\partial F}{\partial N}\Delta N + \frac{\partial F}{\partial M}\Delta M + \frac{\partial F}{\partial S}\Delta S + \frac{\partial F}{\partial P}\Delta P + \frac{\partial F}{\partial V}\Delta V. \tag{2}$$

here ΔN is the increment in the number of standard situations, ΔM is the increment in the number of contingencies, ΔS is the increment in the ergonomics level of the simulator, ΔP is the increment in the theoretical knowledge of the learner, ΔV is the increment in creative skills, and the derivatives in the formula reflect the influence of these increments on the overall efficiency.

On the other hand, the training time is also important. As with the passage of time only the competence of the trainee changes, we will evaluate the effectiveness of the reduced formula

$$Q = \left(\frac{\partial F}{\partial P}\frac{\partial P}{\partial t} + \frac{\partial F}{\partial V}\frac{\partial V}{\partial t}\right)\Delta t. \tag{3}$$

Now added derivatives additionally reflect the speed of mastering knowledge for students and the speed of learning management skills.

It follows that the evaluation of the effectiveness of the simulator can be reduced to a first approximation to assessing the quality of the organization of replenishment of knowledge and creative abilities and skills for students.

3 Data and Methods

In this study, EMCO virtual classroom training simulators were used to develop practical skills in the process of preparing and upgrading teachers, as well as for automated assessment of knowledge and skills for different levels of complexity of work on the milling machine.

4 Methods or Model

The simulators used in the study are a software and hardware complex interacting with the user through interactive dialogue means (mouse, keyboard), a control panel model with visualization of the state of the technological process (See Figure 2), equipment and mechanisms based on modern multimedia technologies for presenting information (text, graphics, sound).

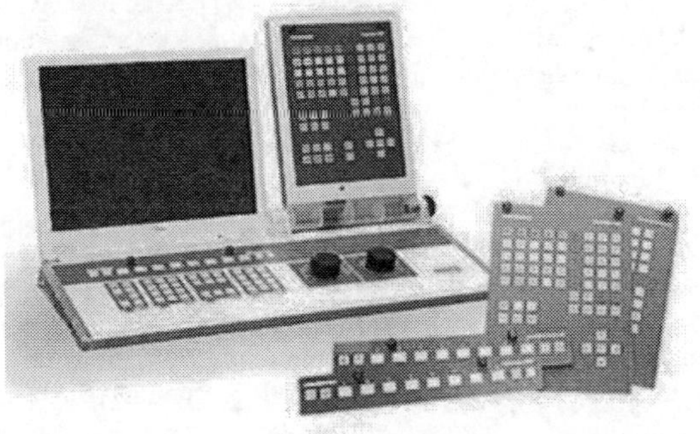

Fig. 2. Appearance of the simulator

The program interface ensures the operation of the complex provided that personnel who are not a qualified user of the PC work there. The training provides an imitation of the technological process applied to the production technology on the machine, provides a concise, easy-to-assimilate form of representation on the main and auxiliary information display. When working with simulators, basic didactic principles of cognitive activity are provided, access to reference and training material, training for developing practical work skills and control over the degree of mastering theoretical and practical knowledge. The software product contains methodological and didactic materials for training on the simulator in accordance with educational programs.

In the development of simulator software, the modular principles of program design were envisaged, providing the possibility of increasing and upgrading the functionality of simulators [16]. The simulator software functions both in stand-alone mode (on a separate local machine) and in a local network. When working in a local network, several (up to 10) training places of the simulator interact with the instructor's workplace.

1. Module "Device of equipment and control panel" is designed to study: - machine tool; - devices of control panels and the purpose of machine controls; - arrangements for setting and resetting the machine; - ways to control the products; - types of product quality mismatch with an indication of the causes and methods of elimination.

2. In the module "Technological process" a virtual model of the machine and control panels is realized, interacting with which the trainee can perform all actions on the machine during the working shift: - selection of the tool in accordance with the current task; - start the machine; - execution of technological operations; - stopping the machine; - quality control of the products; - adjustment of the machine settings depending on the quality mismatch obtained (the type of nonconformity is selected randomly). The module can work in the training mode (prompts and explanations are given during operation), and in the knowledge testing mode (no prompts are given, the correctness of work on the machine is evaluated).

3. The module "Emergency situations" implements a demonstration and description of the main types of emergency situations, methods for their prevention and elimination of consequences in accordance with technological instructions and instructions on safety. For each emergency situation, the following is displayed: - description of the emergency situation; - visual manifestations; - causes of occurrence; - ways of elimination.

4. The module "Testing the knowledge gained in the learning process" consists of a theoretical and practical part. At the end of each part, an assessment is made on a 5-point scale. The theoretical part is questions with the answers, questions are chosen randomly from the base of questions. A mechanism is provided for changing the base of questions with answers from the teacher's workplace. The practical part is the module "Technological process" in the mode of knowledge testing. Knowledge testing is possible on several levels of complexity. A list of difficulty levels and the correspondence of questions to different levels are set from the teacher's place.

The trainee can perform all the actions on the machine: - selecting the tool according to the current job; - start the machine; - execution of technological operations; - stopping the machine; - quality control of the products; - adjustment of the machine settings depending on the quality mismatch obtained (the type of nonconformity is selected randomly). The module can work in the training mode (prompts and explanations in process of work), and in the knowledge testing mode (no prompts are given, the correctness of work on the machine is evaluated). An illustration of the complexity of the material being studied can be the device of a keyboard of controlled equipment (see Figure 3)

Fig. 3. General view of the console

5 Results

The assimilation of knowledge was estimated by the volumes of the acquired textual information, and the creative skills and skills by the number of mastered operations from which the typical actions were composed. The result of assimilation was estimated in points and reflected by the diagram in Figure 4

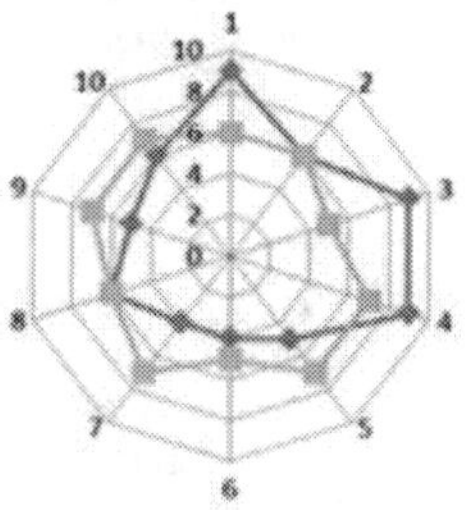

Fig.4. The pie chart of the abilities of students

The result of training as a sum of two aspects of educational activity is reflected by the histogram in Figure 5.

From the first diagram it follows that the cognitive-operational and professional-personal aspects of students in the group differ significantly, but the best results of training were shown by students with a high level of both these aspects.

6 Conclusions or discussion and implication

Thus, the simulator is an effective tool of training to modern equipment management. Moreover, the design

DOI: 10.1051/matecconf/201712906019

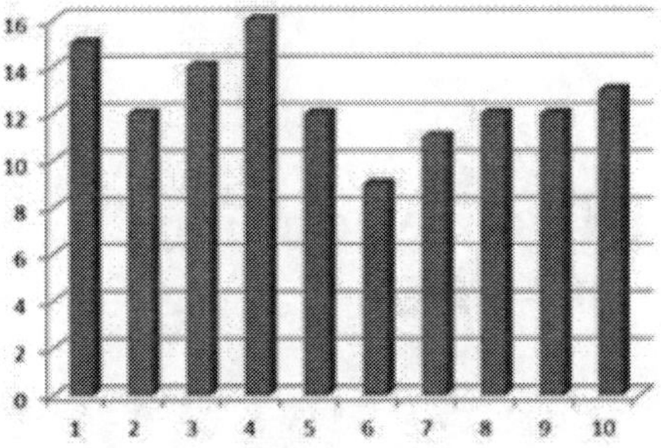

Fig.5. Training Effect Histogram

and algorithmic filling of the simulator, as well as the trainee, must pass their part of the path, only in this case the result is maximized.

Training is provided by a quasi-professional activity that simulates the control of the technological process with respect to the production technology on these machines, provides a concise, easy-to-digest form of representation on the main and auxiliary information display. When working with simulators, basic didactic principles of cognitive activity are provided, access to reference and training material, training for developing practical work skills and control over the degree of mastering theoretical and practical knowledge.

Reference

1. A.M. Pishchukhin, GF.Akhmedyanova, *Automation and technological development are two aspects of the technical equipment of production*, Bulletin of the Orenburg State University, **9(184)**, pp. 170-174 (2015)
2. A.V. Trukhin, *Analysis of existing in Russia simulators and training systems*. Access mode: https://ido.tsu.ru/other_res/pdf/1(29)_2008(32-39).pdf
3. V.V. Belov, I.V. Obraztsov, *Virtualization of physical processes in the theory and practice of building education. The theory and practice of increasing efficiency builds*, Materials: Materials of V Vseros. Conf. Students, graduate students and young scientists. Penza: PGUAS, pp. 186-189 (2010)
4. A.P. Blinov, *Virtual training simulator: effective and safe*, Vocational education and the labor market **2** (2013)
5. A.O. Matlin, D.S. Osintsev, S.A. Fomenkov, *Automated system "interactive virtual simulators"*, Proceedings of the V All-Russian Conference of Students, PhD students and young scientists "Microsoft Technology in the theory and practice of programming." Central region (Moscow: The University Book, p. 75-77, 2008)
6. L. Soyeon, P. Sangjoon, C. Kyoil and C. Choongho *Kinematic Skeleton Based Control of a Virtual Simulator for Military Training*, Symmetry **7(2),** p. 1043-1060 (2015)
7. V.O. Afanasyev, A.G. Brovkin, A.N. Kornievsky, V.P. Podobedov, V.S. Semchenko, A.N. Tomilin, *Research and develop-ment of the interactive observation system of the induced virtual environment (virtual presence systems)*, Cosmonautics and rocket science, 20 (2001)
8. A.O. Matlin, D.S. Osintsev, V.V. Get-mansky, S.A. Fomenkov, *Interactive virtual medical simulators*, Proceedings of the VI All-Russian Conference of 22 students, graduate students and young scientists "Microsoft Technologies in the theory and practice of programming." Central region (Moscow: Moscow. Aviation institute, p. 122-123, 2009)
9. B. Dalgarno, and M. J. W Lee, *What are the Learning Affordances of 3D Virtual Environments?*, British Journal of Educational Technology, **41(1)** (2010), 10–32
10. J. Moskaliuk, J. Bertram & U. Cress, *Training in virtual environments: putting theory into practice*, Ergonomics **56(2),** p. 195-204 (2013)
11. F. D. Rose, E. A.Attree, B. M. Brooks, D. M. Parslow, and P. R. Penn, *Training in Virtual Environments: Transfer to Real World Tasks and Equivalence to Real Task Training*, Ergonomics, **43(4),** p. 494–511 (2000)
12. J. G. Tichon, *Using Presence to Improve a Virtual Training Environment*, CyberPsychology & Behavior, **10(6),** p.781–7. (2007)
13. A.M. Pishchukhin, *Accordance of the components of the system and methods for achieving it*, Bulletin of the Orenburg State University, **1**, p. 87-90 (1999)
14. A.N. Privalov, *Evaluation of the effectiveness of simulators with distributed data processing*, Scientific bulletin **1(56)** (2009)
15. G.F. Ahmedyanova, *Features of development of cognitive-operational and professional-personal competencies of bachelors*, Fundamental research, **2-20**, pp. 4492-4495 (2015)
16. I.P. Norenkov, A.M. Zimin, *Information technologies in education*, (M.: Publishing House of the Moscow State Technical University. N.E. Bauman, 2004)

 DOI: 10.1051/matecconf/201712906020

A setup for electrically controlled liquid-phase sorption of organic pollutants on nanostructured materials

Elena Neskoromnaya[1,*], *Alexander* Babkin[1], and *Alexander* Burakov[1]

[1] Tambov State Technical University, Department "Technology and Methods of Nanoproducts Manufacturing", 392000, 106 Sovetskaya Str., Tambov, Russia

Abstract. The present paper considers the possibility of intensifying the process of liquid-phase sorption of organic molecules of methyl orange dye through exposure to an electromagnetic field. A laboratory setup for water conditioning and an experimental procedure are presented. Activated carbons of different chemical nature (stone carbon – AG-5, charcoal – BAU-A, and NWC coconut shell carbon in its initial form – NWC and nanomodified with carbon nanotubes – NWC-T) were used as sorption materials. Moreover, the possibility of employing "Taunit"-series (Taunit, Taunit-M, Tainit-MD) carbon nanotubes (CNTs), produced at NanoTechCenter Ltd. (Tambov, Russia), in its native form in the process of electrically controlled sorption was also evaluated. In the course of performing comparative studies, the positive influence of the electromagnetic field on the sorption process was established. The adsorption capacity of the sorbents was found to increase 1.5-2 times under the voltage of 100 V. The sorption activity of the materials under study increased in the following sequence: BAU-A<AG-5<NWC <NWC-T <"Taunt"<"Taunit-MD"<"Taunit-M".

1 Introduction

Electrical sorption taking place in a heterogeneous medium under the influence of an external electric field is one of the varieties of sorption technologies. An electric water conditioner represents a water purification system comprising a filter element with porous adsorbents and electrodes that generate an electromagnetic field. This device can be used for treating wastewater to remove dangerous pollutants of different chemical nature [1, 2].

In the present work, the possibility of improving the qualitative characteristics of liquid-phase sorption from aqueous media due to directed electromagnetic exposure in the presence of highly effective sorbents – industrial activated carbons (ACs) and other types of carbon materials – carbon nanotubes (CNTs) – was studied.

Electrochemical polarization of the carbon surface shifts the adsorption equilibrium and, consequently, can be a regulating factor in adsorption systems. The influence of an external electric field on the sorbent causes the polarization of its granules, thereby substantially increasing the electrical sorption of ion-molecular aggregates of the pollutant from an aqueous solution. In adsorption processes involving carbon materials, a change in electric potential occurs at the interface, which appears to be the most pronounced in the case of aqueous solutions. Thus, the controlled electromagnetic influence should contribute to an increase in the sorption activity of the particle [1, 2].

The main groups of physical and physical-chemical factors essential for the electrical sorption are as follows:

- ✓ the distribution of the electric field takes place in a sorption cell with a porous material;
- ✓ the process implies interaction of chemical substances with each other and with water, as well as with adsorbing materials, in the external electric field;
- ✓ the relationship is established between molecules and electrical charges of impurities present in the aqueous solution, in near-electrode zones and directly on the electrodes.

The specific surface area of the ACs is very high and equals to $1\cdot10^5$-$2\cdot10^5$ m^2/kg, and their bulk density is 200-900 kg/m^3. A large number of functional groups have been found on the AC surface: phenol, carbonyl, carboxyl, ester, enol, and lactone. The heteroatoms – nitrogen and sulfur – are often included in six-membered rings. These features of the surface of the ACs frequently determine their ion-exchange and catalytic activity [3, 4].

Since the issue regarding the quality of water consumed is acute during the past years, it is important to develop novel materials capable of extracting different types of pollutants to maximum permissible values, or modify the already existing most widely used adsorbents. At present, many works have aimed at studying standard samples of adsorbing materials and the ability of their surface for activation or modification to improve the key adsorption characteristics.

In the present work CNTs under trade mark "Taunit" three different types (Taunit, Taunit-M, Tainit-

* Corresponding author: lenok.n1992@mail.ru

DOI: 10.1051/matecconf/201712906020

MD) (Table 1), produced at NanoTechCenter Ltd. (Tambov, Russia), were used as a compared material in the adsorption.

CNTs possess a high specific surface area (for single-walled CNTs - up to 3,000 m^2/g) and reactivity. Therefore, adsorption processes are one of the promising trends in their use [5-7].

Table 1. Characteristics of the "Taunit"-series CNTs.

Parameter	"Taunit"	"Taunit-M"	"Taunit-MD"
Outer diameter, nm	20-70	8-15	30-80
Inner diameter, nm	5-10	4-8	10-20
Length, μm	2 and more	2 and more	20 and more
Bulk density, g/cm^3	0.4÷0.6	0.03÷0.05	0.03÷0.05
Specific geometric surface area, m^2/g	120 and more	180-200	300 and more
Adsorption activity regarding methylene blue, mg/g*	110.0	207.5	275.0
Thermal stability (°C)	up to 600	up to 600	up to 600

Thus, in the present study, all three above-described CNTs types were used for electrically controlled liquid-phase sorption, together with the starting (initial) and nanomodified ACs.

2 Experimental

2.1 A Laboratory Setup

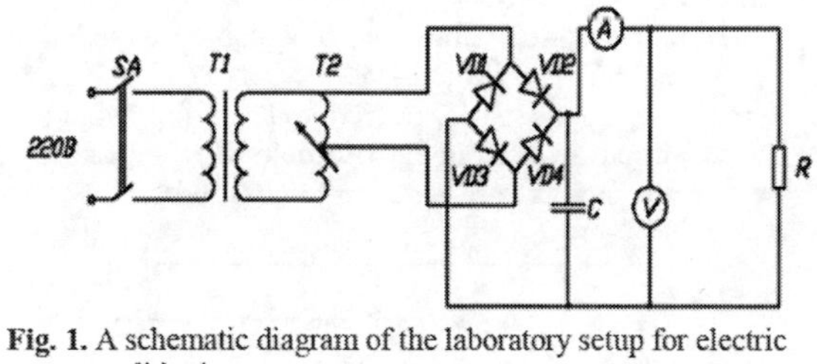

Fig. 1. A schematic diagram of the laboratory setup for electric water conditioning.

The authors have developed a setup for the implementation of the sorption process under the influence of an electromagnetic field. The electric water conditioner represents a closed circuit for circulation of an aqueous solution. The main element is a sorption cell with a sample of the adsorbing material and electrodes for generating an electrostatic field.

The schematic diagram of the laboratory setup is shown in Fig. 1, and its exterior is given in Fig. 2.

The voltage (220 V) is applied to the isolation transformer T1 via the two-pole circuit breaker SA protected against restart and short circuit.

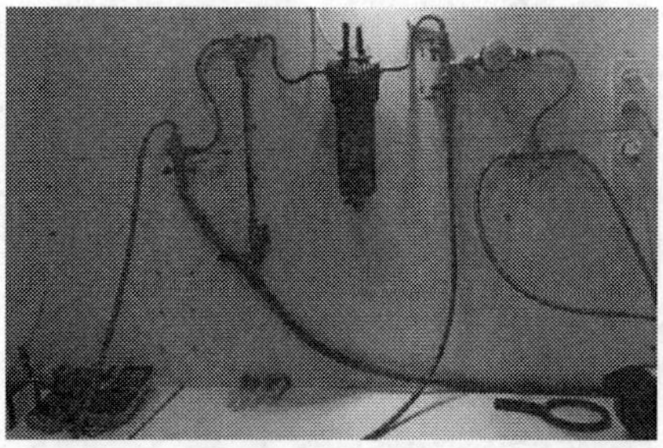

Fig. 2. The laboratory setup for studying the influence of the electromagnetic field on sorption processes.

Since water is an ionic conductor, the transformer is a necessary element in the setup to conduct safe research. Next, from the T1, the voltage is routed to the autotransformer T2 to vary the voltage in the range of 0-220 V.

2.2 Materials

After analyzing a large number of studies in the field of creating new adsorbents and methods for modifying standard sorption materials, the authors of the present paper have developed a technology for obtaining nanomodified sorbents. The sorbent proposed herein was synthesis via a CVD method in the course of growing carbon nanotubes over the surface of coconut shell AC [8].

Electron images (Fig. 3) of nanomodified ACs were obtained by scanning electron microscopy (SEM) on a Neon 40 two-beam instrument (Carl Zeiss, Jena, Germany).

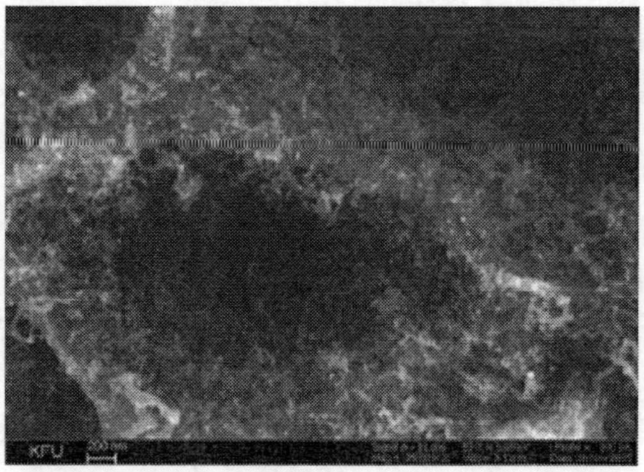

Fig. 3. A micrograph of the surface of the coconut shell carbon modified with a CNTs layer.

Different ACs (BAU-A charcoal, AG-5 stone carbon, and NWC coconut shell carbon) and "Taunit"-series CNTs were used as reference materials. The structure of some of them is presented in Fig. 4.

All the AC-based sorption materials were subjected to preliminary treatment and drying for 2 h at a temperature of 120-150 °C.

DOI: 10.1051/matecconf/201712906020

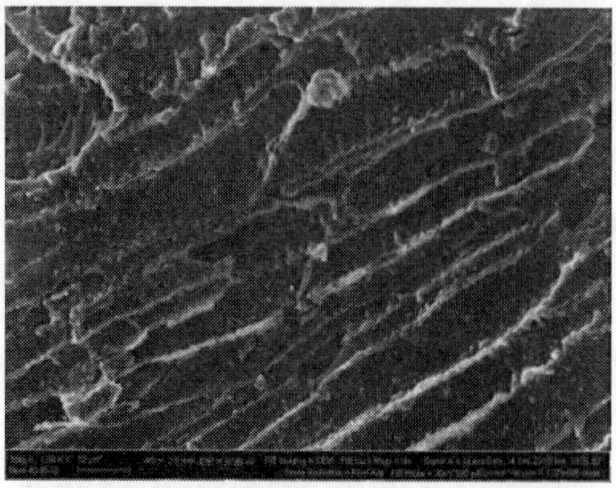

(a)

(b)

Fig. 4. The structure of the NWC AC (a) and "Taunit-M" CNTs (b).

The widely known artificial dye – methyl orange (MO) [other names: helianthin, benzenesulfonate, and sodium p-dimethylaminoazobenzenesulfonate] – was used as model organic pollutant. Its formula is given in Fig. 5.

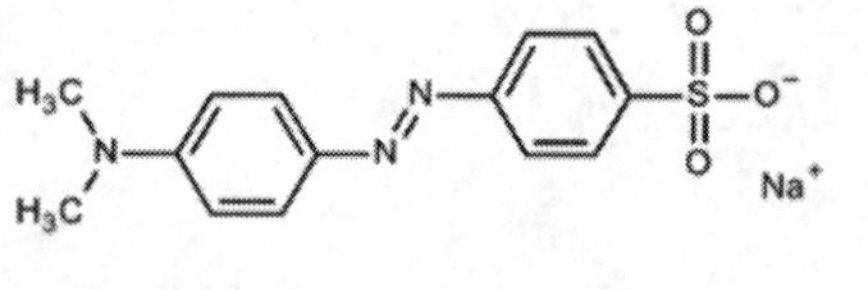

Fig. 5. The chemical formula of methyl orange.

This substance is a sodium salt and represents an acid-base indicator. In addition to the main field of its application as indicator, it is also used in the textile industry for dyeing wool, silk and nylon due to its chemical properties.

2.3 Adsorption studies

To determine the quantitative parameters of the adsorption process, the standard method of spectrophotometric analysis, described in the Russian National Standard (GOST) No. 4453-74 [9], was implemented.

Concentration values of solutions were calculated through optical densities and were obtained spectrophotometrically on a PE 5400 V instrument (Ekros Ltd., St. Petersburg, Russia) at a wavelength of 400 nm. At the preparatory stage of the experiment, a calibration graph was constructed.

A 1500 mg/L stock (initial) solution was prepared by dissolving 1.5 g of the MO in 1 L of distilled water and standing for several hours. From this solution, the samples containing the MO concentrations of 30, 60, 120, 180 and 240 mg/L were obtained using the corresponding dissolutions. The optical density of each sample was measured on the spectrophotometer. Distilled water was used as reference sample. Each sample was replicated 3 times to achieve values with the least error.

To evaluate the effect of the electric field on the adsorption, the experiments were performed on the setup shown in Figs. 1 and 2. The stock solution (volume 4 L) having an initial concentration of 150 mg/L was supplied from the reservoir (using a centrifugal pump) into a sorption flask containing a filter element with the adsorbent and also two electrodes, to which the voltage of 0-150 V was applied. From the sorption flask, the solution was returned to the original reservoir through the reverse loop. The solution flow rate was controlled using a SV-15G meter (Meter, Moscow, Russia). The adsorbent mass was 20 (in the case of the ACs) 2 (in the case of the CNTs) g.

The adsorption capacity of the adsorbent materials was estimated according to the following equation:

$$Q_e = (C_0 - C_e) \cdot V / m \qquad (1)$$

where V is the sample volume (L), m is the adsorbent mass (g), C_0 is the MO initial concentration (mg/L), and C_e is the MO equilibrium concentration in the solution (mg/L).

High adsorption values can be obtained by using the NWC carbon and "Taunit-M" CNTs. To confirm this fact, experiments were carried out on the laboratory setup without use of the electric field. The experimental results obtained are presented in Figs. 6 and 7.

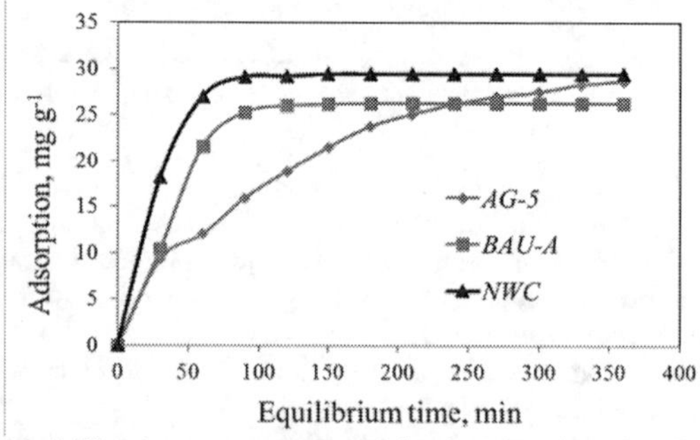

Fig. 6. Kinetic curves constructed for the MO adsorption from aqueous solutions on the ACs: NWC, AG-5, and BAU-A (under normal conditions).

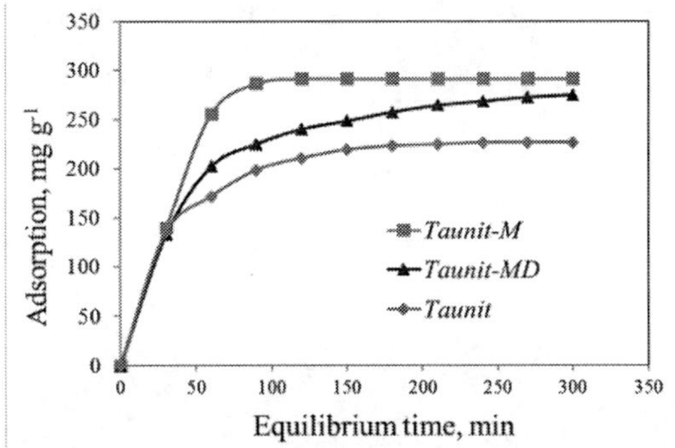

Fig. 7. Kinetic curves constructed for the MO adsorption from aqueous solutions on the CNTs: "Taunit", "Taunit-M", and "Taunit-MD" (under normal conditions).

The analysis of these shows that each adsorbent under study has its own kinetic features of the organic dye sorption, and the NWC AC and the "Taunit-M" CNTs possess the maximum sorption activity (Table 2).

Table 2. Adsorption capacity of the carbon materials under study.

Material	Adsorption capacity, mg/g	Equilibrium time, min
Activated carbons		
NWC	**29**	**90**
AG-5	27	360
BAU-A	26	120
Carbon nanotubes		
Taunit	210	170
Taunit-M	**280**	**110**
Taunit-MD	260	300

Thus, for further studies of the electric field effect, it is advisable to employ the two most sorptionally active materials – NWC AC and "Taunit-M" CNTs.

The effect of the influence of the electromagnetic field on the sorption process on the CNTs (Fig. 8) and the ACs (Fig. 9) was studied at fixed voltage values (0-150 V) in 50 V increments. Samples for diagnosis were taken at 20 min intervals.

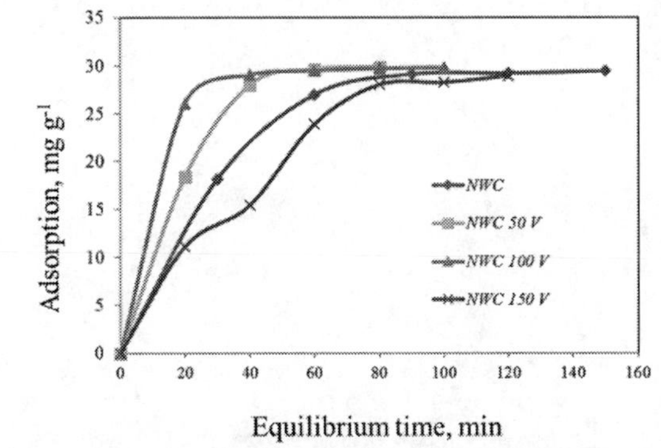

Fig. 8. Kinetic curves constructed for the MO adsorption from aqueous solutions on the initial NWC AC under the voltages of 50, 100 and 150 V.

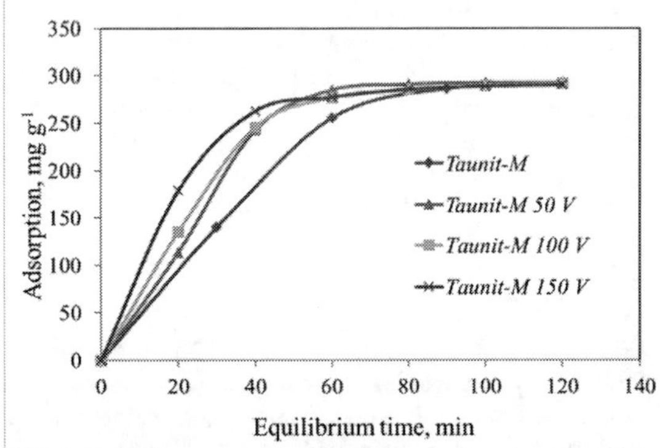

Fig. 9. Kinetic curves constructed for the MO adsorption from aqueous solutions on the "Taunit-M" CNTs under the voltages of 50, 100 and 150 V.

Analyzing the graphs constructed for the "Taunit-M" CNTs, it can be concluded about the opposite results with respect to the standard AC. The equilibrium time for the pollutant recovery process is reached under the influence of the electric field with a voltage of 100 V, which is 1.5 times faster than in the absence of voltage.

The results of the studies on the adsorption characteristics of the nanomodified samples using the electromagnetic field are presented in Fig. 10.

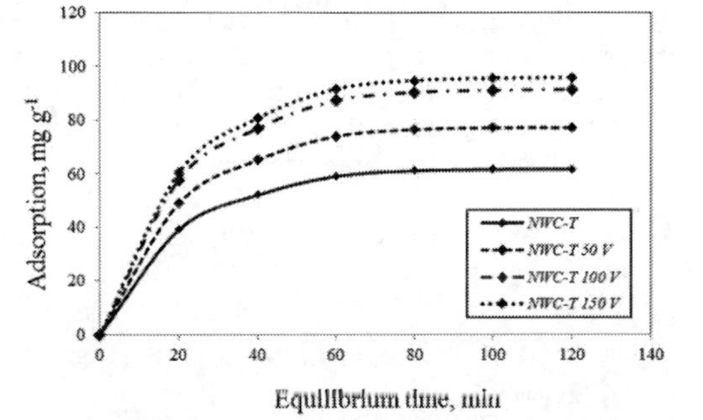

Fig. 10. Kinetic curves constructed for the MO adsorption from aqueous solutions on the NWC-T AC under the voltages of 50, 100 and 150 V.

From this figure, it can be seen that at the same equilibrium time for all the nanomodified AC samples, the maximum adsorption capacity (90 mg/g) corresponds to the voltage of 100 V.

Thus, as a result of the experimental studies, the comparative data shown in Table 3 were obtained.

Table 3. Adsorption capacity of the sorbents under the influence of electromagnetic field.

	NWC		"Taunit-M" CNTs		NWC-T	
Voltage, V	Q_e, mg/g	Time min	Q_e, mg/g	Time min	Q_e, mg/g	Time, min
0	29	90	280	110	60	80
50	29	60	280	80	73	
100	29	37	285	80	90	
150	28	120	275	100	92	

DOI: 10.1051/matecconf/201712906020

Analyzing the results presented in Table 3, it can be clearly seen that the optimum voltage is 100 V for all the samples. Under these conditions, the maximum adsorption capacity is achieved in a shorter period of time. The activity of the sorption materials can be distributed in the following sequence: BAU-A<AG-5<NWC<NWC-T<"Taunt"<"Taunit-MD"<"Taunit-M".

3 Conclusions

The authors of this works have proved that the electric field really intensifies the process of extracting the organic pollutant molecules from aqueous solutions. The studies aimed at carrying out a number of experiments using a special laboratory facility. The influence of the electric field of different strength was elucidated and the degree of the influence of this external factor on the key sorption characteristics of the carbon materials was calculated based on the results obtained.

It was found that the voltage superposition increases the adsorption capacity of the sorbents (ACs, CNTs) while reducing the process time. The sorption activity of the materials under study lies in the following sequence: BAU-A < AG-5 < NWC < NWC-T < "Taunt" < "Taunit-MD" < "Taunit-M".

Thus, the research carried out allows for drawing a conclusion about the positive influence of the electrostatic exposure on the sorption activity of the adsorbents during the extraction of organic pollutants from aqueous media, thereby contributing to the prospects of using the method of electrically controlled sorption on carbon materials on an industrial scale.

The research was funded by the Ministry of Education and Science of the Russian Federation (Project No. 16.1384.2017/PCh).

References

1. M.A. Tsvetnov, *Adsorption and electrical sorption of mono- and semi-functional organic compounds from aqueous and aqueous-organic media on granular and fibrous carbon materials* (PhD (Candidate of Chemical Sciences) Thesis, Vladivostok, 2002)
2. N.I. Gel'perin, *Basic processes and devices of chemical technology. Parts 1 and 2* (Chemistry, Moscow, 1981)
3. A. Mohammad-Khah, R. Ansari Int. J. Chem. Tech. Res., **1**, 4 (2009)
4. Wu Yunhai, Wen Yajun, Zhou Jianxin, Cao Julin, Jin Yanping, Wu Yunying, Environ. Sci. Pollut. Res., **20**, 4 (2013)
5. Xuemei Ren, Changlun Chen, Masaaki Nagatsu, Xiangke Wang, Chem. Eng. J., **170** (2011)
6. T.P. Dyachkova, Yu. A. Khan, N. V. Orlova, S. V. Kondrashov, Vestnik TSTU, **324** (2016)
7. D.V. Tarov, T.V. Gurova, I.N. Shubin, Vestnik TSTU, **361** (2015)
8. A. Burakov, I. Romantsova, A. Kucherova A. Tkachev, Ads. Sci. Tech., **32** (9) (2014)
9. Russian National Standard GOST No. 4453–74 *A Powder of Active Clarifying Charcoal. Technical Conditions* (1996)

DOI: 10.1051/matecconf/201712906021

Implementation of carbon nanomodification for sorption materials

Alexander Babkin[1], *Elena* Neskoromnaya[1], and *Irina* Burakova[1,*]

[1] Tambov State Technical University, Department "Technology and Methods of Nanoproducts Manufacturing", 392000, 106 Sovetskaya Str., Tambov, Russia

Abstract. The article addresses the urgent task of improving the adsorption capacity and expanding the scope of application for commonly used industrial sorbents - activated carbons and synthetic zeolites. Among a variety of methods for modifying these sorbents, more attention is now being given to techniques that employ carbon nanomaterials. This is due to the unique properties of nanostructures - developed surface, availability of active functional groups, etc. In the present work, the classic materials – NWC coconut shell activated carbon and synthetic NaX zeolite – were chosen as initial sorbent samples to be modified. The authors developed a process flowsheet for the carbon nanomodification, which contains the following main stages: preparation of a catalytic mixture solution under given temperature conditions, impregnation of porous materials using the obtained solution, and drying and synthesis of carbon nanotubes via chemical vapor deposition. The proposed technological line consists of a reactor for synthesis of carbon nanotubes, the patented design of which will allow for simultaneously modifying in an effective way different types of materials. As a result, the layer of carbon nanostructures, the quality of which can be varied by changing the conditions of the modification procedure, is formed on the substrate surface.

1 Introduction

The problems of fine purification of aqueous media, regeneration of wastewater and integrated water treatment for high-tech industries are relevant and due to ever more stringent modern requirements. A promising route for developing effective sorbents is their modification with carbon nanomaterials, in particular, carbon nanotubes (CNTs). The most common sorbents are activated carbon (AC) and zeolites, the modification of which substantially increases their adsorption activity [1-18].

As it is known, the AC consisting of a lot of randomly arranged graphite microcrystals is usually employed in the processes of purification and separation of liquids and gases (vapors) for adsorption of organic substances.

The zeolites represent natural or synthetic minerals – aqueous aliminosilicates containing oxides of alkali and alkaline earth metals.

Despite the variety of sorbents used, many of them do not satisfy the entire set of requirements established for materials of this type. In this regard, the search for and development of novel sorption materials are ongoing.

Industrial sorbents must meet the following requirements: have a great sorption capacity and high mechanical strength, possess high selectivity, ability to regenerate and stability of sorption properties under long-term operationconditions, be non-toxic, non-corrosive, and inexpensive.

At present, a number of new sorption materials are being developed, the production of which has not yet been mastered on an industrial scale. Materials with a modified porous space and various kinds of composites may be attributed to such sorbents. The modification of sorbents promotes the expansion of their application areas, as well as an increase in their adsorption selectivity.

Depending on the application field, there exist different methods for modifying the AC: acid/base treatment, ammonium compound treatment, impregnation using polymer solutions; temperature treatment; plasma treatment; exposure to microwave radiation; ozonation; impregnation using metals; and modification using carbon nanomaterials. As a result of the modification, the surface reactivity, chemical, physical and structural properties of the AC change [1-14].

Among the methods for modifying the zeolites, the following ones can be pointed out: composition change during crystallization; ion exchange; introduction of easily adsorbed impurities into the zeolite crystal; decathionization; dealumination; modification using polymers; and modification using carbon nanomaterials [15-18].

As can be seen, for both the AC and zeolites, the modification with carbon nanomaterials is used. The application of CNTs is due to their high adsorption characteristics, the possibility of changing the surface chemistry, and the structural features of the nanometer range which are determined by synthesis method and parameters. Chemical vapor deposition (CVD) with the

* Corresponding author: iris_tamb68@mail.ru

DOI: 10.1051/matecconf/201712906021

use of metal oxide catalysts appears to be a widespread industrial method of CNT production, which makes it possible to obtain CNT structures over substrates. The most important effect on the nanotube structure is provided with the technology of preparation of metal oxide catalysts and chemical nature thereof. The determination of the dependence of the nanotube properties on the parameters of the catalysts used allows to obtain CNTs with adjustable characteristics (length, diameter, degree of defectiveness, etc.) for surface modification of porous substrates [19-25].

Considering the aforementioned, the aim of the present research is to develop a carbon nanomodification process for industrial sorbents used for purification of aqueous media.

2 Materials and Methods

In the modification process, NWC industrial coconut shell AC and synthetic NaX zeolite were used as substrates.

Electron images of nanomodified materials were obtained via scanning electron microscopy (SEM) on a Neon 40 two-beam instrument (Carl Zeiss, Jena, Germany).

3 Modification Procedure

The implementation of the experimental-industrial (pilot) production of nanomodified materials based on the AC and zeolites is possible within the technological line, the flow diagram and elemental composition of which are presented in Fig. 1.

The initial components (nickel, cobalt and magnesium nitrates, and citric acid) enter the reactor-mixer RM1 from the bunkers B1, B2, B3 and B4 through the dispensers D1, D2, D3 and D4. Ethylene glycol comes into the RM1 from the reservoir R1 by means of the dispensing pump DP.

Distilled water is obtained in the distiller DIS and routed into the reservoir R2. From the R2, the distilled water is supplied into the RM1 using the dispensing pump DP, where all the catalyst components are mixed and treated at 80 °C.

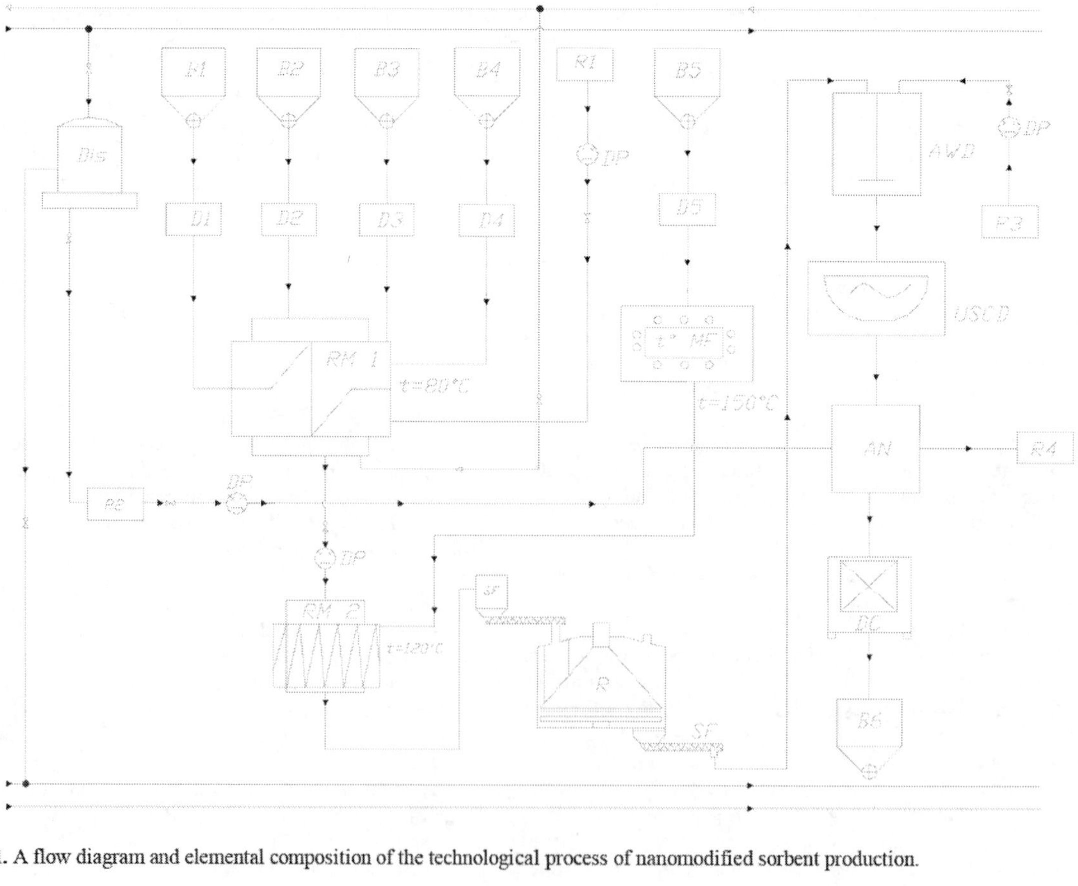

Fig. 1. A flow diagram and elemental composition of the technological process of nanomodified sorbent production.

DOI: 10.1051/matecconf/201712906021

The solution is heated in indirect steam. The temperature inside the reactor is raised up to 140 °C. From this reactor, the xerogel formed is routed through the dispensing pump DP into the auger reactor-mixer RM2.

The sorbent comes into the muffle furnace MF (pre-heated to a temperature of 150 °C) from the bunker B5 through the dispenser D5. From the MF, it is supplied into the RM2, where the porous substrate is impregnated with a catalyst solution at 120 °C. Then, the impregnated sample comes into the batch reactor R, where the carbon nanomaterial is synthesized over the substrate surface via catalytic pyrolysis of hydrocarbons.

After that, the nanomodified material is sent to the acid washing device AWD for washing from impurities of amorphous carbon and catalyst particles. Into this device, nitric acid comes from the reservoir R3 through the dispensing pump DP.

Next, the material passes through the ultrasonic cleaning device USCD, and then it enters the drying chamber DC through the acid neutralizer AN. From the AN, the spent acid is withdrawn into the reservoir R4. The obtained nanomodified material is routed into the bunker B6 for storage.

To implement the CNT synthesis process over the impregnated sorbent samples, the design of an upgraded reactor was developed, which allows for increasing the productivity of this stage due to more rational use of the working area of the device.

In the reaction zone of the reactor, there is a device that makes it possible to modify several types of materials: fibrous, granular, and gradient-porous inorganic plates (Fig. 2). The upgraded reactor comprises a housing consisting of the bottom 1 and the lid 2 connected with the sealing unit 3.

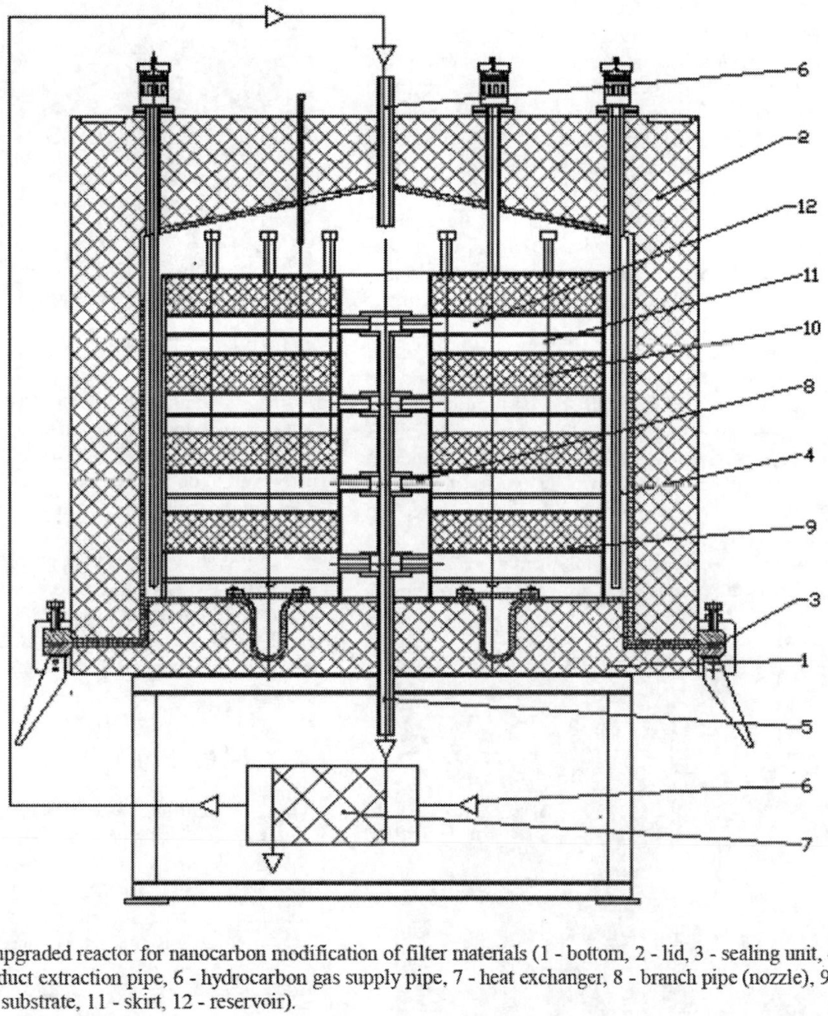

Fig. 2. The design of an upgraded reactor for nanocarbon modification of filter materials (1 - bottom, 2 - lid, 3 - sealing unit, 4 - heaters, 5 - gaseous pyrolysis product extraction pipe, 6 - hydrocarbon gas supply pipe, 7 - heat exchanger, 8 - branch pipe (nozzle), 9 - deposition chamber, 10 - substrate, 11 - skirt, 12 - reservoir).

DOI: 10.1051/matecconf/201712906021

On the lid 2, inside the housing, the heaters 4 and the gaseous pyrolysis product extraction pipe 5 are installed, and on the bottom 1, hydrocarbon gas supply pipe 6 is mounted.

The hydrocarbon gas supply pipe 6 and the gaseous pyrolysis product extraction pipe 5 are connected together with the heat exchanger 7 installed outside the reactor. On the bottom 1, the deposition chambers 9 are mounted. At the outlet, they are connected with the gaseous pyrolysis product extraction pipe through the branch pipes (nozzles) 8. At the inlet of those chambers, the substrates 10 made of a porous material [26] impregnated with a solution of liquid phase precursors of the CNT synthesis catalyst are placed.

The technical result provided with the reactor lies in obtaining nanomodified materials by depositing on the carbon material impregnated with the catalyst solution.

The proposed device operates as follows: when the lid is up, the deposition chambers and the substrates are installed on the bottom. The chambers are equipped with the substrates in the following way - when using the substrates in the form of granules, the latter are loaded into pallets with a mesh bottom. The substrates in the form of disks are mounted in the lid sockets and fixed with rings. The substrates in the form of sheets are laid on the lattice lid and fixed with frames. The nozzles are connected to the gaseous pyrolysis product extraction pipe. Then, the lid is installed on the bottom, and the reactor is hermetically sealed using the sealing units. Before operation, the reactor cavity is purged with argon through the hydrocarbon gas supply pipe, while the reactor air is removed through the gaseous pyrolysis product extraction pipe.

The pyrolysis of the hydrocarbon gas is carried out at a temperature of 600-700 °C. During this process, the hydrocarbon gas passes through the perforation of the skirts 11 and the substrate 10 into the reservoir 12, and is discharged into the gaseous pyrolysis product extraction pipe 5 through the nozzles 8. The exhausted hot gases enter the heat exchanger 7, where the heat is supplied into the hydrocarbon gas supply pipe 6.

The results of the implementation of the nanomodified sample synthesis are presented in Figs. 3 and 4.

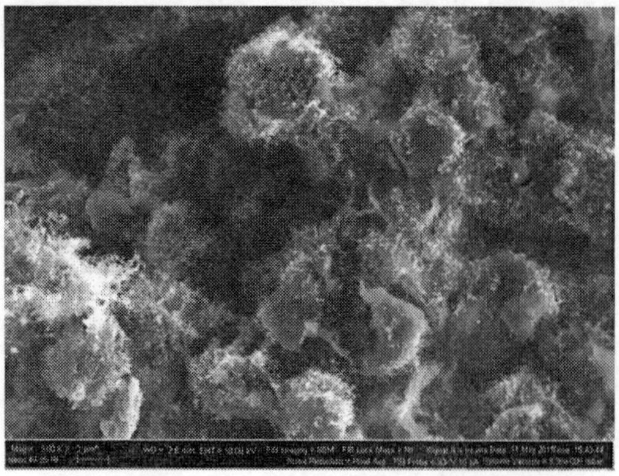

Fig. 3. A modified surface of the NaX zeolite.

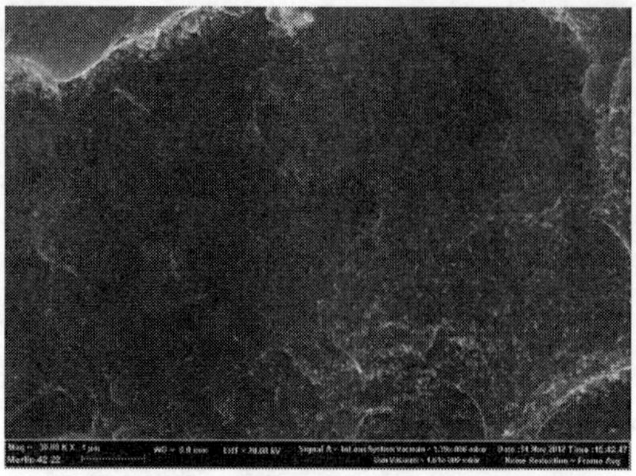

Fig. 4. A modified surface of the NWC coconut shell activated carbon.

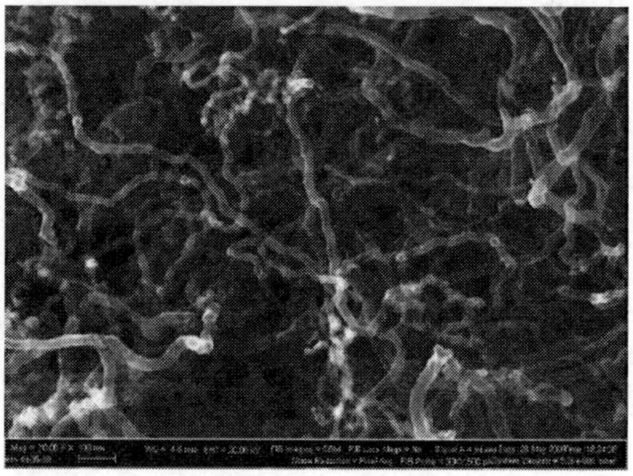

Fig. 5. Morphology of the CNT layer.

Fig. 3 demonstrates the synthetic NaX zeolite, the surface of which is modified with CNTs. In this figure, individual zeolite granules coated with the CNTs can be observed. The nanotube diameter is 15-30 nm. Besides, the catalyst particles with a size of 15-100 nm are available in the material.

Fig. 4 shows the structure of the NWC coconut shell AC with pores of 1-2 microns. It can be seen that the nanomodification resulted in uniform coverage of the CNTs over the AC surface. The nanotube diameter is 15-50 nm, and the catalyst particle size is also 15-50 nm (Fig. 5).

4 Conclusion

Thus, the authors have implemented the formation of a nanotube layer (Patent 106253 RF) in a porous space of commonly used sorbent materials for developing nanostructured sorbents. The proposed technological solution provides obtaining a uniformly distributed array of nanotubes with adjustable parameters.

The adsorption of organic and inorganic compounds from aqueous solutions appears to be the main trend in using modified materials.

By varying the synthesis mode parameters, as well as by changing the chemical composition of the catalytic mixture, it is possible to obtain the CNT layer with a given quality on the surface of porous substrates. This fact can expand the field of application for nanomodified materials.

The research was funded by the Ministry of Education and Science of the Russian Federation (Project No. 16.1384.2017/PCh).

References

1. Yunhai Wu, Yajun Wen, Jianxin Zhou et.al., ESPR, **20**, 4 (2013)
2. Lo Sheng-Fong, Wang Song-Yung, Tsai Ming-Jer, Lin Lang-Dong, Chem. Eng. Res. Des., **90**, 9 (2012)
3. Sato Sanae, Yoshihara Kazuya, Moriyama Koji, Machida Motoi, Tatsumoto Hideki, Appl. Surf. Sci., **253**, 20 (2007)
4. B. Saha, M.H. Tai, M. Streat, Chem. Eng. Res. Des., 2003. **81**, 10 (2003)
5. E. Erdem, N. Karapinar, R. Donat, J. Colloid Interface Sci., **280** (2004)
6. P. Chingombe, B. Saha, R.J Wakeman, Carbon, **43** (2005)
7. Amalina Md. Dali, Ahmmed Saadi Ibrahem, Abdul Hadi, J. Pur. Util. Reac. Environ., **1**, 5 (2012)
8. Chun Yang Yin, Mohd Kheireddine Arou, Wan Mohd Ashri Wan Daud, Sep. Purif. Technol., **52**, 3 (2007)
9. Xuemei Ren, Changlun Chen, Masaaki Nagatsu, Xiangke Wang, Chem. Eng. J., **170** (2011)
10. S. Vinitnantharat, W. Rattanasirisophon, Y. Ishibashi, Water Sci Technol., **55**, 5 (2007)
11. Rachel RibeiroVieira Azzi Rios, Dênio Eduardo Alves, Ilza Dalmázio, Sílvio Fernando Vargas Bento, Claudio Luis Donnici, Rochel Monteiro Lago, Mater. Res., **6**, 2 (2003)
12. W. Shen, Z. Li, Y. Lieu, Rec. Pat. Chem. Eng., **1** (2008)
13. M.S. Shafeeyan, W.M. Wan Daud, A. Houshmand, A. Shamiri, J. Anal. Appl. Pyrolysis, **89** (2010)
14. G.G. Stavropoulos, P. Samaras, G.P. Sakellaropoulos, J. Hazard. Mater., **151** (2008)
15. C.R. Oliveira, J. Rubio, Miner. Eng., **20** (2007)
16. G. Ciobanu, D. Ignat, G. Carja, S. Ratoi, C. Luca, Chem. Bull. "POLITEHNICA" Univ. (Timişoara), **53**, 67 (2008)
17. Yousheng Tao, Hirofumi Kanoh, Lloyd Abrams, Katsumi Kaneko, Prep. Char. Appl. Chem. Rev., **106** (2006)
18. A. H. Janssen, I. Schmidt, C. J. H. Jacobsen, A. J. Koster, K. P. de Jong, Microporous and Mesoporous Mater., **65**, 1 (2003)
19. Ren Xuemei, Chen Changlun, Nagatsu Masaaki, Wang Xiangke, Chem. Eng. J., **170** (2011)
20. Tsu-Wei Chou, Limin Gao, Erik T. Thostenson, Zuoguang Zhang, Joon-Hyung Byun, Compos. Sci. Technol., **70** (2010)
21. Mohamed Abdel Salam, Mohamad S.I. Makki, Magdy Y.A. Abdelaal, J. Alloys Compd., **509**, 5 (2011)
22. M. Bahgat, A.A. Farghali, W.M.A. El Rouby, M.H. Khedr, J. Anal. Appl. Pyrolysis, **92**, 2 (2011)
23. Savage Nora, Diallo Mamadou S., J. Nanopart. Res., **7** (2005)
24. D.V. Tarov, T.V. Gurova, I.N. Shubin, Vestnik TSTU, **361** (2015)
25. A.G. Tkachev, D.V. Tarov, V.P. Tarov, I.N. Shubin, Vestnik TSTU, **503** (2016)
26. A.E. Burakov, Yu. A. Kobcheva, E.A. Burakova, I.V. Ivanova, A.G. Tkachev Patent of the Russian Federation 106253 U1, № 19 (2011)

DOI: 10.1051/matecconf/201712906022

Graphene materials for lead (II) extraction: an equilibrium study

Anastasia Kucherova[1,*], *Irina* Burakova[1], and *Alexandr* Burakov[1]

[1]Tambov State Technical University, Department "Technology and Methods of Nanoproducts Manufacturing", 392000, 106 Sovetskaya Str., Tambov, Russia

Abstract. The present paper describes a study on the adsorption of lead (II) ions on graphene materials obtained via exfoliation of a hydrolyzed expanded graphite intercalation compound. An equilibrium studies were performed implementing the empirical Freundlich and Langmuir isotherm models. It was found that all the isotherms are fitted well by those models, indicating the efficiency of the graphene materials as an adsorbent of heavy metals. The applicability of the two isotherms for sorption processes shows that both monolayer adsorption and heterogeneous energetic distribution of active sites on the adsorbent surface may take place. The maximum experimental lead (II) adsorption capacities of the materials under study – few-layered oxidized graphene suspension, multi-layered graphene paste, and few-layered graphene paste – were found to be 850, 230 and 170 mg/g, respectively. Due to the high sorption characteristics and unique physical and chemical properties of these materials, the adsorption technologies developed herein can act as good sustainable options for the future in heavy metal removal from industrial effluents.

1 Introduction

Rapid industrialization of human society leads to the fact that different contaminants (such as heavy metal ions, organics, bacteria, viruses, etc), which are seriously harmful to human health, are released to wastewater. Among all the water pollutants, heavy metal ions, including Pb(II), Cd(II), Zn(II), Ni(II) and Hg(II), have high toxicity and non-biodegradable properties and can cause severe health problems in animals and human beings. For instance, lead poses several risks to human health, causing insomnia, pain, dizziness, anemia, irritability, muscle weakness, hallucinations, and renal damage [1].

The presence of the above-mentioned heavy metals in aqueous systems poses a serious environmental problem. In this regard, various techniques, such as ion exchange, membrane filtration, coagulation, flocculation, flotation and electrochemical methods, have been proposed to remove and isolate those species [2]. Among them, adsorption appears to be the most efficient way. It has been widely employed in separation, waste effluents treatment, refrigeration, environmental control, life support devices in spacecraft, and heterogeneous catalysis [3-5].

Graphene, a new carbon nanomaterial, possesses unique physical, chemical, electrical and mechanical properties [6-9], and is adequate for removing heavy metals from aqueous solutions. It has demonstrated high adsorption rates and fast adsorption kinetics [10-13].

Considering the aforementioned, the aim of the present research was to develop novel graphene materials for efficiently extracting Pb(II) from aqueous solutions.

Electron images of graphene-based samples were obtained by scanning electron microscopy (SEM) on a Neon 40 two-beam instrument (Carl Zeiss, Jena, Germany) and transmission electron microscopy (TEM) on a JeM-200CX instrument (JEOL, Tokyo, Japan). The equilibrium Pb (II) concentration in aqueous solutions was determined by atomic absorption spectrometry (AAS) on an MGA-915MD instrument (Atompribor Ltd., Saint-Petersburg, Russia).

2 Experimental part

2.1 Adsorbents

1) Graphene samples under study represent aqueous pastes containing multi- or few-layered graphene.

The graphene pastes were synthesized according to the method described in [14]. Graphite was intercalated with an ammonium persulfate solution in sulfuric acid, and then aged for 3 h at 40 ^{o}C for expansion. After that, it was hydrolyzed and washed with water until the acid was removed. The addition of surfactants promotes the formation of few-layered graphene. As a rule, the graphene content in the multi-layered graphene paste (MLGP) was 5 wt.%, and in the few-layered graphene paste – 12 wt.% (based on graphene carbon). The graphene contained about 10 wt.% of oxide groups related to the graphene carbon.

The typical SEM and TEM images of the MLGP are presented in Fig. 1, from which it can be seen that the size of nanoplatelets is on the order of 5-10 nm, and the number of layers is 15-25.

* Corresponding author: anastasia.90k@mail.ru

DOI: 10.1051/matecconf/201712906022

As seen in Fig. 2, the FLGP represents few-layered graphene nanoplatelets with the size of about 5-10 nm and the number of layers of 3-5.

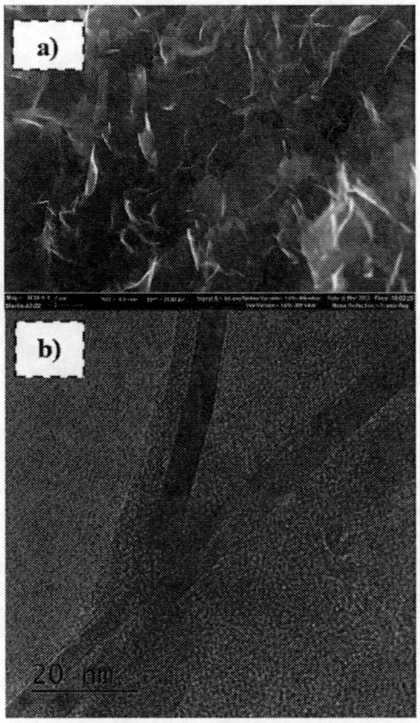

Fig. 1. SEM (*a*) and TEM (*b*) images of the MLGP.

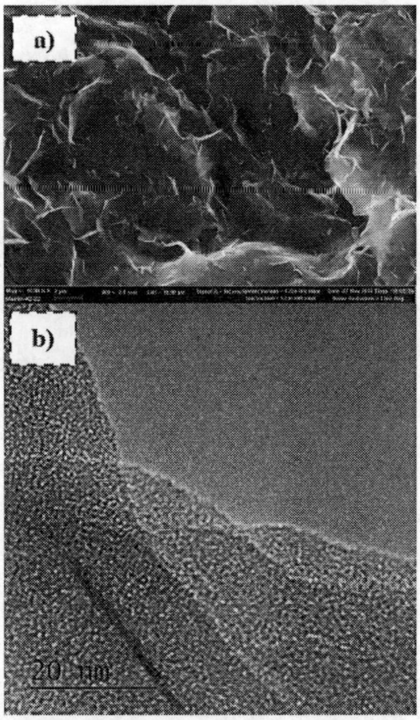

Fig. 2. SEM (*a*) and TEM *(b)* images of the FLGP.

2) Few-layered oxidized graphene suspension (FLOGS) represents an aqueous dispersion of oxidized few-layered graphene nanoplatelets. This dispersion was prepared by forming an expanded graphite compound according to the procedure described elsewhere [15], and then stabilized using a Fenotam GR-326 water-soluble phenol-formaldehyde resole resin (PFR) (Pigment Ltd., Tambov, Russia) containing 50 % of non-volatile substances. GSM-2 natural graphite (Pavna Ltd., Belgorod, Russia), containing ≤0.5 % of mineral impurities, was intercalated by ammonium persulfate in anhydrous sulfuric acid monohydrate (both - reagent grade, Laverna Ltd., Moscow, Russia). After that, the intercalated compound was expanded for 3 h at 40°C, and then filtered and washed with water to neutral pH. The obtained hydrolyzed expanded graphite compound was used for ultrasonic exfoliation without drying. The calculated amount of wet HEGC, obtained after mixing 5 g of the starting graphite and 5 g of the PFR, was placed into a 2-L beaker. Next, water was added to this mixture until a total mass of 1,500 g was obtained. The resulting suspension was sonicated using an IL10 ultrasonic generator (Ultrasonic Technique-INLAB Ltd., Saint-Petersburg, Russia) under continuous stirring for 6 h. The graphene content in the obtained colloidal dispersion was 3.33 g L^{-1}, and the PFR content (on a dry basis) was 1.67 g L^{-1}.

2.1 Research methodology

The experimental study on the Pb(II) adsorption from aqueous solutions was performed with the FLOGS, FLGP and MLGP graphene materials used as adsorbents. Each series consisted of seven 50-mL tubes containing the corresponding adsorbent. 15 mL of $Pb(NO_3)_2$ solutions having concentrations of 200, 500, 700, 1,000, 2,400 and 3,500 mg/L were separately added into those tubes using Pasteur pipettes. Then, the tubes were end-over-end shaken on Bio RS-24 programmable rotators (Biosan, Riga, Latvia) at 100 rpm during the equilibrium time previously found for the adsorbent. After the adsorption process was finished, the solutions were centrifuged using a 5810 R centrifuge (Eppendorf, Hamburg, Germany) at 10,000 rpm for 10 min to separate the phases. Next, the liquid phase was placed into 30-mL tubes with the Pasteur pipettes, preventing ingress of the solid phase and the suspension thereof. Finally, the solutions were measured according to the method of quantitative elemental analysis by atomic absorption spectra on a spectrometer.

The equilibrium Pb(II) adsorption capacity, q_e (mg/g), was evaluated according to Eq. (1):

$$q_e = (C_0 - C_e) \cdot V / m \tag{1}$$

where V is the sample volume (L), m is the mass of the adsorbents (g), C_o is the initial Pb(II) concentration (mg/L), and C_e is the equilibrium Pb(II) concentration in the solution (mg/L).

DOI: 10.1051/matecconf/201712906022

3 Results and discussion

To optimize the design of an adsorption system for adsorbate removal, it is important to establish the most appropriate correlation for equilibrium data. Two adsorption isotherm models – Langmuir and Freundlich were implemented herein to fit the experimental data. The Langmuir model assumes that the adsorption takes place at specific homogeneous sites located on the adsorbent surface. Its form is given in Eq. (2) as follows:

$$q_e = \frac{q_{\max} b C_e}{b C_e + 1} \tag{2}$$

where, q_e is the Pb(II) amount adsorbed at equilibrium per unit weight of adsorbent (mg/g), q_{max} is the saturated monolayer adsorption capacity (mg/g), b is the Langmuir constant related to the binding energy (L/mg) of the adsorption system, and C_e is the Pb(II) equilibrium concentration (mg/L).

The Freundlich model can be applied to non-ideal sorption onto heterogeneous surfaces in multiple layers. It is expressed by Eq. (3) as follows:

$$q_e = k_f (C_e)^{1/n} \tag{3}$$

where q_e is the Pb(II) amount adsorbed at equilibrium (mg/g), C_e is the Pb(II) equilibrium concentration (mg/L), k_f is the equilibrium constant which indicates the adsorption capacity, and n is the equilibrium constant standing for the adsorption intensity (for $1/n$=1, the adsorption is linear; $1/n$<1 implies that the adsorption is of the chemical nature, and if $1/n$> 1, the adsorption is favorable and of the physical nature).

Although many other isotherm models have been developed and employed [16], it can be seen that the Langmuir and Freundlich isotherms still remain the two most commonly used equilibrium adsorption equations due to their simplicity and the ease of estimation of their adjustable parameters.

As seen in Figs. 3 and 4, the isotherms can be well approximated by straight lines, thereby confirming the applicability of both the Langmuir and the Freundlich models for describing the Pb(II) adsorbtion on all the graphene materials under study.

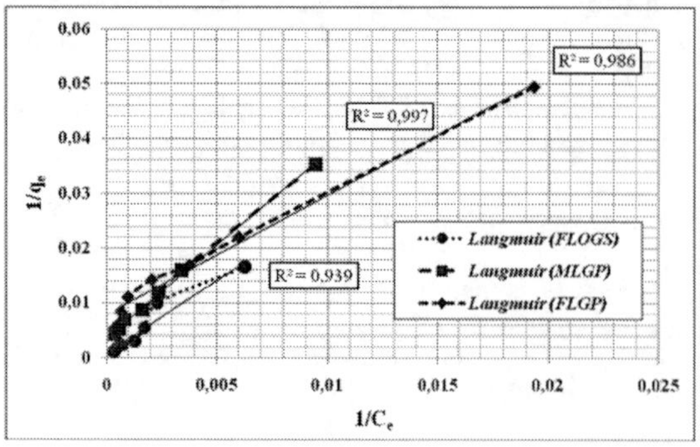

Fig. 3. Langmuir isotherm curves.

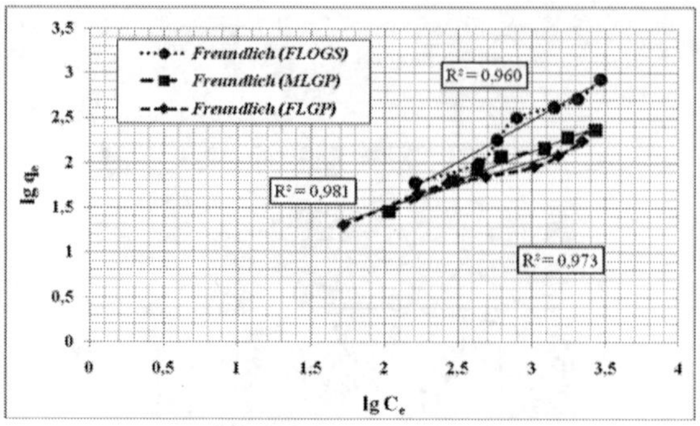

Fig. 4. Freundlich isotherm curves.

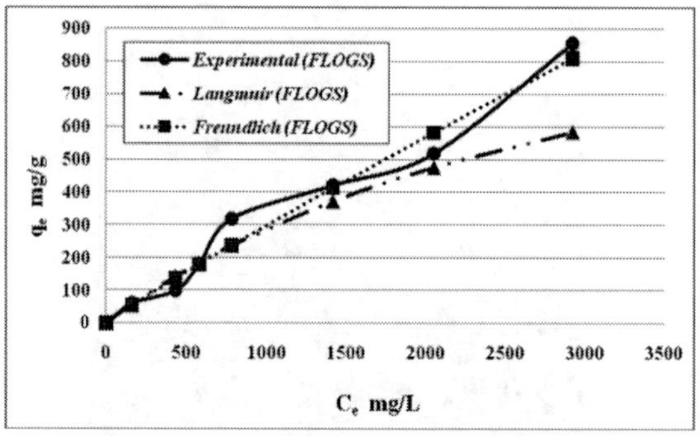

Fig.5. Experimental and theoretical isotherms of the Pb(II) adsorption on the FLOGS.

According to the classification proposed by Smith, the adsorption isotherms constructed for the FLOGS are reminiscent of Type III S-shaped isotherms (Fig. 5), thereby indicating a transitional porous structure of the adsorbent characterized by complicated topology of the pore space and a large variance in the distribution of pore sizes. Furthermore, the nature of the steepness of that isotherm suggests the presence of predominantly very small pores in the adsorbent. The lower part of the S-shaped curve, from the origin to the inflection point, corresponds to the formation of a monomolecular layer, and then turns into multi-molecular adsorption, thereby explaining the further rise of the curve [17-19].

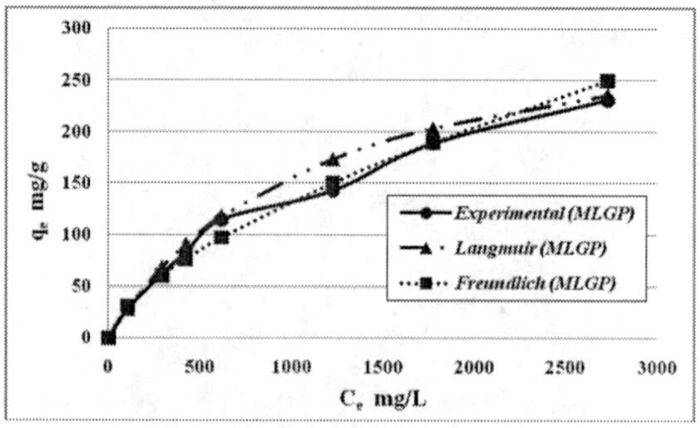

Fig.6. Experimental and theoretical isotherms of the Pb(II) adsorption on the MLGP.

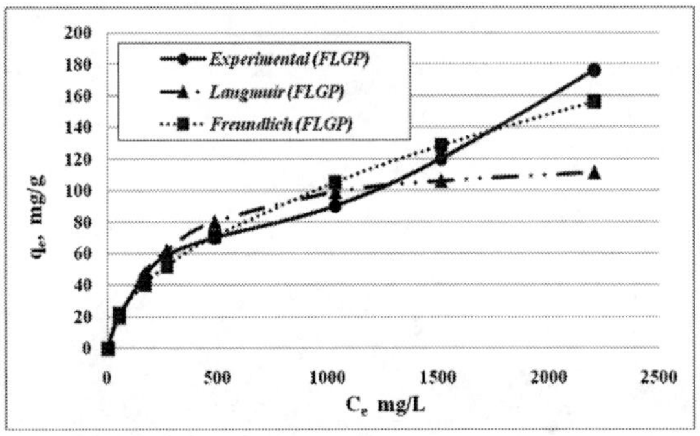

Fig. 7. Experimental and theoretical isotherms of the Pb(II) adsorption on the FLGP.

After studying the adsorption isotherms constructed for the MLGP and FLGP, it can be assumed that they correspond to L-type curves. In Figs. 6 and 7, an initial curved section can be observed relative to the concentration axis, evidencing the formation of a monomolecular layer. The presence of a second plateau after saturation of the adsorption layer is typical for Type IV isotherms and may be due to multi-molecular adsorption [17].

From the isotherm studies (Table 1), it was found that all the isotherms fit well the experimental data, thereby indicating the efficiency of the graphene materials used herein as heavy metal adsorbents.

It should be noted that the maximum experimental adsorption capacity of the materials under study changes in the following sequence: FLOGS > MLGP > FLGP.

Comparing the correlation coefficients (R^2) for the two models used, it can be seen that their values are commensurable. This indicates that the Pb(II) adsorption on the graphene materials has a complex nature, in particular, due to the availability of active sites with different binding energies on the adsorbent surface.

In the range of small concentrations (200...700 mg/L), it is preferable to implement the Langmuir model. Based on the obtained values of the adsorption equilibrium constants of that equation, it can be assumed that the Pb(II) ions in their small concentrations interact with active sorption sites located at the surface of the materials under study and responsible for physical adsorption. The range of high concentrations does not satisfy the boundary conditions for the applicability of the Langmuir model with the monomolecular sorption mechanism, and the active adsorption within this range can serve as the foundation for an indirect confirmation of the applicability of the Freundlich model.

The $1/n$ values obtained can imply favorable conditions for chemical adsorption to occur, since $1/n < 1$. Such an adsorption type is related to the energy heterogeneity of adsorption sites.

Based on the systematization of the above-described results, some recommendations on the use of the graphene materials in the Pb(II) liquid-phase adsorption may be proposed, since the empirical dependencies obtained (Table 2) allow for predicting parameters of the adsorption process.

4 Conclusions

The present study demonstrates the metal adsorption capacity of the graphene materials in aqueous solutions. It confirmed the potential of those materials for employing them as efficient adsorbents to extract metal pollutants at both their lower and higher levels in the liquid phase. This finding may be of great significance for industrial wastewater treatment.

Table 1. Isotherm constants obtained for the Pb(II) adsorption on the FLOGS, MLGP and FLGP materials

Material	Langmuir constants			Freundlich constants		
	q_{max}	b	R^2	k_f	$1/n$	R^2
FLOGS	1250	$3.01\cdot10^{-4}$	0.939	0.488	0.929	0.960
MLGP	333.33	$8.89\cdot10^{-4}$	0.997	1.659	0.634	0.981
FLGP	250	$3.45\cdot10^{-4}$	0.989	2.850	0.520	0.908

Table 2. The Langmuir and Freundlich equations obtained based on the experimental data

Material	Langmuir	Freundlich
	Equations	
FLOGS	$q_e = \frac{1250\cdot 3.01\cdot 10^{-4} C_e}{3.01\cdot 10^{-4} C_e + 1}$	$q_e = 0.488(C_e)^{0.929}$
MLGP	$q_e = \frac{333.33\cdot 8.89\cdot 10^{-4} C_e}{8.89\cdot 10^{-4} C_e + 1}$	$q_e = 1.659(C_e)^{0.634}$
FLGP	$q_e = \frac{250\cdot 3.45\cdot 10^{-4} C_e}{3.45\cdot 10^{-4} C_e + 1}$	$q_e = 2.85(C_e)^{0.52}$

The materials presented high Pb(II) adsorption rates and large adsorption capacity (up to 850 mg/g) under normal conditions. Hence, an extensive research in this area is essential for developing this environmentally friendly process as sustainable technology for the future.

The research was funded by the Ministry of Education and Science of the Russian Federation (Project No. 16.1384.2017/PCh).

References

1. R. Naseem and S. S. Tahir, Water Res. **35** (2001)
2. F. Fu., Q. Wang, J. Environ. Manage., **92**, 3 (2011)
3. Y. Sudhakar, A.K. Dikshit, Int. J. Environ. Poll. **15** (2001)
4. L.R. Radovic, C. Moreno-Catilla, J. Riveral-Utrilla, Chem. Phys. Carbon, **27** (2000)
5. F. Derbyshire, M. Jagtoyen, R. Andrews, A. Rao, I. Martin-Guillon, E.A. Grulke, Chem. Phys. Carbon, **27** (2001)
6. K.S. Novoselov, A.K. Geim, S.V. Morozov, D. Jiang, Y. Zhang, S.V. Dubonos, I.V. Grigorieva, A.A. Firsov, Science, **306** (2004)
7. C.G. Navarro, M. Burghard, K. Kern, Nano Lett., **8** (2008)
8. S. Park, K.S. Lee, G. Bozoklu, W. Cai, S.T. Nguyen, R.S. Ruoff, ACS Nano, **3** (2008)
9. A.A. Balandin, S. Ghosh, W. Bao, I. Calizo, D. Teweldebrhan, F. Miao, C.N. Lau, Nano Lett., **8** (2008)
10. V. Chandra, K.S. Kim, Chem. Commun., **47** (2011)
11. X. Deng, L. Lü, H. Li, F. Luo, J. Hazard. Mater., **183** (2010)
12. S.T. Yang, Y. Chang, H. Wang, G. Liu, S. Chen, Y. Wang, Y. Liu, A. Cao, J. Colloid Interf. Sci., **351** (2010)
13. N. Zhang, H. Qiu, Y. Si, W. Wang, J. Gao, Carbon, **49** (2011)
14. A.V Melezhyk,. A.G. Tkachev, Nanosystems: Physics, Chemistry, Mathematics **5**, 2 (2014)
15. A.E. Kucherova, I.V. Romantsova, A.E. Burakov, M.N. Krasnyansky, N.R. Memetov, Graphene-Based Nano-composites for Enhanced Pb^{2+} Adsorption / NHC, **13** (2016)
16. K.V. Kumar, and S. Sivanesan, J. Hazard. Mater. **136** (2006)
17. G.D. Parfitt and C.H. Rochester. Academic press (1983)
18. T.P. Dyachkova, Yu. A. Khan, N. V. Orlova, S. V. Kondrashov, Vestnik TSTU. **324** (2016)
19. A. G.Tkachev, D. V.Tarov, V. P.Tarov, I. N.Shubin, VestnikTSTU. **503** (2016)

DOI: 10.1051/matecconf/201712906023

Kinematic research of closed planetary gears

Ildar Kutlubaev[1,2], *Elena* Matcko[2], and *Olga* Panfilova[1,*]

[1]Nosov Magnitogorsk State Technical University. Magnitogorsk, Russia
[2]JSC «SPA «Android technics», Moscow, Russia

Abstract. The approach to the kinematic research of planetary gears which are closed by nonparallel circuit is represented. The analytical dependencies of transmission ratio on geometric parameters using graphic method are obtained. The obtained dependencies are the basis for power analysis accomplishment.

Planetary gears are essential part of modern machines and aggregates. Proven production technology of units and workpieces, ability to achieve high transmission ratios determine their dynamic utilizing. Planetary gears (PG) are used for transmission of rotational torque in the wide range. Planetary gears with the torque above or equal of 5 Nm are used in drive units of manipulator links, the rotational torque value of travel mechanisms of building, road, and mining machines reaches 500 Nm. The high level of toughness, smoothing of errors and gaps are achieved by multizone contact of links that leads to decreasing of backlash play.

Planetary gear production keeps steadily growing [1]. Technologies that realize double-flow construction schemes of PG such as biplanetary [2] and complex [3] are advanced as well as planetary-lantern-wheel schemes. Kinematic schemes and linear velocity drafts of double-flow PG with complex dog chart and biplanetary gear are represented correspondingly in figures 1 and 2.

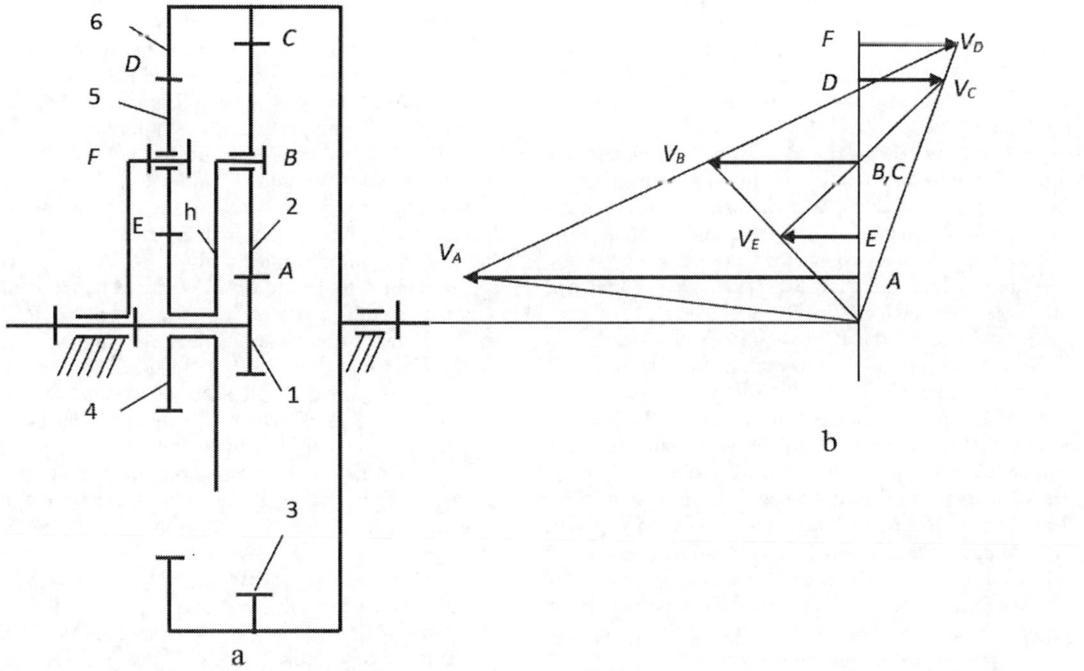

Fig. 1. Kinematic scheme (a) and velocity draft (b) of double-flow PG with complex dog chart

* Corresponding author: halikova@inbox.ru

 DOI: 10.1051/matecconf/201712906023

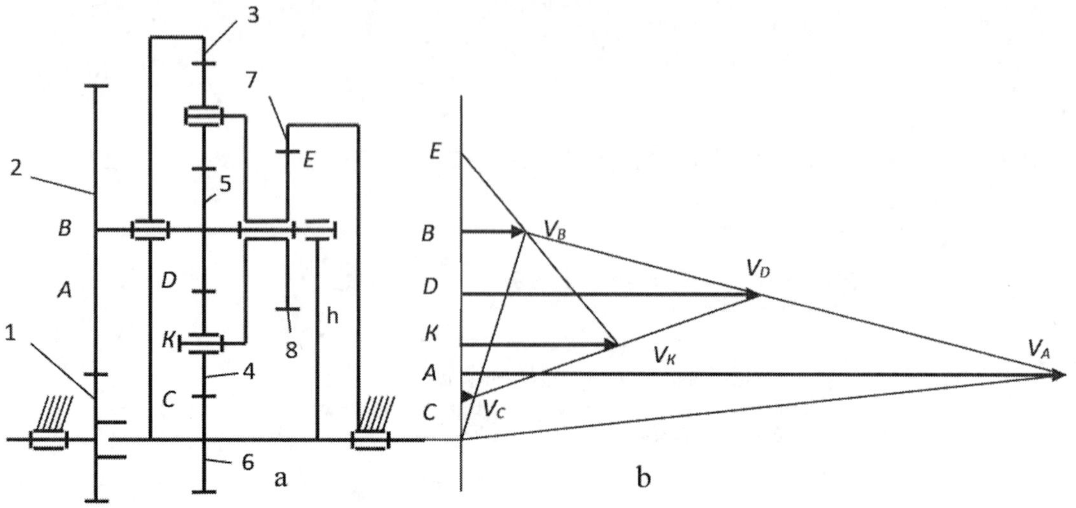

Fig. 2. Kinematic scheme (a) and velocity draft (b) of biplanetary gear

Kinematic chains of represented PGs contain two closed nonparallel circuits. Input power flow falls under two flows in the process of transmission and transformation and after that summarizes in an output link. Transmission of rotational torque through nonparallel flow using tooth gear (figure 1, a) or leverage (figure 2, a) is realized in addition to conventional separation into parallel power flows. This approach decreases load on all tooth wheels and especially on output tooth wheel. As the value of center-to-center distance is determined by transferred torque the separation of rotational torque into two nonparallel flows provides the decrease of transmission size and the additional increase of stiffness.

In a complex scheme (figure 1, a) the rotational torque is transferred to output link 3 (6) through planetary stair (which includes wheels 1, 2, 3 and carrier h) and regular transmission of wheel: 4, 5, 6. At the same time the output link consists of fixedly connected wheels 3 and 6.

In biplanetory gear (figure 2, a) the axle is output and wheel 6 and carrier h are fixedly mounted on it.

It should be noted that closed gears with transmission of several rotational torque flows from two or more input links to one output link are realized [4].

The kinematic research of accepted construction scheme that is presentation of the gear transmission ratio as function of geometric parameters (numbers of teeth) is the foundation of drive unit designing. In relation to PG the research is accomplished on the basis of Willis formula or graphic method. The reversed motion method in regard to closed PG is quite difficult [4]. With the advent of graphic editors graphic methods become more popular in consequence of visibility and simplicity of finite dependences obtainment [5-7].

However the current sequence of construction of velocity diagram starting from the guide link is not workable for closed gears represented in figures 1, 2. It is connected with the fact that the link that engages with the guide link requires the initial information regarding velocity of its two points. Thus complex scheme requires information regarding velocity of points A and C (or B). However only A point velocity is known (given).

Biplanetory gear requires information regarding velocity of points A and B similarly. In such a case only A point velocity is known (given).

The goal of the research performed was the development of approach to determination of transmission ratio of closed PG using graphic method.

The objective of research was obtainment of analytic forms of transmission ratio by geometric parameters by the example of two variants of closed PG.

Kinematic analysis accomplishment in the reverse order (from the output link to the input link) is the basis of the approach. On the basis that links of both flows lock in on output link it appears that linear velocities of links which contact it are certain. This makes it possible to use velocities of two points within each flow of power transmission for kinematic analysis.

In relation to double-flow PG closed by tooth wheels (figure 1, a) the analysis is performed in the following order. Velocity diagram of output link is drawn according to the adopted angular velocity of wheel 3 (6). Then linear velocities of points C and D are determined on basis of this diagram. Point E velocity is determined for the wheel 5 according to pints D and F. Velocity diagram of the wheel 4 (carrier h) is drawn according to the point E velocity. That makes it possible to determine linear velocity of point B that is mutual for carrier h and wheel 2.

As a result velocities of two points (C and B) of wheel 2 are discovered that makes it possible to draw its velocity diagram and determine velocity of point A of input link.

Transmission ratio of a given type planetary gear is determined according to the velocity diagram by means of simple geometrical correlations.

$$u_{13} = 2 \cdot \frac{d_6 \cdot (d_1 + d_2)}{d_4 \cdot d_1} + \frac{d_3}{d_1} \qquad (1)$$

di – diameters of i wheels.

In biplanetory gear locking of flows on an output link is performed by tooth wheel 7 and carrier h, that is PG with wheel-leverage locking. Kinematic analysis is performed in the following order. Velocity diagram of the output link is drawn and velocities of point E (wheel 7), axis of carrier B and point C are determined according to the adopted angular velocity of wheel 7 (carrier h). Velocity diagram of the wheel 8 is drawn and velocity of point K is determined according to known velocities of two points of wheel 8 (E and B).

Velocity diagram is drawn and velocity of point D is determined on the ground of linear velocities of two points of wheel 4 (C and K). That makes it possible to draw velocity diagram of 5-2 wheels according to velocities of two points (D and B) and to determine linear velocity of point A.

Diagram of possible velocities (figure 2, b) is drawn in concordance with the outlined order.

Velocity diagram makes it possible to determine transmission ratio of biplanetory gear according to geometrical representation of the linear velocity vector by the wheels diameters.

$$u_{16} = 2 \cdot \frac{d_2 \cdot (d_1 + d_2) \cdot (d_4 + d_5 + d_8)}{d_1 \cdot d_5 \cdot d_8} - \frac{d_2 \cdot (d_1 + d_2 + d_6)}{d_1 \cdot d_5} + \frac{d_1 + d_2}{d_1} \qquad (2)$$

The result formulas make it possible to accomplish kinematic research and estimate the influence of geometric parameters on transmission ratio of double-flow PG.

Conclusion

The approach is developed and research of kinematics of PG with closed nonparallel power flows is accomplished on the approach basis. Analytical expressions for determination of transmission ratio of PG advanced types are obtained.

The represented approach develops graphic method of PG research and appears as basis for power analysis of PG closed types.

References

1. A.S. Ivanov, V.M. Zyablikov, M.V. Fomin and others, Vestnik mashinostroenija **4**, 15 (2004)
2. J. Drewniak, P. Garlicka, A., Scientific Journal of Silesian University of Technology. Series Transport. **91**, 5 (2016).
3. A.A. Chirkov, Metro and tunnels **2**, 12 (2009)
4. V. Argeşanu, I.S. Borozan, I. Maniu, R.M. Kulcsar, M. Jula., International journal of systems applications, engineering & development **8**, 277 (2014)
5. A.S. Ivanov., M.V. Fomin, M.M. Ermolaev and others, Vestnik mashinostroenija **8**, 22 (2012)
6. Ja.A. Andreeva, I.A. Zhukov, Industrial Engineering **22**, 12 (2012)
7. A.S. Ivanov, M.V. Fomin, M.M. Ermolaev, Higher educational establishment news. Industrial Engineering **13**, 37 (2012)

DOI: 10.1051/matecconf/201712906024

Analysis of the numerical results of the melting paraffin corks, ensuring the safe operation of the pipeline using a mobile source of electromagnetic radiation

Vera Titarenko[1,*]

[1]Moscow Aviation Institute (National Research University), Moscow, Russian Federation

Abstract. For the safe operation of the pipeline, ensuring its smooth operation was conducted analysis of the results of numerical simulation of melting of paraffin jams with using a mobile source of electromagnetic radiation. Carried out varying different parameters of the radiation source, the peculiarities of its movement, the nature of the change of geometry of the boundary of melting of paraffin plugs under different external conditions

Reliable operation of pipelines for the transport of oil and gas condensate products in conditions favorable for the formation of crystalline hydrate and asphalt paraffin plugs requires the development of new methods for destroying such jams, as well as creating ways to prevent their formation. To eliminate paraffin plugs in real oil pipelines, it was proposed [1] to use a mobile source of electromagnetic radiation, called the "electromagnetic mole", in analogy with the "mechanical mole" used in the oil and gas industry for mechanical cleaning of pipelines from paraffin or gas hydrate plugs. The RF power of the electromagnetic wave is absorbed on the inner surface of the pipeline. Treatment of production wells is carried out while moving the source of electromagnetic radiation in the pipe in the longitudinal direction. This source produces an impact on the transported products, hindering the adhesion of the resultant particles of wax, they cannot Deposit on pipe walls and must be carried away by the fluid flow. Using the source of electromagnetic radiation, the turnaround time required cleaning piping (oil) wire increases due to the decrease of crystallization of paraffin's.

The speed of movement of the source was determined by the speed of movement of the boundary between liquid and solid phases in the melting process of asphalt-paraffin deposits under the influence of high-frequency electromagnetic radiation. Paraffin corks in the pipeline will be considered as a cylindrical metallic waveguide completely filled with dielectric. The electromagnetic radiation source located in the plane z = 0 at t = 0. The damping of the electromagnetic wave is caused by volume losses in the dielectric corks and surface losses in the metal walls of a cylindrical waveguide due to the imperfection of the conductivity of its surface The process of heating and melting asphalt-paraffin deposits under the influence of high-frequency electromagnetic radiation was studied on the basis of heat conduction equations with a given external heat source, which moves in the pipe in the longitudinal direction so that the minimum longitudinal distance from it to the surface phase transition was permanent. Considered axisymmetric problem, equation of heat conduction is written in a General form without explicit allocation of the phase [2, 3]:

$$\rho c_T \frac{\partial T}{\partial t} = \frac{1}{r}\frac{\partial}{\partial r}\left(\lambda r \frac{\partial T}{\partial r}\right) + \frac{\partial}{\partial z}\left(\lambda \frac{\partial T}{\partial z}\right) + Q(r,z,t) \quad (1)$$

where ρ - density of high-paraffin oil, c_T — its specific heat λ — thermal conductivity; $Q(r,z,t)$ - the power density of volumetric heat generation, like the thermal conductivity, was considered temperature-independent, and the heat capacity at the phase transition temperature of T_S, has δ - shaped feature: $c_T = c_0 + L\delta(T - T_S)$, where L is the latent heat of the phase transition, $\delta(T - T_S)$ — Delta function of Dirac.

The power density of volumetric heat release takes into account the displacement of the source of electromagnetic radiation according to the law $z = z_0(t)$.

$$Q = Q_0\Theta(z - z_0(t))\exp(-\alpha(z - z_0(t)))$$

$$\Theta(z - z_0) = \begin{cases} 1, & z \geq z_0 \\ 0, & z < z_0 \end{cases} \quad (2)$$

α — power attenuation factor of high frequency electromagnetic wave

The heat equation is complemented by boundary conditions.

1) At the end of the corks at $z = 0$, given in the form of Newtonian convective heat transfer:

$$\lambda \frac{\partial T}{\partial z}\Big|_{z=0} = \varkappa_1[T(r,0,t) - T_0], \quad (3)$$

where T_0 - is the ambient temperature and initial

* Corresponding author: vera_xmel@mail.ru

DOI: 10.1051/matecconf/201712906024

temperature of the paraffin corks, $\varkappa_1$ — is the heat transfer coefficient.

2) At the remote end of the corks $z = H$ there is no heat transfer:

$$\lambda \frac{\partial T}{\partial z}|_{z=H} = 0 \quad (4)$$

3) On the lateral surface of the cylinder $r = R$ the boundary condition is also written as a convective heat transfer law, with a different heat transfer coefficient taking into account the heat release due to the absorption of an electromagnetic wave on the surface of the metal pipeline:

$$\lambda \frac{\partial T}{\partial r}|_{r=R} = \varkappa[T(R,z,t) - T_0] - q(z,t) \quad (5)$$

where $\varkappa = Nu \bullet \lambda / R$ — the coefficient of heat exchange with the external environment, Nu is the Nussle number.

For accuracy control used the law of conservation of energy, which immediately follows from the heat conduction equations with boundary conditions:

$$\lambda\, \rho \frac{\partial}{\partial t}\int dV \int_{T_0}^{T} c(T')dT' = P[1 - \exp(-\alpha(H - z_0(t)))] - \\ -2\pi\varkappa_1 \int_0^R drr[T(r,z=0) - T_0] - 2\pi\varkappa R \int_0^H dz[T(r=R,z) - T_0] \quad (6)$$

The law of displacement z0 (t) of a moving source of electromagnetic radiation, necessary for closing the system of equations determining the process of liquidation of a paraffin plug, was defined as the law of motion of the interface between the liquid and solid phases by the same differential equation of heat conductivity.

Numerical simulation of melting process of paraffin plugs by a moving source of electromagnetic radiation by a wave of the TEM0n type is carried out, which studies various solutions of the heat equation with a movable partition boundary phases by the pass-through account method, using a scheme of alternating directions The calculation was carried out in the operating system MS Windows C++, using libraries: Qt, QCustomPlot, Eigen, Boost. Calculated temperature field displayed in the right part of the window. Because the problem is cylindrically symmetric, it displays only half of the pipeline, and the axis of symmetry is on the left. The position of the radiation source at the process of calculation is dynamically updated and the melting boundary of the paraffin corks is dynamically drawn.

The calculations were limited to consideration of the E_{01} wave. The cut-off frequency of E_{01} waves for the considered cylindrical waveguide is equal to $f_0 \sim 0.97 \cdot 10^9$Hz. Set control input frequencies: if it falls below the cutoff frequency for a given pipeline and types of waves, the entered value is replaced with the value exceeding frequency cutoff at 0.01 GHz. The default values are set in accordance with article [1]. Feature at the point $r = 0$ is the Laplace operator in the heat equation treated in the standard way: δ - function in the expression for the heat capacity was approximated by a step with half-width equal to 04°C.

For the numerical analysis of the process of elimination of asphalt-paraffin corks was used the thermal parameters of high-paraffin oil: density oil and corks ρ = 950kg/m^3, specific heat capacity of oil and corks C_0 = 3kJ/(kg K), the phase transition temperature of T_S = 50°C, the specific heat of melting L = 300kJ/kg, thermal conductivity λ = 0,125 W/(m·K). Real and imaginary part of the dielectric constant ε_0 is chosen for dehydrated oil, in these cases, in a wide range of frequencies, these values are very weakly dependent on frequency, and in the calculations the typical values of the real $\varepsilon_0' \approx 2.3$ and imaginary $\varepsilon_0'' \approx$ 1.276 10- parts of the complex permittivity of paraffin were used: ($\varepsilon = 2{,}3 + i \cdot 0{,}01276$); the heat transfer coefficient (heat transfer from the external environment) from the end $\varkappa_1 = 0.2\text{W}/(\text{m}^2\text{K})$ and from the sides $\varkappa = 1.613\text{W}/(\text{m}^2\text{K})$, and the value of heat transfer coefficient on the lateral surface of the cylinder r = R corresponds to the Nussle number Nu = 1 (pipe in dry ground). The length of the paraffin plug is chosen equal to H = 5 m.

It was believed that the RF power of the electromagnetic wave is absorbed on the inner surface of the pipeline. Because a too small value of the length of penetration complicates the operational control of the movement of the source of electromagnetic radiation and because the depth of skin - layer is much less than the thickness of the pipe, it was believed that the wall thickness is equal to 4111111,110 Cm.

To study and analyze the nature of melting, configuration changes, temperature and speed of movement of the melting boundaries paraffin corks at different possible ambient temperatures (0°C to 20°C) varied values of frequency and power source of electromagnetic radiation. Upon reaching the source of the upper boundary of the corks and if after a specified time interval, the calculation stops automatically.

Enter input parameters is carried out via a graphical interface. The parameters set in the program (including variable ones) include: a) characteristics of oil and paraffin plugs (densities, thermal conductivity coefficients, heat transfer coefficients, dielectric constant, specific heat of melting, ambient temperature, melting point); b) the parameters of the source of electromagnetic radiation (number of mode, conductivity of the pipeline walls, power of the source, frequency); c) the geometry of the tubing (radius, corks length, the distance from the radiator to the corks, the time interval); d) the parameters of the computational grid (the spacing/number of nodes along the radius, the interval/number of nodes along the radius, interval/number of nodes along the length of the corks, the time step (total time steps is calculated automatically depending on the magnitude of the time interval, the half-width of the Delta- function smoothing).

Data entry, management, and dynamic recalculation of the fields associated with the grid, are ensured.

1. Analyze the melting process of paraffin corks HF moving source with a capacity of 5,000 kW with a frequency of 1.4 GHz at the ambient temperature 20 ^{0}C. The first 41 min. the paraffin layer is heated and heated to

50 ^{0}C, the melt boundary starts to move inside the plug, and the phase boundary of the solid and liquid paraffin reaches the pipeline wall after 3 hours, and thus, as the plug penetrates (3 hours), the source begins to flow into the plug, The boundary of the melt (the boundary of the cork) was moved away from the source by a distance of 1.9 m, this distance remains unchanged.

For frequency (f = 1.4 • 10^{9}Hz) paraffin penetration corks coming from its center, which does the initial penetration of the corks. Progresses (center to 215 ^{0}C) the source continues to move, and the boundary of the melt corks reaches a length of 5 m over a period of time of about 10 hours (t =585 min, Fig. 1.1), and the source reaches the border of the corks over 839 min.

2. Let's investigate the process of melting paraffin plug, if the temperature of the external environment is less, namely: 10 ^{0}C. The process of motion of the source occurs with a delay (with respect to item 1) for approximately 0.5 hours, the length from the source to the boundary of the melt also is 1.8 meters, and also the length of the plug is reached by the melt boundary with a delay of just over 0.5 hours and is 649min, the source itself will reach the end of the plug in 913 minutes.

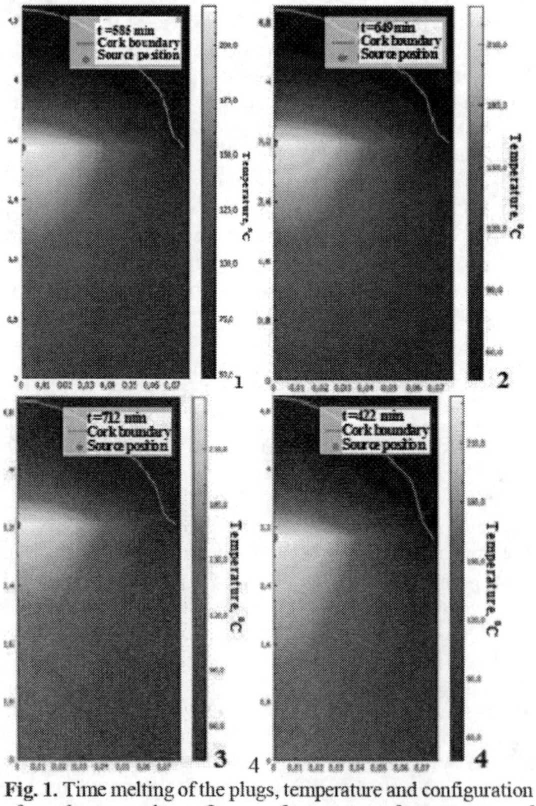

Fig. 1. Time melting of the plugs, temperature and configuration of the melt fuse for a frequency of f = 1.4 • 10^9 Hz when the power source 5 000 kW at an external temperature 1) 20 ^{0}C; 2) 10° C; 3) 0 ^{0}C and 4) at the t = 20 ^{0}C at a power of 7000 kW

3. If the temperature of the external medium drops to 0 °C, the melt of the plug and the achievement of the melt boundary by the walls of the tube also occur with a delay of 3.5 hours, and at the same time starts the movement of the source into the corks (206 min), the length from the source to the boundary of the melt is 1.75 m. Initially, the boundary of the melt reaches the end of the corks (5 m) in the center over a period of time of more than 11.5 hours (712 minutes), the UPS continues to move forward and reach the end of the corks (5m) for 986 min. Temperature in the center (near the source) is 240 ^{0}C.

Thus, with decreasing outside temperature, the source melts the paraffin plug for a longer time.

4. We will analyze the melting process of a paraffin plug with a moving RF source with an increased power to 7,000 kW, but with the same frequency of 1.4 GHz at an ambient temperature of 20 ° C. Already after 0.5 hour (29 min) the melt boundary started to move and after 2 hours (132 min) has reached the inner surface of the pipe, after which the source started to move deeper into the cork, and the distance from the source to the melt boundary was 1.95 m. After 7 hours. (422 min), the melt boundary reached the end of the plug (5 m), and the source advanced to the end of the plug in 608 minutes. The temperature near the source was 230 ^{0}C.

Times of melting of the corks ranges of temperature change and views of configuration of the cork melt for a frequency of f = 1.4 • 109Hz with source capacity of 5 000 kW at different external environment and at different capacities of the radiation source are presented in Fig. 1.(1 - 4).

5. Now consider the same process, only the power of the source was reduced to 3 000 kW and all other parameters are left unchanged. Cork started to melt, about 68 minutes (a little more than 1 hour), and the source started moving source further deep into cork only after 283 min (≈ 5 hour). The distance from the source to the upper limit of the melt amounts to approximately a little less of 1.8 m, and the source melting the entire length of the corks over a period of time more than 16 hours (975 minutes), that is, the lowering of the power source entails a sufficiently large increase in the time of melting of the corks of fixed length and a decrease in the speed of it melting. The source reaches the boundary of the corks only through 1372 min (slightly less than 25 hours). The temperature near the source does not exceed 200 ^{0}C (185 ^{0}C).

The configuration of the pattern boundary melting paraffin (geometry) with the source remains the same in all of these five studies, the difference is observed only in the temperature near the HF source and in the time of reaching the boundary of the when the melt boundary reaches the end of the corks.

6. Let us further investigate the influence of the change in the source frequency on the velocity, the characteristic shape (boundary geometry) of the melt, the time of melting of the corks and the temperature regime inside the pipeline. Analyze the melting process of paraffin corks HF moving source with a capacity of 5,000 kW with a frequency of 1.7 GHz at the ambient temperature 20 ^{0}C. The movement of the melt boundary of the cork begins approximately after 1 hour (56 minutes) and after approximately 2.5 hours (154 minutes) the source starts to move along the length of the cork. The boundary of the melt is separated from the

source by 1.2 m and has a more uniform power density distribution of heat over the cross section of the corks.

Localization of the initial melting is determined by the maximum power density of the heat *Q(r,z)*. Melting the entire length of the cork was carried out at 587 min, and the source itself has moved to the end of the cork over 12 h (733 min). The maximum temperature in the center of the melt was 130 ^{0}C (Fig. 2)

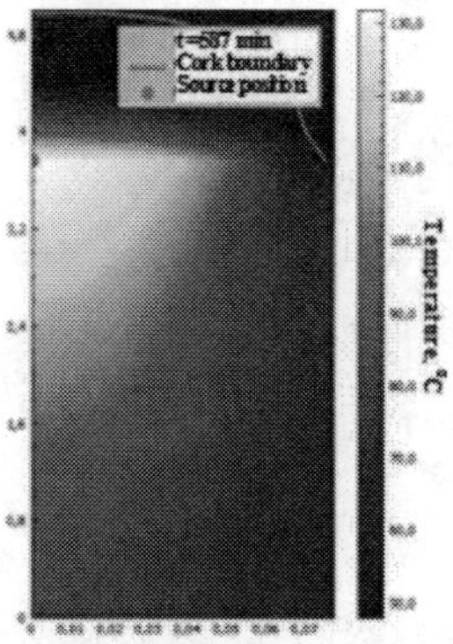

Fig. 2. The appearance of the border of the melt when reaching the end of its paraffin corks with a source frequency f = 1.7 GHz

7. Let us now consider what the situation will occur with the same frequency of 1.7 GHz, but at lower (10 ^{0}C) ambient temperature. After 56 min started melting paraffin corks, also center of melting was at a distance of 0.04 – 0.05 m from the center of the corks (source), and upon reaching the melt of the boundary pipe, source has begun movement (through 121 minutes) at 84 ^{0}C. The center of the border of the melt removed from the source center at a distance of about 0.65 m, moving parallel with the source, reaches the boundaries of the paraffin cork through 562 min, and the source goes to the end of the cork for 658 min. The temperature in the center of the beginning of melting of paraffin plugs (located on concentric circles) does not exceed 92 ^{0}C.

8. Next, change the frequency of the RF source, increasing it to 2.0 GHz and leave all other parameters unchanged. First, the boundary melt paraffin plugs appears not in the center (where is the radiation source), but at a distance of about 0.04 – 0.05 m from the center of the radiation source. Here is the boundary of the melt reaches the maximum value, and then, this by heating boundary continues to move, and when the melt reaches the inner wall of the pipe, starts the motion of the source inside the pipe (Fig. 3). The initial distance from the source center to the center of the border of the melt amounts to little more than 0.65 m. The temperature inside the pipe near the source is close to 90 ^{0}C. After 2 hours, the source starts moving in the direction of the melting corks The boundary of the melt reaches the end of the cork not in its middle, but along the concentric circle at the same distance from the center of the emitter (i.e., distance from the center of the pipe is 0.04-0.05 m) in 564 min, while the source will move until the end of the corks in less than 11 hours (658 minutes).

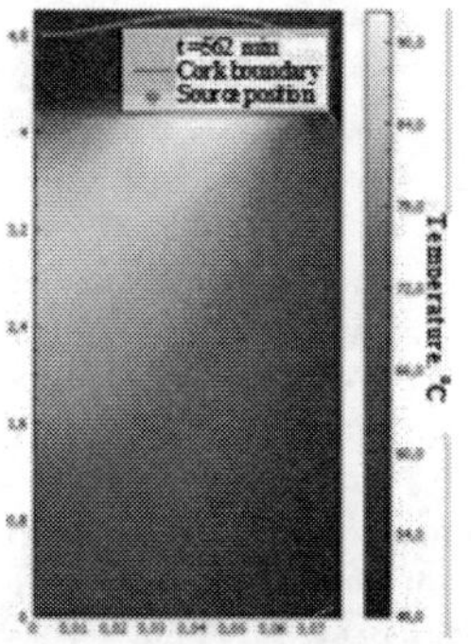

Fig. 3. The appearance of the border of the melt when reaching the end of its paraffin corks with a source frequency f = 2.0 GHz

9. If the ambient temperature will be more low (10 ^{0}C), the overall picture of the geometry of the melt occurring at the same frequency of 2.0 GHz, does not change. The distance between centre of source and center of the border of melting of the cork remains the same (about 0.65 m), the temperature in concentric circles melting is still not superior to 92 ^{0}C, however, to achieve the boundary of the melt throughout the length of the corks required more time: 624 min., and the source would will pass the whole length of the cork for 725 minutes,, and the temperature of the initial melting of the boundaries of the corks will amount t o 98 ^{0}C

10. Increasing further (Fig. 4) the frequency

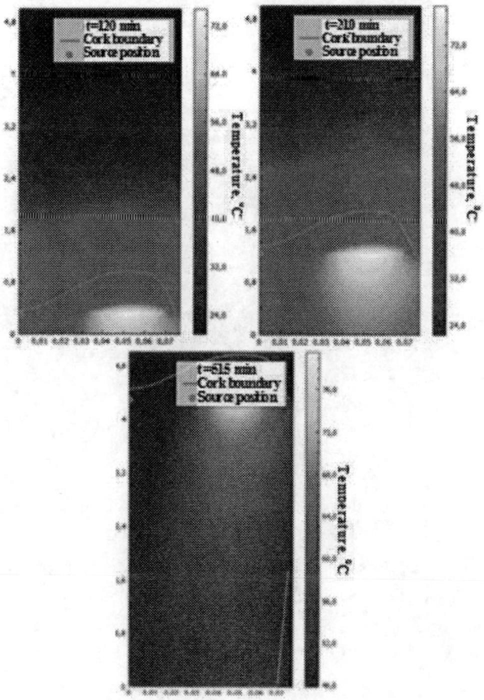

Fig. 4. The Appearance of the border of the melt and place of origin at different points of time at f = 2.5 GHz

of the RF source up to 2.5 GHz at the unchanged all the other parameters, we note that the initial melting of the

DOI: 10.1051/matecconf/201712906024

paraffin corks (at its maximum values) has the form of concentric circles (due to axial symmetry), the initial melting appears at a distance 0,054 m in 47 min and after about 1,5 hour (93min) reaches the inner surface of the pipeline, after which the source of HF energy starts moving, and the boundary of the melt in the center of the corks initially comes directly from the source and then (after 150 min) the boundary of the melt is separated from the source, the distance from it increases gradually and at a distance of 0.2 m it stabilizes. After about 510 minutes (8.5 hours), the boundary reaches the end of the cork, still at a distance of about 0.054 m from the center of the pipeline, and the maximum temperature (in the concentric circle from the center of the pipeline) in the center of the melt does not exceed 78 ° C

11. With further increase of the frequency of the RF source up to 3.0 GHz with unaltered all the other parameters the initial melting of the corks occurs further away from the city center, approximately 0,058 m from the center of the pipeline (Fig. 5).

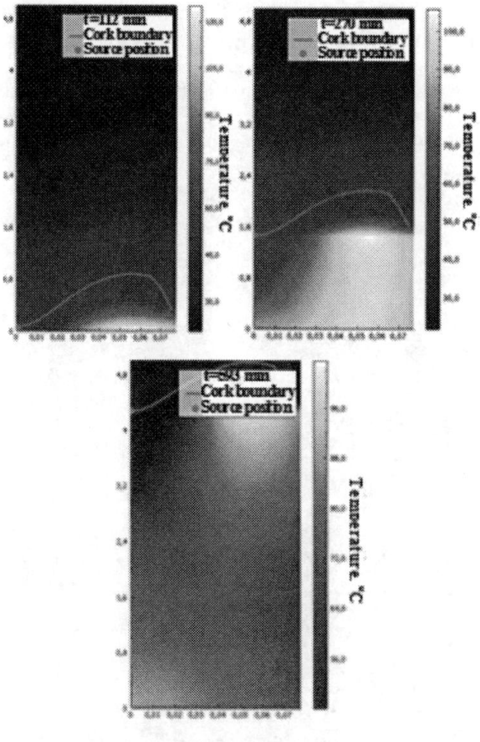

Fig. 5. The border of the melt and source at different points of time at $f = 3.0$ GHz

Through 593 min boundary reaches the end of the corks at the same distance 0,058 m from the center of the pipeline, moreover, the maximum temperature at the epicenter of the melt (obviously located on a concentric circle with a radius 0,058 m) slightly increased, but, nevertheless, at the peak of the melt (in concentric circles) is not greater than 102 ^{0}C.

The boundary of the melt comes from the source and moves with it, anywhere from it not coming off.

Thus, after the variation of these parameters of the RF source at different temperatures of the external environment, we can draw the following conclusions:

- Geometry and configuration of the boundaries of the melt corks depends only on the value of the source's frequency, and if the frequency is less than 2.0 GHz, the melting begins in the Central region. With increasing source frequency, the separation of the melt boundary from the source of RF radiation decreases, the boundary is smoothed
- Neither the ambient temperature nor the power of the radiation source does not have a significant impact on the geometry of the boundary of the melt. Change (increase or decrease) the power of the radiation source has an impact on the duration of melting of paraffin plugs (decrease and increase, respectively), the speed of the source itself (increase or decrease, respectively). The temperature inside the pipe is also affected by the external temperature: with a decrease in external temperatures, a longer warm-up time is required before the plug begins to melt, and the melting rate also decreases. When increasing (decreasing) power of the radiation source significantly increases (decreased) the temperature near the source, the time of melting decreases and Vice versa. As the temperature of the external environment also decreases the temperature near the center of the melting inside of the pipe.
- With an increase in frequency of more than 2.0 GHz, the appearance of the melting surface changes significantly: the melting center is moved away from the center of the pipeline, turning into a concentric circle; the boundary of the melting is moved closer to the source (with increasing frequency) and at f = 3.0 GHz the boundary of the melting is attached to the source and does not detach from it.
- The temperature near the source at a frequency of more than 2.0 GHz oscillates around 90 °C, at f = 2.5 GHz this temperature is minimal (does not exceed 78 °C), and then, with an increase of f = 3.0 GHz, the temperature grows slightly, but not surpasses 102 ^{0}C.

References

1. V. A. Balakirev, G. V. Sotnikov, Yu. V. Tkach, T. Yu. Yatsenko, ZhTF (Thech. Phys.) **71, 9** (2001)
2. A. A. Kislitsyn, R. I. Nigmatulin, J. Appl. Phys. (PMTF) **4** (1990)
3. R. I. Nigmatulin, K. M. Fedorov, Izv. Universities: Oil and gas, **10** (1991)

Simulation of pump performance control schemes with the use of the virtual model of the process

Julia Nikolaevna Zatsarinnaya[1,*], *Nataliya Alexandrovna* Staroverova[2], *Alla Grigorevna* Logacheva[1]

[1]Kazan Power State University, Kazan, Russian Federation
[2]Kazan National Research Technological University, Kazan, Russian Federation

Abstract. The research of pump performance control schemes based on bypassing depending on the moving medium is carried out with the use of the virtual model of the process. Such approach speeds up and simplifies the development and reduces the risk of errors made by the developer in the execution of routine operations. Unlike theoretical methods of research, the virtual model of the process allows the developer to concentrate on the physics of simulated actions or the principles of the functioning of technological systems, without delving into the subtleties of programming. It is established that this simulator will significantly improve the operator's work, provide high-quality process control, and will also allow to ensure safe operation in the real conditions in the chemical-technological production.

1 Current state of the issue

In the modern world, energy saving issues are closely related to the development of the world economy and in the last decade are becoming more relevant for the Russian economy. Most large-scale productions in Russia, such as the oil and chemical industry, thermal power plants and petrochemical industries were built in the former USSR before the 1980s. The Soviet industry can be characterized by the growth of capacity per unit of aggregates, the growth of oil and gas production and the constantly growing demand for electricity. This led to the construction of powerful oil arteries and large power plants. Despite this, the main problem in the operation of oil pipelines up to the 1980s was the search for a reserve to increase the throughput capacity of main oil pipelines. In the mid-1980s, there was a tendency to decline in oil production, but in general, until 1990, the growth in pumping volumes still remained. By the end of the 1990s, due to political and economic instability in Russia, oil pumping volumes in some pipelines declined by 50%. Nowadays, oil pumping volumes have increased by 30% compared to the 90s, but the reduction in pumping volumes below the nominal pumping capacity has led inefficient pump operating [1].

The aim of our work was modeling of pump performance control schemes and increasing the efficiency of main oil pipelines, using a virtual model of the process, which results in energy savings in the operation of pumps in the underload condition. This model can be used for modeling the operating modes of smoke exhausters and fan blowers of boilers, oil-pumping pumps and gas-pumping compressors.

In recent years intensive work is being done on experimental and theoretical studies of the operation of pumping stations in variable modes and methods of control of their operation in such modes by a number of authors such as A.A Veremeenko, E.V. Vinogradova, E.V. Vyazunova, A.I. Golyanova, L.A. Zaitseva, L.G. Kolpakova, Sh.I. Rakhmatullina, N.Z. Aitova, G.S. Salashchenko, In works under consideration, three most common technical schemes for pumps operation control are noted: the control of the rotor speed of the pump, using variable-speed motors; the control of the rotor speed of the pump, using special control couplings; change of the geometry of the flow channels of the pump and the kinematics of the flow at the input to the impeller.

However, some of the technical solutions listed above have not been applied for a number of reasons: the lack of equipment manufactured in Russia and that would comply with requirements for energy-intensive pumping and power equipment and the lack of fundamental research in this area. Conducting of the experiment was restrained because of the lack of mathematical models of the elements of the system, adequately describing their characteristics. We conducted a comparative study of pump performance control schemes based on bypassing and throttling, depending on the moving medium by the virtual process modeling. Using of the virtual model of the process speeds up and simplifies development, lowers the risk of mistakes made by the developer in the execution of routine operations, allows the developer to concentrate on the physics of simulated actions or the principles of the functioning of technological systems, without delving into the subtleties of programming.

The following main research tasks can be distinguished:

* Corresponding author: waysubbota@gmail.com

- Development of mathematical (virtual) models of the process of automatic control of pump performance based on stabilization by throttling on the basis of modern technical automation tools using the Visual Modeler modeling tool.

- Development of an algorithm for calculating the operation modes of the pump and minimizing the energy consumption.

2 Review of modern tools for dynamic modeling

The environments for the technological process simulators development are considered. The following systems were chosen for analysis and comparison:

The CENTUM VP system represents a distributed production control system used to manage and control the operation of plants in a wide variety of industries: oil refining, petrochemical, chemical, pulp and paper, and other industries. It consists of the following components:

- Operator station (HIS) is the operator interface designed for control, monitoring and engineering;

- Control Station (FCS) is a controller that performs automated real-time monitoring and control of the technological process. It allows to perform the following actions:

- Define input / output modules, create switches (Switch);

- Implement logic control schemes using five programming languages.

- Vnet / IP network. All Centum VP system control units are connected via the Vnet / IP bus. It provides data transfer between control stations and operator stations. This bus is a real-time bus and all its traffic is controlled by the system. To connect the operator station to the Vnet / IP network the control bus interface board VI701 is used.

The Trace Mode system is a software package of SCADA HMI class, invented by AdAstra Research Group. It is specialized for the development of software for automated process control systems, telemechanics systems, building automation, systems for electricity, water, gas, heat accounting, and for control of their operation in the present time [2].

A distinctive feature of TRACE MODE is "the technology of a single line programming", that is, the probability of developing all the ACS modules with the support of a single device. The development of a single line programming allows creating human-machine interface tools, resource accounting systems, program industrial controllers and develop a web-interface within the framework of one project. For this reasons special editors are integrated into the TRACE MODE tool system, such as:

- Editor of graphic mnemonic schemes;

- The editor of the screen panels;

- The editor of programs in visual language FBD (standard IEC 6-1131 / 3);

- The editor of programs in visual language SFC (standard IEC 6-1131 / 3);

- The editor of programs in visual language LD (standard IEC 6-1131 / 3);

- The program editor in the procedural language ST (IEC standard 6-1131 / 3);

- Program editor in procedural language IL (IEC standard 6-1131 / 3);

- Document templates editor;

- Builder of links with DBMS;

- Equipment Passport Editor (EAM);

- Staff Editor (HRM);

- The editor of material resources (MES).

The CodeSys system is an instrumental software package for industrial automation. It is developed and distributed by 3S-Smart Software Solutions GmbH (Kempten, Germany).

The CoDeSys programming environment includes a set of tools for preparing and debugging programs, compilers, configurators, visualization editors, etc. If necessary, the functionality of the system is supplemented with optional components. The projects created in CoDeSys can be saved not only on the PC disk, but also in the controller, if it has a large memory size, what allows to avoid the loss of source code or dislocation in projects. For large projects, a version control system (SVN) is envisaged.

In CoDeSys all 5 programming languages defined by the IEC 61131-3 standard are available:

- IL (Instruction List) - assembler-like language;

- ST (Structured Text) - Pascal-like language;

- LD (Ladder Diagram) – Relay-like schemes;

- FBD (Function Block Diagram) - Multi-functional block language;

- SFC (Sequential Function Chart) - The language of state diagrams.

The InSIDE system is a built-in development environment for InSiDE simulators. It runs under the Windows family of operating systems (Windows NT, Windows 2000, Windows XP) and allows the creation of simulators (both analytical and full-scale) for technological objects of various complexity. It allows:

- Automation of the simulators development process, which not only speeds up and simplifies the development, but also reduces the risk of errors made by the developer in the execution of routine operations.

- Involvement in the development process of the engineers who are not professionals in programming languages. Applying CAD systems based on the principle of visual programming, the developer can focus on the physics of simulated actions or the principles of the functioning of technological systems, without delving into the subtleties of programming.

- Recording of the configurations in the simulator by user without involving developers.

Main features of the InSiDE:

- Graphical interface for the development of simulators;

- Multiuser development environment;

- CAD applications for visual programming;

- Tools for debugging and monitoring the work of the simulator;

- Automated integration of simulation modules;

DOI: 10.1051/matecconf/201712906025

- The possibility of using distributed computing (cluster simulators).

The built-in development environment of InSiDE simulators is based on the application of the present-day database InterBase. The simulator database is used to store data according to the simulator being developed and to organize multi-user access to it.

The MATHLAB (MATrix LABoratory) software package is designed for performing scientific and engineering calculations, with the help of which the problems of mathematical modeling are effectively solved. The integrated development environment includes a set of development tools provided through a single user interface.

The Lab VIEW software package is a graphical programming environment used around the world for the rapid creation of complex applications in measurement and automation tasks based on the architecture of data flows.

The Omega Land system represents an integrated environment for dynamic modeling. It contains several "modules", each of which provides an independent function and a mechanism for the systematized connection of these modules (VMspace).

Basic functional modules include the necessary fundamental modules for constructing almost any application system. Optional function modules can be selected depending on the application. Both basic and optional function modules are developed by Omega Simulation Company. The basic function modules consist of the following four modules:

1. The module of the technological installation model (VisualModeler)
2. Execution Control Module (EXEC)
3. Database module (DataBase)
4. Graphic Builder

Each module provides various functions, which provides the flexibility of the system, which allows to model and solve any problems.

Using the "Visual Modeler" software package for dynamic modeling in combination with various modules, it is possible to create a system designed for process design / analysis, for checking control systems, for training, for operating support and so on. This system uses the following simulators:

- A static simulator can quickly model a stable state of equilibrium for certain process conditions. It is used in the following types of situations:
- Analysis of physical properties of chemical components;
- Analysis of design data and border conditions of each block;
- Definition of a steady mode of operation;
- Evaluation of construction and operation costs.
- Dynamic simulator allows to view the change in results depending on the design data provided for each unit, and also depending on the manipulations with the valves, etc. in the process of modeling. It is used in the following situations:
- Checking the start and stop operations;
- Analysis of the limit points of operations;
- Development and research of control systems;
- Analysis of accidents and contingencies.

The aim of both static and dynamic simulators is to get an idea of the process, and also to analyze the operating conditions. Nevertheless, since the main objective of the first one is to extract the design data for the equipment, and for the latter it is to analyze and optimize the operating conditions, it is important to make the right choice in accordance with the purpose of the analysis.

Visual Modeler contains numerous built-in functions, including the function of model editing, which provides the possibility of creating a simulator that works with various mathematical models of processes and devices. The package also includes programs that implement the functions of performing production operations used for the operation of developed simulator and its testing, as well as other functions that technologists and automation specialists may need to create special models for specific tasks.

Visual Modeler offers five standard functions and two libraries:

- Model editing function - this function is used to build a model of petrochemical production, which is the object of modeling, and to create a program that performs calculations;
- Model execution function - this function is used to control the running of model;
- Function of a block model development - this function generates a model of a new block and registers it in the library of unit models;
- Function for calculating of physical properties [S] [C] - The physical property data necessary for the model calculating contains such types of variables as molar mass, equilibrium gas-liquid flow ratio, etc.

Despite the fact that Visual Modeler represents a modular type of simulator, the balance of the pressure stream is not calculated separately for each block, but simultaneously for the whole process model [3].

Visual Modeler breaks the process topology into several "flow lines" for each execution cycle. The flow rate remains the same throughout the single flow line.

As a result of the analysis, it was concluded that Visual Modeler is ideally suited for implementing the process of controlling the operation of the pump. Since this environment has many advantages in comparison with similar systems for the development of visual simulators.

3 Description of the proposed solution

When implementing a process diagram, it is necessary to add a system (in other words, to describe for which mediums the circuit will be used). Since this project provides the implementation of three processes, three systems have been added.

To add a system, the components and their physico-chemical properties are determined. There is a large database of component in the VM, but even if the required component is not available, it is possible to create and describe it. In our case, for the initial settings, the available components can be used.

 DOI: 10.1051/matecconf/201712906025

Once the systems have been added, it is possible to implement the process scheme itself. On the next step, the settings and process sensors are configured.

Using a database of existing equipment, circuits were drawn up, pumps, valves, regulators, and measuring instruments were selected.

The constructed scheme made it possible to carry out research in several directions: 1. studying the behavior of various systems on the equipment used, and how to anticipate possible difficulties in introducing new components or new equipment into those processes. 2. Training of personnel and prevention of emergencies through their modeling. 3. Selection of optimal parameters and control schemes to improve process performance.

4 Conclusions

The result of modeling and conducted research is the solution to the problem of rational loading of main oil pipeline systems. The main part of the research carried out includes complexes of mathematical algorithms for solving the problems of rational operation of main oil pipeline systems and improving the management efficiency of main oil pipelines. This task was solved on the basis of creating a simulator system for automatic regulation of the pump efficiency based on the stabilization method by throttling with the Visual Modeler modeling tool.

As a result of creating of a virtual model of the pump performance control scheme, it became possible to increase the efficiency of the main oil pipelines, resulting in energy savings in the operation of pumps in the underload condition by 12-15%.

The proposed algorithms of the mathematical model make it easier to understand the process of pumps control, enabling future specialists to learn how to manage processes directly before starting working at the site. The use of the developed model and algorithms allowed to cut energy costs on the Surgut-Polotsk and Baku-Tikhoretsk pipelines to 5.5 and 7%, respectively, when solving problems of rational operation and transportation of oil.

References

1. A.A. Kabanov, S.A. Dubovik. Problemy upravleniya i informatiki, **3**, 17 (2010)
2. B.T. Polyak, E.N. Gryazina, A.A. Tremba. Avtomatika i telemekhanika, **12**, 3 (2008)
3. M.V. Astakhov, T.V. Tagantsev. Proc. of the Bauman MSTU “Mathematical Modeling of Complex Technical Systems”, **593**, 125 (2006)

Vibration decrease of electromechanical actuators based on roller screw mechanisms

Valentin Morozov [1], *Pavel* Fadeev [1], *Dmitri* Shtych [1], *Leonid* Belyaev [1], and *Alexey* Zhdanov [1,*]

[1] Department of Mechanical Engineering, Vladimir State University named after Alexander and Nikolay Stoletovs, Vladimir, Russia

Abstract. The reduction of the vibratory activity of the roller screw mechanism (RSM) for converting a rotational movement into a translational movement is discussed. A mathematical model of vibratory activity roller screw mechanisms on the basis of elastic compliances was developed. The resulting equation for calculating the natural frequencies of the basic RSM elements – screw, rollers and nut was obtained. It is proved that the RSM has a low vibroactivity that allows you to create mechatronic actuators, the maximum level of vibration does not exceed 60 to 62 dB, which is 11-16 dB below the level of vibratory activity gear mechanisms

1 Intruduction

Recently, planetary roller-screw mechanisms (RSM) have become widely used among executive mechanisms used for converting rotary movement into linear displacement [1-6]. In papers [2, 3, 6] we have shown methods for calculating the main technical characteristics of those mechanisms. At the same time, an important actuators characteristic, often determining the applicability of a particular mechanism, is its vibroactivity and, as a result, the noise generated. The screw gear smoothness is the cause of decreased RSM vibroactivity and noise and makes those mechanisms promising for use. As for RSM, the closest analogues for the design models are planetary gear mechanisms [4, 5] and rolling bearings [6, 7] which vibrational processes have been studied sufficiently. However, the use of calculation procedures developed here and practical recommendations for the RSM vibrations reduction is unjustified, because of the changing characteristics and difference between screw mechanisms [8,9]. Let us consider the mechanism option where the screw-rollers support interface is made with multistart threads and has end-capping geared rims. The mechanism structure (Fig.1) comprises screw *1* installed in bearings fixed in housing *2*. On the screw and rollers geared rims *6* were cut. The screw is mating the thread rolls *3* through the thread and through the gear rims *6*, and the rollers interact with the nut *4* thread. The casing is fixed to the base *5*. The problems solved during RSM vibroactivity study are as follows: determining vibration sources, their frequencies and amplitudes, calculating the forced vibrations spectrum, determining the audio range vibration levels accounted for by the RSM casing.

2 Materials and methods

Out of the many oscillation theory methods, the most suitable method for solving the tasks in hand is the dynamic compliance method. The method is as follows. The mechanism is divided into a number of simple subsystems. As for the RSM the following subsystems are distinguished: screw, rollers, nuts and housing (Fig. 2).

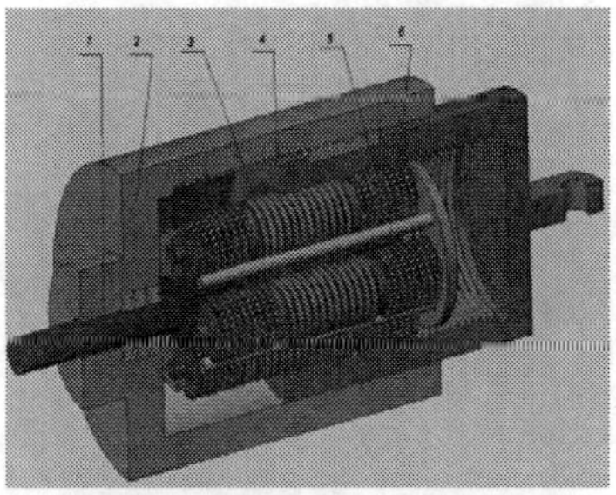

Fig. 1. Structure of Roller Screw Mechanism

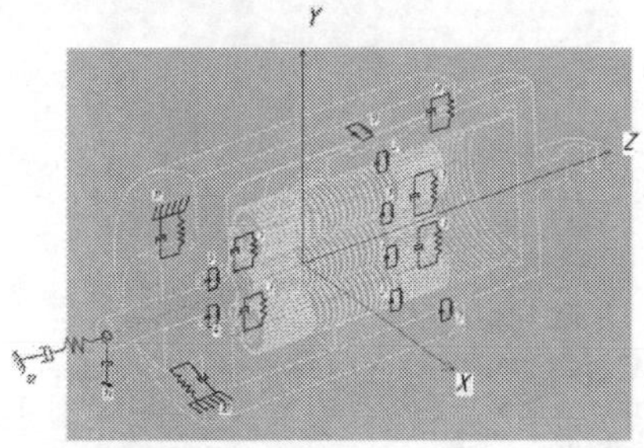

Fig. 2. Diagram of RSM; elastic connection:
1 - screws and clips; 2 - roller and nuts; 3 - screw and toothed roller in the circuit; 4 - screw-on body X axis; 5 - the screw-on housing axis Y; 6 - the screw-on housing axis Z; 7 - a nut body on X-axis; 8 - a nut body on Y-axis; 9 - key - keyway; 10 - body-reliance on the X axis; 11 - body-reliance on the Y axis; 12 - body-reliance on the Z axis; 13 - body-support in angle φ.

* Corresponding author: zhdanov@vlsu.ru

At the points of mechanism dismemberment into simple subsystems constraining forces are applied that vary with some degree of frequency, forced vibrations resulting from the action of those harmonic constraining forces are calculated separately for each subsystem and displacements of points along which the linkages are broken in the linkages of those bonds are located. After that, the strain compatibility equations were set up for each elastic linkage (the number of equations was equal to the number of broken linkages), the frequency equation was solved and displacements value in selected joint coordinates were determined.

3 Vibroactivity calculation of RSM

During RSM operation, the most important sources of vibration are threaded surfaces manufacturing errors. These are zero lead-ins (n = 0) threaded elements eccentricities, ovalities, facetings and undulations, and in case of threaded elements with n> 0 to these must be added periodic pitch errors [2]. The whole set of errors leads to periodic threaded surface generator position motions which we shall hereafter refer to as undulation.

We will make the following assumptions: the deformation in the RSM threaded elements contact follows the Hertz-Belyaev law; the load-side axial load is constant and uniformly distributed over all the contact points; the amount of threaded elements working surfaces undulation is within the elastic deformation limits. We will establishing the exciting force values dependence on the undulation amount.

The normal force at each threaded surfaces contact point P_H causes elastic deformation δ.

The threaded surfaces undulation can be expanded into Fourier series with each i-th undulation harmonic causing additional elastic deformation δ_{zi}.

$\delta = \delta_1 + \sum \delta_{zi}$ The dependence of elastic deformation on the normal force (in case of linear contact of two bodies) is nonlinear [4]. In order to use in a linear equation system additional force caused by the undulation and vibrating the mechanism we will find the linear approximation by way of using direct linearization method.

$$\delta = c \cdot q$$

Where $c = K_1\left(K_2 - lnq_{max} + \frac{1}{3}\right)$; K_1=0,579 E;

$K_2 = \frac{\ln(4E(r_1 + r_2))}{1{,}522^2 \sin\alpha} + 0{,}814$; q_{max} - maximum specific load; E - reduced modulus of elasticity; r_1 and r_2 - the radii of curvature of the contacting surfaces; α- half thread angle.

Given that , where L_{KOH} - the length of the contact line, the variable load is in contact, is the driving force of vibration (P_H)), it is equal to:

$$P_{H_i} = \frac{(a_i \cos iwt + b_i \sin iwt) L_{KOH}}{c}$$

Let us introduce two coordinate systems: one - fixed XOY coordinate system and a X'OY' system rotating relative to the first one with the roller radius vector rotary speed, i.e., connected to the separator, which rotary speed is calculated using the following formula [2]:

$$\omega_P^o = \frac{\omega_B (1 + u)}{2}$$

where ω_B - angular screw speed; $u = d_p/d_B$, d_p, d_B - roller and screw diameters respectively.May the screw rotate at a rotary speed ω_B and have undulation of i-th harmonic. We will consider the mechanism in the XOY coordinate system.

The angle between the j-th roller contact point and the screw in the moving coordinates is

$$_{B_i} = \omega_{OTH} t + 2\pi \frac{(j-1)}{n} \tag{1}$$

where ω_{OTH} - rotor angular velocity in the moving coordinate system (speed of rotation of the screw with respect to the separator) $\omega_{OTH} = -(\omega_B - \omega_P^0)$.The normal force arising from the i-th screw harmonic undulation on all cutting threads (1≤K≤m)

$$F_{nij} = \frac{l_{KOH}}{l}\left(\sum_{K=1}^{m} a_{iK}^b \cos i\varphi_{B_j} + \sum_{K=1}^{m} b_{iK}^b \sin i\varphi_{Bj}\right).$$

Using (1) and turning to a fixed coordinate system we will define after the transformation the joint force along the Z axis, which causes axial vibrations.

$$F_{zi} = \frac{l_{KOH} n \cos\alpha}{c}\left(\sum_{K=1}^{m} a_{iK}^b \cos i\omega_{OTH} t + \sum_{K=1}^{m} b_{iK}^b \sin \omega_{OTH} t\right) \tag{2}$$

where $i = N \cdot n$, *(N=1,2,3,...)*.By finding the normal force projection at the j-th roller and the screw contact point on the Y axis, let's find the forces causing vibrations in radial axis. Similarly, we will determine the constraining forces from the rollers and the nuts threaded surfaces undulation that causes vibrations in radial and axial directions. Frequencies of the constraining forces generated by the threaded elements undulation are used to calculate the RSM vibroactivity/.

Let's separate RSM screw from the elastic system (Fig. 3). It is necessary to determine the dynamic compliance coefficients at the points where it joins the conjugate elements through elastic bonds (Figure 3). To solve this problem, it is necessary to create the selected element (screw) equations of motion under the impact of harmonic forces applied at the bonds cut-off points. In this case, regarding the screw, these forces will act in the normal direction at the screw points of contact with the n rollers on the thread surface, at the screw points of contact with the n rollers in the gear coupling and in the direction of the bearing assembly major rigidity, since,

 DOI: 10.1051/matecconf/201712906026

generally speaking, the bearing assembly design is unknown.

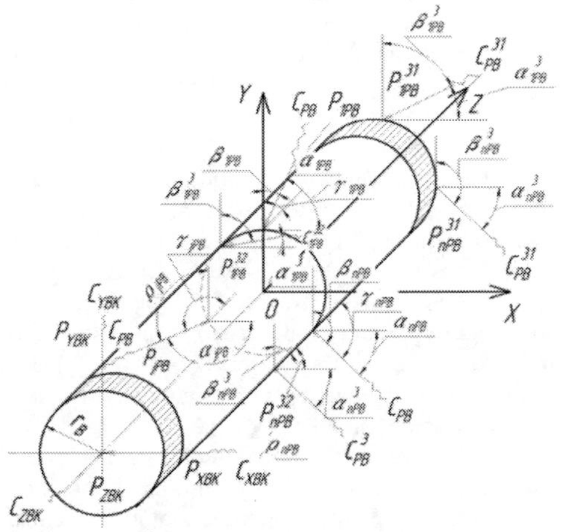

Fig. 3. Scheme of RSM screw

The equations of motion in a plane perpendicular to the screw axis and in the axial direction have the form:

$$
\begin{aligned}
& m_B \cdot \ddot{Y} + b_y \cdot \dot{X} = \\
& \left[\sum_{j=1}^{n} P_{jPB}\alpha_{jPB} + \sum_{j=1}^{n} \left(P_{jPB}^{31} + P_{jPB}^{32} \right) \alpha_{jPB}^{3} + P_{XBK} \cdot 1 \right] \cos \omega t \\
& m_B \cdot \ddot{Y} + b_y \cdot \dot{Y} = \\
& \left[\sum_{j=1}^{n} P_{jPB}\beta_{jPB} + \sum_{j=1}^{n} \left(P_{jPB}^{31} + P_{jPB}^{32} \right) \beta_{jPB}^{3} + P_{YBK} \cdot 1 \right] \cos \omega t \\
& m_B \cdot \ddot{Y} + b_y \cdot Z = \left[\sum_{j=1}^{n} P_{jPB}\gamma_{jPB} + P_{ZBK} \cdot 1 \right] \cos \omega t \\
& j_B \cdot \varphi + b_\varphi \cdot \varphi = \left[\sum_{j=1}^{n} P_{jPB}\varphi_{jPB} + \sum_{j=1}^{n} \left(P_{jPB}^{31} + P_{jPB}^{32} \right) \varphi_{jPB}^{3} \right] \cos \omega
\end{aligned}
\qquad (3)
$$

where m_B - weight screws; j_B - moment of inertia about the axis of the screw; P_{jPB} harmonic amplitudes of the forces acting on the screw at points of contact with the rollers; P_{jPB}^{31} ; P_{jPB}^{32} the amplitude of the harmonic forces acting on the screw at the points of engagement with the toothed rollers; b_x, b_y, b_z, b_φ damping coefficients that take into account the power dissipation in the direction of oscillation of the generalized coordinates; P_{XBK}, P_{YBK}, P_{ZBK} amplitude harmonic force in the contact-screw housing; α_{jPB}, β_{jPB}, γ_{jPB}, φ_{JPB}, α_{jPB}^{3}, β_{jPB}^{3}, φ_{jPB}^{3} points and the direction cosines of the unit vectors directed along the axes *x j*-springs; ω the frequency of the driving force; *X, Y, Z,* φ coordinates of the center of gravity and rotation motion with respect to the screw axis *OZ*.

In the motion equation system (3) the amplitudes of the harmonic forces affecting the screw at the points of gear coupling with the rollers P_{jPB}^{31} ; P_{jPB}^{32} are directed along the coupling line at any given time t. The equations of screw motion were set up in the XOY coordinate system rotating with the angular velocity ω_p^0 . The transition to the fixed XOY coordinate system will take place when determining the vibratory acceleration amplitudes. The same calculation procedure was also applied when considering other RSM subsystems motion. The spot solution of equation (3) will be in the form of:

$$X = A \cos \omega t \sin \omega t \qquad (4)$$

In matrix form the equilibrium equations system has the form of:

$$M_B X_B = R_B P_B \qquad (5)$$

where M_B mass matrix of the screw; X_B vector amplitudes screw displacement of the center of gravity; P_B vector amplitudes forced harmonic forces; R matrix of direction cosines.The movements of J-th points along which the bonds were broken in the direction of those links (in the force line direction

$$U_B = R_B^T X_B \qquad (6)$$

where U_B - motion vector points j-points; - transposed matrix .Solving equation (5) in terms of X_B and multiplying both its parts by a matrix M_B^{-1} inverse to the matrix we obtain:

$$X_B = M_B^{-1} R_B P_B \qquad (7)$$

Let plug (7) into the equation (6) and we obtain a link between the j-th points movements and the forces acting in those directions

$$U_B = R_B^T M_B^{-1} R_B P_B \qquad (8)$$

when $E_B = R_B^T M_B^{-1} R_B$ - the coefficient matrix of the dynamic compliance of the screw.Similarly rollers, nuts and casing dynamic compliance coefficients are determined that are used to work out the system for canonical equations subject to no-movement condition at the points where the whole system is divided into subsystems in the direction of those bonds. In matrix form we will obtain the following equation system:

$$AP = 0 \qquad (9)$$

when P - column matrix of harmonic forces (reaction of links); A - matrix of dynamic compliance RBM system, where each element is the sum of compliances

DOI: 10.1051/matecconf/201712906026

subsystems at the removal of links between them. To find the natural system frequencies of vibration the frequency equation is solved:

$$\det A = 0 \tag{10}$$

When calculating forced vibrations the dynamic compliance method is also used. The general matrix equation for calculating the forced system vibrations has the form of:

$$AP = A^{*}F \tag{11}$$

where A^{*} - matrix dynamic compliance system; P - vector reactions in relationships; F - column matrix driving forces acting on relations.

$$F = \{F_{1PB} \dots F_{nPB}\, F_{1PB}^{31} F_{1PB}^{32} \dots F_{n\mathrm{Pr}}\, F_{1PB}^{31} F_{1PB}^{32} \dots \;\dots F_{n\mathrm{Pr}}$$

When calculating the forced system vibrations, the superposition principle is used due to the linearity of equation (11), i.e., the impact of the system of constraining forces of the like harmonics (frequency) were considered separately, which is fully in line with the spectral analysis' objectives. The obtained RSM element of interest vibratory displacement amplitudes were summed.We would also like to note that when calculating the free vibrations, RSM frequencies that make part of the matrix A are the natural vibration frequencies and are determined from equation (9) and in case of forced vibrations those frequencies are preset (known) frequencies of constraining forces.The general course of forced vibrations calculation by way of dynamic compliance prediction is as follows. In matrix F constraining forces of the same frequency ω_1 were taken into account. This frequency value is plugged into dynamical compliance coefficients expressions in matrices A and A*. From (11) unknown reactions in the bonds caused by the impact of constraining forces are determined. After that each subsystem can be considered independently of each other. From the equilibrium equations of the interesting subsystem, the amplitude of vibratory displacement in the direction of the required generalized coordinate is determined. Then, if necessary, vibration speed, vibration acceleration and vibration log level can be found later. When changing the constraining force frequency to other one, it is necessary to carry out complete recalculation again, because the dynamic compliance coefficients values in matrices A and A* also change.

As already indicated, the most interesting from the practical point of view are vibrations accounted for by the RSM casing and generating its vibroactivity and noise, as well as vibrations overlapping the output motion and determining the drive operation smoothness.

When analyzing the executive units vibration, it is more convenient to use a third-octave spectrum rather than a narrowband one, because in this case the spectrum record is of more compact form and presents a general picture of the vibration process over a wide frequency range. In this case, the volume of vibration amplitude-frequency spectra calculations was significantly reduced.

The methods presented are convenient for the RSM vibrational activity checking calculations when the exciting forces spectrum can be experimentally obtained. For this purpose, the threaded elements undulation is measured. For example, the undulation of zero lead-in elements can be measured using a profilometer.

This results in obtaining an undulation profilogram of all the cutting threads of the nut, screw and rollers being machined. The obtained data arrays are expanded into Fourier series and the exciting forces amplitude-frequency spectrum is formed. This spectrum was used in the technique under discussion, and the vibration amplitude-frequency spectrum of the particular executive unit, which elements undulation has been measured, was determined.

To develop a methodology for vibroactivity forecasting at the design stage when only tolerances for the production of the main members of a mechanism are set, it is necessary to take into account the likely nature of exciting forces.

4 Experimental study of RSM vibroactivity

The item under test was a roller-screw mechanism which structure is shown in Fig. 1. Vibration levels were measured in the frequency range from 20 to 20,000 Hz using a specially manufactured stand shown in Fig. 4. Sample RSM *2* is fixed on the bracket on the foundation *1*. The RSM output rod by a connecting rod *3* is connected to a spring-actuated loading device *4* designed to create forces on the rod of the mechanism. In the loading device a 5.7 kgf / cm force spring with a 100 N preloading was used. In the extreme mechanism rod positions the loading device generated forces of up to 350 N.

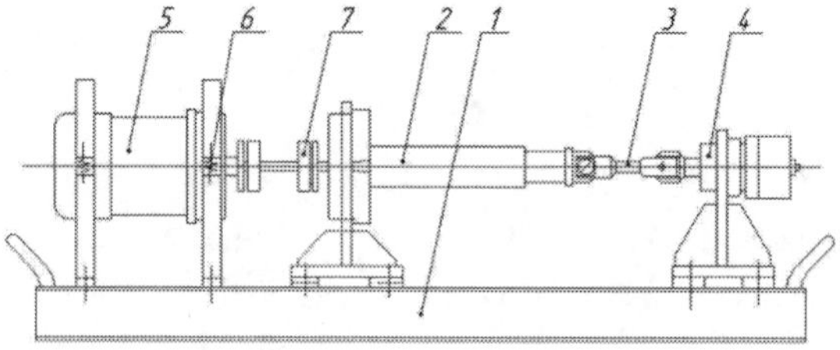

Fig. 4. Test stand for investigation of RSM

Vibration sensors were mounted on the stand basement and on the RSM housing. The vibration parameters were measured in an orthogonal coordinate system in the OY and OZ axes direction. Following vibration characteristics were determined during testing: the RSM vibration level in the radial direction (OY-direction); RSM vibration level in the axial direction (Z^0-direction)Equipment manufactured by Bruel & Kjaer Company was used for testing. Sensors installation, calibration, preparation for testing, measurements and analysis of results were performed in accordance with the above equipment operating instructions.

Values of vibration levels in the radial direction are somewhat lower than in Y-direction and are more evenly distributed over frequency bands with no pronounced peak values. At frequencies lower than 80 Hz vibration levels were not higher than 40 dB, within the range of 250 to 4 kHz their values amounted to 48 to 58 dB. Maximum vibration levels amounted to 65 dB in frequency bands of 6.3 to 12.5 kHz. In case of screw rotation speed increase from 21 to 25 Hz regarding RSM 1, the vibration level in the radial direction increases by 2 to 3 dB. Narrowband RSM vibration analysis spectra within the frequency range of 0 to 2000 Hz shown in Fig. 5 have a high density of vibration frequencies discrete components, which is consequential of a significant number of constraining force harmonics caused by the threaded elements undulation.

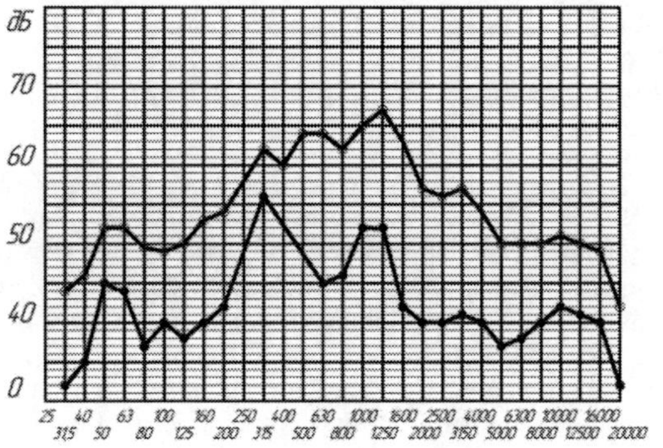

a)

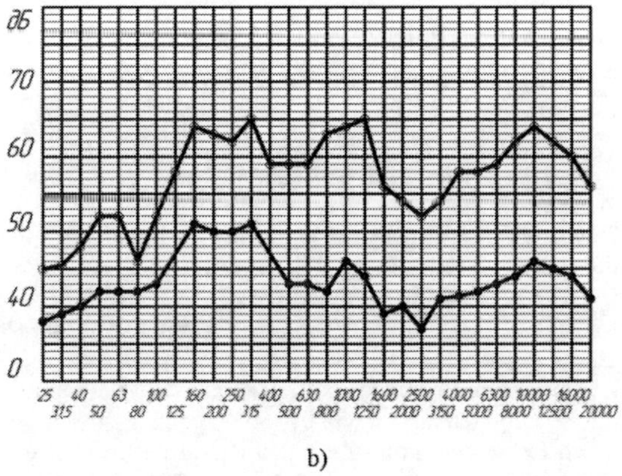

b)

Fig. 5. Third octave spectra RSM vibroactivity:
a) axially vibroactivity: b) radial vibroactivity: ●—●—● - no load, ○—○—○- load 380 N

5 Results and discussion

Third-octave spectra of vibration levels in the radial and axial directions during RSM 1 and RSM 2 testing were obtained in all modes of interest. Third-octave spectra of vibration levels in the radial direction of the RSM testing at regime are shown as an example in Fig. 6. Analysis of the available spectrograms showed that the values of vibration levels in the radial direction were from 36 to 64 dB in different third-octave frequency bands. Vibration level maximums are in the frequency ranges of 250 to 315 Hz, 800 to 1000 Hz and of 8 to 10 kHz.The RSM 1 vibration levels in the axial direction have maximum values in the frequency bands range of 63 to100 Hz, 250 to 400 Hz, 1 to 16 Hz and amount to 53 to 60, 68 to 62 and 58 to 62 dB.Studies allowed us to draw following conclusions:

1. RSM have low vibroactivity which allows developing mechatronic drives based on them with the maximum vibration level not exceeding 60 to 62 dB, which is 11 to 16 dB below the gear mechanisms vibroactivity level.
2. The method developed for calculating RSM vibroactivity allows vibration level predicting at the drives design stage.
3. The study of developed models allows specifying requirements for the accuracy of screw, rollers and nut thread surfaces manufacture. Thus, the vibroactivity of an RSM with the Tolerance Class 1 elements is by 16% less than those with elements of Tolerance Class 2.
4. The given experimental data showed that the maximum RSM vibration level does not exceed 60 to 62 dB. With the increase of axial load from 0 to 760 No, vibration in the radial direction decreases by 2 to 6 dB. When screw rotation speed increases from 21 to 26 Hz, the vibration level increases by 2 to 3 dB.
5. Narrowband vibration analysis showed that the vibration spectrum is continuous with a large number of vibration frequencies discrete components which is due to a significant number of constraining forces harmonics.
6. A comparison of the calculated and experimental vibration levels third-octave spectra showed the adequacy of developed models to the real RSM processes taking place. The difference between calculated and experimental results did not exceed 4 to 10 dB which is 6 to 16%

References

1. M. Shangjun, L. Geng, T. Ruiting, Z. Xiaocai, A Mathematical Problems in Engineering (2012)
2. V. Morozov *Roller screw mechanisms. Kinematic Characteristics* (VlSU, Vladimir, 2005)
3. V. Morozov, A. Zhdanov, *Roller screw mechanisms. Reliability and durability* (Pero, Moscow, 2015)
4. A. Zhdanov, V. Morozov, Procedia Engineering, **150** (2016)
5. V. Morozov, A. Zhdanov, Russian Aeronautics, **59**, 1 (2016)
6. V. Morozov, A. Zhdanov, A. Kosterin, Russian Engineering research, **36**, 9 (2016)
7. V.V. Kozyrev, *Comparison of ball and roller screw-nut transmissions*, Bulletin of Mechanical Engineering., **11**, pp. 31-35 (1983)
8. V.V. Kozyirev, *Design theory and methods of design and study of planetary gear-nut threaded rollers and mechatronic units based on them: monograph* (Vladimir State University, Vladimir, 2011)
9. D.S. Blinov, M.I. Morozov *Promising designs of planetary roller screw mechanisms*, Izvestiya vuzov, Mishinostroenie, **3**, pp. 62-72 (2013)

MATEC Web of Conferences **129**, 06027 (2017) DOI: 10.1051/matecconf/201712906027
ICMTMTE 2017

Investigation of stress-strain state in the flywheel and estimation their specific energy capacity

Dmitri V. Berezhnoi[1,*], *Lejsan R.* Gajnulina[1], and *Andrej A.* Sachenkov[1]

[1]Kazan (Volga Region) Federal University, 18 Kremlyovskaya Street, Kazan, 420008, Russian Federation

Abstract. In this paper, the specific energy intensity of the kinetic energy storage devices, including the flywheel-casing scheme in the potential field, is investigated. The possibilities of using various structural materials in the manufacture of structural elements of a mechanical accumulator are analyzed, the stress-strain state of the flywheel and the casing under quasistatic increase in the rotational speed of the rotor part of the structure is investigated. It is noted that the presence of a potential field in the flywheel-casing system makes it possible to increase the specific energy intensity of the kinetic energy storage.

1 Introduction

Recently, the problem of accumulation and conservation of energy has become increasingly important. The creation of new materials allows not only to improve the existing traditional electrochemical batteries, but also to seek new methods of accumulating and storing energy, including mechanical ones. Mechanical so-called "static" and "dynamic" mechanical energy storage devices can be referred to mechanical energy storage devices.

Static battery stores stored energy in static, in a stationary state. But these batteries have a very small accumulated specific energy. Contrary to this static batteries have provided a solid place in many machines and mechanisms, since they possess such properties - high efficiency, stability of accumulated energy and durability. Simple enough is the technology of energy storage by means of kinetic storage. The flywheel battery is promising because, firstly, flywheels are capable of instantly gaining colossal power; Secondly, they are very reliable, long-lasting, economical (the flywheel has a high coefficient of efficiency, the only one of the batteries capable of using mechanical energy on the descents and braking the vehicles on which they can be used with great efficiency); Thirdly, they are environmentally safe, which in our time is very important. The main limitation in implementing such a technology is the rotational speed of the rotor part of the kinetic storage, which can lead to an increase in the stress level and its possible further destruction. An essential factor determining the specific energy intensity of kinetic energy stores is the ratio of the ultimate destructive stress to the density of the material, which, together with the safety requirements, dictates the use of super flywheels (flywheels made by winding threads or tapes).

For flywheels of arbitrary shape, numerical methods for studying the specific kinetic energy are proposed in the literature [1-9]. In a number of works kinetic energy storage devices made of rubber-like materials are proposed, for the calculation of which special methods are proposed [10-14]. The calculation of super-flywheels, made by winding threads or tapes, presupposes methods for their contact interaction [15-22].

In this paper, we present some results of estimating the specific power consumption of flywheels, including kinetic and potential deformation energy, an algorithm for calculating the stress-strain state and specific energy for flywheels of various shapes and structures.

2 Formulation of the problem

One of the most important indicators of the efficiency of the energy storage device is the specific energy intensity, i.e. The ratio of accumulated energy to mass. Formulas for calculating the specific energy intensity for the kinetic bodies of the simplest form have long been known and are given in many sources, but estimates of the specific energy intensity for potential energy are much less common in the literature. In particular, for a flywheel in the form of a disk with a hole [23, 24], an estimate of the specific energy intensity with respect to the kinetic energy can be written in the form

$$K_m = (\sigma_y(1+k^2))/(\rho(3+\mu+(1-\mu)\,k^2)), \qquad (1)$$

where σ_y and μ are the time resistance and the Poisson's ratio of the flywheel material, k is the ratio of the inner radius of the flywheel to the outer one. An estimate of the specific energy intensity from the potential energy of deformation obtained under the condition that the flywheel is in a plane-stressed state can be written in the form

$$P_m = \alpha\,(\sigma_y)^2 \,/\, (6E\rho(3+\mu+(1-\mu)k^2)), \qquad (2)$$

* Corresponding author: berezhnoi.dmitri@mail.ru

where

$$\alpha = (1+k^4)(7-6\mu+\mu^2) + 2k^2)(17+6\mu+\mu^2), \quad (3)$$

and E is the Young's modulus of the flywheel material. It can be noted that for flywheels made of rubber-like materials, the energy capacity estimate for the potential energy of deformation is much higher than in kinetic energy.

In the framework of the solution of a geometrically nonlinear problem [25-29], the components of the Cauchy-Green tensor are used in the deformation ratio. The global scheme for solving the problem with allowance for geometric nonlinearity is as follows. The general variational equation for the total potential energy is written:

$$\delta W + \delta W_g = 0, \quad (4)$$

where part δW refers to the whole structure, part δW_g describes geometrically nonlinear deformation. Further, within the framework of the finite element method, the global rigidity matrix is assembled, after assembly of which the variational equation takes the form:

$$\{\delta q\}^T (\{R\}+\{R_g\}) = 0, \quad (5)$$

where $\{\delta q\}$ is the variation of the vector of nodal displacements, $\{R\}$ and $\{R_g\}$ - vectors of generalized forces. As a result, the Newton method for the vector function F:

$$F=F(\{q\}) = \{R\}+\{R_g\}. \quad (6)$$

Accordingly, giving the initial approximation ($\{q\}^{(0)}$), we construct an iterative procedure:

$$\{\Delta q\} = - F_u(\{q\})^{-1} F(\{q\}), \quad (7)$$

$$\{q\}^{(n+1)} = \{q\}^{(n)} + \{\Delta q\}^{(n)}. \quad (8)$$

$$F_q = dF_u / d\{q\}. \quad (9)$$

The transition from the variational problem to the algebraic one is done by means of discretization by the finite element method. The relationship between stresses and strains is determined by the well-known Green formula

$$\sigma_{ij} = \partial U / \partial \varepsilon_{ij}. \quad (10)$$

In the case of a linearly elastic material, the elastic potential is taken in the form

$$U= C_{ijkl}\, \varepsilon_{ij}\, \varepsilon_{kl} / 2, \quad (11)$$

where C_{ijkl} are the coefficients of the isothermal elasticity tensor. To study the stress-strain state of flywheels made of elastomers, we will use the Mooney-Rivlin potential in the form

$$U= C_{10} (I_1 - 3) + C_{01} (I_2 - 3) + (J - 1)^2 / d, \quad (12)$$

where C_{10}, C_{01} are the coefficients of the model, material constants; I_1 and I_2 are the invariants of the deviator of deformation; J - determinant of the strain gradient matrix; d is the coefficient of incompressibility of the material. In the subsequent calculation, the Mooney-Rivlin constants for ordinary rubber C_{10}=0.293, C_{01}=0.177 and the material incompressibility coefficient d = 0.0042 were used.

Specific energy intensities for flywheel in the form of a disk with a hole, obtained for a number of materials, whose mechanical characteristics are given in Table 1, are presented in Table 2.

Table 1. Mechanical characteristics of a number of materials used in the manufacture of flywheels.

Materials	Young's modulus [MPa]	Poisson's ratio	Yield stress [MPa]	Density [kg/m³]
High carbon steels	215000	00.3	1155	7900
Titanium alloys	120000	0.3	1245	4800
Boron carbide	472	0.3	5687	2550
Composites polymer CFRP	150000	0.25	1050	1600
Polyurethane elastomers (eiPU)	3	0.48	51	1250
Kevlar aramid fiber 49	170000	0.36	3600	1440

Table 2. Specific energy capacyty of the flywheel [MJ/kg] over the potential (Pm) and kinetic (Km) energies for the thin ring (k = 1).

Materials	P_m	K_m
High carbon steels	0.0004	0.073
Titanium alloys	0.0013	0.129
Boron carbide	0.0134	1.115
Composites polymer CFRP	0.0023	0.328
Polyurethane elastomers (eiPU)	0.3470	0.020

3 Numerical results

Of the potential fields used in engineering, namely gravitational, electrostatic and magnetic, in the flywheel-hull system, the latter can be applied. The use of potential coupling in the mutual repulsion mode allows to unload the outer layers of the flywheel by transferring the load to the body and thus increasing the speed of rotation of the flywheel and the stored energy. Also, due to the inclusion in the work of the body of the kinetic energy storage device, which is in a compressed state due to the difference in internal and external atmospheric pressure, it is possible to reduce its weight by removing the load by the amount created by the potential field and to change the sign of the load from compression to tension, More preferably for modern synthetic materials. To create a magnetic unloading of the outer layers of the flywheel rotor, it is necessary to use a magnetic material, which makes it possible to create repulsive forces between the flywheel rotor and the sturdy body of the magnetic energy store.

When winding the flywheel rim from the magnetized tape, the entire flywheel acquires the properties of a magnet with radial magnetization, with each subsequent winding coil being attracted to the previous one, and, consequently, to the body of the flywheel. The attraction of the subsequent turns prevents the flywheel from breaking off, which occurs at high rotational speeds. The magnetic body is made by winding with a magnetic tape (used earlier for winding the flywheel rim), but with counter magnetization. Under the influence of magnetic forces, the flywheel rim will be repelled by the body magnet. Unloading the rim of the flywheel with the transmission of the centrifugal forces arising and striving to destroy the flywheel on the body of the mechanical energy storage allows substantially increasing the speed of rotation of the flywheel without destroying it.

Consider the simplest axisymmetric mechanical model of the flywheel-casing system (Fig. 1). The flywheel rotates at a constant angular velocity ω inside the casing, a constant potential field acts between the casing and flywheel, which (in the first approximation) generates an excess pressure q in the gap between the flywheel and the casing. The internal radius of the flywheel $r0$, outer - $r1$, the size of the clearance between the flywheel and the casing - t, the inner radius of the casing $r2$, outer - $r3$. The density and time resistance of the material of the flywheel are respectively ρ_m and σ_{ym}, the shell materials are ρ_k and σ_{yk}, respectively. In this case, the solution obtained does not depend on the rotation angle φ, but depends only on the current radius r of the disk.

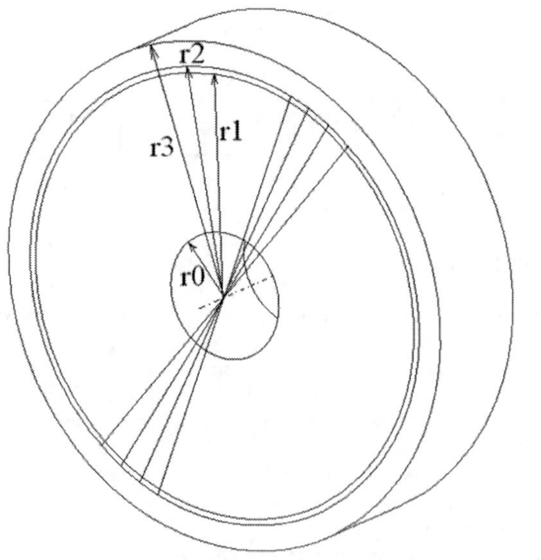

Fig. 1. Model of flywheel-housing system.

Two types of calculations were carried out (for a flywheel without a casing, for a flywheel with a casing in a potential field, in a potential field) to estimate the specific energy intensity of some structural materials whose mechanical characteristics are given in Table 1. Table 3 gives the specific energy intensities of flywheels and flywheels with housings from the boron carbide in the potential field (for kinetic energy *K*m). Table 4 gives the specific energy intensities of flywheels and flywheels with casing from the kevlar aramid fiber 49.

Table 3. Specific energy consumption of flywheels in the form of discs without a hole [MJ/kg] with a casing of the boron carbide and without it in kinetic (*K*m) energy.

Materials	Flywheel scheme	Flywheel-casing scheme
High carbon steels	0.071	0.889
Titanium alloys	0.13	0.871
Boron carbide	0.33	0.909
Composites polymer CFRP	0.020	1.105
Polyurethane elastomers (eiPU)	1.25	0.409

Table 4. Specific energy consumption of flywheels in the form of discs without a hole [MJ/kg] with a casing of the Kevlar aramid fiber 49 and without it in kinetic (Km) energy.

Materials	Flywheel scheme	Flywheel-casing scheme
High carbon steels	0.073	0.849
Titanium alloys	0.13	0.818
Boron carbide	0.328	0.885
Composites polymer CFRP	0.020	1.232
Polyurethane elastomers (eiPU)	1.25	0.409

4 Analysis of results and conclusions

In conclusion, it should be noted that a comparative analysis of the specific power consumption of flywheels (only the kinetic energy of rotation was estimated) with a casing and without it was carried out. It can be noted that when using the flywheel-casing system in the potential field, the specific strength of the casing material and σ_{yk}/ρ_k should be higher than the specific strength of flywheel material σ_{ym}/ρ_m. The presence of a potential field in the flywheel-casing system makes it possible to increase the specific energy intensity for flywheels. The presence of additional winding threads that enhance the power structure of the housing reduces the specific energy intensity of the kinetic energy storage.

The flywheel was considered only as a disk with a hole, but the results can be generalized also to the case of a flywheel in the form of a solid disk. Generally speaking, at high speeds of rotation of a flywheel presence of a casing is obligatory, since In the immediate vicinity of the rotating part of the structure it is necessary to create a vacuum, so taking into account the mass of the casing when calculating the specific energy intensity is mandatory. The presence of a potential field between the flywheel and the casing makes it possible to increase the angular speed of rotation of the flywheel, but we can not increase the infinitely angular velocity, because Thus the mass of the casing begins to grow.

The calculations show that the involvement of not only the flywheel with the geometric and physical-mechanical parameters given in the work, but also the case of the flywheel energy storage, increases the flywheel rotation speed by 1.4 times. Thus, the specific energy intensity of the flywheel energy storage can be increased more than 2 times.

The reported study was supported by RFBR and Government of the Republic of Tatarstan research projects No. 15-41-02555, 15-41-02557.

References

1. T.O. Drum, Energy storage-flywheel, http://www.resilience.org/stories/2011-10-05/energy-storage-flywheel.
2. N.V. Gulia, *Machovichnye dvigately* (Mashinostroenie, Moskva, 1976) 163 (In Russian).
3. N.V. Gulia, *Nakopitely Energii* (Nauka, Moskva, 1980) 152 (In Russian).
4. V.V. Matveev, G.V. Pisarenko, A.P. Yakovlev, *Spravochnik po soprotivleniyu materialov on resistance of materials* (Naukova Dumka, Kiev, 1988) 736 (In Russian).
5. Y. Seong-yeol, L. Wook-ryun, B. Yong-chae, N. Myounggyu, Journal of Mechanical Science and Technology **24**, 231-235 (2010)
6. D.V. Berezhnoi, D.E. Chickrin, A.F. Galimov, Applied Mathematical Sciences **8 (124)**, 6181–6190 (2014)
7. D.V. Berezhnoi, D.E. Chickrin, E.Yu. Kurchatov, A.F. Galimov, Applied Mathematical Sciences **8 (163)**, 8125–8135 (2014)
8. D.V. Berezhnoi, D.E. Chickrin, L.R. Gajnulina, E.Yu. Kurchatov, P.A. Kokunin, I.G. Sozutov, M.I. Shigapov, Contemporary Engineering Sciences **8 (36)**, 1703–1712 (2015)
9. D.V. Berezhnoy, L.R. Gajnulina, L.R. Sekaeva, International Journal of Applied Engineering Research **10 (24)**, 44672–44676 (2015)
10. A.I. Abdrakhmanova, L.U. Sultanov, Materials Physics and Mechanics **26 (1)**, 30–32 (2016)
11. A.I. Golovanov, L.U. Sultanov, International Applied Mechanics **41 (6)**, 614–620 (2005)
12. R.L. Davydov, L.U. Sultanov, Journal of Engineering Physics and Thermophysics **88 (5)**, 1280–1288 (2015)
13. L.U. Sultanov, L.R. Fakhrutdinov, Magazine of Civil Engineering **44 (9)**, 69–74 (2013)
14. L.U. Sultanov, R.L. Davydov, Magazine of Civil Engineering **44 (9)**, 64–68 (2013)
15. O. Sachenkov, L. Kharislamova, N. Shamsutdinova, E. Kirillova, Yu. Konoplev, IOP Conference Series: Materials Science and Engineering **98**, 012015 (2015)
16. O. Sachenkov, R. Hasanov, P. Andreev, Y. Konoplev, IOP Conference Series: Materials Science and Engineering **158 (1)**, 012079 (2016)
17. F.A. Shigapova, R.F. Mustakimova, G.T. Saleeva, O.A. Sachenkov, International Journal of Applied Engineering Research **10 (24)**, 44711–44714 (2015)
18. O.A. Sachenkov, V.I. Mitryaikin, T.A. Zaitseva, Yu.G. Konoplev, Applied Mathematical Sciences **8 (159)**, 7889–7897 (2014)
19. I.B. Badriev, V.V. Banderov, M.V. Makarov, V.N. Paimushin, Applied Mathematical Sciences **9 (78)**, 3887–3895 (2015)
20. I.B. Badriev, V.V. Banderov,O.A. Zadvornov, Applied Mechanics and Materials **392**, 188–190 (2013)
21. I.B. Badriev, G.Z. Garipova, M.V. Makarov, V.N. Paimushin, Research Journal of Applied Sciences **10 (8)**, 428–435 (2015)
22. Y.N. Rabotnov, Soprotivlenye materialov (Fizmatgiz, Moskva, 1962) 456 (In Russian).
23. D.V. Berezhnoi, L.R. Gajnulina, IOP Conference Series: Materials Science and Engineering **158 (1)**, 012017 (2016)
24. D.V. Berezhnoi, D.E. Chickrin, L.R. Gajnulina, E.Yu. Kurchatov, P.A. Kokunin, I.G. Sozutov, M.I. Shigapov, Contemporary Engineering Sciences **9 (32)**, 1591-1598 (2016)
25. D. Kolarov, A. Baltov, N. Boncheva, *Mechanics of plastic mediums* (Mir, Moscow, 1979), 302 (In Russian).
26. A.S. Sakharov, V.N. Kislooky, V.V. Kirichevski, N. Altenbach, U. Gabbert, Y. Dankert, Kh. Kepler, Z. Kochyk, *The finite element method in the mechanics of solids* (Higher School, Kiev, 1982), 480 (In Russian).
27. P.V. Trusov, A.I. Shveikin, *The theory of constitutive relations,* (Perm State Technical University, Perm, 2008), 243 (In Russian).
28. V.I. Levitas, *The large elastic-plastic deformation of materials under high pressure* (Naukova Dumka, Kiev, 1987), 232 (In Russian).
29. A.A. Pozdeev, P.V. Trusov, Y.I. Nyashin, *The large elastic-plastic deformation theory, algorithm, application* (Nauka, Moscow, 1986), 232 (In Russian).

MATEC Web of Conferences 129, 06028 (2017) DOI: 10.1051/matecconf/201712906028
ICMTMTE 2017

The device to improve the working element of the rotary trench excavator in the production process of breaking a main pipeline

Asvar Akhmedov[1,*]

[1]Volgograd State Technical University, 400005, Lenin avenue, 28, Volgograd, Russia

Abstract. The article develops a device for the rotary trench excavator. The device will enable to break an underground major pipeline from one of the side faces and above the top surface of the pipe in one operation. The design of the device enables to use it in any climatic conditions. The application of the improved device attached to the rotary trench excavator enables to enforce the structure of the construction and repair spread.

Today, the construction and the major repairs of the linear section of major pipeline require some "breakthrough" inventions to enhance technical and economic performance. The oil and gas industry has a problem that lies in the fact that the lion's share of the presently exploited pipelines needs immediate overhaul operation. To this end, it is urgent to develop new innovations devices to improve technological processes. Due to the improvement of technological processes and operations it is possible to reach better technical and economic performance. This will enable to improve the situation with the state of main pipelines.

The application of continuous-type machines cuts the duration of stripping works by 10 to 15 times which undoubtedly is profitable for companies doing the overhaul operations.

Today, there is no fleet of rotary trench excavators at the organizations performing the overhaul works for the main pipelines [1-3]. However, our country goes on to produce the rotary trench excavators. The rotary trench excavators are made and sold in this country up to now.

Search for power efficient technologies on the major repairs of main pipelines is being done in many countries and this is connected not only with the fact of increasing volumes of worn-out pipes, but also with the situation when pipeline mains go through various climatic and natural conditions [4-8].

A machine is known for the layer-by-layer stripping of the main pipeline [9] that has milling tools as a working element. To break the pipe using this machine, it is necessary to do several passes which increases the unproductive movements.

The disadvantage of the invention is the fact that after the digging machine operation, it is difficult to do further repair works and there is no access for machines and mechanisms since the soil lifted with the buckets of working elements is falling on the conveyors and brought to the burrows at the both sides of the trench which requires the equipment to trim the excavated soil along the whole length of the stripped section of the pipeline.

A rotary trench excavator is known with a special working element [10] consisting of three rotors with buckets mounted on the frame suitable for movement relative to one another vertically. The present ability to move rotors vertically relative to one another requires a powerful framework and a strong hydraulics which increases the cost of the working element.

A single-bucket excavator is known [11] with a conveyor and a blade on it which enables to remove the soil outside the machine as stripping moves forward. However, the given excavator cannot strip the trench without tip trucks taking the soil from the conveyor.

There is also another technology [12] consisting of the basic machine with the mounted special working element enabling to strip the pipeline from both lateral sides and along the lower moving line. According to the technology, to strip a pipeline, it is necessary to strip the whole pipeline section under repair using any digging machine up to the upper line which makes costs of the earthworks higher. Apart from this, the design of the working element prevents from moving the soil to only one side when stripping the pipeline which involves the need to level one of the trench edges to move a construction train. This results in reducing the technical and economic performance.

From the existing technology level, we know the inventor's certificate [13] lying in the fact that the working element of the machine is two rotors parallel fixed in the vertical plane suitable for changing the distance between them. There is a "dumping knife" between the rotors cutting the soil along the top of the pipe as the stripping moves forward guiding it onto the rotating rotors on both lateral sides; they move the soil to the disposable area via a conveyor.

Despite the fact that the given working element can be applied for stripping pipelines of several diameters due to the rotors' readjustment, the use for all the

* Corresponding author: asvar05@mail.ru

diameter types is limited by the framework sizes. To strip pipes of greater diameters, the change of the framework for another with greater sizes is required. The installation and the following adjustment require the application of lifting facilities and hand labor.

There exists one more invention [14], the working element of the excavator for the pipeline break including a rotor with the buckets and the shortened middle section, and the buckets of the extreme sections are made with the slope angle to prevent the shock impacts on the pipeline walls. The given technological solution is directed at breaking the pipeline from the lateral sides and from the pipe top in one passage of the machine. The drawback of the invention is undoubtedly the inability to use it in all the pipeline diameters which reduces its demand and universal application.

In [15] the stripping of the pipeline is carried out by means of the simultaneous movement of the column of technical facilities and mechanisms connected with one another by cables. There is a tow truck at the front of the column to which a bulldozer is attached by a cable cutting the vegetable mold and moving it to both sides of the stripped pipeline with the help of the blade and the ripper arrangement in the direction of its travel. The second tow truck is fixed to the bulldozer by ropes which pulls the second ripper which does its work and moves apart the mineral soil above the pipeline by its "blades-molds" to both sides from the axis of the stripped trench. The pipe-layers moving in step with the tow trucks and the bulldozer provide the pipeline's lifting with the trolley pipe holders, herewith, the carrying crossbeam is joined to the ripper of the bulldozer which ensures the removal of the soil rests from the upper part of the pipe and its lifting from the foundation level by the trolley pipe holder. Plow-type device behind the multi-bank trolley holder removes the soil to the trench and surfaces the ground onto which the lifted pipeline is laid due to the reverse turning of the plowshares in its forward motion parallel to the ripping and moving the soil apart with the first and the second rippers.

Thus, in the known technique, the break is done due to the simultaneous motion of all the technical facilities and creation of considerable pulling force.

Among the great variety of the existing working elements of the rotary trench excavators for stripping a linear section of main pipelines, there are a multitude of technical solutions with considerable drawbacks. Many of them are outdated and require the upgrading and improving with regard to modern achievements in science and technology.

With regard to the studied problems and drawbacks, a special rotary trench excavator was proposed the working element of which has been modernized. Further, we will consider the proposed device in more details taking into account its applicability.

The device can be used while stripping during the major repairs of main pipeline as well as in disassembling the decommissioned pipes.

The proposed innovative device enables to break a pipeline in the broad range of diameters from 520 to 1620 included. The device is a hitch attachment and is connected to the base machine of the rotary trench excavator that is strengthened with an extra space frame which adds an extra rigidity and stability to the working element.

The developed device can significantly facilitate the break of the main pipeline. The application of the hitch attachment along with the working element of the rotary excavator enables to additionally enhance the productivity and efficiency in the work of the digging machine.

The increase in the values of technical and economic indexes can attract specialists to the developed device. The increase in these indexes and the improvement of technical characteristics of the existing digging machines will enable to save considerable financial means and so their introduction into the production will give a short payback period.

The short payback period is the key to the rapid spread in the market and mainstreaming in those branches where it is required to break a pipeline for repair and maintenance works or to change the pipe's wall.

The proposed device along with the working element of the rotary trench excavator enables to bring the technological process to a new qualitative level.

The device is designed for enlarging the capabilities of the rotary trench excavator due to improvement of its working element. With the improved rotary trench excavator, the pipeline section under repairs can be stripped in one passage.

The device is made as a spatial one, suitable for quick installation, removal due to a special removable connection. Besides, the device can be easily repaired and moved due to its competent and compact performance.

The use of the device enables to bring the technological process of breaking the linear section of main pipelines to a new qualitative level.

Based on the analysis of the existing technical solutions [1-16] for improving the technology of breaking the underground main pipeline using the rotary trench excavator, a special innovative device has been developed. The proposed device is designed on the basis of the common mass-produced rotary trench excavator.

The target of the device is to enlarge technological capabilities of the rotary trench excavator with the simultaneous reduction of labor intensity for earthworks due to the increase of the operating efficiency coupled with the machine's working element.

In order to reach the target, the following problems are determined and put:

• to develop a device which will help to break the main pipeline from the lateral side and above the top plane in one passage;

• to develop a technical solution for the device that will move the soil to one side in the direction of the machine movement;

• to develop a technical solution for the device enabling to use the device for different diameters of main pipelines.

The device consists of a space frame 1that is fixed to the base mounting of the basic machine 2 of the rotary excavator. The space frame 1 consists of the threaded guide rods состоит 3 one each of which the drive components 4 are mounted with the electric motors 5. The drive components 4 with the electric motors 5 are located to move along the rods 3. The milling tools 6 are hardly connected to the drive components 4. Herewith, the number of the milling tools is selected based on the width of the stripped trench 7 and the diameter of the main pipeline 8. The milling tools are mounted rigidly to the rods 3 with their upper and lateral sides which prevents them from shifting in the stripping process. The rods 3 with the milling tools are mounted on the metal frame by which they are connected with the base machine 2 from one lateral side, and from the other side they are connected to the rubber tired base mounting. To move the milling tools 8 in the traverse direction, some extra threaded rods 9 are mounted to the rods 3 on which there are extra drive components 10 with the electric motors 11for the transfer purposes (Fig. 1.- 6).

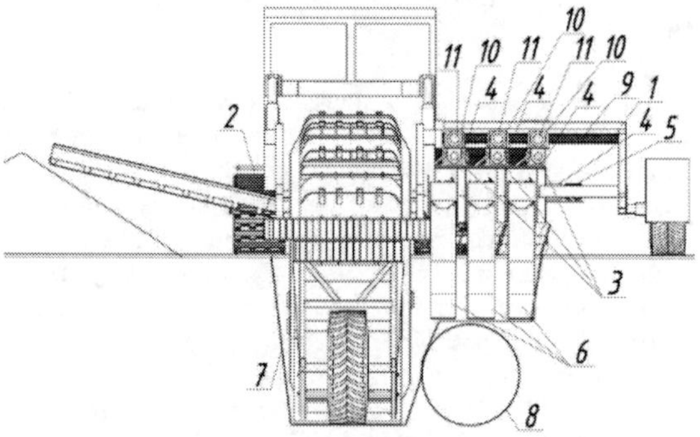

Fig. 1. Caption of the Figure 1. Below the figure: 1 – space frame , 2 – base machine, 3 – threaded guide rod, 4 – drive, 5 – electric motor, 6 – milling tools, 7 – trench, 8 – pipeline, 9 – extra rod, 10 – extra drive, 11 -electric motor

The device operates in coordination with the rotary trench excavator in the following way. At first, the rotary trench excavator is located at the start of the stripping segment of the main pipeline. Hereafter, the device is connected to the lateral side of the base machine. The device is designed so that it can be connected both to the one and to the other lateral side of the rotary trench excavator. Thereafter, the milling tools are moved to the needed location and the break (stripping) starts. The milling tools are located offset relative to one another which enables to get the “effect of the ground failure”. Then the rotor with the buckets is buried so that it is to one side from the stripped segment of the pipe. When the rotor reaches the design reference mark of the trench bottom, the milling tools and the screw conveyers are switched on. While running the milling tools, the screw conveyers are brought to the rotational motion which moves the soil lifted by the drilling tools to the rotating rotor with the buckets. In their turn, the moving rotor’s buckets will take the soil and move it to the stacker conveyer from which it will go to the disposal area. The proposed device enables to carry out the stripping of the main pipeline from one of the lateral sides and from the top side of the pipe in one passage of the rotary trench excavator.

The application of chain milling tools in the device’s design enables to break the pipeline above the top side of the pipe in the wide interval in depth. The adjustment of the milling tools location is done from the remote controller in the operator’s cabin. And the visual control over the location of the device and the working element is carried out by video cameras fixed on the base machine. If necessary to relocate the rotary trench excavator to a new place of the trench break, the device is removed and transferred on the individual basis.

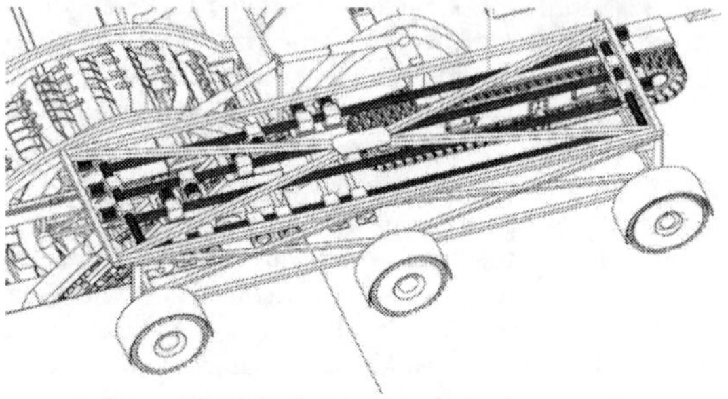

Fig. 2. Space simulation model of the device.

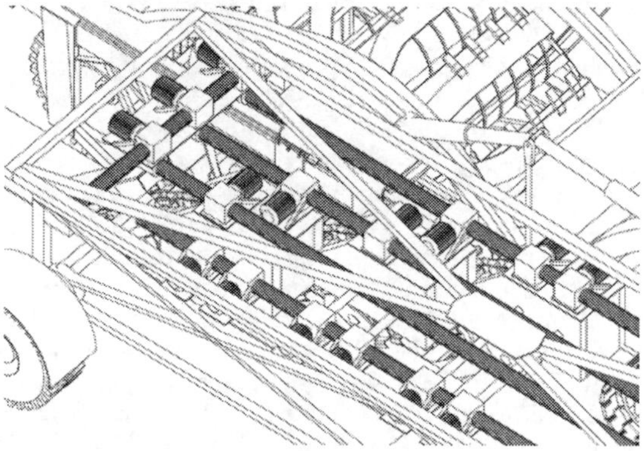

Fig. 3. Dimensional view of the connection for the guide rods with an additional rod.

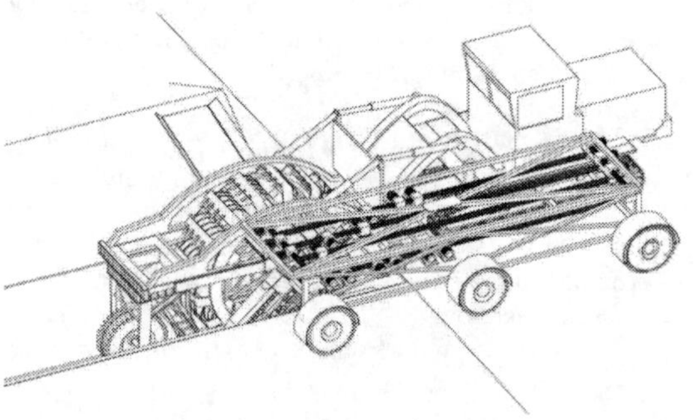

Fig. 4. Lateral dimensional view of the device in the working order while breaking a main pipeline.

DOI: 10.1051/matecconf/201712906028

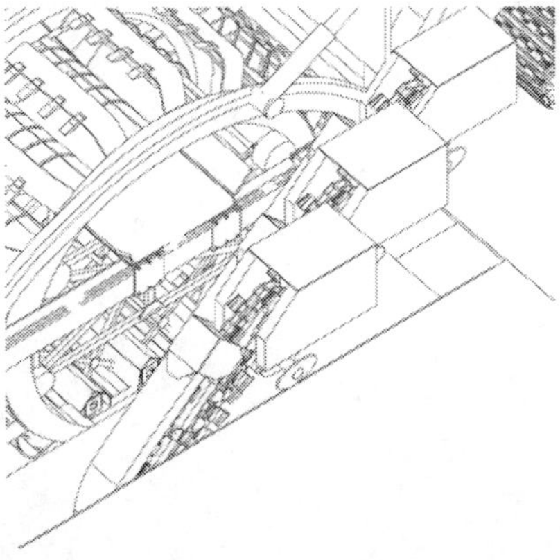

Fig. 5. Milling tools fixed on the devices with the shift relative to one another to get the "effect of the ground failure".

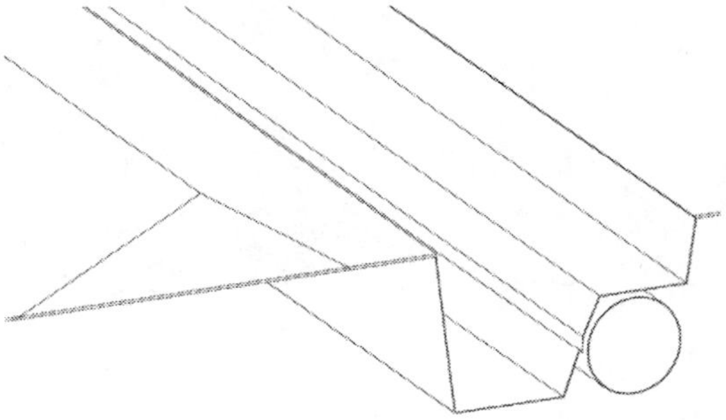

Fig. 6. General space view of the trench after the stripping.

Thus, the device for stripping the main pipeline compared to the known technical solutions enables to reduce the amount of the excavated soil due to the fact that the break is carried out from one lateral side and above the pipeline, coupled with the simultaneous cutting of machines for the pipeline break. Herewith, the time consumption to get the machine with the device into working trim is also cut, that is a new technical result of the proposed technical solution.

To prevent the damage of the main pipeline while breaking it as well as to exclude the chance of milling tools damage, it is proposed to place two fixed stops on the rear wheel assembly and inside the rotor with the buckets.

Both fixed stops 1 and 2 are brought into motion by horizontally mounted jacks. The fixed stops are made of metal based rubber. The fixed stops 1 and 2 are rigidly fixed on the working element of the rotary trench excavator.

The fixed stop 1 located on the rear support is suitable for altitude adjustment, to have a possibility to change its altitude depending upon the diameter of the stripped pipeline. The fixed stops are brought into motion with the horizontally fixed jacks 3 and the guide rods 4 which prevent the shift of fixed stops (Fig. 7-9). Besides, a sensor is mounted on each of the fixed stops that will send a signal to the operator's cabin in case of pressure boost in the hydraulic fluid in the jacks. Since when the fixed stop touches the surface, there appears a pressure loads which can sharply increase the pressure in the hydraulic cylinders. According to the sound signal for the operator he can correct the position of the working element and go on to break the pipeline.

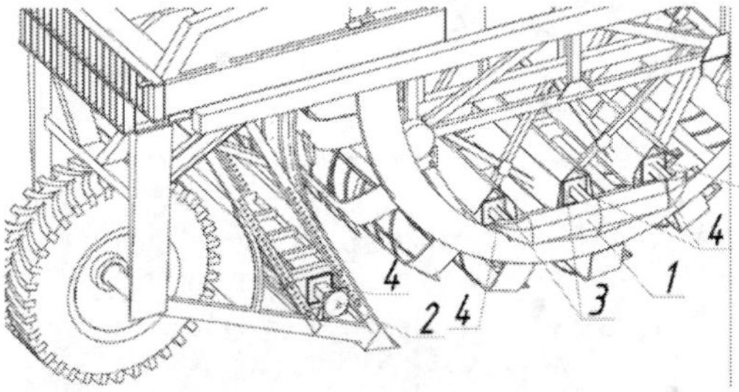

Fig. 7. General space view of the fixed stop system.
1, 2 – fixed stops, 3 - horizontally fixed jacks, 4 - guide rods

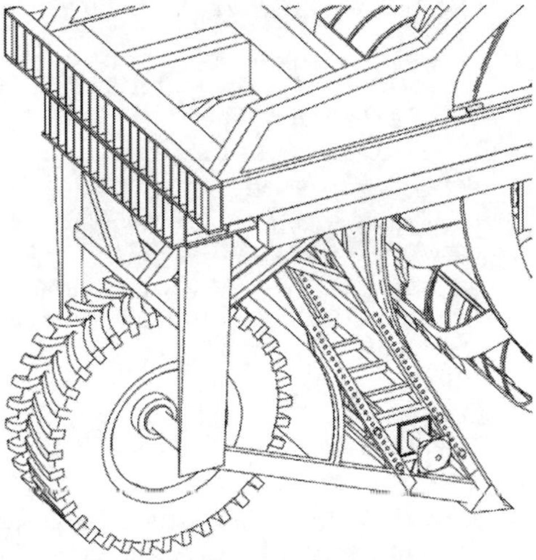

Fig. 8. 3D configuration of a fixed stop mounted on the rear wheel assembly of the rotary trench excavator.

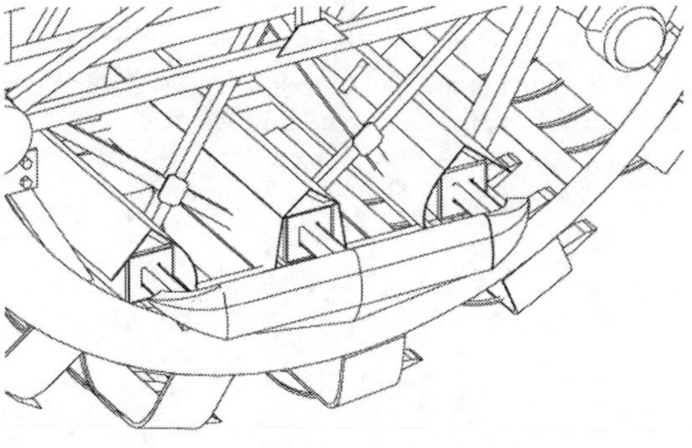

Fig. 9. 3D configuration of a fixed stop placed inside the rotor with buckets.

The proposed system of fixed stops along with the device for stripping the main pipeline help to carry out the break of the main pipeline carefully without damaging it.

DOI: 10.1051/matecconf/201712906028

References

1. A.S. Shatskiy, *On the issue of using rotary trench excavators in Russian pipeline construction,* Truboprobodniy transport, No. **2(36).** pp. 28–31 (2013)
2. A.S. Shatskiy, G.R. Gabelaya, *On the issue of optimization of earthworks building pipelines under the conditions of the Extreme North,* Troboprovodniy transport, No. **3 (31),** pp. 17–21 (2012)
3. S.I.Larin, M.Z. Alshits, *The rotary trench excavator ETR309,* Mashiny i oborudovaniye, No. **1**. P. 21–22 (2009)
4. G. Ferris, S. Newton, M. Porter, Proceedings of the 11th International Pipeline Conference, 2016, 2, UNSP V002T07A001, (2017)
5. F. Mosleh, T. Zayed, MS El-Abbasy, Journal of Performance of Constructed Facilities, 30 (Iss. 6), 04016037 (2016)
6. M. Elchalakani, Journal of Constructional Steel Research, **125**, 26 (2016)
7. R. Shahandeh, H. Showkati, Journal of Constructional Steel Research**, 121,** 237 (2016)
8. V. Ivanov, M. Zykov, V. Rybin, S. Plotnikov, XV International Conference Topical Problems of Architecture, Civil Engineering, Energy Efficiency and Ecology – 2016, 73, UNSP 01024 (2016)
9. D.F. Ihsanov, V.L. Buzhinskij, A. G. Gumerov, A.A. Bikbulatov, V.M. Murov, Z.K. Berdnikova

Patent RU, No. 887740, 2006.

10. Alexandr V. Bykov e.a., Machine for digging into the lower layers of the ground. Patent US, No. 6418646 B1, 2002.
11. Gareth J.T. e.a.,. Patent UK, no. 2435653 (2006)
12. Evgeny P. Kovalev e.a. Patent US, No. 4976051 (1990)
13. Pierluigi Zanin., e.a. Patent US, No. 5601383 (1997)
14. Z.K. Berdnikova, Z.H. Valeev, V.E. Gubin, F.H. Ismagilova, O.N. Kanavin, E.K. Kondrat'eva

Patent SSSR, No. 257355 (1990)

15. A.M. Muhametshin e.a., Patent RU, No. 2389930, (2010)
16. Danny R. MCCARTY e.a., Excavating systems and methods. Patent US, No. 2013/0025169 A1 (2013)

Improving the quality of transient response during automatic control of the turn of a tracked vehicle based on the implementation of structured input shapers

Igor Taratorkin[1,*], *Victor* Derzhanskii[1], and *Alexander* Taratorkin[1]

[1]IES UB RAS, Department of Transportation Vehicles Mechanics, 620049, Ekaterinburg, Russia

Abstract. The article presents the rationale for improving the quality of transients during steering control of a tracked vehicle in order to increase its high speed performance. Based on the comparative analysis of the results of simulation modeling and experimental research of the controlled movement dynamics, it is established that high speed performance is restricted by transient overshoot of a vehicle response to the control input, by transient oscillation and insufficient oscillation damping. The solution of the problem of high speed performance increase is carried out by implementing a modern IT approach – by creating an automated system for stabilizing the vehicle motion trajectory – which applies Input Shaper for adjusting the control input. The research has established that, considering the control object under research, the best results are obtained by Zero Vibration Shaper - (ZV-Shaper), which enables to minimize limiting overshoot, and the transient becomes close to an aperiodic one. The quality of the control process is defined by natural frequency stability of the system, to which the Input Shaper in question is tuned. In this regard, it is appropriate to introduce monitoring and identifying natural frequency into the control system. This creates conditions for increasing vehicle high speed performance, including on soils with bad traction properties..

One of the main operational properties of high-speed tracked vehicles is high speed performance represented by high-speed qualities. Increase in power-to-weight ratio in vehicles up to 25 kW, improvement of transmissions, of motion control systems and data information systems provide for increase in potential high-speed qualities of modern tracked vehicles. However, the manifestation of potential high-speed qualities is limited by a number of dynamic phenomena, which characterize controllability at rectilinear motion and during turning, especially at vehicle robotization.

Severe vehicle operation conditions, limited possibilities for organizing maintenance service, insufficient reliability of electronic devices and hydrostatic drives, as well as cost limits for the steering control system lead to various suggestions on creating simple steering mechanisms and control systems. These systems have their own merits in terms of weight and volume indicators, simplicity, costs, a degree of design and technological perfection of the structural components. Steering devices with discrete properties, with which many pre-developed vehicles as well as the modern advanced TM-140 transport vehicle, intended for severe service conditions of the Arctic region, are equipped, can be referred to such control systems.

Moving along the roads with the hard surface coating (asphalt, concrete, frozen soil), with bad traction properties and intensive change of curvature, the high speed characteristics of the vehicles equipped with the steering control system (SCS) with discrete properties are limited and do not exceed 36 … 38 km/h. It is explained by impossibility of variable (stepless) control of a trajectory curvature, angular acceleration, trajectory deviation compensation. At the same time, the intensity of the driver's control activity, i.e. the number of SCS actuations (steering control) per kilometer of driving increases by a factor of 6 … 9, reaching 96. In this regard, the driver's compensatory control is effective at a speed lower than 36 - 38 km/h, and with application of remote control for robotic centers based on these vehicles, the average speed does not exceed 5 km/h. It determines the urgency of increasing high speed characteristics of tracked vehicles.

The goal (objective) of the research is to improve the quality of transients at controlling the angular speed of tracked vehicle cornering. The degree of implementing potential high-speed characteristics of the tracked vehicle is defined by the dynamic properties in case of coming into a corner (by a response to steering wheel jerk). For tracked vehicles with the steering mechanism providing for the discrete control of the driving direction, the angular speed ω undergoes discontinuity of the second kind, and its derivative $\dot{\omega}$ is the delta function $\delta(t)$, i.e. in absolute magnitude, it can change within an unlimited interval. However, the experimental data show that in an elastic-inertial mechanical system Track Assembly – Transmission, which actuates friction elements slipping and interacts with soil by slipping, the value of the

* Corresponding author: ig_tar@mail.ru

derivative $\dot{\omega}$ changes continuously in terminal, though significant, limits. Even in case of a bounded deviation of angular speed up to 0.1 rad/sec with a frequency 2.5 Hz (due to drift and oscillatory processes in the powertrain), the angular accelerations reach 0.6 rad/sec^2, and in case of cyclic actuation of the mechanism at long corners – up to 1.3 rad/sec^2, and sideways accelerations – up to 10 … 12 m/s^2 [1]. In case of driving on hard surface soil, it is not possible to provide motion stabilization.

The specified features of the nonlinear control system with discrete properties are that at driving along the roads with an intensive change of curvature, the trajectory accuracy is reached by periodic steering brake actuation and its release, i.e. the control is pulsed. To provide a rationale for the ways of transient quality improvement, the track vehicle steering dynamics is simulated on the LMS Amesim platform [2]. The mathematical model of the plane-parallel controlled motion of the tracked vehicle as of a continuous solid body is described by the first two differential equations of the system (1). The third equation allows considering elastic-inertial properties of the dynamic parts of the system from the engine up to the driving sprockets of the vehicle.

$$\begin{gathered} m\ddot{y} + \frac{\dot{y}}{V}\sum K_{s_i} - \left[mV + \frac{1}{V}\sum K_{s_i} l_i\right]\dot{\varphi}_{tv} = 0 \\ J_z\ddot{\varphi}_{tv} + \frac{\dot{\varphi}_{tv}}{V}\sum K_{s_i} {l_i}^2 + \frac{\dot{y}}{V}\sum K_{s_i} l_i + C_{tor}[\varphi_{tv} - \varphi_{ice}] = 0 \\ J_{ice}\ddot{\varphi}_{ice} + C_{tor}[\varphi_{ice} - \varphi_{tv}] = M_{turn} \end{gathered} \quad (1)$$

where $y, \dot{y}, \ddot{y}, \varphi_{tv}, \dot{\varphi}_{tv}, \ddot{\varphi}_{tv} \varphi_{ice}, \dot{\varphi}_{ice}, \ddot{\varphi}_{ice}$ are the generalized coordinates and their corresponding derivatives (at the lateral motion of the vehicle, yaw motion of the vehicle body in reference to the vertical axis and roll motion of the ICE crankshaft reduced to the angular coordinate of the vehicle body); m – vehicle weight; J_z - yaw moment of inertia; J_{ice} - the roll moment of inertia of the engine reduced to the angular coordinate of the vehicle body; K_{s_i} – slipping resistance coefficient of the tracked vehicle; C_{tor} - torsional rigidity of the transmission (main shaft) reduced to the angular coordinate of the vehicle body.

The value of the reduced rigidity C_{tor} can be considered as composition of the following components: rigidity of coupling shafts; driving track runs; suspension unit components of the lower rollers, as well as the stretching (tensioned) of the sagging upper tracks on the upper rollers [3]. This work considers the reduced rigidity as linearized. At track vehicle cornering, the values of the reduced rigidity and reduced moment of inertia of the engine depend on transmission ratio (the number of the engaged gear). The transients have been determined at a driving mode with a constant speed for a vehicle with the following parameters:
vehicle weight $m = 11556$ kg; yaw moment of inertia $J_z = 48200$ kg•m^2; slipping resistance coefficient $K_{s_i} =$ 60 kN/rad; the sum of longitudinal coordinates of the lower roller axes in reference to the center of mass $\sum l_i = 1.412$ м and the sum of their squares $\sum (l_i)^2 = 7.06\, m^2$; the reduced transmission rigidity $C_{tor} = 84000$ Nm/rad. The reduced moment of inertia of the engine is determined considering the transmission ratio between the generalized coordinates of the engine (the third equation) and of the vehicle body in a rotary mode (the second equation), as well as the kinematic peculiarities of the vehicle steering mechanism. For the vehicle in question the reduced moment of inertia J_{ice} varies from 6148 kg•m^2 up to 359589 kg•m^2. The value of the yaw moment was set in the form of the step function imitating the vehicle going into a corner, at the same time, the maximum value was taken to be equal 0.9 from the greatest possible under the motion conditions, which is determined by cornering resistance coefficient.

The simulation results show that the parameters of the transient function significantly depend on a driving speed. The transient is accompanied by oscillations starting with the third gear (5.6 m/sec). The intensity of damping is determined by the soil damping properties. The response overshoot at the top gear reaches 50%, which demands speed restriction according to safety conditions.

The obtained results suggest that the discrete properties of the steering control system restrict high speed characteristics of vehicles because of oscillation transients and significant response overshoot in case of going into a corner. In engineering, such phenomena are described as a Bonanza effect [4, 5], which is observed in the course of controlling transport vehicle mechanical systems. In general, generated oscillations restrict high speed characteristics of vehicles, they are perceived by the driver as disturbing ones, increase structural components load and decrease the comfort level. The suggested measures are necessary to reduce oscillation amplitude by means of frequency detuning, i.e. the maximum amplitudes will be shifted at frequencies which are not critical for the vehicle. The kinetic energy of an oscillatory process depends on moment M squared and is inversely proportional to the doubled value of the reduced rigidity C.

Rigidity increase leads to decrease in oscillation kinetic energy because of the torque impulse created by the steering control system. It results in higher efficiency of damping with all other things being equal, but is, as a rule, infeasible for design, configuration or technological reasons. At invariable design parameters of the vehicle, as an alternative, the impulse which causes oscillations at torque application can be changed to prevent or reduce generation of natural oscillations. The procedure can be explained by means of a practical example described by R Fischer [6].

It appears that the most effective way of restricting oscillations is introduction into the steering and brake control unit of Input Shapers. The main idea of these shapers consists in formation of the controlling signal by convolution of the controlling input with the pulse sequence in the form of the Dirac delta function. Shaper operating algorithm is based on redistribution in time of power impact on a controlled object preserving an invariable value of the total impact. Due to relative increase in time of a system transient from one state into another, compensation of the arising oscillations is provided, and the shift of the control (response, impact) for a fraction of the period of the calculated oscillations of the controlled object provides ideally their full suppression. At the same time, the reference-input signal

DOI: 10.1051/matecconf/201712906029

is accepted such as the resultant effect on the system remained invariable:

$$\sum_{i=1}^{n} A_i = 1 \qquad (2)$$

where n is the number of pulses; A_i is the amplitude of the ith pulse.

The key parameter defining the quality of shaping-algorithm performance is the level of oscillation suppression. It is worthwhile making an estimation of suppression in the form of the ratio of the residual oscillation amplitude in the system in which the input signal was created by the shaping-filter to the amplitude of residual oscillations which would arise in the same system without conversion of an input signal. The degree of oscillation suppression is defined by the following expressions [7]:

$$\begin{cases} V(\omega,\varepsilon) = e^{-\varepsilon\omega t_n}\sqrt{C^2+D^2} \\ C(\omega,\varepsilon) = \sum_{i=1}^{n} A_i\, e^{-\varepsilon\omega t_i} cos\left(\omega t_i\sqrt{1-\varepsilon^2}\right) \\ C(\omega,\varepsilon) = \sum_{i=1}^{n} A_i\, e^{-\varepsilon\omega t_i} sin\left(\omega t_i\sqrt{1-\varepsilon^2}\right) \end{cases} \qquad (3)$$

where ω is an oscillation frequency; ε is a damping coefficient; $t_i$is time of the ith pulse.

Shaping-filter parameters of the A_i and t_i are received by solving the equations (2) and (3), being defined by the maximum allowable level of oscillations $V(\omega,\varepsilon)max$.. At the same time, the values ω and ε are usually selected equal to the natural frequency and to the controlled object damping coefficient, which can be determined by a mathematical model or by the results of the experimental findings. It should be considered that real-life controlled objects (in particular the steering control system of the tracked vehicle) are non-linear that leads to changes of the specified parameters. Thus, there appears a task of shaping such an input signal which will provide robustness in the system in relation to uncertainty of generated oscillation frequency (natural frequency) and to the controlled object damping coefficient. One of the solutions of this task is introduction of an additional condition:

$$\frac{\partial}{\partial\omega}\left(e^{-\varepsilon\omega t_n}\sqrt{C^2+D^2}\right) = 0 \qquad (4)$$

The partial derivative of the expression (4) considering frequency being equal to zero allows for increasing a range of frequencies, in which the desirable degree of oscillation suppression is provided, and by that achieving acceptable Input Shaping performance even in case of a significant error in determination of controlled object natural frequency[7].

Another important property characterizing the quality of Input Shapers performance is high-speed characteristics which is defined by the time of the latest pulse *tn*. For this reason, when calculating the algorithm, a condition of *tn* minimization is usually entered. In such a case, it is necessary to introduce restrictions for pulse minimization amplitude as far as time minimization can result in the necessity of shaping pulses with an infinite amplitude. Execution of the given conditions is defined by the dynamics of the object under research and by the choice of a Shaper-algorithm type.

Shaping-algorithms have found a wide circulation in different industrial applications [8] that allowed for increasing the performance of different devices considerably: of cranes [8,9,10], coordinate measuring machines [11,12], satellite systems [13], milling machines [14], disk drives [15] and others. The above-mentioned references describe the types of Shapers, dependences on their parameter determination, efficiency assessment for different industrial applications and recommendations about their use.

The key parameter according to which an algorithm type is selected is a natural frequency of the system. For the object under study, the frequency value is defined by the number of engaged gear. In this regard, a possibility of applying one algorithm of tracked vehicle steering control for all the gears is assessed. Fig. 1. shows the assessment of oscillation suppression efficiency in case of application of different Shaping-algorithms. The obtained data show that in the range of natural frequencies 1,4 … 3,92 rad/s (in case of tuning the Input Shaper for the frequency corresponding to the 5th gear – direct-driven gear) a satisfactory result is achieved in case of control system tuning according to 3-EI –Shaper-algorithm. In this case, the level of residual oscillations in the range of frequencies under study does not exceed 5% from the amplitude of the residual oscillations which arise in the same system without input signal conversion.

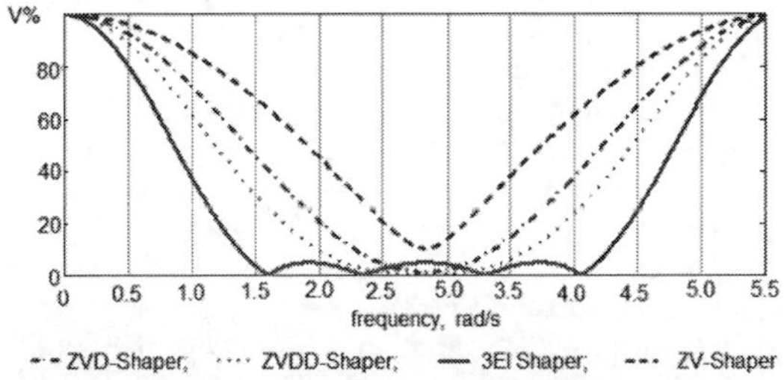

Fig. 1. Assessment of oscillation suppression efficiency by different Input Shaping (residual oscillation)

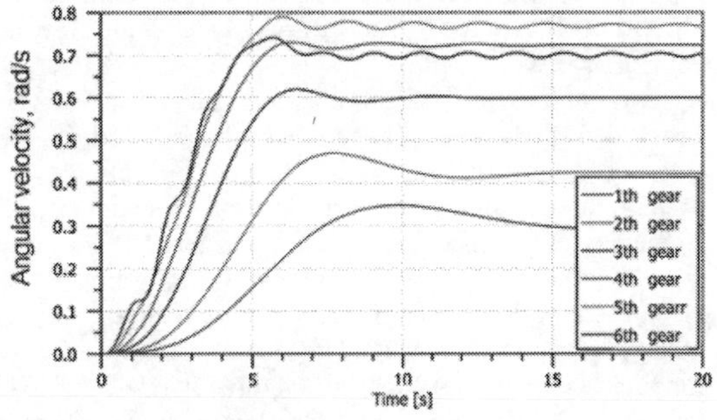

Fig. 2. Assessment of high-speed performance of the system with 3-EI – Input Shaping

However, in this case the required high-speed performance of the vehicle steering control system is not provided as far as the transient duration time increases fourfold (see Fig. 2.) and reaches 5 seconds. In this respect, the use of the simplest ZV, ZVD – Shapers provides for satisfactory high-speed performance (the

DOI: 10.1051/matecconf/201712906029

obtained level of the angular speed is reached for 1.5 … 2.0 sec in comparison with 1.1 sec for the option without Shaping application) in case of tuning up for one frequency. Furthermore, the required robustness level is not provided (see Fig. 3.). Fig. 3. shows that in case of the effective oscillation damping at the fifth gear (ZV-Shaper is tuned up for the frequency of 0.45 Hz and corresponds to natural frequency at the fifth, "direct-driven" gear), the degree of oscillation suppression at other gears is not satisfactory. The similar results are provided by the Shapers working according to a negative algorithm – UM-ZV Shapers, etc.

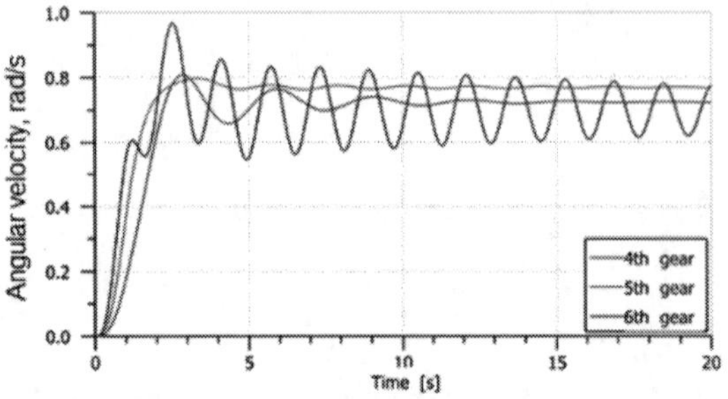

Fig. 3. Assessment of ZV-Shaper Quality on high-speed performance and robustness

In this case, the application of negative algorithms is restricted by the speed change of the controlling (force) input. These algorithms with the specified negative amplitude (SNA) in case of the value of the of the amplitude ratio of two adjacent pulses in the range from 0.5 to 2 allows for reaching a compromise between the requirement of increasing algorithm high-speed performance and reducing the power of the steering control drive. In relation to the research process of tracked vehicle steering, the restriction is caused by impossibility of quick creation of a braking effort at the speeding up vehicle side, owing to the necessity of changing the driving and loose tracks in the track chain. Consideration of this circumstance does not allow for using these algorithms in relation to the research object in the steering control process.

On the basis of the study undertaken, it may be concluded that input Shapers can be applied in the steering control system of the tracked vehicle. In the system under study, the change of system natural frequency at the specified gears in the gearbox can have significant effect on the parameters of a transient. It can be caused both by nonlinearity of the system and by change of the vehicle weight in case of different rates of loading capacity, fuel load, tyre temperature of the road wheels leading to a considerable change of the slipping resistance coefficient, etc.

Comparing different input Shapers, it can be concluded that the best effect considering high-speed performance is reached either by application of ZV and ZVD algorithms, or by algorithms with pulses of different signs UM - ZV, however, the property of robustness decreases. As the key property of the steering control system of the tracked vehicle is high-speed performance, so ZV - Shaper application with introduction into the control system of monitoring and natural frequency identification is preferable. Application of this Input Shaping allows for minimizing the ceiling overshoot up to 5%, and the transient becomes close to aperiodic, including driving with high speed on soils with bad traction properties.

Restrictions

The research is executed on the basis of analyzing the linear model of the dynamic system. The reduced rigidity C_{tor} in the real-life system is significantly nonlinear and asymmetric, which can excite parametrical oscillations [16]. Besides, there is a good reason to consider the longitudinal displacement of the vehicle turning center and nonlinearity of lateral slipping resistance coefficient K_{s_i}. Taking into account the abovementioned, repeatability of the results of the experimental study and simulation, can be increased.

Conclusion

As a result of simulation and experimental study of the dynamics of the controlled movement of the tracked vehicle with the discrete properties of the steering control system, it is established that the extent of realization of potential high-speed qualities is limited to the quality of response to a steering wheel jerk (to a single control input).

To achieve a required transient control quality at tracked vehicle cornering, there is a good reason to apply input Shapers in the control system providing optimum ratio among contradictory properties of high speed performance, an extent of oscillation suppression and robustness.

At driving on the roads with bad traction properties, high-speed qualities can be increased by introduction into the turning brake control unit of a ZV Shaper providing prevention of oscillation generation at response overshoot which does not exceed 5%.

References

1. A. Taratorkin, V. Derzhanskii, I. Taratorkin, Pr. Eng., **136,** 1368-1377 (2016) – DOI 10.1016/j.proeng.2016.07.331
2. LMS Imagine.Lab Model-oriented approach to designing mechatronic systems. URL: https://www.plm.automation.siemens.com/ru_ru/products/lms/imagine-lab/amesim
3. V. Derzhanskii, I. Taratorkin, SAE Technical Papers, **9,** (2013)
4. G. Jurgens, R. Fischer, **697,** VDI-Berichte, Dusseldorf, 233, (1988)
5. H-J Schspf, G Jurgens, R Fischer, **91,** ATZ, 568–575, (1989)
6. R. Fischer, F. Küçükay, G. Jürgens, R. Najork, B. Pollak, DOI 10.1007/978-3-319-05263-2, (2015)
7. N.C. Singer, W.P. Seering, **112,** 76–82, (1990)

8. K.L. Sorensen, A Thesis Presented to The Academic Faculty, (2008)
9. K.L. Sorensen, W.E, Singhose, S. Dickerson, **15,** № 7, 825–837, (2007)
10. A.P. Kuznetsov, A.V. Markov, A.S. Shmarlouski, T.V. Gavrilik, Doklady BGUIR, **6(60),** 5–12, (2011)
11. N. Seth, K. Rattan, R. Brandstetter // IEEE Conf. On Control Apps., 368–373, (1993)
12. S. Jones, A.G. Ulsoy // J. of Dynamic Systems, Measurement, and Control, **121,** № June. 242–247, (1999)
13. W.E. Singhose, L.J. Porter, T.D. Tuttle et al. // J. of Dynamic Systems, Measurement, and Control, **119,** № June, 320–326, (1997)
14. J. Fortgang, W. Singhose, J. Marquez et al., // Proceedings of theAmerican Control Conf., 4531–4536, (2005)
15. W. Singhose, W.P. Seering, N.C. Singer, J. of Dynamic Systems, Measurement and Control, **119,** № June, 198–205, **(**1997)
16. A. Taratorkin, V. Derzhanskii, I. Taratorkin, LNEE, 10, 495-504, (2013)

MATEC Web of Conferences **129**, 06030 (2017) DOI: 10.1051/matecconf/201712906030
ICMTMTE 2017

Problems of wave dynamics of the systems that support moving loads

Sergey I. Gerasimov[1], *Vladimir I.* Erofeev[1,*], and *Alexey O.* Malkhanov[1]

[1]Mechanical engineering research institute RAS, 603024 Nizhny Novgorod, Russia

Abstract. The work is devoted to the study of the stability of mechanical systems with moving loads. We consider lumped objects that move uniformly along the distributed elastic systems. For example, sliding train supports vibrating, excited in the rails elastic waves, which can lead to instability.

1 Flexural and torsional vibrations

To speedup the payload on the rocket track, rocket trains consisting of several stages are used. In some cases, the movement of high-speed stages of rocket trains is associated with the excitement of oscillations in their transverse motion, resulting in wear of the working surfaces and destruction of the clogs of the stage, damage to the rail tracks of the rocket track [1, 2]. Similar problems are also present in testing on US rocket tracks [3].

When carrying out the tests on a rocket track with speeds greater than 1300 m/s, there are effects appear that lead to serious destruction of test objects and rail tracks. After some accidental experiments on rail tracks, residual wave-like deformations in the horizontal plane with an amplitude of up to 2.5 mm were found. That is, in the rail guide under the action of lateral loads, flexural-torsional vibrations can occur.

Based on the equations of motion of small flexural-torsional oscillations

$$
\begin{aligned}
EI_y z_g^{IV} + \rho F z_g + y_s \rho F \varphi_g + H_z z_g = 0, \\
EI_\varphi \varphi_g^{IV} + GI_k \varphi_g'' + \bar{I}_p \ddot{\varphi}_g + y_s \rho F \ddot{z}_g \\
+ H_\varphi \varphi_g = 0
\end{aligned} \qquad (1)
$$

it is possible to obtain a dispersion equation which connects the frequencies of flexural-torsional oscillations with their wave numbers and satisfies the condition phases of the excited waves equality in the guide in the coordinate system associated with the object that moves with velocity V_x. This condition has the form [4]:

$$\omega_z - K_z V_x = \Omega_z, \qquad (2)$$

and, as it has been mentioned above, in the theory of waves is known as a kinematic invariant.

Here $z(x,t)$ is the horizontal deviation of the track rail, $\varphi(x,t)$ is the angle of rotation of the cross section of the track rail around the x axis, E and G are the elasticity and shear moduli of the track rail material ρ, F is the material density and the cross-sectional area of the trackrail, I_y is the moment of inertia of the cross section of the track rail around the y axis, I_φ is the sectorial moment of inertia of the cross section of the track rail, I_k is the torsional moment of inertia of the cross section of the track rail, $\bar{I}_p$ is the polar moment of inertia of the cross of the section track rail, H_z is the relative stiffness of the base in the direction of the axis z, H_φ is the relative stiffness of the base in the direction of angle φ, and y_s is the distance between the centre of gravity and flexural centre of the cross section of the track rail.

The solution to this equation can be represented on the dispersion plane. Parabolas are a graphical portrait of the guide, and an inclined straight line is an object moving along it (Fig. 1).

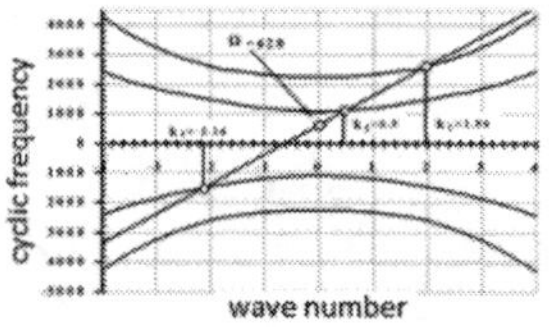

Fig.1. Dispersion plane

To compare the oscillation parameters of the rocket guide (RG), which took place in the experiment and fixed in the form of residual deformations with the calculated, it is necessary to know the frequency of the impact on the RG in the experiment. To determine the frequency of exposure can be as follows.

On the dispersion plane in the coordinates "circular frequency - wave number", where the rail guide is represented by four symmetric curves relative to the coordinate axes, a straight line representing the kinematic invariant is drawn. For the velocity parameter it corresponds to the moving stage of the rocket train (RT), and matches the waveform with the obtained experimental data. The reference values for the definition of the kinematic invariant line will be assumed to be $K_1 = 1.96\frac{1}{m}$ (wave number corresponding to the distance

*Corresponding author: erof.vi@yandex.ru

DOI: 10.1051/matecconf/201712906030

between the supporting clogs of the moving object) and the slope of the straight line corresponding to a speed of $V_x = 1000\frac{m}{s}$ (the speed of the object at the time of the accident) (Fig. 2). The intersection of the kinematic invariant line with the ordinate axis gives the value of the frequency of impact on the RG, which took place in the experiment $\Omega = 620\frac{1}{s}$ (Fig. 1).

In the (Fig. 2) there is shown the dependence of the deviation of the head of the RG from the longitudinal coordinate in the coordinate system associated with the stage (the origin is aligned with the point of contact of the front clog with the RG).

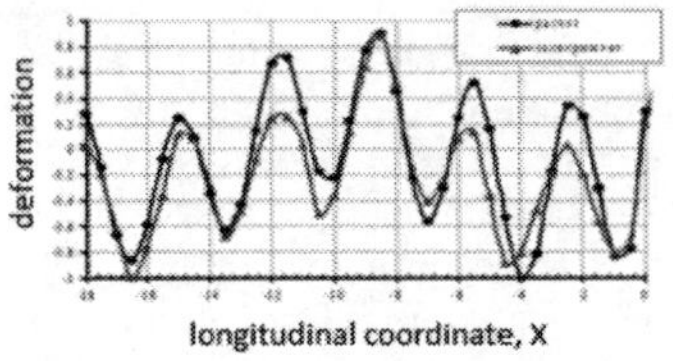

Fig. 2. Comparison of calculated and experimental data for deformation of the RG

The dependence of the measured residual deformations on the longitudinal coordinate is imposed on the calculated dependence (Fig. 2). The imposition is performed in such a way that the main part of the curve coincides. Both relationships are normalized with respect to their maximum values. The calculated and experimental dependences agree satisfactorily in amplitude, wavelength and phase. This is a confirmation of the development of flexural-torsional oscillations in the rail guide and the correctness of the accepted mathematical model.

The occurrence of these oscillations is due to perturbations caused by the object moving along the elastic guide. The dynamic impact of a moving object on a guide causes it to oscillate in the form of traveling waves.

As noted above, the oscillations under consideration can be unstable due to the anomalous Doppler effect, which occurs when the object moves with a velocity exceeding the phase velocity of the wave it excites. In this case, the radiated wave increases the energy of the transverse vibrations of the object, while decreasing the energy that ensures its motion along the guide [4]. A significant influence on the evolution of transverse vibrations of high-speed objects is exerted by dynamic stiffness in the moving contact of the rail tracks of the rocket track. As the simplest model of a moving object, let us consider once again a two-mass oscillator uniformly moving along an elastic guide located on a viscous-elastic base. The small mass m is in continuous contact with the head of the rail guide. A larger mass M is connected with a small mass m by an elastic coupling h_z and a viscous dampener ε_z (Fig. 3)

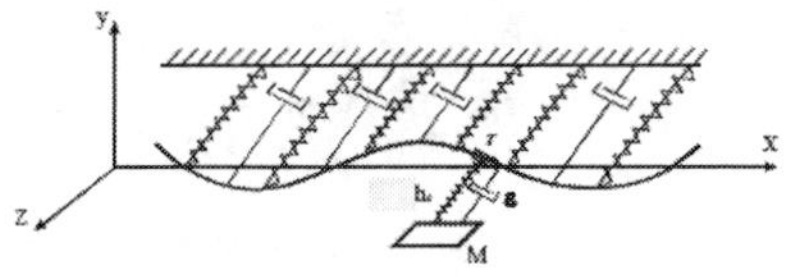

Fig. 3. Uniform continuous motion of a two-mass oscillator along a guide lying on a viscous-elastic base

The oscillations of the guide in the horizontal plane have a flexural-torsional character and are described by the Bernoulli-Euler equations:

$$\rho F\ddot{z} + EI_y z^{IV} + H_z z + v_z\dot{z} + y_s\rho F\ddot{\varphi} = -\delta(X - Vt)\left[m\frac{d^2z^0}{dt^2} + h_z(z^{01} - z^{02})\right],$$

$$\rho I_\varphi\ddot{\varphi} + EI_\varphi\varphi^{IV} - GI_k\varphi'' + H_\varphi\varphi + v_\varphi\dot{\varphi} + y_s\rho F\ddot{z} = -\delta(X - Vt)y_v\left[m\frac{d^2z^0}{dt^2} + h_z(z^{01} - z^{02}) + \varepsilon_z\frac{d}{dt}(z^{02} - z^{01})\right], \quad (3)$$

$$M\frac{d^2z^{02}}{dt^2} + h_z(z^{02} - z^{01}) + \varepsilon_z\frac{d}{dt}(z^{02} - z^{01}) = 0,$$

$$z^{01}(t) = z(x - Vt, t) + y_v\varphi(x - Vt, t), \lim_{x-Vt\to\infty} z(x,t) = 0, \lim_{x-Vt\to\infty}\varphi(x,t) = 0.$$

The mathematical formulation of the problem of the interaction of a moving object with an elastic guide contains partial differential equations describing the flexural-torsional oscillations of the guide and the equation of oscillations of the moving object. These equations are dynamically linked at the moving point of contact. To describe the dynamic conditions in the contact, the generalized Dirac delta function is introduced into the right-hand side of the equations of the elastic guide, describing the position of the moving object and, therefore, depending on time as well as on the spatial variable. Using the above method for studying such equations, which consists of introducing a moving system of coordinates in which the delta function does not depend on time, the application of integral Fourier transforms with respect to spatial variables and Laplace in time, allowing one to move from a system of partial differential equations to a system of algebraic equations, As a result, we can obtain the characteristic equation for the oscillations of an object moving along an elastic guide.

The characteristic equation of the considered oscillatory system, obtained with the help of the proposed technique, has the form:

DOI: 10.1051/matecconf/201712906030

$$\left(-m\Omega^2 + h_z + i\varepsilon_z\Omega + \chi_z(\Omega,V)\right)\left(-M\Omega^2 + h_z + i\varepsilon_z\Omega\right) - (h_z + i\varepsilon_z\Omega)^2 = 0. \quad (4)$$

The integrand in the expression for the horizontal stiffness of the guide is a fraction, in the denominator of which there is a polynomial of the eighth degree. According to the theory of residues, the result of computing such an integral can be written as:

$$\chi_z(\Omega,V) = D_0\left(i\sum_{n=1}^{8}\frac{A_z(\Omega,k_n) - y_v C_z(\Omega,k_n)}{\partial/\partial k\,\left(A_z(\Omega,k_n)B_z(\Omega,k_n) - C_z^2(\Omega,k_n)\right)\big|_{k=k_n}}\right). \quad (5)$$

wherek_n are roots of the equation$A_z(\Omega,k_n)B_z(\Omega,k_n) - C_z^2(\Omega,k_n) = 0$, which lie in the upper half-plane of the complex variable k.

The dynamic stiffness of the guide lying on the elastic-viscous base is equivalent to the reaction of a lumped element whose dynamic rigidity is a complex-valued function of the disturbance frequency and the speed of the motion of the object. The real part of the dynamic stiffness reflects the elastic-inertial properties of the reaction of the guide, and the imaginary - the viscous. The positive imaginary part of the dynamic stiffness reflects the damping properties of the reaction of the guide in the moving contact. The presence of a negative viscosity in a moving contact is a necessary condition for the appearance of instability of oscillations of an object moving along the guide.

For the guide of the rocket track the calculations for the dynamic stiffness in the moving contact based on the expression (3) has been carried out [1, 2]. The results of the calculation are shown in (Fig. 4) and (Fig. 5).

The parameters characterizing the motion of a high-speed object varied within the limits: the speed $V - 0 \ldots 3000\frac{m}{s}$; the cyclic oscillation frequency $\Omega = 0 \ldots 6000\frac{1}{s}$.

The results of the calculation of the real and imaginary components of the dynamic stiffness of the guide in a moving contact are presented in the form of dependences $Re\chi_z(\Omega,V)$, $Im\chi_z(\Omega,V)$ in the (Fig. 4) and (Fig. 5).

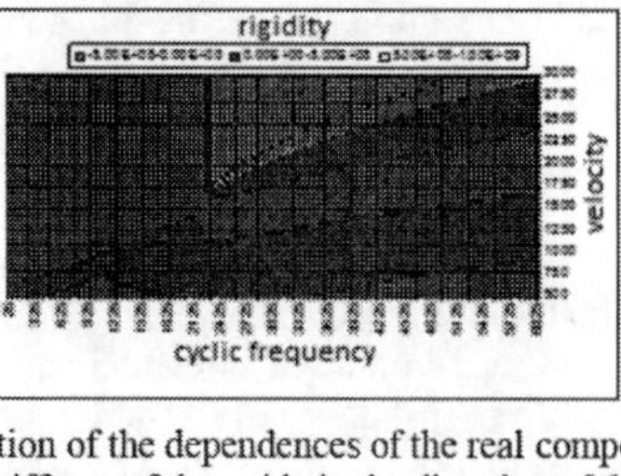

Fig. 4. Projection of the dependences of the real component of the dynamic stiffness of the guide in the direction of the *Z* axis$Re\chi_z$ on the parameter plane — "frequency - speed"

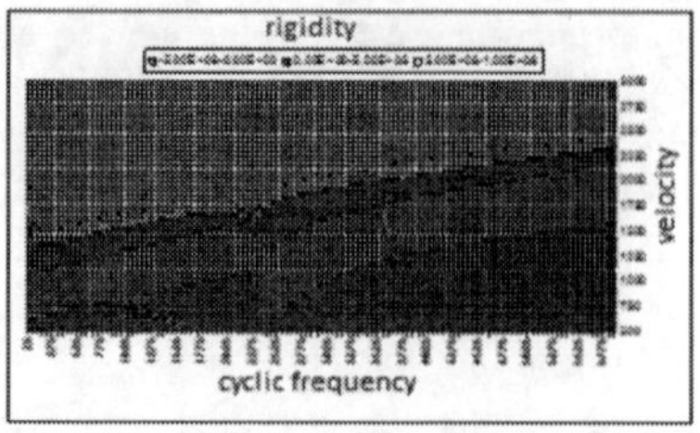

Fig. 5. Projection of the dependences of the imaginary component of the dynamic stiffness of the guide in the direction of the *Z* axis $Im\chi_z$ on the parameter plane — "frequency - speed"

In area A, the imaginary component of the dynamic stiffness of the guide in the direction of the Z axis $Im\chi_z$ is negative, that is, the necessary condition for the appearance of instability is satisfied. To determine its sufficiency, it is necessary to analyze the roots of the characteristic equation (4). The horizontal oscillations of a two-mass oscillator will be unstable if at least one of the roots has a positive real part. However, equation (4) is integral with respect to the variable Ω and it is not easy to find its roots. Therefore, to investigate the roots of such equations, the D-decomposition method is used, the idea of which, as mentioned above, is to map the imaginary axis of the complex plane $i\Omega$ to the plane of the system parameter, which will be temporarily considered as a complex one. The mapping law can be determined from the characteristic equation by expressing the selected parameter explicitly.

The most significant influence on the stability of the transverse motion of a two-mass oscillator is the rigidity of the elastic coupling between the masses of the oscillator h_z, which can be explicitly expressed from equation (4)

$$h_z = \frac{M\Omega^2(m\Omega^2 - \chi_z)}{(M+m)(\Omega^2 - \chi_z)} - i\varepsilon_z\Omega, \quad (6)$$

Using equation (6) as the mapping law, it is possible to vary Ω from $-\infty$ to $+\infty$ and construct the dependences. The line obtained by means of this display divides the parameter space into regions with a different number of roots of the characteristic equation (5) having a positive real part.

For two types of oscillators simulating the motion along the guide of the rocket track of monorail high-speed objects, calculations have been carried out in the ranges of velocities of $500 \ldots 2000\frac{m}{s}$ and circular frequencies of $6000\frac{1}{s}$. The results of the calculations are shown in (Fig. 6) and (Fig. 7).

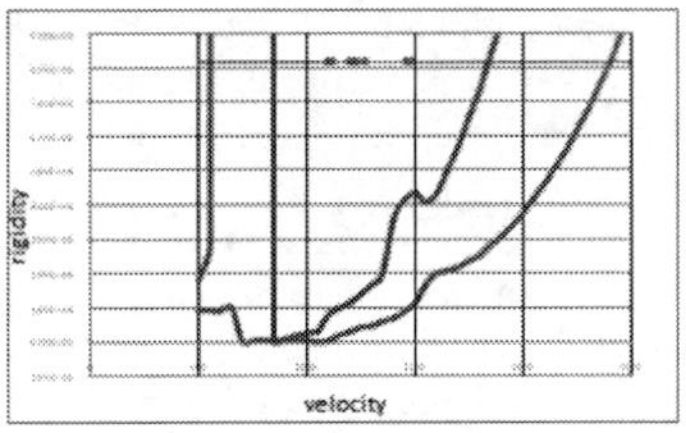

Fig. 6. The boundaries of the region of the transverse motion instability in the direction of the *Z* axis of the oscillator, which simulates a high-speed object (*M = 200 kg, m = 6 kg, ε = 300 N•s/m*)

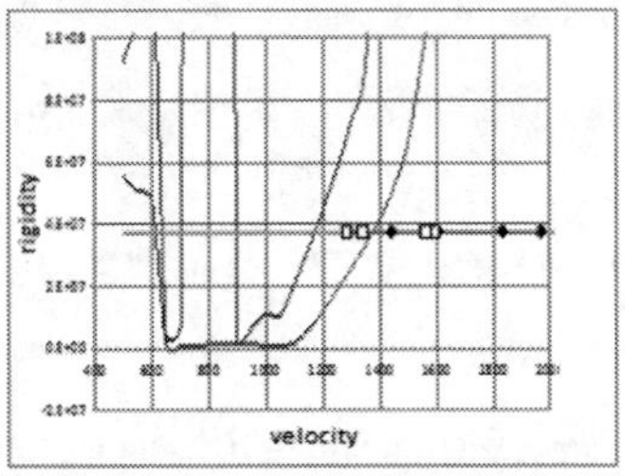

Fig. 7. The boundaries of the region of the transverse motion instability in the direction of the *Z* axis of the oscillator, which simulates a high-speed object (*M = 50 kg, m = 4 kg, ε = 100 N•s/m*)

In (Fig. 6) and (Fig. 7) are shown the boundaries of the regions of transverse motion instability in the *Z*-axis direction of the oscillators simulating the high-speed objects with which the rocket track tests have been carried out. If the icons displaying objects are above the upper boundary or below the bottom, then the movement of such objects is stable, otherwise it is unstable. In (Fig. 6), icons showing experiments with objects weighing $200\ kg$ at velocities of $1080 - 1480\ \frac{m}{s}$are located above the upper boundary of the instability region. All real experiments with these objects have been scored, there nave not been emergency experiments. In (Fig. 7), the icons displaying experiments with objects weighing $50\ kg$ at velocities of $1280 - 1340\ \frac{m}{s}$ are located above the lower and lower upper boundaries of the instability region, i.e., in the region of instability. These experiments have been emergency. The situation is more complicated with the region of stability below the lower boundary of the instability. At speeds of $1440\ \frac{m}{s}$and $1830 - 1970\ \frac{m}{s}$, the stability of their transverse movement is ensured. At speeds of $1560\ \frac{m}{s}$and $1600\ \frac{m}{s}$, experiments were completed abnormally, although at the same speed of $1600\ \frac{m}{s}$ there is also a test experiment. Perhaps the cause of these accidents is not instability, but, for example, resonance. In conclusion, it should be noted that the results of the studies on the stability of the motion of high-speed objects on elastic guides agree satisfactorily with the results of experiments, although certain regimes for the motion of objects require additional study.

2 Flexural vibrations in the vertical plane

One of the main questions in the study of the dynamic behavior of guides under the action of moving loads is the question of critical velocities. At a critical speed (V_*) is meant such a speed of movement of the load, when it exceeds a qualitatively changing picture of wave formation. Examples are the occurrence of waves during the movement of vessels along water, shock waves during supersonic motion in the atmosphere, etc. For high-speed ground roads during their design, it is necessary to correctly calculate the critical speeds, since they are the maximum permissible.

As the reasons why the movement with supercritical velocities is unacceptable, we can note the following [1]: large energy costs in the formation of waves that are comparable to the costs of movement; excited waves lead to the appearance of appreciable braking forces (up to 40% and more total resistance to movement); the energy of the waves, scattered in the guides, can lead to the destruction of the latter; waves, fed into the ground, can cause damage to the environment, etc.

Critical speeds on the ground (in the territory of European countries, where high-speed trains are used) range from 150to 400 $\frac{km}{h}$. In order to exceed them, it is necessary to solve the problem of "decoupling" of railroad tracks with soil, the successful solution of which is impossible without a well-developed theory of the dynamic behavior of structures carrying mobile loads.

Let us consider a single-mass mass oscillator m and rigidity k, uniformly moving along an infinitely long Bernoulli-Euler beam (Fig. 8). It is assumed that the mass and the beam oscillate continuously, and the upper end of the oscillator spring moves strictly horizontally.

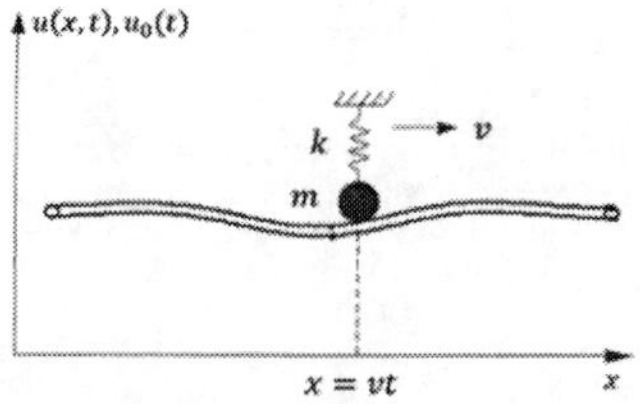

Fig. 8. Uniform motion of a single-mass oscillator along a Bernoulli-Euler beam

The equations of motion and the boundary conditions for this model have the form [5]:

$$\frac{\partial^2 u}{\partial t^2} + \alpha^2 \frac{\partial^4 u}{\partial x^4} = 0, -\infty < x < \infty,$$

$$u_0(t) = u(l(t), t),$$

$$[u]_{x=l(t)} = [u_x]_{x=l(t)} = [u_{xx}]_{x=l(t)} = 0, \quad (7)$$

$$EI[u_{xxx}]_{x=l(t)} = -\left(m\frac{d^2 u_0}{dt^2} + ku_0\right),$$

$$\lim_{|x-l(t)|\to\infty} |u(x,t)| < \infty;$$

where $l(t) = vt$, v – the velocity of the motion of the oscillator, $u_0(t)$ and $u(x,t)$ – the vertical displacement of the mass m and the transverse displacement of the beam; E – Young's modulus; ρF – linear density of the beam material; F – cross-sectional area of a beam; I – moment of inertia of the beam section, $\alpha^2 = \frac{EI}{\rho F}$; square brackets mean the difference between the values of the function standing in them, to the right and to the left of the specified value of the argument:

$$[f(x)]_{x=x_0} = \lim_{x\to x_0+0} f(x) - \lim_{x\to x_0-0} f(x).$$

We will seek the beam bias in the form of a superposition of waves, and the displacement of the mass of the oscillator in the form of oscillations:

$$u(x,t) = \sum_j c_j e^{i(\omega_j t - k_j x)}, \qquad (8)$$

$$u_0(t) = A e^{i\Omega t}, \qquad (9)$$

where the frequencies ω_i and Ω, and also the wave numbers k_i in general, are complex values.

Substituting (8) and (9) into first and second equations of (7), and taking into account that $l(t) = vt$, we obtain a system of two equations for determining the wave numbers and frequencies of the waves radiated by the oscillator:

$$\omega_j^2 = \alpha^2 k_j^4,$$
$$\omega_j = k_j v + \Omega.$$

If we solve this system we get the following:

$$k_1 = \frac{\left(v + \sqrt{v^2 + 4\alpha\Omega}\right)}{2\alpha},$$
$$k_2 = -\frac{\left(v + \sqrt{v^2 - 4\alpha\Omega}\right)}{2\alpha},$$
$$k_3 = \frac{\left(v - \sqrt{v^2 + 4\alpha\Omega}\right)}{2\alpha}, \qquad (10)$$
$$k_4 = \frac{\left(-v + \sqrt{v^2 - 4\alpha\Omega}\right)}{2\alpha}.$$

We will assume that in the case when the oscillator frequency is complex, the branch of radicals in (10) is fixed by the inequalities:

$$\begin{cases} Im\left(\sqrt{v^2 + 4\alpha\Omega}\right) < 0, \\ Im\left(\sqrt{v^2 - 4\alpha\Omega}\right) > 0, \end{cases} \quad \text{if} Im\Omega < 0$$

Such a choice lead to:

$$\begin{cases} Im(k_1, k_2) < 0, \\ Im(k_3, k_4) > 0, \end{cases} \quad \text{if } Im\Omega < 0$$

In connection with the requirement that the beam is bounded at an infinite distance from the oscillator, in accordance with (8), we obtain that as $x \to +\infty$, the imaginary parts of the wave numbers must be nonpositive, and as $x \to -\infty$, nonnegative. Therefore, it follows from (11) that waves with wave numbers k_1 and k_2 propagate ahead of the oscillator *($x > vt$)*, and with k_3 and k_4 – behind *($x < vt$)*.

In the case of a real oscillation frequency ($Im\Omega = 0$), either two or all four wave numbers in (10) are real. In this case, it is necessary to use the radiation condition, according to which the waves emitted by the moving oscillator must divert energy from it. Consequently, for waves propagating in front of the oscillator *($x > vt$)*, the group velocity of the wave must be greater than the speed of the oscillator *($v_i^{gr} = {}^{d\omega_i}/_{dk_i} > v$)*, and for waves propagating behind the oscillator *($x < vt$)* - less *($v_i^{gr} < v$)*.

Thus, the displacements of the beam ahead of the oscillator *(u^+)* and behind it *(u^-)* have the form:

$$u^+ = c_1 e^{i(\omega_1 t - k_1 x)} + c_2 e^{i(\omega_2 t - k_2 x)},$$
$$u^- = c_3 e^{i(\omega_3 t - k_3 x)} + c_4 e^{i(\omega_4 t - k_4 x)}. \qquad (11)$$

The characteristic equation, which determines the natural frequencies of oscillations of a moving oscillator interacting with a beam has the following form:

$$\omega_0^2 - \Omega^2 - i\frac{EI}{2\alpha^3 m}\frac{\sqrt{v^2+4\alpha\Omega}\sqrt{v^2-4\alpha\Omega}\left(v^2+\sqrt{v^2+4\alpha\Omega}\sqrt{v^2-4\alpha\Omega}\right)}{\sqrt{v^2+4\alpha\Omega}-\sqrt{v^2-4\alpha\Omega}} = 0, \qquad (12)$$

where $\omega_0^2 = {}^{k}/_{m}$.

Let us consider the case when the effect of a beam on the oscillations of an oscillator is not small. In this case, we will determine the region of instability of the system using the mentioned above method of D-decompositions. Following this method, we map the imaginary axis of the complex plane Ω to the plane of the complex variable k (only real k has physical meaning) according to the rule:

$$k = m\Omega^2 - i\frac{EI}{2\alpha^3 m}\frac{\sqrt{v^2+4\alpha\Omega}\sqrt{v^2-4\alpha\Omega}\left(v^2+\sqrt{v^2+4\alpha\Omega}\sqrt{v^2-4\alpha\Omega}\right)}{\sqrt{v^2+4\alpha\Omega}-\sqrt{v^2-4\alpha\Omega}} \qquad (13)$$

The resulting D-curve divides the k plane into regions with a different number of roots of equation (12) lying in the lower half-plane of the complex frequency Ω. The oscillations of the oscillator will be stable if equation (12) does not have such roots.

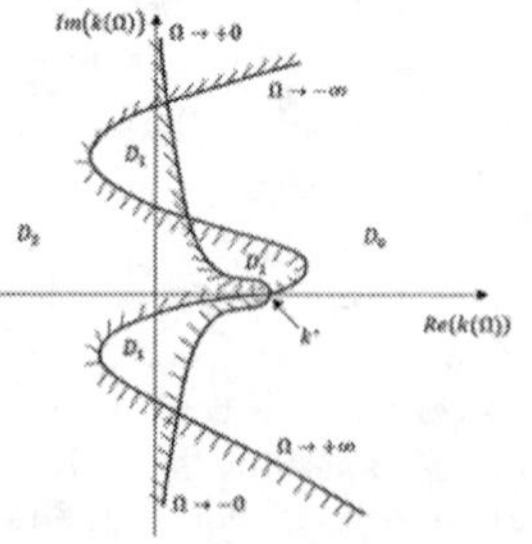

Fig. 9. The partition of the complex k plane into regions with different instability order D_i (i– is the order of instability $k^* = {}^{mv^4}/_{16\alpha^2}$).

DOI: 10.1051/matecconf/201712906030

In (Fig. 9) is shown the result of the mapping for a fixed speed v of the oscillator. The D-decomposition curve has one intersection point $k = k^*$ with the positive semi-axis of the axis $Re(k(\Omega))$. This value can be determined from the relation $Im(k(\Omega)) = 0$. Resolving it we get:

$$k^* = m\frac{v^4}{16\alpha^2}\ or\ \omega_0{}^* = \frac{v^2}{4\alpha}.$$

It is clear that we obtained the same value that was found under the assumption of the weak interaction between the beam and the oscillator. Thus, the transverse oscillations of an oscillator moving uniformly along the Bernoulli-Euler beam are unstable if the following condition is satisfied:

$$\omega_0 < \frac{v^2}{4\alpha} \Rightarrow v > 2\left(\frac{kEI}{m\rho F}\right)^{1/4}.$$

Thus, it has been shown that when the oscillator moves along the beam there is a limiting value of the velocity, the excess of which leads to instability.

The work was carried out with the financial support of the Russian Science Foundation grant # 14-19-01637.

References

1. S.V. Butova, S.I. Gerasimov, V.I. Erofeev et al.,J. Mach. Manuf. Reliab., P.1-6,**44**, 1 (2015)
2. S.I. Gerasimov and V.I. Erofeev, J. Mach. Manuf. Reliab., P.211-213, **45**, 3 (2016)
3. J.L. Lamb, Johns Hopkins APL Technical Digest, P. 448-458, **21**, 3(2000)
4. A.I. Vesnitsky, *Waves in Systems with Moving Boundaries and Loads* (Moscow, Fizmatlit, 2001)

MATEC Web of Conferences **129**, 06031 (2017) DOI: 10.1051/matecconf/201712906031
ICMTMTE 2017

Simulation of heat losses and temperature of blast furnaces tuyeres

Yuriy Tarasov[1,*], *Alexandr* Radyuk[1], and *Sergey* Gorbatyuk[1]

[1]EDP Sciences, National University of Science and Technology «MISiS», Moscow, Russian Federation

Abstract. The calculation technique of heat losses and temperature, adapted to blast furnaces tuyeres is described. Using a linear excel program the effect of gas-thermal coating and insulation lining on thermal state of the blast tuyere was studied. Problems of the blast tuyeres hardening and reducing heat losses through their surface are relevant, but finding ways of the blast tuyeres hardening and reducing heat losses through their surface by simulation of the thermal state of blast furnaces tuyeres are carried out. Key words: blast tuyere, blast furnace, tuyere nose, air passage, internal nozzle, gas-thermal coating, insulation lining.

1 Introduction

Blast tuyere is a key constructional element of blast furnace, determining its running efficiency. For this reason the furnace downtime leads to considerable reduction of iron smelting and increasing coke consumption. Besides, blast tuyeres account for 30% of all heat losses in furnance. So the problems of the blast tuyeres` hardening and reducing heat losses through their surface are relevant. While under normal condition of blast-furnace operation through a tuyere nose, external and internal nozzles of tuyeres flow 18, 36 and 46% respectively [1].

At the present time the given problems are solved by different ways: applying gas-flame sprayed coatings on an external surface of tuyere and lining-up by firebrick from the direction of air passage [2], applying heat-resisting backfill on an external surface of tuyere [3, 4] and others.

For further thinking on the ways of the blast tuyeres` hardening and reducing heat losses through their surface, simulation of thermal state of blast furnaces tuyeres is useful [5-7].

Let us use a calculation procedure of heat losses through blast tuyere [8]. In work the steady heat conduction formulae for flat and cylindrical wall were used [9, 10].

2 Research technique and results

Heat transfer through tuyeres surface is:

$$Q = Q_n + Q_r + Q_d = Q_n + Q_{rn} + Q_{rt} + Q_v + Q_{rv} = K_n^* \cdot (t_g - t_v) \cdot H_n + K_{rn}^*(t_g - t_v) \cdot (H_{rn} - h_{rt}) + K_{rt} \cdot (t_g - t_v) \cdot S_{rt} + K_v^* \cdot (t_d - t_v) \cdot H_v + K_{rv}^* \cdot (t_d - t_v) \cdot (H_{rv} - h_{rt})$$

where Q_n, Q_r, Q_d is the heat transfer through external nozzle, tuyere nose (from outside and cross-cut end) and from side of air passage respectively, W;

Q_{rn}, Q_{rt}, Q_v, Q_{rv} is the heat transfer through external surface of tuyere nose, butt end of tuyere nose, internal nozzle and tuyere nose from side of air passage respectively, W;

K_n^*, K_{rn}^*, K_v^*, K_{rv}^* is the testing heat-transfer rate through external nozzle, external surface of tuyere nose, internal nozzle of tuyere and tuyere nose from the side of air passage respectively, W/(m·K);

K_{rt} is the heat-transfer rate through butt end of tuyere nose, W/(m2·K);

t_g, t_d is tuyere temperature in iron receiver and hot blast temperature respectively, °C;

t_v is water temperature in tuyere, °C;

H_n, H_v, H_{rn}, H_{rv} is the length of external nozzle, internal nozzle, nose outside tuyere and tuyere nose from the side of air passage respectively (fig.1), m;

h_{rt} is the thickness of cross-cut end of tuyere nose (fig.1), m;

S_{rt} is the area of butt end of tuyere nose, m^2.

$$K_n^* = \frac{2\cdot\pi}{\frac{1}{\alpha_{gn}\cdot R_n^*} + \sum_{i=1}^{m}\frac{1}{\lambda_i}\cdot\ln\frac{R_{nj}}{R_{nj+1}} + \frac{1}{\lambda_{Cu}}\cdot\ln\frac{R_n}{R_n-h_n} + \frac{1}{\alpha_v\cdot(R_n-h_n)}}$$

where α_{gn} is coefficient of heat-exchange to external nozzle from gas forge, W/(m2·K);

$$R_l^* = R_l + \sum_{i=1}^{m} h_i\ ,\bar{\imath}$$

$$R_{l1} = R_l^*\ ,\bar{\imath}$$

$$R_{l_j} = R_l^* - \sum_{i=1}^{j-1} h_i\ ,\bar{\imath}$$

* Corresponding author: trsi@mail.ru

DOI: 10.1051/matecconf/201712906031

$$R_{l_{j+1}} = R_l^* - \sum_{i=1}^{j} h_i , i$$

$$j = 1, \dots m$$

R_n is the midradius of external nozzle from the side of furnace, m;

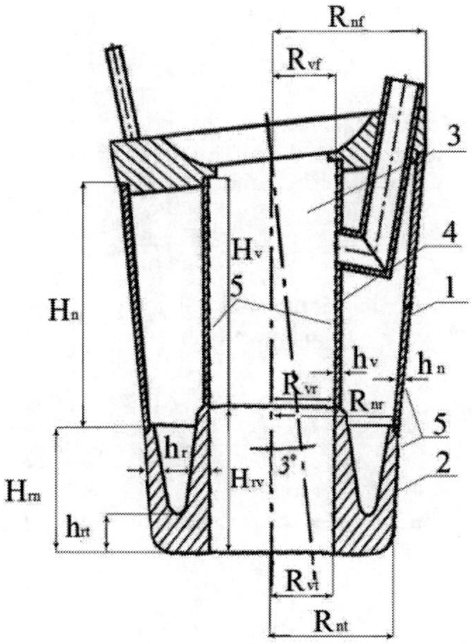

Fig. 1. Longitudinal section of blast tuyere:1- external nozzle, 2- tuyere nose,3- air passage, 4- internal nozzle, 5- covering

$$R_n = \frac{R_{nr} + R_{nf}}{2}$$

R_{nr}, R_{nf} is the radius of external nozzle from the side of furnace against tuyere nose and flange respectively (fig.1), m;

h_n, h_i is the thickness of external nozzle and i- coating material respectively (fig.1), m;

λ_{Cu}, λ_i is the heat conduction coefficient of cuprum M1 and i-coating material respectively W/(m·K);

m is the allotment of covering materials (layers);

α_v is the coefficient of heat-transfer from tuyere wall to water, W/(m²·K).

$$\frac{\mathrm{K}_{rn}^*}{2 \cdot \pi} = \frac{1}{\frac{1}{\alpha_{gr} \cdot R_{rn}^*} + \sum_{i=1}^{m} \frac{1}{\lambda_i} \cdot \ln \frac{R_{rnj}}{R_{rnj+1}} + \frac{1}{\lambda_{Cu}} \cdot \ln \frac{R_{rn}}{R_{rn} - h_r} + \frac{1}{\alpha_v \cdot (R_{rn} - h_r)}}$$

where α_{gr} is the coefficient of heat-transfer of tuyere nose from gas forge W/(m²·K);

$$R_{rn}^* = R_{rn} + \sum_{i=1}^{m} h_i$$

$$R_{rn1} = R_{rn}^*$$

$$R_{rnj} = R_{rn}^* - \sum_{i=1}^{j-1} h_i$$

$$R_{rnj+1} = R_{rn}^* - \sum_{i=1}^{j} h_i$$

$$j = 1, \dots m$$

R_{rn} is the midradius of tuyere nose from the side of furnace, m;

$$R_{rn} = \frac{R_{nr} + R_{nt}}{2}$$

R_{nt} is the radius of tuyere nose against tuyere`s cross-cut end (fig.1), m;

h_r is the average thickness of tuyere nose without butt end thickness considering, m.

$$\mathrm{K}_{rt} = \frac{1}{\frac{1}{\alpha_{gr}} + \sum_{i=1}^{m} \frac{h_i}{\lambda_i} + \frac{h_{rt}}{\lambda_{Cu}} + \frac{1}{\alpha_v}}$$

where h_{rt} is the thickness of cross-cut end of tuyere nose, m.

$$\frac{\mathrm{K}_v^*}{2 \cdot \pi} = \frac{1}{\frac{1}{\alpha_d \cdot R_v^*} + \sum_{i=1}^{m} \frac{1}{\lambda_i} \cdot \ln \frac{R_{vj+1}}{R_{vj}} + \frac{1}{\lambda_{Cu}} \cdot \ln \frac{R_v + h_v}{R_v} + \frac{1}{\alpha_v \cdot (R_v + h_v)}}$$

where α_d is the coefficient of heat-transfer from hot blow to tuyere wall, W/(m²·K);

$$R_v^* = R_v - \sum_{i=1}^{m} h_i$$

$$R_{v1} = R_v^*$$

$$R_{vj} = R_v^* + \sum_{i-1}^{j-1} h_i$$

$$R_{vj+1} = R_v^* + \sum_{i=1}^{j} h_i$$

$$j = 1, \dots m$$

$$R_v = \frac{R_{vr} + R_{vf}}{2}$$

R_{vr}, R_{vf} is the radius of internal nozzle from the side of air passage against tuyere nose and flange respectively (fig.1), m;

h_v is the thickness of internal nozzle (fig.1), m.

$$\frac{\mathrm{K}_{rv}^*}{2 \cdot \pi} = \frac{1}{\frac{1}{\alpha_d \cdot R_{rv}^*} + \sum_{i=1}^{m} \frac{1}{\lambda_i} \cdot \ln \frac{R_{rnj+1}}{R_{rvj}} + \frac{1}{\lambda_{Cu}} \cdot \ln \frac{R_{rv} + h_r}{R_{rv}} + \frac{1}{\alpha_v \cdot (R_{rv} + h_r)}}$$

where,

$$R_{rv}^* = R_{rv} - \sum_{i=1}^{m} h_i$$

$$R_{rv1} = R_{rv}^*$$

$$R_{rvj} = R_{rv}^* + \sum_{i=1}^{j-1} h_i$$

$$R_{rvj+1} = R_{rv}^{*} + \sum_{i=1}^{j} h_i$$

$$j = 1, \dots m$$

$$R_{rv} = \frac{R_{vr} + R_{vt}}{2}$$

R_{vt} is the radius of air passage against cross-cut end of tuyere (fig.1), m.

$$S_{rt} = \pi \cdot (R_{nt}^2 - R_{vt}^2)$$

Methods of temperature calculation on different surfaces of blast tuyere with using steady heat conduction formulae through a cylindrical and flat multiple-layer wall is as follows [10]:

-For external nozzle:

Temperature on the surface of the first coating`s layer from the side of furnace is:

$$t_{g-1} = t_g - \frac{K_n^{*} \cdot (t_g - t_v)}{2\pi \cdot \alpha_{gn} \cdot R_n^{*}}$$

Temperature at the joint of j and j+1 coating`s layers or coating`s layer and external nozzle:

$$t_{j,j+1} =$$

$$= t_g - \frac{K_n^{*} \cdot (t_g - t_v)}{2\pi} \cdot \left(\frac{1}{\alpha_{gn} \cdot R_n^{*}} + \sum_{i=1}^{j} \frac{1}{\lambda_i} \cdot \ln \frac{R_{nj}}{R_{nj+1}} \right)$$

Temperature on the surface of external nozzle from the side of water is:

$$t_{n-v} = t_v + \frac{K_n^{*} \cdot (t_g - t_v)}{2\pi \cdot \alpha_v \cdot (R_n - h_n)}$$

- For tuyere nose from the furnaces side:

Temperature on the surface of the first coating`s layer from the side of furnace is:

$$t_{g-1} = t_g - \frac{K_{rn}^{*} \cdot (t_g - t_v)}{2\pi \cdot \alpha_{gn} \cdot R_{rn}^{*}}$$

Temperature at the joint of j and j+1 coating`s layers or coating`s layer and tuyere nose:

$$t_{j,j+1} =$$

$$= t_g - \frac{K_{rn}^{*} \cdot (t_g - t_v)}{2\pi} \cdot \left(\frac{1}{\alpha_{gr} \cdot R_{rn}^{*}} + \sum_{i=1}^{j} \frac{1}{\lambda_i} \cdot \ln \frac{R_{rnj}}{R_{rnj+1}} \right)$$

Temperature on the surface of tuyere nose from the side of water is:

$$t_{rn-v} = t_v + \frac{K_{rn}^{*} \cdot (t_g - t_v)}{2\pi \cdot \alpha_v \cdot (R_{rn} - h_r)}$$

- For cross-cut end of tuyere nose from the furnaces side:

Temperature on the surface of the first coating`s layer from the side of furnace is:

$$t_{g-1} = t_g - \frac{K_{rt} \cdot (t_g - t_v)}{\alpha_{gr}}$$

Temperature at the joint of j and j+1 coating`s layers or coating`s layer and tuyere nose:

$$t_{j,j+1} = t_g - K_{rt} \cdot (t_g - t_v) \cdot \left(\frac{1}{\alpha_{gr}} + \sum_{i=1}^{j} \frac{h_i}{\lambda_i} \right)$$

Temperature on the surface of tuyere nose from the side of water is:

$$t_{rt-v} = t_v + \frac{K_{rt} \cdot (t_g - t_v)}{\alpha_v}$$

- For internal nozzle:

Temperature on the surface of the first coating`s layer from the side of air passage is:

$$t_{d-1} = t_d - \frac{K_v^{*} \cdot (t_d - t_v)}{2\pi \cdot \alpha_d \cdot R_v^{*}}$$

Temperature at the joint of j and j+1 coating`s layers or coating`s layer and internal nozzle:

$$t_{j,j+1} =$$

$$= t_d - \frac{K_v^{*} \cdot (t_d - t_v)}{2\pi} \cdot \left(\frac{1}{\alpha_d \cdot R_v^{*}} + \sum_{i=1}^{j} \frac{1}{\lambda_i} \cdot \ln \frac{R_{vj+1}}{R_{vj}} \right)$$

Temperature on the surface of internal nozzle from the side of water is:

$$t_{v-v} = t_v + \frac{K_v^{*} \cdot (t_d - t_v)}{2\pi \cdot \alpha_v \cdot (R_v + h_v)}$$

- For tuyere nose from the side of air passage:

Temperature on the surface of the first coating`s layer from the side of air passage is:

$$t_{d-1} = t_d - \frac{K_{rv}^{*} \cdot (t_d - t_v)}{2\pi \cdot \alpha_d \cdot R_{rv}^{*}}$$

Temperature at the joint of j and j+1 coating`s layers or coating`s layer and tuyere nose:

$$t_{j,j+1} =$$

$$= t_d - \frac{K_{rv}^{*} \cdot (t_d - t_v)}{2\pi} \cdot \left(\frac{1}{\alpha_d \cdot R_{rv}^{*}} + \sum_{i=1}^{j} \frac{1}{\lambda_i} \cdot \ln \frac{R_{rvj+1}}{R_{rvj}} \right)$$

Temperature on the surface of tuyere nose from the side of water is:

$$t_{rv-v} = t_v + \frac{K_{rv}^{*} \cdot (t_d - t_v)}{2\pi \cdot \alpha_v \cdot (R_{rv} + h_r)}$$

By the calculation methods of heat losses through tuyere surface and temperature a linear excel program has been developed. With this program it was calculated the heat losses of blast tuyere`s surface and temperature on tuyere`s details for conditions DP-5 OJSC "NLMK". The values of heat-transfer coefficient had been calculated according to experimental data, were used [11].

We take t_g =16000 °C; t_d =11500 °C; t_v = 300 °C; h_n = 7,0 mm; h_r = 16,0 mm; h_{rt} = 45,0 mm; h_v=5,0 mm; h_i= 1,5 mm; λ_{Cu} = 415,6562 W/(m·K); λ_i = 29,8891 W/(m·K); α_{gn} = α_{gr} = 133,745 W/(m2·K); α_d = 465,2 W/(m^2·K); α_b = 5815 W/(m^2·K); R_{nr} = 154 mm; R_{nf} = 185 mm; R_{nt} =

140 mm; R_{vr} = 72,5 mm; R_{vf} = 72,5 mm; R_{vt} = 72,5 mm; H_n = 305 mm; H_m = 145 mm; H_{rv} = 167 mm; H_v = 283 mm.

Taking into account that the heat guard inside of air passage finds wide application [12, 13], it was a study on the thickness effects of insulation lining on heat losses through surface of air passage and it was obtained temperature distribution from water-cooled hole of tuyere to air passage (fig.2-3).

As a comparison, it was calculated heat losses through air passage with aluminum gas-thermal coating. Fig. 2 shows the presence of insulation lining has a significant effect on reduction of heat losses through air passage than application of aluminum coating. Creation of air gap between it and internal nozzle leads to additional reduction of heat losses. Besides, the more thickness of insulation lining, installed without gap in air passage, the less heat losses through air passage. If the insulation lining was installed with gap the thickness has little effect on heat losses through air passage.

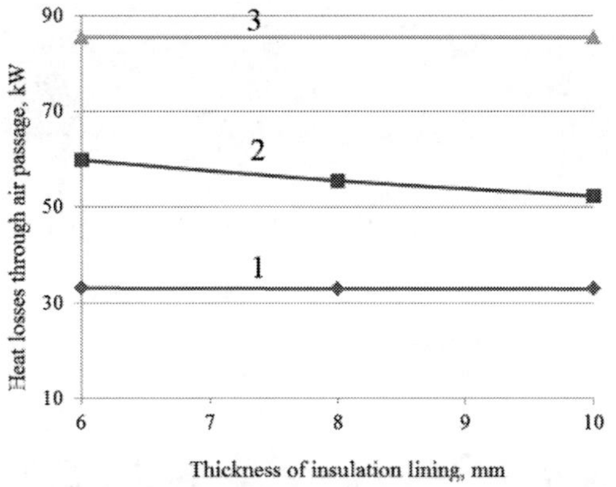

Fig. 2. Dependence of heat losses through air passage on thickness of insulation lining: 1– h_v=0,6 mm; 2– h_v=0 mm; 3– h_{Al} =1,5 mm; λ_{inl}=3 W/(m·K), λ_v=0,0362 W/(m·K), λ_{Al} =29,8891 W/(m·K)

According to figure 3, it is observed the steep gradient of temperature between the insulation lining and internal nozzle in air gap via its low conductivity, so that the temperature of the insulation lining is higher and temperature of the internal nozzle is lower than in the absence of gap between them. In terms of lining`s life the heavy gradient of temperatures on its thickness in the absence of gap may cause of the appearance of high thermoelastic stresses, expediting its destruction [14, 15].

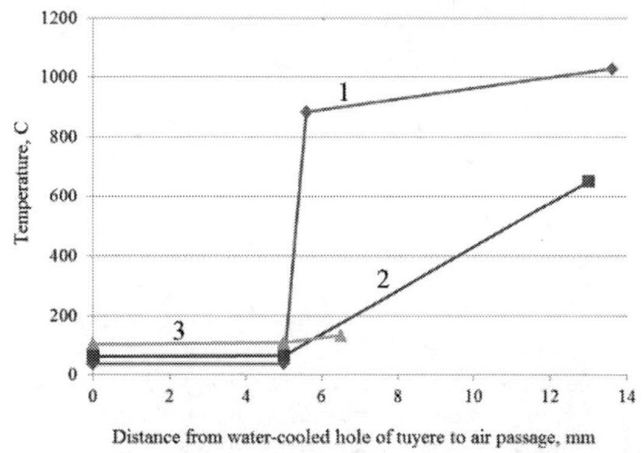

Fig. 3. Temperature distribution from water-cooled hole of tuyere to air passage: h_v =5 mm; 1– h_v=0,6 mm, h_{inl}=8 mm; 2– h_v=0 mm, h_{inl}=8 mm; 3– h_{Al} =1,5 mm; λ_{inl}=3 W/(m·K), λ_v=0,0362 W/(m·K), λ_{Al} =29,8891 W/(m·K)

3 Conclusion

The calculation technique of heat losses and temperature was adapted to blast furnaces tuyeres. Using a linear excel program the effect of gas-thermal coating and insulation lining on thermal state of the blast tuyere was studied.

References

1. A.V. Borodulin, A.P. Vasilev, E.L. Gluschenko and others, *About heat losses of furnace cavity*, The proceedings of the 2nd international research-to-practice conference *"Automatic furnace units and energy saving solutions in metallurgy"*, 424 (2002)
2. V.N. Grigoriev, G.I. Urbanovich, E.G. Urbanovich and others, Pat. № 2215043 RF. IQR C21B7/16. Preparation setting up method of blast furnaces tuyere / - 20011337559/02; ass. 11.12.01, pub 27.10.02 – Bul.№30
3. V.A. Akimov, avtoreferat dis. kand. tekh.nauk, *Issledovanie i sovershenstvovanie tekhnologii izgotovleniia vozdushnykh furm domennykh pechei naneseniem aliumosoderzhashchikh gazoplamennykh pokrytii s posleduiushchei termoobrabotkoi*, 23 (2011)
4. A.G. Radyuk, A.E. Titlyanov, N.L. Kirillova, and others, Pat. 128614 RF, C21B7/16. Tuyere of blast furnace, №2013100964/02, pub. 27.05.13, Bul.№15
5. P.S. Bykovskikh, V.A. Tishchenko, S.V. Bykovskikh, *Matematicheskaia model' teplovoi raboty vozdushnoi furmy domennoi pechi,* Metall i lit'e Ukrainy, **4**, 16 (2015)
6. E.B. Agapitov, V.A. Bolkunova, K.V. Shutov, *Chislennoe modelirovanie teplovogo sostoianiia vodookhlazhdaemoi domennoi furmy pri nalipanii rasplava shlaka na ee poverkhnost,* Vestnik MGTU im. G.I. Nosova, **4**, 73 (2011)
7. V.V. Lavrov, N.A. Spirin, V.Yu Ribalovlev, *Model study of temperature of blast furnace tuyere`s wall*, Not. Of HEI Ferrous metallurgy **2**, 46 (2000)

8. A.G. Manshilin, E.N. Skladanovsky, V.I. Necvetov and others, Pat. № 2235789 RF. IQR C21B7/16. *Blast furnace`s tuyere and technique of barrier coating`s application on blast furnace tuyere,* 2002129285/02, pub..27.05.04 – Bul.№15
9. V.P. Isachenko, V.A. Osipova, A.S. Sukomel, *Teploperedacha*, Energoizdat, 46 (1981)
10. S.S. Kutateladze, V.M. Borischansky, *Heat-transfer reference,* Gosenergopub, 414 (1959)
11. N.L. Kirillova, A.G. Radyuk, S.M. Gorbatyuk and others, *Calculation of coefficients of heat-transfer to blast furnace tuyeres surfaces from forge to hot blow* (Scein.-inform. publishing center "Institute for Strategic Studies", 132, 2012)
12. N.L. Kirillova, A.G. Radyuk, A.E. Titlyanov, *Reducing heat losses through surface of blast furnace tuyere,* Metallurgy **10**, 28 (2013)
13. N.L. Kirillova, A.G. Radyuk, A.E. Titlyanov, S.M. Gorbatyuk, *Improving air-tuyere operation in blast furnaces by means of coatings and sealant*, Steel in Translation **5**, 231 (2013)
14. K. Andreev, Kh. Kharmut, *Raschet formirovaniia treshchin v ogneuporakh*, Fizicheskaia mezomekhanika **4**, 109(2001)
15. L.M. Aksel'rod, A.V. Zabolotskii, *Matematicheskoe modelirovanie razrusheniia futerovok metallurgicheskogo oborudovaniia pod deistviem termoudarov* (Sb. n.tr. Sovremennaia nauka, 2010)

MATEC Web of Conferences **129**, 06032 (2017) DOI: 10.1051/matecconf/201712906032
ICMTMTE 2017

System of information output to electronic maps about vessels and aircrafts movement

Vitaly Vergeichik[1], *Vadim* Kramar[1,*], *Veronika* Dushko[2], and *Andrey* Volodin[1]

[1]Nakhimov Black Sea Higher Naval School, 299035 Dibenko st., 1, Sevastopol, Russian Federation
[2]Sevastopol State University, 299053 Universitetskaya st., 33, Sevastopol, Russian Federation

Abstract The article is devoted to actual problem of development and introduction of the replay system of the current location and direction of movement, in real time vessels and aircrafts, intended for continuous monitoring and obtaining data on the characteristics of vessels.

1 Introduction

Modern technology has caused the increase of technological advancements.

In the past few years it has been seen that the following questions are put in first place: data support, which helps to achieve superiority, real-time information exchange, adjustments of the ship's movements at sea [1].

The research is based on the problem of creation of a possibilities modelling system, which will also aim to technically solve the information field for information reflection via various industrial gauges with a trajectory measurements complex register [3].

At the present time, control of the situation in the area is carried out from various display devices, which cannot ensure that security control is provided to the full extent.

2 Main part

Computer programs used in telemetry are significantly different from those used in less critical applications. Since the telemetry data arrives at the receiving station repeatedly and sometimes even continuously, the hardware and software must be well coordinated with each other. In typical cases, hardware works relatively simple and repetitive tasks (for example, the establishment of synchronization and response to the emergence of an alarming situation); The software performs a primary processing to reproduce the data on the screen. The tasks of the software include setting up all hardware and software, high-speed data entry, possible preliminary hardware testing, high-speed output of specially selected measurement results to display terminals, display formatting and special data processing in accordance with the analysis requirements. The software is also often used to prepare a disk drive to work with all or selected measurement results to retrieve data from the disk for more detailed analysis and to perform a self-diagnosis of the state of the telemetry system before and during data reception.

As a means of operational control [2], each individual electronic map exists only at a certain point of time, as a rule, a short one, while it is visible on a display device. This is their main difference from other visual cartographic documents visualized on a solid surface (paper, plastic) by the means of graphic output (printers, for example).

This meaning agrees best with the word "electronic" [7], i.e. obtained through the movement of electrons, which is what happens in a working electronic device.

In accordance with the standard, a digital terrain model is a logical mathematical representation of the topographic survey data about the objects of the terrain, its characteristics, and includes:

- metric information – geodetic spatial data of the distinguished points of the terrain and situation;
- syntactic information to describe the connections between the points – borders of the buildings, forests, plowlands, water reservoirs, roads, watershed and overflow lines, slope direction between the characteristic points on the slopes, etc.;
- semantic information that characterizes the properties of objects - technical parameters of engineering structures, geological characteristics of the soil, information about the trees in forest areas, etc.;
- structural information that describes the relationship between different objects - the relationship of objects to any set: separate points of the railway line, buildings and constructions of a settlement, buildings and structures of the relevant facilities, etc.;
- general information - name of the site, system of coordinates and heights, nomenclature.

Topographic digital terrain model characterizes the condition and lay of the ground. It consists of a digital terrain model and digital model outline of the (situation) area. In addition, a digital terrain model can be supplemented with a special-purpose engineering model. When it comes to the engineering practice, a combination of digital models is often used; the models feature the situation, terrain, hydrological, engineering, geological, technical, economic and other indicators.

* Corresponding author: kramarv@mail.ru

In seeking to resolve the geodetic engineering problems on a personal computer (PC), a mathematical interpretation of digital models is used, which is called a mathematical terrain model. Computer-assisted design, which is based on a digital terrain model and a mathematical terrain model, reduces labor and time expenditure by the dozens of times, compared to the use of paper topographic maps and plans for this purpose.

In this regard, the electronic map symbolization system includes special fonts, and the classification of the electronic maps corresponds to the general map classification, for example: electronic topographic map, electronic aviation map, electronic geological map, electronic cadastral map, and others.

A digital map is a digital cartographic model, the content of which corresponds to the content of the map of a certain type and size. The digital map is used for the manufacture of paper and electronic maps, it is included in a map database, and it represents a geodetic topographic background for geographic information systems.

An electronic map is a digital cartographic model, visualized and prepared for visualization on the screen of a display device for presentation of information in a special system of symbols, the content of which corresponds to the content of a map of a certain type and size.

The advantages of electronic maps are listed below:

- possibility to change it as soon as new data are processed;
- automatic search and delivery of the information required;
- scaling of an image;
- layered data display, i.e. only of the information the user needs.

The Sirius software (certificate of state registration of computer programs No.2016662511 dated November 11th, 2016) [6] is designed to collect, display and predict the situation in the area with the construction of moving objects trajectory according to the data from a radar; it also helps to control and monitor the air and surface picture for security purposes.

PC with a recording device for trajectory measurements provides a continuous origination of actual real-time coordinates of a product, which is enough to identify it, regardless of geographic application conditions, weather conditions; it also records the instrument errors when determining the angle of azimuth and slant visual range (fig. 1).

In addition, when it comes to determining the characteristics of a product, the software improves the productivity and lowers the cost of production.

The use of electronic maps of the area allows one to outline the air and maritime situation; a block scheme of a hardware component is shown in picture No.1. Data from a radar station reveal the location of ships and unidentified targets; the software displays the coordinates, calculates the speed and direction of a tracked object, indicates the time when the object enters/leaves the area. The air picture is visualized in real-time.

A real-time information output on the electronic map from different technical elements allows one to control the area and, in case of security violation, make decisions concerning the termination of the motion of a sea vessel/aircraft.

It is also possible to save data entered by the user, as well as to open saved files.

On completion of simulation, an output file is being created; it can be used to check real devices for signal processing.

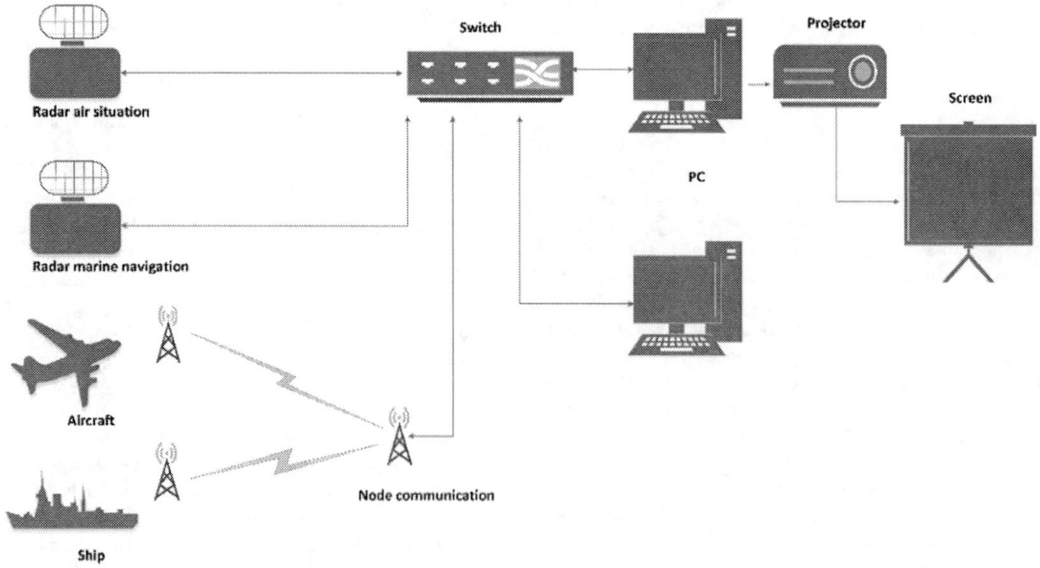

Fig.1. Oscillograms for the seaworthiness evaluation problem

DOI: 10.1051/matecconf/201712906032

Data is imported from the files with a radio-locating situation to the real devices for signal processing.

Determination of available hardware resources, search for drivers for these devices, as well as their initialization, take place at initializing the hardware and software. Data exchange with a limited-purpose device is organized under the control of a previously selected driver.

Data exchange with other applications is organized, the exchange takes the form of graphic information and information that may be used to organize the process of simulation, by the results of simulation itself.

The simulation of a radio-locating situation, based on the data provided by the user, calculates the signals which form a radio-locating situation, as well as converts the curvilinear coordinates system into the Gauss-Kruger plane rectangular coordinates system.

Capability research was carried out, theoretical solutions concerning the way to transfer the information data and technical ways of the information field on the display of information from different technical elements was found.

The software reproduces the current position and direction, in real time [4], as well as displays the most important parameters for control and comparison with the standards in a rapid analysis mode, by means of communication channels it can also transmit digital information to the closely-spaced areas [5].

The following methods have been used at completing general and particular tasks of the research: the methods of matrix theory, linear algebra, curvilinear coordinates system and Gauss-Kruger plane rectangular coordinates system, computer generated simulation. Computer generated simulation was performed with the help of a mathematics software, C ++ Builder 6.0. and the use of matrix algebra numerical computing.

The Sirius software family (fig. 2) solves tasks at three different levels:

- a registration level;
- a collection rate;
- a display level.

Software main features:

- receiving and decoding of information from a radar in real scale;
- calculation of the geographical values of the object tracked by a radar;
- calculation of the distance to the object;
- multicolored visualization of the trajectory of a tracked object on a digital map, with the color depending on the radar tracking quality of the object;
- detailed decoding, display of radar operation, chart output;
- obtainment and display of target indications from a radar;
- collection of objective situation from a radar;
- display of the radar tracking quality of each object;
- the choice of radar, which will calculate the indications;
- display of trajectory and radar tracking quality;
- calculation of the speed and direction of the target movement;
- display of the history sheet and trajectory of an object on a digital map;
- calculation of the time when the object enters/leaves the area.

Digital data can be converted to different formats, as well as transformed from one form to another, it can be visualized on graphic displays, printed, edited, changed, etc.

Electronic map classifier is an independent library of symbols, which contains all possible graphic descriptions of objects.

The external object code, layer, symbol and other characteristics are specified in the description of an object. However, for application of any additional graphic information (descriptive information, split lines, polygons, etc.) it is more convenient to use random graphic primitives without their prior description in a classifier.

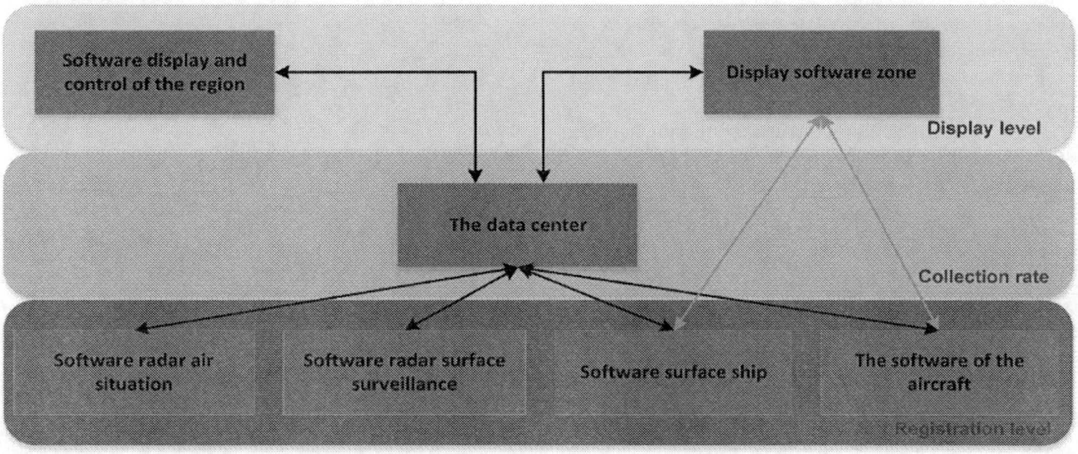

Fig.2 Software architecture of display, collection and recording of information collection Sirius

3 Development of mathematical support

Currently, the modeling of various processes and phenomena is widely used for a variety of research and development in various fields of science and technology. The use of modern computer technology makes it possible to obtain sufficiently accurate models that accurately describe the simulated phenomenon or process. The most widely used mathematical and simulation modeling.

At the heart of mathematical modeling are mathematical models - every process or phenomenon is described by mathematical equations, which requires a serious mathematical apparatus. Implementation of mathematical models on a PC is possible when using numerical methods to implement various mathematical functions and operators, which introduces an error in the calculations.

Simulation modeling involves obtaining and using a model that describes the mechanics of a process or phenomenon, i.e. How the process proceeds in the real situation and what underlies this process. Simulation modeling, in general, relies on physical laws, for the implementation of which the device of mathematical functions is also needed.

The most acceptable way to describe a simulated domain in this paper is to use simulation modeling based on mathematical models. As such an object, one can take a point target, and divide the entire radar model into models of its main modules. In this case, a compromise will be reached between the accuracy of the model and the total computational cost of implementing the model.

4 Conclusions

Therefore, the system of registration, information processing will ensure the increase of security and visibility level, it will also reduce the time of processing and documentation of the results. That will not only comply with all the requirements for the information transfer, but will also reduce the cost of maintenance of the telecommunications infrastructure.

The reliability of the results is confirmed by experimental verification of performance of the developed algorithms, which have confirmed the improved accuracy, in comparison with analogues.

A new technical solution for area monitoring has been found. The software reproduces the current position and direction, in real time, as well as displays the most important parameters for control and comparison with the standards in a rapid analysis mode, by means of communication channels it can also transmit digital information to the closely-spaced areas.

References

1. V. V. Vergeichik,. V. V. Buldakov, *Purpose and performance characteristics of promising measurement tools during the field works*, Scientific readings on aviation dedicated to the memory of N. E. Zhukovsky, No.**4**, Pp.361-370 (2016)

   http://elibrary.ru/item.asp.2id=26570905

2. D. Yu. Polenov, *The Evolution of telemetry in rocketry*, Young scientist, No.**6**, Pp. 216-218 (2014)
3. A. A. Guzinski, V. V. Vergeichik, V. A. Kramar, *Increase of efficiency of use of measuring instruments for the Navy*, Innovative Arsenal youth: proceedings of the seventh scientific - technical conference of young scientists and specialists of FSUE "KB Arsenal", issue No.**7**, Pp. 295-298 (2016)
4. V. I. Pan-Luzin, *Telemetry. The eyes and ears of the Chief designer* (Moscow: OOO "Overlay", 2009)
5. A. V. Nazarov, *Modern telemetry in theory and practice* (St. Petersburg: Science and technology, 2007)
6. A. A. Guzinski, O. V. Hletko, V. V. Vergeichik, *Collection, records and displays the trajectory of the product on an electronic map in real time according to the radar data and telemetry data, control and monitoring of surface and air situation in the area of the experiment to ensure security*, Certificate of state registration of computer programs №2016662511, date of the state registration in the Register of the Federal service for intellectual property (ROSPATENT) computer programs November 11, 2016
7. The Internet: http://dic.academic.ru/dic.nsf/ruwiki/704030 (accessed 22.10.2016), " Electronic map.

AUTHOR INDEX

AUTHOR INDEX

AUTHOR INDEX

AUTHOR INDEX

AUTHOR INDEX